The compounds depicted on the facing page
have been synthesized since the publication
of the second edition in 1964.
Those below have not yet been synthesized.

ORGANIC
CHEMISTRY

ORGANIC CHEMISTRY

THIRD EDITION

JAMES B. HENDRICKSON

Professor of Chemistry

Brandeis University

DONALD J. CRAM

Professor of Chemistry

University of California at Los Angeles

GEORGE S. HAMMOND

Professor of Chemistry

California Institute of Technology

McGRAW-HILL BOOK COMPANY

New York   St. Louis   San Francisco

Düsseldorf   London   Mexico   Panama

Sydney   Toronto

This book was set in Elegante by Graphic Services, Inc., and Black Dot Inc.,
printed and bound on permanent paper by Kingsport Press.
The designer was Barbara Ellwood;
the drawings were done by BMA Associates, Inc.
The editors were Thomas Adams and Sally Anderson.
Peter D. Guilmette supervised the production.

**ORGANIC
CHEMISTRY**

Library of Congress Catalog Card Number 70-109249

28150

1 2 3 4 5 6 7 8 9 0   KPKP   7 9 8 7 6 5 4 3 2 1 0

PREFACE

Organic chemists have seen a dramatic growth of theoretical understanding in the past few decades, and there is little doubt that this understanding—while certainly still imperfect—has nonetheless greatly clarified and simplified our comprehension of the science. If this simplification in learning occurs with development of theory, surely it is the beginning student who needs it most!

The first and second editions of this book were attempts to find a new textbook organization of organic chemistry in which theory could play its proper role of pulling together otherwise unrelated mountains of facts. Since these editions appeared, the advance of theory has made the organizational task much easier. The third edition of "Organic Chemistry" has now been thoroughly revised to take advantage of both the advances in theory and fact, and at the same time to use our own experience and that of others to simplify and refine the presentation initiated in the earlier editions. Our aim is to bring the concepts of organic chemistry directly to the level of the beginning student.

We believe certain major advantages are associated with our organization of the text on lines of theory. Structure is discussed first, with all kinds of structural families introduced and organized at the beginning as natural consequences of the properties of atoms. Then chemical reactions follow, after the structures of both reactants and products are already familiar to the student. The early introduction of stereochemistry allows these concepts to be used throughout the discussions of reactions. The organization of reactions into broad classes not only simplifies their presentation but also makes it possible to consider the mechanism, scope, limitations, and side reactions of whole groups of transformations. Other organizations permit no such generalizations. Subjects such as physical properties, spectra, acid-base theory, orbital and resonance theory, structure elucidation, synthesis, nomenclature, and chemical literature, which are ordinarily scattered in traditional texts, receive integrated treatment here.

The basic philosophy of this unified approach is thus one of gradual and logical development from the properties of single atoms. These properties are shown to lead to specific combinations, yielding the structures of organic molecules, their shapes and their electrical nature. We develop first the structural possibilities for molecules that arise from the properties of the constituent atoms and then show that these properties in turn dictate how molecules behave in reactions with each other.

Students are encouraged to see that they may thus "invent" or predict molecules from the properties and "rules" of atoms, and then in turn that they may predict the nature of chemical reactions from the properties of the molecules. Even though our theoretical knowledge is imperfect for prediction, it creates a unifying conceptual frame which makes learning easier, and more exciting, for the beginning student. Above all, theory invites tests of theory, and students can enjoy the exhilaration of both discovery and invention.

The first section of this book is devoted to structure and its correlation with physical properties. Following an introductory and historical first chapter, the second chapter delineates the properties of atoms and their development into molecules, with specific shapes and electrical natures. The next two chapters survey the classes of possible molecules, along traditional lines. The last three chapters on structure develop three major areas: resonance and related consequences of electronic interactions in structure (Chap. 5); molecular shape, conformation, and stereochemistry (Chap. 6); and the correlations of physical properties with molecular structure (Chap. 7). This last chapter is chiefly involved with methods of deducing molecular structure from spectroscopic observations.

Two transitional chapters lead to the second section on the dynamics of chemical reactions. These include an application of structural concepts to reactivity in the simplest reaction, the acid-base reaction (Chap. 8) followed by a general introduction to the overall nature of reaction dynamics (Chap. 9).

The main section on chemical reactions, Chaps. 10 to 18, deals with ionic reactions, first with nucleophilic agents then with electrophilic ones. In order not to lose the value of the classical organization by functional classes, we have made a special effort in this section to show these reactions as they apply to each traditional class of molecules. Furthermore, each chapter on a major reaction type incorporates a specific section on the practical use of the reaction in synthesis. Beyond this, the chapter on organic synthesis (Chap. 23) has been expanded and reoriented towards the solving of problems in synthesis design.

The third section of the book contains three chapters on nonionic reactions: radical reactions (Chap. 20); thermal pericyclic reactions (Chap. 21); and photolytic reactions (Chap. 22). Special topics include the chemistry of sulfur and phosphorus (Chap. 19), heterocycles (Chap. 24), and polymers, both synthetic and natural (Chap. 25). The natural polymers provide a place to discuss amino acids and nucleotides and thus lead to a chapter (26) which provides an introduction to biochemistry. This chapter is intended primarily to show that the molecules and reactions of biochemistry may be readily understood as further examples of the theory presented in the previous chapters. It offers a capsule survey of the main lines of biochemical research into the operation of the living cell at the molecular level. The chapter provides suitable background for a discussion (Chap. 27) of natural products, since it is organized on biosynthetic lines.

Much material in the present book is new. New compounds, reactions, and theories have been added to the central chapters. Whole new chapters have been added on photochemistry, sulfur and phosphorus chemistry, and the new concept of the pericyclic reaction. The chapter on the dynamics of biochemistry is new, as is most of the material on heterocycles and natural products.

A special feature of this revision is its focus on the methods and protocols of problem solving in organic chemistry. The logic of structure determination by physical and chemical means and of synthesis design is specifically formulated in a stepwise fashion with examples for the student to follow. In a number of other areas such as resonance, stereochemistry, and reaction mechanisms, *detailed procedures* are delineated for attacking the types of problems commonly encountered. Illustrative problems are sprinkled liberally through the text as well as at the ends of chapters; solutions to many of the problems are now incorporated at the end of the book. These problems have been designed for their value not only as exercises but also as pedagogical tools for introducing new concepts. Much of the concept of deducing multistep mechanistic pathways is introduced as problem material.

This book provides more material than can be included in short courses. However, in every chapter except for the first seven, sections of material may be omitted without loss of continuity. We have found that this flexibility allows the teacher to stress whichever of the three general sections he chooses. A course designed for both chemistry and nonchemistry majors could emphasize the material in the first and third sections, whereas a course designed only for chemistry majors might stress the second group of chapters.

In the use of a second color, we apply the simplest of rules—parts of formulas are colored to draw the attention of the reader to a particular feature of the formula in order to make the central features of the illustration or reaction more vivid.

We are indebted to the many users of the previous editions who provided us with many detailed and valuable comments, criticisms and suggestions. These communications were of very great value in preparing the present revision. We should like now to invite and urge users of the present text—teachers and students—to write us of their reactions to it and of their suggestions for later improvement. We are always grateful for this help.

JAMES B. HENDRICKSON
DONALD J. CRAM
GEORGE S. HAMMOND

CONTENTS

DYNAMICS

INTRODUCTION

1-1 THE NATURE OF ORGANIC CHEMISTRY

Chemistry has been called the science of what things are. Its intent is the exploration of the nature of the materials that make up our physical environment, why they possess the different properties that characterize them, how their intimate ultimate structure may be understood, and how they may be manipulated and changed. By far the greatest variety of materials that confront us are organic. At the dawn of the nineteenth century, when chemistry was just beginning, organic materials were understood as substances created by living organisms: wood, bone, cloth, food, medicines, and the intricate materials that made up our own bodies. Inorganic materials from the mineral world—salt, metals, rocks, glass—turned out to have simpler compositions and fitted into the early development of chemical theory more easily. Because of this and of men's natural awe of life, organic materials were believed to be possessed of a mysterious "vital force" and organic chemistry was thus separated in its path of evolution from inorganic chemistry. By the middle of the nineteenth century, however, the vital force theory was largely discredited by repeated preparations from mineral sources of "organic" compounds originally obtained only from living organisms.

The salient feature of organic compounds is that they contain carbon and the best present definition of organic chemistry is the chemistry of carbon-containing compounds. Compounds such as carbonates, carbon dioxide, and metal cyanides, however, are usually excluded as inorganic, and a refined definition, which we shall see more closely describes organic chemistry, would be the chemistry of compounds containing a carbon-carbon bond.

A striking feature of organic chemistry is the vast number of organic compounds known and the fact that the potential number is virtually limitless. Over two million different organic compounds have now been characterized, and every year tens of thousands of new substances are added to the list, either by discovery in nature or by preparation in the laboratory. Commonly a research chemist prepares over a thousand new compounds during his lifetime. The rate of increase in the numbers of known compounds provides an interesting record of the explosive development of organic chemistry: in 1880, the number was approximately 12,000; in 1910, about 150,000; in 1940, about 500,000; at present, several millions. These are pure compounds of known structure, synthesized by chemists in the laboratory. Millions of compounds are also synthesized by living organisms; a single-celled bacterium is itself a sophisticated and highly organized mixture of tens of thousands of organic compounds, and many of these compounds, the stuff of life itself, have also been isolated and scrutinized and their structures unraveled by chemists in this century.

Most natural organic materials which we experience are mixtures, such as wood, rubber, paper, cloth, turpentine, olive oil, vitamins, perfumes, medicines, but even these are often purified or modified by chemical techniques. Many more of the organic materials around us are created by modern chemical knowledge and are substances unknown in past centuries, e.g., polyethylene and other plastics and film, most dyes and pigments, medicinal drugs, gasoline, paints, and surface coatings. Among the pure organic compounds we commonly encounter are found sugar, DDT, lighter fluid, alcohol, penicillin. In general we may distinguish organic from inorganic materials by several common characteristics. Usually organic materials burn, often charring (to elemental carbon), as inorganic substances do not. Organic compounds are either liquids, or solids commonly melting below 300°, whereas mineral solids often sustain red heat; and organic materials are "soft," rarely showing the structural strength of much of mineral matter.

1-2 HISTORY AND PRESENT PRACTICE

Although organic reactions have been deliberately carried out since the discovery of fire, the science of organic chemistry did not evolve until the beginning of the nineteenth century, mainly in France at first, then in Germany, later in England. In this early period many organic compounds were isolated and purified and their compositions accurately determined. Citric and malic acids were crystallized from lemons and apples before 1800, morphine purified from opium, and cholesterol from gallstones.

Berzelius at this time was developing with much success the ideas of electrovalency that worked so well in explaining the composition of salts and inorganic substances, but these complex organic materials implied a bewildering variety of combining modes, or valences, for carbon; the formula of morphine is $C_{17}H_{19}NO_3$ and cholesterol is $C_{27}H_{46}O$! Berzelius tried hard to fit these compounds into his theory but was ultimately obliged to set them aside and, in doing so, he coined the term "organic chemistry" and shared the general view that organic compounds, despite their removal from living organisms, still were invested with "vital force," which presumably caused all the trouble.

In 1828 Wöhler recorded his famous conversion of ammonium cyanate into urea, a substance isolated from animal sources, and this observation has traditionally been taken as a clear demonstration that the vital force theory was incorrect. For Wöhler, however, the importance of the experiment lay in showing that two substances with the same formula (CH_4N_2O) could be physically different materials. A number of such cases came to light about that time and led slowly but inexorably to the idea that the central special feature of organic compounds was not their *composition* but instead the way in which the atoms were combined—in short, their structure.

The vital force conception did not die in 1828. Wöhler seems to have regarded it as unimportant and pointed out himself that the philosophically inclined could argue that ammonium cyanate, obtained by heating bones, might contain residual vital force. We can also object today that urea, despite its suggestively organic name, is only a simple derivative of carbon dioxide, contains no carbon-carbon bond, and if named carbamide instead might always have been viewed as essentially inorganic. In 1845 Kolbe synthesized acetic acid, the principal component of vinegar, in a sequence of reactions starting with carbon itself; the demonstration is clearer since acetic acid ($C_2H_4O_2$) possesses a carbon-carbon bond. Probably the theory of vitalism, like many other scientific theories, disappeared slowly under the weight of accumulated evidence rather than as a consequence of any single brilliant and illuminating experiment. Similar slow change is characteristic of most chemical theories including, no doubt, many which are found most useful today by organic chemists.

The development of the structural theory, about 1860, ushered in the second major period in the growth of organic chemistry. The development of a detailed picture, by purely inductive reasoning, of both the atomic organization and the shapes of molecules stands as a major accomplishment of the human intellect. One of the most fruitful concepts in the history of science was advanced almost simultaneously by Kekulé in Germany and by Couper in Scotland in 1859. They suggested that the atoms in molecules are bound together by bonds and that each kind of atom is characterized by having the same number of bonds in all of its compounds. The key feature of organic compounds, the existence of strong carbon-carbon bonds, was recognized, and used to understand the possibility of large molecules, containing many connected carbon atoms. Now that chemists have used, developed, and verified for a century the idea that organic molecules have definite structural features, with carbon atoms as the main building blocks and chemical bonds as the mortar, it seems so "right" that it is difficult to appreciate the genius of the original suggestion.

It is also curious in retrospect to imagine a structural theory of molecules which did not immediately appreciate that molecules must be three-dimensional physical entities if they are to be understood as the ultimate particles of real matter in space. Nevertheless, it was not until 1875 that van't Hoff and LeBel independently proposed the spatial nature of molecules with the reasonable hypothesis that the four bonds to carbon were located at equal angles to each other in space, thus being directed to the four corners of a regular tetrahedron with the carbon atom at its center. Ample proof of the correctness of this view has long since been supplied but in 1875 it was by no means universally accepted. In fact the chemical establishment of that day looked on the suggestions of van't Hoff and LeBel as speculative nonsense.

The structural theory is not only a keystone of organic chemistry but a remarkably simple idea. In essence it states that by understanding each carbon atom to form four bonds, tetrahedrally arranged in space, we may describe the physical architecture of the most complex molecules. Hence, although molecules are too small to be seen even in the most powerful microscopes, chemists have a clear understanding of their appearance. The truth of their vision has been substantiated literally millions of times by the successful outcome of prediction. The power of the theory is demonstrated by the statement that there has been no chemical observation that cannot be satisfactorily understood in terms of the structural theory. Finally, although structural logic is completely rigorous, it involves no mathematics. Unlike most sciences of comparable sophistication, much of organic chemistry is conducted without the use of any formal mathematics beyond elementary arithmetic.

The third and present period in the history of organic chemistry commenced with the description of chemical bonds as electron pairs, by Lewis in 1917. Although a host of chemical reactions were already known and in active use to transform organic compounds into other compounds, only with this understanding of the nature of the chemical bond did a clear rationale of the nature and mechanism of chemical reactions begin to appear. This will be apparent when one reflects that the transformation of one molecule into another—a chemical reaction—requires the breaking of some bonds and the making of others, and this process could not be understood until one knew what a bond is. Thus if the nineteenth century was devoted to unraveling the static structures of molecules, the twentieth has added the study of the dynamics of their transformations.

The kinds of activities which engage the organic chemist may be grouped in the following way, with the understanding that all three may be involved in a given project:

1 Structure analysis
2 Synthesis
3 Reaction dynamics

In general structure must be determined every time a chemical reaction is carried out and a reaction product isolated. If the product is suspected of being a compound already known, it is then physically compared with the known material to establish its identity. If the product is a new compound, the structure must be proved independently. In this sense structure analysis is common to most organic chemical research. Usually the problem is relatively simple since the molecules used in the reaction and the nature of the reaction itself are already known, and thus offer a strong presumption of the product identity, by analogy. On the other hand specific problems of great complexity usually arise with the isolation of new compounds from nature, the so-called "**natural products.**" The procedure used in structural analysis in either case is broadly the same and is outlined in more detail later in the text. In general the unknown compound is submitted either to known chemical reactions in order to draw structural inferences or examined for certain physical properties which have been correlated with structural features. The latter, or physical, methods have had a phenomenal advance in the past three decades, notably in the development of optical and nuclear magnetic spectroscopy. These methods can yield in minutes extensive information about the structures of unknown compounds that previously could be obtained only by weeks or even years of chemical study. Total determination of structure is also possible by means of x-ray diffraction, so that the classical goal of structural studies— ascertaining the relative spatial positions of all the atoms in any molecule—can be readily and routinely attained. Beyond the gross location of the atoms, however, is the finer understanding of the electrical nature of a molecular structure—the interactions of bond electrons and atomic nuclei—and structural studies in this more modern sense are still under active pursuit.

The synthesis of organic compounds involves conversion of available substances of known structure, through a sequence of particular, controlled chemical reactions, into other compounds bearing a desired molecular structure. By synthesis the chemist can *create* molecules specifically designed to test some theoretical hypothesis or to be tested for medicinal or other commercial value; such synthetic compounds are new substances, having never previously existed in the world. On the other hand it has long been the practice to attempt the synthesis of natural products in the laboratory once their structures had been determined, both as a proof of the deduced structure and as a challenge to the capabilities of chemical knowledge. In the early nineteenth century all organic compounds were natural products, but with the growth of knowledge about chemical reactions it became possible to synthesize many compounds not found in nature. In this way the development of chemical theory was not limited simply to observations made on naturally occurring substances, although these have always provided important instructive examples.

A great many reactions were studied and catalogued in the nineteenth century and used to make many new compounds, but the lack of any understanding then about the mechanisms of these reactions prevented the achievement of real sophistication in the synthesis of complex molecules. The most brilliant exponent of synthetic sequences derived in theoretical terms has been R. B. Woodward, who in recent decades has synthesized such complex natural molecules as quinine, cholesterol, and chlorophyll. If the goal of structural studies is to be able to know the molecular architecture of any substance, then the goal of synthesis is to make it in the laboratory, ultimately by the detailed prediction of the best sequence of reactions and exactly how to carry out each one.

The third broad area of activity involves the study of the intimate details of the course of chemical transformations, the ultimate aim of such work being the ability to predict quantitatively the rate of any given reaction and the nature of its products. This study of reaction mechanisms began in earnest in the 1920s, following Lewis' electronic description of the chemical bond, and was actively pursued in England by Robinson and Ingold, in Germany by Arndt and Meerwein, and in the United States by Hammett, Conant, and Lucas. Extending the methods of physical chemistry these workers and their host of contemporary successors have determined quantitative rates of reaction and the effects on these of temperature, concentration, molecular variations, and other conditions, in an effort to discern the modes and pathways of organic reaction. The considerable success of these studies has not only increased our knowledge of the intimate details of chemical change and greatly expanded the possibilities of synthesis, but has also brought a theoretical unity to the whole field of organic chemistry which has the effect of making its principles easier to teach and to learn.

In addition to these exciting developments, the past few decades have been characterized by a prodigious expansion of the organic-chemical industry. Before World War I a vigorous chemical industry existed, especially in Germany. However, its scope was limited to relatively few fields, notably dyestuffs and explosives and a few preliminary successes with medicines. Since the 1920s, the industry has thrust into many new fields on a massive scale. Plastics, elastomers, fibers, and films are now produced in enormous volumes for myriad purposes. The pharmaceutical industry introduces new drugs so rapidly that it is difficult for the practicing physician to distinguish the new from the obsolete. These industries maintain large laboratories in which organic-chemical research is practiced at a very sophisticated level. A large and thriving industry in fact exists simply to supply research workers with organic compounds, over 50,000 different organic chemicals being commercially available for research purposes. The major industrial sources for organic chemicals are coal, petroleum, and some plant products.

For 30 years we have been in an organic-chemical age. While we may now be thinking in terms of an atomic age, it is unlikely that the impact of organic chemistry on our daily lives will ever be rivaled by that of atomic physics.

Organic chemistry in the future will certainly continue to blossom in the same exciting way that has characterized its recent active past, largely toward the goals suggested in the three areas above. Furthermore, there are areas still virtually undeveloped, such as the organic compounds involving elements in the periodic table other than the half-dozen or so of traditional organic chemistry. The vigorous and fascinating advances in the chemistry of life can only accelerate as we probe more deeply into the organic reactions which are the dynamics of life.

The field of biochemistry, now fashionably labeled as molecular biology, is in fact the study of organic chemistry focused on those molecules—proteins, nucleic acids, polysaccharides—of which living organisms are constructed. In some ways organic chemists freed themselves from dependence on materials from the natural world during the adolescence of chemistry and are now increasingly returning to this chemistry of nature to apply their experience to the subtle and complex molecules of biochemistry. The interaction will work both ways, for as chemists discover the details of the enzyme-catalyzed reactions of life, they will discover ways to emulate these effective catalysts in laboratory reactions as well.

In the foregoing we have sketched a sense of the growth and present activity of organic chemistry. Most of this book will be involved with our understanding of molecular structure and change as deduced over the years from laboratory observations. Before we begin, however, it will be of value to look at the nature of the actual materials which are the subject of organic chemistry and see how they may be purified, examined, and identified, and how we obtain their elemental compositions.

1-3   IDENTITY AND PURITY OF SUBSTANCES

Organic chemistry is conducted at two levels. While chemists do their experiments and observations on actual substances, liquids or solids which can be seen, manipulated, and put into bottles, nevertheless most of our discussion centers on the behavior of the molecules which are the unseen, submicroscopic particles of which every substance is ultimately constituted. Implicit in this molecular theory is the belief that the behavior of substances at the observed level directly reflects the behavior of their constituent molecules. It is important not to confuse these levels, that of physical observation and that of deduced molecular behavior. Discussion commonly passes back and forth between them, and our ideas of molecular properties all have consequences at the observational level with which we may test the truth of theory.

The most general consequences which underlie organic-chemical work are the following:

1   *The characterization of organic compounds.* When we wish a unique description for some object in order to identify it, we commonly specify the details of its shape or color, i.e., its morphology. This is very highly developed in the descriptions of plant and animal species in biology, but it is a relatively undiscriminating approach for describing people, most of whom look approximately alike. To obtain more nearly unique descriptions of individuals we collect a variety of quantitative distinguishing properties such as height, weight, birth date, fingerprints, and residence address, and for convenience summarize these in a name for the individual. The more properties we collect the greater the chance that they uniquely characterize one individual and not several similar ones. We do the same with organic compounds, which are mostly colorless liquids and solids and rarely distinguishable in physical appearance. Physical properties

such as melting and boiling temperatures, adsorption behavior, and several kinds of optical data (index of refraction, intensity of light absorption at different wavelengths, etc.) are commonly used since they offer a collection of discreet numerical data for the purpose of distinguishing one compound from all others. Any quantifiable physical property serves as well, however. The most useful physical properties for this purpose are those with the broadest range of data and hence the greatest capability for discrimination.

These observed physical properties used to characterize a pure compound are a consequence of the properties of its individual characteristic molecules. Therefore, certain features of molecular architecture should be deducible from the variation in physical properties of different substances. In some cases, discussed in Chapter 7, this is certainly true and constitutes a potent approach to structure determination, often a completely definitive one.

Extensive correlations of physical properties with molecular structure have in recent years revolutionized the art of elucidating the constitution of organic molecules. It is also apparent that the more similar the molecules of two compounds the more alike their physical properties will be. Very similar molecules can only be distinguished by very sensitive physical methods, capable of high discrimination.

2   *Criteria of purity.* In molecular terms a pure compound is one in which all the molecules are alike. In order to study the behavior of those molecules we must devise ways of separating out all others first. Then our observations at the macroscopic level are certain to pertain only to a single molecular species, and our conclusions clearly are referent to it and not to some minor impurity. Impure substances are mixtures of the molecules of two or more compounds and mixtures of two compounds will show physical properties different from and usually intermediate between those of the two pure compounds. Therefore, in devising ways of separating the compounds in a mixture we are also devising purification procedures. The criterion that will test the effectiveness of these procedures will be the change in a physical property as purification proceeds. The value of the selected property should pass from the value for the mixture to that of one pure, molecularly homogeneous compound. When the numerical value of a physical property remains constant on further purification this is a demonstration that the substance is a pure compound, molecularly homogeneous.

3   *Comparison of compounds.* Since the argument above also implies that any two compounds with all physical properties identical have identical molecules, we use the comparison of physical properties to compare substances and prove their identity (or lack of it). With many, perhaps most, of the organic compounds that have been described in the chemical literature, physical samples in bottles no longer exist. They may have been lost or discarded or they may have deteriorated with time, but this is no impediment to the advance of chemistry since the originally described experiment can be repeated in order to re-obtain the substance. Proof that the desired substance was obtained can be had by comparing its physical constants with those reported in the original literature. In the same way a new synthesis of a known compound is verified by a comparison of the physical properties of the known with those of the newly synthesized materials.

As a practical example the mixture obtained from a reaction or a plant extraction might be submitted to purification by chromatography (Sec. 1-4) and a partly purified fraction obtained as a solid, of which the melting point is then taken. This solid is purified further by crystallization, and the melting point is found to be higher. Several recrystallizations result in more, but diminishing, increases in melting point until it no longer changes on further recrystallization. The material is now judged to be pure; if its melting point is identical to that of a known compound (the history of the experiment will suggest some), then other physical properties of each substance, such as infrared or ultraviolet spectra, are also examined and compared. If these are completely identical the two substances are themselves judged to be identical also.

Melting behavior has added value as a method of identity comparison since it is almost always true that the melting point of a compound is substantially lowered by the addition of another substance, even a different substance with the same melting point. The melting point (or range) of a mixture of two samples provides a sound basis for identity comparisons, since two different compounds with fortuitously identical melting points will ordinarily show a lower melting point when mixed, whereas with identical compounds, their mixed melting point will remain the same. Such mixed melting point depressions are frequently as large as 20 to 30° and are traditionally the most widely used and powerful test of the identity of two crystalline compounds. Most pure organic compounds are crystalline solids at room temperature.

1-4   PURIFICATION AND SEPARATION METHODS

Methods for separation or purification of compounds all depend on some mechanical procedure allowing a difference of physical properties to be translated into actual movement of different molecules to different places so that they can be physically separated from each other. Many physical properties useful in characterization also constitute a basis for separation. The most important separation methods are crystallization, distillation, extraction (or distribution), and chromatography, and each of these is associated with a physical property which is also used for identity characterization. In no procedure is purification or separation absolutely complete, but the fractions separated from the mixture are enriched in one compound and so purification often involves several repetitions of the procedure. The basic features of these physical separation methods are described in what follows.

Crystallization

If a mixture of (solid) compounds is dissolved in a solvent and the maximum solubility of one compound is then exceeded either by lowering the temperature, adding a poorer solvent, or evaporating some solvent, then the excess of that compound over the maximum soluble amount will come out of solution, depositing as visible crystals. This physical separation of a crystalline fraction much enriched in one component results from the physical property of solubility and is seen in molecular terms as a competition between the attraction of molecules of dissolved compound for each other versus their attraction to solvent molecules. Furthermore, the forces that attract and order molecules of the same kind into a crystal lattice tend to exclude dissimilar molecules, which do not fit, so that the value of crystallization for purification is thereby much enhanced. The related physical property for characterization purposes is the melting point. Crystallization is very widely used for purifying solid substances.

Distillation

For separating liquid compounds advantage is taken of their differences in boiling temperature, a property useful in itself for characterization. The mixture is placed in a vessel which is heated until the boiling point of the lowest-boiling component is reached, whereupon this substance vaporizes and the vapors are carried over into a condenser in which they are cooled and return to a liquid state, running into a different container for separation purposes. At increasing temperatures each component separately distills over in this way and the receiving containers are changed to collect each fraction separately. In practice closely boiling compounds distill together to a large extent and must be more finely separated by use of a fractionating column which in effect provides continuous condensation and revaporization before the final condenser is reached. Thus the lower boiling component is much enriched before entering the condenser and being collected in the receiver. Modern fractional distillation equipment allows the separation of liquid compounds different by as little as 2° in boiling point.

Extraction

When a compound is shaken with two immiscible solvents together, some of it will be found dissolved in one solvent, the rest in the other. Thus the compound distributes itself between the solvents and exhibits a ratio of concentrations in the two (the **distribution coefficient**) which reflects its preferred solubility in one over the other, i.e., the competition of attraction by the two solvents for the molecules of the compound. The distribution coefficient between any pair of immiscible solvents is another physical property useful for characterization of any particular compound and is applicable to separation since the immiscible solvents are physically separable.

Simple extractions are usually carried out after reactions in order to separate the inorganic salts, which are preferentially soluble in water, from the organic compounds, which will stay in an immiscible organic solvent such as ether or chloroform. Advantage is taken of this preferential water solubility of ionic salts for separating organic acids or bases from other organic compounds. This is done by shaking an organic solution of the mixture with aqueous alkali or mineral acid. These will react with organic acids or bases to form their salts and then cause them to distribute preferentially into the water layer.

Many multiple extractions can even serve to separate very similar organic compounds when two immiscible solvents which are nearly alike in solubilizing power are utilized. This repeated distribution, called **countercurrent distribution**, may be done 100 to 1,000 times to achieve fine separations in a modern, automatic apparatus. The procedure is theoretically very similar to that of fractional distillation, for each involves equilibration or passage back and forth of the substances being separated between two separate physical phases, vapor and liquid in fractional distillation and the two immiscible solvent phases in countercurrent distribution.

Chromatography (General)

The ability of organic compounds to become adsorbed on a solid surface varies widely both with the structure of the compound and with the character of the surface. Advantage is taken of these differences in adsorption tendencies both for separation and characterization of compounds.

In general the compounds in a mixture are separated by allowing them to pass in a moving solvent phase through a stationary adsorbent phase so that each mixture component is retarded in its passage by its tendency to adsorb. Each component will move at a rate slower than the solvent, and each at a different rate depending on the tenacity of its adsorption to the stationary phase.

The usual operating technique is known as **chromatography**. In liquid-phase chromatography, a solution of the compounds to be separated is poured through a column of finely divided solid adsorbent, such as alumina (Al_2O_3), silica gel, carbon, or cellulose. As the solution passes down the adsorbent column, the dissolved molecules experience a competition of attraction between the moving solvent molecules and the stationary adsorbent solid such that their rate of progress down the column will be slower than the solvent itself and will differ for each component of the mixture according to its characteristic ratio of attraction for solvent and adsorbent phases. The more powerfully a compound is adsorbed, the slower it travels. This ratio is another physical property useful for both identification and separation and is completely analogous to the distribution coefficient in its utility. All forms of chromatography are characterized by this competition between a moving and a stationary phase. Countercurrent distribution is in fact completely comparable, one of the immiscible solvents remaining stationary in the apparatus and the other moving relatively to it.

Chromatographic separations are carried out in two ways after the mixture is introduced at the head of the column. In one, the moving phase (solvent) is continuously added and passed through the column of stationary phase (adsorbent) so that the components of the mixture are sequentially removed (**eluted**) from the column. For separation, solvent fractions coming off from the bottom of the column are separately collected and evaporated to yield the mixture components. For characterization purposes the time required for elution is a physical characteristic of every compound.

In the other procedure the moving phase is stopped when the first component of the mixture nears the end of the stationary phase. The column is then dried and the positions of the several mixture components are observed. In this case it is the distance each substance has traveled through the stationary phase which is its identifiable physical constant; segments of the stationary phase containing the different components may also be removed for purposes of separating the components they contain. The components are then simply washed off the adsorbent surface.

In order to achieve the optimum separation of any set of compounds it is possible to vary the nature of the moving and stationary phases and sometimes the temperature. Chromatography is by far the most versatile and sensitive method of separation in use today. The various kinds of chromatography can be summarized as follows.

Column Chromatography

In this procedure for separations (Fig. 1-1) the mixture is first introduced at the head of the column and then sequentially eluted by passing a series of solvents through the column. Solvents are chosen in order of increasing affinity for the adsorbing surface, so that the tendency of the solvent to displace the adsorbed compounds continually increases. The more weakly adsorbed compounds are eluted first. Under ideal operating conditions, only a small fraction of the free surface area of the adsorbent is covered with molecules of the mixture, so that the latter are repeatedly adsorbed, desorbed, and re-adsorbed, as they pass down the column. The result is a multiple-fractionation process akin to that in distillation or countercurrent distribution. Such a procedure can be adapted to quantities of material ranging from milligrams to kilograms.

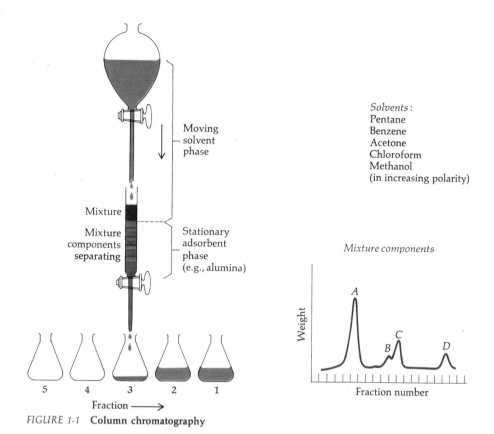

Solvents:
Pentane
Benzene
Acetone
Chloroform
Methanol
(in increasing polarity)

FIGURE 1-1   **Column chromatography**

It is common in organic chemistry to speak of compounds as being graded between polar and nonpolar. The distinction is electrical; charged or ionic compounds such as salts and most inorganic substances are polar; and the nonpolar extreme consists of hydrocarbons such as gasoline and paraffins. This gradation is very inexact and is only very roughly correlated with molecular structure (Chap. 7) but is nonetheless useful in chromatography, for polar compounds are the most strongly adsorbed by most stationary phases and only eluted with polar solvents. The graded series of solvents used for sequential elution (Fig. 1-1) is also applied in order of increasing polarity. However, when carbon is used as the stationary phase the situation is roughly reversed, nonpolar compounds being most adsorbed. When the mixture components are colored they may be seen as bands moving slowly down the column, separating more as they proceed; the term chromatography is in fact derived from the Greek word *chrōmos* for color. With noncolored substances their positions as they move down the column can often be observed with ultraviolet light.

The fractions eluted from the column are collected serially and their contents weighed and examined after solvent evaporation. A typical graph of these weights is shown in Fig. 1-1; the middle peak presumably represents a mixture which will still need rechromatographing, perhaps with other solvents. Automatic fraction collectors are available which move a new collecting tube under the column as each one is filled to a preset volume.

Thin-layer and Paper Chromatography

In these variations the stationary phase is either one of the standard adsorbents, such as alumina or silica, mixed with some plaster and bound as a thin layer onto a glass or plastic sheet, or else simply a strip or sheet of filter paper. For analytical purposes a very small amount (< 1 mg) of the substance is spotted at the bottom of the strip of adsorbent (Fig. 1-2) and the solvent allowed to rise up by capillarity, moving and separating the components. When the solvent front has moved to the top of the strip, this distance from the origin spot is labeled as 1.0. The fractional distance each mixture component has traveled (labeled the R_f **value**) are the identifying physical characteristics of those compounds. The two mixtures shown (rising parallel from two origin spots) have only one compound (R_f 0.21) common to both. The positions of the components can be observed after separation by spraying with a developer which turns color on encountering organic material and shows up each component as a spot partway up the strip. For actual separation of the resolved components the strip may be cut into segments containing the spots and the spots washed from the adsorbent. For purposes of obtaining pure components in usable quantities, preparative thin-layer chromatography allows separation of up to 200 mg of a mixture on an 8 $\times$ 8-in. plate in about half an hour.

> Paper chromatography can be viewed in another way which emphasizes its similarity to countercurrent distribution. Ordinary filter paper, which contains a layer of water adsorbed on its surface, is frequently employed. This water layer forms a stationary phase, and the solute is subjected to innumerable partitions between the water phase and the moving solvent phase as the latter advances. The components of the solute have different solubilities in the two phases and therefore spend different amounts of time in the stationary and moving liquid phases. As a result, each component moves at a different rate along the paper. Thus the process may be viewed as a continuous liquid-liquid extraction rather than a multiple adsorption. The actual physical situation is probably a combination of the two.

Gas or Vapor-phase Chromatography

In gas chromatography the moving phase is a gas; the mixture is vaporized into the column and fractionated by being swept over a solid adsorbent. The surface of the solid is usually coated with a nonvolatile liquid. When the stationary phase is a solid coated with a liquid, the fractionation process consists of multiple partition between liquid and gas phases. An inert gas, such as

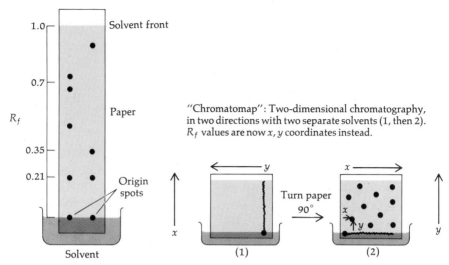

FIGURE 1-2   **Paper and thin-layer chromatography**

helium, nitrogen, or argon, is used as carrier and eluant. The retention time of a particular species depends upon both its vapor pressure and its solubility in the immobile liquid. By varying the nature of the liquid phase, adsorption columns having a wide variety of specialized selectivities can be prepared. For example, polar liquids are most successful in effecting the separation of closely related polar compounds; hydrocarbon greases are especially successful for the separation of hydrocarbons that have very similar boiling points. Use of heated columns and vaporization chambers allows separation by vapor chromatography of compounds that boil at temperatures as high as 400°. The method has most commonly been used for analysis of samples containing 10 mg or less of material. Recently preparative-scale vapor chromatography has been used with increasing frequency for the separation of samples of 1 to 100 g.

A commonly used method for detecting products eluted in a vapor chromatogram is to note the change in thermal conductivity of the eluted gas when a product appears. If helium is used as a carrier, the thermal conductivity falls when some foreign material appears in the gas stream. The output of the thermal conductivity cell is applied to the y axis of a recorder so that the appearance of a product is signified by the appearance of a positive peak on the recorder trace. The apparatus and output graph are shown in Fig. 1-3.

1-5 QUANTITATIVE ELEMENTAL ANALYSIS AND MOLECULAR FORMULAS

When a compound has been purified and its molecular homogeneity assured, the first operation necessary in determining its structure is to know how many atoms of each kind are present in its molecule. This is obtained from the percentage composition of the elements present and leads to an empirical

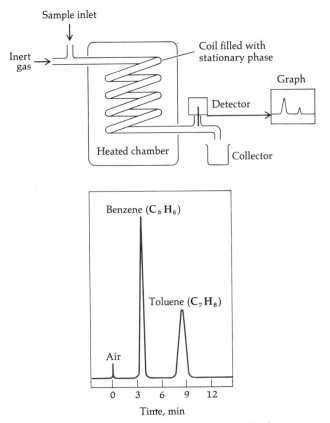

Vapor chromatogram of a mixture of toluene and benzene
on a 6-ft column of Apiezon wax on firebrick ($T = 80°$)

FIGURE 1-3   **Gas- or vapor-phase chromatography**

formula. The elements commonly found in organic compounds, in order of
decreasing frequency of occurrence, are carbon and hydrogen, oxygen, nitrogen,
halogens, sulfur, and phosphorus.

In quantitative combustion analysis, a weighed portion of a compound
is burned in oxygen over hot copper oxide at about 700° (Fig. 1-4). Carbon
dioxide and water are produced (in essentially 100 percent yield) and are
individually captured and weighed in absorption tubes (*A* and *B*). This is
accomplished by first passing the combustion gases through a tube containing
a neutral desiccant (*A*), such as magnesium perchlorate (Anhydrone), which
absorbs the water vapor and converts it to water of hydration; then by leading
them through a tube (*B*) containing finely divided sodium hydroxide (Ascarite),
which absorbs and converts the carbon dioxide to sodium carbonate.

The weight percent of carbon and hydrogen in the unknown compound
can be calculated by the stoichiometric principles of general chemistry. Should

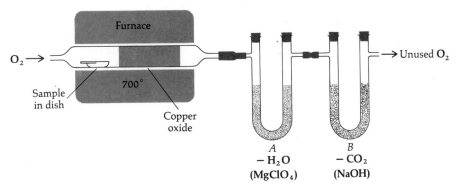

FIGURE 1-4   **Apparatus for combustion analysis**

the sum of these percentages equal approximately 100 percent, no other ele-ments are present in the molecule. If the sum is less than 100 percent and qualitative tests reveal the absence of elements such as nitrogen, sulfur, and halogens, then oxygen is probably present. Frequently the oxygen content is assumed to be the difference between the sum of the percentages of carbon and hydrogen and 100 percent. A better procedure involves the direct determi-nation of oxygen, which is done by a kind of reverse combustion. Since carbon in the sample is isolated by reaction with oxygen, then the oxygen in the sample may be removed by heating to about 1100° with carbon. This forms carbon monoxide, which is assayed by passing the gas into iodine pentoxide and titrat-ing the liberated iodine with thiosulfate.

In order to convert these data into a molecular formula two steps are required: (1) the weight of each element in the sample, and its percentage composition, are calculated and (2) these relative weights in the sample are then converted to a ratio of numbers of atoms by dividing by the relative weight of each atom (atomic weight). The following example will illustrate the pro-cedure with a sample known to contain only carbon, hydrogen, and (perhaps) oxygen; other examples, however, are all done the same way.

1 $\underset{\text{4.337 mg}}{Sample\ wt.}\ \xrightarrow{(+O_2)}\ \underset{\text{10.35 mg}}{wt.\ of\ \mathbf{CO_2}}\ +\ \underset{\text{3.42 mg}}{wt.\ of\ \mathbf{H_2O}}$

Wt. of **C** in sample $= \dfrac{12.01\ \text{(formula wt. of } \mathbf{C})}{44.01\ \text{(formula wt. of } \mathbf{CO_2})} \times 10.35\ (\text{mg } \mathbf{CO_2}) = 2.824\ \text{mg } \mathbf{C}$

Wt. of **H** in sample $= \dfrac{2.018\ \text{(formula wt. of } \mathbf{H_2})}{18.02\ \text{(formula wt. of } \mathbf{H_2O})} \times 3.42\ (\text{mg } \mathbf{H_2O}) = 0.383\ \text{mg } \mathbf{H}$

$\% \ \mathbf{C} = \dfrac{\text{mg } \mathbf{C}}{\text{mg sample}} \times 100\% = \dfrac{2.824}{4.337} \times 100\% = 65.11\% \ \mathbf{C}$

$\% \ \mathbf{H} = \dfrac{\text{mg } \mathbf{H}}{\text{mg sample}} \times 100\% = \dfrac{0.383}{4.337} \times 100\% = 8.83\% \ \mathbf{H}$

Difference $= 100.00\% - 65.11\% - 8.83\% = 26.06\% \ \mathbf{O}$

2 The ratio of numbers of atoms is now obtained by dividing by the atomic weights; the crude ratios (A) so obtained must be converted to whole number ratios first by dividing each crude ratio by the lowest of them so that one number goes to unity (B), then multiplying each by some integer so as to create integers of all the ratio numbers (C).

$$
\begin{array}{cccc}
 & A & B & C \\
\text{C:} & \dfrac{65.11}{12.01} = 5.421 & \div\ 1.628 = 3.34 & \times 3 = 10 \\[2ex]
\text{H:} & \dfrac{8.83}{1.008} = 8.76 & \div\ 1.628 = 5.39 & \times 3 = 16 \\[2ex]
\text{O:} & \dfrac{26.06}{16.00} = 1.628 & \div\ 1.628 = 1 & \times 3 = \ \ 3
\end{array}
$$

The small deviations from whole numbers represent the experimental error, which is commonly acceptable as long as deviations in the percentage composition values are less than ± 0.3.

3 The empirical formula is now written as follows:

$C_{10}H_{16}O_3$

Larger multiples of these numbers of course also satisfy the ratios above so that $C_{20}H_{32}O_6$, $C_{30}H_{48}O_9$, etc., are all equally correct formulas. The general formula determined by combustion analysis is therefore $(C_{10}H_{16}O_3)_n$ where n is an integer. To determine the value of n the approximate molecular weight of the compound must be separately determined.

4 For the formula $C_{10}H_{16}O_3$ the molecular weight and true percentage composition are determined as follows, and compared with the experimental values; in each case the error is less than 0.3 and so deemed acceptable.

$C_{10}H_{16}O_3$:   Mol. wt. $= (10 \times 12.01) + (16 \times 1.008) + (3 \times 16.00) = 184.2$

Experimental

$$\% \ C = \frac{\text{wt. of C}}{\text{mol. wt.}} = \frac{(10 \times 12.01)}{184.2} \times 100\% = 65.20 \qquad 65.11 \qquad (-0.09)$$

$$\% \ H = \frac{\text{wt. of H}}{\text{mol. wt.}} = \frac{16 \times 1.008}{184.2} \times 100\% = \ 8.76 \qquad 8.83 \qquad (+0.07)$$

$$\% \ O = \frac{\text{wt. of O}}{\text{mol. wt.}} = \frac{3 \times 16.00}{184.2} \times 100\% = 26.06 \qquad 26.06 \qquad (\ \ 0.00)$$

The molecular weight of the compound can be most accurately obtained from the mass spectrum of the sample; the method is discussed in Chap. 7.

Other methods, in general use before the advent of mass spectra, include determination of freezing-point lowering (Rast cryoscopic method), vapor density for volatile substances, or variations on Raoult's law such as the Signer isothermal distillation method. These methods will not be discussed.

If a compound contains nitrogen, a Dumas analysis is frequently used to determine the percentage of this element. The unknown substance is mixed with cupric oxide (**CuO**) and brought to a dull-red heat, which results in complete oxidation of the organic material. The resulting gases are passed over a surface of hot copper to reduce nitrogen oxides to nitrogen. The other, more reactive gases are chemically absorbed, and the residual volume of nitrogen is measured at carefully determined temperatures and pressures. The weight of nitrogen can be calculated (ideal-gas law) and is divided by the initial sample weight to give the fraction of nitrogen in the original compound.

Sulfur and halogens are determined by complete destruction of the organic compound with hot fuming nitric acid (Carius method) or sodium peroxide (Parr method). The sulfate or halide ions in the residue are precipitated and weighed as barium sulfate or silver halide.

A striking feature of these elemental analyses is that they can be carried out on a microscale. Not more than 3 to 4 mg of material is required for carbon and hydrogen microanalysis, and not more than 3 to 8 mg is needed for a micro Dumas determination of nitrogen.

1-6   METHOD AND ORGANIZATION OF BOOK

In the history of organic chemistry the first area which had to be mastered was that of molecular structure. The nineteenth century was involved with observing the behavior of the reactive groupings present in organic molecules. For this reason nineteenth-century textbooks were organized as catalogs of these reactive groups (**functional groups**), how they were formed, and how they reacted. In the twentieth century, the mechanisms of reactions began to be studied and chemists began to appreciate broad generalities of reaction class which could be catalogued similarly, but these generalizations make a much simpler catalog. The twentieth century has also witnessed the growth of structural concepts, based on the electrical and spatial nature of molecules, which are far more general than simply their application to particular functional groups.

Nevertheless, textbooks have largely continued to employ a nineteenth-century format, organizing the material in terms of functional groups. The broader structural concepts, such as resonance, tautomerism, stereochemistry, and the universality of reaction mechanisms must then be introduced piecemeal under special functional group headings, often before enough functional groups have been described to appreciate the generality of the concepts.

But theory improves teaching, simplifies learning, and stimulates interest. The underlying general theories, even in their present imperfect state, create a powerful frame for understanding and ordering the vast array of facts in organic chemistry. Practicing research chemists implicitly think in terms of this theoretical frame; it is not only appropriate, but also easier, if students come to the field with the same theoretical guides.

This book does not simply incorporate theory, then; it is completely organized around it. Broad concepts are developed first to orient the student and then their consequences are explored. The book starts with a simple description of atoms and develops logically, step by step, through the consequences of the nature of atoms into all the intricacies of organic chemistry. The student is encouraged to see that we can virtually "invent" most of organic chemistry by following this logical development. Furthermore, care has been exercised in this stepwise presentation not to use examples of ideas before their full discussion. Hence, for example, no names are applied before nomenclature is introduced and virtually no reactions are written until they can be properly presented in context.

The first seven chapters deal broadly with molecular structure. Chap. 2 presents qualitatively the nature of atoms and how they bond together into molecules. In the succeeding chapters the consequences of this behavior are explored: what functional groups are possible (Chap. 3) and a survey of actual compounds (Chap. 4); the electrical (Chap. 5) and spatial (Chap. 6) features of molecules; and finally the physical properties through which we may deduce molecular structure in specific substances (Chap. 7).

The next 15 chapters (8 to 22) deal with molecular dynamics—the reactions of organic molecules—a general discussion of reactions as a consequence of structure (Chaps. 8 and 9). Four chapters (10 to 13) are devoted to reactions of electron-donating molecules (nucleophiles) and the next three (14 to 16) to electron-acceptors (electrophiles). Following this are discussions of other, specialized reaction areas. The remainder of the book deals with several major special topics, including a survey of biochemistry.

Problem solving is an integral part of the presentation. Problems are placed throughout the chapters to help solidify concepts through application to particular cases. More than this, however, a specific attempt has been made to codify and present definite procedures for the solving of the main kinds of chemical problems which daily challenge the practicing chemist. These include protocols for unraveling structure in Chaps. 3 and 7 and for design of synthetic pathways in several chapters, culminating in the discussion in Chap. 23.

PROBLEMS

1-1 A chemist repeated the experimental directions contained in a published account of work in the chemical literature. The final product was stated to be colorless and to melt at 169–170°. On repeating the published experiment, our chemist obtained a yellowish crystalline solid melting at 157–161°. He chromatographed this material and combined several fractions containing colorless solids melting at 165–167°. This substance he recrystallized once to a melting point of 168–169° and twice more without change in melting point. To convince himself that he had in fact prepared the same substance as that published, he took its infrared spectrum, which looked like Fig. 7-5. Account for his procedure and for the melting-point discrepancy. The chemist took the view that an exact identity of his infrared spectrum with that published for the substance is a much more convincing test of the identity of the substances than the melting points. Why?

1-2 When different species of the sage (*Salvia*) family of plants are ground up and distilled, small amounts of odoriferous liquids are obtained. These samples are passed through a gas chromatograph and yield chromatograms such as those shown in Fig. P1-2.

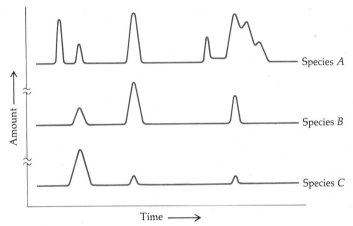

FIGURE P1-2 **Gas chromatograms from** *Salvia* **species**

   a Describe what is happening.
   b What can be deduced about the chemical constituents of the species?
   c When the three liquid samples are heated to a higher temperature and their chromatograms taken again, samples *B* and *C* are unchanged but the first (left-hand) peak in the species *A* trace disappears and the third becomes much bigger. What is happening?

1-3 The identity of two substances is being compared. They both have the same melting point. Compared on a thin-layer chromatography plate using chloroform as a solvent they showed the same R_f value. Using alcohol as a solvent, however, their R_f values were different. Explain and consider what should be done next.

1-4 In order to purify a certain compound the chemist recrystallized it until its melting point was constant. On a thin-layer chromatography plate, however, it showed a major and a minor spot. Explain and consider what he should do next.

1-5 Thin-layer chromatography is relatively quick compared to column chromatography, although only the latter can handle large amounts of material. In separating the components of a mixture by column chromatography it is common to monitor its progress with thin-layer chromatograms of each fraction (taking only a tiny sample of each fraction for this purpose). What is the rationale of this procedure?

1-6 List the materials you can see from where you are sitting that you think might be made of organic compounds.

1-7 a A compound was found to contain 87.8% carbon and 12.2% hydrogen. What is the molecular formula of this substance?

    b A second contained 88.8% carbon and 11.2% hydrogen. What is the formula in this case?

    c What extra factor is involved in the case of solving for 73.1% carbon and 7.4% hydrogen?

1-8 A compound was subjected to combustion analysis. A weight of 3.898 mg of the substance when burned gave 3.172 mg of water and 12.969 mg of carbon dioxide. What is the probable empirical formula of the compound?

1-9 Qualitative analysis of a compound demonstrated the absence of halogens, nitrogen, sulfur, and phosphorus. Combustion analysis of 3.493 mg of the compound gave 2.316 mg of water and 9.394 mg of carbon dioxide. Determine the probable empirical formula of the substance.

1-10 A molecule was known by its mode of synthesis to contain 10 atoms of carbon per molecule, along with unknown numbers of chlorines, hydrogens, and oxygens. Analysis indicated that the substance contained 60.5% carbon, 5.55% hydrogen, 16.1% oxygen, and 17.9% chlorine. Calculate the molecular formula of the substance.

1-11 Compounds were analyzed and their molecular weights determined in the mass spectrometer. The results are given below (* = not determined); solve for the molecular formula in each case.

Compound	Carbon, %	Hydrogen, %	Nitrogen, %	Oxygen, %	Others, %	Mol. wt.
a	52.2	4.38	0.0	*	0.0	184
b	69.5	7.30	*	23.2	*	138
c	40.3	6.04	0.0	0.0	Br = 53.7	149
d	57.8	6.07	16.9	19.3	0.0	166
e	57.1	7.19	16.6	19.0	0.0	*

1-12 The liquid fractions from the distillates of the sage species B or C (Prob. 1-2) were run directly into the mass spectrometer. The three peaks gave values of 136, 138, and 150. Only the third substance contained oxygen and it contained only one oxygen atom per molecule. Other elements were not present (except C and H). What are the molecular formulas of the three substances in these distillates?

1-13 Polyethylene analyzes roughly as 85.6% carbon and 14.4% hydrogen and has a very large molecular weight, variable but in the thousands. How might the formula be written?

1-14 Cholesterol is isolated from gallstones by chromatography and recrystallization to a constant melting point (148.5°). It is known to contain only one oxygen atom besides carbon (83.9%) and hydrogen (12.0%) and no other elements. What can you derive about the formula? Suppose the carbon and hydrogen analyses were each lower by 0.5% in one analysis, will this change the derived formula?

READING REFERENCES

Benfey, O. T., "From Vital Force to Structural Formulas," Houghton Mifflin, Boston, 1964.

Partington, J. R., "A Short History of Chemistry," Macmillan, 1957.

THE CHARACTERISTICS OF CHEMICAL BONDS

2-1 INTRODUCTION

In order to understand the nature and reactivity of organic molecules we shall start with the few pieces from which they are constructed: the atoms and the electron pairs which serve as bonds to hold the atoms together. The point of this chapter will be to focus on the properties of atoms in order to show how they lead naturally and logically to molecule formation and on to an understanding of the nature and reactivity of molecules generally, no matter how complex. We shall see the atoms of our special interest, namely, C, H, O, N, and halogen (**F**, **Cl**, **Br**, **I**), as high-energy unstable particles with a strong urge to fill their outer shells with electrons. Thus they combine or bond together, sharing electrons in pairs between them and so becoming more stable by filling their outer shells while at the same time being attached or bonded to each other by the electron pairs.

  The nature of the bonds formed also derives from the atoms; there are only a few kinds of bonds and the simple rules governing their formation then lead us directly to the *shapes* of the resulting molecules. The atoms involved also determine the electrical nature of the bonds and hence of the molecule. Atomic nuclei are positively charged; electrons are negatively charged. When unlike atoms are bonded to each other, charge imbalances develop in the bonds, leading to positive and negative sites in molecules and hence to their reactivity, since chemical reactions are largely initiated by the attraction of positive for negative charges. In this chapter, therefore, we shall derive the nature of the chemical bond from the qualities of atoms and lay down in consequence several broad concepts about the shapes and the electrical nature of molecules. The exploration of these concepts will then form the rest of the book in which they direct the development of the details of organic chemistry.

2-2 ATOMIC STRUCTURE

Organic chemistry deals with millions of compounds, potentially an infinite number. The molecules of which these compounds are composed are generally constructed from only a few kinds of atoms, viz., hydrogen, carbon, oxygen,

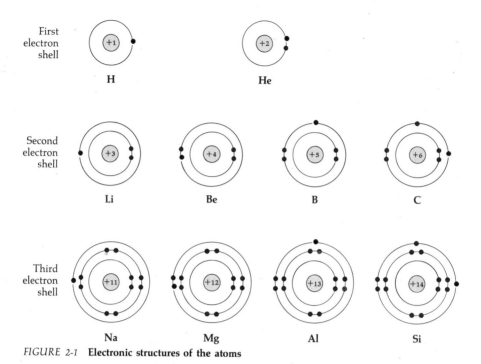

First
electron
shell

H   He

Second
electron
shell

Li   Be   B   C

Third
electron
shell

Na   Mg   Al   Si

FIGURE 2-1 **Electronic structures of the atoms**

nitrogen, and halogens, so that we must look further for the source of this
molecular variety. The first thing to explore is the nature of the atoms themselves.

Although the theory of atomic structure is usually presented in courses
in general chemistry, a brief review is included here. All atoms contain positively charged nuclei, which carry nearly all the atomic mass. The nuclei are
composed of protons and neutrons, each with a mass of 1 on the atomic-weight
scale. The sum of the mass (and also the number) of protons and neutrons
is the atomic weight of the element, while the atomic number is simply the
number of protons alone and hence is the total (positive) nuclear charge. The
lighter and more elusive nuclear particles of atomic physics (mesons, neutrinos,
etc.) do not concern us. Enough electrons (negative charge) are located around
the nucleus to make the whole atom neutral. Since the atomic number gives
the number of positive charges in the nucleus, it is also the number of electrons
in the neutral atom. Hydrogen has one electron, helium two, lithium three, and
so on with increasing atomic numbers of the elements. Differences in the
behavior of the electron "atmospheres" are responsible for the variations in
chemical properties of the elements.

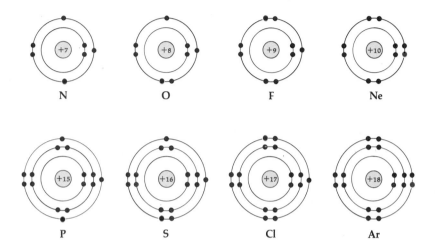

N   O   F   Ne

P   S   Cl   Ar

Both theory and experiment indicate that the electronic structures of the rare gases (helium, neon, etc.) are especially stable; these atoms are said to contain "filled shells." Much of the chemistry of the other elements is understandable in terms of their tendencies to attain "filled-shell" conditions by gaining, losing, or sharing electrons. The first shell holds only 2 electrons; the second holds 8; the third can ultimately hold 18, but a very stable configuration is reached when a shell of 8 is filled (argon structure). Thus hydrogen requires only 2 electrons to fill its outer shell, while the other atoms of organic chemistry require 8. The electronic structures of the atoms of the first 10 elements are schematically represented in Fig. 2-1.

2-3   IONIC AND COVALENT BONDS

The common compounds of inorganic chemistry are generally characterized by **ionic bonding**. This comes about through the gain or loss of one or more electrons by an atom to gain a stable, full outer electron shell. The atom therefore becomes a positively or negatively charged particle called an ion, and two atoms must undergo complementary loss and gain of electrons so that the collection of ions remains electrically neutral overall.

The chemistry of fluorine, which has seven electrons in the second shell, illustrates this. Addition of one more electron gives the very stable negative fluoride ion, which has the same electron configuration as neon. Therefore, reaction of a fluorine atom with a lithium atom involves transfer of an electron from lithium to fluorine. The lithium becomes a positive ion (**cation**) with the helium configuration of electrons, while the fluorine becomes a negative ion (**anion**) with the neon configuration of electrons.

$$\text{Li}\cdot \quad + \quad \cdot\ddot{\underset{\cdot\cdot}{\text{F}}}\text{:} \quad \longrightarrow \quad \text{Li}^{+} \quad + \quad \text{:}\ddot{\underset{\cdot\cdot}{\text{F}}}\text{:}^{-}$$

Lithium ion   Fluoride ion
(helium   (neon
structure)   structure)

Crystals of lithium fluoride and other salts contain ions. Each anion has as nearest neighbors six or more cations and vice versa. No pairs of neighbors can be singled out and called molecules. If lithium fluoride is dissolved in water, free ions are formed.

In this way many ionic species can be constructed with the aid of the periodic table to show how many electrons are lacking from or in excess of a full outer shell. Since sodium and chlorine are in the same columns of the table as lithium and fluorine, respectively, they exhibit identical outer shells and form the same kinds of ionic compounds, i.e., **LiF, NaF, LiCl, NaCl**. Similarly, we can form **BeF$_2$, MgCl$_2$, AlCl$_3$, Na$_2$O**, etc., utilizing elements that gain or lose more than one electron. However, even with more than 100 elements in the periodic table, the number of such ionic compounds is severely limited and cannot account for millions of compounds. Furthermore, a second problem arises with elements in the middle of a row of the period table in that they must take on a sizable charge concentration to acquire a filled shell, and this is a destabilizing circumstance. In order to fill its outer shell in this ionic fashion the carbon atom must either gain or lose four electrons, becoming C^{4+} or C^{4-}, both very unstable ions.

PROBLEM 2-1

Diagram the involved ions in the compounds **BeF$_2$, MgCl$_2$, AlCl$_3$, Na$_2$O, Li$_2$S**.

To counter this difficulty such atoms can form instead a **covalent bond**. The bond itself is composed of two electrons, each one contributed by one of the two bonding atoms. The atoms are then held closely together (bonded) by the electron pair which they share between them such that each atom experiences the full pair as a part of its own outer shell. The covalent bond may be characterized as a mutual deception in which each atom, though contributing only one electron to the bond, finds itself with two electrons contributed to its effort to fill its outer shell. Furthermore, since the electrons are not removed from the atoms, no net charge is generated and no attendant destabilization occurs. This kind of bonding is very widespread throughout the periodic table, especially among the atoms of concern to organic chemistry. Even hydrogen and the halogens, which can easily form ionic bonds by transfer of only one electron, also commonly form covalent bonds. For this reason organic chemistry has often been roughly distinguished as the *chemistry of the covalent bond.*

The simplest molecule is that of hydrogen (H_2) in which two hydrogen atoms lock together and share their two electrons in a covalent bond so that each atom in the molecule simulates a helium configuration, with a full outer shell (two electrons). Similarly, two fluorine atoms share a pair of electrons, giving each a filled shell, in the molecule of fluorine (F_2).

$$H\cdot \quad + \quad H\cdot \quad \longrightarrow \quad H:H$$

$$:\!\overset{\cdot\cdot}{F}\!\cdot \quad + \quad \cdot\overset{\cdot\cdot}{F}\!: \quad \longrightarrow \quad :\!\overset{\cdot\cdot}{F}\!:\!\overset{\cdot\cdot}{F}\!:$$

The central element of organic chemistry is carbon; its atom has four electrons in its outer shell and requires eight to become stable. If it shares one of its electrons with another atom, it now has formed one bond and finds itself counting five electrons in its outer shell. If it engages a second of its original electrons in another such bond, the carbon atom will now have six electrons in its outer shell. If it forms four such bonds by sharing all four electrons, it reaches the stable, full outer shell of eight. In this way it may bond to other carbon atoms, or to hydrogen or fluorine as in the example below in which atoms are successively added to share electrons with the unfilled carbons until the outer shells of all the atoms are filled.

This forms a stable covalent molecule in which each pair of constituent atoms is tightly held together by the electron pair between the atoms. Unlike ionic bonding the covalent bond occurs between two specific atoms, which are not free to separate† as ions are, so that the entire molecule is a rigorously connected unit. The bonding may be simply understood as the attraction of the two positive atomic nuclei for the negative electron pair between them. This simple physical law of *the attraction of unlike charges and the repulsion of like charges* is the most basic force in chemistry and is invoked often to explain chemical phenomena.

---

†Of course covalent bonds can be broken when submitted to the right conditions. This bond breaking is the essence of a chemical reaction, but when a bond is broken, the molecule is changed and the compound on which the reaction is carried out is transformed into another compound.

It should be clear from the operation in the example that the number of electrons needed to fill the outer shell of an atom is equal to the number of covalent bonds it will form. Thus carbon needs four electrons and forms four bonds, while hydrogen and fluorine each needs one electron and forms only one bond. Examination of the periodic table or Fig. 2-1 shows, therefore, that oxygen should similarly form two covalent bonds ·and nitrogen three. Following the same procedure we may bond these atoms into molecules also, for example, adding them to the unfinished fragment (A) instead of fluorine. In each case we may then finish filling the outer shells of the oxygen or nitrogen atoms by creating more covalent bonds, most simply with hydrogen.

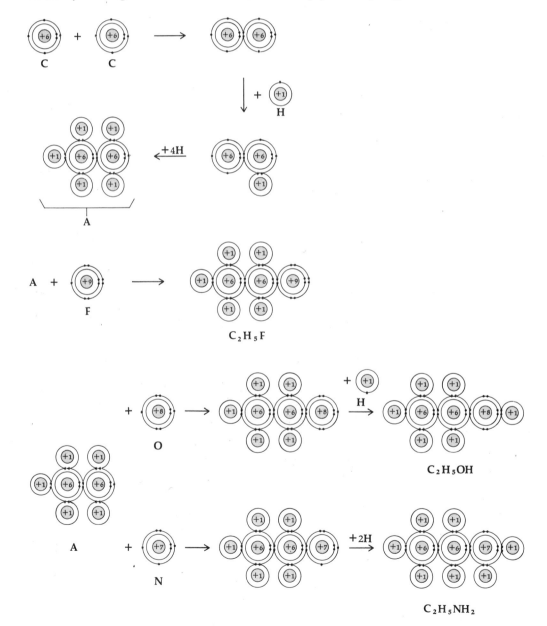

In these examples a carbon-carbon bond is depicted. Clearly, we could have used more carbon atoms instead of hydrogens or fluorine etc. to fill the existing carbon outer-shell vacancies, and the new carbons in their turn would have to be satisfied as well. It is possible to build up long chains of carbon atoms covalently linked in this way, and since carbon must form four covalent bonds, branching chains or even rings of carbon atoms may be constructed also. It is this unique ability of carbon atoms to bond to themselves in long chains which makes possible the essentially limitless number of compounds which contain carbon.

The uniqueness of carbon in forming long chains may seem surprising at first sight, since silicon, which lies below carbon in the periodic table, has the same outer-shell configuration. It does not form stable chains, however. The C—C bond is over 50 percent stronger than the Si—Si bond (83 vs. 53 kcal/mole), and the latter is weak enough to be easily broken. This is in turn caused by the larger size of the silicon atom so that the Si—Si bond must be longer and the electrostatic force holding the atoms together correspondingly weaker. We might also construct a chain of atoms more nearly the same size as carbon by using nitrogen atoms but such a chain would have on each nitrogen one unshared electron pair and the electrostatic repulsion of these adjacent centers of like charge tends to destabilize that N—N bond (39 kcal/mole). Compounds with more than two or three linked nitrogen atoms like

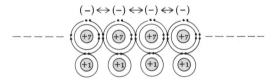

are rare and unstable as are those with more than two linked oxygens, while there is no theoretical limit on the number of carbon atoms which may be linked into a chain; polyethylene contains chains of thousands of carbon atoms in length!

It is also apparent that there is a major role in these compounds for hydrogen in filling up the extra or terminal bonding positions of each carbon; the carbon chains are thus sheathed in bonded hydrogen atoms. In recent years extensive study has been devoted to fluorine-sheathed carbon chains as well, long chains of this kind forming the molecules of teflon.

PROBLEM 2-2

Show the electronic structures of the neutral covalent molecules HCl, H_2S, H_2O, NH_3, CH_4O, CH_5N, H_3NO, HOCl.

2-4 STRUCTURAL FORMULAS AND ISOMERS

The atomic representations used in the previous section to depict molecules are exceedingly cumbersome to construct for any but the simplest molecules, but fortunately they are unnecessary for our purpose and we can simplify them in the several steps of Fig. 2-2. It is only the outer-shell electrons which are involved in bonding and therefore in the nature and behavior of molecules. The nucleus and the electrons of the inner, filled shells may be grouped together and simply represented by the atomic symbol. This inner grouping is called the **kernel** and possesses a net positive charge equal to the number of outer-shell electrons in the neutral atom. For elements other than transition elements this number is also simply the column of the periodic table in which the element is to be found. As an example, the nitrogen kernel includes the nucleus with charge $+7$ and the inner filled shell of two electrons (charge -2); the kernel charge is $+5$. With hydrogen the kernel is simply a proton ($+1$); oxygen is $+6$; halogens are $+7$.

With the kernels represented by the atomic symbol and the outer-shell electrons by dots, in pairs, we evolve the simpler structural formulas of line 2 in Fig. 2-2. Since it is clearer to separate shared and unshared electron pairs, the former (the actual covalent bonds) are written as straight lines connecting the two atomic symbols, as physically the pair of electrons in the bond—which the line represents—connects the two atomic kernels. This is shown in line 3. Finally, to simplify the drawing further, line 4 indicates the atoms (other than hydrogen) of the molecular skeleton bonded together, and each written with the number of its associated hydrogens to the right. Finally (line 5), for typographical convenience we may usually eliminate the bond lines in a linear arrangement without loss of clarity. Lines 4 and 5 represent the conventions most widely in use.

In order to maintain these simple rules of covalent bonding, it is often necessary for one atom to form more than one bond to another. In nitrogen gas (N_2), the two atoms are joined by a triple bond, i.e., $:N:::N:$ or $:N\equiv N:$ and the compound CH_2O can only be $\ddot{O}=CH_2$, with a double bond. Finally it is important to observe that the lines in a structural formula are only notations reporting which atoms are attached to which others. They convey nothing about the geometry of the physical molecule and, conversely, they may be written in any way which does not alter the *sequence* of bonding atoms:

$$
\begin{array}{c}
\text{H} \\
| \\
\text{H}-\text{C}-\text{H} \\
| \quad \ddot{} \\
\text{H}-\text{C}-\ddot{\text{O}}: \\
| \quad | \\
\text{H} \quad \text{H}
\end{array}
\qquad \text{is the same as} \qquad
\begin{array}{c}
\text{H} \quad \text{H} \\
| \quad | \\
\text{H}-\text{C}-\text{C}-\ddot{\text{O}}-\text{H} \\
| \quad | \\
\text{H} \quad \text{H}
\end{array}
$$

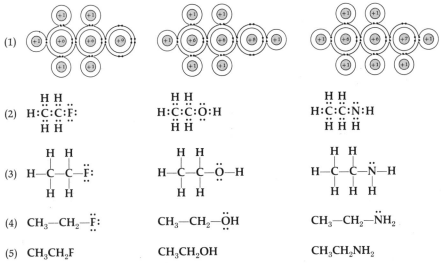

FIGURE 2-2 **Conventions for symbolizing molecular structure**

It should now be clear that we are free to use this simple electron-sharing procedure to create any number of covalent molecules of all sizes, using only the criterion that every atom shall fill its outer shell. If we try this, it soon becomes apparent that more than one structural formula may be written for a given molecular formula. Reexamination of the cases previously built up shows that although CH_3—CH_2—$\ddot{\underset{\cdot\cdot}{F}}$: is the only structural formula for C_2H_5F, there are two structures possible for C_2H_6O,

$$H-\underset{\underset{H}{|}}{\overset{\overset{H}{|}}{C}}-\underset{\underset{H}{|}}{\overset{\overset{H}{|}}{C}}-\ddot{\underset{\cdot\cdot}{O}}-H \quad \text{and} \quad H-\underset{\underset{H}{|}}{\overset{\overset{H}{|}}{C}}-\ddot{\underset{\cdot\cdot}{O}}-\underset{\underset{H}{|}}{\overset{\overset{H}{|}}{C}}-H$$

The two structures represent two quite different molecules and hence two physically distinguishable substances at the level of actual observation, the one at the left being a liquid (ethyl alcohol), the other a gas (methyl ether) at ordinary temperatures. In general, compounds that possess the same molecular formula but different structures are referred to as **isomers**, particularly as **structural isomers**, and the structures as being **isomeric**. The number of possible structural isomers increases very rapidly as the number of atoms increases, and may be calculated. You may check your comprehension by drawing out the structures in the left-hand column and confirming the stated number of isomers.

Number of isomers:

CH_4	1	C_7H_{16}	9
C_2H_6	1	C_8H_{18}	18
C_3H_8	1	C_9H_{20}	35
C_4H_{10}	2	$C_{10}H_{22}$	75
C_5H_{12}	3	$C_{20}H_{42}$	366, 319
C_6H_{14}	5	$C_{40}H_{82}$	62, 491, 178, 805, 831

All 75 of the isomeric compounds from CH_4 through C_9H_{20} have been individually characterized, and many of the larger molecules on the list are also known (up to $C_{120}H_{242}$), while polyethylene consists of a mixture of far larger molecules with thousands of carbons in each molecule.

One of the firm predictions of organic chemistry that has never failed to prove true is that the actual number of isomeric compounds observed is exactly the same as the predicted number of molecular isomers. This is part of the confirmation, at our macroscopic level of observation, that the structural theory of molecules is correct.

Since even the last notation on Fig. 2-2 becomes burdensome when describing molecules of over five to ten carbon atoms in size, a final, more streamlined notation (the bond-line convention) is one in which only the skeleton is shown and bonds to hydrogen are not drawn but are only assumed present as required to make up full bonding at each atom. Each line is then a skeletal bond and is assumed to have carbons at each end unless another element is written in (such as F, O, N; the hydrogens bonded to O and N are written). In order to distinguish one carbon-carbon bond from the next, bond lines are placed at about 120° (hexagon angle), roughly resembling the true physical bond angle.†

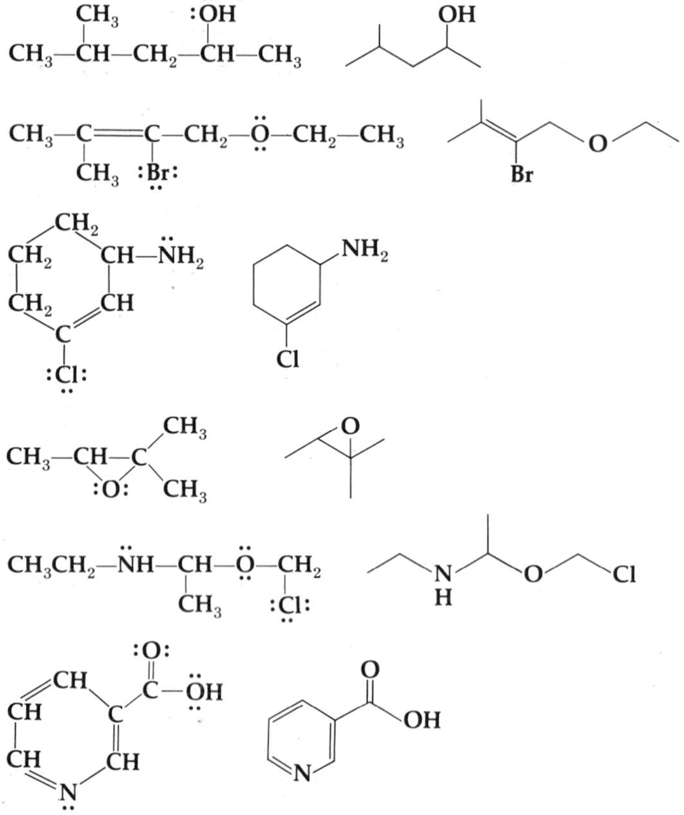

---

† Placement of the unshared pairs (on O, N, halogen) is left to discretion, but it is understood that they are always there as required by the atom.

The end of a single line, otherwise unadorned, is therefore implicitly $-CH_3$ and the end of a double line $-CH_2$, a triple line $-CH$. It is not regarded as an affront to consistency (since the meaning remains clear) if these are written in, e.g.,

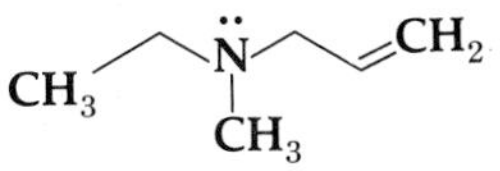

PROBLEM 2-3

Write structures for all the possible isomers of the following, using the convention of line 4 in Fig. 2-2.

a C_2H_6S d C_5H_{12} g C_4H_9Cl j C_3H_6O

b C_2H_7P e C_4H_8O h $C_5H_{11}F$ k $C_3H_4F_2$

c C_3H_9N f $C_3H_8O_2$ i C_3H_9NO l C_5H_8O

PROBLEM 2-4

Rewrite the following structures in the "bond-line" convention.

a $CH_3CHBrCH_2CH_2CH=O$

d

b

e

c $CH_3C\equiv CCH-CH_2-C(CH_3)_3$ (with NH_2 substituent)

f

PROBLEM 2-5

Rewrite the following "bond-line" structures in the notation convention of Fig. 2-2, line 4.

a

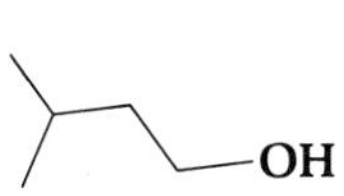

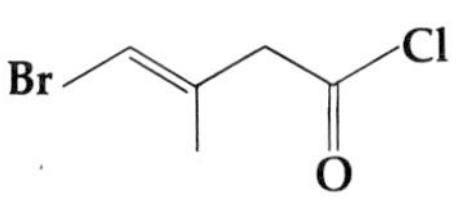

b

c

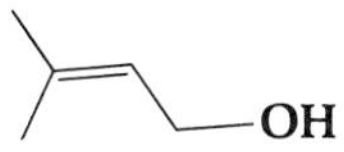

d

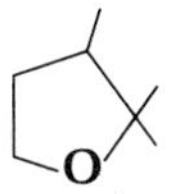

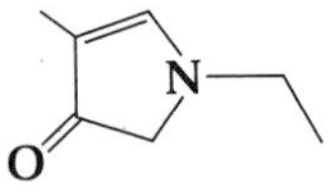

2-5 COVALENT IONS AND ISOELECTRONIC STRUCTURES

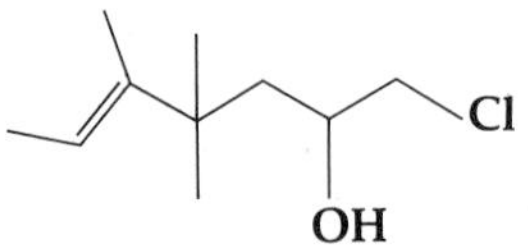

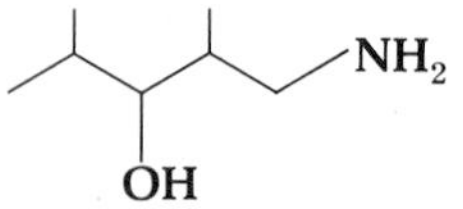

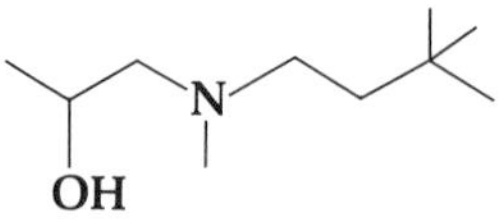

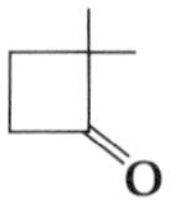

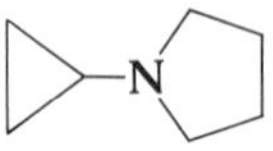

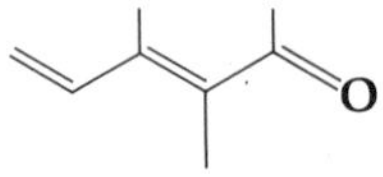

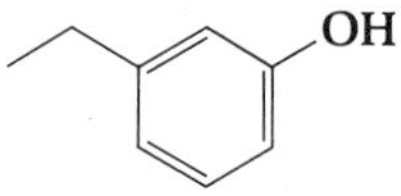

All the covalent molecules considered so far have been electrically neutral, but this need not be the case. Many simple inorganic ions contain covalent bonds and the entire covalent assemblage is then a charged species, or ion. An example is the carbonate ion, as in calcium carbonate;

$$Ca^{++} CO_3^{--} \quad \text{equivalent to} \quad Ca^{++} \quad \overset{..}{:}\overset{-}{\underset{..}{O}} \overset{\overset{:O:}{\|}}{-C-} \overset{-}{\underset{..}{O}}\overset{..}{:}$$

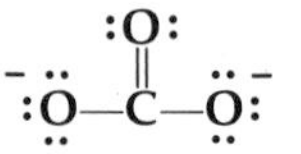

in this as in all cases of ions which are large covalent bound units the ion must be associated with a counter ion (or ions) sufficient to balance the charge. The word **molecule** is usually reserved for neutral covalent species and **ion** for charged species, either monatomic (Ca^{++}) or polyatomic with internal covalent bonding (CO_3^{--}), positive ions being **cations**, negative ions being **anions**.

We can account for the charge on the carbonate ion (-2) by adding the kernel charges of the involved atoms ($6 + 6 + 6 + 4 = +22$) and subtracting the total outer-shell electrons (8 unshared pairs + 4 bonds = 12 electron pairs $= -24$). In order to ascertain with which atoms the two charges are specifically associated, we subtract from the kernel charge of each atom the number of electrons associated with it: two for each of its unshared pairs and only one of each of its covalent bonds (shared pairs). For purposes of this accounting it is unimportant which atom originally contributed the electrons in any bond.

In the present instance we find the carbon with the oxygen doubly bonded to it to be neutral in this accounting and the two oxygens $-\ddot{O}:$ to be -1 each $[+6 -(3 \times 2) -1 = -1]$. It is no surprise that the oxygen with only one single bond should be associated with the charge since the normal, neutral oxygen has two single (or one double) bonds, $-\ddot{O}-$ or $=\ddot{O}$, wherein the accounting is neutral $[+6 -(2 \times 2) - 1 - 1 = 0]$.

In a similar vein, the inorganic compound ammonia ($:NH_3$) is neutral while the ammonium cation ($:NH_4^+ +5 - 4 = +1$) is a charged ion and must be associated with an anion (such as $:\ddot{C}l:^-$) in any compound. However, there is no line of theoretical distinction between inorganic and organic chemistry as the following neutral and ionic compounds show; they are formally derived merely by replacing hydrogens in the inorganic (covalent) compounds with carbons. The same accounting shows the positive charge always associated with nitrogen, as written below.

$$\text{CH}_3-\text{CH}_2-\overset{\cdot\cdot}{\text{N}}-\text{H} \quad\quad \text{CH}_3-\text{CH}_2-\overset{\overset{\displaystyle \text{H}}{|+}}{\text{N}}-\text{H} \quad\quad :\overset{\cdot\cdot}{\underset{\cdot\cdot}{\text{F}}}:^-$$
$$\phantom{\text{CH}_3-\text{CH}_2-}\underset{\text{H}}{|} \quad\quad\quad\quad\quad\quad \phantom{\text{CH}_3-\text{CH}_2-}\underset{\text{H}}{|}$$

$$\text{CH}_3-\text{CH}_2-\overset{\cdot\cdot}{\text{N}}-\text{CH}_2-\text{CH}_3 \quad\quad \text{CH}_3-\text{CH}_2-\overset{\overset{\displaystyle \text{H}}{|+}}{\text{N}}-\text{CH}_2-\text{CH}_3 \quad\quad :\overset{\cdot\cdot}{\text{Br}}:^-$$
$$\phantom{\text{CH}_3-\text{CH}_2-}\underset{\text{H}}{|} \quad\quad\quad\quad\quad\quad\quad\quad \phantom{\text{CH}_3-\text{CH}_2-}\underset{\text{H}}{|}$$

$$\text{CH}_3-\text{CH}_2-\text{CH}_2-\overset{\cdot\cdot}{\text{N}}-\text{CH}_3 \quad\quad \text{CH}_3-\text{CH}_2-\text{CH}_2-\overset{\overset{\displaystyle \text{CH}_3}{|+}}{\text{N}}-\text{CH}_3 \quad\quad :\overset{\cdot\cdot}{\text{I}}:^-$$
$$\phantom{\text{CH}_3-\text{CH}_2-\text{CH}_2-}\underset{\text{CH}_3}{|} \quad\quad\quad\quad\quad\quad\quad \phantom{\text{CH}_3-\text{CH}_2-\text{CH}_2-}\underset{\text{CH}_3}{|}$$

Thus it is clear that $-\ddot{O}:$ will be $-$ and $-\overset{|}{\underset{|}{N}}-$ $+$ in any molecular context in which they appear, and we can easily create and list all the common charged-atom situations of this kind. Furthermore, the singly bonded oxygen anion will be recognized as identical to a singly bonded fluorine ($-F:$) in every atomic respect except that the latter atom has one more proton (and some uninvolved neutrons) and hence one more kernel charge in its nucleus. Hence $-\ddot{F}:$ is neutral while $-\ddot{O}:$ is negative. Bonded atoms, like these, which differ only in nuclear charge, are labeled **isoelectronic**, since their outer-shell electron appearance is identical. (Contrast this with atoms differing only in number of *neutrons*, not charges, which are labeled **isotopic**.) Hence, charged $-\overset{|}{\underset{|}{N}}^{+}$ is isoelectronic with neutral $-\overset{|}{\underset{|}{C}}-$.

It is now a simple matter to construct a chart of the possible neutral and singly charged covalent bonding situations, and this is shown in Fig. 2-3 with the periodic table columns placed vertically and the isoelectronic situations of different atoms seen horizontally. Names are attached to some ions for

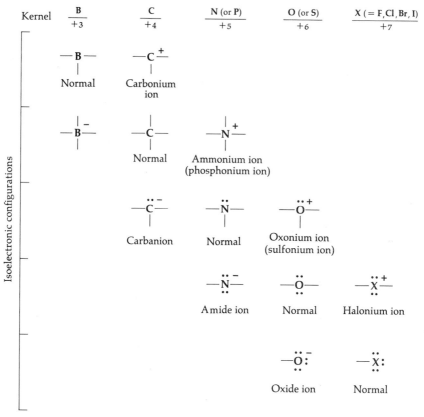

FIGURE 2-3   **Isoelectronic bonding situations**

convenience; the positive ions are all labeled as **-onium ions**. All but the first row possess full outer shells of electrons, while the top have only six electrons in the outer shell and may be expected to be rather unstable and hence reactive (and reactive in the same way: to gain a fourth electron pair). These situations will be recognized as the parts from which all covalent molecules and ions are constructed. They are not altered in any way if the bonds are multiple instead of single bonds.

PROBLEM 2-6

Compose tables comparable to Fig. 2-3 for doubly bonded and triply bonded atoms.

PROBLEM 2-7

Write out structures for the following covalent ions and molecules, showing the bonds and the nonbonded electrons as well as the atoms bearing charges. Write the structures so that all atoms have a full outer shell of electrons (i.e., eight electrons, except two for hydrogen). No structure possesses an oxygen-oxygen bond.

a NO_2^- e ClO_4^- i I_3^-

b HCO_3^- f CO j BH_4^-

c NO_3^- g $(CH_3)_3NO$ k BO_3^{3-}

d PO_4^{3-} h H_3NBH_3 l $AlCl_4^-$

PROBLEM 2-8

Replace the colored atom in each of the following structures with another atom (from the same row of the periodic table) so as to form an isoelectronic structure. Write out the structure and show any charges and all unshared electron pairs.

a $(CH_3)_3CH$ c BF_4^- (two answers)   e $(C_2H_5)_3\ddot{N}$ (two answers)

b $C_2H_5\ddot{\underset{\cdot\cdot}{O}}:^-$ d $(CH_3)_2\ddot{\underset{\cdot\cdot}{O}}$ f $CH_2{=}\dot{\overset{\cdot\cdot}{N}}CH_3$ (two answers)

2-6 ATOMIC ORBITALS

Structural formulas such as those presented in the previous sections, with atomic symbols and bond lines, were in use in the nineteenth century, though without any sense of the electrical nature of the atoms and bonds. The constancy of four bonds to carbon was first predicated simultaneously by Couper in Scotland and Kekulé in Germany in 1859 and their first structural formulas are reproduced here.

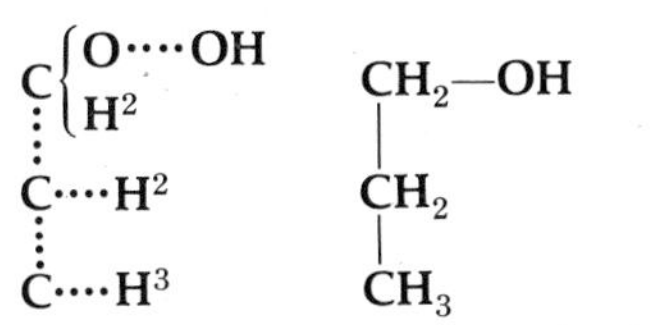

Couper†   Modern equivalent†

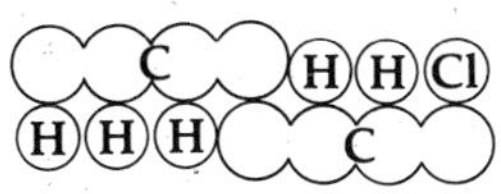

Kekulé

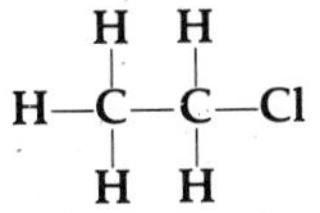

Modern equivalent

As remarked in the first chapter, it is surprising now that these chemists did not recognize that their structures must represent a physical, three-dimensional molecule, but 15 years elapsed before the first projection of structures in space was attempted (correctly) by van't Hoff and, again simultaneously, LeBel. In

---

† Oxygen was always doubled in early formulas owing to an error in understanding atomic weights.

order for us to derive not only the spatial qualities of molecules but also their more detailed electrical nature we must first take a closer look at the atom and its modern physical description.

The first modern theory of atomic structure was proposed by Bohr in 1913: a positive nucleus with a limited number of orbits in which electrons could circulate around the nucleus like planets around the sun, held in place by coulombic attraction opposing centrifugal force. This planetary theory remains the common view of the atom as a simple sphere, but it is not the correct view. The shape of the atom is not spherical but somewhat more complex. The importance of the Bohr atom lay in its recognition that electron orbits were limited in number, or quantized, each one being characterized by a specific energy (and radius) and each given a quantum number, $n = 1, 2, 3, \ldots$, increasing with the energy and distance from the nucleus. In the hydrogen atom the one electron occupied the lowest, $n = 1$, orbit and could be promoted to the next $(n = 2)$ orbit only by absorption of a discrete package of energy ΔE, known as a **quantum**. Radiational energy could achieve this by possessing a frequency ν, such that $\Delta E = h\nu$, where h is Planck's constant.

Electrons as Waves

Bohr's atom explained many but not all of the features of the atomic spectrum of hydrogen despite his introduction of quantum theory. However, in 1924, de Broglie made the important suggestion that the motion of electrons might have the character of wave motion.† This hypothesis was verified experimentally by passing electron beams through thin films of crystals and observing that the beam underwent diffraction in the same way as a beam of light passed through a narrow slit. In other words, shadows cast by electron beams do not have sharp edges, and electrons could exhibit properties of waves as well as particles.

Electron waves may be viewed as analogous to the standing waves in a string held in the hand at one end and tied to a tree at the other. The wave has motion such that it periodically reaches a maximum in one direction, a minimum in the other, with a node, of amplitude zero, halfway between. The sign of the wave is positive on the maximum side of the node and changes to negative on passing through the node. The standing wave in the string shown here moves only in one plane, however, while the electron wave must be in three dimensions.

† The suggestion was in part inspired by the discovery that light, which had long been described as waves, had some of the properties of particles in motion. For example, absorption of light by an electron appears to involve transfer of momentum to the electron; therefore, the photon, or particle of light, behaves as though it has mass.

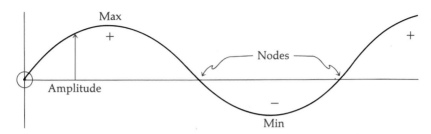

The amplitude of the electron wave is called the wave function (Ψ) and is of course a function of three spatial coordinates (x,y,z).

Almost immediately wave equations were formulated to describe the behavior of electrons in atoms and molecules. Today it is quite conventional to speak and think of electrons in atoms and molecules as though they were minute standing waves of electricity. The new mechanics became known as wave mechanics and is the basis of all modern structural theory. The first successes of the theory were exciting; virtually all the arbitrary postulates of the Bohr quantum theory became axiomatic as soon as the hypothesis of the wave character of the electron was introduced. However, it also soon became apparent that only the very simplest chemical systems could be calculated exactly by wave mechanics. Furthermore, a complete solution to a wave equation for a moderately complex molecule, even if obtainable, would be too complicated to be either tabulated by available methods or comprehended by ordinary human minds. Consequently, myriad methods for finding approximate solutions to wave equations for complex molecules have been sought. The concepts and language usually used in qualitative discussions of molecular structure derive from these methods.

The Hydrogen Atom

In both classical and wave mechanics, the hydrogen atom is treated as a heavy nucleus that stands still and an electron that moves. Hence it is a relatively simple two-body problem. The electron has kinetic energy because it is in motion and potential energy because of the electrostatic interaction between the two particles. The motion of the electron is thought of as a standing wave. A differential equation formulating this concept can be written and solved. A series of solutions (roots), called **wave functions**, are obtained. These wave functions (Ψ) contain the coordinates (x,y,z) of the electron (with the nucleus taken as the origin of the coordinate system). The wave functions are called **orbitals**, by analogy to the orbits, which are solutions to the classical equations of planetary motion. Whereas an orbit is the equation of a line, an orbital is a function that has a finite value everywhere in space; however, the value of the function becomes very, very small at distances greater than a few angstroms (1 Å $= 10^{-8}$ cm) from the nucleus.

The value of the *square* of the wave function at any point is assumed to be proportional to the *probability* that the electron will be there. For example, if we choose two small volume elements of the same size but at different distances from the nucleus and calculate the value of the square of the wave function in those volume elements, a result such as the following might be obtained.

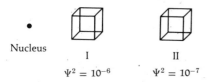

Nucleus   I   II
         $\Psi^2 = 10^{-6}$ $\Psi^2 = 10^{-7}$

Ψ = the wave function (orbital) of the electron

The results would show that the probability of finding the electron in volume element I is 10 times as great as the probability that the electron will be found in volume element II. Since the integral of Ψ^2 over all space is unity, the probability that the electron will be found in II at any arbitrarily chosen time is 10^{-7} to 1. If such calculations were carried out for a large number of volume elements, the results could be plotted by putting different numbers of dots in each element. The resulting figure would look like a cloud of varying density. Values of Ψ^2 become so small a few angstroms from the nucleus that the calculations would surely be terminated because of the trivial value of the results. The resulting figure might look something like the following:

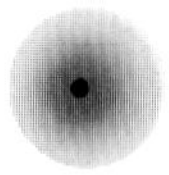

Solutions of the wave equation for the hydrogen atom can be obtained only for certain values of the total energy. The characteristics of the various orbitals provide the basis for discussion of bonding in molecules. The lowest-lying wave function in energy is spherically symmetrical as shown above; the value of the function depends only on distance from the nucleus and not at all on direction. A sphere with a radius of 1.7 Å represents a probability of 0.95 (that is, 19 to 1) of finding the electron within that spherical volume around the nucleus. When we draw an orbital we imply a volume within which there is a high probability, $\Psi^2 = 0.90$ to 0.95, of finding the electron within it. For this most stable (lowest-energy) hydrogen orbital we draw a circle to imply the cross section of a probability surface which is spherical.

Shapes and Energies of Atomic Orbitals

There are only a few characteristic atomic orbitals and only three of these concern us here: the *s*, *p*, and *d* orbitals. The *s* orbital is always a simple sphere

such as the one described above. The higher-energy p orbitals and d orbitals
have the shapes shown, the former with two lobes symmetrically placed on
an axis through the nucleus, the latter with four lobes on two perpendicular
axes intersecting at the nucleus. The number of nodes in the wave functions
are: s = no node, p = 1 node, d = 2 nodes.

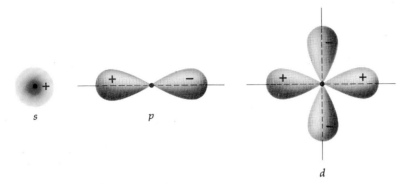

The amplitude signs of the wave functions (Ψ) are shown on the lobes and,
of course, the signs change as one passes through a node. The shapes are in
reality not those of the wave function Ψ, but really of Ψ^2, the probability of
electron presence (0.90 to 0.95) within the indicated space. The signs of Ψ, then,
have no meaning when discussing Ψ^2, which must always be positive, but
these signs will have value later in the formation of molecular orbitals from
atomic orbitals.

There are some more complex, higher atomic orbitals which we shall not
use. In fact basic organic chemistry develops completely from just the s and
p atomic orbitals, since these are the only ones utilized by first-row elements!
The real shapes of atoms are now best pictured as the superposition of these
several atomic orbitals at the nucleus, with electrons in rapid motion in each
one. The number of orbitals occupied depends on the number of electrons
in the atom as set down below.

The atoms of the periodic table may be constructed in sequence by the
successive addition of electrons, one at a time, to a table (Fig. 2-4) of the
possible atomic orbitals arranged in increasing energy. The breakdown of
atomic orbitals is very symmetrical: as one increases in energy each electron
shell contains the same kind of orbitals as the last with a new set added and
the number of orbitals in each set increases regularly.

Shell number	Number of orbitals			
	s	p	d	f
1	1			
2	1	3		
3	1	3	5	
4	1	3	5	7

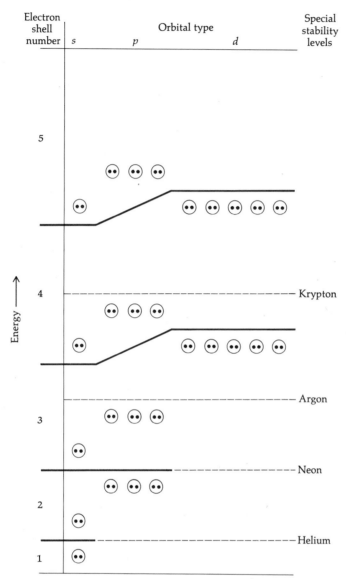

FIGURE 2-4   **Energy levels of atomic orbitals**

The rules for building up the atoms by successive placement of electrons in these orbitals are simple:

1 The next atom is created by addition of one electron in the lowest-energy available orbital. (It is implicit that another proton and some neutrons are also added to the nucleus.)

2 Maximum occupancy of an orbital is two electrons (**Pauli exclusion principle**).

3 Only one electron is placed in each orbital of equal energy (**degenerate** orbitals) until all are full before adding a second in each (**Hund's rule**).

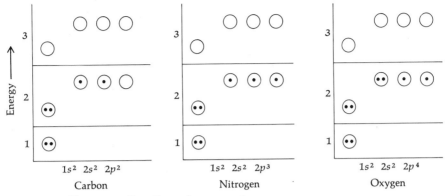

FIGURE 2-5   **Electron configurations of common atoms**

The Pauli exclusion principle also states that the electrons must be "paired" in the sense that they have opposite spins.† We noted previously that all stable molecules and ions had all their electrons in pairs and it is the exclusion principle which gives this physical meaning and implies stability for such pairs.

Hydrogen has one electron, in the 1s orbital (s orbital of shell number 1), and helium has two (paired) electrons in that orbital, thus filling the first shell and achieving a specially stable configuration. Much higher in energy is the (otherwise similar) 2s orbital, in which the lithium atom will have one electron that it readily gives up, forming the lithium ion and so returning to the more stable helium configuration. To pass from lithium to beryllium we place two electrons in the 2s orbital and so on through the periodic table. The configurations of some common organic atoms are shown in Fig. 2-5. A simple shorthand notation of electron configuration is also shown in Fig. 2-5, a simple listing of the filled orbitals with the number of electrons in each written as superscripts.

Wave equations for polyelectronic atoms can be formulated but cannot be solved in exact form, so the exact wave functions are inaccessible. Such an exact wave function would include coordinates of all the electrons with reference to each other as well as to the nucleus. Thus any calculation of the probability of finding an electron in some small volume element would have to take into account the distribution of all other electrons in the atom. Since electrons repel each other strongly,‡ the interaction effects are very large, and details of an exact wave equation for an atom containing a dozen or so electrons are terrifying to contemplate. Approximate methods must be used. It is usually assumed

---

† The electron, regarded as a charged particle, is spinning and like any rotating charge generates a magnetic field in a right-handed direction along the axis of spin. When electrons are paired, their fields point in opposite (antiparallel) directions, often symbolized ↑↓.

‡ This situation is quite different from the problem of calculating orbits of planets in the solar system since the gravitational attraction of planets for each other is always very small compared with the gravitational field of the sun.

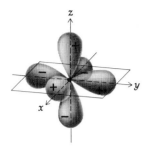

Three mutually
perpendicular
p orbitals

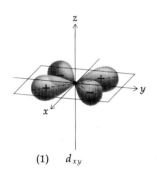

(1) d_{xy}

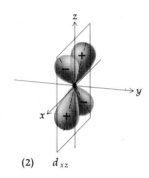

(2) d_{xz}

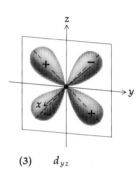

(3) d_{yz}

The five d orbitals

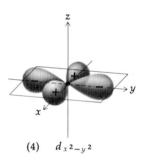

(4) $d_{x^2-y^2}$

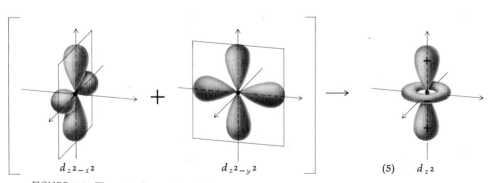

$d_{z^2-x^2}$ $d_{z^2-y^2}$ (5) d_{z^2}

FIGURE 2-6   The sets of p and d orbitals

that each of the electrons in a polyelectronic atom can be described by a hydrogenlike orbital. Approximate corrections to the orbital, and the energy of the individual electron, are then made on the basis of some kind of an averaged distribution of all the other electrons.

PROBLEM 2-9

Write out the electron configurations of the following atoms with both the diagram and shorthand notation of Fig. 2-5; **Be, B, F, Na, P, S, Si.**

There are three equivalent p orbitals, all with the same shape, but they may not occupy the same space. Hence they are placed mutually perpendicular so that they lie along the three coordinate axes (Fig. 2-6). With the shapes of d orbitals we might orient only three along the coordinate axes, and then place three at 45°, or between the three axes, for a total of six. However, the progression of the table (page 43) requires only five d orbitals, so that two of these six possibilities are fused into one, as shown in the tabulation of the d orbitals in Fig. 2-6. It is fortunate that most of organic chemistry does not involve d orbitals since visualization of all five oriented about a single nucleus is quite difficult!

2-7   MOLECULAR ORBITALS

Having seen the appearance of atoms in isolation, we must examine their behavior as they come together, share their electrons in covalent bonds, and form molecules. In the simplest example, let two hydrogen atoms approach each other from infinity. Each nucleus will increasingly attract the other's electron, lowering the energy of the whole system, but the repulsion of the two similarly charged nuclei for each other will predominate if they get too close and then the energy will increase again very rapidly. As a result an optimum or equilibrium internuclear distance, of lowest energy, results and this is the bonding distance. This behavior is general for any two objects possessing both attraction and repulsion and is summarized in a Morse curve, as in Fig. 2-7. It demonstrates the common experience that atoms are more stable in molecules than alone. Hydrogen atoms, produced by input of high energy, have only a transient existence before recombining to H_2.

In the bonding process the two **atomic orbitals (AO)** overlap and are fused into a **molecular orbital (MO)** surrounding both nuclei and containing both electrons (paired). This is pictured in Fig. 2-7 as is the common energy-level diagram for the fusion of the two atomic orbitals of hydrogen atoms, **A** and **B**, into the molecular orbital **AB** of the H_2 molecule; the dots on the orbital energy levels represent electrons.

Mathematically, molecular orbitals (MO) may be obtained by an approximation called the linear combination of atomic orbitals (LCAO). This combination must

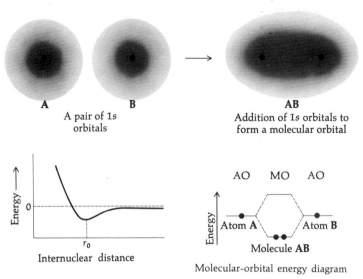

FIGURE 2-7   **Molecular orbital of the H_2 molecule**

produce as many molecular orbitals (Φ) as the number of atomic orbitals (Ψ) originally, in this case two:

$$\Phi_b = N_1(\Psi_A + \Psi_B)$$

$$\Phi_a = N_2(\Psi_A - \Psi_B)$$

The first is the lower-level orbital shown in Fig. 2-7 and is called the bonding orbital (Φ_b); the second is the higher-energy molecular orbital on the diagram in Fig. 2-7 and is called the antibonding orbital (Φ_a). The formation of the two orbitals may be symbolized:

Bonding:

Ψ_A Ψ_B Φ_b

Antibonding:

Ψ_A Ψ_B Φ_a

The antibonding molecular orbital has a node midway between the nuclei and is strongly repelling and of high energy. The electrons available are now added to the molecular orbitals, just as they were to make atoms with atomic orbitals, starting to fill the lowest-energy levels first and using two electrons per orbital as before. In the hydrogen molecule, with two electrons only, the high-energy antibonding orbital need not be used, but if two helium atoms came together similarly, both molecular orbitals would be required to accommodate the four electrons in a He_2 molecule. With both antibonding and bonding orbitals equally filled, the molecule is, in sum, slightly less stable than the atoms (cf. Fig. 2-2). Helium, of course, exists in nature as single atoms and He_2 has never been observed.

Complete solution of wave equations for molecules is, in principle, even more difficult than obtaining exact wave functions for polyelectronic atoms. An exact molecular wave function would contain coordinates of all the electrons and of the nuclei and would include terms sensitive to the interactions of each particle with all others. Such a molecular wave function is unknown, and it is doubtful that anyone could read intelligently the solution for even a simple molecule if it were given to him. A number of methods have been developed for obtaining approximate molecular wave functions (or molecular orbitals). The name "molecular-orbital method" has been attached to a common procedure, which emphasizes the formulation of some molecular orbitals that cover *many* nuclei in a molecule.

The method involves the following steps:

1   Construct a series of molecular orbitals as linear combinations of atomic orbitals and producing the same number of MOs as there were original AOs.
2   Arrange the orbitals in order of increasing energy. This procedure may involve some sort of calculation, or it may be entirely intuitive.
3   Assign the electrons, in pairs with paired spins, to the molecular orbitals, beginning with the orbital of lowest energy and working upward.

The principles of the method are illustrated by the treatment of the hydrogen molecule here and are quite analogous to the buildup of atoms with atomic orbitals as described previously.

2-8   HYBRIDIZATION AND COVALENT BONDS

In coming together to form a molecular orbital, two atomic orbitals (or orbital lobes) *of the same sign* must overlap and fuse. *Maximum overlap affords the strongest bond.* The two s orbitals of hydrogens do not overlap to a major extent (Fig. 2-7) since more overlap would bring the nuclei too close. Because of their greater extension from the nucleus and strong directionality, p orbitals can overlap more effectively.

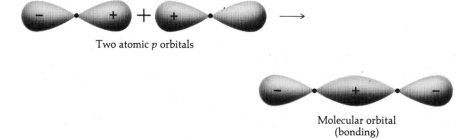

Two atomic *p* orbitals

Molecular orbital
(bonding)

However, even here the uninvolved ($-$) lobes cannot overlap and they represent half of the original orbitals. Pauling originally showed that, if (as in carbon) all the *s* and *p* orbitals in a shell are required for bonding, they may first be combined into **hybrid atomic orbitals** better adapted for the maximum overlap required by bonding. The new set of our atomic orbitals can be produced by taking appropriate combinations of the *s* and *p* orbitals. The new set of orbitals are designated as sp^3, indicating that they are made by mixing one *s* and three *p* orbitals.

An individual sp^3 orbital has the following shape:

— Orbital axis

Nearly all of the orbital is concentrated on one side of the nucleus, although there is a small "tail" which extends in the opposite direction. This distribution of the orbital allows it to overlap more efficiently than either *s* or *p* orbitals with an orbital from another atom placed on or near the orbital axis on the "heavy" ($+$) side. Such overlap gives rise to chemical bonding. The following drawing illustrates bond formation by end-on overlap of two sp^3 orbitals centered at different atoms.

sp^3 atomic orbitals merging

Van't Hoff postulated that the four single bonds to carbon had an arrangement in space such that the four were all at equal angles to each other. This creates a regular tetrahedron, shown in perspective at bottom right in Fig. 2-8. The bond angles can be derived from the geometry of the regular tetrahedron and are found to be ($\cos^{-1}\frac{1}{3}$), or $109°28'$ ($\sim109.5°$). Since we regard these bonds as mutually repelling negative electron pairs, it is reasonable that each should be as far from the others as possible, and this leads to the same geometrical conclusion of tetrahedral angles. This sp^3 hybridization is therefore called **tetrahedral**. In Fig. 2-8 the angle between any two bonds is $109.5°$. Two

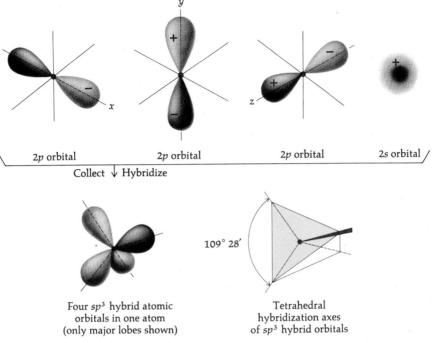

2p orbital   2p orbital   2p orbital   2s orbital

Collect ↓ Hybridize

109° 28′

Four sp³ hybrid atomic
orbitals in one atom
(only major lobes shown)

Tetrahedral
hybridization axes
of sp³ hybrid orbitals

FIGURE 2-8   **Hybridization of atomic orbitals**

bonds are shown in the plane of the page; the other two lie in a plane perpendicular to it (shown in perspective), extending above and below the plane of the page. A single axis bisects both pairs of bonds and lies in both planes.

The tetrahedral arrangement of four separate atoms bound to one carbon atom is universal.† The bonds in such compounds are described by saying that four bonding molecular orbitals are formed by combining each of the four carbon sp³ orbitals with one orbital from the attached atom. The resulting bond (molecular) orbitals are rather like the bonding molecular orbital in the hydrogen molecule. The bond orbitals will be symmetrical about the tetrahedral axes. Orbitals that have this axial symmetry are called **sigma** (σ) **orbitals**. The bonds are called **sigma bonds**.

Molecular σ orbital formed
by carbon sp³ and hydrogen s orbitals

Note that the sigma bonds of **CH₄** are formed by overlapping sp³ orbitals of carbon with s orbitals of hydrogen. Consequently, the molecular orbital resembles an s orbital closely in the region behind the hydrogen nuclei. Figure

†Bond lengths vary somewhat with the nature of the attached substituents, but the tetrahedral bond angle remains nearly constant.

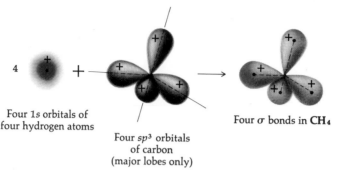

Four 1s orbitals of
four hydrogen atoms

Four sp^3 orbitals
of carbon
(major lobes only)

Four σ bonds in **CH₄**

FIGURE 2-9   **Formation of molecular orbitals of CH₄**

2-9 shows schematically the sigma bond orbitals. The eight bonding electrons are assigned in pairs to these orbitals to give four sigma bonds.

Multiple Bonds

The four sp^3 hybrid atomic orbitals are only useful for bonding to four separate atoms. If bonding is required to only three atoms then only three of the orbitals, one s and two p's are hybridized to form three sp^2 hybrid orbitals. This leaves one p orbital remaining unhybridized. This state of hybridization is called **trigonal**. The simplest example is carbonium ion (Fig. 2-3), where the three bonds are formed from the sp^2 orbitals and the residual p orbital is empty. The sp^2 orbital is essentially the same in appearance as the sp^3 orbital but with not quite the disparity in lobe size and hence not quite the overlap efficiency. The axes of the three sp^2 orbitals are oriented, all in one plane, at mutual angles of 120° (the internal angle of a regular hexagon). The three sigma bonds formed by these three orbitals, therefore, also result in the three attached atoms on carbon being located in one plane with it and at 120° angles to each other. The remaining p orbital is thus perpendicular to this plane with its lobes equally above and below that plane; the plane of the atoms is also the node plane for the remaining p orbital.

The last, or **digonal**, state of hybridization is developed for bonding to only two other atoms. The s and only one p orbital are hybridized to form two sp orbitals, leaving two p orbitals unhybridized. The angle between their axes is now 180°, which is to say that they lie in opposite directions on a straight line as shown in Fig. 2-10. The two remaining p orbitals remain mutually perpendicular and also perpendicular to the line of the hybrid axes so that

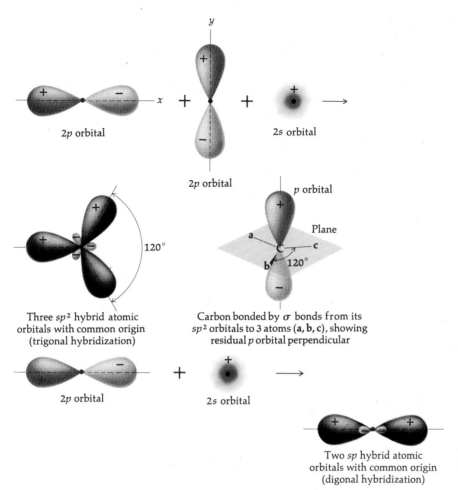

2p orbital

2p orbital

2s orbital

2p orbital

p orbital

Plane

120°

Three sp² hybrid atomic
orbitals with common origin
(trigonal hybridization)

Carbon bonded by σ bonds from its
sp² orbitals to 3 atoms (a, b, c), showing
residual p orbital perpendicular

2p orbital

2s orbital

Two sp hybrid atomic
orbitals with common origin
(digonal hybridization)

FIGURE 2-10   **Hybridization of orbitals of carbon**

a carbon bonded to two atoms (**a**, **b**) by sigma bonds, formed from the *sp*
orbitals, would show its residual *p* orbitals in this fashion:

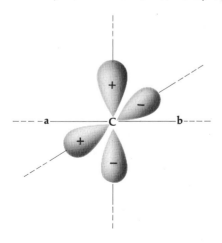

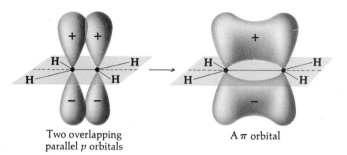

| Two overlapping | A π orbital |
| parallel p orbitals | |

FIGURE 2-11   The carbon-carbon double bond (in C_2H_4)

The most common fate of the residual p orbitals emerges when we examine the bonding of two carbons, both with sp^2 hybridization, to form a carbon-carbon double bond. Let the two carbons form a sigma bond symmetrical about the axis linking their nuclei and the four remaining sp^2 orbitals of the two be bonded to anything else, e.g., four hydrogens, for convenience. *All six atoms must now lie in the same plane.*

The four C—H bonds and one C—C bond account for ten of the bonding electrons necessary for C_2H_4. Two are left for completion of a carbon-carbon double bond. In addition to the three hybridized (sp^2) orbitals, there is the unhybridized p orbital at each carbon atom. These p orbitals are symmetrical about axes perpendicular to the molecular plane. The two parallel p orbitals overlap extensively sideways. Therefore, if added together, the individual p orbitals form a new wave function, considerably different from the original functions. The molecular orbital represented by the new function is usually called a **pi (π) orbital**. The orbital does not have an axis of symmetry and changes sign on passing through the molecular plane. Figure 2-11 shows schematically the combination of two parallel p orbitals (atomic orbitals) to form a π molecular orbital.

The qualitative molecular-orbital procedure for building up the C_2H_4 molecule would be phrased more in this way. We are given four hydrogen atoms with an s atomic orbital and one electron each, and two carbon atoms with an s and three p atomic orbitals hybridized first to three sp^2 atomic hybrids and a p orbital, each carbon having four electrons. Total electrons to place is then 12. We cause an sp^2 atomic orbital on each carbon to overlap, forming a σ molecular orbital, and the remaining four sp^2 atomic orbitals to form four σ molecular orbitals with the hydrogen s orbitals. The two p orbitals (atomic) now overlap sideways to form a π molecular orbital. This generates six bonding molecular orbitals, five σ, and one π, and implicitly six more high-energy antibonding molecular orbitals (five σ and one π). Antibonding orbitals are signified with a superscript asterisk, cf. σ^* and π^*. (Total original atomic orbitals = 12; total molecular orbitals must also be 12). The 12 electrons are now spin-paired and placed in the six lowest (bonding) molecular orbitals. No antibonding orbitals need be used and so the molecule is stable.

As the $2(sp^2 + p)_{AO} \rightarrow (\sigma + \pi)_{MO}$ situation characterizes double bonds, then the $2(sp + 2p)_{AO} \rightarrow (\sigma + 2\pi)_{MO}$ similarly characterizes triple bonds. As the double bond creates a fully coplanar array of six atoms, the triple bond creates a linear orientation of four. The two π orbitals in a triple bond are mutually perpendicular and also perpendicular to the line of the four bonded atoms.

In Fig. 2-12, hybridization of atomic orbitals and formation of molecular orbitals by combination of atomic orbitals centered at different nuclei are shown in color. The p orbitals are shown in grey and s orbitals in red. Since hybrid orbitals are made by mixing s and p orbitals, they are shown in duotone; that is, a combination of grey and red. Molecular orbitals follow the same scheme: those which are a linear combination of two sp^2 hybrids are shown in the same shade as the component atomic orbitals. Similarly, orbitals formed by combination of an s orbital (i.e., from hydrogen) with an sp^2 orbital (from carbon) are also shown in duotone.

In atoms of the second full row of the periodic table (P, S, Cl) the $3d$ orbitals are near enough in energy to the $3p$ orbitals to become involved in some bonding capability, thus changing their chemistry somewhat with respect to their "normal" analogs in the row above (N, O, F). This is discussed in Chap. 19 and referred to in several other chapters. The result is that outer-shell occupancy by more than eight electrons is possible, as implied in these traditional formulas for inorganic ions:

Phosphate ion   Sulfate ion   Perchlorate ion

However, these may be accounted for in a manner consistent with our previous discussion for first-row elements by showing only four bonds (and eight outer-shell electrons) to the second-row atom, in the normal way, but understanding that the four attached oxide anions all (equivalently) donate some of their negative charge back into the positive central atom (into its d orbitals):

All of these ions are isoelectronic, and tetrahedral in geometry.

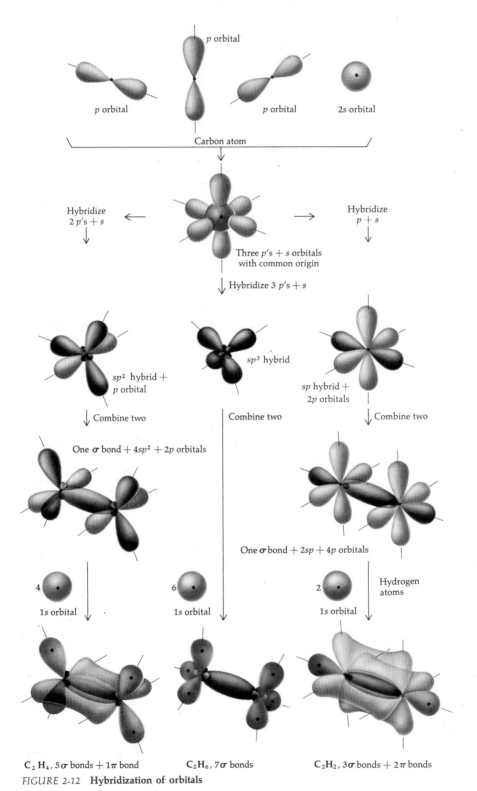

C₂H₄, 5σ bonds + 1π bond   **C₂H₆, 7σ bonds**   **C₂H₂, 3σ bonds + 2π bonds**

FIGURE 2-12 **Hybridization of orbitals**

Rotation Around Bonds

The symmetry of molecular orbitals will on several occasions be an important concept, the first consequence of which shows up here. The σ bond is formed by coaxial, end-on overlap of two orbitals and is symmetrical around its axis, i.e., the line linking the bonded nuclei. If we could grasp the three hydrogens at one end of the CH_3—CH_3 molecule in one hand and the other three in the other hand, we would see that rotation of one CH_3— with respect to the other cannot do any violence to the symmetrical C—C σ bond. Hence we should expect **free rotation** to be possible around sigma bonds and indeed that is the common experience.

The π bond, however, is formed by sideways overlap and is symmetrical by reflection in its molecular plane but certainly not symmetrical with respect to rotation around the internuclear axis like the σ bond. Hence, a similar attempt to rotate one CH_2— with respect to the other in CH_2=CH_2 must break the π bond as one of its constituent p orbitals rotates away from the other. Their parallelism, needed for overlap, is destroyed by rotation. Since the bond formation is accompanied by loss of energy, energy must be put in to effect rotation of a π bond. Since overlap is completely destroyed by a 90° rotation, the full amount of the π-bond energy (see below) must be input to accomplish this. Therefore the ordinary double bond is, in effect, rigid and cannot rotate.

PROBLEM 2-10

Make drawings of the following:

a   A p orbital.

b   An s orbital.

c   Orbitals resulting from hybridization of one p and one s, two p's and one s, and three p's and one s.

d   An orbital resulting from combination of sp^3 atomic orbitals centered at two adjacent atoms.

e   An orbital resulting from combination of two parallel and overlapping p orbitals (two adjacent atoms).

f   An orbital resulting from combination of one sp^3 and one s atomic orbital (two adjacent atoms).

g   The σ orbitals for butadiene.

h   The σ orbitals for benzene.

2-9   BOND LENGTHS AND BOND ANGLES

The bond lengths (internuclear distances) of representative bonds are tabulated in Table 2-1. The generalities about the lengths of bonds are few and simple.

1   The bond length of each kind of bond varies very little from one particular compound to another.

2   Single bonds of first-row elements (**C, N, O, F**) to hydrogen are all about 1 Å.

3   Single bonds between first-row atoms are all about 1.5 Å.

4   Double and triple bonds are shorter: 1.2 to 1.3 Å in first-row elements.

5   Second-row, and higher, atoms (**S, P, Cl,** etc.) form correspondingly longer bonds.

     It is no surprise that bond lengths of first-row atoms are similar since these atoms are all very much alike and use the same orbitals in bonding, as we have seen. The lengths quoted in Table 2-1 for single bonds are for σ bonds composed of sp^3 hybrid orbitals on each atom (except **H**); σ bonds with sp or sp^2 orbitals are slightly shorter, rarely by more than 0.1 Å. Larger atoms (second row and below) must have longer internuclear distances to accommodate their extra, filled inner electron shells. In general the stronger the bond the shorter it is (or vice versa), since greater bonding force (nucleus-electron attractions) presses the nuclei closer against their internuclear repulsion. This accounts for the shorter length of multiple bonds in general.

     Bond angles at carbon are predicted by hybridization and are generally found experimentally to be close to the predictions of 109.5° for four attached atoms (sp^3), 120° for three (sp^2), and 180° for two (sp). When all attached atoms (and bond types) at carbon are identical as in CCl_4, the bond angles are exactly as predicted, but when they differ there is minor variation, as the examples in Fig. 2-13 illustrate. In fact the several methods for determining these angles experimentally do not always exactly agree. Still, it is not unreasonable that this variation should exist in cases of unsymmetrical substitution if we realize that the hybridization at carbon is ultimately only a device for the molecule to attain an optimum configuration energetically. It is possible for the hybridization, or orbital mixing, at carbon to occur in an asymmetric fashion (more s mixed with p in some orbitals than in others) in order best to accommodate different substituents. In the second example in Fig. 2-13 ($CH_3CH_2CH_3$) the central carbon is hybridized so as to form orbitals with about 27 percent s character in the C—C bonds and 23 percent in the C—H bonds instead of 25

TABLE 2-1   **Normal Bond Lengths (in angstroms)**

H—C	1.09 Å	C=C	1.35 Å	C≡C	1.20 Å
H—N	1.00	C=N	1.30	C≡N	1.16
H—O	0.96	C=O	1.22		
C—C	1.54				
C—N	1.47				
C—O	1.43				
C—Cl	1.76				
C—Br	1.94				
C—I	2.14				

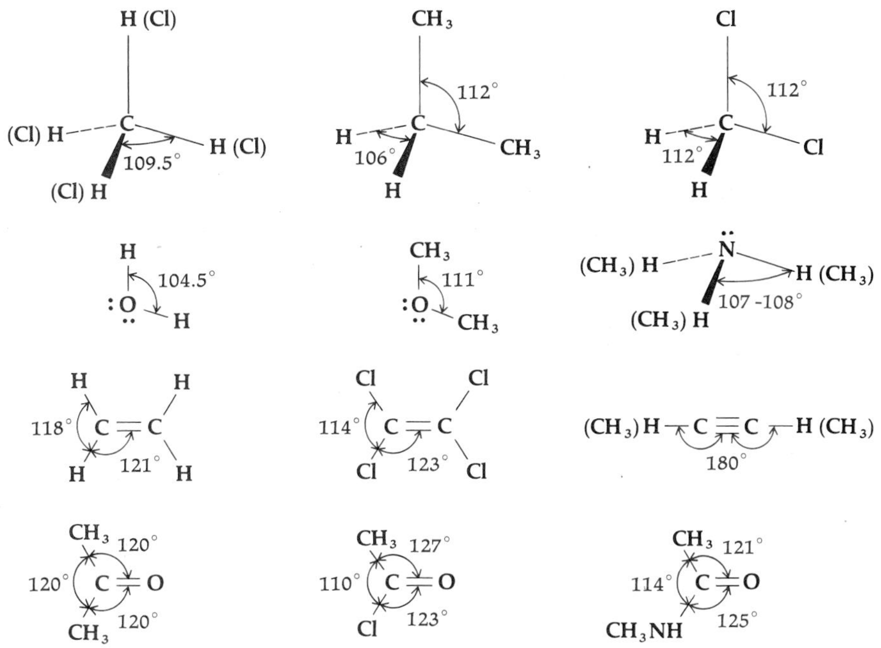

H (Cl)

(Cl) H———C———H (Cl)
      109.5°
   (Cl) H

CH_3
   112°
H———C———CH_3
  106°
   H

Cl
   112°
H———C———Cl
  112°
   H

H
   104.5°
:O
 ·· H

CH_3
   111°
:O
 ·· CH_3

(CH_3) H———N———H (CH_3)
           107–108°
(CH_3) H

H   H
118° C═C
H 121° H

Cl   Cl
114° C═C
Cl 123° Cl

(CH_3) H—C≡C—H (CH_3)
         180°

CH_3 120°
120° C═O
CH_3 120°

CH_3 127°
110° C═O
Cl 123°

CH_3 121°
114° C═O
$CH_3 NH$ 125°

FIGURE 2-13 **Bond angles in selected molecules**

percent in all four as in normal sp^3 hybrids. The **CCC** angle in most compounds containing **C—CH$_2$—C** is 112°, not 109.5°.

Atoms with unshared pairs of electrons apparently follow the same hybridization pattern as asymmetrically substituted carbon, the angles at C—Ö—C and C—N̈—C being close to tetrahedral. This implies that an unshared

pair of electrons occupies a hybrid orbital, roughly sp^3 in these cases, and is like a bonded substituent. Thus the nitrogen atom is roughly tetrahedral like carbon with four substituents, *not* planar ($3sp^2 + 1p$) with its electron pair in a perpendicular p orbital. Where no angle is measurable, as in covalent

halogen (—C̈—Ẍ:) or doubly bonded oxygen (C═Ö), it is not possible to

determine the hybridization experimentally. In the molecular-orbital treatment of C═Ö, however, the oxygen is generally best handled as unhybridized (σ bond and π bond from perpendicular p orbitals, leaving s and p orbitals for the remaining pairs—see Sec. 7-4).

Hybridization and Geometry

The hybrid orbital used for forming a σ bond or holding an unshared electron pair may be characterized as a hybrid of $(s + \lambda^2 p)$ where λ = the mixing coefficient. (λ^2 is called the hybridization index.)

$$\% \text{ of } s \text{ character in each orbital} = \frac{100}{1 + \lambda^2} \qquad\qquad 1$$

Since the total % s character must equal 100% (one original s orbital):

$$\sum_n \frac{100}{1 + \lambda_n^2} = 100 \qquad \text{where } n = \text{no. of orbitals}$$
$$\text{formed (up to four)} \qquad\qquad 2$$

For equivalent substituents:

$n = 4$ orbitals:	sp^3:	$\lambda = \sqrt{3}$	or	25% s character/orbital
$n = 3$ orbitals:	sp^2:	$\lambda = \sqrt{2}$	or	$33\frac{1}{3}\%$ s character
$n = 2$ orbitals:	sp:	$\lambda = 1$	or	50% s character

The angle between two hybrid orbitals, 1 and 2 (with λ_1 and λ_2), is θ_{12}:

$$1 + \lambda_1 \lambda_2 \cos \theta_{12} = 0 \qquad\qquad 3$$

For the angle (θ) between two equivalent orbitals of coefficient λ:

$$1 + \lambda^2 \cos \theta = 0$$

sp^3:	$\cos \theta = -\frac{1}{3}$	or	$\theta = 109.5°$
sp^2:	$\cos \theta = -\frac{1}{2}$	or	$\theta = 120°$
sp:	$\cos \theta = -1$	or	$\theta = 180°$

With C—CH_2—C, the observed CCC angle is 112° ($\cos 112° = -0.375$):

$$\lambda_{CC}^2 = \frac{1}{0.375} = 2.7 \quad \text{and} \quad \% s \text{ character (C—C bond)} = \frac{100}{1 + 2.7} = 27\%$$

This is not exactly an sp^3 ($= sp^{3.0}$) orbital but rather an $sp^{2.7}$ hybrid for the C—C bond. The HCH angle can then be calculated:

$$2\,\frac{100}{1 + \lambda_{CH}^2} + 2\,\frac{100}{1 + 2.7} = 100 \quad \text{and} \quad \lambda_{CH}^2 = 3.35$$

This is then an $sp^{3.3}$ hybrid for the C—H bond.

$$\% s \text{ character/ C—H bond} = \frac{100}{1 + 3.35} = 23\%$$

(Total % s character $= 2 \times 23 + 2 \times 27 = 100\%$)

HCH angle: $\cos \theta_{HCH} = -\dfrac{1}{3.35} = -0.298; \; \theta_{HCH} = 107°$

(observed $= 106°$)

In cases exhibiting bond angles forced to be small, as in a three-membered ring with obligatory CCC angles of 60°, we must distinguish the internuclear

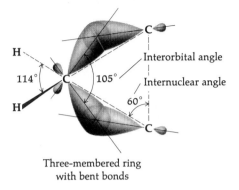

Three-membered ring
with bent bonds

angle (60°) from the interorbital angle, for in such cases it is more stable for the orbitals to project outward from the internuclear axis rather than suffer such a small interorbital angle. This results in a bent bond, not as stable as a normal, coaxially overlapped σ bond since the amount of overlap is less. In this case the HCH angle is measured as 114°, for which the above calculations yield a CCC interorbital angle of 105°.

In summary it is convenient for qualitative purposes to consider sp^3 bonds as forming tetrahedral (109.5°) angles, sp^2 as planar trigonal (120°), and sp as linear, 180°, in accord with orbital theory, despite these observed second-order variations, and to regard N and O with single bonds as essentially tetrahedral also, with the unshared pairs occupying sp^3 orbitals. Furthermore, all the iso-electronic atom situations (Fig. 2-3) may be regarded as possessing nearly identical hybridizations and angles.

PROBLEM 2-11

Calculate the interatomic distances marked in color on these structures.

$$CH_2 \text{---} CH_2$$
$$| \quad\quad\; a \;\;\nearrow | $$
$$CH_2 \text{---} CH_2$$

$$H_3C \text{---} CH_2 \text{---} CH_3 \quad (b)$$

$$Br \text{---} \bigcirc \text{---} Cl \quad (d) \;(c)$$

$$CH_3C \equiv CC \begin{array}{c} \nearrow CH_2 \\ \searrow CH_3 \end{array} \quad (e)\;(f)$$

2-10   BOND ENERGIES

Before discussing the energies involved in holding two atoms together in a covalent bond, we need to establish briefly a general intuitive physical sense of energy on which we may draw in later discussions of making and breaking bonds; i.e., chemical reactions. The various kinds of energy are all equivalent and interconvertible:

1   Kinetic energy (motion) and heat (essentially molecular motion)
2   Potential energy arising from
    a   Gravitational forces
    b   Electrical forces (attraction of unlike, repulsion of like charges)

The latter kind of potential energy (2b) is that of chemical bonds. Interconversion of energy is seen when a ball rolls downhill, losing potential energy (gravitational) to pick up kinetic energy (motion). In a comparable example the chemical potential energy in gasoline is released on combustion and converted to heat which may be converted to kinetic energy of an engine piston and this in turn to kinetic energy of a moving automobile.

It is convenient to regard chemical potential energy—which is "stored" as bond energies in molecules—as analogous to gravitational potential energy, cf., a ball "stored" on a hilltop. Both have a spontaneous tendency to proceed to lower potential energy, i.e., downhill, given the chance. Both spontaneously convert potential energy to heat and/or kinetic energy as they lower potential energy. The lower the potential energy, the more stable the molecule, or ball.

An object or particle (e.g., an atom) in isolation has zero potential energy, but two brought close together interact (gravity, or electrical forces) and their potential energy becomes positive or negative. In the case of molecule formation from isolated atoms, the energy goes down and becomes negative, i.e., more stable. (See diagram in Fig. 2-7.) Bond energies are therefore generally used as negative energies; the strongest bonds are the lowest energies (greatest negative values). In the table, however, it is more convenient simply to list their magnitudes.

In chemical reactions bonds are broken and new bonds are formed. Energy must be added, often in the form of heat, in order to break bonds and (heat) energy is liberated when a bond is formed. The amount of energy in each case is characteristic of the bond being broken or formed and is called the bond energy (Table 2-2). An overall reaction represents a change in (potential) energy, ΔE:

$$\Delta E = E_{\text{products}} - E_{\text{reactants}} = \Sigma E_{\text{bonds formed}} - \Sigma E_{\text{bonds broken}}$$

Reactions in which the energy goes down are proceeding—spontaneously—to lower energy, i.e., to more stable products than reactants. Such reactions usually yield the excess energy (ΔE) as heat and are called **exothermic**. In the opposite case, energy (usually heat) must be input to force the reaction to go (from lower to higher energy) and such reactions (nonspontaneous) are called **endothermic**.

Bond energies unfortunately cannot be measured directly. Those quoted in Table 2-2 are obtained indirectly through thermochemical studies of evolved heat in reactions such as combustion. These studies yield ΔE for overall reactions. The energy per bond must then be extrapolated in various ways and the quoted values are average figures. Also they represent values for reactions in the vapor state at 25°. For ordinary organic reactions in solution bond energies must be corrected with heats of vaporization, solution, etc. Nevertheless, within a few kilocalories per mole they often prove accurate in predictions of reaction energy, as further discussed in Chap. 9.

In order to develop a useful scale of energy units for general use, we shall follow organic chemistry practice in expressing energies in kilocalories per mole (kcal/mole). We can develop a rough scale for the purpose of comparing energies when we appreciate that the energy of thermal molecular motion at room temperature is about 15 to 20 kcal/mole. As most bond energies are substantially more than this, covalent molecules are stable to thermal disruption at room temperature. At higher temperatures, the energy of random molecular motion increases and can often exceed certain bond energies and so cause covalent bond breaking (such thermally induced reactions are called **pyrolyses**).

Single bonds are generally in the range 50 to 100 kcal/mole, with bonds between atoms bearing unshared pairs notably weaker owing to electrostatic repulsions between such pairs, as in O—O, N—O, N—X, O—X bonds. Bonds

TABLE 2-2   **Average Bond Energies (kilocalories per mole at 25°)**

Diatomic molecules

H—H	104	H—F	135	F—F	37
O=O	119	H—Cl	103	Cl—Cl	58
N≡N	226	H—Br	88	Br—Br	46
		H—I	71	I—I	36

Covalent bonds in larger molecules

H—C	99	C—C	83	C=C	146	C≡C	200
H—N	93	C—N	73	C=N	147	C≡N	213
H—O	111	C—O	86	C=O†	179		
H—S	83	C—S	65	C=S‡	128		
N—N	39	C—F	116	N=N	100		
N—O	53	C—Cl	81	N=O	145		
O—O	35	C—Br	68				
S—S	54	C—I	51				

† In **C—CO—C** compounds; **C—CO—H** is 176.
‡ In **S=C=S**, carbon disulfide.

are weaker as one proceeds down columns in the periodic table (e.g., C—F > C—Cl > C—Br > C—I). Double bonds are stronger, though not twice as strong, as single bonds. The difference represents roughly the strength of the π-bond component. Thus, the π bond between carbons is roughly (C=C) − (C—C) = 146 − 83 = 63 kcal/mole, and this is the energy which must be added to cause rotation to occur around a double bond, breaking the π bond. Triple bonds are correspondingly stronger. Bonds to second-row elements (P, S, Cl, etc.) are longer and hence weaker. The π bond, depending on sideways overlap, is especially weakened.

The s atomic orbital is more stable (lower energy; Fig. 2-4) than the p orbital, hence a hybrid orbital having more s character is more stable also. This manifests itself especially in the greater stability (lower energy) of unshared electron pairs occupying hybrid orbitals: $sp > sp^2 > sp^3$.

$$^-:C\equiv C—H \quad > \quad \begin{matrix} & H & H \\ & | & | \\ ^-:C & = & C \\ & & | \\ & & H \end{matrix} \quad > \quad \begin{matrix} & H \\ & | \\ ^-:C & —H \\ & | \\ & H \end{matrix}$$

$$:N\equiv C—H \quad > \quad \begin{matrix} & H & H \\ & | & | \\ :N & = & C \\ & & | \\ & & H \end{matrix} \quad > \quad \begin{matrix} & H \\ & | \\ :N & —H \\ & | \\ & H \end{matrix}$$

Decreasing stability
⟶
(Increasing energy)

PROBLEM 2-12

Calculate the energy released when these molecules are formed from their constituent free atoms.

a CH_4 d CH_3COCH_3 g $H_2C=CHOCH_3$

b CO_2 e C_2H_5Br h $CH_3NHNHCH_3$

$$\qquad\qquad\qquad\qquad\qquad\qquad\qquad\qquad\qquad\qquad\qquad\overset{O}{\overset{\|}{}}$$

c CH_3OH f $HC\equiv C—CN$ i $CH_3—C—OH$

2-11   ELECTRONEGATIVITY AND DIPOLES

Each atomic kernel has a certain affinity for electrons surrounding it; this affinity is called the **electronegativity**. Atoms with greater electronegativity attract electrons more strongly and hold them closer to the nucleus, or kernel; the orbitals bearing unshared electrons on such atoms are in fact smaller, more compressed about the nucleus. This affinity is directly related to the kernel charge as implied in Table 2-3 (a scale of electronegativity devised by Pauling).

Across any row of the periodic table, electronegativity increases with increasing kernel charge. As one proceeds down the table, from first- to second-row elements, etc., electronegativity falls off as the outer electrons are increasingly distant from the positive nucleus and shielded from it by shells of inner electrons.

Covalent bonds between atoms of different electronegativity (A—B) show two effects:

1 The bond energy is increased over the mean of bonds A—A and B—B by an amount roughly proportional to the square of the electronegativity difference.
2 The electrons in the bond orbital are polarized so as to lie closer to the more electronegative atom on the average. This results in partial ionic character in the covalent bond, i.e., to each atom bearing a local partial charge, cf.:

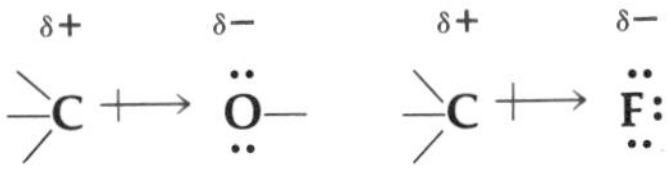

This polarization, or electrical imbalance, in the bonds is important in directing the course of chemical reactions in which these bonds are broken, for they can be said to be already partially broken or ionized, in the molecule and can be induced by appropriate reagents to break, or ionize, completely:

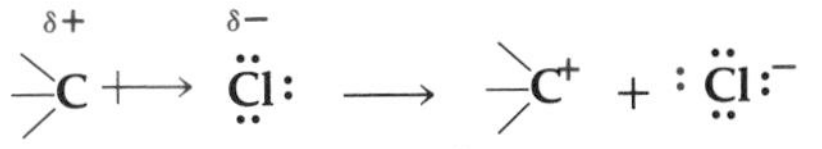

Furthermore, this polarization in the bonds makes each bond between dissimilar atoms a **dipole**. Hence **dipole moments** can exist in the full molecule, and these can be measured. In general bond dipoles faithfully reflect electronegativity differences, except with C—H in which the dipole is reversed (though quite small), directing the bonding electrons toward hydrogen: $-\overset{|}{\underset{|}{C}} \longleftrightarrow H$.

TABLE 2-3 **Electronegativities of Atoms**

Kernel charge:	+1	+4	+5	+6	+7
	H	C	N	O	F
	2.2	2.5	3.0	3.5	4.0
		Si	P	S	Cl
		1.9	2.2	2.5	3.0
					Br
					2.8
					I
					2.5

The orbitals in dipolar bonds are shaped just as previously described but the electrons in it are somewhat bunched up at one end.

2-12   HYDROGEN BONDING

Hydrogen occupies a unique place in our discussion of orbitals and bonding since it alone possesses no inner electron shell isolating its nucleus from the bonding electrons. The hydrogen kernel is a bare proton. Furthermore, when bonded to such electronegative atoms as N, O, and F, the bonding electrons are drawn strongly to the electronegative atom, leaving the proton as a relatively bare positive charge at the outer end of the covalent bond. As such it is peculiarly capable of attracting closely an external negatively charged center, either in the form of an anion or of an orbital on another molecule containing an unshared electron pair, i.e., a localized site of negative charge.

This attraction between a hydrogen covalently bonded to a heteroatom and another molecule (with its unshared electrons) or anion is called a **hydrogen bond**. It is essentially electrostatic or ionic in character and relatively weak (3 to 5 kcal/mole in common cases), but it is nevertheless of enormous importance in ordering the arrangements of molecules in both solution and crystal. In the hydrogen bond are two electronegative atoms with a proton between them. The hydrogen bond is stronger the more electronegative the two atoms and most stable when it is possible to be *linear* (hydrogen bond shown dotted);

$$\overset{\delta-}{\ddot{\text{O}}}\!\!-\!\!\overset{\delta+}{\text{H}}\cdots\ddot{\ddot{\text{F}}}\!:^{-} \quad\text{or}\quad \overset{\delta-}{\ddot{\text{O}}}\!\!-\!\!\overset{\delta+}{\text{H}}\cdots\overset{\delta-}{:\text{O}}:$$

$$\underbrace{\hspace{2.2cm}}_{\text{H bond}} \qquad\qquad \underbrace{\hspace{2.2cm}}_{\text{H bond}}$$

2-13   SUMMARY

In this chapter we have utilized a simple picture of the structure and nature of atoms to derive the nature of their bonding into covalent molecules and so have developed a few simple rules which allow the construction of all kinds of molecules of any complexity, understanding their size and shape, their energies and electrical makeup. These features will allow us to understand, indeed, often to predict, their chemical and physical behavior in the laboratory.

The rules of chemical bonding are simple. Most of chemistry is directed by two premises:

1   Like charges repel; unlike charges attract.
2   Atoms act so as to fill their outer shells with electrons.

These two dominant atomic "motivations" not only direct the details of molecular structure but also the dynamics of chemical reaction.

Atoms of the first row (and often those of the second) all form bonds of the same kind and can all be handled with the same rules, which are summarized in Table 2-4; the hybrid orbitals all look alike and form σ bonds by *end-on* overlap, while the other bond type (π bond) is formed by *sideways* overlap of unhybridized *p* orbitals.

The anatomy of molecular structure involves a **skeleton** made of atoms (**skeletal atoms**) forming two or more bonds, of which carbon is by far the predominant one and the only one capable of forming long chains. Oxygen and nitrogen are the next most abundant and sulfur and phosphorus a poor third. Hydrogen atoms serve to fill up the vacant bond positions on the skeleton, forming an external skin, and halogens, which occur relatively rarely, can serve the same function. Hydrogen is thus clearly the most numerous atom in most molecules.

In the next two chapters we shall see how the rules of chemical bonding allow the definition of all the possible ways of linking atoms other than carbon and hydrogen into organic molecules. These in turn allow organic compounds to be classified into broad structural families.

Following this we shall explore the consequences of more extended orbital overlap in molecules (Chap. 5) and then, in Chap. 6, the results of bonding and orbital geometry in the three-dimensional shapes of organic molecules and the extent of their freedom to change these shapes. Chapter 7 then develops the effects which various atoms and bonds in molecules have on certain physical properties useful for elucidating features of molecular structure in particular substances.

Having in this way determined the structural nature of organic molecules, we may turn, in Chaps. 8 to 22, to their behavior when allowed to interact with each other: the nature of chemical reactions. As we shall see, their behavior in reactions is also largely prefigured by the concepts presented in this chapter.

TABLE 2-4   **Summary of Bonding Characteristics**

No. of bonded atoms or unshared pairs	Bonds to single atoms				Bonding of two atoms			
	Hybrid orbitals (% s character)	Remaining unhybridized orbitals	Interorbital angle	Geometry	Bond	No. of electrons	Bond types	Free rotation
4	$4sp^3$ (25%)	0	109.5°	Tetrahedral	Single	2	σ	Yes
3	$3sp^2$ ($33\frac{1}{3}$%)	1p	120°	Planar	Double	4	σ + π	No
2	$2sp$ (50%)	2p	180°	Linear	Triple	6	σ + 2π	—

PROBLEMS

2-13 What is wrong with the following molecular formulas?

    a C_2H_5 **c** CH_3O **e** C_2H_4Cl

    b C_2H_4N **d** CH_5S **f** C_3H_3NCl

2-14 Write *one* structure which fits each of the following descriptions:

    a $C_{10}H_{10}O_4$ **d** C_4N_2 (no rings)   **g** $C_4H_8N^+$

    b $C_{12}H_{16}N_2O$ **e** C_6N_4 (no rings)   **h** $C_3H_3O^-$

    c C_6F_6 **f** C_3H_5N (no multiple bonds)

2-15 Write all the structures (ions or molecules) which are isoelectronic with CO_2, showing which atoms bear charges. Utilize only first-row atoms.

2-16 Assign charges and add all the unshared electron pairs to the correct atoms in the following structures. (Fill all outer shells of electrons.)

    a $CH_3CH_2N(CH_3)_3$ **e** $O{-}N{=}N{=}O$

    b $CH_3{-}C{\equiv}C$ **f** $CH_2{=}N{=}N$

    c $CH_3{-}C{\equiv}N{-}O$ **g** $(CH_3)_2S{-}O$

    d $CH_3{-}CH{=}N{-}O$ **h** $\overset{\displaystyle Br}{\overset{\displaystyle \diagup \diagdown}{CH_2{-}CH_2}}$

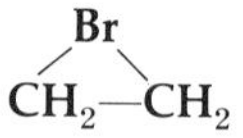

2-17 Point out which atoms do not possess full outer shells of electrons in these structures:

    a $(CH_3)_3C^+$ **d** $CH_3{-}\overset{\displaystyle :O:}{\overset{\displaystyle \|}{C}}{-}\ddot{\underset{..}{N}}$

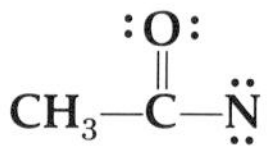

    b $(CH_3)_3B$ **e** Br^+

    c $(CH_3)_3N:$ **f** $AlCl_3$

2-18 Decide whether each reaction shown is exothermic or endothermic. How much energy is released (or must be added) in each reaction?

    a $CH_3\overset{\displaystyle O}{\overset{\displaystyle \|}{C}}{-}OH \longrightarrow CH_4 + CO_2$

    b $(CH_3)_3C{-}N{=}N{-}C(CH_3)_3 \longrightarrow (CH_3)_3C{-}C(CH_3)_3 + N_2$

    c $(CH_3)_3COH \longrightarrow (CH_3)_2C{=}CH_2 + H_2O$

    d $CH_3{-}N{=}O \longrightarrow CH_2{=}N{-}OH$

    e $(CH_3)_2C{=}CH_2 + HBr \longrightarrow (CH_3)_3CBr$

    f $CH_3CN + H_2O \longrightarrow CH_3\overset{\displaystyle O}{\overset{\displaystyle \|}{C}}{-}NH_2$

2-19 Assess the relative stabilities (i.e., resistance to chemical change) of the following molecules. (Note that cleavage of any one bond in a molecule destroys the substance, i.e., converts it to other compounds.)

a $CH_2=CH-CH_2-O-OH$ d N_2

b $(CH_3)_2CO$ e CO_2

c $CH_3\overset{\displaystyle O}{\overset{\|}{C}}-NHOH$

2-20 For each molecular formula below show two different covalent structures: one with and one without charged atoms, all atoms to have filled outer shells.

a CH_2N_2 c HNO_2 e O_3

b $CHON$ d CHN f C_2H_6SO

2-21 Draw each structure below to show its internal hydrogen bond.

a 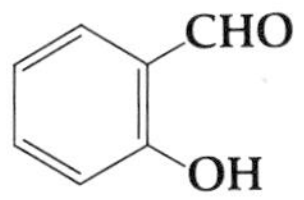   c $CF_3CH_2\underset{\displaystyle CH_3}{\overset{\displaystyle }{CH}}-OH$

b $CH_3OCH_2CH_2CH_2\overset{+}{N}H_3$ d

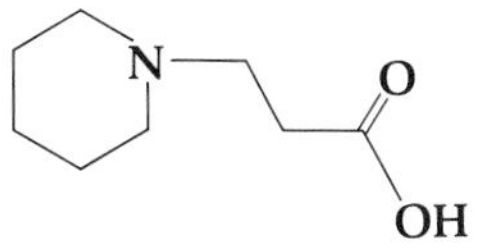

2-22 The following notations are common for certain widespread groupings of atoms. Show each one in full structure, complete with unshared electron pairs, charges, and full outer shells of electrons. (Note: there are no O—O bonds).

a —COOH   c —COCl   e $-N_3$ g $-SO_2H$

b —CHO   d $-CONH_2$ f $-NO_2$ h $-SO_3H$

2-23 Label the hybrid atomic orbitals used to form each colored bond in these molecules. Write in the expected bond angles as well.

a H_3C-H c $CH_3C\equiv CH$ e $(CH_3)_2B-CH_3$

b $H_3C-CH_2CH_3$ d $(CH_3)_3\overset{+}{N}-CH_3$ f $(CH_3)_2C=\overset{..}{N}CH_3$

2-24 Examine the orbitals involved in forming $CH_2=C=CH_2$; label them and determine the orbital geometry required. The three carbons are known to lie in a straight line. Is this reasonable?

2-25 In the following pairs of isomers one is more stable than the other. Explain why for each case.

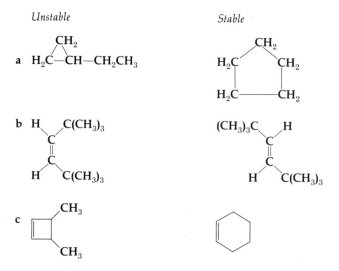

READING REFERENCES

Pauling, L., "The Nature of the Chemical Bond," 3d ed., Cornell, New York, 1960.
Orchin, M., and H. H. Jaffe, "The Importance of Antibonding Orbitals," Houghton
                                                                      Mifflin, Boston, 1967.

THE CONSEQUENCES OF CHEMICAL BONDING: CLASSES OF MOLECULES

THE preceding chapter developed from the separate atoms rules for the construction of organic molecules from atoms of C, H, O, N, S, P, and halogens. Here our purpose is to use these rules to construct the various classes of molecules common in organic chemistry. We are in effect now ready to invent all the possible molecules, and then to explore their chemical and physical behavior.

In general organic molecules are composed of a skeleton of carbon atoms, sheathed in hydrogens, with groupings composed of other atoms attached to that skeleton. These attached groups are called **functional groups**, since they are always the sites of chemical reactivity or function. A molecule composed only of carbon and hydrogen is called a **hydrocarbon**. Its functional groups are only its double and triple (π) bonds, the σ bonds being essentially unreactive. A molecule or group without π bonds is labeled as **saturated** while multiple bonds constitute **unsaturation**. Atoms other than C and H are usefully separated generically as **heteroatoms** (O, N, S, P, halogens), and these are the substance of most functional groups.

In this chapter we shall construct the possible functional groups and then see how the great variety of organic molecules which we can now generate may be logically named to facilitate communication about them. Furthermore, the logical procedures invoked to deduce the molecular structures of unknown compounds will also be examined.

3-1 HYDROCARBONS

As we saw in examining the possibilities of structural isomerism, the potential number of hydrocarbons is limitless. It is therefore necessary to find some kind of order in these possible structures, many thousands of which are known compounds. Hydrocarbons and derived compounds fall into three large classes, which are defined in terms of structural concepts. **Aliphatic** hydrocarbons are composed of chains of carbon atoms not arranged in rings. Substances belonging to this group are sometimes referred to as **open-chain** compounds. In **alicyclic** hydrocarbons, the carbon chains form rings. Aside from a few exceptional ring

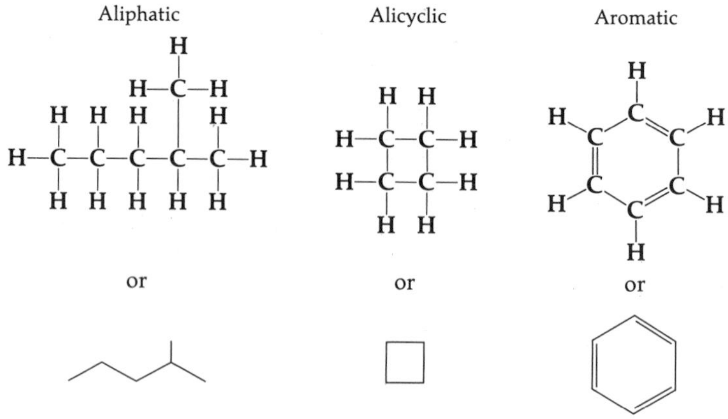

FIGURE 3-1   **Examples of classes of hydrocarbons**

compounds, aliphatic and alicyclic hydrocarbons of like molecular weight are similar in both physical and chemical properties. The third group consists of the **aromatic** hydrocarbons, which contain six-membered rings into which are fitted three carbon-carbon double bonds. Special physical and chemical properties are associated with this arrangement of double bonds in aromatic systems (Chap. 5). Figure 3.1 shows examples of the three classes of hydrocarbons.

The **alkanes (paraffins)** are aliphatic hydrocarbons that contain the maximum number of hydrogen atoms compatible with the requirement that carbon always possesses four bonds and hydrogen one. Thus alkanes are said to be saturated, and they are all represented by the general formula C_nH_{2n+2}, where n is an integer. All compounds corresponding to the structural formulas that can be written for alkanes with one to nine carbon atoms are known. Some substances with n higher than 50 have been prepared in a pure state. This general formula describes what is known as a **homologous series**, members of which are referred to as **homologs** of one another. Each member of a homologous series differs from its immediate neighbors by a **methylene** group (CH_2), i.e., CH_4, C_2H_6, C_3H_8, C_4H_{10}, etc.

Index of Hydrogen Deficiency

The molecular formula of a compound gives important information concerning its structural possibilities. Comparison of the hydrogen content of a compound with that of the alkanes containing the same number of carbon atoms shows the index of hydrogen deficiency of the substance of unknown structure. The number of hydrogen atoms in a molecule is decreased by two for each new ring or double bond introduced into the structure. The **index of hydrogen deficiency** is the number of *pairs* of hydrogen atoms that must be removed from the saturated alkane formula (C_nH_{2n+2}) to give the molecular formula of the compound at hand. Table 3-1 contains a number of examples.

The index† is, therefore, a number equal to the sum of the rings and multiple bonds (triple bonds require 2 index units) in the molecule. Thus, the compound C_6H_{10} in the table (index = 2) has one of the following four combinations:

Examples

Compound = C_6H_{10} 2 double bonds $CH_2{=}CH{-}CH{=}CH{-}CH_2CH_3$

Satd. alkane = C_6H_{14}

$\Delta H = 14 - 10 = 4$ 1 ring and
                              1 double bond
Index = $\frac{4}{2}$ = 2

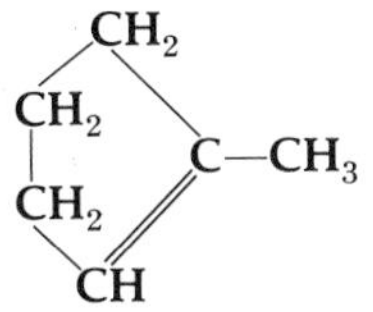

2 saturated rings

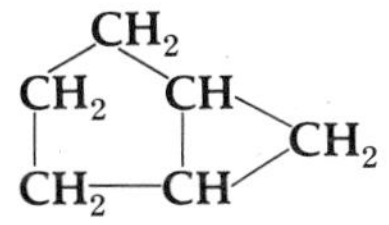

1 triple bond $CH_3{-}C{\equiv}C{-}CH_2CH_2CH_3$

In general the possible hydrocarbon formulas require the following structural features:

C_nH_{2n+2} Saturated alkane
C_nH_{2n} One ring or double bond
C_nH_{2n-2} Two rings and/or double bonds (as C_6H_{10} above)
C_nH_{2n-4} Three rings and/or multiple bonds etc.

No hydrocarbon can have an odd number of hydrogens.

† The index of hydrogen deficiency is sometimes referred to as the "number of sites of unsaturation," which is a clear description but unsatisfactory in that it implies that the hydrogen deficiency is caused only by multiple bonds (unsaturation), not rings as well. Carbon skeletons with rings show hydrogen deficiency but may often be saturated.

TABLE 3-1   **Index of Hydrogen Deficiency**

Molecular formula	Corresponding alkane (C_nH_{2n+2})	Index
C_6H_{14}	C_6H_{14}	0
C_6H_{10}	C_6H_{14}	2
C_5H_{10}	C_5H_{12}	1
C_6H_6	C_6H_{14}	4

PROBLEM 3-1

Derive the index of hydrogen deficiency for each formula and write both a cyclic and an acyclic structure to fit each formula.

a C_5H_{10} d C_3H_4

b C_6H_{10} e C_6H_6

c C_7H_{10} f $C_{10}H_{10}$

3-2   THE FUNCTIONAL GROUPS

The few functional groups which may be generated from the heteroatoms are the same and behave similarly no matter to what hydrocarbon skeleton they may be attached. For this reason it is often convenient to generalize a hydro-carbon group with one free bond for attachment as R—. Thus R— can mean CH_3—, CH_3—CH_2—, CH_3—CH_2—CH_2—, $(CH_3)_2CH$—CH_2—, or any other hy-drocarbon group with one available bonding site. In the sections that follow we shall bond heteroatoms to hydrocarbons in all the possible ways allowed by Chap. 2, writing their structures out once in enough detail to allow de-scription of all bonding and orbital detail, and then in most cases writing a shorthand form as well. These are all summarized in Table 3-2.

In order to invent the functional groups we need simply select one carbon atom of a skeleton and bond various heteroatoms to it.† It will be convenient to group the functional groups formed by the number of bonds from the carbon of the skeleton to heteroatom(s), group I having a single bond from the carbon of the skeleton to a heteroatom, group II having two such bonds to carbon, and group III having three.

Group I

This group incorporates functional groups with only one bond from the skeletal carbon to a heteroatom. The simplest example is the attachment of a halogen to give —C—X: or R—X, the X being used throughout this book specifically to denote the chemically very similar halogen atoms: **F, Cl, Br, I.** This class of compounds is called **halides (fluorides, chlorides, bromides, iodides)** and includes compounds such as CH_3—Cl, CH_3—CH_2—Br, $(CH_3)_2CH$—I, CH_3—CH_2—CH_2—F, etc.

---

†We shall limit ourselves at first to allowing no bonds between two heteroatoms, for simplicity. We shall also limit ourselves to the major elements of organic chemistry: C, H, O, N, S, P, halogen.

If we attach oxygen instead we must then select something else for the second oxygen bond. With no bonds between heteroatoms, this will be either H or C. In the first instance we generate $-\overset{|}{\underset{|}{C}}-\ddot{O}-H$ or R—OH, a class of compounds called **alcohols**, of which common beverage alcohol, CH_3-CH_2-OH, is an example. The other oxygen compounds are $-\overset{|}{\underset{|}{C}}-\ddot{O}-\overset{|}{\underset{|}{C}}-$ or R—O—R', which are called **ethers**, CH_3-O-CH_3 being an ether isomer of the alcohol shown (both are C_2H_6O), and $CH_3-CH_2-O-CH_2-CH_3$ the ether of anesthesia. These classes may also be regarded as deriving from successive replacement of the hydrogens of water by hydrocarbon groups: H—O—H; R—O—H; R—O—R'. The analogous sulfur compounds, R—SH, are called **mercaptans, thio-alcohols**, or more commonly **thiols**, while the R—S—R' class are **sulfides (thio-ethers)**.

Finally, attachment of nitrogen leads to a class of compounds known as **amines**. These represent substitutions of ammonia (replacement of H by R), similar to those of water above and clearly will be of three kinds:

$H-\ddot{N}-H$	$R-\ddot{N}-H$	$R-\ddot{N}-R'$	$R-\ddot{N}-R'$
H	H	H	R''
Ammonia	Primary amines (R—NH$_2$)	Secondary amines (R—NH—R')	Tertiary amines

The **phosphines** (R—PH$_2$ etc.) are analogously obtained from use of phosphorus in place of nitrogen.

These are all the Group I functional groups, or classes of compounds we can obtain in this way. By allowing bonds between heteroatoms, many more are possible, but with two exceptions they are not of common occurrence and are often unstable. Thus R—O—OH (**hydroperoxides**) and R—O—O—R' (**peroxides**) are known, but R—O—O—O—H(R') are too unstable to exist (viz., chains of heteroatoms, page 31). Similarly, **hydroxylamines** are formed by replacing one or more of the hydrogens of NH$_2$—OH with R groups, as **hydrazines** are obtained from NH$_2$—NH$_2$, but R—NH—NH—NH$_2$ and like compounds are exceedingly rare and very unstable.† The two exceptions mentioned are two functional groups with only one bond to carbon but rather complex bonding among the heteroatoms, which stabilizes these groups. The **nitro** and, to a lesser extent, the **azide** group are not only common enough to warrant their inclusion as important functional groups but also serve to show the variety possible on utilizing more involved bonding among the heteroatoms in a functional group.

$-\overset{|}{\underset{|}{C}}-N\overset{\overset{\ddot{O}\cdot}{\diagup}}{\underset{\underset{+}{\diagdown}}{\cdot\ddot{O}\cdot}}$ (R—NO$_2$)	$-\overset{|}{\underset{|}{C}}-\ddot{N}\overset{+}{=}N\overset{-}{=}\ddot{N}$ (R—N$_3$)
Nitro group	Azide group

---

† The **hypohalides**, R—O—X, and **haloamines**, R—NH—X, are also very unstable.

TABLE 3-2 **The Major Functional Groups**

Group I	*Group II*	*Group III*

Group I

Halides
(R—X)

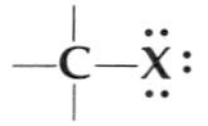

Alcohols
(R—OH)

Ethers
(R—O—R')

Amines

Primary
(R—NH$_2$)

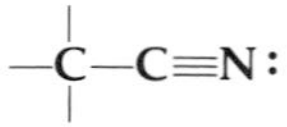

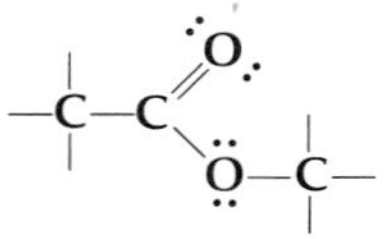

Secondary
(R—NH—R')

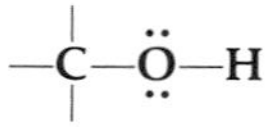

Tertiary
(R—N—R')
   R''

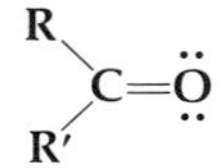

Group II

Aldehydes
(R—CHO)

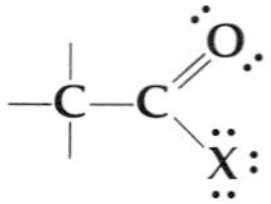

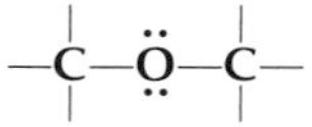

Ketones
(R—CO—R')

Imines
(R—CNR''—R')

Group III

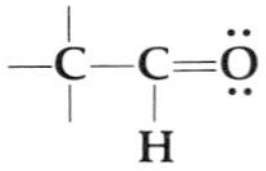

Nitriles (Cyanides)
(R—CN)

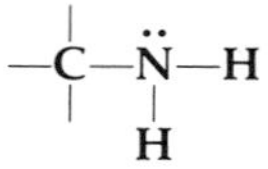

Acid halides
(R—COX)

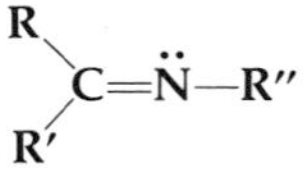

Carboxylic acids
(R—COOH)

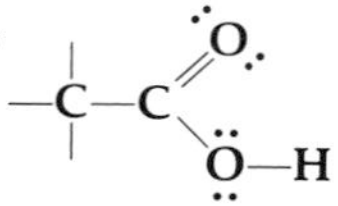

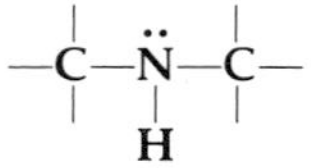

Esters
(R—COOR')

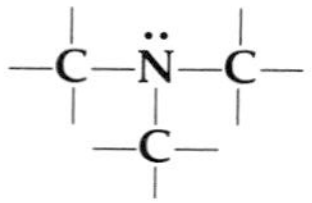

Anhydrides
(R—CO—O—CO—R')

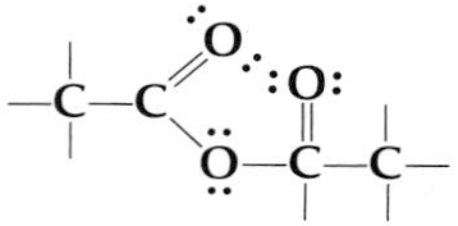

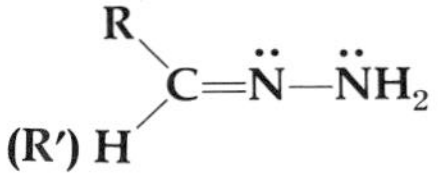

Amides
$(R\text{—}CONH_2)$
$(R\text{—}CONHR')$
$(R\text{—}CO\text{—}N\text{—}R')$
$\phantom{(R\text{—}CO\text{—}N\text{—}}R''$

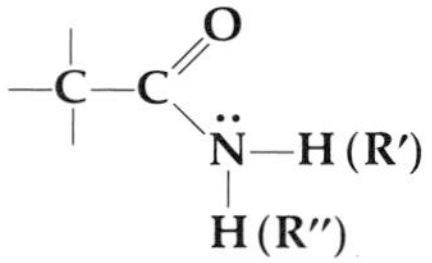

Imides
$(R\text{—}CO\text{—}NH\text{—}CO\text{—}R')$

Acetals

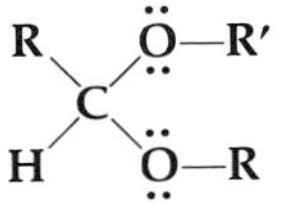

Ketals

Oximes

Hydrazones

Nitro compounds
$(R\text{—}NO_2)$

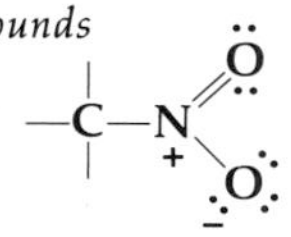

Azides
$(R\text{—}N_3)$

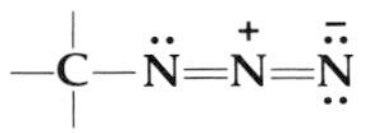

Hypohalides $(R\text{—}O\text{—}X)$
Hydroperoxides $(R\text{—}O\text{—}OH)$
Peroxides $(R\text{—}O\text{—}O\text{—}R')$
Haloamines $(R\text{—}NH\text{—}X)$
Hydroxylamines $(R\text{—}NH\text{—}OH)$
Hydrazines $(R\text{—}NH\text{—}NH_2)$

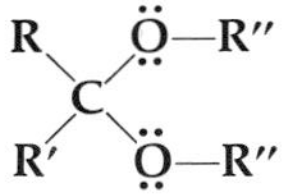

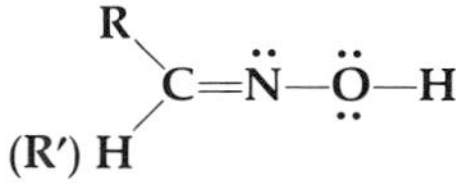

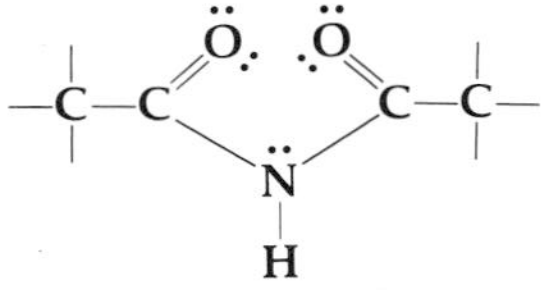

It is certainly possible to put together more functional groups in this way, using the rules of the preceding chapter, and most of these are known, but the important ones are here.

Group II

This group is most obviously exemplified by placing any two of the group I functionalities (functional groups) singly bonded to a single carbon (e.g., $-CX_2-$), but the behavior of such groups is easily encompassed in any description of the single groups. The only compounds of this kind to be singled out will be the **acetals** and **ketals**:

Acetal   Ketal   (General)

Their chemistry is virtually identical but no generic term for the group is in common use. (Stable compounds with two —OH groups on a single carbon are very rare.)

There are two functional groups in group II which are important and unique, both having very similar behavior and deserving of an inclusive generic term. These are the **aldehyde** group and the **ketone** group. The particular assemblage of the carbon-oxygen double bond, of which they are composed, is called a **carbonyl group**,† which also shows up in the group III functions.

A third group, the **imine**, can be analogously constructed with nitrogen,

Carbonyl   Aldehyde   Ketone   Imine
group   (R—CHO)

but is of minor importance, while **thioketones** and **thioaldehydes (thials)** are rare and unstable, the phosphorus analogs more so (cf. page 64). All of these lesser groups, including several shown in Table 3-2, are regarded as a family, being derivatives of the corresponding aldehydes and ketones, to which they may usually be transformed chemically by hydrolysis (reaction with water).

Group III

Leaving aside triplications of group I such as $-CX_3$ and $-C(O-R)_3$ there is really only one unique group III functionality, the **nitrile** or **cyanide (cyano group)**, $R-C\equiv N:$ or R—CN, with three bonds to the only heteroatom (nitrogen) that accepts three. The phosphorus analogs are virtually unknown and certainly unstable (cf. page 64).

The *major* functional groups with three bonds to heteroatoms are con-

---

† The term "double bond," as usually employed by organic chemists, is taken to mean a carbon-carbon double bond, as distinguished from a "carbonyl," or carbon-oxygen double bond.

structed by allowing two of the bonds to carbon to be a carbonyl group and then reserving the third bond for any of the functionalities listed as group I. (Let these be designated for the moment as $-Z$, cf., $-Z{\equiv}-Cl$, $-Br$, $-OH$, $-NH_2$, etc.) Thus, in group I these are attached to a saturated hydro-

carbon group, $R-Z$, but in group III they are attached to a carbonyl, $R-C\overset{\ddot{O}}{\underset{Z}{\diagup}}$.

These important carbonyl derivatives are all listed with their names and short-hand designations in Table 3-2.

All of the group III functional groups are considered to be derivatives of the parent, **carboxylic acid**, $R-COOH$, since they are all chemically converti-ble to it by hydrolysis. Not shown in the table are the logically comparable series constructed from **imino** groups ($C{=}\ddot{N}-$) in place of carbonyl groups or from **thiocarbonyl** groups ($C{=}\ddot{S}$). Such group III functionalities are mostly known but are excluded here because of their minor importance. They are, however, part of the carboxylic acid family because of their chemical hydrolysis to $R-COOH$. Imino compounds are usually labeled as **imino-halides, imino-ethers,** etc.

There are a number of functional groups, somewhat out of the main stream, which involve higher oxidation states of second-row atoms, especially sulfur and phosphorus, and these are mentioned in Chap. 4 and 5 and developed in Chap. 19. The most important is a family of sulfur derivatives of sulfonic acids, analogous to carboxylic acids, in which the group $-SO_2-$ replaces the carbonyl ($-CO-$) of the carboxylic acid derivatives.

$$R-\overset{\overset{\displaystyle :\ddot{O}:^-}{|{++}}}{\underset{\underset{\displaystyle :\ddot{O}:^-}{|}}{S}}-Z \qquad \textbf{Sulfonic acid derivatives}$$

$R-SO_2-OH$ **Sulfonic acids**

$R-SO_2-Cl$ **Sulfonyl chlorides**

$R-SO_2-OR'$ **Sulfonate esters**

$R-SO_2-NR'_2$ **Sulfonamides**

Group IV

These compounds, with four heteroatom bonds to carbon, are of less interest since they leave no bonds for attachment of that carbon to the other carbons of a carbon skeleton. They may be regarded as derivatives of carbon dioxide (CO_2) or its inorganic hydrate, carbonic acid ($HO-CO-OH$, or H_2CO_3), from which the carbonate salts of Sec. 2-5 are derived by removal of the hydrogens as protons. Table 3-3 lists a number of the common compounds and functional groups which make up group IV. More can easily be derived as an inventive exercise, following the patterns of previous groups.

It is important to appreciate that the bonding rules allow only a specific and limited number of functional groups and completely define the orbitals and shape in each. We shall later see that their reactivity is also predetermined by the nature of bonding. It is also important to see that the functional groups are just logical variations on a few themes, and may be examined in different ways to facilitate comprehension and familiarity. For example, there is a formal analogy between the relation of R—OH and R—O—R′ and that among carboxylic acid, ester, and anhydride; it comes about by replacing H— in water

TABLE 3-3 **Group IV Compounds and Functional Groups**

$$
\begin{array}{c}
\quad :\!\ddot{O}\!: \\
\quad \| \\
H\!-\!\ddot{O}\!-\!C\!-\!\ddot{O}\!-\!H
\end{array}
$$
Carbonic acid

$$\ddot{O}\!=\!C\!=\!\ddot{O}$$
Carbon dioxide

$$
\begin{array}{c}
\quad :\!\ddot{O}\!: \\
\quad \| \\
:\!\ddot{C}l\!-\!C\!-\!\ddot{C}l\!:
\end{array}
$$
Phosgene

$$
\begin{array}{c}
\quad :\!\ddot{O}\!: \\
\quad \| \\
R\!-\!\ddot{O}\!-\!C\!-\!Cl
\end{array}
$$
Chloroformates

$$
\begin{array}{c}
\quad :\!\ddot{O}\!: \\
\quad \| \\
R\!-\!\ddot{O}\!-\!C\!-\!\ddot{O}\!-\!R'
\end{array}
$$
Carbonates

$$
\begin{array}{c}
\quad :\!\ddot{O}\!: \\
\quad \| \\
R\!-\!\ddot{N}\!-\!C\!-\!\ddot{O}\!-\!R' \\
\quad | \\
\quad H
\end{array}
$$
Urethans
(carbamates)

$$R\!-\!\ddot{N}\!=\!C\!=\!\ddot{O}$$
Isocyanates

$$R\!-\!\ddot{N}\!=\!C\!=\!\ddot{S}$$
Isothiocyanates

$$
\begin{array}{c}
\quad :\!\ddot{O}\!: \\
\quad \| \\
R\!-\!\ddot{N}\!-\!C\!-\!\ddot{N}\!-\!R' \\
\quad | \quad\quad | \\
\quad H \quad\quad H
\end{array}
$$
Ureas

$$R\!-\!\ddot{N}\!=\!C\!=\!\ddot{N}\!-\!R'$$
Carbodiimides

$$
\begin{array}{c}
\quad :N\!-\!R' \\
\quad \| \\
R\!-\!\ddot{N}\!-\!C\!-\!\ddot{N}\!-\!R'' \\
\quad | \quad\quad | \\
\quad H \quad\quad H
\end{array}
$$
Guanidines

(H—O—H) by either **R**— or **R**—**CO**—. A similar analogously related group may be seen in the replacement of **H**— in either **H**—**X** or **NH**$_3$ by either **R**— or **R**—**CO**—.

PROBLEM 3-2

Draw up a table of the possible compounds derived by replacing **H** in **NH**$_3$, **H**$_2$**O**, and **HX** with both **R**— and **RCO**—, separately and in combination. Label the functional groups so obtained wherever possible.

PROBLEM 3-3

Invent some more group IV compounds and functional groups not shown in Table 3-3.

The point of all this is that the classes of organic compounds are logical and simple. A last way of looking at organic molecules is that they contain a carbon skeleton with the following functionalities hooked on (here only the major ones are noted):

—X	—COX
—OH	—COOH
—OR	—COOR
—NH$_2$	—CONH$_2$
—NR$_2$	—CONR$_2$
—CHO	—CN
=O	—NO$_2$

It is often convenient to define the skeleton as including not just carbon-carbon links (single or multiple) but also intimately involved ether (—O—) or amine (—NH— or —N—) linkages, as the following randomly generated structures
         |

imply (respectively, a carboxylic acid, an ester, a cyanoalcohol).

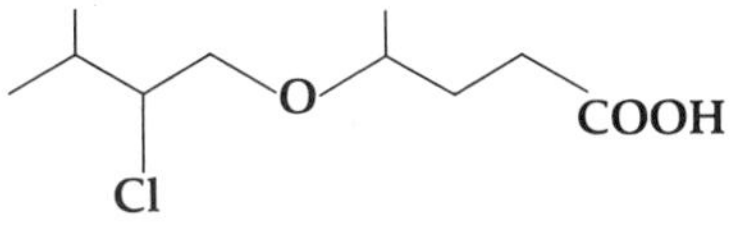

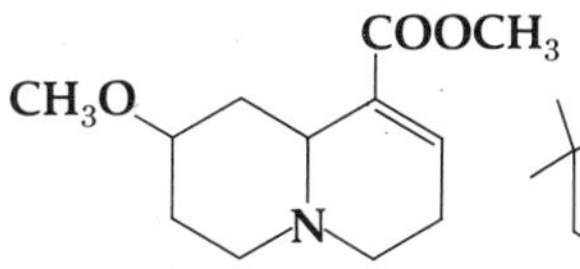

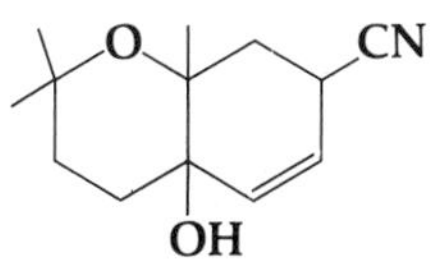

PROBLEM 3-4

Draw structures of particular examples of the following:

a An unsaturated alcohol   i An aromatic amine

b A cyclic anhydride   j A tertiary amine-ether (no N—O bond)

c A cyclic ester   k An alicyclic nitrile

d A cyclic paraffin   l An acyclic dialdehyde

e An amino acid   m A cyclic N-bromoimide

f A saturated trichloride   n A homolog of CH_3OH

g A cyclic thioether   o A secondary phosphine

h A thioamide   p An unsaturated nitroalcohol

The index of hydrogen deficiency may now be expanded to compounds other than hydrocarbons, by making use of the following rules:

1 The presence of oxygen (or sulfur) makes no change in the index of hydrogen deficiency.

2 Each halogen is regarded as the equivalent of one hydrogen.

3 Each nitrogen in the molecule raises the number of hydrogens in the corresponding saturated parent compound by one.

As an example, $C_5H_7NO_3Br_2$ can be first written as $C_5H_9NO_3$, replacing its halogens (bromines) by hydrogens, and the parent compound (saturated and noncyclic) must be $C_nH_{2n+2+1}NO_3 = C_5H_{13}NO_3$, the single nitrogen adding one H to C_nH_{2n+2} and the oxygens being uninvolved. Formulas for the parent saturated, acyclic compounds can alternatively be written formally:

$$C_nH_{2n+2}O_m$$
$$C_nH_{2n+2-m}X_m$$
$$C_nH_{2n+2+m}N_m$$

The relationships are illustrated by the examples found in Table 3-4.

3-3 THE RATIONALE OF STRUCTURE DETERMINATION

As noted earlier, a major task of many organic chemists is the determination of the structure of an unknown compound. The task arises most often simply because he has prepared a new, previously undescribed compound in his laboratory and must confirm what it is. Much more complex problems of structure determination, undertaken for their own sake, arise with the isolation from natural sources (plants, fungi, insects, animals, etc.) of new, pure compounds of unknown structure (Chap. 27). The history of molecular structure assignments in organic chemistry is one of the most impressive feats of deductive logic in the history of mankind.

TABLE 3-4   **Indices of Hydrogen Deficiency of Representative Compounds**

Structure	Formula	Formula of parent compound	Index of hydrogen deficiency	Identification of hydrogen-deficient structures
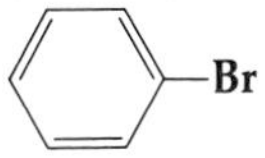 ⬡—Br	C_6H_5Br	$C_6H_{13}Br$	4	3 C=C 1 ring
$Cl_2CHCH_2C{\equiv}CH$	$C_4H_4Cl_2$	$C_4H_8Cl_2$	2	2 from C≡C
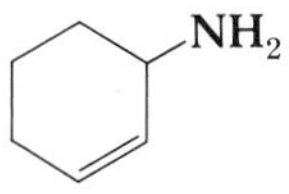 ⬡—NH₂ (cyclohexene)	$C_6H_{11}N$	$C_6H_{15}N$	2	1 C=C 1 ring
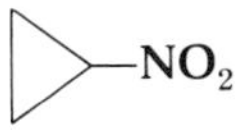 △—NO₂	$C_3H_5NO_2$	$C_3H_9NO_2$	2	1 N=O 1 ring
$CH_3C{\equiv}N$	C_2H_3N	C_2H_7N	2	2 from C≡N
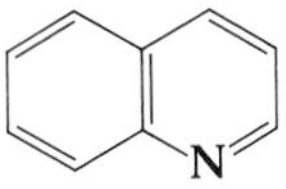 (quinoline)	C_9H_7N	$C_9H_{21}N$	7	4 C=C 1 C=N 2 rings
 (furan)	C_4H_4O	$C_4H_{10}O$	3	2 C=C 1 ring
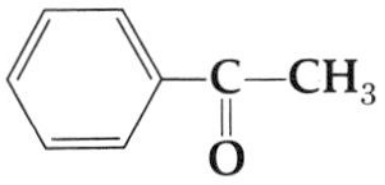 ⬡—C(—CH₃)=O	C_8H_8O	$C_8H_{18}O$	5	3 C=C 1 C=O 1 ring
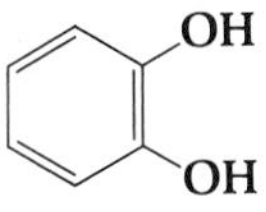 ⬡(OH)(OH)	$C_6H_6O_2$	$C_6H_{14}O_2$	4	3 C=C 1 ring
$CH_2{=}CHCH_2SH$	C_3H_6S	C_3H_8S	1	1 C=C
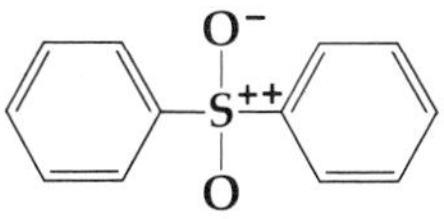 (diphenyl sulfone)	$C_{12}H_{10}SO_2$	$C_{12}H_{26}SO_2$	8	6 C=C 2 rings
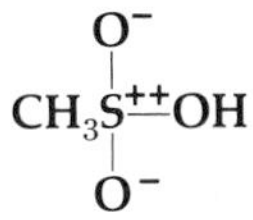 $CH_3\overset{O^-}{\underset{O^-}{S^{++}}}OH$	$C_1H_4SO_3$	$C_1H_4SO_3$	0	
$(CH_3)_3\overset{+}{N}{-}\bar{O}$	C_3H_9NO	C_3H_9NO	0	

The first step in any such problem, once the unknown compound has been purified (Sec. 1-4), is the determination of its molecular formula. This in effect defines the problem by

1 Establishing the size of the molecule
2 Implying the extent of functionality by the number of heteroatoms
3 Indicating the number of rings and/or multiple bonds from the index of hydrogen deficiency

After establishment of the molecular formula, two general avenues of approach may be used (usually profitably combined) to define the structural detail: chemical reactions and physical methods.

Chemical methods were historically the first and until very recently the major avenue of structural investigation. In principle they always involve submitting the unknown compound to chemical reactions and deducing structural features from its responses. Since the functional groups are the reactive sites of organic molecules, the compound is first submitted to chemical reactions which are particular for certain functional groups. In the process the fundamental logic lies in the *expectation of molecular change*. In a certain reaction known to effect a conversion of one functional group to another, a specific change is expected in the molecular formula between the unknown compound and its reaction product. Should this occur in the case under investigation, one could conclude that the presence of a particular functional group had been revealed by the reaction.

As an example, the chemical conversion of an aldehyde or ketone to an oxime (Table 3-2) requires a change in molecular formula which amounts to an overall addition of NH ($C=O \xrightarrow[(-H_2O)]{+NH_2OH} C=N-OH$). An unknown compound $C_8H_{12}O_2$, which yielded $C_8H_{13}NO_2$ in this reaction would be suspected of containing one ketone or aldehyde function. If it yielded $C_8H_{14}N_2O_2$, it would instead be believed to contain two. The continued monitoring of molecular formula after each reaction is the key to chemical procedure in structure determination. In the case in hand, for example, it is clear that—with either result noted—there is no need to test the unknown compound for the presence of a carboxylic acid or ester since the molecular formulas make clear that those groups are not possible. (There are two oxygens in $C_8H_{12}O_2$, but if one is a ketone there are not two oxygens remaining to make up —COOH or —COOR also.)

PROBLEM 3-5

A compound, $C_3H_6O_2$, is submitted to chemical conditions which convert ketones (but not esters or carboxylic acids) to oximes. The product is $C_3H_7NO_2$. What are the structures of the two compounds?

PROBLEM 3-6

There are only two possible kinds of functional groups in $C_8H_{18}O_2$. What are they? Write three possible isomeric structures incorporating these groups.

PROBLEM 3-7

It is known that boiling in aqueous acids converts an amide to a carboxylic acid and an amine ($R—CO—NH—R' + H_2O \xrightarrow[\text{acid}]{\Delta} R—COOH + R'—NH_2$).

   a What can we deduce from the fact that compound A ($C_6H_{13}NO_2$) gives a carboxylic acid, $C_4H_8O_2$, and a primary amine, C_2H_7N, under these conditions?

   b What structural deductions arise from the observation that compound B ($C_6H_{11}NO_2$) similarly yields a carboxylic acid, $C_3H_4O_2$, and a primary amine, C_3H_9NO?

   c Under the same conditions, compound C ($C_6H_{13}NO_2$) does not react. Write a possible structure for compound C.

   d Under the same conditions, compound D ($C_6H_{11}NO_2$) reacts to give only one compound, $C_6H_{13}NO_3$, which has chemical characteristics of both a carboxylic acid and an amine. Write a possible structure for compound D.

When all the functional groups allowed by the molecular formula are established, then chemical reactions are utilized to break down the skeleton into smaller molecules and, ultimately, usually after a sequence of transformations, into known compounds. The identity of these is established by the identity comparisons with physical properties discussed in Sec. 1-3. When enough of such reactions have been examined, it is possible to deduce an unambiguous molecular structure for the original unknown compound by working back through this history of sequential degradative reactions. The deductive process here derives from the expectation of molecular change in each reaction.

The elaborate but rigorous structural logic to be found in the histories of many of these structural studies presents some fascinating intellectual challenges. Some of the great classical structure problems are natural products (Chap. 27) first isolated in the early nineteenth century. Some of these structure elucidations spanned almost the entire history of organic chemistry, involving many chemists and the publication of long series of papers. The structures of quinine, morphine, cholesterol, and strychnine are major examples of this kind and their structures are shown here partly for an appreciation of their complexity and partly for an appreciation that our command of structure, after Chap. 2, allows the details of these structures to be understood; many of the functional groups of the preceding section will be found in them.

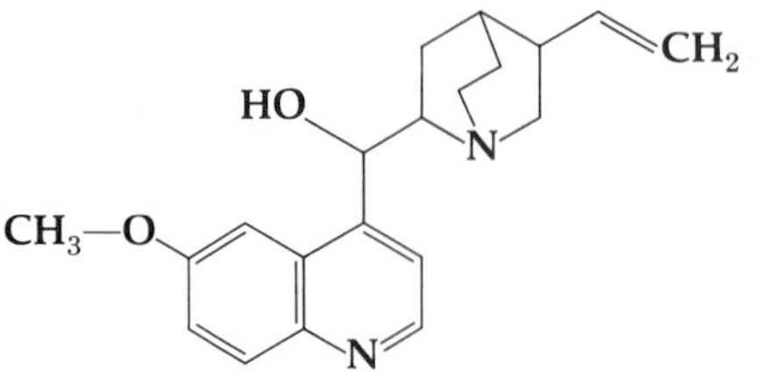

Quinine

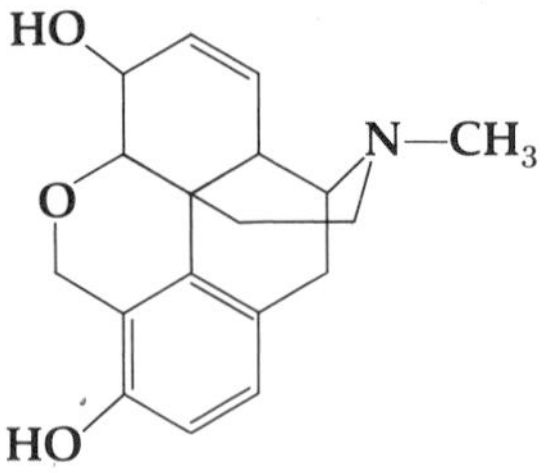

Morphine

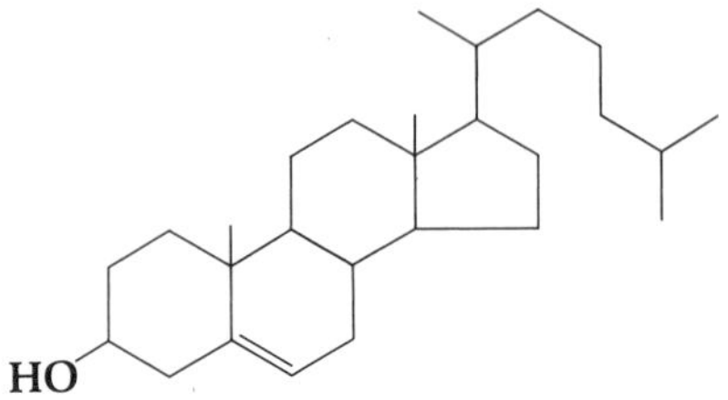

Cholesterol

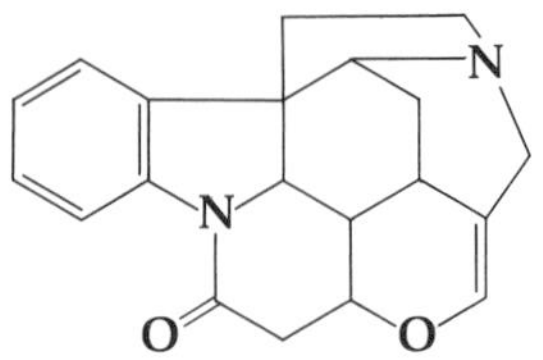

Strychnine

Physical methods of structure determination only became possible after the 1940s. In the last decade, however, they have virtually supplanted the chemical approach. Physical methods have several advantages. Observations can be carried out on very small samples (∼1 mg) of material and these are not affected by the observation, as they are in chemical reactions, and hence may be used over. Furthermore, the observations and their interpretation take far less time.

In the nineteenth century, of course, considerable effort was expended in tabulating physical constants of known compounds in the hope that such lists might be rewarded by some insight into correlations with structural features. Such attempts were made with all physical measurements available, such as melting point, density, index of refraction, etc., and perhaps most extensively with tabulation of crystal facet angles (goniometry). However, these searches all failed to yield significant structural correlations since the physical datum in each case was a *single number*, which encompassed a complex mixture of responses to many structural features at once and these could not be unraveled. Structure correlation became possible only with methods that produced a rich spectrum of separate responses from a single compound such that each particular response might be caused by a single structural feature. The contemporary physical methods of value are all of this kind:

Ultraviolet and visible light spectroscopy (UV and VIS)
Infrared spectroscopy (IR)
Nuclear magnetic resonance spectroscopy (NMR)
Mass spectra (MS)

These methods, as well as the more traditional physical properties, are discussed in Chap. 7 along with protocols for their use in solving structural problems. These physical methods are so powerful for elucidating molecular structure that not only have they largely supplanted chemical procedures but they also now allow the solution to structural problems to be accomplished in a far briefer time. Finally, we may note here that X-ray diffraction is an independent and potent method for complete structure solutions. Although the technique evolved in the last several decades entirely in the hands of crystallographic specialists, it is coming increasingly to be applied by organic chemists to their own problems and with compelling success.

Although physical methods are described in Chap. 7, it is useful here to examine some chemical examples of structure proof, and more will be found in subsequent chapters dealing with various chemical reactions. We shall need to introduce two simple chemical reactions here for use in our examples. These are reactions widely used for structural studies and both are reactions of carbon-carbon multiple (π) bonds.

In **hydrogenation**, the compound is allowed to react with hydrogen (H_2) in the presence of a finely divided metal catalyst (usually **Pt, Pd, Ni**). The hydrogen attacks the π bonds and forms new **C—H** σ bonds in their place. The reaction is said to **saturate** (with hydrogen) the unsaturated groups. As usually practiced, hydrogenation saturates **C=C** and **C=N** (also **C≡C** and **C≡N**) groups but generally leaves **C=O** groups unaffected. Hydrogenation for structural purposes is carried out quantitatively, weighing the compound first and measuring the uptake volume of absorbed hydrogen gas during the reaction. This makes it possible to ascertain the number of moles of hydrogen reacting with one mole of unsaturated compound and thus the number of sites of unsaturation in its molecule.

$$CH_3(CH_2)_5-CH=CH_2 + H_2 \xrightarrow{\text{Pt catalyst}} CH_3(CH_2)_5-CH_2-CH_3$$

$$(CH_3)_2C=CH-CH=CH_2 + 2H_2 \xrightarrow{\text{Pt catalyst}} (CH_3)_2CH-CH_2-CH_2-CH_3$$

In the equations above the functional part of the starting molecule is the carbon-carbon double bond. This bond, along with the atoms attached to it, is shown in color; the same carbon atoms are shown in color in the formulas of the products. The value of hydrogenation in structure proof is that it allows a distinction to be made between multiple bonds (which react) and rings (which do not react†), the sum of which is already known from the index of hydrogen deficiency.

The second reaction of unsaturated compounds, with ozone (O_3), is selective in the sense that saturated centers are almost never attacked. It is useful in cleaving a molecule specifically at its double bonds. **Ozonolysis** (cleavage by ozone) is actually a two-stage process. First, ozone is added and carbon-carbon double bonds are split, forming unstable products known as **ozonides**.

---

†Some three-membered rings can be opened by hydrogenation so that the distinction is not always rigorous.

Ozonides are often explosive and hence are not usually isolated. They are cleaved by hydrolysis in the presence of a reducing agent. The overall procedure produces two new molecules having carbonyl groups in place of the carbon-carbon double bond of the original molecule; the products are aldehydes or ketones.

$$CH_3CH{=}\overset{\overset{\displaystyle CH_3}{|}}{C}CH_3 + O_3 \longrightarrow CH_3\overset{O}{CH}\diagdown\diagup \overset{CH_3}{\underset{CH_3}{C}} \xrightarrow{Zn + H_2O}$$

An ozonide

$$CH_3\overset{\overset{\displaystyle H}{|}}{C}{=}O + CH_3\overset{\overset{\displaystyle O}{\|}}{C}CH_3$$

Assume that a new hydrocarbon, C_4H_8, has just been discovered. The molecular formula shows that the molecule must contain either one ring or a double bond. Only five isomeric structures, I to V, can be written to accommodate the formula.

$$\underset{I}{CH_3CH_2CH{=}CH_2} \qquad \underset{II}{CH_3\overset{\overset{\displaystyle CH_3}{|}}{C}{=}CH_2} \qquad \underset{III}{CH_3CH{=}CHCH_3}$$

$$\underset{IV}{\overset{\displaystyle CH_2{-}CH_2}{\underset{\displaystyle CH_2{-}CH_2}{|\qquad|}}} \qquad \underset{V}{CH_3CH\diagup\diagdown\overset{CH_2}{\underset{CH_2}{|}}}$$

If the compound reacts rapidly with hydrogen in the presence of a palladium catalyst, structures IV and V are eliminated. The three remaining unsaturated hydrocarbons can be distinguished by ozonolysis. If the compound is III, the identification is especially easy. Since the compound is symmetrical, only a single product will be obtained. Note that the two steps in ozonolysis are indicated in a condensed equation by numbers placed over and under the reaction arrow.

$$CH_3CH{=}CHCH_3 \xrightarrow[\text{2) } H_2O, \text{ Zn}]{\text{1) } O_3} 2CH_3\overset{\overset{\displaystyle H}{|}}{C}{=}O$$

If two compounds are obtained on ozonolysis, it will be necessary to isolate the C_3H_6O compound and study its structure, particularly as to whether it is an aldehyde or a ketone.

$$CH_3CH_2CH{=}CH_2 \xrightarrow[\text{2) } H_2O, \text{ Zn}]{\text{1) } O_3} CH_3CH_2\overset{\overset{\displaystyle H}{|}}{C}{=}O + CH_2{=}O$$

$$CH_3\overset{\overset{\displaystyle CH_3}{|}}{C}{=}CH_2 \xrightarrow[\text{2) } H_2O, \text{ Zn}]{\text{1) } O_3} CH_3\overset{\overset{\displaystyle CH_3}{|}}{C}{=}O + CH_2{=}O$$

Consider another molecule of unknown structure having the formula C_8H_{12}. Examination of the formulas for all the possible isomeric compounds of the stated composition is too tedious a job to be practical. Likewise, consideration of all possible results of chemical experiments would require too much time and space. Therefore, experimental results that lead directly to the correct structure will be considered. The original molecule has an index of hydrogen deficiency of 3. Exhaustive hydrogenation gives C_8H_{16}, but careful hydrogenation over a palladium catalyst of low activity gives another compound, C_8H_{14}. The results indicate that the original compound contains one ring and that the other hydrogen deficiencies must be due either to a triple bond† or to two double bonds that differ widely in their reactivity to hydrogenation.

$$C_8H_{12} + 2H_2 \xrightarrow{\text{Pt}} C_8H_{16}$$

$$C_8H_{12} + H_2 \xrightarrow{\text{Pd}} C_8H_{14}$$

Ozonolysis of C_8H_{14} gives a single compound, $C_8H_{14}O_2$. This result shows that the eight carbon atoms must be attached in a ring which contains the double bond, since otherwise the product could not include all eight original carbons. The following is one possible solution:

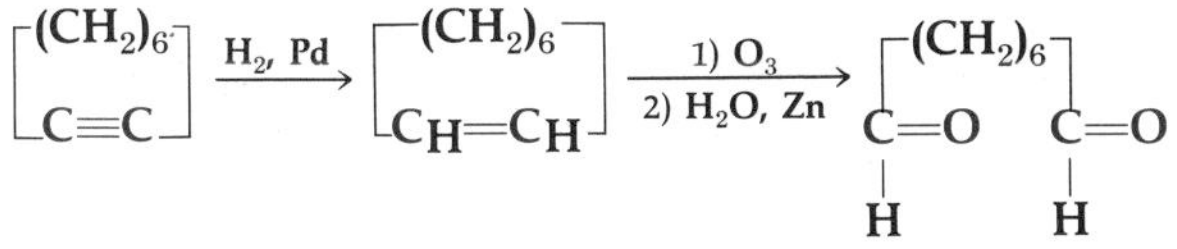

$$\begin{bmatrix} (CH_2)_6 \\ C \equiv C \end{bmatrix} \xrightarrow{H_2,\ Pd} \begin{bmatrix} (CH_2)_6 \\ CH = CH \end{bmatrix} \xrightarrow[\text{2) H}_2\text{O, Zn}]{\text{1) O}_3} \begin{bmatrix} (CH_2)_6 \\ C=O \quad C=O \\ \ \ \ | \qquad | \\ \ \ H \qquad H \end{bmatrix}$$

A number of other structures are still possible, and a further reaction which determines the number of terminal CH_3— groups would be useful. Should it show none in the $C_8H_{14}O_2$ the above structure for C_8H_{14} is uniquely determined.

Using no three-membered ring structures, deduce structures from the following evidence. Examine whether a unique solution is possible in each case.

PROBLEM 3-8

The compound C_6H_6 gives *only* $C_2H_2O_2$ on ozonolysis.

PROBLEM 3-9

The compound $C_{10}H_{16}$ absorbs two moles of H_2 (with catalysis) and gives on ozonolysis two diketones, $C_6H_{10}O_2$ and $C_4H_6O_2$.

PROBLEM 3-10

The compound C_6H_{10} reacts with one mole of H_2 catalytically and gives a single ketoaldehyde $C_6H_{10}O_2$ on ozonolysis.

† Triple bonds usually react faster than double bonds and the resulting double bond product can often be isolated before it reacts further.

PROBLEM 3-11

Catalytic hydrogenation of $C_{10}H_{20}$ requires only one mole of H_2 and ozonolysis yields only a single ketone, $C_5H_{10}O$.

PROBLEM 3-12

A ketone $C_9H_{14}O$ absorbed one H_2 on catalytic hydrogenation to give a new ketone $C_9H_{16}O$ and on reaction with ozone yielded a ketone C_3H_6O and a diketone $C_6H_8O_2$ which was found to have one —CH_3 group.

3-4 NOMENCLATURE

Even though the rules for generating molecular structures are few and simple, there results a fabulous wealth of molecular variety from the process, and with it arises a formidable problem of nomenclature for these structures. Yet nomenclature is vitally important since, for reasons of linear communications either auditory or printed, chemists commonly require verbal descriptions of compounds and their molecules. On the other side it must also be observed that organic chemists rarely talk for long without paper or blackboard available for writing molecular structures, even in the most informal discussion!

The nomenclature problem is in many ways the hardest part of learning organic chemistry and is the major contributor to its deplorable reputation of being only a massive burden of memorization. Various textbooks cope with it in various ways, usually trying to minimize the learning chore by introducing it piecemeal throughout the book. However, the founding fathers of organic nomenclature did devise a system† as neat and logical as structural groups themselves and it seems a pity not to present this in one place, at least in grand outline, for in that way it is more quickly and integrally comprehended. Therefore, the main points of nomenclature are laid down in this section, followed by a survey of the main classes of compounds (Chap. 4), with many named as examples, and finally, for fine points of detail, the entire system is presented in the Appendix. However, the language of organic chemistry, like every other language, grew up first in response to need and so grew up with rather little logic. Therefore, although the system indicated here forms the main core of current practice, there remains in use today a number of "trivial" names, retained from the days before systemization because of their convenience.

The first and last criterion for the name of a structure is that *it must be uniquely definitive*. A name must clearly direct the writing of one molecular structure and one only. The nomenclature system arises parallel to our description of structure: it names a basic skeleton chain of carbons and then designates the names and positions of attached groups on that chain. The basic rules are listed first, with examples following the list.

---

†The systematic nomenclature is called IUPAC nomenclature since its tenets were set by, and are continuously reexamined by, a committee of the International Union of Pure and Applied Chemistry.

1   *The basic unit* in the system is a series of word roots which designate linear chains of carbon atoms of any desired length (Table 3-5). From C_5 upward the roots are the familiar Greek number roots. The generic root for any carbon chain is **alk-**.

2   *A primary suffix* is added to the root to designate unsaturation in the carbon chain:

		General
-ane	Saturated hydrocarbon chain	**Alkane**
-ene	C=C double bond	**Alkene**
-yne	C≡C triple bond	**Alkyne**
-yl	Designates a group (R—)	**Alkyl**

The first three are used to designate whole molecules. If followed by a second suffix they drop the final **-e**. The designation **-yl** is used to indicate a group with a free single bond, named as a unit for attachment to the molecule name (see prefixes below). The generic term **alkyl** is then the term for any saturated hydrocarbon group R—, while **alkenyl** indicates a group R— containing a double bond and **alkynyl** a group with a triple bond.

3   *A secondary suffix*, following the primary suffix, is added to indicate certain functional groups:

-ol	Alcohol (**—OH**)
-one	Ketone (**—CO—**)
-al	Aldehyde (**—CHO**)
-oic acid	Carboxylic acid (**—COOH**)
-oic amide	Amide (**—CONH$_2$**)
-oate	Salt (**—COO⁻**) or ester (**—COOR**) of carboxylic acid
-onitrile	Nitrile (**—CN**)
-oyl	**Acyl** group = generic term for **R—CO—**

4   *A primary prefix*, used immediately *before* the root, is **cyclo-** to indicate a ring compound.†
Without this prefix compounds are assumed to be acyclic.

   †In which all members of the ring are carbon (**carbocyclic** ring).

TABLE 3-5   **Systematic Nomenclature: Word Roots for Carbon Chain Length‡**

C_1	meth-	C_4	but(a)-	C_7	hept(a)-	C_{10}	dec(a)-
C_2	eth-	C_5	pent(a)-	C_8	oct(a)-	Generic:	alk(a)-
C_3	prop-	C_6	hex(a)-	C_9	non(a)-	Higher numbers: see Appendix	

‡The extra (**a**) is used before suffixes beginning with a consonant.

5 *A secondary prefix* is added before the root to designate attached functional groups:

hydroxy-	Alcohol (—**OH**)
alkoxy-	Ether (—**OR**) with simple attached **R**— (alkyl) group
halo-	Halide (—**X**) (**fluoro-, chloro-, bromo-, iodo-**)
keto- or **oxo-**	Ketone (=**O**)
amino-	Primary amine (—**NH₂**)
N-alkylamino-	
or **alkylamino-**	Secondary amine (—**NHR**) with simple attached **R**— (alkyl) group
nitro-	Nitro group (—**NO₂**)
cyano-	Cyano group (—**CN**)

Hydrocarbon branches off the main chain implied by the root are also designated by prefixes **alkyl-**.

6 *Multiple functionality* or branched alkyl groups are labeled with the Greek numerical prefixes:

di-	Two
tri-	Three
tetra-	Four
penta-	Five

7 *Position of attachment* on the carbon chain (designated by the word root) is indicated by numbering the carbons of the chain for identification and placing before each prefix or suffix in the name the number corresponding to its place of attachment on the chain. Multiple bonds are numbered simply by the lower number of the two carbons involved in the bond. In a complex molecule the longest chain bearing the unsaturation and functional groups is chosen as the basic chain and its word root used as a foundation for building the name. The chain is then numbered from the end which gives the lowest numbers to the attached functional groups or multiple bonds. All groups in paragraph 3 which *must* terminate a chain (i.e., all but **-ol** and **-one**) are invariably numbered as 1, the beginning of the chain themselves. If branching alkyl groups must contain multiple bonds or functional groups, separate numbering of the alkyl branch is chosen such that the carbon in the alkyl group which is itself bonded to the main chain is designated as 1. Wherever possible this is avoided by choosing the main chain so as to include all functionality. Groups attached to nitrogen are designated by the letter *N*- instead of a chain number; it is implied that such groups replace hydrogen on the nitrogen of the corresponding primary amine (or amide) for naming purposes.

Finally, it is clear that in numbering several groups on one molecule there will be priorities for naming the groups in order, but it is unnecessary to learn these now when beginning nomenclature study. Such priorities are all included in the more detailed account of nomenclature in the Appendix. For the present, the single criterion will be that the name unambiguously describes a single molecular structure.

The rules may now be illustrated with examples. Unbranched hydrocarbons need only suffixes and numbers to achieve unique definition of structure in the name. In order to clarify the parts of each name, the roots are printed in boldface in each instance.

$$\overset{1}{CH_3}\overset{}{CH_2}\overset{}{CH_2}\overset{}{CH_3} \quad \overset{1}{CH_2}{=}\overset{2}{CH}\overset{3}{CH_2}\overset{4}{CH_3} \quad \overset{1}{CH_3}{-}\overset{2}{CH}{=}\overset{3}{CH}{-}\overset{4}{CH_3} \quad \overset{1}{CH_3}{-}\overset{2}{C}{\equiv}\overset{3}{C}{-}\overset{4}{CH_3}$$

Butane   1-**Butene**   2-**Butene**   2-**Butyne**

$$\overset{1}{CH_2}{=}\overset{2}{CH}{-}\overset{3}{CH}{=}\overset{4}{CH}{-}\overset{5}{CH_3} \quad CH_2{=}CHCH_2CH{=}CH{-}CH_3$$

1,3-**Penta**diene   1,4-**Hexa**diene
(*not* 2,4-pentadiene)   (*not* 2,5-hexadiene)

$$CH_3{-}CH{=}CHCH_2{-}C{\equiv}C{-}CH_3$$

2-**Hept**en-5-yne

Branching in the hydrocarbon may now be illustrated, taking for the base name (root), the longest linear carbon sequence (chain) containing all the multiple bonds. Names for the appended alkyl groups are also generated from Table 3-5 with added -yl: CH_3- (**methyl**); CH_3CH_2- (**ethyl**); $CH_3CH_2CH_2-$ (**propyl**); the alkyl roots are also printed in boldface.

$$CH_3\overset{CH_3}{\overset{|}{C}}{=}CHCH_3 \qquad CH_3CH_2\overset{CH_3}{\overset{|}{C}}{=}CH_2 \qquad CH_3CH_2CH_2\overset{CH_2CH_2CH_3}{\overset{|}{C}}HC{\equiv}CH$$

2-**Methyl**-2-**butene**   2-**Methyl**-1-**butene**   3-**Propyl**-1-**hexyne**

$$CH_3{-}CH_2{-}\underset{\underset{CH_3}{|}}{\overset{\overset{CH_3}{|}}{C}}{-}CH_2{-}\overset{CH_2{-}CH_3}{\overset{|}{CH}}{-}CH_2{-}CH_3$$
$$\quad 7 \quad\;\; 6 \quad\; 5\; \quad 4 \quad\; 3 \quad\; 2 \quad\; 1$$

3-**Ethyl**-5,5-**dimethyl**-**heptane**

3-**Propyl**-4-**ethyl**-1-**heptene**
(*not* 3-**ethenyl**-4-**ethyl**-octane)

4-(1,2-**Dimethylbutyl**)-
1,3,5-**heptatriene**

Examples of the incorporation of functional groups can now be added to the list. The variations shown should be studied with care in order to consolidate a useful understanding of the system.

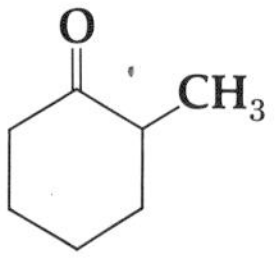

2-Methylcyclohexanone
(2-methyl-1-cyclohexanone
or 2-methyl-cyclohexan-1-
one are redundant)

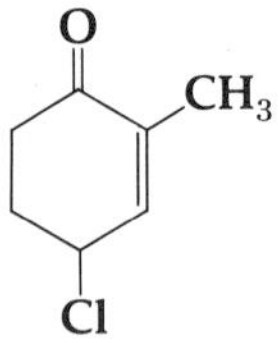

2-Methyl-4-chloro-
2-cyclohexenone

$$\overset{6}{C}H_3\overset{5}{C}H\overset{4}{C}H=\overset{3}{C}\overset{2}{C}H_2\overset{1}{C}HO$$
with CH₂CH₃ on C-3 and OH on C-5

3-Ethyl-5-hydroxy-3-hexenal
(*not* 3-ethyl-3-hexen-5-ol-al)

$$CH_3CH_2NHCH_2\overset{CH_3}{\underset{|}{C}}HOH$$

N-Ethyl-1-amino-2-propanol
(or 1-ethylamino-2-propanol)

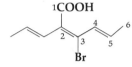

2-(1-Propenyl)-3-bromo-
2,4-hexadienoic acid

$$CH_3\overset{CH_3}{\underset{|}{C}}CH-\overset{O}{\overset{\|}{C}}-CH_2-NH-\overset{O}{\overset{\|}{C}}CH=CH_2$$
with F on the CH

N-Propenoyl-1-amino-3-fluoro-
4-methyl-2-pentanone

CN
$$CH_2CHCH_2COOCH_2CH_3$$
OCH₃

Ethyl-3-methoxy-
4-cyano-butanoate

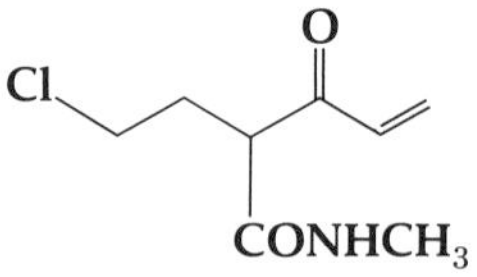

N-Methyl-3-keto-
2-(2-chloroethyl)-4-
pentenoic acid amide

$$CH_3CH_2O-C\equiv CCH_2CCH=CHCH_3$$
$$\overset{\|}{O}\quad\overset{\|}{O}$$

1-Ethoxy-oct-6-en-1-yn-3,5-dione

PROBLEM 3-13

Write systematic names for the following structures.

a $$CH_3CH_2\overset{CH_3}{\underset{|}{C}}HCH_2OH$$

b $$CH_3CH_2\overset{CH_3}{\underset{|}{C}}HCHO$$

c $$CH_3CH=\overset{CH_3}{\underset{|}{C}}-COOH$$

d $$CH_3CH_2-\overset{CH_2}{\overset{\|}{C}}-COOH$$

e $$CH_3CH=CHC\equiv CCH_3$$

f $$(CH_3)_2C=CHCH=CHCH(CH_3)_2$$

g   CH$_3$CHBrCCOOCH$_3$ (with CH$_3$ on top and OH below the central C)

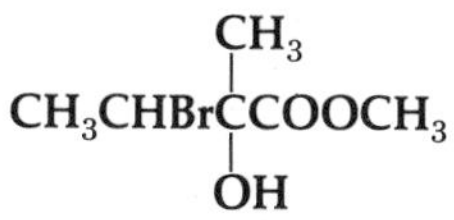

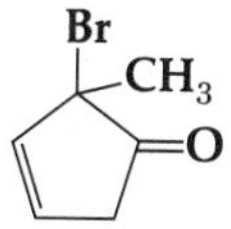

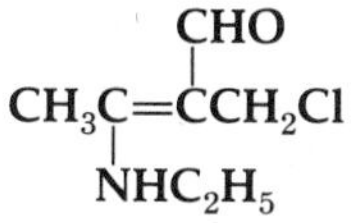

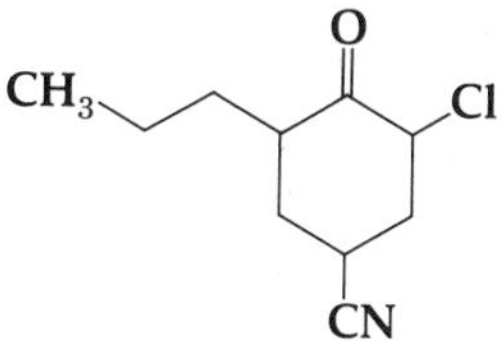

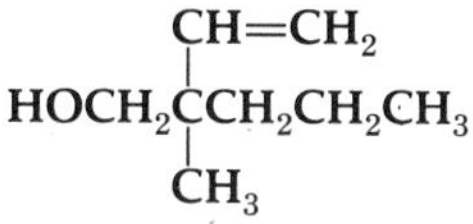

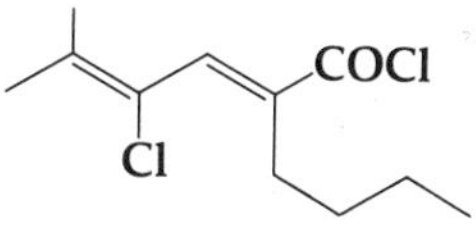

k

h   CH$_3$C=CCH$_2$Cl (with CHO on top and NHC$_2$H$_5$ below)

l

i   HOCH$_2$CCH$_2$CH$_2$CH$_3$ (with CH=CH$_2$ on top and CH$_3$ below)

m

j   HC≡CCH=CCHCHCONHCH$_3$ (with CH$_3$ on top, Br and OH below)

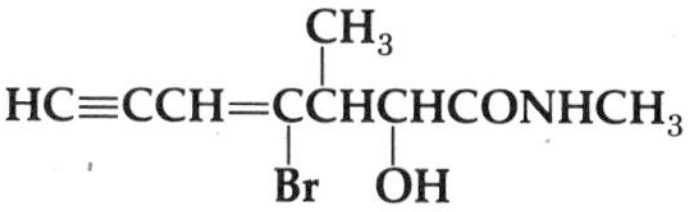

n

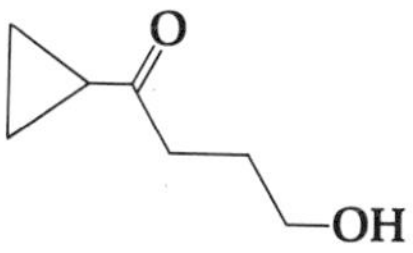

PROBLEM 3-14

Write the structures of the following compounds.

a 4-Hydroxy-3-pentenoic acid
b N,3,3-Trimethyl-butanoamide
c 6-Methyl-1,3,5-heptatriene
d Octa-2,4-dien-6-yn-al

e 4-Cyano-4-ethyl-3-heptanone
f 3,4-Dibromo-4-ethyl-cychlohexanone
g 4,4-Dimethyl-2,5-cyclohexadienone
h 4-Cyclopropyl-3-chloro-1,2-cyclo-butanediol

The IUPAC system was of course grafted onto nineteenth century no-menclature practices and it has fortunately now taken a prime place in practical usage. Nonetheless, certain names were so thoroughly familiar that chemists continue to use them. First among these is the list of names for the common carboxylic acids and their derivatives listed in Table 3-6; aldehydes are here also named as acid derivatives. Greek-letter symbolism for indicating chain position is used with these acid names as well as in many other compounds; the carbons next to a carbonyl group (acid, ketone, etc.) are labeled in order: α, β, γ, δ, etc., as in γ-hydroxybutyric acid (HOCH$_2$CH$_2$CH$_2$COOH) or

β-bromoketones (Br—C—C—C—C—) (with O double bonded to third C).

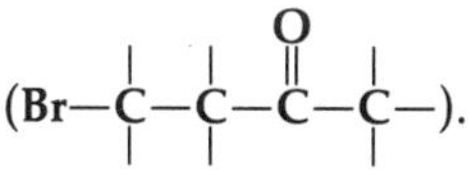

TABLE 3-6　**Naming of Common Carboxylic Acids and Their Derivatives**

Acid (RCOOH)	*Formula*	*Derivatives*
Formic acid	HCOOH	Salts or esters: root + **-ate**
Acetic acid	CH_3COOH	Amides: root + **-amide**
Propionic acid	CH_3CH_2COOH	Nitriles: root + **-onitrile**
Butyric acid	$CH_3CH_2CH_2COOH$	Aldehydes: root + **-aldehyde**
Isobutyric acid	$(CH_3)_2CHCOOH$	Acyl group: root + **-yl**
Valeric acid	$CH_3CH_2CH_2CH_2COOH$	
Acrylic acid	$CH_2{=}CH{-}COOH$	
Benzoic acid	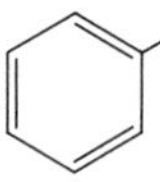 COOH	

Finally, a group of widely used common names is collected in Table 3-7 and is presented without apology. The benzene ring (Table 3-7) is responsible for a great variety of chemistry and the three possible isomeric disubstituted derivatives (two groups attached to the ring) are commonly labeled *ortho-* (*o-*), *meta-* (*m-*), and *para-* (*p-*), referring to 1,2-, 1,3-, and 1,4-orientations of two substituents, respectively:

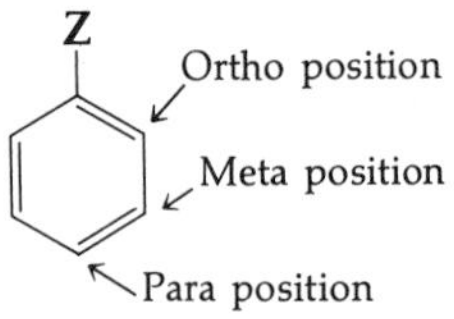

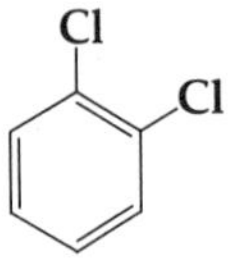

o-Dichlorobenzene
(1,2-dichlorobenzene)

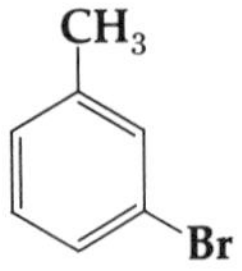

m-Bromotoluene
(1-bromo-3-methyl-
benzene)

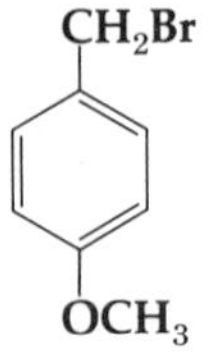

p-Methoxybenzyl
bromide
(4-methoxyphenyl
bromomethane)

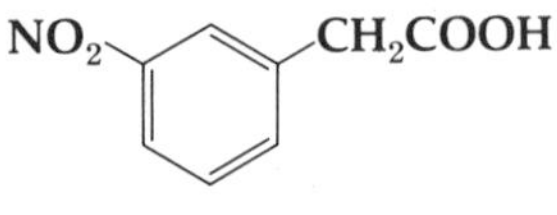

m-Nitrophenyl-
acetic acid
(3-nitrophenyl-acetic acid)

Examples

Methyl formate (**HCOOCH$_3$**)

Ethyl valerate (**CH$_3$CH$_2$CH$_2$CH$_2$COOCH$_2$CH$_3$**)

Sodium isobutyrate (**(CH$_3$)$_2$CHCOO$^-$Na$^+$**)

Acetyl chloride (**CH$_3$COCl**)

Dimethylformamide (**HCON(CH$_3$)$_2$**)

N-Benzoyl-dipropylamine (**(CH$_3$CH$_2$CH$_2$)$_2$N—CO)**)

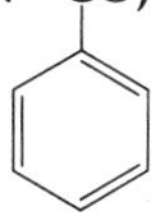

Propionaldehyde (**CH$_3$CH$_2$CHO**)

Butyronitrile (**CH$_3$CH$_2$CH$_2$CN**)

N-Ethyl-acrylamide (**CH$_2$=CH—CONHCH$_2$)**
                                                    CH$_3$

TABLE 3-7   **Common Trivial Names**

Hydrocarbons		*Groups*	
Olefin	Alkene	Methylene	—CH$_2$—
Ethylene	CH$_2$=CH$_2$	Methine	—CH—
Acetylene	HC≡CH	Alkylidene	RCH=
Allene	CH$_2$=C=CH$_2$	(Ethylidene	CH$_3$CH=)
		Isopropyl	(CH$_3$)$_2$CH—
Benzene		Isobutyl	(CH$_3$)$_2$CHCH$_2$—
		sec-Butyl	CH$_3$CH$_2$CHCH$_3$
Toluene		tert-Butyl	(CH$_3$)$_3$C—
		Vinyl	CH$_2$=CH—
		Allyl	CH$_2$=CH—CH$_2$—
Others		Phenyl	
Acetone	CH$_3$COCH$_3$		
Epoxide		Benzyl	
(added to alkene name)			

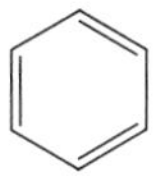

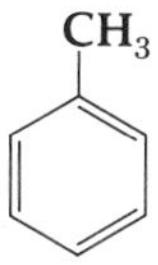

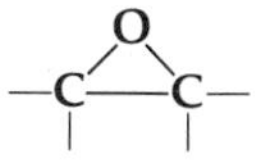

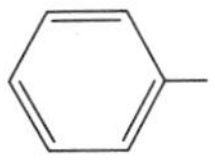

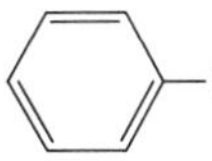

In Table 3-8 are presented a number of common abbreviations used in this book and elsewhere for various molecular groupings.

The procedure to follow in naming a structure is the following:

1 Identify the longest chain (or ring) to contain—if possible—all the functional groups.
2 Select the corresponding word root (Table 3-5), prefixing *cyclo-* for rings.
3 Number the chain from one end, choosing the end
   a so that —CHO, —COOH, —COOR, —COCl, etc., bear number 1, or
   b so that the numbers given to functionalities are as small as possible.
4 Add to the name root a suffix expressing the number and kind of carbon-carbon multiple bond (-ane, -ene, -yne, -diene, -triene, etc.) and number the positions of these in the name.
5 Add either prefixes or suffixes, with appended numbers, to express the number and position of each attached functionality or branched alkyl group.

TABLE 3-8 **Abbreviations for Groups in Common Use**

Abbreviation	Group	Examples
R—	Alkyl	R—Cl (CH_3—Cl, CH_3CH_2Cl, $CH_3CH_2CH_2Cl$, $(CH_3)_2CHCl$, etc.)
Ar—	Aryl	$ArNO_2$ (p-$CH_3C_6H_4NO_2$)
X—	Halide (F, Cl, Br, I)	R—X (R—F, R—Cl, R—Br, or R—I)
Me—	Methyl (CH_3—)	CH_3COOMe; 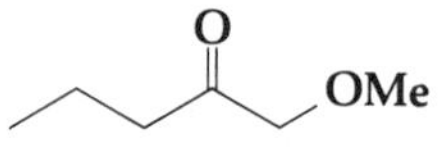
Et—	Ethyl (C_2H_5—)	$HCONEt_2$; $Et_3NH^+Cl^-$
Pr—	Propyl (C_3H_7—)	PrCl; ClCOOPr
i-Pr—	Isopropyl	*i*-PrOH (($CH_3)_2CH$—OH)
i-Bu—	Isobutyl	*i*-BuNHCOPr (($CH_3)_2CHCH_2$—NH—CO—$CH_2CH_2CH_3$)
s-Bu—	*sec*-Butyl	*s*-Bu_2O ($CH_3CH_2\underset{CH_3}{CH}$—O—$\underset{CH_3}{CH}$—$CH_2CH_2$)
t-Bu—	*tert*-Butyl	*t*-BuO^-K^+ (($CH_3)_3C$—O^-K^+)
ϕ— (or Ph—)	Phenyl (C_6H_5—)	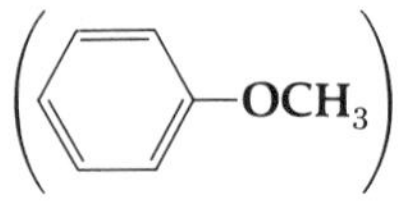 ϕOMe
Ac—	Acetyl (CH_3CO—)	 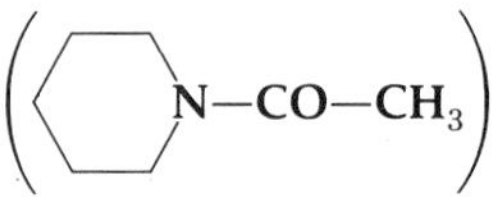

The procedure to follow in reading a name is the following:

1 Locate the word root (cf. Table 3-5), write down a linked straight chain of that many carbons, and number it.
2 Locate the suffix(es) following the root which indicates the number, kind, and position of multiple bonds in the chain, and fill these in on the structure at the numbered carbon atoms.
3 Append the functional groups and branched alkyl groups indicated by prefixes and suffixes at the carbon atoms correspondingly numbered.

The problems at the end of the chapter offer some practice in nomenclature. A much more intimate grasp of the nomenclature procedure is probably obtained by writing down structures at random and then attempting to put proper names to them. In many cases this will not be possible without consulting the Appendix, but the practice is useful and will serve to define the problems of nomenclature more sharply.

PROBLEMS

3-15 Write common and/or systematic names for the following compounds.

a $CH_3-CH_2-CH-C{\equiv}C-CH_3$
    $\quad\quad\quad\quad CH_2-CH_2-CH_3$

e

b $CH_3-CH-CH_2-\overset{CH_3}{\underset{CH_3}{C}}-OH$
    $\quad\quad NO_2$

f

c $Cl-CH_2-CH_2-O-CH-CH-CH_3$
    $\quad\quad\quad\quad\quad CH_3\ Br$

g

d $CH_3-CH=CH-\overset{OH}{\underset{CH_3}{C}}-CH_2-\overset{OH}{\underset{CH_3}{C}}-CH_3$

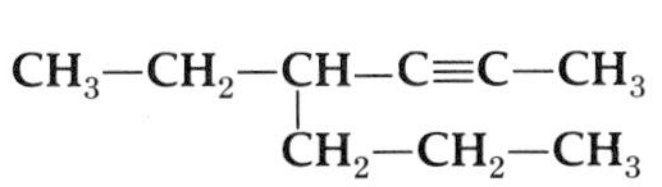

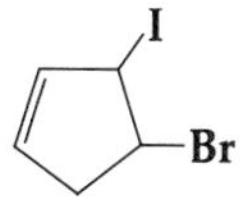

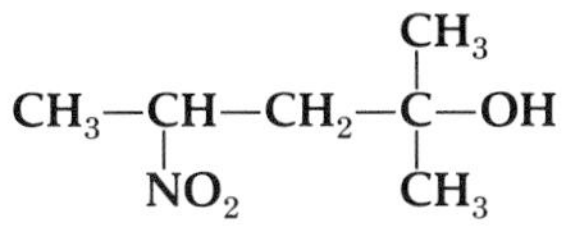

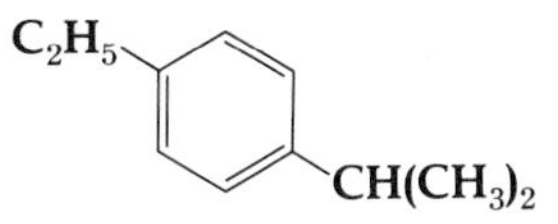

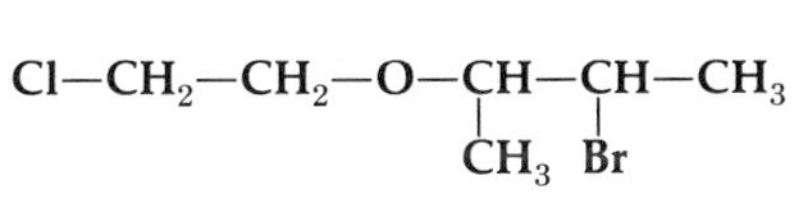

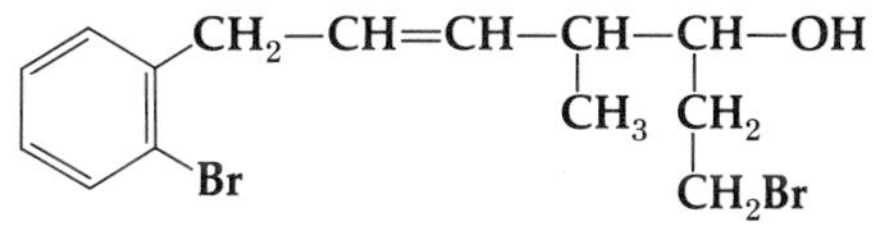

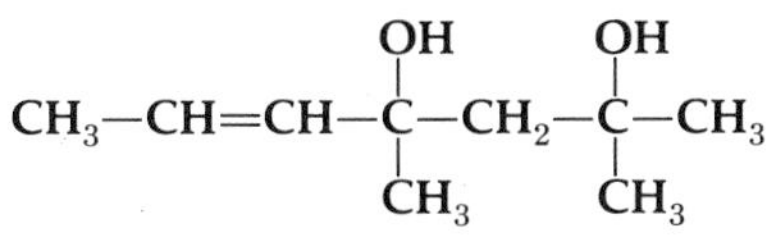

h

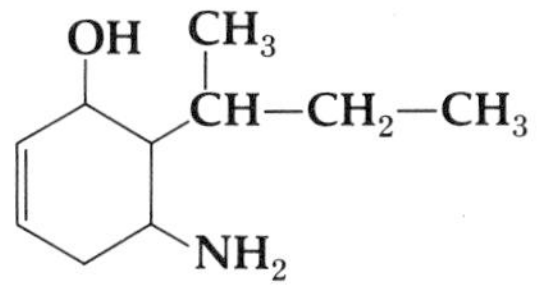

3-16 Write structures for the following compounds:

a Triisopropylamine
b Methyl benzyl sulfide
c Vinyl allyl ether
d 1,3,5-Trimethyl-1,4-cyclo-octadiene

e 2-N-Methylamino-3,4-pentadiene
f p-Aminophenol
g o-Phenylnitrobenzene
h 1-Hydroxymethyl-3-ethyl-4-iodobenzene

3-17 In each case below show: (1) the index of hydrogen deficiency; (2) a structure that fits the given molecular formula; (3) an unambiguous name for the structure.

  a $C_5H_{10}O_2$ d $C_{12}H_{11}NO$

  b $C_8H_{13}N_3O_2$ e $C_{10}H_{17}O_4Br$

  c $C_6H_9NOCl_2$ f $C_7H_{11}NOClBr$

3-18 What is wrong with the following structure names?

  a 3-Methyl-3-pentanone   f 1-Cyclohexenone
  b Propenone •   g *para*-Tribromobenzene
  c Pentanone   h 2,2,2-Tribromobutanal
  d Cyclopentanoic acid   i Tribromobutanol
  e 2-Butynoic acid ester   j Hexenedioic acid

3-19 What should be the principal product of the ozonolysis of natural rubber?

$$(-CH_2\overset{\overset{\displaystyle CH_3}{|}}{C}=CHCH_2-)_x$$

   Natural rubber

3-20 A substance with a molecular formula C_7H_{12} absorbed 1 mole of hydrogen on hydrogenation over platinum. Ozonolysis of the substance yielded one compound only, and this showed no aldehyde groups by chemical test. Trace these reactions with structural formulas.

3-21 A substance C_6H_8 absorbs two moles of hydrogen on hydrogenation over platinum. Ozonolysis yielded 1.8 moles of a single compound. Trace these reactions with structural formulas.

3-22 Compound A ($C_9H_8O_2$) absorbed two moles of hydrogen over an inactive palladium catalyst to yield compound B. Hydrogenation of compound B with active platinum resulted in two more moles of hydrogen uptake to yield compound C, which is also formed from A with platinum-catalyzed hydrogenation. All three compounds react under conditions for converting ketones to oximes, compound C yielding an oxime, $C_9H_{18}N_2O_2$. Compound B on ozonolysis yields 2 moles of one substance and 1 mole of another. Trace these reactions with structural formulas.

A SURVEY
OF THE
CLASSES
OF COMPOUNDS

UP to this point our involvement has been with the theory of covalent molecules, how they are constructed, how they may be classified, and how they may be named. In order to consolidate this information and particularly now to see this structural theory in the light of some real organic compounds in common use, this chapter undertakes a modest excursion among the major classes of organic compounds. Some of the common nomenclature of simple, widely used compounds is incorporated as well since these trivial names are still in general use.

This chapter must be regarded as different from most of the others in that it does not add to the development of the central theory of organic chemistry. The survey of common organic substances conducted in this chapter serves as a pause, primarily for acquaintance with the actual stuff from which the theory has grown. Since the presentation in this book is intended chiefly to develop a theoretical framework from which most facts can be deduced, or at least understood in a context, the student should not look upon the many facts presented in this chapter as candidates for rote memorization. With this view the chapter need not be as formidable as it might first appear; it is simply a tour through the laboratory.

PROBLEM 4-1

Write systematic IUPAC names for every underlined example in this chapter, which is given another (trivial) name.

4-1 HYDROCARBONS

Alkanes

Hydrocarbons in which all the carbon atoms occur in a continuous sequence are known as normal hydrocarbons. They are sometimes called linear or straight-chain hydrocarbons, but these names are misleading because the carbon chains are actually kinked and twisted and are linear only in our symbolic representations. The names and some of the physical properties of a group of normal saturated hydrocarbons are listed in Table 4-1. Much of organic nomenclature is based upon the names in this table (cf. Table 3-5). The common

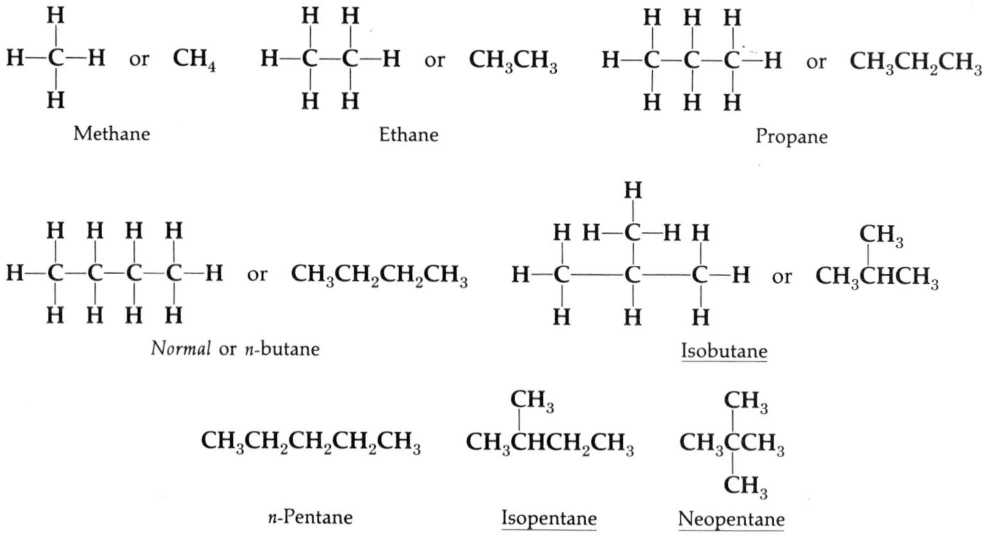

Methane   Ethane   Propane

Normal or n-butane   Isobutane

$$CH_3CH_2CH_2CH_2CH_3 \qquad CH_3CHCH_2CH_3 \qquad CH_3CCH_3$$

n-Pentane   Isopentane   Neopentane

FIGURE 4-1 **Formulas of the eight simplest alkanes**

names of the branched hydrocarbons containing four or five carbon atoms†
are found in Fig. 4-1. These names illustrate some of the early problems of
organic chemical nomenclature. For example, only two butanes exist, one in

† Note the absence of branches in a three-carbon chain. The two formulas

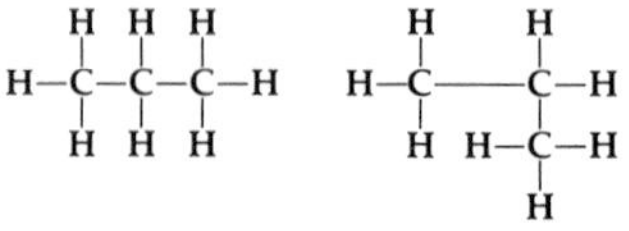

cannot represent different molecules since they indicate the same *sequence* of atoms. *This kind of formula*
never attempts to show the real geometry of molecules.

TABLE 4-1 **Normal Saturated Hydrocarbons**

Name	Formula	Mp	Bp	Sp. gr. (liquid)	State under atmospheric conditions
Methane	CH_4	−183	−162	0.4240	Gas
Ethane	C_2H_6	−183	−89	0.5462	Gas
Propane	C_3H_8	−187	−42	0.5824	Gas
n-Butane	C_4H_{10}	−138	0	0.5788	Gas
n-Pentane	C_5H_{12}	−130	36	0.6264	Liquid
n-Hexane	C_6H_{14}	−95	69	0.6594	Liquid
n-Octane	C_8H_{18}	−57	126	0.7028	Liquid
n-Decane	$C_{10}H_{22}$	−30	174	0.7298	Liquid
n-Eicosane	$C_{20}H_{42}$	36	343	0.7777	Solid
n-Pentacontane	$C_{50}H_{102}$	92		0.7940	Solid

which the four carbon atoms occur in a single, continuous sequence and the other which contains a branch. The first is called *n*-butane, and since the second is an isomeric butane, it is called isobutane. On the same basis the first two pentanes to be characterized were called *n*-pentane and isopentane. Many years later the third isomer was prepared. The new pentane was named neopentane. The prefix iso- is still occasionally used in nomenclature to indicate the presence of the isopropyl group, $(CH_3)_2CH-$ (Table 3-7) in a structure.

Consideration of the common names of the five possible isomeric hexanes shows that while the above principles are repeated, new problems are also encountered. Obviously, the nomenclature devices established with the pentanes are consumed in naming only three of the five hexanes. Furthermore, as the number of carbon atoms in the molecule increases, the number of isomers corresponding to any one composition will multiply rapidly (remember the 62,491,178,805,831 structural formulas for $C_{40}H_{82}$, page 33). Clearly, a systematic procedure for assigning names was needed. The answer of course lies in the systematic nomenclature system developed in Sec. 3-4.

$$CH_3CH_2CH_2CH_2CH_2CH_3 \qquad CH_3\overset{\overset{\displaystyle CH_3}{|}}{C}HCH_2CH_2CH_3$$

$$\text{\textit{n}-Hexane} \qquad\qquad\qquad \text{Isohexane}$$
$$\text{or}$$
$$\text{2-methylpentane}$$

$$CH_3\overset{\overset{\displaystyle CH_3}{|}}{\underset{\underset{\displaystyle CH_3}{|}}{C}}CH_2CH_3 \qquad CH_3CH_2\overset{\overset{\displaystyle CH_3}{|}}{C}HCH_2CH_3$$

$$\text{Neohexane} \qquad\qquad \text{3-Methylpentane}$$

$$CH_3\overset{\overset{\displaystyle CH_3}{|}}{C}H-\overset{\overset{\displaystyle CH_3}{|}}{C}HCH_3$$
$$\text{2,3-Dimethylbutane}$$

Alkenes

The alkenes are aliphatic (open-chain) hydrocarbons containing one or more carbon-carbon double bonds as functional groups. These compounds are said to be *unsaturated* since they do not have the maximum numbers of hydrogen atoms. They are also referred to as **olefins**, an old term that derives from the fact that the simplest member of the series, ethylene (C_2H_4), was called olefiant gas (Latin: *oleum*, oil + *facio*, make) because the gaseous hydrocarbon reacts with chlorine to form an oil ($C_2H_4Cl_2$). Since addition of chlorine to carbon-carbon double bonds is a general reaction of that functional group, the term "olefin" provides both structural description and chemical characterization of the group of compounds containing carbon-carbon double bonds.

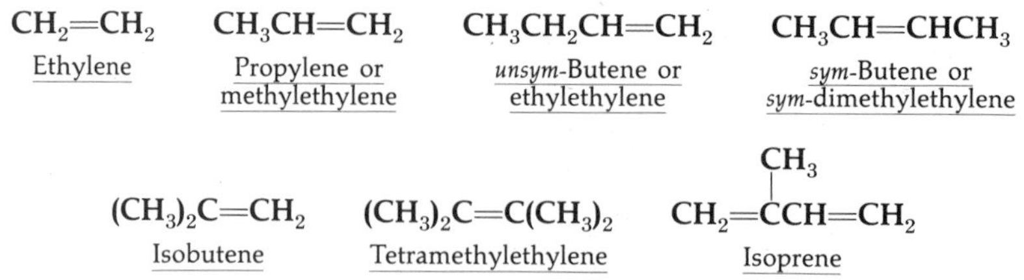

$$CH_2{=}CH_2 \qquad CH_3CH{=}CH_2 \qquad CH_3CH_2CH{=}CH_2 \qquad CH_3CH{=}CHCH_3$$

Ethylene   Propylene or   unsym-Butene or   sym-Butene or
            methylethylene   ethylethylene   sym-dimethylethylene

$$CH_2{=}\overset{\overset{\displaystyle CH_3}{|}}{C}CH{=}CH_2$$

$$(CH_3)_2C{=}CH_2 \qquad (CH_3)_2C{=}C(CH_3)_2 \qquad CH_2{=}\overset{\overset{\displaystyle CH_3}{|}}{C}CH{=}CH_2$$

Isobutene   Tetramethylethylene   Isoprene

FIGURE 4-2 **Common and partially systematic names of the simple alkenes**

The simpler alkenes are often referred to by common names, as shown in Fig. 4-2. These include both names of entirely trivial origin, such as propylene and isobutene,† and partially systematic derived names. In the latter group of names ethylene is often chosen as the parent structure and compounds are named by substitution of alkyl groups for hydrogens in ethylene. Figure 4-2 also illustrates another common nomenclature device, the use of symmetry properties in assigning names. Two compounds can be called dimethylethylenes (sym-butene and isobutene). In one compound both methyl groups are attached to the same carbon atom, but in the other they are symmetrically distributed at each end of the parent structure, ethylene.

The following are examples of IUPAC nomenclature of simple alkenes:

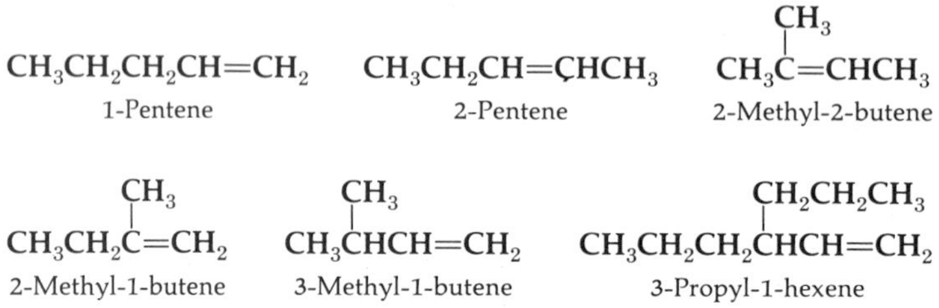

$$CH_3CH_2CH_2CH{=}CH_2 \qquad CH_3CH_2CH{=}CHCH_3 \qquad CH_3\overset{\overset{\displaystyle CH_3}{|}}{C}{=}CHCH_3$$

1-Pentene   2-Pentene   2-Methyl-2-butene

$$CH_3CH_2\overset{\overset{\displaystyle CH_3}{|}}{C}{=}CH_2 \qquad CH_3\overset{\overset{\displaystyle CH_3}{|}}{C}HCH{=}CH_2 \qquad CH_3CH_2CH_2\overset{\overset{\displaystyle CH_2CH_2CH_3}{|}}{C}HCH{=}CH_2$$

2-Methyl-1-butene   3-Methyl-1-butene   3-Propyl-1-hexene

Alkenes that contain two or more double bonds are well known, and some are among the most important organic chemicals. Doubly unsaturated hydrocarbons are known as **dienes**, those containing three double bonds are called **trienes**, and so on. As a group the multiply unsaturated hydrocarbons are called **polyenes**.

Compounds in which a single carbon atom is doubly bonded to two other carbon atoms ($C{=}C{=}C$) are called **allenes**, since allene is the name of the simplest member of the series ($CH_2{=}C{=}CH_2$). Because allenes are both difficult to prepare and very reactive, they are not commonly encountered and only a few members

† Note, however, the similarity between the pairs of names propylene and propane and isobutene and isobutane.

of the series are known. Both IUPAC and common nomenclature are illustrated by the following examples.

$CH_2=C=CH_2$ $CH_3CH=C=CH_2$

    Allene   Methylallene
      or   or
  propadiene   1,2-butadiene

In contrast to allenes, dienes in which two double bonds are separated by one single bond ($C=C-C=C$) are a very important group of chemicals. The best-known example is 1,3-butadiene. The compound is a principal building block for certain kinds of synthetic rubber and is commonly called simply butadiene. Interaction between the two double bonds in 1,3-dienes influences both their chemical and their physical properties. For this reason such dienes are set apart as a subclass, called **conjugated dienes**. Higher polyenes in which the alternate arrangement of double and single bonds are found are called **conjugated polyenes**. Dienes and polyenes in which the double bonds are separated by one or more saturated carbon atoms are said to be **unconjugated**. The special features of conjugated double bonds are explored in Chap. 5.

$CH_2=CH-CH=CH_2$ $CH_3\overset{\overset{\displaystyle CH_3}{|}}{C}=CHCH=CHCH=\overset{\overset{\displaystyle CH_3}{|}}{C}CH_3$

    1,3-Butadiene   2,7-Dimethyl-2,4,6-octatriene
 (a conjugated diene)   (a conjugated triene)

$CH_2=CHCH_2CH=CH_2$ $CH_2=CHCH_2CH_2CH=CH_2$

    1,4-Pentadiene   1,5-Hexadiene
(an unconjugated diene)   (an unconjugated diene)

The two most common *alkenyl groups* are vinyl ($CH_2=CH-$) and allyl ($CH_2=CH-CH_2-$), listed in Table 3-7.

Alkynes

The alkynes contain one or more carbon-carbon triple bonds. The simplest member of the class is acetylene ($HC\equiv CH$), and the class as a whole is frequently referred to as the **acetylenes**. Acetylene itself is very easily prepared and has long been one of the most important raw materials of industrial organic chemistry.

The simple alkynes are often named as derivatives of acetylene. Numbers need not be used in such names to locate substituents on an acetylenic linkage since the maximum possible number of substituents on a triple bond is two. A disubstituted acetylene must have one substituent at each end of the acetylenic linkage, replacing the hydrogens of the parent compound, acetylene. The following examples illustrate the use of both the IUPAC system and the derived names based on acetylene.

$$HC\equiv C-\qquad HC\equiv CCH_2-$$

Ethynyl   Propargyl

FIGURE 4-3   **Common alkynyl groups**

$$CH_3CH_2C\equiv CH\qquad CH_3C\equiv CCH_3$$

Ethylacetylene   Dimethylacetylene
or   or
1-butyne   2-butyne

$$\overset{\displaystyle CH_3}{\underset{\displaystyle |}{CH_3CHC\equiv CCH_3}}\qquad \overset{\displaystyle CH_3}{\underset{\displaystyle |}{CH_3CHCH_2CH_2C\equiv CH}}$$

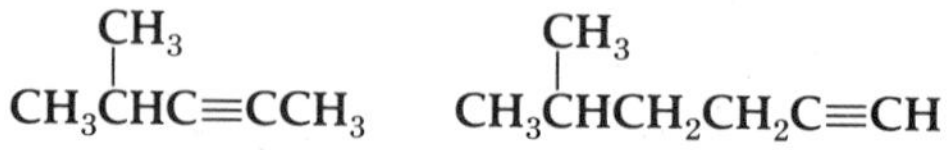

Isopropylmethylacetylene   Isopentylacetylene
or   or
4-methyl-2-pentyne   5-methyl-1-hexyne

$$CH_3C\equiv CC\equiv CC\equiv CCH_3$$

2,4,6-Octatriyne
(a conjugated triyne)

The only two *alkynyl groups* (hydrocarbon groups containing triple bonds) encountered frequently enough to be given common names are shown in Fig. 4-3.

Aromatic Hydrocarbons

Benzene (C_6H_6), the simplest aromatic hydrocarbon, illustrates the characteristic structural features of aromatic compounds. The compound was first discovered by Faraday in 1825, and the controversy over its structure, which has existed since that year, has provided one of the most interesting and important stories of organic chemistry.

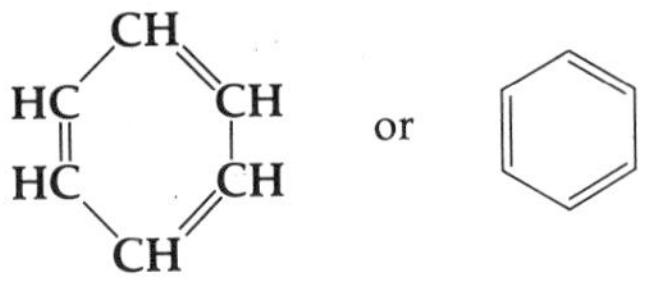

Benzene

The special nature of benzene and other aromatic compounds—molecules containing benzene rings—is discussed in Chap. 5. These compounds are characterized by a special stability and resistance to chemical reaction which is not shown by other unsaturated molecules. For this reason aromatic compounds have always been regarded separately from olefins, despite the fact that it is most convenient to write benzene as if it contained three double bonds. Since we shall see that all six bonds in benzene are equivalent, it is unimportant which positions are written as double bonds. The two possible structures are called **Kekulé forms.**

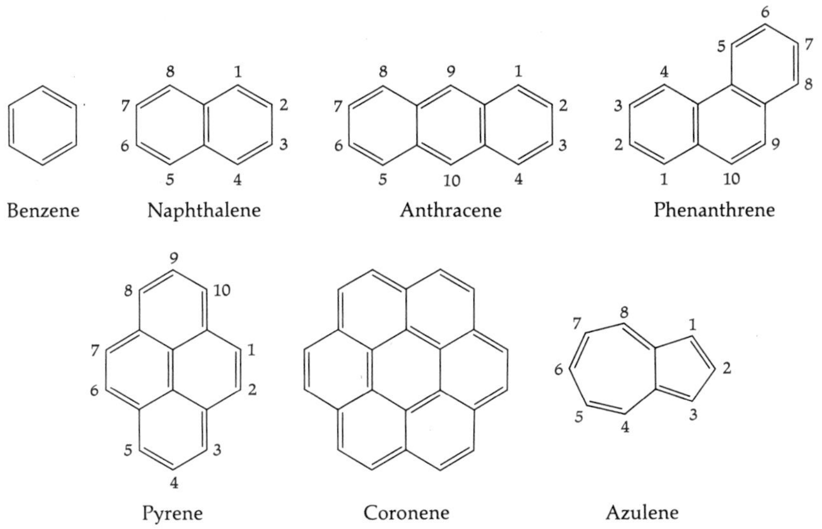

Benzene   Naphthalene   Anthracene   Phenanthrene

Pyrene   Coronene   Azulene

FIGURE 4-4   **The simple aromatic ring systems**

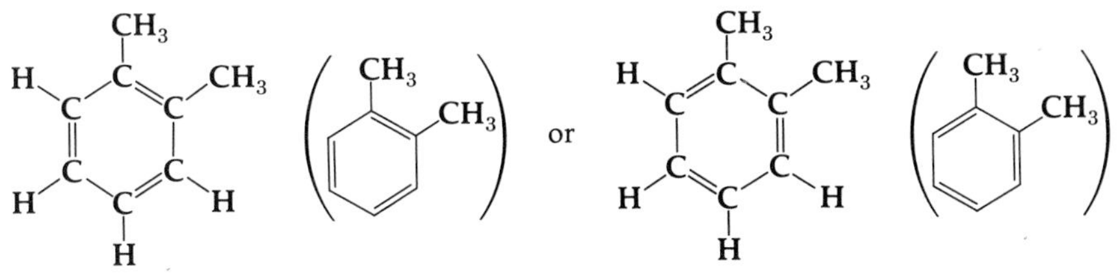

Kekulé forms of 1,2-dimethylbenzene

An important extension of the benzene structure is found in **fused-ring systems**. These are found in compounds containing two or more rings that have two carbon atoms in common. Comparison of the structures of benzene and naphthalene illustrates the principle.

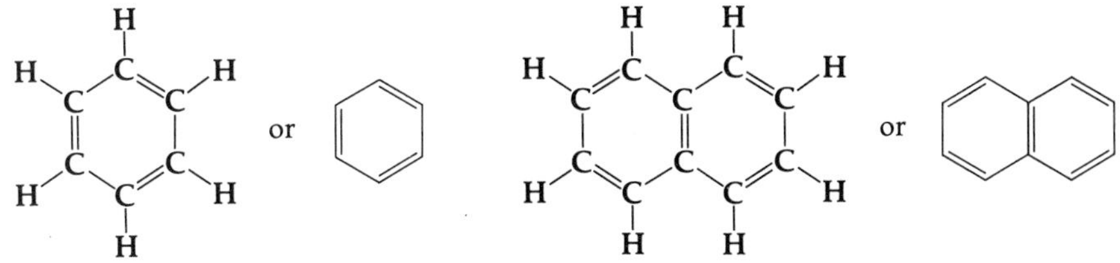

Benzene   Naphthalene

The first six formulas in Fig. 4-4 illustrate the common types of aromatic structure. The last example, azulene, shows that the six-membered ring is not a unique characteristic of aromatic compounds. *All aromatic hydrocarbons contain ring systems in which all carbon atoms are trigonal (sp^2)*, i.e., attached to only three other atoms. Two or more Kekulé formulas can always be written for any aromatic hydrocarbon. For example, four Kekulé formulas can be written for anthracene.†

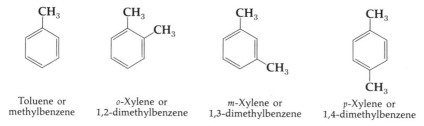

Kekulé formulas for anthracene

Many hydrocarbons contain both aromatic and aliphatic parts and aromatic groups are distinguished from aliphatic by use of the word **aryl** in place of **alkyl**. Particularly important are the alkyl derivatives of naphthalene and benzene. Some of these compounds have been known for a long time and are usually referred to by their common names. They can also be named as derivatives of benzene. In IUPAC nomenclature numbers are assigned to positions on the benzene nucleus and used to locate substituents. Alternatively, the older ortho, meta, para designations are widely used (page 97).

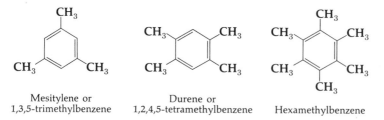

| Toluene or methylbenzene | o-Xylene or 1,2-dimethylbenzene | m-Xylene or 1,3-dimethylbenzene | p-Xylene or 1,4-dimethylbenzene |

The following are other important alkylbenzenes.

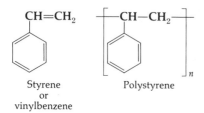

| Mesitylene or 1,3,5-trimethylbenzene | Durene or 1,2,4,5-tetramethylbenzene | Hexamethylbenzene |

An especially important benzene derivative is styrene (vinylbenzene), which can be converted to polystyrene, a familiar plastic.

Styrene
or
vinylbenzene   Polystyrene

† The significance of the double-headed arrow that connects the various Kekulé formulas will be explained in Chap. 5. Briefly, the symbol indicates that more than one formula can be written equally well for a single compound.

Three different kinds of positions are found in the naphthalene nucleus. Those at the junction of the two rings cannot bear substituents if the aromatic structure is maintained. Substituents can, however, be attached to the other two positions. Positions adjacent to the ring junction are called α-positions and those more remote from the ring junction are called β-positions. In IUPAC nomenclature one of the α-positions is always given the number 1.

α-Methylnaphthalene β-Methylnaphthalene

Sources and Uses of Hydrocarbons

Natural gas and petroleum are chief sources of aliphatic hydrocarbons at the present time, and coal is one of the major sources of aromatic hydrocarbons. Coke and carbon monoxide, which are also obtained from coal, can be converted to either aliphatic or aromatic hydrocarbons, and the direct chemical reaction of hydrogen with coal has been used to prepare aliphatic and alicyclic compounds. Most of the volume of hydrocarbons produced commercially is used for fuel, but large amounts are also diverted to the chemical industry for conversion to other useful chemicals.

Natural gas and petroleum are found near the surface of the earth trapped by rock structures. When a well is drilled through the protective cap, the gas is released and gushes forth, carrying with it some of the more volatile liquids (low-boiling petroleum). After the flow of gas ceases, residual petroleum is removed by a variety of pumping procedures. The gaseous fraction is "scrubbed" by passing it through an oil that absorbs the less volatile components. The scrubbed product is called *natural gas* and contains methane, the principal component, plus ethane, propane, butane, carbon dioxide, nitrogen, and hydrogen sulfide. Vast networks of pipelines then carry the natural gas throughout the continent.

Liquid petroleum is a complex mixture of hydrocarbons in which alkanes predominate. It is separated into fractions by distillation or extraction. Gasoline is the petroleum fraction that boils between 40 and 200°. (See Table 4-1.) The other principal fractions of petroleum are kerosine (bp 175–325°, C_8 to C_{14}), gas oil (bp above 275°, C_{12} to C_{18}), lubricating oils and greases (above C_{18}), and asphalt or petroleum coke, depending on the source of the petroleum.

The yield and quality of natural gasoline are improved by processes known as **isomerization, cracking,** and **alkylation**. Isomerization changes straight-chain hydrocarbons into isomeric branched compounds, which are better fuels. Cracking splits alkanes into smaller fragments, lower alkanes and alkenes. In the alkylation reaction, selected cracking products are put back together to form premium-grade motor fuels. Such processes more than double the yield of gasoline from petroleum. Valuable by-products, such as ethylene, propene, and the butenes, are also produced by the cracking process and used in the chemical industry to make other chemicals, plastics, etc.

If coal is heated to 1000 to 1300° in the absence of air, a liquid distillate, *coal tar*, and a solid residue, *coke*, are obtained. Coal tar is an exceedingly complex mixture, which can be separated to yield, among other things, large numbers of hydrocarbons. Benzene is a principal constituent, and alkylbenzenes, naphthalene, anthracene, phenanthrene, and many other aromatic compounds are also produced.

Coke, which is mostly carbon, may also be converted to hydrocarbons by indirect routes. Calcium carbide is formed by heating a mixture of coke and calcium oxide to 2000°. Calcium carbide reacts with water to form acetylene, a very cheap material.

$$3C + CaO \longrightarrow CaC_2 + CO$$

Coke   Calcium
                    carbide

$$CaC_2 + 2H_2O \longrightarrow HC{\equiv}CH + Ca(OH)_2$$

If coke is treated with steam at a high temperature, a mixture of carbon monoxide and hydrogen, known as *water gas*, is produced.

$$C + H_2O \longrightarrow CO + H_2$$

Largely through the efforts of German chemists, a process has been developed for hydrogenation of carbon monoxide to produce mixtures of hydrocarbons. In the Fischer-Tropsch process, more hydrogen is added to water gas and the mixture is passed over a solid catalyst at an elevated temperature. Carefully controlled operations lead to the production of alkanes and alkenes boiling in the gasoline range.

$$nCO + (2n + 1)H_2 \xrightarrow[\text{200--300 atm}]{\text{Co cat., 200°}} C_nH_{2n+2} + nH_2O$$

Alkanes

In the above equation the space above and below the reaction arrow has been used to indicate important reaction conditions. Such information should be carefully noted. Control of conditions is often very critical to the success of organic reactions. For example, the reaction of hydrogen with carbon monoxide can be guided, by varying the reaction conditions, to produce mostly methane or to form oxygen-containing compounds.

The major part of the enormous volume of hydrocarbons produced in the world is used directly as fuel. Petroleum is now the cheapest source of motor fuels, and every shift in the production and distribution of petroleum products involves a potential international crisis. In addition to their use as fuels, *hydrocarbons are the principal raw materials for the synthesis of all organic chemicals.* Virtually all large-scale industrial organic chemical processes are based, at least indirectly, on hydrocarbons (including acetylene from calcium carbide) and carbon monoxide.

Although ages of geological action have been required to convert plant tissue to coal, petroleum, and natural gas, many plants produce hydrocarbons directly as metabolic products. The amounts of such products are usually small. A number of high-molecular-weight alkanes have been isolated from the leaves of waxy plants. Alkenes and cycloalkenes are produced in great abundance in nature. Included are natural rubber, turpentine, essential oils, and some beautifully colored plant pigments. Rubber consists of large molecules composed of long sequences of isoprene units linked together through the 1- and 4-positions (Chap. 25).

$$CH_2=\underset{\underset{CH_3}{|}}{C}-CH=CH_2 \qquad -(\underset{\underset{CH_3}{|}}{CH_2C}=CHCH_2-)_x$$

Isoprene   Rubber

Hydrocarbons are not common products of animal metabolism, but a few examples are known. For example, beeswax contains about 10% hydrocarbons, including $n\text{-}C_{27}H_{56}$ (heptacosane). A few polyacetylenic compounds have been isolated from plants and fungi.

PROBLEM 4-2

Write all reasonable names for the following compounds:

a $CH_3\underset{\underset{CH_3}{|}}{C}HCH_2\underset{\underset{\underset{\underset{CH_3}{|}}{CH}}{|}{CHCH_3}}{C}HCH_2\underset{\underset{CH_3}{|}}{C}HCH_2CH_3$

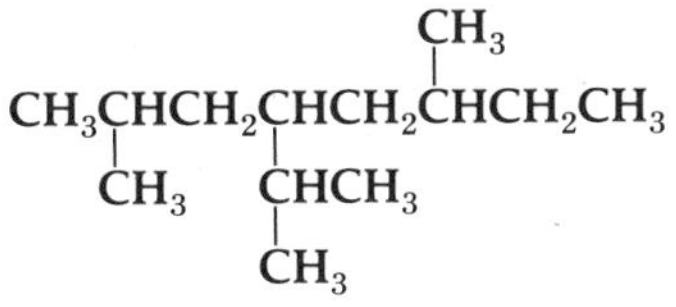

f $CH_3CH_2\underset{\underset{CH_3}{|}}{\overset{\overset{CH_2CH_3}{|}}{C}}CH_2C\equiv CH$

g $CH_3CH=CHCH=CHCH_3$

b $CH_3CH_2\underset{\underset{C_6H_5}{|}}{C}HCH=CHCH_3$

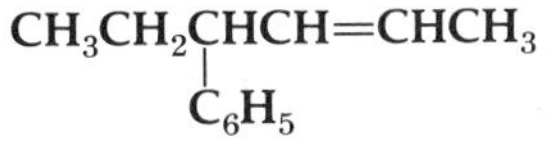

c $(CH_3)_3CC\equiv CC(CH_3)_3$

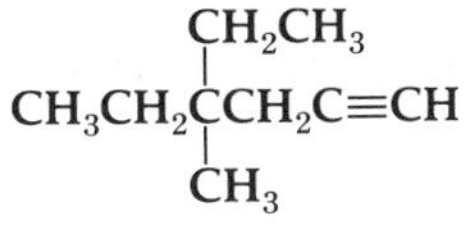

h

d

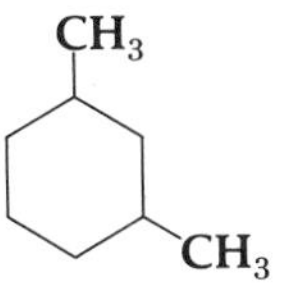

i

e $(CH_3)_3CC_6H_5$

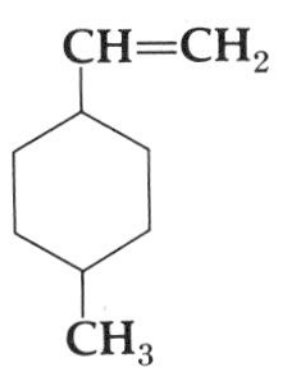

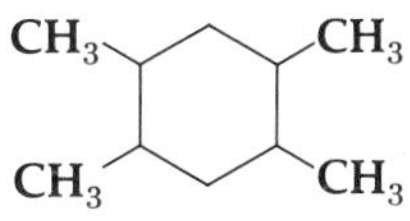

PROBLEM 4-3

Write structural formulas for the following compounds:

a *sec*-Butylbenzene

b 1,2-Dimethyl-1-cyclopentene

c Vinylacetylene

d *sym*-Dimethylethylene

e *p*-Methylstyrene

f Isobutylmethylacetylene

g Allylbenzene

h *o*-Diethylbenzene

PROBLEM 4-4

Write out all possible structures that fit the following molecular formulas. Include cyclic structures *only* if they are aromatic. Name all compounds.

a C_4H_9Cl d C_4H_7Br g C_8H_{10} (ten structures are sufficient)

b C_4H_6 e C_5H_{10} h $C_{10}H_8$ (ten structures are sufficient)

c C_6H_6 f C_5H_8

4-2 COMPOUNDS CONTAINING SIMPLE FUNCTIONAL GROUPS

The compounds to be discussed in this section are all of those containing group I functional groups, i.e., with only one single bond linking carbon to the heteroatom of the functional group.

Organic Halides

Almost any hydrogen in any hydrocarbon can be replaced with a halogen atom to give a stable compound. Since a carbon atom may be bonded to from one to four halogen atoms, an enormous number of organic halides can exist. Completely fluorinated compounds are known as **fluorocarbons**. The stability of fluorocarbons is even greater than that of hydrocarbons. Completely chlorinated compounds are rather unstable if they contain more than three carbon atoms. The following examples represent interesting and important structural types:

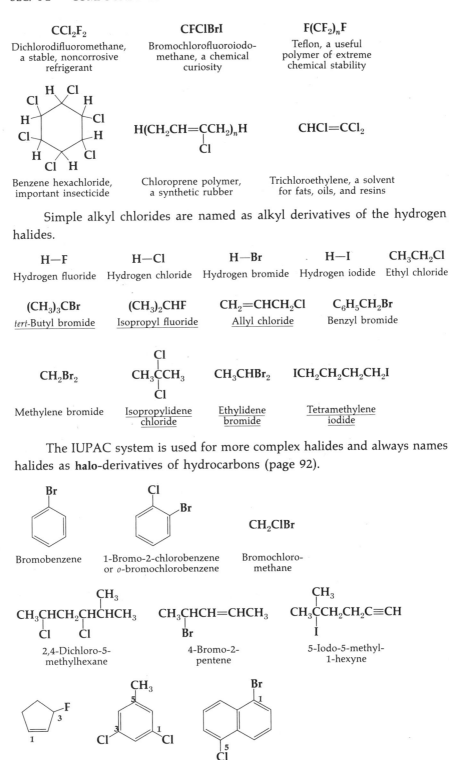

CCl_2F_2

Dichlorodifluoromethane,
a stable, noncorrosive
refrigerant

CFClBrI

Bromochlorofluoroiodo-
methane, a chemical
curiosity

$F(CF_2)_nF$

Teflon, a useful
polymer of extreme
chemical stability

Benzene hexachloride,
important insecticide

$H(CH_2CH{=}CCH_2)_nH$
$\quad\quad\quad\;\;\; Cl$

Chloroprene polymer,
a synthetic rubber

$CHCl{=}CCl_2$

Trichloroethylene, a solvent
for fats, oils, and resins

Simple alkyl chlorides are named as alkyl derivatives of the hydrogen halides.

$H{-}F$

Hydrogen fluoride

$H{-}Cl$

Hydrogen chloride

$H{-}Br$

Hydrogen bromide

$H{-}I$

Hydrogen iodide

CH_3CH_2Cl

Ethyl chloride

$(CH_3)_3CBr$

tert-Butyl bromide

$(CH_3)_2CHF$

Isopropyl fluoride

$CH_2{=}CHCH_2Cl$

Allyl chloride

$C_6H_5CH_2Br$

Benzyl bromide

CH_2Br_2

Methylene bromide

$\qquad Cl$
$CH_3\overset{|}{\underset{|}{C}}CH_3$
$\qquad Cl$

Isopropylidene
chloride

CH_3CHBr_2

Ethylidene
bromide

$ICH_2CH_2CH_2CH_2I$

Tetramethylene
iodide

The IUPAC system is used for more complex halides and always names halides as **halo**-derivatives of hydrocarbons (page 92).

Bromobenzene

1-Bromo-2-chlorobenzene
or *o*-bromochlorobenzene

CH_2ClBr

Bromochloro-
methane

$\qquad\qquad\;\; CH_3$
$CH_3CHCH_2\overset{|}{C}HCHCH_3$
$\;\;\;\;\, \underset{|}{C}l \qquad \underset{|}{C}l$

2,4-Dichloro-5-
methylhexane

$CH_3CHCH{=}CHCH_3$
$\;\;\;\;\; \underset{|}{B}r$

4-Bromo-2-
pentene

$\;\;\; CH_3$
$CH_3\overset{|}{\underset{|}{C}}CH_2CH_2C{\equiv}CH$
$\;\;\; I$

5-Iodo-5-methyl-
1-hexyne

3-Fluoro-
cyclopentene

1,3-Dichloro-5-
methylbenzene

1-Bromo-5-chloro-
naphthalene

The terms **geminal (gem-)** and **vicinal (vic-)** are frequently used to show the relative positions of substituents, such as halogens, in 1,1- and 1,2-disubstituted compounds, respectively. The following examples illustrate the use of the terms.

CH_3CHBr_2

1,1-Dibromoethane
or *gem*-dibromoethane

$BrCH_2CH_2Br$

1,2-Dibromoethane
or *vic*-dibromoethane

Most of the common organic halides are liquids with specific gravities (densities) higher than those of most other organic compounds. They are miscible in all proportions with liquid hydrocarbons and are, in general, good solvents for many organic substances. Consequently, many of the stable and cheap chlorides enjoy extensive use as solvents for purposes such as extraction of fatty materials from animal tissue, for separation of substances by liquid-liquid extraction (Sec. 1-4, e.g., water–carbon tetrachloride), and as media for organic reactions. Increase in the halogen content of organic compounds decreases their flammability. Carbon tetrachloride is used in certain simple fire extinguishers. The physical properties of a number of typical organic halides are recorded in Table 4-2.

TABLE 4-2   **Physical Properties of Organic Halides**

Name	Formula	Mp	Bp	Sp. gr. (liquid)
Methyl chloride	CH_3Cl	−97	−24	0.920
Methyl bromide	CH_3Br	−93	4	1.419
Methyl iodide	CH_3I	−66	42	2.279
Methylene chloride	CH_2Cl_2	−96	40	1.326
Chloroform	$CHCl_3$	−64	62	1.489
Carbon tetrachloride	CCl_4	−23	77	1.594
Ethyl chloride	C_2H_5Cl	−139	12	0.898
Ethyl bromide	C_2H_5Br	−119	38	1.461
Ethyl iodide	C_2H_5I	−111	72	1.936
n-Propyl chloride	$CH_3CH_2CH_2Cl$	−123	47	0.890
Isopropyl chloride	$(CH_3)_2CHCl$	−117	36	0.860
n-Butyl chloride	$CH_3(CH_2)_3Cl$	−123	78	0.884
tert-Butyl chloride	$(CH_3)_3CCl$	−25	51	0.842
Vinyl chloride	$CH_2{=}CHCl$	−154	−14	0.911
Allyl chloride	$CH_2{=}CHCH_2Cl$	−136	45	0.938
Chlorobenzene	C_6H_5Cl	−45	132	1.107
Iodobenzene	C_6H_5I	−29	189	1.824
o-Dichlorobenzene	$o\text{-}C_6H_4Cl_2$	−17	180	1.305
p-Dichlorobenzene	$p\text{-}C_6H_4Cl_2$	53	174	1.247

Alcohols and Phenols

These compounds may be considered as derivatives of water in which one hydrogen atom is replaced by a hydrocarbon group. Substitution of an alkyl group for one hydrogen atom of water gives an alcohol. Aryl substitution for one hydrogen gives a phenol. Both hydrogens of the water structure are replaced by aryl or alkyl groups in ethers. Note that the unshared electrons (Chap. 2) in the oxygen outer shells, a chemically important feature of the structures, are often included in the formulas.

The older names of the simpler alcohols include the name of the alkyl group followed by the word "alcohol."

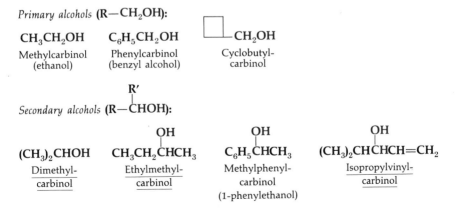

$$CH_3OH$$
Methyl alcohol

$$\overset{\displaystyle OH}{CH_3CH_2CHCH_3}$$
sec-Butyl alcohol

$$CH_2{=}CHCH_2OH$$
Allyl alcohol

$$C_6H_5CH_2OH$$
Benzyl alcohol

$$(CH_3)_3COH$$
t-Butyl alcohol

Cyclopropyl alcohol

Alcohols are sometimes named as derivatives of methyl alcohol, the simplest member of the series, by substitution of alkyl groups for the methyl hydrogens. However, the word **carbinol** is usually used in place of the words *methyl alcohol* in such names. Logically, the system implies that methyl alcohol should be called carbinol, but the name is rarely used. The system is often useful because it leads to easy visualization of the formulas.

The scheme also emphasizes the subclassification of alcohols as **primary**, **secondary**, and **tertiary**. Primary alcohols have only one alkyl group attached to the hydroxyl-bearing carbon atom (i.e., they always contain the $-CH_2OH$ group). Secondary alcohols have two alkyl groups attached to the functional center, and in tertiary alcohols all the C—H bonds of methanol are replaced by alkyl or aryl groups. The following examples illustrate both this classification and the carbinol nomenclature system.

Primary alcohols ($R-CH_2OH$):

$$CH_3CH_2OH$$
Methylcarbinol (ethanol)

$$C_6H_5CH_2OH$$
Phenylcarbinol (benzyl alcohol)

$-CH_2OH$
Cyclobutyl-carbinol

Secondary alcohols ($R-\overset{\displaystyle R'}{C}HOH$):

$$(CH_3)_2CHOH$$
Dimethyl-carbinol

$$\overset{\displaystyle OH}{CH_3CH_2CHCH_3}$$
Ethylmethyl-carbinol

$$\overset{\displaystyle OH}{C_6H_5CHCH_3}$$
Methylphenyl-carbinol
(1-phenylethanol)

$$\overset{\displaystyle OH}{(CH_3)_2CHCHCH{=}CH_2}$$
Isopropylvinyl-carbinol

Tertiary alcohols $\left(R'-\underset{\underset{R}{|}}{\overset{\overset{R''}{|}}{C}}OH\right)$:

$(CH_3)_3COH$	$(C_6H_5)_3COH$	$(CH_3)_2\overset{\overset{OH}{|}}{C}C_6H_5$	$[(CH_3)_3C]_3COH$
Trimethyl-carbinol	Triphenyl-carbinol	Dimethylphenyl-carbinol	Tri-*tert*-butyl-carbinol
(*tert*-butyl alcohol)			

Compounds containing several hydroxyl groups are known as **polyhydric** alcohols. Those which contain only two hydroxyl units are called **glycols** or **diols**. Trihydric alcohols are also known as **triols**. The class includes some important members, such as glycerol (or glycerine), ethylene glycol, and the carbohydrates (Chap. 27).

$HOCH_2CH_2OH$	$HOCH_2CH(OH)CH_2OH$	$HO(CH_2)_3OH$
Ethylene glycol or 1,2-ethanediol	Glycerol or 1,2,3-propanetriol	Trimethylene glycol or 1,3-propanediol

1,1-Glycols, with two —OH groups on one carbon, can almost never be isolated. An alcohol group attached to an unconjugated double bond is also virtually unknown in stable compounds. These vinyl alcohols or **enols** (= **ene** + **ol**), HO—C=C, are important reaction intermediates, however.

Unstable:

$\underset{R'}{\overset{R}{\diagdown}}C\underset{OH}{\overset{OH}{\diagup}}$	$\underset{R'}{\overset{R}{\diagdown}}C=C\underset{R''}{\overset{OH}{\diagup}}$
1,1-Glycol	Enol

Table 4-3 summarizes the properties of some common alcohols. The compounds are much higher-boiling than hydrocarbons containing the same numbers of carbon atoms. The lower alcohols are miscible with water. Alcohols become more waterlike in their physical properties as the ratio of hydroxyl groups to C—H bonds is increased. Note, for example, the high boiling points of ethylene glycol and glycerol, both of which are hygroscopic (water-absorbing) liquids. The high boiling points and miscibility with water arise from considerable intermolecular association of molecules in the liquid state by hydrogen bonding.

Most of the compounds listed in Table 4-3 have considerable commercial importance. *Methanol*, which was once produced by destructive distillation of wood (wood alcohol), is now manufactured in large quantities by catalyzed reaction of hydrogen with carbon monoxide. The compound is much used as a solvent and chemical intermediate. *Ethanol* (ethyl alcohol) is "alcohol" in nonscientific language, the intoxicating constituent of alcoholic beverages.

Ethanol is produced both by fermentation of grains such as corn (which contain carbohydrates) and by catalyzed addition of water to ethylene. At the present time industrial alcohol (nonbeverage use) is produced almost exclusively from ethylene. However, increased demands on the petroleum industry for fuels (as occurred during World War II) can cause diversion of large amounts of fermentation alcohol to industrial use.

Ethanol is widely used as a solvent and is one of the principal raw materials of chemical industry. *n-Butyl alcohol* (butanol) is also produced by fermentation and is widely used as a solvent. *Isopropyl alcohol* (2-propanol) can be prepared from propene and can in turn be oxidized to acetone (CH_3COCH_3). This combination of cheap synthesis and cheap conversion to a compound having varied and useful chemical properties makes isopropyl alcohol a key industrial compound.

Ethylene glycol and *glycerol* can be prepared from ethylene and propene, respectively. Glycerol (or glycerine) is also obtained by hydrolysis (cleavage with water) of fats. Both are used in the manufacture of synthetic resins (Chap. 25). Ethylene glycol is familiar to most of us as the principal ingredient in permanent antifreeze, a use made possible by its high boiling point, low melting point, and miscibility with water.

Phenols are sometimes placed in a different class from alcohols because the two groups of compounds have rather different chemical properties. The simplest member of the class, hydroxybenzene, is known as *phenol*. The most important phenols are hydroxybenzene derivatives and the naphthols (hydroxynaphthalenes). Most of the compounds have been known for a very long time and are, therefore, usually referred to by common names.

TABLE 4-3   **Physical Properties of Alcohols**

Name	Formula	Mp	Bp	Sp. gr.
Methyl alcohol	CH_3OH	−98	65	0.792
Ethyl alcohol	C_2H_5OH	−114	78	0.789
n-Propyl alcohol	$CH_3CH_2CH_2OH$	−126	97	0.804
Isopropyl alcohol	$(CH_3)_2CHOH$	−89	82	0.781
n-Butyl alcohol	$CH_3(CH_2)_3OH$	−90	118	0.810
Isobutyl alcohol	$(CH_3)_2CHCH_2OH$	−108	108	0.798
sec-Butyl alcohol	$CH_3CHOHCH_2CH_3$	−115	100	0.808
tert-Butyl alcohol	$(CH_3)_3COH$	26	83	0.789
Cyclohexyl alcohol	$C_6H_{11}OH$	25	161	0.962
n-Octyl alcohol	$CH_3(CH_2)_7OH$	−17	195	0.827
Allyl alcohol	$CH_2{=}CHCH_2OH$	−129	97	0.855
Benzyl alcohol	$C_6H_5CH_2OH$	−15	205	1.046
Ethylene glycol	$HOCH_2CH_2OH$	−12.6	198	1.113
Glycerol	$HOCH_2CHOHCH_2OH$	18	291	1.260

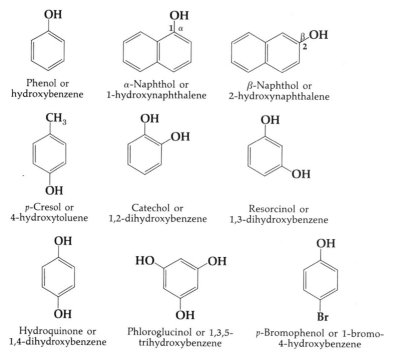

Phenol or
hydroxybenzene

α-Naphthol or
1-hydroxynaphthalene

β-Naphthol or
2-hydroxynaphthalene

p-Cresol or
4-hydroxytoluene

Catechol or
1,2-dihydroxybenzene

Resorcinol or
1,3-dihydroxybenzene

Hydroquinone or
1,4-dihydroxybenzene

Phloroglucinol or 1,3,5-
trihydroxybenzene

p-Bromophenol or 1-bromo-
4-hydroxybenzene

Compared with alcohols, phenols are high-melting, high-boiling compounds. Phenol and the cresols are obtained from coal tar (page 110), but demands of industry for the compounds are so large that they are now manufactured by indirect means from aromatic hydrocarbons. Phenol and its derivatives are consumed in many processes such as synthesis of dyes and plastics (Chap. 25). Phenols and their derivatives are widely distributed in nature. Included are several naturally occurring amino acids (Chap. 25), tanning agents, the component of wood known as lignin (Chap. 25), and a large group of beautiful flower pigments (Chap. 27).

Ethers

Aryl and alkyl ethers are usually grouped together as one class of compounds. Chemically, ethers are far less important than either alcohols or phenols, although some ethers are widely used as solvents. The ether structure is common in natural products. Simple ethers are ordinarily named by first naming the two hydrocarbon groups attached to oxygen and then the word **ether**, as in the examples given. If the two hydrocarbon groups are identical, the group need be named only once. The prefix *di-* is often used to emphasize the identity of the two groups.

CH_3OCH_3 $CH_3CH_2OCH_2CH_3$ $CH_3OCH_2CH_2CH_3$

Methyl ether
or dimethyl ether

Diethyl ether
or "ether"

Methyl *n*-propyl ether

$C_6H_5OCH_3$ $(CH_3)_2CHOC(CH_3)_3$

Methyl phenyl ether
or anisole

tert-Butyl isopropyl ether

Ethers containing one small and one large group are often named as **alkoxy** (RO—) derivatives of the more complicated of the two hydrocarbon groups.

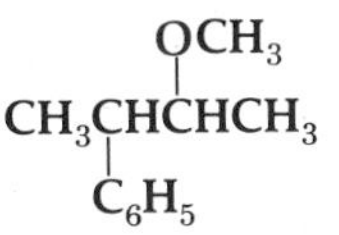

$$
\underset{\substack{| \\ C_6H_5}}{\overset{\overset{\displaystyle OCH_3}{|}}{CH_3CHCHCH_3}}
\qquad\qquad
CH_3OCH_2CH_2OCH_3
$$

2-Methoxy-3-phenylbutane   1,2-Dimethoxyethane or
                                  ethylene glycol dimethyl ether

Since the oxygen is divalent, it can be incorporated into rings. Cyclic systems containing heteroatoms are called **heterocycles** (Chap. 24), as opposed to carbocycles. Some oxygen heterocycles are named as **oxides** and others have common names; **epoxides** are three-membered cyclic ethers.

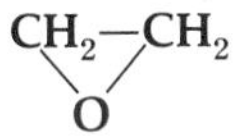

Ethylene oxide† 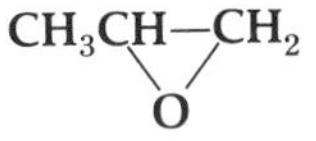   Propylene oxide† 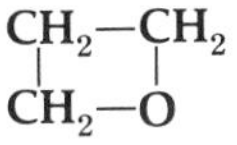   Trimethylene oxide

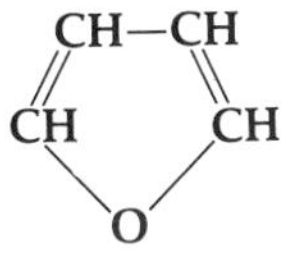

Furan 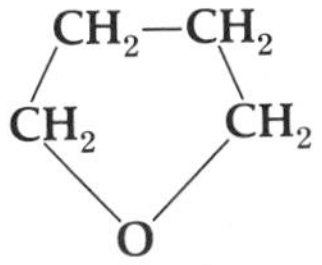   Tetrahydrofuran 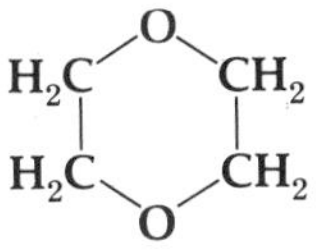   1,4-Dioxan
                                        or "dioxan"

Table 4-4 records physical properties of the most important ethers. Note that the boiling and melting points of ethers, in contrast to those of alcohols and phenols, are very similar to those of hydrocarbons of comparable molecular weight since they exhibit no intermolecular hydrogen bonding.

†So named because the compound can be prepared by oxidation of the corresponding alkene.

TABLE 4-4 **Physical Properties of Ethers**

Name	Formula	Mp	Bp	Sp. gr.
Methyl ether	CH_3OCH_3	−142	−25	0.661
Ethyl ether	$C_2H_5OC_2H_5$	−116	35	0.714
n-Butyl ether	$(CH_3CH_2CH_2CH_2)_2O$	−98	141	0.769
Phenyl ether	$C_6H_5OC_6H_5$	27	259	1.072
Anisole	$C_6H_5OCH_3$	−37	154	0.994
Ethylene oxide	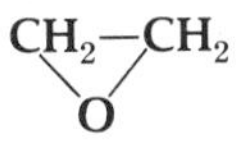	−111	14	0.887
Tetrahydrofuran	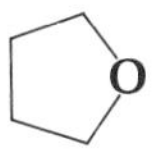	−108	65	0.888
Dioxan	$O(CH_2CH_2)_2O$	12	101	1.033

Ethyl ether, which can be made from ethylene, is a valuable solvent that dissolves both hydrocarbons and organic compounds containing polar functional groups. The compound is commonly called "ether" and was well known as a general anesthetic. Ethylene oxide has also been used as an anesthetic. Tetrahydrofuran, which can be manufactured in two steps from agricultural waste products, has become an important solvent for organic reactions.

Sulfur Compounds

The sulfur analogs of alcohols and phenols are known as *mercaptans*, or *thiols*. Nomenclature usually parallels that of alcohols; in fact, some mercaptans are simply named as *thioalcohols*.

CH_3SH Compare with CH_3OH

Methyl mercaptan   Methyl alcohol

C_6H_5SH Compare with C_6H_5OH

Phenyl mercaptan   Phenol
or thiophenol

Sulfur analogs of ethers are called *sulfides*, or *thioethers*. Analogs of peroxides have the structure **RSSR** and are simply called *disulfides*. Nomenclature of these compounds follows the same rules as that of ethers.

$CH_3CH_2SCH_2CH_3$ $CH_3CH_2SSCH_2CH_3$

Ethyl sulfide   Ethyl disulfide
or diethyl sulfide

$CH_3SC_6H_5$ $C_6H_5SSC_6H_5$

Methyl phenyl sulfide   Phenyl disulfide

Classes of sulfur compounds that have no oxygen analogs are those in which other atoms are formally bound to the unshared pairs of sulfur electrons in mercaptans and sulfides. In most of such compounds the "extra" atoms are oxygen, and a number of classes more or less analogous to sulfurous and sulfuric acid are known (Chap. 19). The following examples illustrate the relationships. The compounds may be formulated as having sulfur-oxygen double bonds. Such electronic configurations put 10 or 12 electrons in the outer electron shell of sulfur. The elements in the second row of the periodic table are capable of expanding their outer electron shells in this way and the bonding in these compounds is discussed in Chap. 5. The compounds may also be formulated with normal eight-electron outer shells as shown.

$$
\underset{\text{Sulfuric acid}}{HO-\overset{\overset{O}{\|}}{\underset{\underset{O}{\|}}{S}}-OH \quad \text{or} \quad HO-\overset{\overset{O^-}{|}}{\underset{\underset{O^-}{|}}{S^{++}}}OH}
\qquad
\underset{\text{Sulfurous acid}}{HO-\overset{\overset{O}{\|}}{\underset{\underset{..}{}}{S}}-OH \quad \text{or} \quad HO-\overset{\overset{O^-}{|}}{\underset{\underset{..}{}}{S^{\pm}}}-OH}
$$

$$R-\overset{\overset{O}{\|}}{\underset{\underset{O}{\|}}{S}}-OH \quad \text{or} \quad R-\overset{\overset{O^-}{|}}{\underset{\underset{O^-}{|}}{S^{++}}}OH$$

Alkanesulfonic acid

$$R-\overset{\overset{O}{\|}}{\underset{..}{S}}-OH \quad \text{or} \quad R-\overset{\overset{O^-}{|}}{\underset{..}{S^{\pm}}}-OH$$

Alkanesulfinic acid

$$R-\overset{\overset{O}{\|}}{\underset{\underset{O}{\|}}{S}}-R' \quad \text{or} \quad R-\overset{\overset{O^-}{|}}{\underset{\underset{O^-}{|}}{S^{++}}}R'$$

Alkyl sulfone

$$R-\overset{\overset{O}{\|}}{\underset{..}{S}}-R' \quad \text{or} \quad R-\overset{\overset{O^-}{|}}{\underset{..}{S^{\pm}}}-R'$$

Alkyl sulfoxide

Sulfones and **sulfoxides** are named by a system analogous to that used for ethers. The prefix *di-* is usually included in the names of symmetrical sulfones and sulfoxides, although the device is not approved by nomenclature experts.

$$C_6H_5-\overset{\overset{O}{\|}}{\underset{\underset{O}{\|}}{S}}-C_6H_5 \qquad C_6H_5SOCH_2CH_2CH_3$$

Diphenyl sulfone   Phenyl *n*-propyl sulfoxide

Sulfonic and **sulfinic acids** are always named as derivatives of hydrocarbons. However, since acidic functional groups are always considered to be the principal functional groups in molecules, the names **-sulfonic acid** and **-sulfinic acid** appear as suffixes rather than prefixes. They are formally analogous to carboxylic acids in replacement of CO by SO_2 and exhibit analogous derivatives:

RCOOH	Carboxylic acid	RSO_2OH	Sulfonic acid
RCOCl	Acid chloride	RSO_2Cl	Sulfonyl chloride
RCOOR'	Ester	RSO_2OR'	Sulfonate ester
$RCONH_2$	Amide	RSO_2NH_2	Sulfonamide

$$C_6H_5\overset{\overset{O}{\|}}{\underset{\underset{O}{\|}}{S}}OH \qquad\qquad \underset{\underset{SO_3H}{|}}{CH_3CH_2CHCH_3}$$

Benzenesulfonic acid   Butane-2-sulfonic acid

$$C_6H_5SO_2H \qquad\qquad \underset{\underset{SO_2H}{|}}{CH_3CH_2CHCH_3}$$

Benzenesulfinic acid   Butane-2-sulfinic acid

Sulfonium salts (Fig. 2-3) are compounds containing cations in which three carbon groups are bound to one sulfur atom. No well known analogies to the structures are found in the inorganic chemistry of sulfur, but a close analogy to the structure of ammonium ions exists.

$$R-\overset{..}{\underset{..}{S}}-R \qquad \underset{\underset{H}{|}}{H-N-H} \qquad R-\overset{\overset{R}{|}}{\underset{..}{S^+}}-R \qquad \overset{\overset{H}{|}}{\underset{\underset{H}{|}}{H-N^+-H}}$$

Sulfide   Ammonia   Sulfonium ion   Ammonium ion

In sulfonium salts some anion is always present to maintain electrical neutrality. However, the anions are not bound to the cations by covalent bonds, and the physical properties of sulfonium salts are like those of inorganic salts, such as sodium chloride. The following are typical sulfonium salts.

$(n\text{-}C_4H_9)_3S^+ \; I^-$ $(CH_3)_2S^+C_2H_5 \; Cl^-$

Tri-*n*-butylsulfonium iodide   Ethyldimethylsulfonium chloride

Mercaptans, sulfides, and disulfides are moderately important chemicals. Mercaptans boil at lower temperatures than the corresponding alcohols. The compounds have very objectionable odors; the well-known scent of the skunk is due mainly to *n*-butyl mercaptan. Some mercaptans are used to regulate industrial polymerization of unsaturated compounds (Chap. 25). Certain proteins contain **sulfhydryl** (mercaptan) (—SH) groups and disulfide linkages, and penicillin contains a sulfide group. The pungent flavors and odors of onion and garlic are due to organic sulfides.

Sulfonic acids, on the other hand, are very important industrial chemicals. They are very strong acids, comparable to sulfuric acid. This property, along with high solubility in many organic solvents, makes the compounds useful as acidic catalysts in many organic reactions. Sulfonic acid groups are also often used to increase the water solubility of large organic molecules. Many dyes and detergents are sulfonic acids or their salts.

Amines and Related Compounds

Simple amines are often named by specifying the groups attached to nitrogen and adding the word *amine* as a suffix. Note that the name is written as one word, in contrast to the nomenclature of alcohols. Also in contrast to alcohols, the structure of attached alkyl groups has no bearing on the classification of amines, which are classified with respect to the number of alkyl groups attached to nitrogen. Thus, *tert*-butylamine is a primary amine even though it contains a *tert*-butyl group.

Primary amines:

CH_3NH_2 $(CH_3)_3CNH_2$

Methylamine   *tert*-Butylamine

Secondary amines:

$(CH_3)_2NH$ $(CH_3)_2CHNHCH_3$

Dimethylamine   Isopropylmethylamine

Tertiary amines:

$$\overset{\displaystyle C_2H_5}{\underset{\displaystyle }{CH_3NCH_2CH_2CH_3}}$$

$(CH_3)_3N$ $CH_3NCH_2CH_2CH_3$

Trimethylamine   Ethylmethyl-*n*-propylamine

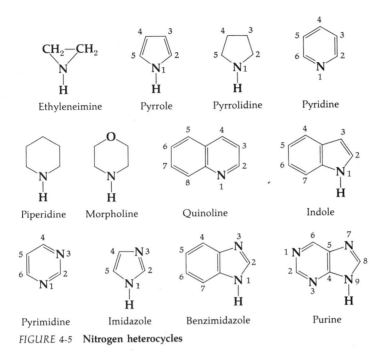

Ethyleneimine   Pyrrole   Pyrrolidine   Pyridine

Piperidine   Morpholine   Quinoline   Indole

Pyrimidine   Imidazole   Benzimidazole   Purine

FIGURE 4-5   **Nitrogen heterocycles**

The most important aromatic amine is aniline ($C_6H_5NH_2$), and many compounds are named as derivatives of aniline.

Aniline   N-Methylaniline   N,N-Dimethylaniline   N,N,3,5-Tetramethylaniline

Heterocyclic amines are of great importance. Nitrogen-containing ring systems are found in hundreds of natural products. The parent ring systems are known by common names, which are widely used as generic structures in the nomenclature of the more complex compounds. Names of the ring systems shown in Fig. 4-5 are all fundamental to a knowledge of organic chemistry. The ring systems are assigned rigidly defined numbering schemes in which nitrogen is usually given the number 1. Further discussion of these heterocyclic compounds is available in Chap. 24.

Amines resemble ammonia in their basic character and ability to form salts by reactions with acids. The salts contain substituted ammonium ions. Salts can also be prepared in which ammonium ions are completely substituted by carbon groups. The latter class are called **quaternary ammonium** compounds. (See Fig. 2-3.) The cations are all usually named as derivatives of ammonium ions. Cations derived from amines with common names are named by adding the ending -*ium* to the name of the amine (i.e., pyridine $\longrightarrow$ pyridin*ium*). Typical ammonium salts are formulated below.

$$
\begin{array}{cc}
\underset{\substack{|\\H}}{\overset{\substack{H\\|}}{H-N^{\pm}-H}} \ Cl^- & \underset{\substack{|\\H}}{\overset{\substack{H\\|}}{CH_3-N^{\pm}-H}} \ Cl^- \\
\end{array}
$$

Ammonium   Methylammonium
chloride   chloride

$$
\underset{\substack{|\\C_2H_5}}{\overset{\substack{CH_3\\|}}{CH_3-N^{+}-C_6H_5}} \ Br^-
$$

Ethyldimethylphenylammonium
bromide (a quaternary
ammonium salt)

N-Methylpyridinium
hydrogen sulfate

At one time the nature of the interaction between acids and amines was not clearly understood, and the formulas of ammonium salts were written in a noncommital way as $RNH_2 \cdot HCl$. A vague nomenclature accompanied this formulation. Ammonium chlorides were called amine hydrochlorides, bromides were known as amine hydrobromides, etc. Both the names and formulas are widespread in the older chemical literature, and the names still appear frequently, although the "dot" formulas have almost disappeared.

Amines are among the two or three most important classes of organic compounds. They undergo many and varied organic reactions, thereby giving them a prominent role both in organic synthesis and in the metabolism of living things. In nature amines are widely distributed in the form of *amino acids*, *proteins*, and *alkaloids*. Some of these substances are principal building blocks of animal tissue, and minute amounts of others have spectacular physiological effects, both harmful and beneficial. Many vitamins, antibiotics, drugs, and poisons contain amino groups. Discussion of these natural compounds is found in Chaps. 25 to 27.

The boiling points of primary and secondary amines are intermediate between those of alcohols and hydrocarbons containing the same number of carbon atoms. Tertiary amines have boiling points very close to those of the corresponding hydrocarbons.† Low-molecular-weight amines are water-soluble, and nearly all amines dissolve in aqueous acid (by salt formation).

Compounds Containing Oxidized Nitrogen

Four classes of compounds can be considered as derived from nitric and nitrous acids. Nitrites and nitrates are regarded as esters of these acids. Note

---

† The properties are to be expected from the weaker hydrogen bonding of nitrogen.

that nitrites, which contain C—O bonds, are isomers of nitro compounds, which contain C—N bonds.

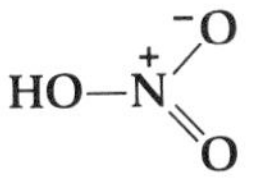

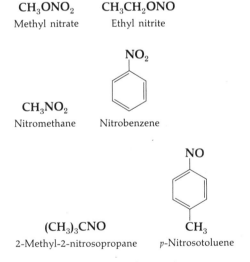

Nitric acid   Nitrous acid

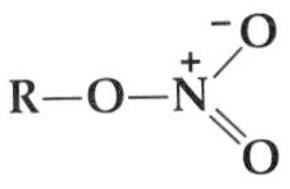

Alkyl nitrate   Alkyl nitrite

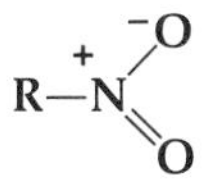

Nitroalkane   Nitrosoalkane

Nomenclature of the individual compounds follows the class names. Nitrates and nitrites are named by combining the name of the alkyl group and the word **nitrate** or **nitrite**. Nitro and nitroso compounds are *always* named as derivatives of some parent structure with *nitro-* and *nitroso-* used as prefixes. The following examples illustrate the rules:

CH_3ONO_2 CH_3CH_2ONO
Methyl nitrate   Ethyl nitrite

NO_2

CH_3NO_2
Nitromethane   Nitrobenzene

NO

$(CH_3)_3CNO$ CH_3
2-Methyl-2-nitrosopropane   *p*-Nitrosotoluene

Nitro compounds are the most important of these groups. Nitrobenzene is manufactured in large amounts from benzene and nitric acid. Most of the product is converted to aniline, but some is diverted to other chemical uses. The compound 2,4,6-trinitrotoluene (TNT) is a well-known high explosive.

Amine oxides are an interesting class of compounds formed by oxidation of tertiary amines with suitable reagents (Chap. 18). Electronic theory demands that these compounds be formulated with *separated positive and negative charges.*

Nitrogen in amine oxides is in the same condition electronically as the nitrogen in ammonium ions.

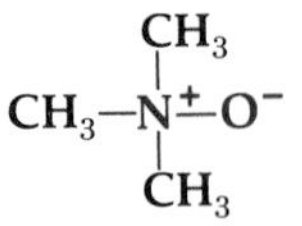

$$CH_3-\overset{\overset{\displaystyle CH_3}{|}}{\underset{\underset{\displaystyle CH_3}{|}}{N}}{}^{\pm}-O^-$$

Trimethylamine oxide 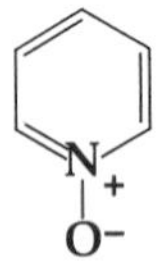   Pyridine oxide

Organometallic Compounds

An important group of organic reagents contain carbon-metal bonds. The compounds are alkyl and aryl derivatives of metals. Compounds containing the alkali metals and magnesium are seldom isolated. Lithium and magnesium compounds are prepared and used in ether solutions (usually ethyl ether or tetrahydrofuran). Sodium and potassium compounds react with ethers and are, therefore, prepared as suspensions in hydrocarbon solvents. The compounds are named as alkyl and aryl derivatives of metals.

CH_3Li C_6H_5Na $(C_2H_5)_2Mg$
Methyllithium   Phenylsodium   Diethylmagnesium

Mixed organomagnesium halides are known as **Grignard reagents**. They are prepared in ether solution and the real structures of the materials in solution have never been determined, although the overall composition corresponds to **RMgX**.

CH_3MgCl $CH_3CH_2CH_2MgBr$
Methylmagnesium chloride n-Propylmagnesium bromide

The above types of organometallic compounds must be protected from the atmosphere since they all react readily with both water and carbon dioxide. The high reactivity of organometallic compounds is due to the partially ionic character of carbon-metal bonds. Even though lithium and magnesium compounds are only slightly ionized in ether solutions, the compounds have many of the properties expected of ionic substances, $R^-{:}M^+$, containing a carbanion. Such properties include low solubility in hydrocarbons and high boiling and melting points.

PROBLEM 4-5

Write systematic, and where possible, common names for the following compounds:

a $C_6H_5CHCl_2$ b $CH_3\underset{\underset{\displaystyle Br}{|}}{CH}CH{=}CH_2$ c $(CH_3)_3CCH_2OH$

d $CH_2{=}CHCH_2OCH_3$

h 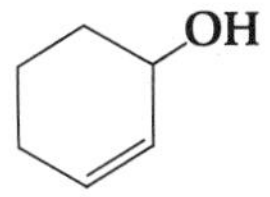

l 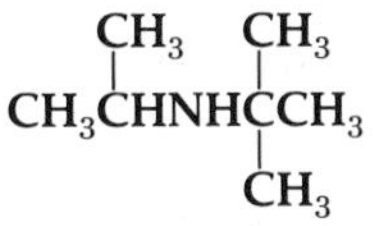

e 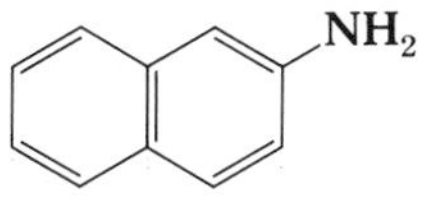

i 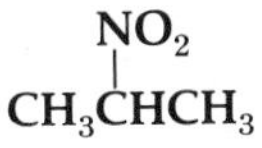

m 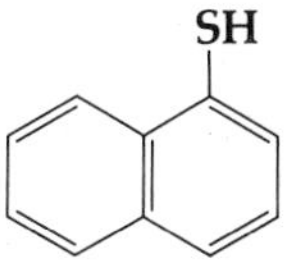

f 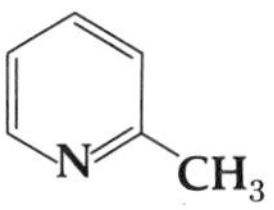

j 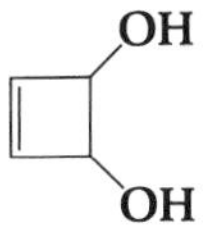

n $C_6H_5CH_2SCH_3$

g 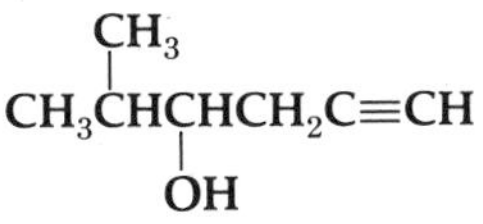

k 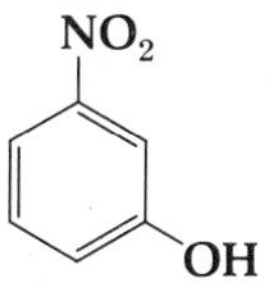

o 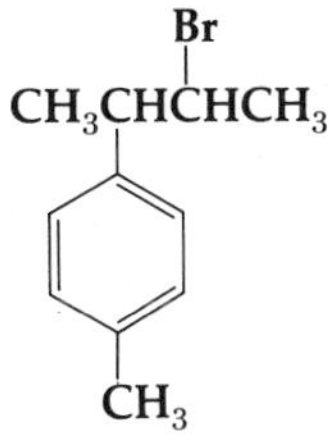

PROBLEM 4-6

Write structural formulas for the following compounds:

a Isopropyldimethylamine
b 3-Methoxycyclopentanol
c 3,5-Dinitrophenol
d 6-Bromo-1-naphthol
e 3-Methylpyridine
f Tetramethylene fluoride
g 3-Phenylquinoline

PROBLEM 4-7

Write structural formulas for and give names to compounds that illustrate each of the following:

a A tertiary alcohol
b An alkyl nitrite
c A nitroalkane
d A secondary amine containing a tertiary alkyl group
e A sulfoxide
f A hydroperoxide
g An alkyl nitrate
h An unstable enol
i A stable polyhydric alcohol
j An amino alcohol
k A heterocyclic amine
l A heterocyclic ether
m A cyclic sulfide
n A cyclic amine oxide

4-3 COMPOUNDS CONTAINING UNSATURATED FUNCTIONAL GROUPS

The compounds in this section all contain functional groups from groups II and III, having multiple bonds from the skeletal carbon to heteroatom(s). In most instances these functional groups incorporate the important and ubiquitous carbonyl group (C=O), but in a few the analogous C=N and C≡N functions are involved instead.

Aldehydes

Two systems are used for the nomenclature of aldehydes. Common names are derived from the names of the corresponding carboxylic acids (Table 4-8). The ending -*ic* (or -*oic*) in the name of the acid is replaced by the word *aldehyde*.

HCOOH	H$_2$CO	CH$_3$COOH	CH$_3$CHO
Formic acid	Formaldehyde	Acetic acid	Acetaldehyde

The following examples illustrate the systematic IUPAC nomenclature.

$$CH_3CH_2CHO \qquad CH_3CH_2CH_2CHO \qquad \overset{\displaystyle CH_3}{\overset{|}{CH_3CHCHO}}$$

Propanal or propionaldehyde	Butanal or n-butyraldehyde	2-Methylpropanal or isobutyraldehyde

$$\overset{\displaystyle CH_3}{\overset{|}{CH_3CH{=}CCHO}} \qquad CH_3C{\equiv}CCHO$$

2-Methyl-2-butenal   2-Butynal

When the name of an aldehyde is used to designate a parent structure, substituents are often located by means of Greek letters. The system has one confusing feature. The lettering is started at the carbon atom *adjacent* to the carbonyl group; no letter is assigned to the carbonyl carbon atom, since attachment of a substituent at that position would produce a compound of some other class.

$$\overset{\gamma \quad \beta \quad \alpha}{-C-C-C-\underset{\parallel}{\underset{O}{CH}}} \qquad \overset{\displaystyle Cl}{\overset{|}{CH_3CHCH_2CHO}}$$

β-Chloro-n-butyraldehyde

Table 4-5 lists the names of the most important aldehydes and summarizes their physical properties. Aldehydes of low molecular weight have unpleasant pungent odors. Formaldehyde is familiar as a 40% aqueous solution, called *formalin*, used for preservation of biological specimens. Acetaldehyde is an important industrial chemical since it can be made by hydration of acetylene. Furfural is produced by the decomposition of agricultural wastes (oat hulls) in the presence of acids, and so finds use as an inexpensive starting material for the synthetic chemist.

Ketones

In the common names of simple ketones, the names of the alkyl and/or aryl groups attached to the carbonyl group are coupled with the word "ketone." A few ketones also have trivial names, which are related to names of acids used in preparation of the ketone.

CH_3COCH_3

Dimethyl ketone
or acetone

$CH_3COC_2H_5$

Ethyl methyl ketone

$CH_3COCH(CH_3)_2$

Isopropyl methyl ketone

$C_6H_5COCH_3$

Methyl phenyl ketone
or acetophenone

Greek letters are frequently used to locate substituents in ketones; the lettering begins with positions adjacent to the carbonyl group as described above for aldehydes. If both **R—** groups bear substituents, the chains are both lettered away from the carbonyl group using prime signs to distinguish one chain from the other. However, ketones are most easily named in the systematic (IUPAC) way.

$$\overset{\gamma}{C}-\overset{\beta}{C}-\overset{\alpha}{C}-\underset{\underset{O}{\|}}{C}-\overset{\alpha'}{C}-\overset{\beta'}{C}-\overset{\gamma'}{C}$$

$$\underset{Cl}{\overset{|}{C}}H_3\overset{}{C}HCOCH_2CH_3$$

α-Chlorodiethyl ketone

$$\underset{Br}{\overset{|}{C}}H_3\overset{}{C}HCOCH_2CH_2Cl$$

α-Bromo-β'-chloro-
diethyl ketone

$CH_3CH_2COCH_2CH_3$

3-Pentanone

$$CH_3CH_2CO\underset{\underset{CH_3}{|}}{C}HCH_3$$

2-Methyl-3-pentanone

$$\underset{\underset{CH_3}{|}}{C}H_3C=CHCO\underset{\underset{CH_3}{|}}{C}HCH_3$$

2,5-Dimethylhex-4-
ene-3-one

$$CH_3\underset{\underset{C_6H_5}{|}}{C}HCOCH_3$$

3-Phenyl-2-butanone

TABLE 4-5 **Physical Properties of Aldehydes**

Name	Formula	Mp	Bp	Sp. gr.
Formaldehyde	CH_2O	−117	−19	0.815
Acetaldehyde	CH_3CHO	−123	21	0.778
n-Butyraldehyde	$CH_3(CH_2)_2CHO$	−97	75	0.801
Glyoxal	OHCCHO	15	51	1.26
Acrolein	$CH_2=CHCHO$	−87	53	0.841
Benzaldehyde	C_6H_5CHO	−56	179	1.046
Furfural	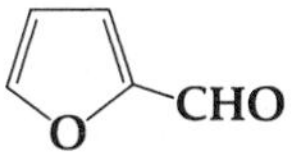 CHO	−37	162	1.156

Table 4-6 lists the names and physical properties of a number of important ketones. As a class, ketones rank among the most important organic compounds. The reactivity of carbonyl groups (Chaps. 12 and 13) is such as to make ketones almost uniquely useful in building branched-carbon skeletons. Special electronic properties of the carbonyl group also render the α-positions unusually reactive. In fact, given a reasonable assortment of ketones as starting materials, a good organic chemist can design syntheses of most carbon skeletons that can be written on paper. In addition low-molecular-weight ketones, such as acetone and ethyl methyl ketone, are used in large volume as solvents. Acetone is a common solvent for coatings ranging from nail polish to outdoor enamel paints.

Many natural products are ketones (Chap. 27), including many perfumes and flavoring materials. Such compounds often have both exotic properties and chemical structures.

Muscone, a perfume extracted from the scent glands of male musk deer

β-Ionone, fragrant scent of the violet

Nitrogen Analogs

Imines are analogs of aldehydes and ketones which have nitrogen in place of oxygen. Few are stable; they readily yield the parent carbonyl on reaction with water. Much more stable are the family of N-substituted imines, regarded as derivatives of aldehydes and ketones because of their easy preparation from

TABLE 4-6   **Physical Properties of Ketones**

Name	Formula	Mp	Bp	Sp. gr.
Acetone	CH_3COCH_3	-95	56	0.792
Ethyl methyl ketone	$CH_3COCH_2CH_3$	-86	80	0.805
Diethyl ketone	$C_2H_5COC_2H_5$	-39	102	0.814
2-Hexanone	$CH_3CO(CH_2)_3CH_3$	-57	127	0.830
tert-Butyl methyl ketone	$CH_3COC(CH_3)_3$	-53	106	0.807
Cyclohexanone		-31	156	0.942
Biacetyl	$CH_3COCOCH_3$	-2.4	88	0.980
Acetylacetone	$CH_3COCH_2COCH_3$	-23	138	0.972
Acetophenone	$CH_3COC_6H_5$	20	202	1.024
Benzophenone	$C_6H_5COC_6H_5$	48	306	1.145

them with inorganic substituted amines. They are named as derivatives of the parent carbonyl compounds, as shown in italic type below.

$$CH_3CHO + NH_2OH \longrightarrow CH_3CH{=}NOH$$

Acetaldehyde Hydroxyl- *Acetaldehyde*
amine *oxime*

$$CH_3COCH_3 + NH_2NH_2 \longrightarrow CH_3\overset{\overset{\displaystyle CH_3}{|}}{C}{=}NNH_2 \xrightarrow{(CH_3)_2CO} CH_3\overset{\overset{\displaystyle CH_3}{|}}{C}{=}N{-}N{=}\overset{\overset{\displaystyle CH_3}{|}}{C}{-}CH_3$$

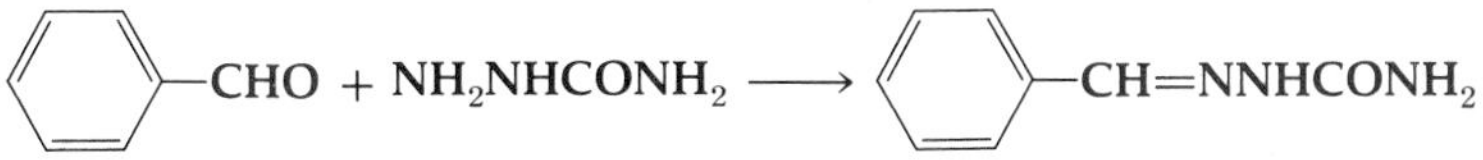

Acetone Hydrazine *Acetone* *Acetone azine*
hydrazone

Benzaldehyde Semicarbazide *Benzaldehyde*
semicarbazone

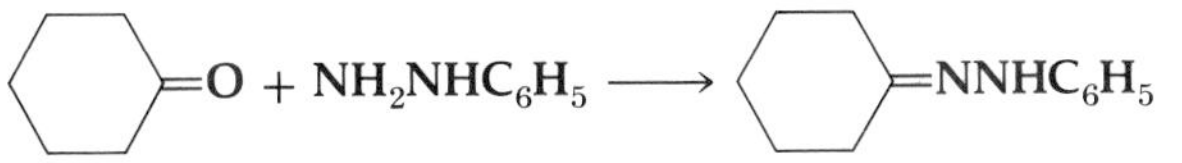

Cyclo- Phenyl- *Cyclohexanone*
hexanone hydrazine *phenylhydrazone*

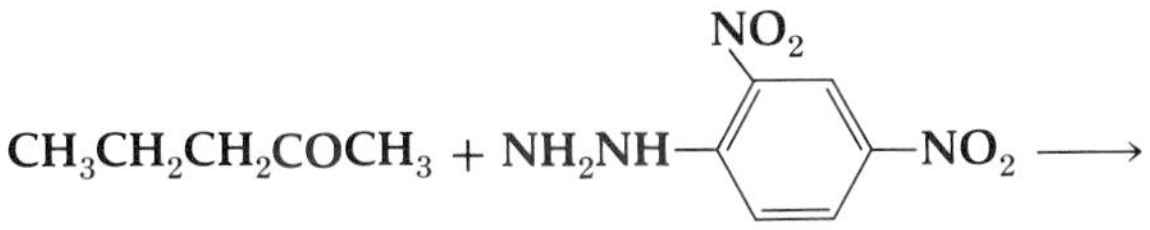

2-Pentanone 2,4-Dinitrophenyl-
hydrazine

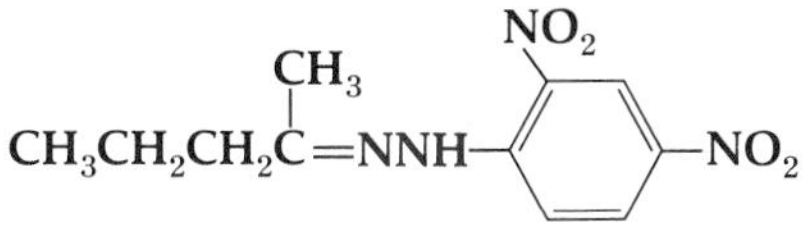

2-Pentanone-
2,4-dinitrophenyl-
hydrazone

Carboxylic Acids

The carboxyl group, $-\overset{\overset{\displaystyle O}{\|}}{C}-OH$, is formally made by combining carbonyl and hydroxyl groups. However, interaction between these two parts so modifies their chemical properties that the entire group is considered as a new function with its own characteristic properties.

The chemical importance of carboxylic acids is illustrated by the fact that their names provide roots for many other compounds. Names and physical properties of many important carboxylic acids are listed in Table 4-7. Although the list is formidable, a serious student should learn the names of all the acids listed in the table. (Division lines are included for convenience.)

Positions of substituents on aliphatic acid structures are frequently designated by the use of Greek letters, as was done with aldehydes and ketones.

CH$_3$CHCOOH   CH$_3$CCH$_2$COOH
    | ‖
    Cl   O

α-Chloropropionic acid   β-Ketobutyric acid

The simple aromatic acids have common names, which are used to designate parent structures in naming complex aromatic acids.

COOH   COOH   COOH
                                         COOH |
                                                             COOH

                                         NO$_2$ COOH   NO$_2$

Benzoic acid   m-Nitrobenzoic acid   Phthalic acid   3-Nitrophthalic
                or 3-nitrobenzoic acid   acid

TABLE 4-7 **Physical Properties of Carboxylic Acids (R—COOH)**

Name	Formula	Mp	Bp	Sp. gr.	pK$_a$
Formic	HCOOH	8.4	101	1.220	3.77
Acetic	CH$_3$COOH	17	118	1.049	4.76
Propionic	CH$_3$CH$_2$COOH	−22	141	0.992	4.88
n-Butyric	CH$_3$(CH$_2$)$_2$COOH	−5	163	0.959	4.82
Isobutyric	(CH$_3$)$_2$CHCOOH	−47	154	0.949	4.85
n-Valeric	CH$_3$(CH$_2$)$_3$COOH	−35	187	0.939	4.86
Caproic	CH$_3$(CH$_2$)$_4$COOH	−2	205	0.929	4.85
Glycolic	HOCH$_2$COOH	79			3.90
Lactic	CH$_3$CHOHCOOH	17		1.249	3.87
Acrylic	CH$_2$=CHCOOH	13	141		4.26
Oxalic	HOOCCOOH	180		1.90	1.27†
Malonic	HOOCCH$_2$COOH	135			2.85†
Succinic	HOOCCH$_2$CH$_2$COOH	183			4.21†
Glutaric	HOOC(CH$_2$)$_3$COOH	98	303		4.33†
Adipic	HOOC(CH$_2$)$_4$COOH	153	338		4.43†
Tartaric	HOOCCHOHCHOHCOOH	174			3.02†
Maleic	HOOCCH=CHCOOH (cis)	133			1.91†
Fumaric	HOOCCH=CHCOOH (trans)	276			3.10†
Benzoic	C$_6$H$_5$COOH	122	249	1.316	4.20
Phthalic	o-C$_6$H$_4$(COOH)$_2$	206–208		1.593	2.95†
Salicylic	o-HOC$_6$H$_4$COOH	158	256		2.94
Anthranilic	o-H$_2$NC$_6$H$_4$COOH	147			4.80
Cinnamic	C$_6$H$_5$CH=CHCOOH (trans)	133	300		4.43

†Based on the dissociation of the first carboxyl group. (See Chap. 8.)

In IUPAC nomenclature monocarboxylic acids are named as *alkanoic acids,* with the carboxylic group itself numbered 1 on the chain.

$$CH_3CH_2CH_2COOH \qquad CH_3CH_2\overset{\overset{\displaystyle CH_3}{|}}{C}H\overset{\overset{\displaystyle }{|}}{C}HCOOH$$

$$\underset{C_6H_5}{|}$$

Butanoic acid
(*n*-butyric acid)

2-Methyl-3-phenylpentanoic acid

$$CH_3CH=CHCH_2\overset{\overset{\displaystyle NH_2}{|}}{C}HCOOH \qquad HOOCCH_2\overset{\overset{\displaystyle Cl}{|}}{C}HCOOH$$

2-Amino-4-hexenoic acid

2-Chlorobutanedioic acid

When the carboxyl group is considered as a substituent on some parent structure, it may be located either by naming the substituent **carboxy** or by naming the parent hydrocarbon structure and following the name with the words **carboxylic acid**.

2-Carboxycyclohexanone

Cyclopentane-carboxylic acid

Examination of the melting- and boiling-point data in Table 4.7 shows that, compared with the other classes of compounds studied thus far, carboxylic acids tend to be high-melting and high-boiling. This shows that the molecules of acids tend to cling together by hydrogen-bonding.

Many of the acids listed in Table 4-7 are found in nature either free, as esters, or as amides. Many acids play important roles in the metabolism of plants and animals. Acetic acid, for example, is the fundamental unit used by living organisms in the **biosynthesis** of such widely diversified classes of natural products as long-chain fatty acids, natural rubber, and steroid hormones (Chap. 27). Acetic acid (vinegar) is the end product of fermentation of most agricultural products.

Most of the simpler carboxylic acids are also readily available commercial chemicals. The acids and their derivatives (see following sections) are valuable synthetic intermediates. Acetic acid is a powerful solvent and is used in both laboratory and industrial chemical procedures. Since pure acetic acid, known as *glacial acetic acid,* is somewhat corrosive, it is not ordinarily used as a solvent in chemical preparations intended for public distribution.

Dibasic acids (acids containing two carboxyl groups) are widely used in synthesis and are fundamental units in the construction of a number of synthetic high polymers (Chap. 25). The latter include such well-known products as the synthetic fibers Dacron and Nylon.

Acid Halides and Anhydrides

Replacement of the —OH group of a carboxylic acid by a chlorine atom produces an **acid chloride.**† This is both a formal structural procedure and a chemical reaction.

$$R-\overset{\overset{\text{O}}{\|}}{C}-OH \qquad R-\overset{\overset{\text{O}}{\|}}{C}-Cl$$

Carboxylic   Acid chloride
acid

Acid anhydrides contain two acyl groups bound to an oxygen atom. Both structurally and chemically anhydrides are produced from carboxylic acids by removal of one molecule of water from two molecules of acid. Both symmetrical and mixed anhydrides are known.

$$RC\overset{\overset{\text{O}}{\|}}{}\boxed{O-H + H}-O-\overset{\overset{\text{O}}{\|}}{C}R \longrightarrow RC\overset{\overset{\text{O}}{\|}}{}-O-\overset{\overset{\text{O}}{\|}}{C}R$$

An anhydride

$$CH_3\overset{\overset{\text{O}}{\|}}{C}-O-\overset{\overset{\text{O}}{\|}}{C}CH_3 \qquad C_6H_5\overset{\overset{\text{O}}{\|}}{C}-O-\overset{\overset{\text{O}}{\|}}{C}C_2H_5$$

Acetic anhydride,   Benzoic propionic
a symmetrical   anhydride, a mixed
anhydride   anhydride

Acid halides (or **acyl halides**) are named by combining the names of the **acyl group (RCO—)** and the particular halide. Names of the acyl groups are made from the stems of the names of carboxylic acids by replacing the ending -*ic* with -*yl*.

† Carboxylic acid chloride is sometimes used as a name, but the shorter form is commoner.

TABLE 4-8 **Physical Properties of Esters**

Name	Formula	Bp	Mp	Sp. gr.
Methyl formate	$HCOOCH_3$	32	−99	0.974
Methyl acetate	CH_3COOCH_3	57		0.959
Ethyl acetate	$CH_3COOCH_2CH_3$	77	−84	0.925
n-Butyl acetate	$CH_3COO(CH_2)_3CH_3$	126	−74	0.902
Methyl propionate	$CH_3CH_2COOCH_3$	80	−88	0.939
Methyl n-butyrate	$CH_3(CH_2)_2COOCH_3$	103	−85	0.920
Ethyl benzoate	$C_6H_5COOCH_2CH_3$	212	−38	1.065
Methyl salicylate (oil of wintergreen)	(benzene ring with OH and COOCH₃)	223	−8.6	1.199
Acetylsalicylic acid (aspirin)	(benzene ring with COOH and OCOCH₃)		143	

$$CH_3\overset{O}{\underset{\|}{C}}- \qquad CH_3\overset{O}{\underset{\|}{C}}-F \qquad C_6H_5COCl \qquad ClCOCH_2CH_2COCl$$

| Acetyl group | Acetyl fluoride | Benzoyl chloride | Succinyl chloride |

Symmetrical anhydrides are named with the name of the parent acid and the word *anhydride* replacing the word *acid*. Both parent acids are mentioned in naming mixed anhydrides.

$$CH_3\overset{O}{\underset{\|}{C}}-O-\overset{O}{\underset{\|}{C}}CH_3 \qquad (CH_3CH_2CH_2CO)_2O \qquad C_6H_5\overset{O}{\underset{\|}{C}}O\overset{O}{\underset{\|}{C}}CH_3$$

Acetic anhydride   *n*-Butyric anhydride   Acetic benzoic anhydride

Cyclic anhydrides are very easily formed from those dibasic acids which can cyclize with formation of five- or six-membered rings.

| Succinic anhydride | Glutaric anhydride | Phthalic anhydride |

Acid chlorides and anhydrides are generally reactive liquids, used principally as sources of acyl groups in chemical syntheses.

Esters and Lactones

Both in a formal sense and in actual chemical practice an ester is derived from an alcohol and an acid with elimination of water. Esters of carboxylic acids are named as *alkyl* or *aryl carboxylates*. Carboxylate groups are named by substituting the ending -*ate* for the ending -*ic* in the name of the acid.†

$$RC\overset{O}{\underset{\|}{}}\lbrack OH + H \rbrack OR' \longrightarrow RCOR'$$

An ester

$$CH_3\overset{O}{\underset{\|}{C}}-OH \qquad CH_3\overset{O}{\underset{\|}{C}}-O- \qquad CH_3\overset{O}{\underset{\|}{C}}-O-CH_3$$

Acetic acid   Acetate (or acetoxy group)   Methyl acetate

$$CH_3COOC_6H_5 \qquad C_6H_5COOCH_2C_6H_5 \qquad CH_3CH_2COOC(CH_3)_3$$

Phenyl acetate   Benzyl benzoate   *tert*-Butyl propionate

† The names of carboxylate groups really originate from the names of salts of carboxylic acids. For example, $CH_3CO_2^- Na^+$ is sodium acetate. At one time the reaction of an alcohol with an acid was considered to be analogous to the reaction of a metallic hydroxide (MOH) with an acid. Despite the dissimilarity in the structures of salts (ionic compounds) and esters (covalent compounds), the similarity in nomenclature of the two classes has been maintained.

Sometimes as a convenience esters are given names in which the entire ester group is considered as a substituent. This procedure is rarely used unless the ester group is either **carbomethoxy** ($-COOCH_3$) or **carboethoxy** ($-COOC_2H_5$).

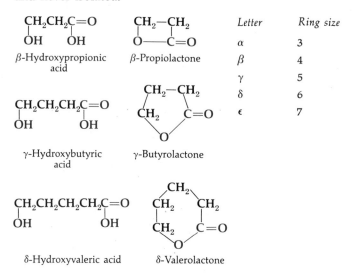

Carbomethoxycyclobutane   2-Carboethoxycyclohexanone
or methyl cyclobutanecarboxylate

The IUPAC names of esters are compounded from the IUPAC names of the alkyl (or aryl) groups attached to oxygen and the IUPAC names of the acids, with the -*ic* ending changed to -*ate*, and the alkyl name placed first.

$$CH_3COOCH_3 \qquad CH_3CH{=}CHCH_2COOCHCH_2CH_3$$
$$\overset{CH_3}{|}$$

Methyl ethanoate   *sec*-Butyl 3-pentenoate

Cyclic esters are known as **lactones**. They are designated α-, β-, γ-, or δ-lactones depending on the ring size, the α-lactones being exceedingly unstable and never isolated.

		Letter	Ring size
β-Hydroxypropionic acid	β-Propiolactone	α	3
		β	4
		γ	5
γ-Hydroxybutyric acid	γ-Butyrolactone	δ	6
		ϵ	7
δ-Hydroxyvaleric acid	δ-Valerolactone		

Esters usually have lower boiling and melting points than either acids or alcohols of comparable molecular weight (see Table 4-8), again as a consequence of their lack of hydrogen bonding. They are usually noncorrosive, nontoxic liquids with pleasant fruity odors. As solvents, esters are somewhat like the lower ketones and are extensively used in lacquers and other commercial preparations. In one class of high polymers, the polyesters, the fundamental units are held together by ester linkages. The most familiar example is polyethylene terephthalate, the polyester from terephthalic acid and ethylene glycol. The polymer can by spun into fibers (Dacron) or made into beautifully clear films (Mylar).

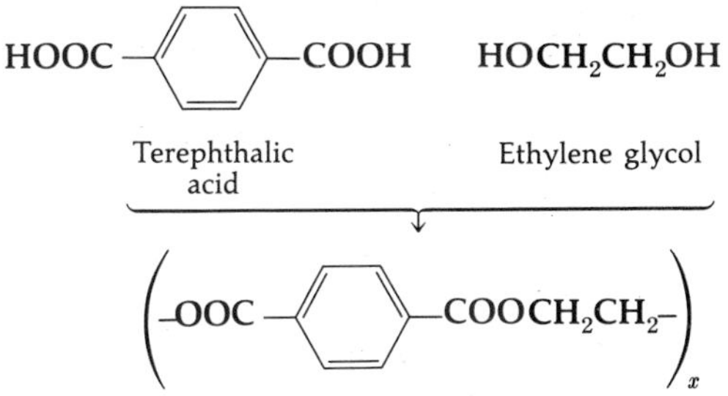

$$HOOC-\left\langle\bigcirc\right\rangle-COOH \qquad HOCH_2CH_2OH$$

Terephthalic acid   Ethylene glycol

$$\left(-OOC-\left\langle\bigcirc\right\rangle-COOCH_2CH_2-\right)_x$$

Polyethylene terephthalate

Esters are widely distributed in nature. Perhaps the most important are *fats* and *vegetable oils,* which are esters of long-chain fatty acids and glycerol (page 116).

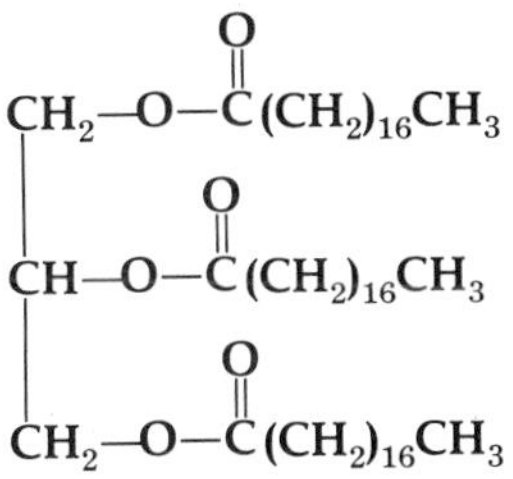

$$\begin{aligned} &\overset{O}{\overset{\|}{CH_2-O-C}}(CH_2)_{16}CH_3 \\ &\overset{O}{\overset{\|}{CH-O-C}}(CH_2)_{16}CH_3 \\ &\overset{O}{\overset{\|}{CH_2-O-C}}(CH_2)_{16}CH_3 \end{aligned}$$

Tristearin, a fat

Amides, Lactams, and Nitriles

Replacement of the **OH** group of an acid by a nitrogen atom, which may in turn bear substituents, produces amides. The class can also be described as acyl derivatives of ammonia and amines.

$$\overset{O}{\overset{\|}{R-C}}-OH \qquad \overset{O}{\overset{\|}{R-C}}-NH_2 \qquad \overset{O}{\overset{\|}{R-C}}-NHR' \qquad \overset{O}{\overset{\|}{R-C}}-NR'_2$$

Carboxylic acid   Amides

Names of amides are based upon the names of the related acids. The word "amide" is attached to the root of the name as an ending. If the nitrogen atom bears substituents, they are located by placing a capital *N* before the name of the substituent.

$$\overset{O}{\overset{\|}{CH_3-C}}-NH_2 \qquad \overset{O}{\overset{\|}{C_6H_5-C}}-NHCH_3 \qquad \overset{Cl}{\overset{|}{CH_3CHCH_2}}-\overset{O}{\overset{\|}{C}}-NHC_2H_5$$

Acetamide   *N*-methyl-benzamide   3-Chloro-*N*-ethyl-butyramide

A few amides, especially *N*-acyl derivatives of aniline, are known by special names.

$$CH_3CONH\,C_6H_5 \qquad CH_3CH_2CONH\,C_6H_5$$

Acetanilide   Propionanilide
(*N*-phenylacetamide)   (*N*-phenylpropionamide)

Lactams are cyclic amides. Greek letters are used to show the ring size in the same way as in lactone nomenclature.

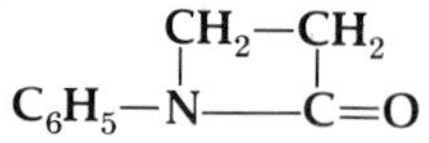

$$\underset{\substack{\text{N-Phenyl-}\beta\text{-}\\\text{propiolactam}}}{C_6H_5-\overset{\overset{\displaystyle CH_2-CH_2}{|}}{N}\underset{}{\overset{|}{\underline{\qquad}}}C=O}$$

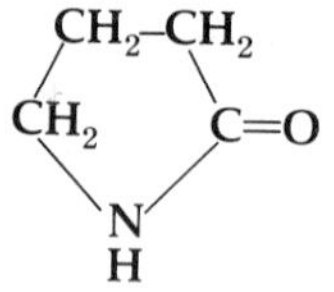

γ-Butyrolactam

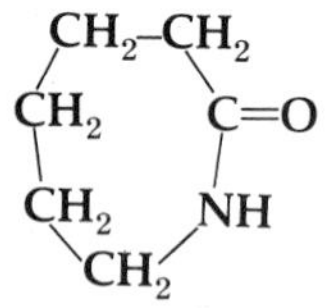

ϵ-Caprolactam

Imides contain two acyl groups attached to the same nitrogen. The structure corresponds to that of the anhydrides with an —NH— rather than an —O— bridge. Only a few cyclic imides have attained much chemical significance.

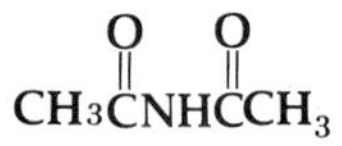

$$CH_3\overset{\overset{\displaystyle O}{\|}}{C}NH\overset{\overset{\displaystyle O}{\|}}{C}CH_3$$

Acetylimide
(acetimide)

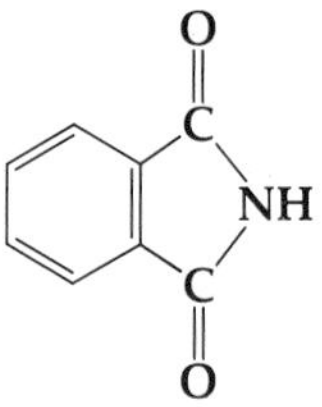

Phthalimide

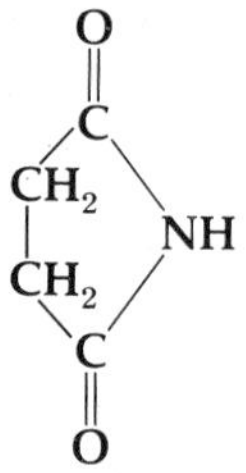

Succinimide

With the exception of formamide, a high-boiling liquid, the unsubstituted amides are all crystalline solids at room temperature. Apparently, amides are strongly hydrogen-bonded in both liquid and solid states. Dimethylformamide is a very powerful solvent for both polar and nonpolar compounds and is often used as a reaction medium both in the research laboratory and in industrial processes.

Nitriles, which contain the *cyano group*, —C≡N, are alkyl and aryl derivatives of hydrogen cyanide.

H—CN   R—CN Ar—CN

Hydrogen cyanide   Nitriles

Although nitriles are quite stable, the —C≡N group can usually be converted to —COOH by appropriate hydrolysis. This structural relationship between nitriles and acids is the basis for the nomenclature of nitriles. The ending -*onitrile* is attached to the root of the name of the corresponding carboxylic acid. Sometimes polyfunctional compounds are named by considering the cyano group as a substituent.

$$CH_3C\equiv N \qquad C_6H_5C\equiv N \qquad CH_3CH_2\overset{\overset{\displaystyle Cl}{|}}{C}HC\equiv N$$

Acetonitrile   Benzonitrile   α-Chlorobutyronitrile

$$CH_3CH_2\overset{\overset{\displaystyle C_6H_5}{|}}{C}HCH_2CN \qquad N\equiv CCH_2COOH \qquad CH_3\overset{\overset{\displaystyle}{|}}{C}HCH_2\overset{\overset{\displaystyle}{|}}{C}HCH_3$$
$$\qquad\qquad\qquad\qquad\qquad\qquad\qquad\qquad\qquad\qquad\qquad C\equiv N \quad Cl$$

3-Phenylpentano-   Cyanoacetic   2-Cyano-4-chloro-pentane
nitrile   acid
(3-phenyl valeronitrile)

Most of the common nitriles are liquids with boiling points slightly higher than those of alcohols of comparable molecular weight. Despite their structural relationship to hydrogen cyanide, nitriles are only moderately toxic. Acetonitrile is a common solvent. Like acetone and dimethylformamide, acetonitrile dissolves surprisingly large amounts of inorganic salts as well as most organic compounds.

PROBLEMS

4-8 Write all possible structural formulas for each of the following:

   a Amines having the formula $C_4H_{11}N$

   b Alcohols having the formula C_3H_6O

   c Compounds containing no carbon-oxygen double bonds and having the formula $C_3H_8O_2$

   d All the derivatives of hydroxylamine having the formula $C_4H_{11}NO$

   e All tertiary amines containing four or fewer carbon atoms

4-9 Two compounds have the formula C_2H_6O. What are their structural formulas? Make the following predictions. Give your reasons.

   a Which is higher boiling?

   b Which is more soluble in water?

   c Which can be made from ethylene?

   d Which will react with elemental sodium with the evolution of hydrogen? (Reason by analogy to the chemistry of water.)

4-10 If water is added to a solution of methyllithium in ether, a gas is produced and the water solution contains a strong base. What reaction do you think has occurred?

4-11 Calculate the indices of hydrogen deficiency in the following compounds:

   a C_4H_7N **d** $C_7H_{10}SO_3$

   b $C_6H_{14}N_2$ **e** C_8H_6NCl

   c $C_6H_4N_2O_4$ **f** $C_{10}H_8SCl_2$

4-12 Identify the hydrogen-deficient parts of the following structures. Indicate the index of hydrogen deficiency for each structure.

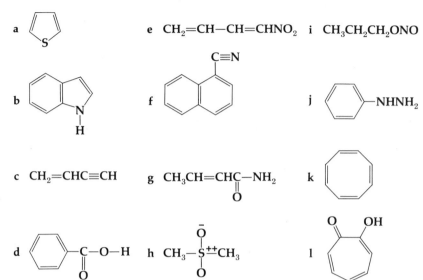

a (structure: thiophene, labeled S)

e $CH_2=CH—CH=CHNO_2$

i $CH_3CH_2CH_2ONO$

b (indole structure, labeled N, H)

f (naphthalene with $C≡N$)

j (phenyl—$NHNH_2$)

c $CH_2=CHC≡CH$

g $CH_3CH=CHC—NH_2$ (with O double bond)

k (cyclooctatetraene structure)

d (phenyl—$C—O—H$ with O)

h $CH_3—\overset{\bar{O}}{\underset{\underset{-}{O}}{S^{++}}}CH_3$

l (tropone with OH, O)

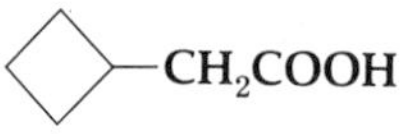

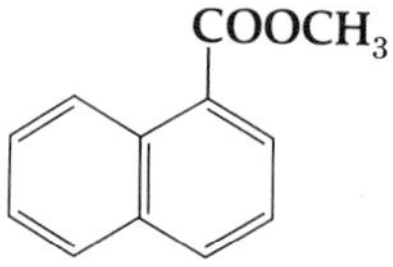

4-13 Name the following compounds in as many ways as you can.

a $CH_3CH_2\underset{\underset{NH_2}{|}}{C}HCOOH$

e $CH_3\underset{\underset{}{}}{\overset{\overset{Cl}{|}}{C}}HCH_2COOCH_3$

b (cyclobutyl)—CH_2COOH

f $CH_3\overset{\overset{CH_3}{|}}{C}=CHCH_2COOH$

c (naphthalene with $COOCH_3$)

g $HOOC(CH_2)_6COOH$

d $(CH_3)_2C=CH\underset{\underset{Cl}{|}}{C}HCH_2COOH$

h $C_6H_5CH_2CH_2COO\overset{\overset{CH_3}{|}}{C}HCH_2CH_3$

4-14 Write structures for the following compounds:

a   *p*-Chlorobenzoic acid
b   *tert*-Butyl cyclohexyl ketone
c   Dibenzoylmethane
d   Hexamethylacetone
e   2-Butyl-4-methylpentanoate

f   3-Carbethoxycyclobutene
g   α-Methyl-β-aminobutyric acid
h   4-Methoxy-2-butenoic acid
i   N-Methyl-N-phenylacetamide
j   2,4,6-Trichlorononanonitrile

4-15 Write structures for compounds that belong in the following classes:

a A β-γ-unsaturated δ-lactone   g The oxime of an aromatic ketone

b A cyclic unsaturated anhydride   h γ-Diketone

d An acyl fluoride   j An imide

e An N-substituted γ-lactam   k An N-substituted urea

f An α-ketoacid   l A semicarbazone

4-16 Write structural formulas for at least half a dozen compounds having each of the following empirical formulas. Give each compound a name and indicate the index of hydrogen deficiency for each group of structures.

a $C_4H_8O_2$ c C_4H_7Cl

b C_3H_5N d $C_3H_7NO_2$

4-17 Write the formulas indicated by these names.

a 2-Methyl-3-propyl-2-pentene

b 3-Methyl-3-ethyl-1-pentyne

c 2,4-Hexadiene

d Hexa-1-ene-4-yne

e 1-Chloro-1-cyclohexene

f 3-Amino-1-cyclopentene

g 2-Methyl-1-cyclobutanone (= 2-methylcyclobutanone)

h 2-Hydroxy-3-pentanone

i Hex-1-en-3-on-4-ol

j 2-Bromo-3-pentenoic acid

k 1-Bromo-2-hydroxy-5-methyl-benzene

l 2-Iodo-3-amino-pentanal

m 4-(1-propenyl)-3,4-dimethyl-2-hexanone (write a better name.)

4-18 Write names for each of these structures.

a $CH_3-CH-CH-C-CH-CH_3$
       | | || |
      CH_3 Cl O CH_2-CH_3

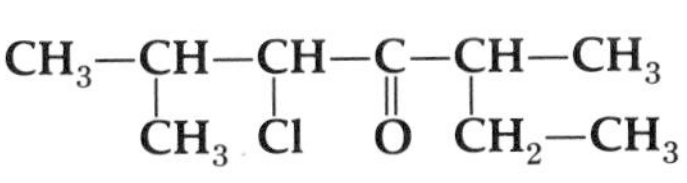

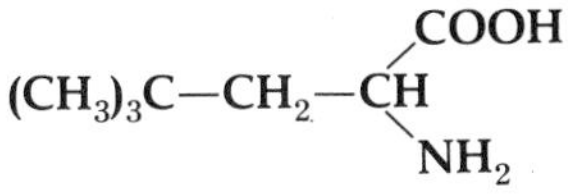

d

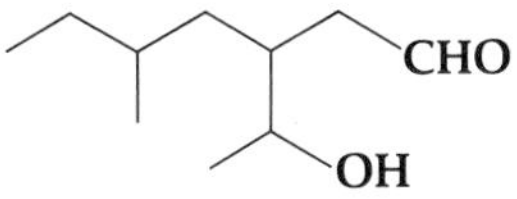

b

e

c $(CH_3)_3C-CH_2-CH$ with COOH and NH$_2$ substituents

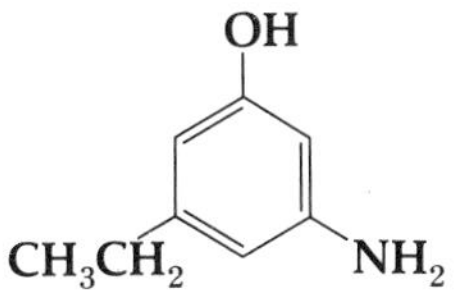

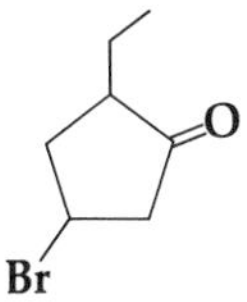

f

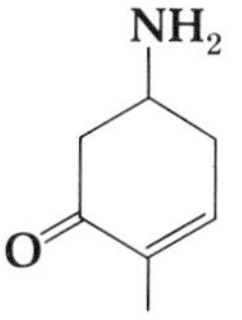

g $CH_3CH_2NH-CH_2CH_2-\overset{\overset{\displaystyle CH_3}{|}}{\underset{\underset{\displaystyle CHO}{|}}{C}}-CH_2CH_3$ k

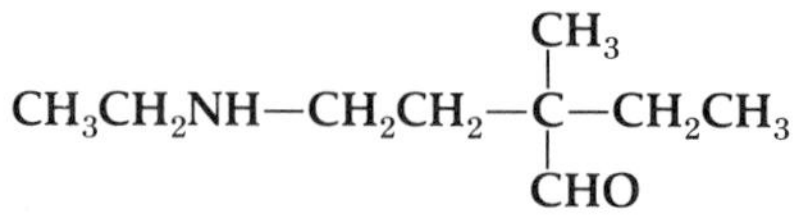

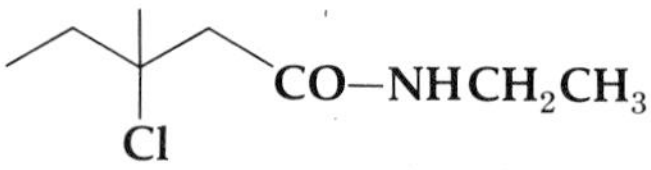

h 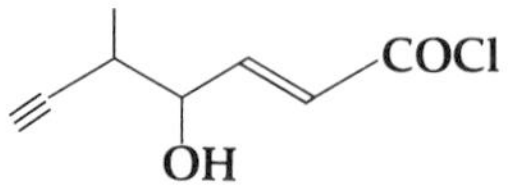   l

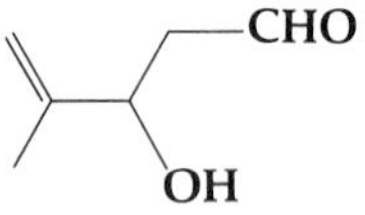

i 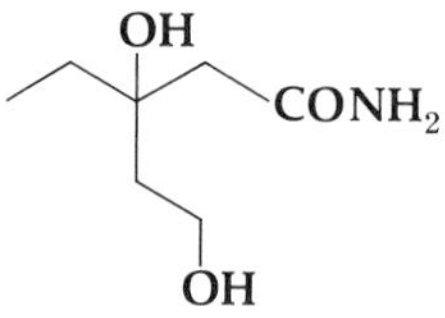   m

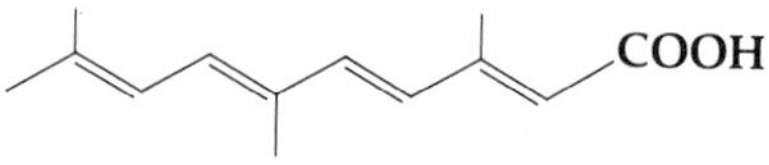

4-19 Write all the isomers fitting each of the following names and give more definitive names to each; if some names are impossible molecules, indicate this.

a   Butanone
b   Propenal
c   Ethyl butanol
d   3-Methyl-3-heptanone
e   Pentenone
f   δ-Bromopropionic acid

g   Cyclodecanoic acid
h   Cyclopropyl-butanone
i   Propanetriol
j   N-Ethyl-amino-pentane
k   4-Hydroxy-heptadiyne

4-20 Write the structure for an example of each of the following and name your choice.

a   A cyclic diketone
b   A saturated secondary amine
c   An aromatic acid chloride
d   A bicyclic ether

e   An unsaturated amino acid
f   An amino amide
g   An N-acyl aromatic compound
h   A tricyclic hydroxy nitrile

CONJUGATED BONDS: RESONANCE, TAUTOMERISM, AND AROMATICITY

WITH respect to covalent bonds we have seen the two major bond types, σ and π, formed between two atoms by overlap of their atomic orbitals: $s + p$ hybrid atomic orbitals for σ bonds; p atomic orbitals for π bonds. In both cases the electron pair of the bond is specifically associated only with the two bonded atoms; such a pair is said to be **localized**. With two orbitals (sp^3, sp^2, sp) overlapping end-to-end to form a σ bond lying directly between the nuclei, there is little chance for the bonding electrons to be anything but localized.

With two p orbitals standing perpendicular to the σ-bonded skeleton and overlapping sideways, however, they may easily encounter more adjacent p orbitals on their opposite sides and overlap further with these. Such a situation could create a more extended π orbital in which the electrons in the orbital are no longer confined between two atoms but rather are **delocalized**, over a greater number of nuclei. For this to occur each successive atom bearing a p orbital for overlap must be adjacent to the last.†

This delocalization of π electrons is of central importance to the chemical and physical properties of unsaturated molecules and is the subject of the present chapter. The particular unsaturated molecules involved are those which have formal structures exhibiting double bonds adjacent to each other (or to unshared electron pairs which can be hybridized into p orbitals) and therefore capable of further π overlap with each other. Such unsaturated molecules are said to be **conjugated** and to have conjugated double bonds.

The special properties of conjugated molecules were only understood and unified with the advent of quantum mechanics in the 1920's and 1930's. Although in principle a complete solution of the wave equation for any molecule would predict all of its properties, in practice these solutions are of such horrendous complexity that nothing more complex than the H_2 molecule has yielded to complete solution. Nevertheless, a judicious (and extensive) use of

---

†Sideways overlap of p orbitals is not significant for atoms more than a bond length apart, that is, ~ 1.5 Å.

simplifying assumptions does afford approximate solutions in the more complex molecules which interest us here. These solutions provide a bridge between theoretical physics and organic chemistry, offering focal points for correlating data and theory and providing a basis for prediction and test. In the solutions for conjugated molecules the main assumption is that the σ-bonded skeleton, or σ frame, of the molecule is fixed and electrically uninvolved with the conjugated π-electron system. An orbital description of the independent π-electron system may then be calculated as a separate and simpler entity. There are two traditional approaches to solving the wave equation for π-electron systems: the **valence-bond (VB)** or **resonance** method and the **molecular-orbital (MO)** method, and they differ essentially in the simplifying assumptions which they utilize. Each will have value for us in particular contexts, but it is important to be able to think of any conjugated molecule in terms of both descriptions.

5-1   THE MOLECULAR-ORBITAL METHOD

Because it develops as a natural extension of our previous discussion of combining atomic orbitals to form molecular orbitals, this approach is offered first. We treat the σ frame as already bonded and bearing, on a series of adjacent atoms, p atomic orbitals, suitably placed for overlap sideways into extended molecular orbitals. The σ frame must be coplanar so that the p orbitals are all parallel for sideways overlap. We must now create as many molecular orbitals as we have p atomic orbitals to combine (LCAO method, Sec. 2-7). Hence the number of molecular orbitals equals the number of atoms with p orbitals to overlap.

The simplest example is 1,3-butadiene, with four parallel p orbitals in a row (Fig. 5-1). Linear combination of all four atomic p orbitals gives a molecular orbital that covers the entire molecule. In the first approximation, the energy of electrons in this π orbital has a value considerably lower than that of any of the other π electrons. Therefore, two of the four π electrons are described by this orbital. The other two must go into one of the higher molecular orbitals. Three more π orbitals are formed by addition and subtraction of the four atomic orbitals. The next lowest in energy is also shown in Fig. 5-1. The second pair of π electrons in butadiene must be assigned to this orbital.

The lowest orbital is the combination of the four atomic orbitals (ψ) into a molecular orbital or wave function (ϕ). The mixing coefficients C are like those used previously in Sec. 2-7. The first (lowest-energy) molecular orbital adds all four atomic orbitals (all positive):

$$\phi_1 = C_1\psi_1 + C_2\psi_2 + C_3\psi_3 + C_4\psi_4$$

The second has one sign change:

$$\phi_2 = C_1\psi_1 + C_2\psi_2 - C_3\psi_3 - C_4\psi_4$$

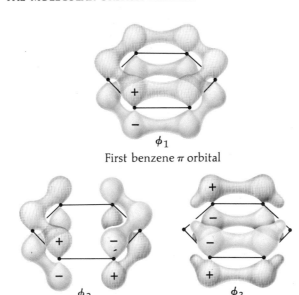

ϕ_1

First benzene π orbital

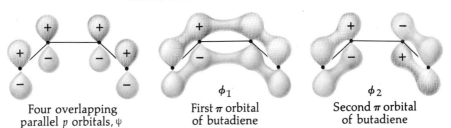

ϕ_2 ϕ_3

Second and third benzene π orbitals

Four overlapping   First π orbital   Second π orbital
parallel p orbitals, ψ of butadiene   of butadiene

ϕ_1 ϕ_2

FIGURE 5-1 **The orbitals of benzene and butadiene on their carbon σ frames (showing the signs of their lobes)**

The higher-energy orbitals are these:

$$\phi_3 = C_1\psi_1 - C_2\psi_2 - C_3\psi_3 + C_4\psi_4$$

$$\phi_4 = C_1\psi_1 - C_2\psi_2 + C_3\psi_3 - C_4\psi_4$$

The four orbitals are shown schematically and arranged in a typical energy-level diagram below. The signs of the coefficients in the LCAO equation dictate the positive and negative amplitudes (+ and −) of the wavefunctions shown in the drawings and Fig. 5-1. These + and − signs are *not* electrical charges but only mathematical signs of wave function amplitude.

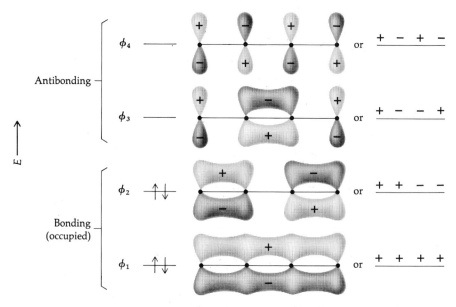

Atomic orbitals with coefficients of *like sign* are **bonding**, those with *unlike sign* **antibonding**. Note that in Fig. 5-1 the positive and negative lobes, shown by dark and light shading, are inverted at carbon atoms 2 and 3 in the second π orbital (ϕ_2). The orbital ϕ is said to be *bonding* between atoms 1 and 2 and between atoms 3 and 4 but *antibonding* between 2 and 3. There is a node between the various antibonding orbital lobes; there can be no electrons at the node. Hence there is antibonding since the electrons are "squeezed out" of the volume between two nuclei and this condition allows unshielded nuclear repulsion. The whole molecule of butadiene is the sum of its two occupied orbitals, described as having much *double-bond character* between atoms 1 and 2 and atoms 3 and 4, as in the classical formula, and a little double-bond character at the central bond between atoms 2 and 3. The molecular-orbital method allows quantitative assessment of double-bond character at each bond.

The lowest-energy (most stable) molecular orbital in a conjugated system is the fully extended one. Molecules in which this orbital is linear are said to have **linear conjugation**. When the extended orbital is branched, **cross-conjugation** is the description.

$$CH_2{=}CH{-}CH{=}CH{-}CH{=}CH_2$$
1,3-Hexatriene = linear conjugation

$$\begin{matrix} CH_2{=}CH \\ {\diagdown} \\ \quad\quad C{=}CH_2 \\ {\diagup} \\ CH_2{=}CH \end{matrix}$$
2-Vinyl-1,3-butadiene = cross-conjugation

The molecular-orbital treatment of butadiene has several attractive features. It leads to the expectation that the conjugated π bonds will behave as a unit chemically and provides a ready explanation of the fact that the ultraviolet absorption spectra of dienes are quite different from those of simple alkenes (Chap. 7). However, the method overemphasizes the delocalization concept, and the stability of conjugated dienes relative to their unconjugated isomers is overestimated. The difficulty arises because the π-molecular-orbital method does not take account of the fact that electrons repel each other. The best picture of 1,3-butadiene is intermediate between the freely delocalized molecular-orbital picture and the traditional model of a molecule containing two localized double bonds.

The π-molecular-orbital method is perhaps at its best when applied to benzene. Since the molecule is completely symmetrical, the six π electrons must be distributed symmetrically with respect to the six carbon nuclei. Each of the six carbon atoms is prepared in the sp^2 state of hybridization in the σ frame and contributes a p orbital to the π system. Linear combination of the six p orbitals generates six π molecular orbitals. The six π orbitals are arranged in order of increasing energy, and the three lowest are filled with two electrons each. The lowest filled orbital, (ϕ_1) is completely symmetrical, as shown at the top of Fig. 5-1. The next two (ϕ_2 and ϕ_3) are a **degenerate pair,** i.e., they have the same energy. One representation of this pair is found in Fig. 5-1. Initially, it appears that the orbitals do not reproduce the symmetry of benzene. However, if the two orbitals are taken together, as would be the case if the total electron density were calculated after two electrons had been assigned to each of the orbitals, they give sixfold symmetry. One of the degenerate orbitals has two nodes at opposite bond midpoints (ϕ_2), the other (ϕ_3) at opposite atoms.

One of the important properties of a molecule is its energy. The wave function allows a calculation of the energy by a procedure in which a mathematical derivative of the wave function ϕ is integrated over all of space. The energy is then minimized with respect to the coefficients (C_1, C_2, etc.). The minimum energy of the system is then obtained by solving the complex determinant (the **secular determinant**) resulting from this minimization process. The proper coefficients for the wave function ϕ with this minimum energy may then be obtained as well. The mathematical procedures involved will not concern us here, however. They are too complex for solution at all unless various simplifying assumptions are made, but with suitable choices for these assumptions which have been developed in practice, rough solutions to these equations can be obtained without either a computer or special mathematical background.

One important result of such calculations is that the energy of a conjugated π-electron system is lower—i.e., it is more stable—than the calculated energy of the same molecule with each pair of π electrons *localized* in simple π bonds between pairs of atoms. Therefore, butadiene is more stable than it would be if it were simply regarded as the sum of two independent ethylenes, and benzene is more stable than a theoretical unconjugated "cyclohexatriene." The difference in energy between the real conjugated molecule and its theoretical unconjugated analog is called the **delocalization energy** or **resonance energy**.

PROBLEM 5-1

Draw the lowest-energy molecular orbital for each of the following (cf. Fig. 5-1):

a 2-Butenal   **c** 2-Methyl-1,3,5-hexatriene

b Cyclobutadiene   **d** Vinylbenzene

5-2 THE RESONANCE METHOD

In the last section a particular approach to finding a molecular wave function was described. The two steps were (1) to form molecular orbitals and (2) to assign electrons, in pairs, to the molecular orbitals. The approach is by no means unique. Another scheme begins at the outset with the notion of localized bonds and then feeds in delocalization at the end. This procedure, the **resonance method**, has inspired a very useful approach to qualitative representation of systems in which electrons are extensively delocalized.

In the molecular-orbital method the molecular wave function ϕ was created by a linear combination of atomic orbitals (ψ) of the involved atoms. In the valence-bond (resonance) approach another linear combination of the same form is used to approximate ϕ, but in this case the individual additive functions ϕ are not atomic orbitals but represent various possible structures of the whole molecule with localized bonds only.

$$\phi = K_1\phi'_1 + K_2\phi'_2 + K_3\phi'_3 + \cdots +$$

In this linear combination we approximate the authentic structure by a mixing of the properties of two or more full structures which contain the same π electrons localized in various ways. The true molecule is then regarded as a **hybrid**, or **resonance hybrid**, of the several localized π-bond molecular structures. The coefficients (K_1, K_2, etc.) in the equation represent the weight, or relative importance, of each contributing structure (ϕ') in the resonance hybrid, the actual form of the molecule.

The resonance approach is more difficult mathematically than the LCAO but is usually easier and clearer for qualitative discussion of delocalization, or resonance. The whole problem of conjugated systems is simply that no single classical structure (localized bonds) adequately describes the molecule and several structures must sometimes be written. None of these is a unique or completely "true" description of the molecule but each contributes to the total behavior of the molecule. Using the resonance method, then, we find we can express the actual molecule as a combination or hybrid of written structures and thereby qualitatively parallel the mathematical expression. The classic case is seen in the two localized-bond, or Kekulé, structures for benzene. Any particular bond (note color) is single in one structure, double in the other.

Neither structure alone fully or correctly expresses benzene, and neither structure alone represents a real molecule; but a resonance hybrid of the two (expressed by the double-headed arrow†) properly represents benzene. Hence the actual bond between every pair of nuclei is intermediate between double and single. Since each of the two Kekulé structures is identical, each contributes equally to the hybrid. ($K_1 = K_2$ in the equation for ϕ.)

> There is a close analogy in a description of a certain person as being a cross between Don Quixote and Sherlock Holmes with a touch of Sir Galahad. This analogy points up the fictional character of the contributors to the real hybrid as well as the possibility of more than two contributors and these in different relative importance ("but he is really mostly like Don Quixote").

Many molecules and ions are poorly described by any single electronic formula of the conventional type. The carbonate ion (CO_3^{--}) is a good example. Each structure follows the rules of valence, since each of the oxygen atoms and the carbon atom have octets of valence electrons.

However, physical evidence shows that the carbonate ion is symmetrical, i.e., all C—O bonds are the same length. In the π-molecular-orbital method six electrons would be put into three π orbitals that spread over the entire ion. The resonance method describes delocalization of the π electrons and accounts for the fact that some electrons can be bonding between all pairs of adjacent atoms with a set of three equivalent classical formulas connected with a double-headed arrow.

Resonance Energy

One of the "peculiarities" of benzene, compared with alkenes, is the stability of the former. A quantitative measure of the stabilization may be obtained by measuring the heat evolved when hydrogen is added to benzene and to cyclohexene. The same product, cyclohexane, is formed in both cases.

---

†The symbol ⟷ must be carefully distinguished from symbols such as ⇌, which indicates a chemical reaction which is reversible. *The double-headed arrow does not have any dynamic significance;* it only implies that a molecule or ion is better represented by several classic formulas than by one alone. The use of the word resonance to refer to the mixing of formal structures is unfortunate since it implies a rapid physical interconversion between them. This is not the case and π-electron delocalization is a less confusing description of the same phenomenon.

If benzene contained three —CH=CH— groups identical to the one in cyclo-hexene, the heat evolved during hydrogenation of benzene should be about three times the heat of hydrogenation of cyclohexene. Actually, there is a large discrepancy between the predicted and measured heats of hydrogenation.

$$+ 3H_2 \longrightarrow \quad + 49.8 \text{ kcal/mole}$$

$$3 \quad + 3H_2 \longrightarrow 3 \quad + 86.4 \text{ kcal/3 moles}$$

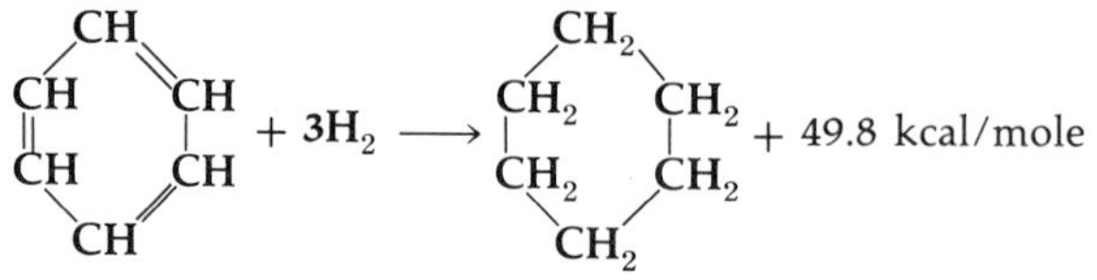

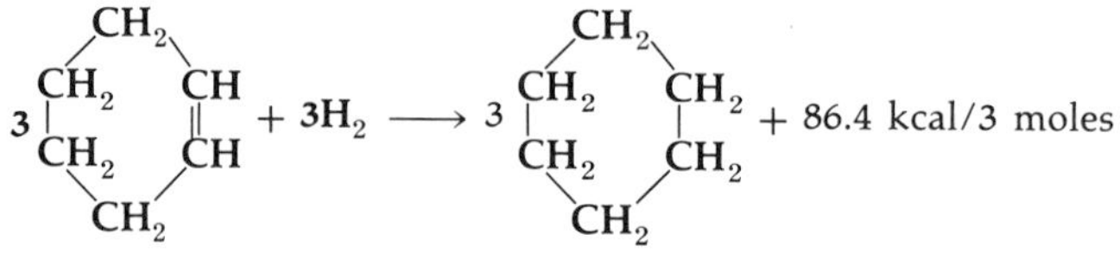

86.4 − 49.8 = 36.6 kcal/mole, resonance energy for benzene

Although other factors may contribute, much of the stabilization of ben-zene apparently results because it contains delocalized electrons. Energy is required to hold the electrons in localized bonds in cyclohexene. Benzene is said to be stabilized by **resonance**, or **delocalization, energy**. The term should be used carefully, since resonance energy is negative; i.e., the molecule has a *lower* energy content than might have been expected by comparison with simple alkenes.

Resonance energies of many conjugated systems have been estimated by comparison with model compounds. Semiempirical methods of calculating resonance energies have also been developed (page 147). Table 5-1 shows a few selected values. Those listed as calculated can be no more accurate than the measured values, since the calculations make use of parameters obtained by comparing measured values with each other. The choice of model compounds is somewhat arbitrary, and factors other than electron delocalization contribute somewhat to the unusual stability of conjugated, unsaturated compounds.

TABLE 5-1 **Representative Resonance Energies**

Name	Formula	Resonance energy, kcal/mole	Source
Benzene		37	Heats of hydrogenation
		42	Heat of combustion
Naphthalene	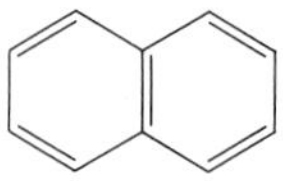	75	Heat of combustion
Acetic acid	CH_3COOH	15	Heat of combustion
Allyl cation	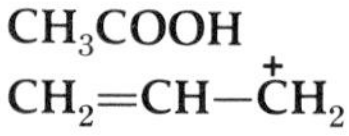 $CH_2=CH-\overset{+}{C}H_2$	25	Calculated

A substance that has a considerable resonance energy is said to be **reso-nance-stabilized**. The concept is particularly important to the understanding of many complex reaction mechanisms. Many reactions involve formation of reactive intermediates, such as cations or anions, and variation in the resonance stabilization of such intermediates is frequently responsible for large variations in reaction rates. The values given in Table 5-1 must be accepted with some reservation as to quantitative significance, however.

PROBLEM 5-2

Write the principal resonance forms of the following ions and molecules.

a NO_3^-

c $R-\overset{\overset{+NH_2}{\|}}{C}-NH_2$

e CH_3- $:^-$

b CH_3NO_2

d NO_2^-

f $(CH_3CO)_3C:^-$

5-3 QUALITATIVE USE OF RESONANCE

As mentioned previously, the resonance method of describing π-electron delocalization by a resonance hybrid of localized-bond structures is usually easier and more satisfactory for qualitative discussion of the sort most commonly in use in organic chemistry. Nevertheless, the detailed orbital picture generated by the molecular-orbital method should always be in mind when discussing any resonance hybrid. The orbital picture serves to emphasize that a resonance hybrid is only a single molecule with delocalized electrons, *not* an interconverting or "resonating" mixture of separate structures. Furthermore, it serves to point up the important geometrical facts, of the π-orbital lobes above and below a planar σ skeleton, which will be of signal importance in discussions of the spatial nature and requirements of reactions in π-electron systems.

Isovalent Resonance Structures
The examples discussed in the last section involved symmetrical resonance hybrids. Electron delocalization, however, is not restricted to symmetrical systems, and good evidence is available for delocalization of electrons in many unsymmetrical molecules and ions. For example, in *o*-xylene the position of two methyl groups on a benzene ring destroys the benzenelike symmetry.

o-Xylene

One of the resonance structures contains a double bond between the two carbon atoms that hold the two methyl groups; the other does not. Since the two structures are not equivalent, they probably do not contribute the same amount to the actual structure of the resonance hybrid. The hybrid will resemble one structure slightly more than the other, although in this case the difference is so slight that it is difficult to decide which structure should be given the greater weight in describing the molecule.

The case of the carboxylic acid presents a more interesting example of delocalization of electrons in an unsymmetrical system. The character of the interaction is suggested by comparing the acid with the carboxylate ion, a closely related symmetrical system, similar to the carbonate ion. The acid is

made from the ion by attaching a proton to the charged oxygen atom. Structure A below is an obvious first approximation to the structure of the acid. However, the oxygen of the —OH group still has four unshared electrons. These electrons are involved with the adjacent carbon atom and contribute something to bonding in the molecule, just as they do in carboxylate ion. The interaction is shown by adding a second resonance structure, B, which may be obtained by attaching a proton to the other oxygen of the ion.

Electron delocalization in carbonate ion, benzene, carboxylate ion, and the carboxylic acid can be described by sets of resonance structures, each of which contain the *same number of bonds*. Such a phenomenon is described as **isovalent resonance**. Isovalent resonance always involves unsaturated molecules and describes some of the most important electron delocalization in organic molecules. A structure such as B, in which separation of positive and negative charges occurs, is called a *dipolar* structure. Charge separation will give the molecule an electrical dipole, which can be determined experimentally (Sec. 2-11). Involvement of the nonbonding electrons of nitrogen, oxygen, sulfur, and halogen atoms in isovalent resonance with adjacent unsaturated systems always gives rise to dipolar structures. Several examples are shown in Fig. 5-2.

When isolated, the unshared electron pairs on these heteroatoms normally occupy sp^3 hybrid orbitals (Sec. 2–8), but conjugation to a double bond offers greater stability if the electrons are delocalized. Therefore, the heteroatom becomes hybridized sp^2 with the conjugating electron pair in the remaining p orbital suitable for overlap and delocalization. This of course changes the geometry of attachment at the heteroatom from tetrahedral to trigonal and planar.

$$CH_2{=}CH{-}\overset{..}{N}(CH_3)_2 \longleftrightarrow :\overset{-}{C}H_2{-}CH{=}\overset{+}{N}(CH_3)_2$$

Dimethylvinylamine

$$CH_2{=}CH{-}\overset{..}{\underset{..}{O}}CH_3 \longleftrightarrow :\overset{-}{C}H_2{-}CH{=}\overset{+}{O}CH_3$$

Methyl vinyl ether

$$CH_2{=}CH{-}\overset{..}{\underset{..}{F}}: \longleftrightarrow :\overset{-}{C}H_2{-}CH{=}\overset{+}{F}:$$

Vinyl fluoride

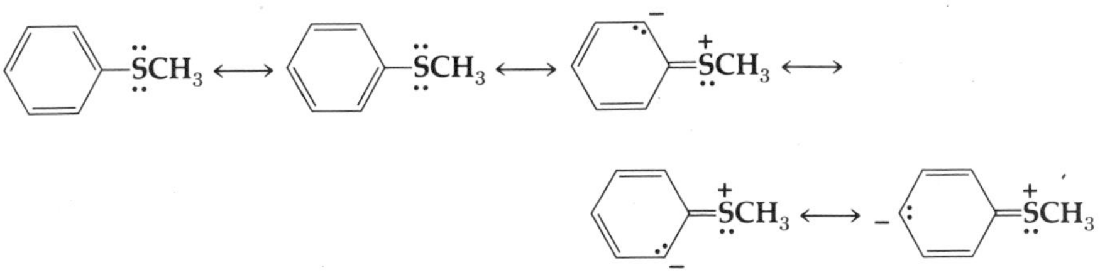

Methyl phenyl sulfide

FIGURE 5-2   Isovalent resonance involving nonbonding electrons

Heterovalent Resonance Structures

Bonds between unlike atoms are usually *polarized* because the nuclei involved have different electronegativities. Such polarization may be described by the resonance method. For example, the following is a resonance description of hydrogen fluoride:

$$H{-}F \longleftrightarrow H^+ \ F^-$$

The two structures are not isovalent since the dipolar structure is represented *formally* as having no H—F bond. Such structures are classified as **heterovalent resonance** structures.

Figure 5-3 describes a number of polar molecules by heterovalent resonance structures, of which the carbonyl resonance (cf. CH_2O) is of primary importance.

Bonds between unlike atoms are of course electrically imbalanced and exhibit dipoles (Sec. 2-11), and heterovalent resonance is simply an application of the resonance method as an alternative description of this partial shift of the bonding electrons towards the more electronegative atom.

Evidence for delocalization of electrons in unsaturated systems containing heteroatoms is good. In a molecule such as acrolein the difference between the electronegativities of carbon and oxygen develops a considerable negative charge at oxygen. In the language of the resonance method, structures *A*, *B*, and *C* for acrolein are given considerable weight in the description of the molecule since they shift negative charge toward electronegative oxygen.

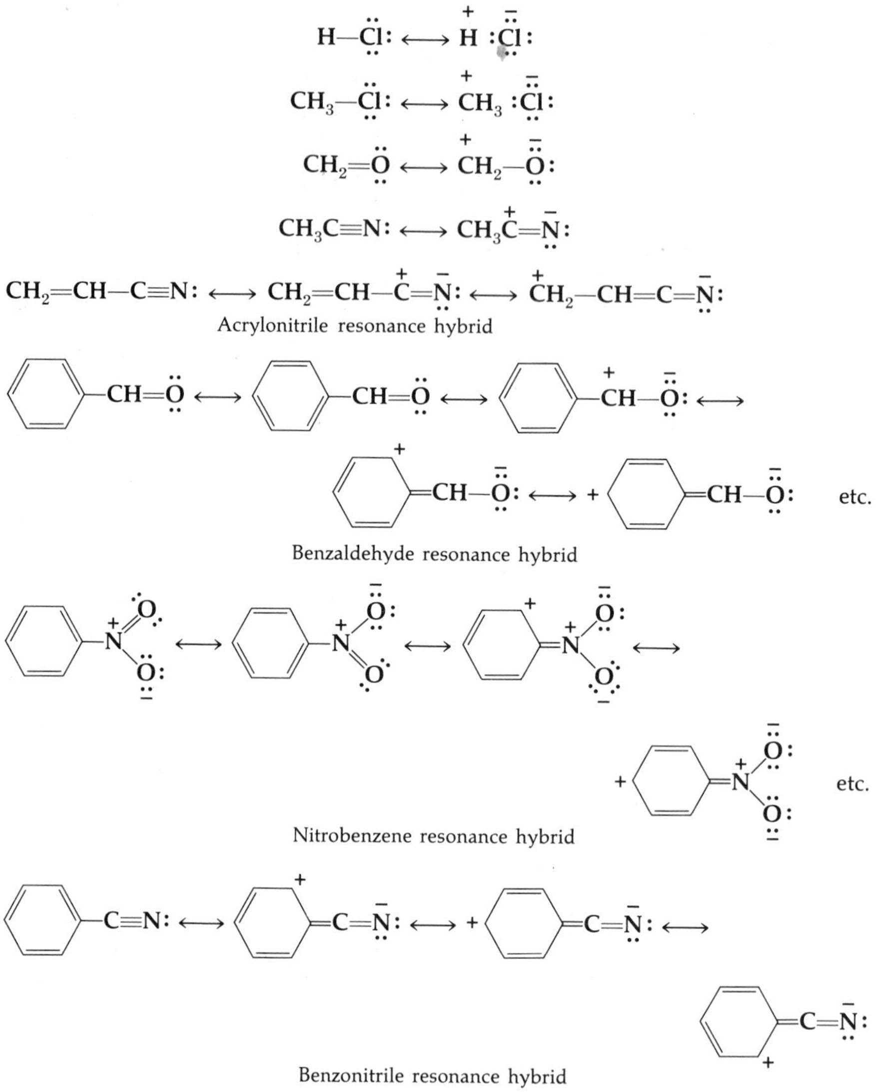

$$H\!-\!\ddot{\underset{..}{Cl}}: \longleftrightarrow \overset{+}{H} \; :\!\overset{..}{\underset{..}{\overset{-}{Cl}}}:$$

$$CH_3\!-\!\ddot{\underset{..}{Cl}}: \longleftrightarrow \overset{+}{CH_3} \; :\!\overset{..}{\underset{..}{\overset{-}{Cl}}}:$$

$$CH_2\!=\!\overset{..}{\underset{..}{O}} \longleftrightarrow \overset{+}{CH_2}\!-\!\overset{-}{\underset{..}{O}}:$$

$$CH_3C\!\equiv\!N: \longleftrightarrow CH_3\overset{+}{C}\!=\!\overset{-}{N}:$$

$$CH_2\!=\!CH\!-\!C\!\equiv\!N: \longleftrightarrow CH_2\!=\!CH\!-\!\overset{+}{C}\!=\!\overset{-}{\underset{..}{N}}: \longleftrightarrow \overset{+}{CH_2}\!-\!CH\!=\!C\!=\!\overset{-}{\underset{..}{N}}:$$

Acrylonitrile resonance hybrid

Benzaldehyde resonance hybrid

Nitrobenzene resonance hybrid

Benzonitrile resonance hybrid

FIGURE 5-3 **Resonance involving heterovalent, ionic structures**

Structures D and E are entirely negligible since they involve resonance polarization of negative charge *away* from the more electronegative atom.

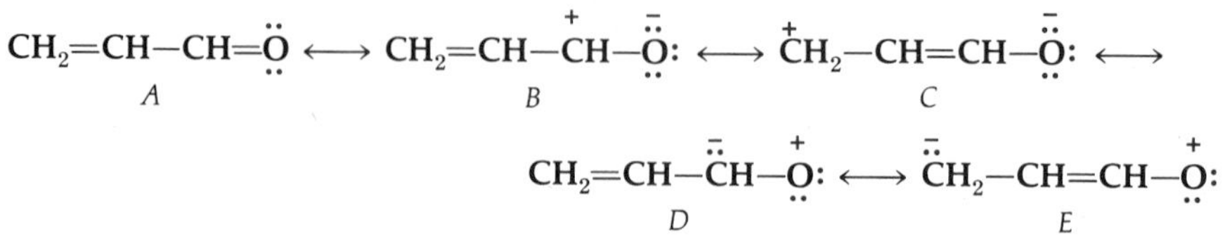

$$CH_2\!=\!CH\!-\!CH\!=\!\overset{..}{\underset{..}{O}} \longleftrightarrow CH_2\!=\!CH\!-\!\overset{+}{CH}\!-\!\overset{-}{\underset{..}{O}}: \longleftrightarrow \overset{+}{CH_2}\!-\!CH\!=\!CH\!-\!\overset{-}{\underset{..}{O}}: \longleftrightarrow$$
$$\underset{A}{} \qquad\qquad\qquad \underset{B}{} \qquad\qquad\qquad \underset{C}{}$$

$$CH_2\!=\!CH\!-\!\overset{-}{\underset{..}{CH}}\!-\!\overset{+}{O}: \longleftrightarrow \overset{-}{\underset{..}{CH_2}}\!-\!CH\!=\!CH\!-\!\overset{+}{O}:$$
$$\underset{D}{} \qquad\qquad\qquad \underset{E}{}$$

Acrolein resonance hybrid

Similar interaction occurs between double and triple bonds and between double or triple bonds and aromatic nuclei, as in the examples of Fig. 5-3. The effects of polar unsaturated substituents are strongest at ortho and para positions in a benzene nucleus.

Understanding the difference between isovalent and heterovalent conjugation is critical to understanding the two different roles played by atoms such as nitrogen, oxygen, and sulfur in unsaturated systems. In isovalent resonance it is the unshared pairs on these heteroatoms (hybridized into p orbitals) which are delocalized; in heterovalent resonance the unshared pairs are perpendicular to the π orbitals and uninvolved in the resonance. If other factors remain constant, isovalent conjugation is much more important than heterovalent conjugation. Consider two different ways of incorporating oxygen in an unsaturated system of methyl vinyl ether and acrolein.

The oxygen atom in methyl vinyl ether is relatively positive, since it contributes nonbonded electrons to the adjacent vinyl group in isovalent conjugation.†

$$CH_2=CH\ddot{O}CH_3 \longleftrightarrow \,:\bar{C}H_2-CH=\overset{+}{\ddot{O}}CH_3$$
Methyl vinyl ether

No analogous contributing structure (using unshared electrons from oxygen) can be written for acrolein. The only way in which oxygen can assume a formal positive charge is by way of an *improbable heterovalent* interaction, structure A, in which bonding electrons are given to carbon. By way of contrast, heterovalent structure B is very important because of the high affinity of the electronegative oxygen atoms for electrons.

$$CH_2=CH-\overset{+}{\underset{..}{C}}H-\overset{+}{\ddot{O}}: \longleftrightarrow CH_2=CH-CH=\ddot{O}: \longleftrightarrow \overset{+}{C}H_2-CH=CH-\overset{-}{\ddot{O}}:$$
A \qquad\qquad\qquad Acrolein \qquad\qquad\qquad B

Atoms of the second row in the periodic table,‡ especially sulfur and phosphorus, exhibit a particular form of heterovalent resonance in many compounds. The two structures involved always include: one with four bonds (or unshared pairs) to the second-row atom bearing positive formal charge; and one with unshared pairs on an adjacent atom donated back into d orbitals of the second-row atom, which then shows more than four bonds. The double bonds shown in these resonance forms are not typical π bonds since they are formed using d orbitals which have a different and more complex shape (Sec. 2-6). This involvement of d orbitals and consequent expansion of the outer shell to more than eight electrons does not occur with first-row elements and is largely responsible for the special chemical properties of sulfur and phos-

---

† The oxygen atom in methyl vinyl ether is not actually positive, since donation of nonbonding electrons is compensated by the inherent polarity of C—O single (σ) bonds towards oxygen. However, oxygen in methyl vinyl ether is more positive than oxygen in a saturated ether.

‡ Since **H** and **He** properly constitute the first "row" of the periodic table, **S** and **P** should be described as third-row elements, but it is common to ignore the **H/He** row and regard **C, N, O, F** as first-row elements and **S, P, Cl** as second-row elements.

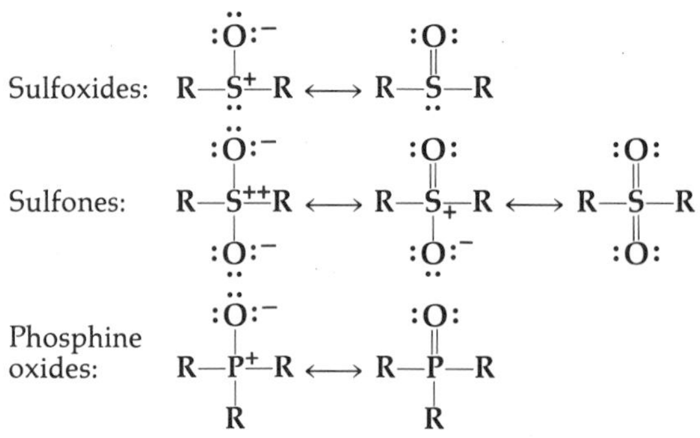

Sulfoxides: $R-\overset{+}{\underset{\cdot\cdot}{S}}-R \longleftrightarrow R-\overset{\overset{:\ddot{O}:^-}{|}}{\underset{}{S}}-R$ $R-\overset{\overset{:O:}{||}}{\underset{}{S}}-R$

Sulfones: $R-\overset{++}{\underset{\cdot\cdot}{S}}R \longleftrightarrow R-\overset{}{\underset{+}{S}}-R \longleftrightarrow R-\overset{}{\underset{}{S}}-R$

Phosphine oxides: $R-\overset{+}{\underset{|}{P}}-R \longleftrightarrow R-\overset{}{\underset{|}{P}}-R$

FIGURE 5-4 **Heterovalent resonance of second-row atoms**

porus discussed in Chap. 19. Examples are shown in Fig. 5-4. Note that amine oxides, $R_3N^+-O^-$, have no pentacovalent resonance form like that of phosphine oxides.

Rules for Resonance Structures

Some of the rules for writing meaningful resonance structures are based upon wave mechanics; others depend upon evaluating the relative energy levels of the various members of a set of structures. The following rules are presented without theoretical justification and serve as an empirical guide.

1 *All resonance structures in a set must indicate the same number of paired electrons.*

$CH_2{=}CH_2 \quad \longleftrightarrow\!\!\!\!\!/ \quad \uparrow\!\cdot CH_2{-}CH_2\,\cdot\!\uparrow$

Paired (π) electrons   Unpaired electrons

2 *A series of resonance structures must never imply movement of nuclei (Sec. 5-4).*
3 *If other factors are equal, isovalent resonance is more important than heterovalent resonance.*

Heterovalent: $CH_2{=}CH{-}CH{=}CH_2 \longleftrightarrow$
                              Major contribution

$\overset{+}{C}H_2{-}CH{=}CH{-}\overset{..}{C}H_2 \longleftrightarrow \overset{..}{C}H_2{-}CH{=}CH{-}\overset{+}{C}H_2$
                           Minor contributions

Isovalent: $CH_2{=}CH{-}\overset{..}{\underset{..}{O}}{-}CH_3 \longleftrightarrow \ :\!\overset{-}{C}H_2{-}CH{=}\overset{+}{\underset{}{O}}{-}CH_3$
                              Significant contribution

4 *The more isolated charges there are in a structure, the less important a contributor it is. Rarely are more than two isolated charges significant (cf. item 3 above).*
5 *Heterovalent resonance increases the electron density (negative charge) at atoms with higher electronegativity.*

Heterovalent: $R_2C{=}\overset{..}{\underset{..}{O}} \longleftrightarrow R_2\overset{+}{C}{-}\overset{-}{\underset{..}{O}}\!: \longleftrightarrow R_2\overset{-}{\underset{}{C}}{-}\overset{+}{\underset{..}{O}}$
                                        Negligible contribution

6 *Structures that crowd like charges close together are usually of minor significance.*

$$\underset{R-\overset{\displaystyle :O:}{\overset{\|}{C}}---\overset{\displaystyle :O:}{\overset{\|}{C}}-R' \longleftrightarrow R-\overset{\displaystyle :\overset{..}{\overset{-}{O}}:}{\underset{+}{C}}---\overset{\displaystyle :\overset{..}{\overset{-}{O}}:}{\underset{+}{C}}-R'}{}$$

Minor contribution

7 *Structures that require the expansion of the outer electron shells of first-row elements are negligible* (page 155).

8 *Delocalization of electrons in an unsaturated system is at a maximum when the σ skeleton is completely planar.*

9 *The resonance energy is greater:*

  a *When the contributing structures are all equivalent, as in benzene or* **RCOO⁻**.

  b *The more contributing structures there are, of roughly comparable energy.*

Most of the rules are self-explanatory or are examined in more detail in other sections, as indicated. The necessity for planarity of the molecule (item 7) obviously arises from the need for the atomic p orbitals to stand parallel for maximum overlap sideways. In a molecule which is constrained by other factors to have adjacent p orbitals skewed to each other the resonance energy is diminished. No interaction occurs between perpendicular unsaturated units. For example, in allene the terminal methylene groups are perpendicular to each other, so the molecule contains two localized double bonds. Similarly, no interaction occurs between the two *perpendicular* π orbitals of a triple bond.

Allene

Generation of Resonance Forms

In the usual case a single obvious structure can be written for any conjugated molecule, and it is then useful to apply some operation which will generate the other important resonance forms from it so as to picture more clearly the nature of the hybrid. The operation used introduces a notation convention for indicating the movement of electron pairs by the use of curved arrows. This notation will be very useful later as a shorthand for the movements of electron pairs in the bond breaking and bond making of chemical reactions. It is useful here in descriptions of resonance to indicate the *formal* shift of electron pairs in structure A which serves to generate structure B of an $A \longleftrightarrow B$ resonance hybrid. This formal shift of electron pairs in resonance descriptions has, again, no physical reality, any more than the separate contributing structures A and B do, but it serves to illustrate the overall delocalization of the π electrons.

The notation is illustrated in these examples:

Heterovalent: $CH_2=CH-CH=CH_2 \longleftrightarrow \overset{+}{C}H_2-CH=CH-\overset{..}{C}H_2$
                              A   B

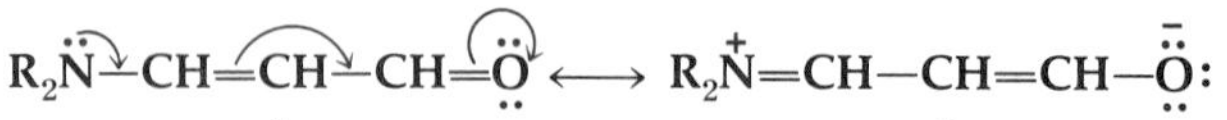

Isovalent: $R_2\overset{..}{N}-CH=CH-CH=\overset{..}{O} \longleftrightarrow R_2\overset{+}{N}=CH-CH=CH-\overset{..}{\underset{..}{O}}:$
                       A   B

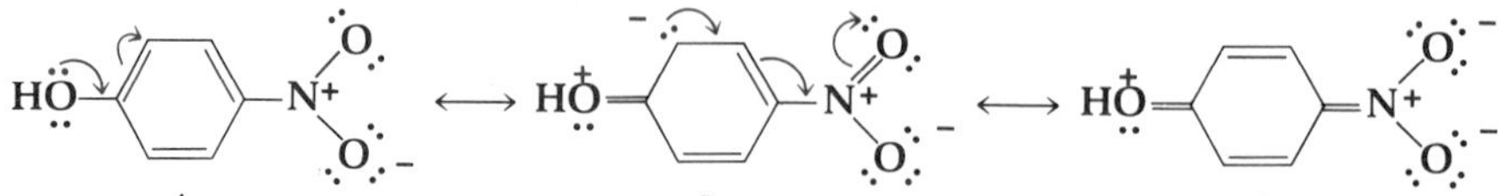

The arrow convention must follow these rules:

1 The curved arrow indicates movement of one electron pair at the foot of the arrow (either a double-bond π pair or an unshared pair) to an adjacent position indicated by the head of the arrow (either between two atoms, becoming a new double bond, or onto one atom, becoming an unshared pair).

2 The atom at the foot of the arrow, or series of arrows, becomes more positive by one charge and that at the head more negative by one.

3 If an electron pair moves in on a new atom, another pair must leave that atom so that the atom does not exceed a full outer shell of eight electrons. There are two obvious exceptions when an atom can accommodate added electrons without one of its present bonds having to leave.

a When the atom already has an incomplete shell (cf. R_3B, R_3C^+);

$R_2\overset{..}{N}-BR_2 \longleftrightarrow R_2\overset{+}{N}=\overset{-}{B}R_2$

b With second-row atoms expanding their valence shells, accepting a new pair into d orbitals. (In this case, of course, the double bond so generated is not a true π bond, as noted on page 155.)

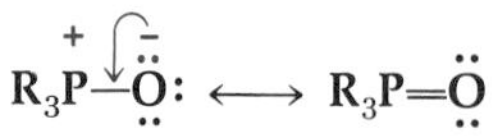

$R_3\overset{+}{P}-\overset{..}{O}: \longleftrightarrow R_3P=\overset{..}{O}$

4 The arrows completely dictate the product structure, and arrows may be written on the product structure which cause it to revert to the original. If two resonance structures cannot be interconverted by use of this notation, one of them is incorrect.

5 The natural polarization of a double bond between unlike atoms is in the direction of the more electronegative atom, and this will be the more important direction of electron movement in generating resonance structures. Similarly, important electron movements will proceed toward positive sites and away from negative ones.

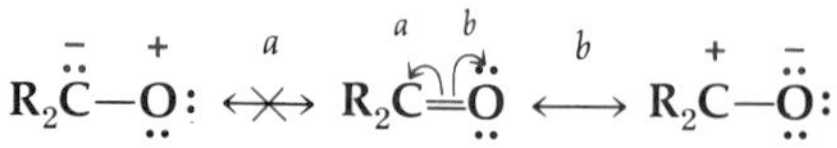

$R_2\overset{-}{C}-\overset{+}{O}: \overset{a}{\longleftrightarrow} R_2\overset{a}{C}\overset{b}{=}\overset{b}{O} \longleftrightarrow R_2\overset{+}{C}-\overset{-}{O}:$

     Negligible   Important

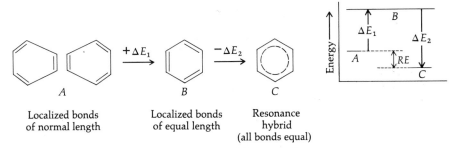

FIGURE 5-5   **Comparison of benzene with "cyclohexatriene"**

PROBLEM 5-3

With the use of curved arrows on the given structure, show the generation of the principal resonance form with the furthest charge separation, in each case.

a 3-Hydroxy-2-butenal

b 4-Dimethylamino-vinylbenzene

c 4-Aminobenzonitrile (p-$H_2NC_6H_4CN$)

d p-Hydroxy-nitrobenzene (p-nitrophenol)

e 4-Methylamino-3-penten-2-one

5-4   RESONANCE VS. TAUTOMERISM

The delocalization of π electrons, or resonance, is described with reference to a fixed σ-bonded skeleton bearing the conjugated π electrons. The fixed σ frame must therefore not be violated when writing resonance structures to express this delocalization. In other words, atomic nuclei must be in identical relative positions in all resonance structures written (rule 2, page 156). The case of benzene itself can serve to elaborate this idea, for if a Kekulé structure of benzene is accepted as written, as a localized "cyclohexatriene," the double bonds would be shorter than the single bonds (Sec. 2-9), and an exaggerated view of its geometry would show the less symmetrical form (A) in Fig. 5-5. We may formally pass to benzene itself in two stages: first, deform all bonds to an equal, intermediate length†; and second, allow delocalization to occur. The first requires that energy be put in (ΔE_1) to change the bonds from their normal, most stable lengths; the second releases energy by delocalizing the π electrons. Since the energy release, ΔE_2, is greater than that required to deform bond lengths (ΔE_1), there is a net resonance energy overall ($RE \simeq 40$ kcal/mole). Resonance stability, or resonance energy (RE), is therefore observed only when the true delocalization energy (ΔE_2; not directly measurable) is enough to compensate for the energy cost (ΔE_1) of deforming the σ skeleton

---

† Ordinary double bond length, 1.34 Å; single bond between sp^2 carbon atoms, $\sim$1.48 Å; observed bonds in benzene, 1.40 Å. That this is closer to double-bond than single-bond length reflects the resonance energy: the stronger a bond the shorter it is. The 1.48-Å length is that between sp^2-hybridized carbons with negligible resonance energy.

to the optimum geometry for resonance. When we write resonance structures, we imply this optimum *geometry*, the two Kekulé forms of benzene being B, not A, so that both have identical geometry and there is no relative movement of nuclei between them.

> Benzene resonance has been described as including minor contributions from three "Dewar forms" each containing a central bond spanning the 2.8-Å distance across the ring, which is too long for significant bonding. If this bond is contracted to normal length (1.5 Å), the isomeric bicyclohexadiene results. The necessary movement of nuclei is too great to be compensated by any delocalization energy; therefore, benzene and bicyclohexadiene are two separate isomeric compounds. The latter is much higher in energy and is easily converted into benzene. Bicyclohexadiene may be described in turn with a very minor resonance contribution of a very deformed Kekulé structure. This illustrates that involvement of σ bonds in resonance hybrids is rarely significant.

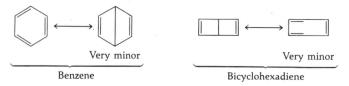

Benzene   Bicyclohexadiene

> Other high-energy isomers of benzene have recently been observed, cf.

> As these three-dimensional views imply, these isomers and bicyclohexadiene all differ from benzene in being nonplanar.

Two isomeric structures which differ significantly in the relative positions of their atoms are not resonance forms but are in fact separate isomeric compounds, called **tautomers**. In most cases tautomers are of similar energy and are interconvertible by the appropriate movement of atoms; this is a chemical reaction (**tautomeric change** or **tautomerism**), not a description of resonance, and is distinguished symbolically by the familiar equilibrium double arrow ⇌. *Tautomerism always involves the making and breaking of single (σ) bonds* in the course of this change in geometry. We may for convenience distinguish two kinds of tautomerism:

1 **Valence tautomerism**
2 **Proton tautomerism**

Valence Tautomerism

This is generally understood as a shift in interatomic distances within a molecule without the separation of any atom from the rest as an intermediate

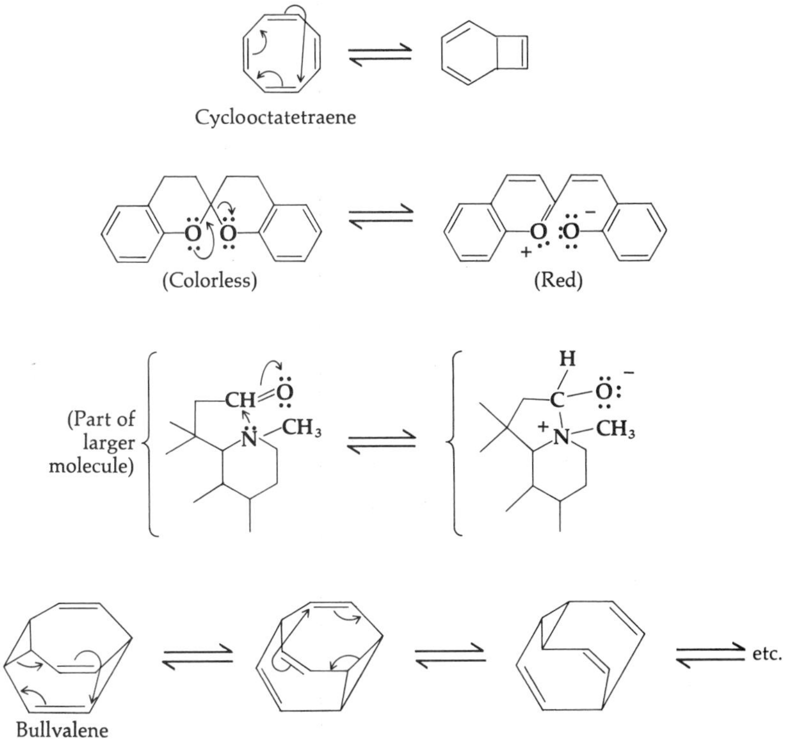

Cyclooctatetraene

(Colorless)   (Red)

(Part of larger molecule)

Bullvalene

FIGURE 5-6   **Examples of valence tautomerism**

stage. The valence tautomers of benzene, discussed previously, are an example; others are shown in Fig. 5-6. Physical evidence (Chap. 7) makes it clear that cyclooctatetraene exists normally as the monocyclic tautomer. However, the valence tautomerism of cyclooctatetraene is postulated to account for reactions of that molecule which give products bearing the bicyclic skeleton of its tautomer. Similarly the second compound in Fig. 5-6 is a colorless solid which goes red on warming, as its tautomer is formed; the red compound reverts to the colorless one on cooling again. The arrow convention signifies movements of electrons although it is also convenient, as described previously, for the formal generation of one resonance form from another. It is equally convenient, and describes real electron shifts, when applied to describe chemical reactions. In the latter cases—which include tautomerism—atomic nuclei also move relative to each other; in resonance they do not. It is appropriate to use the convention here to describe the conversion of one tautomer to another, as illustrated in Fig. 5-6. The bond-making processes are easily pictorialized by this convention, which shows graphically where each pair of electrons goes in a reaction.

If the movement of electrons on one side of a symmetrical molecule is balanced by a comparable reverse movement on the other, the two tautomers can even be physically identical molecules. This can easily be seen if we consider the following two trapezoid shapes for analogy. They are identical in shape (as shown by turning one over from left to right so that it becomes superimposable on the other) but the left one may be converted to the right one by shifting its corner positions (cf., nuclei) as marked by colored arrows, i.e., changing the relative positions of the nuclei.

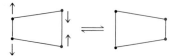

The last example in Fig. 5-6 is one of a group of recently discovered compounds with **fluxional** structures, undergoing rapid tautomeric interconversions among identical molecules. The number of possible tautomeric shifts in bullvalene, discernible by distinguishing all the carbons from each other in some way, is very large, only three being shown in Fig. 5-6. All the carbons become equivalent in this complex tautomeric equilibrium. Before quantum mechanics benzene was looked upon as two Kekulé structures in rapid tautomeric equilibrium between identical molecules. We now know that resonance intervenes and that the separate Kekulé structures are not true descriptions of benzene. However, there is no *a priori* reason why such an equilibrium should not be possible if it does not describe the particular case of benzene. Fluxional molecules now represent the phenomenon of rapid tautomeric interconversion between identical molecules. To identify it more clearly, put two adjacent substituents on the Kekulé benzene structures: if they were tautomers, the substituents would be closer together in one form than the other and would therefore oscillate during tautomeric interconversion. Two adjacent substituents on benzene do not oscillate, but on bullvalene they do.

Proton Tautomerism

Unlike valence tautomerism proton tautomerism is distinguished by removal of a proton from a molecule and then replacement of it (or another) onto a different atom in the molecule. Such a reaction is generally deemed a tautomerism only if the proton is readily and rapidly shifted as in tautomers with the proton bound to oxygen or nitrogen. In the carboxylate anion (page 152) the two resonance forms exhibit the negative charge on each oxygen atom (the actual hybrid being an anion with half a charge on each oxygen). In the carboxylic acid we may consider two tautomers with the hydrogen on either oxygen, as shown in Fig. 5-7; again in this case the two acid tautomers are physically identical molecules (they would not be if the two oxygens were different isotopes). This tautomerism is very facile and rapid as is the closely analogous **keto-enol tautomerism** compared with it in Fig. 5-7.

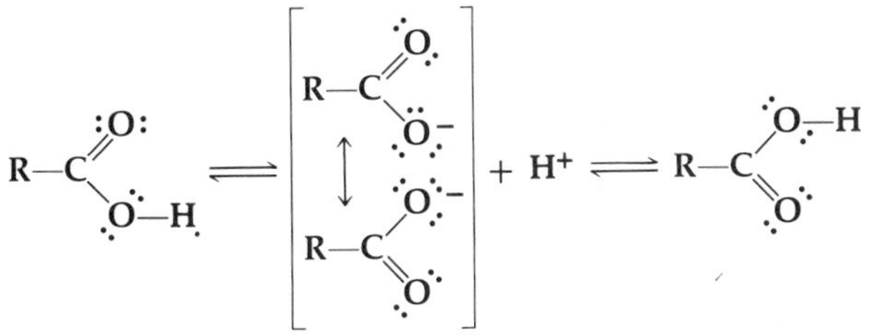

Carboxylic acid   Carboxylate anion   Carboxylic acid

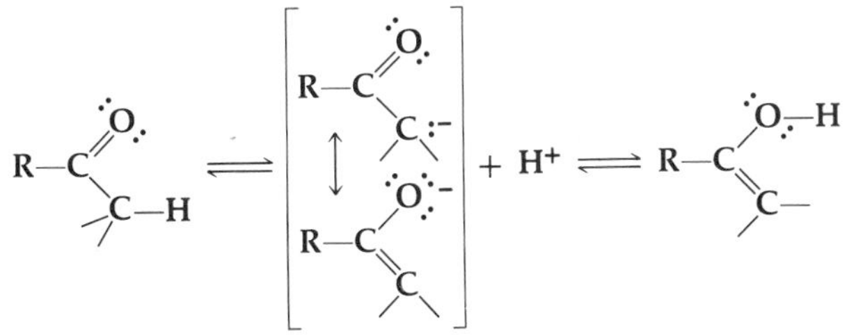

Ketone   Enolate anion   Enol

FIGURE 5-7   **Carboxylic acid and keto-enol tautomerism compared**

In the latter case either the ketone or its tautomeric **enol** (hydroxyl group attached to a double bond $=$ $-$**ene** $+$ $-$**ol**) may lose a proton and form a single intermediate **enolate anion** (with its two resonance forms reflecting both of the initiating tautomers) which may then re-bond the proton to yield the other tautomer. Keto-enol tautomerism is of enormous importance in organic reactions and will be referred to often in that connection. Other forms of proton tautomerism which are all very similar to keto-enol tautomerism are shown in Table 5-2. Much of their chemistry is related to that of ketones for this reason, and all have a delocalized intermediate anion comparable to the enolate anion.

The relative stabilities, or energies, of two tautomers may be roughly calculated from the bond energies of the bonds which change in the tautomerism (Table 2-2). In this way we find that the simple ketone is substantially more stable than its corresponding enol, and in fact simple (unconjugated) enols have never been isolated and are known only as reaction intermediates. When the enol double bond is conjugated, however, and enjoys resonance stabilization, it can be more stable than its corresponding ketone: hydroxy-benzenes (phenols) are the most dramatic examples. Here the possible ketone tautomers are unknown.

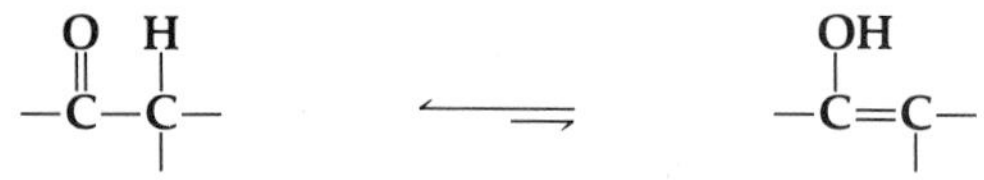

Bond energies:

Keto			Enol		
C=O	−179		C—O	− 86	
C—C	− 83		C=C	−146	
C—H	− 99		O—H	−111	

−361 kcal/mole = more stable than −343 kcal/mole
by $\Delta E = -18$ kcal/mole

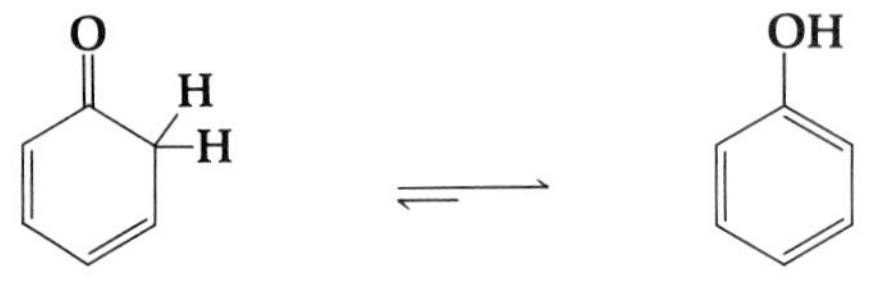

−361   −343

$RE \simeq - \quad 5$ $RE \quad - \quad 36$

−366 kcal/mole = *less* stable than −379 kcal/mole
by $\Delta E = +13$ kcal/mole

TABLE 5-2   **Major Examples of Proton Tautomerism**
(★ = most stable tautomer)

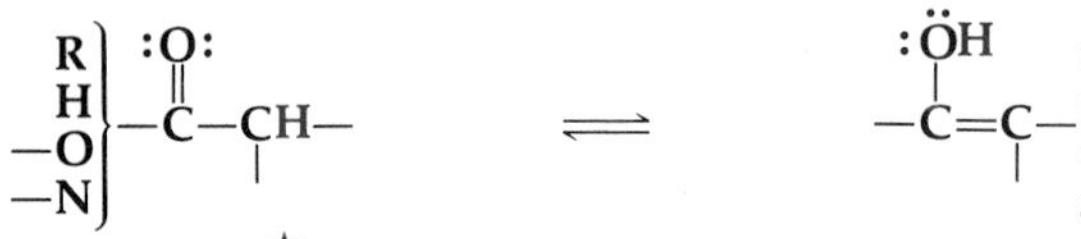

Carbonyl ★   Enol

Keto-enol tautomerism: occurs with all carbonyl-containing functional groups (esters, amides, etc.)

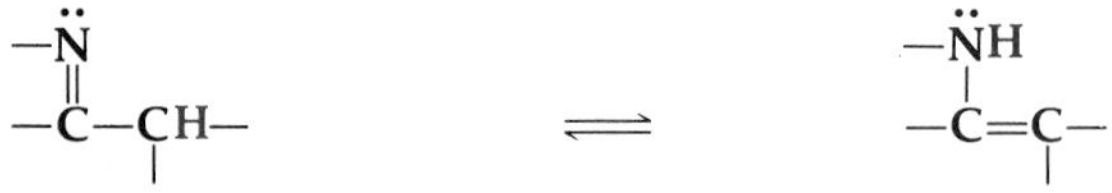

Imine   Enamine ★

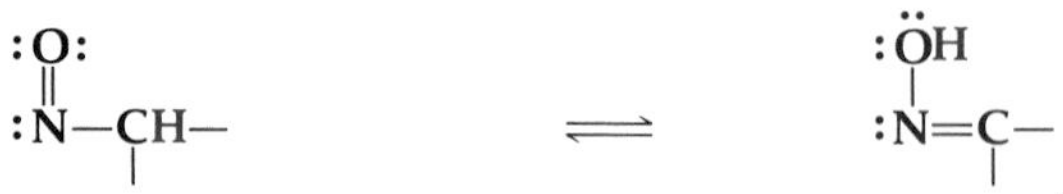

Nitroso   Oxime ★

Nitro ★   *aci*-Nitro

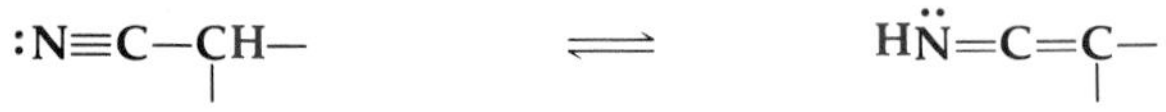

Nitrile ★

These energy calculations are crude, as the known preferred stabilities of Table 5-2 will show on comparison with Table 2-2. Since the bond energies are only for simple average cases and these in the gas phase, there is little surprise if they are off by ± 5 kcal/mole for comparisons in solution. The minor resonance of the enol should improve its standing by a few kcal/mole relative to simple bond energies. If more enols are incorporated in one benzene ring the benzene resonance is less able to overcome the energy difference, so that resorcinol, while enolic, shows more ketonic activity than phenol, and phloroglucinol is just at the borderline and behaves chemically like a triketone in a number of reactions.

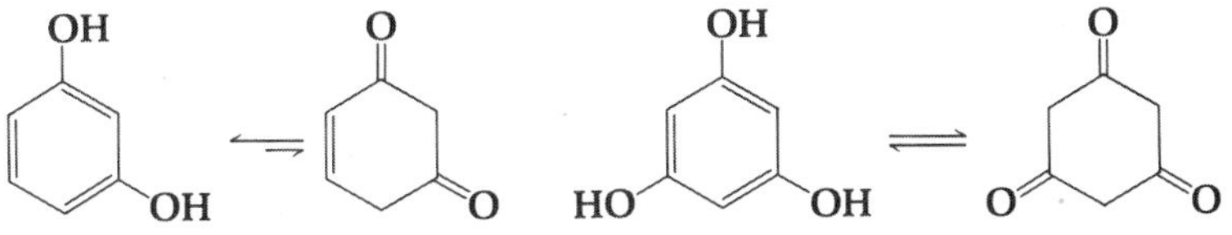

Resorcinol   Phloroglucinol

Another kind of tautomerism, usually combining proton and valence types, is called **ring-chain tautomerism**. This kind can occur when one functional group of a bifunctional molecule (e.g., an acyclic, or chain molecule) attacks the other with its pair of electrons, and thus forms a ring. The simplest example is the third case in Fig. 5-6. These are not resonance structures (though nearly so) since nuclei move. The nitrogen moves into bonding distance of the carbonyl carbon and that shifts from sp^2 to sp^3 hybridization, moving its attached groups (CH_2, H, O) from $\sim 120°$ to $\sim 109°$ angles. The two tautomers can be separately isolated by crystallization from different solvents. In more common examples there is also a proton shift as in the cases below, and ring tautomers of five and six members are usually favored over the corresponding open-chain tautomers. Tautomerism is discussed as an acid-base reaction in Sec. 8-9.

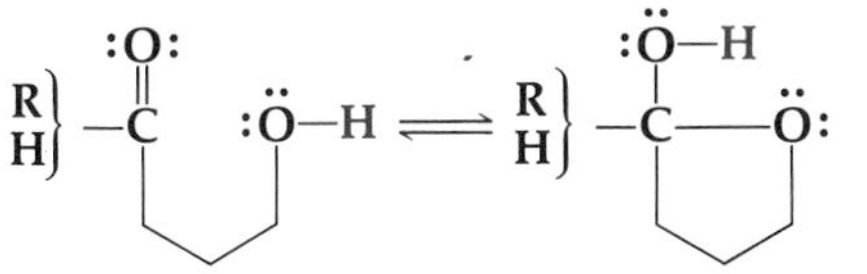

Ketoalcohol or   Hemiketal or   (Usual only with five-
hydroxy-aldehyde   hemiacetal   and six-membered rings)

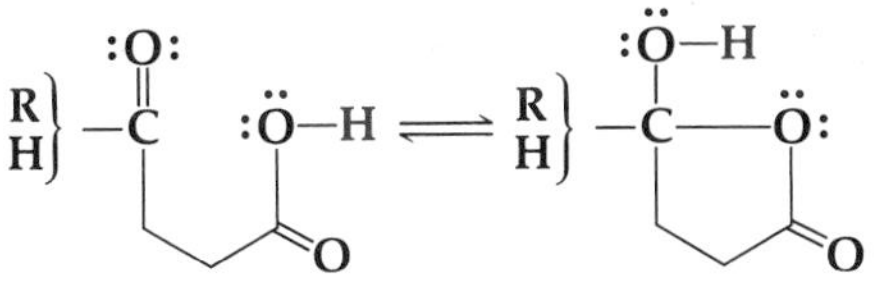

Ketoacid or   Lactol
aldehyde-acid

Hyperconjugation

Before leaving the subject of resonance vs. tautomerism, it is appropriate to point out that a number of effects observed in the last three decades have been explained by involvement of resonance stabilization due to delocalization of a σ bond with the π electrons of an adjacent double bond. This is heterovalent resonance, requiring breaking a σ bond in one structure, and called **hyperconjugation**. The effect, also called "no-bond resonance," is commonly invoked only for protons. It is understood that the proton does not move from bonding distance to carbon in such resonance: *hyperconjugation is not proton tautomerism.*

$$\underset{\begin{array}{c}|\\ \text{H}\end{array}}{}\underset{\begin{array}{c}\\ \overset{+}{\text{C}}\end{array}}{}\overset{\begin{array}{c}\text{H}\\ |\end{array}}{-\text{C}-\text{C}-} \quad \longleftrightarrow \quad \overset{\text{H}^+}{-\text{C}=\text{C}-}$$

$(\text{C}^+ = p \text{ orbital})$ Hyperconjugation

Certain phenomena involving extra stabilization of carbonium ions are frequently ascribed to hyperconjugation, but the whole concept has been surrounded by controversy for years. Authentic hyperconjugation in stable neutral molecules is hard to demonstrate, and the effect must be a minor one in such cases. The stability of the *tert*-butyl cation, $(\text{CH}_3)_3\text{C}^+$, is very great, however, with respect to other saturated carbonium ions. This stability is probably reasonably attributed to hyperconjugation (with 10 resonance forms), but other factors are also certainly involved, and defining relative contributions is very difficult.

PROBLEM 5-4

Are the following pairs tautomers or resonance forms?

a $\underset{\text{CH}_3-\overset{\overset{\displaystyle \text{OCH}_3}{|}}{\text{C}}=\overset{+}{\text{N}}(\text{CH}_3)_2}{}$ $\text{CH}_3-\overset{\overset{\displaystyle \overset{+}{\text{OCH}_3}}{||}}{\text{C}}-\text{N}(\text{CH}_3)_2$

b $\text{CH}_3-\overset{\overset{\displaystyle \text{OCH}_3}{|}}{\text{C}}=\text{NCH}_3$ $\text{CH}_3-\overset{\overset{\displaystyle \text{O}}{||}}{\text{C}}-\text{N}(\text{CH}_3)_2$

c   (cyclohexane-1,3-dione, both C=O)   (enol form: HO and =O)

d $\phi_2\text{CHNO}_2$ $\phi_2\text{C}=\overset{+}{\text{N}}\overset{\text{OH}}{\underset{\text{O}^-}{<}}$

e   (bicyclic lactam with C=O and N–H)   (bicyclic enamine with OH and N)

f

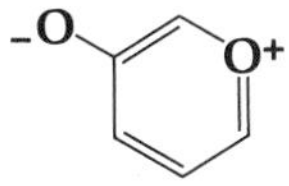

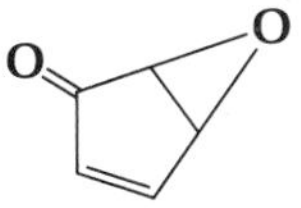

PROBLEM 5-5

a Write curved arrows to show the electron movement which converts the several benzene tautomers to benzene, and the right-hand tautomers to the left-hand ones in Fig. 5-6.

b Write several major resonance forms of each tautomer in the second example of Fig. 5-6.

c Write a second resonance form of the enol; why is this resonance less important than that in the enolate anion?

d Write two tautomers of the amide group (R—CONH$_2$) and two resonance forms of each tautomer. Assess the importance of resonance in each tautomer. Compare bond energies of the two tautomers and comment on the predominance of the normal amide tautomer.

e Two interconvertible forms are tautomers, not resonance forms, if nuclei must move in the interconversion. How far must the acid proton move in passing from one oxygen to the other in the carboxylic acid (see Table 2-1, Fig. 2-14).

f Calculate the relative bond energies of each pair of tautomers in Fig. 5-6 and Table 5-2 using values from Table 2-2.

5-5 AROMATICITY

The most significant quality about structures containing the benzene ring is that they are unusually *unreactive* compared to acyclic polyenes. Recognition of this special stability is almost as old as organic chemistry itself and, indeed because of it, much nineteenth-century chemistry was performed on benzene derivatives. These compounds were said to be **aromatic** or to have **aromatic character**. The quality which renders aromatic rings especially stable is conveniently referred to as **aromaticity**. Of course, aromaticity is only resonance stabilization, or delocalization, but it particularly refers to the situation in which the delocalized π electrons are contained in an orbital ring (doughnut-shape as in benzene, ϕ_1, Fig. 5-1). The resonance energy in these rings can be especially large, as benzene demonstrates.

The amount of aromaticity in unsaturated rings is assessed in various ways:

1 Chemical stability in excess of acyclic unsaturated analogs
2 Thermochemical measurement of RE by heats of formation or hydrogenation
3 Physical measurement, e.g., diamagnetic susceptibility or, more commonly, nuclear magnetic resonance (Sec. 7-6)

The third criterion, currently the most actively employed, derives from the physical expectation of a magnetic field generated by any ring of electrons, circulating as an electric current. This **ring current** in a π-electron circle is therefore a special feature of aromaticity (Sec. 7-6).

In 1931 Hückel presented molecular-orbital calculations which predicted that a monocyclic ring of $(4n + 2)$ π electrons, where n is any integer, would have special stability and that one of $(4n)$ π electrons would not. Recent more refined calculations predict that those with $(4n)$ π electrons should be *less* stable than their acyclic analogs and the $(4n)$ rings have been called **antiaromatic**. This special effect $(+$ or $-)$ in energy falls off with increasing values of n.

Aromatic: $\quad\quad 4n + 2 = 2, 6, 10, 14, 18, \ldots \pi$ electrons

Antiaromatic: $\quad\quad 4n = 4, 8, 12, 16, 20, \ldots \pi$ electrons

Hückel's prediction has stimulated considerable research into the synthesis of various fully unsaturated rings to test their aromaticity and in general the prediction has been substantiated. As with benzene (six π electrons), all aromatic molecules must have enough delocalization energy to force them into essentially planar geometry for full π overlap. The major examples are shown in Fig. 5-8. In the first row the carbonium ions are the most stable known, existing as stable salts in water solution.† In cyclopropenyl ion the resonance energy overcomes the considerable instability associated with squeezing three sp^2 carbon angles from their normal 120° bond angles to 60° (Sec. 6-4). The six-membered heterocyclic rings in the first two rows are isoelectronic, with six π electrons. A perpendicular sp^2 orbital on the heteroatom bears an unshared pair, in the same orientation as the C—H bonds in benzene, coplanar with the ring. The five-membered heterocycles are also planar and isoelectronic (except for the N—H proton) and make up their six π electrons by involving one pair on the heteroatom (in a p atomic orbital). These heterocycles are discussed in Chap. 24.

The 10-electron anions are both planar and relatively stable despite the instability associated with widening their normal 120° angles to 135° and 140°, respectively. There is also strain on the normal bonding angles (Sec. 6-4) in the 10-electron bicyclo-undecapentaene, but the delocalization energy is adequate to overcome it and leave the molecule still observably aromatic. This is also true with the tetracyclic 14-electron hydrocarbon. (The methyl groups are located one above and one below the ring.) The [14]-annulene is plagued with destabilization caused by pressing the inner hydrogens (colored) against each other but still shows aromatic stability. Both the [14]-annulene and the [18]-annulene $(4n + 2)$ are more stable than the [16]- and [20]-annulenes $(4n)$, but it is clear that the stabilization (and destabilization) is diminished in these larger rings, which are all increasingly more similar to acyclic polyenes as their ring size increases.

---

†Cycloheptatrienyl cation—known as "tropylium ion"—was actually first made in 1893 but never recognized since it was thrown out, as a salt, with the water extract of the reaction! Only in 1954, seeking to test Hückel's prediction, did Doering repeat the experiment and isolate the tropylium salt.

Aromatic

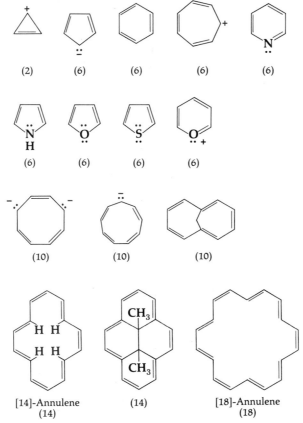

(2)   (6)   (6)   (6)   (6)

(6)   (6)   (6)   (6)

(10)   (10)   (10)

[14]-Annulene   (14)   [18]-Annulene
   (14)   (18)

Nonaromatic

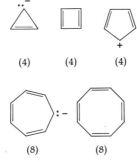

(4)   (4)   (4)

(8)   (8)

FIGURE 5-8   **Aromaticity in unsaturated monocyclic ring systems (number of π electrons in parentheses)**

Only fleeting glimpses of the first row of nonaromatics can usually be obtained owing to their unusual instability; they appear to be truly anti-aromatic. Cyclooctatetraene is not planar and behaves like an acyclic polyene with neither resonance stabilization or destabilization.

PROBLEM 5-6

Comment on the aromaticity or antiaromaticity of the following structures. (Some may have aromatic character owing to another resonance form.)

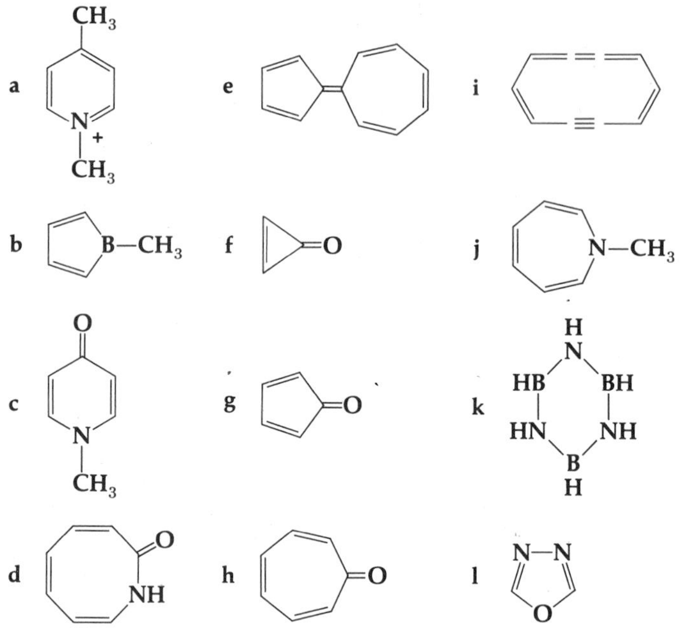

5-6 SUMMARY

The π and p orbitals above a planar σ skeleton in a conjugated unsaturated molecule can overlap into more extensive molecular π orbitals. The resultant electron delocalization is a stabilizing influence, lowering the energy of the molecule by an amount called the delocalization or resonance energy. This is about 40 kcal/mole in a strongly stabilized case like benzene but only a few kilocalories per mole in conjugated polyenes, other common examples being intermediate. The concept of stabilization by resonance will be very important later in assessing acidity and basicity, the stability of charged reaction intermediates, and the spectral behavior of π electron systems.

The molecular-orbital approach affords an important qualitative picture of this delocalization in clearly showing the extended π orbital system perpendicular to the coplanar σ frame. The resonance approach complements this in affording a set of contributing localized-bond structures which are more useful in qualitatively determining charge distribution. These contributing structures are selected and weighted according to the rules in Sec. 5-3. The arrow convention for electron movement allows resonance structures to be generated from each other with confidence by following simple rules (see page 158).

Both visualizations are employed together, but the latter is usually employed for the actual writing of structures. It will be much more convenient (and safer) to write benzene in one of its Kekulé forms when discussing reactions; either Kekulé form written at random will serve for reaction mechanism depictions, it being remembered always that the true picture of the molecule is the symmetrical one with delocalized orbitals as a ring or doughnut above and below the planar skeleton (Fig. 5-1).

When only π electrons are involved, with no movement of atomic nuclei, the various possible structures that can be written for a conjugated molecule are indeed resonance forms collectively describing a single resonance hybrid molecule. When major movement of nuclei and the making and breaking of σ bonds are involved, the several possible structures virtually always represent different molecules (tautomers) rather than a single hybrid. Valence tautomerism describes the shift in geometry in a molecule going to an isomer (tautomer) by breaking and making internal σ bonds (and also shifting π bonds usually) without loss of any atoms. Proton tautomerism arises by breaking, and reforming at another site, a bond to hydrogen; one of the tautomers generally has a heteroatom bonded to the hydrogen.

An aromatic system is a fully conjugated ring. It enjoys special stability (aromatic stabilization is resonance energy) as long as the cycle contains $(4n + 2)$ π electrons. The cycle of π electrons generates a ring current discerned by its effects in the NMR spectrum (Sec. 7-6).

PROBLEMS

5-7 Write sets of resonance structures for the following compounds and indicate only the more important contributors to each of the hybrids.

a CH_3CONH_2	i $CH_2{=}CHBr$	q Naphthalene
b CH_3COOCH_3	j $CH_2{=}CHN(CH_3)_2$	r Phenanthrene
c CH_3CN	k $CH_2{=}CHNO_2$	s α-Naphthol
d $(CH_3CO)_2O$	l C_6H_5Cl	t $C_6H_5CH{=}CH_2$
e CH_3COCl	m $C_6H_5OCH_3$	u C_6H_5CN
f CH_3COOH	n $C_6H_5NO_2$	v $C_6H_5CH{=}CHC_6H_5$
g CH_3NO_2	o $C_6H_5NH_2$	w $p\text{-}CH_3OC_6H_4COCH_3$
h $CH_2{=}CHCHO$	p $C_6H_5COCH_3$	x Furan

5-8 The following is a formula for a well-known stable free radical (a species containing an odd number of electrons), which is also a cation. By writing resonance structures, investigate the true location of the odd electron.

5-9 Write the principal resonance forms stabilizing the following structures:

a 4-Amino-nitrobenzene

b The aromatic C_4H_5N

c

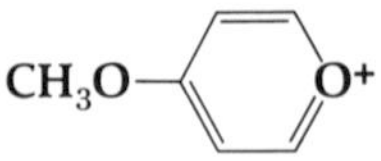

d The violet dye, pinacyanol:

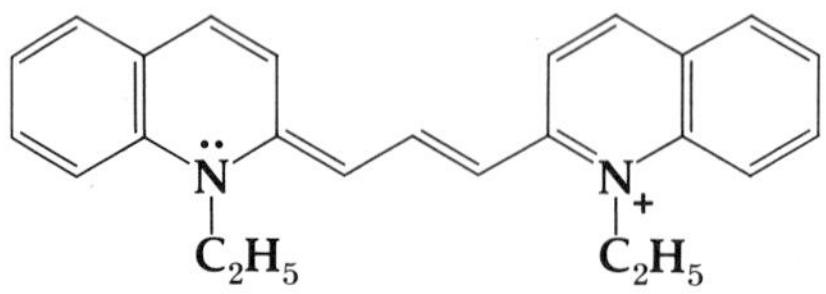

e The natural dye, indigo:

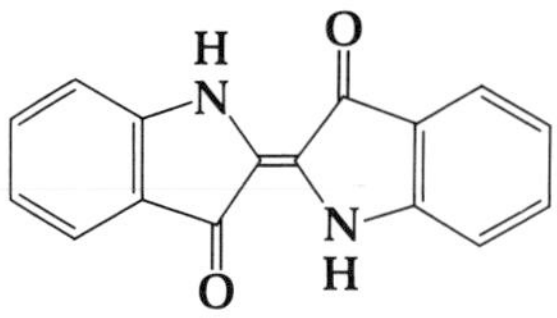

f

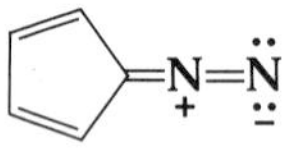

g

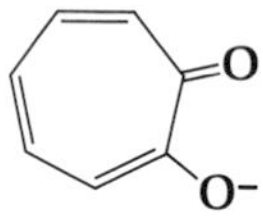

5-10 Criticize the following representations of *resonance.*

a $HO-CH_2-CH=CH-CH=O$

b $CH_3N=CH-CH=CH_2$

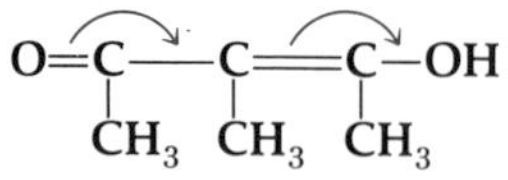

c

d $O=CH-CH=CH_2$

e

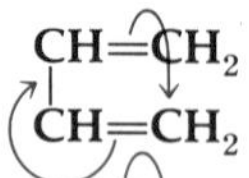

f

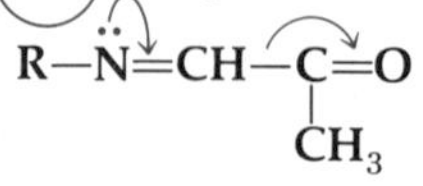

5-11 Write the structure of a proton tautomer of each of the following. Show resonance forms (if any) for the intermediate anion in the tautomerism.

a $HOCH_2CH_2CCl_2COC_6H_5$

b $C_6H_5COCOCH_3$

c $CH_2=CH-CH=N-OH$

d

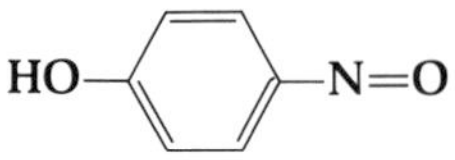

e 2-Cyclohexenone (two ways)

f $CH_3NH-CH=CH-CH_3$

g CH_3NO_2

h *p*-Nitrotoluene

i Cyclopentyl cyanide

j $C_6H_5CH_2CONHCH_3$ (two ways)

5-12 Use curved arrows to generate uncharged resonance forms of the following:

a

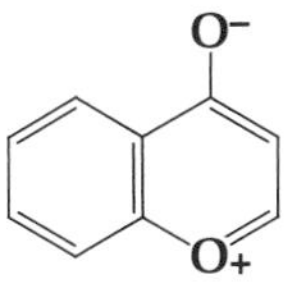

c $(CH_3)_2\bar{B}\!=\!\!\!\left\langle\!\!\!\begin{array}{c}\end{array}\!\!\!\right\rangle\!\!\!=\!\!\overset{+}{N}(CH_3)_2$

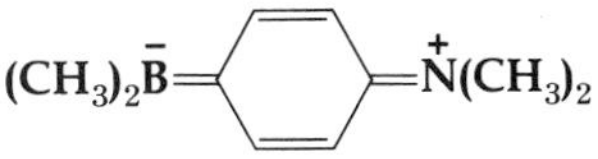

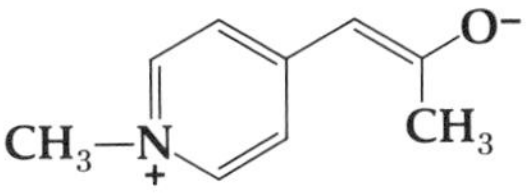

b

d $(C_6H_5)_2\overset{..}{\underset{.}{C}}\!-\!\!\!\left\langle\!\!\!\begin{array}{c}\end{array}\!\!\!\right\rangle\!\!\!-\!\!C\!\!\!\begin{array}{c}\overset{+}{N}(CH_3)_2\\ \underset{..}{N}(CH_3)_2\end{array}$

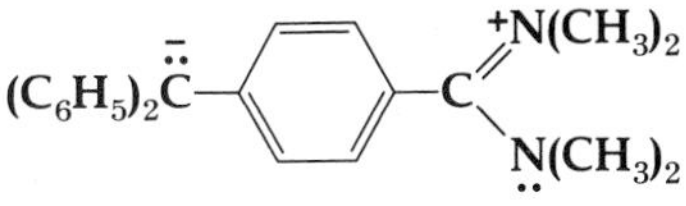

5-13 Criticize certain of the following structures as significant contributors to the resonance hybrid of the molecule. Classify each of the indicated sets as isovalent or heterovalent interactions.

a $CH_3C\equiv N \longleftrightarrow CH_3\,\overset{+}{C}\!=\!\overset{-}{N} \longleftrightarrow CH_3\overset{+}{C}\!-\!\overset{+}{\underset{=}{N}}$

b $C_6H_5\!-\!Cl \longleftrightarrow C_6\bar{H}_5\overset{+}{Cl}$

c $CH_2\!=\!CH\!-\!CH\!=\!O \longleftrightarrow \bar{C}H_2\!-\!CH\!=\!CH\!-\!\overset{+}{O}$

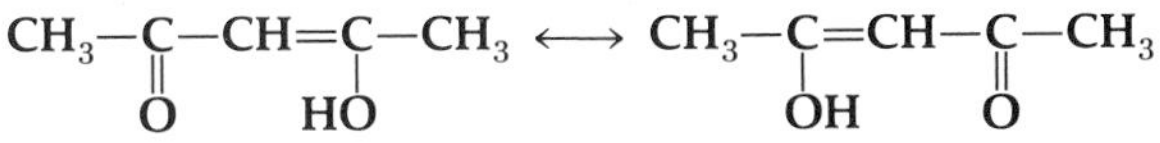

d $CH_2\!=\!CH\!-\!\underset{\underset{O}{\parallel}}{C}\!-\!CH_3 \longleftrightarrow \downarrow CH_2\!-\!CH\!=\!\underset{\underset{O\downarrow}{|}}{C}\!-\!CH_3$

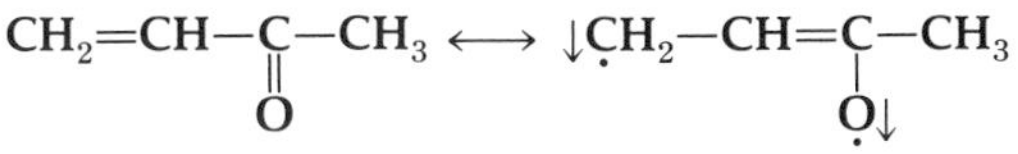

e $CH_3\!-\!\underset{\underset{O}{\parallel}}{C}\!-\!CH\!=\!\underset{\underset{HO}{|}}{C}\!-\!CH_3 \longleftrightarrow CH_3\!-\!\underset{\underset{OH}{|}}{C}\!=\!CH\!-\!\underset{\underset{O}{\parallel}}{C}\!-\!CH_3$

f $CH_3\!-\!\overset{+}{N}\!\!\!\begin{array}{c}\diagup O\\ \diagdown \underset{-}{O}\end{array} \longleftrightarrow CH_3\!-\!N\!\!\!\begin{array}{c}\diagup O\\ \diagdown O\end{array}$

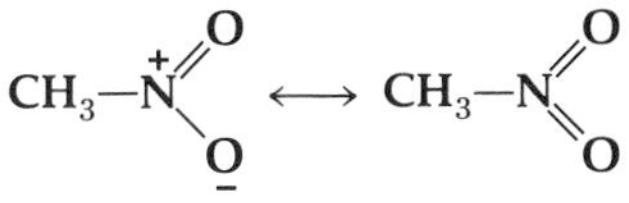

g

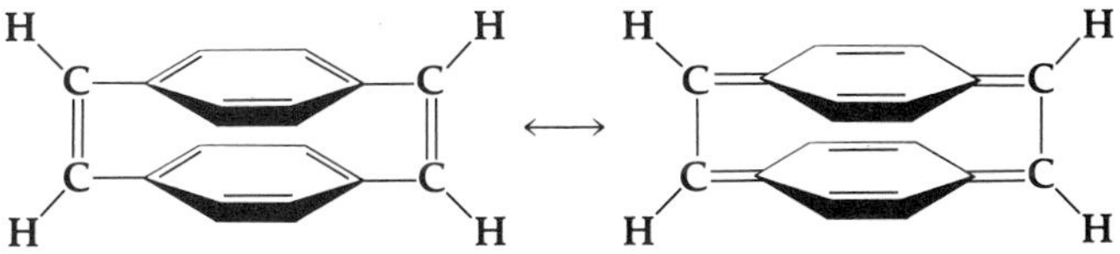

h $CH_2\!=\!C\!=\!O \longleftrightarrow \bar{C}H_2\!-\!C\!\equiv\!\overset{+}{O}$

i $O\!=\!C\!=\!C\!=\!C\!=\!O \longleftrightarrow \bar{O}\!-\!C\!\equiv\!C\!-\!C\!\equiv\!\overset{+}{O}$

j $H\!-\!N\!=\!\overset{+}{N}\!=\!\bar{N} \longleftrightarrow H\!-\!N\!=\!N\!\equiv\!N$

5-14 2,3-Dimethyl-2-cyclohexenone and 2,3-dimethyl-3-cyclohexenone are much more readily interconvertible than 1,2-dimethylcyclohexene and 2,3-dimethylcyclohexene. Explain.

5-15 A ketone may be converted to its enol under acidic as well as basic conditions. The first step involves forming an O—H bond by attaching a proton (H+) to an unshared electron pair of oxygen. Show the steps in converting ketone to enol this way and the parallel with base-catalyzed enol formation.

5-16 Compare the stability of each of the following pairs of organic ions.

a $C_6H_5CH{=}CH{-}CH_2^+$ $CH_3CH{=}CH{-}CH_2^+$

b 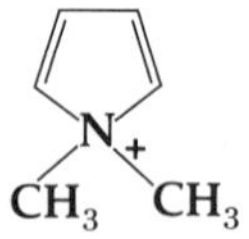 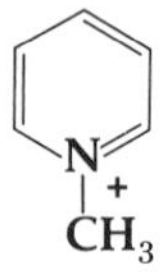

c $CH_3{-}CH_2{-}O^-$ $CH_2{=}CH{-}O^-$

d 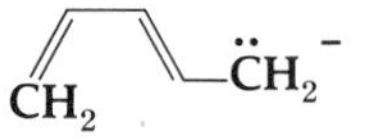 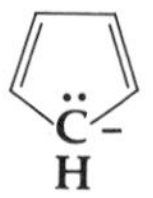

e $(CH_3)_2\ddot{N}{-}CH{=}CH{-}CH{=}\overset{+}{N}(CH_3)_2$ $CH_3\ddot{O}{-}CH{=}CH{-}CH{=}\overset{+}{N}(CH_3)_2$

f $R{-}C({=}O){-}\ddot{N}H^-$ $R{-}C({=}NH){-}\ddot{O}{:}^-$

g $(CH_3)_2\ddot{C}^-{-}CH{=}O$ $(CH_3)_2\ddot{C}^-{-}CH{=}CH_2$

READING REFERENCES

Orchin, M., and H. H. Jaffe, "The Importance of Antibonding Orbitals," Houghton Mifflin, Boston, 1967.

Roberts, J. D., "Molecular Orbital Calculations," W. A. Benjamin, New York, 1961.

Dewar, M. J. S., "The Molecular Orbital Theory of Organic Chemistry," McGraw-Hill, New York, 1969.

THE SHAPES OF MOLECULES— CONFORMATION AND STEREOCHEMISTRY

IN previous chapters most formulas were written in ways not intended to show the detailed shapes of molecules. This chapter is devoted to examining the architectural features of three-dimensional organic molecules. Knowing the bond angles at each atom in an organic molecule allows us to construct the molecular shape in space. Usually, however, such a construction will not result in a single rigid shape since the coaxial symmetry of single bonds allows them to rotate easily. Consideration of even a simple lower alkane shows that it can have many shapes as the several carbon-carbon bonds are allowed to rotate. The **conformations** of a molecule are these various shapes it can take up by rotations about its single bonds, without breaking any bonds. Rotations around double bonds, however, are severely restricted since such rotation must break the π bond and must therefore require substantial energy input. (See page 57).

Although most molecules can assume many conformations by single-bond rotation, there are restrictions on the process. As we shall see, rotation about single bonds is not quite free but exhibits favored orientations. Furthermore, rotations that force one atom too close to a neighboring but nonbonded atom are restricted. Finally, changes in shape or conformation arising from deforming bond angles and lengths from their natural values are also restricted. Each of these four kinds of molecular deformation—**bond stretching**, **bond-angle bending**, **bond rotation**, and **compression of nonbonded atoms**—is subject to restrictions which may be assessed in terms of their energies, called **deformation** or **strain energies**. Thus, it costs energy (makes the molecule higher in energy, or less stable) to deform a molecule, in each of these four ways, from its natural or optimum geometry. The total strain energy of any conformation is the sum of these separate strains at each of its bonds, and the molecule tends to take up the conformation(s) which minimize this strain energy. Most molecules have only a single important conformation, fortunately.

We can often apply certain rules (**conformational analysis**) to see in a qualitative way how a particular molecule will minimize strain, and thus determine its preferred shape(s) or conformation(s). This is particularly true of cyclic saturated molecules since rings tend to limit the conformational possibilities. It will be of importance to determine preferred conformations, not just for a clearer view of the spatial appearance of molecules, but also because their shapes are of salient importance in their chemical reactivity.

Stereochemistry ("spatial chemistry") is a second area of importance, beginning with Sec. 6-8, which is concerned not with the optimum shape of a single molecule but with the fact that a two-dimensional representation of a molecular structure, such as we have previously written, can actually represent two or more different three-dimensional molecules, with different shapes no matter what rotational conformations they assume. This is possible because, with four groups bonded to any carbon, it is important how they are placed in space on the four tetrahedral bonds of that carbon. This placement can lead to more than one molecule.

The idea is easily seen in 1,3-dimethylcyclobutane. In two dimensions, on paper, a single structure (*A*) defines it, but in three dimensions there are two possible real molecules corresponding to structure *A* and these are shown as *B* and *C* in the perspective drawings. Thus structure *A* is not enough to define a real molecule of 1,3-dimethylcyclobutane. We also need to specify whether the two methyls stand on the same (*B*) or opposite (*C*) sides of the ring.

There is no doubt that the two 1,3-dimethylcyclobutanes are different molecules—and so represent physically different chemical substances—since they have different shapes and could only be interconverted by breaking the bonds to methyl and hydrogen on one carbon and remaking them in a reverse sense. This is a chemical reaction, changing one substance into another. The two molecules, *B* and *C*, therefore are two separate isomeric compounds. They are called **stereoisomers** ("spatial isomers") to distinguish them from the **structural isomers** we have defined previously. There are two stereoisomers of 1,3-dimethylcyclobutane but these are in turn *structural isomers* of the (two) 1,2-dimethylcyclobutanes. Structural isomers are distinguishable by two-dimensional structures such as *A*, while stereoisomers are not. Any stereoisomer can of course have a number of conformations.†

In dealing with **stereoisomerism** we shall also be involved with the concept of symmetry in molecular shape, and considerations of symmetry simplify the understanding of, and greatly diminish the number of, stereoisomers for most molecules. The importance of stereoisomerism can be appreciated in considering reactions forming a structure capable of existing as more than one stereoisomer: we need to define and control which stereoisomer is produced in the reaction and be able to separate the product stereoisomers from each other.

---

†There are possible conformations of *B* and *C* with rings which are not square, or are bent. Rotation around the ring bonds is virtually impossible, but the methyl groups can take up a full circle of possible conformations by rotating.

FIGURE 6-1   **Space-filling model of 2-hexenoic acid**

6-1   MODELS AND PICTURES OF MOLECULES

Since it is important to represent molecules for visual inspection, organic chemists have devised various mechanical models as well as conventions for perspective pictorialization on paper. The models are of two kinds, those which show only the framework (bonds + nuclei) and those which show the full bulk of each atom, so-called space-filling or scale models. The latter have some value for determining the strain associated with compression of nonbonded atoms, but do not show the skeleton at all clearly. Inexpensive framework models are available for students and are of enormous assistance in visualizing the spatial relations of molecular groups. Examples of these models are shown in Figs. 6-1 through 6-4.

FIGURE 6-2   **Framework model of 2-hexenoic acid**

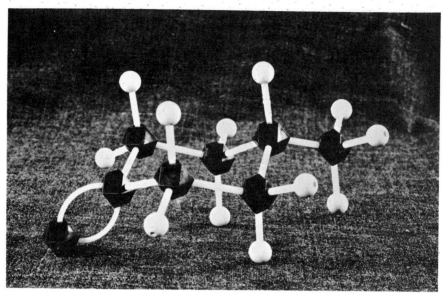

FIGURE 6-3 "Ball-and-stick" model of 4-methylcyclohexanone

Despite (or because of?) their great value for visualizing molecules, models can be deceptive. As we learn more about the responses of real molecules to deformation forces, we will recognize that models do not respond in parallel ways and thus can lead users into false attitudes about molecular strains. In general models are too stiff in resisting angle bending, too loose in rotation about single bonds, and the space-filling variety are pessimistic about the tolerance of nonbonded atom compressions. Rarely will any model fall into the correct conformation by itself, as the molecule it represents does. The first two figures show 2-hexenoic acid in two kinds of models, the first (Fig. 6-1) showing the space occupied by the atoms but a poor sense of the skeletal connections and angles. The second shows the skeleton only and more clearly. In both the coplanarity of the groups attached to the conjugated double bonds should be apparent. The second two figures illustrate cyclohexane derivatives in models which show the skeleton clearly. The cyclohexane ring is not planar, but the benzene ring in Fig. 6-4 is quite flat.

In writing perspective formulas it is common to approximate the tetrahedral angle with the simpler 120° or hexagon angle, and to show two groups in the plane of the paper and the other two above and below as wedges and dashed lines, respectively, as in the structure of hexanol shown here. Heavy and dashed bonds are also used as a convention for showing a ring substituent as above (heavy) or below (dashed) the plane of the ring, as in the two cyclohexanols shown. Conventions for drawing ring compounds in perspective are outlined in Sec. 6-6.

FIGURE 6-4   **Dreiding model of 4-phenyl-1-methylcyclohexylamine**

Figure 6-5 illustrates another convenient device for showing the three-dimensional arrangement of atoms about any single bond. The "end-on" or **projection formulas** show the molecule as it would appear to a viewer looking down the carbon-carbon bond. Consequently, one sees only the front carbon atom and the groups attached to both of the carbon atoms. In the projection formulas the six bonds are arranged like spokes of a wheel. The three bonds that intersect in the center of the circle show which groups are attached to the near carbon atom. Bonds to the more distant carbon atom are drawn only to the edge of the circle, since their intersection at the remote carbon atom is hidden. Rotation around the central bond may be shown by keeping the three bonds at one carbon atom fixed and rotating the group of three bonds at the other carbon atom, as in the two ethane conformations in Fig. 6-5.

6-2   RESTRICTED ROTATION: DOUBLE BONDS

The geometry of ethylene is much simpler than that of ethane, since all six atoms of ethylene must lie in one plane (page 54). The energy barrier to rotation about any carbon-carbon double bond is high enough to hold the four attached substituents in a rigid planar conformation except at high temperatures.

Projection formulas

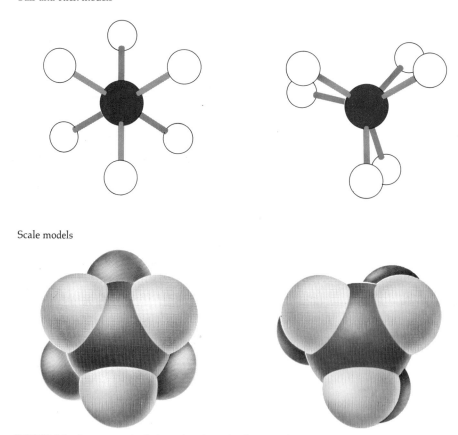

Ball-and-stick models

Scale models

FIGURE 6-5 **Arrangement of atoms in ethane (end-on view)**

Examination of the π orbitals of ethylene explains the rigidity associated with the carbon-carbon double bond. (See Fig. 6-6.) If the plane occupied by one CH_2 group is rotated 90° about the axis for the carbon-carbon σ bond, the π molecular orbital is destroyed. The two atomic p orbitals are now at right angles to each other, and no orbital overlap can occur. Twisting the molecule

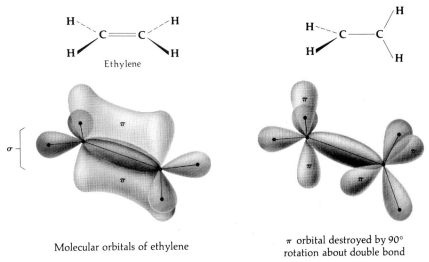

Molecular orbitals of ethylene

π orbital destroyed by 90°
rotation about double bond

FIGURE 6-6   **Rotation about double bonds**

to the perpendicular configuration requires energy input sufficient to break the π bond, i.e., on the order of 60 kcal/mole for the carbon-carbon double bond. (Compare double and single bond energies in Table 2-2.)

A consequence of restricted rotation about double bonds is **geometric isomerism**. For example, two different compounds contain the atomic sequence **HOOCCH=CHCOOH**; one is maleic acid and the other is fumaric acid. These substances have widely different physical properties. The former melts at 130°, the latter at 270°. When heated to 140° in an open flask, maleic acid gives maleic anhydride and water. In contrast, fumaric acid remains unchanged when subjected to the same treatment. However, at 275°, fumaric acid loses water to form maleic anhydride. Reversing the reaction by opening the anhydride ring with water gives only maleic acid.

Maleic acid
mp 130°

Maleic
anhydride

Fumaric acid
mp 270°

The existence of these two compounds and the difference in their behavior are consequences of restricted rotation about double bonds. In maleic acid, the two carboxyl groups are attached on the same side of the double bond; in fumaric acid, the functional groups are on opposite sides. Maleic acid loses water to form a cyclic anhydride because its two carboxyl groups are close together and can interact with each other. In fumaric acid, the carboxyl groups are too far apart to interact directly. If the compound is heated strongly, it slowly *isomerizes* (by rotation around the double bond) to maleic acid; the latter is converted to the anhydride very rapidly at the temperature required for isomerization.

Pairs of geometric isomers are normally referred to as **cis** and **trans** isomers. If the two carbon atoms of a carbon-carbon double bond possess a pair of like substituents, the cis isomer is the compound in which the like substituents are on the same side of the molecule. Thus, maleic acid is *cis*-butenedioic acid and fumaric acid is *trans*-butenedioic acid. If four different groups are attached to a carbon-carbon double bond, the cis isomer is the one in which the groups containing the longest carbon chains are on the same side of the double bond, as illustrated by the nomenclature of the 3-methyl-3-heptenes.

$$
\begin{array}{cc}
\mathrm{CH_3}\diagdown \quad \diagup \mathrm{H} & \mathrm{CH_3CH_2}\diagdown \quad \diagup \mathrm{H} \\
\mathrm{C}=\mathrm{C} & \mathrm{C}=\mathrm{C} \\
\mathrm{CH_3CH_2}\diagup \quad \diagdown \mathrm{CH_2CH_2CH_3} & \mathrm{CH_3}\diagup \quad \diagdown \mathrm{CH_2CH_2CH_3} \\
\textit{cis}\text{-3-Methyl-3-heptene} & \textit{trans}\text{-3-Methyl-3-heptene}
\end{array}
$$

The words cis and trans are used to describe the disposition of pairs of groups with respect to each other. For example, in *cis*-3-methyl-3-heptene the hydrogen atom and the methyl groups are *cis to each other*, and the *hydrogen atom is trans to the ethyl group*. The distinction between words used in naming a molecule as a whole and in identifying geometric relationships between parts must be clearly recognized.

Conventions for representation of unsaturated molecules are illustrated by the formulas for *cis*- and *trans*-2-butene, which are shown in Fig. 6-7, and in the photograph of the models of *trans*-2-hexenoic acid in Figs. 6-1 and 6-2.

In order to be capable of geometrical isomerism a double bond must have two substituents different from each other at either end, as in the cases below:

$$
\begin{array}{ccc}
\mathrm{A}\diagdown \quad \diagup \mathrm{X} & \mathrm{A}\diagdown \quad \diagup \mathrm{A} & \mathrm{A}\diagdown \quad \diagup \mathrm{Y} \\
\mathrm{C}=\mathrm{C} & \mathrm{C}=\mathrm{C} & \mathrm{C}=\mathrm{C} \\
\mathrm{A}\diagup \quad \diagdown \mathrm{Y} & \mathrm{B}\diagup \quad \diagdown \mathrm{B} & \mathrm{B}\diagup \quad \diagdown \mathrm{X}
\end{array}
$$

$$
\begin{array}{cc}
\mathrm{A}\diagdown \quad \diagup \mathrm{B} & \mathrm{A}\diagdown \quad \diagup \mathrm{X} \\
\mathrm{C}=\mathrm{C} & \mathrm{C}=\mathrm{C} \\
\mathrm{B}\diagup \quad \diagdown \mathrm{A} & \mathrm{B}\diagup \quad \diagdown \mathrm{Y}
\end{array}
$$

One isomer   Two isomers   Two isomers

The number of geometrical isomers that have the same basic structure increases with the number of double bonds that fulfill the above condition, and is given by the relation:

Number of geometrical isomers $= 2^n$

$(n =$ no. of double bonds with different substituents at each end$)$

In a diene example, the four $(2^2 = 4)$ isomers of 1-phenyl-1,3-pentadiene are shown in the following formulas.

Formulas

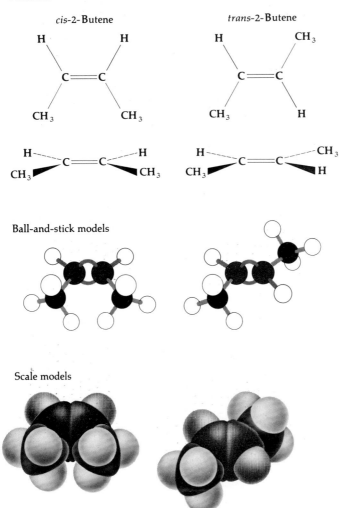

FIGURE 6-7   **Representations of the isomeric 2-butenes**

Geometric isomerism can also arise in the chemistry of compounds containing carbon-nitrogen and nitrogen-nitrogen double bonds. Although interconversion of stereoisomers by rotation about C=N and N=N bonds is easier than isomerization about C=C bonds, many examples are known in which both members of isomeric pairs have been isolated and characterized. A special nomenclature has developed for the description of isomeric oximes. The terms **syn** and **anti** replace **cis** and **trans**, respectively. For example, the syn isomer of an oxime derived from an aldehyde (RCHO) is the compound in which the hydrogen atom and the hydroxyl group are adjacent to each other. In the anti isomer the hydroxyl group is cis to the other group.

syn-Oxime　　　　　*anti*-Oxime
(H and **OH** cis)　　(H and **OH** trans)

As indicated by the above formulas, unshared pairs of electrons at nitrogen occupy fixed positions, corresponding to the positions of one of the groups attached to carbon in unsymmetrical compounds containing carbon-carbon double bonds.

Steric Hindrance

The relative stability of two geometric isomers is usually determined simply by considerations of *steric hindrance*. The term refers to the compression which a group suffers by being forced too close to its (nonbonded) neighbors. The cis positions on a double bond are close enough in space (and cannot rotate away) so that if the groups attached cis are of moderate size they will suffer strain due to crowding each other; this strain is not experienced in the trans isomer.

In general the isomer with the two larger groups at each end of the double bond cis is the less stable of the two. Thus *cis*-2-butene is slightly less stable than *trans*-2-butene, while with the bulkier phenyl groups the difference in energy between *cis*- and *trans*-stilbene (1,2-diphenylethylene) is so great that an equilibrium mixture contains 10^4 times as much trans as cis isomer at room temperature. When the groups in question are not hydrocarbon, however, the relation is less clear, for all the 1,2-dihaloethylenes except diiodo are more stable in the cis form, apparently owing to a counterbalancing electrical attraction between two halogen atoms. This attractive effect is not at present well understood.

6-3　RESTRICTED ROTATION: SINGLE BONDS

Since the single bond, unlike the double bond, consists of an axially symmetrical σ orbital, it should in principle be capable of completely free rotation which would carry the attached atoms through successive **staggered** and **eclipsed** conformations, as shown in Fig. 6-5 for the simplest case, ethane. However, even though the substituents on the single bond of ethane are only hydrogens and very small, the eclipsed conformation is less stable than the staggered by 3 kcal/mole.

This constitutes an **energy barrier** to free rotation around the single bond. It implies that most molecules at any moment are in the most stable staggered orientation but that any one which receives 3 kcal/mole from its environment may rotate to the eclipsed **conformer** (conformational isomer) and then give up the 3 kcal/mole again by rotating on to the next stable staggered conformer. It receives this energy by collision.

All molecules have thermal or kinetic energy owing to their being in constant motion, and this energy is a function of temperature, rising as temperature rises and the molecules travel faster. At room temperature this represents sufficient energy to allow processes requiring 15 to 20 kcal/mole to proceed, activated by molecular collision. This is not enough energy to break bonds (50 to 150 kcal/mole, see Table 2-2) or cause double bonds to rotate (~ 60 kcal/mole), but it is enough to allow rotation around single bonds to occur with ease (only 3-kcal/mole barrier). As we saw with fumaric acid, *sufficient* heat provides enough energy to overcome even the 60-kcal/mole barrier of double bonds.

The basic energy barrier to rotation around single bonds† is not yet fully accounted for but is generally understandable in a useful qualitative way as the mutual repulsion of the negative electrons in the six bonds attached to the central bond when they rotate closest to each other, i.e., in the eclipsed conformation. When the substituents on the single bond become larger than hydrogen their *steric hindrance* in the eclipsed conformation will increase the value of the barrier of rotation above the basic 3 kcal/mole.

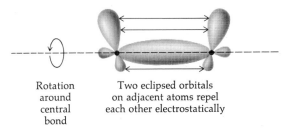

Rotation around central bond

Two eclipsed orbitals on adjacent atoms repel each other electrostatically

An important consequence of the rotation barrier arises in ring compounds in which certain bonds are forced by the restrictions of ring making to have conformations other than the staggered one. In such cases the nearer each bond is to being eclipsed the more strain energy of this kind it contributes to the overall strain energy of the molecule. Preferred conformations of ring molecules are usually characterized by their relative freedom from this **torsional** or **eclipsing strain** (Sec. 6-6).

As shown in Fig. 6-8, the question of free rotation about single bonds assumes major importance when the structure of a compound such as *n*-butane

---

† This barrier is 3 kcal/mole for bonds between sp^3 carbons and somewhat less for other single bonds.

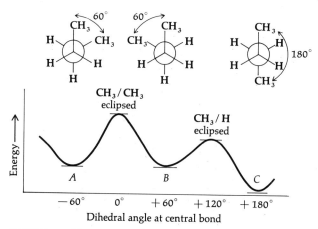

FIGURE 6-8   **Three staggered conformers of *n*-butane**

is considered. Of the three staggered arrangements, *A* and *B* resemble each other but are different from *C*. If the molecule could not pass easily from *C* to *A* and *B*, it would be possible to isolate two isomeric modifications of *n*-butane. Isomeric modifications of any saturated compound in which the isomerism is due solely to restriction of rotation about single bonds have never been isolated, however, owing to the low barrier to free rotation.

Although **rotamers** (rotational isomers) cannot be isolated, rotational arrangement of groups about single bonds sometimes has a profound influence on chemical reactivity. Conformations in which large groups are as far apart as possible are ordinarily most stable. For example, *C* should be the most stable of the *n*-butane rotamers since it has no steric hindrance between the methyl groups, and in a macroscopic sample more molecules have arrangement *C* than have the isomeric conformation *A* or *B*. Straight-chain hydrocarbons tend to take up the conformation with all single bonds in conformation *C*, the bonds to carbon from each single bond being oriented at 180° to each other as in the hexanol conformation shown on page 179. A curve of the changing strain energy with rotation is included in Fig. 6-8.

In saturated compounds generally, then, staggered conformations are preferred and the one with the groups at 180°—farthest from each other—is the best staggered conformation. When a single bond is connected to a double bond, however, the double bond at one end of the single bond eclipses a *hydrogen* at the other end in the preferred conformation. With respect to rotation around the single bond this conformational preference is around 2 kcal/mole, compared to 3 kcal/mole for the saturated compound.

PROBLEM 6-1

In each of the following pairs of compounds, which one will be the more stable and why?

a   *cis*- or *trans*-2-pentene

b   *cis*- or *trans*-2-phenyl-2-butene

c   *cis*- or *trans*-1,2-dimethyl-
cyclopentane

d   Isomers of 1-phenyl-2,3,3-trimethyl-
1-butene

e   *cis*- or *trans*-cyclooctene

PROBLEM 6-2

In planar cyclopentane the carbon bond angles are all 108°, very nearly tetrahedral. However, the preferred conformation is nonplanar. Explain by utilizing a projection diagram of one bond.

6-4   ANGLE-BENDING STRAIN

The great reactivity of cyclopropane compared with other cycloalkanes was explained in terms of steric strain by Baeyer as early as 1885. He reasoned that cyclopropane rings could be formed only by compressing the C—C—C bond far below the "natural" tetrahedral angle of 109°28′. The notion that strain makes unstable the small-ring compounds, such as cyclopropane and cyclobutane, remains essentially unmodified today. Extension of the theory predicts greater strain and higher reactivity for cyclopropene since the normal 120° bond angles at the sp^2 carbons are even more compressed in being forced into a 60° triangle. Cyclopropene is known, but it is so reactive that it explodes at room temperature. Angle compression, with resultant strain, is also observed when a double bond (or more exactly, any sp^2 carbon, with 120° angle requirements) is placed in a four- or five-membered ring. Chemical reactions which convert sp^2 to sp^3 carbon in a small ring (3, 4, 5) are enhanced owing to release of this angle strain.

Cyclopropane   Cyclopropene   Cyclobutane

Ethylene oxide, ethylene imine, and ethylene sulfide are common compounds that are unstable compared with their open-chain counterparts, but they are easily handled at room temperature and are widely used. Their four-membered homologs are also known. As expected, the latter compounds have intermediate stability.

$$CH_2\text{---}CH_2 \qquad CH_2\text{---}CH_2 \qquad CH_2\text{---}CH_2$$
$$O \qquad\qquad N \qquad\qquad S$$
$$\qquad\qquad\quad H$$

| Ethylene oxide | Ethylen- imine | Ethylene sulfide |

In molecules like cyclopropane which force severe angle strain, the molecule is faced with two choices: either keep the *interorbital* angles at about 109° to minimize interelectron repulsions and sacrifice maximum overlap of the bonding orbitals; or achieve maximum overlap by placing the bonding orbitals coaxial with the line between nuclei in the normal way and tolerating the higher interelectron repulsions of orbitals only 60° apart. This choice is apparently resolved, in the usual manner of physical systems, by adopting an intermediate orbital geometry which minimizes the overall energy. The actual cyclopropane molecule which results is the compromise of Fig. 6-9, with **bent bonds**. (See page 61).

6-5   CONFORMATIONAL ANALYSIS

It is possible to assess the strain energy of different conformers in order to decide which is the lowest in energy and hence most favored or preferred in the actual molecule. To do this we have only to examine each possible conformer by adding the separate strain energies due to angle bending, rotation, steric hindrance, and bond-length stretching, and then to select the conformer with lowest strain energy as the preferred one.

Each of these geometric factors is associated with its corresponding strain energy, and the molecule tends to adjust its geometry to minimize its total strain energy accordingly. Angle-bending strain increases with the deviation of each bond angle from the normal. Torsional or rotation strain is a function of the torsion or dihedral angle about a single bond seen in projection (Fig. 6-8) and increases as the angle passes from 60 to 0° by rotation. Steric hindrance is the strain due to mutual compression of groups which are crowded close but not bonded to each other; it increases rapidly as the distance between the nonbonded atoms decreases. Bond stretching is a very minor mode of strain accommodation since the energy necessary to stretch a bond from its optimum length is very much more than that associated with the other modes. A molecule tends to mix these accommodations to minimize its total strain energy, bending angles somewhat to relieve some steric hindrance, accepting some steric hindrance to achieve a more favorable torsional angle, etc. The process of assessing preferred conformations in this way is called conformational analysis and has its major value in cyclic molecules since the restriction of ring formation is usually accompanied by strains of these kinds but also limits the number of possible conformations.

Although conformational analysis is commonly—and very effectively—applied in a simple qualitative manner (Sec. 6-6), it is possible in principle to apply detailed mathematical calculations of the strain energy for a series

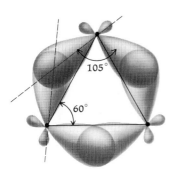

FIGURE 6-9   **Bent bonds in cyclopropane**

of varying conformers in a search for the energy minimum corresponding to the preferred form. Such calculations have been carried out by computer with fair success in matching experimental data.

6-6   CONFORMATION IN CYCLIC COMPOUNDS

Cyclic molecules are commonly designated as in Table 6-1, the five- and six-membered rings being by far the most common since they are easily formed in chemical reactions and have minimal strain energy. The small rings were discussed in Sec. 6-4. Since each carbon on a saturated ring bears two bonds to substituents (hydrogen in the parent rings) one is regarded as above and one below the plane of the ring, as in the cyclohexanols shown on page 179. Two substituents on the same side of the ring are described as cis, on opposite sides as trans, in analogy to double bonds (page 182). The compounds on page 176 are therefore designated as *cis-* and *trans-*1,3-dimethylcyclobutanes.

Cyclopentane is a stable, easily made compound, and Baeyer reasoned that the compound should be a planar stable molecule since the internal angles in a regular pentagon are 108°, a value very close to the tetrahedral angle. However, planar cyclopentane is not the most stable conformation for the molecule. Such a conformation brings all five ring bonds into the eclipsed arrangement. The cumulative effect of the resultant torsional strain (5×3 kcal/mole) is enough to distort cyclopentane from the planar to a slightly *puckered* configuration, in which the bond angles are more strained ($\sim$105°)

TABLE 6-1   **Strain in Cyclic Compounds**

Ring size	Designation	Strain
3–4	Small rings	Large ($3 > 4$); mostly angle-bending strain
5–7	Common rings	Small
8–10	Medium rings	Large (max = 10); mixed strains
12–∞	Large rings	Small

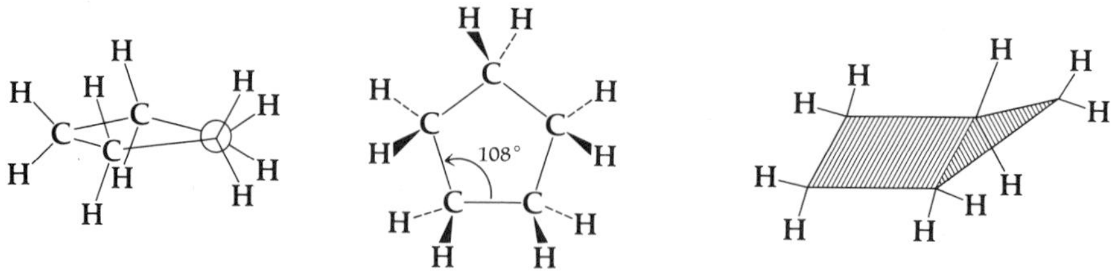

Cyclopentane (carbons
all occupy one plane)

Cyclopentane (puckered)

FIGURE 6-10   **Representations of cyclopentane**

in order to release the torsional strain of eclipsed bonds; only one bond is now eclipsed in the preferred puckered conformation, which has a net strain lower than the planar form. Both planar and puckered cyclopentane are represented in Fig. 6-10. A low-energy vibrational motion continually moves the distortion, or pucker, around the cyclopentane ring. Derivatives of cyclopentane may retain a more permanent distortion so that a carbon atom bearing a bulky substituent is permanently displaced from the plane of the other four carbon atoms.

Cyclohexane

Cyclohexane derivatives occupy a position of unique importance in organic chemistry and their conformational analysis is the easiest and most effective of all the rings. The uniqueness arises from the fact that a conformation for cyclohexane may be constructed with all bond angles tetrahedral and all six bonds in the staggered conformation. Since this ring also exhibits no internal steric hindrance between nonbonded atoms, it is essentially strain-free, unlike any other ring (up to about 12 atoms). This is called the **chair conformation** of cyclohexane, illustrated in Fig. 6-11 with a perspective drawing meant to convey the actual three-dimensional appearance.† The puckered chair conformation may also be seen in the models in Figs. 6-3 and 6-4. In the latter it should be compared with the planar six-membered benzene ring to which it is attached. Other conformations of cyclohexane are possible, as exemplified in the **boat conformation** (Fig. 6-11), but these are all less stable than the chair form by about 6 kcal/mole, largely owing to torsional strain (the boat form has two eclipsed bonds). Hence cyclohexane, and in fact virtually all substituted cyclohexanes, exist in the chair conformation preferentially.

Examination of the 12 substituent positions on a cyclohexane ring (occupied by hydrogens in the parent cyclohexane itself, Fig. 6-11) shows that they are of two kinds with six positions each. The **equatorial** (*e*) positions, or bonds, project out roughly in the plane of the ring, forming a belt around the perimeter. The six **axial** (*a*) positions or bonds, in two groups of three, extend up and

---

†The structure of diamond is a three-dimensional matrix of sp^3 carbons, each bonded to four others. Examined in models it exhibits an infinite array composed entirely of chair cyclohexane rings. This accounts for its remarkable stability. Any fracturing of a diamond must rupture carbon-carbon bonds.

Chair form

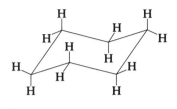

Ball-and-stick model of chair form

Boat form

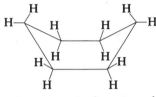

Ball-and-stick model of boat form

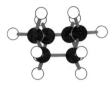

FIGURE 6-11   **Conformations of cyclohexane**

down perpendicular to the ring and are all parallel to each other. The conventional **perspective representation** of the chair form of cyclohexane and its derivatives is easily drawn if a few points are kept in mind.

1   Bonds on opposite sides of the rings are parallel to each other.
2   The axial bonds are all parallel to each other and are perpendicular to a plane which bisects each of the carbon-carbon bonds (the plane of the ring).
3   Each equatorial bond is drawn so as to be parallel to the two ring bonds *once removed* from its point of attachment. The parallel ring bonds for the upper and lower equatorial positions are colored in the example. With these simple rules it is easy to draw cyclohexane rings in correct perspective.

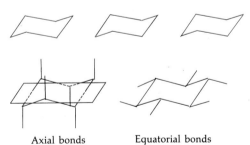

Axial bonds   Equatorial bonds
            Cyclohexane

A substituted cyclohexane may have its substituent in an equatorial position or an axial one, as indicated in Fig. 6-12 for methylcyclohexane. The axial and equatorial conformers have different energies, the equatorial being the more stable since there is steric hindrance between an axial substituent and its neighboring two axial substituents, even hydrogen. In Fig. 6-3 the more

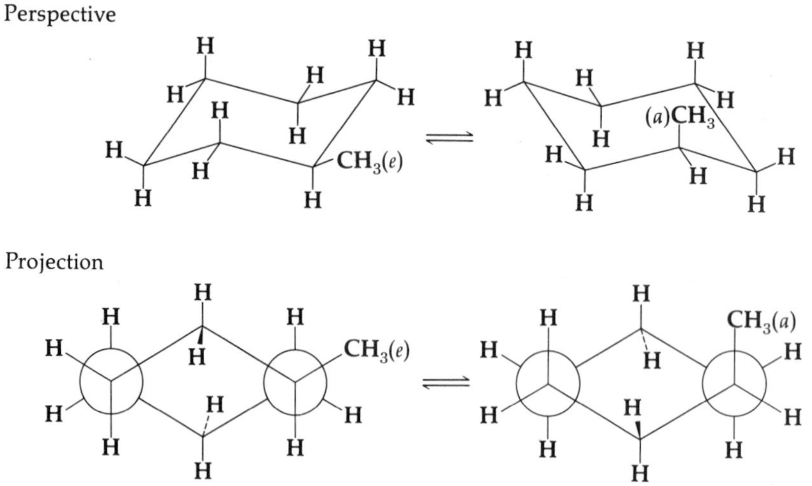

Perspective

Projection

Methyl group in equatorial (e)
position (more stable conformer)

Methyl group in axial (a) position
(less stable conformer)

FIGURE 6-12   **Conformers of cyclohexane**

stable, equatorial methylcyclohexanone is illustrated. In Fig. 6-4 the phenyl and amino substituents are equatorial and the methyl is axial. Furthermore, the axial- and equatorial-substituted conformers are interconvertible, since deformation of a chair cyclohexane can convert it into a boat form and this may in turn be transformed into another chair form in which the six axial positions become equatorial and vice versa. Thus, in this conformational transformation all substituent positions become reversed, as the methyl group does in Fig. 6-12. The interconversion may most easily be seen in this sequence:

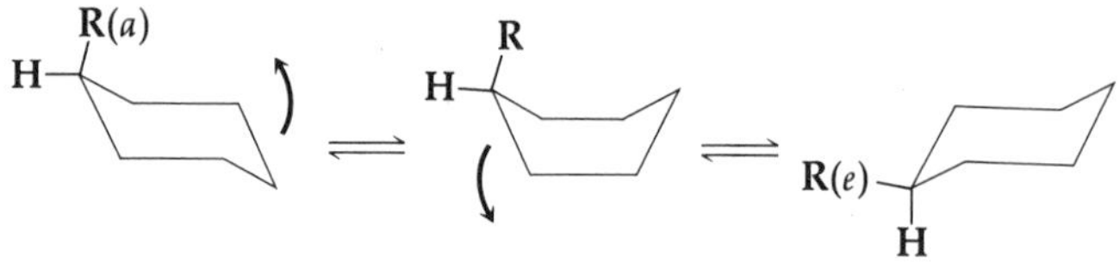

Chair   Boat   Chair

An equatorial substituent becomes axial (and the reverse) and the barrier to the process is about 11 kcal/mole, necessary to deform the chair for transformation to a boat. This means that the process has a very rapid rate at room temperature for it is below 15 to 20 kcal/mole.

PROBLEM 6-3

Draw carefully two chair forms each for the six possible dimethylcyclohexane molecules and indicate which is the more stable of each pair. Then arrange the six molecules (**a** to **f**) in an order of decreasing stability.

a *cis*-1,2-Dimethylcyclohexane

b *trans*-1,2-Dimethylcyclohexane

    c *cis*-1,3-Dimethylcyclohexane

    d *trans*-1,3-Dimethylcyclohexane

    e *cis*-1,4-Dimethylcyclohexane

    f *trans*-1,4-Dimethylcyclohexane

PROBLEM 6-4

Which of these isomers is more stable and why? About what energy difference is to be expected?

        A   *B*

Most methylcyclohexane molecules will have the methyl in the more stable equatorial orientation, but a few at any given moment will be axial since the energy difference is small. Energy differences for selected substituents are given in Table 6-2. The *t*-butyl group is so large that it will not tolerate the conformational flip into an axial orientation; hence it is often used to "lock" a substituted cyclohexane in just one of its two possible chair conformations: that with *t*-butyl equatorial.

These properties allow a basis for conformational analysis of polysubstituted cyclohexanes. In brief, any cyclohexane can take up two chair forms, with substitutent positions at each atom reversed, axial/equatorial, between the two. The conformer with the most equatorial substituents is then the preferred one. In cases of doubt, calculations can be made by adding energies of axial substituents from Table 6-2 for each of the two possible chair forms. In this connection, the *t*-butyl group must always be placed equatorial. The energies of Table 6-2 for axial substituents are valid only if the other two axial substituents on the same side of the ring are hydrogens; cis-diaxial substituents have much more steric hindrance (several kilocalories per mole) and consequent strain energy. It is important to note that, in flipping the ring from one chair form to the other, all substituents which are above the ring stay above when

TABLE 6-2 **Axial-Equatorial Energy Differences for Substituents on Chair Cyclohexane**

Substituent	ΔF, kcal/mole	Substituent	ΔF, kcal/mole
—CH$_3$	1.7	Cl, Br, I	0.5
—CH$_2$CH$_3$	1.8	OH, OR	0.7
—C(CH$_3$)$_3$	Very large	COOR(H)	1.1
—C$_6$H$_5$	3.1	CN	0.2

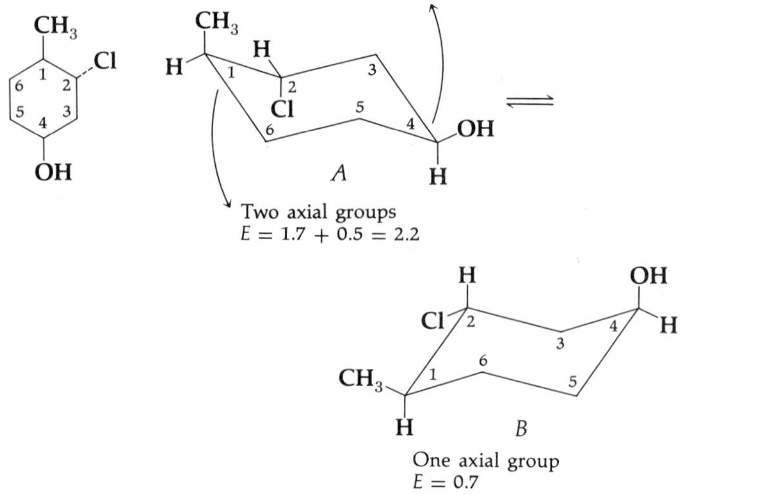

Two axial groups
$E = 1.7 + 0.5 = 2.2$

One axial group
$E = 0.7$

Preferred conformation (by $2.2 - 0.7 = 1.5$ kcal/mole)

FIGURE 6-13   **Conformational analysis of a substituted cyclohexane**

the ring flips even though they interchange axial and equatorial environments; substituents below the ring stay below, cis substituents stay cis, etc. The analysis is exemplified in Fig. 6-13.

PROBLEM 6-5

Draw carefully the preferred conformation of each molecule and indicate its approximate strain energy.

a *cis*-1-Ethyl-3-propylcyclohexane

b *trans*-1-Phenyl-3-methylcyclohexane

c 1,1,4-Trimethylcyclohexane

d 1,3,4-Cyclohexanetriol (all cis)

e 1,3,5-Cyclohexanetriol (all cis)

f *cis*-3-Methyl-*trans*-4-chlorocyclohexane-carboxylic acid

g 3-*t*-Butyl-4-phenylcyclohexanol (all cis)

h

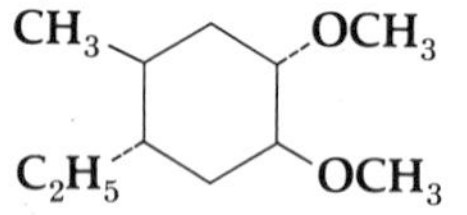

i

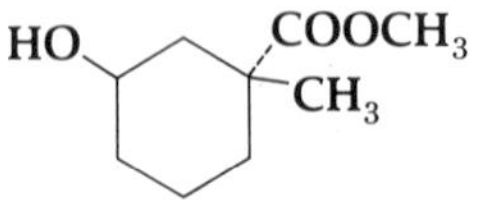

Bicyclic Compounds

When more than one ring is present, this conformational mobility may be frozen out and only one, rigid conformation be possible. This may be con-

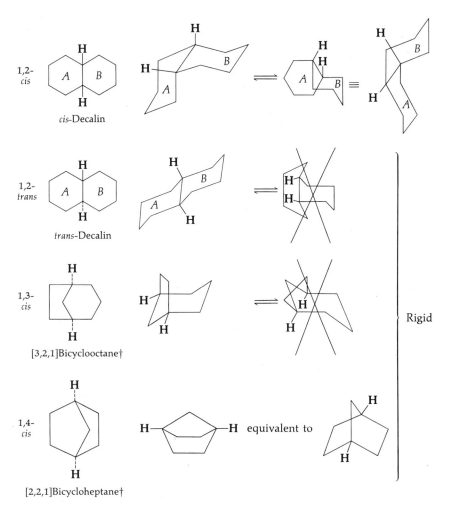

† Nomenclature: See Appendix.

FIGURE 6-14   **Bicyclic compounds and conformation**

veniently seen by considering the six possible disubstituted cyclohexanes and regarding the two substituents as linked together, forming a second ring. Some of these linkages will be recognized as manifestly impossible, and are crossed out or so marked in the chart, which refers to both chair forms of each (except 1,4-cis-linked, only in boat form):

Positions	Cis	Trans
1,2	1-eq, 2-ax ⇌ 1-ax, 2-eq	1-eq, 2-eq ⇌ ~~1-ax, 2-ax~~
1,3	~~1-eq, 3-eq~~ ⇌ 1-ax, 3-ax	Both impossible
1,4	Rigid boat ring only	Both impossible

Of the four possible bicyclic combinations (examples in Fig. 6-14), three are conformationally rigid. Only a cyclohexane 1,2-cis-fused to a second ring shows the conformational flipping typical of monocyclic cyclohexanes. The common example is *cis*-decalin and the chart shows a convenient way of drawing both possible conformers in useful perspective, with both rings visibly in the chair form. *cis*-Decalin is less stable than *trans*-decalin since it has an axial substituent on each ring in either conformation; *cis*- and *trans*-decalins, furthermore, are stereoisomers of each other—different compounds—and not just different conformers (see the analogous 1,3-dimethylcyclobutanes on page 176) of a single compound.

PROBLEM 6-6

Problems involving the flexible conformations of *cis*-decalin are most easily solved with these two chair forms: all substituent positions convert from axial to equatorial and vice versa on passing from one form to the other as the same R and R' groups show.

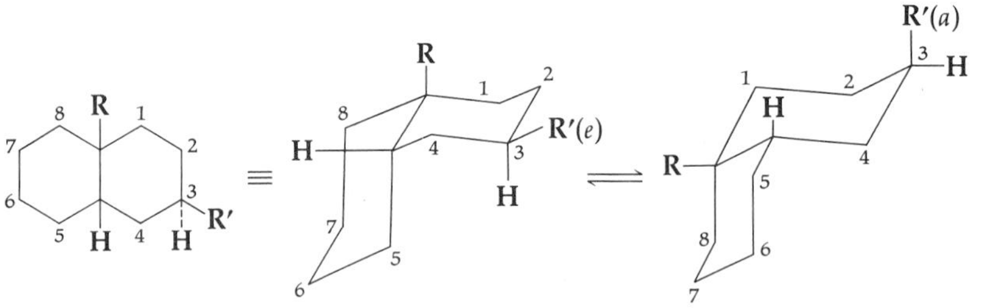

a Show the most stable conformation for each of the following isomers, and place them all in an order of decreasing stability (increasing strain energy).

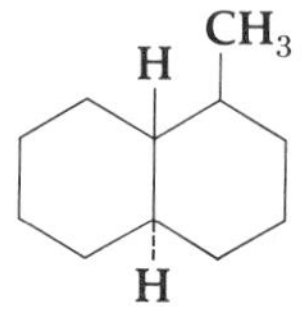

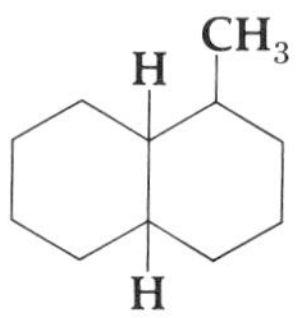

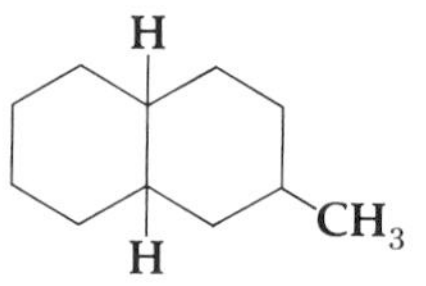

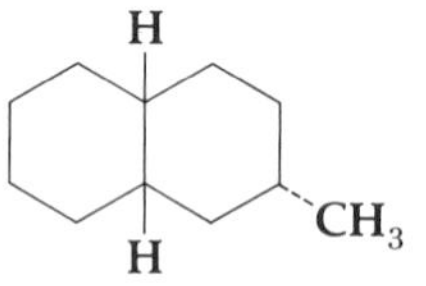

b Draw the most stable *cis*- and the most stable *trans*-decalins from part **a** in a *projection* formula, with projection down the central bond.

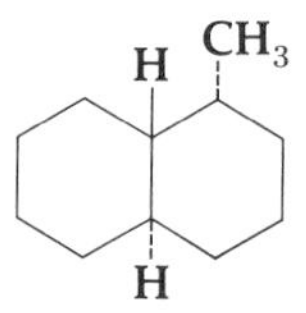

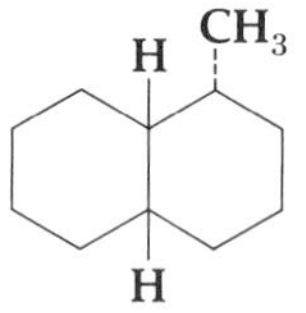

Medium Rings

Larger rings are more strained, the strain order in the whole series being $3 > 4 > 5 > 6 < 7 < 8 < 9 < 10 > 11 > 12$, with cyclohexane unstrained and cyclodecane the most strained among larger rings. In the larger rings steric hindrance between certain hydrogens across the ring becomes inevitable and, to minimize this, the rings find optimum conformations with larger bond angles (up to 118°), some steric crowding of hydrogens across the ring, and imperfect

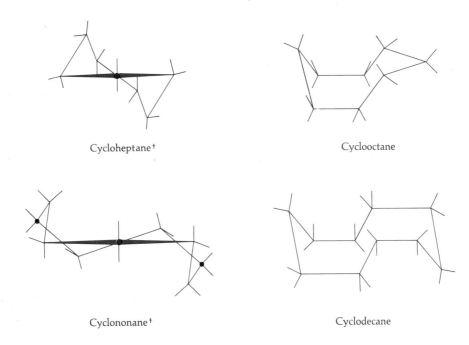

Cycloheptane†   Cyclooctane

Cyclononane†   Cyclodecane

† Seen in projection down the axis of symmetry (colored).

FIGURE 6-15 **Preferred conformations of saturated rings 7 to 10**

staggering of torsional angles. In each case the preferred conformation is symmetrical, however, and has the least torsional strain, but in no ring is the best conformation so much preferred over its competitors as in cyclohexane. In cyclooctane, in fact, there are five conformations within about 1 kcal/mole of each other. As in cyclohexane the energy barriers to interconversion are small so that free equilibration of the various conformers occurs at room temperature. For these reasons conformational analysis of these rings is far more difficult. The favored conformations of these larger rings are shown in Fig. 6-15. The odd-numbered rings have conformations containing only an axis (no plane) of symmetry and are best seen in projection looking down that axis. The even-numbered rings have planes of symmetry as in cyclohexane chair; they are shown with the plane of the page acting as the symmetry plane in the perspective drawings. Still larger rings can take up preferred conformations with normal bond angles and all-staggered torsional arrangements. They are essentially strain-free and behave like open-chain compounds.

Unsaturated Rings

Double bonds can be introduced into even small rings. Cyclopropene, cyclobutene, and cyclopentene are all strained, but cyclohexene is nearly strain-free. However, the double bonds in these rings must have the cis configuration. The smallest cycloalkene stable enough to exist in the trans configuration is *trans*-cyclooctene.† Since acetylenes are normally linear compounds (page 55), it is not surprising that small-ring cycloalkynes have not yet been prepared.† Cyclooctyne is the smallest cycloalkyne known.

Cyclohexenes adopt the chair (or half-chair, since one side is flat, not puckered), not the boat, conformation and the saturated carbons retain their equatorial-axial substituent distinction although the energy differences between them are less, especially at the carbons adjacent to the double bond.

Cyclohexene

New factors are now introduced into the conformational analysis since substituents on the double bond nearly eclipse adjacent equatorial substituents. In such cases, for substituents of moderate size, the adjacent *axial* orientation becomes the *less hindered* and preferred position. A second unsaturated case in which this happens is provided by a double bond projecting from the ring (**exocyclic** double bond). Substituents on the outer end of such a double bond offer so much hindrance to the adjacent eclipsed equatorial position as to cause the axial position to be preferred. These ideas are summarized in Fig. 6-16.

6-7 SYMMETRY, ASYMMETRY, AND CHIRALITY

In order to build a clear basis for understanding stereoisomerism we must first explore the nature of symmetry and apply it to molecular shapes. The most important kind of symmetry is the **plane of symmetry**. An object is symmetrical if, on passing a plane through the center of the object, the reflection of one side of the object in that plane is identical with the other side. Such a plane of symmetry is often called a mirror plane. (If one half of the object mirrors the other half, the object is symmetrical.) Symmetrical objects exhibiting mirror planes include eggs, funnels, books, persons (idealized), and trigonally hybridized atoms, for which the sp^2 molecular plane is a mirror plane. A tetrahedral carbon with *at least two identical substituents* also has a plane of symmetry lying between the identical substituents (Fig. 6-17).

An object without a plane of symmetry is asymmetric.‡ Most trees and houses are asymmetric, hands are asymmetric but simple mittens have a plane of symmetry since back and front are the same and they can be worn on either

---

†Evidence has been obtained for the transient existence of cyclopentyne and *trans*-cycloheptene as short-lived reaction intermediates.

‡More accurately, an object with no mirror plane is **dissymmetric**, whereas an object with no symmetry of any kind is **asymmetric**. (All asymmetric things are dissymmetric but not vice versa.) For our purposes we shall speak only of asymmetry, since virtually all symmetric molecules have a plane of symmetry. The conformations of cycloheptane and cyclononane in Fig. 6-15 are exceptions, having only axes, but no planes, of symmetry.

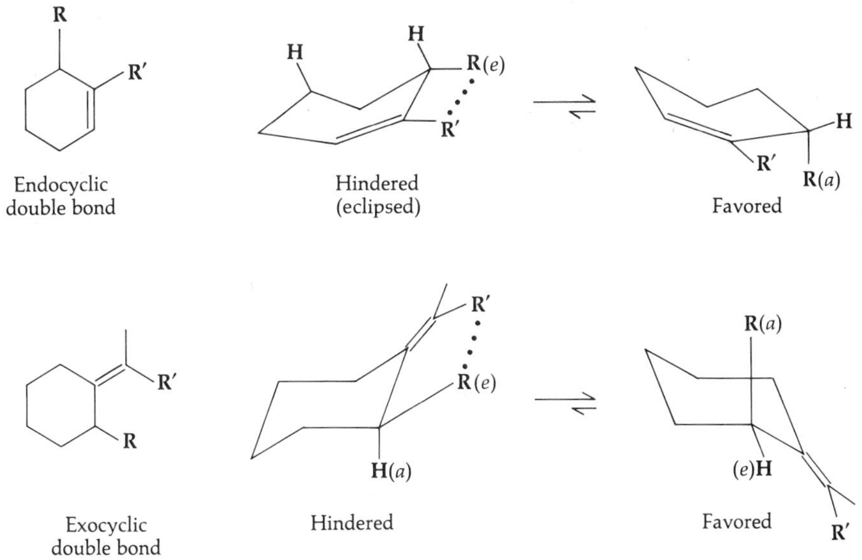

FIGURE 6-16   **Conformations of unsaturated six-membered rings**

hand. A tetrahedral atom with *all four attached groups different* is asymmetric also; such carbons are called **asymmetric carbons** (Fig. 6-17).

Right and left hands illustrate our recognition that two objects can be alike in every description except one: they are mirror images of each other. They are not identical since they cannot be exactly superimposed on each other. In fact for any object (or molecule) *the ultimate test of asymmetry is that it cannot be superimposed on its mirror image.* An asymmetric molecule can exist in either of two mirror-image forms, which are not identical. Such a pair of nonidentical, mirror-image molecules (cf. Fig. 6-17) are called **enantiomers** and represent a major kind of stereoisomerism. The several possible situations can be tabulated as follows:

One object = *symmetric*   Two objects = *identical*
        = *asymmetric*   = *enantiomeric (mirror image)*
                                = *different*

Chirality is the distinction applied to discriminate between two enantiomers, i.e., the right or left handedness of an asymmetric object or molecule. In general we can only distinguish chirality by applying some arbitrary asymmetric standard. For visible objects this will most commonly be recognition by the human eye, as in right-handed screws or gloves, left-hand drive in cars, etc. At the molecular level we distinguish the chirality of asymmetric molecules in two such ways. We can apply an asymmetric kind of physical observation, namely measurement of optical activity with dissymmetric, polarized light, described in Sec. 6-9. Most physical observations (melting point, spectra, etc.), however, distinguish only between *different* molecules but not *enantiomeric* ones,

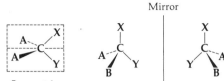

Symmetric:
mirror plane = plane
of paper
bisecting molecule   Asymmetric: two enantiomers

FIGURE 6-17   **Symmetry and asymmetry in tetrahedral carbon**

since there is not intrinsic asymmetry in the physical property observed in such cases.

The second method is to couple the molecule with a standard asymmetric molecule in a chemical reaction. The rationale here is that coupling two enantiomers with something of known chirality gives two new species which are *different*, no longer *enantiomeric*, and physically distinguishable like any two different objects. All these approaches to determining chirality may be summarized as follows ("standard" = human eye, polarized light, or asymmetric molecule):

Two enantiomers + one standard $\longrightarrow$ two different entities

    S   S   SS

    Я   S   ЯS

Asymmetric compounds in nature are virtually always present as only one enantiomer, but in the laboratory reactions which create asymmetric structures are not discriminating and a 50:50 mixture of both enantiomers always results.† Such a mixture of enantiomers is called a **racemic mixture** or **racemate** and may be separated by reaction with a standard asymmetric substance, as explained above, followed by separation of the different products, as described in Sec. 6-13. The separation of a racemic mixture into its two constituent enantiomers is called **resolution**. Most physical properties of racemates are the same as those of its two component enantiomers, except optical activity and melting point. The melting points of the two enantiomers must be the same, but that of the racemate may be higher or lower.

As an example (Fig. 6-18), 2-butanone is symmetrical, having a plane of symmetry, but reduction (addition of H_2 to the double bond) creates 2-butanol with an asymmetric carbon and no plane of symmetry. The hydrogen coming in to bond can approach equally well from either above or below the plane of the sp^2 carbon and so yields a 50:50 (racemic) mixture of the two enantiomers of 2-butanol.

---

†If asymmetric reagents are used, one enantiomer may be favored in the reaction product. Also, when other asymmetric carbons are already present and a new one is created, the original chirality may also cause one enantiomer to predominate at the new center.

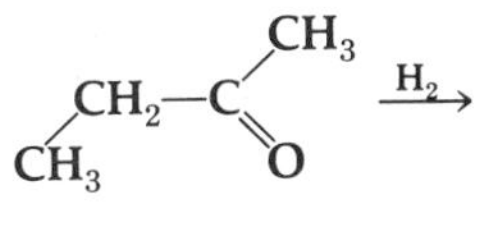

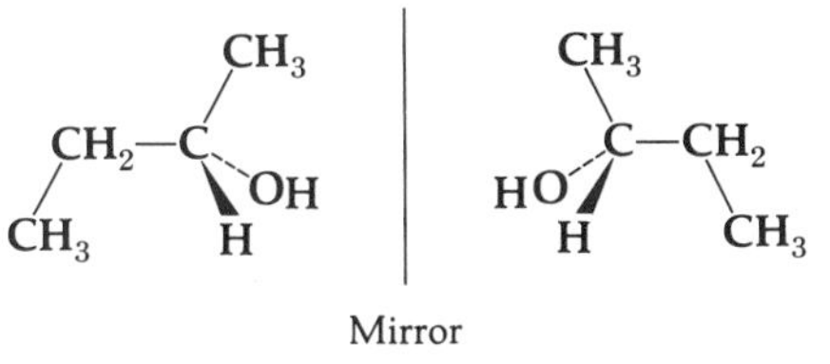

Formulas

Mirror

FIGURE 6-18   **2-Butanone and the enantiomers of 2-butanol**

6-8   OPTICAL ACTIVITY

It has been asserted that enantiomers are isomers, implying that there are differences in their properties. Differences exist, but they can be detected only by using measuring devices that are themselves chiral, or asymmetric. Thus, a pair of enantiomers will have identical melting points, boiling points, etc., and will show identical reactivity toward most chemical reagents. However, enantiomers show different chemical reactivity toward other asymmetric compounds and will respond differently toward asymmetric physical disturbances.

The commonest asymmetric tool is polarized light. If a beam of ordinary light is passed through a Nicol prism, only part of the light emerges. The oscillating electric vector of the emerging beam is oriented in one plane, and the beam is described as **polarized**. If a polarized beam is passed through an asymmetric material, the plane of polarization is rotated, and this may be observed. Hence, the asymmetric material is said to be **optically active**. The angle between the original and final planes of polarization is known as the **optical rotation**. The changing plane of polarization of the light may be visualized as a twisted ribbon passing down the tube of asymmetric material. The optical rotation of any sample depends upon the wavelength of the light, the length of the light path, the number of optically active molecules per unit length of light path, the temperature, the nature of the solvent (if any), and the structure of the compound. Measurement of the optical rotation of asymmetric compounds under carefully specified conditions provides a valuable physical constant for use in characterization and identification of the material.

Rotations of pure samples of enantiomers are equal in magnitude but opposite in sign. When the plane of polarization is rotated clockwise,† the sign of rotation is taken as positive and the compound is said to be **dextrorotatory** (i.e., rotation to the right). **Levorotatory** compounds rotate the plane of polarization to the left, or counterclockwise. The letters *d* and *l*, standing for *dextro-* and *levo-*, refer to the sign of rotation. In the older literature the letters were often placed before the names of optically active compounds. Racemic mixtures

†The observer *faces* the emerging beam of light.

are often referred to as d,l pairs. Current practice is to use $(+)$ and $(-)$ signs to distinguish between enantiomers.

Quantitative measurements of the optical activity of asymmetrical compounds are usually reported as **specific rotations**. The quantity is defined by

$$\text{Specific rotation} = [\alpha]^{\text{temp.}}_{\text{wavelength}} = \frac{\text{observed rotation,}^{\circ}}{\text{length of sample (dm)} \times \text{conc. (g/ml)}}$$

Concentrations are reported on a weight basis to permit data for compounds of unknown molecular weight to be reported. The report should also specify the conditions under which the measurement was made, i.e., solvent and concentration or neat (pure) liquid. The following examples will illustrate general practice in reporting results. The subscript D indicates that the light used to measure the rotation was the sodium D line (587 nm wavelength); $[\alpha]_D$ measurements are by far the most common.

Enantiomers of 1-phenylethanol:

$$[\alpha]^{27^{\circ}}_D = +42.9^{\circ} \text{ (neat)} \qquad [\alpha]^{27}_D = -42.9^{\circ} \text{ (neat)}$$

Enantiomers of atrolactic acid:

$$[\alpha]^{15}_D = +52.0^{\circ} \text{ (C 2\%, H}_2\text{O)} \qquad [\alpha]^{15}_D = -52.0^{\circ} \text{ (C 2\%, H}_2\text{O)}$$

Neither the sign nor magnitude of the rotation, however, can tell us anything about the structure of the molecule, or even its chirality if the gross structure is known. The correlation of structure with optical rotation is very complex and has not been successful to date. Rotations, like melting points, are chiefly used for identification purposes with asymmetric compounds. A substance with an optical rotation of zero, however, is either a racemic mixture or composed of symmetrical molecules, and this fact is often useful in structure determination.

Other measurements of optical activity do afford information about structure and chirality of asymmetric molecules, however. Since the optical rotation is a function of the wavelength of light used, a graph may be constructed consisting of optical rotation plotted against the wavelength at which it is observed. Such a curve is called the **optical rotatory dispersion (ORD)** of the substance and frequently exhibits a sine-wave form, such as B and C in Fig. 6-19, which is referred to as a **Cotton effect**. Such a curve has a **peak** and **a trough** and is labeled positive if the former is at longer wavelength and vice versa. Two enantiomers exhibit mirror-image curves, one negative and one positive, and in many instances these Cotton-effect signs can be correlated with the chirality of the enantiomer, as discussed in more detail on page 221. Struc-

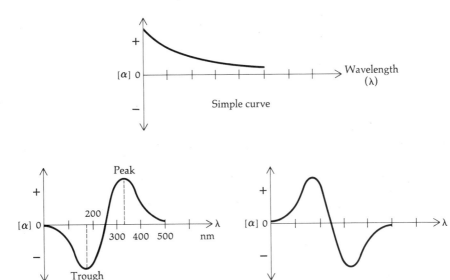

FIGURE 6-19   **Optical rotary dispersion curves**

tural information is also available from measuring the light absorption using asymmetric light which is circularly rather than plane-polarized. Such measurements are called **circular dichroism**.

PROBLEM 6-7

A naturally occurring alcohol, $C_6H_{12}O$, was found to exhibit an optical rotation, $[\alpha]_D^{25} = +49.5°$. On catalytic hydrogenation the alcohol absorbed 1 mole of hydrogen to form a new alcohol with $[\alpha]_D = 0°$, i.e., optically inactive. Write a structure for the natural alcohol which explains this fact. Does this structure determination yield a unique answer?

6-9 THE ASYMMETRIC CARBON: CONFIGURATION

The commonest sort of asymmetry in organic molecules derives from the presence of one or more asymmetric carbons, as in the example of 2-butanol, Fig. 6-18. The **configuration** of an asymmetric carbon is the specification of the relative spatial placement of the four groups attached. In particular the **absolute configuration** specifies their order in such a way as to distinguish the two enantiomers and so define their chirality. **Relative configuration** describes the relation of two asymmetric atoms to each other (viz., cis or trans substituted on a ring) and is discussed in subsequent sections.

The two enantiomeric configurations at any asymmetric atom can only be interconverted by breaking bonds and remaking them in the reverse sense. This process is called **inversion** of the configuration.

The system for specification of absolute configuration is as follows. If a molecule contains an asymmetric carbon atom, the four *atoms* attached to that carbon are first arranged in a sequence of decreasing atomic number. If two or more of these *first atoms* have the same atomic number, one is chosen by comparing the atomic numbers of the *second* group of atoms attached to the first atoms. If ambiguity still persists, the *third, fourth,* etc., sets (working outward from the asymmetric carbon atom) are compared until a selection can be made. In the second group of atoms, those atoms with the highest, next-highest, etc., atomic numbers are always ranked in that order.

Multiple bonds are treated as separate single bonds, so that the carbon of —CHO is regarded as attached to 2O + 1H. Orders of decreasing priority are these.

Atoms: I, Br, Cl, S, P, F, O, N, C, H, unshared electron pair (isotopic atoms by decreasing mass)

Groups†:   —C(CH$_3$)$_3$, —CH(CH$_3$)$_2$, —CH$_2$CH$_3$, —CH$_3$;

—COOCH$_3$, —COOH, —CONH$_2$, —COCH$_3$, —CHO ;

—C≡N, —C$_6$H$_5$ (φ), —C≡CH, —CH=CH$_2$;

Examples:
$$
\begin{array}{c}
\text{CH}_3 \\
| \\
\text{H}-\text{C}-\text{NH}_2 \\
| \\
\text{COOH}
\end{array}
\qquad
\begin{array}{c}
\text{COOH} \\
| \\
\text{CH}_3-\text{C}-\text{C}\equiv\text{CH} \\
| \\
\text{C}_6\text{H}_5
\end{array}
$$

Sequence: NH$_2$, COOH, CH$_3$, H   COOH, C$_6$H$_5$, C≡CH, CH$_3$
              1   2   3   4   1   2   3   4

A three-dimensional model of the isomer to be named is now viewed from the side opposite the group of lowest priority, and the sequence (decreasing priority) of the other three groups is noted as being clockwise or counterclockwise. When clockwise, the symbol R (for Latin, *rectus*, right) is used to denote the configuration. When the sequence is counterclockwise, the symbol S (for Latin, *sinister*, left) is employed. The figure may be viewed as a steering wheel with the lowest-priority atom down the steering column (this is usually hydrogen) and the higher three groups clockwise (R-) or counterclockwise (S-) around the wheel, as in Fig. 6-20.

As an example, let us apply this sequence rule to 2-butanol. (Note that the asymmetric carbon is commonly indicated by means of an asterisk.)

$$
\begin{array}{ccccc}
\text{H} & \text{H} & \overset{*}{\text{OH}} & \text{H} \\
| & | & | & | \\
\text{H}-\text{C}-&\text{C}-&\text{C}-&\text{C}-\text{H} \\
|4 & |3 & |2 & |1 \\
\text{H} & \text{H} & \text{H} & \text{H}
\end{array}
$$

2-Butanol

Sequence of groups at C-2: **OH, C$_2$H$_5$, CH$_3$, H**

1   2   3   4

†Note that a higher atomic number always takes priority in secondary groups even though the sum of atomic numbers on the three substituents may be lower than a competitor; hence —CH$_2$OH has higher priority than —C(CH$_3$)$_3$.

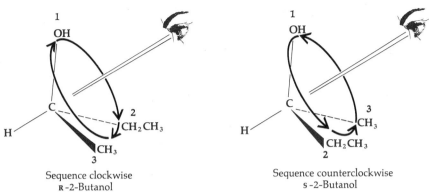

Sequence clockwise
R-2-Butanol

Sequence counterclockwise
S-2-Butanol

FIGURE 6-20 **Conventional views of 2-butanol**

Clearly oxygen has the highest and hydrogen the lowest atomic-number priority. To differentiate between C-1 and C-3, the atoms attached to C-1 (H, H, H) must be compared with those of C-3 (C, H, H). Obviously, C-3 has the higher priority and the groups can be ordered in the sequence O, C-3, C-1, H. This priority is now found to read clockwise (Fig. 6-20) in one enantiomer and so that enantiomer has the R configuration (shown at left). The enantiomer on the right has the sequence order counterclockwise and thus the S configuration.

Other atoms than carbon can be asymmetric, but since such asymmetric centers are usually tetrahedral, the same behavior and nomenclature is to be found in these analogous cases. Quaternary ammonium compounds have been resolved, as have amine oxides. Examples of these types of compounds are formulated below.

$$\bar{Cl} \qquad\qquad \bar{Cl}$$

$$C_6H_5-\overset{+}{N}\underset{CH_3}{\overset{CH_2C_6H_5}{\diagup}}C_2H_5 \qquad C_6H_5CH_2\underset{CH_3}{\overset{}{\diagdown}}\overset{+}{N}-C_6H_5 \qquad CH_3-\overset{\overset{C_2H_5}{|}}{\underset{|}{\overset{+}{N}}}-\bar{O}$$

$$C_2H_5 \qquad\qquad\qquad\qquad C_6H_5$$

R—   S—

| |

Racemate   An asymmetric
                                              amine oxide

Although the arrangement of bonds in amines approaches a tetrahedral configuration (the unshared pair of electrons is equivalent to one bond), no ordinary tertiary amine has been resolved. In molecules of this type, the two forms of the amine undergo very rapid interconversion; the molecule flips inside out, like an umbrella in a high wind.

$$\underset{R_1}{\overset{R_3}{R_2{\diagdown}}}N: \rightleftharpoons :N\underset{R_1}{\overset{R_3}{\diagup R_2}}$$

Unshared pairs of electrons on sulfur or phosphorus (unlike those of nitrogen) can hold configuration at ordinary temperatures, as has been demonstrated by resolution of compounds such as the sulfonium salt and sulfoxide shown.

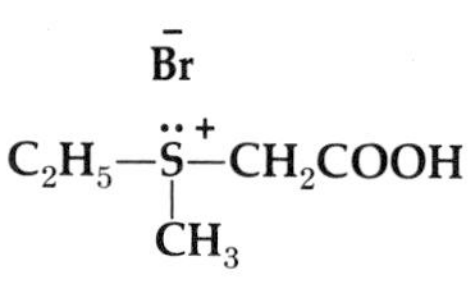

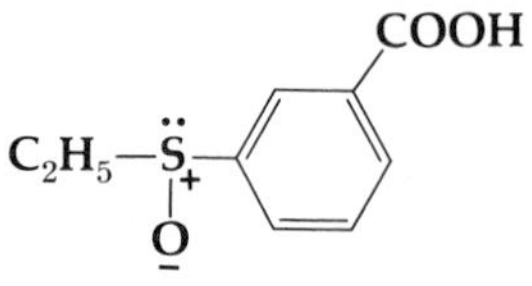

With molecules containing more than one asymmetric center, each center is examined and assigned a configuration, and the terms R or S are then incorporated into the systematic name by simply inserting the number of the asymmetric carbon atom in question in front of the letter, as in the following example. The system has been expanded to embrace asymmetric molecules without asymmetric centers (Sec. 6-12), as well as compounds with asymmetric centers other than carbon.

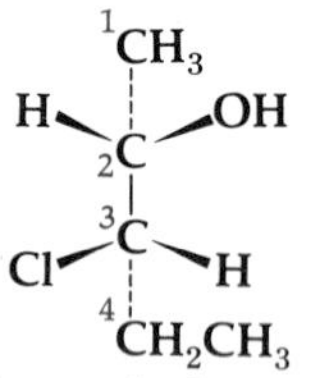

3-s-Chloro-2-s-hydroxypentane 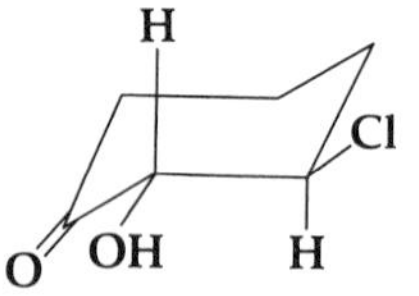   2-s-Hydroxy-3-R-chloro-cyclohexanone

PROBLEM 6-8

Label the following structures R- or S-.

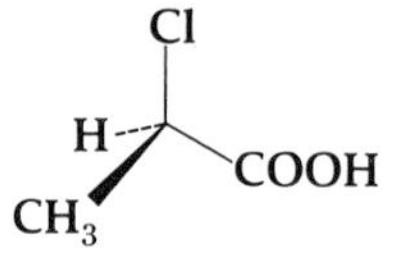

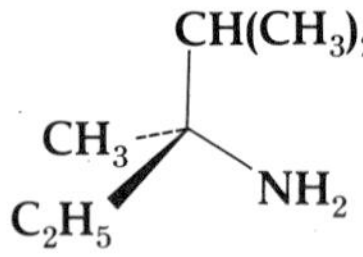

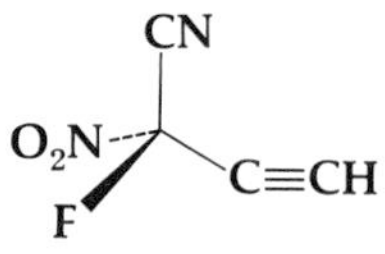

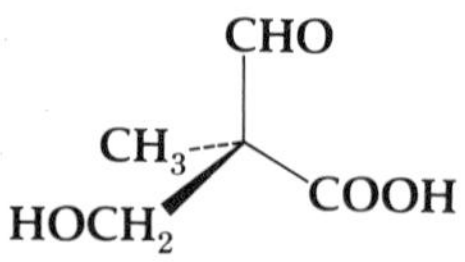

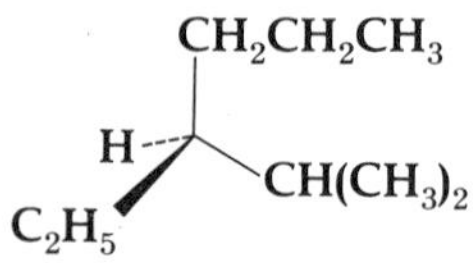

PROBLEM 6-9

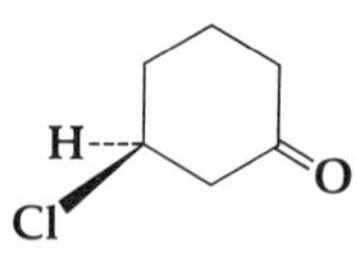

Draw the following structures in correct asymmetric perspective (cf. as in Prob. 6-8).

a R-2-Hydroxypropanoic acid
b R-2,3-Dibromopropanal
c s-3-Methyl-3-methoxy-4-
   hexen-2-one
d R-3-Cyanocyclopentanone

e s-2-Ethyl-hept-1-en-5-yne
f R-2-Deuteropropanoic acid
g s-α-Methylsuccinimide
h s-Methylethylbenzylammonium
   chloride

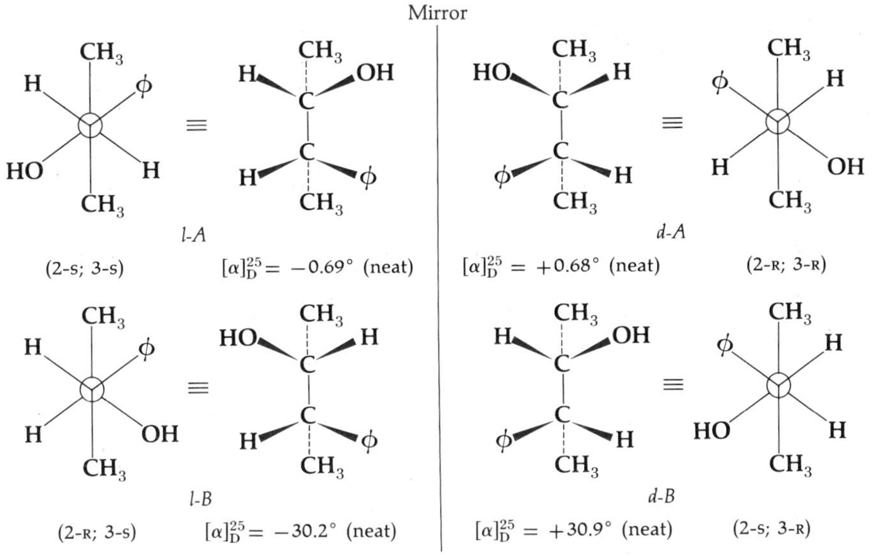

Mirror

l-A

(2-s; 3-s) $[\alpha]_D^{25} = -0.69°$ (neat) $[\alpha]_D^{25} = +0.68°$ (neat)   (2-r; 3-r)

d-A

l-B

(2-r; 3-s) $[\alpha]_D^{25} = -30.2°$ (neat) $[\alpha]_D^{25} = +30.9°$ (neat)   (2-s; 3-r)

d-B

FIGURE 6-21   Stereomers of 3-phenyl-2-butanol

6-10   SEVERAL ASYMMETRIC CENTERS

Structures that contain two or more asymmetric centers can exist in more than two stereoisomeric modifications. Some are pairs of optically active enantiomers; others may be symmetric and therefore optically inactive. Many natural products contain 2 to 10 asymmetric centers per molecule, and molecules such as starch and proteins contain hundreds.

The number of possible stereoisomers having the same structural formula is called the **isomer number** for that structure. Open-chain structures without special symmetry properties have isomer numbers equal to 2^n, where n is the number of asymmetric carbon atoms in the structure. The same is true of cyclic compounds in which steric restrictions from ring formation do not make certain isomers sterically impossible (cf. a 1,3- or 1,4-trans-bridged ring across a cyclohexane, page 195). Thus for almost all compounds the isomer number is 2^n.

Since any asymmetric structure must have an enantiomer, the total number (2^n) of stereoisomers may be broken down into $\frac{1}{2}(2^n) = 2^{n-1}$ *pairs of enantiomers*, i.e., 2^{n-1} possible *racemates* for a structure with n asymmetric atoms. With only one asymmetric carbon the molecule has $2^1 = 2$ stereoisomers, i.e., one pair of enantiomers only. The substance 3-phenyl-2-butanol contains two asymmetric carbon atoms; therefore, four stereoisomers, each with a unique configuration, have this structure. The isomer number of 3-phenyl-2-butanol is 4. All are known compounds and have the configurations shown in Fig. 6-21.

There are two pairs of enantiomers; the two enantiomers of any pair have equal and opposite rotations. Stereoisomer *d-A* is dextrorotatory and is the enantiomer of levorotatory *l-A*, and *d-B* is the enantiomer of *l-B*. Taken together in equal amounts, *d-A* and *l-A* form one racemate (*d,l-A*) and *d-B* and *l-B* form another (*d,l-B*). Stereoisomer *l-A* or *d-A* is a **diastereomer** of *d-B* and *l-B*.

Diastereomers are stereoisomers that are *not* mirror images of each other; they have the same configuration at at least one asymmetric center and, at the same time, different configurations at at least one asymmetric center. Diastereomers are physically different molecules and have different physical and chemical properties. In general two stereoisomers are *enantiomers only if all the asymmetric centers in one are inverted in the other*. All other pairs of stereoisomers are diastereomeric. Thus, if a molecule with four asymmetric centers has configurations RSRR, then its enantiomer is SRSS; while SSSR, SSRR, etc., are diastereomers. A special case of diastereomers is distinguished for convenience: two diastereomers that differ in configuration at *only one* asymmetric center are called **epimers**. One epimer of RRRR would be RSRR and another would be RRRS, but RSSR is a diastereomer which is not an enantiomer or epimer of RRRR.

The formulas in Fig. 6-21 show the difference between enantiomeric and diastereomeric relationships. Comparison of *d-A* and *d-B* shows that this pair cannot be arranged to make all nonbonded interatomic distances the same. As they are drawn, the distances between the hydrogen atoms attached to the two central carbon atoms are different, and if they are made the same by rotation, the methyl distances will then become different. Hence, *d-A* and *d-B* are diastereomers. Since they differ in absolute configuration at only one center (C-2), they are also epimers and are said to be epimeric at C-2. Comparison of *d-A* and *l-A* shows that the two molecules can be arranged so that all nonbonded interatomic distances are the same; the compounds differ only in chirality and are therefore enantiomers (2-s, 3-s and 2-R, 3-R).

The conformations the molecules assume have no bearing on their configurations (only bond breaking changes configuration): no rotation around the bonds will make one stereoisomer identical to another. Each example in Fig. 6-21 is shown in two conformations, depicted differently; the wedge-dot depictions are all eclipsed, and a 180° rotation of the central bond leads to the staggered projections shown adjacent to each. (The equivalency, ≡, implies a change in the conformation but not in the identity of the molecule.) The absolute configurations of the asymmetric carbons (2 and 3), derived from the priority rules, are shown beneath each projection formula; thus, *l-B* is s-3-phenyl-R-2-butanol, etc.

Sometimes substances have special symmetry properties, and the 2^n formula breaks down. The tartaric acids illustrate the point.

$$\text{HOOC}-\overset{*}{\underset{\underset{\text{OH}}{|}}{\text{CH}}}-\overset{\overset{\text{OH}}{|}}{\underset{*}{\text{CH}}}-\text{COOH}$$

Tartaric acid

$$\begin{array}{cccc}
\text{COOH} & \text{COOH} & \text{COOH} & \text{COOH} \\
\text{HO} \diagdown \overset{|}{\underset{C}{}} \diagup \text{H} & \text{H} \diagdown \overset{|}{\underset{C}{}} \diagup \text{OH} & \text{H} \diagdown \overset{|}{\underset{C}{}} \diagup \text{OH} & \text{HO} \diagdown \overset{|}{\underset{C}{}} \diagup \text{H} \\
\end{array}$$

(−)-Tartaric acid, mp 170°	(+)-Tartaric acid, mp 170°	Two different ways of representing *meso*-tartaric acid, mp 140°; substance is optically inactive

Racemic or *d,l*- or (±)-tartaric acid, mp 206°

FIGURE 6-22   **The three stereoisomers of tartaric acid**

Three stereoisomers are known. Two of these are an enantiomeric pair, mp 170°; the third is symmetric. The two formulas on the right in Fig. 6-22 represent the same compound. It can be readily seen that if either formula is rotated 180°, in the plane of the page, about the center of the central carbon-carbon bond, that formula can be superimposed on the other. Hence they are identical; each structure is identical with its mirror image and so possesses symmetry. The fact that a plane bisecting the carbon-carbon bond and perpendicular to it is a mirror plane also indicates that the molecule is symmetric and cannot exist in enantiomeric modifications. The compound is called *meso*-tartaric acid. Symmetric compounds containing asymmetric carbon atoms are called **meso compounds**. The meso condition almost invariably results, as here, from having the asymmetric atoms of opposite configuration and paired equivalently on each side of a mirror plane. The result is a kind of internal racemate with the two asymmetric atoms compensating each other on each side of the plane of symmetry. *meso*-Tartaric acid is a diastereoisomer of either (+)- or (−)-tartaric acid. The isomer number of tartaric acid is 3, therefore, and not $2^n = 4$.

The energy, or stability, of two enantiomers (and their racemate) is of course the same, but diastereomers, being different molecules, differ in non-bonded distances and interactions and will have different stabilities. In general the situation with two adjacent asymmetric carbons can be reduced to a projection and the relative energies of the two diastereomers assessed on grounds of steric hindrance. The molecule having the following configuration is the more stable:

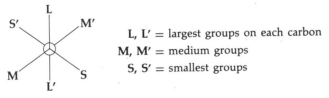

L, L′ = largest groups on each carbon
M, M′ = medium groups
S, S′ = smallest groups

Occasionally electrostatic attractions or hydrogen bonding can upset this simple steric order of stability. If the three groups are the same on each carbon, this will be the meso isomer, and it will be more stable than the *d,l* isomer.

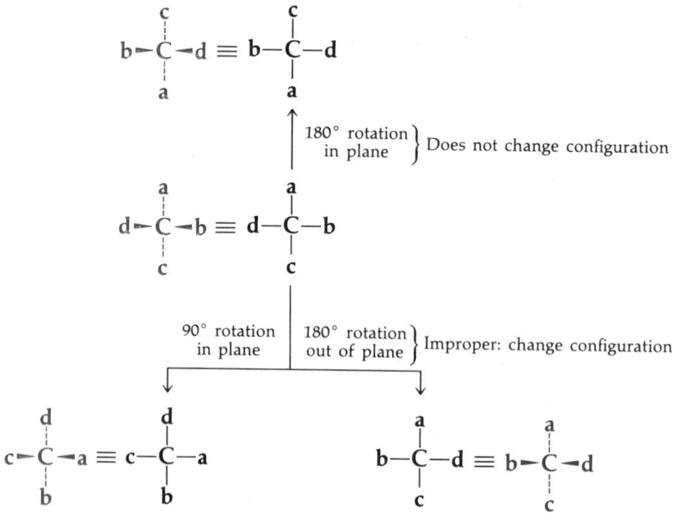

FIGURE 6-23 The Fischer projection convention

PROBLEM 6-10

Examine the relative stabilities of the isomers of tartaric acid by drawing them in projection. If one acid group were a methyl ester instead, how many stereoisomers would there be and what relative stabilities would they show?

The difficulty of drawing understandable three-dimensional formulas of molecules increases as the number of asymmetric centers increases. The classical substitute for a three-dimensional representation of an open-chain molecule is found in **Fischer projection formulas,** the use of which grew up for, and is largely (but not necessarily) used in, the descriptions of sugars. Sugars are widespread natural compounds characterized by a linear chain of carbons (three to nine) with a hydroxyl group on each carbon except one which is carbonyl: an aldehyde at the chain end or a ketone at C-2. Hence nearly every carbon is asymmetric and the number of stereoisomers is large, glucose being but one of 16 possible six-carbon aldehyde sugars (the four central carbons are asymmetric: $2^4 = 16$).

In the Fischer convention (Fig. 6-23) the backbone chain is written straight and vertical with the substituents to right and left at each central carbon.

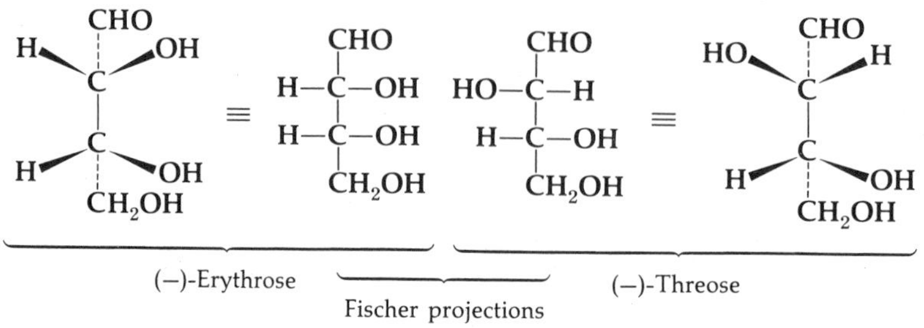

(−)-Erythrose   (−)-Threose

Fischer projections

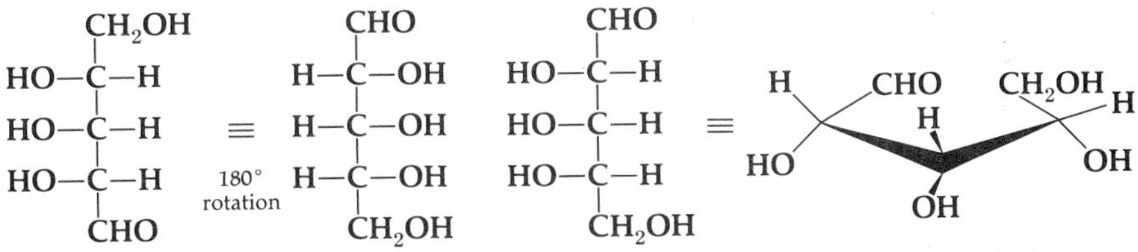

D-Ribose L-Ribose (Perspective view)

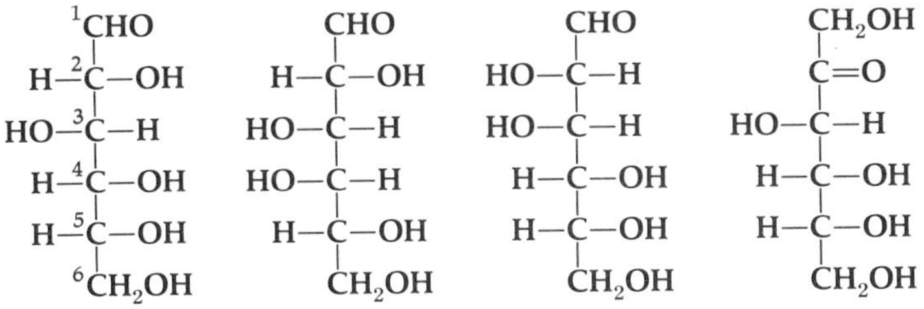

D-Glucose D-Galactose D-Mannose D-Fructose

FIGURE 6-24 **Fischer projection formulas of sugars**

It is understood that this represents the molecule fully eclipsed so that substituents at right and left angle up from the paper and the two end carbons are bent below; with longer chains than four each central bond is seen as if it were in the plane of the paper with the carbons at its ends pointing down (or as if the chain were pulled out straight). Formulas of this type cannot be manipulated as easily as perspective drawings, the only allowed change being a 180° rotation in the plane of the paper (Fig. 6-23). Interchange of any two substituents in a Fischer projection inverts the configuration. Examples of Fischer projection formulas are shown in Fig. 6-24; galactose is an epimer of glucose at C-4, mannose epimeric at C-2. Most of the sugars exist in fact as cyclic tautomers which are discussed in the next section.

An interesting set of symmetry properties is associated with the system represented by the sugarlike penta-ol shown below, which contains three potentially asymmetric centers. Carbon atoms 2 and 4 are ordinary asymmetric centers, but carbon 3 is said to be *pseudoasymmetric,* since its symmetry properties depend on the configurations about carbon atoms 2 and 4. Four stereoisomeric structures can be written for this alcohol. Two of these are symmetric or meso forms, which contain mirror planes that pass through carbon 3 and its hydrogen and hydroxyl substituents. The two optically active stereoisomers are enantiomerically related, and taken together, they represent a racemate. These relationships are shown in Fischer projection formulas as follows:

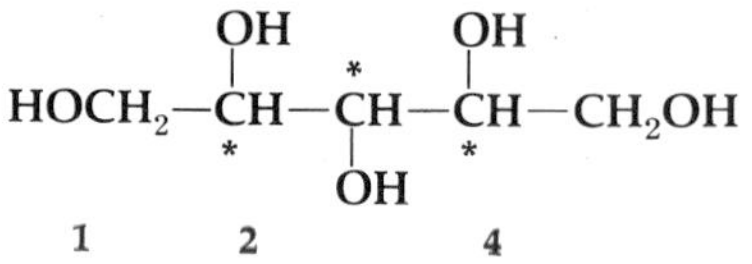

$$HOCH_2-\overset{OH}{\underset{*}{\overset{|}{CH}}}-\overset{OH}{\underset{|}{\overset{*}{\overset{|}{CH}}}}-\overset{OH}{\underset{*}{\overset{|}{CH}}}-CH_2OH$$

1   2   4

1,2,3,4,5-Pentan-penta-ol

	Represent same isomer:
CH$_2$OH CH$_2$OH CH$_2$OH CH$_2$OH	H OH

CH$_2$OH   CH$_2$OH   CH$_2$OH   CH$_2$OH

H—C—OH   H—C—OH   HO—C—H   H—C—OH

Mirror plane ----H—C—OH----HO—C—H----   H—C—OH   HO—C—H

H—C—OH   H—C—OH   H—C—OH   HO—C—H

CH$_2$OH   CH$_2$OH   CH$_2$OH   CH$_2$OH

Meso   Meso   Optically   Optically
A   B   active C   active D

Racemate

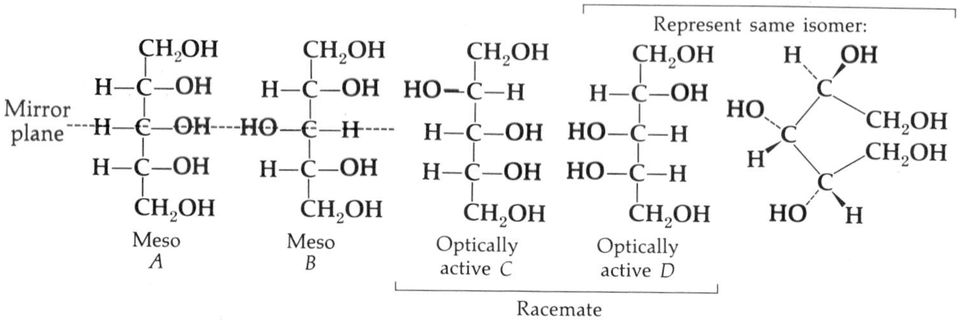

Interconversion of Conventions

In a number of cases sequence-rule designation of an asymmetric center can be made without resorting to models. In the Fischer projection, if the lowest-priority atom (usually H) is not at the bottom, interchange it with the atom (group) at the bottom and note that any interchange of two groups reverses the absolute configuration. The new configuration is now R- or S- if the top three groups are in sequence clockwise or counterclockwise, respectively (and H— is at the bottom).

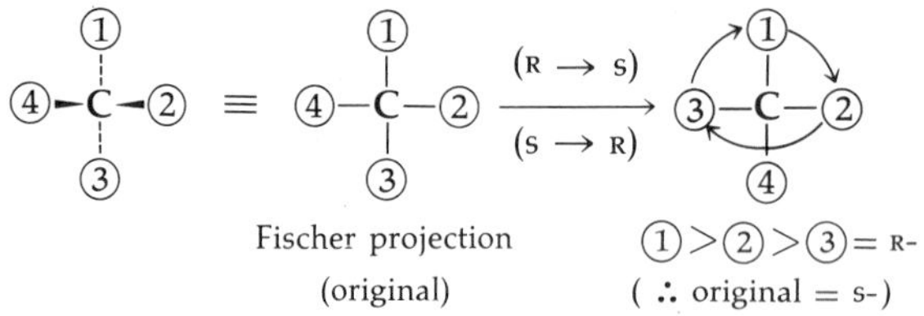

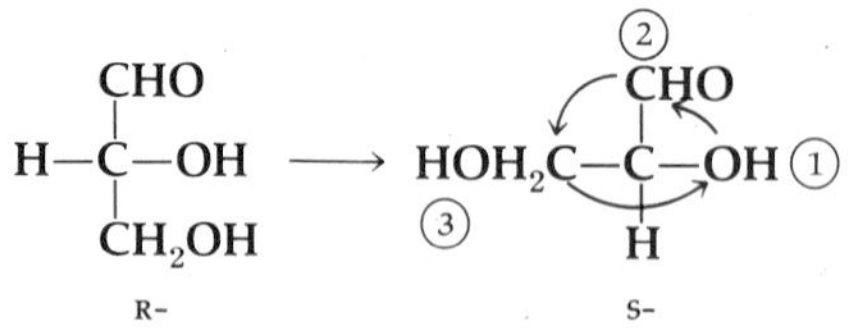

Fischer projection   ①>②>③= R-
(original)   (∴ original = S-)

R-   S-

Similarly, with ring designations, the pattern below shows absolute configuration of R- if the three groups are in sequence clockwise, S- if counterclockwise, the lowest-priority —H not being counted. The ring must be drawn to the left of the asymmetric center and the hydrogen "down"—below the ring plane.

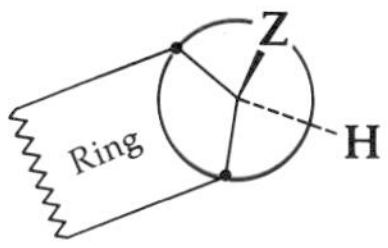

r = clockwise

s = counterclockwise

(ring drawn to the left of the asymmetric center)

Example:

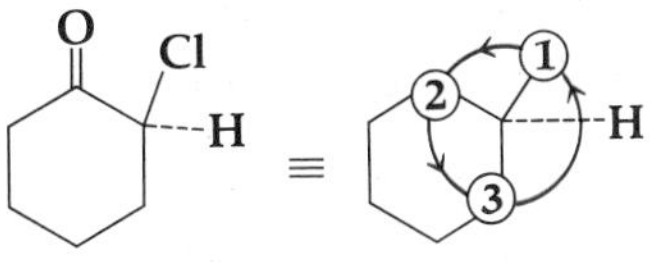

s configuration

PROBLEM 6-11

Each of the following formulas represents two diastereomeric compounds. In each case assess which is the more stable on steric grounds, draw its favored conformation in projection, and label the asymmetric carbons r- or s-.

a 3-Chloro-2-hexanol   **c** 3,4-Dihydroxy-2,6-octadiene

b 2,3-Dimethyl-4-hexenoic acid   **d** α,β-Diphenyl-butyronitrile

PROBLEM 6-12

Draw Fischer projections for all the stereoisomers of 2,3,4-trihydroxypentanoic acid and label each asymmetric carbon r- or s-.

6-11 STEREOISOMERISM IN CYCLIC COMPOUNDS

The same stereochemical principles apply to both cyclic and open-chain compounds. A single substituent on a saturated carbocyclic ring (e.g., methylcyclohexane) cannot make a molecule asymmetric, whereas two substituents properly situated can give rise to stereoisomerism. For instance, the isomer number of 1,2-cyclopropanedicarboxylic acid is 3, and all three compounds are known.

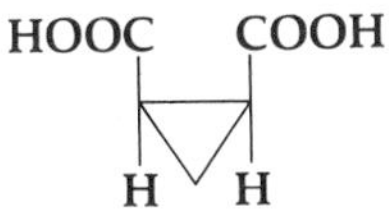

Meso or cis isomer, mp 130° 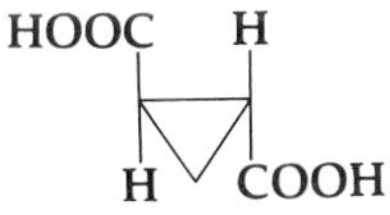   Optically active trans (−) isomer, mp 175°

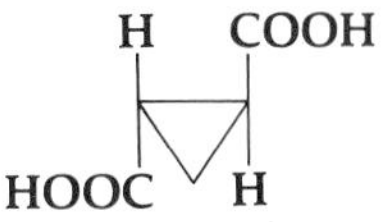

Optically active trans (+) isomer, mp 175°

The major interest lies in cyclohexane derivatives, for which we can discuss both conformation and configuration. Examination of symmetry and isomer number in cyclohexanes need not be confused with considerations of conformation, however, for they are simply examined first on the flat ring formula. This is possible since matters of configuration are not affected by conformation, and any molecule can pass through the planar conformation, hence cannot have less symmetry than that conformation.

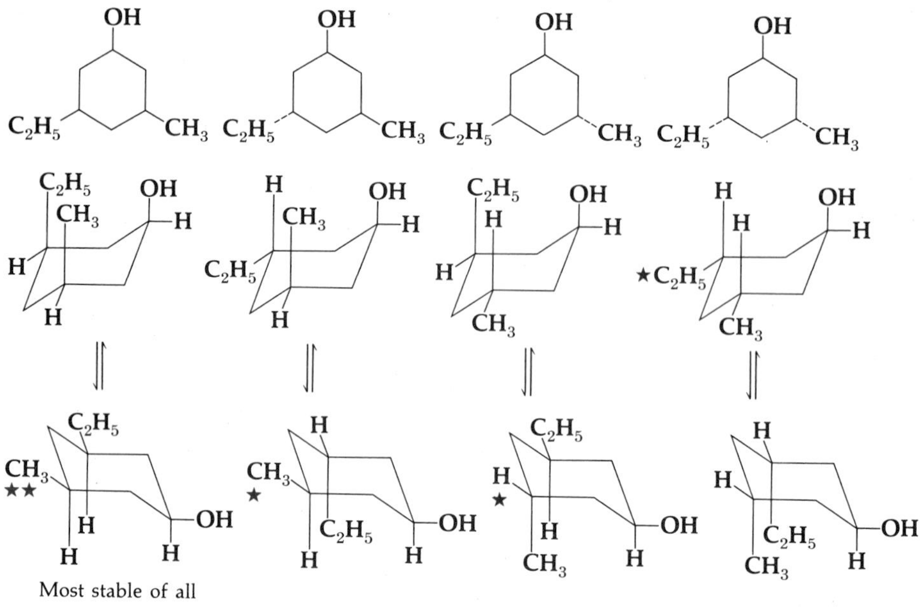

FIGURE 6-25   Analysis of 3-methyl-5-ethylcyclohexanol

The complete analysis of substituted cyclohexanes takes this course:

1 Determine the number of possible racemates $= 2^{n-1}$ where n is the number of asymmetric ring carbons.

2 Write these out with flat hexagon formulas and substituent bonds dotted if below the plane, solid if above. Select one substituent to have the same configuration in each structure.

3 Mark each as symmetric or asymmetric by seeking planes of symmetry—those with planes will be meso; the others, d,l pairs (racemates).

4 Check the asymmetric ones to see if any are enantiomers of each other, and use only one of such a pair.

5 Write the two possible chair forms of each, and decide which is the preferred conformation (via number of axial groups, Fig. 6-13).

6 Compare all racemates to find the most stable—that with the most equatorial substituents.

As an example, consider 3-methyl-5-ethylcyclohexanol (Fig. 6-25). There are three substituents on three asymmetric carbons, hence $2^{n-1} = 2^2 = 4$ possible racemates. These are written out by holding one substituent in the same configuration on each ring and varying the others systematically; this yields one set of diastereomers, and their enantiomers are implied as the comparable set with the chosen substituent in the opposite configuration. In Fig. 6-25 the hydroxyl was drawn as above the ring plane in each formula. This yields four formulas, all diastereomers, the enantiomers of which would be the analogous four more with —OH below the ring plane. None of these is found to have a plane of symmetry. Next we draw each in its two chair conformations; there is no need to do this for their enantiomers which are the same in every relevant conformational particular. The more stable of each pair is assessed and starred; of course, cis-diaxial substituents are worse than simply two axial substituents (page 193). The all-cis diastereomer (A), with all substituents equatorial, is the most stable stereoisomer of this group; its enantiomer is of course the mirror image with the same energy and conformation and is not pictured. Using Table 6-2 the other three racemates could be placed in order of increasing strain energy.

In 3,5-dimethylcyclohexanol, examination of the four corresponding diastereomers shows two with planes of symmetry, hence meso, and the other two asymmetric but actually enantiomers of each other as a result of the 3- and 5- substituents being equal. Hence there are a total of only four, not eight, stereoisomers:

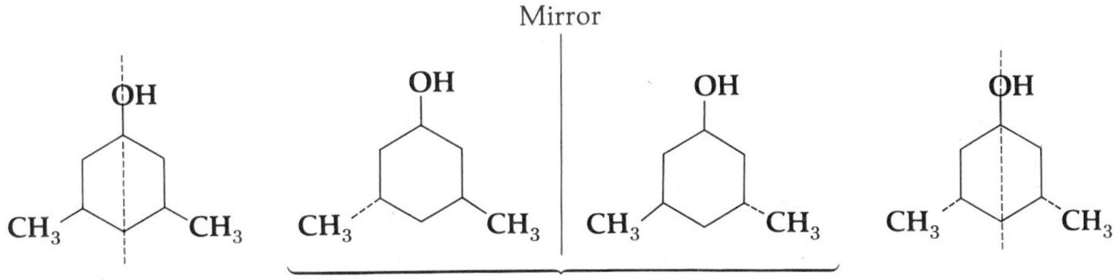

Mirror

| Meso | Enantiomeric = d,l pair | Meso |

Sugars

The sugars in fact are cyclic, not open-chain, molecules and many common ones have six-membered rings; these occupy chair conformations and are subject to the same conformational analysis. The sugars represent a prime example of ring-chain tautomerism (page 165), the chains shown in Sec. 6-10 in equilibrium with the cyclic forms shown below. The equilibrium strongly favors the ring, which in turn seeks the most stable chair conformation. In all of this tautomeric and conformational equilibration, however, the *configurations* at the asymmetric carbons (depicted in the chains in Fig. 6-25) are totally unaffected and so are transferable from the chain formula to the ring without change. A new asymmetric center is created at the carbonyl carbon (C-1), however, and if equilibration is complete, the configuration will be that with the more stable, or equatorial, hydroxyl. Glucose is the most stable six-carbon sugar since all of its substituents are equatorial.

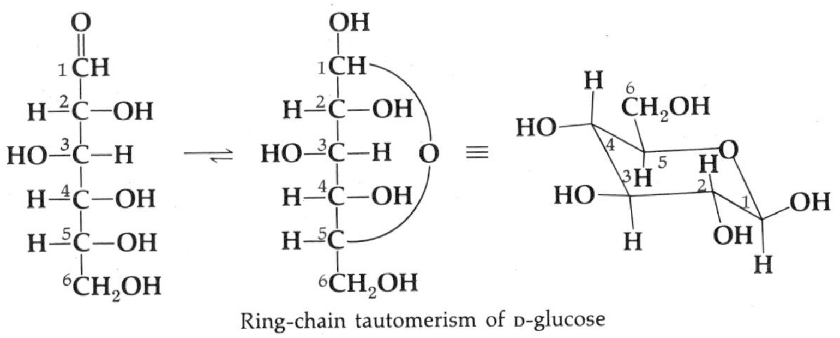

Ring-chain tautomerism of D-glucose

Transfer of the correct configurations from one representation to the other is best accomplished by examining each single bond between two asymmetric carbons in projection, as shown for the 2-3 bond in glucose (C-3 in front). Practice in manipulating representations is readily afforded by generating the possible sugar isomers (that is, 16 for glucose, etc.).

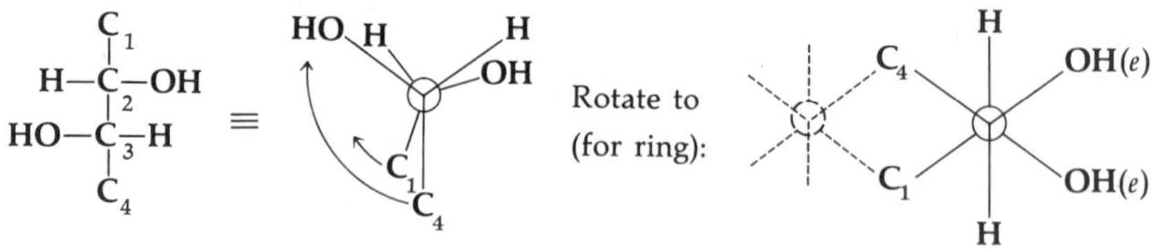

From Fischer projection   Chair ring projection

PROBLEM 6-13

For each example: (1) delineate the number and kind of stereoisomers, (2) draw the most stable conformation of each, and (3) determine the most stable of the stereoisomers.

a 3-Hydroxy-5-methylcyclohexanecarboxylic acid
b 3-Hydroxy-3-methylcyclohexanecarboxylic acid
c 4-Hydroxy-4-methylcyclohexanecarboxylic acid
d 1,3,5-Cyclohexanetriol
e 1,2,3-Cyclohexanetriol

6-12 MOLECULAR ASYMMETRY

In a few uncommon instances, a compound may be optically active even though its molecule has no asymmetric atoms. The most important way in which this can happen arises from the possibility that two rotational conformers might be each asymmetric and incapable of interconversion at room temperature because of too high an energy barrier.† No *saturated* compound with such a high barrier to free conformational interconversion has yet been observed, but a number of unsaturated cases are known. The hindered biphenyls illustrate this idea very clearly, as exemplified in the top row of Fig. 6-26. If the two ortho positions on each ring are differently substituted, and with groups too large to pass each other (steric hindrance) when the rings are rotated about the central bond, then two asymmetric rotational (torsional) isomers can exist.

---

† Conformers *A* and *B* of *n*-butane in Fig. 6-8 are asymmetric objects, mirror images of each other, but they are easily interconverted by rotation and cannot be isolated as enantiomeric isomers, even at low temperature, owing to the low barrier to rotation.

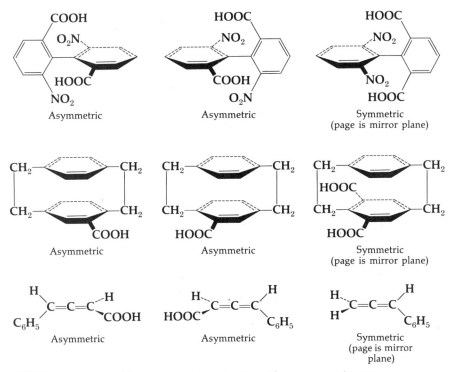

FIGURE 6-26 **Asymmetric and symmetric molecules with no asymmetric atoms**

These isomers will be mirror images and exhibit equal and opposite optical rotations. The third example in the top row is a structural isomer with a plane of symmetry and so cannot be asymmetric even though its rotation is equally restricted.

In the other rows of Fig. 6-26 are two comparable examples. Rotation is restricted in the aromatic molecules since neither benzene ring can turn over through the large ring. In fact the molecule is rigidly held with the rings face to face and somewhat strained even without turning the benzene rings. The allenes offer another common case of molecular asymmetry of this kind (sometimes called **torsional stereoisomerism**), rotation here being restricted by the resistance of double, not single, bonds to rotation. Remember that double-bond geometry requires the pairs of substituents at each end to define perpendicular planes so that the situation is parallel to that of the four substituents on the tetrahedral carbon. In fact an allene would become a tetrahedral carbon if we could compress all three allenic carbons together into one.

PROBLEM 6-14

Which of the following structures are symmetric and which are asymmetric?

a

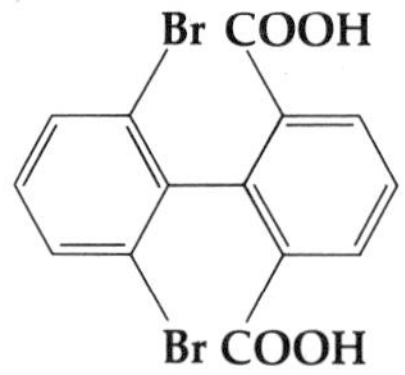

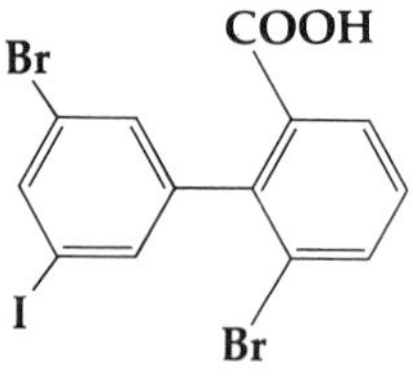

b

c

d

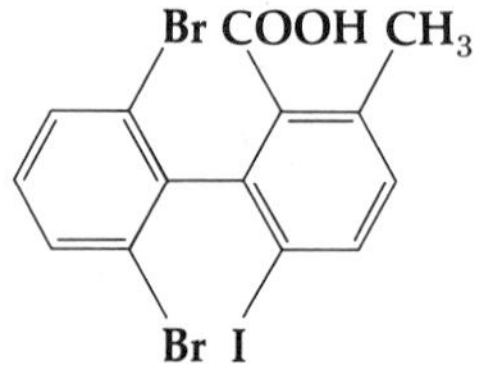

e $CH_3CH=C=CHCH_3$

f $CH_3CH=C=C(CH_3)_2$

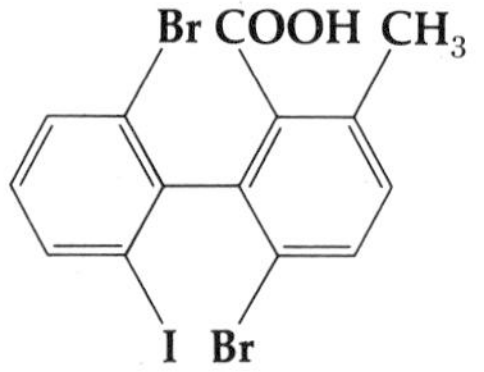

g $CH_3CH=C=CHC_2H_5$

6-13 SEPARATION, DETERMINATION, AND
INTERCONVERSION OF STEREOISOMERS

The separation of diastereomers is no different from separation of any similar but different molecules and can be affected by distillation, crystallization, chromatography, etc. The separation of enantiomers (resolution) can only be effected in three ways, each requiring application of an external chiral standard:

1 A few racemic mixtures crystallize in such a way that molecules of like configuration gather into one kind of *visibly* asymmetric crystal and enantiomeric molecules into a second, the first crystal being the mirror image of the second. These crystals may be sorted by hand and collected into two piles, one rich, or pure, in one enantiomer and the other in the second. Pasteur performed the first resolution of this sort in 1848 in a separation of the (+) and (−) forms of sodium ammonium tartrate, but the opportunity is rare since racemates usually crystallize in only one racemic crystalline form.

2 In the second method, enzyme systems (either inside or outside an organism) are allowed to consume or chemically modify one enantiomer of a pair, while the other is rejected. A case has been recorded in which a racemate was fed to a dog which metabolized one enantiomer and passed the other in its urine. This procedure has not been widely used to resolve enantiomers. Most cases have been accomplished with isolated enzymes.

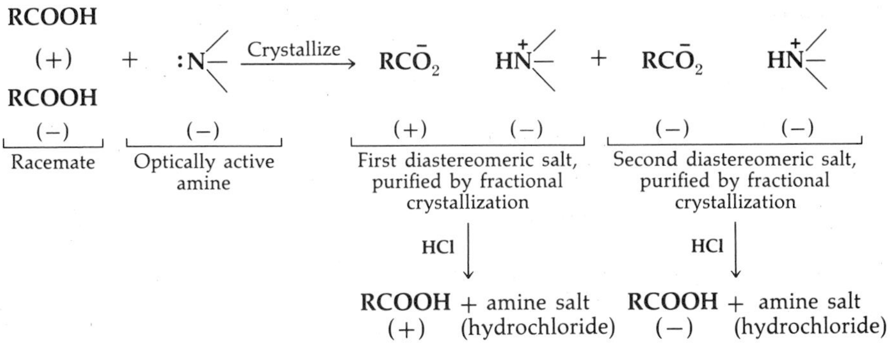

RCOOH

(+)   +   :N— $\xrightarrow{\text{Crystallize}}$ RCŌ₂   HN⁺—   +   RCŌ₂   HN⁺—

RCOOH

(−)	(−)	(+)	(−)	(−)	(−)
Racemate	Optically active amine	First diastereomeric salt, purified by fractional crystallization		Second diastereomeric salt, purified by fractional crystallization	

HCl ↓   HCl ↓

RCOOH + amine salt   RCOOH + amine salt
(+)   (hydrochloride)   (−)   (hydrochloride)

FIGURE 6-27   **Resolution of a racemic acid**

3   The third, and by far most generally useful, method involves a chemical procedure. A racemate is allowed to react with a second, standard asymmetric molecule (Sec. 6-7). This procedure essentially creates, from two enantiomers + and − , two diastereomers + + and − +, which are now separable by common physical means. This requires a chemical reaction to attach the standard and, after separation, its reverse to remove the pure enantiomers from the standard compound.

The simplest and most common reactions used in the third procedure involve salt formation with optically active amines (bases) or acids. Both classes of compounds are abundantly available in nature in optically active form. If the racemate is an acid, an optically active amine such as quinine, brucine, or strychnine is used to resolve the enantiomeric pair. The racemate is mixed with the amine, and *diastereomerically related and optically active salts crystallize*. Since these two salts have different solubility properties, they can be separated by fractional crystallization to give homogeneous substances. Each of the salts is treated with mineral acid (e.g., HCl) to regenerate the original organic acids, now in optically active forms. If the separation is meticulously carried out, **optically pure** (enantiomerically homogeneous) stereoisomers may be prepared.

If the starting racemate is an amine, an optically active acid is used as resolving agent. Compounds such as (+)- and (−)-tartaric acid and its derivatives are frequently employed. Figure 6-27 traces the steps involved in the resolution of a racemic carboxylic acid.

If the racemate is neither an acid nor a base, a chemical "handle" must be attached to the molecule by an appropriate reaction. After the racemate is resolved, the "handle" is removed. Derivatives of the racemate are often made so as to contain free carboxylic acid groups on the derivative portion of the molecule to allow resolution in the same fashion.

Absolute and Relative Configurations

No purely chemical method exists for determining the **absolute configuration** of an optically active molecule, i.e., whether a given enantiomer is R- or S-. However, optically active compounds may be interconverted by chemical routes without disturbing their asymmetric centers, and as a result, series of compounds whose configurations are known in relation to one another have accumulated.

Chemical means are also available for determining the **relative configurations** of two or more asymmetric carbon atoms within the same molecule. Commonly, for example, ring formation between two nonadjacent substituents already attached to one ring can demonstrate their relative configuration as cis (page 195), in analogy with the maleic-fumaric acid case for determining geometrical isomerism (Sec. 6-2). Also spectroscopic evidence can often be used to distinguish cis and trans substituents (Chap. 7). If the absolute configuration is determined for any single member within a series of compounds of known relative configurations, absolute configurations can then be assigned to all compounds of the series.

Lacking means to determine absolute configurations, investigators before 1951 used *assumed configurations* for the enantiomeric glyceraldehydes as a standard. The (+) isomer was assumed to have the structure shown, and this particular arrangement of atoms was named the D configuration of glyceraldehyde.

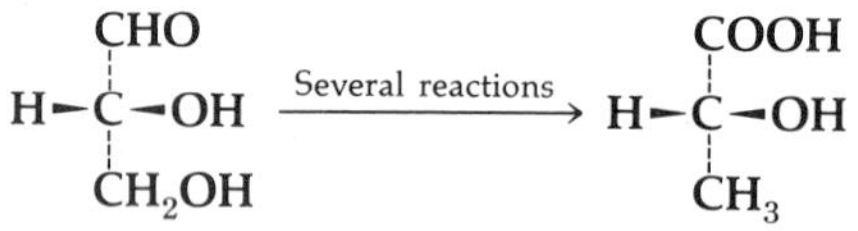

$$
\begin{array}{ccc}
\text{CHO} & & \text{COOH} \\
| & & | \\
\text{H}-\text{C}-\text{OH} & \xrightarrow{\text{Several reactions}} & \text{H}-\text{C}-\text{OH} \\
| & & | \\
\text{CH}_2\text{OH} & & \text{CH}_3 \\
\text{D-(+)-Glyceraldehyde} & & \text{D-(-)-Lactic acid}
\end{array}
$$

The *relative configuration* of natural lactic acid has been determined by converting (+)-glyceraldehyde into (−)-lactic acid, *without breaking any bonds to the asymmetric carbon atom.* Observe that the asymmetric carbon, therefore, must still maintain the same absolute configuration but that the sign of rotation has reversed.† Similarly, relative configurations of many compounds have been established. An imperfect nomenclature system was devised for families of asymmetric compounds, like sugars (Sec. 6-10) and amino acids (Chap. 25), which designated them as D- or L- according to the configurational similarity of one asymmetric carbon with D- or L-glyceraldehyde defined in some arbitrary way for each family.

This system becomes ambiguous in designating configurations of any but the simplest molecules. Different authors, on the basis of different selections of groups similar to one another, have designated (+)-tartaric acid as both D and L. When molecules contain two or more asymmetric centers, authors usually resort to trivial names for diastereomers, except where the terms cis, trans, meso, D, and L suffice. The notation system outlined in Sec. 6-9 is the preferred one since it is systematic and relates all compounds with only one set of rules. All the correlations of configurations of asymmetric molecules with

---

†This observation demonstrates that optical rotation is no guide to the chirality of even very similar optically active molecules (page 202).

the assumed configuration shown for glyceraldehyde can be put in terms of the R- and S- notation; D-glyceraldehyde is R-glyceraldehyde.

The chances were even that the glyceraldehyde convention would couple the sign of rotation with the correct absolute configuration. Fortunately, the selection was demonstrated to be correct by experiment in 1951. Normally, X-ray diffraction photographs of crystalline enantiomers are identical. However, by the use of special X-rays, the absolute configuration of the sodium rubidium salt of (+)-tartaric acid was determined. Now absolute configurational assignments can be given to compounds whose configurations have been established in relation to (+)-tartaric acid, and virtually all optically active compounds have now been so correlated.

Optical rotatory dispersion **(Sec. 6-8) of asymmetric cyclic ketones has been correlated with their absolute configurations and may now be used to determine absolute configuration in new compounds quite easily. In principle this tool can be extended to other functional groups as well. For a ketone** *in a conformationally fixed (cyclic) structure* **we divide the space around it into eight octants by looking down the carbonyl as shown below. Four octants are in front of the oxygen (toward the observer) and usually can be ignored. The four behind it usually contain all the atoms in the molecule, the view of these four as seen**

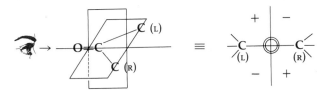

by the observer being shown at the right. Such a view of a ketone molecule can be examined easily with a model of its preferred conformation. The octant containing the bulk of the atoms in this projection determines the sign (+ or −) of the Cotton effect in the ORD.† If the opposite sign is actually observed, the molecule has the opposite (enantiomeric) absolute configuration from that of the model used.

This Octant Rule is most easily applied to six-membered ring ketones without using models. The conformation is drawn as shown at left so that its octant projection is that shown at right. When the substituents are put in one absolute configuration, the Cotton effect is predicted by the sign of the most occupied octant. The example is the natural oil, dihydrocarvone, which has a positive Cotton effect and the absolute configuration shown.

---

†Actually not only the bulk but the proximity and electrical qualities of the atoms enter in, close atoms weighing more than far ones and halogens more than carbons, but most cases are obviously + or − without these second-order considerations.

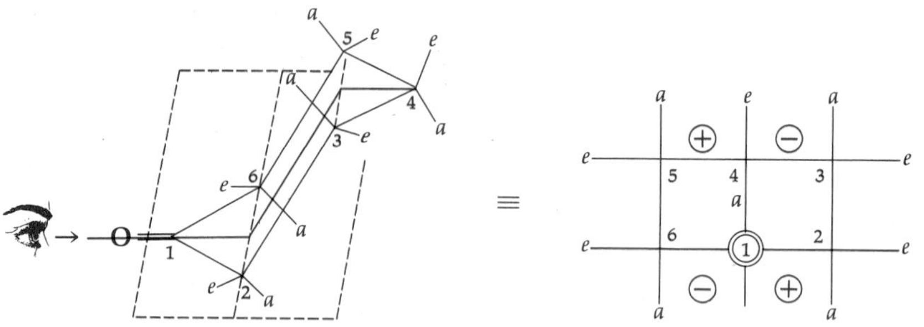

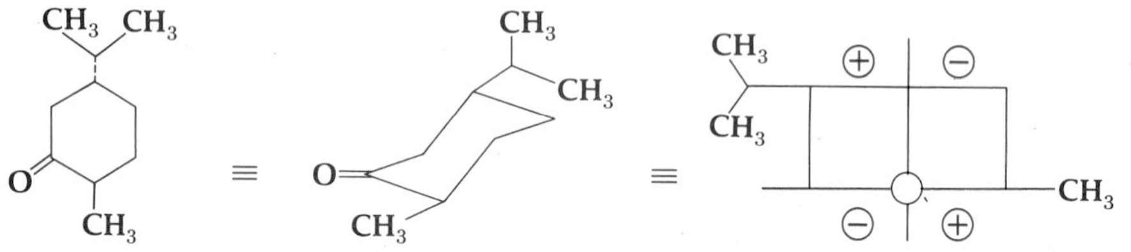

Dihydrocarvone

Chemical Changes in Configuration

In order to convert an asymmetric carbon to its enantiomer, it is necessary to break and remake bonds. It can usually be effected by breaking one bond and allowing the remaining three to become coplanar (as an sp^2 carbon) with a plane of symmetry. The group removed is then replaced from the opposite side to produce the enantiomer (**inversion**); returning the group to the same side results in **retention** of configuration. Such a chemical reaction is illustrated for 2-bromobutane:

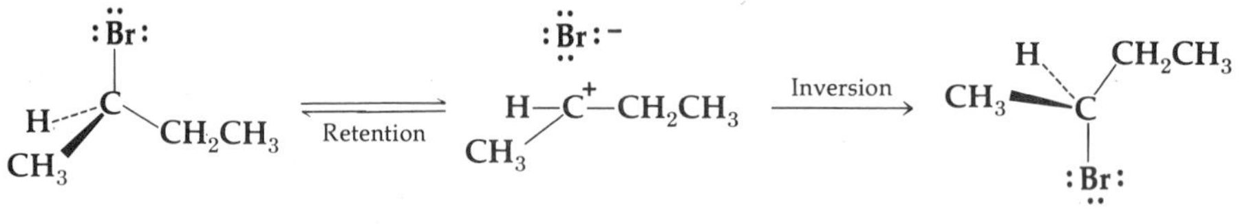

R enantiomer   Symmetrical   s enantiomer

Since the symmetrical intermediate (a carbonium ion) will re-bond equally well from either side, both enantiomers are formed equally in the reaction. In molecules with only one asymmetric center the product is the racemate, and this conversion of one enantiomer to its racemate is called **racemization**. If several asymmetric centers are present but *only one* is involved in the reaction, the process is an **epimerization**, converting one optically active molecule to its epimer, also optically active and epimeric (inverted) at the site of the reaction. Since the rest of the molecule is still asymmetric and unchanged in the reaction, it may influence the epimerization by favoring one side over the other for re-bonding (favoring *retention* or *inversion*, usually by steric hindrance on one side of the symmetrical intermediate). If the remaining asymmetry of the molecule has little influence at the site of epimerization, a mixture of the starting compound and its epimer will be formed and needs to be separated.

Finally, if several asymmetric centers are present, racemization of an optically active compound can *only* occur if the reaction inverts *every* asymmetric center in the molecule (a rare circumstance).

   With the discussion of reactions and their mechanisms in subsequent chapters we shall see what kinds of reactions are capable of racemizing asymmetric atoms. For the present it will suffice to look at one other simple example, which illustrates a very widespread mode of racemizing asymmetric carbons via keto-enol tautomerism (Sec. 5-4). When α-methylbutyric acid is heated, its optical activity is slowly lost, and racemic material is ultimately obtained. This transformation involves equilibration of each enantiomer through a planar and symmetrical enol intermediate in which hydrogen on the α-carbon has shifted (tautomerized) to the carbonyl oxygen of the carboxyl group. In shifting back from oxygen to carbon, hydrogen comes in with equal probability from either above or below the plane of the enol double bond to give equal amounts of the two enantiomers.

R isomer   Planar and symmetric enol   s isomer

   The racemization of optically active hindered biphenyls (Sec. 6-12) can also often be achieved by heating. If the steric hindrance of the ortho substituents to rotation represents a barrier of less than about 15 to 20 kcal/mole (room-temperature thermal energy), then enantiomers could only be separated at lower temperatures. If the barrier is somewhat above 20 kcal/mole, the separated enantiomers would racemize on heating to a higher temperature, at which the increased thermal motion energy is adequate to exceed the rotation barrier. A whole range of such biphenyls has been examined.

6-14   SUMMARY

Conformation

   Rotation about the single bonds of molecules allows them to have many shapes, or conformations, the actual conformation(s) taken up being that (or those) with minimum strain energy. The strain energy of any conformation is the sum of angle-bending, rotational, and steric strain. Simple acyclic molecules with one key single bond are best assessed from projection formulas down that bond. Conformational analysis is mostly applied to six-membered rings, which can exist in two chair conformations, the one with the least axial substituents preferred. Other conformations, such as boat shapes, represent only a negligible proportion of the total molecule population, owing to their higher energy (≥ 6 kcal/mole). The chair conformation is depicted in two ways, perspective and projection, most commonly the former.

Cyclohexane derivatives

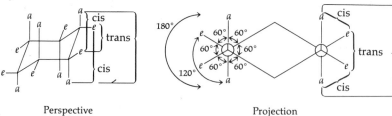

| Perspective | Projection |

Stereoisomerism

The properties of kinds of isomers may be tabulated as follows.

1 Structural isomers   } Asymmetry
2 Geometrical isomers = cis and trans double-bond isomers   } unnecessary
3 Stereoisomers: Number = 2^n (n = number of asymmetric atoms)   Asymmetry
   a *Enantiomers:* $[\alpha] \neq 0°$ necessary

      Mirror images

      All asymmetric centers inverted

      Identical properties except optical activity

      Equal and opposite ($+$, $-$) values of $[\alpha]$

   b *Racemates (racemic mixtures):* $[\alpha] = 0°$

      Equal mixture (50:50) of enantiomers

      Number = 2^{n-1}

   c *Diastereomers:* $[\alpha] \neq 0°$

      Not mirror images

      Some but not all asymmetric centers inverted

      Different properties, including optical activity

   d *Epimers:* $[\alpha] \neq 0°$

      Diastereomers differing at only one asymmetric center

   e *Meso compounds:* $[\alpha] = 0°$

      Even number of asymmetric centers paired and internally compensated in optical activity across a plane of symmetry

    A molecule is asymmetric if it is not superimposable on its mirror image. It is usually easier to look for a plane of symmetry; if such a plane can be found, *in any accessible conformation* the molecule is symmetrical and has no optical rotation. Ring compounds may most easily be examined in flat conformations for this purpose. A compound whose molecules contain asymmetric atoms will be

1 *Asymmetric* and optically active or
2 *Racemic* and capable of resolution into optically active enantiomers or
3 *Meso*, optically inactive and exhibiting a plane of symmetry

A set of the stereoisomers of a single structural isomer may be analyzed in terms of the number of stereoisomers and their mutual relationships as enantiomeric and diastereomeric. The favored conformations of each may then be evaluated and the order of stabilities of the various diastereomers determined. (See page 214.) Epimerization reactions which go to equilibrium will yield the more stable epimer in a predominance determined by the energy difference between the epimers (cf., Table 6-2).

In general the considerations of conformation and stereochemistry here allow an understanding of the isomers and shapes of molecules and their energy preferences. This becomes of major importance in assessing the course and consequences of chemical reaction in the succeeding chapters. It is in any event an enormous step forward in understanding structure to expand our understanding from simple structural formulas to detailed three-dimensional shape, as this example vividly demonstrates:

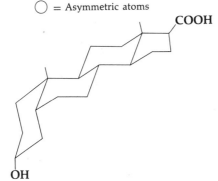

Simple structure
$2^8 = 256$ stereoisomers possible
◯ = Asymmetric atoms

Natural lithocholic acid from bile and gallstones
(absolute configuration)

Conformation of lithocholic acid

PROBLEMS

6-15 Write out structures for the following substances.

    a The three *conformations* of 2-methylbutane with end-on formulas. Which formulas are superimposable?

    b Three conformations of 2-butanone with three-dimensional formulas (the groups attached to any unsaturated carbon atom all lie in one plane).

    c The two conformations of *cis*-4-isopropyl-1-methylcyclohexane.

    d The two conformations of *trans*-4-isopropyl-1-methylcyclohexane.

    e The two conformations of *cis*- and the two conformations of *trans*-1,3-dimethylcyclohexane.

6-16 Write out structural formulas for the following compounds.

  a *cis*-2,3-Dichloro-2-pentene   d *trans*-4-Methyl-3-octene
  b *trans*-2-Phenyl-2-butene   e *anti*-Oxime of α-naphthaldehyde
  c *cis,trans*-1,4-Diphenyl-1,3-butadiene f *cis*-1,3-Dichlorocyclobutane

6-17 State which of the following pairs of compounds you expect to be the more stable and why.

  a *cis*- or *trans*-2-butene
  b *cis*- or *trans*-2-phenyl-2-butenoic acid
  c *cis*- or *trans*-1,4-diphenyl-cyclohexane
  d *cis*- or *trans*-1,3-dibromocyclohexane
  e *cis*- or *trans*-1,2-dimethylcyclopentane
  f *cis*- or *trans*-1-methyl-4-butylcyclohexane

6-18 Write out three-dimensional formulas for all the isomers of the following compounds. Indicate the enantiomeric and the diastereomeric relationships. Indicate all isomers that are optically active.

a $CH_3-CH-CH-CH_3$
        $\;\;\;\;\;\;\;\;|\;\;\;\;|$
        $\;\;\;\;\;\;\;\;Cl\;\;\;Cl$

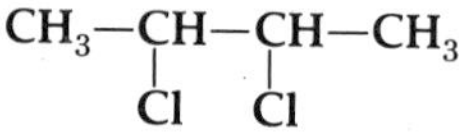

b

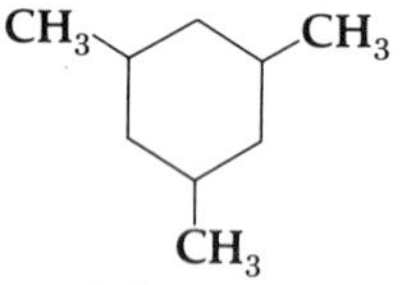

c

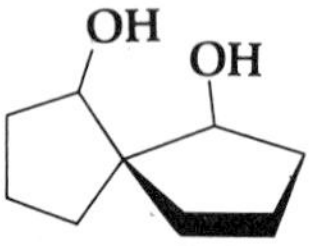

d $CH_3CH{=}CHCH_2CH_2CH{=}CHCH_3$
e $CH_3CH{=}CHCHCH_2CH{=}CHCH_3$
                    $\;\;\;\;\;\;\;\;|$
                    $\;\;\;\;\;\;\;\;OH$

f $CH_3CHCHCHCH_3$
       $\;\;\;\;|\;\;\;\;|$ ... Cl above; $C_6H_5\;\;\;C_6H_5$

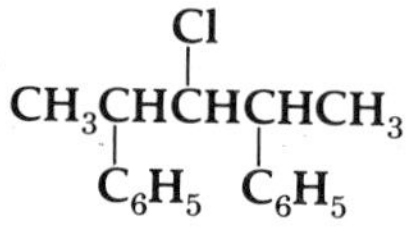

g

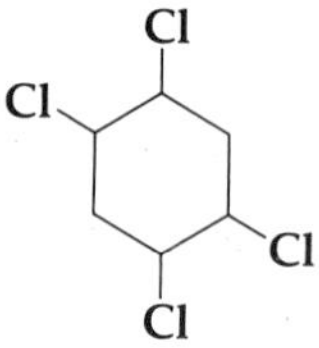

h

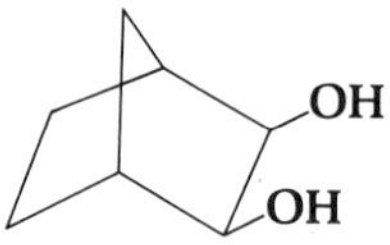

6-19 Write out three-dimensional structures for *all the unique ways* of substituting two carboxyl groups on the two aromatic rings of the compound shown. State which are optically active. Indicate which structure contains a center but not a plane of symmetry.

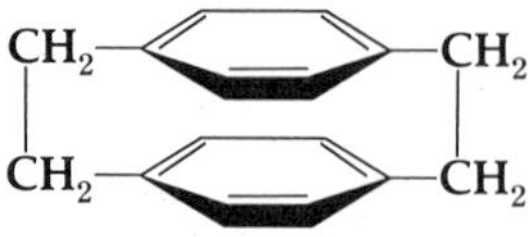

6-20 Define the following terms:

  a Racemization   d Center of symmetry   f Diastereoisomerism
  b Mirror plane   e Resolution   g Equatorial bond
  c Epimerization

6-21 Write out three-dimensional formulas for the following compounds:

a **CHClBrF** (R configuration)

b $CH_3CH_2CH_2-\overset{\overset{\displaystyle CH_3}{|}}{\underset{\underset{\displaystyle OH}{|}}{C}}-CH_2CH_3$ (s configuration)

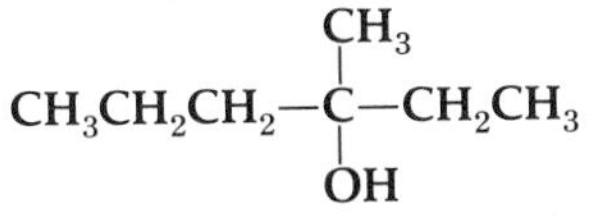

c 3-R-4-s-Dimethylhexane

6-22 Write systematic configurational names for the following structures:

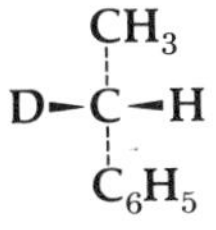

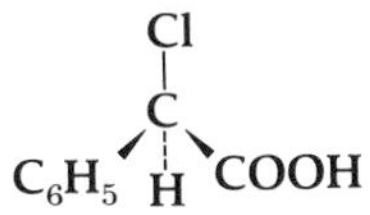

a $H-\overset{\overset{\displaystyle OH}{|}}{\underset{\underset{\displaystyle CH_3}{|}}{C}}-COOH$ c $D-\overset{\overset{\displaystyle CH_3}{|}}{\underset{\underset{\displaystyle C_6H_5}{|}}{C}}-H$

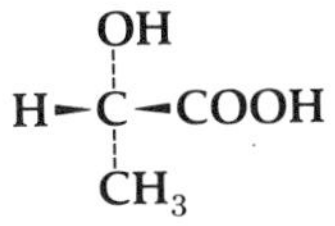

b $\underset{C_6H_5 \ \ H}{\overset{\overset{\displaystyle Cl}{|}}{C}}COOH$ d $\begin{array}{c} H-\overset{\overset{\displaystyle CH_3}{|}}{C}-C_6H_5 \\ C_6H_5-\underset{\underset{\displaystyle CH_3}{|}}{C}-H \end{array}$ Fischer formula

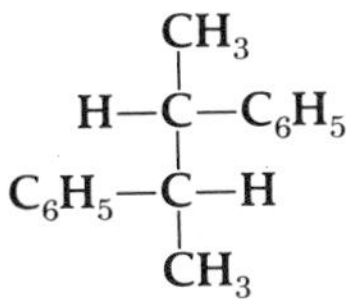

6-23 Differentiate the terms *structure, configuration,* and *conformation.* In ordinary molecules, which of these terms apply to states subject to interconversion without making or breaking bonds?

6-24 What are the isomer numbers for each of the following structural formulas?

a $CH_3CHClCHOHCH_3$ d $OHCCHOHCHOHCHOHCHOHCH_2OH$

b $CH_3\underset{\underset{\displaystyle O}{\diagdown \ \diagup}}{CH-CH}C_2H_5$ e $HOOCCHClCHClCHClCOOH$

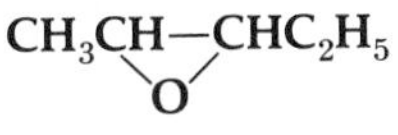

c $CH_3CHOHCH=CHCH_3$

6-25 a How many stereoisomers are possible for 2-bromo-4-hydroxycyclohexane-carboxylic acid?

   b Indicate the breakdown of stereoisomers into enantiomers, racemates, and diastereomers.

   c Draw the two enantiomers of the most stable diastereomer in perspective, and point out why it is the most stable.

   d Draw the 2-epimer of the most stable diastereomer in perspective and indicate the cis-trans relation of the pairs of ring substituents (**Br, OH, COOH**).

   e Draw the other, less stable chair conformation possible for this 2-epimer.

6-26 The most stable acid above (Prob. 6-25) was synthesized in the laboratory from simple symmetrical starting materials. It was neutralized with an asymmetric base in an effort to resolve the acid. Explain how this works in a stereochemical sense.

6-27 The acid in Probs. 6-25 and 6-26 can be chemically converted to 2-bromo-cyclohexan-4-one carboxylic acid without inverting the configuration at carbons 1 and 2. One resolved enantiomer of the most stable diastereomer was converted to this ketone and its optical activity (rotation) measured at different wavelengths and found to be as follows. Draw in perspective the absolute configuration of this enantiomer and show in detail how you arrived at the conclusion.

Wavelength: λ	Rotation: $[\alpha]$
587	$-10°$
500	$+62°$
400	$+98°$
350	$+160°$ (max)
300	$+5°$
250	$-138°$

6-28 **a** How many stereoisomers of 2,5-dimethylhexanedioic acid are possible?
   b Characterize and name them in some definitive way.
   c How many isomers are possible for 2,5-dimethyl-3-hexendioic acid?
   d Is the situation different with the monoesters of each and, if so, explain.

6-29 A derivative of the steroid, cholic acid, found in human gallstones, is shown below:

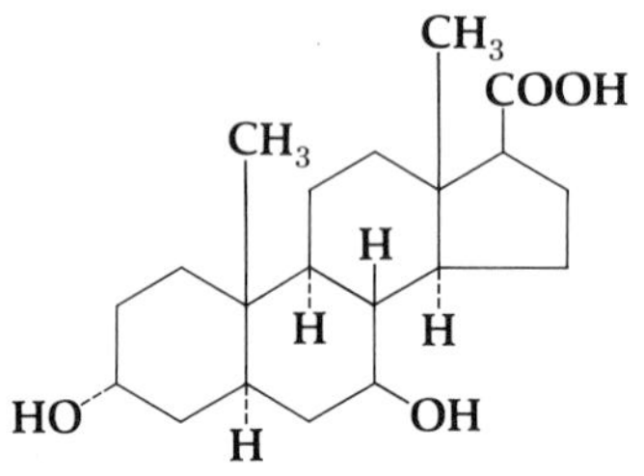

   a Draw the molecule in its most stable conformation. Does it have more than one?
   b Are the epimers at the carbons bearing —OH more or less stable?
   c How many stereoisomers are possible if the asymmetric carbon configurations are *not* specified as they are in this representation?

6-30 **a** Do a stereochemical analysis of 4-methyl-3-hydroxy-cyclohexanecarboxylic acid (number and kind of stereoisomers, the preferred conformations of each drawn in perspective, indication of most stable isomer).
   b Under reaction conditions which convert γ-hydroxy acids like these to lactones

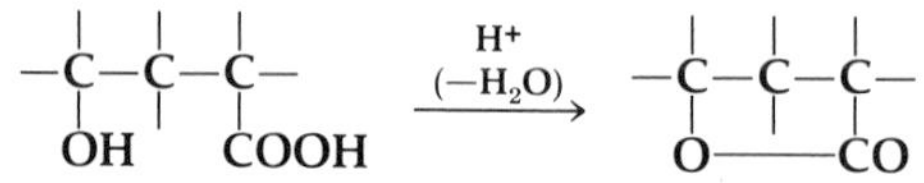

   the most stable isomer gives a lactone. Draw this in perspective. Some of the isomers will not give lactones in this reaction; which ones do and which do not?

c The most stable isomer can be converted to the corresponding 3-ketone (4-methyl-cyclohexan-3-one-1-carboxylic acid). Draw the possible enol tautomer(s) of this ketone.

d The enol-ketone equilibrium is established in acid. When both the above ketone (part **c**) and its 4-methyl epimer are put in acid and later recovered, they both give but one ketone. Explain.

6-31 Consider the isomers of 2-cyano-3-methyl-5,5-dichloro-4-cyclohexanone-1-carboxylic acid.

a How many racemates are possible? How many stereoisomers?

b How many asymmetric carbons must be inverted in configuration to convert one stereoisomer into its enantiomer?

c How many must be inverted to form an epimer?

d Is the isomer with the carbon-containing-ring substituents all cis the most stable? Draw this isomer and, if another is the most stable, draw it also (in perspective).

e Show which enantiomer of the all-cis isomer has a positive Cotton effect.

f What Cotton effect will an equimolar mixture of the enantiomers of the all-cis compound show?

g What is the *source* of instability in the less stable stereoisomers?

6-32 Draw a structure which satisfies each description below.

a A saturated hydrocarbon, $C_{14}H_{26}$, with more than fifty possible stereoisomers.

b A linear symmetric hydrocarbon with four possible geometric isomers.

c A meso hydrocarbon, C_9H_{20}.

d A meso hydrocarbon, $C_{10}H_{20}$, which absorbs one mole of hydrogen over platinum.

6-33 The barrier to free rotation in ethane is 3 kcal/mole; in ethylene, 63 kcal/mole. Qualitatively only what would you expect of the barriers at the three bonds marked below?

$$(CH_3)_2N\downarrow \underset{C}{\overset{H}{\underset{|}{\downarrow C \downarrow}}} \overset{O}{\underset{C}{\overset{\|}{}}}$$

$$\underset{H \quad CH_3}{}$$

6-34 Why is *t*-butyl energetically so *much* worse in an axial position of cyclohexane than ethyl or isopropyl (which are very similar to each other)? What is the likely conformational fate of *cis*-1,4-di-*t*-butyl-cyclohexane?

6-35 Is the cis or trans isomer of 3-methylcyclobutanecarboxylic acid the more stable? (*Hint:* Like cyclopentane, cyclobutane is a nonplanar, or bent, ring.)

6-36 Draw *projections*, seen down the most substituted bond, for the following in their most stable conformations.

    a *cis-* and *trans-*decalin

    b The most stable isomer of 2,3-diphenyl-2,3-diiodo-butane

    c The most stable isomer of 2-bromo-4-methyl-cyclohexanecarboxylic acid

    d The least stable isomer of 2-bromo-4-methyl-cyclohexanecarboxylic acid

READING REFERENCES

Mislow, K., "Introduction to Stereochemistry," W. A. Benjamin, New York, 1965.

Eliel, E. L., "Stereochemistry of Carbon Compounds," McGraw-Hill, New York, 1962.

Eliel, E. L., N. L. Allinger, S. J. Angyal, and G. A. Morrison, "Conformational Analysis,"
Wiley, New York, 1965.

PHYSICAL PROPERTIES AND MOLECULAR STRUCTURE

THE main thrust of the material in the book thus far has been a development of the possibilities of molecular structure from the nature of covalent bonding. We have seen how the rules of bonding allow a variety of carbon skeletons and functional groups to be constructed, and we have endeavored to cope with the main features of naming them. We have also explored the peculiarities of delocalization in π bonds. We have come to a position in which we can say what molecular structure is all about. Now we must examine how these structural and bonding features manifest themselves in terms of the observable physical properties they confer on real compounds. In turn we shall see how to use these physical properties to deduce the molecular structure when we are confronted with unknown compounds.

In the first part of the chapter are discussed the more classical physical properties, such as melting points and refractive indices, which are of practical value chiefly in characterizing or identifying compounds. The nature of inter-molecular forces, which are much involved in these properties, is explored. Following this is an examination of those physical properties, mostly spectroscopic, which form the primary basis for physical methods of structure elucidation: ultraviolet (UV), infrared (IR), nuclear magnetic resonance (NMR), and mass spectra.† Following the introduction to these primary physical tools is a section devoted to the procedures for solving problems of unknown structure from this physical evidence.

7-1 TRANSITION POINTS AND SOLUBILITY

The **melting point** of a solid is that temperature at which it becomes a liquid. Pure crystalline solids normally have *sharp* melting points, or undergo the transition over a temperature range of 1° or less. On the other hand, impure crystalline solids melt over much wider ranges of temperature. Most crystalline

---

†Mass spectra are only listed with the other physical methods for convenience in grouping the problem-solving methods. Any description of mass spectra is really an exposition of molecular reactions or fragmentations in which a high-energy electrical potential is the "reagent."

organic compounds have characteristic melting points, which are easily determined and reproduced. These facts make melting points the most widely used
physical constants for identification purposes.

The melting range of a crystalline compound is frequently used as a
measure of its purity. When two different pure crystalline compounds are
mixed together, the resulting mixture usually melts over a wide range of temperatures and well below the melting point of either pure component. Such
mixed melting-point depressions are frequently as much as 20 or 30°. Determination of a mixed melting point is a common procedure in establishing the
identity of compounds obtained by synthesis, by degradation of complex substances, or by fractionation of naturally occurring materials. If the melting point
of the compound of unknown structure is *not* depressed by admixture with
an *authentic sample* of a compound of known structure, the two substances are
presumed to be identical.

In a crystal, molecules are usually packed together in orderly and rigid
repetitive patterns. Intermolecular forces, referred to as *lattice forces*, hold the
crystal together and limit molecular motions. For example, although rotations
about carbon-carbon single bonds occur readily in the liquid and vapor phases,
these movements are usually frozen in crystals. When brought to its melting
point, the crystal disintegrates; the thermal energy of the molecules within the
crystal becomes great enough to overcome the lattice forces, and the molecules
break out of their prisons.

The **boiling point** of a liquid is that temperature at which the vapor
pressure of the liquid becomes equal to the pressure on the system. Boiling
point depends very markedly on pressure, since volume changes in passing
from the liquid to the gaseous state are very large. Thus, when boiling points
are reported, the pressure must be specified; whenever possible, they are reported at 760 mm of mercury (one atmosphere of pressure). Boiling points
vary widely with structure and provide useful constants for identification and
characterization of liquids. Differences in boiling point allow liquids to be
separated by fractional distillation. Pure compounds have narrow boiling
ranges, and mixtures usually boil over ranges limited by the boiling points
of the pure constituents. Occasionally mixtures may have sharp boiling points
that differ from those of either pure component. Such constant-boiling mixtures
are called **azeotropes**. For example, a mixture containing 4.5% water (bp 100.0°)
and 95.5% ethyl alcohol (bp 78.4°) boils at 78.1° at atmospheric pressure.

In the liquid state, strong intermolecular forces still hold molecules close
together but in a manner that allows rapid molecular movement. Not only do
molecules as a whole move about, but their parts flap and rotate in all but
the most rigid structures. When their thermal energy becomes great enough,
molecules overcome attractive forces and pass into the vapor state, where such
forces are usually negligibly small.

Solubility

Dissolution of solid in liquid in some respects resembles the melting of a solid, since the ordered crystal structure is destroyed. Interactions in a dilute solution are for the most part between solute and solvent, although in some instances solute molecules may show a considerable tendency to cluster together. Since the characteristic crystal energies are lost in both melting and dissolution, one can safely generalize that among groups of closely related substances, the higher-melting compounds will usually be the less soluble. This generalization is really applicable to nonpolar compounds only.

Mixing of two miscible liquids involves the exchange of attractive forces between like molecules for interactions between dissimilar molecules. The familiar generalization "like dissolves like" is based upon the observation that attractive interactions tend to be greater between similar molecules than between structurally dissimilar pairs. Thus, the lower alcohols are soluble in water because the attractive forces between the hydroxyl groups of water and those of an alcohol are similar to those between individual water molecules and individual alcohol molecules (hydrogen bonds). Conversely, water does not dissolve in hexane because the combined attractive forces between water molecules and between hexane molecules are greater than those between water and hexane molecules.

Sometimes high solubility is due to rapid chemical reactions, so that new compounds are formed during the process of solution. Such chemical action can lead to solubilities that appear anomalously high. The most striking and generally useful examples involve acid-base reactions (Chap. 8) that convert neutral compounds to highly water-soluble ionic species. For example, aniline is very sparingly soluble in water, but it dissolves readily in aqueous acid by virtue of its conversion to anilinium ion. Neutralization of the solution with strong base regenerates the weakly basic amine, which separates from solution.

$$\overset{+}{H_3O} + C_6H_5\overset{..}{N}H_2 \rightleftharpoons C_6H_5\overset{+}{N}H_3 + H_2O$$

$$\text{Aniline} \qquad\qquad \underset{\text{ion}}{\text{Anilinium}}$$

7-2 INTERMOLECULAR INTERACTIONS

The nature of intermolecular interactions, which dictate the physical properties discussed in the previous section, will be considered briefly in this section. Although all chemical interactions arise from the same fundamental sources, i.e., interactions of negative electrons and positive nuclei with each other, the forces involved in intermolecular interactions are much smaller than those responsible for chemical bonding.

Dipole Moment

The dipole moment of a compound measures the concentration of positive and negative charges in different parts of a molecule. This physical constant is determined by observing the extent to which molecules of a substance orient themselves when placed between the plates of a charged condenser. Molecules with high dipole moments become highly oriented, with their dipoles parallel to the direction of the field. Dipole moments (μ) are reported in Debye units (D), since values in electrostatic units always include a large negative exponent (1 D $= 10^{-18}$ esu).

Separation of charge in organic molecules is associated with the presence of either formal charges or polarized bonds. The high dipole moment of tri-methylamine oxide illustrates the former, and the smaller dipole moment of methyl iodide the latter, as is indicated in Fig. 7-1.

Many molecules with polar bonds do not possess molecular dipoles, because individual **bond moments** are so oriented that they cancel one another. Any molecule that has a center of symmetry (Sec. 6-7) will not have a dipole moment. The two geometric isomers of 1,2-dichloroethylene illustrate the principle (see Fig. 7-1). Both molecules are planar, and both contain two polar carbon-chlorine bonds. The cis isomer has an appreciable dipole moment, since in this molecule the bond moments are so oriented as to provide overall elec-trical dissymmetry. The trans isomer has a zero dipole moment, since the two individual bond moments are aligned so as to cancel each other.

Resonance effects frequently influence dipole moments. For example, the dipole moment of nitromethane is smaller than that of nitrobenzene. Migration of π electrons from the benzene ring into the nitro group increases charge separation in the aromatic compound.

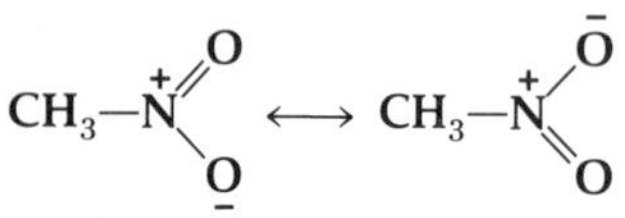

Nitromethane, $\mu = 3.50$ D

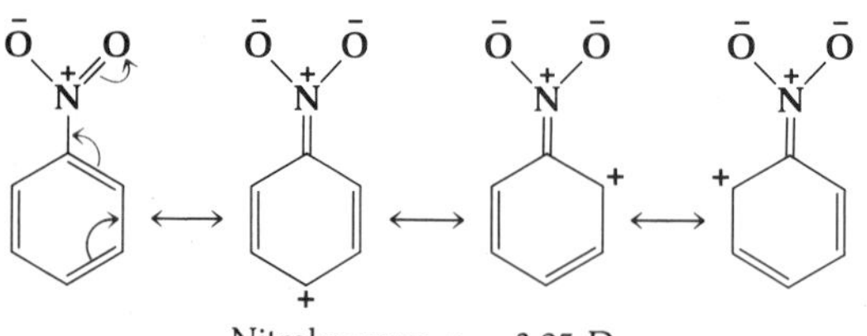

Nitrobenzene, $\mu = 3.95$ D

In contrast, isovalent resonance involving nonbonding electrons of oxygen and nitrogen may make the dipole moments of compounds such as aromatic

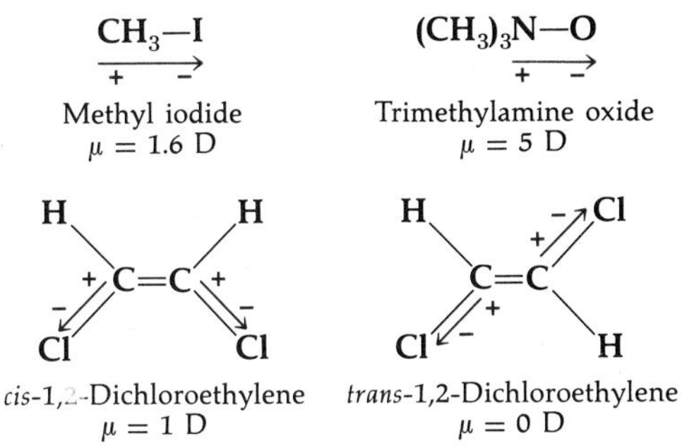

$$CH_3\text{—}I$$
$$\overset{+\quad\;\; -}{\longrightarrow}$$

Methyl iodide
$\mu = 1.6$ D

$$(CH_3)_3N\text{—}O$$
$$\overset{+\quad\; -}{\longrightarrow}$$

Trimethylamine oxide
$\mu = 5$ D

cis-1,2-Dichloroethylene
$\mu = 1$ D

trans-1,2-Dichloroethylene
$\mu = 0$ D

FIGURE 7-1   **Dipole moments**

amines, phenols, and aryl ethers either larger or smaller than the dipoles of analogous aliphatic compounds.

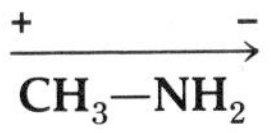

$$\overset{+\qquad\quad -}{\longrightarrow}$$
$$CH_3\text{—}NH_2$$

Methylamine, $\mu = 1.33$ D

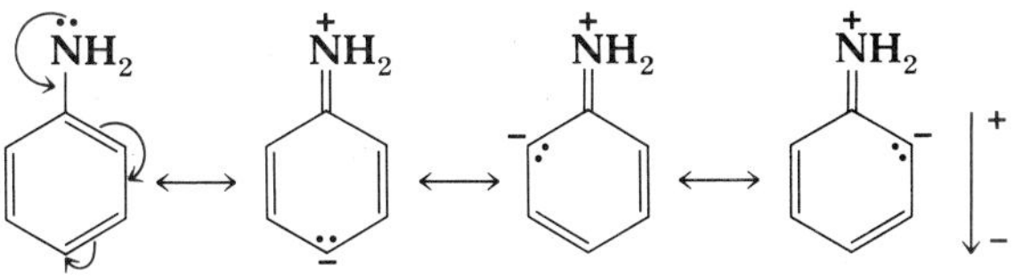

Aniline, $\mu = 1.53$ D

The very large dipole moment of p-nitroaniline is striking evidence for transfer of electrons from the amino group to the nitro group by isovalent resonance through the entire conjugated system.

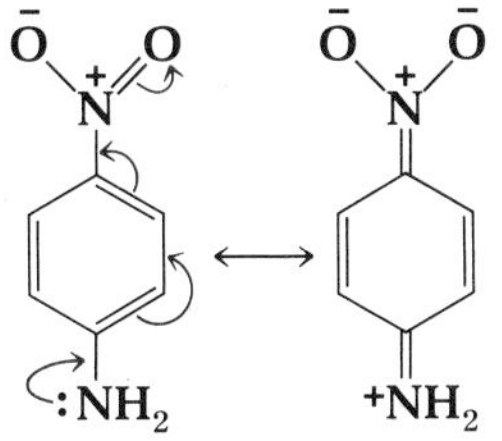

p-Nitroaniline, $\mu = 6.10$ D

Polarizability

Even when molecules of a compound have no permanent dipoles, the substance can be polarized by an applied electric field. Field-induced polarization measures the **polarizability** of molecules. When a molecule is polarized, the average electron distribution is changed in comparison with that of an identical molecule not subjected to an externally applied field.

In molecules, polarization involves deformation of both bonding and nonbonding electron clouds. Unsaturated molecules are much more easily deformed than saturated systems, and conjugated unsaturated systems are highly susceptible to polarization. Weak bonds are usually highly polarizable, since the electrons involved are comparatively weakly bound to the molecule.

A physical property directly related to polarizability is the **refractive index**. Light rays travel more slowly through transparent materials than through a vacuum. The ratio of the speed of a beam in air to its speed in another medium is the refractive index of the latter. Retardation is due to the interaction of light with electrons, and *high refractivity* indicates *easy polarizability*. Refractive indices vary with wavelength of the transmitted light; hence, a notation to indicate wavelength must be included in a report of a refractive index. Since the refractive index is strongly temperature-dependent, temperature is also always included in the notation. Thus, refractive indices are usually reported: $n_D^T = $ ———. They are one of the most easily obtained physical constants that can be found for pure liquids. Table 7-1 lists the refractive indices of typical organic compounds.

Intermolecular Polarization

Interactions of polarizable molecules with one another and with dipolar molecules give rise to intermolecular attractive forces. These forces are responsible for the crystal energies of nonpolar solid compounds and for the existence of liquid phases of such materials.

The attraction between a dipole and an easily polarizable molecule that has no permanent dipole is readily understood, since the dipole supplies a field

TABLE 7-1   **Refractive Indices of Pure Liquids**

Compound	T, °C	n_D^T
H_2O	20	1.3330
$CH_3C \equiv N$	17	1.3460
$(CH_3)_2CHCH_2CH_3$	20	1.3549
$CH_3(CH_2)_4F$	20	1.3562
CH_3COCH_3	20	1.3591
$CH_3CH_2CH_2C \equiv CH$	18	1.4079
$CH_3(CH_2)_4Cl$	20	1.4119
$CH_3CH = CHC \equiv N$	20	1.4156
$HC \equiv CCH_2CH_2C \equiv CH$	20	1.4414
$CH_3(CH_2)_4Br$	20	1.4444
$CH_3(CH_2)_4I$	20	1.4955
C_6H_6	20	1.5017
$C_6H_5NO_2$	20	1.5529
$(C_2H_5)_2Cd$	18	1.5680

NH_2 NH_2

	99	1.7083

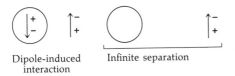

Dipole-induced   Infinite separation
interaction

FIGURE 7-2   **Induced polarization**

to polarize its deformable partner and a small dipole is induced in the latter. Dipole-induced interactions are pictured in Fig. 7-2.

Hydrogen Bonding

Compounds containing oxygen-hydrogen or nitrogen-hydrogen bonds show evidence of association that exceeds all expectations based upon their molecular weights, dipole moments, and molecular polarizability. For example, water, which contains no highly polarizable atoms, melts at 0° and boils at 100°, whereas methyl ether and methyl alcohol (which are compounds of higher molecular weight) boil at −24 and 65°, respectively, and melt at very low temperatures. There is no large difference in the dipole moments of these substances. Accordingly, the remarkably high association of water must be due to some short-range interactions that are not reflected in all the gross properties of water molecules. Comparison of physical properties within the series water, methyl alcohol, and methyl ether indicates that the hydroxyl group has a pronounced specific effect upon physical properties.

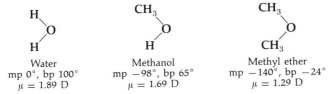

Water	Methanol	Methyl ether
mp 0°, bp 100°	mp −98°, bp 65°	mp −140°, bp −24°
$\mu = 1.89$ D	$\mu = 1.69$ D	$\mu = 1.29$ D

Comparison of the structures and physical properties of water and methyl ether is particularly instructive. In liquid water, molecules are held together by dipole-dipole interactions between positive hydrogen and negative oxygen atoms. The shape of the water molecule allows many different arrangements in the molecular aggregates, since these partial charges are sterically exposed to those of neighboring molecules. In molecules of methyl ether, these partial charges are deeply buried beneath a canopy of relatively nonpolar carbon-hydrogen bonds. As a result, the dipole-dipole interactions must cover longer distances, and their effect is diminished accordingly. Models for aggregates of water and of ether molecules are drawn in Fig. 7-3.

Another important factor contributes to the highly associated character of liquids whose molecules contain hydrogen bonds. These bonds are highly polar, and the hydrogen atom possesses some of the character of a naked proton. The charge of this nucleus is much less shielded by electrons than the charge of other nuclei. As a result, the intense local field associated with

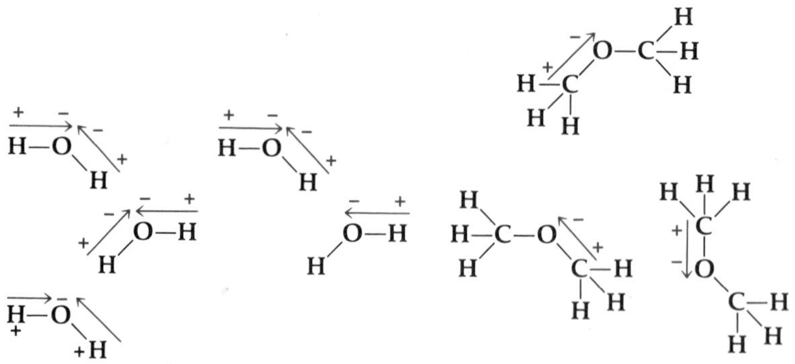

Hydrogen bonding in
liquid water

Weaker dipole-dipole interactions
in liquid methyl ether

FIGURE 7-3   **Intermolecular interactions in liquids**

hydrogen bound to oxygen or nitrogen produces strong dipole-dipole interactions.

Typical hydrogen bonds are shown below; they are most stable and effective when the three involved nuclei are in a straight line.

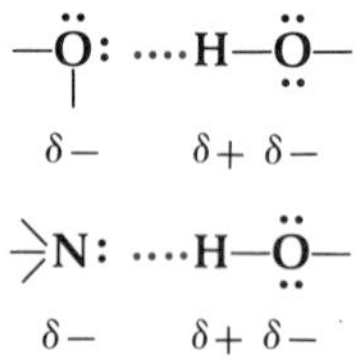

Polar Compounds

Polar molecules cling together tenaciously, as is illustrated by their high boiling and melting points. Such compounds may, however, be rather soluble in polar solvents. The strong intermolecular interactions that hold the molecules together are replaced by similar polar interactions between solvent and solute. Polar substances are also adsorbed very strongly on solids, such as alumina and silica, which are commonly used as adsorbents in liquid-phase chromatography (Sec. 1-4).

The dominant influence of hydrogen bonding on the physical properties of compounds that contain oxygen-hydrogen or nitrogen-hydrogen linkages is illustrated in Table 7-2. Compounds containing these bonds are listed next to related compounds in which these linkages are absent. The most dramatic effects are associated with molecules that contain two functional groups containing polar hydrogens, such as ethylene glycol ($HOCH_2CH_2OH$). The boiling point of this substance is over 100° higher than that of 1,2-dimethoxyethane ($CH_3OCH_2CH_2OCH_3$).

The "like dissolves like" principle requires a special comment in connection with polar solvents. No doubt, an individual water molecule would polarize a hydrocarbon molecule, producing stronger interactions than those between two hydrocarbon molecules. However, water and hydrocarbons are insoluble in each other, because solution can be accomplished *only by separating water molecules*, thereby *sacrificing* the very large association energy of liquid water. The principal exceptions to the general insolubility of these compounds in each other occur with molecules that do not have oxygen-hydrogen or nitrogen-hydrogen bonds but do contain oxygen or nitrogen that can serve as the second partner in a hydrogen-bonded molecular aggregate with water. Table 7-3 provides examples.

Interesting consequences sometimes result from internal interactions of polar groups with each other. Internal hydrogen bonding gives particularly dramatic effects. For example, *o*-nitrophenol is a low-melting solid and can be distilled normally; *p*-nitrophenol has a much higher melting point, and the liquid cannot be distilled without decomposition. (Slow heating of the solid in a vacuum results in sublimation.)

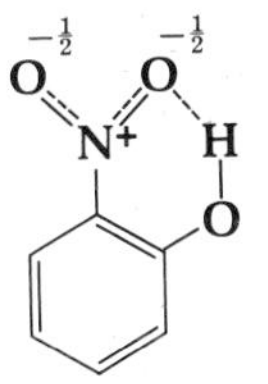

o-Nitrophenol
mp 45°
bp 214°

p-Nitrophenol
mp 113°
sublimes

Ionic Compounds

If an organic molecular species contains a charge it is then an ion and must be associated with another ion of opposite charge, just as with inorganic

TABLE 7-2   **Influence of Hydrogen Bonding on Physical Properties**

Compound	Mp, °C	Bp, °C	Solubility in water
$HOCH_2CH_2OH$	−16	197	Miscible
$CH_3OCH_2CH_2OCH_3$	−58	84	Miscible
$H_2NCH_2CH_2NH_2$	9	117	Miscible
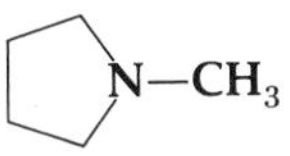		79	Miscible
p-$HOOCC_6H_4COOH$	300		0.001 g/100 g (100°)
p-$CH_3OOCC_6H_4COOCH_3$	140	300	0.3 g/100 g (100°)
o-$HOOCC_6H_4COOH$	191	Sublimes	0.54 g/100 g (14°)
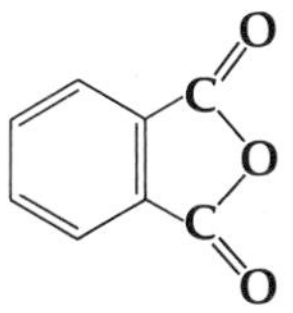	131	285	Slightly soluble

salts like Na$^+$Cl$^-$. The most common stable organic salts (ionic compounds) are the carboxylate anions and the ammonium cations. These ionic compounds are typical salts, dissolving in water because of hydrogen bonding and strong dipole interactions that stabilize the ions in solution. Such stabilization by solvent attraction is called **solvation**. These salts dissolve less well in other organic polar solvents and are usually virtually insoluble in nonpolar solvents. The most common exceptions are quaternary ammonium salts with large alkyl groups. These cations can be fairly soluble in nonpolar solvents since, despite the positive charge buried at the center, they present a hydrocarbon surface to the solvent.

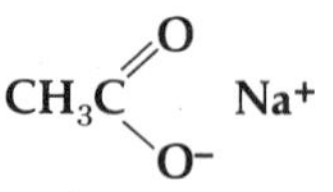

$$CH_3C \overset{O}{\underset{O^-}{\diagup\diagdown}} \quad Na^+ \qquad (C_2H_5)_3\overset{+}{N}-H \; Cl^- \qquad \phi CH_2\overset{+}{N}(CH_3)_3 \; OH^-$$

| Sodium acetate | Triethylammonium chloride | Benzyltrimethylammonium hydroxide |

TABLE 7-3   **Hydrogen Bonding and Solubility**

Compound	Solubility in water
CH_3OCH_3	7.6 g/100 g (18°)
$C_2H_5OC_2H_5$	7.5 g/100 g (19°)
CH_3COCH_3	Miscible
CH_3CN	Miscible
$(CH_3)_3N$	41 g/100 g (19°)
Compare with	
C_2H_5Br	0.9 g/100 g (30°)
$n\text{-}CH_3(CH_2)_3CH_3$	Insoluble

PROBLEM 7-1

Indicate which of the following compounds should have dipole moments. For compounds without a dipole moment, explain why they can possess polar groups and not have a dipole moment. For compounds that have a dipole moment, draw an arrow in the direction of the dipole, with the head of the arrow as the negative end.

a $\overset{H}{}\diagdown \overset{COOH}{\underset{\|}{C}}$ $\overset{C}{}\diagup\diagdown \overset{COOH}{\underset{H}{}}$

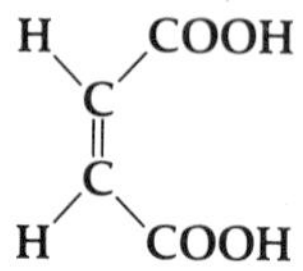

b $\overset{HOOC}{}\diagdown \overset{H}{\underset{\|}{C}}$ $\overset{C}{}\diagup\diagdown \overset{COOH}{\underset{H}{}}$

c (1,4-dinitrobenzene with NO$_2$ groups)

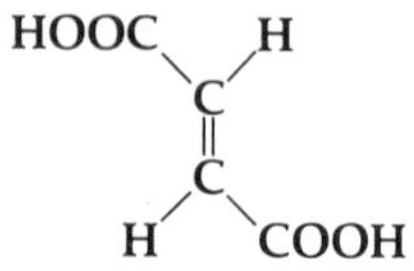

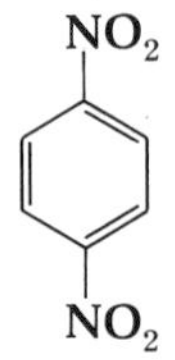

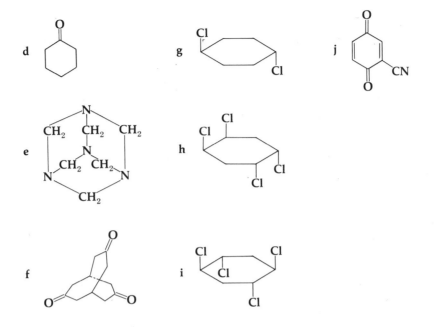

PROBLEM 7-2

Write formulas for each of the following compounds and show the probable orientation of their dipoles.

a The dicyanobenzenes

b The 1,4-dichlorobutadienes

c *trans*-1,2-Dihydroxycyclopentane

d *p*-Nitrobenzylamine and *p*-nitroaniline (which will have the larger dipole moment?)

e Succinonitrile ($NCCH_2CH_2CN$); consider relative stabilities of all staggered conformations.

PROBLEM 7-3

Arrange the following compounds in order of decreasing solubility in water. Give your reasoning.

a $C_6H_5CHCH_3$
       OH

b $C_6H_5CH_2OCH_3$

c $C_6H_5CH_2CH_2CH_3$

d $p\text{-}HOC_6H_4C_2H_5$

e $p\text{-}CH_3C_6H_4COOH$

f $p\text{-}HOC_6H_4CH_2OH$

PROBLEM 7-4

Explain the following sequences of boiling points.

a CH_3OH, H_2O, $HOCH_2CH_2OH$
    65°   100°   197°

b CH_3CH_3, CH_3CH_2Br, CH_3CH_2I, CH_3CH_2OH
     $-89°$ $38°$ $72°$ $78°$

c *n*-Hexane, benzene, cyclohexane
     $69°$ $80°$ $81°$

PROBLEM 7-5

A mixture of naphthalene, ethylene glycol, benzoic acid, and aniline was separated by the following scheme. Show the composition of each phase in every step of the procedure.

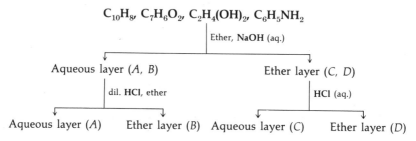

$C_{10}H_8$, $C_7H_6O_2$, $C_2H_4(OH)_2$, $C_6H_5NH_2$

Ether, **NaOH** (aq.)

Aqueous layer (*A, B*)   Ether layer (*C, D*)

dil. HCl, ether   HCl (aq.)

Aqueous layer (*A*)   Ether layer (*B*)   Aqueous layer (*C*)   Ether layer (*D*)

7-3   ABSORPTION SPECTRA: GENERAL

Light, or more generally, electromagnetic radiation, may be described in terms either of the frequency (number per second) ν of its waves or of the wavelength λ between them. These are related to each other by the constant velocity of light, c, through the relation $\lambda\nu = c$. Light rays are composed of individual units of energy, called **quanta**. The energy of a quantum of light is proportional to its frequency, or inversely proportional to its wavelength [Eq. (1)]. In familiar kilocalorie per mole terms the energy of light can be calculated from the wavelength. The wavelength of light is commonly measured in **microns** (μ = micrometers) or in **millimicrons** ($m\mu$) or the equivalent **nanometers** (nm), depending on convenience in the spectral region†:

$$E = h\nu = \frac{hc}{\lambda} = \frac{28.6}{\lambda} \text{ kcal/mole} \qquad \text{for } \lambda \text{ in } \mu \text{ units} \qquad\qquad 1$$

$$= \frac{28,600}{\lambda} \text{ kcal/mole} \qquad \text{for } \lambda \text{ in } m\mu \text{ or nm units}$$

$$1\mu = 10^{-6} \text{ m} = 10^{-3} \text{ mm}$$

$$1 \; m\mu = 1 \text{ nm} = 10^{-9} \text{ m} = 10^{-6} \text{ mm} = 10^{-3} \; \mu$$

Visible light is just that small portion of the entire electromagnetic radiation spectrum (Fig. 7-4) which is registered by the human eye and constitutes the

---

† In this book we shall use the more contemporary nanometer (nm) notation. It is in any case a unit equivalent to the more traditional millimicron (mμ) for the ultraviolet spectral region. Angstrom units are occasionally used in the literature to express wavelengths of light (1 Å = 0.1 nm = 10^{-8} cm).

SPECTRAL REGION

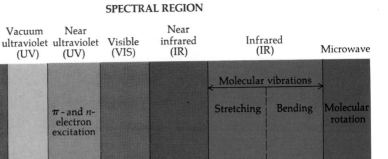

FIGURE 7-4 Correlation between spectroscopic regions and types of molecular excitation

wavelength region from about 400 to 800 nm. At the lower end (400 nm) is violet light and it passes through the rainbow of blue, green, yellow, to red at the long (800-nm) wavelength end. When we observe color with the eye we observe in effect the wavelength(s) of the observed light. Below visible light is the ultraviolet region, above it the infrared region, as symbolized in Fig. 7-4.

Light of a particular energy (a **photon** or quantum) may be absorbed by a molecule if the molecule can by absorbing it pass from one of its quantized energy levels to a higher one by just the energy amount (ΔE) corresponding to the photon energy ($E = h\nu$). There are two kinds of molecular energy jumps which will concern us.

1   *Electronic excitations* in the energy range of chemical bonds (35 to 150 kcal/mole), that is, 200 to 800 nm, involving the promotion of an electron to a higher-energy molecular orbital (usually an antibonding orbital), result in ultraviolet and visible absorption spectra.

2   *Nuclear vibrations* (also quantized) in the range of 1 to 15 kcal/mole are caused by infrared absorption in the spectral region 2 to 25 μ.

In either of these spectral ranges the act of light absorption at a particular wavelength is a direct function of and indication of the presence of a molecular structural feature possessing the capability of absorbing that specific discrete energy package. This is most easily observed instrumentally by passing light of continuously changing wavelength through a sample of the compound and measuring the light intensity before and after passage through the sample. This procedure records those wavelengths at which light absorption occurs and those others which are largely transmitted freely. A record of the amount of light absorbed by a sample as a function of the wavelength of the light is called an **absorption spectrum**. In principle, all absorption spectra should consist of

ULTRAVIOLET

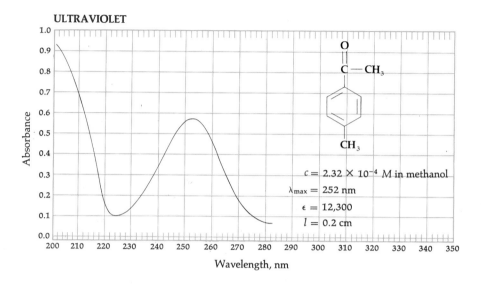

Wavelength, nm

INFRARED

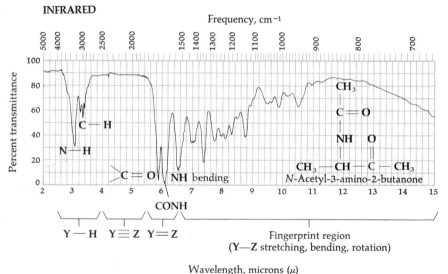

Wavelength, microns (μ)

FIGURE 7-5 **Typical absorption spectra**

lines, but in practice one observes absorption **bands**. A spectrum has a series
of maxima and minima connected by smooth curves. Typical absorption spectra
are shown in Fig. 7-5.

Ultraviolet and visible (electronic) spectra commonly have the appearance
of the upper record in Fig. 7-5 and are plotted with absorption bands showing
as maxima. Infrared (vibrational) spectra commonly resemble the lower graph
with sharper absorption bands, and these actually recorded as minima (peaks
hanging down like stalactites); here it is **transmission**, the opposite of absorbance, of the light that is usually recorded.

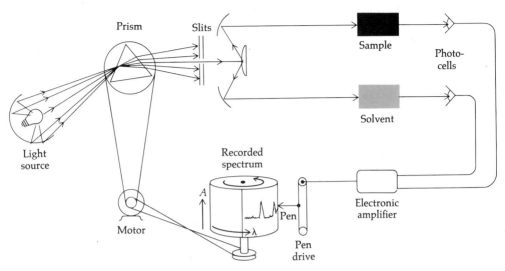

FIGURE 7-6   **Spectrophotometer for recording absorption spectra**

Quantitative spectrophotometric measurements are usually reported in terms of the Beer-Lambert absorption law [Eq. (2)].

$$\frac{I}{I_0} = e^{-\epsilon cl} \qquad\qquad 2$$

where I = intensity of transmitted light (leaving sample)

I_0 = intensity of incident light (entering sample)

ϵ = molar absorption coefficient

c = concentration of the absorbing species, moles/liter

l = sample thickness (usually = 1 cm)

Equation (3) is a useful form of Eq. (2):

$$\text{Optical density or absorbance:} \qquad A = \log\frac{I_0}{I} = \epsilon cl \qquad\qquad 3$$

The intrinsic intensity ϵ is a function of the molecule, and this intensity as well as the wavelength of a given absorption maximum (λ_{max}) are the features which are used for structure correlations. The ultraviolet spectrum at the top of Fig. 7-5 may serve as an example; the intensity ϵ was calculated from the concentration and the observed absorbance.

A generalized instrument for obtaining spectra such as those in Fig. 7-5 is a double-beam spectrophotometer, schematically outlined in Fig. 7-6. The light source generates a continuum of wavelengths which are separated by a prism and then selected by slits such that rotation of the prism causes a continuously increasing series of wavelengths to pass through the slits for recording purposes. The selected **monochromatic** (one wavelength) beam is then split into two beams, one for passage through the unknown sample dissolved in a solvent and the other for passage through solvent alone. The two beams are each measured for intensity by a photoelectric cell and the latter subtracted electronically from the former to give an intensity (absorbance or transmission) signal which is characteristic of the compound alone. This signal is automatically recorded on a graph as a function of the wavelength being utilized. In each spectral region the light source must be a continuous generator of wavelengths in that region and the materials used for prisms and solution container must be transparent to the wavelengths used. The equipment employed is as follows:

Spectral region	Light source	Prism and solution cell construction
UV	Hydrogen lamp	Quartz
VIS	Tungsten lamp	Glass (or quartz)
IR	Ceramic glower	Rock salt ($NaCl$)

While organic chemists have mostly agreed in practice to recording ultraviolet and visible spectra on a wavelength scale in millimicron or nanometer units, there has been no unanimity concerning the scale to be used in infrared spectra. Many chemists report IR bands (peaks) in frequency units which are really reciprocal centimeters (cm^{-1}) obtained as $\nu \times c$, and which are the inverse of wavelengths, on the rationale that these units are directly proportional to energy. The range of common utility for these units is from 5000 cm^{-1} (2 μ) to 500 cm^{-1} (20 μ); the units are frequency units related to the wavelength scale in microns:

$$\nu(\text{cm}^{-1}) = \frac{10^4}{\lambda(\mu)} \quad \text{or} \quad \lambda(\mu) = \frac{10^4}{\nu(\text{cm}^{-1})}$$

The numbers used thus carry four digits of accuracy which are harder to learn and imply more accuracy than is usually warranted. The other scale in common use is the wavelength scale, in microns, 2 to 25 μ, which is easier to master and also reflects a consistent usage through the entire UV, VIS, and IR spectral range, but is inversely proportional to energy. Indeed this proportionality to energy is rarely invoked, but in subsequent sections in this text both scales are recorded since both are in active contemporary use.

7-4 ULTRAVIOLET AND VISIBLE SPECTRA

Electrons in σ bonds require for their excitation energies which are too high to be available through ordinary UV radiation. Hence UV (and VIS) spectra are only caused by the higher-energy, more labile π and n (nonbonded) electrons in a molecule. Saturated molecules (and groups in molecules) are transparent to UV and VIS light. The two major kinds of electronic absorptions in unsaturated molecules are illustrated from the molecular-orbital energy-level diagram for a carbonyl group shown in Fig. 7-7. The oxygen atom is shown initially hybridized with $2sp$ and $2p$ atomic orbitals. One sp orbital then forms a σ molecular orbital with an initial sp^2 atomic orbital of the carbon atom. The

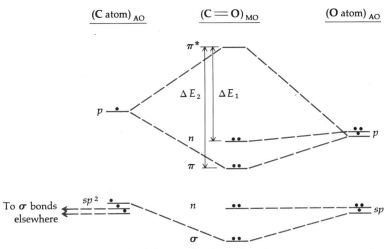

FIGURE 7-7   Molecular-orbital diagram for the carbonyl group

other sp orbital of oxygen is left to provide a nonbonding molecular orbital (n). Then p atomic orbitals on each atom overlap to give a π molecular orbital, and the remaining p orbital of oxygen becomes a second (and higher-energy) n orbital (nonbonding).

The lowest *unoccupied* molecular orbital is the antibonding π^* orbital which is separated from the nearest (n) orbital by energy ΔE_1 and from the π orbital by ΔE_2.

Should light of frequency $\nu_1 = \Delta E_1/h$ be absorbed it would cause one electron to jump from the n level of energy to the π^* level. This is called an $n \longrightarrow \pi^*$ (n to pi star) transition and is geometrically a poor one since the n and π^* orbitals are perpendicular to each other. Hence absorption occurs with carbonyl compounds in this fashion but with low intensity ($\epsilon = 10$ to 100; this is a "forbidden," or infrequent, transition). The $n \longrightarrow \pi^*$ transition corresponds to about 100 kcal/mole since it occurs in simple ketones (and aldehydes) at about 280 nm.† A transition of higher energy, the $\pi \longrightarrow \pi^*$ (pi to pi star) transition (ΔE_2 in Fig. 7-8 occurs at about 180 nm† and is much more intense ($\epsilon \sim 10,000$) owing to the favorable geometry (π and π^* orbitals lie in the same plane). This transition (but not of course the $n \longrightarrow \pi^*$) also occurs in carbon-carbon double bonds, at about the same wavelength.

In general the λ_{max} of absorption indicates the energy difference between the ground and excited states of the molecule. Changes in the molecular environment of the absorbing electrons (i.e., by changes in $\mathbf{R}$ and $\mathbf{R}'$) will cause small shifts in λ_{max} and hence in ΔE between the two states, but it is always important in assessing the cause of these shifts to compare the environmental (i.e., strain and substituent) effects on both ground and excited states since *both* energy levels are involved in the observed ΔE. Conjugated systems of π

---

† The spectra of ketones will show a large maximum at 180 nm and a small, low-intensity peak at 280 nm. These peaks or maxima are designated as λ_{max}.

electrons are stabilized by resonance, and the stabilization is apparently greater in the excited than the ground state since addition of increasing numbers of double bonds in conjugation raises the wavelength (i.e., lowers ΔE) of the absorption maximum.

Absorption of $\mathbf{H}{-}(\mathbf{CH}{=}\mathbf{CH})_n{-}\mathbf{H}$ ($\pi \longrightarrow \pi^*$)

Excited (π^*)

Ground (π) ΔE_1 ΔE_2 ΔE_3 ΔE_4

$n =$ 1   2   3   4

$\lambda_{max} =$ 165   217   258   286   nm

Conjugated π-electron systems are called **chromophores** and some largely empirical but quite successful rules have been developed for calculation of the absorption maxima of a number of kinds. Major shifts to longer wavelength (**bathochromic shifts**) occur on extension of the conjugated system, and minor bathochromic shifts usually accompany increased substitution (of **R**, **X**, etc., for **H**) on the chromophore skeleton. To a remarkable extent the wavelength is proportional to the distance between the terminal atoms of the chromophore, an all-trans multiple double bond (polyene) chromophore absorbing at longer wavelength than its partially cis geometric isomers.

Shifts in absorption position

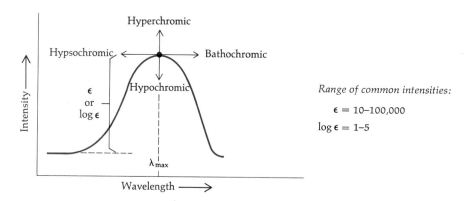

Range of common intensities:

$\epsilon = 10{-}100{,}000$

$\log \epsilon = 1{-}5$

Empirical Calculations

Extensive correlations between UV spectra and structure in known compounds have led to certain simple empirical additive rules for the calculation of the wavelength of maxima in some classes of chromophores, notably dienes and unsaturated carbonyl compounds. These rules are very useful in the solution of structure problems. The general procedure is to begin with a base wavelength for the parent chromophore and add to it values corresponding to each substituent found attached to the parent π-electron skeleton.

	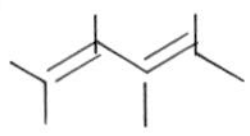 Dienes	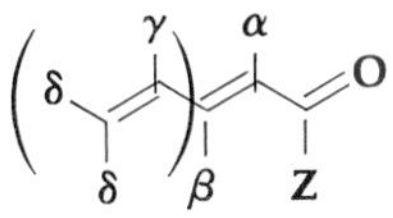 Enones †	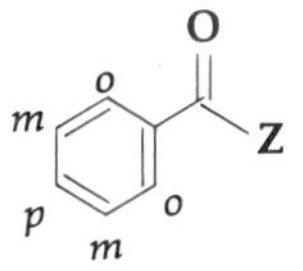 Aromatic carbonyl
Parent	215 nm	**Z = C:** 215 nm (acyclic or 6-ring) **Z = C:** 202 (5-ring) **Z = H:** 207	**Z = H:** 250 nm **Z = C:** 246 **Z = OH, OR:** 230
Increments Extra conjugated **C=C**	+30	+30	—
S-cis **C = C — C = C†**	40	40	
Exocyclic position†	5	5	—
Substituents: **H**	0	0	0
R	5	α 10 β 12 γ , δ 18	o, m 3 p 7
Cl	5	α 15 β 12	o, m 0 p 10
OH, OR	5	α 35 β 30	o, m 7 p 25
OCOR	0	α , β , δ 6	—
O⁻	—	β 75	o, m 15 p 80
SR	30	β 85	—
NR₂	60	β 95	o, m 20 p 85

†Features:

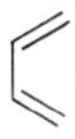

S-cis diene   *S*-trans diene   >C=C (Ring)

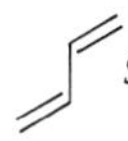

(*S* = single bond)   Exocyclic double bond position

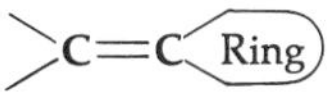

Note that the second column is only for unsaturated ketones and aldehydes, and that acids, esters, and amides follow the same *trends*, but at lower wavelengths.

FIGURE 7-8   **Calculation of UV maxima for conjugated chromophores**

Although a fairly complete empirical table of these increments is listed in Fig. 7-8, the common knowledge of the research chemist who is only infrequently concerned with UV spectra may be summarized here. He is likely to recall that dienes and unsaturated ketones absorb at 215 nm with 30 nm added

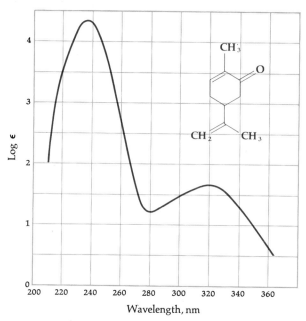

FIGURE 7-9   Ultraviolet spectrum of carvone in ethanol

for another linearly conjugated C=C† and another 40 nm for any two carbon-carbon double bonds held in an *S*-cis alignment. Any C=C which is exocyclic to a ring adds a 5-nm increment, as do carbon, halogen, or ether substituents on a diene. On the unsaturated ketone carbon substituents create about an 11-nm increment on the first double bond next to C=O and an 18-nm increment on the second double bond, if there is one. The ultraviolet spectrum of carvone, the chief component of natural spearmint oil, is shown in Fig. 7-9 and may serve as an example.

For the other values chemists are likely to consult tables such as Fig. 7-8. It may be noted that substituents allowing strong resonance interaction with carbonyl give spectra of much longer wavelength when oriented for direct conjugation, as with β-amino-α,β-unsaturated ketones or oxide anions para to aromatic carbonyls. Values calculated from Fig. 7-8 are usually accurate to better than 5 nm, and are only really valid for spectra taken in alcohol since λ_{max} values are somewhat solvent-dependent.

Aromatic compounds are generally characterized by an intense band around 200 nm. This π-π* transition is of high energy owing to the resonance stabilization and is little affected by substituents. A second aromatic band of low

---

†Cross-conjugated systems usually exhibit spectra with λ_{max} simply that of the longer-wavelength chromophore involved.

Calculated: $A = 215 + 12 = 227$ nm
$B = 215 + 12 + 12 + 5 = 244$ nm
Observed: 244 nm

($\epsilon < 1000$) intensity appears at 255 to 280 nm in most simple benzene derivatives. The same two-peak pattern also characterizes the heterocyclic aromatics (Chap. 24), which are also six-electron monocyclic aromatics. In aromatic carbonyl compounds these characteristic aromatic bands also occur, but the second is often submerged by the overlaid and intense carbonyl absorption maximum of column 3 in Fig. 7-8. The example of p-methylacetophenone in Fig. 7-5 shows an actual spectrum.

The best familiarization with these useful rules will come with practice on the examples shown below.

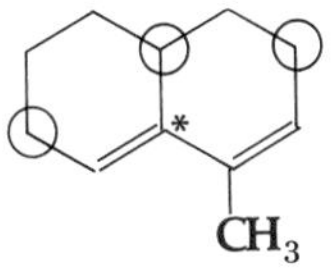

Base = 215 nm
+4 C = 20
+1 exocyclic
  position (*) =   5
———————————
Calc   240 nm
Obs   238 nm

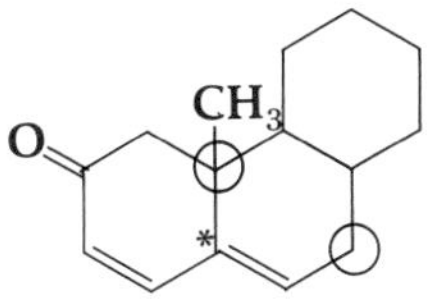

Base = 215 nm
+extra C=C = 30
+2 subst. ($\gamma + \delta$)
  (◯ = 2 × 18) = 36
+1 exocyclic
  position (*) =   5
———————————
Calc   286 nm
Obs   290 nm

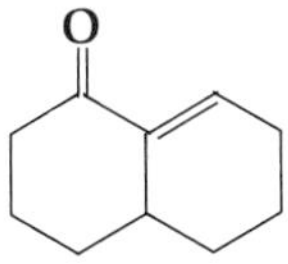

Obs 242 nm

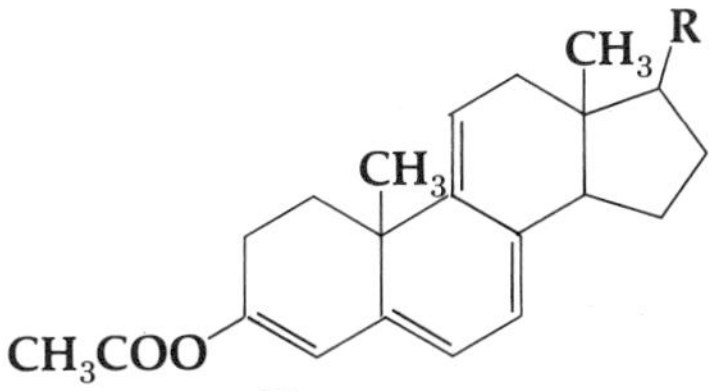

Obs 355 nm

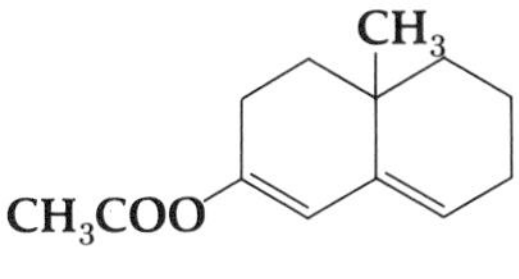

Obs 235 nm

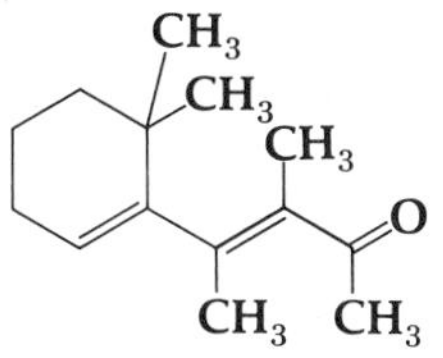

Obs 278 nm

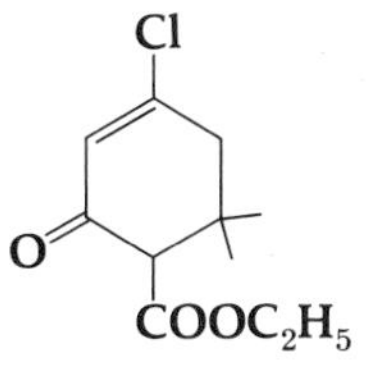

Obs 239 nm

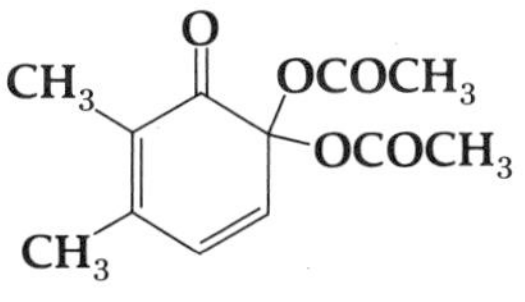

Obs 327 nm

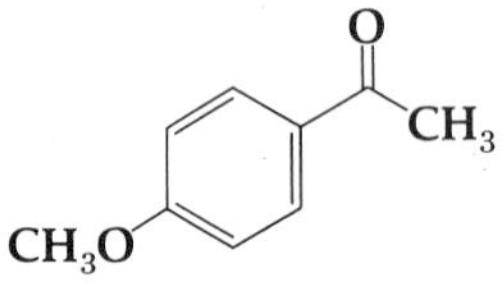

Base = 246
p-OCH$_3$ =   30
Calc   276 nm
Obs   277 nm

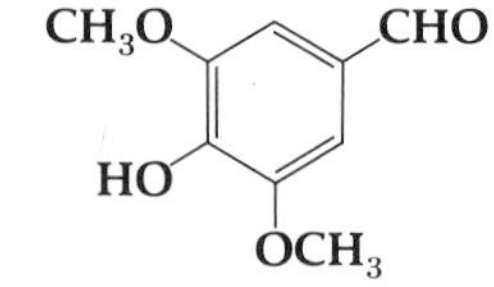

Base = 250
p-OH =   30
2 × m-OCH$_3$ =   20
Calc   300 nm
Obs   308 nm

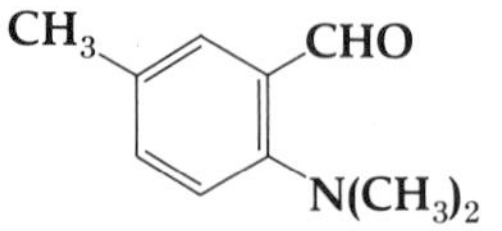

Obs 270 nm

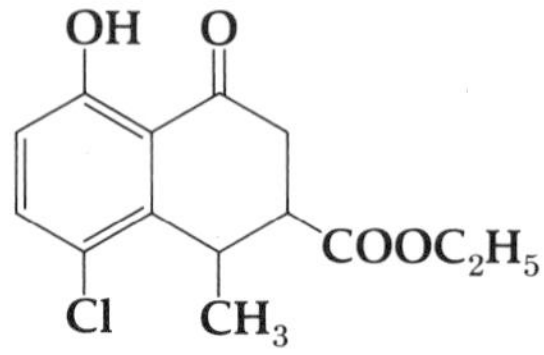

Obs 257 nm

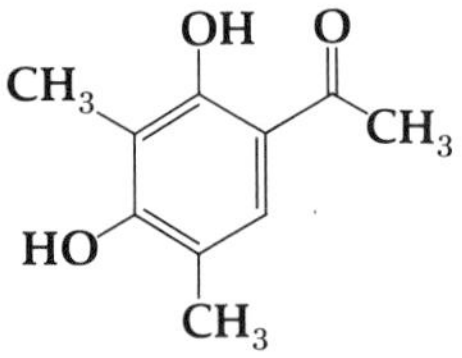

Obs 286 nm

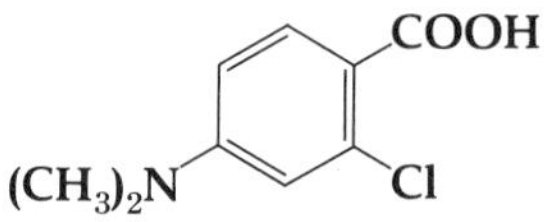

Obs 315 nm

Model Compounds and Additivity

Owing to the fact that the UV spectrum arises only from the chromophore (conjugated π and n electrons), unaffected by the rest of the molecular skeleton, different molecules having the same chromophoric system show identical UV spectra. Thus spectra of the following compounds (A and B) are the same, the chromophore and its substitution pattern being the same in each, as indicated in C:

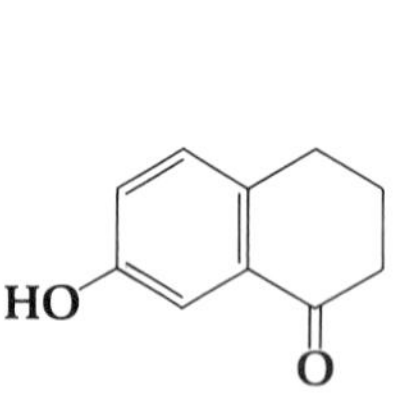

A 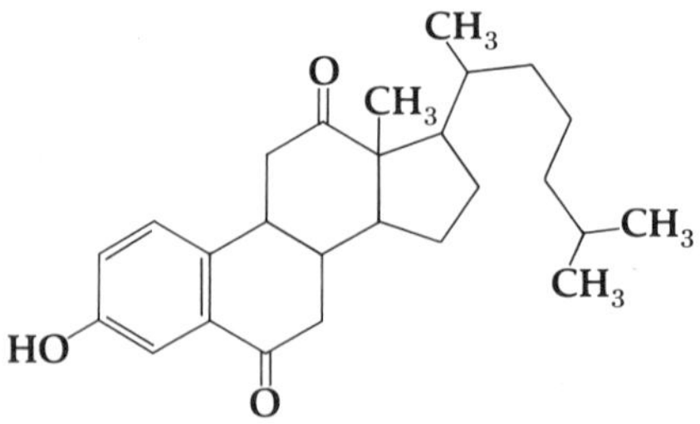   *B* 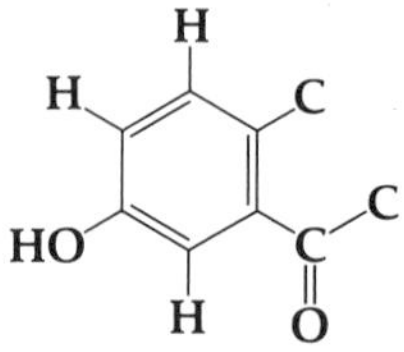   *C*

Thus, if one suspects an unknown compound of having structure B, one could establish at least its chromophore part by synthesizing A and establishing that the UV spectra of A and B are identical. This use of model compounds is common practice in natural-product studies and is often invaluable for establishing the skeleton of a large part (the chromophore part) of the molecule of an unknown compound. Since many UV spectra are now already tabulated in the chemical literature, it is usually not necessary to undertake a synthesis of the model compound as in this case.

When two chromophores in a molecule are insulated from each other by a break in conjugation (more specifically, a break in π-orbital overlap), they give rise to a spectral curve which is the sum of the two chromophores. Again, such a curve can be built up for comparison by adding the curves from two model compounds. A break in conjugation commonly occurs when one (or more) saturated carbon intervenes between two chromophores. In Fig. 7-9 the same spectrum would appear if the isolated double bond were saturated, but it would change if that double bond were moved into conjugation. Another instance involves an allene (C=C=C) in which the π electrons are perpendicular and do not overlap with each other.

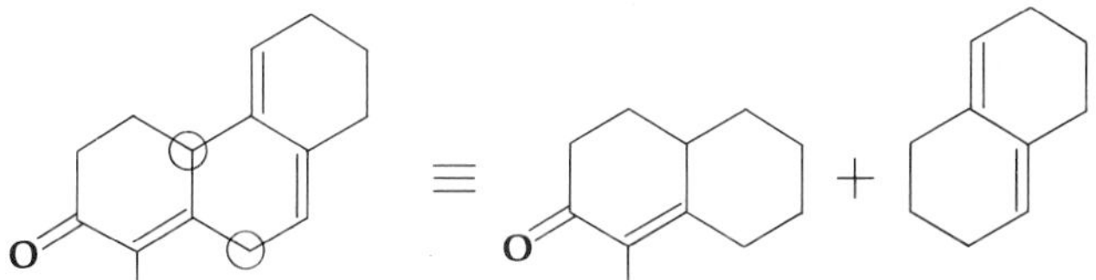

Insulating saturated   Model compounds
carbon shown in color
                          Calc 254 nm   Calc 245 nm

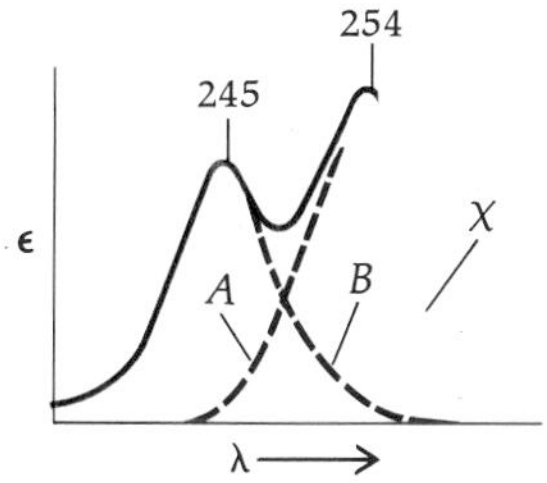

The UV of this antibiotic (mycomycin) is equal to the sum of the two chromophores shown below since the π-electron planes in the allene are perpendicular.

$$\text{HC}\equiv\text{C}-\text{C}\equiv\text{C}-\text{CH}=\text{C}=\text{CH}-\text{CH}=\text{CH}-\text{CH}=\text{CH}-\text{CH}_2\,\text{COOH} \quad \equiv$$

$$\text{C}\equiv\text{C}-\text{C}\equiv\text{C}-\text{C}=\text{C}-\text{C} \;+\; \text{C}-\text{C}=\text{C}-\text{C}=\text{C}-\text{C}=\text{C}-\text{C}$$

It is possible to use models to determine any UV spectrum, and the following examples will illustrate their use further. In practice only absorptions above 200 nm are observed since oxygen in the air absorbs at 200 nm and obscures peaks below this. Hence, simple unconjugated π or n electrons will not show in spectra above 200 nm (cf. ketones and aldehydes, olefins, carboxylic acid derivatives, halides, alcohols and ethers, amines, etc.). The model examples in Fig. 7-10 offer nomenclature practice and should be written out with structural formulas. It should be appreciated how much simpler the models are compared to the original molecule.† The value of these models for structure elucidation should also be apparent here and is explored in more detail in Sec. 7-8.

----

† As shown in the section on calculating spectra, σ-bonded substituents on the chromophore—even alkyl groups with no n electrons—afford slight bathochromic shifts in λ_{max}. Hence truly accurate models reproduce the alkyl substitution pattern on the chromophore (usually simply with added methyl groups). However, these bathochromic shifts are usually so slight that an unsubstituted chromophore model is often close enough.

Ultraviolet spectrum of   is the sum of the spectra of these models

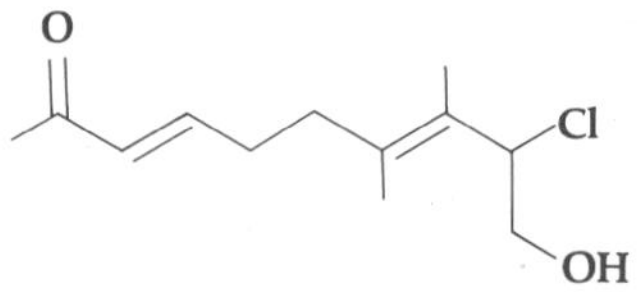

3-Penten-2-one   (calc 227 mμ)

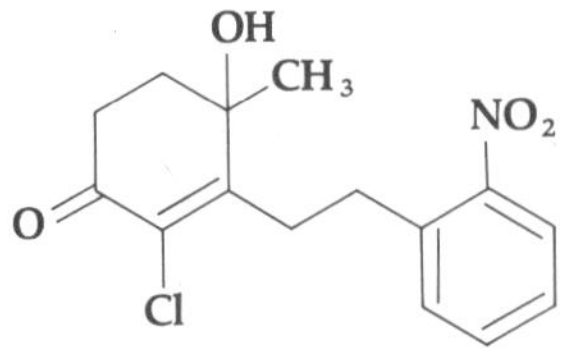

2-Chloro-3-methyl-2-cyclohexenone
(or 3-chloro-4-methyl-3-penten-2-one)
(calc 249 mμ) and o-nitrotoluene

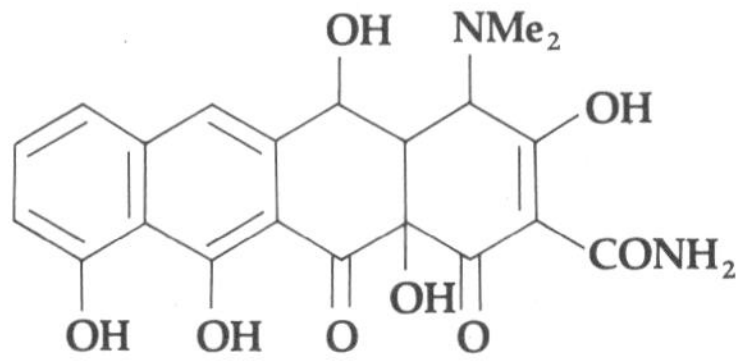

Anhydroterramycin
(from the antibiotic terramycin)

2-Acetyl-3-methyl-1,8-dihydroxynaphthalene
and 2-acetyl-3-hydroxy-2-butenoic amide

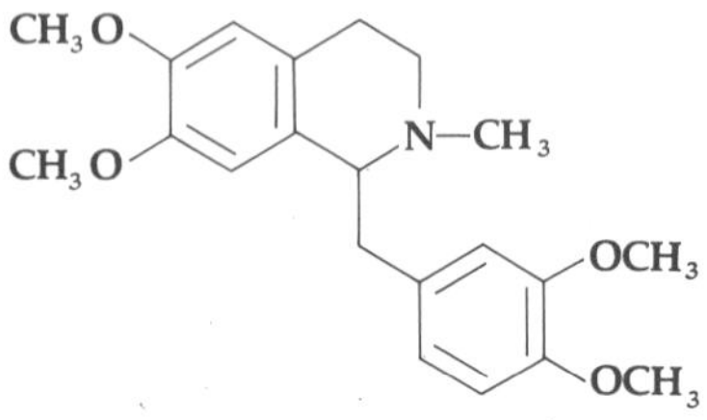

Laudanosine, an opium constituent

3-Methyl-and 3,4-dimethyl-
1,2-dimethoxybenzenes

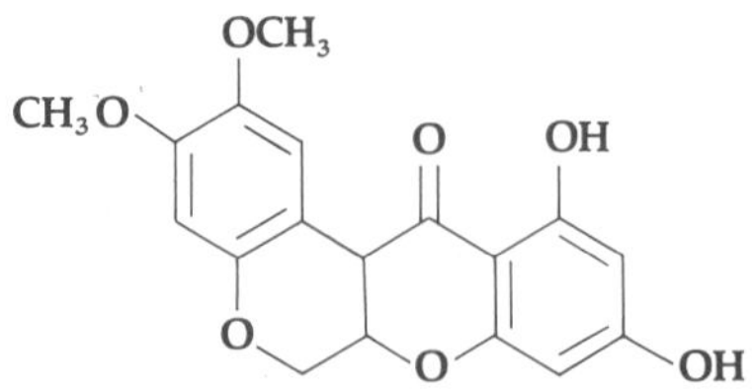

Apotoxicarol, a natural insecticide

1,4,5-Trimethoxytoluene
and 1-acetyl-2,4-dihydroxy-6-methoxy-benzene

FIGURE 7-10 **Examples of model compounds for ultraviolet spectra**

PROBLEM 7-6

Write structures consistent with these observations.

a An acid, $C_7H_4O_2Cl_2$, shows a UV maximum at 242 nm.
b A ketone, $C_8H_{14}O$, absorbs in the UV with λ_{max} 248 nm.
c An aldehyde, $C_8H_{12}O$, absorbs in the UV with λ_{max} 244 nm.

PROBLEM 7-7

Devise simple model compounds for duplicating the UV spectrum of the following complex molecules.

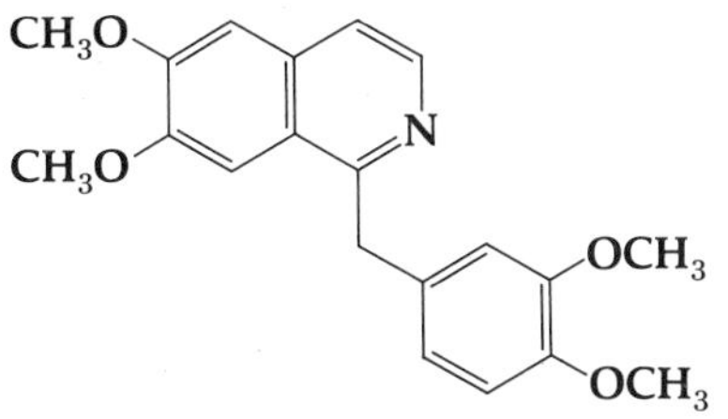

Papaverine
(from opium)

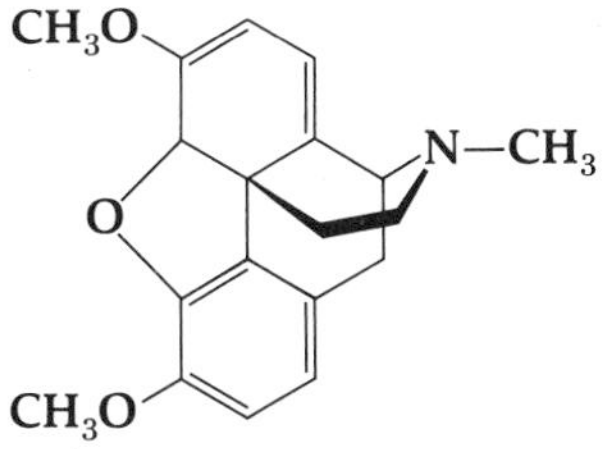

Thebaine
(from opium)

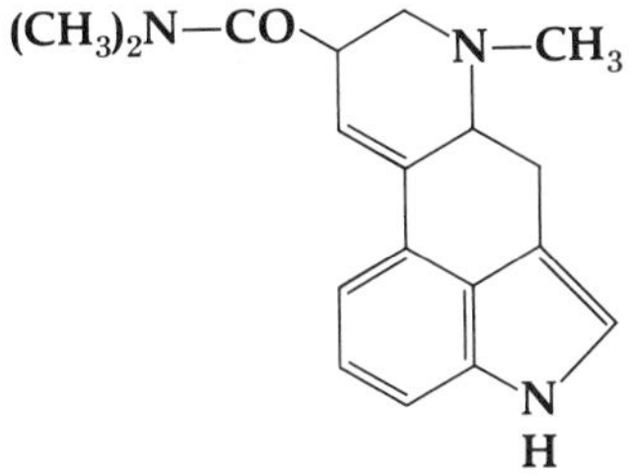

Lysergic acid diethylamide (LSD)

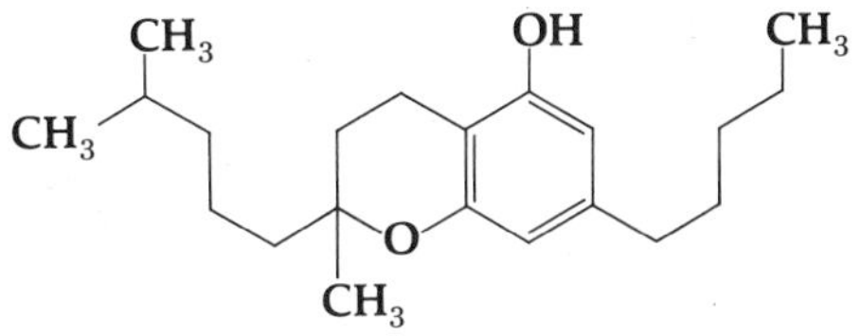

A derivative of marijuana

Pigments and Dyes

In order for the human eye to observe color in an object, the object must absorb light in the visible (400- to 800-nm) region so that some wavelengths are thus removed from ordinary white light (which is a continuum of all wavelengths) and only the remainder reach the eye, as indicated in this approximate table.

λ, nm	Light color	Color of substance absorbing at λ
~400	Blue	Yellow
500	Green	Red
600	Yellow	Blue
700	Red	Green

Organic compounds will thus appear colored if they have λ_{max} of absorption in the visible region and this in turn generally implies a rather extensively conjugated chromophore. Dyes and pigments are accordingly usually rather complex molecules with extensive π-electron systems. A simple diene like

1,3-butadiene absorbs at 215 nm and appears colorless to the eye but increasing the chromophore to the natural pigment, β-carotene, yields $\lambda_{max} = 450$ nm and an orange color; β-carotene is present in carrots, tomatoes, autumn leaves, and many other plant sources.

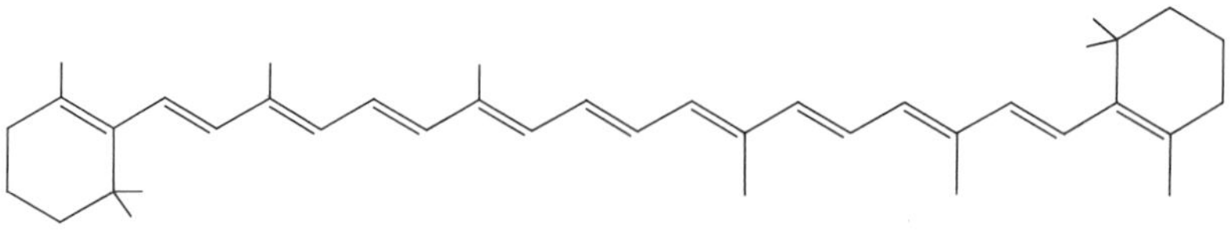

β-Carotene

With heteroatoms involved, chromophores need not be so extended in order for their absorptions to reach the visible region. The **cyanines** are variously substituted chromophores of the general form:

They exhibit strong resonance since the two resonance forms (arrows) are identical, and are intensely colored, only $n = 0$ being colorless. Various cyanines are very widely used as dyes and as pigments for color film, such as the ones shown.

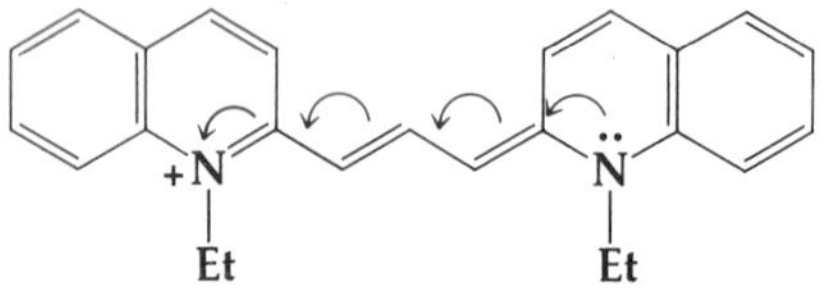

Pinacyanol (violet) 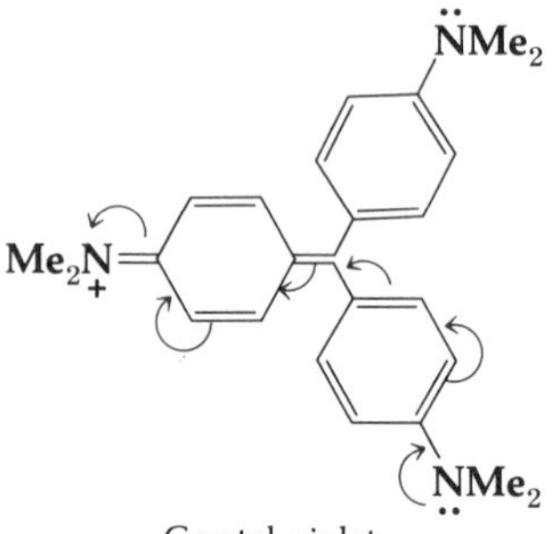   Crystal violet

7-5 INFRARED SPECTRA

The atoms in a molecule are not still. They vibrate and rotate in a variety of ways, always however in certain quantized energy levels. Changes in these energy levels are caused by absorption of infrared light. Considering any two atoms which form a bond, these are the kinds of motions and the infrared region in which they are excited:

1   Stretching: 2.5 to 15 μ (650 to 4000 cm^{-1})
2   Bending: 6.5 to 20 μ (500 to 1550 cm^{-1})
3   Rotation: $\gtrsim$ 12 μ ($\lesssim$ 800 cm^{-1})

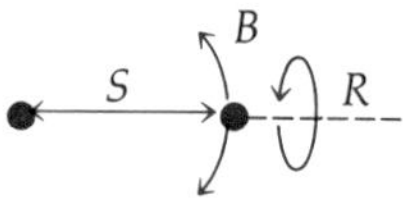

The stretching modes are the most important to structure determination and may conveniently be divided into four kinds, of which the first three show

up most clearly in IR spectra since they occur in regions uncomplicated by any other absorptions.

2.5–4.5 μ (2200–4000 cm^{-1}): bonds to hydrogen (or **D**): **OH, NH, CH, SH, OD**, etc.
4.5–5.0 μ (2000–2200 cm^{-1}): triple bonds (**C≡C, C≡N, A=B=C**)
5.5–6.5 μ (1550–1800 cm^{-1}): double bonds (**C=O, C=N, C=C, C=S**)
6.5–15 μ (650–1550 cm^{-1}):   single bonds (**C—C, C—N, C—O, C—X**, etc.)

The fundamental feature of value here is that most covalent bonds have *characteristic* and *essentially invariant* absorption wavelengths, no matter what their molecular environment, so that the presence of a band in the spectrum (particularly below 6.5 μ) indicates the presence of a bond in the molecule and the absence of an absorption peak guarantees the absence of its corresponding bond. Hence, in the spectrum recorded in Fig. 7-5 the N—H and C—H peaks show clearly near 3 μ, while the absence of triple bonds is implied by the lack of peaks in the 4- to 5-μ region. The two carbonyl groups appear near 6 μ.

Hooke's Law. The stretching vibration of two bonded atoms may be likened to the vibration of two balls at the end of a spring, a situation for which Hooke's law applies:

$$\nu = K\sqrt{\frac{f}{M_0}} \quad \text{or} \quad \lambda = K'\sqrt{\frac{M_0}{f}}$$

$$\longleftrightarrow \;(M_1)\!\!\!\begin{array}{c}\text{\small 000000000}\end{array}\!\!\!(M_2)\; \longleftrightarrow$$

where $M_0 = \dfrac{M_1 M_2}{M_1 + M_2}$ (reduced mass)

f = force constant

Thus the vibration frequency ν is directly related to the force constant f of the bond, which is a measure of its resistance to vibration or roughly to the bond energy, and is inversely related to the combined mass (reduced mass, M_0) of the two vibrating nuclei. From this comparison we may expect stronger bonds (among atoms of comparable mass—i.e., in the same row of the periodic table) to produce absorptions at lower wavelength (higher frequency) and this is confirmed in the bond strengths: triple > double > single. Also the absorptions of bonds including a light atom like hydrogen should also occur at lower wavelength (higher frequency), and this is confirmed in the region for bonds to hydrogen. It may be noted that S—H bonds, as well as bonds to deuterium are in the longer-wavelength (lower frequency) part of this region (4 to 4.5 μ), as expected from Hooke's law.

In the region above 7 μ (<1400 cm^{-1}) there are commonly far too many peaks to interpret since this is a lower-energy region. Accordingly, many more complex polyatomic vibrations as well as single-bond stretchings and bond bending frequencies occur together in this region. This spectral range, while dangerous for particular peak interpretations, is very useful for characterizing

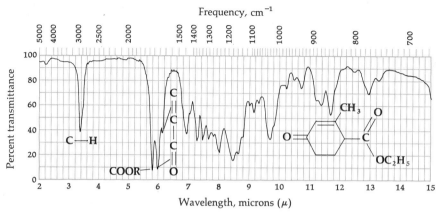

FIGURE 7-11   **Infrared spectrum of a ketoester**

compounds and accordingly has been dubbed the **fingerprint region**. The relative complexity in this part of the spectrum is apparent in Figs. 7-5 and 7-11.

The most valuable aspect of IR spectra is the double-bond region which shows up the absorptions due to the ubiquitous carbonyl group in its many functional variations. Fortunately, the specific position of a carbonyl peak tells two important things about the corresponding carbonyl group:

1   The extent of its resonance involvement
2   The ring size if it is located in a small ring

These two effects on λ_{max} are independent and additive. As to ring size, λ_{max} reflects the sp^2 bond angle and compression of this, as in a three, four-, or five-membered ring, relative to an acyclic carbonyl, lowers the wavelength (raises the energy and frequency) as the values indicate in Table 7-4 for cyclic ketones, esters (lactones), and amides (lactams).

The various functional groups in which C=O appears all differ in the nature and extent of the involvement of the carbonyl in resonance. Resonance can confer more triple-bond character or more single-bond character on the normal carbonyl (itself a hybrid of double- and single-bond character: $C{=}O \longleftrightarrow \overset{+}{C}{-}\overset{-}{O}$). This results in shifts of λ_{max} to, respectively, lower or higher wavelength (higher or lower frequency), i.e., *toward the regions in which triple or single bonds absorb.* These two kinds of resonance involvement may be sum-

TABLE 7-4   **Effect of Ring Size on C=O Absorptions**†

Ring size	Ketones, μ (cm⁻¹)	Lactones	Lactams	Anhydrides
4	5.65 (1770)	~5.5 (1820)	5.72 (1750)	—
5	5.75 (1740)	5.65 (1770)	5.85 (1710)	5.40 (1850) and 5.60 (1786)
6	5.85 (1710)	5.75 (1740)	5.95 (1680)	5.50 (1820) and 5.70 (1754)

† Data are roughly ±0.02 μ (±5 cm⁻¹) and refer to simple unconjugated systems.

marized as follows, with the norm (simple acyclic ketones and aldehydes) in the middle of two opposing resonance contributors,† A and B:

$$\overset{..}{\underset{..}{Y}}: \quad \underset{R}{C}\!\equiv\!\overset{+}{\underset{..}{O}} \longleftrightarrow \overset{..}{Y}\!-\!\underset{R}{C}\!=\!\overset{..}{\underset{..}{O}} \longleftrightarrow \overset{+}{Y}\!=\!\underset{R}{C}\!-\!\overset{..}{\underset{..}{O}}:$$

$$\underset{\to \text{ Lower } \lambda}{A} \quad \underset{5.85\ \mu\ (1710\ \text{cm}^{-1})}{\textit{Normal}} \quad \underset{\to \text{Higher } \lambda}{B}$$

Since neither A nor B is significant for simple aldehydes and ketones, these absorb at about 5.85 μ (1710 cm^{-1}).

Form A, by creating a resonance hybrid with more triple-bond character in the C—O bond, tends to lower the wavelength while form B, with more single-bond character, tends to raise the carbonyl wavelength. A qualitative assessment of how much one resonance form outweighs the other will then serve for a rough prediction of carbonyl absorption peak position in particular functional groups.

The order of groups Y— for form A is an order reflecting the ability of Y— to hold a negative charge and may be assessed by the relative stabilities of the anions Y:$^-$ as exhibited in the acidities of acids H—Y. (See Chap. 8.) The more acidic H—Y, the more stable Y:$^-$, and hence the greater the contribution of form A and the lower the wavelength of the carbonyl in R—CO—Y. Hence the order of common attached groups —Y, aligned with the pK of the corresponding acids H—Y, is listed below:

R—CO—Y	λ/C=O, μ (v, cm^{-1})	pK/H—Y
Y = —Cl	5.5 (1820)	~0
—OCOR	Av. ~5.6 (1790)	5
—Oϕ	5.70 (1750)	10
—OR(H)	5.75 (1740)	15
—NH$_2$(R$_2$)	~6 (1670)	30

On the other hand the greater the availability of electrons on Y for resonance into the carbonyl the greater the importance of form B and a shift to longer wavelength. Conjugation of C=O with ϕ—, C=C—, R$_2\overset{..}{N}$—, or $^-\!:\!\overset{..}{O}$— will have this effect, as the tabulation (Table 7-5) reveals, and comparisons can profitably be made between carbonyl compounds. Thus, conjugation of C=O with phenyl is less than with a simple double bond, owing to aromaticity, and this is evidenced in less shift of λ_{max}, while the cross-conjugation of oxygen on one side and unsaturation on the other diminishes such a shift for unsaturated carboxylic acids and esters relative to ketones. The full charge available for conjugation in —COO$^-$ causes the greatest shift [to 6.25 μ (1600 cm^{-1})], as would be anticipated. Assessments made in this way are of great value even though they are only very roughly quantitative.

---

†As a picture of resonance this diagram violates previously defined notions (Chap. 5) about the non-involvement of σ bonds in ordinary resonance. However, it is so useful a unifying conception with such predictive power that it is introduced here despite this. This example should *not* serve as a precedent for invoking "no-bond" resonance in other examples.

TABLE 7-5 **Infrared Absorptions of Diagnostic Value**

I *Bonds to hydrogen*

λ, μ	ν, cm^{-1}		*Notes*
2.8– 3.1	3200– 3600	O—H	H bonding broadens the band and raises its maximum to longer wavelength; very strong *internal* H bonding may eliminate the band from view, as in aromatic *o*-hydroxy aldehydes. In some cases of *external* H bonding, some OH groups are bonded, some not, and both bands may show. This is common in oximes.
2.9– 3.1	3300– 3500	N—H	Amides, and other nonbasic N—H show a very sharp, clean band here; —CONH$_2$ will show two distinct peaks. In ordinary amines the band is often diffuse or broad and sometimes not visible at all. Hydrogen bonding is often observed here as with alcohols.
3.0–3.7	2700– 3300	C—H	Present in all organic molecules; hence seldom useful.
∼3.5	∼2900	C—H saturated	
3.0–3.1	3200– 3300	C—H acetylenic	
3.1–3.5	2900– 3200	C—H aromatic and ethylenic	
3.6–3.7	2700– 2800	C—H on CHO;	Sharp band distinctive for most aldehydes but often weak in intensity.
∼4.0 ± 0.3	∼2500	S—H	Not much studied, but since few other groups absorb in this region; a band here (and sulfur present) probably indicates S—H.
2.8–3.5	2800– 3600	COOH	A broad scalloped band with no single maximum.
∼4.2	∼2400	$\overset{+}{\underset{}{\diagdown}}$N—H	Either an absorption maximum or a tail in this region indicates ammonium salts.

TABLE 7-5   **Infrared Absorptions of Diagnostic Value** (*Continued*)

II   *Triple-bond region*

λ, μ	ν, cm⁻¹		Notes
4.5 ± 0.2	~2200	—C≡C— —C≡N	Weak band unless conjugated.
4.60 ± 0.05	2200	—C=C—C≡N	Conjugated and aromatic nitriles: strong, sharp band.
4.7–4.8	~2100	—C=C=C— —C=N=N⁻ —N=C=O	Cumulated double bonds, allenes, etc.: strong, sharp band.

III   *Double-bond region*

A   Carbonyl groups (values roughly ±0.03 μ or ±10 cm⁻¹)

λ, μ	ν, cm⁻¹		Notes
5.50	1820	R—COX	Acid halides
5.50 and 5.70	1820 and 1750	(RCO)₂O	Acyclic anhydrides and six-membered-ring anhydrides; a separation of 0.20 μ is always observed between the two peaks of anhydride absorption; one peak is usually more intense than the other but there is no consistency about which peak.
5.70	1750	R—COOϕ	Phenyl (or vinyl) esters; lower wavelength in this range with more electron-withdrawing substituents on the phenyl ring, cf. —NO₂.
5.75	1740	R—COOR′ R—COOH	Esters and six-membered lactones, carboxylic acids.
5.85	1710	>C=O	Cyclohexanones (six-membered-ring) and simple acyclic ketones and aldehydes.
5.95	1680	—CO—N<	Six-membered lactams and simple amides; alkyl substituents on nitrogen raise wavelength to 6.1 or 6.2 μ.

TABLE 7-5 **Infrared Absorption of Diagnostic Value** (*Continued*)

III *Double-bond region* (*Continued*)

λ, μ	v, cm^{-1}		Notes

A Carbonyl groups (*Continued*)

6.25	1600	—COO⁻	Carboxylic acid salts.
			General: Conjugation of the carbonyl group with double bonds or aromatic rings tends to raise the absorption maximum 0.05–0.15 μ, rather less in acids and esters; double bonds have a greater effect than aromatic rings. α-Diketones show two bands, one lower than normal by 0.1–0.2 μ. Acyclic amides with N—H show a second intense band $\sim$6.5 μ, due to N—H *bending.*

B Other double bonds

6.0–6.2	1610–1670	C=C	The double-bond band is very weak, often too small to see, in unconjugated cases, but becomes very intense and at relatively longer wavelength in conjugated cases, more so when conjugated with ketones than with other double bonds; tetrasubstituted double bonds often do not show, while those conjugated to oxygen or nitrogen are very intense.
6.4	1560	R_2N—C=C—C=O RO—C=C—C=O	Exceptional conjugation raises double-bond band to 6.4 μ (ketone at 6.2 μ).
$\sim$6.0	1670	C=N	Intensity usually weak like double bonds. Oximes and imines show this band, which is not much affected by conjugation.
$\sim$6.5–7.5	1330–1540	—NO₂	This pair of strong bands shown by nitro compounds is relatively unaffected by conjugation.
$\sim$7.4 8.6	1160–1350	—SO₂⁻	Two strong bands in these positions are characteristic of sulfonyl chlorides, esters, and amides.

Since infrared spectra are convenient and rapid to obtain, it is common practice to observe the IR spectrum of a crude reaction product even before purification to ascertain that the spectral change from starting material is in accord with expectation. IR spectra are also conveniently used to monitor the course of a chemical reaction by observing spectra of aliquots (portions of the reaction) after several partial-reaction intervals. The great value of IR in structure elucidation is explicitly explored in Sec. 7-8. Meanwhile, examples of IR peaks in particular molecules are appended here and in Fig. 7-11 for acclimation. The C—H absorptions are usually complex and are simply reported here as about 3.5 μ (2860 cm^{-1}); only absorptions below 6.5 μ are shown.

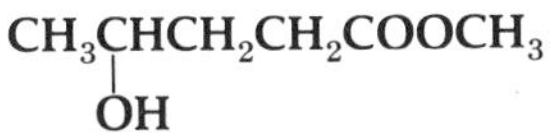
CH3CHCH2CH2COOCH3
OH

2.80 μ (3570 cm^{-1})
~3.5 (2860)
5.76 (1736)

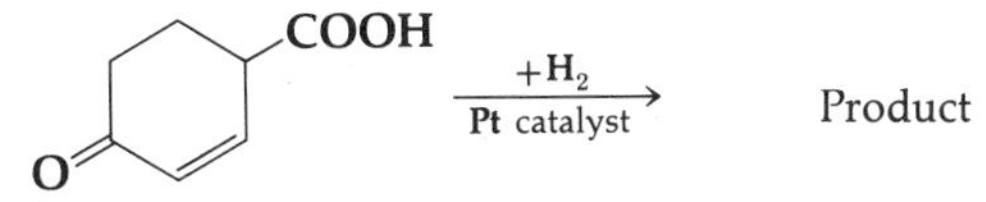
COOH
O
+H2
Pt catalyst
Product

2.9–3.5 μ (2860–3450 cm^{-1}) (broad)
~3.5 (2860)
6.02 (1660)
6.21 (1610)

Product:
2.9–3.5 μ (2860–3450 cm^{-1})
~3.5 (2860)
5.85 (1712)
(Write the structure.)

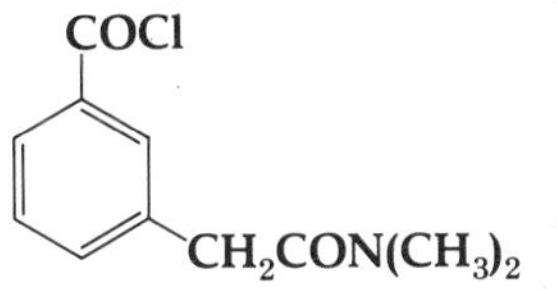
COCl
CH2CON(CH3)2

3.2–3.5 μ (2860–3125 cm^{-1})
5.62 (1780)
6.13 (1630)

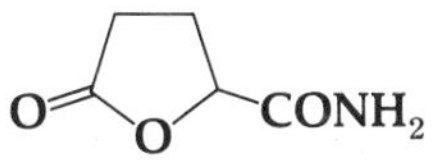
O
O
CONH2

2.95 μ (3390 cm^{-1})
3.00 (3330)
~3.5 (2860)
5.65 (1770)
5.96 (1680)

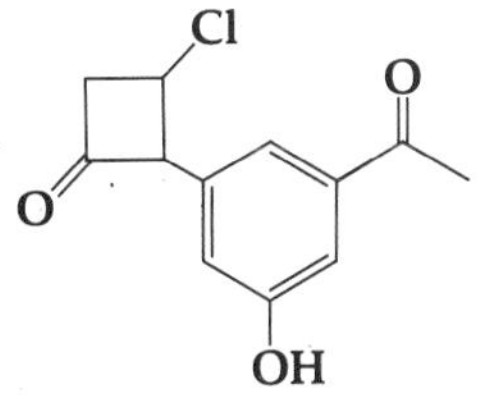
Cl
O
O
OH

2.89 μ (3460 cm^{-1})
3.2–3.5 (2860–3125)
5.62 (1780)
5.95 (1680)

PROBLEM 7-8

Indicate the expected IR peaks for the following compounds.

a Benzamide

b 4-Hydroxy-3-penten-2-one

c Dimethylcyclopropanone

d γ-Butyrolactone

e γ-Butyrolactam

f *p*-Nitrobenzonitrile

g 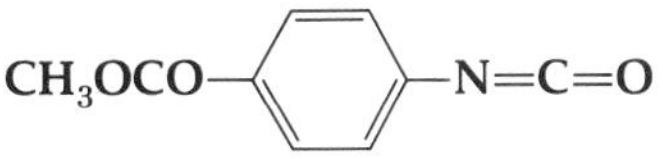
CH3OCO
N=C=O

h Penicillin:

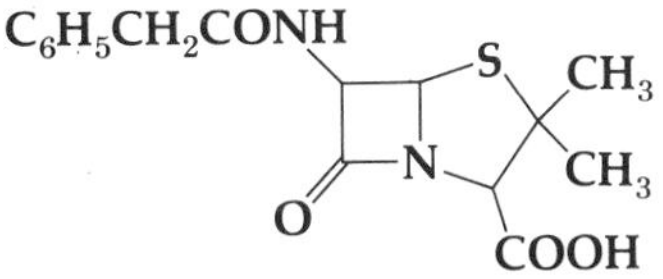
C6H5CH2CONH
S
CH3
N
CH3
O
COOH

i 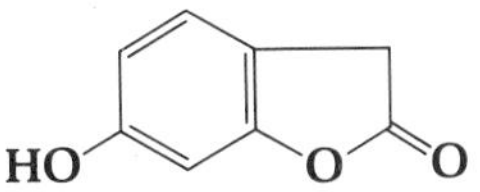
HO
O
O

j 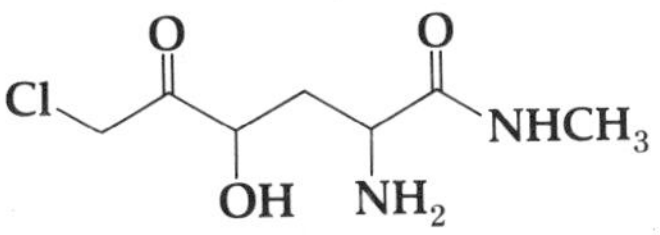
Cl
O
O
NHCH3
OH
NH2

k O=⬡⬡=O

l $(CH_3)_2NCOOCH_3$
   (Examine cross conjugation.)

PROBLEM 7-9

Arrange the following structure groups in order of increasing wavelength of carbonyl absorption in the IR spectrum.

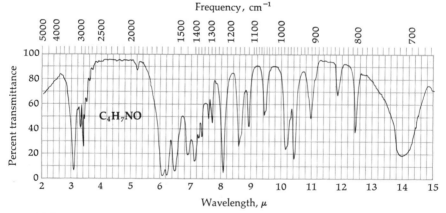

PROBLEM 7-10

Derive structures that fit the following spectral facts. Are any of these unique solutions?

a C_4H_8O shows no bands below about 6.7 μ (above 1500 cm^{-1}) except one at about 3.5 μ (2860 cm^{-1}).

b C_4H_7NO shows one sharp IR band at 3.08 μ (3310 cm^{-1}) and two strong ones at 6.00 and 6.20 μ (1610 and 1670 cm^{-1}), as shown in Fig. P7-10.

Frequency, cm^{-1}

[IR spectrum plot: Percent transmittance vs Wavelength, μ (2 to 15). Labeled C_4H_7NO.]

FIGURE P7-10

c $C_5H_3NO_2$: 4.4, 4.5, 5.8 μ (1725, 2210, 2280 cm^{-1}).
d $C_5H_6O_4$: 2.8, 5.65, 5.8 μ (1725, 1770, 3600 cm^{-1}).

Raman Spectra

Raman spectra are closely related to infrared spectra. Both arise from the excitation of vibrational motions. Since different methods of excitation are used, the selection rules are different. Raman spectrometers rivaling infrared instruments in convenience of operation have not yet been developed, but the newer models utilizing lasers show promise. Since Raman data complement those from infrared, further development of Raman spectroscopy is anticipated.

When a beam of light is passed through a transparent material some of the light is scattered at wide angles to the direction of incidence. In the spectrum of the scattered light, there are new lines, superimposed on the original spectrum. The *shifts* in the frequency correspond to molecular vibrational frequencies in the scattering material and constitute the Raman spectrum of the transmitting medium. Raman intensity depends upon the change in *polarizability* of the bond accompanying the excitation. As a consequence, symmetrical vibrations, which are "forbidden" in the infrared, give rise to the strongest Raman lines.

7.6   NUCLEAR MAGNETIC RESONANCE SPECTRA

The functional groups in a molecule are largely responsible for UV and IR spectra, but the major portion of a molecule is usually its hydrocarbon skeleton. Hence it would be useful for structure elucidation to have a physical method which specifically observes characteristics of the skeleton. Fortunately, the magnetic properties of the hydrogen nucleus allow us a powerful tool for observing the environments of the hydrogen nuclei which are so commonly bonded all over a molecular skeleton. These magnetic properties are observed in the nuclear magnetic resonance (NMR) spectra of organic compounds. These spectra allow the most detailed interpretation of any of the four discussed here and offer the most value for structure elucidation.

Physical Background

The hydrogen nucleus, or proton, may be regarded as a spinning, positively charged unit and so, like any rotating electric charge, it will generate a tiny magnetic field H' along its spinning axis (Fig. 7-12). If this nucleus is placed in an external magnetic field H_0 it will line up either parallel A or antiparallel B (Fig. 7-12) to the direction of the external field, the former A being of lower energy (more stable). The energy difference ΔE between the two states will be absorbed or emitted as the nucleus flips from one orientation to the other. Since $\Delta E = h\nu$, ΔE is related to the applied field H_0 and also to a radiation frequency ν. This small ΔE is in the radio-frequency (r.f.) range (longer wavelengths than Fig. 7-4). If the correct frequency is now applied to the sample containing hydrogen nuclei and the sample is placed in the external field H_0, the low-energy nuclei A absorb $\Delta E = h\nu$ and flip to B; on flipping back down they reemit $h\nu$ as a radiation signal that may be picked up instru-

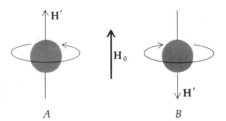

A   *B*

FIGURE 7-12   **Two spin states of the hydrogen nucleus in a magnetic field H**

mentally. The following equations give the fundamental relationships:

$$\Delta E = \frac{\delta h \mathbf{H}}{2\pi} \qquad\qquad 4$$

$$\Delta E = h\nu \qquad\qquad 5$$

$$\nu = \frac{\delta \mathbf{H}}{2\pi} \qquad\qquad 6$$

where ΔE = energy difference between the two spin states
δ = the **gyromagnetic ratio**, a constant characteristic of the particular nucleus
h = Planck's constant
$\mathbf{H}$ = the strength of the applied magnetic field at the nucleus
ν = frequency of the resonance absorption

All of this physical condition for exciting a signal from a hydrogen nucleus would be structurally valueless if it were not for the important fact that the magnetic field $\mathbf{H}$ which a nucleus *truly experiences* is not simply the external field applied $\mathbf{H_0}$, but rather the external field modified by the local environment of that nucleus, electrical and magnetic. The modified field $\mathbf{H}$ is that actually experienced by the hydrogen nucleus (proton) in a molecule and is the $\mathbf{H}$ in Eqs. (4) to (6) above. The external field $\mathbf{H_0}$ is modified by the **magnetic shielding** exhibited by the surrounding bonding electrons. The observed value of $\mathbf{H}$ is therefore a function of the molecular environment of the proton affording the signal.

In the NMR spectrometer the sample is placed in a tube in the midst of a magnetic field $\mathbf{H_0}$ and irradiated with radio waves of frequency ν. One of these applied influences is variable over the required range. Since the field $\mathbf{H}$ is proportional to the frequency ν, the sample may be scanned either by varying the external field $\mathbf{H_0}$ or the applied frequency. As each proton receives the particular applied field and frequency corresponding to its resonant frequency for its environment, it will broadcast an *output* frequency signal which is then picked up by an antenna coil (placed perpendicular to the input coil to avoid receiving simple input frequency). Current instruments, such as the diagram in Fig. 7-13, vary the external magnetic field $\mathbf{H_0}$ (which is usually around 14,000 gauss) and yield corresponding frequencies (for protons) around 60 megacycles/sec (60 million cycles), the variation due to different proton environments constituting a range of only about a thousand cycles per second.

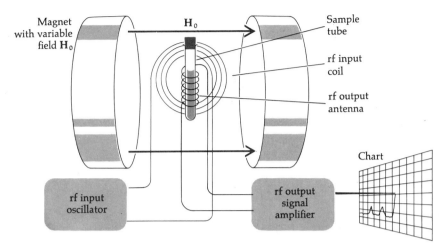

FIGURE 7-13   **Nuclear magnetic resonance spectrometer**

In fact a number of other nuclei also have these magnetic properties, and their molecular environments may similarly be studied. Such nuclei include C^{13}, F^{19}, and P^{31}, and they all resonate in different regions owing to their different characteristic gyromagnetic ratios (δ). More important to our purpose is the fact that the common organic nuclei, C^{12}, N^{14}, and O^{16}, do *not* have these magnetic properties and do not interfere with our observations on protons. A very important consideration is that deuterium has no magnetic properties in this range either. This has two useful consequences: (1) solvents with deuterium replacing hydrogen may be used without causing signals of their own; and (2) replacement of certain hydrogens in test molecules by deuterium causes their signals to disappear and allows such hydrogens to be located by comparing the spectra of compounds with and without the deuterium.

Chemical Shifts

Spectra of proton resonance, our sole concern here, are recorded as a series of sharp peaks at various frequencies relative to that of standard compounds. The position of the peak relative to a standard is called the **chemical shift** and reflects the shift in magnetic environment of the proton relative to the proton in the standard. The intensity of the peak† at a given chemical shift indicates the *number* of protons of identical environment which have that same chemical shift.

---

†The intensity is expressed by the area under the peak. This is electronically integrated automatically in modern instruments and may be read off the graph directly, as in the overlaid line shown in Fig. P7-12. A scale line for automatic integration is shown on a number of the NMR spectra recorded here. It is a horizontal line which shows a vertical displacement each time it encounters an absorption peak. The height of the vertical displacement is proportional to the number of protons causing the peak. The integration line is shown in Fig. P7-12 and in several problems. The spectrum for Prob. 7-12*c* shows three peaks integrating for one, six, and three protons, respectively (from left to right).

Even though it is the magnetic field which is experimentally varied, chemical shifts may be and usually are expressed in frequency units since frequency and field are related by $\nu = \delta H/2\pi$. Thus the chemical shift may be expressed as $\Delta \nu = \nu - \nu_s$, the frequency difference (in cycles per second = cps) from the frequency of the standard (ν_s), which is usually tetramethylsilane [tetramethyl silicon, $(CH_3)_4Si$, or TMS]. Tetramethylsilane yields an intense single peak since it has 12 protons which are structurally identical and all yield the same resonant frequency.

Chemical shifts expressed in cycles per second ($\Delta \nu$) are a function of the magnetic field and input frequency used in the instrument and, since these are different (40, 60, and 100 megacycles/sec instruments are common), it is more universal to apply units in which the *instrumental* frequency has been cancelled out. Such units are dimensionless and are a function only of the proton in the molecule. Two kinds are in common use:

δ *units:* $\delta = \dfrac{\nu - \nu_s}{\nu_0}$

τ *units:* $\tau = 10-\delta$

where ν = observed proton frequency, cps

  ν_s = tetramethylsilane frequency, cps

  ν_0 = input frequency of instrument, megacycles/sec

Both units of chemical shift are such that a range of 0 to 10 accommodates most observed chemical shifts, the standard $[(CH_3)_4Si]$ protons appearing at 0 on the δ scale and 10 on the τ scale. The standard instruments are scaled directly in both δ and τ units, as is Fig. 7-14.

A general view of the major kinds of chemical shifts is offered in Fig. 7-14, with a specific spectrum reproduced above it as an example. The particular proton(s) causing a given signal is indicated in color for identification. Chemical shifts corresponding to more detailed environments are listed in Fig. 7-15. Certain rough generalizations may be made to organize these empirical chemical shifts.

1 In otherwise equivalent local environments the more hydrogens on one carbon the greater the shielding (higher τ).

$\tau:$ $-\overset{\displaystyle |}{\underset{\displaystyle |}{C}}-H \; < \; -\overset{\displaystyle |}{C}H_2 \; < \; -CH_3$

2 In saturated C—H cases electron withdrawal (σ-bond dipoles) by attached electronegative atoms also causes deshielding, and the τ values are lower the greater the electronegativity. The same kind of deshielding is produced when the electronegative atom is one carbon further removed but the magnitude is then very small ($\tau < 1$).

$\tau:$

$H-\overset{\displaystyle |}{\underset{\displaystyle |}{C}}-F \; < \; H-\overset{\displaystyle |}{\underset{\displaystyle |}{C}}-O \; < \; H-\overset{\displaystyle |}{\underset{\displaystyle |}{C}}-N$

$H-\overset{\displaystyle |}{\underset{\displaystyle |}{C}}-F \; < \; H-\overset{\displaystyle |}{\underset{\displaystyle |}{C}}-Cl \; < \; H-\overset{\displaystyle |}{\underset{\displaystyle |}{C}}-Br \; < \; H-\overset{\displaystyle |}{\underset{\displaystyle |}{C}}-I$

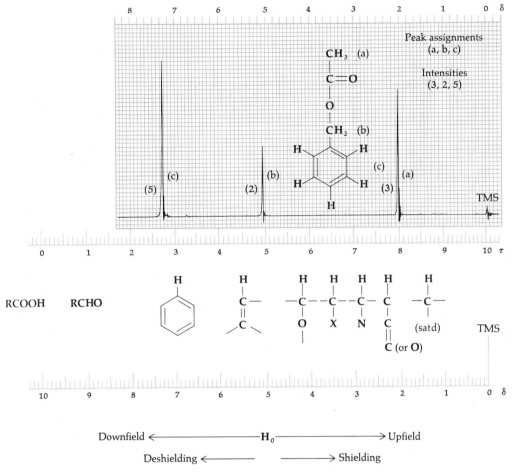

FIGURE 7-14 **Sample NMR spectrum and general chemical-shift regions**

3 These effects are partially additive, $Cl_2CH\underline{C}HCH_3$ absorbing downfield (lower τ) of $ClC\underline{H}(CH_3)_2$ but not twice as far.

4 Ring size in alicyclic compounds causes more shielding the smaller the ring, cyclopropanes having the highest hydrocarbon τ values.

5 Unsaturation lowers τ values, although alkynes ($C{\equiv}C{-}H$) are out of line (Fig. 7-15).

$$\phi{-}H \ < \ C{=}C{-}H \ < \ \phi{-}\overset{|}{C}{-}H \ < \ C{=}C{-}\overset{|}{\underset{|}{C}}{-}H \ < \ {-}\overset{|}{\underset{|}{C}}{-}\overset{|}{\underset{|}{C}}{-}H$$

$$O{=}\underset{|}{C}{-}H \ < \ O{=}\overset{|}{\underset{|}{C}}{-}\overset{|}{\underset{|}{C}}{-}H \ < \ {-}O{-}\overset{|}{\underset{|}{C}}{-}\overset{|}{\underset{|}{C}}{-}H$$

6 Protons on heteroatoms ($H{-}O, H{-}N, H{-}S$, etc.) show highly variable chemical shifts and sometimes broad peaks which can be so broad as not to be visible. These protons are rarely of value for structure correlations.

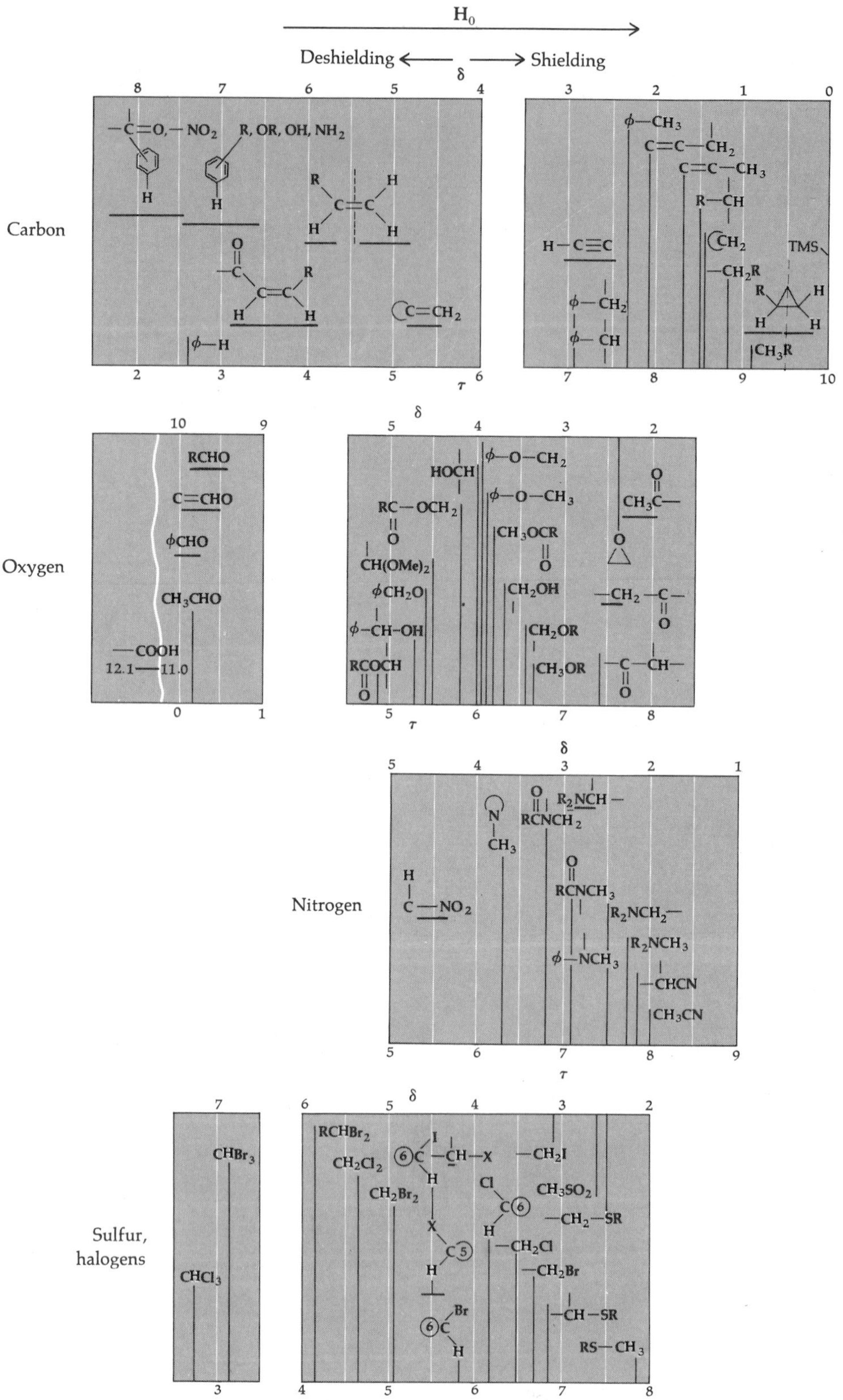

FIGURE 7-15 **NMR chemical shifts**

All hydrogens with identical environments in a molecule have the same chemical shift, i.e., absorb together at the same τ value. Hydrogens which are identical are of two kinds:

1 All hydrogens on a single carbon are identical if
 a They are the three protons of a methyl $—CH_3$.
 b They are the two protons of a methylene $—CH_2—$ which is free to rotate so that they become equivalent. Methylene protons which are not equivalent are those:
    In rigid cyclic systems;
    In cases of hindered rotation; and
    In positions adjacent to an asymmetric atom.
2 Hydrogens on different carbons yield the same absorption signal if they are structurally indistinguishable such as those marked in color in the examples.

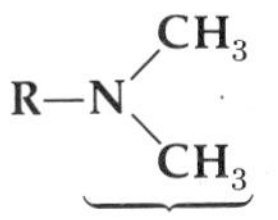

One signal

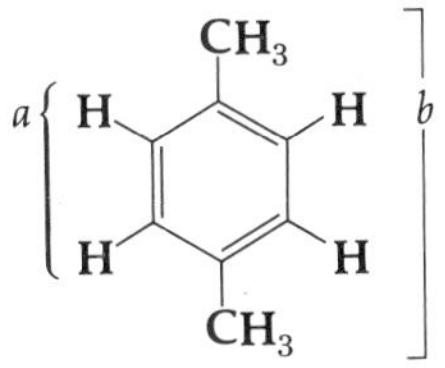

Two signals: 2.7 τ (a)
7.7 τ (b)

In a given compound, however, many protons may be so very *similar* in chemical shift that their signals overlap too much to be distinguishable. This is most commonly the case with saturated C—H signals, of which there are often a large number in a molecule. The total intensity of their combined peaks, however, does serve to ascertain the *number* of protons involved.

PROBLEM 7-11

Assign chemical shifts and relative intensities to the expected NMR peaks for these molecules.

a β-Benzoylpropionic acid
b 5,5-Dimethyl-2-cyclopentenone
c Methyl isopropyl ether
d Diethyl malonate $[CH_2(COOC_2H_5)_2]$
e Ethyl allyl ether

f Dimethylamino acetaldehyde
g 2,3-Butanediol monoacetate
h Propenal dimethyl acetal
i 1,1-Diphenyl-3-chloropropane
j 3-Methoxy-4-pentyn-2-one

PROBLEM 7-12

a An aromatic compound shows signals of the following chemical shifts in the NMR spectrum. Write a structure which fits these observations and indicate the bands expected in the IR spectrum. τ 1.0 (1H); 1.80 (2H); 2.32 (2H); 7.52 (3H).

b A compound $C_6H_{10}O_2$ with an IR peak at 5.65 μ (1770 cm^{-1}) shows three peaks in the NMR spectrum, at 5.8, 7.5, 9.1 τ, with relative intensities of 1:1:3. What is it?

c The three illustrated spectra (Fig. P7-12), represent mesityl oxide, a liquid with a molecular weight of 98. What is its structure?

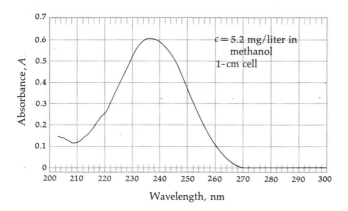

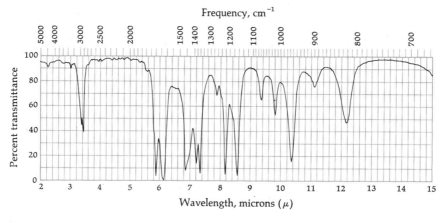

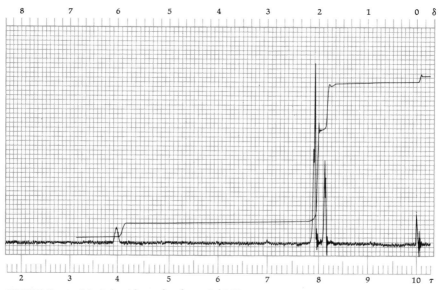

FIGURE P7-12   Mesityl oxide; molecular weight 98

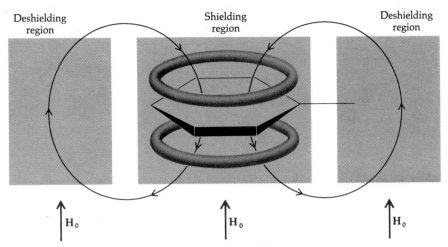

FIGURE 7-16   **Induced aromatic ring current and shielding**

Aromaticity

The shielding situation with respect to the magnetic effects of unsaturated groups is not so simple as presented in rule 5, but a clear and dramatic shielding effect of aromatic rings has been of enormous assistance in assessing aromaticity. An aromatic ring contains a cycle of π electrons, the lowest-lying molecular orbital being in fact the entire π-electron loop (Sec. 5-5). If we apply a magnetic field to this electron loop, electron circulation (**ring current**) is induced which in turn produces a counteracting magnetic field. This field acts either to shield or deshield neighboring protons found within the field according to whether they experience it parallel or antiparallel to the applied field H_0 as illustrated in Fig. 7-16.

The deshielding effect is seen in protons on the ring or on carbons directly

attached ($\phi{-}\overset{|}{\underset{|}{C}}{-}H$), both of which are at lower field than those in correspond-

ing olefinic examples. Furthermore, only if the ring is truly aromatic does the ring current occur, the common example being cyclooctatetraene, which is neither planar nor aromatic and shows no aromatic shift. Large aromatic ($4n + 2 \, \pi$ electrons; Sec. 5-5) systems have protons in both the shielding and deshielding regions, and NMR measurements have confirmed the aromaticity of ($4n + 2$) examples.

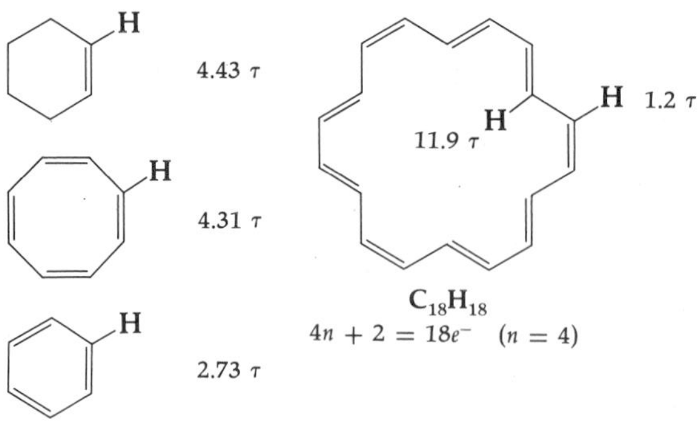
H
4.43 τ
H
4.31 τ
H
2.73 τ
H 1.2 τ
H
11.9 τ
$C_{18}H_{18}$
$4n + 2 = 18e^-$ $(n = 4)$

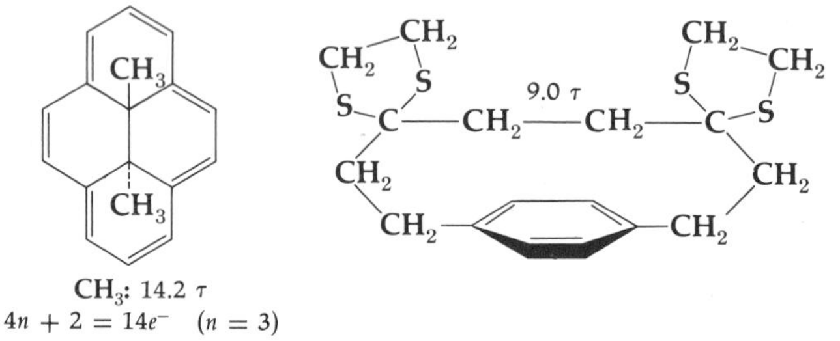
CH_3
CH_3
CH_3: 14.2 τ
$4n + 2 = 14e^-$ $(n = 3)$
CH_2
CH_2
S
9.0 τ
S
C
CH_2
CH_2
C
S
CH_2
CH_2
CH_2
CH_2
CH_2
CH_2

Spin Coupling or Splitting

NMR spectra are in fact both more complex and more informative than implied so far just by chemical shifts and intensities of proton signals. This comes about because the actual magnetic field (**H**) experienced by one proton is modified by the spin states (↑ or ↓)—and their attendant fields ±**H′**—of neighboring protons. This is illustrated in Fig. 7-17, in which we examine the signal given off by the colored proton in three different structural situations.

The neighboring protons (at left) can add or subtract their fields **H′** from the external field $\mathbf{H}_0$. The colored proton experiences each of the modified fields **H** thus created and gives a signal proportional to each and with an intensity ratio proportional to the probability of each field **H**. The effect is that a given proton exhibits an absorption signal which is **split** into several peaks because of **coupling** with its neighboring protons. The chemical-shift position of the given proton is taken as the center of its split peaks. Several rules characterize this splitting of proton signals.

1 Splitting of a proton signal is caused only by neighboring protons with different chemical shifts from the observed proton.

2 Splitting of one proton by another on the same carbon is very uncommon as is splitting of protons separated by more than two atoms. For our purposes here only the case of protons on adjacent carbons (or other atoms), as in Fig. 7-17, need be considered (**vicinal protons**).

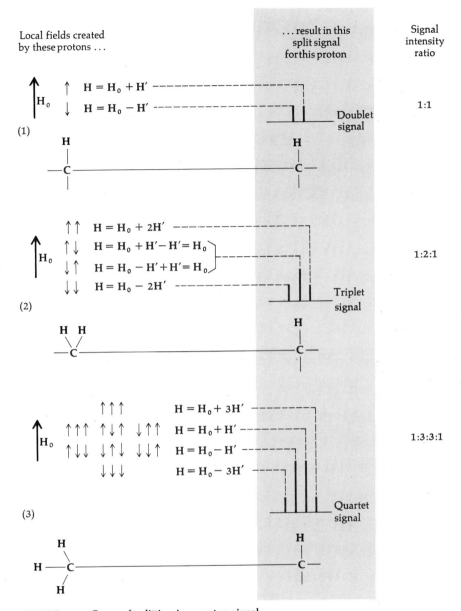

FIGURE 7-17 **Cause of splitting in a proton signal**

3 The number of peaks (N) into which a proton signal is split equals one more than the number of vicinal protons (n);

$$N = n + 1 \qquad N = 2 \text{ (one vicinal H)} = \textbf{doublet (d)}$$
$$= 3 \text{ (two vicinal H's)} = \textbf{triplet (t)} \qquad \text{as in Fig. 7-17}$$
$$= 4 \text{ (three vicinal H's)} = \textbf{quartet (q)}$$

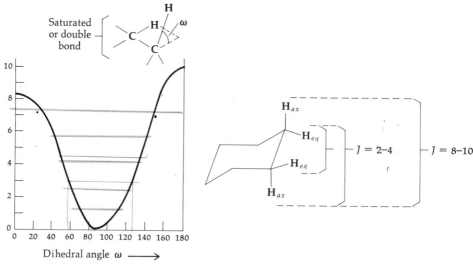

FIGURE 7-18   **Variation of coupling constant with geometry**

More accurately, doublets, triplets, and quartets are produced only by *identical* vicinal hydrogens, whereas two nonidentical vicinal hydrogens (cf. $-\overset{|}{C}-\overset{|}{C}-\overset{|}{C}-$) will yield two
 H H H
superimposed doublets, etc.

4   The *magnitude* of the splitting (distance between peaks) is independent of external field, a function solely of the molecular structure, as Fig. 7-17 should make clear. Accordingly, these splittings or **coupling constants** (J) are expressed in cycles per second and are commonly in the range 0 to 15 cps.

5   The size of the coupling constant J is a function of the geometric relation of the protons to each other. In freely rotating groups (H—C⊖C—H) the coupling constant is $J \sim 7$ cps. In rigid systems (cf. rings), the coupling constant is a function of the dihedral angle ω between the hydrogens as shown in Fig. 7-18. Some olefinic proton couplings are listed below.

	J, cps			J, cps
cis	7–12		ortho	6–9
trans	13–18		meta	1–3
			para	0–1

6   *Each* of a pair of coupled protons is split by the other and their coupling is represented by a single *coupling constant J*. Both signals show this same splitting. It will, however, only be clear in a spectrum if the chemical shifts of the two protons are enough different from each other (and from other protons) that their split signals stand out clear and unencumbered.

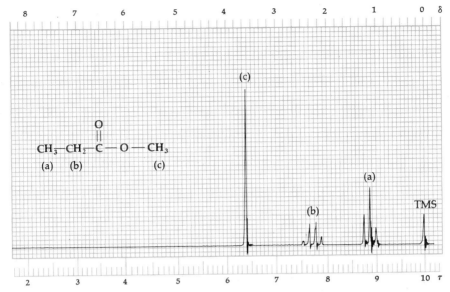

FIGURE 7-19   NMR spectrum of methylpropionate

7   In a technique called **spin decoupling** (or **double resonance**) it is possible to saturate the sample with a second radio frequency (apart from the main radio frequency input used to obtain the overall spectrum—Fig. 7-13). This second frequency is selected to be exactly that of one of the proton signals in the spectrum and has the effect of keeping all those protons effectively in their higher-energy spin state. Hence they only contribute a single field to their vicinal neighbors, and the coupling or splitting seen at the neighbor proton signal disappears. As an example, if the sample of methyl propionate in Fig. 7-19 is irradiated at the frequency (chemical shift) of the β-CH_3, then the α-CH_2 signal will collapse to a singlet, or if protons **e** in Fig. 7-20 are irradiated, the signal for proton **d** will collapse to a simple doublet (split only by **c**).

The several rules are exemplified in the actual spectra of Figs. 7-19 and 7-20. In the former the simple singlet, quartet, and triplet are seen in a clear example with well-separated chemical shifts. In fact the ethyl group is easily recognized in general by its coupled quartet (of two-proton intensity) and triplet (of three-proton intensity). In Fig. 7-20 the somewhat more complex spectrum of p-isopropylcinnamic acid analyzes the makeup of each peak from its splittings by neighboring protons, e.g., proton **b** is a doublet (due to **e**, $J = 16$ cps), and the same splitting is seen in the signal for adjacent proton **e**. The full septet of the **CH** split by two methyls is also visible. A simpler schematic representation of spectra is used with the other examples in Fig. 7-21 which illustrate the correlations between NMR spectra and molecular structure.

Proton-exchange Reactions

In addition to providing a highly informative tool for the study of the structures of organic compounds, NMR spectroscopy permits study of the rates

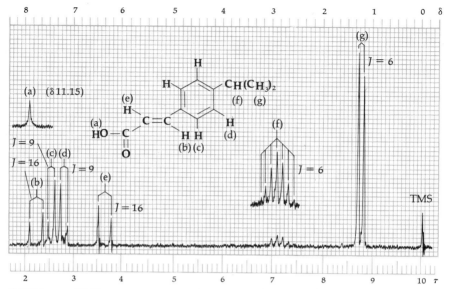

FIGURE 7-20   **NMR spectrum *p*-isopropylcinnamic acid**

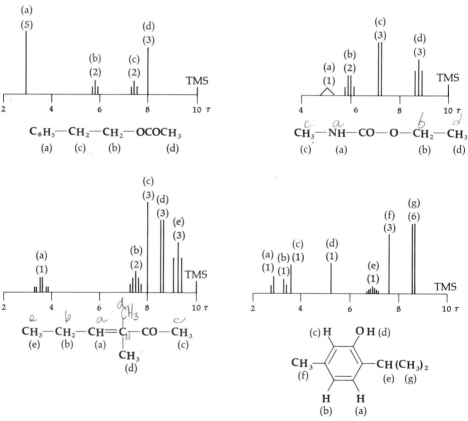

FIGURE 7-21   **Schematic representations of NMR spectra showing splittings**

of certain reactions that are too fast for measurement by ordinary techniques. Such reactions include fast proton transfers. If a proton is shuttled back and forth between two different magnetic environments at a rate that is fast in comparison with NMR transition times (the time involved in the actual excitation), the resonance observed will be simply that of the *averaged* effective field in the two environments. Such a system will show only one resonance for two different kinds of protons. For example, the **OH** protons in acetic acid and those in water have very different chemical shifts, which can be observed by measuring the spectra of the pure liquids. However, mixtures of the two substances show only a single **OH** resonance at a position intermediate between the two parent resonances. The position of the single peak depends upon the composition of the mixture; accurate determination of the position provides a means for quantitative analysis of water-acetic acid mixtures. The significance of the result is that the rate at which water and acetic acid trade protons is faster than the transition times. Specifically, results can be expressed in the form of an uncertainty principle.

$$\tau = \frac{1}{2\pi\Delta\nu}$$

where τ = shortest residence time that allows two distinct states to be distinguished
  $\Delta\nu$ = separation between the two resonances, cps

The two proton resonances are completely merged when measured with a resonance frequency of 40 megacycles/sec. The chemical-shift difference $\Delta\nu$, indicated by the resonance bands for the two pure liquids, is 240 cps. Consequently, the average time required to exchange protons must be less than about 7×10^{-4} sec. A contrasting example is provided by solutions of ethyl alcohol containing small amounts of water (20% or less). Such solutions show separated **OH** proton resonances at room temperature. The two signals can be collapsed to one averaged resonance by raising the temperature. Increasing the fraction of water in the sample causes collapse of the separate signals even at room temperature, showing that proton exchange becomes faster as the water content is increased.

Another method for detection of variations in the rates of a proton-exchange reaction is illustrated by comparison of Figs. 7-22 and 7-23. The latter figure is an NMR spectrum of pure anhydrous ethanol. The low-field resonance due to the **OH** proton is split into a triplet, and the group of bands due to the **CH₂** group is a poorly resolved octet. Clearly, there is strong coupling between the two groups; since the **CH₂** group is also coupled to the protons of the methyl group, the methylene resonance is split into a doublet of quartets. The simplified spectrum shown in Fig. 7-22 was obtained by addition of a trace of sulfuric acid. Coupling between the hydroxyl and methylene protons is destroyed because the exchange of protons between **OH** groups has become very rapid in comparison with the coupling constant (~5 cps) between the

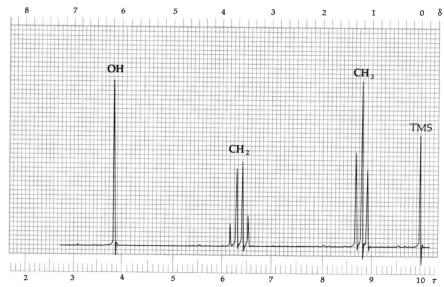

FIGURE 7-22 **NMR spectrum of ethanol with a trace of sulfuric acid at 60 megacycles/sec**

two groups. During the time of one pulse of the radio-frequency oscillator every oxygen has been host to many protons having spins of both $-\frac{1}{2}$ and $+\frac{1}{2}$; the net effect is that the methylene group acts as though it were not coupled to the OH proton. The data show very clearly that the following "symmetrical" exchange reaction is acid-catalyzed.

$$C_2H_5OH\alpha + C_2H_5OH\beta \overset{H^+}{\rightleftharpoons} C_2H_5OH\beta + C_2H_5OH\alpha$$

On the other hand when it is desired to observe the coupling of the —OH proton to those adjacent (cf., to distinguish —CH—OH from —CH$_2$OH), it is

only necessary to dissolve the compound in a solvent to which the —OH proton is strongly hydrogen-bonded. This causes the proton to stay on the oxygen, not exchanging but stabilized in place by its hydrogen bond to the solvent molecule. Dimethylsulfoxide $(CH_3)_2S{=}O$ is commonly used for this purpose.

Another technique of great value is **deuterium exchange**, in which advantage is taken of the fast proton exchange reaction to replace protons on heteroatoms in any molecule by deuterium. This is simply achieved by adding D_2O to the compound in solution (commonly in CCl_4 or $CDCl_3$) in the NMR cell, whereupon all —OH, —NH, —SH groups become —OD, —ND, —SD and their signals (and couplings) disappear. As will be discussed later, certain weakly acidic hydrogens bound to carbon can also be exchanged with deuterium under acidic or basic conditions (Chap. 12).

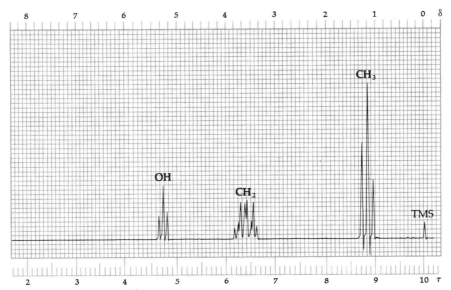

FIGURE 7-23   **High-resolution NMR spectrum of anhydrous ethanol at 60 megacycles/sec**

7-7   MASS SPECTRA

A rather different but very informative kind of spectrum of an organic molecule may be obtained by observing the masses of the fragments produced by destructive bombardment of a molecule with electrons in a vacuum. The electron impact dislodges an electron from the sample molecule, leaving an **ion-radical**, a positively charged species missing one electron (usually from an n or π orbital). This ion-radical is unstable and breaks up in various ways to produce more stable radicals and positive ions.

These fragmentations are bond-breaking processes and may be considered as chemical reactions. In practice many different fragmentations occur on electron bombardment and a **mass spectrum** is a sorted collection of the masses of all the charged molecular fragments (ions) produced. Its first importance is in registering the mass of the **parent ion** ($[A\cdot]^+$) and hence the *molecular weight* of the sample. Furthermore, the most intense peaks in the spectrum correspond to ions produced by the most favored bond-breaking processes. These are usually found to be the more-stabilized, or lower-energy, cations which can be produced from the parent molecule.

$$A: \xrightarrow{-e^-} [A\cdot]^+ \longrightarrow B\cdot + C^+ \xrightarrow{\text{etc.}} D^+ \longrightarrow E^+ \longrightarrow \text{etc.}$$
$$\searrow F\cdot^+ \longrightarrow G \longrightarrow \text{etc.}$$

The apparatus, schematically illustrated in Fig. 7-24, consists of a bombardment chamber in which the positive ions are produced by electron stripping and then accelerated (by attraction to a negative plate) down a tube (colored paths). Here a magnetic field is applied with the effect of bending

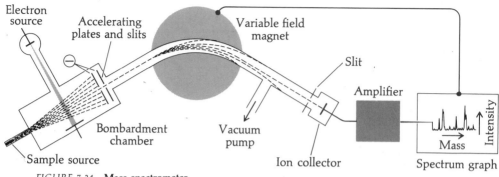

FIGURE 7-24 **Mass spectrometer**

the paths of the flying ions into circles, tangential to the entering path and having a radius proportional to the mass of the ion.† The curved tube has a fixed radius of curvature but the magnetic field can be varied continuously. Hence at any given magnetic field only particles of a single mass will have exactly the correct path curvature to reach the collector slit at the end. By varying the field, therefore, an ordered progression of ion masses is fanned past the collector slit, and their intensities are separately and sequentially measured to yield a spectrum of mass vs. intensity. Ions which do not reach the collector at any time are discharged on the tube walls, and the molecular debris (including neutral bombardment fragments) is swept out by the vacuum pump.

In the fragmentation process only the cationic fragments are collected and measured, not the neutral molecular or free-radical ($R \cdot$) species. The masses of these species may be discerned by subtraction, however. If a molecule of molecular weight (mass) P, observed as the parent ion, i.e., highest mass in the spectrum, fragments to a smaller ion of mass m, then $P - m$ is the mass of the neutral fragment produced. Stabilized cationic species are likely to appear as intense peaks, as in these resonance-stabilized fragments produced by bond breaking as indicated from the parent molecules:

Parent molecule	*Neutral radical*	+	*Stabilized cation*

$$R-\overset{\displaystyle |}{\underset{\displaystyle |}{C}} \! + \! \overset{\displaystyle |}{\underset{\displaystyle |}{C}}-\overset{\displaystyle ..}{\underset{\displaystyle ..}{O}}-R'(H) \xrightarrow{-e^-} \quad R-\overset{\displaystyle |}{\underset{\displaystyle |}{C}}\cdot \quad + \quad \overset{\displaystyle |}{C}=\overset{\displaystyle +}{\underset{\displaystyle ..}{O}}-R'(H)$$

$$R-\overset{\displaystyle |}{\underset{\displaystyle |}{C}} \! + \! \overset{\displaystyle |}{\underset{\displaystyle |}{C}}-\overset{\displaystyle ..}{\underset{\displaystyle |}{N}}- \xrightarrow{-e^-} \quad R-\overset{\displaystyle |}{\underset{\displaystyle |}{C}}\cdot \quad + \quad \overset{\displaystyle |}{C}=\overset{\displaystyle +}{\underset{\displaystyle |}{N}}-$$

$$R-\overset{\displaystyle |}{\underset{\displaystyle |}{C}} \! + \! \overset{\displaystyle |}{C}=\overset{\displaystyle ..}{\underset{\displaystyle ..}{O}} \xrightarrow{-e^-} \quad R-\overset{\displaystyle |}{\underset{\displaystyle |}{C}}\cdot \quad + \quad C\equiv\overset{+}{\underset{\displaystyle ..}{O}}$$

$$R-\overset{\displaystyle |}{\underset{\displaystyle |}{C}} \! + \! \overset{\displaystyle |}{\underset{\displaystyle |}{C}}-\overset{\displaystyle |}{C}=\overset{\displaystyle |}{C}- \xrightarrow{-e^-} \quad R-\overset{\displaystyle |}{\underset{\displaystyle |}{C}}\cdot \quad + \quad \overset{+}{\underset{\displaystyle |}{C}}-\overset{\displaystyle |}{C}=\overset{\displaystyle |}{C}-$$

†The proportionality is actually to the mass/charge (m/e) ratio but in actual instruments the ions are essentially all singly charged so that in effect only the masses are observed.

In each case the original removed electron may be one from a π or n orbital, in

the first instance yielding $\left[R{-}\overset{|}{\underset{|}{C}}{-}\overset{|}{\underset{|}{C}}{-}\overset{..}{\underset{..}{O}}{-}R' \right]^{+}$ which can fragment to the prod-

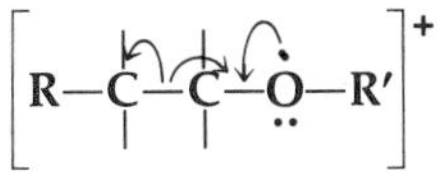

ucts above by movement and re-pairing of single electrons as implied by the arrows. (Single-barbed arrows imply the movement of single electrons.) A common source of fragmentation is a synchronous re-pairing of electrons around a six-membered ring as illustrated in the usual major mode of fragmentation of carbonyl groups bearing γ-hydrogens:

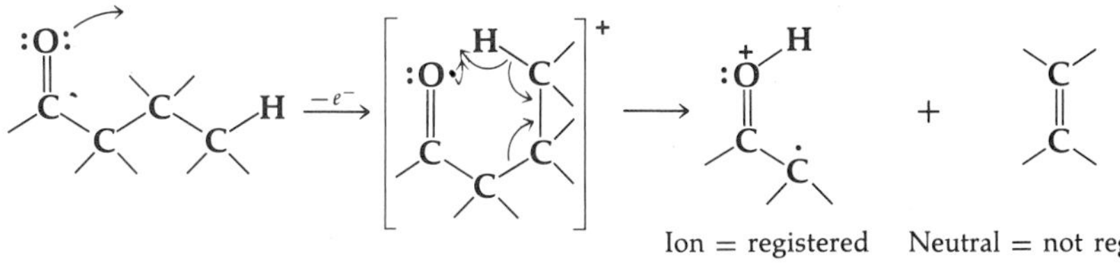

Ion = registered   Neutral = not registered

In order to appreciate the effect of these fragmentations on the mass spectra it will be useful to examine some particular examples so as to see specific fragment masses. In 3-methyl-4-phenyl-2-butanone (molecular weight 162) the cyclic carbonyl fragmentation is impossible since there is no γ-hydrogen. The expected fragments are these.

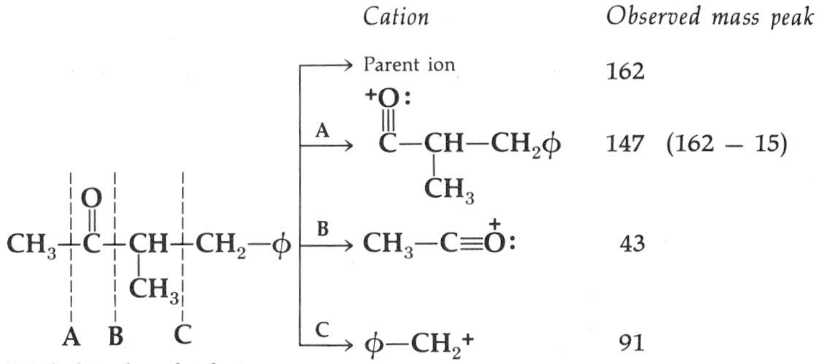

	Cation	*Observed mass peak*
	Parent ion	162
A	$\overset{+O:}{\underset{\|\|}{C}}{-}CH{-}CH_2\phi$ with CH_3	147 (162 − 15)
B	$CH_3{-}C{\equiv}\overset{+}{O}:$	43
C	$\phi{-}CH_2{}^+$	91

3-Methyl-4-phenyl-2-butanone
(mol. wt. = 162)

In 5-dimethylaminopentanoic acid methyl ester (molecular weight 159) we expect fragments from all possible losses of single n or π electrons and the sum of that expectation (in the above terms) is this:

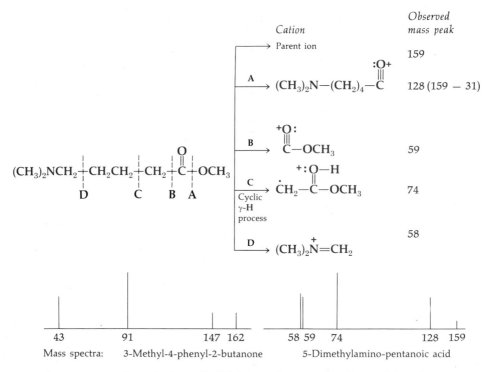

Mass spectra: 3-Methyl-4-phenyl-2-butanone   5-Dimethylamino-pentanoic acid

An easy way to orient one's thinking about molecular weights is to appreciate that the saturated hydrocarbon C_nH_{2n+2} has a molecular weight of $(14n + 2)$. Thus in general saturated compounds with C, H, O, N have molecular weights which are multiples of 14 with an extra two units and one or two extra units for each nitrogen and oxygen present. With unsaturated compounds, of course, the molecular weight is reduced by two for each ring or double bond. In most cases, then, it is easy to count the number of C, N, O atoms by dividing the molecular weight by 14 and considering the integer on each side of the answer. Thus, for a molecular weight of 160 ($160 \div 14 = 11 + \frac{6}{14}$), the number of C, N, O atoms is either 11 or 12, as in $C_9H_{20}O_2$, $C_6H_{16}O_3$, $C_7H_{16}N_2O_2$ with 11 "heavy" atoms or $C_{12}H_{16}$, $C_{11}H_{12}O$, $C_{10}H_{12}N_2$, etc. Always remember that odd numbers of odd-valent heteroatoms (N, P, halogen) result in odd-numbered molecular weights. All other (most) molecular weights are even numbers.

PROBLEM 7-13

Assign the masses of the fragments (positive ions) you might expect to be registered with these compounds in the mass spectrometer.

a 2-Hexanone

b 6-Hydroxy-2-hexanone

c 5-Methoxy-2-hexanone

d 4-Dimethylaminobutanoic
   acid

e 7-Dimethylamino-3-heptenoic acid

f All the previous five with deuteriums
  replacing all hydrogens at the 3-position
  in each

PROBLEM 7-14

Derive a structure consistent with the following peaks in a mass spectrum.
$m/e = 43, 57, 91, 105, 147, 162$

Identity of Fragments

The resolution of ordinary mass spectra is usually about $1:1,000$, which means that individual integer masses are distinguishable up to molecular weights of about 1,000. For most purposes this accuracy (fragment masses known to 1 mass unit) is adequate but it will not serve to differentiate between fragments of the same nominal mass but different compositions, such as C_2H_2O and C_3H_6 ($m = 42$) or C_3H_6N and C_3H_4O ($m = 56$), since CH_4, NH_2, and O are all nominally masses of 16. However, their *true* masses do differ enough to make clear distinctions in the *compositions* of fragments if the spectra can be obtained to an accuracy of 0.001 mass unit.

Molecular unit	Accurate mass
H	1.008
C	12.000
N	14.007
O	15.999
C_2H_2O	42.015
C_3H_6	42.048
C_3H_6N	56.055
C_3H_4O	56.031

Such high-resolution mass spectra can be obtained, although the number of peaks becomes enormous and is best handled by direct introduction of digital output data into a computer, which can then publish for a given sample spectrum a list of the fragments in terms of their *molecular constitutions* (empirical formulas), and ordered by their intensities in the spectrum. The method also serves to yield the exact molecular constitution of the parent molecule, of course, and so can displace the traditional quantitative elemental analysis (Sec. 1-5). With complex molecules this specificity with respect to the composition of each ion is very valuable for analyzing the fragmentation process.

Another method of examining the nature of the fragments involves replacement of an atom by an isotope (cf., deuterium for hydrogen) and observing which of the fragments change appropriately in mass—implying presence of the isotope in the fragment—and which remain unchanged.

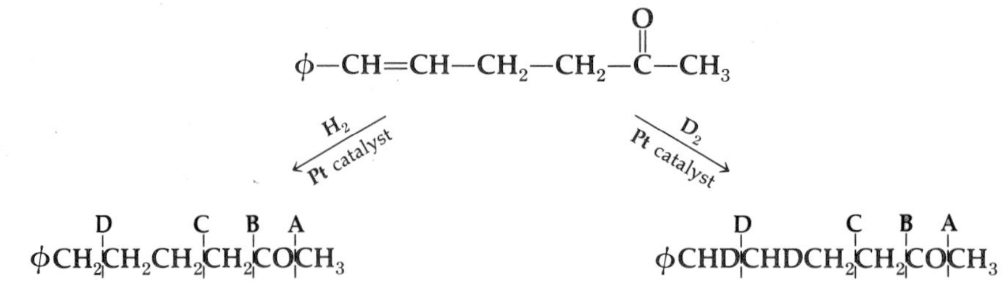

$$\phi-CH{=}CH-CH_2-CH_2-\overset{\overset{\displaystyle O}{\|}}{C}-CH_3$$

$\overset{H_2}{\underset{Pt\ catalyst}{\swarrow}}$ $\overset{D_2}{\underset{Pt\ catalyst}{\searrow}}$

D	C	B	A
$\phi CH_2CH_2CH_2CH_2COCH_3$			

Mass spectrum: Parent = 176
(- - - - -)   A = 161   (176 − 15)
                  B = 43
                  C = 57
                  D = 91

D	C	B	A
$\phi CHDCHDCH_2CH_2COCH_3$			

Parent = 178
   A = 163   (178 − 15)
   B = 43
   C = 57
   D = 92

7-8   SPECTROSCOPIC SOLUTION OF STRUCTURE PROBLEMS

In the previous sections we have explored how features of molecular structure give rise to spectroscopic manifestations and we have learned what spectra to expect for given structures. Now we must invert the process and learn to deduce structural features from given spectra, since this is the basis for the determination of the structures of unknown compounds. The course of reasoning continues from the discussion of Sec. 3-3 and in general proceeds via the following steps. In order to ensure rigor in the deductive process it is important to proceed stepwise by a regular protocol, checking each new deduction against the limits of possibility already established.

1 The molecular formula establishes the overall molecular size and is confirmed by the molecular weight obtained from the mass spectrum. The formula also provides:
   a The number (and kind) of heteroatoms which define the limits on the number and kind of functional groups.
   b The index of hydrogen deficiency, which, taken with hydrogenation or other evidence of double bonds, reveals the number of rings.
2 The data from spectra are separately analyzed, and each feature of structure implied is separately listed first. It is important to list features which are proved absent as well as those proved present.
3 Spectral evidence is taken in the following order: UV, IR, MS, NMR, and simple qualitative evidence on acidity or basicity is included (cf. Chap. 8).
4 Partial structures are written to summarize the structural features deduced at each step. Thus the whole sequence of deductions is ordered and illustrated by a stepwise progression of partial structures of increasingly detailed definition, culminating in a final complete structure.
5 Each structural feature and partial structure is compared with the molecular formula to be certain it does not exceed the number of atoms available, and the number of *unplaced* atoms of each kind is noted at each stage.

6   If several partial structures or families of structures can be drawn at one stage and none easily eliminated, each should be independently and separately followed through to see if later evidence eliminates one.

7   It may not be possible from the data to arrive at a *unique* structural solution, in which case the final solution is still a set of possible structures. Commonly, a test is then devised so as to distinguish among the set: usually one or more chemical reactions followed by analyses and spectra of the products.

8   The final structure obtained must be reexamined to check that it fits all the evidence. If a unique solution is obtained, the evidence should be retraced with a view to querying the uniqueness.

9   The stereochemistry of the molecule is often not fully defined by spectral evidence and must be separately sought after the nonstereochemical structure is determined.

Functional-group Evidence (UV, IR, MS)

The UV (and visible) spectrum shows whether any conjugated chromophores exist. The absence of any λ_{max} above ~210 nm precludes the presence of conjugation. Ketones and benzene rings without conjugating substituents do absorb at 280 and 260 nm, respectively, but at such low intensity ($\epsilon < 100$) that the peak will be missed unless it is specially sought out with high concentration solutions. Aromatic compounds, however, always show (at least) very large absorption ($\epsilon = 5{,}000$ to $10{,}000$) around 210 to 220 nm, usually as a tail from an intense maximum ~200 nm (which is not reliably recorded in air). Such a strong tailing absorption at the lower end of the spectrum is good evidence for an aromatic ring. A single symmetrical peak with λ_{max} below ~320 nm (and $\epsilon = 5{,}000$ to $15{,}000$, that is $\log \epsilon \sim 4.0$) is likely to be a polyene or enone susceptible to calculation (Fig. 7-8). More complex spectra (more than one λ_{max}) require a search for literature models of similar appearance.

While the UV spectrum shows conjugated chromophores, the IR shows functional groups. In examining an IR spectrum of an unknown compound one checks only the three uncomplicated regions for definite assignments:

Bonds to H: *2.8 to 3.2 μ (3100 to 3600 cm^{-1}):* either **OH** or **NH**; hydrogen bonding will broaden the peak and move it to longer wavelength (still <3.4 μ); amine **N—H** is often weak and unreliable; —**NH$_2$** in amides and ϕ—**NH$_2$** give two sharp bands; —**COOH** shows a broad scalloped band across the whole region and can obscure other peaks. *~3.5 μ (~2900 cm^{-1}):* CH bands in all compounds and so not useful and usually not quoted. *3.5 to 4.3 μ (2300 to 2900 cm^{-1}):* broad band usually $\overset{+}{\text{N}}$—H; otherwise, **S—H** or deuterium.

Triple Bonds: Triple bonds rather invariant around 4.5 μ (2200 cm^{-1}) and often weak; cumulated bonds (**X=Y=Z**) usually ~4.8 μ (2100 cm^{-1}) and strong. See Table 7-5.

Carbonyl: *5.5 to 6.4 μ (1550 to 1800 cm^{-1}):* strong band is almost always carbonyl, never **C=C** unless conjugated; absence of band means no carbonyl; in general one band equals one carbonyl, but two carbonyls of similar absorption may show as only one peak. The number of carbonyls cannot be less than the number of peaks. For detailed assignments consider both ring size (Table 7-4) and conjugation (Table 7-5) as roughly additive.

Other bands which may be picked out in an unknown spectrum are the nitro and sulfonyl pairs of absorptions listed in Table 7-5; obviously the presence of these groups must also depend on the presence of nitrogen or sulfur as well as sufficient oxygen in the molecular formula. A strong band at 6.4 to 6.6 μ (1500 to 1550 cm^{-1}) often arises from N—H bending, especially in acyclic amides. Aromatic rings are characterized by a weak band at 6.2 to 6.3 μ ($\sim$1600 cm^{-1}) and a strong one at 6.8 to 6.9 μ ($\sim$1450 cm^{-1}) but so many C—H bending vibrations occur between 6.7 and 7.3 μ (1350 and 1500 cm^{-1}) that this region, like the C—H stretch region ($\sim$3.5 μ), is rarely interpretable.

With the mass spectrum the simplest information is obtained by observing peaks near the parent (i.e., within 50 to 100 mass units) and noting the size of small fragments lost by the mass of the $(P - x)$ fragment, e.g., indications such as loss of 15 for CH$_3$, 17 to 18 for —OH (H$_2$O), 31 for —OCH$_3$, etc. These usually indicate units attached by only one bond and unequivocally identifiable by mass alone.

It should be emphasized that all these spectral interpretations of functional groups should be written down separately and checked against the possibilities allowed by the molecular formula. One hopes to find some evidence of each functionality and so to have compiled a list of functional groups adding up to the molecular formula and its index of hydrogen deficiency. With respect to the formula, ethers are virtually the only functional group affording no evidence in these spectra and are often assumed by default in cases where an oxygen occurs in the formula but is not accounted for spectrally.

The index of hydrogen deficiency must of course equal or exceed the number of carbonyls and conjugated double bonds found in the spectra. The difference will equal isolated double bonds (deduced from molar uptake of H$_2$/catalyst) or rings. Rings are also implicit in carbonyl ring-size evidence (Table 7-4) and must also be accounted for with the index of hydrogen deficiency.

Some simple chemical evidence is commonly adduced and is included here for convenience. Hydrogenation is used to detect carbon-carbon double and triple bonds (Sec. 3-3) and these are readily distinguished from benzene rings, which require especially vigorous conditions and react slowly. The other principal chemical evidence which is readily obtained concerns acidity and basicity, amines being differentiated from all other nitrogenous functions by their basicity, observed by their salt formation and hence solubility in strong aqueous mineral acids (HCl, H$_2$SO$_4$, etc.). Amines are the bases of organic chemistry, and carboxylic acids are the main acidic functions (Chap. 8). Like most organic compounds carboxylic acids are usually only slightly soluble in water but they commonly will dissolve in aqueous bicarbonate (pH $\sim$ 7) since they then form carboxylate anions. Since virtually no other organic compounds are acidic enough to be neutralized by this weak base (HCO$_3^-$), carboxylic acids are often recognized by this simple solubility test (taken with the molecular formula and IR evidence, of course).

Finally, there are certain functional-group combinations which are virtually always disallowed owing to their chemical instability, described in later chapters. A brief list of the most common would include:

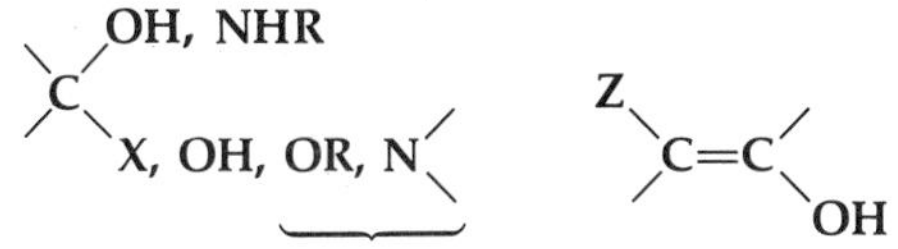

| (Except cyclic, see page 165) | (Except with C=O at Z) |

Carbon Skeleton Evidence (NMR)

The NMR spectrum provides three kinds of information:

1 The number of hydrogens causing each signal (**intensity**)
2 The environment of the hydrogens (**chemical shift**)
3 The number of vicinally adjacent hydrogens (**splitting**)

Each signal can be summarized for interpretation in a convenient format:

9.1 τ (3H) d **($J = 7$):** three hydrogens at 9.1 τ as a doublet with $J = 7$ cps splitting
4.62 τ (1H) s: one hydrogen at 4.62 τ as a singlet (unsplit)

When presented with NMR spectra of unknown compounds it is common to find the saturated CH absorptions at the right end ($>8\ \tau$) often cluttered with several absorptions together. Aromatic hydrogens (2 to 3 τ) may similarly appear as unreadable **multiplets** (abbreviation "*m*") unless they are polysubstituted. In such cases the *number* of hydrogens involved is frequently all that can be read in those regions, but this is still a significant datum. Hydrogens on heteroatoms (—OH, —SH, —NH) have highly variable shifts and may be broad peaks, but they may be counted and removed (to leave only CH signals) by running the spectrum again with added D_2O to cause exchange.

Of the CH peaks those, usually below 8 τ, which are well separated from their neighbors are first picked out and tabulated and their splittings noted (s = singlet, d = doublet, dd = pair of doublets with different splittings, t = triplet, q = quartet, m = multiplet). Next, their chemical shifts are judged with respect to the environment they imply (cf. —COOH, ϕH, C=C—H, O—C—H, X—C—H, O=C—C—H, saturated C—H, as in Fig. 7-14). Finally, the number of hydrogens *adjacent* to the one(s) giving the signal is noted from the splitting. It is then important to see whether both coupled hydrogens yielding the same splitting may be discerned, i.e., pairs which are therefore structurally vicinal. Methyl groups are especially easy to locate, even in the saturated region, since their three identical hydrogens create an intense single signal; also methyls are simple in splitting, being obliged to be singlets (◯N—CH₃, —O—CH₃,

$-\overset{|}{C}-CH_3$) or doublets ($>CH-CH_3$) or triplets ($-CH_2-CH_3$), with $J \simeq 7$ cps as they are free to rotate.

In proceeding to draw partial structures from the information it is useful to build up pieces of carbon skeleton with NMR-observed hydrogens specifically drawn and those bonds to carbon which are as yet uncertain but are *known not to be hydrogen* marked in some other clear way (here we use a small x†). Thus if a hydrogen signal is a doublet of two hydrogens in intensity, we can write

$$x-\overset{\overset{\displaystyle H}{|}}{\underset{\underset{\displaystyle H}{|}}{C}}-\overset{\overset{\displaystyle H}{|}}{\underset{\underset{\displaystyle x}{|}}{C}}-x.$$

(The two colored H's caused the signal; the other one split it.) It should be observed that this single fact has considerable deductive power since the bonds labeled x *cannot be hydrogens*. The number of structural possibilities is circumscribed now that it becomes necessary to fill those bonds with some of the carbons or heteroatoms in the molecular formula, since these atoms are generally in short supply. In general the NMR data has this great potency because of its ability to limit structural possibilities severely, even at a distance of some atoms away from the hydrogen giving a particular signal.

Examples

A few examples are carried through here to illustrate the generalities and procedures outlined above.

1   *A compound,* $C_6H_{10}O$, *shows a single* λ_{max} *of 240 nm (log* $\epsilon = 4.1$). This statement goes far to solve the problem: the index of hydrogen deficiency is 2 and the UV chromophore must be an α,β-unsaturated ketone (or aldehyde) since a diene cannot be created with enough substitution (using C_6 only) to yield λ_{max} so high (Fig. 7-8). There cannot be any rings since the chromophore already uses up the index of 2. Calculation of an α,β-unsaturated ketone shows two β-substituents yielding $215 + 2(12) = 239$ nm. Hence only three formulas are possible; the NMR spectrum with three 3H singlets $\sim 8 \ \tau$ proves the first one correct.

2   $C_5H_9NO_3$ *shows no important UV absorption above 200 nm and has IR peaks at 2.80, 2.95, 2.97, 5.85, 6.01, 6.45 μ (3570, 3390, 3367, 1710, 1664, 1550* cm^{-1}). [*The 3.5-μ (2860-*cm^{-1}) *peak of* CH *stretching is not normally quoted though present.*]

There is an index of 2 and it is all accounted for by two carbonyl groups (5.85, 6.01 μ); so the compound is acyclic. It has no C=C also (index) and no conjugation (UV). The three oxygens are two carbonyls and one O—H, the latter being one of

---

†This x cannot be confused with generalized hàlogen (X) here since, beginning with the molecular formula, we always know which, if any, halogen is present: Cl, Br, etc., not X in general.

three **N—H** or **O—H** peaks $<3\ \mu$; the other two must be **N—H**, and specifically **—NH$_2$** since only one **N** is present. The 6.01-μ (1664-cm^{-1}) peak must be an amide and so must be **—CONH$_2$**. The 5.85-μ (1710-cm^{-1}) peak can only be a ketone or aldehyde (Tables 7-4, 7-5), not **—COOH**. The following partial formulas can be written from the four possible **C$_5$** acyclic skeletons:

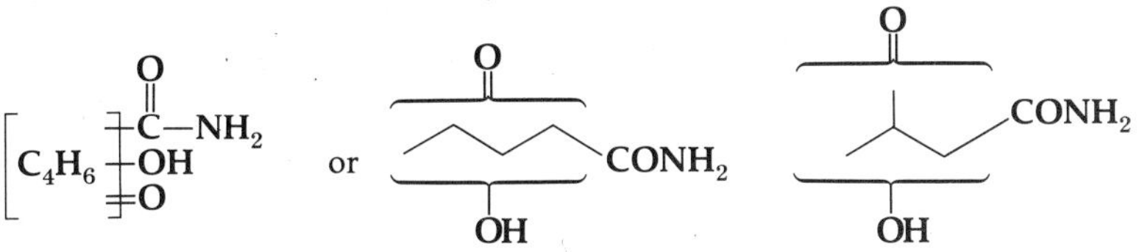

Possible number of isomers: (12)   (5)

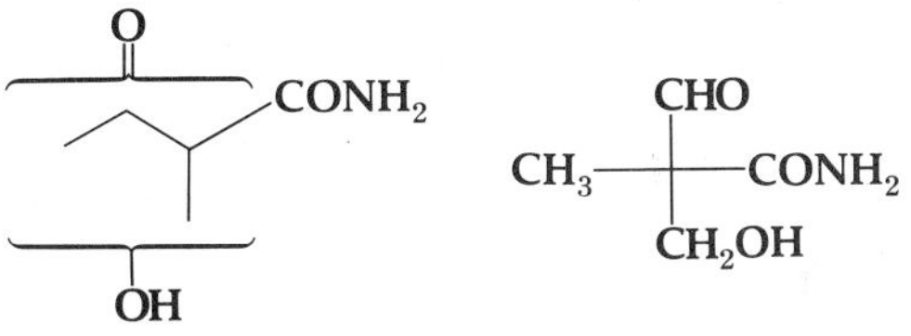

(9)   (1)

*The NMR (after **D$_2$O** exchange which removes **3H** as expected for **OH** + **NH$_2$**) shows only two singlets of equal intensity* so that *A* must be the structure; if the NMR showed only one singlet, it would then be *B*, with the methyl groups equivalent.

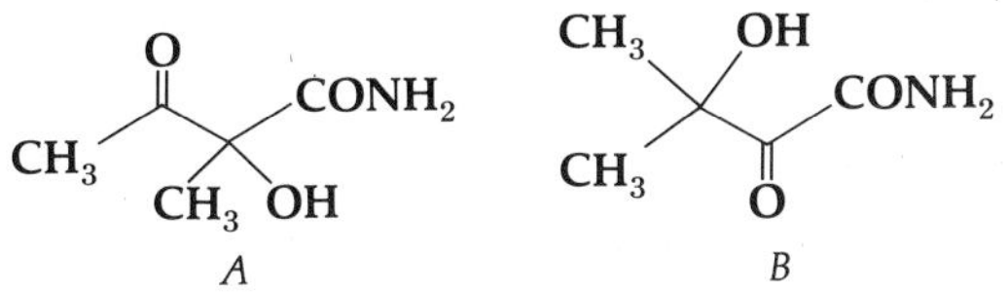

A   *B*

3 **C$_8$H$_{13}$OCl** *shows no important UV* > *200 nm. The IR has only **CH** peaks* $\leq 3.5\ \mu$ *and a peak at 5.73 μ (1740 cm^{-1}). The NMR exhibits three singlets of **3H** each at 8.95, 9.20, 9.28 τ as well as broad absorption (**4H**), 8.7 to 9.5 τ.*

The index is 2 and the carbonyl can only be a ketone in a five-membered ring since there is only one oxygen and the IR wavelength is too high for **—COCl** and too low for **—CHO**. This sets up a partial structure *A*.

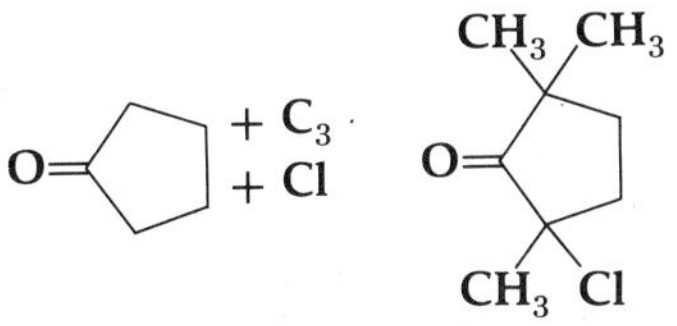

A   *B*

The NMR shows three methyls attached to carbons bearing no hydrogens on any of them (no splitting). This can only be done by putting two CH_3 on one ring carbon and CH_3 and Cl on another. The remaining four ring hydrogens absorb too high to be adjacent to carbonyl (Fig. 7-14). Hence the α-positions to $C{=}O$ are fully substituted as in the second and final (unique) structure B.

4 .*The problem*   *The deductive steps*

$C_{10}H_{16}O_2$; $H_2/Pt = 1$ mole   Index $= 3$ including only one $C{=}C$ double bond

UV: λ_{max} 236 mμ (log $\epsilon = 4.1$) Chromophore cannot be $C{=}C{-}C{=}C$

Must be $C{=}C{-}C{=}O$ and calc $H{-}\underset{\underset{C}{|}}{C}{=}\underset{\underset{C}{|}}{C}{-}C{=}O$

(If acyclic or 6-ring)

IR: Nothing $< 5\ \mu$ (except CH) No $-OH$

6.00, 6.18 μ (1667, 1620 cm^{-1}) Confirms $C{=}C{-}C{=}O$ } No indication/other oxygen

$\therefore$ Ether?

(Also index now implies *one* ring present)

NMR: 9.02 τ (**3H**) t ($J = 7$ cps) $-CH_2-CH_3$

8-9 τ (**6H**) m 6 satd **CH** overlapping signals

7.65 τ (**3H**) s $-CO-CH_3$

6.82 τ (**2H**) q ($J = 7$) $-O-CH_2-CH_3$

5.21 τ (**1H**) t ($J = 4$) $x{-}\underset{\underset{x}{|}}{\overset{\overset{x}{|}}{C}}{-}\underset{\underset{CH_2}{|}}{\underset{\underset{x}{|}}{CH}}{-}OR$ and $\underset{\underset{x}{|}}{\underset{H_2C}{}}C{=}C\overset{\overset{H}{}}{\underset{\underset{x}{|}}{}}$

3.82 τ (**1H**) t ($J = 3$)

Only one vinyl H; hence coupling to satd CH_2 adjacent

Note that 5.21 τ cannot be a second vinyl proton because it would then have to be coupled with the vinyl proton at 3.82 τ and it is not. Partial structure A now incorporates both the collected unsaturated ketone pieces and the ether environment, but the two pieces total 13 carbons and so must be put together such that 2 carbons from each are identical and their merger into one skeleton results in a total of only 10 carbons. The circled pairs are the only ones that can be merged, as trial will show. The final structure is thus B and should be checked back to see that it fits all the data acceptably.

PROBLEMS

In all the following problems assemble the data in a table and indicate the significance of each separate piece of data. Proceed stepwise to put inferences together, writing the inferences and structural limitations as you go. When you have arrived at one possible structure, recheck it to be sure it fits *all* the data given. Then assess whether your solution is a unique solution or not.

7-15 A compound, $C_7H_{12}O$, shows a UV spectrum exhibiting a single λ_{max} at 250 nm (log ϵ = 4.2).

7-16 C_7H_{10}; λ_{max} 237 nm (log ϵ = 4.1).

7-17 **a** $C_5H_8O_2$ shows no intense UV above 200 nm and important IR absorption only at 5.75 μ (1740 cm^{-1}). The NMR shows a doublet (3H) at τ 8.85 and a triplet (2H) at τ 7.65; the remaining protons appear as a multiplet around τ 6.0.

b An isomer of the substance shows IR = 5.6 μ (1780 cm^{-1}) and only two NMR peaks, both singlets, in a ratio of 3:1, the latter at lower field.

7-18 The mass spectrum of a certain solid shows a parent peak at 206 and its largest peak at 91. The IR has a broad absorption from 3.0 to 3.5 μ (2900 to 3300 cm^{-1}) as well as bands at 5.78 and 5.84 μ (1710 and 1730 cm^{-1}); the UV is largely a tailing absorption out from 210 nm. The NMR shows these peaks: τ −1.2 (1H) *s*; 2.7 (5H) *s*; 7.4 (2H) *s*; 7.8 to 8.0 (3H) *m*; 9.0 (3H) *d*, J = 7 cps.

7-19 An oil obtained by distillation of ground leaves from a *Santolina* species is purified by vapor-phase chromatography to a liquid, with bp 180–182°, shown to contain no elements other than C, H, and perhaps O. The mass spectrum showed a parent peak at m/e 152 and the UV a λ_{max} 238 nm (log ϵ = 4.05). In the IR there were no peaks around 3 μ and two at 5.97 and 6.12 μ (1630 and 1670 cm^{-1}). The NMR data are tabulated:

τ	(H)
8.82	(6) *s*
8.11	(3) *s*
7.90	(3) *s*
3.8–5.1	(4) *m*

7-20 A pure liquid with no intense UV absorption, and a parent peak of m/e 113 in the mass spectrum, shows IR bands at 4.5 and 5.8 μ (1730 and 2220 cm^{-1}) and two NMR singlets at τ 6.25 and 7.32 (intensity ratio 3:4, respectively). The second NMR signal surprised the investigators by appearing as a clear singlet. Why?

7-21 The following data characterize a certain solid, $C_{12}H_{15}NO_3$.
UV: λ_{max} <210 nm (ϵ > 10,000); 280 nm (ϵ = 900)
IR: 3.0, 5.84, 6.0 μ (3330, 1710, 1670 cm^{-1})
NMR: τ 2.82 (2H) *d*, J = 8 cps
3.20 (2H) *d*, J = 8 cps
5.66 (2H) *d*, J = 6 cps
6.24 (3H) *s*
6.63 (2H) *s*
7.79 (3H) *s*
(One proton does not show owing to exchange.)

7-22 **a** A hydrocarbon, mass spectral parent peak at m/e 102, showing two NMR peaks: τ 2.6 m and 6.92 s in a ratio of 5:1, respectively.

     b Three NMR singlets characterize an oil, $C_{10}H_{18}O$: τ 7.94 (4H); 8.47 (2H); 8.98 (12H).

     c A liquid with IR 5.5 μ (1820 cm^{-1}), two NMR triplets at τ 5.71 and 6.44 with a splitting of about 5 cps, and a parent peak of m/e 72 in the mass spectrum.

     d A hydrocarbon (C_9H_{12}) with three NMR peaks: τ 2.75 singlet, 7.10 septet, and 8.75 doublet in a ratio of 5:1:6, respectively.

7-23 Mass spectrum: m/e 73, 91, 149, 164 (and others below 149)

     IR: 5.78 μ (1730 cm^{-1})

     NMR: τ 2.7 (5H) $\sim s$

           5.70 (2H) t, $J = 7$ cps

           7.07 (2H) t, $J = 7$ cps

           8.00 (3H) s

7-24 The accompanying NMR spectrum (Fig. P7-24) is that of an oil containing nitrogen and showing a molecular weight of 131 in the mass spectrum. IR: 5.75 μ (1730 cm^{-1}).

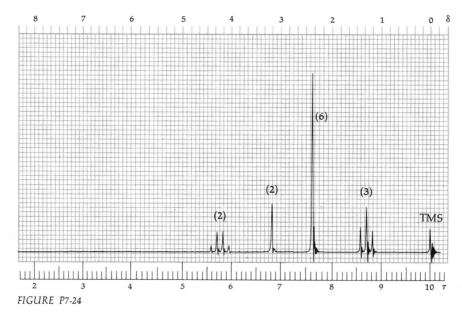

FIGURE P7-24

7-25 The substance $C_7H_{10}N_2O_3$ shown in the accompanying NMR spectrum (Fig. P7-25) also exhibits IR bands at 3.0, 4.5, 5.8, and 6.0 μ (1670, 1720, 2220, 3350 cm^{-1}). When the sample is shaken with D_2O the rounded peak at about τ 3.3 disappears from the NMR spectrum.

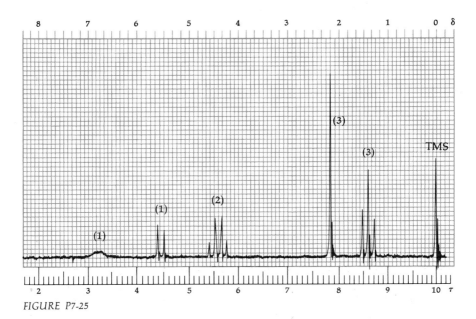

FIGURE P7-25

7-26 The hydrocarbon C_9H_{10} shows the accompanying NMR spectrum:

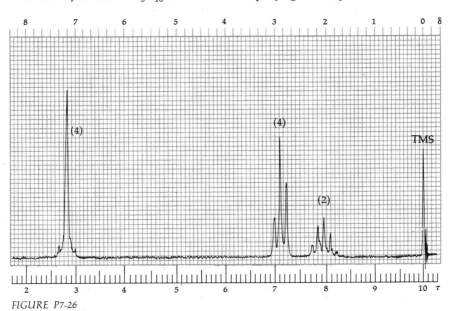

FIGURE P7-26

7-27 The accompanying NMR spectrum (Fig. P7-27) was taken on $C_{12}H_{14}ClNO_2$ which also showed IR peaks at 3.05, 6.03, and 6.20 μ.

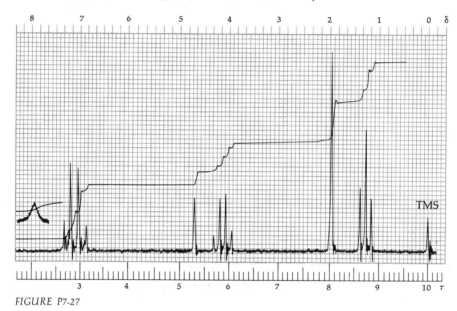

FIGURE P7-27

7-28 A pleasant-smelling liquid, $C_9H_{10}O_2$, shows three singlets in the NMR, at τ 2.69 (5H), 4.92 (2H), and 7.94 (3H), and an IR peak at 5.75 μ (1730 cm^{-1}) but none near 3 μ (3350 cm^{-1}).

7-29 The familiar anesthetic *Novocaine*, $C_{13}H_{20}N_2O_2$, is a synthetic material developed out of studies on the analgesic properties of cocaine. Novocaine is commonly administered as its hydrochloride salt. The NMR spectrum is shown in Fig. P7-29; when D_2O is added, the diffuse peak (2H) at τ 5.87 disappears without affecting other peaks. The UV spectrum is the same as that of ethyl *p*-aminobenzoate. (*Hint:* The five bands just above τ 7 are the result of two superimposed signals, a triplet $\sim\tau$ 7.18 and a quartet $\sim\tau$ 7.38.)

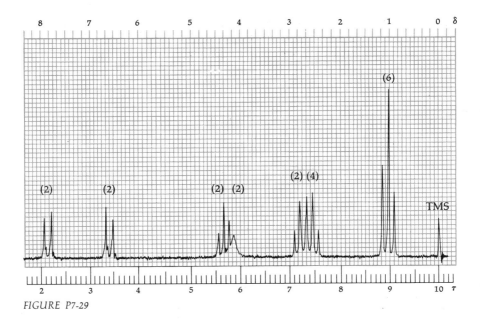

FIGURE P7-29

7-30 The analgesic *Phenacetin*, $C_{10}H_{13}NO_2$, is often mixed with aspirin in mild pain-killer tablets sold without prescription. The NMR spectrum is illustrated (Fig. P7-30); the IR shows peaks at 3.0, 6.0 μ (3350, 1670 cm^{-1}).

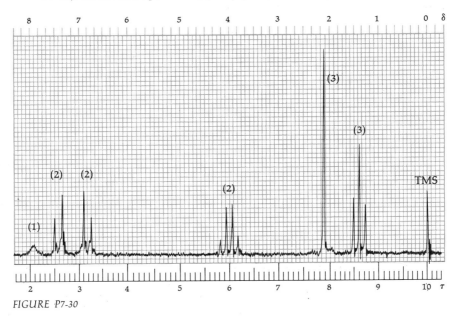

FIGURE P7-30

7-31 Both compound A and compound B give compound C on catalytic hydrogenation.

NMR:	A	B	C
	3.72 (2H) s	3.17 (2H) s	7.38 (4H) s
	5.72 (4H) q, J = 7	5.73 (4H) q, J = 7	5.85 (4H) q, J = 7
	8.70 (6H) t, J = 7	8.68 (6H) t, J = 7	8.75 (6H) t, J = 7

7-32 How many aromatic isomers of $C_{10}H_{13}NO$ will fit this NMR, which shows no change on exchange with D_2O?

NMR: τ 2.6–2.9 (4H) m τ 7.75 (3H) s

              6.80   (3H) s 8.22 (3H) s

7-33 $C_{10}H_{12}O_2$; IR 5.78 μ (1730 cm^{-1}).

NMR: τ 2.71 (5H) s

              5.70 (2H) t, $J = 7$

              7.07 (2H) t, $J = 7$

              7.98 (3H) s

7-34 From the leaves of *Magnolia salicifolia* may be obtained by distillation an optically inactive oil (mp 21°) with the NMR spectrum tabulated below. Analysis shows a formula of $C_{10}H_{12}O$ and its dihydro derivative (from H_2/Pd) shows a simple aromatic UV spectrum. Neither compound exhibits definitive IR peaks.

NMR: · τ 2.77 (2H) d, $J = 9$

              3.20 (2H) d, $J = 9$

              3.72 (1H) s

              3.92 (1H) d, $J = 5$

              6.25 (3H) s

              8.17 (3H) d, $J = 5$

7-35 $C_9H_{11}NO$; NMR: τ 0.30 (1H) s

                    2.30 (2H) d, $J = 9$

                    3.30 (2H) d, $J = 9$

                    6.95 (3H) s

READING REFERENCES

See the more complete list in Sec. 28-1. Only introductory texts and paperbacks are listed here.

Silverstein, R. M., and Bassler, G. C., "Spectrometric Identification of Organic Compounds," Wiley, New York, 1964.

Dyer, J. R., "Applications of Absorption Spectroscopy of Organic Compounds," Prentice-Hall, Englewood Cliffs, N. J., 1965.

Nakanishi, K., "Infrared Absorption Spectroscopy," Holden-Day, San Francisco, 1962.

Rao, C. N. R., "Ultraviolet and Visible Spectroscopy," Butterworth, New York, 1961.

Jackman, C. M., "Applications of Nuclear Magnetic Resonance Spectroscopy in Organic Chemistry," Pergamon, New York, 1960.

McLafferty, F. W., "Interpretation of Mass Spectra," W. A. Benjamin, New York, 1966.

CHEMICAL REACTIVITY AND MOLECULAR STRUCTURE

WITH the first seven chapters we have completed our introduction to that area of organic chemistry which is devoted to the structure of molecules. We have learned the nature of isomerism, the factors in molecular shape, the electrical composition of molecules, and the physical methods for observing these features. In effect we have looked at the single molecule, alone and at rest. Now we must examine the consequences of allowing different molecules to interact with each other. We shall find the same physical forces at work here as those invoked previously to explain structural observations. In short, we shall now explore chemical reactions, how they happen, and how to predict and control their products. Having studied static structure, we proceed to the dynamics of chemical change.

A chemical reaction is characterized by the attack of one molecule on another resulting in the breaking of bonds and the remaking of new bonds and hence new molecules. Commonly the initiating attack stems from the attraction of a site of negative charge in one molecule for a positive charge in another. In most reactions the changes result from the subsequent movement of electrons in pairs, from one place to another, but breaking and making bonds, especially σ bonds, so that the starting molecules are changed. *At the molecular level the essence of chemical reaction is charge attraction and electron movement.*†

We have already seen some chemical reactions briefly described, especially in tautomerism (Sec. 5-4) and racemization (Sec. 6-13). Proton tautomerism and many cases of racemization involve the simplest reaction, a **proton transfer**, the major example of the **acid-base reaction**. In this chapter we shall examine this acid-base reaction, which serves as a smooth transition from structure to dynamics. The reaction offers a clear case for study of the effects of structure on reactivity and provides a cogent, simple model for the variations in more complex reactions which are the subject of subsequent chapters.

---

†Electron movement is conveniently described by the curved-arrow convention introduced in Sec. 5-3. It was used there to describe the interlocking of formal structures into a resonance hybrid and usually only involved π bonds over the molecular skeleton. In reactions, by contrast, σ bonds are usually broken in the electron shifts, and molecules therefore come apart, into separate pieces.

That part of the molecular structure which surrounds the site of reaction often has a profound effect on the reactivity, for reasons of structure which derive from the previous discussions: resonance, steric hindrance, dipoles, etc. Appropriately, then, we enter the subject of reactivity with a discussion of the correlation of reactivity with structure, both as a review of our understanding of structural features and as a demonstration of their involvement in the dynamics of chemical change.

8-1 PROTON ACIDS AND BASES

Consider an unshared pair of electrons on one molecule ($B:^-$) in the same vessel with another (HA) bearing a proton on a more electronegative atom. The attraction of negative electrons for a positive proton can cause the electron pair to attack the proton and form a bond to it. The original bonding electrons of the proton are now free to collapse onto the electronegative atom and become an unshared pair there. This is illustrated in Eq. (1).

$$\bar{B}: \overset{\frown}{+} H \overset{\frown}{\underset{\delta+}{\quad}} A \underset{\delta-}{\rightleftharpoons} B{-}H + :\bar{A} \qquad\qquad 1$$

The initial lineup of the three involved nuclei is just the same as that in the hydrogen bond (Sec. 7-2), which may be regarded as simply an uncompleted proton transfer. The two new molecules (**BH** and **A**$:^-$) which are formed in Eq. (1) separate in solution but can return to reverse the reaction in identical fashion at any later time.†

A molecule with an unshared pair of electrons (cf. **B:**, **A:**) is a **base**, or **proton acceptor**, while an **acid** is a **proton donor** (cf. **HA, HB**). Thus each acid has a related base, called the **conjugate** base, and each base has its conjugate acid. Conjugate acid-base pairs (**B:** and **H—B; H—A** and **:A**) are exemplified in carboxylic acid and their carboxylate anions (**RCOOH** and **RCOO⁻**) or in bases like amines (**RNH₂**) and their salts (**RNH₃⁺**). These two functional groups are the most important acids and bases of organic chemistry.

Organic acids:

$$R{-}C\underset{\ddot{O}{-}H}{\overset{\ddot{O}}{\big\langle}} \rightleftharpoons R{-}C\underset{\ddot{O}:^-}{\overset{\ddot{O}}{\big\langle}} + H^+$$

Organic bases:

$$R{-}\underset{R'',H}{\overset{R',H}{N}}: + H^+ \rightleftharpoons R{-}\underset{R'',H}{\overset{R',H}{N^{\pm}}}H$$

---

†The charges shown represent only one example, for **B:** can be neutral also (⟶ **BH⁺**) or **H—A** could be positive (**HA⁺** ⟶ **:A**, neutral). In each case the electron movement is the same, independent of charges.

The essence of the acid-base reaction is the dissociation of an acid to its conjugate base and a proton, but since protons never exist in solution as separate entities, there must always be a base molecule present to *remove* the proton from the conjugate acid, as indicated in Eq. (1). The base serving to remove the proton is a solvent molecule, usually water, when the simple dissociation of a particular acid is being studied. The acid-base reaction is therefore a **competition reaction** between two bases for a single proton. This is an equilibrium process, as Eq. (1) indicates and the rate of proton transfer, for ordinary proton acids and bases, is exceedingly fast so that equilibrium is attained virtually instantaneously.

The **law of mass action** allows the concentrations at equilibrium to be related with an **equilibrium constant** (K) as exemplified by the dissociation of acetic acid in water, the solvent acting to remove the proton:

$$CH_3COOH + H_2O: \underset{}{\overset{K'_a}{\rightleftharpoons}} CH_3COO:^- + H_3O^+ \qquad K'_a = \frac{[CH_3COO^-][H_3O^+]}{[CH_3COOH][H_2O]} \qquad 2$$

$$\text{HA} \qquad\quad \text{B:} \qquad\qquad \text{A:} \qquad\quad \text{HB}$$

Since the concentration of solvent is essentially constant, the **ionization** or **dissociation constant** of acetic acid is given as K_a, which is also called the **acidity constant**.

$$K_a = K'_a[H_2O] = \frac{[CH_3COO^-][H_3O^+]}{[CH_3COOH]} = 1.75 \times 10^{-5} \qquad 3$$

Dissociation of acids and bases in solvents other than water requires proton transfer either to or from the solvent:

$$HA + CH_3C\equiv N: \rightleftharpoons CH_3C\equiv \overset{+}{N}H + \overset{-}{A}$$
Acid ionization in acetonitrile

$$B: + C_2H_5OH \rightleftharpoons \overset{+}{B}H + C_2H_5\overset{-}{O}$$
Base ionization in ethanol

The expression for acidity constant may then be generalized:

$$HA + solvent \rightleftharpoons H^+ \cdot solvent + :A \qquad K_a = [H^+ \cdot solvent] \times \frac{[:A]}{[HA]} \qquad 4$$

It is frequently convenient to express dissociation constants in logarithmic units. So that positive rather than negative numbers may be compared, the unit is the *negative logarithm,* (pK_a) of the acidity constant, as indicated in Eq. (5). For acetic acid in water, $K_a = 1.75 \times 10^{-5}$ and $pK_a = -\log (1.75 \times 10^{-5}) = 4.76$.

$$pK_a = -\log K_a \qquad 5$$

It is important to keep in mind when using pK_a values that they represent *powers of ten* of the acidity constants, that *one* pK_a *unit* difference between two acids means that one is *ten times as acidic* as the other.

Since every base has its conjugate acid, it is convenient to compare all acids and bases on a single scale: the pK_a of the conjugate acid for any conjugate pair. What we are comparing is, therefore, either the case of proton dissociation (i.e., the acidity) or the avidity of the unshared pair for a proton (i.e., the basicity). The equilibria above make it clear that

The lower the pK_a the stronger the (conjugate) acid
The higher the pK_a the stronger the (conjugate) base

Since every acid-base reaction is a competition, a known acid (or base) may be used to test the basicity (or acidity) of another, measuring the competition equilibrium (K_c):

$$B: + HA \xrightleftharpoons{K_c} HB + :A \qquad \text{(Charges ignored)} \qquad\qquad 6$$

$$K_c = \frac{[:A][HB]}{[HA][:B]} = \frac{K_{a(A)}}{K_{a(B)}} \qquad \text{or} \qquad pK_c = pK_{a(A)} - pK_{a(B)} \qquad 7$$

Comparison of acid-base relationships in terms of pK_a units is particularly easy because the scale is linear. The difference between two pK_a values gives a measure of the difference in the strengths of two acids or bases [Eq. (7)]. A further useful relationship is that a solution containing equal concentrations of an acid and its conjugate base (a buffered mixture) will have a **pH** equal to the value of pK_a for the acid [Eq. (4)].

The degree of dissociation of acids and bases is very solvent-dependent. For example, solutions of hydrogen chloride in benzene will not conduct an electric current, despite the fact that hydrogen chloride is a strong acid in aqueous solution. Although water is superior to other media as an **ionizing solvent**, reactions of many acids and bases can be studied only in nonaqueous solvents, because of the insolubility of the un-ionized species of the pair in water.

Solvent effects in acid-base equilibria are due to these principal causes:

1   Solvents are ordinarily involved in proton equilibria. Accordingly, ionization constants depend upon the *acidity and (or) basicity of the solvent.*

2   All ions in solution strongly polarize solvent molecules near them. The strength of such interactions is enormous in water solutions of ionic compounds and much smaller in nonpolar, difficult-to-polarize solvents such as hydrocarbons. Hydrogen bonding of ions represents a strong interaction.

3   The electrostatic energy of a charged body decreases as the **dielectric constant** ϵ of the surrounding medium is increased. The dielectric constant measures the relative effect of the medium on the force with which two opposite charges attract each other. The dielectric constant of a liquid is determined readily by measuring the electrical capacitance of a condenser when empty and when filled with the liquid.

The effect of the dielectric constant makes solvents such as water ($\epsilon = 80$) and acetronitrile ($\epsilon = 39$) much better solvents for ions than are low dielectric media such as acetone ($\epsilon = 21$) and benzene ($\epsilon = 2.3$). Pure dielectric effects are subordinate in importance to the other specific solvation effects mentioned above.

Table 8-1 illustrates the effect of solvents on the acidity constant of acetic acid.

8-2   SCALE OF ACIDITY AND BASICITY

In order to see clearly the relative strengths of organic acids and bases it is useful to arrange a pK_a scale of significant compounds, and this is shown as Table 8-2. This single scale offers a measure of basicity as well as acidity since bases are compared as their conjugate acids. Unfortunately, the values at the two extremes of the scale (beyond 0 to 14 each way) are not directly measurable in water since water is itself both an acid and a base and therefore only allows direct measurement of pK_a's between about 0 and 14. Stronger acids ($pK_a < 0$) react virtually completely with water to form H_3O^+ and leave no undissociated acid to measure, whereas very strong bases ($pK_a > 14$) merely convert water to its base, OH^-.

However, although solvent (and other) troubles plague these determinations, strong acids and bases can be compared with those measured in water by using nonaqueous competition experiments [Eq. (7)] and in this way the scale of Table 8-2 has been obtained. It must be remembered that values between 0 and 14 are obtained in water and that values <0 and >14 are less accurate.

The scale offers value beyond its obvious usefulness for showing the extent to which a given acid protonates a given base [Eq. (7)].

TABLE 8-1   **Effect of Solvent on Dissociation of Acetic Acid at 25°**

Solvent	K_a	pK_a
Water	1.75×10^{-5}	4.76
20% Dioxan–80% water	5.11×10^{-6}	5.29
45% Dioxan–55% water	4.93×10^{-7}	6.31
70% Dioxan–30% water	4.78×10^{-9}	8.32
82% Dioxan–18% water	7.24×10^{-11}	10.14
10% Methanol–90% water	1.25×10^{-5}	4.90
20% Methanol–80% water	8.34×10^{-6}	5.08
100% Benzene	Too small to measure	

TABLE 8-2 **Scale of Acidities**

Conjugate acid	pK_a	Conjugate base
Cyclohexane	45	$C_6H_{11}^-$
CH_3-CH_3	42	$CH_3CH_2^-$
CH_4	40	$:CH_3^-$
Benzene	37	$C_6H_5^-$
Ethylene	36	$CH_2=CH:^-$
NH_3	36	$\overset{..}{N}H_2^-$
ϕCH_3	35	$\phi CH_2:^-$
$CH_2=CH-CH_3$	35	$CH_2=CH-CH_2:^-$
ϕ_3CH	32	$\phi_3C:^-$
ϕNH_2	27	$\phi\overset{..}{N}H^-$
$HC\equiv CH$	25	$HC\equiv C:^-$
ϕ_2NH	23	$\phi_2\overset{..}{N}:^-$
CH_3COCH_3	20	$CH_3COCH_2:^-$
$t\text{-BuOH}$	19	$t\text{-BuO}^-$
 $O_2N-\!\!\langle\!\!\rangle\!\!-NH_2$	18.5	 $O_2N-\!\!\langle\!\!\rangle\!\!-\overset{..}{N}H^-$
C_2H_5OH (ROH)	17	$C_2H_5O^-$
RCONHR′	~16	$RCO\overset{..}{N}R'^-$
CH_3OH	16	CH_3O^-
H_2O	15.7	HO^-
	15	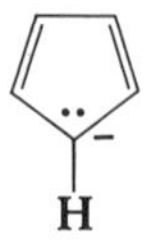
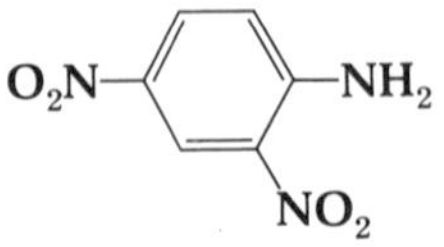	15	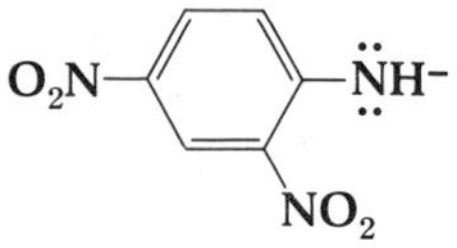
$(ROOC)_2CH_2$	13.5	$(ROOC)_2\overset{..}{C}H-$
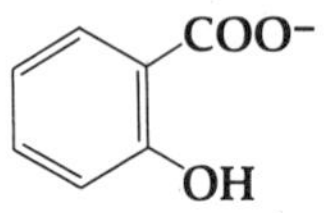	13.4	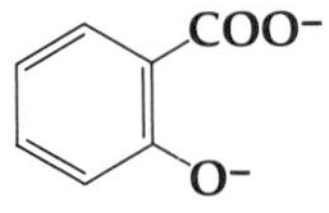

TABLE 8-2 **Scale of Acidities** (*Continued*)

Conjugate acid	pK_a	Conjugate base
H_2N $\quad$ C=$\overset{+}{N}H_2$ H_2N	13.4	H_2N $\quad$ C=NH H_2N
$(NC)_2CH_2$	11.2	$(NC)_2\ddot{C}H^-$
CH_3COCH_2COOR	10.2	$CH_3CO\overset{=}{C}HCOOR$
RNH_3^+ $R_2NH_2^+$ R_3NH^+	~10	$R\ddot{N}H_2$ $R_2\ddot{N}H$ $R_3\ddot{N}$
CH_3NO_2	10.2	$:\bar{C}H_2NO_2$
HCO_3^-	10.2	CO_3^{--}
ϕ—OH	10	ϕ—O$^-$
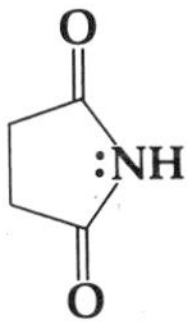	9.6	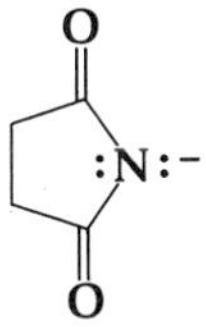
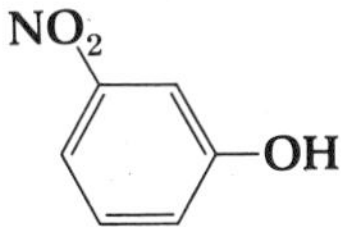	9.3	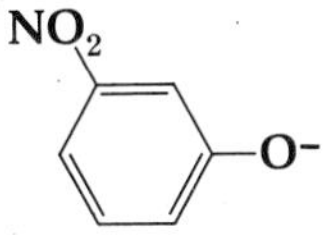
NH_4^+	9.2	$:NH_3$
HCN	9.1	$:CN^-$
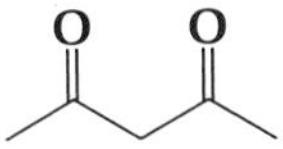	9.0	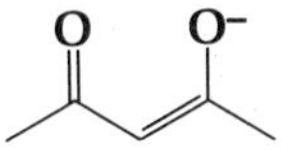
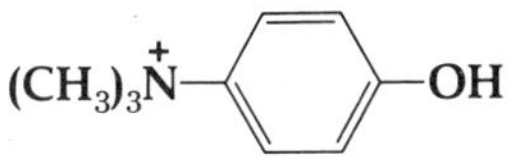 $(CH_3)_3\overset{+}{N}$—〈〉—OH	8.0	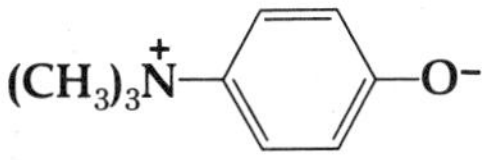 $(CH_3)_3\overset{+}{N}$—〈〉—O$^-$
 O_2N—〈〉—OH	7.2	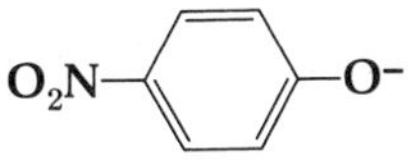 O_2N—〈〉—O$^-$
H_2CO_3	6.5	HCO_3^-
$O_2NCH_2COOCH_3$	5.8	O_2N—$\overset{=}{C}H$—$COOCH_3$
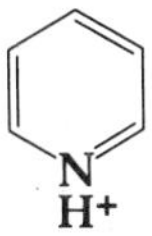	5.2	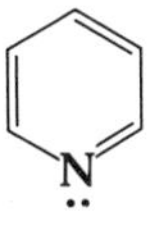

TABLE 8-2 **Scale of Acidities** (*Continued*)

Conjugate acid	pK_a	Conjugate base
$\phi-\overset{H}{N}(CH_3)_2{}^+$	5.1	$\phi\ddot{N}(CH_3)_2$
$\phi-NH_3{}^+$	4.6	$\phi-\ddot{N}H_2$
RCOOH	4.5 ± 0.5	RCOO$^-$
2,4-Dinitrophenol	4.0	$(NO_2)_2\phi-O^-$
HCOOH	3.7	HCOO$^-$
$CH_2(NO_2)_2$	3.6	$^-{:}CH(NO_2)_2$
$ClCH_2COOH$	2.8	$ClCH_2COO^-$
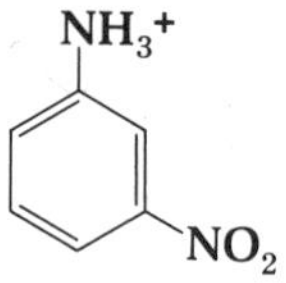 (m-nitroanilinium, $NH_3{}^+$ / NO_2)	2.5	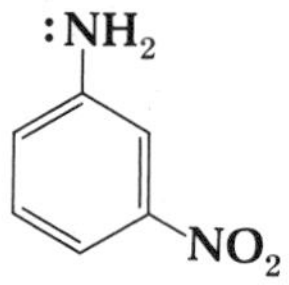 (m-nitroaniline, $:NH_2$ / NO_2)
$R-CH \overset{COOH}{\underset{NH_3{}^+}{}}$	2.4	$R-CH \overset{COO^-}{\underset{NH_3{}^+}{}}$
$Cl_2CHCOOH$	1.3	$ClCHCOO^-$
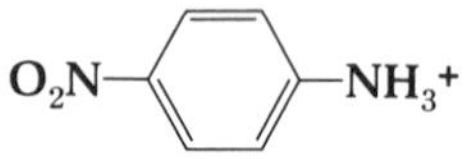 $O_2N-\langle\rangle-NH_3{}^+$	1.0	 $O_2N-\langle\rangle-\ddot{N}H_2$
$\phi_2NH_2{}^+$	1.0	$\phi_2\ddot{N}H$
Cl_3CCOOH	0.9	Cl_3CCOO^-
2,4,6-Trinitrophenol	0.4	$(NO_2)_3\phi O^-$
CF_3COOH	0	CH_3CONH_2
$CH_3CONH_3{}^+$	0.3	CF_3COO^-
HNO_3	-1.4	$NO_3{}^-$
$\phi CONH_3{}^+$	-2	$\phi CONH_2$
$CH_3OH_2{}^+$	-2	CH_3OH
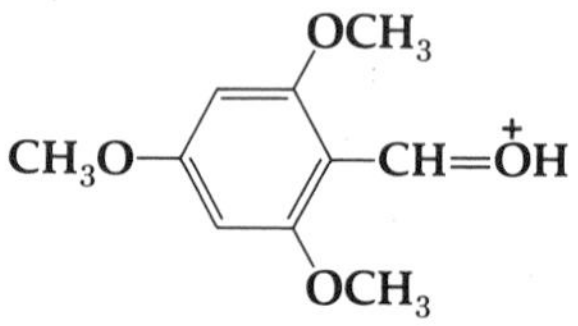 $CH_3O-\langle\rangle-CH{=}\overset{+}{O}H$ (OCH$_3$ ortho, OCH$_3$ para)	-2.1	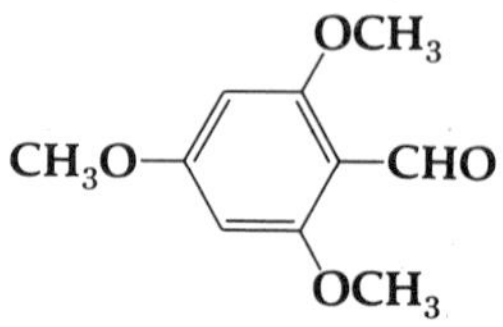 $CH_3O-\langle\rangle-CHO$ (OCH$_3$, OCH$_3$)
$(CH_3)_2OH^+$	-3.8	$(CH_3)_2O$

TABLE 8-2 **Scale of Acidities** (*Continued*)

Conjugate acid	pK_a	Conjugate base
$t\text{-BuOH}_2{}^+$	-4	$t\text{-BuOH}$
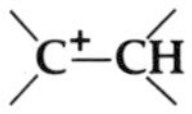 $\underset{\diagup}{\overset{\diagdown}{C}}{}^{\pm}\!-\!\underset{\diagup}{\overset{\diagdown}{C}}H$	$\sim -4\ (?)$	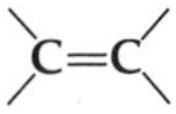 $\underset{\diagup}{\overset{\diagdown}{C}}\!=\!\underset{\diagup}{\overset{\diagdown}{C}}$
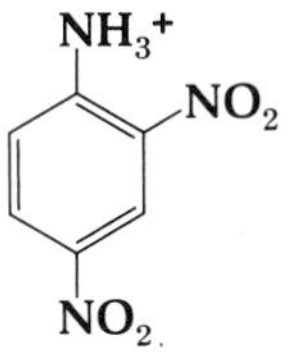 2,4-dinitroaniline ($NH_3{}^+$)	-4.5	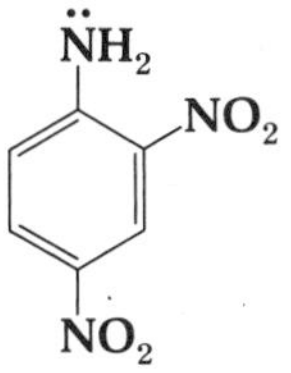 2,4-dinitroaniline ($\ddot{N}H_2$)
$(CH_3)_2SH^+$	-5.2	$(CH_3)_2S$
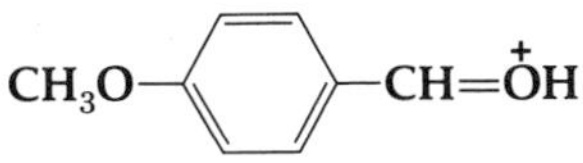 $CH_3O\!-\!\phi\!-\!CH\!\overset{+}{=}\!\overset{..}{O}H$	-5.5	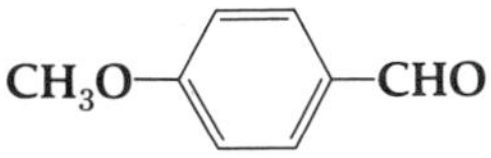 $CH_3O\!-\!\phi\!-\!CHO$
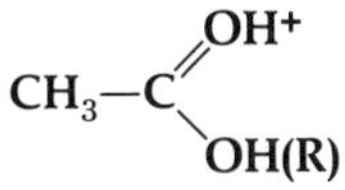 $CH_3\!-\!\underset{OH(R)}{\overset{OH^+}{C}}$	-6.2	$CH_3\!-\!COOH(R)$
$\phi OH_2{}^+$	-6.7	ϕOH (or ϕOR)
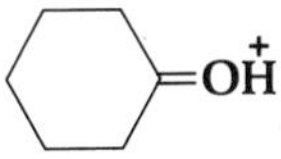 cyclohexanone $=\overset{+}{O}H$	-6.8	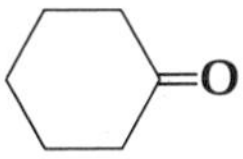 cyclohexanone $=O$
$\phi CH\!=\!OH^+$	-7.1	ϕCHO
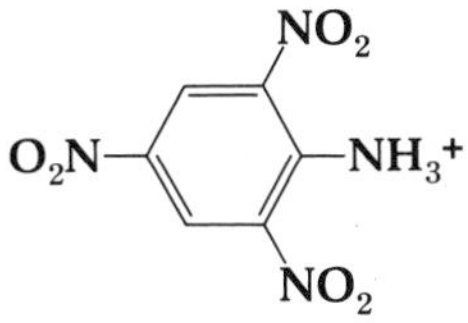 2,4,6-trinitroaniline $NH_3{}^+$	-9.4	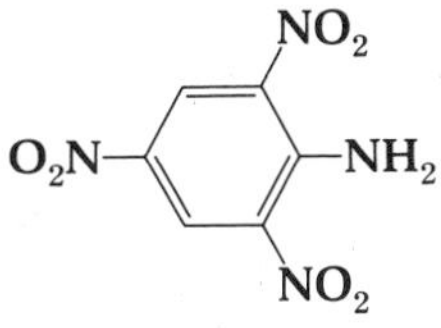 2,4,6-trinitroaniline NH_2
$R\!-\!C\!\equiv\!NH^+$	~ -10	RCN
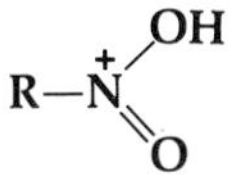 $R\!-\!\overset{+}{N}\!\underset{O}{\overset{OH}{<}}$	~ -11	$R\!-\!NO_2$
H_2SO_4	?	$HSO_4{}^-$
HBF_4	?	$BF_4{}^-$
FSO_3H	?	$FSO_3{}^-$
$HClO_4$	~ -20	$ClO_4{}^-$
HPF_6	-20	$PF_6{}^-$
$SbF_5 \cdot FSO_3H$ (strongest acid)	< -20	$SbF_5 \cdot FSO_3{}^-$

1 Most other reactions involve attack of unshared electron pairs on positive sites, often on carbon instead of hydrogen, as here. Hence the order of basicity is a rough guide to reactivity in such reactions since it catalogs the relative avidity of electron pairs on different bases for positive sites of all kinds.

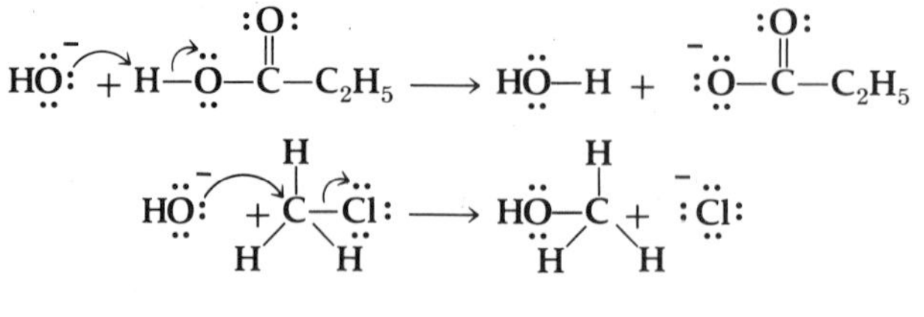

or

$$HO^- + C_2H_5COOH \longrightarrow H_2O + C_2H_5COO^-$$

$$HO^- + CH_3Cl \longrightarrow CH_3OH + Cl^-$$

2 If a base is planned for use in a reaction in which it is meant to attack a positive site, the scale can tell us whether the base would instead remove a proton from some acidic site (of lower pK_a) in the molecule and so perhaps be neutralized and worthless for the planned attack. Acid–base reactions involving O—H and N—H acids are generally faster than others.

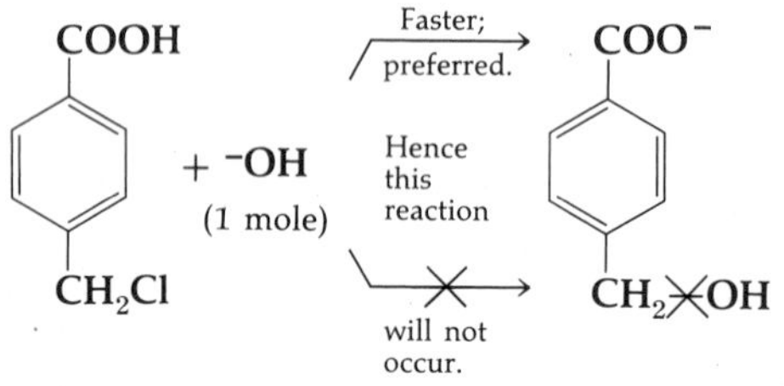

3 A great many organic reactions occur in several steps of which the first is often a protonation of a weakly basic site, to be followed by collapse of the intermediate cation formed; alternatively the first step may be deprotonation by added base followed by collapse of the resultant intermediate anion. Almost all reactions operate optimally within some acidity-basicity range, and the scale can assist in finding appropriate reagents and conditions for these reactions.

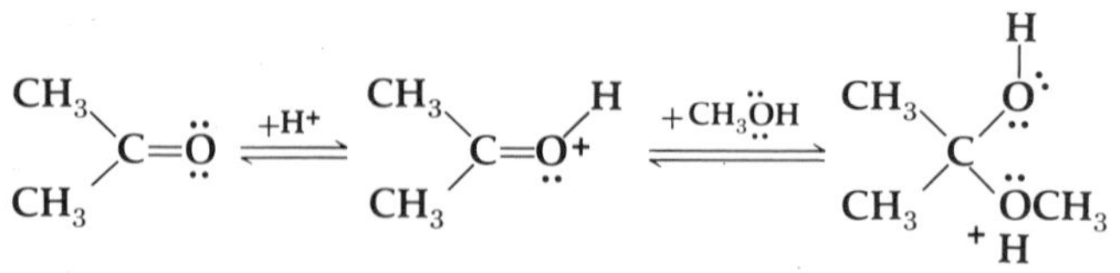

In general we are interested in distinguishing three families of acids and the (roughly parallel) effects of structural environment on acidity in each family. The families are carbon acids, nitrogen acids, and oxygen acids, i.e., dissociation of H—C, H—N, H—O. Furthermore, in the latter two, which can have both electron pair and proton and can be both acids and bases (like water), we must always distinguish two subfamilies, representing two different ionizations, and keep them separate.

Parent pK$_a$ *Parent acid ⇌ base*

~43 $H-\underset{|}{\overset{|}{C}}-$ $-:\underset{|}{\overset{|}{C}}-$

~35 $H-\underset{|}{\overset{..}{N}}-$ $^-:\underset{|}{\overset{..}{N}}-$ Two different equilibria:

$$R_2NH_2{}^+ \overset{pK_1}{\rightleftharpoons} R_2\overset{..}{N}H \overset{pK_2}{\rightleftharpoons} R_2\overset{..}{N}:{}^-$$

~10 $H-\underset{|}{\overset{+|}{N}}-$ $:\underset{|}{N}-$

~18 $H-\overset{..}{\underset{..}{O}}-$ $^-:\overset{..}{\underset{..}{O}}-$

~-2 $H-\overset{+}{\underset{|}{O}}-$ $:\overset{..}{\underset{..}{O}}-$

$$ROH_2{}^+ \overset{pK_1}{\rightleftharpoons} R\overset{..}{\underset{..}{O}}H \overset{pK_2}{\rightleftharpoons} R\overset{..}{\underset{..}{O}}:{}^-$$

$pK_2 > pK_1$

$$C_6H_5-NH_3{}^+ \underset{+H^+}{\overset{-H^+}{\rightleftharpoons}} C_6H_5-\overset{..}{N}H_2 \underset{+H^+}{\overset{-H^+}{\rightleftharpoons}} C_6H_5-\overset{..}{N}H^-$$
$$pK_1 = 5 \qquad\qquad pK_2 \simeq 27$$

$$CH_3CH_2-C\overset{\overset{+}{O}H}{\underset{OH}{\big\langle}} \underset{+H^+}{\overset{-H^+}{\rightleftharpoons}} CH_3CH_2-C\overset{O}{\underset{OH}{\big\langle}} \underset{+H^+}{\overset{-H^+}{\rightleftharpoons}} CH_3CH_2-C\overset{O}{\underset{O^-}{\big\langle}}$$
$$pK_1 \simeq -6 \qquad\qquad pK_2 = 5$$

Each of the families is fundamentally different because each involves ionization of hydrogen from a different basic atom, C, N, or O, or from the latter two in different charge states. Each family, however, exhibits in rough parallel the effects on acidity provided by structural variation in the molecule near the site of dissociation. These effects will be described in terms of ΔpK_a from the simple parent acid (base). The families differ primarily in the electronegativity of the parent atom: the more electronegative it is (larger kernel charge) the more tightly it holds electrons and so the more stable (less basic) the conjugate base is than the conjugate acid.

Basicity: $R_3C: > R_2\overset{..}{N}^- > R\overset{..}{\underset{..}{O}}:^- > :\overset{..}{\underset{..}{F}}:^-$

$R_3N: > R_2\overset{..}{\underset{..}{O}} > R\overset{..}{F}:$

PROBLEM 8-1

The following bases are commonly used in the laboratory: $CH_3COO^-Na^+$, $(C_2H_5)_3N$, $C_2H_5O^-Na^+$, $(CH_3)_3C-O^-Na^+$, $(C_6H_5)_3C:^-Na^+$. Which of these bases would suffice, in each case below, to remove one colored proton, in order to initiate reactions (removal of $\geqslant 50\%$ of one proton)?

a $C_6H_5NH_2$

e CH_3—⟨benzene ring⟩—OH

b $CH_3CH=CH—CH_3$

f (structure: five-membered ring with two C=O groups and NH)

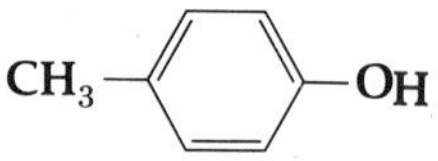

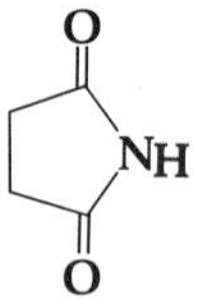

c $C_2H_5COCH_3$

d $(NC)_2CH_2$

g O_2N—⟨benzene ring⟩—NH_2

PROBLEM 8-2

Complete the following acid-base equilibrium reactions and compute their equilibrium constants. Which ones go essentially to completion ($\geq 99\%$ products)?

a $C_2H_5NH_2 + CH_3COOH$

b $CH_3O^- + CH_3COCH_2COOCH_3$

c $CO_3^- + HCN$

d $CH_3NO_2 +$ ⟨cyclopentadienyl ring⟩ : $-$

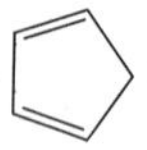

e $HC\equiv CH + :NH_2^-$

f $HCO_3^- + HCOOH$

g $Cl_2CHCOOH + C_6H_5NH_2$

h $C_6H_5OH +$ ⟨pyridine ring⟩

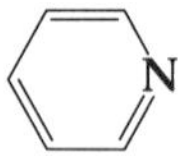

8-3 ENERGY AND THE ACID-BASE REACTION

Any reversible reaction comes ultimately to a position of equilibrium which directly reflects the relative free energies of starting materials and products. The more stable (lower energy) of the two predominates in the equilibrium mixture. The relation is reflected in the standard free-energy equation, in which ΔF represents the difference in free energy of starting material and product.

$$\Delta F = -2.30\ RT \log K \qquad\qquad 8$$

or

$$\Delta F \simeq 1.4\ pK \qquad \text{in kcal/mole at room temperature} \qquad\qquad 9$$

The dissociation of an acid is no exception, and ΔF is the energy difference between the conjugate acid and its conjugate base. Thus a strong base (high pK_a) is one of high energy which loses $1.4 \times pK_a$ kilocalories per mole on picking up a proton. Similarly, a strong acid is characterized by a negative pK_a and loses $1.4 \times pK_a$ kilocalories per mole on giving up its proton. For an exothermic reaction—in which the equilibrium favors the products—the reacting base must have a higher pK_a (properly, pK_a of its conjugate acid) than the reacting acid. The example

$$CH_3NH_3^+ \rightleftharpoons CH_3NH_2 + H^+ \qquad pK_1 = 10$$

$$CH_3COOH \rightleftharpoons CH_3COO^- + H^+ \qquad pK_2 = 5$$

$$CH_3NH_2 + CH_3COOH \rightleftharpoons CH_3NH_3^+CH_3COO^- \qquad pK_c = 5 - 10 = -5$$
$$\Delta F = -7.0 \text{ kcal/mole}$$

shows the reaction of an amine with a carboxylic acid to proceed exothermically, by 7 kcal/mole, to the salt. Finally, it must be remembered that these are equilibria: even if a base is added to an acid a few pK_a units higher, there will be some net reaction producing very small concentrations of their two conjugates. In many cases in which these in turn initiate a reaction, that reaction can still proceed, fed by the tiny but continuously replenished concentrations produced in the unfavorable acid-base equilibrium.

We can derive the criterion for assessing the effects of structural variation on the acid-base reaction from considerations of energy. The pK_a value on the scale is proportional to the energy difference between the conjugate acid and the conjugate base. Hence, if some variation in structure affects both conjugate acid and base forms *equally*, it will not change the pK_a. This is, however, an unlikely happenstance, since it is more common for a variation in the molecule to stabilize (lower the energy)—or destabilize (raise the energy)—one conjugate species more than the other, thus changing ΔF and pK_a. The effects are:

1 *Stabilization of* **B:** *more than* **HB** $=$ *lower* pK_a
2 *Stabilization of* **HA** *more than* **:A** $=$ *higher* pK_a

This is the central criterion we shall use in discussing structural effects on pK_a. It may be rendered qualitatively that any effect that tends to make the conjugate base (**B:**) less avid for protons, or make its electron pair less available to pick up a proton, will lower the pK_a. Any effect that causes the conjugate acid (**HA**) to be more reluctant to lose its proton will raise the pK_a. The first case makes the base less basic (lower pK_a); the second makes the acid less acidic (higher pK_a).

In the next sections we shall examine the four major structural effects listed below. In any given conjugate pair more than one effect may be operating, but usually one is predominant. The principal effects are these:

1 Resonance effects: the most potent—almost always stabilize the base by delocalizing the electron pair, and so lowering pK_a
2 Inductive and electrostatic effects: can stabilize either acid or base
3 Steric effects: can stabilize either acid or base
4 Hydrogen bonding (internal): usually stabilizes the acid, raising pK_a

8-4 RESONANCE EFFECTS

Since *delocalization of electrons always stabilizes a system*, it follows that localization of electrons makes a system less stable. Frequently the attachment of a proton to a base to form the conjugate acid restricts an electron pair that in the free base was able to spread over an unsaturated system. Protonation of the carboxylate anion provides an example. The addition of a proton localizes a pair of electrons (into a bond) and destroys the perfect symmetry of the ion. The acid still has some delocalization energy because of mixing of the unshared electrons on the oxygen of the hydroxyl with the π orbitals of the carbonyl group. However, this effect in the acid is smaller than in the carboxylate anion since it leads to charge separation in the acid, while in the anion the resonance forms are identical. Thus, resonance stabilizes the anion (conjugate base) *more* than the acid and the pK_a is lowered, making the acid more acidic than a simple "parent" —O—H would be, i.e., in alcohols (R—OH).

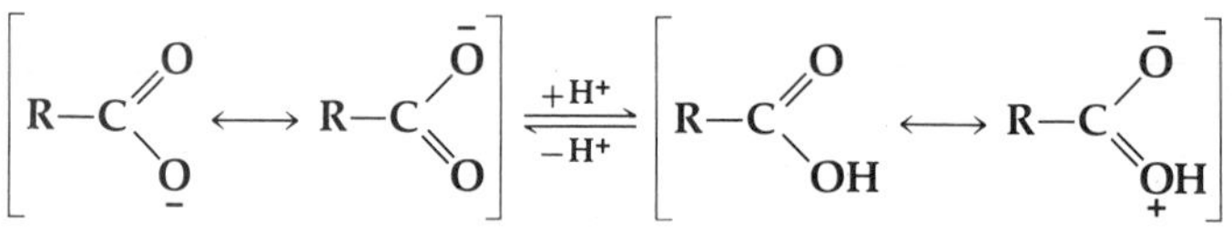

This **resonance effect** is presumed to be the principal reason for the much greater acidity of carboxylic acids compared with alcohols. The acidity of ethanol is undetectable in water solution, but it is estimated that the ionization constant is about 10^{-17}. The factor of 10^{11} between the acidity constants of ethanol and acetic acid is due largely to delocalization of charge in acetate ion.

The comparable effect, of substituting a carbonyl group for an alkyl adjacent to the proton-bearing atom, is seen in all the atom families of acids:

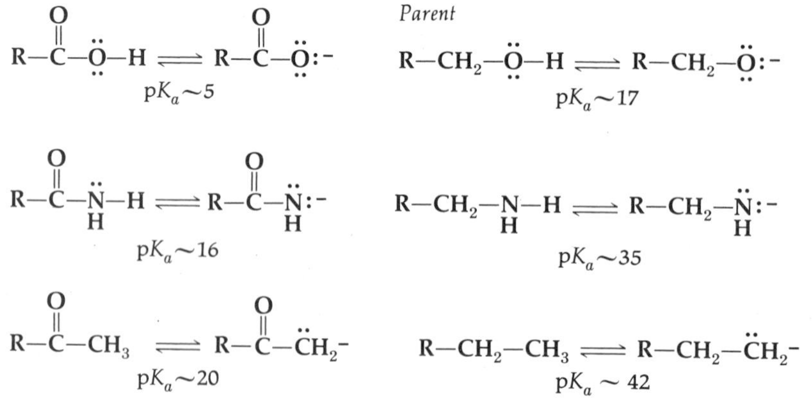

The basicity of amines is very sensitive to resonance effects, as is shown by the fact that aniline is a weaker base than the aliphatic amines by 10^6 ($\Delta pK_a = 6$; see Table 8-2). This dramatic effect is due largely to the delocalization energy of aniline, which is lost when a proton is added to nitrogen.

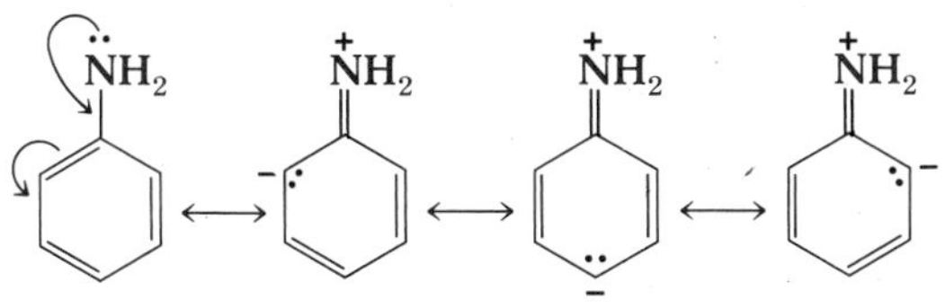

Aniline, electron pair of nitrogen distributed in the benzene ring

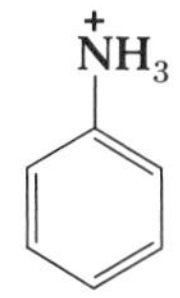

All electrons of nitrogen localized in single bonds

Anilinium ion

In the above electronic formulas the π-electron system, which is involved in resonance, is shown in color to emphasize it in the present discussion.

The influence of nitro groups in further reducing the basicity of aromatic amines illustrates the idea further. An additional base-weakening effect in p-nitroaniline (not present in the meta isomer) results from dispersal of the unshared pair of electrons on nitrogen into the nitro group (colored form) in addition to the resonance seen in aniline or the meta isomer (four forms only). Both nitro isomers are less basic than aniline due to a superimposed inductive effect (Sec. 8-5).

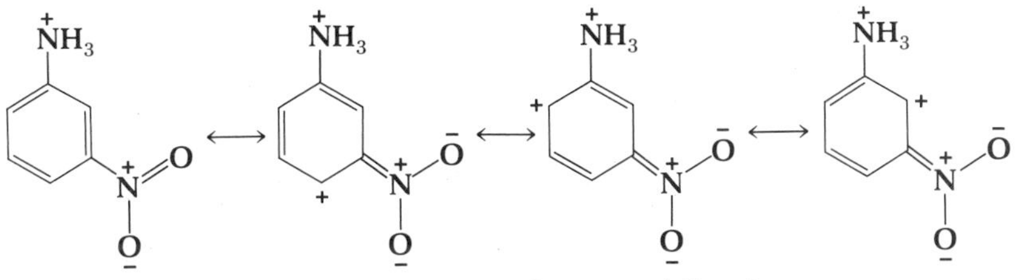

Resonance in m-nitroaniline cation (pK_a 2.5)

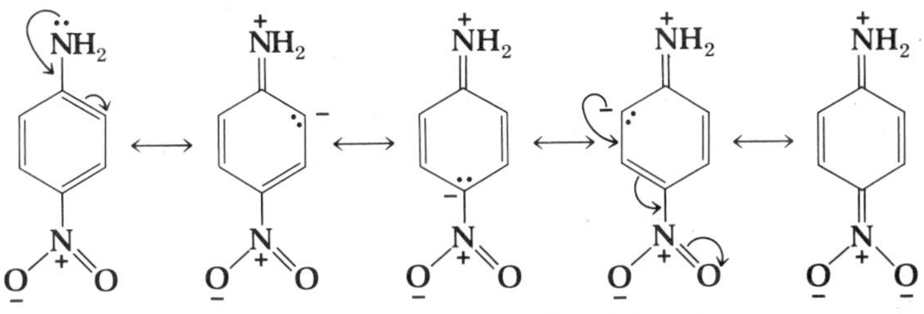

More resonance in p-nitroaniline (pK_a 1.0)

Phenols are much more acidic than alcohols for entirely analogous reasons and substitution with ortho or para nitro groups notably increases this acidity as the examples in Table 8-2 show.

Amides, in contrast to amines, are not detectably basic in water solution, largely because of delocalization of the electrons of nitrogen, this time into carbonyl rather than phenyl unsaturation. The equilibrium discussed here is *not* that discussed above for the *acidity* of amides (loss of —H); but rather the second pK_a of amides, describing their *basicity* (adding —H), and the two must be distinguished.

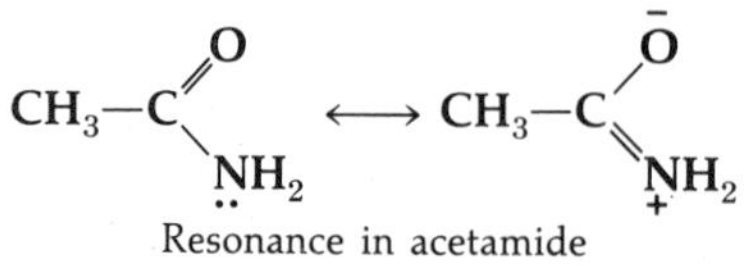

Resonance in acetamide

When nitrogen is flanked by two carbonyl groups, as in the cyclic imides, the influence of the two groups is strong enough to render the compounds not at all basic and even observably acidic in water solution. Succinimide serves as an example. Both the neutral succinimide molecule (as acid) and its conjugate base are stabilized, and the same number of significant resonance structures can be formulated for both species. Delocalization is more effective in the anion, since removal of the proton decreases the electron affinity of the nitrogen atom. *Delocalization of electrons is most effective when excess charge is dispersed throughout a system without charge separation.* Comparable stabilization of a conjugate acid is illustrated in Fig. 8-1.

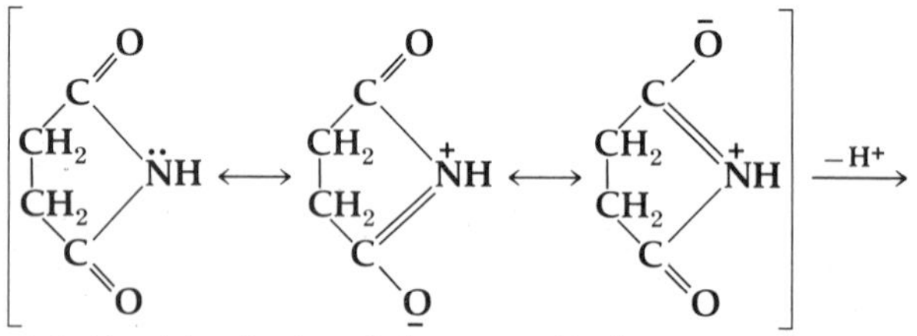

Succinimide—delocalization of electrons, with charge separation

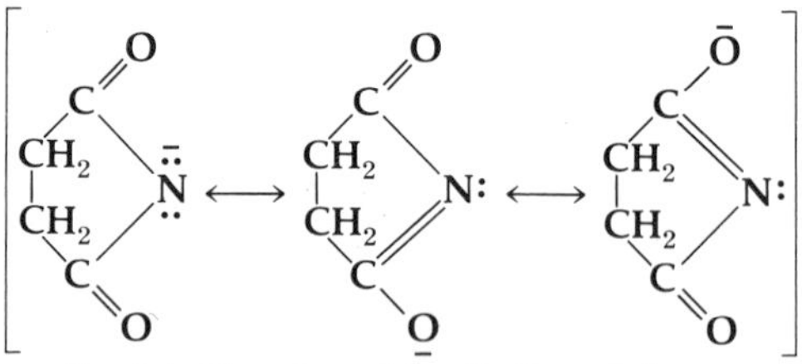

Succinimide anion—delocalization of charge,
without added charge separation

PROBLEM 8-3

It may be argued that the amide acting as a base is protonated on oxygen, not nitrogen, in which case it is properly compared in basicity to carbonyl compounds, not amines. Develop this reasoning with resonance structures and the pK_a values of Table 8-2.

In general whenever a basic site (unshared electron pair) is adjacent (conjugated) to unsaturation, the electron pair is strongly delocalized and sta-

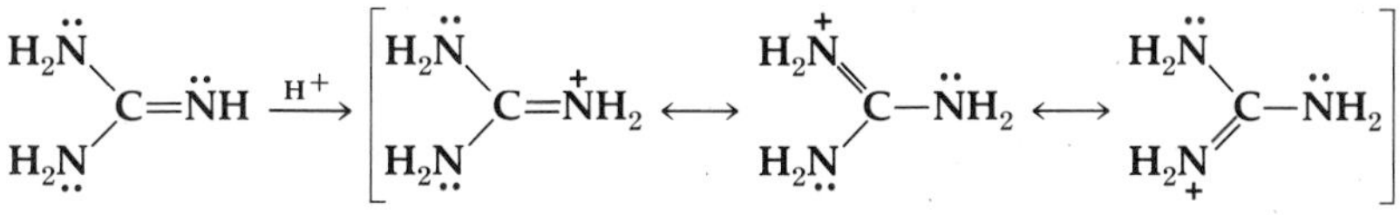

$$H_2\overset{\cdot\cdot}{N}\diagdown C=\overset{\cdot\cdot}{N}H \xrightarrow{H^+} \left[H_2\overset{\cdot\cdot}{N}\diagdown C=\overset{+}{N}H_2 \longleftrightarrow H_2\overset{+}{N}\diagdown C-\overset{\cdot\cdot}{N}H_2 \longleftrightarrow H_2\overset{\cdot\cdot}{N}\diagdown C-\overset{\cdot\cdot}{N}H_2 \right]$$

Guanidine, basicity
too large to measure
accurately in water

Guanidinium ion
resonance-stabilized

FIGURE 8-1 **Resonance in guanidinium ion**

bilized. This lowers the pK_a (relative to saturated analogs) since its conjugate acid derives little stabilization in this fashion. The more unsaturated groups attached, the more the pK_a is lowered (see ϕCH$_3$ vs. ϕ_3CH and others in Table 8-2) and the more electronegative the atom(s) supporting the negative charge in resonance the more effective the resonance stabilization will be.

Stabilization of the conjugate bases of carbon acids is dramatic and very important in organic reactions. The acidity of simple aldehydes and ketones is not measurable in water, but their conversion to enolate anions by bases is a critical step in many of their reactions. On the other hand, β-dicarbonyl compounds are measurably acidic in water since their anions are doubly stabilized by two adjacent carbonyl groups.

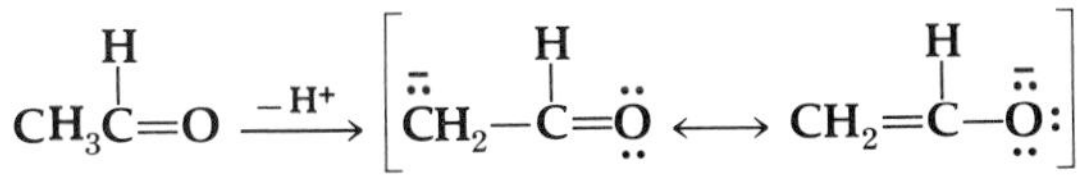

$$CH_3\overset{H}{C}=O \xrightarrow{-H^+} \left[\overset{H}{\underset{}{CH_2}}-\overset{}{C}=\overset{\cdot\cdot}{O} \longleftrightarrow CH_2=\overset{H}{C}-\overset{\cdot\cdot}{O}: \right]$$

Acetaldehyde   Acetaldehyde enolate anion

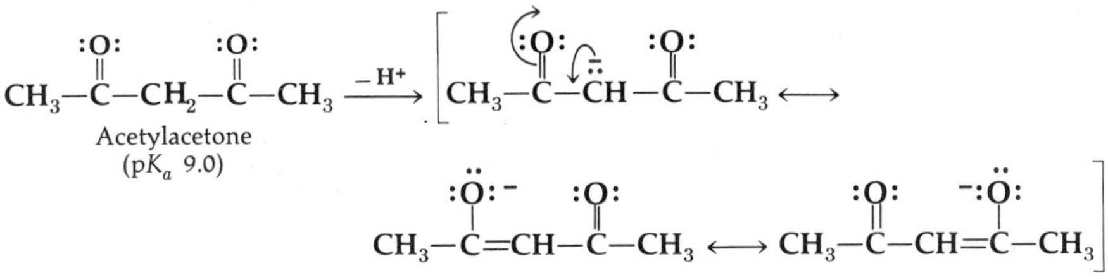

$$CH_3-\overset{:O:}{\overset{\|}{C}}-CH_2-\overset{:O:}{\overset{\|}{C}}-CH_3 \xrightarrow{-H^+} \left[CH_3-\overset{:O:}{\overset{\|}{C}}-\overset{}{CH}-\overset{:O:}{\overset{\|}{C}}-CH_3 \longleftrightarrow \right.$$

Acetylacetone
(pK_a 9.0)

$$\left. CH_3-\overset{:\overset{\cdot\cdot}{O}:^-}{\overset{\|}{C}}=CH-\overset{:O:}{\overset{\|}{C}}-CH_3 \longleftrightarrow CH_3-\overset{:O:}{\overset{\|}{C}}-CH=\overset{-:\overset{\cdot\cdot}{O}:}{\overset{\|}{C}}-CH_3 \right]$$

Acetylacetonate anion

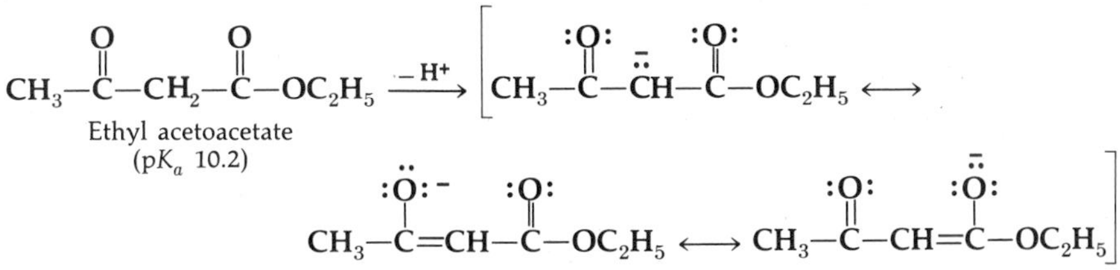

$$CH_3-\overset{O}{\overset{\|}{C}}-CH_2-\overset{O}{\overset{\|}{C}}-OC_2H_5 \xrightarrow{-H^+} \left[CH_3-\overset{:O:}{\overset{\|}{C}}-\overset{}{CH}-\overset{:O:}{\overset{\|}{C}}-OC_2H_5 \longleftrightarrow \right.$$

Ethyl acetoacetate
(pK_a 10.2)

$$\left. CH_3-\overset{:\overset{\cdot\cdot}{O}:^-}{\overset{\|}{C}}=CH-\overset{:O:}{\overset{\|}{C}}-OC_2H_5 \longleftrightarrow CH_3-\overset{:O:}{\overset{\|}{C}}-CH=\overset{:\overset{\cdot\cdot}{O}:}{\overset{\|}{C}}-OC_2H_5 \right]$$

Ethyl acetoacetate anion

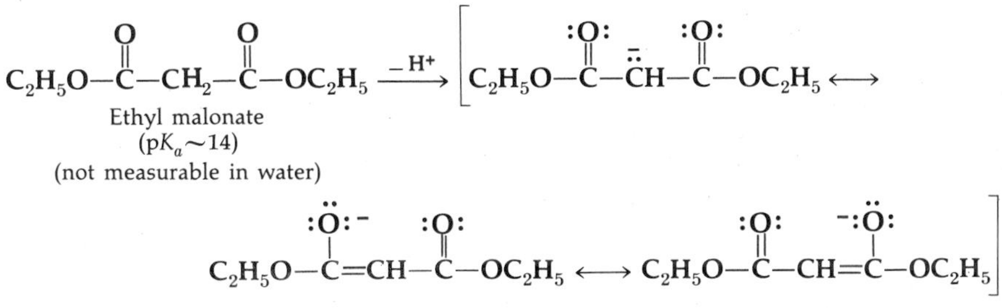

$$C_2H_5O-\overset{O}{\overset{\|}{C}}-CH_2-\overset{O}{\overset{\|}{C}}-OC_2H_5 \xrightarrow{-H^+} \left[C_2H_5O-\overset{:O:}{\overset{\|}{C}}-\overset{}{CH}-\overset{:O:}{\overset{\|}{C}}-OC_2H_5 \longleftrightarrow \right.$$

Ethyl malonate
(p$K_a \sim 14$)
(not measurable in water)

$$\left. C_2H_5O-\overset{:\overset{\cdot\cdot}{O}:^-}{\overset{\|}{C}}=CH-\overset{:O:}{\overset{\|}{C}}-OC_2H_5 \longleftrightarrow C_2H_5O-\overset{:O:}{\overset{\|}{C}}-CH=\overset{-:\overset{\cdot\cdot}{O}:}{\overset{\|}{C}}-OC_2H_5 \right]$$

Ethyl malonate anion

The above examples show that aldehyde and ketone groups are superior to the ester function in their ability to disperse negative charge. Delocalization of electrons within the ester group itself decreases the ability of the carbonyl group to accept electrons from external sources, such as an adjacent carbanion.

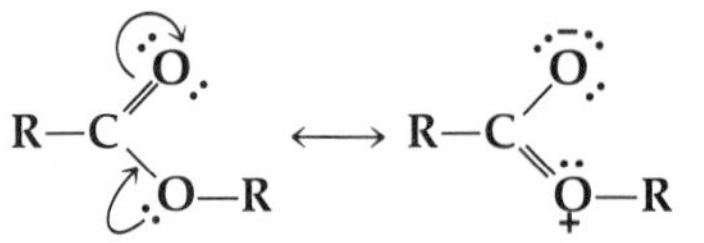

Delocalization of electrons in an ester group

Other unsaturated groups are capable of making adjacent carbon-hydrogen bonds relatively acidic. These groups arrange themselves in the order $NO_2 > CN \cong SO_2R > (C_6H_5)_3$ in their ability to disperse negative charge. Ionization of compounds containing these groups is illustrated.

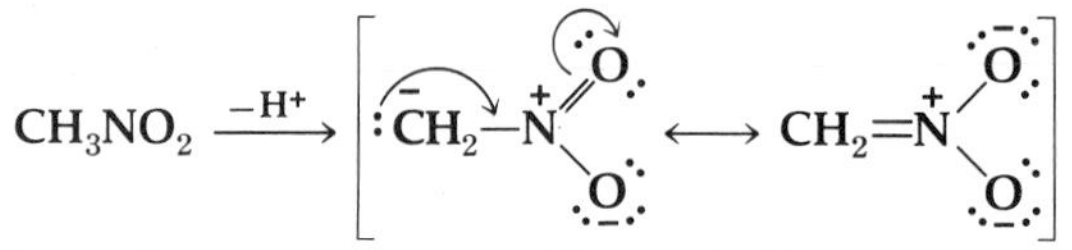

Nitromethane   Nitromethide ion $pK_a = 10.2$

Acetonitrile   Acetonitrile conjugate base $pK_a \sim 25$

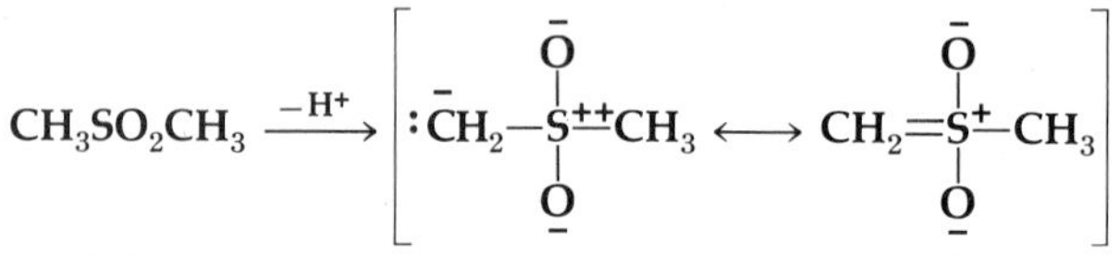

Dimethyl sulfone   Dimethyl sulfone conjugate base $pK_a \sim 23$

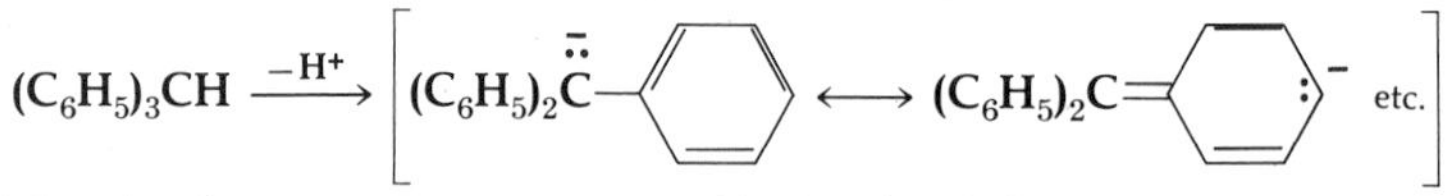

Triphenylmethane   Triphenylmethide ion $pK_a \sim 32$

The cumulative effect of several cyano groups on a carbon-hydrogen bond is illustrated by tricyanomethane and hexacyanoisobutene, which are as strong as the mineral acids ($pK_a \ll 0$). The phenomenal effects of resonance on the acidity of carbon acids is illustrated here in that nearly the entire range of acidities is available, increasing acidity from the parent hydrocarbons by 40 powers of 10 (10^{40} times more acidic)!†

†This resonance effect is more pronounced in carbon acids because the lower electronegativity of carbon supports its anion poorly and resonance forms bearing the charge elsewhere are of correspondingly greater weight, with more attendant stabilization (Sec. 5-3).

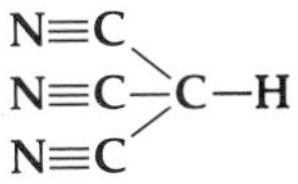

Tricyanomethane 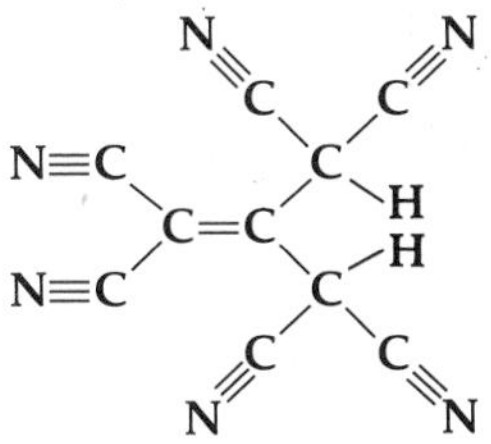   Hexacyanoisobutene

Study of acidity of some weakly acidic carbon-hydrogen linkages is complicated because equilibria are sometimes established slowly. In contrast, the rate of equilibration of protons among various oxygen, nitrogen, and sulfur atoms is too fast for measurement by ordinary methods.

Hybridization Changes and Acidity

A multiple bond attached to the basic atom reduces its basicity (lowers pK_a). For example, ketone oxygens are weaker bases than ethers, and nitrogen compounds, shown below, demonstrate the same effect. Unsaturated carbon acids can also be found in Table 8-2 to illustrate this.

$(CH_3)_3N:$ $CH_3C\equiv N:$

Trimethylamine   Pyridine   Acetonitrile
pK_a 9.8 pK_a 5.2 $pK_a < 0$

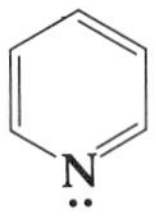

The base-weakening effect of multiple bonding is associated with changes in the hybridization of the orbitals of the heteroatoms. In an aliphatic amine the nonbonding pair of electrons is in an sp^3 orbital; in a compound containing a double bond ($=\ddot{N}-$) the nonbonding pair occupies an sp^2 orbital; and in a nitrile ($-C\equiv N:$) the nonbonding pair is in an sp orbital. The amount of s character increases with the unsaturation. Since s electrons are bound more firmly to the nucleus than p electrons, it is reasonable that electrons in orbitals that have a large amount of s character should be fairly tightly held and so relatively unavailable for bonding to external reagents, such as protons. Alternatively, we can say that since bond formation is weaker, more s character is present; the parent acid—with a bond to hydrogen—is destabilized in these unsaturated cases.

Despite the above generalization some doubly bonded heteroatoms may be very strongly basic, because of resonance effects. A striking example is guanidine, shown in Fig. 8-1. Aqueous solutions of guanidine are essentially completely ionized to guanidinium ions and hydroxide ions. The effect is due to the very great resonance stabilization of the symmetrical guanidinium ion.

Another example of resonance effects resulting in increased base strength is found in the greater basicity of 2,6-dimethyl-δ-pyrone compared with cyclohexanone. The benzenelike resonance of the conjugate acid is undoubtedly responsible for the enhanced basic character of the pyrone, since its aromatic resonance stabilization is especially large. Although ordinary ketones, esters, ethers, and alcohols all show basic properties (i.e., they can be protonated) when treated with strong acids, their base strengths are considerably lower than those of their nitrogen analogs. The conjugate acids of the oxygen compounds are important intermediates in organic reactions (Chaps. 10 and 12). The following examples illustrate the basic character of oxygen-containing compounds:

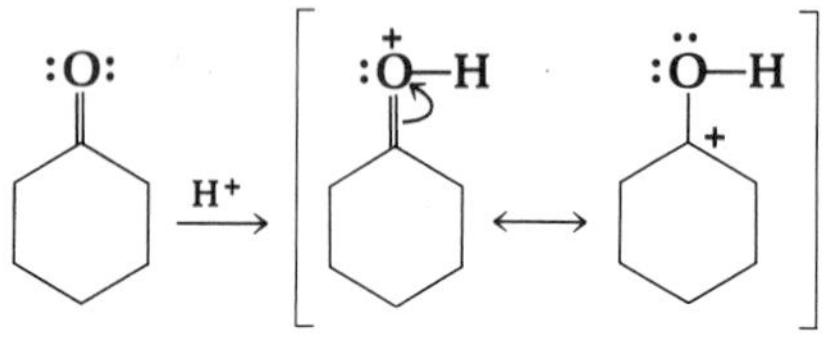

Cyclohexanone as a base

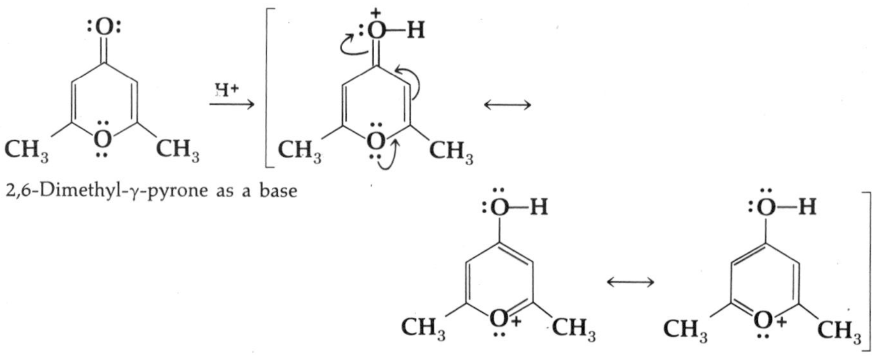

2,6-Dimethyl-γ-pyrone as a base

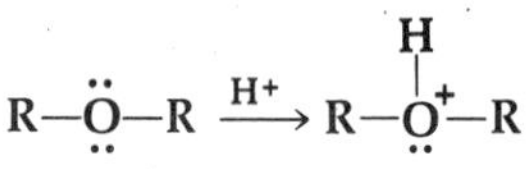

$$R-\overset{\cdot\cdot}{\underset{\cdot\cdot}{O}}-R \xrightarrow{\text{H}^+} R-\overset{\overset{\text{H}}{|}}{\underset{\cdot\cdot}{O}}{}^+-R$$

Ether as a base

$$R-\overset{\cdot\cdot}{\underset{\cdot\cdot}{O}}-H \xrightarrow{\text{H}^+} R-\overset{\overset{\text{H}}{|}}{\underset{\cdot\cdot}{O}}{}^+-H$$

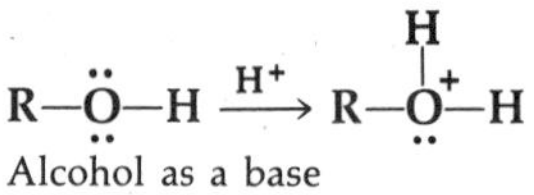

Alcohol as a base

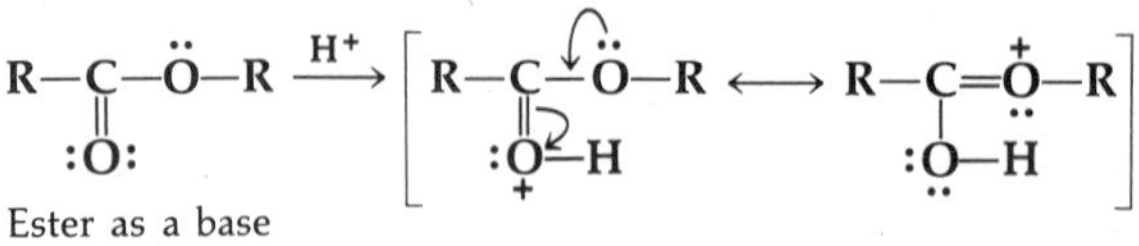

Ester as a base

The ester example deserves comment, for if the sp^3 oxygen is more basic than the sp^2 (carbonyl) oxygen, we should expect protonation there, whereas in fact it appears to occur on the less basic carbonyl oxygen. The important factor upsetting this prediction is that the normal ester resonance is retained to stabilize carbonyl protonation but is lost if the sp^3 oxygen is protonated (see Prob. 8-3 also).

PROBLEM 8-4

Explain why one member in each pair is a stronger acid than the other.

a   *m*- and *p*-cyanophenol

f   *m*- and *p*-amino-acetophenone ($H_2N-C_6H_4-COCH_3$)

b $(CH_3)_2C=\overset{+}{O}H$ and $(CH_3)_2\overset{+}{O}H$

g $CH_2=CHCH_2OH$ and $CH_3CH=CHOH$

c $R-CONH_2$ and $R-C\overset{\displaystyle NH}{\underset{\displaystyle NH_2}{<}}$

h $O_2NCH_2COCH_3$ and $O_2NCH_2COOCH_3$

d $\bigcirc\overset{+}{N}H_2$ and $\bigcirc\overset{+}{N}H$

i $NCCH_2CN$ and $NCCH_2CH_2CN$

e $C_6H_5CONHCH_2C_6H_5$ and $C_6H_5CH_2CONHC_6H_5$

j   Cyclopentadiene and 1,4-pentadiene

8-5   INDUCTIVE AND ELECTROSTATIC EFFECTS

If nearby substituents possess a dipole it will usually be directed away from carbon and will place partial positive character on the carbon. By electrostatic attraction this tends to stabilize a nearby anion and thus usually stabilizes the conjugate base more than the acid in a conjugate pair, and lowers pK_a. (Acids become more acidic.) This is borne out by the compounds collected in Table 8-3. Substitution of iodine on acetic acid is less acid-strengthening than substitution of chlorine since iodine is less electronegative. Increasing the number of halogens increases the effect. The range of these dipole effects can be seen in the acidity of trifluoroacetic acid ($pK_a = 0$), about 10^5 (100,000) times more acidic than acetic acid. It may be noted here that *external resonance effects* are virtually negligible for carboxylic acids since the carboxylate resonance is itself so very strong. This may be seen by comparing the small difference in the *m*- and *p*-nitrobenzoic acids (Table 8-3) with the large differences in the *m*- and *p*-nitroanilines noted previously.

Destabilized by dipoles   Stabilized by dipoles

The dipoles introduced also destabilize the conjugate acid since carboxyl has a dipole also. Since the two are bound together with their positive dipole ends adjacent, they repel each other electrostatically. Substituent effects due to the permanent polarity or polarizability of groups are called **inductive effects** and are perhaps best understood in terms of dipole-dipole and charge-dipole interactions. The data of Table 8-3 provide examples. For instance, only very small dipoles are associated with methyl-carbon bonds, and K_a's for acetic and propionic acids are very close to each other. In contrast, the cyano group has a strong dipole with its positive charge oriented toward the carboxyl group, and as a result pK_a's for acids containing this substituent possess decidedly lower values than those which do not:

$$:N\equiv CCH_2COOH \longleftrightarrow :\ddot{N}=\overset{+}{C}CH_2COOH \text{ or } \overset{- \; +}{N\equiv CCH_2COOH}$$

That the effect drops off as the dipole is moved farther from the carboxyl is shown by the increase in values of pK_a in the series α-, β-, and δ-chlorobutyric acid.

The following groups have strong **electron-withdrawing** inductive effects: NR_3^+, NO_2, ONO_2, CN, CO_2H, CO_2R, $C=O$, F, Cl, Br, I, NO, and ONO.

Electron-donating inductive effects require that substituent groups either carry a negative charge or have a dipole with the negative end directed toward carbon. The latter cases are very rare owing to the relatively low electronegativity of carbon. The effect of charged groups is illustrated by values of the *second* ionization constants of dibasic acids (Table 8-4). The large difference between the first and second ionization constants is mainly due to the repulsion

TABLE 8-3 **Inductive Effects on Acid Strength**

Acid	Structure	$pK_a(H_2O)$ at 25°
Acetic	CH_3COOH	4.8
Propionic	CH_3CH_2COOH	4.9
Iodoacetic	ICH_2COOH	3.9
Chloroacetic	$ClCH_2COOH$	2.8
α-Chlorobutyric	$CH_3CH_2CHClCOOH$	3.8
β-Chlorobutyric	$CH_3CHClCH_2COOH$	4.1
γ-Chlorobutyric	$ClCH_2CH_2CH_2COOH$	4.5
Trichloroacetic	Cl_3CCOOH	0.9
Trifluoroacetic	CF_3COOH	0.3
Methoxyacetic	CH_3OCH_2COOH	3.5
Cyanoacetic	$NCCH_2COOH$	2.4
Phenylacetic	$C_6H_5CH_2COOH$	4.3
Benzoic	C_6H_5COOH	4.2
p-Nitrobenzoic	$p\text{-}NO_2C_6H_4COOH$	3.4
m-Nitrobenzoic	$m\text{-}NO_2C_6H_4COOH$	3.5

between like charges in the second anion. The stepwise ionization of oxalic acid is shown below.†

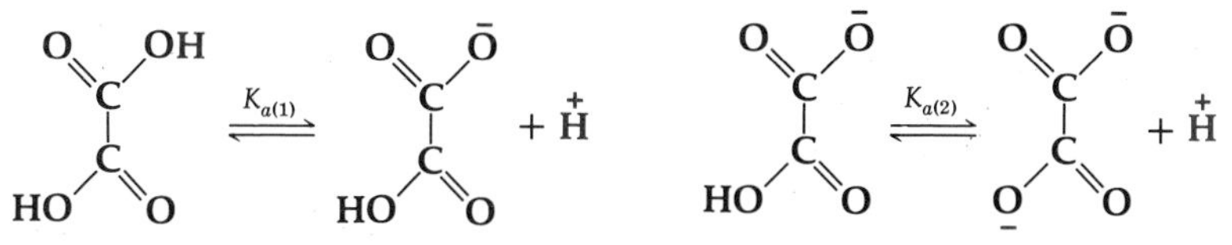

The effect of the first anion on formation of the second is clearly seen in maleic and fumaric acids, in which geometrical isomerism holds the first anion near the second in the cis acid and the forced proximity creates a larger discrepancy in pK_a than in the trans where the groups are held farther apart.

Molecules can have both basic and acidic groups, as is illustrated by the existence of a large number of amino acids in nature (Chap. 25). An example is glycine (H_2NCH_2COOH), which normally exists as an **inner salt** or **zwitterion**, in which the proton of the carboxyl group is transferred to the basic nitrogen atom. The conjugate acid of glycine loses its first proton from the carboxyl group. The substance is a rather strong acid owing to the electron-withdrawing effect of the formal positive charge on nitrogen. The zwitterion can also lose a proton, but it is a rather weak acid because of the formal negative charge now present on the carboxyl group.

$$\overset{+}{N}H_3CH_2COOH \underset{2.4}{\overset{pK_a}{\rightleftharpoons}} \overset{+}{H} + \overset{+}{N}H_3CH_2CO\bar{O} \underset{8.8}{\overset{pK_a}{\rightleftharpoons}} NH_2CH_2CO\bar{O} + \overset{+}{H}$$

Conjugate acid of glycine — Glycine inner salt (zwitterion) — Glycine anion

Inductive effects of single dipoles are usually small, but observable, in other acids and bases. *meta*-Nitrophenol is a somewhat stronger acid than phenol (9.3 vs. 9.8) and the relatively large effect of a trifluoromethyl group is seen in β-trifluoroethyl amine, pK_a 7.5, compared to ethylamine pK_a 10.5.

† Formation of an internal hydrogen bond (page 66) in the first anion may also help to stabilize that species.

TABLE 8-4 **Acidity of Dibasic Acids**

Name	*Structure*	pK_1	pK_2
Oxalic	HOOCCOOH	1.5	?
Malonic	$HOOCCH_2COOH$	1.9	5.7
Succinic	$HOOC(CH_2)_2COOH$	4.2	5.6
Glutaric	$HOOC(CH_2)_3COOH$	4.4	5.4
Maleic	*cis*-HOOCCH=CHCOOH	2.0	6.3
Fumaric	*trans*-HOOCCH=CHCOOH	3.0	4.5

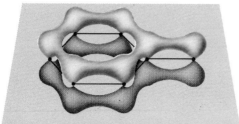

I
Sterically feasible

II
Not sterically feasible

III
Sterically feasible

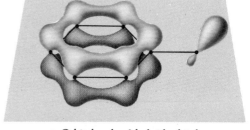

π Orbital planar aromatic amine

π Orbital and sp^3 hybrid orbital
of nonplanar aromatic amine

FIGURE 8-2   **Steric inhibition of resonance**

8-6   STERIC EFFECTS AND HYDROGEN BONDING

Proton acid-base reactions are not particularly sensitive to steric compression, since a proton is so small that its absence or presence in a molecule does not usually affect the volume very much. However, steric effects are commonly observed as a side effect in resonance stabilization. This occurs because resonance stabilization of the electron pair (in the base) requires the involved group at the basic site to become coplanar with the rest of the conjugated system. No such requirement exists for the unconjugated acid of the pair and so a differential *destabilization* of the base is possible if steric compression acts against the σ frame becoming coplanar.

The base *N,N*-dimethylaniline is six times weaker than *N,N*-dimethyl-*o*-toluidine. In both amines, the electron pair on nitrogen is delocalized through resonance, which effect makes both compounds weaker bases than methylamine. However, *N,N*-dimethyl-*o*-toluidine is the stronger base.

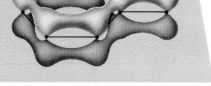

N,N-Dimethylaniline
pK_a 5.1

etc.

N,N-Dimethyl-*o*-toluidine
pK_a 5.9

These facts are explained with formulas in Fig. 8-2. The conformation needed for maximum interaction of the benzene ring π electrons with the unshared electrons of nitrogen is one in which all atoms of the molecule are in one plane except the hydrogens of the methyl groups (structure I of Fig. 8-2). This conformation results in no steric compression for N,N-dimethylaniline, because there is enough space to accommodate both the volumes occupied by the methyl groups attached to the nitrogen and by the ortho hydrogen atoms. In N,N-dimethyl-o-toluidine, a planar conformation compresses the ortho methyl and one of the methyl groups attached to the nitrogen, as indicated in structure II. To avoid this strain, the molecule is twisted toward the conformation shown in III. As a result, resonance structures such as II are less important; the electron pair on nitrogen is more localized, and the substance is a stronger base. In Fig. 8-2 there is also a molecular-orbital representation of planar aromatic amines and an orbital picture of nonplanar aromatic amines. A similar example is found in the dimethyl-p-nitrophenols, in which proximity of the methyls to nitro sterically inhibit its becoming coplanar to stabilize the phenolate anion, while their presence ortho to —O^- is not sterically troublesome for resonance.

X	Y	pK_a
CH_3	H	8.2
H	CH_3	7.2

Conjugate base of dimethyl-p-nitrophenols

A second type of steric compression is associated with the decrease in acidity on transition from the simple carboxylic acids to those which possess highly ramified structures. In the ionization of acetic acid, the charged anion must require more stabilization by solvation than does the uncharged acid. With acids such as A, below, the charge on the anion B is somewhat shielded from solvent molecules by the surrounding methyl groups. As a result, the anion is less stable, and the acid is weaker. This may be called steric hindrance to solvation. In a simpler case this is also seen in the lower acidity of t-butanol than n-butanol ($pK_a \sim 19$ vs. 17).

$$CH_3COOH \rightleftharpoons CH_3COO^- + H^+$$

Acetic acid   Acetate anion
$pK_a = 4.6$ (50% H_2O, 50% CH_3OH)

$pK_a = 7.0$ (50% H_2O, 50% CH_3OH)

Hydrogen Bonds. Internal hydrogen bonding which involves an acidic hydrogen tends to stabilize that acid, and this stabilization is of course lost when the hydrogen is ionized away to produce the conjugate base. This effect causes acidity changes of one to three pK_a units, and can be seen in operation in the appended pair of similar cases, one with and one without internal hydrogen bonding. The large raising of the phenolic pK_a (to 13.4 from 9.3) in the second case is partly a function of the inductive effect of the neighboring anion.

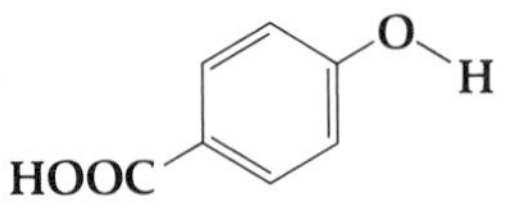

HOOC
$pK_a = $ —OH 9.3, —COOH 4.5

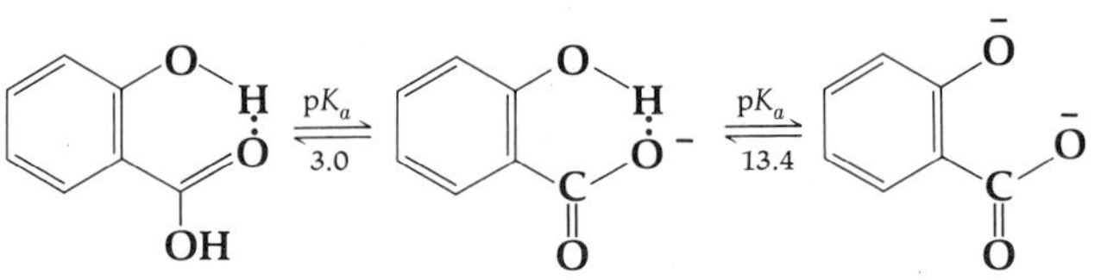

With amines, it is hydrogen bonding of the conjugate acid—usually to a nearby hydroxyl—which is commonly observed. This has been used to distinguish cis and trans relative configurations of cyclic amino alcohols.

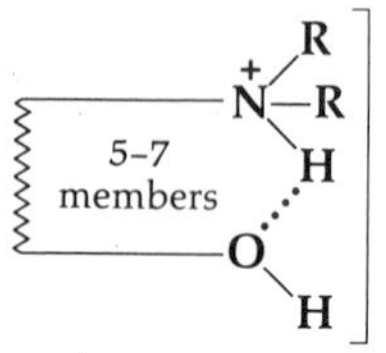

5–7 members

PROBLEM 8-5

Explain why one member in each pair is a stronger acid than the other.

a

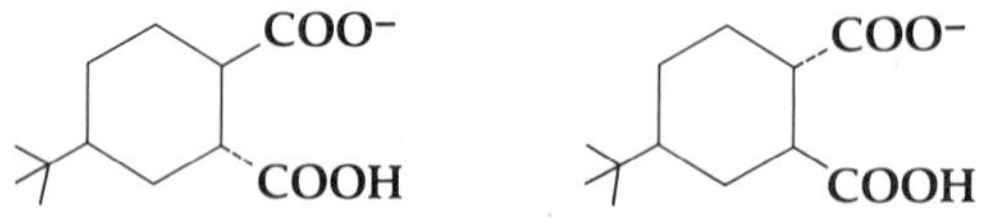

b

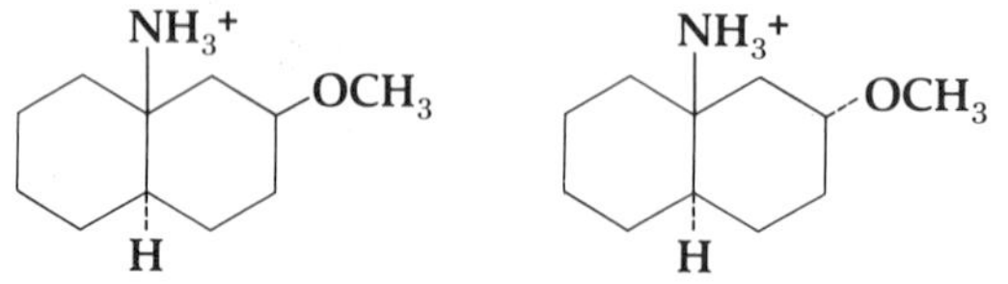

c CF_3CH_2OH CH_3CF_2OH

d   *o*- and *p*-nitrophenol

8-7 LEWIS ACIDS AND BASES

A general set of definitions of acids and bases was proposed by G. N. Lewis:

1 All substances containing unshared electron pairs are bases
2 All those containing an element that is two electrons short of having a complete valence shell are acids

The bases are as defined before but the definition of acids is broader, the *proton* being only one case of a "Lewis acid." Boron (and aluminum) compounds are common examples of **Lewis acids**. The reaction of a Lewis acid with a Lewis base produces an addition compound, called a salt if the Lewis acid is a proton acid. Addition compounds, including salts, are always ionic, inter- or intramolecularly.

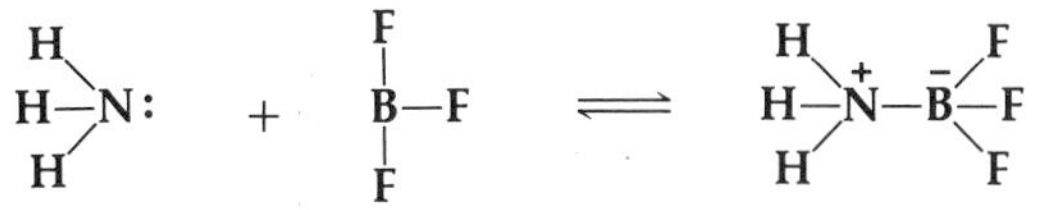

Ammonia, a   Boron trifluoride,   Addition
Lewis base   a Lewis acid   compound

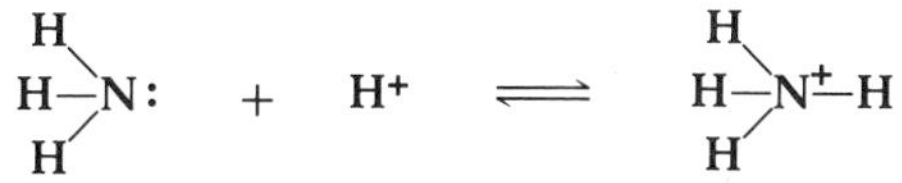

Salt

Lewis acids such as aluminum chloride, boron trifluoride, stannic chloride, zinc chloride, and ferric chloride are extremely important acid catalysts for certain organic reactions, which are initiated by reaction of these Lewis acids with basic sites (electron pairs) on the reactant molecule.

BF_3	$AlCl_3$	$SnCl_4$	$ZnCl_2$	$FeCl_3$
Boron trifluoride	Aluminum chloride	Stannic chloride	Zinc chloride	Ferric chloride

Important Lewis acids

Systematic study of Lewis acid-base reactions has been particularly useful in establishing principles of chemical reactivity. Equilibrium constants have been evaluated for the reversible reactions between alkyl-boron compounds and amines of varying steric requirements, as in Table 8-5.

Comparison of dissociation constants of trimethylboron adducts shows that replacement of two hydrogen atoms of ammonia by methyl groups is base-strengthening. Such a response is expected. Since the electron affinity of carbon is less than that of hydrogen, a methyl group will have a greater electron-donating effect than hydrogen. However, the trimethylamine adduct is much more highly dissociated than the adduct from dimethylamine. The methyl groups must be more compressed in trimethylamine-trimethylboron than are the groups in the dimethylamine adduct, but this steric effect is negligible when the acid is only a proton.

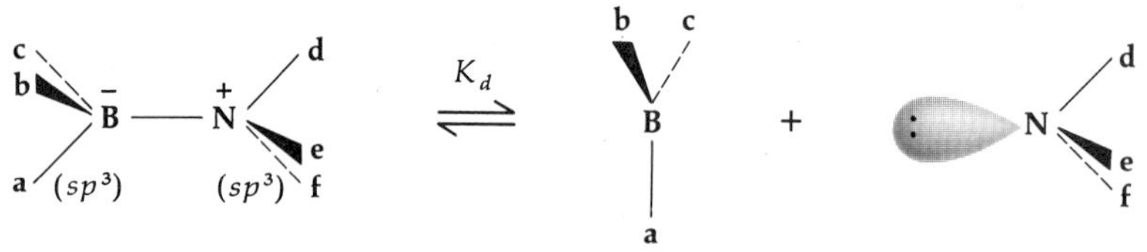

| Alkylborane-amine addition compound | Alkylborane (sp^2; planar) | Amine (sp^3; pyramidal) |

Triethylamine gives no detectable adduct with trimethylboron, since ethyl require more space than methyl groups. At least one of the three ethyl groups of triethylamine must be folded toward, and partly covering, the unshared pair of electrons. In quinuclidine, the carbon atoms are held back by the ring system, and consequently the substance forms a very stable adduct. Configurations of these two compounds are indicated in Fig. 8-3.

An interesting example of the steric requirements of Lewis acids is afforded by 2,6-di-t-butyl-pyridine, the pK_a of which is essentially the same as that of unsubstituted pyridine (reaction with a small proton), but the space available at the base pair is too small for any other Lewis acid and no addition compounds are observed with various boranes (even BH_3), SO_3, etc. The compound is a selective base for protons only among Lewis acids.

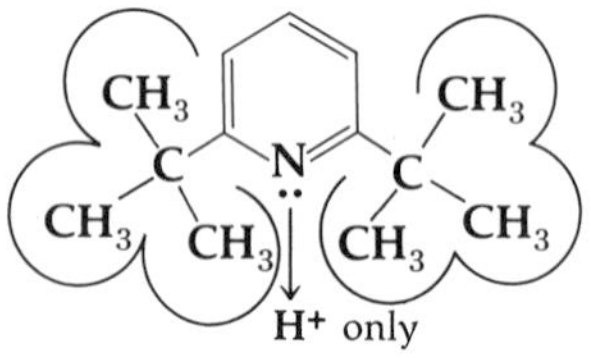

TABLE 8-5 **Dissociation Constants of Trialkylboron-Amine Compounds at 100°**

Acid	Base	K_d
$(CH_3)_3B$	$\ddot{N}H_3$	4.6
$(CH_3)_3B$	$CH_3\ddot{N}H_2$	0.0350
$(CH_3)_3B$	$(CH_3)_2\ddot{N}H$	0.0214
$(CH_3)_3B$	$(CH_3)_3\ddot{N}$	0.472
$(CH_3)_3B$	$(C_2H_5)_3\ddot{N}$	No compound formed
$(CH_3)_3B$	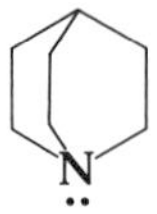	0.0196

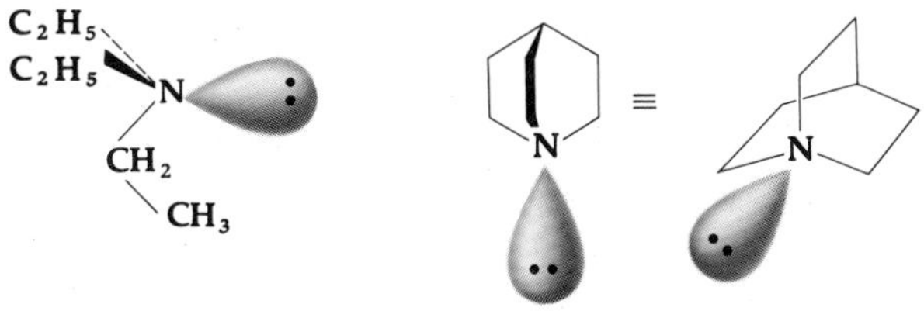

Triethylamine   Quinuclidine

FIGURE 8-3   Configurations of triethylamine and quinuclidine

8-8 CHARGE-TRANSFER COMPLEXES

Bases were defined previously in terms of unshared electron pairs, but the electrons in a π bond are also mobile enough to attack strong acids, as the low pK_a's of alkenes (~ -4) testify. The protonation of carbon-carbon double bonds is an important process in initiating many alkene reactions. Aromatic π electrons are less easily attacked—less basic—owing to their extra resonance stabilization, but they can react with both protons and Lewis acids to form addition compounds. In some cases, these adducts can be isolated in a crystalline state. In general, the greater the number of aromatic rings per molecule, the more basic the hydrocarbon and the more readily are addition compounds formed. Anthracene is more basic than naphthalene, which in turn is more basic than benzene.

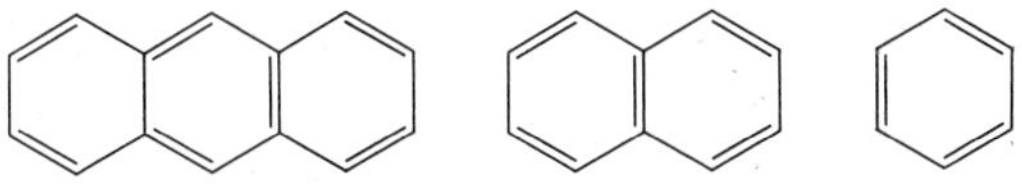

Anthracene   Naphthalene   Benzene

Order of decreasing basicity →

Liquid hydrogen fluoride, tetranitromethane, picric acid, and tetracyanoethylene are examples of Lewis acids capable of forming adducts with aromatic hydrocarbons.

Hydrogen   Tetranitromethane   Picric acid   Tetracyano-
fluoride   ethylene

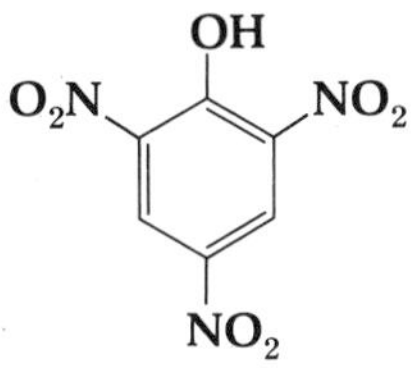

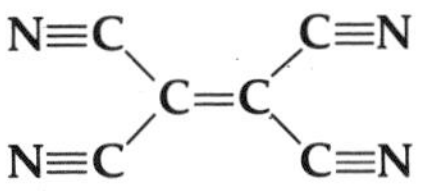

The exact structures of these addition compounds and the nature of the bonds involved are not well understood. For instance, anthracene and tetracyanoethylene react to form a brilliant green complex. Tetracyanoethylene is a special kind of Lewis acid, because the four cyano groups withdraw electrons from the central carbon-carbon double bond and these two carbons become relatively electron-deficient. On the other hand, the π electrons of anthracene are relatively loosely bound. In the complex some of the anthracene π electrons "leak" into tetracyanoethylene. Consequently, the complexes are called **charge-transfer complexes**, or π **complexes**. In each part of the complex the transferred charge is spread throughout a conjugated system, as indicated by the use of dashes for partial bonds in the following formulas.

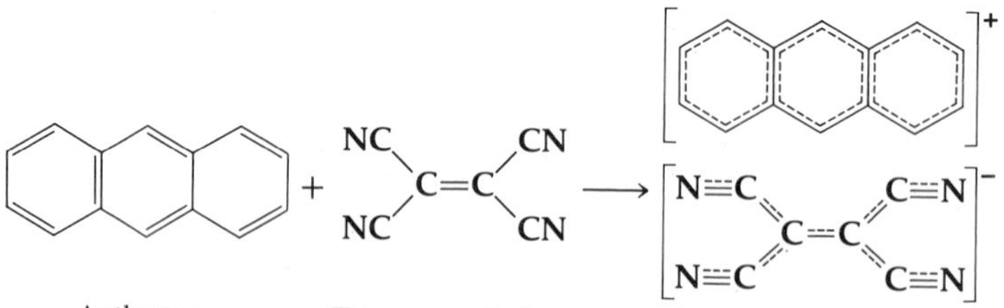

Anthracene   Tetracyanoethylene   Adduct

8-9 TAUTOMERISM

Tautomeric shifts of protons (Sec. 5-4) are internal acid-base reactions. When a proton is completely removed from either of two tautomers, the same resonance-stabilized anion is produced, as illustrated in the conversion of both *carbonyl* and *enol* forms of acetaldehyde into the same enolate anion.

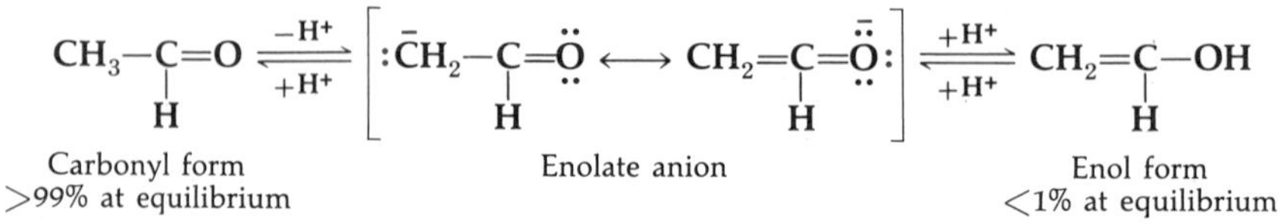

Tautomerism in acetaldehyde

Interconversion of carbonyl and enol forms of aldehydes and ketones is measurably slow, since at some stage a carbon-hydrogen bond must be broken. Many studies of equilibrated systems have been carried out, and tautomerization constants measured. These constants are the ratio of the acidity constants of the carbonyl and enol forms, as indicated in Eq. (10), and recorded in Table 8-6.

$$\text{Tautomerization constant} = K_T = \frac{K_a \text{ of carbonyl form}}{K_a \text{ of enol form}} \qquad 10$$

or $pK_T = pK_{a(CO)} - pK_{a(enol)}$

The enol forms of simple aldehydes and ketones have not been isolated in pure form because they are so easily converted into the more stable carbonyl forms. Conversions of this type occur very readily and are catalyzed by both acids and bases, and even by polar surfaces (such as glass!). Both tautomeric forms of β-dicarbonyl compounds have been isolated. The enol and carbonyl forms of these compounds are close to each other in stability, and in some of the examples shown in Table 8-6 the enol form is the more stable of the two. β-Ketoaldehydes commonly enolize almost 100% at the aldehyde carbon.

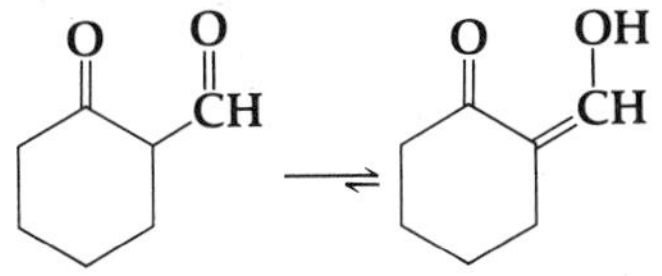

Two factors contribute to the increased stability of the enol form in β-dicarbonyl compounds. The enols are stabilized by a significant resonance energy, similar to that of carboxylic acids. In noncyclic compounds, the enol forms are also stabilized by internal hydrogen bonding.

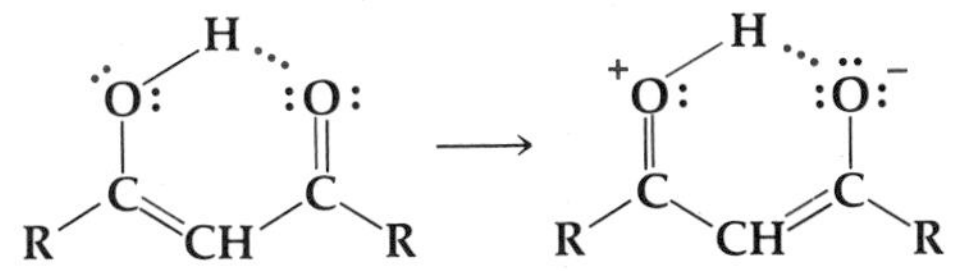

Resonance and hydrogen bonding in enols of β-diketones

TABLE 8-6   **Tautomerization Constants of Carbonyl-containing Compounds**

Carbonyl form	Enol form	K_T	pK_T
CH_3CCH_3 (C=O)	$CH_2=CCH_3$ (OH)	2.5×10^{-6} (in water)	5.6
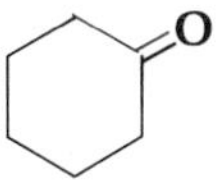 cyclohexanone (C=O)	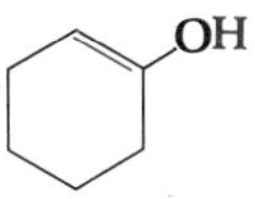 cyclohexenol (OH)	2.0×10^{-4} (in water)	3.7
$CH_3CCH_2COC_2H_5$	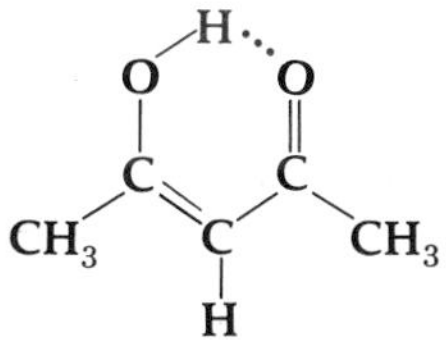 enol form	6.2×10^{-2} (pure liquid)	1.2
$CH_3CCH_2CCH_3$	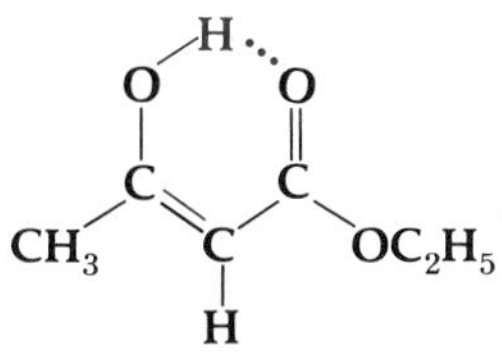 enol form	3.6 (pure liquid)	0.6
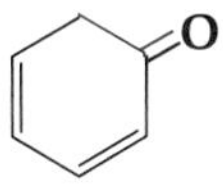 cyclohexadienone (C=O)	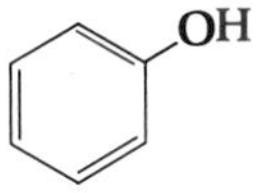 phenol (OH)	Too large to measure	(~20)

8-10   USES OF ACIDS AND BASES

A common application of organic acids and bases is found in indicators. These are compounds which have spectral absorption in the visible region but different λ_{max} (Sec. 7-3) for the acid and its conjugate base. Thus when the pH of the medium changes and passes through a value of pH = pK_a (when [B:] = [HB]) the color of the indicator changes. The common basis for this occurrence is that the conjugate base represents an extended conjugated system (with long-wavelength absorption) which is broken by protonation. A simple example is found in p-nitrophenol, the conjugate acid of which is yellow, the base (anion) being dark red. Phenolphthalein is one member of a large class of structurally similar common indicators, including thymol blue, cresol purple, and brom-cresol green. These have two ionizable hydrogens and both often ionize in water giving two color transitions.

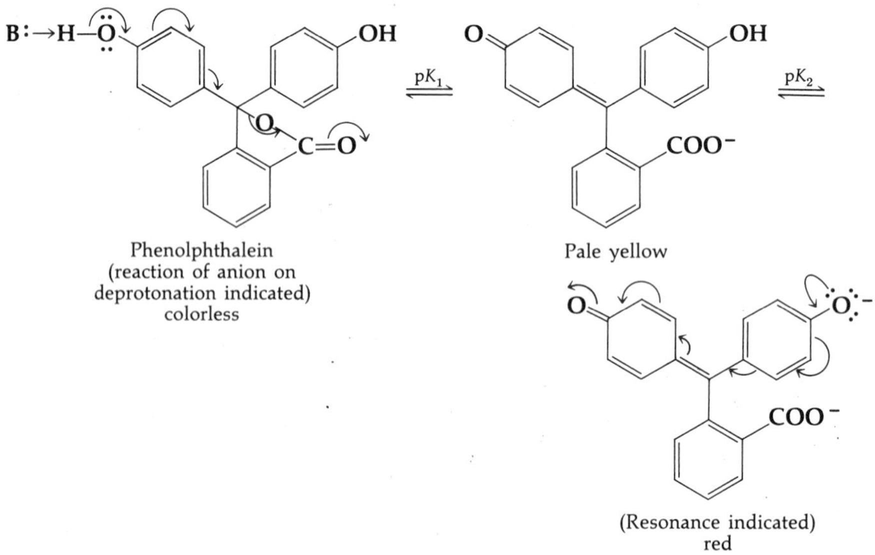

Phenolphthalein
(reaction of anion on
deprotonation indicated)
colorless

Pale yellow

(Resonance indicated)
red

PROBLEM 8-6

a Write several resonance forms of the dianion of phenolphthalein.

b Create variants on the phenolphthalein structure which will have different colors (cf., longer absorption wavelengths) for both the anions.

In general, of course, when an acid or base is part of a chromophore absorbing in the UV or visible region, the spectrum will change on passing over to the conjugate form. From Fig. 7-8 it is clear that hydroxyl conjugated to carbonyls goes to a longer wavelength absorption in basic solution since the group becomes oxide anion with more effective conjugation. Similarly, p-nitroaniline absorbs at longer wavelength than its conjugate acid, in which the nitrogen electrons are removed from conjugation by the protonation.

Accordingly, UV spectra of many systems afford more structural information when recorded in acid and base, as well as neutral, solution in order to discern these wavelength shifts.

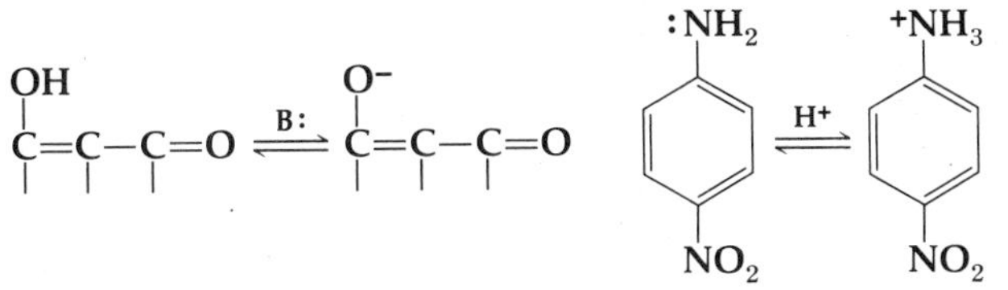

$$\underset{\substack{|\\ C\\ \|}}{OH} \underset{\substack{|\\ C\\ |}}{} \underset{\substack{|\\ C\\ |}}{} =O \overset{B:}{\rightleftharpoons} \underset{\substack{|\\ C\\ \|}}{O^-} \underset{\substack{|\\ C\\ |}}{} \underset{\substack{|\\ C\\ |}}{} =O$$

Many organic reactions are initiated by a first step involving either protonation or deprotonation of a reactant which is a base or an acid, respectively. This is done by applying an acid or base as reagent. This initiating acid or base should be chosen so as to have sufficient strength that the reaction is exothermic, i.e., the equilibrium lies towards the products. This is assessed by comparing the reagent and reactant pK_a's on the scale of Table 8-2. If a reagent is chosen that causes the equilibrium to favor starting materials and hence yield only a small proportion of products, the choice will be acceptable only if this small equilibrium supply does not choke off the subsequent desired reactions of those products.

Reagent bases must also be chosen so as not to attack other positive sites than acidic protons in the molecule or else side reactions will occur. In the list of common reagents in Table 8-7 the bases NaH, $(CH_3)_3C:^-$, $\phi_3C:^-$, and

TABLE 8-7   **Common Acid and Base Reagents Used
to Initiate Organic Reactions**

Bases	Acids
NaH	$FSO_3H \cdot SbF_5$
t-BuLi	FSO_3H
n-BuLi	HBF_4
$LiNEt_2$	H_2SO_4
$NaNH_2$	BF_3
ϕ_3CNa	$C_6H_5SO_3H$
$KOC(CH_3)_3$	HCl
$NaOCH_3$	CF_3COOH
NaOH	CH_3COOH
Et_3N	
CH_3CO_2Na	
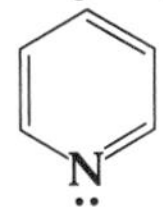 Pyridine	

Decreasing basicity

Decreasing acidity

t-BuO⁻ are commonly used for selective attack on protons rather than other positive sites, the last three because their steric hindrance prohibits attack on sites bulkier than protons. Acids must sometimes be chosen so that they offer no mild base counterion which is capable of attacking the protonated reactant; $FSO_3H \cdot SbF_5$, HBF_4, and BF_3 are common for this purpose.

As an example, consider a reaction in which a ketone is to be converted to its enolate anion so that this may undergo a subsequent reaction (Chap. 12). The pK_a of a simple ketone is about 20. Use of the base triethylamine ($pK_a \sim 10$) may not be adequate to catalyze the reaction, but sodium ethoxide ($pK_a \sim 17$) may be sufficient even though it can enolize very little. The $pK_c = pK_2 - pK_1 = 20 - 17 = 3$, or $K_c = 10^{-3}$ which implies a concentration of enolate about 1/1,000 that of the ketone at equilibrium. However, the rate of attainment of the equilibrium is fairly rapid so that as enolate is used up in the subsequent reaction, it is regenerated rapidly by the ethoxide-ketone equilibrium. If the rate of proton abstraction from the ketone is slow, however, it would be better to use a base which more fully converts ketone to enolate at equilibrium, such as *t*-butoxide (pK_a 19) or, better, $NaNH_2$ ($pK_a \sim 35$) or $\phi_3C{:}^-$ ($pK_a \sim 32$).

Acidic and basic properties can be of great value in structure determinations as well. Compounds that are basic (and are not salts like $CH_3O^-Na^+$) are almost invariably amines, while acidic compounds with pK_a below about 7 are almost always carboxylic acids. Most amines and carboxylic acids, like other neutral organic molecules, do not dissolve in water. However, amines dissolve in aqueous acids (pH 1 to 2) since their ionic conjugate acids are formed. Similarly, carboxylic acids dissolve in aqueous bicarbonate (pH ~ 7), and carbon dioxide is seen bubbling out from the carbonic acid liberated. Phenols ($pK_a \sim 10$) will dissolve in strong aqueous alkali (pH ~ 13) but not in aqueous bicarbonate. Such simple solubility tests have long been used to determine the acidity or basicity of organic substances in structure determination, and so imply the presence or absence of these functional groups.

The detailed measurement of pK_a values, especially in the easily accessible aqueous range of about 1 to 13, can often throw more light on the presence or absence of perturbing influences on normal acidity or basicity, as in the examples of the previous sections. In structural problem-solving the presence of nitrogen and basicity indicate an amine. A lowered pK_a (from ~ 10) implies resonance with the nitrogen electrons usually, and nitrogen without basicity usually implies the nonbasic functionality of *amide*, *nitrile*, or *nitro* although most of the rarer functional groups containing nitrogen are also nonbasic (oximes, azides, etc.).

A number of problems are provided to illustrate the value of acidity-basicity information in simple chemical determinations of structure.

8-11 SUMMARY

Most organic acids are proton acids with protons attached to N, O, and S. The acidity of C—H is negligible and may be ignored unless the carbon is adjacent to one or more groups which stabilize the conjugate base by delocalization of its long pair ($NO_2 > C{=}O > CN$, $SO_2R > \phi$). Every acid is related to a conjugate base with a pair of electrons in place of the acidic proton, and one charge more negative. The acidic proton is removable by another base in an equilibrium reaction, a base of higher pK_a being necessary to force the reaction to proceed substantially to the right. Table 8-2 is a scale of pK_a values which allows rough comparisons of virtually all possible organic acids and bases to be made. Bases are assessed in terms of the pK_a of their conjugate acids in order to use only one scale for all acids and bases.

Bases are molecules with unshared electron pairs at the basic site (or π electrons as very weak bases), either anions at C, N, O, S or neutral amines, $R_3N{:}$. They attack acidic protons on acid molecules with lower pK_a, forming their own conjugate acid (which with amines will be a salt) in the process. Bases also often attack other positive sites, such as certain carbons, in molecules and this dual potential must be dealt with in practical considerations of reactions, but the pK_a scale provides a rough assessment of relative reactivity of bases towards other positive sites as well as protons (discussed more in subsequent chapters).

It is important to keep straight which acid-base equilibrium is involved in cases with two possible proton-transfer steps, particularly amines and alcohols. A brief table of the major organic acids and bases follows.

Acids	pK_a	Bases	pK_a
R—COOH	~5	Satd. $R{:}^-$	~40
ϕ—OH	~10	$-\overset{\displaystyle O}{\overset{\displaystyle \|}{C}}-\underset{\displaystyle \|}{C}{:}^-$	~20
β-Dicarbonyl and related compounds	5–15	$RO{:}^-$	~17
		$R_3N{:}$	~10
		$\phi{-}\ddot{N}R_2$	~5

The general relation of pK_a to reactivity is

High pK_a = weak acid, strong conjugate base (very basic, high-energy base)

Low pK_a = weak base, strong conjugate acid (very acidic, high-energy acid)

Acids and bases with pK_a in the range 0 to 14 undergo their interconversion equilibrium in water.

The energy relationships state that a base is of higher energy than its conjugate acid by $\sim 1.4 \times pK_a$ (and hence of lower energy if the pK_a is negative). An acid and a base will react in an equilibrium which goes substantially to products if $pK_c = pK_{a(A)} - pK_a$ = negative, where $pK_{a(A)}$ is the pK_a of the reacting acid **A** and $pK_{a(B)}$ that of the reacting base **B**.

The major effects of structure on reactivity arise from resonance, which usually acts when the base site (e^- pair) is conjugated with an unsaturated system. (See, however, Fig. 8-1.) This lowers basicity and pK_a, raises acidity, and is the strongest effect, especially in carbon acids. Carboxylic acids, however, are virtually unaffected by resonance. The inductive effect is caused by the electrostatic repulsion or attraction of full charges (or the nearest charge in a dipole) on the reactive site: positive charges and ordinary dipoles stabilize the base and lower the pK_a; negative charges raise it. Both effects (resonance and inductive) are electron-withdrawing usually, and in most cases stabilize the base by withdrawing its electrons. The pK_a of the acid is correspondingly lowered.

Steric effects are less common; they usually act by inhibiting resonance (raising pK_a) or solvation (lowering pK_a) and almost never inhibit the approach of proton to the electron pair, since the proton is small (sterically undemanding). Steric effects are very important with other Lewis acids, which are larger than protons. Finally, internal hydrogen bonding stabilizes acids and raises the pK_a in most cases. All or some of these four effects can operate together, independent and additive, and in some cases their disentanglement can be complex and difficult.

Lewis acids and charge-transfer complexes are extensions of the acid-base reaction beyond protons (acids) and unshared pairs (bases) to include as well electron-deficient species as acids and π electrons as bases. Acid-base reactions are important both as initiators of subsequent reactions of many kinds and as models for more complex reactions.

PROBLEMS

8-7 Arrange the following compounds in order of decreasing acid strength.

CH_3COOH, $[(CH_3)_3C]_3CCOOH$, $CH_3SO_2CH_2COOH$, C_6H_5OH, $p\text{-}CH_3C_6H_4OH$, CH_3NO_2, $n\text{-}C_4H_{10}$, $(C_6H_5)_2CH_2$

8-8 Classify the following compounds as Lewis acids, Lewis bases, neither, or both.

$AlCl_3$, NI_3, $NaOH$, dioxan, pyridine, CH_3NO_2, $FeCl_3$, CH_3Br, CH_3COOH, C_6H_6, $(NC)_2C{=}C(CN)_2$

8-9 When possible, dipoles within a molecule tend to line up in configurations that allow like charges to be as far as possible from each other. An example is found in glyoxal:

more stable than

Two configurations for glyoxal

Keeping the above idea in mind, provide an interpretation for the following facts:

$$K_T = 670 \times 10^{-3}$$

$$CH_3COCOCH_3 \rightleftharpoons CH_3COC\!\!=\!\!CH_2 \qquad K_T = 5.6 \times 10^{-3}$$
$$\qquad\qquad\qquad\qquad\quad\ \ \overset{|}{OH}$$

8-10 In 100% sulfuric acid, acetic acid behaves as a base, taking a proton from sulfuric acid. Write what you think is the structure of the conjugate acid of acetic acid, and justify your choice of this structure over any alternative structures.

8-11 Quinuclidine is a stronger base than triethylamine toward trimethylboron by an immeasurably large amount, but the basicity of the two compounds toward protons is about the same. Explain this fact.

8-12 Explain the order of acidity and basicity observed for the following compounds.

a A solution of p-$CH_3C_6H_4SO_3H$ (or any other strong acid) is somewhat less dissociated in acetic acid solution than in aqueous solution. However, the acidity of the acetic solution appears to be the higher of the two when estimated by measurement of their abilities to protonate bases: acid + **B:** $\rightleftharpoons$ **BH$^+$** + conjugate base.

b
$$\begin{matrix} H-C-COOH \\ \| \\ H-C-COOH \end{matrix} \quad \text{is a stronger acid than} \quad \begin{matrix} H-C-COOH \\ \| \\ HOOC-C-H \end{matrix}$$

while
$$\begin{matrix} H-C-COONa \\ \| \\ HOOC-C-H \end{matrix} \quad \text{is a stronger acid than} \quad \begin{matrix} H-C-COOH \\ \| \\ H-C-COONa \end{matrix}.$$

c $(NO_2)_3CH$ is a stronger acid than $(NO_2)_2CH_2$, which in turn is a stronger acid than NO_2CH_3.

8-13 Which of the two enols would you expect to be the stronger base, and why? Which is the stronger acid, and why?

$$CH_3-\overset{\|}{\underset{O}{C}}-\overset{|}{\underset{OH}{C}}=CH_2 \qquad CH_3\overset{\|}{\underset{O}{C}}-CH=\overset{|}{\underset{OH}{C}}-CH_3$$

Enol of an α-diketone   Enol of a β-diketone

8-14 Arrange the following acids in decreasing order of acid strength:

$$CH_3\overset{+}{N}H_3 \qquad NH_2-\overset{\|}{\underset{\overset{|}{NH_2}}{C}}-NH_2 \qquad C_6H_5\overset{+}{N}H_3 \qquad p\text{-}NO_2C_6H_4\overset{+}{N}H_3$$
$$\qquad\qquad\qquad\qquad\quad \underset{+}{NH_2}$$

8-15 Presume that, in the formation of the addition compound between tetracyano-ethylene and naphthalene (page 328), an electron is completely transferred from hydrocarbon to cyano compound to produce a naphthalene radical-cation and a tetracyanoethylene radical-anion. Draw representative resonance forms for these two species.

8-16 Write all possible tautomeric structures for the following compounds. Applying the principles of resonance, pick what you think would be the most stable tautomer.

a $C_6H_5CH(COOC_2H_5)_2$

b

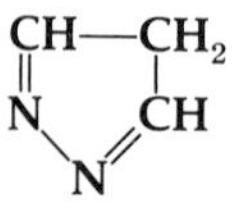

c

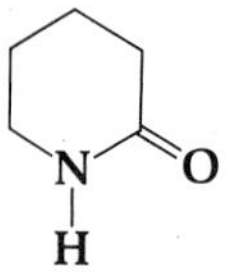

d

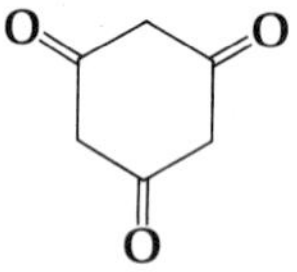

e

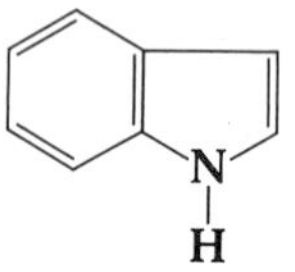

f

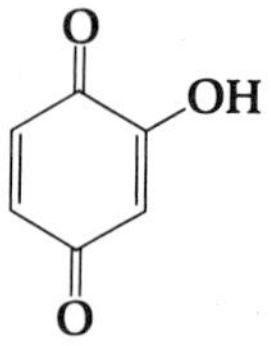

g

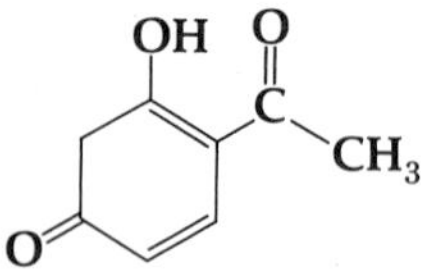

h

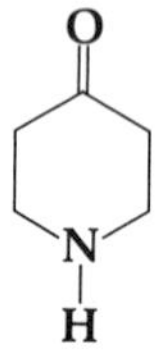

i

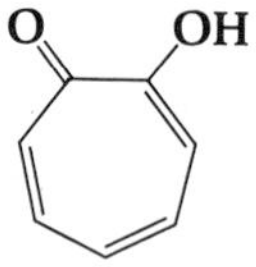

j $CH_3-\underset{O}{\overset{\|}{C}}-CH_2-NO_2$

8-17 Interpret the following facts:

a Although $n\text{-}C_4H_9Na + (C_6H_5)_3CH \longrightarrow (C_6H_5)_3CNa + n\text{-}C_4H_{10}$,

$n\text{-}C_4H_9Na +$ 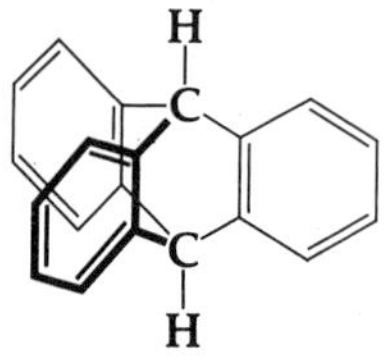
$\longrightarrow$ no reaction.

b In the homologous series of compounds

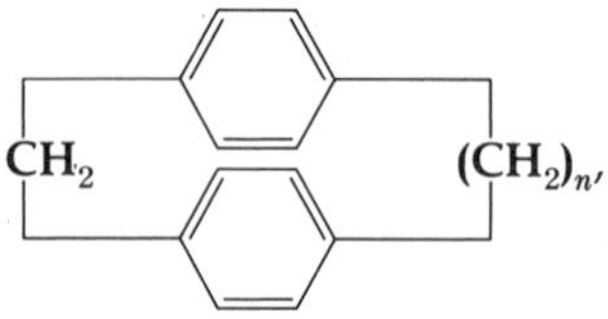

those with higher values of n are more acidic.

c In the homologous series of compounds the

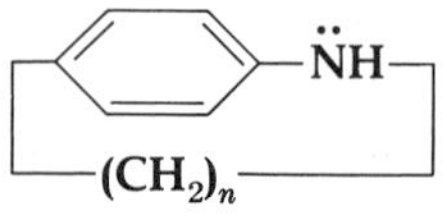

those with smaller values of n are the stronger bases.

d In acid strength,

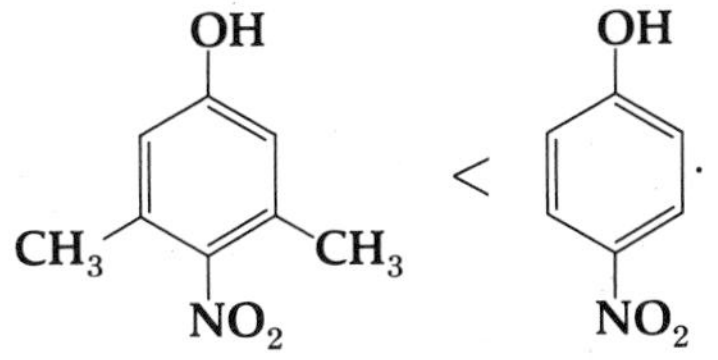

$<$

8-18 **a** Few neutral carboxylic acids show a higher pK_a (lower acidity) than acetic acid. One exception is α-furoic acid. Rationalize this observation, considering resonance effects.

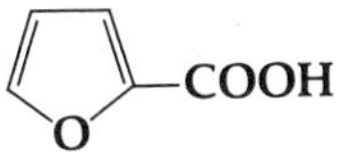

α-Furoic acid
pK_a 6.7

b Account for these acidity differences.

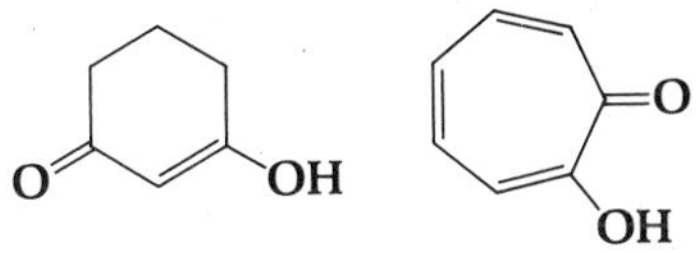

pK_a 10 pK_a 7

c Account for this increase in *acidity* (loss of proton).

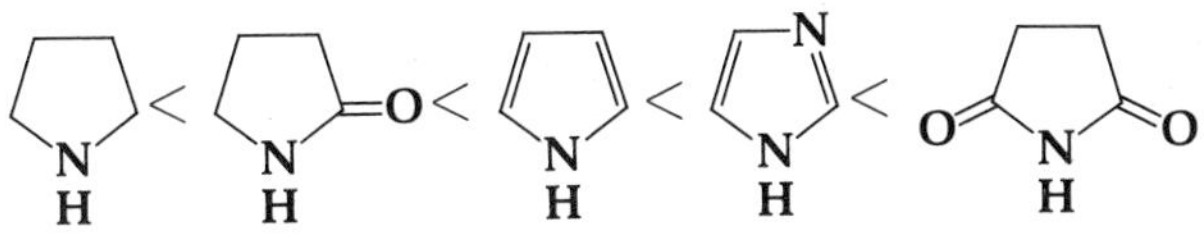

8-19 When 3-chlorocyclohexanone is treated with base, the elements of **HCl** are eliminated. The first step is removal of a proton in base, as shown, forming the enolate anion.

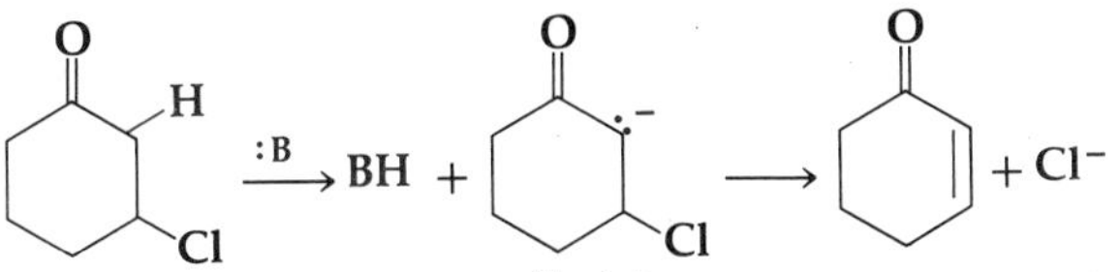

Enolate

a Assuming the first reaction is much faster than the second and that the pK_a of simple ketones is ~20 (cf. **CH$_3$COCH$_3$**, Table 8-2), what is the immediate result of mixing 3-chlorocyclohexanone and potassium *t*-butoxide in equimolar amounts?

b Why is the overall reaction slower when triethylamine is used in place of potassium *t*-butoxide?

c Would you expect elimination of **HCl** to be faster or slower with the following?

Chlorocyclohexane   2-Phenyl-3-chlorocyclohexanone

4-Chlorocyclohexanone   2,2-Dimethyl-3-chlorocyclohexanone

8-20 An unknown compound $C_8H_{10}O_2$ was said to give a basic reaction to wet litmus or pH paper. What response can be made to this claim?

8-21 What structural information can be deduced from the information given in each case about the following substances of unknown structure? Write a structure for each unknown and comment on whether other structures are possible or not.
a A base, C_7H_9N, with pK_a ~5.
b A base, C_7H_9N, with pK_a ~10.
c A base, $C_4H_{11}N$, with pK_a ~10 and optically active.
d A neutral compound (no basicity or acidity observed in water), $C_{10}H_{11}NO_3$.
e A base, $C_{10}H_{11}NO_3$, with pK_a ~5.
f A base, $C_{10}H_{11}NO_3$, with pK_a ~10.
g A weak acid, $C_{10}H_{11}NO_3$, with pK_a ~10.
h An acid, $C_{10}H_{11}NO_3$, with pK_a ~5.
i A compound, $C_{10}H_{11}NO_3$, is insoluble in water (pH ~7) but dissolves in aqueous acid below pH 3 to 4 and also dissolves in strong alkaline solution with pH above 12.

8-22 Which of these two isomeric bases is more basic, and why?

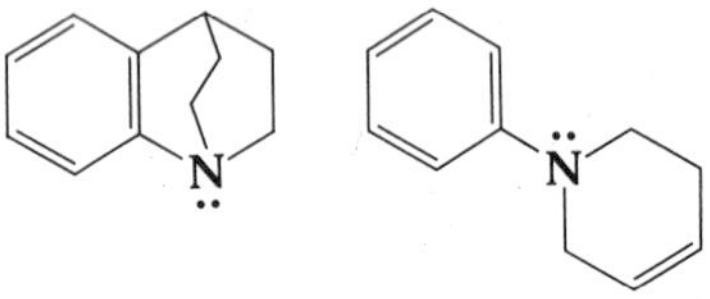

pK_a 8 pK_a 5

8-23 Deduce structures which fit the descriptions of unknown compounds given below.
a A water-insoluble compound, $C_8H_{14}O_3$, dissolves in aqueous $NaHCO_3$ solution (with bubbling). The compound is unaffected by catalytic hydrogenation.
b An acid, $C_7H_{10}O_4$, pK ~5, is optically active. On ozonolysis it produces 1 mole each of two different acids, only one of which is optically active.
c Of two isomeric liquids, C_3H_5N, only one dissolves in aqueous **HCl**.
d A compound, $C_3H_8N_2O_2$, is dissolved in ether and dry **HCl** gas bubbled in. A white crystalline solid, $C_3H_9N_2O_2Cl$, precipitates.
e A compound, $C_9H_3N_3O$, dissolves in aqueous $NaHCO_3$.

8-24 The two very similar substances below have very different properties. Compound *A* is neutral and absorbs in the IR at 6.0 μ (1670 cm^{-1}) while compound *B* dissolves in 5% hydrochloric acid and absorbs in the IR at 5.8 μ (1720 cm^{-1}). Explain the difference.

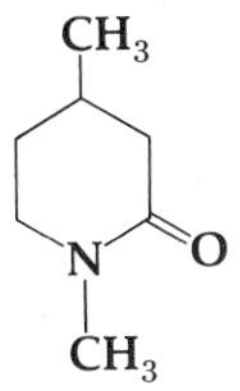

A 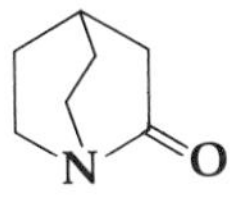   *B*

8-25 The diamine shown below is a stronger base than ordinary *aliphatic* tertiary amines although it is very inactive to Lewis acids like trimethylboron. In accounting for its behavior, examine its geometry, resonance, steric hindrance, and hydrogen bonding.

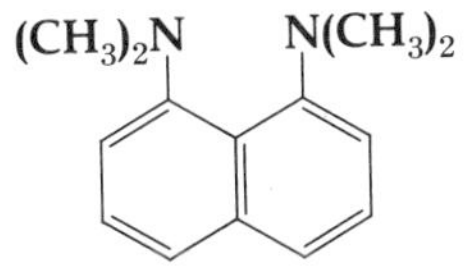

8-26 When 1,2-dimethylcyclopentadiene is treated with sodium methoxide and then reisolated, three isomers (of C_7H_{10}) are isolated. Explain.

8-27 Extraction of the bark of the boola-boola bush yields a pure crystalline compound named odorific acid, $C_8H_{13}NO_3$, mp 162°, optical rotation $[\alpha]_D = 0°$, and the acid could not be resolved into enantiomers. The pK_a of odorific acid was similar to that of acetic acid, and the sodium salt could be isolated by neutralizing with sodium hydroxide. The IR spectrum of the acid showed a strong band at 5.8 μ, that of the salt at 6.2 μ, while both showed a strong band at 6.0 μ. (They each had only two strong bands in the region around 6 μ.) The salt also showed one band at 3.0 μ, and neither showed UV absorption. The NMR spectrum showed one unsplit, very-low-field proton, only one proton (singlet) on a double bond, and one peak of three protons in a sharp unsplit singlet in the hydrocarbon (high-field) region; the other protons showed as unclear multiplet signals. Try to assign structural meaning to these facts, writing out your reasoning stepwise. Then combine these separate interpretations into a structure for odorific acid which is consistent with all the observations above. (*Note:* Odorific acid was not *amphoteric,* i.e., it showed no *basic* properties in addition to its acidic character.)

8-28 The substance $C_{10}H_{10}O_4$ dissolves in aqueous bicarbonate and exhibits NMR peaks at τ 0.12, 2.58, 4.05, and 7.82, all singlets in an intensity ratio of 1:5:1:3, respectively.

ORGANIC REACTIONS

THE second unit in the book (Chaps. 9 to 22) deals largely with organic reactions. Reactions, allowing interconversion of the millions of organic compounds and permitting the synthesis of materials of great value to humanity, are the real stuff of organic chemistry. The number of reactions that have been carried out (or attempted unsuccessfully) with organic compounds far exceeds the number of compounds. Nevertheless, with our present insight into the mechanisms of these reactions we shall see that they mostly fit into a few basically simple patterns. Classification of compounds by structural types and functional groups is indispensable to any discussion of the structure of organic substances. Similar advantages of clarity and organization arise from the classification of organic reactions.

We shall find that organic reactions take place entirely in accord with the expectations derived from the fundamental properties and energies of the molecules as developed in Chap. 2. In the past chapters we developed these properties to see their effects on molecular structure. In this chapter, introducing the general concepts of reactions, we shall develop the same properties to understand the ways in which molecules react with each other to create new molecules.

9-1 TERMINOLOGY AND CLASSIFICATION

A chemical reaction—at the molecular level—is an event in which two molecules collide† in such a way as to break one or more of their bonds and make one or more new bonds, and hence new molecules. The *sequence* and *timing* of the bond-breaking and -making processes will be important to our understanding of reactions. They may occur as separate, discrete steps; they may overlap in the sense that a new bond begins to form before an old one is completely broken; or they may both occur together, the new bond being formed at the same time as (and as a direct consequence of) the cleavage of another. The last process is called a **concerted** or **synchronous** reaction. If an observed chemical transformation is found to consist of several steps—and this is common—the several intermediate compounds can sometimes be isolated, with care and control of reaction conditions, or sometimes observed in fleeting existence by physical methods if too unstable to isolate. The detailed course of an overall reaction—its sequence of steps and the details of electron move-

---

†Even though a few reactions occur by initial disruption of only a single molecule (via input of energy), most reactions are initiated by collision of two, and the definition in this way is more useful for the development of the central theory.

ment, bond breaking and making, and timing—is described as the **mechanism** of the reaction.

A simple generalized reaction would be

$$A + B \longrightarrow C + D$$

in which A and B are **starting materials** and C and D are **products**. A reaction occurring in several steps might proceed as follows, taking a course in which the **intermediates** D and E are formed and destroyed again en route to the products F and G. (Intermediates are often written in brackets to emphasize their transient existence.)

$$A + B \longrightarrow C + [D] \xrightarrow{+Z} [E] \longrightarrow F + G$$

It is not uncommon to find that intermediates can go on to react in more than one way, affording **competing reactions** which lead to **side products** or **by-products** (H and J) as in this example:

$$A + B \longrightarrow C + [D] \xrightarrow{+Z} [E] \longrightarrow F + G$$
$$\phantom{A + B \longrightarrow C + [D] \xrightarrow{+Z} [E] } \xrightarrow{+A} H + J$$

Most reactions involve conversion of one functional group into another at one site on an otherwise unchanged molecular skeleton. The saturated hydrocarbon portion of the molecular skeleton is rarely affected. This organic starting material, on which the attention is thus focused, is called the **substrate** or **reactant**, which is acted on or attacked by the **reagent**. The reagent is very commonly an inorganic or very simple organic substance and is used to create the desired transformation in the substrate. In the special but synthetically important reactions which combine two organic reactants to create new carbon-carbon bonds and hence larger molecules, the designation of either starting material as substrate or reagent is arbitrary and often meaningless. In general,

Substrate + reagent $\longrightarrow$ [intermediate(s)] $\longrightarrow$ products
$$ \dashrightarrow \text{by-products}$$

Stereochemical Features

For new bonds to form from orbitals in the reacting molecules the orbitals must approach and overlap. Since orbitals and their mutual overlap to form bonds have strong directionality, there will be geometrical constraints on the direction and orientation in which reactants can come together for successful reaction. Ordinary hybrid orbitals (sp^3, sp^2, sp) extend along an axis and must be aligned on that axis if they are to bond successfully when they approach each other. These constraints are called the **stereoelectronic** requirements of the reaction.

Other stereochemical features of reactions include the stereospecificity and the stereoselectivity.

A **stereoselective** reaction is one which proceeds (reacts) preferentially with one stereoisomeric reactant instead of another.

A **stereospecific** reaction is one which *produces* one stereoisomeric product preferentially over another.

A simple example is available in the maleic and fumaric acids on page 181. The cis diacid forms an anhydride with more facility than the trans in a *stereoselective* reaction while the hydrolysis of the anhydride is *stereospecific* in that it yields only maleic acid and no fumaric.

Fundamental Classes

Classification of organic reactions very naturally emphasizes the changes that occur in the bonding to carbon atoms at the site of reaction. The first level of classification defines three great classes of reactions: **substitution, addition**, and **elimination**.

In a **substitution reaction** an atom or a group attached to a carbon atom is removed and another enters in its place. No change in the degree of unsaturation at the reactive carbon occurs.

An **addition reaction** involves an *increase* in the number of groups attached to carbon. The molecule becomes more nearly saturated.

An **elimination reaction** involves a *decrease* in the number of groups bound to carbon. The degree of unsaturation increases. The following are examples of each class.

Substitution:

$$HBr + CH_3CH_2OH \longrightarrow CH_3CH_2Br + H_2O$$

Reagent   Substrate   Products

Addition:

$$Br_2 + CH_2{=}CH_2 \longrightarrow CH_2BrCH_2Br$$

Elimination:

$$NaOH + CH_3CH_2Br \longrightarrow CH_2{=}CH_2 + H_2O + NaBr$$

Virtually all organic reactions fall into one of these three classes. The acid-base reaction (proton-transfer, Chap. 8) may be regarded as a fourth class, although it makes no fundamental change in the substrate structure. **Rearrangements** (Chap. 17) are often listed as a different class. These are reactions in which the carbon skeleton of the molecule is internally rearranged, but they may be seen as a sequence of steps, all of which are reactions of the first three classes. Reactions which cleave carbon-carbon bonds and so break up the molecular skeleton are called **fragmentation** reactions; these are usually variations on the elimination class of reactions.

PROBLEM 9-1

Into what class of reactions does each of the following transformations fall?

a $CH_3CH_2CH_2\overset{+}{S}(CH_3)_2 + CH_3NH_2 \longrightarrow CH_3CH_2CH_2\overset{+}{N}H_2CH_3 + (CH_3)_2S$

b $CH_3CH_2CH_2\overset{+}{S}(CH_3)_2 + OH^- \longrightarrow CH_3CH=CH_2 + (CH_3)_2S + H_2O$

c $C_6H_6 + HNO_3 \longrightarrow C_6H_5NO_2 + H_2O$

d $CH_3CH=CH_2 + Cl_2 \longrightarrow CH_3CHClCH_2Cl$

e $CH_3CH=CHCH_2COOH \xrightarrow{\text{Heat}} CH_3CH_2CH=CH_2 + CO_2$

f $(CH_3)_2\underset{OH}{C}-\underset{OH}{C}(CH_3)_2 \xrightarrow{H_2SO_4} CH_3COC(CH_3)_3 + H_2O$

g $CH_3CH=CHBr + {}^-OH \longrightarrow CH_3C\equiv CH + H_2O + Br^-$

Reaction Intermediates

Many organic reactions involve formation of transient intermediates that play a crucial role in the overall processes. The lifetimes of these intermediates range upward from 10^{-12} sec. These intermediates may be formed by attack of various reagents on substrates, by dissociation of organic compounds, or by promotion of molecules to excited states by absorption of light or interaction with high-energy radiation, such as α-, β-, and γ-rays. The following summary describes the important intermediates.

Carbonium ions are carbon cations. They may be thought of as fragments of molecules in which a group *and* the pair of bonding electrons have been removed from one of the carbon atoms. The positively charged carbon atom of a carbonium ion is in the sp^2 state of hybridization,† so that the ion is planar:

$$\overset{a}{\underset{b \diagup \quad \diagdown c}{C^+}}$$

Carbonium ion
(sp^2 planar)

Carbanions are carbon anions. They are formed by removal of one of the groups attached to a carbon atom *without* removing the bonding pair of electrons. Like amines, with which they are isoelectronic, carbanions are nonplanar. The simplest formation of a carbanion is the removal of a proton from its conjugate carbon acid.

$$\overset{a}{\underset{\underset{c}{b}}{\diagdown}} C:^-$$

Carbanion
(sp^3 pyramidal)

Carbon radicals are formed from normal carbon compounds by removal of an attached group along with *one* of the two bonding electrons. Most carbon

---

†If a carbonium ion assumes the trigonal (sp^2) state of hybridization, an unhybridized p orbital is left vacant. Such a condition is energetically desirable, since the six bonding electrons are then accommodated in orbitals that "use up" all the carbon $2s$ orbital. The energy of the system is minimized because electrons in $2s$ orbitals have lower energy than electrons in $2p$ orbitals.

radicals are uncharged, although both anion and cation radicals are known. Radicals are intermediate between carbanions and carbonium ions. The methyl radical is known to be planar, and by inference, this is believed to be the "natural" configuration of free radicals.†

Carbon radical
(probably planar)

Carbenes are fragments of molecules in which two groups attached to carbon have been removed along with only one pair of bonding electrons. Carbenes are neutral divalent carbon compounds which typically have only transient existence. Carbon monoxide is stable, however, as the carbene of the carbonyl, $:C=O$. One interesting feature of the carbene structure is that it may exist in two distinct states, one in which the spins of the unshared electrons are paired and one in which the spins are unpaired.‡ Little information is available concerning the configuration of carbenes.

Carbene
(configuration uncertain)

Excited states of molecules are produced by absorption of light. Visible and ultraviolet light induce states in which electrons occupy high-energy orbitals (Sec. 7-4). Such species are highly reactive and are involved in photochemical reactions of organic compounds (Chap. 22).

Unstable molecules are formed as intermediates in some reactions. Such molecules are unstable and highly reactive, often because of extreme geometric distortion (strain energy), and are converted rapidly to other, more stable products. Because of their instability and reactivity they are classed in the same high-energy category as the preceding ions and radicals. A single example will be cited here. α-Haloacids undergo facile substitution reactions in aqueous alkaline solutions. The products are α-hydroxy-acids. Careful study of the reactions has shown that they do not occur in one step. Highly reactive α-lactones are formed as transient intermediates.

---

†Electronic arguments lead to no decisive conclusion as to the preferred configuration of radicals. Perhaps the planarity of the methyl radical arises from the fact that repulsion between the attached groups is minimized in the planar configuration.

‡Carbon has four low-energy orbitals (one $2s$ and three $2p$). Two of these are used for bonding; therefore, there are two low-energy orbitals available to accommodate the unshared electrons. If both electrons go into one orbital, the spins must pair. If the electrons go into different orbitals, they will have parallel (unpaired) spins. A carbene in the latter state would have a permanent magnetic moment and would exist in three closely grouped energy states if it were placed in a magnetic field. Such a state is called a **triplet state**. The state in which the electrons are paired has no magnetic moment and is called a **singlet state**. It is known that the triplet state of $H_2C:$ has a lower energy than the singlet state. These states are discussed in Chap. 22.

$$Br-CH_2C\overset{O}{\underset{:O:^-}{\diagup}} \longrightarrow CH_2-C=O + Br^-$$
$$\qquad\qquad\qquad\qquad\qquad \overset{}{\underset{O}{\diagdown}}$$

α-Bromoacetate ion   α-Acetolactone
                            (glycololactone)

$$CH_2-C=O + HO^- \longrightarrow HOCH_2C\overset{O}{\underset{O^-}{\diagup}}$$

Glycolate (hydroxyacetate) ion

Nomenclature

Authoritative rules for nomenclature of unstable intermediates have not yet been laid down. There seems to be no confusion concerning names of free radicals and carbenes. Radicals are designated by use of group names followed by the word *radical* or are named as derivatives of the methyl radical. Thus, $CH_3CH_2\cdot$ is called the *ethyl radical* and $(C_6H_5)_3C\cdot$ the *triphenylmethyl radical*. Current custom also seems to be adopting the satisfactory practice of naming carbenes as derivatives of the parent species, $H_2C:$. Thus, $(C_6H_5)_2C:$ is called *diphenylcarbene*. No such harmony exists in the nomenclature of anions and cations. The following systematic scheme accords with good nomenclature practice.†

1 If group names are used, they should be followed by the words *anion* or *cation*. Thus, $CH_3CH_2CH_2:^-$ is the *n*-propyl anion and $CH_3CH_2CH_2{}^+$ the *n*-propyl cation.

2 The words *carbanion* and *carbonium ion* should be reserved for use in derived names. The system is exactly analogous to the nomenclature of carbenes and to the nomenclature of alcohols as carbinols (page 115). The following examples illustrate the system:

Cation	Anion	Carbinol
$(CH_3)_3C^+$, trimethyl-carbonium ion	$(CH_3)_3C:^-$, trimethyl-carbanion	$(CH_3)_3COH$, trimethyl-carbinol
$(C_6H_5)_2CH^+$, diphenyl-carbonium ion	$(C_6H_5)_2CH:^-$, diphenyl-carbanion	$(C_6H_5)_2CHOH$, diphenylcarbinol

†In one respect the practice is bad; the suggested scheme raises some conflict with names already in the chemical literature. For example, $(CH_3)_3C^+$ has often been called *t*-butyl carbonium ion. However, "carbonium ion" in this sense seems pointless, since the simple word "cation" fulfills the purpose perfectly. The present authors urge adoption of the system outlined, but also recommend caution in the interpretation of names found in the chemical literature.

Types of Bond Change

In principle (and practice) a covalent bond can be broken in two ways: **homolytic** cleavage refers to the process in which the bond is broken symmetrically, with one electron remaining on each of the originally bonded atoms; in **heterolytic** cleavage the bond is broken so that the electron pair stays united and attached to only one of the two originally bonded atoms:

$$X\!-\!Y \xrightarrow[\text{cleavage}]{\text{Homolytic}} X\cdot + \cdot Y \qquad \text{(radicals)}$$

$$X\!-\!Y \xrightarrow[\text{cleavage}]{\text{Heterolytic}} X^+ + :Y^- \qquad \text{(ions)}$$

The latter process separates charge and produces ions; hence heterolytic reactions are often called **ionic reactions**. Homolytic reactions produce neutral radicals and are often called **radical** or **free-radical** reactions. The great preponderance of common organic reactions are ionic reactions, and these will be our central preoccupation here, while radical reactions are discussed in Chap. 20.

PROBLEM 9-2

Name the following species:

a $CH_3CH_2{}^+$ d $CH_3\overset{..}{\overset{..}{C}}HCH_3$

b $C_6H_5\overset{..}{C}C_6H_5$ e $(C_6H_5)_3C^+$

c $CH_3CH\!-\!C\!=\!O$ f $(C_6H_5)_2\overset{.}{C}H$
$\qquad\;\;\diagdown\!\!\diagup$
$\qquad\;\;\; O$

9-2   THE NATURE OF IONIC REACTIONS

Since ionic reactions involve charged species, electrostatic forces are of major importance. The attraction of unlike and the repulsion of like charges are the central forces initiating ionic reactions and determining their course. Reagents may thus be broadly divided into two great classes. Those which are electron-rich, and have an unshared pair of electrons acting as a reactive site, seek a positively charged or electron-deficient site in another molecule. The former are called **nucleophiles** ("nucleus loving"). Their opposites are the **electrophiles** ("electron loving"), which are electron-deficient and offer a positive rather than negative reactive site. A positive (electron-deficient) reactive site is either an atom without a full outer electron shell or one at the positive end of a dipole, which may polarize further in the course of reaction. Dipoles are recognized as any bond between different atoms, the atom of lower electronegativity (commonly, carbon) being the positive site. The C—H bond is rarely an important dipole. Reactions are described as **nucleophilic** or **electrophilic** in terms of the nature of the *reagent*.

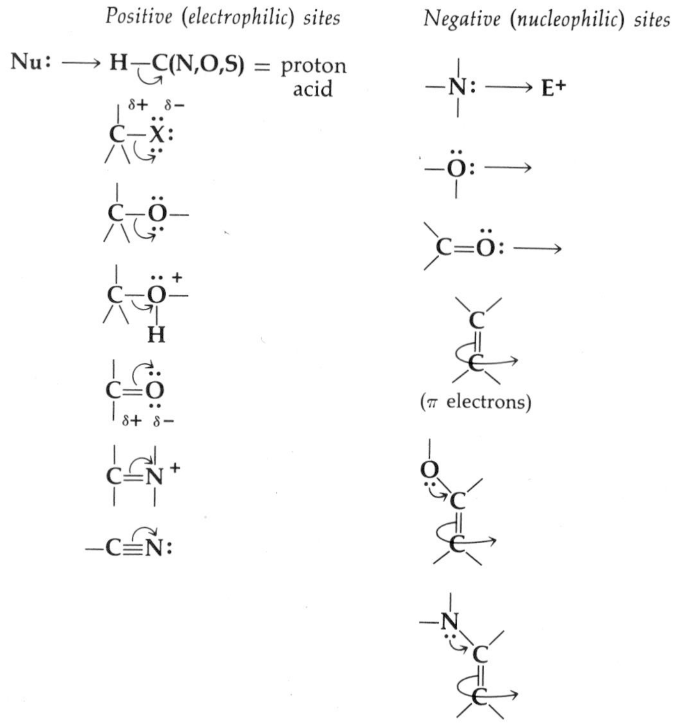

Positive (electrophilic) sites   Negative (nucleophilic) sites

Nu: $\longrightarrow$ H$-$C(N,O,S) = proton acid

$\overset{\delta+}{C}-\overset{\delta-}{\ddot{X}}:$

$C-\ddot{O}-$

$C-\overset{+}{\underset{H}{\ddot{O}}}-$

$\underset{\delta+ \ \delta-}{C=\ddot{O}}$

$C\equiv\overset{+}{N}$

$-C\equiv N:$

$-\overset{|}{\underset{|}{N}}: \longrightarrow E^+$

$-\ddot{O}: \longrightarrow$

$C=\ddot{O}: \longrightarrow$

C (π electrons)

FIGURE 9-1   **Common reactive sites in organic molecules**

The two classes are essentially the same as Lewis bases and acids, but the terms base and acid are usually reserved for the making and breaking of bonds *to hydrogen* while nucleophile and electrophile refer to reactions *at carbon*. Thus, the same compound can be called either a base or a nucleophile depending on whether it attacks a proton or a carbon, respectively. The distinction is important, however, since the order of reactivity of Lewis bases (i.e., pK_a) does not always correspond to their order of reactivity as nucleophiles (to carbon), and the same can be said of Lewis acids/electrophiles. The two classes of compounds may be summarized as shown.

Nucleophile = "negative," electron-rich, anionoid, electron-donating,

Lewis base **B:** or **Nu:**

Common nucleophiles: $H\ddot{O}:^-$, $R\ddot{O}:^-$, $RS:^-$, $:CN:^-$, $H_2\ddot{O}:$, $R\ddot{O}H$, $R_3N:$

Electrophile = "positive," electron-deficient, cationoid, electron-accepting,

Lewis acid, E^+

Common electrophiles: H^+, Br^+,† NO_2^+,† BF_3, $AlCl_3$

A nucleophile or Lewis base has an unshared electron pair (or, less reactive, π electrons) and may be neutral or negative in charge (cf. $R_3N:$ and $H\ddot{O}:^-$). An electrophile or Lewis acid may be neutral or positive in charge (cf. BF_3 and H^+). Carbanions are nucleophiles, while carbonium ions and carbenes are electrophiles since they have unfilled outer shells of electrons.

Mirroring the electrical nature of the attacking reagent, the receptive

---

† As will be seen later, many electrophiles are actually reactive species generated by ionization; halogens afford the reactive electrophiles $:\overset{..}{\ddot{X}}{}^+$ by $X_2 \rightleftharpoons X^+ + X^-$ and $HNO_3 + H^+ \rightleftharpoons NO_2^+ + H_2O$.

	Initiation	*Intermediate(s)*	*Product formation*

A   Electrophilic
attack of **E⁺** on
(**H⁺** = common)
	(*a*) Unshared
	        e^- pair
	(*b*) π electrons
		followed by collapse
		of resultant
		cation via
			(*a*) Addition of nucleophile
			        (often solvent = **R—ÖH**,
			        etc.)
			(*b*) Loss of **H⁺** or small
			        stable fragment (cf.,
			        H₂O)
			(*c*) Rearrangement (Chap. 17)

	Initiation	*Intermediate(s)*	*Product formation*

B   Nucleophilic
attack of
B: (Nu:) on
	(*a*) **H—**(proton
	        acid)
	(*b*) **C—**
	(*c*) Others (rare)
		followed by collapse
		of resultant
		anion via
			(*a*) Addition of **H⁺** (from
			        H—B′ = solvent, etc.)
			(*b*) Loss of more stable
			        anion (like **Cl⁻**, etc.)

FIGURE 9-2   **The general course of ionic reactions**

reactive site in the substrate molecule will also be either electron-rich or -poor. This electrical imbalance can only arise in the functional groups. The saturated-hydrocarbon portion of a molecular skeleton is nonpolar and generally unreactive in ionic reactions; reactions occur at the functional groups. Hence it is possible to catalog the various functional groups with respect to their positive or negative qualities: for reaction with nucleophilic and electrophilic reagents, respectively. The common reactive functionalities are grouped in Fig. 9-1 showing their electrical imbalance, often as a dipole between atoms of differing electronegativity. Their general reaction with nucleophiles (**Nu:**) or electrophiles (**E⁺**) is indicated with arrows, underlining the central fact of ionic reactions: the movement of pairs of electrons—from nucleophiles to electrophiles.

In this fashion we can make an "electronic reactivity map" of any molecule, locating all of its potential positive sites for attack by nucleophiles and all of its negative sites, electron pairs which can attack electrophiles. This allows us to form a basis for the prediction of the kind of reactivity that will be shown by any substrate for a given reagent. The details of these processes and the order and extent of reactivities are developed in the following chapters.

The general course of reactions occurring in several steps is outlined in Fig. 9-2. In many cases the chain of electron movement can be promoted from both ends: an electrophile "pulling" at one end and a nucleophile "pushing" at the other:

Nu: ⌒ ⤳ **W—X—Y—Z—E⁺**

It is very common for the electrophile to be a proton, so that electrophilic initiation of a reaction is simply a protonation, i.e., a proton-transfer or acid-base reaction (Chap. 8), converting the substrate (base) to its conjugate acid. Proton transfer is also seen in nucleophilic attack of bases on substrate protons, forming a conjugate base of the substrate as an intermediate anion. Simple acid-base reactions are very common as steps in multistep reactions. Protonation or deprotonation (often by solvent) serves to prepare the molecule for its central reaction or to afford collapse of intermediates to stable products. These steps are marked "(PT)," for proton transfer, in the examples of Fig. 9-3. In considering mechanisms in future chapters it is very important to recognize the rapid interconvertibility of the conjugate acid and base pairs, for protonation and deprotonation as required are common and rapid steps in many mechanisms.

The interaction of nucleophiles and electrophiles with the positive and negative reactive sites in substrate molecules can be examined by putting together the reagents listed above (page 348) with the appropriate reactive functional groups (Fig. 9-1), and examples illustrating the major ionic reactions are collected in Fig. 9-3.† The reactions are labeled as additions and eliminations, and as substitution reactions at saturated or unsaturated carbons, and then subclassified as nucleophilic and electrophilic. Reaction (5) is a nucleophilic substitution since the overall reaction is a replacement of —OH by the nucleophile Br⁻ even though it is initiated by the electrophilic attack of a proton. This reaction also represents a possible side reaction, forming the by-product diethyl ether by nucleophilic attack of unreacted ethanol instead of Br⁻. Reaction (6) shows the enolate anion—in both its major resonance forms—as an intermediate. As shown, either resonance form may be used to show the subsequent electron movement leading to the product.

PROBLEM 9-3

Compile an extensive list of nucleophiles and electrophiles; include examples not found in this chapter.

PROBLEM 9-4

Consider attack on the following molecules by **OH⁻**. Using a qualitative knowledge of bond dipoles (cf. electronegativity, Sec. 2-11) and of relative acidities of different protons (cf. Table 8-2), predict the *atoms* most likely to be attacked by the **OH⁻** ion.

a $CH_3COC_6H_5$

b $ClCH_2COC_6H_5$

c $CH_3CHBrCH_2COOCH_3$

d $CH_3CHBrCH_2COOH$

$\quad\quad\quad\quad\quad\quad\quad\quad OH$
$\quad\quad\quad\quad\quad\quad\quad\quad |$
e CH_3CHCN

f $CH_3CH{=}CHCH_2COCH_3$

g $CH_3CONHCH_3$

h $H_2N{-}CH{=}\overset{+}{N}H{-}CH_2CH_2OH$

---

†Note again here the rules for describing electron movements by the curved-arrow convention as outlined in Sec. 5-3. The examples in Fig. 9-3 illustrate the important rule that, correctly applied, the arrows exactly predict the product structures.

(1) Electrophilic addition (Chap. 15)

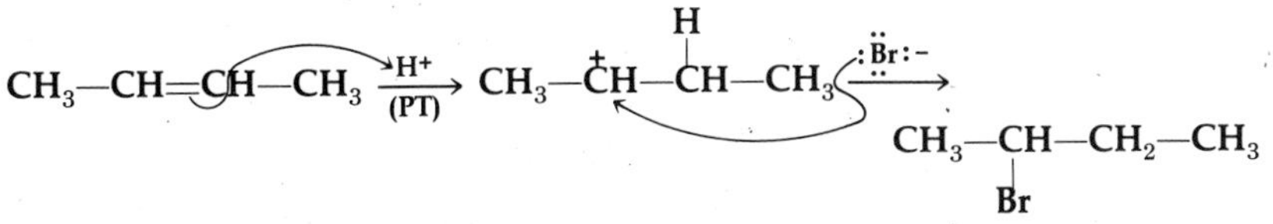

$$CH_3-CH=CH-CH_3 \xrightarrow[\text{(PT)}]{H^+} CH_3-\overset{+}{C}H-CH-CH_3 \cdots \overset{..}{\underset{..}{Br}}:^- \longrightarrow$$

$$CH_3-CH-CH_2-CH_3$$
$$\qquad\quad |$$
$$\qquad\quad Br$$

(2) Electrophilic substitution (at unsaturated carbon) (Chap. 16)

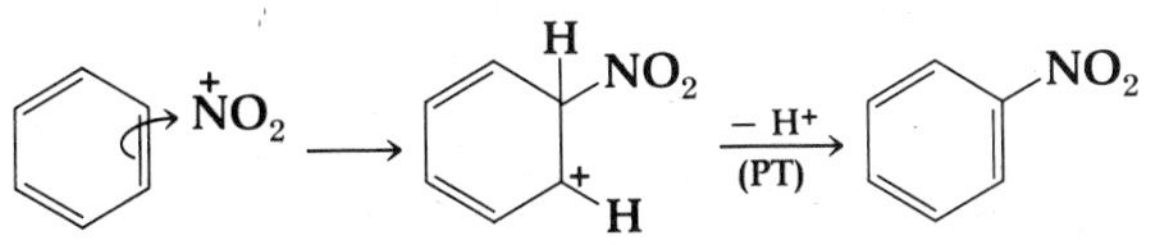

$$\bigcirc \overset{+}{N}O_2 \longrightarrow \overset{H}{\underset{H}{\bigcirc}} NO_2 \xrightarrow[\text{(PT)}]{-H^+} \bigcirc NO_2$$

(3) Nucleophilic addition (Chap. 12)

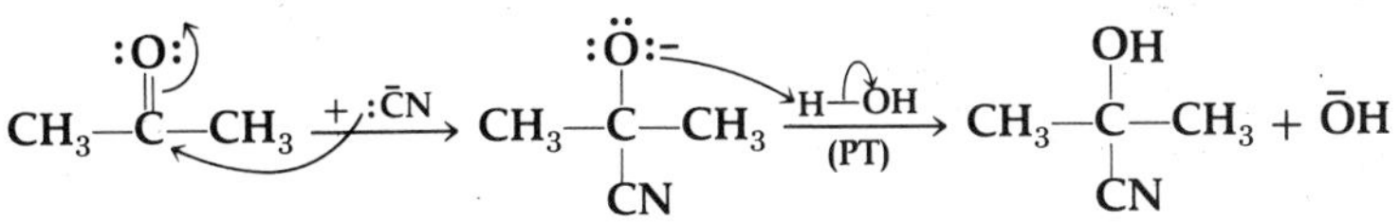

$$CH_3-\overset{\overset{\displaystyle :O:}{\|}}{C}-CH_3 \xrightarrow{+:\bar{C}N} CH_3-\overset{\overset{\displaystyle :\overset{..}{O}:-}{|}}{\underset{CN}{C}}-CH_3 \xrightarrow[\text{(PT)}]{H-OH} CH_3-\overset{\overset{\displaystyle OH}{|}}{\underset{CN}{C}}-CH_3 + \bar{O}H$$

(4) Nucleophilic substitution (at unsaturated carbon) (Chap. 13)

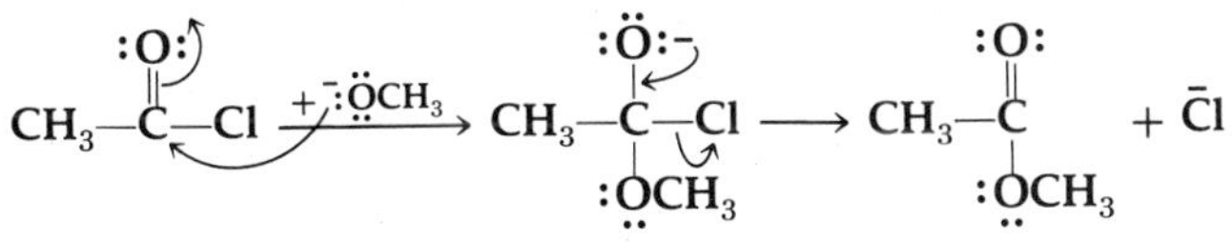

$$CH_3-\overset{\overset{\displaystyle :O:}{\|}}{C}-Cl \xrightarrow{+:\overset{..}{\underset{..}{O}}CH_3} CH_3-\overset{\overset{\displaystyle :\overset{..}{O}:-}{|}}{\underset{:\overset{..}{O}CH_3}{C}}-Cl \longrightarrow CH_3-\overset{\overset{\displaystyle :O:}{\|}}{\underset{:\overset{..}{O}CH_3}{C}} + \bar{C}l$$

(5) Nucleophilic substitution (at saturated carbon) (Chaps. 10, 11)

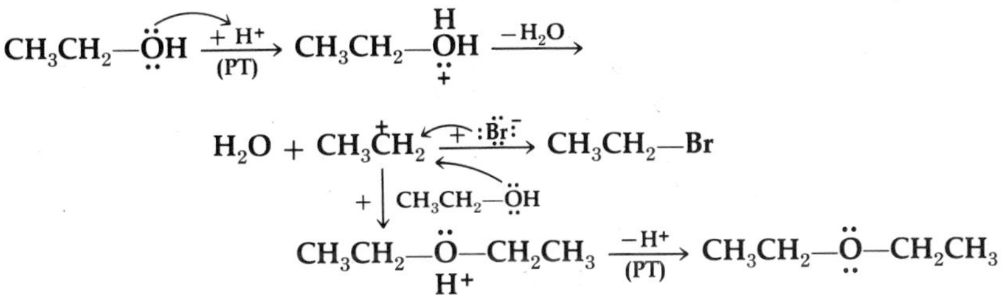

$$CH_3CH_2-\overset{..}{\underset{..}{O}}H \xrightarrow[\text{(PT)}]{+H^+} CH_3CH_2-\overset{H}{\underset{+}{O}}H \xrightarrow{-H_2O}$$

$$H_2O + CH_3\overset{+}{C}H_2 \underset{\longleftarrow}{\overset{+:\overset{..}{\underset{..}{Br}}:^-}{\longrightarrow}} CH_3CH_2-Br$$

$$+ \Big| CH_3CH_2-\overset{..}{O}H$$

$$CH_3CH_2-\overset{..}{\underset{..}{O}}-CH_2CH_3 \xrightarrow[H^+]{-H^+} CH_3CH_2-\overset{..}{\underset{..}{O}}-CH_2CH_3$$
$$\qquad\qquad\quad H^+ \quad \text{(PT)}$$

(6) Elimination (Chap. 14)

$$\left[\cdots \right]$$

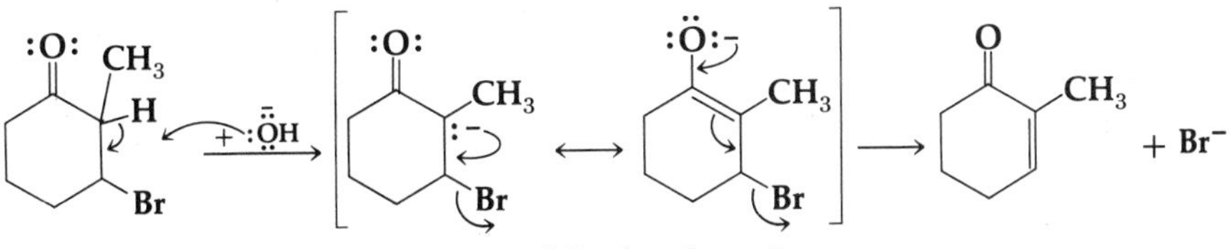

Resonance-stabilized enolate anion

FIGURE 9-3   **Examples of major reaction mechanisms (PT = proton transfer, Chap. 8)**

PROBLEM 9-5

a Similarly, consider the sites in the same molecules which would be susceptible to initial attack by a proton (H^+).

b In each list consider what next step might be possible following the various first steps suggested for either OH^- or H^+.

PROBLEM 9-6

a The reaction below can be observed by ultraviolet spectroscopy, since the product B has a distinctive (yellow) chromophore of known intensity (λ_{max} 362 nm, $\epsilon = 30{,}000$). When a 1 M solution of A in aqueous acetic acid has come to equilibrium it shows $\epsilon = 1{,}000$ at 362 nm. Calculate the equilibrium constant and the energy difference between A and B in this medium.

$$A \qquad\qquad\qquad\qquad B$$

b Write a reasonable acid-catalyzed mechanism for the reaction.

9-3 REACTION ENERGIES AND EQUILIBRIUM

All reactions are in principle equilibria:

$$A + B \rightleftharpoons C + D \qquad\qquad\qquad\qquad\qquad \cdot 1$$

The conditions of concentration obtaining at the time the equilibrium condition is reached are expressed by the familiar equation (2) for the equilibrium constant K:

$$K = \frac{[C][D]}{[A][B]} \qquad \begin{array}{l} K > 1 \text{ favors } \textit{products} \text{ at equilibrium} \\ K < 1 \text{ favors } \textit{reactants} \text{ at equilibrium} \end{array} \qquad 2$$

A very large value of K implies a reaction that goes for practical purposes to completion. These are the reactions commonly used for practical, preparative purposes.

The energy of a chemical reaction is equal to the difference in energy between the sum of starting materials and the sum of products. Most of this is usually the difference in energy between the bonds broken and the bonds formed (Sec. 2-10). Energy must be put in to break bonds (in A and B) but is released again—in greater or less amount depending on the individual bond energies involved—when the new bonds are formed (in C and D). If the net difference is negative, i.e., energy is given up in the reaction, the reaction is said to be **exothermic**. Such a reaction in principle proceeds spontaneously since the whole system goes to a lower-energy, or more stable, state. The energy

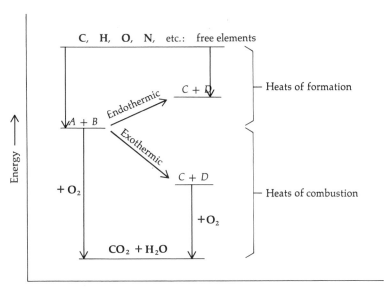

FIGURE 9-4   Energy relations of organic compounds and reactions

given up usually emerges as heat, quite dramatically in the combustion reaction (burning) of organic compounds and oxygen.

An **endothermic** reaction, by contrast, results in products less stable than reactants and can only be achieved by putting energy into the reaction (i.e., with applied heat). These ideas are summarized in Fig. 9-4, showing the exo-thermic nature of the combustion of organic materials generally to the more stable $CO_2 + H_2O$, as well as the greater stability of the organic covalent molecules than their constituent free elements (**C, H, O, N,** etc.).

Figure 9-4 also implies that if a conversion of a more stable to a less stable (higher-energy) molecule is desired, then the *reagent* used must be of high energy since it is the *sum* of reagent and reactant energies $(A + B)$ which characterizes the energy level at which reaction starts. A high-energy reagent B can then initiate an exothermic reaction converting A into a less stable product C. Use of high-energy reagents is common in the laboratory since they serve to drive equilibria to completion.

The difference in free energy of reactants and products (ΔF_0) is related to the equilibrium constant, as recalled in Sec. 8-3, by Eq. (3).

$$\Delta F_0 = -2.3\, RT \log K \qquad \text{or} \qquad \Delta F_0 = 1.4\, pK \qquad\qquad 3$$
$$\text{in kcal/mole at room temperature, with } pK = -\log K$$

The interrelations of energy, equilibrium constant, and relative popula-tions of reactants and products at equilibrium are conveniently summarized in Table 9-1. This table is useful not only for ordinary equilibrium reactions in general but for acid-base reactions $(K = K_a)$ and for conformational equilib-rium mixtures. Thus, axial-methyl cyclohexane is 1.7 kcal/mole less stable than the equatorial isomer (see Sec. 6-6) and will be present in equilibrium with it to the extent of only about 3.6%. Table 9-1 also points out the important fact that very small energy differences are enough to produce substantial yields of product, a 90% yield requiring only that the products be 1.3 kcal/mole more stable than reactants.

Any free-energy difference, ΔF, may be broken down into three component energy terms.

1 The first, and usually major, component is the difference in individual bond energies, ΔE_{bond}, between the bonds broken and those formed (cf. example in Sec. 5-4); this term (ΔE_{bond}) includes resonance energies in reactants and products as well.

2 A second component is the difference in total conformational strain energy, ΔE_{strain} (Sec. 6-5), between reactant and product.

3 The third energy component is the entropy term, $T \Delta S$, which is a measure of the energy expended in the reaction to create more *order or organization* in products than was present in reactants, or, conversely, the energy released if the products represent more disorder or freedom. The total free-energy change in a reaction is therefore†

$$\Delta F = \Delta E_{bond} + \Delta E_{strain} - T \Delta S \qquad \Delta = \text{products} - \text{reactants} \qquad 4$$

The first two components have been discussed before and are fairly straightforward, but the entropy term is a somewhat elusive idea (discussed at greater length in physical chemistry texts under the second law of thermodynamics). For our present purposes, it is enough to visualize the entropy (S) of a molecule (or molecules) as a measure of its freedom of motion, or disorder: the more disorder, the larger the entropy.‡ Work must be done (i.e., energy

---

†This energy relation is commonly seen in the form $\Delta F = \Delta H - T \Delta S$, with ΔH the heat content, or **enthalpy**, term. Hence the bond energies and strain energies here are the parts of the enthalpy change in reaction: $\Delta H = \Delta E_{bond} + \Delta E_{strain}$.

‡Entropy (freedom of motion, or disorder), for example, increases with solid $\longrightarrow$ liquid $\longrightarrow$ gas.

TABLE 9-1 **Relations of Population and Energy for Simple Equilibria ($A \rightleftharpoons B$)**

at 25°. $\Delta F_0 = -1.36 \log K$; $K = \dfrac{[B]}{[A]}$

		Population at equilibrium	
K	ΔF_0 kcal/mole	% reactant A	% product B
0.01	2.72	99	1
0.11	1.30	90	10
0.33	0.65	75	25
1.00	0	50	50
3.00	−0.65	25	75
9.00	−1.30	10	90
99.00	−2.72	1	99
999.0	−4.09	0.1	99.9
9999	−5.46	0.01	99.99

expended) to increase the constraints on its freedom, increasing its organization or order, decreasing its entropy. The work involved is the energy term, $T \, \Delta S$, requiring more energy to accomplish at higher (more disordering) temperatures. Thus a reaction which must bring some order in the products from more disorder in the starting materials cannot be as exothermic as a comparable reaction without this added requirement. In general, however, the entropy term is a small component in the free energy and rarely reverses or markedly affects the overall course of reaction. It does, however, increase in importance with temperature.

A common example of entropy effects is found in *cyclizations,* reactions which create rings by the internal, or *intramolecular,* reaction between two functional groups on the same molecule. The cyclization can be compared to the identical reaction between two separate molecules, which is called *intermolecular.* Both reactions have the same ΔH† but differ in entropy. In the intermolecular case the two molecules must be brought together before they can react, and this constraint lowers the entropy and costs energy, whereas in the cyclic molecule the reactants are already together in the same molecule and the entropy cost is small in their reacting ($\Delta S \sim 0$).

$$\left. \begin{array}{l} \text{Intramolecular:} \quad \Delta F_1 = \Delta H - T \, \Delta S_1 \\[2mm] \text{Intermolecular:} \quad \Delta F_2 = \Delta H - T \, \Delta S_2 \end{array} \right\}$$

$$\Delta F_1 - \Delta F_2 = T \, (\Delta S_2 - \Delta S_1)$$

$$\simeq T \, \Delta S_2 = \text{negative}$$

$$\Delta F_1 = \text{more negative,}$$
$$\text{more exothermic, than } \Delta F_2$$

The effect is dramatic in most cyclizations, which are much more favored reactions than their intermolecular counterparts. The equilibrium between acid and alcohol to form ester ($+ H_2O$) usually has $K \simeq 1$, with all components being present at equilibrium. However, in the cyclic example, of an hydroxy-acid forming a lactone, the hydroxy-acid often cannot even be isolated, the internal reaction to a lactone being so very exothermic.

$$CH_3CH_2COOH + CH_3CH_2OH \; \xrightleftharpoons{K \simeq 1} \;$$

$$CH_3CH_2COOCH_2CH_3 + H_2O \qquad \text{(intramolecular)}$$

$$\begin{array}{l} CH_2CH_2COOH \\ | \\ CH_2CH_2OH \end{array} \; \xrightleftharpoons{K \gg 1} \; \begin{array}{l} CH_2CH_2CO \\ | \qquad\quad | \\ CH_2CH_2O \end{array} + H_2O \qquad \text{(intermolecular)}$$

Similarly, the cyclic hemiacetals in ring-chain tautomerism (page 165) are usually totally favored at equilibrium over the acyclic hydroxy-aldehydes, but the reverse is true in intermolecular cases and no *acyclic* hemiacetals have ever been isolated.

---

†This assumes the cyclization does not form a strained ring.

9-4   REACTION RATES AND TRANSITION STATE THEORY

A knowledge of the equilibrium is not enough to tell whether a reaction will go. All organic molecules in the presence of oxygen are unstable relative to CO_2 and H_2O (Fig. 9-4), and yet they do not spontaneously ignite. If they did, all living organisms (including ourselves)—since they are composed of organic molecules—would be instantly immolated in the very air we breathe! In fact most organic reactions are slow, and combustion, at least at ordinary temperatures, is fortunately one of the very slowest.

Consider the **bimolecular reaction**†

$$A + B \longrightarrow C + D$$

1   Although collision of A with B is an obvious prerequisite for reaction, most collisions between A and B do not result in reaction.

2   Analysis of the way in which reaction rates increase with increasing temperature indicates that reaction occurs only if the collisions occur between molecules of A and B that have more than the average energy content. This *excess energy* required for reaction is called the **activation energy ($\Delta F\ddagger$)**.

3   However, even collisions between molecules having the requisite energy content do not often result in reaction. Unfruitful collisions result if the molecules of A and B collide in the wrong way or if the excess energy in the molecules is not associated with the appropriate internal motions of the molecule. The *probability* of reaction occurring even in a collision between activated molecules may be rather small if the stereoelectronic requirements of the reaction are stringent.

A simplified illustration of the probability factor can be based upon consideration of a simple (nucleophilic) substitution reaction:

$$HO^- + H_3C-Cl \longrightarrow HOCH_3 + Cl^-$$

Since the hydroxide ion becomes bonded to carbon, a collision in which the nucleophile encountered the chlorine atom of methyl chloride would not be likely to lead to reaction. The mutually repelling electron pairs on chlorine and hydroxide also tend to block any contact there.

If the excess energy (activation energy) of the methyl chloride molecule is associated with internal vibrations that stretch the C—Cl bond, the reaction should be assisted. However, if the excess energy is associated with stretching motions of the C—H bonds, the substitution reaction will not be aided.

---

†A **bimolecular reaction** is one that involves two molecules of reactants.

The rate at which the reaction proceeds, therefore, will be a function of all three factors:

$$\text{Rate} = (\text{no. collisions/sec}) \times (\text{\% of molecules with } \Delta F\ddagger) \times (\text{probability factor}) \qquad 5$$

The number of collisions in Eq. (5) is proportional to the concentrations of reactants and is also a function of temperature, increasing with increasing temperature, since this increases molecular motion and thus the frequency of collision. It is common for the rates of organic reactions to double with every 10 to 15° rise in temperature.

The concentration dependence is expressed in Eq. (6):

$$\text{Rate} = k[A][B] \qquad \text{where } k \text{ is the \textbf{specific rate constant} at temperature } T \qquad 6$$

Thus the rate is proportional to the concentration of *each* species involved in reaching the transition state.

Transition States

The entire rate problem can be formulated by defining a state of the reacting system, called the **transition state** or **activated complex**, in which all the stringent requirements for an effective collision have been met. The transition state is a molecular complex in which A and B have been forced together in such a way that they have paid the price (in energy currency) required for reaction and are ready to collapse into products. The change in free energy ($\Delta F\ddagger$) required to make transition states from reactant molecules is a direct measure of the ease of the reaction.

The model is analogous to the physical problem of climbing over a mountain pass from one valley to another. A large number of people trapped in one valley might wish to migrate to the other valley. In order to do so, they would have to climb to the top of the pass between the two valleys. The rate of migration would depend upon the rate at which people arrived at the top. Only the strongest migrants would have sufficient energy to make the climb; furthermore, there would be a high probability that some of the strong ones would expend their energy by climbing paths that did not lead to the pass.

The relation of the reaction rate to the free energy of activation ($\Delta F\ddagger$) is similar to that for the relation of equilibrium to free energy, Eq. (3). The specific rate constant in Eq. (6) (k = rate with reactant concentrations of unity; $[A] = [B] = 1$) is used in these relations in Eqs. (7) and (8), the actual rate in a given case being a function of reactant concentrations as well as k [Eq. (6)].

$$k = Ae^{-\Delta F\ddagger/RT} \qquad \text{or.} \qquad \Delta F\ddagger = -2.3\, RT \log k + \text{const} \quad (\text{at a given temperature}) \qquad 7$$

For reaction at room temperature (25°) this is

$$\Delta F\ddagger = -1.4 \log k + 17.6 \text{ kcal/mole} \qquad 8$$

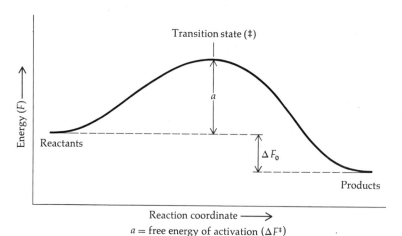

$a = $ free energy of activation ($\Delta F^{\ddagger}$)

FIGURE 9-5 **Free-energy diagram for one-stage reaction**

Here we can see that

1 *Raising activation energy lowers rate and vice versa.*
2 *Raising temperature raises rate.*

As for equilibria, the free energy here has an entropy component ($\Delta F^{\ddagger} = \Delta H^{\ddagger}$ $- T \Delta S^{\ddagger}$), and reactions go slower if some activation energy must be spent forcing constraints on the freedom of the reactants. In bimolecular or more complex reactions, requiring two or more molecules to be coalesced into the activated complex, the reaction will be slower, due to this entropy effect, than those involving only a single molecule. Accordingly, cyclization reactions are *faster* than their bimolecular counterparts, as well as possessing more favorable equilibria.

Free-energy Diagrams

The energy changes in the course of reaction are usually illuminated in the form of a graph (Fig. 9-5) which plots the free energy F, on the vertical axis, against a horizontal coordinate which represents the progress of reaction (**reaction coordinate**). The reaction coordinate may be understood as the total collection of distances among all the particles (atoms and electrons) in the involved molecules at any given moment during the course of reaction. More visually it represents the continuous change in geometry as the two molecules collide, compressing and bending groups, forcing apart the atoms whose bonds break, and drawing together those which are forming new bonds. The molecules start, at the left of the diagram, separate and apart, and proceed along the reaction coordinate as they collide and deform. The potential energy F of their coalescing increases (using up their kinetic energy) until the transition state ($\ddagger$) is reached. The energy then falls off as the transition-state complex begins to regroup into the stabler forms of the product molecules.

Multistage reactions may also be represented by free-energy diagrams, as illustrated in Fig. 9-6. The minimum in the curve represents a reaction intermediate in a two-step reaction. Each step of the reaction involves passage

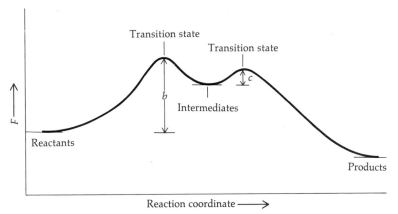

Reactants

Intermediates

Products

Transition state

Transition state

Reaction coordinate ⟶

b = free energy of activation for first stage $(\Delta F_1^{\ddagger})$

c = free energy of activation for second stage $(\Delta F_2^{\ddagger})$

FIGURE 9-6   **Free-energy diagram for two-stage reaction**

through a transition state. However, the overall rate of the reaction is determined only by the free-energy difference between the reactants and the *highest* point on the curve; *that point represents the transition state for the rate-controlling step of the reaction.* In Fig. 9-6 the first maximum is the highest point, so the first step is rate-controlling. The fact that an intermediate is produced before the ultimate products appear will therefore have no measurable influence on the overall rate. The intermediate will pass over to the products more rapidly than the reactants produce the intermediate. The diagram could represent qualitatively the first reaction of Fig. 9-3, the intermediate being the carbonium ion.

Any *minimum* on such curves (reactants, intermediates, products) represents a real molecule, with normal bonds, and even the intermediates can often be isolated and examined despite their higher energy (greater instability). The height of the lowest **energy barrier** (transition state) adjacent to any such molecule is therefore a measure of its reactivity. If that height is less than the 15 to 20 kcal/mole of normal thermal energy, the molecules are unlikely to be isolated at room temperature (but might be at very low temperature).

The transition-state complexes (*maxima* on the curve) have, on the other hand, only fleeting, transient existence; they do not have normal bonds at the reactive site since some are partially broken, others only partially formed at this highest-energy stage of the progress of reaction.

We can speak with confidence of the geometry, resonance, strains, etc., in an *intermediate,* for it is a normal covalent molecule (or ion), but we cannot describe a *transition state* with any accuracy. However, it is generally believed that if a real molecule (energy minimum on curve: reactant, intermediate, product) lies *close in energy* to a transition state (vertical scale), it will also be *close in geometry* (horizontal scale). This is a useful hypothesis for it allows us in many cases to approximate the geometry, resonance, strains, etc., in the transition state by saying that it closely resembles either the reactant or the intermediate (or product) to which it is closest in energy.

In Fig. 9-5 the transition state is similar to neither the initial nor the final molecules in energy and so lies about half-way between them in geometry. In Fig. 9-6, however, the transition state for the first step is close in energy to the intermediate so that we may say it closely resembles this in shape. For example, in the first reaction shown in Fig. 9-3 ($CH_3CH=CHCH_3 + H^+ \longrightarrow CH_3\overset{+}{C}HCH_2CH_3$), at the transition state the π electrons have nearly coalesced into a σ bond to the attacking proton. This σ bond is almost fully formed and the proton almost arrived at full bonding distance (and energy) to its carbon, while the carbonium ion has nearly developed its full positive charge.

We may divide our consideration of any reaction into its **thermodynamics** and its **kinetics**. The former (Sec. 9-3) represents the equilibrium situation and is concerned *only with the energy difference between reactants and products*. The *kinetics* of a reaction refers to its rate and is concerned *only with the energy difference between reactants and the (highest) transition state*. The thermodynamic description is ΔF_0 and the related equilibrium constant K of Eq. (3), while the kinetic description is the activation energy $\Delta F\ddagger$ and the rate constant k of Eq. (7).

The free-energy diagram describes both considerations. It shows the free-energy difference ΔF_0 between reactants and products, which determines the *equilibrium constant K* and whether the reaction is exothermic or endothermic (labeled in Fig. 9-5). However even if the diagram shows more stable products than reactants, the reaction will have a negligible rate if the *energy barrier* ($\Delta F\ddagger$) is too high, and will only be forced to proceed at a sensible rate by raising the temperature.

The energy diagrams shown for conformational change serve as simple analogies to those for reactions. In many conformational change diagrams, the horizontal coordinate of change in geometry can be related to a single simple change in one angle, as in Fig. 6-8 for rotation of *n*-butane around its central bond. Here the geometrical change coordinate is simply the dihedral angle of rotation. The staggered conformations (*A, B,* and *C*) are the stable states in equilibrium, akin to reactant, intermediate, and product. Conversion of the 60° to the 180° isomer is exothermic, the latter predominating in the equilibrium mixture. The barriers (maxima) are the strained transition states of rotation: the eclipsed forms. The rate of the interconversion is determined by the energy height of the barrier ($\Delta F\ddagger$), which in the butane case is much lower than the 15 to 20 kcal/mole of thermal energy. Hence the reaction is very fast at room temperature. The equilibrium table (9-1) is very convenient in assessing conformational change as well as chemical reaction equilibria.

Effects on Reaction Rates

A single mechanism may generalize a whole family of reactions which are all the same in the nature of change at the reactive site but differ in other particulars, of which these are important:

1 The particular nucleophile or electrophile selected as reagent
2 The nature of the groups adjacent to the reactive site in the substrate
3 The nature of the solvent

The nucleophilic substitution reaction mentioned above between methyl chloride and hydroxide ion, will have the same basic mechanism if methoxide ion or ammonia is the nucleophilic reagent or if ethyl or benzyl chlorides, or bromides, are the substrates, and the use of various solvents will not change the generic mechanism. While these major families of reactions are grouped and discussed individually in the succeeding chapters, we can see in broad outline how the factors of structure will influence their rates in much the same way that we compared the effects of these factors on equilibrium constants in the acid-base reaction.

In examining equilibria in Chap. 8 we considered whether an effect would stabilize (or destabilize) one member of the equilibrium (i.e., conjugate acid or base) with respect to the other in order to assess whether ΔF_0 between them would become greater or less than the standard. We can do an analogous comparison with reaction rates, observing the effect on the energy of the reactant compared with that of the transition state, since their energy difference $\Delta F\ddagger$ determines the rate [Eq. (7)]. In general these effects are compared to a standard reaction between specific reactants in a specific solvent.

The four possible cases (A to D) illustrated in Fig. 9-7 are amplified below with examples of different kinds of effects. The transition state here closely resembles the intermediate (this is common) so that the latter is used to approximate the nature of the transition state in discussion.

A The reactant has resonance stabilization which is broken up and lost during reaction, leaving little in the intermediate, hence in the transition state. The rate is decreased.

B The intermediates are ions whereas the reactants were neutral, and the reaction is conducted in a polar solvent which stabilizes the ions by hydrogen bonding. This stabilization only affects the ionic intermediates and the closely similar ionic transition state but not the neutral reactants. The rate is increased.

C The reactant contains a small ring, with angle strain destabilizing it. This ring is opened in the reaction (compare the α-lactone on page 346 to a "standard" six-membered, unstrained lactone), affording a largely unstrained intermediate (and transition state). The rate is increased.

D The collision of reactants leading to the transition state is sterically hindered by large groups on the substrate near the reactive site. These cause no steric strain in the substrate molecule itself but are compressed by the entering reagent molecule as it attacks the active site, thus creating strain energy in the transition state. The rate is decreased.

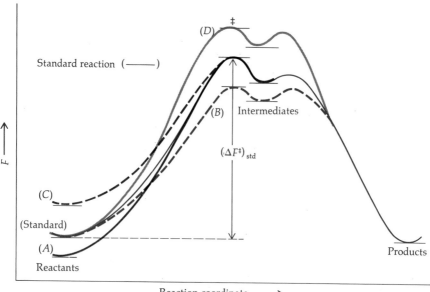

An effect which:

A   stabilizes reactant (more than transition state), increases $\Delta F^{\ddagger}$, and lowers rate;

B   stabilizes transition state (more than reactant), decreases $\Delta F^{\ddagger}$, and raises rate;

C   destabilizes reactant (more than transition state), decreases $\Delta F^{\ddagger}$, and raises rate;

D   destabilizes transition state (more than reactant), increases $\Delta F^{\ddagger}$, and lowers rate.

FIGURE 9-7   **Effects on reaction rate**

Comparisons of this kind are general in assessing the reasons for varying rates of a single general reaction in different particular cases, and will be invoked often for particular cases in succeeding chapters. The observed variations in rate can often be very dramatic: many powers of 10 between certain cases. One reaction may be complete in seconds at room temperature but another may require high temperature for many hours. Furthermore, the kinds of effects which cause these variations are the same structural factors which have been reviewed before, and all have energy terms related to them, i.e., conformational strain (bent angles, bad torsion, steric hindrance), resonance, inductive effects, hydrogen bonding.

PROBLEM 9-7

Consider these protonations of weak bases to yield high-energy carbonium-ion intermediates: the reaction is a simple acid-base reaction from Chap. 8. Compare the *rates* of protonation qualitatively among the examples given in terms of the effects in Fig. 9-7.

a $Z-CH=CH_2 + H^+ \longrightarrow Z-\overset{+}{C}H-CH_3$

b   (cyclohexene ring) $+ H^+ \longrightarrow$ (protonated ring with $\overset{+}{~}$ H, H)

$Z = CH_3-, CH_3O-, CH_3CO-$

c $Z-\underset{\underset{\displaystyle CH_2-CH_2}{|\quad\quad|}}{C}=CH + H^+ \longrightarrow Z-\underset{\underset{\displaystyle CH_2-CH_2}{|\quad\quad|}}{\overset{+}{C}}-CH_2$

(*Hint:* Note angle strains)

9-5   LABORATORY STUDY OF REACTIONS

The previous discussion has all been in terms of an intimate acquaintance with individual molecules, their motivations and behavior. As was the case with questions of molecular structure previously, however, this intimate knowledge of molecular dynamics must be gained through indirect evidence, for these molecules are far too small to be observed directly. Now we must survey the means whereby reaction mechanisms may be deduced from experiment.

The study of reaction rates is called **kinetics** and is a very widely practiced specialty in organic chemistry at present since it is a most powerful tool for illuminating reaction mechanism. Basically, the experimenter must mix potential reactants in known concentration at constant temperature and then determine the rate of decrease of these concentrations with time, thus measuring in fact the rate of reaction (rate of disappearance of starting material). He will observe these concentrations at specific time intervals either by taking spectra of the mixture or by removing a measured sample (**aliquot**), stopping (**quenching**) its reaction, and determining some concentration in the arrested sample, e.g., by titration. Spectroscopic observation of reactions is easiest as in the example of monitoring the reaction of a diene with bromine by observing the rate of decrease of the ultraviolet absorption of the diene with time, the intensity of absorption being a function of diene concentration (Sec. 7-4).

This observed dependence of the rate on reactant concentrations is called the **order** of the reaction. In a **first-order reaction** the rate varies only with the concentration of *one* reactant, in a **second-order reaction** the rate is proportional to *two* reactant concentrations, etc. When the experimental kinetics show these simple dependencies, we may usually correctly deduce that at a molecular level only one molecule goes to the transition state in a first-order reaction, that two combine into a transition state in a second-order reaction, etc. These deductions about the molecular involvement are called the **molecularity** of the reaction. The three most common kinetic situations observed are tabulated in Table 9-2.

Studies of the rate of reaction of 2-butene and bromine [Br_2 for **HBr**, reaction (1), Fig. 9-3] show its proportionality to the concentrations of each reactant, and the specific rate constant (k) for this second-order (and presumed bimolecular) reaction may be determined.† In the bromination of acetone, however, it is found that the rate is independent of the bromine concentration (assuming a reasonable concentration is present) since the **rate-controlling step** is the prior enolization of acetone to its enol. The overall rate is simply that (k_1) of the first, or slow, step:

$$CH_3-\overset{\overset{O}{\|}}{C}-CH_3 + H^+ \xrightarrow[\text{slow}]{k_1} CH_3-\overset{\overset{OH}{|}}{C}=CH_2 \xrightarrow{Br_2} CH_3-\overset{\overset{O}{\|}}{C}-CH_2-Br + HBr$$
$$+ H^+$$

It is clear that such kinetic studies only define reaction progress to the transition state of the rate-controlling step; it is also necessary to determine the products of reaction. Many reactions produce two or more products. Not infrequently the product that *predominates* initially is *not* the most stable; it is said to be the product of the **kinetically controlled reaction**. Sometimes, the various products can be brought to equilibrium by prolonged incubation under the reaction conditions; the most stable product will then ultimately predominate in the resulting **equilibrium-controlled reaction**. Thus, kinetic control rapidly produces a first product, which is, however, then slowly consumed if an equilibrium is slowly reached which favors a more stable (equilibrium-controlled) product. The simple situation is summarized in Fig. 9-8. A somewhat more complex example from actual practice is provided by careful studies of the elimination reaction that occurs when 2-phenyl-2-butyl acetate is heated in

†The energy of activation ($\Delta F\ddagger$) may also be calculated by Eq. (7). Since the entropy component ($T \Delta S\ddagger$) is a function of temperature, determinations of k and $\Delta F\ddagger$ at different temperatures allow deduction of the entropy change.

TABLE 9-2   **Kinetic Results and Mechanistic Deductions**

Kinetic observations		Mechanistic deductions	
Order	Rate [Eq. (6)]	Probable reaction	Molecularity
First	$k[A]$	$A \longrightarrow [\ddagger]$	Unimolecular
Second	$k[A][B]$	$A + B \longrightarrow [\ddagger]$	Bimolecular
Third	$k[A][B][C]$	$A + B + C \longrightarrow [\ddagger]$	Termolecular

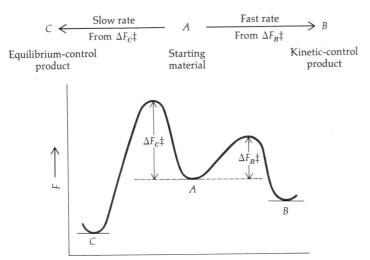

FIGURE 9-8 Kinetic and equilibrium control in reactions

acetic acid solution. If the reaction is carried out with no added catalyst, the products are the following mixture:

$$CH_3-\overset{\displaystyle OCOCH_3}{\underset{\displaystyle C_6H_5}{C}}-CH_2CH_3 \xrightarrow[50°]{CH_3COOH}$$

$$\overset{CH_3}{\underset{C_6H_5}{}}C=C\overset{H}{\underset{CH_3}{}} \quad + \quad \overset{CH_3}{\underset{C_6H_5}{}}C=C\overset{CH_3}{\underset{H}{}} \quad + \quad \overset{CH_2}{\underset{C_6H_5}{}}C-CH_2-CH_3$$

trans-2-Phenyl-2-butene, 2%   cis-2-Phenyl-2-butene, 53%   2-Phenyl-1-butene, 45%

If a small amount of strong acid, such as p-toluenesulfonic acid, is added to the reaction mixture, the alkenes produced are converted to an equilibrium mixture in which the more stable cis isomer predominates over the trans isomer by a ratio of 5 to 1, and the trans over the 1-alkene by a ratio of 5 to 1. The added acid acts as a catalyst to speed up the establishment of equilibrium.

$$\overset{CH_3}{\underset{C_6H_5}{}}C=C\overset{H}{\underset{CH_3}{}} \underset{50°}{\overset{CH_3COOH/H^+}{\rightleftharpoons}} \overset{CH_3}{\underset{C_6H_5}{}}C=C\overset{CH_3}{\underset{H}{}} \underset{50°}{\overset{CH_3COOH/H^+}{\rightleftharpoons}}$$

16%   81%

$$\overset{CH_2}{\underset{C_6H_5}{}}C-CH_2-CH_3$$

3%

If the elimination reaction is originally carried out in the presence of a strong acid, the equilibrium-controlled product distribution is obtained.

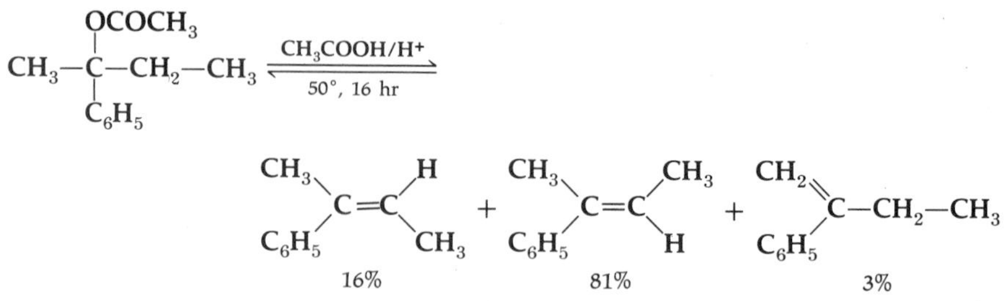

$$
\begin{array}{c}
\text{OCOCH}_3 \\
| \\
\text{CH}_3-\text{C}-\text{CH}_2-\text{CH}_3 \\
| \\
\text{C}_6\text{H}_5
\end{array}
\quad \xrightleftharpoons[\text{50}^\circ,\ 16\ \text{hr}]{\text{CH}_3\text{COOH/H}^+}
$$

$$
\underset{16\%}{\overset{\text{CH}_3}{\underset{\text{C}_6\text{H}_5}{}}\!\!\!\text{C}=\text{C}\!\!\!\overset{\text{H}}{\underset{\text{CH}_3}{}}}
\ +\
\underset{81\%}{\overset{\text{CH}_3}{\underset{\text{C}_6\text{H}_5}{}}\!\!\!\text{C}=\text{C}\!\!\!\overset{\text{CH}_3}{\underset{\text{H}}{}}}
\ +\
\underset{3\%}{\overset{\text{CH}_2}{\underset{\text{C}_6\text{H}_5}{}}\!\!\!\text{C}-\text{CH}_2-\text{CH}_3}
$$

Besides the kinetics of a reaction there are a number of other ways of studying reactions. The final mechanism deduced for a reaction must explain the following:

Products, and side products
Intermediates observable where possible
Kinetics
Stereochemical results (stereospecificity and stereoselectivity)
Isotope studies
Relative reactivity of different reagents
Relative reactivity of different substrates
Variations in rate (and products) with different solvents

Intermediates can sometimes be observed spectroscopically, or even isolated under special conditions. Alternatively, they may be synthesized by an alternate path to see if they collapse to the same products, or they may be trapped by the addition of another substance which will react preferentially with the intermediate to give a new and characteristic product.

Isotopes are used to distinguish two otherwise identical atoms (H and D; O^{16} and O^{18}; etc.) and thus trace their course to particular products, where they are identified by radioactivity or mass spectra. Relative reactivities can be determined by separate rate studies on reactions with different reagents (or substrates). Alternatively, they can be seen by competition experiments in which two reagents are offered to a limited amount of one substrate (or vice versa). The faster reaction is then assessed by the predominance of its product in the final mixture.

Hence a variety of procedures exist for producing information which illuminates the intimate molecular details of a reaction mechanism, and a number of these will appear in discussions of specific reactions in the pages ahead. It is valuable to note here, however, that all of these studies which clarify reaction mechanism are of enormous help in the synthesis of new compounds, for the greater the detail known about a reaction the more certain will be its predictive value for the new cases required in synthesis. This practical value is emphasized in Chap. 23 on synthesis. In another area, knowledge of reaction mechanisms is of great value in deducing the nature of chemical transformations which occasionally take a complex and unexpected course. The unraveling of the details of mechanism in these unusual and abnormal reactions can be a very stimulating intellectual challenge.

9-6   SUMMARY AND PREVIEW

In an ionic reaction an unshared pair of electrons on one molecule or anion (the **nucleophile**) attacks an electron-deficient atom (unfilled outer shell or positive end of a dipole) in another molecule or cation (the **electrophile**) and forms a new bond to that atom. In this attack the potential energy of the system rises to a maximum as the colliding molecules give up kinetic energy and coalesce to a strained, imperfectly bonded **transition state**. This then collapses with full formation of the new bond(s) between the original centers. The molecule (or ion) thus formed may be the final product or only an intermediate which then passes on to other molecules via a second transition state. The energy height ($\Delta F\ddagger$) of the (highest) transition state determines the rate of reaction, which is faster the lower that barrier. Reactions which are reversible (i.e., the reverse reaction has a reasonable rate) come to equilibrium, at which time the thermodynamically most stable compound in the equilibrium mixture will predominate and be isolated. This is an equilibrium-controlled product, whereas a product formed by virtue of its faster rate of formation rather than its relative stability is said to be kinetically controlled.

Reaction rates are related to **activation energy** ($\Delta F\ddagger$)—the energy difference between reactants and transition state. Equilibria are related to the energy difference (ΔF_0) between reactants and products. Both energy terms are differences so that effects on either rate or equilibria may be assessed by considering the energy changes those effects have on *both* members of the pair in order to find the effect on ΔF itself. The kinds of effects which change rates or equilibrium positions are the structural ones already examined: resonance effects, inductive and electrostatic effects, conformational strains (angle bending, torsional or steric strain), hydrogen bonding, and of course combinations of these. Free-energy diagrams are of great value in clarifying the energy relations in reactions.

Reagents are often inorganic or simple organic compounds, and they are often ionic. Nucleophiles are usually anions, and the associated cation is usually uninvolved in reaction, metal ions being stable and unreactive for the most part. To improve solubility of nucleophilic anions in organic solvents, Li^+ or R_4N^+ are often used. Anions are bases if they attack hydrogen (acid-base reaction) and nucleophiles if they attack carbon. The two functions sometimes compete undesirably. Reagents which usually act only as bases and not nucleophiles are therefore valuable:

$$NaH \qquad (CH_3)_3C:^- Li^+ \qquad \phi_3C:^- Na^+ \qquad (CH_3)_3C\!-\!O^- K^+$$

Similarly, when protons (or other electrophiles) are required for reaction it is often important to use a reagent offering an unreactive nucleophile as its associated anion (if any):

$$H^+ BF_4^- \qquad E^+ BF_4^- \text{ or } PF_6^- \qquad H^+ HSO_4^- \text{ or } \phi SO_3^- \qquad BF_3$$

The reactivity of organic molecules lies almost exclusively in their functional groups, and the primary distinction made among functional groups is between their reactivity as nucleophiles or electrophiles (Fig. 9-1), particularly at the carbon atom(s) to which the functionality is attached. We can qualitatively predict how they will react from this first knowledge, but will amplify it with discussion of their relative rates of reactivity to appropriate reagents and the affects of their molecular environment and solvent on these rates.

The bulk of organic reactions can be seen in terms of the reagent as nucleophile or electrophile and the carbon skeleton substrate as its opposite:

Nucleophilic reagent	+	*Electrophilic substrate*		*Electrophilic reagent*	*Nucleophilic substrate*

$$\text{B:} \longrightarrow \text{H} \overset{\frown}{-} \text{Z} \begin{Bmatrix} \text{C} \\ \text{N} \\ \text{O} \end{Bmatrix} \text{Acid-base reaction}$$

or

$$\text{Nu:} \longrightarrow \underset{/\,\,\backslash}{\overset{|}{\text{C}}} \overset{\frown}{-} \text{Y} \quad \text{Saturated carbon}$$

$$\underset{/}{\overset{\backslash}{\text{C}}} \overset{\frown}{=} \text{Y} \quad \text{Unsaturated carbon}$$

$$\text{E}^+ \qquad \overset{-\,\,|}{\underset{|}{:\text{C}-}} \quad \text{Saturated carbon}$$

$$\underset{/}{\overset{\backslash}{\text{C}}} = \underset{\backslash}{\overset{/}{\text{C}}} \quad \text{Unsaturated carbon}$$

There are reactions in which one organic molecule acts as a nucleophile and the other as an electrophile, both acting at carbon and forming a carbon-carbon bond. In such reactions the distinction between "reagent" and "substrate" becomes meaningless. Reactions which form carbon-carbon bonds are of central importance to synthesis as the means of building up more complex carbon skeletons. Such reactions are much less easily reversed than comparable reactions forming unsymmetrical bonds since the carbon-carbon bond has no dipole (electrical imbalance) and so is harder to polarize and break.

For synthetic purposes we may distinguish reactions which interconvert functional groups at one molecular site and those which create carbon-carbon bonds and enlarge the molecular skeleton. Each of the next chapters on reactions is divided into a preliminary theoretical discussion, allowing us to understand the mode and scope of a reaction, and a second practical or synthetic discussion showing how the reaction is actually used.

The major ionic reactions of organic chemistry, discussed in Chaps. 10 to 16, are listed next with generalized mechanisms. It must be appreciated here that the curved arrows serve only to show where the electrons ultimately go. They are not meant to imply that all the bond breaking and making is necessarily concerted.

1   *Nucleophilic substitution at saturated carbon* (Chaps. 10 and 11)

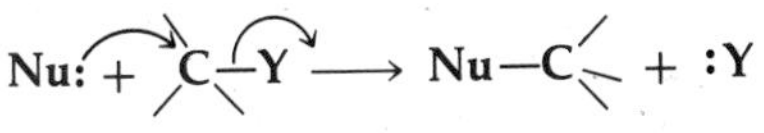

$$\text{Nu:} + \underset{\diagup}{\overset{\diagdown}{C}} - Y \longrightarrow \text{Nu} - \underset{\diagup}{\overset{\diagup}{C}} + \text{:Y}$$

$$(\text{Nu: } + R - Y \longrightarrow \text{Nu} - R + \text{:Y})$$

2   *Nucleophilic addition to unsaturated carbon* (Chap. 12)

3   *Nucleophilic substitution at unsaturated carbon* (Chap. 13)
These reactions are predominantly those of carbonyl groups and are conveniently considered together since both are initiated by the same attack of nucleophile on carbonyl carbon. Some other functional groups react by the same mechanism and are considered in Chaps. 12 and 13 also. The reaction can also be acid-catalyzed (H$^+$ on carbonyl oxygen first), but the mechanistic course is the same.

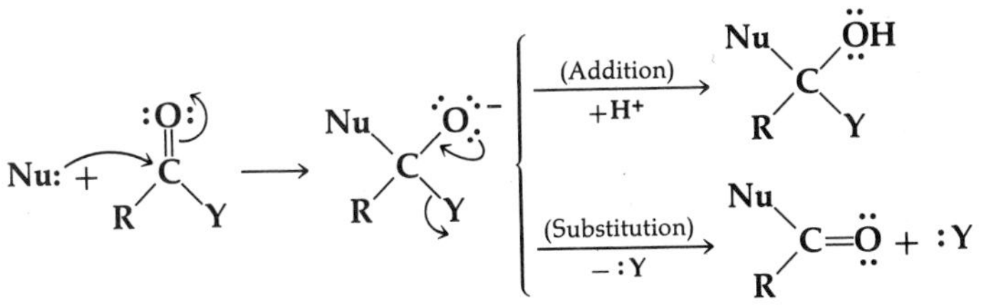

4   *Electrophilic addition* (Chap. 15)

5   *Electrophilic substitution* (Chap. 16)
These reactions are predominantly those of carbon-carbon multiple bonds and also are alike in their initial (and rate-determining) step. Among unsaturated compounds the substitution reaction is peculiarly characteristic of aromatic rings.

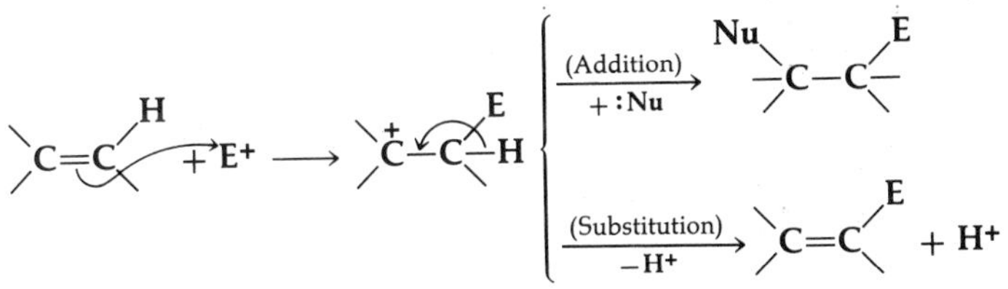

6   *Elimination* (Chap. 14)
The major mechanism for eliminations is shown below. The same elimination mechanism can also create multiple bonds between atoms other than carbon. Because of its similarity in conditions to nucleophilic substitutions the elimination discussion directly follows those reactions in chapter sequence.

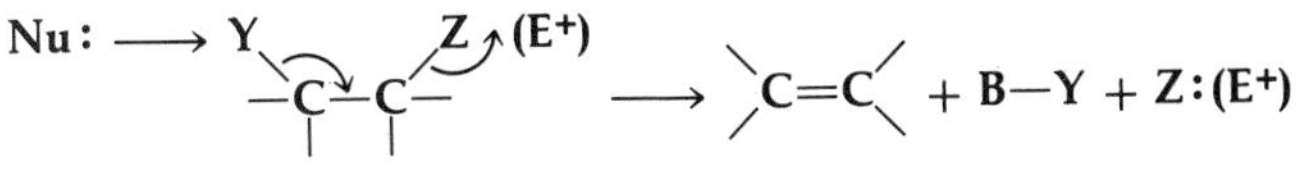

Following the fundamental reaction classes in Chaps. 10 to 16, Chap. 17 deals with molecular rearrangements, or movements of atoms in the skeleton. In these rearrangements the skeletal bonds are shifted as well as the functional-group bonds as seen in this simple example of a migration of a methyl group.

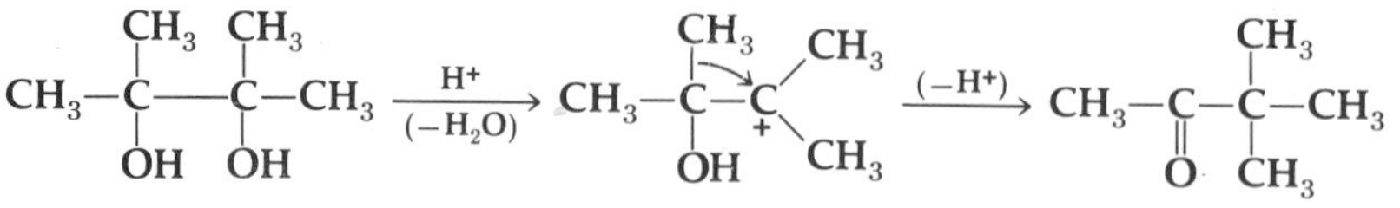

$$CH_3-\underset{\underset{OH}{|}}{\overset{\overset{CH_3}{|}}{C}}-\underset{\underset{OH}{|}}{\overset{\overset{CH_3}{|}}{C}}-CH_3 \xrightarrow[(-H_2O)]{H^+} CH_3-\underset{\underset{OH}{|}}{\overset{\overset{CH_3}{|}}{C}}-\overset{\overset{CH_3}{\diagup}}{\underset{\diagdown CH_3}{C^+}} \xrightarrow{(-H^+)} CH_3-\underset{\underset{O}{\|}}{C}-\underset{\underset{CH_3}{|}}{\overset{\overset{CH_3}{|}}{C}}-CH_3$$

In Chap. 18 we deal with the reactions which allow us to change oxidation state in organic molecules, via treatment with oxidizing or reducing agents, many of them familiar from general chemistry ($KMnO_4$, $H_2Cr_2O_7$, Zn metal, etc.). The oxidation state at any given carbon atom can be most easily appreciated by counting the number of its bonds to heteroatoms. *The more heteroatom bonds the more oxidized the carbon*[†]:

$$-CH_3 \qquad -CH_2OH \qquad -CHO \qquad -COOH$$

Increasing oxidation state

In general, *reductions are reactions that add hydrogen* (as in direct catalytic hydrogenation), and *oxidations are reactions that remove hydrogen, or else add oxygen (or both)*, as in the conversion of R—CHO to R—COOH.

The special properties of elements in the second row of the periodic table are considered in Chap. 19. The major examples are sulfur and phosphorus compounds. These elements exhibit some different reactivity from their first-row counterparts owing to their ability to expand their outer shell of electrons, from the standard capacity of 8 to 10 or 12. The idea was alluded to briefly in Chap. 2.

The fundamental and most common organic reactions are the ionic, or heterolytic, reactions of Chaps. 10 to 19. In Chap. 20, however, are gathered the homolytic or free-radical reactions. Following this are two chapters (21 and 22) which deal solely with reactions which occur without external reagents, simply by heating or by irradiation with light. Both processes afford an input of energy which can break bonds and cause new bonds to be formed. Such reactions may proceed by way of free-radical intermediates or they may simply be characterized by a concerted regrouping of certain bonding orbitals in the molecule. The latter reactions are called **pericyclic reactions.**

One final area of organic reactions is that of polymerization, discussed in Chap. 25. **Polymerization** is a reaction repeated many times with the same kind of molecule such that they link together successively into a long-chain molecule of repeating units. The units linked together are called **monomers** and the resultant long chain of many units is labeled a **polymer** (or merely two units, a **dimer**; three units, a **trimer**; etc.):

[†] Double bonds are assessed for oxidation state by formally adding the elements of water (H—OH) across the double bond first.

$$A \xrightarrow{+A} A-A \xrightarrow{+A} A-A-A \xrightarrow{+A} A-A-A-A \xrightarrow{\cdots} \xrightarrow{+nA} A-(A)_m-A$$

Monomer   Dimer   Trimer   Polymer

The reactions themselves are simply those of Chaps. 10 to 16, repeated many times as in the amide formation from acid and amine that converts natural amino acids into their corresponding polymers, the proteins.

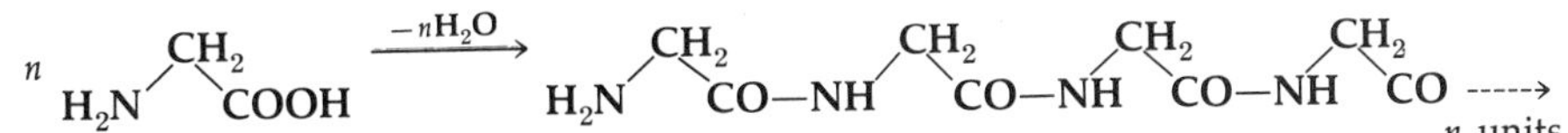

$$n \underset{\underset{\text{Glycine, an amino acid}}{H_2N \quad COOH}}{CH_2} \xrightarrow{-nH_2O} \underset{\underset{\text{Polyglycine, a protein polymer}}{H_2N \quad CO-NH \quad CO-NH \quad CO-NH \quad CO}}{CH_2 \quad CH_2 \quad CH_2 \quad CH_2} \longrightarrow$$
$$n \text{ units}$$

The reactions examined to this point now serve as tools for organic synthesis, which is treated *as a concept* in Chap. 23. The remaining chapters are concerned with special topics, much of Chaps. 25 to 27 being devoted to outlining the state of our present knowledge about the organic molecules and reactions which characterize the phenomena of life.

PROBLEMS

9-8 Identify all the energy terms ΔF_0 and $\Delta F\ddagger$ which are involved in the competitive reactions in Fig. 9-8 and show how each reaction is kinetically or thermodynamically controlled.

9-9 Classify the following reactions as completely as possible:

a $C_6H_5SH + C_6H_5CH_2Br \longrightarrow C_6H_5SCH_2C_6H_5 + HBr$

b $Br_2 + CH_2=CH_2 \longrightarrow BrCH_2CH_2Br$

c $O_2 + (CH_3)_3CH \xrightarrow[\text{Catalyst}]{CoBr_2} (CH_3)_3C-OOH$

d $O=C=O + C_6H_5Li \longrightarrow C_6H_5\overset{\overset{\displaystyle O}{\|}}{C}-O-Li$

e $H_2O + (CH_3)_2C=CH_2 \xrightarrow{H^+} (CH_3)_3COH$

f $Zn + C_6H_5CHClCHClC_6H_5 \longrightarrow C_6H_5CH=CHC_6H_5 + ZnCl_2$

g $CH_2=CH-CH=CH_2 + $ [structure of maleic anhydride] $\xrightarrow[\text{solution}]{\text{Benzene}}$ [structure of Diels-Alder adduct]

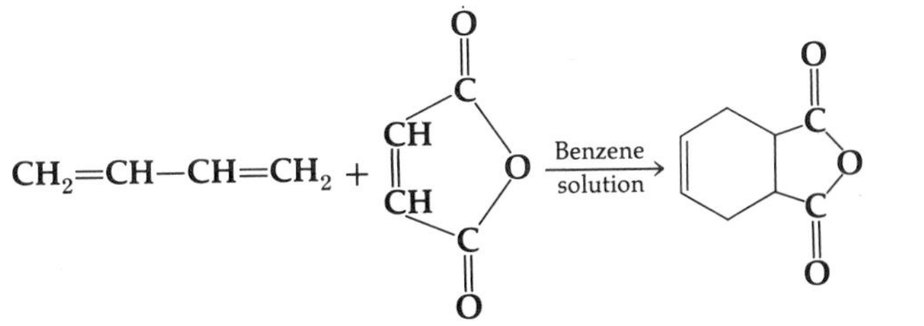

h $H_2SO_4 + C_6H_6 \longrightarrow C_6H_5SO_3H + H_2O$

i $HCN + CH_3CHO \xrightarrow{NaOH} CH_3\overset{\overset{\displaystyle CN}{|}}{C}HOH$

j $C_2H_5OH + CH_2=CHCN \xrightarrow{NaOH} C_2H_5OCH_2CH_2CN$

k $C_2H_5OCOCH_3 \xrightarrow[\text{Gas phase}]{450°} CH_2=CH_2 + CH_3COOH$

9-10 Define the following terms:

- **a** Electrophilic addition to unsaturated carbon
- **b** Multistage reaction
- **c** Transition state
- **d** Rate-controlling step
- **e** Carbene
- **f** Nucleophile
- **g** Lewis acid
- **h** Equilibrium-controlled reaction product

9-11 The reaction of optically active *sec*-butyl chloride with sodium hydroxide gives optically active *sec*-butyl alcohol of configuration opposite to that of the original substrate:

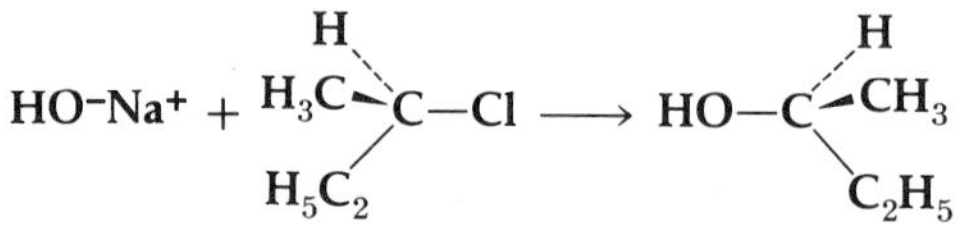

$$HO^-Na^+ + \underset{H_5C_2}{\overset{H}{H_3C-C-Cl}} \longrightarrow HO-\underset{C_2H_5}{\overset{H}{C-CH_3}}$$

The reaction involves only one stage. What do you consider to be the likely configuration of the transition state? How would rate observations be made in practice?

9-12 Draw free-energy diagrams for reactions that possess the following features. Label the activation energies and transition states for each stage and the intermediates in the reaction.

- **a** A one-stage reaction
- **b** A two-stage reaction in which the second stage is slower than the first
- **c** A three-stage reaction in which the first stage has the highest activation energy and the second intermediate is more stable than the first

d $(CH_3)_3CCl \xrightarrow{Slow} (CH_3)_3\overset{+}{C} + \overset{-}{Cl} \xrightarrow[\text{Fast}]{H_2O} (CH_3)_3C\overset{+}{O}H_2 \xrightarrow[\text{Very fast}]{-H^+} (CH_3)_3COH$

9-13 Assume that the solvolysis (cleavage by solvent) of *t*-butyl bromide in acetic acid involves the following mechanism:

$$(CH_3)_3CBr \longrightarrow (CH_3)_3C^+ + Br^-$$

$$(CH_3)_3C^+ + CH_3COO^- \longrightarrow (CH_3)_3C-O-\overset{\overset{\displaystyle O}{\displaystyle \|}}{C}CH_3$$

Addition of sodium acetate to the reaction mixture does not increase the rate of reaction appreciably. Explain. Which step is rate-controlling?

9-14 Irradiation of gaseous azomethane (CH_3—N=N—CH_3) with ultraviolet light produces ethane and nitrogen. Formulate a reasonable mechanism for the reaction.

9-15 In each pair of comparable reactions below indicate which one you expect to proceed with a faster rate, and why.

a CH_2—CH_2 + OH^- $\longrightarrow$ CH_2—CH_2OH

   CH_2 CH_2
   CH_2—CH_2 + OH^- $\longrightarrow$ $CH_2CH_2CH_2CH_2OH$

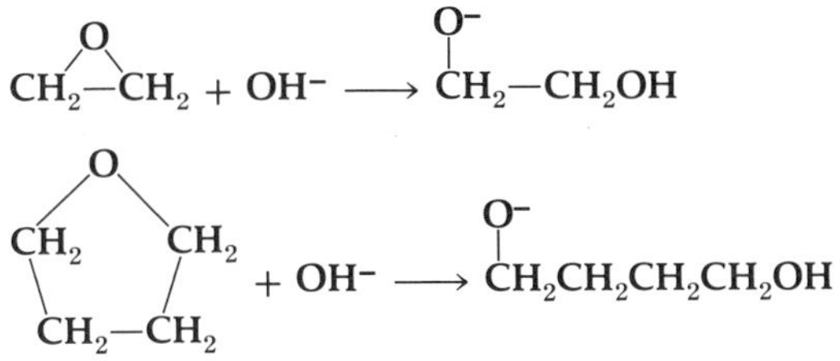

b CH_3Br + OH^- $\longrightarrow$ CH_3OH + Br^-

   $(CH_3)_3CBr$ + OH^- $\longrightarrow$ $(CH_3)_3COH$ + Br^-

c CH_2CH_2Br $\longrightarrow$ CH_2—CH_2 + Br^-

   $CH_2CH_2CH_2CH_2Br$ $\longrightarrow$ CH_2 CH_2 + Br^-
                                       CH_2—CH_2

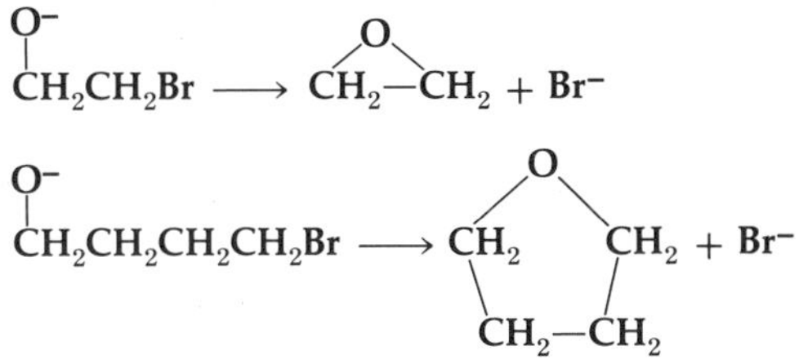

d CH_2=CH—$CHCH_3$ + H^+ $\longrightarrow$ CH_2=$CHCH$=CH_2 + H_2O

   $CH_3CH_2CH_2CH_2OH$ + H^+ $\longrightarrow$ CH_3CH_2CH=CH_2 + H_2O

9-16 In each pair of comparable equilibrium reactions below indicate which equilibrium will lie more to the right, and why.

a $CH_2CH_2NH_2$ $\rightleftharpoons$ CH_2CH_2NH + H_2O
    CH_2CH_2COOH CH_2CH_2CO

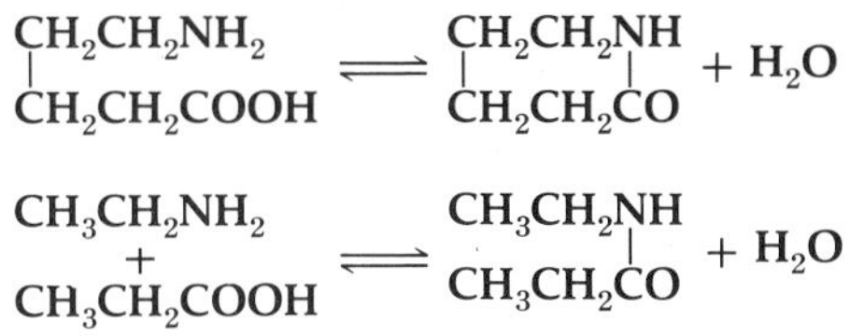

   $CH_3CH_2NH_2$
   + $\rightleftharpoons$ CH_3CH_2NH + H_2O
   CH_3CH_2COOH CH_3CH_2CO

b   —OH $\rightleftharpoons$ —O
    —CHO   OH

   —OH $\rightleftharpoons$ —O
    —CHO   OH

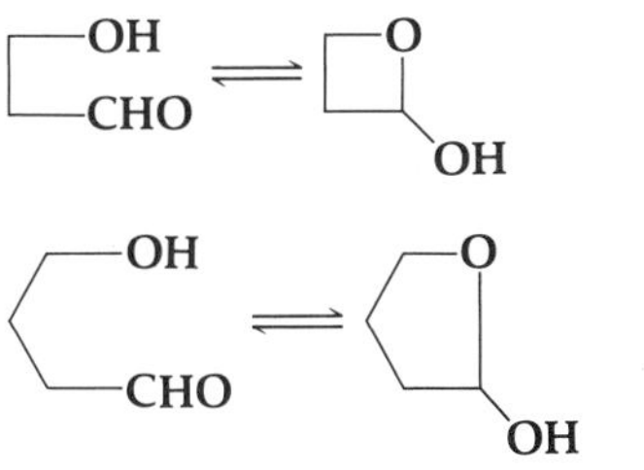

c

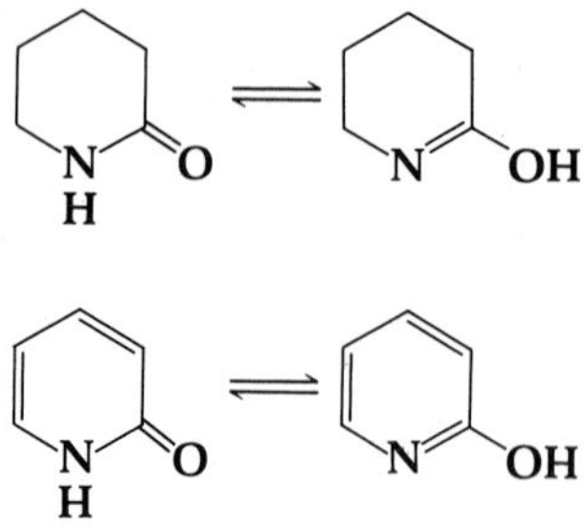

READING REFERENCES

Breslow, R., "Organic Reaction Mechanisms," W. A. Benjamin, New York, 1965.

NUCLEOPHILIC SUBSTITUTION AT SATURATED CARBON

IN the previous chapter we took our first look at the nature of chemical reactions in a general survey. Here we shall examine a particular reaction with the intent of observing the previous generalities in practice. The **nucleophilic substitution** reaction is an ideal selection, both because it has been studied deeply and because it is a broadly useful synthetic reaction capable of a wide variety of interconversions of many kinds of organic molecules. We shall approach the reaction with an effort to visualize events at the molecular level. We shall see again that the detailed nature of the reaction can easily be understood and predicted from the general steric and electronic nature of the involved molecules.

The reaction is simple, involving attack of a **nucleophile (Nu:)** at a saturated carbon bearing a substituent L which may be displaced by that nucleophile. The displaced substituent is called a **leaving group** (—L). In the overall transformation, the C—L bond is ruptured in such a way that the pair of electrons which compose the bond becomes associated with L, so that it leaves as :L. The nucleophile Nu:, which possesses an unshared pair of electrons, uses these to form a new bond to carbon. The general reaction and a specific example are shown in Fig. 10-1. The specific reaction of Fig. 10-1 can be described by the statement that a nucleophilic hydroxide ion attacks a methyl chloride molecule and displaces a chloride ion as the leaving group.

The attacking nucleophile and the leaving group possess the same general character: Whether anion or neutral they contain an unshared electron pair. Four charge types are therefore possible depending on whether they are anionic or neutral:

$$\text{Nu:}^- + \text{R—L} \longrightarrow \text{Nu—R} + \text{:L}^- \qquad (\text{CN}^- + \text{CH}_3\text{I} \longrightarrow \text{CH}_3\text{CN} + \text{I}^-)$$

$$\text{Nu:} + \text{R—L} \longrightarrow \text{Nu}^+\text{—R} + \text{:L}^- \qquad (\text{NH}_3 + \text{CH}_3\text{Br} \longrightarrow \text{CH}_3\text{NH}_3^+ + \text{Br}^-)$$

$$\text{Nu:}^- + \text{R—L}^+ \longrightarrow \text{Nu—R} + \text{:L} \qquad (\text{Cl}^- + \text{CH}_3\overset{+}{\text{O}}\text{H}_2 \longrightarrow \text{CH}_3\text{Cl} + \text{H}_2\text{O})$$

$$\text{Nu:} + \text{R—L}^+ \longrightarrow \text{Nu}^+\text{—R} + \text{:L} \qquad [(\text{CH}_3)_3\text{N} + (\text{CH}_3)_3\text{S}^+ \longrightarrow (\text{CH}_3)_4\text{N}^+ + (\text{CH}_3)_2\text{S}]$$

This variation in charge type is rarely of importance to the nature of the reaction, and the charges are ignored in the general cases discussed next. Most important nucleophiles are anions, the only important neutral ones being the amines.

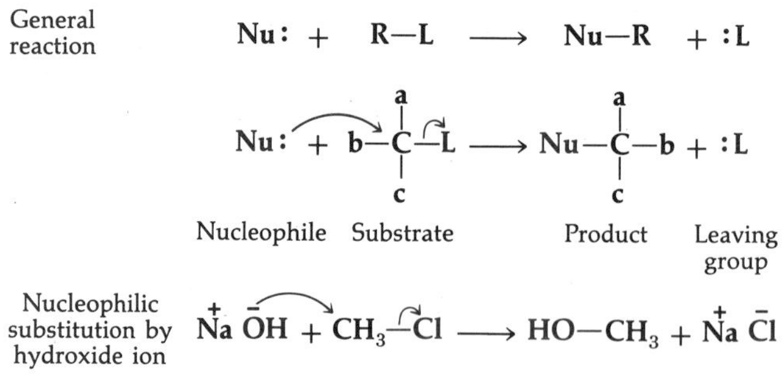

General
reaction

$$\text{Nu:} + \text{R—L} \longrightarrow \text{Nu—R} + \text{:L}$$

$$\text{Nu:} + b\text{—}\overset{a}{\underset{c}{\text{C}}}\text{—L} \longrightarrow \text{Nu—}\overset{a}{\underset{c}{\text{C}}}\text{—b} + \text{:L}$$

    Nucleophile Substrate   Product   Leaving
                                               group

Nucleophilic
substitution by $\text{Na}^+ \ \overset{-}{\text{OH}} + \text{CH}_3\text{—Cl} \longrightarrow \text{HO—CH}_3 + \overset{+}{\text{Na}} \ \overset{-}{\text{Cl}}$
hydroxide ion

FIGURE 10-1   **Nucleophilic substitution at saturated carbon**

        In the course of our discussion we shall examine the following features
of the general reaction:

1 The detailed mechanistic course of events and the observed kinetics of the reaction
2 The scope of substitution in terms of the possible nucleophiles and leaving groups
3 The stereochemical consequences of substitution
4 The relative effects on reactivity which result from variations in
   nucleophile
   leaving group
   the nature of the *carbon site* in the substrate
   the *solvent*
5 The influence of neighboring groups near the reactive site in the substrate
6 The possibility of competitive side reactions, other than substitution, which may occur
   under the reaction conditions.

Following this general clarification of the course of reaction, which will provide
a basis for prediction of results in particular cases, we can take up the practical
uses of the reaction in Chap. 11, showing for each of the major functional
groups with a single bond to carbon how they are created and how they react
in the substitution reaction.

10-1   THE MECHANISM AND SCOPE OF SUBSTITUTION

Available Mechanisms

        In a nucleophilic substitution, two changes occur: breaking of the old
bond and formation of the new. The principal mechanistic variations are associated with changes in the timing of the two processes. Three possibilities can
be imagined.

1 Two-step reaction: (*a*) break the old bond and (*b*) make the new bond.
2 Two-step reaction: (*a*) make the new bond and (*b*) break the old bond.
3 One-step reaction in which bond making and bond breaking are simultaneous or
   concerted.

A Prior ionization (S_N1):

(1) $R-X \xrightarrow{\text{Slow}} \overset{+}{R} + \overset{-}{X}$

(2) $\overset{-}{Nu}: + \overset{+}{R} \xrightarrow{\text{Fast}} Nu-R$

B One-step substitution (S_N2):

(3) $\overset{-}{Nu}: + R-X \longrightarrow [Nu \cdots R \cdots X]^- \longrightarrow Nu-R + \overset{-}{X}$

$\qquad\qquad\qquad\qquad\qquad\qquad\qquad$ Transition
$\qquad\qquad\qquad\qquad\qquad\qquad\qquad$ state

FIGURE 10-2   **Mechanisms of nucleophilic substitution reactions**

Both mechanisms 1 and 3 are observed in nucleophilic substitutions at saturated carbon atoms. To the best of our knowledge, 2 is not.

Saturated carbon has its outer electron shell full and so cannot make a new bond before it breaks one of the old. With trivalent boron, however, substitution can occur by addition of the nucleophile to the boron first, filling its outer shell (and affording boron a formal charge of -1), followed by breaking of the boron bond to the leaving group: $Nu: + BR_2L \longrightarrow Nu-\overset{-}{B}R_2L \longrightarrow Nu-BR_2 + :L$. With second-row atoms, which can expand the outer electron shell, formation of the bond to nucleophile can also precede loss of the leaving group.

In mechanism 1 the first, and rate-determining, step is the breaking of the bond to the leaving group which leaves the carbon site behind as a carbonium ion. This is in effect an ionization reaction, and is often effected by heating the compound in an appropriate solvent. In a second step the high-energy carbonium ion is attacked by a nucleophile to form the new bond. In many such reactions, the nucleophile is actually the solvent molecule (cf. $H_2\overset{..}{O}$, $R\overset{..}{O}H$, $RCO\overset{..}{O}H$), in which case the substitution reaction is called a **solvolysis**. The two steps are shown in Fig. 10-2*A*.

The speed of these ionization reactions depends upon the stability of the fragments formed—the carbonium ion and the displaced leaving group—as well as on the *ionizing power* of the solvent, which must stabilize the ionic fragments formed by solvation. In this mechanism the ionization is the step which involves the high-energy transition state and hence is rate-determining. The capture of a nucleophile by the high-energy carbonium ion thus formed is a rapid, facile process with low activation. These ideas are summarized in Fig. 10-3*A*. The implication of this prior ionization is that only the concentration of the ionizing substrate will affect the rate of the reaction. Hence the *rate will be independent of nucleophile concentration*. If two nucleophiles are present to compete for the carbonium ion, the *reaction rate* is still independent of their concentrations but the *product ratio* will reflect both their relative nucleophilic activity and their relative concentration. When no nucleophiles are present the intermediate carbonium ion can often be isolated, or at least spectroscopically observed in solution. The carbonium ion in such cases must be paired with a non-nucleophilic anion, as in $(CH_3)_3C^+BF_4^-$.

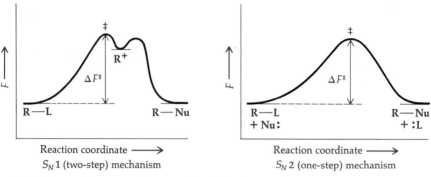

FIGURE 10-3 **Free-energy diagrams for substitution reactions**

In the one-step mechanism 3, by contrast, the nucleophile forces its way in toward the positive substrate carbon, simultaneously pushing out the leaving group (Fig. 10-2B). Hence both the nucleophile and substrate combine in the transition state, the reactivity of both being necessary to attaining the activation energy. The rate of reaction will therefore depend on the concentrations of both reactants. This one-step mechanism, dependent on two reactants, is diagrammed in Fig. 10-3B. This is a concerted mechanism. The nucleophile actively displaces the leaving group, and this mode of substitution is often called **displacement**.

Kinetic Characteristics

Solvolytic reactions of tertiary halides are usually not accelerated by addition of strong nucleophiles. For example, the rate of hydrolysis of *tert*-butyl chloride is not increased by the addition of sodium hydroxide, even though hydroxide ions are consumed in the reaction. Since the rate of the reaction depends only on the concentration of one reactant, the halide, the first-order rate law is followed.

$$NaOH + (CH_3)_3CCl \xrightarrow{\ H_2O,\ C_2H_5OH\ } (CH_3)_3COH + NaCl$$
$$Rate = k[(CH_3)_3CCl]$$

The **first-order** rate law is consistent with the carbonium-ion mechanism for substitution, since the first and rate-controlling step of the reaction does not involve the nucleophilic reagent. The name that has been given to this two-step mechanism is S_N1, which means **substitution, nucleophilic, unimolecular**.

The kinetic behavior of primary halides is very different from that of *tert*-butyl chloride and other tertiary halides. Ethyl chloride, for example, does not hydrolyze at an appreciable rate in neutral solution. However, if sodium hydroxide is added, reaction occurs at a rate that is proportional to the concentrations of both the halide and the base. Such a rate law is **second-order**.

$$NaOH + CH_3CH_2Cl \longrightarrow CH_3CH_2OH + NaCl$$
$$Rate = k[CH_3CH_2Cl][NaOH]$$

The one-step mechanism is consistent with the second-order rate law, although other mechanisms could give the same kinetics. The one-step process is called the S_N2 (**substitution, nucleophilic, bimolecular**) mechanism.

Scope of the Reaction

In order to survey which functional groups can undergo substitution reactions it is convenient to list nucleophiles and leaving groups according to the kind of atom which bonds to the saturated carbon, i.e.: H, C, N, O, X, S, (P). Any of these may act as nucleophiles, given an unshared electron pair. The main examples are listed in Table 10-1. In contrast, there are only a few active leaving groups (Table 10-2). The bonded atom of the leaving group must bear the pair of electrons when leaving (:L) and is virtually never hydrogen or carbon since these would be such high-energy products (:H⁻ and :CR₃⁻). Only in very special cases do nitrogen and phosphorus act as leaving groups. This implies that C—H, C—C, C—N and C—P bonds are almost never broken in substitution reactions, and this is in fact the observed situation. By far the

TABLE 10-1   **Nucleophiles for Nucleophilic Substitution Reactions**
(Nu : + R—L $\longrightarrow$ R—Nu + :L)

		Nucleophile	*Product*
1	*Halogen*	Cl⁻	R—Cl (alkyl chlorides)
		Br⁻	R—Br (alkyl bromides)
		I⁻	R—I (alkyl iodides)
2	*Oxygen*	H₂O, OH⁻	R—OH (alcohols)
		R'OH, R'O⁻	R—OR' (ethers)
		R'COŌH, R'COO⁻	R—O—C(=O)—R' (esters)
		⟩C=C⟨ O⁻ ↓ ⟩C—C⟨ =O	⟩C=C⟨ O—R (enol ethers)† or —C(R)—C(=O)— (ketones)†
		HONO₂, NO₃⁻	R—O—NO₂ (alkyl nitrates)
3	*Sulfur*	H₂S, SH⁻	R—SH (thiols, mercaptans)
		R'SH, R'S⁻	R—SR' (thioethers, sulfides)
		R'₂S	R—S⁺R'₂ (sulfonium ions)

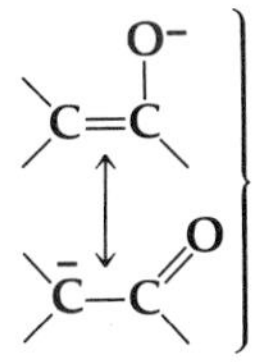

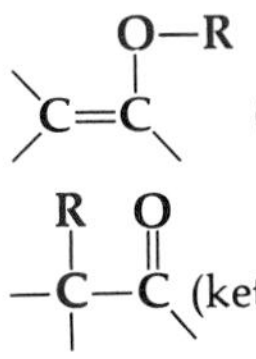

TABLE 10-1   **Nucleophiles for Nucleophilic Substitution Reactions**
(Nu: + R—L $\longrightarrow$ R—Nu + :L) (*Continued*)

	Nucleophile	Product
4 *Nitrogen*	NH_3	R—NH_2 (primary amines)
	R'NH_2	R—NHR' (secondary amines)
	R'$_2$NH	R—NR'$_2$ (tertiary amines)
	R'$_3$N	R—$\overset{+}{N}$R'$_3$ (quarternary ammonium ions)
	NH_2NH_2	R—NHNH$_2$ (alkyl hydrazines)
	NO_2^-	R—NO_2 (nitroalkanes)
	N_3^-	R—N=N=N (alkyl azides)

$$\underset{\displaystyle R\!-\!\overset{\displaystyle O}{\overset{\displaystyle \|}{C}}\!-\!\overset{\displaystyle \cdot\cdot}{N}\!-\!R'}{}\qquad \underset{\displaystyle R'\!-\!\overset{O}{\overset{\|}{C}}\!-\!\underset{\underset{R}{|}}{N}\!-\!R'\ (N\text{-alkyl amides})}{}$$

5 *Phosphorus*	R'$_3$P:	R—$\overset{+}{P}$R'$_3$ (alkyl phosphonium ions)
6 *Carbon†*	R'$_3$C:$^-$	R—CR'$_3$ (hydrocarbons)
	R'C≡C:$^-$	R—C≡C—R (alkynes)
	CN^-	R—C≡N (nitriles)

$$\underset{\displaystyle -\overset{O}{\overset{\|}{C}}\!-\!\overset{\cdot\cdot}{C}\!-\!\overset{O}{\overset{\|}{C}}-}{|}\qquad \underset{\displaystyle -\overset{O}{\overset{\|}{C}}\!-\!\underset{\underset{|}{|}}{\overset{R}{C}}\!-\!\overset{O}{\overset{\|}{C}}-\ (\text{diketones})\dagger}{}$$

7 *Hydrogen*	$LiAlH_4$	R—H (hydrocarbons)

† Carbon nucleophiles are all necessarily carbanions, of all possible kinds. Only the common examples are shown, the β-diketones (and related β-dicarbonyls) being especially common. These ketonic carbanions are enolate anions and can react at oxygen or carbon, forming enol ethers or ketones, respectively.

most common leaving groups are halides and sulfonates (S_N1 or S_N2), and protonated alcohols or ethers (S_N1 only).

Many reactions can be written simply by combining the two lists, although some of the conceivable combinations do not give experimentally feasible processes. The reactivity of the reagents shown in Table 10-1 varies enormously, and some "weak" nucleophiles will not displace any but the most labile leaving groups. Furthermore, the coexistence of some pairs in the same medium is not possible. For example, the first neutral leaving group in Table 10-2 is H_2O. In order to obtain appreciable amounts of the corresponding substrate species, R—OH_2^+, an alcohol must be brought in contact with strong acid, a condition that would destroy many of the nucleophiles listed in Table 10-1 owing to their protonation by the fast acid-base reaction in an acidic medium.

$$R-\overset{..}{\underset{..}{O}}-H + \overset{+}{H} \rightleftharpoons R\overset{+}{\underset{..}{O}}H_2 \quad \left.\begin{array}{l} \\ \\ \end{array}\right\} \begin{array}{l} \text{Acid-base} \\ \text{reactions} \end{array}$$

$$\underset{\substack{\text{Nucleophile}}}{Nu: + \overset{+}{H}} \rightleftharpoons \underset{\substack{\text{Conjugate acid} \\ \text{of nucleophile}}}{\overset{+}{Nu}-H}$$

In general the S_N1 reaction involves nucleophiles of low reactivity since prior carbonium-ion formation is the high-energy step. As implied above these are often substitutions *under acid conditions* (cf. $R'OH + R-X \longrightarrow R'O-R + HX$). By contrast, S_N2 reactions require reactive nucleophiles and these are commonly strong bases (OH^-, $R'_3N:$, $R:^-$) so that the conditions are usually *basic* for this reaction.

TABLE 10-2   **Leaving Groups in Nucleophilic Substitution**

Leaving group (:L)	Substrate (R—L)	Leaving group (:L)	Substrate (R—L)
Cl^-	Alkyl chloride, $R-Cl$	$\bar{O}SO_3H$	Alkyl hydrogen sulfate, $R-OSO_3H$
Br^-	Alkyl bromide, $R-Br$		
I^-	Alkyl iodide, $R-I$	$\bar{O}SO_2R'$	Alkyl alkanesulfonate, $R-OSO_2R'$
H_2O	Alcohol, conjugate acid, $R-\overset{+}{O}H_2$	$\bar{O}SO_3R'$	Alkyl sulfate, $R-OSO_3R'$
ROH	Ether, conjugate acid, $R-\overset{+}{\underset{H}{O}}R$	$\bar{O}SOCl$	Alkyl chlorosulfite, $R-OSOCl$
		$\bar{O}PCl_2$	Alkyl chlorophosphite, $R-OPCl_2$
$\bar{O}-\overset{\overset{\text{O}}{\|\|}}{C}R'$	Ester, $R-O\overset{\overset{\text{O}}{\|\|}}{C}R'$	$\bar{O}PBr_2$	Alkyl bromophosphite, $R-OPBr_2$
$HO-\overset{\overset{\text{O}}{\|\|}}{C}R'$	Ester, conjugate acid, $R-\overset{+}{\underset{H}{O}}-\overset{\overset{\text{O}}{\|\|}}{C}R'$	$N\equiv N$	Alkanediazonium ion, $R-\overset{+}{N}\equiv N$
		SR'_2	Trialkylsulfonium ion, $R-\overset{+}{S}R'_2$

PROBLEM 10-1

   a On the basis of Fig. 9-7 assess the various rates in these S_N1 reactions.

$$Z-CH_2OH + HBr \longrightarrow Z-CH_2Br \quad Z = CH_3, CH_3O, CH_3CO, C_6H_5$$

Substitution with inversion of configuration

Substitution with retention of configuration

Substitution with racemization

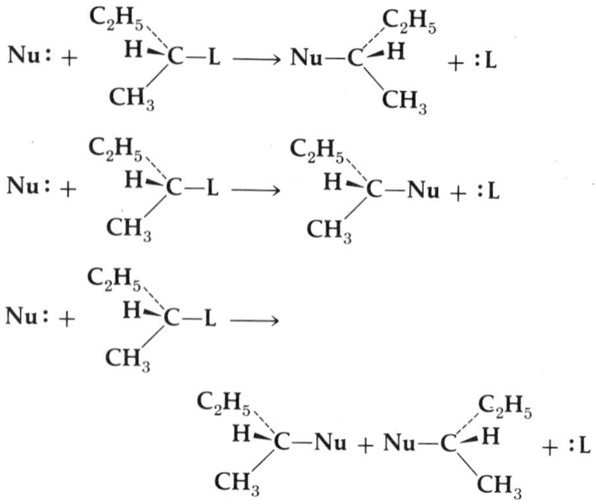

FIGURE 10-4   Possible stereochemical consequences of nucleophilic substitutions

b Draw an energy diagram to illustrate the rate variations.

c Do the same for the rates at which these compounds liberate **HCl** in boiling acetic acid (S_N1 reaction).

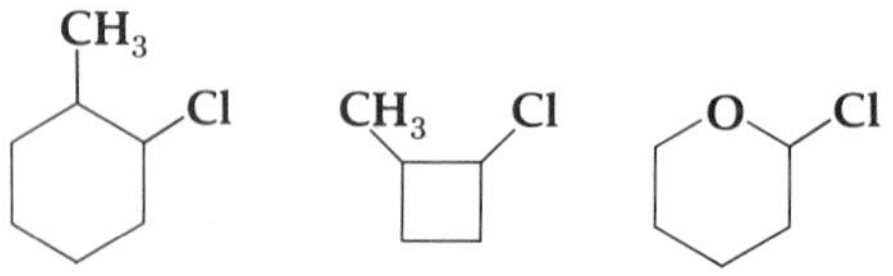

10-2   STEREOCHEMISTRY

Study of the stereochemical course of substitutions at asymmetric carbon has provided a clear picture of the mechanisms of the reactions. In Fig. 10-4 various stereochemical possibilities are illustrated with an optically active *sec*-butyl group. The asymmetric carbon can either retain its configuration in the product or become inverted, and this can be experimentally verified with a polarimeter (Sec. 6-8). If half the product is inverted and half retained in configuration, the optical result is racemization ($[\alpha] = 0°$).†

Stereochemistry of S_N1 Reactions

Consider an asymmetric halide molecule such as that in Fig. 10-5. During ionization the chloride ion pulls away from the asymmetric carbon. The pyramidal arrangement of the other three substituents on that sp^3 carbon atom flattens until it reaches the planarity of the final sp^2 carbonium ion. The two unequal lobes of the sp^3 orbital on the central carbon, which had held the

---

†If other uninvolved asymmetric centers are present in the molecule the result will be partial epimerization, not racemization.

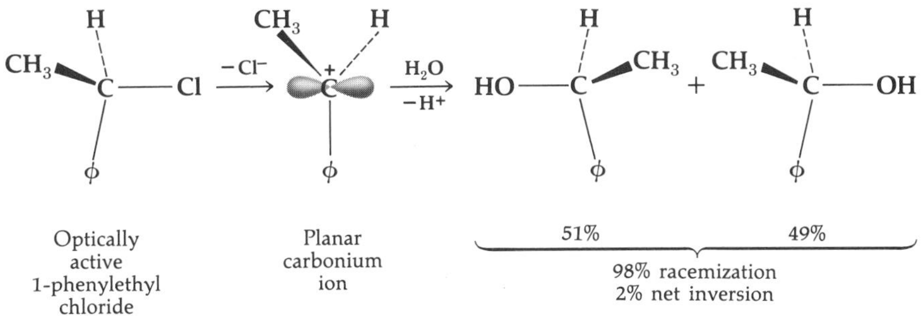

FIGURE 10-5   Hydrolysis of 1-phenylethyl chloride

chloride, smoothly shift to the two equal lobes of the empty p orbital of the carbonium ion, on each side of the molecular plane and perpendicular to it. Asymmetry is lost. The carbonium ion formed is now flat, sp^2, with a plane of symmetry. Into either of the empty p-orbital lobes comes the electron-pair orbital of the nucleophile (H_2O: in Fig. 10-5), overlapping to form the product bond. Since either side is equally accessible owing to the symmetry, both inversion and retention of the original configuration are equally likely, leading to net racemization.

The carbonium ion is a very high-energy species and can only form if its energy is dissipated by **solvation**, i.e., by complexing with a solvent molecule. The solvent molecule moves up below the rear, small lobe of the sp^3 orbital on carbon as it develops into an empty p orbital and loosely complexes it with an electron pair, not forming a true bond. As the leaving group moves away another solvent molecule can come in to replace it on top in the same way. If the ion is stable enough it will have time to become *symmetrically solvated* in this way before going on to bond formation, either with solvent or another nucleophile. This gives true racemization. In many cases some solvent forms a bond opposite the departing leaving group before symmetrical solvation can occur, leading to mostly inverted product. The more stable (and long-lived)

FIGURE 10-6   Methanolysis of 1-phenylethyl bromide

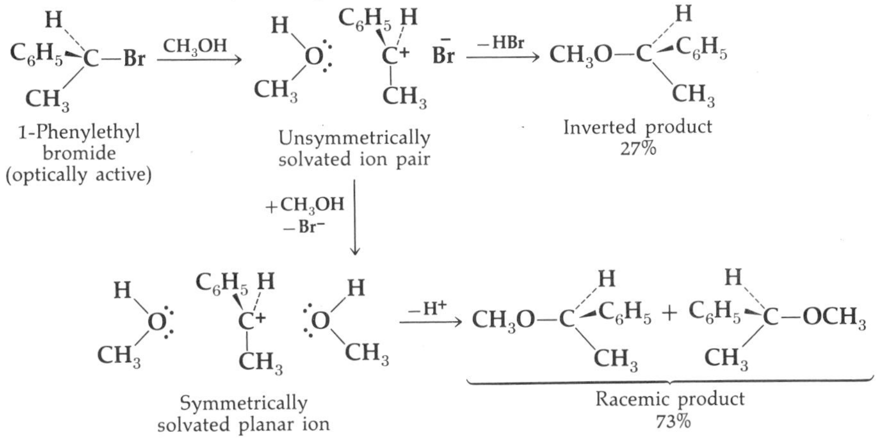

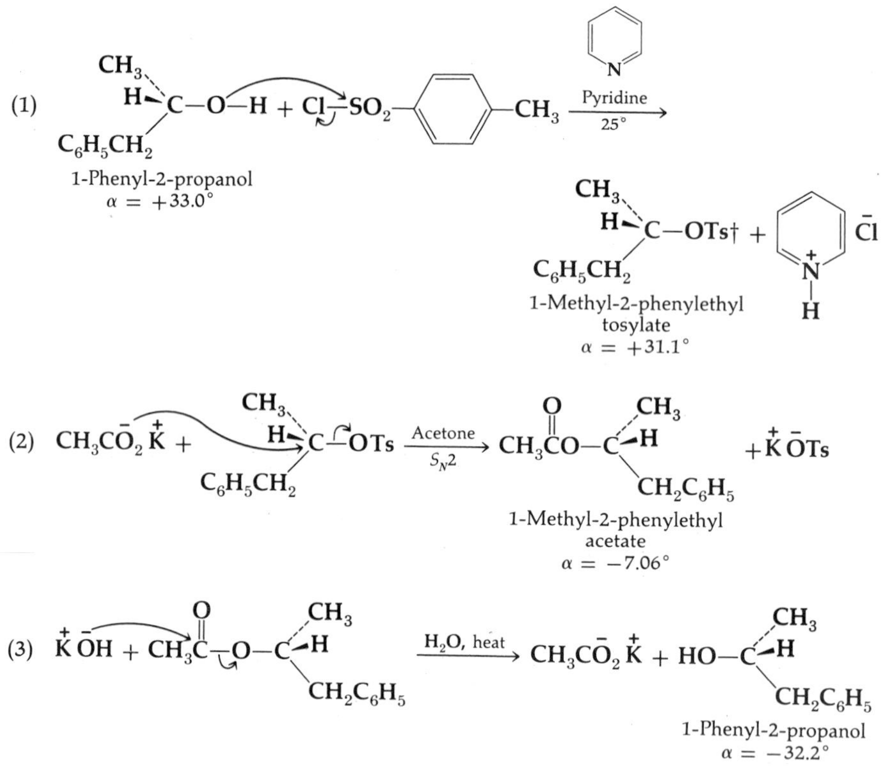

(1) 1-Phenyl-2-propanol $\alpha = +33.0°$

1-Methyl-2-phenylethyl tosylate $\alpha = +31.1°$

(2) 1-Methyl-2-phenylethyl acetate $\alpha = -7.06°$

(3) 1-Phenyl-2-propanol $\alpha = -32.2°$

$\dagger$—OTs $= -OSO_2C_6H_4CH_3$-p.

FIGURE 10-7   Demonstration of the steric course of an S_N2 reaction

the carbonium ion the more complete the racemization will be. The extent of racemization is also dependent on the leaving group, as seen in the imperfect racemization in the case of Fig. 10-6. Thus the steric course of the S_N1 reaction is largely racemization in most cases.

Stereochemistry of S_N2 Reactions

The bimolecular nucleophilic substitution (S_N2) reaction occurs with clear-cut inversion of configuration. This fact was established by study of the stereo-chemistry and kinetics of substitution reactions of sulfonate esters. The equations in Fig. 10-7 indicate a classic set of interlinked reactions that establish the relative configurations of reactants and products in a second-order substitution. The alcohol is first converted to a tosylate (p-toluenesulfonate) by a reaction† which does not disturb the asymmetric C—O bond of interest. The second reaction is the S_N2 displacement, one oxygen nucleophile (acetate anion) displacing the other (tosylate anion). Finally, the acetate product is hydrolyzed† to yield an alcohol, again without disruption of the asymmetric C—O bond.

†The first and third reaction of Fig. 10-7 and a few in subsequent pages are introduced without comment since they are not nucleophilic substitutions at saturated carbon. They are all discussed in context in subsequent chapters and their premature introduction here is held to the very minimum required.

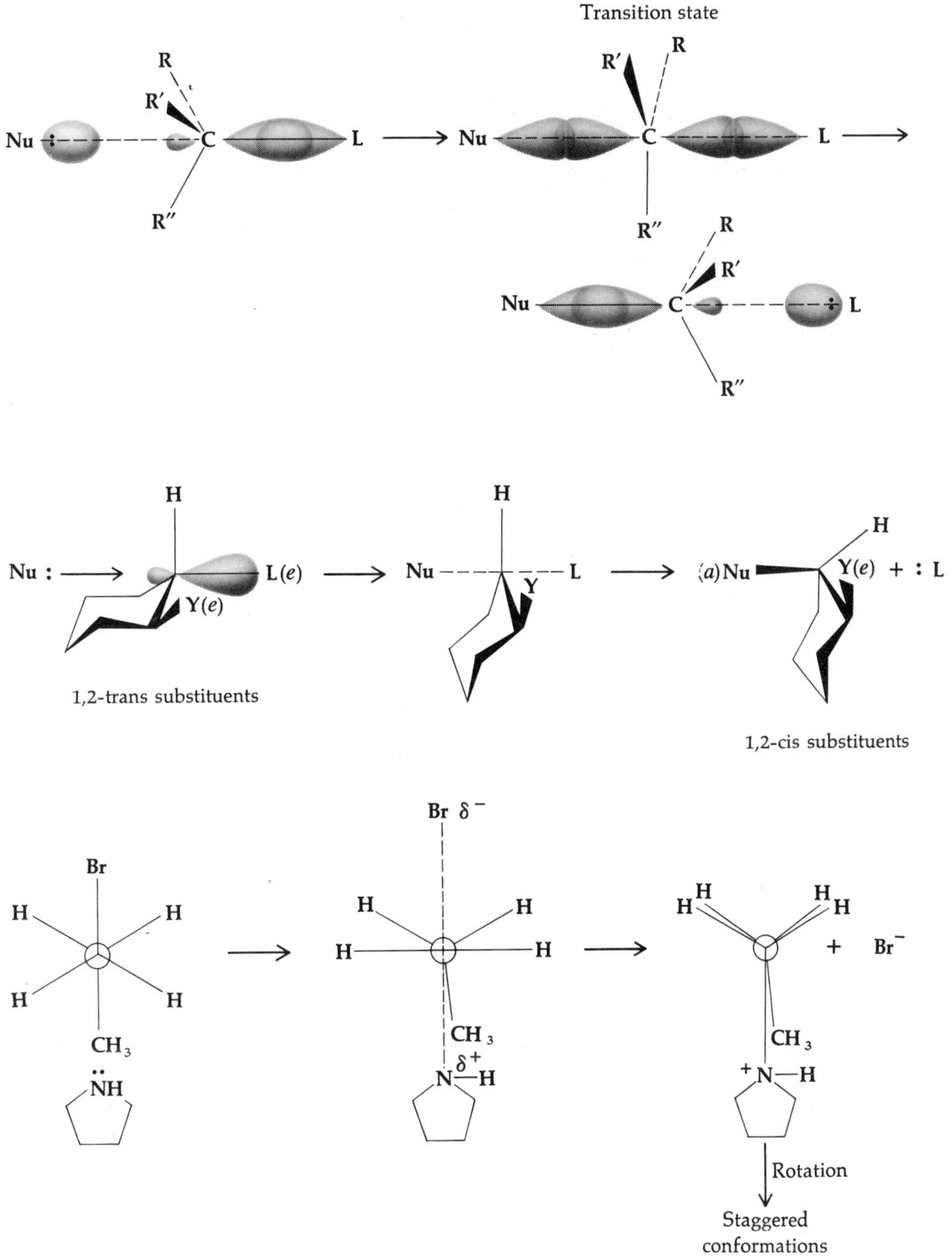

FIGURE 10-8 Mechanism of S_N2 inversion

The key to the demonstration lies in the equal and opposite optical rotation of the initial and final alcohols. Since only the middle, S_N2, reaction involves bonds to the asymmetric carbon, this must have proceeded with complete inversion.

These stereochemical results, observed in many cases, are viewed at the molecular level (Fig. 10-8) as an approach of the nucleophile, its electron-pair orbital foremost, on a line with the axis of the orbital bearing the leaving group and from the backside. As the nucleophile approaches, the leaving group begins to be pushed away and the small lobe of the central carbon orbital holding it grows to meet the incoming nucleophile and shrinks away from the leaving group. The three groups on the central carbon are pushed out into a plane with the central atom perpendicular to the Nu—C—L bond axis in the transition state. The transition state is assumed to be a half-way point wherein the nucleophile and leaving group are each about half-bonded to the central carbon. Overall, the S_N2 mechanism, or **backside displacement**, causes the three attached groups at the central carbon to *invert*, rather like turning an umbrella inside out in a high wind. This can also be seen in the illustration of inversion on a cyclohexane substituent, or in the projection of *n*-propyl bromide displacement, both in Fig. 10-8.

Many similarities between the S_N2 and S_N1 mechanisms can be seen. The geometry of the half-way point in the S_N2 reaction is essentially the same as that of the symmetrically solvated carbonium ion in Fig. 10-6. Indeed the 27% of observed inversion in that case proceeds by a mechanism that is geometrically identical to the S_N2 mechanism. The difference in the two mechanisms is in fact only one of the timing of bond breaking and making. The S_N1 reaction has two steps: passage over the first transition state involves in the main only bond-breaking; and the new bond to nucleophile is made in a discrete second step. In the S_N2 reaction, the two processes are simultaneous, or concerted. It is also possible to envision free-energy diagrams intermediate between the two in Fig. 10-3, so that intermediate mechanisms are possible. This possibility sometimes makes interpretation of borderline cases complex.

PROBLEM 10-2

a What is the expected product (or products), in stereochemical terms, from R-1-phenyl-1-bromobutane in boiling acetic acid? In acetone solution with sodium acetate?

b What is the expectation in similar cases using R-1-phenyl-1-bromo-s-2-methylbutane?

c Similarly examine the tosylates of optically active *cis-* and *trans*-4-phenyl-cyclohexanols.

PROBLEM 10-3

Show with projection formulas the course of S_N1 and S_N2 reactions of 2-bromopentane with azide ion. Draw the intermediate or transition state in this form also.

PROBLEM 10-4

    a The optical rotation of a solution of (+)-2-phenyl-2-pentanol goes to 0° when boiled in formic acid. Explain.

    b The optical rotation of a solution of sodium bromide and (+)-2-bromo-pentane in acetone also goes slowly to 0°. Is there a different explanation?

10-3 RELATIVE REACTIVITY IN SUBSTITUTION

In order to achieve the ultimate goal of quantitative prediction in organic chemistry we should be able to say at what rate substitution will proceed when a certain nucleophile and substrate are allowed to react in a given solvent. The variations in rate observed for each mechanism (S_N1 and S_N2) are enormous and arise from four main effects:

1 *Changes in medium (solvent)*
2 *Substrate structure*
3 *Nature of the nucleophile*
4 *Nature of the leaving group*

    None of these are completely independent, but even if they were, a complete study of the rates of perhaps fifty representative substrates with solvents, leaving groups, and nucleophiles all varied independently would require literally millions of separate quantitative studies of rates and products.

    While this research is far from complete, most of the trends are clear in a semiquantitative way. Tables of relative rates are shown in order to impress a general sense of the *magnitudes* of the variations, but it must be appreciated that these magnitudes often change dramatically with changes in others of the variables: solvent, leaving group, etc. Still the *order* of relative activities is usually consistent even if the numbers change. The data offered should be approached first to see the qualitative trends and how they follow quite reasonably from our previous understanding of structural features, and second to appreciate that several interlocking effects can create large variations in rates with complex rationales. The latter serve to instill caution in making predictions of a quantitative kind.

Solvent Effects

    The nature of the medium in which a heterolytic reaction is conducted often has a profound influence on the reaction rate. The largest medium effects can be anticipated by considering whether ions are formed or destroyed, or neither, in the rate-determining step of the reaction. Good **ionizing solvents** are those in which ions are stabilized by solvation. Reactions which *produce ions* from neutral species are *accelerated* by good ionizing solvents (cf. Fig. 9-7B). Those in which ions collapse to neutral molecules actually go more slowly in ionizing than in nonpolar solvents (Figs. 10-9 and 9-7D).

Mechanism	Reaction	Effect of increasing solvent ionizing power
S_N1	$R{-}Cl \longrightarrow \overset{+}{R} + \bar{Cl}$	Large acceleration
S_N2	$H\bar{O} + R{-}Cl \longrightarrow ROH + \bar{Cl}$	Small effects
S_N2	$R_3N + R'{-}Cl \longrightarrow R_3\overset{+}{N}R' + \bar{Cl}$	Large acceleration
S_N2	$HO^- + R_3S^+ \longrightarrow ROH + R_2S$	Deceleration

FIGURE 10-9   **Medium effects on the rates of nucleophilic substitution reactions**

The most useful tabulation of general solvent activity is probably that of **dielectric constant** (Table 10-3), the more polar, better ionizing solvents having higher dielectric constants. However, the effects of solvents are more particular to their structures and solvation function than such a list of simple numbers can convey. The two particularly important kinds of solvation are the solvation of electron-deficient species (notably cations) by coordination with electron pairs on the solvent, and solvation of anions (or electron-rich atoms) by hydrogen bonding.

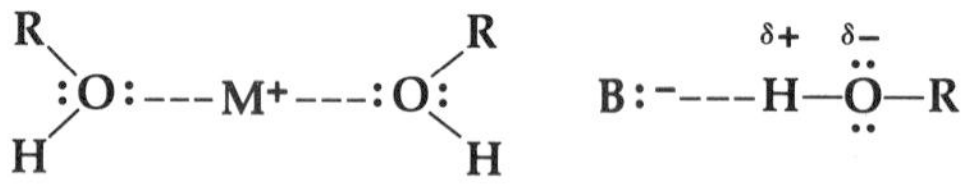

Cation solvation   Anion solvation

Hydrogen bonding by solvent is very important in substitution reactions because of its stabilization of nucleophiles. When a nucleophile reacts, part of the energy required to reach the transition state must be used to break such a hydrogen bond (Fig. 9-7C) and thus solvents which cannot hydrogen-bond to nucleophiles will allow faster rates (lower $\Delta F\ddagger$), often thousands of times faster. Hydrogen-bonding solvents are called **protic** and those with no proton for hydrogen bonding are **aprotic**; Table 10-3 distinguishes the two groups. S_N2 reactions are especially favored in aprotic solvents.

TABLE 10-3   **Polarity and Dielectric Constants of Common Solvents**

	Protic	Dielectric constant	Aprotic
Polar	H_2O	81	
	HCOOH	59	
		45	$(CH_3)_2\overset{+}{S}{-}\bar{O}$
		38	$HCON(CH_3)_2$; CH_3CN
	CH_3OH	33	
	ROH	20–24	CH_3COCH_3
Nonpolar	CH_3COOH	<6	$(C_2H_5)_2O$; $CHCl_3$; hydrocarbons

(Increasing ionizing power)

Variations at the Carbon Site

In the S_N1 mechanism the substrate ionizes to a high-energy carbonium ion. Since the transition state will be similar in energy and structure to that ion, the assessment of effects on the carbonium ion will serve to approximate effects on the transition state. These effects are listed in Fig. 9-7. Resonance stabilization of the carbonium ion lowers $\Delta F\ddagger$ (Fig. 9-7B) and is the most dramatic effect in solvolysis, causing reactions to proceed thousands or even millions of times faster than comparable saturated cases. The principal examples in order of decreasing stabilization (i.e., rate acceleration) are these.

Ether:

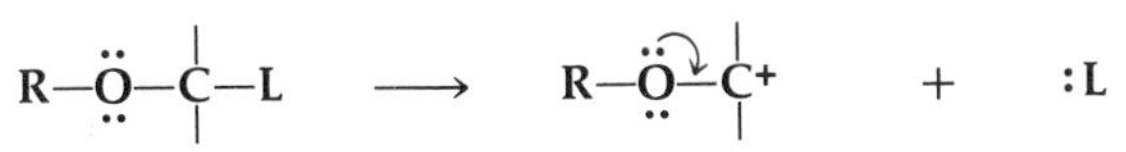

Benzyl:

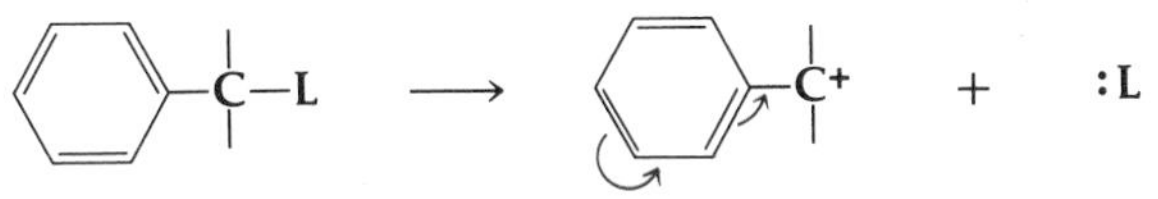

Allyl:

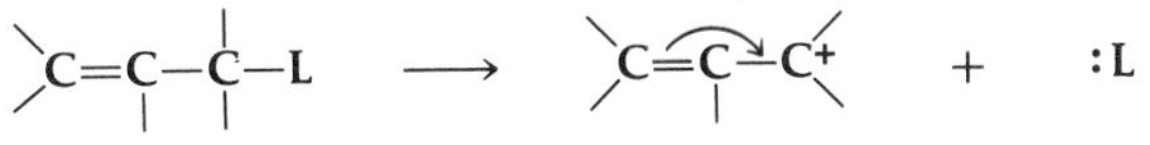

Propargyl:

Particular cases are cited in Table 10-4. The operation of resonance stabilization is also clear in the substitution effects on the benzene ring in benzylic cases: while a meta methoxyl has little effect, a para methoxyl will enhance the S_N1 rate 5,000 to 10,000 times and a para nitro group will reduce it almost as much.

TABLE 10-4 **Relative Solvolysis (S_N1) Rates for Different Substrates**

R—Cl + H_2O/dioxan		R—Br + H_2O		R—Cl + C_2H_5OH	
		CH_3-	1		
$CH_3CH_2CH_2CH_2-$	1	CH_3CH_2-	1	$CH_2=CH-CH_2-$	0.04
$CH_3CH_2\ddot{O}CH_2-$	1×10^9	$(CH_3)_2CH-$	12	$\phi-CH_2-$	0.08
				$\phi-\underset{CH_3}{\overset{\mid}{CH}}-$	1
$CH_3\ddot{O}CH_2CH_2-$	0.2	$(CH_3)_3C-$	1.2×10^6	$(CH_3)_3C-$	1
				ϕ_2CH-	300
				ϕ_3C-	3×10^6

In saturated substrates there is, however, also a dramatic difference in the order:

tert $\gg$ *sec* $>$ *pri* $>$ CH_3 S_N1 reactions

Several causes for this have been advanced:

1 Hyperconjugation (page 166)
2 An inductive effect owing to the greater electronegativity of **H** than **C**, which stabilizes a carbon-substituted carbonium ion over a hydrogen-substituted one
3 *Steric acceleration,* in which some steric strain of three groups separated by a 109.5° angle (sp^3) is released as they go to the 120° separation of the sp^2 carbonium ion.

All three are probably valid but the extent of each is not known. Nevertheless, the rate acceleration is very real, as shown in Table 10-4 (middle list). As the table shows, it is common in S_N1 reactions for two methyl groups to produce a rate equivalent to one phenyl when substituted on the carbonium ion site.

The S_N2 displacement is less affected by electrical effects (resonance, inductive) since there is relatively little charge on carbon in the transition state. It is very subject to *steric hindrance* around the carbon site, however, for the nucleophile must push in among the three carbon substituents to effect successful displacement. Here, therefore, the reverse order of rates holds for saturated substrates:

CH_3 $>$ *pri* $>$ *sec* $\gg$ (*tert*) S_N2 reactions

No clear S_N2 displacement has been demonstrated for a tertiary substrate, since when forcing conditions are used (e.g., raising temperature) ionization of the substrate occurs instead, affording either S_N1 substitution or other reactions of the carbonium ion (Sec. 10-5). Steric hindrance is also clear in the great sluggishness of neopentyl [$(CH_3)_3CCH_2$—] derivatives (Table 10-5), even though their site of reaction is a primary one.

TABLE 10-5 **Relative Displacement Rates of Halides (S_N2)**

CH_3—	30
CH_3CH_2—	1
RCH_2CH_2—	0.4
$(CH_3)_2CH$—	0.02
$(CH_3)_3C$—	~0
$(CH_3)_3C$—CH_2—	0.00001
Neopentyl	
$CH_2{=}CH{-}CH_2$—	40
ϕ—CH_2—	120
CH_3COOCH_2—	100
CH_3OCH_2—	400

Resonance does apparently play some part even in S_N2 reactions but not nearly so spectacularly as in S_N1. The S_N2 transition state is actually an sp^2 hybrid of the carbon site and although the nucleophile and leaving group largely occupy the p lobe with "half-bonds" there is apparently some role for resonance stabilization by adjacent unsaturation since allylic, benzylic, and ether examples do show rates increased by 50 to 500 times (Table 10-5).

Other cases of interest include carbonyls α to the reactive carbon in the substrate. In the S_N1 reaction these inhibit the rate by placing the positive carbon of the carbonyl adjacent to the carbonium ion and so raising its energy (Fig. 9-7D). In the S_N2 reaction, however, α-halocarbonyls are often the most reactive substrates, apparently since the entering nucleophile offers some of its charge to the carbonyl as well as to the adjacent reaction site (cf., Fig. 9-1):

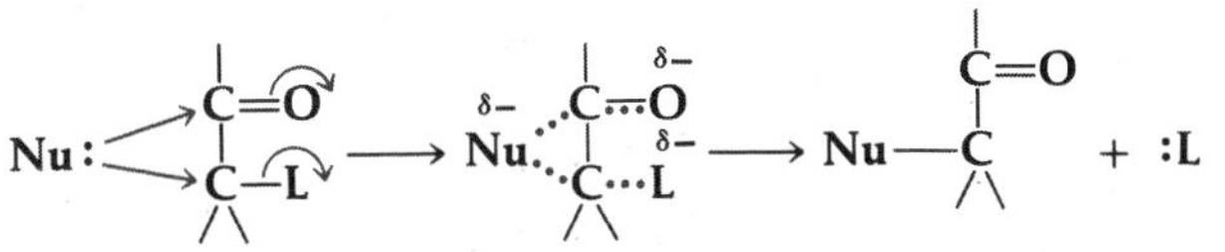

$$\text{Nu:} \overset{\displaystyle|}{\underset{\displaystyle|}{\overset{C=O}{\underset{C-L}{\bigwedge}}}} \longrightarrow \text{Nu:} \overset{\delta-}{\underset{\delta-}{\overset{\overset{\displaystyle|}{C \cdots O}}{\underset{\overset{\displaystyle|}{C \cdots L}}{}}}} \longrightarrow \text{Nu}-\overset{\overset{\displaystyle|}{C=O}}{\underset{\displaystyle\bigwedge}{C}} \quad + \; \text{:L}$$

α-Haloketones are even faster than the α-haloesters of Table 10-5. Finally, it may be added that multiple halogens on one carbon stimulate the S_N1 reaction (10 to 50 times per X) but depress the S_N2 reaction somewhat.

PROBLEM 10-5

Two common tests for the primary, secondary, or tertiary nature of an unknown alkyl bromide or chloride are to allow it to react with $AgNO_3/C_2H_5OH$ and with NaI/acetone. (NaI is soluble in acetone while NaBr and NaCl are not.) In each case one observes whether a precipitate forms rapidly, slowly, or not at all.

a Explain the rationale for this test based on the premise that silver ion complexes with halides, causing them to ionize with precipitation of AgX.

b Predict the expected behavior of each of the following with each test.

C_2H_5Br $(CH_3)_2CHBr$ $(CH_3)_3CCl$

c What behavior would you expect in each test with the following halides: cyclohexyl iodide, allyl chloride, benzyl bromide, triphenylmethyl chloride.

PROBLEM 10-6

In each pair of similar substitution reactions below write the structures of the products of each one; indicate which reaction is likely to have the faster rate and why.

a Phenylmethyl chloride (= benzyl chloride) and 2-phenylethyl chloride with silver acetate in methanol

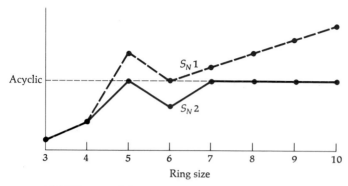

FIGURE 10-10   **Substitution rates of cycloalkyl derivatives**

b Sodium cyanide in acetone with 1-methyl-1-iodomethyl-cyclopentane or 2-cyclopentylethyl iodide

c 2-phenyl-2-propanol or 3-phenyl-2,4-dimethyl-3-pentanol on warming in concentrated **HBr**

d Sodium salt of methyl malonate and ethyl iodide in methanol or in acetonitrile (CH_3CN)

e Lithium aluminum hydride on $CH_3-\overset{\displaystyle O}{CH}-CH_2$ and $CH_3-\overset{\displaystyle O}{CH}-CH-CH_3$

f Heating 2-chloro-3-pentene and 2-chloro-2-pentene in formic acid

Ring compounds exhibit expected characteristics. Since in both S_N1 and S_N2 reactions the angles at the reactive carbon must open to 120° ($sp^3 \longrightarrow sp^2$), the process creates severe angle strain in the small rings and lowers their rates of substitution in the order of ring size: $5 > 4 > 3$. Cyclopentyl derivatives, by a combination of strain relief factors in going to the transition state, are the fastest rings for substitution, faster than comparable acyclic derivatives (5 to 50 times), while cyclohexane is comparable to acyclic derivatives for S_N1 but slower (100 times) in S_N2 owing largely to steric hindrance by the axial hydrogens. In larger rings the S_N2 reaction is similar to that in acyclic cases, but the S_N1 reaction is enhanced ($\longrightarrow$ 500 times) in the strained medium rings (7 to 10 members) owing to relief of steric strain in the parent ring (Fig. 9-7C) as one carbon goes planar and so removes its axial hydrogen from steric compression. Substitution rates in cycloalkyl cases:

$3 < 4 < 5 > 6 < 7\text{--}10$ (see Fig. 10-10)

In bridged bicyclic compounds, a leaving group at the **bridgehead** is totally inaccessible to backside attack for S_N2 displacement and almost equally unreactive in S_N1 since ionization to an sp^2 carbonium ion (planar) requires severe angle-bending strain throughout the molecule.

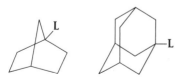

Relative Reactivity of Nucleophiles

All nucleophiles are bases, and the major trend in **nucleophilicity** (reactivity as nucleophiles) is to *parallel base strength*. Differences occur since the electron pair attacks carbon instead of hydrogen. Comparison of nucleophiles *having the same attacking atom* does show that within such a restricted series increased basicity results in increased nucleophilic reactivity. Thus, the following series of decreasing reactivities is observed and parallels decreasing pK_a (Table 8-2):

$$C_2H_5O^- > HO^- > C_6H_5O^- > CH_3CO_2^- > H_2O$$

However, if one stays within one family in the periodic table, reactivity of nucleophiles in series of analogous ions or molecules increases sharply with increasing atomic number, at least in protic solvents. Thus, the following series of decreasing reactivities is usually observed, in protic solvents:

$$I^- > Br^- > Cl^- \gg F^- \quad \text{and} \quad RS^- > RO^-$$

In the larger atoms the external electrons are less tightly bound by the nucleus and thus more **polarizable**, i.e., free to distort to accommodate the electron-deficient carbon in the S_N2 transition state. Finally, solvation is important to nucleophilicity, protic solvents depressing the rates through their relative deactivation of the electron pair in the nucleophile by hydrogen bonding.

Thus, basicity, polarizability, and solvation are all involved in nucleophile reactivity and no quantitative relationship among them has yet been successfully formulated. The list in Table 10-6 is from older studies in protic solvents (alcohols, aqueous acetone, etc.) on the S_N2 reaction. It is now known that rates are thousands of times faster in polar aprotic solvents such as dimethylformamide (Table 10-3). It is also becoming clear that even the *order* of nucleo-

TABLE 10-6   **Average Nucleophilicities in Protic Solvents (Relative Rates for S_N2)**

$\phi-S^-$	500,000
CN^-	5,000
I^-	4,000
$HO^-, CH_3O^-, C_2H_5O^-$	1,000
Br^-	500
ϕO^-	400
Cl^-	100
CH_3COO^-	20
C_5H_5N (Pyridine)	20
NO_3^-	1
RSO_3^-, ClO_4^-	<1

philicities may change drastically in these solvents. Furthermore, a nucleophilic anion must have an associated cation, which has no effect on the substitution reaction as long as it is freely dissociated in the solvent. But in poor ionizing solvents the ions frequently exist as an **ion pair**, tightly bound electrostatically, and this will seriously inhibit the nucleophilicity of the anion.

Relative nucleophilicity is measured in the table by rates of S_N2 reaction, since here the nucleophile is involved in the transition state. In the S_N1 reaction only substrate ionization is involved in the transition state, and added nucleophiles do not change the rate but only the product composition. Thus, the solvolysis of benzyl tosylate in methanol gives methyl benzyl ether. If the better nucleophile lithium bromide is added, the rate is unchanged but the product is benzyl bromide. If equal amounts of lithium chloride and bromide are added, benzyl chloride and bromide are formed in a ratio reflecting their relative nucleophilicities, *but still at the same overall rate.*

Steric hindrance is also exhibited in nucleophiles, as exemplified by the *tert*-butoxide ion. Here the basicity (to H) is comparable to that of ethoxide, but so much steric hindrance is involved when *tert*-butoxide attacks carbon sites that virtually no displacements by *tert*-butoxide are known. This makes it an excellent practical *base* since it will attack only protons, not carbon.

Certain nucleophiles contain more than one atom bearing active electron pairs and so can react at either site and produce more than one product. Such molecules (or ions) are called **ambident** nucleophiles. The enolate anion in Table 10-1 is an important example. Ambident nucleophiles are of two kinds, those in which resonance affords electron density from the active electron pair at either of two (or more) sites ($\ddot{X}-Y=\ddot{Z} \longleftrightarrow X=Y-\ddot{Z}$) and those with two different nucleophilic sites of similar reactivity. The enolate anion is a fundamental example of the first, nitrite anion of the second.

Enolate   C-alkylation

Nitrite   O-alkylation   N-alkylation

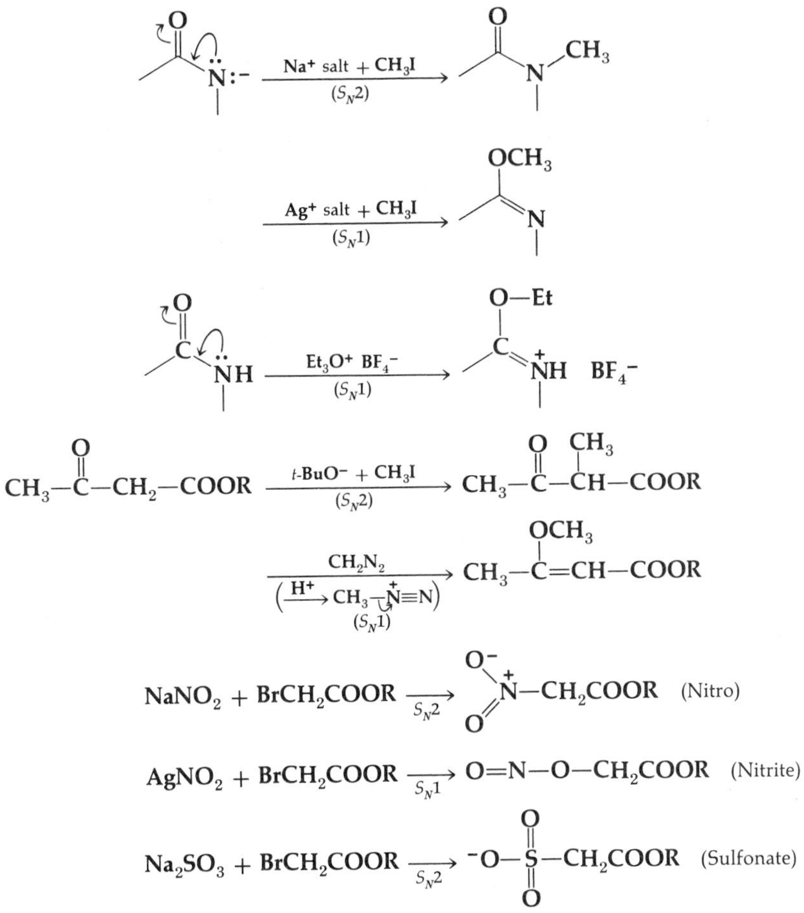

FIGURE 10-11 **Alkylations with ambident nucleophiles**

When attention is focused on the nucleophile instead of the substrate in a substitution reaction, it is frequently called an **alkylation** reaction, implying attachment of an alkyl group (R, from R—L) to the molecule (nucleophile) of interest. The molecule R—L is called an **alkylating agent**. The atom of attachment on the nucleophile can be specified (C, O, N, S alkylations) and is important in clarifying ambident alkylations. Many factors, often subtle, govern the product mixture in ambident alkylation.

The most important feature controlling the alkylation site with ambident nucleophiles (Fig. 10-11) is the relative electronegativity of the possible sites. In attack on a carbonium ion (S_N1) the more electronegative atom ($O > N > C$) is alkylated since the major electron density is available there. By contrast, in the S_N2 reaction the more polarizable and less electronegative atom is alkylated. When one electron pair is on sulfur, sulfur is always the site of alkylation due to its superior polarizability. The latter preference can usually be enhanced by carrying out the reaction in a protic solvent, which preferentially hydrogen bonds to the more electronegative atom, leaving the other site free for displacement. Steric hindrance and ion pairing are also factors in the preference of one ambident site over another.

PROBLEM 10-7

Most nucleophiles are anions but some anions are not nucleophiles. A particular example is BF_4^-. Explain the inertness of BF_4^- as a nucleophile.

PROBLEM 10-8

a The lactam (cyclic amide) of 5-aminopentanoic acid can be converted into a sodium or a silver salt (with appropriate sodium or silver bases). These each react with a molar equivalent of ethyl iodide but give different products. Explain.

b Thiocyanate (SCN^-) and cyanate (OCN^-) each give two products when allowed to react with allyl bromide. In the former case one product predominates very much. Show the products in each reaction, and account for this predominance.

PROBLEM 10-9

Diazomethane ($CH_2\!=\!\overset{+}{N}\!=\!\overset{\cdot\cdot}{\underset{\cdot\cdot}{N}} \longleftrightarrow \overset{\cdot\cdot}{\underset{\cdot\cdot}{C}}H_2\!-\!\overset{+}{N}\!\equiv\!\overset{\cdot\cdot}{N}$) reacts with carboxylic acids cleanly to yield their methyl esters. The reaction is acid catalyzed (by the carboxylic acid itself in fact). Mechanistically, how do you suppose this reaction proceeds?

Relative Leaving-group Activity

Leaving groups have an inverse relation to nucleophiles in substitution mechanisms and so often show inverse orders of activity. Thus, very strong bases, which are good nucleophiles, are very hard to displace and weak, stable bases are easy. As before, with the same atom bearing the leaving electrons, the weaker the base the better the leaving group (e.g., $RSO_2O^- > RCOO^- > \phi-O^-$). Apparently the most effective oxygen leaving group known is fluoro-

TABLE 10-7   **Average Relative Rates
for Leaving Groups**

ϕSO_2-O-	6
$I-$	3
$Br-$	1
H_2O^+-	1
$(CH_3)_2S^+-$	0.5
$Cl-$	0.02
$F-$	0.0001

sulfonate (FSO_2O^-), in accord with the very low pK_a of fluorosulfonic acid, FSO_2OH (Table 8-2). However, *with respect to polarizability the order is the same as with nucleophiles.*† (See Table 10-7.)

$$I^- > Br^- > Cl^- \gg F^- \quad \text{and} \quad Z-S^- > Z-O^-$$

Use is made of the high reactivity of iodide ion, both as a nucleophile and as a leaving group, in the iodide-catalyzed hydrolysis of primary alkyl chlorides. In such reactions potassium iodide is simply added to the reaction to speed up the overall conversion rate but is itself not consumed in the process.

$$H_2O + RCH_2Cl \xrightarrow{\text{Very slow}} RCH_2OH + HCl$$

$$I^- + RCH_2Cl \xrightarrow{\text{Fast}} RCH_2I + Cl^-$$

$$H_2O + RCH_2I \xrightarrow{\text{Fast}} RCH_2OH + HI$$

These strong bases are never leaving groups in simple displacements:

$$H:^- \quad R_3C:^- \quad R_2\ddot{N}:^- \quad R-\ddot{O}:^- \quad H\ddot{O}:^-$$

Hence, the —OH group of alcohols, for example, is never displaced as such in a nonacidic medium. However, oxygen can become a leaving group in two ways. For S_N2 reactions, where the group must leave as an anion, strong resonance stabilization of the anion makes it more stable, less basic, and more susceptible to displacement.

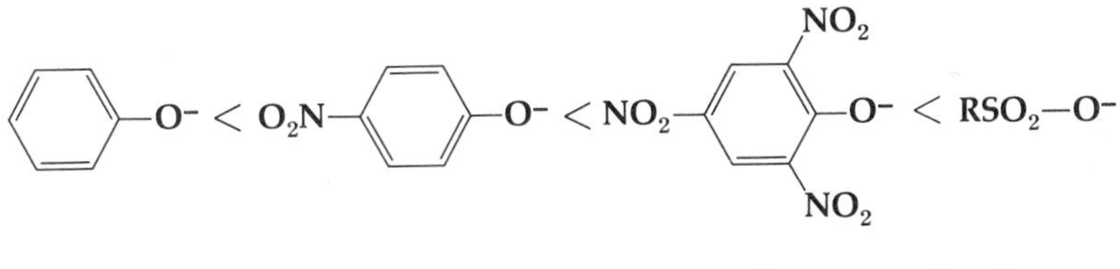

In both S_N1 and S_N2 reactions the ether oxygen becomes a leaving group when it is compressed into an epoxide and releases this strain on C—O bond cleavage.

For S_N1 reactions the oxygen atom is first complexed (on its unshared pairs) by a proton or Lewis acid (BF_3, $AlCl_3$) so that the *actual* leaving group is neutral (H_2O or ROH; Table 10-2):

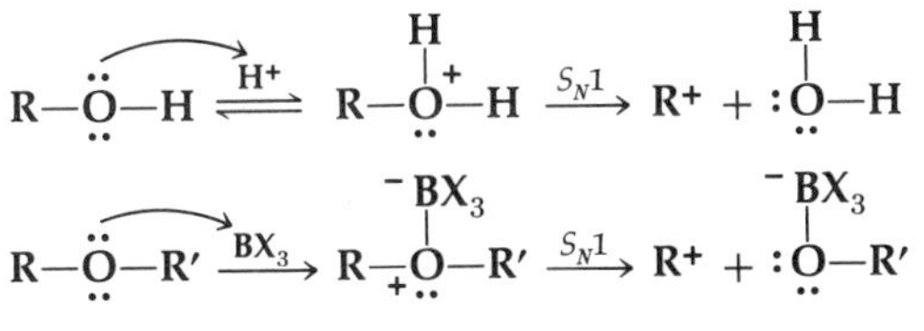

_____

†The order of ease of halide leaving groups is also the order of their C—X bond energies (Table 2-2). On this rationale the C—I bond is easier to cleave than the C—Cl bond and hence iodide is a better leaving group.

The latter procedure is often called **electrophilic catalysis** and occurs usually only with S_N1 reactions. The electrophilic catalyst (H^+, BX_3, etc.) serves to speed up the ionization enormously.† Besides the ionization of alcohols and ethers by Lewis acids, the other important use of electrophilic catalysis applies the silver cation to ionize halides in the same fashion, forming insoluble silver halides and accelerating the ionization rate so that the S_N1 mechanism usually predominates. (See examples in Fig. 10-11.)

$$R-\overset{..}{\underset{..}{Br}}: \overset{\frown}{+} Ag^+ \longrightarrow R^+ + AgBr\downarrow$$

10-4 NEIGHBORING-GROUP PARTICIPATION AND CYCLIZATIONS

If a second functional group is present in the substrate molecule it may participate in the substitution reaction if it fulfills two requirements. The group must offer a pair of electrons, and the atom bearing the pair must be at a favorable distance from the carbon site of substitution: three, five, or six atoms away, generally.

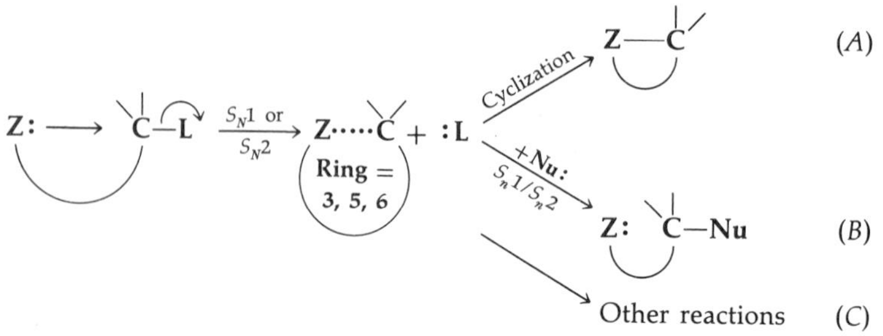

The pair of electrons on the second functional group ($Z = -\overset{|}{\underset{|}{C}}:^-$, $-\overset{..}{O}:^-$, $-\overset{..}{O}H$, $-\overset{..}{O}R$, $-\overset{..}{\underset{..}{X}}:$, $C=C$, ϕ) acts as an internal nucleophile for an *intramolecular* displacement, leading to:

1 A permanent new bond (case A; cyclization) or
2 An internally stabilized or solvated carbonium ion which is then attacked by an external nucleophile (case B) or
3 Some other reaction (case C), as in Sec. 10-5

†Since oxonium ions are such excellent leaving groups it is reasonable that one of the most active *alkylating agents* should be the trialkyl oxonium salts. Some of these can be isolated for use as reagents (see Fig. 10-11) but only when the associated anions are completely nonnucleophilic, for oxonium salts are destroyed by *any* nucleophile.

$$Nu: \longrightarrow CH_3 \overset{\frown}{\underset{\underset{CH_3}{\overset{..}{O}}}{\overset{+}{O}}} \overset{CH_3}{}\quad BF_4^- \longrightarrow Nu^+{-}CH_3 + CH_3\overset{..}{O}CH_3 + BF_4^-$$

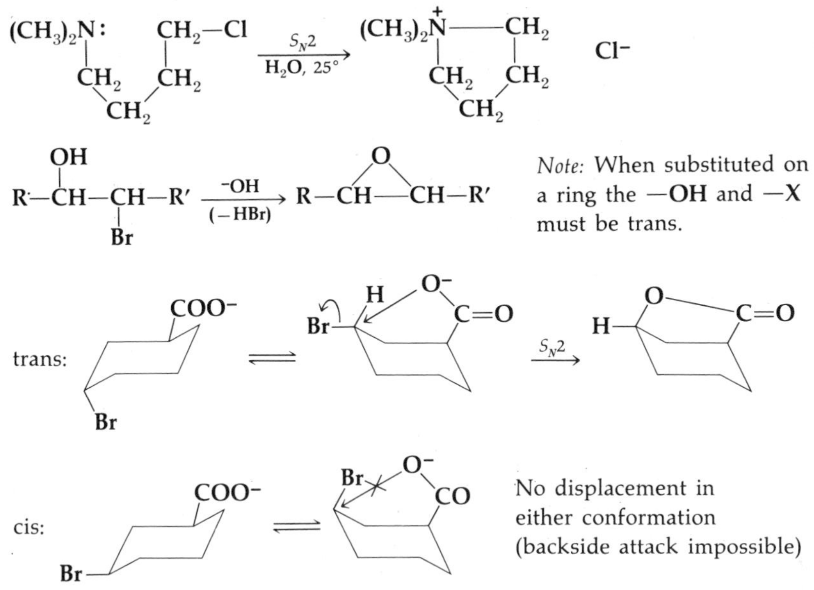

$(CH_3)_2N:$ $\quad$ $CH_2{-}Cl$ $\xrightarrow[\text{H}_2\text{O},\,25°]{S_N2}$ $(CH_3)_2\overset{+}{N}{-\!-\!-\!}CH_2$ $\quad$ Cl^-

$\underset{CH_2}{\overset{|}{C}H_2}$ $\underset{}{\overset{|}{C}H_2}$ $\qquad\qquad$ $\underset{CH_2}{\overset{|}{C}H_2}$ $\overset{|}{C}H_2$

$$R{-}\underset{Br}{\overset{OH}{\overset{|}{C}H}}{-}CH{-}R' \xrightarrow[(-HBr)]{-OH} R{-}\overset{\displaystyle O}{\overset{\displaystyle /\,\backslash}{CH{-\!-\!}CH}}{-}R'$$

Note: When substituted on a ring the —OH and —X must be trans.

trans:

cis:

No displacement in either conformation (backside attack impossible)

FIGURE 10-12   **Internal displacement reactions: cyclizations**

The cyclization (case *A*) necessarily inverts the substituted carbon, while case *B* represents two consecutive substitutions, both obliged geometrically to be inversions, so that the net result is *overall retention* of configuration. The same result occurs if product *A* is unstable to external nucleophiles present and reacts in a second displacement to give *B*. Examples are collected in Figs. 10-12 and 10-13.

Internal substitutions are much favored in rate because of the proximity of the reactive centers (see page 355), commonly thousands of times faster than the comparable *intermolecular* reactions. Formation of three-membered rings is the most favored by this probability-of-reaction factor, and this entropy advantage offsets the energy unfavorability involved in closing a three-membered ring (angle strain). In many cases three-membered-ring formation is even faster than five- and six-membered rings.† The four-membered ring, however, does not have a probability advantage and, as it is also strained, its formation is very rarely observed. All ring formations are made even more rapid the more substituted the ring becomes, probably because the closed ring experiences less steric crowding than the acyclic reactant.

A number of examples are shown in Figs. 10-12 and 10-13.

†The 1,3 interatomic distance is about 2.5 Å and is unchanged by any conformational rotations. Hence the two atoms have only to close up by 1 Å to go to bonding distance. The 1,5 interatomic distance can extend to as much as twice this amount and in an acyclic chain is often in conformations sterically unacceptable for internal displacements.

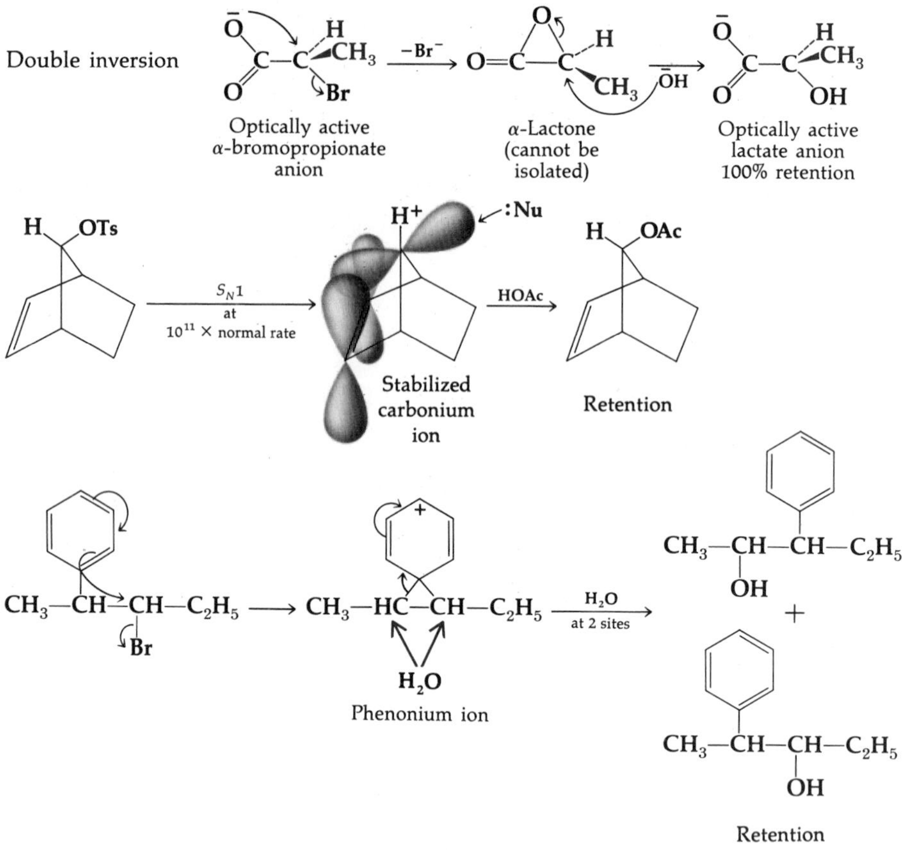

Double inversion

Optically active
α-bromopropionate
anion

$-Br^-$

α-Lactone
(cannot be
isolated)

$\overline{O}H$

Optically active
lactate anion
100% retention

H   OTs

S_N1
at
10^{11} × normal rate

H$^+$:Nu

Stabilized
carbonium
ion

HOAc

H   OAc

Retention

$CH_3-CH-CH-C_2H_5$
                |
               Br

$\longrightarrow$

$CH_3-HC-CH-C_2H_5$

H_2O

Phenonium ion

H_2O
at 2 sites

$CH_3-CH-CH-C_2H_5$
            |
           OH

+

$CH_3-CH-CH-C_2H_5$
            |
           OH

Retention

FIGURE 10-13   **Neighboring-group participation in nucleophilic substitution**

PROBLEM 10-10

Write the expected products of these reactions.

a   2-Bromomethyl-benzoic acid and **NaHCO$_3$**

b   *cis-* and *trans*-3-bromocyclohexyl-acetic acid and **NaOH**

c   *cis-* and *trans*-2-bromocyclohexyl-acetic acid and **NaOH**

d   *cis-* and *trans*-2-bromocyclohexanol and **NaOH**

PROBLEM 10-11

Predict the products and stereochemical outcome of boiling R-2-phenyl-R-3-bromobutane in acetic acid.

10-5   COMPETITIVE REACTIONS

The S_N1 reaction begins with a rate-determining ionization step, which forms a carbonium ion. This in turn reacts with an external or internal nucleophile to afford substitution. However, as implied in Fig. 9-2A, the carbonium ion may also lose a proton to give a stable neutral compound with a double bond, or an adjacent group may rearrange, to satisfy the positive center. The first alternative is an **elimination** reaction, the second a **rearrangement**. Both are treated more extensively in their own right in later chapters (14 and 17) but must be recognized here as possible competing reactions for substitution. Although the three competing courses of reaction (substitution, elimination, rearrangement) yield different products, they all proceed with identical rates since this is determined only by the initial substrate ionization and not by the subsequent mode of collapse elected by the carbonium-ion intermediate.

The S_N2 reaction has only one alternative: the approaching nucleophile can either attack carbon to afford normal substitution, or adjacent hydrogen, to give *elimination*. In the latter instance, the nucleophile acts as a base. Rearrangements virtually never occur under S_N2 conditions.

Elimination Reactions

The chief competition in nucleophilic substitution comes from elimination reactions in which the nucleophile attacks a β-hydrogen rather than carbon. Elimination cannot occur, of course, without a β-hydrogen, and therefore alkylations (substitutions) using such substrates as methyl or benzyl are favored in practice where possible since they bear no β-hydrogens. Elimination will be relatively more favored if steric crowding renders displacement slow. In tertiary cases elimination is usually the only reaction under S_N2 conditions.

Elimination:

$$\text{Nu:} + -\overset{|}{\underset{|}{C}}\overset{H}{\underset{L}{C}}- \longrightarrow \text{Nu:H} + _{/}^{\diagup}C{=}C_{\diagdown}^{\diagup} + \text{:L}$$

Substitution:

$$\text{Nu:} + -\overset{H}{\underset{L}{\underset{|}{C}}}-\overset{|}{\underset{|}{C}}- \longrightarrow -\overset{H}{\underset{|}{C}}-\overset{Nu}{\underset{|}{C}}- + \text{:L}$$

Strong bases, such as HO^-, RO^-, and H_2N^-, are very reactive in hydrogen abstraction and tend to cause elimination by preference over substitution, especially for those with $pK_a > 12$. Highly polarizable nucleophiles, such as I^-, RS^-, and $S_2O_3^{--}$, tend to give nearly exclusive attack on carbon (substitution), and little elimination.

$$C_2H_5O^- \text{ Na}^+ + CH_3-\underset{\underset{\textstyle Br}{|}}{CH}-CH_3 \xrightarrow[55°]{C_2H_5OH} (CH_3)_2CHOC_2H_5 + CH_2{=}CH-CH_3$$
$$\phantom{C_2H_5O^- \text{ Na}^+ + CH_3}\qquad\qquad\qquad\qquad\qquad\qquad 21\% \qquad\qquad\quad 79\%$$

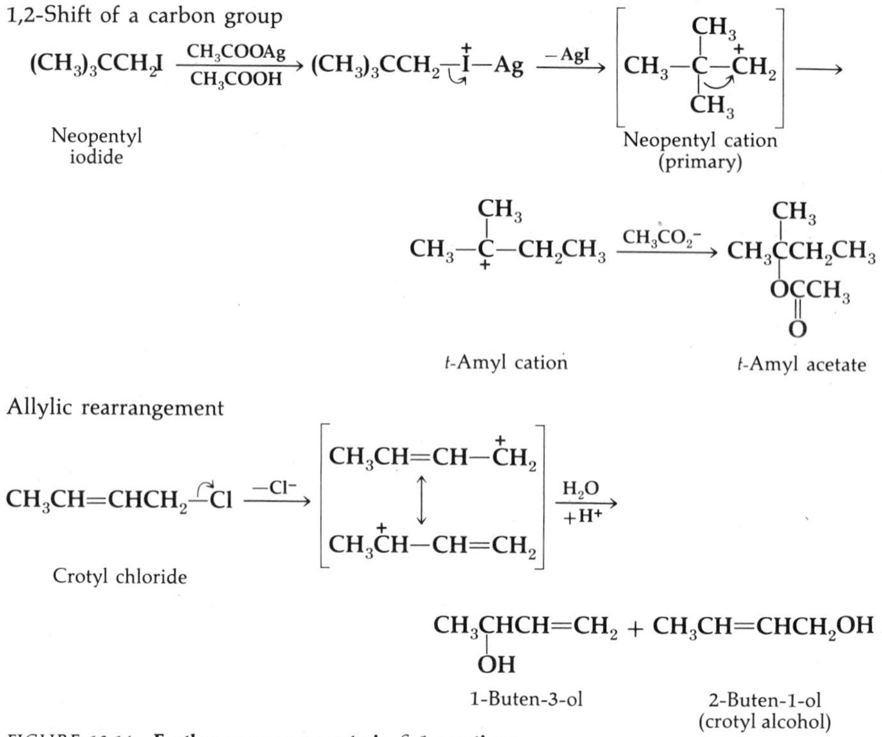

1,2-Shift of a carbon group

$$(CH_3)_3CCH_2I \xrightarrow[CH_3COOH]{CH_3COOAg} (CH_3)_3CCH_2\overset{+}{I}-Ag \xrightarrow{-AgI} \left[CH_3-\overset{CH_3}{\underset{CH_3}{\overset{|}{C}}}-\overset{+}{C}H_2 \right] \longrightarrow$$

Neopentyl   Neopentyl cation
iodide   (primary)

$$CH_3-\overset{CH_3}{\underset{+}{\overset{|}{C}}}-CH_2CH_3 \xrightarrow{CH_3CO_2^-} CH_3\overset{CH_3}{\underset{\overset{|}{O}\overset{\;}{\underset{\overset{\|}{O}}{C}}CH_3}{\overset{|}{C}}}CH_2CH_3$$

                                            t-Amyl cation   *t*-Amyl acetate

Allylic rearrangement

$$CH_3CH=CHCH_2\overset{\frown}{}Cl \xrightarrow{-Cl^-} \left[\begin{array}{c} CH_3CH=CH-\overset{+}{C}H_2 \\ \updownarrow \\ CH_3\overset{+}{C}H-CH=CH_2 \end{array} \right] \xrightarrow[+H^+]{H_2O}$$

Crotyl chloride

$$CH_3\underset{\overset{|}{O}H}{C}HCH=CH_2 + CH_3CH=CHCH_2OH$$

                                1-Buten-3-ol   2-Buten-1-ol
                                                           (crotyl alcohol)

FIGURE 10-14   **Further rearrangements in S_N1 reactions**

Elimination also competes with substitution in S_N1 reactions, since the intermediate carbonium ion may lose a proton to a basic solvent molecule.

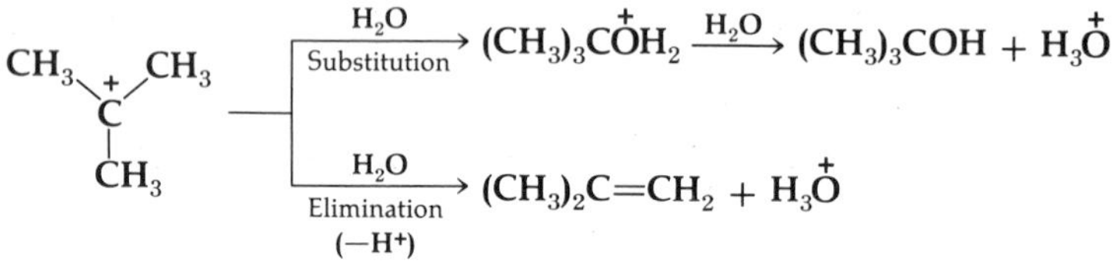

$$\underset{\underset{CH_3}{\overset{|}{C}}}{CH_3\diagdown \overset{+}{} \diagup CH_3} \begin{cases} \xrightarrow[\text{Substitution}]{H_2O} (CH_3)_3C\overset{+}{O}H_2 \xrightarrow{H_2O} (CH_3)_3COH + H_3\overset{+}{O} \\ \\ \xrightarrow[\substack{\text{Elimination} \\ (-H^+)}]{H_2O} (CH_3)_2C=CH_2 + H_3\overset{+}{O} \end{cases}$$

Highly branched substrates give a good deal of elimination under S_N1 reaction conditions. The amount of elimination is decreased by lowering the reaction temperature in both S_N1 and S_N2 reactions.

Carbonium-ion Rearrangements

Reactions that proceed by the S_N1 mechanism frequently give rearrangements. The migrations of an alkyl and a functional neighboring group are illustrated in Fig. 10-14. The first type is usually governed by a simple thermodynamic consideration. *If a carbonium ion can be converted to a more stable ion by migration of an alkyl or aryl group from an adjacent carbon atom, rearrangement will almost inevitably occur.* A common type of functional group migration or rearrangement is that of an adjacent double bond. Allylic cations contain two electron-deficient centers to which nucleophiles can be attached. Therefore, allylic rearrangements may be expected in the solvolysis of allylic substrates, leading to two different products, as in Fig. 10-14. The entire subject of molecular rearrangements is treated more extensively in Chap. 17.

PROBLEM 10-12

For these reactions write possible products which do not arise via normal substitution.

a 2,2,2-Triphenylethanol in hot formic acid
b 2-Phenyl-2-butanol in hydrochloric acid
c *cis*-2-Methyl-1-bromocyclopentane and potassium acetate in acetonitrile
d 1-Chloro-4-methyl-2-pentene in hot aqueous tetrahydrofuran

TABLE 10-8

	S_N1	S_N2
Mechanism	Two step $R—L \longrightarrow R^+ \xrightarrow{+Nu:} R—Nu$	One step $Nu: + R—L \longrightarrow Nu—R + :L$
Common reaction conditions	Acidic	Basic
Kinetics	1st order (~unimolecular) Rate $= k[R—L]$ Ionization $=$ rate determining (2d step does not affect rate)	2d order (bimolecular) Rate $= k[Nu:][R—L]$
Stereochemistry	Racemization $\longrightarrow$ inversion	Inversion
Solvent effects	Much favored by ionizing (polar) solvents	Polarity of solvent rarely important; H-bonding solvents disfavor by deactivating nucleophile
Carbon site in R (must be sp^3)	Resonance stabilization favors reaction Satd.: *tert* $\gg$ *sec* $>$ *pri* Rings: 3 $<$ 4 $<$ 5 $>$ 6 $<$ 7-10	Steric hindrance control $CH_3 >$ *pri* $>$ *sec* (*tert* $= 0$) Same
Competitive reactions	Elimination Rearrangement	Elimination by strongly basic nucleophiles
Catalysis	H^+ (or BF_3): R—OH, R—OR Ag^+: R—X	None

TABLE 10-8 (*Continued*)

	S_N1	S_N2
Leaving groups (L)		
X	$I > Br > Cl \gg F$	Same
O	$-OH, -OR, -OCOR$ 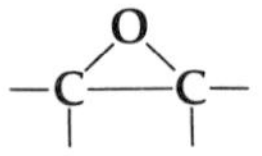 $\left.\begin{array}{c} \\ -\overset{\mid}{C}\overset{\overset{O}{\diagup\diagdown}}{}\overset{\mid}{C}- \end{array}\right\}$ H^+ (or BF_3)	$-OH, -OR$ never $-OH \longrightarrow -OTs$ first 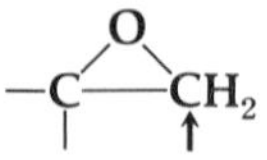 $-\overset{\mid}{C}\overset{\overset{O}{\diagup\diagdown}}{}CH_2$ $\uparrow$
N	$-\overset{+}{N}{\equiv}N$ $-NR_3, -NCOR'$ never	Same
C	Never	Same
Nucleophiles (Nu :)		
X:	$I^- > Br^- > Cl^- \gg F^-$	Same
O: (S:)	$H_2O, ROH, RCOOH$ (Solvolysis)	$RO^- > HO^- > \phi O^- > RCOO^-$ $-S: > -O:$
N:	$NO_2^- \longrightarrow R-O-NO$ N_3^- Amines rarely $[R_3N: \overset{H^+}{\longrightarrow} R_3\overset{+}{N}H,$ inactive$]$	$NO_2^- \longrightarrow R-NO_2$ N_3^- Amines: problems with multiple alkylation 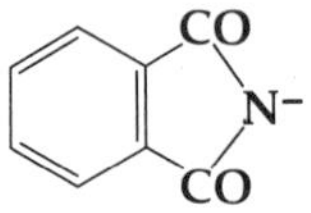 (phthalimide structure with CO, N⁻, CO)
C:	None	Weak bases best to avoid elimination: $Y-\overset{..}{C}H-Z \begin{cases} Y \text{ and } Z = CO, \\ CN, NO_2, SO_2 \end{cases}$ $^-{:}CN, {}^-{:}C{\equiv}C-R$
H:	None	$LiAlH_4$
Ambident	More electronegative site	Less electronegative site

10-6 SUMMARY

The present chapter had as its intent the focusing of our structural knowledge on the problem of a broad and general reaction. We have seen that the observed behavior of many kinds of molecules fits qualitatively quite well into our expectations from the operation of effects previously discussed. Hence we have developed many useful generalizations about substitution reactions which we shall apply to specific cases of synthetic value in the next chapter.

A summary of these generalizations is shown in Table 10-8.

PROBLEMS

10-13 In each case decide whether the reaction is likely to proceed by S_N1 or S_N2 mechanisms, what the product will be (including stereochemistry where appropriate), and what side-products are possible.

    a 3-Phenyl-1-bromopropane + **NaCN** in dimethylformamide

    b R-3-Phenyl-1-bromobutane + **NaCN** in dimethylformamide

    c s-1-Phenyl-1-bromobutane + **NaCN** in dimethylformamide

    d s-1-Phenyl-1-bromobutane + **AgOAc** in ethanol

    e γ,γ-Dimethylallyl tosylate + $(C_2H_5)_3N$ in methanol

    f (+)-4-Chloropentanoic acid + **NaNO$_2$** in water

    g (+)-4-Chloropentanoic acid in boiling water

    h Benzyl chloride + **Na$_2$CS$_3$** in acetone

    i (−)-1,3-Dibromo-1-phenylpropane in hot aqueous acetic acid

    j (−)-1,3-Dibromo-1-phenylpropane + **KOAc** in dimethylsulfoxide

10-14 When p-(2-bromoethyl)-phenol is carefully treated with base a compound, C_8H_8O, can be isolated. This compound shows no IR bands above 3200 cm^{-1} (below 3.1 μ) but a strong band at 1670 cm^{-1} (6.0 μ) and a UV maximum around 230 nm. On treatment with **HBr** the compound reverts easily to the original phenol. Account for these observations.

10-15 When 2-formylcyclohexanone (formyl = —CH=O) is mixed with sodium methoxide in methanol and n-propyl iodide is added, the isolated product shows no intense maximum in the UV above 200 nm (a weak band at 280 nm appears). When isopropyl iodide is used instead, an isomeric product is obtained with an intense UV maximum at about 250 nm. Explain.

10-16 When the *cis*- and *trans*-3,3-dimethyl-2-bromo-cyclohexanols are treated with strong bases, they each yield a single product. The products are isomeric, halogen-free, and only the one from the *cis*-bromo alcohol shows an IR band at 5.85 μ (1710 cm^{-1}). Neither product shows an IR band around 3 μ (3350 cm^{-1}) like the starting materials. Explain.

10-17 In the laboratory, treatment of butanol with sulfuric acid can afford a useful preparation of dibutylether, but if sodium bromide is added very little ether is formed. Explain.

10-18 Give examples of the following reactions:

    a Two examples of reactions that would be expected to give retention of configuration

    b Three examples of reactions that should largely give inversion of configuration

    c Three examples of reaction that should largely give racemization

    d Three examples of reactions that are inhibited by steric hindrance

    e Three examples of reactions that are slow because of adverse electronic effects

10-19 Arrange the following groups of compounds in order of decreasing reactivity in the reactions indicated.

    a The acetolysis reaction:

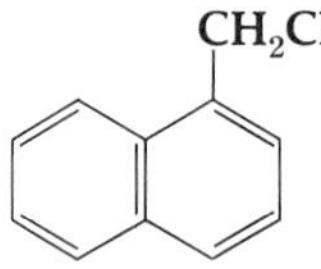

$(CH_3)_2CHCl$ $CH_3CH_2CH_2Cl$ $C_6H_5CH_2Cl$

    b The S_N2 substitution with potassium iodide in dry acetone:

CH_3OTs $CH_2=CHCH_2OTs$

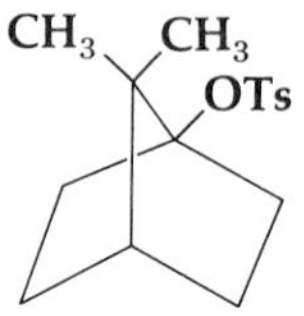

$(CH_3)_2CHOTS$

    c The methanolysis reaction with acid anion as the leaving group:

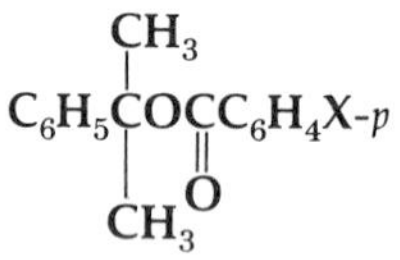

$X = H, I, F, NO_2, OCH_3, CH_3$

10-20 Trace the following interconversions with three-dimensional formulas. Show appropriate reagents, catalysts, and solvents.

    a R-2-Butanol $\longrightarrow$ tosylate $\xrightarrow[\text{Acetone}]{\text{Br}^-}$ bromide $\longrightarrow$ alcohol + olefin

    b s-1-Phenylethanol $\xrightarrow[\text{Ether}]{\text{SOCl}_2}$ alkyl chloride $\longrightarrow$ optically active acetate $\longrightarrow$

        optically active alcohol $\longrightarrow$ tosylate $\longrightarrow$ s-1-phenylethyl iodide

10-21 State concisely the kinetic and stereochemical differences between S_N1 and S_N2 reactions. Why is benzyl chloride more reactive than ethyl chloride in both reactions?

10-22 For each numbered step in the synthetic sequence below, select the best reagents, solvent, and conditions for the desired reaction to proceed. Comment on the mechanism and stereochemistry (if any) in each, whether the reaction would be fast or slow, and what side products might be formed. (*Note:* If you think any individual step in the sequence is poor or impossible, say why.)

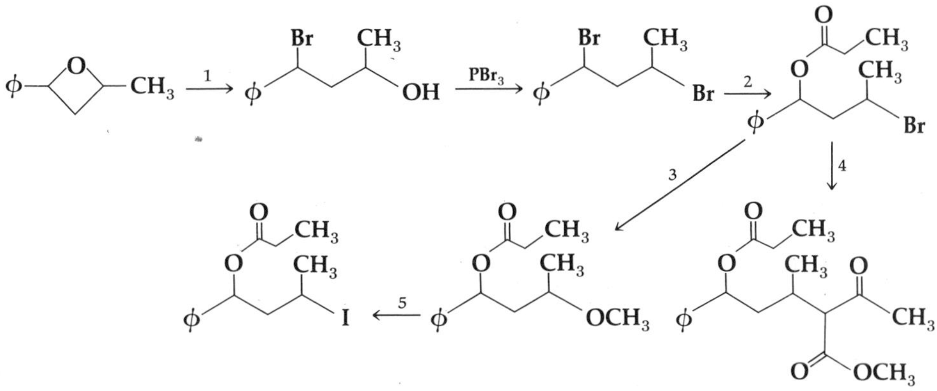

10-23 Provide a single specific structure that satisfies each description below.

    a An isomer of cyclohexyl chloride with a slower rate of displacement by iodide ion in acetone

    b An isomer of 1-methylcyclopentyl chloride which solvolyzes faster in hot formic acid

    c An alcohol more likely to rearrange than substitute or eliminate in acidic solutions

    d An optically active compound with two asymmetric centers giving an optically inactive compound with **$AgNO_3$**

    e An isomer of cyclopentyl bromide reacting faster with **KOAc** in **$HCON(CH_3)_2$**

    f An isomer of s-1-phenyl-1-pentanol which would racemize faster in hot aqueous acid

    g An isomer of s-1-phenyl-1-pentanol which would *not* racemize in hot aqueous acid

    h An isomer of s-1-phenyl-1-pentanol which changes optical rotation in hot aqueous acid but does *not* go to a rotation of 0°

10-24 Write equations for the following:

    a Five examples of reactions in which the nucleophile and substrate are in the same molecule. Try to invent some new examples of reactions you would expect to be successful.

    b Two examples that show the alkylation of ambident nucleophiles in two different ways.

    c Five reactions in which the leaving group is neutral.

    d Five reactions involving substitution by a neutral nucleophile.

10-25 In displacements of halides, **R—X**, the two amines below showed the following *ratios* of rates, with triethylamine always the slower. Account for this rate difference and the fact that it increases in the series shown.

R—X	Rate ratio = quinuclidine: triethylamine
CH_3-	57
C_2H_5-	252
$(CH_3)_2CH-$	706

(quinuclidine = [structure])

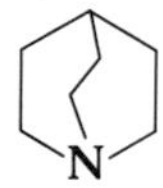

10-26 Explain the following observations:

    a Benzylmagnesium bromide can be made in good yield in dilute solutions but not in concentrated solutions.

    b Under conditions that lead to the hydrolysis of *t*-butyl chloride, $(CH_3)_2CClCOOC_2H_5$ is stable.

    c Neopentyl tosylate reacts slowly with lithium iodide but resists reaction with lithium chloride, whereas *n*-butyl tosylate reacts at reasonable rates with either nucleophile.

    d The yields of substitution product given in the reactions of *sec*-butyl chloride with $(CH_3)_2N^-$, $(CH_3)_3CO^-$, CH_3O^-, and I^- increase in the order listed.

    e Tri-*tert*-butylcarbinol solvolyzes much more rapidly in acid solutions than *tert*-butyl alcohol.

    f Optically active *sec*-butyl chlorosulfite, when warmed in dioxan, gives *sec*-butyl chloride with 97% retention, whereas the same reaction conducted in isooctane gives 43% inversion.

    g 1-Phenylethyl chloride solvolyzes much faster than 2,2-dimethyl-1-phenyl-propyl chloride.

    h In reactions with aqueous silver nitrate

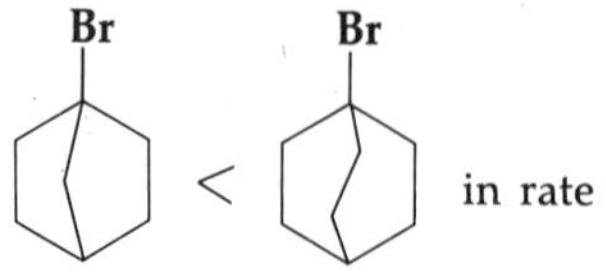

[structure] < [structure]   in rate

10-27 Write structures for the by-products of the following reactions:

a $NH_3 + (CH_3)_3CCl \longrightarrow (CH_3)_3CNH_2$ (trace) +

b $H_2O + (CH_3)_3CCH_2OTs \longrightarrow (CH_3)_3CCH_2OH$ +

c $HI + (CH_3)_2CHOCH(CH_3)_2 \xrightarrow{H_2O} (CH_3)_2CHI$ +

d $C_2H_5\bar{S}\overset{+}{Na} + CH_3CH{=}CHCH_2Cl \longrightarrow CH_3CH{=}CHCH_2SC_2H_5$ +

e $\overset{+}{Na}\bar{S}CN + C_6H_5CH_2Cl \longrightarrow C_6H_5CH_2SCN$ +

SYNTHETIC USES OF NUCLEOPHILIC SUBSTITUTION

THE previous chapter was oriented to theory so as to provide a basis for making predictions about new reactions. In this section attention is turned to the practical features of devising substitution reactions for synthetic purposes. The chapter is focused separately on the several major functional groups with single bonds to carbon, in order to show how they are prepared, and in turn how they react, by substitution reactions.

The two major uses of substitution reactions lie in the interconversion of singly bonded functional groups and in the forging of carbon-carbon bonds in order to build up a carbon skeleton. The two mechanisms, S_N1 and S_N2, have their special features and areas of suitability. The S_N1 reaction is used for tertiary, and sometimes secondary, sites and for allylic, benzylic and α-ether sites which afford resonance-stabilized carbonium ions. It is common for S_N1 reactions to be carried out under acidic conditions. The S_N2 reaction is used for primary and secondary carbon sites and usually employs basic conditions since the common nucleophiles are basic anions. S_N2 reactions always give inversion of configuration but S_N1 conditions usually give mixed racemization and inversion. In all cases it must be emphasized that the carbon site of reaction is saturated (sp^3). Leaving groups attached to simple double bonds (C=C) and to benzene rings are inert to the normal range of substitution conditions. Carbon-carbon bonds are always formed in nucleophilic substitution by S_N2 reactions since the nucleophile must be a basic carbanion.

It is important in assessing a particular reaction to take the entire molecule into account. We must make sure that no neighboring group with an electron pair three, five, or six atoms away is present to usurp a desired external substitution with a faster internal one, or that uninvolved functional groups elsewhere are not also changed by the reaction conditions. Possible competitive reactions of elimination and rearrangement must be borne in mind. Only in the S_N2 reaction has one much control over side reactions, by lowering the basicity of the nucleophile employed so as to damp the undesired elimination side reaction.

11-1 ALCOHOLS AND RELATED C—O COMPOUNDS

This section constitutes a survey of the main functional groups with C—O single bonds and the means of making and breaking this C—O bond. The groups

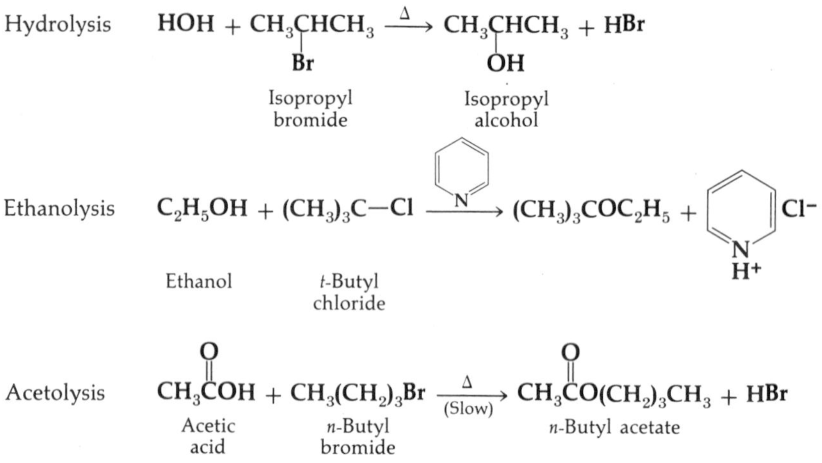

Hydrolysis $HOH + CH_3CHCH_3 \xrightarrow{\Delta} CH_3CHCH_3 + HBr$

                           Br   OH

                         Isopropyl   Isopropyl
                         bromide   alcohol

Ethanolysis $C_2H_5OH + (CH_3)_3C-Cl \xrightarrow{\text{N}} (CH_3)_3COC_2H_5 +$ Cl^-

              Ethanol   *t*-Butyl   N
                             chloride   H+

                   O   O

Acetolysis $CH_3\overset{\text{||}}{C}OH + CH_3(CH_2)_3Br \xrightarrow[\text{(Slow)}]{\Delta} CH_3\overset{\text{||}}{C}O(CH_2)_3CH_3 + HBr$

                 Acetic   *n*-Butyl   *n*-Butyl acetate
                  acid   bromide

FIGURE 11-1   **Solvolysis of alkyl halides (S_N1)**

in question are primarily alcohols (R—OH), ethers (R—OR′), and esters (R—OCOR′). These compounds are frequently prepared through reactions in which oxygen-containing nucleophiles attack saturated carbon and displace a leaving group, utilizing either S_N1 or S_N2 conditions. In the first, the reaction is often a solvolysis, in which a substrate reacts with a nucleophilic solvent molecule. Several examples of solvolysis reactions are found in Fig. 11-1. Alkyl halides are the most generally available substrates for solvolysis reactions. Use of water as solvent leads to alcohols, use of alcohols as solvents leads to ethers, and use of carboxylic acids leads to esters as products. As illustrated in the ethanolysis of *tert*-butyl chloride, a tertiary amine may be added to the reaction mixture. The amine neutralizes the acid produced and makes the hydrochloric acid unavailable for possible reverse reaction.

    The direct displacement (S_N2) is usually preferable to solvolysis as a method for preparing alcohols, ethers, and esters. Salts of the weakly nucleophilic O—H compounds are prepared by reaction with a strong base. These salts are dissolved, either in the parent hydroxylic compound or in some inert solvent. This solution of the negatively charged nucleophile (oxyanion) is then allowed to react with the substrate. The specific examples of such reactions shown in Fig. 11-2 are chosen to parallel the examples of solvolysis of Fig. 11-1.

    Halide ions are the most commonly encountered leaving groups, but sulfonates (R—SO_2O^-) and, to a lesser extent, sulfates (HO—SO_2—O^- and RO—SO_2—O^-) are also employed. Elimination is an undesirable side reaction which usually accompanies substitution to some extent, but is most serious with high temperatures or with nucleophiles which are strong bases. Thus, sodium hydroxide is used in the preparation of alcohols which contain no β-hydrogens, e.g., benzyl alcohol from benzyl bromide, but otherwise hydroxides more commonly cause elimination.

Hydroxide as
the nucleophile

$$\overset{+}{Na}\,\overset{-}{O}H + CH_3CHCH_3 \xrightarrow[\text{solvent}]{H_2O-C_2H_5OH} CH_3CHCH_3 + \overset{+}{Na}\,\overset{-}{Br}$$

with Br on the first substrate and OH on the product.

Alkoxide as
the nucleophile

$$(CH_3)_2CH\overset{-}{O}\,\overset{+}{K} + C_2H_5-Br \xrightarrow{\text{Benzene}} (CH_3)_2CHO-C_2H_5 + \overset{+}{K}\,\overset{-}{Br}$$

Potassium
isopropoxide

Carboxylate ion
as the
nucleophile

$$CH_3\overset{O}{\overset{\|}{C}}\overset{-}{O}\,\overset{+}{Na} + CH_3(CH_2)_3-Br \xrightarrow{C_2H_5OH} CH_3\overset{O}{\overset{\|}{C}}O(CH_2)_3CH_3 + \overset{+}{Na}\,\overset{-}{Br}$$

Sodium
acetate

FIGURE 11-2 **Nucleophilic substitution by oxyanions (S_N2)**

Elimination is usually avoided by using the weaker base sodium acetate to give an ester first, by substitution. This is then cleaved by a hydrolysis, described in Chap. 12, which removes the acetyl group and leaves the alcohol.

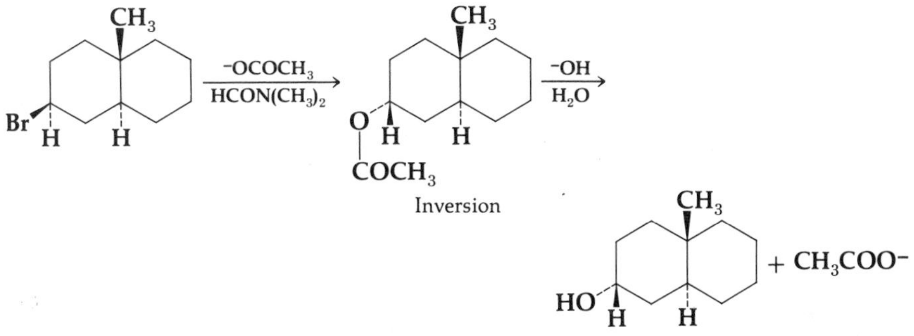

Inversion

The conditions used for S_N2 reactions reliably cause total inversion in configuration, as illustrated in the example above; the ester hydrolysis does not affect this configuration at the substituted carbon. When other asymmetric centers are present, the displacement effects an *epimerization*, inverting only at the site of substitution.

An interesting, and occasionally useful, type of substitution involves molecular nitrogen as a leaving group. Primary amines react with nitrous acid to give diazonium salts (Sec. 12-8). Although aromatic diazonium ions are reasonably stable, aliphatic diazonium salts have never been isolated, presumably because they decompose almost instantly to the parent carbonium ions and molecular nitrogen. Alcohols and alkenes of both rearranged and unrearranged structures are formed, and this multiplicity of products usually prevents its use as a practical reaction. Even the relatively stable aromatic diazonium salts undergo similar substitution to hydroxy-aromatics (phenols) when heated.

$$CH_3CH_2CH_2NH_2 + HNO_2 \xrightarrow{HCl} \left[CH_3CH_2CH_2 - \overset{+}{N} \equiv N \ \bar{C}l \right] + 2H_2O$$

n-Propylamine   Diazonium chloride

$$H_2O + CH_3CH_2CH_2 - \overset{+}{N} \equiv N \xrightarrow[(S_N1)]{-N_2} CH_3CH_2CH_2OH + CH_3CH=CH_2$$

Preparation of Ethers

The substitution reaction is the most common route to ethers. Preparation of ethers by the reaction of metal alkoxides with alkyl halides, sulfates, or sulfonates is known as the **Williamson synthesis.** Since tertiary substrates undergo elimination virtually exclusively in their reactions with strong bases, tertiary ethers can be made via the Williamson synthesis only by utilizing tertiary alkoxides with primary or secondary substrates.

$$C_6H_5CH_2\bar{O}\overset{+}{Na} + CH_3-Br \longrightarrow C_6H_5CH_2OCH_3 + \overset{+}{Na}\ \bar{Br}$$

$$C_6H_5CH_2\bar{O}\overset{+}{Na} + (CH_3)_3C-Br \longrightarrow C_6H_5CH_2OH + CH_2=C(CH_3)_2 + \overset{+}{Na}\ \bar{Br}$$

Sodium benzyloxide

Cyclic ethers have become important commercial materials for use both as synthetic intermediates and as solvents. The rings are closed by a cyclization which is essentially the same reaction as the Williamson synthesis. **Halohydrins** (2-haloalcohols) are readily made from alkenes (Chap. 15) and cyclize with facility in the presence of a weak base to form epoxides.

$$CH_2=CH_2 + HOCl \longrightarrow \underset{\substack{|\\HO}}{CH_2} - \underset{\substack{|\\Cl}}{CH_2} \underset{K_2CO_3}{\rightleftharpoons} CH_2 - CH_2 \longrightarrow CH_2 - CH_2 + \bar{Cl}$$

Ethylene
chlorohydrin
(2-chloroethanol)

Ethylene
oxide

$$\overset{+}{Na}\ \bar{O}H + HOCH_2CH_2CH_2CH_2Cl \longrightarrow \langle \overset{\bigcirc}{O} \rangle + H_2O + \overset{+}{Na}\ \bar{C}l$$

Tetrahydrofuran

Alkylation of phenols with reagents such as methyl iodide or dimethyl sulfate in the presence of an aqueous or alcoholic base is another example of the Williamson synthesis. Phenols, being more acidic than alcohols, can be converted to anions by aqueous base, or even by potassium carbonate in acetone.

$$\langle \rangle - \bar{O}\overset{+}{Na} + CH_3 \overset{\frown}{OSO_2OCH_3} \longrightarrow \langle \rangle - O-CH_3 + \overset{+}{Na}\ \bar{O}_3SOCH_3$$

Dimethyl
sulfate

Anisole
(methyl phenyl
ether)

Diazomethane is an interesting alkylating agent that forms methyl ethers (O-alkylation) with weakly acidic hydroxy compounds such as carboxylic acids, phenols and the enol forms of β-diketones and β-ketoesters. The latter reaction is of particular interest because the ambident anions derived from β-dicarbonyl compounds tend to be alkylated on carbon, rather than on oxygen, in conventional S_N2 displacements. The reaction of diazomethane with acidic compounds probably involves C-protonation to form the methyl diazonium ion as an intermediate, but the latter, if formed, must be very short-lived and readily ionizes to $CH_3^+ + N_2$ for an S_N1 reaction. (See Fig. 10-11.) Alcohols are not acidic enough to protonate themselves and thus do not react with diazomethane spontaneously. The reaction can be carried out if a catalytic amount of non-nucleophilic acid is added (BF_3 or HBF_4; HCl simply yields CH_3Cl).

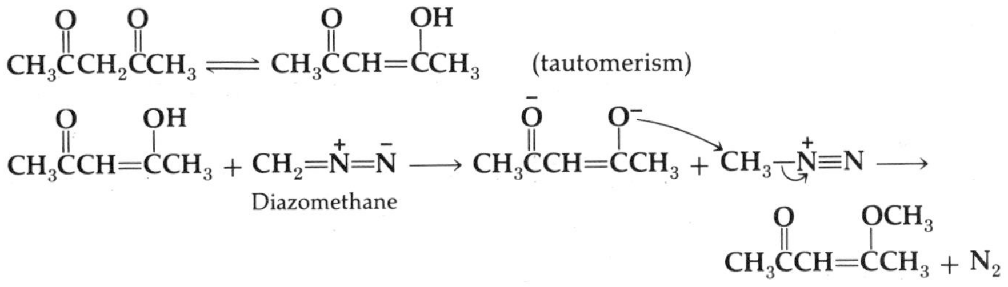

$$CH_3\overset{O}{\overset{\|}{C}}CH_2\overset{O}{\overset{\|}{C}}CH_3 \rightleftharpoons CH_3\overset{O}{\overset{\|}{C}}CH{=}\overset{OH}{\overset{|}{C}}CH_3 \qquad \text{(tautomerism)}$$

$$CH_3\overset{O}{\overset{\|}{C}}CH{=}\overset{OH}{\overset{|}{C}}CH_3 + CH_2{=}\overset{+}{N}{=}\overset{-}{N} \longrightarrow CH_3\overset{\overset{-}{O}}{\overset{\|}{C}}CH{=}\overset{O^-}{\overset{|}{C}}CH_3 + \overset{+}{CH_3{-}N}{\equiv}N \longrightarrow$$

Diazomethane

$$CH_3\overset{O}{\overset{\|}{C}}CH{=}\overset{OCH_3}{\overset{|}{C}}CH_3 + N_2$$

$$\phi CH_2OH + CH_2N_2 \xrightarrow{\;HBF_4\;} \phi CH_2OCH_3 + N_2$$

Diazomethane, which has many uses as a research chemical, is a toxic, yellow, explosive gas at room temperature and is usually handled in ether solution. The structure of diazomethane is unusual. The molecule is linear, and no proper Kekulé structure can be formulated that does not involve separation of charge although the compound itself is neutral. The cyclic isomer, diazirine, is also known.

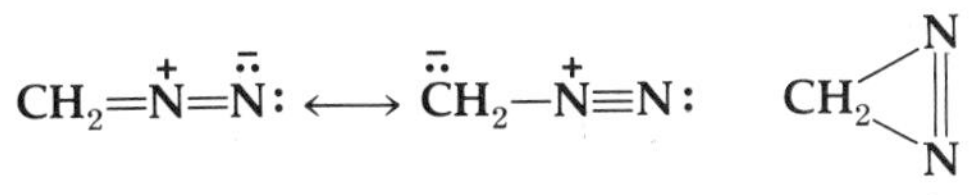

$$CH_2{=}\overset{+}{N}{=}\overset{..}{N}: \longleftrightarrow \overset{..}{C}H_2{-}\overset{+}{N}{\equiv}N: \qquad CH_2\overset{\displaystyle N}{\underset{\displaystyle N}{\Big\langle}}$$

Diazomethane   Diazirine

Preparation of Esters

Esters are usually prepared by the reactions in Chap. 12 but may be prepared by:

1 Solvolysis of halides, sulfonates, etc., in the corresponding carboxylic acid as solvent (Fig. 11-1)
2 S_N2 displacement via the carboxylate anion in a polar, aprotic solvent as on page 000
3 Electrophilic catalysis utilizing the silver salt of the carboxylic acid (S_N1)

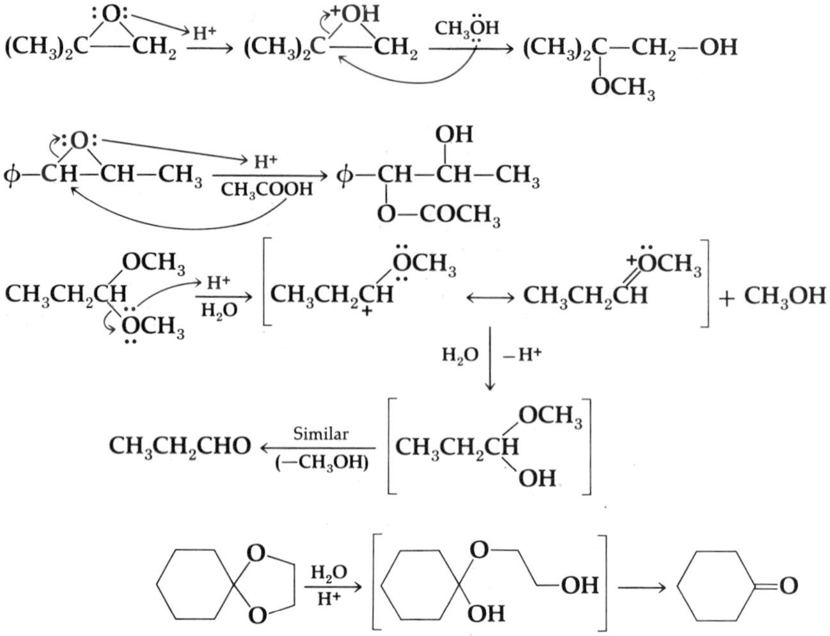

$$(CH_3)_2C\overset{:O:}{\underset{}{\diagdown}}CH_2 \xrightarrow{H^+} (CH_3)_2C\overset{\overset{+}{O}H}{\underset{}{\diagdown}}CH_2 \xrightarrow{CH_3\ddot{O}H} (CH_3)_2C-CH_2-OH$$
$$\underset{OCH_3}{|}$$

$$\phi-\overset{\overset{:O:}{\diagdown}}{CH}-CH-CH_3 \xrightarrow[CH_3COOH]{H^+} \phi-CH-\overset{\overset{OH}{|}}{CH}-CH_3$$
$$\underset{O-COCH_3}{|}$$

$$CH_3CH_2\overset{OCH_3}{\underset{\ddot{O}CH_3}{CH}} \xrightarrow[H_2O]{H^+} \left[CH_3CH_2\overset{\ddot{O}CH_3}{\underset{+}{CH}} \longleftrightarrow CH_3CH_2\overset{+\ddot{O}CH_3}{CH} \right] + CH_3OH$$

$$H_2O \Big| -H^+$$

$$CH_3CH_2CHO \xleftarrow[(-CH_3OH)]{Similar} \left[CH_3CH_2\overset{OCH_3}{\underset{OH}{CH}} \right]$$

FIGURE 11-3   Acid-catalyzed cleavages of epoxides and ketals/acetals

Once formed an ester may be regarded as the source of its corresponding alcohol by hydrolysis (R—OCOR′ ⟶ R—OH).

$$CH_3CH_2-\ddot{Br}: + \overset{+}{Ag} \longrightarrow CH_3CH_2-\overset{+}{\ddot{Br}}-Ag$$

$$CH_3COO^- + CH_3CH_2-\overset{+}{Br}-Ag \longrightarrow CH_3COOCH_2CH_3 + AgBr$$

Methylation with diazomethane provides an elegant, simple, and quantitative synthesis of methyl esters. The method is very useful in synthesis on a small scale and in the preparation of esters that are sensitive to acids and bases.

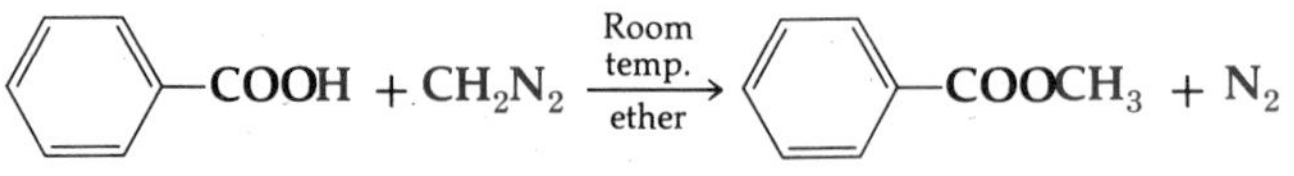

Benzoic
acid

Methyl
benzoate

Reactions of Alcohols, Ethers, and Esters

The *reactions* of the C—O compounds by substitution can only be effected in a direct manner by employing acid catalysis, so that the leaving group is the more stable ROH (H₂O), not RO⁻ or HO⁻. These are S_N1 conditions and proceed best on tertiary, secondary, or conjugated substrates. The simplest and most common examples employ **HI**, **HBr**, and **HCl** to convert alcohols and ethers to halides. When a poorer nucleophile than X⁻ is used, elimination and rearrangement are the more likely reactions (e.g., using $HClO_4$ or HBF_4). The cleavage of aryl alkyl ethers by **HI** is a general procedure, substitution occurring only at the saturated alkyl side of the ether, not the aryl. Diaryl ethers are inert under ordinary acidic conditions.

$$\phi CH_2OH \xrightarrow{\text{HBr}} [\phi CH_2 - \overset{+}{O}H_2] \longrightarrow [\phi CH_2{}^+] \longrightarrow \phi CH_2Br + H_2O$$

$$\phi - OCH_3 \xrightarrow{\text{HI}} \phi - OH + CH_3I$$

Ethers are generally impervious to all but the strongest acid conditions (e.g., $HI > 100°$). Perhaps the smoothest and most useful cleavage of ethers is effected by the Lewis acid BBr_3, at room temperature or below:

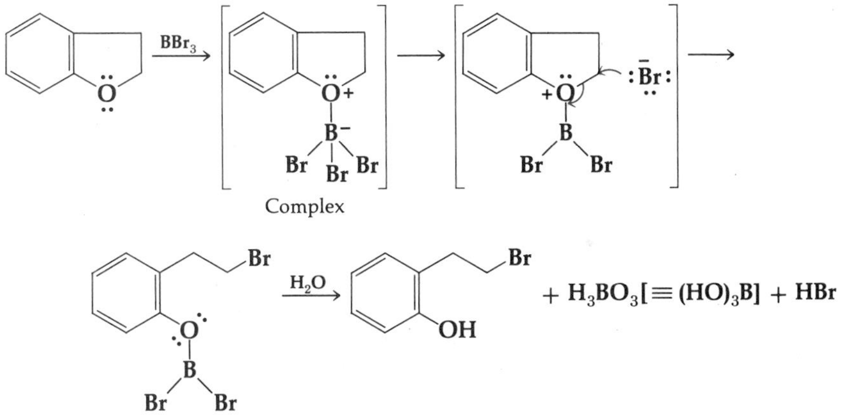

Complex

$+ H_3BO_3[\equiv (HO)_3B] + HBr$

Two cases are exceptions (Fig. 11-3) to the rule of general unreactivity of ethers. Epoxides, being severely strained, open rapidly with mild acids (protonation and S_N1). Of the two possible carbon sites, the one forming the better carbonium ion is the site of cleavage owing to its faster ionization rate. The angle strain in epoxides is so great that even S_N2 displacements can occur, opening the ring and releasing the strain. The carbon attacked is of course the least hindered of the two, as in any S_N2 reaction, and displacement goes with inversion. It is therefore frequently possible to open either one of the epoxide C—O bonds by selection of S_N1 or S_N2 conditions.

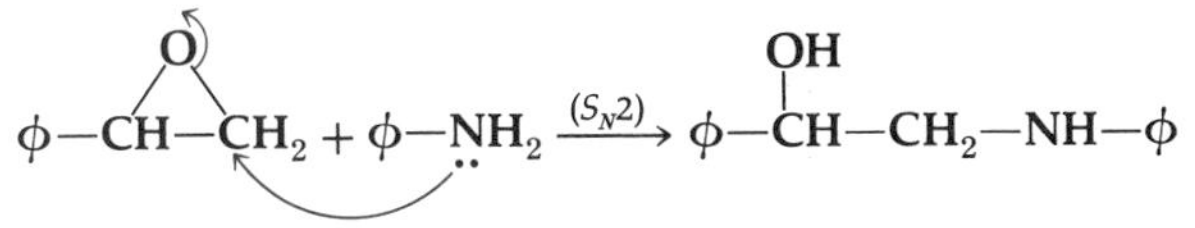

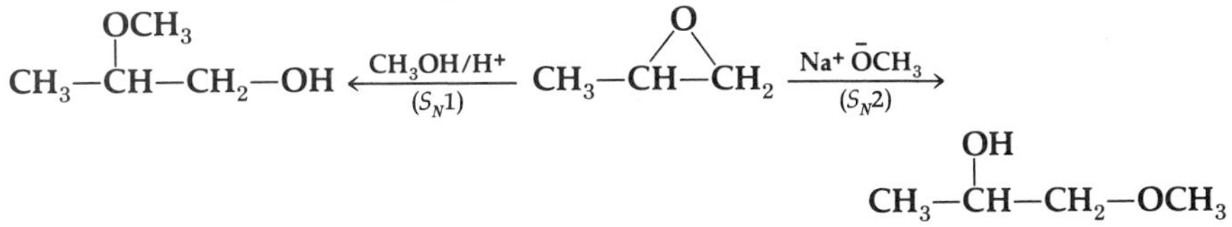

The other exception is that of acetals and ketals, with two ether linkages sited on a single carbon. The one ether so stabilizes the carbonium ion formed by protonation and ionization of the other that cleavage also occurs easily with mild acids, and these compounds are unstable to dilute aqueous acid even at room temperature. They are hydrolyzed to the parent ketone or aldehyde.

Even though the S_N2 reaction cannot be performed on alcohols or ordinary ethers themselves, alcohols at least can be first converted to derivatives (**ROH** ⟶ **ROZ**) in which the —**OH** group has been transformed into a leaving group which is less basic and more stable than **OH**$^-$. Such a group is then easily displaced by nucleophilic anions.

The most common procedure is reaction with toluene- or methanesulfonyl chloride to afford a sulfonate ester. This does not change the configuration at the carbon site since the reaction cleaves only the **O—H**, not the **C—O** bond of the alcohol.

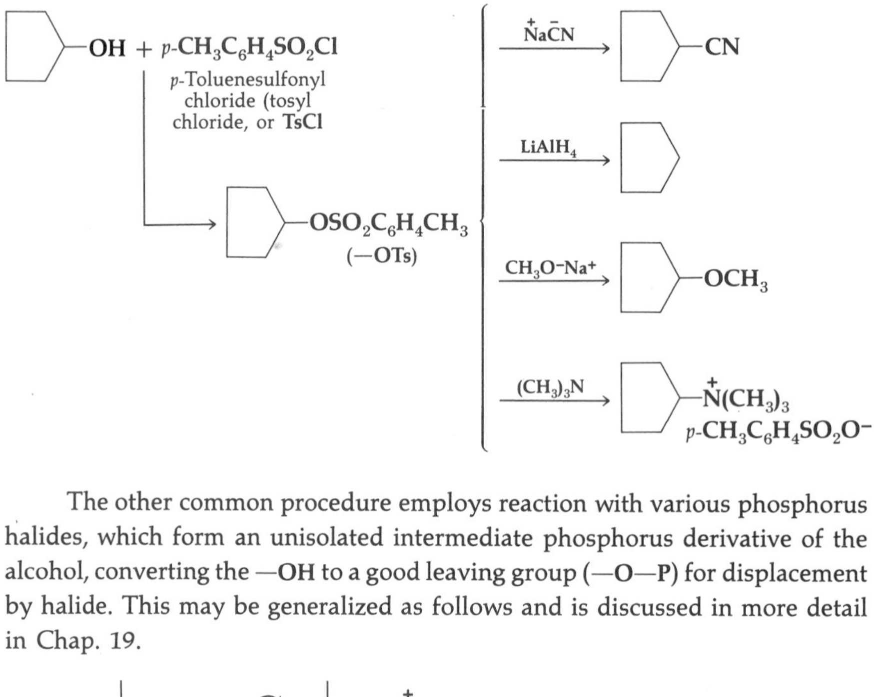

The other common procedure employs reaction with various phosphorus halides, which form an unisolated intermediate phosphorus derivative of the alcohol, converting the —**OH** to a good leaving group (—**O—P**) for displacement by halide. This may be generalized as follows and is discussed in more detail in Chap. 19.

$$\mathbf{R{-}OH} + \mathbf{P{-}X} \longrightarrow [\mathbf{R{-}O{-}\overset{+}{P}{-}}] + \mathbf{\overset{+}{H}X^-} \longrightarrow \mathbf{R{-}X} + \mathbf{P}\text{-products}$$

The best reagent for creating chlorides and bromides is apparently ϕ_3PX_2 ($\phi_3P^+XX^-$) and that for iodides $(\phi O)_3P^+CH_3I^-$. In the older literature PCl_5, PCl_3, $POCl_3$, and PBr_3 were commonly used.

$$(CH_3)_3C{-}CH_2{-}OH + (\phi O)_3\overset{+}{P}CH_3\,I^- \longrightarrow (CH_3)_3CCH_2I + \phi OH + CH_3\overset{+}{P}(O^-)(O\phi)_2$$

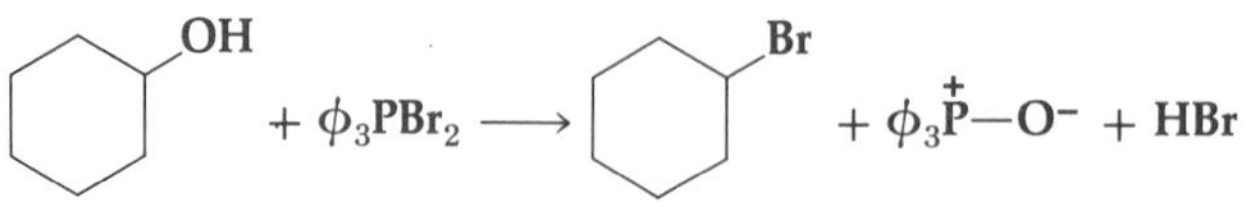

Thionyl chloride ($SOCl_2$) is the other useful reagent for converting alcohols to chlorides, frequently for the purpose of providing a better leaving group for a further displacement. The special practical appeal of this reagent is that all of the other products are gases. Phosgene ($COCl_2$) may be used in the same way but is often avoided because of its toxicity.

$$R-OH + SOCl_2 \longrightarrow R-Cl + SO_2 + HCl$$

$$R-OH + COCl_2 \longrightarrow R-Cl + CO_2 + HCl$$

11-2   THIOLS AND RELATED C—S COMPOUNDS

Sulfur compounds show greater nucleophilic reactivity than their oxygen analogs, and substitutions by sulfur-containing nucleophiles often occur under very mild conditions and give high yields. The important reagents that contain nucleophilic sulfur include hydrogen sulfide, bisulfide ion (HS^-), bisulfite ion ($:SO_2-O^-$), mercaptans (RSH and $ArSH$), mercaptide ions, thiocyanate ion (SCN^-), and disulfide ion (S_2^{--}). The common leaving groups include halide, sulfonate, and sulfate ions, as usual, and water (from conjugate acids of alcohols).

$$CH_3\overset{\frown}{SH} + (CH_3)_2CHCH_2-Br \longrightarrow CH_3SCH_2CH(CH_3)_2 + HBr$$
Isobutyl methyl
sulfide

$$C_6H_5SH + \langle \text{hexagon} \rangle -OSO_2C_6H_4CH_3\text{-}p \longrightarrow \langle \text{hexagon} \rangle -S-C_6H_5 + p\text{-}CH_3C_6H_4SO_3H$$

Thiophenol   Cyclohexyl tosylate   Cyclohexyl
                                                         phenyl sulfide

$$\overset{+}{Na}\ \overset{-}{S}-\overset{-}{S}\overset{+}{Na} + 2CH_3CH_2I \longrightarrow CH_3CH_2SSCH_2CH_3 + 2NaI$$
Ethyl disulfide

$$\overset{+}{Na}\ \overset{-}{S}CN + (CH_3)_2CHBr \longrightarrow (CH_3)_2CHSCN + NaBr$$
Isopropyl
thiocyanate

Ethylene oxide is readily opened by nucleophilic sulfur compounds. The reaction between hydrogen sulfide and the substrate gives 2,2'-dihydroxydiethyl sulfide. The latter compound is converted, by substitution, to 2,2'-dichlorodiethyl sulfide, a powerful vesicant (blistering agent) known as "mustard gas."

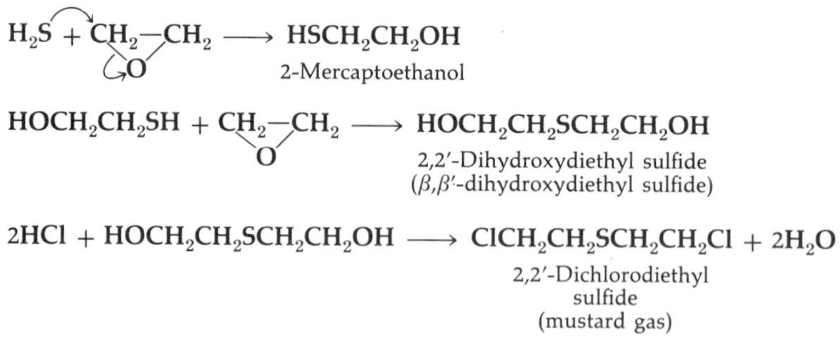

$$H_2S + CH_2\!\!-\!\!CH_2 \longrightarrow HSCH_2CH_2OH$$
$$\underset{O}{}$$
2-Mercaptoethanol

$$HOCH_2CH_2SH + CH_2\!\!-\!\!CH_2 \longrightarrow HOCH_2CH_2SCH_2CH_2OH$$
$$\underset{O}{}$$
2,2'-Dihydroxydiethyl sulfide
(β,β'-dihydroxydiethyl sulfide)

$$2HCl + HOCH_2CH_2SCH_2CH_2OH \longrightarrow ClCH_2CH_2SCH_2CH_2Cl + 2H_2O$$
2,2'-Dichlorodiethyl
sulfide
(mustard gas)

11-3 ALKYL HALIDES: C—X BONDS

Alkyl Halides

Alkyl halides are made by three principal methods:

1 Nucleophilic substitution reactions with alcohols as substrates
2 Addition of halides and hydrogen halides to carbon-carbon double bonds (Chap. 15)
3 Photochemical chlorination of hydrocarbons as an industrial process (Chap. 20)

Most of the methods for converting alcohols to halides were listed in the preceding paragraphs and may be summarized here:

RCl: HCl, $ZnCl_2$,† $SOCl_2$, $COCl_2$, ϕ_3PCl_2, PCl_5, PCl_3, $POCl_3$

RBr: HBr, ϕ_3PBr_2, PBr_3

RI: HI, $(\phi O)_3P^+CH_3I^-$

Halide ions increase both in nucleophilic reactivity and in leaving-group facility as the size (and polarizability) of the ions increases. Thus, while iodide is both an excellent nucleophile and leaving group, fluoride is weak in each role. Alkyl fluorides cannot usually be made by ordinary nucleophilic substitution and they are very unreactive to displacement.

> A classical test for the nature of saturated alkyl bromides or chlorides was devised from the opposite priorities of alkyl substitution in S_N1 and S_N2 reactions. A sample of a halide is treated with $AgNO_3$ in ethanol and another with NaI in acetone. If a rapid precipitate forms in the first (AgBr or AgCl), the halide is tertiary, while a rapid precipitate of NaBr or NaCl (insoluble in acetone) in the second indicates a primary halide. The first test promotes S_N1 conditions by electrophilic catalysis with silver ion and a poor nucleophile (nitrate) for any competing S_N2 displacement, while the second test utilizes a good nucleophile (iodide) for S_N2 reaction and a poor ionizing solvent to depress S_N1 reaction. In each test secondary halides react much more slowly, often requiring heating to yield a precipitate. See Prob. 10-5.

†$ZnCl_2$ acts by electrophilic catalysis, complexing with —OH as a Lewis acid. The **Lucas test** for distinguishing tertiary, secondary, and primary alcohols is based on the speed with which the alkyl chloride is seen to precipitate as a second phase (tertiary = immediate, primary = only slowly with heating).

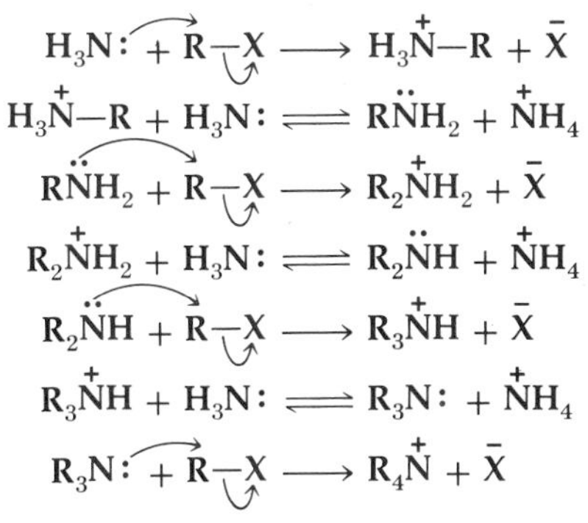

$$H_3N: \overset{\frown}{+ R - X} \longrightarrow H_3\overset{+}{N} - R + \bar{X}$$

$$H_3\overset{+}{N} - R + H_3N: \rightleftharpoons R\overset{..}{N}H_2 + \overset{+}{N}H_4$$

$$R\overset{..}{N}H_2 + \overset{\frown}{R - X} \longrightarrow R_2\overset{+}{N}H_2 + \bar{X}$$

$$R_2\overset{+}{N}H_2 + H_3N: \rightleftharpoons R_2\overset{..}{N}H + \overset{+}{N}H_4$$

$$R_2\overset{..}{N}H + \overset{\frown}{R - X} \longrightarrow R_3\overset{+}{N}H + \bar{X}$$

$$R_3\overset{+}{N}H + H_3N: \rightleftharpoons R_3N: + \overset{+}{N}H_4$$

$$R_3N: \overset{\frown}{+ R - X} \longrightarrow R_4\overset{+}{N} + \bar{X}$$

FIGURE 11-4 Alkylation of ammonia and amines

11-4 AMINES AND RELATED C—N COMPOUNDS

Alkylation of ammonia to form amines occurs in stages to form primary, secondary, and tertiary amines and quaternary ammonium salts. If a particular product is desired, it can sometimes be made to predominate by manipulation of the concentrations of the reactants and other reaction conditions. In each of the first three alkylation steps the initial products are ammonium ions, which are equilibrated by proton transfers with all other bases in the mixture. Since ammonium ions are not nucleophilic, two equivalents of nucleophile are required for each alkylation step, as shown in Fig. 11-4. The anions, $R_2\overset{..}{N}:^-$, are very poor nucleophiles but very strong bases.

Commercial synthesis of the methylamines is based upon the reaction of ammonia with methyl chloride. The products are separated by large-scale fractional distillation. As a matter of practical convenience, procedures that give mixtures of products or require rigid control of reaction conditions are usually avoided in the research laboratory. However, primary amines can often be easily made by using a large excess of ammonia. The alkylation reaction can also usually be stopped at the tertiary amine stage in good yield. Secondary amines are ordinarily made by other methods. Quaternary ammonium salts are virtually always made by alkylation of tertiary amines.

$$BrCH_2CH(OC_2H_5)_2 + 2NH_3 \longrightarrow H_2NCH_2CH(OC_2H_5)_2 + NH_4Br$$

1,1-Diethoxy-2-
bromoethane   2,2-Diethoxyethylamine

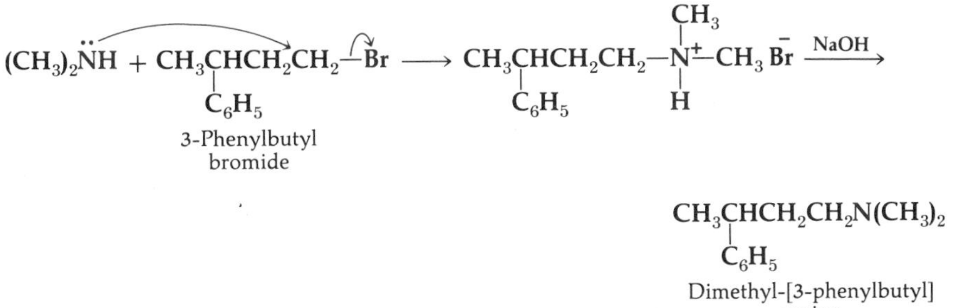

$$(CH_3)_2\overset{..}{N}H + \overset{\frown}{CH_3\overset{\frown}{CHCH_2CH_2}-Br} \longrightarrow CH_3CHCH_2CH_2-\overset{\overset{\displaystyle CH_3}{|}}{\underset{\underset{\displaystyle H}{|}}{N^+}}-CH_3 \ \bar{Br} \xrightarrow{NaOH}$$

$\underset{\displaystyle C_6H_5}{|}$ $\underset{\displaystyle C_6H_5}{|}$

3-Phenylbutyl
bromide

$$CH_3\underset{\underset{\displaystyle C_6H_5}{|}}{CH}CH_2CH_2N(CH_3)_2$$

Dimethyl-[3-phenylbutyl]
amine

In the **Gabriel synthesis**, a special device is employed to allow primary amines to be prepared uncontaminated by more highly alkylated products. Phthalimide is sufficiently acidic to be converted to a potassium salt by treatment with concentrated potassium hydroxide. The phthalimide ion is a good nucleophile, especially in polar aprotic solvents, in which it is reasonably soluble. It is alkylated to give *N*-alkyl phthalimides, which cleave with aqueous base (Chap. 12) to give primary amines.

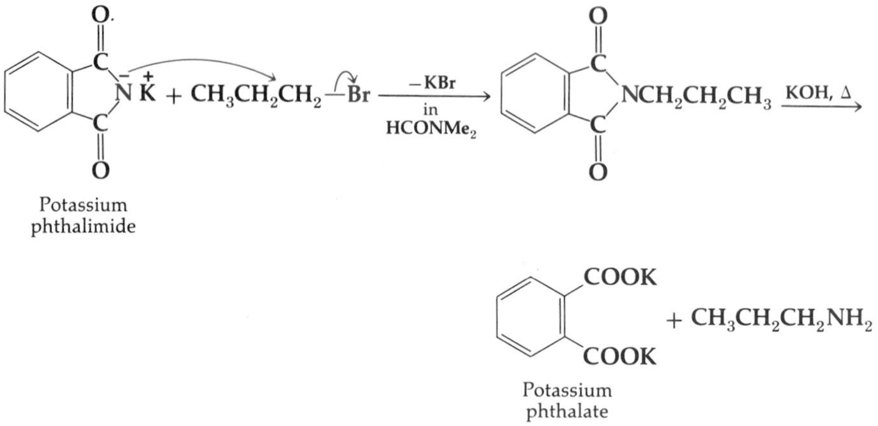
CH3CH2CH2—Br
−KBr
in
HCONMe2
NCH2CH2CH3
KOH, Δ
Potassium
phthalimide
COOK
COOK
+ CH3CH2CH2NH2
Potassium
phthalate

The strained small ring compounds, ethylene oxides and β-propiolactones, react readily with ammonia and amines by an S_N2 mechanism.

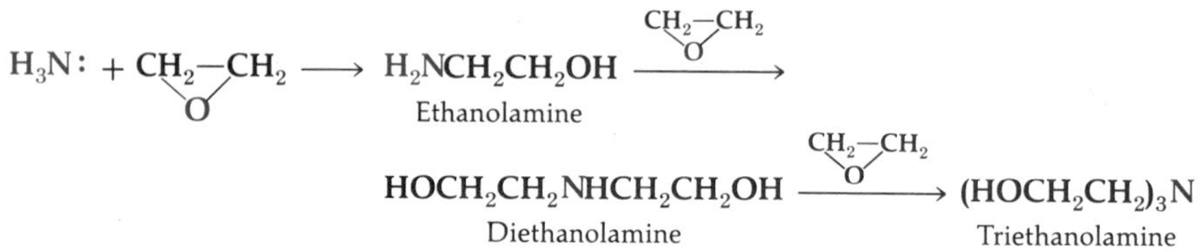
H3N: + CH2—CH2
H2NCH2CH2OH
Ethanolamine
HOCH2CH2NHCH2CH2OH
Diethanolamine
(HOCH2CH2)3N
Triethanolamine

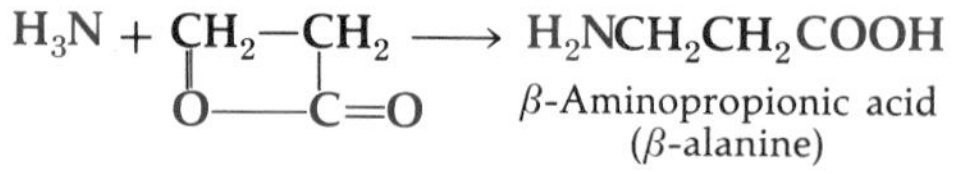
H3N + CH2—CH2
O——C=O
H2NCH2CH2COOH
β-Aminopropionic acid
(β-alanine)

Elimination reactions often become predominant in the amination of secondary substrates, and tertiary substrates generally give only elimination products.

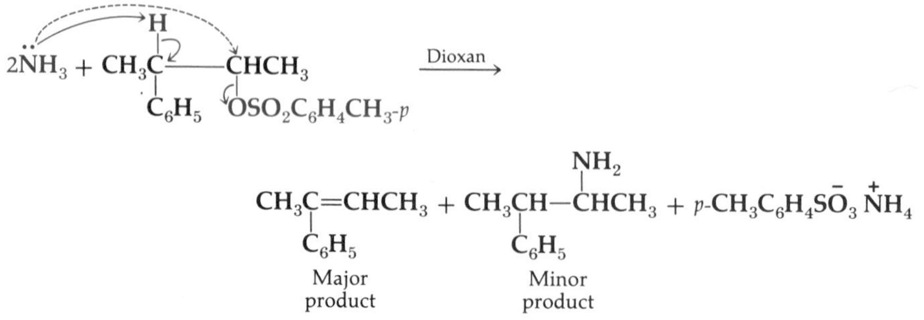
2NH3 + CH3C——CHCH3
C6H5 OSO2C6H4CH3-p
Dioxan
CH3C=CHCH3 + CH3CH—CHCH3 + p-CH3C6H4SO3 NH4
C6H5
NH2
Major
product
Minor
product

The azide ion is a powerful nucleophile, giving alkyl azides in nucleophilic substitution reactions. Azides may be reduced to primary amines with hydrogen and a metallic catalyst. Hydrazine is a weak nucleophile, but it will react satisfactorily with some primary and secondary halides. Multiple alkylation occurs if the alkyl halide is present in excess, and undesirable mixtures are obtained.

$$\overset{+}{K}\ \overset{-}{N}=\overset{+}{N}=\overset{-}{N} + \underset{\underset{Br}{|}}{CH_3CHCOO_2C_2H_5} \xrightarrow{-KBr} \underset{\underset{|}{CH_3CHCOO_2C_2H_5}}{\overset{\overset{N=\overset{+}{N}=\overset{-}{N}}{|}}{}} \xrightarrow[-NH_3]{H_2,\ Pt} \underset{\underset{|}{\dot{C}H_3CHCOO_2C_2H_5}}{\overset{\overset{NH_2}{|}}{}}$$

| Potassium azide | Ethyl α-bromo-propionate | Ethyl α-azido-propionate | Ethyl α-amino-propionate |

$$NH_2NH_2 + CH_3CH_2CH_2I \longrightarrow CH_3CH_2CH_2\overset{+}{N}H_2NH_2\ \overset{-}{I} \xrightarrow{Na_2CO_3}$$

Hydrazine

$$CH_3CH_2CH_2NHNH_2$$
n-Propylhydrazine

Nitroalkanes can often be prepared by S_N2 reactions on alkyl halides with the ambident nitrite anion, usually as its sodium salt, but with secondary or tertiary halides, silver nitrite yields largely nitrite esters (R—O—N=O), by S_N1 reaction.

$$\overset{+}{Na}\ \overset{-}{O}-\overset{..}{N}=O + BrCH_2COOCH_3 \xrightarrow{S_N2} \underset{-O}{\overset{O}{\diagdown}}{}\overset{+}{N}-CH_2COOCH_3$$

Methyl α-nitroacetate

$$\overset{+}{Ag}\ \overset{-}{O}-\overset{..}{N}=O + (CH_3)_3C-Cl \xrightarrow{S_N1} (CH_3)_3CNO_2 + (CH_3)_3CONO + CH_2=C(CH_3)_2$$

| | 2-Methyl-2-nitropropane (trace) | *tert*-Butyl nitrite 60% | |

The cleavage of the C—N single bond in amines is exceedingly difficult, and displacements of amines or even quaternary ammonium salts are virtually unknown. As with —OH, however, —NH₂ and ⟩NH can be converted to derivatives which undergo substitution, but here it is much more difficult and the reagents used on alcohols do not suffice.

The conversion of primary amines to transient diazonium ions was discussed above (page 413). The cleavage of the C—N bond is certainly facile with stable N₂ as the leaving group, but the reaction commonly suffers from yielding an impractical welter of substitution, elimination, and rearrangement products, via the S_N1 ionization route.

The diazonium group, with a C—N single bond, is also formed by protonation of the diazo group. The diazonium ion intermediate then ionizes to a carbonium ion and nitrogen gas with such readiness that the diazonium salt cannot be isolated.

$$CH_2 = \overset{+}{N} = \overset{-}{N} + HCl \longrightarrow [CH_3 - \overset{+}{N} \equiv N]\, Cl^- \longrightarrow CH_3Cl + N_2$$
Diazomethane

$$\phi - \overset{\displaystyle O}{\overset{\|}{C}} - CH = \overset{+}{N} = \overset{-}{N} + CH_3COOH \longrightarrow [\phi - \overset{\displaystyle O}{\overset{\|}{C}} - CH_2 \overset{+}{N} \equiv N]\, CH_3COO^- \longrightarrow$$
α-Diazoacetophenone

$$\phi - \overset{\displaystyle O}{\overset{\|}{C}} - CH_2 - O - COCH_3 + N_2$$

Two more practical variants have been devised, however, which lead more cleanly to substitution. In one the amine is converted first to an amide (page 512) and this is treated with nitrous acid to afford an *N*-nitroso ($\diagup$N—N=O) derivative (page 540) which collapses on warming to an ester and molecular nitrogen. In the second, the amine is allowed to react with an aryl (more stable) diazonium salt to yield a **triazene** (page 490) which can be substituted with a carboxylic acid.

$$R - NH_2 + R'COCl \longrightarrow R - NH - \overset{\displaystyle O}{\overset{\|}{C}} - R' \xrightarrow{HONO} R - \overset{\displaystyle N=O}{\overset{|}{N}} - \overset{\displaystyle}{\underset{\displaystyle O}{\overset{\|}{C}}} - R' \xrightarrow{\Delta}$$

$$R - O - \overset{\displaystyle}{\underset{\displaystyle O}{\overset{\|}{C}}} - R' + N_2$$

$$R - \overset{..}{N}H_2 + \phi - \overset{+}{N} \equiv N: \longrightarrow R - \overset{..}{N}H - \overset{..}{N} = \overset{..}{N} - \phi \rightleftharpoons R - \overset{..}{N} = \overset{..}{N} - \overset{..}{N}H\phi \xrightarrow[\Delta]{R'CO\overset{..}{O}H}$$
Tautomeric triazenes

$$R - O - \overset{\displaystyle}{\underset{\displaystyle O}{\overset{\|}{C}}} - R' + N_2 + \phi NH_2$$

Secondary and tertiary amines can often be cleaved by heating with **BrCN**. This converts the amine to a cyanamide, which is a better leaving group by virtue of resonance stabilization.

$$\phi CH_2 - NH - CH_2\phi \xrightarrow{BrCN} \left[\phi - CH_2 - \overset{\displaystyle C \equiv N}{\overset{|}{N}} - CH_2 - \phi \atop + Br^- \right] \longrightarrow$$

$$\phi CH_2 - \overset{\displaystyle C \equiv \overset{..}{N}:}{\overset{|}{N}:}\;\; + \phi CH_2Br$$
Stabilized
cyanamide

Phosphines are similar to amines and are readily alkylated by methyl and primary halides, less easily by secondary ones. Most reactions of phosphorus compounds are, however, separately discussed in Chap. 19 since they possess a number of peculiarities which arise from the expansion of the phosphorus outer electron shell.

$$\phi_3P: + CH_3CH_2Br \longrightarrow \phi_3\overset{+}{P}-CH_2CH_3\ Br^-$$

Ethyl triphenyl
phosphonium bromide

11-5   DISPLACEMENT BY HYDRIDE TO C—H BONDS

The development of the metal hydrides since 1945 has provided a number of reagents that can donate hydride ($H:^-$) to a wide variety of organic compounds, among which are certain reagents capable of undergoing nucleophilic substitution at saturated carbon. The most commonly used hydride donor is lithium aluminum hydride.

$$Li^+\ H-\overset{\displaystyle H}{\underset{\displaystyle H}{Al^-}}\!\!\widehat{-H} + R\!-\!\widehat{L} \longrightarrow Li^+\ H-\overset{\displaystyle H}{\underset{\displaystyle H}{Al}} + H-R + :L$$

The reagent is remarkably soluble in ether. It has been widely used in addition-reduction reactions (Chap. 12) and has a more limited application in simple nucleophilic substitution. The reagent attacks the more reactive alkyl halides and sulfonates and opens epoxide rings. The displacement is S_N2, for these hydrides are strong bases. Surprisingly, sodium hydride (NaH) does not act in this nucleophilic manner and so is of special importance used as a base, reacting only with hydrogen, never at carbon.

Synthetically, the displacement of leaving groups by LiAlH$_4$ is important since it serves to remove a functional group from the skeleton altogether, affording a hydrocarbon product. It does not react at tertiary sites and rarely causes elimination. All four hydrogen atoms are available as H^-. Lithium aluminum deuteride, LiAlD$_4$, has been used to introduce deuterium into organic molecules.

$$LiAl\ H_4 + 4(CH_3)_2CHCH_2\!-\!Br \xrightarrow{\text{Ether}} 4(CH_3)_3C\ H + \overset{+}{Li}\overset{-}{Al}Br_4$$

$$LiAl\ H_4 + 4CH_2\!\!\overset{\diagdown\ \diagup}{\underset{O}{-}}\!\!CHC\ H_3 \xrightarrow{\text{Ether}} 4C\ H_3CHCH_3 \xrightarrow{H_3O^+} 4CH_3CHCH_3$$
                         OM†   O   H
                           − +

$$LiAl\ H_4 + 4C_6H_5\underset{OCH_3}{CH}CH_2\ OTs \xrightarrow{\text{Ether}} 4C_6H_5\underset{OCH_3}{CH}C\ H_3 + \overset{+}{Li}\overset{-}{Al}(OTs)_4$$

2-Methoxy-2-
phenylethyl
p-toluenesulfonate

$$LiAl\ D_4 + C_6H_5\underset{Br}{CH}CH_3 \xrightarrow{\text{Ether}} C_6H_5\underset{D}{CH}CH_3$$

Ethylbenzene-1-d

---

†The symbol **M** represents any metal. The ether solution contains mixed lithium and aluminum alkoxides which might best be represented as **LiAl(OR)$_4$**.

11-6 FORMATION OF C—C BONDS: GENERAL

The forging of new carbon-carbon bonds is essential in any synthesis, and the substitution reaction is one of the major methods. The reaction requires a carbanion as a nucleophile and, since these are strong bases, proceeds by S_N2 displacement. When the focus of attention is on the carbanion component, the reaction is called an alkylation. Reference to Chap. 8 will show wide variations in the basicity of carbanions. The very strong bases, R:⁻, such as organolithium compounds (R—Li ⟷ R:⁻ Li⁺) and Grignard reagents (R—Mg X ⟷ R:⁻ Mg⁺⁺ X⁻, Chap. 12) are generally poor nucleophiles and rarely used for displacements. This is apparently because they too readily cause eliminations, acting as bases, and because the carbanion electron pair is too encumbered by its associated cation and solvent to act as a displacing nucleophile.

The principal carbon nucleophiles used for substitution are

⁻:C≡N: Cyanides

⁻:C≡C—{R, H}   Acetylides

⁻:C—C— ⟷ C=C—   Enolates (Sec. 11-7)

Although the stronger carbanions (R:⁻ as in RLi, etc.) are of little use as nucleophiles generally, in the case of displacement of strained cyclic ethers and epoxides they do function well, attacking at the less hindered carbon. Carbon chains may be extended by two or three carbon atoms by the reactions with ethylene oxide and trimethylene oxide. The four-membered ring in trimethylene oxide is somewhat strained, but the compound behaves much more like an ordinary ether than does ethylene oxide. As a consequence, opening the larger ring by nucleophilic substitution is often a slow and difficult reaction.

$$\phi MgBr + CH_2—CH_2 \xrightarrow{\text{Ether}} \phi CH_2CH_2OMgBr \xrightarrow{H_3O^+} \phi CH_2CH_2OH$$

2-Phenylethanol

$$C_2H_5MgBr + CH_2CH_2 \xrightarrow[\text{2) } H_3O^+]{\text{1) Ether}} C_2H_5CH_2CH_2CH_2OH$$

Trimethylene oxide   1-Pentanol

Cyanides (Nitriles)

One of the most useful methods of synthesis involves use of cyanide ion as a nucleophile. Substitution by cyanide ion provides a method of extending carbon chains by one unit and preserving a functional group at the end of the sequence. Primary halides and sulfonates are suitable substrates in such syntheses. Secondary compounds give poor yields of substitution products, and tertiary substrates undergo only elimination reactions when treated with metal cyanides. Polar solvents such as ethanol, acetone, dimethyl formamide, acetonitrile, and nitromethane are used in order to dissolve the ionic cyanide.

$$\overset{+}{K}\overset{-}{C}N + CH_3CH_2CH_2CH_2\!-\!Br \xrightarrow{\text{Acetone}} CH_3CH_2CH_2CH_2CN + KBr$$
$$\text{n-Valeronitrile}$$

$$\overset{+}{K}\overset{-}{C}N + C_6H_5CH_2\!-\!Cl \xrightarrow{\text{Ethanol}} C_6H_5CH_2CN + KCl$$
$$\text{Phenyl-}$$
$$\text{acetonitrile}$$

Neopentyl halides, and other substrates in which the approach to the functional carbon atom is sterically hindered, do not react with cyanide ion.

$$\overset{+}{K}\overset{-}{C}N + CH_3\overset{\overset{\displaystyle CH_3}{|}}{\underset{\underset{\displaystyle CH_3}{|}}{C}}CH_2Cl \longrightarrow \text{no reaction}$$
$$\text{Neopentyl}$$
$$\text{chloride}$$

The synthesis of nitriles is important primarily because of the ease with which the functional group can be converted to other groups of more general importance. Carboxylic acids are formed by the hydrolysis of nitriles (Chap. 13) and a number of reducing agents convert nitriles to amines.

$$CH_3CH_2CH_2CH_2C\!\equiv\!N + 2H_2O \xrightarrow{\text{HCl}} CH_3CH_2CH_2CH_2COOH + NH_4Cl$$
$$\text{n-Valeric acid}$$

$$\overset{+}{Li}\overset{-}{Al}H_4 + C_6H_5CH_2C\!\equiv\!N \xrightarrow{\text{Ether}} \left[(C_6H_5CH_2CH_2N)_4\overset{-}{Al}\overset{+}{Li} \right]$$

$$\text{or} \xrightarrow[\text{Pt}]{2H_2} C_6H_5CH_2CH_2NH_2$$
$$\text{β-Phenylethylamine}$$

with H^+, H_2O arrow.

Acetylenes

Acetylide ion is isoelectronic with cyanide and similarly has the anion stability associated with the electron pair in an *sp* orbital (page 317). It is more basic than cyanide, but both these anions have so little steric hindrance that displacement at carbon is relatively favored and attack on hydrogen (elimination) less common. Acetylene and monosubstituted acetylenes are sufficiently acidic to be converted to metal acetylides by treatment with strong bases. Sodamide, prepared by dissolving sodium in liquid ammonia in the presence of a trace of ferric ion, is customarily used in the preparation of acetylides.

$$2Na + 2NH_3 \xrightarrow{FeCl_3, NH_3} 2\overset{+}{Na}\overset{-}{NH_2} + H_2$$

$$NaNH_2 + RC{\equiv}CH \longrightarrow RC{\equiv}\overset{-}{C}\;\overset{+}{Na} + NH_3$$
$$\text{A sodium}$$
$$\text{acetylide}$$

Frequently acetylides are made by exchange reactions with more reactive organometallic compounds acting as bases.

$$\overset{-}{CH_3}\overset{++}{Mg}\overset{-}{Br} + H{-}C{\equiv}CR \longrightarrow RC{\equiv}CMgBr + CH_4$$
Grignard reagent

Use of acetylide anions as nucleophiles is a versatile means of extending carbon chains by two or more carbon atoms. Frequently both ends of the acetylene molecule can be used as nucleophiles.

$$HC{\equiv}\overset{-}{C}\;\overset{+}{Na} + CH_3CH_2{-}Br \longrightarrow CH_3CH_2C{\equiv}CH + NaBr$$
$$\text{1-Butyne}$$

$$CH_3CH_2C{\equiv}CH \xrightarrow{NaNH_2} CH_3CH_2C{\equiv}\overset{-}{C}\;\overset{+}{Na} \xrightarrow{(CH_3)_2CHI} CH_3CH_2C{\equiv}CCHCH_3$$
$$\overset{|}{CH_3}$$
$$\text{2-Methyl-3-hexyne}$$

The following sequence illustrates an exploitation of the difference between bromide and chloride as leaving groups. If a symmetrical dihalo compound had been used in the first step, a considerable amount of the diacetylene would have been produced.

$$HC{\equiv}\overset{-}{C}\;\overset{+}{Na} + BrCH_2CH_2CH_2Cl \xrightarrow[-30°]{NH_3} HC{\equiv}CCH_2CH_2CH_2Cl + NaBr$$
$$\text{5-Chloro-1-pentyne}$$

$$ClCH_2CH_2CH_2C{\equiv}CH \xrightarrow{NaNH_2, NH_3} ClCH_2CH_2CH_2C{\equiv}\overset{-}{C}\;\overset{+}{Na} \xrightarrow[-30°]{Br(CH_2)_3Cl}$$

$$ClCH_2CH_2CH_2C{\equiv}CCH_2CH_2CH_2Cl$$
$$\text{1,8-Dichloro-4-octyne}$$

$$2KCN + ClCH_2CH_2CH_2C{\equiv}CCH_2CH_2CH_2Cl \xrightarrow[Reflux]{Acetone}$$

$$NCCH_2CH_2CH_2C{\equiv}CCH_2CH_2CH_2CN$$

$$NCCH_2CH_2CH_2C{\equiv}CCH_2CH_2CH_2CN \xrightarrow[Reflux]{H_2SO_4, H_2O}$$

$$HOOCCH_2CH_2CH_2C{\equiv}CCH_2CH_2CH_2COOH$$
$$\text{5-Decynedioic acid}$$

A variety of reactions can be used to convert alkyl acetylenes to a host of other compounds, and these are discussed in subsequent chapters, but mention may be made of simple catalytic hydrogenation, which can be controlled to give either cis alkenes $(+1H_2)$ or saturated chains $(+2H_2)$.

$$RC\equiv CR + H_2 \xrightarrow{Pd} \underset{\text{Cis alkene}}{\overset{R}{\underset{H}{\diagdown}}C=C\overset{R}{\underset{H}{\diagup}}} \xrightarrow[Pd]{H_2} RCH_2CH_2R$$

11-7 FORMATION OF C—C BONDS: ENOLATE ALKYLATION

The most versatile and important alkylations are those of carbanions resonance-stabilized by adjacent electron-withdrawing groups, especially carbonyl. These carbanions are enolate anions and generally react by S_N2 displacement to afford alkylation on carbon. Simple enolates tend to be basic enough $(pK_a \sim 20)$ to promote elimination in many cases, but conjugated enolates, i.e., with the carbanion stabilized by two groups, are less basic $(pK_a = 5 \text{ to } 15)$ and are excellent nucleophiles.

The groups which stabilize adjacent carbanions, in order of their effectiveness, are

$$\underset{\text{Nitro}}{-\overset{O}{\underset{O^-}{\overset{\displaystyle\|}{N^+}}}} > \underset{\text{Ketone}}{-\overset{O}{\overset{\displaystyle\|}{C}}-R} > \underset{\text{Sulfone}}{-\overset{O}{\underset{O}{\overset{\displaystyle\|}{\underset{\displaystyle\|}{S}}}}-R} > \underset{\text{Ester}}{-\overset{O}{\overset{\displaystyle\|}{C}}-OR} \quad \underset{\text{Nitrile}}{-C\equiv N} > \underset{\text{Phenyl}}{\phi} \quad \underset{\text{Alkene}}{C=C}$$

A number of representative pK_a values for such compounds are shown in Table 8-2 and offer a measure of the relative stability of these carbanions.

When a simple ketone is treated with a strong base, the weakly acidic α-hydrogen is removed to form an enolate anion.

$$B: \longrightarrow H \overset{:O:}{\underset{\displaystyle\|}{\underset{\displaystyle|}{-C-C-}}} \rightleftharpoons \left[-\overset{:O:}{\underset{\displaystyle|}{\overset{\displaystyle|}{C}}}-\overset{\displaystyle|}{C}- \longleftrightarrow -\overset{:\overset{..}{O}:}{\underset{\displaystyle|}{C}}=\overset{\displaystyle|}{C}- \right] + BH$$

Ketone   Enolate anion
(conjugate acid)   (conjugate base)

The reaction is an acid-base equilibrium. For the enolate to be completely formed (equilibrium well to the right) bases above $pK_a = 20$ must be used, the usual choices being **NaH**, **NaHN$_2$**, **LiNEt$_2$**, ϕ_3**CNa**, or *t*-**BuOK** in solvents like 1,2-dimethoxyethane, liquid ammonia, or dimethylformamide.

When a compound with two stabilizing groups on one carbon (Table 11-1) is used, in some instances the compounds can be converted to salts even by aqueous base. Most compounds in the group, notably malonic esters, β-keto-esters, and β-diketones, are converted to sodium enolates by sodium alkoxides in alcoholic solutions. The dry salts can be isolated, but they are generally used directly in the solutions in which they are prepared. Even though the metal ions may be tightly bound to the anions as **metal chelates**, the chemistry of the compounds is usually discussed in terms of the reactions of the nucleophilic enolate ions.

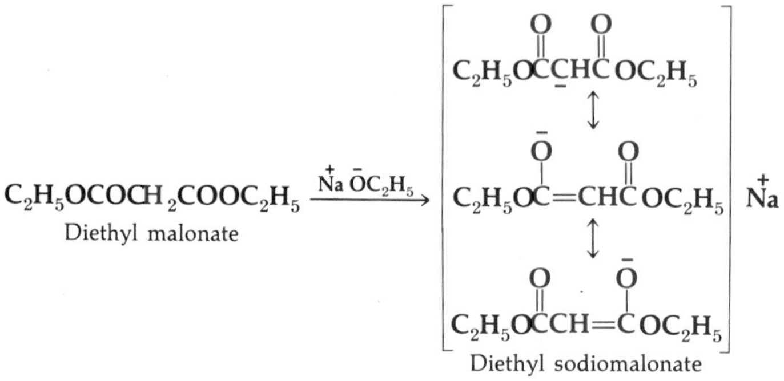

Diethyl malonate → Diethyl sodiomalonate

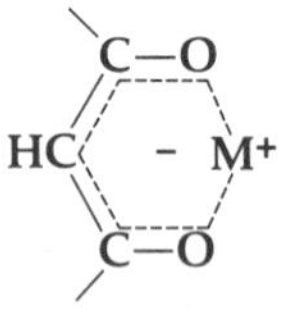

Metal chelate

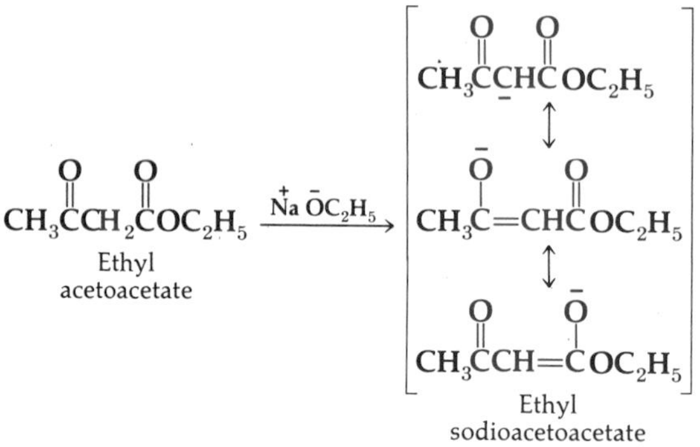

Ethyl acetoacetate → Ethyl sodioacetoacetate

Enolates can be prepared directly from sodium and the dicarbonyl compounds if a trace of alcohol is added as a catalyst. Sodium ethoxide is probably an intermediate in the reaction. Various other stronger bases can also be used in making these enolates. The thallium salts are reported to give especially clean monoalkylation reactions; these salts are quite insoluble chelates and easily isolated and purified before use.

The compounds of Table 11-1 are usually prepared by the reactions in Chap. 13 that yield β-dicarbonyl compounds, but some are made by substitution reactions:

$$K^+C\bar{N} + BrCH_2COOR \longrightarrow NC-CH_2COOR + KBr$$

$$Na^+ NO_2^- + BrCH_2COOR \longrightarrow O_2N-CH_2COOR + NaBr$$

$$\phi SO_2^-Na^+ + BrCH_2CO\phi \longrightarrow \phi SO_2-CH_2-CO\phi + NaBr$$

Alkylations of these various enolates proceeds at carbon when primary, unhindered halides (or sulfonates) are used. Secondary halides may also be used if their carbon chains are not branched, although the yields are not high since elimination reactions and O-alkylation of the ambient anions are undesired side reactions. Tertiary substrates give no C-alkylation.

In the alkylation of simple ketones two problems of control arise. The first is the question of preference between one or the other of the two α-sites adjacent to the carbonyl. The side bearing the more acidic hydrogen is enolized more rapidly and will therefore be the site of alkylation. With saturated α-alkyl groups the order of acidity—and hence preference for enolization—is $CH_3 > CH_2 > CH$, but the differences are small and mixtures of products from alkylation on each side of the carbonyl are commonly observed. With any unsaturation to favor the enolate (by resonance) on one side, however, alkylation proceeds on that side exclusively.

TABLE 11-1   **Complex Enolate Ions For Alkylation**

General carbanion stabilization:

$$R\!=\!C\!\overset{Y=Z}{\underset{Y'=Z'}{\diagdown}} \longleftrightarrow R\!-\!C\!\overset{Y-Z:^-}{\underset{Y'=Z'}{\diagdown}} \longleftrightarrow R\!-\!C\!\overset{Y=Z}{\underset{Y'-Z':^-}{\diagdown}}$$

Parent compounds† to form carbanion by removal of **H** by bases:

| Malonic esters | Cyanoacetate esters | Acetoacetic esters | Acetyl-acetone | Nitroacetate esters |

| Malononitrile | β-Ketosulfones; sulfone-esters | β-Keto-esters | β-Diketones | β-Aldehydo-ketones |

† All compounds are shown with an "active methylene" between the two activating groups but the replacement of one methylene hydrogen by alkyl still leaves a compound that can be converted to its enolate and alkylated.

$$\phi-\overset{\underset{1}{\uparrow}}{CH_2}-\overset{\overset{O}{\|}}{C}-\overset{\underset{2}{\uparrow}}{CH_3} \xrightarrow[\text{1,2-dimethoxy-ethane}]{\phi_3C:^-} \text{two enolates} \xrightarrow{CH_3I}$$

$$\phi-\overset{\overset{CH_3}{|}}{CH}-\overset{\overset{O}{\|}}{C}-CH_3 + \phi-CH_2-\overset{\overset{O}{\|}}{C}-CH_2-CH_3$$
$$\quad\quad 93\% \quad\quad\quad\quad\quad\quad <1\%$$

$$CH_3-\overset{\overset{O}{\|}}{\underset{\underset{2}{\uparrow}}{C}}-\overset{}{\underset{\underset{1}{\uparrow}}{CH_2}}-COOC_2H_5 \xrightarrow[\text{EtOH}]{\text{EtO}^-} CH_3-\overset{\overset{O^-}{|}}{\underset{1}{C}}=CH-COOC_2H_5 \xrightarrow{n\text{-}C_4H_9Br}$$

$$\downarrow \begin{matrix} KNH_2 \\ \text{liq. } NH_3 \end{matrix} \qquad\qquad CH_3-\overset{\overset{O}{\|}}{C}-\overset{\overset{}{|}}{\underset{n\text{-}C_4H_9}{CH}}-COOC_2H_5$$
$$\qquad\qquad\qquad\qquad\qquad 70\%$$

$$\begin{bmatrix} :\overset{..}{C}H_2-\overset{\overset{O^-}{|}}{C}=CH-\overset{\overset{O}{\|}}{C}-OC_2H_5 \\ \updownarrow \\ CH_2=\overset{\overset{O^-}{|}}{C}-CH=\overset{\overset{O^-}{|}}{C}-OC_2H_5 \end{bmatrix} \xrightarrow{CH_3I} \overset{\overset{CH_3}{|}}{CH_2}-\overset{\overset{O^-}{|}}{C}=CH-\overset{\overset{O}{\|}}{C}-OC_2H_5$$
$$\qquad\qquad\qquad\qquad\qquad\qquad\qquad \downarrow H^+$$
$$\qquad\qquad\qquad\qquad\qquad\qquad \overset{\overset{CH_3}{|}}{CH_2}-\overset{\overset{O}{\|}}{C}-CH_2-COOC_2H_5$$
$$\qquad\qquad\qquad\qquad\qquad\qquad\qquad \text{Predominant product}$$

In the case of highly stabilized enolates like acetoacetic ester, strong base (KNH_2) converts the compound to a *dianion* which now alkylates on the *less* acidic side of the ketone because, *once formed*, the enolate there is more basic, hence more nucleophilic and reactive.

The second problem is *dialkylation* at α-methylene positions. Once the enolate is discharged by the first alkylation, unreacted enolate can act as a base for rapid enolization of the monoalkylated ketone. The two enolates now compete for substrate, and product mixtures are formed corresponding to mono- and dialkylation.

$$\underset{\text{Monoalkylated}}{} \qquad\qquad \underset{\text{Dialkylated}}{}$$

Dialkylation is a serious, usually unavoidable problem with simple ketones but with the doubly activated methylenes (Table 11-1), of lower basicity, this is rarely a problem. Hence with the latter compounds it is possible to monoalkylate the active methylene with one substrate (**R—L**) and then alkylate a second time with a second substrate, with complete control.

Dialkylation†:

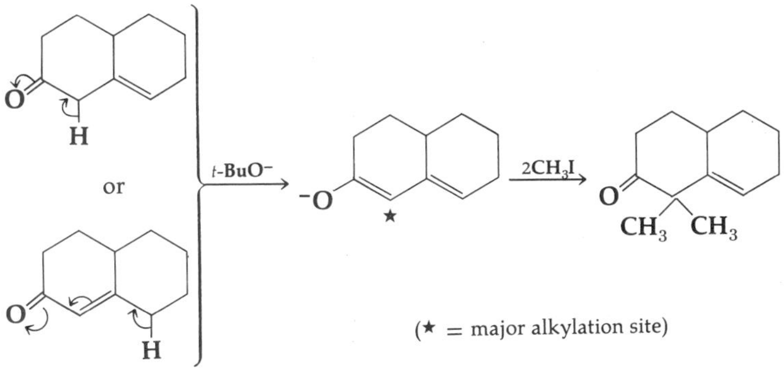

(★ = major alkylation site)

Monoalkylations:

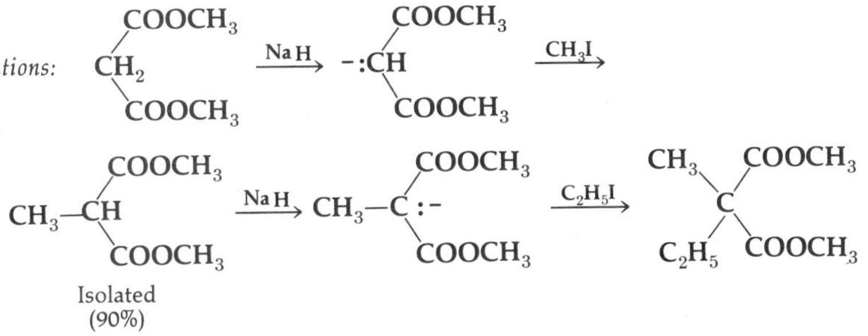

Isolated
(90%)

This very versatile selective double alkylation of doubly activated methylenes (Table 11-1) is a major method of building up branched-chain carbon skeletons. The method is often called the **malonic ester** or **acetoacetic ester synthesis** after the two most commonly used enolates. These enolates also react with epoxides at the less hindered carbon. An interesting variation of the malonic ester synthesis is used in the preparation of small rings (three, four, five, or six membered). The second alkylation, being internal, is now faster and dialkylation proceeds in one step, as expected.

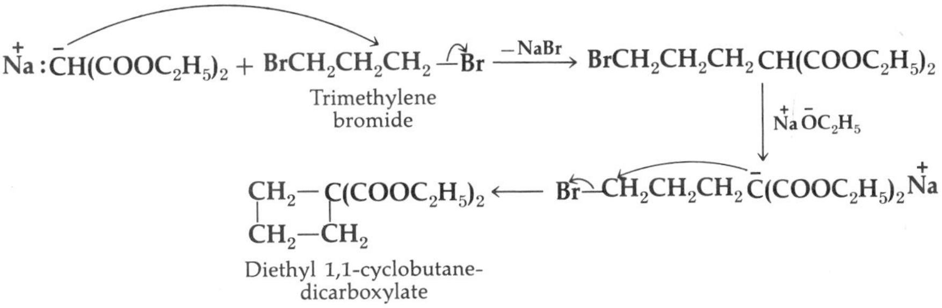

Trimethylene
bromide

Diethyl 1,1-cyclobutane-
dicarboxylate

---

†The lower case illustrates vinylogous enolization (Chap. 12): bond π electrons extend the activation of C—H to the γ-carbon and the enolate formed is the same in each case. Such complex anions commonly alkylate at the *center* carbon of the enolate.

The usefulness of these alkylations would be limited, however, if it were not possible to simplify the complex product by certain cleavages after it had served its purpose in creating new carbon-carbon bonds. These cleavages are discussed in the next two chapters but the main examples may be introduced here to show the utility of these syntheses for creating carboxylic acids and ketones.

$$R—CH_2Br + ^-:CH(COOR')_2 \longrightarrow R—CH_2—CH(COOR')_2 \xrightarrow[\Delta]{H_3O^+}$$
$$R—CH_2—CH_2—COOH + 2R'OH + CO_2$$

$$R—CH_2—Br + ^-:CH\begin{smallmatrix}COCH_3 \\ COOR'\end{smallmatrix} \longrightarrow$$

$$R—CH_2—CH\begin{smallmatrix}COCH_3 \\ COOR'\end{smallmatrix} \xrightarrow[\Delta]{H_3O^+} R—CH_2—CH_2—COCH_3 + R'OH + CO_2$$
$$\xrightarrow[\Delta]{^-OH} R—CH_2—CH_2—COOH + CH_3COO^- + R'OH$$

Two other examples may be briefly introduced. Phenols are enols but usually alkylate only on oxygen since alkylation on carbon must break up the benzene ring aromaticity in the transition state and so is relatively disfavored. Under certain conditions, however, C-alkylation does occur, as when a strongly hydrogen-bonding solvent is used to deactivate the more electronegative oxygen anion.

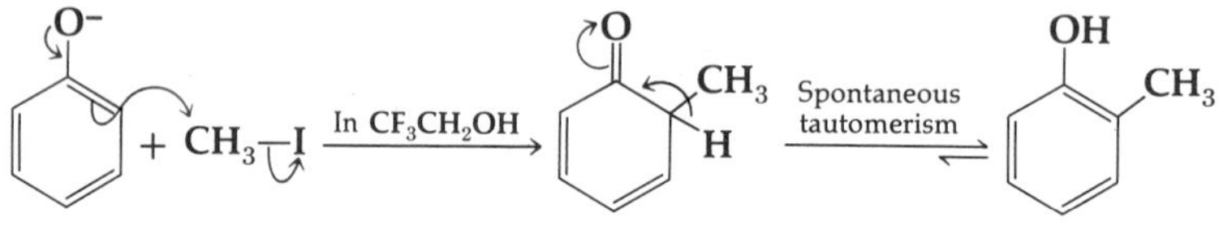

In the second example the nitrogen analogs of enols and enolates are employed instead. Preparation of these **"enamines"** (-ene + -amine = C=C—N) from the corresponding ketones is detailed in Chap. 12. Alkylation of enamines is facile but often not useful because they are ambident nucleophiles and prefer N-alkylation. On the other hand the magnesium salts apparently alkylate well on carbon, probably because the magnesium and solvent are strongly held by the nitrogen anion and deactivate it as a nucleophilic center.

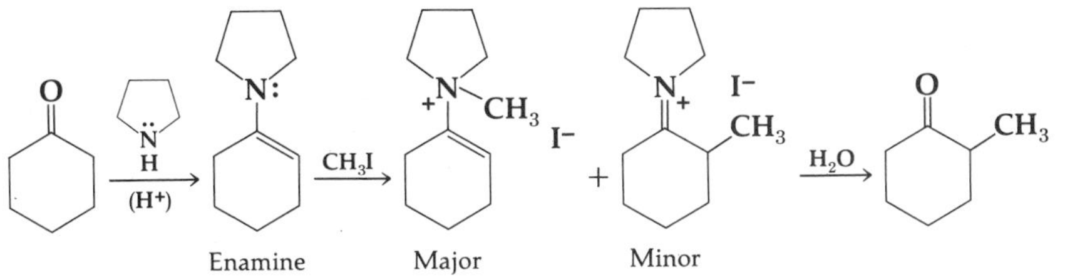

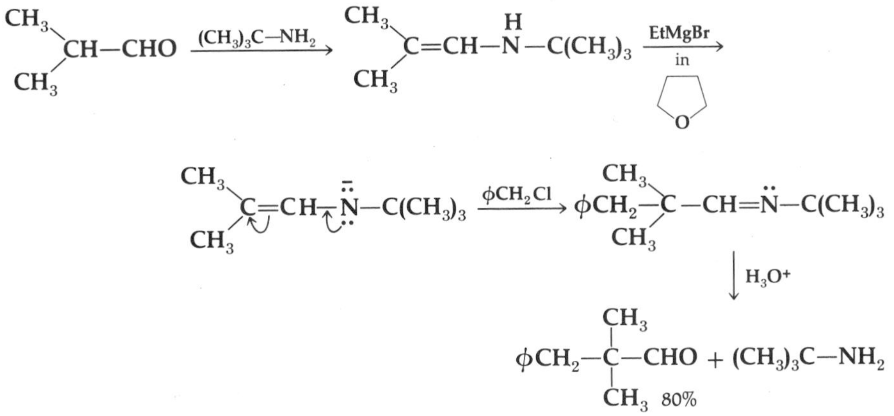

$$CH_3\text{-}CH\text{-}CHO \xrightarrow{(CH_3)_3C\text{-}NH_2} CH_3\text{-}C=CH\text{-}N\text{-}C(CH_3)_3 \xrightarrow[\text{in}]{EtMgBr}$$

$$CH_3\text{-}C=CH\text{-}N\text{-}C(CH_3)_3 \xrightarrow{\phi CH_2Cl} \phi CH_2\text{-}C\text{-}CH=N\text{-}C(CH_3)_3$$

$$\downarrow H_3O^+$$

$$\phi CH_2\text{-}C\text{-}CHO + (CH_3)_3C\text{-}NH_2 \quad 80\%$$

Among these several variants and a few more to be found in Chap. 12, we have a wide variety of control of carbon-carbon bond formation via enolate alkylation. The major carbon-carbon bond formations utilizing carbonyl groups as substrates are outlined in the next chapter, at the end of which is a section dealing with the use of all these procedures for synthetic purposes.

PROBLEMS

11-1 Write out syntheses of the following compounds, starting with the appropriate alkyl or aryl halides and other needed reagents. Indicate the side reactions you would expect, if any.

a sec-Butyl alcohol

b Methylacetylene

c Propyl-benzene

d Isopropyl n-propyl ether

e tert-Butyl mercaptan

f 2-Acetoxyhexane (2-hexyl acetate)

g 1-[1-Naphthyl]-1-dimethylaminoethane

h n-Hexane

i 1,2-Propanediol

j 1,3-Dimethoxybenzene (use a phenol as a starting material)

k Phenylacetic acid

11-2 Write out specific examples of the following reactions:

    a Two independent methods of extending a carbon chain by one carbon atom

    b Two independent methods of extending a carbon chain by two carbon atoms

    c Two independent methods of extending a carbon chain by three carbon atoms

11-3 In the following synthetic sequences, the structures of the intermediates are omitted. What are the missing links?

    a $HC{\equiv}CH \xrightarrow{NaNH_2} ? \xrightarrow[\substack{CH_2-CH_2\\ |\quad\quad|\\ O\text{---}CH_2}]{} ? \xrightarrow{SOCl_2} ? \xrightarrow{KCN}$

                                 $? \xrightarrow{H_3O^+,\ heat} HC{\equiv}CCH_2CH_2CH_2COOH$

    b $CH_3CH_2Br \xrightarrow{KN_3} ? \xrightarrow{H_2,\ Pt} ? \xrightarrow[excess]{CH_3I} CH_3CH_2\overset{+}{N}(CH_3)_3\overset{-}{I}$

    c $CH_2(COOC_2H_5)_2 \xrightarrow{Na} ? \xrightarrow{CH_3I} ? \xrightarrow{Na} ? \xrightarrow{CH_3(CH_2)_3Br} ?$

    d $CH_2{=}CHCH_2OH \xrightarrow{Na} ? \xrightarrow[\substack{CH_2-CH_2\\ \diagdown O \diagup}]{} ? \xrightarrow{PBr_3} ? \xrightarrow[2)\ H_3O^+]{1)\ LiAlH_4,\ ether} ?$

11-4 Starting with ethyl acetoacetate or dimethyl malonate and (or) any compounds containing not more than four carbon atoms, write out syntheses for the following compounds.

    a $CH_3CH_2\underset{\underset{CH_3}{|}}{C}HCOCH_3$ **d** $CH_3(CH_2)_4CH_3$

    b $(C_2H_5)_2CHCOOH$ **e** $(CH_3)_3C\underset{\underset{O}{||}}{C}CH_3$

    c $(CH_3)_2CH\underset{\underset{CH_3}{|}}{C}H\underset{\underset{O}{||}}{C}CH_3$ **f** $CH_3(CH_2)_5\underset{\underset{CH_3}{|}}{C}HCOOH$

11-5 Write the structure of all the products that might be expected from the following reactions.

    a $C_6H_5\overset{-}{S}\ \overset{+}{Na} + CH_3CH{=}CHCH_2OTs \longrightarrow$

    b $CH_3OH + (CH_3)_3CCH_2OTs \longrightarrow$

    c $CH_3CH_2\underset{\underset{NH_2}{|}}{C}HCH_3 \xrightarrow[H_2SO_4]{NaNO_2}$

    d $CH_3CH_2\underset{\underset{O}{\diagdown\diagup}}{CH\text{---}CH}CH_3 + HCl \xrightarrow{H_2O}$

    e $\overset{+}{Ag}\ \overset{-}{N}O_2 + CH_3CH{=}CHCH_2Br \longrightarrow$

f $CH_3CH_2Br + CH_3CH_2CH_2CH_2Br \xrightarrow{\text{Na}}$

g $CH_2N_2 + CH_3\underset{\underset{O}{\|}}{C}CH_2\underset{\underset{O}{\|}}{C}CH_3 \longrightarrow$

11-6   Write out synthetic schemes for the following substances using any two-carbon compounds as the only carbon source. Organometallic carbanions such as **R—Li** may be prepared from halides, $R-X + 2Li \longrightarrow R-Li + LiX$, and nonsubstitution reactions briefly introduced in this chapter for their synthetic value may be utilized. Any molecule once made may serve in another synthesis. Also note that hydration of double bonds occurs as $RCH=CH_2 + H_2O \longrightarrow$ **$RCH(OH)CH_3$**.

a	Propionitrile	**n**	α-Methylpropionitrile
b	Methylpropionate	**o**	3-Methyl-1-pentanol
c	1-Butanol	**p**	5-Methyl-1-heptyne
d	3-Hydroxypropanoic acid	**q**	n-Hexane
e	Succinic (butanedioic) acid	**r**	2-Methylhexane
f	Dimethyl malonate	**s**	Methylethylamine
g	Thiomethylacetic acid	**t**	n-Butylamine
h	1-Butene	**u**	2-Methylpentanoic acid
i	2-Butene	**v**	3-Methylpentanoic acid
j	Isopropyl methyl ether	**w**	2-Aminopentane
k	Diethyl methylmalonate	**x**	$CH_3\overset{*}{C}H_2CH_3$ (C* is carbon-14 which is available as $KC^{14}N$)
l	2-Chloropropane		
m	2-Nitropropane	**y**	1,6-Dinitro-3-hexyne
		z	2,5-Dimethylhexanedioic acid

11-7   Write down, and try to synthesize, five different molecules with five carbons or more and various functional groups. Use the reactions of Prob. 11-6. Aim for some variety in skeleton and for the fewest possible reactions. Look for more than one synthetic route and compare them for economy (number of steps).

11-8   As a means of forming a cyclic compound, 5-bromo-1-pentyne was treated with several bases, such as **KNH_2** and **t-BuOK**, in order to initiate cyclization by removal of the acetylenic hydrogen.

   a   No cyclization occurred. What other reactions intervened?

   b   When **NaH** is used, the acetylenic carbanion is formed, but no cyclization occurs. Consider the stereoelectronic requirements of the reaction in order to explain this.

11-9   Problem 11-8 illustrates some of the obstacles faced in efforts to make cyclic
compounds by substitutions. What is the likelihood that these cyclizations
will occur? Is the base used the best choice?

a   4-Chloro-1-pentanol + NH_3 ⟶ + ?

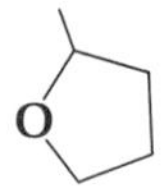

b $BrCH_2CH_2CH_2CH(COOCH_3)_2$ + $^-OCH_3$ ⟶ ?

c $BrCH_2C{\equiv}CCH(COOCH_3)_2$ + $^-OCH_3$ ⟶ ?

d *cis*- and *trans*-5-bromo-2-pentenyl dimethylammonium bromide +
$NaHCO_3$ ⟶ ?

11-10 Account for the following observations.

a

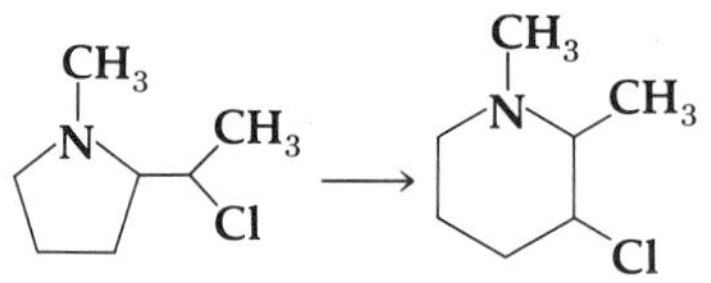

b $BrCH_2CH_2CHBrCH_2CH(COOCH_3)_2$ $\xrightarrow{\text{NaH}}$ $BrCH_2CH_2CH{-}C(COOCH_3)_2$ over CH_2

11-11 Devise synthetic routes to convert the following starting materials into the
products shown.

Starting materials	Products
a $(CH_3)_2C{-}CH_2$ (with epoxide O)	*t*-Butyl chloride; 1-nitro-2-methyl-2-chloropropane; 3-methyl-3-propenonitrile
b $(CH_3)_3CCH{-}CHCH_3$ (with epoxide O)	3-Iodo-2,2-dimethylpentane; 3-bromo-2,2,4-trimethylpentane
c $C_6H_5CH{-}CH_2$ (with epoxide O)	1-Iodo-2-phenyl-2-methoxyethane; ethylbenzene
d $(CH_3)_2C{-}CHCH_3$ with OH Cl	$(CH_3)_2C{-}CHCH_3$ with Cl OH
e $CH_3COCH_2COOCH_3$	3-Methyl-5-hydroxy-2-hexanone

11-12 a Consider the stereochemical requirements and results in the following interconversions. Fill in the necessary reagents, and draw the molecules in perspective.

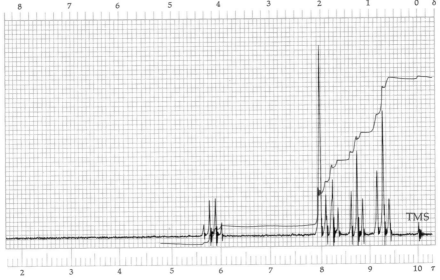

b What are the stereochemical requirements of 2-bromo-4-chlorocyclohex-anone which are necessary for its conversion into the following molecule? Show the steps in the conversion.

11-13 One of the possible pentachloropropanes has an NMR spectrum with two peaks, a doublet at τ 3.93 and a triplet at 5.48, in a ratio of 2:1, respectively. Which isomer is it?

11-14 A certain alkylation reaction yielded a product with IR peaks at 5.75 and 5.85 μ (1710 and 1740 cm^{-1}) and the NMR spectrum shown in Fig. P11-14. The mass spectrum showed a parent peak at m/e 186 and peaks at 171, 158, 73, and 43 (among others).

FIGURE P11-14

a Derive a structure for the alkylation product.

b Deduce the nature of the reactants used in the alkylation.

c Write the mechanism of the alkylation reaction.

d Rationalize the observed peaks in the mass spectrum.

11-15 Consider, as an example, the reaction between an alkyl halide and **NaOH** in a mixture of water and ethanol. In a table, with one column for S_N2 and another for S_N1, compare the two mechanisms with regard to:

a Stereochemistry

b Kinetic order

c Occurrence of rearrangements

d Relative rates of CH_3X, C_2H_5X, iso-C_3H_7X, t-C_4H_9X

e Relative rates of **RCl**, **RBr**, and **RI**

f Effect on rate of a rise in temperature

g Effect on rate of doubling **[RX]**

h Effect on rate of doubling **[OH⁻]**

i Effect on rate of increasing the water content of the solvent

j Effect on rate of increasing the alcohol content of the solvent

11-16 R-1-Phenyl-1-bromo-butane may be converted to the acetate (ester) of 1-phenyl-1-butanol either by refluxing in acetic acid (first case) or by the action of sodium acetate in an inert solvent such as dimethylformamide (second case).

a Write a detailed mechanism for each procedure and draw a qualitative energy diagram for each one.

b What stereochemistry would you expect in the product ester in each case and, briefly, why?

c What effect on the rates of reaction would be observed on adding silver salts in the first case; or adding excess potassium acetate in the second case?

d What effect on the rate of reaction would be occasioned by use in the first case of (i) the p-nitrophenyl analog; (ii) the p-hydroxyphenyl analog?

e What effect on the two cases would be caused by substitution of extra methyls at the 1-position or the 2-position?

f If 1-phenyl-1-Butanol itself were desired (instead of the ester), could sodium hydroxide be used instead of sodium acetate in the second case, and what effect would you expect on rate, stereochemistry, and yield?

CHAPTER TWELVE

CARBONYL AND RELATED GROUPS: NUCLEOPHILIC ADDITION

THE carbonyl group is the centerpiece of organic chemistry. Not only is it present itself in most of the main functional groups with more than one bond from carbon to heteroatom (Table 3-2), but it serves as a model for the reactions of all functionalities with π bonds between dissimilar atoms. Furthermore, its modes of reaction are basically simple, but they are very versatile in terms of synthetic utilization. In keeping with their importance, carbonyl compounds require two chapters of presentation, one primarily on addition reactions and the other on substitution, with a summary for both. However, it is important to appreciate that both chapters simply explore variations on a single central mode of reactivity which unifies all carbonyl reactions. This first chapter opens with a general exposition of these reactions applicable to the substance of both chapters. This is designed to provide an introductory overall survey and to focus on the simple principles which motivate all carbonyl reactivity and allow reliable predictions to be made.

12-1 GENERAL CONSIDERATIONS OF CARBONYL REACTIVITY

Carbonyl reactivity is all presaged by the electron imbalance in the π bond between carbon and a more electronegative atom, or indeed between any two dissimilar atoms, as illustrated in these resonance forms.

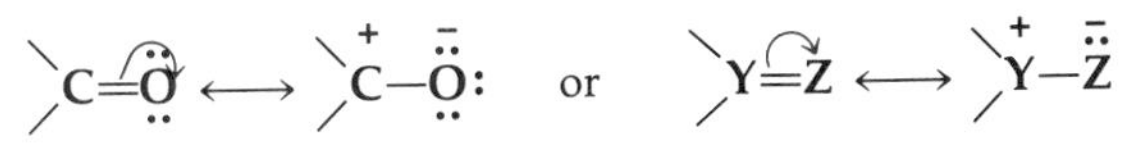

$$\underset{\text{Carbonyl group}}{\Large \sum C=\ddot{O}} \longleftrightarrow \overset{+}{\underset{}{\sum}}C-\overset{-}{\underset{..}{\ddot{O}}}: \quad \text{or} \quad \underset{\text{Related groups}}{\Large \sum Y=\ddot{Z}} \longleftrightarrow \overset{+}{\underset{}{\sum}}Y-\overset{-}{\underset{..}{\ddot{Z}}}$$

The consequences of this electron imbalance for reactivity are depicted in Fig. 12-1, which summarizes the possible first steps initiating almost all carbonyl reactions. The carbonyl dipole can do two things in the presence of a nucleophile/base:

1 It can enolize by attack of base on the α-hydrogen.
2 It can be attacked directly by a nucleophile in a nucleophilic addition.

In the presence of a Lewis acid (commonly H^+), the negative oxygen is first bonded, leaving a stabilized ($\sum\overset{+}{C}-\ddot{O}H \longleftrightarrow \sum C=O^+H$) carbonium ion which can then do the same two things. Since the protonated carbonyl is even

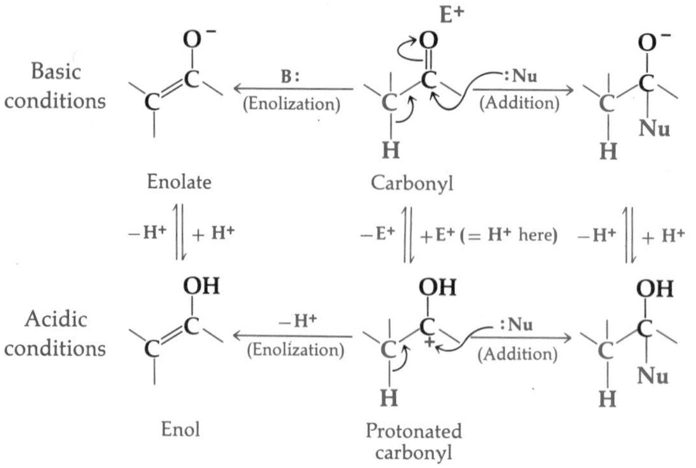

FIGURE 12-1   **The central reactivity of carbonyl and related groups: enolization and nucleophilic addition**

more reactive (higher energy), weaker bases (i.e., solvent) suffice for enolization and weaker nucleophiles for nucleophilic addition.

These reactions may be compared directly with those introduced in the preceding two chapters for groups with only a single bond to carbon, and they frequently occur under comparable acidic or basic conditions with comparable reagents. The comparison is closest with a singly bonded oxygen:

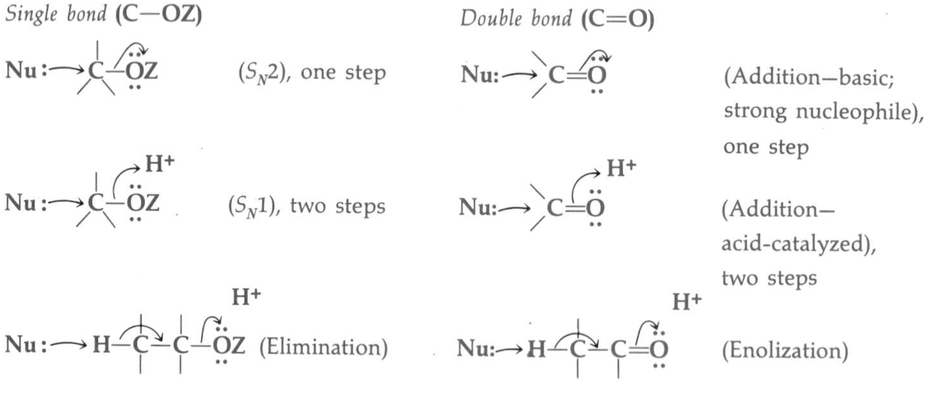

Single bond **(C—OZ)**   *Double bond* **(C=O)**

The corresponding reactions of carbonyl groups differ generally from those at saturated carbon in that carbonyl reactions are reversible equilibria, rapidly established, whereas substitutions at saturated carbon are usually practically irreversible. Therefore, where we were concerned with *rates* in substitution reactions, we shall now be concerned with *equilibria* in the reversible carbonyl reactions.

Subsequent Reaction Steps

The major first steps for carbonyl reactions are presented in Fig. 12-1, but often their products are reactive for some second step. Enolization was

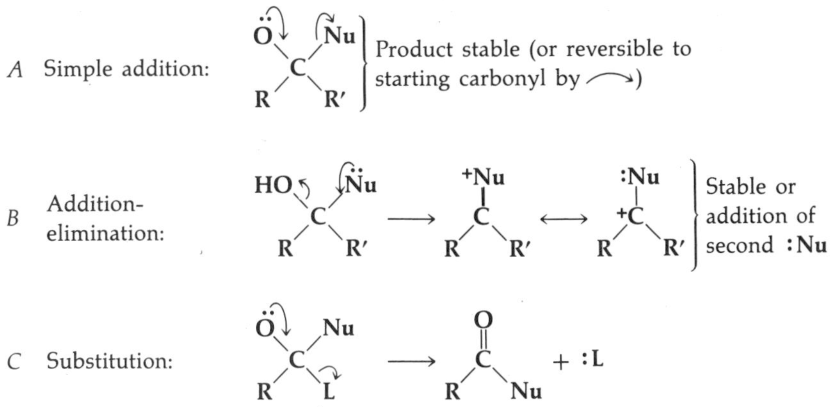

A Simple addition:

$$\overset{\overset{\cdot\cdot}{O}\downarrow\quad \overset{\frown}{Nu}}{\underset{R\quad R'}{C}}\;\Bigg|\;\text{Product stable (or reversible to starting carbonyl by } \frown\!\!\!\rightarrow)$$

B Addition-
     elimination:

$$\overset{HO\curvearrowright \;\overset{\cdot\cdot}{Nu}}{\underset{R\quad R'}{C}}\longrightarrow \overset{^{+}Nu}{\underset{R\quad R'}{\underset{|}{C}}}\longleftrightarrow \overset{:Nu}{\underset{R\quad R'}{^{+}C}}\;\Bigg|\;\begin{array}{l}\text{Stable or}\\ \text{addition of}\\ \text{second }:Nu\end{array}$$

C Substitution:

$$\overset{\overset{\cdot\cdot}{O}\downarrow\quad Nu}{\underset{R\quad L}{C}}\longrightarrow \overset{O}{\underset{R\quad Nu}{\overset{\|}{C}}}\; +\; :L$$

FIGURE 12-2 **Possible modes of reaction for initial addition product**

discussed in this connection in Chaps. 10 and 11, for the enol or enolate anion readily react further in substitution (alkylation) reactions, the enol usually giving O-alkylation by an S_N1 mechanism, the enolate anion giving C-alkylation by S_N2. Thus enolization produces nucleophiles, or electron donors, while the carbonyl group itself is an electron acceptor (at carbon). This suggests that enols and enolate anions may react with other carbonyl compounds—or even their own parents. This is indeed the case and forms the subject of much of these two carbonyl chapters.

The initial product of nucleophilic addition to a carbonyl group (Fig. 12-1), on the other hand, is either stable itself or has three courses of action open to it. All three are analogous to the simple reverse reaction which reforms the parent carbonyl. In this reverse reaction the entered nucleophile leaves again, ionizing off to leave a carbonium ion (i.e., S_N1 reaction) which is stabilized by the electron pair on oxygen (*i*). Since this is a powerful stabilization it is equivalent to regard the nucleophile as eliminated by the oxygen electron pair (*ii*). The illustration is an acid-catalyzed example but the base-catalyzed case is the same without the oxygen protonated.

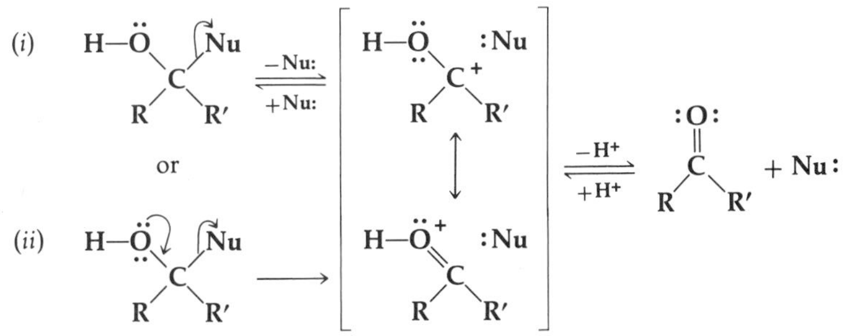

(*i*) $H-\overset{\cdot\cdot}{O}\;\overset{\frown}{Nu}$ ...

(*ii*) $H-\overset{\cdot\cdot}{O}\downarrow\;\overset{\frown}{Nu}$ ...

$$\overset{-H^{+}}{\underset{+H^{+}}{\rightleftharpoons}}\quad \overset{:O:}{\underset{R\quad R'}{\overset{\|}{C}}}\; +\; Nu:$$

The three courses, then, are simply the same elimination applied with the different possible groups leaving, i.e., —OH, Nu, or R', and these are summarized in Fig. 12-2 and on the following page.

A The addition product is stable. When the original carbonyl was bonded only to carbon and/or hydrogen (**R, R'** = **C, H**)—i.e., aldehydes and ketones—and the attached nucleophilic atom (**Nu**) has no other electron pair, the only course open to the addition product is the reverse reaction. The addition product is then stable if the nucleophile is a poor leaving group and the equilibrium favors the addition product. This is common with hydride and carbanion nucleophiles (Secs. 12-3, 12-4).

$$C_6H_5-\overset{\overset{\displaystyle O}{\|}}{C}-CH_3 + LiAlH_4 \longrightarrow C_6H_5-\overset{\overset{\displaystyle OH}{|}}{\underset{\underset{\displaystyle H}{|}}{C}}-CH_3$$

B If the nucleophilic atom (**Nu**) has another electron pair (or can develop one by easy loss of proton) this pair can equally serve to facilitate elimination and it is then the **—OH** (or **—OH$_2^+$**) which is eliminated. With oxygen nucleophiles (e.g., **Nu** = **R—OH**) the resultant oxonium ion adds a second nucleophile (Sec. 12-2) but with nitrogen and enolate nucleophiles, loss of a proton gives a stable doubly bonded product, discussed in Sec. 12-4.

$$CH_3CH_2CHO \begin{cases} \xrightarrow{+CH_3OH} & CH_3CH_2CH(OCH_3)_2 + H_2O \\ \xrightarrow{+H_2NOH} & CH_3CH_2CH{=}N{-}OH + H_2O \\ \xrightarrow[-OH]{+CH_3COCH_3} & CH_3CH_2CH{=}CHCOCH_3 + H_2O \end{cases}$$

C If the starting carbonyl bears a heteroatom (**R'** = **N, O, S, X**), as in the carboxylic acid class of functional groups, this can be the leaving group from the addition product. This results in a net effect of **nucleophilic substitution at unsaturated carbon** in the two steps. The relative facility of leaving groups parallels that in the nucleophilic substitution reaction at saturated carbon, discussed in the previous two chapters. Carbonyl substitution is discussed in Chap. 13.

$$C_6H_5COCl \begin{cases} \xrightarrow{+H_2O} & C_6H_5COOH + HCl \\ \xrightarrow{+C_2H_5OH} & C_6H_5COOC_2H_5 + HCl \\ \xrightarrow{+2CH_3NH_2} & C_6H_5CONHCH_3 + CH_3NH_3^+Cl^- \end{cases}$$

It should be noted that the nucleophilic substitution here is mechanistically different from that at saturated carbon since with carbonyl substitution, addition to the C=O double bond is the first step, not direct substitution. Direct substitution (e.g., S_N2) on double-bonded sites has rarely been observed either in O=C—L or C=C—L compounds. The latter compounds (leaving group on alkene) do *not* undergo substitution by the carbonyl two-step route either, since the alkene is not polarized for an initial addition. Substituents on alkenes or aromatic rings are virtually never substituted by nucleophiles.

Finally, the enolization step of Fig. 12-1 suggests that other α bonds besides those to hydrogen may be broken in the same way to supply electrons to the electron deficiency of the carbonyl carbon. We shall encounter examples of such reactions in the sections to come. If the bond broken at the α-position is a single bond the reaction is a **fragmentation**; if it is double, it is a **vinylogous enolization**,† or **vinylogous** (or **conjugate**), **addition**. The latter cases are those of the conjugated unsaturated ketone, acid, ester, etc. The mechanism is fundamentally the same but the reactivity at the double bond is somewhat diminished.

Fragmentation:

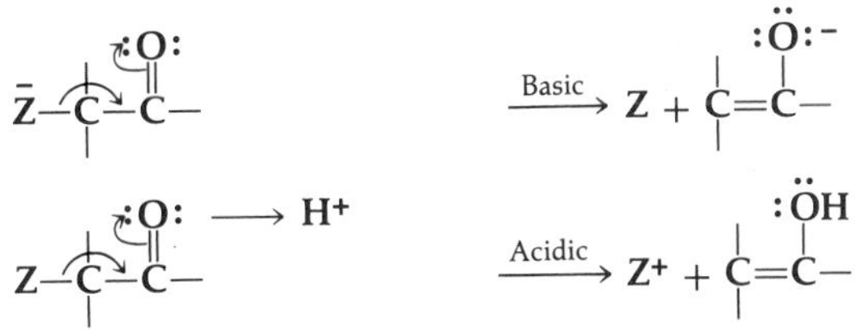

Vinylogous enolization:

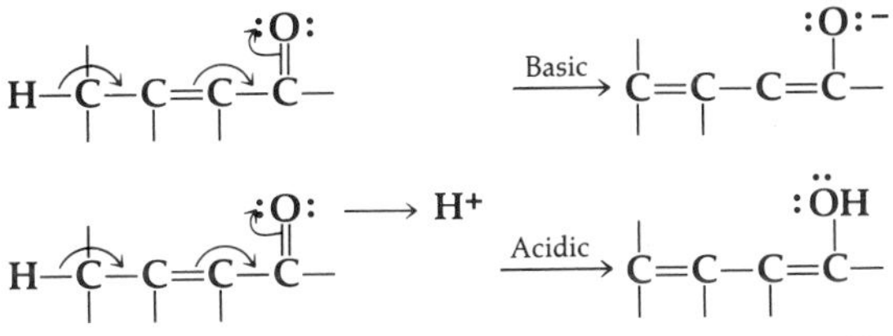

Vinylogous addition:

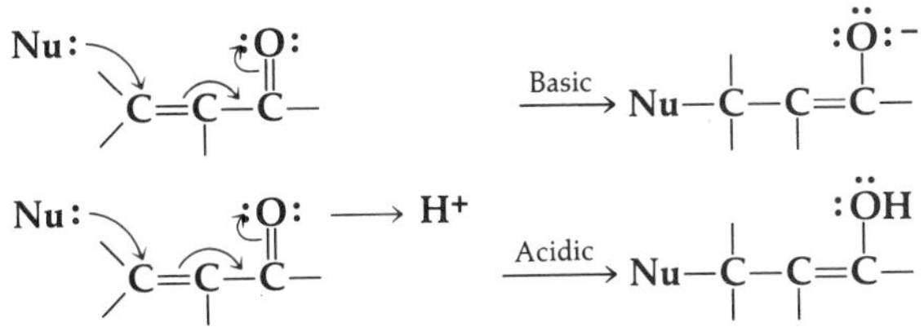

†A **vinylogous** reaction or a vinylogous site of reactivity is one in which the same reaction can occur at the β-carbon of a double bond conjugated to the reaction site. An unsaturated ketone is a **vinylog** of a ketone:

Examples of these processes are indicated here and discussed in Secs. 12-6, 13-5, 13-6, and 13-8.

Fragmentation: $CH_3-\overset{\overset{\textstyle O}{\|}}{C}-CH_2-\overset{\overset{\textstyle O}{\|}}{C}-OH \xrightarrow{\Delta} CH_3COCH_3 + CO_2$

Conjugate addition: $CH_2{=}CH-\overset{\overset{\textstyle O}{\|}}{C}-C_6H_5 \xrightarrow{+HCN} NC-CH_2-CH_2-CO-C_6H_5$

In all of these fundamental reactions we see a complete formal parallel between modes of carbonyl reaction with base (nucleophile) and with acid, the latter modes simply operating, with the same mechanisms, on the conjugate acids (protonated carbonyls) corresponding to the former. It is true of many carbonyl reactions, in fact, that they may proceed to the same products under either acidic or basic conditions.

PROBLEM 12-1

Write two specific examples that seem reasonable to illustrate each of the generalized reactions in this section. Wherever possible, show the steps involved in both acidic and basic catalysis of the example.

Stereochemistry

Like substitutions at saturated carbon, carbonyl reactions have stereoelectronic requirements, but they appear to be less critical. Carbonyl reactions occur in the π-electron plane, hence perpendicular to the molecular plane. In addition reactions the nucleophile approaches carbon in the π-electron plane so as to maximize the developing overlap of its own electron-pair orbital with the electron-deficient π orbital over the carbonyl carbon. In enolization the αC—H bond should roughly parallel the carbonyl π orbital, so that the orbitals may best overlap sideways to yield the enol or enolate anion.

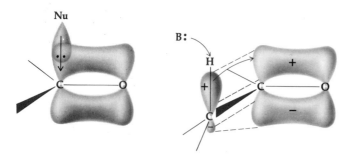

The stereochemistry of addition reactions may be considered since they generate a new asymmetric carbon from the carbonyl group. When no other asymmetric atoms are present in the molecule, the addition occurs equally well from either side and yields a racemic mixture. When other asymmetry is present in the molecule, this will influence which side of the carbonyl π orbital is more accessible (less hindered) and mixtures will result which are often not readily predictable. Carbonyl substitution reactions, on the other hand, involve no asymmetry in the product and hence no stereochemical problem; they yield only one product.

Enolization serves to racemize an asymmetric α-carbon to carbonyl since the enol has no asymmetry. The most stable configuration at the α-carbon will result, from equilibration by enolization.

Less stable epimer equilibrated   Enolate anion (or enol)   More stable epimer isolated

Related Functional Groups

Although carbonyl groups are the most studied examples of these reactions, other functional groups with π bonds between dissimilar atoms react in a similar way in most cases and these are collected in Fig. 12-3 together with their enolic analogs.

It represents a very important simplification for us to recognize that all of these different functions *react in the same fashion as carbonyl groups and their corresponding enols.* This is apparent in the collection of general classes of carbonyl reactions summarized in Fig. 12-4.

The reactivity of the parent group is always electron-accepting, the enolic form electron-donating. Only the relevant forms for basic media are shown in Fig. 12-3—the conjugate acids of each constitute the parallel functions for reaction in acidic media. All of these also show vinylogous enolization and addition when conjugated to double bonds. In the succeeding discussion a section (12-8) is devoted to addition reactions of these functional groups, but in the main their activity as enolates is not separately discussed since all of these simply represent stabilized carbanions and act as carbanion nucleophiles no matter which group stabilizes the carbanion.

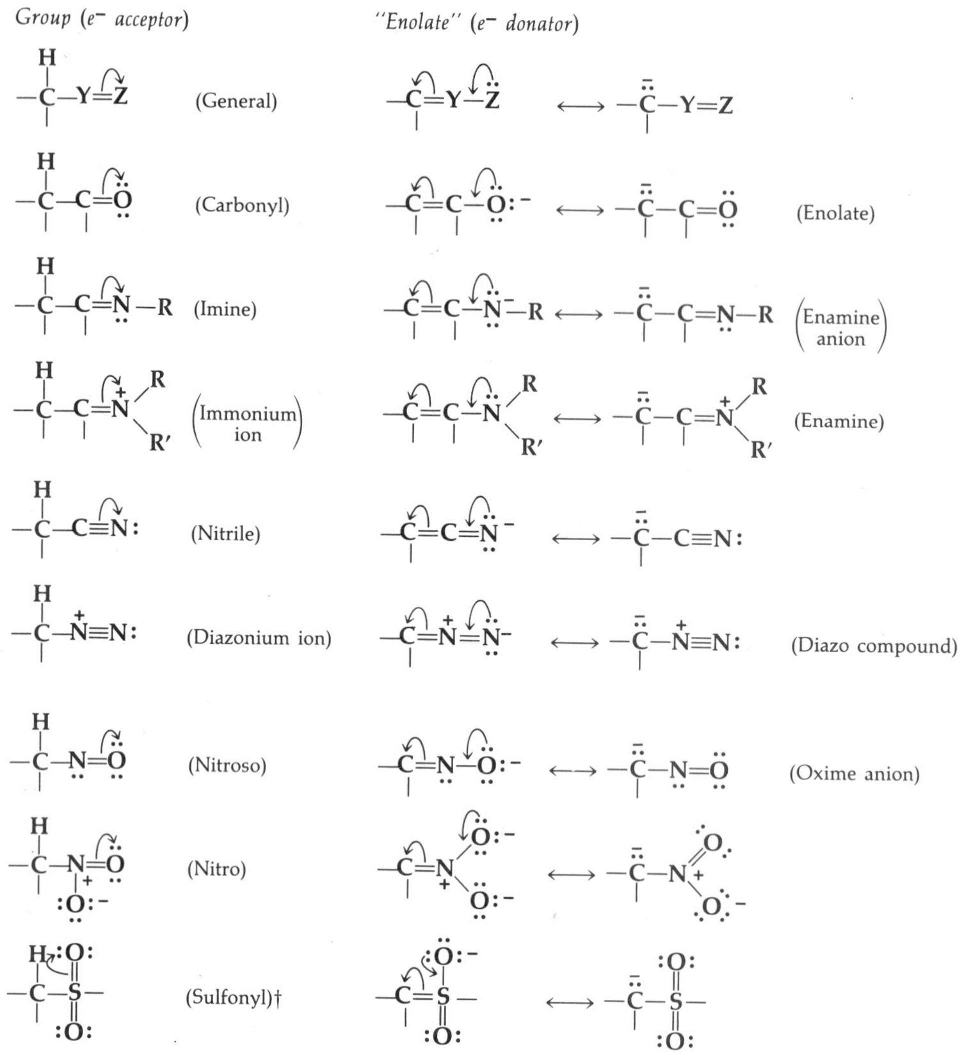

Group (e⁻ acceptor)   *"Enolate" (e⁻ donator)*

†The sulfonyl group often acts like carbonyl but its enolic form is not strictly parallel since the "enol" double bond here is not a π bond, as in the other cases, but a more complex bond involving *d* orbitals on sulfur. The two forms may also be written as follows:

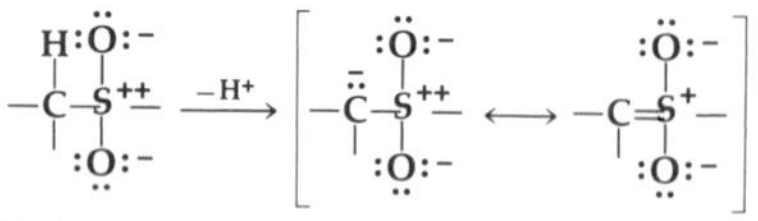

FIGURE 12-3   **Related functional groups and their enolic forms**

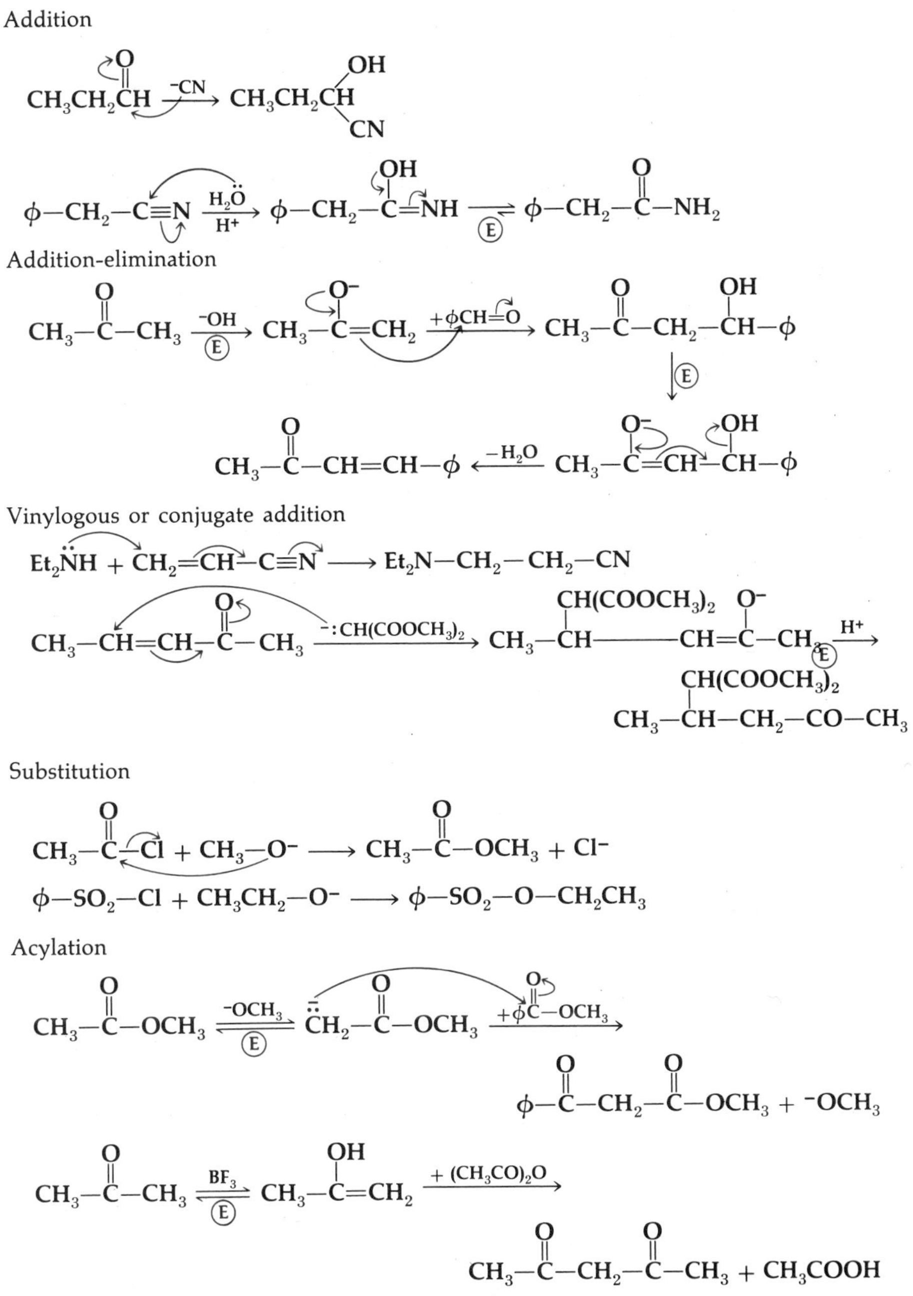

Addition

$$CH_3CH_2\overset{O}{\overset{\|}{C}}H \xrightarrow{\ ^-CN\ } CH_3CH_2\overset{OH}{\underset{CN}{\overset{|}{C}}H}$$

$$\phi-CH_2-C\equiv N \xrightarrow[H^+]{H_2\ddot{O}} \phi-CH_2-\overset{OH}{\overset{|}{C}}=NH \underset{(E)}{\rightleftharpoons} \phi-CH_2-\overset{O}{\overset{\|}{C}}-NH_2$$

Addition-elimination

$$CH_3-\overset{O}{\overset{\|}{C}}-CH_3 \xrightarrow[(E)]{\ ^-OH\ } CH_3-\overset{O^-}{\overset{|}{C}}=CH_2 \xrightarrow{+\phi CH=O} CH_3-\overset{O}{\overset{\|}{C}}-CH_2-\overset{OH}{\overset{|}{C}}H-\phi$$

$$\downarrow (E)$$

$$CH_3-\overset{O}{\overset{\|}{C}}-CH=CH-\phi \xleftarrow{-H_2O} CH_3-\overset{O^-}{\overset{|}{C}}=CH-\overset{OH}{\overset{|}{C}}H-\phi$$

Vinylogous or conjugate addition

$$Et_2\ddot{N}H + CH_2=CH-C\equiv N \longrightarrow Et_2N-CH_2-CH_2-CN$$

$$CH_3-CH=CH-\overset{O}{\overset{\|}{C}}-CH_3 \xrightarrow{^-:CH(COOCH_3)_2} CH_3-\overset{CH(COOCH_3)_2}{\overset{|}{C}H}\text{———}CH=\overset{O^-}{\overset{|}{C}}-CH_2 \xrightarrow[(E)]{H^+}$$

$$\overset{CH(COOCH_3)_2}{\underset{|}{}}$$

$$CH_3-CH-CH_2-CO-CH_3$$

Substitution

$$CH_3-\overset{O}{\overset{\|}{C}}-Cl + CH_3-O^- \longrightarrow CH_3-\overset{O}{\overset{\|}{C}}-OCH_3 + Cl^-$$

$$\phi-SO_2-Cl + CH_3CH_2-O^- \longrightarrow \phi-SO_2-O-CH_2CH_3$$

Acylation

$$CH_3-\overset{O}{\overset{\|}{C}}-OCH_3 \underset{(E)}{\overset{^-OCH_3}{\rightleftharpoons}} \ddot{C}H_2-\overset{O}{\overset{\|}{C}}-OCH_3 \xrightarrow{+\phi\overset{O}{\overset{\|}{C}}-OCH_3}$$

$$\phi-\overset{O}{\overset{\|}{C}}-CH_2-\overset{O}{\overset{\|}{C}}-OCH_3 + {}^-OCH_3$$

$$CH_3-\overset{O}{\overset{\|}{C}}-CH_3 \underset{(E)}{\overset{BF_3}{\rightleftharpoons}} CH_3-\overset{OH}{\overset{|}{C}}=CH_2 \xrightarrow{+ (CH_3CO)_2O}$$

$$CH_3-\overset{O}{\overset{\|}{C}}-CH_2-\overset{O}{\overset{\|}{C}}-CH_3 + CH_3COOH$$

FIGURE 12-4   **Examples of carbonyl reactivity (enolization reactions and their reverse are noted with Ⓔ)**

It is interesting to consider all the possible cases of the general function meeting "enolization" requirements. These may be created by substituting C, N, and O for X, Y, and Z in the general case below in all possible ways and include some nominally inorganic compounds like hydrazoic acid (HN_3) and cyanic acid (HNCO). There are many combinations possible and something is known about almost all of them although some are unstable and cannot be isolated.

$\overset{H}{\underset{\|}{X}}-Y=Z$ $\xrightarrow{-H^+}$	$\ddot{X}-Y=\overset{\curvearrowleft}{Z}$	*General*
$\overset{H}{\underset{\|}{R-N}}-\overset{\|}{C}=O$ $\longrightarrow$	$R-\overset{-}{N}-\overset{\|}{C}=O$	Amide
$\overset{H}{\underset{\|}{R-N}}-N=O$ $\longrightarrow$	$R-\overset{-}{N}-N=O$	Unstable N-nitroso $\rightarrow$ diazo-oxide
$\overset{H}{\underset{\|}{R-N}}-N=N-R'$ $\longrightarrow$	$R-\overset{-}{N}-N=N-R'$	Triazine
$\overset{H}{\underset{\|}{R-N}}-\overset{+}{N}\equiv N$ $\longrightarrow$	$R-\overset{-}{N}-\overset{+}{N}\equiv N$	Azide

Relative Reactivities of Carbonyl Groups

Three dominant influences control the reactivity of carbonyl groups, and more or less equally for both initial addition or enolization. These factors are summarized for common substrate types in Fig. 12-5.

1 *Resonance* of adjacent *p*- or *π*-electron pairs stabilizes the carbonyl and makes it *less reactive*. The order is generally the same as that observed for resonance shifts of infrared absorption maxima of carbonyls (Table 7-5) and for the same reasons.
2 *Inductive effects* due to polar substituents on adjacent carbons make the carbonyl *more reactive* owing to unfavorable dipole-dipole repulsions (i.e., the presence of two adjacent partial positive charges). This repulsion is lost on addition or enolization.

3 *Steric effects:* bulky groups adjacent to C=O cause more steric strain in the addition product than in the parent carbonyl and *reduce reactivity* toward addition, though they have little effect on enolization.

Aldehydes are more reactive than ketones largely because of the last two factors.

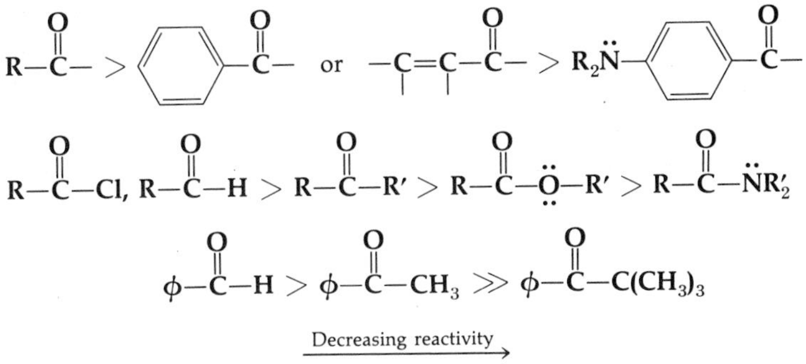

$$R-\overset{\overset{\displaystyle O}{\|}}{C}- \; > \; \phi-\overset{\overset{\displaystyle O}{\|}}{C}- \quad \text{or} \quad -C=C-\overset{\overset{\displaystyle O}{\|}}{C}- \; > \; R_2\ddot{N}-\phi-\overset{\overset{\displaystyle O}{\|}}{C}-$$

$$R-\overset{\overset{\displaystyle O}{\|}}{C}-Cl, \; R-\overset{\overset{\displaystyle O}{\|}}{C}-H \; > \; R-\overset{\overset{\displaystyle O}{\|}}{C}-R' \; > \; R-\overset{\overset{\displaystyle O}{\|}}{C}-\overset{\cdot\cdot}{\underset{\cdot\cdot}{O}}-R' \; > \; R-\overset{\overset{\displaystyle O}{\|}}{C}-\ddot{N}R'_2$$

$$\phi-\overset{\overset{\displaystyle O}{\|}}{C}-H \; > \; \phi-\overset{\overset{\displaystyle O}{\|}}{C}-CH_3 \; \gg \; \phi-\overset{\overset{\displaystyle O}{\|}}{C}-C(CH_3)_3$$

Decreasing reactivity $\longrightarrow$

FIGURE 12-5   **Relative reactivity of carbonyl groups**

Ketones can offer a choice of enolization in two directions. The more resonance-stabilized enol or enolate is always preferred. In saturated cases, where this is not a factor: *in base* the enolate from the less substituted side is favored; *in acid* the more substituted enol is preferred.

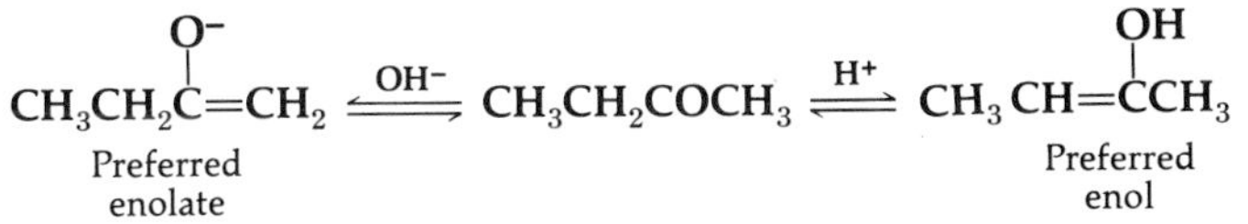

$$\underset{\substack{\text{Preferred}\\\text{enolate}}}{CH_3CH_2\overset{\overset{\displaystyle O^-}{|}}{C}=CH_2} \; \overset{OH^-}{\rightleftharpoons} \; CH_3CH_2COCH_3 \; \overset{H^+}{\rightleftharpoons} \; \underset{\substack{\text{Preferred}\\\text{enol}}}{CH_3\,CH=\overset{\overset{\displaystyle OH}{|}}{C}CH_3}$$

PROBLEM 12-2

In each of the following molecules, write the product expected from reaction with one equivalent of cyanide ion.

a $CH_3COCH_2CH_2CHO$ e *m*-Acetylbenzoyl chloride

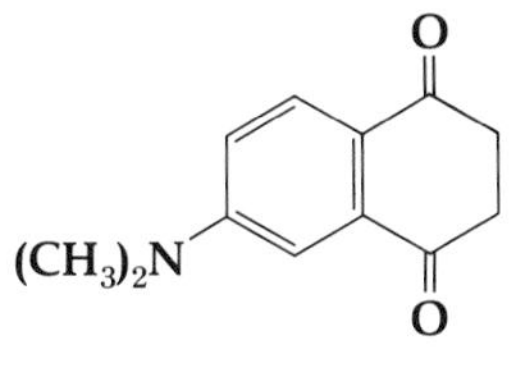

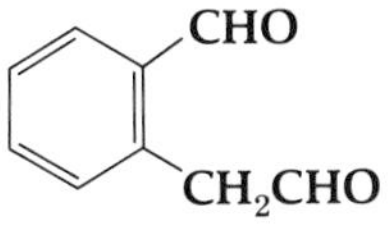

b $C_6H_5COCH_2CH_2COCH_3$ f Methyl *m*-acetylbenzoate

c

(CH_3)_2N

d

g

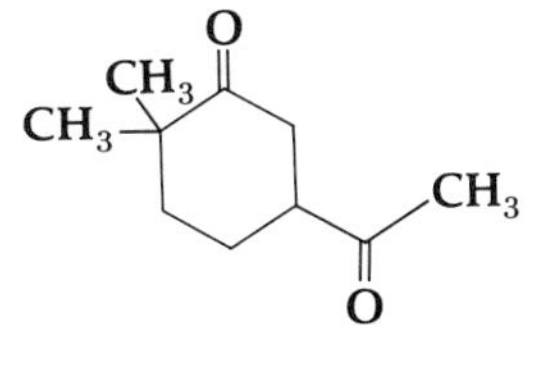

12-2   NUCLEOPHILIC ADDITION WITH HETEROATOMS (N, O, S, X)

Nucleophilic addition to carbonyl groups, as summarized in Fig. 12-1, can occur in general with the same set of nucleophiles considered in previous chapters. The reaction is reversible and occurs in basic media, or in acidic media which allow the existence of the free nucleophile (solvents and weak bases).

Halohydrins and Dihalides

Hydrogen halides add reversibly to aldehydes and ketones to give 1,1-halohydrins. These products cannot be isolated in pure form, because equilibrium is established too rapidly and the adducts reverse on isolation.

$$CH_3\overset{O}{\overset{\|}{C}}CH_3 + HCl \rightleftharpoons CH_3\overset{\overset{+}{O}H}{C}CH_3 + \bar{C}l \rightleftharpoons CH_3\overset{OH}{\underset{Cl}{C}}CH_3$$

2-Chloro-2-propanol

In alcohol solutions, α-haloethers are formed from 1,1-halohydrins by nucleophilic substitution reactions. Since equilibrium here can be established rapidly only in the presence of acid, α-haloethers may be isolated after the acid catalyst is neutralized, thus stopping the reverse reaction.

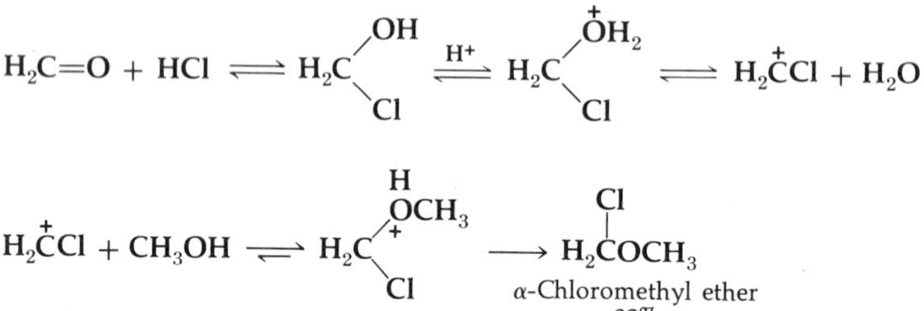

$$H_2C=O + HCl \rightleftharpoons H_2C\overset{OH}{\underset{Cl}{\diagup}} \overset{H^+}{\rightleftharpoons} H_2C\overset{\overset{+}{O}H_2}{\underset{Cl}{\diagup}} \rightleftharpoons H_2\overset{+}{C}Cl + H_2O$$

$$H_2\overset{+}{C}Cl + CH_3OH \rightleftharpoons H_2C\overset{\overset{H}{\underset{}{O}CH_3}}{\underset{Cl}{\diagup+}} \longrightarrow H_2\overset{Cl}{\underset{}{C}}OCH_3$$

α-Chloromethyl ether
89%

Aldehydes and ketones can be converted into *gem*-dihalides† by treatment with reagents such as phosphorus pentachloride or tribromide. The reactions probably involve addition of the halogen reagents to the carbonyl group, followed by a nucleophilic substitution reaction by halide ion.

---

†*gem-*, from *gemini* (twins), signifies two substituents on the same carbon.

$$PCl_5 + CH_3CH_2\underset{\underset{O}{\parallel}}{C}CH_3 \rightleftharpoons \left[CH_3CH_2\underset{\underset{OPCl_4}{|}}{\overset{\overset{Cl}{|}}{C}}CH_3 \right] \xrightarrow[(S_N1)]{Cl^-} CH_3CH_2\underset{\underset{Cl}{|}}{\overset{\overset{Cl}{|}}{C}}CH_3 + POCl_3$$

(PCl$_4^+$Cl$^-$)

Ethyl methyl ketone   (Not isolated)   2,2-Dichloro-butane 50%

$$PCl_5 + (CH_3)_2CHCH_2CHO \longrightarrow \left[(CH_3)_2CHCH_2\underset{\underset{OPCl_4}{|}}{\overset{\overset{Cl}{|}}{C}}HOPCl_4 \right] \longrightarrow$$

3-Methylbutyr-aldehyde   (Not isolated)

$$(CH_3)_2CHCH_2CHCl_2 + POCl_3$$

1,1-Dichloro-3-methylbutane 34%

The dihalides can be hydrolyzed to regenerate aldehydes or ketones or submitted to base-catalyzed elimination reactions to give vinyl halides and acetylenes (Chap. 14).

$$H_2O + \phi CHCl_2 \xrightarrow{\text{Heat}} \phi CHO + 2HCl$$

Benzal chloride   Benzaldehyde

$$CH_3CH_2CHO \xrightarrow{PBr_3} CH_3CH_2CHBr_2$$

Acetals, Ketals, and Hydrates

Addition of water or alcohols to aldehydes and ketones is similarly reversible and acid-catalyzed. The full sequence of events is outlined stepwise in Fig. 12-6. The initial addition products (hemiacetals, hemiketals, *gem*-diols) are unstable, the equilibria generally favoring the parent ketone or aldehyde.† With alcohols the subsequent steps are completely analogous to the first two, protonation of the —OH allowing it to ionize as in an S_N1 reaction, the carbonium ion being stabilized by the electrons on the adjacent oxygen (and being completely analogous to the original protonated carbonyl) for addition of a second alcohol molecule. The stabilization of the carbonium ion ensures a rapid rate for the ionization and for establishment of the equilibrium. The following examples show the overall reaction in particular cases.

$$CH_3CH_2CHO + C_2H_5OH \xrightarrow{H^+} CH_3CH_2CH\underset{OC_2H_5}{\overset{OC_2H_5}{<}} + H_2O$$

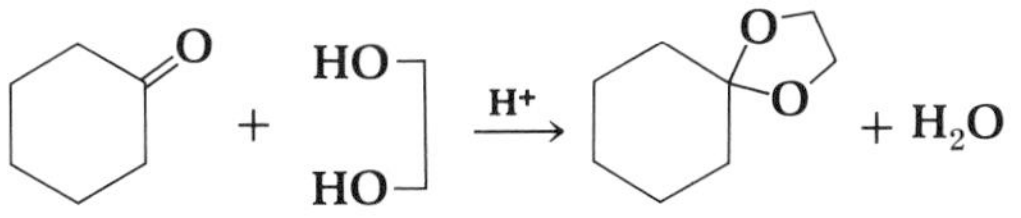

Ethylene glycol   Cyclic ketal

† Cyclic hemiacetals and hemiketals are exceptions, as discussed on page 355.

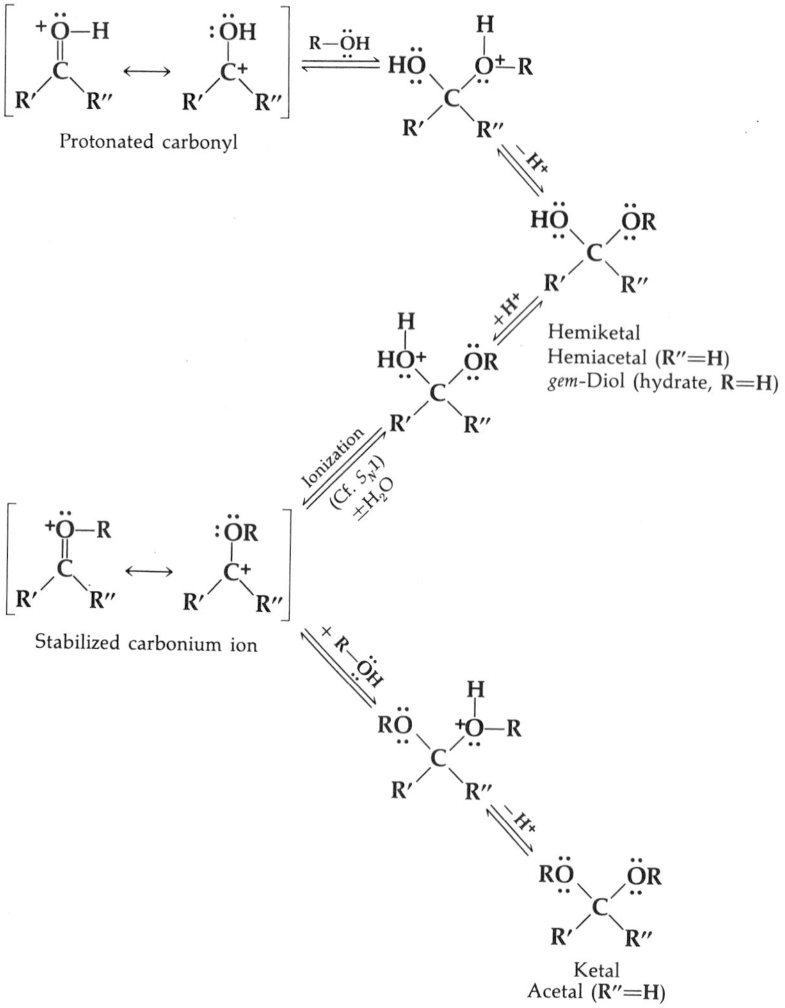

FIGURE 12-6 **Formation of ketals and acetals**

Only when the carbonyl compound is heavily substituted with electron-withdrawing groups are the *gem*-diols isolable. The stability of such diols is probably due to unfavorable dipole-dipole repulsion in the parent carbonyl compounds. Such hydrates are rare.

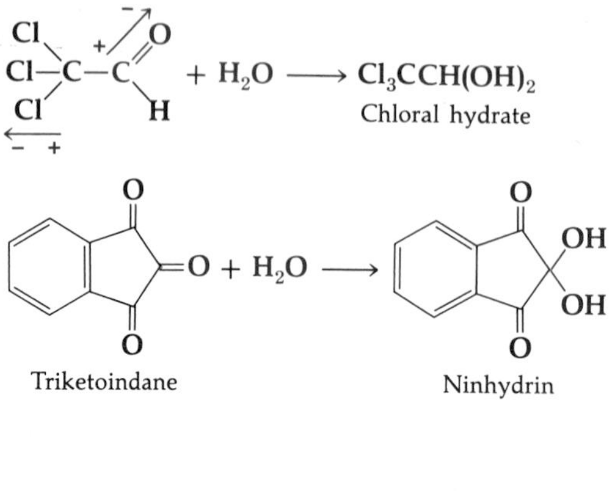

$$Cl_3CCH(OH)_2$$
Chloral hydrate

Triketoindane   Ninhydrin

Other carbonyl compounds are extensively converted to *gem*-diols in water solution, but attempts to isolate the diols fail because they are rapidly dehydrated during isolation. An example is *formalin*, the solution of 40% formaldehyde in water, which is used to preserve biological specimens. The amount of free formaldehyde in solution is almost undetectably small, but the diol cannot be isolated in an anhydrous condition.

$$H_2C=O + H_2O \rightleftharpoons H_2C\begin{array}{c} OH \\ OH \end{array}$$

Formaldehyde   Methanediol

Equilibrium between carbonyl compounds and acetals or ketals is rapidly established under acidic conditions. The substitution step is specifically acid-catalyzed. The diethers can be isolated by making the solution neutral or alkaline. Since the ether linkage resists attack by nucleophilic reagents under basic conditions, acetals and ketals are useful as **protective groups** in synthesis. A polyfunctional aldehyde or ketone is first converted to an acetal or a ketal. Reactions requiring basic conditions that would destroy the carbonyl function may then be safely carried out on *another* group in the molecule. Finally, the original carbonyl group is regenerated by hydrolysis of the acetal (ketal) group with dilute aqueous acid. Cyclic ketals from ethylene glycol, as in the example above, are often used synthetically for this purpose, commonly to prevent undesired enolization of a ketone when basic conditions are employed for some reaction elsewhere in the molecule.

$$CH_2=CHCHO + HC(OC_2H_5)_3 \xrightarrow{NH_4^+NO_3^-} CH_2=CHCH(OC_2H_5)_2 + HCOOC_2H_5$$

Acrolein   Ethyl ortho-   Acetal   Acrolein diethyl
                formate   exchange   acetal
                                              73%

$$Br_2 + CH_2=CHCH(OC_2H_5)_2 \xrightarrow{Bromine\ addition} BrCH_2CHBrCH(OC_2H_5)_2$$

$$2NaOH + BrCH_2CHBrCH(OC_2H_5)_2 \xrightarrow{Elimination} CH\equiv CCH(OC_2H_5)_2 + 2NaBr + 2H_2O$$

$$CH\equiv CCH(OC_2H_5)_2 + H_2O \xrightarrow{H^+} CH\equiv CCHO + 2C_2H_5OH$$

Propiol-
aldehyde

The above equations illustrate the use of ethyl orthoformate as a reagent for acetal synthesis under anhydrous conditions. This common procedure is an **acetal exchange** and involves the same mechanisms as in Fig. 12-6. Ammonium nitrate is used as a weakly acidic catalyst in the exchange reaction.

Cyclic hemiacetals and acetals are formed from 1,4- and 1,5-hydroxy-aldehydes and hydroxyketones, usually spontaneously. Ring formation of this type is common in carbohydrate chemistry (Secs. 25-3 and 27-3).

$$
\underset{\substack{\text{4-Hydroxybutanal}}}{\overset{\displaystyle\begin{array}{c}\text{CH}_2\text{CH}_2\text{CH}_2\\ \text{HO}\qquad\text{HC=O}\end{array}}{}} \rightleftharpoons \underset{\substack{\text{2-Hydroxytetra-}\\\text{hydrofuran}\\\text{(a cyclic hemiacetal)}}}{\overset{\displaystyle\bigcirc\!\!\!-\!\text{CHOH}}{}} \xrightarrow[\text{HCl}]{\text{CH}_3\text{OH}} \underset{\substack{\text{2-Methoxytetrahydrofuran}}}{\overset{\displaystyle\bigcirc\!\!\!-\!\text{CHOCH}_3}{}}
$$

A rough idea of the equilibrium position between these derivatives and the parent carbonyl can be obtained from consideration of bond energies:

$$
\underset{\text{H}}{\overset{\text{R}'}{>}}\!\!\text{C=O} + \text{R-O-H} \rightleftharpoons \underset{\text{H}}{\overset{\text{R}'}{>}}\!\!\text{C}\!\underset{\text{O-R}}{\overset{\text{O-H}}{<}}
$$

Bonds broken

C=O	+176 kcal/mole
O—H	+111
$\Sigma = +287$	

Bonds made

C—O	− 86 kcal/mole
C—O	− 86
O—H	−111
$\Sigma = -283$	

$$\Delta E = +4 \text{ kcal/mole}$$

The value of 4 kcal/mole favors the ketone slightly at equilibrium as is usually observed with the hydration of ketones. However, with values of ΔE so close to zero and other factors like entropy not considered, we can only say that the equilibrium does not favor either substance significantly, except in cyclic cases, in which the cyclized forms is generally favored (page 355). Ordinary ketals are formed by using an excess of an anhydrous alcohol with a catalytic amount of added acid, and the reverse reaction, hydrolysis to the carbonyl, is easily effected by warming in *aqueous* acids. In each case the excess of one reagent (ROH or H_2O) forces the equilibrium one way by the law of mass action.

Addition Compounds with Sulfur and Nitrogen

Thiols react even more rapidly than alcohols with aldehydes and ketones to give thioacetals and thioketals.

$$
2C_2H_5SH + C_6H_5CH_2COCH_3 \xrightleftharpoons{\text{Dry HCl}} \underset{\substack{\text{1-Phenyl-2,2-diethyl-}\\\text{mercaptopropane}}}{\overset{\substack{\text{SC}_2\text{H}_5\\|\\C_6H_5CH_2CCH_3\\|\\\text{SC}_2\text{H}_5}}{}}
$$

Benzyl methyl ketone

Aldehydes and ketones, such as acetone and cyclohexanone, which are not branched near the functional group add bisulfite ions in aqueous solution to form **bisulfite addition compounds**. These products are α-hydroxysulfonates, which can be crystallized as sodium salts.

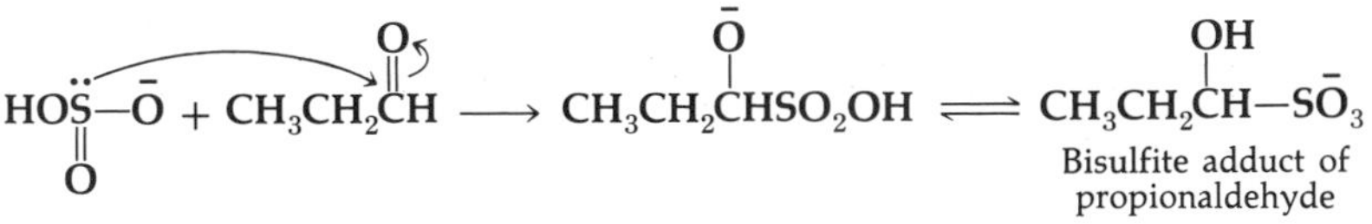

$$HO\overset{..}{\underset{\underset{O}{\parallel}}{S}}-\bar{O} + CH_3CH_2\overset{\overset{O}{\parallel}}{CH} \longrightarrow CH_3CH_2\overset{\overset{\bar{O}}{|}}{C}HSO_2OH \rightleftharpoons CH_3CH_2\overset{\overset{OH}{|}}{C}H-\bar{S}O_3$$

Bisulfite adduct of
propionaldehyde

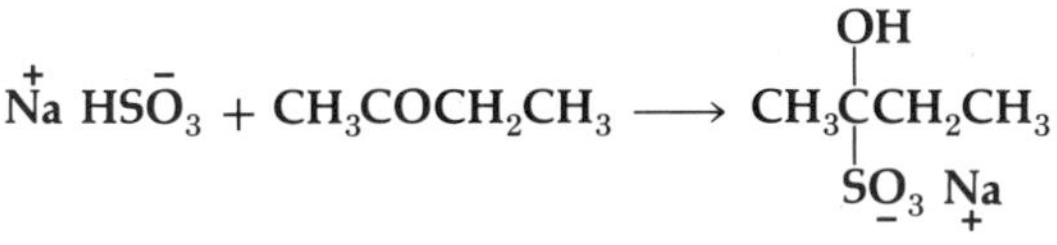

$$\overset{+}{Na}\ H\bar{S}O_3 + CH_3COCH_2CH_3 \longrightarrow CH_3\overset{\overset{OH}{|}}{\underset{\underset{\bar{S}O_3\ \overset{+}{Na}}{|}}{C}}CH_2CH_3$$

Carbonyl compounds are easily regenerated from their bisulfite addition compounds by treatment with either acid or base.

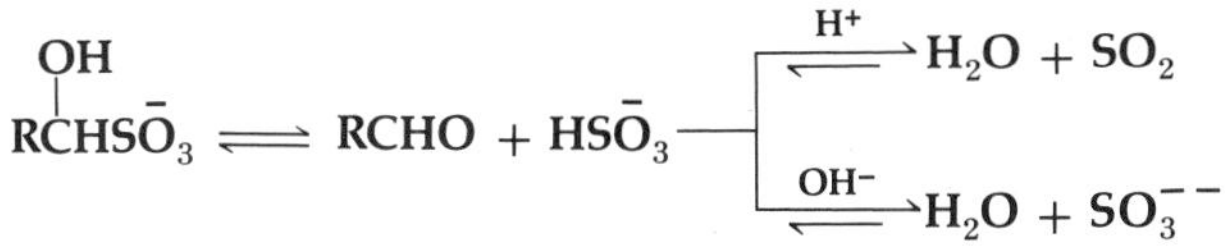

$$\overset{\overset{OH}{|}}{RCH\bar{S}O_3} \rightleftharpoons RCHO + H\bar{S}O_3 - \begin{cases} \overset{H^+}{\rightleftharpoons} H_2O + SO_2 \\ \overset{OH^-}{\rightleftharpoons} H_2O + SO_3^{--} \end{cases}$$

Aldehydes and ketones also add primary amines, secondary amines, and ammonia. The adducts, like *gem*-diols, are usually too labile to be isolated. Ammonia addition is typical of reactions involving uncharged nucleophiles. The neutral reactants first combine to give a charged intermediate that requires a proton shift to complete the addition.

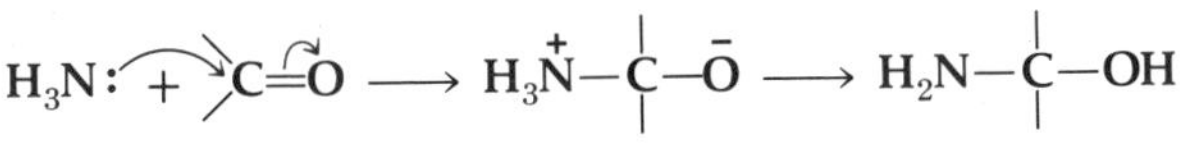

$$H_3N: + \ \ \overset{}{\underset{}{C}}\overset{\frown}{=}O \longrightarrow H_3\overset{+}{N}-\overset{|}{\underset{|}{C}}-\bar{O} \longrightarrow H_2N-\overset{|}{\underset{|}{C}}-OH$$

While such addition compounds are rarely stable themselves, they do easily eliminate water to form imines ($R_2C=N-R'$), and their enolic counterparts, enamines, as discussed in Sec. 12-5.

PROBLEM 12-3

Show preparations of each of the following from carbonyl compounds.

a C_6H_5 — (1,3-dioxane ring structure)

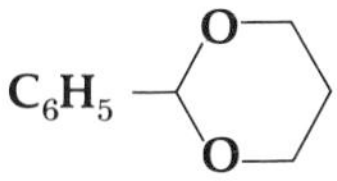

d $C_6H_5CH_2CCl_2CH_3$

b $CH_3CH_2CH_2\overset{\overset{OH}{|}}{\underset{\underset{CH_3}{|}}{C}}SO_3H$

e   (benzene ring fused with CHCl—O—C=O ring structure)

c CH_3O — (cyclic structure with O) — OCH_3

f   (cyclobutane ring with S—S spiro structure)

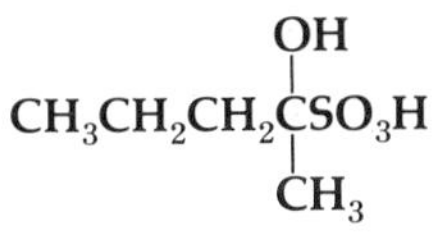

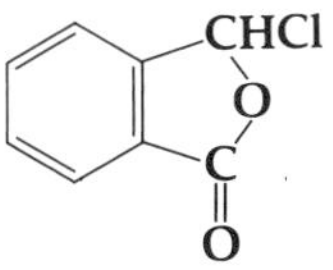

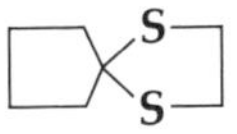

12-3   HYDRIDE DONORS AS NUCLEOPHILES

Metal Hydrides

Lithium aluminum hydride ($LiAlH_4$) serves as a **hydride donor** toward electron-deficient groups, mainly the carbonyl family. All four hydrogen atoms are available as negative hydrogen or hydride anion ($H:^-$). Metal (Li, Al) oxides are first produced from carbonyls and $LiAlH_4$, and these are hydrolyzed, when reaction is finished, to form alcohols, using dilute aqueous or alcoholic acid. Anhydrous ethers are commonly used as solvents for the reaction. The reaction constitutes a reduction of doubly bonded carbonyls to singly bonded alcohols and is the most important method of effecting this conversion. Unlike most other carbonyl reactions hydride addition is essentially irreversible owing to the high energy of the reagent. Enolization caused by hydride acting as a base is slower than carbonyl attack and almost never observed.

$$\overset{+}{Li}\overset{-}{Al}H_4 + \phi CHO \longrightarrow (\phi CH_2O)_4\overset{-}{Al}\overset{+}{Li} \xrightarrow{4HCl} 4\phi CH_2OH + AlCl_3 + LiCl$$

Lithium
aluminum
hydride   Benzaldehyde

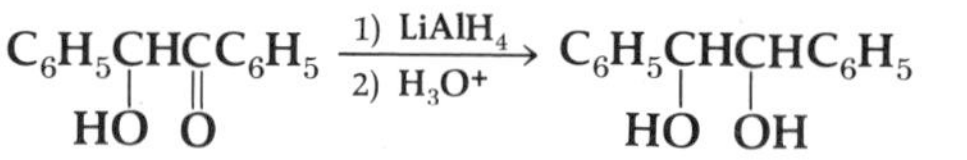

$$\underset{\underset{HO\ \ O}{|\ \ \ ||}}{C_6H_5CHCC_6H_5} \xrightarrow[\text{2) } H_3O^+]{\text{1) } LiAlH_4} \underset{\underset{HO\ \ OH}{|\ \ \ |}}{C_6H_5CHCHC_6H_5}$$

Benzoin   Hydrobenzoin
(1,2-diphenylethyleneglycol)

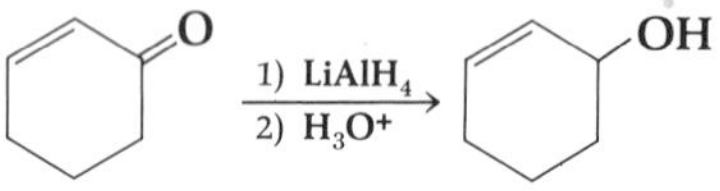

Hydride addition may most simply be regarded as an attack of $H:^-$. However, sodium hydride ($Na^+H:^-$) does not reduce carbonyls and it is presumed that the released AlH_3 acts as a Lewis acid to complex the carbonyl initially (E^+ in Fig. 12-1) for hydride attack. The reaction is often viewed as a concerted one in which the $H:^-$ attacks carbon as the oxygen is complexing with the released AlH_3:

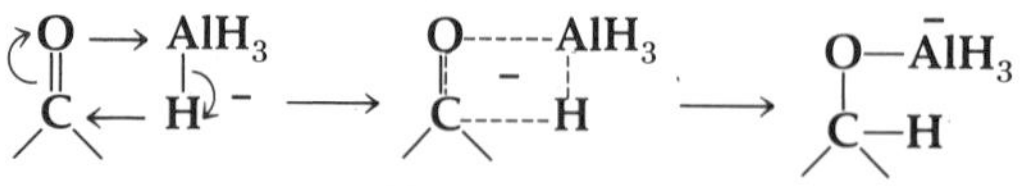

Transition state

Lithium aluminum hydride attacks all the many functional groups containing carbonyls as well as the related functions of Fig. 12-3. Reductions of these other functional groups are discussed in later sections (12-8 and 13-7) and also in Chap. 18.

Sodium borohydride is a less vigorous and more selective reagent than lithium aluminum hydride, although it acts in the same way. The former reduces aldehydes and ketones but does not react with esters, amides, nitriles, or nitro compounds under ordinary conditions. Water and alcohols may even be used as solvents for sodium borohydride reductions whereas they react vigorously with lithium aluminum hydride (yielding H_2).

$$CH_3CCH_2CH_2COOC_2H_5 \xrightarrow[\text{2) } H_3O^+]{\text{1) } NaBH_4} CH_3CHCH_2CH_2COOC_2H_5$$
$$\overset{\|}{O} \qquad\qquad\qquad\qquad\qquad\qquad \overset{|}{OH}$$

Ethyl 4-ketopentanoate   Ethyl 4-hydroxypentanoate

$$NCCH_2CH_2CHO \xrightarrow[\text{2) } H_3O^+]{\text{1) } NaBH_4} NCCH_2CH_2CH_2OH$$

3-Cyano-   4-Hydroxybutyronitrile
propionaldehyde

A number of partially substituted hydrides [e.g., $LiAl(OR)_3H$] have been used for special purposes in carbonyl reduction, and diborane ($B_2H_6 \rightleftharpoons 2BH_3$) reduces carbonyls as well as carbon-carbon double bonds. The latter is taken up as an electrophile in Chap. 15, and the various capabilities of all these hydride-donor reducing agents are explored further and summarized in the chapter on reduction (Sec. 18-6).

In converting an unsymmetrical ketone to a secondary alcohol these hydrides create a new asymmetric center. It is therefore of interest to inquire into the stereochemistry of this reduction. There are two opposed factors at work, however, which usually make prediction of the stereochemical result uncertain and very often result in mixtures of epimers. The first of these is steric hindrance favoring approach of the reagent from one side or another of the carbonyl. The second is steric hindrance in the developing product, which favors the more stable product. The mixture of epimeric alcohols which results in any given case will depend on the predominance of either this steric-approach control or product-development control in the transition state. If the transition state more closely resembles starting materials, steric-approach control predominates, but if it is more similar to the products, then product-development control is ascendant. In most cases mixtures are formed and in cyclohexanones the more stable equatorial alcohol usually predominates. In many cases prediction is risky since the position of the transition state on the (horizontal) reaction coordinate in the free-energy diagram cannot be accurately assessed.

4-tert-Butylcyclohexanone   Equatorial approach favored   Equatorial product favored
                                    (steric approach control)   (product development control)
                      10%   90%

Hydride Transfer from Carbon

Aldehydes that have no *α-hydrogen for enolization* disproportionate in the presence of strong base, giving equal amounts of the corresponding alcohol and carboxylic acid (**Cannizzaro reaction**). One molecule of aldehyde acts as a hydride donor and another functions as an acceptor.

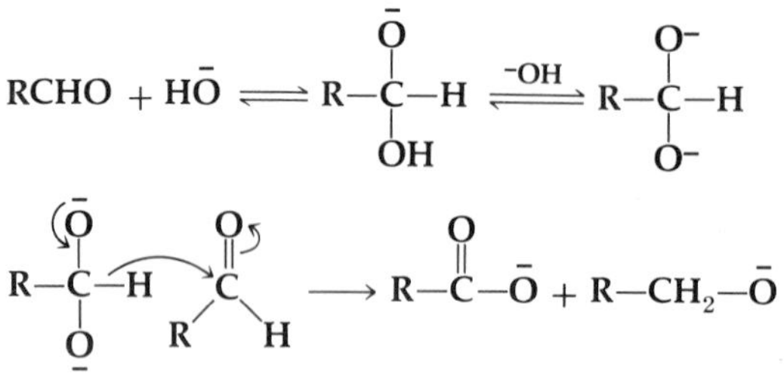

$$RCHO + H\bar{O} \rightleftharpoons R-\underset{\underset{OH}{|}}{\overset{\overset{\bar{O}}{\|}}{C}}-H \xrightarrow{-OH} R-\underset{\underset{O^-}{|}}{\overset{\overset{O^-}{|}}{C}}-H$$

$$R-CH_2-\bar{O} + H_2O \longrightarrow R-CH_2-OH + \bar{O}H$$

$$2(CH_3)_3CCHO + NaOH \longrightarrow (CH_3)_3CCH_2OH + (CH_3)_3CC\bar{O}_2\overset{+}{N}a$$

Trimethyl-   2,2-Dimethyl-1-   Sodium pivalate
acetaldehyde   propanol

Ketones and aromatic aldehydes react with ammonium formate at high temperatures to give primary amines (**Leuckart reaction**). The first step is probably a condensation of ammonia with the carbonyl compound to produce an imine (Sec. 12-5). Formate ion then serves as a hydride donor and reduces the imine to an amine.

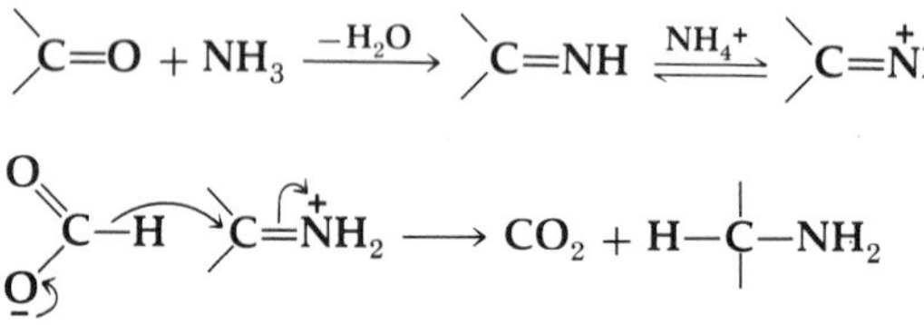

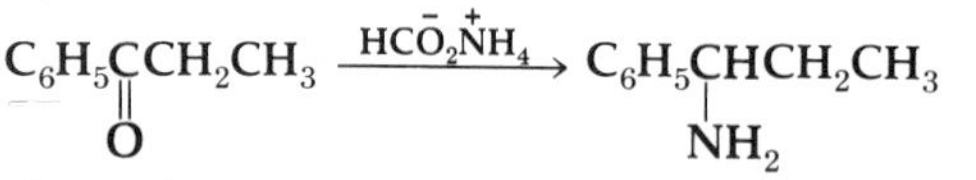

$$\overset{\diagdown}{\underset{\diagup}{C}}=O + NH_3 \xrightarrow{-H_2O} \overset{\diagdown}{\underset{\diagup}{C}}=NH \underset{}{\overset{NH_4^+}{\rightleftharpoons}} \overset{\diagdown}{\underset{\diagup}{C}}=\overset{+}{N}H_2$$

$$C_6H_5CCH_2CH_3 \xrightarrow{HC\bar{O}_2\overset{+}{N}H_4} C_6H_5CHCH_2CH_3$$

Propiophenone   1-Phenyl-1-propylamine
(ethyl phenyl
ketone)

Metal salts of primary and secondary alcohols can also transfer hydride to carbonyl groups of aldehydes and ketones. Aluminum is the metal most commonly used (**Meerwein-Ponndorf reduction**).

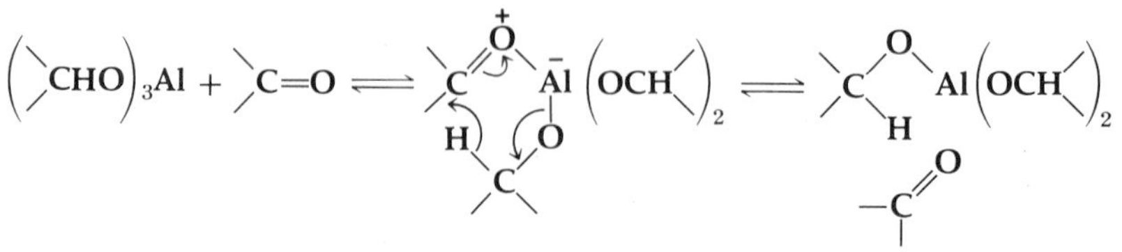

Since metal alkoxides undergo rapid acid-base exchange with their corresponding alcohols, the alcohol may be used as the source of hydrogen in the presence of a catalytic amount of aluminum alkoxide. Isopropanol and aluminum isopropoxide are frequently used. Reactions can be forced to high conversion by continuous removal, via distillation, of the resultant acetone from the isopropanol.

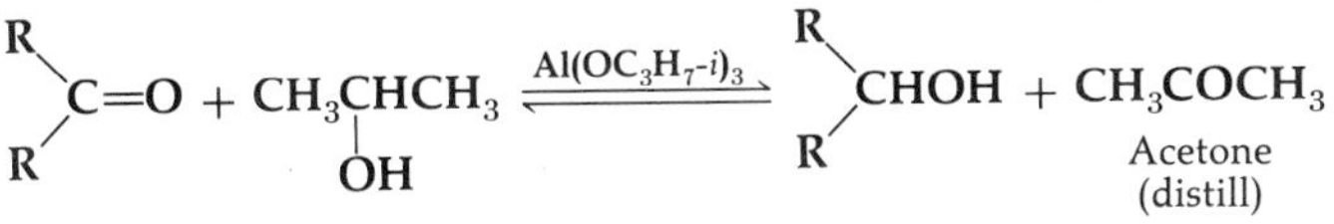

$$\underset{R}{\overset{R}{>}}C{=}O + CH_3CHCH_3 \underset{}{\overset{Al(OC_3H_7\text{-}i)_3}{\rightleftharpoons}} \underset{R}{\overset{R}{>}}CHOH + \underset{\substack{\text{Acetone}\\\text{(distill)}}}{CH_3COCH_3}$$

where the isopropanol bears an OH.

$$CH_3COCH_2CH_2CH_2Br + (CH_3)_2CHOH \xrightarrow{Al(OC_3H_7\text{-}i)_3}$$
5-Bromo-2-pentanone

$$CH_3\underset{OH}{CH}CH_2CH_2CH_2Br + CH_3COCH_3$$
5-Bromo-2-pentanol

$$(CH_3)_2CHOH + \underset{\text{Crotonaldehyde}}{CH_3CH{=}CHCHO} \xrightarrow{Al(OC_3H_7\text{-}i)_3}$$

$$\underset{\text{Crotyl alcohol}}{CH_3CH{=}CHCH_2OH} + CH_3COCH_3$$

The same reaction in reverse becomes a preparative method for oxidation of secondary alcohols to ketones if a large excess of acetone is used as a hydride-acceptor (**Oppenauer oxidation**). Aluminum *tert*-butoxide is then usually employed as the catalyst in Oppenauer oxidations. Cyclohexanone and benzoquinone have also been used as hydrogen acceptors in order to allow reactions to be run at temperatures above the boiling point of acetone.

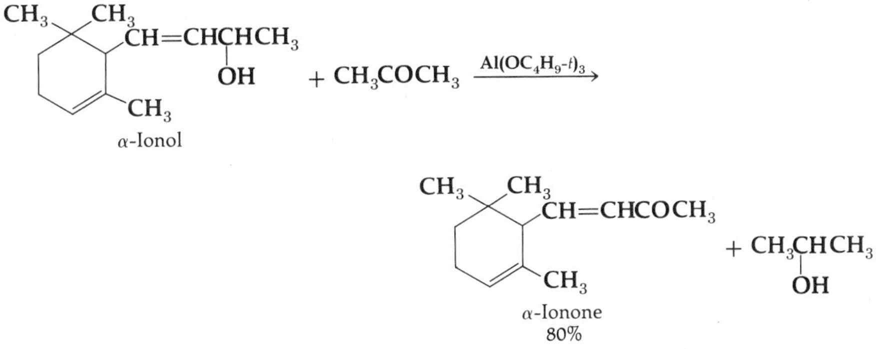

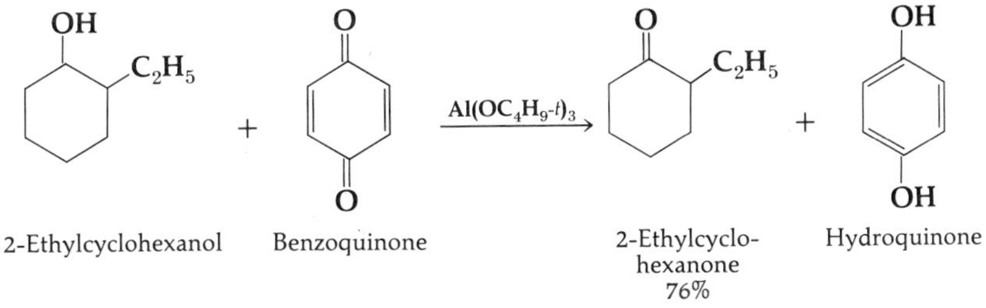

α-Ionol + CH_3COCH_3 $\xrightarrow{Al(OC_4H_9\text{-}t)_3}$

α-Ionone
80%
+ $CH_3\underset{OH}{CH}CH_3$

2-Ethylcyclohexanol   Benzoquinone $\xrightarrow{Al(OC_4H_9\text{-}t)_3}$ 2-Ethylcyclo-hexanone 76%   Hydroquinone

Both Ponndorf reductions and Oppenauer oxidations are particularly useful with polyfunctional molecules containing sensitive groups that are destroyed by the conditions of many other oxidations and reductions (Chap. 18).

PROBLEM 12-4

Fill in the missing blanks.

a   4-Cyano-2-pentanone + $NaBH_4 \longrightarrow$?

b   1-Chloro-3-pentanone + $NaBH_4 \longrightarrow$? $\xrightarrow{\text{LiAlH}_4}$?

c $C_6H_5CH{=}CHCOCH_3$ + ? $\longrightarrow C_6H_5CH{=}CHCHOHCH_3 \xrightarrow[\text{Pd}]{H_2}$?

d $CH_3CH{=}\underset{\underset{\displaystyle CH_3}{|}}{C}CHO + \bar{O}H \longrightarrow$?

e   ⬡=O + $Al[OCH(CH_3)_2]_3 \xrightarrow{+?}$? + ?

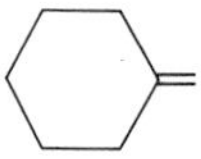

f $(C_6H_5CH_2)_2CO$ + ? $\longrightarrow$ 1,3-diphenyl-2-propylamine

PROBLEM 12-5

Explain with structures and curved arrows the fate of benzoquinone used in an Oppenauer oxidation

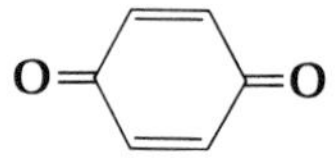

O=⬡=O

Benzoquinone

12-4   CARBANION ADDITIONS; ORGANOMETALLIC REAGENTS

Carbanions were discussed in Secs. 11-6 and 11-7 as important nucleophiles for carbon-carbon bond formation by substitution. The same carbanion reagents may attack ketones and aldehydes to form new skeletal bonds by nucleophilic addition.

Cyanides and Acetylides

Hydrogen cyanide addition to aldehydes and many of the more reactive ketones is catalyzed by traces of base. Need for a basic catalyst indicates that it is cyanide ion which makes the first attack on the substrate.

The α-hydroxynitriles produced in the reaction are commonly known as **cyanohydrins**. They can be converted to α-hydroxy acids by hydrolysis of the nitrile and are *dehydrated* to form α,β-unsaturated nitriles, acids, and acid derivatives. One of the commercial syntheses of methyl methacrylate, which polymerizes (Chap. 25) to form a transparent plastic known as Plexiglas or Lucite, is based upon the cyanohydrin reaction of acetone.

$$HCN + C_6H_5CH_2CHO \xrightarrow{\text{Aqueous NaOH}} \underset{\underset{OH}{|}}{C_6H_5CH_2CHCN}$$

Phenylacetaldehyde   2-Hydroxy-3-phenylpropionitrile
67%

$$HCN + CH_3COCH_3 \xrightarrow{\text{Aqueous NaOH}} \underset{\underset{CN}{|}}{\overset{\overset{OH}{|}}{CH_3CCH_3}} \xrightarrow{CH_3OH, H_2SO_4} \underset{}{\overset{\overset{CH_3}{|}}{CH_2{=}CCOOCH_3}}$$

Acetone   Methyl methacrylate
cyanohydrin   90%
78%

Reactivity of carbonyl compounds in cyanohydrin formation has been studied carefully, and is controlled by the three factors discussed on page 450. Most aldehydes react to give high conversions to products, although benzaldehyde reacts much more slowly than most aliphatic aldehydes. However, *p*-dimethylaminobenzaldehyde does not form a cyanohydrin because resonance interaction between the amino and carbonyl groups stabilizes the carbonyl and so disfavors the equilibrium to cyanohydrin. Steric hindrance is shown clearly by the comparison of acetophenone and *t*-butyl phenyl ketone. Under conditions that give 91% conversion of the methyl ketone, the *t*-butyl ketone gives only 46% of the addition compound.

Acetylene salts ($R-C{\equiv}C{:^-}\ M^+$) similarly add to ketones and aldehydes. Here the carbanion is more basic (and so higher in energy) than cyanide so that the equilibrium always strongly favors the addition products. The reaction is essentially irreversible in practice. One of the antiovulation hormones used in birth-control pills is synthesized by this reaction.

$$2CH_2O + HC{\equiv}CH \xrightarrow[90°]{CuC{\equiv}CCu} HOCH_2C{\equiv}CCH_2OH$$

2-Butyne-1,4-diol
92%

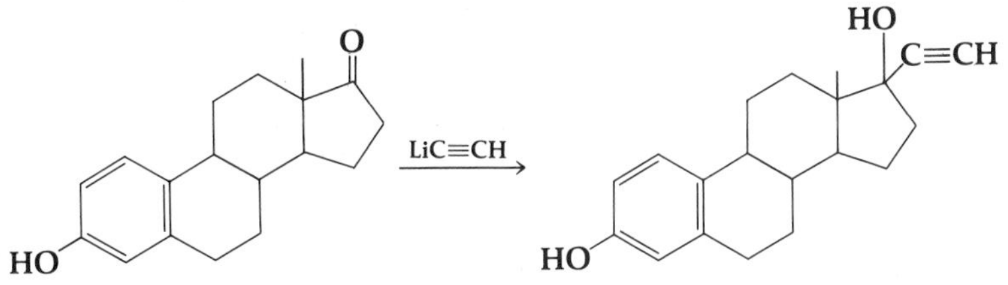

Ethinyl estradiol

Grignard and Related Reagents

Alkyl- and arylmagnesium halides (**Grignard reagents**) and **organolithium compounds** are decidedly nucleophilic and are very widely used in synthetic procedures. Although the compounds are not really ionized, much of their chemistry is easily predicted by regarding them as **carbanion donors**. While rarely useful in nucleophilic substitution at saturated carbon, they attack virtually all carbonyl groups with ease. They do not, however, generally attack carbon-carbon double bonds.

$$R—Li \longleftrightarrow \bar{R}: \overset{+}{Li} \qquad R—\overset{+}{Mg}\bar{Br} \longleftrightarrow \bar{R}: \overset{++}{Mg}\bar{Br}$$
Alkyllithium   Grignard reagent

Many methods have been developed for making magnesium and lithium compounds, but the most general is to treat an organic halide with a metal. Ethers are unique as solvents for preparation and use of the reagents. The ethers complex with empty orbitals on the magnesium and stabilize the combination.

$$CH_3I + Mg \xrightarrow{C_2H_5OC_2H_5} CH_3MgI$$
Methylmagnesium iodide

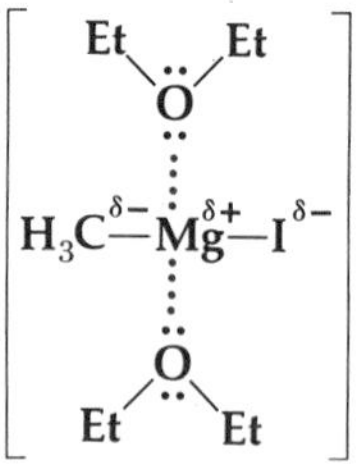

$$CH_3CH_2CH_2CH_2Cl + 2Li \xrightarrow{C_2H_5OC_2H_5} CH_3CH_2CH_2CH_2Li + LiCl$$
n-Butyllithium

$$C_6H_5Cl + Mg \xrightarrow{O} C_6H_5MgCl$$
Phenylmagnesium chloride

Lithium and magnesium compounds are very reactive and must be protected carefully from the atmosphere, because they react with oxygen, carbon dioxide, and water vapor. Since these carbanion derivatives are also strong bases, they readily react with weakly acidic hydrogens such as —OH and even $\rangle$NH, forming the conjugate base (alkoxides, etc.) and a hydrocarbon derived from the carbanion derivative.†

$$R—MgX + R'—OH \longrightarrow RH + R'—O^- Mg^{++} X^-$$

†This reaction represents a useful alternative mode to **LiAlH$_4$** for removing halide from a carbon skeleton and is more general since Grignard reagents are formed from almost any halide, which contains no other active functional group.

$$R—X \xrightarrow[Et_2O]{Mg} R—MgX \xrightarrow{H_2O} R—H + Mg(OH)X$$

Consequently, organometallic compounds cannot be used in hydroxylic solvents. Organometallics react with most other functional groups by addition or substitution; therefore, *halides that contain a second functional group cannot usually be converted to stable organometallic derivatives.* The only groups compatible with organometallic formation are the tertiary amines, ethers, alkenes, aromatic rings, and a few inert halides. Since a carbonyl group cannot be present in a Grignard reagent, ketones and aldehydes are commonly converted to cyclic ketals (acetals) as protecting groups to render them safe for Grignard formation,

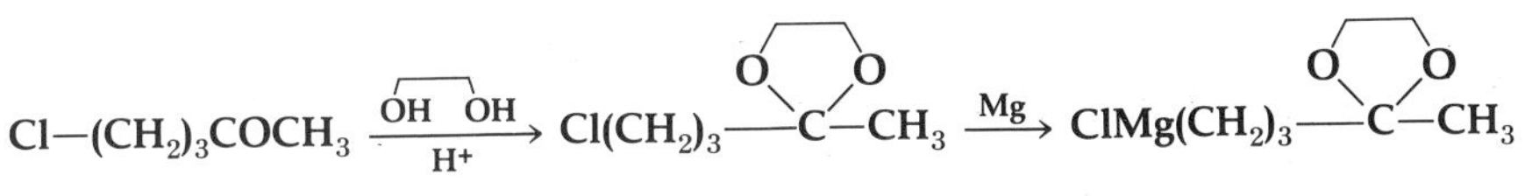

$$Cl-(CH_2)_3COCH_3 \xrightarrow[H^+]{OH \quad OH} Cl(CH_2)_3\overset{O \quad O}{-\overset{|}{C}-CH_3} \xrightarrow{Mg} ClMg(CH_2)_3\overset{O \quad O}{-\overset{|}{C}-CH_3}$$

The preparation of these organometallics can also be carried out by an acid-base reaction, sometimes called metalation in this context, in which a more basic carbanion is used to remove a proton from a more acidic hydrocarbon. Such reactions afford the best measure of the relative acidities of hydrocarbons. Grignard reagents are not sufficiently reactive to function well in most metalation reactions, probably because the carbanion is so strongly bound to the magnesium:

$$C_6H_6 + C_2H_5Na \xrightarrow{Pentane} C_6H_5Na + C_2H_6$$
Ethyl sodium   Phenyl sodium

$$C_6H_5Na + C_6H_5CH_3 \xrightarrow{Pentane} C_6H_5CH_2Na + C_6H_6$$
Phenyl sodium   Benzyl sodium

$$C_6H_5CH_2Na + (C_6H_5)_2CH_2 \longrightarrow (C_6H_5)_2CHNa + C_6H_5H_3$$
Benzyl sodium   Diphenylmethyl
                                       sodium

$$(C_6H_5)_2CHNa + (C_6H_5)_3CH \longrightarrow (C_6H_5)_3CNa + (C_6H_5)_2CH_2$$
Diphenylmethyl   Triphenylmethyl
sodium   sodium

The above reactions reflect the sequence of hydrocarbon acidities (Table 8-2): $Ar_3CH > Ar_2CH_2 > ArCH_3 > ArH > RH$. Comparisons of the reactivity of metal alkyls indicate the following stability order: primary > secondary > tertiary. Terminal acetylenes, in which the hydrogen is bound by an *sp* orbital, are so acidic that the compounds are easily metalated by Grignard reagents.

$$CH_3C{\equiv}CH + CH_3MgBr \longrightarrow CH_3C{\equiv}CMgBr + CH_4$$

In the metalation (or metal-hydrogen exchange) above, the acid-base mechanism required nucleophilic attack on hydrogen. An analogous reaction (metal-halide exchange) can be used with equal facility and involves nucleophilic attack on halide instead of acidic hydrogen:

$$R{:}^- + H{-}R' \longrightarrow R{-}H + {}^-{:}R'$$

$$R{:}^- + X{-}R' \longrightarrow R{-}X + {}^-{:}R'$$

The same requirement that the product carbanion be lower in energy (less basic, lower pK_a) than the starting carbanion prevails in each reaction:

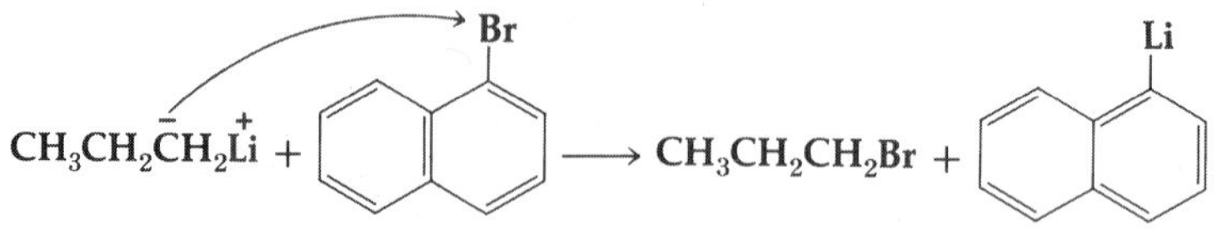

α-Bromonaphthalene α-Naphthyllithium

Grignard reagents and organolithium compounds add to carbonyl compounds to produce metal alkoxides, which, on hydrolysis, give alcohols. Formaldehyde gives primary alcohols, other aldehydes give secondary alcohols, and ketones give tertiary alcohols. The reaction provides one of the most versatile and reliable methods for constructing branched carbon skeletons.

$$(CH_3)_2CHCH_2MgBr + H_2C{=}O \xrightarrow{\text{Ether}} (CH_3)_2CHCH_2CH_2OMgBr \xrightarrow{H_3O^+} (CH_3)_2CHCH_2CH_2OH$$

Isoamyl alcohol

$$C_6H_5MgBr + CH_3CH_2CHO \xrightarrow[\text{2) } H_3O^+]{\text{1) Ether}} C_6H_5\underset{\displaystyle OH}{\underset{|}{C}}HCH_2CH_3$$

1-Phenyl-1-propanol

$$CH_3C{\equiv}CMgBr + CH_3\underset{\displaystyle O}{\underset{\|}{C}}CH_3 \xrightarrow[\text{2) } NH_4Cl,\ H_2O]{\text{1) Ether}} (CH_3)_2\overset{\displaystyle OH}{\overset{|}{C}}C{\equiv}CCH_3$$

2-Methyl-3-pentyn-2-ol

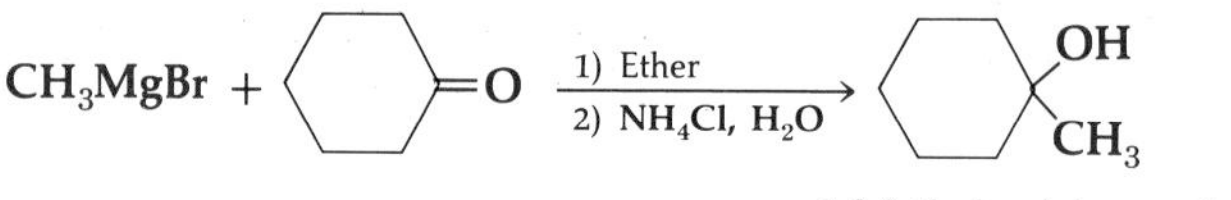

1-Methylcyclohexanol

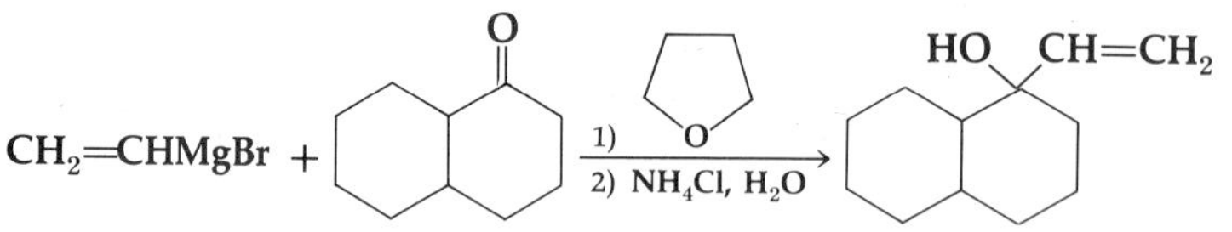

There is a very close parallel in reactivity and mechanism of reaction between metal hydrides and organometallics. Except that one acts to deliver hydride and the other a carbanion, they may conveniently be regarded as closely similar reagents. The organometallic reagents also react with carbonyl compounds other than ketones and aldehydes, as discussed later (Secs. 12-8 and 13-2).

Side reactions sometimes occur with sterically hindered ketones as substrates. These are viewed as involving cyclic mechanisms.

Enolization:

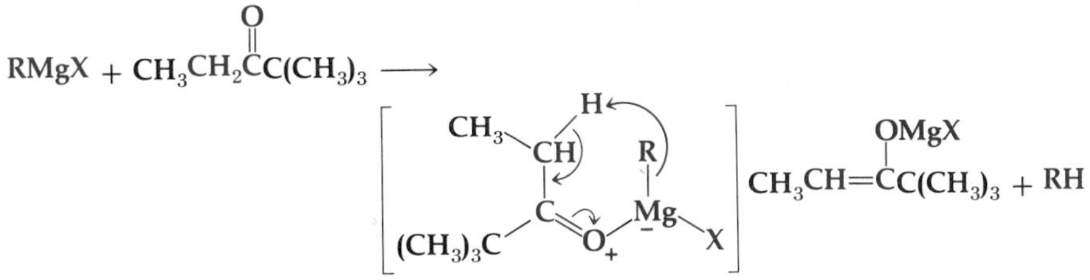

$$RMgX + CH_3CH_2\overset{\overset{\displaystyle O}{\|}}{C}C(CH_3)_3 \longrightarrow$$

$$\left[\begin{array}{c} CH_3 \\ \end{array} \right] \quad CH_3CH=\overset{\overset{\displaystyle OMgX}{|}}{C}C(CH_3)_3 + RH$$

Reduction:

$$-\overset{|}{\underset{H}{C}}-\overset{|}{\underset{|}{C}}-MgX + R\overset{\overset{\displaystyle O}{\|}}{C}R \longrightarrow \left[\begin{array}{c} R \\ R \end{array} \right] \longrightarrow R_2\overset{|}{\underset{H}{C}}-OMgX + \;\; \diagdown C=C\diagup$$

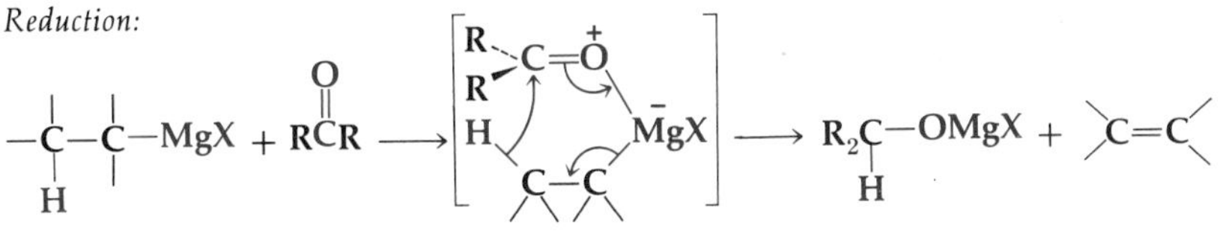

Coordination of a metal atom with the carbonyl oxygen, shown above in the reduction reaction, is undoubtedly also involved in the simple addition reaction since an important feature of Grignard attack is complexation of the very electronegative carbonyl oxygen with the magnesium atom. In general, low reaction temperatures and substitution of lithium compounds for Grignard reagents favor addition in preference to enolization and reduction. Steric hindrance in either the carbonyl compound or the metallic reagent decreases yields of addition product.

$$(CH_3)_3CLi + (CH_3)_3CCOC(CH_3)_3 \xrightarrow[-60°]{\text{Ether}}$$

Hexamethylacetone

$$[(CH_3)_3C]_3COLi + [(CH_3)_3C]_2CHOLi + (CH_3)_2C=CH_2 \xrightarrow{\text{H}_2\text{O}}$$

$$[(CH_3)_3C]_3COH + [(CH_3)_3C]_2CHOH$$

Tri-*t*-butyl-   Di-*t*-butylcarbinol
carbinol
81%

Grignard and lithium reagents cannot be made from compounds that contain reactive functional groups such as carbonyl, ester, amide, and many others. However, zinc compounds are less reactive, and zinc reagents can be prepared from α-haloesters and added to aldehydes and ketones. This is called the **Reformatsky reaction**, and the carbanion of the organometallic reagent is here an enolate, as it is α to carbonyl.

$$Zn + BrCH_2COOC_2H_5 \xrightarrow{\text{Ether}} BrZnCH_2COOC_2H_5 \xrightarrow{(CH_3)_2CO}$$

Zinc enolate

$$(CH_3)_2\overset{\overset{\displaystyle OZnBr}{|}}{C}CH_2COOC_2H_5 \xrightarrow{\text{H}_3\text{O}^+} (CH_3)_2\overset{\overset{\displaystyle OH}{|}}{C}CH_2COOC_2H_5$$

Ethyl 3-methyl-3-hydroxybutyrate

PROBLEM 12-6

Utilizing as starting materials ketones or aldehydes of fewer carbons, write syntheses of the following compounds.

a   3-Pentanol

b   1-Pentyn-3-ol

c   1-Pentanol

d   *t*-Butyl bromide

e   2-Methyl-2-hydroxy-
    butanonitrile

f C_2H_5 ⟨structure⟩ O   (from a chloroketone)

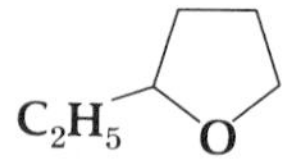

g   2-Hexyn-1,4-diol

h   Propylbenzene

i   Benzyl chloride

j   (1-Hydroxycyclobutyl)-propionic
    acid methyl ester

12-5   ADDITION ELIMINATION; THE ALDOL CONDENSATION

Addition to carbonyl initially affords a hydroxyl group. In cases of simple addition this alcohol represents the product. When the added nucleophilic group bears an adjacent electron pair—or can easily lose a proton to get one—this serves to eliminate the hydroxyl, i.e., assist its ionization and removal. In the cases where the nucleophile is oxygen, this leads to an oxonium ion ($\overset{+}{>}C{-}\ddot{O}R \longleftrightarrow >C{=}\overset{+}{\ddot{O}}R$) which must add a second nucleophile (**ROH**), as in Sec. 12-2. When it is nitrogen or carbon, it can be stabilized by loss of a proton to form a new π bond which replaces the original C=O π bond. The net reaction is then loss of water. Reactions in which two components combine with loss of water (or simple alcohols) are called **condensations**.

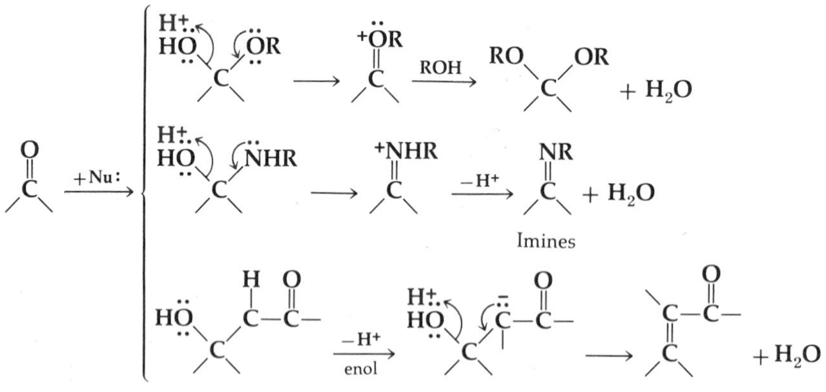

Condensations with Nitrogen Compounds

Compounds bearing a primary amino group ($Z{-}NH_2$) react with aldehydes and ketones to replace the carbonyl oxygen by nitrogen ($>C{=}O \longrightarrow >C{=}N{-}Z + H_2O$). The major condensations are these, all proceeding by the same general mechanism (Fig. 12-7):

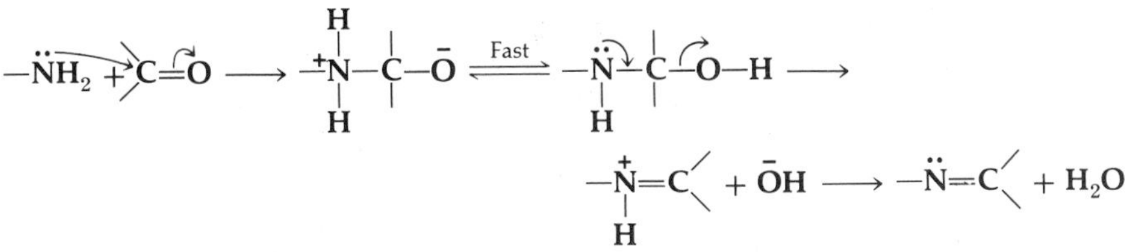

$$-\ddot{N}H_2 + \overset{\frown}{C}\!\!=\!\!\overset{\frown}{O} \longrightarrow \ \overset{H}{\underset{H}{\overset{|}{\overset{+}{N}}}}\!\!-\!\!\overset{|}{\underset{|}{C}}\!\!-\!\!\bar{O} \ \underset{\text{Fast}}{\rightleftharpoons} \ \overset{H}{\underset{H}{\overset{|}{\ddot{N}}}}\!\!\overset{\curvearrowleft}{}\!\!\overset{|}{\underset{|}{C}}\!\!\overset{\frown}{O}\!\!-\!\!H \longrightarrow$$

$$-\overset{+}{\underset{H}{\overset{|}{N}}}\!\!=\!\!C\!\!\overset{/}{\underset{\backslash}{}} + \ \bar{O}H \longrightarrow \ -\ddot{N}\!\!=\!\!C\!\!\overset{/}{\underset{\backslash}{}} + H_2O$$

FIGURE 12-7 **General path for replacement of carbonyl by imino**

$R_2C\!\!=\!\!O \ +$

    $H_2N\!\!-\!\!OH$ (hydroxylamine) $\longrightarrow R_2C\!\!=\!\!N\!\!-\!\!OH$ (**oximes**)

    $H_2N\!\!-\!\!NH_2$ (hydrazine) $\longrightarrow R_2C\!\!=\!\!N\!\!-\!\!NH_2$

    $H_2N\!\!-\!\!NHR'$ (substituted hydrazines) $\longrightarrow R_2C\!\!=\!\!N\!\!-\!\!NHR'$ $\Big\}$ (**hydrazones**)

    $H_2N\!\!-\!\!NHCONH_2$ (semicarbazide) $\longrightarrow R_2C\!\!=\!\!N\!\!-\!\!NHCONH_2$ (**semicarbazones**)

    $H_2N\!\!-\!\!R'$ (primary amines) $\longrightarrow R_2C\!\!=\!\!N\!\!-\!\!R'$ (**imines**)

The products of such reactions are usually sharp-melting solids, useful for the characterization of their parent aldehydes or ketones.

$$\underset{\overset{||}{O}}{CH_3\overset{}{C}CH_3} + NH_2OH \xrightarrow{\text{Weak base}} (CH_3)_2C\!\!=\!\!NOH + H_2O$$

       Hydroxyl-   Acetoxime
       amine   (an oxime)

$$C_6H_5COCH_3 + C_6H_5NHNH_2 \xrightarrow{H_3O^+, \ CH_3COOH} \underset{\overset{|}{CH_3}}{C_6H_5\overset{}{C}\!\!=\!\!NNHC_6H_5} + H_2O$$

Acetophenone   Phenylhydrazine   Acetophenone
                                        phenylhydrazone

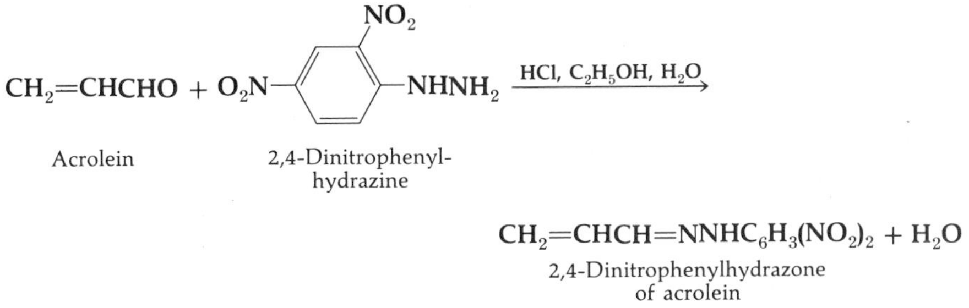

$$CH_2\!\!=\!\!CHCHO + O_2N\!\!-\!\!\underset{}{\overset{NO_2}{\bigcirc}}\!\!-\!\!NHNH_2 \xrightarrow{HCl, \ C_2H_5OH, \ H_2O}$$

    Acrolein   2,4-Dinitrophenyl-
                    hydrazine

$$CH_2\!\!=\!\!CHCH\!\!=\!\!NNHC_6H_3(NO_2)_2 + H_2O$$

                2,4-Dinitrophenylhydrazone
                   of acrolein

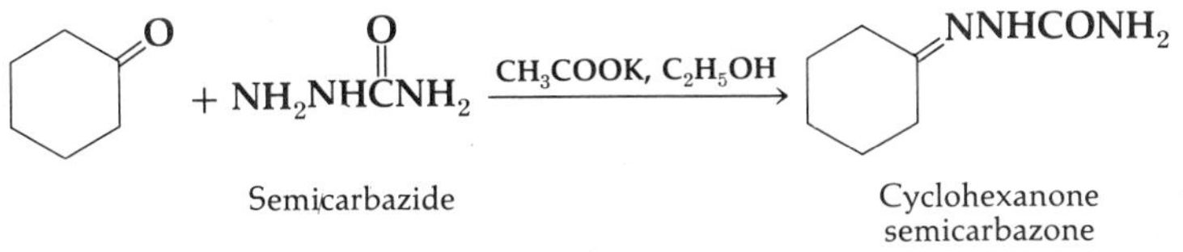

$$\underset{}{\overset{O}{\bigcirc\!\!=\!\!}} + NH_2NH\overset{\overset{O}{||}}{C}NH_2 \xrightarrow{CH_3COOK, \ C_2H_5OH} \underset{}{\overset{NNHCONH_2}{\bigcirc}}$$

          Semicarbazide   Cyclohexanone
                                              semicarbazone

It is difficult to prepare hydrazones from hydrazine, because the former tend to react a second time, giving **azines**.

$$CH_3CHO + NH_2NH_2 \longrightarrow CH_3CH{=}NNH_2 \xrightarrow{CH_3CHO} CH_3CH{=}NN{=}CHCH_3$$

Acetaldehyde   Acetaldehyde
hydrazone   azine

All condensations of carbonyl groups with nitrogen compounds are acid-catalyzed. Catalysis by *strong* acids is not very effective, however, because of protonation of the nucleophiles to form inactive derivatives of the ammonium ion.

$$R_2C{=}O + \overset{+}{H} \rightleftharpoons R_2C{=}\overset{+}{O}H$$

$$NH_2OH + \overset{+}{H} \rightleftharpoons \overset{+}{N}H_3OH$$

Inactive
hydroxyl-
ammonium ion

These reactions illustrate general acid catalysis. In a general acid-catalyzed reaction, the rate is dependent on the concentration of *all acid species in the solution* rather than on just the hydrogen-ion concentration. In the formation of oximes, and possibly for many other condensation reactions, dehydration of the intermediate adduct is the rate-determining step. The following mechanism accounts for the general acid catalysis.

$$R_2C{=}O + H_2\ddot{N}OH \rightleftharpoons R_2\overset{OH}{\underset{|}{C}}{-}NHOH$$

$$R_2\overset{OH}{\underset{|}{C}}{-}NHOH + \overset{+}{H} \rightleftharpoons R_2\overset{\overset{+}{O}H_2}{\underset{|}{C}}{-}NHOH$$

$$R_2\overset{\overset{+}{O}H_2}{\underset{|}{C}}{-}NOH + :\bar{B} \longrightarrow R_2C{=}NOH + H_2O + HB$$
$$\underset{H}{}$$

General acid-catalyzed reactions can be strongly catalyzed by massive concentrations of weak acids, such as acetic acid, which do not tie up large amounts of the nucleophile. Formation of oximes is also base-catalyzed. This fact is readily accommodated by a mechanism involving an elimination of the neutral addition compound.

$$R_2\overset{OH}{\underset{|}{C}}{-}N{-}OH + :\bar{O}H \longrightarrow R_2C{=}NOH + H\bar{O} + H_2O$$
$$\underset{H}{}$$

Regeneration of an aldehyde or a ketone from an oxime, a semicarbazone, or a hydrazone by direct hydrolysis is difficult, because the equilibrium constants favor the condensation products very strongly.

$$\ce{C=O} + NH_2Z \overset{K}{\rightleftharpoons} \ce{C=NZ} + H_2O \qquad K \text{ is large}$$

Recovery of a carbonyl compound can be accomplished by equilibrating the derivative with a reactive carbonyl compound, such as pyruvic acid, and in this way achieving an exchange reaction yielding the more stable free ketone.

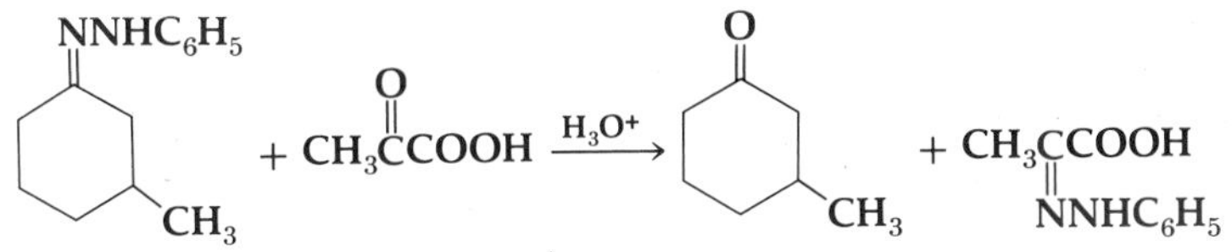

3-Methyl-
cyclohexanone
phenylhydrazone   Pyruvic
                               acid

In contrast to oximes, hydrazones, and semicarbazones, unhindered imines are difficult to isolate and readily hydrolyzed.

$$H_2O + C_6H_5CH{=}NC_6H_5 \longrightarrow C_6H_5CHO + C_6H_5NH_2$$

The Aldol Condensation

An important reaction for carbon-carbon bond formation occurs when an enol (or enolate) of one carbonyl compound adds to the carbonyl of an aldehyde or ketone. This fundamental condensation is used under various names which relate to the choice of carbonyl components, but is illustrated first with the original aldol example. All variations on this condensation may conveniently be included in the name **aldol condensation.**

Aldehydes undergo self-condensation reactions when treated with catalytic amounts of aqueous base. The basic catalyst removes a proton from the α-position of one molecule, and the resulting enolate ion then adds to the carbonyl group of a second molecule. The hydroxy-aldehyde formed from acetaldehyde was originally named **aldol** and gives its name to the general reaction.

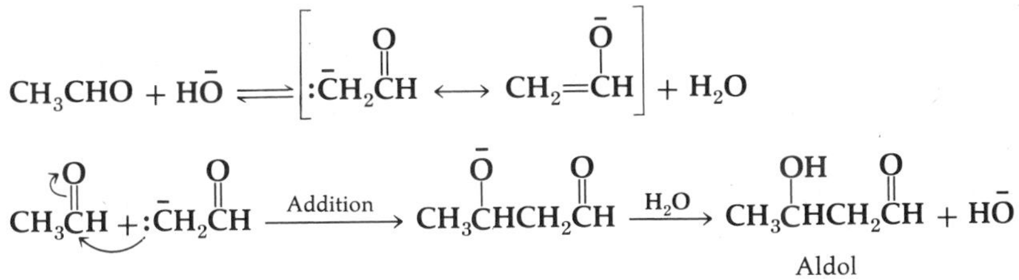

Aldols (aldehyde-alcohol) may be dehydrated either by heating the basic reaction mixture or by a separate, acid-catalyzed reaction. This is the elimination step and is very facile since the proton adjacent to —OH is α to the carbonyl and readily removed. Dehydration usually occurs under conditions required for the condensation of unreactive aldehydes such as sterically hindered aliphatic or aromatic aldehydes.

Acid-catalyzed:

$$CH_3\overset{\overset{\displaystyle OH}{|}}{C}HCH_2CHO \xrightarrow{H^+} CH_3-\overset{\overset{\displaystyle H^+ \ddot{O}H}{}}{CH}-CH=CH-\overset{\displaystyle \ddot{O}H}{} \xrightarrow{-H_2O} CH_3CH=CHCHO$$

                                          Enol   Crotonaldehyde

Base-catalyzed:

$$CH_3\overset{\overset{\displaystyle OH}{|}}{C}HCH_2CHO \xrightarrow{HO^-} CH_3\overset{\overset{\displaystyle OH}{}}{CH}-\overset{\displaystyle \ddot{\ }}{C}HCHO \longrightarrow CH_3CH=CHCHO$$

                               Enolate ion   Crotonaldehyde

Crossed condensations between two aldehydes that both have α-hydrogens give complex reaction mixtures and are not used for synthesis. However, an excess of an aldehyde that has *no α-hydrogen* (such as benzaldehyde or formaldehyde) can serve as a substrate for a second enolizable aldehyde. Acrolein can be made by carrying out the condensation of 1 mole of formaldehyde with one of acetaldehyde at a high enough temperature to effect dehydration of the intermediate aldol, but with base catalysis in solution successive additions occur with no elimination.

$$CH_2O + CH_3CHO \xrightarrow[300°]{\underset{\text{silicate}}{\overset{\text{Sodium}}{}}} [HOCH_2CH_2CHO] \longrightarrow CH_2=CHCHO$$

                                                              Acrolein
                                                               75%

$$3CH_2O + CH_3CHO \xrightarrow{Ca(OH)_2} (HOCH_2)_3CCHO$$

An aldehyde that bears no α-hydrogen may also be condensed with a ketone or an ester that can enolize. The conditions are too mild to allow attack on the less reactive ketone or ester carbonyl in competition with the aldehyde carbonyl.

$$C_6H_5CHO + CH_3COCH_3 \xrightarrow{10\% \text{ NaOH}} C_6H_5CH=CHCOCH_3$$

                                                    Benzalacetone

$$C_6H_5CHO + CH_3COOC_2H_5 \xrightarrow{NaOC_2H_5, \ 0°} C_6H_5CH=CHCOOC_2H_5$$

                                                       Ethyl cinnamate

For both electronic and steric reasons, ketones are less reactive than aldehydes. Acetone undergoes self-condensation, giving diacetone alcohol, but only a small amount of the product is formed at equilibrium. High yields of diacetone alcohol are obtained only by continuously cycling acetone vapor over a mild basic catalyst, with the small amount of product formed in each pass being retained as a high-boiling residue in a reservoir.

$$2CH_3COCH_3 \underset{}{\overset{Ba(OH)_2}{\rightleftharpoons}} CH_3\overset{\overset{\displaystyle OH}{|}}{\underset{\underset{\displaystyle CH_3}{|}}{C}}CH_2COCH_3$$

                              Diacetone alcohol
                              70% by recycling

Base-catalyzed aldol cyclizations to five- and six-membered rings are especially favored by entropy, however, and occur with ease.

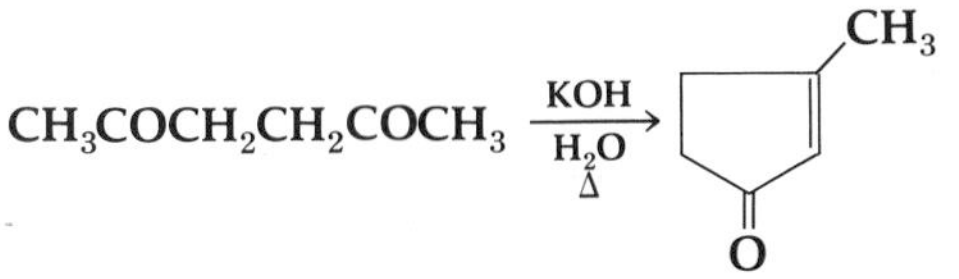

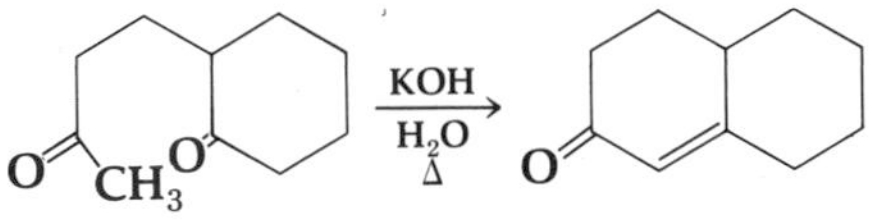

$$\text{CH}_3\text{COCH}_2\text{CH}_2\text{COCH}_3 \xrightarrow[\text{H}_2\text{O}]{\text{KOH} \atop \Delta}$$

Several valuable synthetic procedures are based upon condensations of aldehydes and ketones with enolates and similar carbanions. The elimination step following addition can often be suppressed by using mild basic catalysts. On the other hand the elimination is usually desirable since the conjugated unsaturated product is more stable and tends to drive the equilibria toward the product, thus countering and displacing other unfavorable side-reaction equilibria that may occur previously.

$$\phi-\text{CHO} + \phi\text{CH}_2\text{CN} \longrightarrow \phi-\text{CH}=\text{C}\begin{smallmatrix}\phi\\ \text{CN}\end{smallmatrix}$$

$$\text{CH}_3\text{CH}_2\text{CH}_2\text{CHO} + \text{CH}_3\text{NO}_2 \xrightarrow{\text{KOH}} \text{CH}_3\text{CH}_2\text{CH}_2\overset{\text{OH}}{\underset{}{\text{CH}}}-\text{CH}_2\text{NO}_2$$
$$71\%$$

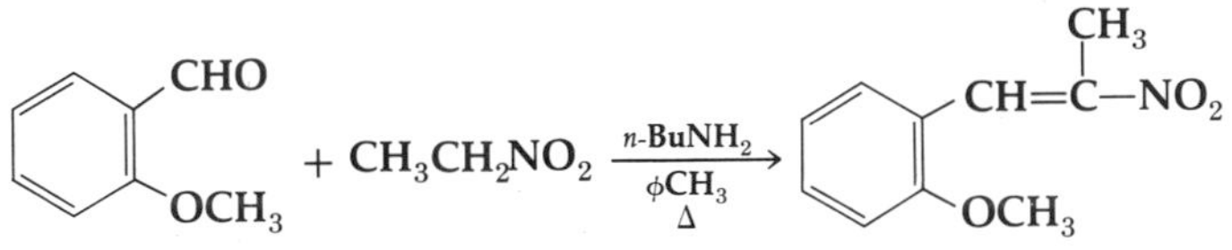

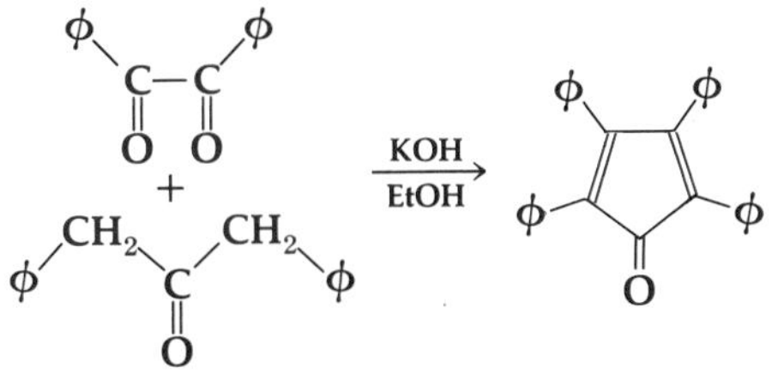

The key intermediate in the commercial synthesis of vitamin A is β-ionone, which is prepared from the natural oil, citral, via a condensation with acetone (the second reaction is treated in Chap. 15):

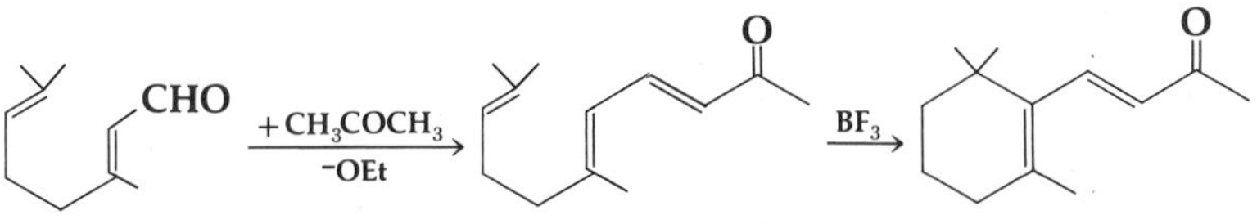

Citral   β-Ionone

A number of special cases of aldol condensations well-known through long usage and synthetic value are collected in the following numbered paragraphs.

1 Active methylene compounds (with two carbonyls or other electron-withdrawing groups on methylene) react with aldehydes and many ketones in an addition-elimination known as the **Knoevenagel condensation**. Here the acidity of the active methylene compound allows enolate formation under basic conditions so mild that self-condensation of aldehydes is rarely a hazardous side reaction. The use of both weak base (**R$_2$NH**) and weak acid (**RCOOH**) *together* provides excellent catalysis in this reaction.

$$\phi-CHO + CH_2 \overset{COOC_2H_5}{\underset{COOC_2H_5}{\big<}} \xrightarrow[\Delta - H_2O]{R_2NH, \, \phi COOH} \phi-CH=C \overset{COOC_2H_5}{\underset{COOC_2H_5}{\big<}}$$

Ethyl malonate   90%

$$CH_3CHO + CH_2 \overset{CN}{\underset{COOC_2H_5}{\big<}} \xrightarrow{R_2NH} CH_3CH=C \overset{CN}{\underset{COOC_2H_5}{\big<}}$$

Ethyl cyanoacetate

$$\phi-\overset{O}{\overset{\|}{C}}-CH_3 + CH_2(COOC_2H_5)_2 \xrightarrow[\phi H/\Delta]{NH_4^+OAc^-} \phi-\underset{CH_3}{\overset{|}{C}}=C(COOC_2H_5)_2$$

2 When the reaction is run in hot pyridine and cyanoacetic or malonic acid is used as the enol component, it loses **CO$_2$** on condensing (**Doebner condensation**).

$$CH_3CH=CHCHO + CH_2 \overset{COOH}{\underset{COOH}{\big<}} \xrightarrow[100°]{Pyridine}$$

$$CH_3CH=CHCH=CH-COOH + CO_2 + H_2O$$

30%

3 The **Perkin reaction** employs condensation of an aromatic aldehyde with an aliphatic anhydride and its corresponding carboxylate salt as a preparation of cinnamic acids.

$$\phi CHO + (CH_3CO)_2O + CH_3COO^- \xrightarrow{180°, \, H_2O} \phi-CH=CH-COOH$$

Cinnamic acid
60%

4 The **benzoin condensation** is a remarkable self-condensation of aromatic aldehydes. The reaction is *specifically catalyzed by cyanide ions*. Three properties of the catalyst combine to make cyanide a unique catalyst:
   a Sufficient nucleophilic reactivity to add to a carbonyl function
   b A marked acid-strengthening effect on hydrogen atoms attached to carbon atoms adjacent to the —C≡N function (Fig. 12-3)
   c The ability of cyanide to depart from the cyanohydrin structure in the last intermediate

$$ArCHO + C\bar{N} \rightleftharpoons Ar\overset{\overset{\bar{O}}{|}}{C}HCN$$

$$Ar\overset{\overset{\bar{O}}{|}}{C}HCN \longrightarrow \left[Ar\overset{\overset{OH}{|}}{\underset{..}{C}}{-}CN \longleftrightarrow Ar\overset{\overset{OH}{|}}{C}{=}C{=}\bar{N} \right]$$

$$Ar\overset{\overset{OH}{|}}{\underset{..}{C}}CN + ArCHO \longrightarrow Ar\overset{\overset{OH}{|}}{C}{-}\underset{\underset{CN}{|}}{\overset{}{C}}H Ar \rightleftharpoons Ar\overset{\overset{\bar{O}}{|}}{C}{-}\underset{\underset{CN}{|}}{\overset{}{C}}HAr$$

with subscripts CN $\underset{O}{}$ and CN OH

$$Ar\overset{\overset{\bar{O}}{|}}{C}{-}CHAr \longrightarrow Ar\underset{\underset{O}{\|}}{C}{-}\underset{\underset{OH}{|}}{C}HAr + C\bar{N}$$

with $\underset{CNOH}{}$

A benzoin
(or acyloin)

Although the reaction gives good yields with many aromatic aldehydes, it is unsuccessful with aliphatic aldehydes. Probably, competing reactions involving enolization of α-hydrogens cause failure with the latter compounds.

$$CH_3O{-}\bigcirc{-}CHO \xrightarrow[\substack{95\% \ C_2H_5OH \\ reflux}]{KCN} CH_3O{-}\bigcirc{-}\underset{\underset{O}{\|}}{C}{-}\underset{\underset{OH}{|}}{C}H{-}\bigcirc{-}OCH_3$$

4,4'-Dimethoxybenzoin
60%

5 When the **—OH** group of the intermediate aldol addition product can do a facile internal reaction (cyclization), this will intercede and normal elimination of water will not occur. The **Darzens synthesis** involves an internal substitution following addition of the enolate from α-chloroesters. The intermediate ester epoxide and the epoxy-salt can be easily isolated. The reaction is useful for preparation of certain aldehydes and ketones.

$$ClCH_2COOC_2H_5 \xrightarrow{NaOC_2H_5} Na\overset{+}{C}\overset{\overset{Cl}{|}}{\overset{}{\bar{H}}}COOC_2H_5 + C_2H_5OH$$

$$Na\overset{+}{\underset{..}{C}}\overset{\overset{Cl}{|}}{\bar{H}}COOC_2H_5 + \phi{-}\overset{\overset{CH_3}{|}}{\underset{\underset{O}{\|}}{C}} \xrightarrow[-5°]{Hexane} \phi{-}\overset{\overset{CH_3}{|}}{\underset{\underset{O}{}}{C}}{-}\overset{\overset{Cl}{}}{C}H{-}COOC_2H_5 \longrightarrow$$

$$\phi{-}\overset{\overset{CH_3}{|}}{C}\underset{\underset{O}{\diagdown\diagup}}{}CH{-}COOC_2H_5 \xrightarrow{KOH} \phi{-}\overset{\overset{CH_3}{|}}{C}\underset{\underset{O}{\diagdown\diagup}}{}CH{-}C\bar{O}_2\overset{+}{K} \xrightarrow[\Delta]{H_2SO_4}$$

$$\phi{-}\overset{\overset{CH_3}{|}}{C}\underset{\underset{\overset{+}{\underset{H}{|}}O}{}}{}CH{-}C\overset{}{\underset{\underset{O}{\|}}{}}O{-}H \xrightarrow{-CO_2} \phi{-}\overset{\overset{CH_3}{|}}{C}{=}CH{-}O{-}H \xrightarrow{Tautomerizes} \phi{-}\overset{\overset{CH_3}{|}}{C}H{-}CHO$$

Enol   Hydratropaldehyde
55% overall

6   In the **Stobbe condensation**, the anion from diethyl succinate adds to a ketone. The first adduct cyclizes by an ester-lactone interchange reaction (Sec. 13-1) that involves the more remote ester group and the alkoxide generated in the addition reaction. The five-membered lactone ring is then opened by a base-catalyzed elimination, which generates a free carboxyl group. The base thus generates the stable carboxylate anion, which renders the reaction irreversible because of its stability (low pK_a):

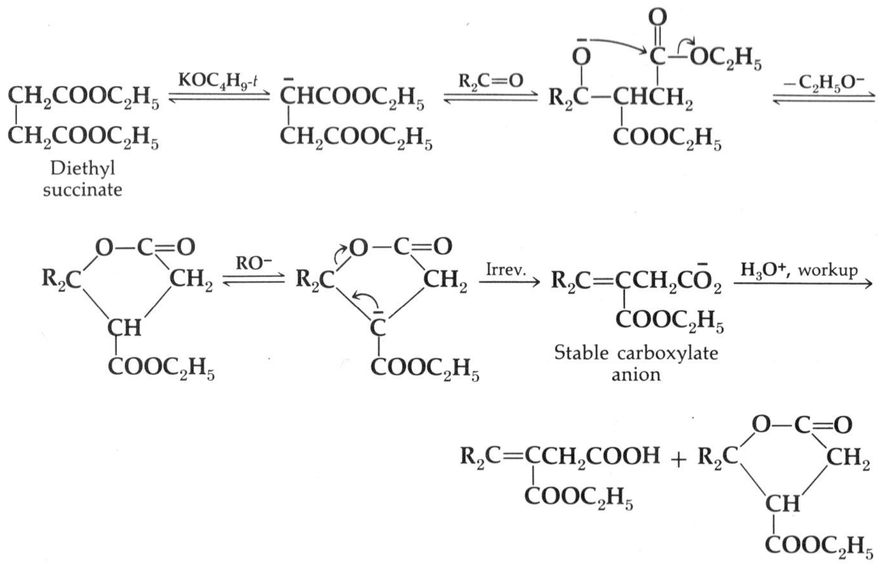

Diethyl
succinate

Stobbe condensations have been used extensively in syntheses of polycyclic ring systems. Potassium t-butoxide and sodium hydride are commonly used as basic condensing agents.

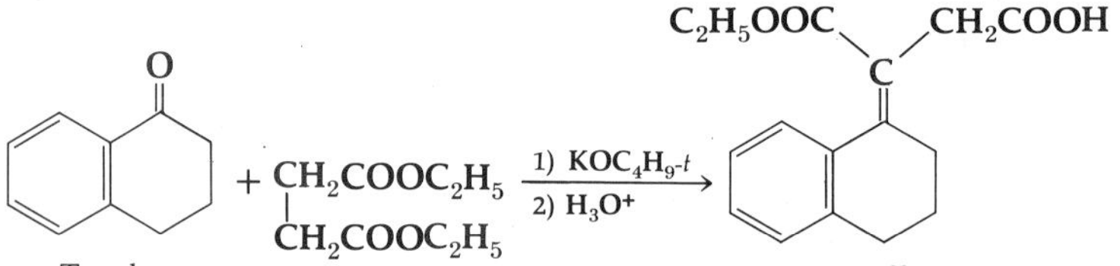

α-Tetralone

90%

7   In the **Wittig reaction** the carbanion is stabilized by resonance into the d orbitals of an adjacent phosphonium cation.† The phosphorus has a special affinity for the carbonyl oxygen which it removes as a phosphine oxide. (See Chap. 19.) The reaction is largely free from side reactions and is an especially valuable synthetic device for creating alkenes.

†This is not properly an aldol condensation as the carbanion is not an enolate, but it achieves so closely parallel a purpose as to justify including it here.

$$\phi_3 P: + \underset{\underset{Br}{|}}{CH_2} - R \xrightarrow{S_N 2} \overset{+}{\phi_3 P} - CH_2 - R \xrightarrow{NaNH_2 \text{ (base)}} \overset{+}{\phi_3 P} - \overset{\cdot\cdot}{\overset{-}{C}H} - R \longleftrightarrow \phi_3 P = CH - R$$
$$Br^- \qquad\qquad\qquad\qquad\qquad \text{Wittig reagent}$$

$$\overset{+}{\underset{\text{:CH}}{\overset{\displaystyle C=O \ \ P\phi_3}{}}} \quad \overset{\displaystyle C \overset{O^-}{\underset{\underset{R}{CH}}{\overset{+}{P\phi_3}}} \longrightarrow \quad \overset{\displaystyle C=CH-R}{} + \overset{+}{\phi_3 P} - \bar{O} \longleftrightarrow \phi_3 P = O$$
$$\qquad\qquad\qquad\qquad\qquad\qquad\qquad \text{Triphenylphosphine oxide}$$

$$(C_6H_5)_3\overset{+}{P} - \bar{C}H_2 + \underset{}{\bigcirc}\!\!=\!\!O \longrightarrow \underset{}{\bigcirc}\!\!=\!\!CH_2 + (C_6H_5)_3\overset{+}{P} - \bar{O}$$

$$[(C_6H_5)_3\overset{+}{P} - \bar{C}H]_2CH_2 + 2(CH_3)_2C=O \longrightarrow$$
$$(CH_3)_2C=CHCH_2CH=C(CH_3)_2 + (C_6H_5)_3\overset{+}{P} - \bar{O}$$

Summary

The overall reaction for the aldol and related condensations may be generalized as follows:

| Enol component (or Z—CH₂—; Z=CN, NO₂, etc.) | Substrate or acceptor (ketone; aldehyde) | Addition product | Condensation (elimination) product |

1   The final elimination is desirable to give an unsaturated product, which is usually more stable; this is often spontaneous. The substrate carbonyl becomes one end of that double bond and its oxygen emerges as water.

2   Two α-hydrogens on the enol component are required for addition-elimination.

3   Elimination following addition is favored by acid or strong base and often suppressed with weak base catalysis.

4   Use of a substrate with no α-hydrogens is preferable in reducing by-products arising from mixed and self-condensations.

5   Use of a very acidic enol component, such as active methylene compounds, usually avoids any enolization of the acceptor carbonyl, even with aldehydes, and is therefore desirable for avoiding side reactions. These condensation products usually eliminate water spontaneously.

6 If cyclization of the —OH group in the addition product can occur the reaction takes a different subsequent course (Stobbe and Darzens condensations).

7 Intramolecular aldol condensations leading to five- and six-membered rings always take precedence over bimolecular competition.

PROBLEM 12-7

Complete the equations for these transformations and indicate what added acid or base catalyst is needed wherever appropriate.

a $CH_3CH=CHCHO + CH_2(COOCH_3)_2 \longrightarrow$?

b ? $\longrightarrow (C_6H_5CH_2)_2C=NOCH_3$

c ? $\longrightarrow$ 1-phenyl-2-methyl-1-penten-3-one

d α-Methylpropionaldehyde + cyclopentanone $\longrightarrow$?

e ? $\longrightarrow (C_2H_5)_2C=C(CN)COOC_2H_5$

f 1,4-Cyclohexanedione + $C_6H_5NHNH_2 \longrightarrow$?

g ? $\longrightarrow$ 3-methyl-2-cyclopentenone $\xrightarrow{+H_2NNHCONH_2}$?

h β-Ionone + benzaldehyde $\longrightarrow$?

i Phenylnitromethane + acetone $\longrightarrow$?

j ? $\longrightarrow$

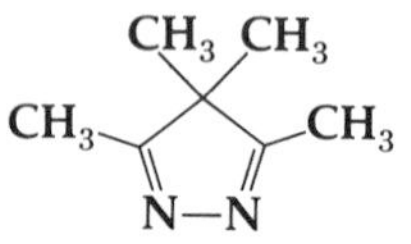

k Benzaldehyde (excess) + $H_2NNH_2 \longrightarrow$?

l ? $\longrightarrow$ p-methoxy cinnamic acid

m ? $\longrightarrow$ Br—⟨ ⟩—COCHOH—⟨ ⟩—Br

n Citral + malonic acid $\longrightarrow$ (monoacid) ?

o ? $\longrightarrow$ 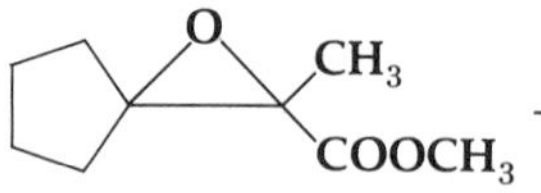
$\xrightarrow{H_3O^+}$?

p ? $\longrightarrow C_6H_5C=CCOOC_2H_5$
 $\qquad\qquad\quad\ \ \underset{CH_3}{|}\ \underset{CH_2COOH}{|}$

q Acetone + $(C_6H_5)_3P=CHCH_2CH_2CH=P(C_6H_5)_3 \longrightarrow$?

r ? $\longrightarrow CH_3CH=$⟨ ⟩$=CHCH_3$

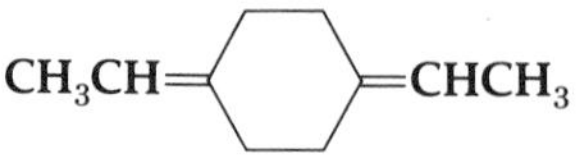

s ? $\longrightarrow CH_3CH=CHCH=CHCN + CO_2$

12-6   VINYLOGOUS OR CONJUGATE ADDITION

Carbon-carbon double bonds conjugated with carbonyls or the related groups of Fig. 12-3 act as vinylogs, or extensions, of carbonyl reactivity in accepting nucleophiles. Such conjugated double bonds may serve as substrates for nucleophilic addition, but *ordinary isolated double bonds are unreactive*.

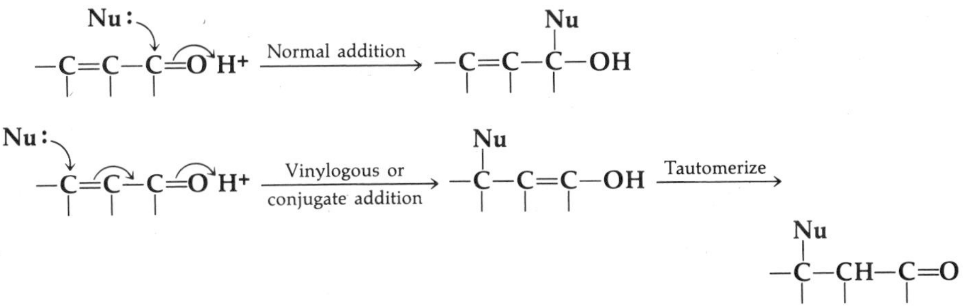

$$
\underset{\text{Nu:}}{-C=C-C=O^+H^+} \xrightarrow{\text{Normal addition}} \underset{\text{Nu}}{-C=C-C-OH}
$$

$$
\underset{\text{Nu:}}{-C=C-C=O^+H^+} \xrightarrow[\text{conjugate addition}]{\text{Vinylogous or}} \underset{\text{Nu}}{-C-C=C-OH} \xrightarrow{\text{Tautomerize}}
$$

$$
\underset{\text{Nu}}{-C-CH-C=O}
$$

In vinylogous addition the initial enol formed reverts to the more stable ketone tautomer.

Addition of Hydrogen Halides

Hydrogen halides add readily to the carbon-carbon double bonds of α,β-unsaturated aldehydes, ketones, esters, acids, and so forth. The products revert to unsaturated carbonyl quite readily with base.

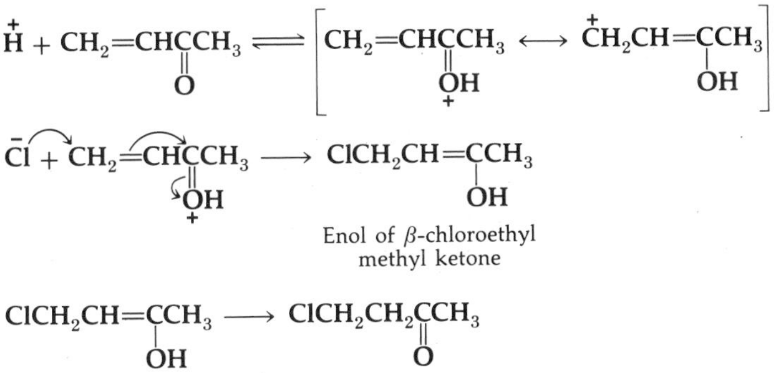

$$
\overset{+}{H} + CH_2=CHCCH_3 \rightleftharpoons \left[CH_2=CHCCH_3 \longleftrightarrow \overset{+}{C}H_2CH=CCH_3 \right]
$$
$$
\underset{O}{} \qquad \underset{\overset{OH}{+}}{} \qquad \underset{OH}{}
$$

$$
\overset{-}{Cl} + CH_2=CHCCH_3 \longrightarrow ClCH_2CH=CCH_3
$$
$$
\underset{\overset{OH}{+}}{} \qquad\qquad \underset{OH}{}
$$

Enol of β-chloroethyl
methyl ketone

$$
ClCH_2CH=CCH_3 \longrightarrow ClCH_2CH_2CCH_3
$$
$$
\underset{OH}{} \qquad\qquad \underset{O}{}
$$

Net reaction: $HCl + CH_2=CHCOCH_3 \longrightarrow ClCH_2CH_2COCH_3$

Methyl vinyl
ketone

β-Chloroethyl
methyl ketone
67%

$$\mathrm{HBr} + \mathrm{CH_2{=}C(CH_3)CN} \longrightarrow \mathrm{BrCH_2CH(CH_3)CN}$$

α-Methacrylonitrile → β-Bromoisobutyronitrile 72%

H—Cl +
CHCl3
O
OH
Cl
H
Benzoquinone
Chlorohydroquinone
74%

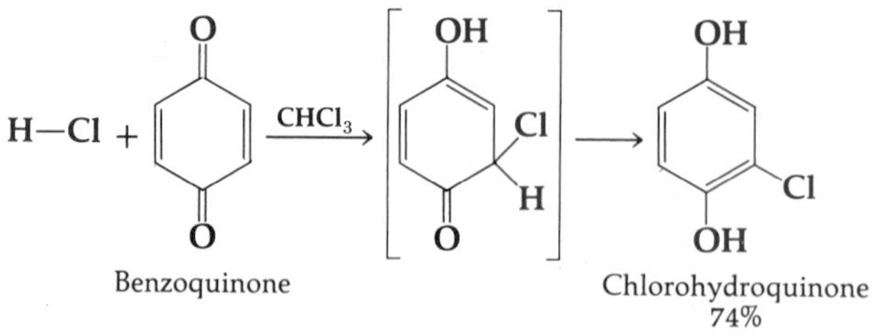

Addition of Alcohols, Thiols, Amines, and Carboxylic Acids

These four classes of organic compounds add readily to highly polarized double bonds in conjugated systems. Such additions are catalyzed by both acids and bases, although acid conditions are often chosen to avoid base-catalyzed side reactions. Acrylonitrile is widely used to cyanoethylate alcohols and amines.

$$\underset{\text{Methyl acrylate}}{\mathrm{CH_2{=}CHCOCH_3}} \overset{+H^+}{\rightleftharpoons} \left[\mathrm{CH_2{=}CH\overset{\overset{+}{O}H}{\overset{\|}{C}}OCH_3} \longleftrightarrow \mathrm{\overset{+}{C}H_2CH{=}\overset{OH}{\overset{|}{C}}OCH_3}\right] \xrightarrow{C_2H_5OH} \underset{\text{Methyl }\beta\text{-ethoxypropionate, 91\%}}{\mathrm{C_2H_5OCH_2CH_2\overset{O}{\overset{\|}{C}}OCH_3}}$$

$$\underset{\text{Thiophenol}}{\mathrm{C_6H_5SH}} \underset{}{\overset{\text{NaOCH}_3\text{ (trace)}}{\rightleftharpoons}} \mathrm{C_6H_5\bar{S}} \xrightarrow{\mathrm{CH_2{=}CH\overset{O}{\overset{\|}{C}}OCH_3}} \underset{\text{Methyl }\beta\text{-phenylmercaptopropionate, 96\%}}{\mathrm{C_6H_5SCH_2CH_2\overset{O}{\overset{\|}{C}}OCH_3}} + \mathrm{C_6H_5\bar{S}}$$

$$\mathrm{C_2H_5OH} + \underset{\text{Acrylonitrile}}{\mathrm{CH_2{=}CHCN}} \xrightarrow{H_2SO_4} \underset{\beta\text{-Ethoxypropionitrile, 89\%}}{\mathrm{C_2H_5OCH_2CH_2CN}}$$

ϕNH2 +
O
Via:
O H
NHϕ
OH
1,4-Naphthoquinone
1,4-Dihydroxy-2-anilinonaphthalene

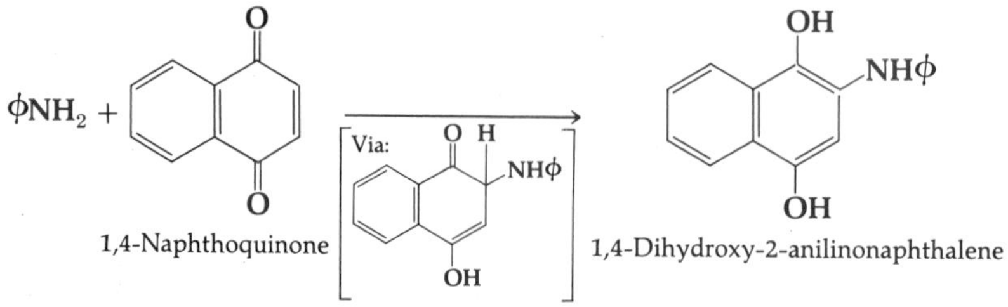

Hydrides

Hydrides almost never undergo conjugate addition. This makes it possible to reduce conjugated carbonyls without loss of the double bond through the use of hydrides, while alternatively the double bond may be selectively reduced by catalytic hydrogenation without disturbing the carbonyl group.

$$CH_3-CH_2-CH_2-\overset{\overset{\textstyle O}{\|}}{C}-CH_3 \xleftarrow{\ \ H_2\ \ \atop Pt} CH_3-CH=CH-\overset{\overset{\textstyle O}{\|}}{C}-CH_3 \xrightarrow{\ \ NaBH_4\ \ }$$

$$CH_3-CH=CH-\overset{\overset{\textstyle OH}{|}}{C}H-CH_3$$

Michael Additions

A family of conjugate additions of stabilized carbanions is known as the **Michael addition** reaction, a name originally applied only to transformations that involved acetoacetic or malonic ester anions as nucleophiles. Michael reactions are catalyzed by bases such as sodium hydroxide, sodium ethoxide, and amines (usually piperidine). These additions have special value since they serve to form carbon-carbon bonds.

$$\overset{-}{C}N + (CH_3)_2C=CHNO_2 \xrightarrow[CH_3OCH_2CH_2OCH_3]{KOH} (CH_3)_2\overset{\overset{\textstyle }{|}}{\underset{\underset{\textstyle CN}{|}}{C}}CH_2NO_2$$

2-Methyl-1-
nitro-1-propene

2,2-Dimethyl-3-nitropropanonitrile
75%

$$\overset{-}{C}H(COOC_2H_5)_2 \xrightarrow{C_6H_5CH=CHCOC_6H_5} C_6H_5\overset{\overset{\textstyle }{|}}{\underset{\underset{\textstyle CH(COOC_2H_5)_2}{|}}{C}}HCH_2COC_6H_5$$

$$CH_3NO_2 + CH_3CH=CHCOOC_2H_5 \xrightarrow[C_2H_5OH]{C_2H_5ONa} CH_3\overset{\overset{\textstyle }{|}}{\underset{\underset{\textstyle CH_2NO_2}{|}}{C}}HCH_2COOC_2H_5$$

Ethyl crotonate

Ethyl 3-methyl-4-nitrobutyrate
55%

$$CH_2(COOC_2H_5)_2 + CH_3CH=CHCH=CHCOOCH_3 \xrightarrow[CH_3OH]{CH_3ONa}$$

Methyl sorbate

$$CH_3\overset{\overset{\textstyle }{|}}{\underset{\underset{\textstyle CH(COOC_2H_5)_2}{|}}{C}}HCH_2CH=CHCOOCH_3 + CH_3CH=CH\overset{\overset{\textstyle }{|}}{\underset{\underset{\textstyle CH(COOC_2H_5)_2}{|}}{C}}HCH_2COOCH_3$$

72%   8%

$$CH_3COCH_3 + 3CH_2=CHCN \xrightarrow[t\text{-}C_4H_9OH]{KOH} CH_3COC\overset{\diagup CH_2CH_2CN}{\underset{\diagdown CH_2CH_2CN}{-CH_2CH_2CN}}$$

1,1,1-Tris(2-cyanoethyl)acetone

Enamines have special value in their facile conjugate additions. The enamine acts simply as a nitrogen enol. (See Sec. 11-7.)

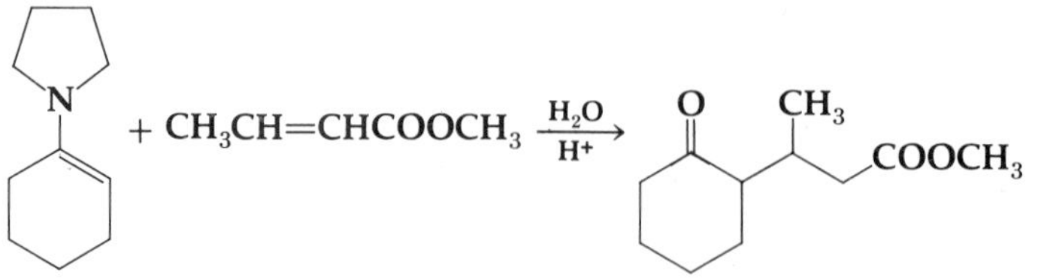

The **Robinson annellation** ("ring-forming") reaction has been widely used for cyclohexenone synthesis; the unsaturated ketone is prepared by aldol or Mannich (page 487) condensation first, or the simplest available reagent, methyl vinyl ketone, is used.

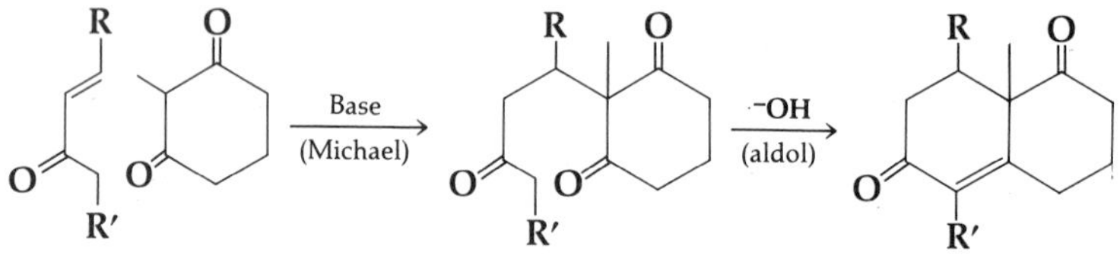

Conjugate Grignard Additions

When the carbonyl group of an α,β-unsaturated ketone is sterically hindered, the system is susceptible to conjugate addition reactions with hindered Grignard reagents. In ordinary unsaturated substrates, simple addition to the carbonyl is the rule but conjugate addition can usually be favored by adding copper salts which apparently break up the complex involved in the more rapid simple addition and allow conjugate addition to proceed instead by default. A cyclic intermediate is probably often involved where it is sterically possible:

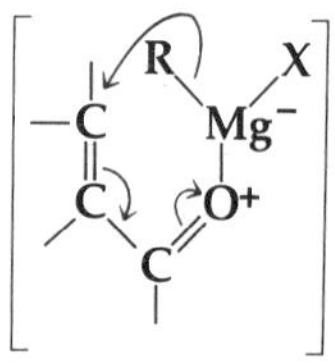

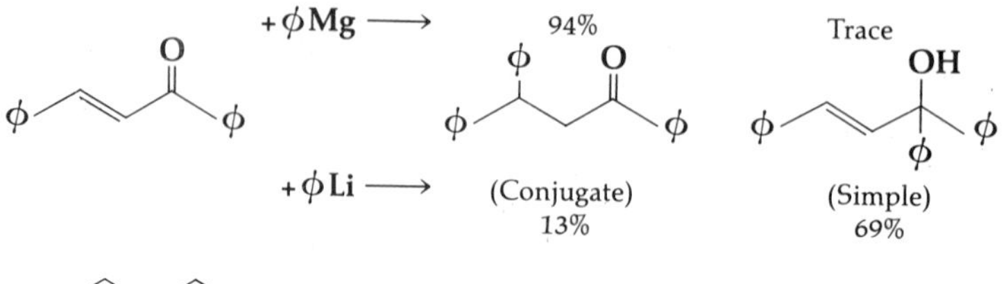

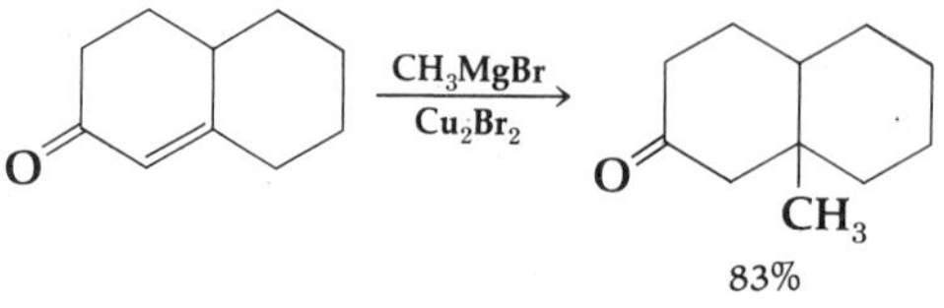

PROBLEM 12-8

In attempting to make the following ketone from 2-methylcyclohexanone, other products were obtained. Show the route selected and the source of the trouble.

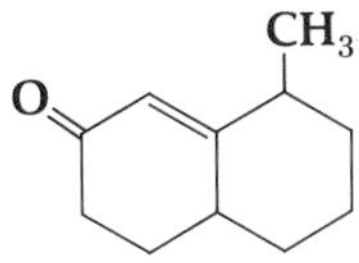

PROBLEM 12-9

Show how conjugate additions may be used to create these substances.

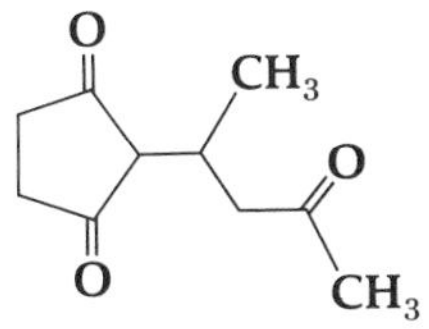

b $(CH_3)_2NCH_2CH_2CN$

c $(CH_3OOC)_2CHCH_2CH_2COOCH_3$

d 3-Cyanocyclopentanone

e 3-Phenyl-3-thiophenyl-2-nitropropane

f $N(CH_2CH_2CN)_3$

g 3-Methoxy-1,4-dihydroxybenzene

h $(C_2H_5)_2NCH{=}CHCOCH_3$

i

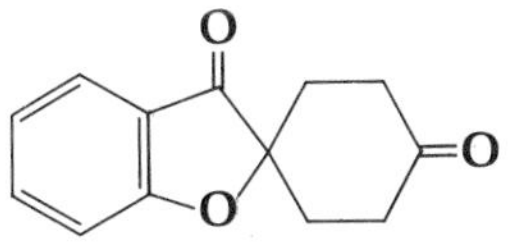

j 4-Phenyl-2-pentanone

12-7 CARBON DIOXIDE DERIVATIVES

The functional groups which have four heteroatom bonds to carbon are a class which may be considered to be derivatives of carbon dioxide ($O{=}C{=}O$). Almost all of these compounds (except a few such as CCl_4) contain a carbonyl group, or $C{=}N$ or $C{=}S$ of similar reactivity, and their reactions are all simply derived from those of the carbonyl compounds considered so far.

These functional groups may be considered in two classes, those with sp hybridization at the central carbon, like carbon dioxide, and those with sp^2 hybridization like carbonic acid. The two classes are listed in Table 12-1.

The carbon dioxide class may all be regarded as internal anhydrides (like P_2O_5 or SO_3). Virtually the only reaction of these compounds is simple carbonyl addition, which they generally undergo with great facility. Carbon dioxide is by far the most stable and least reactive in the left column. The additions yield analogs of carboxylates or enolates in base, and their conjugate acids in acid media.

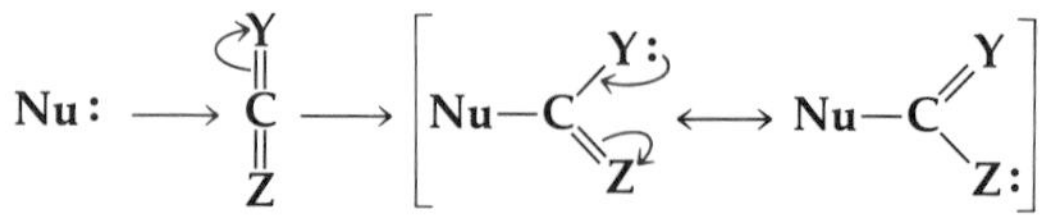

$$\text{Nu:} \longrightarrow \underset{Z}{\overset{Y}{\underset{\|}{C}}} \longrightarrow \left[\text{Nu} - C \overset{Y:}{\underset{Z}{\Bigg\langle}} \longleftrightarrow \text{Nu} - C \overset{Y}{\underset{Z:}{\Big\langle}} \right]$$

From this description it is easy to write the primary products of addition of water, alcohols, and amines to these compounds and in most cases the products are the stable derivatives listed in the second column of Table 12-1. They are rapidly formed at room temperature.

TABLE 12-1   Compounds with Four Heteroatom Bonds to Carbon

Carbon dioxide class (sp hybridization)		Carbonic acid class† (sp² hybridization)	
O=C=O	carbon dioxide	$HO-\overset{O}{\overset{\|}{C}}-OH$	carbonic acid
R—N=C=O	isocyanates	$Cl-\overset{O}{\overset{\|}{C}}-Cl$	phosgene
R—N=C=S	isothiocyanates	$Cl-\overset{O}{\overset{\|}{C}}-OR$	chlorocarbonates (chloroformates)
R—N=C=N—R	carbodiimides	$Cl-\overset{O}{\overset{\|}{C}}-NR_2$	chlorocarbamates
O=C=S	carbon oxysulfide	$RO-\overset{O}{\overset{\|}{C}}-OR'$	carbonate esters
S=C=S	carbon disulfide	$RO-\overset{S}{\overset{\|}{C}}-SR'$	xanthate esters
(R₂C=C=O	ketenes)	$R_2N-\overset{O}{\overset{\|}{C}}-OR'$	urethans (carbamates)
		$R_2N-\overset{O}{\overset{\|}{C}}-NR'_2$	ureas
		$R_2N-\overset{NR'}{\overset{\|}{C}}-NR''_2$	guanidines

(Between the two columns, vertically: Decreasing reactivity ↓)

†—NR₂ is taken to mean —NH₂, NHR, —NRR′, —NR₂

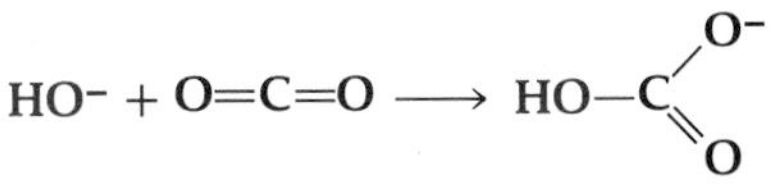

$$HO^- + O{=}C{=}O \longrightarrow HO{-}\overset{O^-}{\underset{O}{C}}$$

Bicarbonate

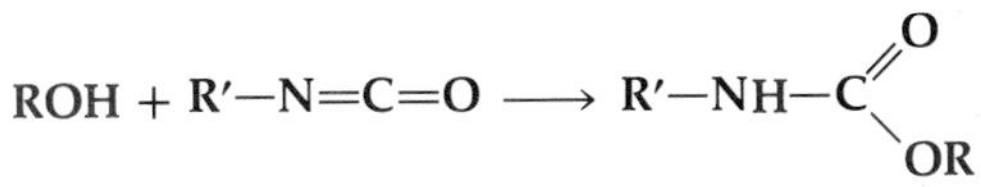

$$ROH + R'{-}N{=}C{=}O \longrightarrow R'{-}NH{-}\overset{O}{\underset{OR}{C}}$$

Urethans

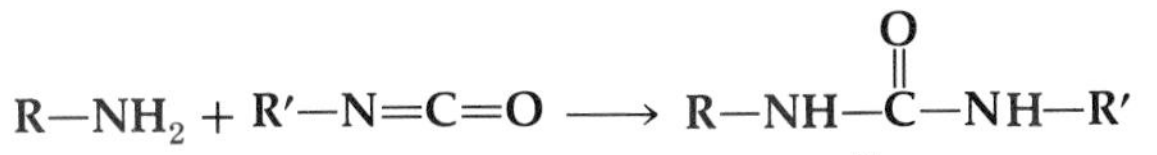

$$R{-}NH_2 + R'{-}N{=}C{=}O \longrightarrow R{-}NH{-}\overset{O}{\overset{\|}{C}}{-}NH{-}R'$$

Ureas

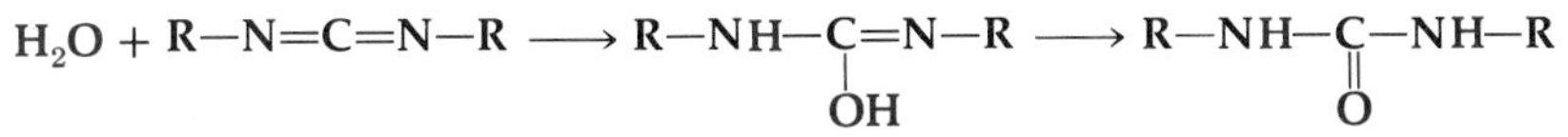

$$H_2O + R{-}N{=}C{=}N{-}R \longrightarrow R{-}NH{-}\underset{OH}{\overset{|}{C}}{=}N{-}R \longrightarrow R{-}NH{-}\underset{O}{\overset{\|}{C}}{-}NH{-}R$$

Ureas

$$R_2NH + CS_2 \longrightarrow R_2\overset{H}{\underset{+}{N}}{-}\overset{S}{\underset{S^-}{C}}$$

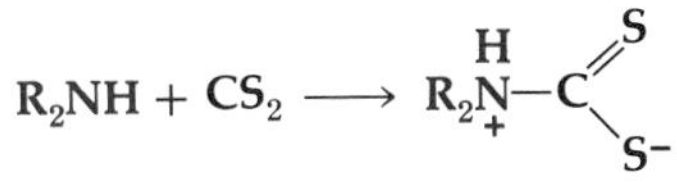

Dithiocarbamates

PROBLEM 12-10

To each of the substances in the left-hand column of Fig. 12-2 write the products expected from adding each of the following.

a Methanol
b Aniline
c Phenylmagnesium bromide

With the stable CO_2 molecule addition occurs only with strong bases (HO^-, RO^-, R_2N^-) and forms a resonance-stabilized carboxylate anion. On acidification, however, the equilibrium is reversed and CO_2 gas is liberated.

The isocyanates are formed from phosgene and arylamines or from several rearrangements like that of acyl azides ($R{-}CO{-}N_3 \overset{\Delta}{\longrightarrow} R{-}N{=}C{=}O$), described in Chap. 17. They are the common precursor for urethans and often for ureas.

The carbodiimides do not form stable addition products with alcohols and carboxylic acids. The initial adduct transforms the —OH group added into a good leaving group since a stable urea is formed on its removal. Carbodiimides thus act as dehydrating agents and are often employed in amide and ester formation (pages 512 and 510):

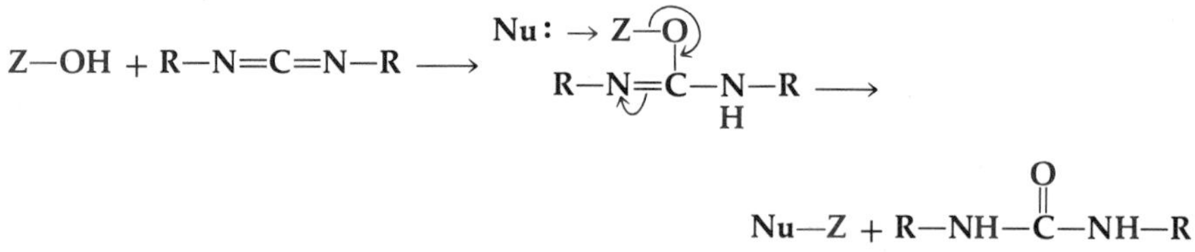

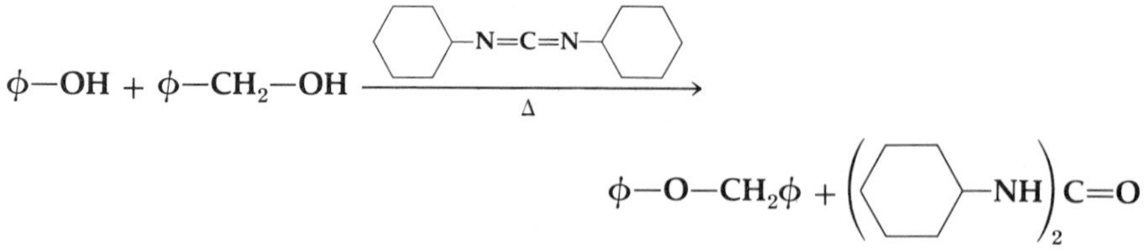

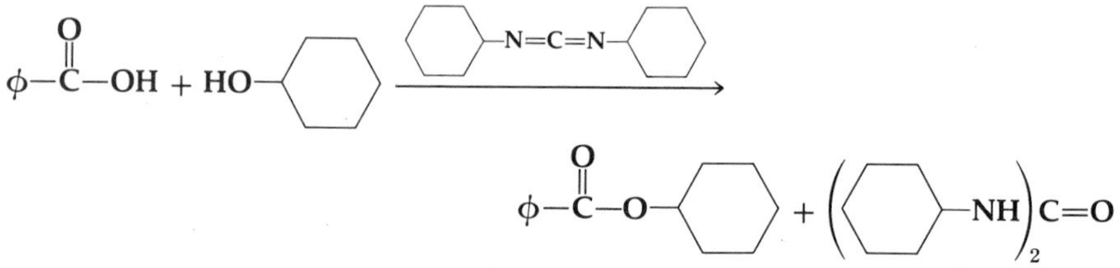

$$Z—OH + R—N{=}C{=}N—R \longrightarrow \quad \underset{\underset{H}{\overset{|}{N}}}{\overset{\overset{Nu: \rightarrow Z—O}{|}}{R—N{=}C—N—R}} \longrightarrow$$

$$Nu—Z + R—NH—\overset{\overset{O}{\|}}{C}—NH—R$$

$$\phi—OH + \phi—CH_2—OH \xrightarrow[\Delta]{\quad \text{(—N=C=N—)} \quad}$$

$$\phi—O—CH_2\phi + \left(\!\!\left\langle \right\rangle\!\!—NH\right)_2 C{=}O$$

$$\phi—\overset{\overset{O}{\|}}{C}—OH + HO—\!\!\left\langle \right\rangle \xrightarrow{\quad \text{(—N=C=N—)} \quad}$$

$$\phi—\overset{\overset{O}{\|}}{C}—O—\!\!\left\langle \right\rangle + \left(\!\!\left\langle \right\rangle\!\!—NH\right)_2 C{=}O$$

Ynamines (R—C≡C—N̈R′$_2$, Sec. 13-7) protonate very readily on carbon and yield quaternary intermediates, which do the same kind of dehydrative reaction but are more reactive.

$$\phi—C{\equiv}C—\overset{..}{N}(CH_3)_2 \underset{\quad}{\overset{H^+}{\rightleftarrows}} \phi—CH{=}C{=}\overset{+}{N}(CH_3)_2 \xrightarrow{2CH_3(CH_2)_2COOH}$$

$$CH_3(CH_2)_2CO—O—CO(CH_2)_2CH_3 + \phi CH_2CON(CH_3)_2$$

Grignard reagents give addition products which are resistant to further attack because of the resonant anion formed. Carboxylation of Grignard reagents (or alkyllithiums) is an excellent procedure for introducing the carboxyl group in synthesis (R—X $\longrightarrow$ R—MgX + CO$_2$ $\longrightarrow$ RCOOH). The reagent is usually formed and poured over crushed dry ice, and the resulting carboxylic acid is liberated by acidification of the carboxylic salt.

$$p\text{-}CH_3C_6H_4MgBr + CO_2 \longrightarrow p\text{-}CH_3C_6H_4CO_2MgBr \xrightarrow{HCl} p\text{-}CH_3C_6H_4COOH$$

p-Toluic acid
90%

Aryl isocyanates add organometallic compounds very easily to form amides, and the others react analogously.

$$(CH_3)_2CHMgCl + C_6H_5N{=}C{=}O \xrightarrow[2)\ H_3O^+]{1)\ Ether} C_6H_5NH\overset{\overset{O}{\|}}{C}CH(CH_3)_2$$

Isobutyranilide

Lithium aluminum hydride, however, proceeds beyond addition to hydride substitution (cf. Sec. 13-2), CO_2 giving methanol and isocyanates going to N-methyl amines ($R-NCO + LiAlH_4 \longrightarrow RNHCH_3$). In each case the reaction follows the pattern expected for the initial addition products, formic acid and formamides, respectively, as discussed in Sec. 13-2.

12-8   NUCLEOPHILIC ADDITIONS TO RELATED GROUPS

Imines and Immonium Cations

Of the related functions listed in Fig. 12-3, imines and immonium ions undergo additions quite analogous to those of carbonyls. The ions are faster, being more electrophilic by reason of charge, but they are exceedingly easily "enolized" to **enamines** by most bases. Hydrides reduce them to amines, as expected by analogy. All the imines or immonium salts are readily hydrolyzed to the parent ketone (aldehyde), either in aqueous acid or base. Immonium ions are the reactive species undergoing addition by enolates in the **Mannich reaction**. An amine (usually secondary to minimize the side reactions) is mixed with an aldehyde (formaldehyde has been most common) and an enolizable ketone. The amine and aldehyde first condense to form the immonium salt. Then the ketone enolate—formed in small concentration by the amine acting as a base on the ketone—then adds to the active immonium intermediate, forming a "**Mannich base**."

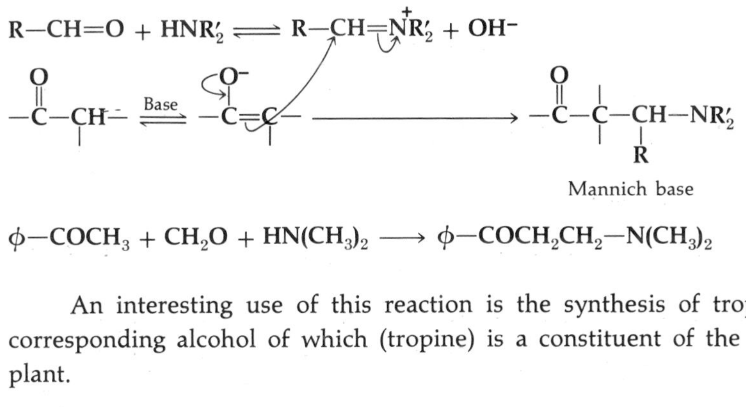

$$R-CH=O + HNR'_2 \rightleftharpoons R-CH=\overset{+}{N}R'_2 + OH^-$$

Mannich base

$$\phi-COCH_3 + CH_2O + HN(CH_3)_2 \longrightarrow \phi-COCH_2CH_2-N(CH_3)_2$$

An interesting use of this reaction is the synthesis of tropinone, the corresponding alcohol of which (tropine) is a constituent of the Belladonna plant.

Tropinone

The Mannich reaction is a mild procedure for obtaining unsaturated ketones (usually $-CO-C=CH_2$) since elimination of the amine is facile:

$$\phi-CO-CH_2CH_2-N(CH_3)_2 \xrightarrow{\Delta} \phi-CO-CH=CH_2 + HN(CH_3)_2$$

Nitriles as Substrates

Nitriles add water when heated with either aqueous acid or base. The addition products are amides, which can in turn be hydrolyzed by continued heating. The reaction can ordinarily be stopped at the amide stage.

$$RC\equiv N: \overset{H^+}{\rightleftharpoons} \left[RC\equiv \overset{+}{N}H \longleftrightarrow R\overset{+}{C}=\ddot{N}H \right] \xrightarrow[-H^+]{H_2O} \overset{OH}{RC=NH} \rightleftharpoons \overset{O}{R\overset{\parallel}{C}-NH_2}$$

A nitrile   An amide

$$H\bar{O} + RC\equiv N \longrightarrow \overset{OH}{RC=\bar{N}} \xrightarrow{H_2O} \overset{OH}{RC=NH} \rightleftharpoons \overset{O}{R\overset{\parallel}{C}-NH_2} \xrightarrow[\Delta]{H_2O} RCOOH + NH_3$$

$$H_2O + CH_3CH_2CH_2CN \xrightarrow{H_3O^+} CH_3CH_2CH_2CONH_2$$

n-Butyronitrile   n-Butyramide

Addition of some hydrogen peroxide accelerates the hydration of nitriles in alkaline solutions and is therefore usually the method of choice for mild hydrolysis.

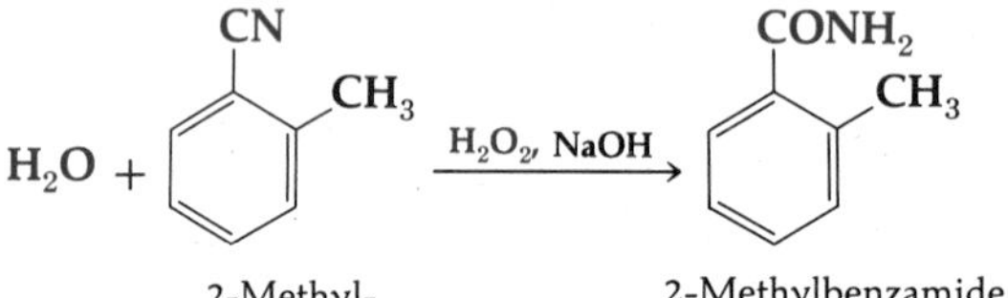

$$H_2O +$$

2-Methyl-
benzonitrile
(o-tolunitrile)

2-Methylbenzamide
92%

Nitriles also add alcohols in anhydrous alcoholic solutions of hydrogen chloride. The products are salts of **imino ethers**. These salts are readily hydrolyzed to esters by refluxing in aqueous solutions.

$$C_2H_5OH + CH_3(CH_2)_3CN \xrightarrow{HCl} CH_3(CH_2)_3\underset{\overset{\parallel}{H_2\overset{+}{N}} \overset{-}{Cl}}{C}OC_2H_5 \xrightarrow{H_2O, \Delta}$$

$$CH_3(CH_2)_3COOC_2H_5 + NH_4Cl$$

Ethyl valerate

Lithium aluminum hydride adds to nitriles to give imines, which add again to yield primary amines ($R—C\equiv N \longrightarrow [R—CH=NH] \longrightarrow R—CH_2—NH_2$).

$$CH_3CH_2CH_2CN \xrightarrow[\text{ether}>]{\text{LiAlH}_4} \xrightarrow{\text{H}_2\text{O}} CH_3CH_2CH_2CH_2NH_2$$

Butyronitrile   *n*-Butylamine

In a completely analogous fashion, but much more sluggishly, Grignard reagents add to nitriles yielding an imine anion as its magnesium salt. This is transformed to a ketone on hydrolysis.

$$R—C\equiv N + R'—MgX \longrightarrow R—\overset{\overset{\displaystyle N^-Mg^{++}X^-}{\|}}{C}—R' \xrightarrow{H_3O^+} R—\overset{\overset{\displaystyle O}{\|}}{C}—R'$$

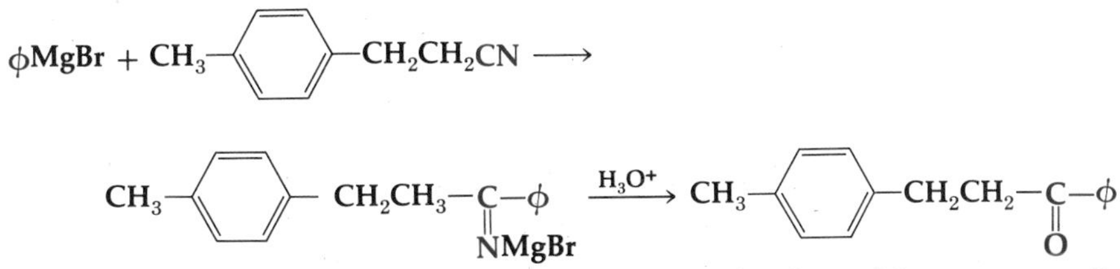

$$\phi MgBr + CH_3—\langle\ \rangle—CH_2CH_2CN \longrightarrow$$

$$CH_3—\langle\ \rangle—CH_2CH_3—\overset{\overset{\displaystyle }{\underset{\displaystyle NMgBr}{\|}}}{C}—\phi \xrightarrow{H_3O^+} CH_3—\langle\ \rangle—CH_2CH_2—\overset{\overset{\displaystyle }{\underset{\displaystyle O}{\|}}}{C}—\phi$$

1-Phenyl-3-*p*-tolyl-1-propanone 70%

Unsaturated nitriles are excellent substrates for vinylogous addition (page 480).

Diazonium Ions, Diazoketones, and Azides

Closely related to the triply bonded nitrile is the much less stable and more reactive diazonium cation ($R—\overset{+}{N}\equiv N:$) which can only be isolated when attached to a benzene ring. The same group appears as in the charged resonance form of certain neutral diazo compounds such as α-diazoketones and azides. In all contexts it is capable of acting as substrate for nucleophiles ($Nu: + :N\equiv\overset{+}{N}—Z \longrightarrow Nu—\ddot{N}=\ddot{N}—Z$). When attention is focused upon carbon as electrophile in these reactions, the diazonium salt is regarded as an electrophilic reagent.

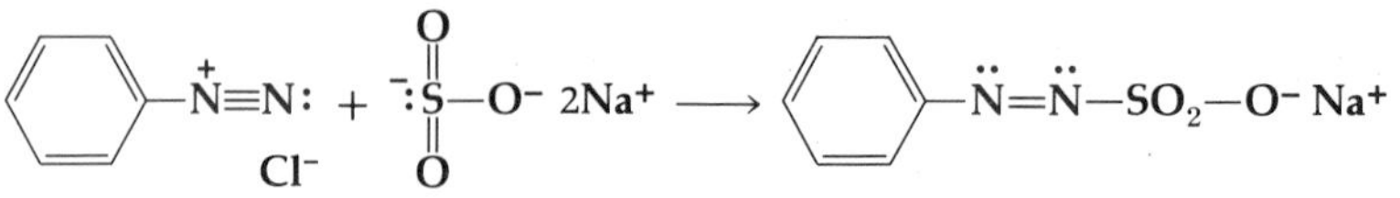

Benzenediazonium   Sodium   Sodium benzenediazonium sulfonate
     chloride   sulfite

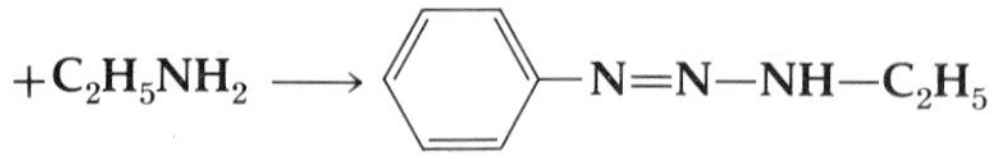

A triazene (page 424)

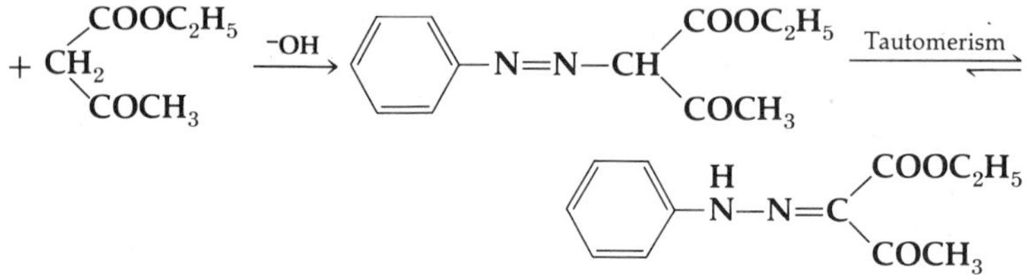

A phenylhydrazone

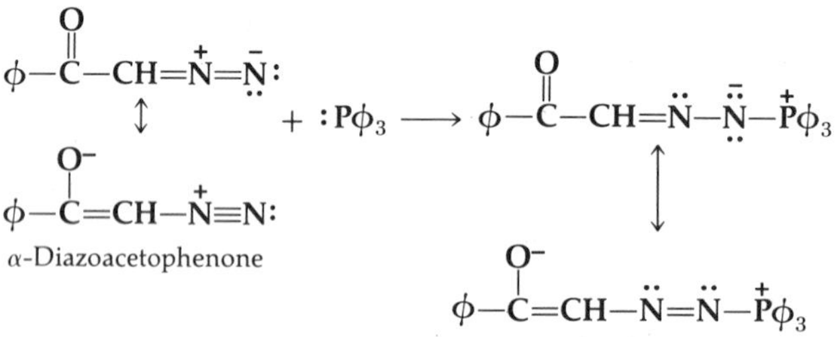

α-Diazoacetophenone

A phosphorazine

Diazonium ions may also lose nitrogen by a simple substitution reaction $(R-\overset{+}{N}\equiv N \longrightarrow R^+ + N_2 \xrightarrow{+Nu:} R-Nu)$, discussed for aromatic diazonium ions in Sec. 13-9 and for protonated (aliphatic) diazo compounds in Sec. 11-4, or by spontaneous extrusion of that stable molecule from an addition product.

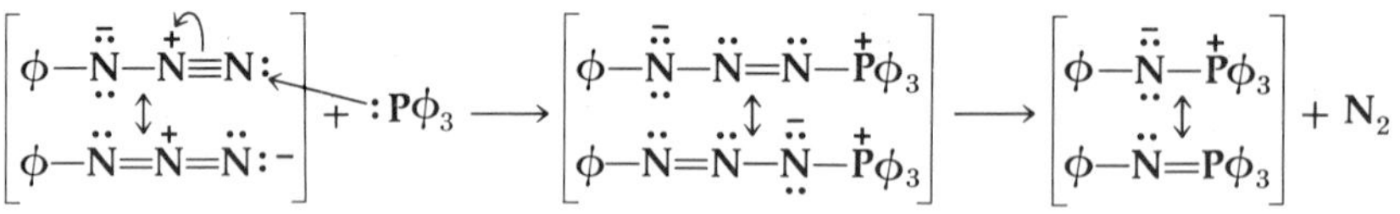

Nitro Groups

The nitro group is very stable and resistant to direct addition although very potent in stabilizing adjacent anions (readily enolized). Accordingly, it is very active in conjugate addition, acting to stabilize the immediate carbanion product.

$$\text{Nu:} + \text{C=C-N} \overset{O}{\underset{O^-}{\Big\langle}} \longrightarrow \left[\text{Nu-C-C-N} \overset{O}{\underset{O^-}{\Big\langle}} \longleftrightarrow \text{Nu-C-C=N} \overset{O^-}{\underset{O^-}{\Big\langle}} \right]$$

The resistance to direct addition is probably involved with the resonance stabilization in the nitro group which is isoelectronic with the unreactive carboxylate anion:

$$-\text{N}\overset{O}{\underset{O^-}{\Big\langle}} \longleftrightarrow -\text{N}\overset{O^-}{\underset{O}{\Big\langle}} \qquad -\text{C}\overset{O}{\underset{O^-}{\Big\langle}} \longleftrightarrow -\text{C}\overset{O^-}{\underset{O}{\Big\langle}}$$

Both are reduced by lithium aluminum hydride, however. The initial reaction is a nucleophilic addition of hydride in each case, and the product from the nitro group is a primary amine ($-NO_2 \longrightarrow -NH_2$). Aromatic nitro groups, however, often give mixed and more complex products with hydrides, but all nitro groups are smoothly reduced to primary amines by catalytic hydrogenation.

The aliphatic nitro compounds (with α-hydrogen) exhibit a unique character in that protonation of the enolate form gives the *aci*-nitro group, a tautomer which can be isolated in many cases, and this "enol" grouping is amphoterically reactive, being able either to accept or deliver electrons:

$$\text{CH-N}\overset{O}{\underset{O^-}{\Big\langle}} \underset{-H^+}{\overset{}{\rightleftharpoons}} \text{C=N}\overset{O^-}{\underset{O^-}{\Big\langle}} \underset{+H^+}{\overset{}{\rightleftharpoons}} \text{C=N}\overset{OH}{\underset{O^-}{\Big\langle}}$$

Nitro   *aci*-Nitro

aci-Nitro as nucleophile: $\text{C=N}\overset{O^-}{\underset{OH}{\Big\langle}}$

aci-Nitro as electrophile: $\text{Nu:} \longrightarrow \text{C=N}\overset{OH}{\underset{O^-}{\Big\langle}}$

Most of our discussion centers around the "enolate" form (*aci*-nitro anion) as a stabilized carbanion acting as nucleophile but the *aci*-nitro form can also be a nucleophile acceptor, reacting with water for example to give ketones or acids.

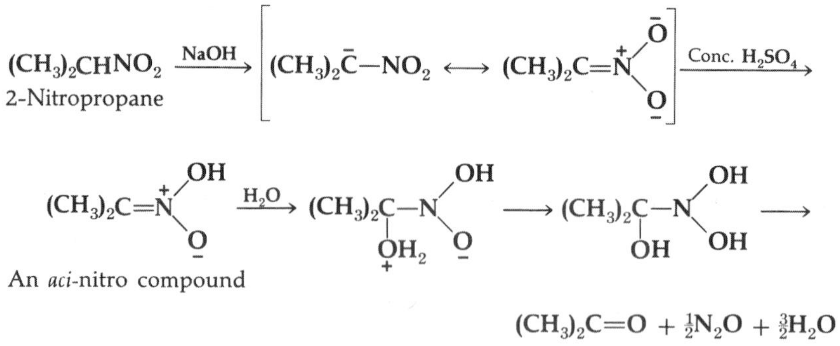

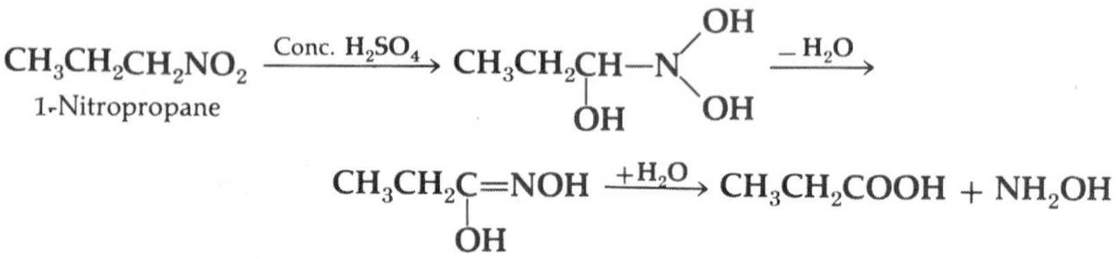

$$(CH_3)_2CHNO_2 \xrightarrow{NaOH} \left[(CH_3)_2\bar{C}-NO_2 \longleftrightarrow (CH_3)_2C=\overset{+}{N}\overset{\bar{O}}{\underset{O}{}} \right] \xrightarrow{Conc.\ H_2SO_4}$$

2-Nitropropane

$$(CH_3)_2C=\overset{+}{N}\overset{OH}{\underset{\underset{-}{O}}{}} \xrightarrow{H_2O} (CH_3)_2C-N\overset{OH}{\underset{\underset{+}{O}H_2\ \ \underset{-}{O}}{}} \longrightarrow (CH_3)_2C-N\overset{OH}{\underset{OH\ \ OH}{}} \longrightarrow$$

An *aci*-nitro compound

$$(CH_3)_2C=O + \tfrac{1}{2}N_2O + \tfrac{3}{2}H_2O$$

$$CH_3CH_2CH_2NO_2 \xrightarrow{Conc.\ H_2SO_4} CH_3CH_2CH-N\overset{OH}{\underset{OH\ \ OH}{}} \xrightarrow{-H_2O}$$

1-Nitropropane

$$CH_3CH_2\underset{\underset{OH}{|}}{C}=NOH \xrightarrow{+H_2O} CH_3CH_2COOH + NH_2OH$$

In some cases the group may serve in both capacities so that the study of aliphatic nitro compounds, while rich in complexity, has not been pursued extensively.

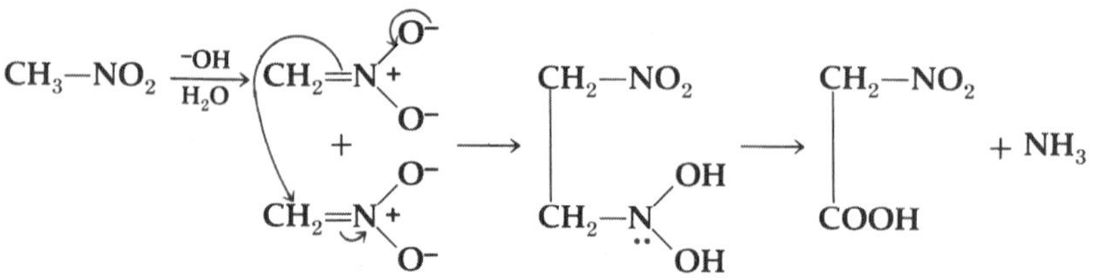

PROBLEM 12-11

Synthesize each of the following from benzyl cyanide, $C_6H_5CH_2CN$

a 1-Phenyl-2-butanone

b β-Phenylethylamine

c $C_6H_5CH_2C(OCH_2C_6H_5)=NH_2^+$

d Phenylacetamide

e 2,3-Diphenyl-4-nitro-pentanonitrile

f α-Phenylglutarimide

g β-Phenylpropionaldehyde

Carbon-Carbon Multiple Bonds

Nucleophiles in general do not add to simple alkenes except when they are conjugated to the groups of Fig. 12-3 and so act for those groups as vinylogous substrates (Sec. 12-6).

Alkynes are much more reactive toward nucleophiles than are alkenes, a fact utilized in the preparation of certain vinyl ethers and vinyl thioethers. The reaction assumes a predominately trans steric course, and addition proceeds in that orientation which serves best to stabilize the initial vinyl carbanion formed on addition ($Nu\!: + -C\equiv C- \longrightarrow Nu-C\!=\!\overset{..}{C}-$).

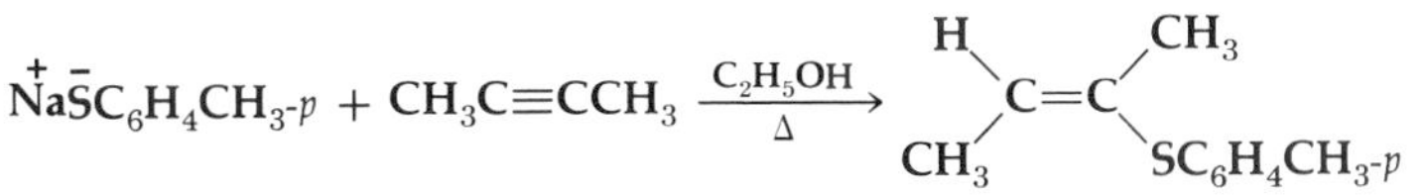

$$\overset{+}{Na}\overset{-}{S}C_6H_4CH_3\text{-}p + CH_3C\equiv CCH_3 \xrightarrow[\Delta]{C_2H_5OH}$$

$$\underset{\substack{\text{2-}p\text{-Tolylmercapto-}trans\text{-}\\ \text{2-butene}\\ 65\%}}{\overset{\displaystyle H}{\underset{\displaystyle CH_3}{>}}C\!=\!C\overset{\displaystyle CH_3}{\underset{\displaystyle SC_6H_4CH_3\text{-}p}{<}}}$$

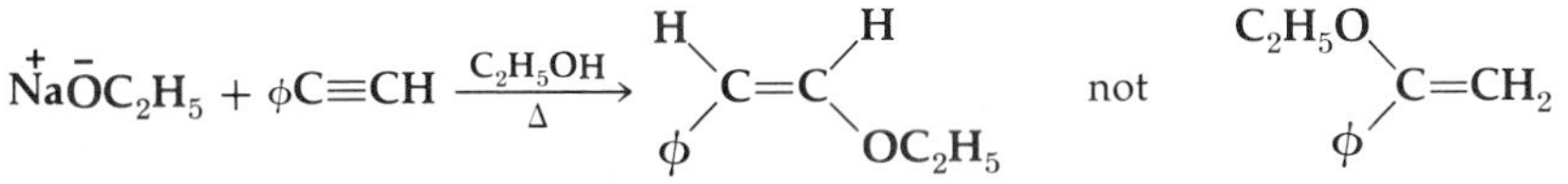

$$\overset{+}{Na}\overset{-}{O}C_2H_5 + \phi C\equiv CH \xrightarrow[\Delta]{C_2H_5OH} \underset{\substack{cis\text{-}\beta\text{-Ethoxystyrene}\\ 75\%}}{\overset{\displaystyle H}{\underset{\displaystyle \phi}{>}}C\!=\!C\overset{\displaystyle H}{\underset{\displaystyle OC_2H_5}{<}}} \quad not \quad \overset{\displaystyle C_2H_5O}{\underset{\displaystyle \phi}{>}}C\!=\!CH_2$$

PROBLEMS

Mechanism Problems. In a number of problems we must attempt to deduce the mechanistic pathway taken by certain known compounds which yielded unusual or unexpected products under standard reaction conditions. In considering these problems the following questions will help to find a likely mechanism. In most such problems the route taken by the molecules consists of a number of reaction steps. The proper use of the curved-arrow convention for electron movements is of great assistance in solving such problems.

a Is the overall molecular change an isomerization, or the change expected for the conditions used (e.g., A + B − H₂O, etc.)? If not, the problem may involve an unexpected oxidation or reduction and so be more difficult.

b What bonds are broken and what bonds are formed, on the basis of the given structures of starting material and product? Carbon-carbon bonds are of central importance in these considerations.

c What are the expected initial reactions considering the reactive functions present? The carbons which nucleophiles can attack and the protons which bases can abstract, in initial reversible reactions, must all be examined. One of these should be the first step in the actual case. With carbonyl reactions, consider reverse reactions also since carbonyl reactions are reversible. Thus, *reverse* aldol and Michael reactions may serve to *break* carbon-carbon bonds.

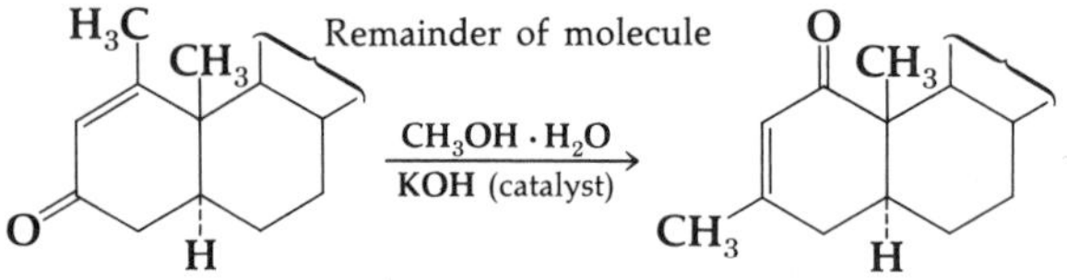

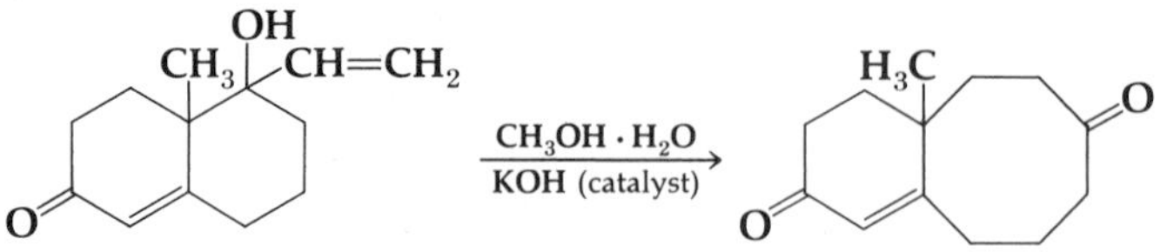

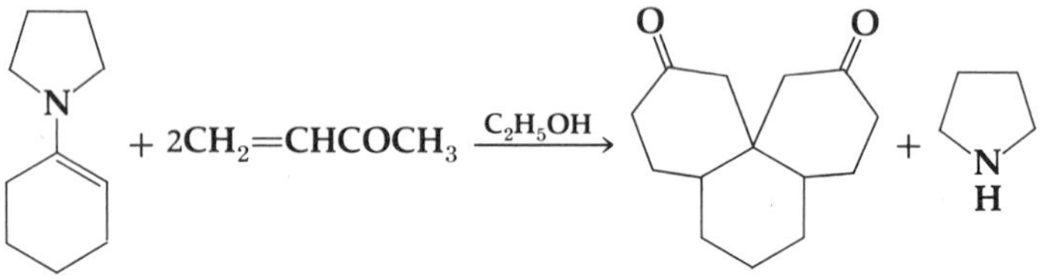

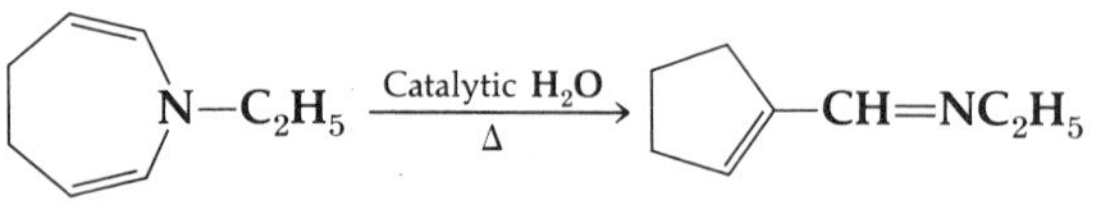

H3C
CH3
Remainder of molecule
CH3OH · H2O
KOH (catalyst)
O
H
O
CH3
CH3
H

OH
CH3
CH=CH2
CH3OH · H2O
KOH (catalyst)
O
H3C
O
O

N
+ 2CH2=CHCOCH3
C2H5OH
O
O
+
N
H

N—C2H5
Catalytic H2O
Δ
CH=NC2H5

12-12 Write specific examples of the following.

a Meerwein-Ponndorf reaction.

b An acid-catalyzed condensation reaction.

c A Michael reaction involving nitromethane.

d A crossed Cannizzaro reaction.

e Vinylogous enolization.

f The synthesis of a secondary alcohol by a Grignard reaction.

g An equilibrium reaction driven to completion by continuous removal of one of the products. Explain.

h A stable hydrate of an aldehyde. Explain the factors which make the substrate stable.

i Three addition reactions to a carbon-nitrogen triple bond.

j Conjugate addition.

k A reaction that increases the length of the carbon chain by three atoms.

12-13 Explain the following facts:

a Ethyl pyruvate $CH_3COCOOC_2H_5$ forms a ketal more readily than acetone.

b If one attempts to carbonate a Grignard reagent by passing a stream of carbon dioxide through an ether solution of the reagent, alcohols, rather than carboxylic acids, are produced.

c An elimination reaction usually accompanies the Perkin reaction.

d Ethyl acetoacetate and Grignard reagents bubble when mixed, but working up the reaction leads to recovery of ethyl acetoacetate.

12-14 How do you account for the fact that conjugate addition to quinones is more facile than to ordinary α,β-unsaturated ketones?

12-15 The following ester is formed in good yield when formaldehyde, acetone and ethyl acetoacetate are allowed to react with weak bases. Explain the steps in this conversion with structures.

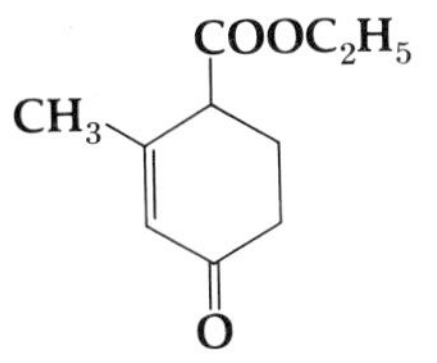

12-16 The following interconversion can be carried out in two steps. What are they?

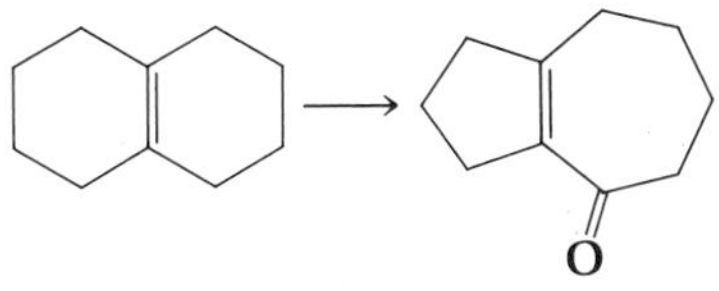

12-17 Indicate the products expected from treatment of each of these three aldehydes with the reagents listed: m-chlorobenzaldehyde, phenylacetaldehyde, α-methyl-cinnamaldehyde (2-methyl-3-phenyl-propenal).

a $NaBH_4$

b KOH

c C_2H_5OH/H^+

d Dry HCl

e $LiAlH_4$

f CH_3NO_2/OH^-

g Cyclohexanone/OH^-

h CH_3MgI

i Hot ammonium formate

j $CH_3CH{=}P(C_6H_5)_3$

k CN^-

l Ethyl acetoacetate/triethylamine

m $HSCH_2CH_2SH/BF_3$

n $CH_3C{\equiv}C{:}^-\ Na^+$

o $ClCH_2COOCH_3/OCH_3^-$

p Cyanoacetic acid/pyridine

q Phenylhydrazine

r Aluminum isopropoxide/isopropanol

s Ethyl orthoformate/H^+

12-18 Utilizing only one- or two-carbon compounds as carbon sources, devise syntheses of the following substances. Make a particular compound or intermediate only once in the problem set.

a $(CH_3)_2CHOH$

b $CH_3CH_2CH_2CH_2OH$

c $(CH_3)_2CHCH_2NH_2$

d $CH_3CH(SC_2H_5)_2$

e $(CH_3)_2CCOOH$
    $\overset{|}{O}H$

f $CH_3CH_2CONH_2$

g $CH_3COOC_2H_5$

h $CH_3SCH_2CH_2CN$

i $CH_3CHCH_2COOC_2H_5$
    $\overset{|}{O}H$

j $CH_3CH=CHCH_2OH$

k $(CH_3)_2C=CHCHCH=C(CH_3)_2$
    $\overset{|}{O}H$

l $CH_3CCH_2CH_3$
   (bicyclic acetal: O O bridged by CH_2-CH_2)

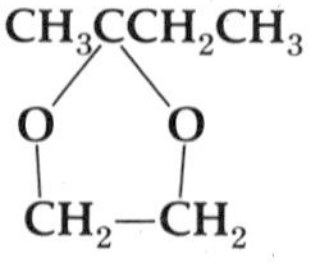

m $CH_3CHClCH_2COOH$

n NH_2CH_2COOH

o $(CH_3)_2\overset{+}{N}HCH_2CH_2CN$ $\bar{Cl}$

12-19 Indicate the best reaction for forming the multiple bonds in each case below.

a $CH_2=C(CH_2CH_2CH_3)_2$

b (ortho-methoxyphenyl)$CH=CHCOOH$ with OCH_3

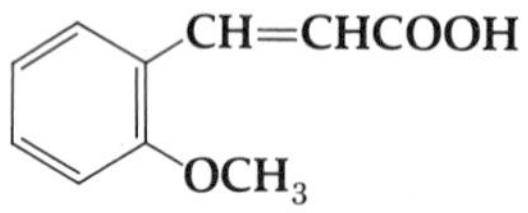

c 2-Methyl-2-pentenal

d CH_3 / CH_3 ... CH_3 / CH_3

e (cyclohexylidene) $C(COOCH_3)=$ ... $COOH$

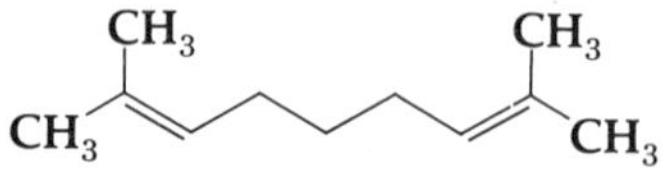

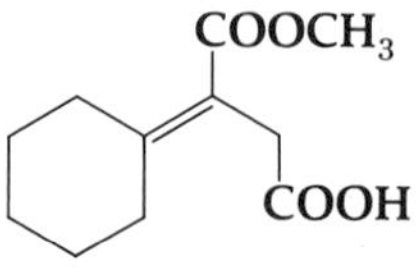

f $C_6H_5CH=N-N=CHC_6H_5$

12-20 Indicate the mechanistic steps in the reaction of benzaldehyde with an appropriate substance to yield the following amino-lactam.

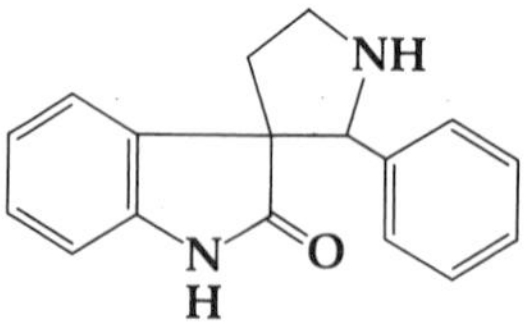

12-21 The first two steps in a synthesis of the male hormone, androsterone, are shown below. Show what reagents were used and the mechanism in each.

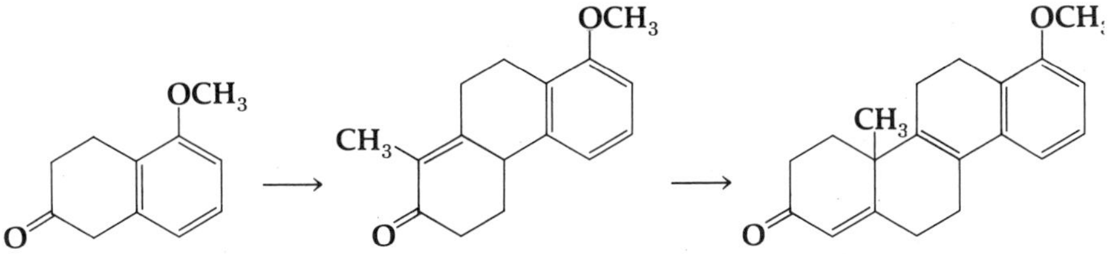

12-22 Write out syntheses of the following compounds, starting with materials one can buy such as acetophenone, benzaldehyde, acrylonitrile, nitromethane, acetone, ethyl acetate, bromobenzene, acetaldehyde, and cyclohexanone.

a $C_6H_5CH=CHCH=CHCH_2OH$

b $(C_6H_5)_2C\underset{\underset{OH}{|}}{}\!\!-\!\!\underset{\underset{OH}{|}}{C}HC_6H_5$

(*Hint:* Grignard additions can be used with hydroxy ketones if an excess of the reagent is used.)

c $CH_3\underset{\underset{C_6H_5}{|}}{C}H\underset{\underset{OH}{|}}{C}HCH_3$

d $C_6H_5\underset{\underset{CH_3}{|}}{C}HCH_2CH_2COOH$

e $CH_3OCH_2CH_2COC_2H_5$

f $C_6H_5CH_2CH_2CH_2CH_2OH$

g $HO(CH_2)_8OH$

h $CH_3(CH_2)_3\underset{\underset{H}{|}}{C}=\underset{\underset{H}{|}}{C}(CH_2)_3CH_3$

i $-CH_2CHO$

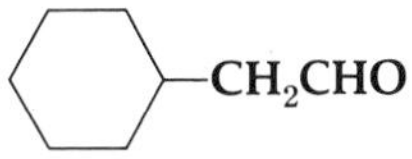

12-23 Account for the following transformations with stepwise mechanisms.

a (+)-4-Methyl-3-hexanone, benzaldehyde, and alkali give a racemic product.

b (+)-4-Methyl-3-hexanone with formaldehyde-diethylamine gives two racemic products.

c 3-Cyclohexenone is partially converted to an isomer with alkali or acid. Which of the two isomers do you expect to predominate in the finally isolated mixture?

d Benzaldehyde and acrylonitrile with **KOH** yield, among other products, 3-phenyl-2-hydroxymethyl-acrylonitrile.

e Another reaction product in part **d** is $C_{13}H_{12}N_2O$ with the same UV spectrum.

f

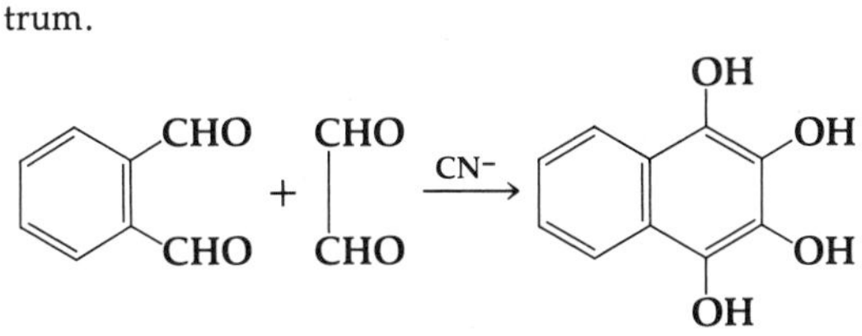

g Amino-acetone and dimethyl acetylenedicarboxylate yield

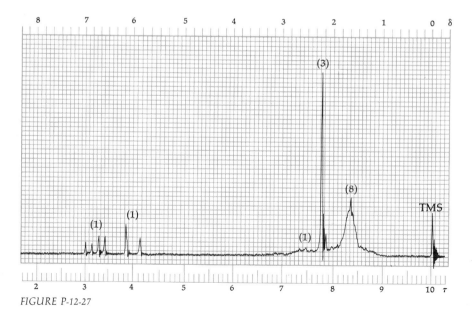

h 2-Hydroxycyclohexanone easily forms a dimer with no carbonyl absorption in the IR spectrum.

12-24 Account for these observations.

   a 5-Ketoheptanal, unlike most simple aliphatic aldehydes, gives but one product in good yield when treated with alkali.

   b Biacetyl (2,3-butanedione) forms 2,5-dimethylphenol under a variety of mild conditions. What is the role of aromatic stabilization in this conversion? Show the steps in the mechanism.

12-25 Write out the mechanistic steps in the formation of an acetal with ethyl ortho-formate in acids:

$$R-CHO + HC(OC_2H_5)_3 \xrightarrow{H^+} R-CH(OC_2H_5)_2 + HCOOC_2H_5$$
Ethyl orthoformate

12-26 An unknown ketone, $C_9H_{10}O_2$, was treated with O-methyl-hydroxylamine to form a derivative showing in the NMR spectrum two doublets of two protons each at τ 2.42 and 3.13 and three singlets of three protons each at τ 6.03, 6.14, and 7.82.

12-27 A compound, $C_9H_{14}O$, with a UV spectrum showing a strong maximum (log $\epsilon > 4$) at 227 nm, exhibited the NMR spectrum illustrated in Fig. P12-27. Write a reaction for its synthesis.

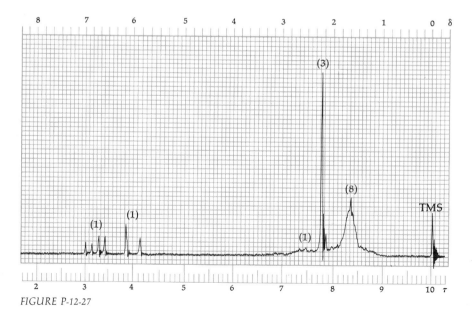

FIGURE P-12-27

12-28 On dissolving a certain ketone in concentrated aqueous ammonia one observes warming, and soon a rapid growth of crystals descends from the solution. These crystals ($C_5H_9NO_2$) show in the NMR three singlets, at τ 5.47 (1H), 6.38 (3H), and 8.08 (3H), as well as a broad signal at about τ 4 (2H) which disappears with D_2O. The crystals, fairly insoluble in water, dissolve easily in aqueous acid and soon revert therein to the starting ketone. What are the structures?

12-29 Two aldehydes (A and B) are allowed to react under catalysis with potassium cyanide. One of the several products formed (C) is isolated by chromatography and found to contain an alcohol group, by IR. This substance (C) is treated with thionyl chloride and the chlorine-containing residue reduced with zinc, replacing **Cl** by **H**. The product (D) shows a parent peak in the mass spectrum at m/e 226 and the NMR spectrum is shown in Fig. P12-29. Derive structures for all four substances and write a mechanism for the first reaction showing the other expected products, besides C.

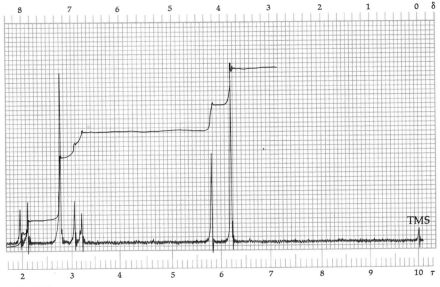

FIGURE P-12-29

12-30 If levulinic acid (4-oxo-pentanoic acid) is slowly distilled, two products are formed, both of which show parent peaks in the mass spectrum at m/e 98. (The acid provides some acid catalysis by itself.) The IR spectrum of each shows one strong peak, at 5.70 μ (1750 cm^{-1}) in one, 5.55 μ (1800 cm^{-1}) in the other. The former exhibits four NMR peaks in a ratio of 1:1:1:3 (reading upfield); the latter has only three (3:2:1). What are the structures and what mechanistic pathway accounts for their formation? What differences might you expect in their UV spectra?

CARBONYL AND RELATED GROUPS: NUCLEOPHILIC SUBSTITUTION

THE reactions of carbonyl compounds in this chapter take up the second major reaction indicated in the preamble to the previous chapter—nucleophilic substitution at unsaturated carbon. This two-stage reaction is recapitulated in Fig. 13-1. The intermediate addition compounds usually have only a transient existence and have not been isolated. Basic conditions are implicit in Fig. 13-1, but the reaction works equally well in many cases under acidic conditions (initial protonation of carbonyl).

13-1 THE CARBOXYLIC ACID FAMILY

It is clear that compounds of the type R—CO—L, where L is a heteroatom and leaving group, are all potentially interconvertible by the substitution reaction and therefore are conveniently regarded as a unified family of compounds. It is common to designate the parent compound as the carboxylic acid and the others as its derivatives. Arranged in order of decreasing activity the main members are listed in Table 13-1. The decreasing reactivity order is a function of increased resonance with carbonyl and decreasing leaving group activity. This order may be roughly correlated with the carbonyl absorptions in the infrared spectra (page 476) and with the order of decreasing stability (increasing pK_a) of the leaving group anions, $L:^- = Cl^-, \overline{O}COR', \overline{O}\phi, \overline{O}R, \overline{N}R_2, O^{--}$.

The conditions for substitution are similar to those for substitution at saturated carbon but without the complication of elimination. They may be carried out in acid or base as appropriate, and in general products must be formed from starting compounds which lie above them in the reactivity order (Table 13-1). Hence acid chlorides are the most common precursors for preparing the others.

When the focus of attention is on the nucleophile, substitution at saturated carbon is called *alkylation*. Substitution at carbonyl is called **acylation**.

FIGURE 13-1 Nucleophilic substitution at unsaturated carbon

$$\overline{Nu}: + \ \overset{Z}{\underset{|}{\overset{\|}{C}}}-L \longrightarrow Nu-\overset{\overline{Z}}{\underset{|}{C}}-L \longrightarrow Nu-\overset{Z}{\underset{|}{\overset{\|}{C}}}- \ + \ :\overline{L}$$

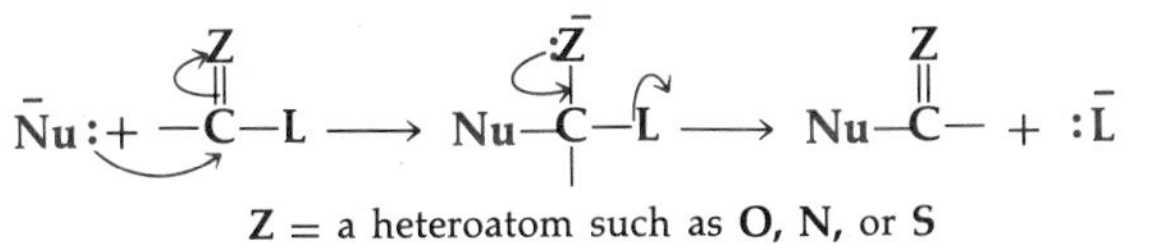

Z = a heteroatom such as O, N, or S

Carboxylic acid derivatives are made by acylating heteroatoms (e.g., —OH, —SH, >NH) while acylation at carbon is an important device for carbon-carbon bond formation and is usually applied to enolate nucleophiles.

Acid Halides

Acid chlorides are the only important compounds in which halogen is attached directly to a carbonyl group. Acid bromides and acid iodides may be prepared, but they are less stable and have no properties that make them superior to acid chlorides for any known chemical purpose. Since all acid halides are highly reactive, their principal use is as reagents for introduction of acyl groups in the course of syntheses.

Acids and their salts are the only common starting materials for the synthesis of acid chlorides, and several reagents are utilized as sources of nucleophilic chloride. All the reagents form intermediate reactive derivatives so that the leaving group in the substitution is not —OH but a derivative thereof. These are the same reagents used to convert alcohols to saturated chlorides (list on page 420).

$$CH_3\overset{O}{\underset{}{C}}{-}OH + SOCl_2 \xrightarrow{-HCl} CH_3\overset{O}{\underset{}{C}}OSOCl \xrightarrow{Cl^-}$$

Thionyl chloride

$$CH_3\overset{\bar{O}}{\underset{Cl}{\overset{|}{C}}}{-}O{-}\overset{O}{\underset{}{S}}{-}Cl \longrightarrow CH_3\overset{O}{\underset{}{C}}Cl + SO_2 + \bar{C}l$$

$$(CH_3)_3C\overset{O}{\underset{}{C}}OH + PCl_5 \xrightarrow{-HCl} (CH_3)_3C\overset{O}{\underset{}{C}}{-}OPCl_4 \xrightarrow[\text{2) } -POCl_3,\ Cl^-]{\text{1) } +Cl^-} (CH_3)_3C\overset{O}{\underset{}{C}}Cl + POCl_3$$

Pivalic acid   Pivaloyl chloride

$$CH_3CH_2CH_2\overset{O}{\underset{}{C}}OH + PCl_3 \xrightarrow{-HCl} CH_3CH_2CH_2\overset{O}{\underset{}{C}}OPCl_2 \longrightarrow CH_3CH_2CH_2\overset{O}{\underset{}{C}}Cl + (POCl)$$

TABLE 13-1   **The Carboxyl Family of Functional Groups**

R—CO—Cl	Acid (acyl) chlorides
R—CO—SR'	Thiol esters
R—CO—OCOR'	Acid anhydrides
R—CO—Oφ	Esters
R—CO—OR'	
R—CO—OH	Carboxylic acids
R—CO—N⟨R'(H) R''(H)	Amides
R—CO—O⁻	Carboxylate anions

(Decreasing reactivity — vertical arrow at left)

Each reagent has its unique characteristics. Thionyl chloride is often the reagent of choice for laboratory preparations, because the only by-products, SO_2 and HCl, are volatile.

Hydrogen chloride *cannot* be used as a reagent in acid chloride preparations because of the unfavorable equilibrium relationship

$$R—COOH + HCl \rightleftharpoons R—COCl + H_2O$$

Anhydrides

Anhydrides are made by displacing chloride from an acid chloride by a carboxylate ion or by heating an acid with an acidic dehydrating agent, such as phosphorus pentoxide or acetic anhydride. Acetic anhydride is a convenient reagent, since the by-product, acetic acid, can be removed continuously by distillation to force high conversion in this reversible reaction.

$$CH_3(CH_2)_5COOH + CH_3(CH_2)_5COCl \xrightarrow{\text{Pyridine}} [CH_3(CH_2)_5CO]_2O + HCl$$

    Heptanoic acid   Heptanoyl   Heptanoic anhydride
                chloride

$$(CH_3CO)_2O + 2C_6H_5COOH \underset{\text{Excess }(CH_3CO)_2O}{\rightleftharpoons} (C_6H_5CO)_2O + 2CH_3COOH$$

                                     Benzoic   Distill
                                    anhydride   out
                                        74%

Mixed anhydrides are also known, but they are usually of little importance. Although formic anhydride has never been prepared, mixed anhydrides that involve a formyl group are known and used. Cyclic anhydrides containing five- and six-membered rings form particularly easily, in many cases simply by heating the parent dibasic acid and driving off the water formed. This is self-acid-catalyzed and favored by being cyclic.

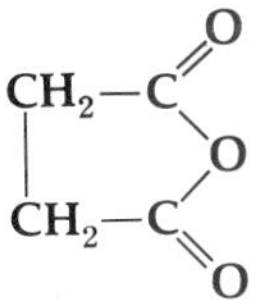

Succinic   Maleic   Glutaric
anhydride   anhydride   anhydride

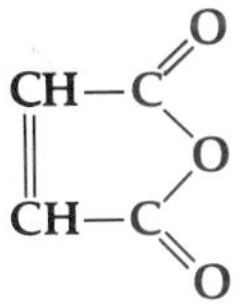

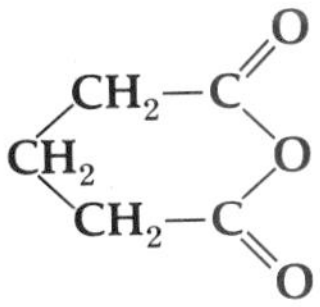

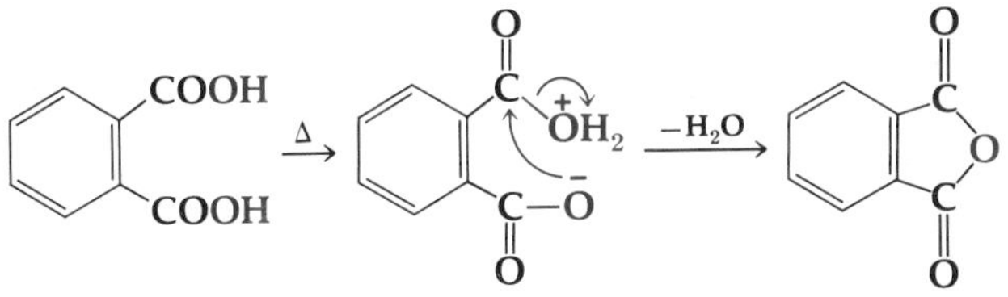

Phthalic   Phthalic
acid   anhydride

Carboxylic Acids

All the derivatives of carboxylic acids may be hydrolyzed under one or another set of conditions. Aliphatic acid chlorides and anhydrides usually react so rapidly with water that they must be protected from atmospheric moisture during storage.

$$CH_3\overset{O}{\overset{\|}{C}}Cl + H_2O \xrightarrow[\text{in air}]{\text{Fumes}} CH_3COOH + HCl$$

$$CH_3\overset{O}{\overset{\|}{C}}O\overset{O}{\overset{\|}{C}}CH_3 + H_2O \xrightarrow[\text{Exothermic}]{\text{Room temperature}} 2CH_3COOH$$

The mechanism of acid chloride hydrolysis is typical. The first step, a *nucleophilic addition* to a polarized carbonyl group, is followed by *elimination* of hydrogen chloride.

$$C_6H_5\overset{:O:}{\overset{\|}{C}}Cl + H_2\overset{..}{\overset{..}{O}}: \longrightarrow C_6H_5\overset{:\overset{..}{\overset{-}{O}}:}{\underset{:\overset{+}{O}H_2}{C}}-Cl \underset{}{\overset{-H^+}{\rightleftharpoons}} C_6H_5\overset{:\overset{..}{O}H}{\underset{:\overset{..}{O}H}{C}}-Cl \xrightarrow{-Cl^-}$$

$$C_6H_5\overset{:\overset{..}{O}H}{\underset{:\overset{..}{O}H}{C}}+ \xrightarrow{-H^+} C_6H_5\overset{:O:}{\overset{\|}{C}}-\overset{..}{\overset{..}{O}}H$$

The intermediate adduct, $C_6H_5C(OH)_2Cl$, is an **orthoacid chloride**, a species that has never been isolated. Loss of chloride ion from the intermediate is rapid because a stable cation (which is actually the conjugate acid of benzoic acid) is formed.

Esters are hydrolyzed to carboxylic acids when heated with catalytic amounts of acid in aqueous solution. The reaction is reversible. In many simple cases, equilibrium constants are about unity. If a large excess of water is used, hydrolysis may be made essentially quantitative. On the other hand, if water is removed by slow distillation, the reaction can be used to convert an acid and an alcohol to an ester (**Fischer esterification**). The steps are outlined in Fig. 13-2.

Esters can also be transformed into acids by an S_N1 reaction, in which the protonated ester becomes a leaving group. This is common with esters of tertiary alcohols.

$$\phi-\overset{\overset{H^+}{\curvearrowleft}}{\overset{O}{C}}\overset{\curvearrowright}{-O}-C\overset{CH_3}{\underset{CH_3}{\overset{CH_3}{-}}} \xrightarrow{HCl} \phi-C\overset{OH}{\underset{O}{\diagdown}} + \left[^+C\overset{CH_3}{\underset{CH_3}{\overset{CH_3}{-}}} \right] \xrightarrow{Cl^-} Cl-\overset{CH_3}{\underset{CH_3}{\overset{CH_3}{\overset{|}{C}}}} CH_3$$

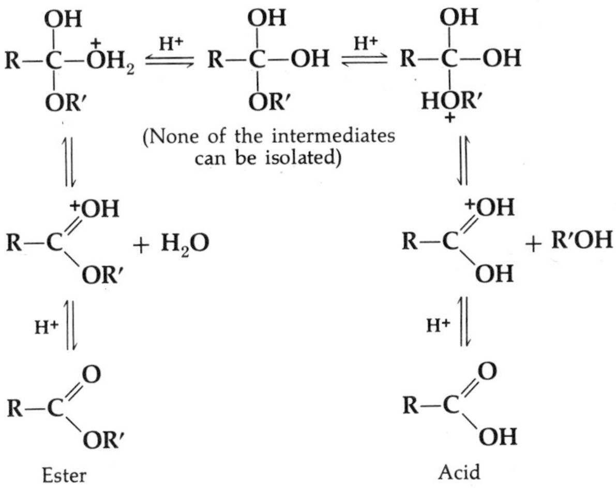

$$
\begin{array}{ccc}
\overset{\displaystyle OH}{\underset{\displaystyle OR'}{R-\overset{|}{\underset{|}{C}}-\overset{+}{O}H_2}}
\;\overset{H^+}{\rightleftharpoons}\;
\overset{\displaystyle OH}{\underset{\displaystyle OR'}{R-\overset{|}{\underset{|}{C}}-OH}}
\;\overset{H^+}{\rightleftharpoons}\;
\overset{\displaystyle OH}{\underset{\displaystyle \overset{+}{H}OR'}{R-\overset{|}{\underset{|}{C}}-OH}}
\end{array}
$$

(None of the intermediates can be isolated)

$$
R-C\!\!\begin{array}{c}\overset{+OH}{\diagup}\\[-2pt]\diagdown\,OR'\end{array} + H_2O
\qquad\qquad
R-C\!\!\begin{array}{c}\overset{+OH}{\diagup}\\[-2pt]\diagdown\,OH\end{array} + R'OH
$$

$$
H^+ \Big\Updownarrow \qquad\qquad\qquad\qquad H^+ \Big\Updownarrow
$$

$$
R-C\!\!\begin{array}{c}\overset{O}{\diagup}\\[-2pt]\diagdown\,OR'\end{array}
\qquad\qquad\qquad
R-C\!\!\begin{array}{c}\overset{O}{\diagup}\\[-2pt]\diagdown\,OH\end{array}
$$

Ester   Acid

FIGURE 13-2   **Acid-catalyzed hydrolysis and esterification**

Normal hydrolysis (Fig. 13-2) is by far the more common, proceeding faster than S_N1 substitution in all cases except those in which the alcohol moiety of the ester is tertiary or otherwise stabilizing for a carbonium ion. The S_N1 mechanism proceeds by **alkyl-oxygen cleavage** and normal hydrolysis by **acyl-oxygen cleavage**. The difference in mechanism was demonstrated by the use of an isotopic oxygen (O^{18}) tracer: O^{18} initially in the alcohol is found after hydrolysis either in the acid or the alcohol, according to the mechanism.

$$
\phi-\overset{\displaystyle O}{\overset{\|}{C}}-Cl + HO^{18}-CH_3 \longrightarrow \phi-\overset{\displaystyle O}{\overset{\|}{C}}-O^{18}-CH_3 \xrightarrow{H_3O^+} \phi-\overset{\displaystyle O}{\overset{\|}{C}}-OH + HO^{18}-CH_3
$$

$$
\phi-\overset{\displaystyle O}{\overset{\|}{C}}-Cl + HO^{18}-C(CH_3)_3 \longrightarrow \phi-\overset{\displaystyle O}{\overset{\|}{C}}-O^{18}-C(CH_3)_3 \xrightarrow[(S_N1)]{H_3O^+}
$$

$$
\phi-\overset{\displaystyle O}{\overset{\|}{C}}-O^{18}H + HO-C(CH_3)_3
$$

Alkali also accelerates hydrolysis, but an equivalent amount of base must be used, since the acidic product is converted to carboxylate ions. This formation of the low-energy (low pK_a) carboxylate anion from the initial less stable hydroxide ion drives the equilibrium to the right in this instance and makes the reaction essentially irreversible. Alkaline hydrolysis of esters is sometimes termed **saponification** from the special case of soap manufacture by alkali treatment of natural fatty esters.

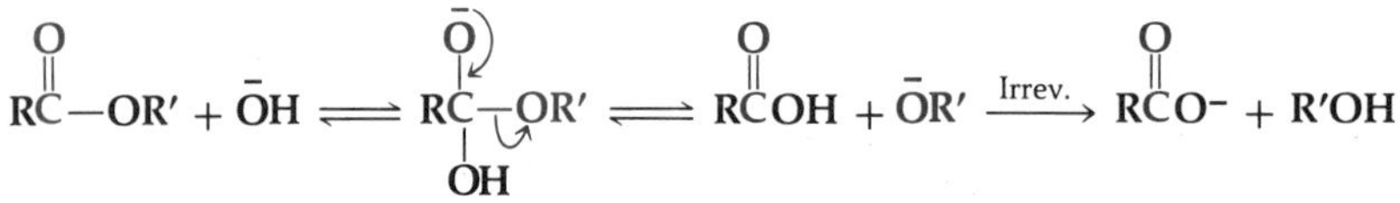

$$RC{-}OR' + {}^-OH \rightleftharpoons RC{-}OR' \rightleftharpoons RCOH + {}^-OR' \xrightarrow{\text{Irrev.}} RCO^- + R'OH$$

$$CH_3COOC_2H_5 + \overset{+}{Na}\ {}^-OH \xrightarrow{20\%\ NaOH} CH_3C\bar{O}_2\ \overset{+}{Na} + C_2H_5OH$$

100%
(by titration)

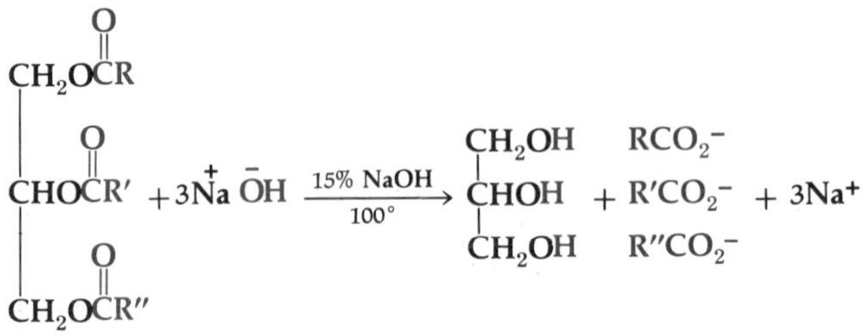

$$\begin{array}{l}
CH_2OCR \\
\quad \\
CHOCR' + 3\overset{+}{Na}\ {}^-OH \xrightarrow[100°]{15\%\ NaOH} \\
\quad \\
CH_2OCR''
\end{array}
\quad
\begin{array}{ll}
CH_2OH & RCO_2^- \\
CHOH & + R'CO_2^- + 3Na^+ \\
CH_2OH & R''CO_2^-
\end{array}$$

A fat; R, R', and   Glycerol   Soaps
R'' are long-   100%
chain saturated
and unsaturated
alkyl groups

The reactivity of esters in hydrolysis or of acids in esterification is controlled by the necessity of passing through an intermediate with a tetrahedral carbon atom in the functional group. If a carbonyl group is surrounded by bulky groups, hydrolysis may become very slow.

$$CH_3COOH \qquad \begin{array}{l} CH_3CH_2 \\ \qquad\qquad CHCOOH \\ CH_3CH_2 \end{array}$$

1   0.0004   Too slow to measure

Relative rates of acid-catalyzed
esterification with ethanol

Conjugated esters lose a certain amount of resonance energy when converted to addition products and are consequently less reactive than their saturated analogs. Alkaline hydrolysis, but not acid-catalyzed esterification or hydrolysis, is very sensitive to inductive effects of polar substituents. Trifluoroacetates, which are very easily hydrolyzed, illustrate the effect. In the ester, the dipoles of the carboxyl and trifluoromethyl groups repel each other. In the transition state for hydrolysis, the negative charge is stabilized by charge-dipole interaction (a case of B + C in Fig. 9-7).

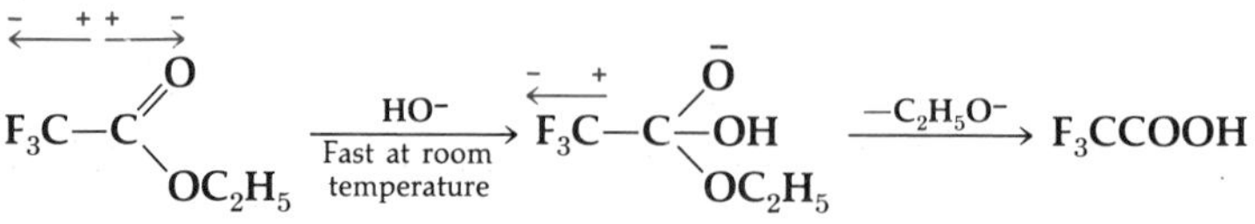

$$F_3C{-}C\overset{O}{\underset{OC_2H_5}{\big<}} \xrightarrow[\substack{\text{Fast at room} \\ \text{temperature}}]{HO^-} F_3C{-}C\overset{\bar{O}}{\underset{OC_2H_5}{\big<}}{-}OH \xrightarrow{-C_2H_5O^-} F_3CCOOH$$

Sterically hindered aromatic acids and esters may be interconverted by the **Newman procedure**. All acids and esters act as bases and are protonated by concentrated sulfuric acid, but steric hindrance promotes further ionization and produces less hindered acyl cations instead of normal conjugate acids.

Normal protonation:

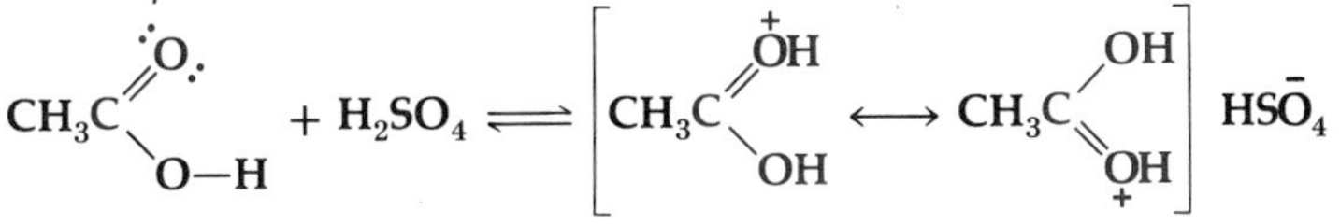

Further ionization:

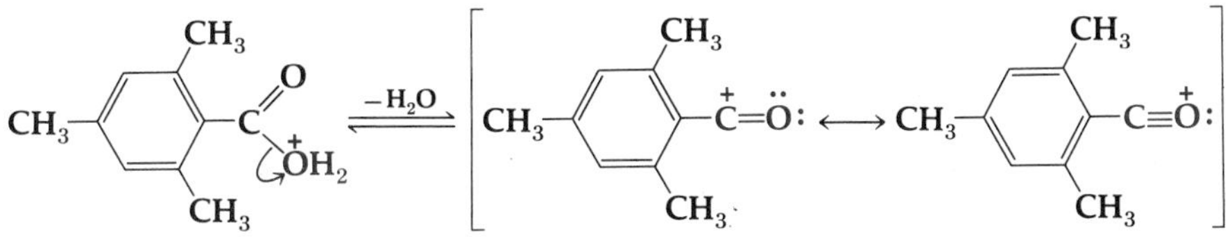

Mesitoic acid (protonated)

The $-C\equiv O:^+$ group is linear and has essentially the same π orbitals as nitriles or acetylenes. When attached to an unsaturated system, charge is highly delocalized, much more than in the corresponding carboxylic acid. The ionization of mesitoic acid relieves steric strain. Esters of hindered acids produce acyl cations by analogous reactions. If a sulfuric acid solution containing acyl cations is diluted with excess water or alcohol, the cations are captured immediately. Since the method is not applicable to unhindered acids and esters, it is an example of *steric acceleration* of a reaction.

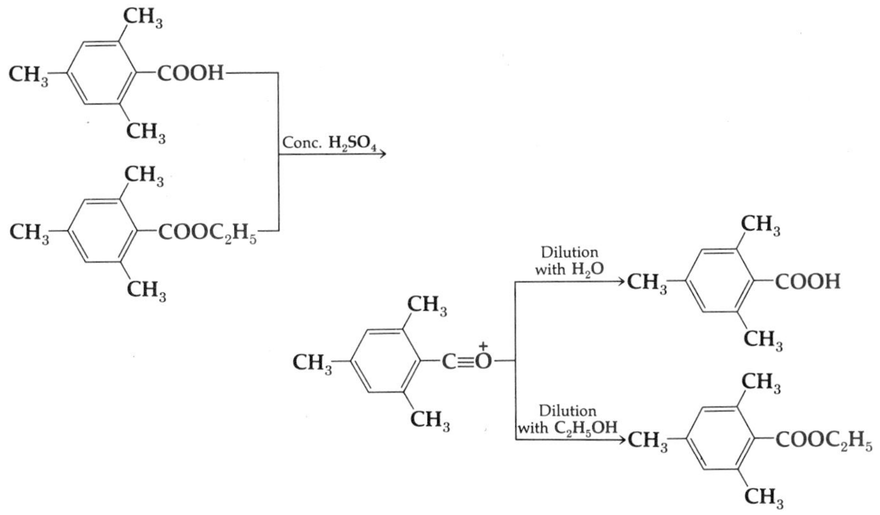

Although not formed in acid from normal esters, acyl cations may be prepared from acyl halides and have occasionally been isolated with a non-nucleophilic anion.

$$CH_3-\overset{\overset{\displaystyle O}{\|}}{C}-F + BF_3 \rightleftharpoons CH_3-C\equiv O^+ \ BF_4^-$$

The following example of stepwise saponification demonstrates variation in ester reactivity. The aliphatic ester group is hydrolyzed readily. The other two carboxyl functions are both conjugated and sterically hindered, but under forcing conditions the less hindered of the two is selectively attacked. The third could probably be cleaved with sodium hydroxide in a high-boiling solvent such as ethylene glycol.

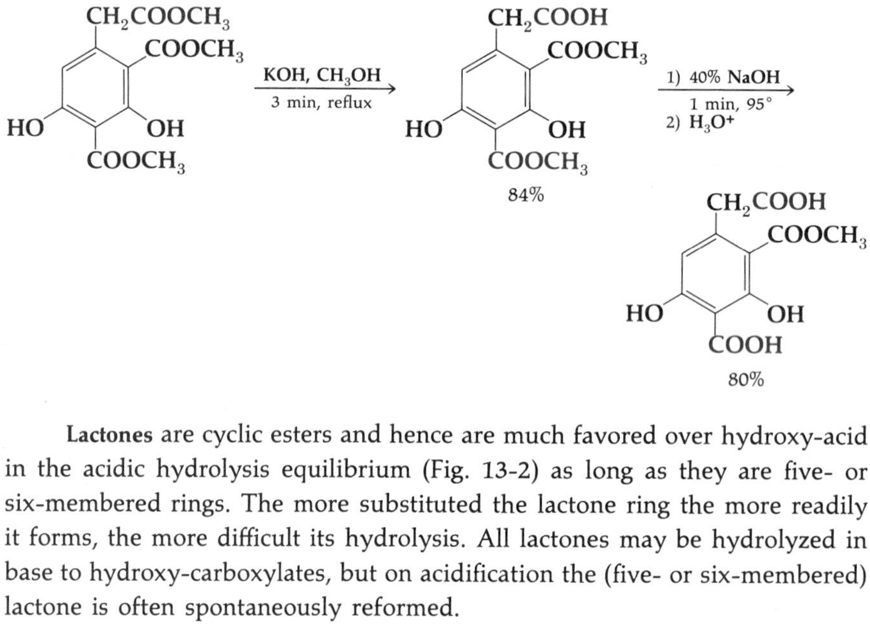

Lactones are cyclic esters and hence are much favored over hydroxy-acid in the acidic hydrolysis equilibrium (Fig. 13-2) as long as they are five- or six-membered rings. The more substituted the lactone ring the more readily it forms, the more difficult its hydrolysis. All lactones may be hydrolyzed in base to hydroxy-carboxylates, but on acidification the (five- or six-membered) lactone is often spontaneously reformed.

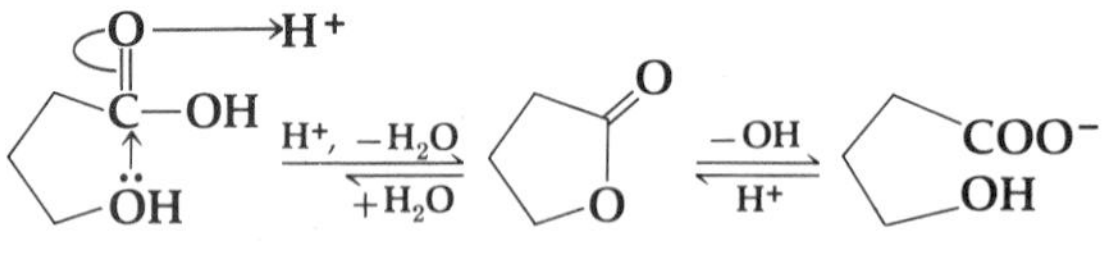

Carboxylic Esters

Direct esterification of an acid with an alcohol was discussed above, because of the intimate relationship with ester hydrolysis. Useful ester syntheses by substitution reactions were treated in Sec. 11-1. However, the most general laboratory procedures for esterification involve *acylation of alcohols by acid derivatives.* Acid chlorides and anhydrides are usually used for the purpose. The most convenient procedure for converting small samples of acids to esters involves intermediate formation of an acid chloride. Aryl esters constitute a prime example, since they cannot be made by direct esterification of phenols, owing to unfavorable equilibrium constants. Even acid chlorides derived from highly hindered acids are converted to esters easily. A base like pyridine is used to neutralize the HCl formed.

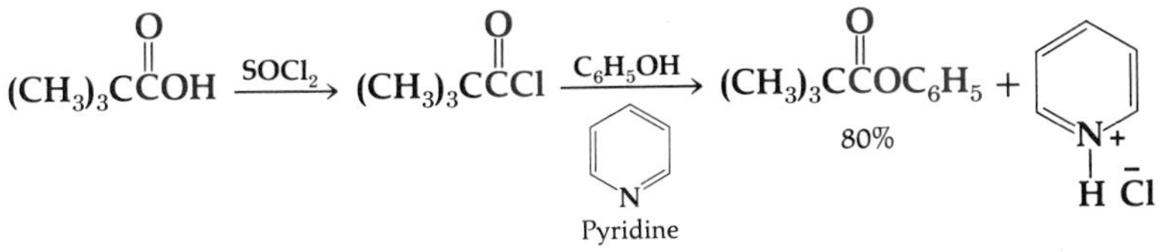

$$(CH_3)_3CCOH \xrightarrow{\ SOCl_2\ } (CH_3)_3CCCl \xrightarrow{\ C_6H_5OH\ } (CH_3)_3CCOC_6H_5 \ + $$

80%

Pyridine

A common application of the procedure is the conversion of alcohols to characteristic solid derivatives. A sample of the alcohol to be identified is heated with an excess of an acid chloride such as benzoyl chloride, p-nitrobenzoyl chloride, or 3,5-dinitrobenzoyl chloride. Excess acid halide is removed by treatment with aqueous base.

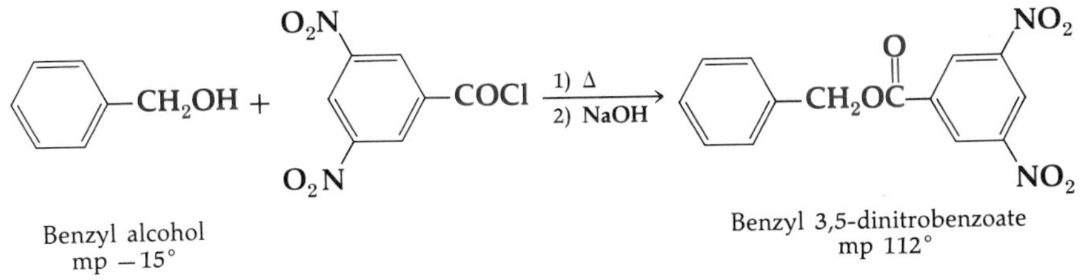

Benzyl alcohol
mp −15°

Benzyl 3,5-dinitrobenzoate
mp 112°

Acetic anhydride is often used to esterify alcohols. Tertiary amines catalyze the reactions and probably act by converting the anhydrides to transient acyl-ammonium ions. Acetylation by heating in acetic anhydride/pyridine is common practice.

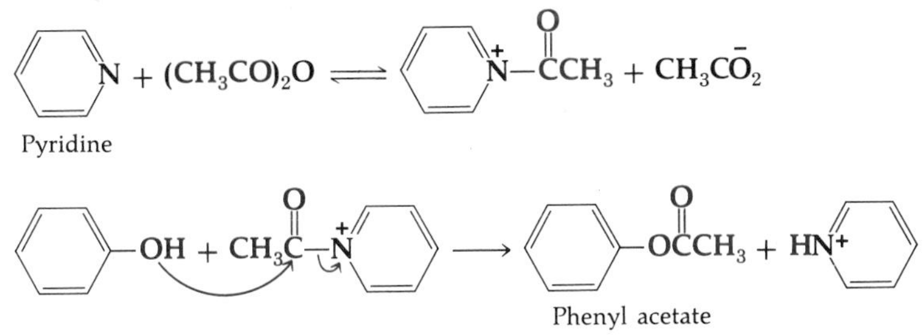

Pyridine

Phenyl acetate

Cyclic anhydrides, such as phthalic anhydride, react easily with alcohols, giving half-esters. An interesting application of the reaction is the conversion of racemic alcohols to acidic derivatives, which may be resolved into enantiomers by means of optically active bases. After the diastereomeric salts have been purified by recrystallization, the optically active alcohols may be recovered by hydrolysis of the active esters.

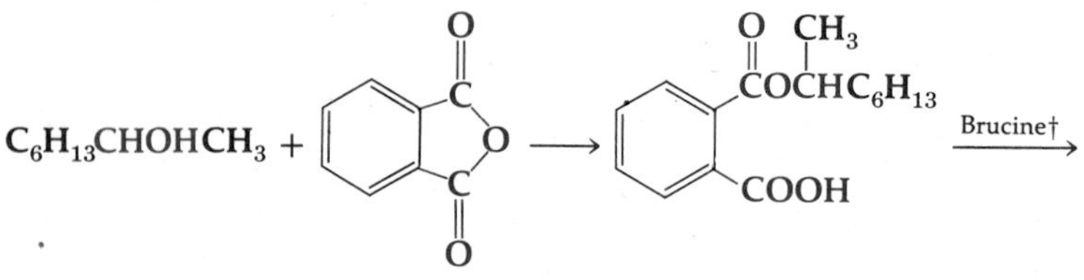

Racemic 2-octanol	Phthalic anhydride	2-Octyl hydrogen phthalate (1-methylheptyl hydrogen phthalate)

Brucine d-2-octyl phthalate + brucine l-2-octyl phthalate

Brucine salts, separated by fractional recrystallization

Brucine d-2-octyl phthalate $\xrightarrow{\text{HCl}}$ brucine HCl + d-2-octyl hydrogen phthalate

$+ CH_3\overset{*}{C}HOHC_6H_{13}$

d-2-Octyl hydrogen phthalate	Sodium phthalate	d-2-Octanol

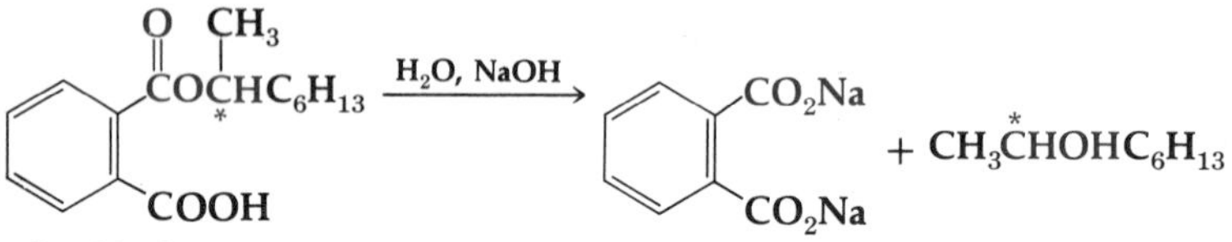

Various condensing agents may be used to cause rapid esterification of alcohols by acids without prior conversion of acid to acid chloride. Mixed anhydrides usually form first and serve as substrates in the acylation step. Trifluoroacetic anhydride is particularly effective, and carbodiimides (Sec. 12-7) have been much used recently, especially in biochemistry. The carbodiimides often serve to esterify even in neutral aqueous solution.

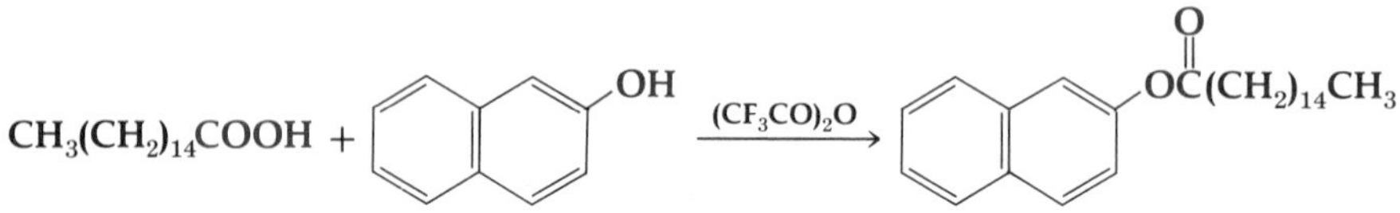

$CH_3(CH_2)_{14}COOH +$ $\xrightarrow{(CF_3CO)_2O}$

Palmitic acid	β-Naphthol	β-Naphthyl palmitate

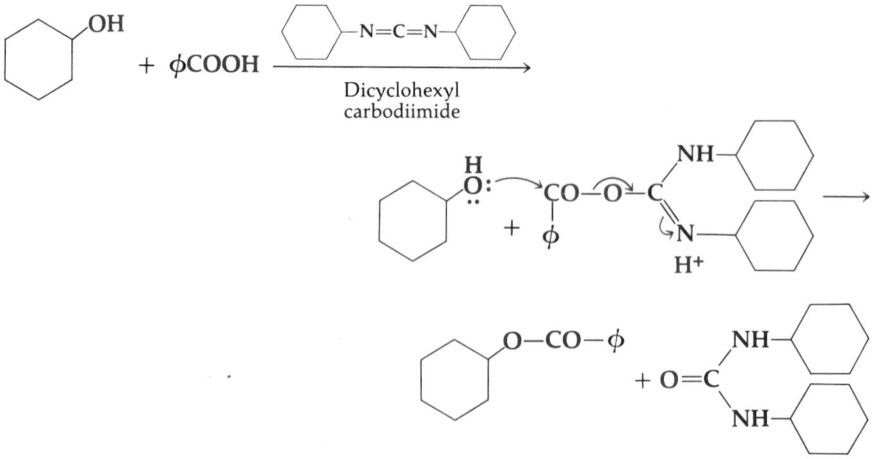

$+ \phi COOH \xrightarrow{\text{Dicyclohexyl carbodiimide}}$

Transesterification, or **ester interchange,** occurs when an ester is heated with an alcohol in the presence of acids or bases. The reaction is useful for the

†Brucine is an optically active, naturally occurring amine.

preparation of esters derived from insoluble acids. The mechanism is completely analogous to hydrolysis-esterification (Fig. 13-2). Internal transesterification converts γ- and δ-hydroxy-esters to lactones spontaneously with mild acidic or basic catalysis, in just the same manner as the corresponding hydroxy-acids,† cf. the second step of the Stobbe condensation (page 476).

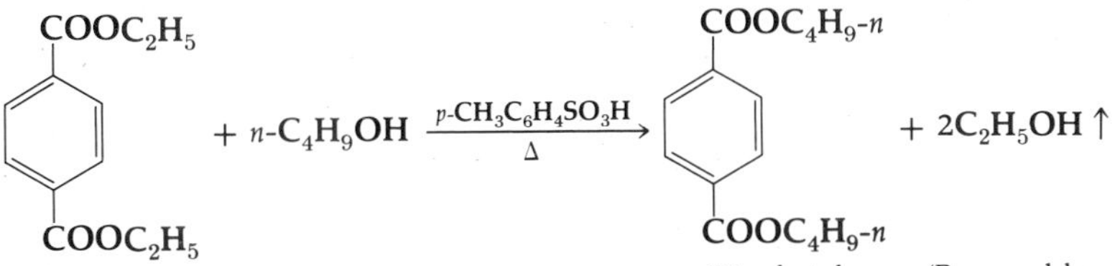

Diethyl terephthalate $+ n\text{-}C_4H_9OH \xrightarrow[\Delta]{p\text{-}CH_3C_6H_4SO_3H}$ Di-n-butyl terephthalate $+ 2C_2H_5OH \uparrow$ (Removed by distillation)

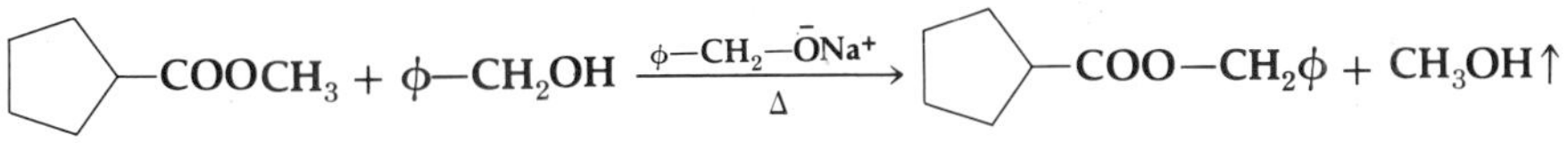

$\bigcirc\!\!-COOCH_3 + \phi-CH_2OH \xrightarrow[\Delta]{\phi-CH_2-\bar{O}Na^+} \bigcirc\!\!-COO-CH_2\phi + CH_3OH\uparrow$

Sulfur Derivatives of Acids

Thiol acids and thiol esters are made by various nucleophilic substitution reactions as illustrated below.

$$H_2S + \text{acylating agent } (RCO-L) \longrightarrow \text{thiol acid } (R-CO-SH)$$

$$H_2S + CH_3\overset{O}{\overset{\|}{C}}O\overset{O}{\overset{\|}{C}}CH_3 \longrightarrow CH_3\overset{O}{\overset{\|}{C}}SH + CH_3COOH$$

Thiolacetic acid

$$\text{Mercaptan } (R-SH) + \text{acylating agent } (R'CO-L) \longrightarrow \text{thiol acid ester } (R'CO-S-R)$$

$$CH_3SH + CH_3COCl \xrightarrow{\text{Pyridine}} CH_3\overset{O}{\overset{\|}{C}}SCH_3 + \text{pyridine} \cdot HCl$$

Methyl thiolacetate

†Of course in some cases a rigid skeleton disallows closure of a five- or six-membered lactone from the hydroxy-acid or ester; 1,3- and 1,4-trans-substituted cyclohexanes are examples.

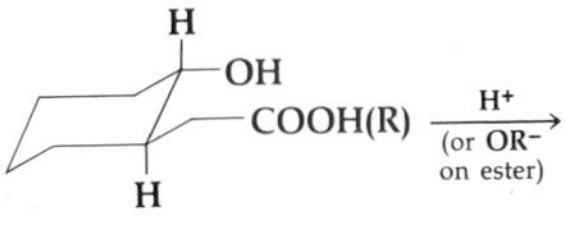

1,2-trans

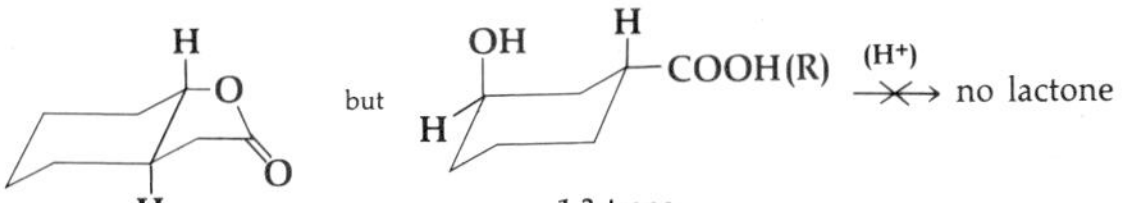

but

1,3-trans

Amides

Amides are prepared by acylation of ammonia or amines with acid chlorides, anhydrides, esters, or acids. The reaction with acids is of limited interest, since pyrolysis of ammonium salts is required and is a poor practical reaction. Anhydrides and acid chlorides react violently with ammonia unless the reactions are moderated by cooling, dilution, or slow mixing of the reactants. Reaction with esters is generally very slow. The following preparations are typical. As with esterification a common acylating procedure for amide formation employs acid chlorides or anhydrides in pyridine.

$$(CH_3\overset{O}{\overset{\|}{C}})_2O + (CH_3)_2NH \longrightarrow CH_3\overset{O}{\overset{\|}{C}}N(CH_3)_2 + CH_3COOH$$

N,N-Dimethyl-
acetamide

$$(C_2H_5)_2CHCOCl + NH_3 \xrightarrow{Benzene} (C_2H_5)_2CHCONH_2 + [(C_2H_5)_2CHCO]_2NH$$

Diethylacetyl   Diethylacetamide   bis-Diethyl-
chloride   91%   acetimide
                                                    9%

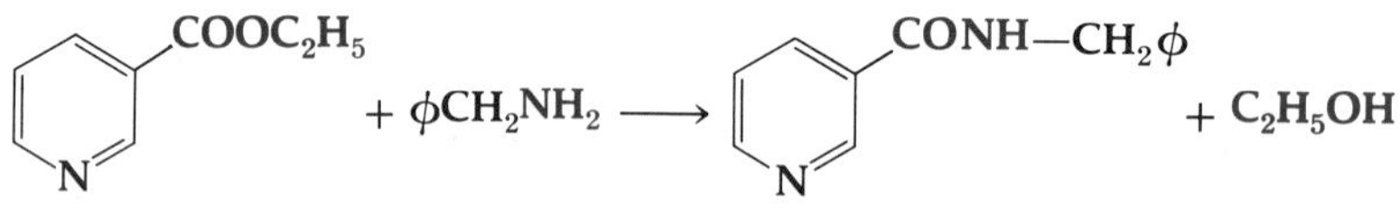

Ethyl nicotinate   N-Benzyl nicotinamide

The following example demonstrates the selective reaction of an anhydride group in preference to an ester group.

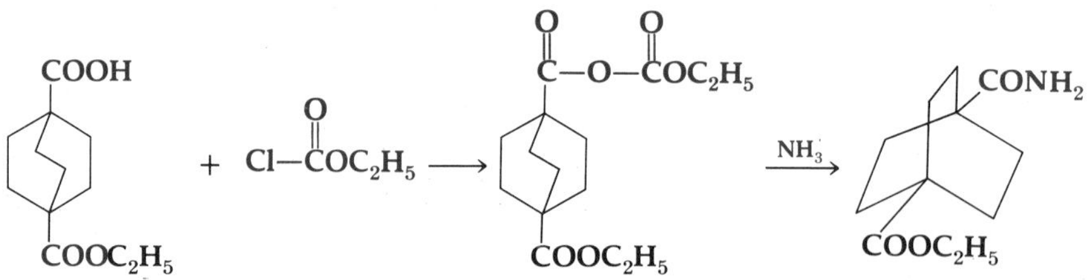

Ethyl   Ethyl   (Not isolated)
bicyclo[2,2,2]-   chloroformate
octane-1,4-dicarboxylate

As with acylation of alcohols, direct one-step conversion of amine and carboxylic acid to amide may be achieved by initial activation of the acid so that the —OH becomes a better leaving group. Carbodiimides again have been widely used, especially in forming the amide bonds of peptides from amino acid derivatives (Chap. 25).

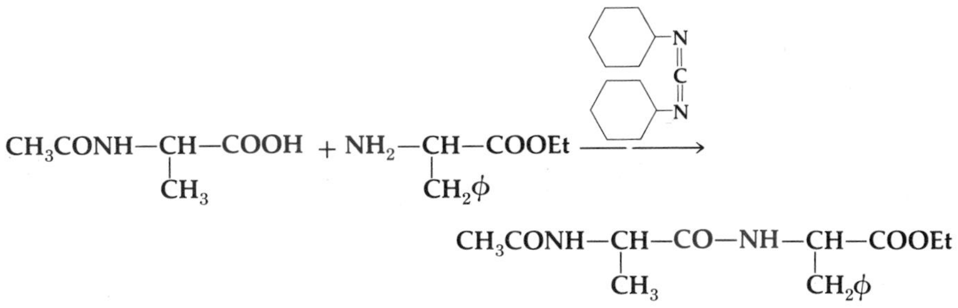

$$CH_3CONH-\underset{\underset{CH_3}{|}}{CH}-COOH + NH_2-\underset{\underset{CH_2\phi}{|}}{CH}-COOEt \longrightarrow$$

$$CH_3CONH-\underset{\underset{CH_3}{|}}{CH}-CO-NH-\underset{\underset{CH_2\phi}{|}}{CH}-COOEt$$

Lactams are cyclic amides and completely analogous to lactones in being highly favored and usually spontaneously formed, when five- or six-membered, from appropriate amino acids or esters.

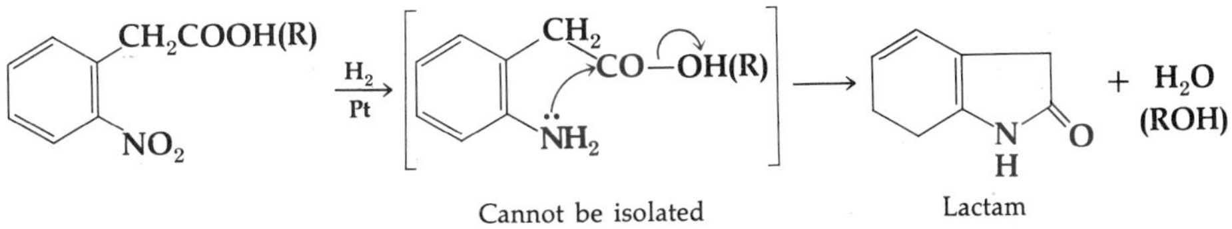

Cannot be isolated   Lactam

Cyclic anhydrides are easily converted to cyclic imides, although it is usually necessary to isolate the intermediate half-amides.

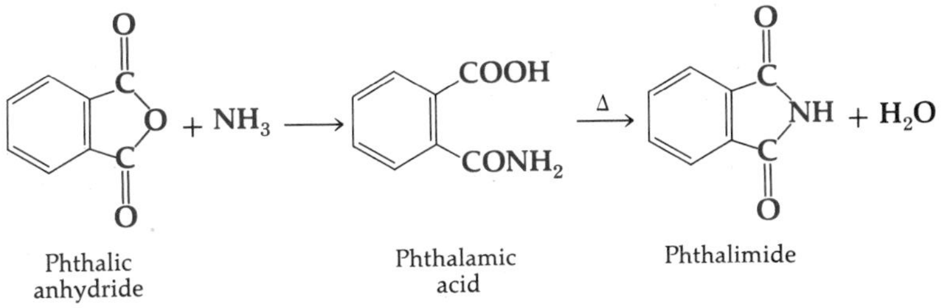

Phthalic   Phthalamic   Phthalimide
anhydride   acid

Amides are much harder to hydrolyze to acids than esters are. Prolonged heating under reflux with strong aqueous acid or base is required to effect cleavage, and even moderately hindered compounds are inert toward hydrolysis for all practical purposes. Simple primary amides are often hydrolyzable in the mildest manner by treatment with nitrous acid. This serves to convert $-NH_2$ to $-N_2^+$ as with aliphatic amines (page 413). Amides rather than acids may be produced in high yield from nitriles by hydrolysis (page 488).

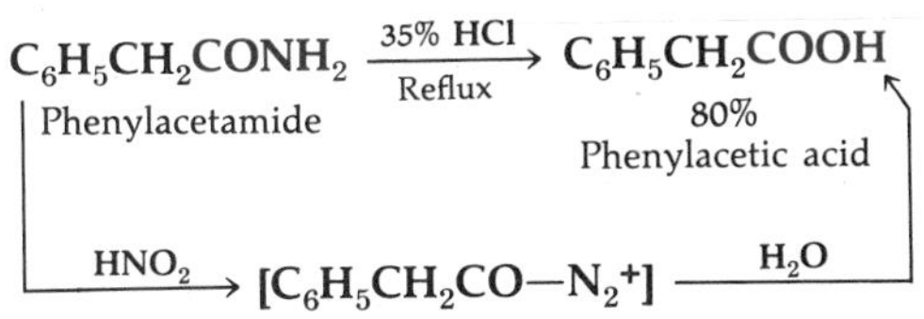

$$C_6H_5CH_2CONH_2 \xrightarrow[\text{Reflux}]{35\% \text{ HCl}} C_6H_5CH_2COOH$$

Phenylacetamide   80%
                    Phenylacetic acid

$$\xrightarrow{HNO_2} [C_6H_5CH_2CO-N_2^+] \xrightarrow{H_2O}$$

Treatment of reactive acylating agents such as acid chlorides, anhydrides, and even esters with hydroxylamine produces hydroxamic acids. The latter are usually not isolated but are converted to derivatives or isolated as salts. Hydroxamic acids are easily detected in solution because they form deeply colored complexes with ferric ions. Active acylating agents are often characterized by conversion to hydroxamic acids.

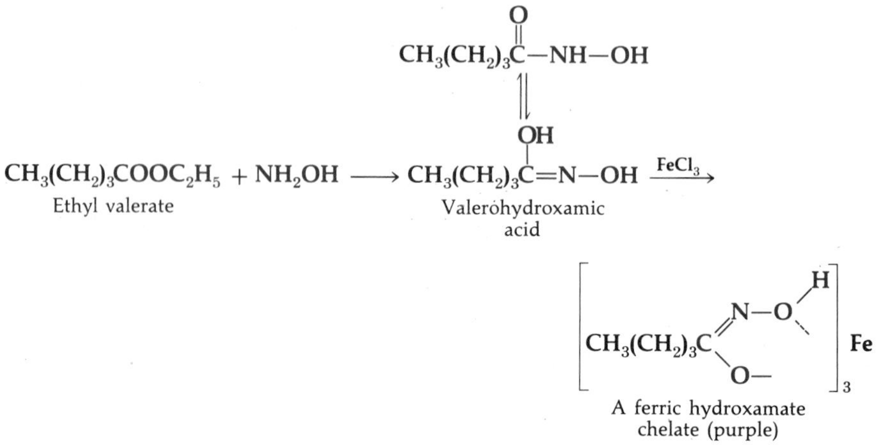

$$CH_3(CH_2)_3\overset{\overset{\displaystyle O}{\|}}{C}-NH-OH$$

$$\Updownarrow$$

$$CH_3(CH_2)_3COOC_2H_5 + NH_2OH \longrightarrow CH_3(CH_2)_3\overset{\overset{\displaystyle OH}{|}}{C}=N-OH \xrightarrow{\ FeCl_3\ }$$

Ethyl valerate   Valerohydroxamic
                                        acid

$$\left[CH_3(CH_2)_3C \overset{N-O}{\underset{O-}{\diagdown}} \,^{H} \right]_3 Fe$$

A ferric hydroxamate
chelate (purple)

Hydrazides, acylhydrazines, and azides are formed by the acylation of hydrazine and metal azides. Acyl azides may also be prepared by oxidation of hydrazides with nitrous acid.

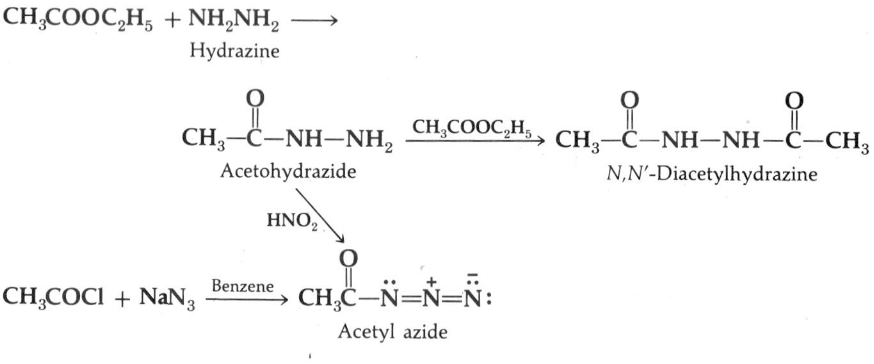

$$CH_3COOC_2H_5 + NH_2NH_2 \longrightarrow$$
    Hydrazine

$$CH_3-\overset{\overset{\displaystyle O}{\|}}{C}-NH-NH_2 \xrightarrow{\ CH_3COOC_2H_5\ } CH_3-\overset{\overset{\displaystyle O}{\|}}{C}-NH-NH-\overset{\overset{\displaystyle O}{\|}}{C}-CH_3$$

Acetohydrazide   N,N'-Diacetylhydrazine

$$\overset{HNO_2}{\searrow}$$

$$CH_3COCl + NaN_3 \xrightarrow{\ Benzene\ } CH_3\overset{\overset{\displaystyle O}{\|}}{C}-\overset{..}{N}=\overset{+}{N}=\overset{..}{N}:$$

Acetyl azide

Nitriles may be regarded as part of the carboxylic acid family since they are hydrolyzable to amides (Sec. 12-8) and have similar reactivity.† Primary

---

† Nitriles may be regarded as amides in which the single nitrogen serves the function of both oxygen and

nitrogen in amides: $R-C\overset{\nearrow O}{\underset{NH_2}{\diagdown}}$ and $R-C{\equiv}N$. Addition reactions of nitriles (Sec. 12-8) are thus analogous

to substitution reactions of amides.

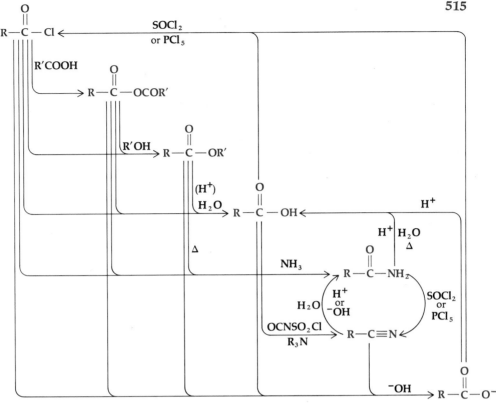

FIGURE 13-3 Interconversions of carboxyl derivatives (order of reactivity top to bottom, most reactive at top)

amides are also converted into nitriles by the same series of dehydrating agents that convert acids to acid chlorides ($SOCl_2$, PCl_5, $POCl_3$, etc.). An analogous mechanism is at work.†

$$R-\overset{\overset{\displaystyle O}{\|}}{C}-OH + SOCl_2 \longrightarrow \left[R-\overset{\overset{\displaystyle O}{\|}}{C}-\overset{\frown}{O}-SO-\overset{\frown}{Cl} \right] \longrightarrow R-\overset{\overset{\displaystyle O}{\|}}{C}-Cl + SO_2 + HCl$$
Acid

$$+ Cl^-$$

$$R-\overset{\overset{\displaystyle NH_2}{|}}{C}=O \rightleftharpoons R-\overset{\overset{\displaystyle NH}{\|}}{C}-OH + SOCl_2 \longrightarrow \left[R-\overset{\overset{\displaystyle NH}{\|}}{C}-O-SO-Cl \right] \longrightarrow$$
Amide

$$R-\overset{\overset{\displaystyle NH}{\|}}{C}-Cl \xrightarrow[\text{(elim.)}]{-HCl} R-C\equiv N + SO_2 + HCl$$

† Most acidic electrophiles attack the ambident nucleophile of the amide group at the more electronegative oxygen instead of nitrogen, analogous to the behavior of S_N1 alkylation (Fig. 10-11):

$$R-C\overset{\displaystyle \ddot{O}:}{\underset{\displaystyle \ddot{N}H_2}{\diagdown}} \xrightarrow{E^+} \left[R-C\overset{\displaystyle \ddot{O}-E}{\underset{\displaystyle \ddot{N}H_2}{\diagdown}} \longleftrightarrow R-C\overset{\displaystyle \ddot{\ddot{O}}-E}{\underset{\displaystyle \overset{+}{N}H_2}{\diagdown}} \right]$$

Carboxylic acids may be converted directly to nitriles by the interesting conversion shown below which makes use of commercially available chlorosulfonylisocyanate.

$$R-COOH + O=C=N-SO_2Cl \xrightarrow{-CO_2} R-CO-NH-SO_2Cl \xrightarrow{Et_3N}$$

$$RCN + Et_3NH^+ClSO_3^-$$

The interrelating reactions of the carboxylic acid family are united in Fig. 13-3.

PROBLEM 13-1

Write the products of the following transformations.

a Succinic anhydride + methanol

b Benzoyl chloride + dibenzylamine

c Benzoyl chloride + tribenzylamine in methanol

d Sodium benzoate + phosphorus pentachloride

e o-Aminophenylacetic acid heated

f Propionamide + thionyl chloride

g N-Methylbutyramide + phosphorus pentachloride

h cis-4-Hydroxycyclohexane-carboxylic acid + dicyclohexylcarbodiimide

i Benzoic acetic anhydride + isopropyl alcohol

j Ethyl heptanoate + benzyl alcohol boiled with a trace of sulfuric acid

k Hydroxylamine + γ-butyrolactone (γ-hydroxybutyric acid lactone)

l Tetrabutylammonium azide + phenylacetyl chloride

m 2-Acetoxy-2-pentene + hot ethanol

PROBLEM 13-2

Fill in the missing compounds in this synthetic scheme.

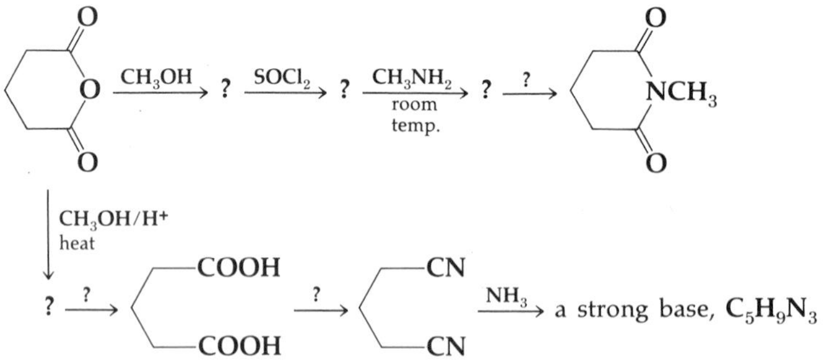

13-2   SUBSTITUTION BY HYDRIDES AND ORGANOMETALLIC CARBANIONS

Reactions of Carboxylic Derivatives with Hydrides and Organometallics

The normal reaction of carboxylic acid derivatives with hydrides and Grignard and alkyllithium reagents is one of *substitution followed by addition.* These reagents may be considered together since they are all high-energy, irreversible nucleophiles of comparable reactivity and complex structure which must be used in aprotic solvents. They differ mainly in delivering H:⁻ or R:⁻ as nucleophile. Each reacts, as noted above, by substitution first but the carbonyl product so formed is then attacked again. The net result, with all but amides and nitriles is to form an alcohol. *Hydrides yield primary* alcohols and *organometallics yield tertiary* alcohols with two attached groups (R′) the same.

General reaction:

$$R-\overset{\overset{\displaystyle O}{\|}}{C}-X(OR) \begin{cases} \xrightarrow[(+2H:^-)]{LiAlH_4} & R-CH_2-OH \\ \xrightarrow[\text{or } R'Li]{R'MgX} & R-\overset{\overset{\displaystyle R'}{|}}{\underset{\underset{\displaystyle R'}{|}}{C}}-OH \end{cases} + X^-(OR^-)$$

$$CH_3MgBr + CH_3CH_2\overset{\overset{\displaystyle O}{\|}}{C}OCH_3 \xrightarrow{\text{Ether}} CH_3CH_2\overset{\overset{\displaystyle \bar{O}\overset{+}{M}gBr}{|}}{\underset{\underset{\displaystyle OCH_3}{|}}{C}}-CH_3 \xrightarrow{-Mg(OCH_3)Br}$$

$$CH_3CH_2\overset{\overset{\displaystyle O}{\|}}{C}CH_3 \xrightarrow{CH_3MgBr} CH_3CH_2\overset{\overset{\displaystyle \bar{O}\overset{+}{M}gBr}{|}}{\underset{\underset{\displaystyle CH_3}{|}}{C}}CH_3 \xrightarrow{H_3O^+} CH_3CH_2\overset{\overset{\displaystyle OH}{|}}{\underset{\underset{\displaystyle CH_3}{|}}{C}}CH_3$$

2-Methyl-2-butanol
(*tert*-amyl alcohol)

$$C_6H_5COOH + LiAlH_4 \longrightarrow C_6H_5C\bar{O}_2\overset{+}{Li} + H_2 + AlH_3$$

Benzoic
acid

$$AlH_3 \downarrow$$

$$C_6H_5\overset{\overset{\displaystyle OAlH_2}{|}}{\underset{\underset{\displaystyle H}{|}}{C}}\bar{O}\overset{+}{Li} \xrightarrow{-LiOAlH_2} C_6H_5\overset{\overset{\displaystyle O}{\|}}{C}H \xrightarrow[2) H_3O^+]{1) LiAlH_4} C_6H_5CH_2OH$$

Benzyl
alcohol

$$2\,C_6H_5MgBr + CH_3CH_2\overset{\overset{\displaystyle O}{\|}}{C}Cl \xrightarrow[2) NH_4Cl]{1) Ether} CH_3CH_2\overset{\overset{\displaystyle OH}{|}}{C}(C_6H_5)_2$$

1,1-Diphenyl-1-propanol

$$C_6H_5\overset{\overset{\displaystyle CH_3}{|}}{C}HCOOC_2H_5 \xrightarrow[2) H_3O^+]{1) LiAlH_4} C_6H_5\overset{\overset{\displaystyle CH_3}{|}}{C}HCH_2OH + C_2H_5OH$$

Ethyl 2-phenyl-   2-Phenylpropanol
propanoate

Lactones behave just like open-chain esters and undergo ring opening to give diols.

$$2CH_3CH_2MgBr + \underset{\text{γ-Butyrolactone}}{\underset{\displaystyle H_2C\diagdown_O\diagup C\diagdown_O}{\overset{\displaystyle H_2C-CH_2}{}}} \xrightarrow[\text{2) NH}_4\text{Cl}]{\text{1) Ether}} \underset{\text{4-Ethyl-1,4-hexanediol}}{HO(CH_2)_3\underset{\underset{\displaystyle OH}{|}}{C}(CH_2CH_3)_2}$$

$$\underset{\displaystyle H_2C\diagdown_O\diagup C\diagdown_O}{\overset{\displaystyle H_2C-CH_2}{}} \xrightarrow{\text{LiAlH}_4} \underset{\text{1,4-butanediol}}{HOCH_2CH_2CH_2CH_2OH}$$

The amides are rarely used with Grignard reagents; they tend to be slow to react and to stop with the substitution step. With LiAlH$_4$, however, amides, lactams, and nitriles are all reduced without substitution to afford amines (C=O $\longrightarrow$ CH$_2$); this constitutes an important synthesis of amines.

$$\textit{General reaction:}\quad \left.\begin{array}{l} R-\overset{\displaystyle O}{\overset{\|}{C}}-NH_2 \\[2mm] R-C\equiv N \end{array}\right\} \xrightarrow{\text{LiAlH}_4} R-CH_2-NH_2$$

$$R-\overset{\displaystyle O}{\overset{\|}{C}}-\underset{\underset{\displaystyle R'}{|}}{N}-R'' \longrightarrow R-CH_2-\underset{\underset{\displaystyle R'}{|}}{N}-R''$$

$$\underset{\text{Butyronitrile}}{2CH_3CH_2CH_2CN} \xrightarrow{\text{LiAlH}_4} (CH_3CH_2CH_2CH_2N)_2\overset{-\ +}{AlLi} \xrightarrow{H_2O} \underset{\text{n-Butylamine}}{2CH_3CH_2CH_2CH_2NH_2}$$

$$\underset{\substack{\text{N,N-Dimethyl-}\\\text{isobutyramide}}}{(CH_3)_2CHCON(CH_3)_2} \xrightarrow{\text{LiAlH}_4} (CH_3)_2CH\underset{\underset{\displaystyle H}{|}}{\overset{\overset{\displaystyle \overset{-\ +}{OLi}}{|}}{C}}N(CH_3)_2 \xrightarrow[-\text{Li}_2O]{\text{LiAlH}_4} \underset{\text{Dimethylisobutylamine}}{(CH_3)_2CHCH_2N(CH_3)_2}$$

Aldehydes by Substitution with Hydride

Aldehydes are not regarded as belonging to the carboxylic acid family since a reduction (+2H) is required to prepare them from carboxyl derivatives. This reduction, however, can be carried out by nucleophilic substitution, with hydride ion as nucleophile, and this leads to a variety of aldehyde syntheses.

Preparation of aldehydes by reduction with metal hydrides involves rather delicate balancing of reactivities because of the great ease with which aldehydes are themselves reduced to alcohols by hydride addition. The use of a very reactive acylating agent, such as an acid chloride as substrate, coupled with the sterically hindered lithium tri-t-butoxyaluminum hydride as a hydride donor gives good results. The relatively unreactive reducing agent selects the more reactive acid chloride and leaves the desired aldehyde intact. Reduction of acyl imidazoles is comparable but much more successful in practice; the imidazole ring is a good leaving group (resonance stabilization of $\ddot{N}^-$). Lithium triethoxyaluminum hydride has been successful in reduction of nitriles and disubstituted amides to aldehydes. The reaction in this case probably stops at the intermediate addition salt and is only finally completed on aqueous workup of the reaction.

$$M-H + R-\overset{O}{\overset{\|}{C}}-X \longrightarrow \overset{+}{M} + R-\overset{O}{\overset{\|}{C}}-H + \bar{X}$$

$$LiAlH(OC_4H_9\text{-}t)_3 + C_6H_5COCl \xrightarrow{CH_3OCH_2CH_2OCH_3} C_6H_5CHO + LiCl + Al(OC_4H_9\text{-}t)_3$$

Lithium tri-*tert*-butoxyaluminum hydride   Aluminum tri-*tert*-butoxide

$$LiAlH(OC_2H_5)_3 + C_6H_5CON(CH_3)_2 \xrightarrow[0°]{\overset{O}{\bigcirc}} \left[C_6H_5\overset{OLi \; CH_3}{\overset{|}{CH}}-\overset{|}{NCH_3} \right] \xrightarrow{H_3O^+}$$

Lithium tri-ethoxyaluminum hydride   N,N-Dimethyl phenylbenzamide

$$\left[C_6H_5\overset{OH \; CH_3}{\overset{|}{CH}}-\overset{|}{NCH_3} \right] \longrightarrow C_6H_5CHO + HN(CH_3)_2$$

$$\tfrac{1}{4} LiAlH_4 + (CH_3)_2CHCO-N\diagup\diagdown N \longrightarrow (CH_3)_2CHCHO + HN\diagup\diagdown N$$

Isobutyryl imidazole   Isobutyraldehyde   Imidazole

Among other methods for reduction of carboxylic acid derivatives to aldehydes are the **Rosenmund hydrogenation** of acid chlorides and the **McFadyen-Stevens decomposition** of sulfonhydrazides (aromatic aldehydes only).

Rosenmund reduction:

$$CH_3O-\!\!\langle\bigcirc\rangle\!\!-\overset{\overset{\displaystyle O}{\|}}{C}-Cl + H_2 \xrightarrow{\textbf{Pd catalyst}} CH_3O-\!\!\langle\bigcirc\rangle\!\!-CHO + HCl$$

p-Anisoyl chloride p-Anisaldehyde
81%

McFadyen-Stevens reaction:

$$C_6H_5COCl + NH_2NH_2 \longrightarrow C_6H_5CONHNH_2 \xrightarrow{C_6H_5SO_2Cl}$$

Benzoylhydrazide
90%

$$C_6H_5CO-NH-NH-SO_2C_6H_5 \xrightarrow[Na_2CO_3,\ 160°]{HOCH_2CH_2OH}$$

1-Benzoyl-2-benzene-
sulfonylhydrazide

$$\left[C_6H_5\overset{\overset{\displaystyle O}{\|}}{C}-\underset{\curvearrowright}{N}-NH-\overset{\curvearrowleft}{S}O_2C_6H_5\right] \xrightarrow{-C_6H_5SO_2^-} \left[C_6H_5\overset{\overset{\displaystyle O}{\|}}{C}-N=NH\right] \xrightarrow{-N_2} C_6H_5CHO$$
70%

Ketones by Substitution with Carbanion Organometallics

Although Grignard reagents and alkyllithium compounds react with acyl halides and esters to give ketones initially, the ketones thus formed then react rapidly with the organometallic reagent, and tertiary alcohols are obtained. If equimolar quantities of reagents are employed, mixtures of starting material, ketone, and tertiary alcohol are produced and the reaction is not synthetically valuable for ketone preparation.

Three methods allow ketones to be prepared satisfactorily by substitution at unsaturated carbon. In the first, the lithium salt of a carboxylic acid is treated with an organolithium reagent. The initial adduct does not decompose to produce ketone until the reaction mixture is treated with water.

$$C_6H_5Li + C_6H_5\overset{\overset{\displaystyle CH_3}{|}}{\underset{\underset{\displaystyle C_2H_5}{|}}{C}}-C\bar{O}_2\overset{+}{Li} \xrightarrow{Ether} C_6H_5\overset{\overset{\displaystyle CH_3}{|}}{\underset{\underset{\displaystyle C_2H_5}{|}}{C}}-\overset{\overset{\displaystyle \bar{O}\overset{+}{Li}}{|}}{\underset{\underset{\displaystyle \bar{O}\overset{+}{Li}}{|}}{C}}-C_6H_5 \xrightarrow{H_2O} C_6H_5\overset{\overset{\displaystyle CH_3}{|}}{\underset{\underset{\displaystyle C_2H_5}{|}}{C}}-\overset{\overset{\displaystyle O}{\|}}{C}-C_6H_5 + 2LiOH$$

65%

The second method involves the reaction of Grignard or alkyl lithium reagents with nitriles or amides. The reason for success here is the same as that in the previous procedure, i.e., formation of an intermediate complex. The reactions are much slower than most Grignard reactions, and the yields vary considerably with the nature of the substrate. However, it is often possible to accelerate Grignard reactions greatly by admixture of a dipolar aprotic solvent, especially hexamethylphosphoramide $[(CH_3)_2N]_3PO$, which apparently gives the reagent more free carbanion character. The reaction of organolithium compounds with dimethylformamide produces aldehydes ($RLi + HCONMe_2 \longrightarrow R\text{—}CHO$).

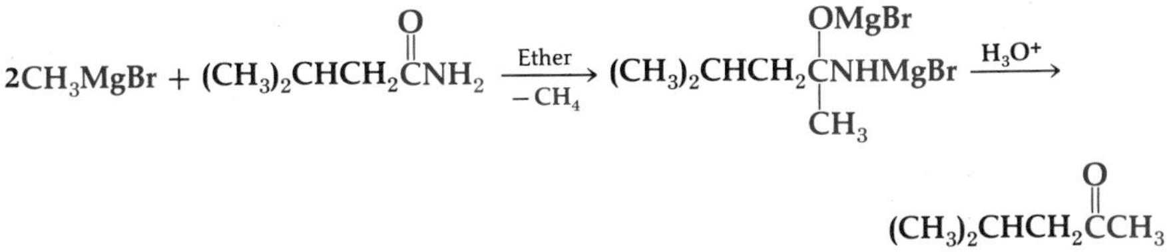

$$2CH_3MgBr + (CH_3)_2CHCH_2\overset{\overset{O}{\|}}{C}NH_2 \xrightarrow[-CH_4]{Ether} (CH_3)_2CHCH_2\underset{\underset{CH_3}{|}}{\overset{\overset{OMgBr}{|}}{C}}NHMgBr \xrightarrow{H_3O^+}$$

$$(CH_3)_2CHCH_2\overset{\overset{O}{\|}}{C}CH_3$$
4-Methyl-2-pentanone
60%

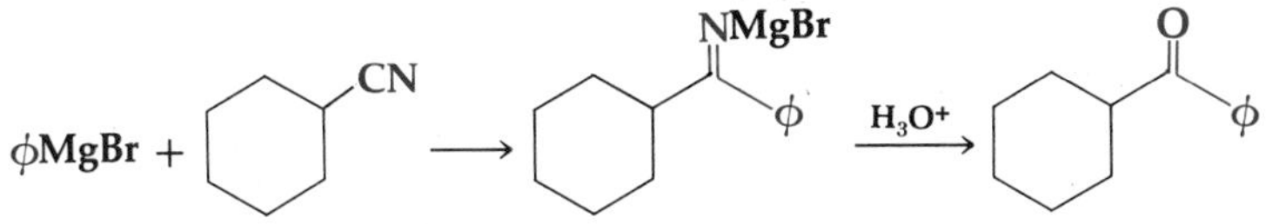

In a third method, a Grignard reagent is converted to an organocadmium reagent with cadmium chloride. Alkyl- and arylcadmium compounds are much less reactive than their Grignard or organolithium counterparts, and although they react with acyl halides, they fail to react rapidly with the initially produced ketone. The method is limited by the fact that only aryl- and primary alkyl-cadmium are stable enough for practical use.

$$2C_6H_5MgBr + CdCl_2 \xrightarrow[40°]{Benzene} (C_6H_5)_2Cd + MgBr_2 + MgCl_2$$

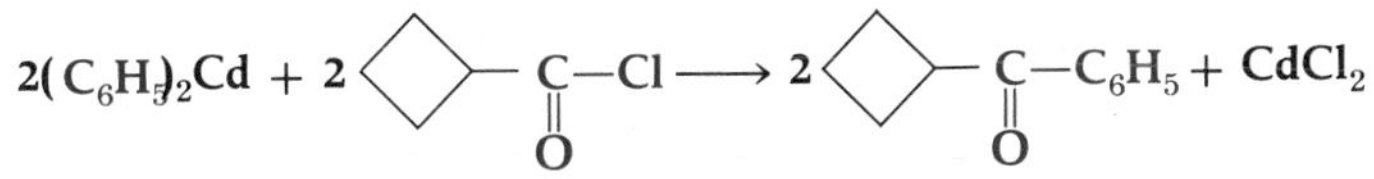

Cyclobutyl phenyl
ketone, 40%

PROBLEM 13-3

Write the products of reaction of (i) excess lithium aluminum hydride and (ii) excess ethylmagnesium iodide on the following.

a Ethyl phenylacetate   c γ-Butyrolactone
b 6-Ketononanoic acid chloride d N-Phenylbutyramide

13-3 ACYLATION OF CARBON

Acylation of carbon ranks as one of the most important synthetic procedures in organic chemistry. The reaction is that of an acylating agent (acid chloride, anhydride, or ester) with an enolate anion (basic media) or enol (acidic media) as nucleophile, the enolate anion being the most usual choice.

Ester Condensations

Aliphatic esters that have at least two α-hydrogens undergo self-condensation (**Claisen ester condensation**) to form β-ketoesters. The classical synthesis of ethyl acetoacetate is a prototype for the group of reactions.

$$2CH_3COOC_2H_5 \xrightarrow{\ C_2H_5\bar{O}\overset{+}{N}a\ } CH_3CO-CH_2COOC_2H_5 + C_2H_5OH$$

Ethyl acetoacetate
(acetoacetic ester)
75%

The reaction is analogous to base-catalyzed aldol condensations (Sec. 12-5). Esters are less acidic for enolization than aldehydes and ketones (because of resonance stabilization in the ester group), but ethoxide ions are sufficiently basic to convert a small amount of ethyl acetate to an enolate ion in a rapid equilibrium. Attack of the enolate on an ester molecule is then a typical nucleophilic substitution reaction.

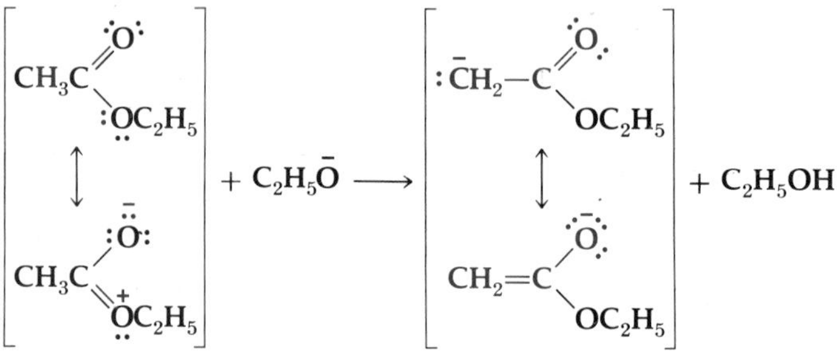

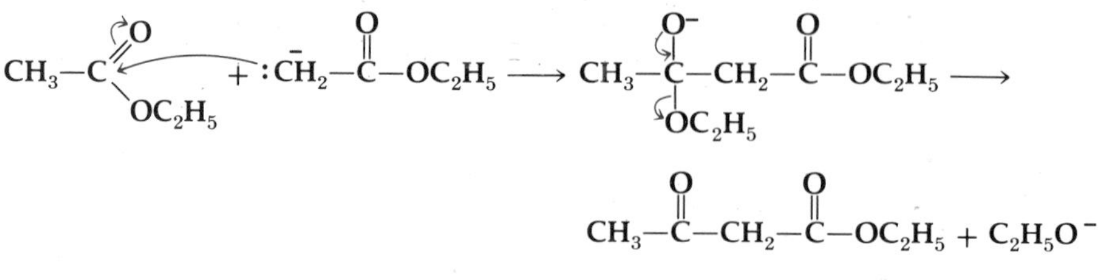

One equivalent of sodium ethoxide is consumed, because the β-ketoester formed is acidic enough to neutralize ethoxide ion.

$$CH_3\overset{O}{\overset{\|}{C}}-CH_2-\overset{O}{\overset{\|}{C}}OC_2H_5 + C_2H_5\bar{O} \rightleftharpoons \left[\begin{array}{c} CH_3\overset{O}{\overset{\|}{C}}-\overset{\cdot\cdot}{\underset{}{CH}}-\overset{O}{\overset{\|}{C}}OC_2H_5 \\ \updownarrow \\ CH_3\overset{\bar{O}}{\overset{|}{C}}=CH-\overset{O}{\overset{\|}{C}}OC_2H_5 \\ \updownarrow \\ CH_3\overset{O}{\overset{\|}{C}}-CH=\overset{\bar{O}}{\overset{|}{C}}OC_2H_5 \end{array}\right] + C_2H_5OH$$

After condensation is completed, the sodium salt of the β-ketoester is neutralized by addition of mineral acid. Formation of the salt or anion (conjugate base) of the β-ketoester is critical for the success of the condensation. The equilibrium constant for the condensation reaction itself is small; but equilibrium is displaced in favor of the product by subsequent conversion to the β-ketoester enolate anion since this is *more stable than the initial alkoxide base* (see Table 8-2).

The situation is analogous to substitution of RO⁻ and HO⁻ on esters. Alkoxide ions simply set up an ester exchange equilibrium, not especially favoring products. With hydroxide ion, however, the product in equilibrium is carboxylic acid which can react with base to form a much more stable carboxylate anion and so displace the equilibrium to the right (toward products).

$$R-COOR' + R''O^- \rightleftharpoons R-COOR'' + R'O^-$$

$$R-COOR' + HO^- \rightleftharpoons R-COOH + R'O^- \xrightarrow{\text{Irrev.}} R-COO^- + R'OH$$

Need for an acidic hydrogen in the product (and hence *two* α-hydrogens in the original enol component) is demonstrated by the behavior of ethyl isobutyrate, which does not undergo condensation in the presence of sodium ethoxide. However, this and other esters with only a single α-hydrogen will condense in the presence of either very strong bases or bases that react irreversibly because the second product of the acid-base reaction escapes as a gas.

$$(CH_3)_2CHCOOC_2H_5 + \left\{\begin{array}{l} (C_6H_5)_3\overset{\cdot\cdot}{\overset{+}{C}}\overset{+}{Na} \\ \overset{+}{Na}\bar{N}H_2 \\ \overset{+}{Na}\bar{H} \end{array}\right. \longrightarrow (CH_3)_2C=\overset{\bar{O}}{\overset{|}{C}}\overset{+}{Na}OC_2H_5 + \left\{\begin{array}{l} (C_6H_5)_3CH \\ NH_3 \\ H_2 \end{array}\right.$$

$$(CH_3)_2\overset{\cdot\cdot}{\overset{}{C}}-COOC_2H_5 + (CH_3)_2CHCOOC_2H_5 \longrightarrow$$

$$(CH_3)_2CH\overset{O}{\overset{\|}{C}}-\overset{CH_3}{\underset{CH_3}{\overset{|}{\underset{|}{C}}}}-COOC_2H_5 + C_2H_5\bar{O}$$

Ethyl 2,2,4-trimethyl-
3-ketopentanoate
60%

In the last example, the driving force for condensation is provided by formation of *ethoxide ion, the weakest and most stable base in the system.*

The general Claisen condensation involves an ester with an α-methylene which acts as enol component and a second ester to act as substrate for the substitution:

$$R-COOR' + \underset{\substack{| \\ CH_2-COOR'}}{\overset{\substack{R'' \\ |}}{}} \xrightarrow{-OR'} R-CO-\underset{\substack{| \\ \ddot{C}-COOR'}}{\overset{\substack{R'' \\ |}}{}} + 2R'OH$$

<table>
<tr><td>Substrate or
acceptor</td><td>Enol
component</td><td>β-Keto-ester
enolate anion</td></tr>
</table>

Crossed condensations between two different esters with α-methylenes are not ordinarily of synthetic value, because a mixture of the four possible keto esters is produced. However, an ester having no α-hydrogen but possessing high carbonyl reactivity can often be condensed with a second ester in good yield. The common acceptors with no α-hydrogens are these esters:

$$HCOOR \qquad \underset{\substack{| \\ COOR}}{\overset{\substack{COOR \\ |}}{}} \qquad RO-CO-OR \qquad \phi-COOR$$

<table>
<tr><td>Formates</td><td>Oxalates</td><td>Carbonates</td><td>Benzoates</td></tr>
</table>

$$\underset{\substack{| \\ COOC_2H_5}}{\overset{\substack{COOC_2H_5 \\ |}}{}} + CH_3CH_2COOC_2H_5 \xrightarrow{C_2H_5ONa} \underset{\substack{| \\ CH_3CH COOC_2H_5}}{\overset{\substack{COCOOC_2H_5 \\ |}}{}} + C_2H_5OH$$

<table>
<tr><td>Diethyl
oxalate</td><td>Ethyl
propionate</td><td>Diethyl α-keto-
β-methylsuccinate
70%</td></tr>
</table>

$$C_2H_5O\overset{\overset{\displaystyle O}{\|}}{C}OC_2H_5 + C_6H_5CH_2COOC_2H_5 \xrightarrow{C_2H_5ONa} C_6H_5CH(COOC_2H_5)_2 + C_2H_5OH$$

<table>
<tr><td>Diethyl
carbonate</td><td>Ethyl
phenylacetate</td><td>Diethyl
phenylmalonate
65%</td></tr>
</table>

$$HCOOC_2H_5 + CH_3COOC_2H_5 \xrightarrow{C_2H_5ONa} HCOCH_2COOC_2H_5 + C_2H_5OH$$

Even when a mixed ester condensation cannot be carried out, desired *ketoesters* can often be made by acylation of enolate ions with acid chlorides. Enolates must be preformed by treatment of an ester with a strong base, such as triphenylmethylsodium, sodium hydride, or sodamide. An acid chloride is then added to the enolate and the reaction is essentially irreversible.

$$(CH_3)_2CH\,COOC_2H_5 \xrightarrow[\text{Ether}]{(C_6H_5)_3\bar{C}\overset{+}{N}a} \underset{\substack{ \\ Na^+}}{(CH_3)_2\ddot{C}^{\,-}-COOC_2H_5} \xrightarrow{C_6H_5COCl}$$

$$\underset{\substack{| \\ COOC_2H_5}}{\overset{\substack{(CH_3)_2C-COC_6H_5 \\ |}}{}} + Na^+Cl^-$$

<table>
<tr><td align="center">Ethyl dimethyl-
benzoyl acetate
79%</td></tr>
</table>

Acylation of Other Carbanions

Enolate ions from ketones and nitriles serve as nucleophiles in useful substitutions on esters and acid chlorides. Because of their high aptitude for self-condensation, aldehydes are not used in this manner. Condensation of an ester with a ketone is a classic method for the preparation of 1,3-diketones. As before, the driving force for the reaction is usually the formation of the final conjugated and stabilized enolate anion of the 1,3-dicarbonyl compound.

$$CH_3CN \xrightarrow[\text{Ether}]{\text{NaNH}_2} \left[:\bar{C}H_2CN \longleftrightarrow CH_2{=}C{=}\bar{N} \right] \xrightarrow{CH_3CH_2COOC_2H_5}$$

$$CH_3CH_2CO{-}CH_2CN$$

β-Ketovaleronitrile
(propionylacetonitrile)

$$CH_3COCH_2COOC_2H_5 \xrightarrow{Mg(OC_2H_5)_2} CH_3CO\ddot{C}HCOOC_2H_5 \xrightarrow{C_6H_5CH_2COCl}$$

$$CH_3CO{-}\underset{\underset{C_6H_5CH_2CO}{|}}{CH}{-}COOC_2H_5$$

3-Carbethoxy-1-
phenyl-2,4-pentanedione

$$(CH_3)_2CHCOOC_2H_5 + CH_3\overset{\overset{O}{\|}}{C}CH(CH_3)_2 \xrightarrow{\text{NaNH}_2} (CH_3)_2CHCOCH_2COCH(CH_3)_2$$

Ethyl isobutyrate   Isopropyl methyl   Diisobutyrylmethane
                           ketone   70%

2-Formyl-   2-Hydroxymethylene
cyclohexanone   cyclohexanone

All β-dicarbonyl compounds are somewhat enolic (Sec. 8-9), but their structures are generally represented by diketo formulas, although formyl ketones are sometimes named by reference to their enol structures. α-Formyl-ketones are essentially 100% enolic and are often named as α-hydroxy-methylene-ketones. The hydroxyl groups in these enolized β-dicarbonyl compounds are strongly hydrogen-bonded to the keto group when sterically possible.

Since enolization of saturated ketones proceeds to the less substituted side, Claisen condensations usually occur there by preference in cases with choice (page 000).

Application of the Claisen condensation to a diester of a six- or seven-carbon diacid or to an appropriate ketoester is called a **Dieckmann reaction.** This is a cyclization reaction and, being reversible, can usually be used only to form rings of five to seven members.

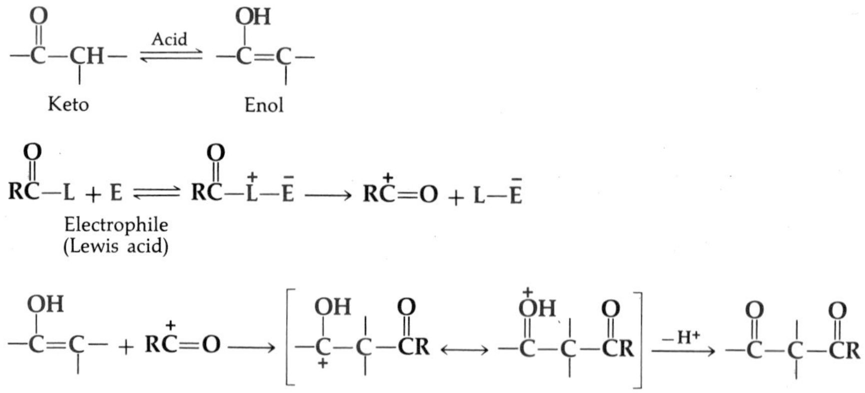

$$\underset{\text{Keto}}{-\overset{\overset{\displaystyle O}{\|}}{C}-CH-} \quad \underset{\text{Acid}}{\rightleftharpoons} \quad \underset{\text{Enol}}{-\overset{\overset{\displaystyle OH}{|}}{C}=\overset{|}{C}-}$$

$$\underset{\substack{\text{Electrophile}\\ \text{(Lewis acid)}}}{\overset{\overset{\displaystyle O}{\|}}{R C}-L} + E \rightleftharpoons \overset{\overset{\displaystyle O}{\|}}{R C}-\overset{+}{L}-\overset{-}{E} \longrightarrow \overset{+}{R C}=O + L-\overset{-}{E}$$

$$-\overset{\overset{\displaystyle OH}{|}}{C}=\overset{|}{C}- + \overset{+}{R C}=O \longrightarrow \left[-\overset{\overset{\displaystyle OH}{|}}{\underset{+}{C}}-\overset{|}{C}-\overset{\overset{\displaystyle O}{\|}}{C}R \longleftrightarrow -\overset{\overset{\displaystyle \overset{+}{O}H}{|}}{C}-\overset{|}{C}-\overset{\overset{\displaystyle O}{\|}}{C}R \right] \xrightarrow{-H^+} -\overset{\overset{\displaystyle O}{\|}}{C}-\overset{|}{C}-\overset{\overset{\displaystyle O}{\|}}{C}R$$

FIGURE 13-4   Acid-catalyzed acylation

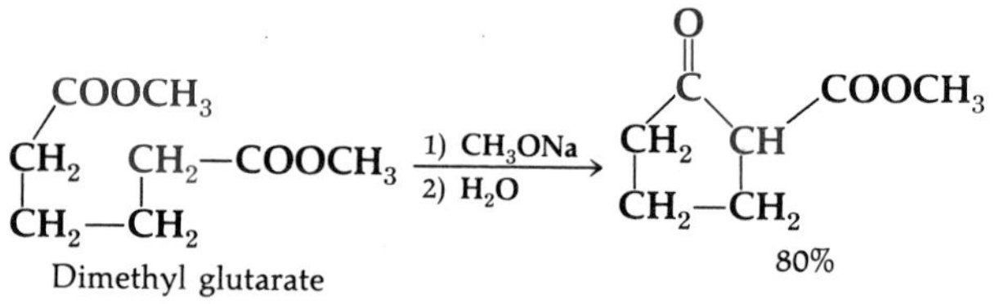

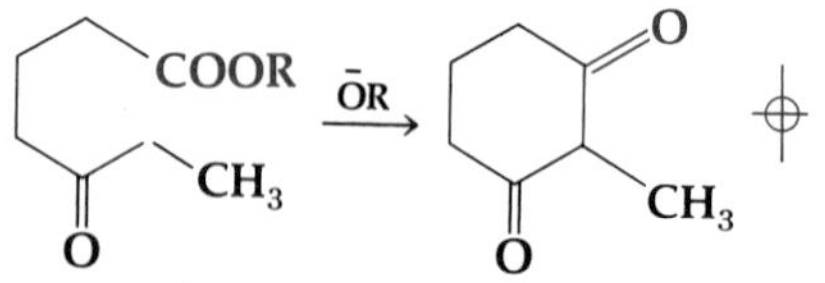

Dimethyl glutarate

$$\xrightarrow[\text{2) H}_2\text{O}]{\text{1) CH}_3\text{ONa}}$$

80%

$$\xrightarrow{\bar{O}R}$$

Acylation of Enamines

A good β-diketone synthesis utilizes the nucleophilic character of enamines. The reaction is especially useful since, unlike the Claisen condensation,
the acylating agent can possess an α-methylene without creating by-products
from cross- or self-condensations.

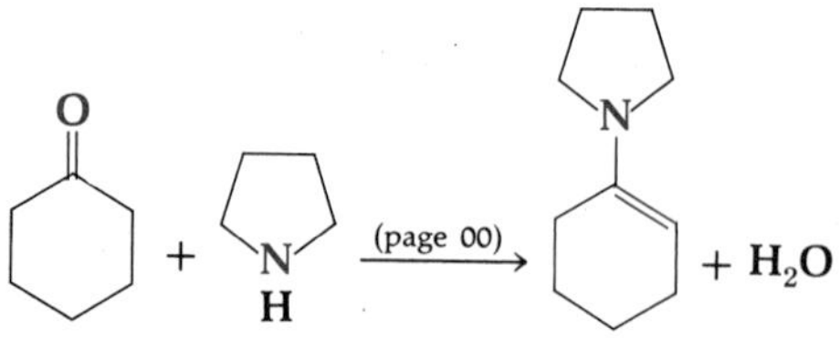

$$+ \quad \xrightarrow{\text{(page 00)}} \quad + H_2O$$

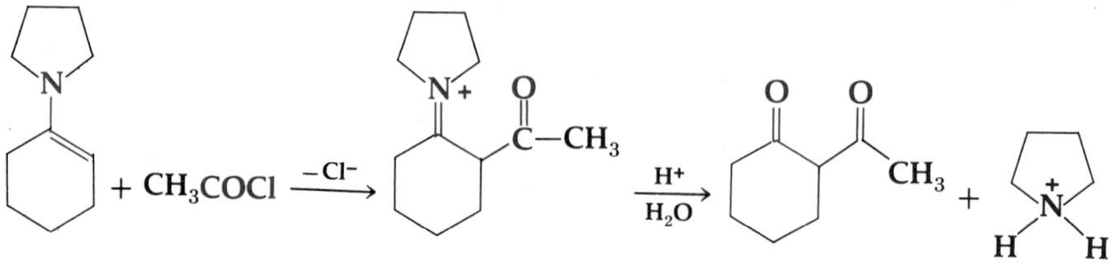

$$+ CH_3COCl \xrightarrow{-Cl^-} \xrightarrow[H_2O]{H^+} +$$

Acid-catalyzed Acylation

The general course of acid-catalyzed acylation reactions at carbon is outlined in Fig. 13-4. The reaction is used much less than the base-catalyzed examples. Unsymmetrical saturated ketones frequently give products in acid-catalyzed condensations that are different from those formed under basic conditions since the enol is formed predominantly on the more substituted side, in contrast to the enolate ion.

$$CH_3CH_2COCH_3 + (CH_3CO)_2O \xrightarrow{BF_3} CH_3CO-\underset{\underset{CH_3}{|}}{CH}COCH_3$$

3-Methylacetylacetone
(3-Methylpentane-2,4-dione)
30%

$$CH_3CH_2COCH_3 + CH_3COOC_2H_5 \xrightarrow{C_2H_5ONa} CH_3CH_2COCH_2-COCH_3$$

Propionylacetone
60%

Enols are sufficiently nucleophilic to be acylated by very reactive substrates such as anhydride-boron trifluoride addition compounds.

$$CH_3\overset{O}{\overset{||}{C}}CH_2CH_3 \underset{\text{acid}}{\overset{\text{Fast with}}{\rightleftharpoons}} CH_3\overset{OH}{\overset{|}{C}}=CHCH_3$$

$$CH_3\overset{:\ddot{O}H}{\overset{\curvearrowright}{C}}=CHCH_3 + CH_3-\overset{O}{\overset{||}{C}}-O\overset{\overset{+}{O}-\bar{B}F_3}{\overset{||}{C}}CH_3 \longrightarrow$$

$$CH_3\overset{\overset{+}{O}H}{\underset{\underset{CH_3}{|}}{\overset{|}{C}}}\overset{O}{\overset{||}{C}}CH\overset{O}{\overset{||}{C}}CH_3 + CH_3CO_2\bar{B}F_3 \longrightarrow CH_3\overset{O}{\overset{||}{C}}\underset{\underset{CH_3}{|}}{CH}\overset{O}{\overset{||}{C}}CH_3 + CH_3COOH + BF_3$$

Acylation of enol esters is a strictly analogous reaction.

$$\phi-\overset{\overset{O}{\overset{||}{OCCH_3}}}{\underset{}{C}}=CH_2 + (CH_3CO)_2O \xrightarrow{BF_3} \left[\phi-\overset{\overset{+}{O}-\overset{O}{\overset{||}{C}}-CH_3}{\underset{}{\overset{||}{C}}}-CH_2-COCH_3\right] + \bar{O}\overset{O}{\overset{||}{C}}-CH_3 \longrightarrow$$

$$\phi-CO-CH_2-COCH_3$$

Benzoylacetone
68%

Related Condensations

The **Thorpe reaction** involves nucleophilic addition to cyano groups. The reaction is best considered in conjunction with the Claisen ester condensation, since the two reactions accomplish essentially the same synthetic objectives and operate by basically similar mechanisms. Highly hindered bases are used to catalyze Thorpe reactions, in order to favor proton abstraction in preference to addition of the base to nitrile groups. The reaction produces imino-nitriles, which are usually hydrolyzed to ketonitriles before products are isolated. The only extensive application of the reaction has been in syntheses of cyclic ketones (Sec. 13-4), similar to the Dieckmann condensation.

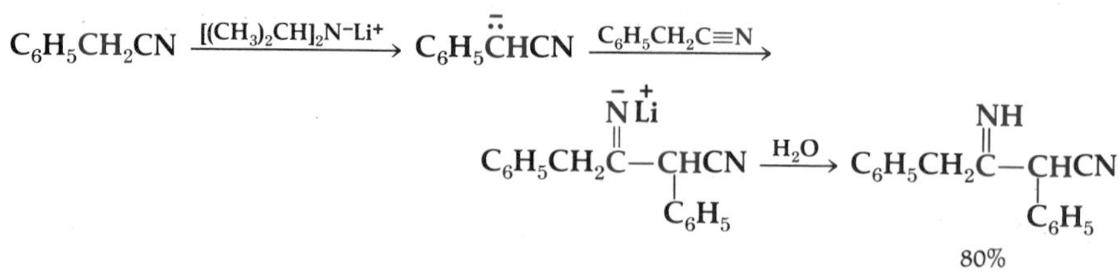

$$C_6H_5CH_2CN \xrightarrow{[(CH_3)_2CH]_2N^-Li^+} C_6H_5\overset{..}{C}HCN \xrightarrow{C_6H_5CH_2C \equiv N}$$

$$\underset{\overset{||}{N}Li}{\overset{-\ +}{\underset{|}{N}Li}} \atop C_6H_5CH_2\overset{||}{C}-\underset{\underset{C_6H_5}{|}}{C}HCN \xrightarrow{H_2O} C_6H_5CH_2\overset{\overset{NH}{||}}{C}-\underset{\underset{C_6H_5}{|}}{C}HCN$$

$$80\%$$

$$(CH_2)_n\overset{CH_2CN}{\underset{CN}{<}} \xrightarrow{(C_2H_5)_2N^-Li^+} (CH_2)_n\underset{\underset{C=N\overset{-\ +}{Li}}{|}}{CHCN} + \xrightarrow{H_3O^+} (CH_2)_n\underset{\underset{C=O}{|}}{CHCN}$$

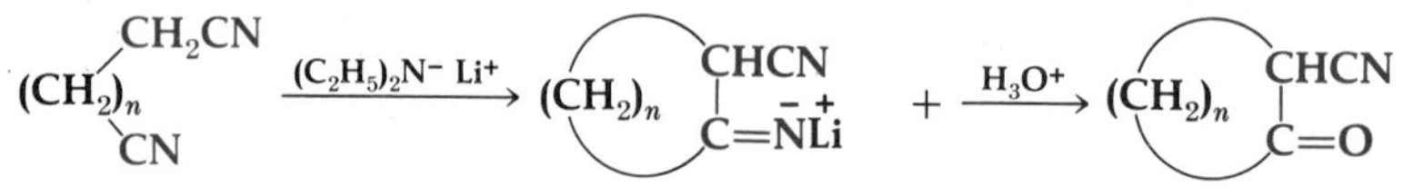

$$1-70\%$$

Pyrolysis of heavy metal salts of carboxylic acids produces ketones. One carboxyl group is converted to a carbonyl group, and the other is lost as carbon dioxide (carbonate). Although acetone was first discovered as a product of the pyrolysis of calcium salts, the method is now seldom used except for synthesis of cyclic ketones. It is mechanistically analogous to the ester condensations.

$$(CH_3COO)_2Ca \xrightarrow{350°} CH_3COCH_3 + CaCO_3$$

$$C_6H_5CH_2COOH + CH_3COOH \xrightarrow[\text{tube, }450°]{\text{ThO}_2\text{-packed}} C_6H_5CH_2COCH_3 + CO_2$$
$$60\%$$

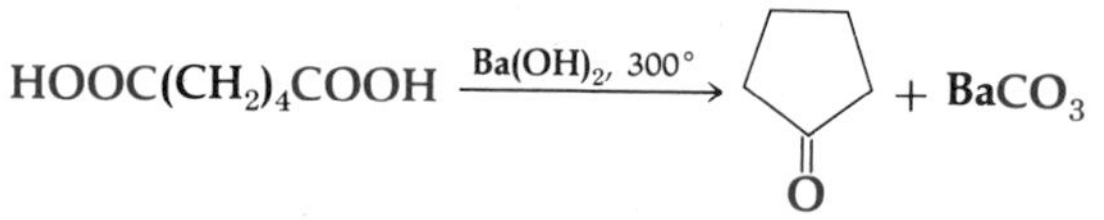

$$HOOC(CH_2)_4COOH \xrightarrow{Ba(OH)_2,\ 300°} \text{(cyclopentanone)} + BaCO_3$$

The **acyloin condensation** results from treatment of an ester or a diester with sodium. The reaction is carried out in an aprotic solvent, such as ether or a high-boiling hydrocarbon. Acyloin condensations have been used with spectacular results in the synthesis of cyclic compounds, considered in the next section. The reaction is not, however, analogous to the foregoing reactions and is discussed as a reduction in Chap. 18.

$$Na\cdot + R\overset{\overset{\displaystyle \ddot{O}:}{\|}}{C}OR' \longrightarrow R\overset{\overset{\displaystyle :\ddot{\ddot{O}}:}{|}}{\underset{\displaystyle \cdot}{C}}-OR' + \overset{+}{Na}$$

A radical
anion

$$2R\overset{\overset{\displaystyle :\ddot{\ddot{O}}:}{|}}{\underset{\displaystyle \cdot}{C}}-OR' \longrightarrow R\overset{\overset{\displaystyle :\ddot{O}:)}{|}}{\underset{\displaystyle \overset{|}{O}R'}{C}}-\overset{\overset{\displaystyle :\ddot{O}:)}{|}}{\underset{\displaystyle \overset{|}{O}R'}{C}}R \longrightarrow R\overset{\overset{\displaystyle :\ddot{O} \quad \ddot{O}:}{\| \qquad \|}}{C}-CR + 2R'\overset{-}{O}$$

$$R\overset{\overset{\displaystyle (:\ddot{O} \quad \ddot{O}:)}{\| \qquad \|}}{C}-CR + 2Na\cdot \longrightarrow R\overset{\overset{\displaystyle :\ddot{O}: \quad :\ddot{\ddot{O}}:}{| \qquad |}}{C}=CR + 2\overset{+}{Na} \xrightarrow{\ H^{+}\ } R\overset{\overset{\displaystyle HO \quad OH}{| \qquad |}}{C}=CR \rightleftharpoons R\overset{\overset{\displaystyle O \quad OH}{\| \qquad |}}{C}-CHR$$

An acyloin

PROBLEM 13-4

Write the expected products in each case when the substances shown are treated with sodium ethoxide.

a   Ethyl propionate.

b   Ethyl 2-methylpentanoate.

c   Ethyl 3-methylpentanoate.

d   Ethyl cyanoacetate.

e   2-Methylcyclohexanone and ethyl formate.

f   Acetophenone and ethyl isobutyrate.

g   Diethyl glutarate (pentanedioate).

h   Ethyl cyclobutylacetate and ethyl carbonate.

i   Ethyl benzoate and γ-butyrolactone.

j   The three monomethyl heptandioic acid diethyl esters. Show that two of these give one product each and the other yields two products. Show the stereo-chemistry of all products.

13-4   CARBONYL CYCLIZATION REACTIONS

Four reactions that effect cyclization of difunctional aliphatic molecules have been presented in this chapter. Another, the intramolecular aldol condensation, was discussed in the previous chapter. Much attention has been centered on development of cyclization reactions for both practical and theoretical purposes. Alicyclic ring systems are very common in natural products, and dozens of syntheses of five- and six-membered rings are used for these compounds. Interest in synthesis of larger rings was stimulated by the discovery by Ruzicka in 1926 that the active principles in two exotic perfume bases, musk and civet, are large-ring ketones. The Himalayan musk deer and the African civet cat produce the compounds, but not in sufficient quantity to satiate the world-wide demands for perfume ingredients. As a consequence, methods were developed for the synthesis of large-ring ketones.

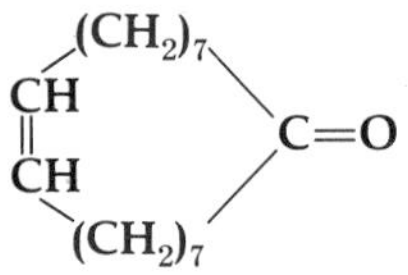

$$
\begin{array}{ll}
\underset{\text{Civetone}}{
\begin{array}{l}
\;\;\diagup(CH_2)_7\diagdown \\
CH \\
\| \qquad\qquad\quad C=O \\
CH \\
\;\;\diagdown(CH_2)_7\diagup
\end{array}} &
\underset{\text{Muscone}}{
\begin{array}{l}
CH_3 \\
| \\
CH-CH_2 \\
| \qquad\qquad\; C=O \\
(CH_2)_{12}\diagup
\end{array}}
\end{array}
$$

Civetone 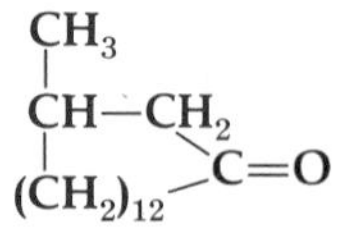   Muscone

The net conclusions of studies of cyclization reactions are as follows:

1 Three- and four-membered rings are not produced in high yields in any intramolecular condensation reaction (i.e., equilibrium reaction).

2 Five-, six-, and seven-membered rings are very easily made by almost any condensation reaction.

3 The yield of cyclization product drops sharply with the rings in the C_8–C_{14} range, with a minimum at C_{10}.

4 Yields again increase and become more or less constant at about C_{15} or C_{16}.

Development of theories to account for the above observations has been slow but interesting. At the present time, the results are attributed to a combination of internal-angle strains and the steric interactions between nonbonded groups attached to the rings (Secs. 6-3 to 6-6). The conclusions can be summarized as follows:

1 The small rings, C_3 and C_4, are badly strained. Condensation reactions, which are for the most part reversible, do not produce them in appreciable yield, because the equilibrium constants for cyclization are too small. There is evidence for the rapid passage of materials through intermediate cyclic C_4 stages in the course of some condensations. Three- and four-membered rings must be made by internal alkylation and addition reactions that are essentially irreversible.

2 C_5 and C_6 rings are formed rapidly (perhaps not as rapidly as C_3) and are stable when formed.

3 The *stability* of large rings favors their formation, but the *rates* of formation fall off. The most important factor in this decrease in rate is the improbability of finding the two ends of a randomly coiled chain close together. In all cases, cyclization competes with intermolecular condensation to form polymeric materials.

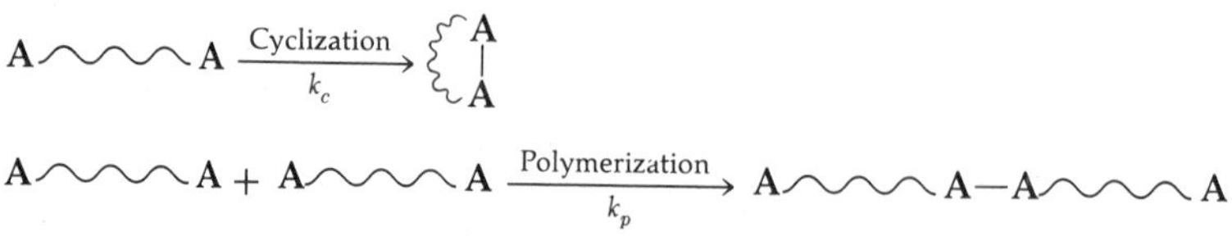

$$
A\!\sim\!\sim\!\sim\!A \xrightarrow[k_c]{\text{Cyclization}} \left\{
\begin{array}{c}
A \\
| \\
A
\end{array}
\right.
$$

$$
A\!\sim\!\sim\!A + A\!\sim\!\sim\!\sim\!A \xrightarrow[k_p]{\text{Polymerization}} A\!\sim\!\sim\!\sim\!A\!-\!A\!\sim\!\sim\!A
$$

4 The minimum yield at C_{10} in the medium rings is apparently caused by the steric strain between nonbonded hydrogens as well as the poor torsional strain which characterizes these rings (page 196).

Various techniques have been developed to overcome unfavorable competition with intermolecular reactions in the synthesis of medium and large rings. The first success with the pyrolysis of salts of divalent metals is probably due to the fact that some cyclic salt structures are formed in the solid mixture, thus bringing two functional ends together.

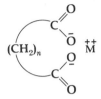

Success with the Thorpe cyclization is due to the choice of a reaction that can be run at very *high dilution* in homogeneous solutions. The intermolecular reaction of a cyano carbanion is bimolecular, so that the rate of polymerization depends upon the concentration of both the anion and the dinitrile.

Rate of polymerization $= k_P[RCHCN][RCN]$

On the other hand, the rate of cyclization is first-order and unimolecular.

Rate of cyclization $= k_C[RCHCN]$

Therefore, the ratio of the two reaction rates depends on the concentration of the nitrile, and cyclization is favored by high dilution (low [RCN]).

$$\frac{\text{Rate of cyclization}}{\text{Rate of polymerization}} = \frac{k_C}{k_P[RCN]}$$

Despite the success of the high-dilution technique, a dramatic improvement was made with the cyclic acyloin condensation. The reaction occurs in a *dilute layer* of ester molecules adsorbed on the surface of sodium metal. Not all surface reactions simulate high-dilution conditions, but in the present case, the ester is only weakly adsorbed on the surface. As a result, ester molecules are too far apart to permit intermolecular coupling to compete with cyclization. Yields in the cyclic acyloin condensation fall to a minimum at C_{10}, but the minimum is a respectable 60%. Large rings, such as C_{16}, can be made in almost quantitative yield.

$$(CH_2)_8 \begin{matrix} -COOCH_3 \\ \\ -COOCH_3 \end{matrix} \xrightarrow[\text{2) } CH_3COOH]{\text{1) } Na, C_6H_5CH_3} (CH_2)_8 \begin{matrix} C=O \\ | \\ CHOH \end{matrix}$$

Dimethyl
sebacate

Sebacoin
60%

PROBLEM 13-5

In each case below a three- or a five-membered ring product is possible, starting with base-catalyzed enolization of the ketone. Each case proceeds exclusively to only one product; one case gives a three- and one a five-membered ring. Account for this behavior and show what the products are.

$$\underset{\underset{\text{Br}}{|}}{CH_3CHCH_2CH_2COCH_2CH_3} \quad \text{or} \quad CH_3COCH_2CH_2COOCH_3 \quad \text{with} \quad NaOCH_3$$

13-5 CLEAVAGE OF β-DICARBONYL COMPOUNDS

The β-dicarbonyl and related compounds prepared by these acylations are of course just those which prove so useful synthetically for alkylations (Sec. 11-7) and aldol condensations (Sec. 12-5). They are made even more versatile for synthetic use in that they may be hydrolytically cleaved again by alkali. Their cleavages are mechanistically the reverse of their formation.

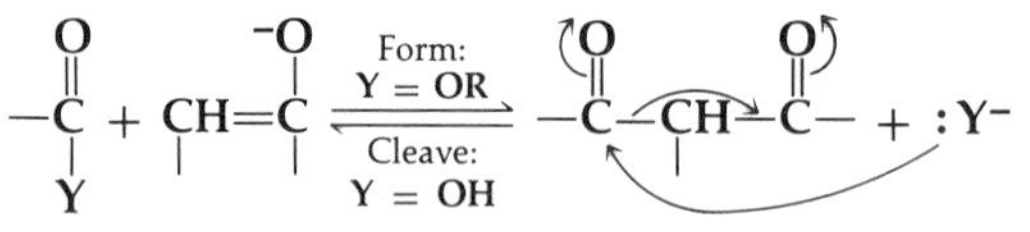

In the formation of β-dicarbonyl molecules the equilibrium is shifted to the right by the formation of the stable enolate of the β-dicarbonyl product. In cleavage by alkali the equilibrium is shifted the other way by formation of the even more stable carboxylate anion. The β-dicarbonyl compounds with no α-hydrogen have no competitive possible enolization and so cleave much

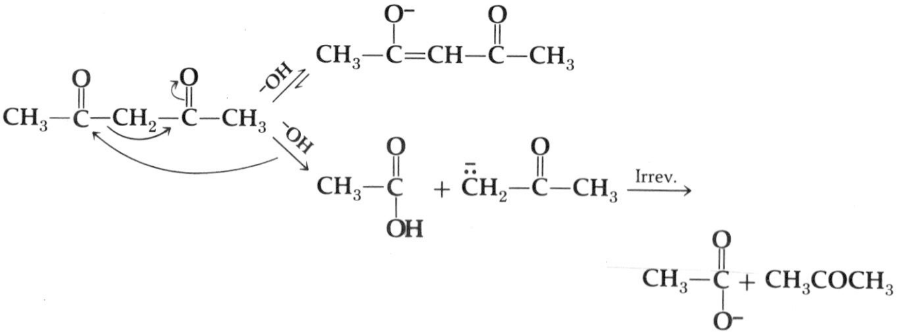

faster in base and may also be cleaved by alkoxides (RO⁻). The cleavage may be regarded as a nucleophilic substitution by hydroxide, with enolate as the leaving group. A comparable substitution at saturated carbon does not occur.

The cleavage of α-acylcyclohexanones and α-acylcyclopentanones occurs at the ring ketone generally and allows a useful synthesis of long-chain acids.

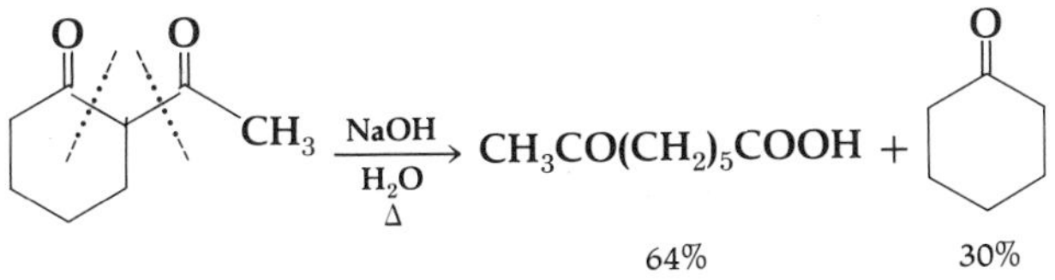

$$\text{(cyclohexanone ring with two C=O and CH}_3\text{)} \xrightarrow[\substack{H_2O \\ \Delta}]{\text{NaOH}} CH_3CO(CH_2)_5COOH + \text{(cyclohexanone)}$$

$$\underset{64\%}{} \qquad \underset{30\%}{}$$

Of the two possible carbonyls the hydroxide attacks the more reactive (or less hindered). In β-ketoesters like acetoacetic esters, it is therefore the ketone which is the site of attack since it is more reactive than ester. Weak base or acid hydrolysis on the other hand simply hydrolyzes the ester instead of cleaving between the two carbonyls. This represents a useful alternative as noted in the next section but summarized here.

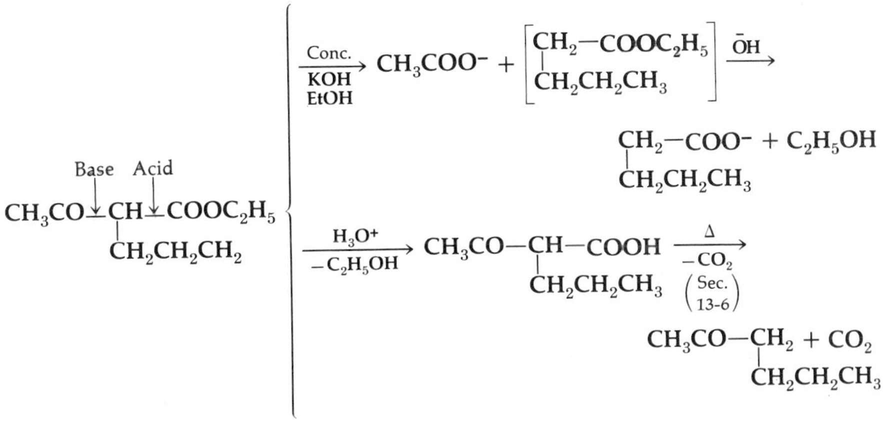

$$
\begin{array}{l}
\xrightarrow[\substack{\text{KOH} \\ \text{EtOH}}]{\text{Conc.}} CH_3COO^- + \begin{bmatrix} CH_2-COOC_2H_5 \\ | \\ CH_2CH_2CH_3 \end{bmatrix} \xrightarrow{\bar{O}H} \\[3em]
\qquad\qquad\qquad\qquad CH_2-COO^- + C_2H_5OH \\
\qquad\qquad\qquad\qquad | \\
\qquad\qquad\qquad\qquad CH_2CH_2CH_3 \\
\end{array}
$$

Base Acid

$$CH_3CO\overset{\downarrow}{-}CH\overset{\downarrow}{-}COOC_2H_5$$
$$\qquad\quad |$$
$$\qquad\quad CH_2CH_2CH_2$$

$$\xrightarrow[-C_2H_5OH]{H_3O^+} CH_3CO-CH-COOH \xrightarrow[\substack{-CO_2 \\ \left(\substack{\text{Sec.} \\ 13\text{-}6}\right)}]{\Delta}$$
$$\qquad\qquad\qquad\qquad\qquad | $$
$$\qquad\qquad\qquad\qquad\qquad CH_2CH_2CH_3$$

$$\qquad\qquad\qquad\qquad\qquad CH_3CO-CH_2 + CO_2$$
$$\qquad\qquad\qquad\qquad\qquad\qquad\quad |$$
$$\qquad\qquad\qquad\qquad\qquad\qquad\quad CH_2CH_2CH_3$$

PROBLEM 13-6

Use methyl acetoacetate, dimethyl malonate, or methyl cyanoacetate as starting materials for these syntheses, as well as any other compounds required.

a 3-Phenyl-2-methylpropionitrile

b Cyclopentanecarboxylic acid

c 2-Methylpentylamine

d 3-Ethyl-2-pentanol

e 2-Methylcyclopentanone

PROBLEM 13-7

Cleavages of α-disubstituted β-diketones with alkali are much faster than those of α-monosubstituted β-diketones. Explain.

13-6 CARBONIC ACID DERIVATIVES AND DECARBOXYLATION

Compounds with four heteroatom bonds to a single sp^2 carbon may be classed as derivatives of carbonic acid and are listed in Table 12-2. These derivatives undergo nucleophilic substitution, closely following the carboxylic acid family in mode and order of reactivity (Tables 12-2 and 13-1). In these compounds with two heteroatom substituents on the carbonyl the more reactive one is substituted first.

$$\phi\text{—CH}_2\text{—OH} + \text{COCl}_2 \longrightarrow \phi\text{—CH}_2\text{—O—}\overset{\overset{\text{O}}{\|}}{\text{C}}\text{—Cl} \xrightarrow{\phi\text{—NH}_2} \phi\text{—CH}_2\text{—O—}\overset{\overset{\text{O}}{\|}}{\text{C}}\text{—NH—}\phi$$

$$(\text{CH}_3)_2\text{N—}\overset{\overset{\text{O}}{\|}}{\text{C}}\text{—Cl} + \text{HO—CH}_2\text{CH}_3 \longrightarrow (\text{CH}_3)_2\text{N—}\overset{\overset{\text{O}}{\|}}{\text{C}}\text{—O—CH}_2\text{CH}_3$$

Phosgene ($COCl_2$) is therefore the primary precursor for most of these compounds. The compound carbonyl diimidazole acts in a completely analogous way but is preferable in giving cleaner reactions and not producing a strong acid (HCl) in the reaction, as well as lacking the toxicity of phosgene.

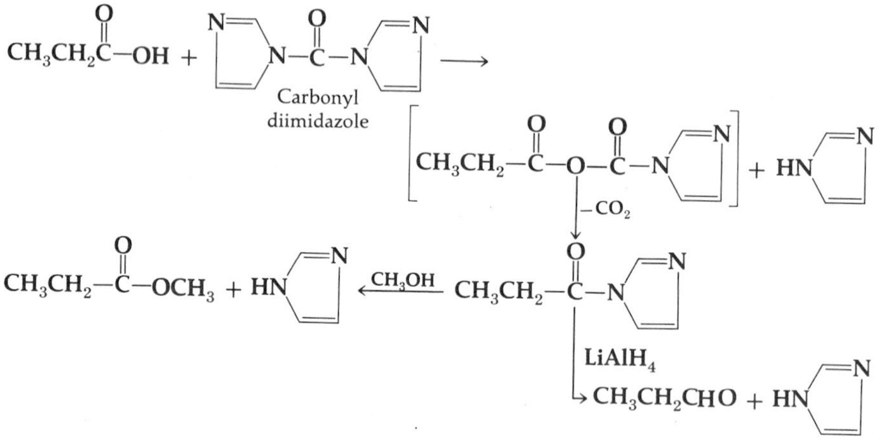

Grignard reagents and hydrides react as expected from analogy with carboxylic acid derivatives.

$$\phi\text{—NH—}\overset{\overset{\text{O}}{\|}}{\text{C}}\text{—OCH}_3 \xrightarrow[-\text{CH}_3\text{OH}]{\text{LiAlH}_4} \left[\phi\text{—NH—}\overset{\overset{\text{O}}{\|}}{\text{C}}\text{—H}\right] \longrightarrow \left[\phi\text{—NH—CH}_2\text{—OH}\right] \xrightarrow{-\text{H}_2\text{O}}$$

$$\left[\phi\text{—N=CH}_2\right] \longrightarrow \phi\text{—NHCH}_3$$

Decarboxylation

The carboxyl group may be regarded as an addition product of CO_2 with a carbanion. Accordingly, in certain circumstances CO_2 may be lost from carboxyl compounds by the reverse reaction, which is called **decarboxylation**. This is an elimination reaction and is thermodynamically favored by the very low energy of carbon dioxide. Decarboxylation can occur if a moderately good electron-acceptor exists next to the carboxyl.

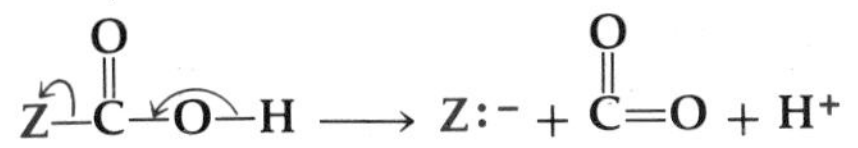

$$Z-\overset{\overset{\displaystyle O}{\|}}{C}-O-H \longrightarrow Z:^- + \overset{\overset{\displaystyle O}{\|}}{C}{=}O + H^+$$

Carbamic acids ($R_2N-COOH$) and half-esters of carbonic acid ($RO-CO-OH$) exist only as their salts, the free acids readily losing carbon dioxide and forming amines and alcohols, respectively. The salts are stable since decarboxylation must release anions, RO^- or R_2N^-, which are less stable than $RO-CO_2^-$ or $R_2N-CO_2^-$.

In true carboxylic acids decarboxylation can occur when the adjacent carbon skeleton contains a good electron acceptor. The most usual case is that of malonic or β-ketoacids. The β-carbonyl accepts the electrons from the bond to **COOH** which is breaking. This occurs on warming the acid and proceeds by a cyclic mechanism where this is sterically feasible. The initial enol formed immediately reverts to the more stable ketone tautomer.†

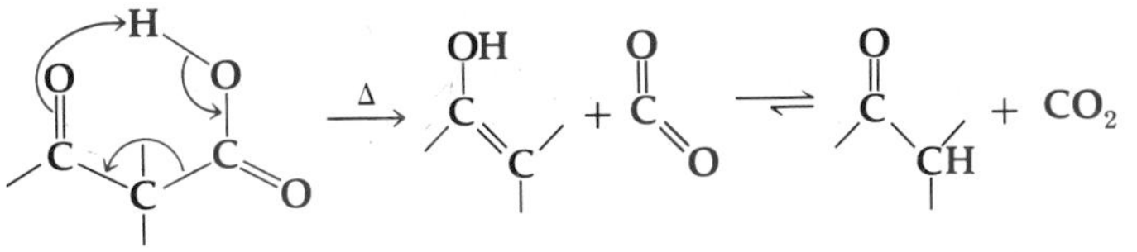

The reaction also occurs in warm acidic media and is the basis for an alternative cleavage of acetoacetic esters (page 533). The ester group in these compounds is the first to hydrolyze in aqueous acid, and the resultant β-ketoacid then decarboxylates. Malonic esters similarly hydrolyze and decarboxylate, malonic acid itself losing carbon dioxide at 50°.

---

† This decarboxylation may be regarded as a fragmentation reaction as described in Sec. 12-1, in which a bond to the α-carbon of a carbonyl is broken, analogous to the breaking of an α C—H bond.

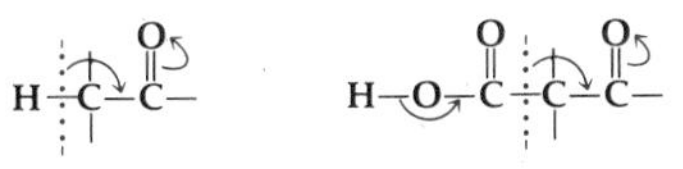

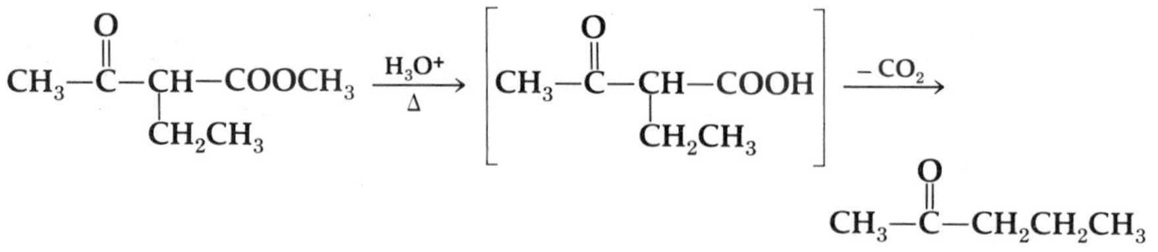

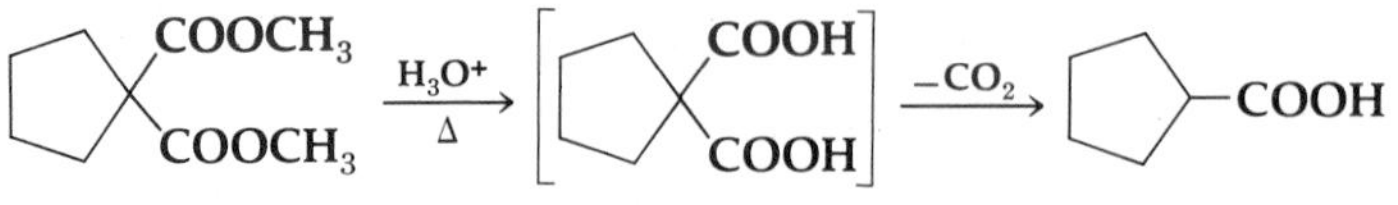

$$CH_3-\overset{\overset{O}{\|}}{C}-\underset{\underset{CH_2CH_3}{|}}{CH}-COOCH_3 \xrightarrow[\Delta]{H_3O^+} \left[CH_3-\overset{\overset{O}{\|}}{C}-\underset{\underset{CH_2CH_3}{|}}{CH}-COOH \right] \xrightarrow{-CO_2}$$

$$CH_3-\overset{\overset{O}{\|}}{C}-CH_2CH_2CH_3$$

$$\overset{COOCH_3}{\underset{COOCH_3}{\diamond}} \xrightarrow[\Delta]{H_3O^+} \left[\overset{COOH}{\underset{COOH}{\diamond}} \right] \xrightarrow{-CO_2} \diamond-COOH$$

Even in highly enolic compounds like the aromatic phenols, the enol can first ketonize in a highly unfavorable equilibrium, which is, however, adequate to provide a ketone for decarboxylation. Supplying their own acid catalysts, *o*- and *p*-hydroxybenzoic acids decarboxylate on heating to the melting point.

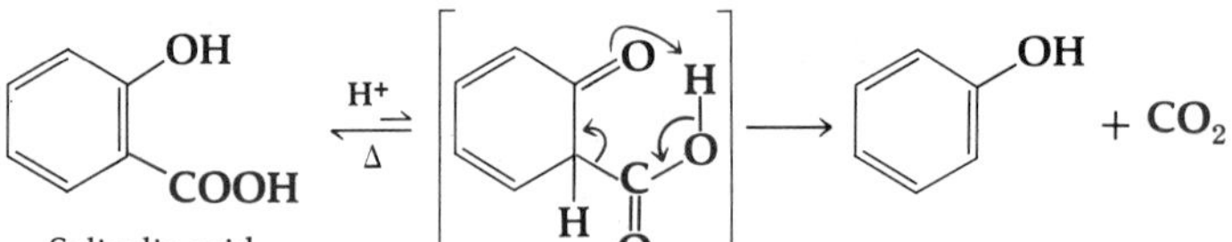

Salicylic acid

Other groups which promote decarboxylation are adjacent three-membered rings, which relieve angle strain by opening, and β,γ-double bonds. When α,β-unsaturated acids can isomerize to β,γ via enolization, they also decarboxylate. Cyclic mechanisms again probably predominate in all these decarboxylations unless prevented by structural restrictions.

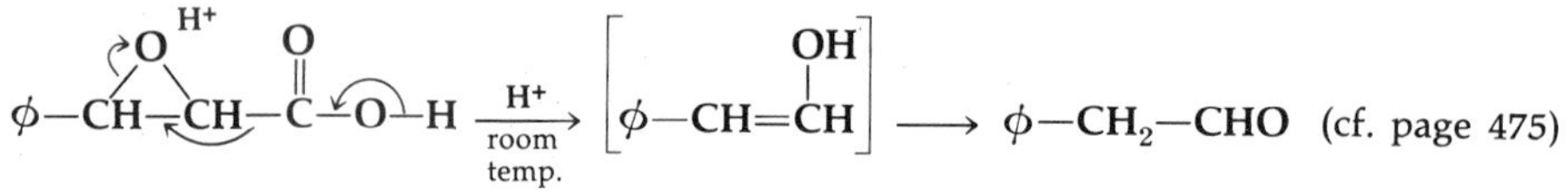

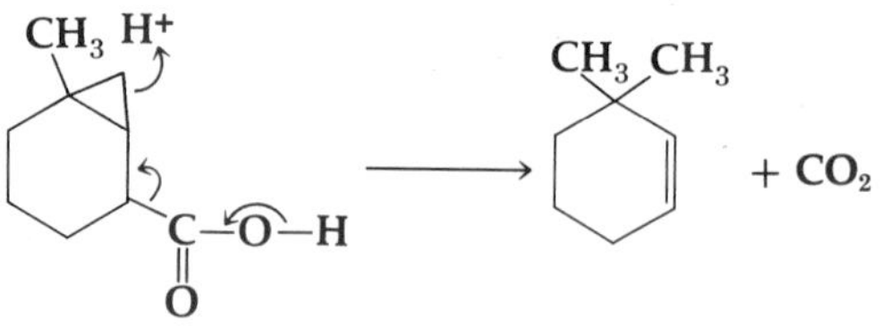

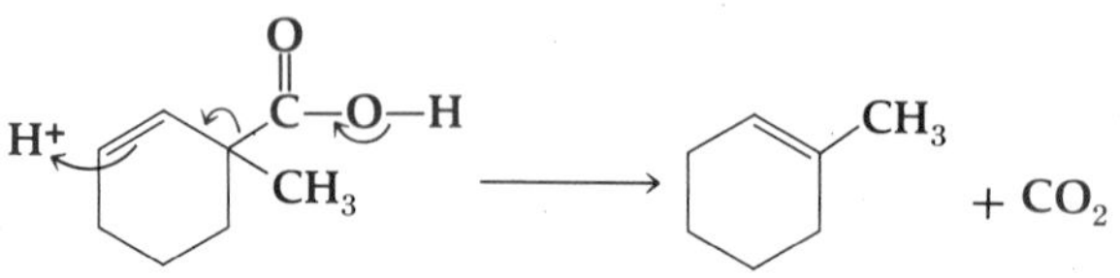

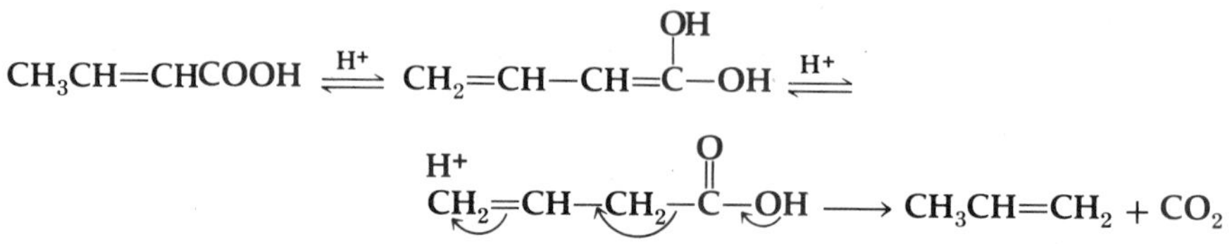

$$\phi-CH-CH-\overset{\overset{O}{\|}}{C}-O-H \xrightarrow[\text{room temp.}]{H^+} \left[\phi-CH=\overset{\overset{OH}{|}}{CH} \right] \longrightarrow \phi-CH_2-CHO \quad (\text{cf. page 475})$$

$$CH_3CH=CHCOOH \rightleftharpoons CH_2=CH-CH=\overset{\overset{OH}{|}}{C}-OH \xrightarrow{H^+}$$

$$\overset{H^+}{CH_2=CH-CH_2-\overset{\overset{O}{\|}}{C}-OH} \longrightarrow CH_3CH=CH_2 + CO_2$$

PROBLEM 13-8

Rank these substances in an order of increasing ease of decarboxylation.

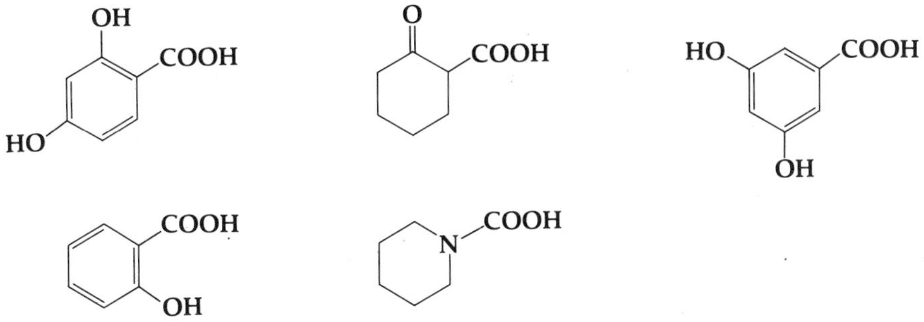

13-7 NUCLEOPHILIC SUBSTITUTION IN RELATED UNSATURATED GROUPS

The list of related functional groups in Fig. 12-3 is adaptable to substitution as well as simple addition when a leaving group is attached. The possibilities are summarized in Fig. 13-5. Each family of derivatives (with various groups L) is characterized by a parent compound, often an inorganic acid, just as in the carboxylic acid family. Similarly, the derivatives in each family are in principle interconvertible like the acyl derivatives of Sec. 13-1, but in practice this has not been explored so deeply and other special reactions of the family may intervene. The principal interconversions are surveyed here.

Imino Compounds

The imino-ethers and -chlorides, corresponding to esters and acyl chlorides, are comparably reactive, more so in acid owing to protonation of nitrogen, less in base since the C=N dipole is less pronounced than C=O (nitrogen less electronegative).

Imino-chlorides   Imino-ethers   Amidines

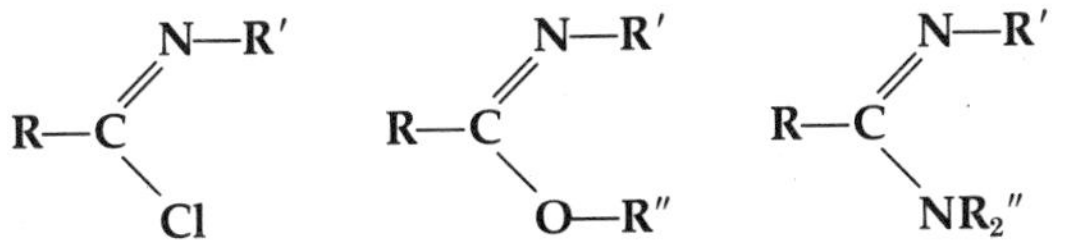

Both derivatives are usually hydrolyzed readily to amides simply with water, unless stabilized by conjugation. The imino-ethers are commonly prepared by addition of alcohols to nitriles (page 488) or by O-alkylation of amides (page 395) both in anhydrous acid media.

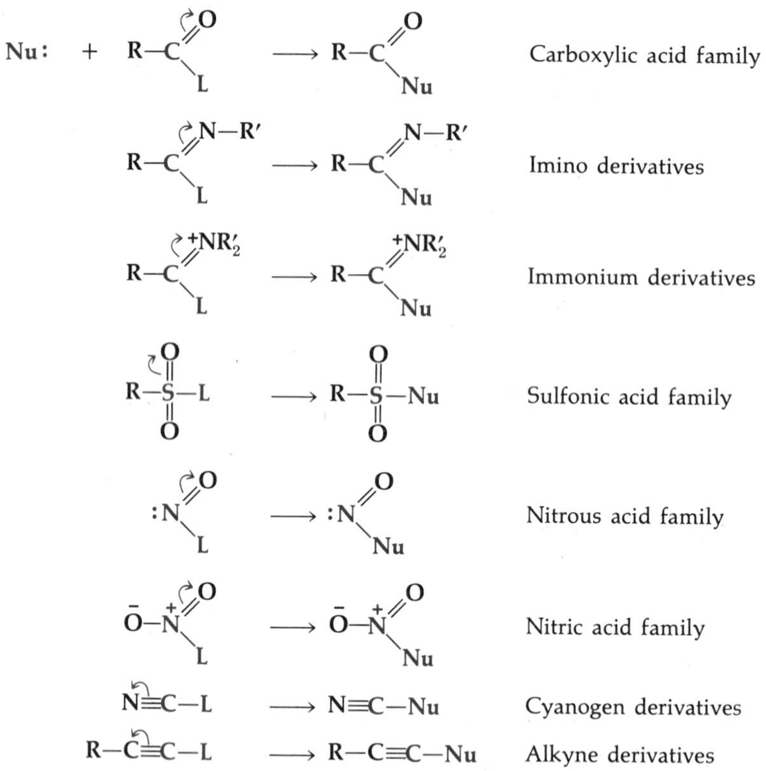

FIGURE 13-5   Nucleophilic substitution in unsaturated functional group

The imino-chlorides arise from attack of PCl_5 and similar dehydrative chlorides on N-substituted amides. (Primary amides simply dehydrate to nitriles by this mechanism.) The reaction is generally an attack on oxygen by the electrophile, followed by nucleophilic substitution with chloride ion, as discussed for the reactions of amides in Sec. 13-1.

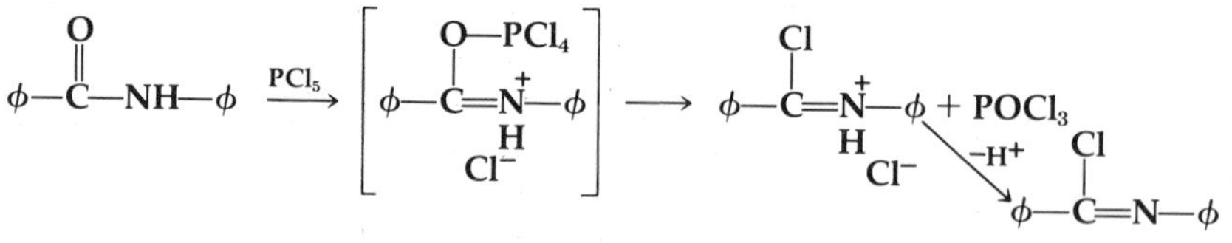

The immonium chlorides and ethers are even more reactive and rarely isolated. The amidines, however, are unusually resistant to substitution owing to their very high resonance stabilization. The immonium ion here has extra stability in that the positive charge can be equally distributed between two like atoms, like the carboxylate anion or guanidinium ion (Fig. 8-1).

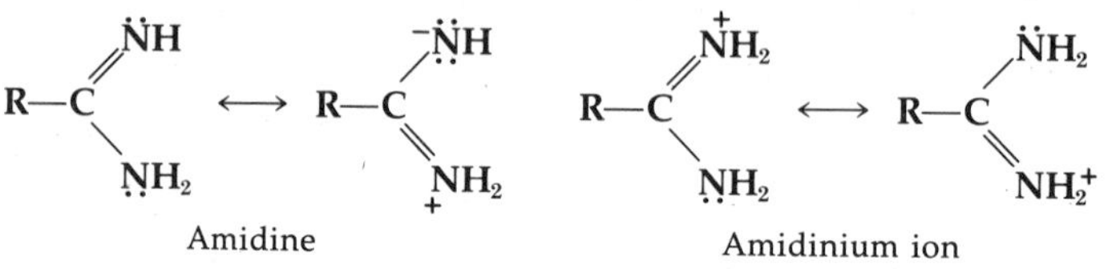

Amidine   Amidinium ion

Sulfonic Acid Derivatives

This family is closest to the carboxylic acid family in the simple parallelism of its derivatives and interconversion reactions. In general the comparable interconversions here are slower than those of acyl derivatives. Special reactions peculiar to sulfonic acid derivatives (and caused by d orbital availability on sulfur) are discussed in Chap. 19.

R—SO_2—Cl	R—SO_2—OH	R—SO_2—OR$'$	R—SO_2—NH_2 (—NHR$'$, NR_2')
Sulfonyl chlorides	Sulfonic acids	Sulfonic esters (alkyl sulfonates)	Sulfonamides

The use of sulfonyl chlorides (usually p-toluenesulfonyl = "tosyl," and methanesulfonyl = "mesyl") in pyridine to convert alcohols to sulfonic esters is a simple sulfonation, analogous to acylation with carboxylic acid chlorides in pyridine. It is primarily used to convert alcohols to good leaving groups for aliphatic substitutions (page 396). The same sulfonation of amides yields the nonbasic sulfonamides, analogous to amide formation.

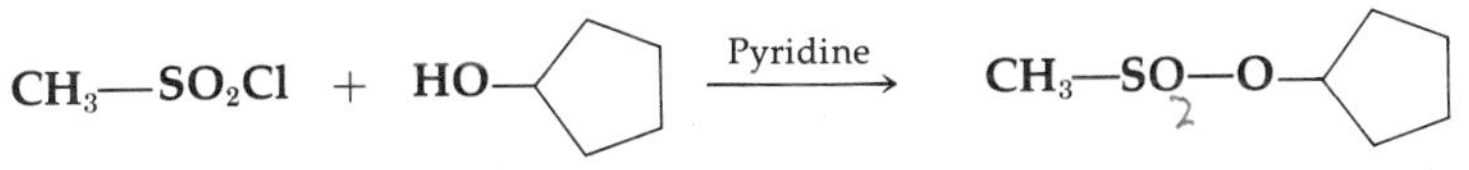

$$CH_3—SO_2Cl \ + \ HO-\!\!\bigcirc \ \xrightarrow{\text{Pyridine}} \ CH_3—SO_2—O-\!\!\bigcirc$$

Methanesulfonyl chloride   Cyclopentanol   Cyclopentyl methanesulfonate

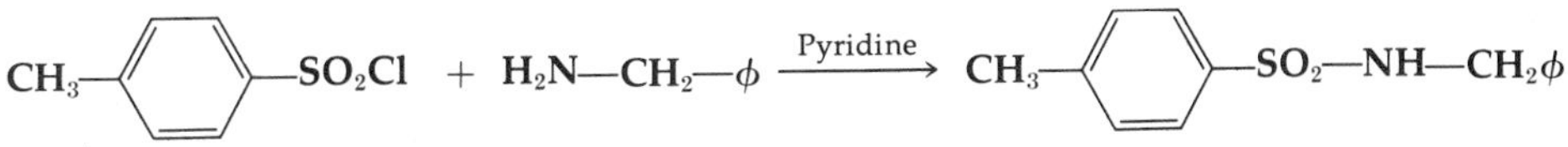

$$CH_3-\!\!\bigcirc\!\!-SO_2Cl \ + \ H_2N—CH_2—\phi \ \xrightarrow{\text{Pyridine}} \ CH_3-\!\!\bigcirc\!\!-SO_2—NH—CH_2\phi$$

p-Toluenesulfonyl chloride   Benzylamine N-Benzyl p-toluenesulfonamide

The sulfonyl chlorides and sulfonic esters hydrolyze in acid or base to sulfonic acids (R—SO_2—OH) but more slowly than acyl chlorides and esters. On the other hand aliphatic substitution is facile (alkyl-oxygen cleavage with sulfonate leaving group) and hence is usually the predominant course of hydrolysis with aliphatic sulfonic esters. Sulfonamides are very stable to hydrolysis, often requiring vigorous conditions such as hot concentrated hydriodic acid for success.

Nitric and Nitrous Acid Derivatives

Substitution occurs in these compounds in a manner quite analogous to that in acyl derivatives, but since the reaction involves substitution at nitrogen instead of carbon, it is common to regard these compounds as electrophilic reagents, focusing instead upon the nucleophilic carbon with which they react. In this connection they appear again in Chap. 16, but it is still valuable to see here some of their reactions which closely parallel those of acylating agents. Nitrous acid derivatives are shown here. The comparable nitric acid derivatives are much less commonly encountered.

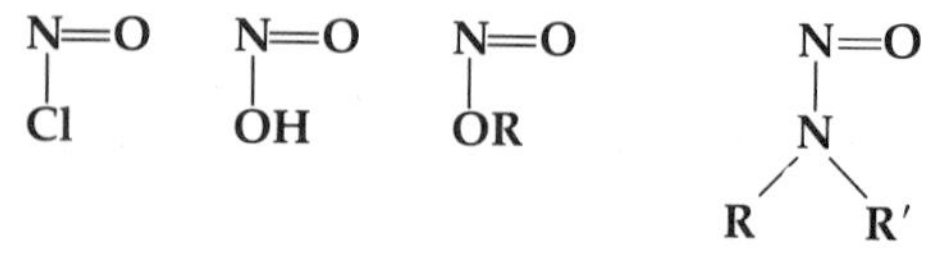

Nitrosyl chloride | Nitrous acid | Nitrites (nitrite esters) | N-Nitroso compounds

Treatment of primary or secondary amines with nitrous acid (or nitrosyl chloride in pyridine, or nitrites) affords the N-**nitroso** derivative as the first step. These are reasonably stable compounds when formed from secondary amines. The primary N-nitroso compound, however, immediately tautomerizes and dehydrates to a diazonium ion since it retains a hydrogen on nitrogen to make this possible.

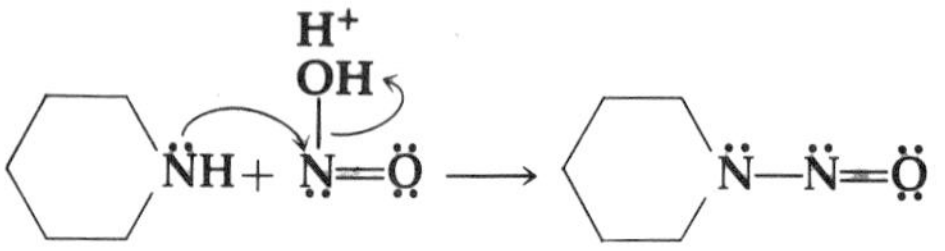

Piperidine   N-Nitrosopiperidine

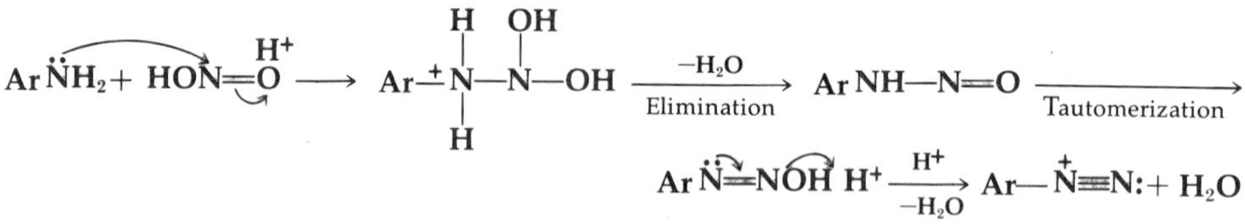

As has been mentioned previously (page 413), aliphatic diazonium ions are unstable. Solutions of aromatic diazonium salts can be preserved reasonably well at 0°, but they are hydrolyzed to phenols on warming (page 546). Aryl diazonium salts with nonnucleophilic ions such as fluoroborate (BF_4^-) can be isolated as ordinary salts.

Nitrosation of enols and enolates is comparable to their acylation and may be carried out on ketones or esters with either acidic or basic catalysis. The C-nitroso compounds with an α-hydrogen always tautomerize preferentially to their enolic tautomers (Table 12-3), the oximes. The oximes are more stable since bonds between heteroatoms are always weak (Table 2-2), and the oxime has only one such bond while nitroso has two.

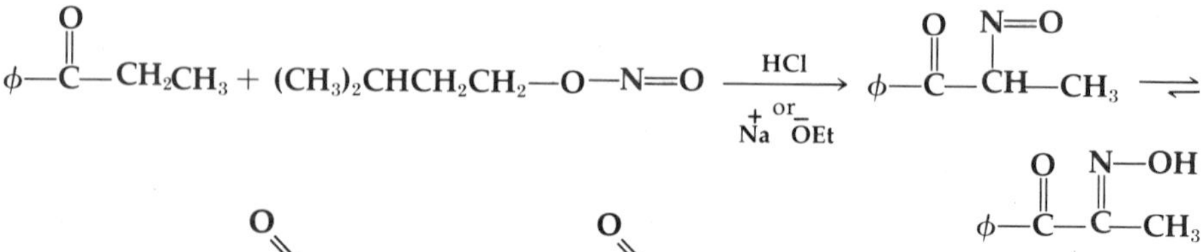

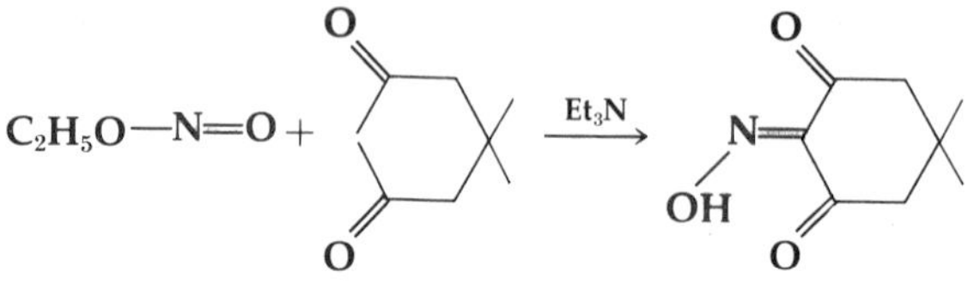

Triple-bonded Substrates

Cyanogen derivatives (L—$C\equiv N$) belong formally to the carbon dioxide class, having four heteroatom bonds to carbon. The only common example is cyanogen bromide (Br—$C\equiv N$) which is attacked by amines to replace the bromine. The products are cyanamides with a large measure of resonance stabilization. Tertiary amines are cleaved (**von Braun reaction**), forming first a quaternary cyanamide which is in turn displaced (in an S_N2 reaction probably) by bromide ion. The displacement is favored by release of the cyanamide resonance not present in the quaternary intermediate. Reaction of cyanogen derivatives with enolates has seldom been examined, but eneamines react to place —CN on carbon.

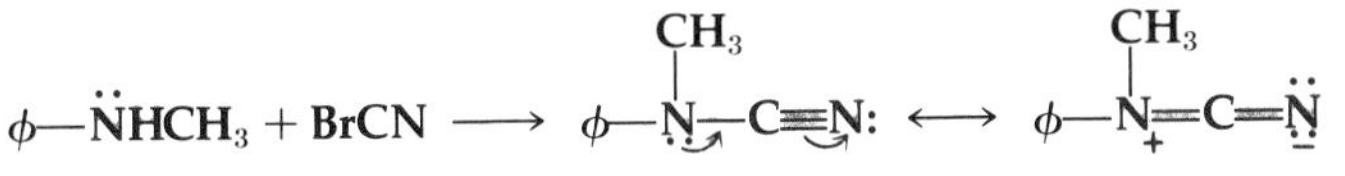

Phenyl-methyl-cyanamide

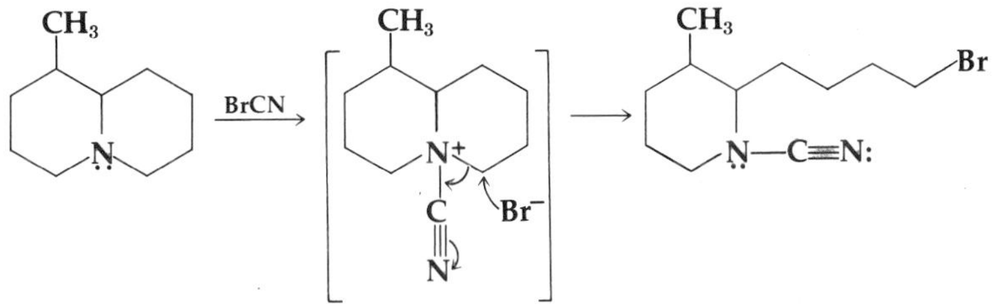

Haloacetylenes undergo substitution with nucleophiles like RS^- and R_2N^- since acetylenes are subject to nucleophilic addition, the initial reaction. Vinyl halides ($-C\overset{|}{=}\overset{|}{C}-X$) are not subject to addition, and so not to substitution, unless conjugated with electron-withdrawing (carbonyl type) functional groups (Sec. 13-8). In the acetylenes, the order of leaving groups is $F > Cl > Br$, the reverse of their activity in substitutions at saturated carbon. (See also Sec. 13-8.) Acetylene ethers also undergo substitution.

$$\phi-C\equiv C-Cl + Li^+ \ \bar{N}CH_3)_2 \longrightarrow \phi-C\equiv C-NCH_3)_2 + Li^+ \ Cl^-$$

An ynamine (87%)

$$F-C\equiv C-Cl + Li^+ \ \bar{N}C_2H_5)_2 \longrightarrow (C_2H_5)_2 \ N-C\equiv C-Cl + Li^+ \ F^-$$

$$C_2H_5-C\equiv C-OC_2H_5 + Li^+ \ \bar{N}C_3H_7)_2 \longrightarrow C_2H_5-C\equiv C-NC_3H_7)_2 + Li^+ \ \bar{O}C_2H_5$$

The amino-acetylenes, or **ynamines**, are very reactive compounds, especially for dehydrations (Sec. 12-7), and are currently being actively studied, as is the whole field of alkyne chemistry, about which we have until recently known far less than about other common functional groups.

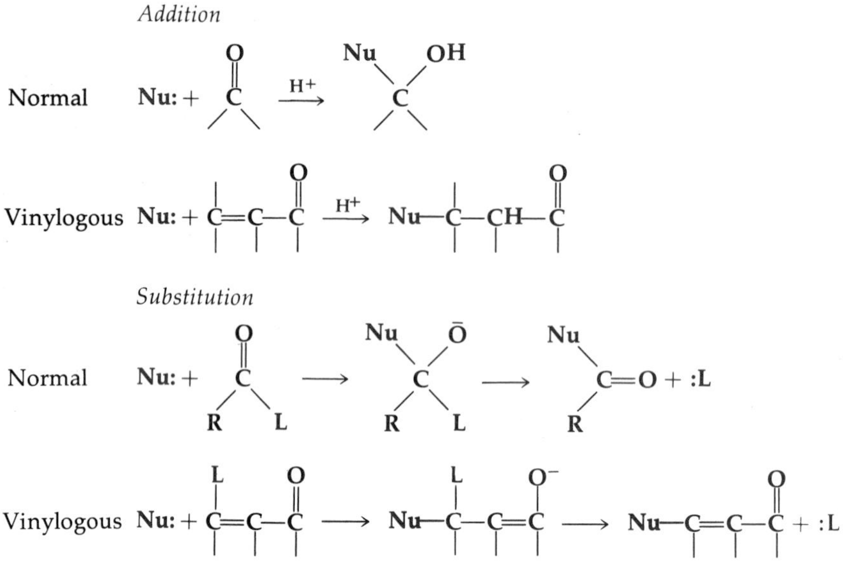

Addition

Normal $Nu: + \underset{/\;\backslash}{\overset{O}{\overset{\|}{C}}} \xrightarrow{H^+} \underset{/\;\backslash}{\overset{Nu\quad OH}{\overset{\backslash\;/}{C}}}$

Vinylogous $Nu: + \underset{|\;|\;|}{C=C-\overset{O}{\overset{\|}{C}}} \xrightarrow{H^+} \underset{|\;|\;|}{Nu-C-CH-\overset{O}{\overset{\|}{C}}}$

Substitution

Normal $Nu: + \underset{R\quad L}{\overset{O}{\overset{\|}{C}}} \longrightarrow \underset{R\quad L}{\overset{Nu\quad \bar{O}}{\overset{\backslash\;/}{C}}} \longrightarrow \underset{R}{\overset{Nu}{\overset{\backslash}{C}}}=O\; +\; :L$

Vinylogous $Nu: + \underset{|\;|\;|}{\overset{L}{\overset{|}{C}}=C-\overset{O}{\overset{\|}{C}}} \longrightarrow \underset{|\;|\;|}{Nu-\overset{L}{\overset{|}{C}}-C=\overset{O^-}{\overset{|}{C}}} \longrightarrow \underset{|\;|\;|}{Nu-C=C-\overset{O}{\overset{\|}{C}}\; +\; :L}$

FIGURE 13-6   **Vinylogous substitution in unsaturated ketones**

13-8   VINYLOGOUS SUBSTITUTION

If α,β-unsaturated ketones and analogous compounds undergo addition to the double bond by reason of its vinylogous extension of the electron deficiency of the carbonyl carbon, then we should also expect similar addition followed by elimination to result in the overall substitution of β-substituents on α,β-unsaturated ketones. This analogy (or vinylogy) is outlined in Fig. 13-6 and is in fact a facile reaction. A second mole of nucleophile could then add to the double bond, but this is uncommon with oxygen and nitrogen nucleophiles because the double-bond reactivity is lowered by the resonance in the initial substitution product.

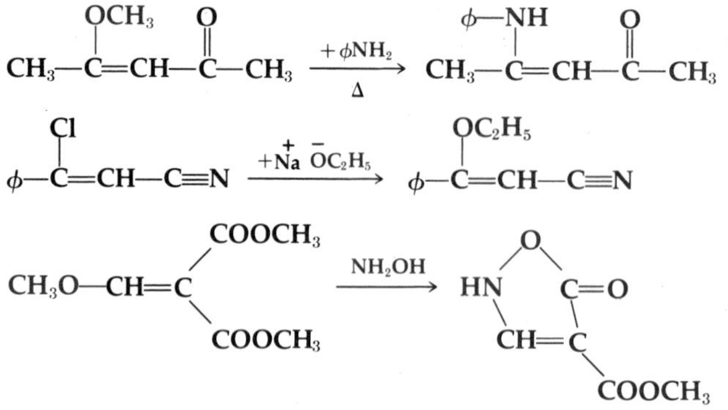

$$CH_3-\underset{}{\overset{OCH_3}{\overset{|}{C}}}=CH-\overset{O}{\overset{\|}{C}}-CH_3 \xrightarrow[\Delta]{+\phi NH_2} CH_3-\underset{}{\overset{\phi-NH}{\overset{|}{C}}}=CH-\overset{O}{\overset{\|}{C}}-CH_3$$

$$\phi-\underset{}{\overset{Cl}{\overset{|}{C}}}=CH-C\equiv N \xrightarrow{+\overset{+}{Na}\;\bar{O}C_2H_5} \phi-\underset{}{\overset{OC_2H_5}{\overset{|}{C}}}=CH-C\equiv N$$

$$CH_3O-CH=\underset{\backslash COOCH_3}{\overset{/COOCH_3}{C}} \xrightarrow{NH_2OH} \underset{CH=\underset{\backslash COOCH_3}{\overset{}{C}}}{\overset{O}{HN\qquad C=O}}$$

The β-chloro compounds useful for this reaction may often be prepared by treating the β-keto analog with PCl_5, which presumably attacks the enol or causes enolization by attack on the keto oxygen directly.

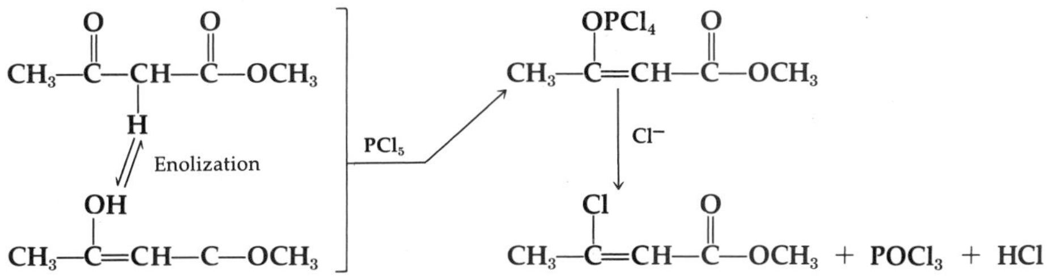

$$CH_3-\overset{\overset{O}{\|}}{C}-\overset{\overset{\displaystyle H}{|}}{CH}-\overset{\overset{O}{\|}}{C}-OCH_3$$

Enolization

$$CH_3-\overset{\overset{\displaystyle OH}{|}}{C}=CH-\overset{\overset{O}{\|}}{C}-OCH_3$$

PCl$_5$

$$CH_3-\overset{\overset{\displaystyle OPCl_4}{|}}{C}=CH-\overset{\overset{O}{\|}}{C}-OCH_3$$

Cl$^-$

$$CH_3-\overset{\overset{\displaystyle Cl}{|}}{C}=CH-\overset{\overset{O}{\|}}{C}-OCH_3 + POCl_3 + HCl$$

The same vinylogous nucleophilic substitution reaction may be extended to aromatic compounds with strong electron-withdrawing substituents and a leaving group in the ortho or para position. Nitro groups are particularly effective, and picryl chloride is as reactive as a carboxylic acid chloride.

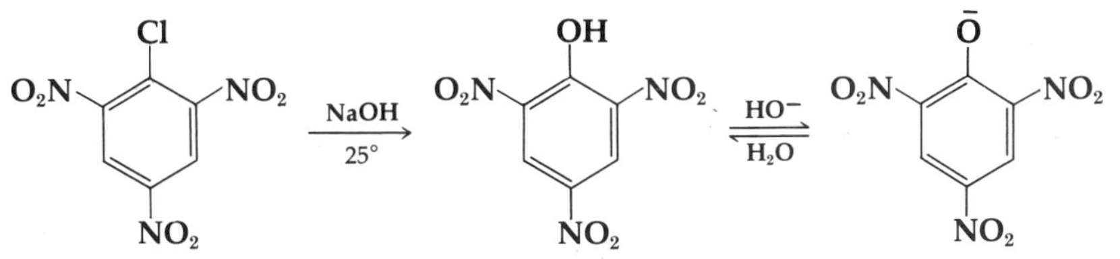

Picryl chloride
(2,4,6-trinitrochlorobenzene)   Picric acid

Activation by nitro groups is due to their ability to absorb the negative charge brought into the molecule by the entering nucleophile, which initially performs the same conjugate addition reaction as in the previous examples. In a few cases the intermediate has been isolated.

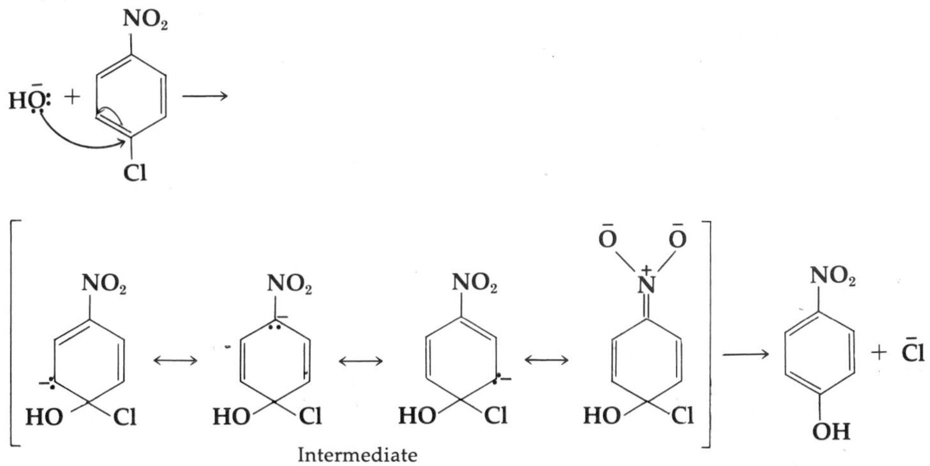

Intermediate

Other strong nucleophiles such as sodium alkoxide or primary and secondary amines also react readily with aryl halides that contain nitro groups in the ortho or para positions. When nitro groups are present in both positions, weaker nucleophiles such as aniline can be used. Yields are ordinarily high.

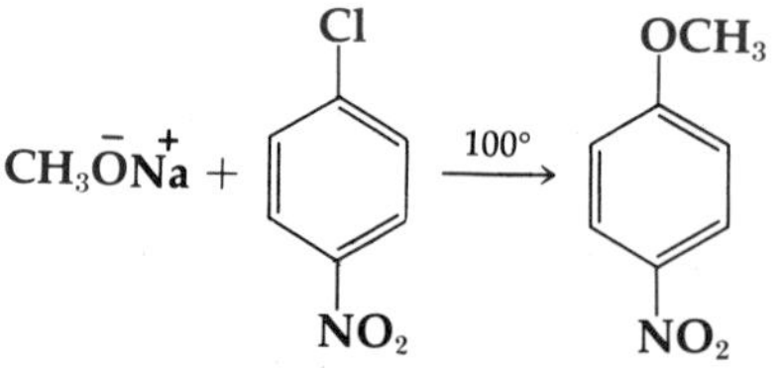

$$CH_3\bar{O}Na^+ + \quad \xrightarrow{100°}$$

p-Nitroanisole

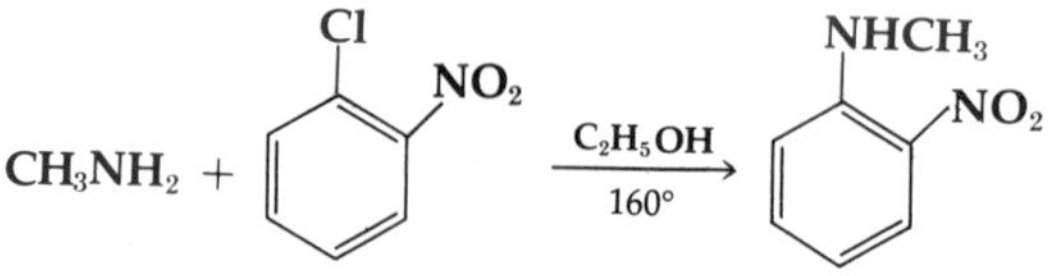

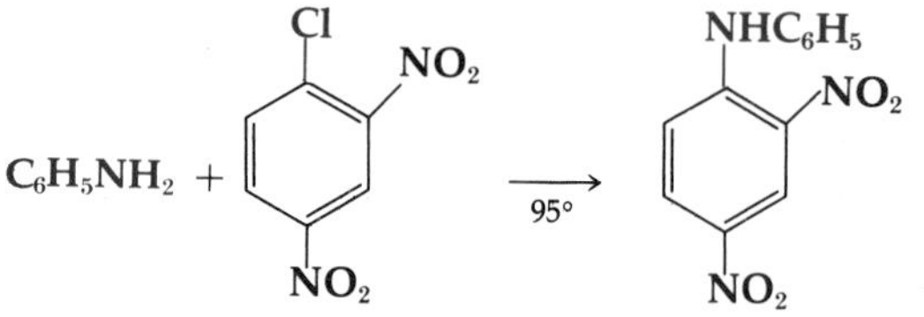

$$CH_3NH_2 + \quad \xrightarrow[160°]{C_2H_5OH}$$

$$C_6H_5NH_2 + \quad \xrightarrow{95°}$$

N-Phenyl-2,4-dinitroaniline
98%

The order of reactivity of halides as leaving groups in aromatic nucleophilic substitution reactions is usually $F \gg Cl > Br > I$. Clearly, the factors that govern reactivity in nucleophilic substitution at *unsaturated* carbon are quite different from those observed for nucleophilic substitution at *saturated* carbon (page 397).† Unlike reactions at saturated centers, polarizability has relatively little influence on the reactivity of nucleophiles and leaving groups in substitution at unsaturated centers.

The superiority of fluoride over the heavier halides as a leaving group has led to the use of 2,4-dinitrofluorobenzene as a specific means for characterization of primary and secondary amines. A particularly valuable application lies in marking the terminal units of proteins and polypeptides (Chap. 25). The peptide is treated with DNFB (dinitrofluorobenzene), and the product is hydrolyzed to cleave the amide linkages which bind the amino acid units together in the peptide. Isolation and identification of the amino acid fragment bearing the dinitrophenyl group indicate the N-terminal unit of the peptide.

---

†If the rate-determining step is formation of the intermediate, rather than its collapse, then of course the "leaving" facility of the halide is inconsequential for its bond is not broken.

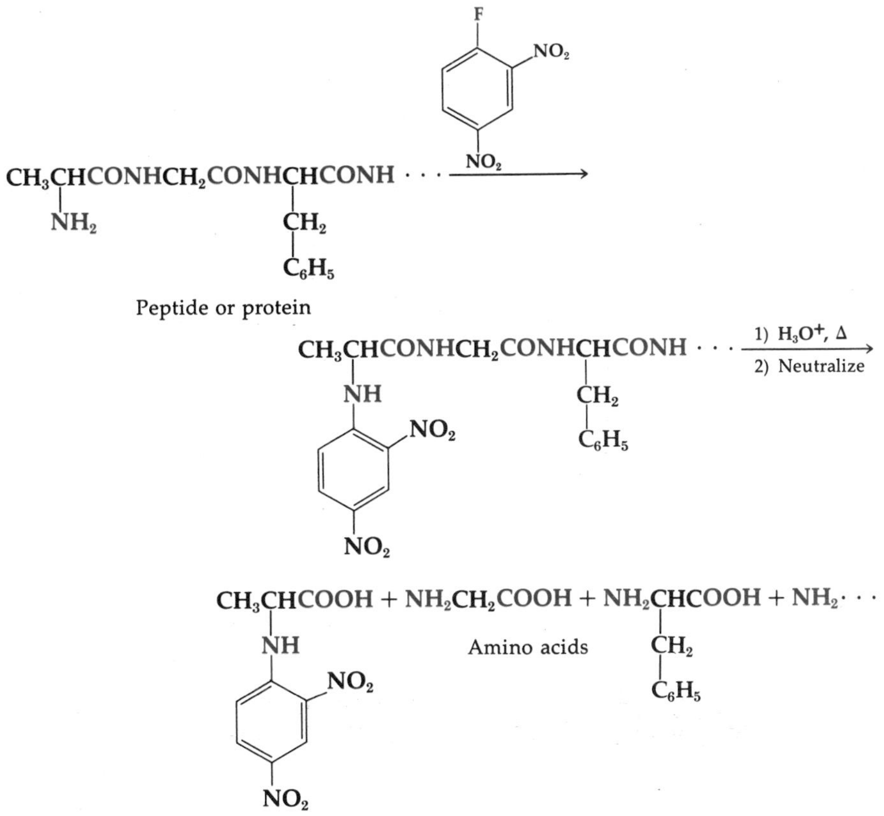

$$CH_3CHCONHCH_2CONHCHCONH \cdots \quad \longrightarrow$$

(with F and two NO_2 groups on benzene ring)

$CH_3\underset{|}{C}HCONHCH_2CONH\underset{|}{C}HCONH \cdots$

$\underset{NH_2}{}$ $\underset{\underset{C_6H_5}{|}}{CH_2}$

Peptide or protein

$$CH_3CHCONHCH_2CONHCHCONH \cdots \xrightarrow[\text{2) Neutralize}]{\text{1) } H_3O^+, \Delta}$$

$\underset{NH}{|}$ $\underset{\underset{C_6H_5}{|}}{CH_2}$

(with NO_2 and NO_2 on benzene ring)

$$CH_3CHCOOH + NH_2CH_2COOH + NH_2CHCOOH + NH_2 \cdots$$

$\underset{NH}{|}$ Amino acids $\underset{\underset{C_6H_5}{|}}{CH_2}$

(with NO_2 and NO_2 on benzene ring)

13-9   OTHER AROMATIC NUCLEOPHILIC SUBSTITUTIONS

In other aromatic cases in which a potential leaving group on the benzene ring is not activated for vinylogous substitution, the attack of nucleophiles is very sluggish as it always is in direct displacement at alkene (sp^2) carbons. Many apparent substitutions in fact proceed by initial elimination to **benzyne** followed by readdition to the transient triple bond, and these reactions are discussed in the next chapter as eliminations.

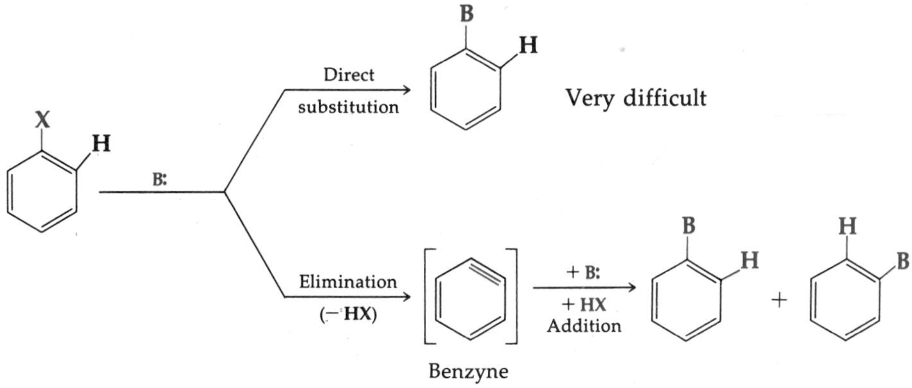

Direct substitution — Very difficult

Elimination ($-HX$) — Benzyne — $+B:$, $+HX$ Addition

Nucleophilic Substitution of Aryl Diazonium Salts

Simple substitution can apparently occur with an exceptional leaving group like the diazonium ion, which irreversibly vacates as stable molecular nitrogen (N_2). The detailed mechanism is not clear here, however, since copper ion catalysis is common and this often implies free-radical mechanisms.

Aryl diazonium salts are prepared at low temperatures from aromatic amines, sodium nitrite, and acids in either aqueous solutions or mixtures of water and organic solvents (Sec. 13-7).

When the associated anion is a nucleophile the salts can be isolated only with great care and are often explosive when dry, but with nonnucleophilic anions like tetrafluoroborate (BF_4^-) they are usually quite stable and easily prepared.

Diazonium salts undergo solvolytic reactions when warmed in hydroxylic solvents, and presumably a very short-lived aryl cation is an intermediate in the reaction ($\phi-N_2^+ \longrightarrow \phi^+ + N_2$). If other nucleophilic anions (either added or from starting salt) are present, the substance formed by coupling of the carbonium ion and the anion arises as by-product.

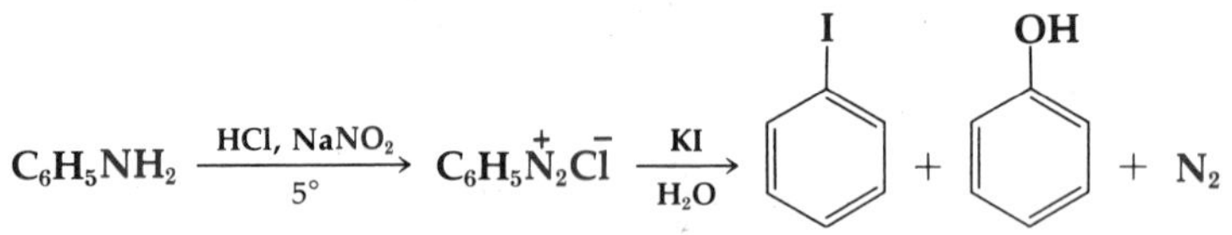

Benzenediazonium
chloride

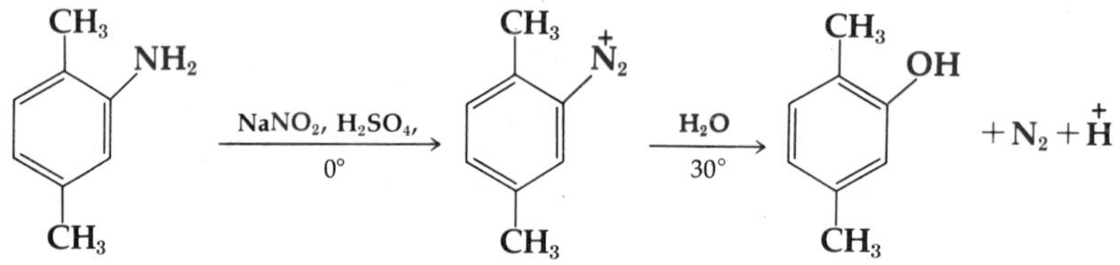

p-Xylidine
(2,5-dimethylaniline)

p-Xylenol
70%

When the solvolysis is carried out in ethanol, both solvolysis and reduction products are formed. Testimony for the reactivity of aryl cations is found in the fact that Ar^+ abstracts a hydride ion from ethanol to give acetaldehyde in that portion of the reaction which gives the reduction product. Better reducing agents for aryl diazonium salts are hypophosphorous acid or sodium borohydride, both of which substitute —H for —N_2^+.

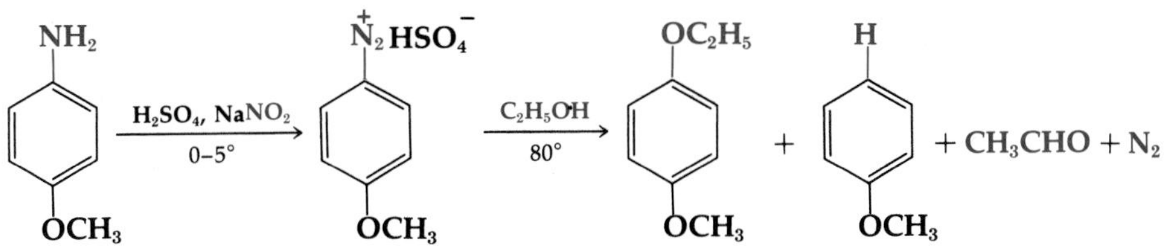

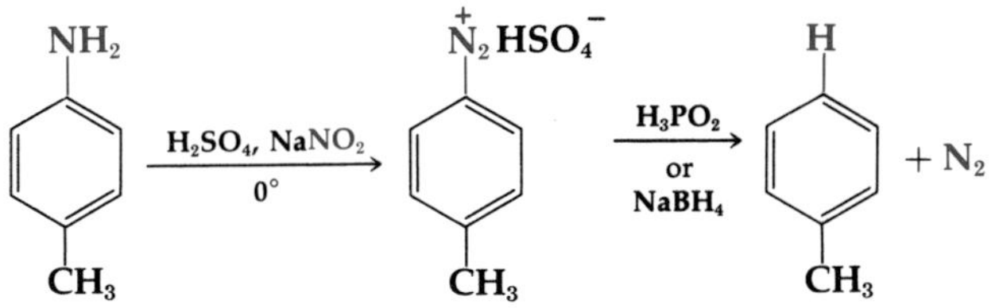

p-Methylaniline   p-Toluenediazonium   Toluene
                   hydrogen sulfate

A number of other reactions involve substitution of nucleophiles for the nitrogen of diazonium salts. As in other substitutions of the same type, the yields depend very much on the exact structure of the starting material and on reaction conditions.

$$C_6H_5NH_2 \xrightarrow{HBF_4, NaNO_2} C_6H_5\overset{+}{N}_2\overset{-}{B}F_4 \xrightarrow[150°]{\overset{\Delta}{\text{Dry salt}}} C_6H_5F + N_2 + BF_3$$

Benzenediazonium   Fluorobenzene
fluoroborate

In the **Sandmeyer reaction**, cuprous salts are employed as reactants in the decomposition of aryl diazonium salts. The anion of the cuprous salt replaces the nitrogen of the diazonium salt at least in a formal sense. Some of the reactions may involve phenyl radicals as intermediates. The transformation is highly versatile and useful in syntheses, since a large variety of cuprous salts are effective.

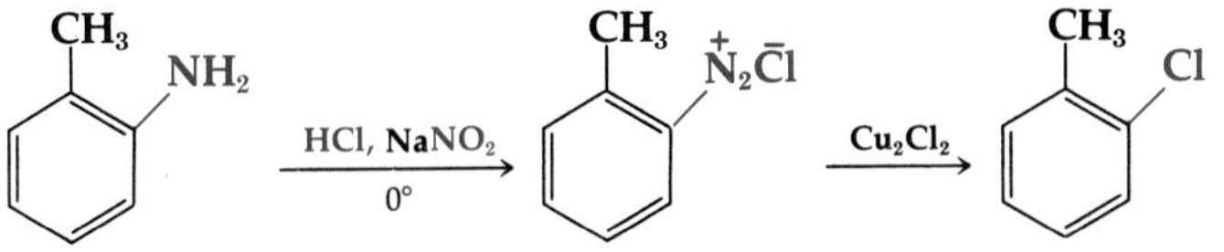

o-Methylaniline   o-Toluenediazonium
                   chloride

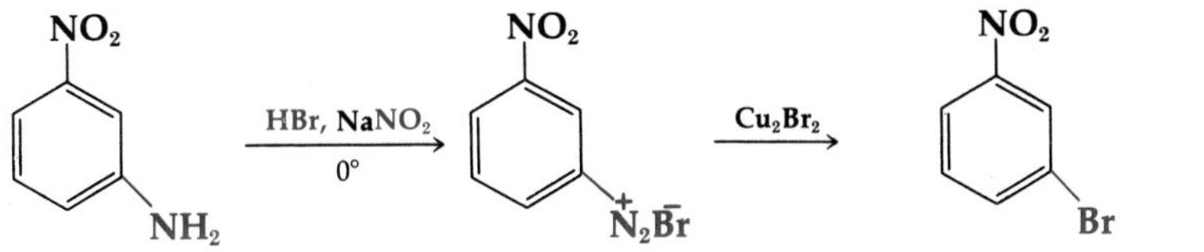

m-Nitroaniline   m-Nitrobenzene-   m-Nitrobromobenzene
                  diazonium bromide

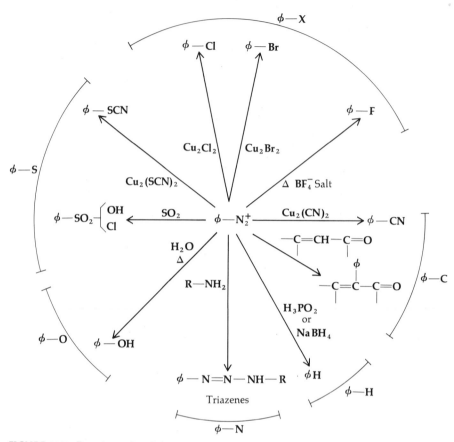

FIGURE 13-7   **Reactions of aryl diazonium salts**

$$C_6H_5\overset{+}{N}_2\bar{C}l \xrightarrow{\text{Cu}_2\text{(SCN)}_2} C_6H_5SCN$$

Benzenediazonium   Phenylthiocyanate
chloride

Diazonium salts add amines to form triazenes and under controlled conditions react with sulfur dioxide to form the ϕ—S bond as in sulfonic acids and sulfonyl chlorides. Formation of a ϕ—C bond is generally a poor reaction unless it is a cyclization. Unsaturated carbonyls react at the α-position with aryldiazonium salts with copper catalysis. A range of reactions of diazonium salts is summarized in Fig. 13-7.

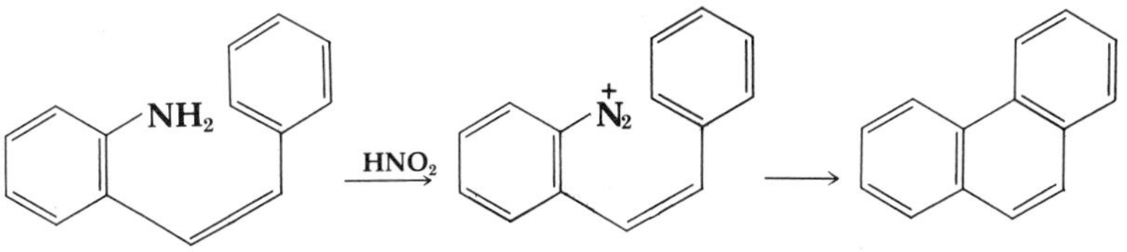

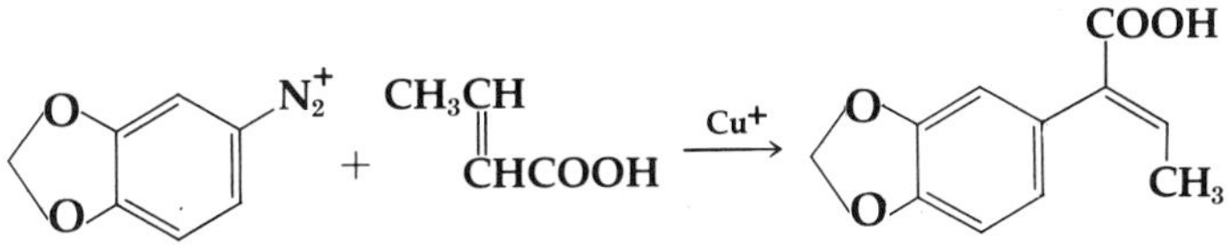

Other nucleophilic substitutions on aromatic rings are few and not very versatile. The major substitution reactions of aromatics by far are electrophilic and are discussed in Chap. 16. Synthetic uses of the nucleophilic substitutions are incorporated there in a discussion of the practical modes of interconversion of aromatic compounds for synthesis.

PROBLEM 13-9

From chlorobenzene as the sole organic starting material with more than one carbon, write out syntheses of the following compounds. Don't make any given intermediate any more than once.

a C_6H_5F

b $C_6H_5S\!-\!\underset{\underset{O}{\|}}{C}\!-\!C_6H_5$

c C_6H_5CHO

d $C_6H_5\!-\!C_6H_5$

e $C_6H_5CH_2NH_2$

f $C_6H_5\underset{\underset{NOH}{\|}}{C}C_6H_5$

g $C_6H_5O\!-\!\underset{\underset{O}{\|}}{C}\!-\!C_6H_5$

h $(C_6H_5)_2\underset{\underset{OH}{|}}{C}\!-\!\underset{\underset{OH}{|}}{C}(C_6H_5)_2$

i $(C_6H_5)_3COH$

j   (cyclohexanone oxime, $=NOH$)

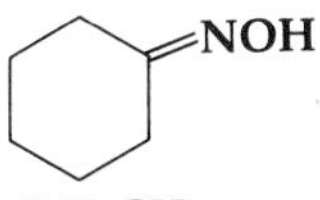

k $C_6H_5CH_3$

l $C_6H_5CH_2CH_2OH$

m $C_6H_5CH\!=\!NC_6H_5$

13-10   NUCLEOPHILIC ADDITIONS AND SUBSTITUTIONS IN SYNTHESIS

Although the reactions of carbonyl compounds are simple in principle, the several variations and combinations that are possible offer a considerable number of possibilities for practical synthetic use. It is appropriate now to shift from an emphasis on theory to one on practice, and to collect the various carbonyl reactions in a manner organized to show their usefulness. For this purpose the alkylations of Chaps. 10 and 11 are incorporated in the present discussion along with carbonyl reactions.

A general discussion of the methods of developing synthetic routes to specific molecular structures is offered in Chap. 23 following the presentation of all the major organic reactions, but it is useful to introduce the subject here since the reactions presented so far form the major collection of methods for creating carbon-carbon bonds and thus building up molecular skeletons. The synthesis of vitamin A in Fig. 23-1 is an excellent example of this.

The reactions used in a synthetic sequence may be divided into reactions which form carbon-carbon bonds and those which interconvert functional groups in preparation for this.

Carbon-Carbon Bond Formation

The reactions explored so far (Chaps. 10 to 13) are all nucleophilic substitutions or additions. Those useful for forming carbon-carbon bonds all involve relatively strong nucleophiles (carbanions) and so are generally reactions under basic conditions. Electrophilic reactions are examined in Chaps. 15 and 16. These occur under acidic conditions wherein the reagents are strong electrophiles, usually aimed at reaction with π electrons. Some of these reactions are merely strong acid variations on the reactions of the present chapters.

In carbon-carbon bond formation it is important to bear in mind that one reactant is always the nucleophile ($R:^-$, enolate, etc.) and the other a substrate, electron acceptor, or electrophile ($C=O$, $R—L$, etc.). Generally in a synthetic plan attention is focused upon a molecular skeleton, to which are sequentially added small, simple units which often have only a few carbons and are commercially available. Hence in forming a new bond and extending the existing skeleton, the reactive site on the skeleton may be either the nucleophile or the substrate for an added nucleophilic reagent.

	Skeleton as nucleophile	(Added unit as substrate)	New skeleton as substrate	(Added unit as nucleophile)

New skeleton

Consideration of the reactions presented so far also makes it clear that reactions on the skeleton can only occur at a functional-group carbon or at a saturated carbon α to carbonyl (as enolate). Other saturated carbons away from functional group sites are not activated for reaction and new bonds cannot usually be formed to such sites. Operations on a given molecular skeleton may then be divided into these reactions, which are summarized for synthetic purposes in the accompanying tables (13-2 to 13-4).

	Skeleton	*Added unit*	*Table*
1	Substrate, or electrophile, at functional group site (C=O or R—L)	Nucleophile (R:⁻)	13-2
2	Nucleophile		
a	At functional group site (RMgX, RLi)	Electrophile (C=O; R—L)	13-3
b	α to carbonyl (enolate, enamine)	Electrophile (C=O; R—L)	13-4

The tables summarize the reactions presented so far which have major synthetic value. They are only general summaries and caution must be exercised in their use, for many reactions are limited by side products (i.e., elimination), steric hindrance, and other factors which are discussed in the text. Only primary halides, sulfonates, and epoxides are listed for aliphatic substitution (S_N2) since secondary sites are less reactive or reliable and tend to eliminate instead. Furthermore, to maintain simplicity in the main forms, only carbonyl acceptors or enols are usually shown, although many reactions are possible using related functional groups (C=N, C≡N, NO_2, etc.) as acceptors or enolate nucleophiles.

These tables, or their alternative presentation in Table 23-2, constitute the basic tools for devising synthetic routes since they summarize the principal modes of carbon-carbon bond formation used to build carbon skeletons.

The main reactions are simply these:

Alkylation: $-\overset{\overset{\displaystyle O}{\|}}{C}-CH- \;+\; R-L \longrightarrow -\overset{\overset{\displaystyle O}{\|}}{C}-\overset{|}{\underset{|}{C}}-R$

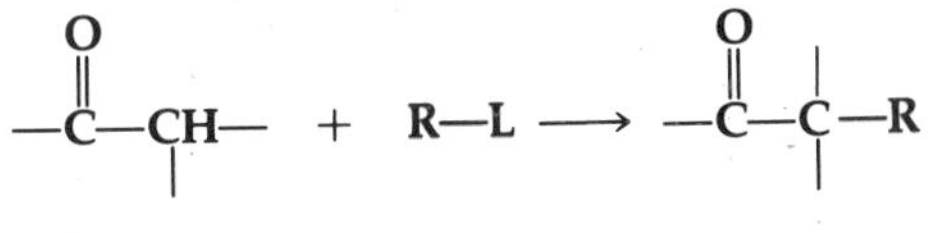

Acylation: $-\overset{\overset{\displaystyle O}{\|}}{C}-CH- \;+\; R-COOR'\,(R-COCl) \longrightarrow -\overset{\overset{\displaystyle O}{\|}}{C}-\overset{|}{\underset{|}{C}}-\overset{\overset{\displaystyle O}{\|}}{C}-R$

Addition: $-\overset{\overset{\displaystyle O}{\|}}{C}- \;+\; RMgX\,(RLi) \longrightarrow -\overset{\overset{\displaystyle HO\quad R}{\diagdown\diagup}}{C}-$

Aldol: $-\overset{\overset{\displaystyle O}{\|}}{C}-CH_2 \;+\; O=C\diagdown \longrightarrow -\overset{\overset{\displaystyle O}{\|}}{C}-\overset{|}{\underset{|}{C}}=C\diagdown$

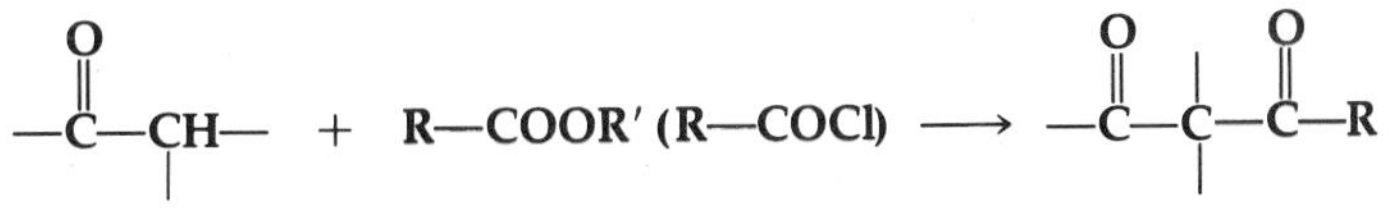

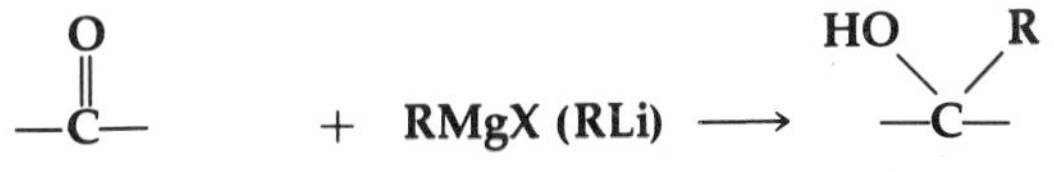

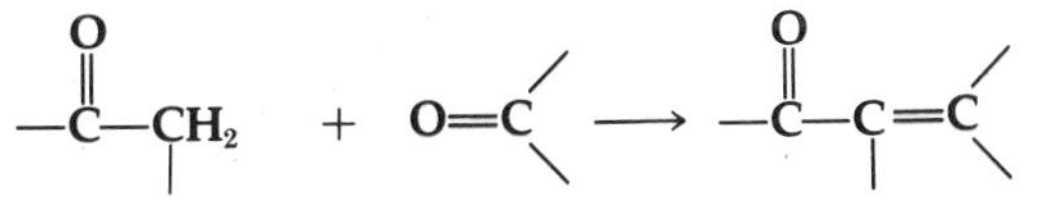

TABLE 13-2 **Carbon-skeleton Extension at Functional-group Site by Nucleophiles**

Initial skeleton	Carbanion nucleophile or enol component	Product and comments	Page
—CHO	$:\overset{-}{C}N$	$-CH{\overset{OH}{\underset{CN}{<}}}$	462
	$:\overset{-}{C}\equiv C-R$	$-\overset{OH}{CH}-C\equiv C-R$	462
	$RMgX$ or RLi	$-\overset{OH}{CH}-R$	464
	$BrZnCH_2COOR$	$-\overset{OH}{CH}-CH_2COOR$	467
	$CH_2{\overset{COCH_3}{\underset{COOR}{<}}}$ ($+ R_2NH + HOAc$)	$-CH=C{\overset{COCH_3}{\underset{COOR}{<}}}$ Knoevenagel	474
	$CH_2{\overset{COOR}{\underset{COOR(CN)}{<}}}$ ($+ R_2NH + HOAc$)	$-CH=C{\overset{COOR}{\underset{COOR(CN)}{<}}}$ Knoevenagel	474
	$CH_2{\overset{CN}{\underset{COOH}{<}}}$ or $CH_2{\overset{COOR}{\underset{COOH}{<}}}$ (pyridine)	$-CH=CH-CN(COOR)$ Doebner	474
	$CH_2{\overset{COCH_3}{\underset{COOH}{<}}}$	$-CH=CH-\overset{O}{\overset{\|}{C}}-CH_3$ Doebner	474
	$-\underset{R}{\overset{..}{C}H}-\overset{+}{P}\phi_3$ (from RCH_2X)	$-CH=CH-R$ Wittig	476

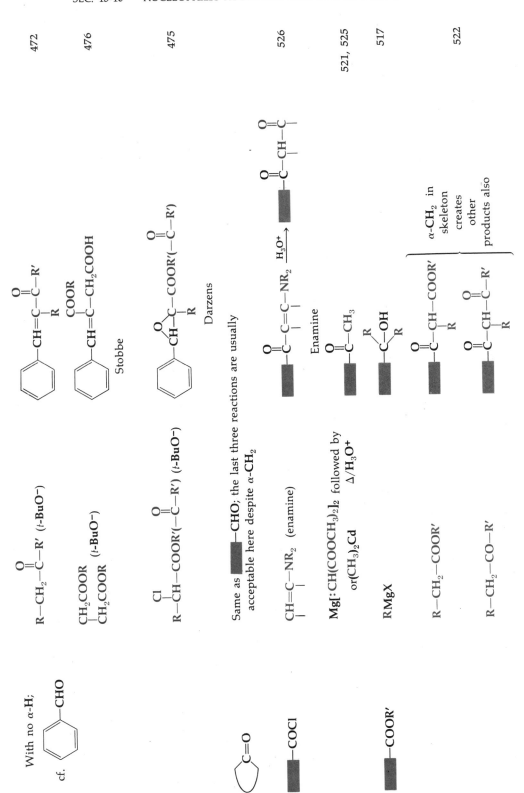

TABLE 13-2 **Carbon-skeleton Extension at Functional-group Site by Nucleophiles** (*Continued*)

Initial skeleton	*Carbanion nucleophile or enol component*	*Product and comments*	*Page*
■—CN	RLi or RMgX	■—C(=O)—R	489
■—CH_2—X (OSO_2R)	$^-$:CN	■—CH_2—CN	428
	$^-$:CH($COCH_3$)(COOR)	■—CH_2—CH($COCH_3$)(COOR) $\xrightarrow[^-OH]{H_3O^+}$ ■—$CH_2CH_2COCH_3$	434
	$^-$:CH(COOR)(COOR)	■—CH_2—CH(COOR)(COOR) $\xrightarrow[\text{or } OH^-]{H_3O^+}$ ■—CH_2CH_2COOH	434
—C=C—C=O	$^-$:CN	—C^3(—CN^4)—CH^2—C^1=O (functions = 1,4 on skeleton)	481
	$^-$:CH(COOR)(COOR) (etc.)	—C(—CH(COOR)(COOR))—CH—C=O $\xrightarrow{H_3O^+}$ —C^3(—CH_2^4—$COOH^5$)—CH^2—C^1=O (functions = 1,5 on skeleton)	481

TABLE 13-3   Carbon-skeleton Extension at Functional-group Site by Electrophiles

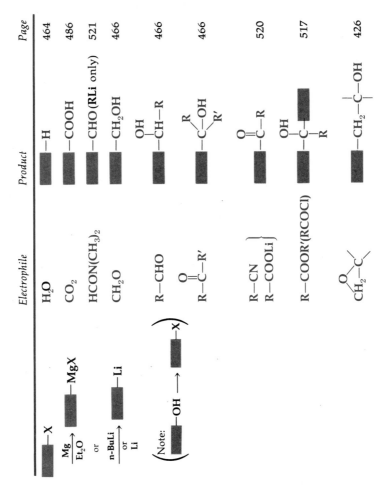

Electrophile	Product	Page
H_2O	—H	464
CO_2	—COOH	486
$HCON(CH_3)_2$	—CHO (**RLi** only)	521
CH_2O	—CH_2OH	466
R—CHO	—CH(OH)—R	466
R—C(=O)—R′	—C(OH)(R)(R′)	466
R—CN, R—COOLi	—C(=O)—R	520
R—COOR′(RCOCl)	—C(OH)(R)(R)	517
CH₂—C (epoxide)	—CH_2—C—OH	426

Reagent formation:

—X $\xrightarrow{\text{Mg, Et}_2\text{O}}$ —MgX   or   —X $\xrightarrow{\text{n-BuLi or Li}}$ —Li

(Note: —OH → —X)

TABLE 13-4 **Skeleton Extension α to Carbonyl**

Initial skeleton	Base	Electrophile (alkylating or acylating agent)	Product and comments	Page
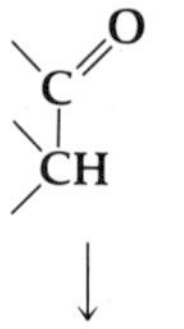 (structure: C=O / CH) → Enamine	t-BuO⁻ ⁻NEt₂ ⁻NH₂ φ₃C⁻	R—CH₂X (OSO₂R′)	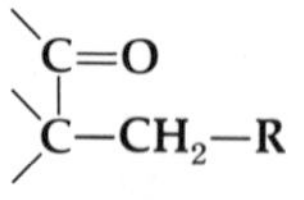 (structure: C=O / C—CH₂—R)	432
		R—COOR′ (no α-CH₂) RCOCl	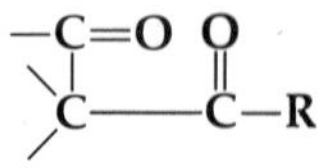 (structure: —C=O O ‖ / C——C—R)	523
—C=O / —CH₂	⁻OH	R—CHO (no α-H) { R=H, φ, C=O }	—C=O / —C=CH—R { R=H, φ, C=O } Also Mannich procedure	472 487
	⁻OR′	R—COOR′ (no α-H) { R=H, φ, OR, COOR }	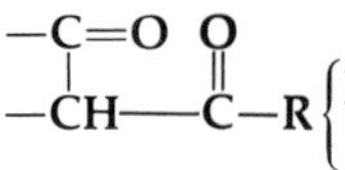 —C=O O ‖ / —CH——C—R { R=H, φ, OR, COOR }	552
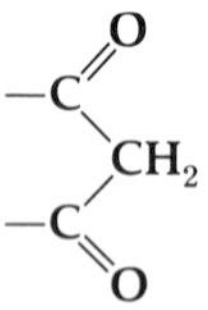 (structure: —C(=O) CH₂ / —C(=O))	R₂NH (HOAc)	R—CHO	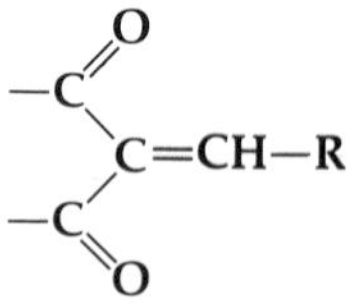 (structure: —C(=O) / C=CH—R / —C(=O))	474
		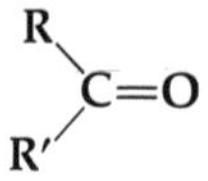 (structure: R / C=O / R′)	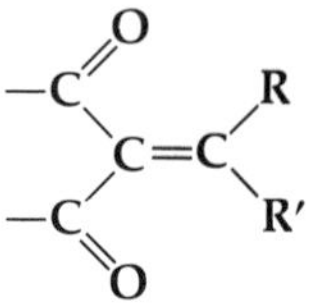 (structure: —C(=O) / C=C with R, R′ / —C(=O))	474
		R—CH=CH—C=O / R′	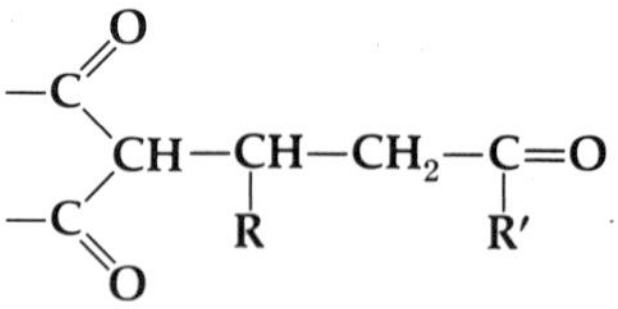 (structure: —C(=O) / CH—CH—CH₂—C=O / —C(=O), R, R′)	481
	NaH ⁻OR	R—CH₂X (OSO₂R′)	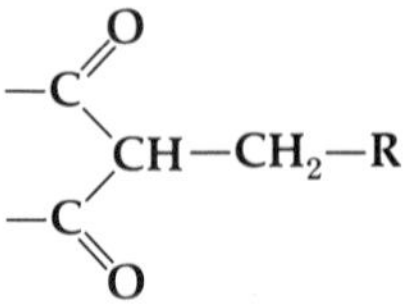 (structure: —C(=O) / CH—CH₂—R / —C(=O))	433
	NaH ⁻OR	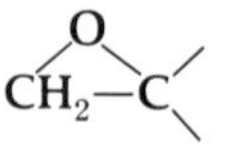 (structure: O / CH₂—C)	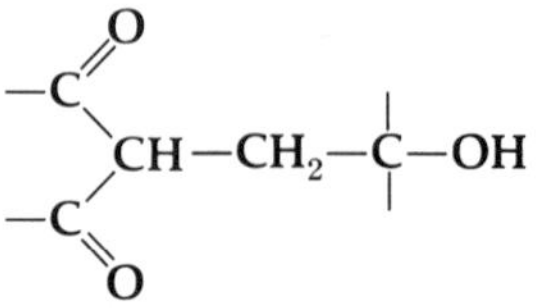 (structure: —C(=O) / CH—CH₂—C—OH / —C(=O))	426

Of special interest among the small units (synthons) used to extend a carbon skeleton are those which are bifunctional and thus allow further chain extension from themselves·after they are added to the skeleton. These include the Reformatsky, Stobbe, and Darzens condensations and bifunctional Wittig and Grignard reagents, discussed below, as nucleophiles as well as electrophiles like ethylene oxide or halides (tosylates) with functions elsewhere in the molecule. Ethylene oxide yields lactones with malonate esters and similar activated ester enolates, owing to a second step of internal ester exchange.

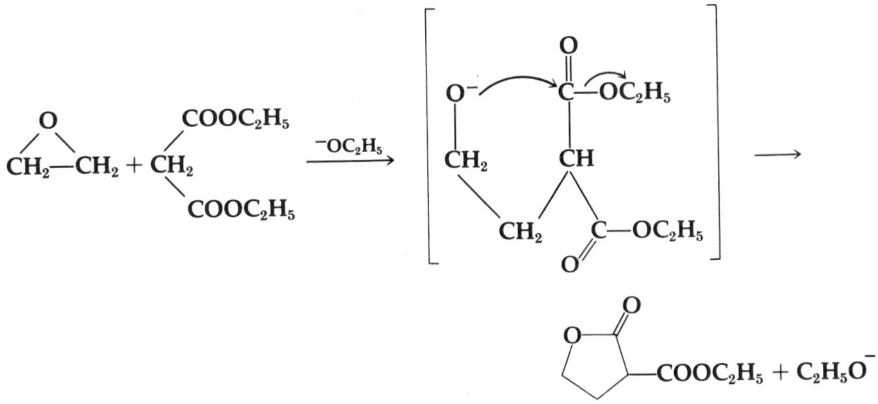

Interconversions of Functional Groups

After any skeleton extension (C—C bond formation) the remaining functional groups must be transformed into those required for the next skeleton extension, or at the end into those required in the final product. The important interconversions are those of carbonyls to alcohols by hydrides (reduction) and the reverse oxidations, which are presented in Chap. 18. These procedures allow easy access to any of these functions from any other as listed in Fig. 13-8. Since alcohols result from addition (e.g., Grignard), these conversions are important as are the conversions of alcohols to sulfonate esters for S_N2 displacement or to halides for Grignard or organolithium formation.

Formation of Grignard (and organolithium) reagents is not compatible with most functional groups (page 465). Ethers, tertiary amines, and carbon-carbon multiple bonds are the main exceptions. In a synthetic scheme employing a Grignard reagent and requiring another function in that molecule for later use, it is occasionally useful to protect the other function, as shown here for ketones (aldehydes) and halides. Wittig reagents (like Reformatsky) can be prepared with ester groups elsewhere in the molecule.

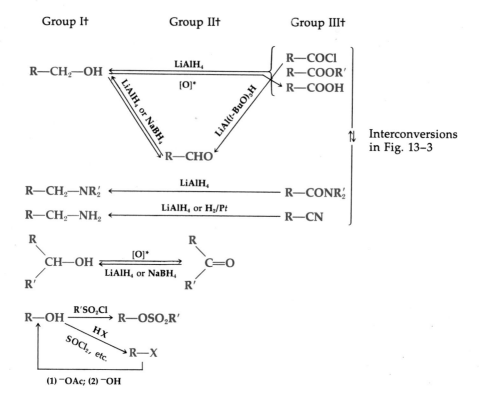

*Specific oxidation procedures [O] are presented in Chap. 19.
†See Table 3-2.

FIGURE 13-8   Important interconversions of functional groups for synthesis

Finally, some functional groups which are needed for directing and orienting formation of the carbon skeleton do not appear in the final product. Accordingly, it is important to have ways of reducing such functional groups to saturated hydrocarbon. The three methods we have encountered so far are these:

$$R\text{—}OH \xrightarrow{R'SO_2Cl} R\text{—}OSO_2R' \xrightarrow{LiAlH_4} RH$$

$$R\text{—}X \xrightarrow[Et_2O]{Mg} R\text{—}MgX \xrightarrow{H_3O^+} RH$$

$$\left.\begin{array}{c} \diagdown\!\!\diagup \\ C{=}C \\ \diagup\;\;\;\diagdown \\ \\ \text{—}C{\equiv}C\text{—} \end{array}\right\} \xrightarrow[\text{Catalyst (Pd or Pt)}]{H_2} \begin{array}{c} \diagdown\;\;\;\diagup \\ CH\text{—}CH \\ \diagup\;\;\;\diagdown \\ \\ \text{—}CH_2\text{—}CH_2\text{—} \end{array}$$

Selectivity

In polyfunctional compounds it is clear that reaction conditions chosen to effect a change in one function should leave the other function(s) untouched. In many cases this selectivity control can be subtle and difficult, but a few examples of broad value out of the present reactions will be helpful in synthesis design.

1 *Hydride reduction:* **NaBH₄** generally reduces only ketones, aldehydes, and acid chlorides, while **LiAlH₄** reduces these as well as all carboxylic acid derivatives and nitro groups.

2 *Catalytic hydrogenation:* carbon-carbon double and triple bonds and nitriles can be hydrogenated without affecting aromatic rings or carbonyls. Triple bonds may be hydrogenated to double bonds with "poisoned" catalysts.

3 *Carbon sites:* the reactivity differences in primary, secondary, and tertiary groups which are discussed in Chap. 10 are adequate to allow selectivity in most cases. Tertiary alcohols are not esterified (acylated) under normal conditions. ′

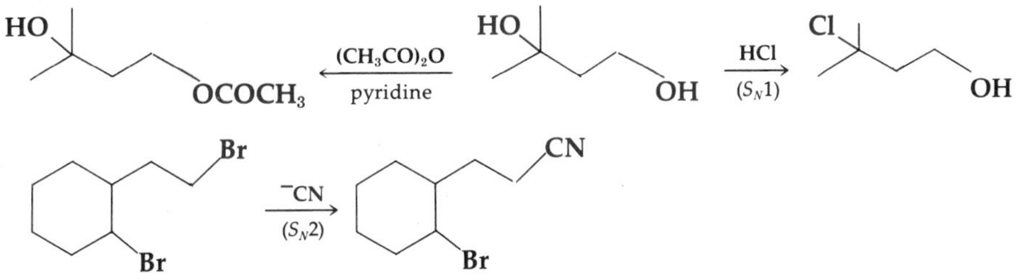

4 *Cyclic reactions* to form five- and six-membered rings take precedence over their inter- molecular counterparts.

5   *Carboxyl derivatives:* the order of reactivity (Table 13-1) allows selectivity in substitutions generally among each group (e.g., page 508), except with organometallics and lithium aluminum hydride.

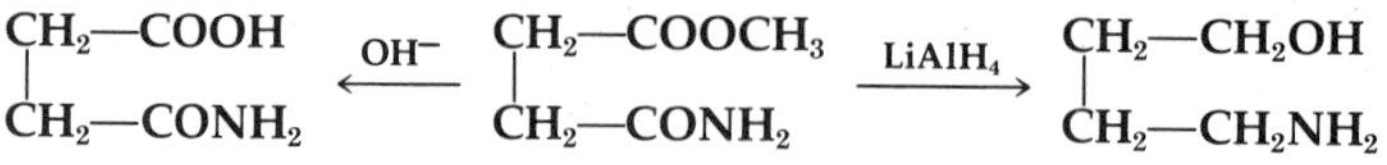

$$\begin{array}{c} CH_2-COOH \\ | \\ CH_2-CONH_2 \end{array} \xleftarrow{OH^-} \begin{array}{c} CH_2-COOCH_3 \\ | \\ CH_2-CONH_2 \end{array} \xrightarrow{LiAlH_4} \begin{array}{c} CH_2-CH_2OH \\ | \\ CH_2-CH_2NH_2 \end{array}$$

Ketones usually present two sides (α-positions) capable of enolization and selectivity here can be complex, but a few simple rules and devices arise from carbonyl chemistry.

1   Aldol condensations with dehydration and reversible acylations (ester condensations) can occur only at *methylene (or methyl).*

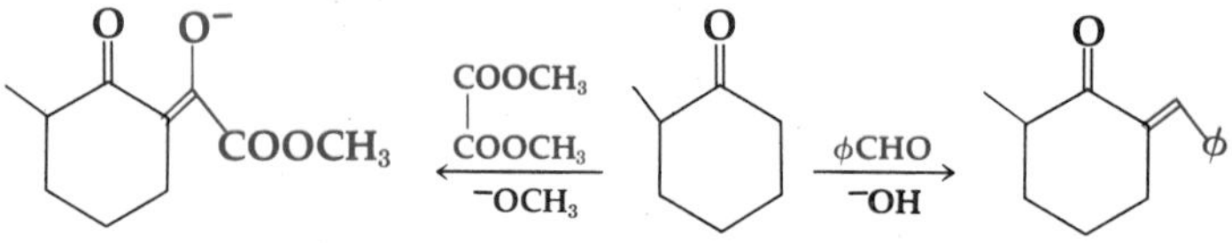

2   The more acidic methylene (more stabilized carbanion) is preferred for enolization in base if a choice exists.

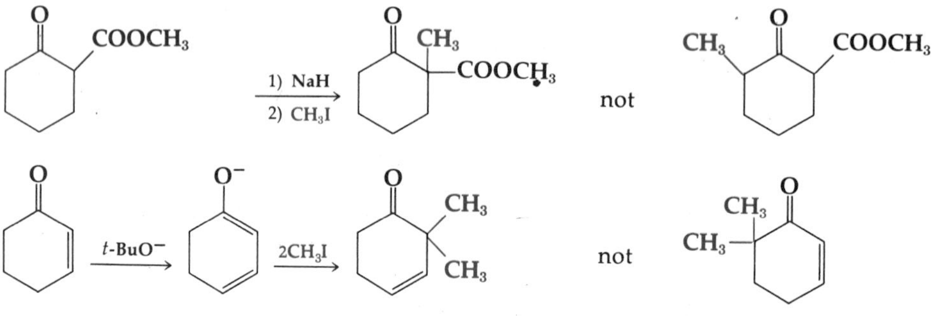

3   Alkylation of simple enolates rarely stops with the monoalkylated product. Monoalkylation can usually be achieved only with β-dicarbonyl enolates.

It is possible to take advantage of these rules in two ways. To achieve monoalkylation on the side which preferentially acylates, the ketone is acylated first. The resultant β-dicarbonyl enolate is then alkylated and the added acyl group removed by alkaline cleavage (Sec. 13-5). Formate and oxalate esters are preferred for this acylation since they are very reactive for cleavage afterwards. To achieve alkylation on the less preferred side, the acylation is carried out as before on the preferred side and then rendered unreactive to alkylation so that only the other side of the ketone can be alkylated. Here the added acyl groups acts as a **blocking group.**

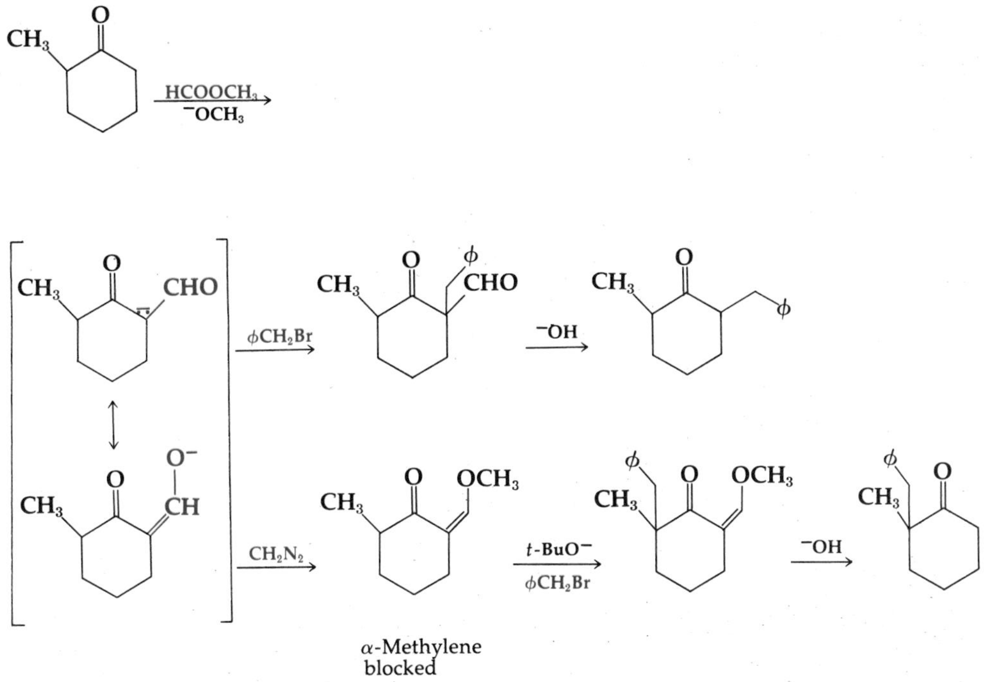

α-Methylene
blocked

Synthesis Design

There is no simple set of rules for devising a synthesis of a given molecule nor is any single synthesis necessarily the only or the best one. Nevertheless, there are certain criteria which guide the design process, and these are discussed in more detail in Chap. 23. However, in the main a good synthesis will be as short as possible, will result in respectable yields of the desired product, and will consist of reliable reactions with a clear and single outcome.

In general the structure of the synthetic goal is first seen as a carbon skeleton and dissected to see which carbon-carbon bonds should be created in a sequence from smaller units. The units used in the synthesis are called **synthons**, and their functionality is deduced from the nature of the bond-forming reactions they are required to undergo. When two reacting molecules contain functional groups other than those required for the reaction, it is important that these others are not affected by the reaction conditions (or caused to cyclize by reaction intermediates—cf. Stobbe condensation, page 476).

In dissecting the final skeleton it is useful to focus upon chain branching sites (and rings); these will be sites for C—C bond formation since simple starting materials and synthons are usually linear. It is then necessary to select reactions from Tables 13-2, 13-3, and 13-4, which create the necessary branching and yield products with useful functions for the next step. These steps are then taken in sequence in an overall plan. It is also necessary, of course, to develop functional groups in the several steps in such a way that they are convertible at the end to the functionality required in the product. The dissection of the desired product in terms of its branching and functional groups allows design of the synthesis by working backward from the product.

There are a great number of commercially available starting materials or synthons, discussed in Chap. 23, but for our trial purposes here the reagents gathered in Table 13-5 are common and should generally suffice.

While it is true that several syntheses of a single product may often be about equally good (at least on paper), certain functional-group collections should be recognized as implying synthesis by particular reactions:

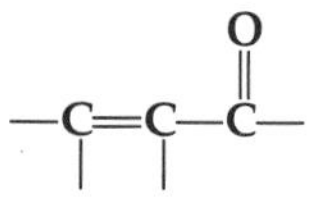

Unsaturated carbonyls by aldol condensations

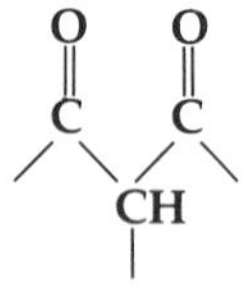

β-Dicarbonyl compounds from ester condensation (C-acylation)

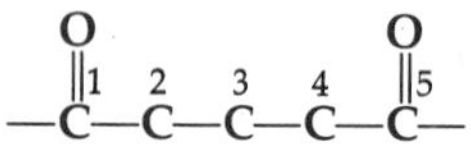

1,5-Dicarbonyl compounds by Michael addition

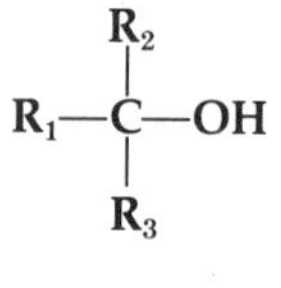

Tertiary alcohols from ketones by organometallics
$(R_2 \neq R_3)$
from esters by organometallics
$(R_2 = R_3)$

As an example, the ketone below ("Goal") is recognized as an aldol product of 2-methyl-3-pentanone and benzaldehyde:

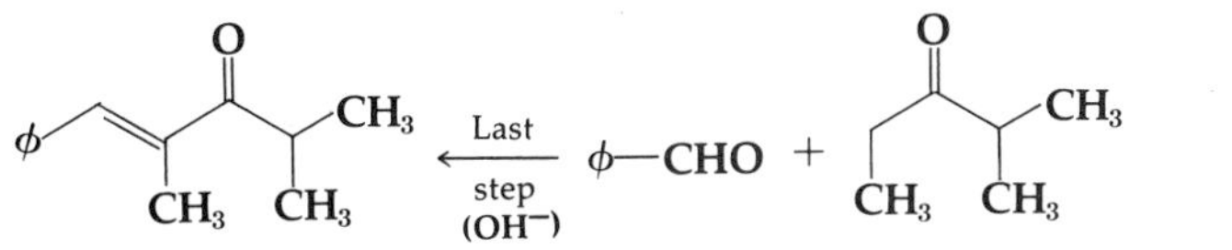

This will be the last step in the synthesis. Benzaldehyde is an acceptable and available synthon, and the reaction shows no problems of selectivity or side reactions (no α-H in ϕCHO, no choice of α-methylene in ketone). Synthesis of the 2-methyl-3-pentanone may be considered next and should be formed by creation of one of the bonds to the carbonyl, i.e., by Grignard attack on nitrile, or on aldehyde followed by oxidation of the secondary alcohol. Any of the necessary starting materials are now included in Table 13-5.

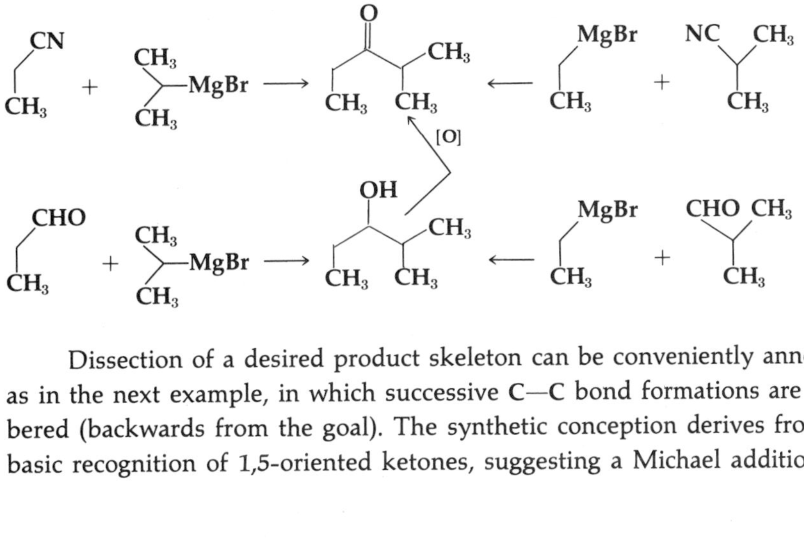

Dissection of a desired product skeleton can be conveniently annotated as in the next example, in which successive C—C bond formations are numbered (backwards from the goal). The synthetic conception derives from the basic recognition of 1,5-oriented ketones, suggesting a Michael addition.

TABLE 13-5   **Some Standard Synthetic Starting Materials (R = H, CH$_3$, C$_2$H$_5$)**

Monofunctional: Any stable compound with four carbons or less; also
             cyclohexanone and cyclopentanone

Bifunctional:	CH_3COCH_2COOR	$\phi-COOR$	XCH_2COOR
	$CH_2(COOR)_2$	$\phi-CHO$	
	$CH_3COCH_2COCH_3$	$\phi-Br$	CHOCOOR
	$CNCH_2COOR$	$\overset{\mid}{\underset{\mid}{COOR}}_{COOR}$	$CH_2=CHCOCH_3$
	$ROOC(CH_2)_{2-4}COOR$		$CH_2=CHCN$

Dissection.

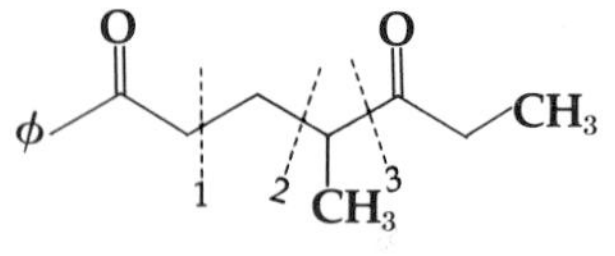
O
O
ϕ
CH3
1
2
3
CH3

Goal

Derived synthesis:

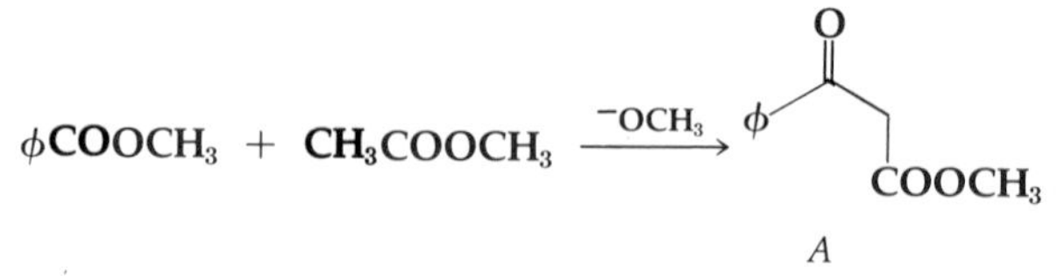
ϕCOOCH3 + CH3COOCH3
⁻OCH3
ϕ
O
COOCH3
A

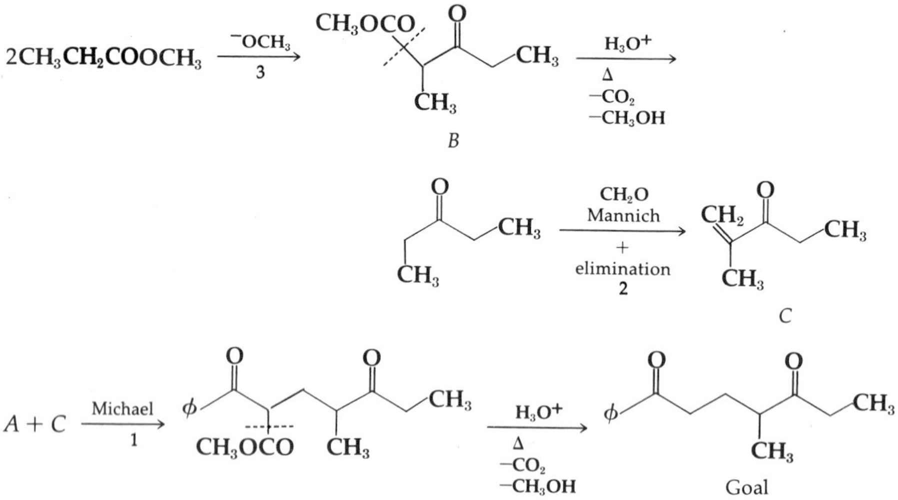
2CH3CH2COOCH3
⁻OCH3
3
CH3OCO
O
CH3
CH3
B
H3O+
Δ
−CO2
−CH3OH
O
CH3
CH3
CH2O
Mannich
+
elimination
2
CH2
O
CH3
CH3
C
A + C
Michael
1
ϕ
O
O
CH3
CH3OCO
CH3
H3O+
Δ
−CO2
−CH3OH
ϕ
O
O
CH3
CH3
Goal

Other syntheses of each example are also valid. In the first case, ester condensation of ethyl propionate and ethyl isobutyrate followed by acidic hydrolysis-decarboxylation of the β-ketoester would afford 2-methyl-3-pentanone. In the second case the Michael addition could proceed in reverse, using $\phi COCH{=}CH_2$ (from a Mannich reaction) as a substrate for Michael addition and β-ketoester *B* as the nucleophile. It is common that a number of synthetic routes to a given molecule are possible.

In cases with no functionality near branched sites which need to be formed, a functional group must be employed and later removed. The following example is offered as an exercise, one dissection being shown; but other reasonable schemes may also be advanced. As a further exercise one may place a methyl group anywhere on the skeleton and see whether the scheme is still effective or needs serious modification.

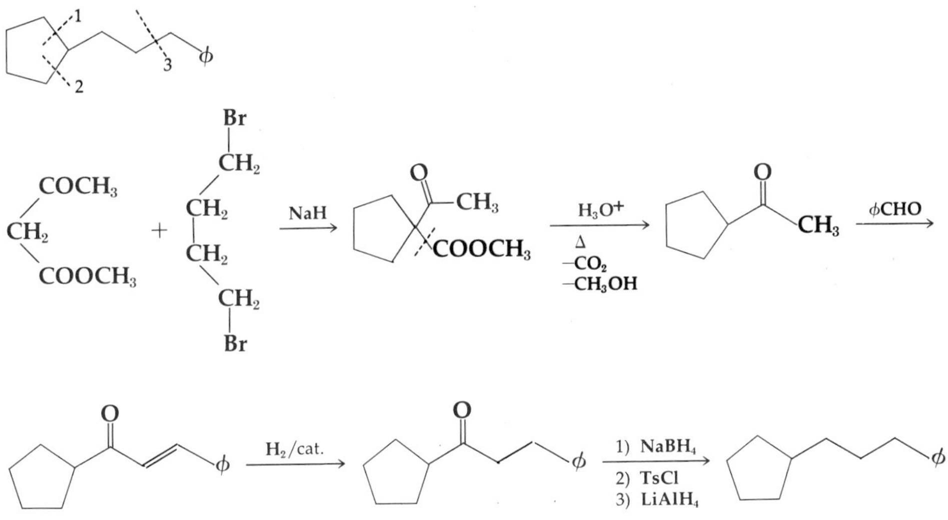

13-11 SUMMARY OF CARBONYL REACTIONS

Carbonyl compounds in general may either enolize or undergo nucleophilic addition. Either reaction may occur under acidic or basic conditions and, unless the nucleophiles are very active (high-energy) like hydrides or organometallic carbanions, these reactions are freely reversible. They are also usually fast. In these addition equilibria, with heteroatom nucleophiles or enolates (stabilized carbanions as nucleophiles), the most stable product is therefore formed, as it is in enolization equilibria which cause epimerization at an asymmetric α-carbon.

If the original carbonyl is not bonded to a leaving group, i.e., in aldehydes and ketones, and the nucleophile offers no further electron pair, then the initial addition product is stable, but another electron pair on the nucleophile can cause the addition product to eliminate water (from the original carbonyl oxygen, protonated). The third possibility occurs when the original carbonyl is bonded to a leaving group (carboxylic acid derivatives), which is spontaneously eliminated from the addition product—causing an overall (two-step) nucleophilic substitution. These various main reactions are summarized in Fig. 13-9.

We have now seen the three major kinds of reactions which involve attack of nucleophiles (bases). These may attack protons (Chap. 8), saturated carbon (Chaps. 10 and 11), and unsaturated carbon of the carbonyl family (Chaps. 12 and 13). These three possible modes of nucleophile/base attack may usefully be compared with each other as in Fig. 13-10. This summary shows carbonyl reactions to be intermediate between proton removal and substitution at saturated carbon (S_N) in facility of attack by nucleophiles/bases.

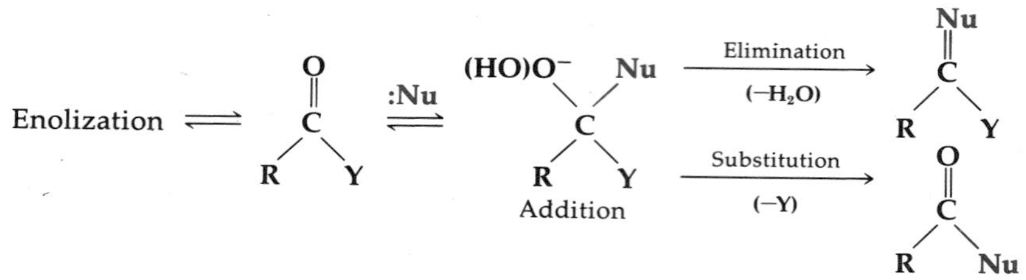

Enolization $\rightleftharpoons$

$$\underset{\substack{R}}{\overset{O}{\|}}C-Y \quad \overset{:Nu}{\rightleftharpoons} \quad (HO)O^- -\!\!\!\underset{\substack{| \\ R}}{\overset{Nu}{|}}C-Y \quad \underset{\text{Addition}}{}$$

Addition

$$\xrightarrow[\text{(}-H_2O\text{)}]{\text{Elimination}} \quad \underset{R}{Nu}C=\!\!\!C$$

$$\xrightarrow[\text{(}-Y\text{)}]{\text{Substitution}} \quad \underset{\substack{R}}{\overset{O}{\|}}C-Nu$$

Addition:*

Substrate	Nucleophile (Nu:)	Product	
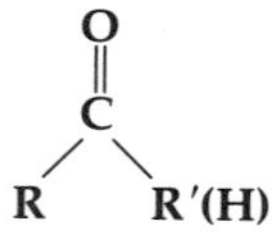 Aldehydes Ketones $\underset{\substack{R}}{\overset{O}{\|}}C-R'(H)$	R:⁻ (Grignard, organolithium) (⁻:CN, ⁻:C≡C—R)	$HO-\underset{\substack{| \\ R}}{\overset{R}{|}}C-R'(H)$	Alcohols (sec + tert)
	H:⁻ (NaBH₄, LiAlH₄)	$HO-\underset{\substack{| \\ R}}{\overset{H}{|}}C-R'(H)$	Alcohols (pri + sec)
	RÖH (H⁺)	$RO-\underset{\substack{| \\ R}}{\overset{OR}{|}}C-R'(H)$	Acetals, ketals

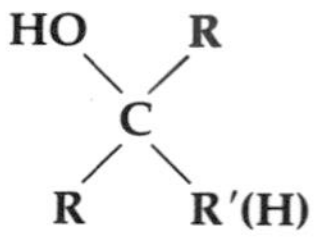

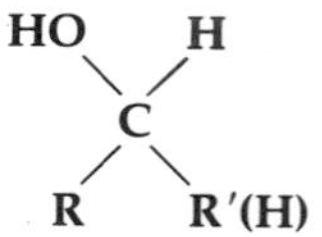

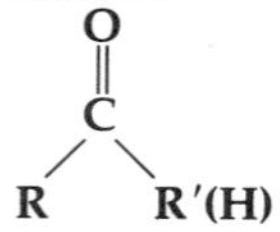

Addition-elimination:

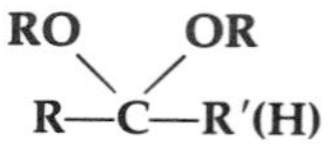

Aldehydes Ketones $\underset{\substack{R}}{\overset{O}{\|}}C-R'(H)$	:NH₂—Z (H⁺)	$\underset{\substack{R}}{\overset{N-Z}{\|}}C-R'(H)$	Imino derivatives
	$\underset{\substack{|}}{\overset{|}{}}CH_2-CO-$ (acid or base)†	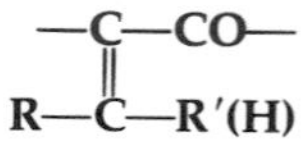 $\underset{\substack{R}}{\overset{-C-CO-}{}}\!\!C-R'(H)$	Aldol products
	⁻:CH—Pϕ₃ (Wittig)†	$\underset{\substack{R}}{\overset{CH-}{\|}}C-R'(H)$	Alkenes

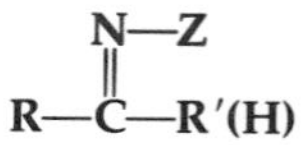

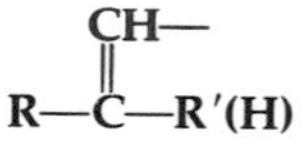

Substitution (acylation).*

$$\underset{R-C-L}{\overset{O}{\|}}$$ $\quad$ L = Cl > OR' $\quad$ R:⁻ (Grignard, organolithium) $\quad\longrightarrow\quad$ $\underset{R}{\overset{R}{R-C-OH}}$ $\quad$ Alcohols (tert)

Carboxylic acid derivatives

$\qquad\qquad$ H:⁻ (LiAlH₄) $\quad\longrightarrow\quad$ R—CH₂—OH $\quad$ Alcohols (pri)

$\qquad$ L = NR'₂ $\qquad$ H:⁻ (LiAlH₄) $\quad\longrightarrow\quad$ R—CH₂—NR'₂ $\quad$ Amines

$\qquad$ L = Cl > OR' > NR'₂ $\qquad$ RÖH, H₂Ö, R₂N̈H (acid or base) $\quad\longrightarrow\quad$ $\underset{R-C-Nu}{\overset{O}{\|}}$ $\quad$ Carboxylic acid derivatives

$\qquad$ L = Cl > OR' $\qquad$ $\underset{|}{CH_2-CO-}$ (+ base, ⁻OR')† (or enamine) $\quad\longrightarrow\quad$ $\underset{R-C-CH-CO-}{\overset{O}{\|}}$ $\quad$ β-Dicarbonyl compounds

*Both addition and substitution can occur vinylogously also:

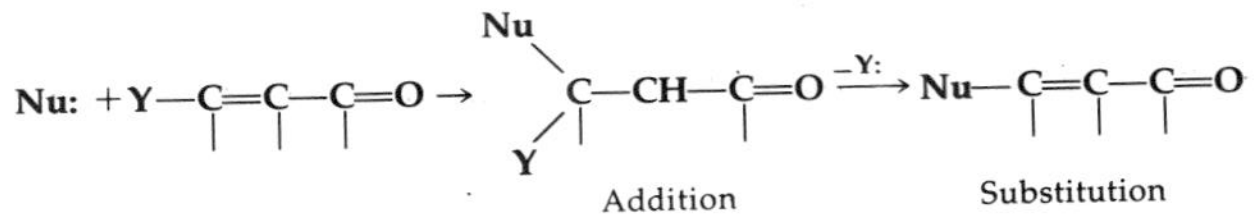

Nu: + Y—C=C—C=O ⟶ [C—CH—C=O with Nu and Y] —Y:⁻⟶ Nu—C=C—C=O

$\qquad\qquad\qquad\qquad$ Addition $\qquad\qquad\qquad\qquad$ Substitution

†Other groups (Fig. 12-3) can also stabilize carbanions in an analogous manner.

FIGURE 13-9 **Summary of major carbonyl reactions**

	Initial reaction	Rate	Steric hindrance	No. of attached atoms at site	Side reaction
Y: + H—Z	Proton transfer (acid-base)	Fast	Low	1	—
Y: + C=O	Addition	Medium	Medium	3	Enolization
Y: + C—L	Substitution (S_N1 or S_N2)	Slow	High	4	Elimination

FIGURE 13-10 **Comparative reactions of bases/nucleophiles**

The basic modes of carbonyl reactivity—enolization and addition—may be extended in two ways. A conjugated double bond extends the electron-deficiency of the carbonyl carbon to the β-carbon of the double bond by resonance. This extension allows the opportunity for vinylogous addition and, with a β-leaving group, vinylogous substitution. These reactions may be seen as exactly parallel to simple carbonyl reactions by regarding the entire unsaturated carbonyl system as if it were a simple carbonyl. Vinylogous enolization to the γ-carbanion is also completely analogous.

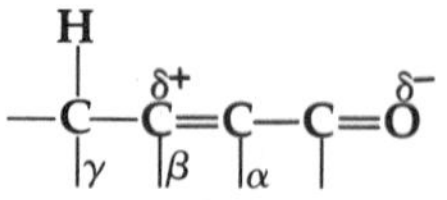

Vinylogous carbonyl

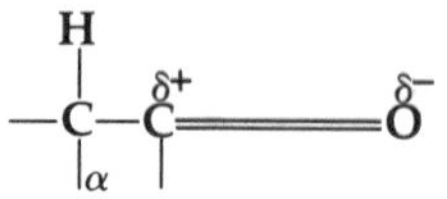

Simple carbonyl

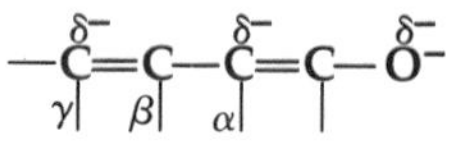

Vinylogous enolate

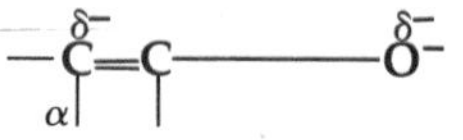

Simple enolate

Vinylogous reactions are generally less rapid but also less freely reversible than simple carbonyl reactions since the β-carbon is not so electrophilic as the carbonyl carbon.

The other extension of carbonyl reactivity involves other multiply bonded groups with a charge imbalance (—Y=Z) which behave analogously to the carbonyl group for enolization and addition (Fig. 12-3) as well as for vinylogous reactions. With enolization and vinylogous reactions these related groups are very closely parallel to carbonyl in mode and reactivity, but addition reactions sometimes take different courses owing to special features in the related functional group (Secs. 12-8 and 13-7). The functional groups with four heteroatoms to carbon—carbon dioxide and carbonic acid derivatives—usually possess carbonyl groups and exhibit reactivity entirely in accord with expectation (Secs. 12-7 and 13-6).

The entire schema of carbonyl reactions is of vast importance in synthesis and therefore the reactions are reorganized for emphasis on synthetic utility in Sec. 13-10. Most of the reactions considered so far invoke strongly basic nucleophilic reagents and the major functional groups which are inert to these reagents are isolated double bonds and aromatic rings, ethers, and tertiary amines. As π electrons are weak nucleophiles themselves, reagents to attack these functions must be strongly electrophilic and hence strongly acidic in most cases. These reactions are considered in Chaps. 15 to 17 following a chapter (14) on elimination, which serves to create double bonds, usually employing the nucleophilic reagents already familiar.

PROBLEMS

13-10 Organic syntheses are often described as proceeding from "coal, air, and water." The phrase generally implies starting with the simplest, primitive inorganic materials, or even the elements themselves, and synthesizing complex organic molecules by a sequence of reactions. Thus the first examples illustrate some of these steps, and the later ones are more complex molecules that can be prepared by continuing the process. Any product once made serves as a starting material if required. Assume NH_3 and any other inorganic reagents.

a $CO_2 + PCl_5 \longrightarrow COCl_2$

b $Ca + 2C \longrightarrow$

   $CaC_2 \xrightarrow{H_2O} HC\equiv CH$

c $HC\equiv CH + H_2 \longrightarrow$

   $CH_2\!\!=\!\!CH_2 \xrightarrow{HOCl} HOCH_2CH_2Cl$

d CH_3OH

e CH_3COOCH_3

f $ClCOOCH_3$

g $HCOOCH_3$

h CH_3CHO

i $CH_2\!\!-\!\!CH_2$

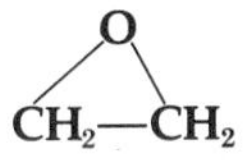

j HCN

k CH_3CN

l C_2H_5OH (four ways from **e, h, i, k**)

m Acetone

n Isopropyl iodide

o Diethyl malonate

p t-Butanol

q Ethyl 2-isopropylacetoacetate

r 3-Ethyl-3-pentanol

s Acrylonitrile

t Succinonitrile

u γ-Butyrolactone

v Methyl vinyl ketone

w 3-Methyl-2-cyclohexenone

x Ethyl 2-methyl-3-ketopentanoate

y 1,3,3-Trimethyl-cyclohexanol

z

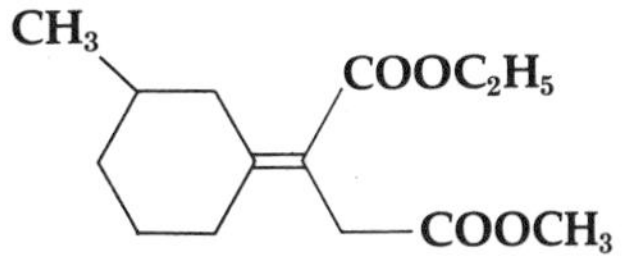

13-11 Account for these observations.

a Dipropylmaleic anhydride dissolves in alkali and is reprecipitated on acidification.

b γ-Butyrolactone is unaffected by heating with acidic methanol.

c On boiling in acidic methanol, levulinic acid (4-ketopentanoic acid) forms two isomeric neutral compounds readily distinguished by their infrared spectra.

d Very little α-ketoacid is formed by adding cyanide ion to an acid chloride and then hydrolyzing the product.

13-12 Acetone and aqueous sodium cyanide yield a neutral compound $C_7H_{12}O_3$ with IR peaks at 2.8 and 5.7 μ (about 3600 and 1760 cm^{-1}) and four close NMR singlets in an intensity ratio of 2:3:3:3. What is the structure?

13-13 Diketene is a useful reagent for introducing acetoacetyl groups. Account for this typical reaction in mechanistic terms.

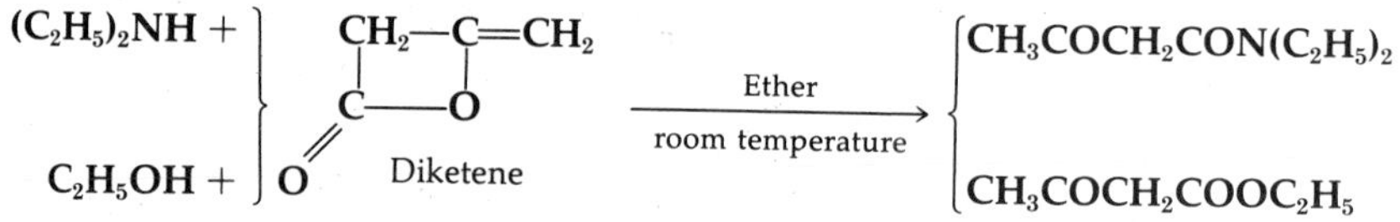

The reaction has been put to more extended synthetic use as in the reactions shown below. Write out a detailed stepwise mechanism for each case.

a 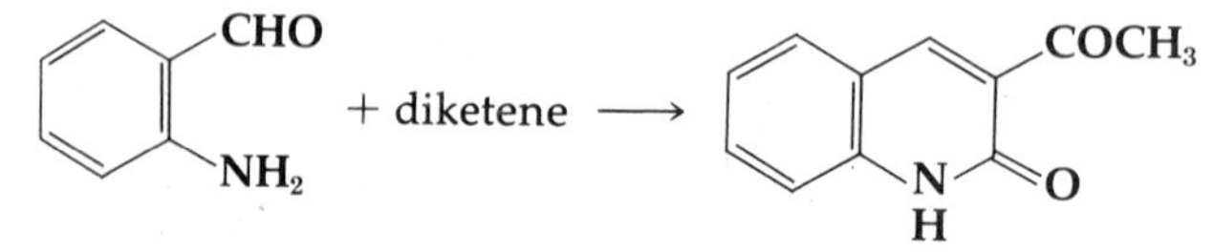

b 2-Hydroxycyclohexanone + diketene ⟶ 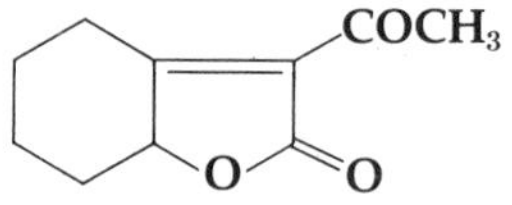

c $(CH_3)_2C(OH)CH_2COCH_3$ + diketene ⟶ $(CH_3)_2C(OCOCH_2COCH_3)CH_2COCH_3$ $\xrightarrow{\bar{O}H}$

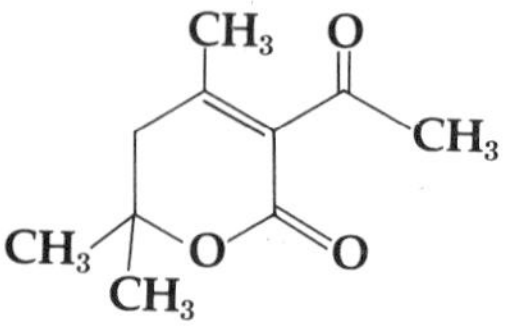

d The product from part **b** on treatment with $HCl/HOAc/H_2O$ yields the more stable aromatic compound shown below. Considering steps such as enolization, ester hydrolysis, and aldol dehydration, construct a mechanism for this isomerization.

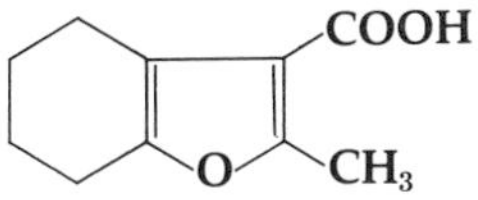

13-14 Write equations for the following transformations:

    a Benzamide is hydrolyzed, and the acid produced is then esterified with phenol, trifluoroacetic anhydride being used as a condensing agent.

    b n-Butyl mercaptan is acylated with propionic anhydride.

    c Diethyl adipate is subjected to the Dieckmann condensation, and the product is alkylated with ethyl bromide.

    d n-Valeronitrile is condensed in the Thorpe reaction, and the product is converted to a ketone.

    e p-Chlorotoluene is made by a Sandmeyer reaction and then aminated by treatment with sodamide in liquid ammonia.

    f Methyl n-propyl ketone is converted to several characteristic derivatives.

    g Dimethyl pimelate is cyclized by the acyloin condensation. The product is reduced to a mixture of stereoisomeric diols, which are converted to dibenzoates.

    h n-Butyl methyl ketone is acetylated under acidic and basic conditions.

    i Nitromethane is acylated with p-nitrobenzoyl chloride.

    j 4-Methylcyclohexanone is recovered from its semicarbazone.

13-15 Devise syntheses of the following compounds from the starting materials indicated and compounds containing no more than three carbon atoms.

    a $CH_3CHOHCH_2CH_2OH$ from $CH_3COOC_2H_5$

    b $C_6H_5COCH_2COCH_3$ from C_6H_5Cl

    c

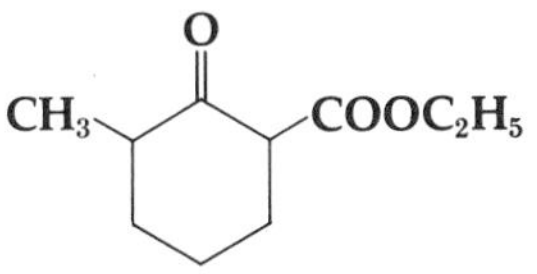

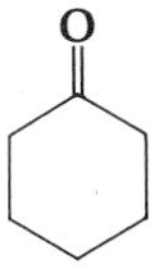

    (*Hint:* The ring is opened and closed again.)

    d $(CH_3CO)_3CH$

    e $(CH_3)_3CC\overset{O}{\overset{\|}{C}}-\overset{O}{\overset{\|}{C}}C(CH_3)_3$

f 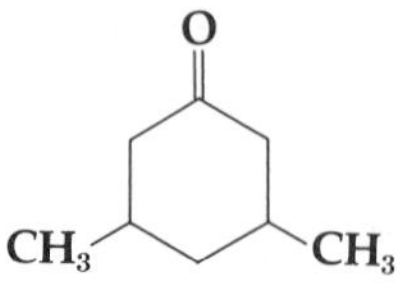

g $CH_3CH_2COCH(CH_3)_2$

h 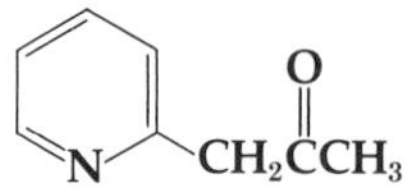 from 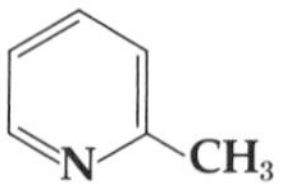

(*Hint:* The hydrogens of the methyl group of the starting material are acidic.)

i 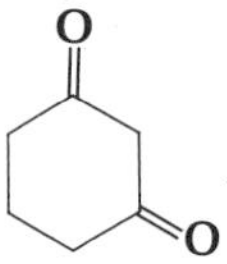 from $CH_2{=}CHCN$ and CH_3COCH_3

j $CH_3CH_2COCH_2COCH_3$ (at least three different syntheses).

13-16 Write mechanisms for each of the following reactions:

a $CH_3COOH \xrightarrow{HCl,\ C_2H_5OH} CH_3COOC_2H_5$

b $C_6H_5COCl + (CH_3)_2Cd \longrightarrow C_6H_5COCH_3$

c $CH_3COCl + KCN \longrightarrow CH_3COCN + KCl$

d $2CH_3\overset{O}{\overset{\|}{C}}CN \xrightarrow[\text{of } CN^-]{\text{Trace}} CH_3\overset{O}{\overset{\|}{C}}O\underset{CN}{\overset{CN}{C}}CH_3$

e $CH_3CH{=}CHCOOC_2H_5 + CH_3CH(COOC_2H_5)_2 \xrightarrow{NaOC_2H_5} \begin{matrix} CH_3CHCH(COOC_2H_5)_2 \\ | \\ CH_3CHCOOC_2H_5 \end{matrix}$

(A cyclobutanone derivate is an intermediate)

f $(CH_3CO)_2O + CH_3COCH_2COOC_2H_5 \xrightarrow{BF_3} (CH_3CO)_2CHCOOC_2H_5$

g $(CH_3)_3COH + CH_3COCH_2COOC_2H_5 \xrightarrow{BF_3} CH_3CO\overset{C(CH_3)_3}{\overset{|}{C}}HCOOC_2H_5$

h $CH_3COCH_2COOC_2H_5 + NH_2OH \longrightarrow$ 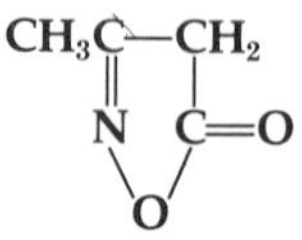

13-17 When chlorobenzene, labeled in the C-1 position with C^{14}, was treated with sodamide, aniline was produced labeled to an equal extent in the C-1 and C-2 positions. Explain this result.

13-18 Using as starting materials any substance from Table 13-5, devise syntheses for these compounds.

a

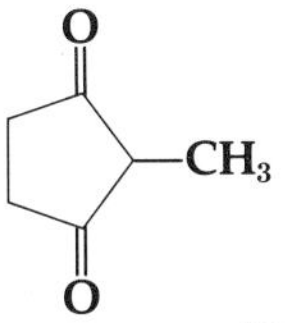

b HOOC— =O

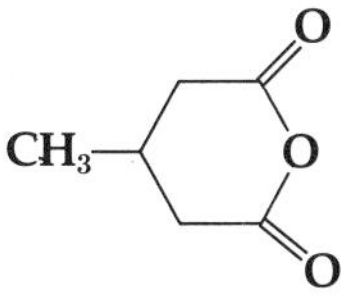

c

d CH$_3$—

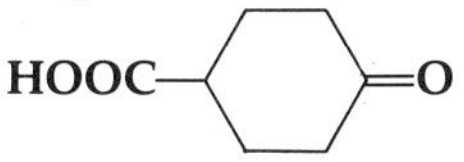

e (CH$_3$)$_2$CHCOCH=CHCH$_2$CH$_3$

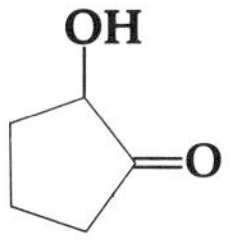

f

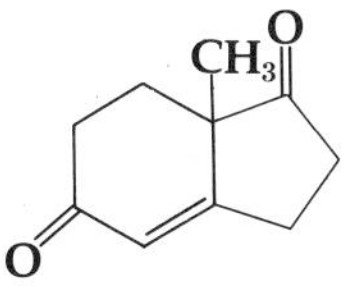

13-19 Convert 2-methylcyclopentanone into each of the following substances, using any other organic materials that may be required.

a

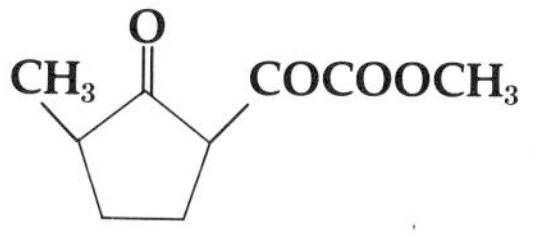

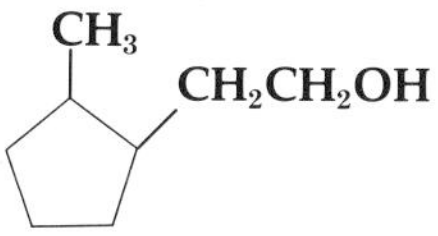

b

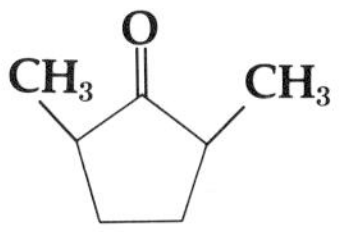

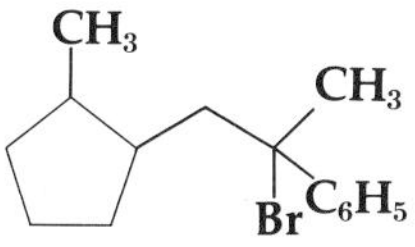

c

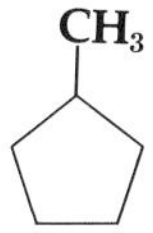

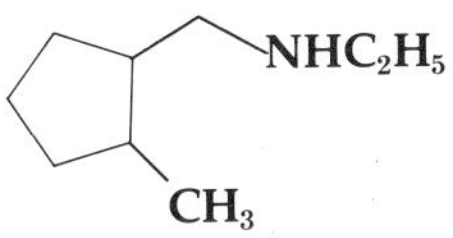

d

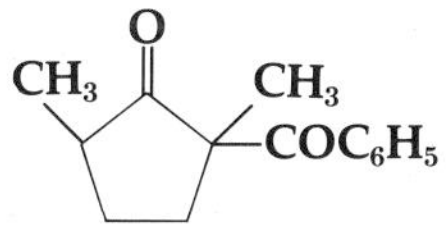

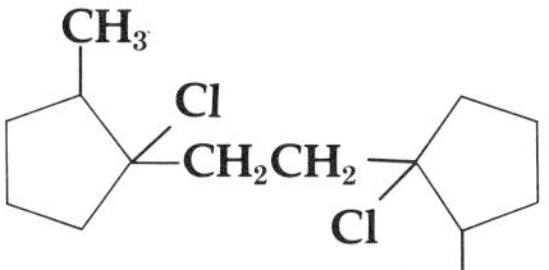

e

f

g

h

i

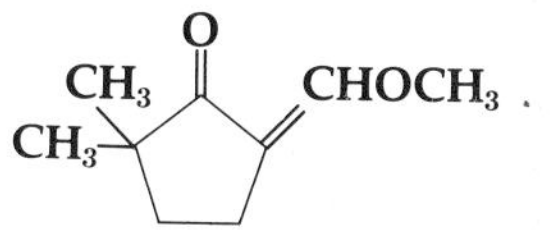

j

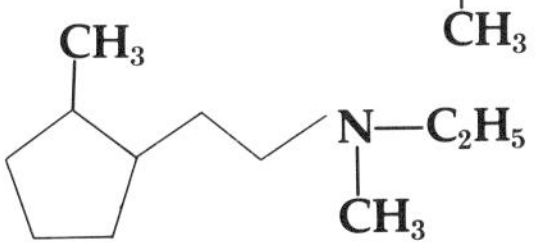

13-20 Convert cyclohexanone into each of the following compounds, using any other organic compounds that are required.

a

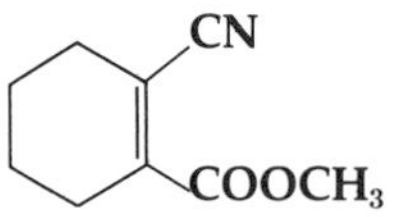

b

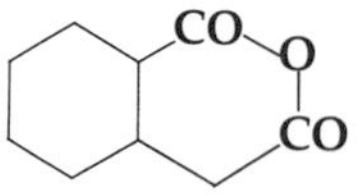

(What stereochemistry?)

c

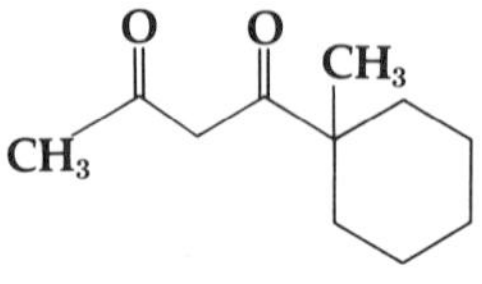

d

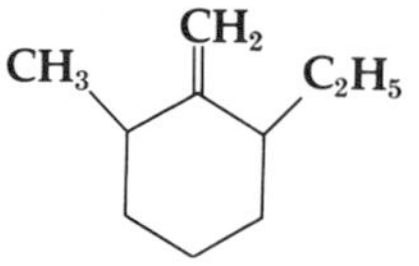

(What stereochemistry?)

e

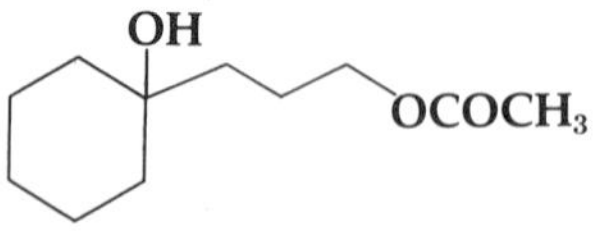

f

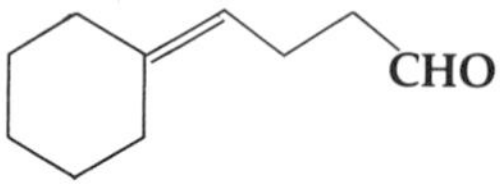

g

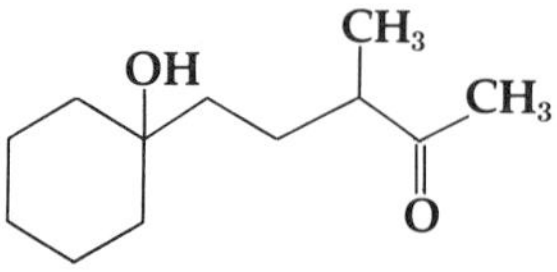

h

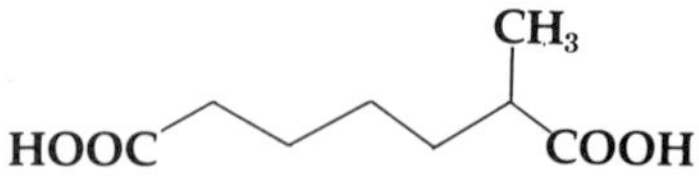

i $CH_3CH_2C(CH_2)_5COOH$

j

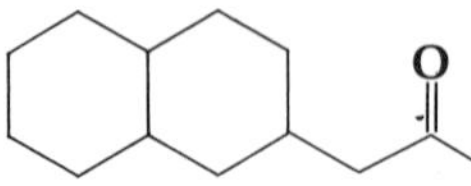

(Disregard stereochemistry)

k

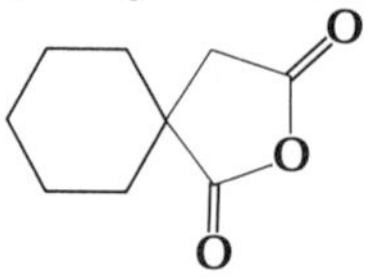

l

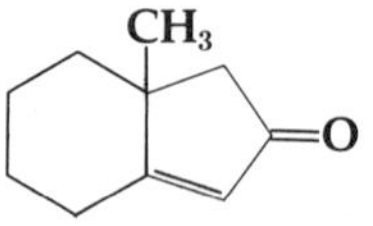

m

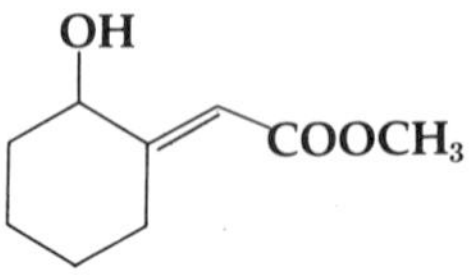

n

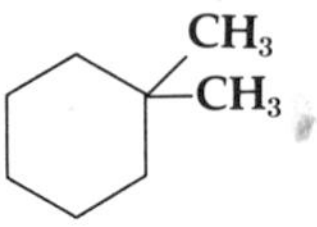

13-21 Account for the following observations.

a

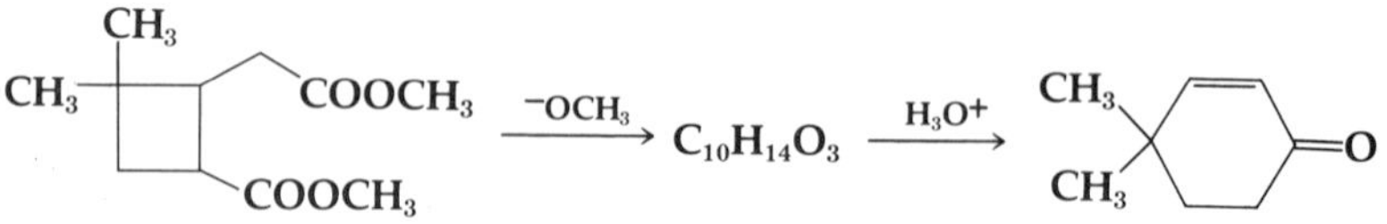

$\xrightarrow{\ ^-OCH_3\ } C_{10}H_{14}O_3 \xrightarrow{\ H_3O^+\ }$

b

$\xrightarrow{\ -OCH_3\ }$

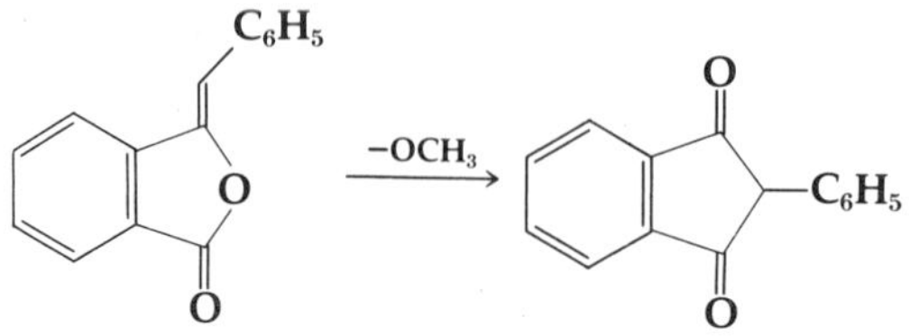

c C_6H_5 [structure] $\xrightarrow{\text{Dilute OH}^-}$ [structure]

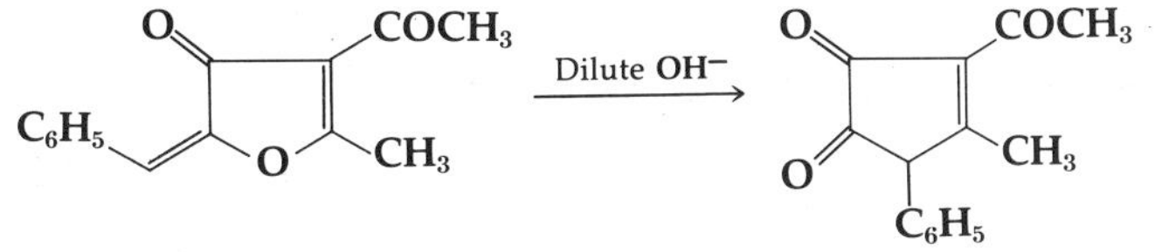

d The product from part **c,** on further treatment with mild alkali, goes on to form 2-phenyl-3-methyl-5-hydroxybenzoic acid. Consider the first step to be a reversible addition of water to the double bond, i.e., the reverse of the dehydrative second step of the aldol condensation. Then examine the carbon skeleton of the desired aromatic product and see how it may be formed from the given carbon skeleton by carbonyl reactions such as cleavage and aldol cyclization.

e Diethyl malonate, formaldehyde, and sodium ethoxide react to form the following ester (A). In order to deduce the chain of events, consider that the anion of the malonate is the first intermediate and that a Knoevenagel reaction is rapid and yields a reactive, conjugated double bond. Also it is known that acyl-substituted malonates may be cleaved by alkoxide ions to form dialkyl carbonate and a stable β-ketoester anion. The reaction is shown below and should be examined mechanistically to demonstrate that it fits the same pattern and criteria as acylation and acyl cleavages previously shown.

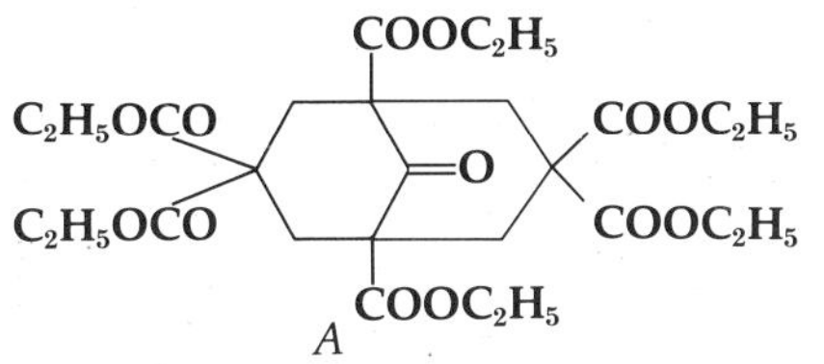

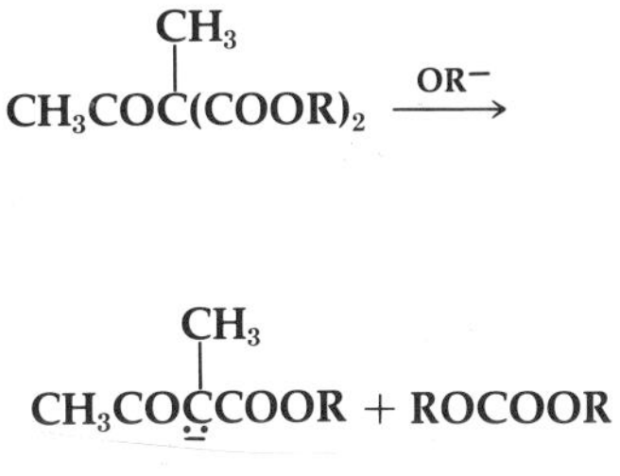

[structure A with $COOC_2H_5$, C_2H_5OCO, C_2H_5OCO, $COOC_2H_5$, $COOC_2H_5$, $COOC_2H_5$ groups]

$$CH_3COC(COOR)_2 \text{ (with } CH_3 \text{ substituent)} \xrightarrow{OR^-}$$

$$CH_3CO\underset{..}{C}COOR \text{ (with } CH_3 \text{ substituent)} + ROCOOR$$

f Acetone $\xrightarrow{H^+}$ $C_6H_{10}O$ $\xrightarrow[\text{metal (Et}_2\text{O)}]{\text{Lithium}}$ $C_{12}H_{20}O_2$ $\xrightarrow{H^+}$ [structure with CH_3, CH_3, CH_3, CH_3, $COCH_3$]
Mesityl oxide
UV: λ_{max} 238 nm
(log ϵ 4.0)
No UV

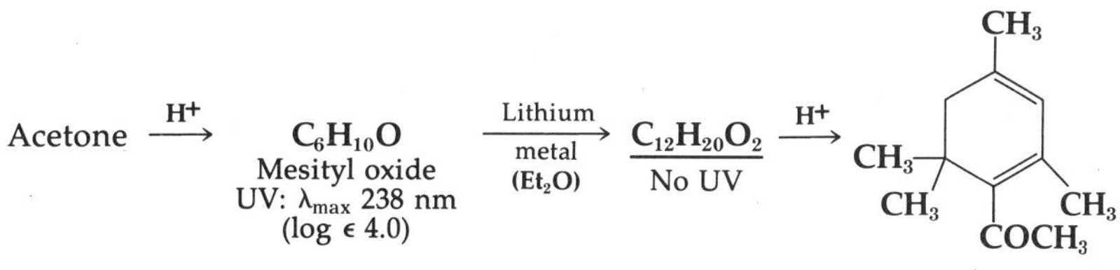

13-22 The compound 5-phenyl-6-oxo-heptanoic acid methyl ester was treated with methoxide in methanol with a view to obtaining a seven-membered ring by Dieckmann condensation. However, the isolated product was 2-acetyl-5-phenyl-cyclopentanone. Explain how this product may have been formed.

13-23 When the benzoate ester of 4-hydroxycyclohexanone is treated with potassium t-butoxide in t-butanol, a carboxylic acid was isolated (after a work-up procedure of pouring the reaction mixture into iced aqueous acid). The rather unexpected product was found to be β-(2-benzoylcyclopropyl)-propionic acid. Try to rationalize the course of this reaction. In doing this, consider that the base serves only to enolize, that the benzoyl group must become attached to carbon instead of oxygen in the product, and that the last step is one which displaces the carboxylate as a stable anion, thus finally consuming the base and stopping reaction.

13-24 Delineate a mechanistic pathway for this observed transformation. (*Hint:* it is not necessary to invoke rearrangements.)

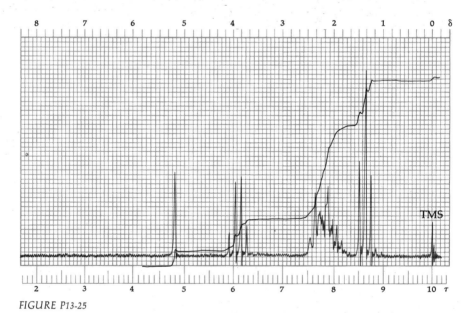

13-25 Compound A is formed by methoxide treatment of a certain ketoester. It is converted to compound B by ethylation with diazoethane. The NMR spectrum of compound B is shown in Fig. P13-25. Its mass spectrum yielded a parent peak at m/e 140. Derive structures for the ketoester and compounds A and B, and write a mechanism for each reaction.

FIGURE P13-25

13-26 Rationalize the molecular events in these observed conversions.

a

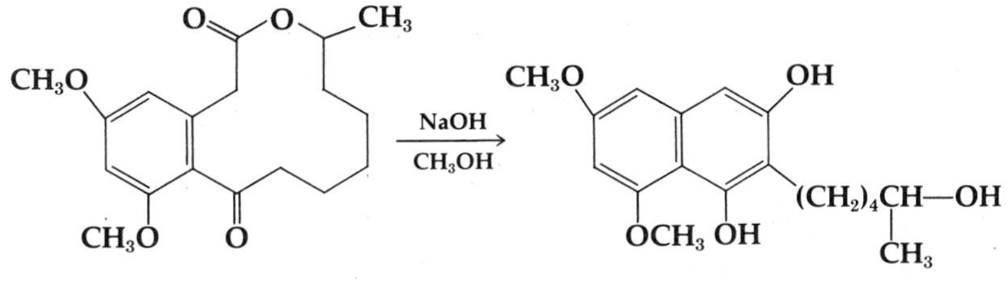

b

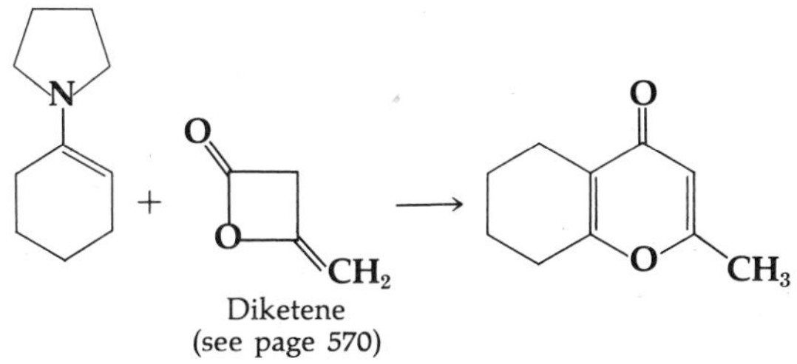

Diketene
(see page 570)

c

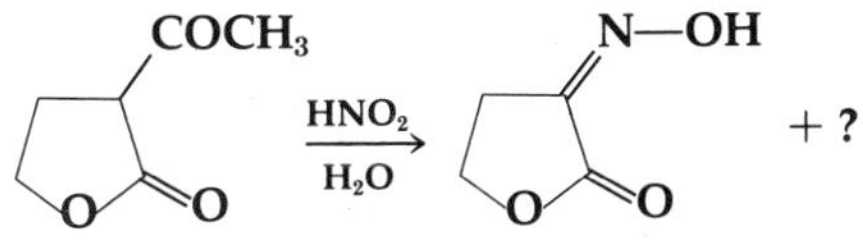

ELIMINATION REACTIONS

14-1 INTRODUCTION

ELIMINATION reactions are formally the reverse of addition reactions, some of which have been discussed in Chap. 12 (others will appear in Chap. 15). Elimination reactions themselves have also been introduced previously since they appear as competitive reactions in substitutions at saturated carbon. They also appear in carbonyl reactions such as the aldol condensation. The most common eliminations are 1,2-eliminations in which substituents on two adjacent atoms are removed, leaving a π bond behind, as generalized in Fig. 14-1. The multiple bonds formed by elimination include $C=C$, $C\equiv C$, $C=N$, $C\equiv N$, $C=O$, $C=S$, and $N=N$, but formation of alkenes and alkynes are the most important uses. The two atoms in the substrate are labeled α and β to distinguish them, α being the atom bearing the leaving group (L). Much less common are the 1,1- (or α-) eliminations in which the two substituents are removed from the same atom to form a highly reactive, electron-deficient species (carbene) with only six electrons remaining on the atom. The main discussion in this chapter concerns 1,2-eliminations.†

Of the two groups removed in the elimination, one (E) is an electrophile, usually a proton, and the other (L) is a leaving group just like those in nucleophilic substitutions. There is a close analogy in eliminations with the substitution reactions at saturated carbon and also with the enolization reaction of carbonyls and related groups. The entire collection of reactions surveyed so far involves attack of a nucleophile and cleavage of a heteroatom bond. The latter cleavage either separates the heteroatom if it is singly bonded or breaks a π bond in multiply bonded cases (e.g., carbonyl). The similarities in all these reactions are emphasized in Fig. 14-2. Eliminations represent the final family of these nucleophilic reactions to be discussed.

The timing of bond-breaking (E—Y and Z—L) events in 1,2-elimination is not specified in Fig. 14-1 (or 14-2) but we have already seen two kinds of mechanism distinguished in this way, and they are summarized in Table 14-1. Similarly, eliminations proceed in two ways. With active nucleophiles (usually strong bases) elimination is a concerted reaction in which there is only one transition state. This is called an *E2* **elimination reaction** since it is bimolecular, like S_N2 substitution, its rate being dependent on both base and substrate concentrations. On the other hand, if ionization is the first step for a leaving group, the subsequent carbonium ion may either be substituted by a nucleophile (S_N1) or lose a proton to yield a double bond (*E1* or **unimolecular elimination reaction**). The rate should depend only on substrate (R—L) concentration,

---

† 1,3-Eliminations, etc., are included as cyclic substitution reactions in Chaps. 10 and 11.

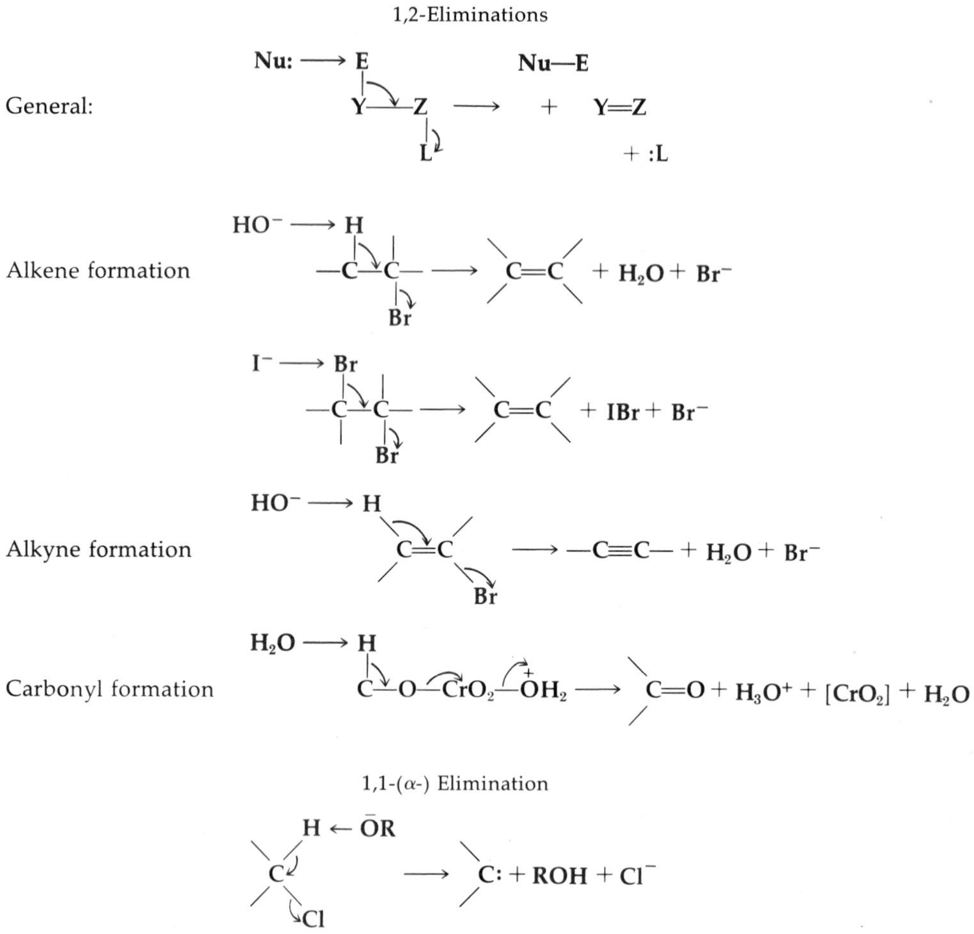

1,2-Eliminations

General:

Alkene formation

$$\quad \longrightarrow \quad \diagdown C=C \diagup + H_2O + Br^-$$

$$\quad \longrightarrow \quad \diagdown C=C \diagup + IBr + Br^-$$

Alkyne formation

$$\longrightarrow -C \equiv C- + H_2O + Br^-$$

Carbonyl formation

$$\longrightarrow \diagdown C=O + H_3O^+ + [CrO_2] + H_2O$$

1,1-(α-) Elimination

$$\longrightarrow \diagdown C: + ROH + Cl^-$$

FIGURE 14-1 **Elimination reactions**

since the initial carbonium-ion formation is rate-determining, and it should be the same for substitution or elimination. As with substitutions, a fair range of rate dependence is observed between the two extremes for elimination reactions, but it is convenient to discuss the two mechanisms, *E*1 and *E*2, as if they were clearly separated, just as we did for substitutions in Chap. 10.

Accordingly, our discussion will deal with the following effects on elimination reactions, similar to our discussion of effects on substitutions.

1 Direction of elimination (choice of hydrogens yields different double bonds)
2 Stereoelectronic requirements and stereospecificity
3 Relative activity of nucleophiles and leaving groups
4 Competition with substitution

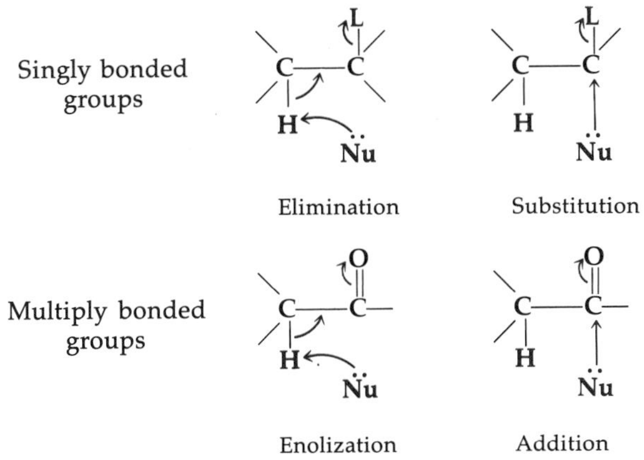

Singly bonded groups

Elimination   Substitution

Multiply bonded groups

Enolization   Addition

FIGURE 14-2   **Similarities in nucleophilic reactions**

The primary synthetic value of elimination reactions is their provision of the main routes to double and triple bonds between carbon atoms.

The 1,1-eliminations produce very reactive neutral **carbenes** ($R_2C:$) or **nitrenes** ($R\ddot{N}:$). These are electron-deficient species and instantly attack any available electron source (hence they are electrophiles). They are too reactive to be isolated and are always generated in the presence of the substrate needed for reaction. For this reason, in Chap. 15 they are primarily treated as electrophilic additions. Carbenes and nitrenes may also accept internal electrons, thus causing rearrangement and these cases are found in Chap. 17. The most commonly used carbene is dichlorocarbene ($Cl_2C:$) which arises by attack of base (^-OH, $t\text{-}BuO^-$, etc.) on chloroform:

$$HCCl_3 + {}^-OR \longrightarrow [{}^-:CCl_3] \longrightarrow :CCl_2 + Cl^-$$
$$\text{Dichlorocarbene}$$

TABLE 14-1   **Summary of Nucleophilic Reaction Mechanism Types**

One-step	Two-step
Concerted	Rate-determining ionization
Bimolecular	Unimolecular
Single transition state	Two transition states
Usually basic	Usually acidic
S_N2 substitution	S_N1 substitution
Enolization in base	Enolization in acid
Carbonyl addition	Carbonyl addition
$E2$ elimination	$E1$ elimination

In a similar reaction, N-haloamides react with strong base to give nitrenes which rearrange.

$$R\text{—}CO\text{—}NHBr \xrightarrow{^-OH} R\text{—}CO\text{—}N \xrightarrow{\text{Rearrangement}} O\text{=}C\text{=}N\text{—}R$$
$$\text{(Nitrene)}$$

PROBLEM 14-1

Using 4-chloro- and 5-chloro-2-hexanones as examples, describe base-catalyzed 1,2- and 1,3-eliminations in terms of orbital changes, mechanism, kinetics and relative rates, and stereoelectronic requirements. Note the differences in the two reactions and consider why 1,3-eliminations are not specially discussed in the present chapter.

PROBLEM 14-2

The elimination reaction has already been invoked as an intimate part of the mechanisms of a number of carbonyl reactions. Write the elimination step in five different reactions, selecting a specific molecule to exemplify.

14-2 ALKENES BY ELIMINATIONS WITH PROTON LOSS

The most widely studied and synthetically useful eliminations are those which form alkenes by loss of proton and leaving group (cf. first example in Fig. 14-1). The base-catalyzed ($E2$) elimination is more commonly used since it exhibits more specific requirements and so is more predictable in terms of products. Bases in the amine class and stronger (higher pK_a) cause elimination reactions with substrates that have a reactive leaving group and a β-hydrogen. In general, the elimination reaction is favored by

1 High reaction temperatures
2 Increase of the basic strength of the nucleophile
3 Increase of steric hindrance in the substrate or base

Attachment of electron-withdrawing substituents to C_β increases the acidity of the proton to be removed and accelerates elimination very greatly.

Bases used in elimination reactions include R_3N, $CH_3CO_2^-$, HO^-, RO^-, H_2N^-, and R_2N^-. The reactivity of bases in elimination reactions parallels their basicity in proton-transfer reactions (Table 8-2). As a consequence, low-basicity nucleophiles such as RS^-, which owe much of their reactivity to high polarizability, tend to bring about substitution rather than elimination. Very strong bases, such as H_2N^-, give elimination predominantly, even with primary substrates. Some of the trends are illustrated by the following examples.

The leaving groups include all those involved in nucleophilic substitutions (Table 10-2), such as Cl^-, Br^-, I^-, RSO_3^-, RCO_2^-, R_3N, R_3P, R_2S, and others such as RO^- and enolate ions. Variation in the reactivity of leaving groups is qualitatively the same as in nucleophilic substitution; reactivity increases with increasing polarizability and decreasing basicity of the group. (See Table 10-7.)

		Substitution	*Elimination*

$$C_2H_5\bar{O}\overset{+}{N}a + CH_3CH_2Br \xrightarrow[-NaBr]{C_2H_5OH,\ 55°} CH_3CH_2OCH_2CH_3 + CH_2{=}CH_2$$
$$\qquad\qquad\qquad\qquad\qquad\qquad\qquad 90\% \qquad\qquad\quad 10\%$$

$$C_2H_5\bar{O}\overset{+}{N}a + CH_3\underset{\underset{Br}{|}}{C}HCH_3 \xrightarrow[-NaBr]{C_2H_5OH,\ 55°} (CH_3)_2CHOC_2H_5 + CH_2{=}CHCH_3$$
$$\qquad\qquad\qquad\qquad\qquad\qquad\qquad 21\% \qquad\qquad\quad 79\%$$

$$C_2H_5\bar{O}\overset{+}{N}a + (CH_3)_3CBr \xrightarrow[-NaBr]{C_2H_5OH,\ 55°} \qquad\qquad CH_2{=}C(CH_3)_2$$
$$\qquad\qquad\qquad\qquad\qquad\qquad\qquad\qquad\qquad\qquad 100\%$$

$$C_2H_5\bar{O}\overset{+}{N}a + C_2H_5\overset{+}{S}(CH_3)_2 \xrightarrow[-(CH_3)_2S]{C_2H_5OH,\ 45°} C_2H_5OC_2H_5 + CH_2{=}CH_2$$
$$\qquad\qquad\qquad\qquad\qquad\qquad\qquad\ 88\% \qquad\qquad\quad 12\%$$

$$C_2H_5\bar{O}\overset{+}{N}a + C_2H_5\overset{+}{S}(CH_3)_2 \xrightarrow[-(CH_3)_2S]{C_2H_5OH,\ 64°} C_2H_5OC_2H_5 + CH_2{=}CH_2$$
$$\qquad\qquad\qquad\qquad\qquad\qquad\qquad\ 85\% \qquad\qquad\quad 15\%$$

Eliminations with halide † and sulfonate ions as leaving groups are usually encountered as side reactions in nucleophilic substitution. Synthetically, dehydrohalogenation† is carried out with strong bases which are not very nucleophilic at carbon, such as NH_2^-, R_2N^-, t-BuO^-, or often OH^-, in order to suppress substitution. Primary halides are more readily substituted, but elimination is common with secondary sites and is the only course with tertiary halides. Sulfonates are somewhat more prone to undergo substitution than elimination.

$$ClCH_2CH_2Cl \xrightarrow[Reflux]{Na\ OH,\ C_2H_5OH} CH_2{=}CHCl$$
$$\qquad\qquad\qquad\qquad\qquad \text{Vinyl chloride}$$

$$CH_3CH_2\underset{\underset{Br}{|}}{C}HCOOH \xrightarrow[Reflux]{Pyridine} CH_3CH{=}CHCOOH$$
$$\qquad\qquad\qquad\qquad\qquad \text{Crotonic acid}$$

---

† Halide eliminations (—HX) are often called **dehydrohalogenation reactions**.

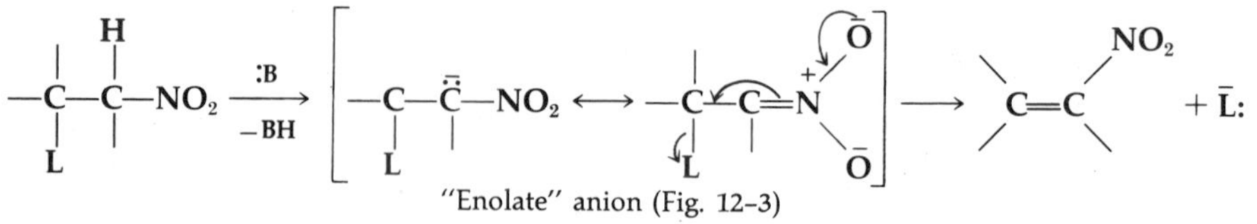

FIGURE 14-3   **Stepwise β-elimination**

An interesting halide elimination is found in the preparation of ketenes from acid chlorides.

$$(C_6H_5)_2CHCOCl + (CH_3CH_2CH_2)_3N: \xrightarrow{\text{Ether}} (C_6H_5)_2C{=}C{=}O + (CH_3CH_2CH_2)_3\overset{+}{N}H\ \bar{C}l$$

Diphenylacetyl chloride   Diphenylketene

Elimination rather than substitution is the rule, even at primary carbon, when the leaving group is positively charged *onium ion* such as quaternary ammonium salts [e.g., $R{-}N^+(CH_3)_3$] or sulfonium salts [e.g., $R{-}S^+(CH_3)_2$]. Pyrolysis of quaternary ammonium hydroxides leads to elimination if β-hydrogens are available, the hydroxide acting as the base for the elimination. This is called the **Hofmann elimination**. The hydroxides are usually prepared by adding silver oxide to an aqueous solution of a quaternary ammonium halide or by passing the quaternary halide through a hydroxide ion-exchange resin. The mixture is then heated to distill off the water. Continued heating of the residue causes elimination to occur. The reaction has been much used in the stepwise degradation of naturally occurring amines in the course of structure determination. Amines to be degraded in this manner are first **quaternized** by alkylation with methyl iodide.

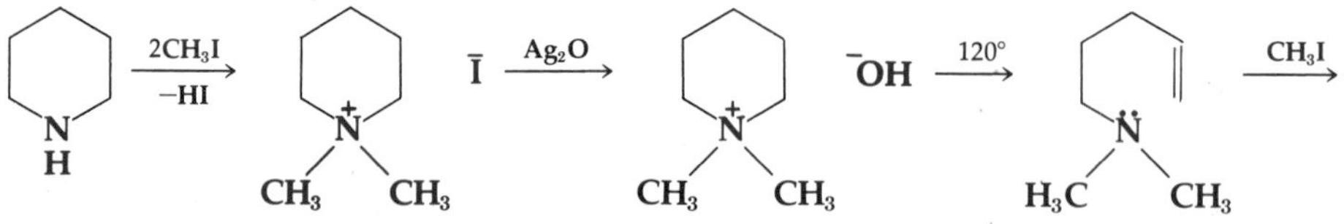

Piperidine   N,N-Dimethyl-
                  piperidinium
                  iodine

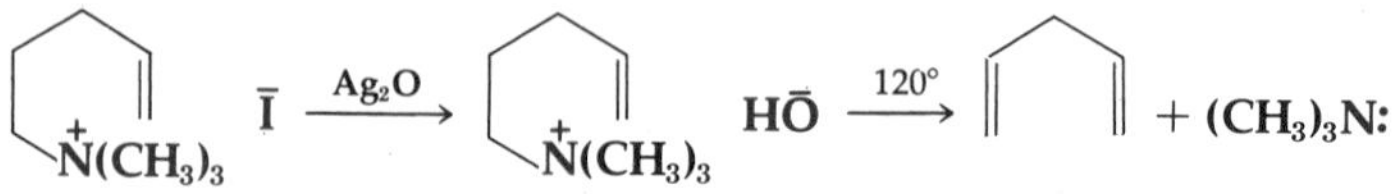

Elimination reactions of β-substituted esters, acids, ketones, aldehydes, and nitro compounds are called β-**eliminations** and are very rapid. The electron-withdrawing groups have strong acid-strengthening effects on the α-proton, which is removed by bases during the reaction. The reactions may be formulated as involving anionic intermediates (Fig. 14-3), although it is doubtful that such intermediates are *always* involved.

Even leaving groups such as amino, hydroxyl, and enolate ions, which are not ordinarily lost in either normal nucleophilic eliminations or substitutions, leave readily in β-elimination. These reactions are the reverse of conjugate additions to α,β-unsaturated systems (Sec. 12-6). The first example is an elimination of a Mannich base (page 487), and the dehydration in the aldol condensation is another example (Sec. 12-5). The β-eliminations, especially of β—OH or —OR, are also facile under acid conditions, via the enol. The β-elimination of carbon, in the form of the enolate carbanion, is a **reverse Michael reaction**. Such reactions can sometimes be brought about by vigorous heating with dry potassium carbonate, but rarely occur under ordinary reaction conditions.

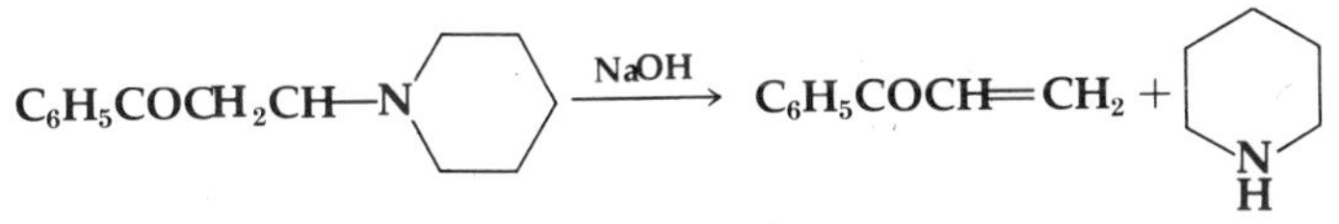

$$C_6H_5COCH_2CH-N \xrightarrow{\text{NaOH}} C_6H_5COCH{=}CH_2 +$$

β-Piperidinopropiophenone   Phenyl vinyl ketone   Piperidine
(a *Mannich* base)

$$\text{(cyclohexanone with Cl)} \xrightarrow[\text{(C}_5\text{H}_5\text{N)}]{\text{Pyridine}} \text{(cyclohexenone)} + C_5H_5NH^+ Cl^-$$

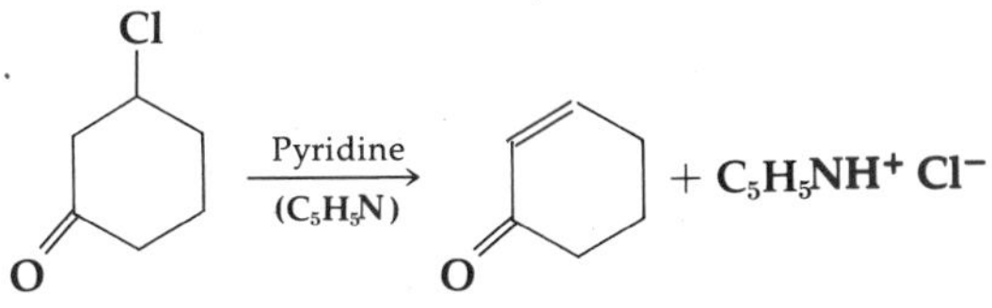

PROBLEM 14-3

Using projection formulas, show why a trans double bond is produced in the aldol condensation of benzaldehyde and acetone.

Stereochemistry

Most E2 reactions appear to be stereospecific. To achieve maximum overlap in the developing bond orbitals, the orbitals must all lie in one plane and the two groups eliminated (e.g., H and X) must be trans to each other across the central bond which becomes the π bond. This arrangement of groups is called **trans antiparallel** or **trans periplanar**. The relation to S_N2 substitution may be seen here, for in both reactions the leaving group is displaced from the backside, either by external nucleophile in S_N2, or by released bond electrons (to E, or H) in E2.

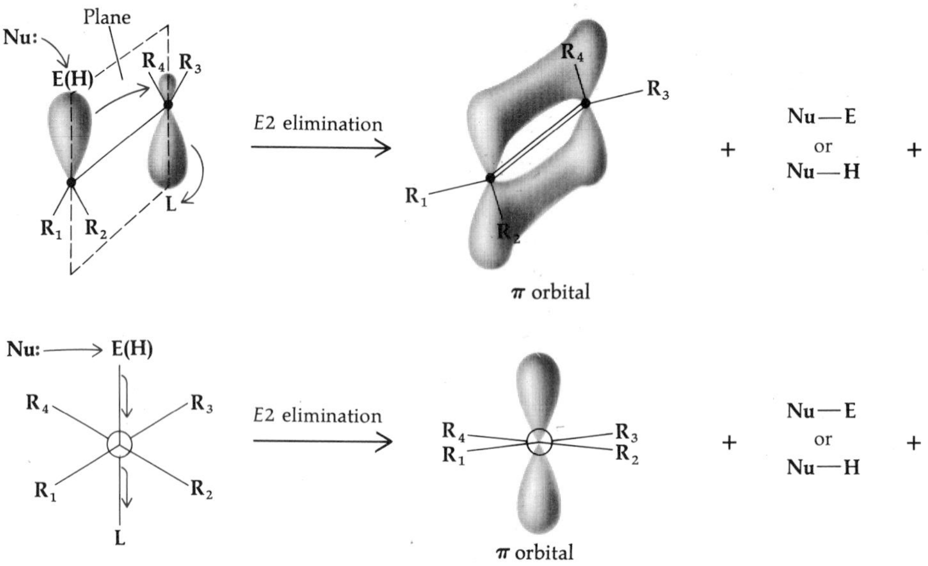

Plane

Nu:

E(H) R_4 | R_3

    E2 elimination → R_4 R_3 Nu—E

R_1 | R_2 L R_1 R_2 + or +

            Nu—H

         π orbital

Nu:→ E(H)

R_4 R_3 *E2 elimination* → R_4——R_3 Nu—E

            R_1——R_2 + or +

R_1 R_2 Nu—H

L

        π orbital

The process is known as **trans elimination**, with the term *trans* referring to the *orientation of the leaving groups* in the substrate at the time of reaction, and *not to the configuration of the alkene product.* Elimination is facile when the molecule can rotate so as to allow trans elimination and will be fastest for cases in which large groups (R_1, R_2, etc.) are not forced together on one side in the required conformation, as with the large phenyls in the example below. The projection notation allows a clear view of the stereoelectronic requirements.

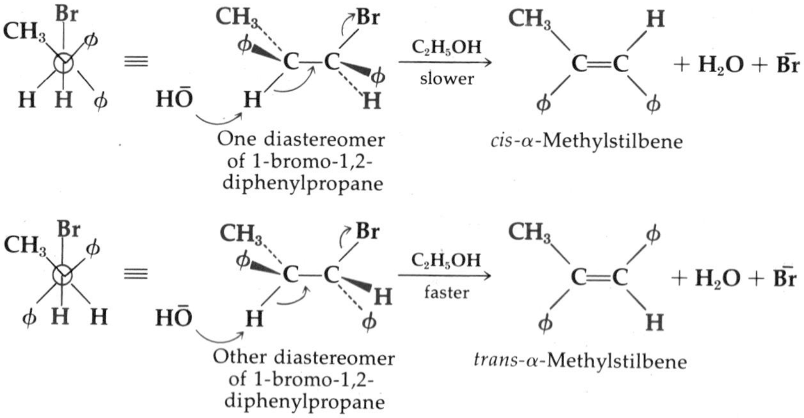

CH_3 Br ϕ CH_3 Br CH_3 H

    $\equiv$ ϕ—C—C—ϕ $\xrightarrow[\text{slower}]{C_2H_5OH}$ C=C + H_2O + $\bar{B}r$

H H ϕ H$\bar{O}$ H H ϕ ϕ

      One diastereomer   *cis*-α-Methylstilbene
      of 1-bromo-1,2-
      diphenylpropane

CH_3 Br ϕ CH_3 Br CH_3 ϕ

    $\equiv$ ϕ—C—C—H $\xrightarrow[\text{faster}]{C_2H_5OH}$ C=C + H_2O + $\bar{B}r$

ϕ H H H$\bar{O}$ H ϕ ϕ H

      Other diastereomer *trans*-α-Methylstilbene
      of 1-bromo-1,2-
      diphenylpropane

This orientation rule, when applied to cyclohexane derivatives means that both leaving groups must be in or be able to enter *axial* positions for a facile elimination (**trans-diaxial orientation**). The preference is shown by the diastereomeric 4-*t*-butylcyclohexyl *p*-toluenesulfonates, in which the bulky *t*-butyl group always assumes an equatorial position, locking the ring in a single conformation. The cis isomer, in which the tosylate group is axial, undergoes bimolecular elimination very easily. Under the same reaction conditions, the trans isomer undergoes no bimolecular reaction, even in the presence of strong base.

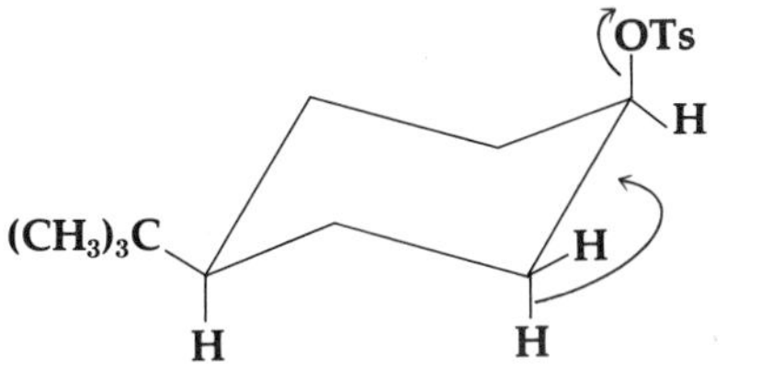

cis-4-Butylcyclohexyl tosylate
(undergoes *E*2 reaction with 0.02 *M* C₂H₅ONa in C₂H₅OH at 75°)

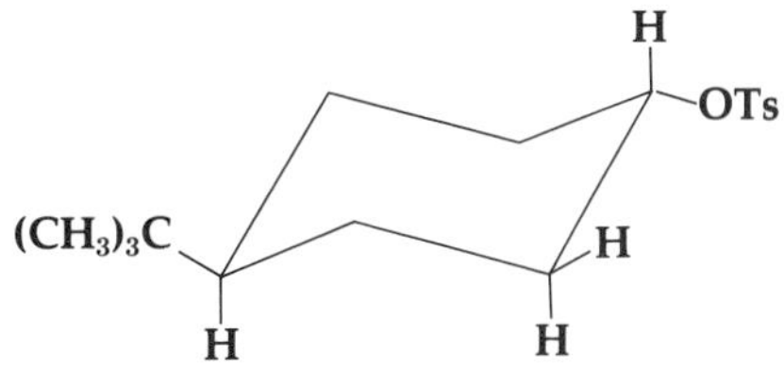

trans-4-Butylcyclohexyl tosylate
(undergoes only solvolytic elimination, *E*1, solvolysis with 0.02 *M* C₂H₅ONa in C₂H₅OH

When the leaving group is axial but has no trans, axial β-hydrogen, or when a rigid ring prevents the trans-periplanar arrangement, no E2 elimination occurs or else it is excessively slow. The first instance is provided by the eliminations of menthyl chlorides (related to the natural product, menthol, the corresponding alcohol) shown in Fig. 14-4. Elimination only occurs where a trans-diaxial orientation is possible, giving two products in one case (neo-menthyl) but only one in the other (menthyl). The corresponding diastereomeric menthyltrimethylammonium hydroxides show parallel behavior.

Similarly, the (conformationally rigid) epimeric chlorohydrins below either eliminate (to an enol) or undergo internal S_N2 displacement (epoxide formation) depending on trans-diaxial orientation of the reacting groups.

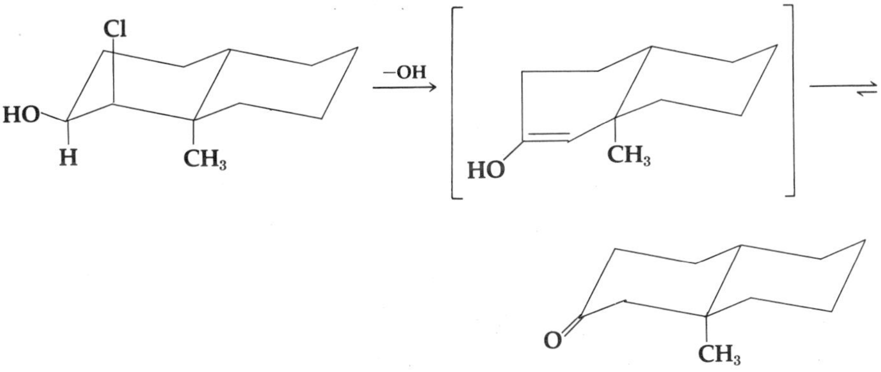

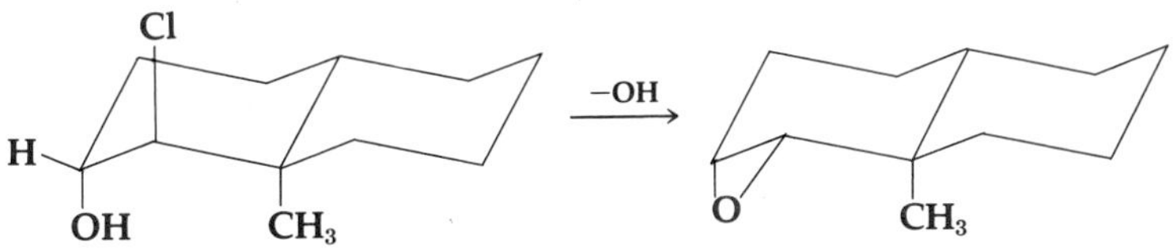

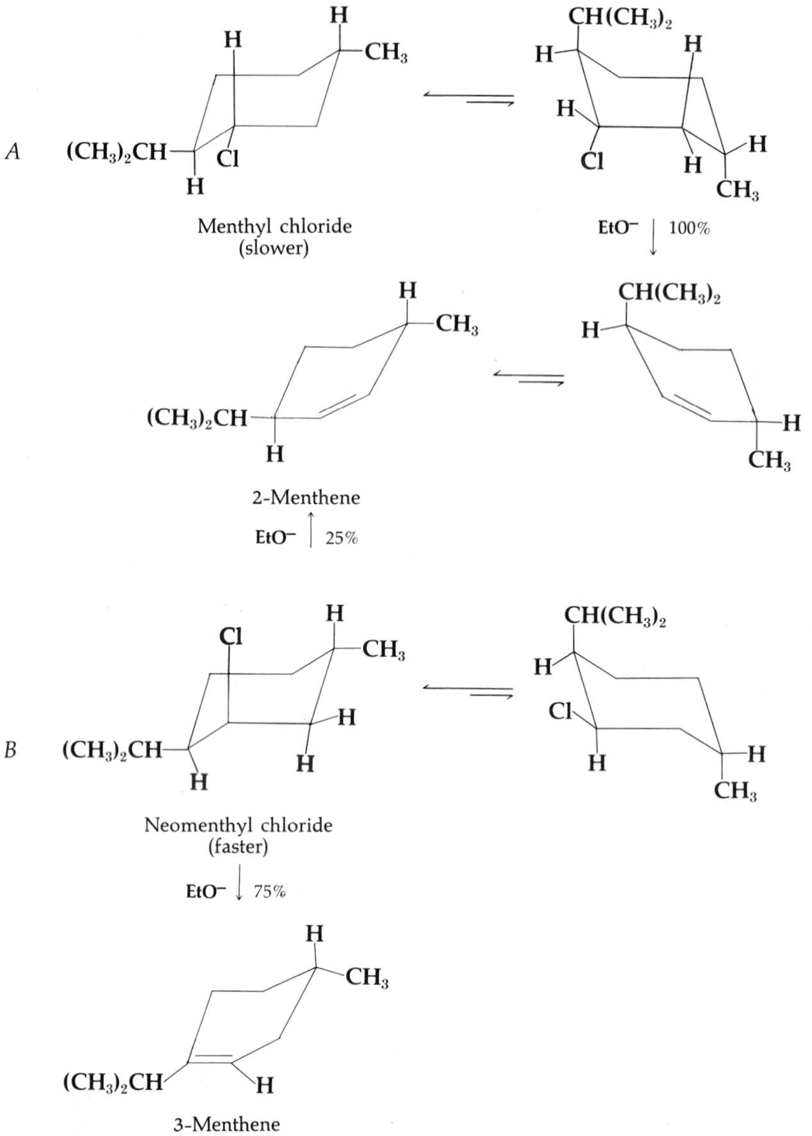

A   (CH$_3$)$_2$CH

Menthyl chloride
(slower)

EtO$^-$ | 100%

2-Menthene

EtO$^-$ | 25%

B   (CH$_3$)$_2$CH

Neomenthyl chloride
(faster)

EtO$^-$ | 75%

3-Menthene

FIGURE 14-4   **Stereochemistry and direction of elimination in menthyl chlorides**

The rigid epimers shown below, however, cannot become trans periplanar and are very unreactive to base.

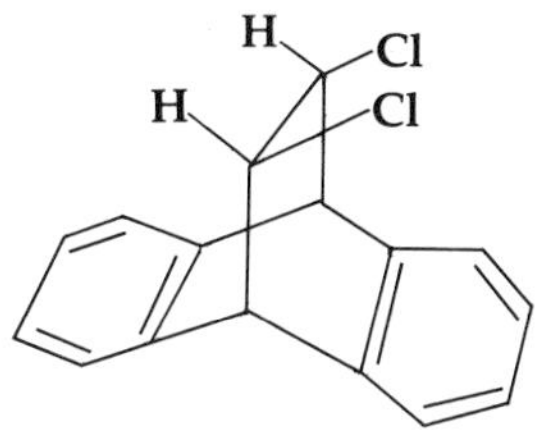

This strong stereoelectronic requirement for *E2* elimination is useful for product prediction but is also strong evidence for the concerted nature of the reaction, along with the second-order kinetics commonly observed. An alternate mechanism (nonconcerted) would involve prior deprotonation to a carbanion before the actual elimination step. This is unlikely because of the very weak acidity of these carbon acids to the bases which cause elimination.

It should be pointed out that the necessary orbital overlap for concerted elimination can be geometrically achieved if the two groups are *cis* and *coplanar* (dihedral angle of 0° instead of 180°). A few such cases are known, but trans-periplanar elimination appears to be the general rule.

Direction of Elimination

Unsymmetrical secondary and tertiary substrates give mixtures of elimination products, as long as the necessary protons on the β-carbons can become trans periplanar. An example is provided by neomenthyl chloride in Fig. 14-4. Two empirical rules govern the direction of elimination:

1 The **Saytzeff rule:** Neutral substrates (halides and sulfonates) yield predominantly the *most substituted* alkene; the predominance is often only moderate.
2 The **Hofmann rule:** Charged substrates (quaternary ammonium and sulfonium salts) yield predominantly the *least substituted* alkene; the predominance here is strong.

The mechanistic reasons for these rules are certainly complex and subtle and have not yet been properly unraveled, but they must involve steric factors (base and substrate) in the transition state, relative stabilities of the developing alkenes, and solvation and interaction of both base and leaving group.

$$C_2H_5\bar{O}\ Na^+ + CH_3CH_2\underset{\underset{Br}{|}}{C}HCH_3 \xrightarrow[25°]{C_2H_5OH}$$

$$CH_3CH{=}CHCH_3 + CH_3CH_2CH{=}CH_2 + CH_3CH_2\underset{\underset{CH_3}{|}}{C}HOC_2H_5$$

$$\text{4 parts} \qquad\qquad\qquad \text{1 part}$$

$$(CH_3)_3C\bar{O}\ K^+ + CH_3CH_2CH_2\underset{\underset{Br}{|}}{C}HCH_3 \longrightarrow CH_3CH_2CH{=}CHCH_3 + CH_3CH_2CH_2CH{=}CH_2$$

$$\text{3 parts} \qquad\qquad\qquad \text{1 part}$$

$$(CH_3)_3C{-}\underset{\underset{CH_3}{|}}{\overset{\overset{CH_3}{|}}{N^+}}{-}CH_2CH_3\ \bar{O}H \xrightarrow{150°} CH_2{=}CH_2 + (CH_3)_2C{=}CH_2$$

$$\qquad\qquad\qquad\qquad\quad 99\% \qquad\quad 1\%$$

$$CH_3CH_2\underset{\underset{^+N(CH_3)_3}{|}}{C}HCH_3\ \bar{O}H \xrightarrow{150°} CH_3CH{=}CHCH_3 + CH_3CH_2CH{=}CH_2$$

$$\qquad\qquad\qquad\qquad\quad \textit{cis}\text{-2-Butene 3\%} \qquad \text{1-Butene}$$
$$\qquad\qquad\qquad\qquad\quad \textit{trans}\text{-2-Butene 2\%} \qquad 95\%$$

$$CH_3CH_2\underset{\underset{^+S(CH_3)_2}{|}}{C}HCH_3\ \bar{O}C_2H_5 \longrightarrow CH_3CH{=}CHCH_3 + CH_3CH_2CH{=}CH_2$$

$$\qquad\qquad\qquad\qquad\quad 26\% \qquad\qquad 74\%$$

Resonance stabilization for the developing double bond facilitates the elimination and dominates the product choice.

$$C_6H_5CH_2CH_2{-}\underset{\underset{CH_3}{|}}{\overset{\overset{CH_3}{|}}{N^+}}CH_2CH_3\ \bar{O}H \xrightarrow{150°} C_6H_5CH{=}CH_2 + CH_2{=}CH_2$$

$$\qquad\qquad\qquad\qquad\qquad\quad \text{Styrene} \qquad 0.4\%$$
$$\qquad\qquad\qquad\qquad\qquad\quad 93\%$$

$\Delta^{1,2}$[2.2.1]Bicycloheptene
(has never been prepared)

$\Delta^{1,2}$[3.3.1]Bicyclononene
(known, but strained and
relatively unstable)

Elimination reactions designed to produce double bonds to the **bridgehead atoms** in bridged bicyclic systems (like the ones above) are either entirely unsuccessful or very slow. The $\Delta^{1,2}$[2.2.1]bicycloheptene† system has never been prepared. This failure is due to the fact that a rigid ring system requires an acute twist from the coplanar configuration required for strong π bonding between the bridgehead and adjacent carbons. The smallest ring system that has been prepared with a bridgehead double bond is $\Delta^{1,2}$[3.3.1]bicyclononene.

† In the names of alkenes, Δ is sometimes used as a prefix in conjunction with numbers (superscripts) which indicate the position of the double bond in the carbon chain.

The failure of reactions to produce compounds with double bonds at a bridge-head position is known as **Bredt's rule**. This double-bond twisting also accounts for the inability of ketones to enolize to an α-carbon which is a bridgehead, as well as the instability of bridgehead amides in which the resonance form with a C—N double bond violates Bredt's rule.

PROBLEM 14-4

Account for the following observations.

a Three isomers of bromocyclohexanol, on treatment with strong bases, give three separate major products (one each), an allylic alcohol, a ketone, and an epoxide.

b In part **a,** if the starting bromocyclohexanols are all optically active, what will the products be?

c The methyl ester of 2-bromo-3-hydroxycyclohexanecarboxylic acid readily eliminates **HBr** in mild bases, but the lactone does not.

d Treatment of 4-bromo-2-hexanone with aqueous **KCN** affords $C_8H_{12}N_2O$ with IR bands at 2.8 and 4.5 μ (about 3600 and 2900 cm^{-1}).

e 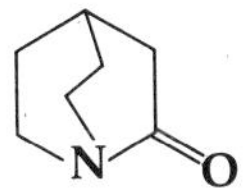 is basic but 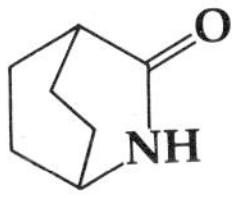 is not.

f $CH_3\overset{\displaystyle{}^+N(CH_3)_3}{\underset{\displaystyle CH_3}{CHCHCOCH_3}}$ gives the more substituted alkene on treatment with base.

PROBLEM 14-5

Indicate the primary product in each transformation and explain the basis for your choice. Discuss the likelihood of cis or trans products.

a $C_6H_5CH_2CHClCH_3$ + KOH

b $C_6H_5CH_2CHClCH_2CH_3$ + KOH

c $C_6H_5CHBrCH_2C_6H_5$ + C_2H_5OK

d Exhaustive methylation and Hofmann elimination of 3-amino-3-methylhexane

e Exhaustive methylation and Hofmann elimination of 3-ethylamino-3-ethylhexane

f Two diastereomers of 2-bromo-1-ethylcyclohexane + potassium *t*-butoxide

g All diastereomers of 2-bromo-1,3-diethylcyclohexane heated with triethylamine

Solvolytic Elimination Reaction (E1)

Carbonium ions formed in solvolysis reactions often undergo elimination reactions as well as nucleophilic substitution, as shown in Fig. 14-5. The slow step in the reaction is the formation of the carbonium ion, and the term $E1$, which is applied to the transformation, refers to the monomolecular character of the rate-determining first stage of the reaction.

Since the same types of carbonium ions intervene in both the $E1$ and S_N1 reactions (Chap. 10), the general reactivity relationships apply to both transformations.

Competition between elimination and substitution is determined largely by the basicity of the solvent and the reactivity of nucleophiles present in solution. For example, solvolysis in pyridine, a basic solvent, usually gives a good deal of elimination, and solvolysis in the presence of thiosulfate ($S_2O_3^{--}$) or azide (N_3^-) ions usually leads to capture of the carbonium ion to form a substitution product. Branching at the β-carbon atom increases the amount of elimination, as does raising the reaction temperature.

As demonstrated in the example below, $E1$ reactions tend to give predominantly the most highly substituted ethylene (Saytzeff rule). Since the leaving group is not involved in the product-determining steps, the product distribution in $E1$ reactions is approximately the same for halides, sulfonates, ammonium ions, and sulfonium ions.

$$CH_3CH_2\underset{\underset{Br}{|}}{C}(CH_3)_2 \xrightarrow[25°]{C_2H_5OH} CH_3CH_2\overset{+}{C}(CH_3)_2 \longrightarrow$$

$$\xrightarrow[-H^+]{C_2H_5OH} CH_3CH_2\underset{\underset{CH_3}{|}}{\overset{\overset{CH_3}{|}}{C}}OC_2H_5 \quad 66\%$$

$$\xrightarrow{-H^+} CH_3CH{=}C(CH_3)_2 \quad 27\%$$

$$+ CH_3CH_2\underset{\underset{CH_3}{|}}{C}{=}CH_2 \quad 7\%$$

$$(CH_3)_3CBr \xrightarrow[65°]{80\% \ C_2H_5OH, \ 20\% \ H_2O} \underset{36\%}{CH_2{=}C(CH_3)_2} + \underset{64\%}{\underline{(CH_3)_3COH + (CH_3)_3COC_2H_5}}$$

$$(CH_3)_3C\overset{+}{S}(CH_3)_2 \xrightarrow{80\% \ C_2H_5OH, \ 20\% \ H_2O} \underset{36\%}{CH_2{=}C(CH_3)_2}$$

$$+ \underset{64\%}{\underline{(CH_3)_3COH + (CH_3)_3COC_2H_5 + (CH_3)_2S}}$$

The $E1$ reactions with halides and sulfonates is little used for preparative purposes, because of the undesirable complexity of the product mixtures. However, a reaction very commonly used in synthesis of alkenes is the acid-catalyzed dehydration of alcohols. Reaction conditions vary widely, depending on the reactivity of the substrate. The role of acid catalysts is the same as in

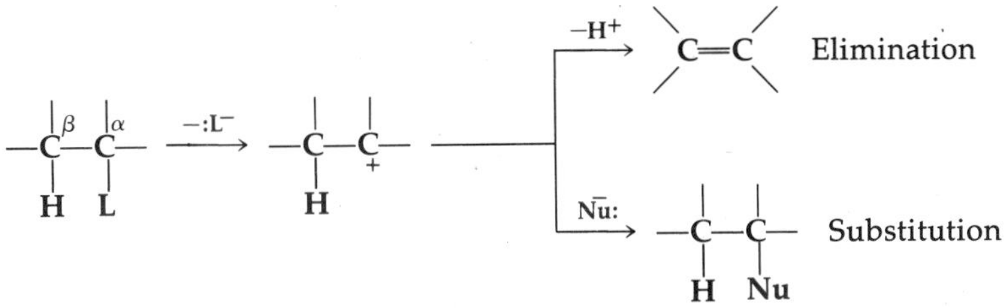

FIGURE 14-5   The relationship between $E1$ and S_N1 reactions

acid-catalyzed nucleophilic substitution reactions of alcohols and ethers. Acids incorporating poor nucleophiles are of course preferred for elimination. The overall reactivity relationships in dehydration reactions are those expected from carbonium-ion reactions.

$$CH_3CH_2OH \xrightarrow[180°]{Conc.\ H_2SO_4} CH_2{=}CH_2$$

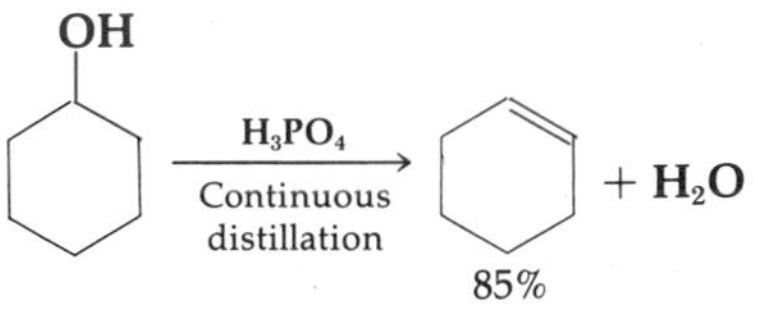

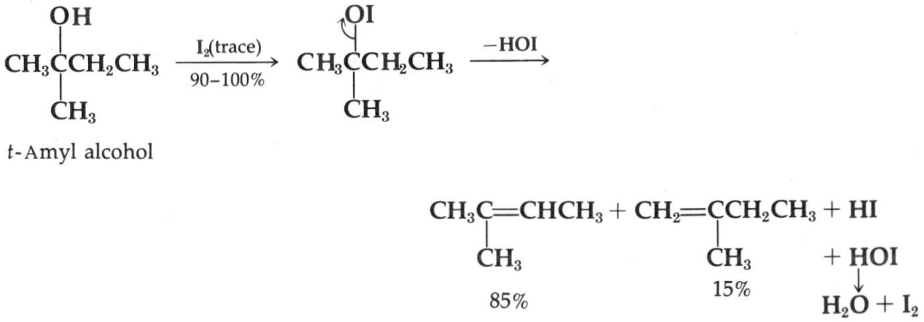

$$\underset{\substack{\\ \\ CH_3 \\ \\ t\text{-Amyl alcohol}}}{\overset{\overset{\displaystyle OH}{|}}{CH_3\overset{|}{C}CH_2CH_3}} \xrightarrow[90\text{--}100\%]{I_2(trace)} \underset{CH_3}{\overset{\overset{\displaystyle OI}{|}}{CH_3\overset{|}{C}CH_2CH_3}} \xrightarrow{-HOI}$$

$$\underset{\substack{CH_3 \\ 85\%}}{CH_3C{=}CHCH_3} + \underset{\substack{CH_3 \\ 15\%}}{CH_2{=}CCH_2CH_3} + HI$$

$$+ HOI \\ \downarrow \\ H_2O + I_2$$

A carbonium ion formed in the course of $E1/S_N1$ reactions can also rearrange especially if a more stable ion can be formed by the migration of a group from an adjacent atom (Chap. 17).

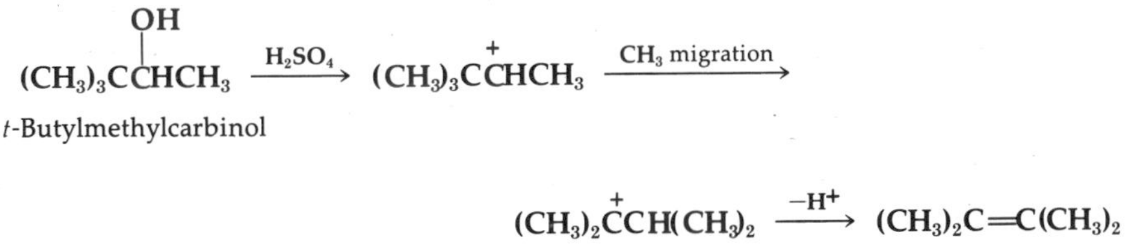

$$\underset{\substack{\\ \\ t\text{-Butylmethylcarbinol}}}{\overset{\overset{\displaystyle OH}{|}}{(CH_3)_3C\overset{|}{C}HCH_3}} \xrightarrow{H_2SO_4} (CH_3)_3C\overset{+}{C}HCH_3 \xrightarrow{CH_3\ migration}$$

$$(CH_3)_2\overset{+}{C}CH(CH_3)_2 \xrightarrow{-H^+} (CH_3)_2C{=}C(CH_3)_2$$

The t-butyl group undergoes elimination especially easily since it forms a stable tertiary carbonium ion. This has been put to use in several synthetic contexts. t-Butyl esters yield carboxylic acids by elimination (alkyl-oxygen cleavage, cf. page 505) under relatively mild acidic (nonhydrolytic) conditions. Hence these esters can be cleaved without affecting other, normal esters and are useful variants in ester reactions (Chaps. 12, 13).

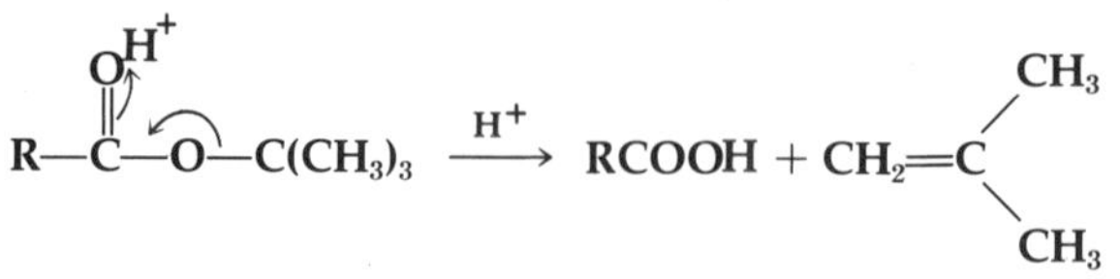

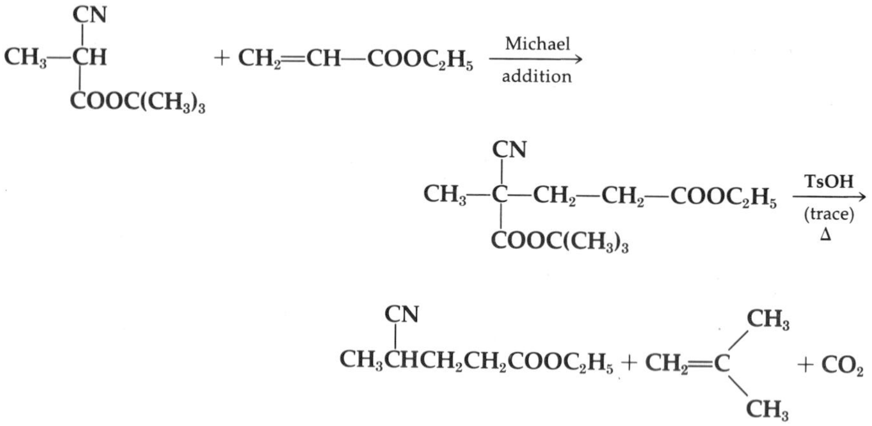

In a similar vein, t-butyl-urethans decompose rapidly to amines on dissolving in trifluoroacetic acid. The initial $E1$ elimination to isobutylene is followed by decarboxylation of the carbamic acid (R—NHCOOH). This "t-butyloxycarbonyl" group is widely used in peptide synthesis for temporary protection of amino groups (Sec. 25-4):

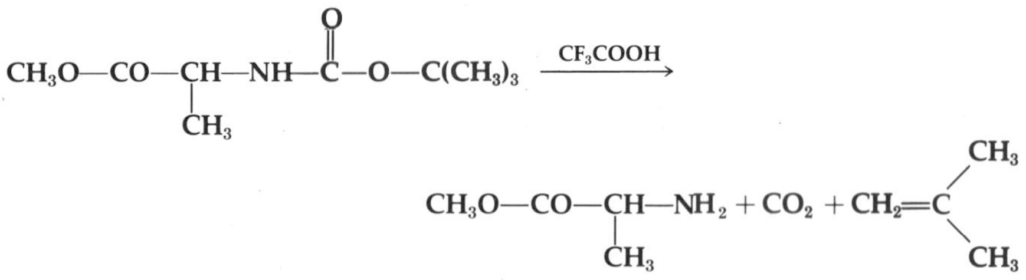

Use is also made of the technique in a modified Gabriel synthesis of primary amines:

$$R—CH_2—X + Na^+ \; \bar{:}N(COO—t\text{-}Bu)_2 \xrightarrow{S_N2} R—CH_2—N(COO—t\text{-}Bu)_2 \xrightarrow[E1]{HCl}$$

$$RCH_2NH_3{}^+Cl^- + 2CO_2 + 2CH_2{=}C(CH_3)_2$$

Thermal Eliminations

In certain eliminations the leaving group also acts as the base, internally attacking the proton in a cyclic mechanism which proceeds in concerted fashion and yields cis elimination. The groups must be oriented **cis coplanar**; they cannot reach if they are trans periplanar. Although cis coplanar they still lie all in one plane for maximum overlap of breaking and forming bond orbitals. The reaction is carried out simply by heating the substrate alone, with no added base, i.e., a **pyrolysis** or **thermolysis** reaction.

In the **Chugaev reaction**, xanthate esters ($R-O-CS-S-R'$; Table 12-2) are pyrolyzed at 140 to 200°. As in nearly all elimination reactions, unsymmetrical substrates give mixtures of products, although there is a tendency for the more highly substituted ethylenes to predominate.

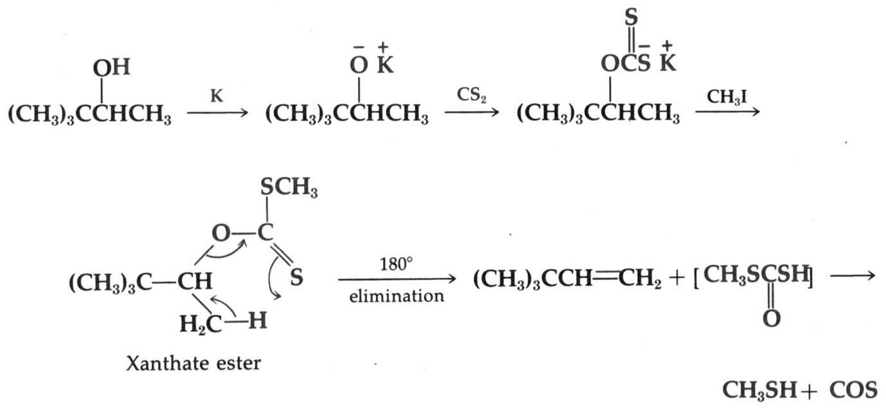

$$(CH_3)_3CCHCH_3 \xrightarrow{K} (CH_3)_3CCHCH_3 \xrightarrow{CS_2} (CH_3)_3CCHCH_3 \xrightarrow{CH_3I}$$

with substituents OH, $\overset{-}{O}\overset{+}{K}$, and $O\overset{S}{\overset{\parallel}{C}}S^- \overset{+}{K}$ respectively

$$(CH_3)_3C-CH \xrightarrow[\text{elimination}]{180°} (CH_3)_3CCH{=}CH_2 + [CH_3SCSH] \longrightarrow$$

Xanthate ester

$$CH_3SH + COS$$

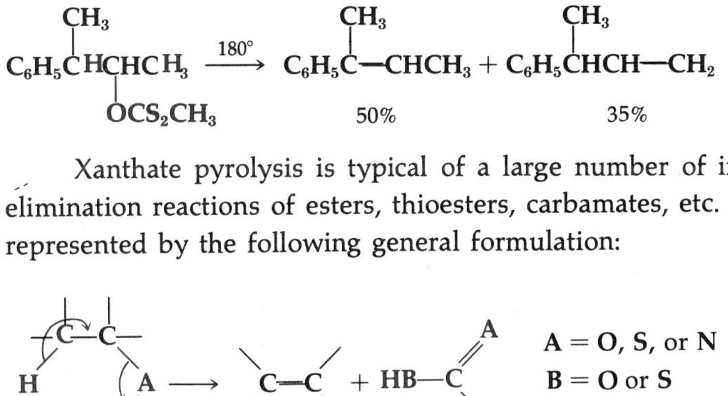

$$C_6H_5\overset{CH_3}{\underset{OCS_2CH_3}{CHCHCH_3}} \xrightarrow{180°} C_6H_5\overset{CH_3}{C}{-}CHCH_3 + C_6H_5\overset{CH_3}{CH}CH{-}CH_2$$

$$\quad\quad\quad\quad\quad\quad\quad\quad\quad 50\% \quad\quad\quad\quad\quad 35\%$$

Xanthate pyrolysis is typical of a large number of internal (cyclic) cis elimination reactions of esters, thioesters, carbamates, etc. The group can be represented by the following general formulation:

$$\rightarrow C{-}C + HB{-}C$$

A = O, S, or N
B = O or S
Z = R, Ar, SR, OR, NHC$_6$H$_5$, or Cl

Pyrolysis of acetate esters, which requires temperatures of 200 to 500°, has been much used in olefin synthesis, because the starting materials are readily prepared. Although there have been reports of highly selective reactions that give the least substituted olefin, the commoner experience is that the abstraction of various β-hydrogen atoms is nearly a random process. Since the reactions are often carried out by passing the vapor of the ester through a hot tube packed with glass helices or beads, it is possible that subtle changes in the surface of the packing material will have a profound influence on the selectivity of the reaction.

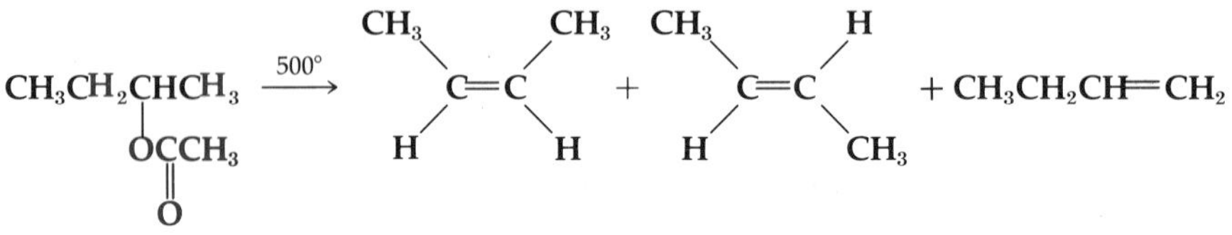

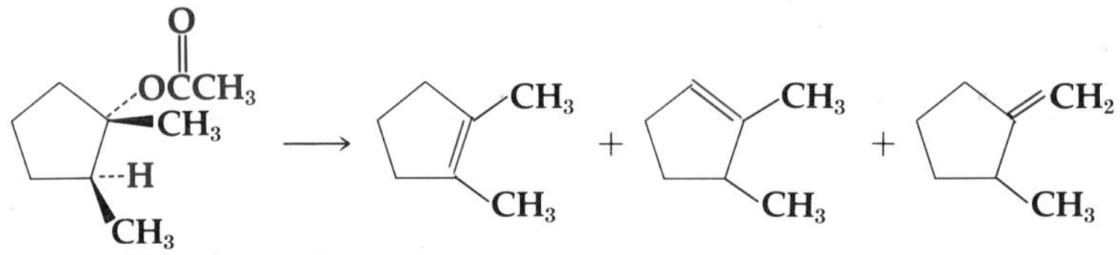

$$CH_3CH_2\underset{\underset{\overset{\|}{O}}{\overset{|}{OCCH_3}}}{CH}CH_3 \xrightarrow{500°} \underset{H}{\overset{CH_3}{C}}=\underset{H}{\overset{CH_3}{C}} + \underset{H}{\overset{CH_3}{C}}=\underset{CH_3}{\overset{H}{C}} + CH_3CH_2CH{=}CH_2$$

cis-1,3-Dimethylcyclopentyl acetate

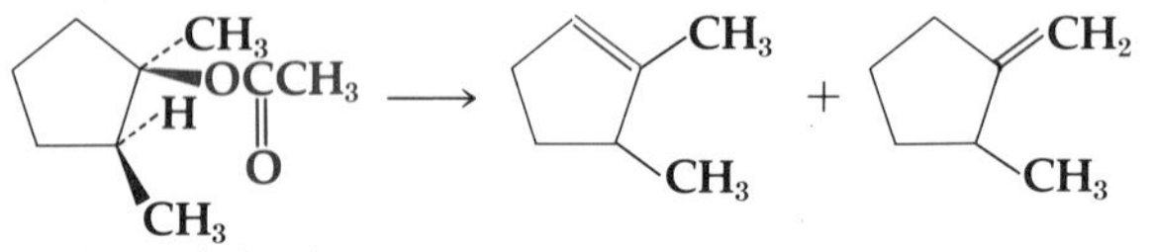

trans-1,2-Dimethylcyclopentyl acetate

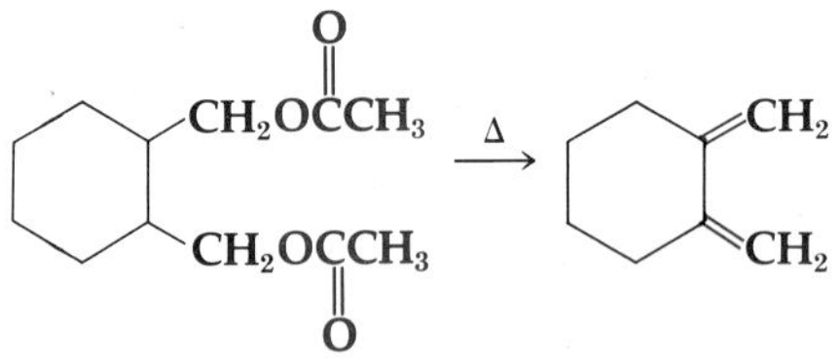

Pyrolysis of amine oxides affords an elegant alkene synthesis. The reaction occurs at relatively low temperatures but, as usual, gives mixtures when the elimination can take two alternative paths, unless there is a pronounced stereochemical advantage for some particular route. The mechanism is probably a cyclic, concerted cis elimination in every case, however:

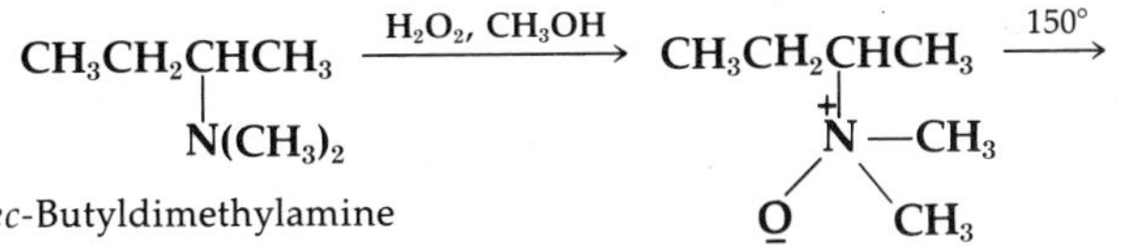

$$CH_3CH_2\underset{\underset{N(CH_3)_2}{|}}{C}HCH_3 \xrightarrow{H_2O_2,\ CH_3OH} CH_3CH_2\underset{\substack{+\\ N-CH_3\\ \underset{O}{|}\ \ \ CH_3}}{C}HCH_3 \xrightarrow{150°}$$

sec-Butyldimethylamine

$$CH_3CH{=}CHCH_3 + CH_3CH_2CH{=}CH_2 + (CH_3)_2NOH$$

21% trans   67%
12% cis

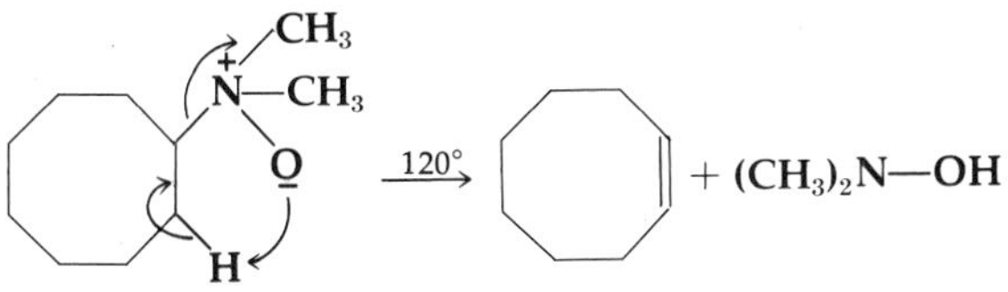

Cyclooctyl-
dimethylamine
oxide

cis-Cyclooctene
81%

Compare with:

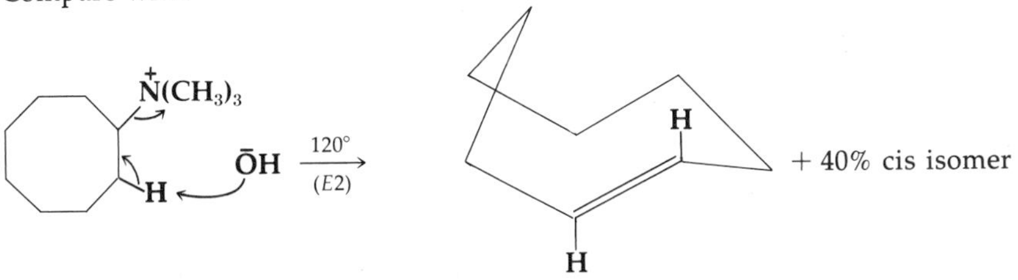

+ 40% cis isomer

Cyclooctyl-
trimethylammonium
hydroxide

trans-Cyclooctene
60%

Pyrolysis of sulfoxides ($R_2S^+{-}O^-$) provides an elimination reaction that at low temperature follows a predominantly cis steric course also but becomes stereochemically indiscriminate at higher temperatures.

PROBLEM 14-6

Select the best procedure to perform the following transformations.

a   Dehydration of a primary alcohol to avoid rearrangement (two ways).

b   Conversion of a nitrile to a terminal olefin.

c   Elimination of an amine without hydrolysis of other functional groups (e.g., esters).

d   Chain extension of primary halides, R—X, to RCH_2COOH without hydrolytic conditions.

14-3 ALKENES BY OTHER ELIMINATIONS

Dehalogenation

Eliminations can occur by similar mechanism with other atoms than hydrogen being removed by the attacking nucleophile (base). Iodide ion causes elimination of 1,2-dihalides in a concerted trans elimination entirely analogous to the $E2$ mechanism described above. The reaction is stereospecific; eliminations that appear to be cis may occur as a consequence of halide exchange (S_N2) reactions that give trans compounds. Similarly 1,2-diols may be converted to disulfonates and eliminated by iodide, probably also by initial displacement.

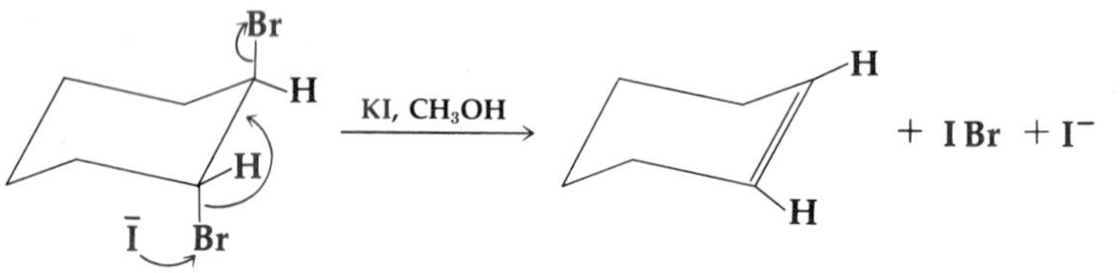

trans-1,2-Dibromocyclohexane

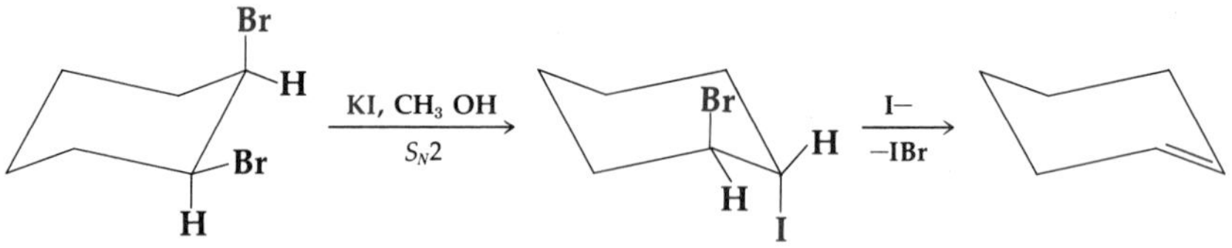

cis-1,2-Dibromocyclohexane   trans-1-Bromo-
                                             2-iodocyclohexane

Reactive metals also remove halogens from 1,2-dihalides and cause similar elimination reactions with related compounds, such as halohydrins or their esters and 1,2-haloethers. Zinc is generally used in preparative work, although other metals effect the same reaction. Grignard and lithium reagents cannot be made from 1,2-dihalides because of this reaction. As in $E2$ reactions, elimination occurs preferentially in a trans fashion, but the metals appear to be less selective since cis 1,2-halohydrins eliminate almost as easily as trans.

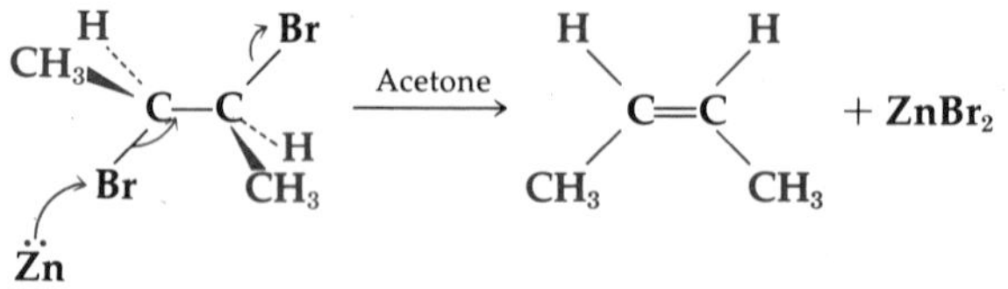

threo-2,3-Dibromobutane   cis-2-Butene

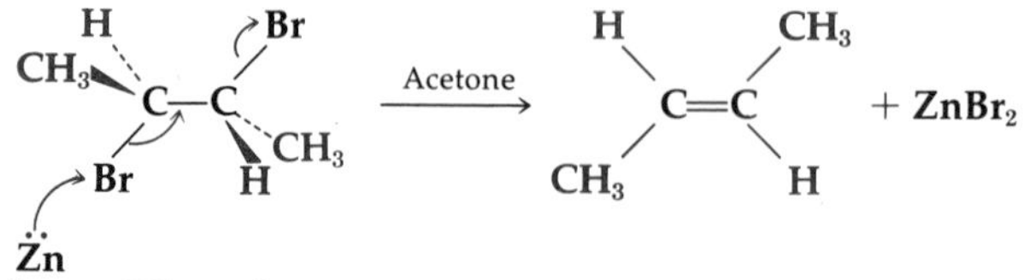

erythro-2,3-Dibromobutane   trans-2-Butene

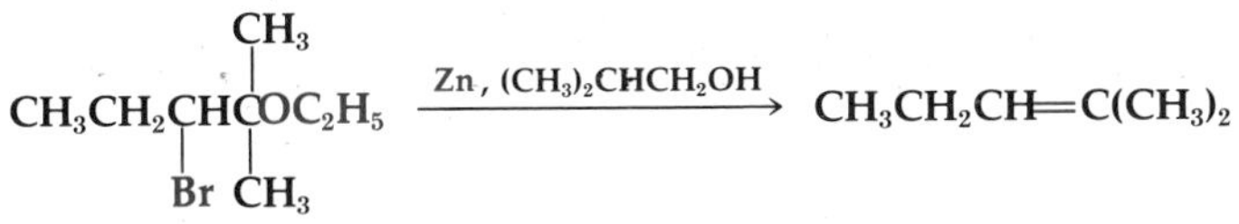

$$\underset{\substack{| \\ Br \quad CH_3}}{CH_3CH_2CHCOC_2H_5} \xrightarrow{\text{Zn, }(CH_3)_2CHCH_2OH} CH_3CH_2CH\!=\!C(CH_3)_2$$

with CH₃ above the chain

$$Mg + BrCH_2CH_2Br \xrightarrow{\text{Ether}} MgBr_2 + CH_2\!=\!CH_2$$

A reaction which bears a close mechanistic resemblance (related as elimination to enolization) is the zinc reduction of α-haloketones, as well as other α leaving groups (—OR, —OCOR, etc.). The procedure creates the zinc enolate, which may be used in subsequent reaction for alkylations or other nucleophilic reactions, the Reformatsky reaction (page 467) being one example.

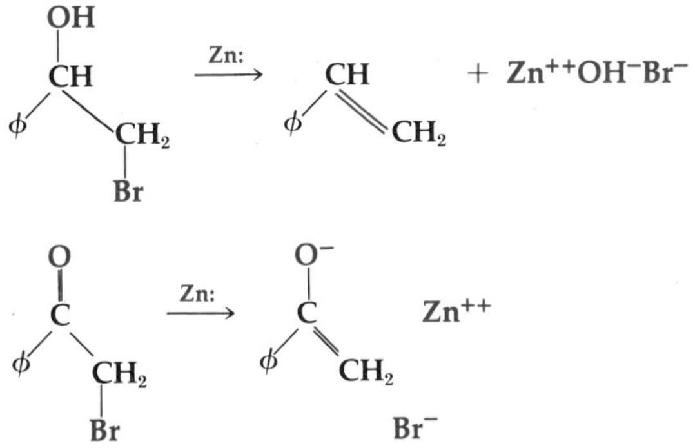

Decarboxylative Elimination

Salts of β-haloacids may undergo decarboxylation with elimination. Like other decarboxylations (Sec. 13-6) this represents loss of CO_2 in place of proton in the normal elimination. The reaction usually occurs with di-α-substituted acids, in which no proton is available for normal β-elimination.

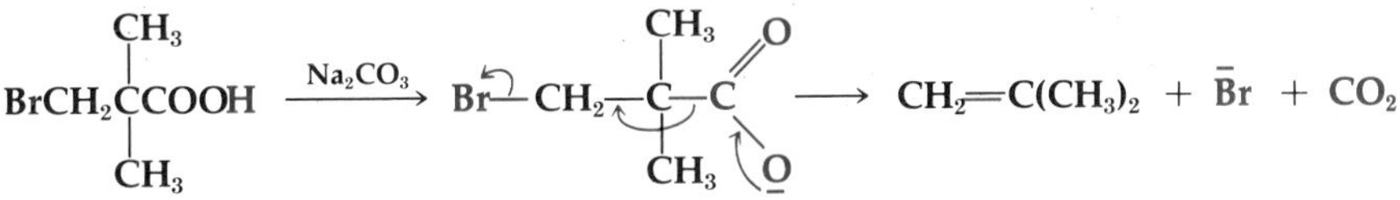

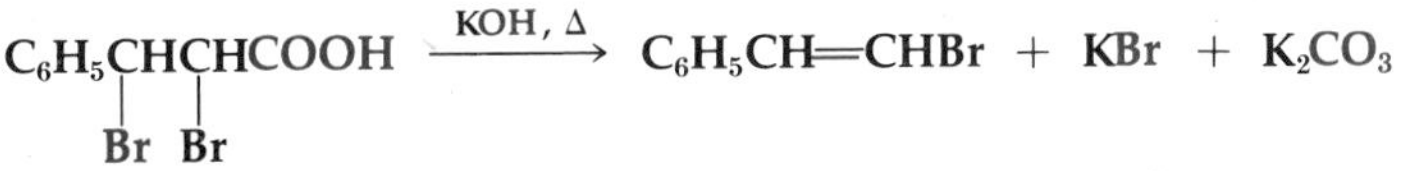

$$C_6H_5\underset{\substack{| \quad | \\ Br \quad Br}}{CHCHCOOH} \xrightarrow{\text{KOH, }\Delta} C_6H_5CH\!=\!CHBr + KBr + K_2CO_3$$

Cinnamic acid dibromide β-Bromostyrene

Fragmentations

In decarboxylative elimination a carbon-carbon bond (C—COO⁻) is broken in the elimination reaction, so that this may be described as a **fragmentation** reaction, since the carbon skeleton as a whole is cleaved. The decarboxylation reactions of β-ketoacids, epoxy-acids, cyclopropane acids, and unsaturated acids (Sec. 13-6) may also be viewed as eliminations, since they form double bonds, or as fragmentations, since they cleave the carbon skeleton.

A few other fragmentation reactions are also known, fitting a general pattern with the leaving group γ to the atom bearing the electron pair.

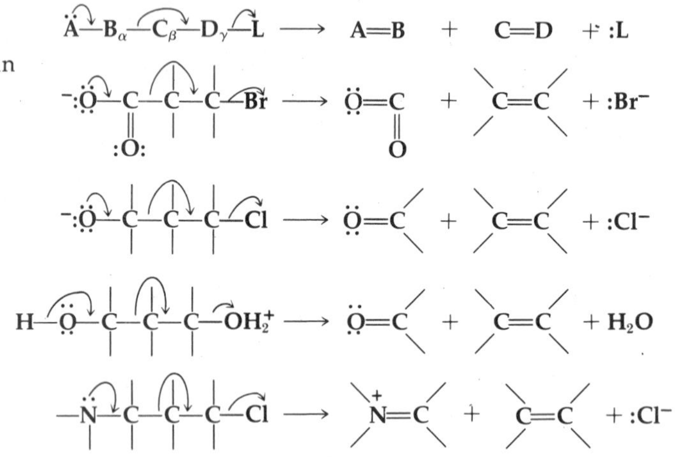

as in

$$\overset{\frown}{A}-B_\alpha\overset{\frown}{-}C_\beta-D_\gamma\overset{\frown}{-}L \longrightarrow A{=}B \quad + \quad C{=}D \quad + \quad :L$$

$$-:\ddot{O}-C\overset{\frown}{-}C-\overset{\frown}{C}-\overset{\frown}{Br} \longrightarrow \ddot{O}{=}C \quad + \quad C{=}C \quad + :Br^-$$

$$-:\ddot{O}-C-C-C\overset{\frown}{-}Cl \longrightarrow \ddot{O}{=}C \quad + \quad C{=}C \quad + :Cl^-$$

$$H-\ddot{O}-C-C-C-\overset{\frown}{O}H_2^+ \longrightarrow \ddot{O}{=}C \quad + \quad C{=}C \quad + H_2O$$

$$-\ddot{N}-C-C-C\overset{\frown}{-}Cl \longrightarrow \overset{+}{N}{=}C \quad + \quad C{=}C \quad + :Cl^-$$

These reactions are not common, since carbon-carbon bond cleavage is disfavored (lack of polarity), and usually more ordinary substitutions or eliminations occur under the conditions employed. The second example below illustrates the trans-periplanar orientation of the two groups (colored bonds) required in the central elimination reaction.

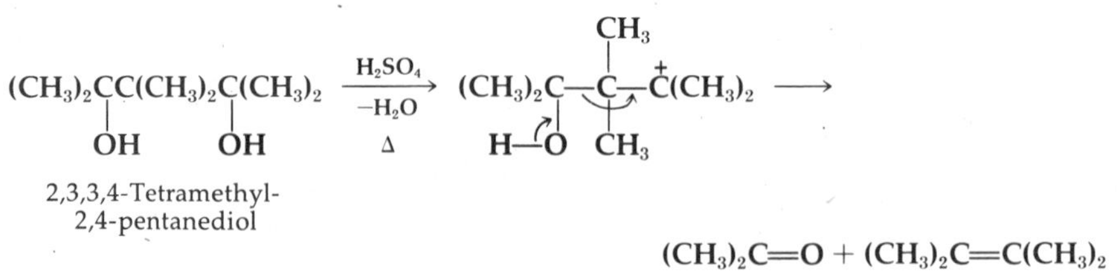

$$(CH_3)_2\underset{\underset{OH}{|}}{C}C(CH_3)_2\underset{\underset{OH}{|}}{C}(CH_3)_2 \xrightarrow[\substack{-H_2O \\ \Delta}]{H_2SO_4} (CH_3)_2C-\underset{\underset{H-O}{\nearrow}}{\overset{\overset{CH_3}{|}}{C}}-\overset{+}{C}(CH_3)_2 \longrightarrow$$

2,3,3,4-Tetramethyl-
2,4-pentanediol

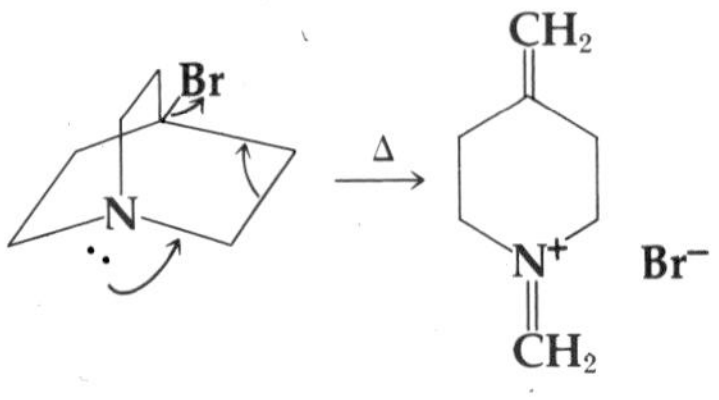

$$(CH_3)_2C{=}O + (CH_3)_2C{=}C(CH_3)_2$$

Phosphine Eliminations

Phosphines (R_3P:) and phosphites [$(RO)_3P$:] show a strong tendency to react by bonding to oxygens (and sulfurs) since the phosphorus-oxygen (and phosphorus-sulfur) bond is very strong, as in R_3P^+—O^- and $(RO)_3P^+$—O^-. This may be seen in the second, or elimination, step of the Wittig reaction (page 476). Phosphines remove the oxygen from amine oxides by direct attack on oxygen. They also convert epoxides into their corresponding alkenes by oxygen removal but in this case the process is not a direct attack on oxygen but is several steps, initiated by phosphine attack on carbon.

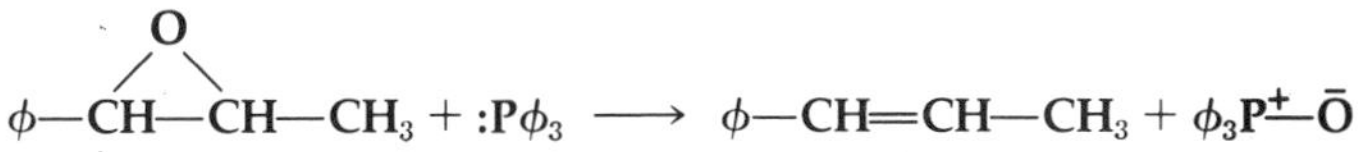

$$\phi\text{—}\overset{\displaystyle \overset{O}{\diagup \backslash}}{CH\text{—}CH}\text{—}CH_3 + :P\phi_3 \longrightarrow \phi\text{—}CH{=}CH\text{—}CH_3 + \phi_3P^{\pm}\bar{O}$$

A synthetically useful elimination of 1,2-diols to alkenes involves phosphine (or phosphite) treatment of a cyclic thiocarbonate of the diol. Cyclic cis diols yield cis alkenes and trans diols yield trans alkenes. In this way the very reactive *trans*-cycloheptene could be observed briefly before it decomposed.

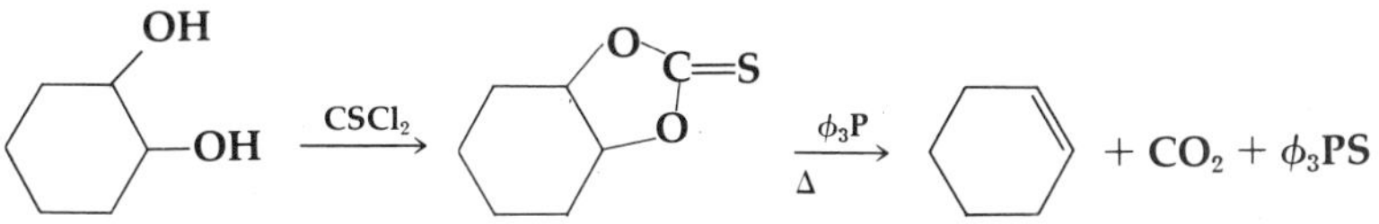

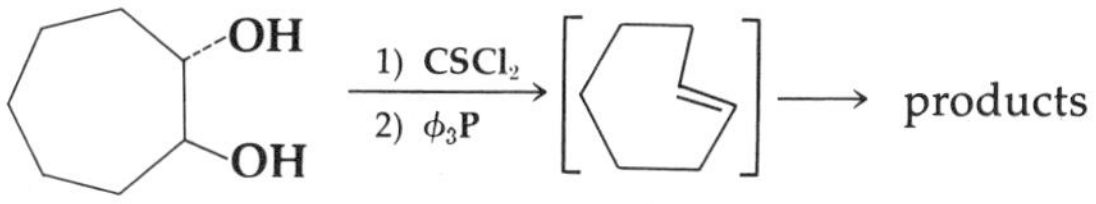

trans-Cycloheptene

PROBLEM 14-7

Write specific examples of the following descriptions.

a   An elimination of a saturated dibromo compound with zinc to give a conjugated diene

b   An alkene formed in base with loss of CO_2

c   A saturated lactone giving an unsaturated acid with zinc

d   A saturated monoalcohol yielding formaldehyde and an olefin

e   A cyclohexylamine which fragments to an acyclic salt hydrolyzable to an aldehyde

f   An epimer of **e** which does not fragment ͻ

PROBLEM 14-8

The reverse aldol (**retroaldol**) reaction and the reverse Michael reaction are fragmentation reactions. Show how they parallel the fragmentation reactions discussed above in terms of conditions, mechanism, etc. The reverse Michael reaction commonly requires high temperatures as well as base, as in the case of compound *A* heated to 450° with K_2CO_3 and distilled. Explain, and show the products.

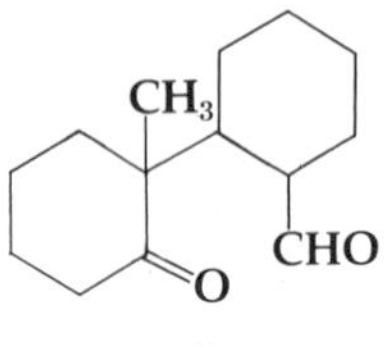

A

14-4   OTHER DOUBLE BONDS (C=N; C=O) BY ELIMINATION

Application of the same principle of elimination to form heteroatom multiple bonds allows two possibilities owing to the asymmetry of the **C=Z** bond: the leaving group may initially be bonded to carbon or to heteroatom, as shown in Fig. 14-6.

The first of these will be recognized as the reverse of carbonyl addition (of **L:** + **C=Z**; Chap. 12) and is found in the addition-elimination reactions discussed in Secs. 12-2 and 12-5. The second type is discussed here. It is characterized by attachment of a leaving group to the heteroatom, i.e., by a bond between two heteroatoms. Such bonds are relatively unstable (Table 2-2) and often undergo elimination as soon as they are formed. All of these reactions involve oxidation and are generalized as the addition of a reagent, as *electrophile*, to the heteroatom and its leaving with the electron-pair as a leaving group (i.e., *nucleophile*). The reagent is thus reduced. This generalization is discussed further in the chapter (18) on oxidation, but a few examples are offered here.

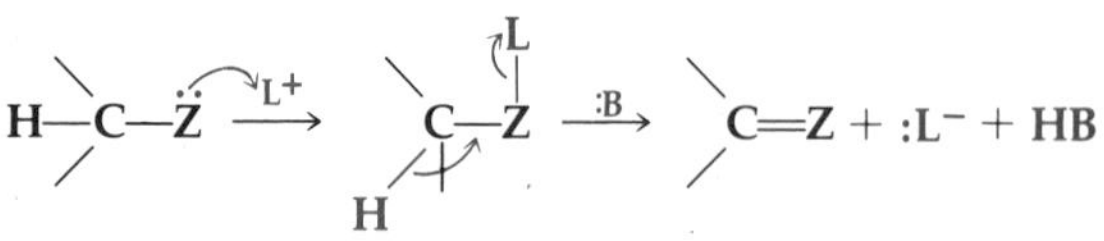

Tertiary amines can be converted to immonium salts by reaction with mercuric acetate, the first step apparently being formation of an **N—Hg** bond, from which mercury acts as a leaving group.

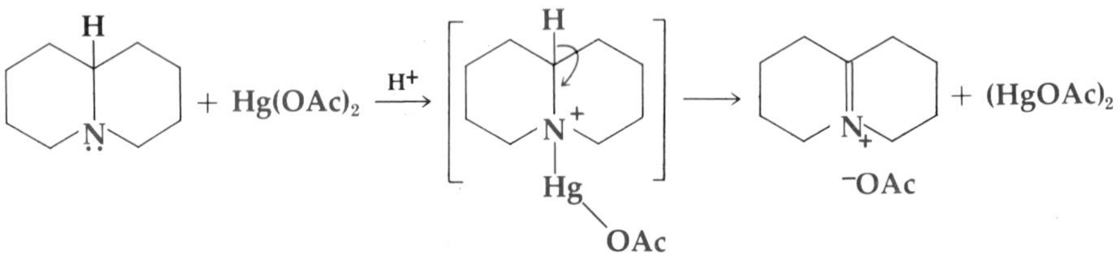

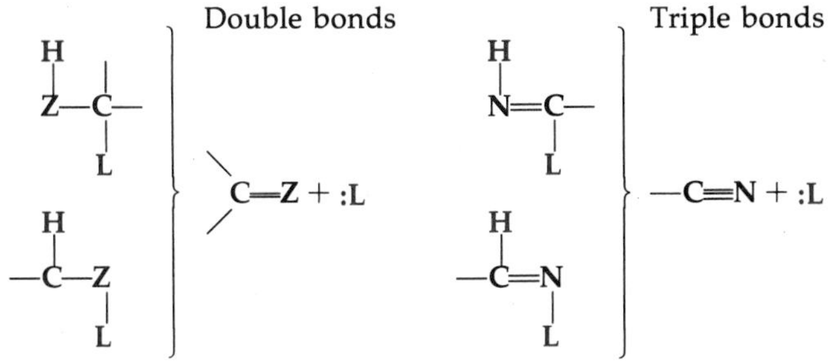

Double bonds   Triple bonds

$$\left.\begin{array}{c} \overset{H}{\underset{L}{Z-C-}} \\[2em] \overset{H}{\underset{L}{-C-Z}} \end{array}\right\} \quad \overset{\diagdown}{\diagup}C=Z + :L \qquad \left.\begin{array}{c} \overset{H}{\underset{L}{N=C-}} \\[2em] \overset{H}{\underset{L}{-C=N}} \end{array}\right\} \quad -C\equiv N + :L$$

$(Z = N, O, S)$

FIGURE 14-6 **Two modes of elimination from unsymmetrical bonds**

Primary and secondary alcohols are treated with chromic acid to give aldehydes and ketones, respectively. With secondary alcohols the reaction affords high yields, whereas with primary alcohols the aldehyde produced is subject to further oxidation to a carboxylic acid or to self-condensation reactions, both of which reduce the yield of aldehyde unless proper care is exercised. The reactions are carried out in the presence of a proton acceptor such as water or pyridine, and thus may be done under acidic or basic conditions. This chromic oxidation is a primary practical procedure for oxidation of alcohols (tertiary alcohols are inert).

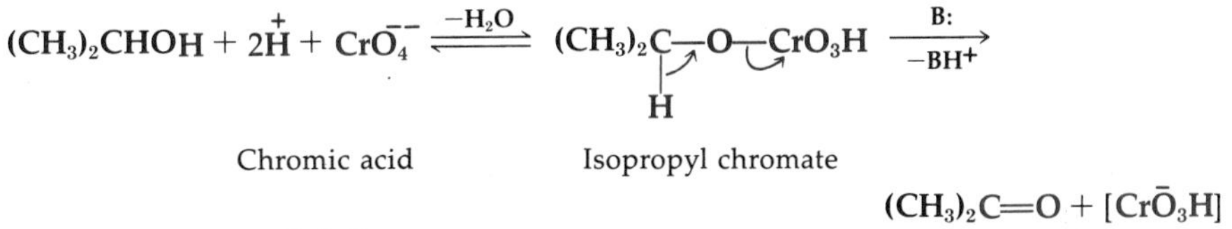

$$(CH_3)_2CHOH + 2\overset{+}{H} + Cr\overset{--}{O_4} \underset{-H_2O}{\rightleftharpoons} (CH_3)_2\underset{H}{C}-O-CrO_3H \underset{-BH^+}{\overset{B:}{\longrightarrow}}$$

Chromic acid   Isopropyl chromate

$$(CH_3)_2C=O + [Cr\bar{O}_3H]$$

$$CH_3CH_2CH_2OH \underset{H_2O}{\overset{H_2Cr_2O_7}{\longrightarrow}} CH_3CH_2CHO$$

Removed by continuous distillation

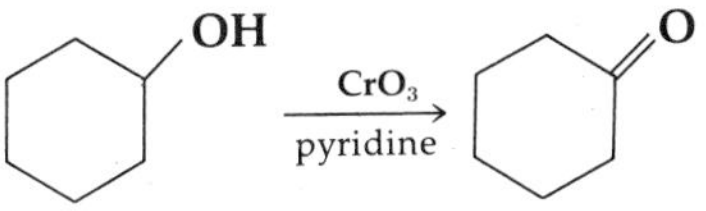

OH → CrO₃ / pyridine → O

The Oppenauer oxidation may also be regarded as an elimination, in which the leaving group is *hydride* (H:⁻) ion. The reverse reaction is thus hydride addition to carbonyl (Sec. 12-3), particularly as in the Meerwein-Ponndorf reduction (page 461). The Canizzaro reaction is another similar hydride transfer which constitutes an elimination (page 460).

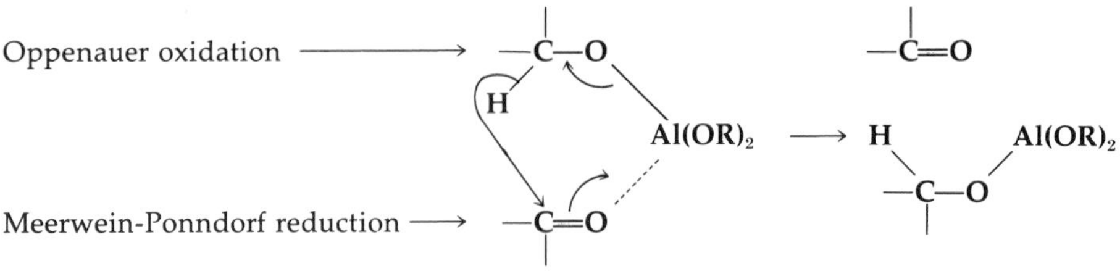

Oppenauer oxidation

Meerwein-Ponndorf reduction

PROBLEM 14-9

Write simple ionic mechanisms to account for these conversions.

a Dimethylbenzylamine + mercuric acetate $\longrightarrow$ X $\xrightarrow{H_2O}$ benzaldehyde + ?

b 1,2-Dicarbomethoxyhydrazine + lead tetraacetate $\longrightarrow$ dimethyl azodicarboxylate + ?

c 2-Hydroxybutyric acid + lead tetraacetate $\longrightarrow$ propionaldehyde + ?

d Cyclopentanol + aluminum *t*-butoxide + acetone $\longrightarrow$ cyclopentanone + ?

e 1-Phenyl-1-butanol + chromic anhydride (CrO_3) in pyridine $\longrightarrow$ butyrophenone ($C_6H_5COC_3H_7$) + ?

14-5   TRIPLE BONDS BY ELIMINATION

Alkynes

Alkynes can be readily prepared by a number of base-initiated elimination reactions resembling reactions that lead to alkenes. In general, somewhat more drastic conditions with respect to base and temperature are needed to produce alkynes, but the prevalence of trans elimination as a stereoelectronic requirement is the same.

Treatment of vinyl halides or *gem*- or *vic*-dihalides† with strong bases effects eliminations, with the consequent formation of acetylenes. Nonterminal acetylenes are isomerized to terminal acetylenes with strong base. Therefore, direct synthesis by elimination is usually restricted to the terminal type or to structures in which rearrangement is blocked by phenyl or vinyl groups.

†Ketones can be converted to *gem*-dihalides with phosphorus pentachloride (page 452).

$$CH_3CH_2CH_2CCl_2CH_3 + 3NaNH_2 \xrightarrow[\substack{-2NaCl \\ -2NH_3}]{Toluene} CH_3CH_2CH_2C\equiv\overset{-}{C}\overset{+}{Na} \xrightarrow{H_3O^+}$$

$$CH_3CH_2CH_2C\equiv CH$$
1-Pentyne

$$CH_3(CH_2)_4CBr=CHCH_3 \xrightarrow[\text{2) } H_3O^+]{\text{1) } NaNH_2} CH_3(CH_2)_4CH_2-C\equiv CH$$

$$C_6H_5CHBrCH_2Br \xrightarrow[\text{2) } H^+]{\text{1) } NaNH_2} C_6H_5C\equiv CH$$
Styrene dibromide   Phenylacetylene
50%

$$C_6H_5CHBrCHBrCOOH \xrightarrow{KOH} C_6H_5C\equiv C\overset{-}{C}\overset{+}{O}_2 K \xrightarrow{H^+} C_6H_5C\equiv CCOOH$$
Cinnamic acid   Phenylpropiolic
dibromide   acid
80%

The base-catalyzed rearrangement involves allenes as intermediates. Rearrangement is not complete with bases such as sodium ethoxide, and mixtures are obtained as reaction products. The driving force for rearrangement to terminal acetylenes derives from the low acidity of the C≡C—H group. With very strong bases such as sodamide, acetylide salts are formed as the most stable anionic end product.

$$\bar{B} + RC\equiv CCH_3 \rightleftharpoons BH + \begin{bmatrix} RC\equiv C-\overset{..}{\overset{-}{C}}H_2 \\ \updownarrow \\ R\overset{..}{\overset{-}{C}}=C=CH_2 \end{bmatrix} \underset{B^-}{\overset{BH}{\rightleftharpoons}} RCH=C=CH_2 \underset{BH}{\overset{B^-}{\rightleftharpoons}}$$
An allene

$$\begin{bmatrix} RCH=C=\overset{..}{\overset{-}{C}}H \\ \updownarrow \\ R\overset{..}{\overset{-}{C}}H-C\equiv CH \end{bmatrix} \underset{B^-}{\overset{BH}{\rightleftharpoons}} RCH_2C\equiv CH \xrightarrow{B^-} RCH_2C\equiv \overset{..}{C}: + BH$$

By combination of the synthesis of terminal acetylenes with use of acetylide ions as nucleophiles in substitution reactions, nonterminal acetylenes can be made. Starting materials for the elimination are obtained by addition of halogens to double bonds (Chap. 15) or by the reaction of phosphorus pentachloride with aldehydes or ketones.

The formation of acetylenes is an *E2* reaction and shows the same preference for trans eliminations as the corresponding olefin-producing eliminations.

$$\left.\begin{array}{c}
\underset{\text{H}}{\overset{\text{CH}_3}{\diagdown}}C=C\underset{\text{CH}_3}{\overset{\text{Br}}{\diagup}} \\
\text{Fast} \\
\underset{\text{H}}{\overset{\text{CH}_3}{\diagdown}}C=C\underset{\text{Br}}{\overset{\text{CH}_3}{\diagup}} \\
\text{Slow}
\end{array}\right\} \xrightarrow{\text{KOH, C}_2\text{H}_5\text{OH}} \text{CH}_3\text{C}\equiv\text{CCH}_3 + \text{CH}_3\text{CH}_2\text{C}\equiv\text{CH}$$

Active metals dehalogenate 1,2-dihaloethylenes with ease. The reaction has little preparative value, however, because of the inaccessibility of the dihalides. Decarboxylative eliminations to give acetylenes are also known. Both reactions are facilitated if trans elimination can occur.

$$\left.\begin{array}{c}
\underset{\text{Br}}{\overset{\text{CH}_3}{\diagdown}}C=C\underset{\text{CH}_3}{\overset{\text{Br}}{\diagup}} \\
\textit{trans-2,3-Dibromo-2-butene} \\
\underset{\text{Br}}{\overset{\text{CH}_3}{\diagdown}}C=C\underset{\text{Br}}{\overset{\text{CH}_3}{\diagup}} \\
\textit{cis-2,3-Dibromo-2-butene}
\end{array}\right\} \xrightarrow{\text{Zn}} \text{CH}_3\text{C}\equiv\text{CCH}_3 + \text{ZnBr}_2$$

(Fast / Slow)

$$\underset{\text{Br}}{\overset{\text{CH}_3}{\diagdown}}C=C\underset{\text{CH}_3}{\overset{\overset{\displaystyle\text{O}}{\|}}{\overset{\displaystyle\text{C}-\text{O}}{}}\overset{+}{\text{Na}}} \longrightarrow \text{CH}_3\text{C}\equiv\text{CCH}_3 + \text{CO}_2 + \text{NaBr}$$

Benzyne (Arynes)

When treated with strong bases such as sodamide, aryl halides undergo successive elimination and addition reactions in which **arynes** (aromatic ynes) appear as discrete intermediates. These eliminations are necessarily cis coplanar, not trans. The aryne intermediates are very unstable owing to the serious deformation of the bond angles involved (from 180 to 120°), and they readily undergo addition reactions. Although no arynes have been isolated, spectroscopic evidence for the transient existence of **benzyne** has been obtained. As would be expected, mixtures of products are obtained from substituted halobenzenes.

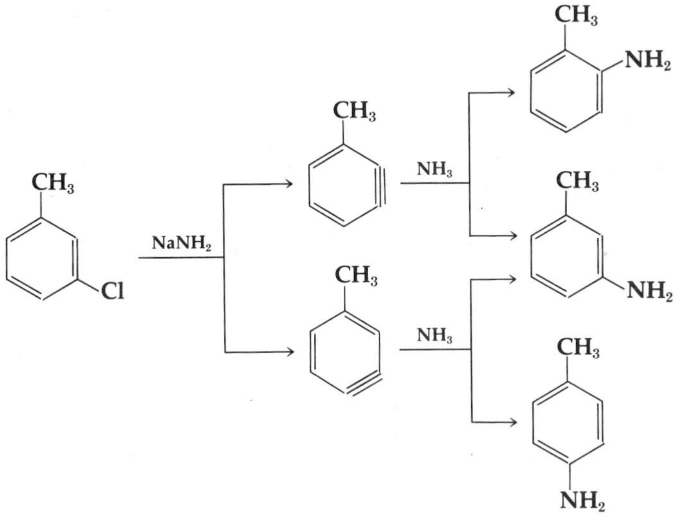

By generation of arynes through different reactions and in different media, a number of syntheses can be carried out.

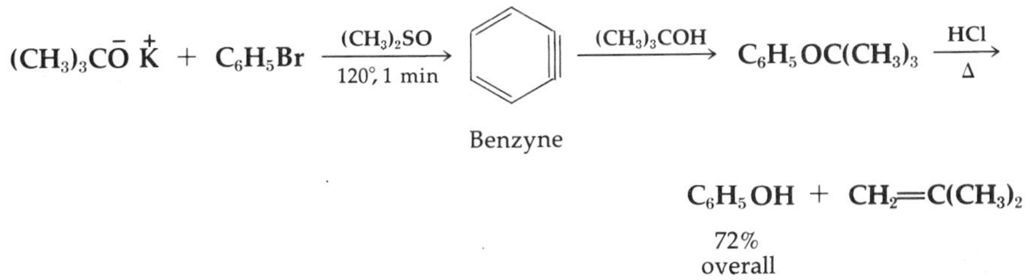

Benzyne

$$C_6H_5OH + CH_2\!=\!C(CH_3)_2$$

72%
overall

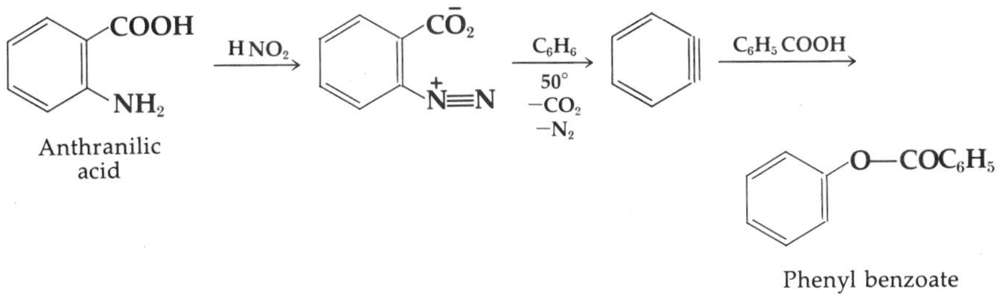

Anthranilic
acid

Phenyl benzoate

Nitriles

As with the generation of double bonds containing heteroatoms, eliminations which form nitriles can also occur by two different routes (Fig. 14-6), and again the first one has been seen before in the dehydration of amides with thionyl chloride, phosphorus pentachloride, and other electrophiles which attack amide oxygen and convert it to a good leaving group for elimination (page 515).

The second elimination type, as before, requires a bond between hetero-atoms, i.e., from N to N, O, Cl, etc. The oximes (C=N—OH) have such a bond and are quite stable compounds. However, any of the reagents which convert —OH to a good leaving group (Sec. 11-1) will readily dehydrate aldoximes (oximes of aldehydes) to nitriles, i.e., PCl_5, $SOCl_2$, RSO_2Cl, Ac_2O, or strong acid. This reaction, when coupled with the hydrolysis of the nitrile, provides a reaction sequence for oxidation of an aldehyde to an acid without recourse to the use of strong oxidizing agents.

$$C_6H_5\underset{\underset{CH_3}{|}}{C}HCHO \xrightarrow[-H_2O]{NH_2OH} C_6H_5\underset{\underset{CH_3}{|}}{C}HCH{=}NOH \xrightarrow[-CH_3COOH]{(CH_3CO)_2O}$$

2-Phenylpropanal   2-Phenylpropanald-
           oxime

$$C_6H_5\underset{\underset{CH_3}{|}}{C}H{-}\overset{\overset{H}{\downarrow}}{C}{=}N{-}O{-}\underset{\underset{O}{\parallel}}{C}CH_3 \longrightarrow C_6H_5\underset{\underset{CH_3}{|}}{C}HCN + CH_3COOH$$

    2-Phenylpropanaldoxime   2-Phenylpropionitrile
      acetate

$$CH_3CH_2CH_2CH{=}NOH \xrightarrow{(CH_3CO)_2O} CH_3CH_2CH_2CN$$

  n-Butyraldoxime   n-Butyronitrile

The oximes also exhibit the expected decarboxylative fragmentation when bonded to carboxyl. Thus, the oximes of α-ketoacids readily fragment on heating to yield nitriles:

$$\phi{-}CH_2{-}\underset{\underset{O}{\parallel}}{C}{-}COOH \xrightarrow{NH_2OH} \phi{-}CH_2{-}\underset{\underset{\underset{OH}{|}}{N}}{C}\underset{\underset{H}{|}}{\overset{\overset{O}{\parallel}}{\underset{C}{\diagup}}}O \xrightarrow{\Delta} \phi{-}CH_2{-}CN + CO_2 + H_2O$$

PROBLEMS

14-10 Write the products of the following transformations.

  a   o-Chlorotoluene + t-**BuOK** $\xrightarrow{\Delta}$ products $\xrightarrow{CF_3COOH}$?

  b   Di-t-butyl-α-ketoglutarate + $NH_2OH \longrightarrow A \xrightarrow[\Delta]{TsOH} B + ?$

  c   Di-t-butyl-α-ketosuccinate + $NH_2OCH_3 \longrightarrow A \xrightarrow[\Delta]{TsOH} B + ?$

  d   Benzaldehyde + $NH_2OH \longrightarrow A \xrightarrow{SOCl_2} B + ?$

  e $C_6H_5CO\underset{\underset{P(C_6H_5)_3}{|}}{C}C_6H_5 \xrightarrow{\Delta}$ hydrocarbon + ?

f 1-Bromo-1-cyclododecene $\xrightarrow{\text{NaNH}_2}$?

g 1-Bromo-1-cyclohexene $\xrightarrow{\text{NaNH}_2}$?

h 2-Bromo-2-hexene $\xrightarrow{\text{NaNH}_2}$?

i 2,3-Dibromohexane $\xrightarrow{\text{NaNH}_2}$?

j Dimethyl malonate sodium salt + 4-iodotoluene $\xrightarrow{\Delta}$?

14-11 Write out 10 base-catalyzed olefin-forming eliminations, each of which leads to a different olefin and in which the same combination of base and leaving groups is never employed.

14-12 Indicate the major olefinic products of each of the following reactions, and state the reasons for your choices.

a $CH_3CH_2CH_2\underset{\underset{\displaystyle Br}{|}}{C}(CH_3)_2 \xrightarrow{C_2H_5ONa}$

b $CH_3CH_2CH_2\underset{\underset{\displaystyle (CH_3)_3\overset{+}{N}\overset{-}{I}}{|}}{C}(CH_3)_2 \xrightarrow[\Delta]{Ag_2O}$

c $CH_3CH_2\underset{\underset{\displaystyle OTs}{|}}{C}HCH_3 \xrightarrow{\text{Acetic acid, } \Delta}$

d $CH_3CH_2\underset{\underset{\displaystyle (CH_3)_2\overset{+}{N}-\overset{-}{O}}{|}}{C}HCH_3 \xrightarrow{\text{Acetic acid, } \Delta}$

Wait, c has $(CH_3)_3\overset{+}{N}\overset{-}{I}$

c $CH_3CH_2\underset{\underset{\displaystyle OTs}{|}}{C}HCH_3 \xrightarrow{\text{Acetic acid, } \Delta}$

$(CH_3)_3\overset{+}{N}\overset{-}{I}$

d $CH_3CH_2\underset{\underset{\displaystyle (CH_3)_2\overset{+}{N}-\overset{-}{O}}{|}}{C}HCH_3 \xrightarrow{\text{Acetic acid, } \Delta}$

e $C_6H_5CH_2\overset{}{C}HCH_3 \xrightarrow{\Delta}$

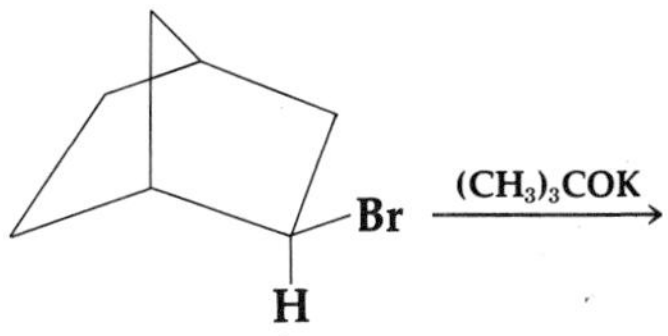

f $\xrightarrow{(CH_3)_3COK}$

g $\xrightarrow{(CH_3)_3COK}$

h $\xrightarrow{C_2H_5OK}$

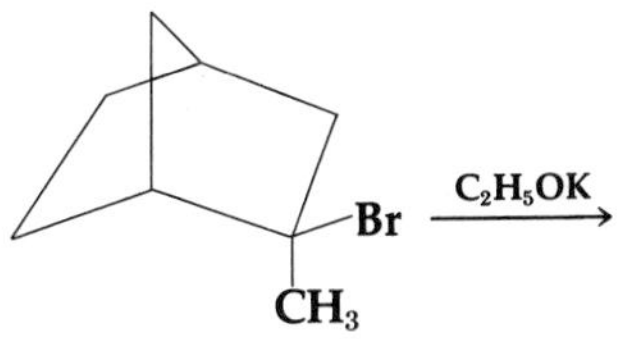

i $(C_6H_5)_2\underset{\underset{\displaystyle OH}{|}}{C}-\underset{\underset{\displaystyle CH_3}{|}}{\overset{\overset{\displaystyle CH_3}{|}}{C}}-\underset{\underset{\displaystyle OH}{|}}{C}(CH_3)_2 \xrightarrow{H_2SO_4}$

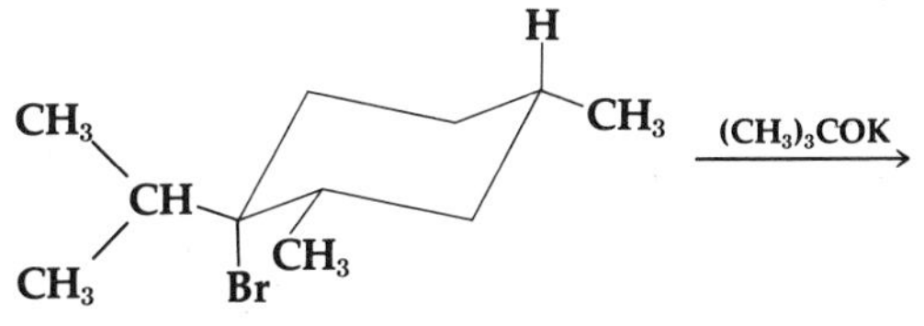

14-13 Write out formulas for the following conversions. Note the useful reaction of bromine with double bonds discussed in the next chapter.

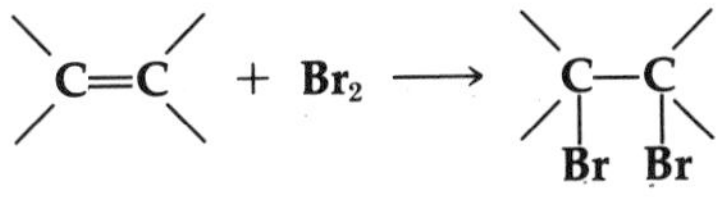

$$\overset{\diagdown}{\underset{\diagup}{C}}{=}\overset{\diagup}{\underset{\diagdown}{C}} + Br_2 \longrightarrow \underset{\underset{\displaystyle Br}{|}}{\overset{\diagdown}{C}}-\underset{\underset{\displaystyle Br}{|}}{\overset{\diagup}{C}}$$

a Propionaldehyde and other organic materials to *pure* 2-butene

b Propionaldehyde and other organic materials to 2-methyl-2-butene

c *n*-Butanol to 1-butyne

d Benzophenone [$(C_6H_5)_2CO$] and other materials to 1,1-diphenyl-1-butene

    e Stilbene to α-bromostilbene (1-bromo-1,2-diphenyl-ethene)

    f *o*-Methoxychlorobenzene to *m*-methoxyaniline

    g Methyl-*n*-propyl ketone to *n*-propylacetylene

    h Acetaldehyde to crotononitrile

    i Propionic acid to propionitrile (without using $ClSO_2NCO$)

    j Bromobenzene and other materials to benzylphenyl ether

    k 1,6-Heptadiene from 1,3-dibromopropane and other materials

    l Benzylphenylcarbinol to diphenylacetylene

    m 1,4-Pentadiene from piperidine $[(CH_2)_5NH]$

    n 1,1-Dimethoxypentane to 1-hexyne

    o Cycloheptanone to cycloheptene

    p Acetophenone and other materials to 3-phenyl-1-butene

    q 1-Hexyne from 3-hexene

    r Cyclohexanone from 1-butylcyclohexanol

    s Diphenylacetylene from 2,3-diphenylpropenoic acid

    t 1-Octyne from 4-octyne

14-14 Deduce the structures of the following compounds, and write out the transformation involved.

    a A compound when treated with acid gave 1 mole of benzaldehyde and 1 mole of α-methylstilbene.

    b A compound when heated with base gave 1 mole of carbon dioxide and 1 mole of 1-methylcyclohexene

    c A compound when treated with acid gave 2 moles of acetic acid and 1 mole of 3,5-dimethylphenol.

14-15 With three-dimensional formulas, work out the probable steric course of the following reaction for each of the two diastereomeric starting materials.

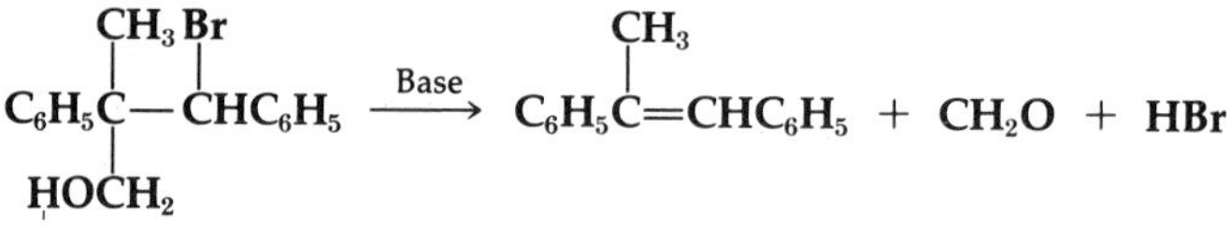

$$C_6H_5\overset{\overset{\displaystyle CH_3}{|}}{\underset{\underset{\displaystyle HOCH_2}{|}}{C}}-\overset{\overset{\displaystyle Br}{|}}{C}HC_6H_5 \xrightarrow{Base} C_6H_5\overset{\overset{\displaystyle CH_3}{|}}{C}=CHC_6H_5 + CH_2O + HBr$$

14-16 Suggest mechanisms to explain the following transformations:

b $C_6H_5\underset{\underset{\displaystyle Br}{|}}{C}HCH_2CH_2\underset{\underset{\displaystyle Br}{|}}{C}HC_6H_5 \xrightarrow{Zn} 2C_6H_5CH{=}CH_2 + ZnBr_2$

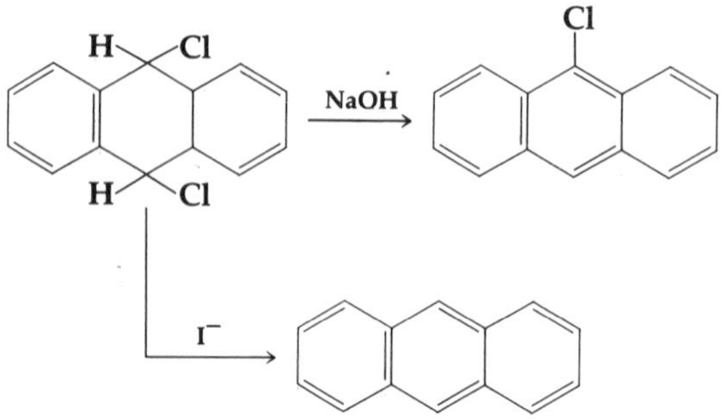

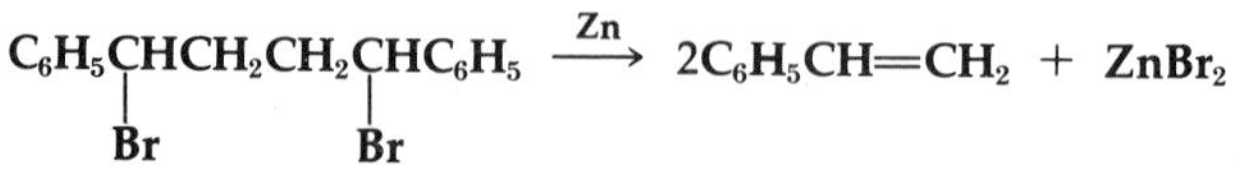

14-17 With three-dimensional structures to show stereochemistry trace the following transformations:

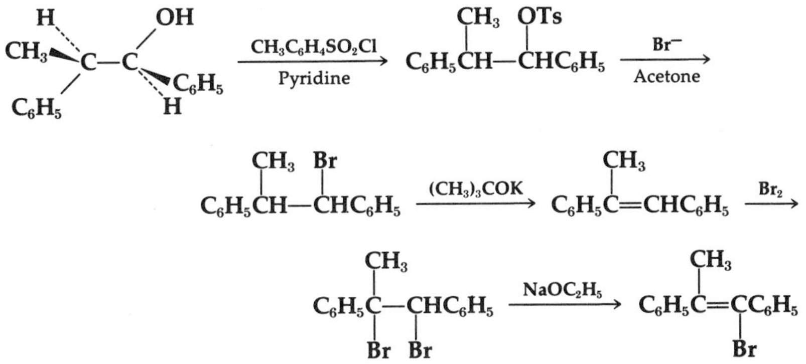

$$C_6H_5\overset{\overset{\displaystyle CH_3}{|}}{C}H-\overset{\overset{\displaystyle Br}{|}}{C}HC_6H_5 \xrightarrow{(CH_3)_3COK} C_6H_5\overset{\overset{\displaystyle CH_3}{|}}{C}=CHC_6H_5 \xrightarrow{Br_2}$$

$$C_6H_5\overset{\overset{\displaystyle CH_3}{|}}{\underset{\underset{\displaystyle Br}{|}}{C}}-\overset{\underset{\displaystyle Br}{|}}{C}HC_6H_5 \xrightarrow{NaOC_2H_5} C_6H_5\overset{\overset{\displaystyle CH_3}{|}}{C}=\overset{}{\underset{\underset{\displaystyle Br}{|}}{C}}C_6H_5$$

14-18 With three-dimensional formulas, write out the structures for one enantiomer of each of the four racemates of

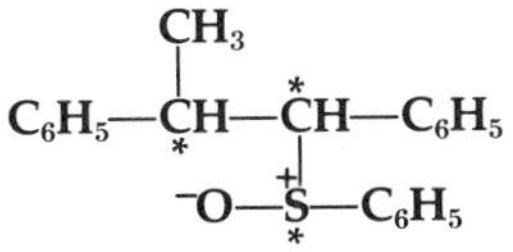

$$C_6H_5-\overset{\overset{\displaystyle CH_3}{|}}{\underset{*}{C}}H-\overset{*}{\underset{\underset{\displaystyle \overset{\displaystyle -O-\overset{+}{\underset{*}{S}}-C_6H_5}{|}}{}}{C}H-C_6H_5$$

14-19 An unknown compound, $C_7H_{12}O_2$, showed IR bands at 2.8 and 5.8 μ (3600, 1710 cm^{-1}) and no intense UV absorption above 200 nm. On heating with thionyl chloride it produced two isomers, $C_7H_{10}O$, the major one with an intense UV maximum at 258 nm, the minor one with an intense band at 220 nm. Write structures for the three compounds.

14-20 When acetophenone is treated with conc. H_2SO_4 the sulfate of the aromatic pyrilium salt is formed.

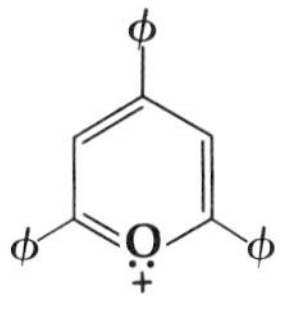

At first it was believed that the other product must be methane because of the net stoichiometry implicit in the simplest reaction. No methane was found, however. A more sophisticated view of the mechanism implied that the other product must be 2-phenylpropene (or the products formed from it in sulfuric acid). Demonstrate that this fits a revised stoichiometry and then examine the mechanism, first linking all the necessary acetophenones by aldol reactions, with or without dehydrations, as required. What is the key step in creating the correct carbon skeletons?

14-21 Account for the following reaction, and show how the stereochemistry in the starting material is important to the reaction by drawing it in perspective.

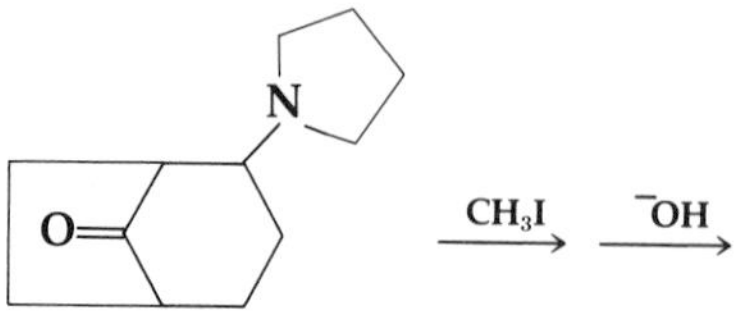

$\xrightarrow{CH_3I}$ $\xrightarrow{^-OH}$ 4-cycloheptene-1-carboxylic acid

14-22 Delineate a mechanism for the following reaction. Would you expect it to proceed the same way on the epoxide of 3-cyclohexenone?

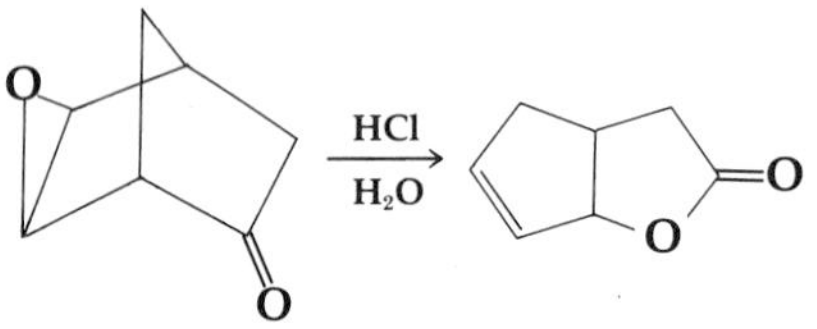

$\xrightarrow[H_2O]{HCl}$

14-23 Write a stereochemically correct structure for the product in this reaction and show a mechanism for the reaction.

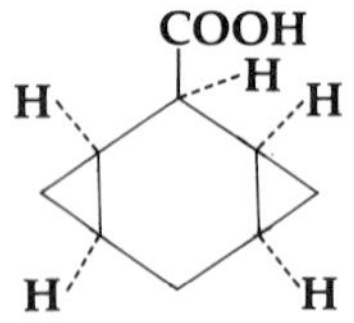

$\xrightarrow{\Delta}$ $C_8H_{12} + CO_2$

14-24 Predict the major elimination product from the following reactions.

a $CH_3CH_2CH_2CH_2\underset{\underset{Br}{|}}{C}(CH_3)_2$ $\xrightarrow{NaOH}$

b 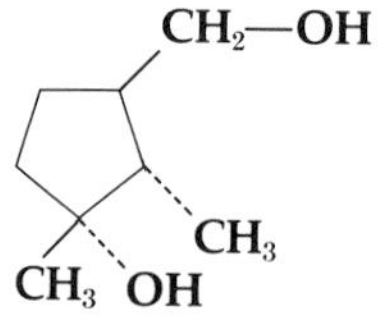 $\xrightarrow{H_2SO_4}$

c $CH_3CH_2\underset{\underset{^+N(CH_3)_3}{|}}{C}HCH_3$ OH^- $\xrightarrow{\Delta}$

d $\xrightarrow[Acetone]{NaI}$

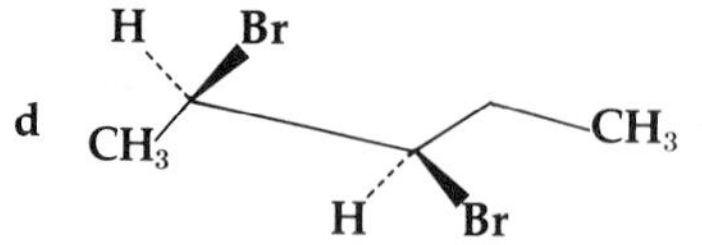

ELECTROPHILIC ADDITIONS TO MULTIPLE BONDS

IN the reactions considered so far the carbon skeleton of interest has generally acted as an electron-deficient (electrophilic) substrate, $R^{\delta+}—L^{\delta-}$, $C^{\delta+}=O^{\delta-}$, for attack by nucleophilic reagents. In Chaps. 15 and 16 we shall change focus and examine strong electrophilic reagents acting on a mild electron source in the carbon skeleton of interest. The electron source here is primarily the very weakly basic (nucleophilic) π electrons of the double and triple bond. These are inactive except to strong electrophilic reagents, and are not generally attacked by nucleophiles.

Many electrophilic reactions are only mechanistic variations on reactions we have seen already, but are initiated by a high-energy electrophile rather than a nucleophile. While most of the nucleophilic reagents and conditions examined so far have been strongly basic, the electrophilic reactions are usually acidic, and the proton (H^+) is of course a common and powerful electrophile. After initial attack of the electrophile, the intermediate carbonium ion formed has two courses open to it: addition of weak nucleophile (as in S_N1) or loss of proton. Thus the two electrophilic reactions are addition and substitution, respectively.

Electrophilic reagents are characterized by an atom with an incomplete outer electron shell (usually six electrons only and an empty orbital) and a desire to acquire two electrons, as from a π bond, to fill this shell. Such configurations are listed for the common atoms in Fig. 15-1 (cf. Fig. 2-3) and include both positive (cf. H^+) and neutral (cf. BF_3) species. Since these configurations are intrinsically unstable, they are of course very reactive and in many cases do not commonly exist in the form listed but must be created as transient, high-energy species in the reaction mixture. The major electrophiles† are listed in Fig. 15-1 and discussed in detail in these two chapters.

---

† Some possible species are not useful electrophilic reagents since they combine with themselves too easily (oxygen atom) or rearrange internally ($R—\ddot{O}:^+$).

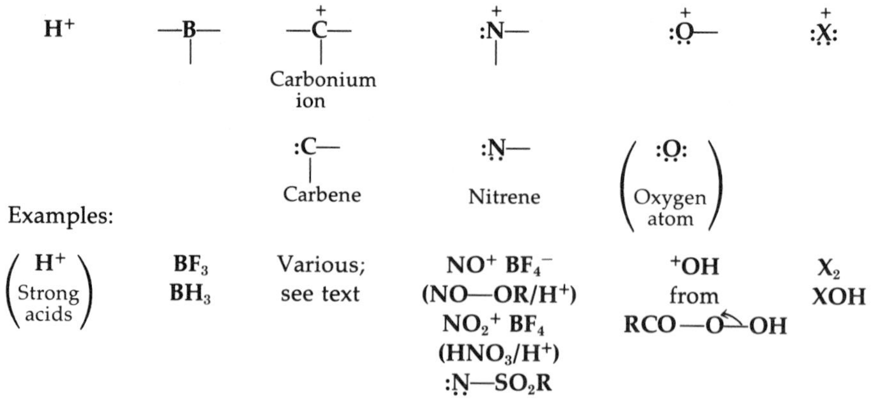

H^+ $-\overset{\displaystyle +}{\underset{\displaystyle |}{B}}-$ $-\overset{\displaystyle +}{\underset{\displaystyle |}{C}}-$ $:\overset{\displaystyle +}{\underset{\displaystyle |}{N}}-$ $:\overset{\displaystyle +}{\underset{\displaystyle \cdot\cdot}{O}}-$ $:\overset{\displaystyle +}{\underset{\displaystyle \cdot\cdot}{X}}:$

Carbonium
ion

$:\underset{\displaystyle |}{C}-$ $:\underset{\displaystyle \cdot}{N}-$ $\left(\begin{array}{c}:\overset{\cdot\cdot}{O}:\\ \text{Oxygen}\\ \text{atom}\end{array}\right.$

Carbene   Nitrene

Examples:

$\left(\begin{array}{c}H^+\\ \text{Strong}\\ \text{acids}\end{array}\right)$ BF_3 Various; $NO^+ BF_4^-$ ^+OH X_2

BH_3 see text $(NO{-}OR/H^+)$ from XOH

$NO_2^+ BF_4$ $RCO{-}O{\overset{\frown}{}}OH$

(HNO_3/H^+)

$:\underset{\displaystyle \cdot}{N}{-}SO_2R$

FIGURE 15-1   **Electronic configurations in electrophiles**

The present chapter deals with electrophilic additions, the next with electrophilic substitutions. Addition is generally the reaction of simple double and triple bonds, while substitution is characteristic of aromatic rings. The large resonance energy of the aromatic π-electron cycle makes these electrophilic reactions more difficult (larger $\Delta F\ddagger$) and also encourages substitution over addition, since proton loss from the intermediate carbonium ion regenerates the resonance-stabilized aromatic system.

Electrophilic substitution is generally a *substitution for hydrogen*, which leaves as H^+, in contrast to nucleophilic substitutions, in which leaving groups are negative. Alternatively stated, the group substituted in nucleophilic substitution leaves with its original bonding pair of electrons, while in electrophilic substitution it leaves without that electron pair.

15-1   DIRECTION AND STEREOCHEMISTRY OF ADDITION

The various reagents to be discussed for electrophilic addition to alkenes act by a single mechanism and so a general introduction can serve for them all. Furthermore, the mechanism is both a simple and a reasonable one and can practically be discerned by a consideration of the stereoelectronic necessities. These in turn make three kinds of predictions about the orientation of substituents, which are usually valid in practice.

1   The geometrical orientation of the two added substituents to each other and the rest of the molecule

2   The choice of which of the two double-bond carbons is ultimately bonded to the electrophilic atom (**E**) and which to the secondary nucleophile (**Nu**) in unsymmetrical double bonds

3   The side of the double bond which is attacked by the electrophile.

Since the six atomic nuclei involved in a double bond are all coplanar and the π orbital is perpendicular to this plane on both sides, we expect approach of an electrophilic atom to be perpendicular also as its empty orbital comes in for bonding overlap with the π orbital. This initial overlap is often called a π complex, and may or may not be an actual intermediate on the free-energy diagram.

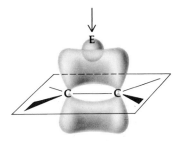

Initial π complex formed by attack of π electrons on vacant orbital of an electrophile, **E**

Most of the electrophiles of Fig. 15-1 have another pair of electrons on the electrophilic atom and so may form a true three-membered ring. This may be only a reactive intermediate as with halogens (X⁺) or it may be the final product, as with carbenes and nitrenes. The proton electrophile has no other electrons and so full bonding in a three-membered ring intermediate is not possible with H⁺.

Product

Product

Orientation of Added Substituents

The π complex or three-membered ring intermediate is now roughly above the original molecular plane of the double bond, and so must be opened by approach of a nucleophile from *below* that plane. This amounts to a displacement, exactly analogous to that for opening opoxides (page 417) and creates an initial addition product with the nucleophile oriented trans periplanar to the electrophilic atom.

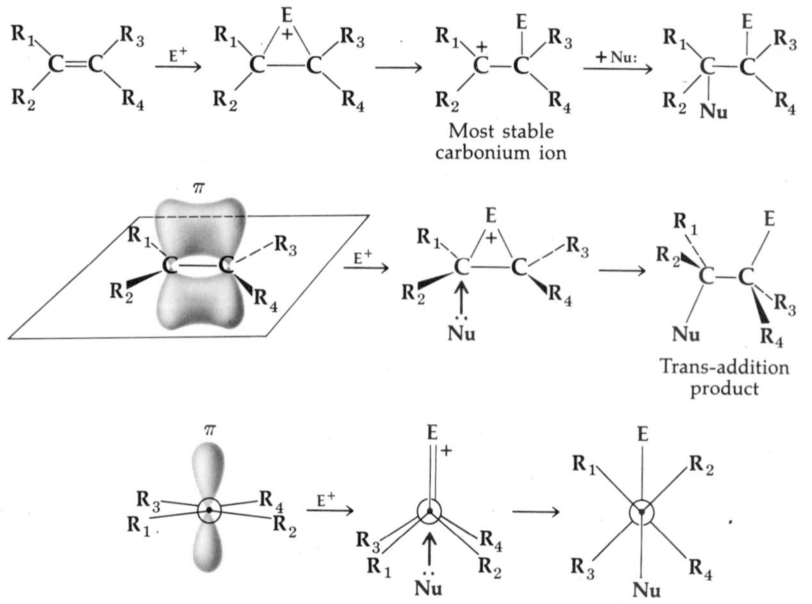

FIGURE 15-2   Steps in electrophilic addition

This overall mechanism is called **trans addition** to the double bond and is depicted in several views in Fig. 15-2. The overall hybridization change of the two original sp^2 carbons is to fold one down to its sp^3 configuration and the other up, resulting in a final staggered conformation about the two carbons, as the projection formulas most clearly indicate.

Trans addition is the common experience in additions to double bonds. It is of course a pathway which is exactly the reverse of the elimination reaction requiring trans periplanar substituents to form a double bond (page 585). The stereoelectronic requirements are the same for reaction in either direction. Eliminations, however, are usually base-initiated while additions commonly proceed under acidic conditions.

Position of Added Substituents

There remains the question of which of the two double-bond carbons is chosen for attack of the nucleophile, for with unsymmetrical double bonds two products are possible in principle (Fig. 15-2). We encountered this choice before with the opening of epoxides, and the choice was made on the basis of whether the substitution occurred by direct displacement (S_N2; basic conditions) or by initial ionization of the protonated epoxide (S_N1; acidic conditions). The acidic conditions of electrophilic addition imply only the latter course and indeed this is observed quite generally.

Hence we may envision the primary bond-breaking in the three-membered-ring intermediate (or π complex) as occurring at the carbon yielding the *more stable carbonium ion,* followed *directly* by nucleophilic attack from below. The transition state, then, has a lot of carbonium ion character and so the route

More hindered side

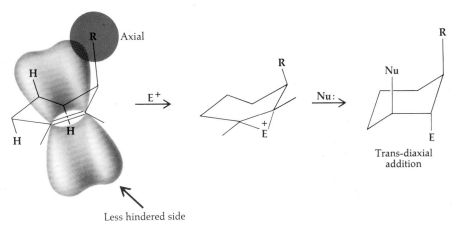

FIGURE 15-3   **Trans addition to cyclohexenes**

via the more stable carbonium ion is the energetically preferred one. This is also indicated in Fig. 15-2 and will be apparent in the examples of subsequent sections. The mechanistic course of opening the three-membered intermediate or π complex here is precisely the same as that in S_N1 opening of epoxides. Relative stability of carbonium ions was discussed in Chap. 10.

Steric Hindrance and Cyclic Olefins

Since the substituents on a double bond are all in one plane, it is unlikely in an *acyclic* molecule that they could induce any preference for electrophilic attack from above or below that plane. In cyclic molecules, however, substituents can sterically hinder reagent approach from one side, especially if they are axial. Double bonds in six-membered rings (cyclohexenes) exhibit a molecular plane which is roughly parallel with the plane of the ring so that the π orbitals are then perpendicular to it and hence parallel to the axial substituents. An axial substituent larger than hydrogen—such as methyl—will then sterically hinder approach of the electrophile on its side of the ring, and the attack will occur preferentially from the opposite, less hindered side.† This is shown in Fig. 15-3.

Attack of the nucleophile on the three-membered intermediate (or π complex) now has no choice and must approach from the hindered side for backside displacement, giving a product which is trans diaxial. It is a general observation with cyclohexenes that they yield trans-diaxial substituents on electrophilic addition (or epoxide opening). However, unless the ring is conformationally rigid it may undergo a *subsequent* conformational change to the other chair form in which the trans substituents have become diequatorial.

---

† In contrast to this steric effect, however, hydrogen bonding by an axial —**OH** to the incoming electrophile has been observed in some cases to direct attack on the *same* side of the ring as the axial —**OH**.

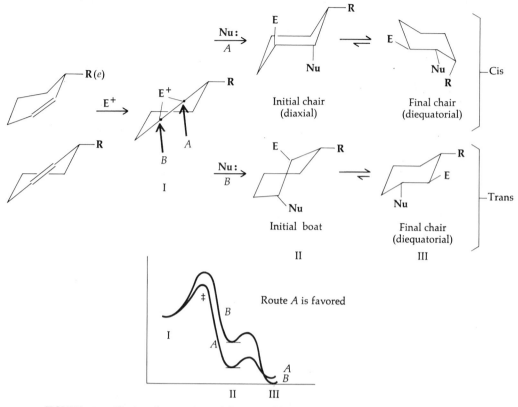

FIGURE 15-4   Choice of routes in cyclohexene addition

In the mechanism shown the two added substituents in cyclohexene addition must complete the addition in a trans-periplanar orientation, but it is not necessarily clear from this that they must become trans diaxial on a chair cyclohexane. In the diagrams of Fig. 15-4 attack on the initial three-membered intermediate is examined at both possible carbons. In each case, of course, inversion of the carbon attacked by the nucleophile must occur. Each situation results in a trans-periplanar arrangement of the added substituents, but one (*A*) results in a chair and the other (*B*) a boat conformation. Since the latter is less stable, the transition state leading to it is also less stable (higher energy) and so relatively disfavored. Each initial product must undergo a subsequent conformational change to a more stable form with the added groups still trans, but diequatorial on chair rings. However, it is the transition to the *initial* product which determines the favored route, and route *A*, to an initial chair ring with diaxial substituents, is thus the pathway which is virtually always observed.

PROBLEM 15-1

   a When *trans*-1-phenyl-1-octene is brominated, two asymmetric centers are
   formed. Two diastereomers are possible; are they both formed? Are the
   products optically active? Explain.
   b Discuss the stereochemical result of brominating optically active *cis*- and
   trans-2-pentene-4-ol.
   c Discuss the stereochemical result of brominating optically active *cis*- and
   trans-4-bromo-2-pentene. (It will be useful to draw these in projection for
   clarification; be careful to show the correct stereochemistry.)
   d Similarly discuss the bromination of optically active R-4-*t*-butyl-1-cyclo-
   hexene, drawing the stereoisomers involved in perspective with the correct
   absolute stereochemistry.

Relative Rates

Electrophilic attack on a double bond implies the availability of an electron pair in the π orbital. Hence it is reasonable that if that electron density
is somewhat diluted by resonance with an electron-withdrawing group the rate
of attack will be lowered. Conversely substituents on the double bond which
can donate electrons to it increase the rate. More specifically, if substituents
can stabilize the carbonium ion which develops in the transition state, they
lower the activation energy and increase the rate (they also determine which
carbon ultimately bears the positive charge). Similarly, electron-withdrawing
substituents raise the transition-state energy by destabilizing the developing
carbonium ion, and they decrease the rate.

Electron-donating	*Electron-withdrawing*
(*Rate increase*)	(*Rate decrease*)
$R_2\ddot{N}$— (enamine)	—NO_2
$R\ddot{O}$— (enol ether)	—CO—
C=C—	—CN
ϕ—	—SO_2—
R—	

   These effects are of course the same ones that we have seen before for
stabilizing and destabilizing positive charge (cf. S_N1 reaction). A double bond
stabilized by —**OH** is the familiar enol, and its attack on an electrophile like
the carbonyl has already been treated in the acid-catalyzed carbonyl reactions
of Chaps. 12 and 13. Electron-withdrawing groups on double bonds confer
on them the receptivity to *nucleophilic* attack treated in conjugate additions (Sec.
12-6), but make them either very sluggish or even quite unreactive to most
of the *electrophilic* reagents treated in this chapter.
   Alkenes are much more susceptible to electrophilic attack than alkynes,
which react slowly or not at all with many reagents that readily attack carbon-
carbon double bonds. The π electrons are much more tightly bound in the

carbon-carbon triple than in the double bond. The shorter distance between nuclei in the triple bond allows better p-orbital overlap and makes the π electrons less available to electrophilic reagents.

Summary

The mechanism of electrophilic addition is generally straightforward and in accord with our theoretical expectations. The stereoelectronic requirements are critical enough usually that only one product is formed, and we may predict that product with four rules:

1 Attack of electrophile proceeds from the less hindered side of the molecular plane of the double bond and perpendicular to it.
2 The added substituents in the initial product are oriented trans periplanar to each other (trans diaxial in cyclohexane products).†
3 The carbon of the original double bond which can form the better carbonium ion receives the nucleophile.†
4 Electron-donating groups enhance the rate; electron-withdrawing groups reduce it. The range of rates is very large among these groups.

$$R_2\ddot{N} > R\ddot{O} > C{=}C > \phi > R > H > SO_2, CN > CO > NO_2$$
Decreasing rate $\longrightarrow$

15-2 ALKYL HALIDES FROM ADDITIONS

Both halogen halides and halogens add readily to alkenes; in the former case the electrophile is H^+, in the latter X^+ (X_2 acting as $X^+ X^-$). Examples of both reactions are shown; these usually proceed in a short time at room temperature. In the case of hypohalous acids, X—OH, the active electrophile is also X^+.

† In saturated cyclohexanes, rule 2 usually predominates over rule 3 when conflicting results are predicted:

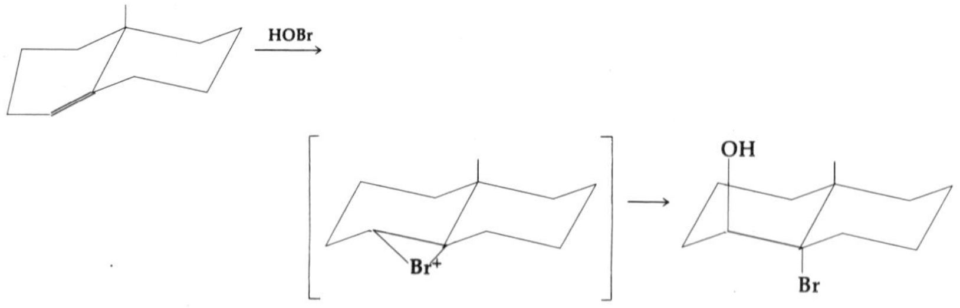

Hydrogen halide addition (**E$^+$ = H$^+$**):

$$\text{HBr} + \text{CH}_3\text{CH}_2\text{CH}{=}\text{CH}_2 \xrightarrow{\text{Acetic acid}} \text{CH}_3\text{CH}_2\underset{\underset{\text{Br}}{|}}{\text{CH}}\text{CH}_3$$

sec-Butyl bromide
80%

$$\text{HCl} + (\text{CH}_3)_2\text{C}{=}\text{CH}_2 \xrightarrow{\text{Conc. HCl}} (\text{CH}_3)_3\text{CCl}$$

tert-Butyl chloride
100%

Halogen and hypohalous acid addition (**E = X$^+$**):

$$\text{Br}_2 + \text{CH}_2{=}\text{CH}_2 \longrightarrow \text{BrCH}_2\text{CH}_2\text{Br}$$

Ethylene dibromide
100% (by titration)

$$\text{Br}_2 + \text{CH}_2{=}\text{CHCH}_2\text{Br} \xrightarrow[-20°]{\text{CHCL}_3} \text{BrCH}_2\text{CHBrCH}_2\text{Br}$$

1,2,3-Tribromopropane
98%

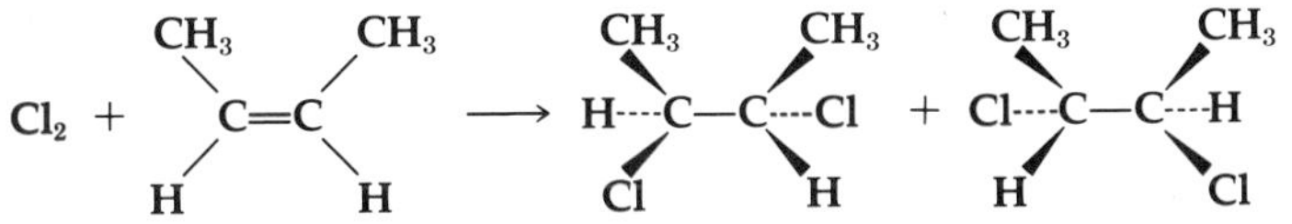

cis-2-Butene   (±)2,3-Dichlorobutane

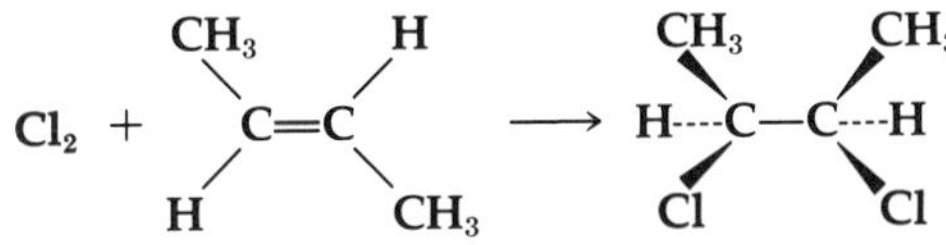

trans-2-Butene   meso-2,3-Dichlorobutane
61%

$$\text{I—Cl} + (\text{CH}_3)_2\text{C}{=}\text{CH}_2 \xrightarrow{\text{HgCl}_2} (\text{CH}_3)_2\underset{\underset{\text{Cl}}{|}}{\text{C}}\text{CH}_2\text{I}$$

Iodine   2-Chloro-1-iodo-
monochloride   2-methylpropane 67%

$$\text{HOCl} + \text{CH}_3\text{CH}{=}\text{C}(\text{CH}_3)_2 \longrightarrow \text{CH}_3\underset{\underset{\text{Cl}}{|}}{\text{CH}}{-}\underset{\underset{\text{OH}}{|}}{\text{C}}(\text{CH}_3)_2$$

2-Methyl-2-butene   3-Chloro-2-methyl-2-butanol

    Addition of hydrogen halides to alkenes has some preparative values in the conversion of unsaturated natural products and petroleum alkenes to halides. The reaction is not often used in synthesis, since the best synthetic route to most olefins leads through alcohols, which can be converted directly to halides by substitution reactions. If a synthetic sequence leads to a primary alcohol, dehydration (page 592) to the corresponding olefin may precede conversion to the nonterminal halide.

$$CH_3CH_2CH_2CH_2CH_2OH \quad \underbrace{\qquad}$$

$$\xrightarrow[\text{ZnCl}_2]{\text{Conc. HCl}} \quad CH_3CH_2CH_2CH_2CH_2Cl$$

$$\xrightarrow[-H_2O]{P_2O_5} \quad CH_3CH_2CH_2CH=CH_2 \quad \xrightarrow{HCl}$$

$$CH_3CH_2CH_2\underset{\underset{\displaystyle Cl}{|}}{C}HCH_3$$

The above examples illustrate the points made in the last section concerning the direction of additions. **Markownikoff's rule**, which was first stated on empirical grounds in 1871, predicts the direction of addition to alkenes. The rule simply states that in the addition of HX to a double bond the hydrogen becomes attached to that carbon atom which already bears the largest number of hydrogen atoms. This rule is of course only another statement of rule 3 in the last section, which was derived from modern theory. It is safer to apply it by considering carbonium ion stability than numbers of hydrogens, however.

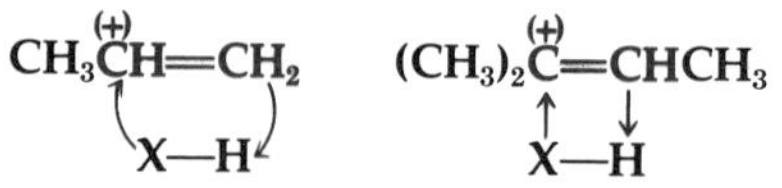

$$CH_3\overset{(+)}{C}H=CH_2 \qquad (CH_3)_2\overset{(+)}{C}=CHCH_3$$
$$\underset{X-H}{\Big\downarrow} \qquad\qquad \underset{X-H}{\overset{\uparrow\;\downarrow}{}}$$

Addition of hydrogen halides to conjugated systems of carbon-carbon double bonds is very rapid because of the stability of the intermediate stabilized carbonium ions. When the system is a diene, the intermediate allylic cation usually reacts with halide ion at both positive centers to give a mixture of products. With benzylic cations, however, the nucleophile adds only to the aliphatic α-carbon and does not disrupt the aromatic ring resonance.

$$\overset{+}{H} + CH_2=CH-CH=CH_2 \longrightarrow [CH_3\overset{+}{C}H-CH=CH_2 \longleftrightarrow CH_3CH=CH-\overset{+}{C}H_2] \xrightarrow{Cl^-}$$

$$CH_3CHClCH=CH_2 \;+\; CH_3CH=CHCH_2Cl$$

Methylvinylcarbinyl   Crotyl chloride
chloride

Styrene   Two more structures   1-Phenylethyl chloride

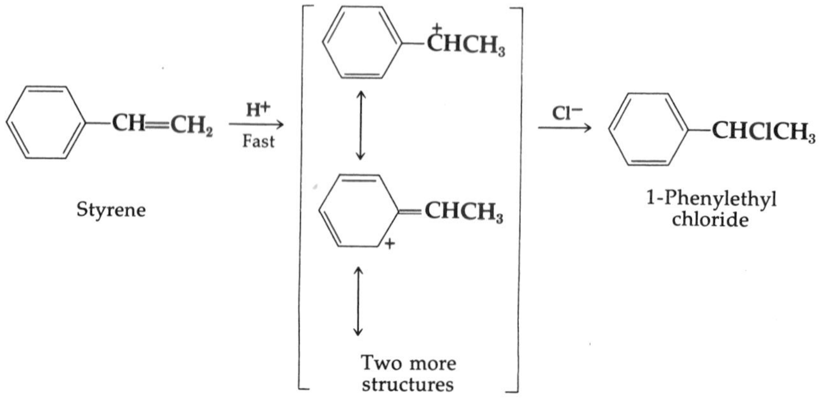

In the common addition of hydrogen halides (usually **HCl** and **HBr**) the proton is the electrophile and first forms a **protonium ion** as a π complex which is in turn attacked from behind by halide ion, at the better carbonium ion site.

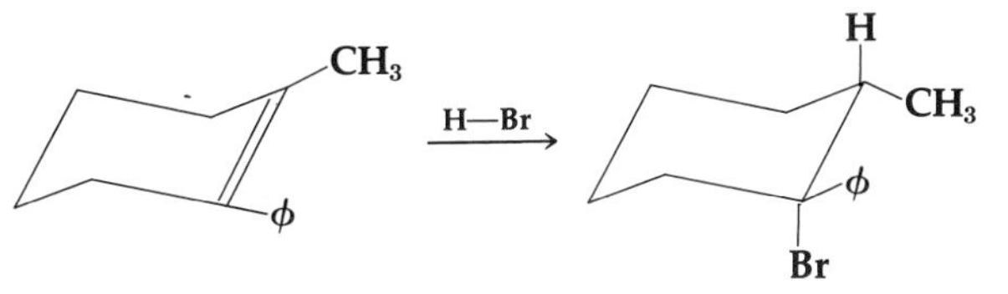

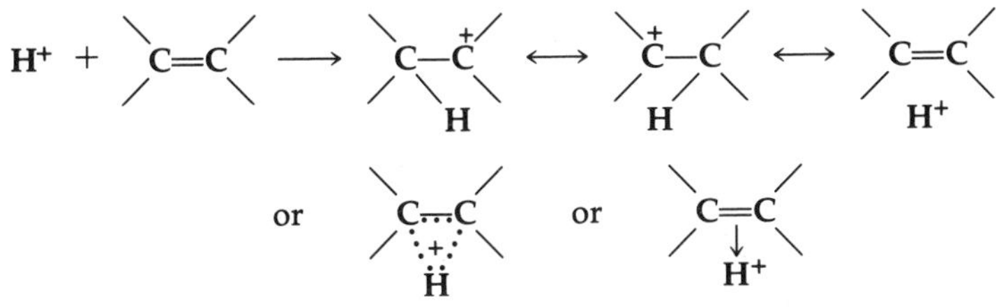

Different representations of protonium ion

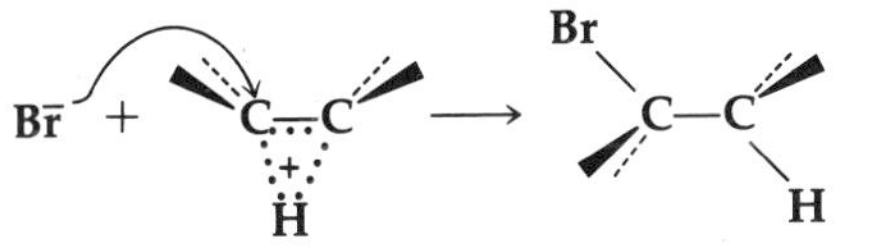

In the reaction of bromine with alkenes, the first step is the transfer of a halogen cation to the hydrocarbon and a cyclic **bromonium ion** is formed. Attack of bromide ion on a bromonium ion occurs with inversion of configuration giving trans addition of **Br$_2$** as specified in the last section. Nucleophilic attack on unsymmetrical bromonium ions also occurs at that carbon atom which is best able to support a positive charge (most highly substituted position in saturated molecules). Chlorine acts in analogous fashion, but iodine does not add since it is a weak electrophile and the sterically crowded diiodide product is not favored in the equilibrium.

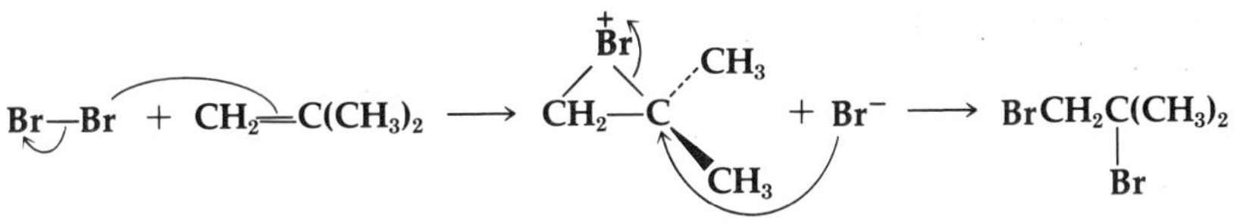

Halogens give products of both 1,2- and 1,4-addition with conjugated dienes. However, 1,2-addition predominates since it arises from the best carbonium ion site in the allylic resonant ion.

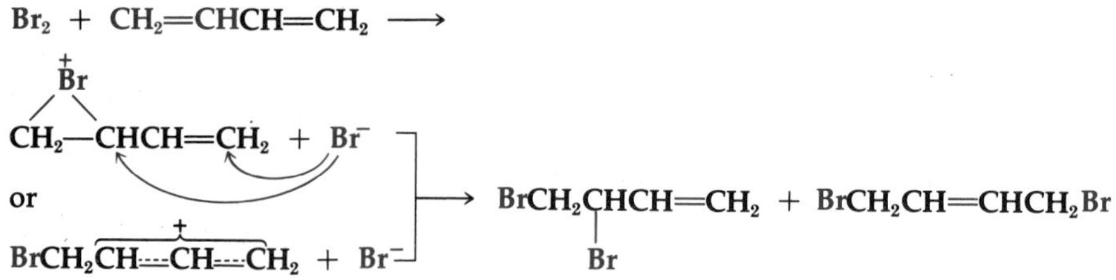

Br2 + CH2=CHCH=CH2
Br
CH2—CHCH=CH2 + Br−
or
BrCH2CH---CH---CH2 + Br−
BrCH2CHCH=CH2 + BrCH2CH=CHCH2Br
Br

Other nucleophiles may complete the addition. Dilute solutions of halogens in water attack alkenes with the formation of **halohydrins** ($X-\overset{|}{\underset{|}{C}}-\overset{|}{\underset{|}{C}}-OH$). The reaction is one of initial halogen attack to form the halonium ion, which then captures solvent water. Dihalides are usually formed as by-products, less so when preformed hypohalous acids, **HOX**, are used as reagents.

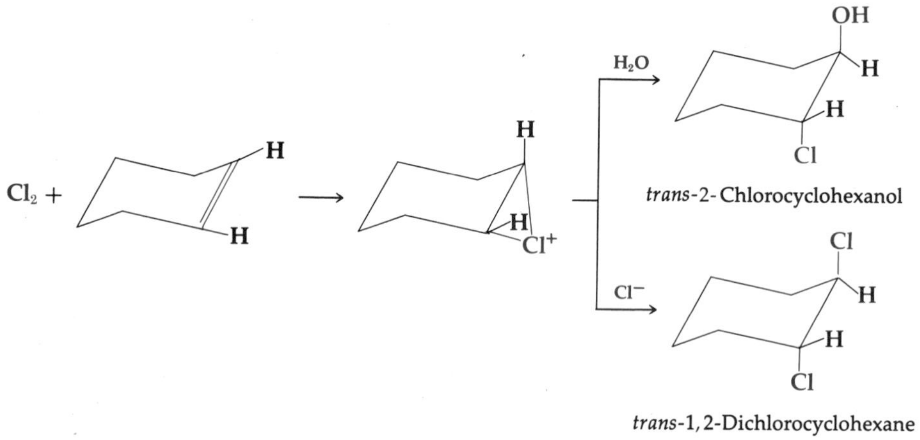

Cl2 +
H
H
H
H
Cl+
H2O
OH
H
H
Cl
trans-2-Chlorocyclohexanol
Cl−
Cl
H
H
Cl
trans-1,2-Dichlorocyclohexane

An alternative and preferable method of adding the elements of **HOX** or **ZOX** to an olefin is through the use of compounds, other than molecular halogen, which are **positive halogen donors**. The *N*-haloamides have been used for this purpose, the solvent providing nucleophile (**HO**, **ROH**, **RCOOH**). Alternatively, halogen is used with the silver salt of a carboxylic acid so that halide ion is removed from nucleophilic competition as insoluble silver halide. (See page 398.)

N-Bromosuccinimide

trans-2-Bromocyclohexyl
acetate

Since hypohalous acids are unsymmetrical and the halogen is the electrophile, the —OH will be found bonded to the better carbonium ion site of the original double bond. An example which shows the effects of the orientation rules in the last section would be the following.

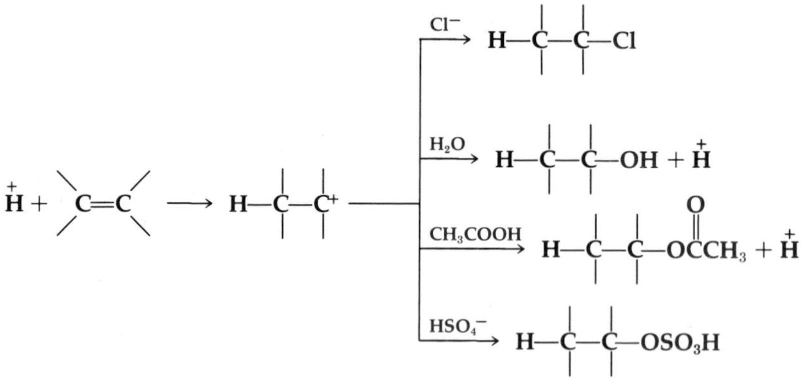

FIGURE 15-5   Competition in alkene addition

PROBLEM 15-2

Write the products of the following transformations. Show the stereochemistry and conformation of the products in two different ways for each case.

a   trans-1-Phenyl-propene + **Br$_2$**

b   trans-1-Phenyl-propene + **HBr**

c   trans-1-Phenyl-butene + **HOCl**

d   1,3-Butadiene + **HOBr**

e   2-Methoxy-2-butene + dry **HCl**

f   2-Methoxy-2-butene + **Br$_2$/H$_2$O**

g   4-Butylcyclohexene + **Br$_2$**

h   1-Bromo-1-ethyl-2-cyclohexene + **Br$_2$**

i   1-Bromo-1-ethyl-2-cyclohexene + **HOCl**

j   1,3,3-Trimethyl-cyclohexene + N-bromosuccinimide in acetic acid

k   2,3,3-Trimethyl-cyclohexene + N-bromosuccinimide in methanol

PROBLEM 15-3

Show the predominant product of chlorinating these dienes, with one mole of **Cl$_2$**.

a   3-Methoxy-2,5-heptadiene

b   3-Ethyl-2,5-heptadiene

c   2,6-Heptadienoic acid

d   Allyl acrylate

e   An equimolar mixture of 2-cyclohexenone and cyclopentene

f   1-Cyclohexenyl-1-cyclobutene

g   N-vinyl-3-hexenoamide

15-3 HYDRATION TO ALCOHOLS AND RELATED COMPOUNDS

Virtually all hydroxylic compounds can be added to alkenes under acidic conditions. **Hydration** (addition of water) is a very important process, used commercially for preparation of alcohols from petroleum by-products. The synthesis of esters, especially *t*-alkyl carboxylates, frequently utilizes this re- action, as does the synthesis of diethyl ether by the addition of water and ethanol to ethylene. Conditions required to effect the transformations vary considerably with alkene structure.

$$H_2O + CH_2{=}CH_2 \xrightarrow[140°]{H_2SO_4} CH_3CH_2OH$$

$$H_2O + CH_2{=}CH_2 \xrightarrow[180°]{H_2SO_4} [CH_3CH_2OH] \xrightarrow{CH_2{=}CH_2} (CH_3CH_2)_2O$$

Diethyl ether

$$H_2O + CH_2{=}C(CH_3)_2 \xrightarrow[25°]{Dil.\ H_2SO_4} (CH_3)_3COH$$

$$Cl_3CCOOH + CH_2{=}C(CH_3)_2 \xrightarrow[0°]{BF_3} Cl_3C\overset{\displaystyle O}{\overset{\|}{C}}{-}O{-}C(CH_3)_3$$

Butyl trichloroacetate
80%

The reactions are proton-initiated (or BF_3) and therefore follow Markow- nikoff's rule. Since capture of the intermediate carbonium ion (or protonium ion depending on the system) is a matter of competition among the various nucleophiles in the system, hydrogen halides cannot be used as catalysts when the object is to add a molecule of a weakly nucleophilic hydroxylic solvent. Sulfuric acid produces only weakly nucleophilic bisulfate ions, which are not good competitors. In any case, acid sulfates formed as intermediates are readily solvolyzed. Figure 15-5 shows the competition for capture of carbonium ions in an acidic solution of an alkene.

Hydration of alkynes produces carbonyl compounds. The reaction is catalyzed by mercuric salts, which are known to form complexes with acety- lenes. Mercuric ion is probably the active electrophile here. The formal similar- ity to alkene hydration may be illusory, but the overall consequences are exactly those expected from normal hydration, followed by tautomerization of the resulting enol.

$$H_2O + C_6H_5C{\equiv}CH \xrightarrow[HgSO_4]{H_2SO_4} \left[C_6H_5\underset{OH}{\overset{\displaystyle}{C}}{=}CH_2 \right] \longrightarrow C_6H_5\underset{O}{\overset{\displaystyle}{C}}CH_3$$

Phenylacetylene   α-Phenylvinyl   Acetophenone
                              alcohol

Coupling of acetylene hydration with acetylene alkylation. (page 428) provides an interesting method for extending carbon chains to give products with nonterminal functional groups.

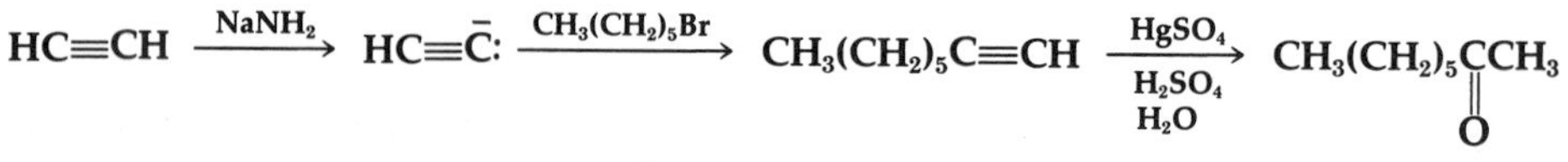

$$HC\equiv CH \xrightarrow{\text{NaNH}_2} HC\equiv \bar{C}: \xrightarrow{\text{CH}_3(\text{CH}_2)_5\text{Br}} CH_3(CH_2)_5C\equiv CH \xrightarrow[\substack{\text{H}_2\text{SO}_4 \\ \text{H}_2\text{O}}]{\text{HgSO}_4} CH_3(CH_2)_5\underset{\underset{O}{\|}}{C}CH_3$$

72%

n-Hexyl methyl ketone
91%

Hydration of the products of condensation of acetylenes with carbonyl compounds is also an excellent synthetic procedure.

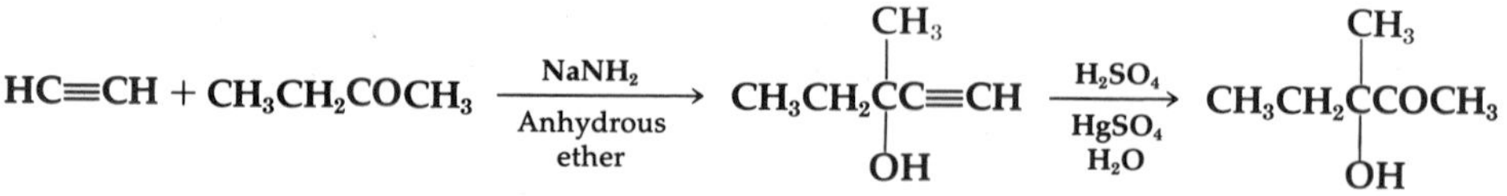

$$HC\equiv CH + CH_3CH_2COCH_3 \xrightarrow[\substack{\text{Anhydrous} \\ \text{ether}}]{\text{NaNH}_2} CH_3CH_2\underset{\underset{OH}{|}}{\overset{\overset{CH_3}{|}}{C}}C\equiv CH \xrightarrow[\substack{\text{HgSO}_4 \\ \text{H}_2\text{O}}]{\text{H}_2\text{SO}_4} CH_3CH_2\underset{\underset{OH}{|}}{\overset{\overset{CH_3}{|}}{C}}COCH_3$$

3-Hydroxy-3-methyl-
2-pentanone

Addition of carboxylic acids to acetylenes leads to alkenyl esters, a versatile class of compound. Vinyl esters undergo polymerization (Sec. 25-1), and all alkenyl esters may be further modified by a host of reactions.

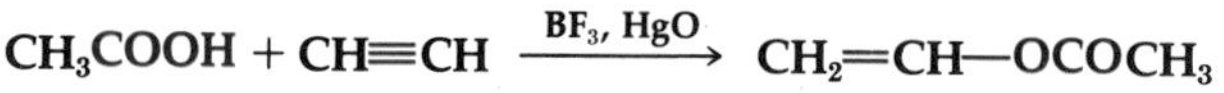

$$CH_3COOH + CH\equiv CH \xrightarrow{\text{BF}_3,\ \text{HgO}} CH_2{=}CH{-}OCOCH_3$$

Vinyl acetate

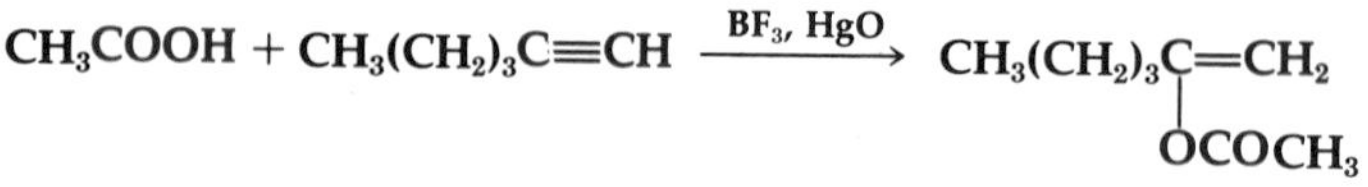

$$CH_3COOH + CH_3(CH_2)_3C\equiv CH \xrightarrow{\text{BF}_3,\ \text{HgO}} CH_3(CH_2)_3\underset{\underset{OCOCH_3}{|}}{C}{=}CH_2$$

31%

$$CH_3(CH_2)_3\underset{\underset{OCOCH_3}{|}}{C}{=}CH_2 \xrightarrow{\text{Br}_2} \left[CH_3(CH_2)_3\underset{\underset{OCOCH_3}{|}}{C}BrCH_2Br \right] \xrightarrow{\text{Spontaneous}}$$

$$\left[CH_3(CH_2)_3\overset{\overset{CH_2Br}{|}}{\underset{:\overset{..}{O}:}{C^+}} \quad Br^{\bar{}} \atop \underset{\underset{O}{\|}}{C}{-}CH_3 \right] \longrightarrow CH_3(CH_2)_3\underset{\underset{O}{\|}}{C}CH_2Br + CH_3\underset{\underset{O}{\|}}{C}Br$$

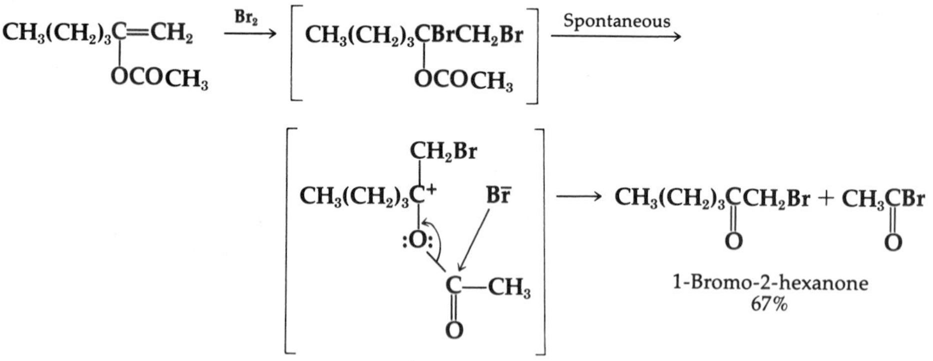

1-Bromo-2-hexanone
67%

Epoxides and Diols

Conversion of an alkene to an epoxide may be readily effected by oxidation with percarboxylic acids ("peracids," **RCO—O—OH**). The reaction shows those variations in reactivity with varying alkene structure which are characteristic of electrophilic addition. Substitution of alkyl or alkoxyl groups for vinyl hydrogens increases substrate reactivity. Peracids that have been investigated fall into the reactivity series $CF_3CO_3H \gg HCO_3H > CH_3CO_3H > C_6H_5CO_3H$. The peracid itself is reduced to carboxylic acid in the reaction. Since carboxylic acids can open oxide rings by nucleophilic substitution reactions, an insoluble weak base is often added to the reaction mixture to neutralize the carboxylic acid formed. Many useful products can be made by opening the three-membered rings of epoxides in a subsequent step with various nucleophilic reagents, or by strong acid catalysis.

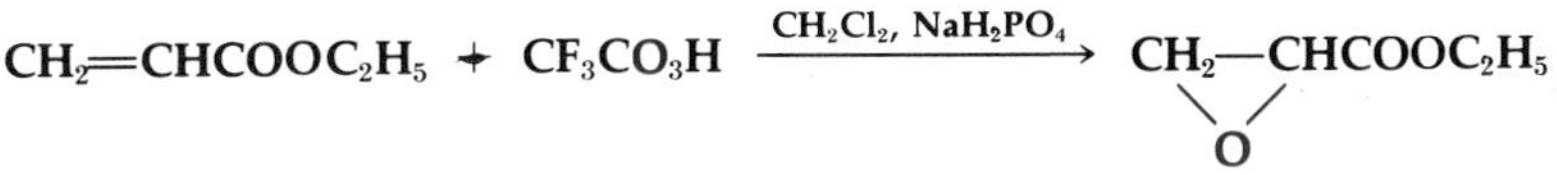
CH2=CHCOOC2H5 + CF3CO3H
CH2Cl2, NaH2PO4
CH2—CHCOOC2H5
O

Ethyl acrylate

Ethyl glycidate 80%

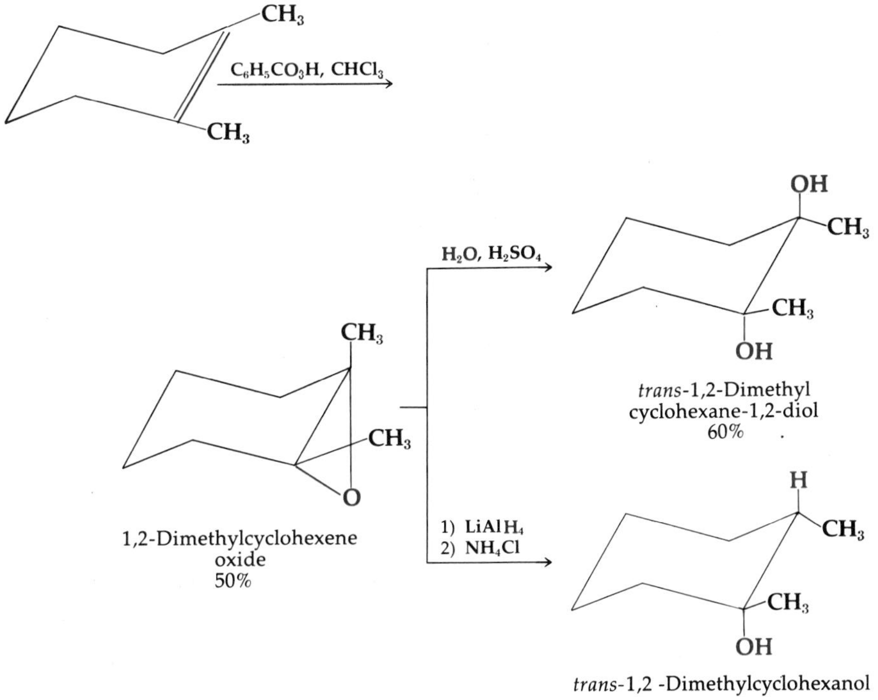
CH3
C6H5CO3H, CHCl3
CH3
CH3
CH3
O
1,2-Dimethylcyclohexene oxide 50%
H2O, H2SO4
OH
CH3
CH3
OH
trans-1,2-Dimethyl cyclohexane-1,2-diol 60%
1) LiAlH4
2) NH4Cl
H
CH3
CH3
OH
trans-1,2 -Dimethylcyclohexanol

In asymmetric cyclic systems, the epoxide is produced on the less hindered side. The epoxide *on the opposite side* may be produced in two steps by forming the halohydrin (or its acetate) first, then treating it with alkali to form the epoxide by internal displacement.

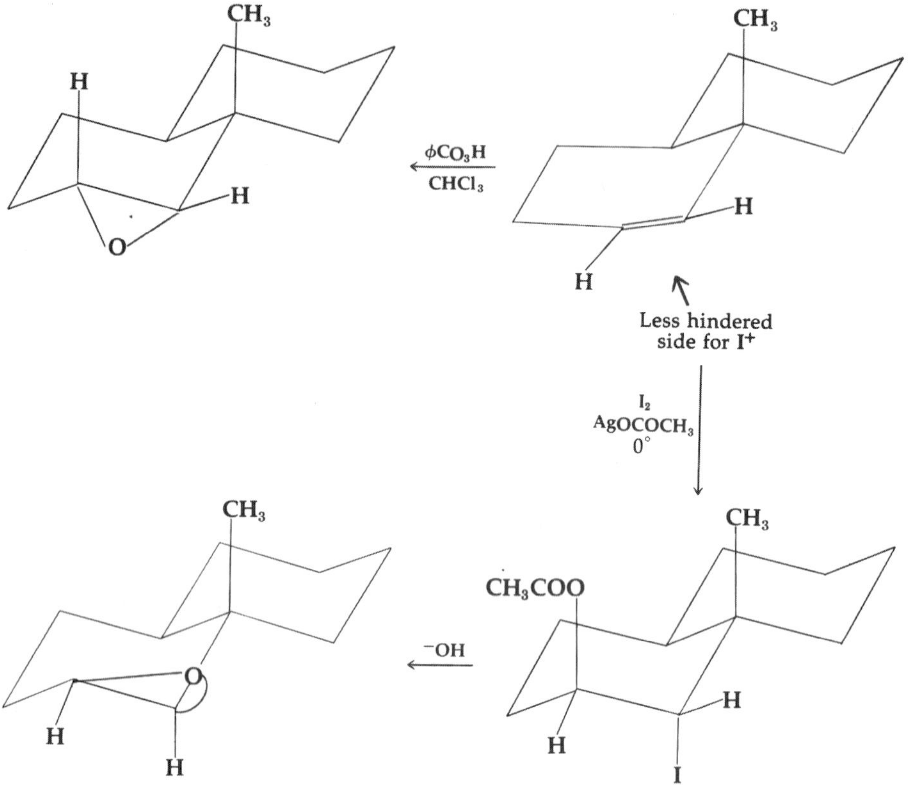

Hydroxylation (diol production) of a double bond via epoxide formation and hydrolysis results in a trans diol. A cis diol can be obtained (on the *more* hindered side) by reaction with silver acetate and bromine in hot wet acetic acid, in which neighboring group participation (Sec. 10-4) of the acetoxy group displaces the original nucleophilic bromine and the intermediate cyclic cation is destroyed by addition of water. Saponification of the product monoacetate produces a cis diol.

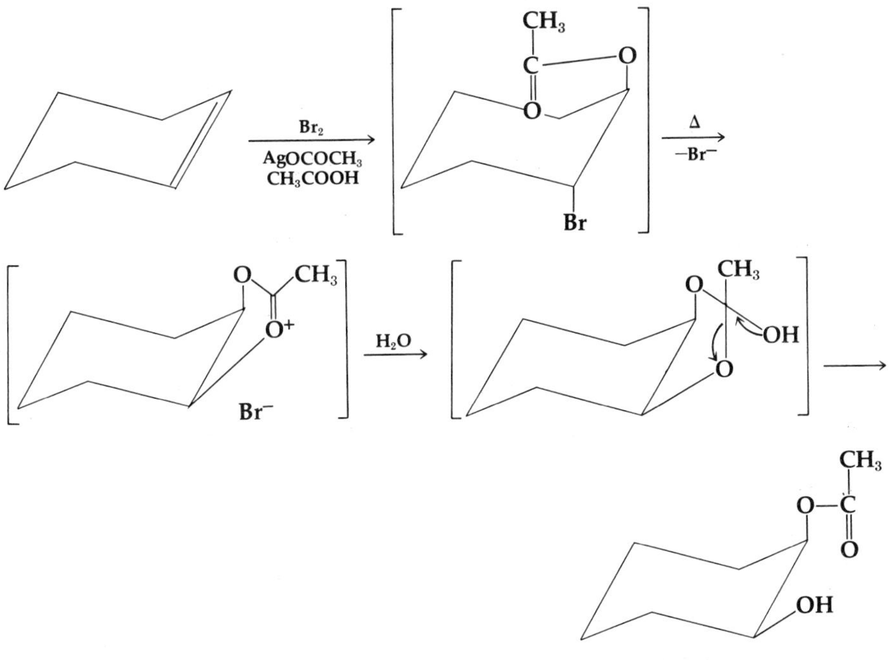

cis-Diol
monoacetate

PROBLEM 15-4

Synthesize the following compounds from appropriate hydrocarbons.

a 1-Methoxy-1-phenyl-propane

b 1-Acetoxy-1-phenyl-1-propene

c 3-Methyl-3-hydroxy-2-hexanone (from 1-pentyne)

d Cyclohexene epoxide

e 4-*t*-Butyl-1,2-cyclohexanediol (all trans)

f A cis 1,2-diol from *cis*-1-methyl-4-*t*-butyl-2-cyclohexene

g *meso*-2,3-Butanediol

h Both epoxides from [2.2.1]bicycloheptene. (*Hint:* The *less* hindered side of the double bond is cis to the methylene bridge.)

PROBLEM 15-5

The compound below yields three different glycols. Which one is formed in each of the following transformations?

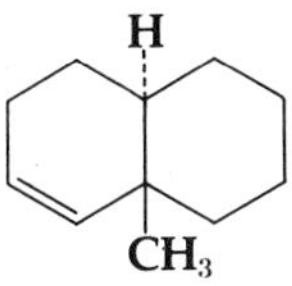

a (1) $C_6H_5CO_3H$ (2) H_3O^+
b (1) CH_3COOAg/I_2 (2) heat/glacial acetic acid (3) ^-OH
c (1) OsO_4 (2) hydrolysis
d (1) CH_3COOAg/I_2 (2) heat/aqueous acetic acid (3) ^-OH
e (1) N-Bromosuccinimide/HOAc (2) ^-OH (3) H_3O^+

15-4 OTHER TRANS ADDITIONS

Hydrocarbons

Addition of one double bond to another can be accomplished if one is protonated by an acid and no other active nucleophile is present to compete with the sluggish nucleophilicity of a second olefin. In practice, sulfuric acid or BF_3 are often used. The reaction is widely used for polymerization, but can only be stopped after the first addition if it is a cyclization since this is so much faster than the *intermolecular* polymerization and so easily controlled. These additions in certain cyclization cases have been shown to have the expected trans orientation.

Cyclization:

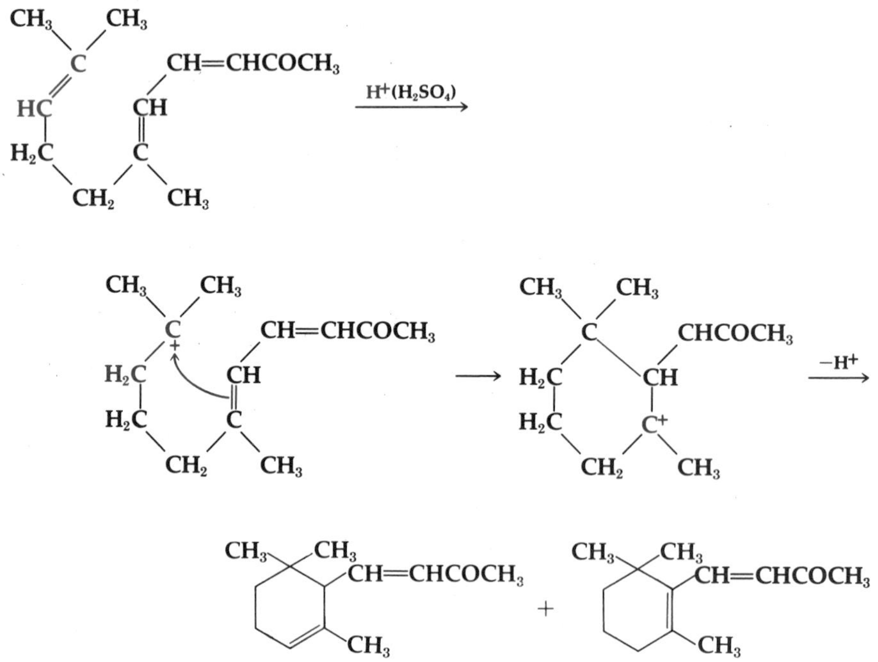

β-Ionone   α-Ionone

High polymerization:

$$C_6H_5CH=CH_2 \xrightarrow[\text{(H}^+\text{BF}_3\text{OH}^-)]{\text{BF}_3, \text{ trace } H_2O} \text{-------}(CHCH_2)_n\text{-------}$$
$$\qquad\qquad\qquad\qquad\qquad\qquad\qquad\qquad\qquad C_6H_5$$

Styrene   Polystyrene

Alkylation and Acylation of Alkenes

In the presence of Lewis acids, alkenes can be alkylated or acylated with alkyl or acyl halides, respectively. The catalyst functions by polarizing and perhaps ionizing the alkylating or acylating agent. Because of the ability of strong Lewis acids to isomerize any but the simplest alkenes (Sec. 17-4), application of these reactions to rational organic synthesis is severely limited. Acylation is, however, widely used in electrophilic substitution on aromatic rings (Chap. 16).

$$C_2H_5COCl \xrightarrow{\text{AlCl}_3} [C_2H_5C\overset{+}{\equiv}\overset{-}{O} \text{ A}l\bar{C}l_4] \xrightarrow{CH_2=CH_2}$$

$$C_2H_5COCH_2\overset{+}{C}H_2 \text{ A}l\bar{C}l_4 \longrightarrow C_2H_5COCH_2CH_2Cl + AlCl_3$$
$$\qquad\qquad\qquad\qquad\qquad\qquad\qquad \beta\text{-Chloroethyl ethyl ketone}$$
$$\qquad\qquad\qquad\qquad\qquad\qquad\qquad 45\%$$

$$(CH_3)_3CCl + CH_2=CH_2 \xrightarrow[0^\circ]{\text{SnCl}_4} (CH_3)_3CCH_2CH_2Cl$$

Nitro and Related Compounds

The introduction of nitro groups (nitration) into organic compounds is of primary importance when applied to aromatic compounds but of limited practical use for alkene additions. Although concentrated nitric acid reacts readily with alkenes, the products are usually mixtures. Like other nitrating agents, concentrated nitric acid produces the electrophilic **nitronium ion** NO_2^+ by complex ionization (Sec. 16-3).

$$(CH_3)_2C=CHCH_3 \xrightarrow[(\equiv NO_2^+)]{\text{Conc. HNO}_3} \left[(CH_3)_2\overset{+}{C}-CH\overset{CH_3}{\underset{NO_2}{}} \right]$$

$$\xrightarrow{-H^+} (CH_3)_2C=C(NO_2)CH_3$$

$$\xrightarrow{+NO_3^-} (CH_3)_2C\text{---}CHCH$$
$$\qquad\qquad\qquad ONO_2\; NO_2$$

$$\xrightarrow{+H_2O} (CH_3)_2C\text{---}CHCH_3$$
$$\qquad\qquad\qquad HO\quad NO_2$$

Addition of nitrosyl chloride (**NOCl**) to olefins has been used to characterize alkenes and, occasionally, for specialized preparative purposes; the nitrosyl group (—N=O) is the electrophile.

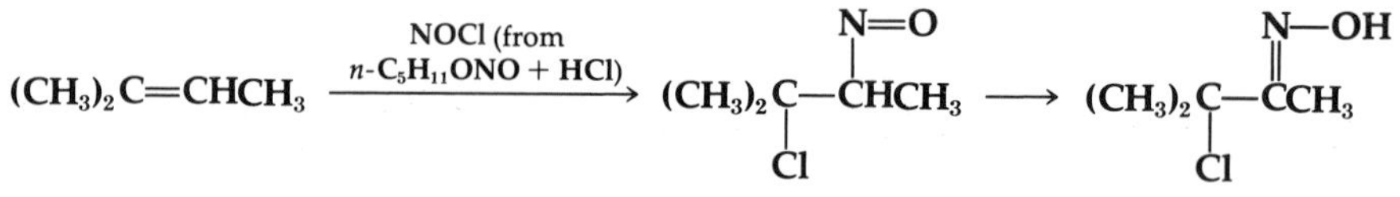

$$(CH_3)_2C{=}CHCH_3 \xrightarrow[\text{n-C}_5\text{H}_{11}\text{ONO} + \text{HCl})]{\text{NOCl (from}}}$$

(with products)

$$(CH_3)_2\underset{\underset{Cl}{|}}{C}{-}\overset{\overset{N{=}O}{\|}}{C}HCH_3 \longrightarrow (CH_3)_2\underset{\underset{Cl}{|}}{C}{-}\overset{\overset{N{-}OH}{\|}}{C}CH_3$$

Trimethylethylene
nitroso chloride
30%

Organomercury Compounds

Alkenes undergo electrophilic addition reactions with mercuric acetate, a process referred to as an **oxymercuration reaction**. The transformation is usually carried out in an alcohol as solvent, and an alkoxyl mercurial is produced. The product is usually isolated as *trans*-alkoxyalkylmercury halide.

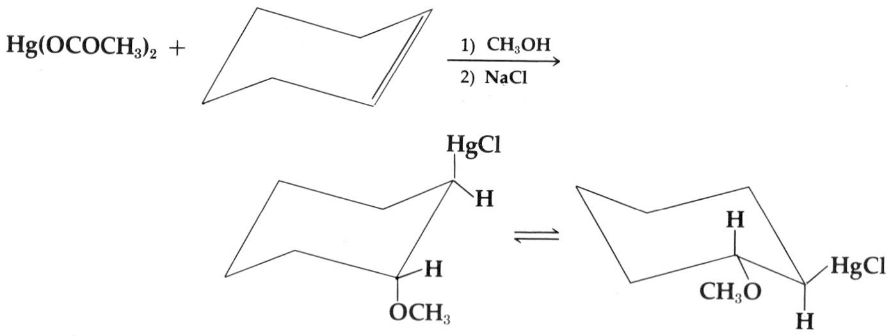

$Hg(OCOCH_3)_2$ +

1) CH_3OH
2) $NaCl$

80%

15-5   CIS ADDITIONS

Epoxidation by peracids necessarily forms a cis adduct, the epoxide. The additions of carbenes and nitrenes are also cis, since they represent merely the first step of normal electrophilic addition—the formation of a three-membered ring, which in these cases is a stable product and does not proceed further. However, if two atoms in an electrophilic molecule can both attack together, this molecule can settle down on the π orbital of the double bond and form two bonds to its two carbons from the same side, affording cis addition. The ring size in the intermediate (or transition state) must then be a reasonable one for optimum orbital overlap and little orbital- or bond-angle strain.

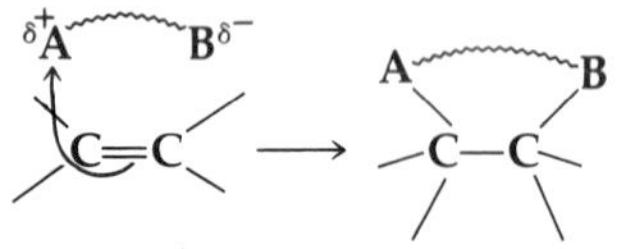

Such reagents include osmium tetroxide (OsO_4) and permanganate ion (MnO_4^-) for cis hydroxylation, ozone (O_3) for ozonolysis (page 87), and boron hydrides (cf. BH_3) for **hydroboration**. Catalytic hydrogenation involves cis addition of hydrogen from the surface of the metal catalyst and is discussed in the chapter on reductions (Sec. 18-5). Cis addition even occurs with other olefins and unsaturated functional groups which act by merging their π electrons with those of the substrate under the influence of heat or light energy and no other catalysis. These latter reactions make up a class called cyclo-addition reactions, which are described in Chap. 21.

Cis Hydroxylation

Both osmium tetroxide and permanganate ion attack alkenes readily, the electrophilic center being the central metal atom which accepts extra electrons and is thus reduced.

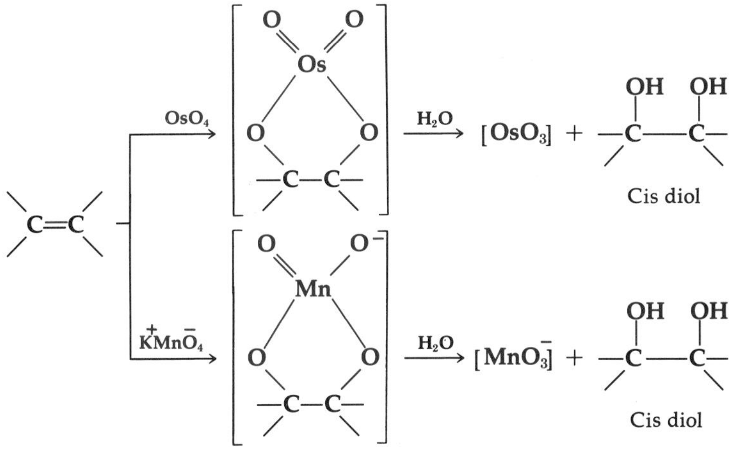

Because of its expense and toxicity, osmium tetroxide is used only in the synthesis of fine chemicals, such as pharmaceuticals, and for degradative studies. To conserve the reagent, it is sometimes used only in minor catalytic amounts, with hydrogen peroxide available to reoxidize osmium in its lower valence states to OsO_4. The reaction is valued since it is mild, quantitative, and highly specific for cis hydroxylation on the less hindered side of the olefin.

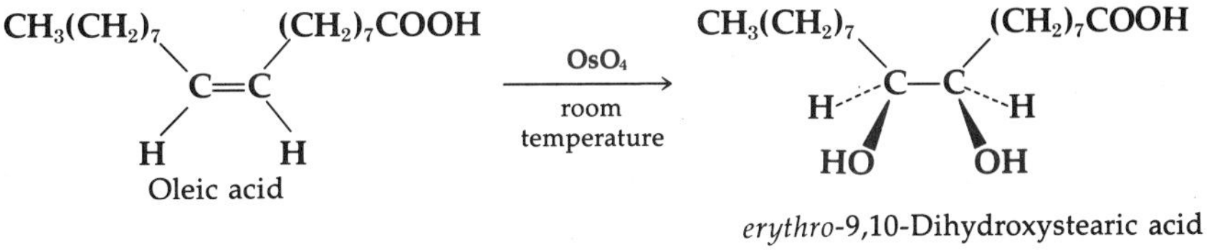

erythro-9,10-Dihydroxystearic acid

Permanganate is very rarely used for this purpose since, unless conditions are very carefully controlled, it reacts further with the initial diol product to cleave the carbon-carbon bond.

Ozonolysis (Ozonization)

Cleavage of alkenes with ozone is a very useful degradative procedure. This potent reagent breaks both carbon-carbon bonds with the formation of cyclic peroxides known as *ozonides*. Decomposition of the ozonide produces carbonyl groups at the carbons initially doubly bonded. Ozonides themselves are usually not isolated because of their explosive character. In addition to degradative uses, ozonolysis has been used synthetically for preparation of aldehydes and ketones from alkenes.

The electrophilic character of ozone is derived from the fact that each of the three oxygen atoms has a considerable affinity for electrons.

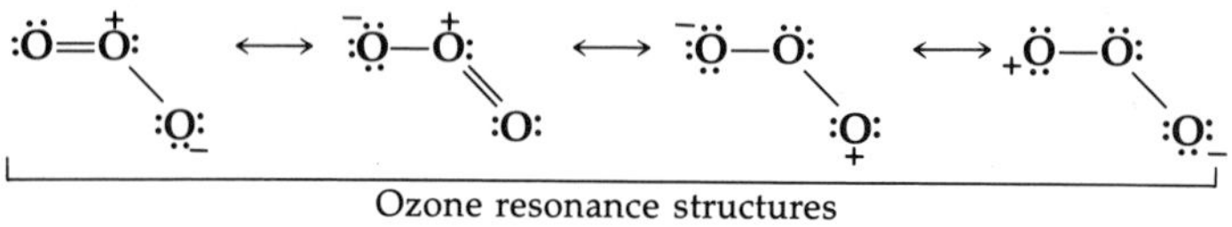

Ozone resonance structures

The mechanism of ozonide formation involves a number of discrete stages, starting with cis addition of O_3 to the alkene.

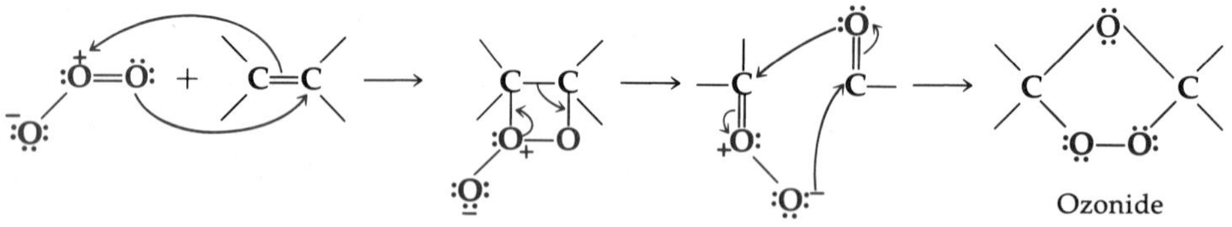

Ozonide

If the original double bond bore two carbon substituents on either of its carbons, that carbon becomes a ketone after decomposing the ozonide. If there were hydrogen on either of the carbons, reductive decomposition of the ozonide would yield an aldehyde but oxidative decomposition would yield a carboxylic acid. Reductive decomposition may be achieved by adding triphenylphosphine ($\phi_3P:$), which takes up the third oxygen as the oxide (ϕ_3P-O^-) or by the use of zinc and water. Oxidative decomposition usually occurs on simple hydrolysis but is assured by adding H_2O_2.

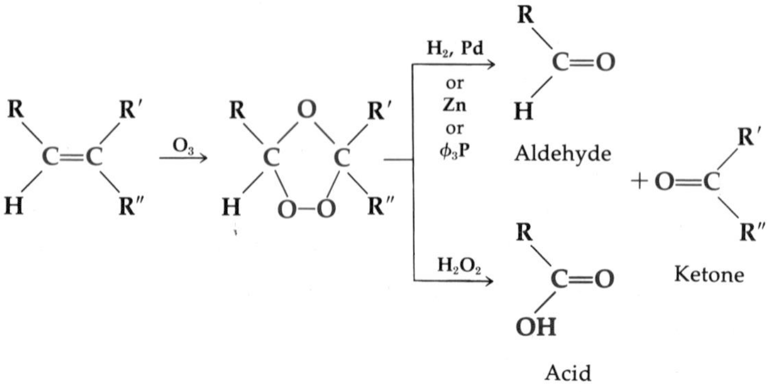

Acid

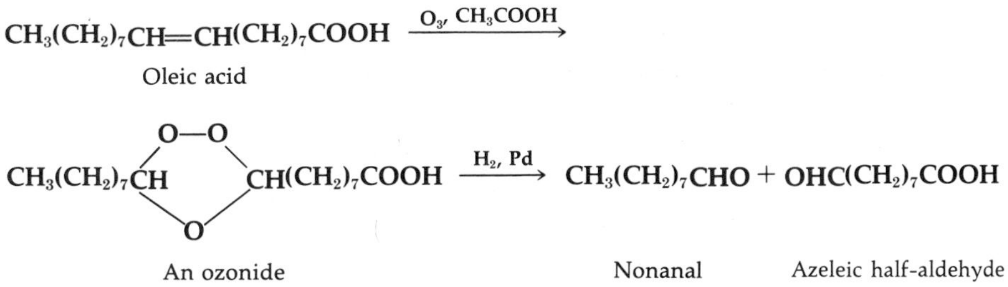

$$CH_3(CH_2)_7CH\!\!=\!\!CH(CH_2)_7COOH \xrightarrow{\;O_3,\; CH_3COOH\;}$$

Oleic acid

$$CH_3(CH_2)_7\overset{O\,-\,O}{\underset{O}{\overbrace{CH}\quad\underbrace{}CH}}(CH_2)_7COOH \xrightarrow{\;H_2,\; Pd\;} CH_3(CH_2)_7CHO + OHC(CH_2)_7COOH$$

An ozonide   Nonanal   Azeleic half-aldehyde

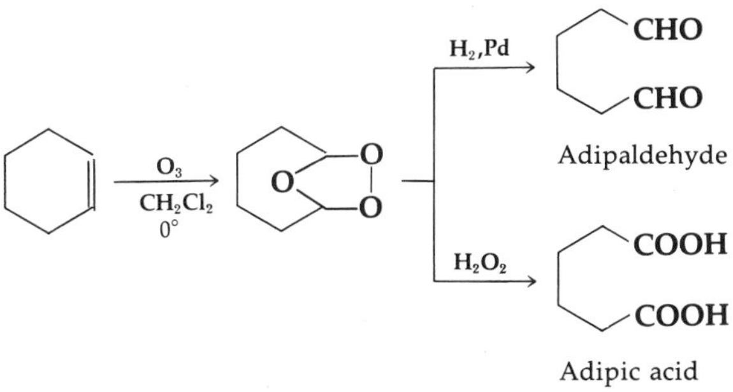

CHO
CHO

Adipaldehyde

COOH
COOH

Adipic acid

Hydroboration

Boron hydrides (BH_3, RBH_2, R_2BH) act as electrophiles because of the vacant orbital on boron (Fig. 15-1). When they attack olefins, however, the hydrogen is bonded cis to the boron. The initial attack on boron proceeds in a direction that leaves the more stable carbonium ion, and it is to this site that the hydrogen bonds in a cis fashion. This may be understood by viewing the reaction stepwise.

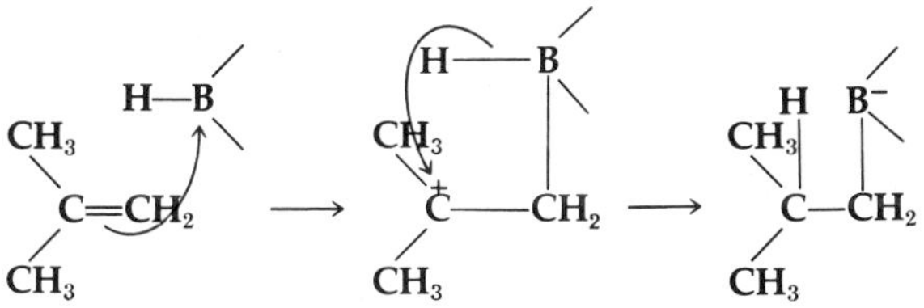

Borane, BH_3, is unknown since it dimerizes to diborane, B_2H_6, but this substance undergoes additions as if it were BH_3. Diborane is prepared from sodium borohydride and boron trifluoride.

$$3NaBH_4 + 4BF_3 \longrightarrow 2(BH_3)_2\!\uparrow + 3NaBF_4$$

Diborane

Addition of BH_3 to $RCH\!\!=\!\!CH_2$ yields $R\!-\!CH_2\!-\!CH_2\!-\!BH_2$, which acts on a second mole in the same way until a trialkylborane, $(RCH_2CH_2)_3B$, is formed. The addition occurs exclusively in the direction indicated, which is that predicted by Markownikoff's rule for attachment of *boron* as the initial electrophile.

With the more highly substituted alkenes, di- and even monoalkylboranes can be isolated. The fact that the hydroboration reaction slows down with hindered olefins indicates the influence of steric effects. An example of the use of steric effects for synthetic purposes involves *bis*-3-methyl-2-butylborane itself as a reagent for addition to less hindered olefins:

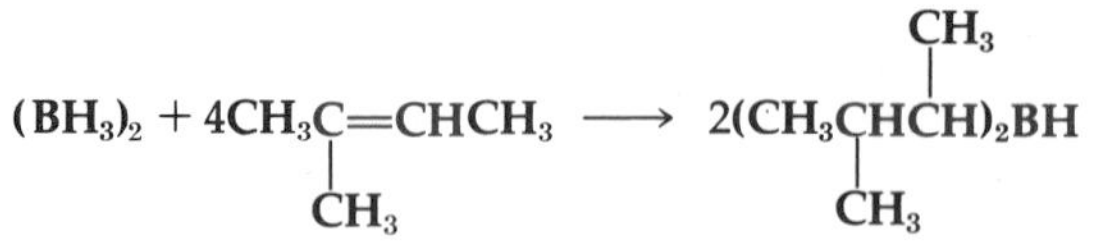

$$(BH_3)_2 + 4CH_3\overset{\underset{\displaystyle CH_3}{|}}{C}=CHCH_3 \longrightarrow 2(CH_3\overset{\underset{\displaystyle CH_3}{|}}{CH}CH)_2BH$$

bis-(1,2-Dimethylpropyl)
borane, or
bis-3-methyl-2-butylborane

The reagent can discriminate between two *differently hindered* double bonds in the same molecule.

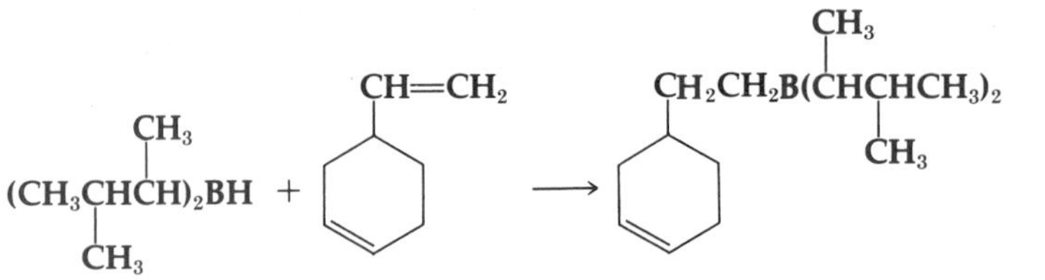

$$(CH_3\overset{\underset{\displaystyle CH_3}{|}}{CH}CH)_2BH \; + \; \text{[cyclohexene with CH=CH}_2\text{]} \longrightarrow \text{[product CH}_2CH_2B(\overset{\underset{\displaystyle CH_3}{|}}{CH}CHCH_3)_2\text{]}$$

Much of the synthetic utility of the hydroboration reaction is derived from the fact that the alkylboranes can be converted to either alcohols or alkanes by treatment with hydrogen peroxide or acetic acid, respectively. Thus, a method is available for preparing alcohols from alkenes by non-Markownikoff addition, the reverse orientation from simple acid-catalyzed hydration. Halides and amines may also be obtained in this powerful procedure, discussed at greater length on page 711.

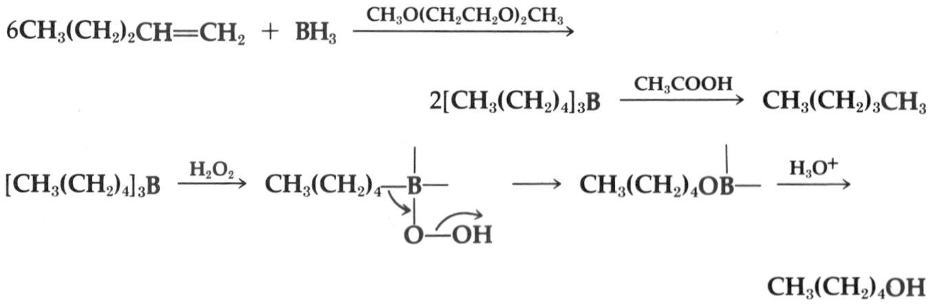

$$6CH_3(CH_2)_2CH=CH_2 \; + \; BH_3 \xrightarrow{\;CH_3O(CH_2CH_2O)_2CH_3\;}$$

$$2[CH_3(CH_2)_4]_3B \xrightarrow{\;CH_3COOH\;} CH_3(CH_2)_3CH_3$$

$$[CH_3(CH_2)_4]_3B \xrightarrow{\;H_2O_2\;} CH_3(CH_2)_4\!-\!\overset{|}{\underset{\overset{\displaystyle \curvearrowright}{O\!-\!OH}}{B}}\!-\; \longrightarrow \; CH_3(CH_2)_4O\overset{|}{B}\!-\; \xrightarrow{\;H_3O^+\;}$$

$CH_3(CH_2)_4OH$

The overall cis stereochemical course of the alkene to alcohol reaction is illustrated by the conversion of *cis*- and *trans*-2-*p*-anisyl-2-butenes to diastereomeric alcohols.

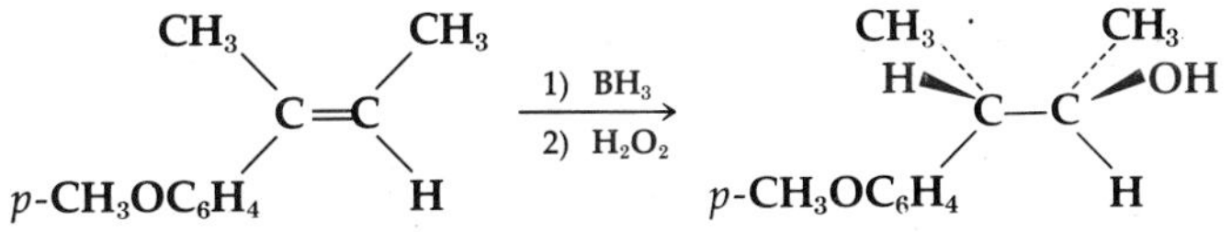

cis-2-p-Anisyl-2-butene

threo-3-p-Anisyl-2-butanol
72%

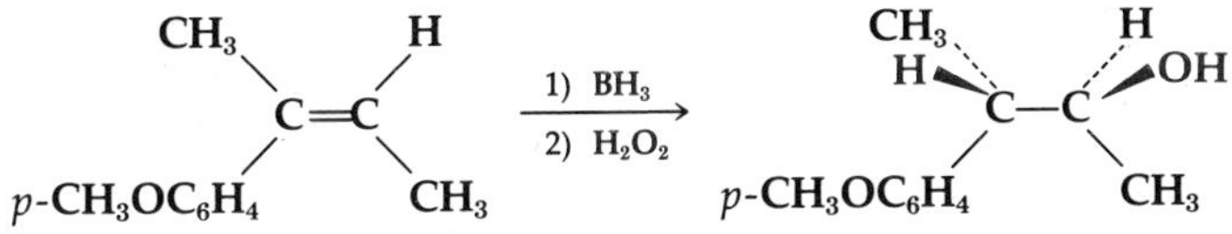

trans-2-p-Anisyl-2-butene

erythro-3-p-Anisyl-2-butanol

Acetylenes undergo the hydroboration reaction *faster* than alkenes, and the products of addition may be readily converted to a variety of compounds. By using stoichiometric amounts, the reaction can be stopped at the *vinylborane* stage, and the vinylborane can be treated with a proton donor to give overall cis addition of the elements of hydrogen to the carbon-carbon triple bond. Alternatively, peroxide oxidation yields a ketone or aldehyde at the site of boron attachment.

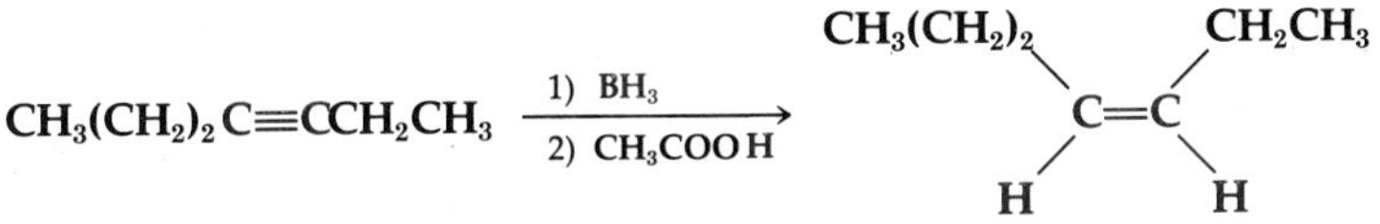

98%

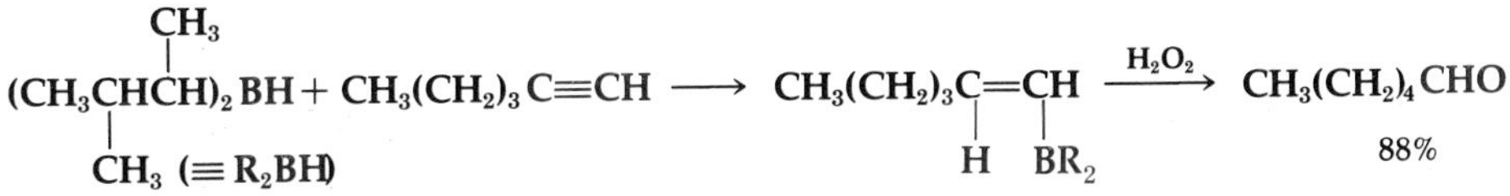

88%

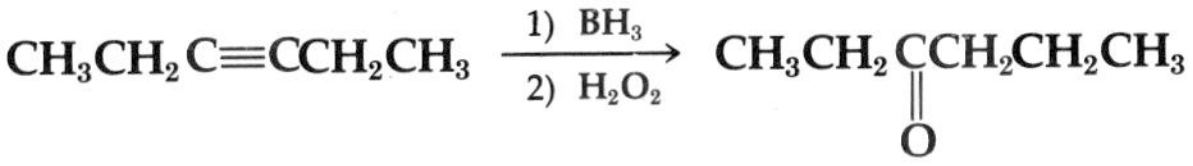

$$CH_3CH_2C\equiv CCH_2CH_3 \xrightarrow[\text{2) } H_2O_2]{\text{1) } BH_3} CH_3CH_2 \underset{\overset{\|}{O}}{C} CH_2CH_2CH_3$$

When heated to high temperatures, the alkylboron compounds regenerate alkenes reversibly. This allows migration of boron along a chain and ultimately the least substituted, least hindered terminal alkylboron compound predominates. This reaction greatly broadens the synthetic scope of the hydroboration reaction.

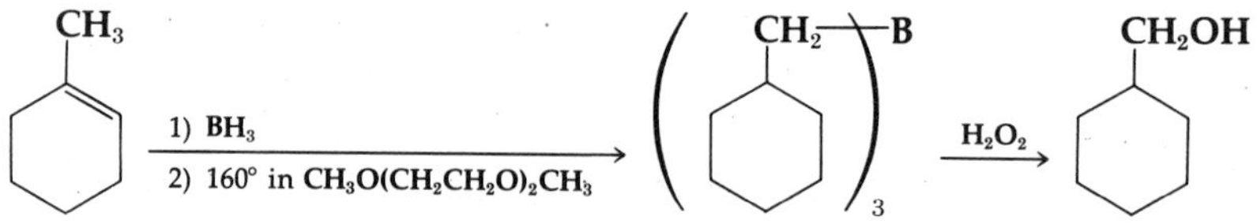

$$CH_3(CH_2)_3C\equiv C(CH_2)_3CH_3 \xrightarrow[\substack{2)\ 160°\ in\ CH_3O(CH_2CH_2O)_2CH_3 \\ 3)\ H_2O_2}]{1)\ BH_3} HO(CH_2)_{10}OH$$

Carbene Additions

A number of reactions involve highly reactive divalent carbon compounds (**carbenes**) as transient intermediates. These species are electrically neutral, are bound to only two substituents, and possess two unshared electrons. The most useful carbene reaction is addition to the carbon-carbon double bond to give cyclopropane and its derivatives. Classification of this electron-deficient carbon as an electrophile is somewhat arbitrary, since in the overall reaction the two unshared electrons of the carbene and the two π electrons of the alkene are converted into two sigma bonds. However, the *electron-deficient nature* of the carbene (Fig. 15-1) causes it to seek π electrons like any other electrophile.

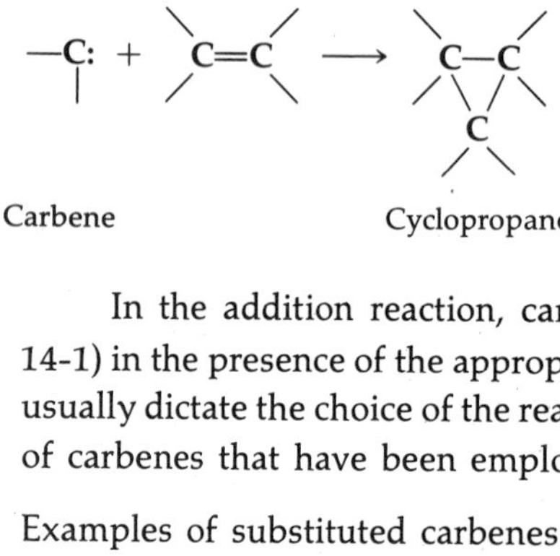

Carbene   Cyclopropane

In the addition reaction, carbenes are usually generated (by base, Fig. 14-1) in the presence of the appropriate alkene. The substituents of the carbene usually dictate the choice of the reaction employed for its production. Examples of carbenes that have been employed in addition reactions are listed.

Examples of substituted carbenes:

$:CH_2$ $:CCl_2$ $:CBr_2$ $:CClC_6H_5$ $:CHCl$ $:C(C_6H_5)_2$ $:CHCOOC_2H_5$

Methylene itself ($:CH_2$) is produced by irradiation of diazomethane (CH_2N_2, page 415) with ultraviolet light. Similarly, ethyl diazoacetate and diphenyldiazomethane give the corresponding carbenes on photolysis. The halocarbenes are produced by base-catalyzed 1,1-elimination reactions in systems unable to undergo the more usual 1,2-elimination reaction.

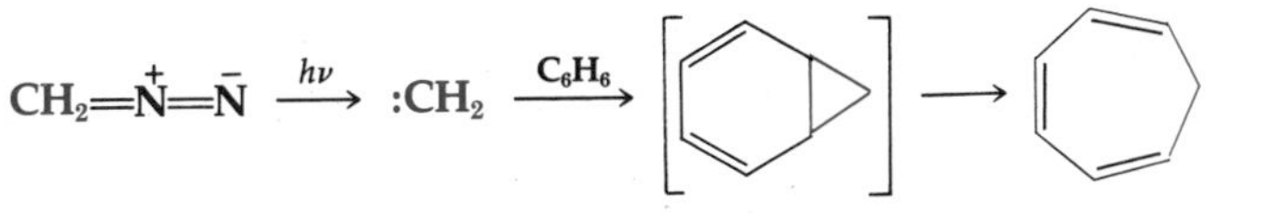

$$CH_2=\overset{+}{N}=\overset{-}{N} \xrightarrow{h\nu} :CH_2 \xrightarrow{C_6H_6} \left[\text{} \right] \longrightarrow \text{}$$

Norcaradiene   Cycloheptatriene
(not isolated)

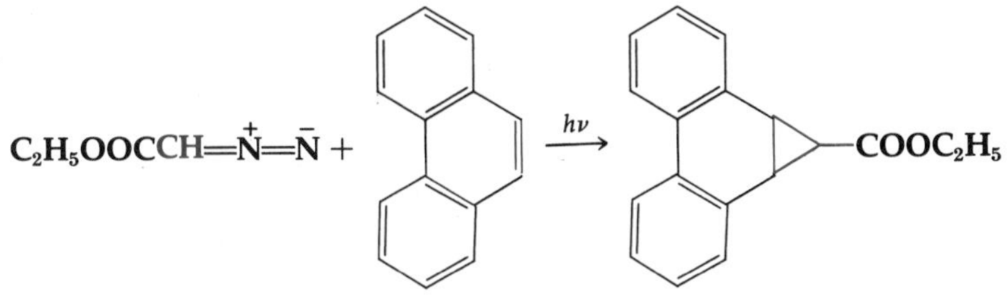

$$C_2H_5OOCCH=\overset{+}{N}=\overset{-}{N} + \text{} \xrightarrow{h\nu} \text{}-COOC_2H_5$$

Phenanthrene

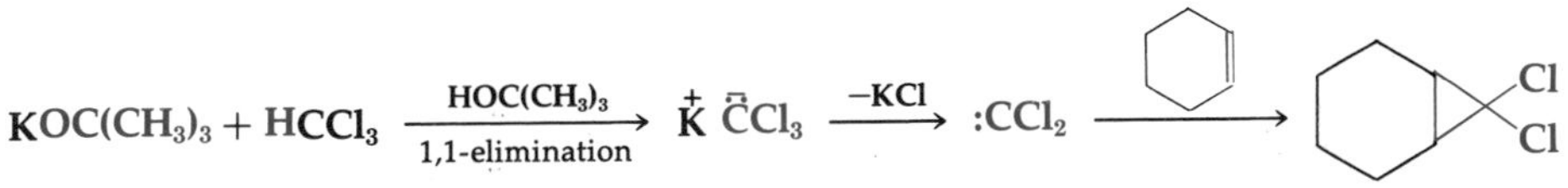

$$KOC(CH_3)_3 + HCCl_3 \xrightarrow[\text{1,1-elimination}]{HOC(CH_3)_3} \overset{+}{K}\ \overset{-}{C}Cl_3 \xrightarrow{-KCl} :CCl_2 \xrightarrow{} \text{}$$

59%

$$n\text{-}C_4H_9Li + CH_2Cl_2 + (CH_3)_2C=C(CH_3)_2 \xrightarrow[+LiCl]{-25°} (CH_3)_2\underset{\underset{CHCl}{}}{C}\text{---}C(CH_3)_2 + n\text{-}C_4H_{10} + LiCl$$

Additions of carbenes to cis and trans olefins are stereospecific and give products of cis addition. A carbenelike reaction that is also stereospecific involves the reaction of methylene diiodide with zinc-copper couple in the presence of an alkene to give a three-membered ring.

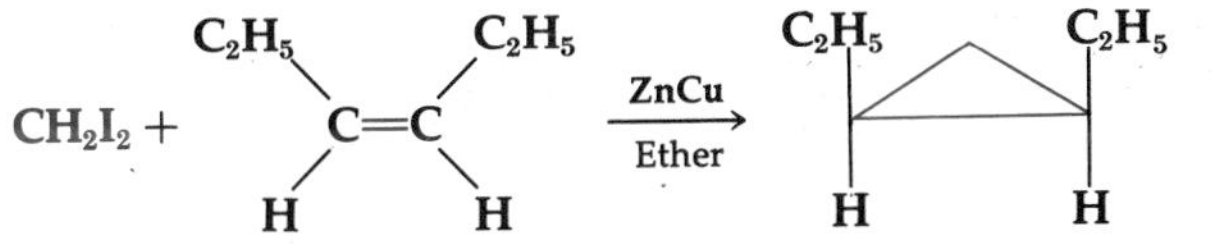

$$CH_2I_2 + \text{} \xrightarrow[\text{Ether}]{ZnCu} \text{}$$

cis-3-Hexene   cis-1,2-Diethylcyclopropane

$$CH_2I_2 + \text{} \xrightarrow[\text{Ether}]{ZnCu} \text{}$$

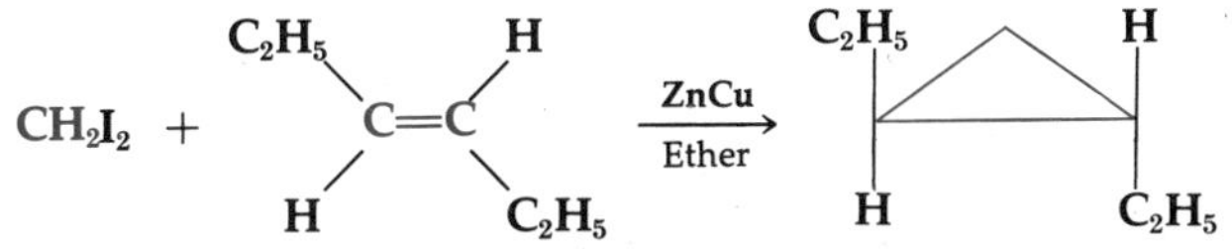

trans-3-Hexene   trans-1,2-Diethylcyclopropane

The highly strained cyclopropene ring system has been synthesized by addition of methylene to dimethylacetylene. Cyclopropene itself has been synthesized by a different route and is very unstable.

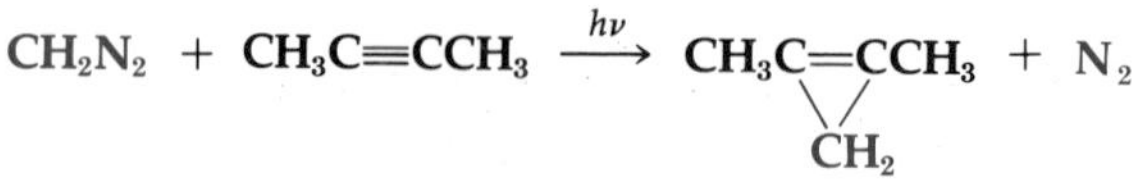

$$CH_2N_2 \; + \; CH_3C{\equiv}CCH_3 \; \xrightarrow{\;h\nu\;} \; CH_3C{=}CCH_3 \; + \; N_2$$

$$\underset{CH_2}{\diagdown\diagup}$$

1,2-Dimethylcyclopropene

Nitrene Additions

Additions of nitrenes have been very little investigated, since the most easily formed nitrenes ($R—CO—\ddot{\underset{..}{N}}$:) undergo rearrangement, not addition (page 704). Sulfonyl nitrenes are formed, just like carbenes from diazo compounds, by heating sulfonyl azides.

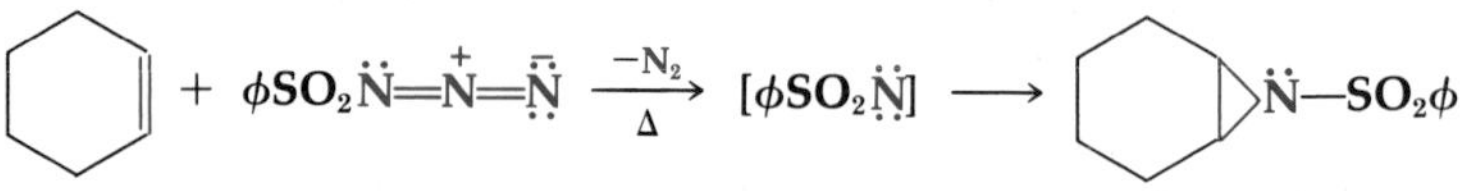

Benzenesulfonyl azide

PROBLEM 15-6

Show the products, and relative stereochemistry where relevant, in the following examples.

a   1-Methylcyclopentene + B_2H_6 followed by H_2O_2/OH^-
b   4-t-Butyl-1-cyclohexene + B_2H_6 followed by $^+NH_3OSO_3{}^-$
c   1-Methylcyclopentene + O_3 followed by H_2O_2
d $C_6H_5C{\equiv}CCH_2C_6H_5$ + B_2H_6 followed by HOAc
e p-$CH_3OC_6H_4C{\equiv}CC_6H_5$ + B_2H_6 followed by H_2O_2
f   1-Methylcyclopentene + $HCCl_3$ + t-BuOK
g   Compound in Prob. 15-5 + B_2H_6 followed by H_2O_2/OH^-
h   Compound in Prob. 15-5 + CH_2I_2 + Zn $\cdot$ Cu

15-6   ADDITIONS AND ELIMINATIONS IN SYNTHESIS

Additions are most commonly used in synthesis to convert an alkene to another function. Eliminations are mechanistically the reverse and serve to *form* alkenes in synthesis. Because of their similarity in mechanism and stereoelectronic requirements, the two reactions can now be reviewed together with the focus on the chief procedures commonly employed in synthetic practice.

Creation of Double Bonds

The chief methods in practical use for introducing carbon-carbon double bonds in synthesis are listed below.

Strong basic conditions (E2):   Loss of:

—HX or —HOTs from halides and alcohols (via tosylates)

—NR$_3$ from amines (via quaternary ammonium salts—Hofmann elimination

Strong acidic conditions (E1):   —H$_2$O from alcohols at good carbonium ion sites

Pyrolysis: —HOAc from alcohols (via acetates)

—COS, —CH$_3$SH from alcohols (via xanthates)

—R$_2$NOH from amines (via tertiary amine oxides)

Although the first category results in trans elimination and the third in cis elimination, none of these methods except the Hofmann elimination affords a high selectivity as to the *direction* of elimination. Even in rigid cyclic systems, while the eliminating groups must be trans (diaxial) or cis, respectively, there may still be two directions of elimination.

In the first category the base must be a strong one which is also a poor nucleophile, in order to suppress competition from the substitution reaction. In the acid-catalyzed reactions, similarly, it is useful to employ an acid embodying a poor nucleophile. Alcohols eliminate best if they are tertiary, allylic or benzylic, and carbon-skeleton rearrangement (Secs. 17-1 and 17-2) is always a potential side reaction.

Eliminations of β-substituents from carbonyls (or other analogously enolizable groups) is always very facile and occurs with almost any heteroatom substituent, under mild acidic or basic conditions.

There are two procedures which are highly selective for creating a double bond at a single location. One is the elimination of vicinal substituents (—X, —OTs) with iodine or (—X + —OH, —OR, —OZ) with zinc. The other is the Wittig reaction (page 476) in which the double bond is created by addition of carbon units at a ketone or aldehyde site.

Reactions of Double Bonds

The major reagents used to add to double bonds (without cleaving the carbon skeleton as in ozonolysis) are the following:

HX $\xrightarrow{\text{Trans}}$ monohalides

H$_2$O/H$^+$ $\xrightarrow{\text{Trans}}$ monoalcohols

BH$_3$ $\xrightarrow{\text{Cis}}$ monoalcohols, halides, and amines (from initial borane)

+

$X_2 \xrightarrow{\text{Trans}}$ dihalides

N-Bromosuccinimide $\xrightarrow{\text{Trans}}$ halohydrins and their ethers or esters

$RCO_3H \xrightarrow{\text{Cis}}$ epoxides, trans diols

$OsO_4 \xrightarrow{\text{Cis}}$ cis diols

Carbenes $\xrightarrow{\text{Cis}}$ cyclopropanes

The additions are cis or trans as discussed in the text and noted above. The active electrophile attacks from the less hindered side of the double bond. In trans additions the subsequent nucleophile, in a two-step reaction, then becomes attached at the better carbonium ion site, or in cyclohexenes, such as to afford trans-diaxial products. A simple rigid and hindered cyclohexene is offered as a sample in Fig. 15-6 to summarize the possible variations in alkene addition.

Carbon-carbon bonds are rarely formed by reaction with alkenes, except in the case of carbene addition. Epoxides can be attacked by carbanions to achieve carbon-carbon bond formation. Certain rearrangements of organoboranes also serve this purpose, but these reactions are new and their synthetic value has not yet been fully assessed.

PROBLEMS

15-7 Write equations for the reactions of 2-methyl-1,3-butadiene with each of the following reagents.

a Br_2

b HCl

c ICl

d O_3 with $(C_6H_5)_3P$

e $C_6H_5CO_3H$ (one equivalent)

f HBr

g D_2 in the presence of finely divided platinum

h A trace of concentrated H_2SO_4

i OsO_4

j A dilute solution of bromine in methanol

k $C_6H_5COCl + AlCl_3$ at $-50°$

15-8 Devise syntheses of the following compounds, utilizing the starting materials indicated and compounds containing no more than three carbon atoms. Do not repeat the preparation of a given intermediate within the problem set.

a $(CH_2)_7 \begin{array}{c} CHOH \\ | \\ CHOH \end{array}$ from $CH_3(CH_2)_7CH{=}CH(CH_2)_7COOH$

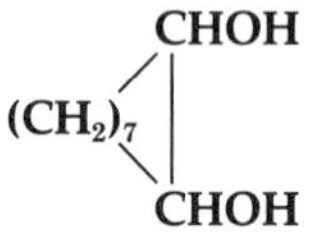

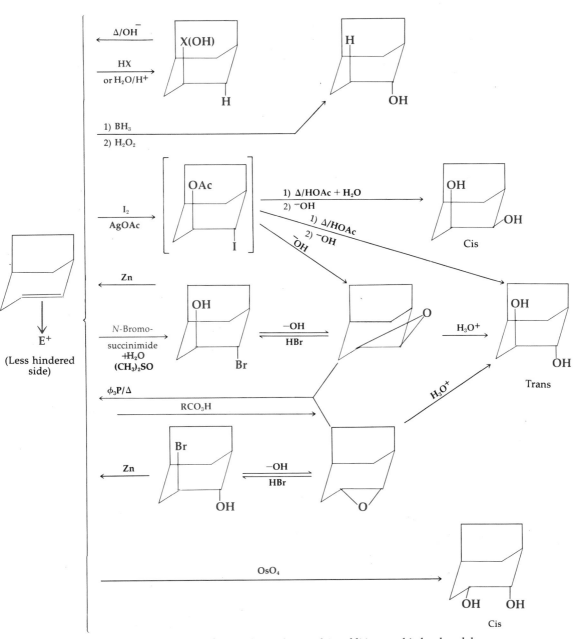

FIGURE 15-6 **Summary of stereochemical control in additions to hindered cyclohexenes**

b $(CH_3)_2CHCH_2COOH$ from CH_3COCH_3

c 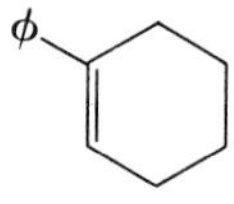   from

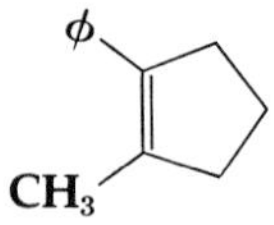

d *trans*-1,2-Cyclohexanediol from cyclohexanone

e 2-Heptanone and heptanal from 1-heptyne

f Acetylcycloheptane from cycloheptanone

g 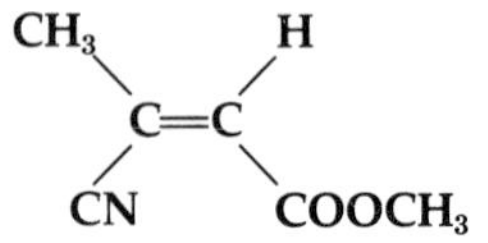   from methylacetylene

h *cis*-Stilbene from *trans*-stilbene

15-9 The natural compound, terrein, is synthesized by certain microorganisms. Terrein is optically active and its absolute stereochemistry was quickly and elegantly determined by conversion to the diacetate with acetic anhydride/pyridine, followed by ozonolysis with a peroxide workup. (*Note:* α-Ketoacids are cleaved to acids by peroxide.) The experiment yielded (+)-tartaric acid diacetate (Fig. 6-22). How does this determine the absolute configuration in terrein, and what is it?

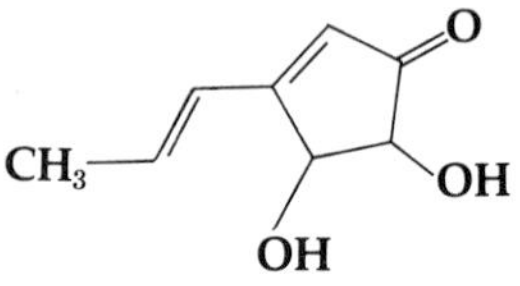

Terrein

15-10 Show synthetic sequences for converting 1-methyl-1-cyclohexene into each of the following compounds.

a

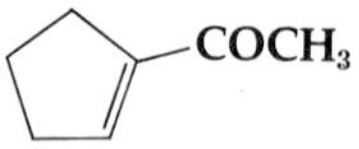

b

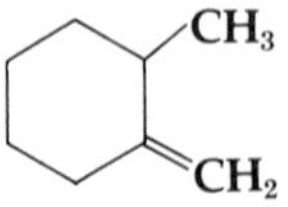

c

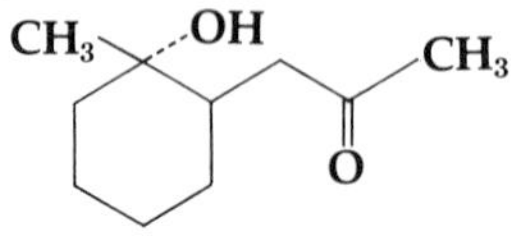

d

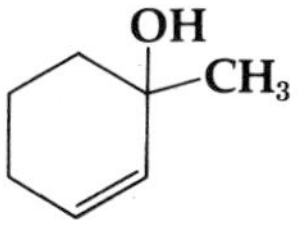

e

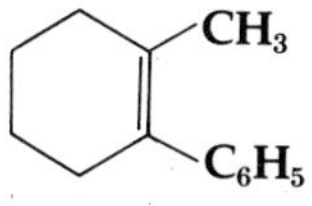

f

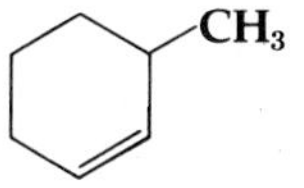

g $CH_2{=}CH(CH_2)_4COOH$

15-11 With three-dimensional formulas, trace the following reactions:

a Compound A was converted to the oxide of its N,N-dimethyl derivative, which was pyrolyzed and the product treated with a mixture of hydrogen peroxide and osmium tetroxide.

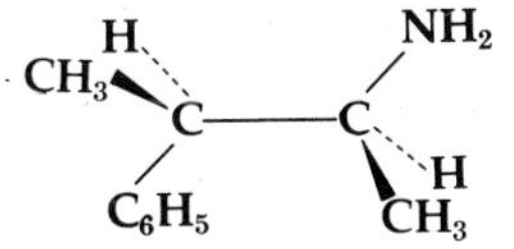

b Cyclopentene was treated with perbenzoic acid, and the product was reduced with lithium aluminum hydride. The resulting substance was converted to a tosylate, and the tosylate was heated with aqueous alkali.

c Dimethylacetylene was reduced with 1 mole of hydrogen over a partially poisoned palladium catalyst. The product reacted with hypochlorous acid, and the resulting compound, when treated with base, gave 1,2-dimethylethylene oxide.

15-12 Write equations for the following reactions, which serve to establish a structure for compound A:

$$\underset{A}{C_7H_{10}} \xrightarrow{\ Br_2\ } C_7H_{10}Br_4$$

$$A \xrightarrow{\ O_3\ } CH_2O + \underset{B}{C_6H_8O_3}$$

$$B \xrightarrow{\ H_2Cr_2O_7\ } \underset{C}{C_6H_8O_5}$$

$$C \xrightarrow{\ 150°\ } \underset{D}{C_5H_8O_3} + CO_2$$

$$D \xrightarrow{\ NaBH_4\ } \underset{E}{C_5H_{10}O_3} \quad \text{(could be resolved)}$$

15-13 Some of the following reactions cannot be carried out; some are possible under conditions other than those indicated; others are reasonable as written. Indicate the category into which each falls, and supply proper conditions where necessary.

a $CH_3CH=CH_2$ $\xrightarrow{\text{HCl, Pt}}$ CH_3CHCH_3
                                                            |
                                                            Cl

b (structure: cyclohexane with $=CH_2$) $\xrightarrow{\text{Br}_2,\text{ conc. }H_2SO_4}$ (structure: cyclohexane with OH and CH_2Br)

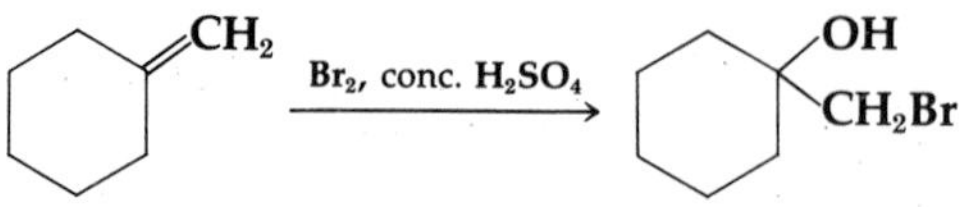

c $CH_3CH=CH_2$ $\xrightarrow{\text{BF}_3,\text{ NaOCH}_3}$ $CH_3CH=CHCH(CH_3)_2$

d (structure: cyclohexene with $=CH_2$) $\xrightarrow[\text{Zn}]{\text{O}_3}$ (structure: cyclohexenone with $=O$)

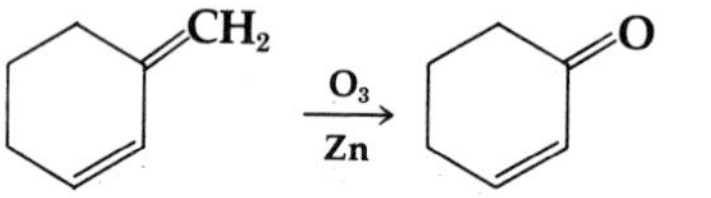

e (structure: cyclopentadiene with CH_3 and COOH) $\xrightarrow{N\text{-Bromosuccinimide}}$ (structure with O—CO, CH_3, Br)

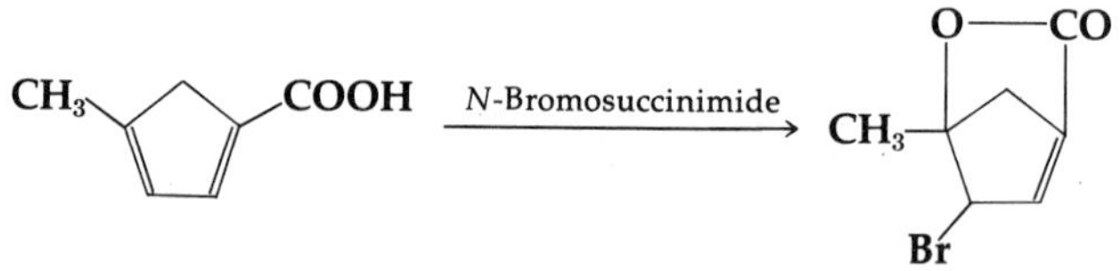

f $(CH_3)_2C=CHCH_3$ $\xrightarrow{\text{Br}_2,\text{ CH}_3OH}$ $(CH_3)_2CCHBrCH_3$
                                                                        |
                                                                        OCH_3

15-14 **a** 2-Cyclohexene-1-carboxylic acid reacts with bromine to form a neutral monobromo compound A [IR 5.65 μ (1770 cm^{-1})]. This compound is exceedingly resistant to treatment with hot triethylamine or quinoline as well as to iodide ion in dimethylformamide. However, in alkali it very rapidly yields an acidic, bromine-free compound B. Explain.

b Compound B in mildly acidic aqueous solution yields a dihydroxy-acid C, which cannot be dehydrated to a lactone. Explain.

c 3-Cyclohexene-1-carboxylic acid reacts analogously with bromine, forming an isomer of A which readily loses **HBr** to yield a single compound on treatment with hot triethylamine or quinoline. Explain the difference.

15-15 Enol ethers are very reactive in aqueous acid. Benzene rings bearing alkoxyl substituents are often usefully converted to cyclohexenones by lithium reduction to enol ethers (Chap. 18) and hydrolysis. Show the mechanism for hydrolysis of dihydro-anisole to a compound with λ_{max} 227 nm.

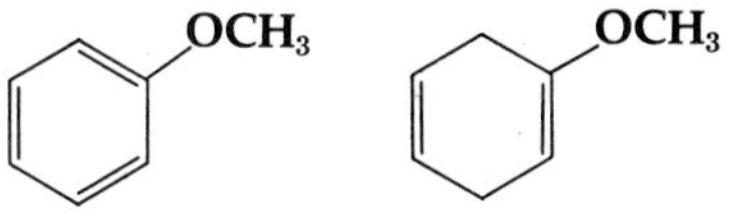

Anisole   Dihydroanisole

15-16 Complete the blanks below to appreciate two other useful synthetic tools applying enol ethers.

a $ClCH_2OCH_3 + P(C_6H_5)_3 \longrightarrow A \xrightarrow{C_6H_5Li} B \xrightarrow{R_2CO} C \xrightarrow{H_3O^+} D$

b 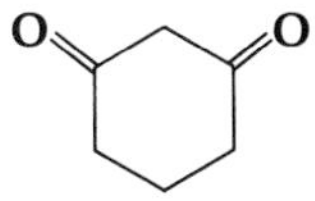 $\xrightarrow{CH_2N_2} A \xrightarrow{LiAlH_4} B \xrightarrow{H_3O^+} C \ (C_6H_{10}O; \lambda_{max}227 \text{ nm})$

15-17 Assign a structure to the following product.

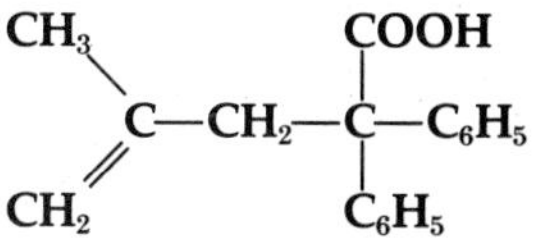

$\xrightarrow{Br_2}$ $(C_6H_5)_2C_6H_7O_2Br$

(Insoluble in HCO_3^-)

ELECTROPHILIC SUBSTITUTION; AROMATIC COMPOUNDS

AROMATIC compounds dominated the early study of organic chemistry, largely because their resonance energy offered extra stability against strong, primitive reagents. Fuming nitric and sulfuric acids often give clean reactions with aromatic compounds. The central reaction of aromatic compounds is that of electrophilic substitution, and this substitution of other groups for hydrogens on aromatic rings was the center of intense study in the nineteenth century. Much of this arose in the search for a rationale for the special stability of the benzene ring (Sec. 5-5).

In the present century the first reaction mechanism to receive extensive study was that of aromatic (electrophilic) substitution, examined in a number of laboratories as early as the 1920's. The salient observation in need of explanation was that a substituent A already present on a benzene ring *directed* the position of substitution of a second added group B. Of the three possible isomeric positions, ortho, meta, and para, available to a substituting group B, some initial substituents A were found to direct substitution primarily to the ortho and para positions, others mainly to the meta location.

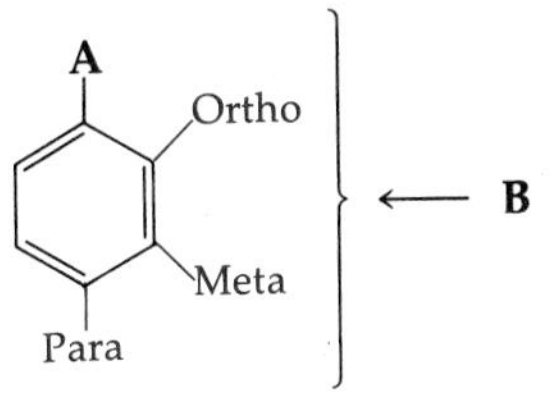

As we shall see, the reagents used for aromatic substitution are all strong electrophiles seeking the electrons in the aromatic π-electron cycle. As we saw in Chap. 5, another substituent already attached to the ring usually interacts strongly with that π-electron cycle by delocalization (resonance) and this is certain to have effects on the entering electrophile.

The actual electrophilic atom in the reagent, the one which bonds to the ring, can be carbon, nitrogen, sulfur, or halogen, and a section is devoted to each in the subsequent discussion so that we may see how the possible aromatic ring substituents may be created by electrophilic substitution. Following this is a section devoted to their various interconversions which allow an appreciation of all the synthetic methods available for making the many possible substituted aromatic compounds.

The single unified mechanism which serves for all aromatic electrophilic substitutions is presented first, however, in generalized form. It follows easily from the previous discussion of electrophilic reactions in Chap. 15 and serves admirably to explain both the rate and orientation of substitution.

The first task of the nineteenth-century chemist was to determine *which* structural isomers were formed when he added a second substituent to an already substituted benzene. Without spectroscopic aids he needed a procedure for identifying ortho, meta, and para products. The method of Körner offered a very elegant use of structural logic. In this method a disubstituted benzene, which must be ortho, meta, or para, is identified by carrying out a further substitution reaction and determining *how many* trisubstituted isomers are formed, as shown in the accompanying table.

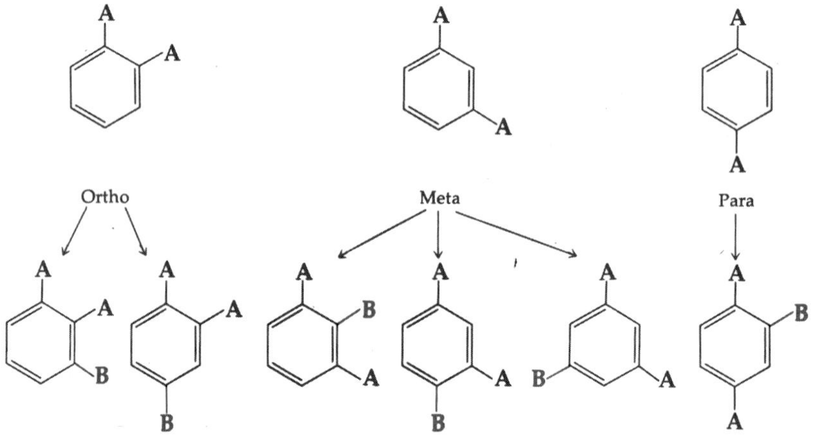

Thus it is rigorously logical that only the ortho isomer can form just two products, the meta, three, and the para, one. Each is therefore identified by the *number* of products it forms.

16-1   THE MECHANISM OF ELECTROPHILIC SUBSTITUTION

The cyclic π orbital of the benzene ring is the electron source sought by the reagent electrophile. Just as in electrophilic addition to alkenes this attack forms a carbonium ion as an intermediate, although with the benzene ring this is

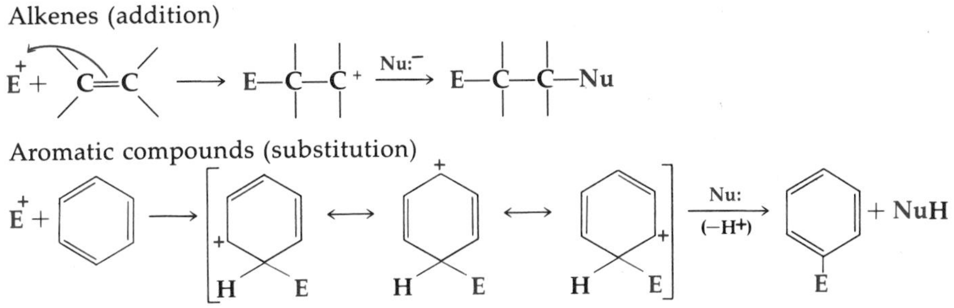

Alkenes (addition)

Aromatic compounds (substitution)

FIGURE 16-1 Reactions of electrophilic reagents with alkenes and aromatic compounds

a doubly conjugated cyclohexadienyl ion. Instead of undergoing nucleophile addition next, as with alkenes, this ion loses a proton to the nucleophile (as base) and so reverts to a stable aromatic π-electron cycle, now substituted with the electrophilic group E. This course of reaction is shown and contrasted with electrophilic additions in Fig. 16-1. Under special conditions, free of nucleophiles/bases, these cyclohexadienyl cations have actually been isolated.

TABLE 16-1 Substituent Effects on Rates of Substitution

	Order of decreasing rate	Electronic effect	Orientation of substitution
	$-\ddot{N}R_2, -\ddot{N}H_2, -\ddot{O}H, -\ddot{O}\!:^-$	Resonance donation	Ortho, para
	$-\ddot{N}HCOR, -\ddot{O}R, -\ddot{O}COR$		
	$-Ar$		
Activating	$-R$	Donation	
	$-H$	(Standard of comparison)	
Deactivating	$-\ddot{X}\!:$	Resonance donation with inductive withdrawal	
	$-\overset{+}{N}R_3, -\overset{+}{S}R_2, -CX_3$	Inductive withdrawal	Meta
	$-\underset{\mid}{C}=O, -C\equiv N, -SO_2, -NO_2$	Resonance withdrawal (Fig. 12-3)	

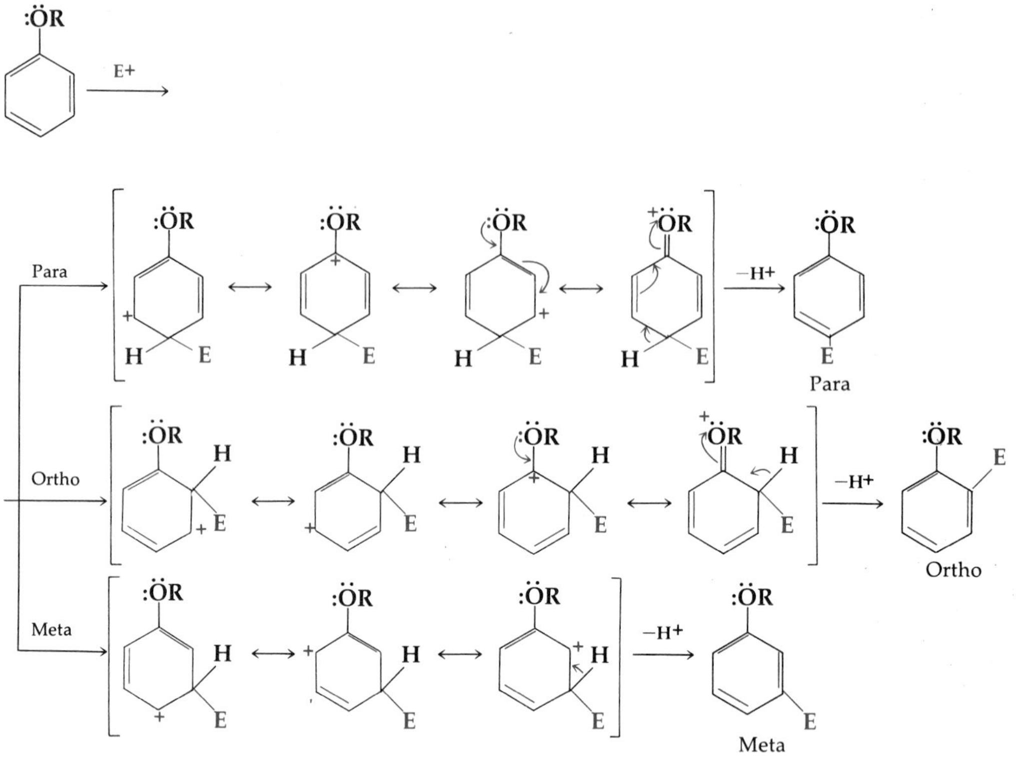

FIGURE 16-2 Effect of an alkoxyl group in an electrophilic substitution reaction

Reaction Rates

The rate of aromatic substitution will be less than that of electrophilic attack on alkenes since the benzene resonance energy is destroyed in the first, rate-determining step to the carbonium ion.† As a consequence electrophilic substitution in benzene is slower, or requires stronger electrophiles, than addition to simple alkenes. (See Fig. 9-7A.) As with alkenes we may expect the rate to be enhanced by electron-donating groups, either through resonance or (less) by alkyl groups through their combined inductive and hyperconjugation effects. These are all called **activating groups**.

Similarly, electron withdrawal by resonance or inductive (dipole) interaction retards the rate (**deactivating groups**). Common substituents are listed in a rough order of rate in Table 16-1, which shows the expected predominance of resonance over inductive effects encountered also in acids and bases (Chap. 8).

---

† The intermediate cyclohexadienyl cation is resonance-stabilized relative to a simple carbonium ion but this stabilization is not nearly as much as that conferred on the original benzene ring by its six cyclic π electrons ($\sim$40 kcal/mole).

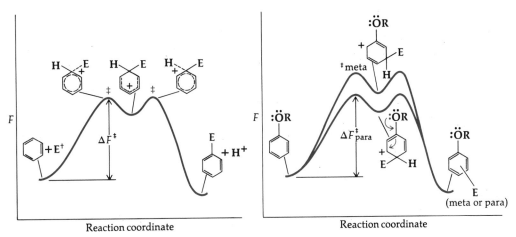

FIGURE 16-3   **Free-energy diagrams for aromatic electrophilic substitution**

Orientation of Substitution

The central test of the effect of an existing ring substituent on the position of electrophile attack is to compare the ability of that substituent to stabilize or destabilize the three transition states leading to ortho, meta, and para substitution, respectively. The favored orientation of attack will then be the one with the lowest activation energy ($\Delta F\ddagger$). As usual, we cannot assess the transition state directly, but must examine the high-energy cyclohexadienyl cation intermediate as a close model for the transition state, which it resembles in energy and geometry.

The procedure may be illustrated with an alkoxyl substituent, as in Fig. 16-2, in which the important resonance forms stabilizing the cationic intermediate are written out for each of the three possible reaction courses. The unshared pair on oxygen can help delocalize the positive charge only in the ortho and para substitutions, which therefore have four instead of only three important resonance forms and more stabilization. This results in a lower-energy transition state for ortho and para than for meta substitution and thus favors ortho and para over meta products.

The free-energy diagrams for substitution of benzene and alkoxybenzene are shown in Fig. 16-3. In the first a rough representation of the transition states is drawn as well as a depiction of the cyclohexadienyl cation with its five-carbon π-electron system shown with dots instead of three resonance forms. Only para and meta substitution paths are shown for alkoxybenzene; ortho

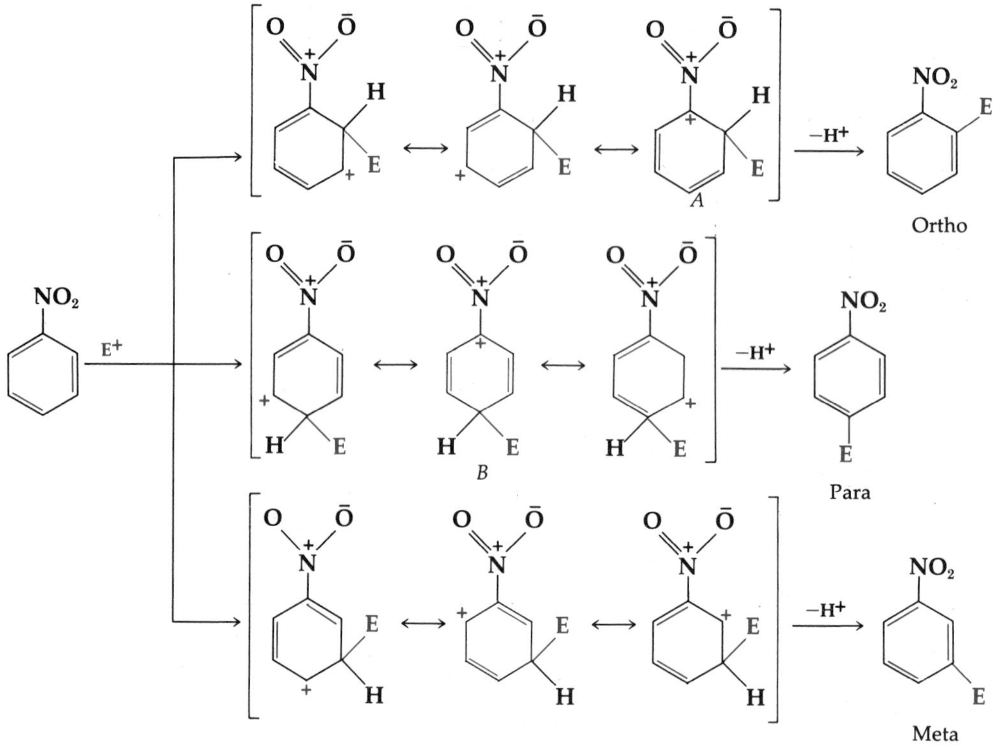

FIGURE 16-4   Effect of a nitro group on electrophilic substitution

is like para, with possibly a slightly higher-energy transition state due to some steric crowding between adjacent substitutions.

Electron-withdrawing substituents deactivate aromatic nuclei, the effect being *smallest* at the meta position. Figure 16-4 shows the principal resonance structures of the intermediate ions that could be formed by electrophilic attack on nitrobenzene. Attack at the ortho and para positions places part of the positive charge adjacent to the electron-withdrawing substituent group. This arrangement gives rise to electrostatic repulsions and destabilization larger than those involved in meta attack. The structures in question are labeled A and B in Fig. 16-4.

A further barrier to attack on nitrobenzene is that formation of the intermediate cation requires sacrifice of some of the resonance energy of the substrate. The interaction between the substituent and the ring removes π electrons from the ring. Since electrophilic substitution also withdraws electrons from the ring, interaction between the ring and the nitro group must be decreased in the transition state for substitution. The overall effect is that nitrobenzene shows a slow substitution rate and yields primarily meta substitution products, rather by default since these are *disfavored least*.

Other substituents may be analyzed in the same way and the resultant orientations they impose on substitution are listed in Table 16-1. With electron-donating groups the sites of substitution are rapidly identified by use of the arrow convention for electron donation in the starting material.

The effects of halogens as substituents were long considered anomalous. Chloro-, bromo-, and iodobenzene are less reactive than benzene itself but give predominantly ortho and para substitution. Fluorobenzene orients ortho-para and is about equal in reactivity to benzene. Most of the other groups that orient ortho-para activate aromatic rings. The curious effect of the halogens arises from the interplay between inductive effects and weak resonance effects. Halobenzenes have dipole moments with halogen negative, but the moments are smaller than those of aliphatic halides. An oversimplified explanation of the effect attributes the result to partial compensation of the polarity of the C—X σ bonds by some resonance feedback of nonbonding halogen p electrons by π bonding, qualitatively just as in the alkoxybenzenes (Fig. 16-2). However, except for fluorine, the larger halogens form poor π bonds (Sec. 19-1) and so exhibit little resonance stabilization. The π-electron feedback is greatest if attack occurs at the ortho and para positions (as with alkoxy substituents), and this effect provides preferential attack at those positions. However, the unfavorable electrostatic interaction between the polar C—X bond and the positive charge is still sufficient to make the halobenzenes less reactive than benzene itself. Although the duality of effects makes the phenomenon of ortho-para orientation by deactivating groups understandable, there is no *a priori* basis for prediction of the actual balance of effects.

With more than one substituent, the more activating substituent usually controls the site of substitution, and steric hindrance usually disfavors ortho over para and particularly disfavors substitution *between* two meta substituents. The major positions of substitution are illustrated by the following examples.

Polynuclear aromatic compounds are more reactive than benzene derivatives. Such a result would be expected from consideration of the extensive delocalization of electrons in the cations formed by adding electrophiles to polynuclear systems. Naphthalenes almost always react in the α-position since aromatic resonance in the other ring is less damaged in the intermediate.

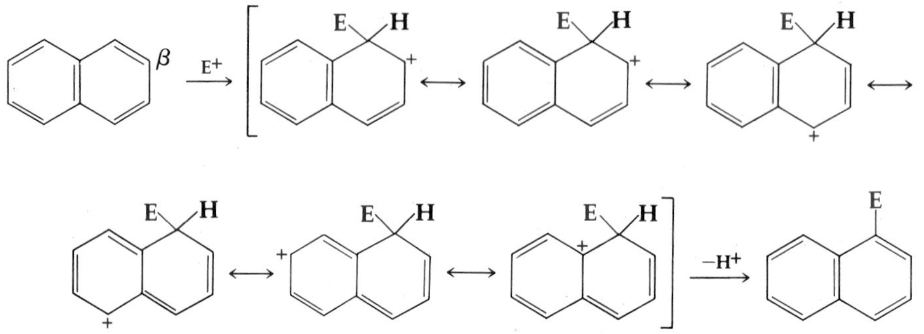

Quantitatively, relative reactivities and orienting influences in aromatic substitutions vary considerably from one reaction to another. Toluene is about 200 times as reactive as benzene to bromination in acetic acid solution but only 30 times as reactive in nitration in nitromethane solution. The reasons for such variations are not reflected in the simplified discussion presented. Error arises in treating a reaction intermediate as though it were actually the transition state in the rate-determining step of the substitution. While the transition states are probably similar to the intermediates, they are not identical with them. In the actual transition states, the extent to which π electrons have been withdrawn from the aromatic systems probably varies with the nature of the incoming reagent and even with the solvent.

PROBLEM 16-1

Place the following compounds in order of decreasing rate of halogenation and draw the structure of the primary monohalo derivative formed.

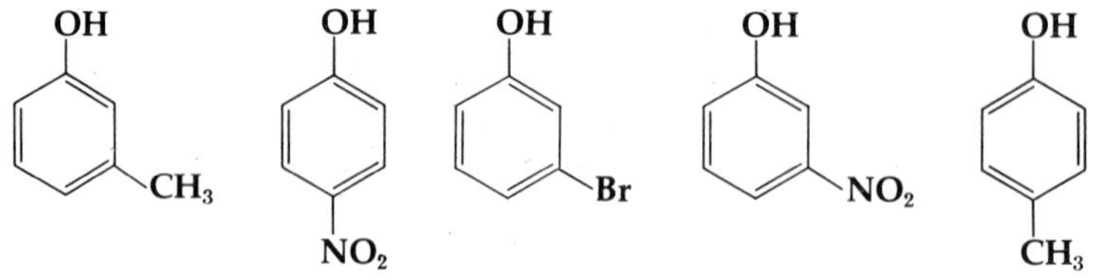

16-2   HALOGENATION; ARYL HALIDES

Halogenation is one of the most useful methods for introduction of substituents in an aromatic system. Reactive substrates are substituted very readily, as shown by the following:

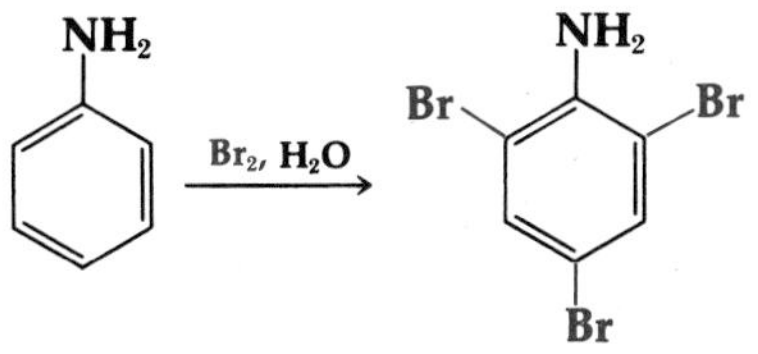

NH$_2$ —Br$_2$, H$_2$O→ 2,4,6-tribromoaniline (NH$_2$ with Br at 2, 4, 6 positions)   100%

OH —Br$_2$, H$_2$O→ 2,4,6-tribromophenol (OH with Br at 2, 4, 6 positions)   100%

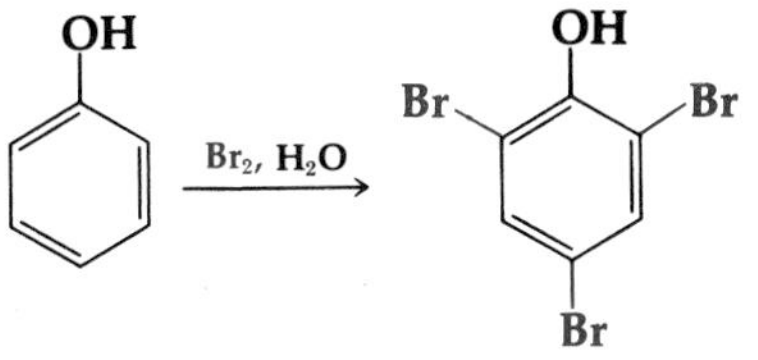

With less reactive substrates, Lewis acid catalysts of varying activity are used to accelerate reactions. The function of the catalysts is to complex with, and polarize, the halogen, rendering it a better electrophile.

$$X_2 + FeX_3 \rightleftharpoons \overset{+}{X}\ \overset{-}{FeX_4}$$

$$X_2 + AlX_3 \rightleftharpoons \overset{+}{X}\ \overset{-}{AlX_4}$$

Lewis acid catalysis is used even with alkylbenzenes to allow a more favorable competition between electrophilic substitution, which gives nuclear attack, and free-radical halogenation, which results in side-chain attack (Chap. 19). In reactions that provide mixtures of compounds, as in the chlorination of toluene, the pure components are sometimes hard to isolate.

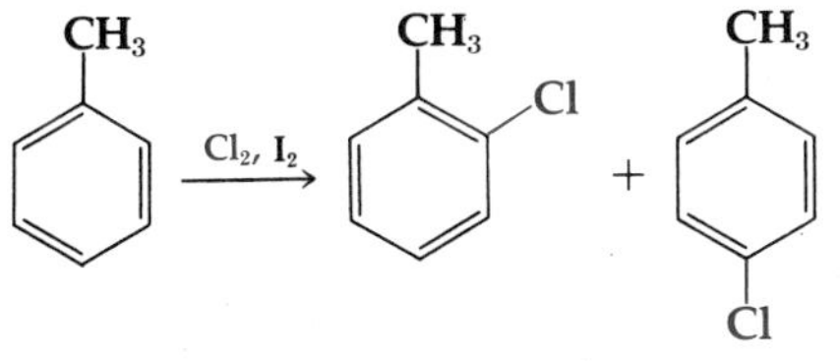

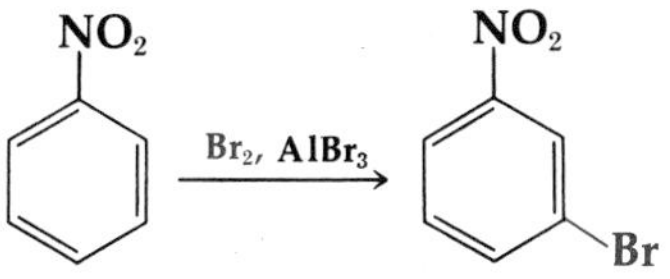

CH$_3$ (toluene) —Cl$_2$, I$_2$→ o-chlorotoluene (CH$_3$ with Cl ortho) + p-chlorotoluene (CH$_3$ with Cl para)

NO$_2$ (nitrobenzene) —Br$_2$, AlBr$_3$→ m-bromonitrobenzene (NO$_2$ with Br meta)   96%

The above examples demonstrate that direct halogenation is not a feasible way to produce all isomers of a given di- or polysubstituted benzene. Various artifices are employed to produce derivatives that cannot be formed by direct introduction of a second substituent. Sometimes different compounds are formed by simply reversing the order of substitution. For example, o- and p-nitrohalobenzenes are formed by nitration of halobenzenes, since the halogens orient ortho-para. Such a procedure would not suffice for the production of an *m*-halotoluene, since both halogen and methyl orient ortho-para.

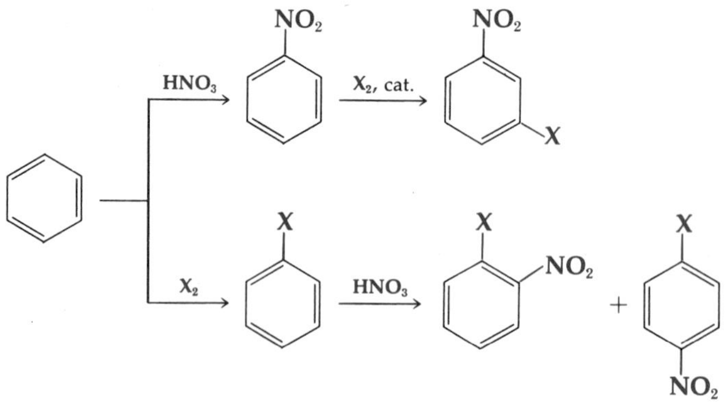

PROBLEM 16-2

Phenol will react under forcing conditions with excess bromine to form a neutral compound, $C_6H_2Br_4O$. On the basis of the mechanism of electrophilic substitution what formula(s) might be written for this compound? What UV spectrum is expected for it?

16-3   NITRATION; ARYL-NITROGEN BONDS

Nitration of aromatic compounds provides the first step in the synthesis of a large number of substances. Polynitro compounds themselves are used as high explosives (TNT is 1,3,5-trinitrotoluene). Reduction of nitro groups leads to a variety of functional groups (Chap. 18), the most important of which is the amino group. Reduction to amines converts a strongly deactivating meta-directing substituent into a powerful ortho-para orienting group. Since the Sandmeyer and related reactions permit replacement of the amino group with many other groups, a synthesis of virtually any polysubstituted aromatic compound can be based upon nitration as a first or early step.

$$HNO_3 + ArH \longrightarrow ArNO_2 + H_2O$$

Conditions for the nitration reaction vary greatly with the reactivity of the aromatic substrate. The nitration mixture required for introduction of a second nitro group into benzene (concentrated nitric and sulfuric acids as 95°) would provide an uncontrollably exothermic reaction with phenol. Nearly all nitrations involve electrophilic attack by the **nitronium ion**, NO_2^+.† Consequently, reactions can be regulated by controlling the concentration of nitronium ion in solution. In 95% sulfuric acid, the ionization of dissolved nitric acid is complete.

† The reactive salt, $NO_2^+BF_4^-$, is available commercially and serves to nitrate aromatics, as expected.

$$H_2SO_4 + HONO_2 \rightleftharpoons H_2\overset{+}{O}NO_2 + HS\bar{O}_4$$

$$H_2\overset{+}{O}NO_2 \rightleftharpoons H_2O + N\overset{+}{O}_2$$

$$H_2O + H_2SO_4 \rightleftharpoons H_3\overset{+}{O} + HS\bar{O}_4$$

$$2H_2SO_4 + HNO_3 \rightleftharpoons H_3\overset{+}{O} + N\overset{+}{O}_2 + 2HS\bar{O}_4$$

In concentrated nitric acid, a small amount of NO_2^+ is present, and even less is formed in solutions of nitric acid in solvents such as acetic acid, acetic anhydride, and nitromethane. By these means, nitration can also be controlled to yield only mononitration products or else dinitro (or trinitro), as desired.

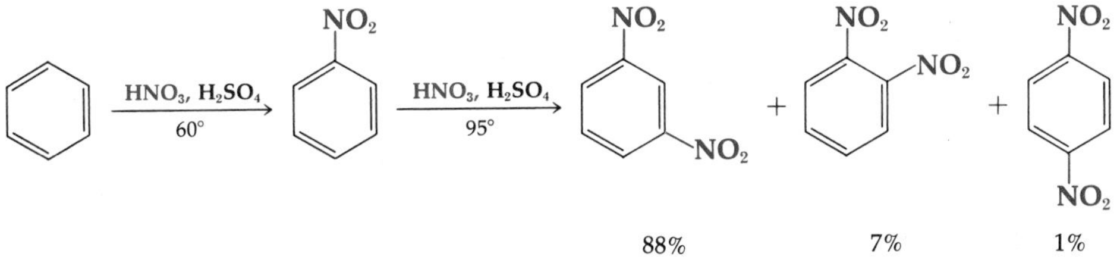

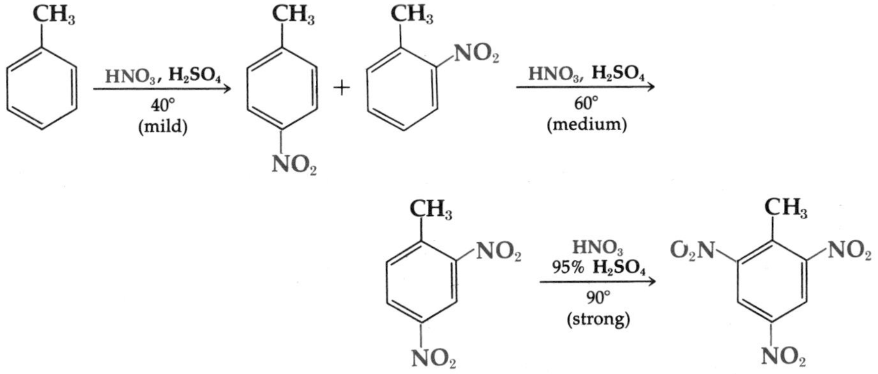

88%   7%   1%

Trinitrotoluene
(TNT)

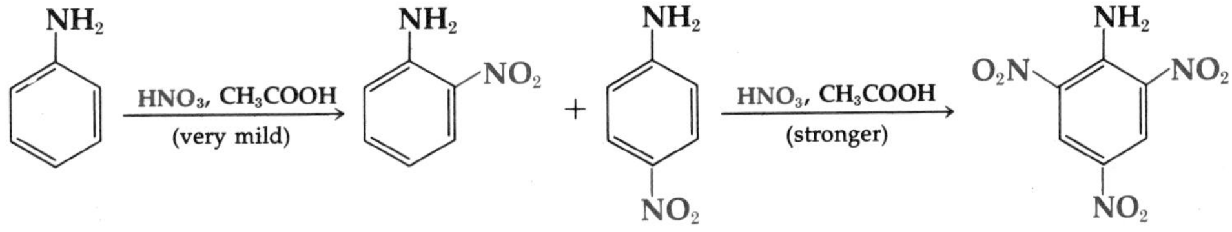

Picramide
(low yield)

Compare

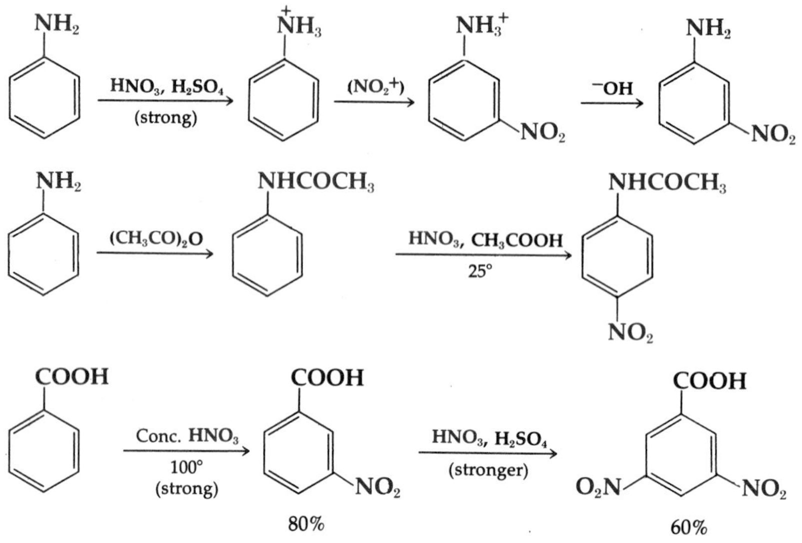

Nitration, as well as other electrophilic substitutions, is subject to steric hindrance, and substitution ortho to a large group may become difficult, as illustrated by the isomer distribution in the mononitration of alkyl benzenes.

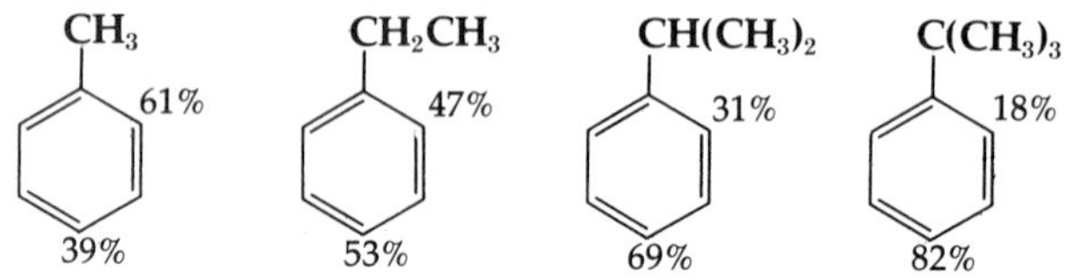

Other nitrogen electrophiles are not so active and so require more activation in the aromatic substrate. Nitrous acid and other nitroso electrophiles (see Sec. 13-7) deliver o- and p-nitroso groups to phenols and aromatic tertiary amines. The same activated substrates will attack the aryl diazonium ion to form brightly colored **azo compounds** used commercially as azo dyes. The second reaction, with β-naphthol, is used as a test for diazonium compound.

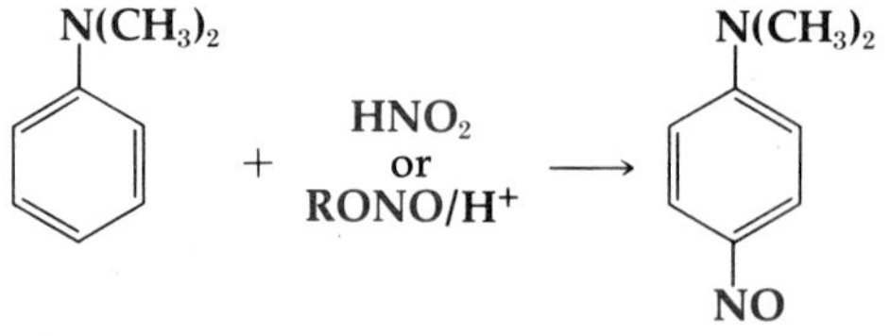

p-Nitrosodimethylaniline

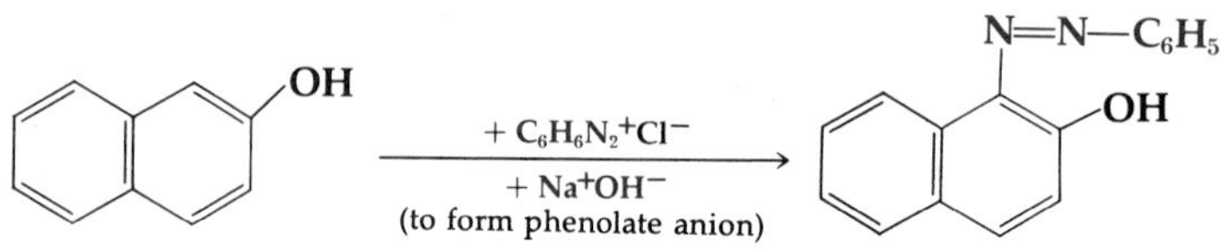

β-Naphthol   Azo compound

PROBLEM 16-3

Nitration of styrene, $C_6H_5CH=CH_2$, affords *three* (mononitro) isomers. What are they?

PROBLEM 16-4

Write the product(s) of mononitration of each of the following and explain the orientation you expect.

a   *m*-Cresol (*m*-hydroxytoluene)
b   *m*-Cyanophenol
c   *o*-Chlorophenol

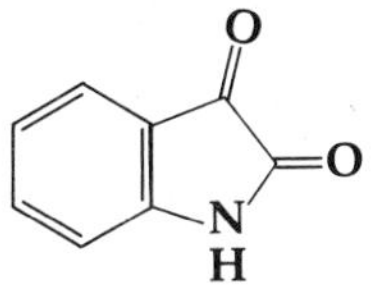

e   *p*-Nitroacetanilide
    (p-$O_2NC_6H_4NHCOCH_3$)

PROBLEM 16-5

a   Show why nitration of naphthalene goes exclusively at the α-position.
b   What products would you expect from mononitration of α-naphthol(1-hydroxynaphthalene)? of β-naphthol?

16-4   SULFONATION; ARYL-SULFUR BONDS

Sulfonation is one of the most important aromatic substitution reactions not only because of the intrinsic value of the sulfonic acids produced but also because sulfonic acids can be converted to many important derivatives.

$$ArH + H_2SO_4 \longrightarrow Ar-SO_3H + H_2O$$

The reactive electrophile in sulfonation is either SO_3 or SO_3H^+, species present in fuming sulfuric acid, the reagent generally used in sulfonation. In the following equations, the yield data in boldface type show *isomer distribution* rather than isolable yields.

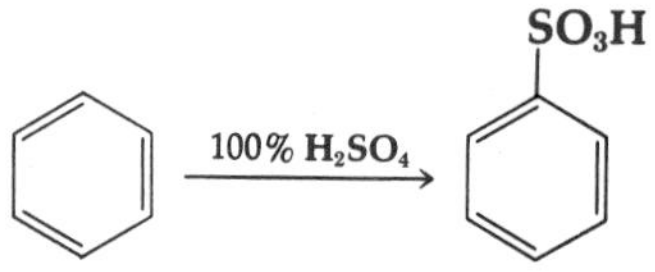

Benzenesulfonic acid

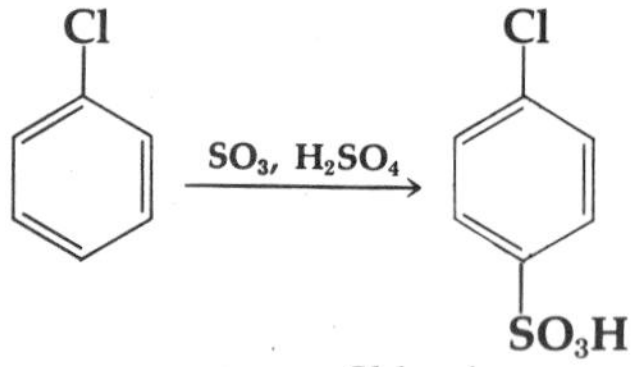

p-Chlorobenzenesulfonic acid
100%

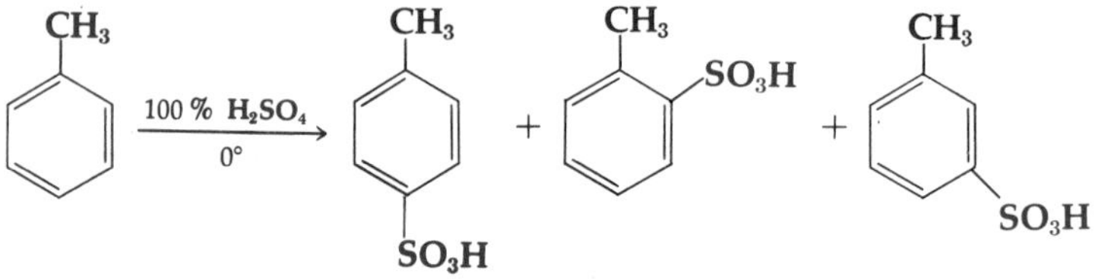

62%   32%   6%

Since sulfonation is reversible, sulfonic acid groups are removed if the acids are heated with aqueous sulfuric acid of only 50 to 60% concentration, for this is merely a proton-donating medium. Because of the powerful meta-directing influence of the sulfonic acid function, sulfonic acid groups are often introduced for blocking and orienting purposes. After other substituents have been introduced, the sulfonate group is removed by hydrolysis in strong acid, replacing $-SO_3H$ by $-H$.

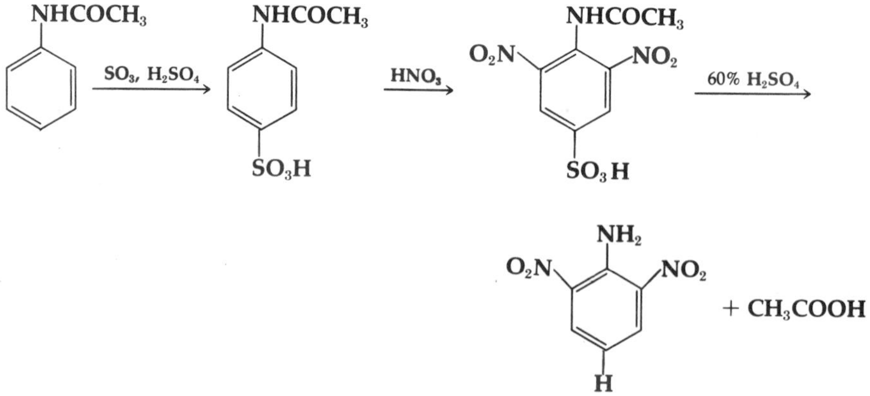

$+ CH_3COOH$

The reversibility of the reaction is also demonstrated by a tendency for initially formed products to revert to more stable isomers at high temperatures.

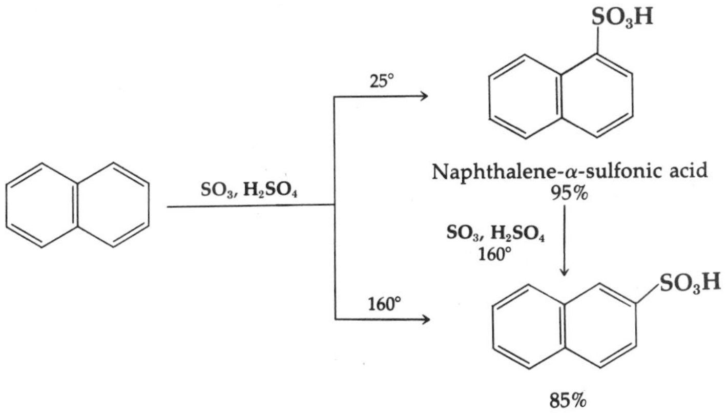

Naphthalene-α-sulfonic acid
95%

85%

Rapid formation of the α-isomer in the sulfonation of naphthalene reflects the greater stabilization of the intermediate carbonium ion. However, the α-sulfonic acid group is sterically hindered by the hydrogen in the 8- (or *peri*) position. The unhindered β-isomer is the more stable of the two compounds.

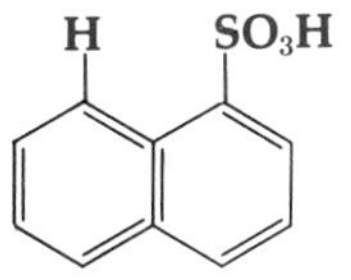

peri-Steric hindrance

Disulfonation is very slow, because of the strong electron-withdrawing influence of SO_3H, and requires forcing conditions. Note the extensive reversion to the stable para isomer under the vigorous conditions in the following example.

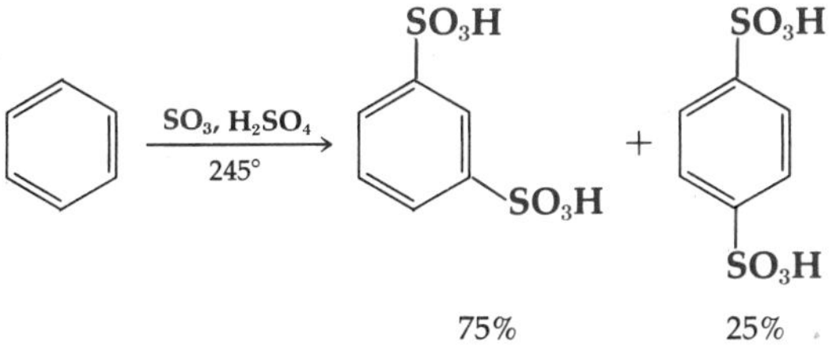

75%   25%

Sulfonic acids are useful as cheap strong acids that are more soluble than strong mineral acids in organic solvents. Another important use is in dyes and detergents, which are frequently water-soluble salts of sulfonic acids.

Sulfonyl chlorides can be made directly by treatment of aromatics with chlorosulfonic acid or by treatment of sulfonic acids with PCl_5, etc. (Sec. 19-6). In the former instance sulfonation probably occurs first and is followed by conversion of the sulfonic acid to the acid chloride by reaction with excess reagent. As is true in most electrophilic substitutions, the actual reagent is produced by a complex initial ionization reaction.

$$2ClSO_2OH \rightleftharpoons HCl + HO\overset{+}{S}O_2 + Cl\overset{-}{S}O_3$$

Chlorosulfonic
acid

$$HO\overset{+}{S}O_2 + ArH \longrightarrow ArSO_2OH + \overset{+}{H}$$

$$ArSO_2OH + ClSO_2OH \rightleftharpoons ArSO_2Cl + H_2SO_4$$

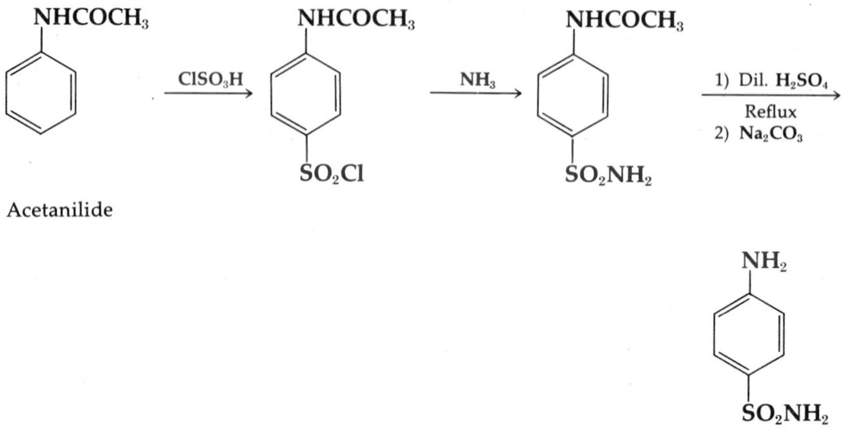

Acetanilide

Sulfanilamide

An important use of arenesulfonic acids involves their conversion to phenols by fusion with sodium or potassium hydroxide. This nucleophilic substitution at unsaturated carbon (only under very vigorous conditions) provides one of the few methods of preparing phenols since there is no oxygen electrophile (like RCO_3H) which is strong enough to attack benzene.

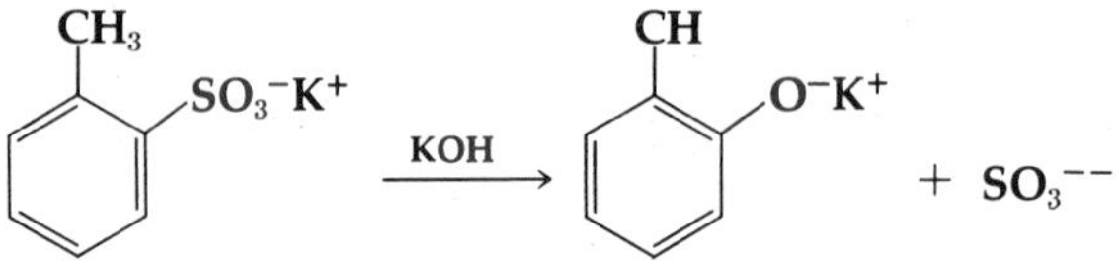

16-5 PROTONS AS ELECTROPHILES

Electrophilic substitution by protons can be observed in isotopic exchange reactions. Treatment of benzene with D_2SO_4 results in slow exchange, and, as would be expected, toluene undergoes fairly rapid exchange in the ortho and para positions. Specific isotopic labeling of a ring is possible by this exchange of proton for deuterium or tritium.

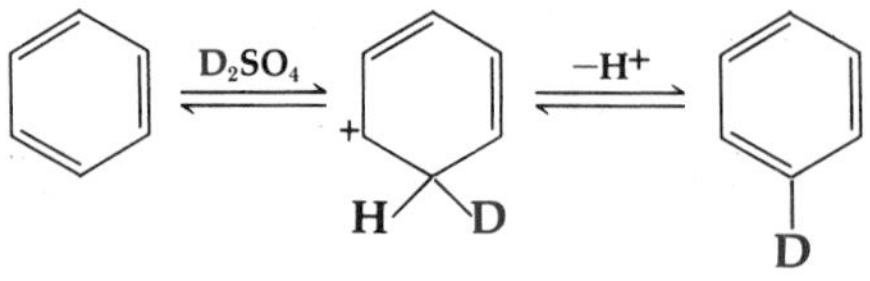

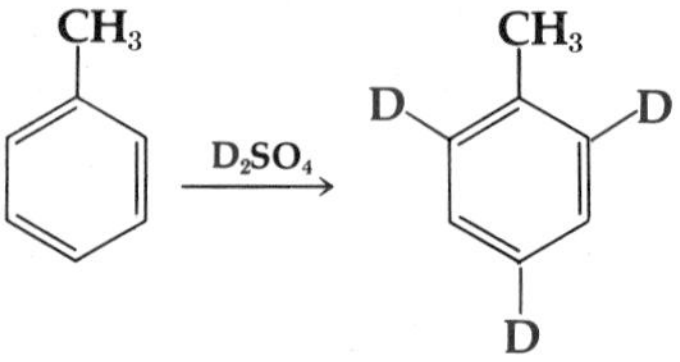

A number of functional groups can be removed from aromatic nuclei by acid cleavage. The most important example is the previously mentioned hydrolysis of aromatic sulfonic acids. Arylcarboxylic acids and aryl ketones that have two alkyl substituents ortho to the functional groups are cleaved by concentrated acids. Less hindered acids and ketones do not undergo the reaction unless there is ortho or para activation, as with a phenolic —OH. With the simple sterically hindered carbonyl compounds, the hindrance probably forces the carbonyl groups out of the plane of the aromatic ring, thus interfering with the normal resonance interaction stabilizing the groups. Cleavage occurs easily, since electrophilic attack does not involve as much loss of resonance energy as would be the case with unhindered carbonyl compounds.

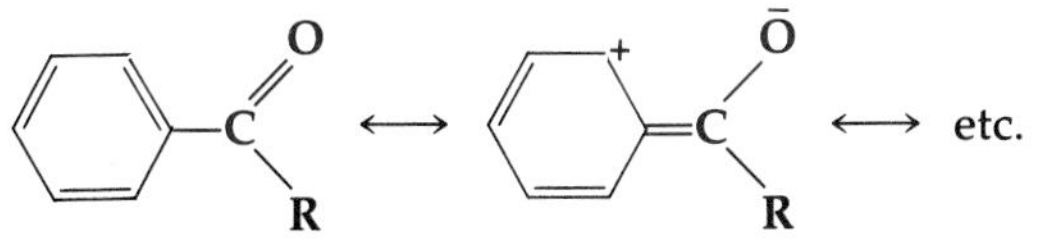

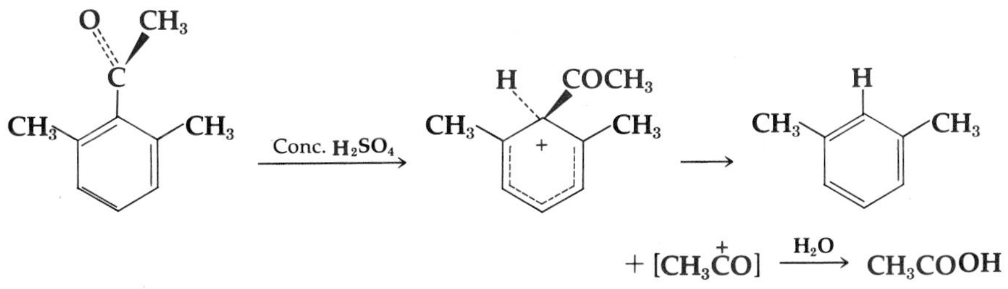

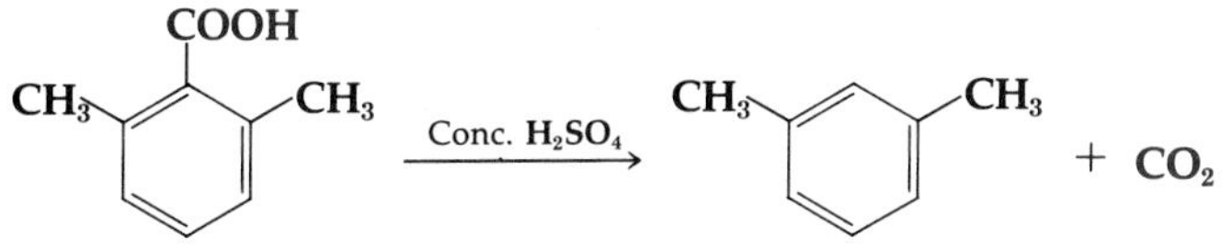

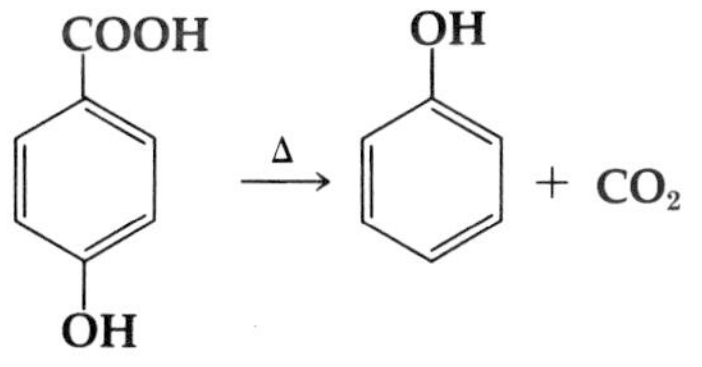

PROBLEM 16-6

Compare the three isomers of resorcylic acid, the dihydroxybenzoic acids with meta hydroxyls. Rank them with respect to ease of decarboxylation in acid. (*Hint:* see page 165.) Decarboxylation in base is much more difficult; what mechanistic feature makes this true? Resorcinol can be *carboxylated* by heating with potassium carbonate in an atmosphere of CO_2. Explain.

16-6 ALKYLATION AND ACYLATION

Generation of carbonium ions, or very reactive carbonium-ion donors, in the presence of reactive aromatic compounds leads to the alkylation and acylation of the aromatic nucleus. The entire group of reactions is known as the **Friedel-Crafts reaction**, although the name was originally applied only to alkylations by alkyl halides catalyzed by aluminum chloride. All the reactions may be formulated in terms of carbonium-ion electrophiles, although the attacking electrophile may often be a highly polarized complex.

Alkylation

A selection of alkylation reactions is illustrated; the carbonium-ion electrophiles are generated in the several ways already seen, in S_N1 ionization or alkene protonation.

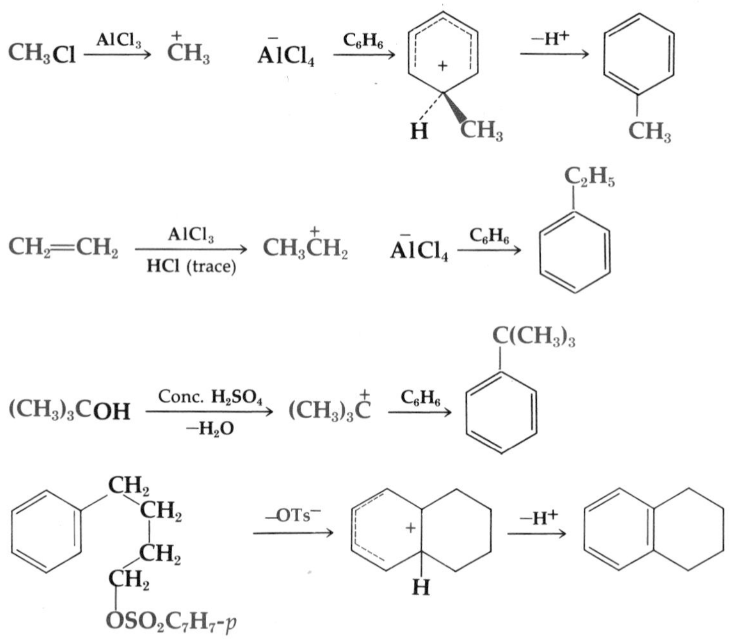

$$CH_3Cl \xrightarrow{AlCl_3} \overset{+}{C}H_3 \quad \overset{-}{A}lCl_4 \xrightarrow{C_6H_6} \quad \xrightarrow{-H^+}$$

$$CH_2{=}CH_2 \xrightarrow[HCl \text{ (trace)}]{AlCl_3} CH_3\overset{+}{C}H_2 \quad \overset{-}{A}lCl_4 \xrightarrow{C_6H_6}$$

$$(CH_3)_3COH \xrightarrow[-H_2O]{Conc. H_2SO_4} (CH_3)_3\overset{+}{C} \xrightarrow{C_6H_6}$$

4-Phenylbutyl
p-toluenesulfonate

Alkylation reactions are complicated by several factors. First, alkyl groups activate aromatic nuclei, so that mixtures of polyalkylated products are formed. This occurs because the rate of alkylation is increased by substitution of the first alkyl group and so the second alkylation proceeds faster than the first (subject to steric hindrance, however).

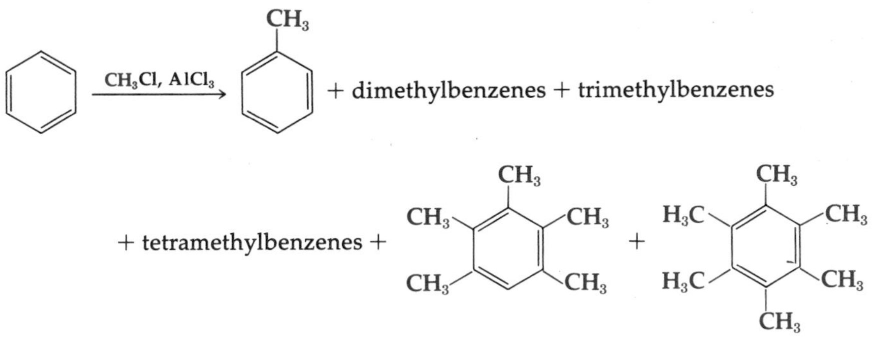

$$\text{(benzene)} \xrightarrow{\text{CH}_3\text{Cl, AlCl}_3} \text{(toluene, } CH_3\text{)} + \text{dimethylbenzenes} + \text{trimethylbenzenes}$$

$$+ \text{ tetramethylbenzenes } + \text{(durene)} + \text{(isodurene)}$$

Relative amounts of products depend upon the ratios of reactants used and contact times. Toluene can be made in high yield by using a high ratio of benzene to methyl chloride. If four equivalents of methyl chloride are used, a significant yield (10%) of durene may be isolated by distillation and freezing of the tetramethylbenzene fraction. Since polyalkylbenzenes rearrange under the reaction conditions, the yield is increased to 25% by treating the liquid mixture of the other tetramethyl benzenes with aluminum chloride and freezing out more durene from the equilibrated mixture.

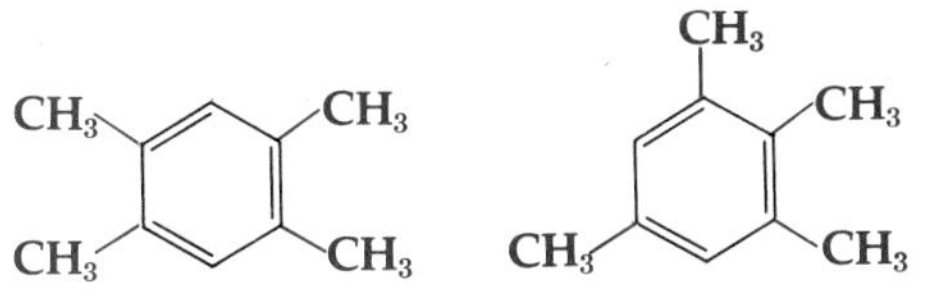

Durene
mp 80°

Isodurene
mp −24°

Mixed xylenes (dimethylbenzenes) and trimethylbenzenes are produced commercially by adjustment of alkylation conditions. Rearrangement of primary alkylation products can be partially avoided by using very gentle reaction conditions.

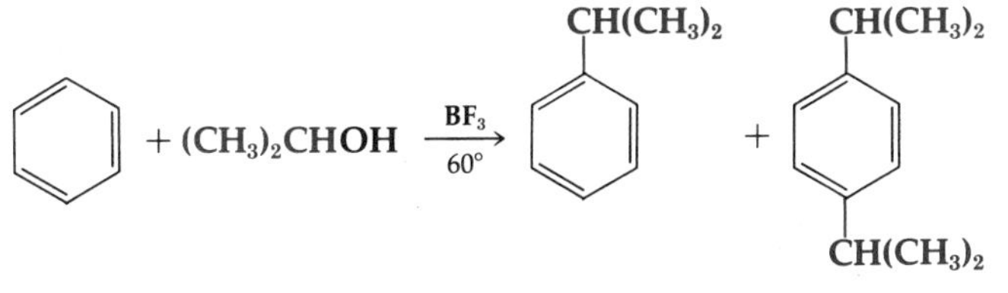

$$\text{(benzene)} + (CH_3)_2\text{CHOH} \xrightarrow[60°]{BF_3} \text{(cumene, } CH(CH_3)_2) + \text{(p-diisopropylbenzene)}$$

Rearrangement of alkyl groups have been observed to produce *sym*-trisubstituted benzenes when alkyl groups larger than methyl are introduced. Under equilibrating conditions, steric hindrance forces the system toward configurations having no two adjacent alkyl groups.

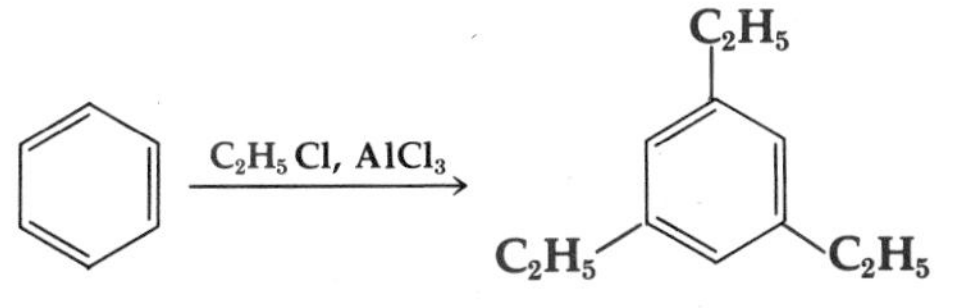

$$\text{(benzene)} \xrightarrow{C_2H_5\text{ Cl, AlCl}_3} \text{(1,3,5-triethylbenzene)}$$

Rearrangement of alkyl groups may occur by way of cleavage followed by realkylation in a less active, meta position. Direct intramolecular transfer from one position to another is also believed to occur.

Dealkylation-realkylation:

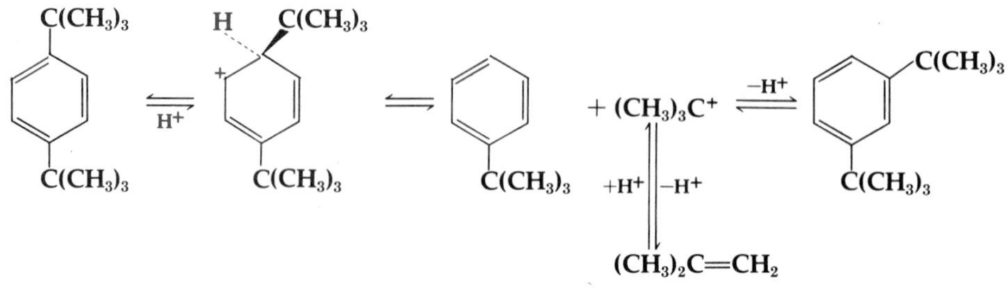

Another undesirable feature of Friedel-Crafts alkylations is rearrangement of alkyl groups themselves during the reaction (Chap. 17). Low yields of *n*-alkyl derivatives may be obtained by using carefully controlled conditions, but usually the products obtained are those derived from the most stable carbonium ion having the same number of carbons as the original alkyl group.

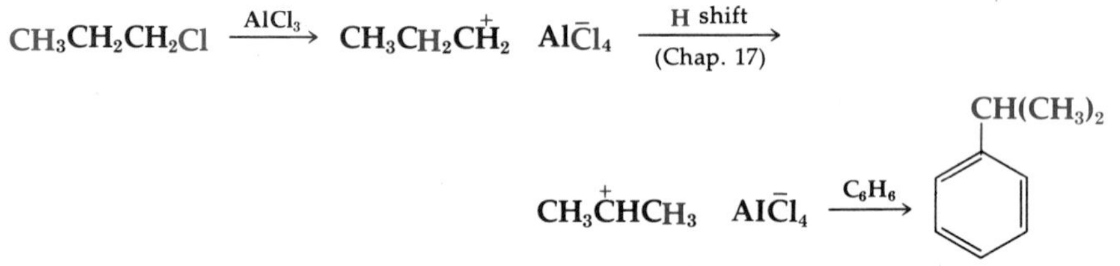

Acylation

The major *useful* reaction for creating carbon-carbon bonds to benzene rings is the Friedel-Crafts acylation reaction. The electrophile is an acyl cation (RCO^+). The reaction involves the aluminum chloride-catalyzed substitution of an acyl group for hydrogen on an aromatic nucleus. In some cases, other Lewis acids and even proton acids are employed as catalysts. The acylating agent is usually an acid chloride or an anhydride. Certain acyl cations, such as $CH_3CO^+BF_4^-$, have actually been isolated as reactive crystalline salts. Common solvents for the reaction include carbon disulfide, nitrobenzene, methylene chloride, and 1,2-dichloroethylene.

$$CH_3\overset{\displaystyle ||}{\underset{\displaystyle O}{C}}\!\!-\!Cl + AlCl_3 \longrightarrow [CH_3\overset{+}{C}\!\!=\!\!\ddot{O} \longleftrightarrow CH_3C\!\!\equiv\!\!\overset{+}{O}]\ AlCl_4^-$$

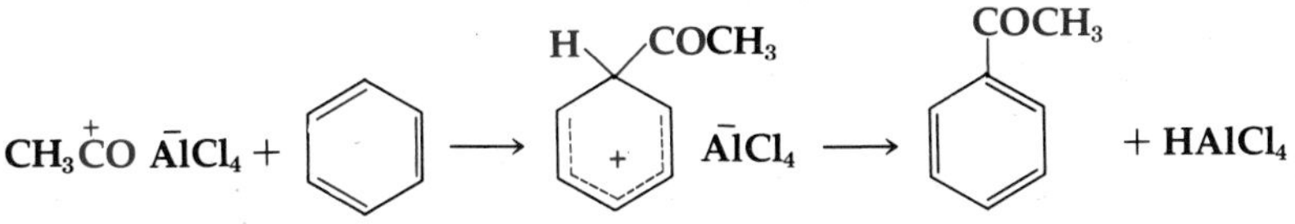

$$CH_3\overset{+}{C}O\ \bar{A}lCl_4 + \bigcirc \longrightarrow \bigcirc_{+}^{\overset{H}{\diagdown}\overset{COCH_3}{\diagup}} \bar{A}lCl_4 \longrightarrow \bigcirc^{COCH_3} + HAlCl_4$$

Acetophenone
90%

$$(CH_3CO)_2O + BF_3 \longrightarrow CH_3\overset{+}{C}O\ \ CH_3COO\ \bar{BF_3}\ \xrightarrow{C_6H_6}\ \bigcirc^{COCH_3}$$

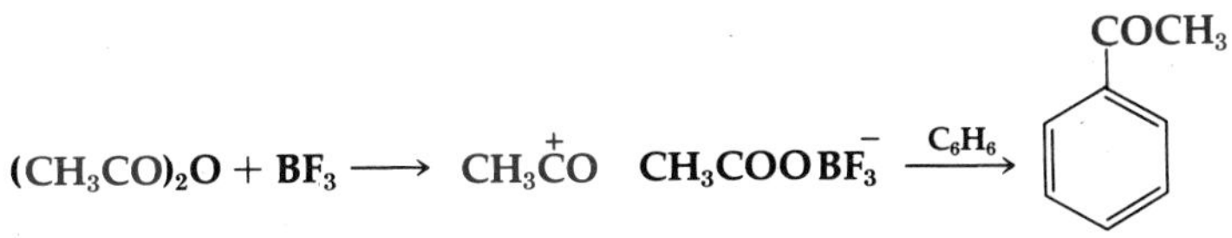

Since an acyl group deactivates an aromatic nucleus toward further electrophilic attack, only a single acyl group can be introduced directly into an aromatic nucleus. This deactivation is in contrast to the situation with alkylation and constitutes the reason for the synthetic value of acylation over alkylation. Acylation can be stopped after introduction of one group. With activating groups acylation can be carried out with mild conditions [cf. $(CF_3CO)_2O$, below].

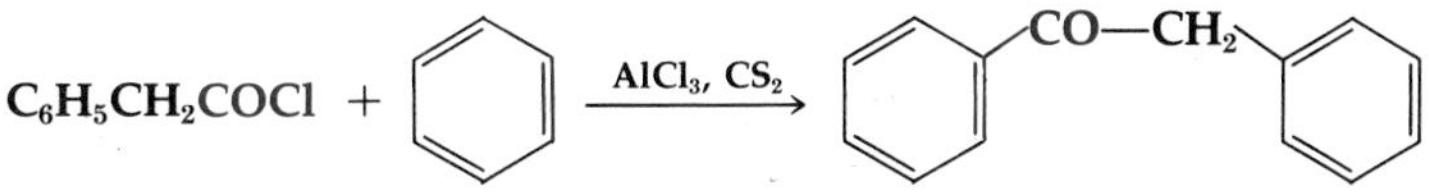

$$C_6H_5CH_2COCl + \bigcirc \xrightarrow{AlCl_3,\ CS_2} \bigcirc\!\!-\!\!CO\!\!-\!\!CH_2\!\!-\!\!\bigcirc$$

Benzyl phenyl ketone
83%

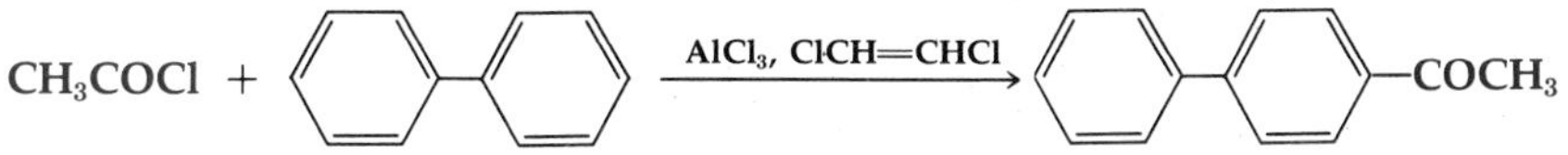

$$CH_3COCl + \bigcirc\!\!-\!\!\bigcirc \xrightarrow{AlCl_3,\ ClCH=CHCl} \bigcirc\!\!-\!\!\bigcirc\!\!-\!\!COCH_3$$

4-Acetylbiphenyl

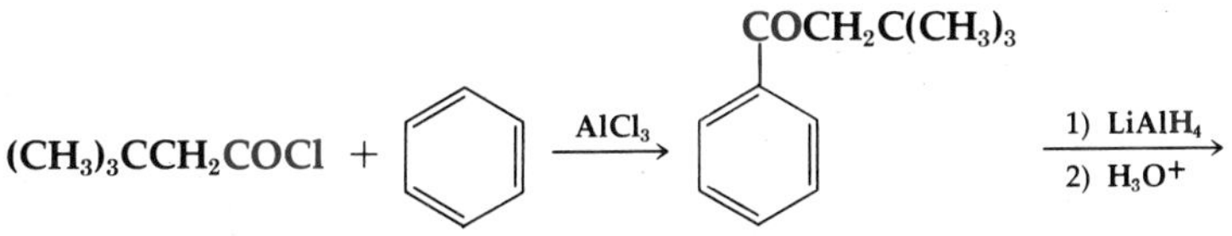

$$(CH_3)_3CCH_2COCl + \bigcirc \xrightarrow{AlCl_3} \bigcirc^{COCH_2C(CH_3)_3} \xrightarrow[2)\ H_3O^+]{1)\ LiAlH_4}$$

87%

$$\bigcirc^{\overset{OH}{\overset{|}{CH}}-CH_2C(CH_3)_3} \xrightarrow[-H_2O]{I_2} \bigcirc^{CH=CHC(CH_3)_3}$$

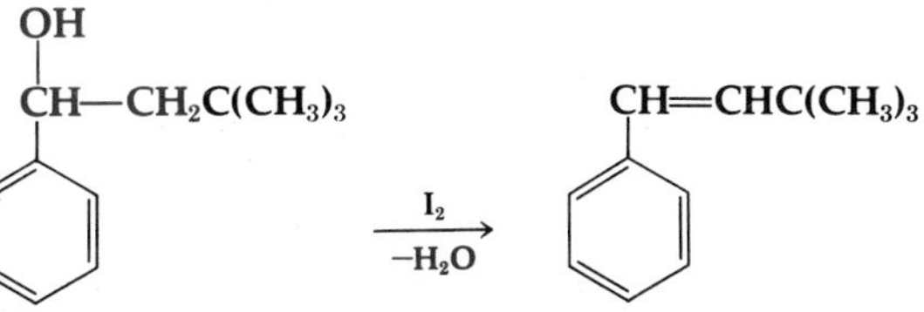

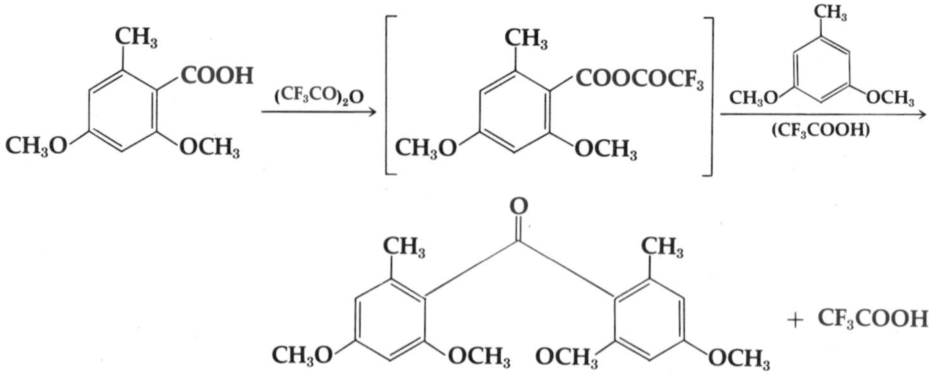

Aromatic rings with meta-directing substituents are not sufficiently re-active to undergo ready acylation; meta-substituted ketones must frequently be made by indirect procedures (Sec. 16-7).

Alkylbenzenes acylate almost exclusively in the para position.

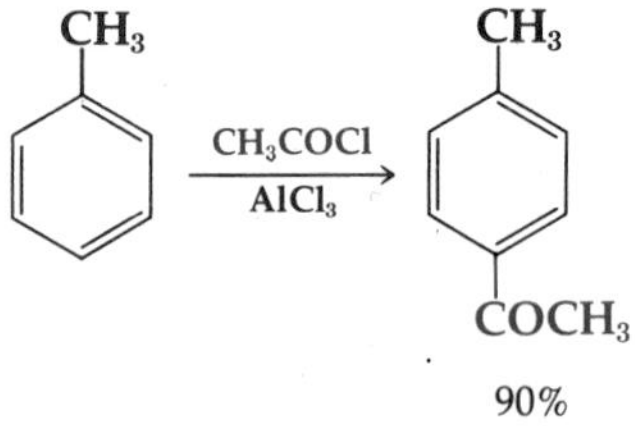

90%

Cyclizations by Acylation

Cyclization of β- and γ-arylalkanoic acids (and acid chlorides) under acid conditions is an elegant method for synthesis of carbocyclic rings fused to aromatic nuclei. A multitude of reaction conditions have been investigated in the course of syntheses of polycyclic systems. Being internal these reactions proceed under milder conditions.

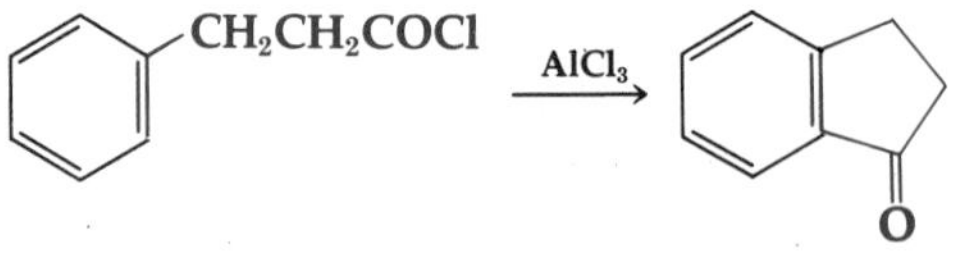

1-Hydrindone
90%

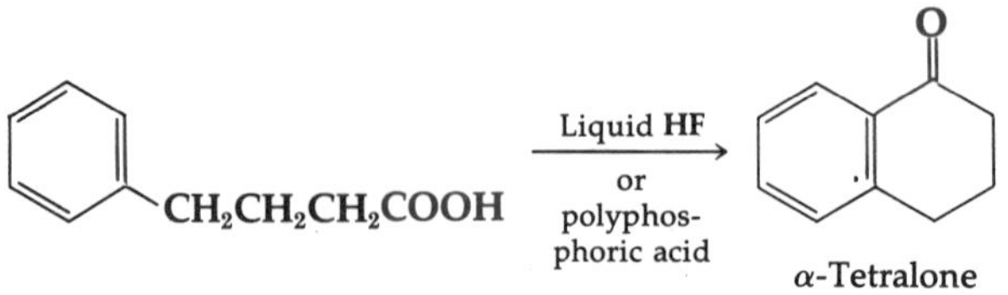

α-Tetralone

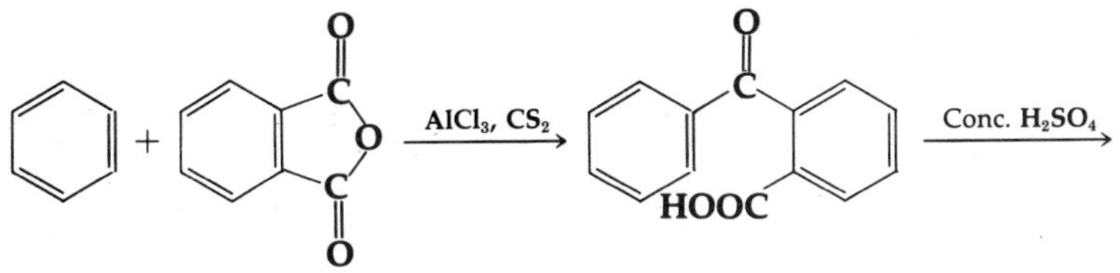

α-Benzoylbenzoic
acid
85%

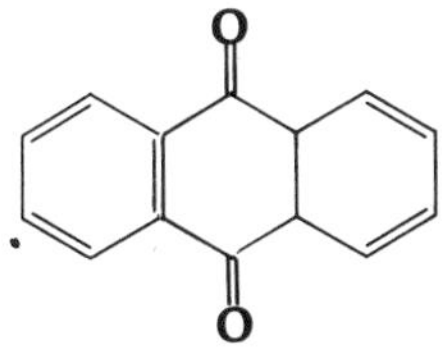

Anthraquinone
100%

Friedel-Crafts cyclizations are useful not only for synthesis of new ring systems but also for preparation of derivatives of common polynuclear hydrocarbons such as naphthalene, phenanthrene, and anthracene. Monosubstituted naphthalenes are nearly all made from naphthalene, but the isomer mixtures obtained in attempting disubstitution are often too complex for practical separation. Even monosubstitution of phenanthrene often gives complex mixtures. The structures of substitution products must be established by unambiguous syntheses, which usually involve cyclization. Some examples of this are presented in Fig. 16-5, mostly as an appreciation of the variety of possibilities. A few reactions in Fig. 16-5 have not yet been presented and these have asterisks (*).

Succinoylation of naphthalene in nitrobenzene gives a separable mixture of α- and β-isomers, both of which are useful in the synthesis of phenanthrene derivatives.

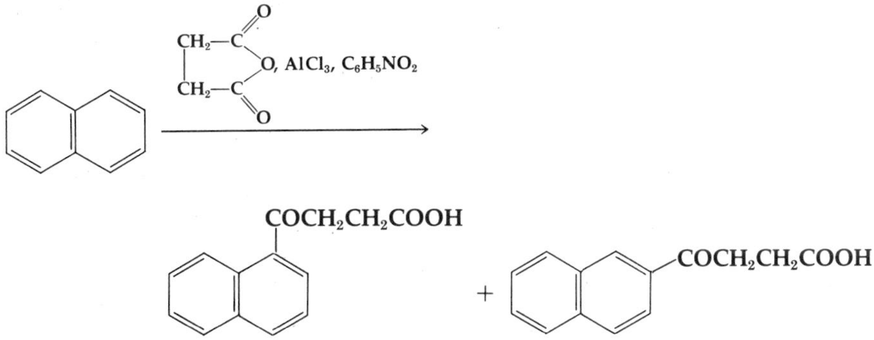

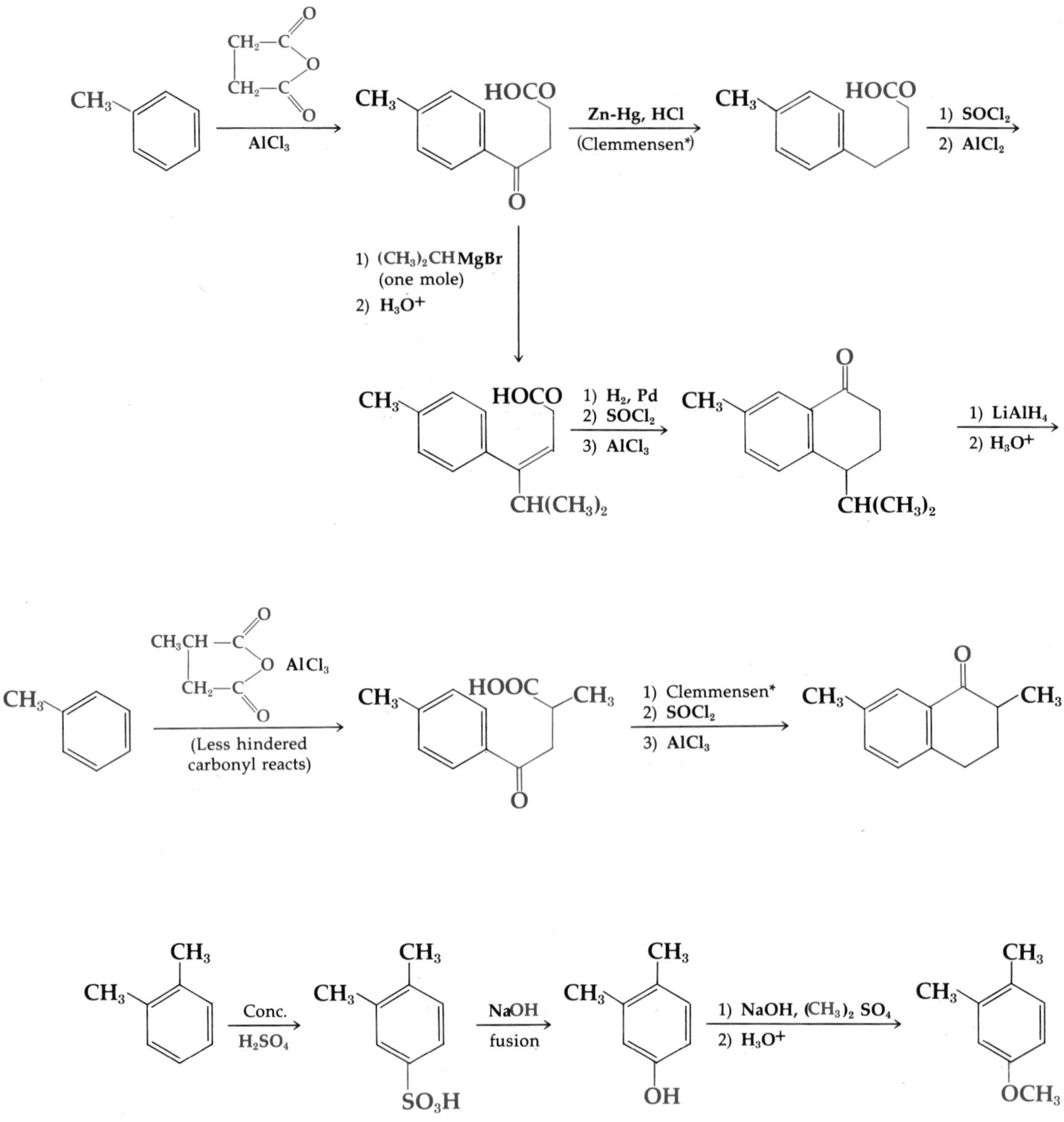

CH_3 —⬡ + (anhydride: CH_2—C=O, O, CH_2—C=O) →[$AlCl_3$] CH_3—⬡—HOCO ... C=O →[Zn-Hg, HCl (Clemmensen*)] CH_3—⬡—HOCO →[1) $SOCl_2$ 2) $AlCl_2$]

1) $(CH_3)_2CHMgBr$ (one mole)
2) H_3O^+

CH_3—⬡—HOCO ... $CH(CH_3)_2$ →[1) H_2, Pd 2) $SOCl_2$ 3) $AlCl_3$] CH_3—bicyclic ketone— $CH(CH_3)_2$ →[1) $LiAlH_4$ 2) H_3O^+]

CH_3—⬡ + (CH_3CH—C=O, O, CH_2—C=O) →[$AlCl_3$ (Less hindered carbonyl reacts)] CH_3—⬡—HOOC ... CH_3 ... C=O →[1) Clemmensen* 2) $SOCl_2$ 3) $AlCl_3$] CH_3—bicyclic ketone—CH_3

CH_3 / CH_3—⬡ →[Conc. H_2SO_4] CH_3 / CH_3—⬡—SO_3H →[NaOH fusion] CH_3 / CH_3—⬡—OH →[1) NaOH, $(CH_3)_2SO_4$ 2) H_3O^+] CH_3 / CH_3—⬡—OCH_3

*Reactions from Chap. 18.

FIGURE 16-5 Syntheses of substituted naphthalenes

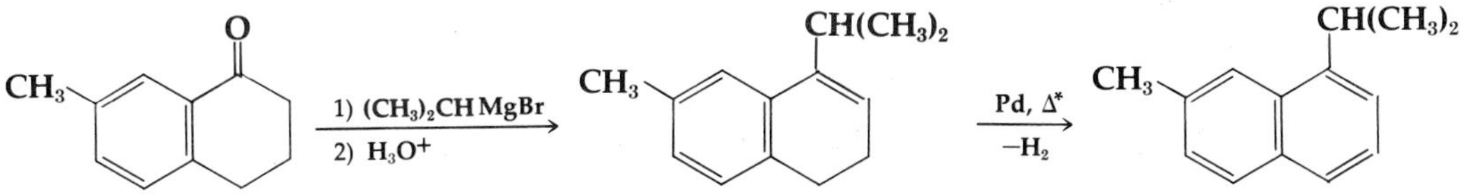
O
CH3
1) (CH3)2CHMgBr
2) H3O+
CH(CH3)2
CH3
Pd, Δ*
−H2
CH(CH3)2
CH3

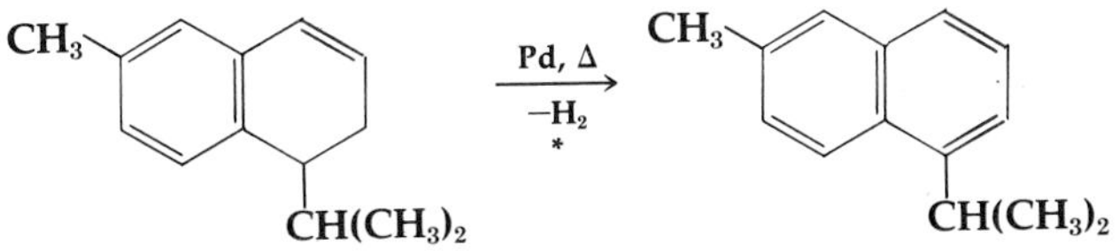
CH3
CH(CH3)2
Pd, Δ
−H2
*
CH3
CH(CH3)2

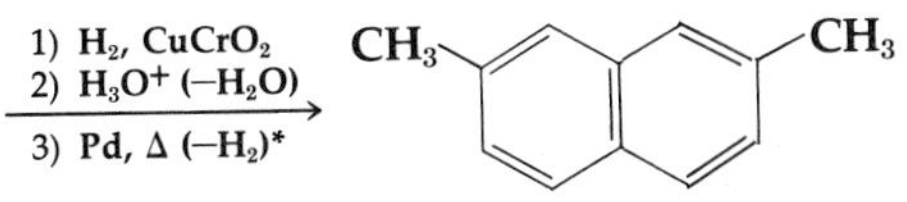
1) H2, CuCrO2
2) H3O+ (−H2O)
3) Pd, Δ (−H2)*
CH3
CH3

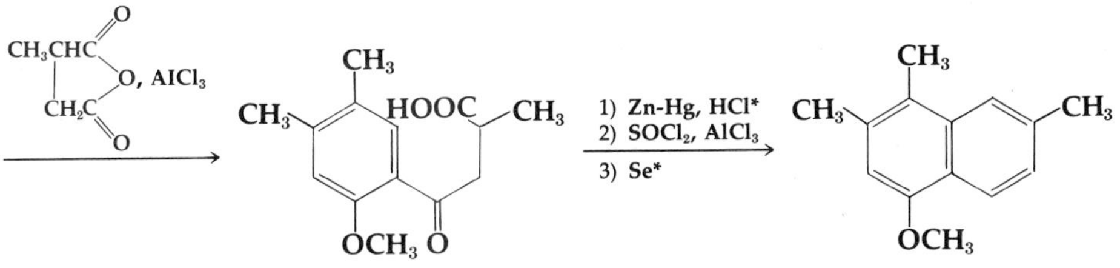
CH3CHC
O
O, AlCl3
CH2C
O
CH3
CH3
HOOC
CH3
OCH3 O
1) Zn-Hg, HCl*
2) SOCl2, AlCl3
3) Se*
CH3
CH3
CH3
OCH3

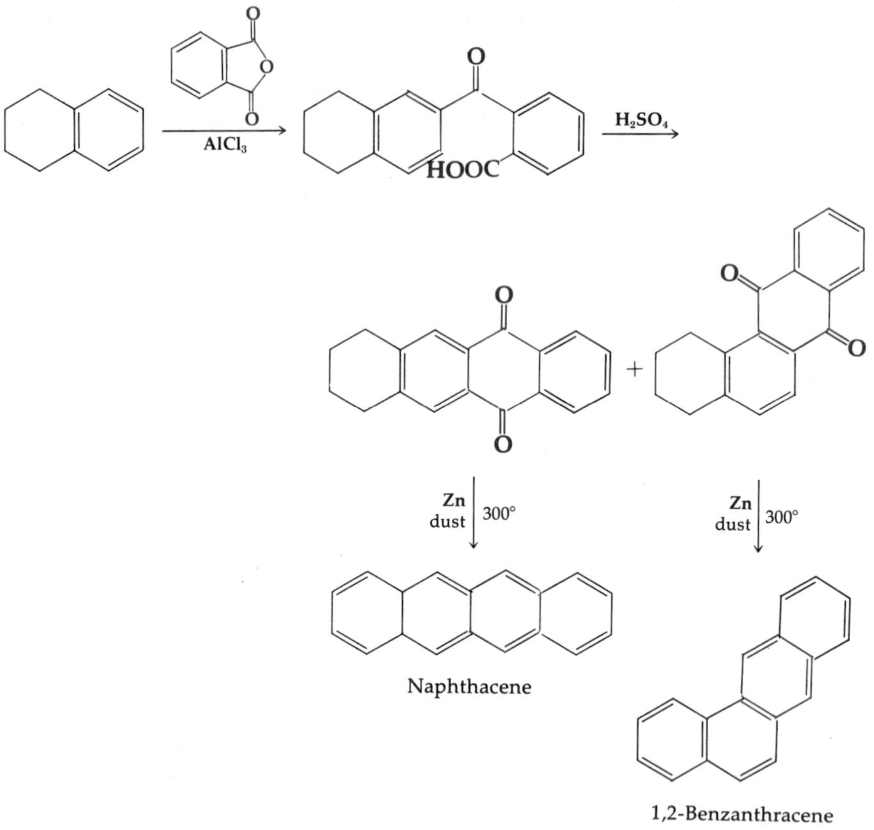

Naphthacene

1,2-Benzanthracene

Other Ketone Syntheses

Phenolic ketones are often prepared by the **Fries rearrangement** of aryl esters. The reaction is an internal self-acylation.

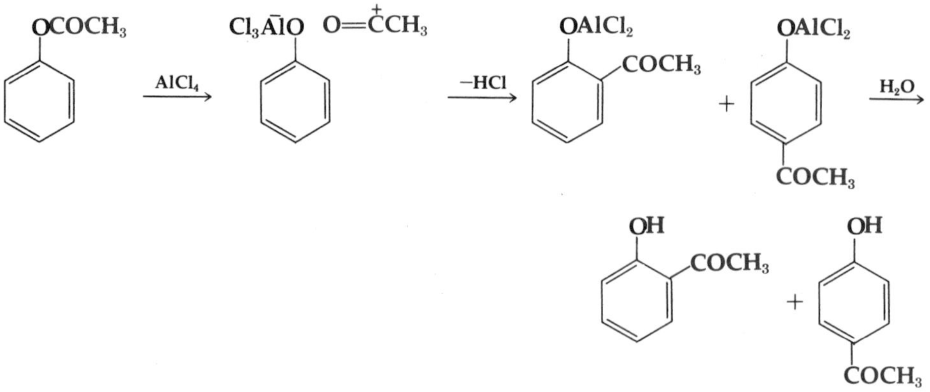

At low temperatures, para product predominates; at high temperatures, ortho isomers are the major product. Solvent variation and aluminum chloride concentration also affect the product distribution.

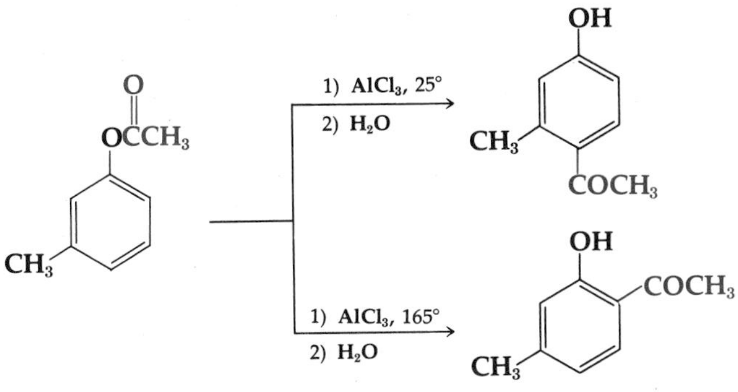

A variation of the Fries rearrangement is available in the **Hoesch reaction**. The process consists of substitution by the conjugate acid of a nitrile. The first product is an imine, which is hydrolyzed during isolation.

$$RC\equiv N + \overset{+}{H} \xrightleftharpoons{ZnCl_2} R\overset{+}{C}=NH$$

$$R\overset{+}{C}=NH + ArH \xrightarrow{-H^+} \underset{NH}{ArCR} \xrightarrow{H_2O} \underset{O}{ArCR}$$

Unfortunately, yields in the Hoesch reaction are not consistently high, and experience indicates that a successful reaction requires a very reactive aromatic substrate. Monohydric phenols are usually attacked almost exclusively at oxygen rather than at nuclear positions.

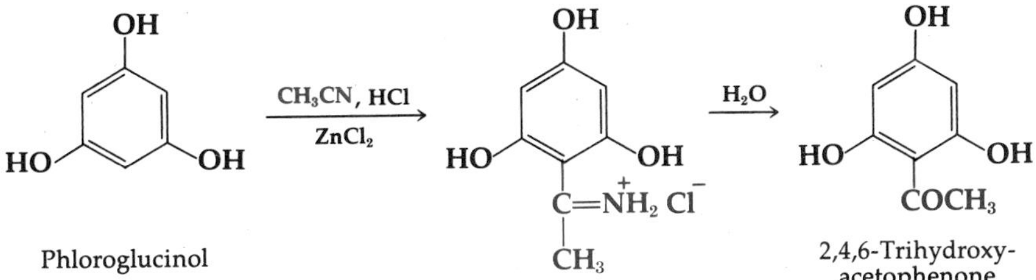

Phloroglucinol

2,4,6-Trihydroxy-acetophenone
87%

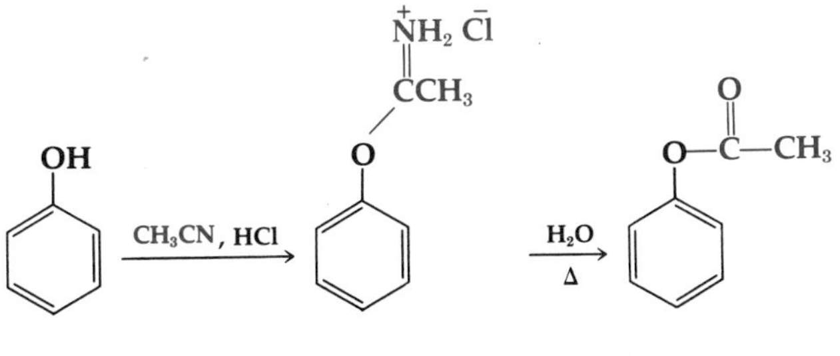

Hydrochloride of
phenyliminoacetate

Aromatic Aldehydes

Although formyl chloride (HCOCl) and formic anhydride [(HCO)$_2$O] are unknown compounds, the ion, HC≡O$^+$, needed for formylation can be produced by protonation of carbon monoxide. In the presence of HCl, CO, and AlCl$_3$, aldehydes are produced from substrates that are ordinarily subject to acylation. Aldehyde synthesis is important because of the versatile reactivity of aldehydes in synthesis.

$$:C{=}O + HCl + AlCl_3 \rightleftharpoons H\overset{+}{C}{=}O \ A\bar{l}Cl_4$$

$$H\overset{+}{C}{=}O + ArH \longrightarrow ArCHO$$

Protonation of HCN produces a similar electrophile. For some reactions, direct use of the very poisonous hydrogen cyanide can be avoided by using hydrogen chloride and zinc cyanide, which generate hydrogen cyanide and zinc chloride (a mild electrophile) in the reaction flask.

$$HC{\equiv}N \xrightarrow{\ ZnCl_2,\ HCl\ } H\overset{+}{C}{=}NH$$

$$H\overset{+}{C}{=}NH + ArH \longrightarrow ArCH{=}\overset{+}{N}H_2 \xrightarrow{\ H_2O\ } ArCHO$$

The aldehyde synthesis with carbon monoxide and hydrogen chloride is known as the **Gattermann-Koch** reaction and that with hydrogen cyanide as the **Gattermann reaction**. Like the Hoesch reaction, the Gattermann reaction requires a very reactive substrate.

Gattermann-Koch:

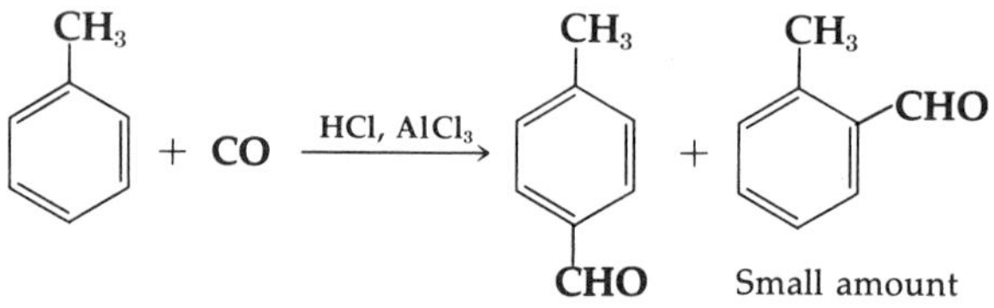

p-Tolualdehyde
51%

Small amount

Gattermann:

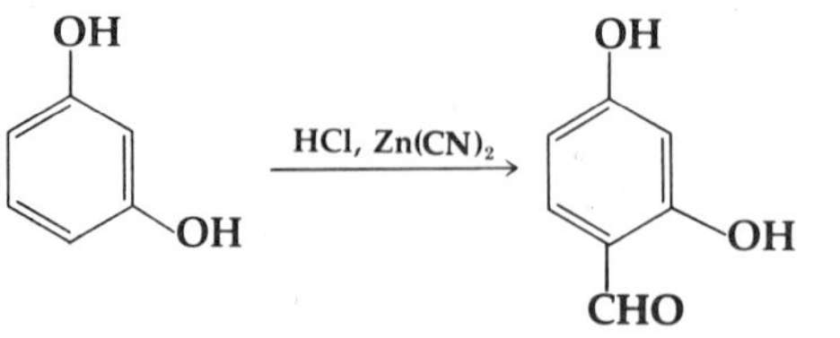

Resorcinol

Resorcyl aldehyde
95%

In the **Vilsmeyer reaction**, moderately reactive aromatic compounds are formylated with an active formyl derivative prepared from dimethylformamide and the electrophilic phosphorus oxychloride. Various other electrophiles have been substituted for $POCl_3$, and other formamides have also been used.

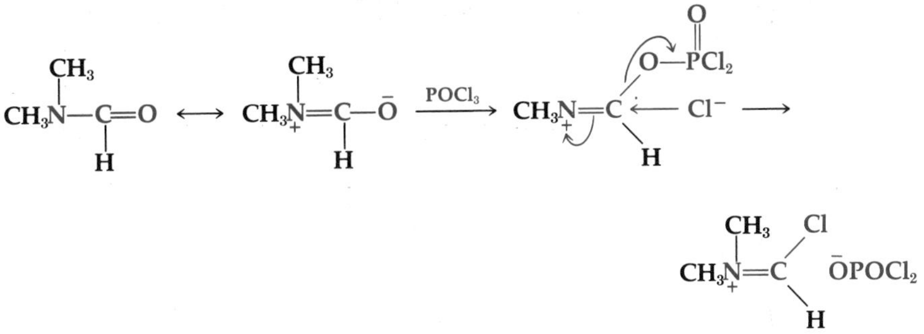

$$CH_3\overset{CH_3}{\underset{H}{N}}-C=O \longleftrightarrow CH_3\overset{CH_3}{\underset{H}{\overset{+}{N}}}=C-\bar{O} \xrightarrow{POCl_3} CH_3\overset{CH_3}{\underset{H}{N}}=C \xleftarrow{} Cl^- \longrightarrow$$

$$CH_3\overset{CH_3}{\underset{+}{N}}=\overset{Cl}{\underset{H}{C}} \quad \bar{O}POCl_2$$

Active electrophile

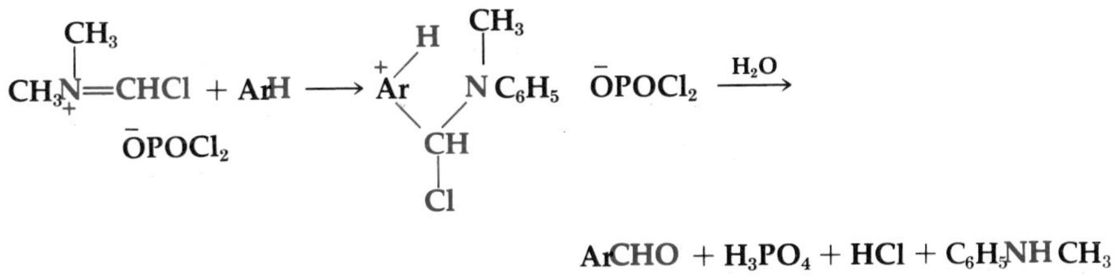

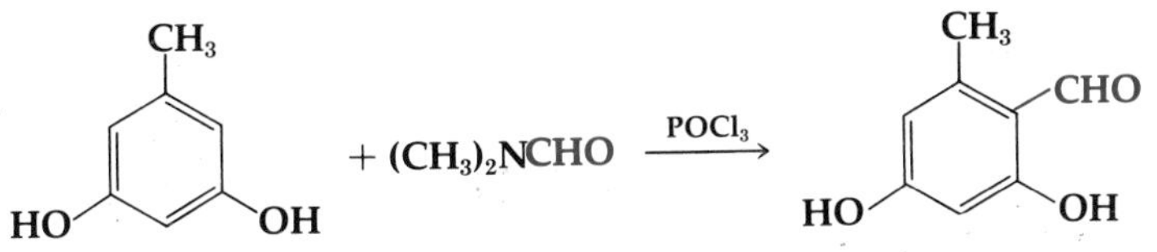

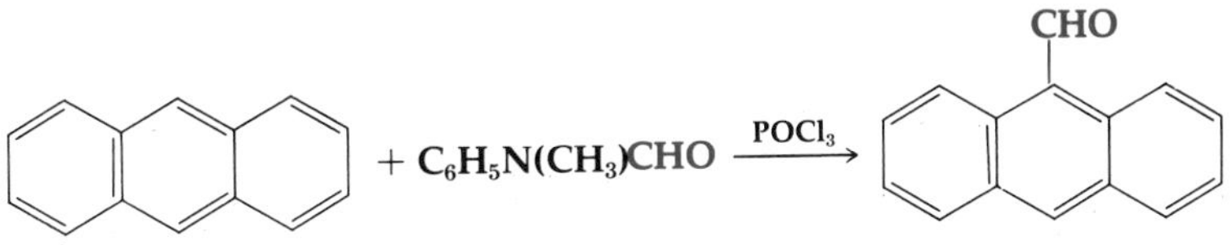

$$CH_3\overset{CH_3}{\underset{+}{N}}=CHCl + ArH \longrightarrow \overset{H}{\underset{\overset{CH}{\underset{Cl}{}}}{Ar}}\overset{CH_3}{\underset{}{N}}C_6H_5 \quad \bar{O}POCl_2 \xrightarrow{H_2O}$$

$$\bar{O}POCl_2$$

$$ArCHO + H_3PO_4 + HCl + C_6H_5NH\,CH_3$$

Aldehydes and Ketones as Electrophiles

Previously we have seen simple carbonium ions as electrophiles for alkylation and the carbonyl group involved as acyl cations for acylation. Aldehydes and ketones also form carbonium ions useful as electrophiles. In the presence of strong acid, carbonyl compounds are converted to their powerfully electrophilic conjugate acids, $>C=\overset{+}{O}H$. Examples of the condensation of these conjugate acids with the enols of carbonyl compounds in acid-catalyzed aldol condensations have been encountered earlier (Chap. 12). In a completely analogous reaction, aldehydes and ketones condense with reactive aromatic nuclei.

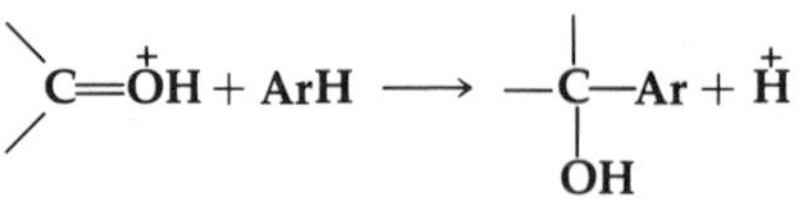

$$\underset{/}{\overset{\backslash}{C}}=\overset{+}{O}H + ArH \longrightarrow -\overset{|}{\underset{\underset{OH}{|}}{C}}-Ar + \overset{+}{H}$$

Since the products are benzyl alcohols, they are readily converted to carbonium ions, which may alkylate another nucleus.

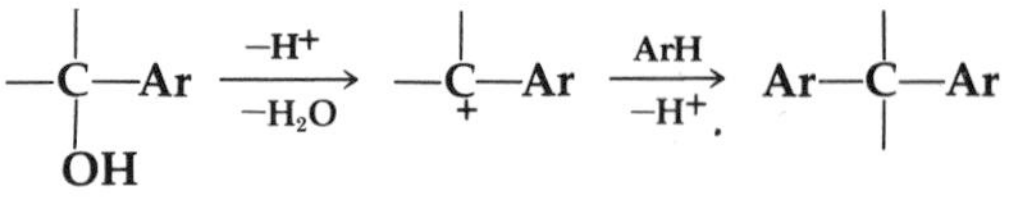

$$-\overset{|}{\underset{\underset{OH}{|}}{C}}-Ar \xrightarrow[-H_2O]{-H^+} -\overset{|}{\underset{+}{C}}-Ar \xrightarrow[-H^+]{ArH} Ar-\overset{|}{\underset{|}{C}}-Ar$$

The activated nucleus is ordinarily still subject to further attack, so that the process generally leads to the formation of high polymers such as **phenol-formaldehyde resins** ("Bakelite").

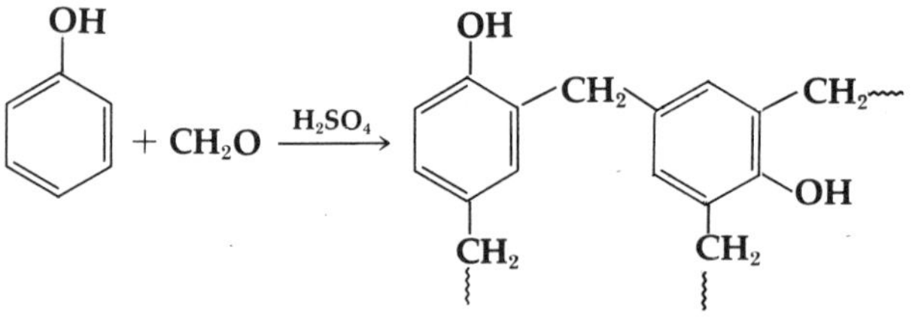

Bakelite structure
(phenol-formaldehyde resin)

If the reaction mixture contains a high concentration of a nucleophilic ion, such as halide when using **HX** as the acid, the benzyl alcohols may be intercepted and converted to halides. The steps include substitution to a benzyl alcohol, acid-catalyzed ionization, and nucleophilic substitution (S_N1) of halide at the benzylic carbon. The combination of formaldehyde with hydrogen chloride introduces the $-CH_2Cl$ group, a process known as **chloromethylation**.

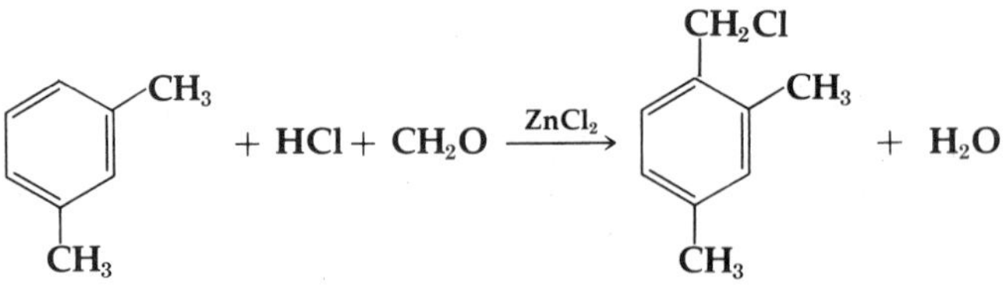

61%

It may be noted here that the reactions of phenols with acyl cations or with the conjugate acids of aldehydes or ketones are completely analogous to acid-catalyzed carbonyl condensations, since phenols are merely enols embedded in benzene rings.

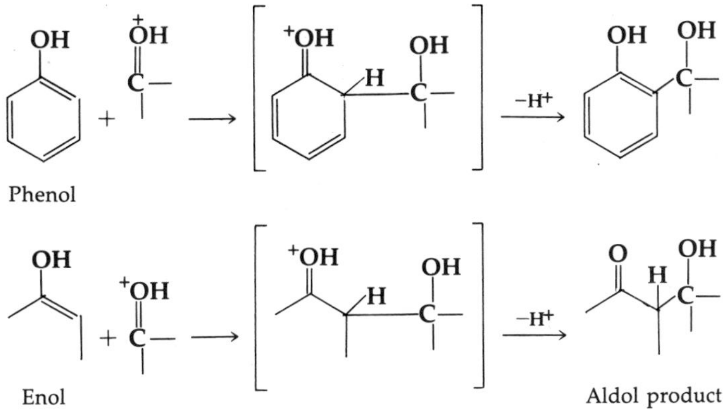

Phenol

Enol   Aldol product

A few reactions of phenols even occur under basic conditions, like the common carbonyl reactions. In the **Kolbe reaction** a phenolate salt is heated with carbon dioxide under pressure to effect carbonation (carboxylation) of the phenolic ring (ortho, usually). The reaction is similar to the reverse of decarboxylation of the anion of a β-ketoacid, and more specifically to the decarboxylation of o- and p-hydroxy-benzoic acids (Sec. 13-6). Polyhydric phenols usually give good yields in the reaction under relatively mild conditions.

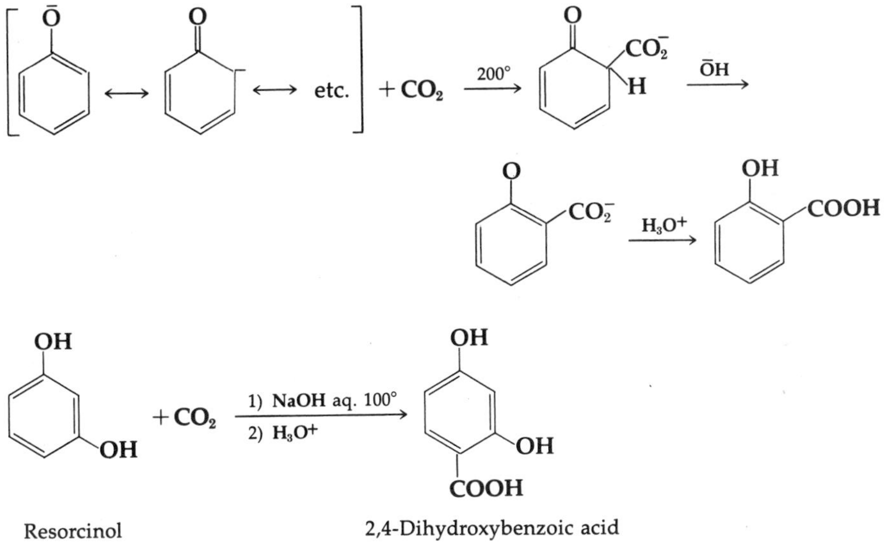

Resorcinol   2,4-Dihydroxybenzoic acid

PROBLEM 16-7

Fill in the missing structures in this synthesis. If there is a choice of more than one structure, write the major one. In the first reaction write the probable structure of the active cation which acts as the electrophile.

Anisole ($C_6H_5OCH_3$) + succinic anhydride $\xrightarrow{AlCl_3}$ A $\xrightarrow[HCl]{Zn \cdot Hg}$

B $\xrightarrow[\text{acid}]{\text{Polyphosphoric}}$ C $\xrightarrow[AlCl_3]{CH_3COCl}$ D $\xrightarrow{NaBH_4}$ E $\xrightarrow[\text{Pyridine}]{C_6H_5SO_2Cl}$

F $\xrightarrow{LiAlH_4}$ G $\xrightarrow{BBr_3}$ H $\xrightarrow{Br_2}$ $C_{12}H_{15}OBr$

PROBLEM 16-8

a Show a procedure for converting p-cresol (p-$CH_3C_6H_4OH$) into 4-hydroxy-3-methyl-benzoic acid.

b Show the steps in the mechanism by which aromatic aldehydes are produced directly in the reaction of aromatic substrates with $Cl_2CHOCH_3/AlCl_3$; use p-cresol as the example. What are the other products? What product might you expect using $HCCl{=}NOH/AlCl_3$?

c In the **Reimer-Tiemann reaction**, phenols are treated with chloroform and alkali. In this reaction p-cresol yields two products, as shown below. Explain. (See page 641.)

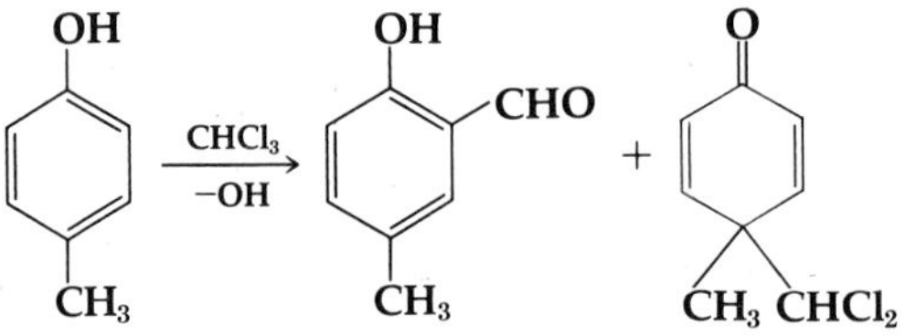

d In the **Pechmann reaction** phenols are allowed to react with acetoacetic ester and acid. The product from p-cresol is shown below. How does the reaction proceed?

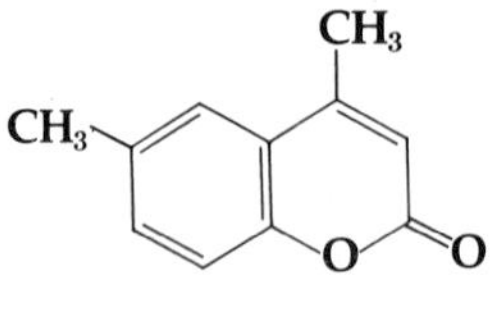

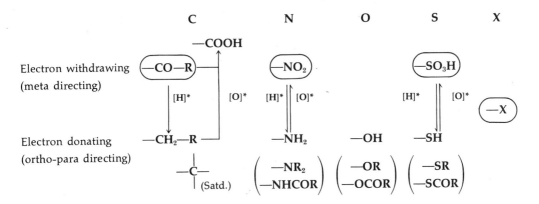

(Circled groups are those which may be introduced by electrophilic substitution.)

*Practical interconversions are available as shown by oxidation-reduction, discussed in Chap. 18.

FIGURE 16-6   **Summary of major aromatic substituents**

16-7   SYNTHESIS OF AROMATIC COMPOUNDS

A little reflection serves to show that all possible disubstituted benzenes cannot be created by electrophilic substitutions alone, despite the versatility of the reaction. The fullest capability in aromatic synthesis comes from combining electrophilic substitution with other reactions (including a few not yet presented) for interconverting substituents. Such reactions are reviewed here in order to provide a practical basis for designing syntheses of any aromatic compounds.

The major reliable substitution reactions for synthesis provide for aryl attachment to the atoms C, N, S, X, but not to oxygen:

X^+ X_2/AlX_3 or FeX_3

NO_2^+ HNO_3, etc.

SO_3H^+ $H_2SO_4\cdot SO_3$; $ClSO_3H$, etc.

RCO^+ $RCOCl/AlCl_3$; $(RCO)_2O$/Lewis acids, etc.

The major aromatic substituents are summarized in Fig. 16-6 and organized so as to see which ones may be introduced by electrophilic substitution (circles), what directive influences they exert on subsequent substitutions, and which ones may be interconverted to the opposite directive influence. Only rate-deactivating substituents can be introduced by substitution since, once introduced, they deactivate the ring and prevent further substitution and undesirable polysubstituted by-products. Hence only the circled substituents can be added; the others (electron-donating) must be created by functional-group interconversions.

The problem of designing a synthesis for a particular polysubstituted aromatic compound is therefore reduced to a choice of which substituents to

TABLE 16-2   **Interconversions of Aromatic Substituents†**

Products ⟶

Starting materials	C	N	O
H	Electrophilic substitution reactions Ar—H + RCOCl $\xrightarrow{AlCl_3}$ Ar—COR	Ar—H + HNO$_3$ ⟶ Ar—NO$_2$	
C	(Fig. 16-6)	Ar—COR + HN$_3$ $\xrightarrow{(18)}$ Ar—NHCOR Ar—COOH ⟶ Ar—CON$_3$ $\xrightarrow{(18)}$ Ar—NCO	Ar—COR + R'CO$_3$H $\xrightarrow{(18)}$ Ar—OCOR
N	Ar—N$_2$⁺ + Cu$_2$(CN)$_2$ ⟶ Ar—CN	(Fig. 16-6)	Ar—N$_2$⁺ + H$_2$O $\xrightarrow{\Delta}$ Ar—OH
O			
S	Ar—SO$_3^-$ + ⁻CN $\xrightarrow{\Delta}$ Ar—CN	Ar—SO$_3^-$ + NH$_3$ $\xrightarrow{\Delta}$ Ar—NH$_2$	Ar—SO$_3^-$ + ⁻OH $\xrightarrow{\Delta}$ Ar—OH
X	Ar—X ⟶ Ar—MgX Ar—MgX + $\overset{\diagdown}{\underset{\diagup}{C}}$=O ⟶ Ar—$\overset{\diagdown}{\underset{\diagup}{C}}$—OH		Ar—MgX + O$_2$ ⟶ Ar—OH

†Reactions discussed in subsequent chapters are included with the chapter number in parentheses.

introduce and in what sequence. Since oxygen cannot be introduced it must either be present in the starting material (cf., phenol) or created from another substituent, as in the hydrolysis of a diazonium salt. In general, other attached atoms also need not be directly introduced by substitution but may often be placed by interconversions of other existing functional groups, as in the conversion of —SO$_3$H to —OH by alkali. In Table 16-2 are listed the chief practical synthetic reactions used to interconvert one functional group to another on a benzene ring. The top row incorporates the chief electrophilic substitution reactions, as conversions of attached —H to various functional atoms.

Armed with these tabular summaries we may approach any synthesis. The problem becomes one of choosing which function to introduce—and in-

S	X	H
$Ar-H + SO_3 \longrightarrow$ $Ar-SO_3H$	$Ar-H + X_2 \xrightarrow{(AlX_3)} Ar-X$	
	$Ar-COO^-Ag^+ + Br_2 \xrightarrow{(18)}$ $Ar-Br + CO_2$	$Ar-COOH \xrightarrow[\substack{or\\ H^+}]{\Delta} Ar-H$
$Ar-N_2^+ + SO_2 \xrightarrow{H_2O}$ $Ar-SO_3H$ $Ar-N_2^+ + Cu_2(SCN)_2 \longrightarrow$ $Ar-SCN$	$Ar-N_2^+ + Cu_2X_2 \longrightarrow$ $Ar-X$	$Ar-N_2^+ + H_3PO_2$ or $NaBH_4 \longrightarrow$ $Ar-H$
		$Ar-SO_3H + H_3O^+ \xrightarrow{\Delta}$ $Ar-H$
(Fig. 16-6)		$Ar-SR + H_2(Ni) \longrightarrow$ $Ar-H$
$Ar-MgX + SO_2 \xrightarrow{(20)}$ $Ar-SO_2H$		$Ar-MgX + H_2O \longrightarrow$ $Ar-H$
		$Ar-X + H_2(Pd) \longrightarrow$ $Ar-H$

terconvert—first so as to provide appropriate directive influence to introduce the next substituent, and so forth. The major synthetic devices employed are the following:

1 Conversion of meta-directing substituents to their ortho-para-directing analogs (Fig. 16-6) in order to change the orientation of introduction of new substituents.
2 Introduction of blocking substituents to occupy an undesired site during substitution. The blocking group is later removed (converted to —H, last column in Table 16-2).
3 Introduction of extraneous activating substituents to provide correct orientation in substitution, followed by final removal of the activating substituents (conversion to —H, Table 16-2).

The conversion of one directive influence to another (device no. 1) is illustrated in this preparation of *m*-bromochlorobenzene, which is unavailable by direct substitution.

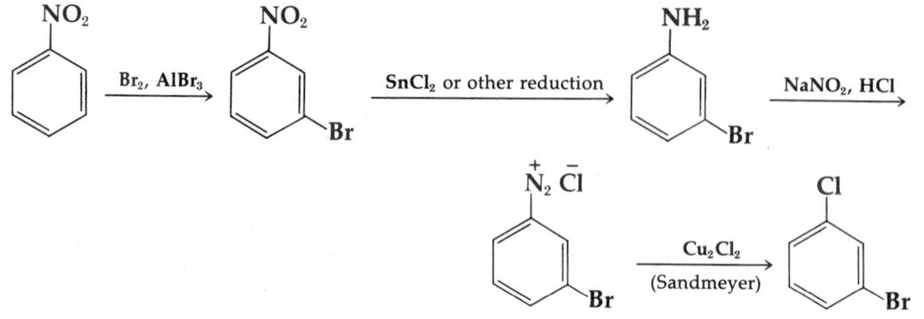

m-Bromochloro-
benzene

The removal of blocking or strongly orienting groups after they have served their synthetic purpose (device nos. 2 and 3) is shown in two examples below. A number of synthetic conceptions are also invoked in the syntheses of naphthalene derivatives shown in Fig. 16-5.

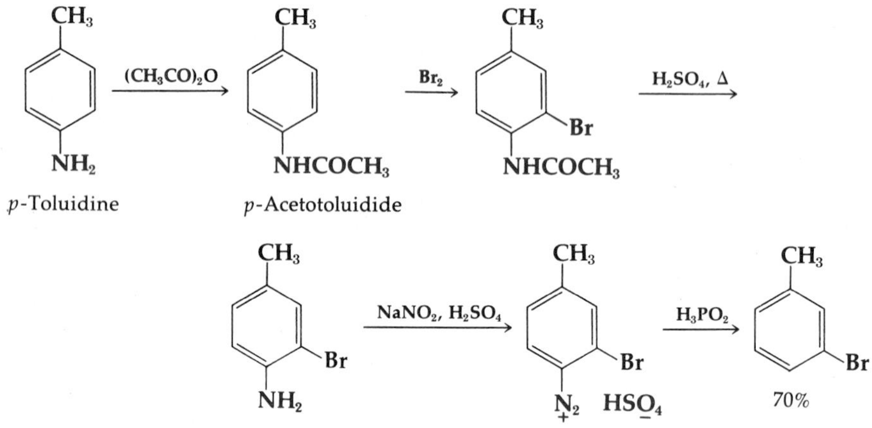

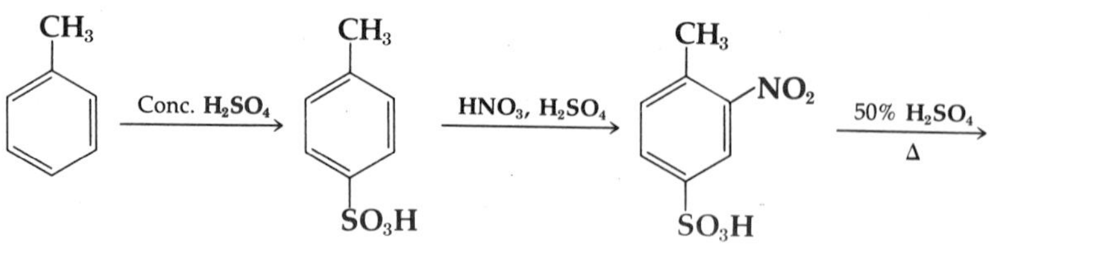

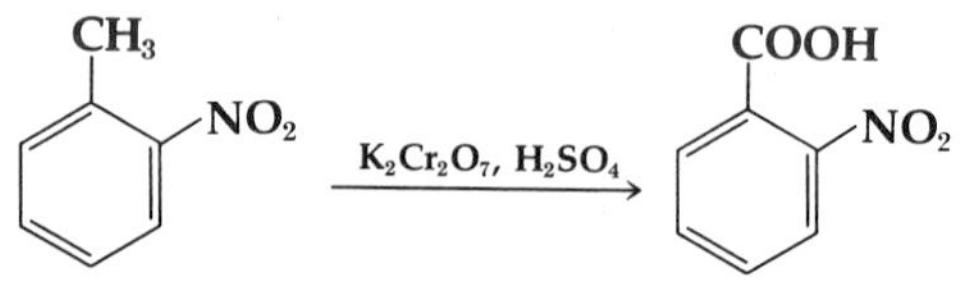

p-Toluidine   *p*-Acetotoluidide

70%

The most important and versatile attached atom is nitrogen since the nitro group is unreactive but strongly deactivating and may be reduced smoothly to amino by several methods discussed in Chap. 18. The chief methods are the use of catalytic hydrogenation or of metals like tin, zinc, or iron in hydrochloric acid. The amino group is not only strongly ortho-para directing but also smoothly convertible to a diazonium salt for replacement of aryl nitrogen by any other attached atom (Table 16-2 and Fig. 13-7).

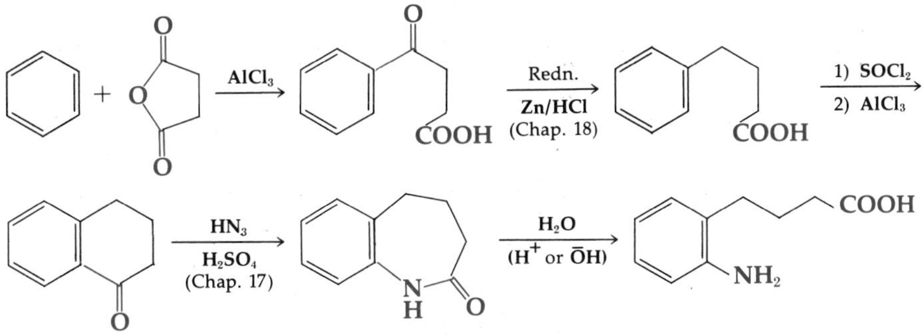

Ortho-para-directing groups with both ortho and para positions free yield mixtures of both the ortho- and para-substituted products. These mixtures commonly favor the para isomer but must be separated. When the ortho isomer is desired, it is made either by first blocking the para position with a group to be removed later, or by using a cyclization reaction, which is obliged to yield a pure ortho attachment.

Acylation provides aryl ketones which are useful for further extension of side chains by the carbonyl reactions of Chaps. 12 and 13. The other important attachment of aryl groups to carbon chains is the use of halogen substituents and their conversion to aryl Grignard reagents.

With this summary of aromatic reactions it should be possible for students to devise syntheses for virtually any polysubstituted aromatic molecule.

PROBLEMS

16-9   Show sequences of reactions that could be used to convert the following substances to benzene.

   a   3-Chloro-4-amino-benzenesulfonic acid

   b   2,6-Dibromoacetophenone

   c   t-Butylbenzene + $AlCl_3$ (Explain.)

16-10   a   The **Bischler-Napieralsky reaction**, illustrated below, is often used for the synthesis of nitrogen heterocycles. Deduce a reasonable mechanism for the reaction.

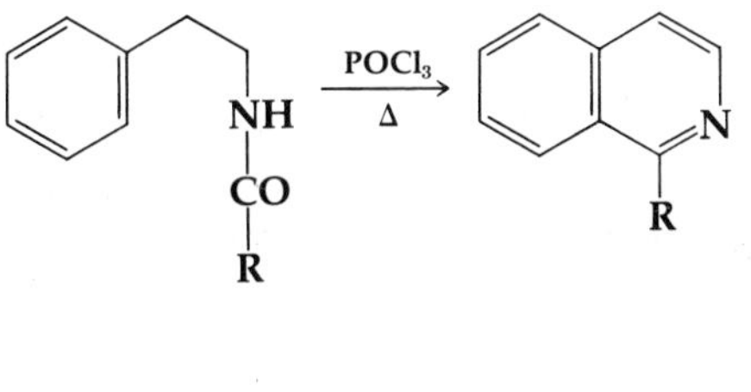

b The Bischler-Napieralsky reaction was used in the synthesis of the natural compound papaverine, which is one of the constituents of opium. The starting material was 3,4-dimethoxybenzaldehyde. One key step was the formation of its cyanohydrin. Formulate the steps in this synthesis (there are nine reactions).

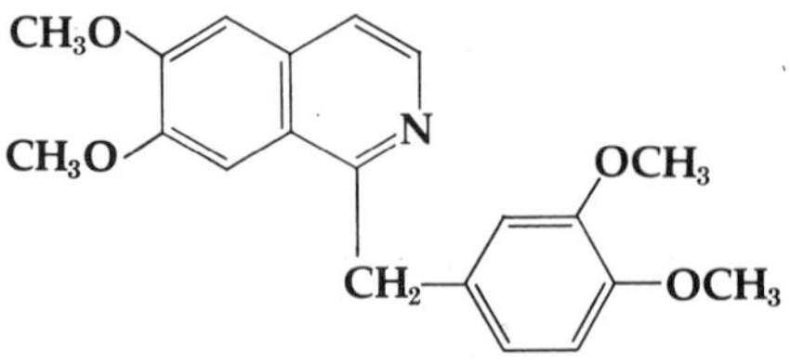

Papaverine

16-11 Show the parallel between the following reaction (known as the **Pictet-Spengler** reaction) and the Mannich reaction (page 487).

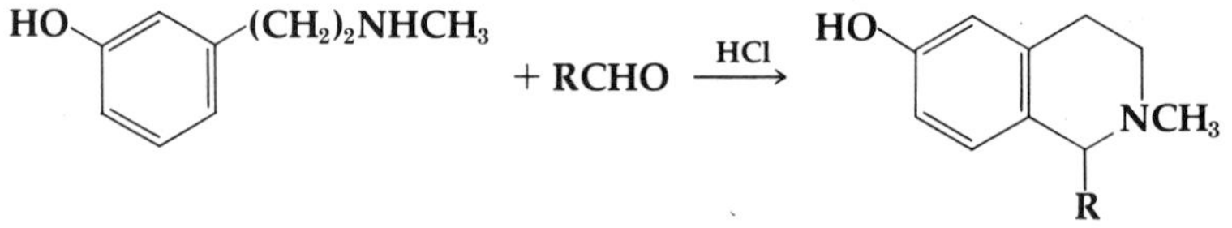

16-12 The orange pigment, fuscin, is synthesized by a species of mold, to which it imparts a characteristic mottled orange appearance. Reductions of various kinds all lead to the colorless aromatic derivative dihydrofuscin. In the structure-determination studies, dihydrofuscin was boiled for several hours in weakly alkaline aqueous solution with a slow stream of nitrogen passed through the solution. The nitrogen gas was then passed through a solution of 2,4-dinitrophenylhydrazine which slowly yielded a yellow precipitate during the course of the reaction. The product isolated on acidification of the alkaline solution was a colorless crystalline acid, soluble in bicarbonate, with a formula $C_{13}H_{16}O_5$. Formulate the reaction in mechanistic terms, and show the structures of the products.

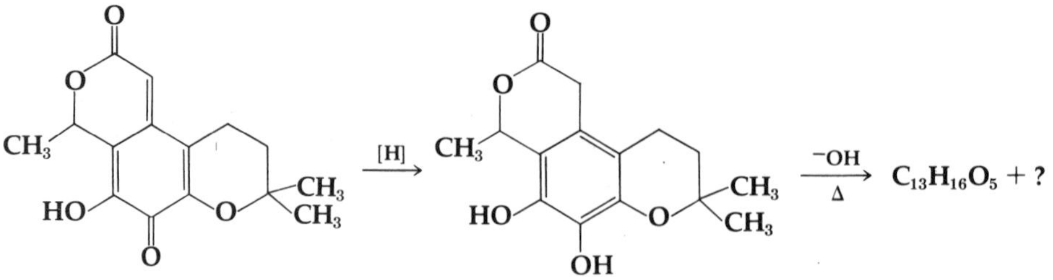

Fuscin   Dihydrofuscin

16-13 The following reactions have been observed. Make a careful analysis of the substitution process in each case, and give an explanation of the observations.

Example:

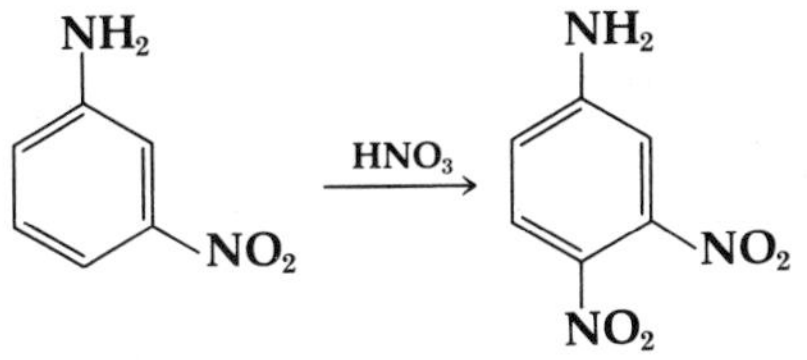

+ smaller amounts of 2,3- and 3,6-isomers

A. The nitro group weakens the basicity of the amine, so that the protonation, $ArNH_2 \longrightarrow ArNH_3^+$, is not as extensive as with aniline. Nitration of the free base can take place even in acidic medium.

B. Nitration para to amino and ortho to nitro involves conflict between two orienting influences. Look at the resonance description of the intermediate ion.

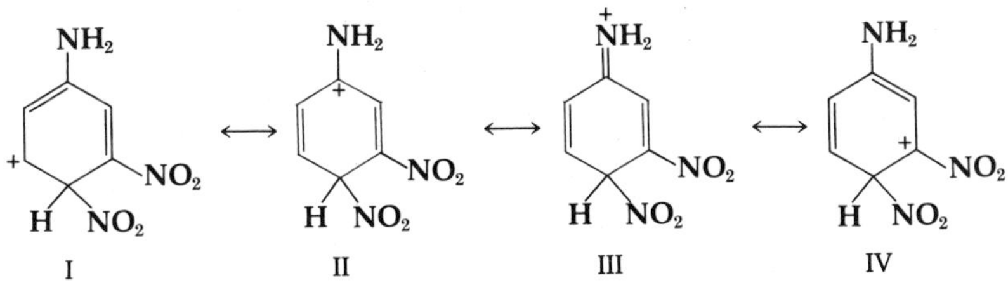

I   II   III   IV

The specific help of NH_2 (see III) must outweigh the specific influence of NO_2 (see IV).†

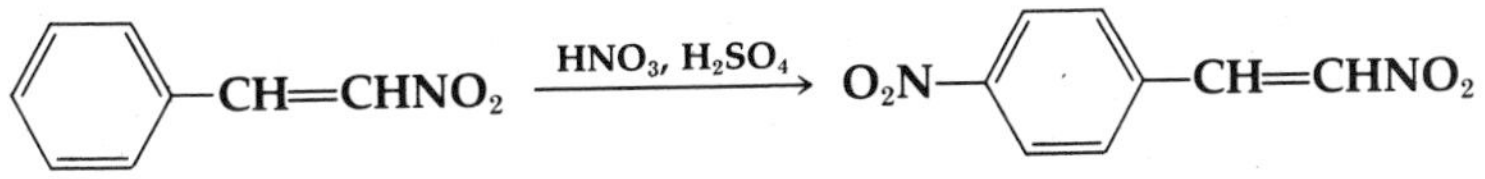

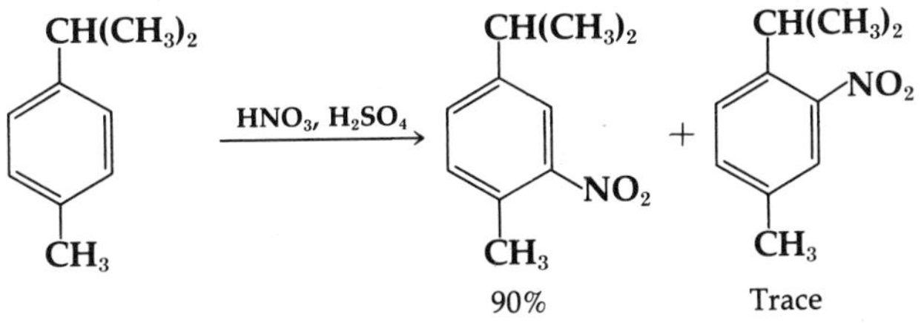

† When they are brought into conflict, the influence of an activating group virtually always outweighs the effect of a deactivating substituent.

c 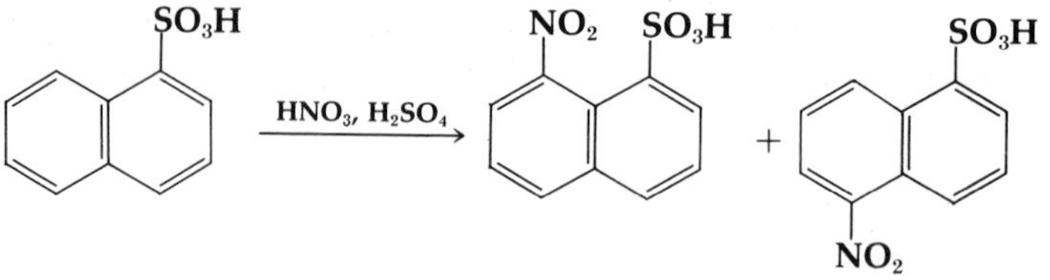
SO₃H
HNO₃, H₂SO₄
NO₂ SO₃H
+
SO₃H
NO₂

d 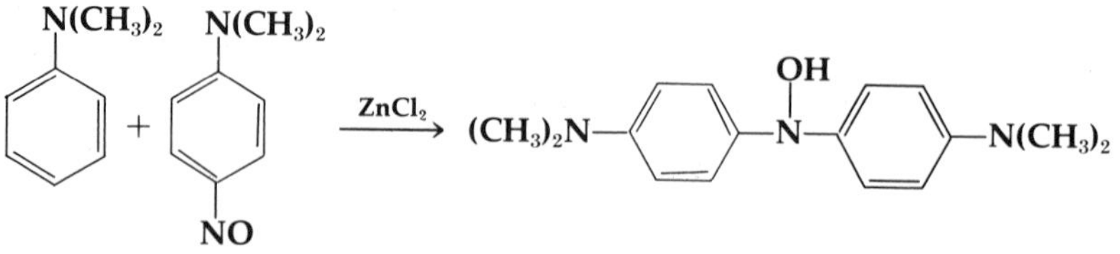
N(CH₃)₂
+
N(CH₃)₂
NO
ZnCl₂
OH
(CH₃)₂N—N—N(CH₃)₂

e Attempts to use esters for Friedel-Crafts acylation,

$$\text{ArH} + \text{RCOOR}' \xrightarrow[\text{or AlCl}_3]{\text{FeCl}_3} \text{ArCOR},$$

always lead to a complex mixture of *alkylated and acylated* products.

f 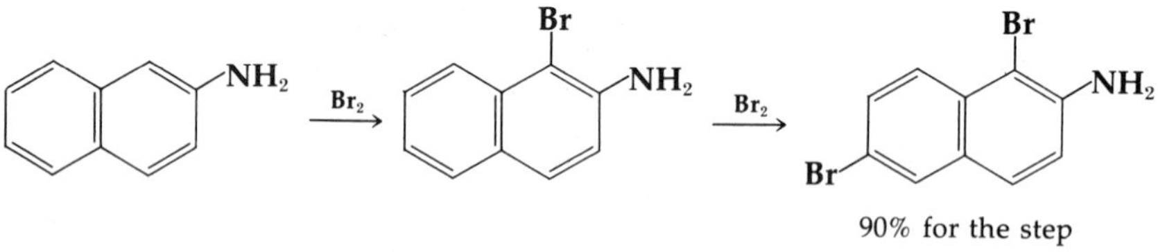
NH₂
Br₂
Br
NH₂
Br₂
Br
NH₂
Br
90% for the step

g 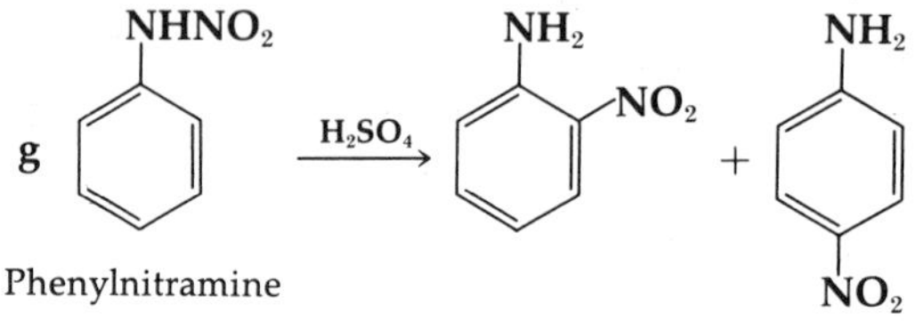
NHNO₂
Phenylnitramine
H₂SO₄
NH₂
NO₂
+
NH₂
NO₂

h 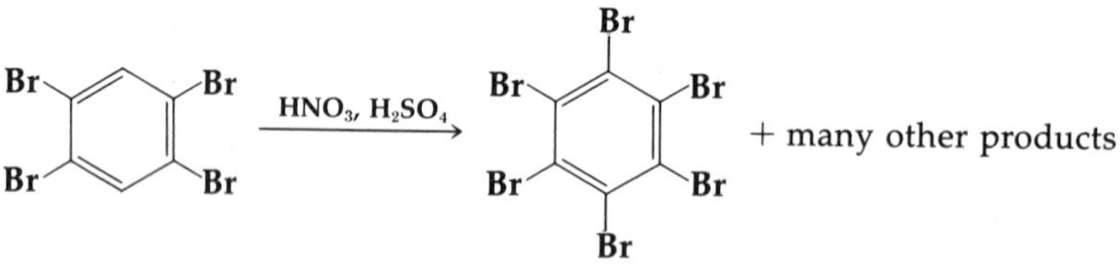
Br Br Br Br
HNO₃, H₂SO₄
Br Br Br Br Br Br
+ many other products

i 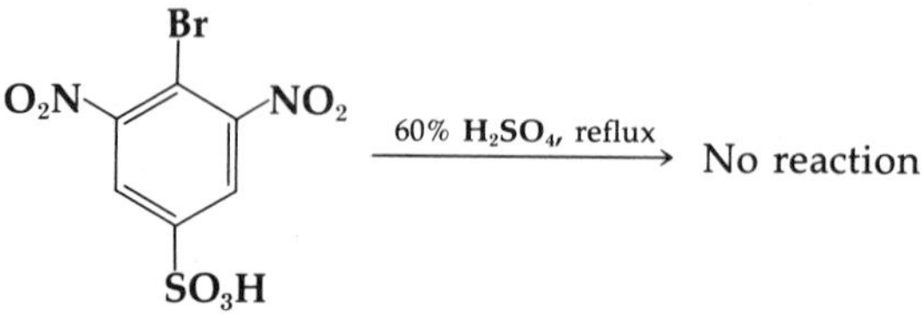
Br
O₂N NO₂
SO₃H
60% H₂SO₄, reflux
No reaction

16-14 Devise suitable synthetic sequences to accomplish the following transformations:

a

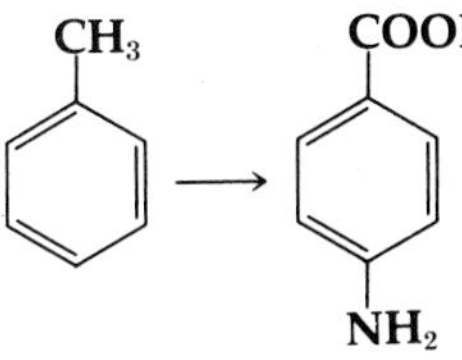

e

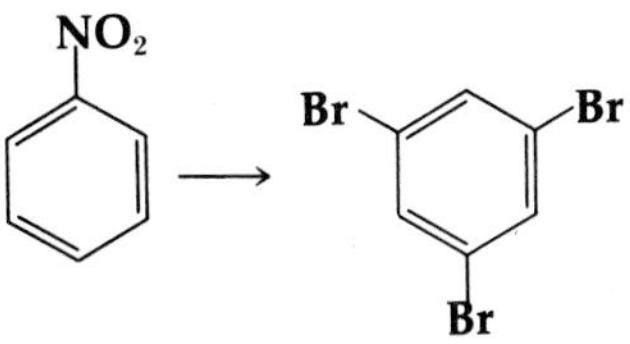

b

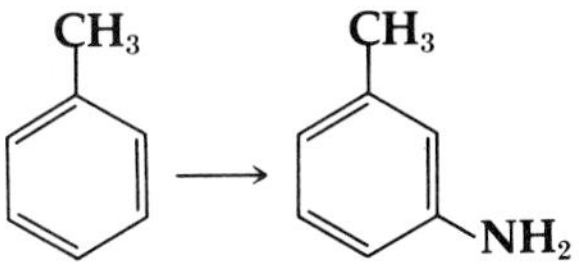

f

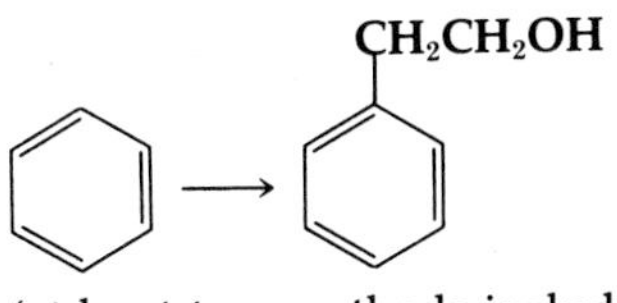

(at least two methods including one that utilizes $CH_2\!-\!CH_2$ with an O bridge)

c

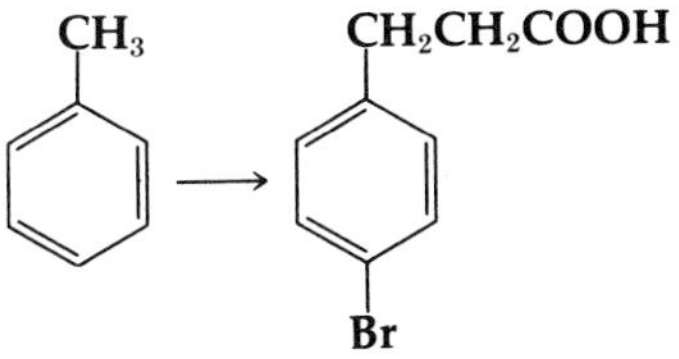

g

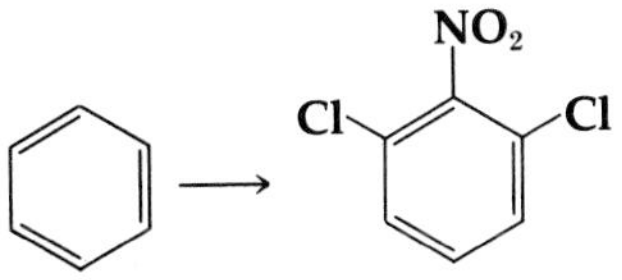

d

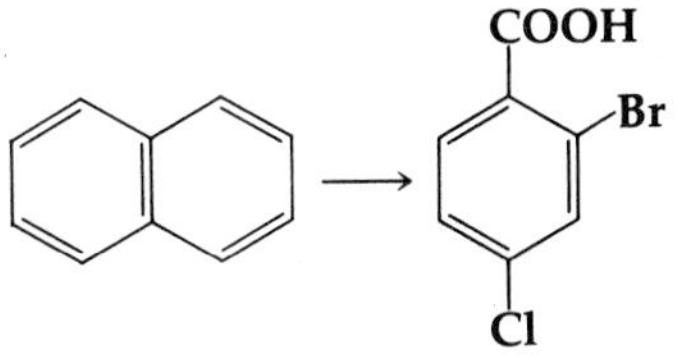

h

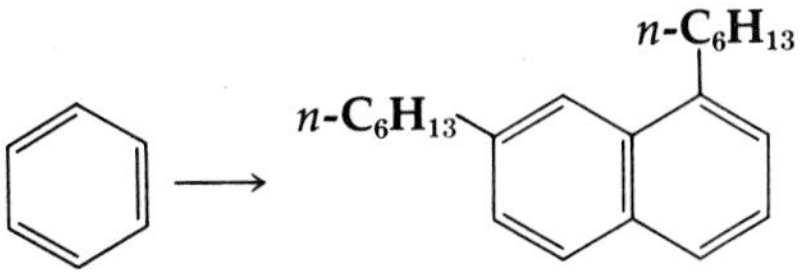

16-15 Formulate reasonable mechanisms for each of the following reactions:

a $(C_6H_5)_2C\!=\!CH_2 \xrightarrow{\ SnCl_4\ }$

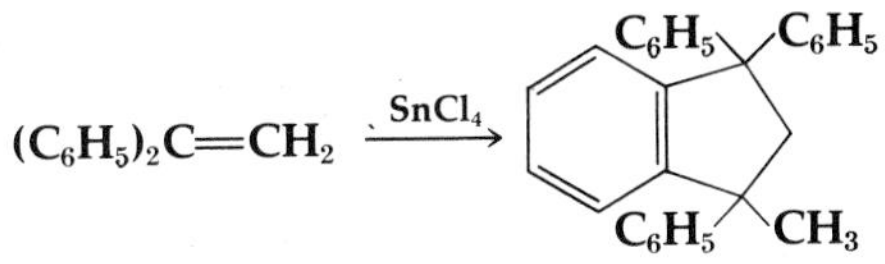

b
$+ Hg(O_2CCH_3)_2 \longrightarrow$

c
$\xrightarrow{\ Br_2\ }$

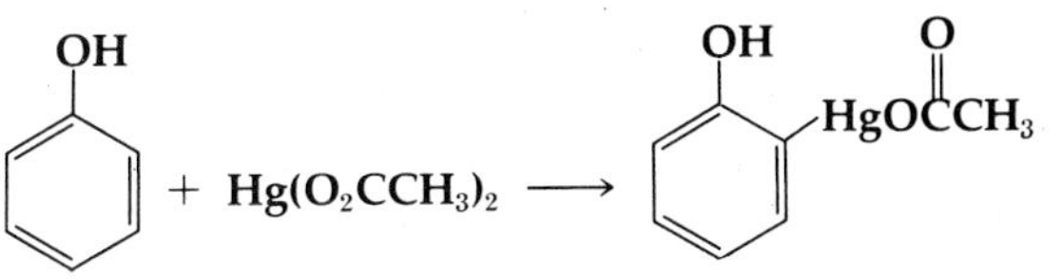

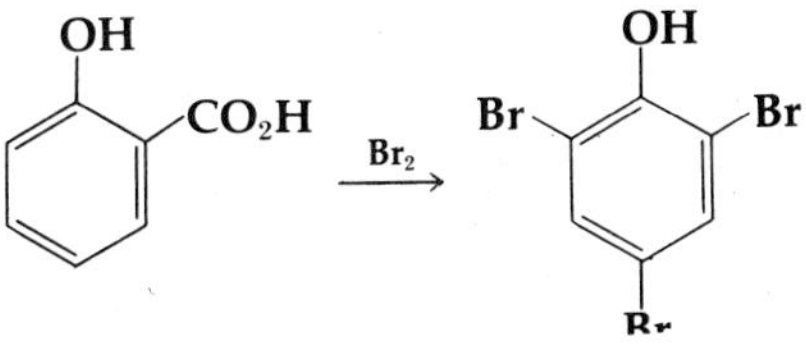

d

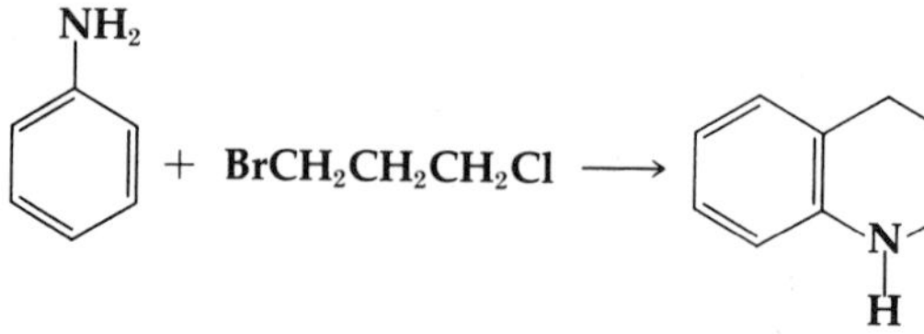

$$\underset{\text{NH}_2}{\text{C}_6\text{H}_5} + \text{BrCH}_2\text{CH}_2\text{CH}_2\text{Cl} \longrightarrow$$

e $\text{C}_6\text{H}_6 + \underset{\text{O}}{\text{CH}_2\text{—CH}_2} \xrightarrow{\text{AlCl}_3} \text{C}_6\text{H}_5\text{CH}_2\text{CH}_2\text{OH}$

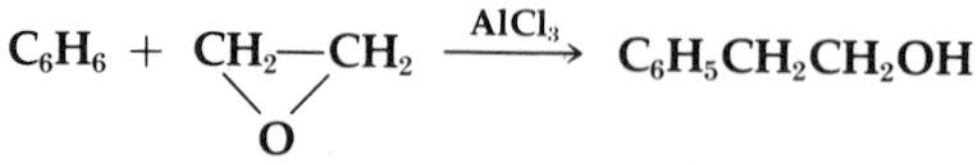

f $\text{C}_6\text{H}_5\text{CHO} + 2\text{C}_6\text{H}_6 \xrightarrow{\text{H}_2\text{SO}_4} (\text{C}_6\text{H}_5)_3\text{CH}$

g (phenol with OH, NO$_2$) $+ \text{CH}_2(\text{OCH}_3)_2 + \text{HCl} \xrightarrow{\text{Conc. HCl}}$ (product with OH, CH$_2$Cl, NO$_2$) $+ 2\text{CH}_3\text{OH}$

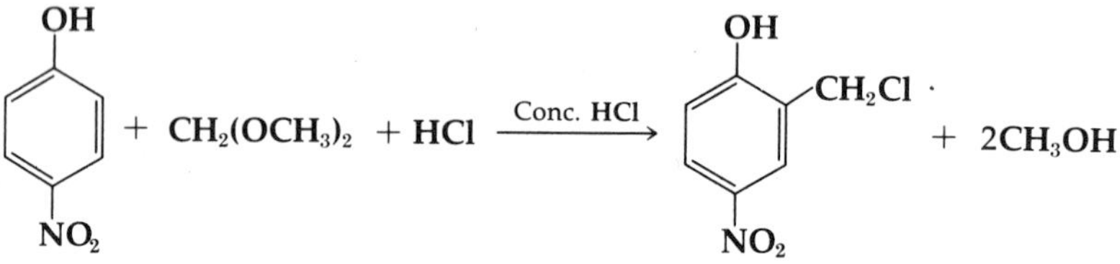

h (3,5-dimethylbenzene, CH$_3$, CH$_3$) $+ \text{Cl}_3\text{CCN} \xrightarrow{\text{ZnCl}_2}$ (imine intermediate with CH$_3$, NH, CCCl$_3$) $\xrightarrow{\text{KOH}}$ (nitrile product with CH$_3$, CN, CH$_3$)

i (dimethoxyaniline with OCH$_3$, NH$_2$, OCH$_3$) $+ \underset{\substack{\text{CH}_2\text{—C}=\text{O} \\ \text{Ketene} \\ \text{dimer}}}{\text{CH}_2=\text{C—O}} \xrightarrow[\text{2) Add H}_2\text{SO}_4]{\text{1) Mix}}$ (quinoline product with OCH$_3$, CH$_3$, OCH$_3$, N, OH)

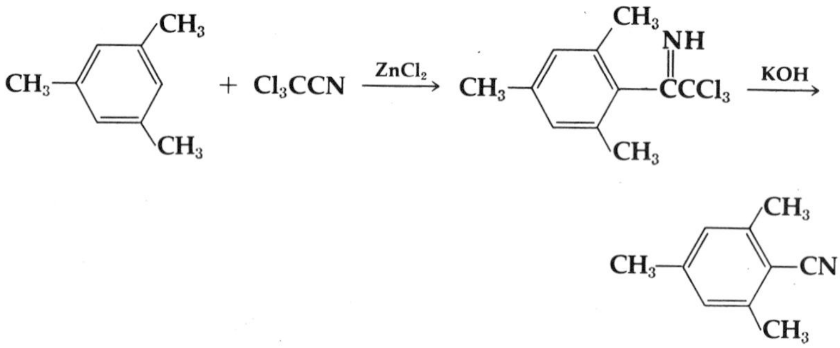

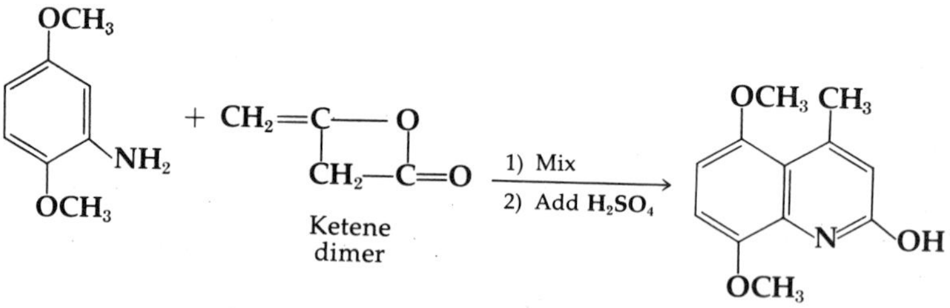

16-16 Devise syntheses of the following compounds from readily available aromatic and aliphatic chemicals.

a

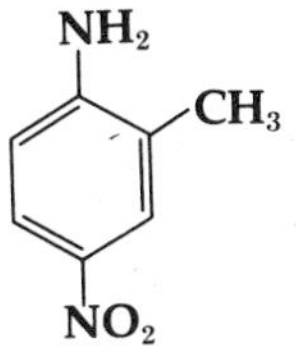

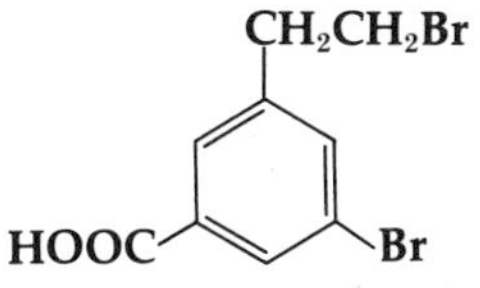

f

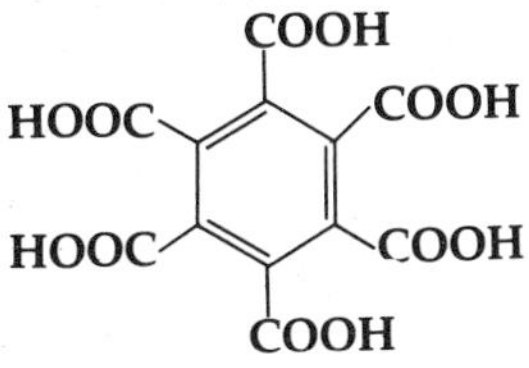

Mellitic acid

b

g

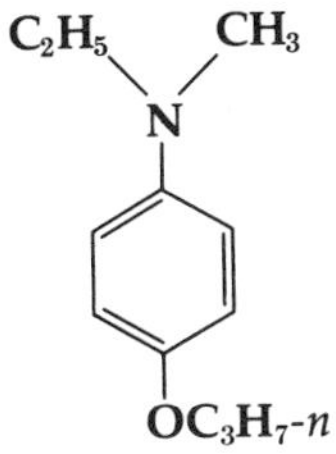

c

h

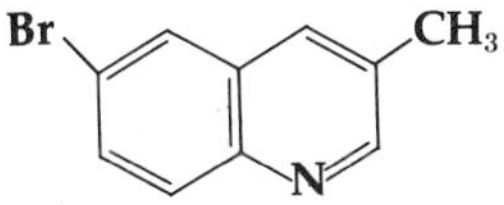

d

i CH_2 $(CH_2)_8$

e

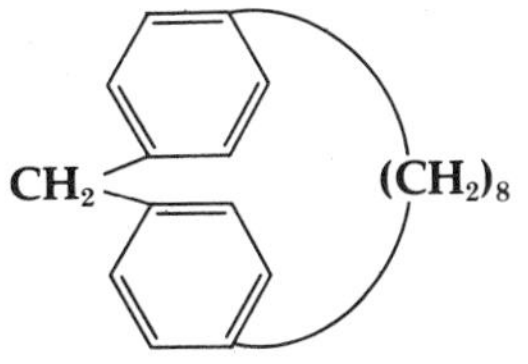

j

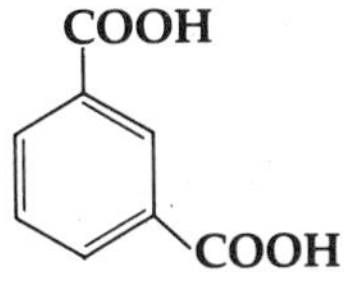

16-17 Substitute formulas for capital letters in the following sequences:

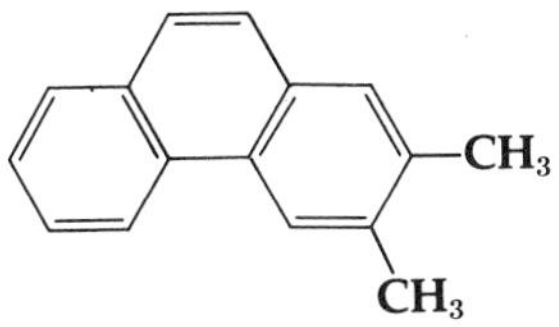

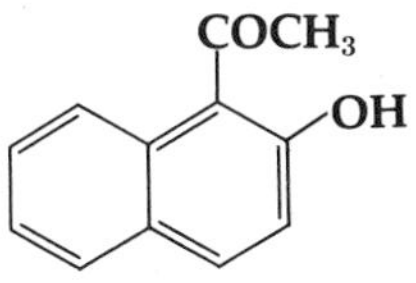

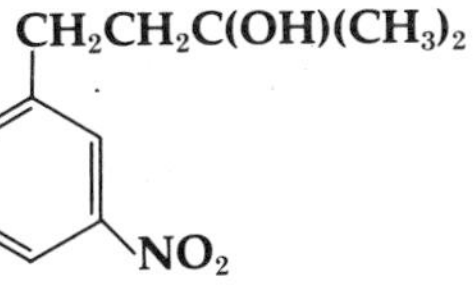

a $2H_2$ + $\xrightarrow{Pt}$ A $\xrightarrow[\text{AlCl}_3]{\text{CH}_3\text{COCl}}$ B $\xrightarrow[\text{NaOH}]{\text{C}_6\text{H}_5\text{CHO}}$ C $\xrightarrow{\text{LiAlH}_4}$ D

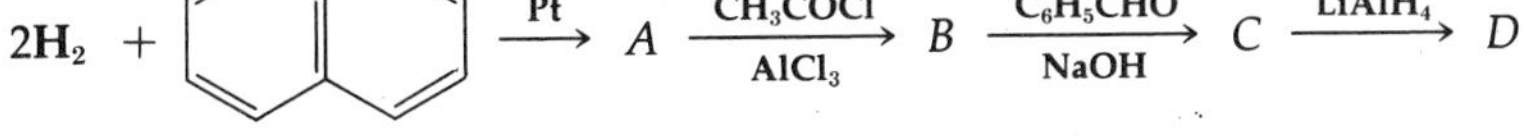

b $-CH_2-CH_2-$ $\xrightarrow[\text{ZnCl}_2]{\text{HCl,CH}_2\text{O}}$ A $\xrightarrow{\text{KCN}}$ B $\xrightarrow{\text{H}_3\text{O}^+}$

C $\xrightarrow{\text{SOCl}_2}$ D $\xrightarrow[\text{AlCl}_6]{\text{C}_6\text{H}_6}$ E

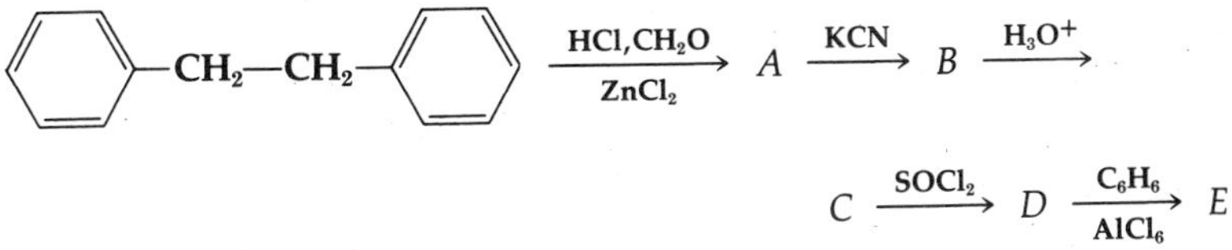

c $C_6H_5C_2H_5 \xrightarrow[\text{AlCl}_3]{\text{CO, HCl}} A \xrightarrow[\text{Heat}]{\text{NaOH}} \xrightarrow{H_3O^+} B + C$

$A \xrightarrow[\text{NaCN}]{\text{NaOH}} D \xrightarrow[\text{2) } H_2O]{\text{1) LiAlH}_4} E + F$

d

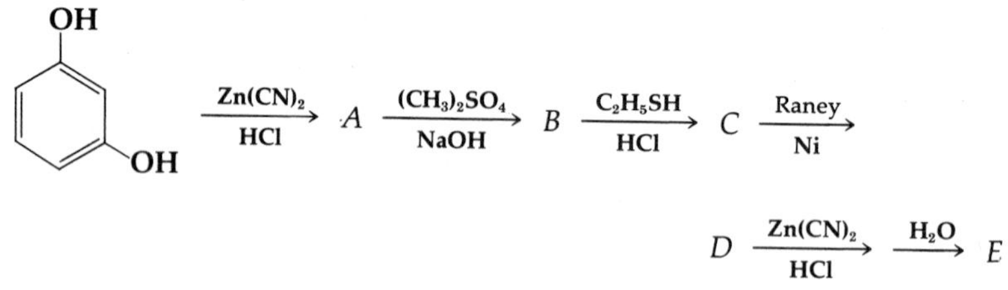

$\xrightarrow[\text{HCl}]{\text{Zn(CN)}_2} A \xrightarrow[\text{NaOH}]{(CH_3)_2SO_4} B \xrightarrow[\text{HCl}]{C_2H_5SH} C \xrightarrow[\text{Ni}]{\text{Raney}}$

$D \xrightarrow[\text{HCl}]{\text{Zn(CN)}_2} \xrightarrow{H_2O} E$

e

$\xrightarrow[\text{CO}_2, \Delta]{\text{NaOH}} A \xrightarrow[\text{NaOH}]{(CH_3)_2SO_4} B \xrightarrow{\text{SOCl}_2} C \xrightarrow[\text{AlCl}_3]{CH_3OC_6H_5} D$

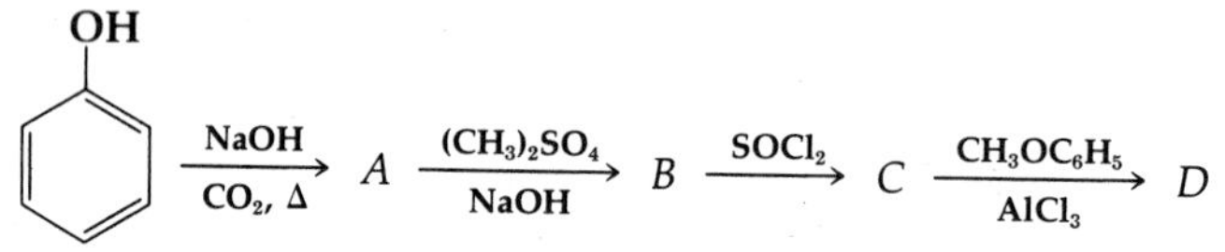

16-18   Provide an explanation for the following phenomena:

a

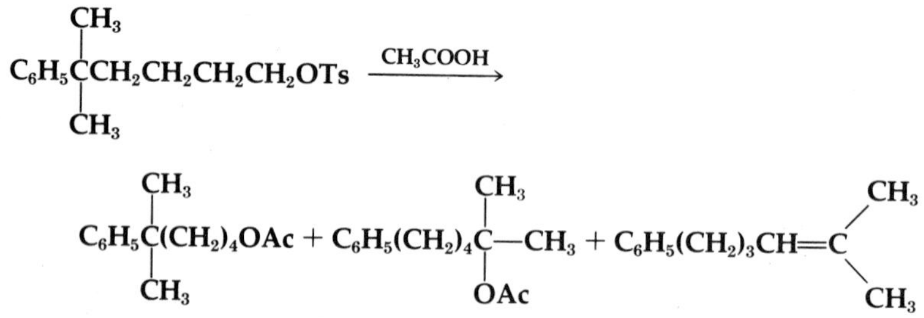

$C_6H_5\overset{\overset{\displaystyle CH_3}{|}}{\underset{\underset{\displaystyle CH_3}{|}}{C}}CH_2CH_2CH_2CH_2OTs \xrightarrow{CH_3COOH}$

$C_6H_5\overset{\overset{\displaystyle CH_3}{|}}{\underset{\underset{\displaystyle CH_3}{|}}{C}}(CH_2)_4OAc + C_6H_5(CH_2)_4\overset{\overset{\displaystyle CH_3}{|}}{\underset{\underset{\displaystyle OAc}{|}}{C}}-CH_3 + C_6H_5(CH_2)_3CH=\overset{\overset{\displaystyle CH_3}{}}{C}\underset{\underset{\displaystyle CH_3}{}}{}$

b

$\xrightarrow{HF}$

c

$\xrightarrow[\Delta]{HBr}$

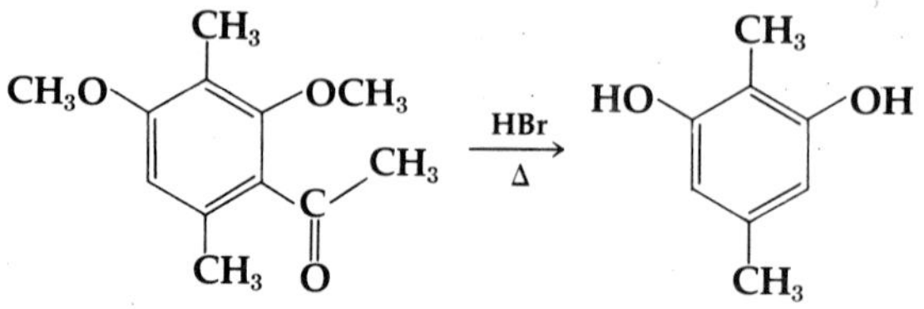

d

$\xrightarrow[\text{AlCl}_3]{(CH_3)_3CCl}$

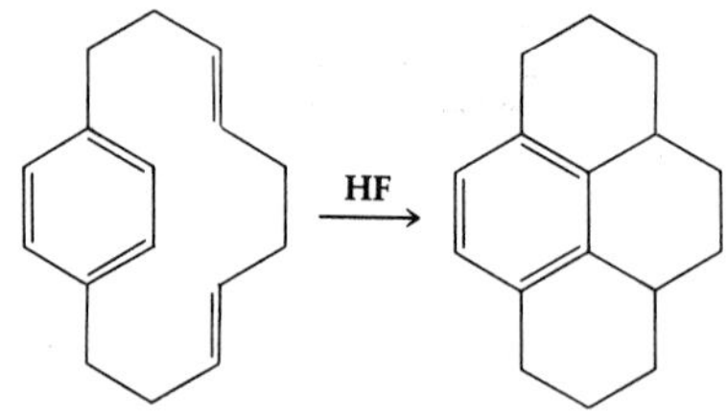

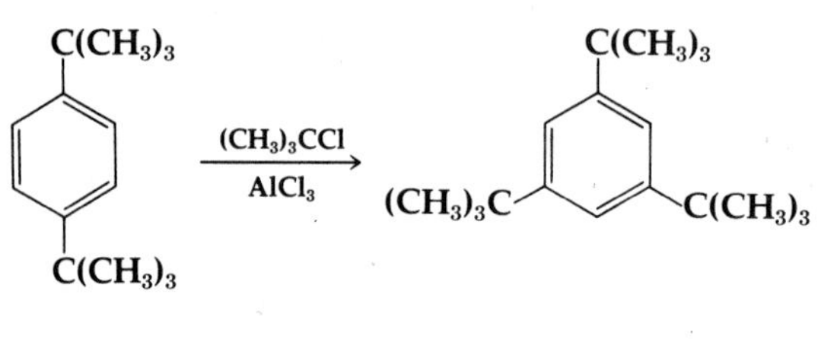

16-19 Presume that the structures of the three dibromobenzenes had not been assigned, although each was available in pure form and had the following physical properties: mp −6.9°; mp 1.8°; mp 88°. Further bromination of each isomer gave a number of discrete tribromobenzenes. The compound with mp −6.9° gave three tribromobenzene isomers, that with mp 1.8° gave two, and that with mp 88° gave only one. Assign structures to the three dibromobenzenes and to all the tribromobenzenes. Show your reasoning.

16-20 Devise syntheses for the following aromatic compounds using benzene as the only aromatic starting material. Any substance once prepared may serve as a further starting material. Do not forget the methods in Sec. 13-9.

a Phenol
b m-Cresol
c p-Cresol
d m-Cyano-acetanilide
   $(CH_3CONHC_6H_4CN)$
e p-Cyano-acetanilide
f m-Bromobenzophenone
   $(BrC_6H_4COC_6H_5)$

g m-Dicyanobenzene
h p-Dicyanobenzene
i α-Phenylbutyric acid
j m-Aminobenzenesulfonic acid
k p-Aminobenzenesulfonic acid
l 1,3,5-Tricyanobenzene
m 2-Bromo-6-chlorophenol

16-21 The major product of a certain aromatic electrophilic substitution reaction yielded the NMR spectrum shown in Fig. P16-21. Write the structure of the compound as well as the reaction involved. In order to check its structure, show an alternate synthesis of the substance, starting with an appropriate hydrocarbon.

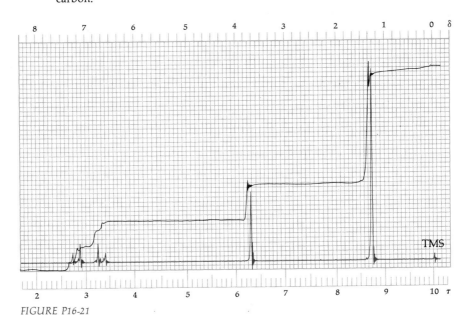

FIGURE P16-21

MOLECULAR REARRANGEMENTS

A vastly simplifying assumption in the study of organic chemistry is that reactions occur at various functional groups and leave carbon skeletons unchanged. However, reactions are encountered in which functional groups migrate within molecules and carbon skeletons are modified. The term **rearrangement** is most commonly used to refer to the migration of a group from one atom to another within a molecule.

17-1 ELECTRON-DEFICIENT SKELETAL REARRANGEMENTS

The most important group of molecular rearrangements involves the **migration (1,2-shift)** of a group from one atom to an adjacent atom that has only six electrons in its valence shell and is therefore electron deficient. The molecular system involved may be either a cation or a neutral molecule. Examples are listed in Fig. 17-1.

The migrating group never leaves the molecule. The migrating group, Z, in 1,2-shifts may be halogen, oxygen, sulfur, nitrogen, carbon, or hydrogen, but by far the most important and common are carbon migrations. When group Z is halogen, oxygen, nitrogen, or sulfur, nonbonding electrons on Z may become involved in the new bond formed during migration. When Z is carbon or hydrogen, nonbonding electrons are not available. In spite of these differences, systems containing both types of migrating groups enter into a similar spectrum of reaction mechanisms. Mechanistically similar electron-deficient rearrangements in which the group Z migrates *farther* than just to the adjacent atom (i.e., 1,3-shifts, 1,4-shifts, etc.) are exceedingly rare.

In Fig. 17-2, the general mechanistic possibilities for the 1,2-shift are outlined. The discrete species involved in the reaction are denoted by A through E. In mechanism $A \longrightarrow B \longrightarrow C \longrightarrow D \longrightarrow E$, the starting material first goes to an **open** or **classical ion**. This then passes to a **bridged** or **nonclassical** ion, which in turn gives the *rearranged* open ion, which produces product. Halfway in the migration of Z from one atom to the next must be a geometrically symmetrical species—a bridged ion. Much attention has been devoted in recent years to very subtle tests of whether this species is simply a transition state as the migrating group slips across from one bonding orbital to the next (route I on the free-energy diagram) or whether it represents a true intermediate, C (route II). In a number of cases the results indicate that this bridged ion (C) is a discrete intermediate and its tenuous existence has been recorded spectrally in some instances. Such an intermediate bridged ion

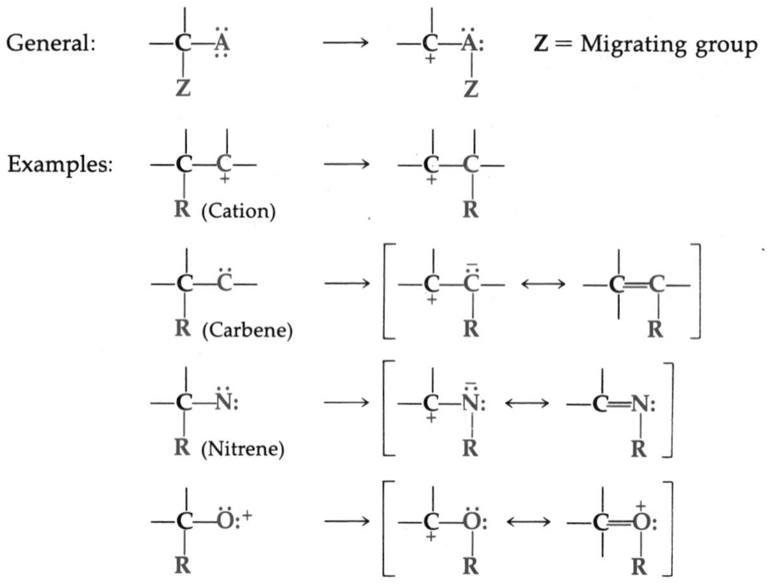

General:

$$-\overset{|}{\underset{Z}{C}}-\ddot{A} \longrightarrow -\overset{|}{\underset{Z}{\overset{+}{C}}}-\ddot{A}{:} \qquad Z = \text{Migrating group}$$

Examples:

$$-\overset{|}{\underset{R}{C}}-\overset{|}{\overset{+}{C}}- \longrightarrow -\overset{|}{\underset{R}{\overset{+}{C}}}-\overset{|}{C}- \qquad \text{(Cation)}$$

$$-\overset{|}{\underset{R}{C}}-\overset{|}{\ddot{C}}- \longrightarrow \left[-\overset{|}{\underset{R}{\overset{+}{C}}}-\overset{\cdot\cdot}{\underset{}{C}}- \longleftrightarrow -C=\overset{|}{\underset{R}{C}}- \right] \qquad \text{(Carbene)}$$

$$-\overset{|}{\underset{R}{C}}-\ddot{N}{:} \longrightarrow \left[-\overset{|}{\underset{R}{\overset{+}{C}}}-\overset{\cdot\cdot}{N}{:} \longleftrightarrow -C=\overset{}{\underset{R}{N}}{:} \right] \qquad \text{(Nitrene)}$$

$$-\overset{|}{\underset{R}{C}}-\ddot{O}{:}^{+} \longrightarrow \left[-\overset{|}{\underset{R}{\overset{+}{C}}}-\overset{\cdot\cdot}{O}{:} \longleftrightarrow -\overset{|}{\underset{R}{C}}=\overset{+}{O}{:} \right]$$

FIGURE 17-1   Rearrangement of electron-deficient systems

is labeled **nonclassical** since the bonding of three atoms with only two electrons is implicit in the structure.

Migration to Carbon

Heretofore we have seen that, when a carbonium ion is formed by ionization or by protonation of a double bond, it has two choices: substitution by a nucleophile as in the S_N1 substitution reaction; or elimination of a proton, as in the $E1$ elimination reaction. The third alternative is that of rearrangement to a new carbonium ion before substitution or elimination (Fig. 9.2). Rearrangements occur regularly in the course of such reactions if the intermediate carbonium ions can be converted to more stable ions by 1,2-shifts of hydrogen atoms or of alkyl or aryl groups.

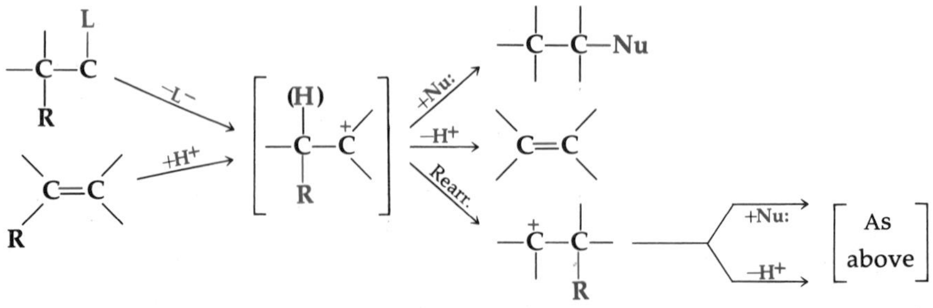

In some instances, rearrangement is initiated before a carbonium ion is formed, and in some solvolytic reactions acceleration of rate is attributed to driving force supplied by the migration. In such cases the reaction takes a

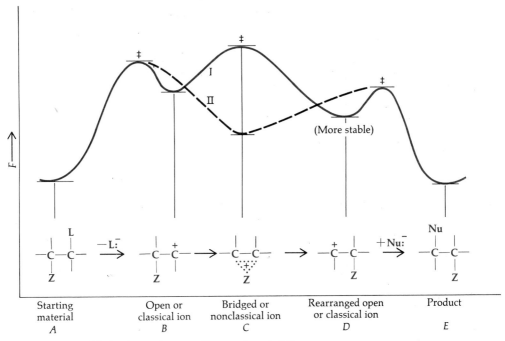

FIGURE 17-2 Mechanism and energy diagram for 1,2-shifts

concerted course and the energy diagram may show only a single transition state between A and D or even between A and E in Fig. 17-2. Rearrangements tend to be promoted by steric crowding of the migrating group in the starting material. The rearrangements of neopentyl compounds under conditions designed to effect substitution or electrophilic additions are representative.

$$
\underset{\substack{\text{Neopentyl alcohol}}}{CH_3\overset{\displaystyle CH_3}{\underset{\displaystyle CH_3}{C}}CH_2OH}
\xrightarrow{HCl}
CH_3\overset{\displaystyle CH_3}{\underset{\displaystyle CH_3}{C}}CH_2\overset{+}{O}H_2
\xrightarrow{-H_2O}
CH_3\overset{\displaystyle CH_3}{\underset{\displaystyle CH_3}{C}}{-}\overset{+}{C}H_2 \ \cdot \longrightarrow
$$

$$
CH_3\overset{+}{\underset{\displaystyle CH_3}{C}}CH_2CH_3
\xrightarrow{Cl^-}
CH_3\overset{\displaystyle Cl}{\underset{\displaystyle CH_3}{C}}CH_2CH_3
$$

tert-Amyl chloride

$$
\underset{\substack{\text{tert-Butylethylene}}}{(CH_3)_3CCH{=}CH_2}
\xrightarrow{HCl}
(CH_3)_2\underset{\displaystyle CH_3}{C}{-}\overset{+}{C}H{-}CH_3 \longrightarrow
$$

$$
(CH_3)_2\overset{+}{C}{-}\underset{\displaystyle CH_3}{CH}{-}CH_3
\xrightarrow{-H^+}
(CH_3)_2C{=}C(CH_3)_2
$$

Tetramethylethylene

PROBLEM 17-1

Write all the possible products which you can expect from the following situations. If you think some product(s) should predominate, indicate which.

a   3,3-Dimethyl-2-butanol heated with aqueous **HBr**

b   3,3-Dimethyl-2-butanol heated with aqueous **HBF$_4$**

c   3,3-Dimethyl-2-butanol treated with concentrated sulfuric acid (small amounts of water are essentially fully protonated in this medium and unavailable to act as nucleophiles)

d   3-Ethyl-4-methyl-2-pentanol heated with aqueous sulfuric acid

e   2,3-Dimethyl-2,3-butanediol treated with concentrated sulfuric acid

Rearrangements of alcohols under acidic conditions were originally known as the **Wagner-Meerwein rearrangement**, but the term has been broadened to include rearrangements involving many other leaving groups. A wide variety of substrates and electrophilic reagents (to generate initial carbonium ions) are involved, all rearranging via carbonium ion routes. A number of these rearrangements are used synthetically for ring enlargement or contraction. These reactions often have particular names, and the main ones are collected in the succeeding paragraphs.

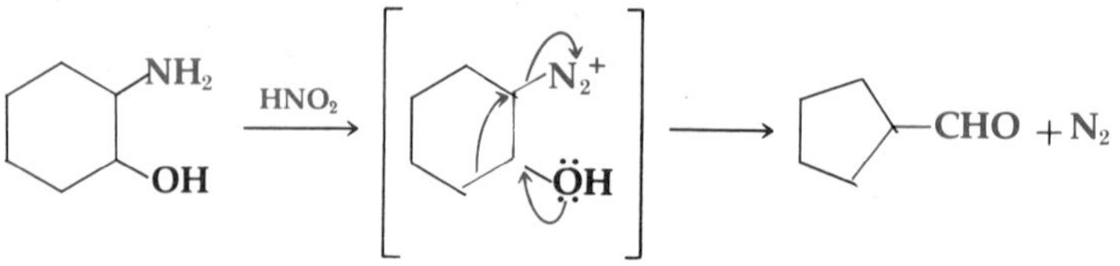

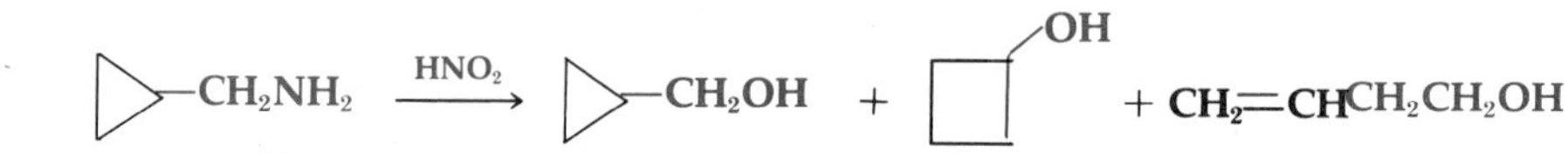

Cyclopropylcarbinylamine   Cyclopropylcarbinol   Cyclobutanol   Allylcarbinol (trace only)

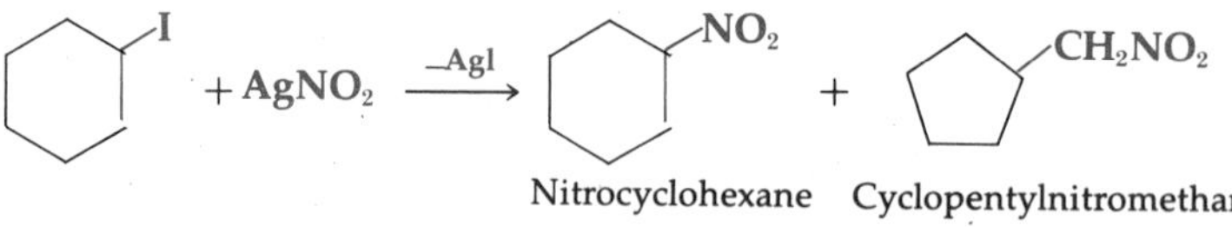

Nitrocyclohexane   Cyclopentylnitromethane

Attempts to dehydrate substituted *vic*-diols (pinacols) usually lead to rearrangements with formation of ketones. This reaction is called the **pinacol rearrangement** and is favored because of the resonance stabilization of the rearranged ion by the attached oxygen. Because the highly branched structures of the products are not easily constructed by other reactions, this rearrangement has found interesting applications in synthesis. Pinacols themselves are made by reductive dimerization of ketones with bivalent metals, usually magnesium, under anhydrous conditions (page 777).

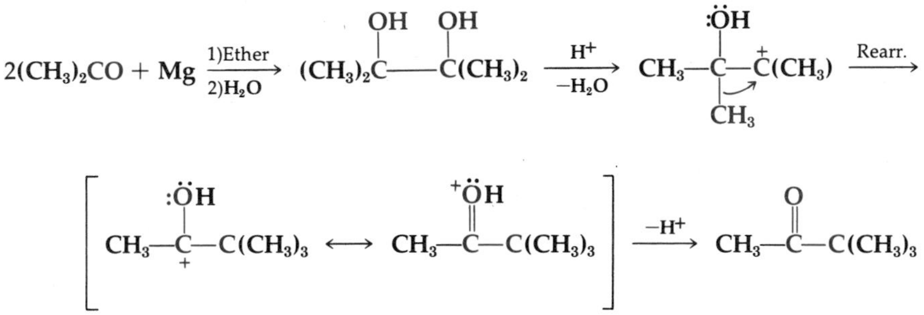

$$2(CH_3)_2CO + Mg \xrightarrow[\text{2)H}_2\text{O}]{\text{1)Ether}} (CH_3)_2\overset{\displaystyle OH}{C}\!\!-\!\!\overset{\displaystyle OH}{C}(CH_3)_2 \xrightarrow[-H_2O]{H^+} CH_3-\overset{\displaystyle :\overset{..}{O}H}{\underset{\displaystyle CH_3}{C}}-\overset{+}{C}(CH_3) \xrightarrow{\text{Rearr.}}$$

$$\left[\quad CH_3-\overset{\displaystyle :\overset{..}{O}H}{\underset{+}{C}}-C(CH_3)_3 \longleftrightarrow CH_3-\overset{\displaystyle +\overset{..}{O}H}{\overset{\|}{C}}-C(CH_3)_3 \quad \right] \xrightarrow{-H^+} CH_3-\overset{\displaystyle O}{\overset{\|}{C}}-C(CH_3)_3$$

Stabilized cation   Pinacolone

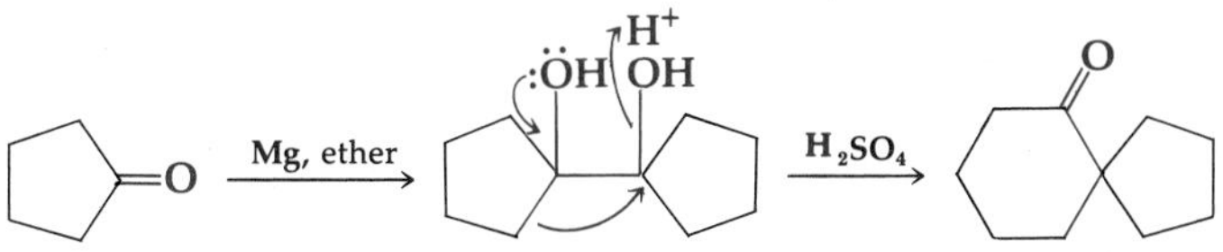

In some rearrangements the electron-deficient carbon is a carbene instead of a carbonium ion. These are usually created by loss of N_2 from a diazo group, $C=N^+=N^-$. In the **Wolff rearrangement**, α-diazoketones lose nitrogen and rearrange to ketenes in the presence of solid silver oxide or by irradiation with light. The reaction is ordinarily carried out in the presence of water or alcohols, which convert the ketenes to carboxylic acids or their esters.

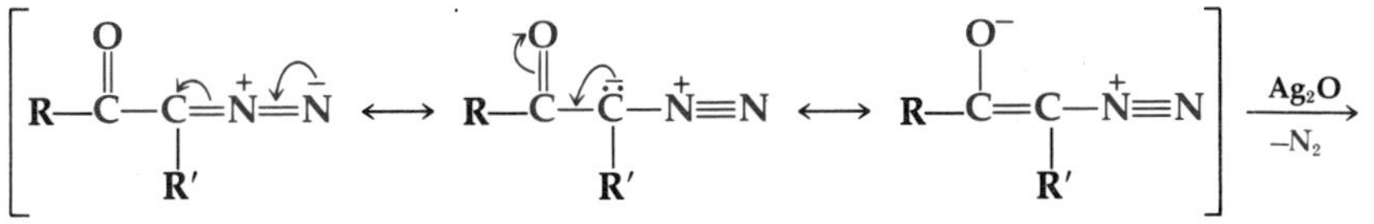

Diazoketones

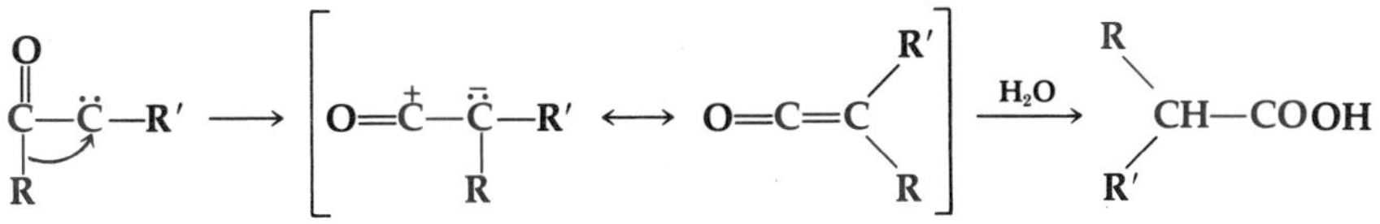

Ketenes

The Wolff rearrangement has been incorporated in a general sequence, the **Arndt-Eistert synthesis**, in which an acid is converted to its next-higher homolog.

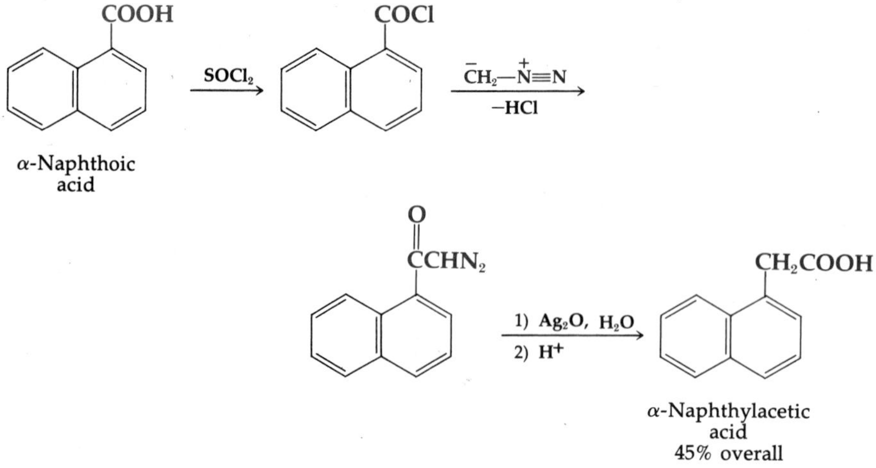

α-Naphthoic
acid

$\overset{-}{C}H_2-\overset{+}{N}\equiv N$
$\xrightarrow{\ -HCl\ }$

$\xrightarrow[\text{2) } H^+]{\text{1) } Ag_2O,\ H_2O}$

α-Naphthylacetic
acid
45% overall

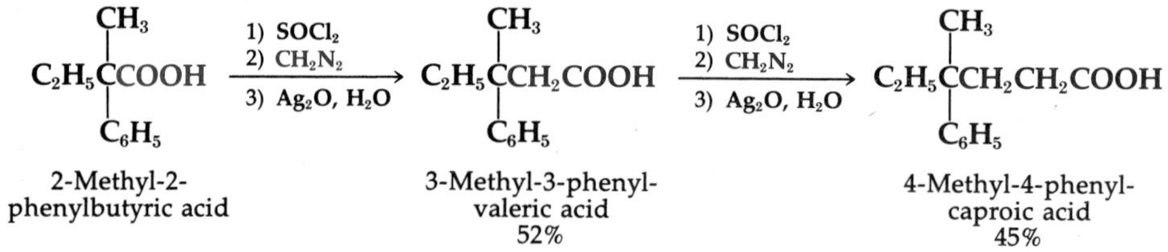

$$C_2H_5\underset{\underset{C_6H_5}{|}}{\overset{\overset{CH_3}{|}}{C}}COOH \quad \xrightarrow[\text{3) } Ag_2O,\ H_2O]{\substack{\text{1) } SOCl_2 \\ \text{2) } CH_2N_2}} \quad C_2H_5\underset{\underset{C_6H_5}{|}}{\overset{\overset{CH_3}{|}}{C}}CH_2COOH \quad \xrightarrow[\text{3) } Ag_2O,\ H_2O]{\substack{\text{1) } SOCl_2 \\ \text{2) } CH_2N_2}} \quad C_2H_5\underset{\underset{C_6H_5}{|}}{\overset{\overset{CH_3}{|}}{C}}CH_2CH_2COOH$$

2-Methyl-2-
phenylbutyric acid

3-Methyl-3-phenyl-
valeric acid
52%

4-Methyl-4-phenyl-
caproic acid
45%

Diazomethane is also used in an analogous reaction that converts ketones to higher homologs. The process has been used for expansion of cycloalkanone rings. A ring closure to form ethylene oxides competes with the rearrangement step and often becomes the principal reaction.

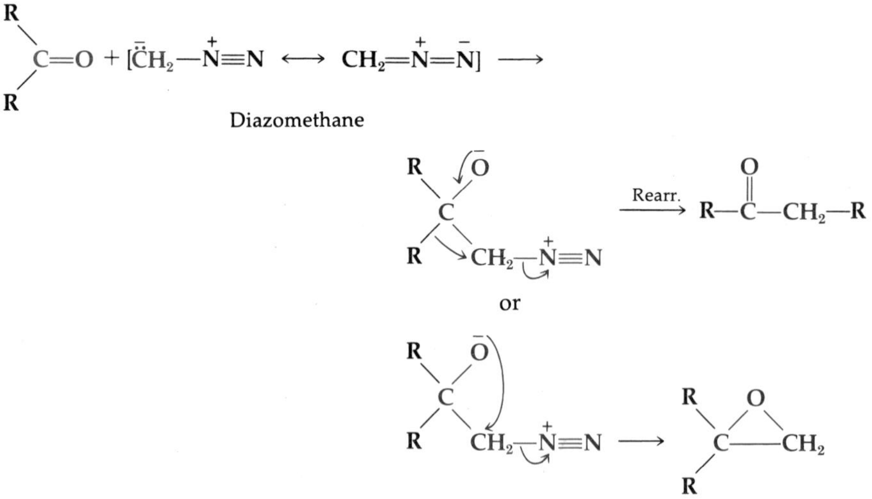

$$\underset{R}{\overset{R}{>}}C=O + [\overset{-}{C}H_2-\overset{+}{N}\equiv N \longleftrightarrow CH_2=\overset{+}{N}=\overset{-}{N}] \longrightarrow$$

Diazomethane

$\xrightarrow{\text{Rearr.}}$ $R-\overset{\overset{O}{\|}}{C}-CH_2-R$

or

$\longrightarrow$

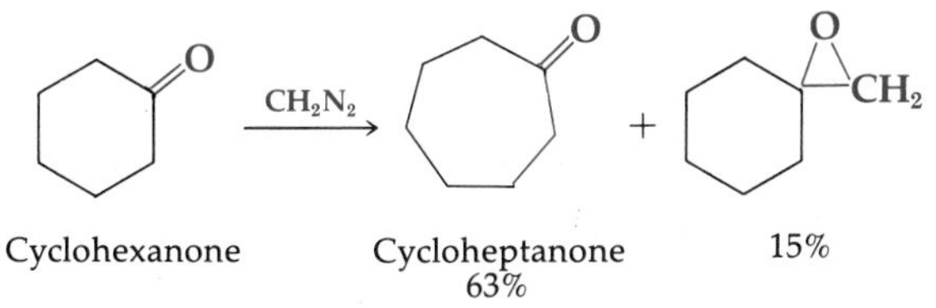

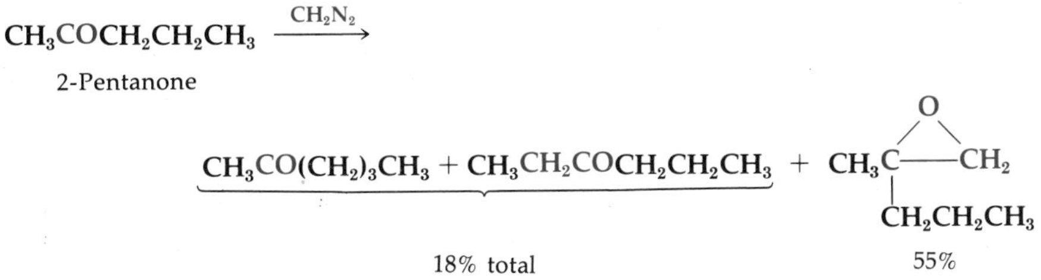

Cyclohexanone   Cycloheptanone   15%
                              63%

$$CH_3COCH_2CH_2CH_3 \xrightarrow{CH_2N_2}$$

2-Pentanone

$$CH_3CO(CH_2)_3CH_3 + CH_3CH_2COCH_2CH_2CH_3 \ + \ \underset{\substack{| \\ CH_2CH_2CH_3}}{CH_3\overset{O}{\overset{\triangle}{C}\!-\!-\!-\!CH_2}}$$

$$\underbrace{}$$

18% total   55%

Migrations of hydrogen instead of carbon pass through the same bridged protonium ion or π complex discussed in Chap. 15 for the electrophilic addition of protons to double bonds. The same sequence of ions then accounts for double-bond protonation, $E1$ elimination (deprotonation), and hydrogen migration.

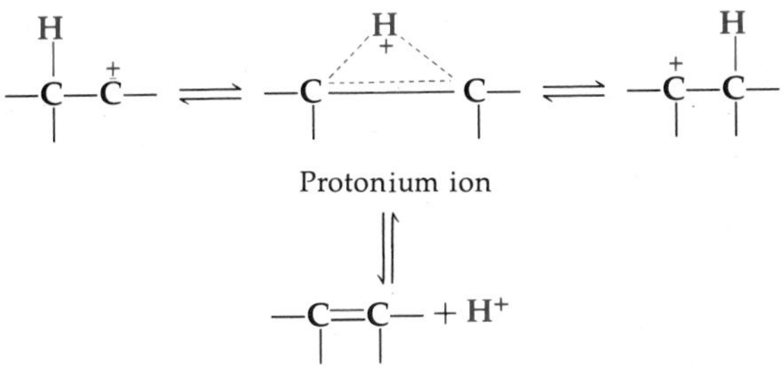

Protonium ion

$$-C\!=\!C- + H^+$$

It is not necessarily clear whether hydrogen migrates internally in a true rearrangement or is merely lost, with formation of an intermediate alkene, which reprotonates on the other carbon. This can, however, be tested by use of isotopic labels (e.g., deuterium or tritium for hydrogen).

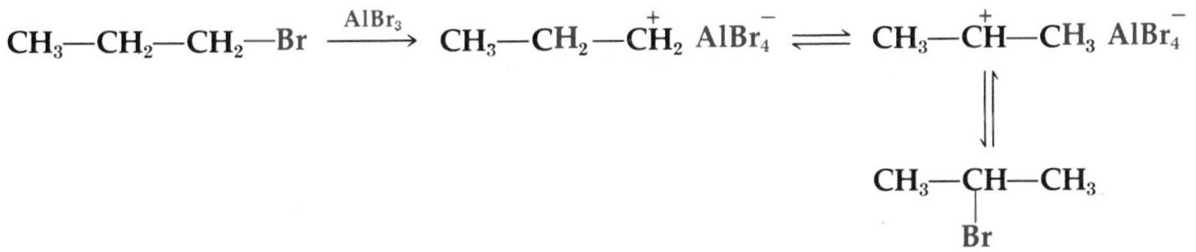

$$CH_3-CH_2-CH_2-Br \xrightarrow{AlBr_3} CH_3-CH_2-\overset{+}{C}H_2 \ AlBr_4^- \rightleftharpoons CH_3-\overset{+}{C}H-CH_3 \ AlBr_4^-$$

$$CH_3-\underset{\substack{| \\ Br}}{C}H-CH_3$$

PROBLEM 17-2

Write the major products expected from the following reactions.

a $+ HNO_2$ in H_2O

b $+ AgBF_4$ in CH_2Cl_2

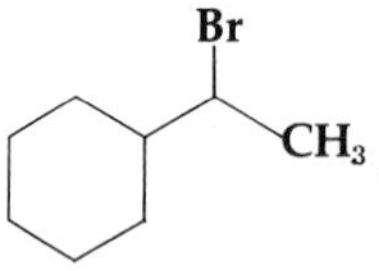

c $+ AlBr_3$ in $C_6H_5NO_2$

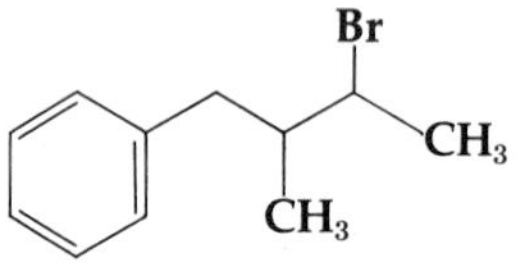

d $C_2H_5COCH(CH_3)_2 + CH_2N_2$

e $C_6H_5CH_2COCl + CH_2N_2 \longrightarrow A \xrightarrow{Ag_2O} B$

PROBLEM 17-3

3,4-Dimethyl-pentane-2,3-diol gives predominantly one carbonyl-containing compound on treatment with strong acids. Discuss which product you might expect to be preferred, considering intermediate energies as in Fig. 17-2.

Migration to Nitrogen

In these rearrangements an electron-deficient nitrogen (**nitrene**, —N:) is first created by loss of a leaving group from the nitrogen atom. This can occur in several ways. There are several reactions which convert N-substituted amides to isocyanates.

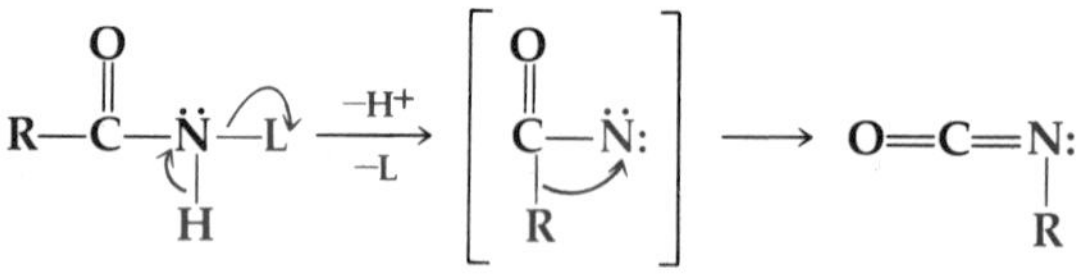

Nitrene   Isocyanates

The **Hofmann rearrangement** of N-haloamides is the commonest of the group. The reactions are useful for conversion of carboxylic acids and their derivatives to amines that contain one less carbon atom than the starting materials. Such a procedure is synthetically useful if the amine cannot be made directly by a nucleophilic substitution reaction.

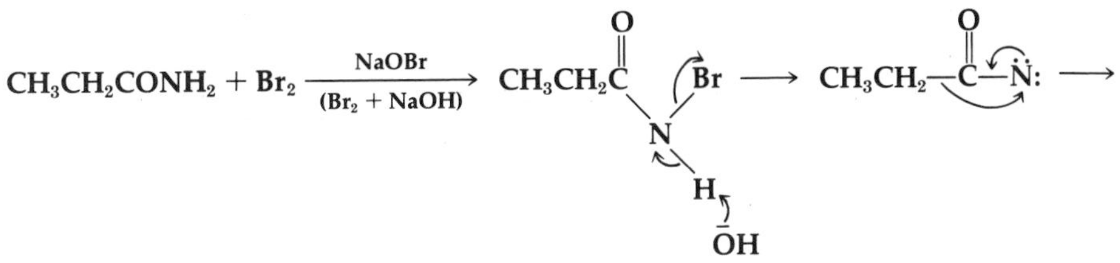

$$CH_3CH_2CONH_2 + Br_2 \xrightarrow[\text{(Br}_2 + \text{NaOH)}]{\text{NaOBr}} \quad CH_3CH_2\overset{\displaystyle O}{\overset{\|}{C}}\underset{\underset{\displaystyle OH}{\underset{|}{H}}}{\overset{Br}{\underset{N}{}}} \longrightarrow CH_3CH_2-\overset{\displaystyle O}{\overset{\|}{C}}-\ddot{N}: \longrightarrow$$

$$CH_3CH_2N{=}C{=}O \xrightarrow{H_2O} CH_3CH_2\overset{\displaystyle OH}{\underset{\displaystyle O}{\overset{\|}{\underset{\|}{NHC}}}} \longrightarrow CH_3CH_2NH_2 + CO$$

Ethyl isocyanate	N-Ethylcarbamic acid	Ethylamine 90%

$$(CH_3)_3CCH_2CONH_2 \xrightarrow{\text{NaOBr}} (CH_3)_3CCH_2NH_2$$

β,β-Dimethylbutyramide	Neopentylamine 94%

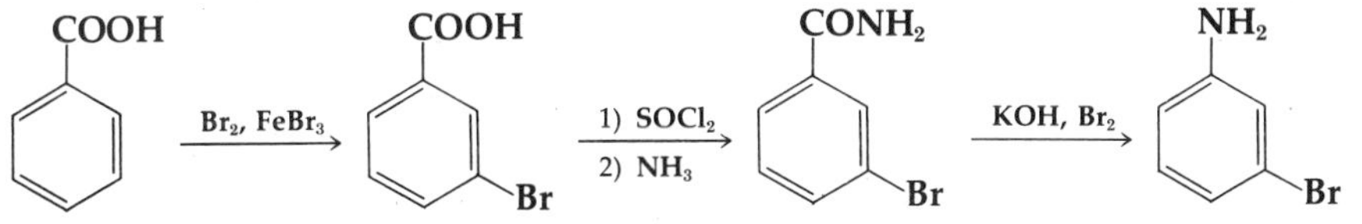

$$\text{COOH} \xrightarrow{Br_2, FeBr_3} \text{COOH}\ (Br) \xrightarrow[\text{2) NH}_3]{\text{1) SOCl}_2} \text{CONH}_2\ (Br) \xrightarrow{\text{KOH, Br}_2} \text{NH}_2\ (Br)$$

		m-Bromobenzamide	m-Bromoaniline 87% (last step)

One of the most important industrial applications of the Hofmann rearrangement is in the synthesis of anthranilic acid, a basic starting material for preparation of ortho-disubstituted benzene derivatives.

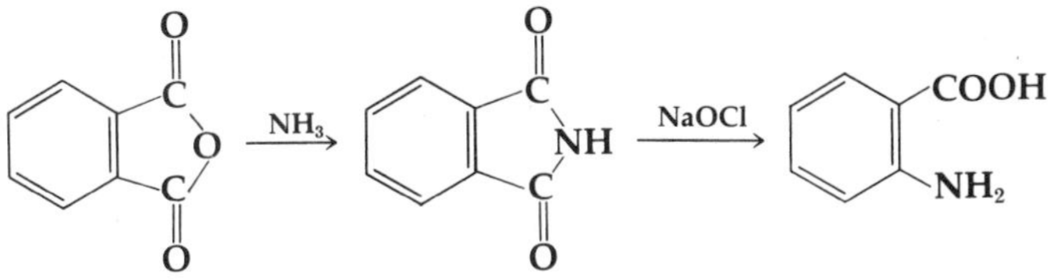

Phthalic anhydride	Phthalimide	Anthranilic acid

In the **Lossen rearrangement**, a salt of a hydroxamic acid or a related compound is decomposed in a manner strictly analogous to N-haloamide decomposition. Isocyanates may be isolated by reaction of the hydroxamic acid with thionyl chloride. The reaction is of limited synthetic value since hydroxamic acids are not readily available.

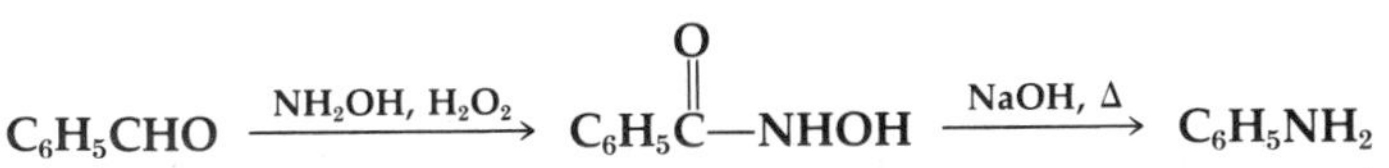

$$C_6H_5CHO \xrightarrow{NH_2OH, H_2O_2} C_6H_5\overset{\displaystyle O}{\overset{\|}{C}}{-}NHOH \xrightarrow{\text{NaOH, }\Delta} C_6H_5NH_2$$

Benzohydroxamic
acid

The **Curtius rearrangement** results from the thermal decomposition of an acyl azide. This reaction is the strict mechanistic analog of the diazoketone rearrangement for carbon, above. The reaction can be carried out in inert solvents, which allow isolation of the isocyanates. The azides required as starting materials may be made by the substitution of acid chlorides with sodium azide or by the reaction of acyl hydrazides with nitrous acid.

$$(CH_3)_2CHCH_2COCl \xrightarrow{NaN_3} (CH_3)_2CHCH_2-\overset{O}{\overset{\|}{C}}-\overset{\ddot{..}}{N}-\overset{+}{N}\equiv\overset{..}{N} \xrightarrow[-N_2]{CHCl_3}$$

Isovaleryl chloride

$$(CH_3)_2CHCH_2-N=C=O \xrightarrow[-CO_2]{H_2O} (CH_3)_2CHCH_2NH_2$$

Isobutylamine
70% (overall)

$$CH_3O-\langle\rangle-COOC_2H_5 \xrightarrow{NH_2NH_2} CH_3O-\langle\rangle-\overset{O}{\overset{\|}{C}}-NHNH_2 \xrightarrow{HNO_2}$$

Ethyl anisate   Anisoylhydrazide
                                   95%

$$CH_3O-\langle\rangle-\overset{O}{\overset{\|}{C}}-N_3 \xrightarrow[\substack{\Delta \\ benzene}]{-N_2} CH_3O-\langle\rangle-N=C=O$$

95%   Anisyl isocyanate
                                  80%

The final amine formed in the Hofmann rearrangement, or by hydrolysis of isocyanates generally, often reacts with unchanged isocyanate to form ureas. To avoid this the acyl azides may be rearranged in hot *t*-butyl alcohol to form *t*-butyl-urethans (carbamates), which in turn are instantly degraded with acid (page 594). Urethans of other alcohols are formed by heating the acyl azide in that alcohol.

$$R-CONH_2 \xrightarrow{NaOBr} R-N=C=O \xrightarrow{H_2O} R-NH_2 \xrightarrow{+RNCO} R-NHCONH-R$$

RCOOH   Isocyanates   Ureas

$$R-CON_3 \xrightarrow[t\text{-BuOH}]{\Delta} R-\overset{\frown}{N}\overset{\frown}{H}-CO-O-\overset{CH_3}{\underset{CH_3}{\overset{\|}{C}}}-CH_2-H \xrightarrow{CF_3COOH}$$

t-Butyl-urethans (carbamates)

$$R-\overset{+}{N}H_3\ CF_3COO^- + CO_2\uparrow + CH_2=C(CH_3)_2\uparrow$$

The reactions of hydrazoic acid with carboxylic acids and ketones in the presence of strong acids are known as **Schmidt rearrangements**. Hydrazoic acid HN_3 ($H\ddot{N}^- \!-\! N^+ \!\equiv\! N\!:$) is structurally analogous to diazomethane, and the Schmidt rearrangement takes essentially the same course as the Wolff rearrangement ($R\!-\!CO\!-\!CHN_2$ vs $R\!-\!CO\!-\!NN_2$) and the rearrangement of ketones with diazomethane. The Schmidt rearrangement is also closely analogous to the Curtius transformation. Yields are usually high, and sterically hindered acids react very smoothly. The rate-determining step in the reaction is the formation of an acyl cation, a process accelerated by bulky groups. Urea formation is not observed since the product amine is fully protonated in the strong acid.

Schmidt reaction with acids:

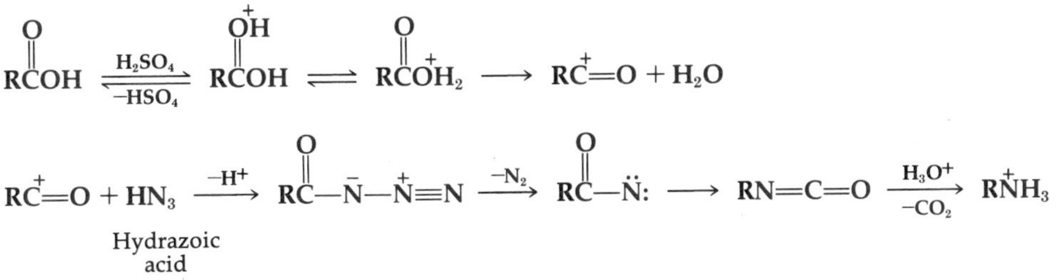

$$RCOH \underset{-HSO_4}{\overset{H_2SO_4}{\rightleftharpoons}} R\overset{+}{C}OH \rightleftharpoons R\overset{+}{C}OH_2 \longrightarrow R\overset{+}{C}{=}O + H_2O$$

$$R\overset{+}{C}{=}O + HN_3 \xrightarrow{-H^+} RC\!-\!\bar{N}\!-\!\overset{+}{N}{\equiv}N \xrightarrow{-N_2} RC\!-\!\ddot{N}: \longrightarrow RN{=}C{=}O \xrightarrow[-CO_2]{H_3O^+} R\overset{+}{N}H_3$$

Hydrazoic
acid

Schmidt reaction with ketones:

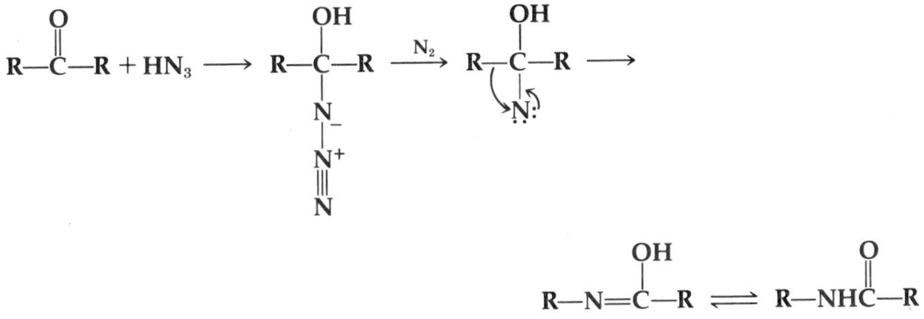

$$R\!-\!C\!-\!R + HN_3 \longrightarrow R\!-\!\underset{\substack{N_-\\ \| \\ N^+\\ \| \| \\ N}}{\overset{OH}{C}}\!-\!R \xrightarrow{N_2} R\!-\!\underset{N:}{\overset{OH}{C}}\!-\!R \longrightarrow$$

$$R\!-\!N{=}\overset{OH}{C}\!-\!R \rightleftharpoons R\!-\!NH\overset{O}{C}\!-\!R$$

$$CH_3(CH_2)_4COOH + HN_3 \xrightarrow{H_2SO_4} CH_3(CH_2)_4NH_2$$

Hexanoic acid

n-Pentylamine
70–75%

$$C_6H_5COCH_3 + HN_3 \xrightarrow{H_2SO_4} C_6H_5NHCOCH_3$$

Acetophenone

Acetanilide
77%

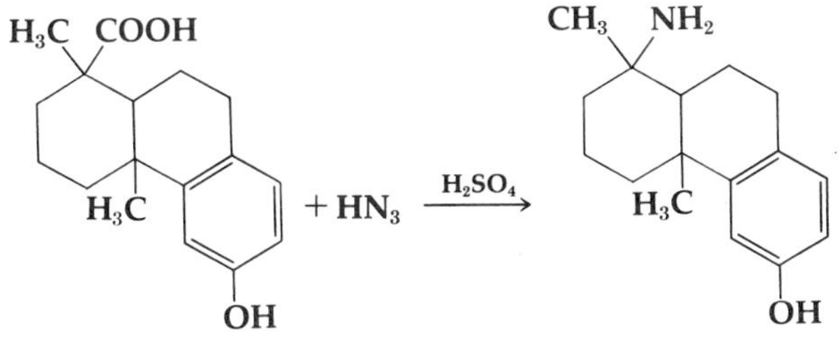

Podocarpic acid

$+ HN_3 \xrightarrow{H_2SO_4}$

73%

The **Beckmann rearrangement** is an acid-catalyzed transformation of a ketoxime to an amide. The reaction is highly stereospecific in that the group anti (trans) to the hydroxyl group of the oxime always migrates.

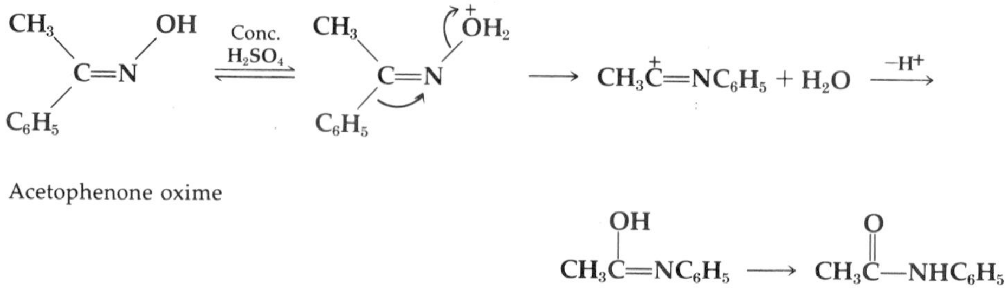

Acetophenone oxime

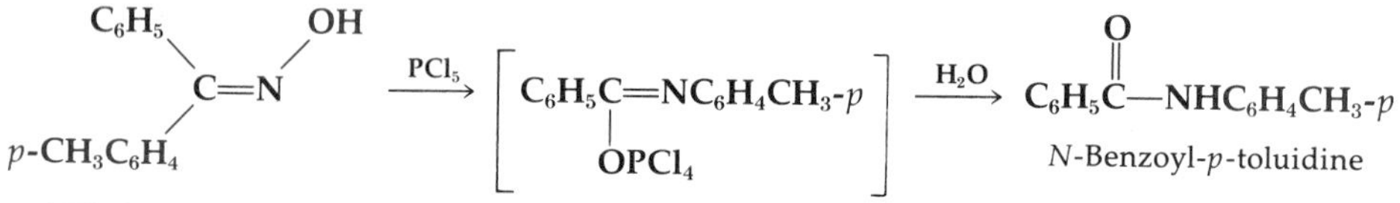

N-Benzoyl-p-toluidine

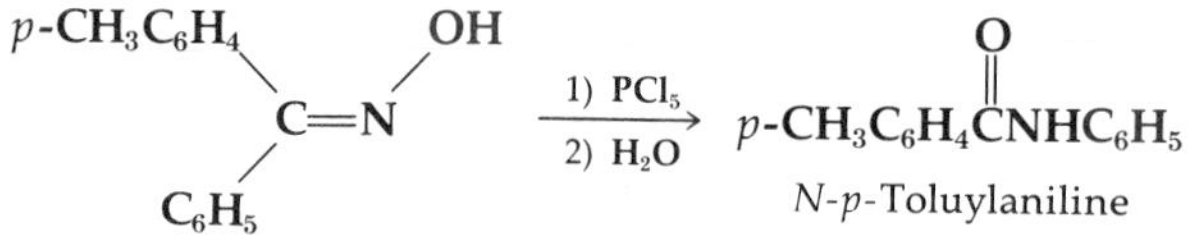

N-p-Toluylaniline

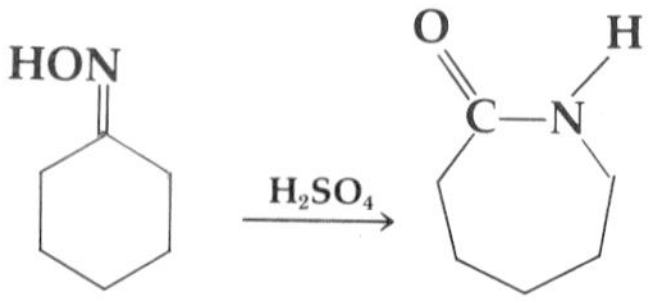

Cyclohexanone   ε-Caprolactam
    oxime   65%

Migration to Oxygen

In the **Baeyer-Villiger oxidation**, a ketone is converted to an ester by reaction with a peracid (RCO_3H). The reaction has been studied carefully, and evidence indicates that the key step in the mechanism is heterolytic dissociation of the O—O bond in an adduct formed from the peracid and a carbonyl compound.

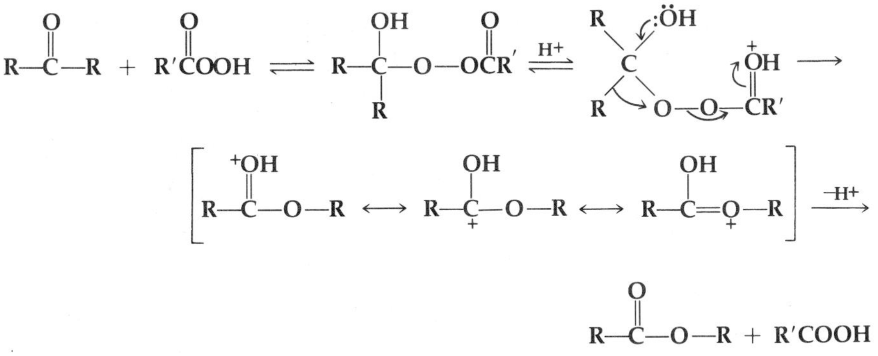

Overall reaction:

$$R—CO—R + R'CO_3H \longrightarrow R—CO—O—R + R'COOH$$

$$C_6H_5COCH_3 + CF_3CO_3H \xrightarrow{CH_2Cl_2,\ CF_3COOH} C_6H_5—O—COCH_3 + CF_3COOH$$

Acetophenone   Peroxytrifluoracetic
                      acid

$$CH_3COC(CH_3)_3 + CF_3CO_3H \xrightarrow{CH_2Cl_2,\ CF_3COOH} CH_3COOC(CH_3)_3 + CF_3COOH$$

Pinacolone
(*t*-butyl methyl
ketone)

Peroxytrifluoroacetic acid is the most reactive peracid known, but both peracetic acid and various substituted perbenzoic acids have been used with good results. A related transformation is the **Dakin oxidation** of *o*- and *p*-hydroxybenzaldehydes or phenyl ketones by the action of alkaline hydrogen peroxide. The esters are usually hydrolyzed under the reaction conditions.

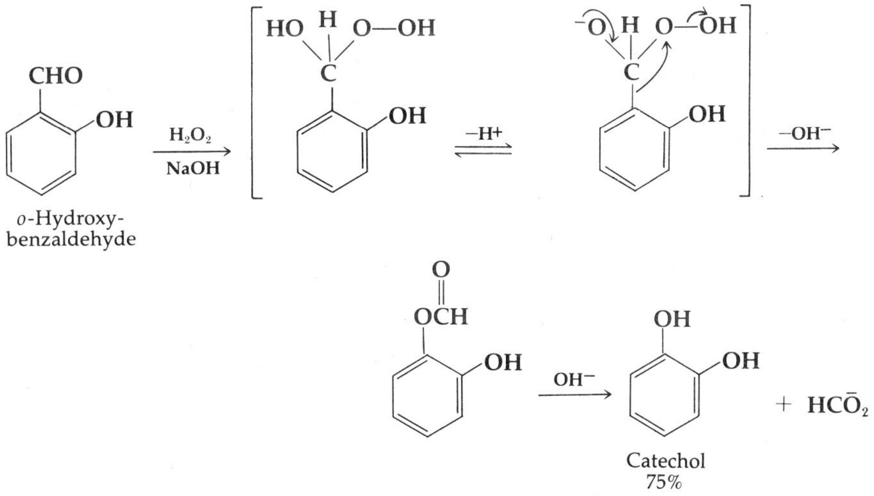

o-Hydroxy-
benzaldehyde

Catechol
75%

All the rearrangements to nitrogen and oxygen serve to insert oxygen or nitrogen into the carbon skeleton.

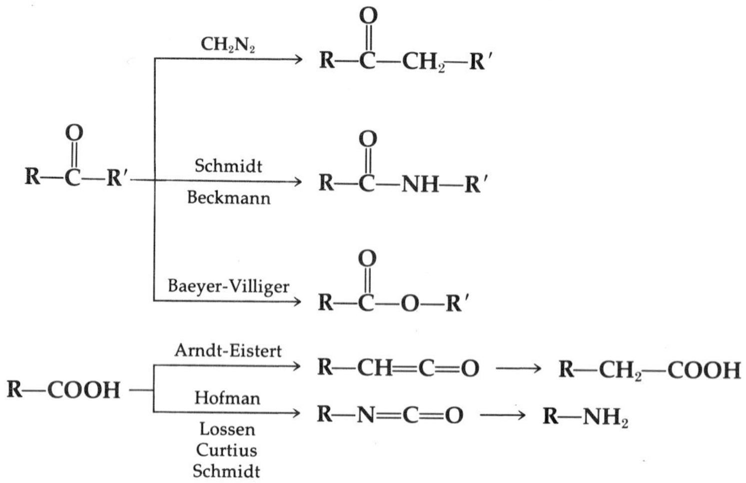

$$R-\overset{\overset{\displaystyle O}{\|}}{C}-R' \quad\xrightarrow{CH_2N_2}\quad R-\overset{\overset{\displaystyle O}{\|}}{C}-CH_2-R'$$

$$\xrightarrow[\text{Beckmann}]{\text{Schmidt}} \quad R-\overset{\overset{\displaystyle O}{\|}}{C}-NH-R'$$

$$\xrightarrow{\text{Baeyer-Villiger}} \quad R-\overset{\overset{\displaystyle O}{\|}}{C}-O-R'$$

$$R-COOH \quad\xrightarrow{\text{Arndt-Eistert}}\quad R-CH=C=O \quad\longrightarrow\quad R-CH_2-COOH$$

$$\xrightarrow[\substack{\text{Lossen}\\\text{Curtius}\\\text{Schmidt}}]{\text{Hofman}}\quad R-N=C=O \quad\longrightarrow\quad R-NH_2$$

PROBLEM 17-4

Complete the following reaction series.

a Each isomer of the oxime of propiophenone $(C_6H_5COCH_2CH_3)$ with PCl_3

b Cyclopentanecarboxylic acid $\xrightarrow{\text{SOCl}_2}$ A $\xrightarrow{\text{NaN}_3}$ B $\xrightarrow[\text{C}_6\text{H}_6]{\Delta}$ C $\xrightarrow[\Delta]{\text{H}_3\text{O}^+}$ D

c Cyclobutanone + CF_3CO_3H $\longrightarrow$ A $\xrightarrow{+\text{NH}_2\text{OH}}$ B $\xrightarrow[\text{NaOH}]{\Delta}$ C

$$\Delta\Big\downarrow\substack{\text{HN}_3\\\text{H}_2\text{SO}_4}$$

$$C_4H_7NO_2 \text{ (neutral)}$$

d [structure: benzene ring with COCl and OCH$_3$ substituents] $\xrightarrow{\text{H}_2\text{NNH}_2}$ A $\xrightarrow{\text{HNO}_2}$ B $\xrightarrow[t\text{-BuOH}]{\Delta}$ C $\xrightarrow{\text{CF}_3\text{COOH}}$ D

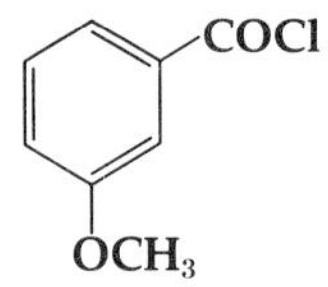

e [structure: phthalic anhydride] $+ HN_3/H_2SO_4$

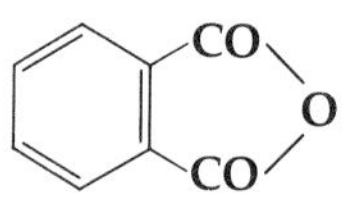

f Benzaldehyde oxime + PCl_5

g Cinnamic acid $(C_6H_5CH=CHCOOH)$ $\longrightarrow$ phenylacetaldehyde

PROBLEM 17-5

Synthesize the following compounds from any desired precursor, utilizing a reliable rearrangement reaction in the sequence.

a Aniline

b N-Phenylbenzamide

c 3-Methylcycloheptanone

d 6-Hydroxyhexanoic acid lactone

e Phenyl propionate

f N-Carbomethoxybenzylamine

g 4-Chloro-1,2-dihydroxybenzene

h Cyclopentanone from succinic acid

Rearrangements from Boron

As discussed in Sec. 15-5, alkyl boranes, R_3B, are readily obtained from alkenes by **hydroboration**. These compounds can be very useful synthetic intermediates because of their ease of rearrangement. Generally, a mild nucleophile with a bond between heteroatoms ($:Z—Y$) adds to the empty orbital of the electron-deficient boron. Then a rearrangement of the alkyl group can occur from boron to heteroatom in a manner analogous to the Baeyer-Villiger mechanism. The boron is subsequently lost by hydrolysis. The reaction with halogen requires the presence of peroxide and may involve **XOH** as the active species. The rearrangement is especially facile since its product is a neutral (though electron-deficient) boron atom and more stable than the resultant (and isoelectronic) carbonium ions in Wagner-Meerwein rearrangements.

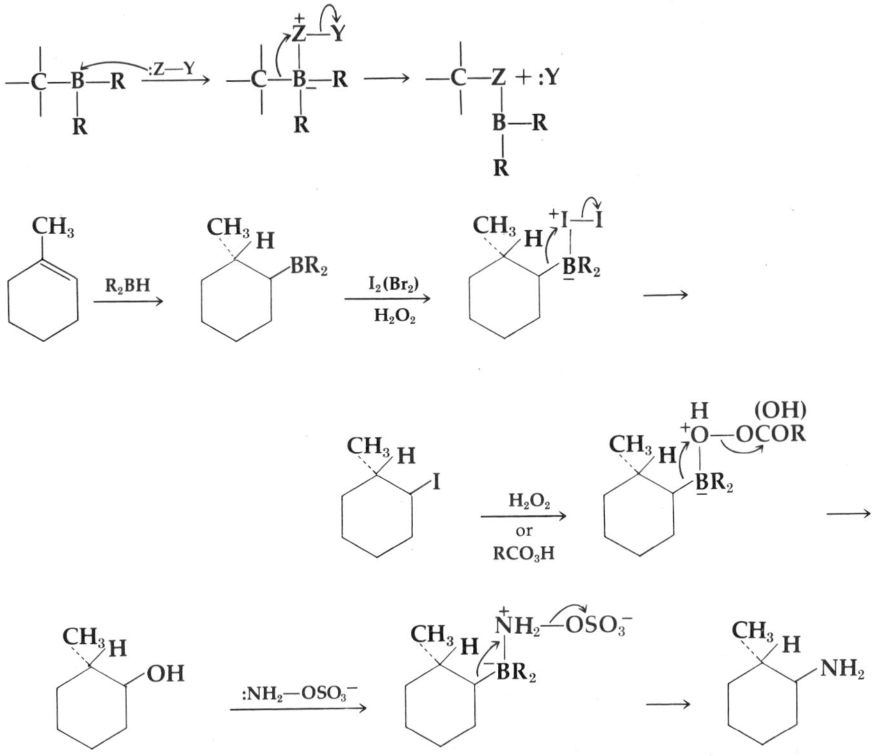

(Note cis addition)

In each case the configuration of the migrating carbon of the skeleton is retained (Sec. 17-2). Hence, for synthetic purposes the overall change is one of *anti-Markownikoff cis addition of* **H—X, H—OH,** and **H—NH₂** to a double bond. This procedure constitutes a very powerful synthetic tool. Furthermore, the reaction is easy since the hydroboration is usually fast and quantitative and the intermediate borane does not require isolation before proceeding to effect rearrangement.

Other rearrangements of organoboranes can create carbon-carbon bonds and so may become potent synthesis tools, but these reactions are still under study to define their scope.

17-2 DIRECTION AND STEREOCHEMISTRY OF REARRANGEMENT

Migratory Aptitudes or Preferences

In a number of the preceding rearrangements there exists a choice as to which group migrates toward the electron-deficient atom. The migrating group passes through a bridged ion which is electron-deficient, and this deficiency is shared by all three centers involved. Hence groups which stabilize carbonium ions lower the barrier and increase the rate. This rule holds for the migrating groups as well, and the relative preference for rearrangement, or **migratory aptitude**, expresses the familiar order of carbonium ion stabilization:

Aryl > *tert*-alkyl > *sec*-alkyl > *pri*-alkyl > **CH₃** > **H**

Aryl groups distribute the charge into their π electrons, creating an intermediate bridged ion (**phenonium ion**) which is a fully bonded cyclopropane and specially stabilized with the charge delocalized in the benzene ring:

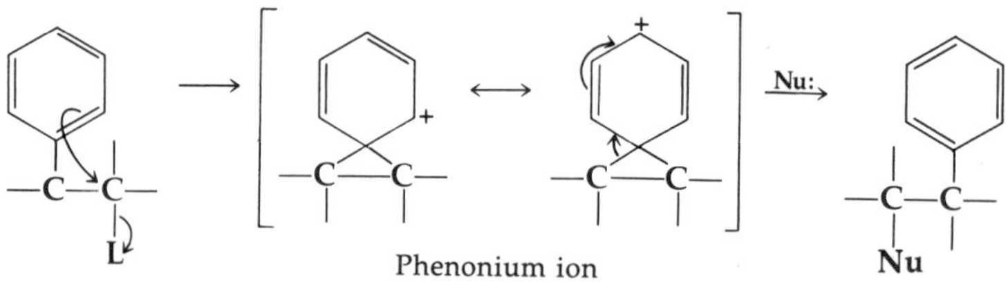

Phenonium ion

The plane of the aromatic ring is *perpendicular* to the three-carbon ring. The rearrangement of phenyl is therefore really an internal electrophilic substitution on the migrating aromatic ring, and the mechanism is quite analogous to that of normal electrophilic substitution (Chap. 16). This pathway is route II in Fig. 17-2, and migratory aptitudes lie in the order expected for electrophilic substitution:

p-$CH_3OC_6H_4$ > p-$CH_3C_6H_4$ > C_6H_5 > p-$O_2NC_6H_4$

Preferred migrating atom starred:

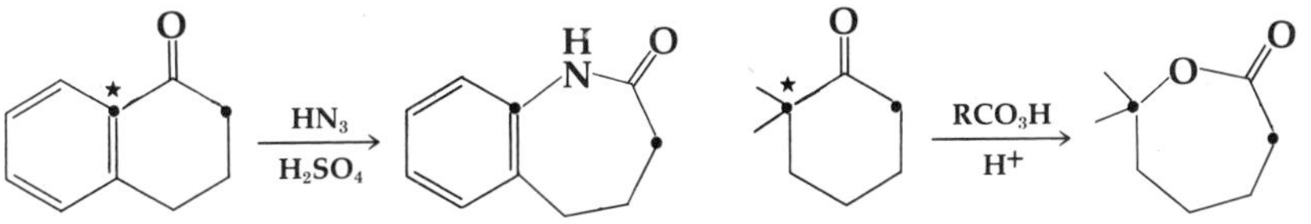

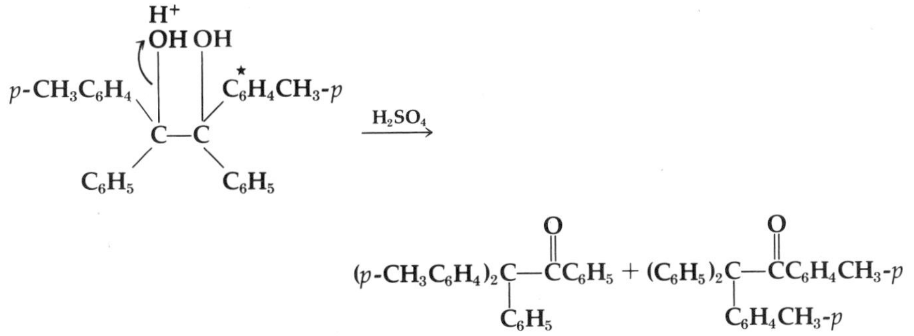

$$(p\text{-}CH_3C_6H_4)_2\underset{\underset{C_6H_5}{|}}{C}\overset{\overset{O}{\|}}{-}CC_6H_5 + (C_6H_5)_2\underset{\underset{C_6H_4CH_3\text{-}p}{|}}{C}\overset{\overset{O}{\|}}{-}CC_6H_4CH_3\text{-}p$$

    94%   6%

In unsymmetrical pinacols, the direction of rearrangement is largely determined by the relative ease of removal of hydroxyl groups from the two possible positions. The hydroxyl group is usually lost from the center that is more easily converted to a carbonium ion.

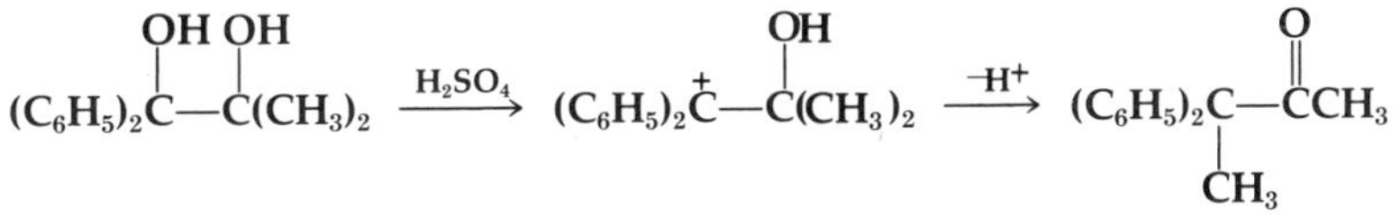

3,3-Diphenyl-2-butanone

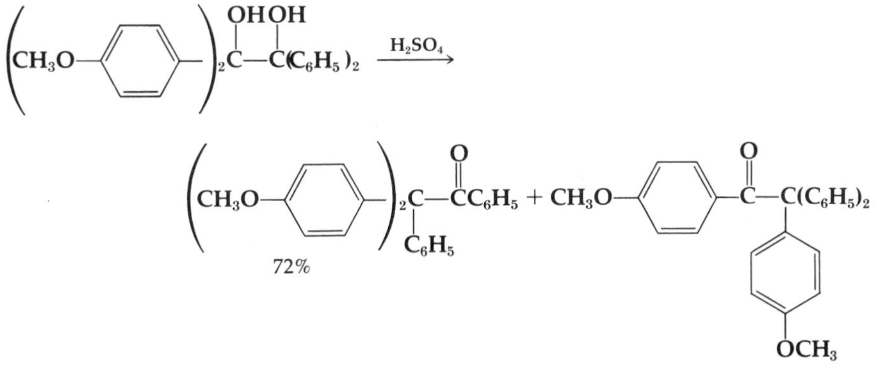

    72%   28%

Stereochemistry of Rearrangement

Consideration of the rearrangement process in terms of orbitals leads to the prediction that the configuration of the migrating group will be retained in the transition, while the two carbons it migrates across should be inverted. This is generally found to be the case as indicated in the examples shown.

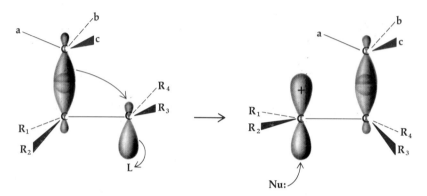

For the two carbons across which the group migrates, their inversion is contingent on some concerted character to the rearrangement, so that migration has begun to some extent before the leaving group is fully ionized. Otherwise, the intermediate carbonium ions can become symmetrical and substitute from both sides. Owing to hindrance the entering nucleophile usually approaches at the opposite side from the migrating group.

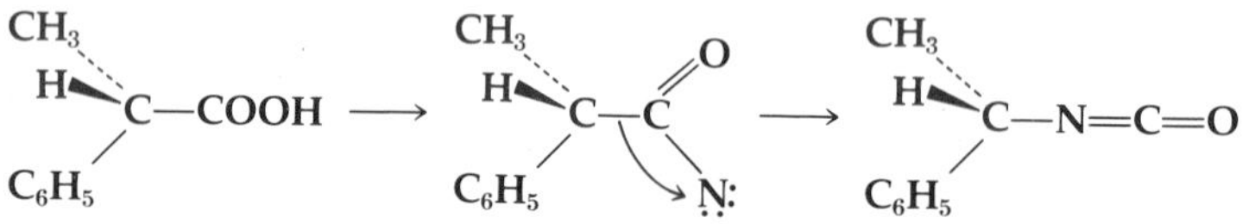

$$
\underset{\substack{\text{(+)-2-Phenylpropionic}\\\text{acid}}}{\overset{\displaystyle CH_3}{\underset{\displaystyle C_6H_5}{H\cdots\overset{|}{C}-COOH}}}
\longrightarrow
\overset{\displaystyle CH_3}{\underset{\displaystyle C_6H_5}{H\cdots\overset{|}{C}-\underset{\searrow\ N:}{C}\overset{\nearrow O}{}}}
\longrightarrow
\overset{\displaystyle CH_3}{\underset{\displaystyle C_6H_5}{H\cdots\overset{|}{C}-N=C=O}}
$$

HO— ... H $\xrightarrow{PCl_5}$ Cl_4P-O ... H $\xrightarrow{\text{Rearr.}}$

H   H

$\xrightarrow{-H^+}$

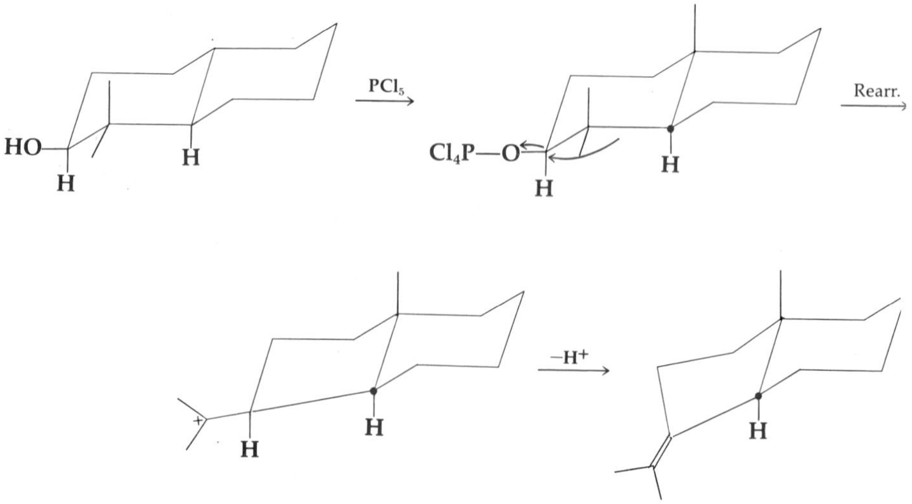

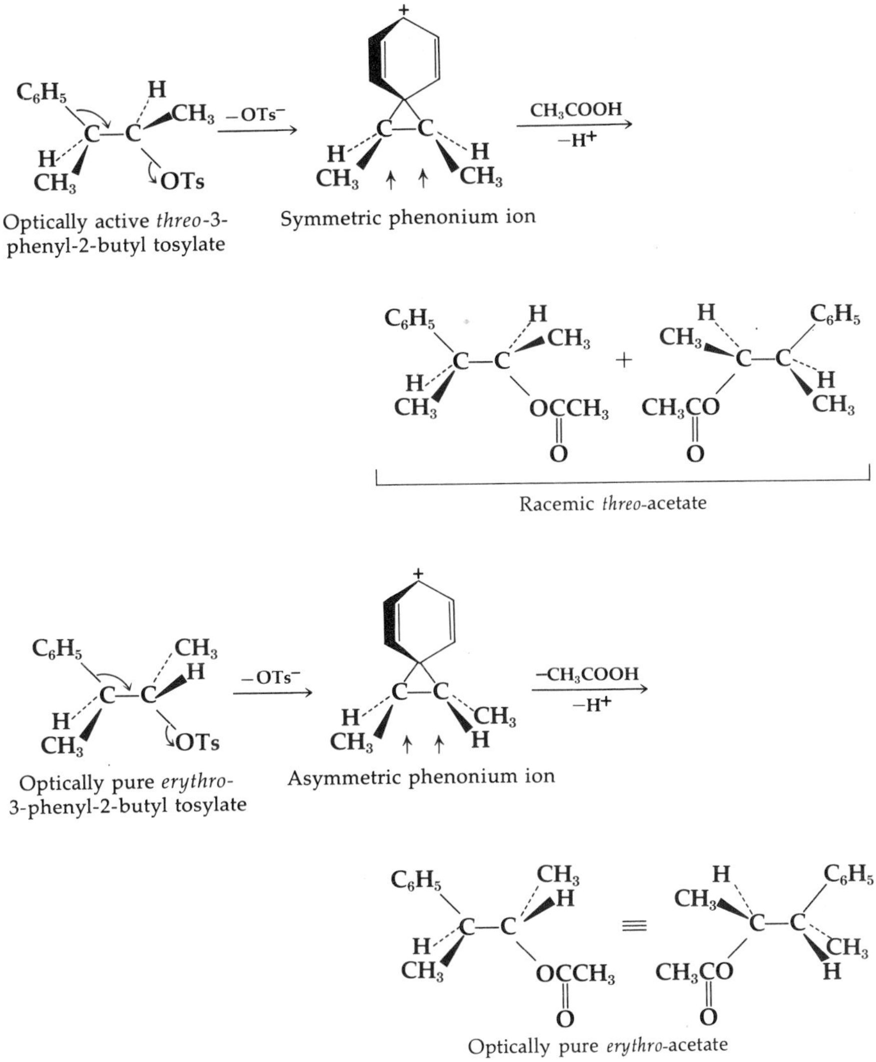
–OTs⁻
CH3COOH
–H+
Optically active threo-3-phenyl-2-butyl tosylate
Symmetric phenonium ion
Racemic threo-acetate
–OTs⁻
–CH3COOH
–H+
Optically pure erythro-3-phenyl-2-butyl tosylate
Asymmetric phenonium ion
Optically pure erythro-acetate

FIGURE 17-3 **Stereochemistry of rearrangement**

In the last example the migrating carbon is shown as a colored spot and retains its configuration. The migrating bond and the leaving group bond are shown in the trans-antiparallel (periplanar) orientation best suited for elimination (Chap. 14), which is also that required for concerted rearrangement since the orbital overlap is better. Another example, involving the phenonium ion, is shown in Fig. 17-3.

PROBLEM 17-6

Indicate the predominant product in each case, stating reasons.

a Reassess Prob. 17-3.

b Acetophenone + HN_3/H_2SO_4.

c p-Methoxybenzophenone + CF_3CO_3H.

d 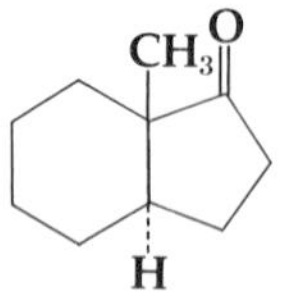 + CF_3CO_3H.

e 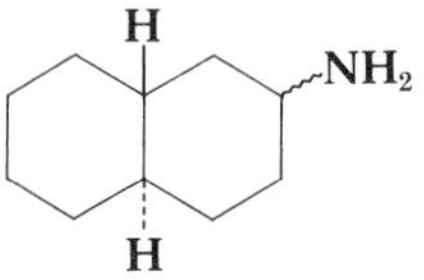 (both amino epimers) + HNO_2/H_2O.

f R-2-Methylbutanoyl azide + C_2H_5OH/Δ.

g Optically active cis-2,3-dimethylcyclobutanone with diazomethane yields an optically inactive product.

h Solvolysis of one optically active 3-phenyl-4-bromobutane in hot acetic acid yields an optically inactive ester.

17-3 ELECTRON-RICH (ANIONIC) SKELETAL REARRANGEMENTS

The rearrangements in the previous two sections are by far the most commonly observed, but they only occur under acid (or neutral) conditions since they require electron-deficient atoms. The rearrangements in this section require strong base. These are three-center reactions like the previous ones, but in detail they more closely resemble internal S_N2 displacement, or cyclization, reactions, as shown in Fig. 17-4. The transition state is electron-rich, with two more electrons than in the previous electron-deficient cases. They are usually initiated by those basic reagents which remove a group or an atom, such as hydrogen. The residual anion then stabilizes itself by rearrangement (Fig. 17-6). An acid-strengthening substituent, which stabilizes the first ionic center Z by conjugation or electrostatic effects, is usually required in order to start the process.

In the **Stevens rearrangement**, keto-quaternary ammonium or sulfonium salts rearrange to amino ketones under the influence of strong base.

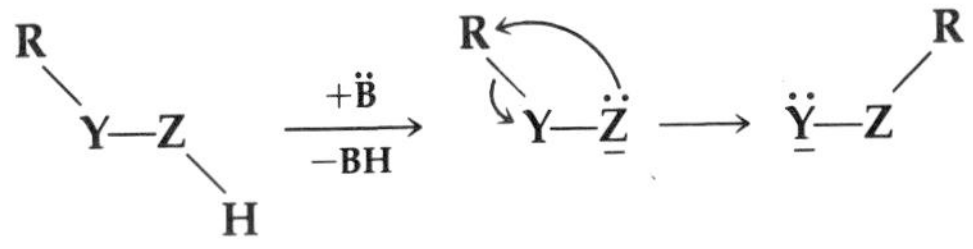

FIGURE 17-4 **Rearrangement of an electron-rich system**

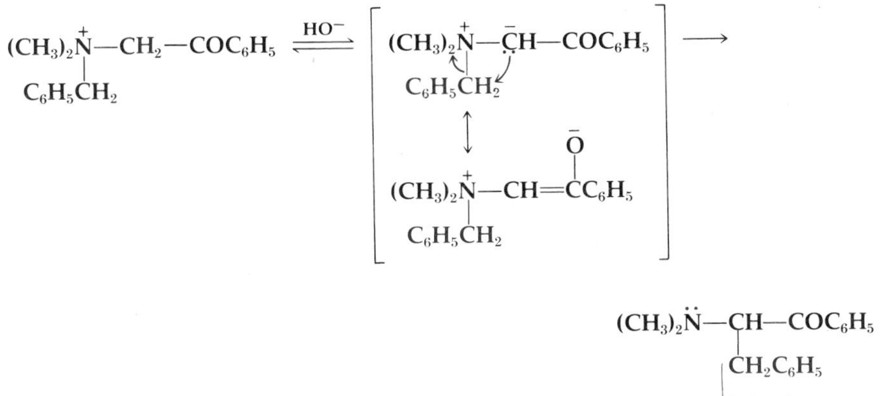

α-Dimethylamino-α-
benzylacetophenone

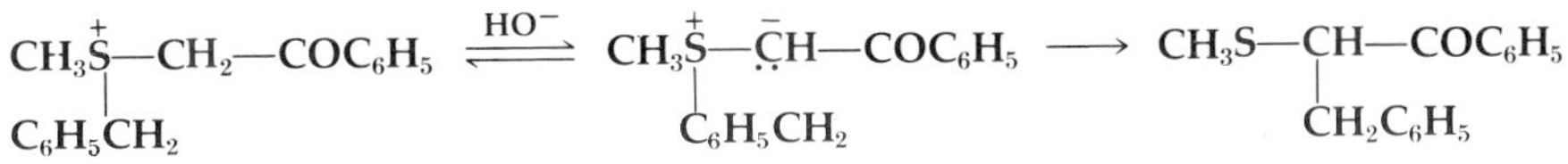

Benzylmethylphenacyl-
sulfonium ion

α-Methylmercapto-
α-benzylacetophenone

Proton removal in the first step of a Stevens rearrangement is facilitated both by the positive charge in the cationic substrates and by virtue of the delocalization energy of the enolate ions. Migrating groups are usually either benzyl or allyl systems.

The **Wittig rearrangement** follows a similar path. Since the substrates are much less acidic than those encountered in the Stevens transformation, powerful basic reagents are required to cause the Wittig reaction.

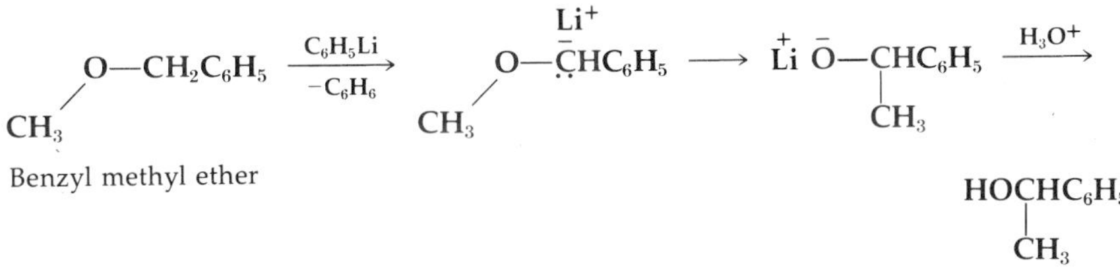

Benzyl methyl ether

1-Phenylethanol

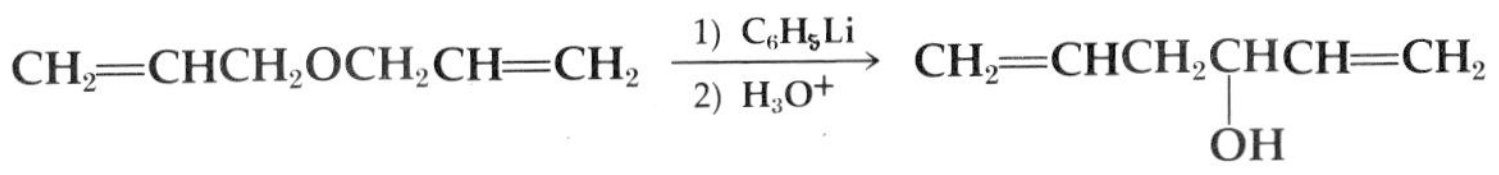

Diallyl ether

1,5-Hexadien-3-ol

Sommelet rearrangements involve the nucleophilic alkylation of the aromatic ring of a benzyltrimethylammonium ion. Protons are first removed from the more acidic benzyl position and subsequently from a methyl group.

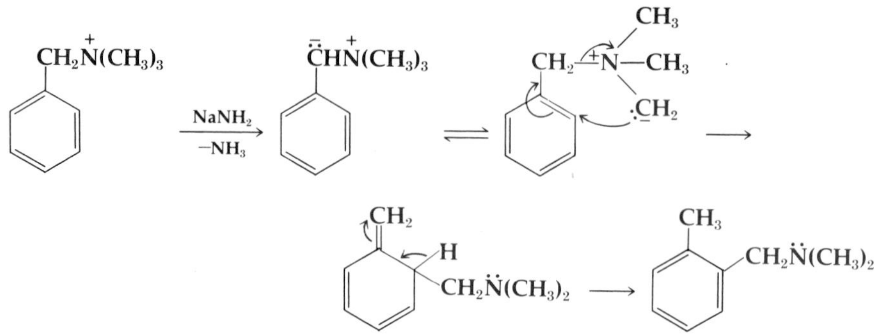

The rearrangement occurs only where none of the alkyl groups of the quaternary ammonium ion carries the β-hydrogen required for an $E2$ reaction (the Hofmann elimination, page 584):

$$C_6H_5CH_2\overset{+}{N}(CH_2CH_2CH_3) \xrightarrow[-NH_3]{NaNH_2} C_6H_5CH_2N(CH_2CH_2CH_3)_2 + CH_3CH{=}CH_2$$

Benzyl tri-*n*-propyl-
ammonium ion   ($E2$)

In the **Favorski rearrangement**, an α-haloketone rearranges by way of intermediate formation of a cyclopropanone. In cyclic systems, the process leads to overall ring contraction.

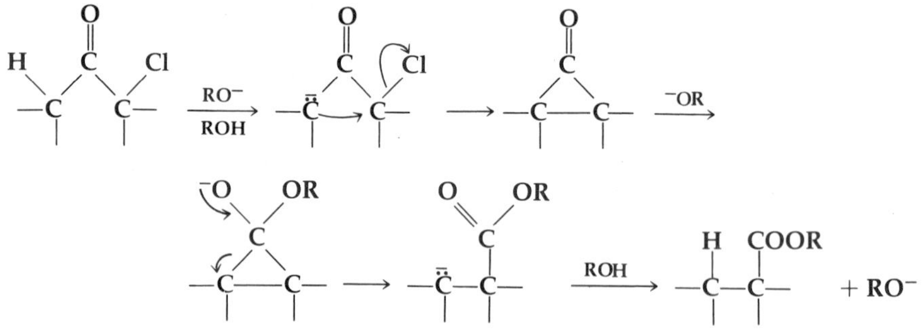

$$C_2H_5\bar{O}\ \overset{+}{N}a + (CH_3)\ CCHCCH_3 \longrightarrow (CH_3)_3CCH_2CH_2COOC_2H_5 + NaBr$$

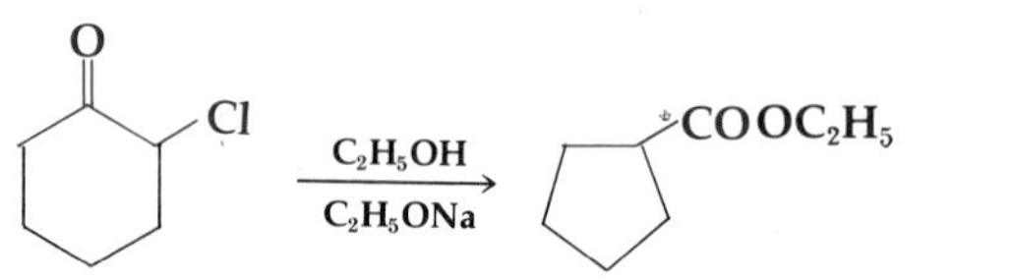

2-Chlorocyclohexanone   Ethyl cyclopentanecarboxylate

The **benzilic acid rearrangement** is the only base-catalyzed rearrangement which closely resembles in mechanism the acid-catalyzed reactions of the last sections. This rearrangement converts certain α-diketones to α-hydroxy-acids on treatment with alkali.

$$C_6H_5\overset{O}{\overset{\|}{C}}\overset{O}{\overset{\|}{C}}C_6H_5 \xrightarrow[\text{Reflux}]{\text{KOH, ethanol}} (C_6H_5)_2\overset{OH}{\overset{|}{C}}COOH$$

Benzil   Benzilic acid

Driving force for the reaction is provided by addition of hydroxide ions to one of the carbonyl groups and by the ultimate formation of a stable carboxylate salt from a strong base.

$$Ar\overset{O}{\overset{\|}{C}}\overset{O}{\overset{\|}{C}}Ar + O\bar{H} \rightleftharpoons Ar\overset{O}{\overset{\|}{\underset{\underset{Ar}{|}}{C}}}\overset{\bar{O}}{\overset{|}{C}}-OH \longrightarrow Ar-\overset{\bar{O}}{\underset{\underset{Ar}{|}}{C}}-\overset{O}{\overset{\|}{C}}OH \longrightarrow Ar_2C\bar{CO_2}$$

Application of the benzilic acid rearrangement has been limited almost exclusively to aromatic diketones. Several examples of the rearrangement of aliphatic diketones are known, but condensations involving α-hydrogens usually compete with rearrangement and keep the yields low.

9,10-Phenanthrenequinone   9-Hydroxy-9-fluorenecarboxylic acid

$$HOOCCH_2\overset{O}{\overset{\|}{C}}\overset{O}{\overset{\|}{C}}CH_2COOH \xrightarrow[\text{2) H}^+]{\text{1) KOH, H}_2\text{O, reflux}} HOOCCH_2\overset{OH}{\overset{|}{\underset{\underset{COOH}{|}}{C}}}CH_2COOH$$

Ketopinic acid   Citric acid

PROBLEM 17-7

Fill in the missing compounds

a Allyldimethylamine + phenacyl bromide ($C_6H_5COCH_2Br$) $\longrightarrow$ A $\xrightarrow{\text{NaOH}}$

B $\xrightarrow{C_6H_5CH_2Br}$ C $\xrightarrow[\Delta]{\text{NaOH}}$ amine D + E ($C_{11}H_{10}O$; λ_{max} 245 nm)

b $C_6H_5COCOC_2H_5$ $\xrightarrow{\text{NaOH}}$ A $\xrightarrow[-CO_2]{\Delta}$ B $\xrightarrow[CH_3COCH_3]{(t\text{-BuO})_3Al}$ C

c Methyl α-methylacetoacetate $\xrightarrow[\text{H}_2\text{O}]{2\text{Br}_2}$ [α,γ-dibromo derivative] $\xrightarrow[\text{H}_2\text{O}]{\text{K}_2\text{CO}_3}$

[intermediates] $\longrightarrow$ α-methylfumaric acid

17-4 MIGRATION OF DOUBLE AND TRIPLE BONDS

A double bond can be caused to migrate by the simple expedient of protonating it and deprotonating from another carbon.

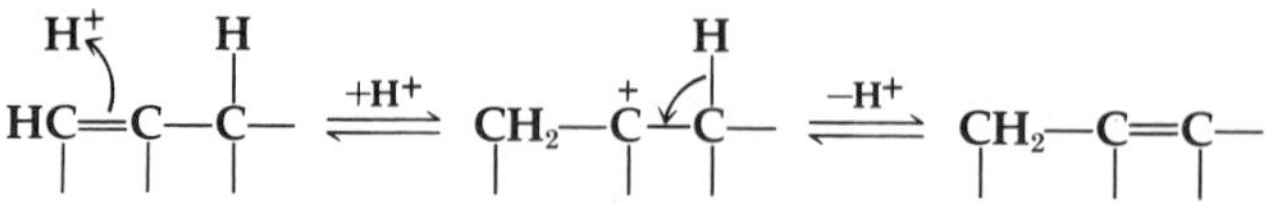

In strong acids such equilibria are established and double-bond migrations within a carbon skeleton are observed. These migrations tend, of course, to yield the most stable double-bond isomer, i.e., one in which the double bond is conjugated or most heavily substituted. However, competition from addition, polymerization, and skeletal rearrangements limits the use of the method. Equilibration of alkene isomers often occurs during acid-catalyzed dehydration of alcohols.

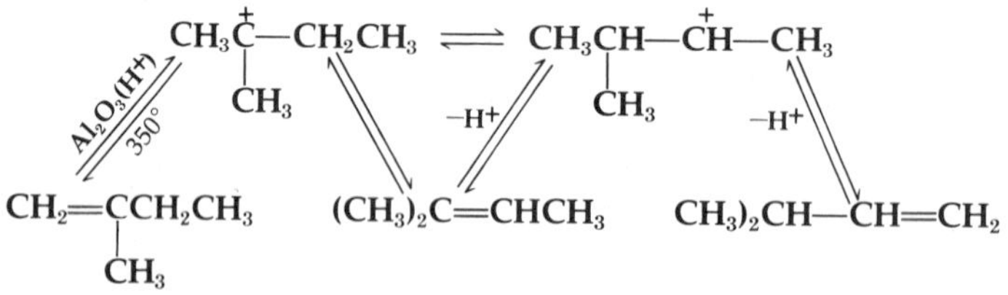

2-Methyl-1-butene
27%†   69%†   4%†

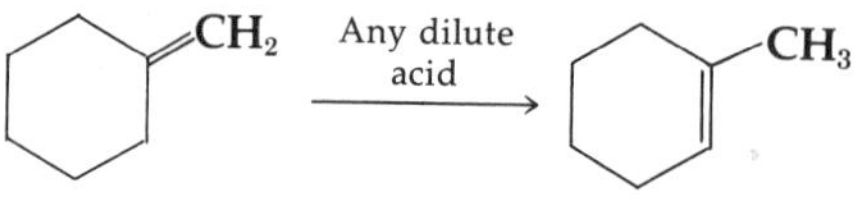

Methylenecyclohexane   1-Methylcyclohexene

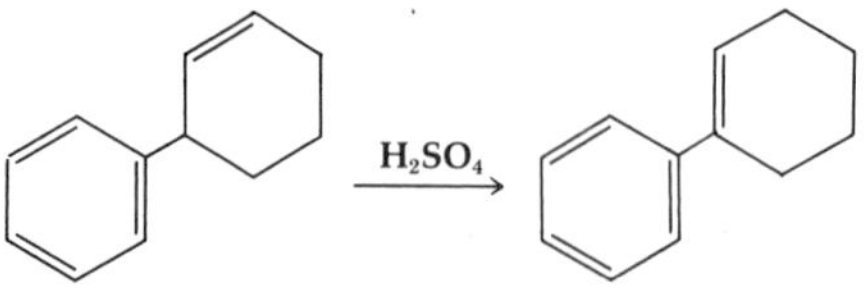

In the presence of strong bases, double bonds will also migrate within carbon skeletons by the removal and readdition of protons. The rigor of the reaction conditions required to cause the change depends upon the effects of other substituents on the acidity of the hydrogen which must be first removed. Again the double bond tends to move to a more stable position. The most common base-catalyzed migration is that of a double bond β,γ to a carbonyl, which readily moves into conjugation (α,β).

† Equilibrated mixture.

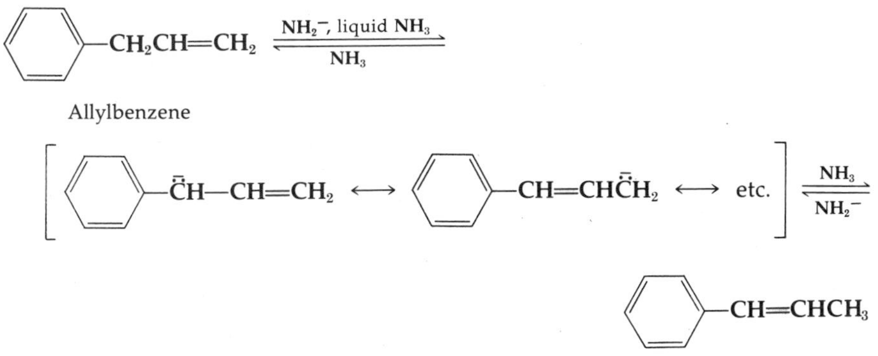

Allylbenzene

1-Phenylpropene
(propenylbenzene

$$CH_2\text{=}CHCH_2COOH \xrightarrow{\text{Dil. NaOH, reflux}} CH_3CH\text{=}CHCOOH$$

3-Butenoic acid   2-Butenoic acid
(vinylacetic acid)   (crotonic acid)

Allylic Rearrangements

Substitution reactions of allylic leaving groups can occur so as to involve migration of the double bond to the leaving group. The net effect is that the double bond and the adjacent substituent change places on the carbon skeleton. Mechanistically, such rearrangements may be regarded as *vinylogous substitutions*, as outlined in Fig. 17-5. Such allylic rearrangements are thus parallel to conjugate addition reactions (see 12-6). When they are bimolecular, they are called S_N2' **reactions.**

Rearrangement by the S_N2' mechanism:

α-Methallyl chloride
(methylvinylcarbinyl
chloride)

Diethylcrotylamine

Crotyl chloride

As illustrated above, S_N2' reactions do not occur regularly but must compete with the normal, and faster, S_N2 processes. Steric hindrance to substitution without rearrangement may channel a bimolecular substitution into the

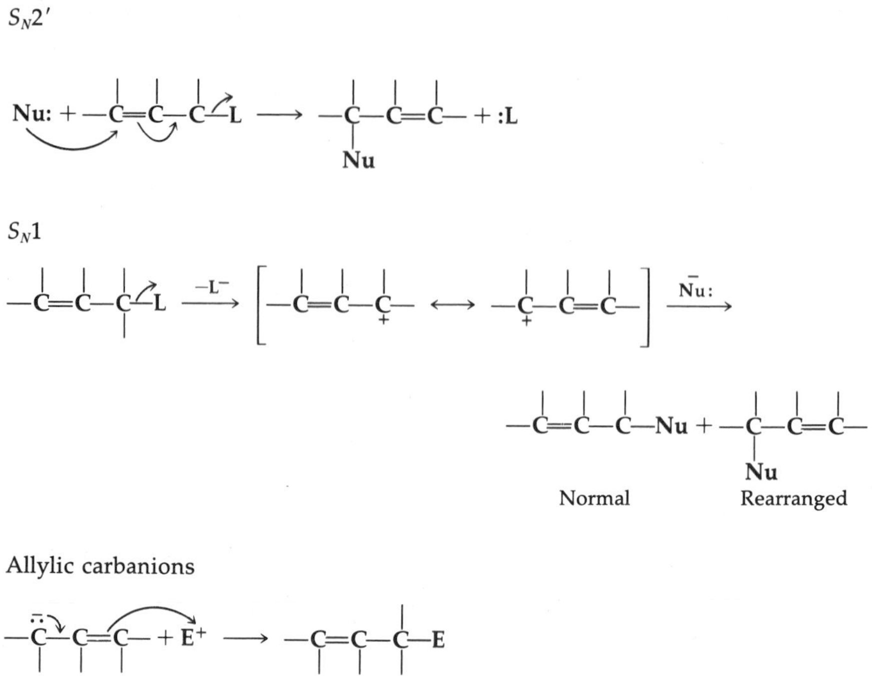

S_N2'

$$\text{Nu:} + -\overset{|}{C}=\overset{|}{C}-\overset{|}{C}-L \longrightarrow -\overset{|}{C}-\overset{|}{C}=\overset{|}{C}- \ + \ :L$$
$$\overset{|}{\text{Nu}}$$

S_N1

$$-\overset{|}{C}=\overset{|}{C}-\overset{|}{C}-L \xrightarrow{-L^-} \left[-\overset{|}{C}=\overset{|}{C}-\overset{+}{\underset{|}{C}} \longleftrightarrow -\overset{+}{\underset{|}{C}}-\overset{|}{C}=\overset{|}{C}- \right] \xrightarrow{\overset{..}{\text{Nu}}:}$$

$$-\overset{|}{C}=\overset{|}{C}-\overset{|}{C}-\text{Nu} \ + \ -\overset{|}{C}-\overset{|}{C}=\overset{|}{C}-$$
$$\overset{|}{\text{Nu}}$$
$$\text{Normal} \qquad\qquad \text{Rearranged}$$

Allylic carbanions

$$-\overset{..}{\underset{|}{C}}-\overset{|}{C}=\overset{|}{C}- \ + \ \overset{+}{E} \longrightarrow -\overset{|}{C}=\overset{|}{C}-\overset{|}{C}-E$$

FIGURE 17-5 **Mechanistic possibilities for allylic rearrangements in substitution**

"abnormal" S_N2' path. The stereochemistry of the S_N2' reaction has been studied in cyclic systems. The results indicate that the leaving group and nucleophile are on the same side of the molecule in the transition state.

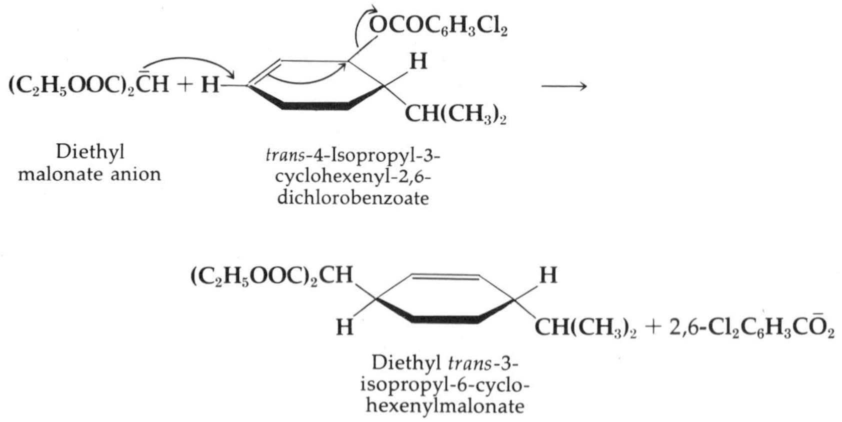

$$(C_2H_5OOC)_2\overset{-}{C}H + H-$$

Diethyl
malonate anion

trans-4-Isopropyl-3-
cyclohexenyl-2,6-
dichlorobenzoate

$$(C_2H_5OOC)_2CH \qquad\qquad H$$
$$H \qquad\qquad CH(CH_3)_2 + 2,6\text{-}Cl_2C_6H_3C\overset{-}{O}_2$$

Diethyl *trans*-3-
isopropyl-6-cyclo-
hexenylmalonate

The approach used in the study shown above is instructive. A compound of known configuration, *trans*-4-isopropyl-3-cyclohexenol, was available as a starting material. For the preparation of a suitable substrate for the substitution reaction, it was necessary to convert the hydroxyl group to some group that could be displaced as a negative ion in nucleophilic substitution reactions. Cyclohexenyl tosylates are so reactive that they are unstable, and they would be quite likely to undergo substitution by the S_N1 mechanism under most reaction conditions. Carboxylate esters are usually not suitable leaving groups in nucleophilic substitutions, because of the preferential attack of nucleophiles at the carbonyl group of the ester function (page 504). The problem was solved by use of the 2,6-dichlorobenzoate in which the carbonyl group is protected by steric hindrance.

The substitutions above are generally those of allylic carbonium ions. The same delocalization occurs with allylic carbanions and similarly can lead to two products in their use as nucleophiles. Reactions with allylic organo-metallic compounds usually give mixtures of products. Furthermore, when organometallic reagents are made from isomeric allylic halides, the same metal derivative is obtained from both isomers.

$$CH_3CH\!=\!CHCH_2Cl \xrightarrow{\ Mg\ }$$

$$CH_3\underset{\underset{Cl}{|}}{C}HCH\!=\!CH_2 \xrightarrow{\ Mg\ }$$

$$\left[\begin{array}{c} CH_3CH\!=\!CH\bar{C}H_2\overset{+}{M}gCl \\[4pt] \updownarrow \\[4pt] CH_3\bar{C}HCH\!=\!CH_2MgCl \end{array} \right] \xrightarrow{\ H^+,\ H_2O\ }$$

Butenyl Grignard
reagent

$$CH_3CH\!=\!CHCH_3 + CH_3CH_2CH\!=\!CH_2$$

43%   57%

Although the reaction of Grignard reagents with *highly hindered ketones* usually takes the trivial course of a proton removal (enolization) rather than an addition, allylic Grignard reagents give addition products in good yield. The results suggest that allylic reagents add by way of a cyclic process.

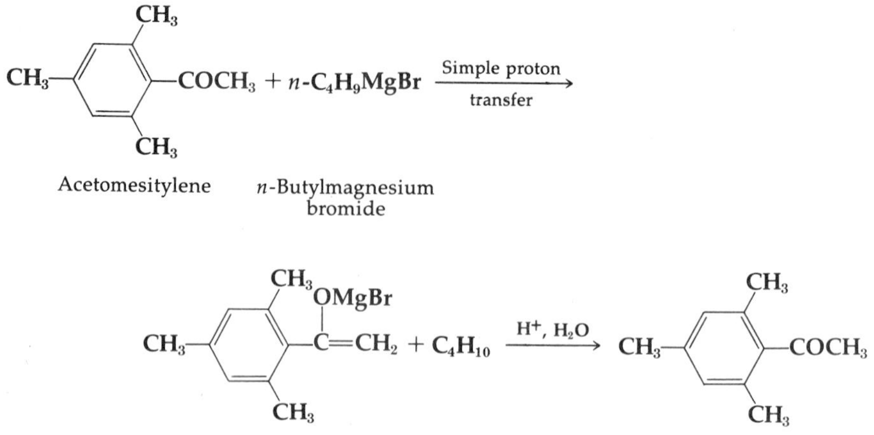

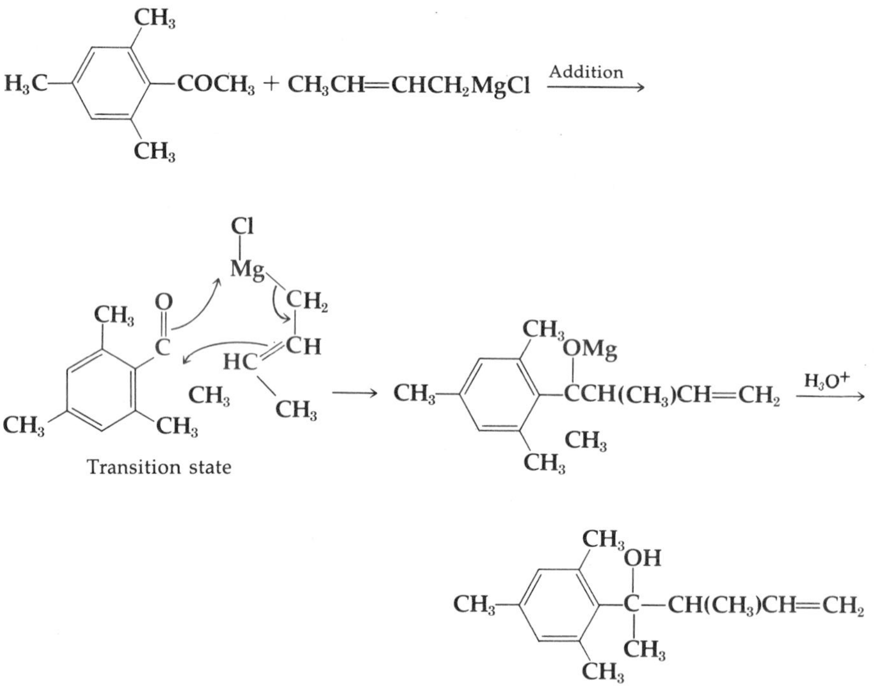

Acetomesitylene n-Butylmagnesium bromide

Transition state

2-Mesityl-3-methylpent-4-en-2-ol

Acetylenic rearrangements are exemplified in the transformations of acetylenic alcohols, which are readily available from acetylene-anion additions to ketones (page 463). Acid-catalyzed rearrangement yields α,β-unsaturated ketones.

$$CH_3CH_2\overset{\overset{\displaystyle O}{\|}}{C}CH_3 + CH\equiv CMgBr \xrightarrow[\text{2) }H^+,\ H_2O]{\text{1) Addition}} CH_3CH_2\overset{\overset{\displaystyle HO}{|}}{\underset{\underset{\displaystyle CH_3}{|}}{C}}C\equiv CH \xrightarrow{HCOOH,\ H^+}$$

$$CH_3CH_2C\overset{\overset{\displaystyle \overset{\displaystyle H}{|}}{O^+}}{\underset{\underset{\displaystyle CH_3}{|}}{\diagup\ \diagdown}}C=CH_2 \longrightarrow CH_3CH_2\overset{+}{\underset{\underset{\displaystyle CH_3}{|}}{C}}\overset{\overset{\displaystyle OH}{|}}{C}=CH_2 \xrightarrow{-H^+}$$

$$CH_3CH=\overset{\overset{\displaystyle OH}{|}}{\underset{\underset{\displaystyle CH_3}{|}}{C}}C=CH_2 \longrightarrow CH_3CH=\overset{}{\underset{\underset{\displaystyle CH_3}{|}}{C}}COCH_3$$

17-5   MIGRATION OF FUNCTIONAL GROUPS

The net effect of migration of a heteroatom from one site to another usually comes about through an initial internal S_N2 displacement, or cyclization, by the unshared electron pair on the heteroatom. This forms an intermediate ring which is subsequently cleaved at the initial site of heteroatom attachment. When this intermediate ring is three-membered, the behavior entirely parallels the chemistry of epoxides (page 417) and bromonium ions (page 623). These principles are illustrated by the following examples.

$$(CH_3)_2\overset{}{\underset{\underset{\displaystyle CH_3\overset{\cdot\cdot}{O}:}{|}}{C}}-\overset{\overset{\displaystyle Br}{|}}{C}HCH_3 \xrightarrow[-AgBr]{Ag^+} (CH_3)_2C-CHCH_3 \xrightarrow[-H^+]{H_2O} (CH_3)_2\overset{}{\underset{\underset{\displaystyle OH}{}}{C}}-\overset{}{\underset{\underset{\displaystyle OCH_3}{}}{C}}HCH_3$$

with center: $\overset{\overset{\displaystyle \diagup\diagdown}{O^+}}{\underset{\underset{\displaystyle CH_3}{|}}{}}$

An ethylene-
oxonium ion

$$CH_3\overset{}{\underset{\underset{\displaystyle C_2H_5S}{|}}{C}}HCH_2OH \underset{}{\overset{HCl}{\rightleftharpoons}} CH_3\overset{}{\underset{\underset{\displaystyle C_2H_5\overset{\cdot\cdot}{S}:}{|}}{C}}H-CH_2^{\overset{\displaystyle \overset{+}{O}H_2}{|}} \xrightarrow{-H_2O} CH_3CH-CH_2 \xrightarrow{Cl^-}$$

with $\underset{\underset{\displaystyle C_2H_5}{|}}{S^+}$

2-[Ethylmercapto]-
1-propanol

An ethylene-
sulfonium ion

$$CH_3\overset{}{\underset{\underset{\displaystyle Cl}{|}}{C}}HCH_2SC_2H_5$$

2-Chloro-1-propyl
ethyl sulfide

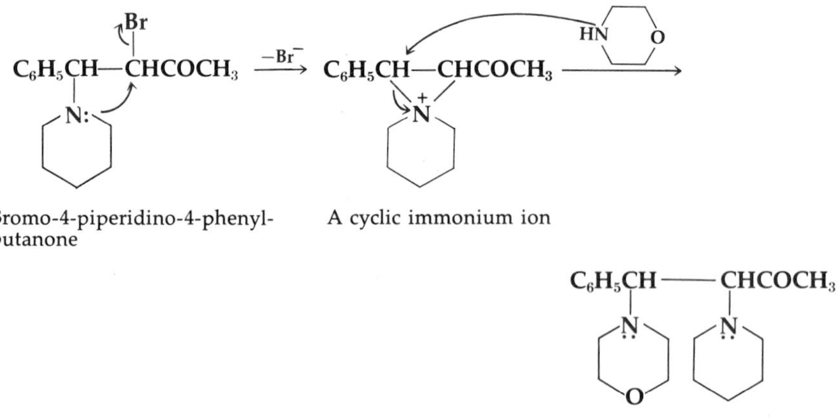

3-Bromo-4-piperidino-4-phenyl-
2-butanone   A cyclic immonium ion

3-Piperidino-4-morpholino-
4-phenyl-2-butanone

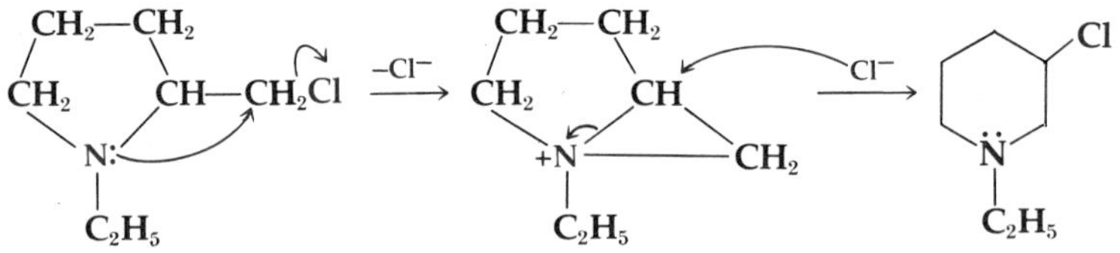

2-[Chloromethyl]-1-ethyl-
pyrrolidine   3-Chloro-1-ethylpiperidine

Acyl and Related Migrating Groups

Rearrangements by way of five-membered cyclic ions are exemplified by the migration of acetoxyl groups in solvolytic reactions (see Fig. 17-6).

The resultant five-membered rings may be opened, with inversion of configuration, at the carbon atom that can best support a positive charge. They are also opened by an entirely different mechanism, which involves attack at the carbonyl carbon. Hence the two mechanisms allow either a cis or trans product depending on whether water is present ($\longrightarrow$ cis) or absent ($\longrightarrow$ trans). The same reaction is to be found on page 630 for hydroxylation of alkenes by silver acetate and halogens.

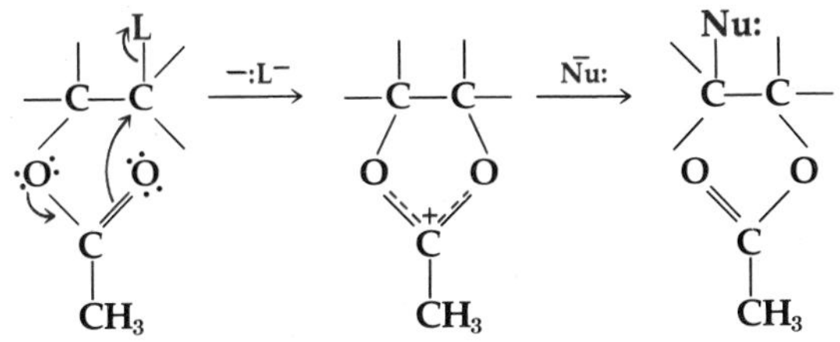
L
−:L⁻
Nu:
Nu:
O
O
C
CH₃

FIGURE 17-6 **Acetoxyl migration**

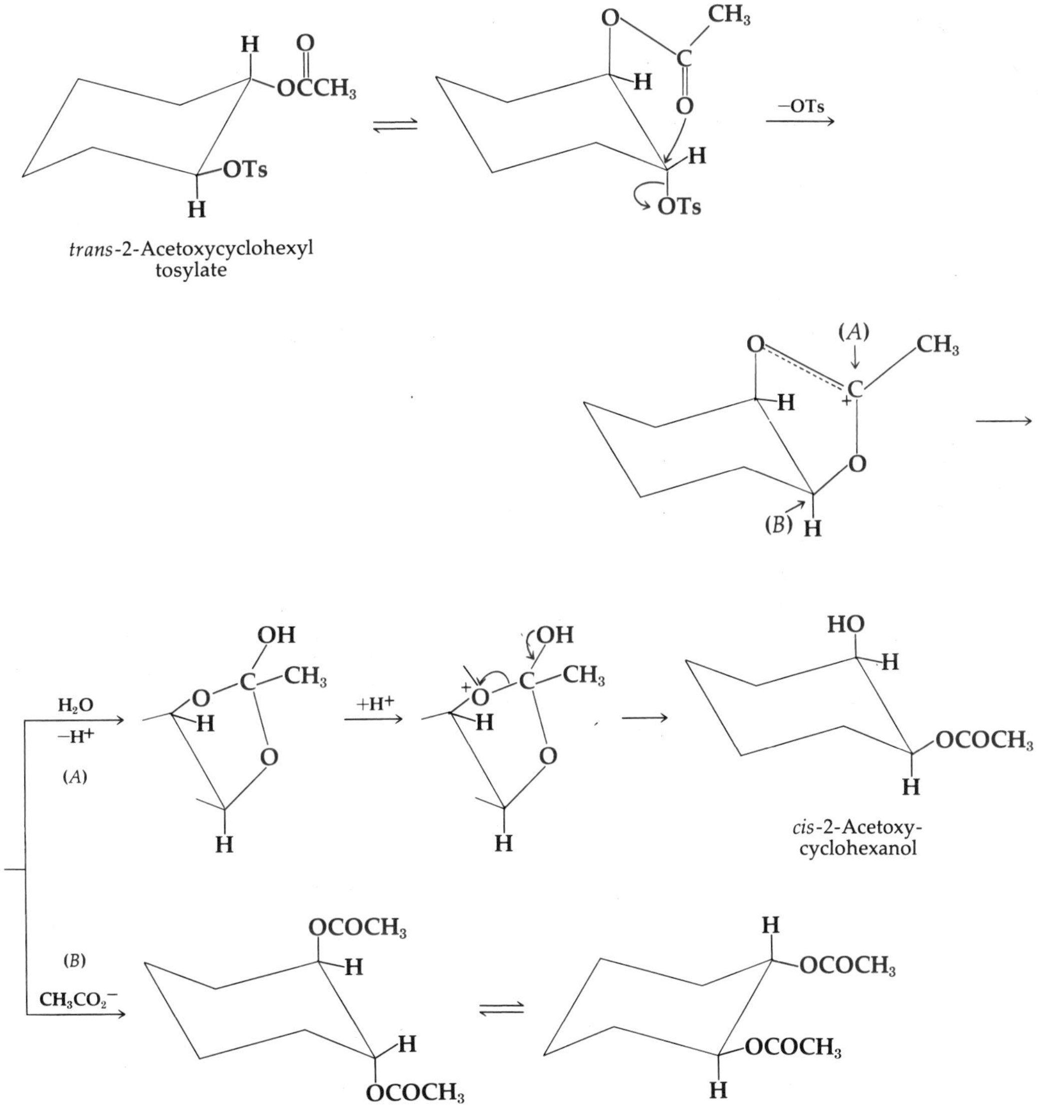
OCCH₃
OTs
H
trans-2-Acetoxycyclohexyl tosylate
−OTs
(A)
(B)
CH₃
H₂O
−H⁺
+H⁺
OH
HO
OCOCH₃
cis-2-Acetoxy-cyclohexanol
CH₃CO₂⁻
trans-1,2-Diacetoxycyclohexane

In a second family of rearrangements, acyl groups migrate from one heteroatom to another. The benzoyl substituent of the hormone ephedrine migrates from nitrogen to oxygen, or in the reverse direction, depending on the acidity of the medium in which the compound is dissolved. In strong acid, the equilibrium is shifted toward the amine salt, whereas in basic media, the amide is favored.

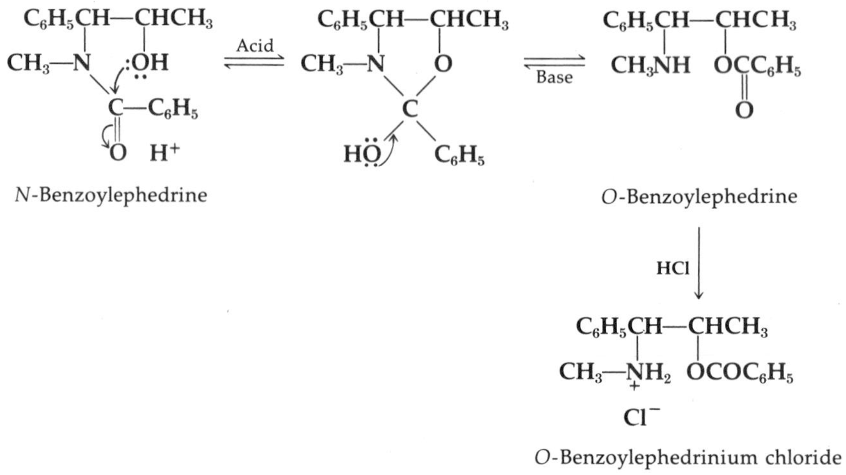

N-Benzoylephedrine

O-Benzoylephedrine

O-Benzoylephedrinium chloride

PROBLEM 17-8

Considering projection formulas of the two diastereomeric N-benzoylephedrines, explain why one has a faster acyl migration rate than the other.

If the geometry of a system is favorable, 1,3-acyl migrations occur. In these cases the cyclic intermediate is a six-membered ring and the rearrangement only occurs if the ring can take up a stable chair conformation. Thus the first stereoisomer shown below (A) undergoes acyl migration whereas its diastereomer (B) does not, since in this case the cyclic intermediate must bear an axial phenyl group.

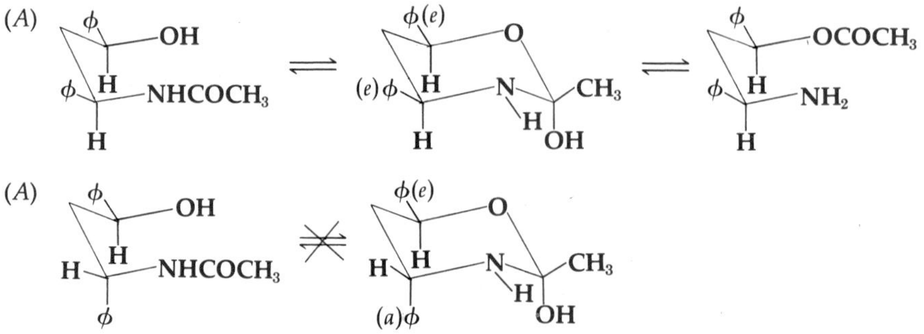

The rearrangements have been carefully studied with the acetyl derivatives of the alkaloid degradation products nortropine and ψ-nortropine. The results have been used to establish the stereochemistry of the parent alkaloids.

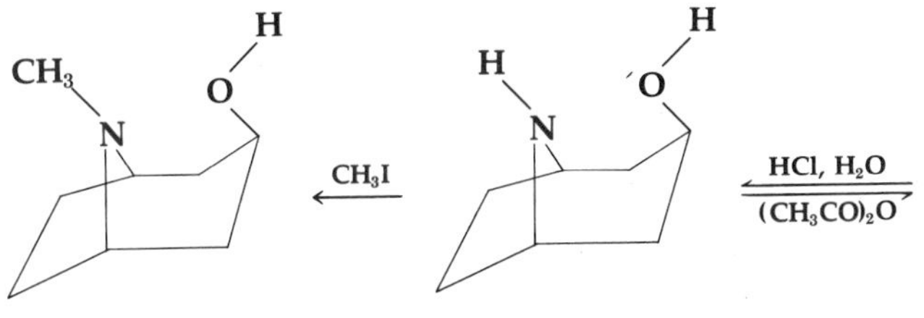
CH₃
N
O
H
CH₃I
H
N
O
H
HCl, H₂O
(CH₃CO)₂O

ψ-Tropine ψ-Nortropine

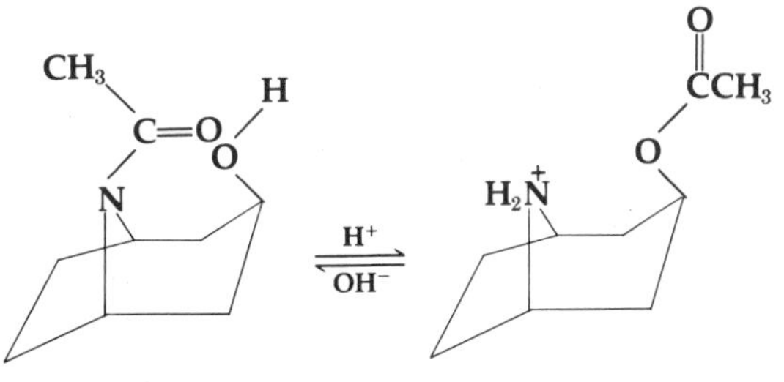
CH₃
C=O
N
O
H
H⁺
OH⁻
H₂N
O
CCH₃

N-Acetyl-ψ-nortropine *O*-Acetyl-ψ-nortropinium cation

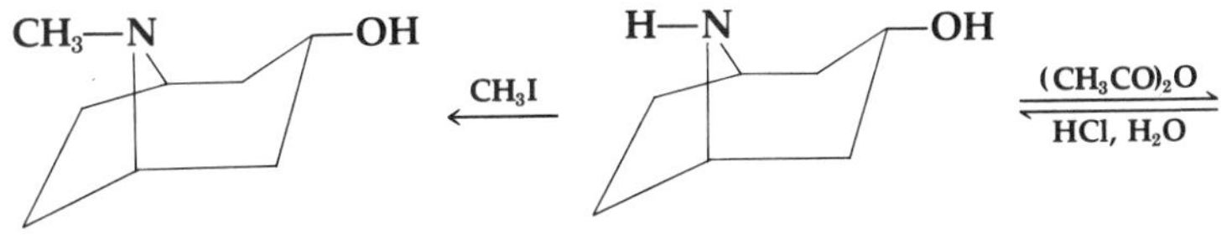
CH₃—N
OH
CH₃I
H—N
OH
(CH₃CO)₂O
HCl, H₂O

Tropine Nortropine

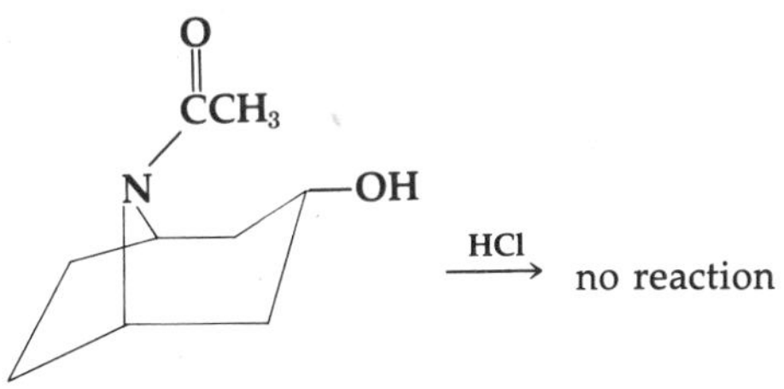
O
CCH₃
N
OH
HCl
no reaction

N-Acetylnortropine

17-6 REARRANGEMENTS ON AN AROMATIC RING

In a large number of reactions a group A migrates from a substituted heteroatom Z and becomes directly attached to an ortho or para position of an aromatic nucleus.

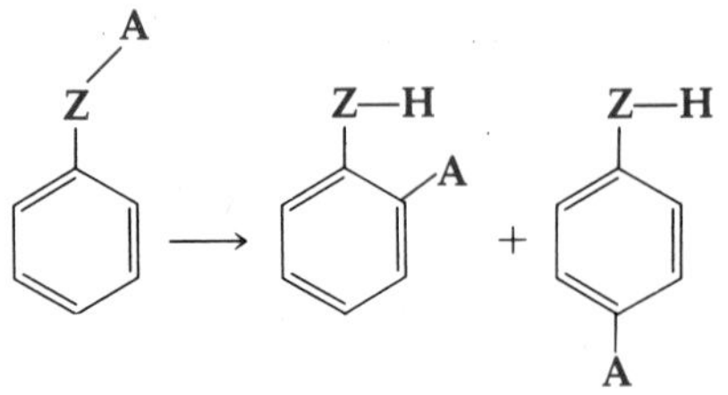

The reactions are nearly all *acid-catalyzed*; hence, attack of **A** on the nucleus has the character of an electrophilic substitution reaction. The reactions of phenolic derivatives include the Fries rearrangement, which is essentially an internal Friedel-Crafts acylation (Sec. 16-6), and the Claisen rearrangement as an example of a pericyclic reaction (Sec. 21-3).

Fries rearrangement:

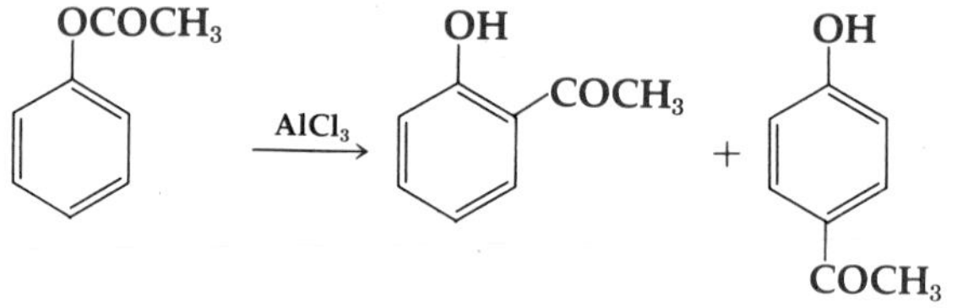

Claisen rearrangement:

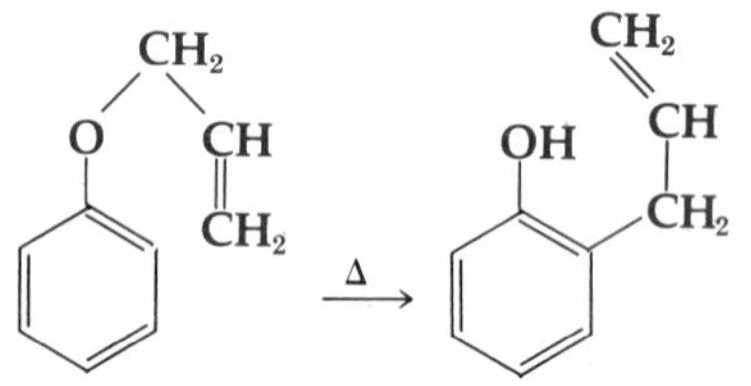

Rearrangements of Derivatives of Aniline

N-Haloacetanilides rearrange to *o*- and *p*-haloanilides when treated with mineral acid. In this reaction, halogen becomes detached from the molecule and then reenters in an ordinary electrophilic substitution reaction, acting as electrophile, X^+.

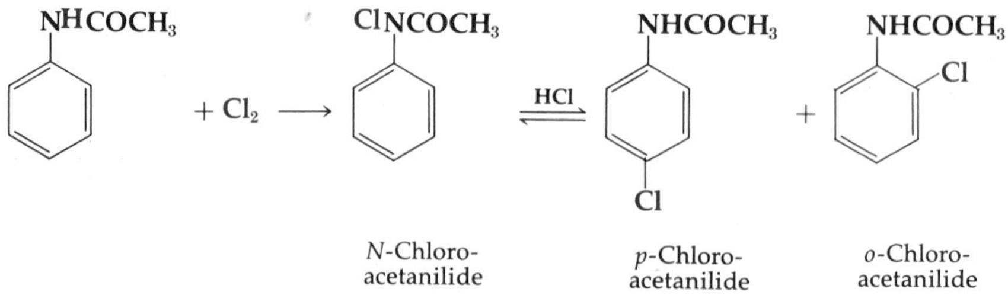

| | N-Chloro-acetanilide | p-Chloro-acetanilide | o-Chloro-acetanilide |

The *N*-haloamide rearrangement is one of a group of formally similar rearrangements. Several of the group are believed to involve *intermolecular* mechanisms, but in three cases (nitramine, *N*-sulfonic acid, and benzidine rearrangements), all available evidence points to *intramolecular* mechanisms in which the migrating group never becomes completely detached from the substrate.

Intermolecular

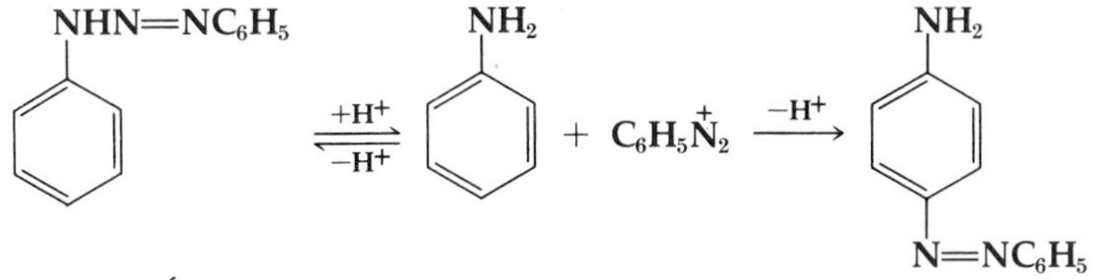

Diazoaminobenzene   *p*-Aminoazobenzene

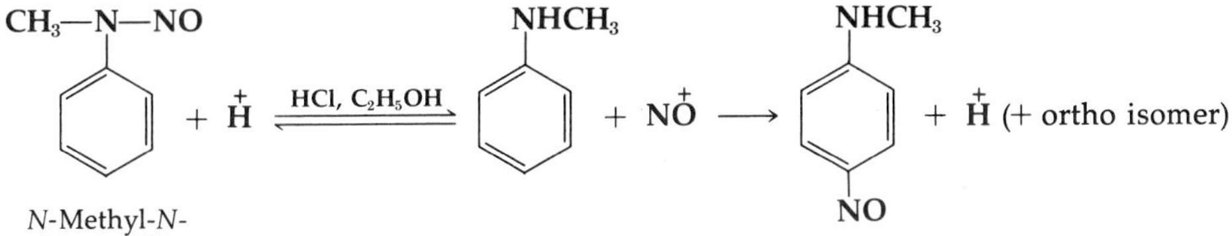

N-Methyl-*N*-
nitrosoaniline

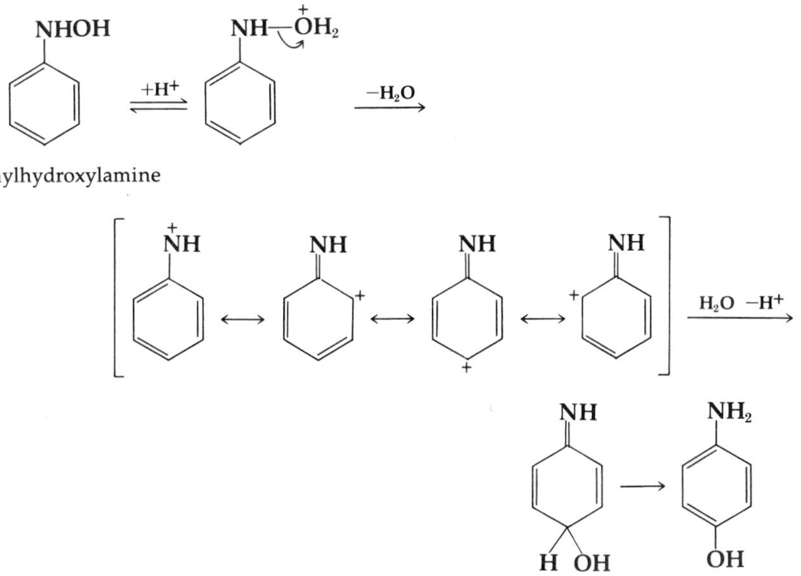

N-Phenylhydroxylamine

p-Aminophenol

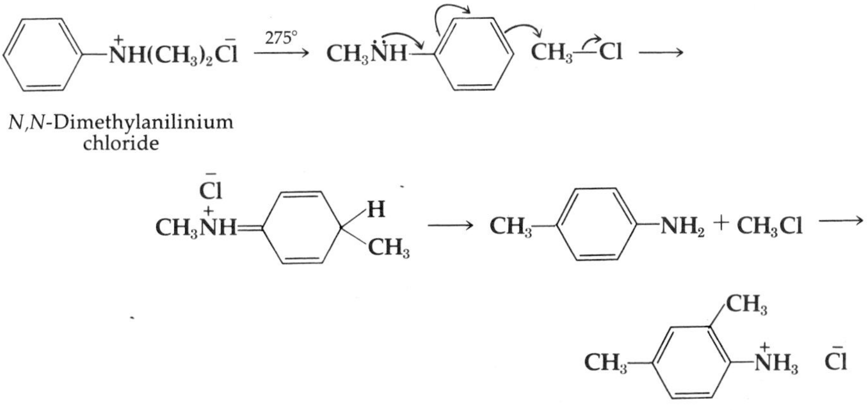

N,N-Dimethylanilinium
chloride

2,4-Dimethylanilinium chloride

Intramolecular:

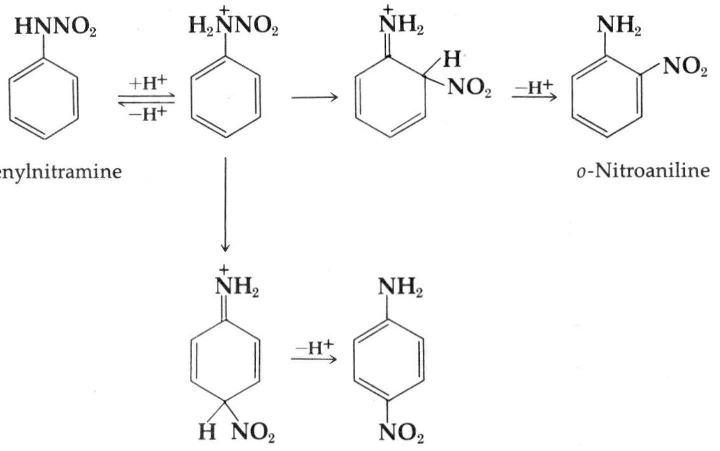

Phenylnitramine   *o*-Nitroaniline

p-Nitroaniline

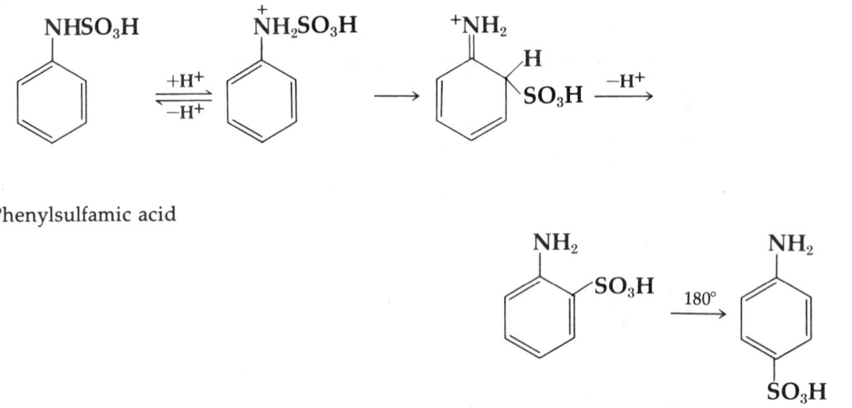

Phenylsulfamic acid

Orthanilic acid   Sulfanilic acid

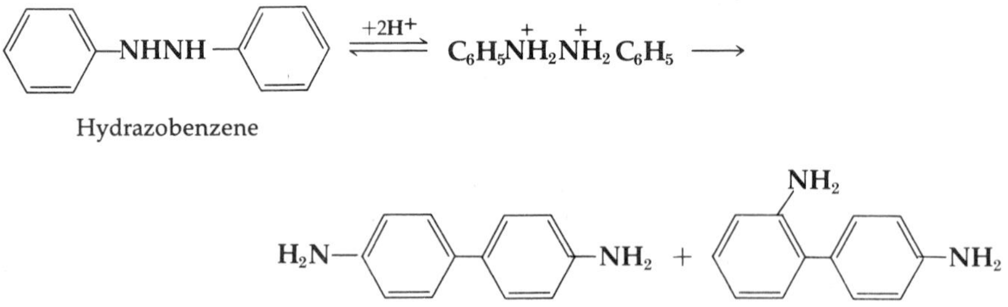

Hydrazobenzene

Benzidine

PROBLEMS

17-9 A useful synthesis of medium ring compounds is implicit in this reaction. Write the products indicated and then consider a synthesis of 1,4-diamino-cyclooctane by a similar process.

Cyclohexanone + 1,4-*bis*-diazobutane $\longrightarrow$ $C_{10}H_{16}O$ $\xrightarrow[\text{2) NaOH}]{\text{1) } CF_3CO_3H}$
neutral,
mp 120°

$C_{10}H_{18}O_3$
acidic, mp 87°

17-10 Account for these observations:

a $\xrightarrow{\text{NaOH}}$

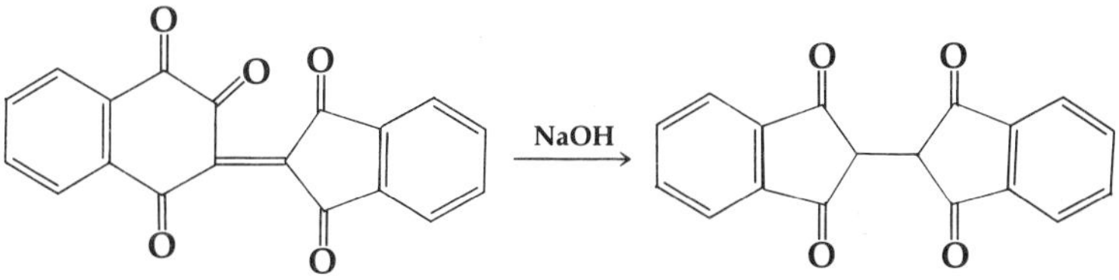

b 2-Cyclohexenone oxime in acetic anhydride does not give a Beckmann rearrangement. The main product is *N*-acetylaniline.

17-11 Devise syntheses of the following compounds, using no organic starting materials except benzene and any substances containing three carbons or less. Make a given compound only once for the entire problem set.

a $(CH_3)_3CCOCH_3$

b $(CH_3)_3CCOOH$

c $(CH_3)_3CCH_2COOH$

d $C_6H_5CH_2CH_2COOH$

e $(C_6H_5)_2CCOOH$
    $\overset{|}{OH}$

f Cycloheptanone

g $NH_2(CH_2)_5COOH$

h $(C_6H_5)_3CCHOHC_6H_5$

i $Cl-$$-Cl$

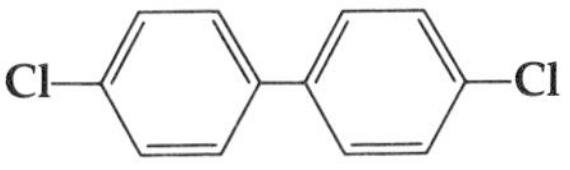

j

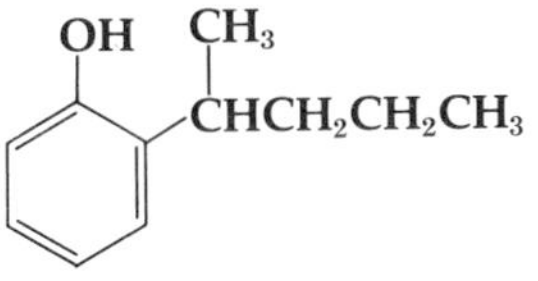

k $C_6H_5CH=\overset{\overset{\displaystyle CH_3}{|}}{\underset{\underset{\displaystyle O}{\|}}{C}}CCH_3$

l

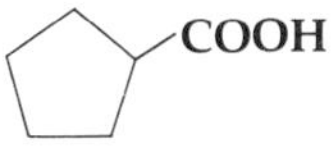

17-12 With three-dimensional formulas, where required, represent structurally the products of the following reactions:

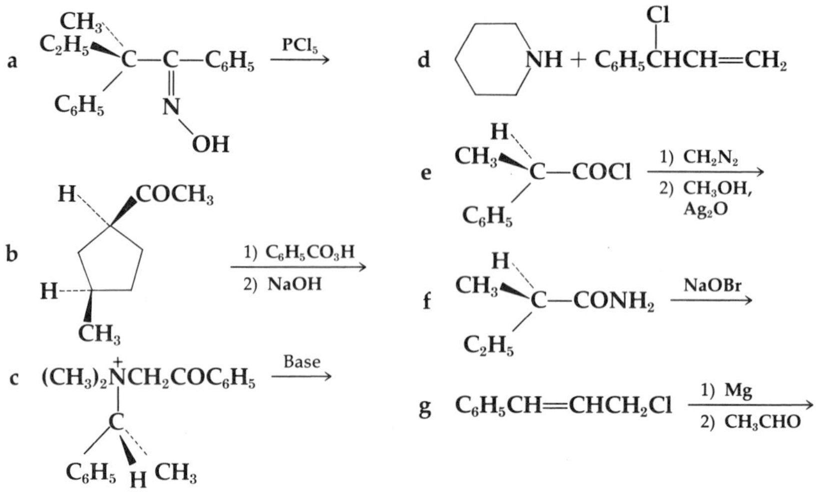

(Predict on geometric grounds.)

17-13 On the basis of the stereochemistry and mechanism of the acetolysis of 3-phenyl-2-butyl-p-toluenesulfonate, predict the stereochemical structures of the isomers of 3-phenyl-2-pentyl-p-toluenesulfonate and 2-phenyl-3-pentyl-p-toluenesulfonate that give rise to the same mixture of acetates.

17-14 Explain the following observations:

    a In the synthesis of 2-acetamido-1-propylamine, the final step involved the treatment of 2-acetamido-1-propyl chloride with ammonia. The product was a mixture of two amino amides instead of the single substance expected.

    b

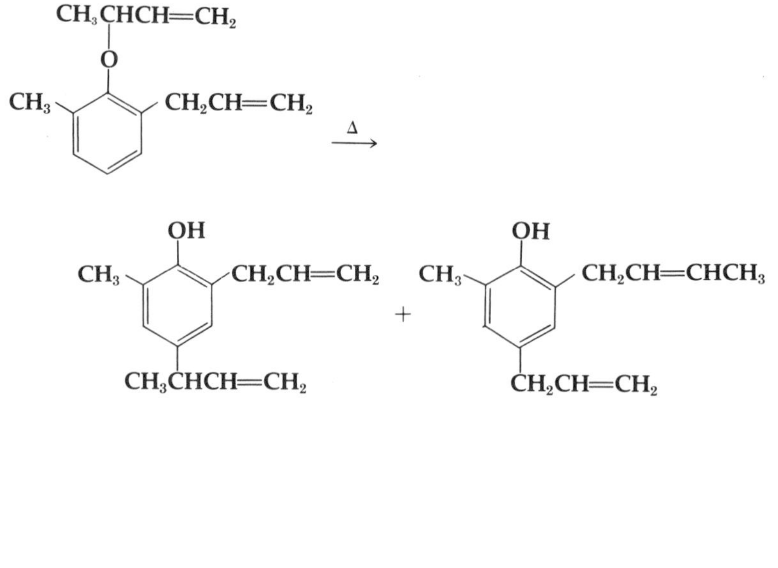

c When subjected to the reaction conditions of a Favorski rearrangement, 1,3-dibromo-3-methyl-2-butanone gave β-methylcrotonic acid.

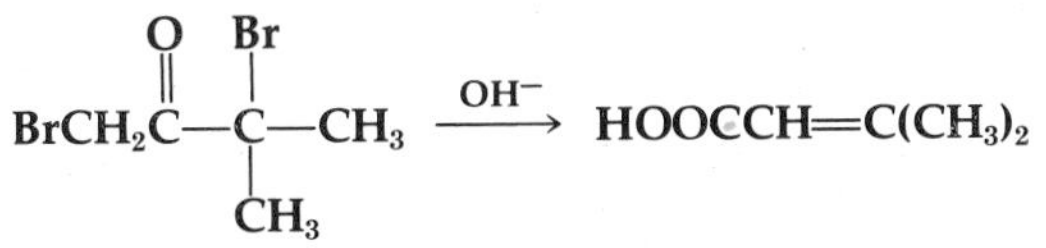

$$BrCH_2\overset{\overset{\displaystyle O}{\|}}{C}-\overset{\overset{\displaystyle Br}{|}}{\underset{\underset{\displaystyle CH_3}{|}}{C}}-CH_3 \xrightarrow{OH^-} HOOCCH=C(CH_3)_2$$

d When treated with base, the oxime tosylate drawn below gave an unstable substance with the molecular formula $C_9H_7N_3O_4$.

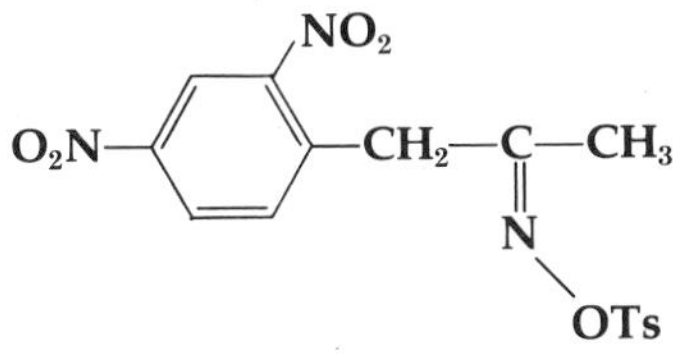

e

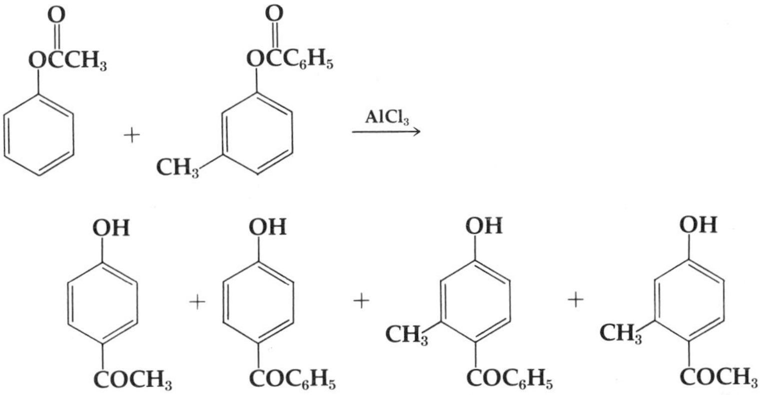

f As the following data show, different solvents give different steric results.

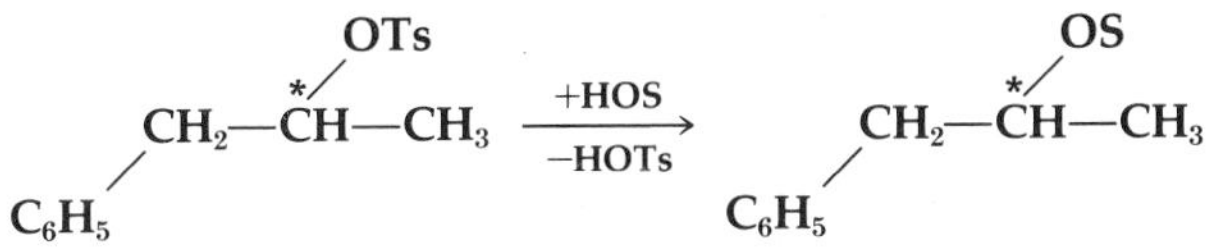

Solvent	Inversion	Retention
CH_3CH_2OH	93%	7%
CH_3COOH	65%	35%
$HCOOH$	15%	85%

17-15 The commercial synthesis of camphor from α-pinene, the major constituent of turpentine, is shown below in part. Every step is a rearrangement. The last step is an oxidation but may be seen formally as a rearrangement in aqueous acid yielding an alcohol which is then oxidized to the ketone. Formulate each step with perspective drawings and show clearly what migrates and with what stereochemistry. What other reactions might you expect to intervene?

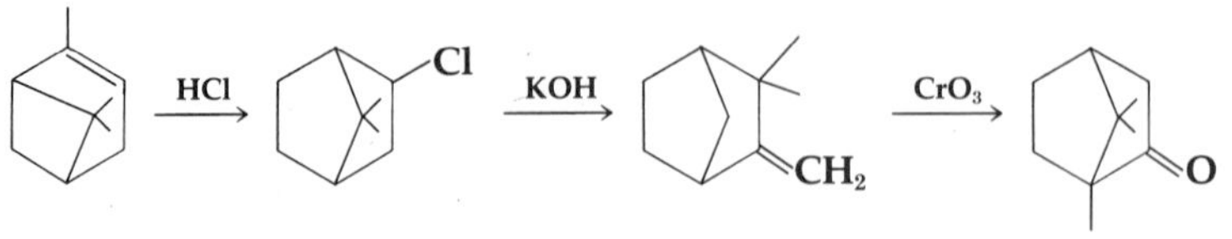

α-Pinene   Bornyl chloride   Camphene   Camphor

17-16 The natural keto-lactone, santonin, is extracted from Indian plants for use as a local medicinal to control intestinal parasites. In sulfuric acid santonin yields desmotroposantonin. Explain.

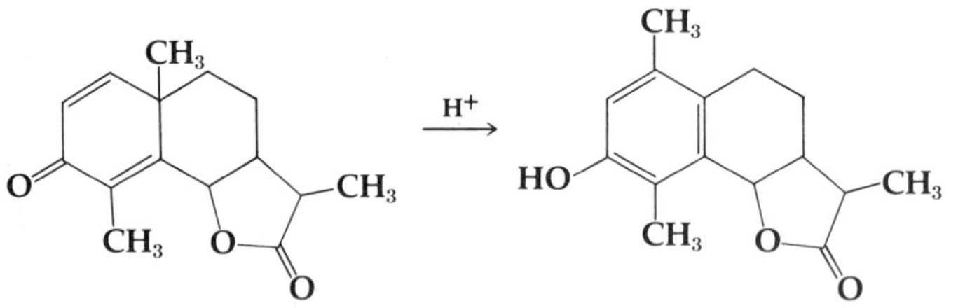

Santonin   Desmotroposantonin

17-17 Considering the mechanism of the santonin change in Prob. 17-16, rationalize the somewhat more complex, but analogous, reaction below. Show every intermediate along the mechanistic path in order to see more clearly what can occur.

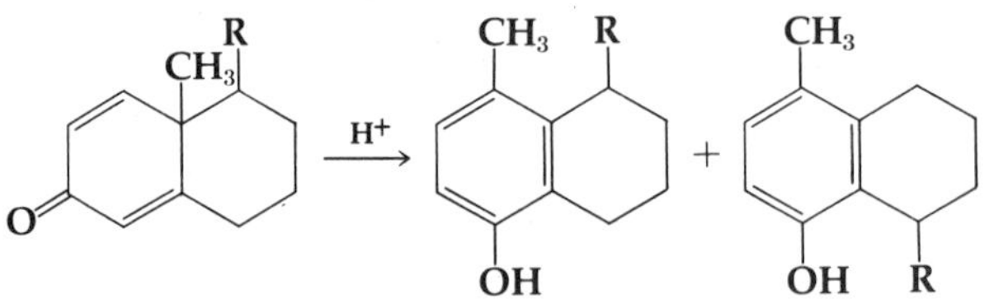

17-18 Only one tosylate epimer gives the following sovolysis reaction. Show which one, and show the stereochemistry of the product.

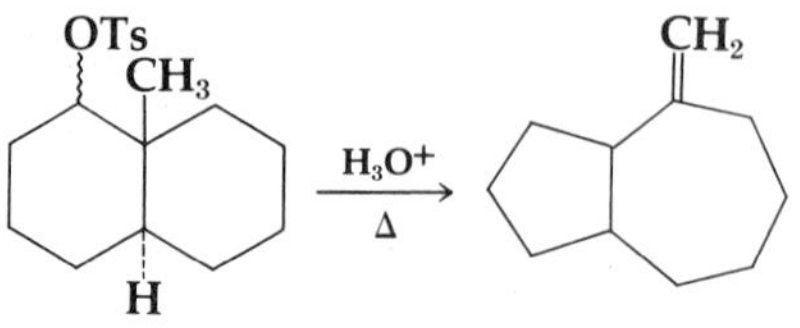

17-19 The following two compounds give halogen-free acids on treatment with alkali. Show the products and comment on the mechanism in the text as applied to these examples (especially the first step). Can you produce a mechanistic variation more suitable for these cases?

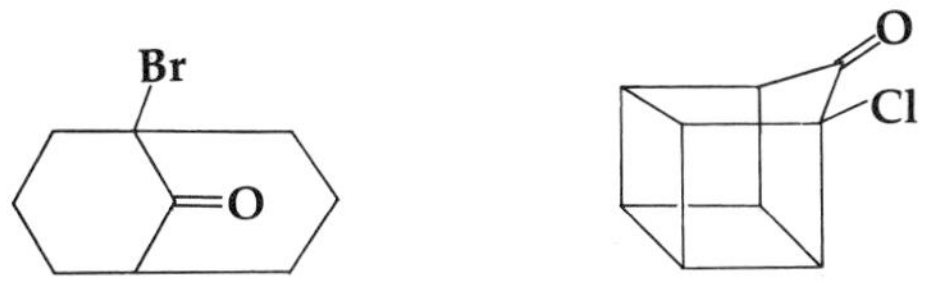

17-20 The oil, caryophyllene, is a natural constituent of cloves. On treatment with sulfuric acid it isomerizes to a substance called clovene. Account for the sequence of mechanistic steps involved. There are several steps, initiated by protonation of the upper double bond.

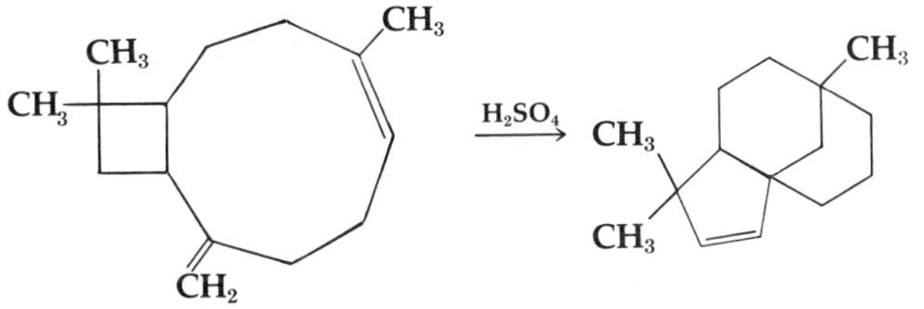

Caryophyllene   Clovene

17-21 The acid-catalyzed isomerization of the natural substance, cinenic acid, to geronic acid, shown below, was considered for years to be a unique example of a 1,5-shift, or rearrangement, of a methyl group. However, this apparent structural change was shown to be specious by labeling the methyl group adjacent to carboxyl in cinenic acid and finding all the label present in the methyl ketone of geronic acid. Considering the most likely initial carbonium ion on protonating cinenic acid, proceed to devise a mechanism incorporating a rearrangement with more precedent, i.e., involving only a 1,2-shift.

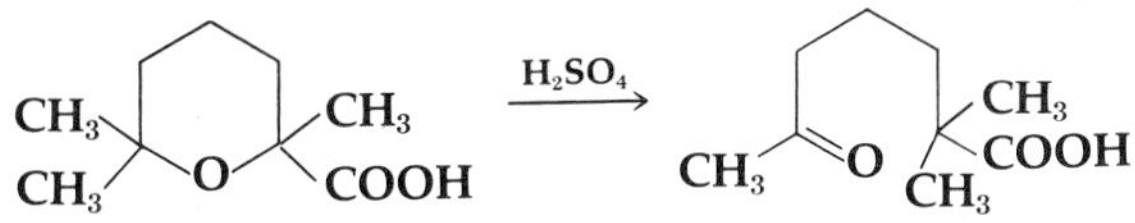

Cinenic acid   Geronic acid

17-22 The rearrangement in Prob. 17-21 was postulated by the investigators to bear a close resemblance to the Baeyer-Villiger rearrangement. To support this they synthesized geronic acid by trifluoroperacetic acid treatment of 2-hydroxy-2,6,6-trimethylcyclohexanone. Explain their synthesis and its relation to the mechanism of Prob. 7-21.

17-23 Deduce the mechanism for the following transformation. How is the starting material synthesized? Why does the product phenol not methylate further?

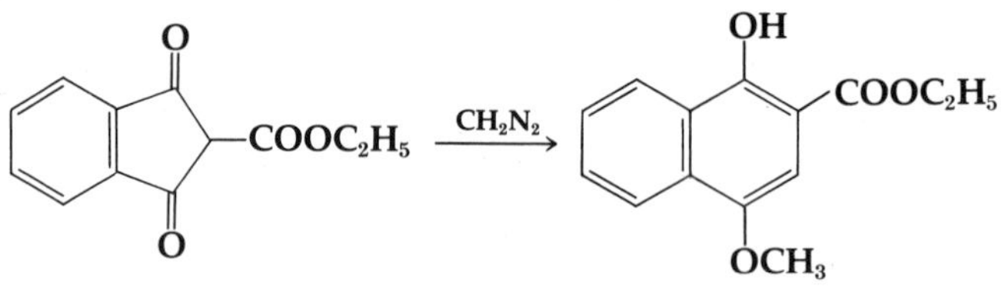

17-24 Show the products of the Favorski reactions on these compounds; both products are halogen-free.

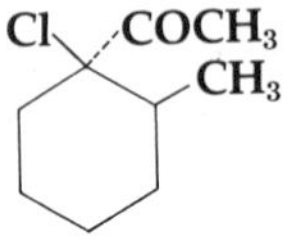

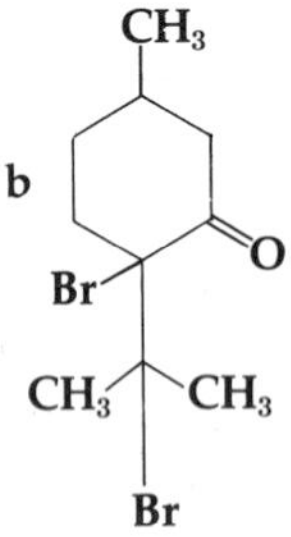

OXIDATION AND REDUCTION

18-1   GENERAL; OXIDATION STATE

The terms oxidation and reduction in inorganic chemistry refer respectively to the loss and gain of electrons at an atom or ion. The elementary atom is assigned an oxidation state of zero, and loss of N electrons is then an oxidation to an oxidation state of $+N$. A gain of electrons is similarly a reduction to an oxidation state lower by an amount equal to the number of gained electrons.†

$$\overset{0}{Zn:} \xrightarrow{\text{Oxidation}} \overset{+2}{Zn}^{++} + 2e^-$$

$$\overset{0}{:\ddot{B}r\cdot} + e^- \xrightarrow{\text{Reduction}} \overset{-1}{:\ddot{B}r:^-}$$

These considerations are based on full electron transfer between ions and so are rarely directly applicable to the carbon atoms *covalently* bonded in an organic molecule. However, common inorganic oxidizing and reducing reagents are often used to effect changes in organic molecules, and these are certainly oxidation-reduction reactions of the covalent molecule.

A rational explanation of the oxidation states of carbon in organic molecules is discussed in the indented section that follows. The conclusions from that analysis are easily summarized for simple application in this rule: *The oxidation state for any single carbon atom is obtained by adding the following values from each of its four bonds:*

-1 *for each* —**H**

0 *for each* —**C**

$+1$ *for each bond to a heteroatom*

The functional group families (Sec. 3-2) are expressions of the oxidation state of the involved carbon inasmuch as they are categorized by the number of bonds from the carbon to heteroatoms. This is exemplified in Table 18-1.

Each carbon in a molecule can be so labeled with its oxidation state, and the sum of these oxidation states in a molecule may be compared to the sum in any reaction product to ascertain whether oxidation (increased oxidation state) or reduction (decreased oxidation state) has occurred. An important criterion of the correctness of an assignment is that any reaction of a molecule

† The sign of the oxidation state of an atom (ion) arises from the charge of -1 on the electron, so that a *gain* of two electrons creates a new oxidation state -2 from the starting state, and a *loss* of negative electrons (oxidation) must yield a more positive (higher) oxidation state.

with water does not change its total oxidation state (sum of the individual carbon states). Examples of oxidation states are shown below:

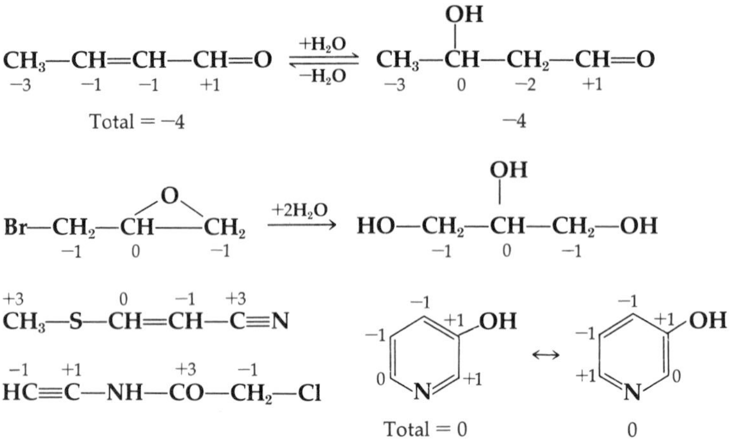

$$\underset{-3-1-1+1}{CH_3-CH=CH-CH=O} \underset{-H_2O}{\overset{+H_2O}{\rightleftharpoons}} \underset{-30-2+1}{CH_3-\overset{\displaystyle OH}{\underset{|}{C}H}-CH_2-CH=O}$$

$$Total = -4 \qquad\qquad\qquad\qquad -4$$

$$\underset{-10-1}{Br-CH_2-\overset{\displaystyle O}{\underset{\diagdown\diagup}{CH-CH_2}}} \overset{+2H_2O}{\longrightarrow} \underset{-10-1}{HO-CH_2-\overset{\displaystyle OH}{\underset{|}{C}H}-CH_2-OH}$$

$$\underset{-1+1+3-1}{\overset{+30-1+3}{CH_3-S-CH=CH-C\equiv N}}$$

$$\underset{-1+1+3-1}{HC\equiv C-NH-CO-CH_2-Cl}$$

$$Total = 0 \qquad\qquad\qquad\qquad\qquad 0$$

TABLE 18-1   **Oxidation States of Carbon in Organic Compounds**

Oxidation state	Primary	Secondary	Tertiary	Quaternary
-4	CH_4			
-3		RCH_3		
-2	CH_3OH	R_2CH_2		
-1		RCH_2OH	R_3CH	
0	CH_2O	R_2CHOH		R_4C
$+1$		$RCHO$	R_3COH	
$+2$	$HCOOH$	R_2CO		
$+3$		$RCOOH$		
$+4$	CO_2			

The number of transferred electrons in a chemical change is then the oxidation-state change from reactant to product, for purposes of balancing oxidation-reduction equations. The oxidation of ethanol to acetic acid by chromic acid is taken as an example, showing the relevance to equation balancing as normally practiced in inorganic chemistry.

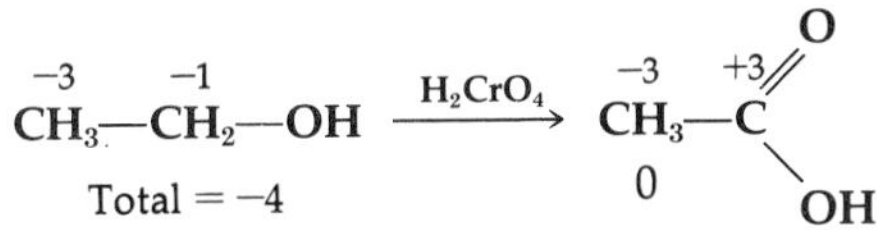

$$\overset{-3}{CH_3}-\overset{-1}{CH_2}-OH \xrightarrow{H_2CrO_4} \overset{-3}{CH_3}-\overset{+3}{C}\overset{\displaystyle O}{\underset{0\qquad OH}{\diagup\diagdown}}$$

Total $= -4$

$\Delta e^- = -4$

$$\overset{+6}{H_2CrO_4} \longrightarrow \overset{+3}{Cr^{3+}} \qquad\qquad \Delta e^- = +3$$

Functional-group families (cf. Table 3-2)

0	I	II	III	IV
CH_4				
RCH_3				
R_2CH_2	CH_3OH (CH_3X)			
R_3CH	RCH_2OH (RCH_2X)			
R_4C	R_2CHOH (R_2CHX)	CH_2O (CH_2X_2)		
	R_3COH (R_3CX)	$RCHO$ $(RCHX_2)$		
		R_2CO (R_2CX_2)	$HCOOH$ (HCX_3)	
			$RCOOH$ (RCX_3) (RCN)	
				CO_2 (COX_2) (CX_4)

Hence

$$3(H_2O + CH_3CH_2OH \longrightarrow CH_3COOH + 4H^+ + 4e^-)$$

$$\underline{4(3e^- + 8H^+ + CrO_4^- \longrightarrow Cr^{3+} + 4H_2O)}$$

$$3CH_3CH_2OH + 20H^+ + 4CrO_4^- \longrightarrow 3CH_3COOH + 4Cr^{3+} + 13H_2O$$

Oxidation States. In order to bridge the gap between full electron transfer in ionic redox reactions and the situation with the shared electrons of covalent bonds, we must formally assign the "possession" of the electron pair in a covalent bond to the *more electronegative atom* of the two bonded atoms. In this way we can then count the electrons on each atom as we would with simple inorganic ions and so determine its oxidation state.

The neutral atom of the element has an oxidation state of zero. The oxidation state of that atom in an ion or molecule is then ascertained by the outer-shell electron change from that elementary condition, electrons added (reduction) making the oxidation state more negative, those lost (oxidation) making it more positive as in the following examples. The circles incorporate electrons "possessed" by the atom circled:

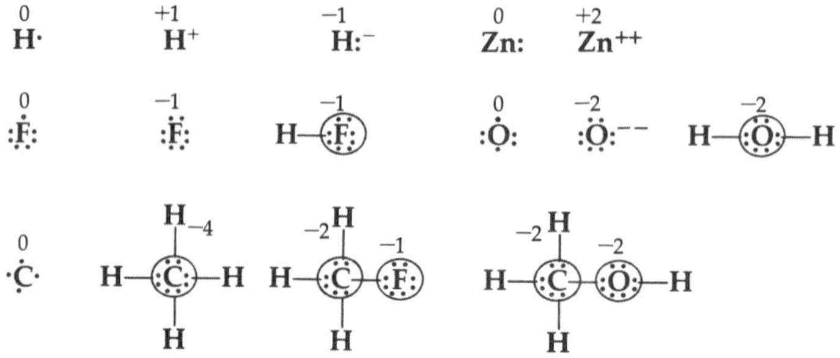

Since carbon is less electronegative than the heteroatoms to which it is commonly bound, the general rule is that, in any bond of carbon to heteroatom, the bond electrons are assigned to the heteroatom. With carbon-hydrogen bonds the electrons are assigned to carbon. With carbon-carbon bonds neither carbon can easily be assigned both bonding electrons and so each is assigned only one. This leads to the simple rule shown above in the main text for determining the oxidation state of any carbon atom. All atoms are treated in the same fashion. As a result oxidation-state change is equivalent to electron transfer and so may be used for balancing redox reaction equations. The electrons in bonds between like atoms (C—C, N—N, O—O, etc.) are assigned either as one to each atom or both to one atom, as convenient. When atoms higher than the first row of the periodic table form covalent bonds, they can generally be regarded as having just four bonds (or unshared pairs) like first-row atoms, for the purposes of assigning electrons and oxidation states. This is exemplified in Chap. 19 for sulfur and phosphorus and may also be seen in chromate ion:

$$\overset{\displaystyle O^-}{\underset{\displaystyle O^-}{\overset{|}{\underset{|}{O-\overset{++}{Cr}-O}}}} \quad = \quad \overset{\displaystyle (\ddot{O})^{-2}}{\underset{\displaystyle (\ddot{O})^{-2}}{\overset{|}{\underset{|}{(\ddot{O})^{-2}\overset{-2}{}\overset{+6}{Cr}(\ddot{O})^{-2}}}}} \quad \text{vs.} \quad :\dot{\underset{.}{C}}r: \quad (\text{element, } 6e^-)$$

For nitrogen compounds the range of oxidation states is -3 to $+5$ but -3 is the normal state of nitrogen in amines and amides (in nitro groups, $N = +3$):

$$\overset{0}{\cdot\ddot{N}\cdot} \qquad \overset{-3}{H-(\ddot{N}:)-R} \qquad R-\overset{-3}{(\overset{\displaystyle H}{\underset{\displaystyle H}{\ddot{N}}})}-R \qquad R-\overset{+3}{(N)}=(:\ddot{O}:)$$

$$\overset{-1}{R-\ddot{N}H-OH} \qquad \overset{+1}{R-\dot{N}=O} \qquad \overset{+3}{R-O-\dot{N}=O} \qquad \overset{+5}{HO-\underset{\underset{\displaystyle -O}{\overset{\displaystyle +}{N}}}{\overset{\displaystyle O}{\diagup\!\!\!/}}}$$

In general for any atom the oxidation state may be computed by adding:

-1 for each bond to a less electronegative atom (or a minus charge)

 0 for each bond to an identical atom

$+1$ for each bond to a more electronegative atom (or a plus charge)

No oxidation-state change in the total organic molecule is observed in reactions like hydrolysis, or comparable reactions with alcohols (alcoholysis) or amines (aminolysis), with addition or elimination of **HX, HOH, HOR,** or with tautomerization. On the other hand oxidation (electron loss) occurs when *hydrogen is removed or oxygen added* to the molecule, or both, as the ethanol oxidation illustrates. Conversely, in reduction *hydrogen is added or oxygen removed,* or both. Addition of other heteroatoms is also an oxidation (unless hydrogen is also added), and their removal (without also removing hydrogen) is a reduction. These changes can often be seen simply from the empirical formula change. Many of the reactions we have already examined are oxidations and reductions and serve to demonstrate these ideas.

Oxidation: $\overset{-2}{C}H_2 = \overset{-2}{C}H_2 + R - \overset{O}{\underset{\|}{C}} - \overset{-1}{O} - \overset{-1}{O}H \longrightarrow \overset{-1}{C}H_2 \overset{\overset{-2}{O}}{\diagup \diagdown} \overset{-1}{C}H_2 + R - \overset{O}{\underset{\|}{C}} - \overset{-2}{O}H$

Total: carbon $= -4$ oxygen $= -2$ carbon $= -2$ oxygen $= -4$

Empirical change: $C_2H_4 \longrightarrow C_2H_4O$

$$\overset{-3}{C}H_3 - \overset{-1}{C}H = \overset{-2}{C}H_2 + :\overset{0}{B}r - \overset{0}{B}r: \longrightarrow \overset{-3}{C}H_3 - \overset{-1}{C}H - \overset{-1}{C}H_2 - \overset{:\overset{-1}{B}r:}{} \overset{|}{\underset{}{}} :\overset{-1}{B}r:$$

Total: $C = -6$ $Br = 0$ $C = -4$ $Br = -2$

Empirical change: $C_3H_6 \longrightarrow C_3H_6Br_2$

Reduction: $\overset{-3}{C}H_3 - \overset{+1}{C}H = O + \overset{-1}{H}:^- \longrightarrow \overset{-3}{C}H_3 - \overset{-1}{C}H_2 - O^- \xrightarrow{H_2O} CH_3CH_2OH$

Total: $C = -2$ $H = -1$ $C = -4$ $H = +1$

Empirical change: $C_2H_4O \longrightarrow C_2H_6O$

$$\overset{-2}{C}H_3 - \overset{-1}{\underset{}{I}} + \overset{0}{Mg} \longrightarrow \overset{-4}{C}H_3 - \overset{+2}{Mg} - \overset{-1}{I} \xrightarrow{H_2O} \overset{-4}{C}H_4 + \overset{+2}{Mg}(OH)\overset{-1}{I}$$

In oxidation reactions the reagent atom(s) accepts electrons from the organic molecule being oxidized. Such oxidations are almost always two-electron changes, the two electrons from some covalent bond being transferred to a reagent atom, often as an unshared pair. Reducing agents by contrast deliver an electron pair, which becomes a covalent bond, as in hydride reduction. Oxidations and reductions at a single carbon atom are changes between groups in any single vertical column in the left-hand arrangement of Table 18-1.

$$R - CH_3 \underset{Red}{\overset{Ox}{\rightleftharpoons}} R - CH_2 - OH \underset{Red}{\overset{Ox}{\rightleftharpoons}} R - CHO \underset{Red}{\overset{Ox}{\rightleftharpoons}} R - COOH$$

$$R_2CH - OH \underset{Red}{\overset{Ox}{\rightleftharpoons}} R_2CO$$

TABLE 18-2   **Common Oxidizing Agents in Organic Chemistry**

O_2	HNO_3	SO_3	Cl_2	Ag_2O	MnO_2
O_3	$RO-NO$	$(CH_3)_2S - O$	Br_2	HgO	MnO_4^-
$HO-OH$	$\phi - N_2^+$	SeO_2	I_2	$Hg(OAc)_2$	CrO_3
$t\text{-}BuO-OH$	H_2NCl		NBS	$Pb(OAc)_4$	CrO_2Cl_2
$RCOO-OH$	$H_3N^+ - OSO_3^-$		(N-bromo-		
	$R_3N^+ - O^-$		succinimide)	$FeCl_3$	OsO_4
			$t\text{-}BuOCl$	$Fe(CN)_6^{3-}$	IO_4^-

Dehydrogenation ($-2H$): **Pt, Pd, S, Se** (with heating)

Substituted quinones (chloranil)

$(t\text{-}BuO)_3Al/R_2CO$ (page 461)

Hydrogenation adds hydrogen to multiple bonds and so is a reduction. **Dehydrogenation** is a removal of two hydrogens to form a multiple bond. The oxidation of alcohols to aldehydes (or ketones) is simple dehydrogenation, but the term is usually reserved for formation of a carbon-carbon double bond from a single bond. Some oxidations actually cleave carbon-carbon bonds.

The reagents for oxidation and reduction are commonly inorganic oxidizing and reducing agents. Many of these are useful and many are highly selective in their actions on various organic functional groups, thus making them valuable for synthesis in polyfunctional molecules. A general summary of useful reagents is collected in Tables 18-2 and 18-3.

PROBLEM 18-1

Examine the oxidation-state changes in each reaction quoted in the remainder of the chapter. This should quickly become second nature in all but very complex cases.

PROBLEM 18-2

Do the following conversions involve oxidation-reduction, either at individual carbons or overall?

a

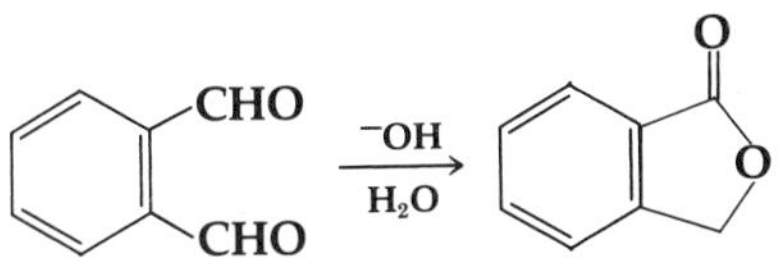

b

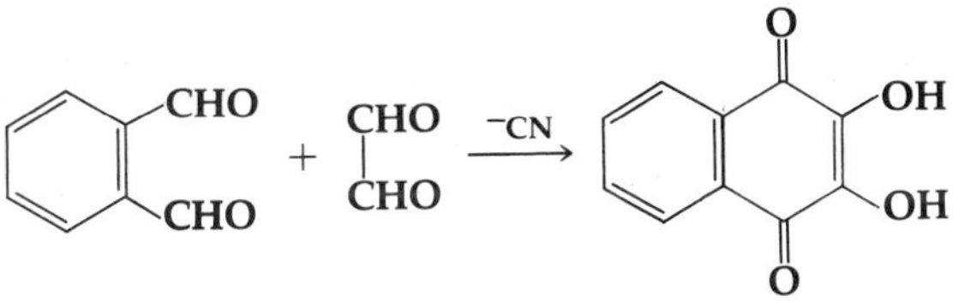

c

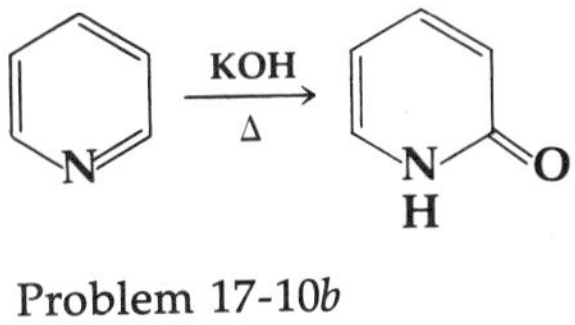

d   Problem 17-10*b*

TABLE 18-3   **Common Reducing Agents in Organic Chemistry**

Catalytic hydrogenation: H_2 + catalyst: **Pt, Pd, Ni**
Hydrides: **LiAlH$_4$, AlH$_3$, NaBH$_4$, BH$_3$, R$_2$BH, ϕ_3SnH**
Metals: **Li, Na, K, Zn, Mg**
Others: **NH$_2$NH$_2$, R$_3$P:, SO$_3{}^{--}$, SnCl$_2$, FeCl$_2$,**
[(CH$_3$)$_2$CHO]$_3$Al (page 460)

A number of oxidizing and reducing agents have already been discussed inasmuch as their mechanisms are generally clear and fall into our previous categories. Hydride addition is a reduction, for example, involving electron-pair delivery by the nucleophilic hydride anion. Also, addition to double bonds by electrophiles like bromine and peracids constitutes an oxidation of the double-bond carbons. The mechanisms of a number of other redox reagents are less clear and some act via free-radical pathways. Nevertheless, many conform in general to a simple electrophile-nucleophile classification.

Reductions are effected by reagents delivering electrons, either as two-electron nucleophiles like H:⁻ and Zn: or as one-electron deliverers like Li· or H· (H₂ with metal catalysts). The former include the complex hydrides (LiAlH₄, NaBH₄, etc.), internal hydride transfer (Meerwein-Ponndorf reduction, page 460), and two-electron metals of the second column of the periodic table. The one-electron transfers are generally from first-column metals or catalytic hydrogenation. These are conveniently grouped in Table 18-3 and discussed in these categories in subsequent sections.

Oxidations usually follow one of two mechanistic patterns. In the first a nucleophilic carbon, either as carbanion or double-bond π electrons, attacks a heteroatom electrophile. Such oxidations have already been encountered in halogenation of alkenes or their attack on peracids (Chap. 15). Aromatic electrophilic substitutions are similarly oxidations of this kind when the electrophile is a heteroatom (Br^+, NO_2^+, SO_3, etc.). Halogenation of enolate anions or enols is another major example, treated in Sec. 18-2.

Carbon nucleophiles:

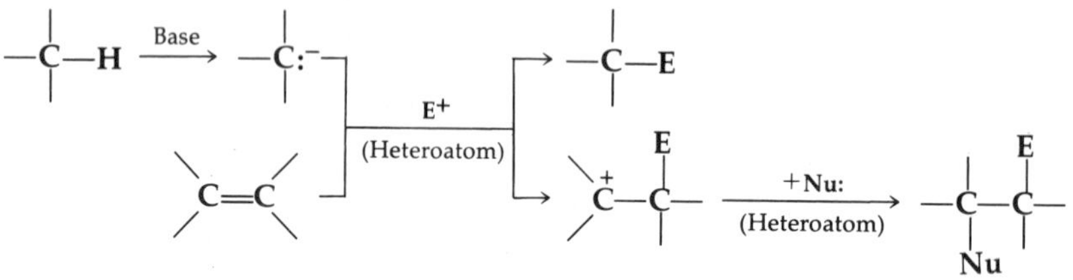

The second mode is characterized by an electrophilic carbon bearing a hydrogen which is converted to an intermediate containing a bond between heteroatoms. This bond then breaks in an elimination step, carrying off the electron pair. The common starting situations will be CH—X, CH—OH, CH=O, and one heteroatom (L) is often a metal, changing oxidation state by -2 ($L^+ + 2e^- \longrightarrow :L^-$).

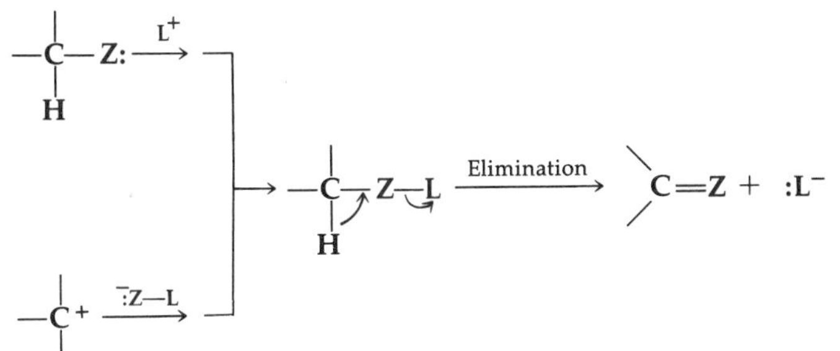

$$\begin{array}{c} -\overset{|}{\underset{H}{C}}-Z: \xrightarrow{L^+} \\[2em] -\overset{|}{\underset{H}{C}}+ \xrightarrow{\ :Z-L} \end{array} \Bigg] \longrightarrow -\overset{|}{\underset{H}{C}}\!\!-\!\!Z\!\!-\!\!L \xrightarrow{\ \text{Elimination}\ } \ \overset{\diagdown}{\underset{\diagup}{C}}\!\!=\!\!Z + \ :L^- $$

$$(Z = O, N, S)$$

PROBLEM 18-3

Identify the changes in oxidation state in all species in each of the following cases and write balanced equations.

a Acetone + **NaBH$_4$**

b Formation of Grignard reagent from iodobenzene and its reaction with acetone

c Acyloin condensation of diethyl glutarate with sodium

d Beckmann rearrangement of benzophenone

e Oppenauer oxidation of **$(C_6H_5)_2$CHOH** with acetone and aluminum t-butoxide

f Nitration of benzene

g Nitrous acid reaction with aqueous α-amino acids

18-2   OXIDATIONS AT A SINGLE CARBON

These reactions are all those which proceed one step down any vertical column in the left-hand presentation in Table 18-1. Hence they involve three kinds of changes:

1 Oxidation of hydrocarbon sites to alcohols (or halides)
2 Oxidation of alcohols (or halides) to aldehydes/ketones
3 Oxidation of aldehydes to acids

These oxidations may also be regarded as conversions of one functional-group family (0, I, II, or III) to the next higher one.

Hydrocarbons

Oxidations at unactivated hydrocarbon sites, C—H (converting to C—O or C—X), are rarely useful since methods like catalyzed air oxidation (or simple combustion!) are too indiscriminate. The cases of practical value involve oxidation at the activated α-CH to a carbonyl, double bond, or aromatic ring.

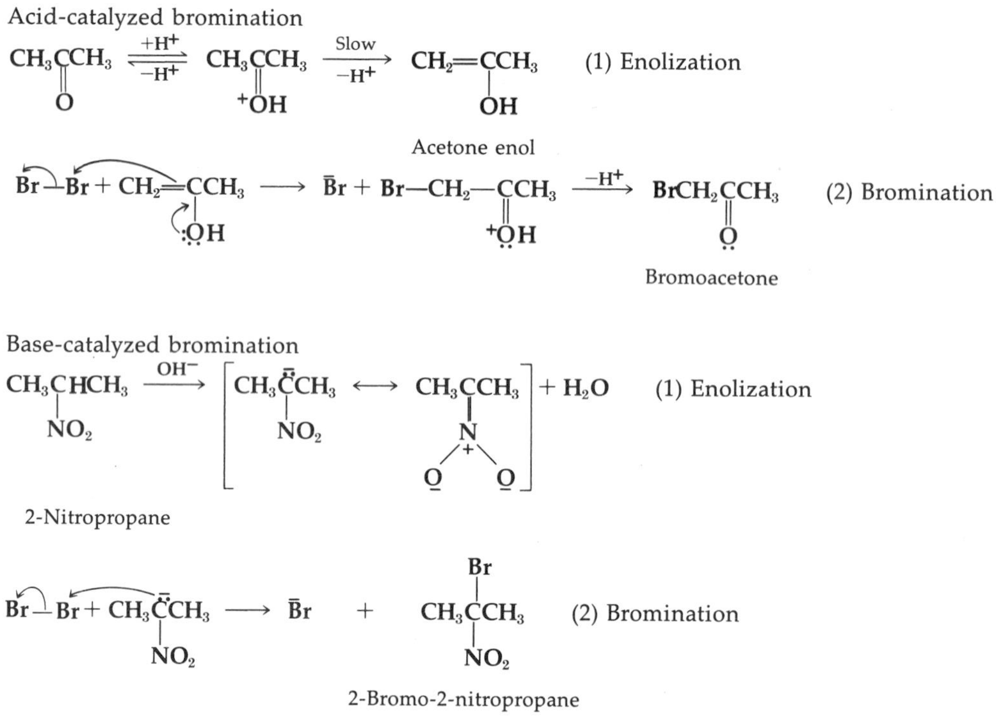

Acid-catalyzed bromination

$$CH_3CCH_3 \underset{-H^+}{\overset{+H^+}{\rightleftharpoons}} CH_3CCH_3 \xrightarrow[-H^+]{Slow} CH_2{=}CCH_3 \quad \text{(1) Enolization}$$
$$\quad \parallel \qquad\qquad \parallel \qquad\qquad\quad |$$
$$\quad O \qquad\qquad +OH \qquad\qquad\quad OH$$

Acetone enol

$$Br{-}Br + CH_2{=}CCH_3 \longrightarrow \bar{B}r + Br{-}CH_2{-}CCH_3 \xrightarrow{-H^+} BrCH_2CCH_3 \quad \text{(2) Bromination}$$
$$\qquad\qquad |\qquad\qquad\qquad\qquad\qquad \parallel \qquad\qquad\qquad \parallel$$
$$\qquad\qquad :OH \qquad\qquad\qquad\qquad +OH \qquad\qquad\quad O$$

Bromoacetone

Base-catalyzed bromination

$$CH_3CHCH_3 \xrightarrow{OH^-} \left[CH_3\bar{C}CH_3 \longleftrightarrow CH_3CCH_3 \right] + H_2O \quad \text{(1) Enolization}$$
$$\quad |\qquad\qquad\qquad\quad |\qquad\qquad\qquad\quad \parallel$$
$$\quad NO_2 \qquad\qquad\qquad NO_2 \qquad\qquad\qquad N$$
$$\qquad\qquad\qquad\qquad\qquad\qquad\qquad\qquad\quad / {\overset{+}{\,}} \backslash$$
$$\qquad\qquad\qquad\qquad\qquad\qquad\qquad\qquad \underline{O} \quad \underline{O}$$

2-Nitropropane

$$\qquad\qquad\qquad\qquad\qquad\qquad\qquad Br$$
$$\qquad\qquad\qquad\qquad\qquad\qquad\qquad |$$
$$Br{-}Br + CH_3\bar{C}CH_3 \longrightarrow \bar{B}r \quad + \quad CH_3CCH_3 \quad \text{(2) Bromination}$$
$$\qquad\qquad\quad |\qquad\qquad\qquad\qquad\qquad\qquad |$$
$$\qquad\qquad\quad NO_2 \qquad\qquad\qquad\qquad\qquad NO_2$$

2-Bromo-2-nitropropane

FIGURE 18-1   α-Halogenation via enolization

Halogenation in the α-position of carbonyl groups proceeds easily via initial activation of that position as a nucleophilic enolate or enol which then attacks the electrophilic halogen, e.g., $X_2 \rightleftharpoons X^+ + :X^-$, as shown in the examples of Fig. 18-1.

Bromination on carbon α to ketones, nitriles, and nitro groups is a facile reaction as is bromination of the doubly activated α-positions of malonic and related acids and esters. Ordinary esters do not satisfactorily α-brominate. Carboxylic acids undergo α-halogenation only in the presence of a small amount of phosphorus trichloride. The acid is converted to an acid chloride, which enolizes more readily than the free acid. Since acids and acid chlorides are readily equilibrated under the reaction conditions, only catalytic amounts of phosphorus trichloride are needed.

Hell-Volhard-Zelinsky reaction:

$$CH_3CH_2CH_2CH_2\,COOH + PCl_3 \longrightarrow CH_3CH_2CH_2CH_2\,COCl$$

$$CH_3CH_2CH_2\,CH_2COCl \rightleftharpoons CH_3CH_2CH_2\,CH{=}CCl \xrightarrow{Br_2} CH_3CH_2CH_2\,CHBrCOCl + HBr$$
$$\qquad\qquad\qquad\qquad\qquad\qquad\qquad\qquad\qquad |$$
$$\qquad\qquad\qquad\qquad\qquad\qquad\qquad\qquad\quad OH$$

$$CH_3CH_2CH_2\,CHBrCOCl + CH_3CH_2CH_2CH_2\,COOH \rightleftharpoons$$

$$\qquad\qquad\qquad CH_3CH_2CH_2\,CHBrCOOH + CH_3CH_2CH_2CH_2\,COCl$$

α-Bromovaleric acid
85% yield

Nucleophilic substitution reactions can then be used to convert α-halo-acids or α-haloketones to useful functional derivatives. The most important examples are α-amino and α-hydroxy-acids, which are of biochemical significance.

Base-catalyzed halogenation of methylene and methyl ketones cannot be stopped at the monohaloketone stage. After complete halogenation occurs, the polyhaloketones are cleaved under the basic reaction conditions. This is called the **haloform reaction** when conducted on methyl ketones, since these are converted to carboxylic acids and haloform (HCX_3) in the overall reaction. This reaction is commonly used as a characteristic test for methyl ketones. In the test, the material of unknown structure is treated with a solution of halogen (X_2) in weak hydroxide solution. Base-catalyzed halogenation occurs. Each halogen introduced increases the acidity of the remaining hydrogens attached to the same carbon atom. Consequently, halogenation of methyl groups does not stop at intermediate stages, and trihaloketones are produced. The latter are cleaved, forming carboxylic acids and haloforms. Iodoform is a yellow solid that is sparingly soluble in aqueous solvents; it is easily detected, therefore, when formed in a positive **iodoform reaction**.

$$RCCH_3 + \bar{O}H \longrightarrow RC\ddot{C}H_2^- + H_2O$$

$$RC\ddot{C}H_2^- + I_2 \longrightarrow RCCH_2I + \bar{I}$$

$$RCCH_2I \xrightarrow{OH^-} RC\ddot{C}HI^- \xrightarrow{I_2} RCCHI_2 \xrightarrow{OH^-} RC\ddot{C}I_2^- \xrightarrow{I_2} RCCI_3$$

$$RCCI_3 + \bar{O}H \longrightarrow \left[RC\overset{\bar{O}}{\underset{OH}{-}}CI_3 \right] \longrightarrow RCOH + :\bar{C}I_3 \longrightarrow RC\bar{O}_2 + HCI_3$$

Iodoform

Net reaction: $RCCH_3 + 3I_2 + 4\bar{O}H \longrightarrow RC\bar{O}_2 + HCI_3 + 3\bar{I} + 3H_2O$

The haloform reaction is given by methyl ketones and other compounds, notably secondary methyl carbinols (CH_3CHOHR) and ethanol, which are readily oxidized to methyl ketones (or CH_3CHO). The reaction is used in synthesis for the preparation of carboxylic acids from methyl ketones.

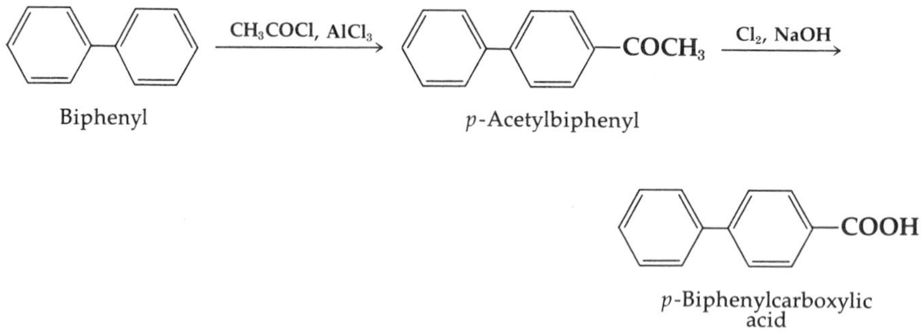
CH3COCl, AlCl3
COCH3
Cl2, NaOH
Biphenyl
p-Acetylbiphenyl
COOH
p-Biphenylcarboxylic acid

Halogenation of ketones at the α-position has also been carried out with advantage using cupric bromide or chloride, or sulfuryl chloride (SO_2Cl_2). Halogenation of other carbanions is facile but rarely of synthetic value. Virtually all organometallic carbanions, such as Grignard reagents and alkyllithium, attack halogens avidly, but the product is only the starting halide used to create the organometallic compound.

$$R{-}MgBr + X_2 \longrightarrow R{-}X + MgBrX$$

Halogenation α to double bonds and aromatic rings is commonly achieved with *N*-bromosuccinimide. This reagent also attacks α to ketones and is selective in its reactivity: $-CH_3 > -CH_2- \gg -\overset{|}{\underset{|}{C}}H$. A free-radical mechanism, involving **Br·**, is apparently involved.

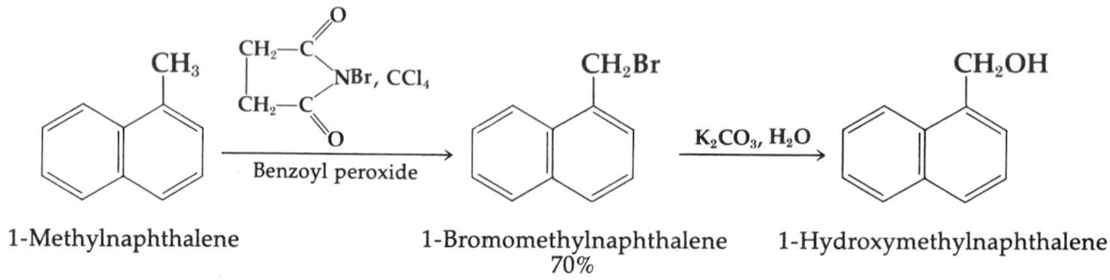
CH3
CH2—C
NBr, CCl4
CH2—C
O
O
Benzoyl peroxide
CH2Br
K2CO3, H2O
CH2OH
1-Methylnaphthalene
1-Bromomethylnaphthalene
70%
1-Hydroxymethylnaphthalene

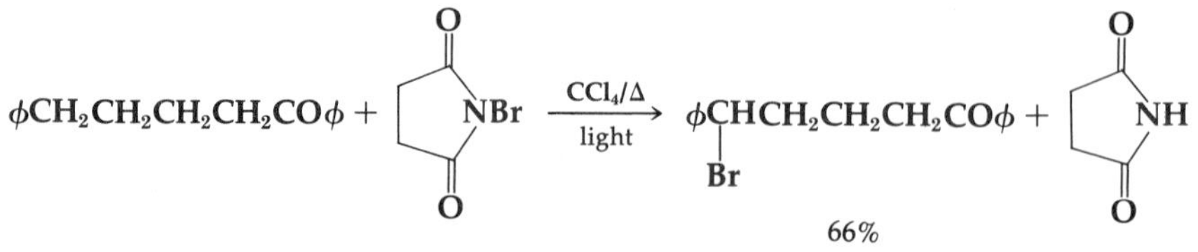
φCH2CH2CH2CH2COφ +
O
NBr
O
CCl4/Δ
light
φCHCH2CH2CH2COφ +
Br
66%
NH

Direct introduction of oxygen at α-CH positions is less easily achieved with selectivity and control. Such C—O bonds are usually created by halogenation followed by substitution as in the example shown above. Low-temperature air oxidation of alkenes or alkyl aromatics occurs at positions adjacent to the double bonds, by a free-radical mechanism. The first products are hydroperoxides, which are seldom isolated but oxidize further with ease.

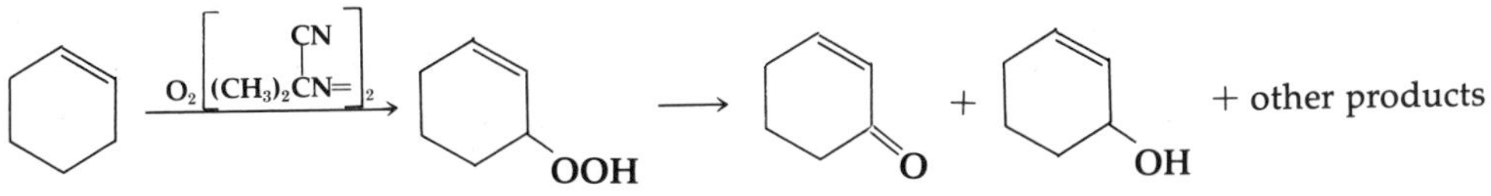

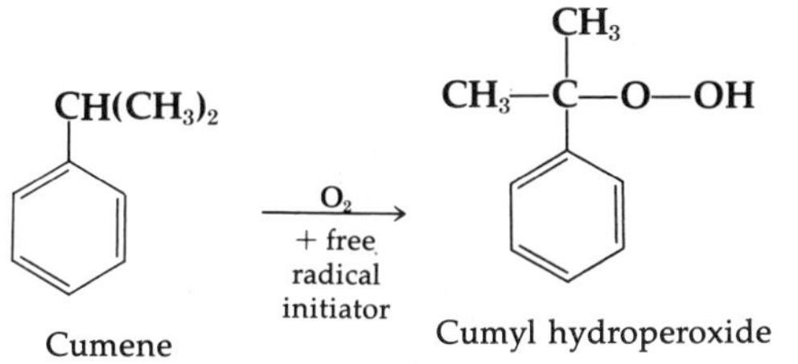

Cumene   Cumyl hydroperoxide

Aliphatic side chains attached to aromatic ring systems are much more readily oxidized than are the rings themselves. Drastic oxidation degrades a side chain to a carboxyl group, a useful step in proof of the structures of certain aromatic compounds.

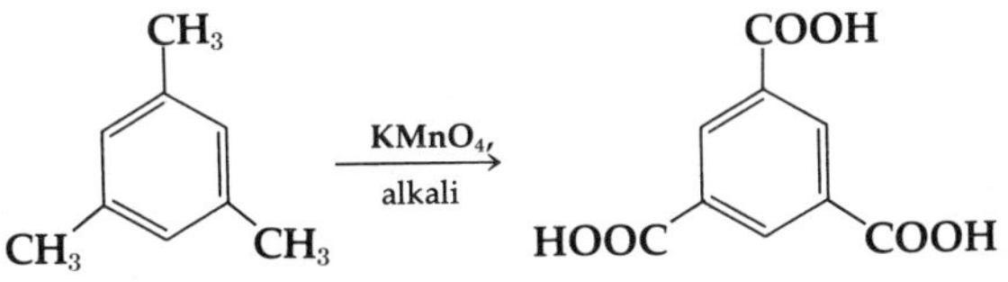

Mesitylene   Trimesic acid

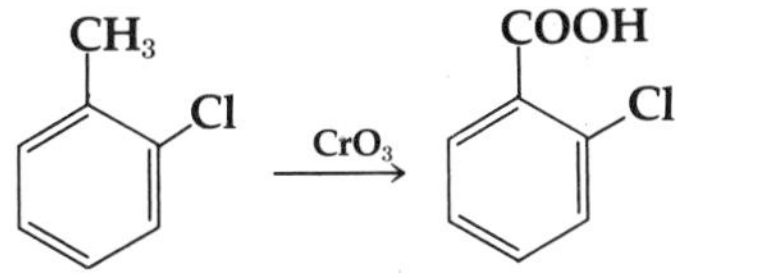

o-Chlorotoluene   o-Chlorobenzoic acid

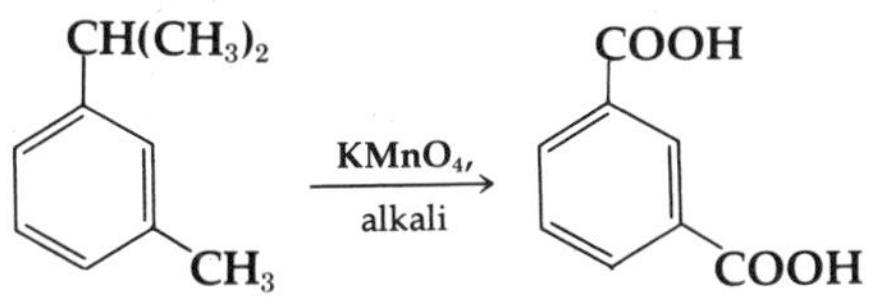

m-Isopropylmethylbenzene   Isophthalic acid

The hydrogen-carbon bond of terminal acetylenes is subject to **oxidative coupling** to produce conjugated acetylenes. This reaction has been used to great advantage as the key step in preparing some interesting and bizarre hydrocarbons. A number of large rings have been prepared by this coupling and converted to monocyclic fully unsaturated hydrocarbons called **annulenes**, $(C_2H_2)_n$ (page 168).

$$C_6H_5C\equiv CH \xrightarrow{Cu(OH)_2} C_6H_5C\equiv CCu \xrightarrow[C_2H_5OH]{O_2,\ NH_3} C_6H_5C\equiv CC\equiv CC_6H_5$$

Diphenyldiacetylene

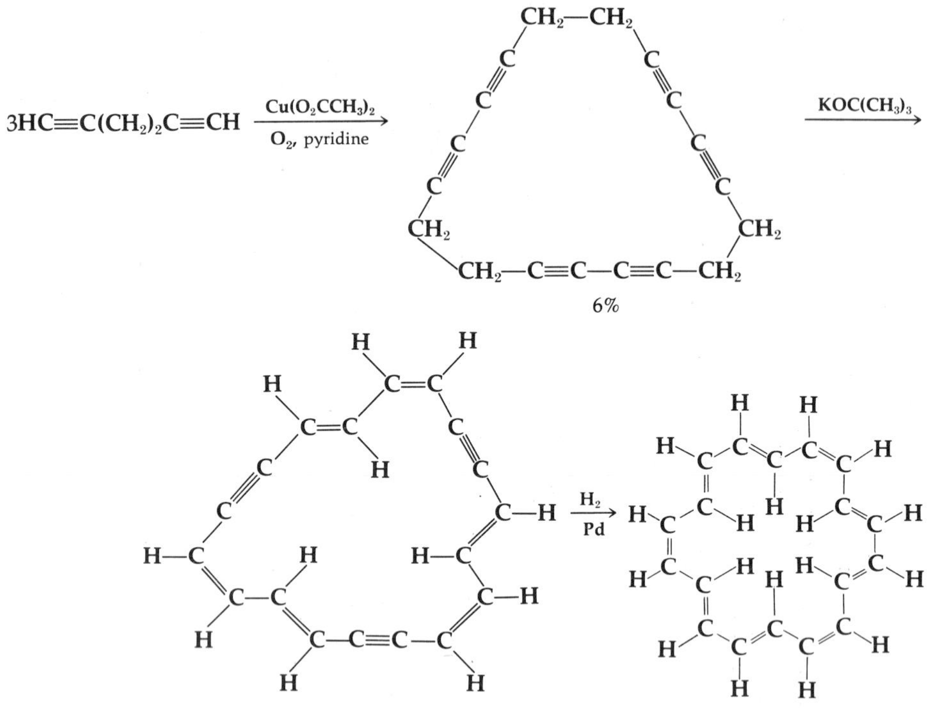

$$3HC\equiv C(CH_2)_2C\equiv CH \xrightarrow[O_2,\ pyridine]{Cu(O_2CCH_3)_2}$$

6%

$$\xrightarrow{KOC(CH_3)_3}$$

$$\xrightarrow[Pd]{H_2}$$

Cyclooctadecanonaene

Analogous to the action of N-bromosuccinimide is that of selenium dioxide, which introduces oxygen α to ketones, double bonds, and aromatic rings. Chromyl chloride has also been used to convert aromatic methyl to aldehyde.

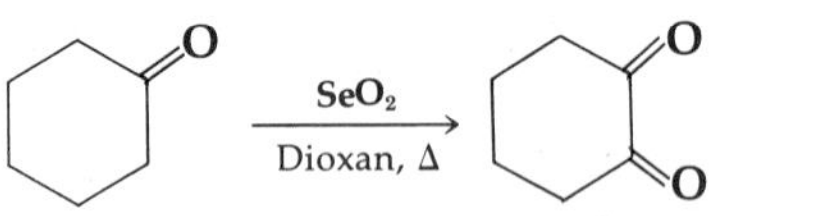

$$\xrightarrow[\text{Dioxan, }\Delta]{SeO_2}$$

1,2-Cyclohexanedione
60%

$$CH_3COCH_3 \xrightarrow{SeO_2} CH_3COCHO$$

Pyruvaldehyde
60%

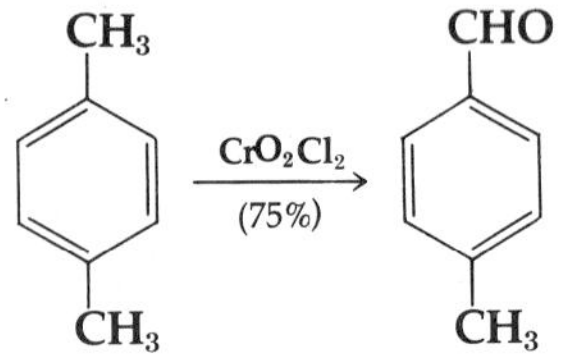

$$\xrightarrow[(75\%)]{CrO_2Cl_2}$$

Introduction of nitrogen is usually achieved α to ketones or esters via acylation with nitrous esters, either base- or acid-catalyzed. Similarly, doubly activated methylenes in base attack diazonium salts. The reagents here act as nitrogen electrophiles, while the carbon nucleophile being oxidized is analogous to the reaction of aromatics with the same electrophiles (page 662).

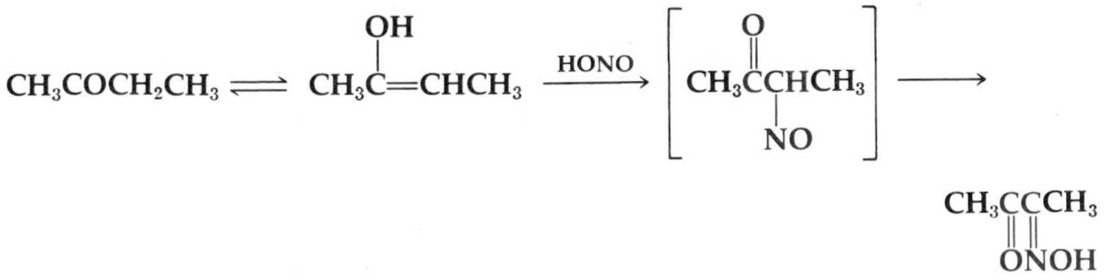

2,3-Butanedione
monooxime

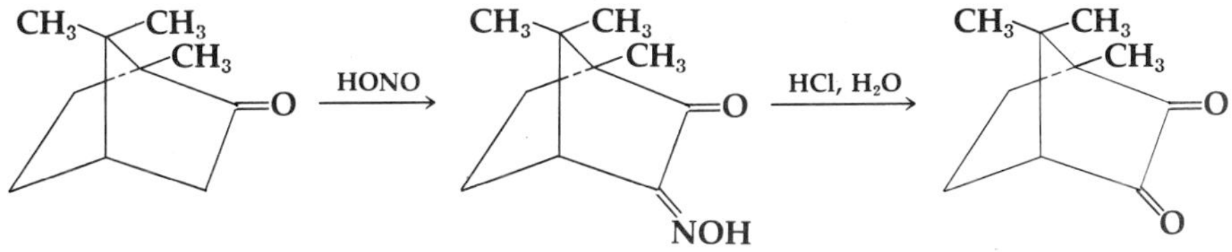

Camphor

3-Ketocamphor
("camphorquinone")

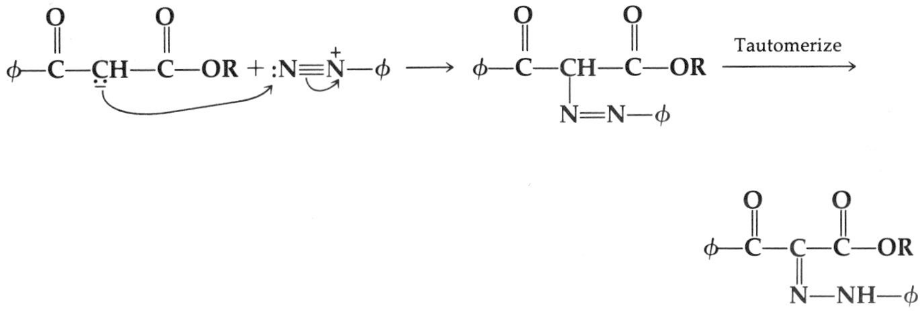

PROBLEM 18-4

List all the methods for introducing heteroatoms at a methylene group. Write their mechanisms.

a Consider the reaction with nitrosobenzene also; is it a reasonable idea and what conditions would be suitable for the reaction?

b In order to create a practical list for synthetic purposes, note for each listed reaction the requirements of substrate structure and experimental conditions.

Alcohols, Halides, and Amines

The usual oxidations of these groups follow the second mechanistic mode outlined in the previous section. Oxidative replacement of α-hydrogen atoms of primary and secondary alcohols is accomplished by many reagents. Chromic acid or anhydride is most frequently used in laboratory procedures (Sec. 14-4). The Oppenauer method, in which an excess of some inexpensive ketone is equilibrated with a secondary alcohol (page 461), is also common. Catalytic dehydrogenation is the usual industrial procedure for conversion of alcohols to aldehydes and ketones.

$$CH_3CH_2OH \xrightarrow{H_2Cr_2O_7} CH_3CHO \xrightarrow{H_2Cr_2O_7} CH_3COOH$$

Isolated only if
removed by continuous
distillation

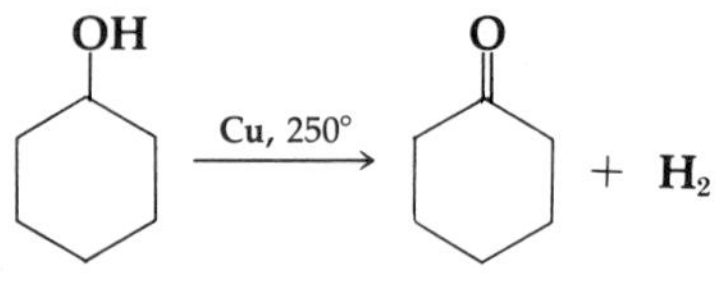

Cyclohexanol   Cyclohexanone

$$(CH_3)_2CHCH_2CH_2OH \xrightarrow[275°]{CuCrO_2} (CH_3)_2CHCH_2CHO + H_2$$

Isoamyl alcohol   Isovaleraldehyde

Oxidation of ethanol by atmospheric oxygen is responsible for formation of the acetic acid in vinegar. The reaction is catalyzed by enzymes produced by the metabolism of certain bacteria. If wine or cider is left exposed to the atmosphere, the beverage is inoculated by airborne cultures of the organism and oxidation occurs, yielding vinegar. Fortified wines, such as sherry, are not oxidized because the enzyme is inactivated by high concentrations of ethanol. Dilute solutions of *pure* ethanol in water are stable because the microorganism needs other nutrients, present in fermentation mixtures.

Chromic acid oxidation of secondary alcohols has been studied in detail. The oxidation step is an elimination reaction of an alkyl ester of chromic acid.

$$R_2CHOH + Cr\bar{\bar{O}}_4 + 2\overset{+}{H} \rightleftharpoons R_2CHOCrO_3H + H_2O$$

Alkyl hydrogen
chromate

(Usually a
weakly basic
solvent)

Di-*t*-butyl chromate is a useful oxidizing agent that can be employed in organic solvents. Ester interchange reactions produce chromates of the alcohols that are to be oxidized. Chromic oxide (CrO_3) is also widely used in pyridine, in acetic acid, in acetone, or in other solvents under either acidic or basic conditions.

A convenient procedure for oxidizing alcohols to aldehydes or ketones involves conversion of the alcohols to sulfonate esters. When heated in basic solutions with dimethyl sulfoxide, the sulfonates then produce the corresponding aldehydes or ketones:

$$CH_3(CH_2)_6CH_2OH \xrightarrow[\text{Pyridine}]{p\text{-}CH_3C_6H_4SO_2Cl} CH_3(CH_2)_6CH_2OTs$$

n-Octyl tosylate
95%

$$(CH_3)_2\overset{+}{S}-\overset{-}{O} + CH_3(CH_2)_6CH_2-OTs \xrightarrow[150°]{NaHCO_3}$$

$$\left[CH_3(CH_2)_6\underset{\underset{\underset{B:}{}}{H}}{CH}-O-\overset{+}{\underset{CH_3}{\overset{CH_3}{S}}} \right] \longrightarrow CH_3(CH_2)_6CHO + (CH_3)_2S$$

Octanal
70%

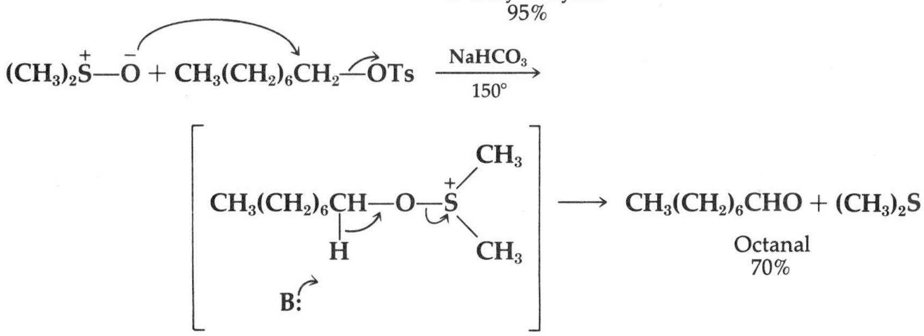

Since the procedure involves prior conversion of the alcohol to a better leaving group, the two steps can be combined into one operation with the following procedure (compare page 510).

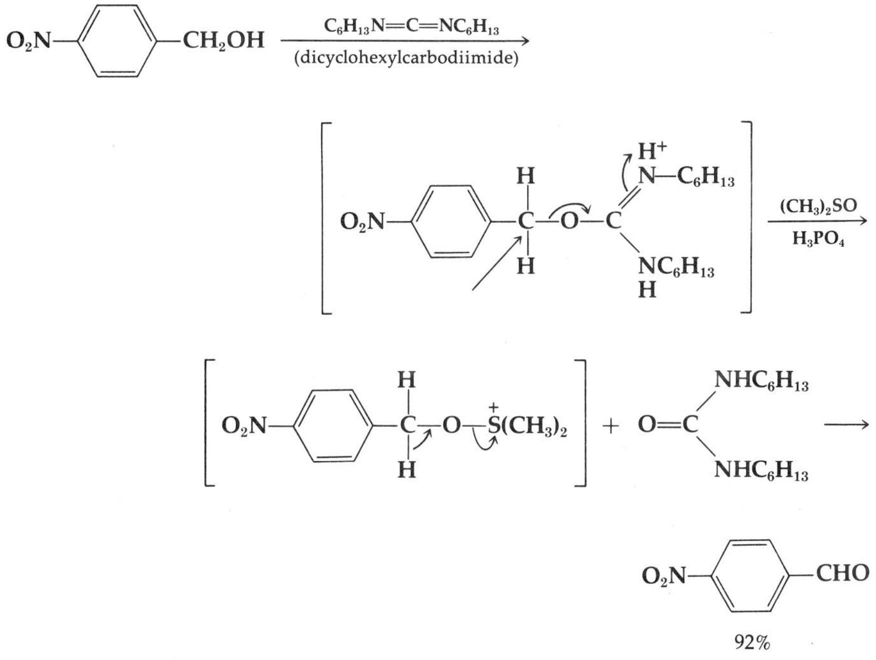

$$O_2N-\langle \text{benzene ring} \rangle-CH_2OH \xrightarrow[\text{(dicyclohexylcarbodiimide)}]{C_6H_{13}N=C=NC_6H_{13}}$$

$$\left[O_2N-\langle \rangle-\underset{H}{\overset{H}{C}}-O-C\overset{\overset{H^+}{N-C_6H_{13}}}{\underset{\underset{H}{NC_6H_{13}}}{}} \right] \xrightarrow[H_3PO_4]{(CH_3)_2SO}$$

$$\left[O_2N-\langle \rangle-\underset{H}{\overset{H}{C}}-O-\overset{+}{S}(CH_3)_2 \right] + O=C\overset{NHC_6H_{13}}{\underset{NHC_6H_{13}}{}} \longrightarrow$$

$$O_2N-\langle \rangle-CHO$$

92%

Epoxides are also opened by displacement with dimethylsulfoxide and thus converted to α-hydroxyketones.

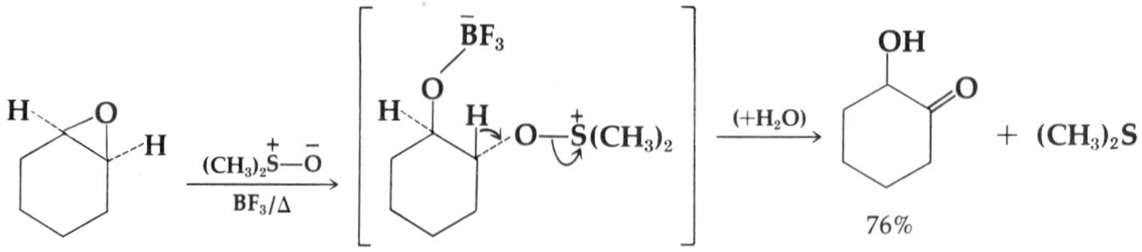

H
O
H
$(CH_3)_2\overset{+}{S}—\overset{-}{O}$
BF_3/Δ
$\overline{B}F_3$
O
H
H
$O—\overset{+}{S}(CH_3)_2$
$(+H_2O)$
OH
O
$+ (CH_3)_2S$
76%

A similar oxidation involves the action of an amine oxide on halides or sulfonate esters of alcohols.

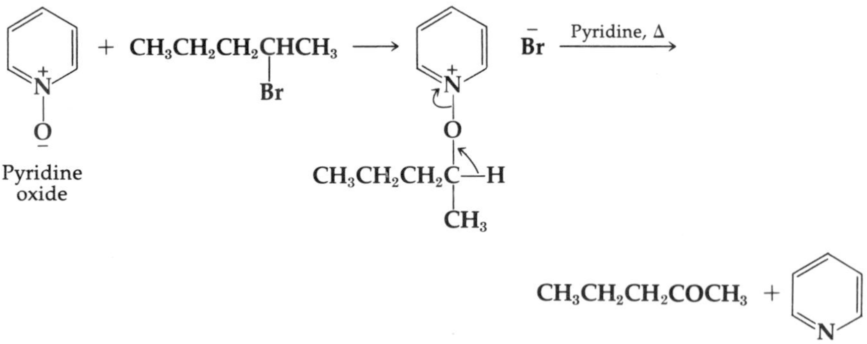

$+ CH_3CH_2CH_2CHCH_3$
Br
N
O
Pyridine oxide
$\overline{B}r$
Pyridine, Δ
$CH_3CH_2CH_2C—H$
CH_3
$CH_3CH_2CH_2COCH_3$ +
N

Drastic conditions are required for oxidation of tertiary alcohols, since carbon-carbon bond cleavage must occur. The products are acids or ketones resulting from oxidative replacement of all hydrogen atoms attached to the carbon atoms involved in the cleavage. Hot chromium trioxide solutions are frequently used in these degradative cleavages. Few applications have been made in synthesis. Under normal (mild) conditions for oxidizing primary and secondary alcohols, tertiary alcohols are unaffected.

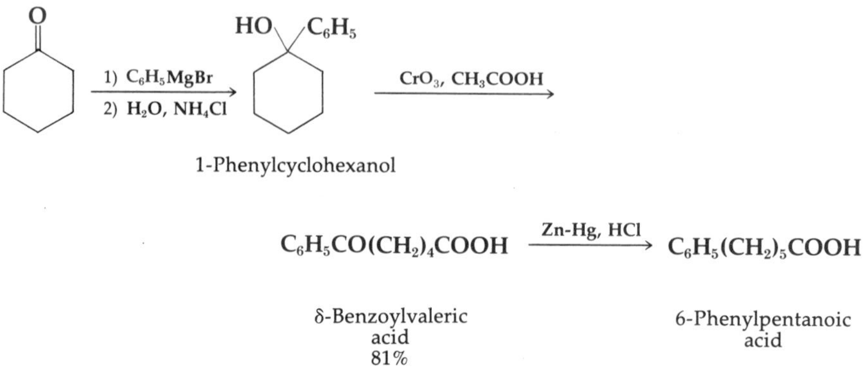

O
1) C_6H_5MgBr
2) H_2O, NH_4Cl
HO
C_6H_5
1-Phenylcyclohexanol
CrO_3, CH_3COOH
$C_6H_5CO(CH_2)_4COOH$
Zn-Hg, HCl
$C_6H_5(CH_2)_5COOH$
δ-Benzoylvaleric acid 81%
6-Phenylpentanoic acid

Oxidation of amines is very facile with most oxidizing agents, but in most cases the products are complex. Tertiary amines are converted to enamines by mercuric acetate in a mechanism parallel to the foregoing examples and lead tetraacetate acts in a comparable manner, although it is more active. The mechanistic analogy to oxymercuration of double bonds (page 634) should be noted.

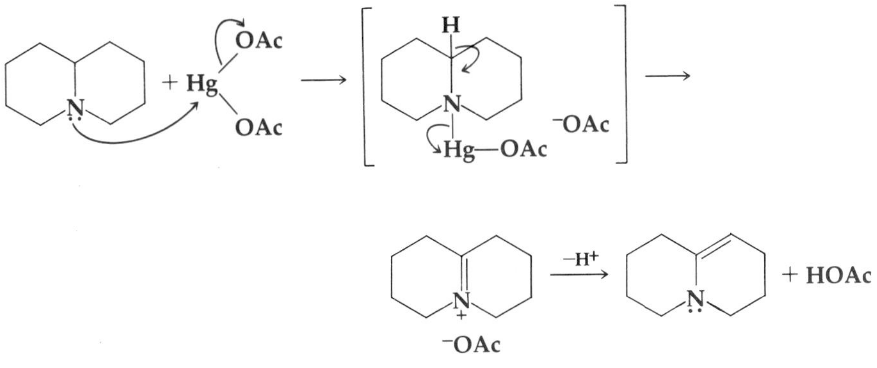

$$[:\overset{-}{\text{H}}\text{gOAc}] \xrightarrow{\text{Hg(OAc)}_2} \text{Hg}_2(\text{OAc})_2 + {}^{-}\text{OAc}$$

Aldehydes and Ketones

The oxidation of aldehydes to carboxylic acids usually follows the second mechanistic pattern as in the oxidation with peracids (RCO—OOH):

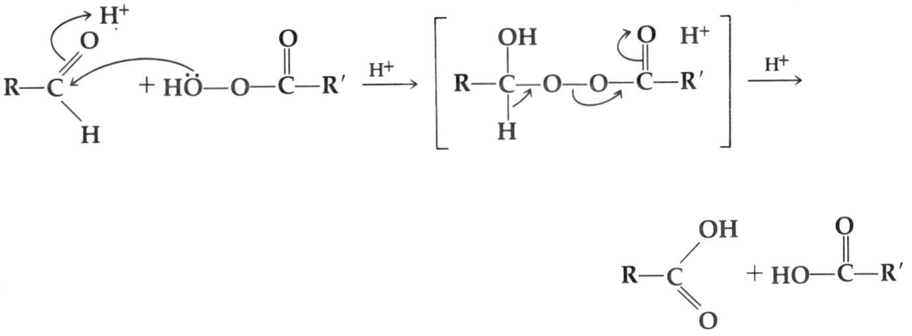

Aldehydes are very easily oxidized by most oxidizing agents, including air, although few reagents give clean conversions. Air oxidation is a chain reaction that proceeds by way of peracids as intermediates. Free-radical traps, such as aromatic amines and phenols, inhibit air oxidation and are added to aldehydes in small amounts to preserve them during storage. Silver oxide, chromic acid, or mild permanganate oxidations are often used in practice to effect the conversion of aldehydes to acids.

The behavior of ketones with respect to oxidation is very similar to that of tertiary alcohols, and must involve carbon-carbon bond cleavage. Such oxidations only occur with vigorous oxidizing conditions and are usually not of preparative value. Hot nitric acid, alkaline or acid permanganate, and chromic acid have all been used to cleave ketones. The commercial preparation of adipic acid by oxidation of cyclohexanone is an example. It is likely that the initial attack is on an enol rather than on the ketone itself.

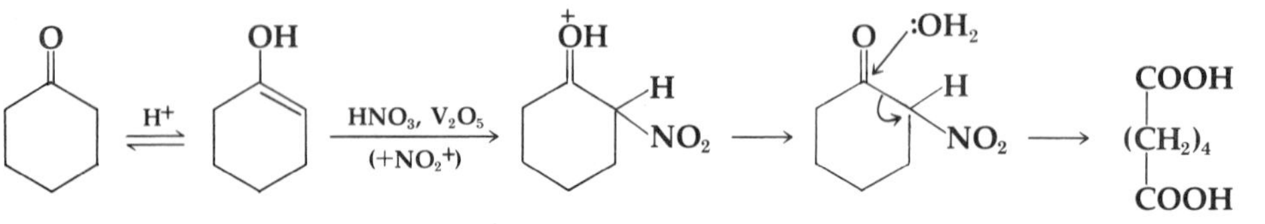

Adipic acid
60%

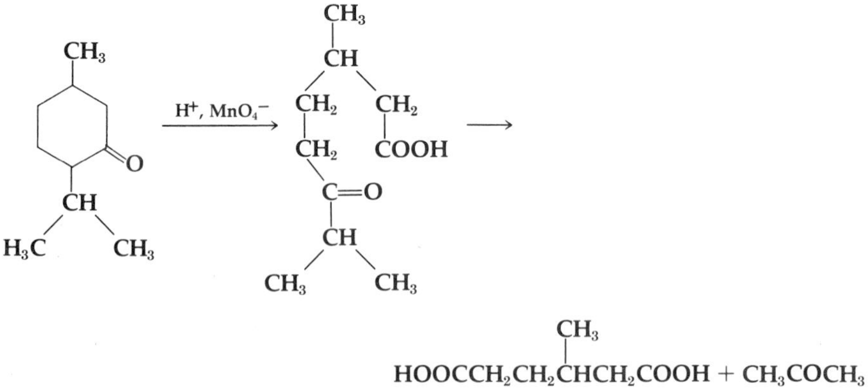

$$CH_3$$
$$HOOCCH_2CH_2CHCH_2COOH + CH_3COCH_3$$

Quinones

Quinones are oxidized forms of aromatic diols and are characterized by two carbonyl groups on a fully unsaturated ring.

Methods for the synthesis of quinones vary with their structure. Benzoquinone is prepared by oxidation of aromatic amines or phenols. Naphthalene and the larger linear polynuclear aromatic hydrocarbons are oxidized directly to form quinones. Orthoquinones are usually prepared from ortho-disubstituted derivatives of the corresponding aromatic systems.

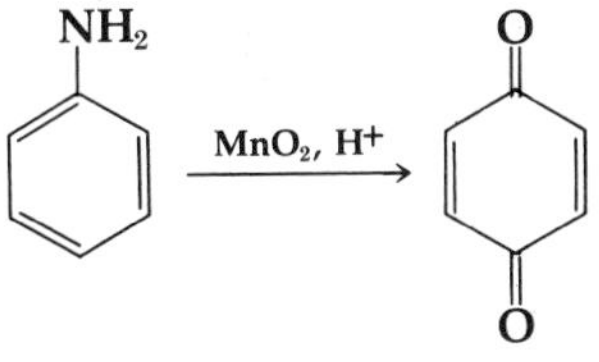

Benzoquinone

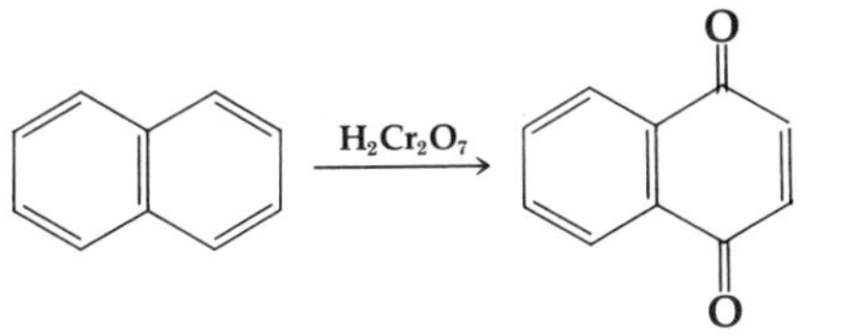

1,4-Naphthoquinone
35%

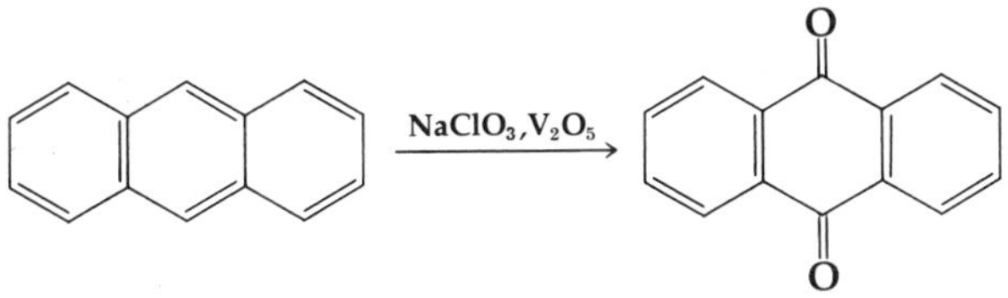

Anthraquinone

The stable radical nitrosyldisulfonate has been used in synthesis of quinones difficult to prepare by other methods.

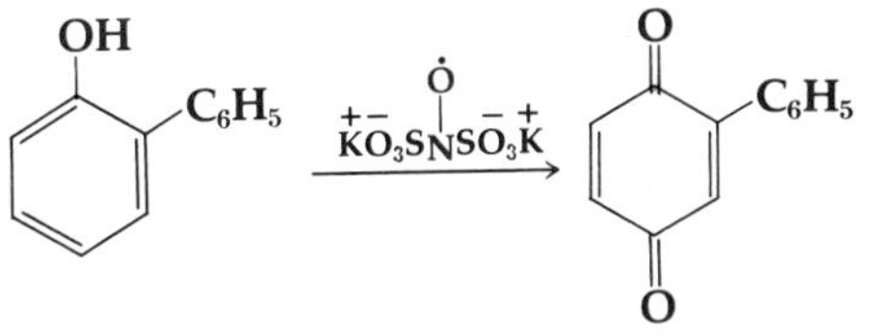

2-Phenylphenol   2-Phenylbenzoquinone

Quinones and the corresponding hydroquinones form oxidation-reduction couples that give reproducible electrode potentials.

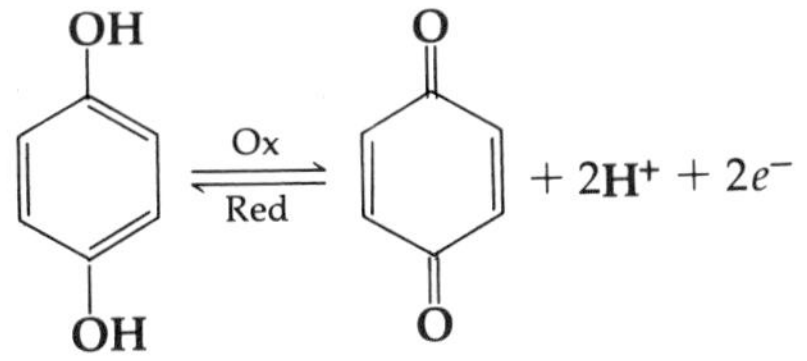

Hydroquinone   Quinone   Catechol   o-Quinone
                  (benzoquinone)

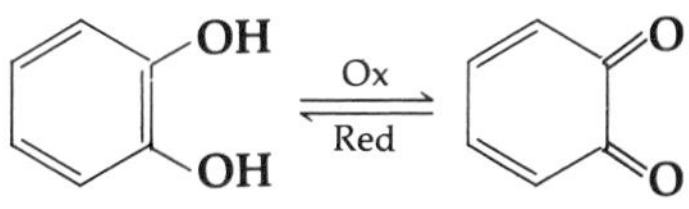

The oxidation potentials of many quinones have been measured by potentiometric titration of the hydroquinones with oxidants of known oxidation potential.

Electron-withdrawing substituents, such as $-NO_2$, $-CN$, $-SO_2Ar$, $-COAr$, $-CO_2H$, *raise the oxidation potential,* making the quinones more powerful oxidants.

Electron-donor substituents, such as $-NHCH_3$, $-NH_2$, $-N(CH_3)_2$, $-OH$, $-OCH_3$, $-CH_3$, $-NHCOCH_3$, C_6H_5, and $-OCOCH_3$, *lower the oxidation potential.*

Since reduction of a quinone involves hydrogen ions, the quinone-hydroquinone system is used as an indicator electrode for measurement of the hydrogen-ion activities of water solutions. The system is known as the **quin-hydrone electrode**, because hydroquinone and quinone combine to form a molecular compound called quinhydrone. The molecular complex has a characteristic black color.

PROBLEM 18-5

Devise ionic mechanisms for the following reactions.

a N-Bromosuccinimide + secondary alcohols ⟶ ketones

b N-Bromosuccinimide + α-hydroxy-acids ⟶ CO_2 + ?

c Benzyl chloride and 2-nitropropane with ethoxide ion yield benzaldehyde and acetone oxime

d ROOCNHNHCOOR + Pb(OAc)₄ ⟶ ROOCN=NCOOR + ?

PROBLEM 18-6

Complete the following interconversions; arrows may represent several steps.

a $C_6H_5COCH_3$ ⟶ $C_6H_5CHOHCH_3$ ⟶ $C_6H_5C_2H_5$ ⟶ C_6H_5COOH

b 2-Cyclohexenone ⟶ cyclohexene

c 2-Cyclohexenone ⟶ cyclohexanone

d C_7H_8O ⟶ $C_6H_5COCH_2OH$ ⟶ C_6H_5COOH

e Cyclohexene ⟶ 1,3-cyclohexadiene

f p-$CH_3C_6H_4CH_2Br$ ⟶ p-$CH_3C_6H_4COCOOH$ ⟶ p-$HOCC_6H_4COCOOH$

18-3 OXIDATIONS AT TWO ADJACENT CARBONS

Double-bond Oxidations

The most important oxidations involving two adjacent carbons are oxidations of double bonds. These are of two kinds, those that cleave the carbon skeleton and those that do not. Most of these reagents have been discussed in Chap. 15 since they all act by an initial electrophilic addition to the double bond. These interrelated oxidations are all summarized in Fig. 18-2, with the

Oxidation
state

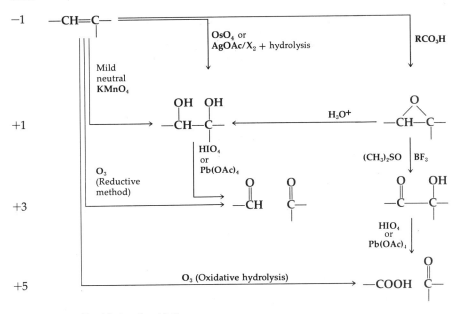

FIGURE 18-2 **Double-bond oxidations**

groups ordered by increasing oxidation state at the two carbons of the original double bond.† These reactions have been very widely used since they are generally selective and specific, the major exception being permanganate, a strong oxidant which attacks many other functions also.

Two reagents, periodic acid and lead tetraacetate, cleave 1,2-diols smoothly to form carbonyl compounds. The reactions have been formulated as passing through cyclic esters as intermediates, but evidence has accumulated that lead tetraacetate oxidations may involve noncyclic, heterolytic mechanisms.

†Since different products arise in some cases depending on whether the double bond bears a hydrogen, the sample in Fig. 18-2 contains one carbon bearing hydrogen and one without, so that both reactions are seen together.

These reagents attack similar compounds which have —OH, —NHR, or C=O on two adjacent carbons:

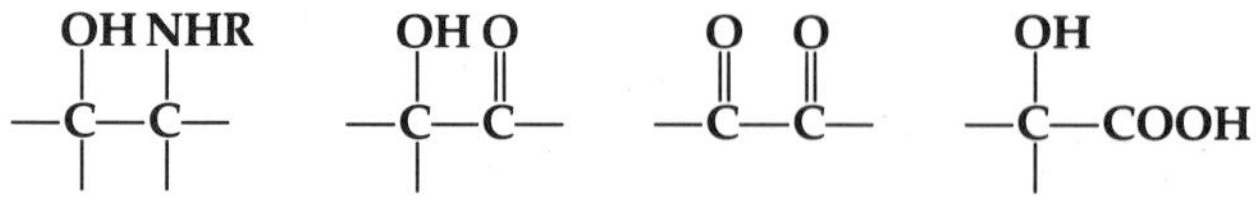

Periodic acid is thus quite selective but lead tetraacetate attacks some other functions (e.g., amines and acids, pages 757 and 764), commonly by a mechanism like the above, with initial displacement, at the lead atom, of an acetate by some nucleophilic center, followed by collapse of the complex with the electrons passing to lead, which leaves as lead diacetate.

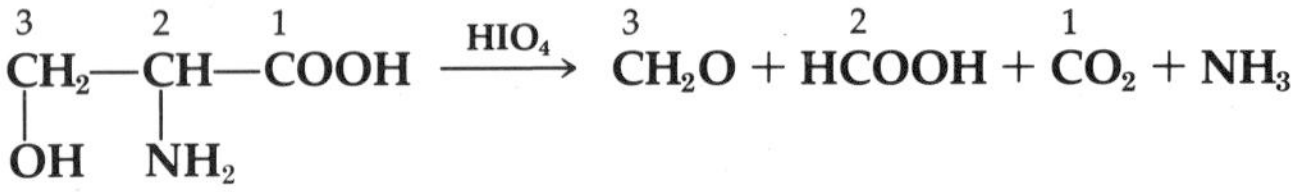

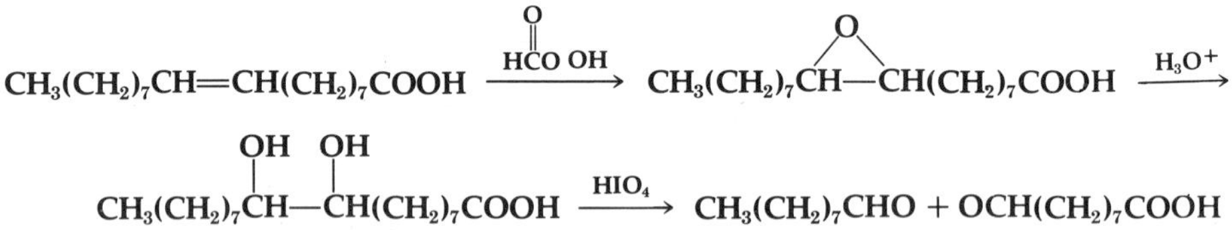

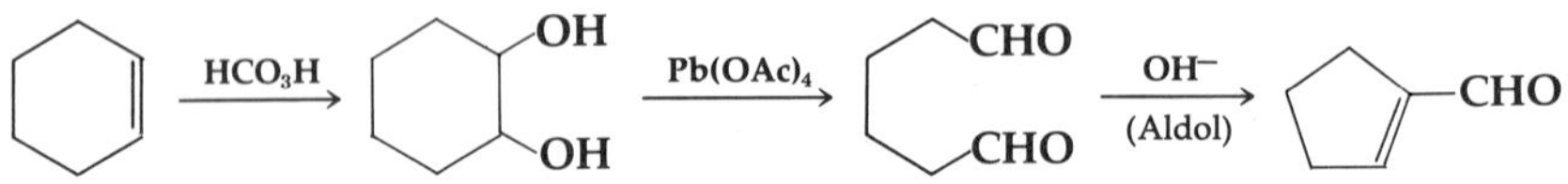

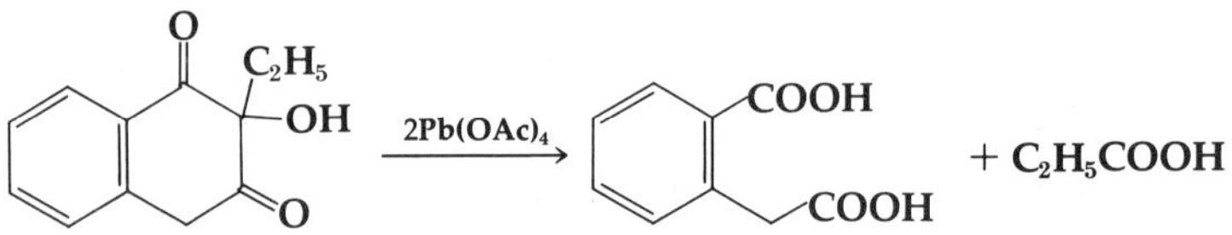

$$\overset{3}{C}H_2-\overset{2}{C}H-\overset{1}{C}OOH \xrightarrow{\text{HIO}_4} \overset{3}{C}H_2O + H\overset{2}{C}OOH + \overset{1}{C}O_2 + NH_3$$
$$\underset{OH}{|}\underset{NH_2}{|}$$

$$CH_3(CH_2)_7CH=CH(CH_2)_7COOH \xrightarrow{\text{HCO OH}} CH_3(CH_2)_7\overset{\displaystyle O}{\overset{\diagup\diagdown}{CH-CH}}(CH_2)_7COOH \xrightarrow{H_3O^+}$$

$$CH_3(CH_2)_7\underset{OH}{CH}-\underset{OH}{CH}(CH_2)_7COOH \xrightarrow{\text{HIO}_4} CH_3(CH_2)_7CHO + OCH(CH_2)_7COOH$$

Chromic acid is sometimes used to cleave double bonds, but it is a strong oxidant also and will react with other functions. Chromic acid oxidations are a key step in the **Barbier-Wieland method** for the stepwise degradation of carboxylic acids.

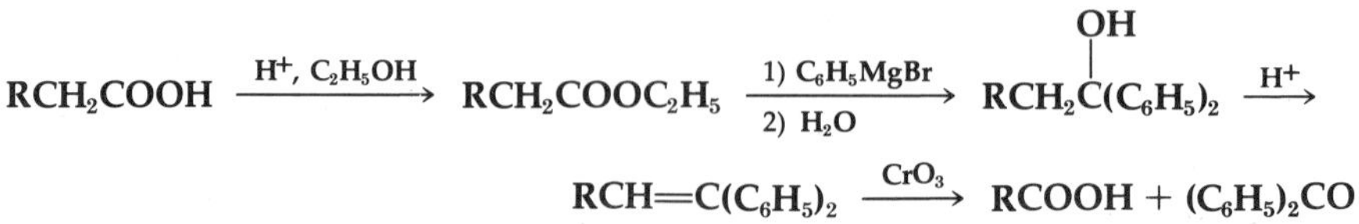

$$RCH_2COOH \xrightarrow{H^+,\ C_2H_5OH} RCH_2COOC_2H_5 \xrightarrow[2)\ H_2O]{1)\ C_6H_5MgBr} RCH_2\overset{\displaystyle OH}{\underset{|}{C}}(C_6H_5)_2 \xrightarrow{H^+}$$

$$RCH=C(C_6H_5)_2 \xrightarrow{CrO_3} RCOOH + (C_6H_5)_2CO$$

Dehydrogenation

Specially prepared solid catalysts are used industrially for the dehydrogenation of alkanes to alkenes. High temperatures are required, and provision must be made for continuous removal of hydrogen. The reactions are frequently carried out by passage of vapor of the substrate over a hot catalyst bed.

$$(CH_3)_2CHCH_3 \xrightarrow[500°]{Cr_2O_3\text{-}Al_2O_3} (CH_3)_2C\!\!=\!\!CH_2$$

$$CH_2\!\!=\!\!CHCH_2CH_3 \xrightarrow[\Delta]{\text{Oxide catalyst}} CH_2\!\!=\!\!CHCH\!\!=\!\!CH_2$$
$$\text{1,3-Butadiene}$$

In the laboratory dehydrogenation is only commonly effected when the product is a resonance-stabilized aromatic ring. Alicyclics that already contain six-membered rings are **aromatized** when heated in the presence of hydrogenation catalysts, such as platinum or palladium, or easily reducible substances such as selenium, sulfur, or chloranil (tetrachlorobenzoquinone).

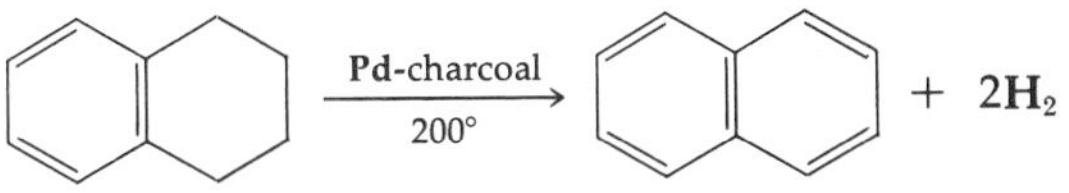

Tetralin

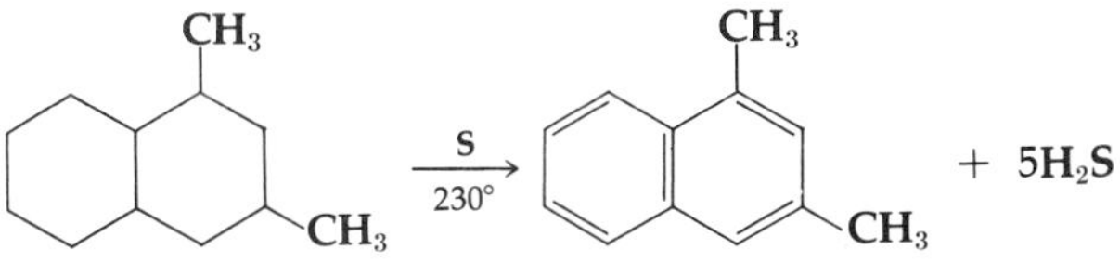

1,3-Dimethyldecalin

Aromatization is frequently the last step in synthetic sequences in which complex aromatic systems are synthesized from aliphatic or alicyclic starting materials.

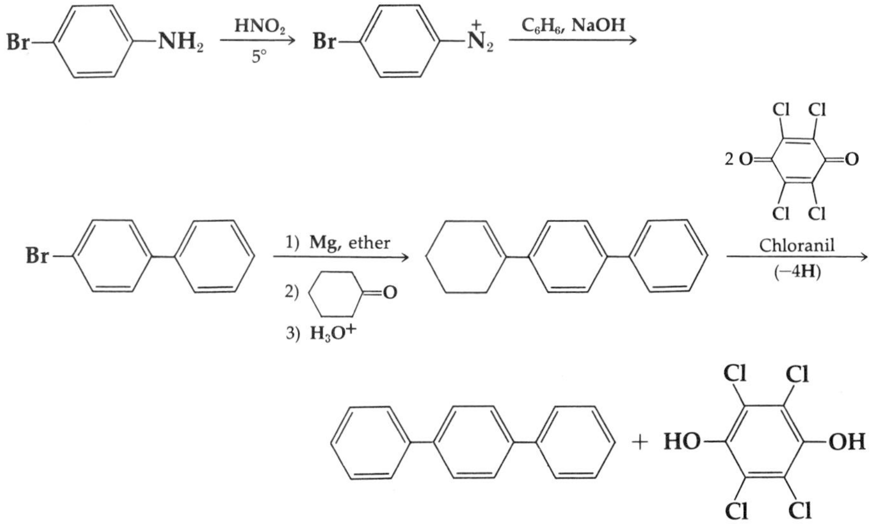

p-Terphenyl   Tetrachloro-
                                              hydroquinone

Other C—C Bond Cleavages

Cleavages of the carbon-carbon single bond occur in certain specific oxidations, several of which have been previously noted:

$$\underset{\overset{|}{\text{OH}}}{-\text{C}}-\text{CH}_2- \xrightarrow{\text{H}_2\text{CrO}_4} \text{C}=\text{O} + \text{HOOC}-$$

$$\underset{\overset{\|}{\text{O}}}{-\text{C}}-\text{CH}_2- \xrightarrow[\text{or HNO}_3]{\overset{\text{H}_2\text{CrO}_4}{\text{or KMnO}_4}} -\text{COOH} + \text{HOOC}-$$

Not selective, rarely useful (vigorous conditions) (p. 758)

$$\underset{\overset{\|}{\text{O}}}{-\text{C}}-\text{R} \xrightarrow{\text{RCO}_3\text{H}} \underset{\overset{\|}{\text{O}}}{-\text{C}}-\text{O}-\text{R}$$

Baeyer-Villiger oxidation (p. 708)

$$\underset{\overset{\|}{\text{O}}}{-\text{C}}-\text{R} \xrightarrow{\text{HN}_3} \underset{\overset{\|}{\text{O}}}{-\text{C}}-\text{NH}-\text{R}$$

Schmidt reaction (p. 707)

$$\underset{\overset{\|}{\text{O}}}{-\text{C}}-\text{CH}_3 \xrightarrow[\text{I}_2]{\text{NaOH}} -\text{COO}^- + \text{HCI}_3$$

Iodoform reaction (p. 747)

The last three reactions are usually selective and of preparative value.

Several procedures for the cleavage of carboxylic acids have been developed, of which the Barbier-Wieland degradation (page 762) is an older, stepwise method. In the **Hunsdiecker reaction**, the silver or thallium salt of a carboxylic acid is treated with bromine. The carboxyl group is lost as carbon dioxide and alkyl or aryl bromides are produced. Evidence points strongly to a homolytic mechanism for the transformation.

$$\text{Br}_2 + (\text{CH}_3)_3\text{CCH}_2\text{C}\bar{\text{O}}_2\overset{+}{\text{Ag}} \xrightarrow[-\text{AgBr}]{\text{CCl}_4} (\text{CH}_3)_3\text{CCH}_2\underset{\overset{\|}{\text{O}}}{\text{C}}\text{O}\text{Br} \xrightarrow{-\text{CO}_2} (\text{CH}_3)_3\text{CCH}_2-\text{Br}$$

Lead tetraacetate has been used in various ways to cleave acids. The carboxyl group is replaced by halide or acetoxy, and the mechanism is probably analogous to that for glycol cleavage.

$$\underset{/}{\overset{\diagdown}{\text{CH}}}-\text{COOH} + \text{Pb(OAc)}_4 \rightleftharpoons \underset{/}{\overset{\diagdown}{\text{CH}}}-\underset{\overset{\|}{\text{O}}}{\text{C}}-\text{O}-\underset{\text{OAc}}{\text{Pb(OAc)}_2} \xrightarrow[\substack{\text{or}\\\text{AcO}^-}]{\text{I}^-}$$

$$\underset{/}{\overset{\diagdown}{\text{CH}}}-\text{I (OAc)} + \text{CO}_2 + \text{Pb(OAc)}_2$$

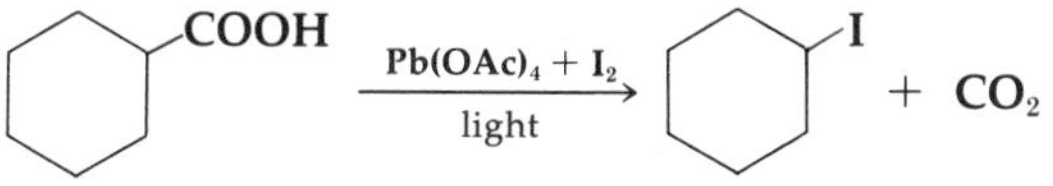

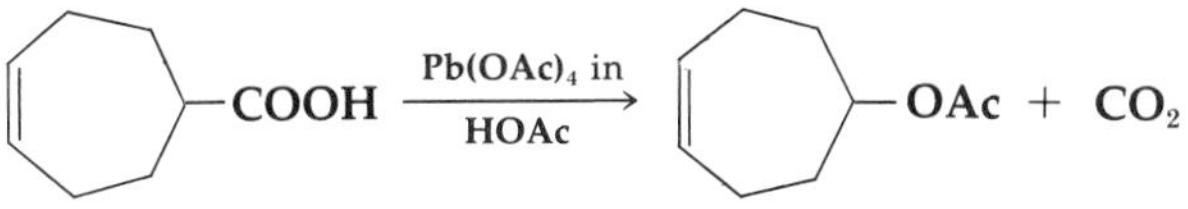

$$\text{cyclohexane-COOH} \xrightarrow[\text{light}]{Pb(OAc)_4 + I_2} \text{cyclohexane-I} + CO_2$$

$$\text{cycloheptene-COOH} \xrightarrow[\text{HOAc}]{Pb(OAc)_4 \text{ in}} \text{cycloheptene-OAc} + CO_2$$

$$CH_3CH_2CH_2COOH \xrightarrow[+LiI]{Pb(OAc)_4} CH_3CH_2CH_2I + CO_2$$

Substituted malonic and succinic acids undergo oxidative bis-decarboxylation with lead tetraacetate.

$$(C_4H_9)_2C\begin{smallmatrix}COOH\\COOH\end{smallmatrix} \xrightarrow[-2CO_2]{Pb(OAc)_4} (C_4H_9)_2C\begin{smallmatrix}OAc\\OAc\end{smallmatrix} \xrightarrow{H_2O} (C_4H_9)C{=}O$$

$$+ 2CO_2 + Pb(OAc)_2 + 2HOAc$$

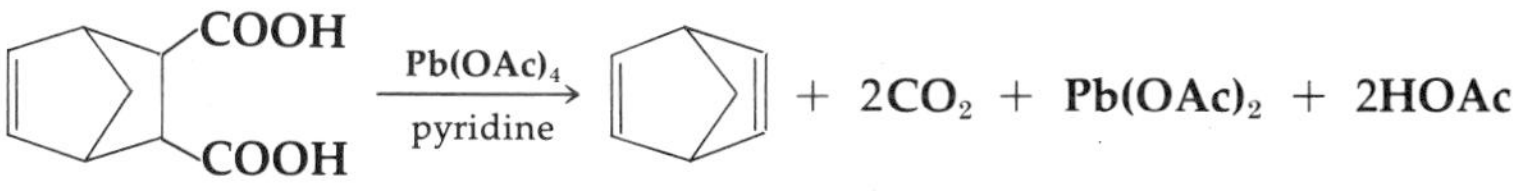

The ease of oxidative cleavage of aromatic rings varies considerably. Benzene is oxidized only under very drastic conditions (page 751), but one ring of naphthalene is much more readily oxidized.

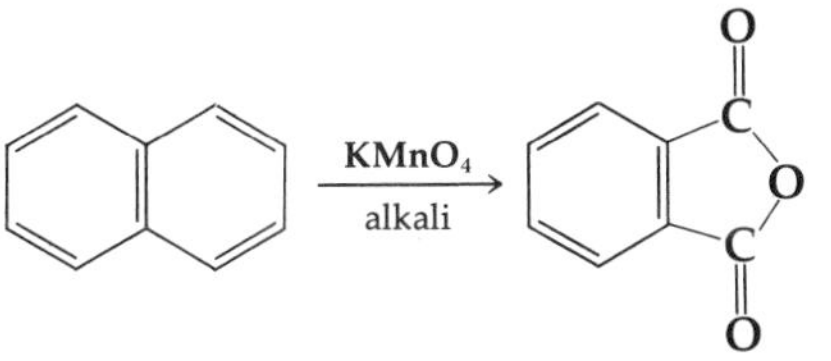

Phthalic anhydride

Electron-withdrawing substituents deactivate aromatic nuclei toward oxidation, and electron-donors have the opposite effect. These facts were used in a classic proof that naphthalene contains two fused aromatic rings.

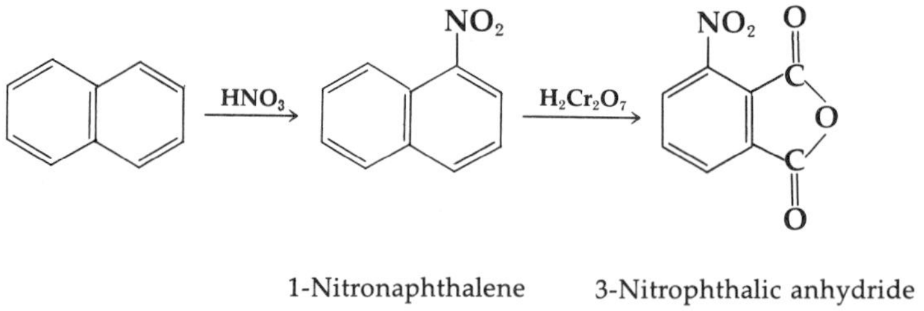

1-Nitronaphthalene   3-Nitrophthalic anhydride

PROBLEM 18-7

a Make a synthetic chart of the possible modes (one or more steps) of converting cyclopentene to cyclopentane derivatives with different oxidation states at one or both of the olefin carbons.

b Make another such chart for modes of converting cyclopentene to acyclic 1,5-derivatives of pentane with various oxidation states at the two terminal carbon atoms.

c Identify the oxidation states on each involved carbon, and write the overall oxidation-state change on each conversion in the two charts.

PROBLEM 18-8

Indicate reagents for the following conversions.

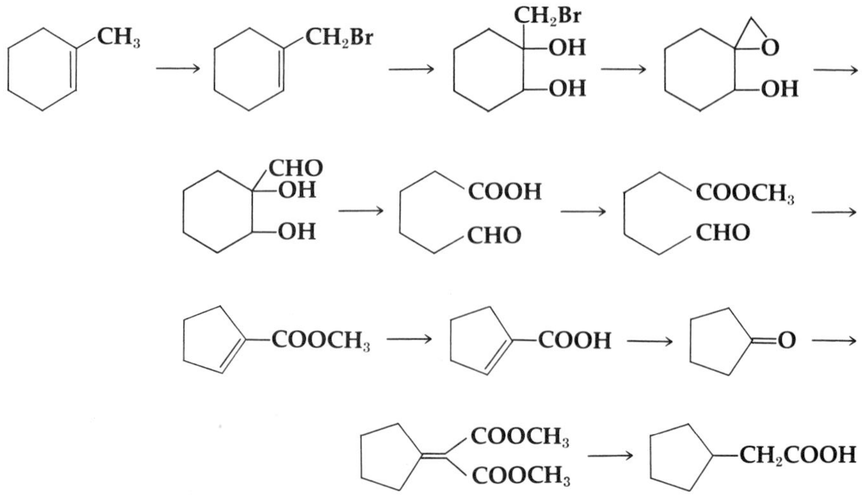

18-4   OXIDATIONS AT OTHER ATOMS

In general amines, and sulfur groups with an unshared electron pair on sulfur, are readily oxidized since their electron pairs attack oxidizing electrophiles with ease. Amines are rendered generally stable to oxidation by making these electrons inaccessible, as occurs on acylation. Nonbasic nitrogenous functional groups are usually not subject to oxidation, e.g., amides, nitriles, nitro groups, and quaternary salts. Among sulfur groups only sulfonyl ($-SO_2-$) bears no electron pair on sulfur and so is generally impervious to oxidation.

Direct oxidation of amines with most oxidants leads to complex mixtures, but peracids and hydrogen peroxide convert tertiary amines to amine oxides, and peroxytrifluoroacetic acid has been used successfully for oxidation of some primary aromatic amines to nitro compounds.

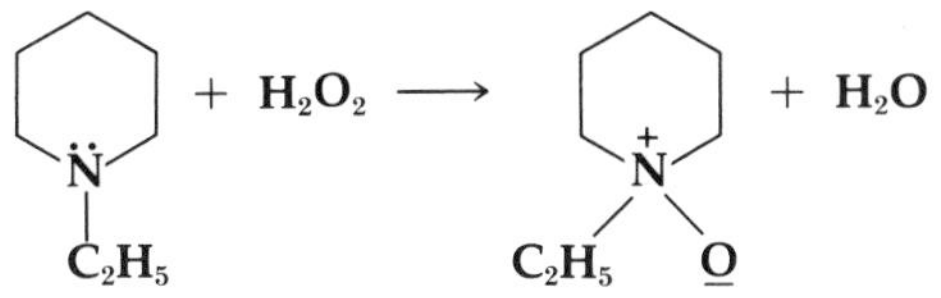

N-Ethylpiperidine

N-Ethylpiperidine oxide

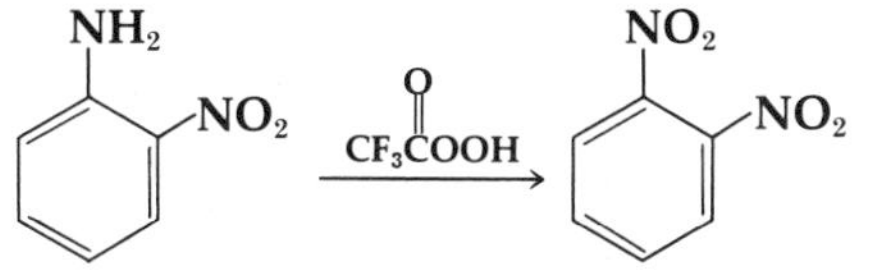

o-Nitroaniline

o-Dinitrobenzene

Amine oxides are of some interest, since they are easily pyrolyzed to give olefins (page 596), and because they have some potential value as specific oxidants (page 756).

Mercaptans are very rapidly oxidized to disulfides by many mild oxidants, including ferric salts, iodine, and oxygen. The reverse reaction is also easily accomplished by reducing agents. This reversible oxidation-reduction equilibrium is very widespread and important in biochemical systems (Chap. 26). Stronger oxidants convert mercaptans and disulfides to sulfonic acids.

$$2C_6H_5SH \xrightarrow[\text{Zn, CH}_3\text{COOH}]{\text{O}_2 \text{ (air)}} C_6H_5SSC_6H_5$$

Thiophenol

Diphenyl disulfide

Sulfides can be oxidized in steps to sulfoxides and sulfones by a variety of strong oxidants. Hydrogen peroxide is usually the reagent of choice. It is difficult to interrupt the oxidation of mercaptans at any stage intermediate between the disulfide and the sulfonic acid. High yields of sulfonic acids are obtained through the action of nitric acid on lead mercaptides, and chlorine water yields sulfonyl chlorides.

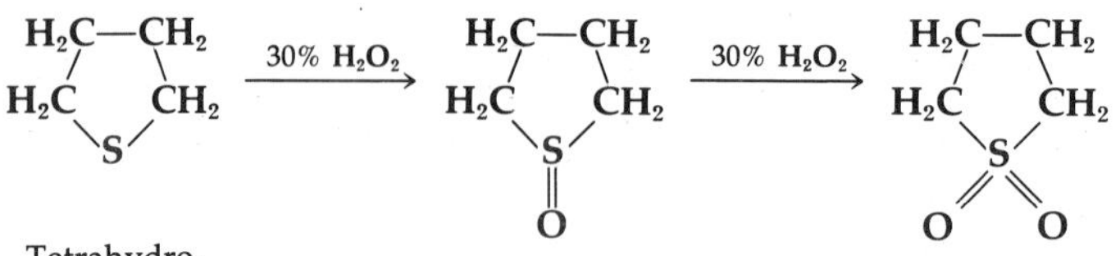

Tetrahydro-
thiophene

$$CH_3(CH_2)_3SH \xrightarrow{Pb(NO_3)_2} [CH_3(CH_2)_3S]_2Pb \xrightarrow{HNO_3} CH_3(CH_2)_3SO_3H$$

Butyl mercaptan
(butanethiol)

Lead
dibutyl mercaptide

1-Butane-
sulfonic acid

18-5　CATALYTIC HYDROGENATION

Nearly all unsaturated compounds add hydrogen in the presence of finely divided metal catalysts, commonly platinum, palladium, and nickel. Although such heterogeneous reactions are exceedingly difficult to study in detail, the mode of action is generally viewed as a bonding of hydrogen atoms to the metal surface first, followed by delivery of the activated hydrogen to π electrons of the organic molecule. Implicit in this view is the prediction that hydrogen addition to a double bond should be cis from the less hindered side, and this is actually the usual result.

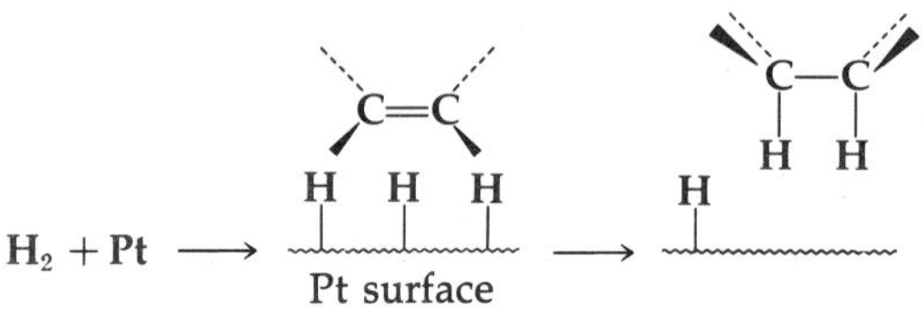

$H_2 + Pt \longrightarrow$　Pt surface

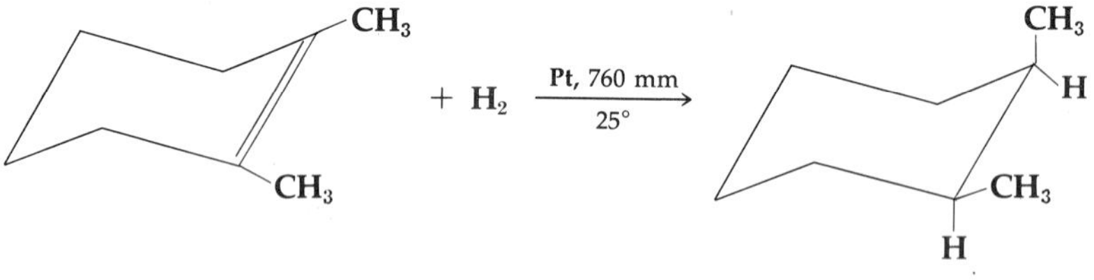

$+ H_2 \xrightarrow[25°]{Pt, 760 mm}$

cis-1,2-Dimethylcyclohexane

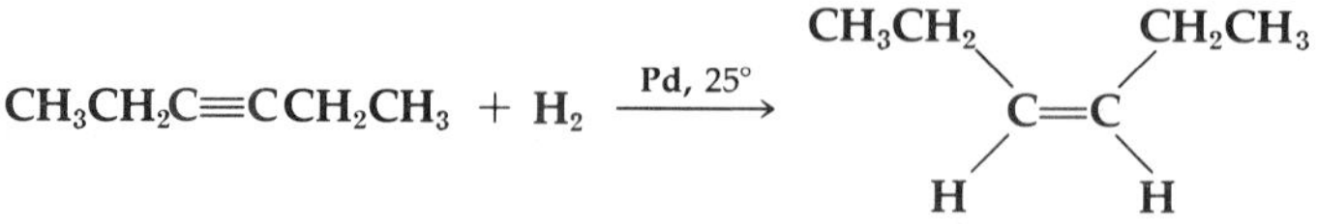

$$CH_3CH_2C{\equiv}CCH_2CH_3 + H_2 \xrightarrow{Pd, 25°}$$

cis-3-Hexene

The reaction is commonly carried out by stirring a solution of the compound with catalyst powder in an atmosphere of hydrogen gas and observing the rate of decrease in hydrogen volume manometrically as a measure of reaction rate. In this way reaction can easily be stopped when the calculated, stoichiometric quantity of hydrogen has been taken up by the sample. Hence when the rates of reduction of different functions are different, the faster one may often be reduced, and the reaction stopped, before any significant amount of the slower group is reduced. The order of rates is shown in Table 18-4 for unsaturated groups. The rates can be vastly different and generally allow hydrogenation of groups at the top of the table without affecting those at the bottom.

Carboxylic acids, esters, and amides are not reduced. Ketones are usually reduced slowly enough that hydrogenation of carbon-carbon double bonds (or other groups above it in the table) may be achieved without affecting the ketone. Benzene rings are inert to normal hydrogenation conditions, owing to resonance stabilization, and normally require high pressure for their reduction.

In the **Rosenmund reduction**, an acid chloride is catalytically reduced with hydrogen to give an aldehyde. The chief drawback of the method is the ease with which aldehydes are reduced. Although results are somewhat erratic, considerable success has attended the use of platinum and palladium catalysts partially inactivated by sulfur or quinoline.

TABLE 18-4   **Order of Reactivity for Hydrogenation**

Group reduced	Product
R—COCl	R—CHO
R—NO$_2$	R—NH$_2$
R—C≡C—R'	R—CH=CH—R' (cis)
R—CHO	R—CH$_2$OH
R—CH=CH—R'	R—CH$_2$—CH$_2$—R'
(more substituted alkenes = slower)	
R—CN	R—CH$_2$NH$_2$
R—CO—R'	R—CHOH—R'

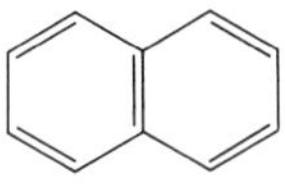

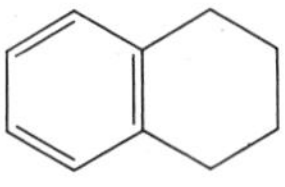

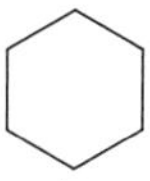

$$C_6H_5CH_2CH_2COCl \xrightarrow[\text{Refluxing xylene}]{H_2,\ Pd\text{-}BaCO_3} C_6H_5CH_2CH_2CHO$$

β-Phenylpropionyl chloride   β-Phenylpropionaldehyde

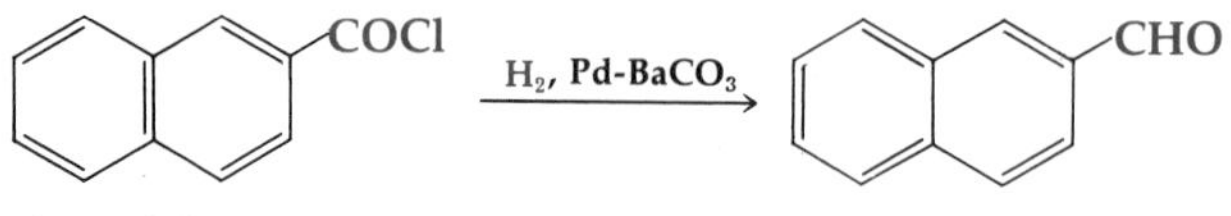

β-Naphthoyl chloride   β-Naphthaldehyde

Hydrogenolysis

Single-bond cleavage on hydrogenation is termed **hydrogenolysis** and occurs in the following situations:

$$C\text{—}X \longrightarrow C\text{—}H$$

$$C\text{—}S \longrightarrow C\text{—}H \quad \text{(only with nickel catalyst)}$$

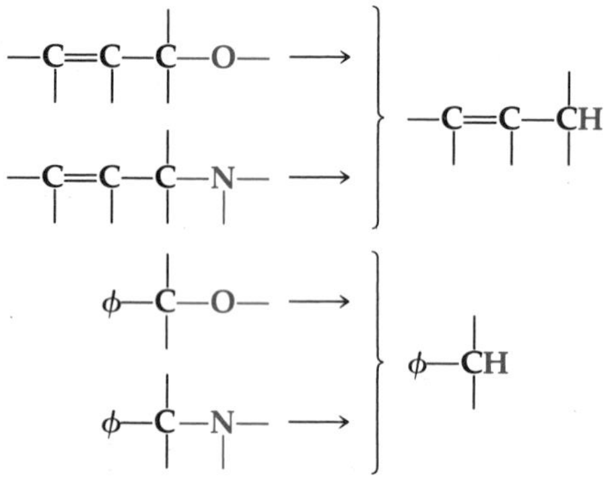

Hydrogenolysis of halides occurs in the presence of platinum or palladium catalysts. The ease of cleavage varies a great deal, aryl halides being rather resistant while allylic and benzylic halides are very reactive.

$$C_6H_5Br + H_2 \xrightarrow[\text{Alcoholic KOH}]{Pd\text{-}CaCO_3} C_6H_6 + HBr$$

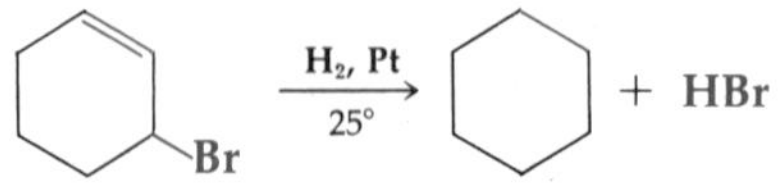

3-Bromocyclohexene   Cyclohexane

Hydrogenolysis of carbon–sulfur bonds (**desulfurization**) is carried out with **Raney nickel**, a finely divided nickel already saturated with hydrogen. Raney nickel is prepared by dissolving the aluminum from a nickel-aluminum alloy with aqueous sodium hydroxide. The hydrogen product in the reaction (of hydroxide with aluminum) adsorbs on the nickel surface. Other catalysts are poisoned and inactivated by sulfur compounds and hence cannot catalyze their hydrogenolysis. The examples include a mild and useful reduction of ketones to methylenes by thioketal desulfurization.

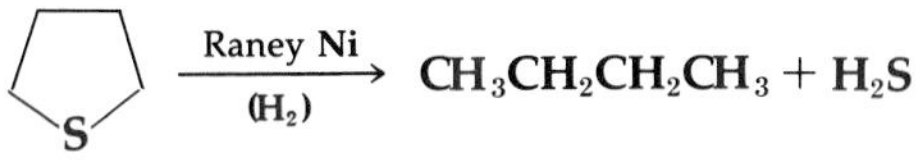

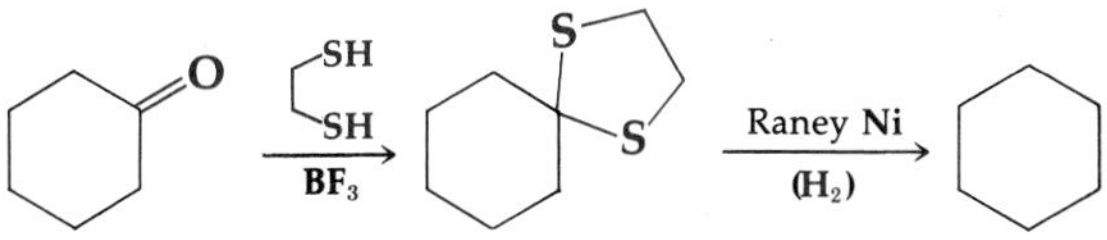

Thioketal

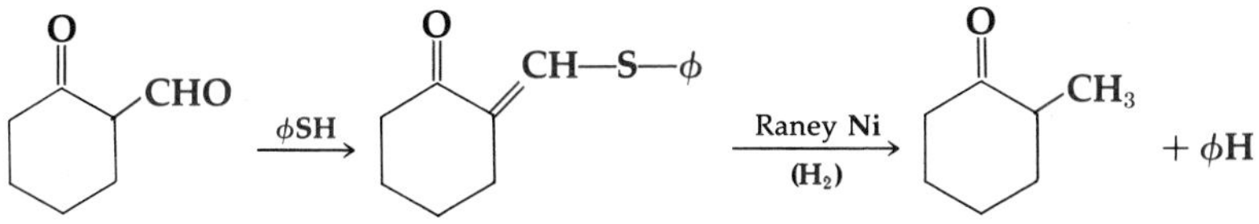

With allylic alcohols, ethers, and amines hydrogenolysis occurs only if its rate is faster than hydrogenation of the double bond. Since they are often comparable, product prediction is difficult. Benzylic groups are smoothly hydrogenolyzed in this order, however:

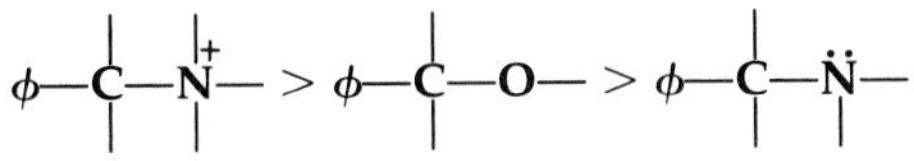

Benzyl esters and ethers are often made in the course of synthetic sequences to protect sensitive functional groups in polyfunctional molecules. The benzyl group may be removed by catalytic hydrogenolysis at the end of the synthesis. The protecting group can be extended to amines by conversion to **carbobenzoxy** derivatives, and these have often been used in peptide synthesis (Sec. 25-4).

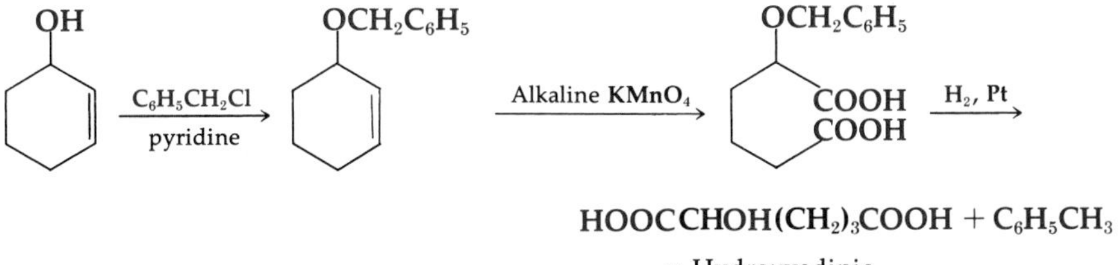

$$HOOCCHOH(CH_2)_3COOH + C_6H_5CH_3$$

α-Hydroxyadipic
acid

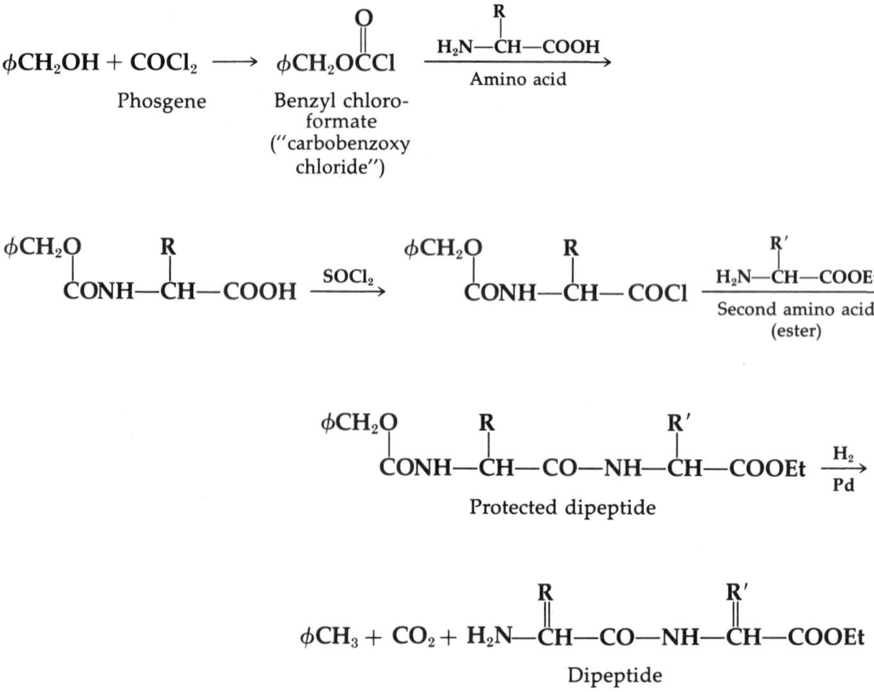

$$\phi CH_2OH + COCl_2 \longrightarrow \phi CH_2O\overset{\overset{\displaystyle O}{\|}}{C}Cl \quad \xrightarrow{\underset{\text{Amino acid}}{\overset{\displaystyle \overset{R}{|}}{H_2N-CH-COOH}}}$$

Phosgene   Benzyl chloro-
                  formate
                  ("carbobenzoxy
                  chloride")

$$\phi CH_2\overset{\overset{\displaystyle O}{|}}{}\quad\overset{\displaystyle R}{|} \qquad\qquad \phi CH_2\overset{\displaystyle O}{|}\quad\overset{\displaystyle R}{|}$$

$$CONH-CH-COOH \xrightarrow{SOCl_2} CONH-CH-COCl \xrightarrow[\text{Second amino acid}]{\overset{\displaystyle \overset{R'}{|}}{H_2N-CH-COOEt}}$$

(ester)

$$\phi CH_2\overset{\displaystyle O}{|}\quad\overset{\displaystyle R}{|}\qquad\qquad\overset{\displaystyle R'}{|}$$

$$CONH-CH-CO-NH-CH-COOEt \xrightarrow[\text{Pd}]{H_2}$$

Protected dipeptide

$$\phi CH_3 + CO_2 + H_2N-\overset{\overset{\displaystyle R}{\|}}{CH}-CO-NH-\overset{\overset{\displaystyle R'}{\|}}{CH}-COOEt$$

Dipeptide

PROBLEM 18-9

Predict the expected products of vigorous catalytic hydrogenation on the following substances. Examine each one stepwise and defend your product predictions
with reasons.

a 2-Cyclopenten-1-ol
b Benzoyl chloride
c Tribenzylamine oxide
d R-2-(p-Nitrophenyl)-2-butanol (optically active)

PROBLEM 18-10

Would you expect the following hydrogenations to be possible under selected
conditions? If not, what products would you expect?

a

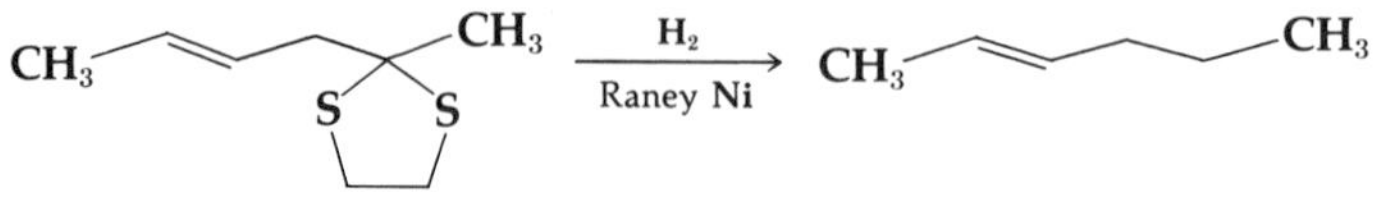

b

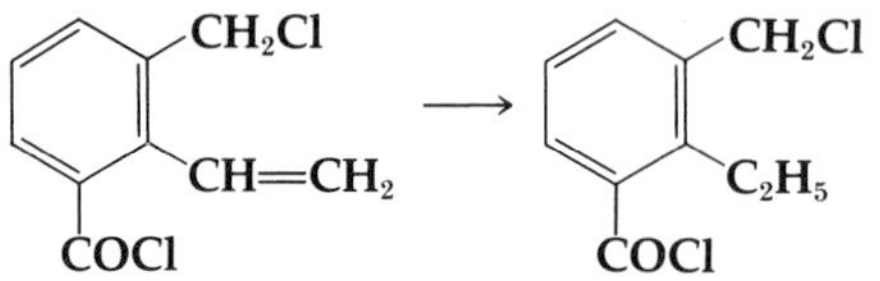

c $CH_3C{\equiv}CCN \longrightarrow CH_3CH{=}CHCN$

d

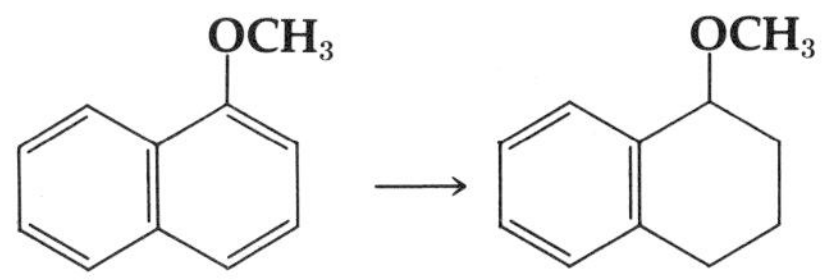

PROBLEM 18-11

Unless nitriles are hydrogenated in acidic medium there is a tendency to form large amounts of amidine. Explain.

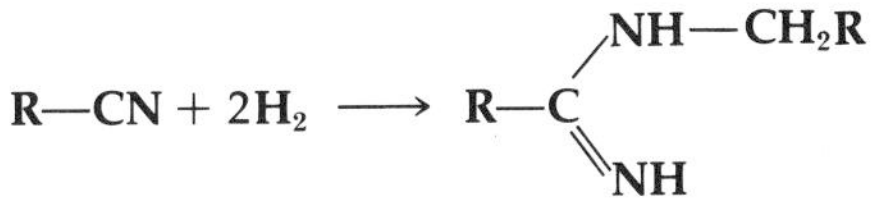

$$R{-}CN + 2H_2 \longrightarrow R{-}C{\big\langle}^{NH{-}CH_2R}_{NH}$$

18-6   REDUCTIONS WITH HYDRIDES

Delivery of hydride ion is usually achieved with complex hydrides like $LiAlH_4$ and $NaBH_4$ or with boranes like BH_3 (actually B_2H_6) and R_2BH. Their reactions have been examined in other contexts, notably substitutions and additions with hydride acting as a nucleophile (Secs. 11-5 and 12-3) and hydroboration of alkenes and alkynes (Sec. 15-5). A summary of their reactions is included in Table 18-5 which shows the variations possible by selecting the correct reducing agent.

Some of the important aspects of hydride reduction selectivity are listed below.

1   $LiAlH_4$ reduces most unsaturated groups except alkenes and alkynes.
2   $NaBH_4$ is much milder and only reduces ketones and aldehydes.
3   The important conversion of **RCOCl** to **RCHO** cannot be achieved with the normal reagents but has been done using a deactivated hydride with three hydrogens replaced by bulky t-butoxy groups: $LiAlH\ [OC(CH_3)_3]_3$.
4   Conversion of nitriles to aldehydes can be effected by using deactivated LiAlH-$(OC_2H_5)_3$.
5   BH_3 (or substituted R_2BH derivatives) is somewhat inverted in activity and so reduces **RCOOH** rapidly and **RCOCl** almost not at all.
6   BH_3 is used for reduction of alkenes and alkynes by hydroboration followed by reaction with acetic acid (page 638).
7   Halides and tosylates are reduced by $LiAlH_4$ only if S_N2 displacement is possible, but all halides are apparently easily reduced (in a free-radical mechanism) by using a hydride of tin, e.g., Bu_3SnH.

TABLE 18-5   **Reactivity of Functional Groups with Reducing Agents**

	Group	Product	H_2/cat	$LiAlH_4$	$NaBH_4$	BH_3 or R_2BH	Li;Na	Others
I	C=C	CH—CH	√	×	×	√	×	
	—C≡C—	—CH=CH—	Fast (cis)	×	×	√ (cis)	√ (trans)	
	R—X { Primary, Secondary, Tertiary } → R—H	R—H	√	√	×	×	√	Bu_3SnH
	Aryl	R—H	√	×	×	×	√	Bu_3SnH
	C—S—	C—H	Raney Ni only	×	×	×	√	
	R—OH R—OR'	R—H	×	×	×	×	×	
	φ—C—OH / —OR / —NR₂	φ—C—H + ROH, H_2O, R_2NH	√	×	×	×	√	
	—C—CH— (epoxide) O	—C—CH₂— OH	√	√	×	√	√	
II	R—NO₂	R—NH₂	Fast	√ †	×	×	√	$Sn;Zn/H^+$
	R—CHO	R—CH₂OH	√	√	√	√	√	Bu_3SnH
	R C=O R'	R CH—OH R'	Slow	√	√	√	√	Bu_3SnH
	C=N—OH	CH—NH₂	Slow	√	×	×	√	$Sn;Zn/H^+$
III	R—COOH	R—CH₂OH	×	√	×	Fast	×	
	R—COOR'	R—CH₂OH	×	√	×	Slow	√	
	R—COCl	R—CHO	Fast	$LiAlH(t\text{-}BuO)_3$	×	×		Bu_3SnH
	R—CO—N<	R—CH₂—N<	×	√	×	×	×	
	R—CN	R—CH₂NH₂	√	√	×	√	√	

† Aromatic—NO₂ yields complex products.

Selectivity in stereochemical result is also possible in ketone reductions if bulky substituted boranes are used, as the following example illustrates. Here the less hindered approach of the reagent is controlling even though it yields (irreversibly) the less stable product:

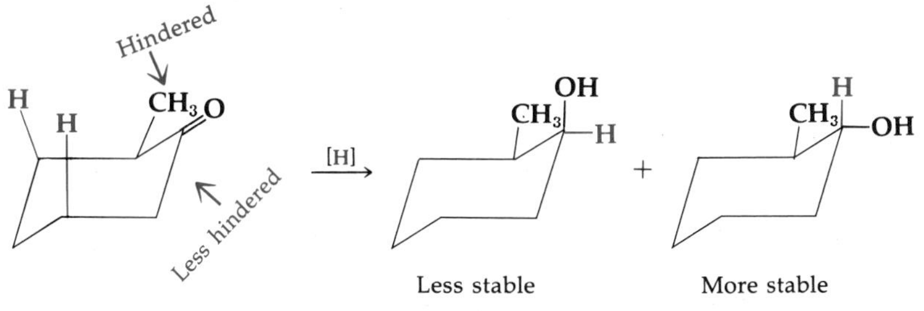

	Less stable	More stable
LiAlH$_4$ or BH$_3$	25%	75%
R$_2$BH ~	100%	~0%

$$R_2BH =$$

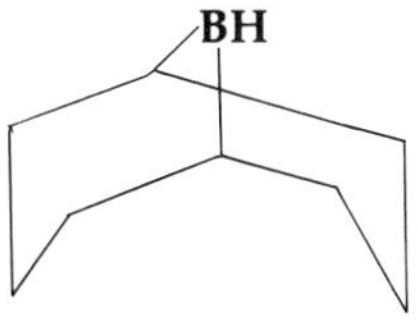

Reduction at nitrogen and sulfur is also quite general with lithium aluminum hydride. In general bonds between these atoms and other heteroatoms are replaced with hydrogen or unshared electron pairs through reductions with LiAlH$_4$.

$$C_6H_5CH_2SO_2Cl \xrightarrow{\text{LiAlH}_4} C_6H_5CH_2SH$$

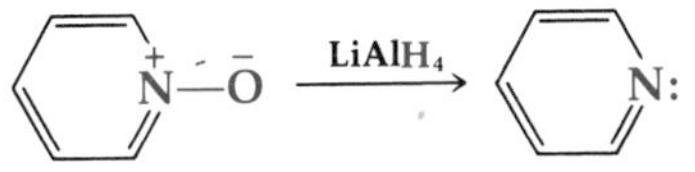

Pyridine N-oxide   Pyridine

It has been shown recently that sodium borohydride is capable of reduction of sulfonylhydrazones of aldehydes and ketones. The reduction proceeds smoothly to yield the hydrocarbon and so constitutes a new and potentially valuable procedure for removal of carbonyl groups in synthesis.

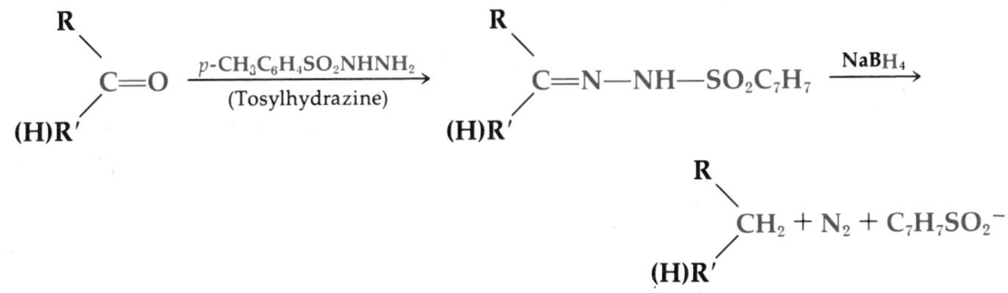

18-7   REDUCTIONS WITH METALS

In general metals are strong reducing agents usually delivering their electrons to a π-electron system. Conjugated alkenes and even aromatic rings are reduced by sodium or lithium but isolated double bonds are rarely affected.

Reduction of aromatic rings with solutions of sodium or lithium in liquid ammonia or amines has become a useful means of producing certain alicyclic compounds (**Birch reduction**). The last double bond is frequently left in the ring system since it reduces with much more difficulty than the others. The first step produces radical-anions, species that are stable only in aprotic solvents such as ether. If a proton source is available (**ROH, RNH$_2$**) the radical-anion is first protonated, then further reduced to another anion. Benzene derivatives yield unconjugated cyclohexadienes, electron-withdrawing substituents remaining on a saturated carbon and electron-donating ones on an unsaturated carbon. The reduction of aryl ethers, followed by hydrolysis, is a very important synthetic device for obtaining substituted cyclohexenones, as illustrated in the fourth example.

Birch reduction:

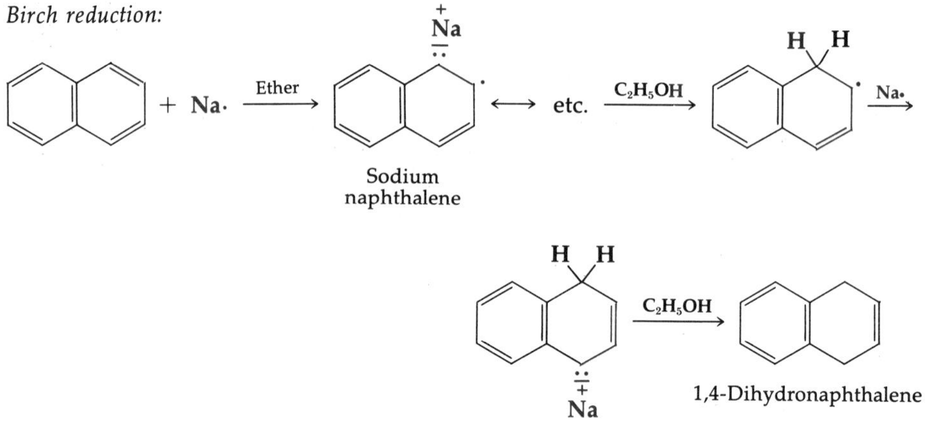

Sodium
naphthalene

1,4-Dihydronaphthalene

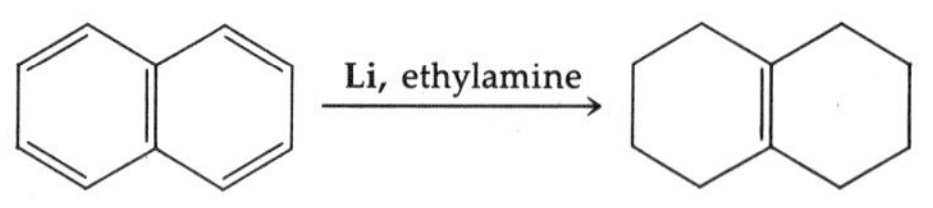

9,10-Octalin
90%

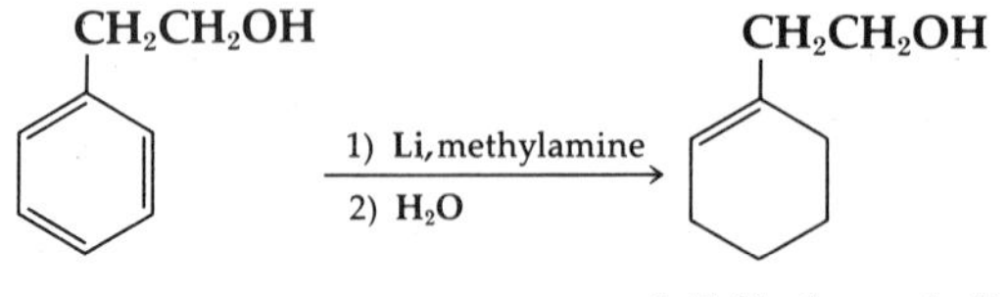

1-(2-Hydroxyethyl)-
cyclohexene

Orientation effects:

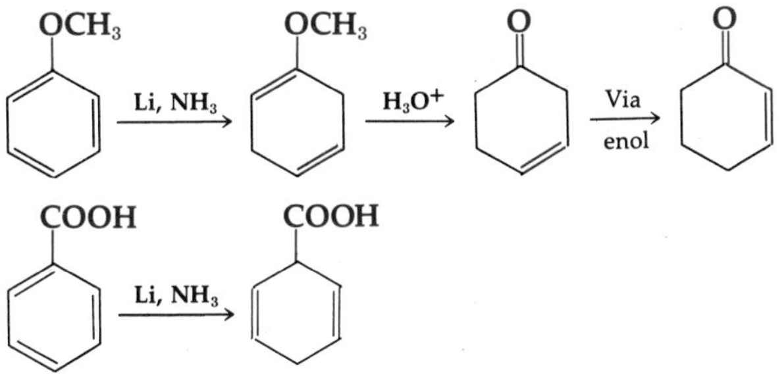

Sodium in liquid ammonia reduces alkynes stereospecifically to trans alkenes. The reaction assumes this course since the negative charges in the dianion intermediate distribute themselves as far from one another as possible. This is in contrast to the cis double bond formed by hydrogenation or hydroboration.

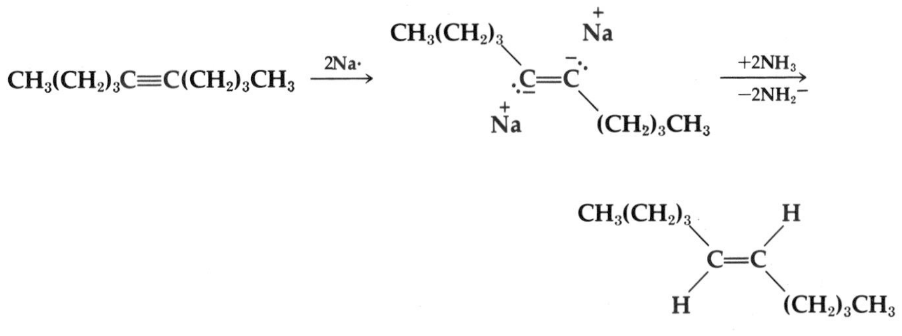

trans-5-Decene
85%

The comparable radical-anion formed by reduction of carbonyls is the first step in several metal reductions which couple two carbon atoms, by linking the radicals. The **acyloin condensation** (page 528) is such a reductive coupling; the **pinacol reduction** is the prototype.

Pinacol reduction:

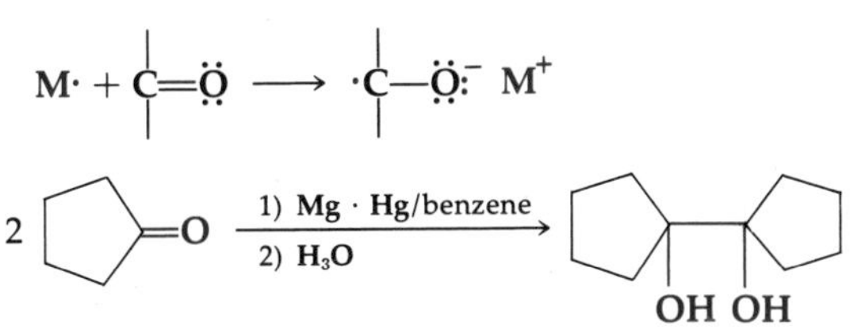

Acyloin condensation:

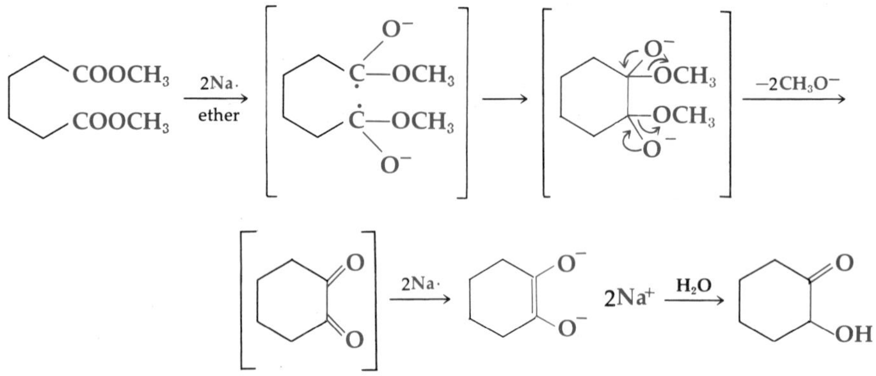

When a proton source intervenes, esters are simply reduced to primary alcohols, and ketones yield secondary alcohols. Carboxylic acids merely form their salts and are unreactive.

$$C_2H_5OCO(CH_2)_8COOC_2H_5 \xrightarrow[C_2H_5OH]{Na} HO—CH_2—(CH_2)_8—CH_2—OH$$

Alkali metals are widely used for reduction of double bonds conjugated with each other, with ketones, or with aromatics. Aromatic rings are slower to reduce.

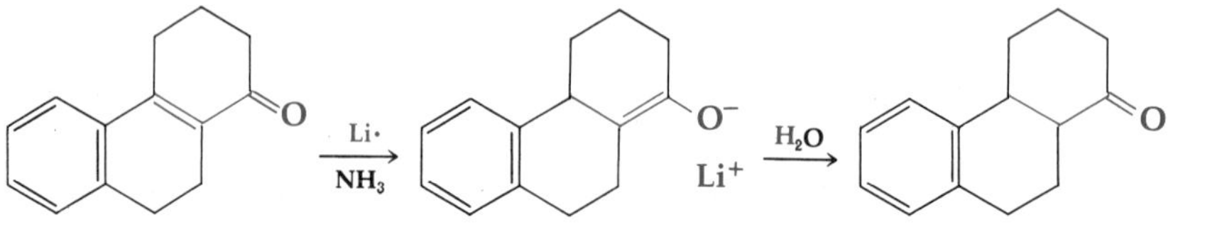

Zinc is used to reduce ketones with oxygen or halogen α-substituents (Sec. 14-3), and under vigorously acidic conditions the ketones themselves are reduced to methylene groups (**Clemmensen reduction**).

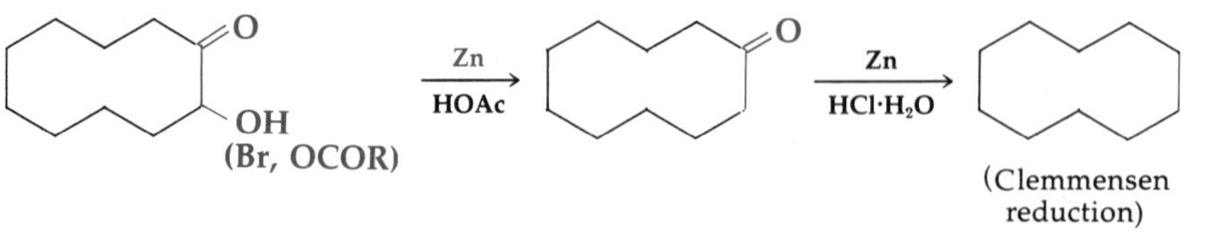

Table 18-5 summarizes the major modes of reduction of the different families of functional groups. Such a table has value not only as a summary but also as a guide to reduction in synthesis, for it allows the selection of a reagent which will effect a desired reduction at one site and leave another function unchanged.

PROBLEM 18-12

a Account for the following reaction, widely observed in certain natural products.

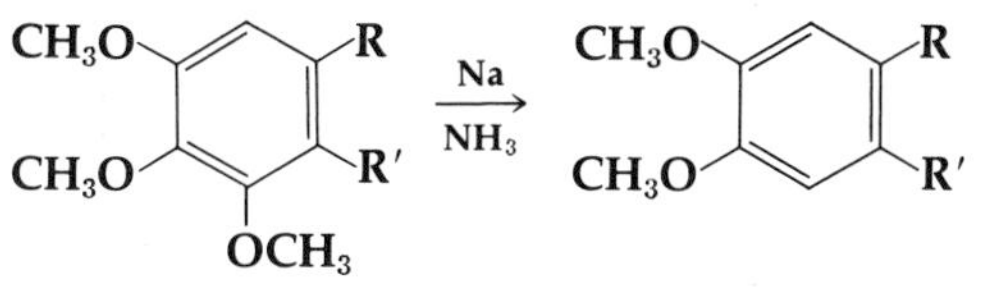

b Oxygen groups attached at the benzylic or α-position to a benzene ring may also be lost in Birch reductions. However, benzoic acids rarely suffer any reduction of the carboxyl group. Explain.

PROBLEM 18-13

Show a mechanism that accounts for the following cyclization reaction.

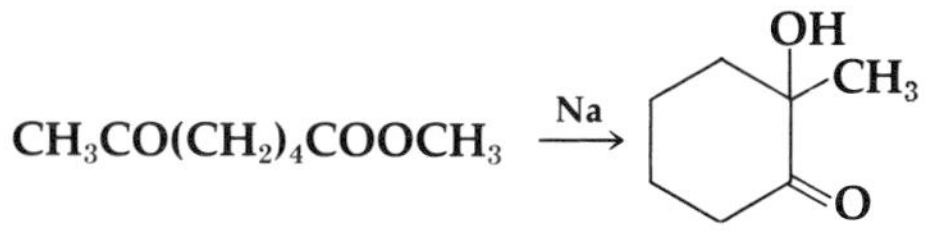

$$CH_3CO(CH_2)_4COOCH_3 \xrightarrow{\text{Na}}$$

18-8 OXIDATION AND REDUCTION IN SYNTHESIS

The previous sections have introduced methods and reagents largely in terms of their modes of action. In the present section we shall gather the chief tools in terms of interconversions between functional groups to emphasize the choices of methods to use in synthesis. These choices are often dictated by the necessity of transforming one functional group without affecting another in the same molecule; Table 18-5 affords a key to this selectivity for reductions.

Hydrocarbons

Saturated hydrocarbon sites are only controllably oxidized when α to unsaturation and thus activated. The chief methods in use are summarized in the following.

$$\left.\begin{array}{c} -\overset{|}{C}=\overset{|}{C}-CH_2- \\[1em] \phi-CH_2- \end{array}\right\} \xrightarrow[\substack{(NBS) \\ (CH_3 > CH_2)}]{N\text{-bromosuccinimide}} \left\{\begin{array}{c} C=C-\overset{|}{C}H-Br \\[1em] \phi-\overset{|}{C}H-Br \end{array}\right.$$

$$\phi-\overset{/}{\underset{\backslash}{C}}\!\!-\ \xrightarrow[OH^-]{Hot\ KMnO_4}\ \phi-COOH$$

$$-CO-\overset{|}{C}H_2 \underset{Zn}{\overset{Br_2\ or\ CuBr_2}{\rightleftharpoons}} -CO-\overset{|}{C}H-Br$$

$$\xrightarrow{SeO_2}\ -CO-CO$$

$$\xrightarrow[H^+\ or\ RO^-]{RONO}\ -CO-\overset{|}{C}=NOH$$

Reduction of alkenes occurs only with hydrogen and with boranes; alkynes are reduced by hydrogen, boranes, or alkali metals. Conjugated alkenes also reduce with metals. Birch reduction is the general choice for partial reduction of aromatic rings although catalytic hydrogenation under pressure yields the fully saturated (and all-cis) cyclohexane derivative.

$$\text{(structure with W, D)} \xrightarrow[\substack{Na\ or\ Li/NH_3}]{Birch\ reduction:} \text{(structure with W, D)} \qquad \begin{array}{l} W = e^-\ withdrawing \\ D = e^-\ donating \end{array}$$

Oxidation of alkenes without cleavage is chiefly executed using halogens (Br_2, Cl_2), N-bromosuccinimide in **ROH** or **RCOOH**, or peracids. It may be noted that epoxides are reconvertible to their parent alkenes by heating with tributyl phosphine.

$$R-\overset{O}{\overset{/\backslash}{CH-CH}}-R' + Bu_3P: \xrightarrow{\Delta} R-CH=CH-R' + Bu_3\overset{+}{P}-\bar{O}$$

Alcohols and Halides

Alcohols are generally stable to reduction and constitute the usual products from reductions of carbonyl functions. Their conversion to ketones and aldehydes is most commonly undertaken with chromic acid or anhydride, under a range of mild conditions ranging from acid to base. Conversion to tosylates or halides is necessary for reduction to hydrocarbon. Halides are fairly easily reduced to hydrocarbons, quite generally with catalytic hydrogenation, tributyl tin hydride, or metal reduction, but only on primary and secondary sites for S_N2 displacement by lithium aluminum hydride.

Ketones and Aldehydes

These groups are both about equally reducible to alcohols by all metal hydrides, sodium borohydride being the choice when other functions are present. Reduction with aluminum isopropoxide (page 460) is also very selective. Complete reduction to hydrocarbon can be effected in four quite different ways:

1 Clemmenson reduction (acidic): $\diagdown$C=O $\xrightarrow{\text{Zn + HCl·H}_2\text{O}}$ $\diagdown$CH$_2$

2 Thioketal reduction (neutral): $\diagdown$C=O $\xrightarrow[\text{BF}_3]{\overset{\text{CH}_2\text{SH}}{\underset{\text{CH}_2\text{SH}}{}}}$ $\overset{\text{S—CH}_2}{\underset{\text{S—CH}_2}{\text{C}}}$ $\xrightarrow[\text{nickel}]{\text{Raney}}$ $\diagdown$CH$_2$

3 Tosylhydrazone reduction (~ neutral): $\diagdown$C=O $\xrightarrow{\text{H}_2\text{NNHSO}_2\text{C}_7\text{H}_7}$

$\diagdown$C=N—NHSO$_2$C$_7$H$_7$ $\xrightarrow{\text{NaBH}_4}$ $\diagdown$CH$_2$ + N$_2$ + C$_7$H$_7$SO$_2{}^-$

4 Wolff-Kishner reduction (basic): $\diagdown$C=O $\xrightarrow{\text{H}_2\text{N—NH}_2}$ $\diagdown$C=N—NH$_2$ $\xrightarrow[\Delta]{{}^-\text{OH}}$

$\diagdown$CH$_2$ + N$_2$

The **Wolff-Kishner reduction** is often carried out by heating the ketone with hydrazine and alkali, not isolating the hydrazone. The mechanism involves base-catalyzed tautomerization of the hydrazone followed by loss of nitrogen.

$$\underset{\text{O}}{\overset{\|}{\text{RCR}}} \xrightarrow[\text{KOH, 100}°]{\overset{\text{NH}_2\text{NH}_2,}{\text{O(CH}_2\text{CH}_2\text{OH)}_2}} \underset{\text{NNH}_2}{\overset{\|}{\text{RCR}}} \underset{\text{H}_2\text{O}}{\overset{\text{HO}^-, 200°}{\rightleftharpoons}} \underset{\underset{*}{\text{NNH}^-}}{\overset{\|}{\text{RCR}}} \underset{\text{OH}^-}{\overset{\text{H}_2\text{O}}{\rightleftharpoons}} \underset{\text{N}=\text{NH}}{\overset{|}{\text{RCHR}}} \rightleftharpoons$$

$$\underset{\underset{\text{N}\rightleftharpoons\text{N}}{}}{\text{RCHR}} \xrightarrow{-\text{N}_2} \text{R}\bar{\text{C}}\text{HR} \xrightarrow{\text{H}_2\text{O}} \text{RCH}_2\text{R}$$

$$\underset{\text{Dibenzyl ketone}}{\text{C}_6\text{H}_5\text{CH}_2\text{COCH}_2\text{C}_6\text{H}_5} \xrightarrow[\text{KOH, 200}°]{\overset{\text{NH}_2\text{NH}_2,}{\text{O(CH}_2\text{CH}_2\text{OH)}_2}} \underset{\underset{93\%}{\text{1,3-Diphenylpropane}}}{\text{C}_6\text{H}_5\text{CH}_2\text{CH}_2\text{CH}_2\text{C}_6\text{H}_5}$$

Oxidation
state

Oxidation
state

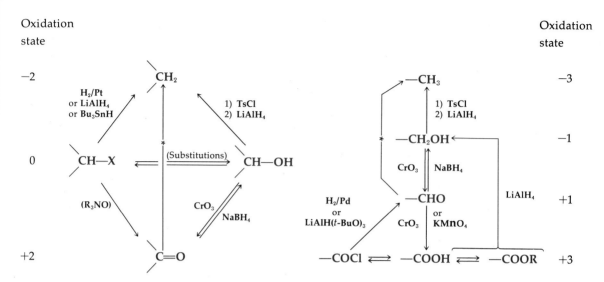

*Clemmensen, Thioketal or Wolff-Kishner reductions.

FIGURE 18-3   Major oxidation-reduction interconversions

Carboxylic Acid Derivatives

Generally these resist hydrogenation but are smoothly converted to primary alcohols by hydrides. Acid chlorides are generally the most reactive, although inert to boranes. Their conversion to aldehydes is often difficult because of the further reactivity of the product aldehyde to reduction.

The interconversions of these groups are summarized in Fig. 18-3.

Nitrogen Compounds

The lowest oxidation state of nitrogen is represented by the amines (-3). Amines bearing hydrogen are easily oxidized but rarely to single, useful products. In synthesis they are usually protected by acylation before oxidations are attempted on other groups. Tertiary amines are converted to enamines by mercuric acetate and to N-oxides by peracids.

Reductions which produce primary amines are the chief routes to these compounds:

$$R—C≡N \xrightarrow[\text{or} \quad \text{LiAlH}_4]{\text{Na·/EtOH} \quad \text{or} \quad \text{H}_2/\text{Pt}/\text{H}^+} \left.\begin{array}{c} \\ \\ \end{array}\right\} R—CH_2NH_2$$

$$R—CONH_2 \xrightarrow{\text{LiAlH}_4 \text{ only}}$$

$$\underset{/}{\overset{\backslash}{C}}H—NO_2 \xrightarrow[\text{or} \quad \text{Zn/H}^+]{\text{H}_2/\text{Pt} \quad \text{or} \quad \text{LiAlH}_4} \underset{/}{\overset{\backslash}{C}}H—NH_2$$

$$\phi—NO_2 \xrightarrow{\text{H}_2/\text{Pt} \quad \text{or} \quad \text{Zn, Sn/H}^+} \phi—NH_2$$

$$\underset{/}{\overset{\backslash}{C}}=NOH \xrightarrow[\text{or} \quad \text{Zn/H}^+]{\text{H}_2/\text{Ni} \quad \text{or} \quad \text{LiAlH}_4} \underset{/}{\overset{\backslash}{C}}H—NH_2$$

$$R—N_3 \xrightarrow{\text{H}_2/\text{Pt} \quad \text{or} \quad \text{LiAlH}_4} R—NH_2$$

Secondary and tertiary amines are commonly produced from reduction of amides or imines:

$$R—CO—N\underset{R''(H)}{\overset{R'}{<}} \xrightarrow{\text{LiAlH}_4} R—CH_2N\underset{R''(H)}{\overset{R'}{<}}$$

$$\left.\begin{array}{c} \underset{/}{\overset{\backslash}{C}}=N—R \\ \\ \underset{/}{\overset{\backslash}{C}}=\overset{+}{N}\underset{R'}{\overset{R}{<}} \end{array}\right\} \xrightarrow[\text{or LiAlH}_4]{\substack{\text{H}_2/\text{Pt} \\ \text{or NaBH}_4}} \left\{\begin{array}{c} \underset{/}{\overset{\backslash}{C}}H—NH—R \\ \\ \underset{/}{\overset{\backslash}{C}}H—N\underset{R'}{\overset{R}{<}} \end{array}\right.$$

Nitro compounds and their relatives are very readily reduced by many reagents. In acidic media, nitro groups are converted smoothly to primary amino groups. Reduction in alkaline solutions is complicated by condensations among the various species present. LiAlH$_4$ reduces aliphatic nitro compounds to amines but creates complex mixtures with aromatic derivatives. Catalytic hydrogenation causes complete reduction to amines very smoothly.

Since aromatic nitro compounds are easily made by nitration reactions, they are used as starting materials in syntheses of virtually all aryl-nitrogen compounds.

$$\phi—NO_2 \text{ (Nitrobenzene)}$$

$$\xrightarrow{\text{Sn, HCl}} \phi—NH_2 \xrightarrow{\text{HNO}_2} \phi—\overset{+}{N}≡N$$
Aniline 86%

$$\xrightarrow[65°]{\text{Zn, NH}_4\text{Cl}} \phi—NHOH \xrightarrow{\text{H}_2\text{Cr}_2\text{O}_7,\ 0°} \phi—NO$$
Phenylhydroxylamine 65%   Nitrosobenzene 77%

$$\xrightarrow{\text{Zn, NaOH, CH}_3\text{OH}} \phi—N=N—\phi \xrightarrow{\text{Zn, NaOH, CH}_3\text{OH}} \phi—NHNH—\phi$$
Azobenzene 85%   Hydrazobenzene 88%

PROBLEMS

18-14 In each case below deduce structures which account for the empirical change noted as well as the reagent used. The equations are not balanced as written; using the oxidation state changes involved, balance each equation.

a $C_7H_{12} + KMnO_4 \longrightarrow C_7H_{12}O_3$ (acidic) $+ Mn^{++}$

b $C_9H_{12} + KMnO_4 \xrightarrow{\Delta} C_8H_6O_4$ (acidic) $+ MnO_2$

c $C_7H_{14}O_2 + H_2Cr_2O_7 \longrightarrow C_7H_{10}O_3$ ^{0}acidic) $[H_2CrO_3]$

d $C_6H_{10}O_4 + Pb(OAc)_4 \longrightarrow C_4H_8 + Pb(OAc)_2$

e $C_{10}H_8O + KMnO_4 \longrightarrow C_8H_6O_4$ (acidic) $+ MnO_2$

f $C_9H_{10}O + I_2/NaOH \longrightarrow C_8H_8O_2$ (acidic) $+ NaI$

g $C_7H_{12}O_4 + NaIO_4 \longrightarrow C_4H_6O_3$ (acidic) $+ NaIO_3 + ?$

h $C_{10}H_{20}O + (CH_3)_2SO \longrightarrow C_{10}H_{20}O_2 + (CH_3)_2S$

i $C_{11}H_{10}O + Li \longrightarrow C_{11}H_{14}O + Li^+$

j $C_{10}H_{12}O_2 + H_2NNH_2/NaOH \longrightarrow C_{10}H_{16}$

k $C_4H_2N_2 + H_2/Pt \longrightarrow C_4H_{12}N_2$

18-15 With starting materials containing no more than six carbon atoms, write equations for syntheses of the following substances:

a Cyclohexanone

b $C_6H_5CH_2CH_2N(CH_3)_2$

c $C_6H_5CH_2CHO$

d 1,2-Cyclodecanedione

e 1-Methylnaphthalene

f m-Aminotoluene

g 1,6-Cyclodecanedione

h

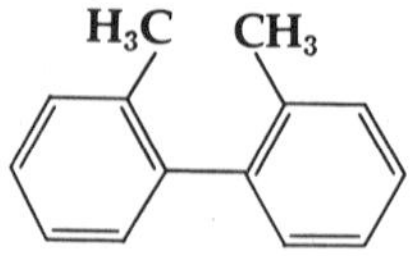

i $C_6H_5CHOHCHNH_2CH_2CH_2CH_3$

j

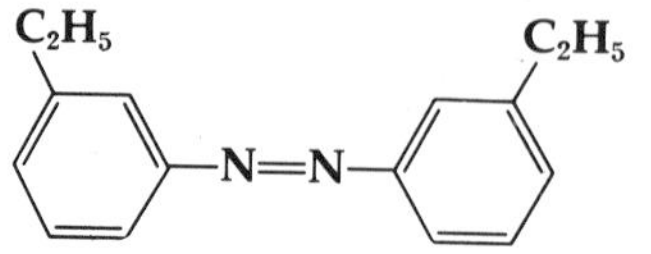

k $CH_3(CH_2)_5CH_2\overset{+}{N}(CH_3)_2$ with $\overset{|}{\underset{}{O}}$

l $C_6H_5CH_2SO_3H$

m $CH_3CH_2CH_2\underset{\underset{NH_2}{|}}{CH}CH_2CH_2CH_3$

n

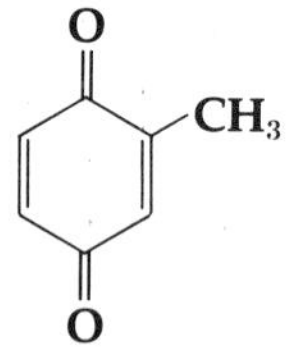

o 4,5-Nonanedione

p $C_6H_5(CH_2)_5CHO$

q $CH_3(CH_2)_{10}COOH$

r $[CH_3(CH_2)_4]_2SO_2$

s $trans$-4-Octene

t Cyclodecane

u

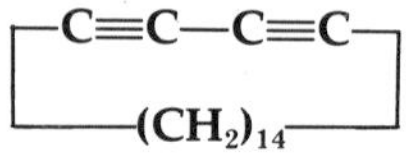

18-16 Formulate examples of the following:

    a A quinone with a higher and one with a lower oxidation potential than methylbenzoquinone

    b Five separate methods for conversion of a ketone to a hydrocarbon

    c Five reactions or processes for degradation of a carbon chain by one carbon atom

    d Three ways of making aromatic compounds from saturated hydrocarbons

    e Three reactions making use of metals in either liquid ammonia or primary amine solutions

    f Six methods of converting a secondary alcohol to a ketone

18-17 An unknown compound, $C_{10}H_{12}O$, when treated with $(NaSO_3)_2NO\cdot$ gave a quinone. The original substance, when heated with zinc, gave a new substance $C_{10}H_{12}$, which when treated with palladium at $210°$, gave two moles of hydrogen and a new compound $C_{10}H_8$. Write equations for the reactions.

18-18 Phenanthrene, when catalytically reduced with only 1 mole of hydrogen, added hydrogen in the 9,10-positions. Careful oxidation of the product with nitric acid introduced a carbonyl group into the 9-position. Write equations for these reactions, and indicate with additional equations how the following compound (A) could be prepared.

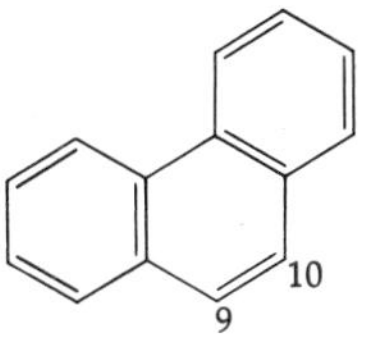

    Phenanthrene 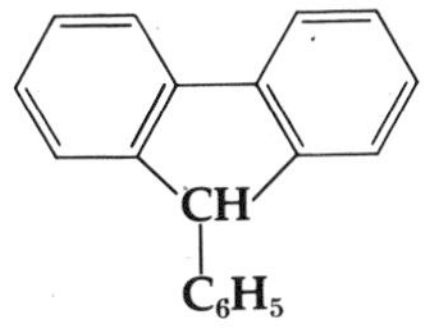 A

18-19 Formulate mechanisms for the following reactions:

    a Na (liquid NH_3) + 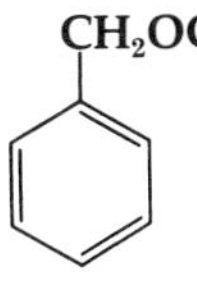 $\longrightarrow$ $C_6H_5CH_3 + NaOCH_3$

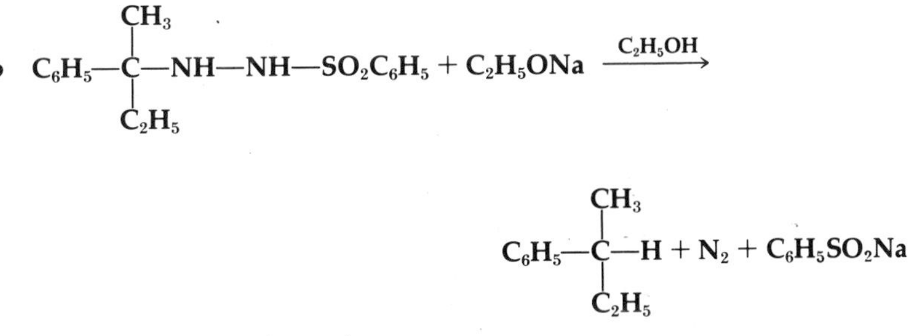

(*Hint:* $C_6H_5SO_2^-$ is a good leaving group.)

c $CH_3CH_2\overset{\overset{\displaystyle O}{\|}}{C}-\underset{\underset{\displaystyle OH}{|}}{CH}CH_2CH_3 + HS(CH_2)_3SH \xrightarrow[\text{HCl}]{\text{ZnCl}_2}$

$CH_3CH_2-\underset{\underset{(CH_2)_3}{\overset{S}{\diagdown}\ \overset{S}{\diagup}}}{C}-CH_2CH_2CH_3 +$ (S–S ring structure with CH₂ CH₂ CH₂)

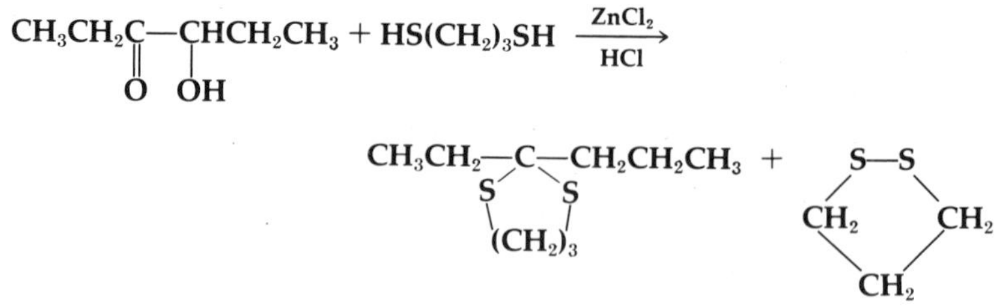

d $H_2O_2 + OsO_4 + CH_3CH{=}CHCH_3 \longrightarrow CH_3\underset{\underset{OH}{|}}{CH}-\underset{\underset{OH}{|}}{CH}CH_3$

1 mole   Trace   1 mole   Meso
                         cis

(*Hint:* H_2O_2 is used in excess of the stoichiometric amount, and only a trace of OsO_4 is used as a catalyst.)

e In Sec. 18-3 a cyclic mechanism for cleavage of 1,2-diols is proposed. This mechanism is consistent with the fact that cis cyclic 1,2-diols in general cleave faster than their trans isomers. Many systems do cleave, however, in which the hydroxyl groups are rigidly trans to one another. Propose a mechanism for the cleavage of such a system with lead tetraacetate.

f In the following overall reaction, an oxidation as well as a hydrolysis has occurred. Write a mechanistic scheme that explains the observation.

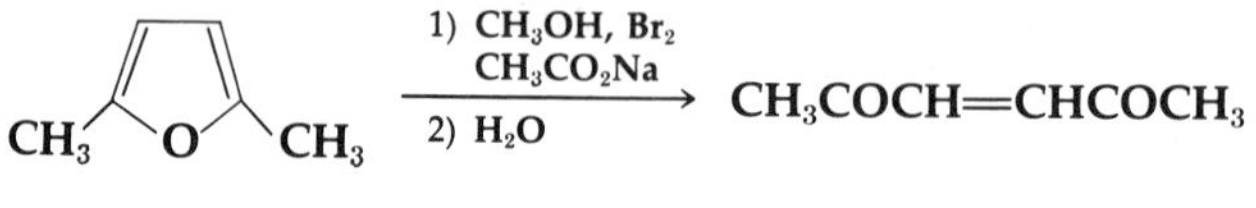

$\xrightarrow[\text{2) } H_2O]{\substack{\text{1) } CH_3OH,\ Br_2 \\ CH_3CO_2Na}}$ $CH_3COCH{=}CHCOCH_3$

18-20 The dark pigment polyporic acid is found in lichens. Lead tetraacetate oxidation of the pigment leads to pulvic anhydride, another natural lichen pigment. It seems likely that such an oxidation *in vivo* is the source of natural pulvic anhydride.

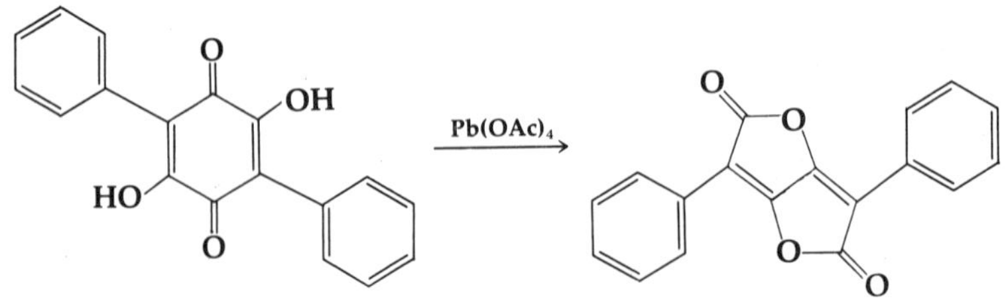

Polyporic acid   Pulvic anhydride

18-21 Reduction of optically active $C_{12}H_{16}O_3$ with $LiAlH_4$ led to optically *inactive* $C_{10}H_{14}O_2$. Upon discerning that the ultraviolet spectrum of each compound was about identical to that of toluene, the investigator decided that the structures of each were *uniquely* determined. Can you agree? Write the structures.

18-22 Devise a mechanism for the following observation and offer a reason for the reaction taking this course.

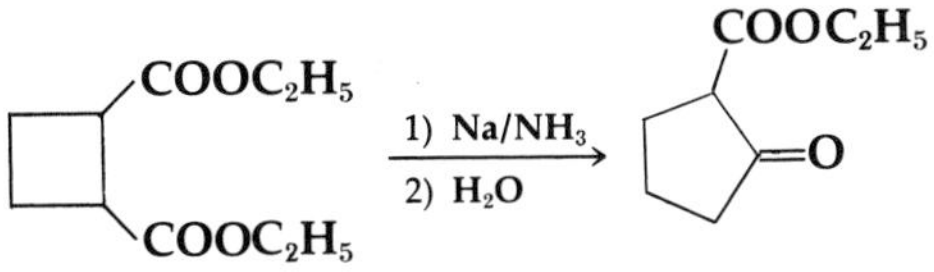

18-23 Provide sequences of reactions to effect the following overall conversions.

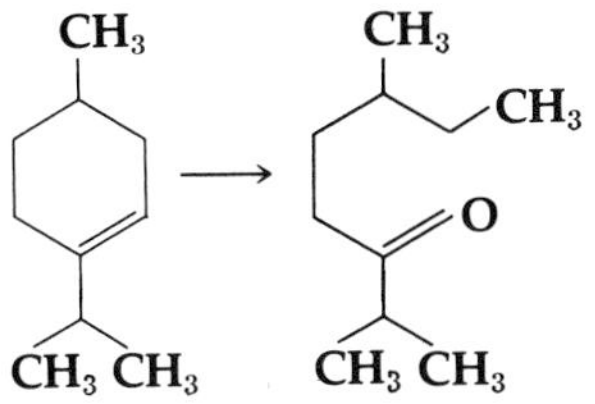

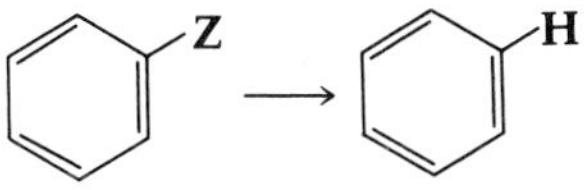

$Z = COOH, NO_2, Br, C_2H_5$

SECOND-ROW ELEMENTS: SULFUR AND PHOSPHORUS

ONE of the remarkable features of organic chemistry is the fact that such a tremendous number of substances and such a wealth and variety of chemical detail can arise from the simple properties of only four elements: carbon, hydrogen, oxygen, and nitrogen. We have also, of course, examined the chemistry of organic halogen and, to some extent, that of sulfur compounds when the context demanded it.

There are, however, organic compounds of virtually all the hundred or so elements in the periodic table. With most of these only the barest start has been made in unraveling their special chemistry. Organic compounds containing transition metals such as iron, nickel, or platinum have been intensively studied in recent years, and this study has revealed an entire new area of different and complex bonding styles involving d and higher atomic orbitals. Although this area is beyond the scope of the present text, we can make a brief survey of the special features of some higher elements as a sample of the kinds of differences to be found. For this purpose we select the relatively common compounds of sulfur and phosphorus as illustrative of the organic behavior of the higher nonmetallic elements.

19-1 THE SPECIAL PROPERTIES OF SECOND-ROW ELEMENTS

As we have seen previously in several places, there are many similarities between the first and second element in a given periodic table column. The behavior of thiols and alcohols, or sulfides and ethers, shows that they have much in common and all obey the general expectations of reaction theory. However, there are several important features which distinguish elements in the second row (Si, P, S, Cl) from their counterparts in the first, (C, N, O, and F, respectively). These distinctions are also usually simply derived from considerations of their atomic structures (Chap. 2).

The similarities arise from the fact that both atoms of a first- and second-row pair (C/Si; N/P; etc.) have the same number of outer-shell electrons and the same kernel charge. Hence they require the same bonding and make the same s- and p-orbital hybrids in order to achieve their stable octet of outer-shell electrons.

The differences arise from the fact that the second-row atoms possess an extra inner electron shell beneath the outer, bonding shell. These atoms are larger, and their outer electrons are farther from the nucleus and more shielded from its attractive force. Furthermore, the outer shell in second-row elements contains not only the s and p orbitals, which dominate bonding by seeking the full octet, but also empty d orbitals into which covalent electrons can be placed. These orbitals are not much higher in energy and can be filled on demand. This is an option first-row elements do not have.

These similarities and differences are illustrated for sample atoms in Fig. 19-1. The differences in second-row atoms create three major effects in the chemistry of their compounds.

1   *Second-row atoms are less electronegative, and their unshared electron pairs are better nucleophiles though weaker bases (to* H^+*).* These effects are apparently caused by the greater size of the atom which allows a greater polarizability of the outer-shell electrons. Less tightly bound by the nucleus, these electrons are more free to move toward external positive centers. Unshared pairs can accommodate better to the stereoelectronic requirements of a reaction, such as nucleophilic displacement. Also the substituents on the larger atom are not crowded in so closely in the transition state and offer less steric hindrance to reaction.

It is generally observed that a molecule or ion bearing an unshared pair on sulfur or phosphorus is a much better nucleophile for displacement reactions than its counterpart with oxygen or nitrogen. Similarly, the order of nucleophilicity for halogens is $I^- > Br^- > Cl^- \gg F^-$, as discussed in Sec. 10-3.

2   *Normal* π*-orbital double bonds are unstable and very rarely occur with second-row atoms.* This effect is also easily understood as a consequence of the greater size and longer bond lengths of these atoms. The requisite sideways overlap of parallel p orbitals necessary to form a stable π orbital is much less extensive if the nuclei are farther apart. Hence the π bond is certain to be very weak.

Thus, while **R—SH** is similar to **R—OH**, the doubly bonded thioketones and thioaldehydes ($R_2C{=}S$, $RCH{=}S$) are rare and far more reactive than corresponding carbonyl compounds. Thione derivatives of carboxylic acids are more stable. The more conjugation or delocalization they enjoy, the more stable thiocarbonyl groups are.

$$\begin{array}{ccc} \overset{\displaystyle S}{\overset{\|}{\text{R—C—OR}'}} & \overset{\displaystyle S}{\overset{\|}{\text{R—C—NR}'_2}} & \overset{\displaystyle S}{\overset{\|}{\text{H}_2\text{N—C—NH}_2}} \\ \text{Thionesters} & \text{Thioamides} & \text{Thiourea} \end{array}$$

True π double bonds to silicon and phosphorus are virtually unknown. In all cases the usual product on attempted preparation is a singly bonded polymer.

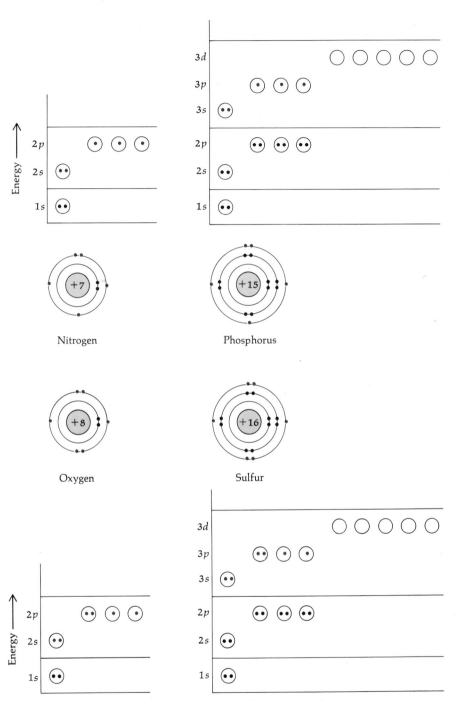

FIGURE 19-1   Electronic configurations of related first- and second-row atoms

$$\longrightarrow \cdot[R_2C\!\!=\!\!PR'] \longrightarrow \left(\!R_2C\!\!-\!\!PR'\!\right)_n$$

$$RCHS \longrightarrow$$ (structure: six-membered ring with alternating S atoms, R groups at carbon positions)

3 *Second-row elements can form more than four covalent bonds.* These atoms can accept outside electron pairs into their empty d orbitals to form a fifth or sixth bond. Thus these atoms can tolerate an expanded outer shell containing 10 or 12 electrons. The effect is seen most simply in the existence of relatively stable compounds containing five or six single bonds to as many atoms:

$$PCl_5 \qquad P\phi_5 \qquad K^+ PF_6^- \qquad SF_6$$

The more common manifestation is found in molecules or ions containing a second-row atom bonded to some atom with an unshared pair. This unshared pair may now be shared with a d orbital on the second-row atom, as symbolized in the resonance descriptions of Fig. 19-2.

Either resonance form may be written for a given case, the right-hand form with the "double bond" being more commonly used. It is unfortunate that the same written convention is used to symbolize both normal π double bonds and these d-orbital stabilizations, referred to as **back-donation** of electrons. The normal π bond is the familiar parallel overlap of two adjacent p orbitals, but this situation special to second-row elements is an overlap of a d orbital on the second-row atom with a p orbital on its neighbor (d_π-p_π **bonding**).

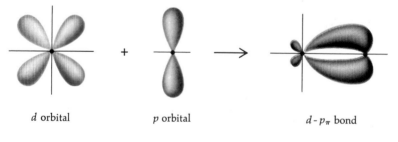

d orbital p orbital d-p_π bond

Atomic orbitals   Molecular orbital

Stable compounds of sulfur and phosphorus almost always possess four substituents (or unshared pairs) attached to the central atom, as the cases in Fig. 19-2 indicate. It is good practice to become accustomed to writing the resonance forms (on the left in Fig. 19-2) with only four bonds (and unshared pairs) on the central atom. This avoids the double-bond confusion and allows the comparisons with tetravalent first-row atoms to be more apparent.†

† In this chapter many instances of back-donation resonance are written with the "double bond," since this is the most common practice in the chemical literature.

General

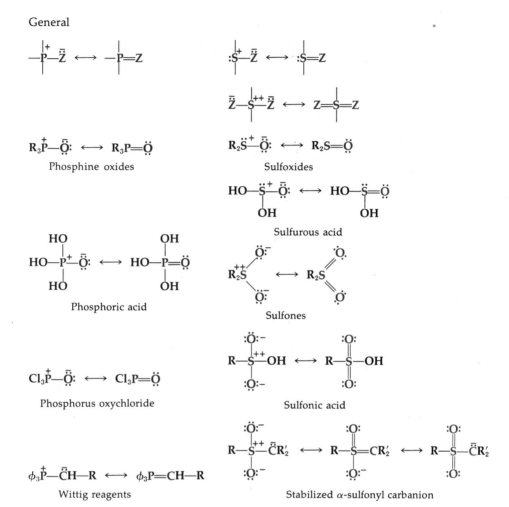

FIGURE 19-2 **Expanded outer shells and *d* orbital bonding in second-row atoms**

The chemistry of sulfur and phosphorus is dominated by the special stability arising from the attachment of oxygen to the central atom. This special stability is presumably caused by the possibility of overlapping *two* of the unshared pairs of oxygen with *d* orbitals on the central atom.† Attached fluorine has similar stability.

One important feature of the delocalization of adjacent electron pairs by back-donation (Fig. 19-2) is the relative stability of adjacent carbanions. The Wittig reagents are carbanions stabilized alpha to phosphorus. Anions adjacent to the isoelectronic positive nitrogen are slightly stabilized by electrostatic charge cancellation but cannot enjoy the much greater stability resulting from the back-donation resonance which characterizes the second-row elements.

---

† Amine oxides are formally similar to phosphine oxides but do not possess the extra stability arising from back-donation, i.e., from the second resonance form in Fig. 19-2.

$$R_3C\text{—}\ddot{C}H_2 < R_3\overset{+}{N}\text{—}\ddot{C}H_2 \ll [R_3\overset{+}{P}\text{—}\ddot{C}H_2 \longleftrightarrow R_3P\text{=}CH_2]$$

Stability increase →

For the same reason the α-hydrogens in sulfonium cations, sulfoxides, and sulfones are also acidic, owing to stabilization of the conjugate anions. This makes possible alkylation and acylation reactions at the α-carbon to these groups in the presence of strong base. Since the C—S bond in the product can be reduced (by metals, cf. **Zn**, **Al·Hg**), this procedure constitutes a valuable method of carbon-carbon bond formation.

$$CH_3\text{—}\overset{\overset{\displaystyle O}{\|}}{S}\text{—}CH_3 + NaH \longrightarrow Na^+ \ \ddot{C}H_2\text{—}\overset{\overset{\displaystyle O}{\|}}{S}\text{—}CH_3 \ \xrightarrow{+\ CH_3COOCH_3}$$

$$CH_3\text{—}\overset{\overset{\displaystyle O}{\|}}{C}\text{—}CH_2\text{—}\overset{\overset{\displaystyle O}{\|}}{S}\text{—}CH_3 + \overset{\textbf{·}}{C}H_3\bar{O}$$

Since double bonds to second-row atoms are poor, these carbanions do not reprotonate on oxygen as enolates can. The tautomers corresponding to enols have never been observed.

$$CH_3\text{—}\overset{\overset{\displaystyle O}{\|}}{C}\text{—}CH_3 \rightleftharpoons CH_3\text{—}\overset{\overset{\displaystyle OH}{|}}{C}\text{=}CH_2$$

$$CH_3\text{—}\overset{\overset{\displaystyle O}{\|}}{\underset{\underset{\displaystyle O}{\|}}{S}}\text{—}CH_3 \rightleftharpoons CH_3\text{—}\overset{\overset{\displaystyle OH}{|}}{\underset{\underset{\displaystyle O}{\|}}{\cancel{S}}}\text{=}CH_2$$

PROBLEM 19-1

Using resonance structures show why the left-hand compound in each pair below is the stronger acid of the two.

a $HCCl_3$ HCF_3

b $H_2C(OCH)_2$ $H_2C(SCH_3)_2$

c RSO_2H $RCOOH$

Geometry of Second-row Atoms

Virtually all of these molecules exhibit tetrahedral geometry at the second-row atom. The back-donation of electrons from an adjacent atom, into a $d\text{-}p_\pi$ bond, *does not change the normal tetrahedral arrangement* of four groups (or unshared pairs) on silicon, phosphorus, or sulfur. In this respect, too, the $d\text{-}p_\pi$ bond differs from a normal π bond, which causes a change from tetrahedral to trigonal geometry. The stereoelectronic demands for back-donation into the

d orbitals are nearly nonexistent, by contrast with the high degree of coplanarity required for normal π-bond stability.

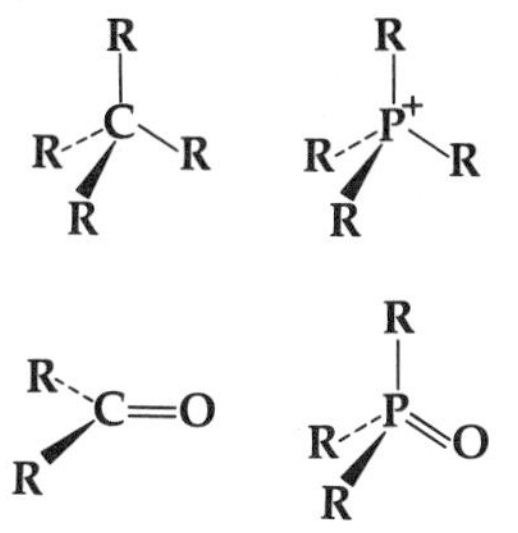

When four different atoms are attached to phosphorus, as to carbon or nitrogen, two mirror images are possible, and examples of optically active enantiomers of quaternary ammonium and phosphonium salts, as well as tertiary amine oxides and phosphine oxides, are all known. In contrast to amines and carbanions, however, phosphorus and sulfur molecules bearing an unshared pair of electrons on the central atom are also capable of resolution into enantiomers. In the isoelectronic cases with first-row atoms, these compounds rapidly invert, but the inversion is very slow (higher-energy barrier) in the analogous second-row atoms. The inversion process racemizes the asymmetric atom since it passes through a trigonal symmetrical configuration.

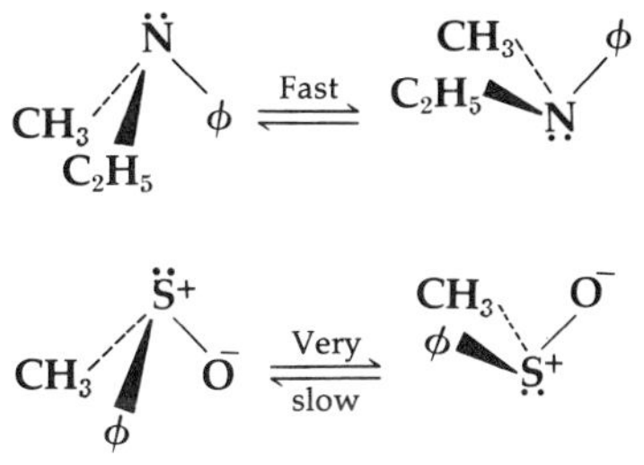

When five atoms are attached to a second-row atom (pentacovalent), the atom usually takes on the geometry of a **trigonal bipyramid**. This form is characterized by three substituents at 120° (trigonal) in a plane and the other two perpendicularly placed above and below that plane. The three trigonal substituents are usually labeled **equatorial**, the other two **apical** (at the apex). This form is the intermediate when nucleophiles add to tetravalent phosphorus. The two most electronegative of the attached atoms always occupy the apical positions. The geometry is the same as that at an sp^2 carbon with the two perpendicular p-orbital lobe positions now constituting true bonds to the apical positions. The molecular orbitals in the trigonal bipyramid are sp^3d hybrids. Pentacovalent phosphorus compounds are called **phosphoranes**.

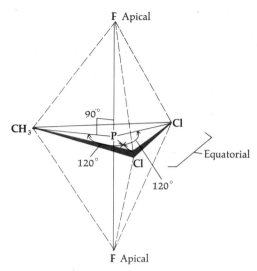

Configuration of $CH_3PCl_2F_2$

PROBLEM 19-2

Explain the following observations.

a   Primary and secondary amines react with CS_2 and SO_2 but not with CO_2.

b   Long chains of silicon atoms alternating with oxygen atoms are very stable while chains composed only of silicon atoms are not.

c   Trimethyl borate $[B(OCH_3)_3]$ is much more stable than the thio analog, $B(SCH_3)_3$.

d   Four-membered-ring intermediates form with facility in reactions at phosphorus although they do not at carbon. The Wittig reaction intermediate is an example:

$$\begin{array}{c}\diagdown\\\diagup\end{array}C{=}O + :\bar{C}H_2{-}\overset{+}{P}\phi_3 \longrightarrow \left[\begin{array}{c} O \\ \diagup\;\diagdown \\ C\qquad P\phi_3 \\ \diagdown\;\diagup \\ CH_2 \end{array}\right] \longrightarrow \begin{array}{c}\diagdown\\\diagup\end{array}C{=}CH_2 + \phi_3PO$$

e   Thioketals are much harder to hydrolyze to ketones than ordinary ketals are.

19-2 OXIDATION STATES AND THEIR INTERCONVERSIONS

The possibility of stabilization by back-donation of electrons allows the existence of a variety of molecular types for covalent compounds of sulfur and phosphorus, differing in oxidation state at the sulfur or phosphorus atom. The lowest oxidation states are -2 and -3, respectively, as for oxygen and nitrogen. The oxidation state may be computed simply by adding to this lowest state $+2$ for each bond to a more electronegative atom (usually N, O, X) but none for bonds to carbon or hydrogen (or unshared pairs). This accounting is the same as that described in Sec. 18-1 and is performed on the tetravalent resonance form, without the expanded outer shell. The major types of compounds are collected in Tables 19-1 and 19-2.

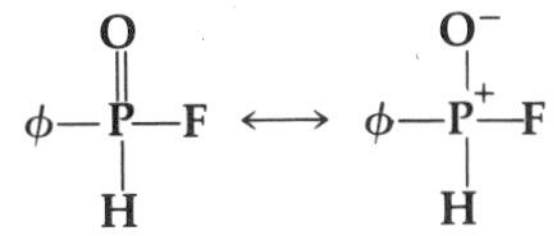

$$CH_3-\overset{\overset{\textstyle O}{\|}}{\underset{..}{S}}-Cl \longleftrightarrow CH_3-\overset{\overset{\textstyle O^-}{|}}{\underset{..}{\overset{+}{S}}}-Cl \qquad\qquad \phi-\overset{\overset{\textstyle O}{\|}}{\underset{\underset{\textstyle H}{|}}{P}}-F \longleftrightarrow \phi-\overset{\overset{\textstyle O^-}{|}}{\underset{\underset{\textstyle H}{|}}{\overset{+}{P}}}-F$$

Oxidation state [O]:	2 bonds to heteroatoms $[O] = -2 + 2(+2) = +2$	2 bonds to heteroatoms $[O] = -3 + 2(+2) = +1$

There is a question of tautomerism in many of the compounds bearing hydrogen, which may be bonded either to oxygen or to the second-row element. The conjugate base is the same from either tautomer, and both forms and their single conjugate base must have the same oxidation state. This tautomerism is analogous to the following familiar keto-enol example.

TABLE 19-1   Oxidation States of Phosphorus in Covalent Compounds

Oxidation state			
-3	$:PH_3,\ R\overset{..}{P}H_2,\ R_2\overset{..}{P}H,\ R_3P:$ Phosphines		
-1	$R_2\overset{\overset{\textstyle O}{\|}}{P}H$	$R_3P{=}O$ Phosphine oxides	
$+1$	$H_2\overset{\overset{\textstyle O}{\|}}{P}{-}OH$ Hypophosphorous acid	$R_2\overset{\overset{\textstyle O}{\|}}{P}{-}OH$ Phosphinic acids	
$+3$	$H\overset{\overset{\textstyle O}{\|}}{P}(OH)_2$ Phosphorous acid	$R\overset{\overset{\textstyle O}{\|}}{P}(OH)_2$ Phosphonic acids	$:P(OR)_3$ Phosphites
$+5$	$O{=}P(OH)_3$ Phosphoric acid		$O{=}P(OR)_3$ Phosphates

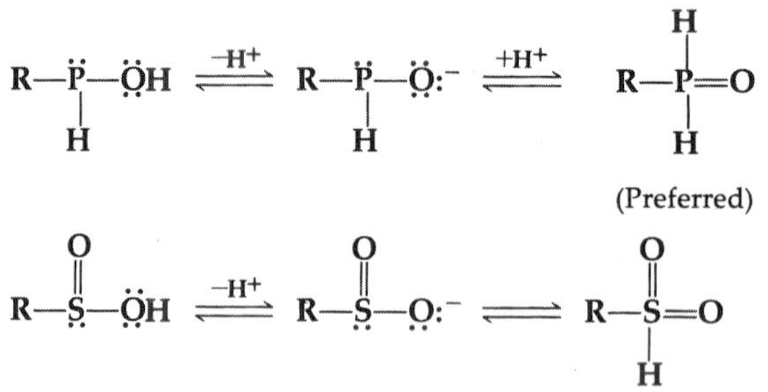

$$R-\overset{..}{\underset{H}{P}}-\overset{..}{O}H \underset{}{\overset{-H^+}{\rightleftharpoons}} R-\overset{..}{\underset{H}{P}}-\overset{..}{O}:^- \underset{}{\overset{+H^+}{\rightleftharpoons}} R-\overset{H}{\underset{H}{P}}=O$$

(Preferred)

$$R-\overset{O}{\underset{..}{S}}-\overset{..}{O}H \overset{-H^+}{\rightleftharpoons} R-\overset{O}{\underset{..}{S}}-\overset{..}{O}:^- \rightleftharpoons R-\overset{O}{\underset{H}{S}}=O$$

All of these compounds are moderate to strong acids. In some sulfur compounds such as the sulfinic acids shown it is not known for certain which tautomer represents the acid. In phosphorus compounds the tautomer with the phosphoryl (P=O ⟷ P⁺—O⁻) group is always the preferred one, owing to the high stability or bond energy for the P=O group.

Many of the types of compounds shown in the tables are interconvertible by adding oxygen to unshared pairs (or in place of hydrogen) attached to the central atom, or the reverse. Such interconversions are two-electron oxidations and reductions, and are generally effected with various of the common reagents of Chap. 18.

TABLE 19-2   **Oxidation States of Sulfur in Covalent Compounds**

Oxidation state			
−2	H_2S	$R-\overset{..}{\underset{..}{S}}-H$ Thiols	$R-\overset{..}{\underset{..}{S}}-R'$ Sulfides
0		$R-\overset{..}{\underset{..}{S}}-OH$ Sulfenic acids	$R-\overset{O}{\overset{\|}{\underset{..}{S}}}-R'$ Sulfoxides
+2		$R-\overset{O}{\overset{\|}{\underset{..}{S}}}-OH$ Sulfinic acids	$R-\overset{O}{\overset{\|}{\underset{O}{S}}}-R'$ Sulfones
+4	HSO_2OH Sulfurous acid	$R-\overset{O}{\overset{\|}{\underset{O}{S}}}-OH$ Sulfonic acids	$RO-\overset{O}{\overset{\|}{\underset{..}{S}}}-OR'$ Sulfites
+6	$HO-SO_2-OH$ Sulfuric acid		$RO-\overset{O}{\overset{\|}{\underset{O}{S}}}-OR'$ Sulfates

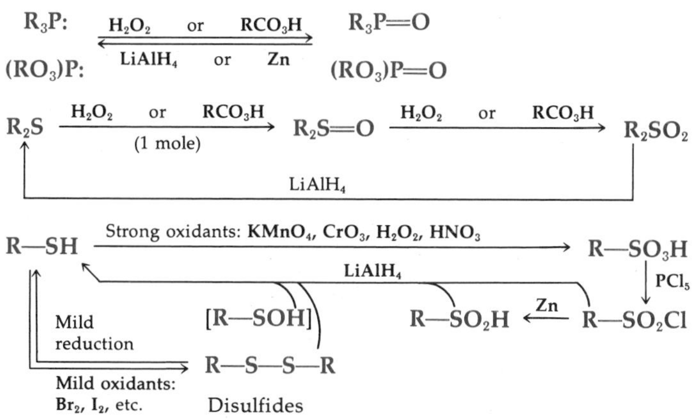

$$R_3P: \quad \xrightleftharpoons[\text{LiAlH}_4 \quad \text{or} \quad \text{Zn}]{\text{H}_2\text{O}_2 \quad \text{or} \quad \text{RCO}_3\text{H}} \quad R_3P{=}O$$

$$(RO_3)P: \qquad\qquad\qquad\qquad\qquad (RO_3)P{=}O$$

$$R_2S \xrightarrow[\text{(1 mole)}]{\text{H}_2\text{O}_2 \quad \text{or} \quad \text{RCO}_3\text{H}} R_2S{=}O \xrightarrow{\text{H}_2\text{O}_2 \quad \text{or} \quad \text{RCO}_3\text{H}} R_2SO_2$$

$$\text{LiAlH}_4$$

R—SH $\xrightarrow{\text{Strong oxidants: KMnO}_4\text{, CrO}_3\text{, H}_2\text{O}_2\text{, HNO}_3}$ R—SO$_3$H

LiAlH$_4$ $\qquad\qquad\qquad$ | PCl$_5$

Mild reduction [R—SOH] R—SO$_2$H $\xleftarrow{\text{Zn}}$ R—SO$_2$Cl

Mild oxidants: R—S—S—R
Br$_2$, I$_2$, etc. Disulfides

FIGURE 19-3 **Redox interconversions of phosphorus and sulfur compounds**

Sulfides and phosphines (like tertiary amines) are converted to their oxides with hydrogen peroxide or peracids. Use of 1 mole of oxidant on sulfides allows the sulfoxide to be prepared, while 2 moles of oxidant carries the oxidation cleanly to the sulfone. Phosphites are generally transformed into the corresponding phosphates by the same reagents. The reverse reactions, reduction back down the oxidation-state scale to sulfides and phosphines, may be effected by lithium aluminum hydride. Zinc reduction is also often effective.

Three oxidation steps (of two electrons each) separate thiols from the corresponding sulfonic acids, and most moderately strong oxidants (KMnO$_4$, CrO$_3$, H$_2$O$_2$, etc.) effect the whole change. It is very difficult to obtain the intermediate sulfenic or sulfinic acid products by partial oxidation. **Sulfenic acids** (RSOH) are unstable with respect to disproportionation (self-oxidation-reduction) and cannot usually be isolated. Sulfenic acid derivatives such as the chlorides (RSCl), esters (RSOR'), and sulfenamides (RSNHR') are known, however. The **sulfinic acids** (RSO$_2$H) may be prepared by zinc reduction of sulfonyl chlorides, but lithium aluminum hydride reduces sulfonyl chlorides or sulfinic acids to thiols. Sulfonic acids are stable to reduction. Finally, mild oxidants such as halogens yield disulfides (R—S—S—R) from thiols. These common transformations are summarized in Fig. 19-3.

PROBLEM 19-3

Construct a table of possible kinds of silicon compounds with respect to oxidation state, analogous to Tables 19-1 and 19-2 for phosphorus and sulfur.

PROBLEM 19-4

The following organic iodine compounds are known. Discuss their detailed structures and oxidation states.

a $(C_6H_5)_2I^+Cl^-$ c $C_6H_5I(OCOCH_3)_2$

　　Diphenyliodonium   　Iodosobenzene
　　　　chloride   　　diacetate

b C_6H_5IO d $C_6H_5IO_2$

　　Iodosobenzene   　Iodoxybenzene

19-3   PHOSPHORUS AS NUCLEOPHILE

Trivalent phosphines and phosphites bear an unshared electron pair and are powerful nucleophiles. They are readily alkylated by primary halides in the S_N2 reactions, yielding phosphonium salts. Removal of α-hydrogen from these salts with strong bases yields Wittig reagents.

$$\phi_3P: + \phi{-}CH_2Br \longrightarrow \overset{+}{\phi_3P}{-}CH_2\,\phi \xrightarrow{\phi Li} \phi_3P{=}CH\phi + LiBr + \phi H$$

$$Br^-$$

Triphenyl-   Triphenylbenzyl-   Wittig reagent
phosphine   phosphonium bromide

$$(\phi O)_3P: + CH_3I \longrightarrow (\phi O)_3\overset{+}{P}CH_3\,I^-$$

Triphenyl   Methyl-triphenoxy-
phosphite   phosphonium iodide

　　When alkyl instead of aryl phosphites are used, a further reaction (**Arbusov reaction**) follows spontaneously. The released nucleophile, usually halide ion, returns to effect displacement of oxygen from one alkyl group. The leaving group in this second displacement is the very stable phosphoryl (P=O) group.

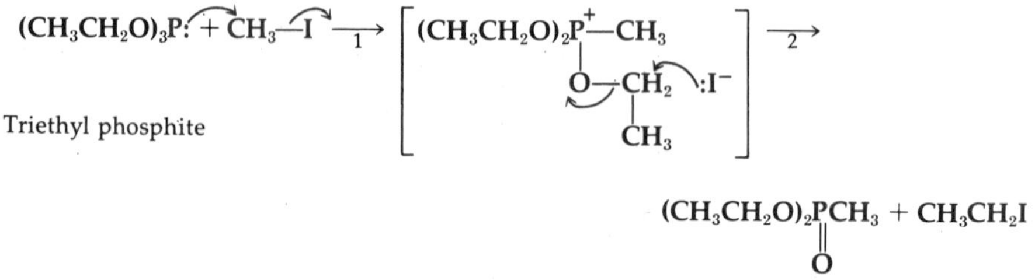

$$(CH_3CH_2O)_3P: + CH_3{-}I \xrightarrow{1} \left[(CH_3CH_2O)_2\overset{+}{P}{-}CH_3 \\ \underset{\underset{CH_3}{|}}{O{-}CH_2}\; :I^- \right] \xrightarrow{2}$$

Triethyl phosphite

$$(CH_3CH_2O)_2\underset{\underset{O}{\|}}{P}CH_3 + CH_3CH_2I$$

Diethyl methylphosphonate

The second displacement in the Arbusov reaction is so rapid that no aliphatic compounds containing $-\overset{|}{\overset{+}{P}}-O-\overset{|}{\underset{|}{C}}$ have ever been isolated. Only when the carbon atom attached to oxygen is part of an aromatic ring and resistant to displacement is the compound stable, as in the methylation of triphenyl phosphite shown above.

Reversible additions of substituted phosphines to carbonyls rarely lead to products since no stabilizing electrons are available adjacent to the phosphorus for back-donation. If an Arbusov reaction can occur as an irreversible second step, however, the addition does go to completion. The reaction of alkyl phosphites with acid chlorides yields α-ketophosphonate esters.

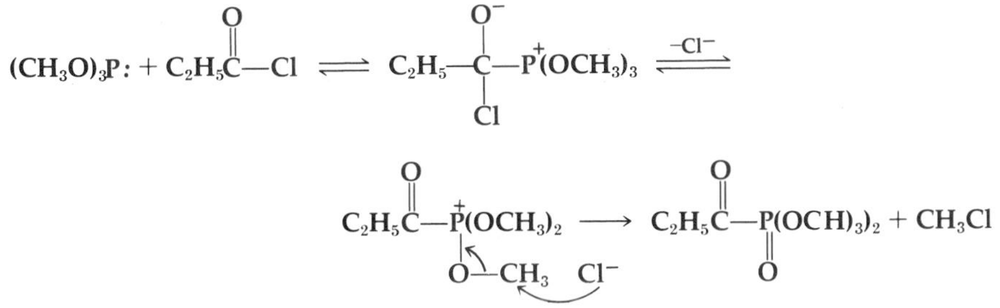

$$(CH_3O)_3P: + C_2H_5\overset{O}{\overset{\|}{C}}-Cl \; \rightleftharpoons \; C_2H_5-\underset{\underset{Cl}{|}}{\overset{\overset{O^-}{|}}{C}}-\overset{+}{P}(OCH_3)_3 \; \xrightarrow{\;-Cl^-\;}$$

$$C_2H_5\overset{O}{\overset{\|}{C}}-\underset{\underset{O-CH_3 \quad Cl^-}{|}}{\overset{+}{P}}(OCH_3)_2 \; \longrightarrow \; C_2H_5\overset{O}{\overset{\|}{C}}-\underset{\underset{O}{\|}}{P}(OCH)_3)_2 + CH_3Cl$$

Stabilization of the added phosphorus occurs on addition to diazonium or azide groups; the products commonly lose nitrogen. The nitrogen phosphorane in the second case is analogous to the Wittig reagent and may be used to convert aldehydes and ketones to imines in a like manner to the Wittig reaction.

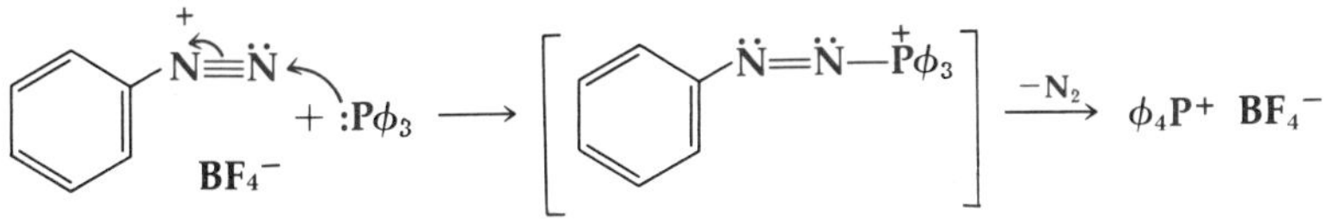

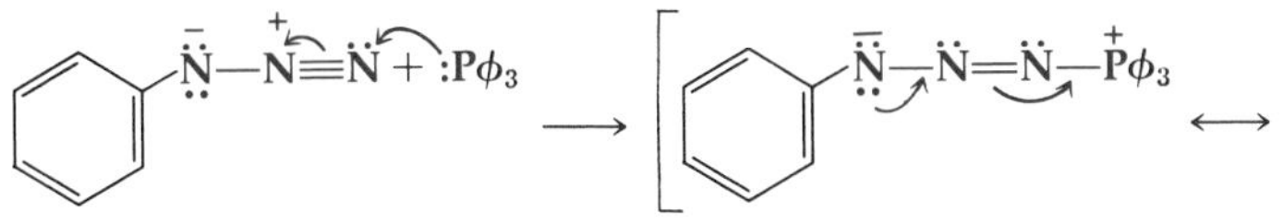

Phenyl azide

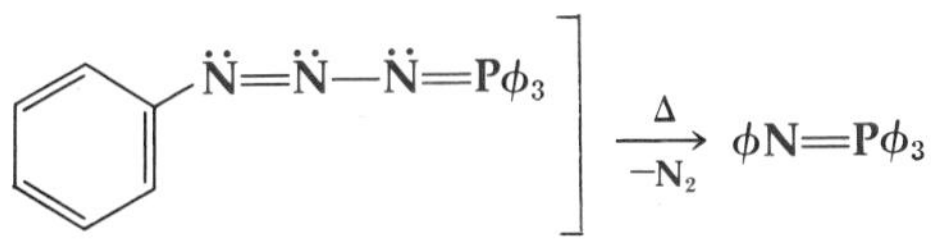

Because of back-donation by unshared electron pairs, phosphorus nucleophiles can attack oxygen and sulfur directly in certain cases, deoxygenating other oxides, such as amine oxides. With 1,2-diketones a pentaoxyphosphorane emerges as a fairly stable product.

$$\overset{+}{R_3N}-\overset{\cdot\cdot}{\underset{\cdot\cdot}{O}}:+ :P\phi_3 \longrightarrow R_3N: + \left[:\overset{\cdot\cdot}{\underset{\cdot\cdot}{O}}-\overset{+}{P}\phi \longleftrightarrow \overset{\cdot\cdot}{\underset{\cdot\cdot}{O}}=P\phi_3 \right]$$

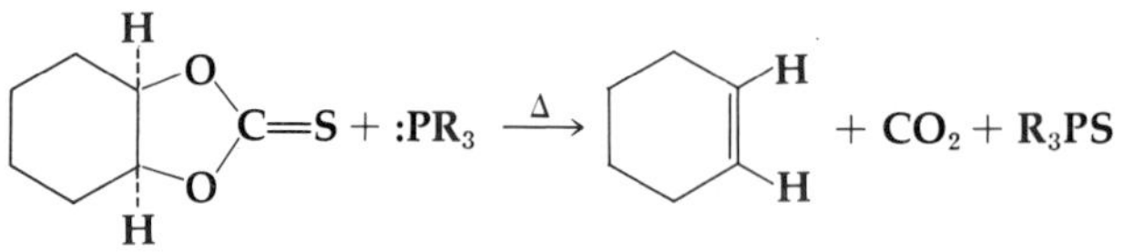

(structure) $C=S + :PR_3 \xrightarrow{\Delta}$ (cyclohexene structure) $+ CO_2 + R_3PS$

Cis   Cis

(structure) $+ :P(OC_3H_7)_3 \longrightarrow$ (trigonal bipyramid structure with OC_3H_7, $90°$)

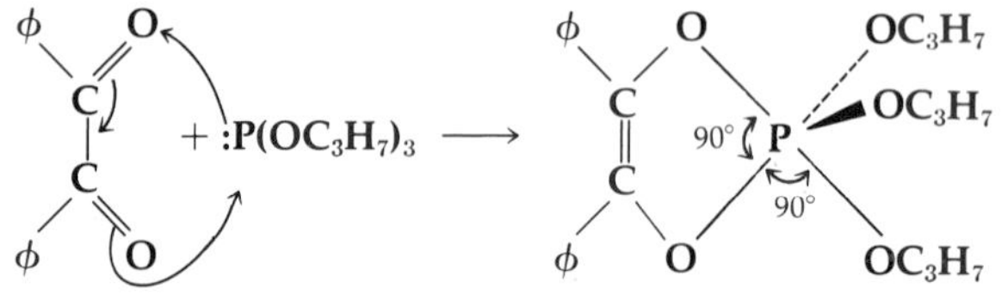

(Trigonal bipyramid)

The reduction of epoxides is apparently *not* a direct removal of oxygen, since *cis*-butene oxide yields *trans*-butene as the major product.

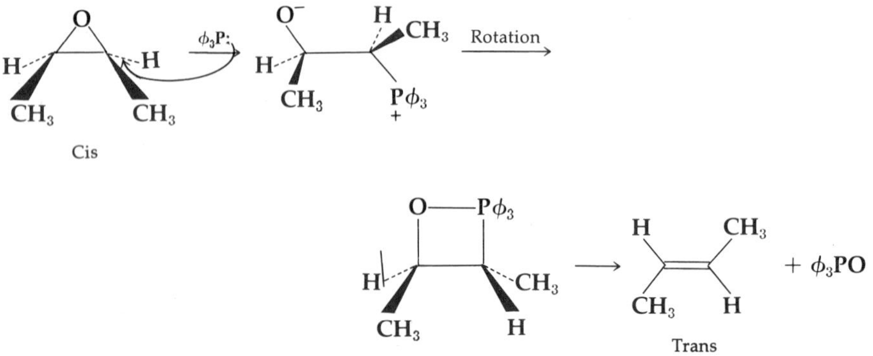

(epoxide structure) $\xrightarrow{\phi_3P:}$ (structure) $\xrightarrow{\text{Rotation}}$

Cis

(structure) $\longrightarrow$ (trans-butene structure) $+ \phi_3PO$

Trans

PROBLEM 19-5

In each of the following known reactions show a reasonable mechanism and indicate wherein the reaction illustrates behavior peculiar to a second-row element.

a $RCON_3 + \phi_3P \xrightarrow{\Delta} RCN + ?$

b $C_6H_5COCH_2Br + \phi_3P \xrightarrow{CH_3OH} C_6H_5COCH_3 + ?$

c $C_6H_5COCH_2Br + (EtO)_3P \longrightarrow C_2H_5Br + C_{12}H_{17}PO_4$

$\qquad C_{12}H_{17}PO_4 + EtOH \xrightarrow{\Delta} C_6H_5COCH_3 + (EtO)_3PO$

d $CCl_3CON(CH_3)_2 + \phi_3P \longrightarrow Cl_2C=CClN(CH_3)_2 + ?$

e $(EtO)_2\overset{\cdot\cdot}{P}-O-\underset{\underset{CH_3}{|}}{CH}-CH=CH_2 \xrightarrow{\Delta} (EtO)_2\overset{\overset{O}{\|}}{P}-CH_2CH=CHCH_3$

f $(EtO)_3P + CH_2{=}CHCHO \longrightarrow (EtO)_2\overset{\displaystyle O}{\overset{\|}{P}}CH_2CH{=}CHOEt$

g $(MeO)_3P + CH_2{=}CHCOOH \longrightarrow (MeO)_2\overset{\displaystyle O}{\overset{\|}{P}}CH_2CH_2COOCH_3$

h $\phi_2C{=}C{=}O + (EtO)_3P \longrightarrow \phi C{\equiv}C\phi + (EtO)_3PO$

PROBLEM 19-6

Show a sequence of reactions which will convert trimethylphosphite to tri-methylphosphine.

19-4   PHOSPHORUS AS ELECTROPHILE

Even when phosphorus enjoys a full octet of electrons in its outer shell, it may easily accept more into its d orbitals, if attacked by a nucleophile. Such attack leads to five attachments to the phosphorus atom (one may be an unshared pair), i.e., an outer shell of 10 electrons. These pentacovalent phosphoranes usually have the trigonal bipyramid geometry. They are also usually unstable and pass on to release a leaving group, returning to tetrahedral form.

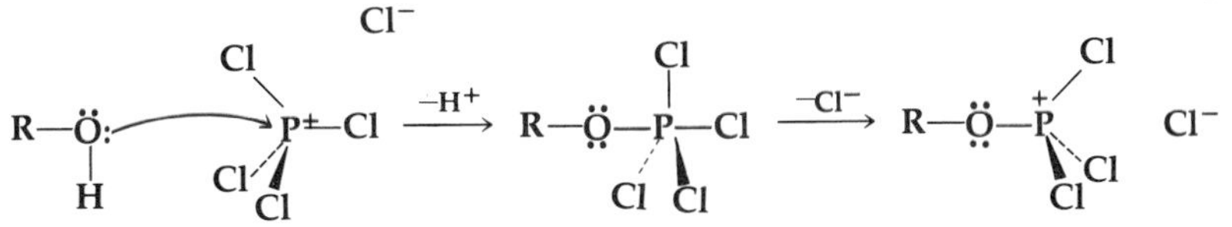

The geometry of the process is the same as that of the S_N2 displacement reaction (see page 385), but the trigonal bipyramid in this instance is a *true intermediate, not just a transition state*. When none of the five groups are good leaving groups, the pentacovalent phosphorane may be isolated. On the other hand, the driving force for formation of the P=O bond is so strong that even phenyl anion can be a leaving group, leaving (as above) from the apical position of a phosphorane intermediate.

$\phi MgBr + \phi_4P_.^+ \longrightarrow \phi_5P + M\overset{+}{g}Br$

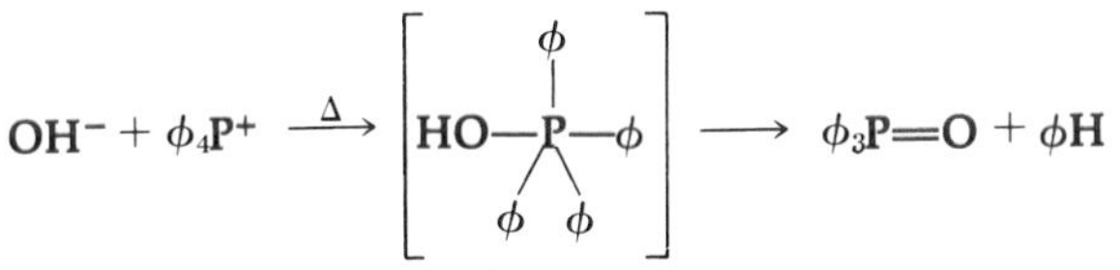

$$OH^- + \phi_4P^+ \overset{\Delta}{\longrightarrow} \left[HO\overset{\displaystyle \phi}{\underset{\displaystyle \phi\ \ \phi}{-P-\phi}} \right] \longrightarrow \phi_3P{=}O + \phi H$$

The reaction of PCl_5† with alcohols, used as illustration above, is typical of the family of reagents used to convert alcohols to halides (see page 420). Each reagent is a phosphonium compound to which the alcohol adds as nu-

† Phosphorus pentachloride may be regarded as pentacovalent PCl_5 with substantial ionic character in the P—Cl bond, so that it reacts as $PCl_4{}^+ Cl^-$.

cleophile. This has the effect of converting —OH in the substrate alcohol to a leaving group. In a second reaction (the same as the Arbusov reaction) halide ion then displaces phosphoryl from the unstable $\geq$P$^+$—O—R intermediate. These reactions are all very facile reactions, proceeding in high yield at room temperature, and surprisingly unaffected by steric hindrance. Triphenylphosphorus dichloride has even been used to convert phenols to phenyl chlorides under vigorous conditions.

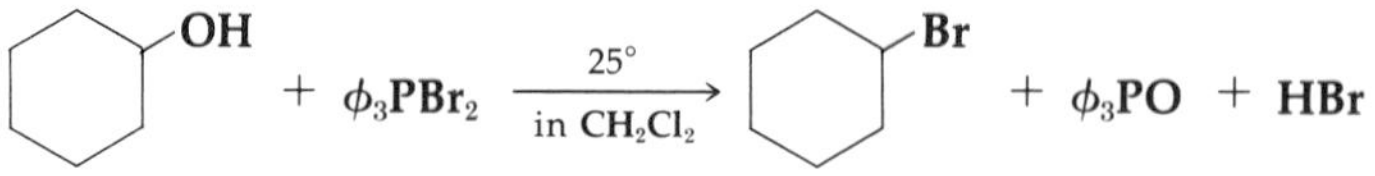

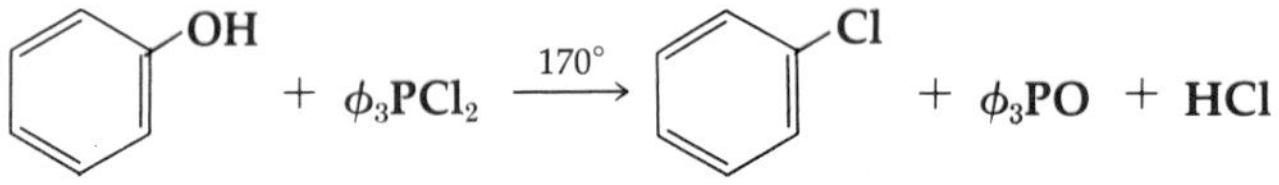

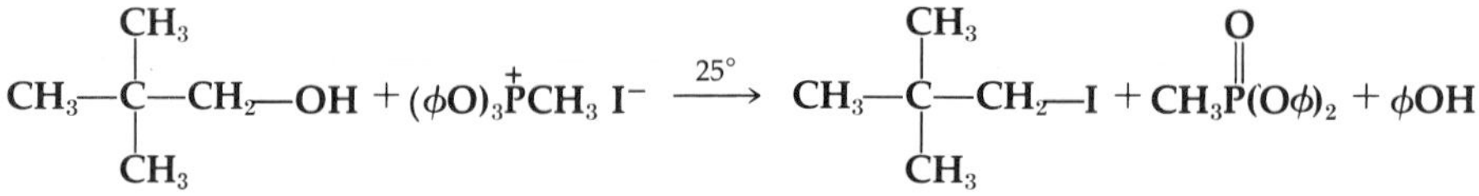

Other nucleophiles are also effective in attack at phosphorus. The examples selected below serve to illustrate the breadth and synthetic utility of the reaction as well as the generality of nucleophilic attack on phosphorus regardless of its oxidation state.

$$\phi\text{MgBr} + (CH_3)_2\ddot{P}\text{—Cl} \longrightarrow (CH_3)_2\ddot{P}\phi + \text{MgClBr}$$

$$3CH_3\text{MgBr} + \text{POCl}_3 \longrightarrow (CH_3)_3\text{P}{=}\text{O} + 3\text{MgClBr}$$

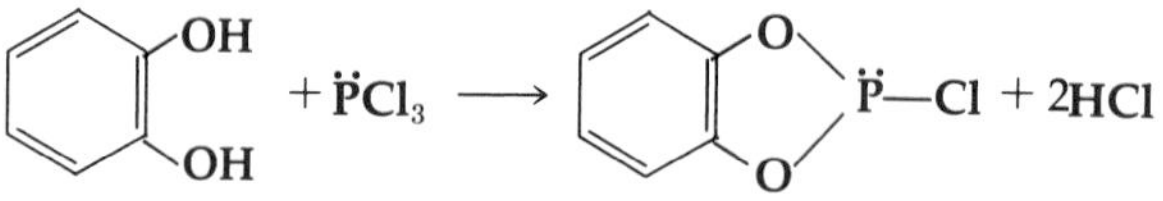

$$2(C_2H_5)_2\dot{\ddot{N}}H + \ddot{P}Cl_3 \longrightarrow (C_2H_5)_2\text{N}\text{—}\ddot{P}Cl_2 + (C_2H_5)_2\overset{+}{\dot{N}}H_2\ \text{Cl}^-$$

$$\phi\text{—C}{\equiv}\text{N} + \text{POCl}_3 + \text{HCl}$$

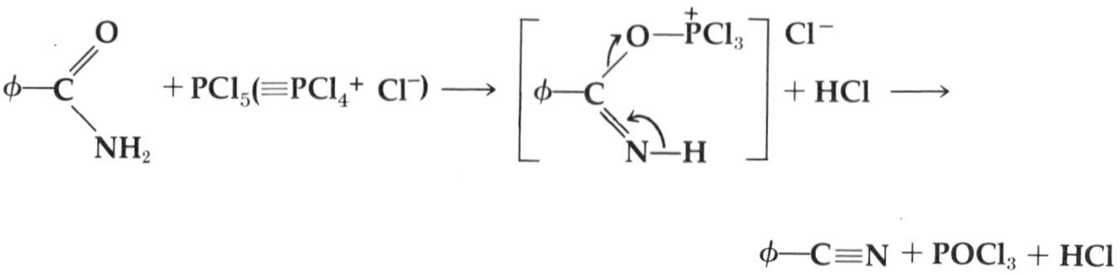

Phosphorylation, the interchange of the alcohol groups attached to phosphates, $O=P(OR)_3$, have essentially the same kind of mechanism as carboxylic ester interchanges. They involve nucleophilic attack of water or alcohols, with acidic or basic catalysis, on $P=O$ instead of $C=O$. Phosphorylations are very important in the chemistry of living systems (Chap. 26).

$$\textbf{ROH} + O=P(OR')_3 \longrightarrow O=P(OR')_2 + R'OH \xrightarrow[+ROH]{etc.} O=P(OR)_3 + 3R'OH$$
$$\overset{|}{OR}$$

In biological chemistry, phosphates are commonly employed to activate alcohols as leaving groups. This comes about because biochemical reactions commonly proceed in aqueous media, in the cell, and the phosphate group possesses other oxygen atoms which exist as solubilizing anions in neutral solution. Phosphorylation occurs more nearly irreversibly when the leaving group is a better anion than simply that of another alcohol in the alcohol interchange illustrated. However, phosphoryl chlorides, used in the laboratory for this purpose, are replaced in biochemical phosphorylation by **pyrophosphates** in which phosphate anion itself (rather than chloride) is the stable anionic leaving group.

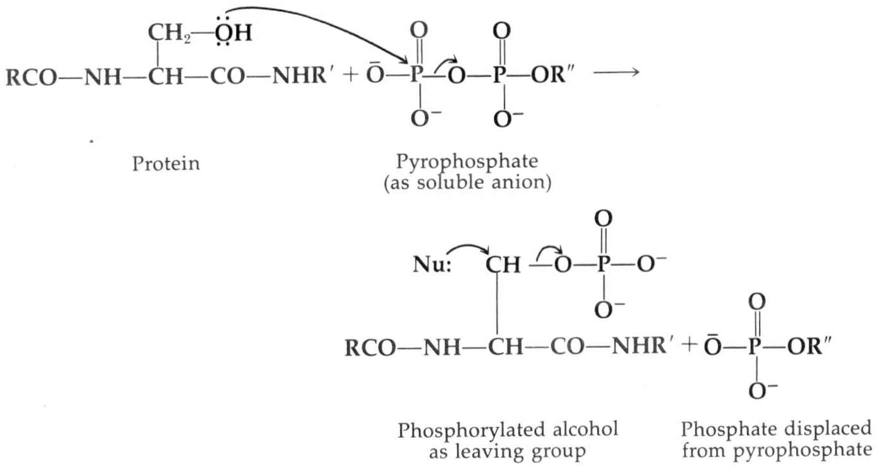

Protein

Pyrophosphate
(as soluble anion)

Phosphorylated alcohol
as leaving group

Phosphate displaced
from pyrophosphate

Protein synthesis:

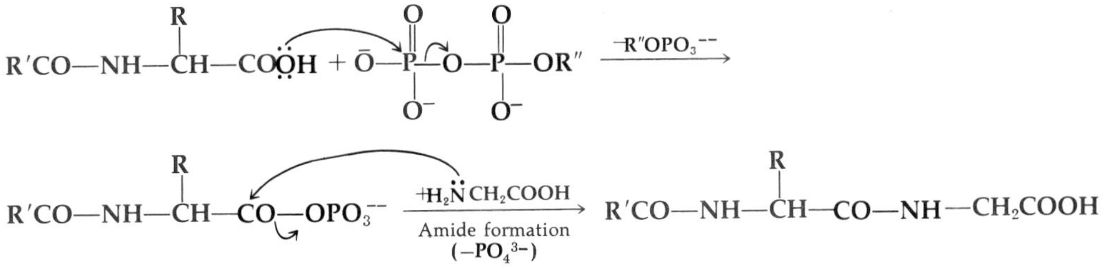

PROBLEM 19-7

A recent mild synthetic method for converting alcohols to their corresponding iodides involves reaction of the alcohol first with the cyclic chlorophosphite shown below, in pyridine, followed by treatment with iodine. Rationalize this reaction. What would be your expectation if *t*-butyl alcohol were used for the reaction?

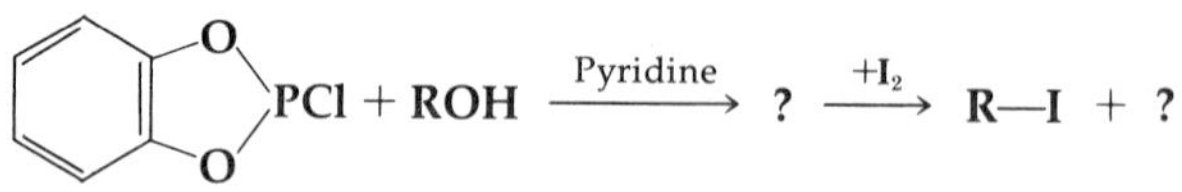

PROBLEM 19-8

Suggest syntheses for the following compounds from phosphorus trichloride as the sole source of phosphorus and any organic substance. You may assume that stepwise substitution of halogens on phosphorus is an allowable procedure.

a   Phosphorus pentachloride

b $POCl_3$

c $(CH_3)_2POCH_3$

d $(C_6H_5)_2PCH_3$ (with O double-bonded to P)

e $(CH_3)_2P—O—C_6H_5$

f $(CH_3)_3P^+—O—C_6H_5$ I^-

g   An asymmetric (resolvable) substance

h $CH_3CO—P(OCH_3)_2$ (with O double-bonded to P)

i $C_6H_5N{=}P(C_6H_5)_3$

j $CH_3P—OCH_3$ with O double-bonded to P and OC_6H_5 below

PROBLEM 19-9

Pyridine *N*-oxide reacts with tributylphosphine but not tributylamine. Give a mechanistic rationale for this difference in behavior.

PROBLEM 19-10

An important synthetic step in the plant cell toward the production of a number of natural substances, such as turpentine, camphor, rubber, and the scent of geraniums, is shown below. Outline the steps in the mechanism of the reaction and write out the products. What is the structure of the phosphorus group on the product and what is a likely kind of biochemical reagent for placing it there?

$$CH_3 \quad OH$$

Mevalonic
acid

$$\text{Mevalonic acid} + \text{an organic pyrophosphate} \longrightarrow CH_2{=}\overset{\underset{\displaystyle CH_3}{|}}{C}{-}CH_2CH_2{-}O{-}(P_2O_6H_3)$$

19-5   SULFUR AS NUCLEOPHILE

Almost any covalent molecule bearing an unshared electron pair on sulfur acts as a powerful nucleophile. Displacements by sulfur are generally fast, and they proceed in high yield owing to the low basicity of sulfur species. Consequently there is negligible elimination as a side reaction. The major types of sulfur nucleophiles are collected in Fig. 19-4.

The reactions of Fig. 19-4 are all nucleophilic substitutions at saturated carbon, irreversible and bimolecular. The reversible substitutions at carbonyl occur only with thiols, which react with acid chlorides to form thiolesters and with ketones and aldehydes to form thioketals and thioacetals. Bisulfite ion adds to aldehydes and some unhindered ketones (Sec. 12-2) but does not afford substitutions.

$$\phi{-}COCl + C_2H_5SH \longrightarrow \phi{-}CO{-}SC_2H_5 + HCl$$

$$\text{(cyclohexanone)}{=}O + C_2H_5SH \xrightarrow{\text{BF}_3} \text{(cyclohexane)}\begin{array}{l} SC_2H_5 \\ SC_2H_5 \end{array} + H_2O$$

While they are good nucleophiles most of the sulfur functions are poor leaving groups, partly because the carbon-sulfur bond is so unpolarized (see Table 2-3). The bond is polarized in the sulfonium salts, and these will undergo displacements on heating.

$$(CH_3)_3S^+{:}\ I^- \xrightarrow{\Delta} (CH_3)_2\ddot{\underset{\cdot\cdot}{S}} + CH_3I$$

The sulfonyl group, $-SO_2-$, is interesting in that it bears an obvious analogy to the carbonyl. The sulfones, like ketones, activate carbanion formation in the α-position, and the sulfonic acid family of derivatives is superficially similar to the carboxylic acid family (page 539) in its general behavior. However, the increased variety of behavior available to sulfur as a second-row element affords some new features, for the sulfinic acid is acidic, unlike its counterpart, the aldehyde, and the corresponding sulfinate anion can act as a sulfur nucleophile (Fig. 19-4), with no real analogy in carbonyl chemistry.

Since it is a stabilized anion, the sulfinate anion is occasionally found as a leaving group as well. The most common situation is found in sulfonyl-hydrazine derivatives, which can eliminate sulfinate anions to yield a nitrogen-

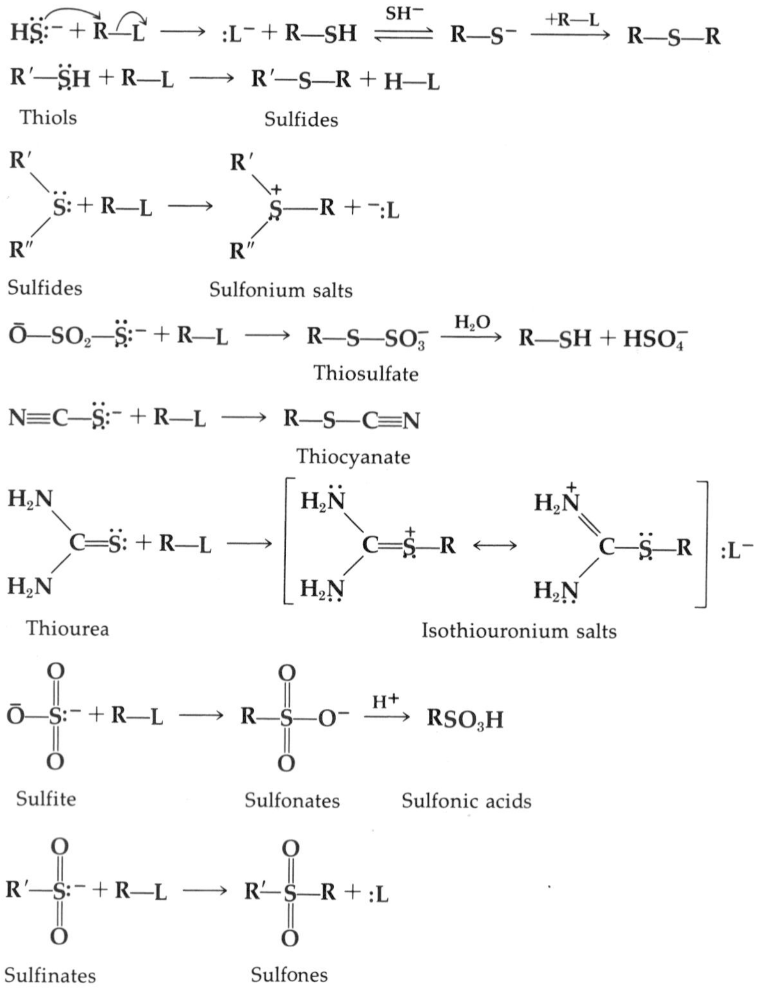

$$HS:^- + R\text{—}L \longrightarrow :L^- + R\text{—}SH \underset{SH^-}{\rightleftharpoons} R\text{—}S^- \xrightarrow{+R\text{—}L} R\text{—}S\text{—}R$$

$$R'\text{—}SH + R\text{—}L \longrightarrow R'\text{—}S\text{—}R + H\text{—}L$$

Thiols   Sulfides

$$\begin{matrix} R' \\ \diagdown \\ & S: + R\text{—}L \longrightarrow \\ \diagup \\ R'' \end{matrix} \qquad \begin{matrix} R' \\ \diagdown \\ & \overset{+}{S}\text{—}R + {}^-:L \\ \diagup \\ R'' \end{matrix}$$

Sulfides   Sulfonium salts

$$\bar{O}\text{—}SO_2\text{—}S:^- + R\text{—}L \longrightarrow R\text{—}S\text{—}SO_3^- \xrightarrow{H_2O} R\text{—}SH + HSO_4^-$$

Thiosulfate

$$N\equiv C\text{—}S:^- + R\text{—}L \longrightarrow R\text{—}S\text{—}C\equiv N$$

Thiocyanate

$$\begin{matrix} H_2N \\ \diagdown \\ & C=S: + R\text{—}L \longrightarrow \\ \diagup \\ H_2N \end{matrix} \left[\begin{matrix} H_2\ddot{N} \\ \diagdown \\ & C=\overset{+}{S}\text{—}R \longleftrightarrow \\ \diagup \\ H_2\ddot{N} \end{matrix} \quad \begin{matrix} H_2\overset{+}{N} \\ \diagup\diagup \\ C\text{—}S\text{—}R \\ \diagdown \\ H_2N \end{matrix} \right] :L^-$$

Thiourea   Isothiouronium salts

$$\begin{matrix} & O \\ & \| \\ \bar{O}\text{—}S:^- + R\text{—}L \longrightarrow \\ & \| \\ & O \end{matrix} \qquad \begin{matrix} & O \\ & \| \\ R\text{—}S\text{—}O^- \xrightarrow{H^+} RSO_3H \\ & \| \\ & O \end{matrix}$$

Sulfite   Sulfonates   Sulfonic acids

$$\begin{matrix} & O \\ & \| \\ R'\text{—}S:^- + R\text{—}L \longrightarrow \\ & \| \\ & O \end{matrix} \qquad \begin{matrix} & O \\ & \| \\ R'\text{—}S\text{—}R + :L \\ & \| \\ & O \end{matrix}$$

Sulfinates   Sulfones

FIGURE 19-4   **Types of sulfur nucleophiles in displacement reactions**

nitrogen double bond. The McFadyen-Stevens reduction of aromatic acylhydrazides to aldehydes is the classic example (page 520), and the conversion of monotosylhydrazones of α-diketones into α-diazoketones is another. The reduction of ketones (or aldehydes) to methylenes via the tosylhydrazone and sodium borohydride is a third, and a useful synthetic tool (page 775).

$$C_6H_5CONHNHSO_2C_6H_5 \xrightarrow[Na_2CO_3,\ 160°]{HOCH_2CH_2OH} \left[C_6H_5\overset{O}{\overset{\|}{C}}N{-}NH{-}SO_2C_6H_5 \right] \xrightarrow{-C_6H_5SO_2^-}$$

1-Benzoyl-2-benzene-
sulfonylhydrazide

$$\left[C_6H_5\overset{O}{\overset{\|}{C}}N{=}NH \right] \xrightarrow{-N_2} C_6H_5CHO$$

70%

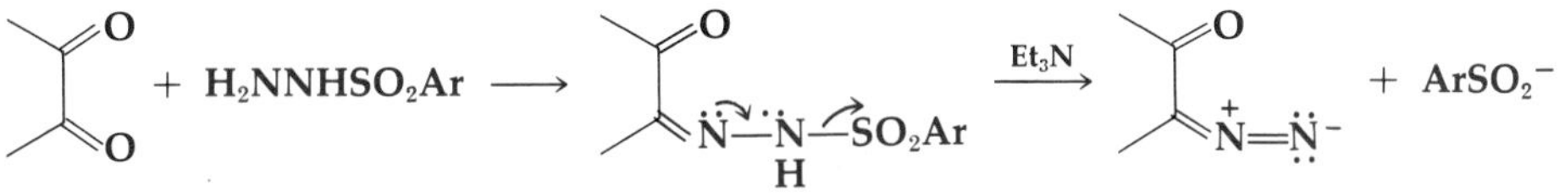

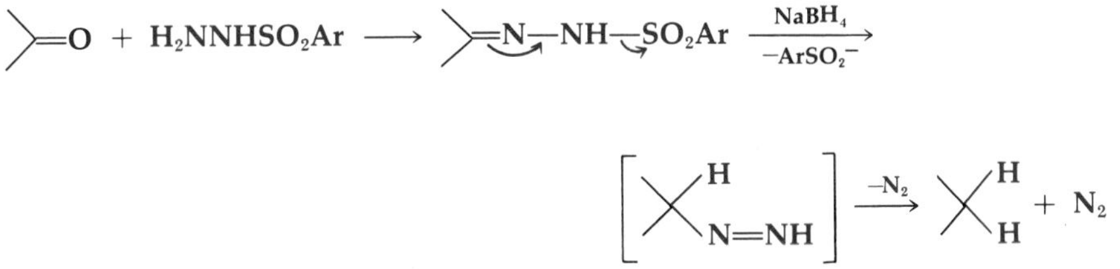

Other such reactions are known. Treatment of tosylhydrazones with bases causes toluenesulfinate to leave and consequent elimination of nitrogen with *migration of a proton*. The net result is an elimination reaction creating an alkene.

$$C_6H_5CH_2\underset{\overset{\|}{NNHSO_2C_7H_7}}{C}C_6H_5 \xrightarrow[-C_7H_7SO_2^-]{Bases} C_6H_5CH_2\underset{\overset{\|}{N_2}}{C}C_6H_5 \xrightarrow[-N_2]{H\ migration} C_6H_5CH{=}CHC_6H_5$$

PROBLEM 19-11

Rationalize the following observations.

a   The chlorides in $(ClCH_2CH_2)_2S$ are solvolyzed very much faster than in $(ClCH_2CH_2CH_2)_2S$, with first-order kinetics.

b   R—CH—CH—R + SCN$^-$ ⟶ R—CH—CH—R + OCN$^-$
       \O/   \S/

c $ClCH{=}CHCOOH + SO_3^{--} \longrightarrow {}^-O_3SCH{=}CHCOOH$

d $HOSO_2CH_2CH_2SO_2OH + PCl_5 \longrightarrow CH_2{=}CHSO_2Cl + ?$

e $CH_3CHBrCOC_6H_5 + CH_3SC_2H_5 \longrightarrow$ two isomers of $C_{12}H_{17}SOBr$

PROBLEM 19-12

Write five synthetic procedures for converting $R-X$ to $R-SH$ without obtaining R_2S as a by-product.

19-6 SULFUR AS ELECTROPHILE

The most obvious example of electrophilic sulfur is the sulfonyl group acting like carbonyl, in the sulfonic acid derivatives and their interconversions. These principal interconversions are summarized in Fig. 19-5. The general observation is that the sulfonic acid family is less readily reactive than the carboxylic acid family. Perhaps their most widespread practical use is in the sulfonation of alcohols in order to convert them to good leaving groups (cf. tosylates), as discussed in Sec. 11-1. This finds analogy in the conversion of alcohols to good leaving groups with phosphorus derivatives, discussed in Sec. 19-4.

The major exception to the analogy of normal nucleophilic substitution on sulfonyl chlorides is found in the behavior of carbanions, which generally attack chlorine instead of sulfur, displacing sulfinate anion.

$$\phi-SO_2-Cl + :CN \longrightarrow \phi-SO_2:^- + Cl-C\equiv N$$

$$\phi-SO-Cl + :CH\begin{array}{c} COOCH_3 \\ \\ COOCH_3 \end{array} \longrightarrow \phi-SO_2:^- + Cl-CH\begin{array}{c} COOCH_3 \\ \\ COOCH_3 \end{array}$$

Elemental sulfur acts as a electrophile owing to its cyclic polysulfide structure (S_8). Grignard reagents attack sulfur, to yield thiols on hydrolysis. Sulfur dioxide and trioxide are also electrophilic. Grignard reagents react smoothly with sulfur dioxide to create sulfinic acids.

$$\phi MgBr + S_8 \longrightarrow \phi-S^- MgBr^+ + S_7^- \xrightarrow{H_3O^+} \phi-SH$$

$$CH_3-\underset{\underset{CH_3}{|}}{\overset{\overset{CH_3}{|}}{C}}-MgCl + SO_2 \longrightarrow CH_3-\underset{\underset{CH_3}{|}}{\overset{\overset{CH_3}{|}}{C}}-SO_2:^-\ M\overset{+}{g}Cl \xrightarrow{H_3O^+} CH_3-\underset{\underset{CH_3}{|}}{\overset{\overset{CH_3}{|}}{C}}-SO_2H$$

Sulfur trioxide is the anhydride of sulfuric acid and a very powerful electrophile. Just as it exothermically reacts with water to form sulfuric acid, so it also reacts with alcohols and amines to form **sulfates** and **sulfamates**. The latter exist as internal salts. Many commercial detergents are the sulfates (as sodium salts) of long-chain alcohols such as n-$C_{18}H_{37}OH$. Such a molecule possesses an ionic end ($-OSO_3^-Na^+$) for water solubility and a hydrocarbon end ($C_{18}H_{37}-$) for attracting and adhering to nonpolar fats and greases.

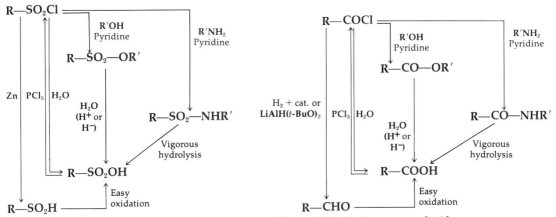

FIGURE 19-5 **Major interconversions of sulfonic acid derivatives compared with carboxylic acid derivatives**

R—OH + $\left[\begin{array}{c} O \\ \| \\ S=O \\ \| \\ O \end{array} \longleftrightarrow \begin{array}{c} ^-O \\ | \\ ^{++}S=O \\ | \\ ^-O \end{array}\right] \longrightarrow$ R—O—$\overset{\overset{\displaystyle O}{\|}}{\underset{\underset{\displaystyle O}{\|}}{S}}$—O⁻ H⁺

Sulfur trioxide   Sulfates

R—NH₂ + SO₃ $\longrightarrow$ R—NH—$\overset{\overset{\displaystyle O}{\|}}{\underset{\underset{\displaystyle O}{\|}}{S}}$—O⁻ H⁺ $\rightleftharpoons$ R—$\overset{+}{N}$H₂—$\overset{\overset{\displaystyle O}{\|}}{\underset{\underset{\displaystyle O}{\|}}{S}}$—O⁻

Sulfamates

The potency of sulfur trioxide as an electrophile is attested by its use in the sulfonation of the very weakly nucleophilic benzene ring (Sec. 16-4).

A comparison of the sulfonyl with the carbonyl groups is made in Fig. 19-5. Both are electrophilic and in the two acid families shown in the figure are reasonably comparable. They are also comparable in that they stabilize carbanions in the α-position. But differences must also be noted. They are to be expected since C=O contains a normal π bond and S=O a d-p_π bond with different characteristics and geometry. Differences were noted in the last section in that R—SO₂:⁻ possesses considerable stability whereas R—CO:⁻ is unknown. Furthermore, sulfonic acids are much stronger acids than carboxylic acids, comparable in strength with sulfuric acid. Toluenesulfonic acid is widely used as an acid catalyst in organic reactions. Similarly, sulfur trioxide is a much more powerful electrophile than carbon dioxide. Finally, sulfones differ from ketones in being unable to form enols and in being totally resistant to nucleophilic attack.

19-7 SUMMARY

There is much in the behavior of second-row elements which mirrors the behavior of their counterparts in the first row of the periodic table. This is most apparent in their saturated, lower-oxidation-state compounds. The key to most of their differences lies in the ability of second-row atoms to expand their outer electron shell, and take into their d orbitals one or two more pairs of electrons (to outer shells of 10 or 12).

The rest of their differences come from the increased size of the second-row atom, with its attendant weakening of bond strength, especially in the inability to form stable π bonds. Thiocarbonyls are very rare functional groups unless they are conjugated. Otherwise they tend to polymerize. No authentic compounds bearing a carbon-phosphorus normal π bond are known and monomeric H—C≡P: (compare stable H—C≡N:) only exists below $-100°C$.

When these atoms bear unshared electron pairs they become especially powerful nucleophiles, as in substitution reactions at saturated carbon.

In all of their oxidation states, sulfur and phosphorus atoms accept added external electrons to expand their outer shells. This may occur as stabilization of an adjacent negative charge on C, N, or O, or it may be manifested in the facility of attack of external nucleophiles on the phosphorus or sulfur atom. This attack of external nucleophiles most often leads to nucleophilic substitution reactions at phosphorus or sulfur, as at carbon, and accounts for most of their characteristic electrophilic reactivity.

Back-donation of electron pairs is so potent in phosphorus (and fairly so in sulfur) when the adjacent donating atom is oxygen that the formation of phosphorus-oxygen bonds (especially phosphoryl, P=O) is a major driving force in many of these substitutions.

The expansion of the outer shell allows for the stability of a variety of higher oxidation states than those found in first-row atoms and their covalent derivatives. A number of reliable procedures exists for interconverting these oxidation states (Fig. 19-3).

The geometry at the sulfur or phosphorus atom is almost invariably tetrahedral except in those few cases with more than four separate attached atoms (e.g., $\phi_5 P$). The tetrahedral geometry of second-row atoms is not altered by the stereoelectronic requirements of the bonding in back-donation resonance (d-p_π bonds). Pentacovalent compounds and reaction intermediates usually exist as trigonal bipyramids.

It is of some value to compare —SO_2— with —CO— in their families of compounds for they are often similar, and their differences are generally derived from the several properties that distinguish second- from first-row elements.

PROBLEMS

19-13   The following reaction creates a carbon-carbon bond and illustrates a group of reactions, called **extrusion reactions**, in which sulfur dioxide is produced.

$$RCH_2SO_2CHClR \xrightarrow[+B:^-]{\text{Base}} RCH{=}CHR + SO_2 + BH + Cl^-$$

a What mechanism can you ascribe to the reaction?

b The chlorine is introduced by simple chlorination of the parent sulfone; how might the sulfone be prepared from RCH_2OH?

c Should the double bond formed be cis or trans?

d When an α,α'-dichlorosulfone is treated with base, what product would you expect?

19-14 Supply the missing information in the following reactions. Write out all the sulfur species, with all the unshared electron pairs, in tetravalent resonance forms and show each mechanism.

a $C_6H_5MgBr + SO_2 \longrightarrow \xrightarrow{H_3O^+} ? \xrightarrow{CH_3I} ?$

b $C_6H_5COCH_3 + SO_3 \longrightarrow ? \xrightarrow{PCl_5} ? \xrightarrow{C_6H_5NH_2} ?$

c $p\text{-}CH_3OC_6H_4COCl + C_6H_5SO_2NHNH_2 \longrightarrow ? \xrightarrow[\Delta]{Na_2CO_3}$

$? \xrightarrow{C_6H_5SO_2NHNH_2} ? \xrightarrow{NaBH_4} ?$

d $CH_3I + SO_3^{--} \longrightarrow ? \xrightarrow{PCl_5} ? \xrightarrow{C_6H_5CH_2NHCH_3} ? \xrightarrow[\Delta]{NaOH}$

$CH_3SO_2^- + ? \xrightarrow{H_3O^+} C_6H_5CHO$

19-15 Synthesize the following compounds using elemental sulfur (S_8) as the sole source of that element.

a CH_3SO_2Cl **f** $C_6H_5SO_2SCH_3$

b $(C_2H_5)_3S^+ Cl^-$ **g** $(CH_3S)_2$

c $(CH_3)_2SO$ **h** $CH_3COCH_2SO_3H$

d C_6H_5SH

e Na_2SO_3

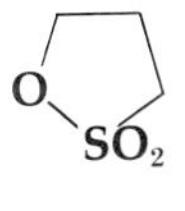

19-16 Although in (irreversible) substitution at saturated carbon the sulfinate anions (RSO_2^-) generally yield sulfones, the situation is different with substitution at unsaturated carbon. The corresponding α-ketosulfones expected from reaction with acid chlorides are not formed (they are in fact unknown compounds). The usual products from acid chlorides and sulfinates are the anhydride, $(RCO)_2O$, of the acid and products from the sulfinate. The latter are believed to form by subsequent reactions on an initial intermediate, $R{-}SO{-}SO_2{-}R$. Explain what occurs and why the α-ketosulfones are not formed. (Note the acidity of RSO_2H.)

19-17 The reaction of N-toluenesulfonyl-hydroxylamine and aldehydes in base affords hydroxamic acids. Explain.

19-18 Consider the reaction of triphenylphosphine with the α-diazoketone, $C_6H_5COCN_2C_6H_5$. A stable crystalline adduct is formed from the two at room temperature. On melting, the adduct bubbles and yields a new crystalline compound and this in turn, on heating more strongly, yields a hydrocarbon and a crystalline solid. Explain and provide mechanisms where possible.

19-19 Protein chains are often connected to each other via disulfide "bridges." It is common practice in protein studies to separate them by using performic acid. What functional change occurs?

19-20 Account for the following conversions in mechanistic terms.

a

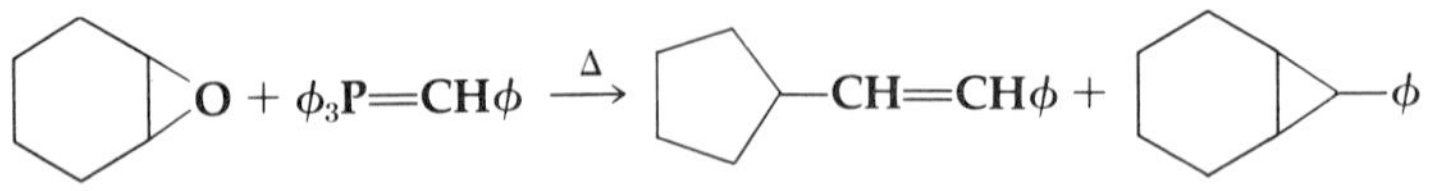

$$\text{(cyclohexane epoxide)} O + \phi_3P{=}CH\phi \xrightarrow{\Delta} \text{(cyclopentane)}{-}CH{=}CH\phi + \text{(bicyclic)}{-}\phi$$

b

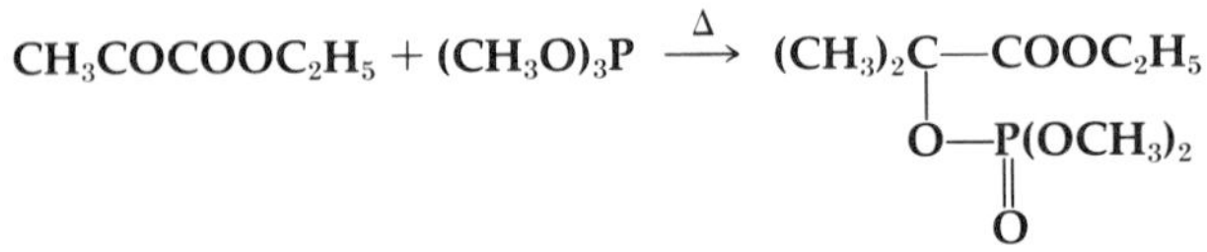

$$CH_3COCOOC_2H_5 + (CH_3O)_3P \xrightarrow{\Delta} (CH_3)_2\underset{\underset{\underset{O}{\parallel}}{\overset{|}{O{-}P(OCH_3)_2}}}{C}{-}COOC_2H_5$$

c

$$\text{(2,5-dichloro-1,4-benzoquinone)} + (C_2H_5O)_3P \longrightarrow$$

$$\text{(product with } O{-}P(OC_2H_5)_2 \text{, Cl, Cl, } OC_2H_5) + \text{(product with } OH, P(OC_2H_5)_2, Cl, OH) + C_2H_5Cl$$

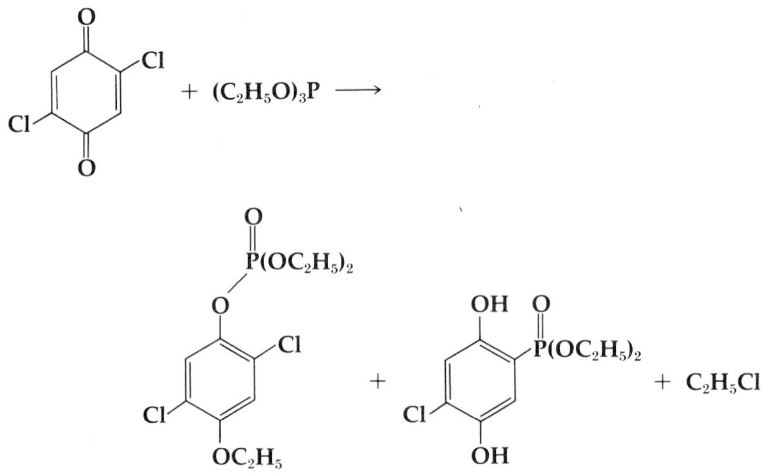

19-21 When the peroxylactone shown below is treated with triphenylphosphine, carbon dioxide is evolved and triphenylphosphine oxide can be isolated, as well as a good yield of C_9H_{10} showing four NMR peaks, all somewhat split by second-order splittings, at τ 2.9 (5H), 4.70 (1H), 5.00 (1H), and 7.90 (3H). Deduce the structure of the product and account for its production.

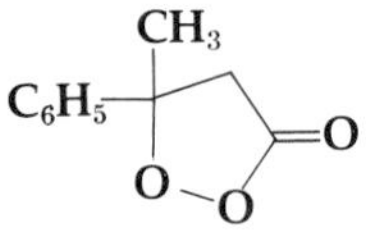

$$C_6H_5\underset{\underset{O{-}O}{\big|}}{\overset{\overset{CH_3}{\big|}}{C}}\cdots{=}O$$

READING REFERENCES

Kirby, A. J., and S. G. Warren, "The Organic Chemistry of Phosphorus," American Elsevier Publishing Company, Inc., New York, 1967.

RADICAL REACTIONS

THUS far, no difficulty has been encountered in the classification of reactions, because the reagents involved had structural characteristics that immediately stamped them as either electrophilic or nucleophilic. However, the operational characteristics of homolytic reactions are more subtle. The reactions are usually recognized as occurring under conditions conducive to the dissociation of reagents or **initiators** into atoms or **free radicals**. In other cases, the key to recognition may be found in the nature of the reaction products. When such guides are lacking, it becomes necessary to *base classification upon detailed studies of reaction mechanism*. Since interpretations of reaction mechanisms are subject to modification, occasionally a reaction is misclassified.

20-1 STABLE FREE RADICALS

In the period between 1840 and 1865, many reports appeared in the literature on the preparation of "free radicals," such as methyl and ethyl. However, after Cannizzaro's suggestions in 1860 led to the development of methods for the assignment of molecular weights, chemists came to recognize that the so-called radicals isolated by earlier workers were invariably dimers, such as ethane (methyl "radical") and butane (ethyl "radical"). The 1896 pronouncement of Ostwald, "It took a long time before it was finally recognized that the very nature of organic radicals is inherently such as to preclude the possibility of isolating them" was accepted and considered final by most workers in the field. Curiously, this dogmatic statement and its acceptance barely preceded Gomberg's announcement in 1900 of the preparation of the free triphenylmethyl radical. Since that time, many compounds have been shown, by magnetic measurements,† to exist as free radicals either in solution or as pure crystalline solids. "Stable" carbon radicals always contain extensive unsaturated systems, so that the unpaired electron is free to spread throughout a large volume, as in triphenylmethyl. Furthermore, the dimers that would be formed by the **coupling** of the radicals are ordinarily subject to great steric strain.

$$(C_6H_5)_3C-C(C_6H_5)_3 \;\overset{\text{Solvents}}{\rightleftharpoons}\; 2(C_6H_5)_3C\cdot \;\longleftrightarrow\; \text{[cyclohexadienyl structure]}=C(C_6H_5)_2 \;\longleftrightarrow\; \text{etc.}$$

Hexaphenylethane   Triphenylmethyl

The classic example of a stable free radical, triphenylmethyl, barely meets the requirements for independent characterization. In solution, the radical exists

† Free radicals have magnetic moments because of the spins of the unpaired electrons. As a consequence, a solution or solid that contains a free radical is drawn into the magnetic field when placed between the poles of a powerful magnet. Such material is said to be **paramagnetic**.

in appreciable concentrations in equilibrium with its dimer. Various reagents, including atmospheric oxygen, react with the radical, and concentration of its solutions leads to the separation of crystals of hexaphenylethane, which are not paramagnetic.

Gomberg's method, which consists in treating a halide with a metal, remains one of the standard procedures for producing stable radicals. If sodium is employed to abstract the halogen, an exactly equivalent amount must be used, because an excess reduces the radical, producing trityl sodium.

$$(C_6H_5)_3CCl \xrightarrow[\text{Na; benzene}]{\text{Ag, Hg, Zn,or}} (C_6H_5)_3C\cdot + AgCl, HgCl_2, ZnCl_2, \text{ or } NaCl$$

Triphenylmethyl
chloride
(trityl chloride)

$$(C_6H_5)_3C\cdot + Na\cdot \longrightarrow (C_6H_5)_3\bar{C}: \overset{+}{Na}$$

Triphenylmethylsodium
(tritylsodium)

Triphenylmethane dyes, which are really substituted triphenylcarbonium-ion salts, are reduced to free radicals by inorganic reductants such as titanous and vanadous salts.

$$(C_6H_5)_3COH \xrightleftharpoons{\text{Conc. } H_2SO_4} (C_6H_5)_3\overset{+}{C} \xrightarrow{\text{VCl}_2} (C_6H_5)_3C-C(C_6H_5)_3$$

A third method of synthesis is used if the related halide cannot be prepared. An ether is cleaved by potassium metal, and the resulting organometallic compound is oxidized by reaction with a *vicinal* dibromide.

$$cyclo\text{-}C_6H_{11}\overset{\overset{\displaystyle C_6H_5}{|}}{\underset{\underset{\displaystyle C_6H_5}{|}}{C}}OCH_3 \xrightarrow{\text{K, ether}} cyclo\text{-}C_6H_{11}\overset{\overset{\displaystyle C_6H_5}{|}}{\underset{\underset{\displaystyle C_6H_5}{|}}{C}}{:}^- \; K^+ \xrightarrow[-KBr,-(CH_3)_2C=C(CH_3)_2]{(CH_3)_2CBrCBr(CH_3)_2}$$

$$cyclo\text{-}C_6H_{11}\overset{\overset{\displaystyle C_6H_5}{|}}{\underset{\underset{\displaystyle C_6H_5}{|}}{C}}\cdot \rightleftharpoons \text{dimer}$$

Diphenylcyclo-
hexylmethyl radical

Oxidation of certain highly substituted phenols, arylamines, and hydrazines produces free radicals that are usually classified as oxygen or nitrogen radicals by reference to the structures of parent compounds. However, the odd electrons in the radicals are spread over extensive conjugated systems.

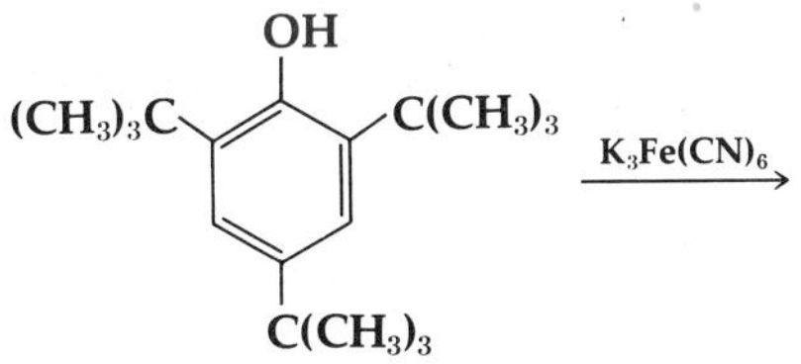

2,4,6-Tri-*tert*-butyl phenol

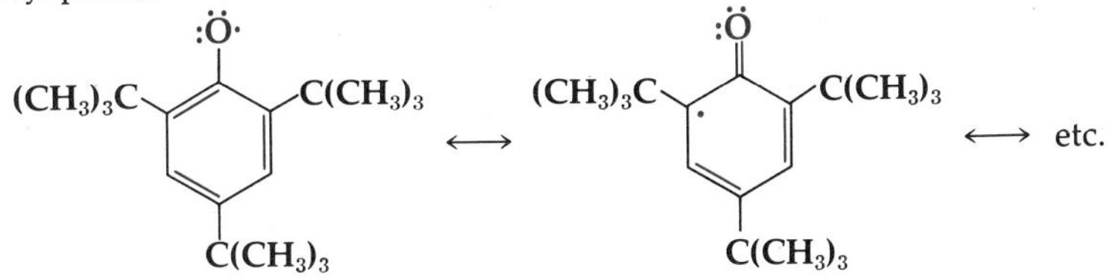

2,4,6-Tri-*tert*-butylphenoxy radical

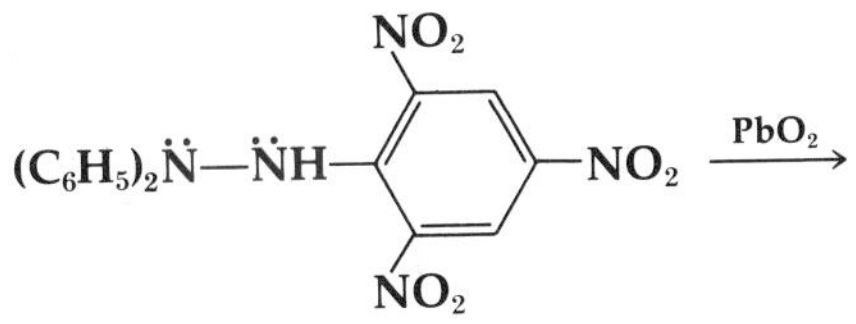

α,α-Diphenyl-β-picrylhydrazine

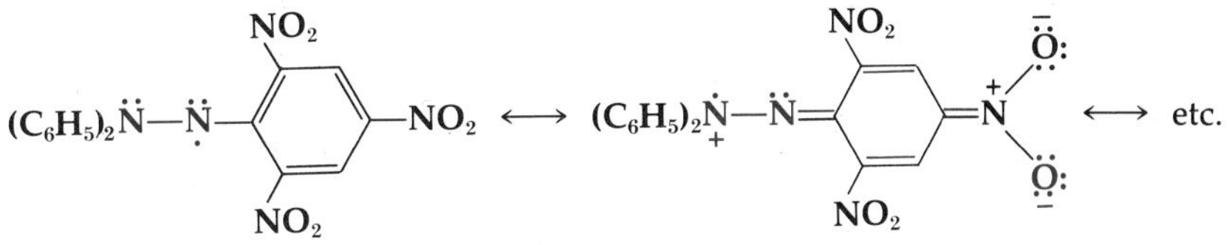

α,α-Diphenyl-β-picrylhydrazyl radical

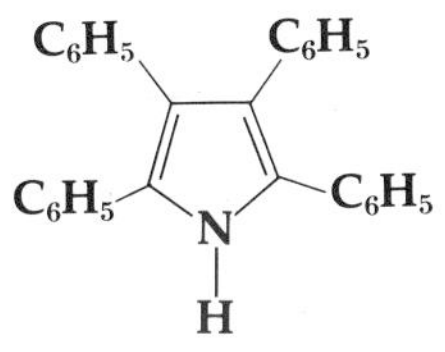

Tetraphenylpyrrole 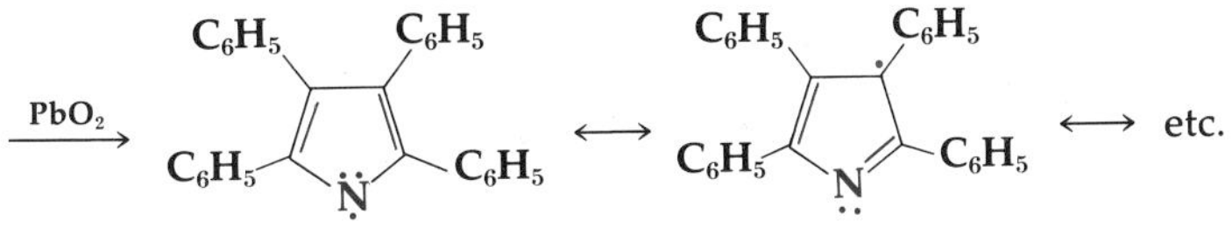   Tetraphenylpyrrole
                                   radical

20-2 GENERATION OF RADICALS

Short-lived free radicals are produced by three principal methods:

1   *Thermal decomposition of compounds containing weak bonds* often gives free radicals. Simple free radicals, such as methyl and ethyl, were first detected in the vapor-phase decomposition of lead and mercury alkyls.

$$(CH_3)_4Pb \xrightarrow{600°} Pb + 4CH_3\cdot \longrightarrow 2C_2H_6$$

Peroxides, which contain weak O—O bonds, and aliphatic azo compounds serve as convenient sources of free radicals at relatively low temperatures.

$$(CH_3)_3C—O—O—C(CH_3)_3 \xrightarrow{100-130°} 2(CH_3)_3C—O \cdot$$

Di-*tert*-butyl   *tert*-Butoxy radical
peroxide

$$2C_6H_5COCl \xrightarrow{Na_2O_2} (C_6H_5COO)_2 \xrightarrow{60-100°} 2C_6H_5COO \cdot$$

Dibenzoyl   Benzoyloxy
peroxide   radical

$$(CH_3)_2\overset{\displaystyle CN}{C}—N{=}N—\overset{\displaystyle CN}{C}(CH_3)_2 \xrightarrow{60-100°} 2(CH_3)_2\overset{\displaystyle \cdot}{C}—CN + N_2$$

α,α'-Azo-*bis*-   2-Cyano-2-
isobutyronitrile   propyl radical

2 *Photochemical reactions* of two types form radicals. Absorption of visible or ultraviolet light gives molecules with sufficient energy to break chemical bonds, and dissociation to give radicals may occur. The wavelength of the light must correspond to an absorption band of the substance to be decomposed.

$$Cl_2 \xrightarrow{Sunlight} 2Cl \cdot$$

Chlorine
atoms

$$CH_3COCH_3 \xrightarrow[\text{Vapor phase}]{\lambda \sim 3000 \text{ Å}} CH_3\overset{\displaystyle \cdot}{C}O + CH_3\cdot$$

Acetyl   Methyl
radical   radical

In solution, acetone and most other ketones do not cleave to give radicals but may produce radicals by the reaction of photochemically excited molecules with solvents.

$$C_6H_5COCH_3 \xrightarrow{\lambda = 3000-3500 \text{ Å}} C_6H_5COCH_3{}^* \xrightarrow{ZH \text{ (solvent)}} Z\cdot + C_6H_5\overset{\displaystyle OH}{\underset{}{C}}CH_3$$

Excited state α-Hydroxy-α-phenyl
of acetophenone   ethyl radical

3 *Oxidation-reduction reactions with inorganic ions that can change their valence state by the gain or loss of a single electron* can be used for generation of radicals.

$$H_2O_2 + Fe^{++} \longrightarrow HO\cdot + Fe(OH)^{++}$$

Ferrous   Hydroxyl   Ferric
ion   radical   ion

$$(CH_3)_3C—O—O—H + Co^{3+} \longrightarrow (CH_3)_3C—O—O\cdot + Co^{++} + H^+$$

tert-Butyl   Cobaltic   *tert*-Butylperoxy   Cobaltous
hydroperoxide   ion   radical   ion

Radicals made by the above procedures undergo rapid reactions in solution. Within their short lifetime, they may initiate important reactions of other constituents of the solution. Sometimes a radical produced in the primary process undergoes a fragmentation reaction to produce a smaller radical and a stable molecule.

$$(CH_3)_3CO \cdot \longrightarrow CH_3COCH_3 + CH_3 \cdot$$
$$C_6H_5CO_2 \cdot \longrightarrow C_6H_5 \cdot + CO_2$$

Oxidation of carbanions with oxygen or aromatic nitro compounds also generates radicals. The anions are prepared by treatment of weakly acidic substrates with strong base in a solvent such as dimethyl sulfoxide or *tert*-butyl alcohol. In the presence of oxygen, the ultimate products are formed by coupling of the radicals or by exhaustive oxidation at the carbon atom, the principal site of the unpaired electron in the radical.

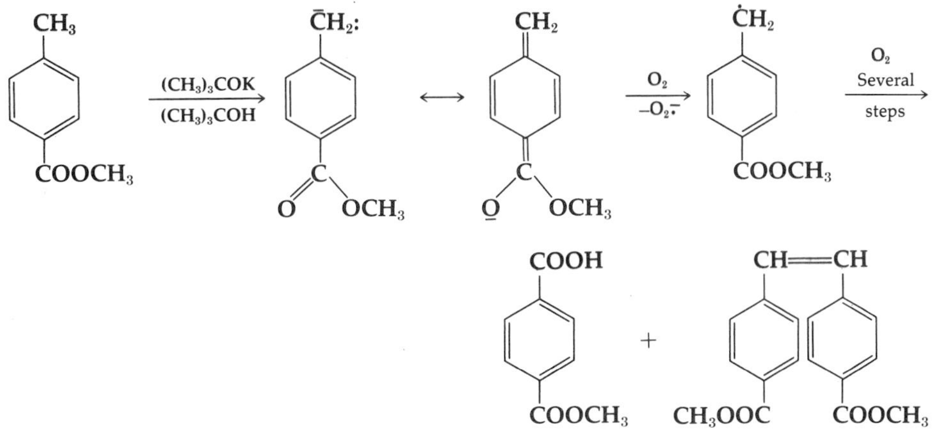

Radical Ions

Both radical anions and radical cations are known. As might be expected, their stability and lifetimes are greatly enhanced by unsaturated systems capable of dispersing both the odd electron and the charge.

Treatment of solutions of diaryl ketones with alkali metals produces deeply colored solutions of **ketyls**, which are salts of radical anions. Ketyls dimerize reversibly, and acidification of an equilibrium mixture of dimer and ketyls leads to *vic*-diols.

$K\cdot + (C_6H_5)_2C{=}O \xrightarrow{\text{Ether}}$

$$\overset{+}{K} \left[C_6H_5{-}\overset{\overset{\bar{O}}{|}}{\underset{\cdot}{C}}{=}\bigcirc \longleftrightarrow C_6H_5{-}\overset{\overset{\bar{O}}{|}}{C}{=}\bigcirc\cdot \longleftrightarrow C_6H_5{-}\overset{\overset{:\ddot{O}:}{|}}{C}{=}\bigcirc :^- \text{ etc.} \right]$$

Diphenyl ketyl

$$2\overset{+}{K}\,\bar{O}{-}\overset{\cdot}{C}(C_6H_5)_2 \rightleftharpoons (C_6H_5)_2\overset{\overset{\overset{-}{O}\overset{+}{K}}{|}}{C}{-}\overset{\overset{\overset{-}{O}\overset{+}{K}}{|}}{C}(C_6H_5)_2 \xrightarrow{H_2O} (C_6H_5)_2\overset{\overset{OH}{|}}{C}{-}\overset{\overset{OH}{|}}{C}(C_6H_5)_2$$

Benzpinacol

Somewhat similar radical anions can be prepared by addition of alkali metals to certain polynuclear hydrocarbons in liquid ammonia. In the case of naphthalene radical anion, a second atom of metal then adds to form a dianion, which abstracts protons from ammonia to give initially 1,4-dihydronaphthalene. Under the influence of the sodamide formed, this material isomerizes to 1,2-dihydronaphthalene (page 720).

$Na\cdot + $ [naphthalene] $\xrightarrow{NH_3}$ [naphthalene radical] $\longleftrightarrow$ [naphthalene radical anion] etc.

Naphthalene radical anion

$Na\cdot + $ [structure] $\xrightarrow{Na}$ [structure] $\xrightarrow{NH_3}$ [structure] $\xrightarrow{NaNH_2}$ [structure]

1,4-Dihydro-naphthalene   1,2-Dihydro-naphthalene

Tetracyanoethylene (page 327) possesses strong electrophilic tendencies which are demonstrated by the fact that it abstracts an electron from potassium metal to give a radical anion.

$K\cdot + (NC)_2C{=}C(CN_2) \xrightarrow[\text{phase}]{\text{Vapor}} \overset{+}{K}$ [radical structure] $\longrightarrow \overset{+}{K}$ [structure] etc.

Careful oxidation of hydroquinones or controlled reduction of quinones in basic solution produces **semiquinones**, which are radical anions of moderate stability in basic media. When acidified, semiquinones disproportionate to mixtures of quinones and hydroquinones, which in many cases form π-molecular complexes (page 328) with one another (**quinhydrones**).

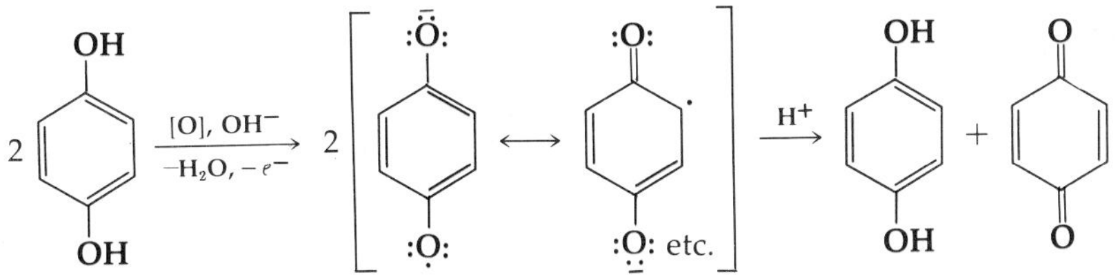

Electron exchange between carbanions and unsaturated organic compounds converts the latter to radical anions. As might be expected, dianions are especially active as electron donors.

$$\bar{A}: \quad + \quad B \quad \rightleftharpoons \quad A\cdot \quad + \quad \bar{B}\cdot$$

Carbanion   Unsaturated   Radical   Radical
              compound   anion

$$A^{--} \quad + \quad B \quad \rightleftharpoons \quad \bar{A}\cdot \quad + \quad \bar{B}\cdot$$

Unsaturated   Unsaturated   Radical   Radical
dianion   compound   anion   anion

An example of this behavior is found in the reactions of o- and p-nitrotoluenes with strong base.

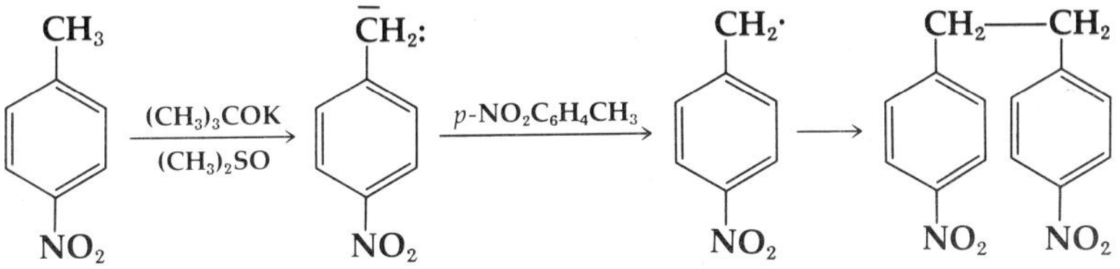

A number of radical cations have also been detected, and their stabilities vary over a wide range. At the high end of the stability scale are the Würster radical cations, in which the positive charge and odd electron are highly delocalized. At the opposite end is CH_4^+, which is produced in a mass spectrograph by knocking an electron out of methane by bombardment with electrons that have been accelerated under moderate potentials.

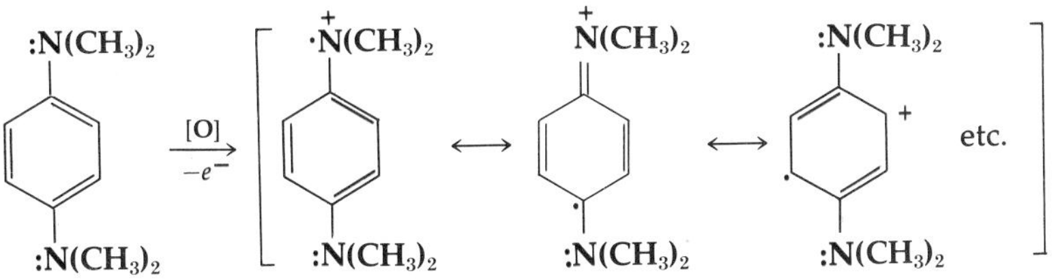

A Würster radical cation

$(C_6H_5)_3N: \xrightarrow[-e]{[O]} (C_6H_5)_2\overset{\cdot}{\underset{+}{N}}\!\!-\!\!\langle \rangle \longleftrightarrow (C_6H_5)_2N\!\!=\!\!\langle \rangle \cdot$ etc.

$H-\overset{\overset{\displaystyle H}{|}}{\underset{\underset{\displaystyle H}{|}}{C}}-H \xrightarrow{-e^-} H-\overset{\overset{\displaystyle H}{|}}{\underset{\underset{\displaystyle H}{|}}{C}}\cdot \overset{+}{H} \longleftrightarrow H\cdot \overset{+}{\overset{\overset{\displaystyle H}{|}}{\underset{\underset{\displaystyle H}{|}}{C}}}-H$ etc.

20-3   RADICAL COUPLING REACTIONS

About the simplest way for radicals to dispose of themselves is to couple with one another to produce a covalent bond. When the radicals are stabilized by extensive delocalization of the electrons, or when the coupled product is destabilized by either steric effects or repulsion of like charges, the coupling reaction is reversible. Examples of reversible coupling reactions were noted with triphenylmethyl and diphenylketyls (page 819). When the radicals are less stable and the coupled products more stable, the coupling reaction occurs very rapidly and is not reversible.

$$C_2H_5OOC(CH_2)_8COOH \xrightarrow[\text{2) Na}_2O_2]{\text{1) SOCl}_2} [C_2H_5OOC(CH_2)_8COO]_2 \xrightarrow[100°]{CH_3COOH}$$

Ethyl hydrogen
sebacate

$$C_2H_5OOC(CH_2)_{16}COOC_2H_5 + 2CO_2$$

Diethyl octadecanedioate
55%

In some cases, the radicals that couple are generated from less stable radicals by hydrogen-atom abstraction.

$$(CH_3COO_2 \xrightarrow{100°} 2CH_3CO_2\cdot \longrightarrow 2CH_3\cdot + 2CO_2$$

$$CH_3\cdot + CH_3COCH_3 \longrightarrow CH_4 + \cdot CH_2COCH_3$$

$$2CH_3COCH_2\cdot \longrightarrow CH_3COCH_2CH_2COCH_3$$

2,5-Hexanedione

$$C_6H_5CH(CH_3)_2 \xrightarrow[120°]{(CH_3)_3COOC(CH_3)_3} C_6H_5\overset{\overset{\displaystyle CH_3}{|}}{\underset{\underset{\displaystyle CH_3}{|}}{C}}\!\!-\!\!\overset{\overset{\displaystyle CH_3}{|}}{\underset{\underset{\displaystyle CH_3}{|}}{C}}C_6H_5$$

$$CH_3COOH \xrightarrow[100°]{(CH_3CO_2)_2} \overset{\displaystyle CH_2COOH}{\underset{\displaystyle CH_2COOH}{|}}$$

In the **Kolbe electrolysis**, salts of carboxylic acids are electrolyzed, and the carboxy radicals generated at the anode lose carbon dioxide and couple. This reaction has found limited synthetic application.

$$RC\overset{-}{O}_2 \xrightarrow[-e]{\text{Anode}} RCO_2 \cdot \longrightarrow CO_2 + R \cdot \xrightarrow{R \cdot} R\!-\!R$$

$$2CH_3(CH_2)_2C\overset{-}{O}_2\overset{+}{K} \xrightarrow[-e^-]{\text{Anode}} CH_3(CH_2)_4CH_3$$

Potassium n-butyrate n-Hexane

$$CH_3OOC(CH_2)_4CO_2Na + NaO_2C(CH_2)_4CH_3 \xrightarrow{\text{Electrolysis}} CH_3OOC(CH_2)_8CH_3$$

Methyl sodium   Sodium caproate   Methyl decanoate
adipate   58%

Oxidative coupling of phenols is a widespread reaction in nature, where enzymes catalyze the formation and coupling of phenoxy radicals. In the laboratory the reaction is less manageable and many products, including polymers, are formed. A number of oxidants have been used successfully, but alkaline ferricyanide is the most common. Resonance allows considerable radical presence at the ortho and para positions as well as on oxygen with the result that six coupling products are theoretically possible. All of these have been isolated except the diphenyl peroxide, which presumably redissociates to radicals.

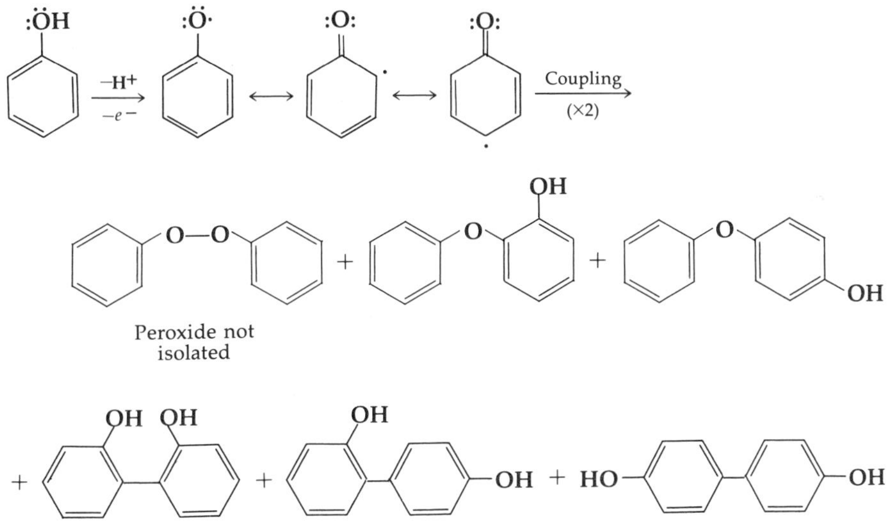

Peroxide not
isolated

Oxidative coupling can occur at a substituted ortho or para position so that rearomatization is not possible. The reaction also often proceeds cleanly in one direction when the coupling is a cyclization of two already linked phenols.

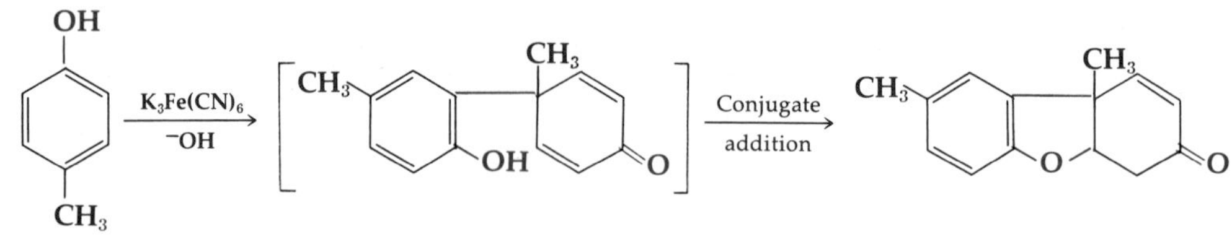

p-Cresol   Pummerer's ketone (52%)

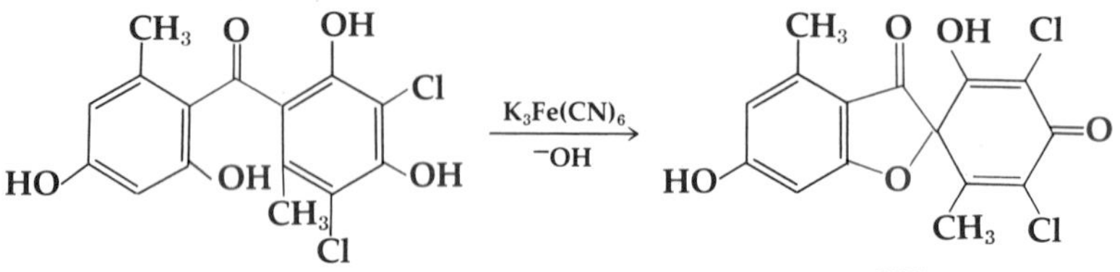

35%

20-4 SUBSTITUTION AT SATURATED CARBON

Homolytic substitution reactions are considerably less important in practice then heterolytic substitutions. In general, ions or polar molecules are the active reagents in heterolytic substitutions, whereas radicals are the active reagents in homolytic substitutions. Of the two types, the polar species are much more selective in their attack on organic compounds. As a result, yields of single products are generally higher in heterolytic than in homolytic substitution reactions. Exceptions to this generalization provide the organic chemist with a few radical substitution reactions of considerable synthetic utility.

Radical reactions commonly take a **chain reaction** course.

1 **Initiation:** Reagent, or a trace of radical initiator, is first converted to radicals in catalytic amounts. The short-lived radicals of Sec. 21-2 are examples.

2 **Propagation:** The initial radicals create radicals from the substrate, often by abstracting a hydrogen atom. The active substrate radical then undergoes a reaction, leaving a radical product which similarly generates another substrate radical for the same reaction, and then another and so on in a chain of identical reaction steps.

3 **Termination:** The chain-carrying radical is destroyed by coupling or reaction with the vessel walls.

Halogenation

The action of light on halogens produces halogen atoms. The latter easily abstract hydrogen atoms from saturated carbon atoms to initiate *chain reactions* that lead to the **photochemical halogenation** of hydrocarbons. The steps of the chain reaction are illustrated with methane as an example.

Initiation $Cl_2 \xrightarrow{\text{Sunlight}} 2Cl\cdot$

Propagation $\begin{cases} Cl\cdot + CH_4 \longrightarrow HCl + CH_3\cdot \\ CH_3\cdot + Cl_2 \longrightarrow CH_3Cl + Cl\cdot \end{cases}$

Termination $\begin{cases} 2Cl\cdot \longrightarrow Cl_2 \\ 2CH_3\cdot \longrightarrow C_2H_6 \\ Cl\cdot + CH_3\cdot \longrightarrow CH_3Cl \end{cases}$

Photochlorination is used industrially for production of mixtures of alkyl halides from petroleum hydrocarbons and for chlorination of aromatic side chains. Attack of chlorine atoms on an alkane shows little discrimination, although *tertiary* C—H bonds are slightly more susceptible to attack than secondary and primary positions. Aromatic side chains, on the other hand, are specifically activated in the positions adjacent to the ring.

$$CH_4 \xrightarrow[-HCl]{Cl_2,\ light} CH_3Cl \xrightarrow[-HCl]{Cl_2} CH_2Cl_2 \xrightarrow[-HCl]{Cl_2} HCCl_3 \xrightarrow[-HCl]{Cl_2} CCl_4$$

Methyl chloride   Methylene chloride   Chloroform   Carbon tetrachloride

$$CH_3CHCH_2CH_3 \xrightarrow{Cl_2,\ light} \left[\begin{array}{c} ClCH_2CHCH_2CH_3 + (CH_3)_2CClCH_2CH_3 \\ | \\ CH_3 \\ + (CH_3)_2CHCHClCH_3 + (CH_3)_2CHCH_2CH_2Cl \end{array} \right]$$

$|$
CH_3
Isopentane

$$C_6H_5CH_3 \xrightarrow[-HCl]{Cl_2,\ light} C_6H_5CH_2Cl \xrightarrow[-HCl]{Cl_2} C_6H_5CHCl_2 \xrightarrow[-HCl]{Cl_2} C_6H_5CCl_3$$

Benzyl chloride   Benzylidene dichloride   Benzochloride

$$C_6H_5CH_2CH_3 + Cl_2 \xrightarrow{Light} C_6H_5CHCH_3 + HCl$$
$|$
Cl

Ethylbenzene α-Phenylethyl chloride

Photochemical bromination occurs in a manner analogous to chlorination, but the reactions are slow because the reaction chains are rather short. Bromination with N-bromosuccinimide is extensively used in the laboratory to introduce halogen in positions adjacent to olefinic, aromatic, and carbonyl groups (page 750). This reaction has the characteristics of a radical-chain process, since it is initiated by light absorbed by the reagent or by the decomposition of labile azo compounds and peroxides. The reaction is also *inhibited* by addition of small amounts of materials, such as benzoquinone, that react readily with free radicals and destroy them.

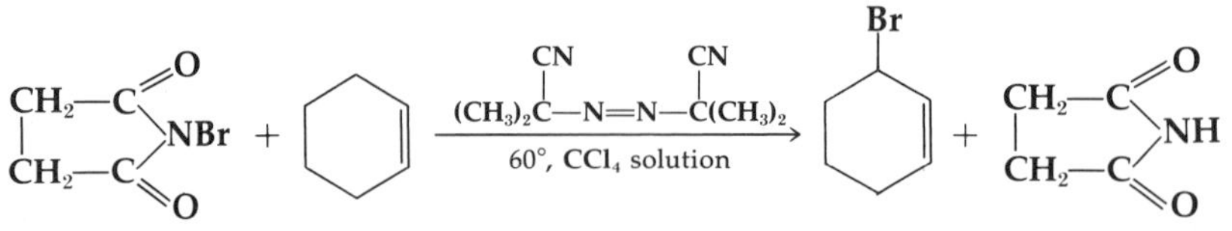

N-Bromosuccinimide
(NBS)

2-Cyclohexenyl
bromide
70%

$$CH_3CH=CHCOOCH_3 \xrightarrow[\text{dibenzoyl peroxide}]{\text{NBS,}} BrCH_2CH=CHCOOCH_3$$

Methyl crotonate   Methyl γ-bromocrotonate

Oxidation

In the presence of free-radical initiators, oxygen attacks saturated hydro-carbon structures, especially at allylic positions. The reactions are chain proc-esses, and the first products are hydroperoxides.

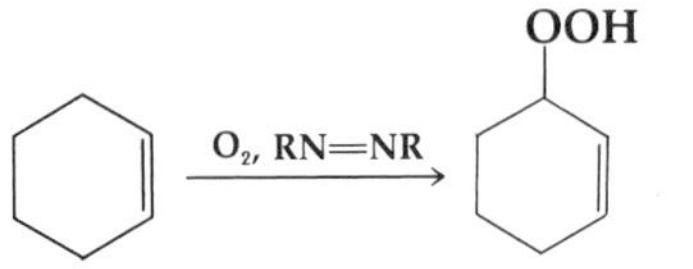

2-Cyclohexenyl
hydroperoxide

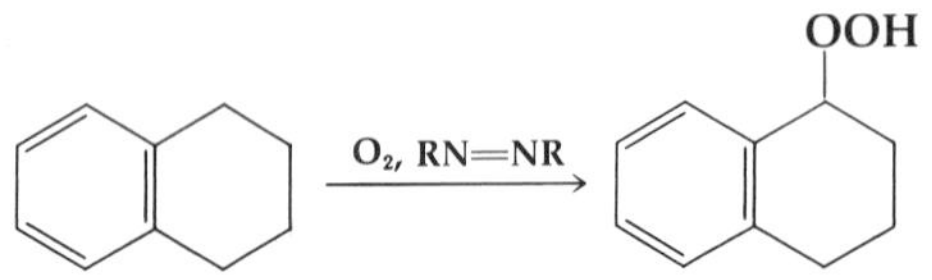

Tetralin   α-Tetralyl
hydroperoxide

The chain-carrying steps (propagation) of the reaction are as follows:

$$RO_2\cdot + RH \longrightarrow RO_2H + R\cdot$$

$$R\cdot + O_2 \longrightarrow RO_2\cdot$$

The slow deterioration of most organic materials when exposed to air and sunlight is due largely to photosensitized air oxidation. Aromatic amines and phenols inhibit oxidation by destroying $RO_2\cdot$ radicals. The inclusion of such **antioxidants** in rubber, gasoline, plastics, etc., prolongs the useful lifetime of such materials.

Activation of allylic and benzylic hydrogen atoms toward abstraction is associated with the fact that removal of hydrogen from such a position leaves a resonance-stabilized allylic radical.

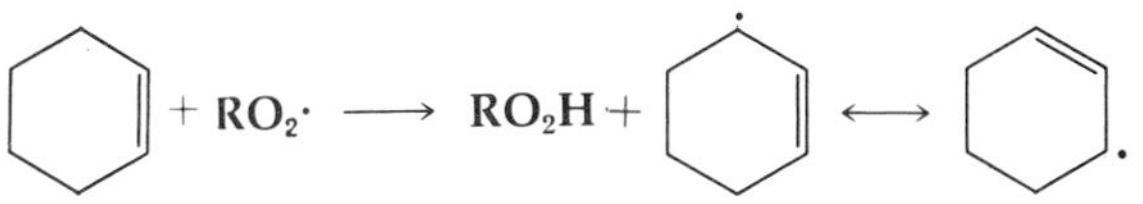

Cyclohexenyl radical

Hydroperoxides decompose to give radicals capable of initiating oxidation when heated to 100° or higher. As a consequence, high-temperature oxidation is autocatalytic and leads to extensive degradation or complete combustion.

Although a polar reagent, chromic acid probably abstracts a hydrogen atom from carbon as the first stage in the oxidation of saturated hydrocarbons. When tertiary, the carbon atom undergoing substitution goes to a tertiary alcohol, which in the case of triphenylcarbinol survives. With other compounds the oxidative process continues and stops at a ketone, carboxylic acid, or carbon dioxide stage, depending on the vigor (temperature, acidity) of the conditions. In many cases, concentrated nitric acid or potassium permanganate can be employed as an oxidizing agent similarly.

$$H_2Cr_2O_7 + (C_6H_5)_3CH \longrightarrow (C_6H_5)_3COH$$

$$HNO_3 + (C_6H_5)_2CH_2 \xrightarrow{\Delta} (C_6H_5)_2C{=}O$$
$$90\%$$

$$KMnO_4 + C_6H_5CH_3 \xrightarrow[100°]{NaOH} C_6H_5COOH$$
$$95\%$$

$$H_2Cr_2O_7 + C_6H_5CH(CH_3)_2 \xrightarrow{\Delta} C_6H_5COOH + CO_2$$

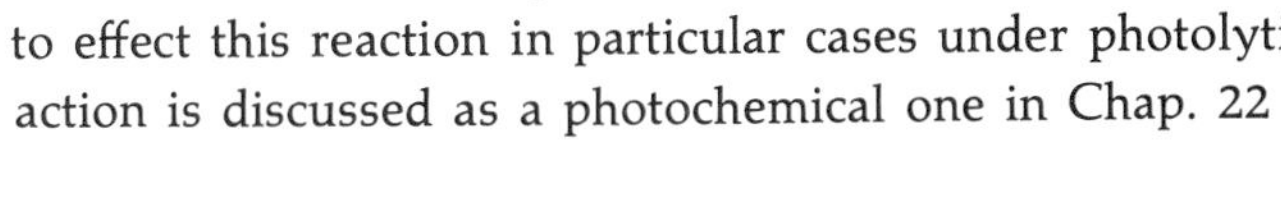

Saturated sites without some particular activation, such as adjacent carbonyl or olefin, cannot usually be caused to undergo specific or selective reactions. One small but synthetically useful family of radical reactions does serve to effect this reaction in particular cases under photolytic initiation. This reaction is discussed as a photochemical one in Chap. 22 (page 894).

20-5 RADICAL ADDITIONS

Many alkene additions can be catalyzed by small amounts of peroxides, by thermally labile azo compounds, or by irradiation by ultraviolet light. Reactions occurring under such conditions are nearly always *free-radical chain* reactions. They are perhaps best illustrated by the *abnormal addition of hydrogen bromide* to alkenes.

Chain
initiation

$$C_6H_5\overset{\displaystyle O}{\overset{\displaystyle \|}{C}}OO\overset{\displaystyle O}{\overset{\displaystyle \|}{C}}C_6H_5 \xrightarrow{\Delta, 60-80°} 2C_6H_5CO_2\cdot$$

Dibenzoyl peroxide   Benzoyloxy radical

$$C_6H_5CO_2\cdot + HBr \longrightarrow C_6H_5COOH + Br\cdot$$

Chain
propagation

$$Br\cdot + CH_3CH{=}CH_2 \longrightarrow CH_3\dot{C}HCH_2Br$$

$$CH_3\dot{C}HCH_2Br + HBr \longrightarrow CH_3CH_2CH_2Br + Br\cdot$$

Chain
termination

$$2Br\cdot \longrightarrow Br_2$$

$$CH_3\dot{C}HCH_2Br + Br\cdot \longrightarrow CH_3CHBrCH_2Br$$

$$2CH_3\dot{C}HCH_2Br \longrightarrow BrCH_2\underset{\overset{\displaystyle |}{CH_3}}{CH}{-}\underset{\overset{\displaystyle |}{CH_3}}{CH}CH_2Br$$

FIGURE 20-1 **Addition of hydrogen bromide to propylene by a peroxide-initiated radical**

$$C_6H_5CH{=}CH_2 + HBr \xrightarrow{\text{Peroxide}} C_6H_5CH_2CH_2Br$$

β-Phenylethyl bromide
85%

The reaction is termed *abnormal* because it does not follow Markownikoff's rule. The steps in addition by a free-radical chain mechanism are illustrated in Fig. 20-1.

Since the chain-propagating reactions recur again and again, the decomposition of a single initiator molecule may produce a large number of molecules of product. The chains eventually end when two of the active chain-carrying intermediates meet and destroy each other.

Hydrogen chloride and hydrogen iodide do not undergo abnormal addition to olefins under normal circumstances. The reasons for failure are different for the two compounds. The hydrogen-chlorine bond is too strong to be broken readily in the *atom-abstraction* reaction. Hydrogen iodide gives up a hydrogen atom very readily, but iodine atoms do not add rapidly to double bonds.

$$R\cdot + HCl \xrightarrow{\text{Slow}} RH + Cl\cdot$$

$$R\cdot + HI \xrightarrow{\text{Fast}} RH + I\cdot$$

$$I\cdot + \overset{\diagdown}{\underset{\diagup}{C}}{=}\overset{\diagup}{\underset{\diagdown}{C}} \xrightarrow{\text{Slow}} I{-}\overset{|}{\underset{|}{C}}{-}\overset{|}{\underset{|}{C}}\cdot$$

The structure of the product is determined in the addition step. The results show that orientation in radical addition to a double bond is the same as in proton addition. The final products of hydrogen bromide addition by ionic and radical reactions are different because in one case hydrogen enters first and in the other bromine is added first.

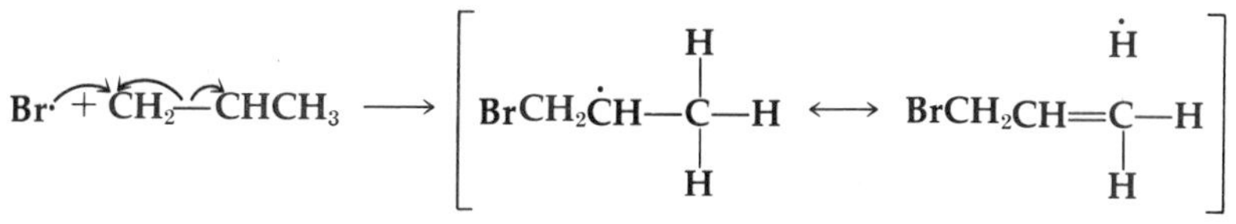

$$\text{Br}^\cdot + \text{CH}_2\text{—CHCH}_3 \longrightarrow \left[\text{BrCH}_2\dot{\text{C}}\text{H—}\overset{\text{H}}{\underset{\text{H}}{\text{C}}}\text{—H} \longleftrightarrow \text{BrCH}_2\text{CH}=\overset{\dot{\text{H}}}{\underset{\text{H}}{\text{C}}}\text{—H} \right]$$

not:

$$\text{Br}^\cdot + \text{CH}_2\text{—CHCH}_3 \longrightarrow \ \cdot\text{CH}_2\text{CHBrCH}_3$$

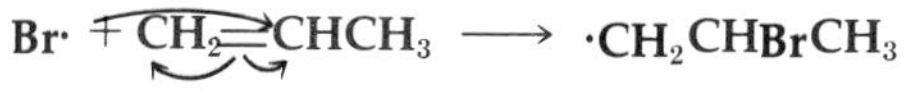

Abnormal addition of hydrogen bromide is a means of preparing primary bromides from terminal olefins, although hydroboration (page 637) is usually preferable.

$$\text{HBr} + (\text{CH}_3)_3\text{CCH}_2\underset{\text{CH}_3}{\text{C}}=\text{CH}_2 \ \xrightarrow{\text{Peroxides}} \ (\text{CH}_3)_3\text{CCH}_2\underset{\text{CH}_3}{\text{CHCH}_2\text{Br}}$$

A host of other radical additions, many of them carbon-carbon bond-making processes, have been carried out successfully. In many cases, it is necessary to use the addend in large excess in order to inhibit polymerization, which arises from the addition of radicals from the alkene to a second alkene molecule (Chap. 25). Table 20-1 shows reagents that have been used, the adding entities, and the structures of the products expected in addition to unsymmetrical olefins.

Benzene undergoes exclusively *addition* of chlorine in photochemical experiments. The resulting mixture of stereoisomers is used as an insecticide, known as Gammexane. The activity is due to the γ-isomer, which constitutes 10 to 12% of the mixture.

TABLE 20-1 **Free-radical Addition to Alkenes**

Reagent	Atom abstracted	Product from $RCH{=}CH_2$ (or $ArCH{=}CH_2$)
R'SH	H	RCH_2CH_2SR'
CCl_4	Cl	$RCHClCH_2CCl_3$
CBr_4	Br	$RCHBrCH_2CBr_3$
$CHBr_3$	Br	$RCHBrCH_2CHBr_2$
CCl_3Br	Br	$RCHBrCH_2CCl_3$
$BrCH_2COOC_2H_5$	Br	$RCHBrCH_2CH_2COOC_2H_5$
R'CHO	H	RCH_2CH_2COR'
$(C_6H_5)_3SiH$	H	$RCH_2CH_2Si(C_6H_5)_3$
$COCl_2$	Cl	$RCHClCH_2COCl$

$$3Cl_2 + \text{(benzene)} \xrightarrow{\text{Light}} C_6H_6Cl_6$$

Benzene
hexachlorides

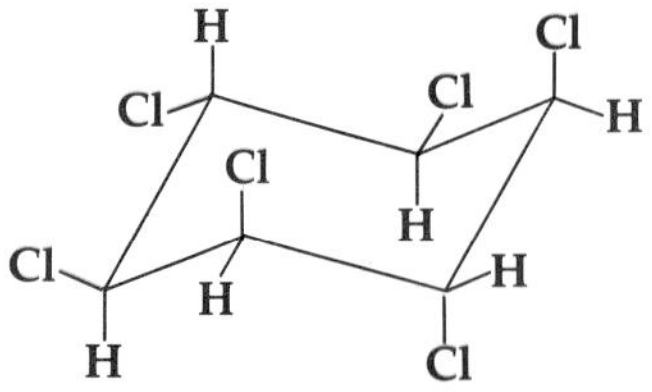

γ-Benzene hexachloride

20-6 AROMATIC SUBSTITUTION

Aromatic compounds undergo substitution by some radicals. Most of the examples involve substitution by an aryl radical with formation of biaryls. Many sources of aryl radicals have been investigated, as illustrated by the following preparations of the phenyl radical. Some details of the decomposition reactions remain uncertain.

$$(C_6H_5CO_2)_2 \longrightarrow 2C_6H_5CO_2 \cdot \longrightarrow C_6H_5 \cdot + CO_2$$

$$C_6H_5N{=}NC(C_6H_5)_3 \longrightarrow C_6H_5 \cdot + (C_6H_5)_3C \cdot + N_2$$

Phenylazotri-
phenylmethane

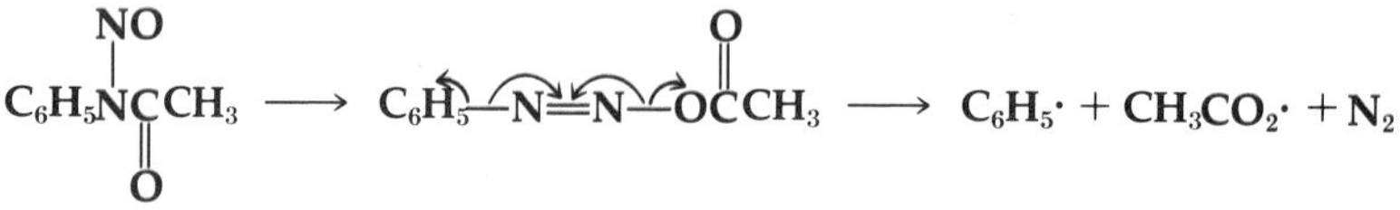

$$C_6H_5\overset{\overset{\displaystyle NO}{|}}{\underset{\underset{\displaystyle O}{||}}{N}}CCH_3 \longrightarrow C_6H_5{-}N{=}N{-}O\overset{\overset{\displaystyle O}{||}}{C}CH_3 \longrightarrow C_6H_5 \cdot + CH_3CO_2 \cdot + N_2$$

N-Nitroso-
acetanilide

Benzene
diazoacetate

$$C_6H_5\overset{+}{N_2}\bar{C}l \xrightarrow{OH^-} C_6H_5N{=}NOH \longrightarrow C_6H_5 \cdot + HO \cdot + N_2$$

Benzenediazonium
chloride

Benzenediazo
hydroxide

$$Pb(OOCC_6H_5)_4 \longrightarrow Pb(OOCC_6H_5)_2 + 2C_6H_5CO_2 \cdot \longrightarrow 2C_6H_5 \cdot + 2CO_2$$

Lead
tetrabenzoate

Although the timing of the substitution reaction is not yet understood, material conservation demands that two radicals must sooner or later be involved in the substitution, one to take the displaced hydrogen atom and one to form the new carbon-carbon bond.

$$Ar \cdot + Ar'H + R \cdot \longrightarrow Ar{-}Ar' + RH$$

Substituent effects are much smaller in aromatic substitutions by free radicals than in electrophilic aromatic substitutions (Chap. 16). Both electron-withdrawing and electron-donating substituents on AR′ have a very mild tendency to direct the incoming substituent to the ortho positions, and unless the substituent is quite large, ortho substitution predominates. Although yields in the reaction are never high (maximum of about 70% on the basis of the stoichiometry indicated above), the reactions have been useful in the synthesis of the derivatives of biphenyl, which are difficult to prepare in other ways.

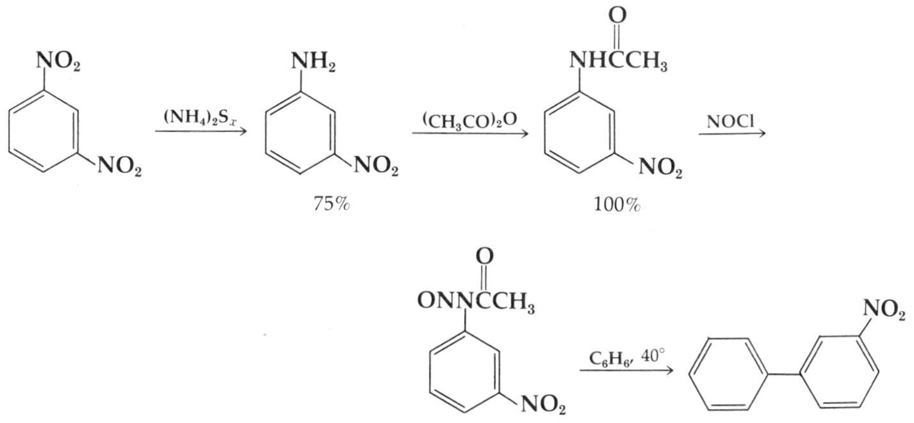

m-Nitro-N-nitroso-
acetanilde
50%

m-Nitrobiphenyl
40%

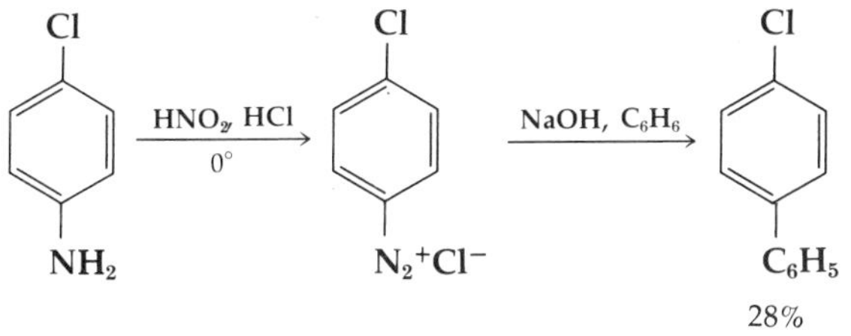

28%

20-7   STEREOCHEMISTRY OF RADICALS

Examination of the configuration of radicals through stereochemical techniques suggests that the three groups attached to carbon are disposed in a planar trigonal arrangement (i.e., sp^2). Substitution reactions carried out with optically active starting materials lead to racemic products in those cases in which free radicals have been demonstrated as intermediates.

Optically active — Racemic

Racemic

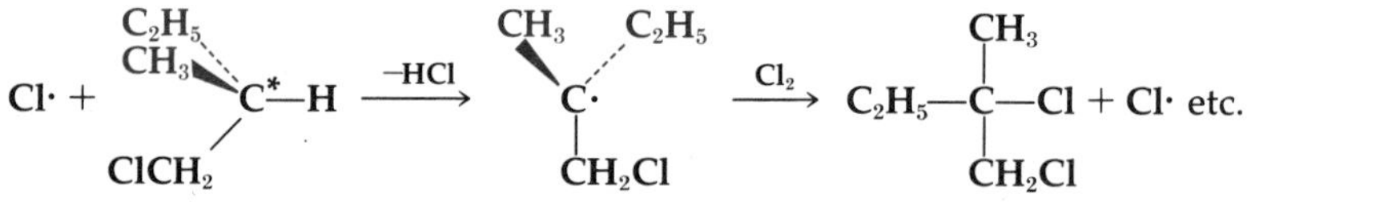

Other experiments demonstrate that radicals can be made in which a tetrahedral configuration is mandatory. In decomposition of apocamphoryl peroxide, the hypothetical intermediate radical is situated at a bridgehead, and an sp^2 configuration is prevented by the geometric requirements of the bicyclic system.

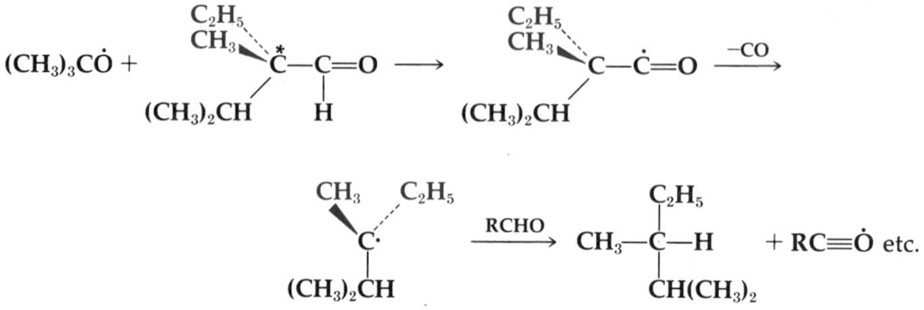

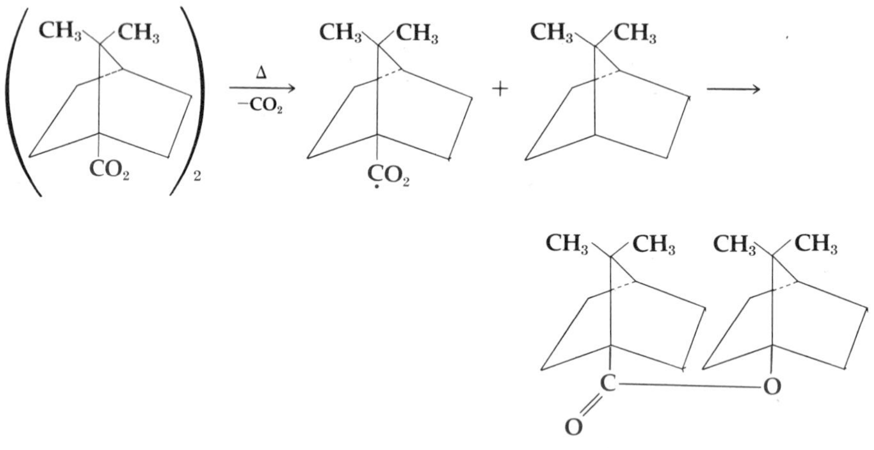

When the same type of reaction was carried out on an open-chain and optically active peroxide, ester was obtained, which when hydrolyzed, gave optically active alcohol.

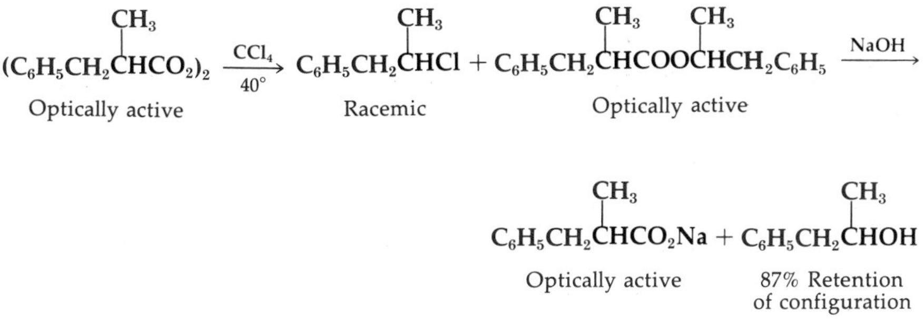

$$(C_6H_5CH_2\overset{\overset{\displaystyle CH_3}{|}}{C}HCO_2)_2 \xrightarrow[40°]{CCl_4} C_6H_5CH_2\overset{\overset{\displaystyle CH_3}{|}}{C}HCl + C_6H_5CH_2\overset{\overset{\displaystyle CH_3}{|}}{C}HCOO\overset{\overset{\displaystyle CH_3}{|}}{C}HCH_2C_6H_5 \xrightarrow{NaOH}$$

Optically active   Racemic   Optically active

$$C_6H_5CH_2\overset{\overset{\displaystyle CH_3}{|}}{C}HCO_2Na + C_6H_5CH_2\overset{\overset{\displaystyle CH_3}{|}}{C}HOH$$

Optically active   87% Retention
of configuration

Two explanations are possible for the observed optical activity of the 1-phenyl-2-propanol obtained on hydrolysis of the ester product. Either the ester was formed by a cyclic one-stage process, or the asymmetric environment of the otherwise symmetric radical led to the final asymmetry in the alcohol portion of the ester. If the latter explanation is correct, the ester must have been formed largely by coupling of the pairs of radicals before they escaped from the "solvent cage" in which they were originally produced.

One-stage process:

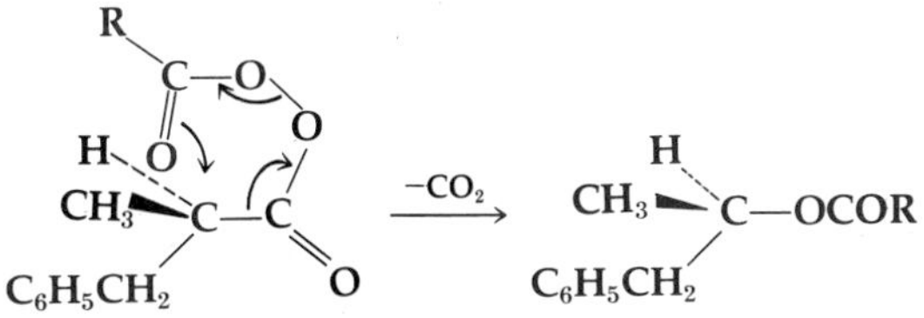

Two-stage process:

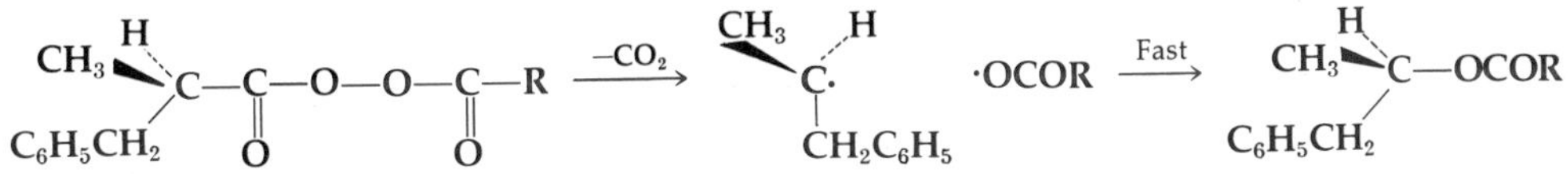

PROBLEMS

20-1 Write equations for the following conversions:

   a Dimethyl ketone to pivalic acid

   b Trityl alcohol to trityl peroxide

   c *p*-Xylene to terephthalic acid

   d Propylene to 1,5-hexadiene

   e 1-Pentene to caproic acid

   f Cyclobutanecarboxylic acid to cyclobutene

   g Glutaric acid to suberic acid

   h 1-Hexene to 2-hexenal

   i 1-Butene to di-*n*-butyl sulfone

    j Benzoic acid to phenylbenzoate

    k Ethane to ethylamine

    l Benzaldehyde and methane to β-nitrostyrene

    m Toluene to β-phenylethylamine

    n Toluene to benzilic acid

    o Diphenylmethane to trityl phenyl ketone

    p Cyclohexene to cyclohexanone

    q Acetone to 2,5-dimethyl-2,4-hexadiene

    r Ethane to 1-amino-3-hydroxybutane

20-2 Formulate reasonable mechanisms for the following reactions:

a

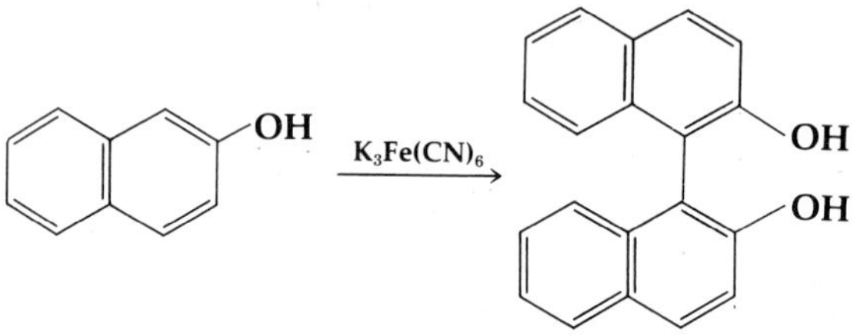

$\xrightarrow{\text{K}_3\text{Fe(CN)}_6}$

b

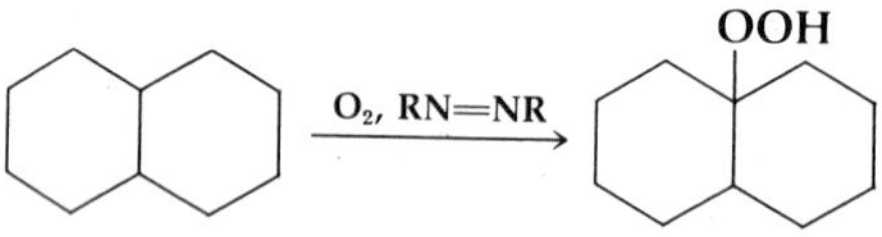

$\xrightarrow{\text{O}_2,\ \text{RN}=\text{NR}}$

c

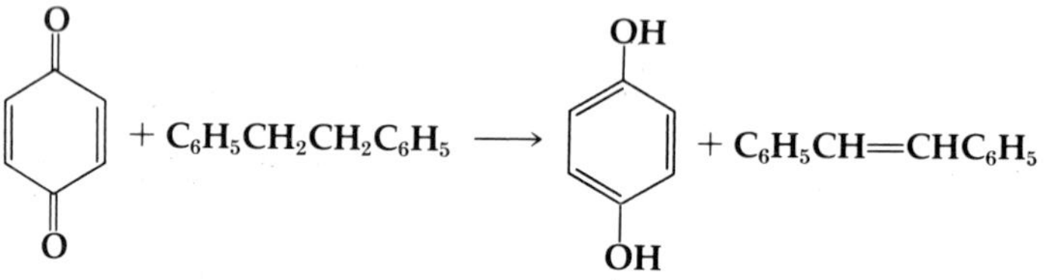

$+ \text{C}_6\text{H}_5\text{CH}_2\text{CH}_2\text{C}_6\text{H}_5 \longrightarrow$

$+ \text{C}_6\text{H}_5\text{CH}=\text{CHC}_6\text{H}_5$

d $\underset{\displaystyle }{(\text{CH}_3)_2\overset{\text{CN}}{\underset{|}{\text{C}}}-\text{N}=\text{N}-\overset{\text{CN}}{\underset{|}{\text{C}}}(\text{CH}_3)_2} \xrightarrow{80°} (\text{CH}_3)_2\text{C}=\text{C}=\overset{\text{CN}}{\underset{|}{\text{N}}}\text{C}(\text{CH}_3)_2 \xrightarrow{80°}$ $(\text{CH}_3)_2\underset{|}{\text{CCN}}$
$(\text{CH}_3)_2\underset{|}{\text{CCN}}$

e $\text{CH}_3\text{CH}=\text{CHCH}_3 + \text{Cl}_2 \xrightarrow{\text{Sunlight}} \text{CH}_3\underset{\underset{\displaystyle \text{Cl}}{|}}{\text{CH}}\text{CH}=\text{CH}_2 +$ other compounds

f

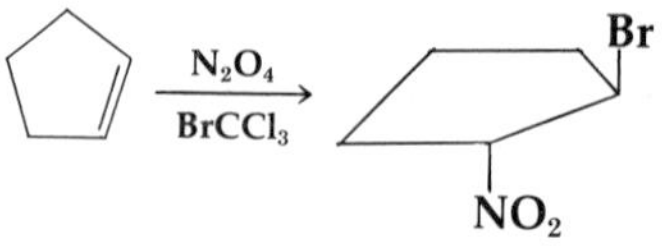

$\xrightarrow[\text{BrCCl}_3]{\text{N}_2\text{O}_4}$

g $\text{CH}_2=\text{CH}-\underset{\underset{\displaystyle \text{Br}}{|}}{\text{CH}}-\text{CH}_3 \xrightarrow[\substack{\text{peroxides} \\ -12°}]{\text{H}-\text{Br}} \text{CH}_3\text{CH}=\text{CH}-\text{CH}_2-\text{Br}$

h $H-Br +$ $\xrightarrow[-80°]{\text{Peroxides}}$

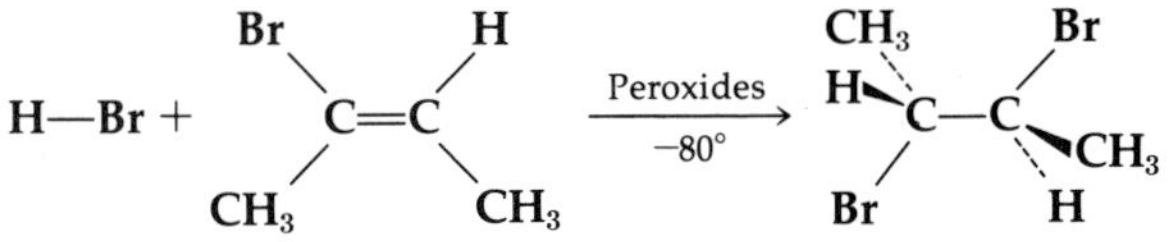

i $\xrightarrow[I_2]{\Delta}$

j $CCl_3Br +$ $\xrightarrow{h\nu}$

k $\underset{\Delta}{\overset{O_2, h\nu}{\rightleftharpoons}}$

l $C_6H_5-\underset{\underset{O}{\|}}{C}-CH_3 + (CH_3)_2C\text{=}CHCH_3$ $\xrightarrow{h\nu}$

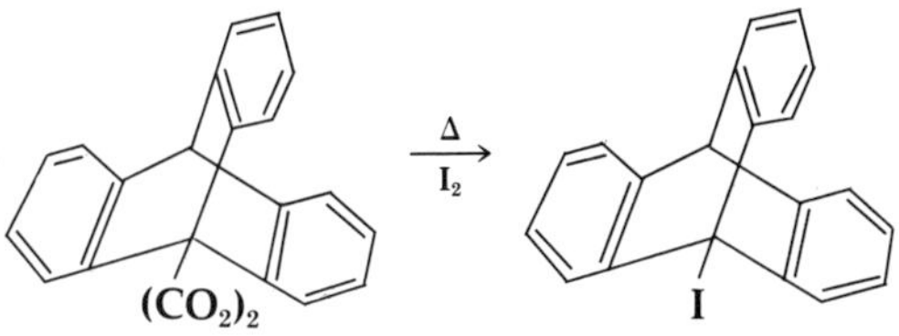

m $(C_6H_5)_3C-O-O-C(C_6H_5)_3$ $\xrightarrow{\Delta}$

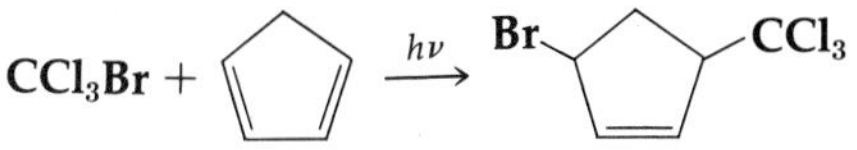

n $C_6H_5SH +$ $\xrightarrow{\text{Peroxides}}$

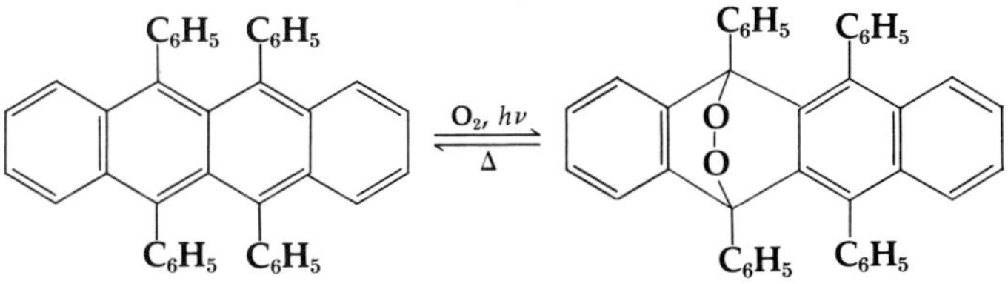

o $C_6H_5-\underset{\underset{CH_3}{|}}{\overset{\overset{CH_3}{|}}{C}}-CH_2-\overset{\overset{O}{\|}}{C}-H$ $\xrightarrow[130°]{[(CH_3)_3CO]_2}$

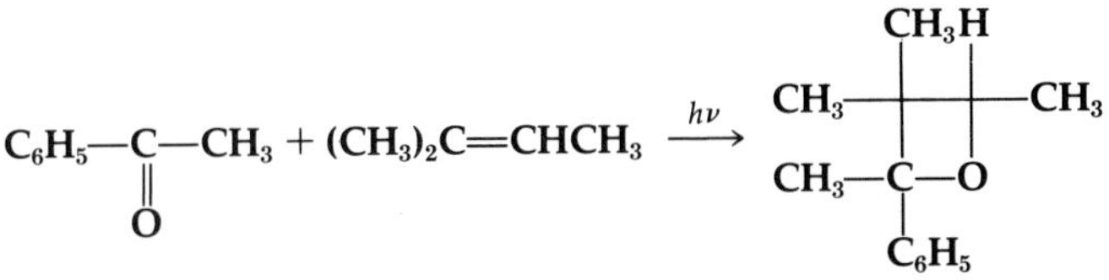

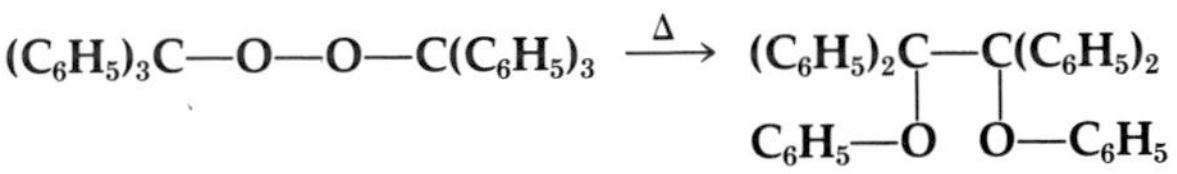

$CO + C_6H_5-\underset{\underset{CH_3}{|}}{\overset{\overset{CH_3}{|}}{C}}-CH_3 + C_6H_5-CH_2-\underset{}{\overset{\overset{CH_3}{|}}{C}H}-CH_3$

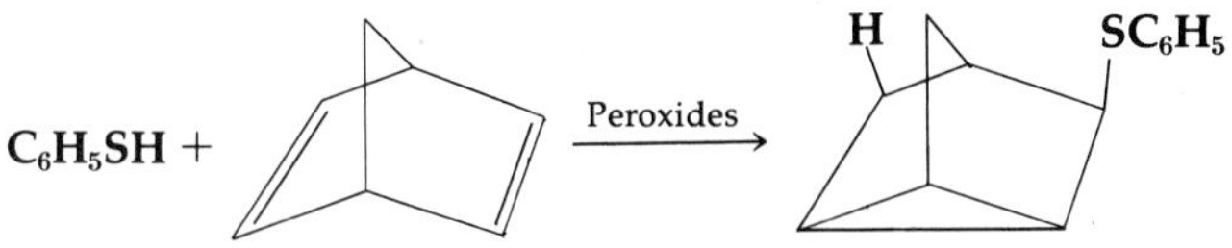

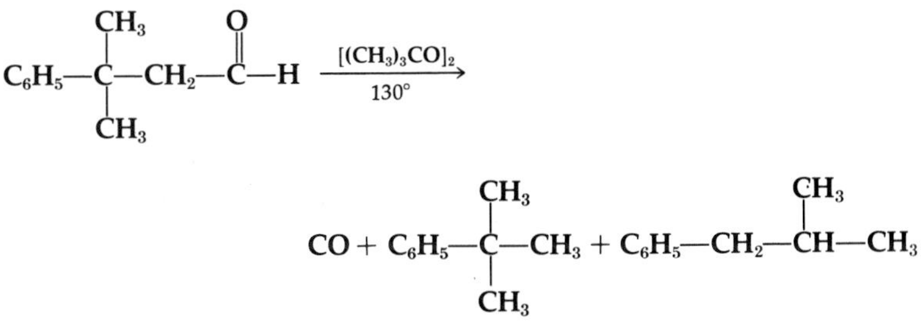

20-3 a The following reaction, carried out at 0°, gave optically pure product; when carried out at 36°, it gave largely racemized product: The optically pure

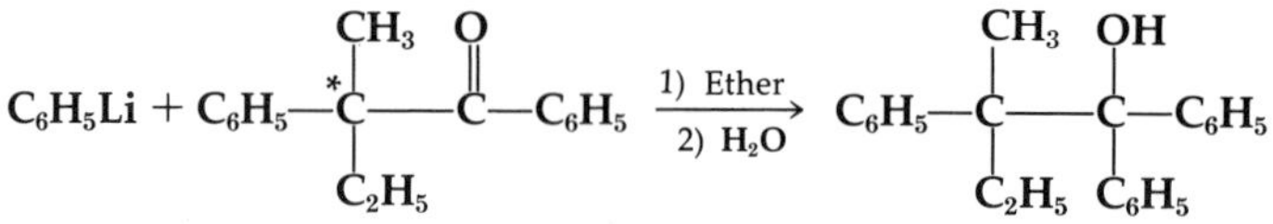

Optically pure

product, when heated in dioxan in the presence of lithium *tert*-butoxide, slowly gave the following products:

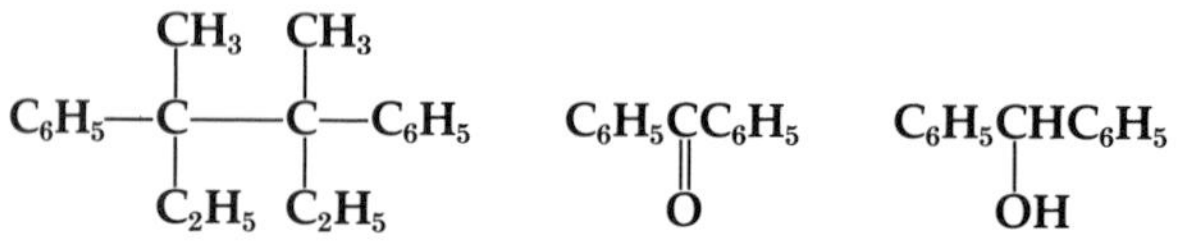

Not optically
active

b Explain the following rate difference:

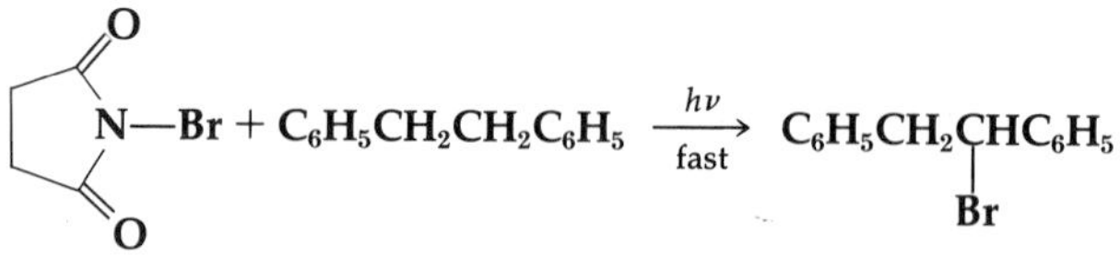

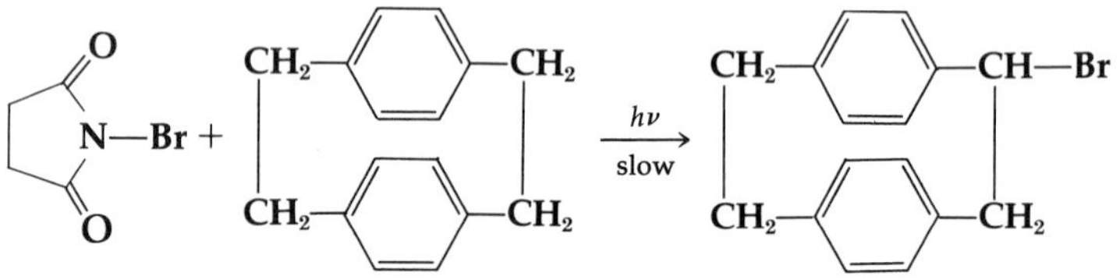

20-4 Predict the products of the following reactions, and indicate the basis of your prediction.

a $(C_6H_5)_3C-C(C_6H_5)_3 +$

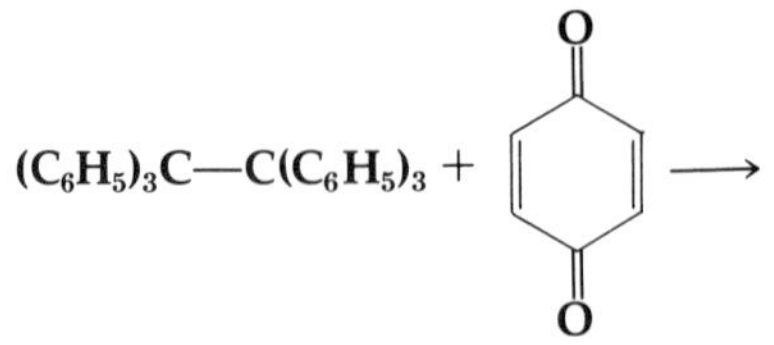

b $CH_3CHO + CH_2{=}CHCH_2OCOCH_3 \xrightarrow[\Delta]{(C_6H_5CO_2)_2}$

c $C_6H_5SH +$

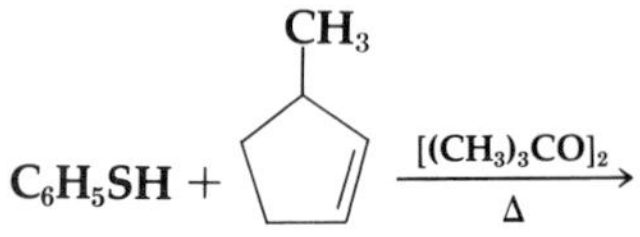

d 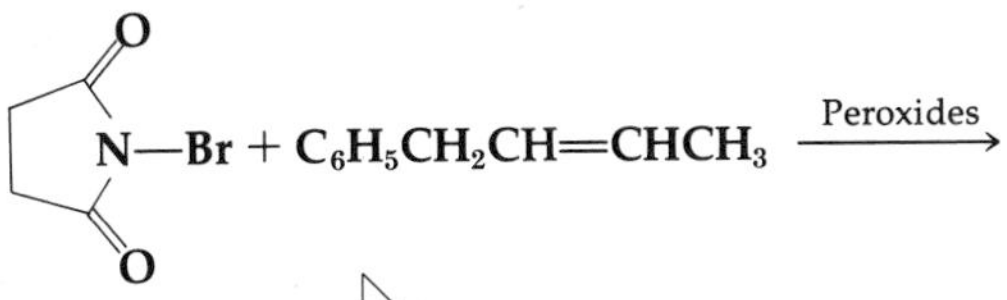
O
N—Br + C6H5CH2CH=CHCH3
O
Peroxides

e 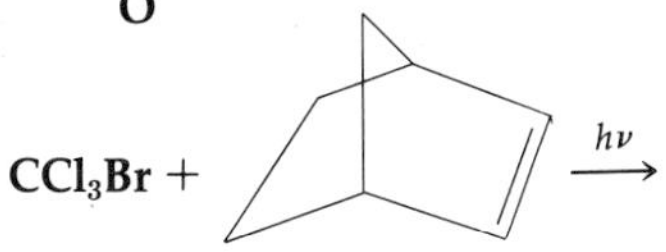
CCl3Br +
hν

f 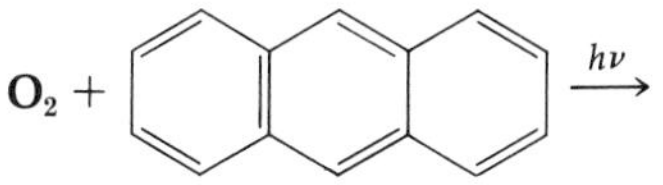
O2 +
hν

g 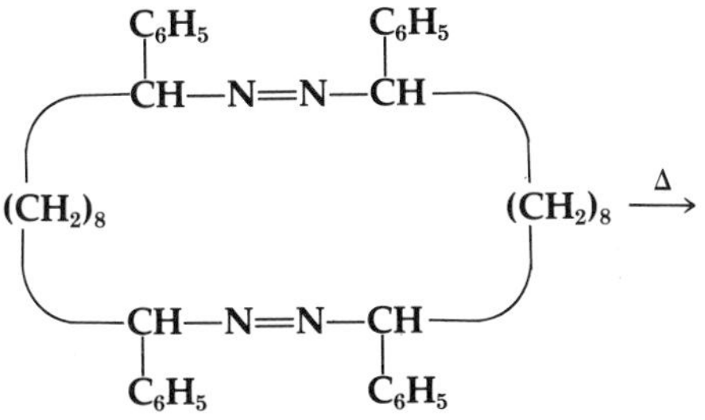
C6H5
C6H5
CH—N=N—CH
(CH2)8
(CH2)8
Δ
CH—N=N—CH
C6H5
C6H5

h 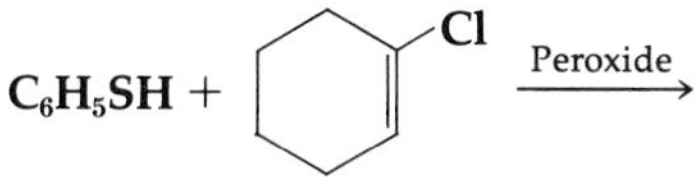
C6H5SH +
Cl
Peroxide

20-5 A hydrocarbon, C_6H_{14}, gives a mixture containing only two monochlorides in photochemical chlorination. One of these compounds solvolyzes very rapidly in ethanol, whereas the other is very slow. What is the hydrocarbon?

READING REFERENCES

Gould, E. S., "Mechanism and Structure in Organic Chemistry," p. 16, Holt, Rinehart and Winston, Inc., New York, 1959.

Walling, C., "Free Radicals in Solution," John Wiley & Sons, Inc., New York, 1957.

Hine, J., "Physical Organic Chemistry," 2d ed., pp. 402–483, McGraw-Hill Book Company, Inc., New York, 1962.

PERICYCLIC REACTIONS

THERE is a large body of chemical reactions which differ from those previously discussed in a number of significant ways:

1 Their mechanisms involve no discernible ionic or free radical intermediates.
2 They involve no electrophilic or nucleophilic reagents.
3 They are largely unaffected by solvent changes or by catalysis.
4 Two or more bonds are made and broken in a single *concerted and cyclic* transition.

Because of their cyclic transition states these concerted reactions are called **pericyclic reactions**. These reactions are equilibria, either side of which may involve one or more organic molecules. The cyclic transition state may be discussed equally well as arising from either side of the equilibrium ("reactants" and "products").

Since no external reagents are used, the only initiation required to effect these reactions is heat or light. As we shall see, some examples occur only on heating (**thermolysis** or **pyrolysis**), others only on **photolysis**. Many of these reactions have been known for a long time but it was only in 1965 that they were assembled by Woodward and Hoffmann into a cogent theoretical framework known as *"the conservation of orbital symmetry."*

The essential feature of pericyclic reactions is a simultaneous overlapping of the participating bond orbitals such that they may pass smoothly over into the bond orbitals of the product. This transformation of reactant orbitals into product orbitals passes through a transition state of merging orbitals. The basic principle guiding these reactions is that the transition state will be of relatively low energy—and the reaction favored—as long as the symmetry of the reactant orbitals is retained, or conserved, in passing to product orbitals. The reaction will be disfavored if the symmetry of the reactant and product orbitals is not the same. This may be understood in an alternative way by noting that only if *some bonding character is retained through the transition* will the activation energy be low and the reaction favored. This retention of bonding character occurs only if orbital symmetry is conserved.

In the course of a thermolysis the participating bond electrons are found in pairs in bonding orbitals, σ or π, which are characterized by phase signs (+ or −) and nodes.

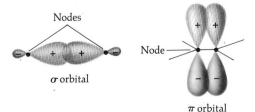

In a favored reaction the collection of participating orbitals must pass to product orbitals with these phases intact so as to maintain some bonding character throughout the transition.

The symmetry of the involved orbitals is a guide to whether this bonding character will be maintained during a given reaction. Hence when the symmetry of reactant and product orbitals is the same, i.e., orbital symmetry is conserved in the reaction, the reaction is said to be **symmetry-allowed**, or energetically favored. When symmetry is not retained, the reaction is **symmetry-forbidden**, or energetically disfavored. Reactions which are symmetry-forbidden can often occur anyway but only under higher-energy conditions, either by a different mechanism—a nonconcerted, free-radical mechanism with several steps—or by investing the reactant(s) first with high energy by irradiation, as discussed further below and in Chap. 22.

The main families of pericyclic reactions will be outlined in the following sections with an examination of which cases are symmetry-allowed. The examples found to be symmetry-allowed proceed smoothly on heating. The other cases, those which are found to be symmetry-forbidden, will usually proceed in photolysis, but not with heat alone. Some simple rules are offered for deducing whether a given reaction is symmetry-allowed, without invoking a rigorous or detailed examination of orbital symmetry. The three main families of pericyclic reactions, detailed in the following sections, are these: electrocyclic reactions, cycloadditions, and sigmatropic rearrangements.

The practical aspects of pericyclic reactions in synthesis are very important for several reasons. In the first place many of these reactions create new carbon-carbon σ bonds and hence are available for carbon-skeleton construction. Secondly, the reactions are generally independent of external influences so that unanticipated effects of solvent, concentration, catalysis, side reactions, etc., which frequently complicate other reactions, rarely deflect the course of these pericyclic reactions. Finally, the stereospecificity of these reactions is extremely high in practice and follows prediction with a consistency rare in ionic reactions.

21-1 ANALYSIS OF ELECTROCYCLIC REACTIONS

Many reactions are known in which a substituted butadiene is converted to a cyclobutene on heating (or the reverse reaction, since it is an equilibrium system). This change is called an **electrocyclic reaction**. In this reaction two π bonds are interconverted to a σ and a π bond.

The reaction is completely stereospecific. Substituents on the bond which breaks in the cyclobutene must rotate to come into the developing plane of the butadiene molecule. There are two rotational ways in which this can occur: The substituents may rotate in the same direction, called **conrotatory**, or they may rotate in opposite directions, called **disrotatory**.

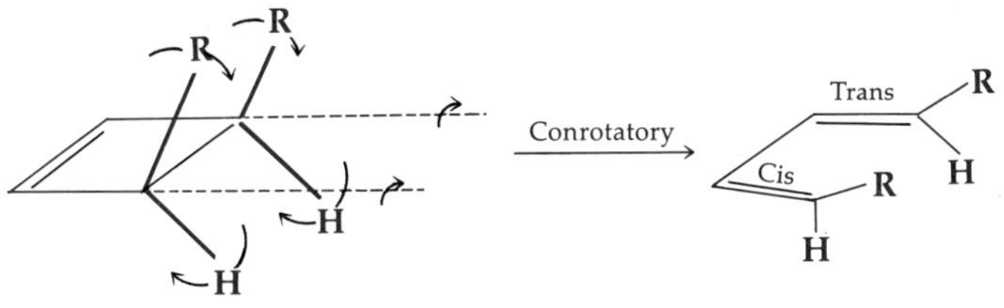

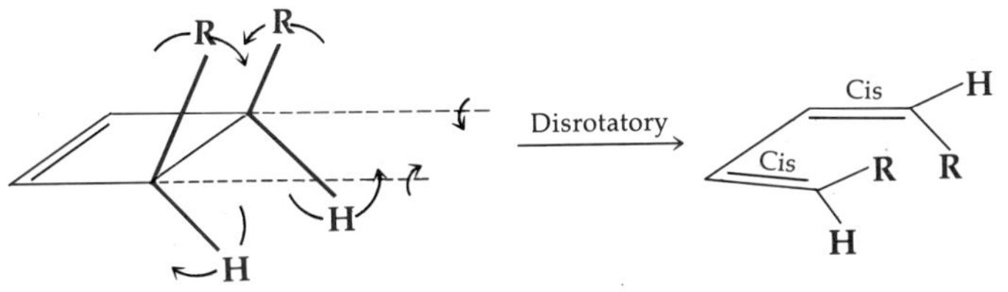

In all cases that have been observed, the *thermolysis of cyclobutenes proceeds with conrotatory opening* of the single bond. Thus, 3,4-dicarbomethoxy-cyclobutene yields only *cis,trans*-1,4-dicarbomethoxy-butadiene and none of the *cis,cis*- or *trans,trans*- isomers.

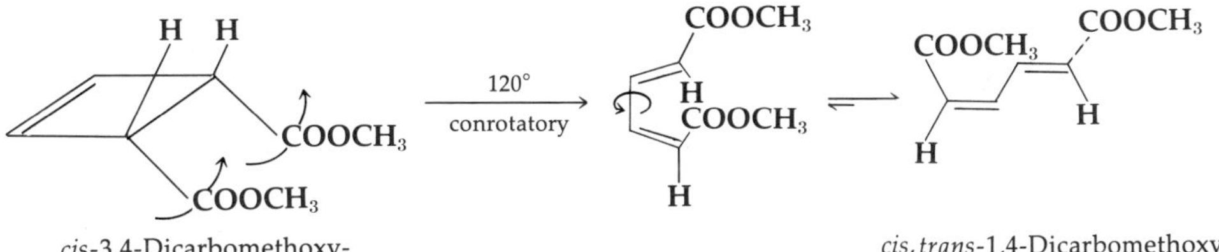

cis-3,4-Dicarbomethoxy-cyclobutene

cis,trans-1,4-Dicarbomethoxy-butadiene

In order to analyze this reaction we need to observe both the orbitals and the symmetry of the reactants and products and we must see which orbitals of product arise from each orbital of reactant. The four molecular orbitals of the involved bonds in each component are listed in order of increasing energy in Fig. 21-1 (compare Fig. 5-1 and page 146 for butadiene). Only the bottom two orbitals, the bonding orbitals, are occupied (filled with electron pairs) in the ground state of each molecule. If we follow the reaction path of the σ orbital in a disrotatory bond opening, we see both the rotation of the two atomic orbitals involved and their rehybridization from sp^3 to p.

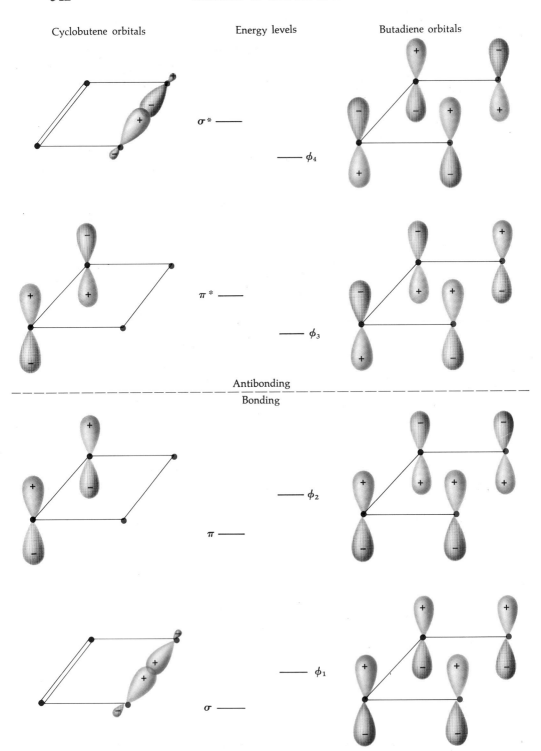

FIGURE 21-1   Orbitals in cyclobutene and butadiene

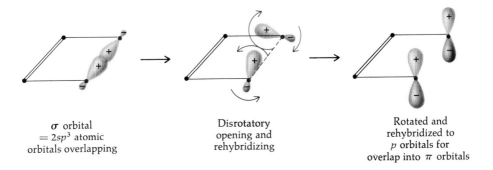

σ orbital
$\equiv 2sp^3$ atomic
orbitals overlapping

Disrotatory
opening and
rehybridizing

Rotated and
rehybridized to
p orbitals for
overlap into π orbitals

The σ orbital is converted into one of the four π orbitals and this must be either ϕ_1 or ϕ_3 since the lobes at the terminal atoms involved (colored dots in Fig. 21-1) have the *same sign*. If we follow the unoccupied σ^* orbital similarly—in a disrotatory opening—it must become the ϕ_2 or ϕ_4 orbital of butadiene, with *opposite-sign* lobes at the terminal atoms (colored dots). On the other hand the π orbital of cyclobutene, although it does not move, can only become ϕ_1 or ϕ_3 in butadiene, the orbitals with the same signs at those two atoms (black dots in Fig. 21-1). The π^* orbital similarly will go to ϕ_2 or ϕ_4, since π^* and ϕ_2 or ϕ_4 all have lobes of opposite sign at those atoms.

The ground state of cyclobutene has only the σ and π (not σ^* and π^*) orbitals filled with electron pairs; these are bonding orbitals. Either of these orbitals could pass to the bonding ϕ_1 orbital, but then the other would have to pass to ϕ_3, which is a high-energy antibonding orbital not occupied in the ground state of butadiene. Hence disrotatory opening of cyclobutene cannot produce stable, ground-state butadiene.

The same analysis for conrotatory opening shows the σ-orbital passing to terminal p lobes of opposite sign, suitable to become ϕ_2 or ϕ_4 butadiene orbitals. This is of course the reverse situation from the disrotatory conversion outlined above and thus allows both ground-state (bonding) orbitals in cyclobutene to yield ground-state (bonding) orbitals in butadiene ($\sigma \rightleftharpoons \phi_2$; $\pi \rightleftharpoons \phi_1$).

σ orbital in conrotatory opening and rehybridization to two p orbitals

The passing of one orbital to another is referred to as a **correlation** of these orbitals (reactant $\rightleftharpoons$ product), and an orbital energy diagram showing the correlations of reactant orbitals with product orbitals is a **correlation diagram**. These correlation diagrams are analogous to those shown in Chap. 5 for corre-

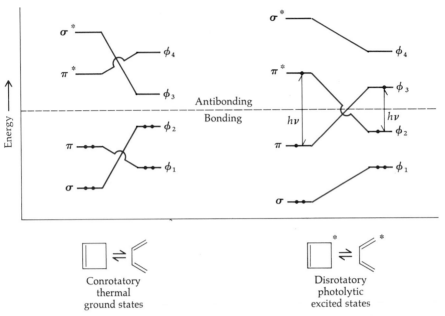

FIGURE 21-2   **Correlation diagrams for cyclobutene-butadiene interconversions**

lating atomic orbitals of atoms with the molecular orbitals of their covalent combinations. In the present instance the correlation is between the molecular orbitals of two covalent molecules which interconvert. In Fig. 21-2 are shown the correlation diagrams for the reaction under discussion, the conrotatory mode at left showing the allowed pathway with no correlations crossing the antibonding energy line.

The correlation diagram does not indicate anything about the transition state energy. However, a correlation diagram showing correlations in which a reactant ground-state (bonding) orbital must correlate with an antibonding product orbital indicates a forbidden, or high-energy, transition, as in the disrotatory mode in the right-hand diagram in Fig. 21-2.

If the reactant, on the other hand, is first converted from the ground state to an excited state, then an antibonding orbital becomes occupied in the starting material. Now a reaction pathway is allowed, for this highest occupied, antibonding orbital decays to a lower-energy bonding orbital. This occurs on photolytic excitation in which a quantum of absorbed light provides the energy to lift one electron from the highest occupied ground-state orbital to the antibonding level above it. The right-hand diagram (Fig. 21-2) illustrates this conversion, which includes the initial promotion of an electron (by $h\nu$) to an antibonding π^* orbital. This highest occupied orbital now reverts to a bonding state (ϕ_2) in the product by disrotatory electrocyclic reaction. We may conclude that photolytic reaction of substituted cyclobutenes yields disrotatory opening to butadienes, and that photolysis of butadienes affords cyclobutenes in a disrotatory ring closure. This is in fact generally and quite specifically observed.

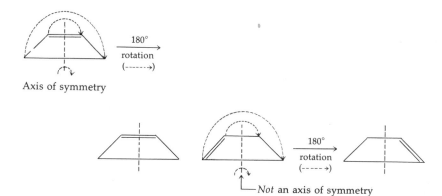

trans,trans-2,4-Hexadiene   *cis*-3,4-Dimethylcyclobutene

In order to analyze the reaction in terms of symmetry we find that both the reactant and product possess a plane of symmetry (mirror plane) and an axis of symmetry. An axis of symmetry† is present if rotation of the molecule around the axis by 180° (halfway around a full rotation) results in a form identical to the original.

We now examine the lobe signs of the several orbitals with respect to a plane or axis of symmetry. The lobe signs will either represent that symmetry or be opposite to it, i.e., symmetric (S) and anti-symmetric (A), respectively. These relations of the lobe signs to the molecular symmetry element (axis or mirror plane) are catalogued in Fig. 21-3. A rotation of the cyclobutene σ orbital halfway around the perpendicular axis (colored circle) through its midpoint regenerates the same σ-orbital form, while similar rotation of σ^* yields a different figure. Reflection in the mirror plane (dotted line) yields the same signs on each side for the σ (or π or ϕ_1 and ϕ_3) orbital but not for the σ^*.

The reactant orbitals are correlated with product orbitals of the same symmetry, and a symmetry-allowed reaction is one in which reactant bonding orbitals all correlate with product bonding orbitals. This occurs only with an axis of symmetry in the present case. The axis of symmetry also implies conrotatory opening of the σ bond since all the motion in conrotatory opening is consistent with the axis of symmetry, as the disrotatory motion fits mirror-plane symmetry.

† Strictly speaking, this is a **twofold axis of symmetry** since duplication of the original occurs two times during one full rotation around the symmetry axis.

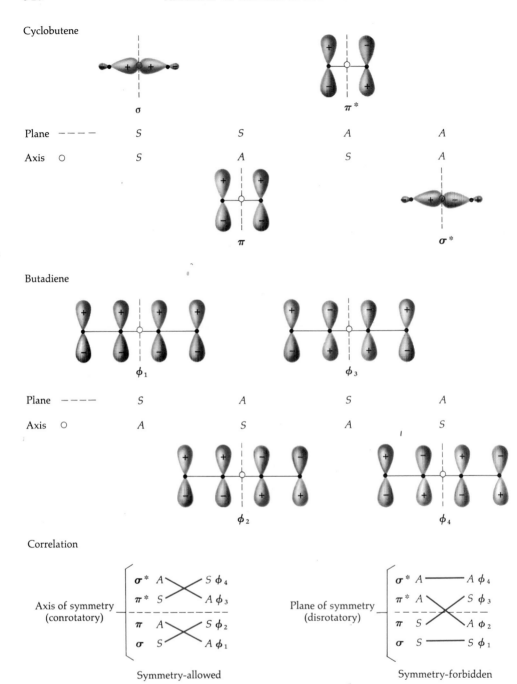

FIGURE 21-3   **Orbital symmetries in cyclobutene-butadiene interconversion**

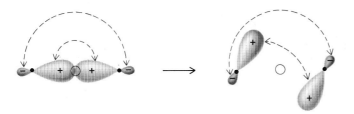

Conrotatory opening with axis (○) of symmetry
showing identity of orbitals (←– – –→) on rotation

Disrotatory opening with plane (– – – – –) of symmetry

If the same analysis is applied to the case with one more double bond
in the electrocyclic system, the result is reversed and cyclohexadienes must
open thermally in a disrotatory rather than conrotatory fashion. Thermal cyclization of hexatrienes, conversely, proceeds in disrotatory fashion to cyclohexadienes.

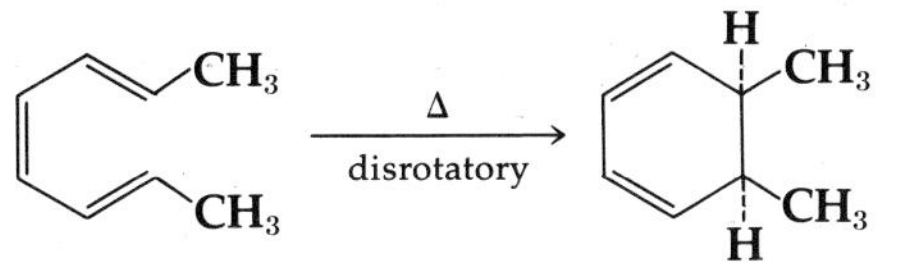

<table>
<tr><td style="text-align:center">

![structure]

trans,cis,trans-
2,4,6-Octadiene

</td><td style="text-align:center">

$\xrightarrow[\text{disrotatory}]{\Delta}$

</td><td style="text-align:center">

![structure]

cis-5,6-Dimethyl-
1,3-cyclohexadiene

</td></tr>
<tr><td style="text-align:center">

![structure]

trans,cis,cis-
2,4,6-Octadiene

</td><td style="text-align:center">

$\xrightarrow[\text{disrotatory}]{\Delta}$

</td><td style="text-align:center">

![structure]

trans-5,6-Dimethyl-
1,3-cyclohexadiene

</td></tr>
</table>

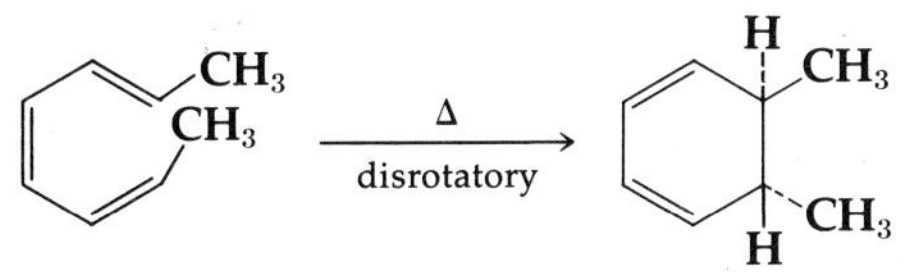

These electrocyclic reactions are easily generalized in terms of the number of electron pairs in the participating bonds:

Number of participating electron-pairs	Thermal reaction course	Photolytic reaction course
Even	Conrotatory	Disrotatory
Odd	Disrotatory	Conrotatory

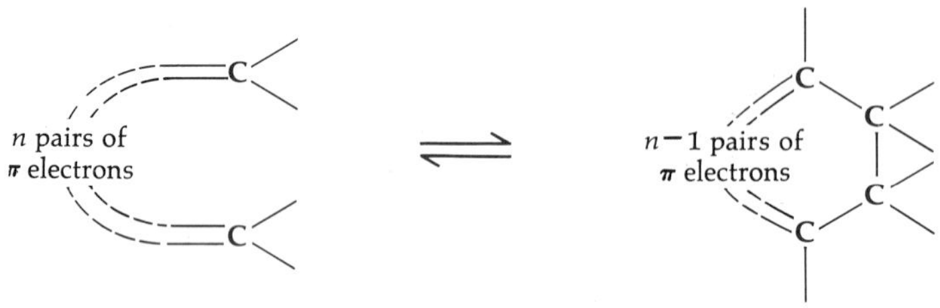

A number of relatively unstable molecules can be created by photolytic electrocyclic reactions since the products cannot revert to the more stable reactants thermally by a symmetry-allowed, or favored, reaction, and also since no agents which might cause other chemical changes need to be present during photolysis. It is often convenient, in examining the rotations in these reactions, to look at models. This practice will offer more assurance that certain conversions are geometrically very strained or virtually physically impossible with reasonable bond lengths.

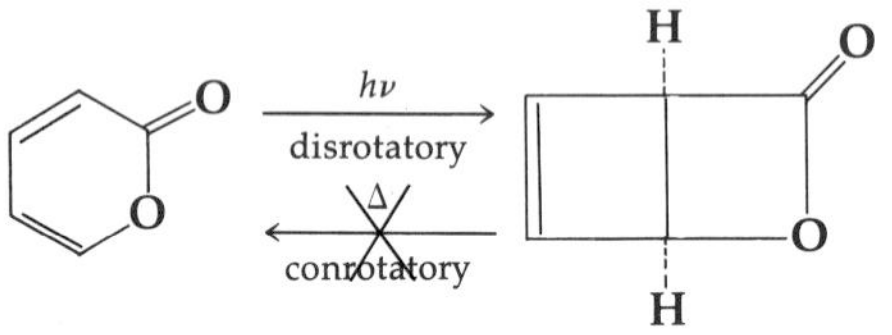

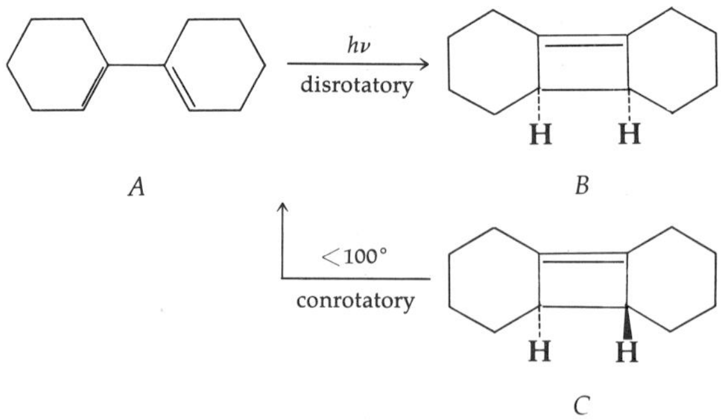

In the second example the cis-fused tricyclic cyclobutene (B) formed by photolysis is stable above 250°, for its conrotatory thermal opening would yield a very unstable *trans*-cyclohexene ring. The trans-substituted cyclobutene (C), however, very easily reverts to the stable diene (A) on warming, although it must be made in another way.

A sequence of two electrocyclic reactions occurs in the following example.

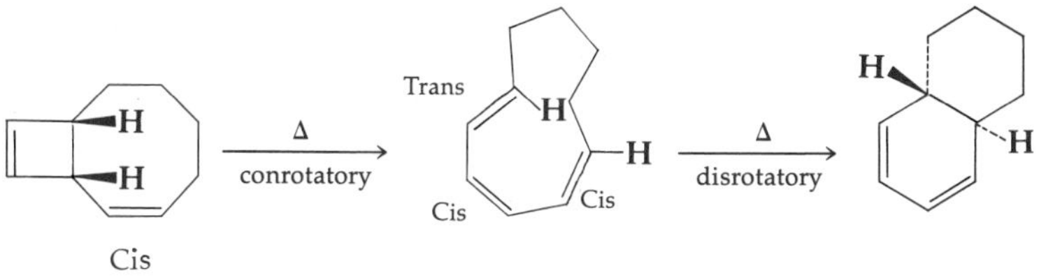

PROBLEM 21-1

Give the products expected from heating the following compounds.

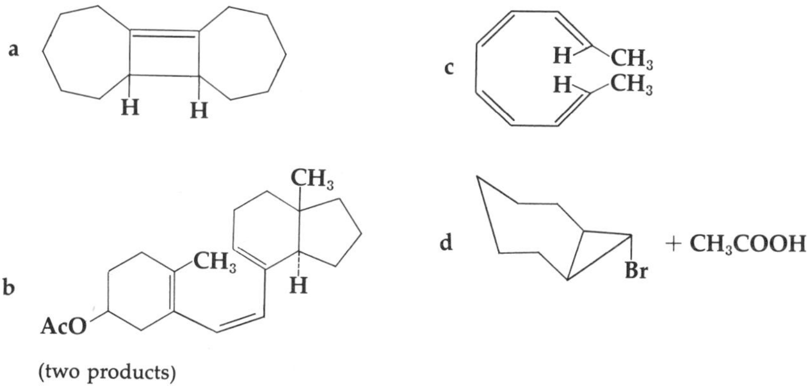

Ions may also participate in electrocyclic reactions, following the same rules. The cyclopropyl and allyl cations can interconvert, and the mode is disrotatory since it involves one electron pair, while the corresponding anions (with two electron pairs) interconvert in a conrotatory fashion.

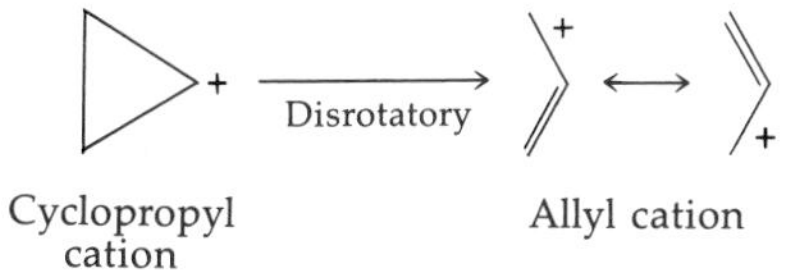

Cyclopropyl cations are often generated by solvolytic ionization of cyclopropanes substituted with leaving groups (—OTS, —X, etc.). The disrotatory cleavage which follows occurs so rapidly that it is concerted with ionization, or nearly so, and the presence and location of the leaving ion affect the direction of disrotatory cleavage. The cleavage occurs in that particular disrotatory fashion which turns the σ electrons downward so that they may be viewed as affording a backside displacement to the leaving group. In the example of Fig. 21-4 this is demonstrated dramatically by linking the rotating substituents into a ring. Only in case B can the resultant ring be strain-free with its three sp^2 atoms cis substituted. The trans-allylic cation (like a trans double bond) cannot exist in a seven-membered ring, as would be required in case A.

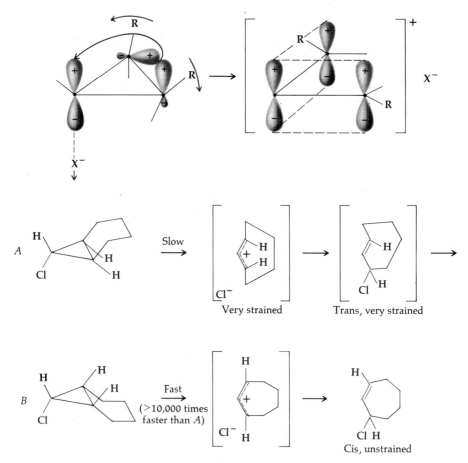

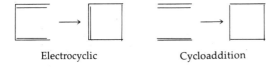

FIGURE 21-4   **Electrocyclic opening of cyclopropyl cations**

21-2   CYCLOADDITION REACTIONS

In the electrocyclic reaction considered in the last section, two double bonds in one molecule interacted to form a π and a σ bond. An analogous system would be that of double bonds in two separate molecules joining together into a single new molecule. The two units would be joined by two σ bonds created from the two original π bonds. Such a reaction is formally an addition reaction and is labeled a **cycloaddition**.

Electrocyclic   Cycloaddition

The general cycloaddition reaction (Fig. 21-5) is the coupling of two unsaturated molecules, each containing one or more (conjugated) double bonds. As labeled in Fig. 21-5 the cycloaddition is designated by a bracket containing

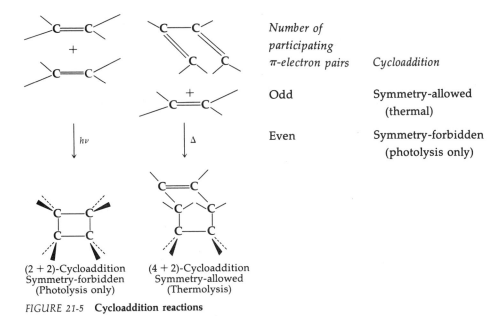

<table>
<tr><td></td><td>Number of
participating
π-electron pairs</td><td>Cycloaddition</td></tr>
<tr><td></td><td>Odd</td><td>Symmetry-allowed
(thermal)</td></tr>
<tr><td></td><td>Even</td><td>Symmetry-forbidden
(photolysis only)</td></tr>
</table>

(2 + 2)-Cycloaddition   (4 + 2)-Cycloaddition
Symmetry-forbidden   Symmetry-allowed
(Photolysis only)   (Thermolysis)

FIGURE 21-5 **Cycloaddition reactions**

the number of π electrons in each component. The molecules meet so that their π orbitals overlap and undergo cycloaddition. The energy required to effect this process will be provided by simple heating for a symmetry-allowed cycloaddition, while only photolysis will effect the symmetry-forbidden cases.

Analysis of the orbitals may be carried out as before. It is convenient to start with the product, i.e., examine the reverse reaction, since this involves only a single starting molecule. The rules will of course be identical for the reaction in either direction.

The two possible bonding molecular orbitals of cyclobutane are shown below.

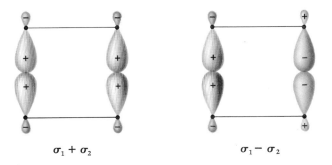

$\sigma_1 + \sigma_2$ $\sigma_1 - \sigma_2$

If these orbitals are pulled apart and allowed sideways overlap to form the two π orbitals of two ethylene product molecules, only $(\sigma_1 + \sigma_2)$ is capable of forming a bonding π orbital, while $(\sigma_1 - \sigma_2)$ must correlate with an antibonding π* orbital. Hence the reaction is symmetry-forbidden. However, if

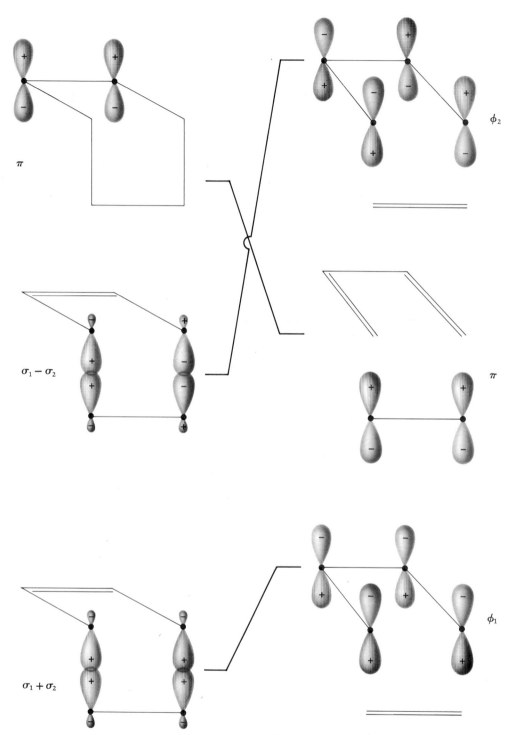

FIGURE 21-6   Correlations in the [4+2]-cycloaddition bonding orbitals

one ethylene is excited first by light absorption to occupancy of its π^* orbital, the fusion of π and π^* in two ethylenes to form a ground-state cyclobutane is symmetry-allowed.

With a [4 + 2] cycloaddition a diene overlaps with a single double bond to form a cyclohexene. Examining the reverse reaction we see the two single bonds of the cyclohexene as before, but now the ($\sigma_1 + \sigma_2$) combination correlates with the ϕ_1 orbital of the product diene while the ($\sigma_1 - \sigma_2$) combination correlates with the ϕ_2 orbital, the other diene bonding orbital. Hence ground-state orbitals are formed and the reaction is symmetry-allowed. The symmetry involved is the mirror plane down the middle and the bonding π orbitals of cyclohexene and products all have the same symmetry with respect to that plane; hence they also correlate in the ground state. These bonding orbitals and their correlations are shown in Fig. 21-6.

Diels-Alder Reaction

The thermal [4 + 2] cycloaddition has been known for almost half a century as the **Diels-Alder reaction**. It has served an important role in many syntheses since it smoothly and stereospecifically unites two carbon skeletons. In the Diels-Alder reaction, a six-membered ring is formed by 1,4-addition of an olefinic unit to a conjugated diene (Fig. 21-7). The **dienophile** is activated by electron-withdrawing substituents (**Z**) such as —COOH, —COOR, —CHO, —COR, —NO_2, —CN, and —SO_2—. Some reactive dienes do not require activated dienophiles, as is illustrated by the dimerization of cyclopentadiene. Dienes are activated by electron-donor substituents. For example 1,3-butadiene is less reactive than most of its mono-, di-, and trimethyl derivatives. However, in the case of the tetramethylbutadienes, steric hindrance decreases reactivity.

FIGURE 21-7   **Diels-Alder reaction or [4+2]-cycloaddition**

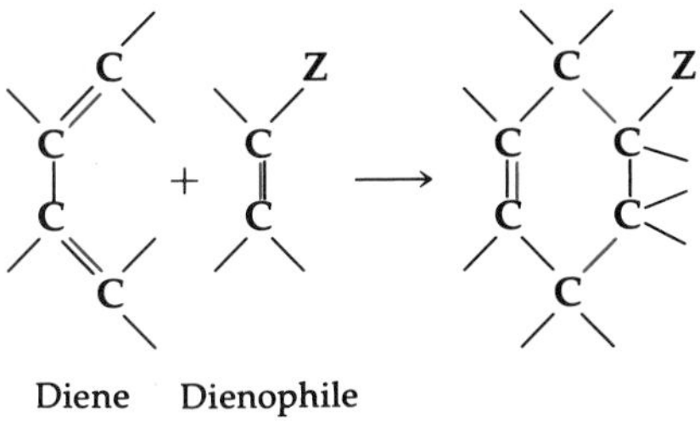

Diene   Dienophile

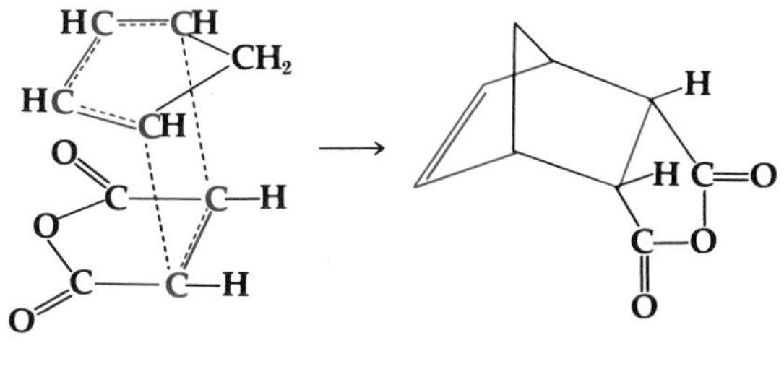

Diels-Alder reactions are highly stereospecific, as illustrated by the reaction of maleic anhydride and cyclopentadiene. The structures of products can be predicted by placing the reactants in two parallel planes in such a way as to obtain maximum overlapping of double-bond π orbitals (Fig. 21-7). Diels-Alder reactions sometimes provide excellent methods for synthesis of aromatic compounds from aliphatic starting materials, as illustrated by some of the following examples. A number of the common dienes and dienophiles are also illustrated in the examples.

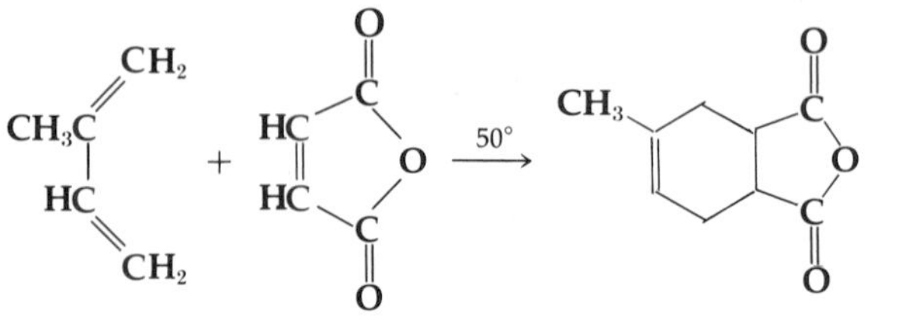

| Isoprene | Maleic anhydride | 4-Methyl-1,2,3,6-tetrahydro-phthalic anhydride 97% |

| 2,3-Dimethyl-1,3-butadiene | Diethyl acetylene-dicarboxylate | | Diethyl 4,5-dimethylphthalate |

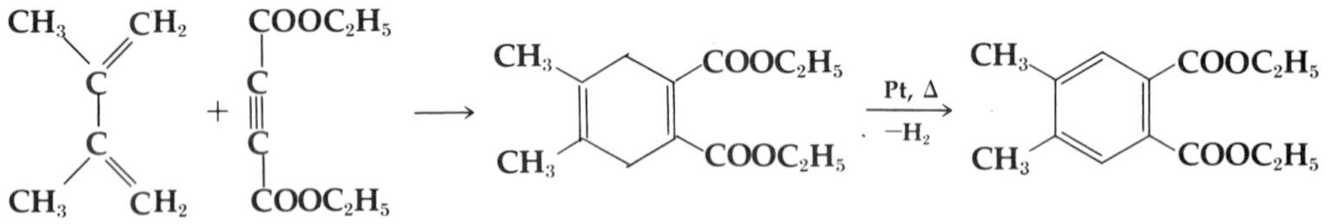

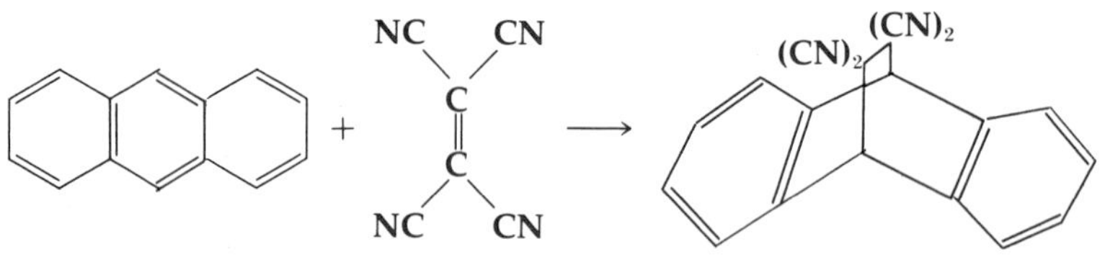

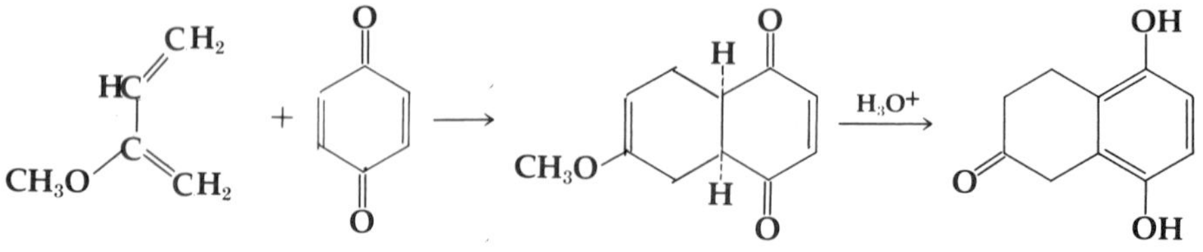

Neither the diene nor the dienophile requires an all-carbon skeleton since basically only the uniting of π orbitals is critical to the mechanism.

The Diels-Alder reaction is reversible, although occasionally the fragmentation assumes a course different from the reverse of the forward reaction when a reasonable mechanistic choice is available.

One advantageous use of the Diels-Alder reaction in synthesis is illustrated in the initial carbon-skeleton construction in the synthesis of reserpine (page 934).

1,3-Dipolar Cycloadditions

A five-membered instead of a six-membered ring should result from a [4 + 2] cycloaddition if the four π electrons in the diene component be compressed into only a three-atom skeleton. This is most commonly seen in the **1,3-dipolar additions,** in which the neutral three-atom, 4π-electron system is isoelectronic with ozone. The term dipolar here refers to the necessary formal charges in the structure of the 4π-electron component. The 2π-electron component, or **dipolarophile**, is usually activated by being strained, highly polar or highly polarizable.

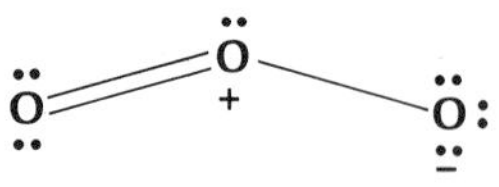

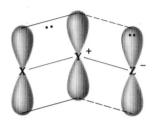

Ozone

General 1,3-dipolar component

Examples of 1,3-dipolar compounds: *Examples of dipolarophiles:*

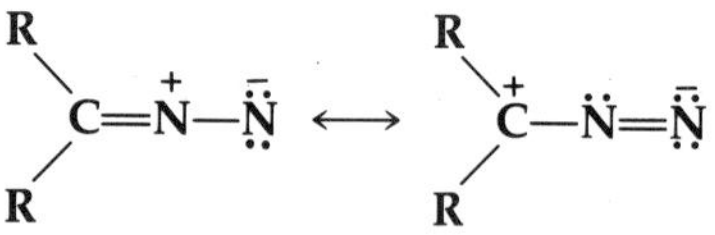

Diazo compounds   Norbornene

Azides   Nitriles

Nitrile oxides   **ROCOC≡CCOOR**

                                                       Acetylene dicarboxylate esters

A large number of ring systems can be synthesized by this reaction. Notable among these is phenylpentazole, which contains a five-membered ring composed entirely of nitrogen atoms. The compound can be prepared and preserved only at very low temperatures.

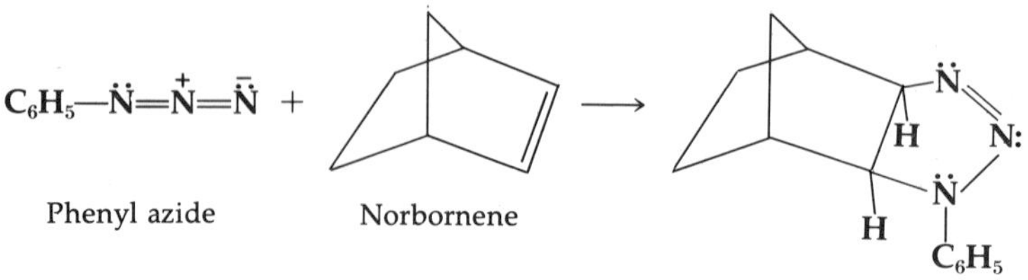

$C_6H_5-\ddot{N}=\overset{+}{N}=\bar{\ddot{N}}$ +

Phenyl azide   Norbornene

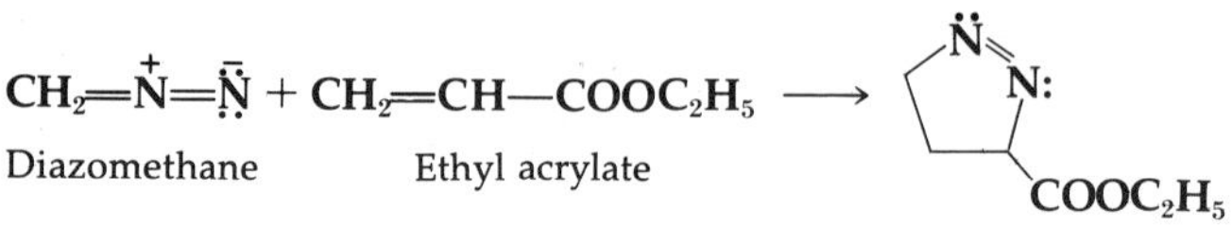

$CH_2=\overset{+}{N}=\bar{\ddot{N}}$ + CH_2=CH—$COOC_2H_5$ ⟶

Diazomethane   Ethyl acrylate

$COOC_2H_5$

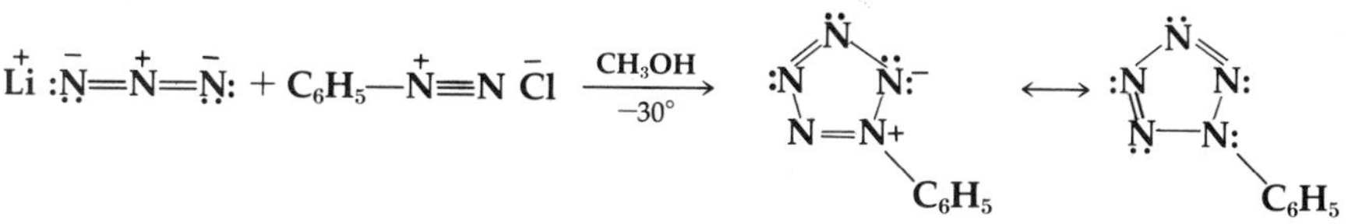

$$\overset{+}{Li} \; :\overset{-}{\underset{\cdot\cdot}{N}}=\overset{+}{N}=\overset{-}{\underset{\cdot\cdot}{N}}: \; + \; C_6H_5-\overset{+}{N}\equiv N \; \overset{-}{C}l \; \xrightarrow[-30°]{CH_3OH}$$

Lithium azide   Phenyldiazonium
chloride

Phenylpentazole

The direction of addition to unsymmetrical dipolarophiles is easily predicted from the natural direction of polarization in the multiple bond of the
dipolarophile and the charges in the 1,3-dipolar molecule.

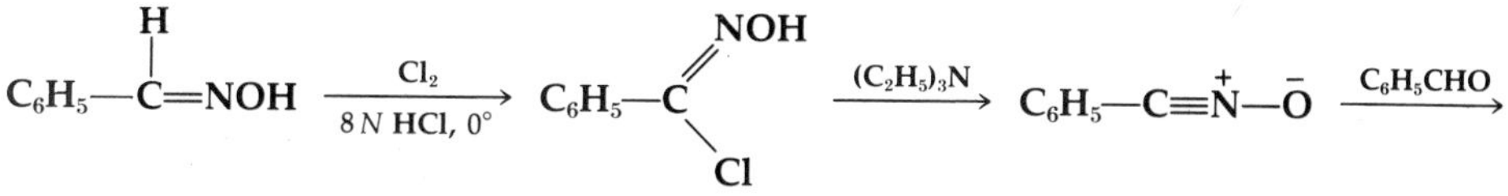

$$C_6H_5-\overset{\overset{\displaystyle H}{|}}{C}=NOH \; \xrightarrow[8\,N\,HCl,\,0°]{Cl_2} \; C_6H_5-\overset{\overset{\displaystyle NOH}{\parallel}}{\underset{\underset{\displaystyle Cl}{\diagdown}}{C}} \; \xrightarrow{(C_2H_5)_3N} \; C_6H_5-C\equiv\overset{+}{N}-\overset{-}{O} \; \xrightarrow{C_6H_5CHO}$$

Benzaldoxime   Benzonitrile
oxide

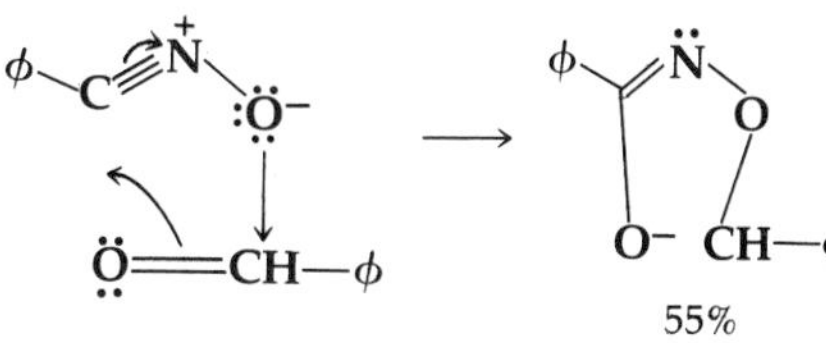

55%

Four-membered Ring Compounds

The photolytic union of two double bonds by [2 + 2] cycloaddition
produces a cyclobutane. Photocycloaddition is caused by irradiation of many
unsaturated compounds with visible or ultraviolet light. A classic example is
the formation of truxinic and truxillic acids by irradiation of cinnamic acid
either in the solid state or in saturated aqueous solution. Under various conditions several stereoisomers of each general structure are formed. The reaction
is complicated by the fact that irradiation also causes cis-trans isomerization
of the cinnamic acids.

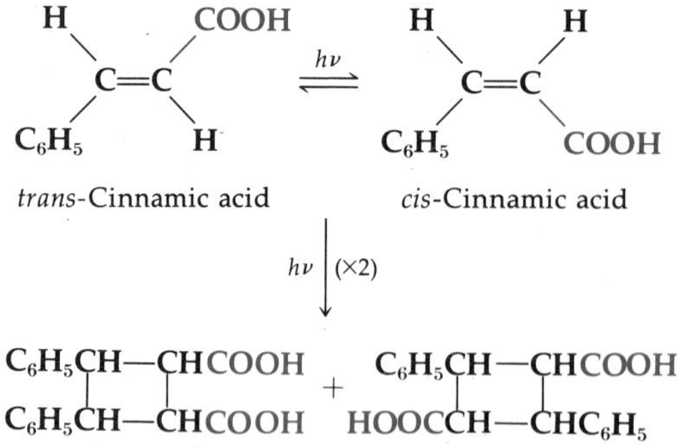

trans-Cinnamic acid   *cis*-Cinnamic acid

$h\nu$ | (×2)

C_6H_5CH—$CHCOOH$ C_6H_5CH—$CHCOOH$
C_6H_5CH—$CHCOOH$ + $HOOCCH$—CHC_6H_5

Truxinic acids   Truxillic acids

Large numbers of crossed and symmetrical photocycloaddition reactions
have been discovered. Intramolecular cyclization reactions can lead to "cage"
compounds.

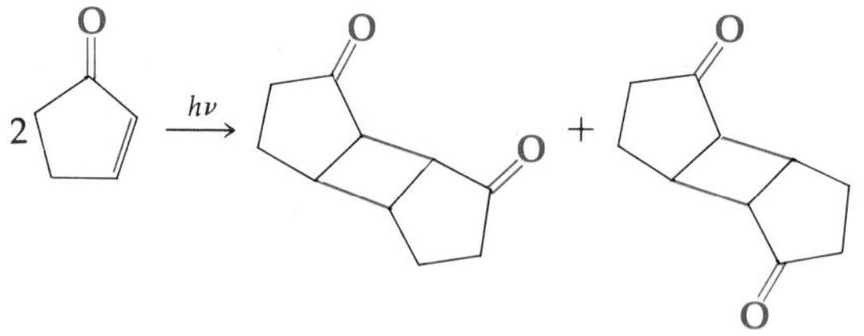

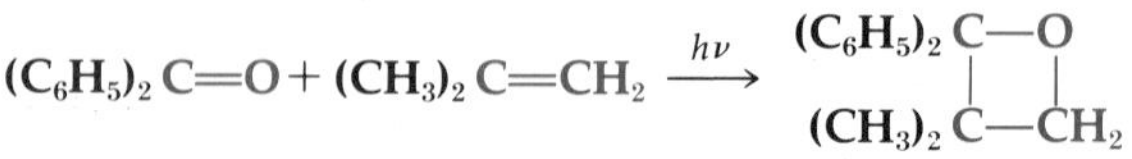

$(C_6H_5)_2\,C$=O + $(CH_3)_2\,C$=CH_2 $\xrightarrow{h\nu}$ $(C_6H_5)_2\,C$—O $\quad$ $(CH_3)_2\,C$—CH_2

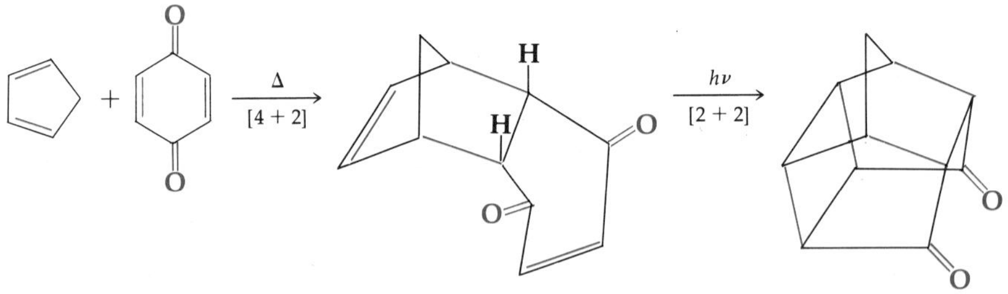

Benzoquinone   "Cage"

Other Ring Sizes by Cycloaddition

The Diels-Alder reaction creates six-membered rings by [4 + 2] cycloaddition. Five- and four-membered rings are generally created by 1,3-dipolar
cycloadditions and photolytic [2 + 2] cycloadditions, respectively. Larger rings
can be created by photolytic [4 + 4] or [6 + 2] cycloadditions or thermal
[6 + 4] or [8 + 2] cycloadditions, although the number of available examples
to date is far fewer than those for the Diels-Alder reaction.

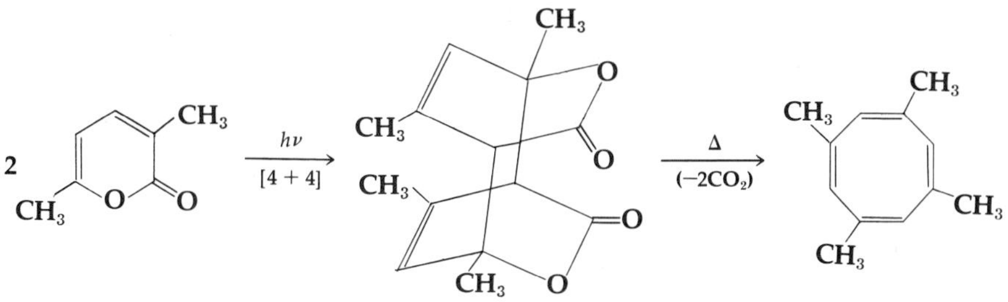

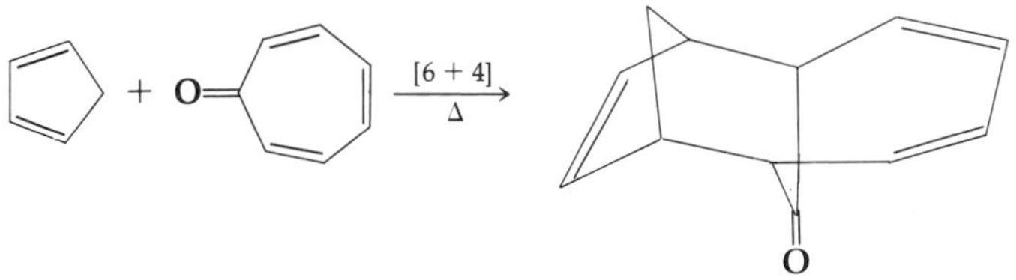

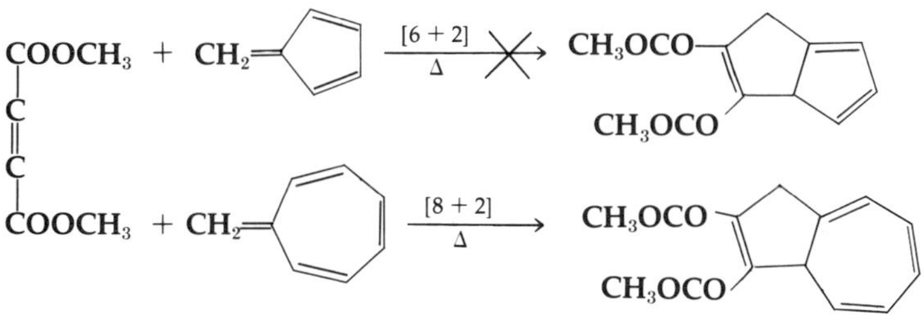

The last example, although it involves a 10-electron cycloaddition, does not form a large ring because of the cyclic nature of the starting tetraene. The new ring formed in this case is five-membered, as it is with 1,3-dipolar cyclo-additions.

PROBLEM 21-2

Write the products expected from allowing these pairs of compounds to react. Show stereochemistry where possible.

a

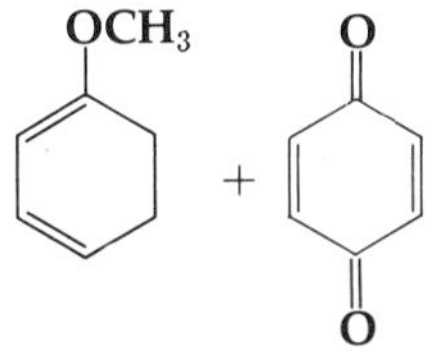

b Dimerization of propenal (acrolein) yields two products.

c

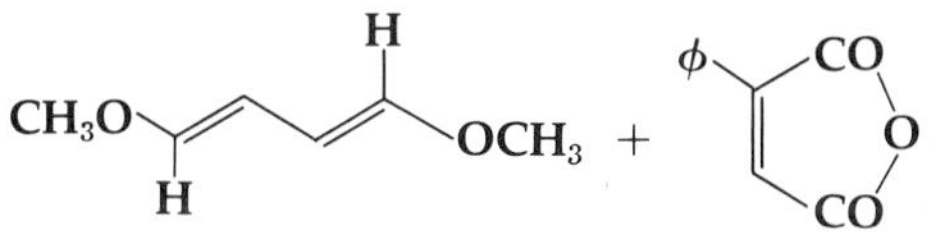

d $C_6H_5N_3 + p\text{-}CH_3C_6H_4CN$

e $CH_3\overset{\displaystyle OCH_3}{\overset{|}{CH}}=CHCOCH_3 + C_6H_5C\overset{+}{\equiv}N-\overset{-}{O}$

      Product loses CH_3OH spontaneously. Why?

f $(CH_3)_2C=\overset{\displaystyle O^-}{\underset{\displaystyle CH_3}{\overset{+}{N}}} \quad + HC\equiv C-COOCH_3$

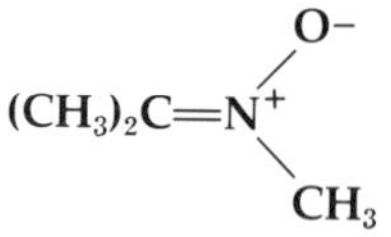

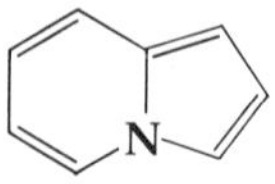

g   (indolizine ring structure) $+ CH_3OOC-C\equiv C-COOCH_3$

h   (cyclopentadienone structure with ϕ, ϕ, CH_3, CH_3 and O) $+$ (cycloheptatrienone structure)

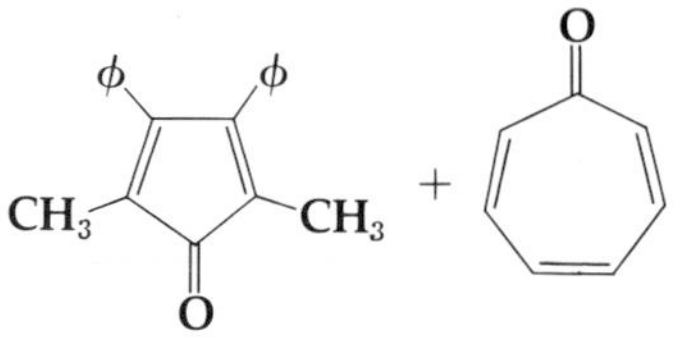

PROBLEM 21-3

The adduct obtained by heating 1,3-cyclohexadiene with maleic anhydride can in principle have one of two structures with respect to the placement of the double bond and the anhydride. To prove the structure, the adduct is dissolved in aqueous bicarbonate and iodine is slowly added. A mole of iodine is consumed and compound A is precipitated on acidification. On boiling the bicarbonate solution dry and heating further, one obtains a crystalline sublimate B. The mass spectra of A and B show parent peaks at 322 and 194, respectively. Both show IR peaks at 5.6 μ (1790 cm^{-1}), and A also shows a peak at 5.8 μ (1730 cm^{-1}). Show the two adduct structures which were possible, which one is correct, and why. Write structures for A and B.

21-3   SIGMATROPIC REARRANGEMENTS

Rearrangements in a molecular skeleton can also occur on heating (or in appropriate cases by photolysis) as long as the π and σ bonds which shift in a concerted fashion conserve orbital symmetry. The usual cases involve one σ bond at the end of a system of conjugated π bonds. The net effect of rearrangement is the reformation of the σ bond at the other end of the π system with a concerted shift of the π bonds.

$$\overset{\displaystyle Z}{\overset{|}{C}}-(C=C)_n \longrightarrow (C=C)_n-\overset{\displaystyle Z}{\overset{|}{C}}$$

Since there is a rearrangement of a σ bond, the reactions are labeled
sigmatropic rearrangements. The σ bond can equally well be in between two
conjugated π-bond systems. Although illustrated here with carbon atoms forming the molecular backbone, there is no special requirement that the backbone
atoms be carbons.

$$
\begin{array}{c}
C-(C{=}C)_m \\
| \\
C-(C{=}C)_n
\end{array}
\longrightarrow
\begin{array}{c}
(C{=}C)_m-C \\
| \\
(C{=}C)_n-C
\end{array}
$$

As a general nomenclature for these rearrangements a particular rearrangement is classified with two numbers, *i* and *j*, set in brackets [*i,j*]. The
numbers refer to the number of the atom to which each end of the migrating
σ bond goes, numbering from the original two atoms forming the original σ
bond.

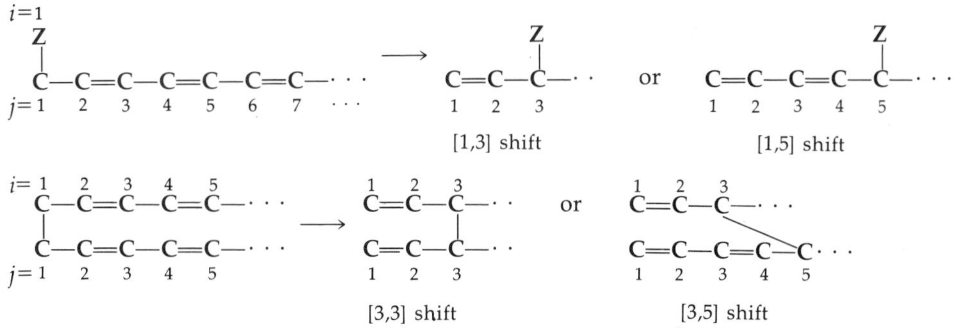

$i=1$

$$
\begin{array}{cccccccc}
& Z & & & & & & \\
& | & & & & & & \\
& C{-}C{=}C{-}C{=}C{-}C{=}C{-}\cdot\cdot\cdot \\
j{=}1 & 2 & 3 & 4 & 5 & 6 & 7 & \cdots
\end{array}
\longrightarrow
\begin{array}{ccc}
& Z & \\
& | & \\
C{=}C{-}C{-}\cdot\cdot \\
1 & 2 & 3
\end{array}
\quad \text{or} \quad
\begin{array}{ccccc}
& & & Z & \\
& & & | & \\
C{=}C{-}C{=}C{-}C{-}\cdot\cdot\cdot \\
1 & 2 & 3 & 4 & 5
\end{array}
$$

[1,3] shift   [1,5] shift

$$
\begin{array}{ccccc}
i{=}1 & 2 & 3 & 4 & 5 \\
C{-}C{=}C{-}C{=}C{-}\cdot\cdot\cdot \\
| \\
C{-}C{=}C{-}C{=}C{-}\cdot\cdot\cdot \\
j{=}1 & 2 & 3 & 4 & 5
\end{array}
\longrightarrow
\begin{array}{ccc}
1 & 2 & 3 \\
C{=}C{-}C{-}\cdot\cdot \\
| \\
C{=}C{-}C{-}\cdot\cdot \\
1 & 2 & 3
\end{array}
\quad \text{or} \quad
\begin{array}{ccc}
1 & 2 & 3 \\
C{=}C{-}C{-}\cdot\cdot\cdot \\
| \\
C{=}C{-}C{=}C{-}C\cdot\cdot\cdot \\
1 & 2 & 3 & 4 & 5
\end{array}
$$

[3,3] shift   [3,5] shift

As with cycloaddition the rearrangement is *symmetry-allowed and proceeds
thermally when the number of participating bonds (electron pairs) is an odd number.*†
Therefore, the [1,5] and [3,3] shifts are favored processes and the ones most
commonly observed. Implicit in this description is the assumption that the
migrating σ bond moves *across the same face* of the π-electron system. Such
rearrangements are called **suprafacial** and are the most common instances for
obvious steric reasons. However, with extended π systems or in certain carefully
designed molecules, the migrating σ bond can be *reformed to the opposite π-electron
face* of the conjugated system. Such a rearrangement is **antarafacial**.

---

† This may also be expressed by the rule that a symmetry-allowed reaction is $i + j = 4n + 2$, where
$n =$ any integer, or as $(i + j)/2 =$ odd number, as with cycloadditions (Fig. 21-5).

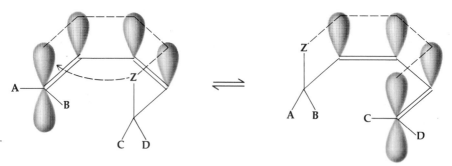

Suprafacial [1,5] rearrangement of **Z**

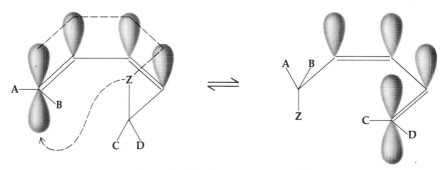

Antarafacial [1,5] rearrangement of **Z**

The rule for symmetry-allowed rearrangements is reversed if the σ bond is reformed in antarafacial fashion on the conjugated π-electron skeleton.

Symmetry-forbidden sigmatropic rearrangements are often observed under photolytic initiation in molecules where the shift is sterically reasonable. Several suprafacial rearrangements, both symmetry-allowed (and pyrolytic) and symmetry-forbidden (occurring on photolysis), are illustrated in Fig. 21-8. An antarafacial, symmetry-allowed [1,7] sigmatropic shift is illustrated below, but it will be clear on inspection of models that antarafacial shifts are rarely physically (sterically) possible.

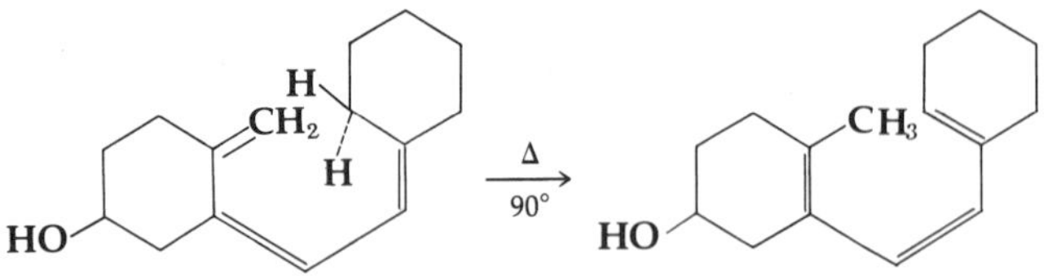

Cope and Claisen Rearrangements

Two thermal rearrangements known for some time are the **Cope** and **Claisen rearrangements**, both of which are [3,3] sigmatropic shifts, essentially differing only in the substitution of oxygen (Claisen) for carbon (Cope) in the molecular framework.

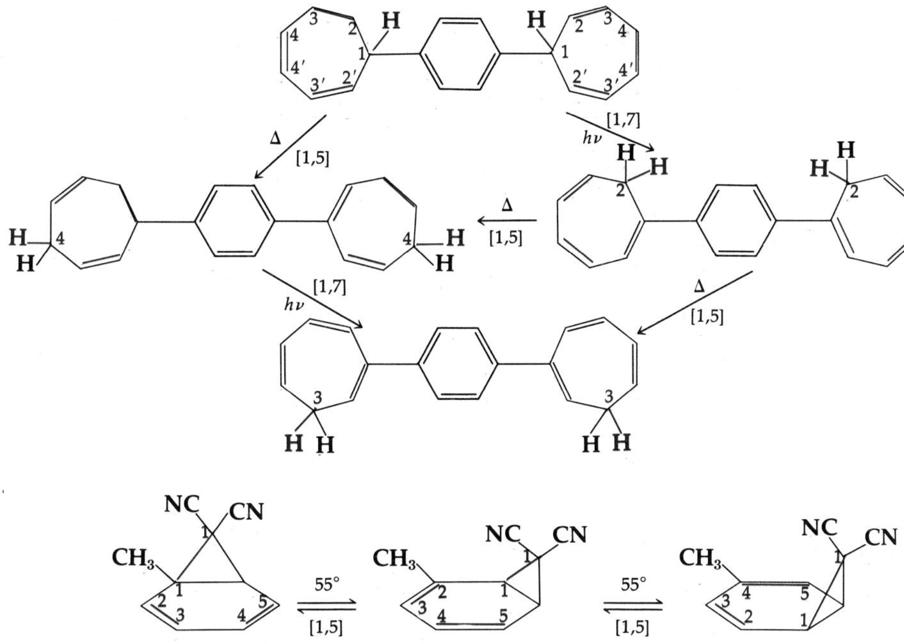

FIGURE 21-8   **Suprafacial sigmatropic shifts**

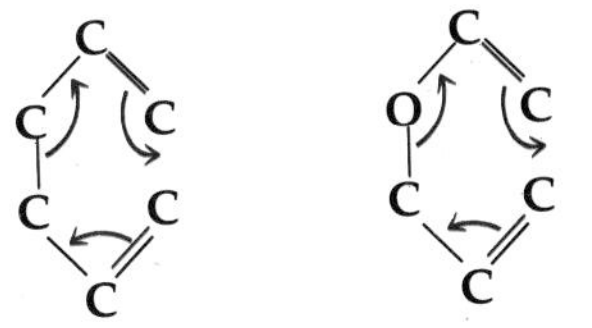

Cope
rearrangement

Claisen
rearrangement

The Cope rearrangement of doubly unsaturated systems of the bis-allylic type is the best known of these transformations. This reaction may be carried out in the vapor phase, and so it is unlikely that the reaction involves dissociation of the starting material into fragments such as radicals or ions. Bonds are probably made and broken in the same transition state, as shown for the rearrangement of *cis*-1,2-divinylcyclobutane to *cis,cis*-1,5-cyclooctadiene.

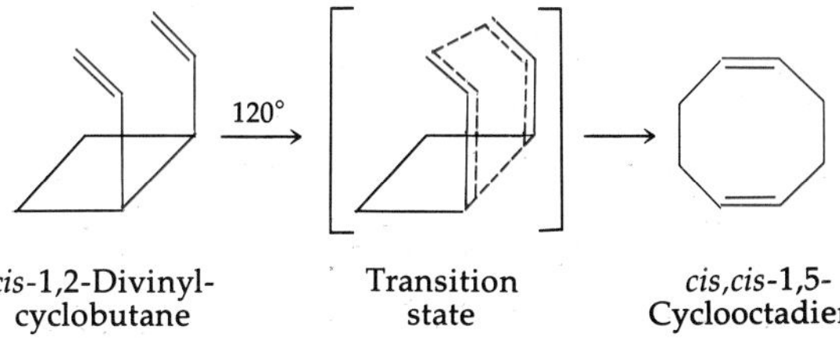

cis-1,2-Divinyl-
cyclobutane

Transition
state

cis,cis-1,5-
Cyclooctadiene

Rearrangement of *trans*-1,2-divinylcyclobutane occurs only at a much higher temperature, and gives a mixture of products. Apparently *cis*-1,2-divinylcyclobutane is a primary product of the latter reaction although it is short-lived under the reaction conditions.

A detailed inference of the preferred stereochemical path of the Cope rearrangement is based on the results of rearrangement of the stereoisomers of 3,4-dimethyl-1,5-hexadiene.

		trans,trans	*cis,cis*	*cis,trans*
Racemic	$\xrightarrow{225°}$	90%	10%	~0%
Meso	$\xrightarrow{180°}$	0.3%	~0%	99.7%

Two kinds of transition states can be visualized for the rearrangements. The first involves a boatlike structure (*A*), the second a chairlike structure (*B*).

A   *B*

The experimental results show that transition state *B* must be energetically preferred. This accounts for all products as suprafacial [3,3] shifts.

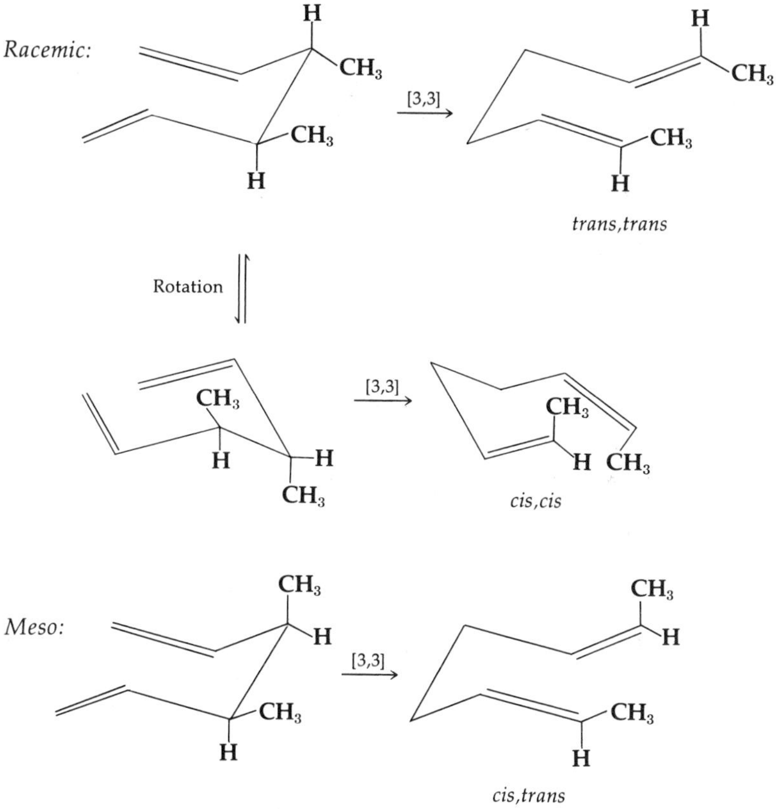

Racemic:

trans,trans

Rotation

cis,cis

Meso:

cis,trans

The Claisen rearrangement of aryl allyl ethers to allylphenols has been studied carefully and can be described with great precision. If there is an open ortho position, only ortho rearrangement occurs. However, if both ortho positions are blocked, rearrangement to an open para position takes place.

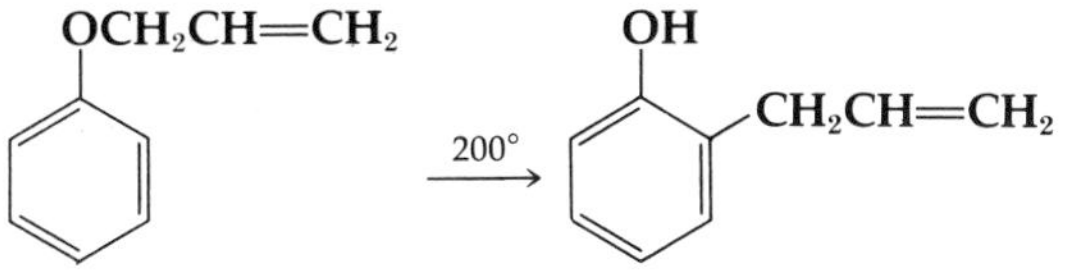

Phenyl allyl ether

o-Allylphenol
90%

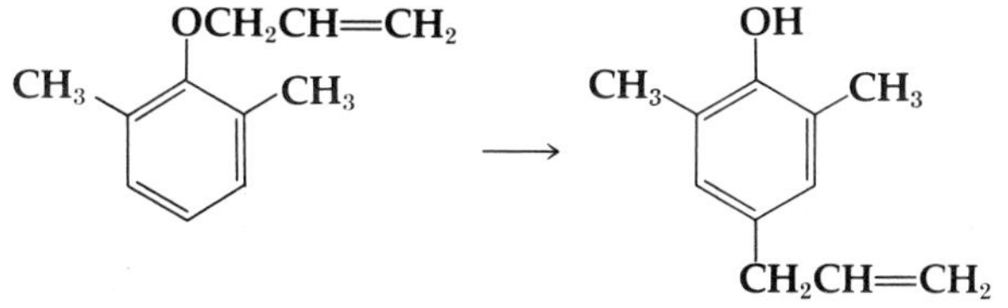

Allyl 2,6-dimethylphenyl
ether

4-Allyl-2,6-dimethylphenol

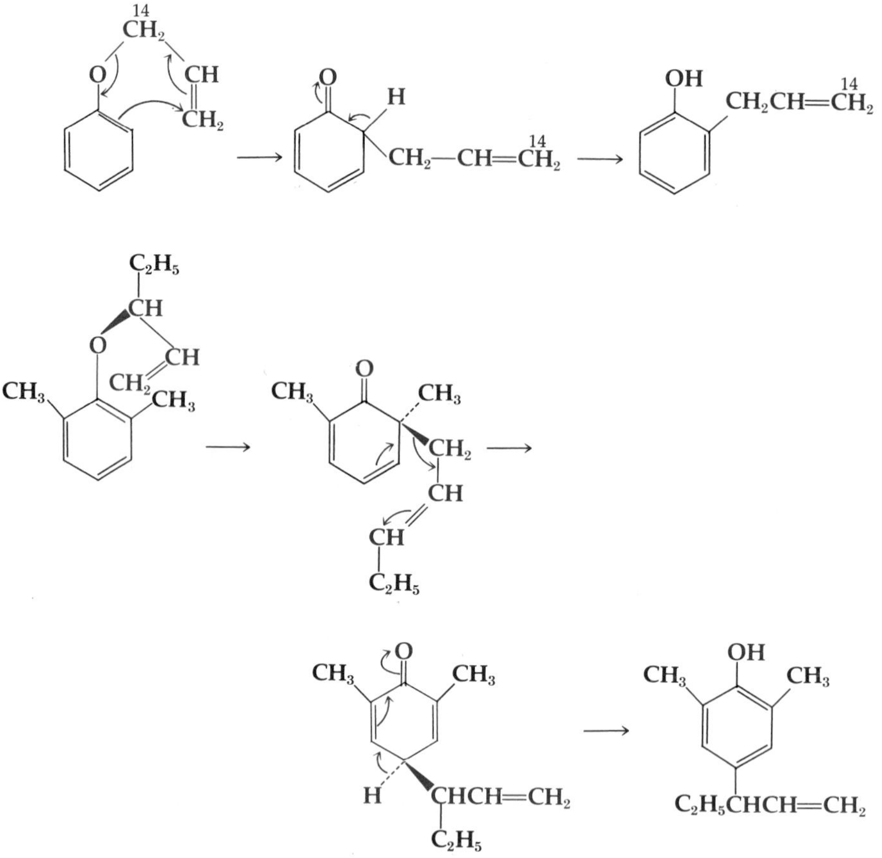

FIGURE 21-9 **Mechanism of the Claisen rearrangement**

Studies with allylic groups labeled with substituents or carbon-14 (Fig. 21.9) show that the allylic group is end-interchanged during the ortho rearrangement but maintains its structure (or is end-interchanged twice) during migration to a para position. These and other experimental data virtually necessitate the view that the rearrangements have cyclic mechanisms and involve dienones as intermediates. The second rearrangement in the para-Claisen isomerization (Fig. 21-9) is a second [3,3] shift, of the all-carbon or Cope variety. Thus, when the ortho position bears hydrogen, re-aromatization intervenes but when it does not—and re-aromatization is blocked—then a second [3,3] sigmatropic rearrangement occurs followed by final aromatization through loss of the original para hydrogen.

An allowed [5,5] version of the Claisen rearrangement has recently been observed. The [3,5] and [5,3] rearrangements are not observed either in this example or in other models.

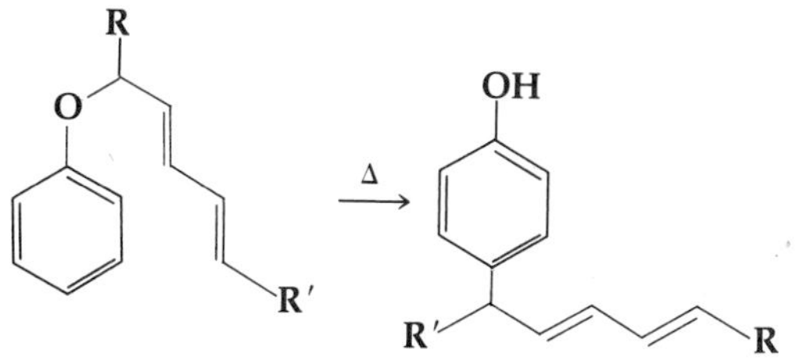

The Claisen rearrangement allows formation of a carbon-carbon bond to the 3-position of an allylic alcohol through its conversion first to a transient vinyl ether and rearrangement. Sigmatropic rearrangements of this sort are valuable in synthesis.

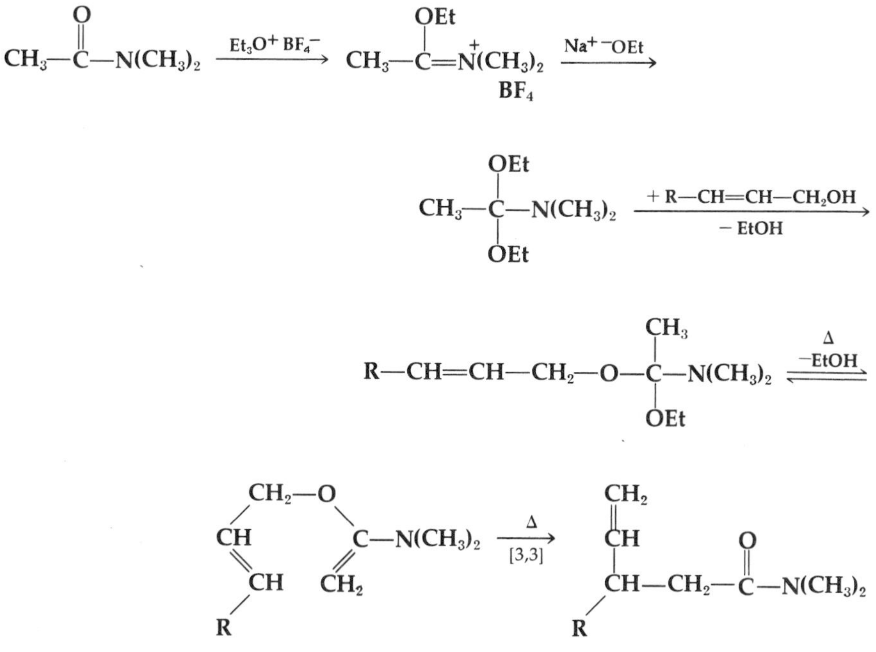

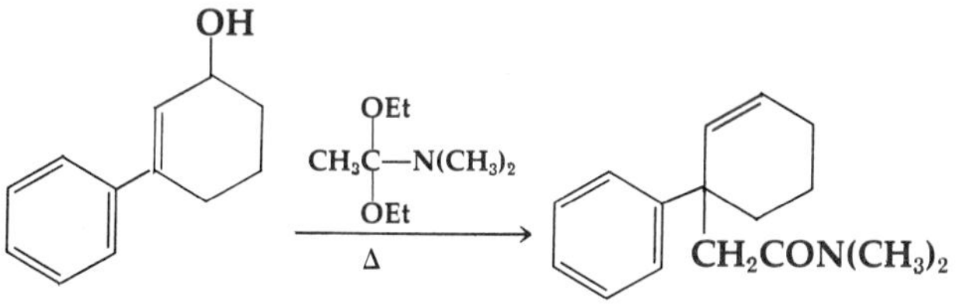

$$CH_3-\overset{O}{\overset{\|}{C}}-N(CH_3)_2 \xrightarrow{Et_3O^+ BF_4^-} CH_3-\overset{OEt}{\overset{|}{C}}=\overset{+}{N}(CH_3)_2 \xrightarrow{Na^+ {}^-OEt}$$
$$BF_4$$

$$CH_3-\overset{OEt}{\underset{OEt}{\overset{|}{\underset{|}{C}}}}-N(CH_3)_2 \xrightarrow[-\ EtOH]{+\ R-CH=CH-CH_2OH}$$

$$R-CH=CH-CH_2-O-\overset{CH_3}{\underset{OEt}{\overset{|}{\underset{|}{C}}}}-N(CH_3)_2 \underset{-EtOH}{\overset{\Delta}{\rightleftharpoons}}$$

$$\xrightarrow[{[3,3]}]{\Delta}$$

The fluxional tautomerism of bullvalene (page 162) is only a series of identical suprafacial and allowed [3,3] sigmatropic rearrangements. These occur rapidly even at room temperature since the transition barrier is so low. At 100° the protons of bullvalene are interchanging their environments so fast that they appear in the NMR spectrum as a single peak. Only by lowering the temperature to −25° does the rearrangement slow down enough to distinguish different hydrogens as separate NMR signals.

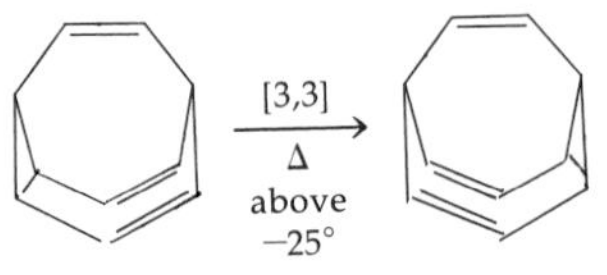

NMR at 100°
   5.8 τ (sharp singlet)
NMR at −25°
   4.3 τ (6H) C=C—H

7.9 τ (4H) —C—H

Bullvalene

[3,3]
Δ
above
−25°

PROBLEM 21-4

Show the expected products from thermolysis of the following compounds. Show stereochemistry where relevant.

a

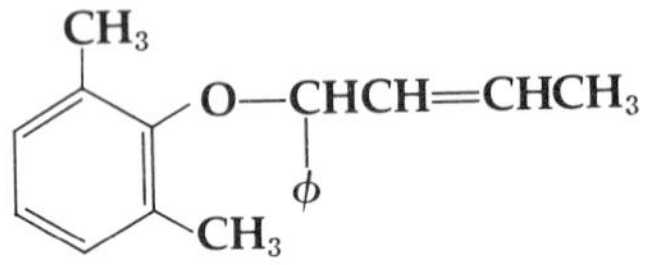

CH$_3$

O—CHCH=CHCH$_3$

φ

CH$_3$

c   (trans,trans)

b

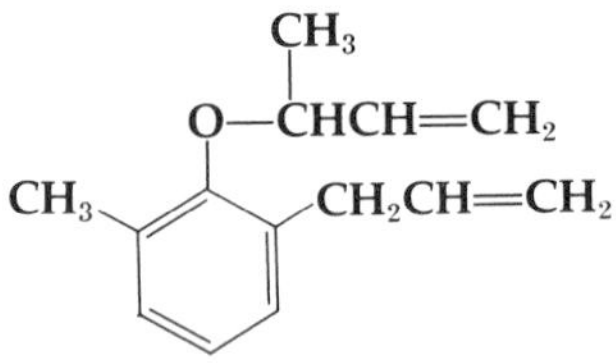

CH$_3$

O—CHCH=CH$_2$

CH$_3$ CH$_2$CH=CH$_2$ (two products)

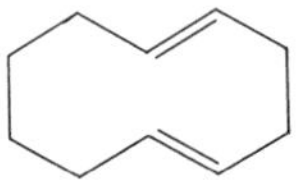

d   (cis,trans)

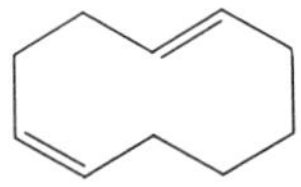

PROBLEM 21-5

The labeled compound below was heated for a period of time and then recovered. The various carbons were separately assessed to see whether all were now partially bonded to deuterium, i.e., complete scrambling of C—H and C—D. Complete scrambling was not observed. In fact deuterium label was found only at certain specific locations. Show how this experiment serves to distinguish [1,3] from [1,5] sigmatropic shifts, and indicate what pattern of deuterium is expected on thermal equilibration.

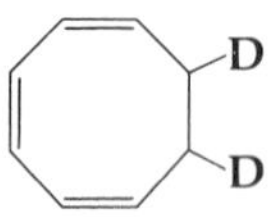

—D

D

21-4   GENERALIZATION AND SUMMARY

There are many variations on the reactions of the last three sections and with selection rules now clear, more will certainly be developed for use in the future. The pericyclic reactions of ions have yet to be explored in any depth and the synthetic possibilities, apart from the Diels-Alder reaction, have not been fully exploited.

All these reactions have in common the cyclic transition from one set of alternating bonds (π or σ) to another.

2 bonds
4 electrons

3 bonds
6 electrons

We may develop a convenient generalization which provides a quick assessment of whether a given example is symmetry-allowed. This generalization treats all participating bonds as separate and is equally applicable to all three of the reaction families: electrocyclic; cycloaddition; and sigmatropic. The generalization depends on the stereochemistry of bond formation, specifically whether it is suprafacial or antarafacial. In the cycloadditions shown in Sec. 21-2 the components always join in a manner suprafacial in both components.

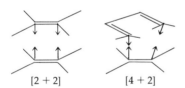

[2 + 2]   [4 + 2]

Suprafacial cycloadditions

The essence of suprafacial addition to a double bond is that the two new bonds are formed on the *same side of the molecular plane* of the double bond. Consider the π bond as two p orbitals combined into a bonding π-molecular orbital. This implies that in cycloaddition the two new bonds are both formed by overlap to p lobes of the same sign, in suprafacial addition. Antarafacial addition proceeds to opposite sides of the π plane or to opposite-signed lobes of the two p orbitals making up the π bond. The two additions are shown at the top of Fig. 21-10.

The same geometrical distinction can be applied to single (σ) bonds participating in pericyclic reactions. When a bonding σ orbital breaks (Fig. 21-10, bottom) either atom can form its new bond (π or σ) with either the $(+)$ or $(-)$ lobe of its sp^3 atomic orbital. If the $(+)$, or overlapped, lobe becomes involved in the new bond, no change in configuration occurs at that atom (retention of configuration), except the minor flattening from tetrahedral to

To π bonds

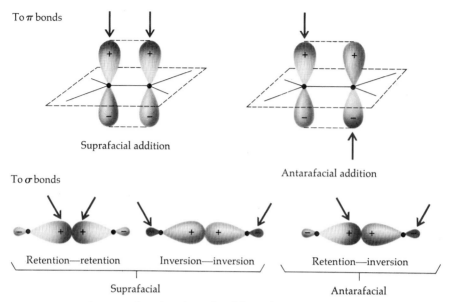

Suprafacial addition

Antarafacial addition

To σ bonds

Retention—retention   Inversion—inversion   Retention—inversion

Suprafacial   Antarafacial

FIGURE 21-10 **Geometrical modes of new bond formation**

trigonal if the new bond created from the orbital is a π bond. If the minor lobe ($-$) expands to become the new bonding lobe, configuration is inverted. The latter case is completely analogous to the S_N2 displacement.

A suprafacial cleavage of a σ bond (Fig. 21-10) is defined as either retention or inversion at both atoms.

An antarafacial cleavage involves inversion at one atom, retention at the other.

We may use a shorthand designating the bonding lobes with short lines simply to see how they are oriented or how they move so as to be suitable for overlap to give the required bonds of the product. Each bond is designated as reacting in a suprafacial (**s**) or antarafacial (**a**) fashion.

Conrotatory:

Disrotatory:

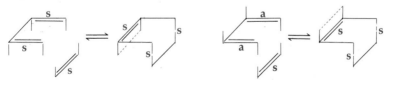

The important observation about this convention is that, no matter how the suprafacial and antarafacial modes are chosen, both the forward and reverse reaction in the conrotatory case involve one suprafacial and one antarafacial regrouping of the two original bonds. In the disrotatory case they are always both suprafacial or antarafacial. The latter case is symmetry-forbidden, the former symmetry-allowed.

The general rule about all pericyclic reactions which proceed in a concerted fashion is the following:

> *The reaction is symmetry-allowed if the total number of suprafacial regroupings of the participating bonds is an odd number.*

This rule allows an assessment of any case without a detailed analysis of symmetry or orbital levels and interactions. The rule then allows us to examine any given case to see whether it is expected to proceed thermally (odd number of suprafacial bond breakings) or photolytically (even number of suprafacial bond breakings).†

Cycloadditions may be taken as examples. The simplest cycloaddition—of two ethylenes—must in any normal unstrained situation be suprafacial in both components and hence, with an even number (two) of suprafacial bond cleavages (or formations), be symmetry-forbidden. The several ways of examining the Diels-Alder reaction all reveal it to be symmetry-allowed.

3 s = Symmetry-allowed   1 s = Symmetry-allowed

---

† When single orbitals are involved, as in the cyclopropyl cation p orbital, they may be included in the general rule with appropriate designations of **s** and **a**, but there are difficulties in these examples and they are not included in the present treatment. The rule quoted therefore pertains only to neutral molecules.

In the sigmatropic rearrangement a rearranging hydrogen must pass over with retention since it possesses only an *s* orbital with only one lobe and no nodes (Sec. 2-6). However, a rearranging carbon can invert during rearrangement. The antarafacial example on page 862 will be seen as [1,7] **sssa** and symmetry-allowed.

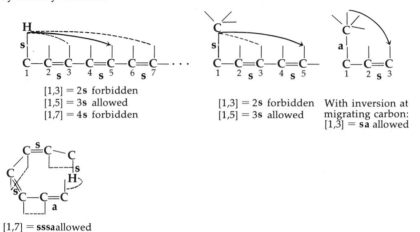

[1,3] = 2s forbidden
[1,5] = 3s allowed
[1,7] = 4s forbidden

[1,3] = 2s forbidden
[1,5] = 3s allowed

With inversion at migrating carbon:
[1,3] = **sa** allowed

[1,7] = **sssa** allowed
(antarafacial overall)

An example of an allowed [1,3] shift with inversion at the migrating carbon is shown below. The steric strain necessary to perform this inversion appears at first sight to be a serious objection. Without the orbital symmetry rules we would expect the sterically easier shift with retention—in which the migrating carbon just slides across the allyl system—to be of lower energy and preferred. However, violation of orbital symmetry places an even larger barrier than angle strain so that the only route available is the strained but symmetry-allowed rearrangement with inversion, as shown.

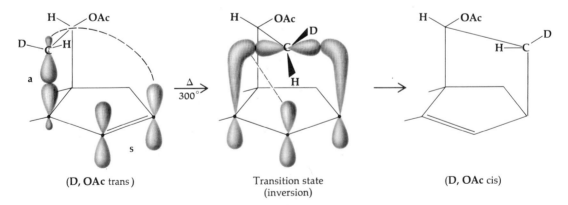

(**D, OAc** trans)

Transition state
(inversion)

(**D, OAc** cis)

Since most bond breaking is suprafacial, it is reasonable in any analysis of a given case to label as many reactant bonds suprafacial as is consistent with the necessary overlap to form product bonds. This will usually reveal either that only one or none need be labeled antarafacial, and then the sum will reveal whether or not the reaction is symmetry-allowed. Two examples are labeled below. The first proceeds easily; the second does not go at all on heating but instead forms the cis isomer (as predicted; by conrotatory cyclization).

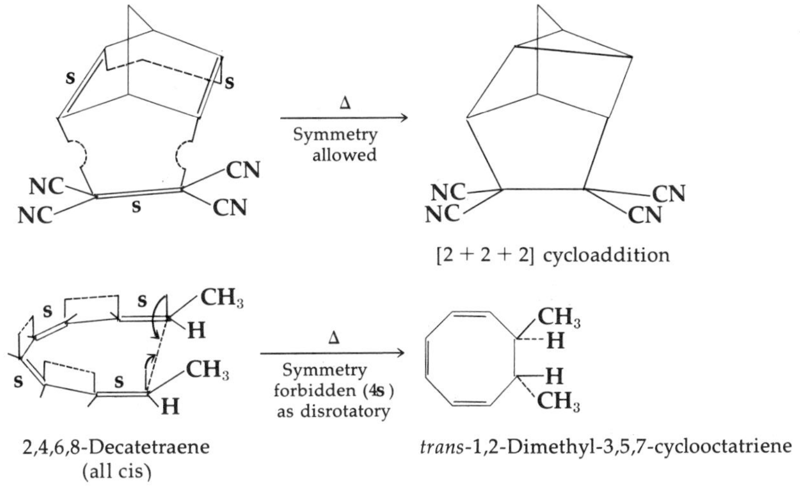

s
s
NC
NC
s
CN
CN
Δ
Symmetry allowed
NC
NC
CN
CN
[2 + 2 + 2] cycloaddition
s
s
s
s
CH3
H
CH3
H
Δ
Symmetry forbidden (4s) as disrotatory
CH3
H
H
CH3
2,4,6,8-Decatetraene (all cis)
trans-1,2-Dimethyl-3,5,7-cyclooctatriene

The most common thermal reactions are those involving three bonds (six electrons) in an all-suprafacial cyclic transition. The central elements of these are summarized below and include the thermal elimination reactions of acetates and xanthates in Chap. 14 (page 595). If the interconversion of the two Kekulé forms of benzene were a tautomerism instead of merely a device for describing the resonance in a single molecule, then that tautomerism would also be an electrocyclic reaction of this same class.

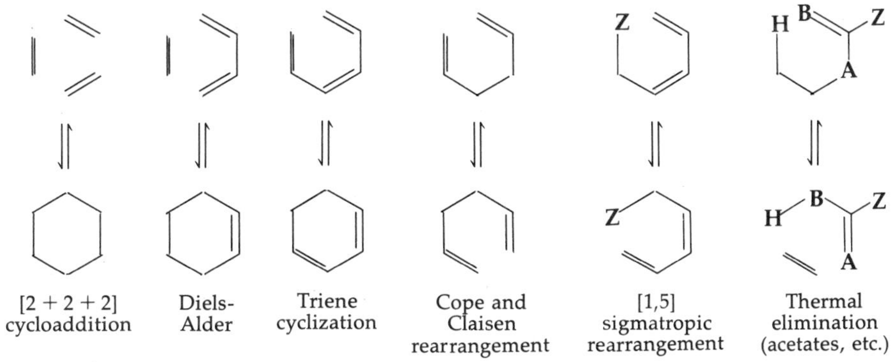

Z
H
B
Z
A
Z
H
B
Z
A
[2 + 2 + 2] cycloaddition
Diels-Alder
Triene cyclization
Cope and Claisen rearrangement
[1,5] sigmatropic rearrangement
Thermal elimination (acetates, etc.)

The pericyclic reactions are all equilibria. The rules tell us about the stereoelectronic requirements, or stereochemical consequences of allowed reactions. The rules tell nothing about the rate or position of equilibrium, i.e., the nature of the free-energy diagram. A reasonable guess about the position of equilibrium—the preferred product—can usually be made from the energies of the involved bonds (Table 2-2). This generally will show the preferred product as that with more σ and fewer π bonds. The equilibria are subject to some control. A reverse Diels-Alder reaction can usually be carried out by distilling away a volatile product and thus driving the equilibrium in one direction.

While the prediction rules have dramatic success for thermolysis reactions (i.e., for symmetry-allowed reactions on heating), photolytic reactions are much more complex. The predicted products from the excited state often form, but the fact that photolysis is initiated by light absorption first to a high-energy excited state implies that such a state may also collapse via other, not necessarily concerted, pathways. Often, indeed, the initial excited state will pass on stepwise through several radical intermediates before affording stable products. These more complex photolysis reactions form the subject of the next chapter.

PROBLEMS

21-6 The two isomeric hydrocarbons below give two isomeric adducts when heated with maleic anhydride. The reaction is believed to proceed in two steps, with maleic anhydride only involved in the second. Indicate the reactions and the stereochemistry of the isomeric adducts in each case.

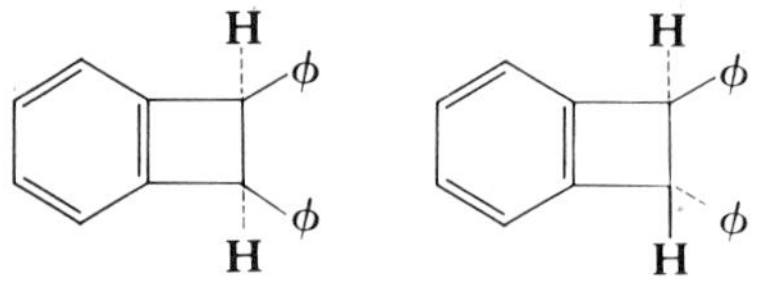

21-7 Explain the following transformation in terms of four successive pericyclic reactions, of which the first to occur does not involve the quinone.

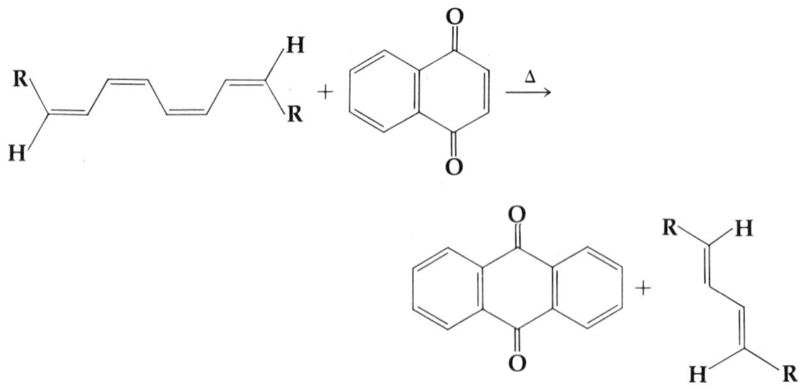

21-8 The following reaction proceeds in good yield:

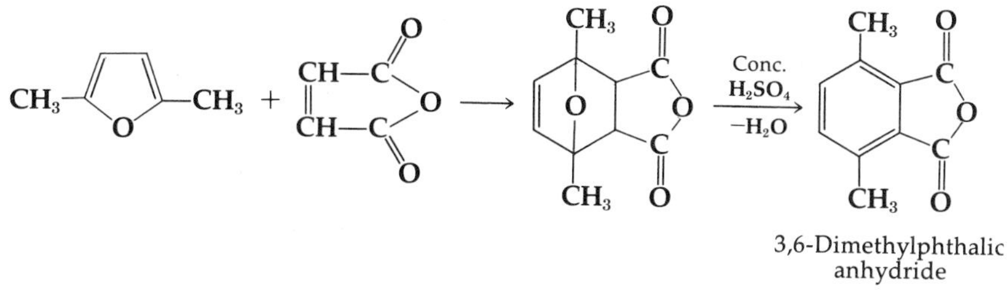

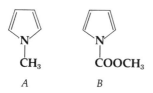

3,6-Dimethylphthalic
anhydride

Using the product as a starting material, devise syntheses of several 1,2,3,4-tetrasubstituted benzenes having four different substituents.

21-9 Devise syntheses of:

a

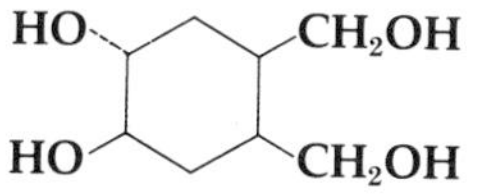

from maleic anhydride

b

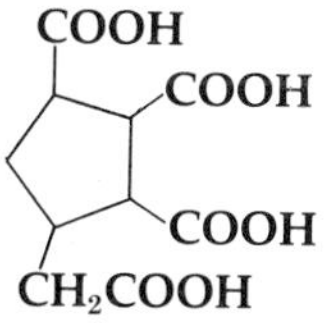

(all cis) from cyclopentadiene

21-10 A simple expedient for preparing substituted furan-3,4-dicarboxylic acids from furans is outlined below. Fill in the missing parts.

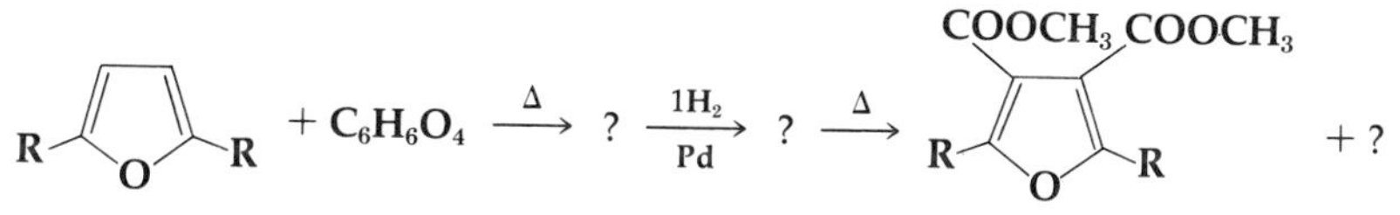

$$R-\text{furan}-R + C_6H_6O_4 \xrightarrow{\Delta} ? \xrightarrow[\text{Pd}]{1H_2} ? \xrightarrow{\Delta} \text{(COOCH}_3,\text{COOCH}_3 \text{ furan)}-R + ?$$

21-11 Compound *A* does not give a Diels-Alder adduct with acetylene dicarboxylic esters but compound *B* does. Explain, after looking again at Fig. 5-8.

A   *B*

CH₃   COOCH₃

21-12 When bromobenzene, anthracene, and sodamide ($NaNH_2$) are allowed to react together there is formed a hydrocarbon, $C_{20}H_{14}$.

a Write a structure for this substance and explain its formation.

b How many such products would you expect in each case if the first starting material were *o*-bromotoluene, *m*-bromotoluene, and *p*-bromotoluene?

21-13 Rationalize the following observed reaction.

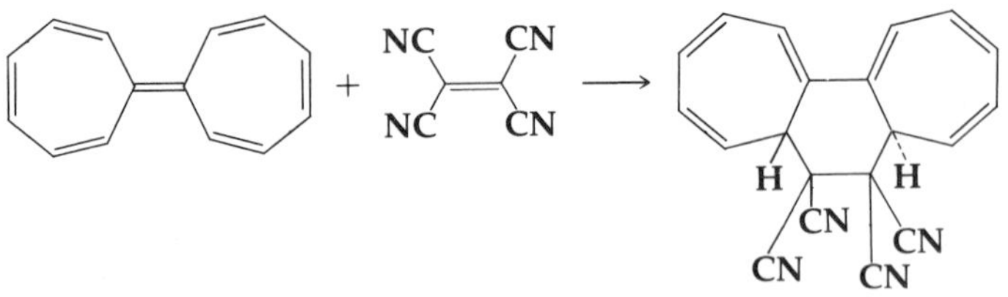

21-14 Devise a mechanism for the following transformation. (*Hint:* the first step is loss of the acetic acid.)

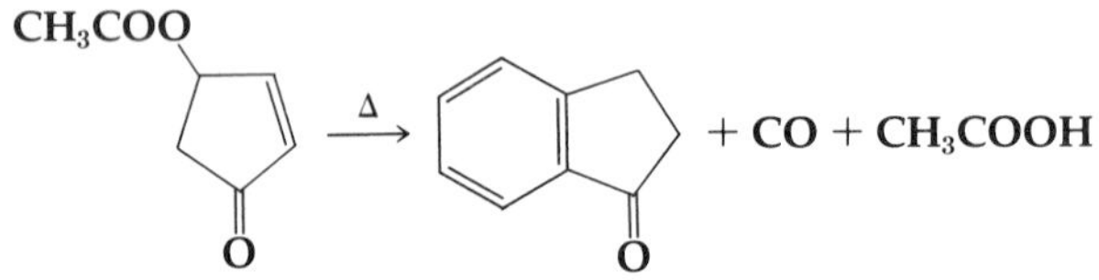

$$+ CO + CH_3COOH$$

21-15 Account for the following observation.

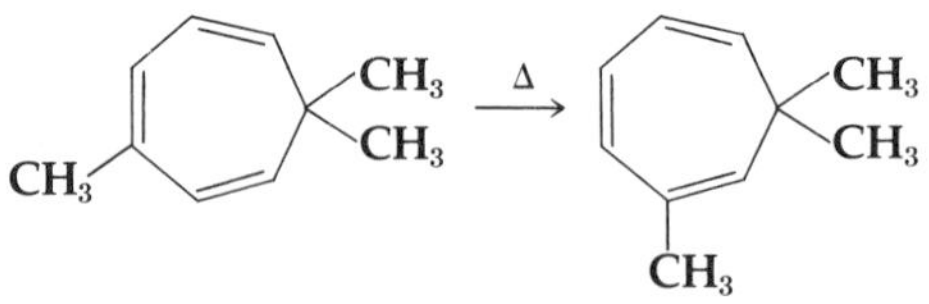

21-16 Isomerization of 3-phenyl-3-acetoxy-1-propene is effected on heating. The isomer has a similar IR spectrum below 6 μ (1670 cm^{-1}) but shows a new UV spectrum with an intense maximum about 250 nm. The NMR spectrum of the isomer shows five peaks, reading upfield, in a ratio of 5:1:1:2:3. What is the isomer and what is the reaction?

READING REFERENCES

Woodward, R. B., and R. Hoffmann, "The Conservation of Orbital Symmetry," Academic Press, New York, 1969.

Orchin, M., and H. H. Jaffe, "The Importance of Antibonding Orbitals," Houghton Mifflin, Boston, 1967.

PHOTOCHEMISTRY

PHOTOCHEMISTRY is the study of reactions caused by absorption of light. Everyone is familiar with some effects of chemical change induced by sunlight. The miracle of photosynthesis, on which all life on this planet depends, is fundamentally a chemical reaction between carbon dioxide and water effected by chlorophyll, the pigment responsible for the color of green plants, and discussed in Sec. 26-4.

$$6CO_2 + 6H_2O \xrightarrow[\text{chlorophyll}]{\text{Sunlight}} \underset{\text{Glucose}}{C_6H_{12}O_6} + 6O_2$$

The response of an eye to light is triggered by a photochemical change of the pigment rhodopsin in the visual rods. Less beneficial photochemical changes are experienced in the form of sunburns and skin cancers, the latter affliction being especially prevalent among fair-skinned people living in tropical regions.

Although Cannizzaro did research in photochemistry in the early nineteenth century, the subject was long considered by many organic chemists as a small, esoteric branch of their field until relatively recently. During the past decade, there has been a great surge of interest in the subject. The change has been caused by a number of factors: some items of simple, but useful, laboratory equipment have become available; the unique value of photochemistry in the synthesis of certain high-energy compounds is attractive; and, probably most important, chemists have come to realize that photochemistry can be studied in a meaningful, systematic way.

22-1 THEORY OF PHOTOCHEMISTRY

The fundamentals of photochemistry are remarkably simple. A molecule becomes energized to an **excited state** by absorption of a photon of light and undergoes a chemical reaction while still in that excited state. The difference between photoreactions and ordinary thermal reactions lies in the fact that in photochemistry individual molecules are promoted to highly excited forms without immediate effect on surrounding molecules. In a thermally activated reaction, energy entering the system in the form of heat is apportioned among all of the molecules in the system according to statistical principles. The unique character of selective excitation of individual molecules in photochemistry is a little like being able to touch a match to a single straw in a haystack and burn it without having the fire spread to the rest of the stack.

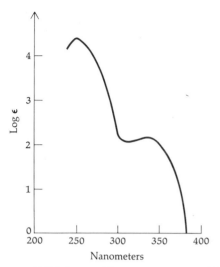

FIGURE 22-1 **Absorption spectrum of benzophenone in ethanol solution**

Another characteristic of photochemistry arises from the fact that the chemical changes occur in energy-rich molecules. Reaction is usually accompanied by loss of energy from the excited species. This is in sharp contrast to thermally activated reactions in which activation energy is gathered in small bits from the environment. In Chap. 9 we discussed elementary theory of the rates of thermal reactions by using an analogy to people wandering over mountain ranges from one valley to another. In photochemistry the problem really starts with a man on a mountain top about to descend. The problem of deciding which of several valleys he will arrive in is not related to the energy that he will spend in getting over passes, but only involves his choice among alternate downhill paths.

Electronic Excitation

Photochemical reactions are caused by absorption of visible and ultraviolet light. Light of these wavelengths causes excitation of electrons in the molecules to higher-energy orbitals producing electronically excited states. In the visible region of the spectrum the photon energies range from 38 kcal/mole (750 nm) to 71 kcal/mole (400 nm), as shown in Fig. 7-5. By using light in the near ultraviolet (to 200 nm), energies up to 143 kcal/mole become available. These energies cover the range of chemical bond energies (Table 2-2). Some photochemistry is done using light of even shorter wavelengths but this is uncommon because strong absorption by atmospheric oxygen below 200 nm necessitates the use of vacuum equipment.

Absorption of light by a molecule is usually discontinuous and the absorption spectrum consists of one or more bands which may show some substructure (Chap. 7). The absorption spectrum of benzophenone (ϕ_2CO) is shown in Fig. 22-1. Two absorption maxima are observed, one at 245 nm and the other at 345 nm. These broad absorption regions are due to different electronic transitions in the molecule. The low-intensity band at long wavelength is due to an $n \longrightarrow \pi^*$ excitation, in which an electron in a nonbonding orbital at oxygen is promoted to the lowest of the unoccupied π orbitals of the benzo-phenone molecule (page 247). The 345-nm transition of benzophenone is more complex than the n,π^* of a normal ketone because the π^* orbital covers the entire molecule rather than being localized in the carbonyl group.

The broad band at shorter wavelength is due to a $\pi \longrightarrow \pi^*$ transition in which an electron is excited from an occupied π orbital to a vacant anti-bonding one. The upper orbital is the same as the one involved in the $n \longrightarrow \pi^*$ transition, but more energy is required because the π electron is more firmly bound in the ground state than are the nonbonding (n) oxygen electrons.[†] The transition is related to the 265-nm bond in the spectrum of benzene. The 245-nm absorption of benzophenone is probably really complex and may consist of two overlapping transitions of very similar energy. We need not be concerned about this spectroscopic complication in most photochemical work. As a matter of fact, the same photochemical results are usually obtained as a result of irradiation in any of the visible and near-ultraviolet bands of most molecules. This is surprising, since the energy contents and electronic structures of the excited states may be quite different. For reasons that are not well understood, the excited π,π^* state of benzophenone decays very rapidly to the lower energy n,π^* state.[‡] This rapid decay of excited states into lower excited states is quite common and has an important leveling effect in photo-chemistry. However, it also introduces a complication because excited states that are not formed by direct absorption of light may be formed by decay and become important in photoreactions.

Singlets and Triplets

Figure 22-2 shows the complication that arises from consideration of electron spins. Nearly all stable molecules have even numbers of electrons with spins paired in their ground states. The pairing means that the magnetic moments of the electrons compensate each other so that the molecule as a whole has no net electronic magnetic moment. These nonmagnetic states are called **singlets** (abbreviated S, S_0, S_1, etc., for different energy singlet states). Excitation of electrons by light absorption *does not change electron spins*, so only excited singlets are formed by absorption. In an excited singlet state, electrons are still spin paired but present in different orbitals (e.g., π and π^*).

Such a state can then be converted to another by inverting one of the electron spins. This creates a new state in which the electronic magnetic moments do not compensate. The new excited state will then have a net

---

[†] This discussion is highly oversimplified. Actually, a change in the electronic configuration of a molecule changes all of the orbitals, because of electron-electron interactions.

[‡] This decay process requires loss of energy, so some energy must be delivered to the environment. In the gas phase at low pressure, the process of decay of the higher excited state to the lower might become slow. This has been shown to be the case with a few molecules, such as benzene, which have been studied carefully.

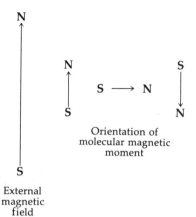

FIGURE 22-2   Orientations of the molecular magnetic moment of a molecule in its triplet state with respect to an external magnetic field. (N and S refer to north and south magnetic poles.)

magnetic moment. This new state is called a **triplet** (abbreviated T, T_1, T_2, etc.). In a magnetic field three states will actually appear because the magnetic moment can be aligned in three different ways with respect to the applied field as shown in Fig. 22-2.

Figure 22-3 shows various possible decay processes. The decay of an excited state to another state of the same **multiplicity** (singlet or triplet)—as in $S_2 \longrightarrow S_1$, $T_2 \longrightarrow T_1$, etc.—is called **internal conversion**. Decay with a change in multiplicity ($S_1 \longrightarrow T_1$, etc.) is called **intersystem crossing**. Although corresponding excited singlets and triplets, for example S_1 and T_1, usually have the same electronic configurations to a first approximation, the triplets lie at lower energies than the singlets because electron-electron repulsion is lower in the triplets.

Decay of S_1 and T_1 to the ground state, S_0, also occurs, but usually much more slowly than decay from higher states to the lowest excited states. As a result, S_1 and T_1 are the longest lived of all the excited states and are responsible for most photochemical changes. T_1 states have especially long lives because decay back to S_0 requires not only energy loss but also reorientation of an electron spin. Because T_1 states decay much more slowly than any other excited states, they are most likely to be involved in photochemical mechanisms, despite the fact that they are the least energetic of all the excited levels.

Fluorescence and Phosphorescence

S_1 and T_1 states of many molecules live long enough to suffer deactivation by *emission* of light, the reverse of light absorption. Emission from S_1 is called **fluorescence** and corresponds to absorption in the first absorption band of a molecule. However, most of this emission lies at longer wavelength and, ideally, the emission band is the mirror image of absorption as shown in Fig. 22-4.

The reason for the shift of the emission to longer wavelength is related to the fact that the absorption is a broad band rather than a single, sharp line.

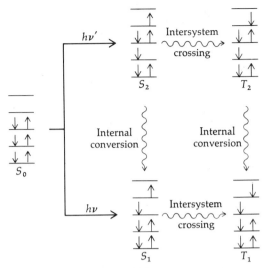

FIGURE 22-3 **Excitation and decay processes;** S_0 = ground singlet state. (S_1, S_2 = excited singlet states; T_1, T_2 = excited triplet states. Colored arrows indicate spin of each electron in its orbital.)

Electronic transitions are accompanied by changes in vibrational states. Since vibrational quantum states are much closer together than electronic levels, they may be run together to form the kind of broad band usually seen in the spectra of polyatomic molecules in solution. Sometimes the vibrational substructure is resolved into a group of fairly sharp submaxima. When an excited molecule is born, it usually has varying amounts of excess vibrational energy. However, this vibrational energy is lost after a few collisions with other molecules. Emission takes place from vibrationally cold molecules, but will produce ground-state molecules with varying amounts of vibrational energy. The only common transition in the absorption and emission spectra will be the O ⟶ O **component**, defined as the transition connecting vibrationally cold-ground and excited-state molecules. These relationships are shown schematically in Fig. 22-5.

The *intensity* of fluorescence depends on the rate of emission compared to the rates of all other decay modes of S_1 states. Radiationless decay paths include intersystem crossing to triplets, internal conversion to S_1 states, and deactivation by energy transfer to other molecules. The rates of all the competitive processes vary widely among systems. However, fluorescence is usually

FIGURE 22-4 **Relationship between first absorption band and fluorescence.**

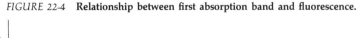

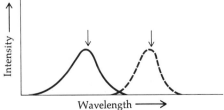

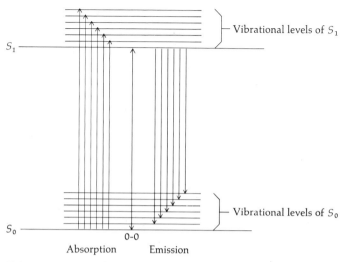

FIGURE 22-5 **Relationship between vibrational contributions to absorption and fluorescence**

either weak or undetectable, except in small molecules (e.g., diatomic molecules) or rigid molecules such as aromatic compounds. Moreover, fluorescence rates are quantitatively related to the intensities of $S_0 \longrightarrow S_1$ transitions in absorption. Many dyes which have intense first absorption bands also exhibit strong fluorescence. A well-known example is the yellow dye, fluorescein, which has a brilliant greenish fluorescence. A principal use of the material lies in aesthetic enhancement of the appearance of lakes and fountains on college campuses.

Fluorescein

The slow emission from long-lived excited states is called **phosphorescence**. In organic molecules phosphorescence arises almost exclusively from triplet states. Whereas fluorescence lifetimes are usually in the range 10^{-7} to 10^{-9} sec, phosphorescence lifetimes are commonly measured in the range from 10 to 10^{-3} sec. This correlates with the fact that $S_0 \longrightarrow T_1$ transitions are almost always too weak to observe in absorption spectra. The fact that some T_1 states live long enough to give observable emission shows that other decay processes are much slower than with S_1 states. In fluid solutions most T_1 states do find nonradiative decay paths, so most phosphorescence studies are made using rigid media, such as glasses formed by rapid cooling of solutions to low temperatures.

22-2   ENERGY TRANSFER

The most basic law of photochemistry is the **Einstein law of photochemical equivalence**, which states in essence that light must be absorbed in order for a photochemical reaction to occur. This may seem to be a truism, but the relationship between chemical action and absorption of light quanta was not at all well appreciated in the early days of photochemistry. Furthermore, some photoreactions occur under conditions that seem at first sight to violate the law.

As an example we can pose a problem. One can irradiate a solution containing butadiene with visible light ($>$400 nm) and observe formation of dimers of the diene, despite the fact that butadiene does not absorb appreciable amounts of light of wavelength longer than 250 nm!

| cis-1,2-Divinyl-cyclobutane | trans-1,2-Divinyl-cyclobutane | 3-Vinylcyclohexene |

The above statement sounds a little less fraudulent if we add the fact that a material, such as biacetyl ($CH_3COCOCH_3$), which does absorb visible light must be included in the solution in catalytic amounts. However, the results are still somewhat puzzling because the light is absorbed by one species and the reaction occurs with another. Clearly excitation energy must be transferred from biacetyl to butadiene. Even this cannot be given a trivial explanation based on the absorption spectra of the compounds. The 0–0 band is not well defined in the spectrum of butadiene, but it is probably close to 250 nm. This represents the minimum energy required to excite butadiene to its S_1 state.

$$\varepsilon = \text{energy per quantum} = h\frac{c}{\lambda}$$

h = Planck's constant = 6.63×10^{-27} erg sec

ν = frequency of light quantum

c = velocity of light = 3.00×10^{10} cm sec^{-1}

λ = wavelength

$$\epsilon = \frac{(6.63 \times 10^{-27})(3.00 \times 10^{10})}{250 \times 10^{-7}} \text{ erg sec cm sec}^{-1}/\text{cm}$$

$$= 7.96 \times 10^{-12} \text{ erg}$$

E = energy per mole = $\epsilon \times 6.02 \times 10^{23}$ ergs

$$= 4.79 \times 10^{12} \text{ ergs}$$

$$= (4.79 \times 10^{12})(2.39 \times 10^{-11}) = 114 \text{ kcal/mole}$$

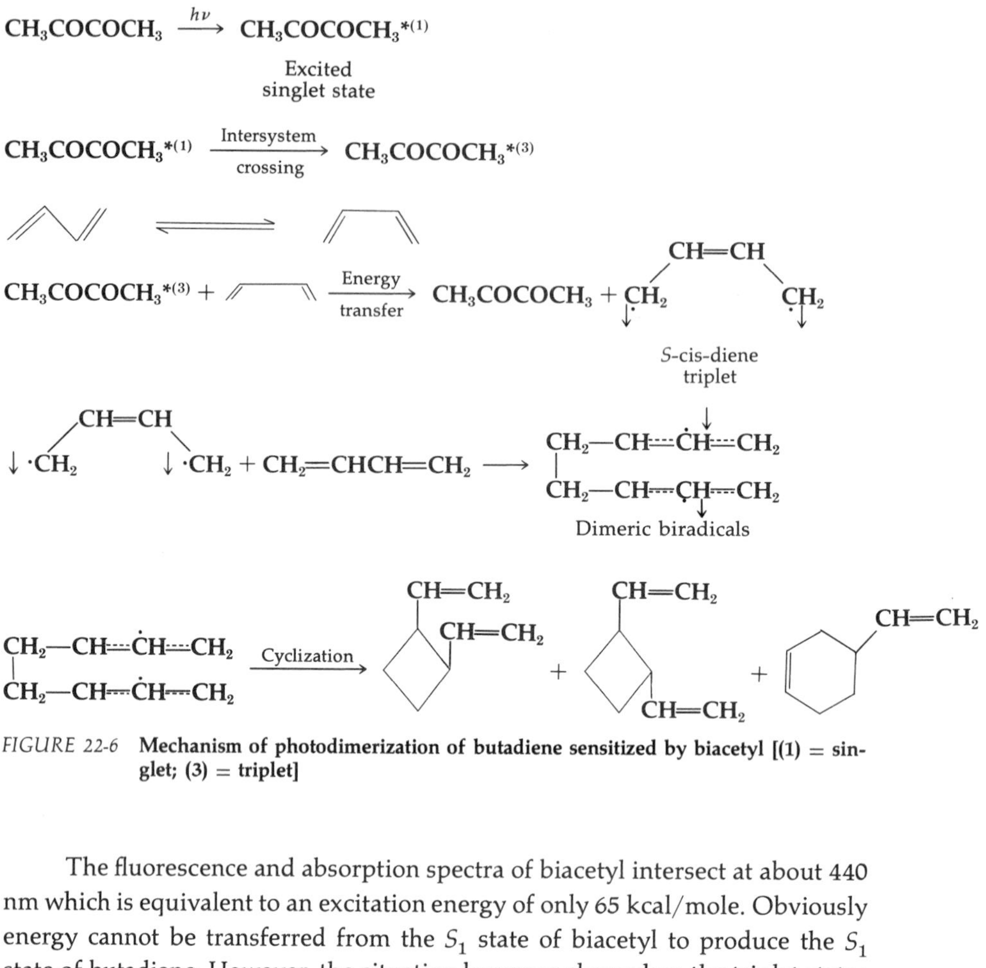

$$CH_3COCOCH_3 \xrightarrow{h\nu} CH_3COCOCH_3^{*(1)}$$

Excited
singlet state

$$CH_3COCOCH_3^{*(1)} \xrightarrow[\text{crossing}]{\text{Intersystem}} CH_3COCOCH_3^{*(3)}$$

$$CH_3COCOCH_3^{*(3)} + \text{(diene)} \xrightarrow[\text{transfer}]{\text{Energy}} CH_3COCOCH_3 + \dot{C}H_2 \cdots CH_2$$

S-cis-diene
triplet

CH₂—CH‑‑‑ĊH‑‑‑CH₂ | CH₂—CH‑‑‑ĊH‑‑‑CH₂ + CH₂=CHCH=CH₂ ⟶ Dimeric biradicals

Cyclization products

FIGURE 22-6   **Mechanism of photodimerization of butadiene sensitized by biacetyl [(1) = singlet; (3) = triplet]**

The fluorescence and absorption spectra of biacetyl intersect at about 440 nm which is equivalent to an excitation energy of only 65 kcal/mole. Obviously energy cannot be transferred from the S_1 state of biacetyl to produce the S_1 state of butadiene. However, the situation becomes clear when the triplet states of the two molecules are considered. The phosphorescence and fluorescence spectra of biacetyl are very close together, showing that there is only a small energy gap between the S_1 and T_1 states. The T_1 excitation energy is 55 kcal/mole.

Butadiene, like most conjugated dienes, has no observable emission spectrum, but the triplet excitation energy can be studied by an unusual absorption technique. If liquid butadiene is sealed in an absorption cell under a pressure of 100 atm of oxygen, weak new absorption is observed in the visible region of the spectrum. This is attributed to the $S \longrightarrow T_1$ transition which becomes observably strong in molecules having paramagnetic oxygen molecules as nearest neighbors.† The long-wavelength edge of the absorption is not

† Even though oxygen molecules contain an even number of electrons, they are paramagnetic. The ground state of the molecule is a triplet rather than a singlet state. This behavior is predicted by molecular-orbital theory. The molecule contains six π electrons. Four are assigned to two bonding π orbitals (π_x and π_y). The last two must be assigned to antibonding, π^* orbitals. Since there are two equivalent orbitals, π_x^* and π_y^*, one electron is assigned to each. Since the two electrons occupy different orbitals they can have parallel spins without violating the Pauli exclusion principle. The configuration with parallel spins minimizes repulsive interaction between the electrons, so the triplet is the most stable state of the molecule.

sharply defined, but the 0–0 band has been assigned at 480 nm (60 kcal/mole). The splitting between the S_1 and T_1 states of butadiene is so large that the triplet energy appears in the same range as that of biacetyl.

Transfer of triplet excitation energy from biacetyl to butadiene has at least become conceivable when triplet energies are considered. However, the triplet state of biacetyl is still deficient by 5 kcal/mole of the energy required to promote butadiene to the triplet state observed spectroscopically. One additional factor comes to the rescue. Butadiene is known to exist primarily in the S-trans form but a small amount ($\sim$2.5%) is in the S-cis form, which is somewhat more strained.

S-trans-	*S-cis-*
1,3-Butadiene	1,3-Butadiene
$\sim$97.5%	$\sim$2.5%

The singlet-triplet absorption spectrum of butadiene must be dominated by the predominant S-trans species, so very weak absorption by the S-cis material is not likely to be resolved. However, 1,3-cyclohexadiene is probably a good model for S-cis-butadiene. Measurement of the absorption spectrum of the cyclic diene by the oxygen-perturbation technique shows singlet-triplet absorption starting at 535 nm corresponding to 52.5 kcal/mole. If the excitation energy of S-cis-butadiene is similar, energy transfer from biacetyl triplets to the cisoid diene can occur without supply of any energy from the environment. Figure 22-6 shows the accepted mechanism for dimerization of butadiene using biacetyl as a **photosensitizer.**

The dimerization mechanism may seem to have been constructed artificially and to be hung together by a slender thread. However, interesting confirmation is obtained by study of the relative amounts of the three dimeric products formed and comparison with results obtained using other sensitizers having higher triplet excitation energies. Typical results are shown in Table 22-1.

TABLE 22-1   **Dimerization of 1,3-Butadiene**

	Triplet energy of sensitizer, kcal/mole	% distribution of products		
Sensitizer		cis-Divinyl-cyclobutane	trans-Divinyl-cyclobutane	Vinyl-cyclohexene
Acetophenone	74	19	78	3
Benzophenone	69	18	80	4
Biacetyl	55	13	52	35

Acetophenone and benzophenone both have triplet excitation energies high enough to excite S-trans-diene molecules to their triplet states. When an energy donor has energy as much as 5 kcal/mole in excess of that required to promote a potential energy acceptor to its lowest excited state, energy transfer apparently occurs on every collision, or nearly every collision. Consequently, high-energy triplets should transfer their energy predominantly to S-trans-diene molecules.

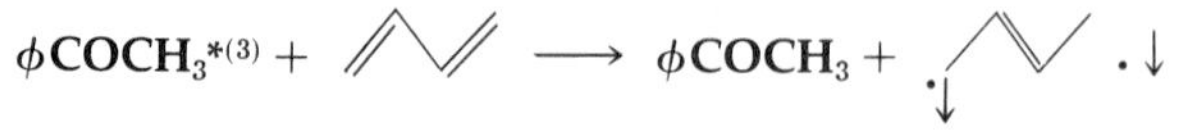

$$\phi COCH_3{}^{*(3)} + \diagup\hspace{-3pt}\diagdown\hspace{-3pt}\diagup\hspace{-3pt}\diagup \longrightarrow \phi COCH_3 + \cdot\diagup\hspace{-3pt}\diagdown\hspace{-3pt}\diagup\cdot\downarrow$$

Acetophenone S-trans-
  triplet   diene triplet

The results in Table 22-1 can be explained by assuming that different product mixtures are obtained from the stereoisomeric diene triplets. Consideration of the results makes this seem very reasonable. Reaction of the trans triplet with an S-trans-diene molecule should preferentially form a biradical in which both of the allylic units have the trans configuration.

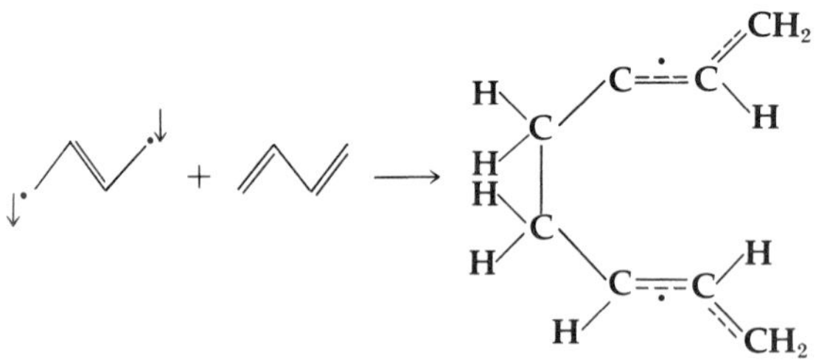

Trans,trans
biradical

The trans,trans biradical cannot cyclize to form a cis-cyclohexene derivative without undergoing a change in the configuration of at least one of the allylic units. Independent study of the behavior of allylic radicals shows that they maintain their steric configuration for appreciable periods of time. Consequently it is not surprising that reaction occurring by way of the trans triplet gives little or no vinylcyclohexene.† Reaction of cis triplets with S-trans diene should give a biradical in which one allylic unit has the cis configuration.

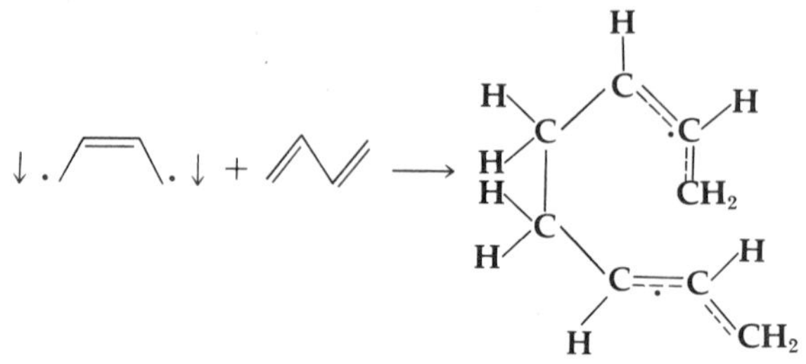

Trans,cis biradical

† The small amount formed with high-energy sensitizers may arise from reaction of cis triplets with S-trans diene and from reaction of trans triplets with S-cis diene.

The trans,cis biradical can close smoothly to vinylcyclohexene, but should also give the vinylcyclobutanes in accordance with the experimental facts.

Energy transfer is often a critical factor in photochemistry. A large number of important reactions can be carried out by **photosensitization**, the name given to reactions in which energy is absorbed by a sensitizer and then transferred to a reactive substrate. Energy acceptors may also **quench** reactions, by accepting energy from excited states before they undergo reactions that would occur in the absence of the quencher.

22-3   CHARACTERISTICS OF PHOTOREACTIONS

The study of photochemical reactions and their use in preparative and synthetic chemistry requires some reorientation of thinking for those who are accustomed to thinking about thermal reactions. In the following paragraphs we shall outline a few of the operational concepts and practices of photochemistry.

Unimolecular Reactions

In a unimolecular reaction an electronically excited state decays chemically without involving other molecules. Two common forms of unimolecular decay are: (1) *photolysis,* in which the molecule fragments into smaller parts such as ions or free radicals, and (2) *molecular rearrangements.* A unimolecular reaction of an excited state can usually be regarded as a nonradiative decay and is fundamentally similar to decay that restores the molecule to its original starting state. For example, chemical decay of a triplet state may be inhibited by the necessity of electron-spin pairing. Dissociative reactions, such as ejection of an electron or fragmentation to free radicals, may turn out to be especially favored paths because spin pairing is not required. In other cases, decay may occur in steps, with the first irreversible change being production of a short-lived intermediate in which electron spins can remain unpaired.

Bimolecular Reactions

In a bimolecular photoreaction an excited state usually reacts with a second molecule in its ground state. Reactions between excited states are only rarely encountered in ordinary photochemical experiments because of their low concentration. They may become important under conditions of high light intensity where relatively high concentrations of excited states can be generated. The ground-state molecule that is the second partner in the bimolecular photoreaction may be an unexcited form of the reactive species, or may be some other constituent of the system.

We expect spin conservation to be an important factor, but hard documentation of the presumption is scanty. The mechanism for sensitized dimerization of butadiene (page 883) illustrates the spin conservation principle. Addition of a diene triplet to a ground-state singlet is believed to result in the formation of only one bond in the first step. This allows spin to be conserved, since the intermediate biradical can be a triplet. Spin inversion can then occur as a separate step unaccompanied by bond formation.

$$
\begin{array}{l}
CH_2-CH{=}\dot{C}H{=}CH_2 \\
| \\
CH_2-CH{=}CH{=}CH_2
\end{array}
\quad \xrightarrow[\text{inversion}]{\text{Spin}} \quad
\begin{array}{l}
CH_2-CH{=}\dot{C}H{=}CH_2 \\
| \\
CH_2-CH{=}\dot{C}H{=}CH_2
\end{array}
\quad \xrightarrow{\text{Cyclization}} \quad \text{products}
$$

Triplet biradical	Singlet biradical

Quantitative Measurements

The quantitative measurement of most importance in characterization of a photochemical reaction is the **quantum yield**.

$$
\Phi = \frac{\text{molecules reacting}}{\text{light quanta absorbed}}
$$

Quantum
  yield

Measurement of the extent of reaction is the same kind of analytical problem as is encountered in quantitative monitoring of any reaction. Measurement of the amount of light absorbed can be done by the same technique as is used in absorption spectrophotometry by using a photoelectric cell placed behind the reaction cell. However, a much more convenient technique involves the use of a chemical **actinometer**. The actinometer is a sample containing known amounts of photosensitive reactants which undergo reaction with a known quantum yield. A frequently used actinometer is the ferrioxalate complex.

$$
\left(\begin{matrix} O \\ \ \end{matrix} C-O \\ O \\ C-O \end{matrix}\right)_3 Fe^{3-} + H_2O \xrightarrow{h\nu} CO_2 + CO + 2HO^- + \left(\begin{matrix} O \\ C-O \\ O \\ C-O \end{matrix}\right)_2 Fe^{--}
$$

Trisoxalatoiron(III)	Bisoxalatoiron(II)

The quantum yield depends on the wavelength of the exciting light, but varies very little in the wavelength region below 400 nm. The quantum yield in this region is 1.20 to 1.25. The value is a little disconcerting because it would appear that the maximum value should be 1.00. One can think of many reasons for the higher-than-expected quantum yield, but most of them can be ruled out because the value is independent of concentration. The most likely explanation is that the stoichiometry of photolysis is not exactly as shown in the equation above. Since the actual measurement is the amount of **Fe(II)** produced, there is probably some other decomposition path that leads to formation of excess **Fe(II)**.

A report of a "yield" in the usual sense has little meaning in photochemistry, unless the conditions of irradiation are very carefully specified. The wavelength and intensity of the light, the absorption spectrum and concentration of the absorbing species, the concentrations of other substances, and the thickness of the sample in which light is absorbed are all important variables. In qualitative work the most important factor is the cleanness of the reaction, that is, the extent to which the reported products predominate over others. This can be ascertained by carrying out a careful material balance when the reaction mixture is worked up. Frequently qualitative reports of photochemical experiments prove difficult to duplicate because of lack of information concerning experimental conditions, or because of the controlling influence of small amounts of impurities which act as sensitizers or quenchers. Confusing data are sometimes the result of control of the reaction by primary photoproducts which either quench excited states or prevent their formation by competitive light absorption. Where this occurs a reaction which seems very clean when studied in dilute solution can become a total failure when scaled up by increasing the concentration of the photoreactive substrate.

22-4   TYPICAL PHOTOREACTIONS

Hundreds of photoreactions of organic compounds are known, and the list grows longer every year. We shall make no attempt to give a comprehensive presentation but shall give representative examples of various types of reactions.

Photoreduction

The excited triplet states of many aldehydes and ketones abstract hydrogen atoms from reactive substrates producing free radicals. The ultimate products of the reactions are those expected from coupling and disproportionation of the radicals. Figure 22-7 shows the entire mechanism for photoreduction of benzophenone by toluene. In simple aldehydes and ketones the lowest excited states have the n,π^* configuration in which a nonbonding oxygen electron has been excited to a π^* orbital (page 247). The electronic configuration around the oxygen is somewhat like that of alkoxy radicals, which are also reactive in hydrogen atom abstraction.

R
$\diagdown$ π^* n
    C═Ö:   R—O̤:
$\diagup$
R

n, π^* state   An alkoxy
of a ketone   radical

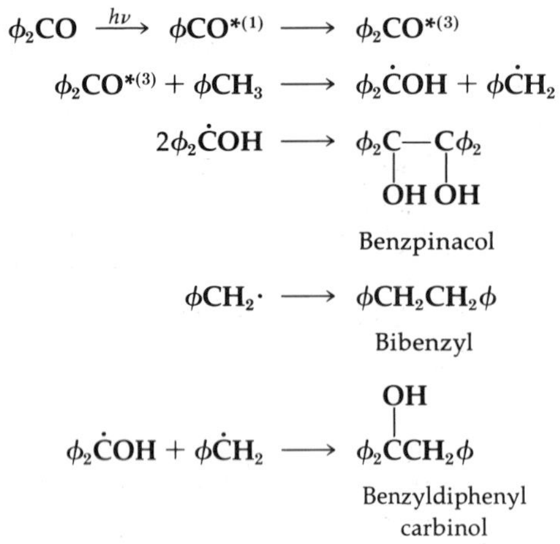

$$\phi_2CO \xrightarrow{h\nu} \phi CO^{*(1)} \longrightarrow \phi_2CO^{*(3)}$$

$$\phi_2CO^{*(3)} + \phi CH_3 \longrightarrow \phi_2\dot{C}OH + \phi\dot{C}H_2$$

$$2\phi_2\dot{C}OH \longrightarrow \phi_2\underset{\underset{OH}{|}}{C}\!-\!\underset{\underset{OH}{|}}{C}\phi_2$$

Benzpinacol

$$\phi CH_2\cdot \longrightarrow \phi CH_2CH_2\phi$$

Bibenzyl

$$\phi_2\dot{C}OH + \phi\dot{C}H_2 \longrightarrow \phi_2\underset{\underset{OH}{|}}{C}CH_2\phi$$

Benzyldiphenyl
carbinol

FIGURE 22-7   Photoreduction of benzophenone by toluene

There are many minor variations in the details of photoreduction reactions. For example, irradiation of a solution of benzophenone in isopropyl alcohol with light of low intensity leads to formation of only benzpinacol and acetone.

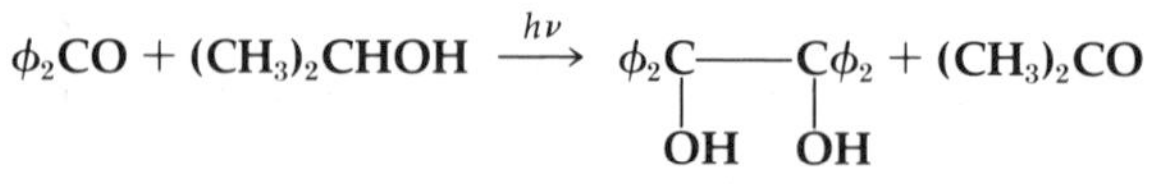

$$\phi_2CO + (CH_3)_2CHOH \xrightarrow{h\nu} \phi_2\underset{\underset{OH}{|}}{C}\!-\!\underset{\underset{OH}{|}}{C}\phi_2 + (CH_3)_2CO$$

A key step in the mechanism is the transfer of hydrogen atoms from 2-hydroxy-2-propyl radicals to benzophenone.

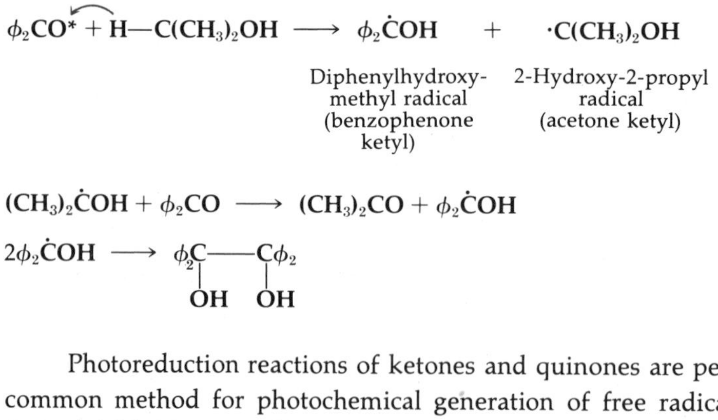

$$\phi_2CO^* + H\!-\!C(CH_3)_2OH \longrightarrow \phi_2\dot{C}OH \quad + \quad \cdot C(CH_3)_2OH$$

Diphenylhydroxy-   2-Hydroxy-2-propyl
methyl radical   radical
(benzophenone   (acetone ketyl)
ketyl)

$$(CH_3)_2\dot{C}OH + \phi_2CO \longrightarrow (CH_3)_2CO + \phi_2\dot{C}OH$$

$$2\phi_2\dot{C}OH \longrightarrow \phi_2\underset{\underset{OH}{|}}{C}\!-\!\underset{\underset{OH}{|}}{C}\phi_2$$

Photoreduction reactions of ketones and quinones are perhaps the most common method for photochemical generation of free radicals in solution. Considerable use is made of the method for initiation of vinyl polymerization (Sec. 25-1) by light. A number of photoimaging processes have also been based on the method.

Not all carbonyl compounds undergo facile photoreduction. For example Michler's ketone and 2-acetylnaphthalene are not reduced with significant quantum yield when irradiated in isopropyl alcohol solution.

$$(CH_3)_2N-\text{[benzene ring]}-\overset{\overset{C}{\underset{O}{}}}{}-\text{[benzene ring]}-N(CH_3)_2$$

Michler's ketone
[*p,p'-bis*(dimethylamino)benzophenone]

COCH$_3$

2-Acetylnaphthalene

The contrast to the behavior of readily reducible ketones such as benzo-phenone and acetophenone is striking. One possible explanation might be the failure of the unreactive ketones to cross over to triplet states. However, this is easily shown to be incorrect because both Michler's ketone and 2-acetyl-naphthalene have strong phosphorescence spectra and **flash kinetic**[†] studies show that both give good high yields of long-lived triplets.

Study of the phosphorescence spectra indicates another explanation. For example, the phosphorescence of 2-acetylnaphthalene is very similar to that of other 2-substituted naphthalenes and not like that of simple ketones.[‡] A straightforward interpretation assigns to the lowest triplet state a π,π^* con-figuration, in which the excitation energy is located in the naphthalene nucleus. Similar comparisons indicate that Michler's ketone triplet also has a π,π^* configuration. We believe that the n,π^* state of Michler's ketone lies at about the same energy as that of benzophenone. However, conjugative interaction of the nonbonding electrons of the amino groups with the aromatic rings lowers the energy of the π,π^* state so that it becomes the lowest triplet of the system. The relationships are shown schematically in Fig. 22-8.

We conclude that the electronic configuration of an excited state, not just the fact that it has plenty of available energy, is an important controlling factor in determining its chemical reactivity. In particular, the electron deficiency created at the oxygen atoms of n,π^* states of carbonyl compounds appears to be important in hydrogen abstraction reactions.

Photolytic Reactions of Ketones

Photochemical cleavage of molecules usually produces free radicals. An interesting example is photolysis of simple aldehydes and ketones. The most thoroughly studied example is acetone which is photolyzed to methyl and acetyl radicals in the vapor phase. At room temperature products derived from the acetyl radical are obtained, but at temperatures above 100° the rate of loss of carbon monoxide becomes so rapid that nearly every acetone molecule decomposed produces two methyl radicals.

$$CH_3COCH_3 \xrightarrow{h\nu} \cdot CH_3 + CH_3\dot{C}O \longrightarrow CO + \cdot CH_3$$

[†] In flash kinetic experiments a solution is irradiated for a short interval, usually of the order of micro-seconds, with an intense pulse of light. Sufficiently high concentrations of triplet states are produced to allow them to be observed by direct measurement of their absorption spectra. By monitoring the decay of the triplet-triplet absorption, the kinetics of disappearance of the triplets can be studied.

[‡] Simple carbonyl compounds which emit from n,π^* excited states usually show vibrational structure characterized by resolvable maxima separated by about 1600 cm^{-1}, the carbonyl stretching frequency (Sec. 7-5).

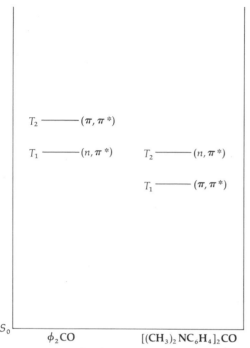

T_2 ——— (π, π^*)

T_1 ——— (n, π^*) T_2 ——— (n, π^*)

T_1 ——— (π, π^*)

S_0

$\phi_2 CO$ $[(CH_3)_2 NC_6H_4]_2 CO$

FIGURE 22-8 **Triplet excitation energies of benzophenone and Michler's ketone**

Irradiation of acetone in solution produces only miniscule amounts of acetyl radicals, an observation that confused early workers. Quantitative study of the gas-phase reaction, however, shows the reasons very clearly. The quantum yield for acetone photolysis increases as the wavelength of the exciting light is decreased. From this we infer that the vibrationally deactivated excited states do not fragment with high efficiency. In solution vibrational deactivation occurs very rapidly. However, other ketones which can fragment to give more stable radicals do photolyze even in solution.

$$\phi COC(CH_3)_3 \xrightarrow{h\nu} \phi \dot{C}O + (CH_3)_3 C\cdot$$

t-Butylphenylketone   *t*-Butyl
  (pivalophenone)   radical

The excitation energy of the lowest triplet state of acetone is 75 kcal/mole, which is just about the energy required to break the carbon-carbon bond in acetone. Consequently, the higher efficiency of bond breaking in vibrationally excited molecules is not surprising. One consequence of this closely balanced energy requirement is very high selectivity in the direction of fragmentation of unsymmetrical ketones, which dissociate to give the more stable of the two possible alkyl radicals.

$$CH_3 COCH_2 CH_3 \longrightarrow CH_3 \dot{C}O + CH_3 CH_2\cdot$$

$$CH_3 COCH(CH_3)_2 \longrightarrow CH_3 \dot{C}O + (CH_3)_2 CH\cdot$$

Another dissociative reaction of ketones is the photoelimination reaction known as the **Norrish type II split**. The reaction occurs with ketones bearing a hydrogen atom attached to a γ-carbon atom, and is analogous to the similar fate of the radical in mass spectral fragmentation shown on page 283.

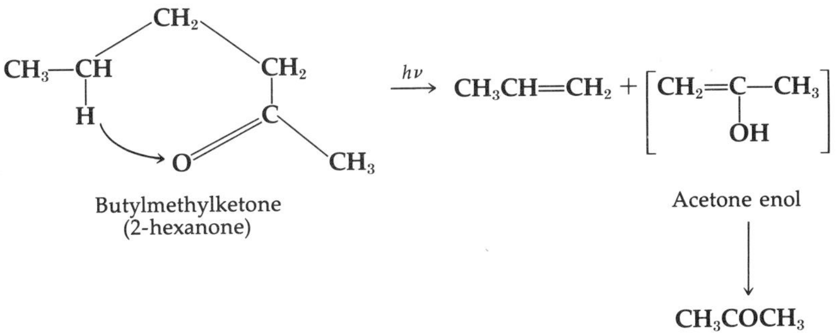

Butylmethylketone
(2-hexanone)   Acetone enol

CH_3COCH_3

Cyclic ketones undergo a number of reactions which are closely related to the photolysis of the open-chain analogs. The principal products can usually be rationalized by a biradical mechanism. Cyclohexanone provides a representative case.

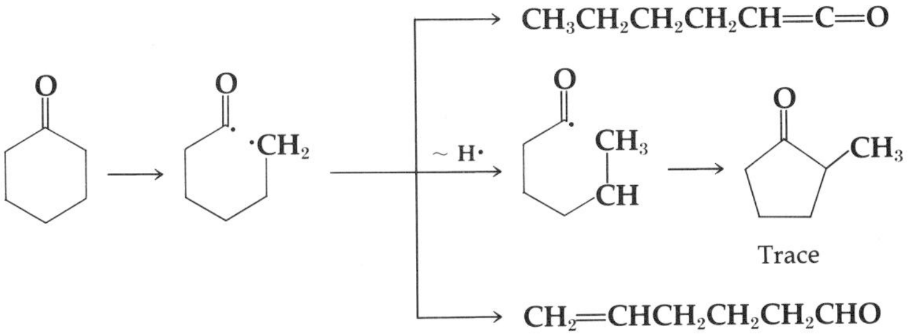

Trace

Compounds containing weak bonds often undergo photolysis also to give free radicals initially. The following are a somewhat random set of examples.

$$(\phi CO_2)_2 \xrightarrow{h\nu} 2\phi CO_2\cdot \longrightarrow \phi\cdot + CO_2$$

Benzoyl   Benzoyloxy
peroxide   radicals

$$(CH_3)_3C\!-\!N\!=\!N\!-\!C(CH_3)_3 \xrightarrow{h\nu} 2(CH_3)_3C\cdot + N_2$$

Azoisobutane

$$\phi I \xrightarrow{h\nu} \phi\cdot \quad + \quad I\cdot$$
$$\downarrow \qquad\qquad \downarrow$$
$$\phi\!-\!\phi \qquad\quad I_2$$

The *fate* of the free radicals formed in photolytic experiments depends upon the other species present in the reaction mixture. In the absence of reactive **radical scavengers** the principal products are those formed in radical coupling and disproportionation reactions. If reactive hydrogen donors are present, reduction products are formed. If reactive unsaturated compounds are

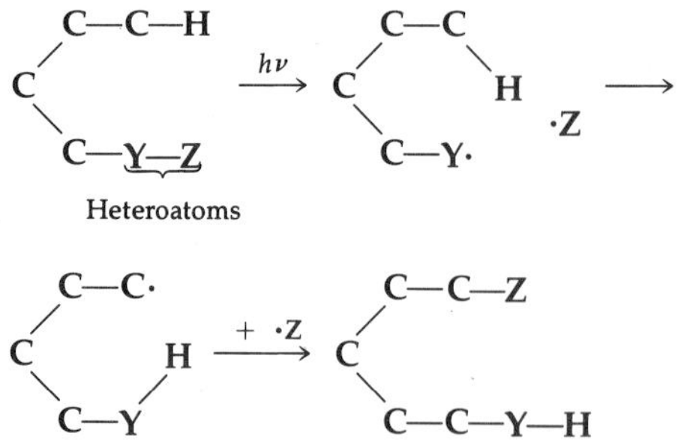

C—C—H   C—C
/   /   \
C $\xrightarrow{h\nu}$ C   H $\longrightarrow$
\   \   /   ·Z
C—Y—Z   C—Y·

Heteroatoms

C—C·   C—C—Z
/   /
C   H $\xrightarrow{+ \cdot Z}$ C
\   /   \
C—Y   C—C—Y—H

FIGURE 22-9 Generalized pathway for photolytic reactions at unactivated sites

present, the radicals add to them to produce new radicals. Essentially all of the reactions described in Chap. 20 can be initiated photolytically.

Reactions at Unactivated Centers

Saturated sites without some particular activation, such as adjacent carbonyl or olefin, cannot usually be caused to undergo reactions except in an indiscriminate and nonuseful fashion. For this reason synthesis design must always accommodate *some* functional group at any skeletal site at which a chemical change is to occur, even if no group appears there in the final product. There is a small group of photolytic reactions, however, which do serve to functionalize an unactivated, saturated site, and which accordingly have considerable synthetic value.

These reactions have certain features in common.

1 The starting material contains a functional group with a weak bond between two heteroatoms.
2 This bond is cleaved homolytically by irradiation (photolysis).
3 The radical produced abstracts a hydrogen from a saturated carbon atom of the skeleton four atoms away, in a cyclic six-membered transition.
4 The other heteroatom radical, released from the skeleton, now bonds to the new carbon radical; other radical scavengers, if present, will also attack the new radical. These steps are pictorialized in Fig. 22-9.

The two most common reactions of this kind are the photolyses of nitrites and of N-haloamines and amides, of which several examples are shown below. In the nitrite photolysis the initial aliphatic nitroso compounds usually tautomerize readily to oximes.

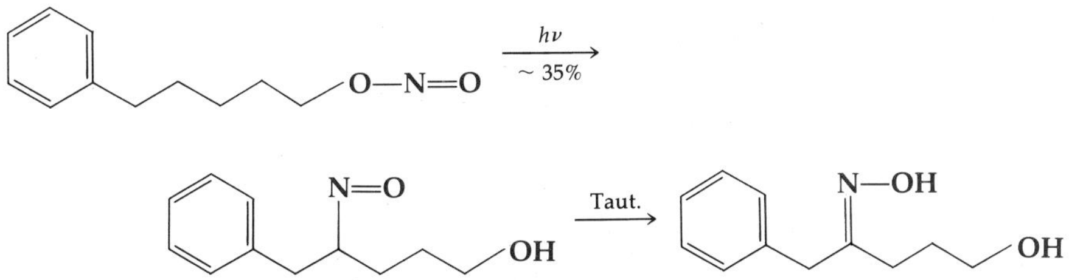

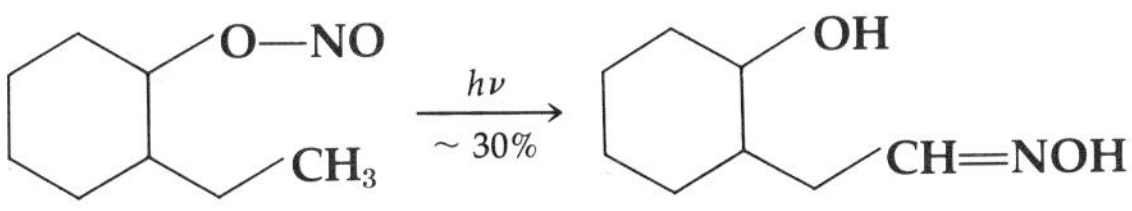

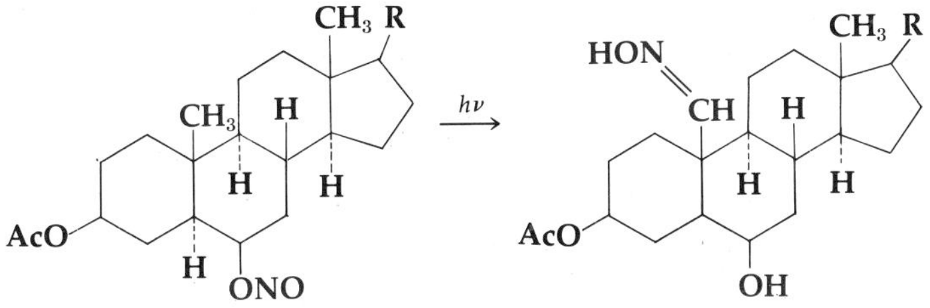

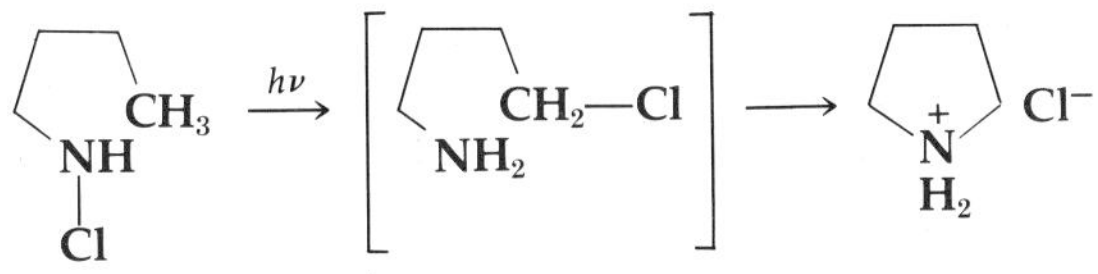

$\xrightarrow[\sim 35\%]{h\nu}$

O—N=O

N=O   Taut.   N—OH

OH $\xrightarrow{\text{Taut.}}$ OH

O—NO   OH

$\xrightarrow[\sim 30\%]{h\nu}$

CH$_3$ CH=NOH

CH$_3$ R   CH$_3$ R

CH$_3$ H $\xrightarrow{h\nu}$ HON   CH   H

H   H   H   H

AcO   AcO

H   ONO   OH

CH$_3$ $\xrightarrow{h\nu}$ $\left[\quad \text{CH}_2\text{—Cl} \right]$ $\longrightarrow$ $\overset{+}{N}$ Cl$^-$

NH   NH$_2$ H$_2$

Cl

As the third example shows, the molecular conformation must allow a feasible cyclic six-membered transition. The identical case has been carried out with the hypochlorite instead of nitrite and may serve to illustrate the ideality of this structure for reaction owing to rigidly cis, diaxial substituents.

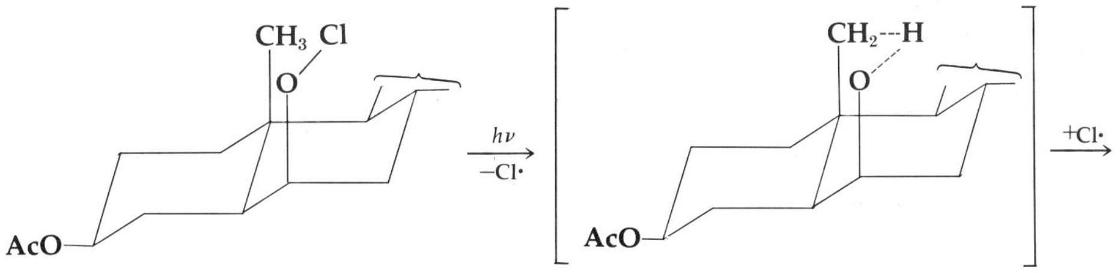

CH$_3$ Cl $\xrightarrow[-\text{Cl}\cdot]{h\nu}$ CH$_2$---H $\xrightarrow{+\text{Cl}\cdot}$

O   O

AcO   AcO

Hypochlorite

Cl   CH$_2$ CH$_2$

H   O

O $\xrightarrow[(-\text{HCl})]{\text{Base}}$

AcO   AcO

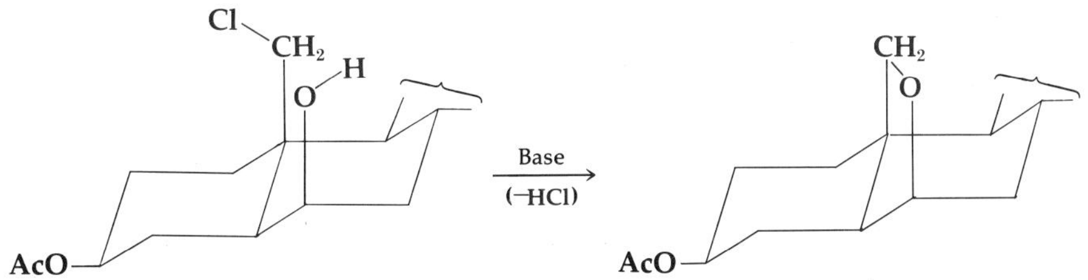

Chlorohydrin   Cyclic ether

PROBLEM 22-1

Rationalize the course of the following reactions.

a

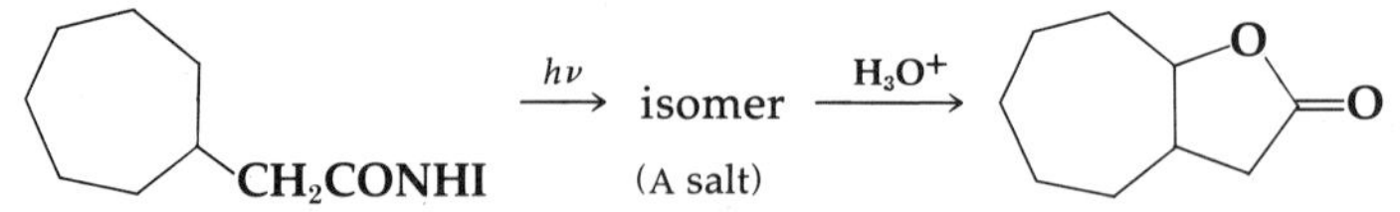

$$\xrightarrow{\quad h\nu \quad} \text{isomer} \xrightarrow{\quad H_3O^+ \quad}$$

(A salt)

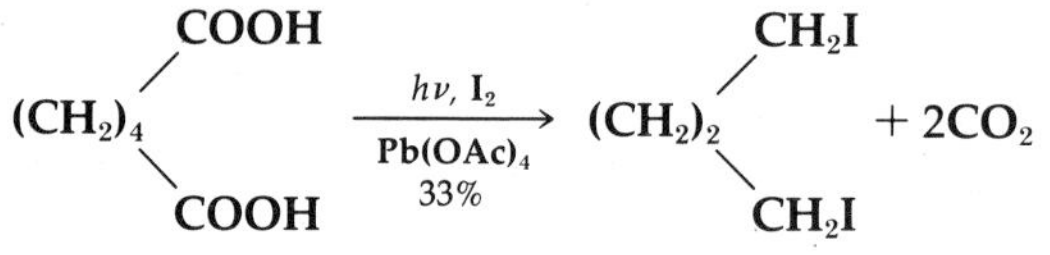

b

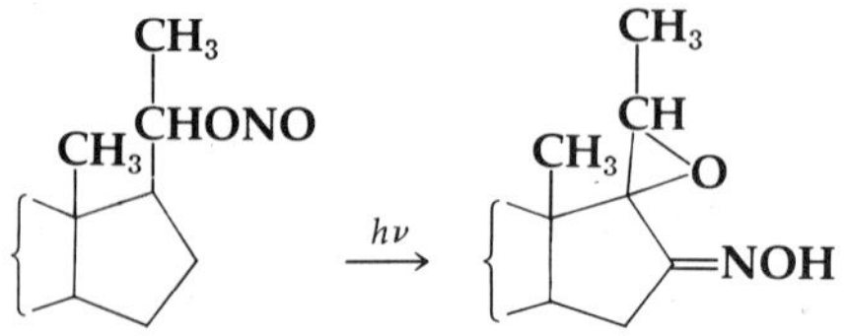

$$\xrightarrow{\quad h\nu \quad}$$

c $(CH_2)_4 \begin{subarray}{l} \nearrow COOH \\ \searrow COOH \end{subarray} \xrightarrow[\substack{Pb(OAc)_4 \\ 33\%}]{\quad h\nu,\ I_2 \quad} (CH_2)_2 \begin{subarray}{l} \nearrow CH_2I \\ \searrow CH_2I \end{subarray} + 2CO_2$

d $CH_3(CH_2)_5CH_2N_3 \xrightarrow{\quad h\nu \quad} CH_3(CH_2)_5CH{=}NH + n{-}C_3H_7{-}$

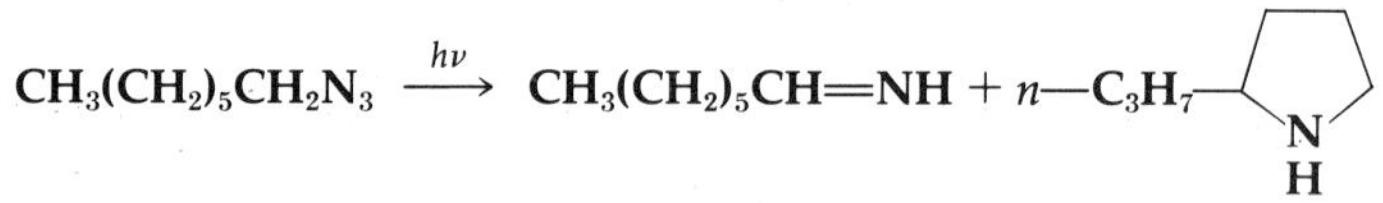

45%   15%

Cycloaddition Reactions

Many unsaturated compounds undergo photochemical cycloaddition reactions. The photosensitized dimerization of butadiene described earlier is a
typical example (page 884). The reaction is observed with many conjugated
dienes, and a considerable spectrum of products is normally produced. In
general, one expects formation of all products that can be produced by cyclization reactions of biradicals containing two allylic radical units. When the
reactions are concerted, however, no radical intermediates are formed and
reaction follows the rules for pericyclic reactions described in Chap. 21.

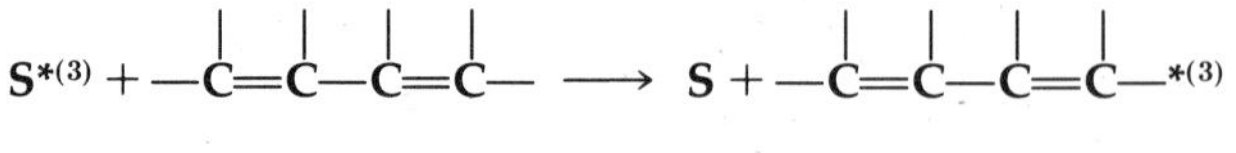

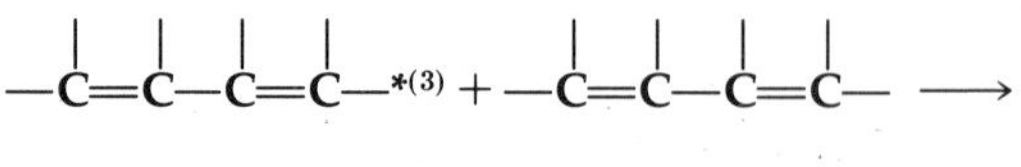

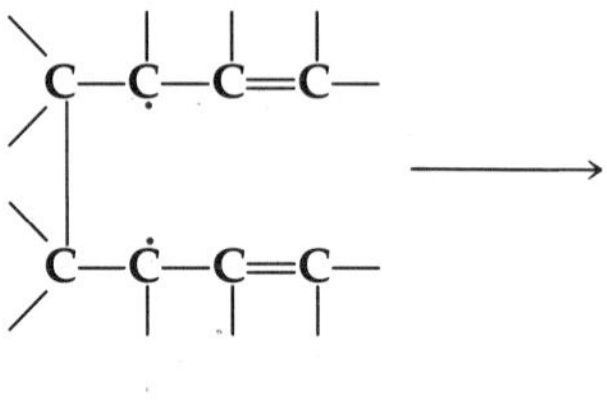

$$S^{*(3)} + {-}\overset{|}{C}{=}\overset{|}{C}{-}\overset{|}{C}{=}\overset{|}{C}{-} \longrightarrow S + {-}\overset{|}{C}{=}\overset{|}{C}{-}\overset{|}{C}{=}\overset{|}{C}{-}^{*(3)}$$

$$-\overset{|}{C}{=}\overset{|}{C}{-}\overset{|}{C}{=}\overset{|}{C}{-}^{*(3)} + {-}\overset{|}{C}{=}\overset{|}{C}{-}\overset{|}{C}{=}\overset{|}{C}{-} \longrightarrow$$

$$\longrightarrow \text{Cyclobutanes, cyclohexenes, and cyclooctadienes}$$

The sensitized dimerization of isoprene and of 1,3-cyclohexadiene are typical photochemical cycloadditions.

$$2CH_2{=}\underset{}{\overset{CH_3}{\overset{|}{C}}}{-}CH{=}CH_2 \xrightarrow[\phi_2CO]{h\nu}$$

[2 + 2]

[4 + 2]

[4 + 4]

[2 + 2]

[4 + 2]

The value of triplet sensitizers is illustrated by diene dimerization. Irradiation of pure liquid, 1,3-cyclohexadiene, with light absorbed by the diene leads to no dimerization. The only photoproduct formed in significant yield is 1,3,5-hexatriene, formed by concerted electrocyclic reaction (Sec. 21-1).

 $\xrightarrow{h\nu} CH_2{=}CHCH{=}CHCH{=}CH_2$

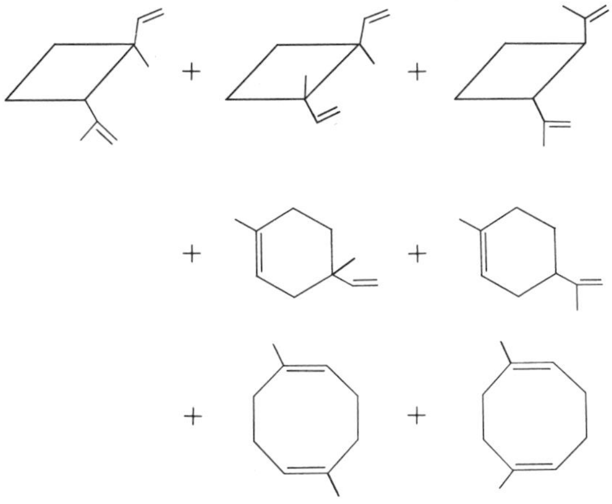

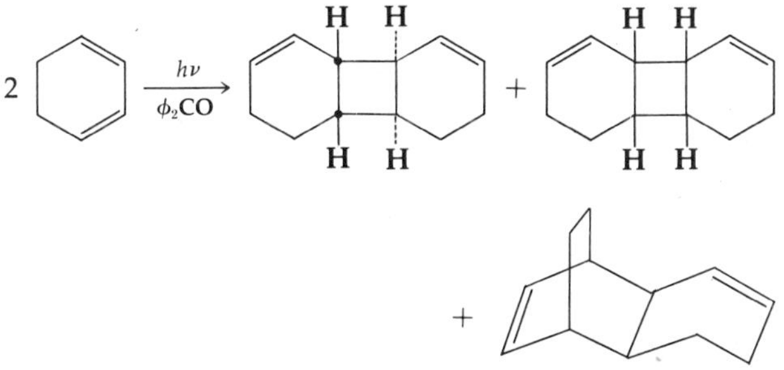

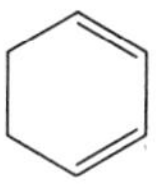

The excited singlet state formed by direct absorption of light has entirely different chemistry than the excited triplets formed by energy transfer from sensitizer triplets. Furthermore, the excited hexadiene triplets must decay rapidly to the ground singlet state without going through triplets because they have no fluorescence and the isomerization to hexatriene has a quantum yield of only about 0.1. In the vapor phase a number of other products are formed in addition to hexatriene.

Another interesting cyclodimerization is observed with 2-cyclohexenone.

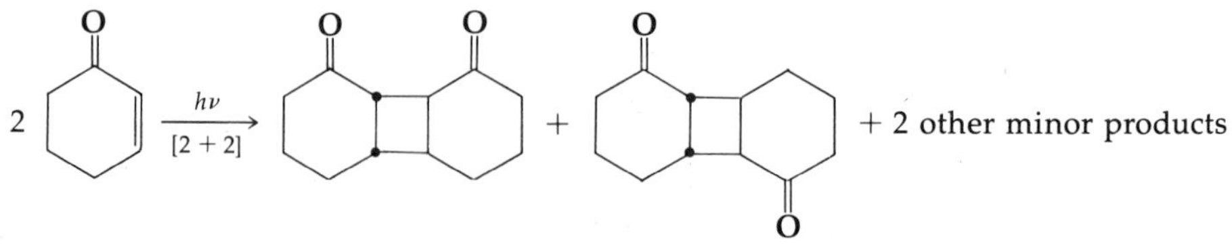

Cyclohexenone   A   B

In this case identical results are obtained in direct excitation and in experiments in which all the light is absorbed by a sensitizer and transferred as triplet excitation to the diene. Furthermore, both the direct and sensitized reactions are quenched in the same way by addition of piperylene (1,3-pentadiene). The diene undergoes cis $\rightleftharpoons$ trans isomerization, just as it does when triplet energy is transferred from a common triplet sensitizer such as benzophenone. A clear inference is that enone triplets are formed very rapidly by intersystem crossing. The mechanism seems to be as follows:

$$K \xrightarrow{h\nu} K^{*(1)} \longrightarrow K^{*(3)}$$

$$\text{or} \quad K + S^{*(3)} \longrightarrow K^{*(3)} + S$$

$$K^{*(3)} + K \longrightarrow \text{dimers}$$

$$K^{*(3)} + CH_2{=}CHCH{=}CHCH_3 \xrightarrow{\text{Quenching}} K + CH_2{=}CHCH{=}CHCH_3{}^{*(3)} \longrightarrow \text{cis and trans}$$

Piperylene

The relative amounts of the two dimers vary with polarity of the medium, with the quantum yield of the unsymmetrical dimer A being greatest in polar solvents. The excited state of cyclohexenone also adds to many other unsaturated compounds.

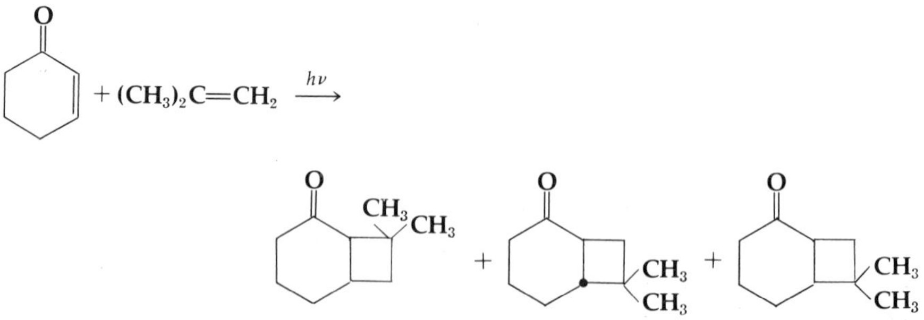

Carbonyl compounds, in which the lowest excited state has the n,π^* configuration, add to simple alkenes to form **oxetanes** (oxa-cyclobutanes).

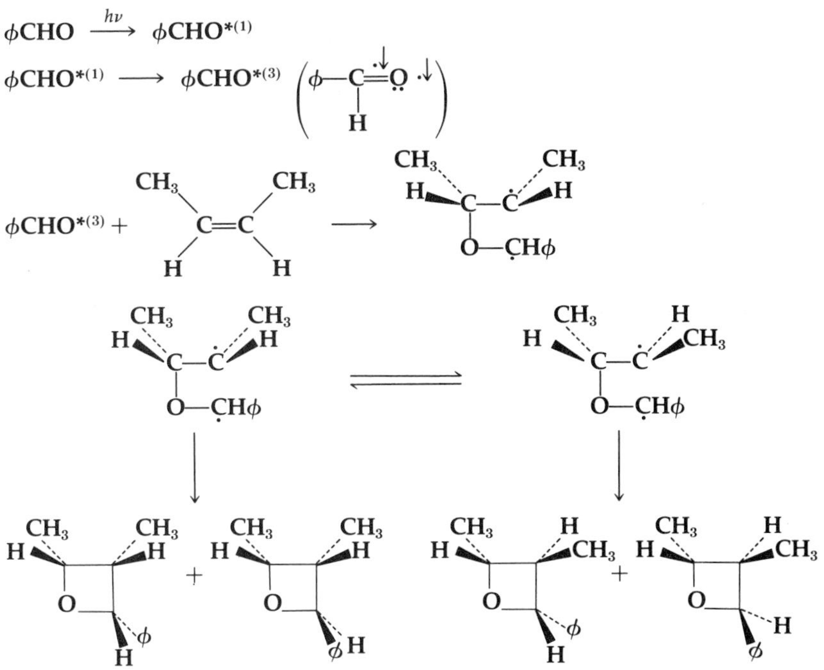

$$\phi CHO \xrightarrow{h\nu} \phi CHO^{*(1)}$$

$$\phi CHO^{*(1)} \longrightarrow \phi CHO^{*(3)} \quad \left(\phi - \overset{\downarrow}{\underset{H}{C}} = \overset{..}{\underset{..}{O}} \cdot \downarrow\right)$$

FIGURE 22-10   **Photocycloaddition of benzaldehyde to *cis*-2-butene**

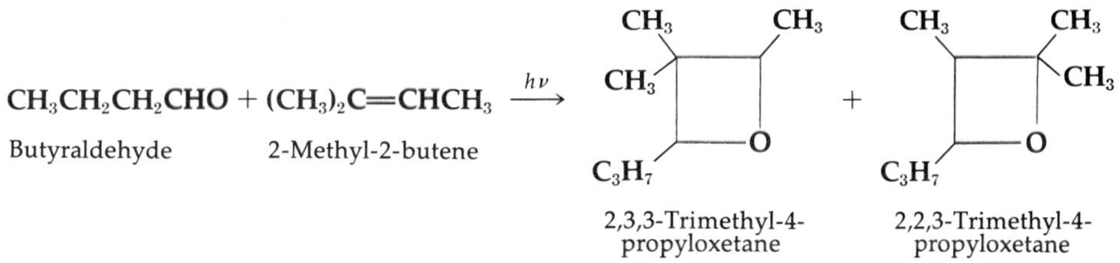

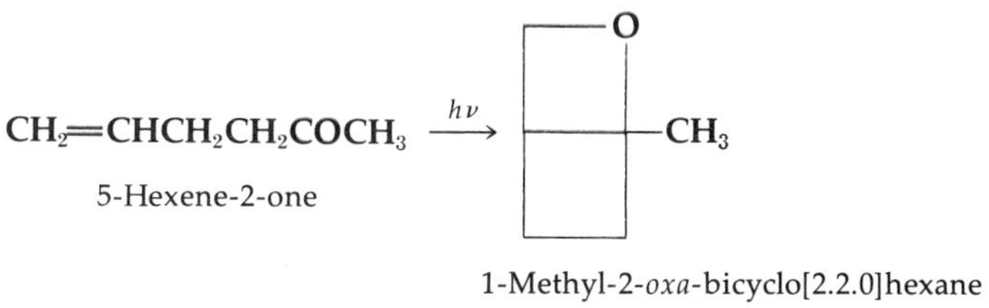

$$\underset{\text{Butyraldehyde}}{CH_3CH_2CH_2CHO} + \underset{\text{2-Methyl-2-butene}}{(CH_3)_2C=CHCH_3} \xrightarrow{h\nu}$$

2,3,3-Trimethyl-4-
propyloxetane

2,2,3-Trimethyl-4-
propyloxetane

$$\underset{\text{5-Hexene-2-one}}{CH_2=CHCH_2CH_2COCH_3} \xrightarrow{h\nu}$$

1-Methyl-2-*oxa*-bicyclo[2.2.0]hexane

Addition to either stereoisomer of 2-butene gives all four possible diastereomeric oxetanes, indicating that a biradical may be an intermediate and live long enough to undergo rotation about bonds at the center of geometric isomerism. The full mechanism believed to occur is shown in Fig. 22-10. The lack of stereochemical discrimination makes clear that this reaction is not concerted and so not subject to the rules of electrocyclic reactions in Chap. 21.

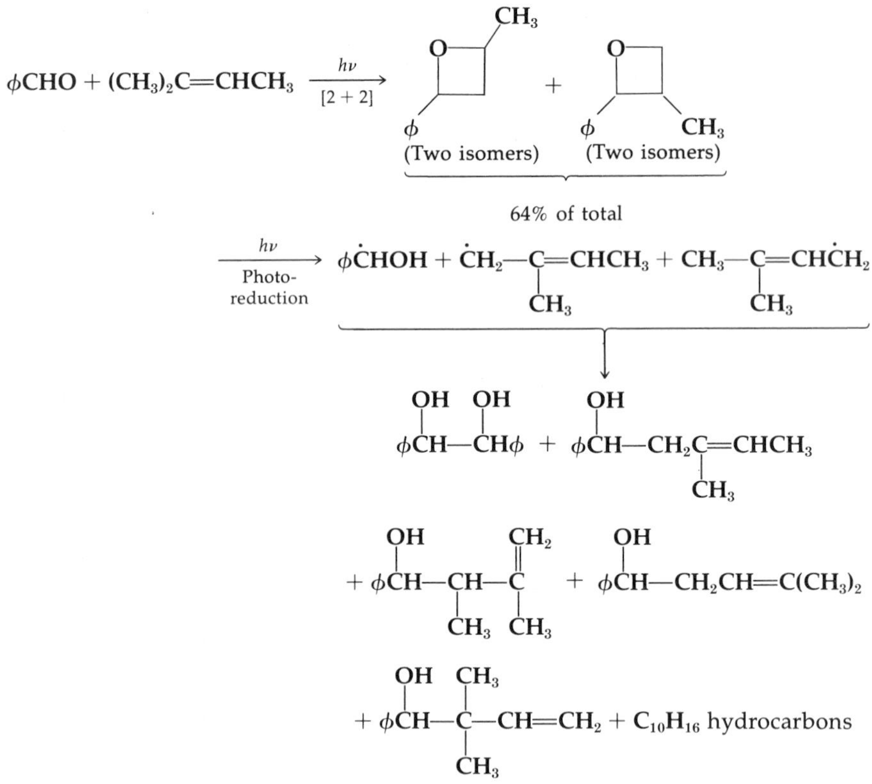

ϕCHO + (CH$_3$)$_2$C=CHCH$_3$ $\xrightarrow[[2+2]]{h\nu}$

(Two isomers)   (Two isomers)

64% of total

$\xrightarrow[\substack{\text{Photo-}\\\text{reduction}}]{h\nu}$ $\phi\dot{C}HOH + \dot{C}H_2$—C=CHCH$_3$ + CH$_3$—C=CH$\dot{C}H_2$

$\overset{|}{C}H_3$ $\overset{|}{C}H_3$

OH OH   OH
ϕCH—CHϕ + ϕCH—CH$_2$C=CHCH$_3$
                              $\overset{|}{C}H_3$

OH   CH$_2$ OH
+ ϕCH—CH—C + ϕCH—CH$_2$CH=C(CH$_3$)$_2$
      $\overset{|}{C}H_3$ $\overset{|}{C}H_3$

OH CH$_3$
+ ϕCH—C—CH=CH$_2$ + C$_{10}$H$_{16}$ hydrocarbons
      $\overset{|}{C}H_3$

FIGURE 22-11 Competitive cycloaddition and photoreduction reactions

Oxetane formation does not occur with conjugated dienes, because energy transfer becomes the predominant reaction between the carbonyl triplet and the unsaturated molecule. This is made reasonable by consideration of the triplet excitation energies of the compounds.

	Triplet excitation energy, kcal/mole
Simple aldehydes and ketones	68–75
Ethylene	~84
Butadiene	60

The results illustrate a general principle: *When energy transfer can occur without addition of energy from the environment, transfer usually takes preference over reactions requiring making and breaking of bonds.*

Photoreduction does become competitive with oxetane formation, since the allylic hydrogen atoms present in most olefinic substrates can be abstracted. The spectrum of reactions observed is summarized in Fig. 22-11.

Molecular Rearrangements

The excited states of many organic compounds decay to isomers of the starting structures, as is partly illustrated by the following examples. These various rearrangements may well have a variety of mechanisms.

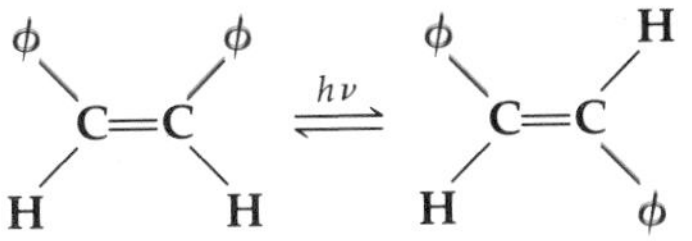
hν

cis-Stilbene *trans*-Stilbene

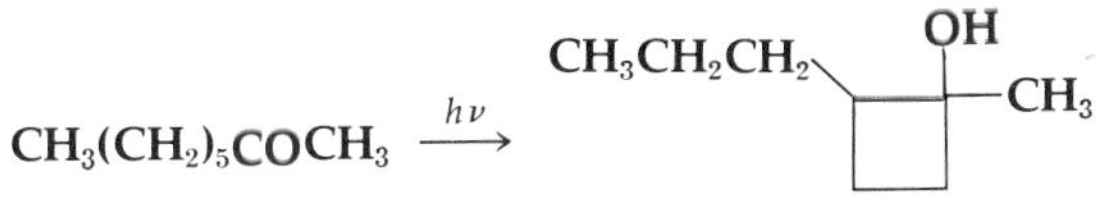
CH3(CH2)5COCH3
hν
CH3CH2CH2
OH
CH3

2-Octanone 1-Methyl-2-propylcyclo- -butanol

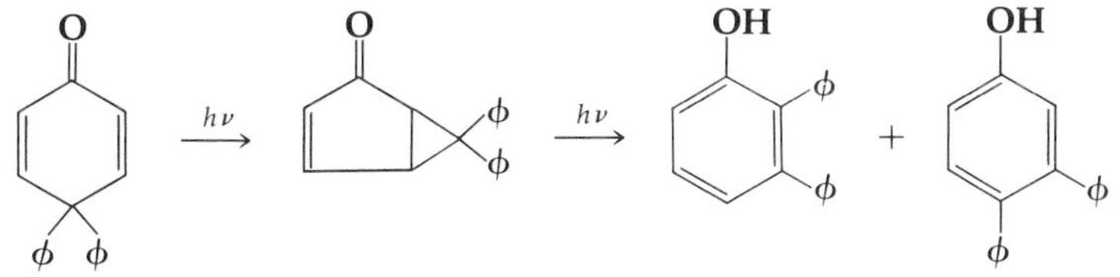
hν
hν
+

4,4-Diphenyl-2,5-cyclohexadienone 6,6-Diphenyl-[3,1,0] bicyclohexanone-2 2,3-Diphenylphenol 3,4-Diphenylphenol

The cis-trans isomerization of stilbene is a typical reaction of many olefinic compounds in which geometrical isomerism can exist. The reaction is effected both by direct light absorption by the isomerizable substrate and in sensitized reactions in which energy is delivered by sensitizers in their triplet states. High-quality theoretical calculations of the potential energies of the excited states of simple alkenes as a function of geometry indicate that both the lowest excited singlets and triplets should be most stable in twisted configurations as shown below. Decay of such states would be expected to give both cis and trans isomers of the planar ground state.

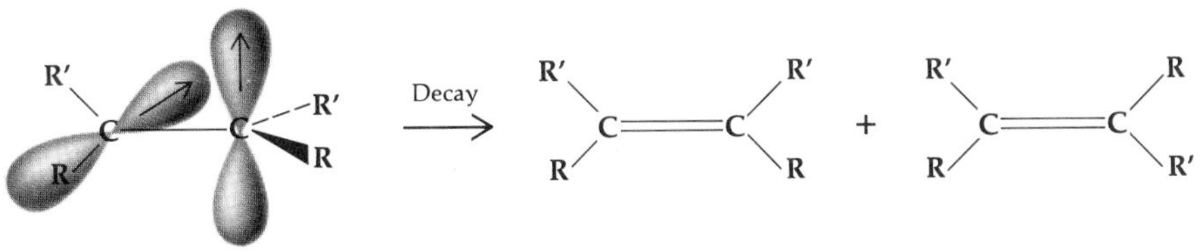
Decay
+

Cyclization of 2-octanone is typical of ketones having hydrogen attached to γ-carbon atoms and is directly competitive with the type II photoelimination reaction (page 893). There is a good deal of evidence to indicate that both reactions involve a common biradical intermediate. The cyclobutane formation is directly related to the bimolecular photoreductions discussed on page 889.

$$CH_3(CH_2)_5COCH_3{}^* \longrightarrow CH_3(CH_2)_2\overset{\overset{\displaystyle OH}{|}}{C}HCH_2CH_2\overset{\overset{\displaystyle OH}{|}}{C}CH_3$$

Type II split   Cyclization

$$CH_3(CH_2)_2CH{=}CH_2$$

$$+ \quad \overset{\overset{\displaystyle OH}{|}}{}$$

$$CH_3COCH_3 \longleftarrow [CH_2{=}\overset{\overset{\displaystyle OH}{|}}{C}CH_3]$$

The intervention of high-energy isomers as intermediates in many photochemical reactions is suspected and in many cases these intermediates are now being detected in photochemical experiments carried out in rigid glasses at low temperatures. An interesting example is found in the much studied photoisomerization reactions of tropolones and related compounds.

α-Tropolone methyl ether   A   B

The first rearrangement is simply an electrocyclic reaction and appears unexceptional. The second looks like a migration of a methoxy group, but use of derivatives shows that the carbon skeleton is actually reshuffled in the process. Various mechanisms have been suggested. Irradiation of first bicyclic ketone (A) in a glass consisting of a mixture of hydrocarbons cooled to liquid-nitrogen temperature (below $-180°$) leads to development of strong new infrared absorption bands, including a band at 2118 cm^{-1} characteristic of ketenes. Other bands in the spectrum are consistent with assignment of a bicyclic ketene structure (Fig. 22-12) to the intermediate. This product could be formed by photolytic [2 + 2] cycloaddition and reversion as outlined in Fig. 22-12.

When samples containing the transient species are warmed to $-70°$ the ketene undergoes rapid rearrangement in the dark to a mixture of the two bicyclic ketones ($A:B = 1:9$). These conversions are both Cope rearrange-

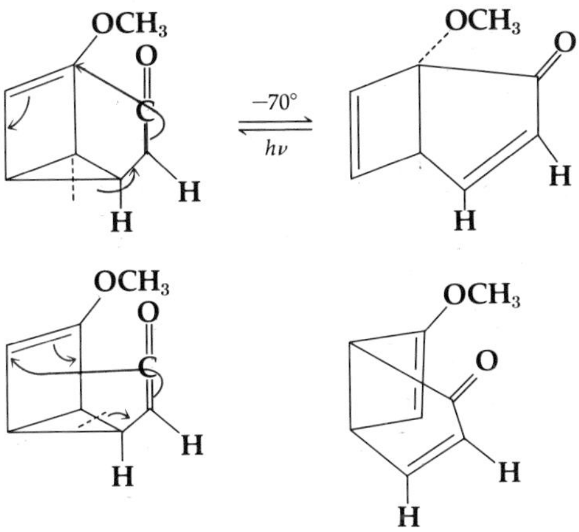

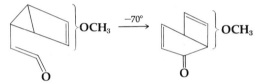

Alternative presentation as sigmatropic shift:

FIGURE 22-12   Interconversion of [3.2.0] bicycloheptadienones by way of an intermediate ketene

ments, i.e., thermal, symmetry-allowed [3,3] sigmatropic rearrangements. As is shown in Figure 22-12, the ketene can be generated photochemically from either bicyclic methoxy ketone (A or B).

Recognition of the role played by high-energy intermediates, such as the ketene in the above example, is certain to be of primary importance in establishment of the pattern of structure-reactivity relationships in photochemistry. The decay of an excited state to a transient species can become a key step in directing the overall course of a reaction. It is not at all surprising that the first "stopping point" in decay is often a very energetic molecule, since this will minimize the amount of electronic excitation energy that must be converted to vibrational energy in the internal conversion process.

PROBLEMS

22-2 Predict the products that you might expect from irradiation of a solution containing the indicated species. The light is absorbed by the material shown in color.

a $\phi COCH_3$ + [cyclohexene] $\xrightarrow{h\nu}$

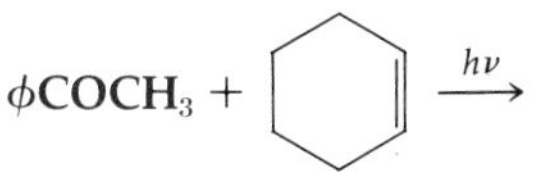

b $\phi COCH_3$ + [cyclohexadiene] $\xrightarrow{h\nu}$

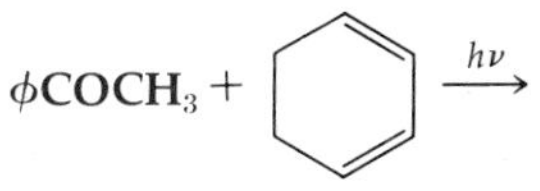

$$\text{c} \quad \text{(cycloheptanone)} \xrightarrow{h\nu}$$

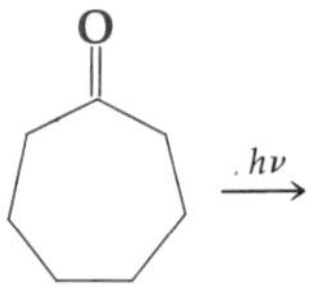

d $CH_3CH_2CH_2CH_2COCOCH_2CH_2CH_2CH_3 \xrightarrow{h\nu}$

e $\phi\overset{O}{\overset{\|}{C}}O-O\overset{O}{\overset{\|}{C}}\phi \xrightarrow{h\nu}$

f $CH_2{=}CHCH_2CH_2CH_2\overset{O}{\overset{\|}{C}}CH_3 \xrightarrow{h\nu}$

22-3 Solutions of *trans*-piperylene and 2-cyclohexenone in *n*-octane are irradiated in hexane solution. In a series of experiments with fixed concentration of cyclohexenone concentration and varying concentrations of piperylene, the following observations are recorded:

a The quantum yield of cyclohexenone dimers falls as the concentration of piperylene is increased.

b Vapor chromatograms show the formation of many new compounds including: (i) *cis*-piperylene; (ii) a group of hydrocarbons having much higher retention times than either the piperylenes and the solvent; and (iii) at least two compounds having retention times intermediate between those of the mixture (ii) and the cyclohexenone dimers.

c In an experiment with xanthone added as a photosensitizer, using light absorbed only by the sensitizer and involving a concentration of piperylene much higher than that of cyclohexenone, very nearly the same products are formed as are produced in experiments with the same concentrations of enone and diene but no sensitizer.

Formulate a mechanism to explain the observations and use it to predict possible structures of some of the compounds produced by piperylene.

22-4 Formulate reasonable mechanisms for the following reactions:

a $\xrightarrow{h\nu}$ $CH_3CH{=}CHCH_2O\overset{O}{\overset{\|}{C}}H + CH_3\overset{CHO}{\overset{|}{C}}HCH_2CHO + CH_3{-}\triangle$

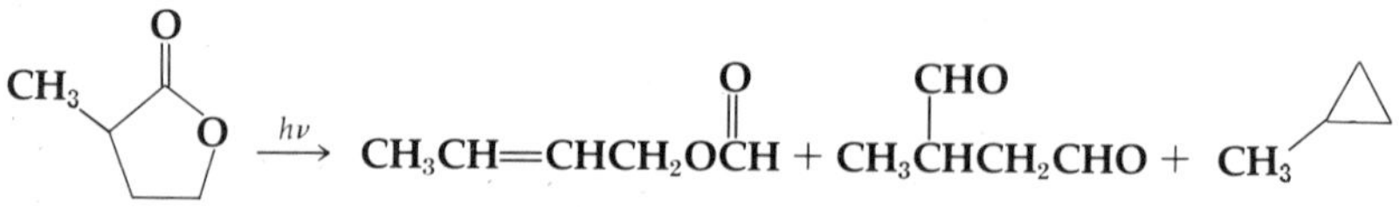

Major   Small   Trace

$+ CH_3CH{=}CH_2 + CH_2O + CO + CO_2$

Trace   Trace

b

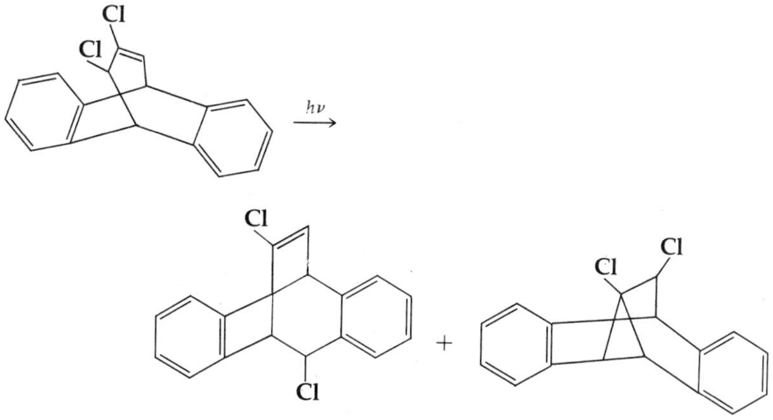

(Two isomers)

c

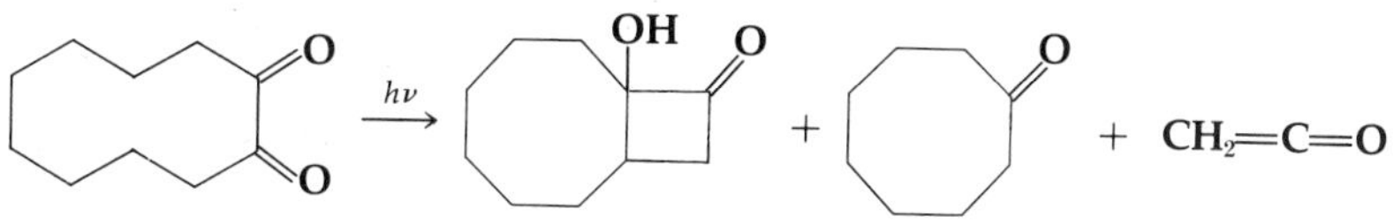

74%

9%

d $\phi COCH_2CH{-}CH\phi \xrightarrow{h\nu} \phi COCH_2CH_2CO\phi$

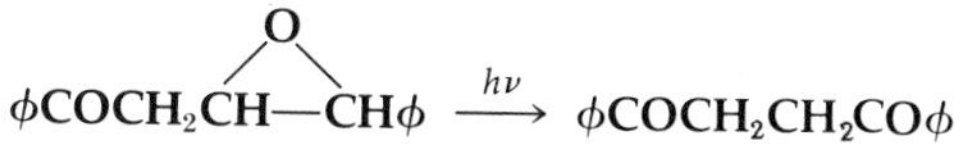

e

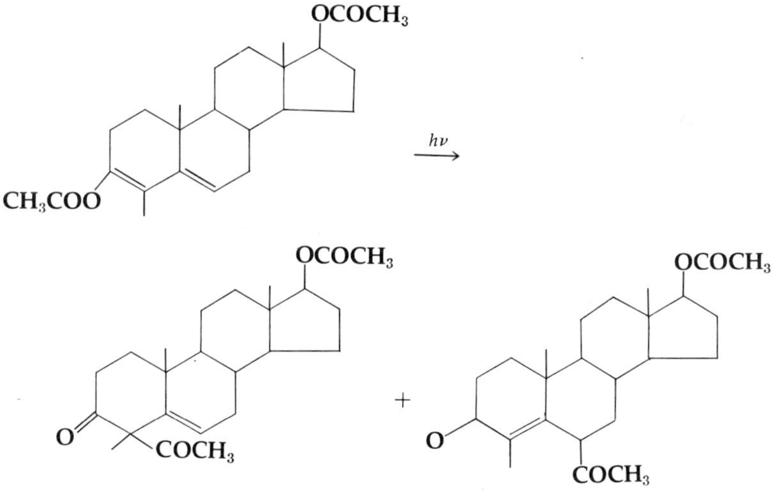

22-5 Dianhydrides are potentially of great use as starting materials for the synthesis of condensation polymers. For example, polyimides derived from pyromellitic anhydride have extraordinary stability at high temperatures, in addition to forming films and fibers with desirable mechanical properties.

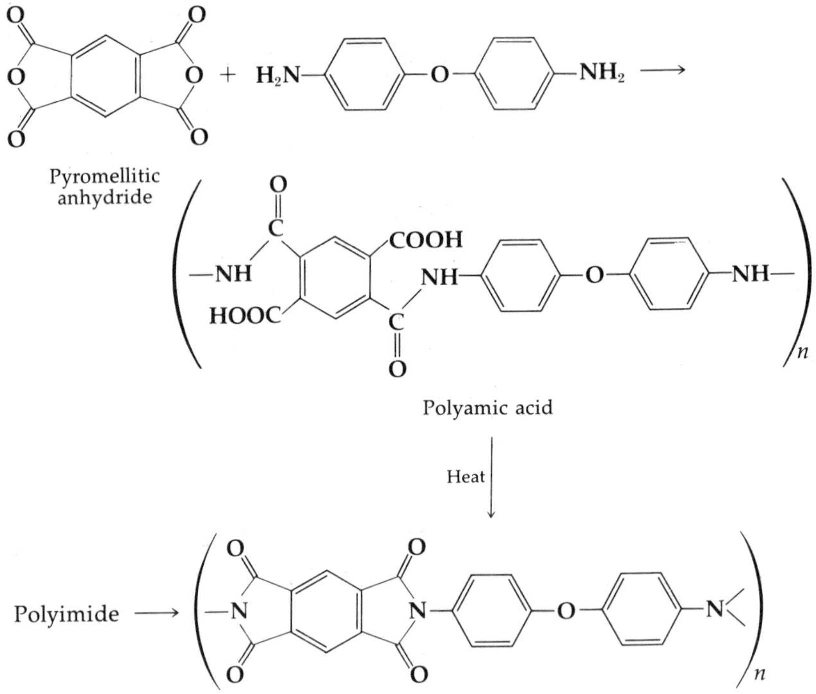

Pyromellitic anhydride

Polyamic acid

Polyimide →

Devise some photosynthetic routes which might be used to convert readily available starting materials into dianhydrides for evaluation as intermediates for the synthesis of condensation polymers.

22-6 By consultation of the chemical literature, evaluate the differences and similarities to be expected of the following ketones, when used as reactants or sensitizers in photochemical experiments:

a Acetophenone

b *m*-Methoxyacetophenone

c Benzophenone

d Naphthyl phenyl ketone

ORGANIC SYNTHESIS

SYNTHESIS of organic compounds lies at the heart of the science. One of the earliest ambitions of chemists was to produce in the laboratory, through rational sequences of reactions, organic compounds discovered in nature. Successful synthesis of a natural product was regarded as a confirmation of the structure of the substance, which had been deduced by degradative reactions. As success at synthesis developed, many compounds not present in nature were also synthesized for a variety of purposes.

At present, a particular organic compound may be synthesized for any of a number of reasons. Rough correlations exist between structures of compounds and their physical, chemical, or biological properties. Research chemists can frequently draw structures of compounds as yet unknown, which, when prepared, have a good chance of possessing a desired property. Such properties are then put to either scientific or commercial use. Many organic compounds are prepared in attempts to substantiate theories, to discover new properties or correlations, or to study organic reaction mechanisms. Syntheses of new drugs and pharmaceuticals or of new pigments and plastics are constantly undertaken in the research laboratories of the chemical industry.

A number of sections have already been specifically devoted to the uses of the major reaction families in synthesis. Within these sections are also groups of useful tables and charts meant to codify information of value for synthesis design. These sections are listed here for more convenient reference.

The great variety of reactions and techniques in organic chemistry make possible a diversity of approaches to synthetic problems. In a sense, organic chemistry is a language, with compounds as words, reactions as grammar, sequences of reacting compounds as phrases, and syntheses as composition. Clearly, solutions of synthetic problems rest squarely on a knowledge of structure and reactions, just as composition depends on vocabulary and grammar. Just as clearly, fluency comes with practice and experience, as much in organic chemistry as in any language.

23-1 FACTORS IN DESIGN OF SYNTHESES

Any given organic structure can be synthesized by several, often many, different routes. A given synthetic route is a sequence of reactions designed to convert commercially available starting materials into the desired substance. The best practice usually is to devise a number of different routes for a given molecule and compare them to select the best one, since the time spent in synthesis design is far less than that spent in the actual laboratory.

Generally, the best synthesis of a substance involves conversion of the most available and cheapest starting materials into the desired product by the least number of steps and in the highest possible overall yield. However, criteria for good syntheses vary with context. In commercial syntheses, costs of starting materials and economy of operations play a dominant role, whereas in many syntheses carried out for research purposes, the dispatch with which a compound can be obtained is more important. Economy is always a central criterion in synthesis, but the relative importance of economy of time, of expense, of material, or of number of operations will depend on the situation.

In many cases, an excellent synthesis in principle has to be abandoned for practical reasons. Sometimes an intermediate product proves too difficult to purify. Possibly an intermediate is too unstable for storage or too insoluble in any medium to permit confinement of the reaction medium to reasonable volumes. In other cases, yields vary because reaction rates are highly sensitive to impurities, reaction conditions, solvent, catalyst grade, or other variables. Difficulties of this sort are hard to anticipate and sometimes difficult to overcome. For this reason a synthesis design containing alternate routes or "detours" will be wiser and safer than one whose success is wholly dependent on one critical reaction.

Conception of Syntheses

Conception of organic syntheses for molecules of any complexity usually involves a stepwise procedure of working backward from the structure of the final product to the structures of available starting materials. Hence possible reactions that might lead to the desired final product are considered first. Compounds needed for these final reactions are next examined and treated in turn as if they were the synthesis problem. This procedure is repeated until available starting compounds are encountered. At every step, reactions are chosen that allow the desired (final or intermediate) compound to be made from the simplest starting materials. Frequently, "blind alleys" are encountered, and other sequences must be envisioned. From this procedure emerges a number of possible synthetic routes, the most attractive of which is selected for trial in the laboratory.

The kinds of reactions involved in any synthesis may be conveniently divided into two categories for separate consideration.

1   *Skeleton construction.* The actual carbon skeleton must be constructed from smaller units (**synthons**) by reactions which form carbon-carbon bonds. This process is the central consideration in synthesis design. The overall conception of the synthesis is basically a dissection of the desired skeleton into the particular synthon units that must be united to create it.

2   *Functional alteration.* Certain of the reactions in the synthetic sequence must be devoted to interconversion of the functional groups on the skeleton. The functionality resulting from one carbon-carbon bond formation must be transformed into that required for the next, or ultimately into the final functional groups of the end product. Furthermore, functional groups must sometimes be protected by conversion to a derivative, so as to survive reactions carried out elsewhere on the molecule. The **protecting group** must be one that can be readily removed later when desired.

The saturated hydrocarbon portion of a molecule is generally unreactive. Reactions occur only at functional group sites or at α-positions to carbonyls or double bonds. This means that synthons must be chosen not just for their carbon skeletons but also with functional groups at the sites requiring later bond formation. Dummy functional groups are often used to provide reactive sites for synthesis. At the end of the synthesis they can then be reduced to hydrocarbon groups if unwanted in the final product. (See Tables 16-2 and 19-5.)

Yields

The criterion of a good synthesis then becomes one of minimizing the number of steps. The number of starting skeletal units and the number of operations adjusting functionality must be kept as small as possible. Furthermore, the reactions chosen should be of high yield wherever possible since the overall yield for the synthesis is dramatically reduced by poor yields in individual steps.

The overall yield is the product of all the yields of the individual steps. The consequence of poor yields is that large and unwieldy quantities of starting material are required, as illustrated in Table 23-1, and that the problem of logistics or supply of intermediates becomes burdensome.

TABLE 23-1   **Yields and Required Starting Materials in Synthesis**

Sequence: $A \longrightarrow B \longrightarrow C \longrightarrow D \longrightarrow$ product

Average yield per step, %	Overall yields, %			Grams of starting material, A, required for 1 g of product†		
	5 steps	10 steps	15 steps	5 steps	10 steps	15 steps
50	3.1	0.1	0.003	16	512	16,384
70	16.8	2.8	0.5	3	18	105
90	59.5	35.4	21.1	0.8	1.4	2.4

†The molecular weight of A is arbitrarily assumed to be one-half of that of the product.

Starting Materials

Knowledge of what starting materials are available is fundamental to the design of synthetic sequences. Companies exist that make, collect, and sell small amounts of the common organic compounds. Some catalogs list thousands of individual substances. A good research laboratory possesses a storeroom of several hundred compounds. Most research chemists have general notions about what compounds can be purchased, but the list is too vast, changes too quickly, and is too dependent on quirks in nature or industrial operations for more precise familiarity. The following generalizations provide an introduction.

1 Most aliphatic hydrocarbons, alkyl bromides or chlorides, alcohols, aldehydes, ketones, and carboxylic acids that contain six carbon atoms or less are articles of commerce. Many polyfunctional compounds with the same number of carbon atoms are also available, although many such compounds which can be envisioned cannot be purchased. Some common bifunctional synthons are listed in Table 13-5. Many of the other simple (six carbon atoms or less) monofunctional compounds such as ethers, esters, amines, nitro compounds, amides, anhydrides, mercaptans, sulfides, sulfones, and sulfonic acids are also available. In general, the lower the number of carbon atoms and the less branched the chains, the greater the possibility that a compound is available.

2 Some alicyclic compounds that contain five- and six-membered rings, not more than six or seven carbon atoms, and a single, common functional group can be purchased. Relatively few compounds that contain small (3-4) or medium (7-12) rings can be purchased.

3 Aromatic hydrocarbons containing up to three aromatic rings are abundant, as are the monoalkylated (alkyl groups of up to four carbon atoms) benzenes and the poly-methylated benzenes. Essentially all monofunctional benzene derivatives and many

di- and trifunctional benzene compounds (the functional groups attached to the benzene ring) can be purchased. Many alcohols and ketones that contain a benzene ring and a carbon chain of four or less carbon atoms are also available.

4   Essentially all the simple five- and six-membered heterocyclic compounds that contain one or two heteroatoms are sold.

5   Virtually the only optically active commercial compounds are those obtained from natural sources; they are relatively few.

6   Compounds that are unstable, difficult to store or ship, or are hardly ever used in synthetic operations are usually not available, even if the substances possess simple structures.

7   Many very complicated compounds can also be purchased, but no correlation exists between their structures and availability.

Case History of a Synthesis

Diketone *A* was required for a stereochemical and spectral study.

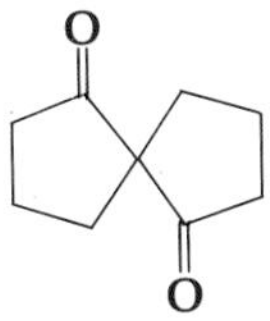

A

Of the ring-closing reactions that lead to ketones, the acyloin (page 528), the Dieckmann (page 525), the Thorpe (page 527), the cyclic anhydride pyrolysis (page 528), and the dicarboxylic acid salt pyrolysis (page 528) were considered. Except for the acyloin reaction all are variations on the enolate acylation theme. The acyloin reaction of *B* was discarded on two counts: when applied to preparation of five-membered rings, the yields are poor; the large number of isomers produced would create difficulties.

Acyloin:

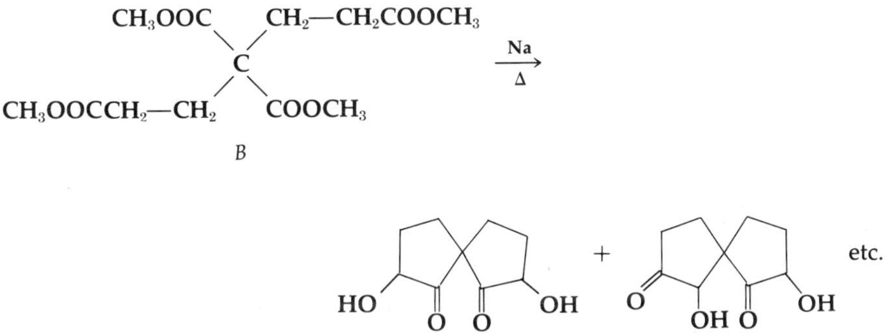

B

The other four reactions were all considered as possible candidates, and they all possessed the virtue that the starting material for each reaction was derivable from a common tetracarboxylic acid (*C*).

Dieckmann:

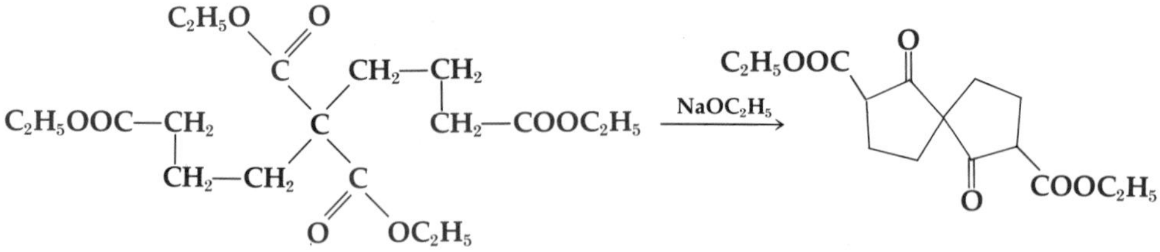

Thorpe:

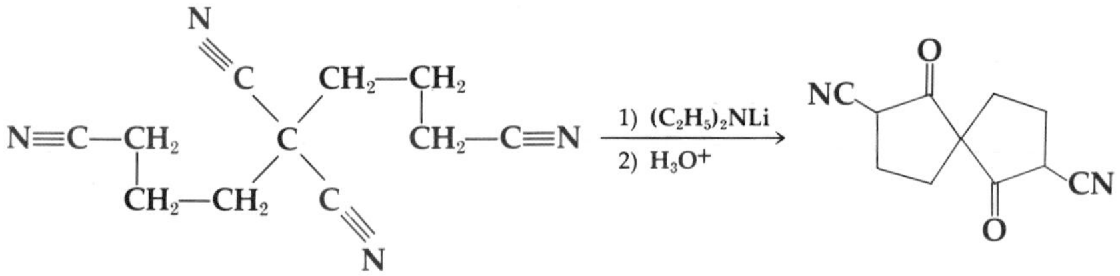

Anhydride pyrolysis:

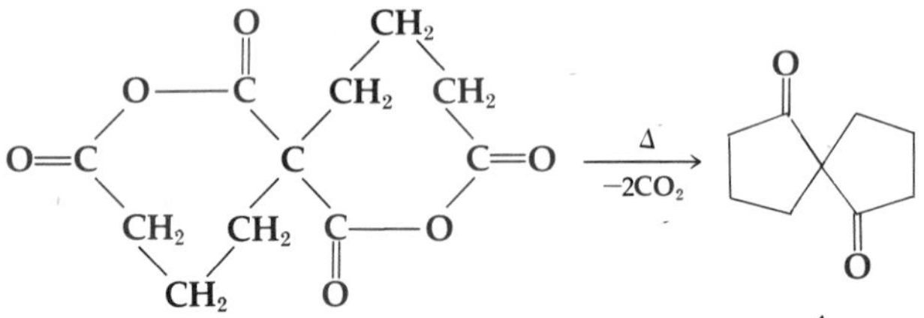

A

Salt pyrolysis:

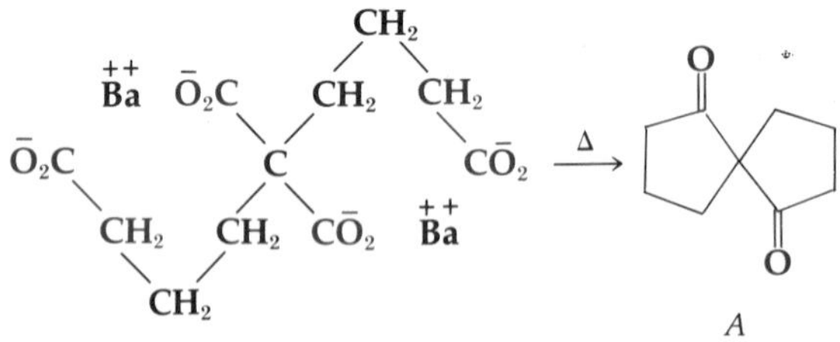

A

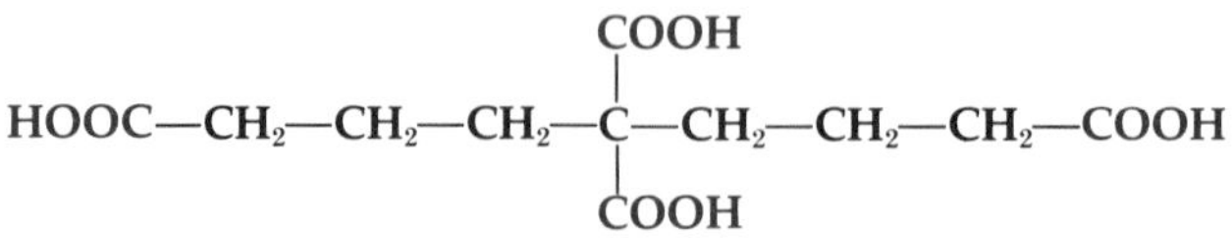

C

The most economical way to determine the feasibility of the critical ring-closing reactions was to try them out on a readily available "model compound." Dicarboxylic acid D was prepared by the sequence indicated. Its ester underwent the Dieckmann reaction in excellent yields. The cyclic anhydride derived from D pyrolyzed to produce ketone in good yield, whereas pyrolysis of the barium salt of D gave the same product in poor yield.

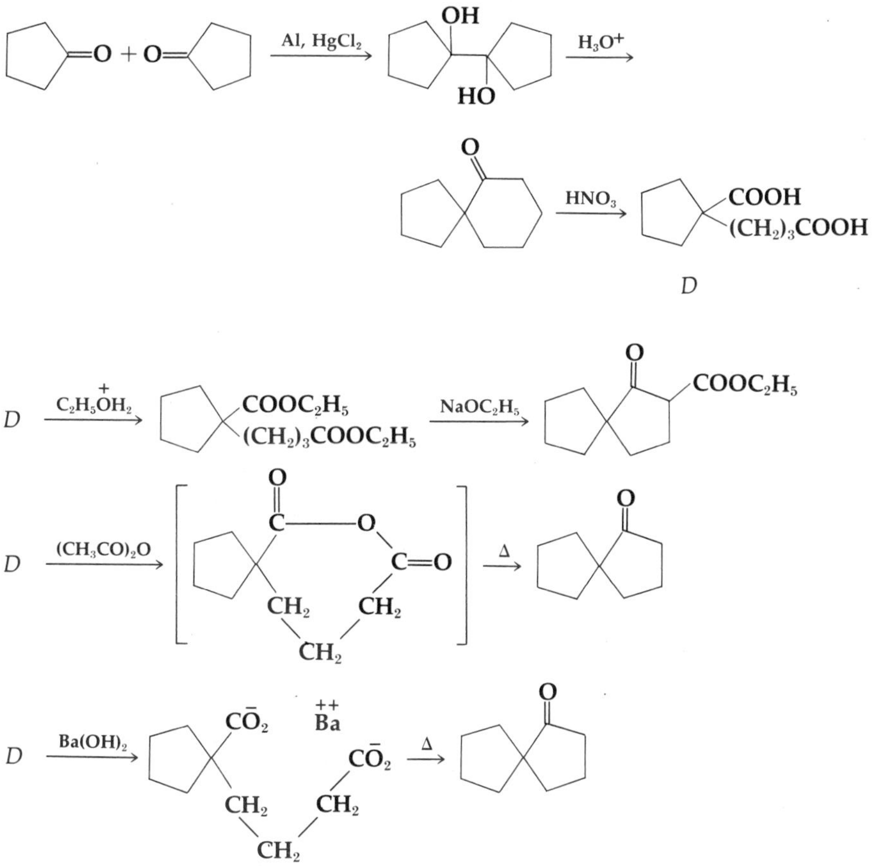

The next step was to design a synthesis of tetraacid C. Since this compound is a derivative of malonic acid, a malonic ester synthesis (page 433) was conceived, using ethyl 4-bromobutanoate as the alkylating agent.

$$\overset{+}{Na}\overset{..}{\overline{C}}H(COOC_2H_5)_2 + Br(CH_2)_3COOC_2H_5 \xrightarrow{C_2H_5OH}$$

$$(C_2H_5OOC)_2CH(CH_2)_3COOC_2H_5 \xrightarrow{NaOC_2H_5}$$

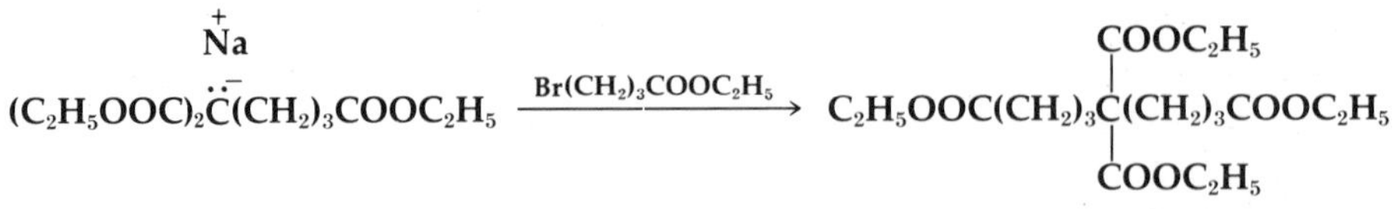

$$\underset{(C_2H_5OOC)_2\overset{..}{\overline{C}}(CH_2)_3COOC_2H_5}{\overset{\overset{+}{Na}}{}} \xrightarrow{Br(CH_2)_3COOC_2H_5} C_2H_5OOC(CH_2)_3\underset{\underset{COOC_2H_5}{|}}{\overset{\overset{COOC_2H_5}{|}}{C}}(CH_2)_3COOC_2H_5$$

Tetraester of *C*

Since ethyl 4-bromobutanoate was not commercially available, a method was required to prepare it. Consultation of the research literature (Chap. 28) indicated the substance had been made by the three syntheses outlined, the last of which was chosen because of its simplicity.

$$C_6H_5\overset{-}{O}\ \overset{+}{Na} + Br(CH_2)_3Br \longrightarrow C_6H_5O(CH_2)_3Br \xrightarrow{NaCN} C_6H_5O(CH_2)_3CN \xrightarrow[2)\ H_3O^+]{1)\ NaOH}$$

$$C_6H_5O(CH_2)_3COOH \xrightarrow{HBr} Br(CH_2)_3COOH \xrightarrow{C_2H_5\overset{+}{O}H_2} Br(CH_2)_3COOC_2H_5$$

$$\overset{+}{K}\ \overset{-}{CN} + BrCH_2CH_2CH_2Br \longrightarrow Br(CH_2)_3CN \xrightarrow{H_3O^+} Br(CH_2)_3COOH \xrightarrow{C_2H_5\overset{+}{O}H_2}$$

$$Br(CH_2)_3COOC_2H_5$$

$$HBr + \underset{O}{\overset{\overset{CH_2-CH_2}{\diagup\quad\diagdown}}{CH_2\qquad C=O}} \xrightarrow{C_2H_5\overset{+}{O}H_2} Br(CH_2)_3COOC_2H_5$$

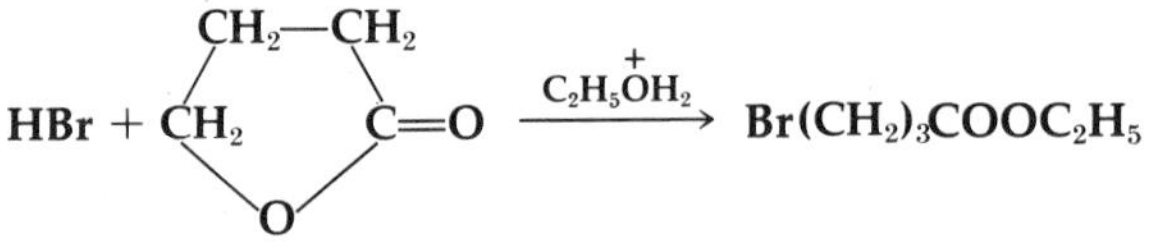

The tetraester of *C* was prepared by the sequence indicated and was subjected to the conditions of the Dieckmann condensation. The reaction failed to give the desired diketone *A* since an unanticipated reaction destroyed starting material at a rate faster than the condensation occurred. The tetraester of

C is a disubstituted malonic ester, which under basic conditions at elevated temperature is subject to a decarboethyoxylation reaction. This reaction belongs to the family of base-catalyzed cleavage reactions discussed previously (page 532). Clearly, the Dieckmann reaction as applied to the diester of diacid D proved to be a poor model for the same reaction on the tetraester of C. This is not an uncommon experience since no model is ever exactly the same as the actual compound of interest.

$$\text{C}_2\text{H}_5\bar{\text{O}}\,\text{Na}^+ + \text{C}_2\text{H}_5\text{OOC(CH}_2)_3\overset{\displaystyle \overset{\text{COOC}_2\text{H}_5}{|}}{\underset{\displaystyle \underset{\text{O} \quad \text{OC}_2\text{H}_5}{\text{C}}}{\text{C}}}(\text{CH}_2)_3\text{COOC}_2\text{H}_5 \xrightarrow[\Delta]{\text{C}_2\text{H}_5\text{OH}}$$

$$\text{C}_2\text{H}_5\text{OOC(CH}_2)_3\overset{\displaystyle \overset{\text{COOC}_2\text{H}_5}{|}}{\underset{\displaystyle \underset{\text{OC}_2\text{H}_5}{\overset{\text{Na}^+\ \bar{\text{O}}\!-\!\text{C}\!-\!\text{OC}_2\text{H}_5}{|}}}{\text{C}}}(\text{CH}_2)_3\text{COOC}_2\text{H}_5 \xrightarrow{-(\text{C}_2\text{H}_5\text{O})_2\text{CO}}$$

$$\text{C}_2\text{H}_5\text{OOC(CH}_2)_3\overset{\displaystyle \overset{\text{COOC}_2\text{H}_5}{|}}{\underset{\displaystyle \underset{\text{Na}}{\bar{\text{C}}^+}}{\text{C}}}(\text{CH}_2)_3\text{COOC}_2\text{H}_5 \xrightarrow{\text{C}_2\text{H}_5\text{OH}}$$

$$\text{C}_2\text{H}_5\text{OOC(CH}_2)_3\overset{\displaystyle \overset{\text{COOC}_2\text{H}_5}{|}}{\underset{\displaystyle \underset{\text{H}}{|}}{\text{C}}}(\text{CH}_2)_3\text{COOC}_2\text{H}_5 \xrightarrow{\text{NaOC}_2\text{H}_5}$$

The next best ring closure of the model system involved pyrolysis of the anhydride of diacid D. When submitted to the same reaction conditions, tetraacid C gave the desired diketone A. The synthetic sequence from available starting materials to final diketone A provided an overall yield of 9% (based on δ-butyrolactone), the product of all the individual yields. The synthesis may be seen to contain three reactions forming carbon-carbon bonds and two transforming functional groups.

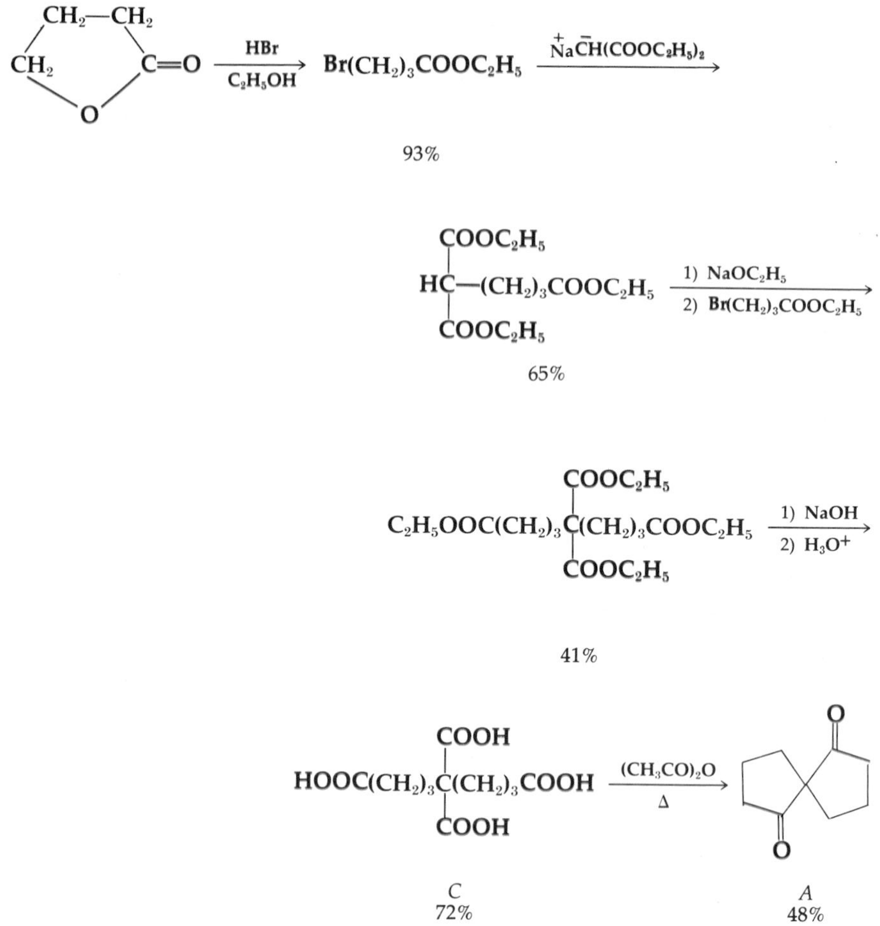

$$\underset{\underset{O}{\diagdown}}{\overset{CH_2\!-\!CH_2}{\overset{\diagup}{CH_2}}}C\!=\!O \quad \xrightarrow[C_2H_5OH]{HBr} \quad Br(CH_2)_3COOC_2H_5 \quad \xrightarrow{Na\overset{+}{C}H(COOC_2H_5)_2}$$

93%

$$\underset{\underset{COOC_2H_5}{|}}{\overset{\overset{COOC_2H_5}{|}}{HC}}\!\!-\!(CH_2)_3COOC_2H_5 \quad \xrightarrow[\text{2) Br(CH}_2)_3\text{COOC}_2\text{H}_5]{\text{1) NaOC}_2\text{H}_5}$$

65%

$$C_2H_5OOC(CH_2)_3\underset{\underset{COOC_2H_5}{|}}{\overset{\overset{COOC_2H_5}{|}}{C}}(CH_2)_3COOC_2H_5 \quad \xrightarrow[\text{2) H}_3\text{O}^+]{\text{1) NaOH}}$$

41%

$$HOOC(CH_2)_3\underset{\underset{COOH}{|}}{\overset{\overset{COOH}{|}}{C}}(CH_2)_3COOH \quad \xrightarrow[\Delta]{(CH_3CO)_2O}$$

C
72%

A
48%

The synthesis illustrates several principles whose recognition can facilitate solution of other synthetic problems. Firstly, the symmetry in the product A allowed both halves of the molecule to be developed simultaneously. Had the desired product been one of the following, the synthesis would necessarily have taken a different course.

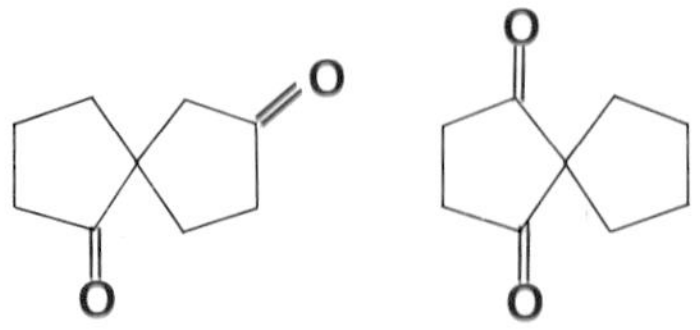

Secondly, the sequence has a key intermediate in the tetraacid C, which could be utilized in several ways for the critical ring closure. It is important that this tetraacid should be made available in quantity to try these reactions if necessary. Hence it may be important to be able to make it in several ways. It is also helpful to have several ways of carrying out the crucial reaction, i.e., C $\longrightarrow$ A in this case.

Recognition of the critical intermediates and reactions in a contemplated synthetic sequence provides focal points for literature work and for investigation of reactions in model compounds. Dominant in developing any organic synthesis is "reasoning by analogy." The chemical literature (Chap. 28) contains a vast inventory of organic syntheses that serve as models for contemplated synthetic schemes. When reactions of model compounds close enough in structure to those envisioned in the synthetic sequence cannot be found in the literature, sometimes actual study of model compounds expedites the overall project. On the other hand, if the model is imperfect its exploration can be a waste of time, as the use of the model D showed in this case.

23-2   CARBON-CARBON BOND FORMATION

Reliable, well-tested reactions for forming carbon-carbon sigma bonds in good yield are of central importance in synthesis. The nature of the available reactions determines the possibilities of dissecting any given structural goal into synthons. The actual number of such reactions is limited, in fact, and may be broken down in the following fashion.

1 *Electrophile-nucleophile reactions* (Chaps. 10–16). These are by far the most common and versatile. Almost all of these reactions involve carbanion nucleophiles and so are carried out under basic conditions. These are reviewed in Sec. 13-10. The only important acid-catalyzed carbon-carbon bond formation is the Friedel-Crafts acylation of aromatic rings. These reactions are discussed below and tabulated in Table 23-2.

2 *Pericyclic reactions* (Chap. 21). The Diels-Alder cycloaddition is the principal reaction in this category. The Cope and Claisen rearrangements also form carbon-carbon bonds but have been used less. All of these reactions are especially valuable for their high stereospecificity. Photochemical cycloadditions (especially [2 + 2]) also belong in this category, as do additions of carbenes to double bonds.

3 *Rearrangements* (Chap. 17). Certain rearrangements make (and break) carbon-carbon bonds in a manner useful for synthesis. Usually these reactions are used for contracting or expanding rings or chains. In general, rearrangements are of limited utility for building up carbon skeletons. Synthetically useful rearrangements are collected in Table 23-3.

4 *Oxidation-reduction reactions* (Chaps. 18, 20). In general these reactions rarely form carbon-carbon bonds; the few exceptions are the oxidative and reductive couplings which follow a free-radical path. In most instances such couplings create symmetrical dimers, i.e., two identical units joined together. This can be appreciated in those most used for synthesis: the acyloin and pinacol reductive couplings of carbonyl groups and the oxidative coupling of copper acetylides.

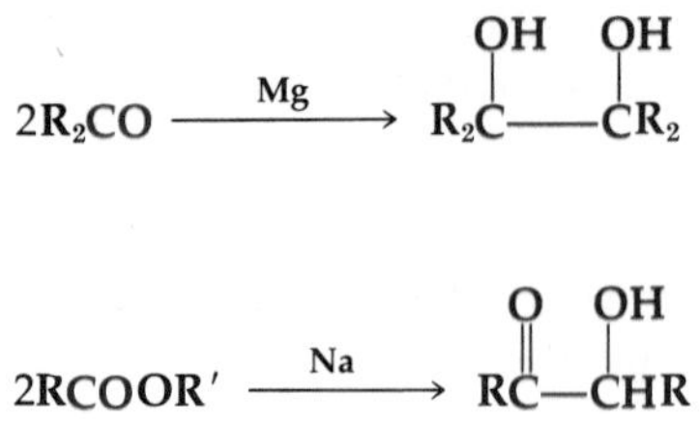

$$2R_2CO \xrightarrow{\text{Mg}} \underset{\overset{|}{R_2C}\text{———}\overset{|}{CR_2}}{\overset{OH\quad OH}{}}$$

$$2RCOOR' \xrightarrow{\text{Na}} \underset{RC\text{—}CHR}{\overset{O\quad OH}{}}$$

$$2RC{\equiv}CH \xrightarrow[+O_2]{\text{Copper salt}} RC{\equiv}C{-}C{\equiv}CR$$

TABLE 23-2 **Major Ionic Reactions for Carbon**

	Electrophiles *I* $RCH_2{-}L$
Nucleophiles	*Halides, tosylates, epoxides* (usually primary)
Organometallic carbanions $-\overset{\|}{\underset{\|}{C}}-MgX(Li)$	(Epoxides only) $-\overset{\|}{\underset{\|}{C}}-CH_2-\overset{OH}{\underset{\|}{C}}-$
Wittig reagent $-\overset{..}{C}H{-}\overset{+}{P}\phi_3$	—
Enolate ion (or enamine) $-CO-\overset{\|}{\underset{\|}{C}}{:}^-$	*Alkylation* (S_N2) (Enolates only) $-CO-\overset{\|}{\underset{\|}{C}}-CH_2R$
sp Anion $\left.-\overset{C}{\underset{N}{}}\right\}{\equiv}C{:}^-$	$-C{\equiv}C-CH_2R$ $NC{-}CH_2R$
Aromatic rings (acidic conditions)  ⬡	—

The most important category is the first one, the electrophile-nucleophile reactions, summarized as ionic reactions in Table 23-2. The number of such reactions is not great when condensed in this general way. The carbon nucleophiles are shown vertically, roughly in an order of decreasing basicity, with the nucleophilic carbon atom colored. The electrophiles are arranged horizontally in order of functional-group families (I, II, III, as in Table 3-2), with vinylogous, or conjugate-addition, cases last.

In deciding how to dissect a product skeleton it is important to consider the placement of the functional groups on that skeleton. This orientation con-

Skeleton Construction

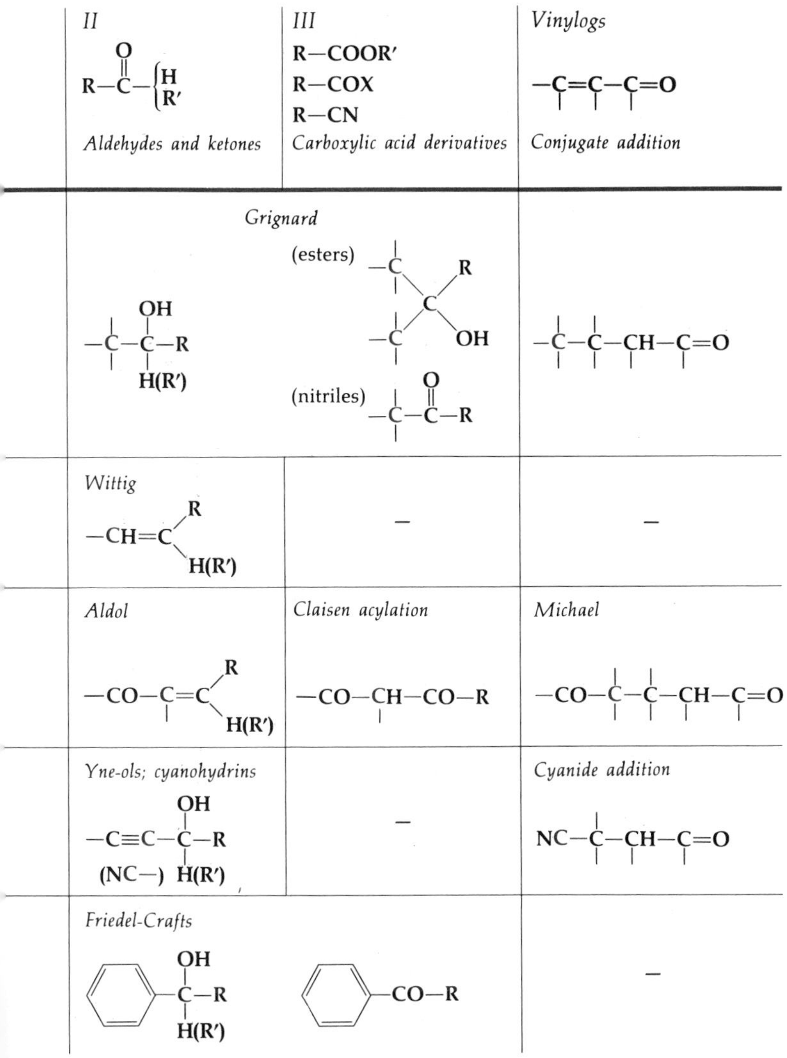

II	III	Vinylogs
O‖ R—C—{H / R′} *Aldehydes and ketones*	R—COOR′ R—COX R—CN *Carboxylic acid derivatives*	—C=C—C=O *Conjugate addition*
OH ‖ —C—C—R ‖ H(R′)	*Grignard* (esters) —C⟨R / C⟨ / —C⟨OH⟩ (nitriles) —C—C—R‖O	—C—C—CH—C=O
Wittig —CH=C⟨R / H(R′)	—	—
Aldol —CO—C=C⟨R / H(R′)	*Claisen acylation* —CO—CH—CO—R	*Michael* —CO—C—C—CH—C=O
Yne-ols; cyanohydrins OH —C≡C—C—R (NC—) H(R′)	—	*Cyanide addition* NC—C—CH—C=O
Friedel-Crafts ⬡—C—R OH H(R′)	⬡—CO—R	—

TABLE 23-3 **Rearrangements Generally Useful in Synthesis**

Carbon insertion

  Diazomethane (p. 702): $R-CO-R' \xrightarrow{CH_2N_2} R-CH_2-CO-R'$

  Arndt-Eistert (p. 701): $R-COCl \xrightarrow{CH_2N_2} R-COCHN_2 \xrightarrow{Ag_2O}$

$$R-CH_2COOH$$

Nitrogen insertion

  Hofmann, Curtius (p. 704): $R-COOH \longrightarrow \begin{Bmatrix} R-CONH_2 \\ R-CON_3 \end{Bmatrix} \longrightarrow$

$$R-N=C=O$$

  Beckmann, Schmidt (p. 707): $R-CO-R' \xrightarrow{\hspace{3cm}} R-NH-CO-R'$

Oxygen insertion

  Baeyer-Villiger (p. 708): $R-CO-R' \xrightarrow{R''CO_3H/H^+} R-O-CO-R'$

Ring or chain contraction

  Wolff (p. 701):

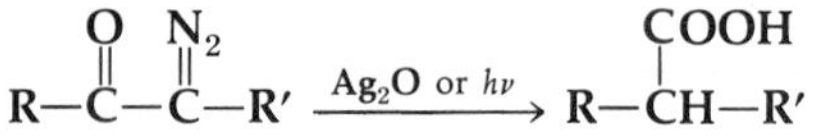

$$R-\overset{O}{\underset{\|}{C}}-\overset{N_2}{\underset{\|}{C}}-R' \xrightarrow{Ag_2O \text{ or } h\nu} R-\underset{\underset{COOH}{|}}{C}H-R'$$

  Favorski (p. 718):

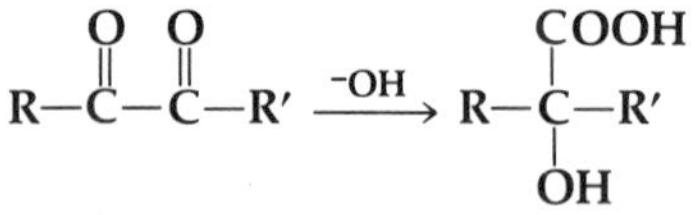

$$R-\overset{O}{\underset{\|}{C}}-\overset{O}{\underset{\|}{C}}-R' \xrightarrow{-OH} R-\underset{\underset{OH}{|}}{\overset{\overset{COOH}{|}}{C}}-R'$$

  Benzilic acid (p. 719):

$$R-\overset{O}{\underset{\|}{C}}-\overset{Cl}{\underset{|}{C}}H-R' \xrightarrow{-OH} R-\underset{\underset{COOH}{|}}{C}H-R'$$

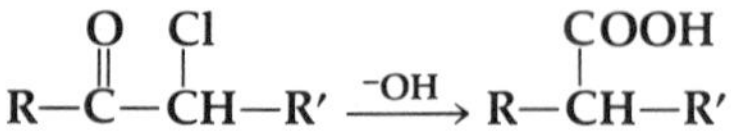

stitutes a clue as to which carbon-carbon bonds may easily be formed so as to result in the product functionality. To this end the nature of the functions in the reaction products is very important. These are shown in the boxes of Table 23-2.

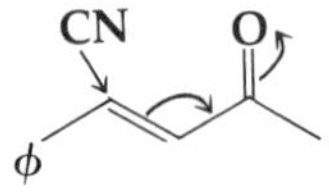

Desired:

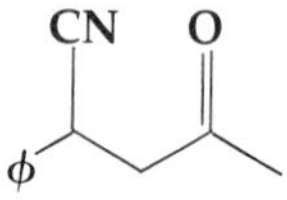

(Cyanide addition)   (Aldol)

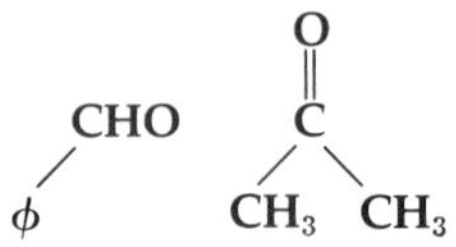

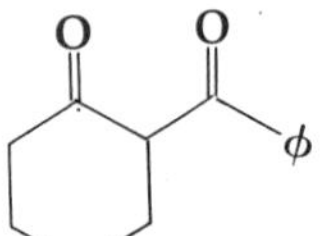

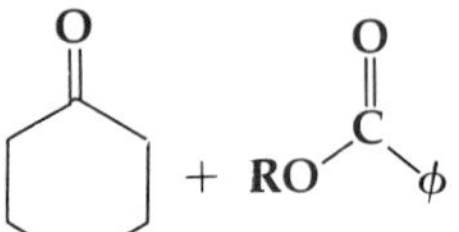

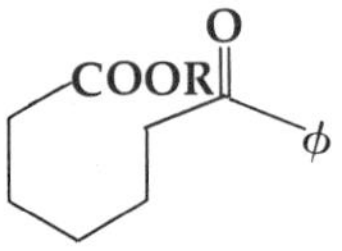

(Claisen acylations)

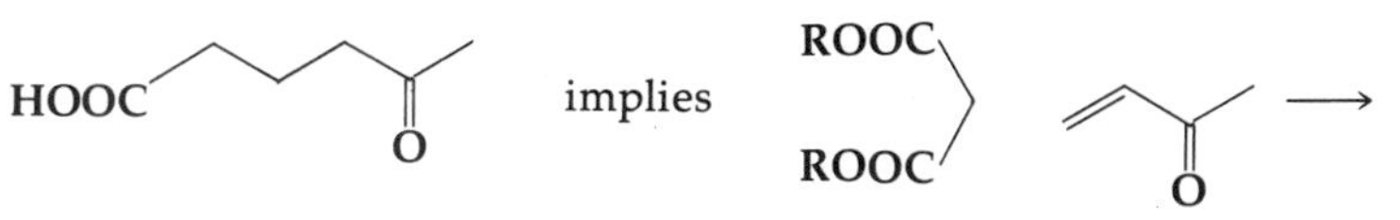

(Michael addition)

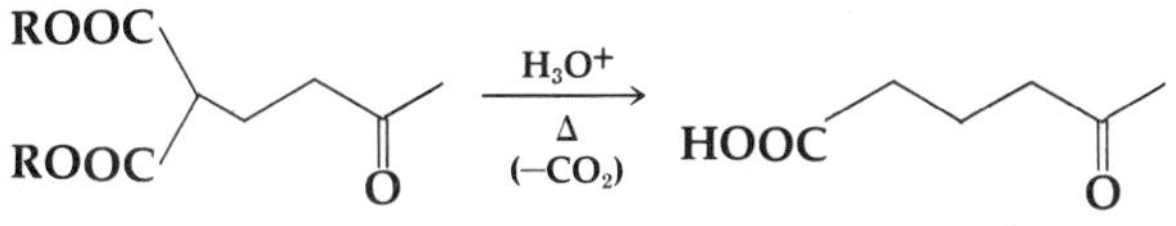

Since starting materials are usually relatively simple molecules (see above), the dissection of the carbon skeleton must also be made in terms of these relatively simple structural units. The synthesis of vitamin A from β-ionone (Fig. 23-1) is a fine example of the use of the reactions in Table 23-2 to build a carbon skeleton from carefully tailored structural units. The dissection of the molecule is shown below. The critical carbon-carbon bond formations are starred in Fig. 23-1.

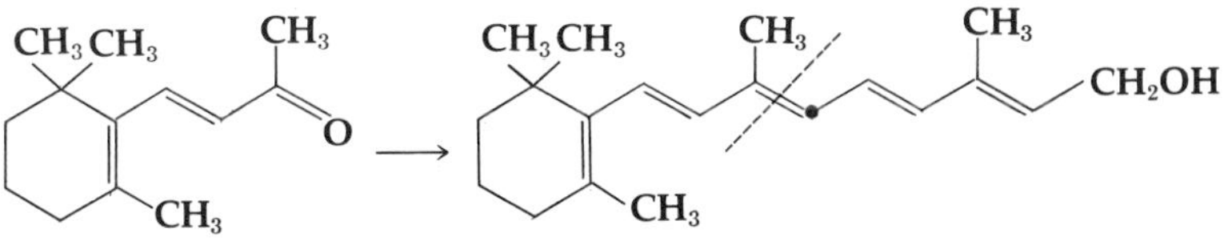

β-Ionone
(p. 473)

Vitamin A

Ring formation is an important part of carbon-carbon bond formation. Cyclizations are simply internal versions of the reactions in Table 23-2 and are generally viable only for enolate reactions and Friedel-Crafts reactions. An important reaction for assembling a ring by addition of an external synthon is the Robinson annellation reaction (page 482).

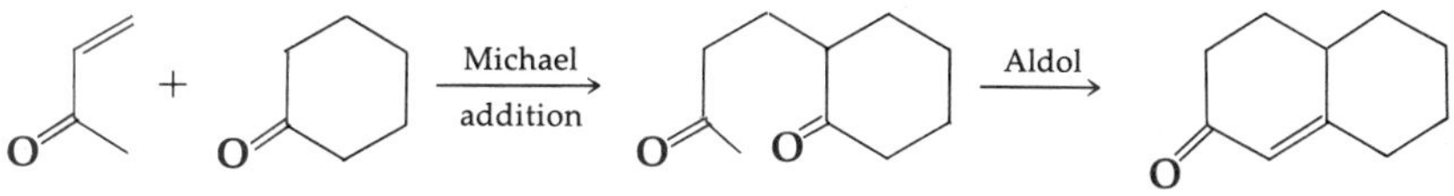

Synthesis of Epiandrosterone

The value of this annellation procedure is clear in a number of syntheses of natural steroid hormones (cf., page 1109), of which that in Fig. 23-2 is the simplest. Of the four rings in the final product two are assembled by the Robinson annellation reaction in the first two steps. (Each step is actually two successive reactions, the Michael addition followed by an aldol cyclization.) It should be noted in each case that the Michael reaction proceeds at that α-position which forms the more stable conjugated enolate of the ketone. In the second annellation this is the complex enolate of the unsaturated ketone.

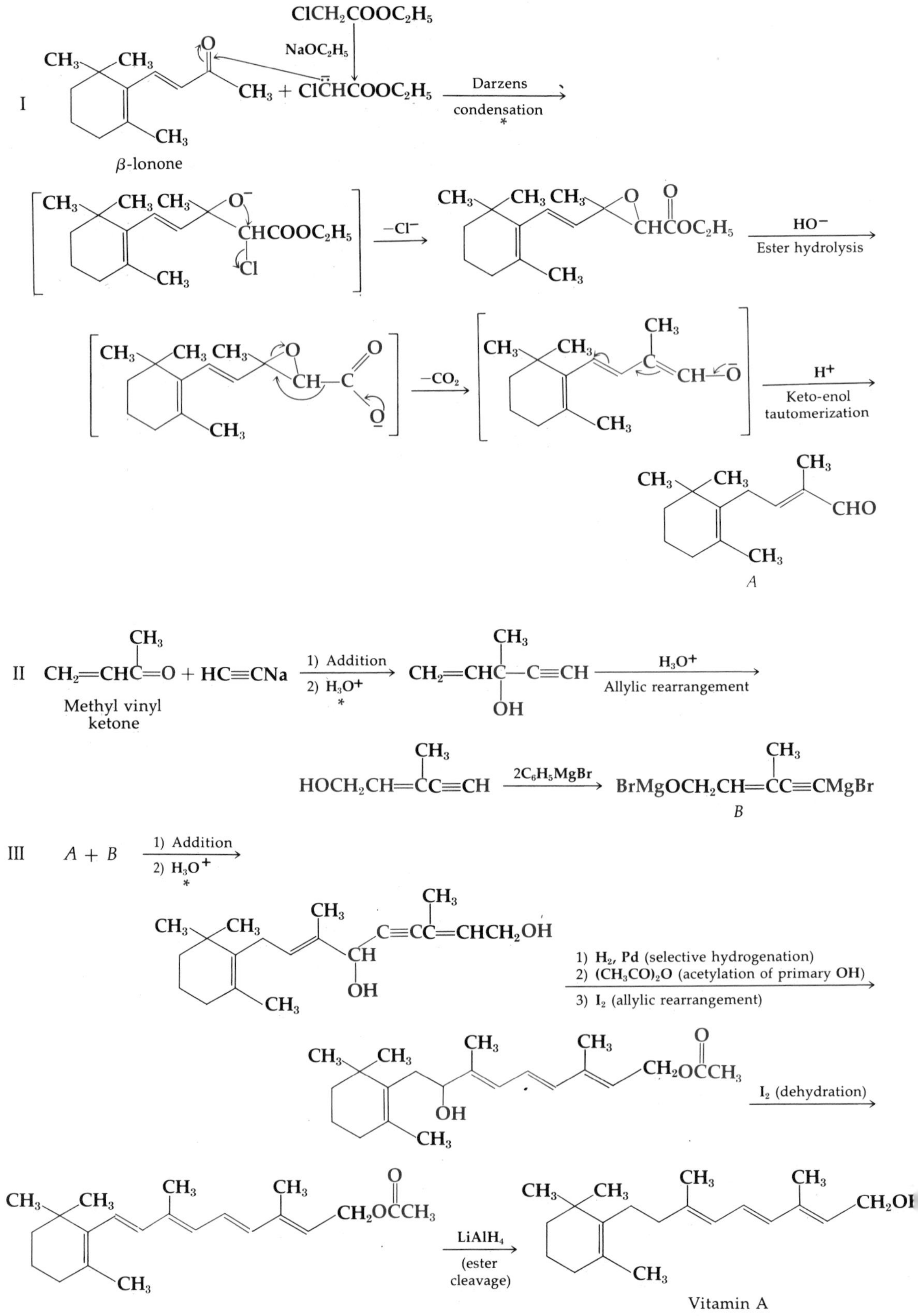

I

CH₃ CH₃ (on cyclohexene ring)

CICH₂COOC₂H₅

NaOC₂H₅

$CH_3 + Cl\ddot{C}HCOOC_2H_5$ → Darzens condensation *

β-Ionone

$\left[\cdots \text{CHCOOC}_2\text{H}_5 \quad \text{Cl} \right]$ —Cl⁻ →

CH₃ CH₃ CH₃ ... CHCOC₂H₅   HO⁻ / Ester hydrolysis →

$\left[\cdots \text{CH—C} \right]$ —CO₂ →

$\left[\cdots \text{CH}—\bar{\text{O}} \right]$ CH₃   H⁺ / Keto-enol tautomerization →

CH₃ CH₃ CH₃ CHO

CH₃

A

II $CH_2{=}CHC{=}O + HC{\equiv}CNa$ 1) Addition / 2) H₃O⁺ *   → $CH_2{=}CHC{-}C{\equiv}CH$ (CH₃, OH)   H₃O⁺ / Allylic rearrangement →

Methyl vinyl ketone

$HOCH_2CH{=}CC{\equiv}CH$ (CH₃)   2C₆H₅MgBr → $BrMgOCH_2CH{=}CC{\equiv}CMgBr$ (CH₃)

B

III   A + B   1) Addition / 2) H₃O⁺ *   →

CH₃ CH₃ CH₃   C≡CC=CHCH₂OH   CH   OH   CH₃

1) H₂, Pd (selective hydrogenation)
2) (CH₃CO)₂O (acetylation of primary OH)
3) I₂ (allylic rearrangement)

→

CH₃ CH₃ CH₃ CH₃   CH₂OCCH₃ (O)   OH   CH₃

I₂ (dehydration) →

CH₃ CH₃ CH₃ CH₃   CH₂OCCH₃ (O)   CH₃

LiAlH₄ (ester cleavage) →

CH₃ CH₃ CH₃ CH₃   CH₂OH

CH₃

Vitamin A

FIGURE 23-1   Synthesis of vitamin A (* = carbon-carbon bond formations)

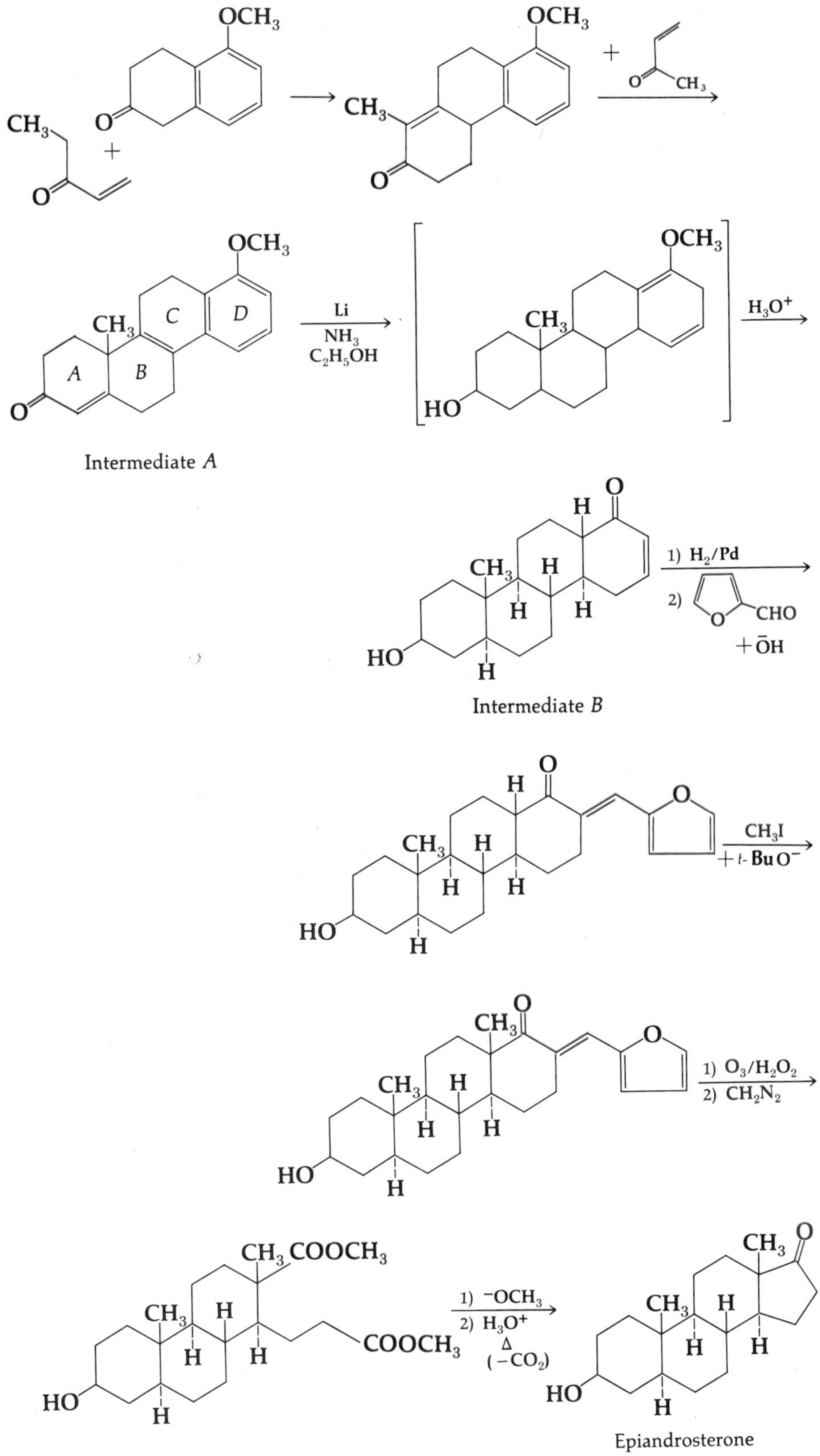

FIGURE 23-2 Synthesis of epiandrosterone

Two other features of the skeleton construction deserve mention. The six-membered ring D, in intermediate A, is introduced as an aromatic ring for easy production of the starting material and must ultimately be contracted to a five-membered ring in the final product. This is accomplished by attaching an aromatic aldehyde, furfural (α-furan-aldehyde, Chap. 24), in an aldol re-action after removing the double bond in B by hydrogenation. Note that there is only one α-methylene for aldol condensation. This affords an unsaturated system amenable to an oxidative cleavage, opening up the six-membered ring at a particular bond and yielding a diacid. Conversion to the diester (with diazomethane) now allows reclosure to the desired five-membered ring by a cyclic acylation (Dieckmann condensation). This is followed by hydrolysis and decarboxylation of the extraneous carboxyl in the resultant β-ketoester. This sequence is a widely used one and illustrates a number of the valuable carbonyl reactions in synthesis.

The second feature is the addition of the final required carbon in the skeleton, namely, the quaternary methyl group. This is introduced via methyl iodide alkylation after the aldol condensation with furfural has blocked the other α-position, allowing the methylation no choice of position except the desired one (page 560).

Synthesis of Cantharidin

Cantharidin is a natural product (isolated from dried beetles called Span-ish fly) that possesses vesicant and rubefacient activity. The preparation of this compound illustrates the use of the Diels-Alder reaction, the second important reaction for creating a ring by adding an external structural unit in one opera-tion.

Cantharidin

The two-step synthesis of cantharidin from 2,3-dimethylmaleic anhydride and furan was first tried, but the initial Diels-Alder reaction failed for steric reasons. Diels-Alder reactions (page 853) are reversible, and the adduct appears to contain enough steric compression to render it thermodynamically less stable than the starting materials.

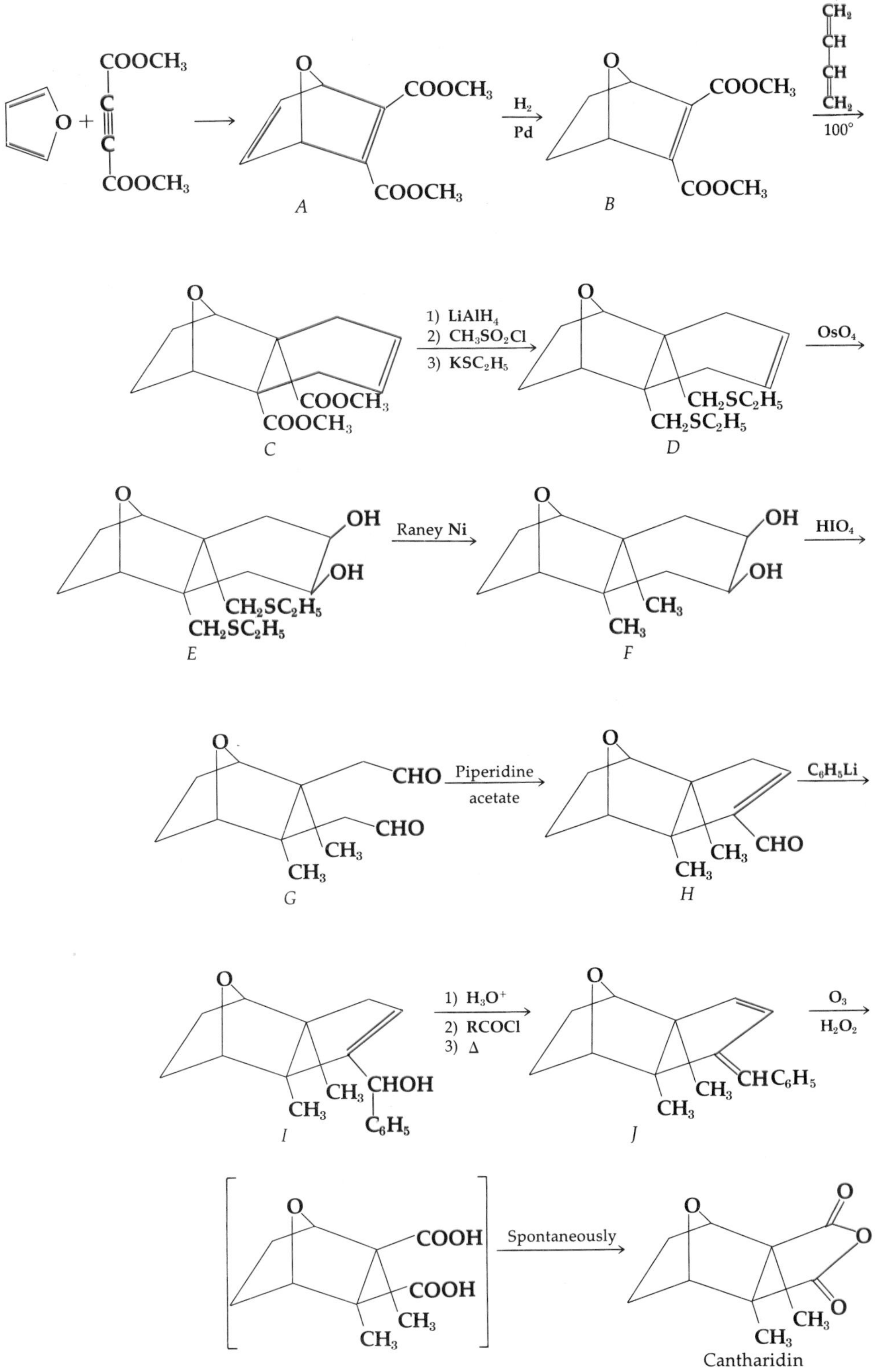

FIGURE 23-3 Synthesis of cantharidin

A much longer synthesis, based on another Diels-Alder reaction, proved successful, as outlined in Fig. 23-3. Except for the methyl groups the skeleton is still constructed with a Diels-Alder reaction. The methyl groups proved impossible to introduce by alkylation alpha to the carbonyls (after hydrogenation). Hence a special synthetic technique was introduced in which a second Diels-Alder reaction was used to introduce the extra carbons. The ester groups are now destined to become the methyls and the carbons added in the Diels-Alder reaction are planned for conversion to the final carboxyls.

In the second Diels-Alder reaction ($B \longrightarrow C$) butadiene became attached to the bicyclic dienophile from the least hindered side of the molecule. This step fixed the configurational relationship for the whole sequence of compounds to the final product. The added cyclohexene ring was made smaller by conversion to a form bearing double bonds at the carbons intended for carboxyls (as their anhydride) in the product. With this preparation, ozonization then created the carboxyls required and swept away the extraneous carbons.

PROBLEM 23-1

Other, even shorter, routes from D to a form suitable for oxidation to the requisite diacid (or anhydride) can be envisioned for the cantharidin synthesis. Some of these were probably tried. Try to write down such an alternate route.

PROBLEM 23-2

In each of the following molecules the functional groups present imply that some particular carbon-carbon bond formation should be involved in the synthesis. Write the equation for that reaction in each case.

a $CH_3CH_2COCHCOOCH_3$
        $|$
        CH_3

b

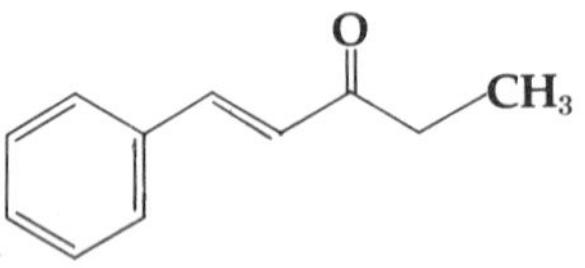

c

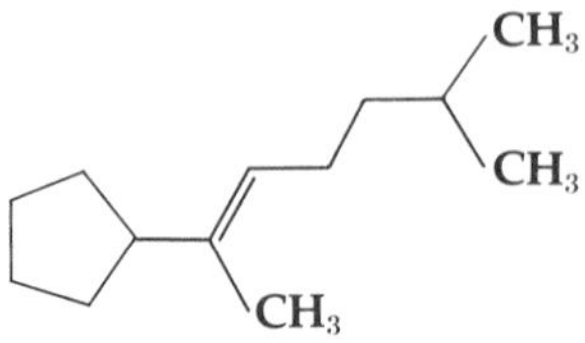

d

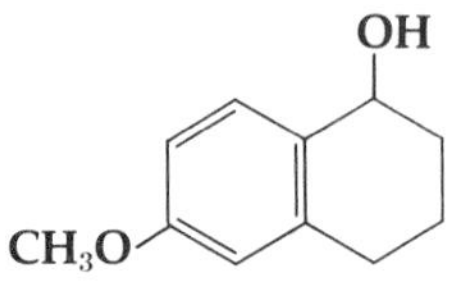

e

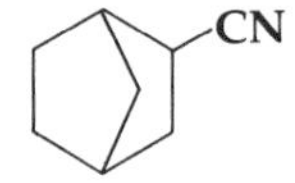

f $(CH_3)_2CCH_2CH_2CHCH_3$
        with OH groups on the indicated carbons

g

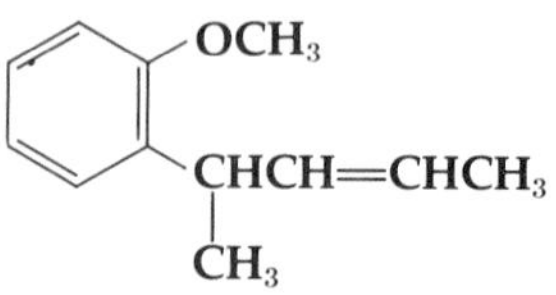

h

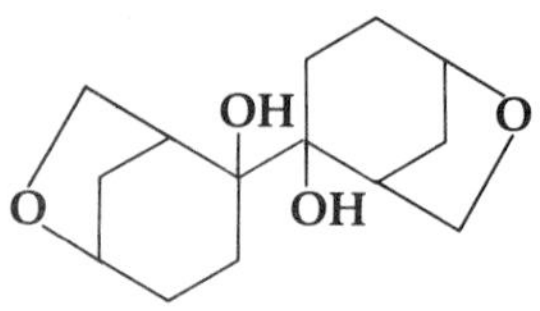

23-3   FUNCTIONAL-GROUP INTERCONVERSIONS

Functional groups are used during synthesis as handles to control the reactions used for carbon-skeleton formation. Ideally, the functions left at the end of skeleton-building operations are those desired in the final product or at least are readily converted to them. The functions left after a given carbon-carbon bond formation are dictated by the nature of the reaction used, as summarized in Table 23-2. These must then be converted to the requisite groups for the next such carbon-skeleton extension, or ultimately to those required for the final product. Economy and efficiency in synthesis therefore demand a proper choice and sequence of skeleton building so as to minimize the number of these functional-group transformations which are required.

Selectivity

More than one functional group will often be present in a synthetic intermediate. It is also often necessary to change one without affecting the other. In selecting reactions to use for some synthetic change it is always important to ascertain that the reaction conditions will not also create an undesired transformation of another functional group in the molecule. Knowledge of the selectivity of reactions is therefore important and has frequently been discussed in previous chapters. (Note especially the question of selectivity in reduction summarized in Table 19-5.)

In the synthesis of vitamin A (Fig. 23-1) two selective reactions are involved. In the first, a carbon-carbon triple bond was reduced with palladium and hydrogen to a carbon-carbon double bond, without reducing the many other reducible multiple bonds in the molecule. A partially poisoned catalyst was employed for the purpose, and the greater reactivity toward catalytic reduction of triple over double bonds (page 769) was utilized. In the second reaction, a primary hydroxyl group was acetylated and a secondary hydroxyl group was left untouched. For steric reasons, the rate of acetylation of the unhindered hydroxyl group at the end of the chain was greater than that of the hindered hydroxyl group toward the middle of the chain.

In the synthesis of epiandrosterone (Fig. 23-2) the preference of enolate formation toward the more conjugated α-position provides the selectivity needed to assure the correct orientation in the two Robinson annellation reactions. The hydroxyl group formed in $A \longrightarrow B$ is unaffected by all the subsequent reactions used.

The cantharidin synthesis (Fig. 23-3) poses a problem of selectivity in that the double bond needs to be oxidized under conditions which leave the oxidation-prone sulfides untouched; osmium tetroxide provided the requisite selectivity. On the other hand the desulfurization of those sulfides in D could not be carried out with Raney nickel without also hydrogenating and thus losing the double bond.

The sequences for converting the esters in *C* to the methyl groups in cantharidin and for revising the aldol product in *H* to the diene needed to produce a diacid from *J* are typical of functional-group transformations required in synthesis.

Protecting groups are often utilized to prevent reaction at an undesired site in the molecule. The aldol condensation with furfural in Fig. 23-2 may be regarded as protecting the α-methylene from being methylated by methyl iodide in the next step. Acetylation of alcohols, as in part III of the vitamin A synthesis, is a common procedure for their protection. Acetylation of amines is also an important protecting device since it renders the amino group nonbasic as well as resistant to oxidation. The protecting acetyl group is then later removed by hydrolysis.

The most common protecting group arises in the conversion of ketones and aldehydes to cyclic ketals and acetals, respectively (page 453), so as to render them unreactive to Grignard reactions and to base-catalyzed carbonyl reactions generally (i.e., Table 23-2). The cyclic ketal (or acetal) readily reverts to the parent ketone (or aldehyde) on mild acidic hydrolysis.

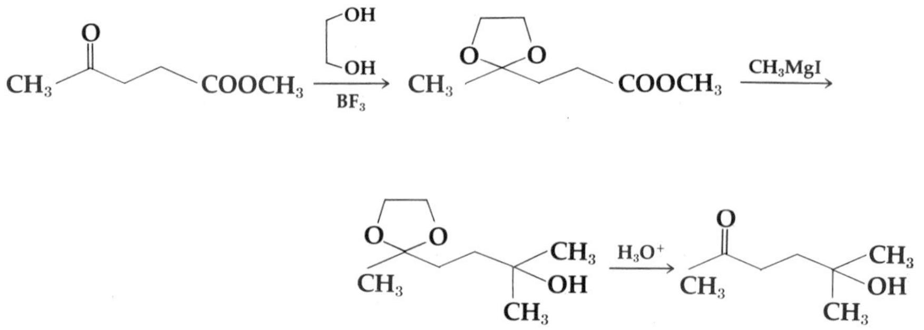

PROBLEM 23-3

Comment on whether these proposed synthetic reactions are likely to be successful. If you think not, try to devise an alternate route to convert the given starting material to the product shown.

a $CH_3COCH_2CH_2CHO \xrightarrow{NaBH_4} CH_3CHOHCH_2CH_2CHO$

b $CH_3OOC(CH_2)_4COOCH_3 + CH_3CH_2COOCH_3 \xrightarrow{^-OCH_3}$

$$CH_3OOC(CH_2)_3\overset{\displaystyle \overset{COOCH_3}{|}}{C}HCOCH_2CH_3$$

c $ClCH_2\overset{\displaystyle \underset{|}{\,}}{C}HCH_2CH_2Cl + NaOH \longrightarrow HOCH_2\overset{\displaystyle \underset{|}{\,}}{C}HCH_2CH_2OH$
    $\quad\quad\;\; OH \quad\quad\quad\quad\quad\quad\quad\quad\quad\quad\quad\quad OH$

d $CH_3COCH_2CH_2COOH + C_6H_5MgBr \longrightarrow$

$$C_6H_5\underset{\underset{CH_3}{|}}{\overset{\overset{O-CO}{|\quad|}}{C}}CH_2CH_2$$

e $H_2NCH_2CH_2CHO + KMnO_4 \longrightarrow H_2NCH_2CH_2COOH$

f $CH_2{=}CHCH_2CH_2CN + H_2/Pd \longrightarrow CH_2{=}CHCH_2CH_2CH_2NH_2$

23-4   STEREOCHEMISTRY

For a molecule with n asymmetric centers there are normally 2^n possible stereoisomers. If a synthesis of the molecule is not under some stereochemical control, all of these isomers will be formed and must be separated. With more than a few asymmetric centers this would be a practically impossible task. Epiandrosterone (Fig. 23-2) contains seven asymmetric carbons, so that $2^7 = 128$ stereoisomers are possible. Fortunately they are not all formed in the synthesis!

Starting materials are usually not asymmetric molecules. When an asymmetric center is formed in a reaction of symmetrical molecules, a racemic mixture is produced. A racemic mixture can only be separated into its two enantiomeric components by resolution, a process involving combination with an optically active material. Resolutions (Sec. 6-13) are usually effected in practice on acids (RCOOH) or bases (amines), by forming optically active salts with optically active organic bases or acids. In any synthetic sequence leading to any number of asymmetric centers (cf. Fig. 23-2) the final product must be racemic if the starting materials are not optically active. To obtain one optically active enantiomer only, a resolution must be performed, either at the end of the synthesis or at some intermediate stage (containing a carboxyl or amino group). In the synthesis of epiandrosterone the product is actually racemic epiandrosterone. The particular natural enantiomer, with the absolute configuration depicted in Fig. 23-2, can be obtained by a resolution only. This would probably best be done by resolving the diacid produced in the ozonolysis step.

Optically Pure 2-Phenylpentane

The hydrocarbon 2-phenylpentane was needed in an optically pure state for use in a study of a reaction mechanism. The racemic material was available, but the molecule lacked a "chemical handle" for use in resolution with optically active acids or bases. The problem was solved by introducing a carboxyl group into the benzene ring and removing it after resolution of the carboxylic acid.

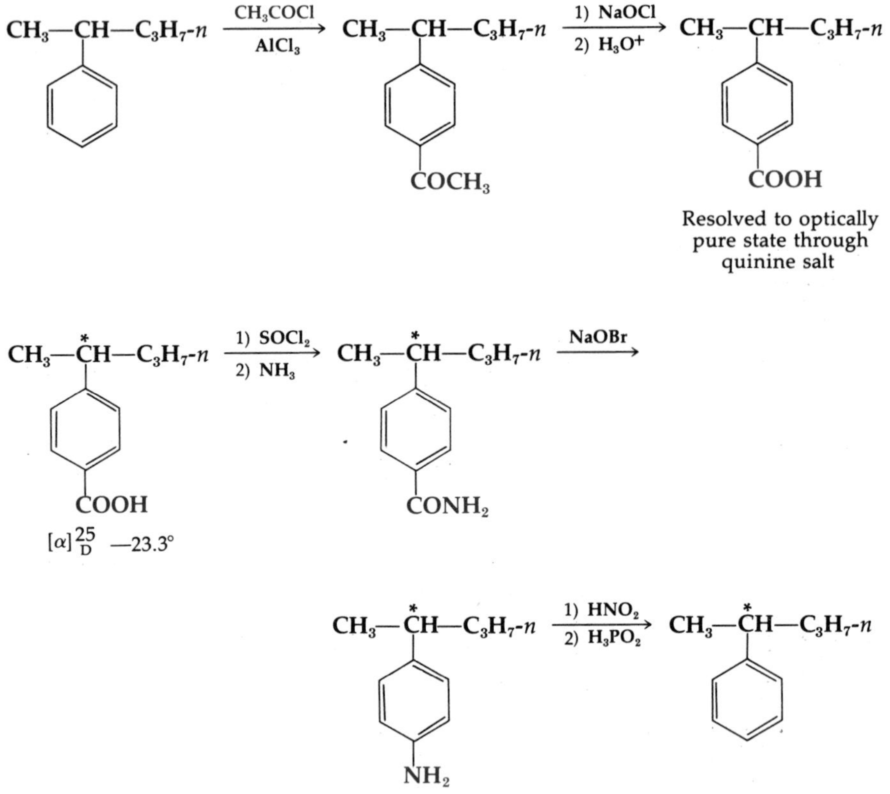

CH$_3$—CH—C$_3$H$_7$-n $\xrightarrow[\text{AlCl}_3]{\text{CH}_3\text{COCl}}$ CH$_3$—CH—C$_3$H$_7$-n $\xrightarrow[\text{2) H}_3\text{O}^+]{\text{1) NaOCl}}$ CH$_3$—CH—C$_3$H$_7$-n

COCH$_3$ COOH

Resolved to optically
pure state through
quinine salt

CH$_3$—*CH—C$_3$H$_7$-n $\xrightarrow[\text{2) NH}_3]{\text{1) SOCl}_2}$ CH$_3$—*CH—C$_3$H$_7$-n $\xrightarrow{\text{NaOBr}}$

COOH   CONH$_2$

$[\alpha]_D^{25}$ —23.3°

CH$_3$—*CH—C$_3$H$_7$-n $\xrightarrow[\text{2) H}_3\text{PO}_2]{\text{1) HNO}_2}$ CH$_3$—*CH—C$_3$H$_7$-n

NH$_2$

A remote possibility existed that, in the reactions employed to remove the carboxyl group, partial racemization of the compounds occurred. This possibility was obviated as follows. The optically active 2-phenylpentane obtained was converted back to the carboxylic acid by the original reaction sequence, and the rotation of the resulting carboxylic acid was identical to that obtained directly from the resolution. Thus, the asymmetric carbon atom in the benzyl position of the molecule maintained its configurational integrity throughout the full cycle of reactions.

Synthesis of Chloramphenicol

An antibiotic of commercial importance was isolated from cultures of certain microorganisms and was found to possess the structure indicated.

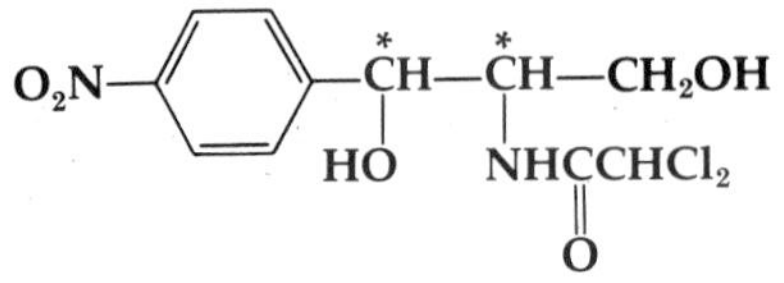

O$_2$N—⟨benzene ring⟩—*CH—*CH—CH$_2$OH
           HO   NHCCHCl$_2$
                        ‖
                        O

Chloramphenicol

The substance possesses two asymmetric carbon atoms, and the synthesis was undertaken without the detailed configuration having been proved.

$$CH_3NO_2 + CH_2O \xrightarrow{NaOH} O_2NCH_2CH_2OH \xrightarrow[NaOCH_3]{C_6H_5CHO} C_6H_5\underset{\underset{OH}{|}}{CH}-\underset{\underset{NO_2}{|}}{CH}-CH_2OH \xrightarrow{H_2 \atop Pd}$$

$$C_6H_5\underset{\underset{OH}{|}}{CH}-\underset{\underset{NH_2}{|}}{CH}-CH_2OH \xrightarrow{(CH_3CO)_2O} C_6H_5\underset{\underset{CH_3COO}{|}}{CH}-\underset{\underset{NHCOCH_3}{|}}{CH}-CH_2OCOCH_3$$

Mixture of diastereomeric
racemates, *A* and *B*; *A*
isolated in pure state
by recrystallization

Mixture of diastereomeric
racemates, *A* and *B*; *B*
isolated in pure state
by recrystallization

$$C_6H_5\underset{\underset{CH_3COO}{|}}{CH}-\underset{\underset{NHCOCH_3}{|}}{CH}-CH_2OCOCH_3 \xrightarrow{HNO_3} O_2N-\text{⟨benzene⟩}-\underset{\underset{CH_3COO}{|}}{CH}-\underset{\underset{NHCOCH_3}{|}}{CH}-CH_2OCOCH_3 \xrightarrow[H_2O]{OH^-}$$

$$O_2N-\text{⟨benzene⟩}-\underset{\underset{HO}{|}}{CH}-\underset{\underset{NHCOCH_3}{|}}{CH}-CH_2OH \xrightarrow{H_3O^+} O_2N-\text{⟨benzene⟩}-\underset{\underset{HO}{|}}{CH}-\underset{\underset{NH_2}{|}}{CH}-CH_2OH$$

Resolved into (+)- and (−)-enantiomers
through fractional crystallization of
d-camphorsulfonate salt

$$O_2N-\text{⟨benzene⟩}-\underset{\underset{HO}{|}}{CH}-\underset{\underset{NH_2}{|}}{CH}-CH_2OH \xrightarrow[\Delta]{Cl_2CHCOOCH_3} O_2N-\text{⟨benzene⟩}-\underset{\underset{HO}{|}}{CH}-\underset{\underset{NHCCHCl_2}{|}}{CH}-CH_2OH$$

(−)-*B* isomer

Chloramphenicol

All four stereoisomers were prepared in a pure state, and one of them proved to be identical with the substance extracted from the microorganism. The other three stereoisomers were biologically inactive. All four possible isomers were formed in the condensation of β-nitroethyl alcohol with benzaldehyde, although the two racemates were produced in unequal proportions. In this particular synthesis, even if the exact configuration of the desired product had been known, little could have been done in this condensation to favor formation of the desired over the undesired racemate.

When mixtures of racemates are produced, they can usually be separated by fractional crystallization of either the isomers themselves or crystalline derivatives. Two different racemates, unlike two enantiomers, are physically different compounds and may be separated by normal physical means, viz., distillation, chromatography, or crystallization. In this particular synthesis, one of the racemates (*A*) crystallized and the other did not. The filtrates from the crystallization, when acetylated, gave the triacetyl derivative. Fortunately, racemate *B* separated as a crystalline compound in this case. If it had not, the acetylated derivative might have been nitrated in the next step and separation attempted on the nitro products.

Because nitric acid oxidizes primary and secondary alcohols as well as primary amines, the aromatic nitration step had to be carried out on a compound in which the two hydroxyl groups and one amino group were protected. The three groups were acetylated, and after the nitration the protective groups were removed.

Stereochemical Control

In order to avoid a multiplicity of stereoisomers as well as to know which stereoisomer is formed, it is important to utilize reactions of known stereospecificity when forming asymmetric centers. When generating a new asymmetric center in a reaction we need to employ conditions which take advantage of the existing asymmetry in the molecule so as to produce the new center in only one relative configuration, i.e., to produce only one epimer.

In the epiandrosterone synthesis the first asymmetric center was formed in intermediate *A*, which is therefore racemic. The lithium reduction of *A* then creates not one, but five, new asymmetriċ centers. The asymmetry of the original quaternary methyl group in *A* dictates the relative configurations of each of the five new centers created. Happily, the relative configurations at all centers are the correct ones for the final product. Clearly, this stereospecific reduction was crucial to the synthesis!

The major reactions useful for their stereospecificity in syntheses are these:

1   Nucleophilic substitutions (S_N2) proceed irreversibly with inversion of configuration. Androsterone can, for example, be made from epiandrosterone by such an inversion.

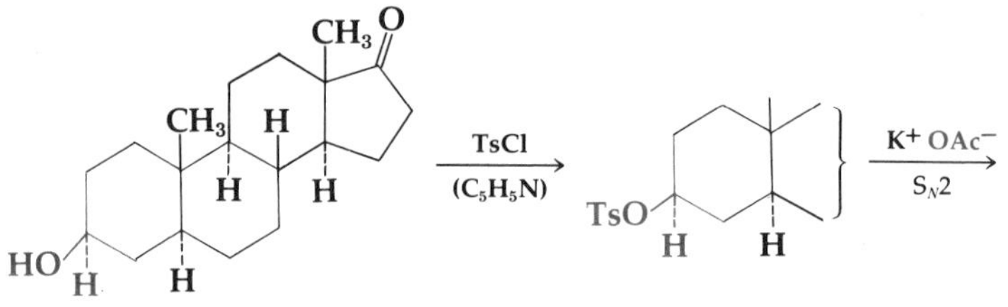

Epiandrosterone

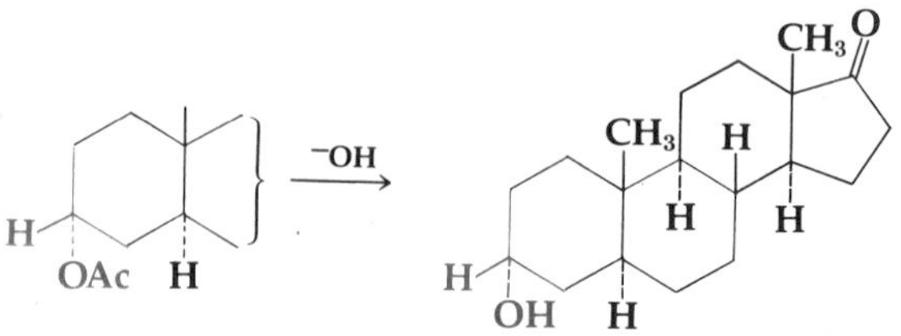

Androsterone

2   Reductions of cyclohexanones usually yield predominantly the more stable, equatorial epimer with sodium borohydride or lithium aluminum hydride, but with hindered boranes the less stable epimer is formed.

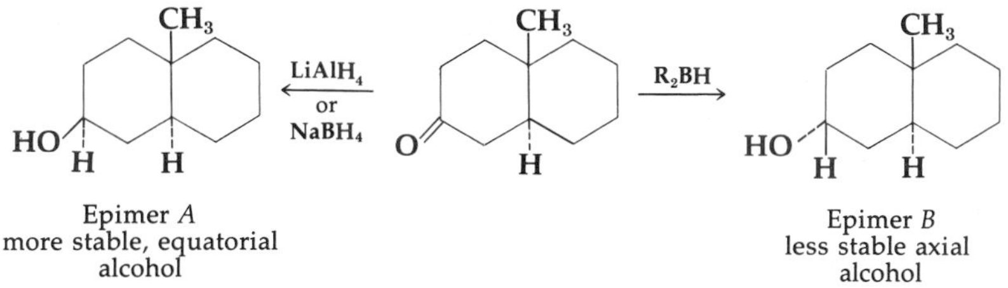

Epimer A
more stable, equatorial
alcohol

Epimer B
less stable axial
alcohol

3 Additions to cyclic double bonds yield initially the trans diaxial product with the electrophilic atom on the less hindered side, as summarized in the examples of Fig. 15-5.

4 Diels-Alder reactions (and other concerted electrocyclic reactions) proceed with predictable stereochemistry. The cis addition of butadiene in the cantharidin synthesis exemplified this, for the configurations at the centers formed in the cycloaddition are all controlled by the geometry of cycloaddition (page 853). Since cantharidin is a meso compound, no final resolution is necessary, or possible, of course.

When the desired relative configuration at an asymmetric center is the more stable of the two possible epimers, it may often be secured by an equilibration at that center, or by use of a reversible reaction (thermodynamic control). This procedure is most commonly applied to asymmetric carbons bearing hydrogen α to a carbonyl in cyclohexane derivatives. The method involves equilibration, in acid or base, via the enol or enolate and leads to the more stable (equatorial) epimer. In the following example if the initial compound were epimer A, the epimer B would be isolated from the reaction.

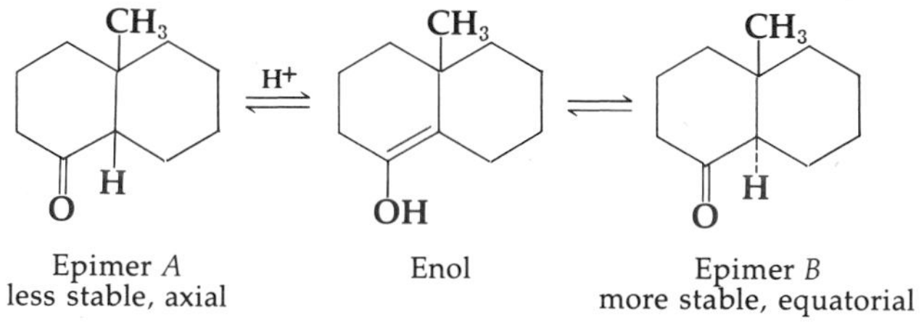

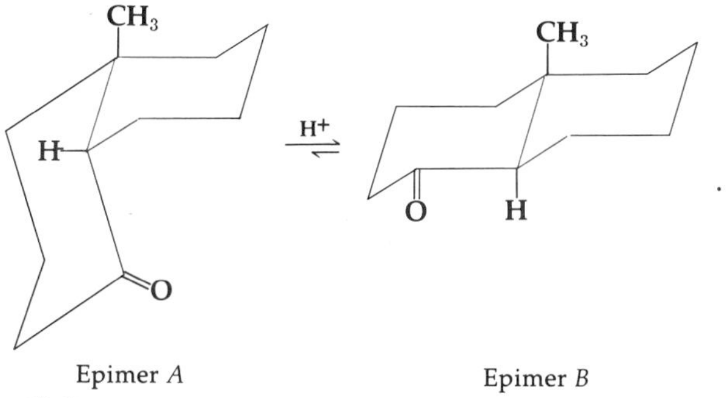

Epimer A
less stable, axial

Enol

Epimer B
more stable, equatorial

Epimer A

Epimer B

Synthesis of Reserpine

Much of the incentive to synthesize natural products has come from the traditional assumption that the structure of a compound is not established with certainty unless the substance is prepared by a rational series of reactions. The synthesis of reserpine by Woodward (1956) provides an instructive example of the construction of an involved ring system. The stereochemical features of the compound were neatly mastered at appropriate stages in the sequence.

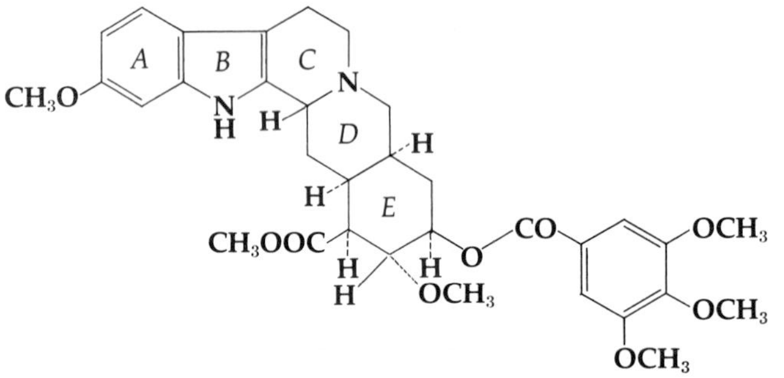

The stereochemical problem may be appreciated from the six asymmetric centers shown with their hydrogens in color. These are not all in the more stable configuration as the perspective views below will demonstrate; two substituents occupy axial positions. The D/E ring junction is cis, as in *cis*-decalin, and hence flexible, capable of two all-chair conformations (see Fig. 6-14).

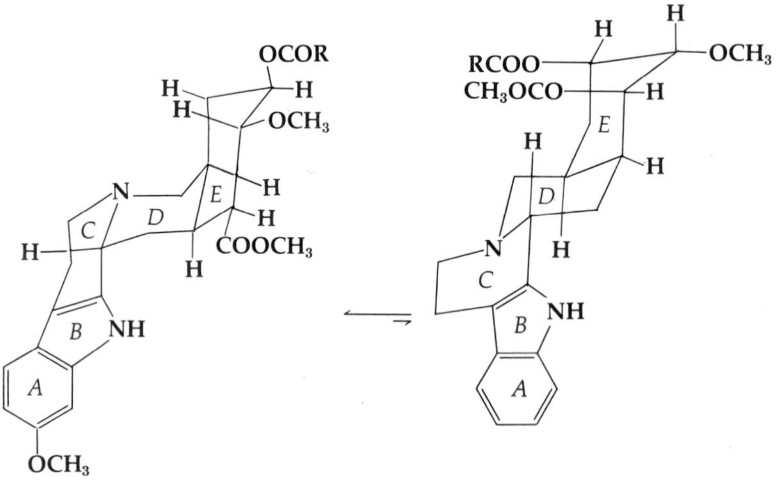

More stable   Less stable

Reserpine conformations

Five of the six asymmetric centers are in ring E of the *cis*-decalin. Hence the major stereochemical problem of synthesis lies there. In order to create a *cis*-decalin stereospecifically a Diels-Alder reaction was selected. A *cis*-decalin was made and then one carbon removed and replaced by nitrogen of an amine linked to the indole heterocycle (page 959). Thus the synthetic conception was the following:

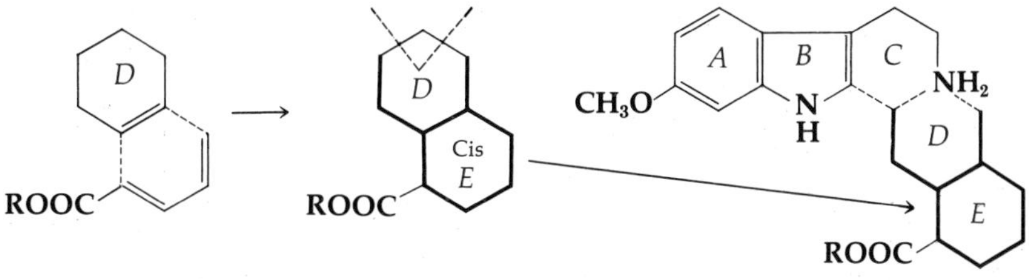

The Diels-Alder reaction used employed benzoquinone and vinylacrylic acid (2,4-pentadienoic acid), yielding the *cis*-decalin adduct (1).

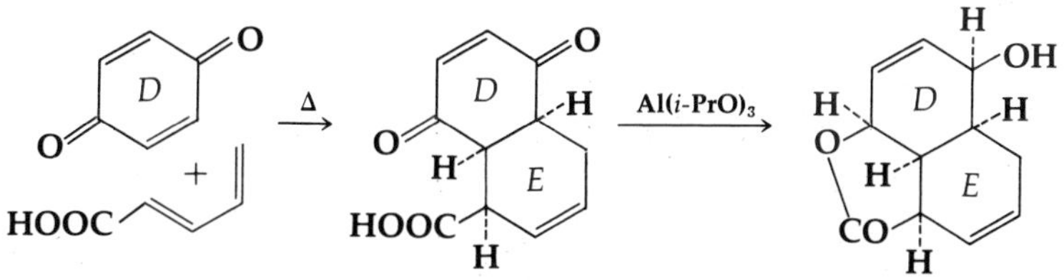

The *cis*-decalin affords stereochemical control owing to its curved, dishlike shape. Approach of reagents is only accessible from the less hindered outer or convex surface, which is the side occupied by the hydrogens at the ring junction. This is clear from the approach of reducing hydrogens (colored) in the aluminum isopropoxide reduction (page 460) of 1 to 2 above. One hydroxyl thus formed cyclizes to a lactone spontaneously.

Perspective drawings make this steric hindrance more apparent, as shown below for the next reaction in which the ring E, and the more reactive, double bond was attacked by bromine from the less hindered side. The intermediate bromonium ion was opened by internal attack of the hydroxyl group, forming a cyclic ether (3).

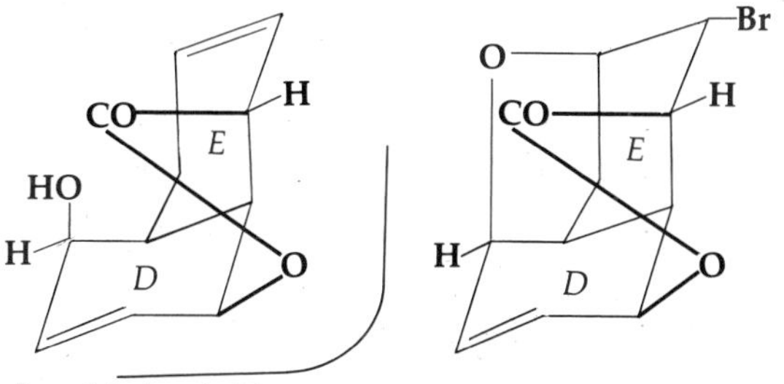

Less hindered side

2   3

The bromine was now eliminated from **3** by methoxide ion. This is a β-elimination, initiated by formation of the enolate of the lactone carbonyl. Methoxide ion immediately added to the double bond formed, in a conjugate addition. The net effect is simply one of replacement of —**Br** by the required —OCH_3 of reserpine.

Product **4** has the same stereochemistry as **3** and has neatly attained correct stereochemistry at all five asymmetric centers in ring E. The next task was to open ring D in order to insert the nitrogen. This was begun by attack on the ring D double bond via bromohydrin formation (**5**). The stereochemistry of the product is predicted as before, with attack of Br^+ on the less hindered side; stereo formula **5** below shows the necessary positions of trans diaxial addition.

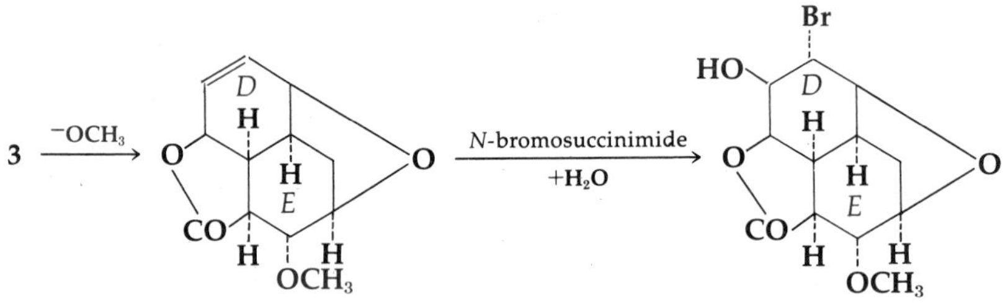

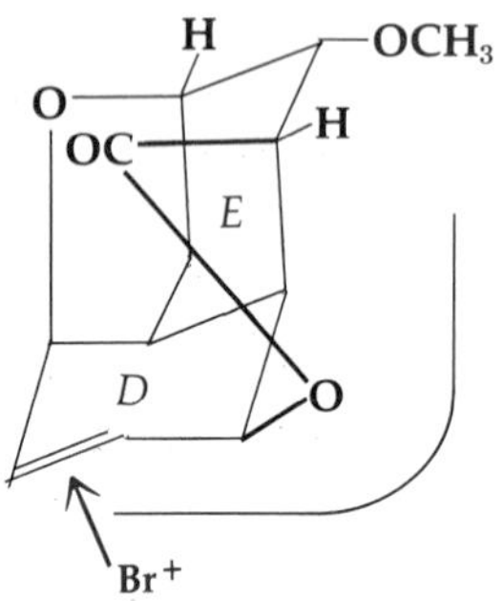

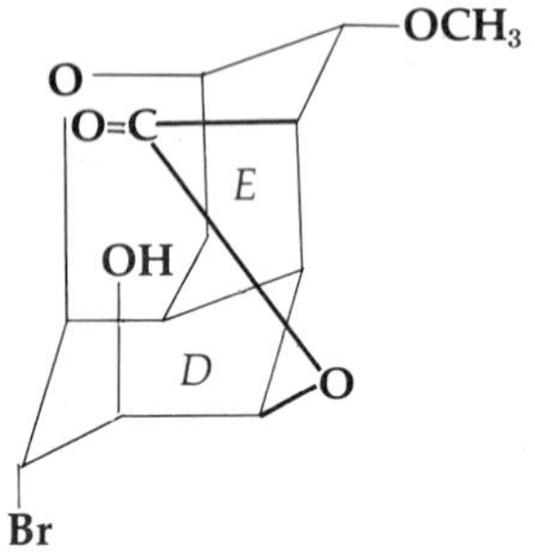

Chromic acid oxidation of the hydroxyl group of **5** gave ketone **6**. When treated with zinc and acetic acid; **6** underwent a twofold reaction.

1 The bromine and ether link trans to each other (axial-axial relationship, page 586) underwent elimination to give an α,β-unsaturated ketone.
2 The hydroxyl originally bound in the lactone function was reductively removed from its site α to the ketone (page 599).

Compound **7** was the product.

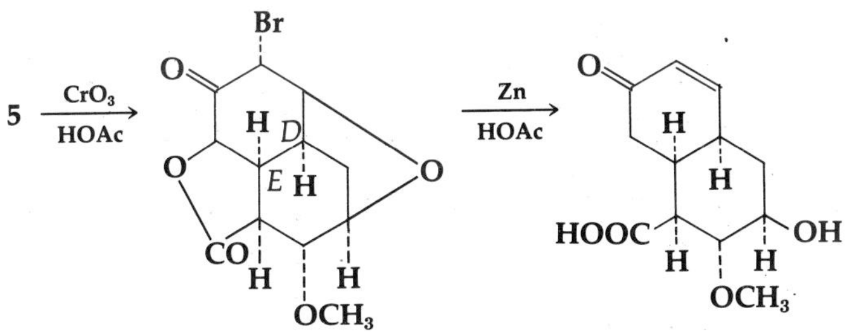

6

7

The carboxyl group of **7** was then esterified with diazomethane and the free hydroxyl group was acetylated (esterified) with acetic anhydride and pyridine to produce **8**. With these groups protected, the double bond of **8** was hydroxylated with osmium tetroxide (page 635) in a cis addition. In this reaction, the oxidizing agent approached the double bond from the least hindered side of the molecule. The resulting diol (**9**) was cleaved with periodic acid (page 761) to produce aldehyde-acid **10**, which was converted to ester **11**.

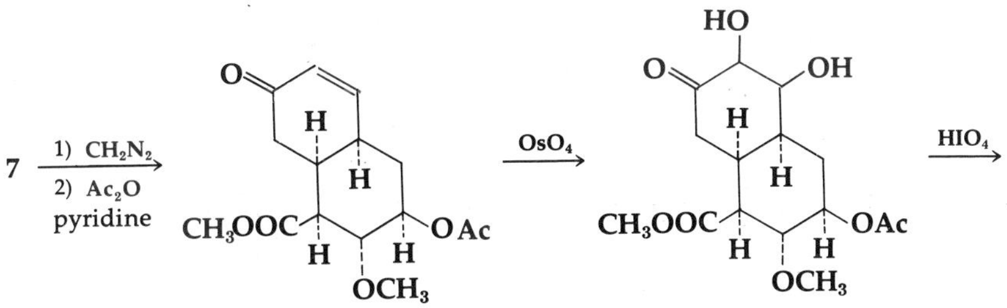

8

9

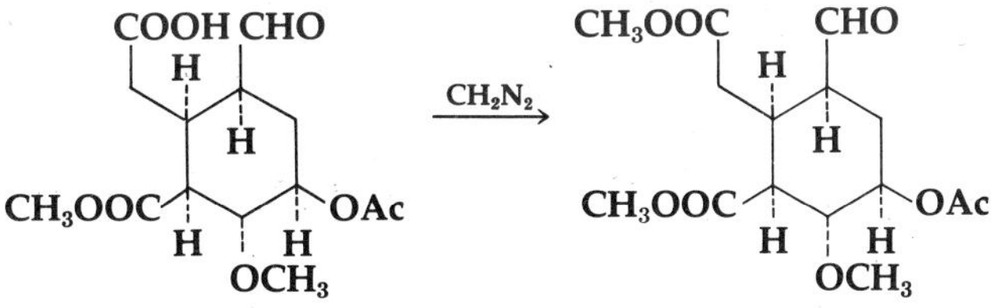

10

11

At this point in the sequence, we are ready for amine **12**. This compound contains two rings, which eventually become rings *A* and *B* of reserpine. In the first step, 6-methoxyindole was converted to its Grignard organometallic anion by removal of its acidic hydrogen with a Grignard reagent (acting as base). The product was employed as a nucleophile in substitution of α-chloroacetonitrile. The resulting nitrile was reduced with sodium and ethanol to give amine **12**.

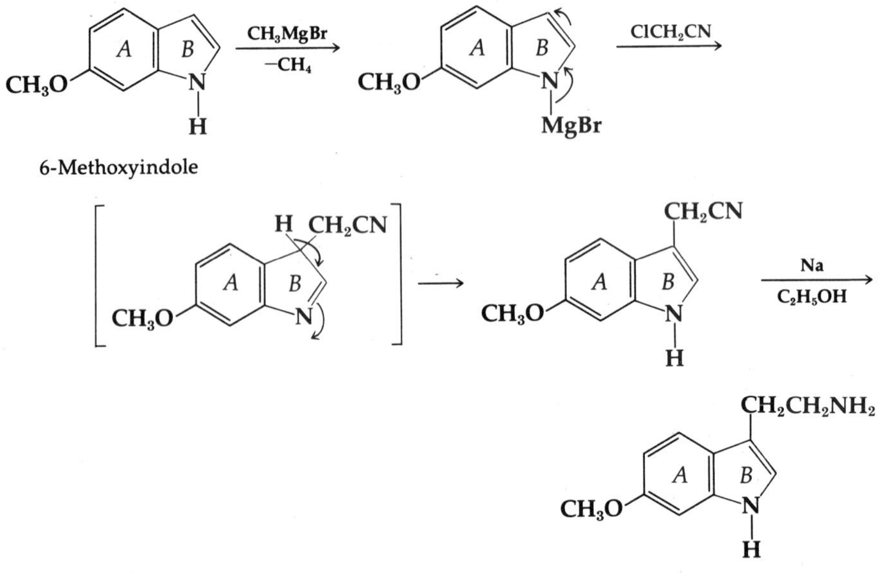

6-Methoxyindole

6-Methoxytryptamine
12

The elements of rings *A*, *B*, *C*, *D*, and *E* were now put together through condensation of amine **12** with aldehyde **11**. The imine linkage in product **13** was reduced selectively with sodium borohydride (page 459). As the reduction proceeded, it was followed by a spontaneous cyclization of the amine and ester that led to lactam **14**. When treated with phosphorus oxychloride, this substance underwent ring closure to give immonium salt **15** via the activated imino-phosphate (page 515). With this cyclization, product **15** now contains the full ring system of reserpine. Reduction of the imine linkage with sodium borohydride led to **16**, formed by approach of the hydride from the least hindered side of the ring system. This compound was resolved, and the (−)-rotating enantiomer was isolated in optically pure form.

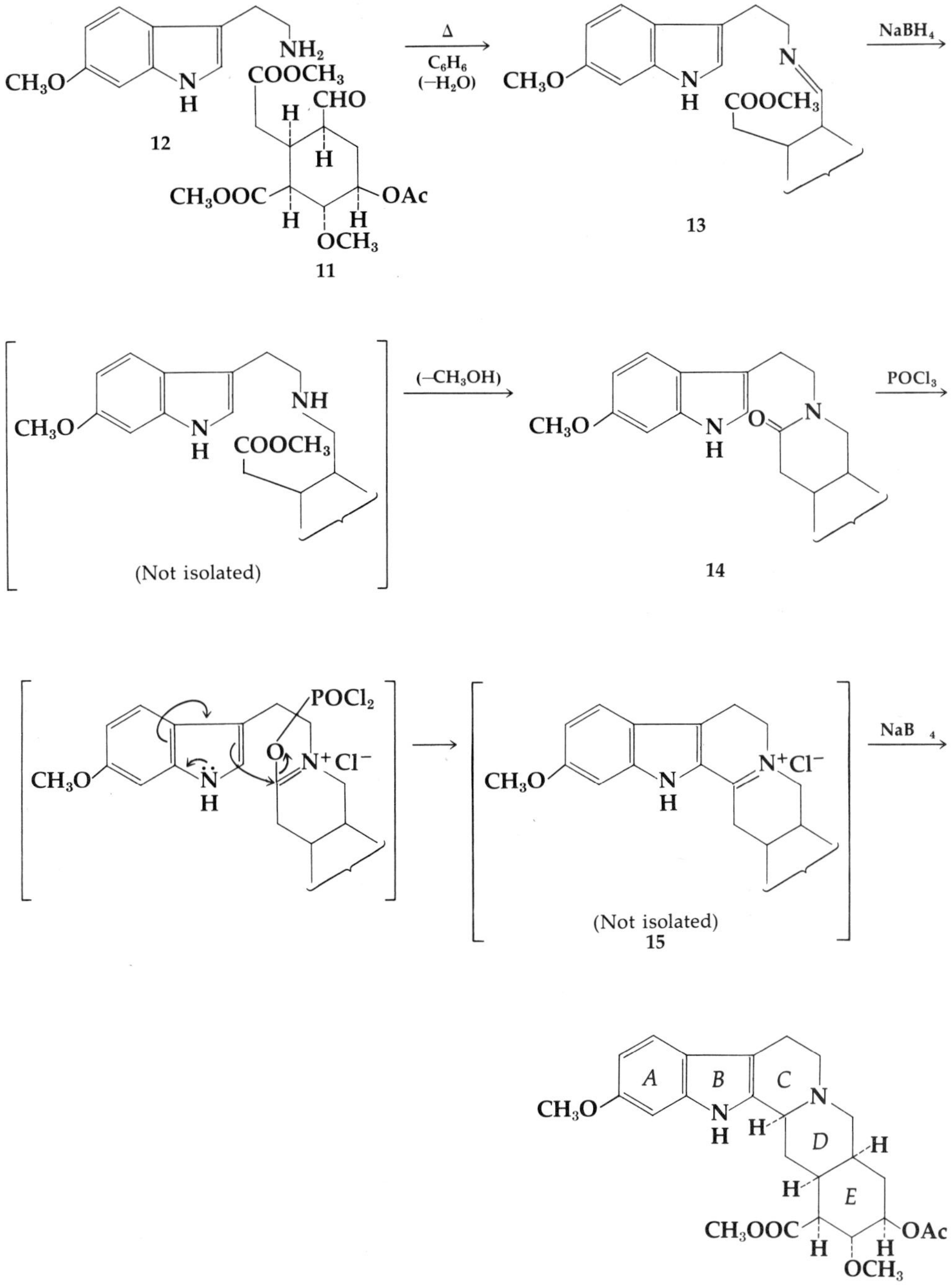

12

11

13

(Not isolated)

14

(Not isolated)
15

16
Resolved through salt
with optically active
di-*p*-toluyl-*l*-tartaric acid

One stereochemical problem remains. The newly introduced hydrogen (colored) in **16** is in the wrong configuration for reserpine. This center can be equilibrated by heating in weak acid, owing to the adjacent indole heterocycle, but structure 16 has the *more* stable configuration at that center while reserpine (page 934) has the *less* stable configuration. However, if the *cis*-decalin ring system could be forced into its less stable conformation (page 934), and fixed there, then the asymmetric center of concern in 16 would be the less stable one and could be equilibrated to the more stable reserpine configuration there.

The two conformations on page 934 suggest how the less stable one might be locked in without a possibility of equilibration. The cis-oriented —COOCH₃ and —OCOR on ring E must be linked into a lactone in order to freeze the *cis*-decalin into its less stable conformation. This reaction was accordingly forced on compound 16 by saponification of the two ester groups followed by dehydrative lactone formation, utilizing N,N'-dicyclohexylcarbodiimide (page 510).

The product is the conformationally rigid lactone 17, which will be seen now to have the asymmetric center of interest (colored) in the less stable configuration. (Compare stereo structures on page 934.) Heating with pivalic acid (trimethylacetic acid) causes epimerization at that center to yield the more stable 18, now bearing the full configuration of reserpine. The synthesis was now completed by methanolysis of the lactone and esterification of the hydroxyl thus freed with trimethoxybenzoyl chloride and pyridine. The final substance so obtained was identical in all respects with natural reserpine. The synthesis represents an especially fine example of full stereochemical control in a series of steps remarkable for economy. The overall yield was above 10%.

17   18

There is no substitute for trial and practice in synthesis. Accordingly, facility can be reached by trying to design syntheses of *any* given molecule. These can be simply invented structures written out solely to serve as synthetic goals. Alternatively, the compounds shown in Chap. 7 (and its structure-proof problems) can serve as challenges for synthesis. More examples are listed in Prob. 23-4.

PROBLEMS

23-4 From readily available starting materials develop synthetic schemes for the following compounds.

a $HOOC-\!\!\!\bigcirc\!\!\!-(CH_2)_3\!\!\!-\!\!\!\bigcirc\!\!\!-COOH$

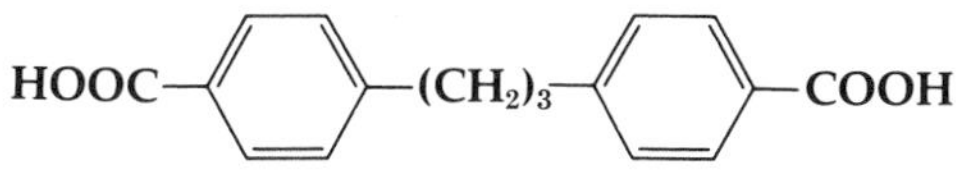

b

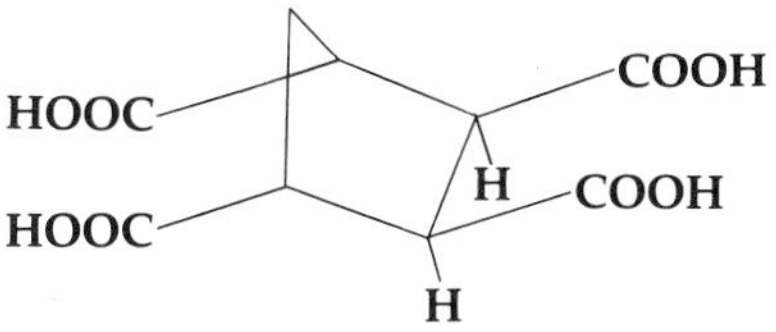

$HOOC$ ... $COOH$
$HOOC$ H $COOH$
H

c

CH_2Br

CH_3

d $CH_3OCH_2CHCHC_6H_5$
        $\overset{OH}{|}$
        $\underset{\underset{O}{\|}}{NHCCH_3}$

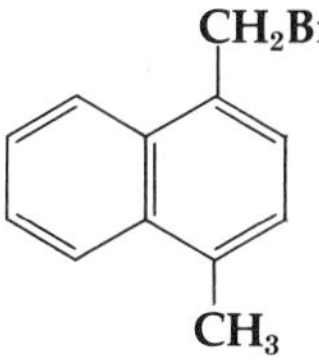

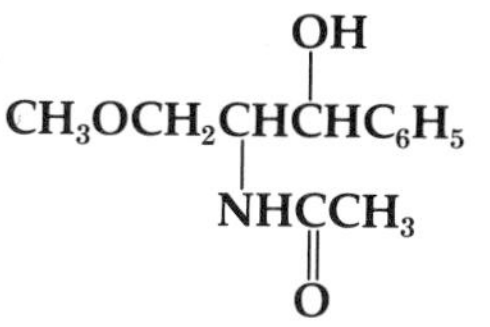

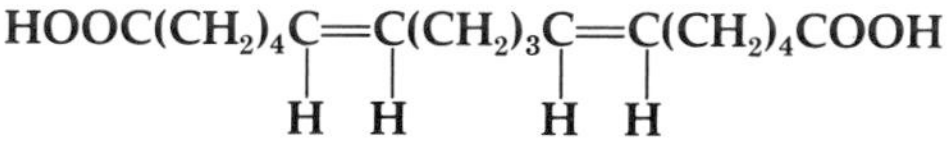

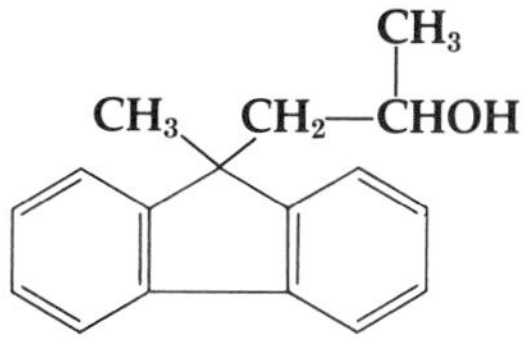

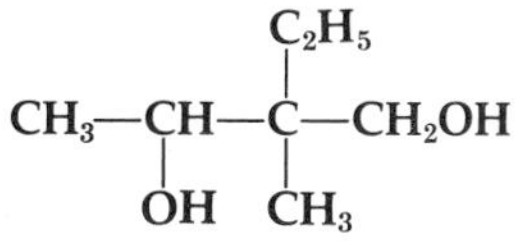

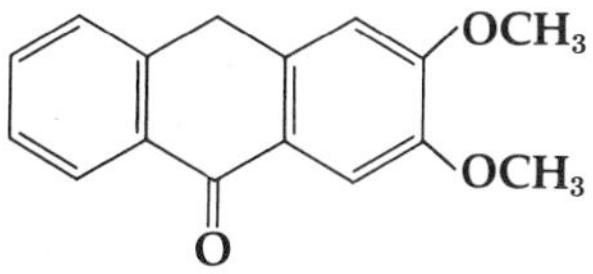

e $HOOC(CH_2)_4C\!\!=\!\!C(CH_2)_3C\!\!=\!\!C(CH_2)_4COOH$
                    $\overset{|}{H}\ \overset{|}{H}$ $\overset{|}{H}\ \overset{|}{H}$

f

CH_3
$CH_3\ \ CH_2\!\!-\!\!CHOH$

g $CH_3\!\!-\!\!CH\!\!-\!\!\underset{\underset{CH_3}{|}}{\overset{\overset{C_2H_5}{|}}{C}}\!\!-\!\!CH_2OH$
          $\underset{OH}{|}$

h

OCH_3
OCH_3
O

i CH_3 OH
    CH_3

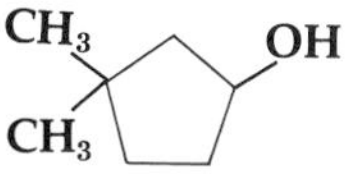

j

H_3C
        H H CH_3
H_3C H H CH_3

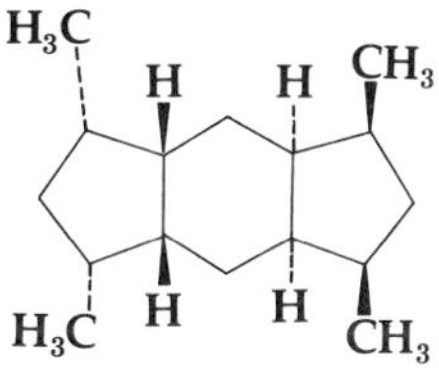

23-5 Consider the synthesis of the hydrocarbon below. Try to analyze it in a systematic fashion so as to allow a number of different syntheses to be devised. Try to select the best one on grounds of economy and utilization of reliable reactions.

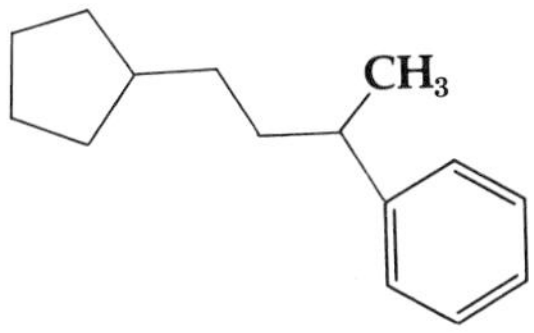

CH_3

23-6 Develop syntheses of the following compounds from coal (carbon only), air, and water. Develop the necessary simple starting materials by synthesizing them. Other inorganic substances may be used as long as they contain no carbon.

a

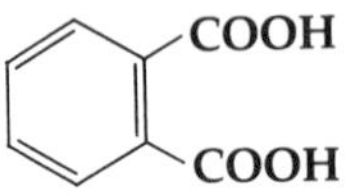

b

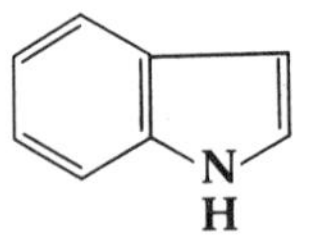

c $(CH_3)_2C=CHCOCH=C(CH_3)_2$

d $(CH_3)_3CCOCH_2CH_2CH_3$

e · Cyclopentanone

f

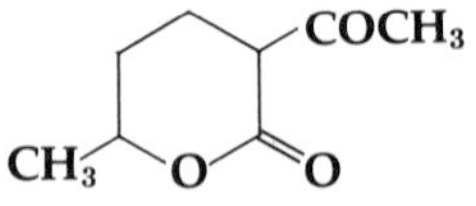

g $C_6H_5CH=CHCOCH_3$

h

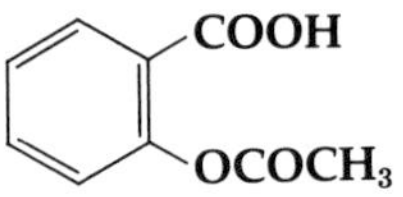

Aspirin

23-7 Show an operation, of one or more steps, for carrying out these general transformations.

a $RCOOH \longrightarrow RCH_2COOCH_3$

b $RCHO \longrightarrow RCH_2CHO$

c

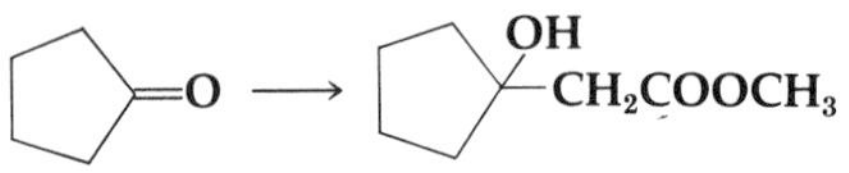

d $\phi Br \longrightarrow \phi CH_2COOCH_3$

e $ROH \longrightarrow RCH_2CH_2OH$

f $R_2CHOH \longrightarrow R_2C=CHCH_3$

g $ROH \longrightarrow RCH_2CH_2CH_2OH$

h

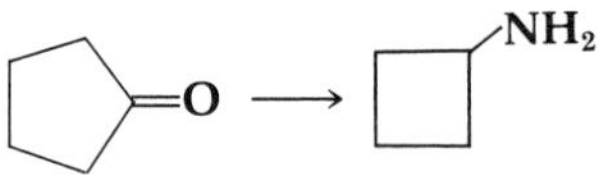

23-8 Synthesize the following compound from o-methoxybenzaldehyde. Comment on the stereochemical problems involved and indicate which isomer or isomers you expect to obtain by your route.

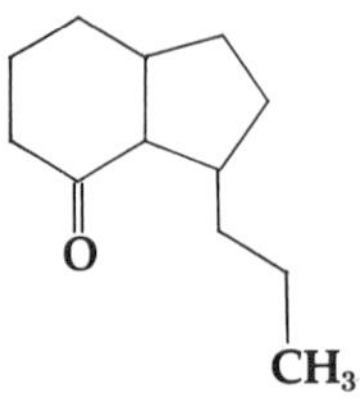

23-9 The synthesis of bullvalene (pages 161 and 867) was carried out in the following fashion [W. Doering et al., *Tetrahedron*, **23**: 3943 (1967)]. Fill in the missing reagents and intermediates.

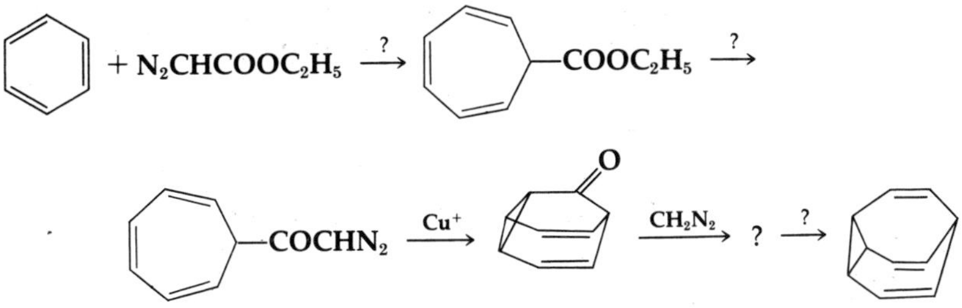

$$+ \ N_2CHCOOC_2H_5 \xrightarrow{?} \quad \text{—COOC}_2\text{H}_5 \xrightarrow{?}$$

$$\text{—COCHN}_2 \xrightarrow{Cu^+} \quad \xrightarrow{CH_2N_2} \ ? \ \xrightarrow{?}$$

Bullvalene

The syntheses of natural products which follow are partially written out. Fill in the missing reagents or intermediates (there may be several steps) and comment on those conversions which have asterisks, with respect either to mechanism and/or stereochemistry. Most of the references are to short communications in the *Journal of the American Chemical Society* (*J. Am. Chem. Soc.*).

23-10 *β-Eudesmol* [J. A. Marshall and M. T. Pike, *Tetrahedron Letters*, **1965**: 3107]

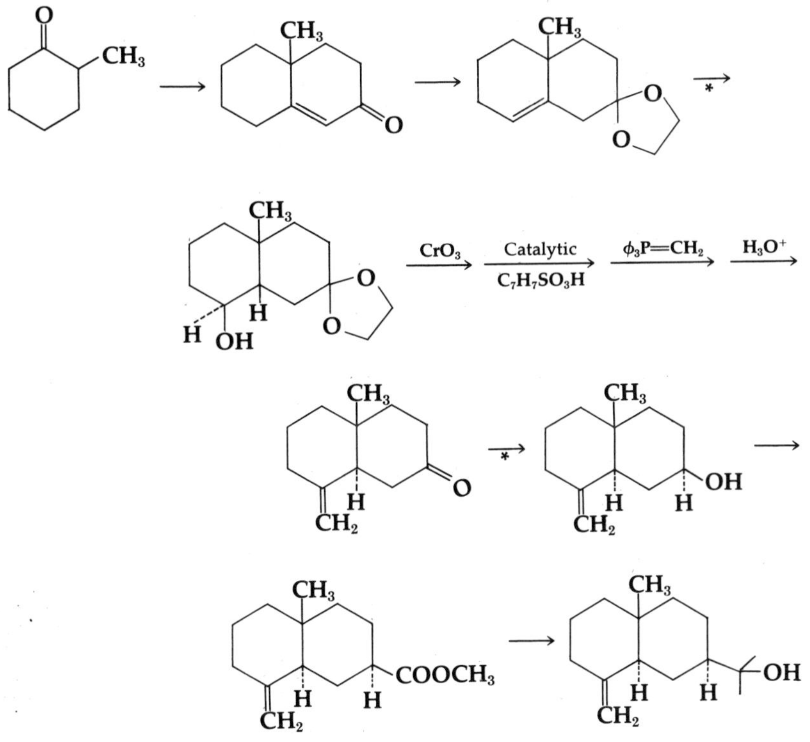

β-Eudesmol

23-11 *Epilupinine* [E. E. van Tamelen and R. L. Foltz, *J. Am. Chem. Soc.*, **82**: 502 (1960)]

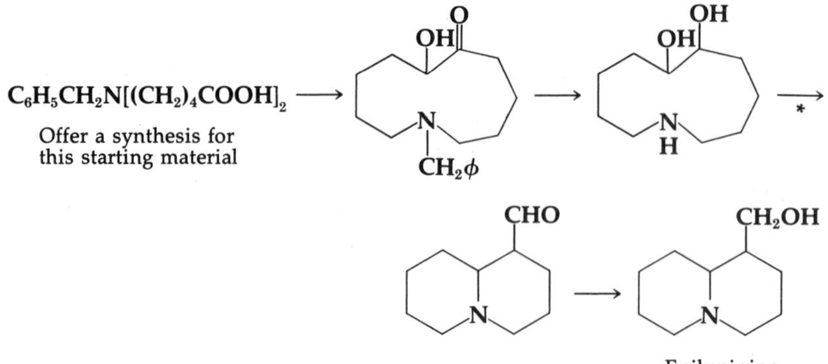
$C_6H_5CH_2N[(CH_2)_4COOH]_2$
Offer a synthesis for this starting material
O
OH
N
$CH_2\phi$
OH
OH
N
H
*
CHO
N
CH_2OH
N

Epilupinine

23-12 *Aromadendrene* [G. Büchi, W. Hofheinz, and J. V. Paukstelis, *J. Am. Chem. Soc.*, **88**: 4113 (1966)]

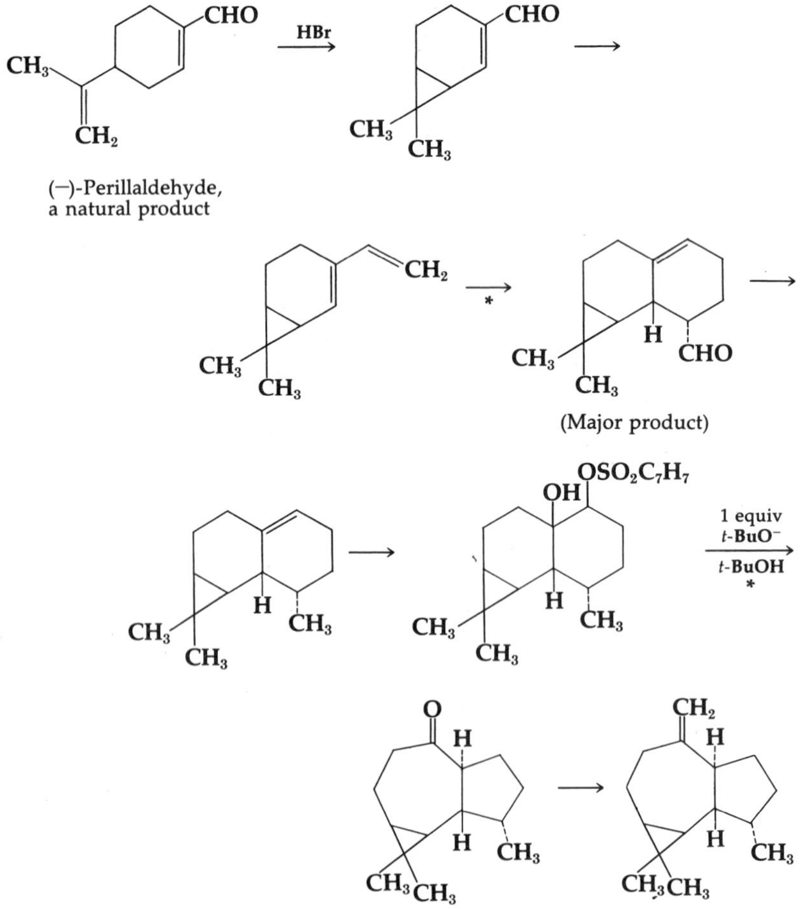
CHO
CH_3
CH_2
(−)-Perillaldehyde, a natural product
HBr
CHO
CH_3
CH_3
CH_2
CH_3
CH_3
*
H
CHO
CH_3
CH_3
(Major product)
H
CH_3
CH_3
CH_3
$OSO_2C_7H_7$
OH
H
CH_3
CH_3
CH_3
1 equiv
t-BuO⁻
t-BuOH
*
O
H
H
CH_3
CH_3
CH_3
CH_2
H
H
CH_3
CH_3
CH_3

(−)-Aromadendrene

23-13 *Aspidospermine* [G. Stork and J. E. Dolfini, *J. Am. Chem. Soc.,* **85**: 2872 (1963)]

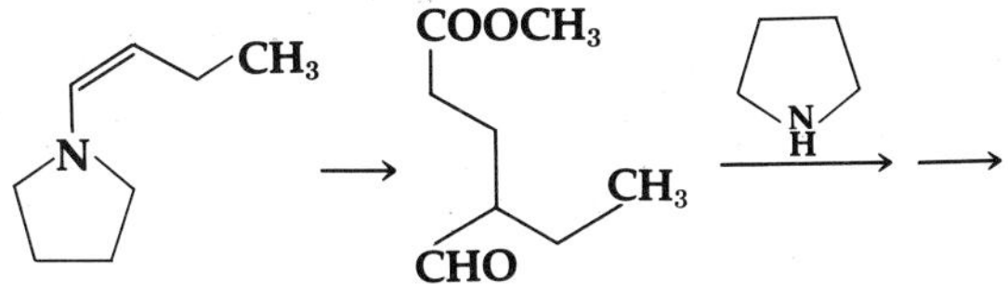

Made from
four-carbon
compounds

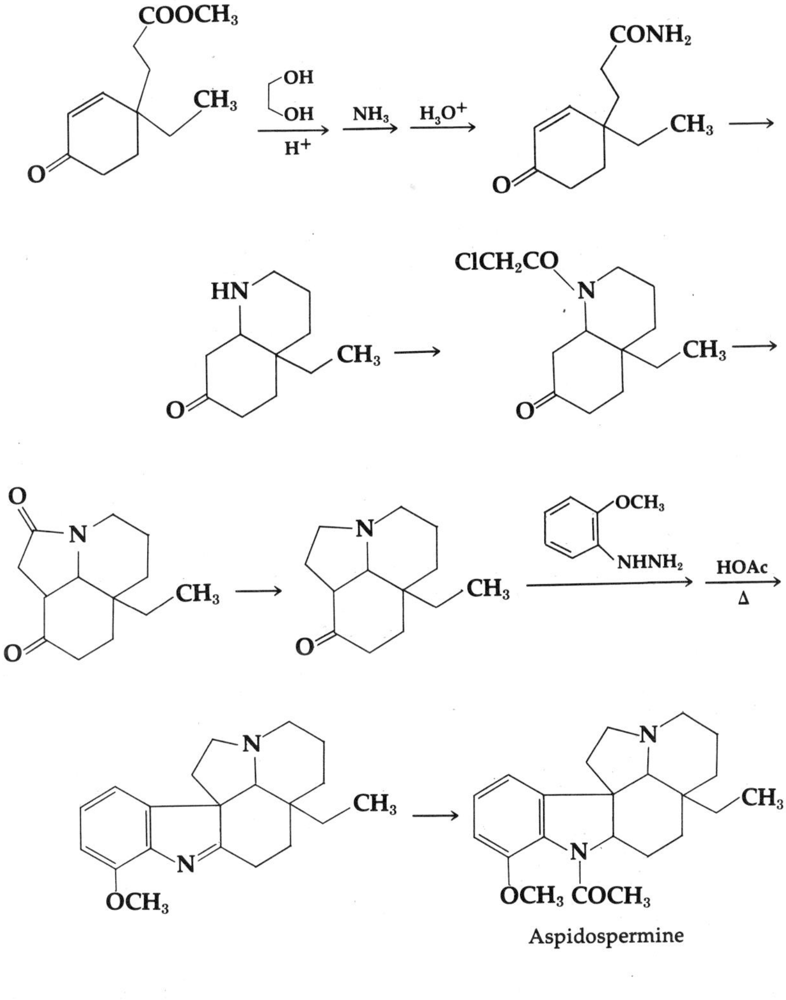

Aspidospermine

23-14   *Ajmalicine* [E. E. van Tamelen and C. Placeway, *J. Am. Chem. Soc.,* **83**: 2594 (1961)]

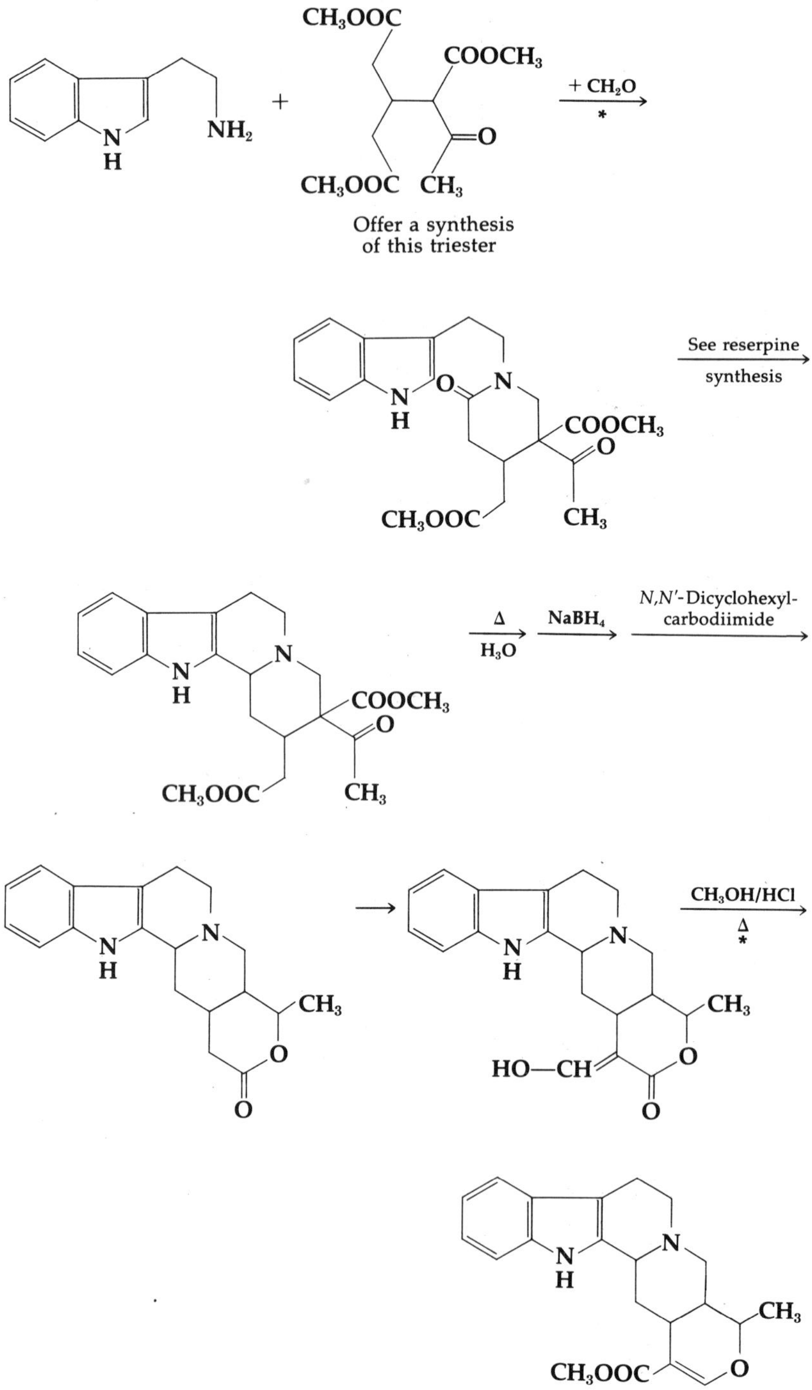

Offer a synthesis
of this triester

See reserpine
synthesis

Ajmalicine

Comment on the stereochemistry
in this synthesis and suggest
the correct stereochemistry in
the product

23-15 *Dihydrocostunolide* [E. J. Corey and A. G. Hortmann, *J. Am. Chem. Soc.*, **87**: 5736 (1965)]

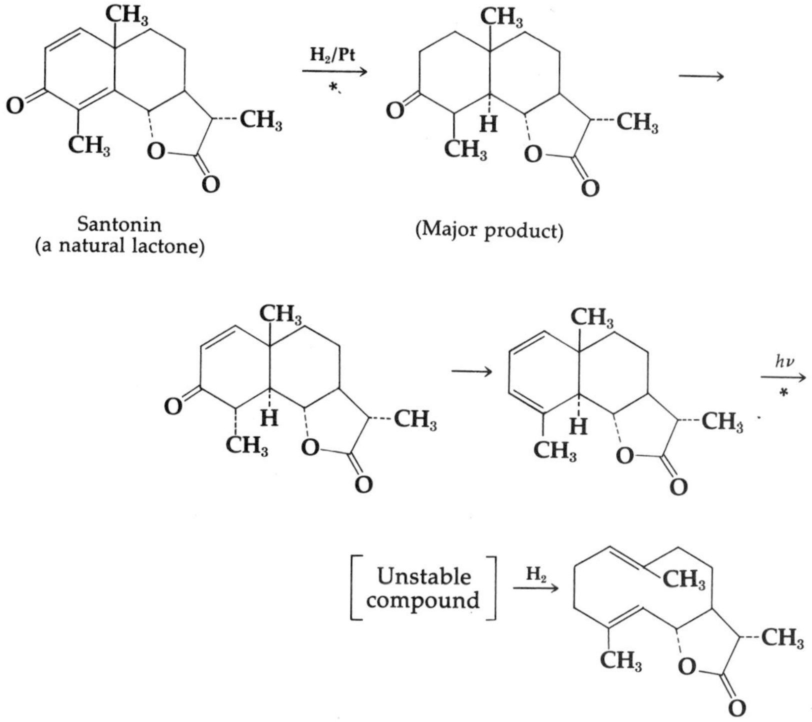

CH3
O
CH3
O
CH3
O
Santonin
(a natural lactone)
H2/Pt
*
H
(Major product)
hν
Unstable compound
H2

Dihydrocostunolide

23-16 Develop a synthesis for (+)-*laudanosine*, one of the constituents of opium.

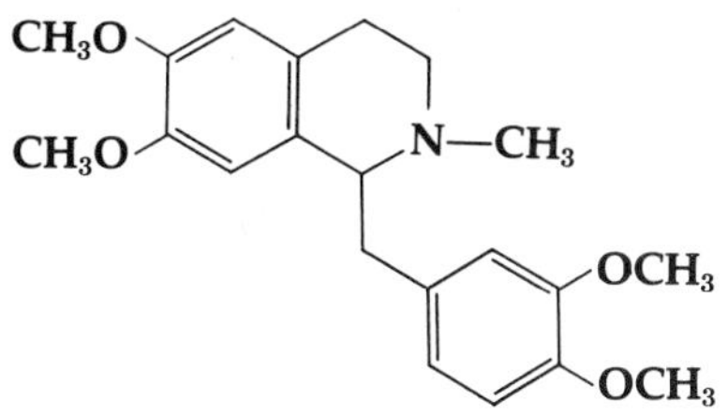

CH3O
CH3O
N—CH3
OCH3
OCH3

Laudanosine

23-17 The structure of the alkaloid *crinine* is shown below. Also shown is the key reaction around which a total synthesis was conceived [H. Muxfeldt, R. S. Schneider, and J. B. Mooberry, J. Am. Chem. Soc., **88**: 3670 (1966)].

a Identify the key reaction.

b Deduce steps to complete the synthesis from the key reaction product.

c Deduce a route to the key reaction from piperonal (3,4-methylenedioxy-benzaldehyde). The authors began by creating compound *A*.

d Comment on the stereochemical problems.

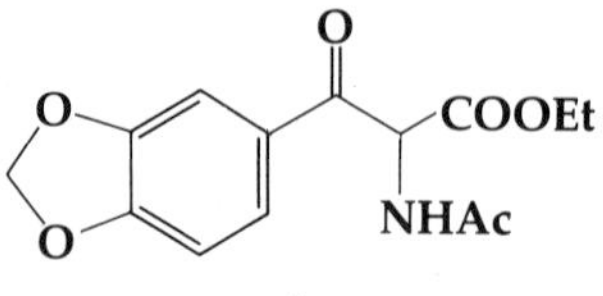

A 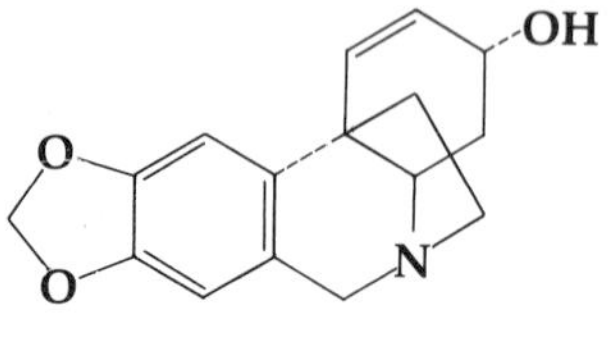   Crinine

Key reaction

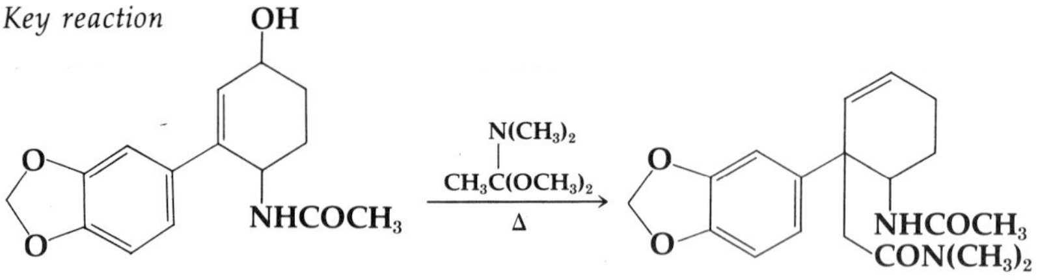

23-18 Outline alternate syntheses for two of the compounds whose preparations are described in this chapter.

READING REFERENCES

House, H. O., "Modern Synthetic Reactions," Benjamin, New York, 1965.
Ireland, R. E., "Organic Synthesis," Prentice-Hall, New York, 1969.

HETEROCYCLIC COMPOUNDS

24-1 INTRODUCTION

A very substantial portion—perhaps over a third—of all the chemical literature deals with **heterocycles**, those molecules having rings composed of both carbon and heteroatoms, chiefly oxygen, sulfur, and nitrogen. The number of such possible rings is obviously enormous and so the variety possible in heterocyclic chemistry is vast. Before the advent of modern reaction theory this led to a large accumulation of unrelated fact which discouraged digestion by any but the most dedicated. Heterocyclic chemistry tended to become a separate field only understood by its experts. Now this has all changed, for the framework of reaction theory in which the previous chapters are cast allows any chemist an easy overall grasp of the behavior of heterocyclic molecules.

Our main concern here will be with the aromatic heterocycles, i.e., the fully unsaturated ones, since it is only these which exhibit special behavior. Saturated heterocyclic rings like cyclic ethers or amines, or lactones behave completely analogously to their acyclic analogs and need no special treatment.

The possible aromatic heterocycles can be seen as two families, created in a logical fashion by two formal procedures from a carbocyclic aromatic, such as benzene:

1 A carbon in the aromatic is replaced by an isoelectronic heteroatom.

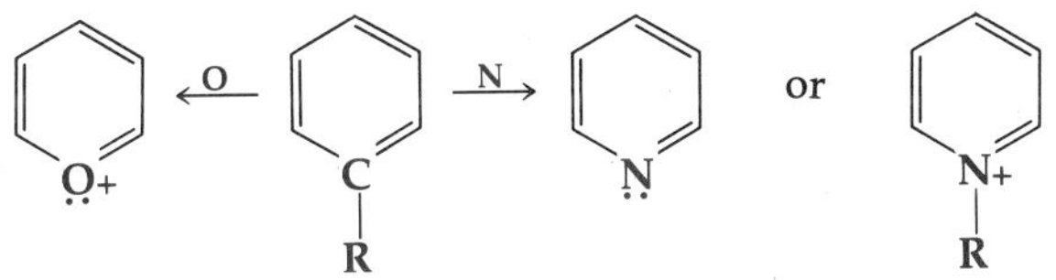

The electron-pair replacing the **C—R** bond occupies the same geometry, in the plane of the ring and perpendicular to the π electrons (not in resonance with them). Examples of this family are shown in Table 24-1.

2 A formal **C=C** double bond is replaced by a heteroatom with an unshared electron pair.

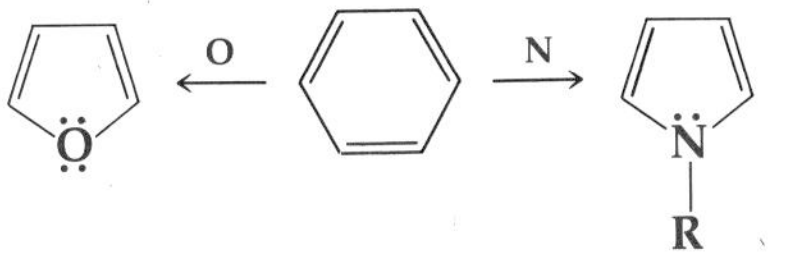

Here one unshared electron pair on the heteroatom is placed in a perpendicular p orbital for resonance with the remaining π electrons (the other heteroatom orbitals are sp^2). Examples of this family are shown in Table 24-2.

The effect in each case is to retain a stabilized, delocalized cycle of π electrons equal in number to those in the original aromatic ring (in these cases, the six of benzene). Examples of each are shown in Tables 24-1 and 24-2, respectively.

Nomenclature

Names for the several aromatic heterocycles so created are logically and simply derived by combining these prefixes and suffixes:

Prefixes		*Suffixes*	
1 *oxa-*	Oxygen	*-ole*	5-membered ring
2 *thia-*	Sulfur	*-ine*	6-membered ring with nitrogen
3 *aza-*	Nitrogen	*-epine*	7-membered ring

When more than one heteroatom is incorporated, the prefixes are placed in the order shown in the list (*oxa-* first, etc.). Ring positions are then designated by number, starting with number 1 at the heteroatom highest on the prefix list. The ring is then numbered in order so that other heteroatoms have the lowest possible numbers. Examples of names and numbers are shown in Tables 24-1 and 24-2. Certain trivial names from a presystematic period of nomenclature are, however, still retained for some common heterocycles. These are shown with their systematic names in parentheses, except for the four most important rings: pyridine, furan, thiophene, and pyrrole.

TABLE 24-1 **Six-membered Aromatic Heterocycles**

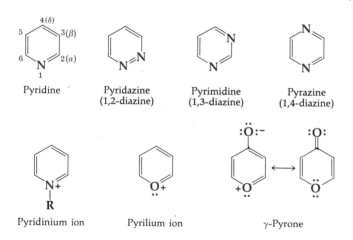

PROBLEM 24-1

Write names for the given structures and structures for the given names.

a

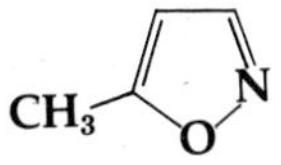

b

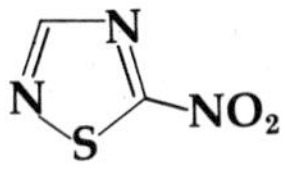

c

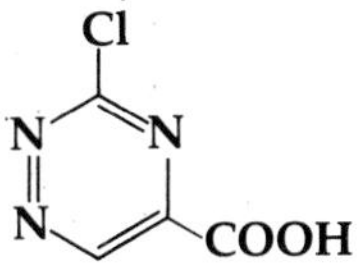

d   3-Phenyl-1,2,4-triazole

e   5-Methyl-1,3,4-thiadiazole-2-carboxylic acid

f   4-Nitro-1,2,3,5-tetrazine

g   Sulfamethizole, an internal antiseptic ("sulfa drug")

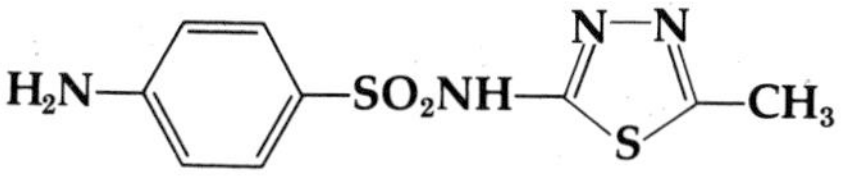

h   4-Acetyl-1,2,3,5-thiatriazole

TABLE 24-2   **Five-membered Aromatic Heterocycles**

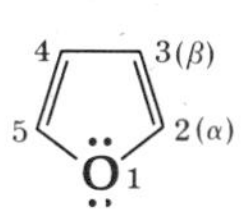

| Furan | Thiophene | Pyrrole | Oxazole (1,3-oxazole) | Thiazole (1,3-thiazole) |

| Pyrazole (1,2-diazole) | Imidazole (1,3-diazole) | 1,2,3-Triazole | 1,2,4-Triazole | Tetrazole |

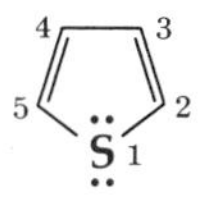

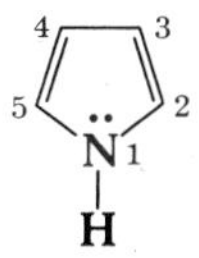

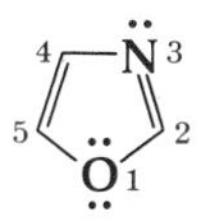

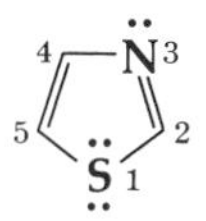

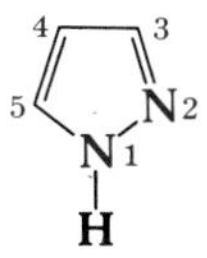

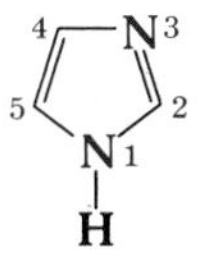

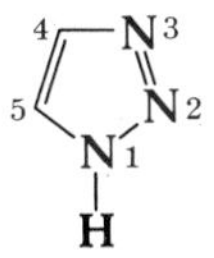

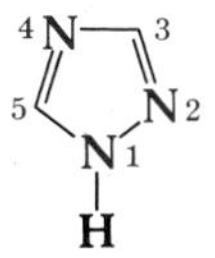

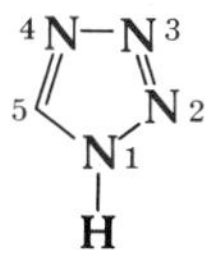

24-2 AROMATICITY AND REACTIVITY

Aromatic Character

Of the two kinds of heterocycles, the six-membered ones (e.g., pyridine) have the same six-electron π-electron cycle as benzene and their delocalization and aromaticity are essentially the same as that in benzene. Because the nitrogen in pyridine is more electronegative than carbon, however, dipolar resonance structures are more important and the molecule has a dipole moment with its negative pole at nitrogen.

Pyridine as a resonance hybrid

The five-membered rings also have six π electrons in a cycle made of $(2\pi + p)$ orbitals and so exhibit the familiar delocalization into a cyclic molecular orbital like two doughnuts above and below the ring plane, similar to that in benzene. The ring is planar and the substituents or unshared pairs also lie in this plane (as they do in benzene), in accord with the requirements of planar sp^2 hybridization geometry. The resonance energies afforded by the π-electron delocalization are somewhat less than in benzene but still quite substantial, pyridine being about the same as benzene and the five-membered heterocycles less, decreasing in the order: thiophene > pyrrole > furan.

Dipolar structures make important contributions to the resonance hybrids of these substances. The charge distribution in five-membered heterocycles provides the carbon portion of the nucleus with partial negative charge and accounts for the dipole moments of these heterocycles.

$Z = \ddot{O}, \ddot{S}, \text{ or } NH$

Basicity

The reactivity of nitrogen as a base in these rings is determined by the availability of its electron pair for bonding to an acid (i.e., H⁺). In pyrrole the electron pair is involved in resonance and can only act as a base (bond to H⁺) by loss of the ring resonance. Since this requires the input of much extra energy, pyrroles are not basic at all to aqueous acids, i.e., they are less basic than water (pK_a of pyrrole $= -0.3$). When protonated by stronger acids, in fact, they do not even protonate on nitrogen, but rather on carbon, as explained below for the general reactivity to electrophiles.

Pyridine is a base, however, since its unshared pair is not involved in π-electron resonance. In aqueous acid pyridines are converted to stable pyridinium salts (Table 24-1, **R** = **H**). They are weaker bases (p$K_a \sim 5$) than saturated amines (p$K_a \sim 10$) for the same reason that ethylene and acetylene carbanions are weaker than saturated carbanions (Sec. 8-4), since the unshared pair has more s character ($sp > sp^2 > sp^3$) and so is less available for forming bonds with protons. Nitriles (**R**—**C**≡**N:**) on the same basis are even less basic than pyridine, with p$K_a < 0$. Pyridine is extensively used as a basic catalyst and solvent, as in ester and amide formation from acid chlorides.

Reactivity

Because of their resonance energy the aromatic heterocycles tend to revert to aromatic form after reaction, as benzene does, and thus characteristically give substitution instead of addition reactions. However, the two types of heterocycles have opposed modes of preferred reaction, the six-membered ones acting as electrophiles, the five-membered ones as nucleophiles.

Virtually the entire reactive behavior of the aromatic heterocycles can be generalized by observing the parallels they display with the familiar reactions of ketones and enols, for the heterocycles all contain analogous functional parts. This is outlined in Fig. 24-1 which shows the six-membered heterocycles exhibiting the reactions of ketones discussed in Chaps. 12 and 13, and the five-membered rings acting analogously to the enols. Since the latter undergo substitution rather than addition, in an effort to revert to aromatic products, their reactions with electrophiles can be most closely compared to the reactions of phenols (aromatic enols). All the reactions of these rings with nucleophiles and electrophiles can be understood in these terms, regarding the rings as analogs of ketones if six-membered, and enols if five membered. Examples are offered in the following sections.†

PROBLEM 24-2

The two "diazoles," pyrazole and imidazole, differ markedly in their basicity, the latter being many times more basic. Explain.

---

† There are of course some rings which exhibit both kinds of reactivity, as indicated in the imidazole example below. Such rings have both an enolic and a ketonic portion and are always five-membered rings with nitrogen and one other heteroatom.

Nucleophilic (enol) ring

Electrophilic (ketone) ring

Imidazole

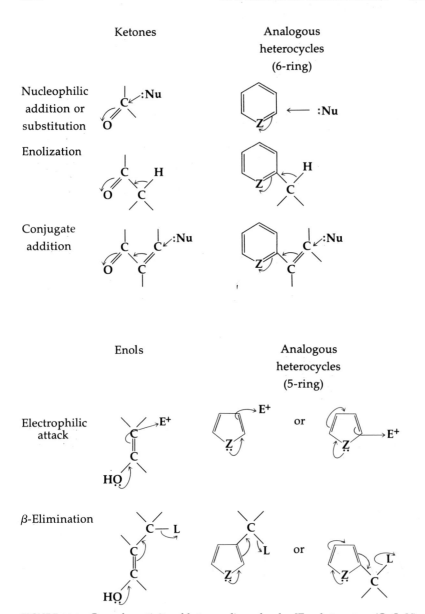

FIGURE 24-1   General reactivity of heterocyclic molecules [Z = heteroatom (O, S, N)]

24-3   FIVE-MEMBERED HETEROCYCLES

The major members of this group are those with only one heteroatom: thiophene, pyrrole, and furan in a rough order of decreasing stability and increasing reactivity. Their common properties arise as expected from the foregoing discussion and may be catalogued as follows:

A Effects of aromatic resonance.

  1 Major reaction is attack of electrophiles (including H^+): on carbon rather than at the heteroatom (Fig. 24-1).

  2 Reversion to aromatic ring occurs after electrophile attack, resulting in substitution.

  3 Hydrogenation of their nominal double bonds is difficult, as in benzene.

B Reactivity to electrophiles.

  1 More reactive than benzene: thiophene about $10^3 \times$ benzene; pyrrole similar to phenol.

  2 Reactive at both α- and β-positions, with α substantially preferred (see below).

  3 Relatively susceptible to oxidation.

  4 Substituents affect rate and position of electrophile attack as in benzene:

   a Electron-donating substituents (—OR, —NR$_2$, etc.) create so much reactivity that such compounds are very unstable and few are known.

   b Electron-withdrawing substituents (—NO$_2$, —COR, —COOR, etc.) deactivate and stabilize the ring by added resonance with the heteroatom:

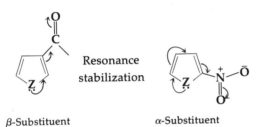

Resonance stabilization

β-Substituent α-Substituent

Reactivity

Reaction at the α-position is preferred since an analysis of the intermediates, as was done with substituted benzenes (Sec. 16-1), implies a transition state with more resonance stabilization in the α- than in the β-substitution.

(fewer stabilizing resonance forms)

With heterocycles not stabilized by electron-withdrawing substituents (—NO$_2$, —COR, —COOH, etc.) their reactivity is so high as to require only mild electrophiles, and even moderately strong acids cause extensive destruction of pyrrole and furan. Initial attack of H^+ as an electrophile and subsequent reaction of the intermediate (acting as a new electrophile) with unchanged heterocycle leads to polymers and complex products.

Halogenation usually may not even be stopped short of substituting all available ring hydrogens by halogen, and may even go further to complex products. Friedel-Crafts reactions are successful only under mild conditions, acid chlorides usually requiring no added Lewis acid (such as $AlCl_3$). Pyrrole reacts even with methyl iodide to form 2,3,4,5-tetramethylpyrrole and other products. Mild nitration and even nitrosation (with HNO_2) are also facile.

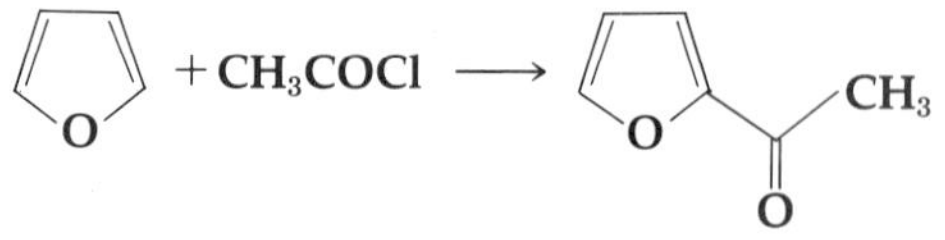

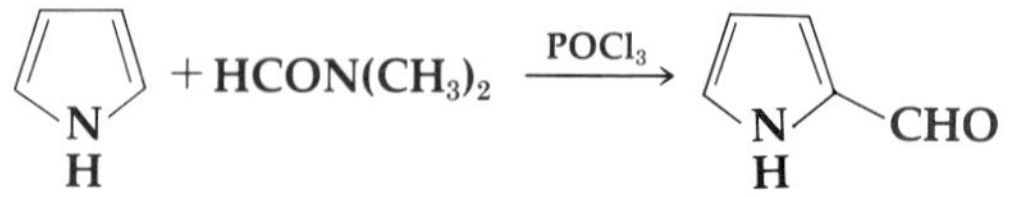

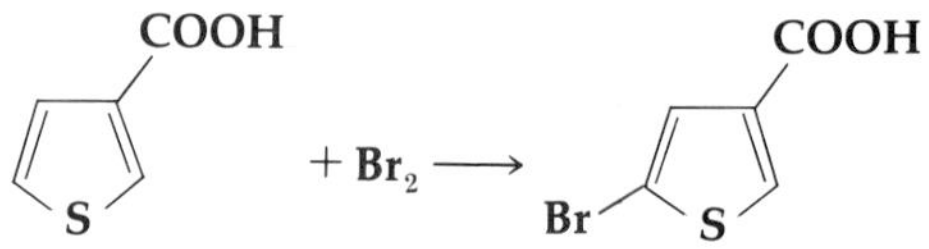

The last example shows the directing effect of substituents. In general, *whenever an α-position is unsubstituted, substitution occurs there.* When other choices exist, the preferred position of attack can be successfully deduced by examining the resonance in the several possible transition states as above for the α-, β- choice. The typical directive effects on electrophilic substitution can be summarized (W = electron-withdrawing; D = electron-donating):

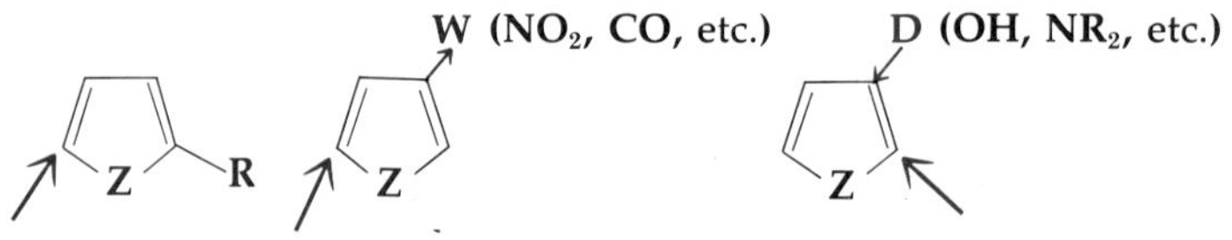

Similar to the Friedel-Crafts acylation is the common reaction with aldehydes and ketones under mild acid catalysis. The intermediate alcohol is readily eliminated (Fig. 24-1) to create a new electrophile for an attack on the unchanged parent heterocycle. As Fig. 24-1 shows, all of these reaction steps are the typical reactions of carbonyls and conjugated carbonyls.

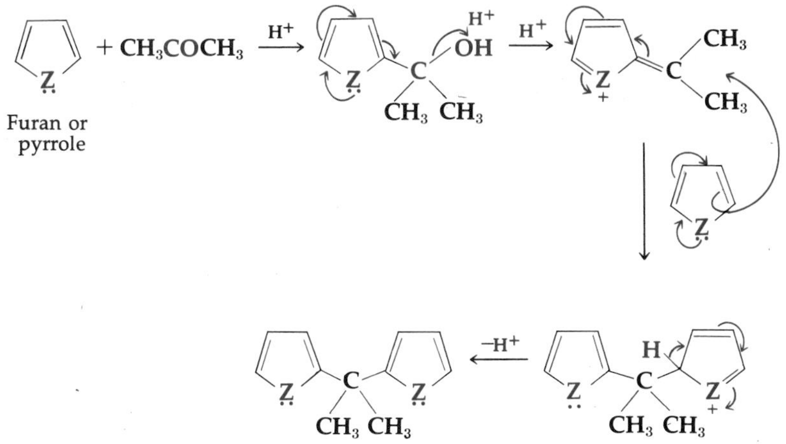

Furan or
pyrrole

Further analogous steps carry this product to polymers as well as the following cyclic tetramer, similar to the natural porphyrins (page 967).

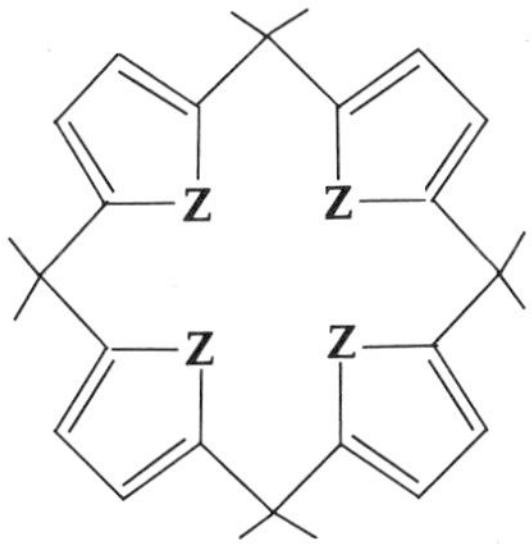

There are many reactions of these heterocycles which lead through many mechanistic steps to complex products. The mechanistic analysis of these transformations is often a fascinating challenge. The several basic modes of carbonyl reactivity, however, always control these mechanisms.

Synthesis

The synthesis of heterocycles can be examined by dissecting the molecule so that every bond to a heteroatom is replaced by bonds to separate oxygens. Thus, all the five-membered heterocycles are formally equivalent to the enols of 1,4-dicarbonyl compounds and in fact can be made by allowing these 1,4-dicarbonyl compounds to react with an appropriate source of the correct heteroatom (H_2O, NH_3, H_2S or S^{--}).

The appropriate 1,4-dicarbonyl compounds in turn are often made by alkylation of an enol with an α-halocarbonyl compound. Synthesis of the 1,4-dicarbonyl can often be combined with its conversion to the heterocycle.

Benz-fused Heterocycles

When a benzene ring is fused to a five-membered heterocycle, the latter largely retains its characteristic behavior, although the β-position now becomes the more reactive since reaction at α must disturb the adjacent benzene resonance. Indole and purine are two fused heterocyclic ring systems of particular importance in biological chemistry.

Indole   Purine

Indole and many of its derivatives are prepared from phenylhydrazones by an interesting and useful reaction termed the **Fischer indole synthesis.**

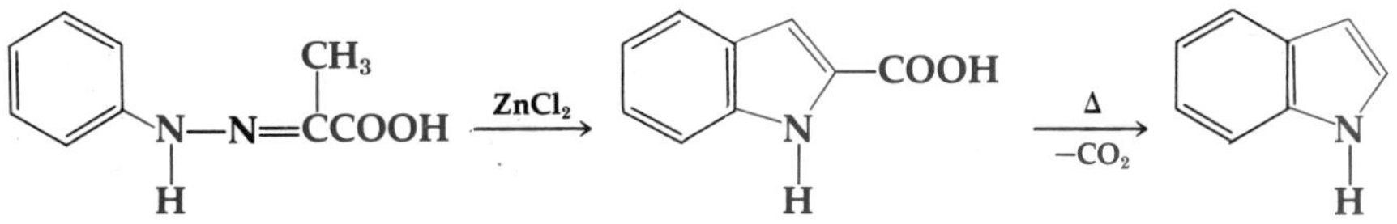

| Phenylhydrazone of pyruvic acid | Indole-2-carboxylic acid | Indole |

PROBLEM 24-3

Analyze the electrophilic substitution of 3-nitropyrrole to show the preferred site for reaction with acetyl chloride.

PROBLEM 24-4

5-Methylfuran-2-carboxylic acid on nitration yields 2-nitro-5-methylfuran as the major product. Account for this result in mechanistic terms.

PROBLEM 24-5

A traditional color test for pyrroles consists in allowing them to react with *p*-dimethylaminobenzaldehyde in acid. A deeply colored substance precipitates immediately. Write the structure of this material.

PROBLEM 24-6

Write the major product of nitration in each of the following compounds.

a 2-Chloropyrrole e 2-Methyl-3-nitrofuran
b 3-Methoxythiophene f 5-Methyl-2-acetylpyrrole
c 2-Acetylthiophene g 5-Methyl-2-methoxythiophene
d 3-Nitrofuran h 2-Phenyl-5-nitropyrrole

PROBLEM 24-7

a Write out the mechanistic steps in the synthesis of methyl 2,5-dimethyl-pyrrole-3-carboxylate shown in the text.
b Write out the mechanistic steps in the Fischer indole synthesis from butanone phenylhydrazone with strong acid.

24-4 SIX-MEMBERED HETEROCYCLES

Ring Reactivity

The chief six-membered heterocycle is pyridine. Its stability and strong tendency to revert to the aromatic ring on substitution parallel the behavior of benzene. The basic electron-pair is the first site attacked by electrophiles and this creates a positive pyridinium ion which now strongly resists any further attack by the electrophile on carbon. When this electrophilic attack is forced, it occurs at the β-position, analogous to the reluctant meta substitution of benzene rings bearing electron-withdrawing groups. For this reason pyridines are quite inert to electrophilic attack and also to oxidation, in vivid contrast to the five-membered heterocycles.

The polarity of electrons toward nitrogen, however, invites attack by nucleophiles, at the α- and γ-positions (α preferred), and this is much enhanced in the pyridinium and pyrilium ions owing to their full positive charge.

Electrophiles:

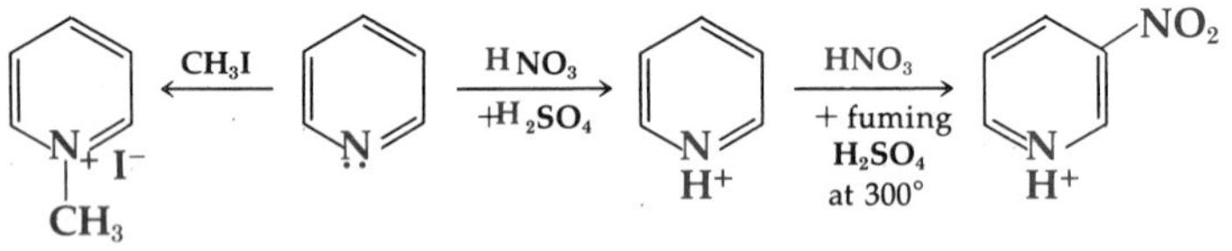

N-Methyl pyridinium iodide

Nucleophiles:

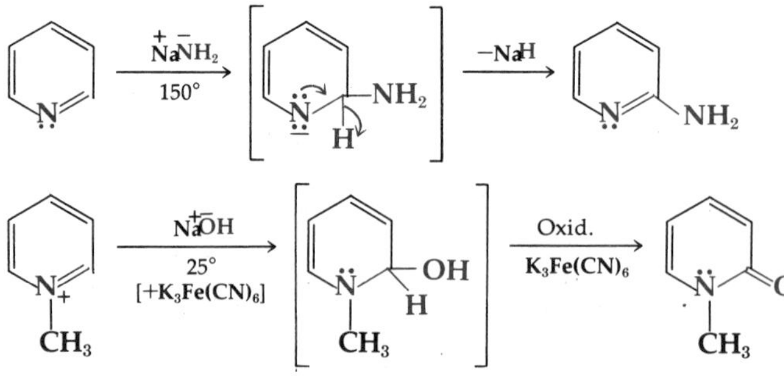

N-Methyl-α-pyridone

The last two reactions show the greater ease with which nucleophiles attack the pyridinium ring and the ready reversion to the aromatic system (second reaction). They also point up the possible tautomerism of α- (and γ-) substituents. Both the α- or γ-hydroxy- and aminopyridines can exist either as ketonic or substituted pyridine tautomers. The amino substituents exist as normal aminopyridines but the α- and γ-hydroxypyridines prefer the ketonic tautomers, called α- and γ-pyridones; these have considerable resonance energy, as implied by the major resonance forms shown. (See also γ-pyrone, Table 24-1). The β-hydroxypyridines are normal phenolic compounds since no stabilized ketonic tautomer is possible.

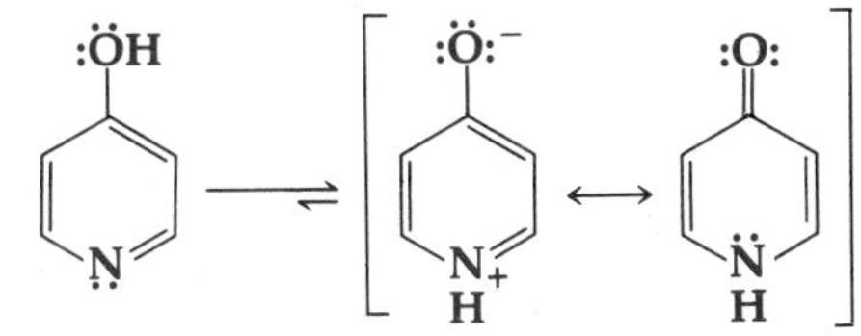

γ-Hydroxy-
pyridine
(negligible)

γ-Pyridone
(preferred)

Electrophilic substitution is of course rendered more facile with electron-donating substituents on the ring. These direct substitution ortho and para to themselves as in benzene. One of the most synthetically useful variations of this occurs with the pyridine-N-oxides, which accept both electrophiles and nucleophiles.

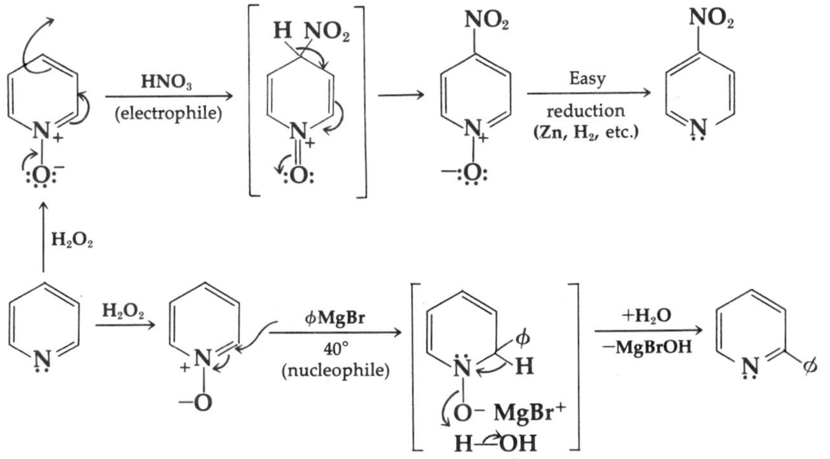

Side Chain Reactivity

Methyl groups attached to the pyridine nucleus in the α- or γ-positions are more acidic than the methyl group in toluene, and their derived anions are nucleophiles that resemble enolate anions in their behavior (Fig. 24-1).

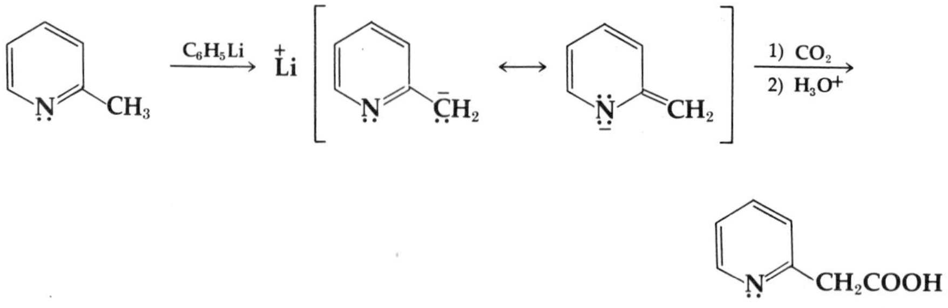

α-Pyridylacetic
acid

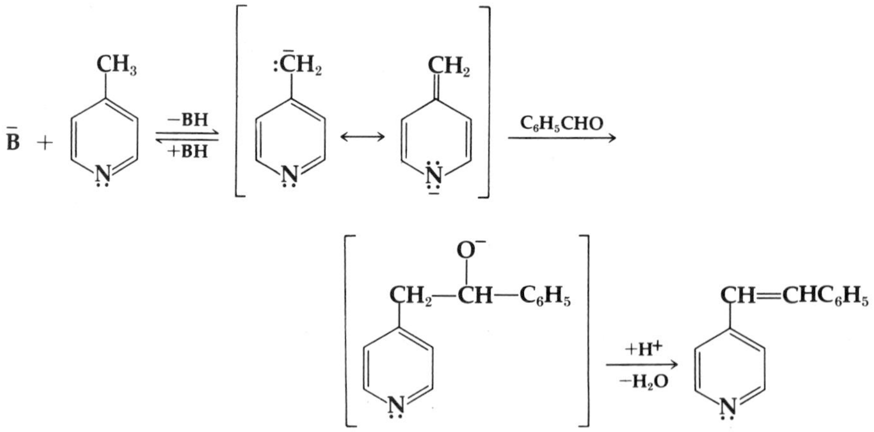

These reactions are analogous to enolization of carbonyls and subsequent reaction of the enol. Other reactions of pyridines which are analogous to those of carbonyls are summarized in Fig. 24-1. Nucleophilic substitution like that of carboxylic derivatives is common, and decarboxylation of α- and γ-pyridyl-acetic acids is the analog of that in β-ketoacids (page 535). Conjugate addition to α- and γ-vinyl-pyridines is also a facile reaction.

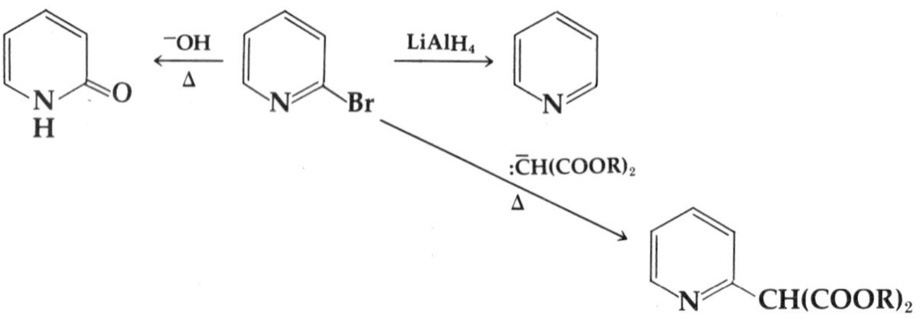

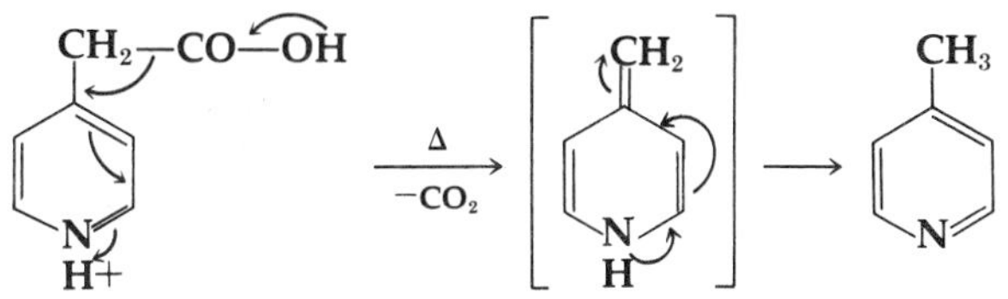

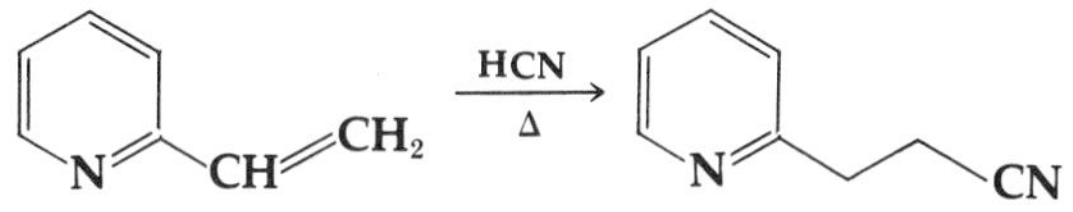

Synthesis

As with the previous heterocycles, dissection of pyridines or pyridones for synthesis yields acyclic dicarbonyl compounds. These in turn can be made by aldol or Claisen condensations, and the aromatic stability of pyridines is utilized to drive the equilibria to completion.

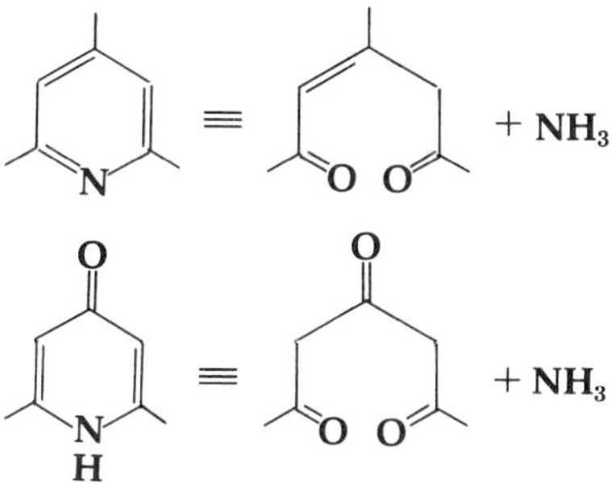

The most widely used synthesis employs this principle, leading to a dihydropyridine, which oxidizes very easily to the aromatic pyridine.

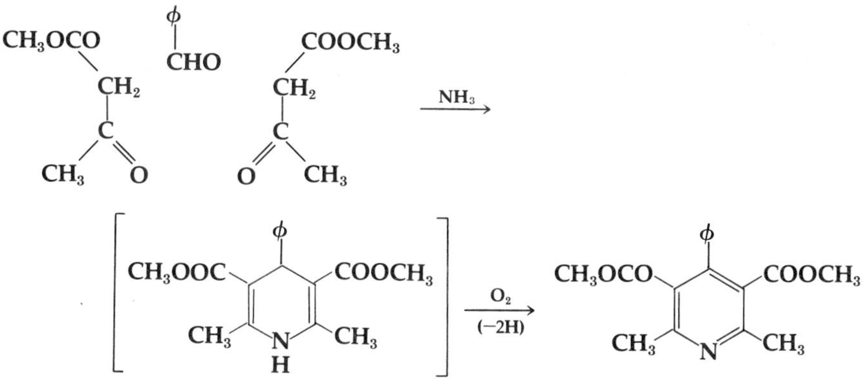

The sum of the oxidation states in the starting materials will equal that in the product (see Sec. 18-1) and will determine whether the primary product is in a dihydropyridine, pyridine, or pyridone oxidation state. The aromaticity provides such a driving force that almost any synthesis devised with the right oxidation states will yield some of the desired pyridine derivative.

Benz-fused Pyridines

Quinoline and isoquinoline are the most common of the pyridines fused to benzene rings (see Table 24-3 for examples of this family). They show largely the same reactivity as pyridine and are usually synthesized from anilines via Friedel-Crafts cyclization.

2,4-Diphenylquinoline

TABLE 24-3 **Fused Six-membered Heterocycles**

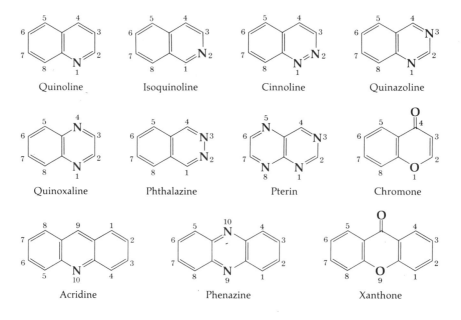

Quinoline   Isoquinoline   Cinnoline   Quinazoline

Quinoxaline   Phthalazine   Pterin   Chromone

Acridine   Phenazine   Xanthone

PROBLEM 24-8

Pyridine forms crystalline 1:1 adducts with boron trifluoride and with sulfur trioxide. Write their structures, with more than one resonance form in each case.

PROBLEM 24-9

a The pyridine-sulfur trioxide complex reacts with sodium hydroxide to form a yellow salt, $C_5H_5O_2^- Na^+$. Write the structure of the salt and the other product of the reaction and show a mechanism.

b By completely analogous mechanism, N-methylaniline hydrochloride reacts with the complex or with the sodium salt, to form a deep red-purple dye, $C_{19}H_{21}N_2Cl$. Formulate these reactions.

PROBLEM 24-10

Write stepwise mechanisms for the following reactions.

a

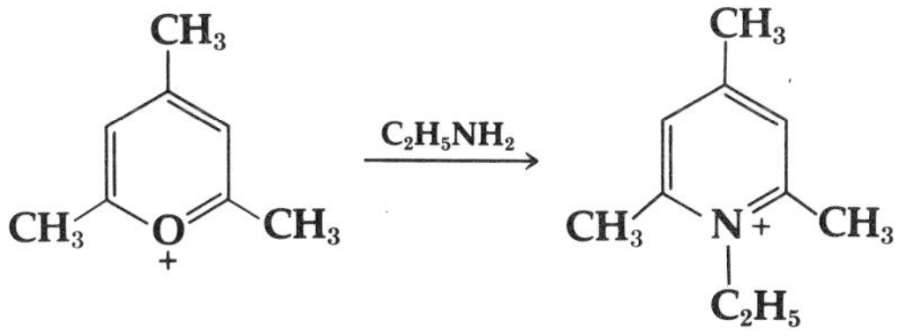

b

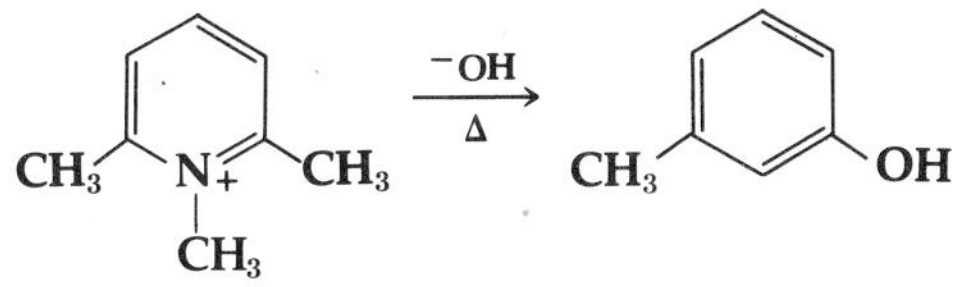

c

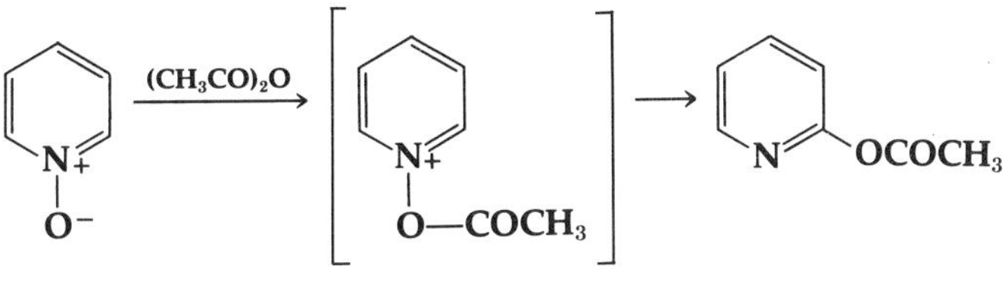

d

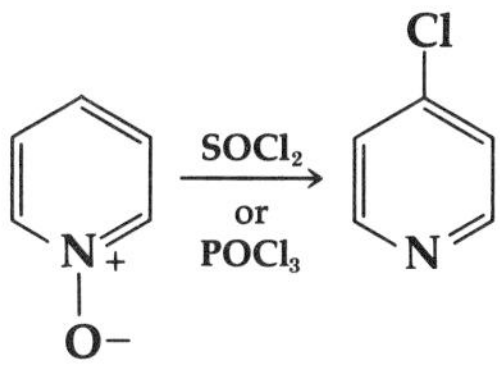

PROBLEM 24-11

On heating acetaldehyde and ammonia in the air, four pyridines are formed, two being isomers of C_6H_7N and two of $C_8H_{11}N$. Write structures for these isomers.

24-5 HETEROCYCLES WITH TWO OR MORE HETEROATOMS

The only aromatic heterocycles with two or more heteroatoms arise by replacing one or more **CH** centers with N in the four simple parent compounds, pyridine, thiophene, pyrrole, and furan. Such formal replacement affords three diazines from pyridine and two azoles from each of the five-membered parent rings.

The six-membered heterocycles so derived behave as expected, being even more reactive to nucleophiles and less to electrophiles. They are also less basic, with pK_a 0–3. They are characterized by somewhat less resonance energy than pyridine but still sufficient to cause them to revert to aromatic cycles after substitution and in general to be quite stable to most reaction conditions.

The azoles exhibit a balance between the reactivity to electrophiles shown by the simple five-membered rings and the ketonic acceptance of nucleophiles characteristic of pyridines and all rings containing C=N. Hence the azoles are usually generally as reactive as benzene to electrophiles. They are somewhat more reactive than pyridine to nucleophiles. The latter reactions commonly open the rings.

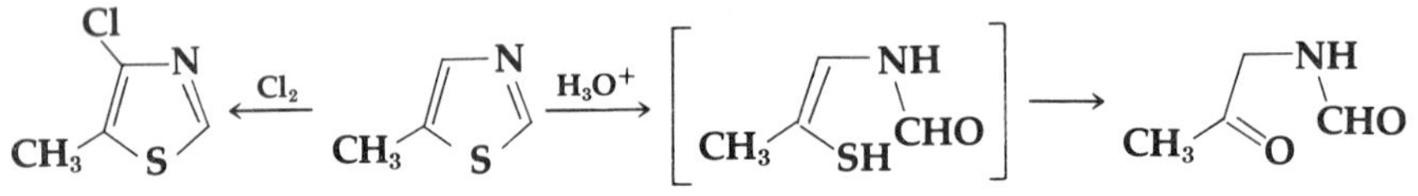

The azoles are usually less basic than pyridine, except for imidazole (Table 24-2), which has $pK_a = 7$ owing to favorable equivalent resonance forms in the conjugate acid (protonated at C=N:).

Synthesis of 1,2-diazoles, oxazoles, and diazines is achieved via NH_2OH or NH_2NH_2 with the appropriate 1,3- or 1,4-dicarbonyl compound, respectively. The 1,3-diazoles and diazines are commonly synthesized from amides, thioamides, or amidines with α-haloketones or 1,3-dicarbonyl compounds.

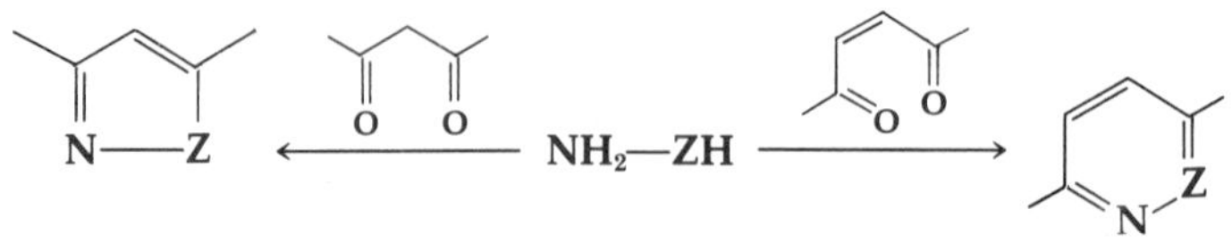

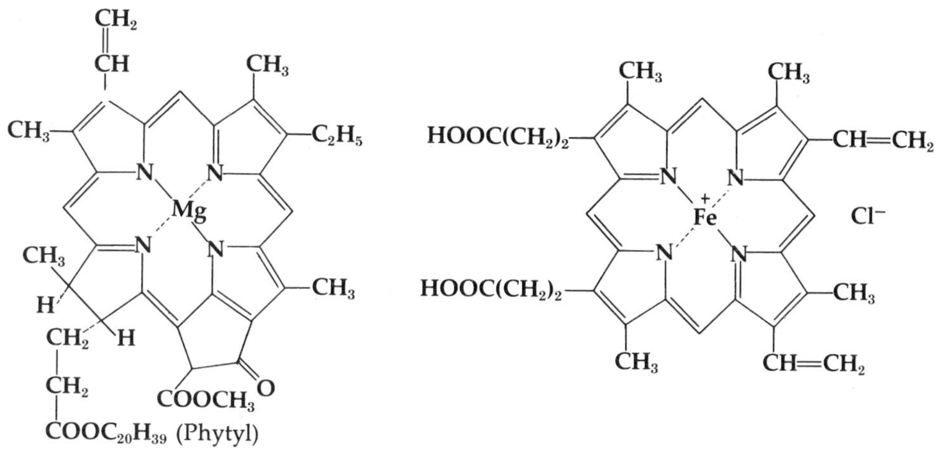

FIGURE 24-2   **Structures of the plant and blood pigments**

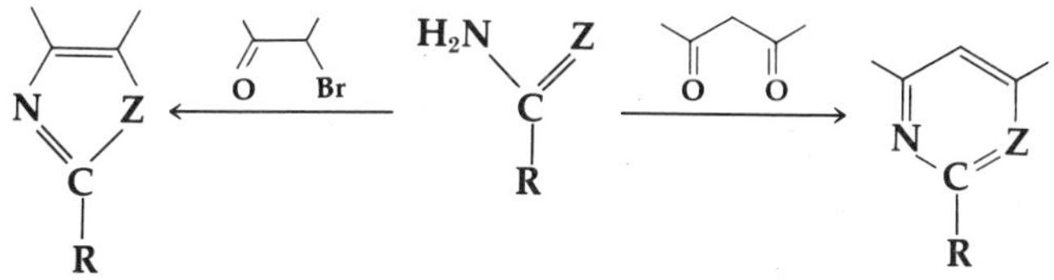

24-6   BIOLOGICALLY IMPORTANT HETEROCYCLES

The wide variations in properties possible with various heterocycles make them very adaptable candidates for the specialized reactivity required in the chemistry of living organisms. Hence they are very widely found in natural products and in biochemistry (Chaps. 26 and 27).

The major pigments of the plant and animal world are the cyclic tetrapyrrole structures shown in Fig. 24-2. Woodward's synthesis of chlorophyll involved extensive pyrrole chemistry since each pyrrole ring is differently substituted. The four pyrroles were each synthesized and then combined into two substituted dipyrrylmethanes which were then joined, in the correct orientation, to afford the cyclic molecule.

A selection of important biochemicals is illustrated in Fig. 24-3. At the top are shown the four bases that constitute the celebrated sequence code of DNA (Sec. 26-6), determining the genetic characteristics of all living things. The first two are pyrimidines, the second purines. The vitamins act in the body usually as the active sites of enzymes which control major metabolic reactions. One synthesis of thiamine is shown in Fig. 24-4 below to illustrate synthetic methods in heterocycle chemistry.

Nucleic
acid
bases

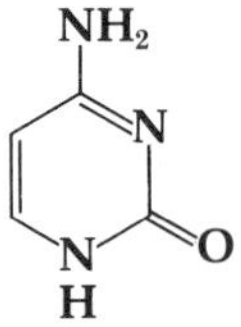

Cytosine

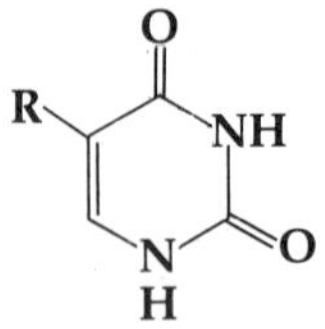

Uracil (R = H)
Thymine (R = CH$_3$)

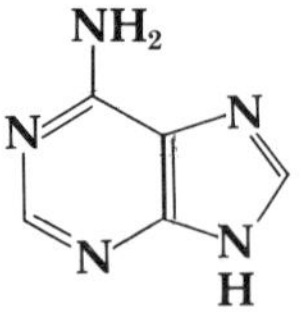

Adenine

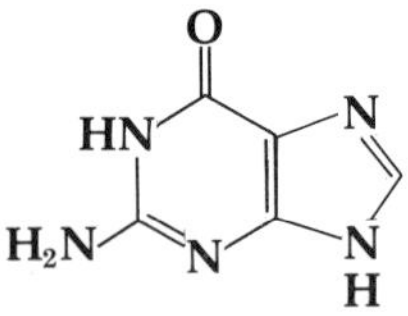

Guanine

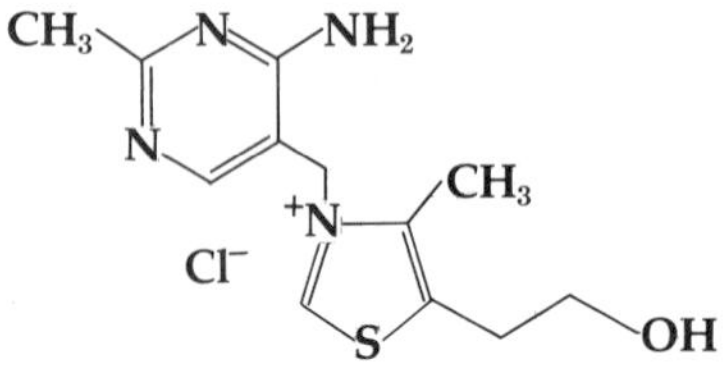

Thiamin
(vitamin B$_1$)

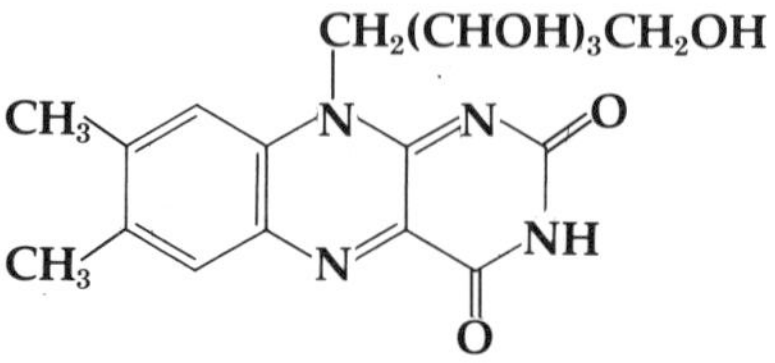

CH$_2$(CHOH)$_3$CH$_2$OH

Riboflavin
(vitamin B$_2$)

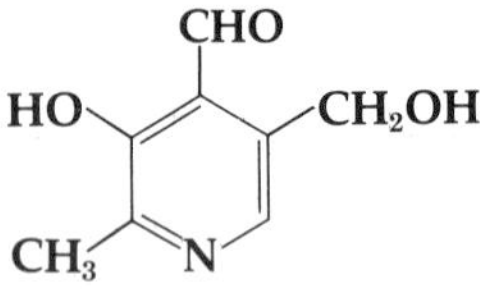

Pyridoxal
(vitamin B$_6$)

FIGURE 24-3   Selected heterocyclic biochemicals

Just as many natural chemicals are heterocycles, so it is no surprise that unnatural, synthetic heterocycles should be found to exhibit physiological responses. A number of these (selections in Fig. 24-5) have unique action in the body and are used as drugs and antibiotics in medicine. Most of the natural heterocycles with medicinal value come from plants and have also been synthesized in the laboratory. Many of these are mentioned in Chap. 27. Others are totally synthetic and have never previously existed in nature. They have all been created in the laboratory, usually in the research departments of pharmaceutical companies, and found to exhibit medically useful drug action. Some, like LSD, are synthetic variations made on naturally occurring compounds.

FIGURE 24-4   **Synthesis of thiamine (vitamin B₁)**

FIGURE 24-5   **Heterocyclic drugs and antibiotics (N = natural product; S = synthetic)**

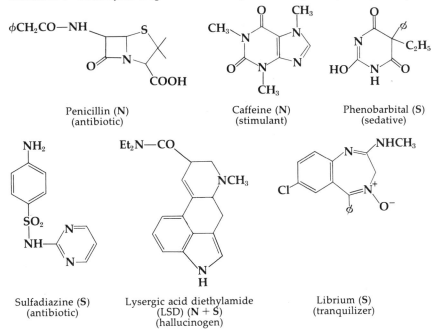

Penicillin (**N**)
(antibiotic)

Caffeine (**N**)
(stimulant)

Phenobarbital (**S**)
(sedative)

Sulfadiazine (**S**)
(antibiotic)

Lysergic acid diethylamide
(LSD) (**N + S**)
(hallucinogen)

Librium (**S**)
(tranquilizer)

PROBLEMS

24-12 Deduce the major product of nitration of the following compounds and show your reasoning by analysis of transition states.

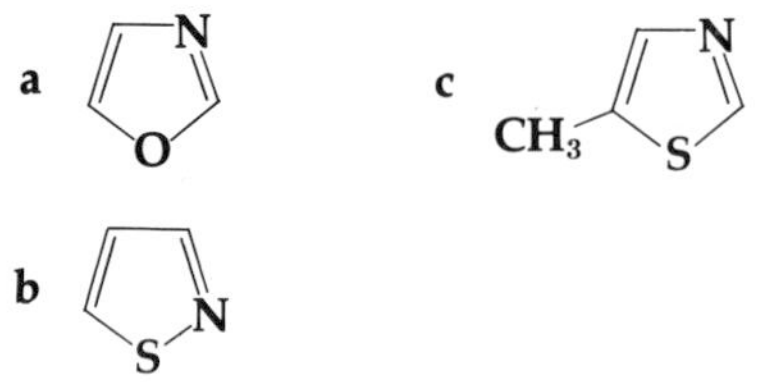

a   c

b

24-13 When ethyl acetoacetate, formaldehyde, and hydroxylamine are heated together in the absence of air, a substituted pyridine may be isolated in good yield. What is it? What is the function of the hydroxylamine?

24-14 Carbon atoms attached to five-membered heterocycles and bearing leaving groups solvolyze very readily but do not always yield the products of direct solvolysis. The two cases listed are typical examples. Show why the solvolysis is rapid and how the subsequent changes occur in these two examples.

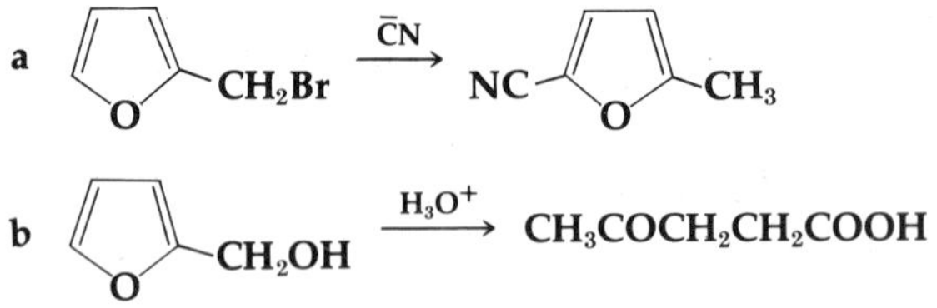

a $\xrightarrow{\bar{C}N}$ NC—⟨furan⟩—CH_3

b $\xrightarrow{H_3O^+}$ $CH_3COCH_2CH_2COOH$

24-15 Rationalize this transformation. The reagents are those commonly used to generate dichlorocarbene.

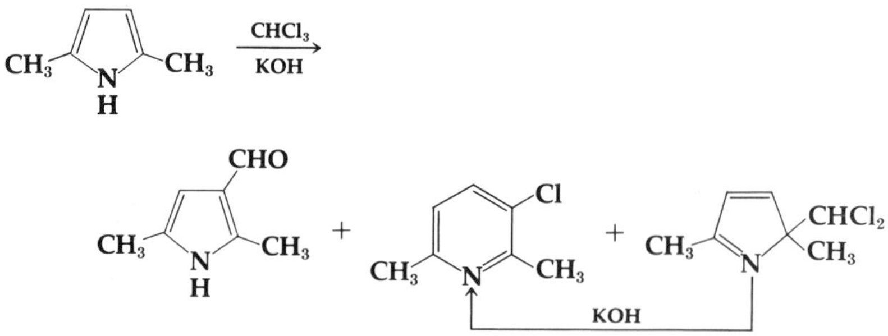

24-16 One of the four pyrrole ring starting materials needed for the chlorophyll synthesis was elaborated in the following manner. Fill in the blanks and write mechanisms for the steps marked with an asterisk.

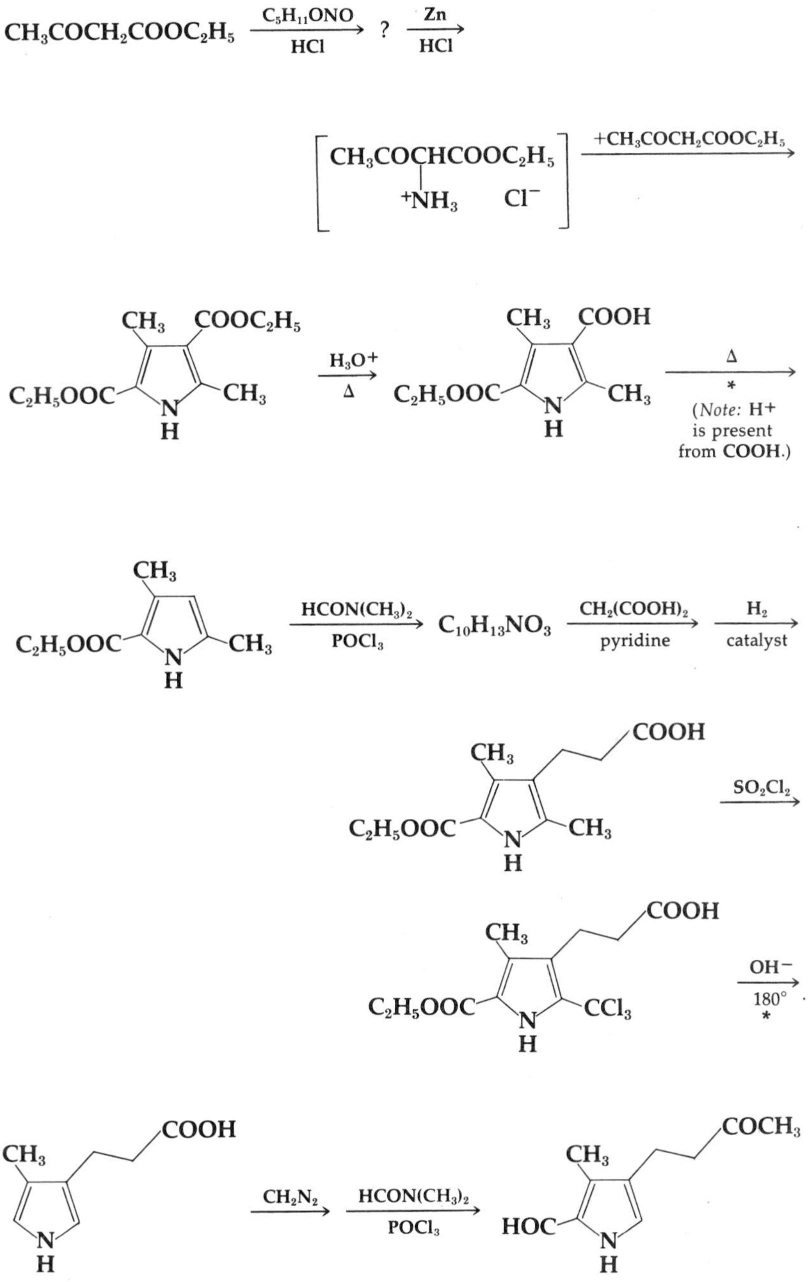

$$CH_3COCH_2COOC_2H_5 \xrightarrow[HCl]{C_5H_{11}ONO} ? \xrightarrow[HCl]{Zn}$$

$$\left[\begin{array}{c} CH_3COCHCOOC_2H_5 \\ {}^+NH_3 \quad Cl^- \end{array}\right] \xrightarrow{+CH_3COCH_2COOC_2H_5}$$

$$\text{(pyrrole: } CH_3, COOC_2H_5, C_2H_5OOC, N-H, CH_3) \xrightarrow[\Delta]{H_3O^+} \text{(pyrrole: } CH_3, COOH, C_2H_5OOC, N-H, CH_3) \xrightarrow[\substack{* \\ (Note: H^+ \\ is\ present \\ from\ COOH.)}]{\Delta}$$

$$\text{(pyrrole: } CH_3, C_2H_5OOC, N-H, CH_3) \xrightarrow[POCl_3]{HCON(CH_3)_2} C_{10}H_{13}NO_3 \xrightarrow[pyridine]{CH_2(COOH)_2} \xrightarrow[catalyst]{H_2}$$

$$\text{(pyrrole with CH_3, CH_2CH_2COOH, C_2H_5OOC, N-H, CH_3)} \xrightarrow{SO_2Cl_2}$$

$$\text{(pyrrole with CH_3, CH_2CH_2COOH, C_2H_5OOC, N-H, CCl_3)} \xrightarrow[\substack{180° \\ *}]{OH^-}$$

$$\text{(pyrrole with CH_3, CH_2CH_2COOH, N-H)} \xrightarrow{CH_2N_2} \xrightarrow[POCl_3]{HCON(CH_3)_2} \text{(pyrrole with CH_3, CH_2CH_2COCH_3, HOC, N-H)}$$

24-17 Devise rational syntheses of the following heterocycles from non-heterocyclic starting materials.

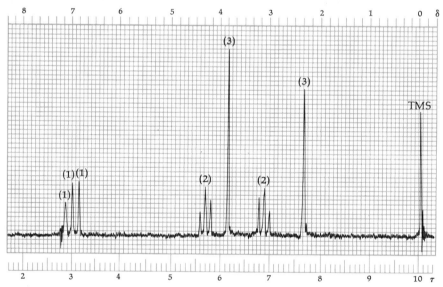

24-18 An indole derivative, $C_{13}H_{13}NO_2$, showed a UV spectrum very similar to that of 5-methoxy-2-acetyl-indole. The NMR spectrum is illustrated in Fig. P24-18. What is the structure?

FIGURE P24-18

24-19 Mild acylation of a certain heterocyclic compound led to $C_8H_{10}O_3$ which shows this NMR spectrum. What are the structures of the involved compounds?

2.92 (1H) d ($J = 5$)
3.87 (1H) d ($J = 5$)
3.22 (2H) q ($J = 7$)
7.62 (3H) s
8.80 (3H) t ($J = 7$)

24-20 A heterocyclic compound, C_7H_9NO, exhibits in the NMR spectrum a singlet at τ 0.62, two doublets ($J = 4$ cps) at τ 3.19 and 3.98, and two singlets at τ 6.13 and 7.73, the five peaks in an intensity ratio of $1:1:1:3:3$. What is the structure?

24-21 Devise a pathway for the following observed conversions of heterocycles. In doing so, note the hidden carbonyl and enol reactivity implicit in each structure and locate the parent carbon skeleton common to each. Consult page 954.

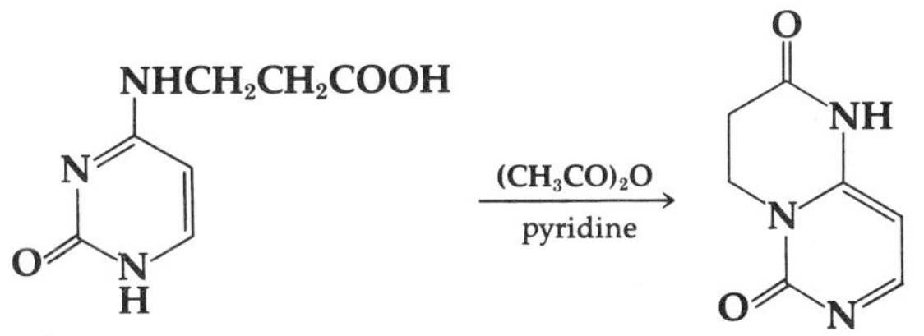

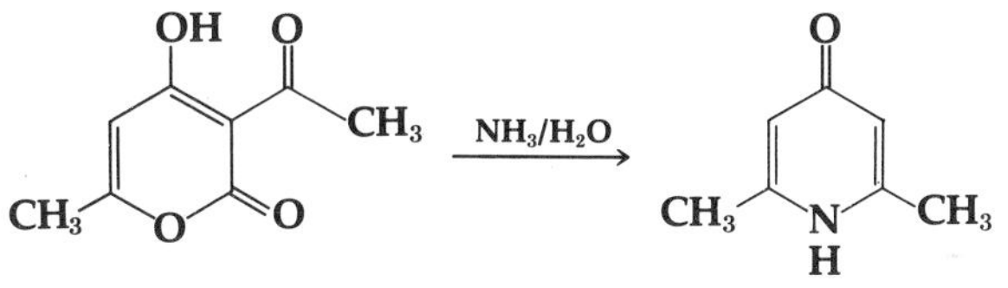

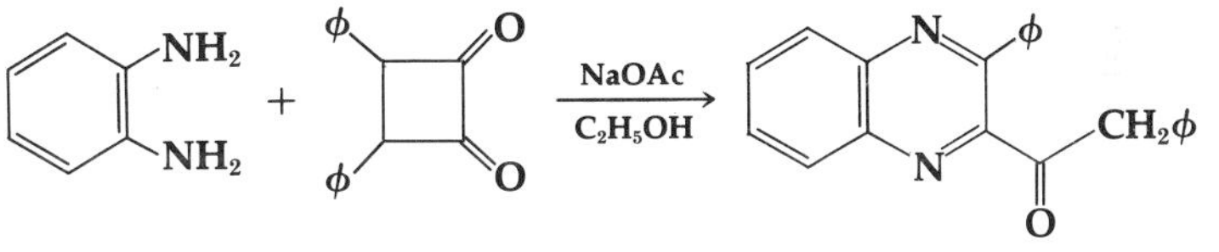

24-22 Deduce a mechanism for the following observed transformation.

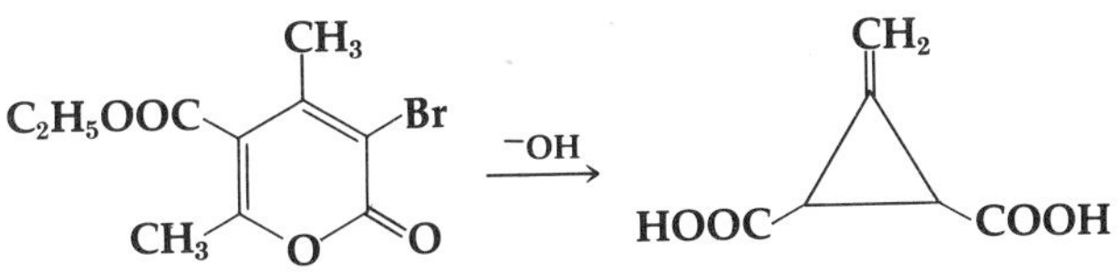

READING REFERENCES

Katrizky, A. R., and J. M. Lagowski, "Principles of Heterocyclic Chemistry," Academic, New York, 1968.

Paquette, L. A., "Principles of Modern Heterocyclic Chemistry," Benjamin, New York, 1968.

CHAPTER TWENTY-FIVE

NATURAL AND SYNTHETIC POLYMERS

A substance made up of long molecules which are characterized by many repeating molecular units in linear sequence is called a **polymer**. Polymers are made by sequential addition of many **monomer** molecules to each other.

$$nA \longrightarrow A—A—A—A— \cdots \equiv A—(A)_{n-2}—A$$

Monomer   Polymer

In many polymers, the fundamental units are not all the same but are two or more similar molecules. Such substances are called **copolymers** to distinguish them from **homopolymers**, which contain only one kind of fundamental unit.

$$nA + mC \longrightarrow (A—C—A—C—C—A)_{n+m}$$

Comonomers   **Copolymer**

The products of linking only small numbers of monomer units are designated by the use of Greek prefixes.

Monomer	A	
Dimer	$A—A$	
Trimer	$A—A—A$	**Oligomers** (short-chain polymers)
Tetramer	$A—A—A—A$	
Pentamer	$A—A—A—A—A$	
	etc.	

Two kinds of polymer may be distinguished: the **addition polymer** is one in which the monomers are molecules with multiple bonds which undergo true addition reactions with each other; the **condensation polymer** is one in which a small molecule (usually water) is eliminated in the condensation of any two monomer units. It is clear, therefore, that monomers for condensation polymers must be (at least) bifunctional molecules in which one function on one monomer molecule reacts with the other function on another molecule. Examples are listed:

Addition polymers:

$$n\text{R---CH}{=}\text{CH---R} \longrightarrow +\!\!\text{CH---CH}\!\!+_n$$
$$\qquad\qquad\qquad\qquad\quad \overset{|}{\text{R}}\quad\overset{|}{\text{R}}$$

Vinyl polymer

$$n\text{CH}_2{=}\text{O} \longrightarrow +\!\!\text{CH}_2\text{---O}\!\!+_n$$

Condensation polymers:

$$n\text{HO---}\rule{0.8cm}{0.3cm}\text{---COOH} \longrightarrow (n-1)\text{H}_2\text{O} + \text{HO}+\!\!\rule{0.8cm}{0.3cm}\text{---CO---O}\!\!+_n\text{H}$$

Polyester

$$n\text{H}_2\text{N---}\rule{0.8cm}{0.3cm}\text{---COOH} \longrightarrow (n-1)\text{H}_2\text{O} + \text{H}_2\text{N}+\!\!\rule{0.8cm}{0.3cm}\text{---CO---NH}\!\!+_n\text{H}$$

Polyamide

$$n\text{HO---}\rule{0.8cm}{0.3cm}\text{---OH} + n\text{HOCO---}\rule{0.8cm}{0.3cm}\text{---COOH} \longrightarrow$$
$$(2n-1)\text{H}_2\text{O} + \text{HO}+\!\!\rule{0.8cm}{0.3cm}\text{---O---CO---}\rule{0.8cm}{0.3cm}\text{---CO---O}\!\!+_n\text{H}$$

Polyester copolymer

The last example implies a variant in which two bifunctional units are present on each of two kinds of monomer to create a condensation copolymer. Polymers which possess only long sequential strands like those listed are called **linear polymers**. It is also possible, with *three* functional groups (or two different monomers, at least one of which is trifunctional), to have long linkage sequences in two (or three) dimensions and such polymers are distinguished as **cross-linked polymers**. Linear polymers are commonly relatively soft, often rubbery substances, and often likely to soften (or melt) on heating and to dissolve in certain solvents. Cross-linked polymers are hard and do not melt, soften or dissolve in most cases.

                 Homopolymers   *Copolymers*

Linear: $A\text{---}A\text{---}A\text{---}A\text{---}\cdots$ or $A\text{---}C\text{---}A\text{---}C\text{---}A\text{---}\cdots$

Cross-linked:

$$\begin{array}{l} A\text{---}A\text{---}A\text{---}A\text{---} \\ \qquad\; \overset{|}{A} \qquad\; \overset{|}{A}\text{---}A \\ A\text{---}A\text{---}A\text{---}A\text{---} \\ \qquad\qquad\qquad\; | \end{array} \quad \text{or} \quad \begin{array}{l} A\text{---}C\text{---}A\text{---}C\text{---}A\text{---}\cdots \\ \qquad\; \overset{|}{A} \qquad\; \overset{|}{A} \\ A\text{---}C\text{---}A\text{---}C\text{---}A\text{---}\cdots \end{array}$$

The importance of polymers in our life is almost breathtaking. Proteins and carbohydrates, which constitute two of the three principal classes of human foodstuffs, are natural polymers of high molecular weight. Nucleic acids are responsible for transmission of genetic characteristics in living organisms. Natural rubber, synthetic elastomers (synthetic rubbers), plastics, synthetic fibers, and resins are all polymers having uses that reach into every part of our lives. The tonnage of polymers marketed by the American chemical industry exceeds by a wide margin the volume of all other synthetic organic chemicals. Most of the structural tissue of living things is composed of polymers. In plants these are chiefly cellulose (a polysaccharide) and lignins. In animals the main structural polymers are proteins, which take different forms as skin, hair, muscle, etc. Highlights of the chemistry of these natural polymers are offered in Secs. 25-3 to 25-5.

25-1 VINYL POLYMERS

The most common of the commercial addition polymers are vinyl polymers. A wide variety of vinyl monomers undergoes *addition polymerization* to produce various high polymers. Familiar products of vinyl polymerization include synthetic elastomers, Plexiglas, polystyrene, polyethylene, Orlon, and most of the ion-exchange resins. Natural rubber is a prototype of the class.

Natural Rubber

Natural rubber is a mixture of hydrocarbons having the empirical composition C_5H_8. Rubber is highly unsaturated, and careful study of the ozonolysis of rubber leaves little doubt that it is a linear polymer of isoprene $[CH_2{=}C(CH_3)CH{=}CH_2]$ in which the residual double bonds are located exclusively between C-2 and C-3 of the isoprene units in the polymer. Although isoprene is not the biological precursor of rubber, the latter may be thought of as the product of the 1,4-addition polymerization of isoprene. Production of a synthetic material identical with natural rubber was first accomplished in 1955.

$$nCH_2{=}\underset{\underset{CH_3}{|}}{C}{-}CH{=}CH_2 \xrightarrow[\text{TiCl}_4]{[(CH_3)_2CHCH_2]_3Al} {+}CH_2{-}\underset{\underset{CH_3}{|}}{C}{=}CHCH_2{+}_n$$

"Natural" rubber

The average molecular weight of rubber is estimated to be in the range of 60,000 to 350,000, which corresponds to a degree of polymerization (number of monomer units per molecule) of 1,000 to 5,000. The steric configuration of the polymer molecules has a critical effect upon their physical properties. All, or nearly all, olefinic linkages in rubber have the cis configuration. *Guttapercha*, a trans isomer of rubber, is horny rather than elastic. Hydrogenation of both polymers gives the same saturated polymer product. The difference in physical properties must, therefore, be associated with differences in the configurations about the double bonds in the natural polymers. Analysis of stretched rubber by X-ray diffraction shows that the polymer has a repeat period of 9.1 Å, whereas α-gutta-percha has a period of 8.7 Å.

Rubber latex, a colloidal suspension of rubber in water, occurs in the interstitial tubules of many plants, including the dandelion, goldenrod, and rubber tree. The tree is native to Brazil but has been introduced to the Far East. During World War II, interruption of the flow of rubber from the plantations of Ceylon, Indonesia, and the countries of the Malay peninsula created a crisis that stimulated fervid effort to develop substitutes for rubber. While the early "synthetic rubbers" were inferior in many respects to natural rubber, synthetic **elastomers** have now been developed that are superior to natural rubber for many applications.

Coagulation of rubber latex by addition of acetic acid and salts gives *crepe* or *gum* rubber. Gum rubber does not have high tensile strength and becomes very brittle at low temperatures. The widespread use of rubber as an elastic structural material was made possible by development of the *vulcanization* process in 1839 by Charles Goodyear, who accidentally discovered that addition of sulfur to hot rubber causes changes that improve the physical properties in a spectacular manner. Vulcanization establishes cross-links between the linear polymer chains of rubber. Cross-linking probably occurs both through saturation of double bonds and by way of coupling and addition reactions of C—H groups α to the double bonds.

$$\{CH_2C(CH_3)=CHCH_2\}_n$$

$$+ S \longrightarrow$$

$$\{CH_2C(CH_3)=CHCH\}_n$$
$$\{CH_2C(CH_3)-CHCH_2\}_m$$
$$|$$
$$S$$
$$|$$

$$\{CH_2C(CH_3)=CHCH_2\}_m$$

Soft rubber contains about 1 to 2% sulfur, and *hard rubber*, which has lost most of its plasticity, contains up to 35% sulfur, enough to effect complete saturation of the rubber if all cross-linking occurred by addition.

Initiation of Polymerization

Polymerization of vinyl monomers occurs in the presence of small amounts of a wide variety of reagents known as **initiators**. Since initiators are often destroyed, it is not proper to refer to the substances as catalysts, although the latter term is sometimes used. Initiators are believed to form some reactive species, such as an ion or a free radical, which can add to carbon-carbon double bonds to form a new ion or radical, which can in turn add to another unit. Various polymerization processes are most easily described in terms of the chemical nature of the *growing polymer chains*.

Chain initiation

$$\text{ROOR} \xrightarrow{100°} 2\text{RO·}$$

$$\text{RO·} + \text{CH}_2{=}\text{CH}{-}\text{OCOCH}_3 \longrightarrow \text{ROCH}_2\overset{·}{\text{C}}\text{H}{-}\text{OCOCH}_3$$

　　　　　　　Vinyl acetate

Chain propagation

$$\text{RO}{-}\!\!\!\!\left(\!\!{-}\text{CH}_2\underset{\underset{\text{OCOCH}_3}{|}}{\text{CH}}{-}\!\right)_{\!\!n}\!\!{-}\text{CH}_2\overset{·}{\text{C}}\text{H}{-}\text{OCOCH}_3 + \text{CH}_2{=}\text{CH}{-}\text{OCOCH}_3 \longrightarrow$$

$$\text{RO}{-}\!\!\!\!\left(\!\!{-}\text{CH}_2\underset{\underset{\text{OCOCH}_3}{|}}{\text{CH}}{-}\!\right)_{\!\!\overline{n+1}}\!\!\text{CH}_2\overset{·}{\text{C}}\text{H}{-}\text{OCOCH}_3$$

                                                                    (P·)

Chain termination

$$2\text{P·} \longrightarrow \text{P}{-}\text{P}$$

FIGURE 25-1   **Peroxide-initiated polymerization of vinyl acetate**

Free-radical Polymerization

The most widely used initiators are those which produce reactive free radicals at a controllable rate. There are three principal classes of free-radical initiators:

1 Compounds, such as peroxides and certain azo compounds, that undergo thermal decomposition at temperatures not far above room temperature (See Fig. 25.1.)

$$(\text{CH}_3)_3\text{COOC}(\text{CH}_3)_3 \xrightarrow{100{-}130°} 2(\text{CH}_3)_3\text{CO·}$$

　　Di-*tert*-butyl　　　　　　*tert*-Butoxy
　　　peroxide　　　　　　　　radical

$$(\text{CH}_3)_2\underset{\underset{\text{CN}}{|}}{\text{C}}\text{N}{=}\text{N}\underset{\underset{\text{CN}}{|}}{\text{C}}(\text{CH}_3)_2 \xrightarrow{60{-}100°} 2(\text{CH}_3)_2\overset{·}{\text{C}}\text{CN} + \text{N}_2$$

α,α'-Azo-*bis*-isobutyronitrile　　　2-Cyano-2-propyl
　　　　　　　　　　　　　　　　　　radical

2 *Photosensitizers,* which, on the absorption of light, decompose or react with other molecules to give radicals.

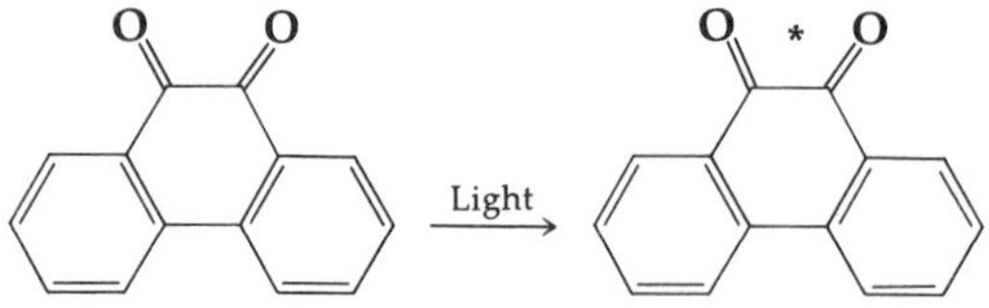

Phenanthrenequinone   Excited state

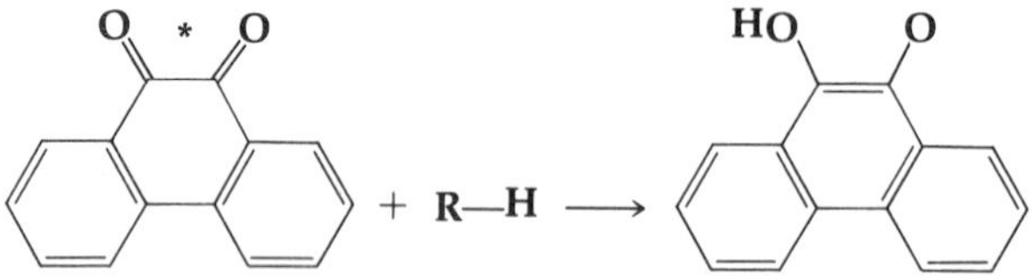

+ R—H ⟶

Hydrogen-
atom
donor

3 *Redox systems,* in which a one-electron transfer reaction, often involving a metal ion, produces reactive radicals.

$$Fe^{++} + H_2O_2 \longrightarrow Fe^{3+} + OH^- + \cdot OH$$

Although free radicals are highly reactive and short-lived, they show a high degree of selectivity in some reactions. For example, a high degree of order is found in the structure of all vinyl polymers, which indicates that addition virtually always occurs in the sense:

$$R \cdot + CH_2=CHX \longrightarrow RCH_2CHX$$

rather than in the opposite sense:

$$R \cdot + CH_2=CHX \longrightarrow RCHXCH_2 \cdot$$

This is, of course, the same observation as was made concerning the addition of reagents such as **HBr** to olefins by the free-radical mechanism (page 828).

Various other reactions of free radicals can compete with the addition reaction. Such side reactions may involve the monomer, the polymer, or foreign additives and impurities. "Side reactions" may have a spectacular influence on the course of a polymerization and are often used to control polymer properties and monomer stability.

Inhibitors are substances that degrade growing radicals to inactive products. Each time such a reaction occurs a growing chain is terminated. Common inhibitors are oxygen, iodine, quinones, and polycyclic aromatic hydrocarbons. Typical inhibitor action is that of benzoquinone.

$$R\cdot \; + \; \underset{\text{Benzoquinone}}{\underset{\displaystyle O}{\displaystyle O}} \longrightarrow \left[\underset{\text{Resonance-stabilized radical}}{\underset{O\cdot}{OR}} \longleftrightarrow \underset{O}{OR} \longleftrightarrow \text{etc.} \right] \longrightarrow \text{inert products}$$

Benzoquinone

Resonance-stabilized radical
incapable of adding to
monomer

Most vinyl monomers are stored with small amounts of added inhibitors to protect them against premature polymerization. Slow oxidation of monomers, which produces peroxides (cf. hydrocarbon oxidation, page 750), is a source of initiators that cause polymerization during storage. As a consequence, antioxidants, which are not effective polymerization inhibitors in the absence of oxygen, are often used to stabilize monomers. Polymerization of a monomer containing an inhibitor is characterized by an **induction period** during which little or no polymerization occurs. After all the inhibitor has been destroyed, polymerization assumes its normal course. Stabilizers are usually removed before polymerization is initiated.

Chain-transfer agents are molecules that can react with growing chains to interrupt the growth of a particular chain. The products, however, are radicals capable of adding to a monomer to start the growth process again. The overall effect is to reduce the average molecular weight of the polymer without reducing the polymerization rate. The chain-transfer agents commonly used to regulate the molecular weight of commercial polymers include carbon tetrachloride, mercaptans, and toluene.

$$R\cdot \; + \; CCl_4 \longrightarrow RCl + Cl_3C\cdot \xrightarrow{\;CH_2=CHX\;} Cl_3CCH_2\dot{C}HX$$

$$R\cdot \; + \; R'SH \longrightarrow RH + R'S\cdot \xrightarrow{\;CH_2=CHX\;} R'SCH_2\dot{C}HX$$

$$R\cdot \; + \; C_6H_5CH_3 \longrightarrow RH + C_6H_5CH_2\cdot \xrightarrow{\;CH_2=CHX\;} C_6H_5CH_2CH_2\dot{C}HX$$

Copolymerization occurs when a mixture of two or more monomers polymerizes, so that units from each monomer enter the same polymer chain. Such a copolymer has properties quite different from those of mixtures of the individual homopolymers (polymer from a single monomer). The high selectivity of free radicals is reflected in the preferences shown by growing radicals in a copolymerization. For example, maleic anhydride, which will not easily form a homopolymer, gives high-molecular-weight copolymers with styrene, butadiene, and vinyl acetate. In the copolymers, there is a nearly perfect alternation of monomer units.

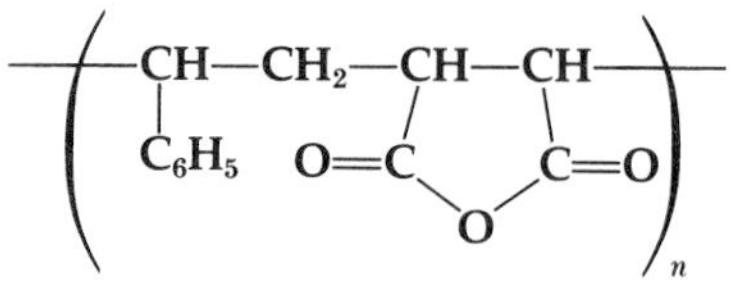

$$\left(\begin{array}{c} -CH-CH_2-CH-CH- \\ \quad | \qquad\qquad | \quad\;\; | \\ \quad C_6H_5 \quad O=C \quad C=O \\ \qquad\qquad\qquad \searrow O \swarrow \end{array}\right)_n$$

Styrene-maleic anhydride copolymer

Growing radicals with maleic anhydride "ends" add preferentially to styrene, and styryl-terminated radicals add more rapidly to the anhydride than to styrene.

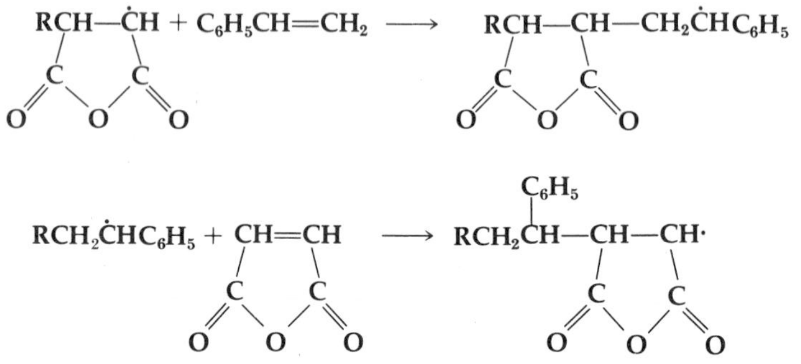

$$RCH\!-\!\dot{C}H + C_6H_5CH\!=\!CH_2 \longrightarrow RCH\!-\!CH\!-\!CH_2\dot{C}HC_6H_5$$

$$RCH_2\dot{C}HC_6H_5 + CH\!=\!CH \longrightarrow RCH_2\overset{C_6H_5}{\underset{\;}{CH}}\!-\!CH\!-\!CH\cdot$$

Cationic Polymerization

Strong acids cause polymerization of a number of vinyl monomers. The growing chains are carbonium ions, and the growing ends are probably closely associated with anions. Most commonly used initiators are Lewis acids, such as boron trifluoride, stannic chloride, or aluminum chloride, assisted by a small amount of water. Polymerization of isobutene to give the product known as *butyl rubber* is a typical example.

$$BF_3 + H_2O \rightleftharpoons HO\bar{B}F_3\overset{+}{H}$$

$$HO\bar{B}F_3\overset{+}{H} + (CH_3)\,C\!=\!CH_2 \longrightarrow (CH_3)_3\overset{+}{C}\ F_3\bar{B}OH \xrightarrow{n(CH_3)_2C=CH_2}$$

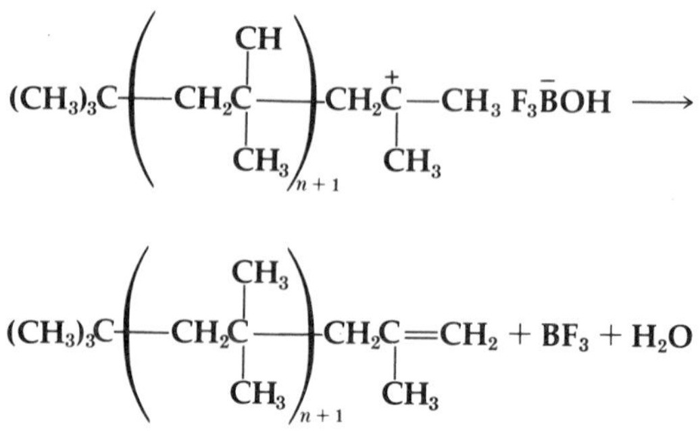

$$(CH_3)_3C\left(\!-\!CH_2\underset{\underset{CH_3}{|}}{\overset{\overset{CH}{|}}{C}}\!-\!\right)_{n+1}\!\!CH_2\overset{+}{\underset{\underset{CH_3}{|}}{C}}\!-\!CH_3\ F_3\bar{B}OH \longrightarrow$$

$$(CH_3)_3C\left(\!-\!CH_2\underset{\underset{CH_3}{|}}{\overset{\overset{CH_3}{|}}{C}}\!-\!\right)_{n+1}\!\!CH_2\underset{\underset{CH_3}{|}}{C}\!=\!CH_2 + BF_3 + H_2O$$

Isobutene, styrene, butadiene, and vinyl ethers are converted to high polymers by acid catalysis, but the process is not nearly as general as radical polymerization. The mechanism of polymerization here is essentially that of electrophilic addition (Sec. 15-4).

Anionic Polymerization

Strong bases initiate polymerization of vinyl compounds that bear electron-withdrawing substituents. The reaction is essentially one of nucleophilic conjugate addition (Sec. 12-6). In the absence of proton-donor solvents, chain growth can occur, following initial conjugate addition of a nucleophilic initiator ($B:^-$). Acrylonitrile ($CH_2=CHCN$) is polymerized by sodium amide in liquid ammonia.

$$\bar{B}: + CH_2=CHCN \longrightarrow BCH_2\bar{C}HCN \xrightarrow{CH_2=CHCN} BCH_2CHCH_2\bar{C}HCN \longrightarrow$$
$$\underset{\displaystyle CN}{|}$$

$$B(CH_2CH)_nCH_2\bar{C}HCN \xrightarrow[\text{termination steps}]{\text{Chain}} \text{polymer}$$

Organometallic Initiators

New types of organometallic initiators for vinyl polymerization have been developed recently. The initiators are complex mixtures formed by admixture of organometallic compounds with various metallic halides. The halides are usually species, such as $TiCl_4$ and $VOCl_3$, containing a metal that can be reduced to one or more lower valence states. The first organometallic "catalyst" of this type was reported by Ziegler and was prepared from aluminum alkyls (R_3Al) and titanium tetrachloride. Ziegler initiators accomplish some very dramatic new polymerizations, including low-temperature polymerization of ethylene and polymerization of propylene. In addition, the new initiators produce crystalline (page 1013) polymers from monomers such as styrene and propylene.

Little is yet known of the mechanism of organometallic polymerization. It is possible that the addition step should be regarded as a Lewis acid–Lewis base attack on alkene groups. In the following equation, the **M**'s refer to metal compounds.

$$M_1{-}R \quad \overset{\displaystyle \diagdown}{\underset{\displaystyle \diagup}{C}}{=}\overset{\displaystyle \diagup}{\underset{\displaystyle \diagdown}{C}} \quad M_2 \longrightarrow \overset{+}{M_1}R{-}\overset{|}{\underset{|}{C}}{-}\overset{|}{\underset{|}{C}}{-}\bar{M}_2$$

Lewis base   Lewis
(R^- donor)   acid

As most Ziegler systems contain suspended solids, addition may well occur through the agency of two metal atoms located on solid surfaces. Almost any terminal alkene ($C=CH_2$) can be polymerized by Ziegler initiators. Functional groups interfere with the action of the catalysts that are most effective in alkene polymerization. Monomers that bear functional groups polymerize in the presence of other organometallic initiators, many of which have not yet been thoroughly described in the literature. It was through the agency of a Ziegler system that the first synthetic all-cis polyisoprene was prepared. More recently, the natural rubber structure has also been reproduced by polymerization of isoprene with metallic lithium.

Ion-exchange Resins

A cross-linked polymer of styrene and divinylbenzene is a hard, insoluble resin which may be formed into tiny beads. Since the molecular network contains pendant phenyl groups (from the styrene units), these may undergo electrophilic substitution reactions and thus produce functionalized polymers. Many chemical reactions may in fact be carried out on functions which are attached to the cross-linked net of the polymer.

When such a polymer is sulfonated there is obtained an insoluble resin with sulfonic acid groups appended to its surface. Such a substance is called an **ion-exchange resin**.

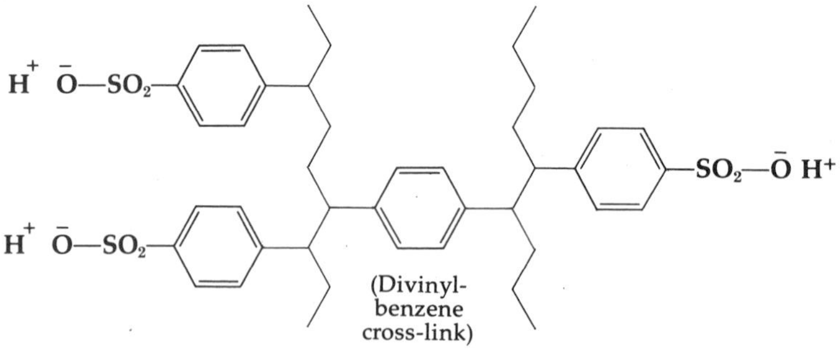

Sulfonic acid ion-exchange resin

While the sulfonate groups are attached to the insoluble resin, the protons are only held electrostatically and may easily migrate away in an adjacent water phase as long as another positive ion from the aqueous solution can move in to take its place and balance the negative sulfonate charge. When packed into a column with water the resin can serve to exchange cations. If a solution of sodium chloride is passed down the column, the passing sodium ions are caught by the resin, releasing protons into the solution so that a solution of HCl emerges from the bottom.

An anion-exchange resin is similarly prepared by chloromethylating the original polymer and displacing chloride with trimethyl amine.

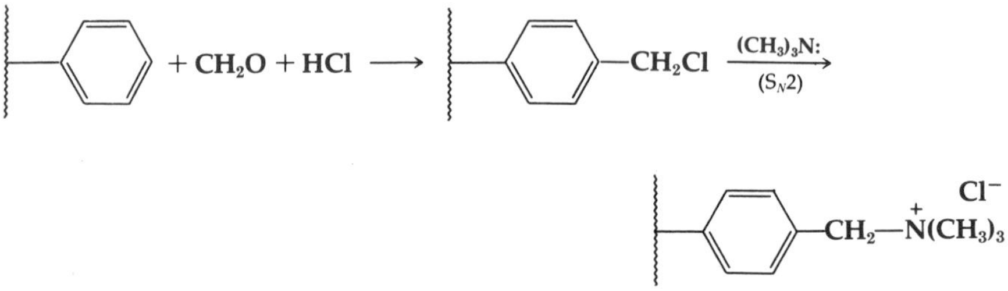

The two kinds of resin are widely used to exchange ions in solution conveniently. When the counterions on the two resins are H^+ or OH^- they are very convenient for providing acid or basic catalysis to a reaction without retaining acid or base in the filtered reaction solution at the end. The resins are also widely used to remove all ions from water as a means of purifying it (**deionizing**). This is the principle of commercial water softeners and works by using an acidic and a basic resin in sequence:

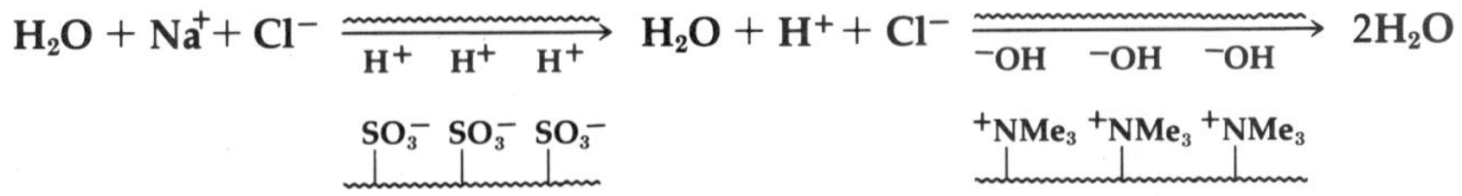

$$H_2O + Na^+ + Cl^- \xrightarrow[\substack{H^+ \quad H^+ \quad H^+ \\ | \quad | \quad | \\ SO_3^- \; SO_3^- \; SO_3^-}]{} H_2O + H^+ + Cl^- \xrightarrow[\substack{^-OH \quad ^-OH \quad ^-OH \\ | \quad | \quad | \\ {}^+NMe_3 \; {}^+NMe_3 \; {}^+NMe_3}]{} 2H_2O$$

25-2 CONDENSATION POLYMERS

Polyesters, polyamides, polyethers, the products of the condensation of form-aldehyde with amines, and reactive aromatic substrates are the common condensation polymers. Not all the products are actually made by condensation reactions.

$$(n + 2)HO(CH_2)_5COOH \xrightarrow{H^+, \; \Delta}$$

$$HO(CH_2)_5\overset{O}{\underset{\|}{C}}-\left[-O(CH_2)_5\overset{O}{\underset{\|}{C}}-\right]_n-O(CH_2)_5COOH + nH_2O$$

$$(n + 1)HOCH_2CH_2OH + (n + 1)HOOC(CH_2)_4COOH \xrightarrow{H^+, \; \Delta}$$

Ethylene   Adipic
glycol   acid

$$HOCH_2CH_2-\left[-O\overset{O}{\underset{\|}{C}}(CH_2)_4\overset{O}{\underset{\|}{C}}OCH_2CH_2-\right]_n-O\overset{O}{\underset{\|}{C}}(CH_2)_4COOH + nH_2O$$

The best known of the polyesters is an ethylene glycol–terephthalic acid polyester, which is spun into a fiber called Dacron or Terylene. The polymer is made by transesterification of diethyl terephthalate with ethylene glycol. Direct esterification is unsuitable, because of the insolubility of terephthalic acid.

$$\text{HOCH}_2\text{CH}_2\text{OH} + \text{C}_2\text{H}_5\overset{\displaystyle \text{O}}{\underset{}{\text{OC}}}\text{—}\left\langle \underset{}{}\right\rangle\text{—}\overset{\displaystyle \text{O}}{\underset{}{\text{COC}}}_2\text{H}_5 \xrightarrow{\text{H}^+}$$

Diethyl terephthalate

$$\text{HOCH}_2\text{CH}_2\text{O}\text{—}\left[\text{—}\overset{\displaystyle \text{O}}{\underset{}{\text{C}}}\text{—}\left\langle\right\rangle\text{—}\overset{\displaystyle \text{O}}{\underset{}{\text{C}}}\text{OCH}_2\text{CH}_2\text{O}\text{—}\right]_n\text{—}\overset{\displaystyle \text{O}}{\underset{}{\text{C}}}\text{—}\left\langle\right\rangle\text{—}\text{COOC}_2\text{H}_5 + 2n\text{C}_2\text{H}_5\text{OH}$$

Dacron

Polyamides, which include nylon, the grandfather of all synthetic fibers, are made by techniques similar to those used in the synthesis of polyesters.

$$(n+1)\text{H}_2\text{N}(\text{CH}_2)_6\text{NH}_2 + (n+1)\text{HOOC}(\text{CH}_2)_4\text{COOH} \longrightarrow$$

$$(n+1) \text{ salt} \xrightarrow{\Delta} \text{H}_2\text{N}(\text{CH}_2)_6\text{NH}\text{—}\left[\text{—}\overset{\displaystyle \text{O}}{\underset{}{\text{C}}}(\text{CH}_2)_4\overset{\displaystyle \text{O}}{\underset{}{\text{C}}}\text{NH}(\text{CH}_2)_6\text{NH}\text{—}\right]_n\text{—}\overset{\displaystyle \text{O}}{\underset{}{\text{C}}}(\text{CH}_2)_4\text{COOH} + 2n\text{H}_2\text{O}$$

Nylon

Polyamides can also be made from ω-amino acids or by a ring-opening reaction of lactams.

$$(\text{CH}_2)_5\overset{\displaystyle \text{O}}{\underset{\displaystyle \text{NH}}{\text{C}}} \xrightarrow{\text{RO}^-,\ \Delta} \text{H}_2\text{N}(\text{CH}_2)_5\overset{\displaystyle \text{O}}{\underset{}{\text{C}}}\text{—}\left[\text{—}\text{NH}(\text{CH}_2)_5\overset{\displaystyle \text{O}}{\underset{}{\text{C}}}\text{—}\right]_n\text{—}\text{NH}(\text{CH}_2)_5\text{COOH}$$

ϵ-Caprolactam

Polyethers are, in principle, derived from glycols with loss of water, but they are usually made by polymerization of cyclic ethers. Ethylene oxide is a particularly important monomer, which undergoes polymerization in the presence of either acids or bases.

$$(n+2)\overset{}{\underset{\text{O}}{\text{CH}_2\text{CH}_2}} \xrightarrow{\text{H}^+ \text{ or } \text{RO}^-} \text{HOCH}_2\text{CH}_2\text{—}\left[\text{—OCH}_2\text{CH}_2\text{—}\right]_n\text{—OCH}_2\text{CH}_2\text{OH}$$

or

$$\text{ROCH}_2\text{CH}_2\text{—}\left[\text{—OCH}_2\text{CH}_2\text{—}\right]_n\text{—OCH}_2\text{CH}_2\text{OH}$$

Polyethylene glycol

Urea and polyfunctional amides condense with formaldehyde to form an important class of copolymer resins. The reactants first condense to a liquid linear polymer, which is then heated in a mold to effect the secondary, cross-linking condensations and yield a hard, infusible product. Such plastics are described as **thermosetting. Thermoplastic** resins are linear polymers which cannot cross-link and these merely soften, without chemical change, when heated.

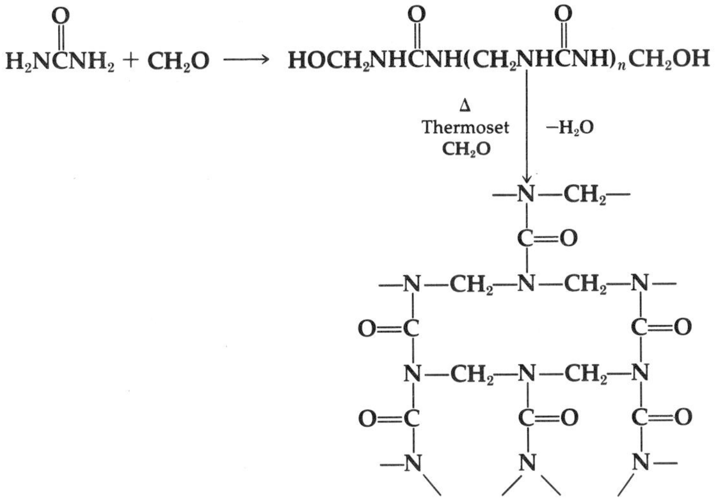

$$H_2N\overset{O}{\overset{\|}{C}}NH_2 + CH_2O \longrightarrow HOCH_2NH\overset{O}{\overset{\|}{C}}NH(CH_2NH\overset{O}{\overset{\|}{C}}NH)_n CH_2OH$$

Urea-formaldehyde resin

Phenol and its derivatives also copolymerize with aldehydes to form useful resins. The best known is Bakelite, produced from phenol itself and formaldehyde. The course of reaction is fairly well understood. Condensation first occurs to give hydroxymethyl phenols.

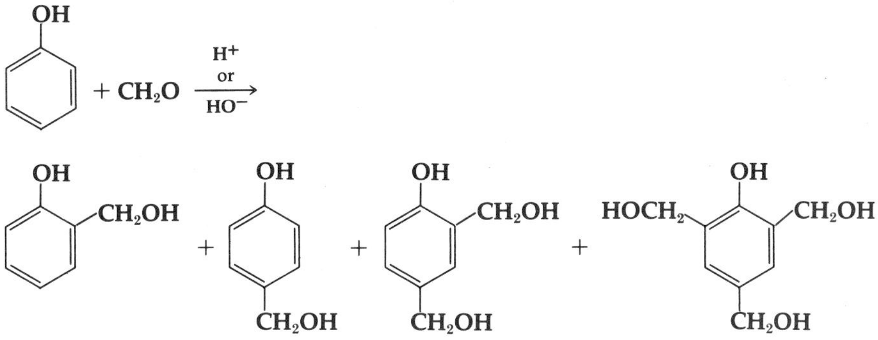

Acid catalysts cause further condensation of the following type:

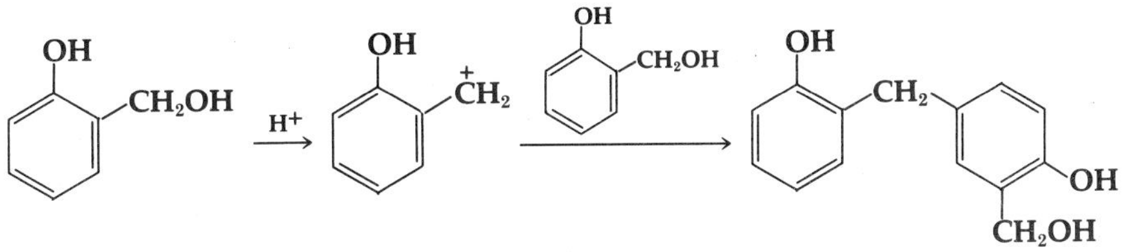

At a low degree of polymerization the fluid material is poured into a mold. The final cure to produce an infinite network is accomplished by heating, which causes further phenol alkylation with loss of water.

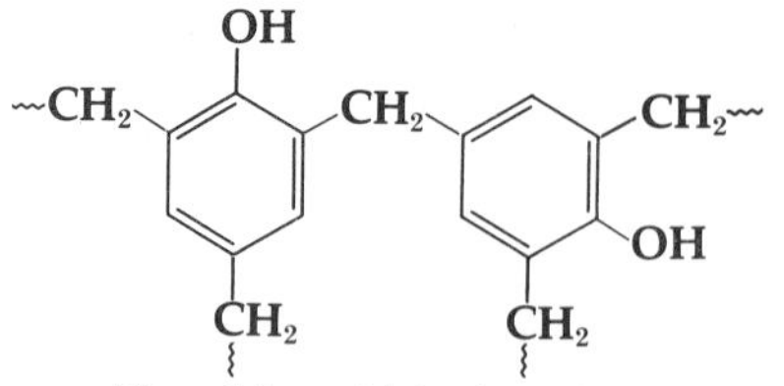

Phenol-formaldehyde resin

PROBLEM 25-1

Indicate the monomers and the polymerization method which are likely to be used in forming each of these commercial polymers.

a $\cdots CH_2CHClCH_2CH_2CH_2CHClCH_2CH_2 \cdots$

   Neoprene

b $\cdots NHCH_2CH_2NHCH_2CH_2NHCH_2CH_2 \cdots$

c $\cdots CH_2OCH_2OCH_2O \cdots$

d $\cdots CH_2CHCH_2CHCH_2CHCH_2CH \cdots$
         $\vert$ $\vert$ $\vert$ $\vert$
         CN   CN   CN   CN

   Orlon

e $\cdots$ OCO—⬡—COOCH$_2$CH$_2$OCO—⬡—COOCH$_2$CH$_2$ $\cdots$

   Dacron

PROBLEM 25-2

Write a structure for the polymer ("Glyptal") formed by heating glycerol (1,2,3-propanetriol) and phthalic anhydride in proportions of 2:3. What physical properties might you expect of this polymer?

PROBLEM 25-3

a One kind of nylon is prepared from cyclohexanone by treating with hydroxylamine, then strong acid. The monomer so produced polymerizes above 250°, especially with a little water present. Show what occurs in this process.

b Another nylon is made from cyclohexanone and butadiene as starting materials. These are first transformed into appropriate monomers and then into the nylon, which has the structure below. How would you make nylon from these monomers?

$\cdots$ —CO—(CH$_2$)$_4$—CO—NH—(CH$_2$)$_6$—NH—CO—(CH$_2$)$_4$—CO—NH— $\cdots$

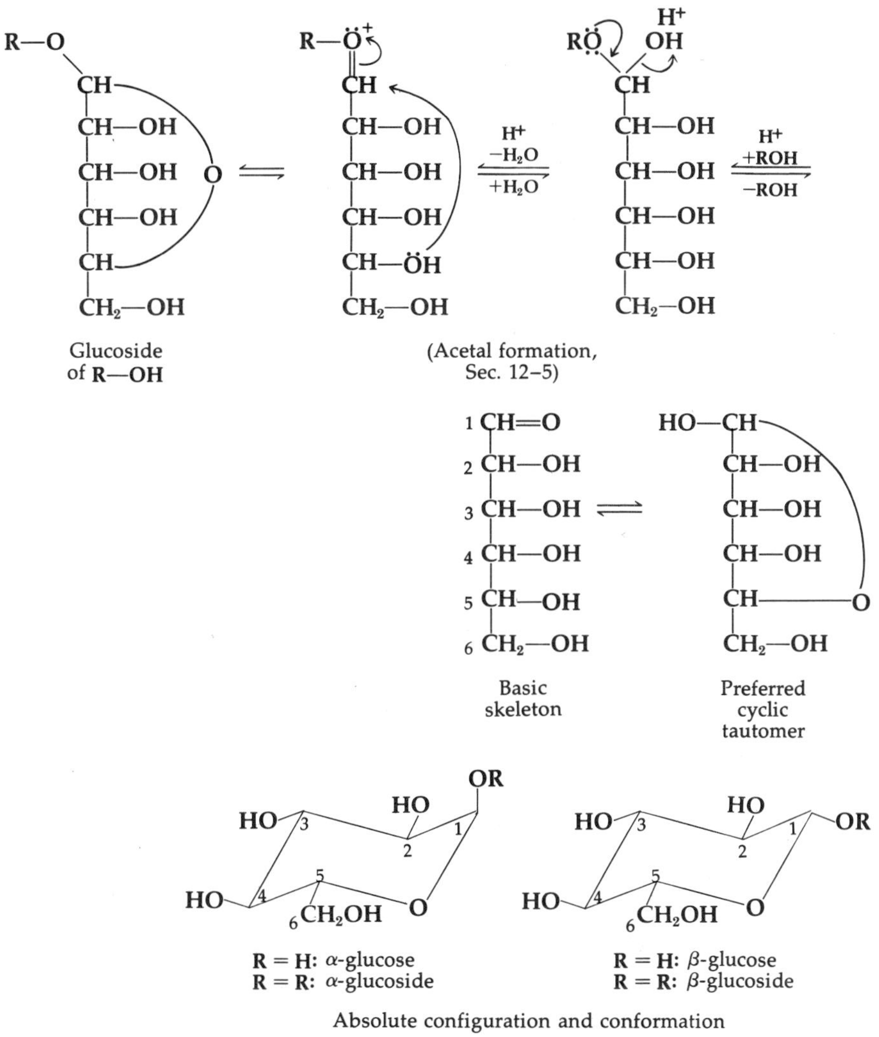

R—O

CH
CH—OH
CH—OH   O
CH—OH
CH
CH_2—OH

Glucoside
of R—OH

$\rightleftharpoons$

R—Ö⁺

CH
CH—OH
CH—OH
CH—OH
CH—OH
CH—ÖH
CH_2—OH

$\xrightarrow[+H_2O]{\begin{array}{c}H^+\\-H_2O\end{array}}$

(Acetal formation,
Sec. 12–5)

H⁺

RÖ   OH

CH
CH—OH
CH—OH
CH—OH
CH—OH
CH_2—OH

$\xrightarrow[-ROH]{\begin{array}{c}H^+\\+ROH\end{array}}$

1 CH=O
2 CH—OH
3 CH—OH $\rightleftharpoons$
4 CH—OH
5 CH—OH
6 CH_2—OH

Basic
skeleton

HO—CH
CH—OH
CH—OH
CH—OH
CH————O
CH_2—OH

Preferred
cyclic
tautomer

R = H: α-glucose
R = R: α-glucoside

R = H: β-glucose
R = R: β-glucoside

Absolute configuration and conformation

FIGURE 25-2   Glucose and the formation of glucosides

25-3   POLYSACCHARIDES AND LIGNINS

The carbohydrates, or saccharides, consist of compounds like sugar, starch, and cellulose. These compounds are of the most fundamental importance in living systems, both plant and animal, and their varied functions are discussed in Chap. 26. The central molecule in all these compounds is a linear carbon chain of four to seven carbons with a hydroxyl group on each carbon and the terminal carbon an aldehyde.† Such molecules spontaneously cyclize (tautomerize) to five- or six-membered hemiacetals (page 215) and these are the monosaccharides, or sugars, of nature.

† In some cases the carbonyl is a ketone on the *second* carbon of the chain instead.

The most common natural example is the six-carbon sugar, glucose, used in Fig. 25-2 to exemplify the nature of all sugars. The stereochemical variety of sugars is outlined in Chap. 6. One of the central modes of reaction is centered at the aldehyde carbon and involves normal formation of a cyclic acetal with an external alcohol, R—OH (Fig. 25-2). The acetals of sugars with alcohols and phenols are called **glycosides**, those of glucose being **glucosides**. Attachment of sugars to alcohols and phenols in this fashion is very common in natural compounds. Since sugars themselves bear hydroxyl groups, it is not surprising that glycosidic linkages between sugar molecules should also be common in nature. Thus, **disaccharides** are formed by joining two sugars, as in the common dimers of glucose shown below.

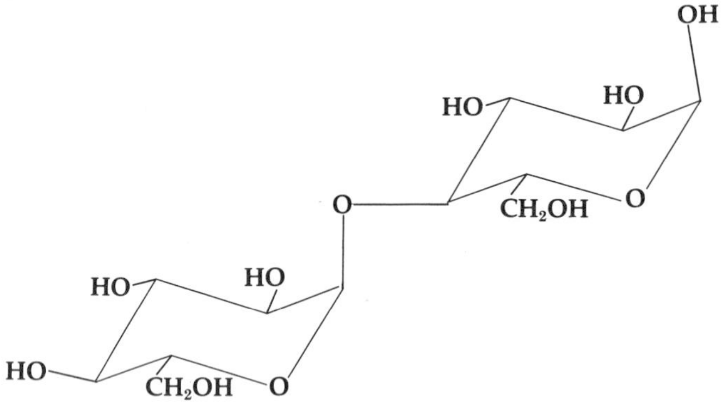

α-Maltose

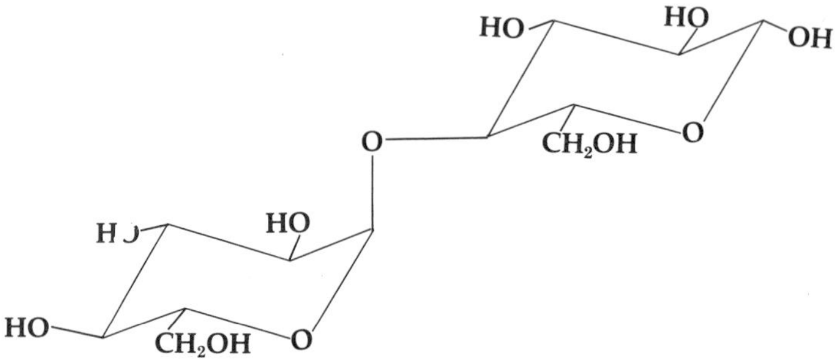

β-Maltose

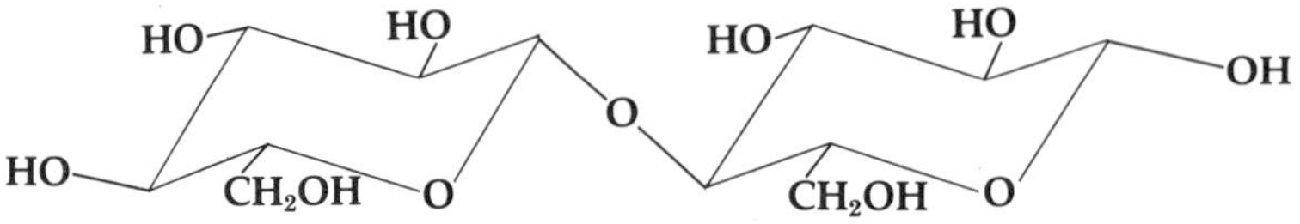

Cellobiose

The simple sugars are called **monosaccharides** (cf. glucose, Fig. 25-2) and serve as monomers for the formation of **polysaccharides**, the corresponding condensation polymers with repeating sugar or monosaccharide units joined with loss of water. Acidic hydrolysis of glycosides and polysaccharides reverses the acetal formation and yields monosaccharides; incomplete hydrolysis may yield di- and trisaccharides.

The glycosidic carbon atom (C-1) is linked by either an α- or a β-linkage to either C-4 or C-6 of the next monosaccharide unit. The commonest polymers contain only glucose units; cellulose is an example.

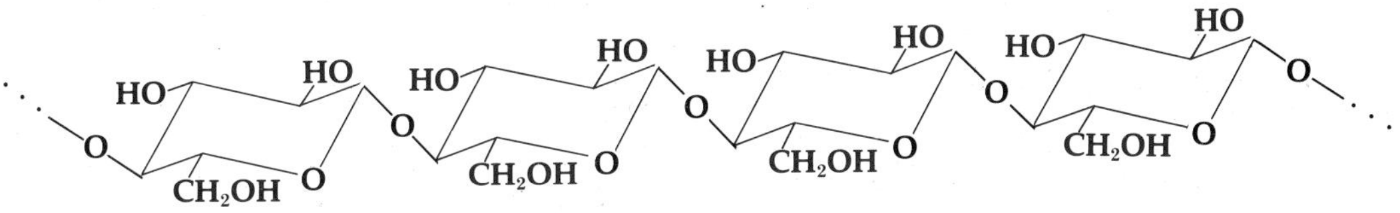

Cellulose

Starch is stored in the roots, seeds, and fruits of plants (e.g., corn, potatoes, wheat, tapioca, and rice). It is one of the principal nutrient materials for the animal world. All starches give only glucose on hydrolysis, but both 1,4- and 1,6-linkages occur in most starches. The glycosidic linkages have the α-configuration, so the links between the repeating units look like those in the maltoses (see above). Hot water separates starch into an insoluble fraction, amylopectin, and a soluble fraction, amylose. Amylose is hydrolyzed exclusively to maltose by the enzyme **maltase**. If amylose is methylated and then hydrolyzed, the major product is 2,3,6-trimethylglucose and small amounts of 2,3,4,6-tetramethylglucose are also found.

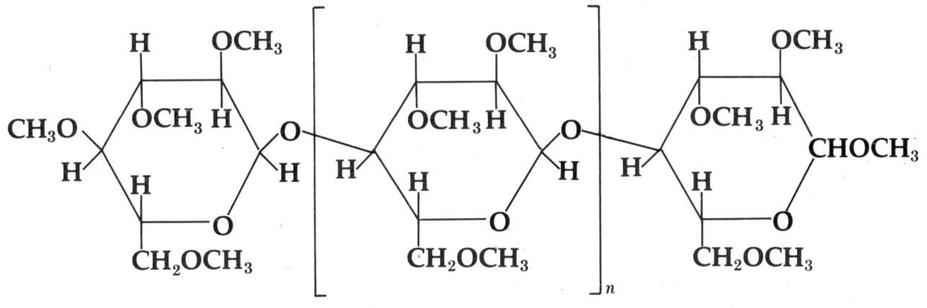

Methylated amylose

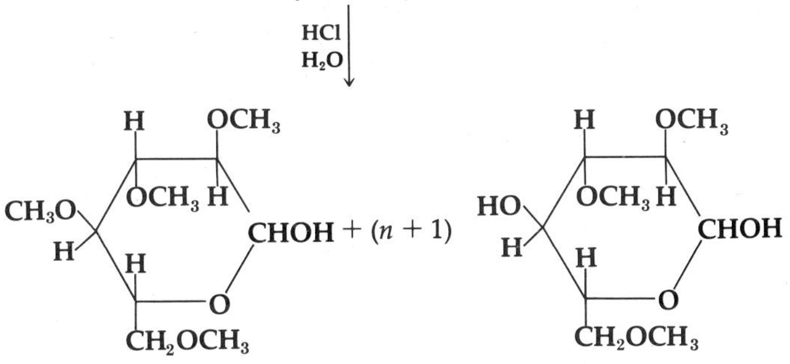

$$\begin{array}{c} \text{HCl} \\ \text{H}_2\text{O} \end{array} \downarrow$$

2,3,4,6-Tetramethylglucose   2,3,6-Trimethylglucose

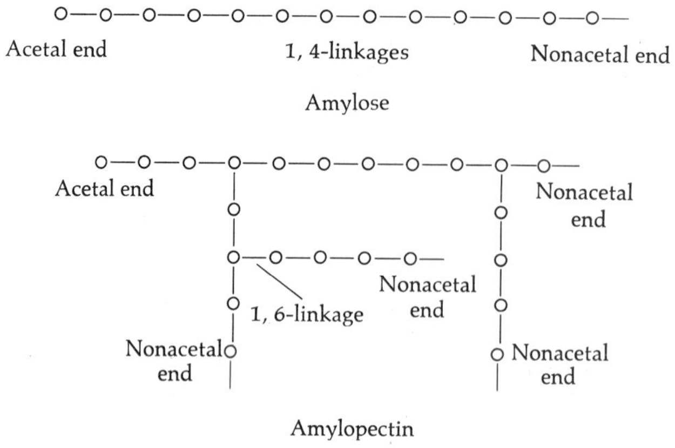

O—O—O—O—O—O—O—O—O—O—O—O—O—

Acetal end   1, 4-linkages   Nonacetal end

Amylose

O—O—O—O—O—O—O—O—O—O—O—
Acetal end | | Nonacetal
                    O   O   end
                    | |
                    O—O—O—O—O—   O
                    | \   Nonacetal |
                    O 1, 6-linkage end   O
                    | |
      Nonacetal O   O Nonacetal
         end | | end

Amylopectin

FIGURE 25-3   **Starch structures**

The free hydroxyl position in the major product indicates that the polymer
is held together by 1,4-linkages, and the amount of the tetramethylglucose
produced gives a measure of the number of nonacetal end groups and, therefore, of the average molecular weight of the amylose. The extra methyl group
at the acetal ends of the polymer chains is lost easily on acidic hydrolysis, since
it is bound by a glycoside (acetal) linkage.

Amylopectin gives only about a 50% yield of maltose when digested by
maltase. Since the enzyme is specific for hydrolysis of 1,4-linkages, there must
be some other links in amylopectin. Exhaustive methylation of amylopectin
followed by hydrolysis gives a higher yield of 2,3,4,6-tetramethylglucose than
is obtained from amylose. However, physical measurements (page 1012) show
that amylopectin has a much higher molecular weight than amylose. Therefore,
there must be a number of nonacetal "ends" on an amylopectin molecule;
hence, the substance must be a branched or cross-linked rather than a linear
polymer. The conclusion is borne out by estimation of the number of acetal
end groups, which is larger for amylose than for amylopectin. The structure
of the two starches can be shown schematically, as in Fig. 25-3. Assignment
of the branch points as 1,6-linkages is indicated by isolation of 2,3-dimethyl-
glucose after methylation and hydrolysis.

Glycogen, the reserve carbohydrate of animals, is stored in the liver and
muscles. On complete hydrolysis, the polymer gives glucose units. Methylation
studies and degradation by enzymatic hydrolysis show that the substance is
linked by β-1,4-glycoside linkages with some 1,6-linkages, such that a cross-
linked structure results.

Cellulose is a glucose polymer of high molecular weight found in all plants.
About 50% of wood and 90% of cotton fiber are cellulose. To the best of our
knowledge, cellulose is completely linear, and the glucose units are joined by
β-1,4-linkages. The monomeric units of cellulose are held together by β-1,4-
linkages, and there are an average of about 3,000 monomer units per cellulose
molecule. Cellobiose (above) is the disaccharide formed on hydrolysis.

Chemical modification of cellulose produces polymeric materials that are familiar commercial products. The principal chemical function of native cellulose is found in the —OH groups that cover the spine of the polymeric molecule. Reaction with acetic anhydride produces cellulose acetate, which can be spun into the synthetic fiber known as *rayon* or formed into photographic film. Nitration produces *nitrocellulose*, which is the explosive binder used in the formulation of many solid propellants for rockets. Perhaps the best known product from cellulose is *cellophane*, a clear sparkling film having approximately the same composition as cellulose, and made from it by a process which brings the cellulose into solution so that it may be reprecipitated while being stretched into film.

Lignin

Wood contains another, noncarbohydrate, polymer known as *lignin*. The structure of lignin has not been entirely elucidated. There is no doubt that soluble "lignins" obtained by digestion of wood with aqueous alkalis and alkaline bisulfites (as is done in the manufacture of paper) are polymers containing a variety of groups. Since there is no unequivocal evidence that native lignin can be dissolved in totally inert solvents, a strong case can be constructed for the view that native lignin may have a simple repetitive structure that is chemically modified during solubilizing processes. Various degradations, such as oxidative cleavage, zinc-dust distillation, and dry distillation of lignin, give guaiacol and its derivatives.

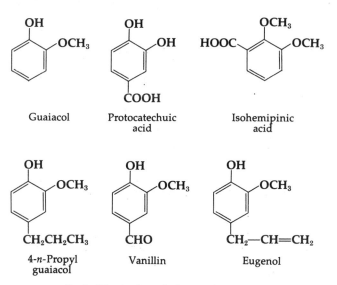

Typical lignin degradation products

Such degradation studies have led to a partial structure of lignin as ortho-dioxygenated benzene rings connected by three-carbon chains. These chains have various attached hydroxyls or carbonyls, generally such that the oxidation state is the equivalent of two hydroxyls per three-carbon chain, as in the hypothetical polymer section shown:

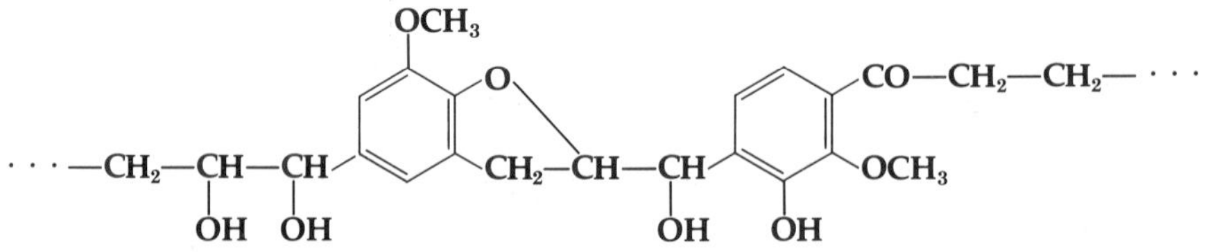

Hypothetical partial lignin structure

Lignin is at present a waste product in the processing of wood during manufacture of paper and other cellulose products. The future will probably see development of significant uses for this source of interesting chemicals. There are two principal processes for the conversion of wood into paper. Both are primarily designed to remove lignin as water-soluble derivatives. In the *alkali process*, lignin is degraded and solubilized by virtue of the acidity of phenolic hydroxyl groups. In the *sulfite process*, an alkaline bisulfite solution sulfonates some of the aromatic nuclei, thus making them water-soluble.

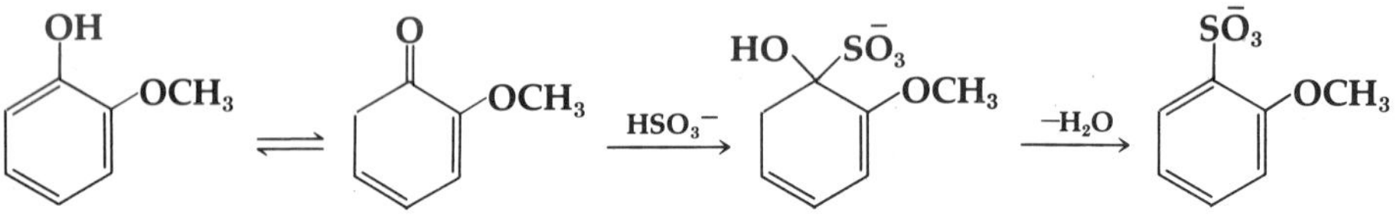

PROBLEM 25-4

Consider glycogen, the animal starch, to be fully water-soluble. How might reaction with periodate be used to determine the proportion of 1,4- to 1,6-linkages? What products of reaction would be expected?

PROBLEM 25-5

Write a detailed mechanism for the hydrolysis of starch to glucose.

PROBLEM 25-6

Consider using glucose as a source for synthesis of higher saccharides. What distinctions among the hydroxyl groups can be made with respect to their reactivity which might allow rational synthetic schemes to be developed? Try to apply these distinctions in a synthesis of the 1,6-linked disaccharide.

25-4   AMINO ACIDS AND PROTEINS

Proteins are omnipresent in living matter. **Fibrous proteins** constitute the structural matter of animals, in the form of skin, muscle, silk fibers, connective tissue, and so forth. **Soluble**, or **globular**, **proteins** play critical roles in all life processes as enzymes, metabolic intermediates, character-determining genetic factors, etc.

Hydrolysis of proteins with acids, bases, or enzymes, produces mixtures of α-amino acids as the principal products. All evidence shows that simple proteins are linear copolymers of α-amino acids in which the units are linked by **peptide** (amide) linkages. Molecules that contain only short sequences of amino acids are known as **peptides** or **polypeptides**.

α-Amino acid units
section of a protein chain

α-Amino acids

There are in nature about twenty different amino acids used in the construction of proteins. Natural proteins commonly contain 100–400 amino acid units per chain, with molecular weights of 15,000–60,000. The possible variety of protein structure available through different numbers and sequences of amino acids is truly staggering.

> The total number of different proteins that can be assembled from these twenty different amino acids is $N = 20^n$, where $n = $ number of amino acid units in the chain. If we could collect only one molecule each of all the possible proteins containing chains which are 20 units in length, this would be 20^{20} molecules and would occupy 25 liters of space, i.e., a sphere of solid protein about the size of a basketball with every molecule in it different! With a 30-unit protein the occupied volume becomes 60 cubic miles, while the possible variants of only a 100-unit protein would fill a space larger than the universe! Fortunately, with this variety we are not obliged to find as food the particular proteins our bodies require to function. In fact since the functional proteins of the organism differ slightly in sequence from one species to another, such a requirement would demand our subsisting on human food. However, the body can ingest all proteins since they are hydrolyzed by stomach enzymes to a simple mixture of the twenty constituent amino acids and these are then resynthesized into the particular proteins which the body requires.

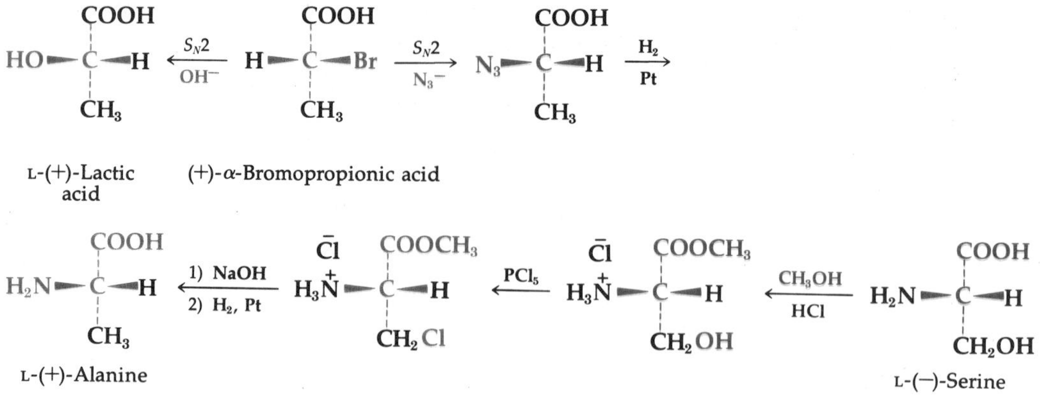

COOH   COOH   COOH
HO—C—H $\xleftarrow[OH^-]{S_N2}$ H—C—Br $\xrightarrow[N_3^-]{S_N2}$ N$_3$—C—H $\xrightarrow[Pt]{H_2}$
CH$_3$ CH$_3$ CH$_3$

L-(+)-Lactic   (+)-α-Bromopropionic acid
acid

COOH $\bar{C}l$ COOCH$_3$ $\bar{C}l$ COOCH$_3$ COOH
H$_2$N—C—H $\xleftarrow[2)\ H_2,\ Pt]{1)\ NaOH}$ H$_3$N$^+$—C—H $\xleftarrow{PCl_5}$ H$_3$N$^+$—C—H $\xleftarrow[HCl]{CH_3OH}$ H$_2$N—C—H
CH$_3$ CH$_2$Cl   CH$_2$OH   CH$_2$OH

L-(+)-Alanine   L-(−)-Serine

FIGURE 25-4   Relative configurations of amino acids

Amino Acids

The chief amino acids from which natural proteins are made all contain an amino group α to the carboxylic acid and are optically active, with the structures shown in Table 25-1. All but two (prolines) are primary amines and all but one (glycine) have the same absolute configuration at the α-carbon. This natural configuration is commonly called L- in the amino acid series and is always the s-absolute configuration, although the optical rotation may be (+) or (−) as in Fig. 25-4.

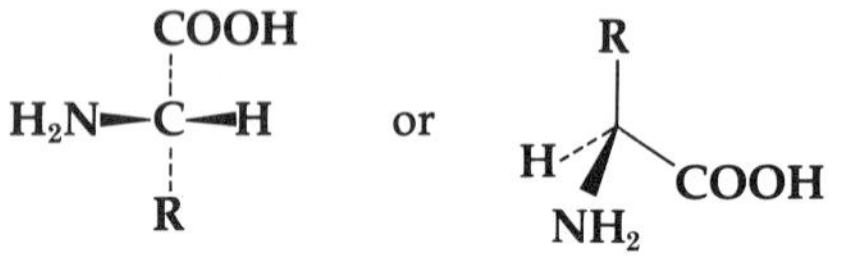

COOH   R
H$_2$N—C—H   or   H···⟋\COOH
R   NH$_2$

Generalized naturally
occurring amino acid
(L series)

The interrelations in configuration of amino acids have been established by the principle that if the configurations of two compounds are established in relation to that of a third, the two become known in relation to each other. The configurations of lactic acid and alanine are key links in the chain of structural relationships recorded in Fig. 25-4. (See also lactic acid on page 220.) From lactic acid, relationships in configuration were also extended to the sugars (page 210). Only two of these reactions involve the making or breaking of bonds to the asymmetric carbon atom. Reactions of (+)-α-bromopropionic acid with either hydroxide or azide ion were found by rate measurements to belong to the S_N2 reaction-mechanism family, and these transformations have been demonstrated to occur with inversion of configuration (Sec. 10-2).

The side chains R contain a variety of functional groups, including extra carboxyl or amino groups which lead to the acidic or basic amino acids of Table 25-1. Most amino acids are neutral, however, since the α-amino and carboxyl groups neutralize each other. Neutral amino acids exist as inner salts with dipolar structures. These saltlike compounds are high-melting (with decomposition), relatively nonvolatile, water-soluble, and ether-insoluble.

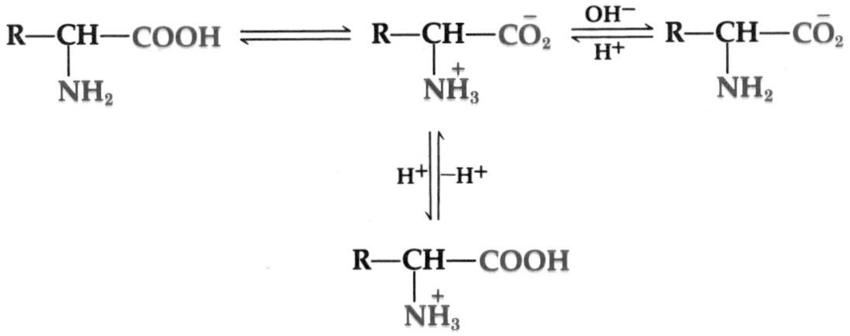

$$R-\underset{\underset{NH_2}{|}}{CH}-COOH \rightleftharpoons R-\underset{\underset{\overset{+}{N}H_3}{|}}{CH}-C\bar{O}_2 \underset{H^+}{\overset{OH^-}{\rightleftharpoons}} R-\underset{\underset{NH_2}{|}}{CH}-C\bar{O}_2$$

$$H^+ \updownarrow -H^+$$

$$R-\underset{\underset{\overset{+}{N}H_3}{|}}{CH}-COOH$$

The lowest solubility of neutral amino acids in water occurs at the acidity level that provides the highest concentration of inner salt. This pH, called the **isoelectric point**, ranges from 4.8 to 6.3 for neutral amino acids and varies according to the relative effect of structural features on the acidity and basicity of the two functions (Chap. 8). Above the isoelectric point, amino acids are converted to anions; below the critical pH, they add protons and form cations.

TABLE 25-1   **Amino Acids from Protein Hydrolysates**

Neutral amino acids

Glycine	$CH_2(NH_2)COOH$	Serine	$HOCH_2CH(NH_2)COOH$
Alanine	$CH_3CH(NH_2)COOH$	Threonine	$CH_3CH(OH)CH(NH_2)COOH$
Valine	$(CH_3)_2CHCH(NH_2)COOH$	Methionine	$CH_3SCH_2CH_2CH(NH_2)COOH$
Leucine	$(CH_3)_2CHCH_2CH(NH_2)COOH$	Cysteine	$HSCH_2CH(NH_2)COOH$
Isoleucine	$CH_3CH_2CH(CH_3)CH(NH_2)COOH$	Cystine	$SCH_2CH(NH_2)COOH$
Phenylalanine	$C_6H_5CH_2CH(NH_2)COOH$		$SCH_2CH(NH_2)COOH$

Tyrosine $HO-\langle\rangle-CH_2CH(NH_2)COOH$

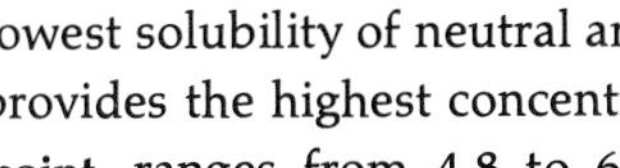

Proline
(pyrrolidine ring with COOH)

Tryptophan
(indole ring with $CH_2CH(NH_2)COOH$)

Hydroxyproline HO (pyrrolidine ring with COOH)

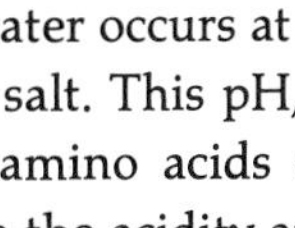

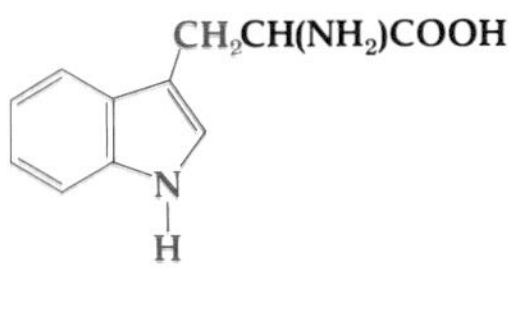

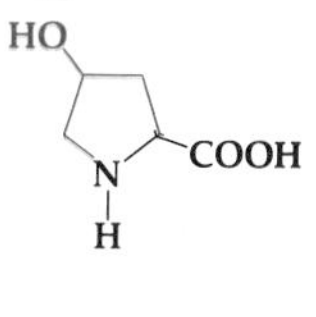

Acidic amino acids

Aspartic acid	$HOCOCH_2CH(NH_2)COOH$	Glutamic acid	$HOCOCH_2CH_2CH(NH_2)COOH$

Basic amino acids

Lysine	$H_2N(CH_2)_4CH(NH_2)COOH$		
Arginine	$H_2N\underset{\underset{NH}{\parallel}}{C}NH(CH_2)_3CH(NH_2)COOH$	Histidine	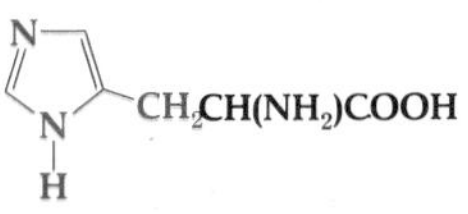 (imidazole ring with $CH_2CH(NH_2)COOH$)

Most reactions of amino acids are those normally expected from the two functions separately, such as esterification of the acid or acylation of the amine. Acid chlorides, of course, can only be prepared if the basic amino group is protected first.

Amino acids may be synthesized in a number of ways, of which two of the most common general routes may serve as examples. In the first, the full carbon skeleton is used and the amino group introduced by α-halogenation followed by reaction with ammonia. A large excess of ammonia is used to suppress further alkylation of the initial primary amines to secondary amines.

$$R-CH_2-COOH \xrightarrow[\substack{PX_3 \\ (p.\ 748)}]{X_2} R-\underset{\underset{X}{|}}{C}H-COOH \xrightarrow[\text{excess}]{NH_3} R-\underset{\underset{NH_2}{|}}{C}H-COOH$$

In the second example, α-aminomalonic ester is used as a source of the $-CH(NH_2)COOH$ portion and is alkylated (S_N2 alkylation with malonate anion, page 430) with R—X, the halide corresponding to the side chain of the amino acid in question.

$$\underset{COOEt}{\overset{COOEt}{>}} \xrightarrow[EtO^-]{EtO-NO} \underset{COOEt}{\overset{COOEt}{>}}=N-OH \xrightarrow[Pd]{H_2} \underset{COOEt}{\overset{COOEt}{\searrow}}CH-NH_2 \xrightarrow{Ac_2O} \underset{COOEt}{\overset{COOEt}{\searrow}}CH-NHAc \xrightarrow{Base}$$

$$\underset{COOEt}{\overset{COOEt}{\searrow}}AcNH-C\colon^- \xrightarrow{R-X} \underset{COOEt}{\overset{COOEt}{\searrow}}AcNH-C-R \xrightarrow[\substack{-2EtOH \\ -CO_2 \\ -AcOH}]{H_3O+} \underset{}{\overset{COOH}{|}}H_2N-CH-R$$

PROBLEM 25-7

If you had just isolated optically active cysteine and serine for the first time, how would you determine their absolute configurations? Assume that the other amino acids in Table 25-1 are all known and available to you.

PROBLEM 25-8

Ordinary amino acids (such as alanine, etc.) show two pK_a values on titration, at about 2.4 and 9.0. Explain this in terms of the species present as an aqueous solution is acidified from pH 12 to pH 1. Predict the pK_a values and species involved in aspartic acid and in arginine.

Determination of Protein Structure

Two problems present themselves in elucidating the structure of a protein. The first is amino acid analysis, from which may be deduced the number and kind of amino acids present. The second problem is concerned with deducing their sequence in the chain. Amino acid analysis is conducted by total hydrolysis of the protein with acid, followed by quantitative chromatography of the component amino acids in the hydrolysate. A completely automatic amino acid analyzer is now commercially available to separate all 20 amino acids chromatographically and record their relative amounts on a graph, as in Fig. 1-1. The amount of amino acid in each fraction is ascertained by reaction with a compound (ninhydrin) which combines with amino acids to produce a colored product, the amount of which is then determined spectrophotometrically.

The end amino acids in a protein or peptide chain are distinguished from the others by possessing either a free carboxyl or amino group. These terminal units may be identified by causing them to react with a tagging reagent so that they are identifiable after hydrolysis. The idea is illustrated by the most common tag for an amino terminal, the dinitrophenyl (DNP) group, attached by reaction of the protein or peptide with 1-fluoro-2,4-dinitrobenzene. The group is colored for easy identification in paper chromatography and is not affected by acid hydrolysis.

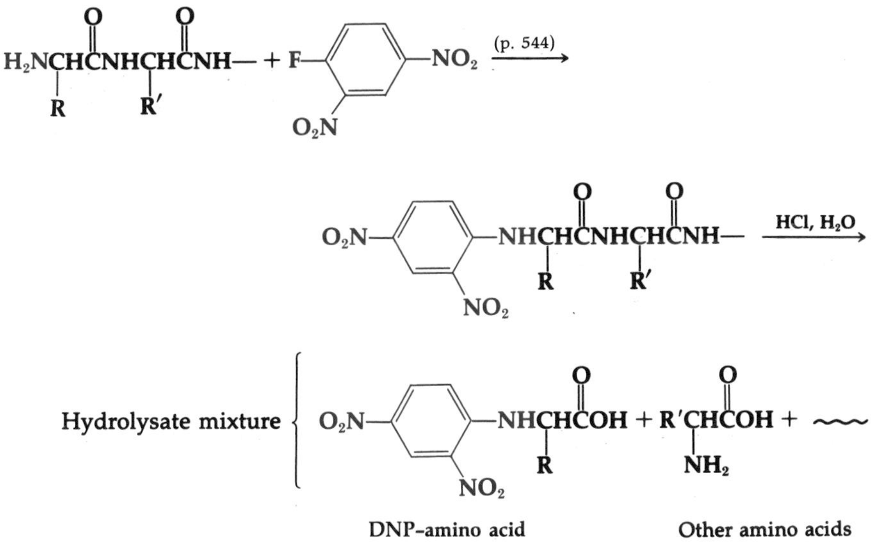

The sequence of amino acids in a protein may be determined by partial hydrolysis of the terminal-tagged protein into peptides of two to eight units in length. These are then chromatographically separated and treated as separate structure problems. The end groups are tagged and the peptide further hydrolyzed and analyzed, the DNP–amino acids marking the amino end of any peptide fragment. The overlapping of fragments from the original chain allows deduction of the sequence.

The approach may be illustrated with two examples of partial hydrolysis of a pentapeptide containing amino acids A, B, C, D, and E.

Peptide $\longrightarrow$ peptide —DNP $\xrightarrow{H_3O^+}$

$$A + B + C\text{—DNP} + D + E + 4 \text{ dipeptides}$$
Amino acids

H_3O^+
- $\longrightarrow$ $A + C$—DNP
- $\longrightarrow$ $B + D$
- $\longrightarrow$ $A + D$
- $\longrightarrow$ $B + E$

Since any nonterminal amino acid can only be linked to two others (and terminals with only one), this establishes the sequence:

$$H_2N—C—A—D—B—E—COOH$$

Alternatively, the hydrolysis might yield the following mixed peptides, which also establish the same sequence:

Peptide—DNP $\longrightarrow$ Peptides
- $\longrightarrow$ C—DNP
- $\longrightarrow$ $A + C$—DNP
- $\longrightarrow$ $A + D + C$—DNP
- $\longrightarrow$ $A + B + D + C$—DNP

Another tool used in connection with sequence studies is that of selective cleavages. Hydrolysis catalyzed by certain enzymes is known to be selective in cleaving only the amide link adjacent to particular amino acids, so that these enzymes may be used to break a protein down into peptide fragments with known terminals. Similarly, certain chemical methods have been devised to achieve the same specific cleavage, as in the oxidative cleavage adjacent to tyrosine by N-bromosuccinimide.

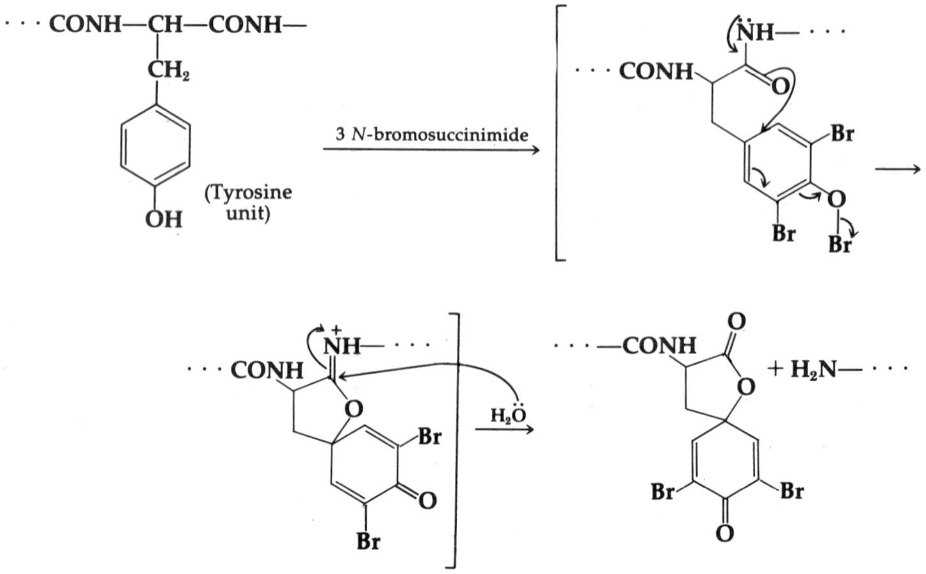

Finally, certain chemical methods have been developed which specifically remove only a single terminal amino acid from the chain. Successive use of such a reaction with identification of the removed amino acid each time provides a specific procedure for determining sequences in proteins. Although the procedure is tedious, it has recently been developed to such an extent that an automatic machine ("amino acid sequence analyzer") has been built to carry out the successive determinations.

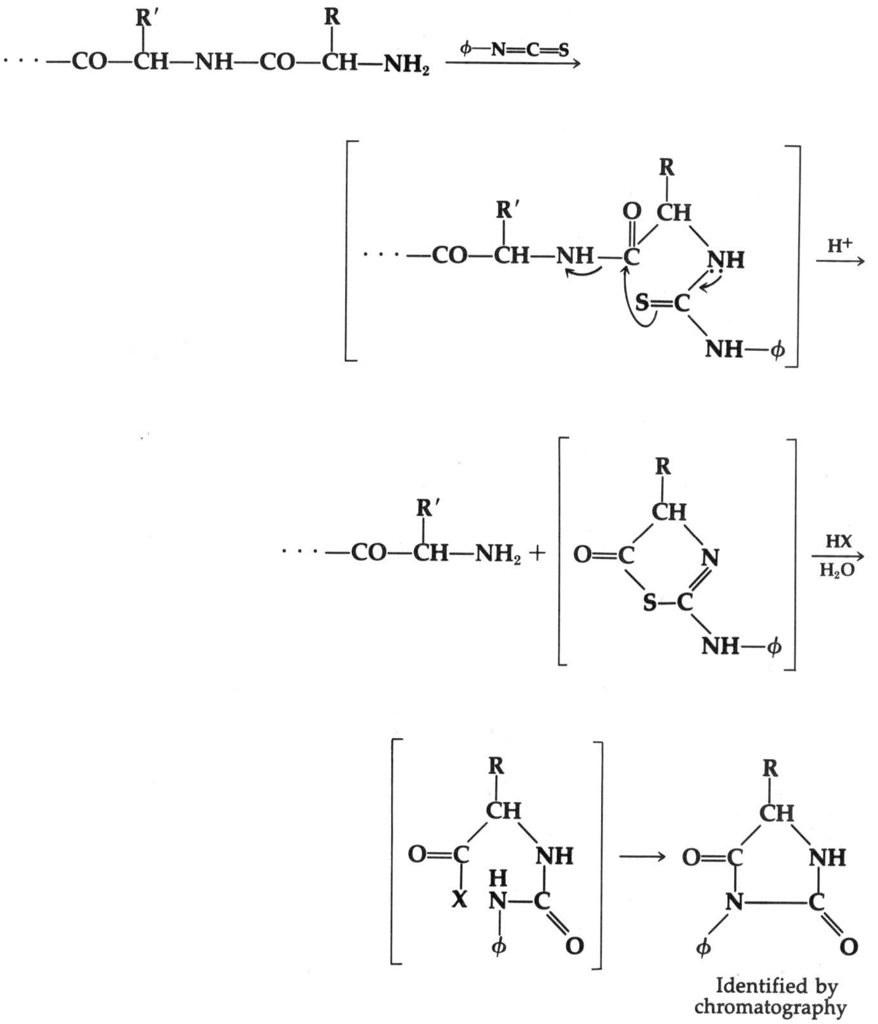

$$\cdots-CO-\underset{\underset{R'}{|}}{CH}-NH-CO-\underset{\underset{R}{|}}{CH}-NH_2 \xrightarrow{\ \phi-N=C=S\ }$$

$$\left[\ \cdots-CO-\underset{\underset{R'}{|}}{CH}-NH-\overset{\overset{O}{\|}}{C}\ \underset{S=C}{\overset{CH(R)}{\diagup}}\ NH\ \diagdown NH-\phi\ \right] \xrightarrow{\ H^+\ }$$

$$\cdots-CO-\underset{\underset{R'}{|}}{CH}-NH_2 + \left[\ O=C\ \underset{S-C}{\overset{CH(R)}{\diagup}}\ N\ \diagdown NH-\phi\ \right] \xrightarrow[H_2O]{\ HX\ }$$

$$\left[\ O=C\ \underset{X\ \underset{\phi}{\overset{H}{N}}-C}{\overset{CH(R)}{\diagup}}\ NH\ \diagdown O\ \right] \longrightarrow\ O=C\ \underset{\underset{\phi}{N}-C}{\overset{CH(R)}{\diagup}}\ NH\ \diagdown O$$

Identified by
chromatography

The detailed sequences of a number of proteins have been determined and more are under investigation. Insulin, with two chains totaling 51 amino acids and linked by disulfide bridges between their cystine units, was the first to be determined. Others include hemoglobin (146 units), enzymes like lysozyme (129 units; Fig. 25-5), and muscle proteins like myoglobin (153 units). Different species produce proteins of similar function which may differ in several of the units in the chain.

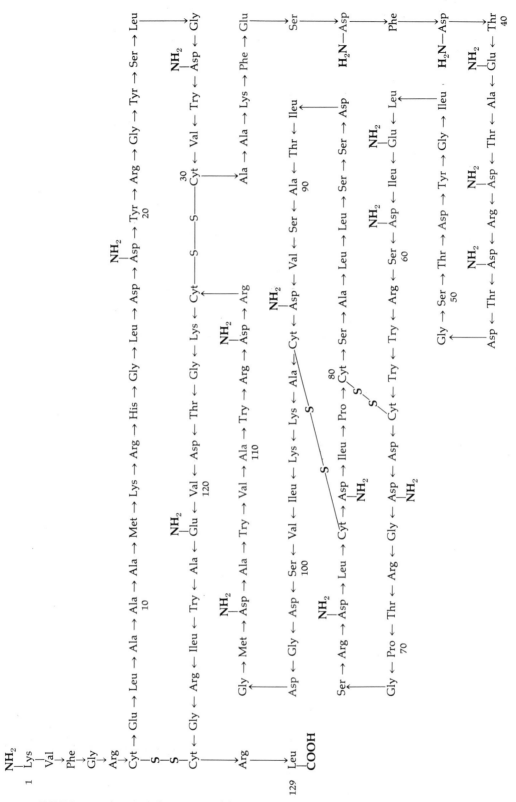

FIGURE 25-5 Amino acid sequence of lysozyme (residues in color are those found in the cleft of the molecule, see page 1034)

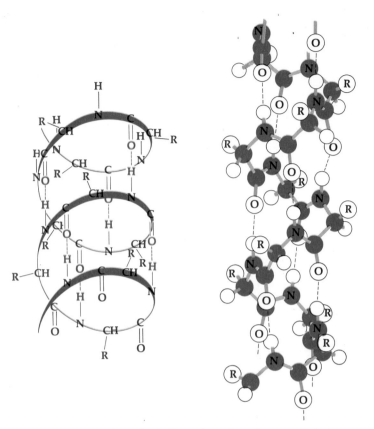

FIGURE 25-6 Right-handed α-helix conformation of protein chain in two representations

Secondary Structure

The structure of proteins may be considered at a secondary level, beyond that of amino acid sequence, and this will concern the shape or conformation of the protein chain. This secondary structure is of enormous importance to the functioning of most proteins in living organisms. The amide link contains much double-bond character owing to resonance. This in turn restricts rotation about the C—N bond and reduces the number of possible conformations of the protein chain.

Barrier to rotation
$\Delta E = 10$ kcal/mole

It has also been shown that the C=O and N—H bonds form strong hydrogen bonds which link one part of the chain to another. These hydrogen bonds are perfectly suited to orient many protein chains into a coiled conformation, the α-**helix**, with about 3.5 amino acid units per turn of the coil (Fig. 25-6). The actual physical appearance of many protein molecules is that of a long coil which is itself bent and folded into a roughly globular shape.

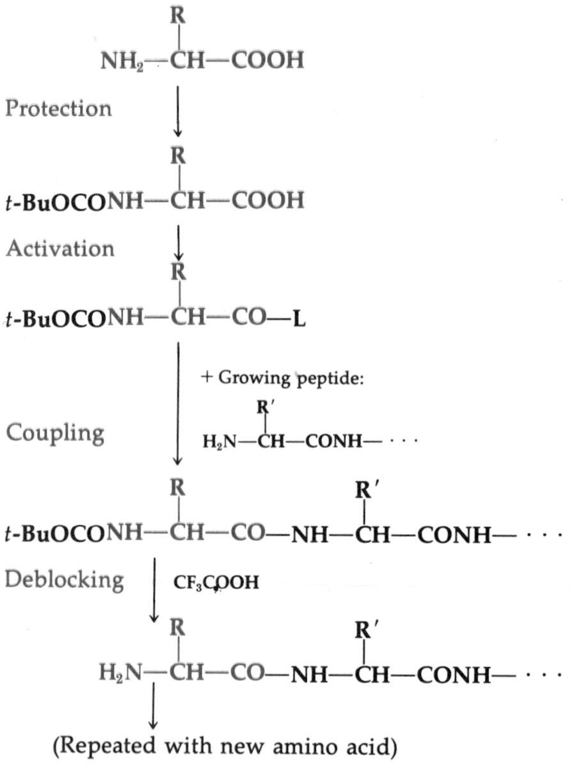

$$\text{NH}_2\text{—}\overset{\displaystyle R}{\overset{|}{\text{CH}}}\text{—COOH}$$

Protection

$$t\text{-BuOCONH}\text{—}\overset{\displaystyle R}{\overset{|}{\text{CH}}}\text{—COOH}$$

Activation

$$t\text{-BuOCONH}\text{—}\overset{\displaystyle R}{\overset{|}{\text{CH}}}\text{—CO—L}$$

Coupling

+ Growing peptide:

$$\text{H}_2\text{N}\text{—}\overset{\displaystyle R'}{\overset{|}{\text{CH}}}\text{—CONH—}\cdots$$

$$t\text{-BuOCONH}\text{—}\overset{\displaystyle R}{\overset{|}{\text{CH}}}\text{—CO—NH—}\overset{\displaystyle R'}{\overset{|}{\text{CH}}}\text{—CONH—}\cdots$$

Deblocking CF_3COOH

$$\text{H}_2\text{N}\text{—}\overset{\displaystyle R}{\overset{|}{\text{CH}}}\text{—CO—NH—}\overset{\displaystyle R'}{\overset{|}{\text{CH}}}\text{—CONH—}\cdots$$

(Repeated with new amino acid)

FIGURE 25-7   Peptide synthesis

Peptide and Protein Synthesis

The problem of protein synthesis is simply stated but not so simply realized in practice. Amide links must be formed to specific amino acids in sequence. For the amino acid being added to the chain the group (amino or carboxyl) which is *not* involved in amide formation must be protected first. Following amide formation the protecting group must be removed so as then to be reactive for addition of the next protected amino acid unit. Furthermore, the carboxyl must be converted to a more reactive acyl form in order to react with the amino of the joining amino acid. Hence the addition of just one amino acid to a growing chain involves several steps. A sample of such a sequence is shown in Fig. 25-7.

Protecting groups are commonly applied to the amino function in the form of easily removed urethans (carbamates), which have enough amide character to render the nitrogen nonbasic:

$$\text{R—NH}_2 + \text{Cl—CO—O—CH}_2\phi \xrightarrow[\text{Protection}]{-\text{Cl}^-} \text{R—NH—CO—O—CH}_2\phi \xrightarrow[\text{Removal}]{\text{H}_2/\text{Pd}}$$

"Carbobenzoxy group"

$$\text{R—NH}_2 + \text{CO}_2 + \phi\text{CH}_3$$

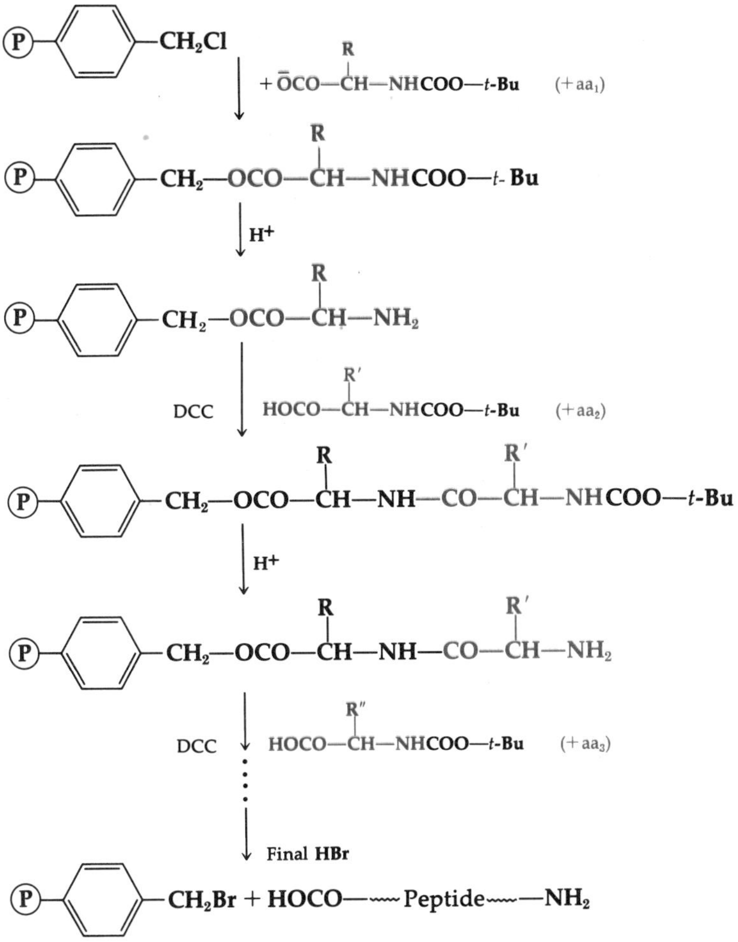

(P)—⟨benzene⟩—CH_2Cl

$\downarrow$ $+ \overset{-}{\text{O}}\text{CO}\overset{\displaystyle \text{R}}{-\text{CH}}\text{—NHCOO}\text{—}t\text{-Bu}$ $(+\text{aa}_1)$

(P)—⟨benzene⟩—$\text{CH}_2\text{—OCO}\overset{\displaystyle \text{R}}{-\text{CH}}\text{—NHCOO}\text{—}t\text{- Bu}$

$\downarrow \text{H}^+$

(P)—⟨benzene⟩—$\text{CH}_2\text{—OCO}\overset{\displaystyle \text{R}}{-\text{CH}}\text{—NH}_2$

$\text{DCC}\ \Big\downarrow\ \ \text{HOCO}\overset{\displaystyle \text{R}'}{-\text{CH}}\text{—NHCOO}\text{—}t\text{-Bu}$ $(+\text{aa}_2)$

(P)—⟨benzene⟩—$\text{CH}_2\text{—OCO}\overset{\displaystyle \text{R}}{-\text{CH}}\text{—NH—CO}\overset{\displaystyle \text{R}'}{-\text{CH}}\text{—NHCOO}\text{—}t\text{-Bu}$

$\downarrow \text{H}^+$

(P)—⟨benzene⟩—$\text{CH}_2\text{—OCO}\overset{\displaystyle \text{R}}{-\text{CH}}\text{—NH—CO}\overset{\displaystyle \text{R}'}{-\text{CH}}\text{—NH}_2$

$\text{DCC}\ \Big\downarrow\ \ \text{HOCO}\overset{\displaystyle \text{R}''}{-\text{CH}}\text{—NHCOO}\text{—}t\text{-Bu}$ $(+\text{aa}_3)$

⋮

$\downarrow$ Final **HBr**

(P)—⟨benzene⟩—$\text{CH}_2\text{Br} + \text{HOCO}\text{—}\backsim\backsim\text{Peptide}\backsim\backsim\text{—NH}_2$

(P) = insoluble polymer
aa$_1$, aa$_2$, aa$_3$ = added, protected amino acids
DCC = dicyclohexyl-carbodiimide, $\text{C}_6\text{H}_{11}\text{N}{=}\text{C}{=}\text{NC}_6\text{H}_{11}$

FIGURE 25-8   **Solid-phase peptide synthesis**

$$\text{R—NH}_2 + \text{N}_3\text{—CO—O—}\overset{-}{\text{C}}(\text{CH}_3)_3 \ \xrightarrow[\text{Protection}]{-\text{N}_3^-}\ \text{R—NH—CO—O—C(CH}_3)_3 \ \xrightarrow[\text{Removal}]{\text{CF}_3\text{COOH}}$$

t-Butyloxycarbonyl group

$$\text{R—NH}_2 + \text{CO}_2 + \text{CH}_2{=}\text{C(CH}_3)_2$$

Activation of carboxyl is usually achieved by formation of an acyl derivative containing a good leaving group, such as acyl azide or esters of phenol.
In other cases carbodiimides are used (page 485) to activate carboxyl and form
amide in one step. Acid chlorides cannot be used since racemization of the
α-carbon via enolization is too facile in the acid chlorides. Racemization is a
severe problem in any case, since it leads to protein mixtures containing some

proportion of D amidst the L amino acid units. In a synthesis of some length this can drastically reduce the total yield of the correct all-L protein in a welter of other diastereomers.

Such syntheses are long and difficult and give low yields at the end of so many steps with the methods currently available. Nevertheless, the peptides produced are pure since purification may be effected at each step. An alternative procedure which gains speed at the sacrifice of stepwise purification is **solid-phase peptide synthesis**. In this procedure (Fig. 25-8) the first amino acid is linked, as an ester, to a benzyl group on an insoluble polymer (often polystyrene). The amino group is then deblocked and a second protected and activated amino acid is added, followed by sequential deblocking and coupling so that the peptide grows on the polymer support.

The resin (polymer support + peptide), in tiny granules, is simply steeped in excess amino acid derivative, then washed, then treated with acid and washed again, and the sequence repeated up to 10 or 20 amino acid additions. Then the resin is treated with stronger acid to remove peptide from the polymer benzyl groups (via S_N1 solvolysis at benzyl carbons). The purity of the final peptide obtained depends on the yields of each step, and these must be extraordinarily good to produce even moderately pure product. However, peptides of 10 to 20 amino acids have been prepared in apparently fair purity by this process, and such syntheses can be performed in a few days.

PROBLEM 25-9

In the hydrolysis of the protein, insulin, a certain peptide was obtained and allowed to react with dinitrofluorobenzene. Hydrolysis then yielded five lesser peptides (*A* to *E*) as shown below. Determine the sequence of the original peptide. (Amino acids are listed by the first three letters of their names, following common practice in peptide chemistry.)

DNP—peptide $\longrightarrow$
$\begin{cases} A \longrightarrow \text{Tyr, Phe, Phe—DNP} \\ B \longrightarrow \text{Thr, Tyr} \\ C \longrightarrow \text{Pro, Ala, Lys} \\ D \longrightarrow \text{Lys, Pro} \\ E \longrightarrow \text{Thr, Pro} \end{cases}$

PROBLEM 25-10

Write out a synthesis of the tripeptide, H_2N—Phe—Ala—Asp—COOH, starting from the amino acids. Do not use a polymer support (solid-phase) synthesis.

25-5 NUCLEIC ACIDS

Nucleic acids are water-soluble high polymers found in high concentration in the nuclei of living cells. These materials have been unequivocally identified as the species responsible for the control of genetic processes, i.e., the replication of cells and whole individuals in a predetermined fashion. Hence the nucleic acids are central to the control of reproduction. Their function in this role is discussed in Sec. 26-6, and only their structures are taken up in this section.

Hydrolysis of nucleic acids produces phosphoric acid, either of two sugars, ribose or deoxyribose, and heterocyclic bases. The latter are derivatives of either purine or pyrimidine.

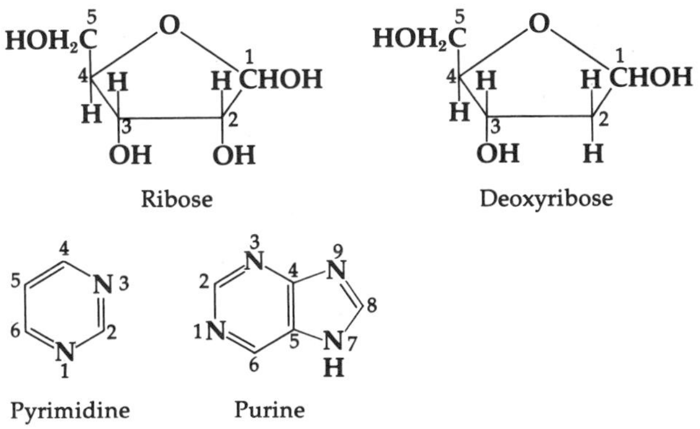

Ribose   Deoxyribose

Pyrimidine   Purine

The four bases produced in the hydrolysis of **ribonucleic acids (RNA)** are adenine (6-aminopurine), guanine (2-amino-6-hydroxypurine), cytosine (4-amino-2-hydroxypyrimidine), and uracil (2,4-dihydroxypyrimidine). The only sugar found in ribonucleic acid is ribose.

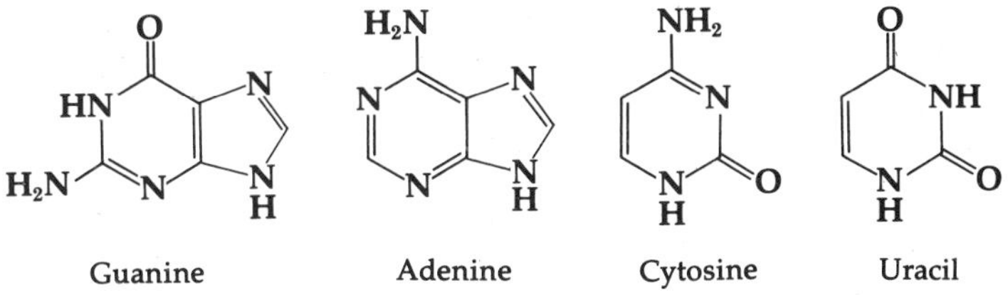

Guanine   Adenine   Cytosine   Uracil

Partial hydrolysis produces phosphoric acid plus **nucleosides**, which contain the carbohydrate residue bound to the various nitrogen bases. The carbohydrate is attached, by way of the glycosidic carbon atom, to nitrogen of the base. In guanosine and adenosine the attachment is made by way of the 7-positions of the purine nuclei. In cytidine and uridine the sugar is attached to the 1-position of the pyrimidine nuclei. Attachment in this position is possible if oxygen in either the 2- or the 4-position tautomerizes to a carbonyl group.

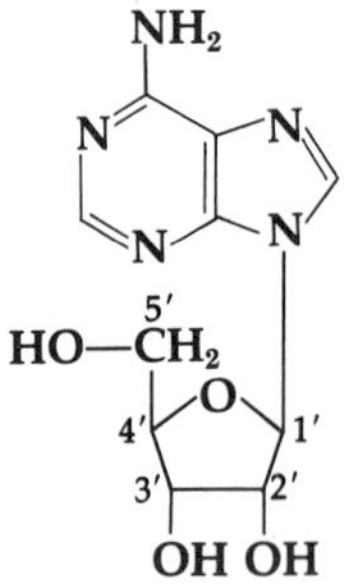

Adenosine

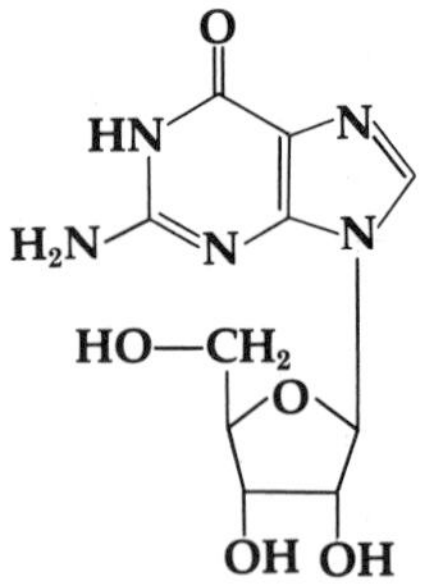

Guanosine

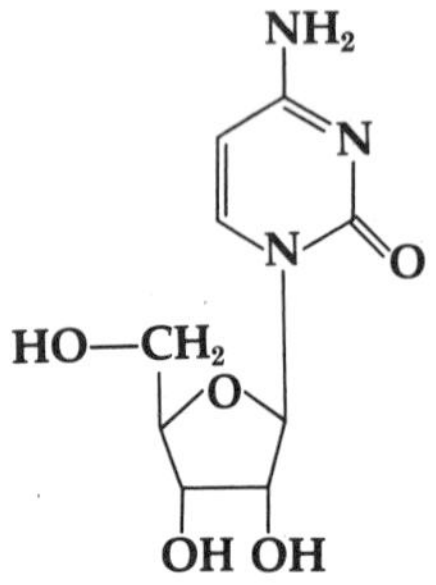

Cytidine

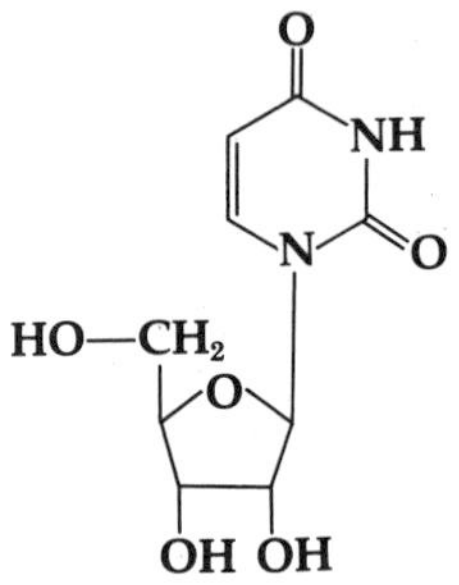

Uridine

Deoxyribonucleic acid (DNA) is hydrolyzed to give deoxyribose instead of ribose. Another difference between RNA and DNA is the fact that the latter yields thymine (2,4-dihydroxy-5-methylpyrimidine), rather than uracil, on hydrolysis. These are the only differences between RNA and DNA.

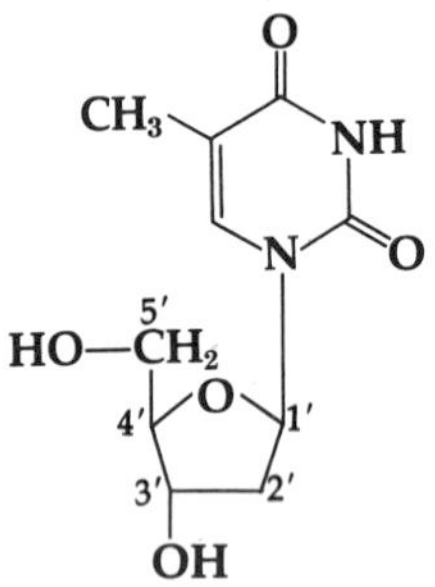

Thymidine, a deoxyribonucleoside

Nucleotides are phosphoric acid esters of nucleosides in which the phosphoric acid residue is usually attached to the 3'- or 5'-position of the ribose (or deoxyribose) nucleus; these are also obtained on partial hydrolysis of nucleic acids.

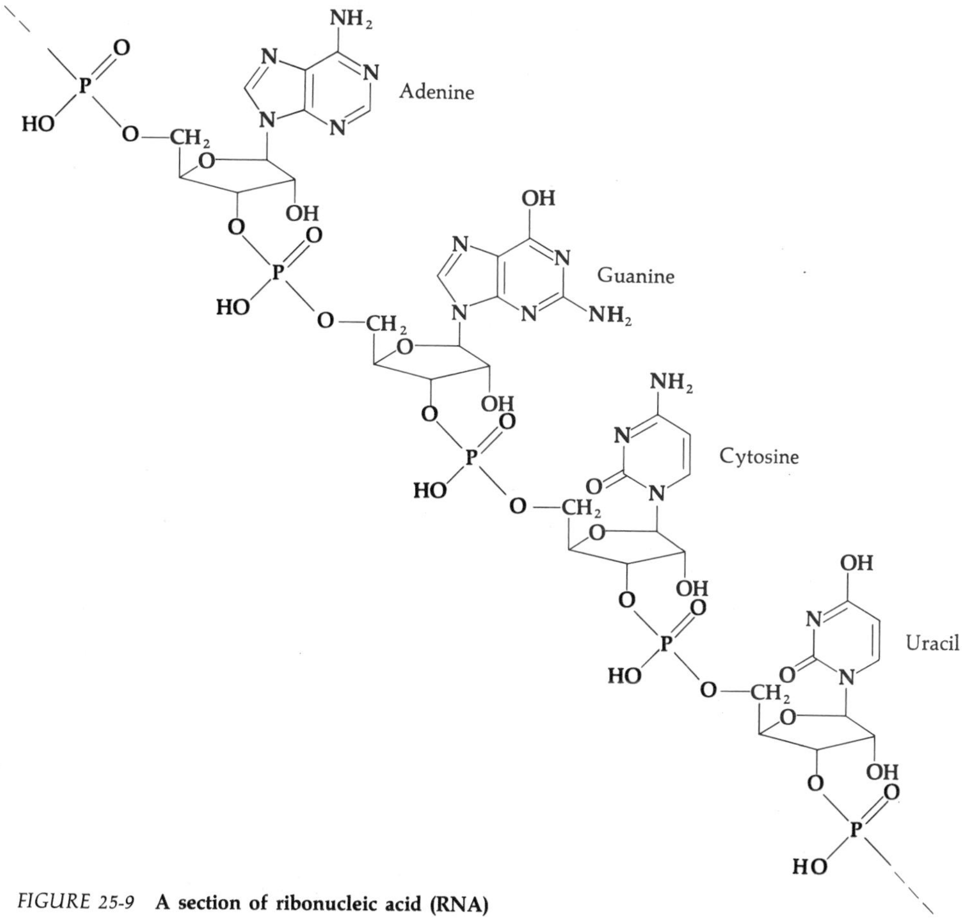

FIGURE 25-9 **A section of ribonucleic acid (RNA)**

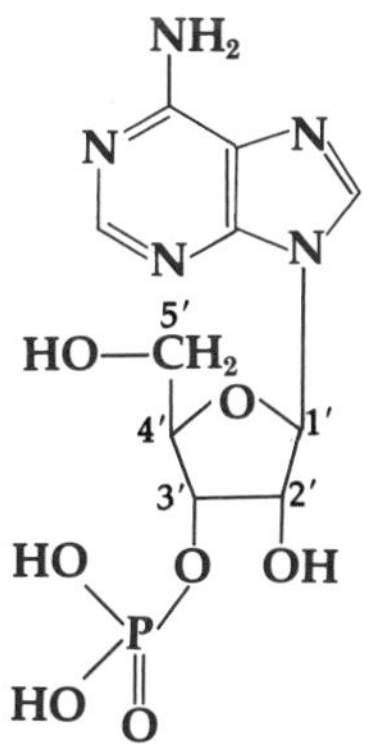

Adenosine-3'-phosphate

The nucleic acids themselves are high polymers in which nucleosides are linked, from the 3'- to 5'-positions, by phosphate ester linkages, as in Fig. 25-9. Hence, two kinds of nucleic acid are ribonucleic acid (RNA), characterized by ribose as the sugar, and deoxyribonucleic acid (DNA), characterized by deoxyribose. Each kind has only four bases, thymine in DNA replacing uracil in RNA. The four bases of nucleic acid sequences correspond to the twenty amino acids of protein sequences. As with proteins, the basic units and description of the nucleic acids are simple, but the *variety* in sequence which is possible is not only enormous but is in fact the very code of life.

The conformation of the nucleic acid polymer is intimately bound up with its reproducing function at the heart of the chemistry of life (Sec. 26-6). The chemistry and synthesis of nucleic acids is outlined in this section. The study of this chemistry is more recent and more difficult than that of proteins and amino acids. Accordingly the art is much less sophisticated. Synthesis has not really proceeded beyond the trinucleotide stage and even trinucleotides are very difficult to synthesize and isolate in pure form. Most of the samples used in biochemical research are separated from nucleic acid hydrolysis mixtures.

While nucleosides behave as normal, rather insoluble organic materials, the presence of the linking phosphates, which are ionized and usually in salt form, makes nucleotides and polynucleotides very intractable to normal organic reaction procedures owing to insolubility in organic solvents. Both nucleosides and mononucleotides are commercially available from nucleic acid hydrolysis, although they have all been totally synthesized in the past.

Synthesis of Polynucleotides

The sequential linking of nucleosides into an oligonucleotide or polynucleotide is basically an exercise in the placement, use, and mild removal of specific protecting groups. The several hydroxyl groups on ribose are a special problem for there is little difference in their reactivity, and so little control is possible in selective blocking. The 5'-OH, being primary, is somewhat less hindered. On acylation all three hydroxyls usually react, but the bulky 5'-monotrityl ether can be prepared selectively with trityl (triphenylmethyl) chloride, and this group is fairly easy to remove with mild acid. Both reactions are facile S_N1 substitutions. The 2'- and 3'-OH groups, on the other hand, are only distinguishable by enzyme catalysis.

The typical synthesis outlined in Fig. 25-10 illustrates the use of protecting groups as well as a shorthand notation common in the field:

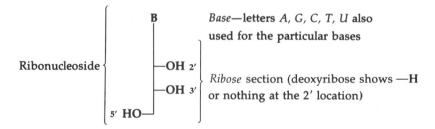

Preparation of nucleosides:

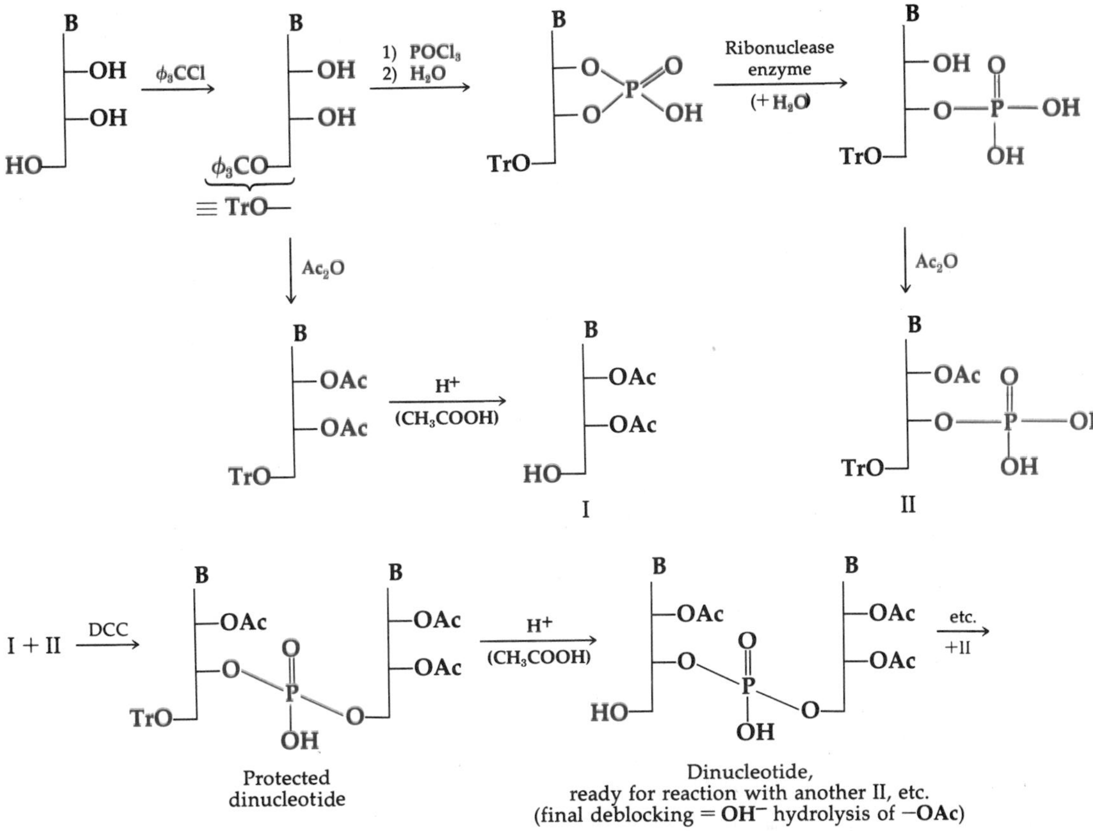

FIGURE 25-10   Typical synthesis scheme for polynucleotides

PROBLEM 25-11

Write possible syntheses of the bases, cytosine and thymine, found in nucleic acids.

PROBLEM 25-12

a Examine the five heterocyclic bases with a view to distinguishing them by spectral means, particularly IR and NMR spectra.

b Which of the five bases will be the most basic, and why?

25-6 PHYSICAL PROPERTIES OF POLYMERS

Polymeric materials vary in their physical properties enough to be used as elastomers, fibers, adhesives, and rigid plastics. Study of the correlation of physical properties with chemical structure is in its infancy, but some reasonable conclusions have already been reached.

Molecular Weight

Conventional methods of molecular-weight measurement, which depend upon colligative properties such as freezing-point depression or boiling-point elevation, are of little value in the high-polymer field. The failure of such methods, in which the number of moles of a solute in a dilute solution are counted, arises from the relatively small number of molecules in a typical sample of a polymer. For example, 1 g of a polymer of molecular weight 100,000 would depress the freezing point of 100 g of benzene by only 5.12×10^{-4}°. Such a minute change is essentially impossible to measure accurately. The one colligative property that has been useful is **osmotic pressure**. If a solution of polymer is separated from pure solvent by a membrane permeable only to the solvent molecules, the solvent molecules pass both ways through the membrane. However, the rate of diffusion from pure solvent is faster than from the solution (which has a slightly lower vapor pressure). As a consequence, a hydrostatic head is created on the solution side of the membrane, which, at equilibrium, compensates for the difference in vapor pressure of the solvent on the two sides of the membrane. Precise determination of the height of the osmotic head gives a delicate method for determination of the small change in vapor pressure due to dilution of the solvent with a relatively small number of solute molecules. Precision osmometry has been much used as a standard for the calibration of other methods of determining molecular weights.

Empirical observations lead to the generalization that the *viscosity* of polymer solutions increases with increasing molecular weight of the solute. No exact theory has been developed to correlate the effect with molecular weight. Furthermore, the magnitude of the effect is sensitive to the shape as well as to the size of polymer molecules. Since viscosity measurements are very easy to carry out, viscosimetry is often used to obtain approximate relative values of molecular weights. If viscosity effects for a particular type of polymer are calibrated by comparison with another method, the viscosity method can be used for routine determinations.

The rate at which molecules undergo sedimentation in an ultracentrifuge is a function of their weight; hence, both **sedimentation** rate and equilibrium are used to measure the molecular weight of polymers. The methods give an estimate of the average weight of solute molecules, as contrasted with osmotic measurements, which give the *average number* of molecules per sample unit. Weight-average and number-average molecular weights for typical samples are usually different, since the former are strongly affected by a relatively small number of very massive particles, and the latter may be dominated by a small number of molecules of low molecular weight.

Light scattering has also been used to determine molecular size. The measurement depends upon the fact that light is scattered by particles that are large in comparison with the wavelength of the light.

The molecular weights of polymers cover a very wide range. Some useful synthetic polymers, such as certain ethylene oxide polymers and some poly-

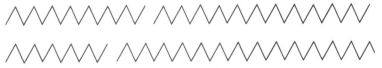

FIGURE 25-11   Schematic representation of a crystalline polymer

amides, have molecular weights of only a few thousand, but many synthetic fibers fall in the range of hundreds of thousands. The amylose fraction of starch is in the 10,000 to 50,000 range, but amylopectin and cellulose are probably as high as 500,000. Proteins show great ranges of molecular weight. Insulin, for example, has a value of 50,000, whereas the tobacco mosaic virus, with molecules that look like boulders to the electron microscope, has a molecular weight of more than 15,000,000. On the other hand many of these large proteins are only loose aggregates of smaller protein chains, not connected by normal peptide links. The tobacco mosaic virus is in fact a core of DNA surrounded by a coating of 2,100 identical protein molecules of 158 amino acids each. It is this entire (and crystalline!) aggregate which behaves as a molecular unit in molecular-weight determinations.

If synthetic polymers are *cross-linked* by bonding between polymer chains, their molecular weights approach infinity. An entire sample might be essentially one molecule.

When the molecular weights of polymers become very high, either by cross-linking or by linear extension, their solubility in all solvents drops essentially to zero. The effect is shown by three-dimensional polymers such as phenolformaldehyde resins, cross-linked ion-exchange resins, and, possibly, by native lignin before chemical degradation to destroy cross-links.

Crystallinity

Since polymers never consist of a single molecular species, a sample cannot be a crystal in exactly the same way as a pure compound in which the molecules pack in a perfectly ordered way. However, some polymers have many of the physical characteristics of crystals and are said to be crystalline. In a crystalline polymer, long segments of linear polymer chains are oriented in a regular manner with respect to one another. The occurrence of an occasional discontinuity at chain ends has only a negligible effect (Fig. 25-11).

The best evidence for crystallinity is the X-ray diffraction pattern of a polymer. The patterns obtained are of the same type as those given by micro-crystalline solids. A crystalline polymer may be oriented by a mechanical stress, such as stretching, and the resulting material will be optically anisotropic; i.e., the refractive index will be different along the direction of the stress than in the perpendicular direction. Such an oriented polymer also has different absorption spectra along different axes.

"Crystalline" polymers tend to be relatively strong and nonelastic. The best examples are synthetic fibers such as nylon and Dacron. The polymers

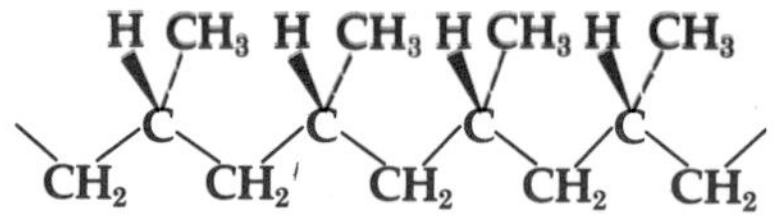

FIGURE 25-12   **Segment of an isotactic polymer chain**

are noncrystalline (*amorphous*) when spun into thread, but crystallization is induced by slow stretching.

Polymers that have polar functional groups show a considerable tendency to be crystalline, and many proteins have been crystallized. Orientation of the chains is aided by alignment of the dipole moments of the groups on different chains. Furthermore, the regular helix structure of proteins, with the order imposed by hydrogen bonds, is a further inducement to crystallinity. However, polar groups are not a necessary prerequisite to crystallization, as has been demonstrated by production of crystalline polyethylene, polypropylene, and polystyrene by polymerization with organometallic initiators. Polypropylene and polystyrene are especially interesting, since preliminary evidence suggests that the molecules owe their crystalline character to the systematic repetition of one configuration of asymmetric centers along a polymer chain. The term **isotactic** is applied to such a configuration (Fig. 25-12).

In each chain, all asymmetric centers have the same configuration, which may be either R or s. Of course, there are equal numbers of R and s chains in any synthetic polymer sample. The polymer is, therefore, a racemate.

Any irregularity, such as introduction of bulky groups or inclusion of comonomer units, will tend to prevent crystallinity. A fine example of disordering effects is found in the polymerization of methyl methacrylate [$CH_2=C(CH_3)COOCH_3$]. The homopolymer is tough and somewhat brittle since it is nearly crystalline. The desirable nonshattering qualities of Plexiglas are obtained by copolymerization of a few percent of ethyl acrylate ($CH_2=CHCOOC_2H_5$) with the methacrylate. The same effect can be obtained by adding a **plasticizer** such as diethyl sebacate [$C_2H_5OCO(CH_2)_8COOC_2H_5$] to polymethyl methacrylate. The long plasticizer molecules intrude between polymer chains and form a solid solution that is disordered. In order to be effective, a plasticizer must be compatible with (soluble in) the polymer.

Elastomers are highly disordered systems that undergo plastic deformation very easily and recover relatively slowly after stretching. The deformation of such "rubbery" materials is accomplished by the uncoiling of randomly piled polymer chains.

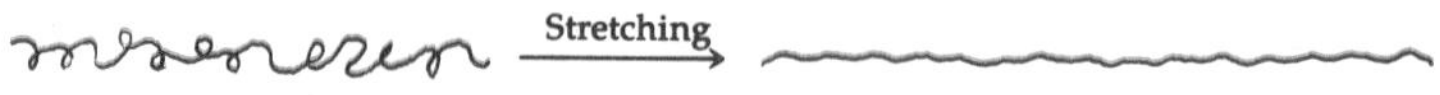

Stretching →

PROBLEMS

25-13 Write out syntheses of the following compounds from substances ordinarily readily available (see Sec. 23-1).

a C$_6$H$_5$CH$_2$CHCOOH
            |
           NH$_2$

b

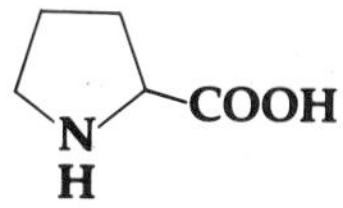

c H$_2$NCH$_2$CH$_2$CH$_2$CHCOOH
                          |
                         NH$_2$

d H$_2$NCH$_2$CONHCH$_2$COOH

e CH$_3$CHCONHCH$_2$COOH
           |
          NH$_2$

f C$_6$H$_5$CH$_2$CHCONHCHCH$_2$CH$_2$COOH
                  | |
                 NH$_2$ COOH

(COOH shown above the CH)

g

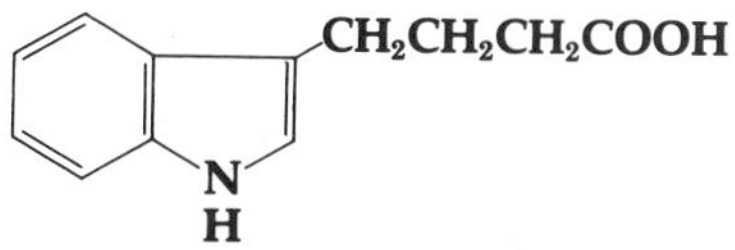

25-14 Assume that the compounds BaC^{14}O$_3$ and KC^{14}N are available and that HOCOCH$_2$CHNH$_2$C^{14}OOH and HOCH$_2$CHNH$_2$C^{14}OOH have to be prepared. Write out syntheses.

25-15 A synthesis of amino acids which has some general value is begun by treating glycine with acetic anhydride to yield 5-oxazolone, C$_4$H$_5$NO$_2$. This condenses with aldehydes in base to yield a derivative which may be hydrolyzed, hydrogenated, and hydrolyzed to yield an amino acid, RCH$_2$CH(NH$_2$)COOH, from aldehydes, RCHO. (The hydrolysis merely adds the elements of water; in the second, acetic acid is produced.) Delineate the course of this synthesis for preparing phenylalanine.

25-16 Cyanogen bromide (BrCN) is a reagent for cleaving a peptide specifically at the peptide bond bearing the carbonyl of methionine. The reagent also reacts with cysteine but causes no peptide bond cleavage. Deduce the reaction mechanism involved.

25-17 What other substances might one add to natural rubber to cause cross-linking? Explain.

25-18 **a** Ion-exchange resins have been used as acid catalysts for certain reactions. What special advantage might they have?

   b A weakly acidic ion-exchange resin uses carboxyl as its active group. How would you synthesize such a resin?

25-19 Polymer foams are widely used to provide very lightweight porous material (e.g., polyurethan foam). These substances create a gas as they polymerize and so "rise" like bread dough to fill molds, or wall spaces if used as insulation. The procedure is to allow ethylene glycol and 2,4-toluene-diisocyanate to react first to form an oligomer as a thick, gluey liquid. This is mixed with some water in order to cause it to rise. The liquid becomes warm, froths, rises, and hardens in minutes. Delineate the chemical processes involved.

25-20 Write equations that illustrate each of the following:
  a Condensation polymerization
  b Addition polymerization
  c Chain transfer
  d Initiation of polymerization by an organometallic compound
  e Photochemical initiation of polymerization
  f Formation of a cross-linked polymer
  g Use of carbobenzoxylation in peptide synthesis
  h Enzymolysis of a polysaccharide
  i Copolymerization
  j End-group marking with DNP

25-21 Each of the following compounds has one or more specialized uses in some branch of polymer chemistry. Speculate as to the use of each.

  a $n\text{-}C_8H_{17}SH$

  b $(C_6H_5)_3CCH(C_6H_5)_2$

  c $(HOCH_2)_4C$ (pentaerythritol)

  d Di-n-butyl phthalate

  e A mercury-vapor lamp

  f

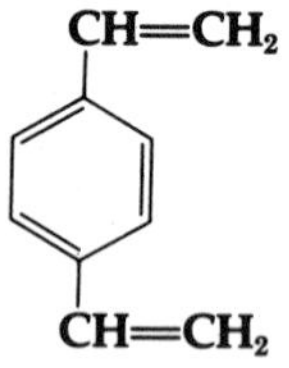

25-22 Explain the following facts:
  a Acrylic anhydride $[(CH_2=CHCO)_2O]$ polymerizes to give a soluble (not cross-linked) polymer that contains no residual unsaturation.
  b Vinylbenzoquinone does not polymerize.
  c Nitroglycerine $[O_2NOCH_2CH(ONO_2)CH_2ONO_2]$ is an excellent plasticizer for nitrocellulose (formed by the action of nitric acid on cellulose).
  d When the compound formulated below is treated with a trace of water, it evolves a gas and forms a polymer. Explain the reaction and offer a synthesis of this compound from alanine.

  e Small amounts of cyclopentadiene, C_5H_6, inhibit the polymerization of butadiene.

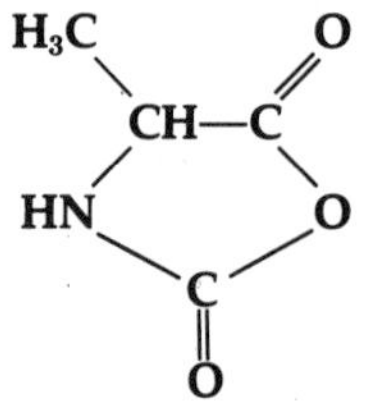

25-23 Consider the general hypothetical structure of lignin. Assuming it were adequately soluble for reactions, what information of a quantitative sort could you expect from a periodate reaction? What monobenzenoid products might you expect? What other chemical reactions might be adapted for probing its structure and what products might they yield?

25-24 One of the standard starting materials for amino acid synthesis is a solid, $C_9H_{15}NO_5$, which shows IR bands at 3.0, 5.8, and 6.0 μ (1670, 1730, and 3330 cm^{-1}) as well as NMR peaks at τ 3.40 (1H) (diffuse); 4.81 (1H) d ($J = 7$); 5.72 (4H) q ($J = 7$); 7.92 (3H) s; 8.70 (6H) t ($J = 7$). In D_2O the first peak disappears and the second becomes a singlet. What is this starting material?

25-25 Chloral hydrate, $CCl_3CH(OH)_2$, is used medically as a sedative and hypnotic for people and as an anesthetic for horses. Under certain conditions of dehydration to the reactive parent aldehyde (chloral, CCl_3CHO) one obtains polymeric material as well as two isomers of $C_6H_3Cl_9O_3$ named α- and β-parachloral. The latter shows one clean singlet in the NMR, at τ 4.54, and the former shows two singlets (2:1) at τ 3.85 and 4.33. What are the structures of the polymer and the parachlorals?

READING REFERENCES

Kopple, K. D., "Peptides and Amino Acids," W. A. Benjamin, Inc., New York, 1966.

THE CHEMISTRY OF LIFE

PERHAPS the most important and far-reaching use of the precepts of organic chemistry lies in the study of living systems, variously called biochemistry, biological chemistry, or molecular biology. This area is of course so vast that there is no chance to treat it in any depth in a book of this size. What we shall attempt in this chapter is simply to sketch in broad outline the major areas and concepts in biochemistry, as presently understood. The chief purpose of the chapter will be to show that the chemical reactions which constitute the living organism are basically just those we have already discussed in the preceding chapters. The understanding of organic chemistry gained already will carry us far in comprehending biochemistry and will serve as an indispensable foundation for further study in this field.

26-1 THE NATURE OF BIOCHEMISTRY

The living cell is a dynamic matrix of organic reactions, continuously occurring and interdependent. This flux of molecular change is the basis and stuff of life. The sum of these reactions is called the **metabolism** of the cell, and broadly consists of two kinds of reactions:

1 Those that produce the energy necessary for cell functions
2 Those that synthesize new cellular components for growth and regeneration

These chemical reactions are mechanistically the same ones we have already studied. In fact they are fewer in kind than the range of reactions of the organic laboratory. There are very few aromatic substitutions, for example, and substitution reactions at saturated carbon are usually limited to phosphate and sulfide leaving groups. The reactions of the carbonyl group (and its imine analogs) are as central to biochemistry as they are to general organic chemistry. The functional groups of the major biological molecules are those with which we are already familiar, but again are somewhat more limited than the range discussed up to now. A rough listing of these functional groups may serve as orientation.

Common in biochemistry	*Uncommon*
—OH	—OR (simple ethers)
—NH$_2$	—X
—SH, —SR, —SSR	—NO$_2$, —N$_3$
=O (aldehydes, ketones and simple imino derivatives)	=N—Z (derivatives of ketones)
—COOH derivatives:	
—COOR	—COCl
—COSR	—COOCOR
—CONHR	—CN
$\diagup$C=C$\diagdown$	—C≡C—
Phosphate esters	—SO$_2$Z (most sulfonic acid derivatives)
Nitrogen heterocycles	

Two central facts about biochemical reactions tend to explain the limitations observed in common functional groups and common reaction types.

1   Biochemical reactions virtually always take place in water as solvent. Hence functional groups which react with water easily, or which are not readily hydrogen-bonded for solubility, are uncommon.

2   All reactions occur at about room temperature.

These are also two of the three salient features which distinguish biological reactions from those carried out in the organic chemical laboratory. The third, probably most important, distinction is that biochemical reactions are catalyzed and controlled by **enzymes** and hence are usually exceedingly rapid, even at room temperature.

Universality of Metabolism

One of the most important generalizations to emerge from the study of biochemistry is that virtually *the same sets of chemical reactions serve to produce the functions of life in all organisms,* whether in primitive bacteria or in man. This is a vastly simplifying truth, for it allows metabolic studies to be carried out in simple organisms with considerable confidence that the same mechanisms operate in higher animals. Not only are the reactions and mechanisms the same, but even the identical fundamental molecules which pass through these metabolic pathways are found in all species. The four bases and 20 amino acids discussed in Secs. 25-4 and 25-5 are examples.

This fundamental identity of the metabolic process in all of the literally millions of known biological species is believed to be evidence for the common origin, and subsequent evolution, of all life forms. In the primitive earth certain interrelated chemical reactions spontaneously developed and assembled into the first monocellular organisms, which in turn evolved into the myriad variants of polycellular creatures. But the basic chemical systems designed to derive, store, and use energy and to synthesize tissue components have virtually not changed at all during this evolution and proliferation of species. The situation may be likened to a variety of automobile (and motorboat) bodies utilizing the same engine.

Two major qualifications should be made to this generalization. The first is that there are two major families of organisms—plants and animals—which are distinguished by using a different source of chemical energy (next paragraph), so that the segments of their metabolism dealing with energy production are different. The second is that there are secondary metabolic sequences particular only to certain families of organisms and that even in the main metabolism there are minor species differences. The remarkable feature is really just how very minor these differences are in the general scheme of life.

Functions of Biochemical Reactions

The first necessity for the metabolic economy of the living cell is a supply of energy to power its other functions and of material to use in synthesizing its needed chemical components. Plants obtain their energy directly from the sun and their organic starting materials principally from carbon dioxide and water. These are combined into **carbohydrates** (sugars and starches) by a process called **photosynthesis**, driven by the energy of sunlight.

Animals take in their food and energy supplies in the form of food, chiefly plant carbohydrates, and oxygen from the air. The food molecules are oxidized by oxygen in a series of reactions which liberate energy and, ultimately, carbon dioxide, which is expelled. The energy is transferred, as chemical bond energy, to labile molecules which transfer their energy in turn into chemical reactions producing body heat, muscular work, electrical nerve impulses and sensory perception, as well as the synthesis of the molecules needed for growth and replacement of cell materials. These labile molecules are similar to the reagents of the laboratory, being high-energy substances suitable for driving their reaction equilibria to completion thermodynamically. These biochemical reagents are remarkably few in number but are utilized in many different reactions.

The energy of all living processes therefore is initially derived from the sun, captured in the photochemical reaction driving photosynthesis in plants, and stored as the bond energies of carbohydrates. These molecules are thermodynamically of higher energy—as all organic molecules are—than the carbon dioxide and water from which they are formed. Animal energy then derives from the controlled, stepwise combustion of plant carbohydrates back to carbon dioxide and water. Plants and animals are thus mutually dependent organisms with respect to the reaction cycle which distributes solar energy into the life process and maintains an oxygen–carbon dioxide balance.

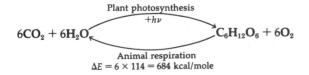

Plant photosynthesis
$+h\nu$

$$6CO_2 + 6H_2O \rightleftharpoons C_6H_{12}O_6 + 6O_2$$

Animal respiration
$\Delta E = 6 \times 114 = 684 \ \text{kcal/mole}$

The sequence of reactions producing energy from the oxidative breakdown of sugars passes through a series of chemical intermediates. These molecules may also be removed from this combustion pathway, as required, to serve as starting materials for a variety of synthetic routes which produce the many chemicals needed in metabolism and in tissue construction. There are 20 intermediates in the complete breakdown of starch, and these intermediates are commonly used in synthetic sequences of 10 to 20 reactions each. The total number of chemical reactions in a simple living cell, such as a bacterium, is estimated to be about 3,000, of which biochemists have identified and studied perhaps a third.

The number of different small molecules and monomers in the cell is also surprisingly small for so complex an entity, but enormous variety in material is possible in the polymeric molecules present (proteins, nucleic acids, polysaccharides; Chap. 25), which constitute the bulk of cell material (apart from water). The possible number of different proteins or nucleic acids that can be made from 20 amino acids or four heterocyclic bases, respectively, is virtually infinite. The small molecules in the cell are primarily sugars, amino acids, purines and pyrimidines, fatty acids (n-alkanoic acids, usually C_{16-20}), as well as various metabolites, vitamins, and hormones, and metabolic breakdown products.

PROBLEM 26-1

How many two-electron steps are involved in the oxidative conversion of a molecule of glucose into six of carbon dioxide (see Table 18-1)? What is the average energy available per oxidation step? Using *specific* oxidation reagents from Chap. 18 show a series of laboratory reactions which would oxidize glucose to carbon dioxide.

Control of Biochemical Reactions

When two reactants are mixed in the laboratory, their reaction may be very slow even when equilibrium favors the products. This slow rate is occasioned by the fact that in order to react the molecules must first find each other through random thermal collision and even then must approach each other in the correct stereoelectronic orientation. This random-collision mode of convergence is very inefficient, and so the rate of reaction is slow (Sec. 9-4). We can increase the frequency of collision by heating but we can rarely improve the probability that collision will lead to successful reaction.

If the two reactant molecules could be held, as we hold their models in our hands, and brought together in the correct orientation, then reaction could occur every time and the rate would be dramatically increased. To an extent we see this happen when a cyclic (intramolecular) reaction is compared with its intermolecular counterpart; cyclization rates are usually orders of magnitude faster simply because the reactive sites are held close together by the molecule itself.

Biochemical reactions occur in just this way. A large protein molecule, called an **enzyme**, picks up the two required reactant molecules, often by means of hydrogen bonds or rapidly reversible covalent bonding. The reactant molecules are attached to the enzyme in a highly specific fashion and so brought together in the correct stereoelectronic orientation for reaction. Since the product molecule is not so strongly bound to the enzyme surface, it is immediately released and the enzyme is free to repeat the process.

Enzymes act only as catalysts. While they vastly increase the rates of chemical reactions, they have no effect on their equilibria or thermodynamic properties. The rates, however, are often increased by factors of 10^8 to 10^{11} over their uncatalyzed counterparts and are often as fast as the diffusion of reactants through the solution to arrive at the enzyme. There are commonly a thousand reactions per second at one enzyme site.

Biochemical reactions are therefore rapid and quantitative at room temperature. The mechanism of the reaction is not altered by the enzyme, but only assisted by it, again as if we demonstrated a mechanism on a molecular model by holding it in our hands. Several-step mechanisms are common for organic reactions and these often involve acid or base catalysts (Fig. 9-2). In related biochemical mechanisms the requisite base catalyst or acidic proton is internally supplied to the substrates by the enzyme, and so need not be captured from the medium. The side chains of the amino acids incorporated into the enzyme protein afford amines and carboxylic acid groups to act as internal catalysts of this kind.

Enzymes are generally quite specific. *There is a single enzyme for each single metabolic reaction.* Its structure is particular to catch a particular substrate and compel it to react. Almost every reaction in metabolism is mediated by its specific enzyme, but most enzymes will capture molecules chemically similar to their intended substrates and cause them to react also, although usually with lower rates.

Nature of the Cell

All living things are composed of cells, varying in size, shape, and function, but all characteristic of the particular organism. An average animal cell is very small, with a diameter of only about 20 μ (0.02 mm). Nevertheless, it is a highly organized structure, surrounded by a double membrane which encloses a great number of morphologically different elements set in **cytoplasmic fluid** (Fig. 26-1). The principal structural material is protein.

Functionally important organization exists at every level in the cell. Amino acids are covalently "organized" in a functionally important order into proteins. Protein molecules—structural or enzyme—are then grouped into ensembles, held together by strong physical forces and having particular functions. These ensembles then make up the structures of subcellular components, which in turn comprise the cell. Cellular processes are controlled to a large extent by their isolation into separate compartments in the cell. Enzymes are generally

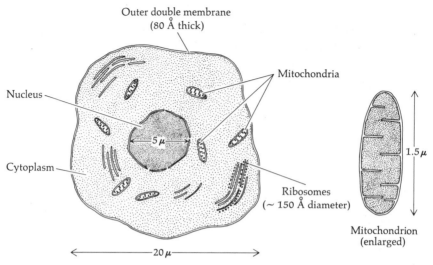

FIGURE 26-1 **Structure of a typical cell**

not free-floating in the cytoplasm but are attached to structural membranes, so that the reactions they catalyze are localized in certain places in the cell. Hence their reactions are moderated by cell structure, which can allow or deny accessibility of substrates to the locale of the enzyme.

Among the separate structures within the cell (Fig. 26-1), the **nucleus** is the largest. Nearly all of the cell's DNA is found in the nucleus, where it is organized into linear structures called **chromosomes**. This DNA contains the genetic information characterizing the cell. The next largest structures are the **mitochondria**, double-membraned ellipsoids which serve as the centers for energy production and transfer. The enzymes involved in the breakdown of sugar molecules for energy are localized on the inner surface of the mitochondrion membrane, which is doubled into the mitochondrion interior, presumably to attain greater surface area. In green plants the photosynthetic reactions are similarly localized in small bodies called **chloroplasts**. Attached to the surfaces of other extended structures in the cytoplasm are tiny granules called **ribosomes**, composed of organized clusters of protein and RNA. The ribosomes are the site of protein synthesis in the cell.

Two major tools are used for the study of cell structure: the electron microscope and the ultracentrifuge. The first allows photography of cell sections by electron beams, for studying the shape and appearance of cells. The available resolution goes down to 20 Å in optimal cases (compare with bond lengths $\sim$1.5 Å). Samples for the electron microscope must be dried, however, so that no observation of the working cell is possible. Separation of cellular components following physical disruption of the cell (grinding, etc.) is carried out with the ultracentrifuge, which separates components as a function of their masses. The procedure allows separation at various levels: single proteins, organized clusters, or cellular components such as mitochondria or ribosomes.

All cells arise from preexisting cells. The cell grows and then divides into two equivalent cells in a process called **mitosis**. This cell self-duplication, or replication, lies at the heart of the accurate reproduction of new individuals and generations of individuals in any species. In order to duplicate all the substances of which it is composed, the cell possesses a record of what all these substances are. This record, called the **genetic code**, specifies the amino acid sequences of all the proteins, structural and enzyme, in the cell and also directs their synthesis when new substance is required for growth, regeneration, or duplication.† The record is contained in the nucleus in the form of molecules of DNA. These molecules are called **genes** and, prior to mitosis, they aggregate into linear chromosomes and cause their own duplication so that two identical records result, one for each daughter cell after mitosis.

Therefore, DNA is the substance in the cell nucleus containing all the information necessary to characterize and duplicate the cell. The mode of action of DNA in replication was brilliantly elucidated in 1953 by Crick and Watson and is surveyed in Sec. 26-6. The related fields of chemical genetics and molecular biology have experienced explosive growth since that date and witnessed the revelation in some detail of the chemical inner mysteries of genetics and reproduction in living organisms.

26-2   THE BASIC MOLECULES OF LIFE

The substances that characterize living matter can be grouped into small molecules or monomers on the one hand and polymers on the other. Many of these have been discussed before in Chap. 25. The three major polymers are the proteins, nucleic acids and polysaccharides. The proteins are utilized for structural purposes and as enzymes. Proteins are synthesized from 20 different amino acids in specific sequences which dictate the nature and function of the proteins. Protein chemistry is outlined in Sec. 25-4 and the 20 amino acids are shown in Table 25-1. The side chains of these amino acids, and hence of proteins, carry virtually every functional group—usually one per side chain —to be found in biochemical reactions.

The nucleic acids are composed of only four heterocyclic bases (Sec. 25-5). Nucleosides contain these bases attached to the sugars ribose (from RNA) or deoxyribose (from DNA). These in turn are phosphorylated to form nucleotides, which condense in long linear chains of particular sequence to form RNA and DNA. These specific sequences of nucleic acids serve to code and to synthesize proteins of specific sequence.

The polysaccharides (Sec. 25-3) are usually composed mostly of glucose. The chief structural (and insoluble) polysaccharide in plants is cellulose, while soluble polysaccharides serve as carbohydrate food storage: starches in plants and glycogen in animals. A series of sugars (monosaccharides) of from three

---

† Since the enzymes in turn dictate the synthesis of smaller molecules and the reactions transforming food into energy, it is only necessary in the record to code the synthesis of the enzymes.

to seven carbons are utilized in the reactions of photosynthesis for capturing carbon dioxide. Sugars are generalized according to their carbon content: C_6 = hexoses; C_5 = pentoses; C_4 = tetroses; etc. Ribose and deoxyribose (pentoses) are integral parts of nucleotides. Sucrose, a common plant disaccharide, is common table sugar.

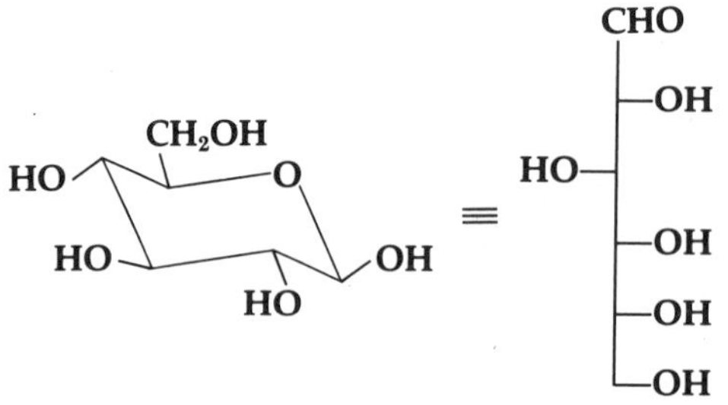

β-Glucose

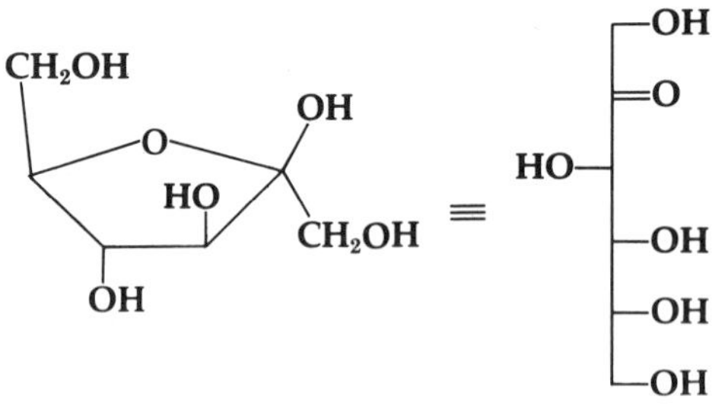

β-Fructose

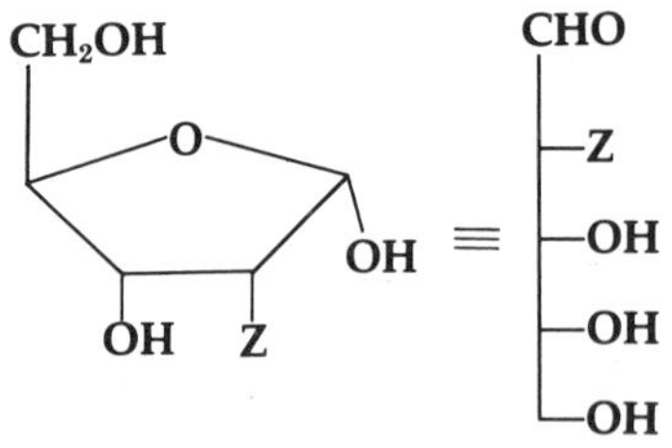

Deoxyribose (Z = H)
Ribose   (Z = OH)

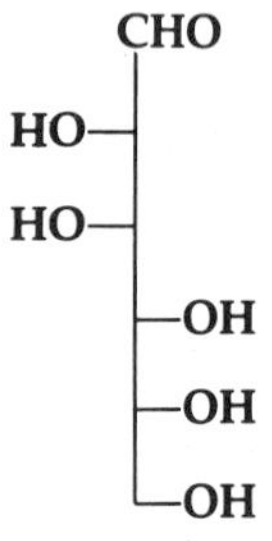

Mannose 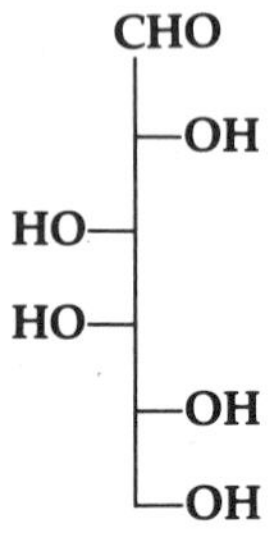   Galactose 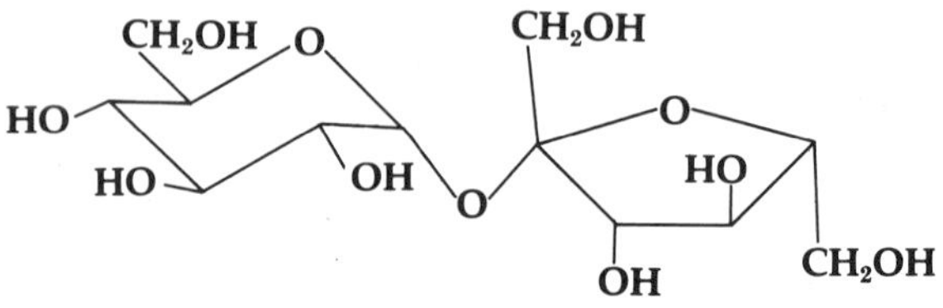   Sucrose
(glucose + fructose)

Nonpolymeric Small Molecules

Apart from the monomers intended for polymer synthesis, the cell contains the small molecules which are synthetic intermediates on the pathways for interconversion and synthesis of amino acids, nucleotides, etc. Some of these syntheses are outlined in Sec. 26-5. The various dicarboxylic acids of the citric acid cycle (Sec. 26-4) are involved in that cyclic pathway both for energy production (in converting pyruvic and acetic acid to carbon dioxide) and for supply of synthetic intermediates for producing amino acids. These diacids are mostly simple molecules like fumaric and α-ketoglutaric acids.†

The **fatty acids** are straight-chain saturated and unsaturated monocarboxylic acids usually with an even number of carbons around eighteen in number. Only about twenty fatty acids are at all common, stearic, palmitic and oleic being by far the most ubiquitous. These fatty acids usually occur as water-insoluble **lipids** (fats and oils). The lipids are triesters of glycerol with fatty acids, or occasionally with another esterifying function at one hydroxyl. The fats constitute high-energy food stored by the organism and utilized for energy by stepwise oxidation to acetic acid. Fatty-acid metabolism is outlined in Sec. 26-5.

$$CH_3(CH_2)_nCOOH \qquad\qquad CH_3(CH_2)_7CH=CH(CH_2)_7COOH$$

Palmitic acid ($n=14$)   Oleic acid
Stearic acid ($n=16$)

$$CH_2-O-CO(CH_2)_{14}CH_3$$
$$CH-O-CO(CH_2)_{14}CH_3$$
$$CH_2-O-CO(CH_2)_{14}CH_3$$

A lipid,
glyceryl tripalmitate

$$CH_2-O-COR_1$$
$$CH-O-COR_2$$
$$\overset{\displaystyle O}{CH_2-O-\underset{\displaystyle O^-}{P}-OCH_2CH_2\overset{+}{N}(CH_3)_3}$$

A phospholipid,
lecithin

A number of molecules with special metabolic functions are also synthesized in the cell. The major compounds in this class are the **coenzymes**, particular reagents which, in conjunction with enzymes, carry out many of the metabolic transformations. There are surprisingly few of these versatile reagents, probably only a few dozen. Many of them are well known as the **vitamins** of nutrition. These are detailed in the next section. Finally, a number of structurally interesting but functionally mysterious molecules also arise in the cell. These are collected as secondary metabolites and include hormones like the steroids, products of detoxification of foreign substances, and a surprisingly large group of substances of no known function. The latter have been isolated

---

† It is common practice in biochemistry to name carboxylic acids as their conjugate bases, the carboxylate anions, since at a metabolic pH ~7 most carboxylic acids are largely ionized. Thus acetic acid is called acetate, pyruvic acid is pyruvate, etc.

(mostly from plants) and studied by organic chemists for decades. They have loosely been referred to as **natural products** and constitute the substance of Chap. 27.

A summary of the interrelations of these compounds in the flux of metabolic interconversions in the cell is depicted in Fig. 26-2.

26-3 ENZYMES AND COENZYMES

An enzyme is a protein molecule specifically tailored to speed up (catalyze) the rate of a certain reaction on a particular substrate. The action of the enzyme is fast and specific, and its catalyzed reaction is quantitative. Furthermore, the action of the enzyme is under a number of forms of control. The actual amount of enzyme present is genetically controlled by the rate of its DNA-directed synthesis. Many enzymes may take up an inactive form and only be activated for catalytic action on some command from the environment. The enzymes are also localized so that their reactions can only occur in particular places in the cell, and supply of substrates can be under control by other cellular mechanisms.

Enzymology

The study of enzymes is concerned with isolating and purifying enzymes, determining their protein structure, and examining their mode of action. Isolation of enzymes is achieved by disrupting the cells and fractionating the solution of soluble proteins obtained. Fractions are then tested for the presence of the desired enzyme by observing their efficacy when used to catalyze the specific reaction *in vitro*.† The most potent tool for fractionating these enzyme proteins is the ultracentrifuge. Enzymes carefully purified by this and other fractionation procedures can often then be crystallized. Once purified their amino acid sequences can be determined. Full structures of some simple enzymes have been determined by X-ray crystallography, which shows not only the sequence ("primary structure") but also the detailed coiling and folding of the chains ("tertiary structure"). Many enzymes fall in a size range of 100 to 400 amino acids.

Enzymes are usually named by adding the suffix "**-ase**" to a root indicating their function or substrate. **Oxidoreductases**, for example, catalyze oxidation-reduction reactions and **transferases** serve to transfer molecular groups; chymotrypsin (below) is an *acyl-transferase*. *Lactate dehydrogenase* catalyzes the oxidation (dehydrogenation) of lactic acid (as lactate anion).

A major strategy in the study of enzyme-catalyzed reactions consists of isolating and purifying the protein components and the involved small organic components, and then reconstructing a working cell-free system *in vitro* with these components. In this fashion the necessary components and cofactors (metal ions, reducing agents, energy-transfer agents, etc., discussed next) can

---

† The term *in vitro* ("in glass") refers to reactions carried out in laboratory apparatus in contrast to those, *in vivo,* which occur within the living cell.

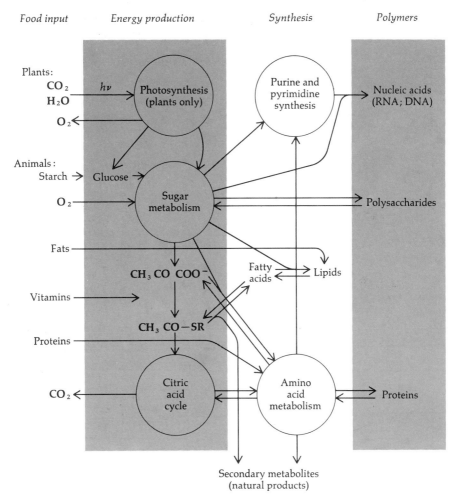

FIGURE 26-2   **General areas of cell metabolism**

be identified. Further, the effects of inhibitors, false substrates, concentrations, etc., on the rate and efficiency of the system can be used as probes to infer the molecular nature of events.

The Active Site

The central question about an enzyme is how does it work. The substrate molecule (or molecules) is loosely held onto the enzyme surface at one site, called the **active site**. The characteristics of the active site must generally be these.

1 The site must offer a perfect steric fit for the shape of the substrate molecule (in the conformation in which it will react). This is the so-called "lock-and-key" hypothesis.

2 The site must also offer some mild attractive binding force, specific for the substrate, in order to catch the molecule from solution and also to hold it in the correct orientation

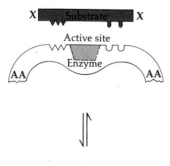

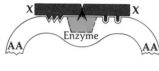

Enzyme-substrate complex

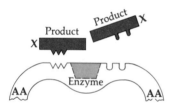

FIGURE 26-3　**Diagrammatic model of enzyme action**

for reaction. Ionic attractions, hydrogen bonds and, less commonly, rapidly reversible covalent bonds serve this function.

3　The site must provide whatever acidic or basic catalyst is needed for reaction.

4　The product molecule must not be so tightly bound to the enzyme that it cannot rapidly leave after reaction.

A diagrammatic presentation of enzyme action is shown in Fig. 26-3. The reaction shown is implicitly a cleavage, or its reverse reaction, a condensation. The enzyme is pictured roughly as a linear protein shape, only the active site region shown. The substrate is drawn from solution by attractive sites—charges, hydrogen bonds, etc.—which are symbolized by the matching shapes on the two surfaces in Fig. 26-3. These align the substrate and hold it in place for catalysis of the required reaction by the enzyme. When reaction is complete, the product molecule(s) leave the enzyme free for another capture and catalysis. Two other features may be illustrated with Fig. 26-3.

1 Variations in substrate structure can occur and still allow successful catalysis but only if they occur in noncritical places on the molecule, such as at the extreme ends (marked **X**) in this diagram. Other molecules may act as substrates in that they bind to the enzyme but will not undergo its directed reaction. Such substrates are enzyme poisons or deactivators since they usually attach irreversibly to the site and so make the enzyme subsequently unavailable for its proper function.

2 Variations in the amino acid sequence of the enzyme are acceptable in other regions of the enzyme protein chain (as at **AA** in the diagram) as long as they do not interfere with the active site structure and conformation. Small differences in sequence are known in different species or mutations and these do not materially change enzymatic activity.

Chymotrypsin

Several enzymes exist in the stomach and upper intestine to catalyze the hydrolysis of amide (peptide) bonds so as to convert proteins taken in as food into their constituent amino acids for resynthesis into native tissue protein. The stomach wall is coated with lipid so that these enzymes do not hydrolyze the protein of the wall itself. When a rupture occurs in the lipid lining, these digestive enzymes do operate on the structural protein beneath, causing an ulcer.

Chymotrypsin is the most studied of these enzymes. Most specifically, chymotrypsin catalyzes the interconversion of carboxylic acid derivatives. The enzyme is described as an acyl transferase, i.e., as an acylating or deacylating catalyst. Hence it catalyzes the hydrolysis, formation, or interconversion of carboxylic acids, esters, and amides. Chymotrypsin is only optimally effective, however, with aromatic amino acids such as phenylalanine and catalyzes the hydrolytic cleavage of their peptide bonds in proteins.

The mechanism of chymotrypsin action is pictured in Fig. 26-4, insofar as it is presently understood. The picture serves best to illustrate the *kind* of mechanistic understanding current about enzyme action in a relatively well understood case. The substrate is an acyl derivative (ester, amide, etc.) of an aromatic amino acid (shown as phenylalanine), either as a separate small molecule as shown or embedded in a protein chain. The enzyme grips it in a specific fashion and offers it the hydroxyl of its serine side chain for conversion to the carbonyl addition product. The reaction is both acid- and base-catalyzed, utilizing a histidine residue and a water molecule.

The reaction is mechanistically no more than the general familiar path for hydrolysis or transesterification of carboxylic acid derivatives (Sec. 13-1), proceeding via the usual two-step substitution mechanism to replace —Z (—OH, —OR′, —NHR′) with the serine oxygen. This forms an acylated chymotrypsin, ready to transfer the attached phenylalanine acyl group to another $\ddot{Z}$H (i.e. H_2O = hydrolysis; R′OH = esterification, etc.) molecule from the medium.

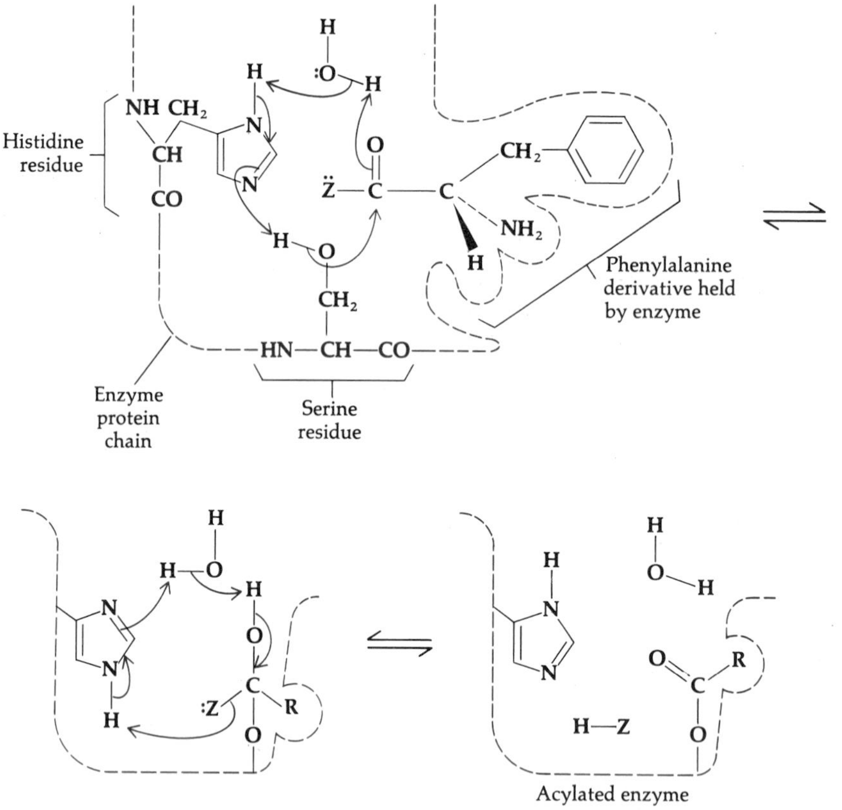

Acyl derivative (of phenylalanine) held by
enzyme in preparation for acyl transfer
to enzyme from —Z (= OH, OR', NHR', etc.)

FIGURE 26-4 **Possible mechanism of chymotrypsin action**

Lysozyme

The important function of hydrolyzing certain polysaccharides is effected by the enzyme lysozyme, the sequence of which is shown in Fig. 25-6. A full X-ray crystallographic analysis of the crystalline protein has resulted in a three-dimensional view of this enzyme. Furthermore, an enzyme-substrate complex has also been studied, allowing a full view of the enzyme with its polysaccharide substrate in position for cleavage. The enzyme is globular in shape with a cleft across one side. The linear polysaccharide slips neatly into this cleft and is held in place by hydrogen bonding, as illustrated in Fig. 26-5; the cleft accommodates and holds six of the sugar units (*A–F* in Fig. 26-5) of the polymer. This particular polysaccharide is made of *N*-acetylglucosamine units and represents the structure of the cell wall of a bacterium which is thus attacked by this enzyme. The *N*-acetylamino groups can be seen in Fig. 26-5 hydrogen-bonded to the enzyme.

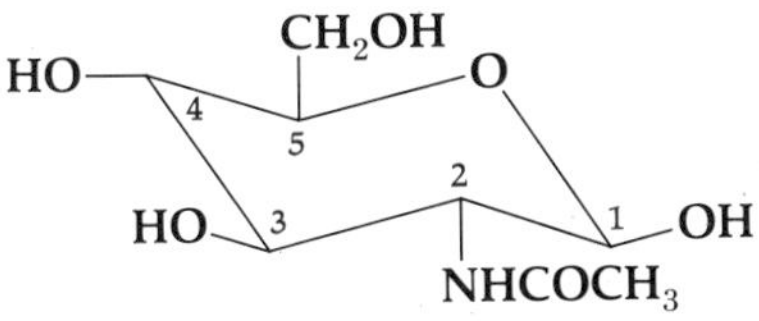

N-acetylglucosamine

The hydrolysis of the C-1 acetal now occurs through initial protonation of the ether link by the carboxyl group of glutamic acid (35 in the sequence, Fig. 25-6). The ether-stabilized carbonium ion at C-1 is further stabilized by an aspartic acid carboxylate ion (52, Fig. 25-6) and is ultimately discharged by capture of a water molecule from solution, thus completing a hydrolytic cleavage.

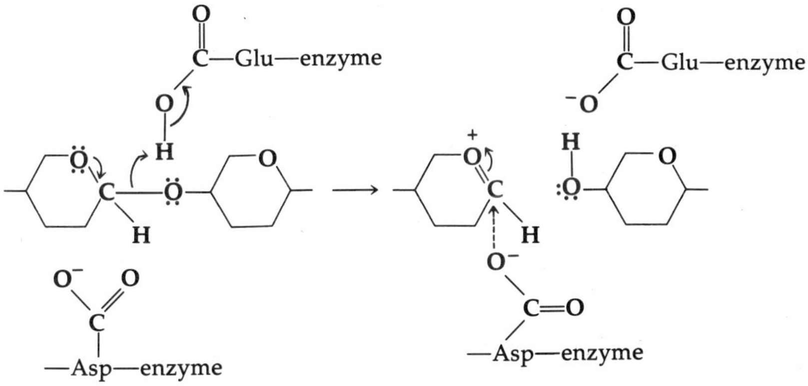

These two acidic amino acids are labeled, above and below the substrate, in Fig. 26-5; they are located in exactly the right place physically to catalyze this hydrolysis. A further important feature of the enzyme which also serves to increase the rate is that the sugar ring attacked (ring *D* in Fig. 26-5) is forced into a *boat conformation* by hydrogen-bonding when it fits into the enzyme crevice. This induces a higher-energy starting material and so lowers $\Delta F\ddagger$ and increases the initial rate of ionization (cf. Fig. 9-7).

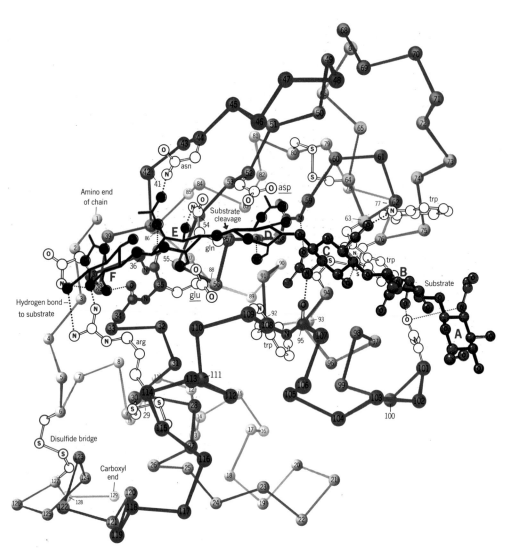

FIGURE 26-5   Enzyme lysozyme with substrate polysaccharide (reproduced, by permission, from "The Structure and Action of Proteins," Harper & Row, New York, copyright 1969 by Richard E. Dickerson and Irving Geis; coordinates courtesy of Dr. D. C. Phillips, Oxford)

PROBLEM 26-2

Pure crystalline chymotrypsin is sold commercially and may be used to catalyze hydrolysis of esters or amides of appropriate substrates. By selection of artificial substrates and comparison of their hydrolysis rates with a standard substrate, such as a phenylalanine ester, one may study the nature of the "fit" of substrate into enzyme required for active catalysis. What can we conclude from the following rates?

Rate

$C_6H_5CH_2CH-COOC_2H_5$
  $|$
  $NHCOCH_3$

L-Phenylalanine derivative	6.2×10^4
D-Phenylalanine derivative	~0

$C_6H_5CH_2CH_2COOC_2H_5$ 15

$L-CH_3-CH-COOC_2H_5$ 2
       $|$
      $NHCOCH_3$

$\qquad\quad CH_3$
$\qquad\quad |$
$L-C_6H_5CH_2-C-COOC_2H_5$ ~0
$\qquad\quad |$
$\qquad\quad NHCOCH_3$

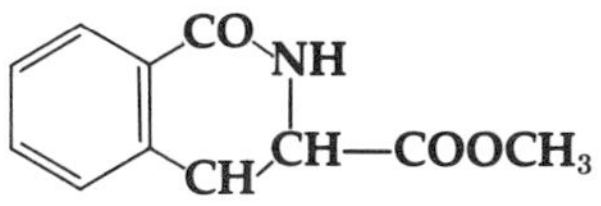	
D isomer	4.3×10^4
L isomer	11

Coenzymes

The 20 amino acid side chains in proteins do not provide enough functional-group variation to catalyze all the reactions needed in metabolism. Accordingly, many enzyme active sites require specially constructed functional groups for their catalytic work. These are supplied to, and bound to, the enzyme as small molecules called **coenzymes**. The lower section of the active site in Fig. 26-3, shown shaded above the dashed line, might be a removable small coenzyme molecule as loosely attached as the substrate. In the oxidoreductase enzymes, the coenzyme is an oxidizing or reducing agent which usually must leave after functioning to be "recharged" at another locale, i.e., reconverted to its oxidizing or reducing form. Vitamins in the diet are usually coenzymes supplied intact in food, since the human body synthesizes some of these molecules poorly and must depend on other organisms to make them.

The most prominent oxidoreductase coenzyme is nicotinamide-adenine dinucleotide (NAD) which can accept two hydrogens to become "NADH" (or

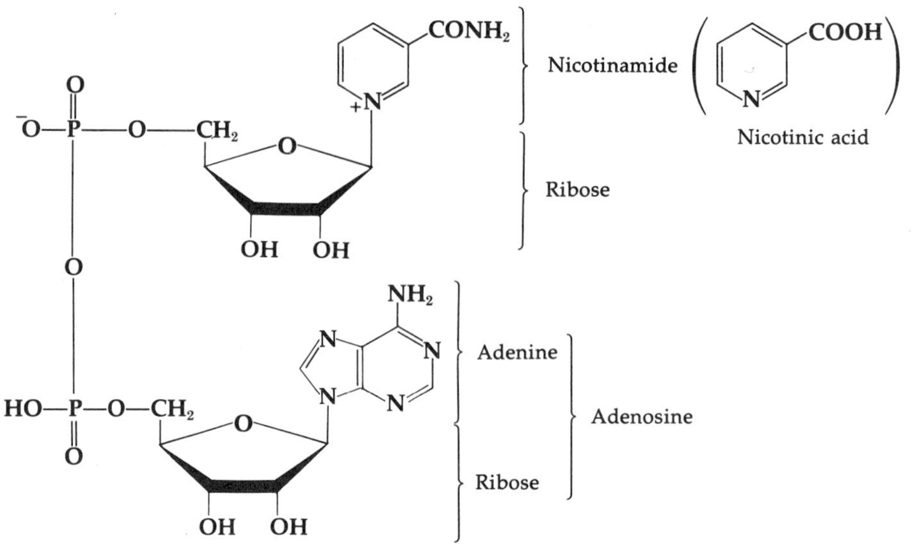

Mechanism of oxidation-reduction:

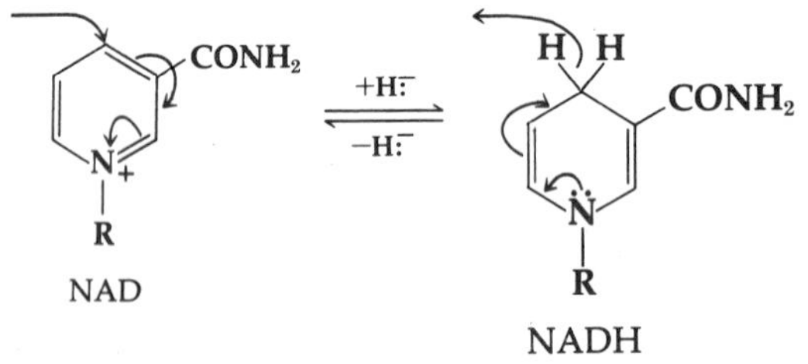

FIGURE 26-6   **Nicotinamide-adenine dinucleotide (NAD)**

NADH$_2$), the reduced form. NAD is thus an oxidizing agent since it dehydrogenates a substrate. Conversely, NADH is a reducing agent which can deliver hydride ion, analogous to the metal hydrides.

Its formula and mode of action are shown in Fig. 26-6. The common format for symbolizing biochemical reactions is shown below for the reduction of pyruvate ion with NADH, analogous to that of sodium borohydride. The delivery of hydride ion is mechanistically analogous to the Cannizzaro reaction (page 460), but is given more impetus by the concurrent aromatization of the pyridine ring.

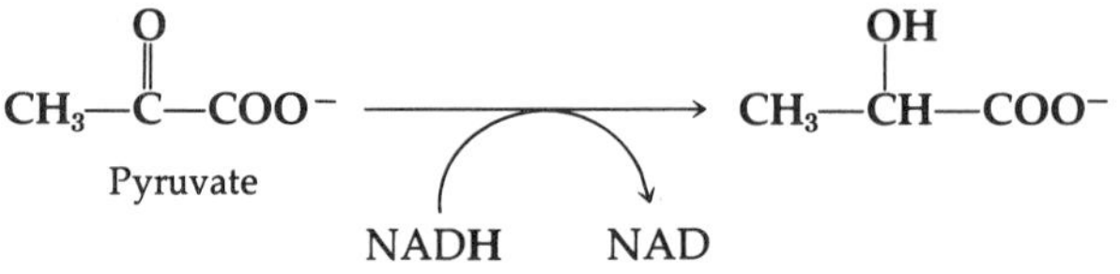

PROBLEM 26-3

The oxidation-reduction process with the NAD-NADH enzyme system is completely stereospecific.

a Outline the *stereochemical* information available in this set of reactions.

b Write the mechanism of the reduction step.

c How may the particular asymmetry of combination of NADH and pyruvate, which leads to asymmetric lactate, be determined experimentally?

d How may the absolute configuration of lactic acid be determined by relating to amino acids?

$$CH_3COCOOH \xrightarrow{\ \text{NaBH}_4\ } \underset{\underset{OH}{|}}{CH_3CHCOOH} \xrightarrow[\substack{\text{dehydrogenase}\\ \text{(with NAD)}}]{\ \text{Lactic}\ } CH_3COCOOH + NADH$$

Lactic acid

$$NADH + CH_3COCOOH \longrightarrow \text{lactic acid} + NAD$$

Vitamin B_1, or thiamine pyrophosphate, is a universal coenzyme, the specific catalytic function of which is mechanistically well understood.

Thiamine pyrophosphate

This coenzyme is responsible for the decarboxylation of α-ketoacids, most particularly in the conversion of pyruvate ion to the centrally important acetic acid thiol-ester, **acetyl-coenzyme A**. This thiol-ester links the glucose breakdown reactions with the final oxidation of acetic acid to carbon dioxide and water in the citric acid cycle (Fig. 26-2).

The key to thiamine action lies in the acidity of the proton shown in color. The conjugate base is stabilized electrostatically by the adjacent N^+ on one side and also by resonance (back-donation into the d orbitals of the adjacent sulfur) on the other side. Hence there is significant anion present for reaction even at pH 7. The anion attacks the very reactive α-ketone in α-ketoacids, following the subsequent pathway shown in Fig. 26-7.

The chief point of Fig. 26-7 for our purposes is to show that the action of enzymes involves no chemical mysteries. The reaction mechanisms involved are all those we have previously seen: additions, substitutions, and eliminations, with their electron movements directed by unshared pairs, charges, and electronegativity as in all the reactions of previous chapters. The five reactions in Fig. 26-7 are these:

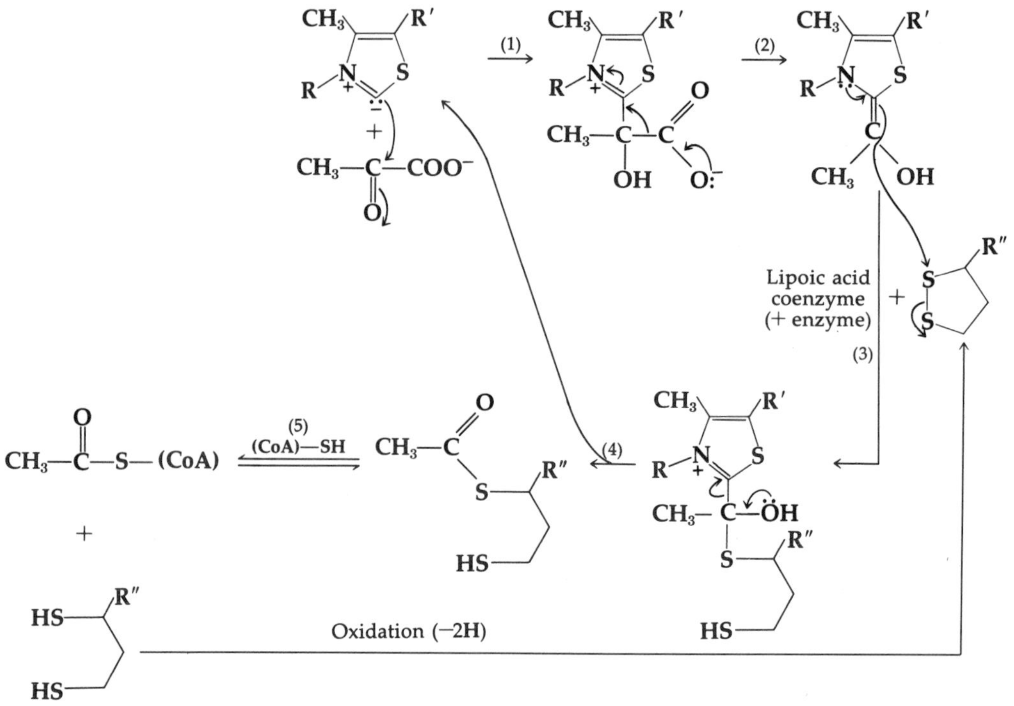

FIGURE 26-7 **Decarboxylation of pyruvate by thiamine**

1 Carbonyl addition by a carbanion nucleophile.
2 Decarboxylation of a β-ketoacid.
3 Nucleophilic displacement by enamine carbon (at a sulfur rather than a carbon atom) with R—SH as the leaving group.
4 Elimination of thiamine carbanion to form a carboxylic acid derivative—the usual second step in substitutions at carbonyl (Fig. 13-1).
5 Transesterification for thiol-esters. The acetyl group is transferred to a thiol of known structure called coenzyme A and abbreviated as **(CoA)—SH**.

Note the importance of the unusual thiamine carbanion for creating the necessary β-ketoacid grouping for decarboxylation; the structure of thiamine is mechanistically well designed for this coenzyme function. The central difference in biochemical from laboratory reactions is simply that the reactions occur on enzyme surfaces which loosely hold and bring together the reactants in optimum orientations for reaction, thus improving the reaction rate.

A second example of coenzyme action is the versatile capability of vitamin B_6, called pyridoxal. This coenzyme acts in the synthesis and degradation of amino acids.

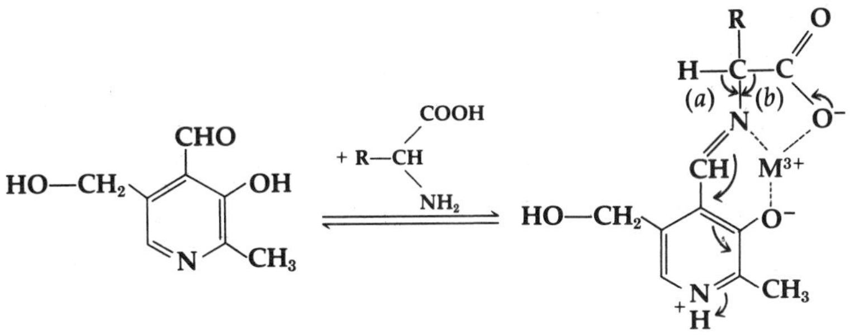

Pyridoxal

Imine complex with amino acid
(M^{3+} = metal ion)

The aldehyde group readily reacts with an amino acid to form an imine, complexed as a chelate with a metal ion. Acceptance of electrons by the pyridinium ring allows it to act as a vinylogous carbonyl, for enolization (*a*) or decarboxylation (*b*), or even cleavage of the C—R bond in appropriate cases. The product in any case will be an imine which on hydrolysis readily yields either an aldehyde, R—CHO, or an α-ketoacid, R—CO—COOH. The reverse reaction is used for synthesis of amino acids from α-ketoacids, such as the conversion of pyruvic acid to alanine. All of these reactions proceed *in vitro* with pyridoxal and metal ions alone but are much faster in the presence of the proper enzyme.

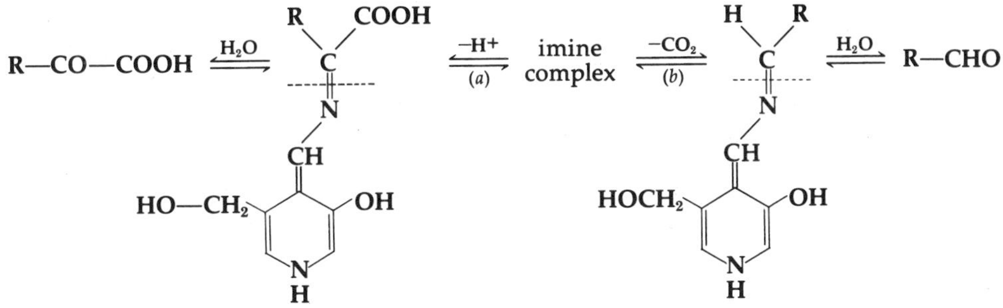

When sodium borohydride is added to the operating enzymatic system and the enzyme is then isolated and hydrolyzed, the reduced imine complex may be isolated from the hydrolysis. Borohydride adds to the imine (colored, above) and so stops enzyme action by rendering the complex stable. This kind of proof of the molecular species involved in enzyme action is used whenever possible in enzyme work.

PROBLEM 26-4

Serine is converted to glycine by pyridoxal. Write a full mechanism for this transformation. Can it occur on a serine unit in a protein?

26-4 BIOLOGICAL ENERGY PRODUCTION

The chief food, or fuel, for animal metabolism is starches, formed in plants by photosynthesis. These polysaccharides are hydrolyzed to glucose first and may be converted to *glycogen* (animal polysaccharide) for storage purposes. Glucose is the primary fuel source for animals. As energy is needed for the cell's work, glucose is fed into a metabolic furnace consisting of a sequence of chemical reactions that oxidize glucose first to pyruvate and then on to carbon dioxide and water. The energy liberated piecemeal in these reactions is used to reduce NAD to NADH as well as to produce ATP, the prime molecule used for energy storage and transfer in metabolism.

Adenosine triphosphate (ATP) is a molecule of central importance in metabolism, acting as a high-energy coenzyme to drive many reactions which would otherwise be thermodynamically unfavorable. The high-energy feature of ATP lies in its pyrophosphate bonds, hydrolysis of which to the di- or monophosphates (ADP and AMP) liberates 7 kcal/mole. However, this hydrolysis is slow unless catalyzed so that ATP has a considerable lifetime in the aqueous medium of the cell.

Attack of alcohols (R—OH) or acids (R—CO—OH) is a common reaction utilizing ATP. These reactions (as well as hydrolysis) are symbolized below as nucleophilic substitution reactions of phosphoric acid derivatives, attacked by Z—OH (cf. page 805). In this reaction alcohols and carboxylic acids have their —OH groups converted to good leaving groups (phosphates) for subsequent nucleophilic substitution reactions used in biosynthesis.

Adenosine triphosphate (ATP)

The first half of the oxidative breakdown of glucose carries it to pyruvate. This reaction sequence is called **glycolysis** and is outlined in Fig. 26-8. The pyruvate is then converted by thiamine (page 1038) into the thiol-ester, acetyl-coenzyme A, shown as CH_3CO—SR in Fig. 26-2. Acetyl-coenzyme A is a central biosynthetic starting material, chiefly for fatty acids (Sec. 26-5), and is the fuel for the chief metabolic furnace, the **citric acid cycle**, which is summarized in Fig. 26-9.

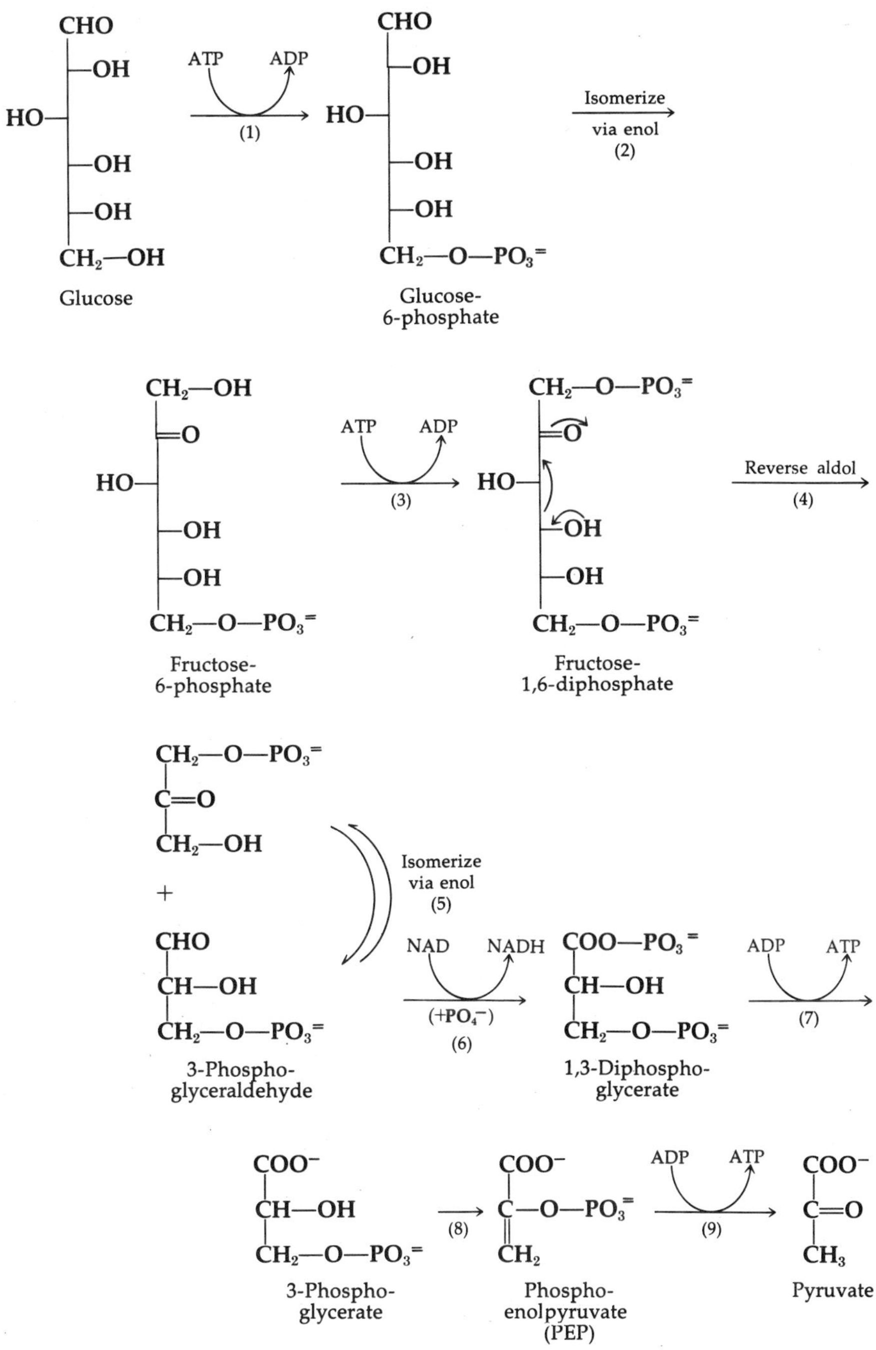

FIGURE 26-8   **The glycolytic pathway**

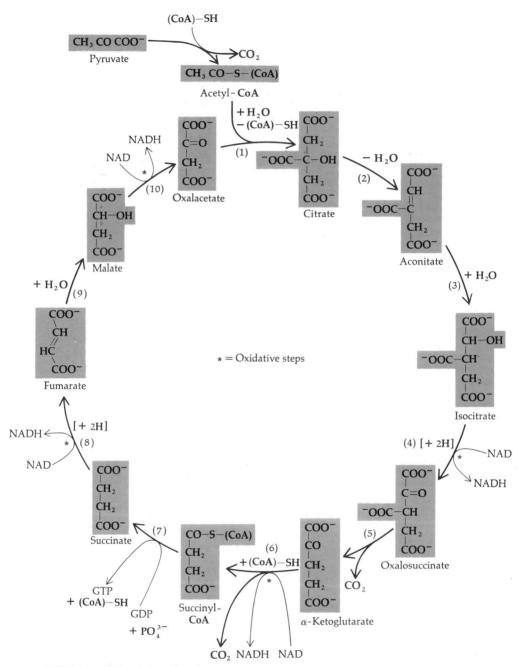

FIGURE 26-9 **The citric acid cycle**

The central degradation in glycolysis (Fig. 26-8) is the reverse aldol cleavage [reaction (4)] of a hexose to 2 moles of triose. Both triose fragments continue along the path to pyruvate owing to isomerization (5), in which the carbonyl enolizes to an ene-diol and then reprotonates in the opposite sense. The same mechanism occurs in reaction (2). Two hexose molecules utilize ATP to phosphorylate alcohol groups [(1) and (3)] and two trioses [(7) and (9)] regenerate ATP by phosphate removal. Since there are twice as many triose molecules this results in a net gain of two ATP molecules to be stored for use as energy sources elsewhere.

In the oxidation of aldehyde to acid derivative [reaction (6)], NAD is reduced and energy is thus made available. This energy allows the stable phosphate ion (PO_4^{3-}) to be transformed into a phosphoric anhydride with sufficient reactivity to transfer a phosphate unit to ADP so as to create more ATP in step (7). The reaction is a common one, called **oxidative phosphorylation**. The enol phosphate (PEP) is also a fairly high-energy compound and serves not only to generate more ATP but also, in biosynthesis, as an active enol for carbon-carbon bond formation via aldol and Michael reactions.

The second part of the oxidative breakdown of glucose is the **citric acid cycle**, also known as the Krebs cycle or the tricarboxylic acid cycle. This is a continuing cyclic reaction sequence into which acetate, as acetyl-coenzyme A, is fed and from which two molecules of carbon dioxide are released per cycle. Four two-electron oxidation steps (Sec. 18-1) are required to convert acetic acid to 2 moles of carbon dioxide. These four oxidations are starred in Fig. 26-9. The 10 reactions of the cycle are summarized below.

1   The readily enolized thiol-ester (**CH₃CO—SCoA**) enolizes and adds to α-ketoacid in an aldol reaction, driven thermodynamically by concurrent hydrolysis of the thiol-ester to a more stable carboxylate anion.

2   β-Elimination of water, activated by carboxyl carbonyl, represents the second step of the aldol condensation.

3   Readdition of water in a conjugate addition to the other carbonyl occurs.

4   Dehydrogenation of the alcohol serves to reduce NAD to NADH.

5   Decarboxylation of a β-ketoacid frees the first molecule of carbon dioxide.

6   The same oxidative decarboxylation as that of pyruvate in Fig. 26-7 releases the second carbon dioxide molecule. This is a five-reaction sequence. The dithiol freed in step (5) of Fig. 26-7 is reoxidized to lipoic acid coenzyme by NAD, producing NADH overall in the sequence.

7   The hydrolysis of the thiol-ester, like that of an acid chloride, is exothermic and here is coupled with phosphorylation of a nucleotide phosphate from the di- to the higher-energy triphosphate. The nucleotides here are guanosine phosphates (GDP $\longrightarrow$ GTP) which serve the same function as the more common adenosine phosphates, i.e., ADP and ATP.

8   Dehydrogenation is accomplished by another reaction which reduces NAD to NADH as in step (4).

9   Conjugate addition of water parallels that of reaction (3).

10 In the final oxidation, NAD dehydrogenates the secondary alcohol to the ketone of oxalacetate [like step (4)], preparing it for condensation with a new activated acetate molecule in reaction (1).

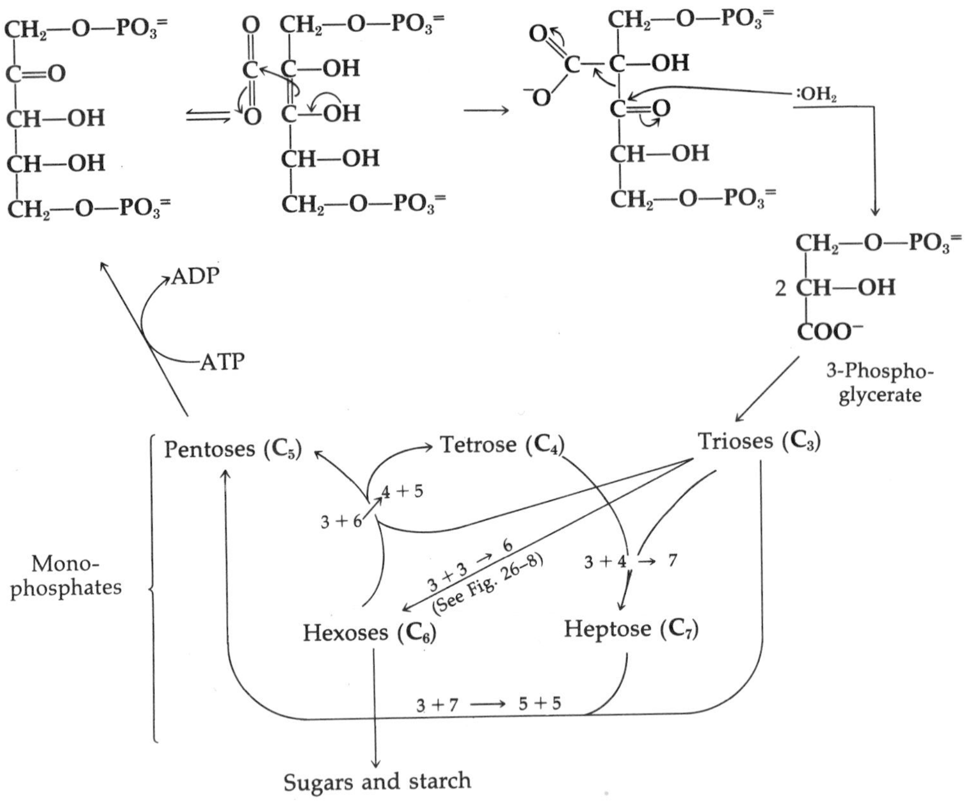

$$
\begin{array}{llll}
CH_2-O-PO_3^= & O\ \ CH_2-O-PO_3^= & O\ \ \ C-C-OH\ \ CH_2-O-PO_3^= & \\
C=O & C-C-OH & & \\
CH-OH & O\ \ C-OH & -O\ \ \ C=O & :OH_2 \\
CH-OH & CH-OH & CH-OH & \\
CH_2-O-PO_3^= & CH_2-O-PO_3^= & CH_2-O-PO_3^= &
\end{array}
$$

$$
\begin{array}{l}
CH_2-O-PO_3^= \\
2\ CH-OH \\
COO^-
\end{array}
$$

3-Phospho-
glycerate

ADP

ATP

Mono-
phosphates

Pentoses (C_5)   Tetrose (C_4)   Trioses (C_3)

4 + 5

3 + 6

$3 + 3 \rightarrow 6$
(See Fig. 26-8)

$3 + 4 \rightarrow 7$

Hexoses (C_6)   Heptose (C_7)

$3 + 7 \longrightarrow 5 + 5$

Sugars and starch

FIGURE 26-10   Carbon dioxide capture in photosynthesis

The oxidative steps represent a free-energy decrease in the substrates. The energy is picked up by NADH or GTP, which may now continue through other, auxiliary reactions to pass this energy on to the phosphorylation of ADP into ATP. Overall, these processes afford 12 molecules of ATP (from ADP) per molecule of acetyl-CoA completely oxidized.

Other foods than glucose can feed the metabolic furnace. Enzymes are available to convert other sugars to glucose or fructose. Proteins serve as food since their initial fate is degradation to amino acids and we have seen (page 1029) how amino acids may be converted, by the reactions of the pyridoxal-enzyme system, into α-ketoacids. These in turn are transformed into the acids of the citric acid cycle and so oxidized to carbon dioxide. Glutamic acid yields α-ketoglutarate directly in this fashion, and alanine goes to pyruvate.

Fats are hydrolyzed to fatty acids which are degraded in turn to acetyl-coenzyme A by successive two-carbon cleavages back from the carboxyl group of the linear fatty acid molecule. This stepwise fatty acid oxidation is very similar mechanistically to the reverse sequence used to synthesize fatty acids from acetyl-coenzyme A (described in Sec. 26-5).

In plants energy is taken directly from the sun to drive all metabolic functions. The initial absorption of light energy occurs by excitation of the chromophores of the universal green pigment, chlorophyll (Fig. 24-2), and several other pigments. The energy is transferred in paths, not yet fully understood, which convert ADP to ATP with remarkable energetic efficiency. In this way light energy is converted into chemical energy in the usable form of ATP.

In a coupled sequence of reactions not requiring sunlight, carbon dioxide from the air is converted into glucose. The carboxylation step is shown in Fig. 26-10 along with a summary of the other sugar conversions making up the cyclic sequence. Aldol reactions are involved to make and break the carbon-carbon bonds necessary to reassemble the required pentoses from the trioses which are formed in the carboxylation step. The same reactions build up hexoses which are stored as food energy. The variety of involved sugars of various sizes arises from the necessity of adding the one carbon of CO_2 to a C_5 sugar, cleaving to C_3, then being obliged to build C_6 for storage and more C_5 to retain the carboxylation cycle. The plant solution to this problem (Fig. 26-10) is an exercise in mathematics!

PROBLEM 26-5

a The enzyme phosphoenolpyruvate carboxykinase catalyzes a reaction between phosphoenolpyruvate and carbon dioxide. Write a mechanism for the reaction which yields H_3PO_4 (or its conjugate base ions, $H_2PO_4^-$, etc.).

b Phosphoenolypyruvate also reacts with certain sugars. Depict its reaction with a tetrose, forming a carbon-carbon bond.

PROBLEM 26-6

Consider the citric acid cycle shown in Fig. 26-9. Each reaction is mediated by a specific enzyme and these have all been isolated and separately characterized. If the isolated enzymes are mixed in aqueous solution with the required cofactors (certain metal ions, NAD, GDP, coenzyme A, etc.), the cycle will proceed to oxidize added pyruvate.

  a Detail the fate of labeled pyruvate added, with C^{14} either in the 2- or 3- positions.

  b If the malic dehydrogenase enzyme for reaction (9) is not added to the mixture, what labeled material(s) will accumulate in the system on addition of either of the labeled pyruvates of part **a**?

  c Suppose that intact mitochondria are isolated and an inhibitor specific to deactivate isocitric dehydrogenase is added. The system is then fed unlabeled pyruvate and also glutamic acid labeled, in separate experiments, at each of its carbons. Where will the labels appear in the compound(s) that accumulate?

26-5 BIOSYNTHESIS

Having examined the utilization of food to produce energy in the last section, we now turn to the reactions which synthesize the necessary molecules required by the cell. These two functions are separated in two columns of the overall metabolic chart in Fig. 26-2. The economy of the cell is such that the same molecules which appear as intermediates in glycolysis and the citric acid cycle are also the starting materials for biosynthesis.

   Amino acids for protein synthesis (Sec. 26-6) may be created by hydrolysis of input food protein, but many may also be created from pyruvate and the citric acid cycle intermediates. The final synthetic step will be the reductive amination of an α-ketoacid, the reverse of the pyridoxal-enzyme system used for their formation (page 1039). Aspartic and glutamic acids and alanine are thus synthesized in one step from the appropriate α-ketoacids of Fig. 26-9, and other amino acids are made from these. Aromatic amino acids arise by cyclization of a seven-carbon sugar (heptose) followed by an aldol addition of phosphoenolpyruvate (C_3) and loss of an original sugar carbon. This yields phenylpyruvate (C_9) first for conversion into phenylalanine, and by oxidation into tyrosine. The sequence is understood in stepwise detail (see Prob. 26-7).

   The nucleotide bases are built up in a sequence of reactions from aspartic acid and glycine, as well as one-carbon units such as formate and carbon dioxide. They are built up from a base of ribose (or deoxyribose) which is first converted to an amine.

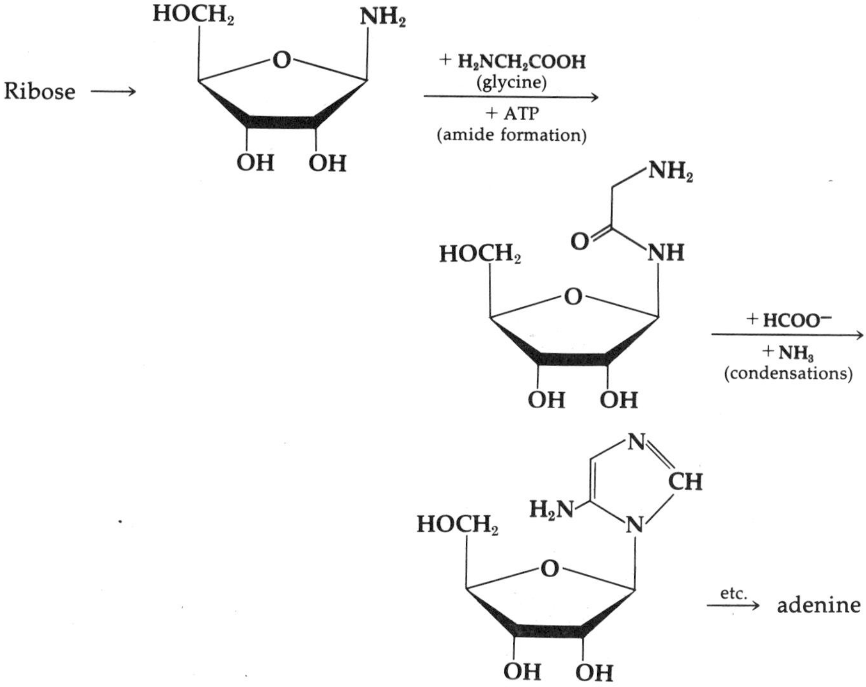

Ribose $\longrightarrow$

(Ribose derivative with HOCH$_2$ and NH$_2$ groups, OH OH)

$\xrightarrow[\substack{+\ ATP \\ \text{(amide formation)}}]{+\ H_2NCH_2COOH \\ \text{(glycine)}}$

(Ribose-amine acylated with glycine, HOCH$_2$, NH$_2$, O, NH, OH OH)

$\xrightarrow[\substack{+\ NH_3 \\ \text{(condensations)}}]{+\ HCOO^-}$

(Ribose purine precursor, HOCH$_2$, H$_2$N, N, CH, N, O, OH OH)

$\xrightarrow{\text{etc.}}$ adenine

For synthetic purposes acids are converted to acyl phosphates ($RCO-O-PO_3^{--}$) with ATP in order to effect substitution reactions (acylations) in much the same fashion that we convert acids to their acid chlorides first for this purpose in the laboratory. An example is implicit in the use of glycine above to acylate the ribose-amine.

The same phosphorylation is used to activate amino acid carboxyls for acylating the amino end of a growing protein chain during protein synthesis. This reaction is somewhat different in detail from the simple phosphorylations by ATP described before (page 1043), since the acyl group here becomes attached to a phosphate which is still also esterified to adenosine instead of to a simple phosphate group. The effect of creating a good leaving group is the same, however.

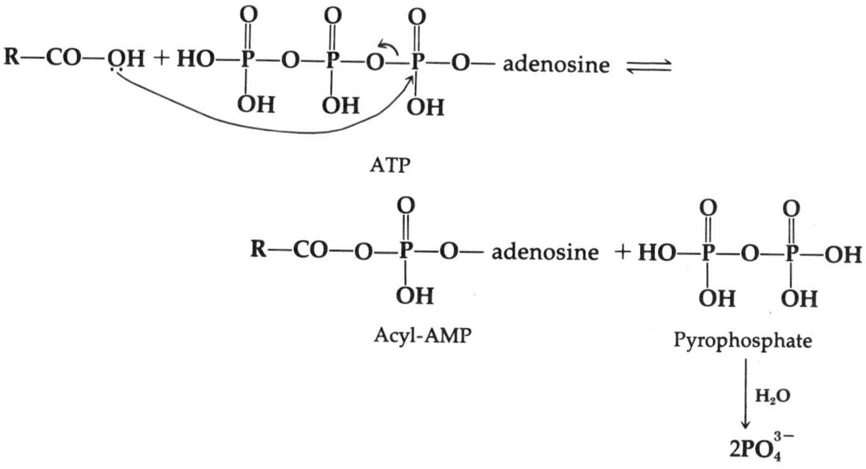

$$R-CO-OH + HO-\overset{\overset{\displaystyle O}{\|}}{\underset{\underset{\displaystyle OH}{|}}{P}}-O-\overset{\overset{\displaystyle O}{\|}}{\underset{\underset{\displaystyle OH}{|}}{P}}-O-\overset{\overset{\displaystyle O}{\|}}{\underset{\underset{\displaystyle OH}{|}}{P}}-O-\text{ adenosine} \rightleftharpoons$$

ATP

$$R-CO-O-\overset{\overset{\displaystyle O}{\|}}{\underset{\underset{\displaystyle OH}{|}}{P}}-O-\text{ adenosine } + HO-\overset{\overset{\displaystyle O}{\|}}{\underset{\underset{\displaystyle OH}{|}}{P}}-O-\overset{\overset{\displaystyle O}{\|}}{\underset{\underset{\displaystyle OH}{|}}{P}}-OH$$

Acyl-AMP   Pyrophosphate

$\downarrow$ H_2O

$2PO_4^{3-}$

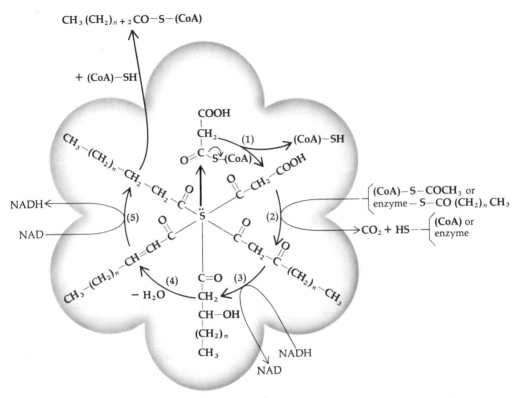

FIGURE 26-11   **The mode of action of fatty-acid synthetase**

The reaction is reversible itself but is rendered irreversible by the immediate removal of the high-energy pyrophosphate ion from the equilibrium via enzyme-catalyzed, exothermic hydrolysis.

Similarly, in the synthesis of nucleic acids the nucleoside triphosphates, like ATP, serve as the activated unit for attachment to the ribose (or deoxyribose) hydroxyl of the growing polynucleotide chain. Here is another instance of cell economy, in which ATP serves not only as a general energy-transfer and phosphorylating agent but also as a synthetic unit for creating nucleic acids.

The synthesis of fatty acids provides an important and interesting example of a biosynthetic pathway. In many organisms this synthesis has the added interest that the whole sequence of reactions apparently occurs while the substrate remains attached to the enzyme. The attachment here is a covalent one, as a thiol-ester of the thiol group of a cysteine residue in the protein chain of the enzyme. The enzyme may then be regarded as an organized cluster of separate enzymes, one for each reaction, around this cysteine site of attachment. The attached substrate may then be regarded as rotating around from one catalytic site to the next, undergoing the successive steps of its whole synthetic sequence while remaining attached to the enzyme. The conception is pictorialized in Fig. 26-11.

There is an important feature of efficiency in this mode of action. The substrate must first be captured from solution by the enzyme. This process will take some time and its costs something to create the covalent CO—S bond which holds the substrate in place. Once captured, the substrate is most rapidly and efficiently transformed through several steps if all the enzymes necessary can operate while it is trapped. In other synthetic sequences the substrate apparently is released from the enzyme after each reaction, but it is very likely that it is merely passed on directly to the next enzyme in the sequence without going loose into solution. This affords virtually the same efficiency and probably occurs because the successive enzymes are all organized in a line on a cellular superstructure and so pass the substrate from hand to hand like a bucket brigade—or, more accurately, an assembly line.

The first step in Fig. 26-11 is the carboxylation of acetyl-coenzyme A to malonyl-coenzyme A in order to create an active methylene for aldol condensation. This malonyl unit, seen at the top in the diagram, is then transferred to the thiol on the synthetase enzyme in step (1). Here it reacts—in a Claisen acylation (2)—with either acetyl-coenzyme A or a short fatty acid attached to another synthetase enzyme on which it was built. Carbon dioxide is lost in the acylation, leaving a β-ketoester still attached to the enzyme and providing a thermodynamic driving force to carry the equilibrium reaction to completion. This β-ketoester now swings around to the next site for reaction (3), a hydride reduction of the ketone by NADH. The substrate thiol-ester continues to move around to accept catalysis for the subsequent β-elimination (4) and hydrogenation (5) which finally yields a fatty acid two carbons longer than the original. This may be the final product (C_{16-18}), and, if so, it is transferred to coenzyme A for removal from the enzyme. If reaction (5) yields a lower product, the entire acyl-enzyme complex is involved in a new acylation reaction (2) at another synthetase enzyme bearing a fresh malonyl unit. Thus each cycle around the synthetase surface adds two carbons to a fatty acid chain. Eight such synthetase cycles convert acetate to stearate (stearic acid, n-$C_{17}H_{35}COOH$), ready to be esterified to glycerol and stored as fat.

The number of synthetic reactions used in metabolism is not large. In this synthesis—and the closely analogous reactions used for fatty acid degradations—we see the same kinds of transformations that were used in the citric acid cycle.

PROBLEM 26-7

The chief biosynthesis of aromatic rings is shown below in the creation of phenylalanine from sugars. Fill in the blanks and identify the mechanism of each step. How might tyrosine be formed?

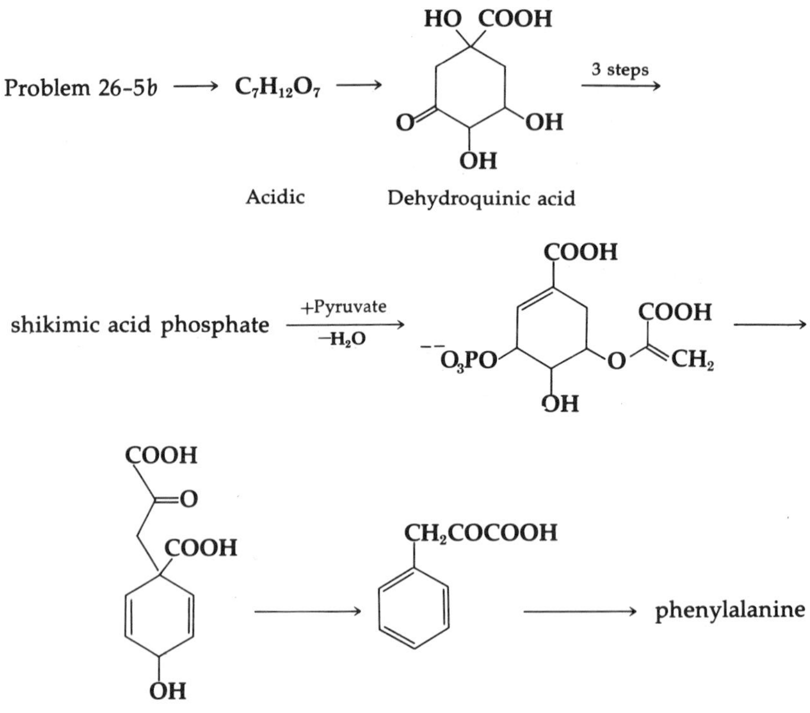

Problem 26-5b $\longrightarrow$ $C_7H_{12}O_7$ $\longrightarrow$ [structure] $\xrightarrow{\text{3 steps}}$

Acidic   Dehydroquinic acid

shikimic acid phosphate $\xrightarrow[\text{$-H_2O$}]{\text{+Pyruvate}}$ [structure] $\longrightarrow$

[structure] $\longrightarrow$ [structure with $CH_2COCOOH$] $\longrightarrow$ phenylalanine

Prephenic acid

PROBLEM 26-8

Propose a reasonable biosynthesis for adenine which fills in the details of the route outlined in the text. Identify the mechanisms and the necessary activating groups probably required.

PROBLEM 26-9

The metabolic route for oxidation of fatty acids to acetic acid is called "β-oxidation." This route is employed in order to utilize fats as food, for energy production.

a Outline a stepwise sequence of reactions to accomplish this conversion. Use fatty acid synthesis as a model for the kinds of reactions the cell is capable of performing.

    b Which of the reactions is most unusual in terms of laboratory precedent?

    c Only one molecule of ATP is required to convert stearic acid (C_{18}) into all nine molecules of acetic acid (as acetyl-coenzyme A). Account for this observation.

26-6 REPLICATION

A keystone in any definition of life must be the ability of living things to reproduce themselves. Furthermore, the offspring must be accurate, functioning replicas of the parents. At the most primitive level this reproduction is observed in the self-duplication of a cell by division (mitosis). In order for a cell to live it must possess a record of the amino acid sequences in its metabolic enzymes and structural protein since these need to be constantly manufactured, both for production of new cells and for replacement of chemically damaged parts in existing cells. These enzymes in turn, as we have seen, define and direct the metabolism which constitutes the life of the cell.

When a cell divides, both new cells must be equipped with the same accurate record of the particular enzyme content in order to function. With higher, multicellular organisms the record, or genetic code, defining all the cells (and their enzyme proteins) for the whole organism must be present in the initial egg cell which will grow into the organism.

The Watson-Crick hypothesis, now extensively supported with experimental evidence, states that the genetic code is contained in the DNA found in the cell nucleus and that this DNA can duplicate itself for cell division. A DNA molecule is a long chain defined by the sequence in which the four nucleotide bases appear along the backbone of ribose and phosphate (Fig. 25-9). This sequence is in turn a code for the sequence of amino acids in an enzyme or structural protein molecule. The DNA molecule holds the information to build protein molecules. This DNA molecule is the **gene** of genetics, holding the key to one characteristic of the organism. Each DNA molecule directs the construction of its characteristic protein (or several proteins), with complete sequence accuracy, whenever this synthesis is required either for replacement in the cell or for cell growth in preparation for mitosis.

Nucleotide Pairing and Self-duplication

The key to both the use and the duplication of the code contained in a sequence of bases in a DNA molecule lies in the concept of the complementary pairing of nucleotide bases by hydrogen bonding. In Fig. 26-12 is shown the strong hydrogen bonding which can hold a pair of nucleic acid bases together. Among the four DNA bases the only pairs that form these strong hydrogen bonds are adenine-thymine and guanine-cytosine, one purine and one pyrimidine base in each pair.

Hence a long DNA molecule, its sequence of bases evenly spaced along the regular backbone, couples with another DNA molecule bearing the com-

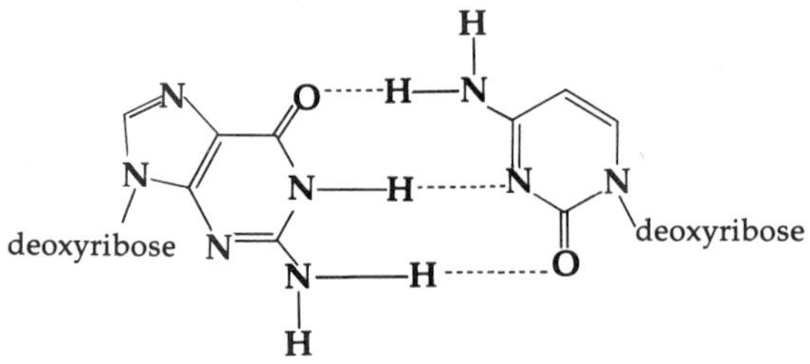

Hydrogen-bonding system for
guanine-cytosine

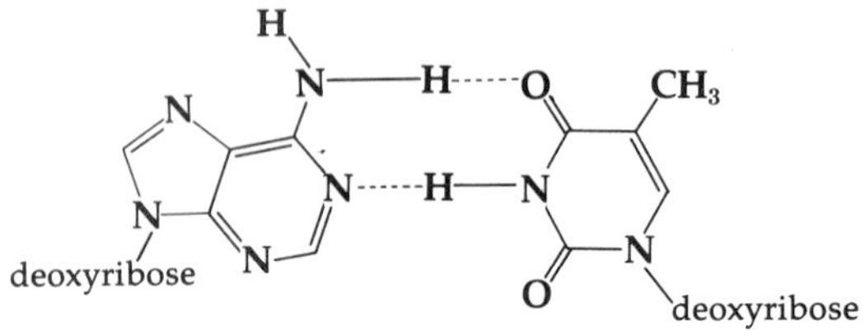

Hydrogen-bonding system for
adenine-thymine

FIGURE 26-12 **Complementary hydrogen bonding of base pairs in DNA**

plementary sequence of bases. The two chains then coil together into a double helix held together by these numerous hydrogen bonds, as shown in Fig. 26-13.

H bonds:

A—C—T—T—A—G—C—G— · · ·   DNA chain (base sequence)

T—G—A—A—T—C—G—C— · · ·   Complementary DNA chain

This is the heart of the Watson-Crick proposal. It is essentially simple but has far-reaching consequences for replication and directed-sequence protein synthesis.

In order accurately to replicate its own sequences of bases a double helix of DNA need only separate into two isolated strands. Each strand then serves as a template for synthesis of another complementary strand. For this synthesis the separate nucleotides from the surrounding medium are first attracted to align themselves by hydrogen bonding only to their opposite numbers in the chain. Then a molecule of DNA-polymerase enzyme links these separate, hydrogen-bonded nucleotides into a new and *necessarily complimentary* strand of DNA, and the two strands coil up together into a new and accurately duplicated double helix. Following this self-duplication of its genetic code material, the DNA now becomes organized as the genes of two identical sets of chromosomes to provide two nuclei on cell division.

Protein Synthesis

The use of the code is to direct the synthesis of cell proteins—mostly enzymes—with the correct amino acid sequences. Since proteins form the structure of the cell and catalyze all of its reactions, their synthesis completely defines the cell. In a primitive cell (e.g., a bacterium) with about 3,000 chemical reactions there must be 3,000 enzymes and, in the nucleus, a corresponding 3,000 genes (DNA molecules) to direct the correct sequential syntheses of these enzymes.

In order to synthesize protein, the parent DNA in the nucleus first causes a strand of RNA of complementary sequence to be created, using the DNA as a template, just as in replication. The RNA differs in that the base uracil (U) replaces thymine (T) but this does not affect the hydrogen bonding. It is this complementary RNA, called **messenger RNA** (*m*-**RNA**), which travels out of the nucleus to the ribosomes to direct protein synthesis there.

The sequence of bases in the DNA, and the complementary sequence in RNA, is composed of only four bases. If this sequence is to record a corresponding sequence composed of 20 amino acids in a protein, there can be no fewer than 3 nucleotide bases in a row to specify 1 amino acid.† Hence a group of 3 bases in a row in the RNA is called a **codon** for a single amino acid. The correspondence can be visualized as follows; the codons shown are those actually utilized in the cell, with the initials (A, C, G, U) of the four ribonucleotide bases used to identify them (see Fig. 25-9).

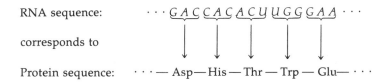

RNA sequence: $\cdots GAC\ CAC\ ACU\ UGG\ GAA \cdots$

corresponds to

Protein sequence: $\cdots$ — Asp—His—Thr—Trp—Glu— $\cdots$

Codons

Bases	*Amino acids*
GAC	Aspartic acid (Asp)
CAC	Histidine (His)
ACU	Threonine (Thr)
UGG	Tryptophan (Trp)
GAA	Glutamic acid (Glu)

The molecular mechanics involved in utilizing a messenger RNA for accurate stepwise protein synthesis are complex and fascinating, but very sensible for the job required. Details of this process are now beginning to be

---

† If two bases were used there would be only $4^2 = 16$ possible combinations, not enough for 20 amino acids. With three bases, $4^3 = 64$ combinations exist and this is ample. Some of the extra triplet combinations are used for auxiliary directions—punctuation of the code—such as "stop synthesis after this amino acid," etc.

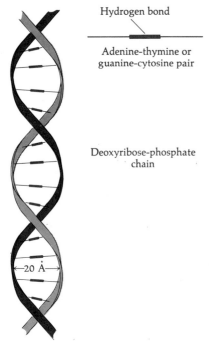

Hydrogen bond

Adenine-thymine or
guanine-cytosine pair

Deoxyribose-phosphate
chain

←20 Å→

FIGURE 26-13   **Hydrogen bonding in double helix of DNA**

clear and are outlined in Fig. 26-14. The essential logic of the process is that
the messenger RNA is a long molecular "sentence" spelling out the amino acid
sequence of the corresponding protein in three-letter words (codons). The
amino acids are brought up in order and in line along the RNA chain by
molecular units bearing at one end the matching three-letter codon and at the
other the amino acid. After these units are snapped into their places on the
messenger RNA (by base pairing), the amino acids at their other ends can be
linked with peptide bonds so as to form the correspondingly ordered protein.

   The matching unit molecules which bring in the amino acids are called
transfer RNA (*t*-RNA). These are relatively small (70 to 80 nucleotides in sequence), soluble, nucleic acid molecules, the nucleotide sequences of some
having been recently deduced. Transfer RNA is most conveniently regarded
as a long loop with the critical codon triplet at one end, free to hydrogen bond
to matching sites along the *m*-RNA chain. Many of the other bases in the *t*-RNA
are hydrogen-bonded across the loop down the center and hence are not
available as triplet codons to bond to the *m*-RNA. At the other end of the
loop the amino acid is attached to an adenosine as an acyl derivative. Here
we recognize the ATP function (page 1047) of activating the carboxyl group
for effective acylation.

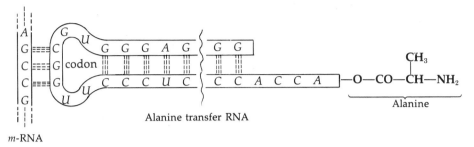

Alanine transfer RNA

m-RNA

The process of protein synthesis (Fig. 26-14) begins with attachment of a ribosome particle at one end of the messenger-RNA chain. In the surrounding medium the 20 kinds of transfer RNAs and their corresponding 20 amino acids are being brought together by 20 amino-acid-specific enzymes. From this pool of amino-acylated *t*-RNA the messenger RNA, with some help from the ribosome, sequentially selects the complementary *t*-RNAs by hydrogen bonding (base pairing) their codons. Then at the other end of the *t*-RNA molecule the amino acids may be linked together by an enzyme. This forms the protein and also frees the *t*-RNA to release itself from the *m*-RNA template and return to the medium to seek another molecule of its particular amino acid.

Some present evidence implies intriguing features of molecular mechanics in protein synthesis such as the view that the ribosome particle rolls along the *m*-RNA chain as synthesis occurs and that a second and more ribosomes may start synthesis of a new protein afresh at the front end of the *m*-RNA chain even before the first protein has been completed at the other end. A mammalian protein of 150 amino acids is synthesized in about 3 minutes.

FIGURE 26-14   **Protein synthesis at the ribosomes**

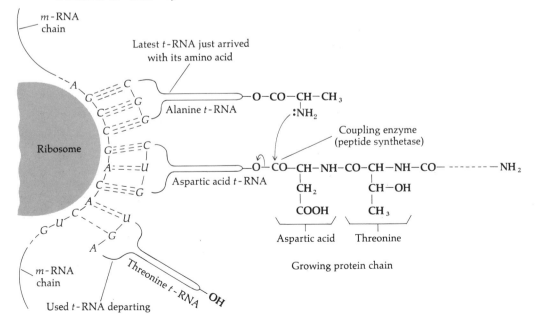

Viruses

A virus may be regarded either as the simplest form of life or as the dividing line between the living and nonliving. A virus may be crystallized like any ordinary organic chemical and is then completely stable and unchanging in the absence of living cells. The virus itself does not have cell structure. In the presence of cells, however, viruses replicate and multiply, using up the material of the living cell to do so. Even simple one-celled bacteria suffer from viral parasites (**bacteriophages**, meaning "bacterium eaters").

A virus particle is simply a chromosome of viral-specific nucleic acid (DNA or RNA) in a limited number of genes, surrounded by a protective coat of protein. This virus particle enters a cell wall (or merely injects its nucleic acid into it) and then simply uses the nucleotides, amino acids, enzymes, and ribosomes within the living cell as components for its own replication and protein synthesis, in competition with those processes of the cell. The virus nucleic acid chain is duplicated many times (until it has used up the cellular supply of nucleotides) and is also used to synthesize its protective-coat protein and a few specific enzymes it may require for this work. For most of its replication the virus uses normal cell enzymes. At least one viral gene also serves to create an enzyme which will break through the cell-wall protein. The material of the living cell is thus reordered into the construction of many virus particles, directed by the nucleic acid of the one original virus particle. When the cell is thus exhausted and its wall broken through, these virus particles swarm out to attack many other cells.

Because of their simplicity the study of viruses is a very useful tool for studying the basic nature of replication, but it also affords a potent probe of the nature of viral diseases. While this study may illuminate the course of these diseases and point to some mechanisms for their containment, it also focusses on the difficult medical problem that the virus and its mode of multiplication are exceedingly similar to the nature of healthy metabolism in the organism attacked. The same medical difficulty applies to cancer, which is a subtle change in the course of normal cell replication, operating by virtually all the same mechanisms as the normal cell. This metabolic similarity to the healthy cell makes selective destruction of cancer or virus an exceedingly subtle and difficult medical problem.

PROBLEMS

26-10 The nature of nucleic acid self-replication implies a procedure for simplifying laboratory synthesis of particular sequences. Suppose a very small amount of a particular oligonucleotide, of about a dozen units in length, were synthesized in the laboratory.

    a How might larger quantities of this substance be obtained?

b If for practical reasons the longest chain available by stepwise chemical synthesis were a dozen units long, how might a 77-unit transfer RNA of specific sequence be made?

26-11 The number of reaction types in primary metabolism is small. List the number of kinds of reaction which occur in the whole system of sugar oxidation to carbon dioxide and fatty acid synthesis and degradation. An example of a kind of reaction would be

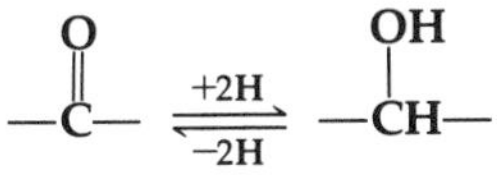

$$\underset{\text{—C—}}{\overset{\text{O}}{\|}} \quad \underset{\xrightarrow{-2\text{H}}}{\overset{+2\text{H}}{\rightleftharpoons}} \quad \underset{\text{—CH—}}{\overset{\text{OH}}{|}}$$

26-12 **a** How many different enzymes are required to synthesize any protein from twenty amino acids. Assume that any reaction that does not demand specificity for a single amino acid may be catalyzed by one enzyme equally well for all twenty amino acids.

b Similarly, how many enzymes are required for DNA synthesis from the four deoxyribonucleotides?

26-13 Examine the life cycle of a virus and roughly determine the minimum number of DNA molecules, or genes, which it requires. Assume one DNA molecule per protein which will be synthesized under its direction.

26-14 Write out the specific reactions necessary to fill in the photosynthesis chart shown in Fig. 26-10.

26-15 In the synthesis of fatty acids the loss of CO_2 in the initial malonate condensation may serve to drive the reaction to completion thermodynamically. Consider the condensation in detail, at pH ~ 7, taking note that it requires the presence of magnesium, and formulate a particular mechanism for the reaction.

26-16 The biosynthesis of proline takes place using glutamic acid as a starting material. Devise such a biosynthesis that occurs following biochemical precedents established in this chapter.

26-17 Mevalonic acid is an important intermediate for biosynthesis of a number of natural compounds, including turpentine, rubber, steroid hormones, and the odors of spices. The sole source of carbon is in acetyl-coenzyme A. Devise a biosynthetic route to mevalonic acid consistent with rational mechanistic precedent.

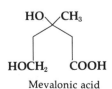

Mevalonic acid

READING REFERENCES

Mahler, H. R., and E. H. Cordes, "Basic Biological Chemistry," Harper & Row, New York, 1966.

Watson, J. D., "Molecular Biology of the Gene," W. A. Benjamin, Inc., New York, 1965.

Lehninger, A. L., "Bioenergetics," W. A. Benjamin, Inc., New York, 1964.

Ingram, V. M., "The Biosynthesis of Macromolecules," W. A. Benjamin, Inc., New York, 1964.

Dickerson, R. E., and Geis, I., "The Structure and Action of Proteins," Harper & Row, New York, 1969.

THE CHEMISTRY OF NATURAL PRODUCTS

27-1 INTRODUCTION

Man's earliest interest in organic compounds goes back to prehistory. Pigments for dying and painting, perfumes, and folk medicines are all organic compounds, and all required some crude extraction from their natural sources, usually plants. Over sixty perfume components were known in the Renaissance, usually prepared by distillation from crushed plant leaves. The extraction of medicinals from leaves and bark by grinding and boiling has a worldwide and ancient history.

Many such organic compounds from natural sources, often crystalline, were isolated and purified in the early nineteenth century. The study of their reactions constituted the beginnings of organic chemistry, and indeed during the whole of that century these **natural products** constituted the main source of organic chemicals and afforded the main problems of organic chemistry. Only with the growth of synthesis was the field freed from a dependence on compounds available from natural sources. However, the variety of structures found in natural products is astonishing, and the most subtle and challenging problems of structure elucidation and of synthesis have been and still are those of natural products.

The common procedure for obtaining natural products involves drying, grinding, and extracting plant material with some solvent like methanol or chloroform. Evaporation of the solvent then yields a mixed gum or oil which must be separated into its component compounds. Chromatography is the usual procedure, although in a number of cases one or two components may be crystallizable from the mixture. There may however be literally hundreds of compounds in the gum. A number of these may be the familiar and common chemicals of general metabolism (i.e., citric acid and others in Fig. 26-9), but more will be compounds which are characteristic of the plant species extracted.

Once a pure natural product is isolated, attention is focused on the elucidation of its molecular structure. In the older tradition—prior to about 1940—this was achieved by deducing structural information from the nature of the products obtained in particular chemical reactions. In current practice the chemical approach has been largely supplanted by the use of physical tools (Chap. 7). The essential logic of structure determination was initially presented in Sec. 3-3 while the protocols for the use of physical methods were offered in Chap. 7. It is appropriate, now that we have explored chemical reactions, to select some which have proved useful in structure determination by chemical means in order to expand the discussion of Sec. 3-3.

Structure Determination

The first focus in structure determination is to establish the nature of the functional groups implied by the number and kind of heteroatoms in the empirical formula. In terms of chemical methods, it is common to determine the number of hydroxyls and (primary and secondary) amino groups by acetylation. Amines are recognized by their basicity (solubility in aqueous acid); primary and secondary amines lose their basicity on acetylation, changing to nonbasic amides, but tertiary amines are unchanged. Primary, secondary, and tertiary alcohols may also be distinguished. Tertiary alcohols are not normally acetylated and do not oxidize with chromic oxide, while primary alcohols give aldehydes or acids with that reagent and secondary alcohols yield ketones.

Carboxylic acids are recognized by their acidity (solubility in aqueous bicarbonate), and the number of acid groups may be determined by titration or by methylation with diazomethane. The presence of methylene groups adjacent to ketones may be determined by aldol condensation with benzaldehyde, yielding a "benzylidene derivative."

In all of these chemical procedures the product is analyzed and its empirical formula compared to that expected from such a procedure on the starting compound.

Consider a possible natural product, $C_7H_{12}O_5$, which is soluble in aqueous bicarbonate. The following reactions are carried out.

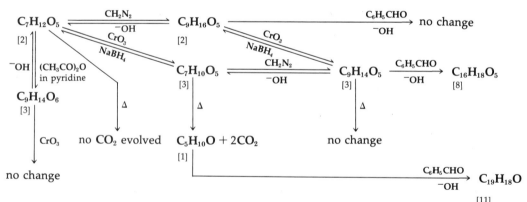

In order to deduce the structure of the acid we first examine each formula to determine the index of hydrogen deficiency; these are shown as colored numbers in brackets. Next we must examine each reaction in terms of the *expectation of chemical change* (see Sec. 3-3), as follows. The initial problem is to define the functionality represented by five oxygens and an index of hydrogen deficiency showing two rings and/or double bonds in $C_7H_{12}O_5$.

1 CH_2N_2/OH^-. Diazomethane yields a methyl ester from an acid, $R—COOH \longrightarrow$ $R—COOCH_3$. The expectation is a net increase in the empirical formula of $+CH_2$ for each carboxyl group.† The observed change is $+C_2H_4$, implying two —COOH groups. The implication is tested by applying the common saponification of esters (with OH^-), which in fact yields again the starting diacid, as expected. The formula allows two carboxyl groups since the four requisite oxygens are present in $C_7H_{12}O_5$. Two carboxyl groups also require an index of 2 (one $C=O$ double bond in each carboxyl) and so we may deduce that the rest of the molecule is saturated and has no rings. There is one more oxygen to uncover, but with no unsaturation this can be only an alcohol or an ether.

2 $(CH_3CO)_2O/OH^-$. Acetylation is applied to distinguish alcohol from ether. Ethers do not acetylate, while acetylation of an alcohol occurs with a formula change of $+C_2H_2O$. This is exactly the change observed in going to $C_9H_{14}O_6$ and the ester (acetate) nature of the product is again confirmed by saponification, which recreates the starting material. As in the previous case this confirmation is important to prove that no unexpected skeletal change (rearrangement) has occurred.

3 CrO_3. The nature of the alcohol is shown to be secondary by oxidation with chromic oxide. A tertiary alcohol would have shown no change. A secondary alcohol proceeds to a ketone with a molecular change of $-2H$ and an increase of one in the index of hydrogen deficiency, as observed here. A primary alcohol, however, under the conditions used would be transformed into a carboxylic acid ($R—CH_2OH \longrightarrow$ $R—COOH$), with a formula change of $-2H + O$; this is not observed. The deduction is confirmed by sodium borohydride reduction to starting material and by the inertness of the acetylated alcohol to CrO_3. The same oxidation is also available to the diester, giving $C_9H_{14}O_5$ as expected.

These reactions establish the original compound as a secondary alcohol as well as a diacid. The formula shows that there are no more functional groups. The original alcohol diacid yields an alcohol diester (CH_2N_2) and both are oxidized (CrO_3) to a keto-diacid and a keto-diester, respectively. We can summarize these functions in the following way and it now remains to locate their relative positions on the acyclic $C_5H_9\equiv$ carbon skeleton.

> †Phenols also yield methyl ethers with diazomethane, but the presence of a phenol would require at least six carbons and an index of hydrogen deficiency of at least four. Since the latter is not available a phenol is ruled out.

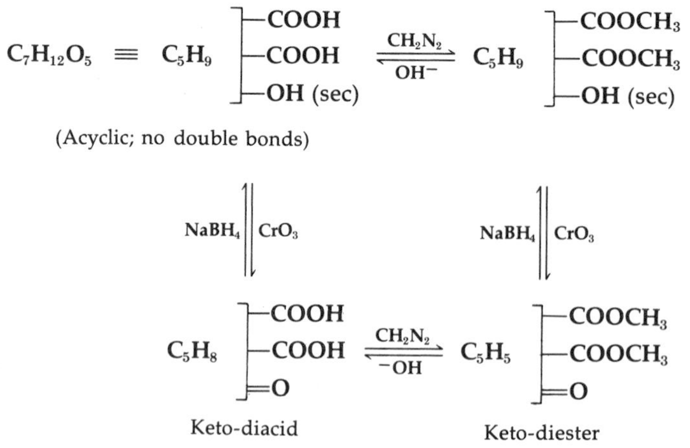

$$C_7H_{12}O_5 \equiv C_5H_9 \begin{bmatrix} -COOH \\ -COOH \\ -OH \text{ (sec)} \end{bmatrix} \underset{OH^-}{\overset{CH_2N_2}{\rightleftharpoons}} C_5H_9 \begin{bmatrix} -COOCH_3 \\ -COOCH_3 \\ -OH \text{ (sec)} \end{bmatrix}$$

(Acyclic; no double bonds)

$$NaBH_4 \updownarrow CrO_3 \qquad\qquad NaBH_4 \updownarrow CrO_3$$

$$C_5H_8 \begin{bmatrix} -COOH \\ -COOH \\ =O \end{bmatrix} \underset{-OH}{\overset{CH_2N_2}{\rightleftharpoons}} C_5H_5 \begin{bmatrix} -COOCH_3 \\ -COOCH_3 \\ =O \end{bmatrix}$$

Keto-diacid   Keto-diester

4   *Pyrolysis.* Mild heating of the keto-diacid yields a neutral compound, $C_5H_{10}O$. It is neutral since to be basic it must contain nitrogen and to be acidic it must contain two oxygens for —**COOH**. Two molecules of carbon dioxide are lost on heating. These must arise by decarboxylation of the two acid groups since when they are esterified no change occurs on pyrolysis. This decarboxylation of ketoacids is characteristic only of β-ketoacids. Hence one of these two groupings must be present:

HOOC—C—CO—C—COOH   HOOC—C—CO—C
                                                         |
                                                       COOH

The second of these would require the initial diacid to be a substituted malonic acid. This would have decarboxylated, in the original diacid with no ketone present, but pyrolysis of the original diacid shows that this is not the case. Hence the malonic acid structure is ruled out.

5   *Aldol condensation.* Benzaldehyde in alkali is now used to examine whether the carbons adjacent to the ketone are methyl or methylene (CH_2) as required by the aldol reaction (Sec. 12-5)

—CO—CH_2 + OCHC_6H_5 $\longrightarrow$ —CO—C=CHC_6H_5

The expected formula change is $+C_7H_4$ and the index of hydrogen deficiency of the product must be increased by 5, i.e., one double bond, one ring (benzene), and the three nominal double bonds of a formal benzene ring. There is no ketone in the initial hydroxy-diester ($C_9H_{16}O_5$) and the ester groups are not adequately activated to support an aldol condensation. hence this diester is unaffected, but the keto-diester reacts with one benzaldehyde to form a benzylidene derivative representing the correct molecular change.

The ketone, $C_5H_{10}O$, remaining after decarboxylation of the keto-diacid, is found to form a *bis*-benzylidene derivative from two molecules of benzaldehyde. From this data we conclude that the keto-diester possesses only one methylene *alpha* to the keto group while the final ketone has two. The following two formulas show the only allowable deductions. At least one α-carbon substituent must be present in the ester, but *only* one is allowed at that α-carbon in the ketone.

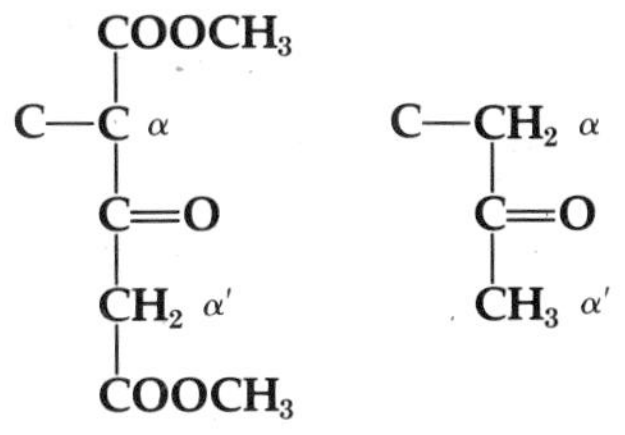

COOCH₃ structures:

$$
\begin{array}{l}
\text{COOCH}_3 \\
|\\
\text{C—C} \; \alpha \\
|\\
\text{C=O} \\
|\\
\text{CH}_2 \; \alpha' \\
|\\
\text{COOCH}_3
\end{array}
\qquad
\begin{array}{l}
\text{C—CH}_2 \; \alpha \\
|\\
\text{C=O} \\
|\\
\text{CH}_3 \; \alpha'
\end{array}
$$

Keto-diester   Ketone ($C_5H_{10}O$)

Of the $C_5H_{10}O$ of the ketone this reasoning accounts for all but CH_5, and these remaining atoms can only be attached at the colored α-carbon substituent, and in only one way—to form 2-pentanone. The final structures are therefore rigorously proved to be the following.

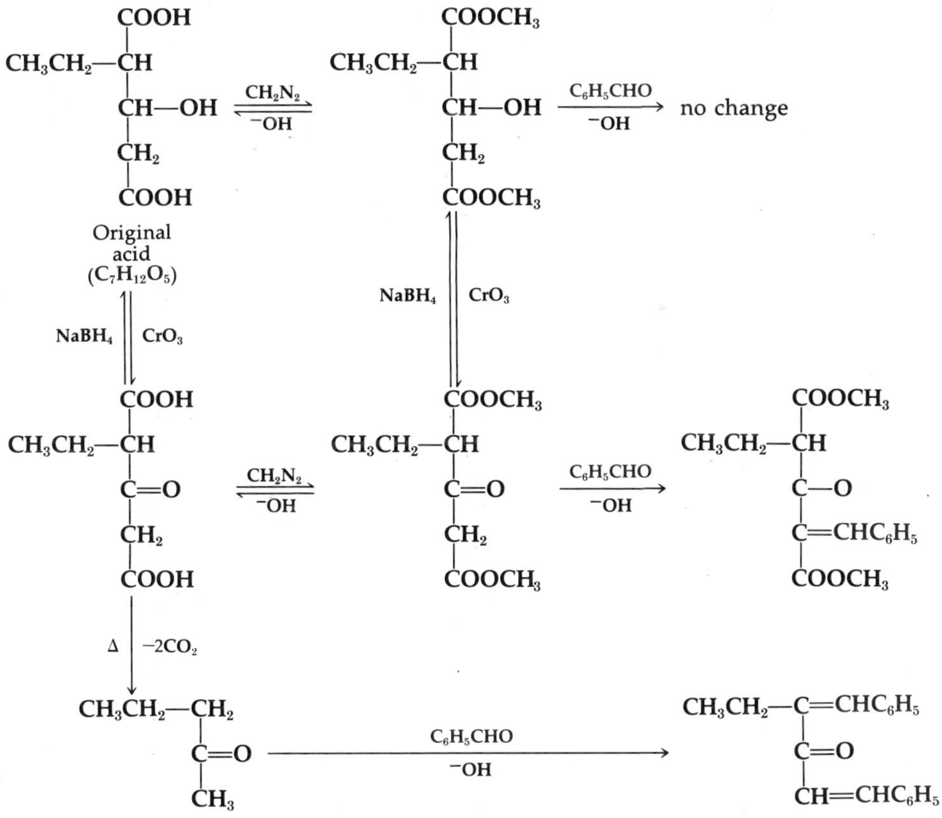

This sort of logical proceeding is used in all chemical determinations of structure, and certain examples from known natural products are presented in the succeeding sections.

Apart from structure elucidation, the chief activities in the natural products field are synthesis of these natural compounds and studies aimed at deducing their bioxynthesis. Our present knowledge of their biosynthesis allows a classification of the thousands of known natural products into several broad families:

Sugars
Acetogenins
Terpenes (and steroids)
Alkaloids

Most, though not all, natural products fall into these categories. Many which do not fit into these groups have specific functions in the central metabolism, such as the citric acid cycle acids or the coenzymes of Chap. 26. The biological purpose or function of most natural products in the last three major families above has in most cases, however, eluded detection. Indeed there is a suspicion that many may serve no function at all in the life of the producing organism, that they may be merely "metabolic accidents." However that may be, many natural products serve mankind as medicines, pigments, flavors, perfumes, solvents, and in other various uses, not least of which is to furnish organic chemists with a fund of fascinating structures and unusual reactions, many of which have provided the springboard for new areas of theoretical advance.

In the sections that follow only the barest outline and suggestion is possible of the enormous variety in the chemistry of natural products. A survey of representative structures and their biosynthesis is offered, followed by a few examples from each family illustrating structure elucidation, synthesis, biosynthesis studies, and stereochemistry.

PROBLEM 27-1

Construct the set of spectral data you would expect from the diacid, $C_7H_{12}O_5$, in the preceding structure problem. Show how they might serve to solve the structure problem without carrying out any chemical reactions.

PROBLEM 27-2

In the structure problem elucidated, certain other facts might have been gathered which need accounting for in the final analysis. Account for these:

a The original diacid and its diester are optically active but their two products from chromic oxide oxidation are not.

b The borohydride reductions of these two oxidation products (ketones) actually afforded starting diacid and diester in less than 50% yield. A second product was formed in each reduction which was very similar in properties.

c The pyrolysis of the initial diacid yielded an optically inactive neutral compound, $C_7H_8O_3$.

d The yield of keto-diacid on saponification of the keto-diester ($C_9H_{14}O_5$) was quite poor, although saponification of the hydroxy-diester ($C_9H_{16}O_5$) was quite free of by-products. What other product(s) might be expected from the saponification of the keto-diester ($C_9H_{14}O_5$)?

27-2 BIOSYNTHESIS OF NATURAL PRODUCTS

Natural products are usually secondary metabolites, synthesized from the primary molecules which constitute the central metabolism of the living cell (Chap. 26). By far the most important starting material is acetic acid (as acetyl-coenzyme A). The other major sources are several amino acids, notably lysine, tryptophan, phenylalanine, and tyrosine (Table 25-1).

The reactions utilized in natural-product biosynthesis are all mechanistically familiar from the previous chapters. The enzymes that catalyze most if not all of these synthetic steps are often quite nonspecific. For example there are enzymes capable of reactions such as oxidation-reduction by transfer of two hydrogens, as is observed in the interconversion of alcohols and ketones, and these may often be quite effective on substrates other than those for which they are intended. Some of the key reactions are capable of proceeding spontaneously in neutral aqueous media without any enzymic mediation.

Each of the three major families, apart from sugars, is characterized by certain structural features which are in turn the result of a common general biosynthetic mode for that family. Hence introduction to this general bio-synthesis pattern for each family offers an opportunity to survey the main characteristic structural types of natural products.

Acetogenins

The biosynthesis of fatty acids was outlined in Sec. 26-5. The sequence consists essentially of an acylation of an acetyl group attached to an enzyme to yield an acetoacetyl group esterified to the enzyme. This is then followed by stepwise reduction of the β-keto group. A very large body of natural products appears to arise from a variation in the normal fatty-acid biosynthesis whereby the further acetyl units are added, by acylation, without the intermediacy of the ketone reduction steps. This leads to a poly-β-keto, or poly-acetyl, linear chain still attached to the enzyme. Natural products derived from further reactions of this linear polyacetyl chain are called **acetogenins**.

$$Enzyme-S-CO-CH_2-CO-CH_2-CO-CH_2-\cdots$$

Many of these natural compounds are characterized by long linear carbon chains with functionality at *alternate* carbons along the chain, reflecting the initial poly-β-keto periodicity in the precursor. In the examples the original carbonyl carbon of the polyacetyl chain is marked in color.

$$CH_3-C\equiv C-C\equiv C-C\equiv C-CH=CH-COOCH_3$$

Matricaria ester

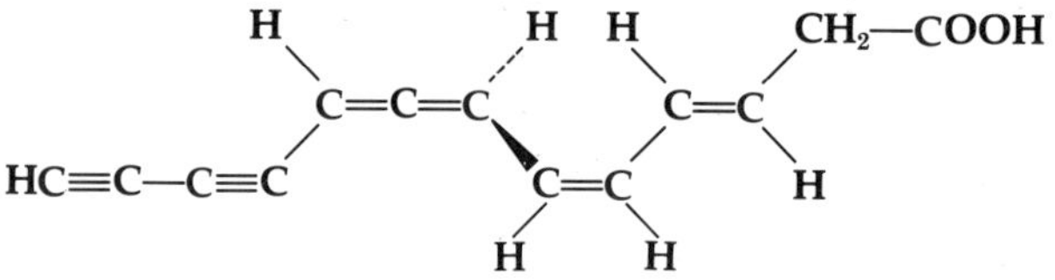

Mycomycin

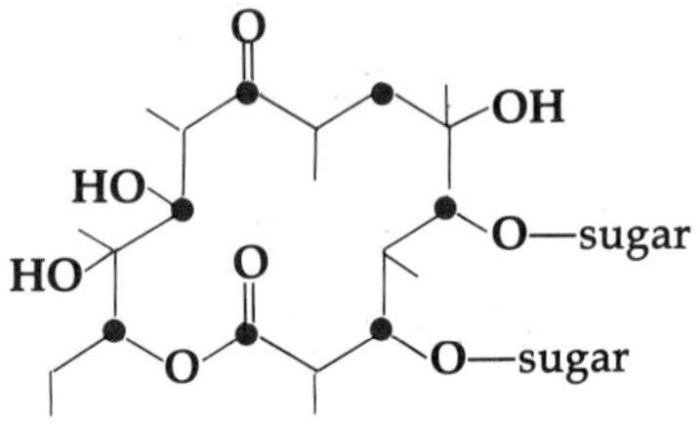

Erythromycin
(an antibiotic
from fungi)

The examples show occasional reduction of the several original keto groups to alcohols and these are sometimes dehydrated to double bonds. One original carbonyl site (top, right) in erythromycin is fully reduced, as in normal fatty acids. The colored lines on the erythromycin periphery indicate methyl groups, always attached to a site α to the original polyacetyl carbonyls. These arise in two ways, either by methylation of the β-diketone (cf. Sec. 11-7) or by substitution of propionic for acetic acid in the original acylations on the enzyme which creates the chain. The alkylation of these β-diketone anions is a very facile process and can occur on the carbon between two carbonyls or on the oxygen of the β-diketone enolate.

Almost always the only groups introduced by alkylation in biosynthesis are methyl groups, and the terpenoid C_5 (isopentenyl, see page 1070) groups. Oxidation of the β-diketone enolate (i.e., "OH^+" or its equivalent) can also produce hydroxyls at the α-position, as seen in two sites on erythromycin.

Accessory changes:

Polyacetyl chain:

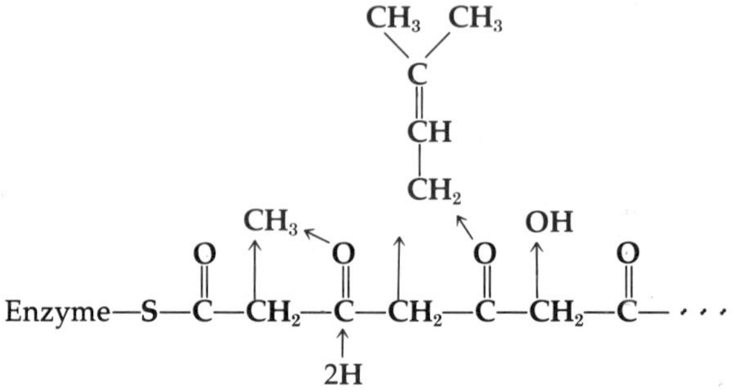

The most common fate of the polyacetyl chain is an aldol-type cyclization. This can occur in two ways to yield aromatic rings with meta-oriented phenolic hydroxyl groups, shown below for a tetraacetyl precursor.

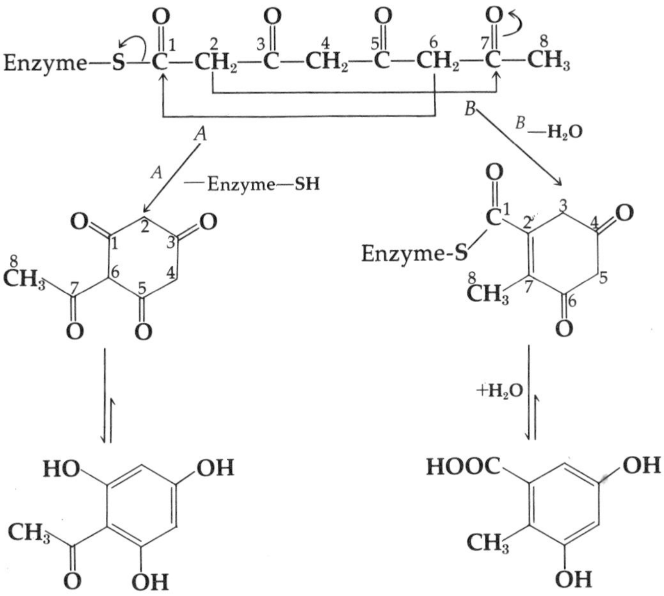

Phloroacetophenone

Orsellinic acid

This biosynthesis is one of the two methods used in nature to produce aromatic rings. The phenols produced all exhibit the meta orientation of hydroxyls. The other route is one from sugars used to synthesize phenylalanine and tyrosine (see Prob. 26-7), and these in turn are the source of ϕ—C—C—C skeletons in many natural products.† These aromatics are characterized by oxygen para to the C_3 chain (as in tyrosine) and sometimes second and third oxygens ortho to that first one. The lignin skeleton (page 993) shows molecular units from this source as do the examples in Fig. 27-1.

The aromatics from polyacetyl cyclization can also exhibit the pattern of methylation (or attachment of isopentenyl groups) on oxygen as in Fig. 27-1, or on the carbons originally between ketone groups in the initial polyacetyl chain. The cyclization can also produce more rings than one, including oxygen heterocycles which are merely the initial poly-β-keto chain in cyclized disguise. Examples of these processes are found in the natural products of Fig. 27-2, in which the linear polyacetyl precursor is traced in color with its original carbonyl sites marked. The pathway for their biosynthesis should be clear from

† The usual first step in plants is elimination of ammonia from the amino acid to yield a cinnamic acid:

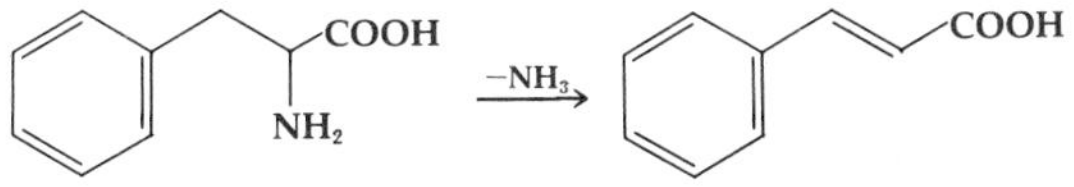

Phenylalanine

Cinnamic acid

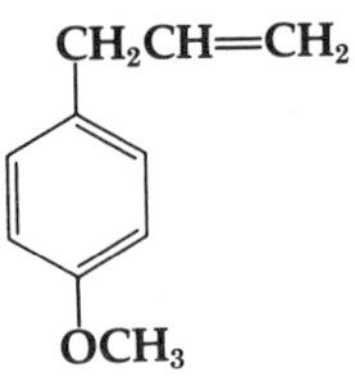

Anethole (from anise seed)

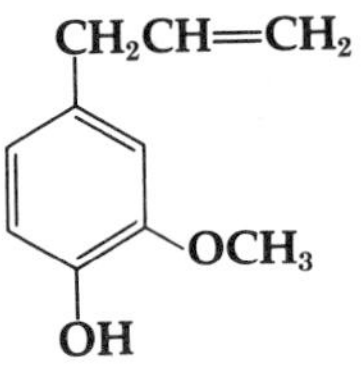

Eugenol (found in cloves)

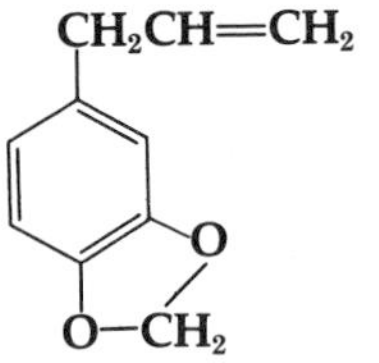

Safrole (chief constituent of oil of sassafras)

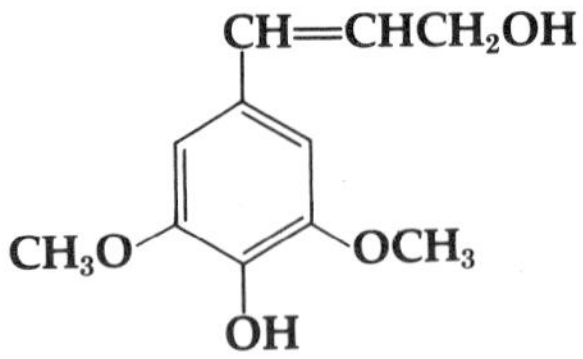

Syringyl alcohol (from syringa blossoms)

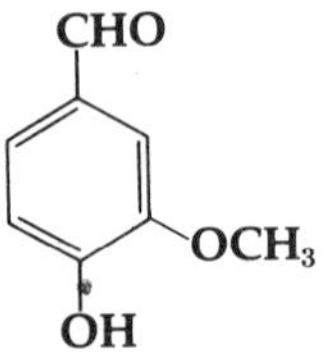

Vanillin (fragrant component of vanilla bean)

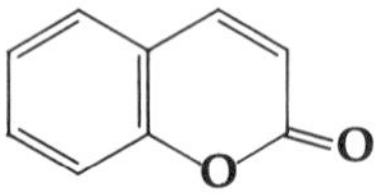

Coumarin (perfume in clover)

FIGURE 27-1   Natural phenolic compounds arising from phenylalanine

their structures. Indeed the original "acetate hypothesis" for their biosynthesis was first proposed because of these suggestive natural structures. This understanding of the biosynthetic process in return has allowed errors in several published structures to be discovered when they did not fit a polyacetyl origin.

The oxidative coupling reaction of phenols (page 823), or of β-diketone enols, is a very widespread biosynthetic reaction which may be illustrated in a synthesis of the common yellow lichen pigment, usnic acid. The two identical coupled units are simple acetogenins, and the biosynthesis of usnic acid is presumably roughly the same as this laboratory synthesis modeled after it.

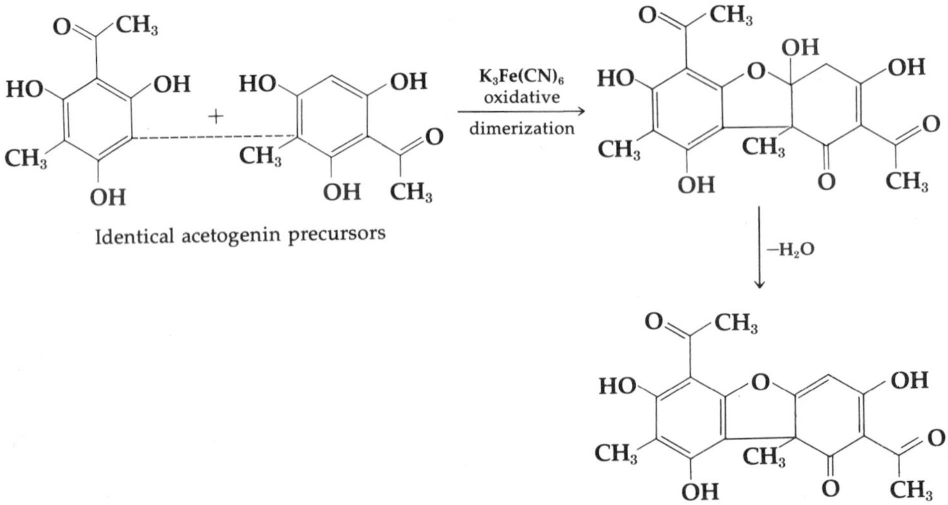

$K_3Fe(CN)_6$ oxidative dimerization

Identical acetogenin precursors

$-H_2O$

Usnic acid

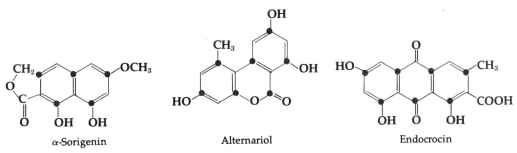

Hydroxy-peonol
(from peonies)

Dehydroangustione

Eugenone

Angustifolionol

Colupulone
(from hops)

Visnamminol

Ustic acid

α-Sorigenin

Alternariol

Endocrocin

FIGURE 27-2   Examples of natural aromatic acetogenins

Terpenes and Steroids

Many of the components of essential oils from fragrant plants were grouped into families very early when it became clear that they all had formulas containing a multiple of five carbons. As structures became known, these compounds exhibited carbon skeletons composed of **isoprene units**.

Isoprene   Isoprene unit
of structure

These compounds (including rubber, a natural polymer of isoprene, page 977) were labeled **terpenes** and subdivided by the number of isoprene units they contained in their structures:

Monoterpenes	C_{10}	Two isoprene units
Sesquiterpenes	C_{15}	Three
Diterpenes	C_{20}	Four
Triterpenes	C_{30}	Six
Tetraterpenes	C_{40}	Eight

The isoprene units can be found, attached to each other in a head-to-tail fashion, by inspecting the representative terpene formulas shown in Figs. 27-3 and 27-4. The isoprene units are outlined in color.

It is clear that this dissection of terpene structures into isoprene units must reflect a general biosynthesis from an isoprenoid precursor which links several times with itself in a head-to-tail fashion. The biosynthesis of terpenes has been extensively studied using compounds containing C^{14} at "labeled" positions in order to trace the path of particular compounds during biosynthesis. The actual biosynthetic isoprene unit is mevalonic acid, which is synthesized from 3 moles of acetic acid (as acetyl-coenzyme A) via normal acylations to a branched skeleton. Formation of the branched skeleton follows a mechanistically sound aldol reaction; this branching constitutes a major biosynthetic departure from the linear fatty acids and acetogenins. Mevalonic acid is then phosphorylated and decarboxylated to isopentenyl pyrophosphate, the actual biosynthetic isoprene unit.

This biosynthesis is outlined in Fig. 27-5 with color to follow the C^{14}-labeled methyl group of acetate, as was actually done in the tracer studies. The isopentenyl pyrophosphate isomerizes and then dimerizes to form geranyl pyrophosphate in an attack of a double bond on the carbonium ion left by ionization of the pyrophosphate, the common leaving group of biological reactions (Chap. 26). Several linkings of isopentenyl pyrophosphate lead to C_{10}, C_{15}, and C_{20} acyclic skeletons, and these may dimerize head to head to form precursors of tri- and tetraterpenes. Squalene is the general precursor of triterpenes and is very widespread in living organisms.

Monoterpenes

Myrcene
(bayberry)

Geraniol
(gingergrass)

Citral
(oil of lemon grass)

Menthol
(Japanese
peppermint oil)

d-Limonene
(lemon, orange)

Δ^3-Carene
(oleoresin of
Pinus longifolia)

α-Pinene
(turpentine oil)

Camphor
(camphor tree)

Sesquiterpenes

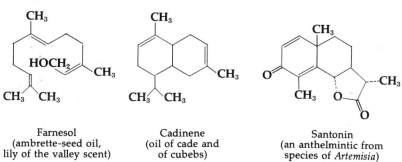

Farnesol
(ambrette-seed oil,
lily of the valley scent)

Cadinene
(oil of cade and
of cubebs)

Santonin
(an anthelmintic from
species of *Artemisia*)

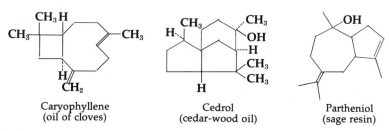

Caryophyllene
(oil of cloves)

Cedrol
(cedar-wood oil)

Partheniol
(sage resin)

FIGURE 27-3   Representative natural mono- and sesquiterpenes

Diterpenes

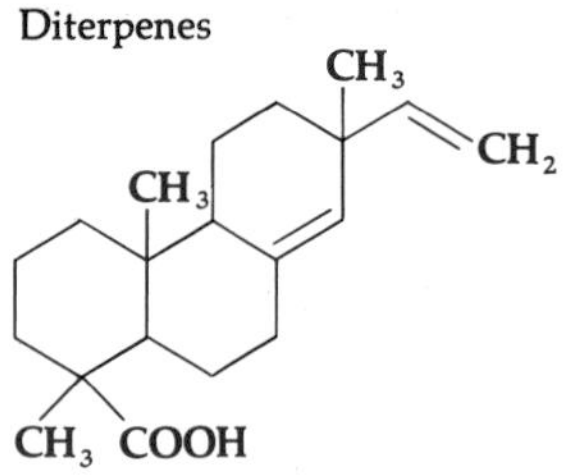

Dextropimaric acid
(pine rosin)

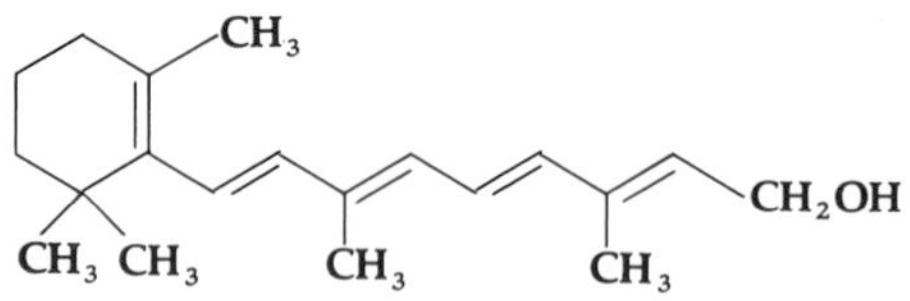

Vitamin A

Triterpenes

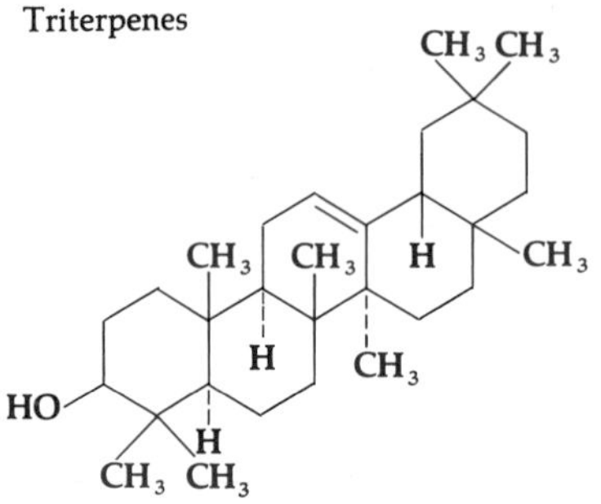

β-Amyrin

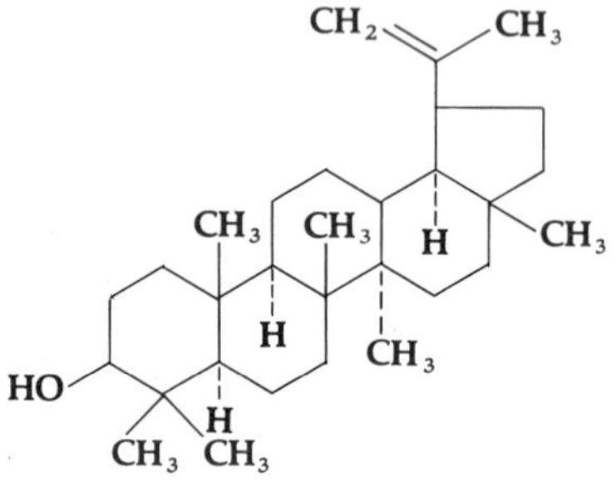

Lupeol

Tetraterpene

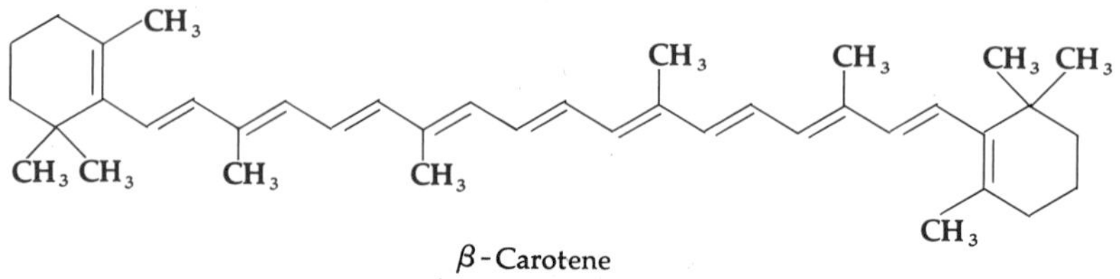

β-Carotene
(carrot pigment)

FIGURE 27-4 Representative higher terpenes

The **steroids** are a group of tetracyclic molecules, ubiquitous in both plants and animals, that have been much studied. The two examples shown indicate how similar they appear to higher terpenes in that they show polycyclic aliphatic structures with several methyl substituents. However, they rarely possess an even 30 carbons; cholesterol has only 27 carbons.

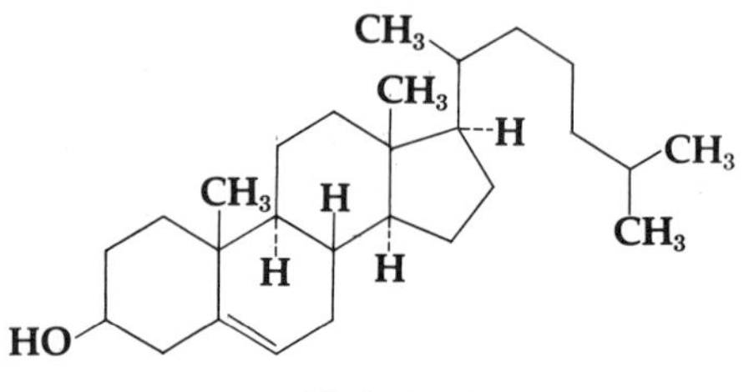

Cholesterol

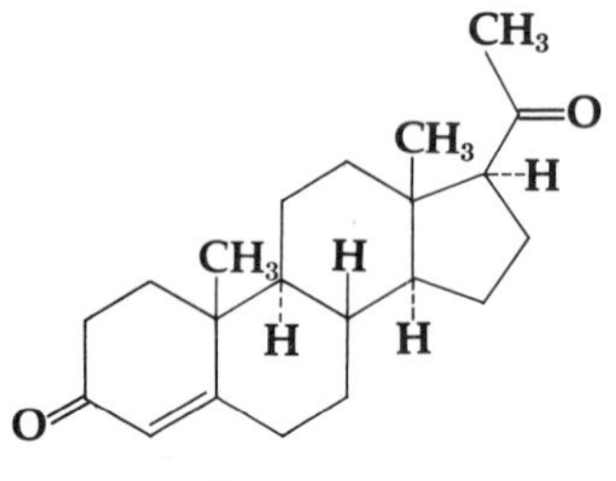

Progesterone

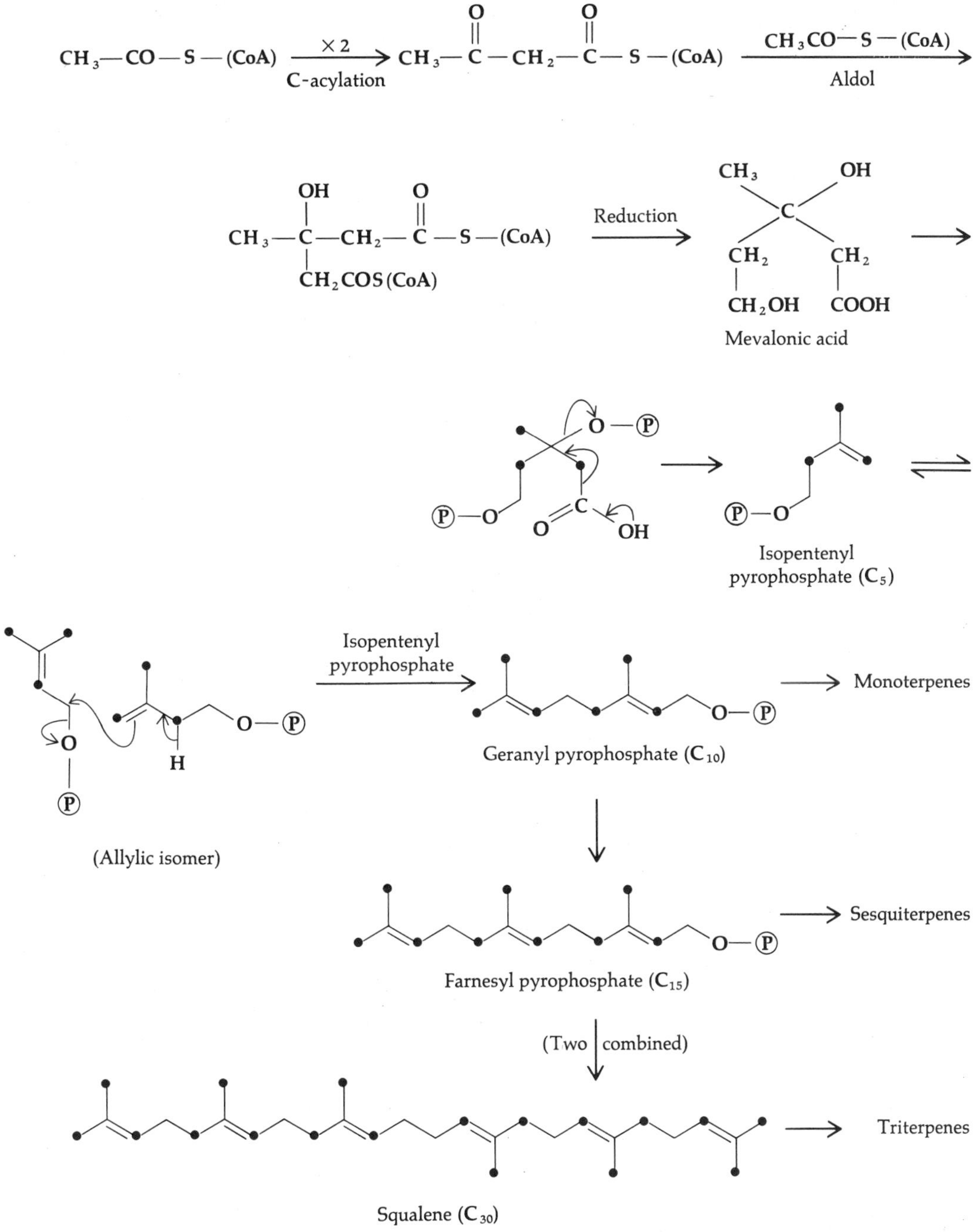

$$CH_3-CO-S-(CoA) \xrightarrow[\text{C-acylation}]{\times 2} CH_3-\overset{O}{\overset{||}{C}}-CH_2-\overset{O}{\overset{||}{C}}-S-(CoA) \xrightarrow[\text{Aldol}]{CH_3CO-S-(CoA)}$$

$$CH_3-\underset{\underset{CH_2COS(CoA)}{|}}{\overset{\overset{OH}{|}}{C}}-CH_2-\overset{O}{\overset{||}{C}}-S-(CoA) \xrightarrow{\text{Reduction}}$$

Mevalonic acid

Isopentenyl pyrophosphate (C_5)

(Allylic isomer)

Isopentenyl pyrophosphate →

Geranyl pyrophosphate (C_{10}) → Monoterpenes

Farnesyl pyrophosphate (C_{15}) → Sesquiterpenes

(Two combined)

Squalene (C_{30}) → Triterpenes

C or ● $= C^{14}$ tracer incorporated in initial acetic acid ($C^{14}H_3COOH$) fed to the organism

$$\textcircled{P} = -\overset{O}{\overset{||}{P}}-O-\overset{O}{\overset{||}{P}}-OH \quad \text{(pyrophosphate)}$$

S—(CoA) = coenzyme A

FIGURE 27-5 **Biosynthesis of terpenes**

In fact the steroids are merely modified triterpenes and all arise by a remarkable cyclization of squalene epoxide followed by a concerted series of four rearrangements, all of which are trans diaxial on the ring system for facile migration. This biosynthesis of cholesterol has been extensively studied. When $CH_3C^{14}OOH$ or $C^{14}H_3COOH$ is fed to experimental animals, the cholesterol produced in the tissues of those animals is labeled with C^{14}. The origin (in terms of CH_3 and $COOH$ groups of acetic acid) of all the carbon atoms of cholesterol has been determined by chemical degradations of the radioactive product steroid in such a way as to isolate specific single carbons and measure their radioactivity. The specific pathway whereby the 2-carbon molecule is converted to the complex 27-carbon substance has been investigated in great detail. The methyl-labeled (C^{14}) squalene of Fig. 27-5 is followed through the biosynthesis of cholesterol below; the C^{14} radioactivity from labeled acetate appears *only* in the marked carbons of the cholesterol.

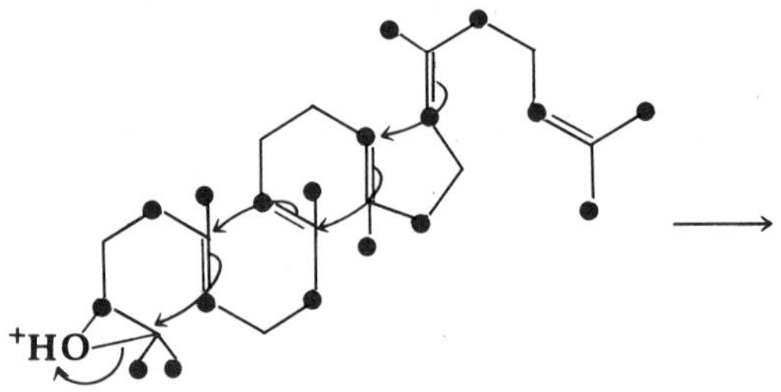

Squalene oxide
(C^{14}-labeled at ●)

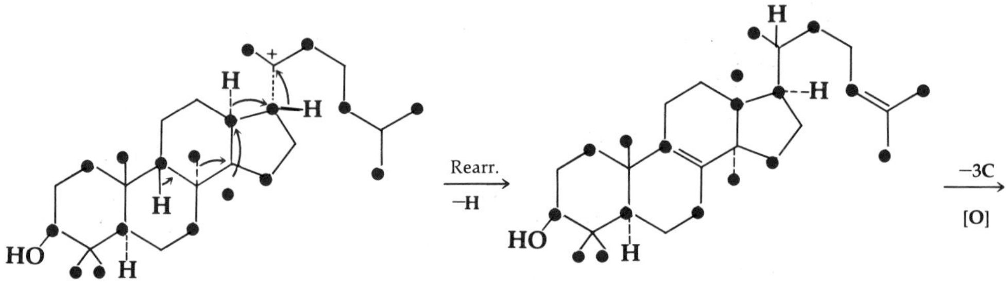

Lanosterol
(high concentration
in wool fat)

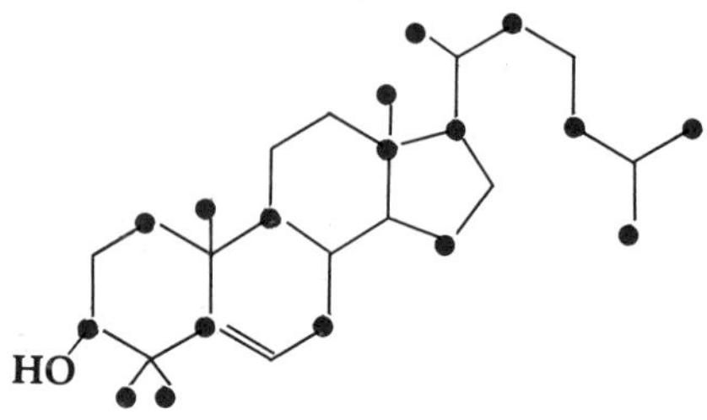

Cholesterol
(source of other steroids)

Alkaloids

The widest structural variety and the greatest number of individual representatives among natural products are to be found in the alkaloids. In fact these thousands of natural products are only grouped together by their common possession of a basic nitrogen atom. The name **alkaloid** refers to their basic (alkaline) properties such as solubility and salt formation in aqueous acid. Nevertheless, the major source for the biosynthesis of alkaloids is just a few amino acids, and one major reaction—a Mannich reaction—is the common pathway for building the many diverse skeletons of the alkaloids. This biosynthetic route was deduced over half a century ago simply from examination of known alkaloid structures.

The Mannich reaction can be generalized as the linking of a carbanion site (enolate or phenolate) with an aldehyde and an amine.

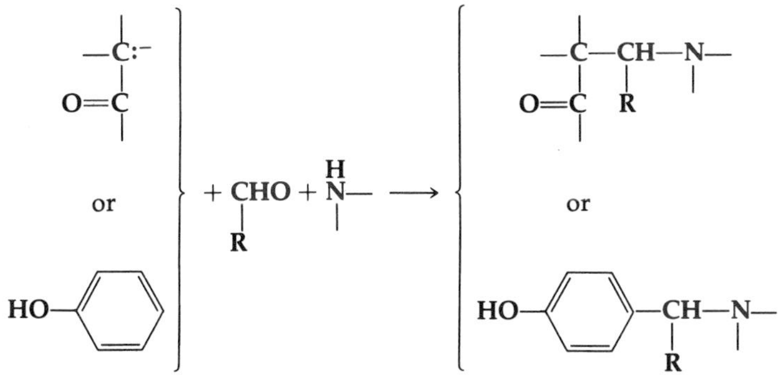

$$\left.\begin{array}{c} -\overset{|}{\underset{|}{C}}:^{-} \\[2pt] O{=}\overset{|}{\underset{|}{C}} \\[10pt] \text{or} \\[10pt] HO{-}\!\!\bigcirc \end{array}\right\} + CHO + \overset{H}{\underset{R}{N{-}}} \longrightarrow \left\{\begin{array}{c} -\overset{|}{\underset{|}{C}}{-}CH{-}N{-} \\ O{=}\overset{|}{C}\;\;R \\[10pt] \text{or} \\[10pt] HO{-}\!\!\bigcirc\!\!{-}\underset{R}{CH}{-}N{-} \end{array}\right.$$

Both the amine and the aldehyde components required can be obtained from amino acids, by decarboxylation (page 1039) or oxidative decarboxylation.

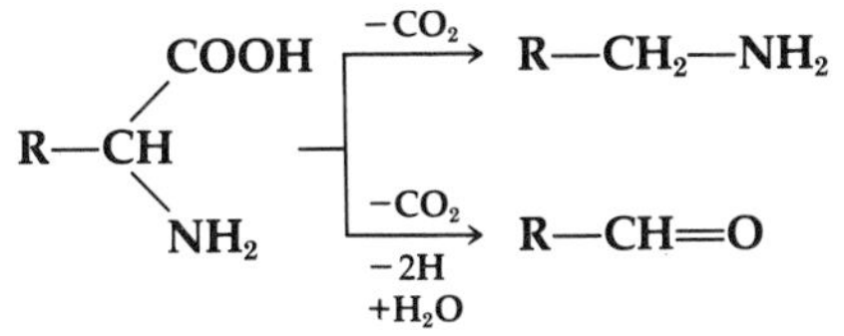

$$\begin{array}{c} \underset{R{-}CH}{\overset{COOH}{\diagup}}\!\!\diagdown_{NH_2} \left[\begin{array}{c} \xrightarrow{-CO_2} R{-}CH_2{-}NH_2 \\[12pt] \xrightarrow[\substack{-2H \\ +H_2O}]{-CO_2} R{-}CH{=}O \end{array}\right. \end{array}$$

In a simple example in which lysine affords both aldehyde and amine and the anionic center is acetoacetate, the following biosynthesis of methylisopelletierine is presumed.

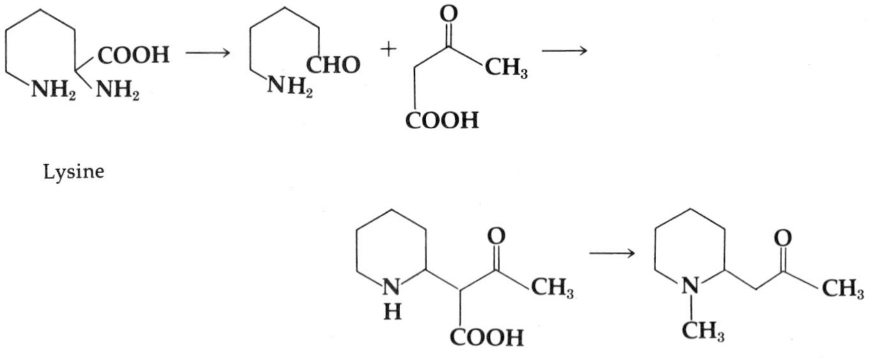

Lysine

Methylisopelletierine

The comparable case for phenols is the biosynthesis of the opium alkaloids from phenylalanine, discussed in more detail in Sec. 27-6.

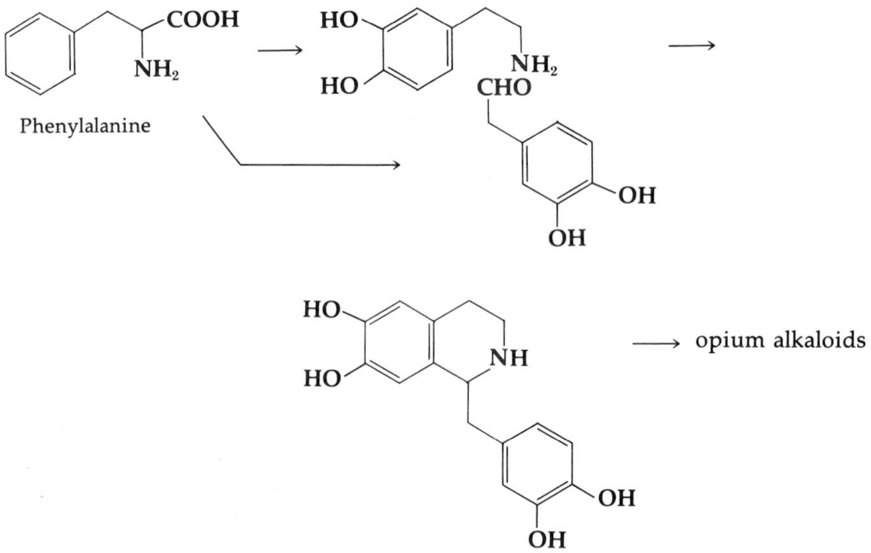

Phenylalanine

→ opium alkaloids

Norlaudanosine

Oxidative coupling of phenol rings is a frequent variant after forging the skeleton in a key Mannich reaction. Methylation of —OH or >NH is also widespread. Both of these features are at work in the further elaboration of the family of opium alkaloids, of which thebaine is an example. The biosynthesis of another (morphine) is discussed on page 1119.

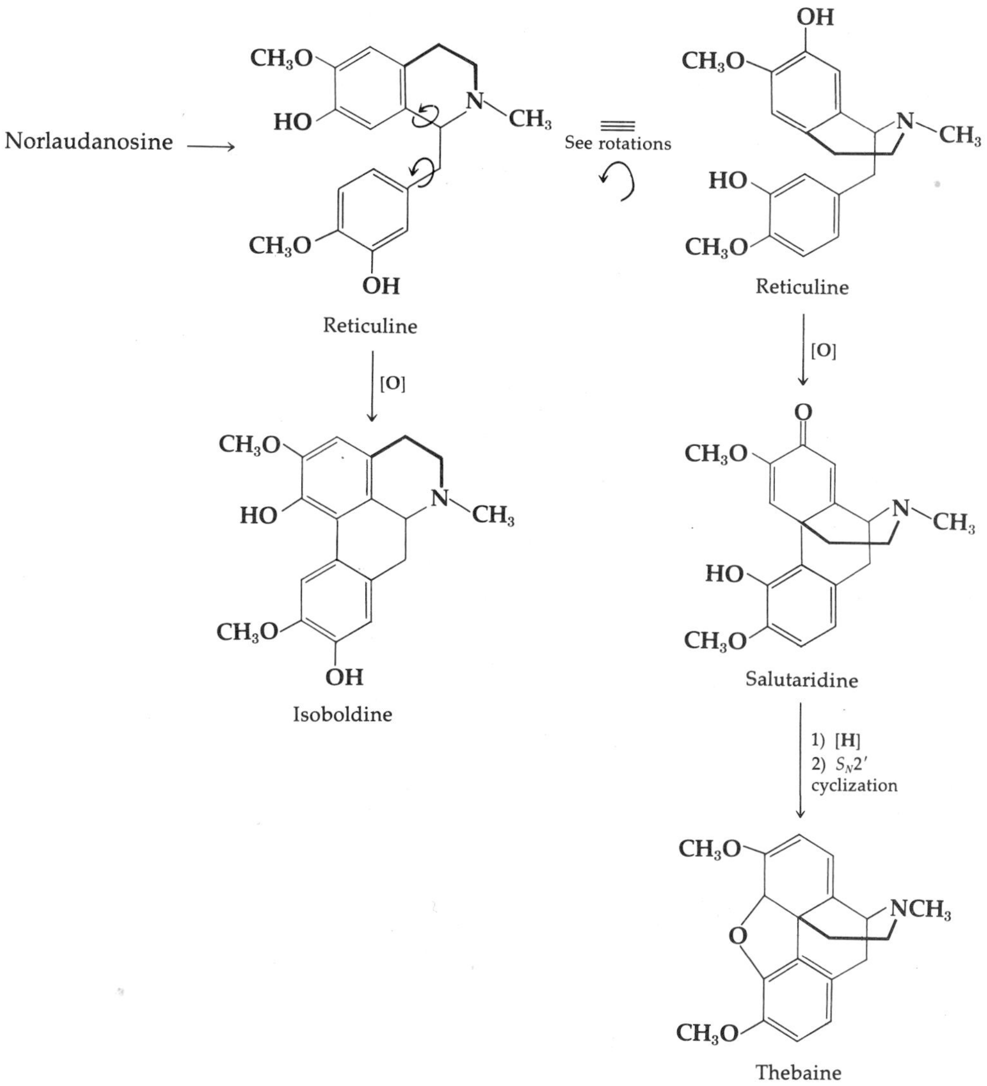
Norlaudanosine
CH3O
HO
N
CH3
CH3O
OH
Reticuline
See rotations
OH
CH3O
N
CH3
HO
CH3O
Reticuline
[O]
[O]
CH3O
HO
N
CH3
CH3O
OH
Isoboldine
O
CH3O
N
CH3
HO
CH3O
Salutaridine
1) [H]
2) SN2′
cyclization
CH3O
NCH3
O
CH3O
Thebaine

A group of representative alkaloids is shown in Fig. 27-6, which illustrates the striking richness of structural diversity which they exhibit. There are over 600 known indole alkaloids all derived from tryptophan. The other 9 to 10 carbons of most of these actually arise from a monoterpene precursor.

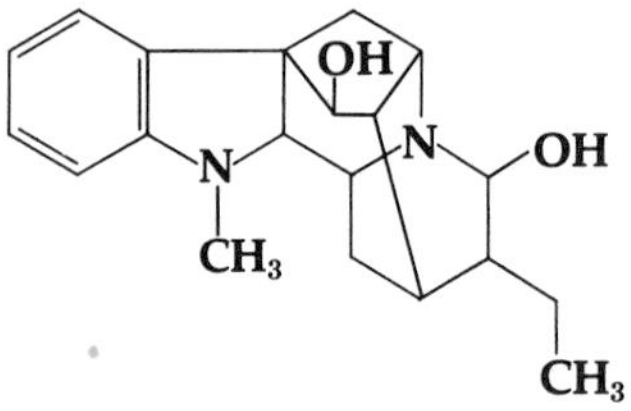

Ajmaline
(*Rauwolfia serpentina*)

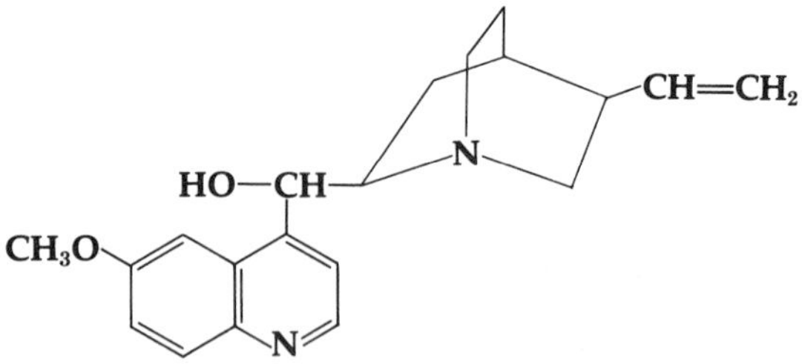

Quinine
(*Cinchona* tree),
antimalarial

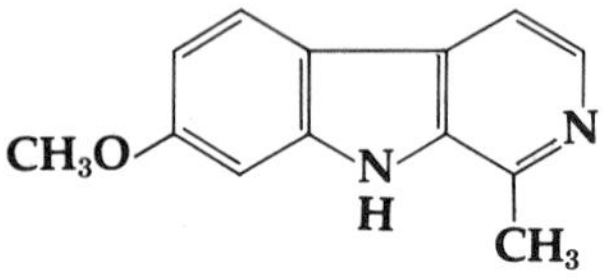

Harmine
(South American liana
extract is native
drug "yage")

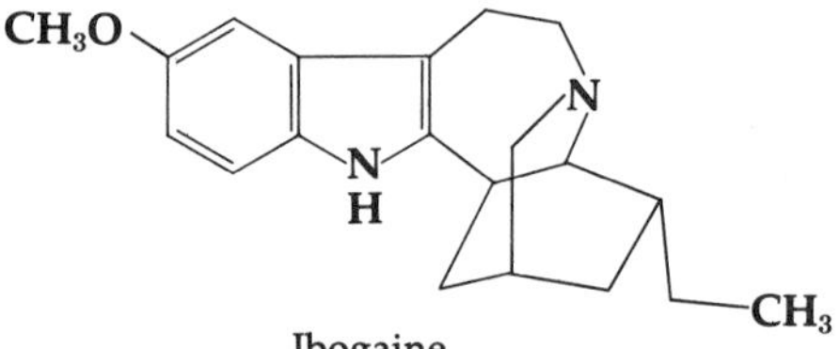

Ibogaine
(root bark of African
medicinal plant)

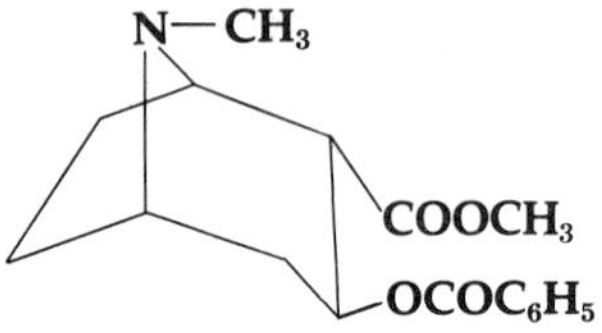

Cocaine
(coca leaves),
powerful anesthetic,
produces fever, stimulant

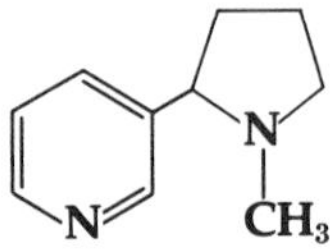

Nicotine
(tobacco plant),
stimulant, increases blood
pressure

FIGURE 27-6 **Representative alkaloids**

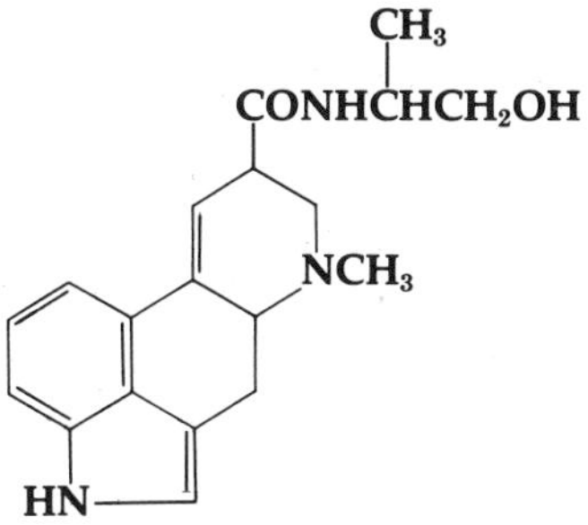

Ergonovine
(parasitic fungus),
uterotonic activity

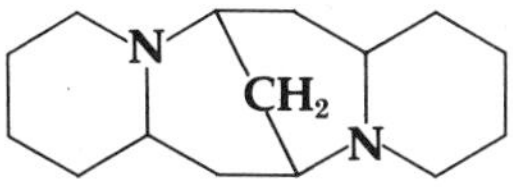

Sparteine
(e.g., *Lupinus arboreus*),
toxic

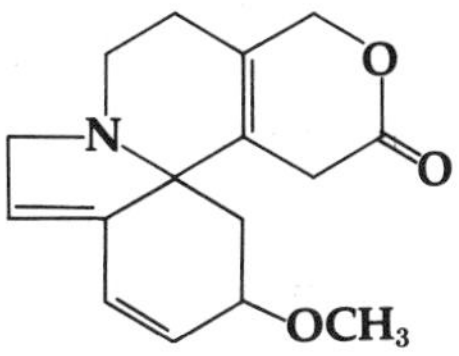

β-Erythroidine
(*Erythrina* genus),
curarizing agent

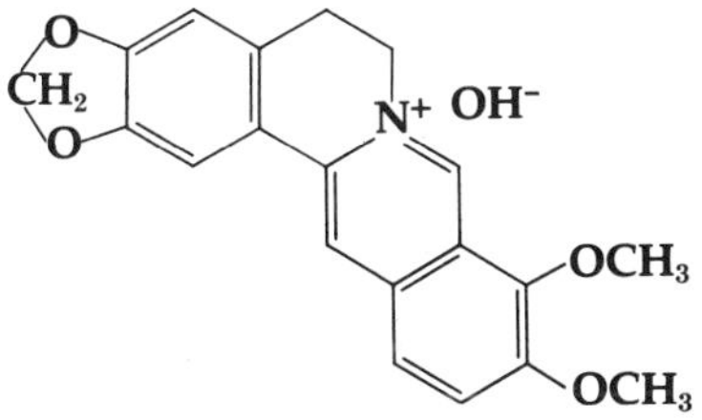

Berberine
(barberry root),
paralytic effect,
increases blood
pressure

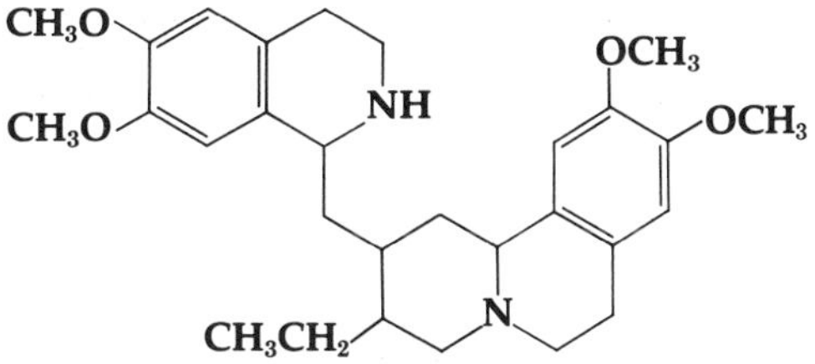

Emetine
(South American creeping
plant), emetic used in
amebic dysentery

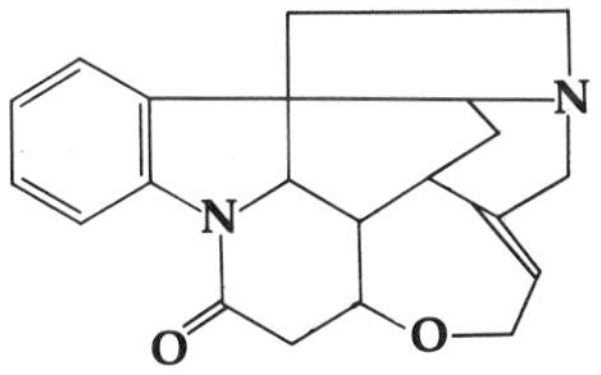

Strychnine
(*Strychnos nux-vomica*),
poison, used to exter-
minate pests

PROBLEM 27-3

Show the steps in the biosyntheses of these compounds from acetate and/or other sources that may be noted.

a   Angustifolionol (Fig. 27-2)

b   Alternariol (Fig. 27-2)

c   Griseofulvin (page 1095)

d   Sparteine (Fig. 27-6), from lysine

e   Cularine, from tyrosine

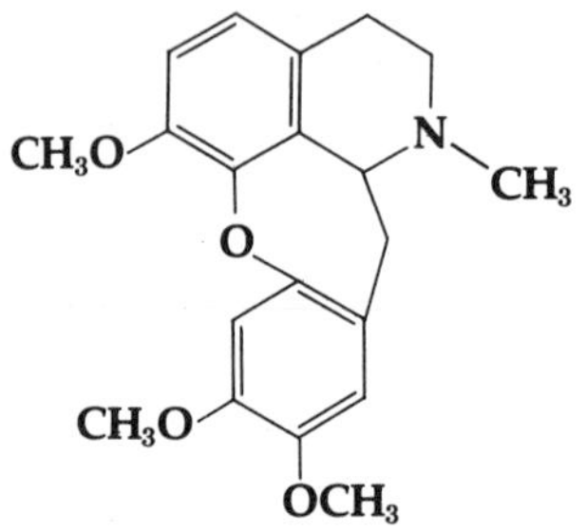

f   Atrovenetin

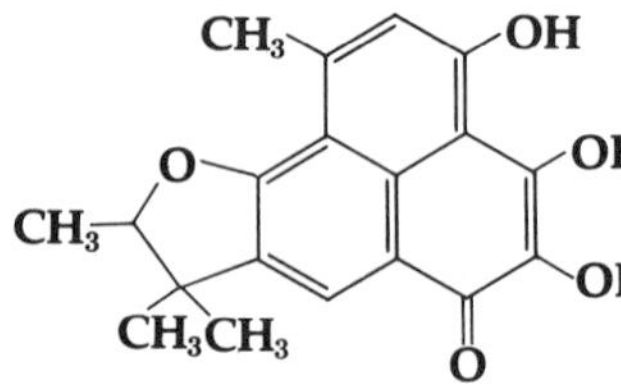

g   Pterostilbene, from tyrosine, etc.

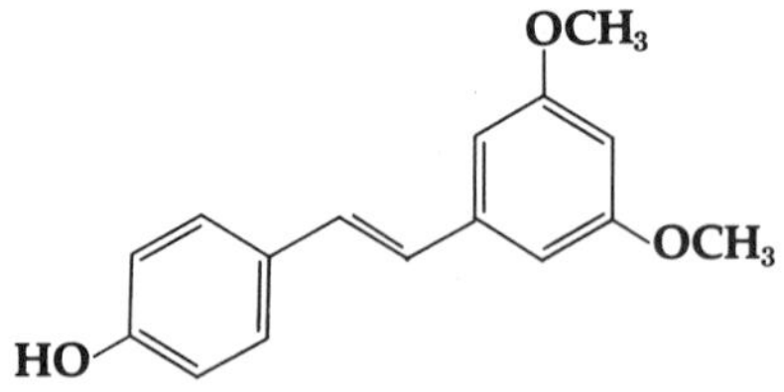

PROBLEM 27-4

One of the original triumphs of the "acetate hypothesis" was Birch's correction of the published structure of eleutherinol, shown below. He revised the locations of the same functions shown into an isomer which was in accord with polyacetyl biosynthesis, and then proved that this isomer had the correct structure. Can you devise the correct isomeric structure?

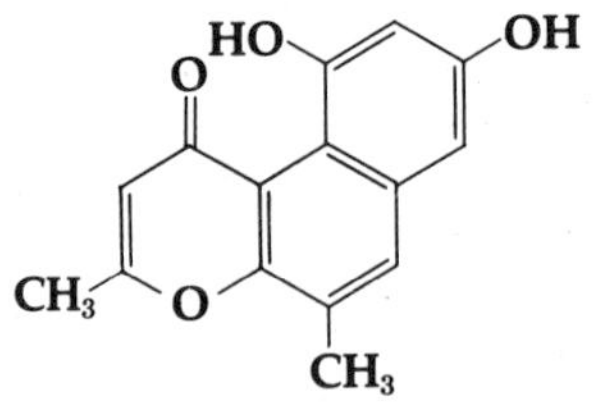

Incorrect structure of eleutherinol

PROBLEM 27-5

Show a reasonable biosynthetic pathway for these terpenes:

a   Menthol from geraniol

b   Cadinene from farnesol

27-3 SUGARS

The term **carbohydrate** arose historically from the observation that a group of compounds isolated from natural sources possessed molecular formulas that could be fitted to the general formula $C_x(H_2O)_y$. After the structures of these compounds were elucidated, many other substances were discovered whose constitution placed them within the carbohydrate family but whose molecular formulas were in conflict with the implications of the term. Carbohydrates are now classified as polyhydroxylated compounds, many of which contain aldehydic or ketonic groups or yield such groups on hydrolysis.

Simple carbohydrates are referred to as sugars, or saccharides, because they are sweet to the taste. Usually their names end in **-ose**. Sugars are classified as **monosaccharides**, **oligosaccharides**, or **polysaccharides**, depending on the number of simple sugar units linked together in the molecule (Sec. 25-3). Monosaccharide units usually consist of chains of five or six carbon atoms and are called **pentoses** and **hexoses**, respectively. If monosaccharides contain an aldehyde function, they are classified as **aldoses**; if a ketonic group, as **ketoses**. Thus, a monosaccharide might be an **aldopentose**, **aldohexose**, **ketopentose**, or **ketohexose**. Monosaccharides that contain from three to eight carbon atoms are found in nature.

The present section focusses on the chemistry of the monosaccharides, while the polysaccharides are discussed as natural polymers in Sec. 25-3. Sugars are unlike the other natural products of this chapter since they do serve a central metabolic function as a source of food for energy production and biosynthesis. The metabolic functions of sugars are a part of Chap. 26, chiefly discussed in Sec. 26-4.

Structures and Configurations of Glucose and Other Monosaccharides

Many of the principles associated with the simple sugars, and many of their reactions, can be illustrated by an argument for the structure of the commonest aldohexose, glucose. This substance is readily obtained by acid hydrolysis of starch, cellulose, cane sugar (sucrose), and a host of other natural products. Glucose, in free or combined state, is one of the most plentiful of all organic compounds.

The molecular formula of glucose is $C_6H_{12}O_6$. The gross structure of the sugar is established by the following facts:

1 Reduction of the substance with hydrogen iodide and red phosphorus gives n-hexane and reveals an unbranched chain of six carbon atoms.
2 Glucose reacts with reagents such as hydroxylamine and phenylhydrazine, which are used to characterize aldehydes and ketones (page 469).
3 Oxidation of the compound with bromine water gives gluconic acid ($C_5H_{11}O_5COOH$), a monocarboxylic acid. This fact and fact **2** establish that glucose is an aldehyde.
4 Reduction of the aldehyde with sodium amalgam gives sorbitol, $C_6H_{14}O_6$, which upon acetylation gives a hexaacetate.

Thus, sorbitol must contain six hydroxyl groups, one due to reduction of the aldehyde and five originating in glucose. Each of the six carbon atoms of sorbitol would appear to be linked to one hydroxyl group.† If the structural formula of glucose is assigned as $HOCH_2(CHOH)_4CHO$, all the reactions can be readily formulated.

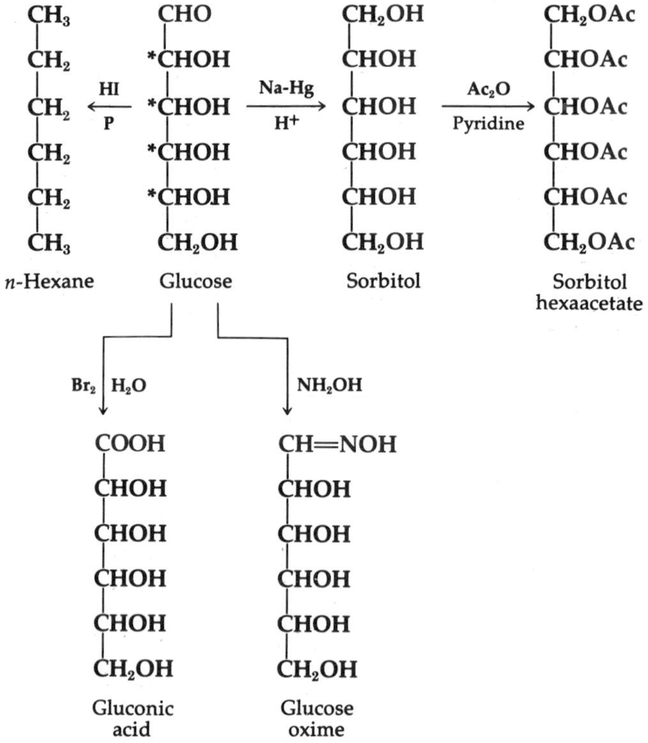

CH_3		CHO		CH_2OH		CH_2OAc
CH_2		*CHOH		CHOH		CHOAc
CH_2	$\xleftarrow[P]{HI}$	*CHOH	$\xrightarrow[H^+]{Na-Hg}$	CHOH	$\xrightarrow[Pyridine]{Ac_2O}$	CHOAc
CH_2		*CHOH		CHOH		CHOAc
CH_2		*CHOH		CHOH		CHOAc
CH_3		CH_2OH		CH_2OH		CH_2OAc
n-Hexane		Glucose		Sorbitol		Sorbitol hexaacetate

$\text{Br}_2 \mid \text{H}_2\text{O}$ NH_2OH

COOH	CH=NOH
CHOH	CHOH
CHOH	CHOH
CHOH	CHOH
CHOH	CHOH
CH_2OH	CH_2OH
Gluconic acid	Glucose oxime

Four asymmetric centers are found in the glucose molecule, and the number of possible stereoisomeric structures is 2^4, or 16. The task of relating the configurations of these four asymmetric centers to one another was completed by Emil Fischer in 1896, and the essential portions of the work in slightly modified form are reproduced here. Although the work was done long before most of the modern methods now available, it still stands as a classic of clear structural reasoning.

---

† Examples have been given of compounds that contain two hydroxyl groups per carbon atom. In each case, however, strong electron-withdrawing groups occupy adjacent carbon atoms, as in chloral hydrate and ninhydrin.

$CCl_3CH(OH)_2$

Chloral hydrate

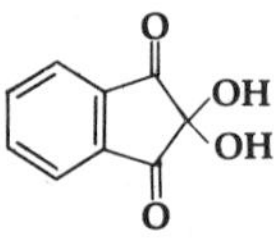

Ninhydrin

At that time, three aldohexoses, (+)-glucose, (+)-mannose, and (+)-gulose, were known, together with the aldopentose, (−)-arabinose, and the ketohexose, (−)-fructose. The gross structures of these substances were known, but their relative configurations were not established. Three general reaction sequences were available for interrelating and characterizing monosaccharides.

1 *Osazone formation.* Compounds containing the group **CHOHCHO**, when treated with three moles of hydrazine or phenylhydrazine, produce 1,2-*bis*-hydrazones, known as **osazones**, which are usually nicely crystalline compounds. One mole of hydrazine is involved in oxidizing a hydroxyl to a carbonyl group, and the other two moles give the osazone.

$$
\begin{array}{c}
\text{CHO} \\
| \\
\text{CHOH} \\
| \\
\text{(CHOH)}_n \\
| \\
\text{CH}_2\text{OH}
\end{array}
\quad \xrightarrow[-\text{H}_2\text{O}]{\text{C}_6\text{H}_5\text{NHNH}_2}
\quad
\left[
\begin{array}{c}
\text{CH}=\text{NNHC}_6\text{H}_5 \\
| \\
\text{CHOH} \\
| \\
\text{(CHOH)}_n \\
| \\
\text{CH}_2\text{OH}
\end{array}
\right]
\quad \xrightarrow[\substack{-\text{C}_6\text{H}_5\text{NH}_2 \\ -\text{NH}_3}]{\text{C}_6\text{H}_5\text{NHNH}_2}
$$

An
aldose

$$
\left[
\begin{array}{c}
\text{CH}=\text{NNHC}_6\text{H}_5 \\
| \\
\text{C}=\text{O} \\
| \\
\text{(CHOH)}_n \\
| \\
\text{CH}_2\text{OH}
\end{array}
\right]
\quad \xrightarrow[-\text{H}_2\text{O}]{\text{C}_6\text{H}_5\text{NHNH}_2}
\quad
\begin{array}{c}
\text{CH}=\text{NNHC}_6\text{H}_5 \\
| \\
\text{C}=\text{NNHC}_6\text{H}_5 \\
| \\
\text{(CHOH)}_n \\
| \\
\text{CH}_2\text{OH}
\end{array}
$$

An
osazone

$$
\begin{array}{c}
\text{CH}_2\text{OH} \\
| \\
\text{C}=\text{O} \\
| \\
\text{(CHOH)}_n \\
| \\
\text{CH}_2\text{OH}
\end{array}
\quad \xrightarrow{3\text{C}_6\text{H}_5\text{NHNH}_2}
\quad
\begin{array}{c}
\text{CH}=\text{NNHC}_6\text{H}_5 \\
| \\
\text{C}=\text{NNHC}_6\text{H}_5 \\
| \\
\text{(CHOH)}_n \\
| \\
\text{CH}_2\text{OH}
\end{array}
$$

A ketose   An osazone

2 *Oxidation of aldoses to glycaric acids.* When treated with nitric acid, aldoses are converted to glycaric acids.

$$
\begin{array}{c}
\text{CHO} \\
| \\
\text{(CHOH)}_n \\
| \\
\text{CH}_2\text{OH}
\end{array}
\quad \xrightarrow{\text{HNO}_3}
\quad
\begin{array}{c}
\text{COOH} \\
| \\
\text{(CHOH)}_n \\
| \\
\text{COOH}
\end{array}
$$

An
aldose

A glycaric
(saccharic)
acid

3 *Chain extension.* Aldoses or ketoses when subjected to the cyanohydrin reaction (page 462) give mixtures of diastereomeric nitriles, which may be hydrolyzed to mixtures of diastereomeric carboxylic acids.

$$
\begin{array}{ccc}
\text{CHO} & & \text{CN} \qquad\qquad \text{CN} \\
| & \xrightarrow{\text{HCN}} & \text{H--C--OH} + \text{HO--C--H} \\
\text{(CHOH)}_n & & | \qquad\qquad\quad | \\
| & & \text{(CHOH)}_n \qquad \text{(CHOH)}_n \\
\text{CH}_2\text{OH} & & | \qquad\qquad\quad | \\
& & \text{CH}_2\text{OH} \qquad\quad \text{CH}_2\text{OH}
\end{array}
$$

$$
\xrightarrow{\text{H}_3\text{O}^+}
\begin{array}{c}
\text{COOH} \qquad\qquad \text{COOH} \\
\text{H--C--OH} + \text{HO--C--H} \\
| \qquad\qquad\quad | \\
\text{(CHOH)}_n \qquad \text{(CHOH)}_n \\
| \qquad\qquad\quad | \\
\text{CH}_2\text{OH} \qquad\quad \text{CH}_2\text{OH}
\end{array}
$$

Diastereomeric nitriles   Diastereomeric acids

At the time Fischer did this work, *no absolute configurations were known,* and he assumed (and by chance correctly) that the absolute configuration about carbon 5 in (+)-glucose was as written.

$$
\begin{array}{l}
^1\text{CHO} \\
^2\text{CHOH} \\
^3\text{CHOH} \\
^4\text{CHOH} \\
\text{H--}^5\text{C--OH} \\
^6\text{CH}_2\text{OH}
\end{array}
$$

(+)-Glucose
(Fischer projection,
page 211)

The compounds (+)-mannose and (+)-glucose were found to yield the same osazone. This fact demonstrates that the two compounds are *epimers* and *differ only in their configurations at carbon 2.* The same osazone was formed from (−)-fructose. This experiment shows that the latter substance has a carbonyl group at carbon 2 and that carbons 3, 4, and 5 possess configurations like those of (+)-mannose and (+)-glucose.

$$
\begin{array}{c}
^1\text{CHO} \\
^2\text{CHOH} \\
^3\text{CHOH} \\
^4\text{CHOH} \\
\text{H--}^5\text{C--OH} \\
^6\text{CH}_2\text{OH}
\end{array}
\xrightarrow{\text{3C}_6\text{H}_5\text{NHNH}_2}
\begin{array}{c}
\text{CH}{=}\text{NNHC}_6\text{H}_5 \\
\text{C}{=}\text{NNHC}_6\text{H}_5 \\
\text{CHOH} \\
\text{CHOH} \\
\text{H--C--OH} \\
\text{CH}_2\text{OH}
\end{array}
\xleftarrow{\text{3C}_6\text{N}_5\text{NHNH}_2}
\begin{array}{c}
^1\text{CH}_2\text{OH} \\
^2\text{C}{=}\text{O} \\
^3\text{CHOH} \\
^4\text{CHOH} \\
\text{H--}^5\text{C--OH} \\
^6\text{CH}_2\text{OH}
\end{array}
$$

(+)-Glucose or   Glucosazone   (−)-Fructose
(+)-mannose

Configurations about carbons 3, 4, and 5 are the same

When (+)-mannose and (+)-glucose were oxidized with nitric acid, *two different optically active* glycaric acids, *mannaric* and *glucaric acids,* were produced. Because these glycaric acids have identical terminal groups, the total number of stereoisomers is reduced to ten, two meso forms and four enantiomeric pairs (see Table 27-1). The fact that mannaric and glucaric acid are optically active indicates that neither one can correspond to structures I and II. Structure III is also eliminated, since mannaric and glucaric acid, like their parents, glucose and mannose, differ in configuration only at carbon 2. Should structure III apply to either mannaric or glucaric acid, the structure of the other would have to correspond to I or II, both of which are optically inactive. Since both mannaric and glucaric acid are optically active, they must possess two of the last three structures, IV, V, or VI.

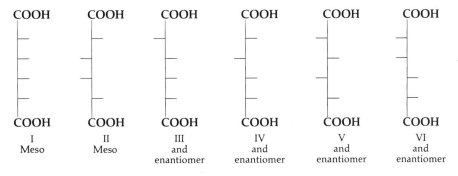

The aldopentose, (−)-arabinose, was subjected to the cyanohydrin reaction, and the cyanohydrins produced were hydrolyzed to give two glyconic acids. These substances were found to be identical with the *two glyconic acids* obtained by oxidation of (+)-glucose and (+)-mannose with bromine water. Thus, the configurations about carbons 2, 3, and 4 of (−)-arabinose must be the same as the configurations about carbons 3, 4, and 5, respectively, of (+)-glucose and (+)-mannose.

TABLE 27-1 **Glycaric Acids†**

COOH	COOH	COOH	COOH	COOH	COOH
COOH	COOH	COOH	COOH	COOH	COOH
I	II	III	IV	V	VI
Meso	Meso	and enantiomer	and enantiomer	and enantiomer	and enantiomer

†Horizontal bonds stand for positions of hydroxyl groups.

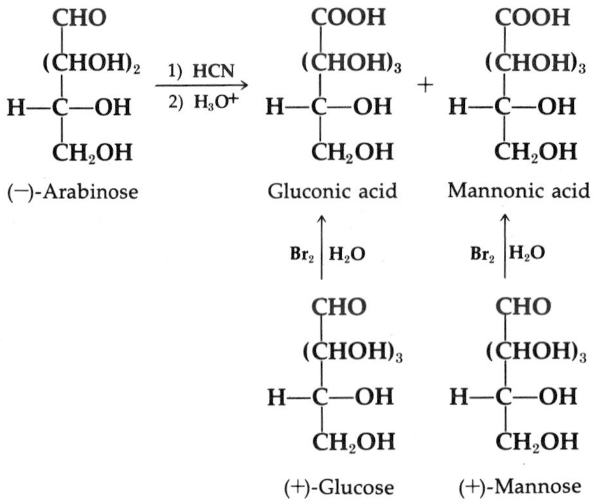

CHO   COOH   COOH

$(CHOH)_2$ 1) HCN $(CHOH)_3$ + $(CHOH)_3$

H—C—OH   2) H_3O^+ H—C—OH   H—C—OH

CH_2OH CH_2OH CH_2OH

(−)-Arabinose   Gluconic acid   Mannonic acid

                 Br_2 | H_2O Br_2 | H_2O

                 CHO   CHO

                $(CHOH)_3$ $(CHOH)_3$

               H—C—OH   H—C—OH

                CH_2OH CH_2OH

              (+)-Glucose   (+)-Mannose

When (−)-arabinose was oxidized to the corresponding dicarboxylic acid, an optically active product was obtained. The two possible meso structures for this acid were thus eliminated, and only the asymmetric structure remained.

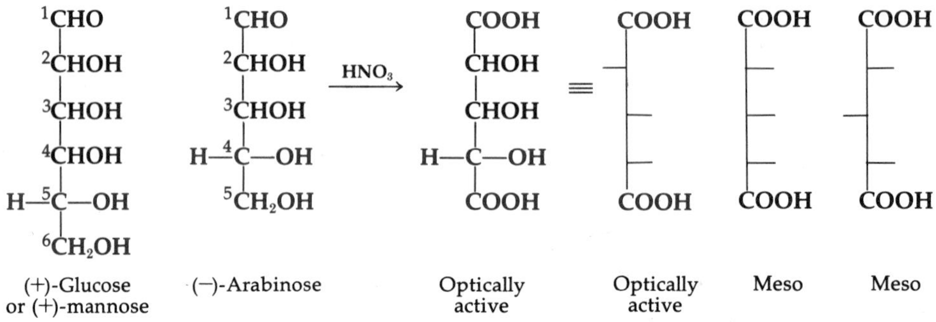

^{1}CHO ^{1}CHO   COOH   COOH   COOH   COOH

^{2}CHOH ^{2}CHOH   HNO_3   CHOH

^{3}CHOH ^{3}CHOH   CHOH   ≡

^{4}CHOH   H—^{4}C—OH   H—C—OH

H—^{5}C—OH $^5CH_2OH$ COOH   COOH   COOH   COOH

$^6CH_2OH$

(+)-Glucose   (−)-Arabinose   Optically   Optically   Meso   Meso
or (+)-mannose   active   active

This chain of interlocking configurational relationships eliminated configuration V for either glucaric or mannaric acid, and limited the configurations available for these two acids to IV and VI. Hence, only two configurations, VII and VIII, remained for (+)-glucose and (+)-mannose, and the structure of (−)-fructose was established.

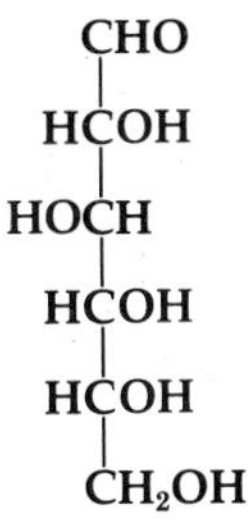

```
        CHO                 CHO                CH₂OH
      HCOH                HOCH                 C=O
      HOCH                HOCH               HOCH
      HCOH                HCOH               HCOH
      HCOH                HCOH               HCOH
      CH₂OH               CH₂OH              CH₂OH
        VII                VIII            (—)-Fructose
```

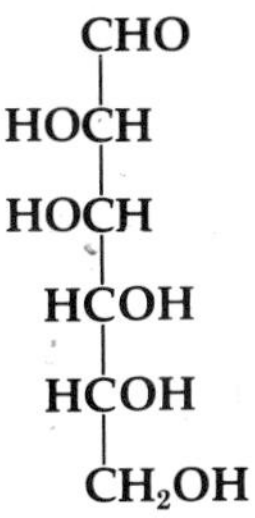

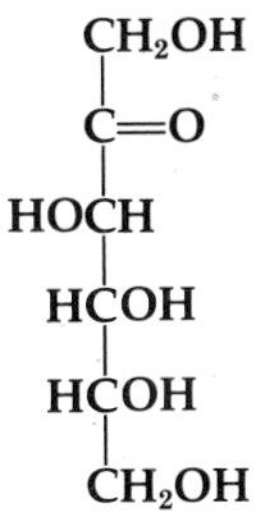

TABLE 27-2 **Configurations of the D-Aldoses**

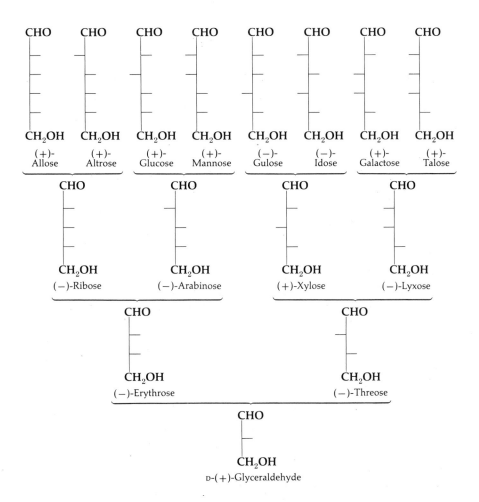

(+)-Allose   (+)-Altrose   (+)-Glucose   (+)-Mannose   (−)-Gulose   (−)-Idose   (+)-Galactose   (+)-Talose

(−)-Ribose   (−)-Arabinose   (+)-Xylose   (−)-Lyxose

(−)-Erythrose   (−)-Threose

D-(+)-Glyceraldehyde

A final experiment distinguished between configurations VII and VIII for
(+)-glucose. The third available aldohexose, (+)-gulose, was not enantiomerically related to either (+)-glucose or (+)-mannose. This substance, when
oxidized with nitric acid, gave glucaric acid, which was identical with a sample
of the substance similarly obtained from glucose. Hence, (+)-glucose and
(+)-gulose must differ only in that their aldehyde and hydroxymethylene
groups are interchanged. Such a relationship is incompatible with structure
VIII for (+)-glucose, and only structure VII is compatible with the evidence
for glucose. Thus, (+)-mannose must have structure VIII.

CHO		COOH		CH$_2$OH		CHO
HCOH		HCOH		HCOH		HOCH
HOCH	$\xrightarrow{HNO_3}$	HOCH	$\xleftarrow{HNO_3}$	HOCH	or	HOCH
HCOH		HCOH		HCOH		HCOH
HCOH		HCOH		HCOH		HOCH
CH$_2$OH		COOH		CHO		CH$_2$OH

(+)-Glucose   Glucaric acid   (+)-Gulose

With experiments of the above type and similar reasoning, investigators
have established the relative configurations of a multitude of sugars, including
all the aldoses (see Table 27-2). Since the absolute configuration of (+)-
glyceraldehyde is now established (page 221), the absolute configurations of
the compounds in Table 27-2 are known to be correct as shown.

Structures of Methyl Glucosides

When treated with hot methanolic hydrogen chloride, glucose gives two
mixed acetals at carbon 1, called methyl **glucosides**. The two methyl glucosides
differ only in their configurations at carbon 1, the new asymmetric center. These
isomers are diastereomers but are also designated as **anomers**.

Formulas written in perspective are superior for configurational designation to the flat Fischer formulas. Ring size is indicated in names by the suffix
-pyranoside, which relates to the parent heterocycle, pyran. This term is necessary, since five-membered acetal rings are also encountered, and these are
designated as **furanosides**. The class name **glycoside** has been given to mixed
acetals of aldoses, irrespective of ring size or the nature of the group attached
to the noncyclic ether oxygen. The principles of conformational analysis (Sec.
6-6) apply to the chemistry of the pyranosides. In the following formulas, the
pyranosides are written in the conformation that places the hydroxymethylene
(largest) group in an equatorial position.

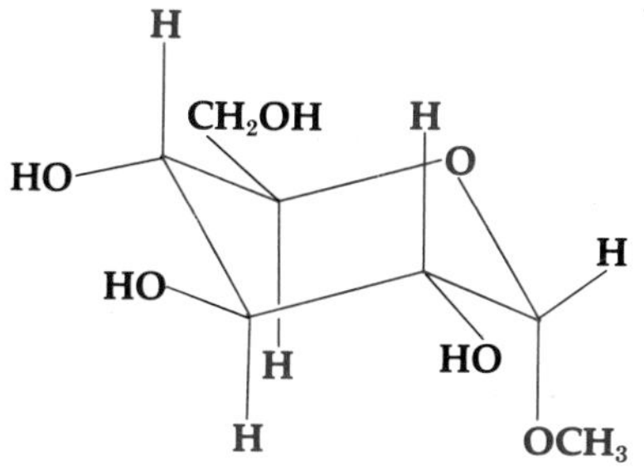

Methyl α-D-gluco-
pyranoside
(α-anomer)

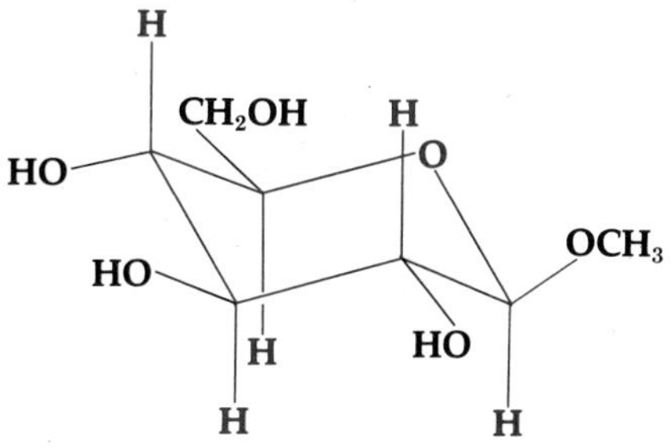

Methyl β-D-gluco-
pyranoside
(β-anomer)

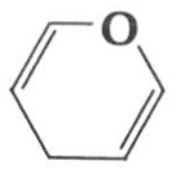

Pyran

Furan

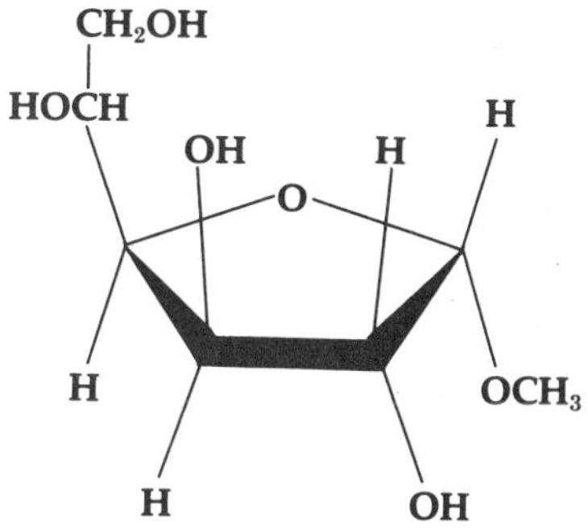

Methyl α-D-gluco-
furanoside

Although glucose itself has been represented as an aldehyde, the more stable form of the substance in either solution or the solid state contains a pyranose structure. (See pages 989 to 993.) Both the α- and β-isomers of glucose have been isolated in crystalline form. When dissolved in water, these anomers equilibrate both with the open aldehyde form and with each other. These changes are followed by observation of change in rotation with time, the equilibrium value being $[\alpha]_D + 52°$. The change in rotation that occurs when either the α- or the β-epimer goes to the equilibrium mixture is called **muta-rotation**. Configurational assignments given to the anomeric carbon in these cyclic hemiacetals and to the methylglucosides depend partly on physical measurements, partly on enzymatic reactions, and partly on chemical syntheses. Although the aldehyde is present in very small amounts at equilibrium, there is enough so that glucose undergoes aldehyde reactions, such as hydrazone formation above. The equilibrium pool rapidly replenishes the aldehyde form as it is used up in reaction.

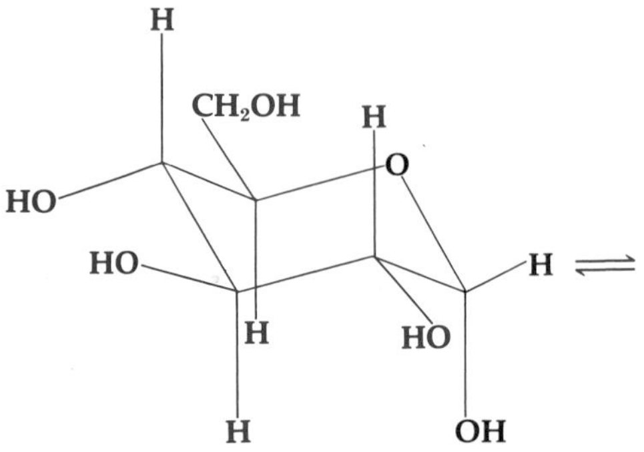

α-D-Glucopyranose
[α]D + 113°

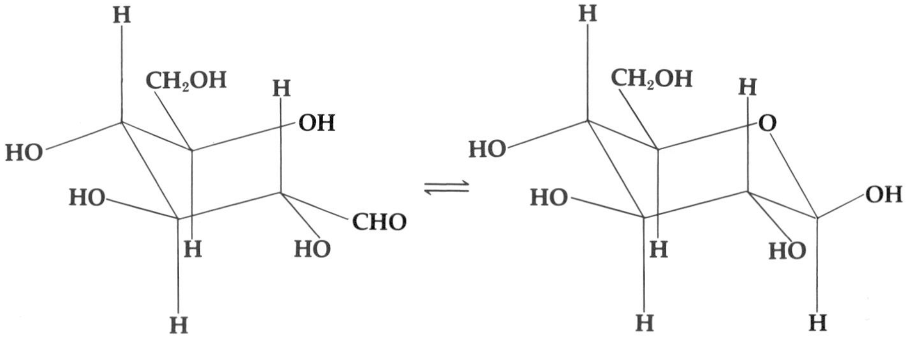

Aldehyde form

β-D-Glucopyranose
[α]D + 19°

In β-D-glucopyranose, all the bulky substituents (hydroxyl and hydroxy-methylene groups) occupy equatorial positions in one of the chair conformations and axial positions in the other. Clearly, the former arrangement is by far the more stable, a conclusion confirmed by X-ray crystal-structure studies. Among the aldohexoses, glucose is probably the most thermodynamically stable diastereomer and is also by far the most widely distributed in nature. This correlation suggests the possibility that the enzymatic reactions finally responsible for the configuration of glucose are reversible and that to some extent equilibria are reached with respect to possible configurations.

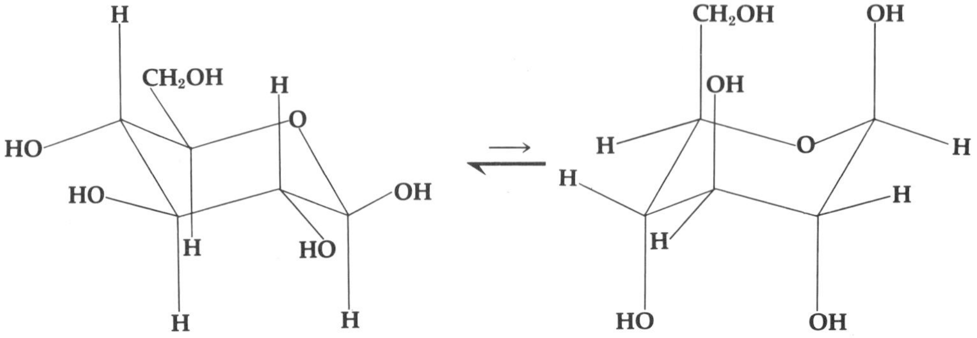

More stable conformation
of β-D-glucopyranose

Reactions of Monosaccharides

When acetylated, glucose forms the cyclic pentaacetate, α-D-glucopyranoside pentaacetate. A special method is employed to prepare the acyclic pentaacetate of glucose. The superiority of sulfur over oxygen as a nucleophile (page 393) is demonstrated by the fact that an open-chain thioacetal (unlike an acetal) of glucose can be prepared directly. This material is then acetylated, and the thioacetal group is removed by hydrolysis.

$$
\text{D-Glucose} \xrightarrow[\text{HCl}]{C_2H_5SH}
\begin{array}{c} CH(SC_2H_5)_2 \\ (CHOH)_4 \\ CH_2OH \end{array}
\xrightarrow[\text{AcONa}]{Ac_2O}
\begin{array}{c} CH(SC_2H_5)_2 \\ (CHOAc)_4 \\ CH_2OAc \end{array}
\xrightarrow[\substack{CdCO_3 \\ H_2O}]{HgCl_2}
\begin{array}{c} CHO \\ (CHOAc)_4 \\ CH_2OAc \end{array}
$$

D-Glucose diethyl mercaptal   Aldehydo-D-glucose pentaacetate

Although simple sugars are unstable in alkaline solution, the methyl glycosides are stable enough to permit formation of ethers from the hydroxyl groups by means of the Williamson synthesis (page 414), using dimethyl sulfate and alkali to obtain the methyl ethers, which are often used for structure determination owing to their relative unreactivity.

Triphenylmethyl chloride (trityl chloride) is a selective reagent for ether formation since it reacts much faster with the less sterically hindered primary hydroxyl groups than with secondary hydroxyl groups. Use of this property is made in reactions of the saccharides in which hydroxymethylene groups (CH_2OH) are selectively converted to some other groups, as in the examples of Fig. 25-10.

Carbohydrates react with such reagents as acetone or benzaldehyde and dry hydrogen chloride to form isopropylidene (ketals of acetone) or benzylidene (acetals of benzaldehyde) derivatives, respectively. In some cases these derivatives are used as protective groups, which are easily and sometimes selectively removed (dilute acid), in others for purposes of obtaining furanosides difficult to prepare by other means. The cyclic ketal or acetal formation is generally specific for cis-vicinal hydroxyls on a ring.

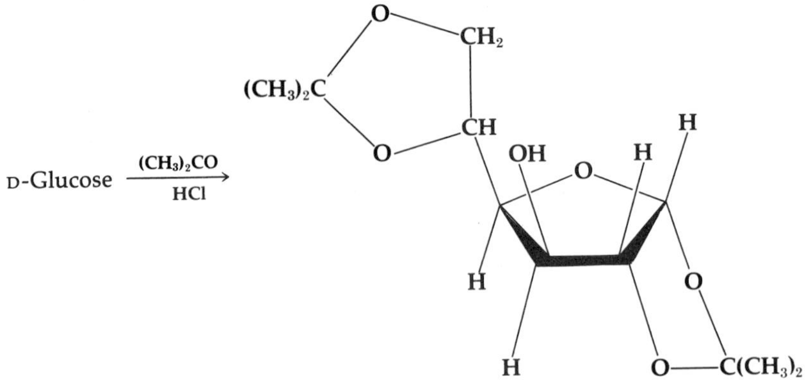

1,2:5,6-Di-*O*-isopropylidene-
α-D-glucofuranose

When treated with either strong acids or bases, the monosaccharides suffer rather profound chemical modification. In strong base, a series of reverse and forward aldol condensations occur that lead to very complicated mixtures. Similarly when formaldehyde, glycolic aldehyde, or glyceraldehyde are treated with strong alkali, complex mixtures of sugars arise from which racemic glucose has been isolated in very low yields.

PROBLEM 27-6

When a sugar is fully methylated the methyl ether at carbon 1 can be removed selectively. Explain why this is so in accounting for the following procedure used to prepare tetramethylglucose. Write out perspective formulas for the series:

$$\text{Glucose} \xrightarrow[\substack{\text{HCl} \\ \Delta}]{\text{CH}_3\text{OH}} \text{methyl glucoside} \xrightarrow[-\text{OH}]{(\text{CH}_3)_2\text{SO}_4}$$

$$\text{pentamethylglucose} \xrightarrow[\text{HCl}]{\text{H}_2\text{O}} \text{tetramethylglucose}$$

PROBLEM 27-7

Periodic acid is a very valuable reagent for distinguishing sugar derivatives. Two observations are commonly made in quantitative determinations: the number of moles of HIO_4 consumed by a mole of sugar, and the number of moles each of formaldehyde and formic acid produced.

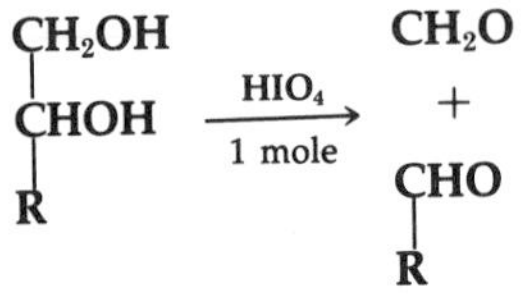

$$\begin{array}{c}\text{CH}_2\text{OH} \\ | \\ \text{CHOH} \\ | \\ \text{R}\end{array} \xrightarrow[\text{1 mole}]{\text{HIO}_4} \begin{array}{c}\text{CH}_2\text{O} \\ + \\ \text{CHO} \\ | \\ \text{R}\end{array}$$

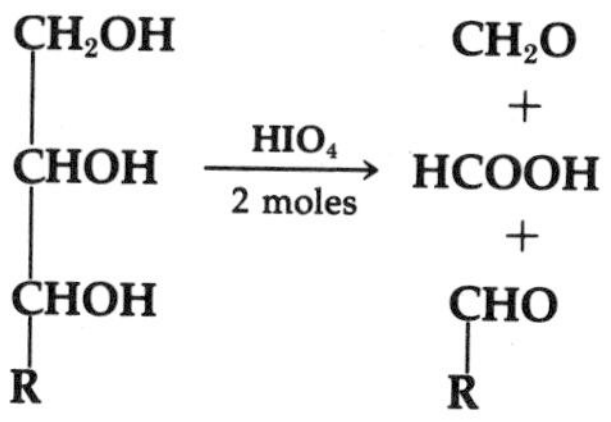

$$
\begin{array}{ccc}
\text{CH}_2\text{OH} & & \text{CH}_2\text{O} \\
| & & + \\
\text{CHOH} & \xrightarrow[\text{2 moles}]{\text{HIO}_4} & \text{HCOOH} \\
| & & + \\
\text{CHOH} & & \text{CHO} \\
| & & | \\
\text{R} & & \text{R}
\end{array}
$$

Show how the following derivatives may be distinguished by periodic acid.

a Methyl glucopyranoside

b Methyl glucofuranoside

c Tetramethyl glucose (Prob. 27-6)

d Methyl fructofuranoside

e The theoretically possible seven-membered cyclic methylglucoside

PROBLEM 27-8

Write a perspective structure for mannose as a pyranoside. How might 4-mono-ethylmannose be prepared, using the several specific protecting groups discussed in this section?

PROBLEM 27-9

Would you expect the sugar altrose (Table 27-2) to exist preferentially as a furanoside or a pyranoside? In making this decision, note that on a five-membered ring the most favored steric situation has all substituents trans. Assume for simplicity that the basic five- and six-membered rings themselves have equal strain energy.

PROBLEM 27-10

One product of treating the several aldopentoses with acid is furfuraldehyde (furan-2-carboxaldehyde), which is also a widespread natural product. Write a reasonable mechanism for this transformation and consider whether the steps might be acceptable for a biosynthetic route also.

27-4 ACETOGENINS

Most of the acetogenins are complex phenols, very often colored substances. They constitute much of the pigmentation of the natural world; flowers, autumn leaves, lichens, insects, and tropical woods all owe their colors to the presence of acetogenins. Virtually the only other major coloring matters in fact are chlorophyll (page 967) and the tetraterpenes like carotene (Fig. 27-4).

Another group of acetogenins are a collection of structurally diverse products of mold or fungus metabolism, many of which are antibiotics, like erythromycin (page 1066). Many pharmaceutical research groups grow various molds in culture media, causing them to mutate, and harvest the acetogenins they produce in order to look for medically active substances. Penicillin is such a mold metabolite (though not an acetogenin†) as are the acetogenins erythromycin, terramycin, griseofulvin, and a number of others.

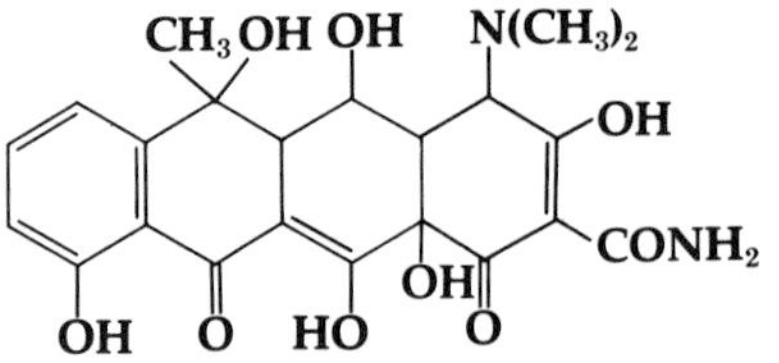

Terramycin 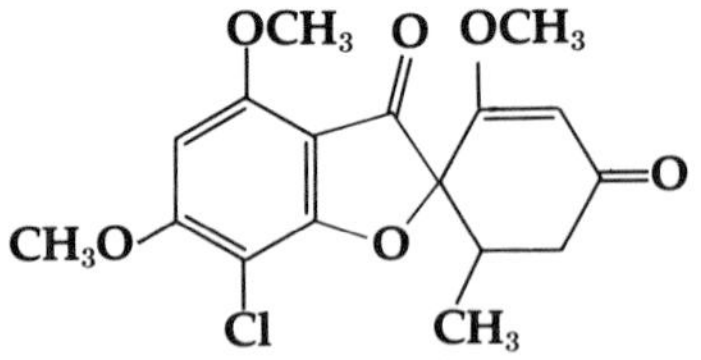   Griseofulvin

Two syntheses of griseofulvin may be examined. The first is modeled after the biosynthesis, the key step of which is an internal oxidative coupling of a phenolic benzophenone, synthesized in the laboratory by a Friedel-Crafts reaction.

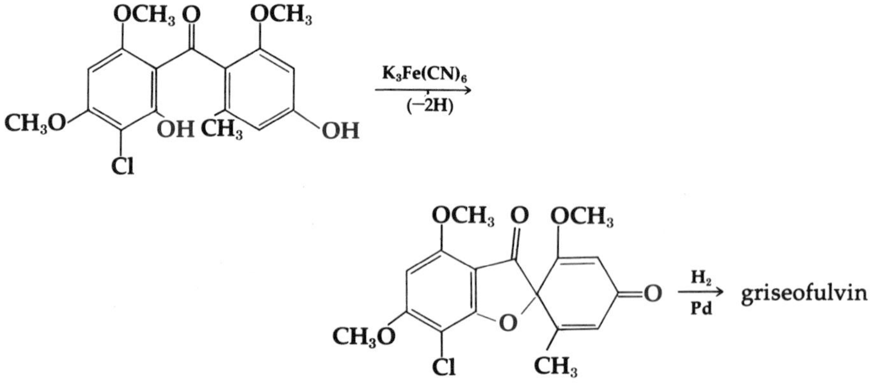

$$\xrightarrow[(-2H)]{K_3Fe(CN)_6}$$

$$\xrightarrow[Pd]{H_2}$$ griseofulvin

The second synthesis was designed to create the cyclohexenone ring in one step by a double Michael addition. Not only was this design successful, but it created the correct relative stereochemistry at the two asymmetric centers in griseofulvin.

† Penicillin is synthesized by the microorganism from amino acids and has the structure shown, in which amino acid groups can be discerned.

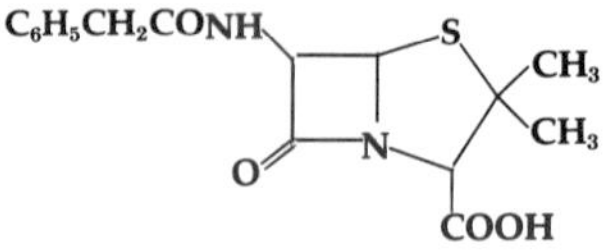

Penicillin G

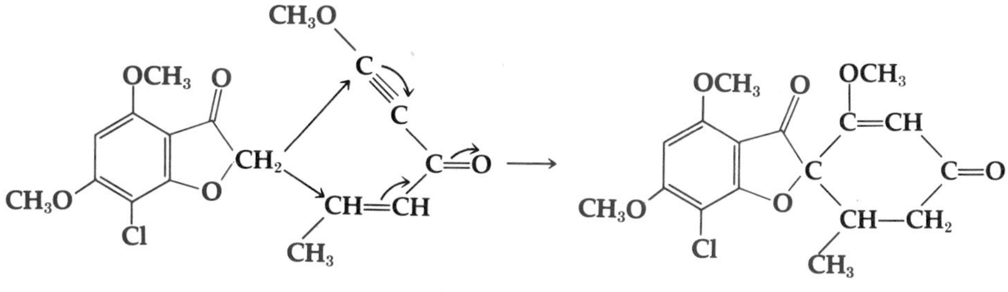

Griseofulvin

Flavonoids

A main family of some 300 flower pigments is based on the flavone skeleton, bearing hydroxyl groups located in accord with polyacetyl biosynthesis. These compounds represent a variation on polyacetyl biosynthesis, for the polyacetyl chain here is terminated in a cinnamic acid, from phenylalanine (or tyrosine), which is the source of ring B in the flavones.

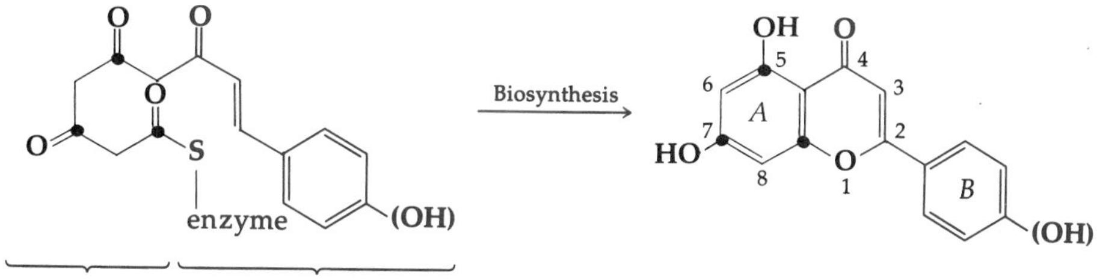

Triacetyl   Cinnamic   5,7-Dihydroxyflavone
chain   unit

Since the heterocyclic ring is a disguised β-diketone, it is cleaved by hot alkali. This has been a standard procedure for structure elucidation since it divides the molecule into two simpler and readily synthesized aromatic acids. Some phenolic hydroxyls will be found methylated in various natural flavonoids. In order to distinguish these and also to protect the phenolic rings themselves from alkaline depredation, the free phenolic groups are first ethylated. A typical procedure is shown below. The whole structure of rhamnetin may be simply deduced from its molecular formula and the structures deduced for the two product acids. The *phenolic* acid must come from ring A. Rhamnetin comes from cactus flowers.

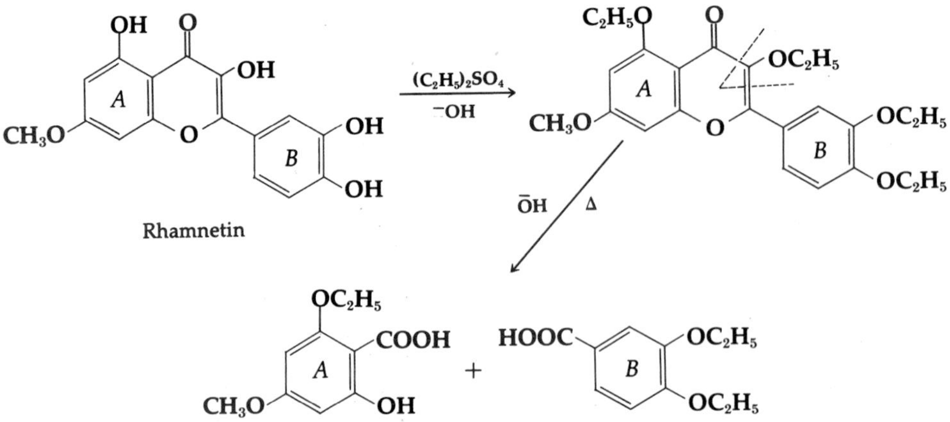

Rhamnetin

Synthesis of flavones is usually effected through condensation reactions of the type involved in the preparation of fisetin, a pigment of yellow cedar and sumac. The steps of the synthesis are recorded in Fig. 27-7.

Anthocyanins are one of the main classes of plant pigments. They occur in flowers and fruit as glycosides (often glucosides), hydrolysis of which provides colored aglycones known as **anthocyanidins**. The vivid blues and reds of anthocyanins are associated with the distribution of positive charge throughout an aryl-substituted chroman ring system.

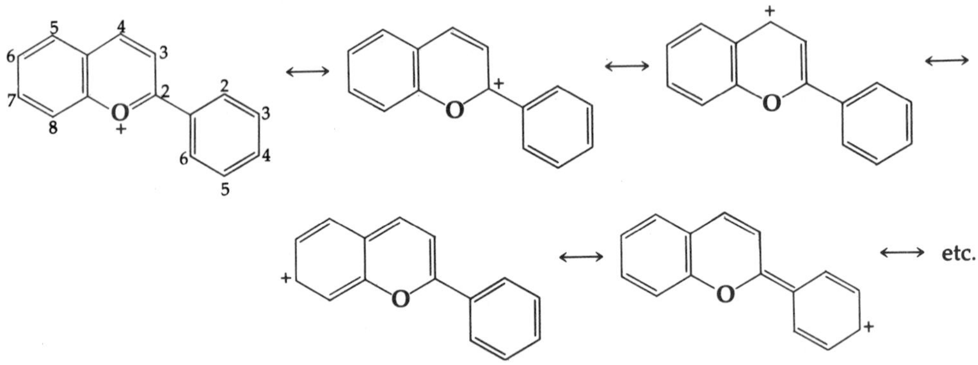

Anthocyanidins are usually isolated in the form of chloride salts and are frequently hydroxylated in the 5-, 7-, 3'-, 4'-, and 5'-positions. The character of the resonating system is seriously affected by the presence of mineral salts and by the pH of the environment. Consequently, colors of flower pigments sometimes vary markedly with type of soil. Thus, cyanin, the pigment of the red rose and blue cornflower, is pale violet in neutral solution, red in dilute acid, and blue in dilute base. The flowers themselves often change color on immersion in water at various pH values. Other anthocyanins differ only in number and position of hydroxyl groups and in the character of the sugars to which they are attached.

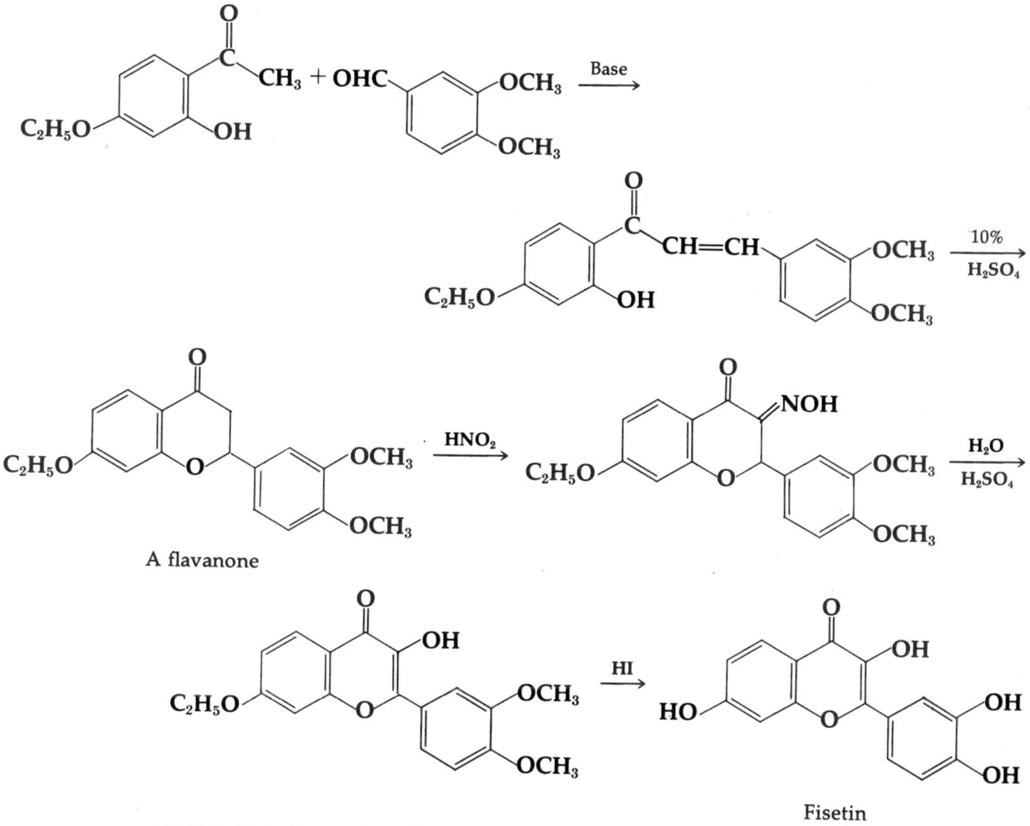

FIGURE 27-7 **Synthesis of fisetin**

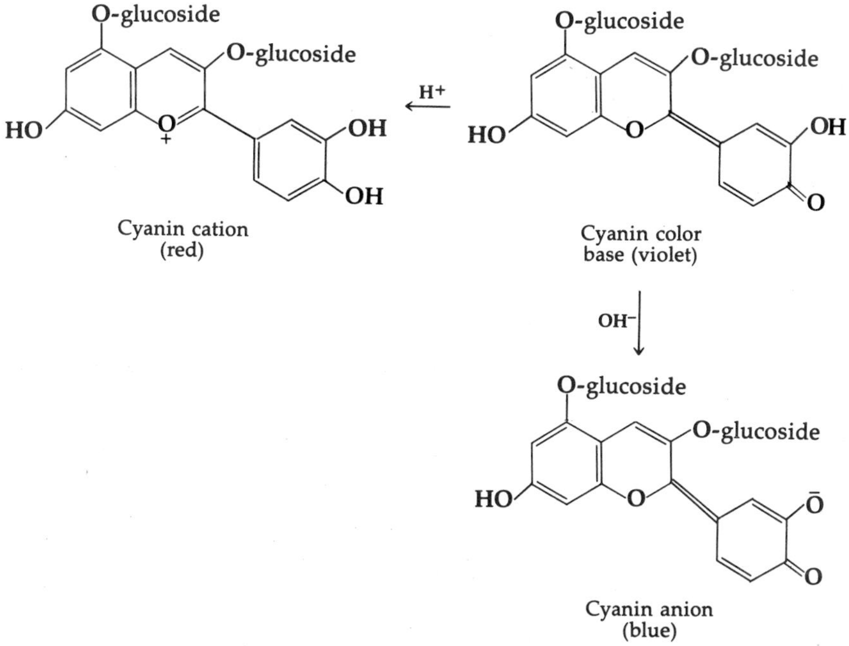

Cyanin cation
(red)

Cyanin color
base (violet)

Cyanin anion
(blue)

The synthesis of cyanin is representative of methods applied to the preparation of a number of these plant pigments.

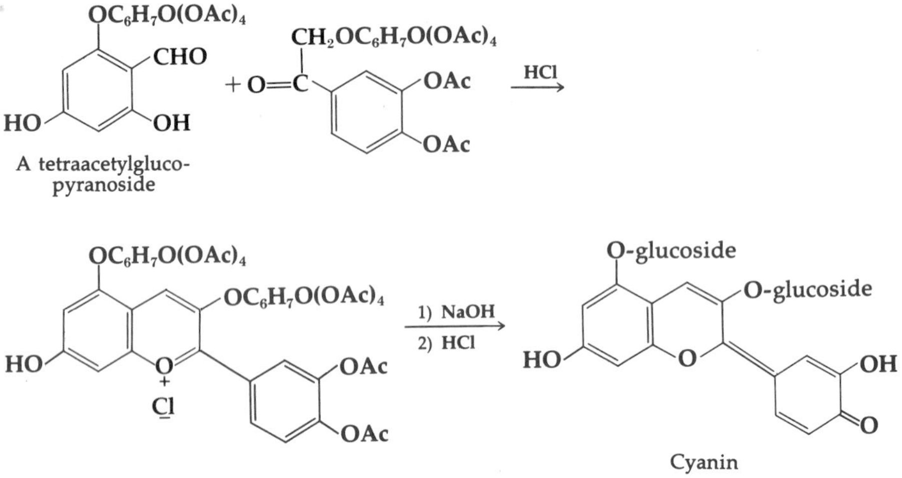

A tetraacetylgluco-
pyranoside

Cyanin

Chemical and Physical Structure Proofs

It is typical of the polycyclic phenols that they react with and are often cleaved by alkali since their structures incorporate the original poly-β-keto chain, itself an entity very labile to alkaline cleavage (page 532). The flavones represent a family in which this works very well. In other cases the variety of possible reactions of these polycarbonyl systems with base can lead to unexpected results and incorrect structural inferences.

On the other hand the power of physical methods for structure elucidation will be clear on inspecting the formula of evodionol, the structure of which was originally determined chemically by oxidation ($KMnO_4$) of the double bond followed by acidic cleavages to two fragments, a phenolic acetophenone and α-hydroxyisobutyric acid. These products were then synthesized as proof of structure. The work involved was considerable.

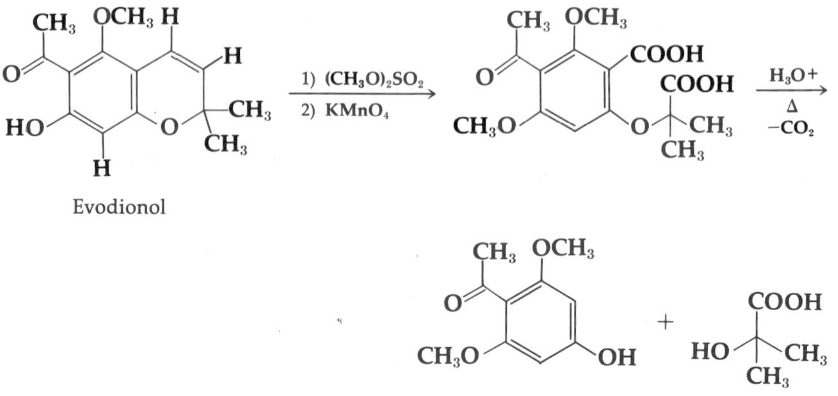

Evodionol

In a contemporary solution of this structure problem, the IR and UV spectra would show the characteristic hydrogen-bonded *o*-hydroxyacetophenone chromophore. Models would furthermore make the 2,4,6-oxygen pattern of the phloroglucinol chromophore clear just from the UV and NMR spectra. The NMR spectrum would appear roughly as shown, with all 16 protons unsplit except the olefinic pair, and all well separated into characteristic chemical shifts.

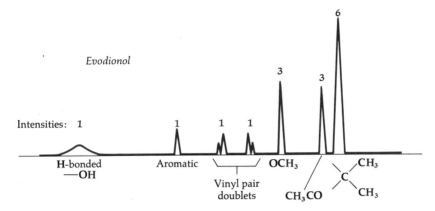

Thus this spectral solution, available in a few hours, immediately reduces the structure problem from $C_{14}H_{16}O_4$ to these three isomeric possibilities, differing only in the placement of functions:

A	B	C

PROBLEM 27-11

    a Structure *A* for evodionol, above, is rendered unlikely since it does not produce any of a new isomeric phenolic ketone (not *B* or *C*) on acid catalysis. Show why such a result would be anticipated for structure *A*.

    b Trace the steps in a *chemical* proof of structure for evodionol.

    c Evodionol is actually structure *B*. How might you distinguish between the three choices *A*, *B*, and *C*?

PROBLEM 27-12

Suggest syntheses of the starting materials used in the two syntheses shown for griseofulvin.

PROBLEM 27-13

A pigment from *Hibiscus sabdariffa* was named hibiscetin. Its formula was shown to be $C_{15}H_{10}O_9$. Full methylation with dimethyl sulfate yielded a derivative, $C_{22}H_{24}O_9$. This derivative was boiled in alkali and afforded two acids, soluble in bicarbonate. Acid *A*, $C_{10}H_{12}O_5$, yielded $C_{11}H_{14}O_5$ with diazomethane. Acid *B*, $C_{10}H_{12}O_6$, yielded $C_{12}H_{16}O_6$ with diazomethane. Acid *A* was sublimed unchanged on heating, while acid *B* yielded $C_9H_{12}O_4$, soluble in strong alkali (but not in bicarbonate), when it was heated. Armed with this information and an acquaintance with the biosynthesis of flower pigments, the investigators could write two very likely structures for hibiscetin. Can you?

PROBLEM 27-14

In some cases the events attendant on hydroxide reaction with acetogenins are more complex and lead to erroneous deductions of structure. The initial incorrect structure and the biosynthetically revised (and later proven) structure for flavosperone, a mold pigment, are illustrated. The wrong structure arose in part from the production of orcinol (3,5-dihydroxytoluene) on heating flavosperone with alkali.

a Write the biosynthetic pathway for flavosperone.

b Taking account of keto tautomers of phenols, vinylogous substitution, and other carbonyl reactions expected in alkali, deduce a pathway for the formation of orcinol in the ill-fated reaction.

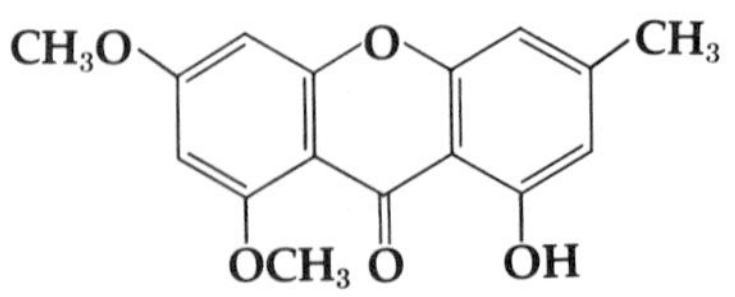

Incorrect structure 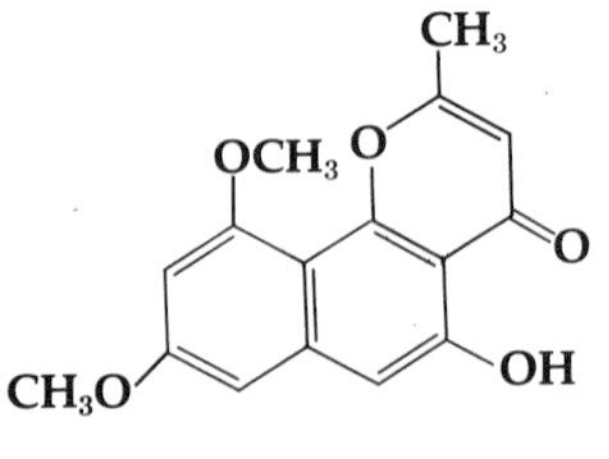   Flavosperone

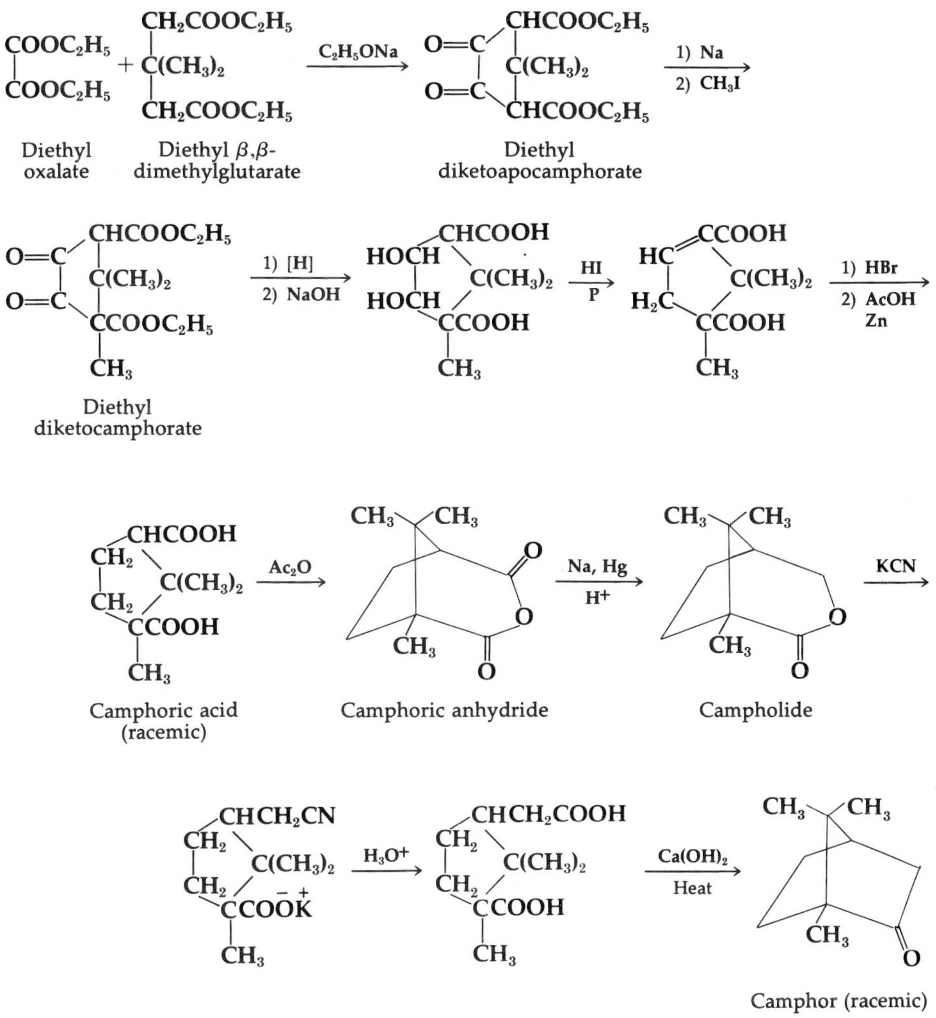

$COOC_2H_5$, $COOC_2H_5$ (Diethyl oxalate) $+$ $CH_2COOC_2H_5$, $C(CH_3)_2$, $CH_2COOC_2H_5$ (Diethyl β,β-dimethylglutarate) $\xrightarrow{C_2H_5ONa}$ Diethyl diketoapocamphorate $\xrightarrow[2)\ CH_3I]{1)\ Na}$

Diethyl diketocamphorate $\xrightarrow[2)\ NaOH]{1)\ [H]}$ (HOCH, HOCH ... $C(CH_3)_2$, CCOOH, CH_3) $\xrightarrow[P]{HI}$ (HC, H_2C ... $C(CH_3)_2$, CCOOH, CH_3) $\xrightarrow[2)\ AcOH\ Zn]{1)\ HBr}$

Camphoric acid (racemic) $\xrightarrow{Ac_2O}$ Camphoric anhydride $\xrightarrow[H^+]{Na,\ Hg}$ Campholide $\xrightarrow{KCN}$

(CHCH_2CN ... $C(CH_3)_2$, $CCOO^-K^+$, CH_3) $\xrightarrow{H_3O^+}$ (CHCH_2COOH ... $C(CH_3)_2$, CCOOH, CH_3) $\xrightarrow[Heat]{Ca(OH)_2}$ Camphor (racemic)

FIGURE 27-8   **Synthesis of camphor**

27-5   TERPENES AND STEROIDS

Monoterpenes (C_{10} Compounds)

These volatile substances (cf. Fig. 27-3) provide plants and flowers with much of their fragrance, and certain of them are used commercially in perfumes and flavors. Some of the terpenes were known in antiquity and were employed as medicines. Camphor (from the camphor tree) and α-pinene (from pines) are among the commercially important terpenes. The former is used as a plasticizer in the manufacture of celluloid and photographic film base, and the latter is the chief component of turpentine (paint thinner). The total synthesis of camphor is recorded in Fig. 27-8.

A number of open-chain terpenes, when heated with acid, undergo ring closure to give other terpenes, as in the following example:

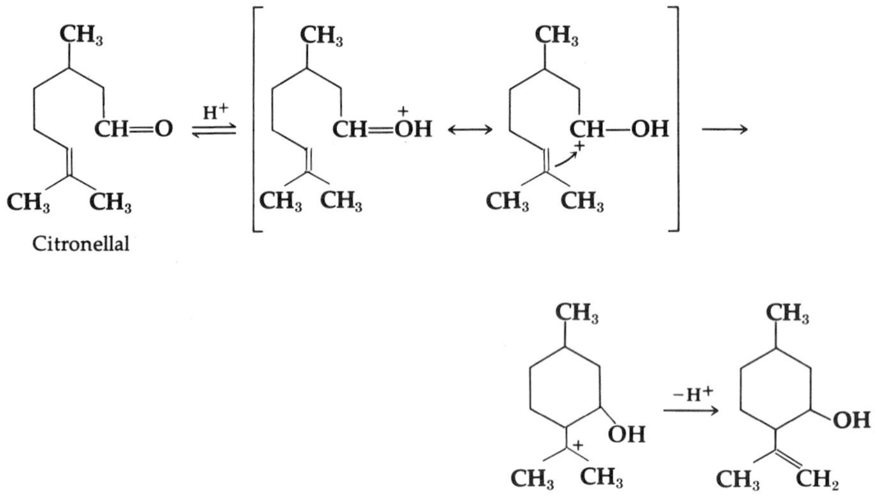

Citronellal

Isopulegol

Aromatization of the more plentiful terpenes provides such compounds as *p*-cymene, which can be oxidized to useful phenols and to terephthalic acid, for plastic manufacture. The device of aromatization was often used in early structure determination also since the resultant aromatics were easier to synthesize for confirmation.

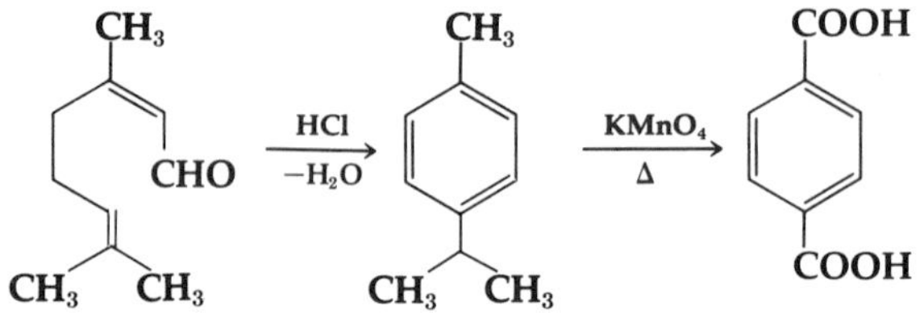

Citral   *p*-Cymene   Terephthalic acid

For many years a controversy existed over the structure of eucarvone, which had been formulated as *A* since it yielded some *cis*-3,3-dimethylcyclopropane-1,2-dicarboxylic acid on ozonolysis and also as *B* since it afforded α,α-dimethylsuccinic acid with permanganate oxidation. The structures are now recognized as readily interconvertible by a six-electron (disrotatory) electrocyclic reaction of their enols. Such a reaction apparently supervenes in certain reactions of eucarvone and leads to the not uncommon situation in which chemical methods will not serve to prove a structure. The NMR spectrum, however, clearly indicates three olefinic protons and so confirms structure *B*.

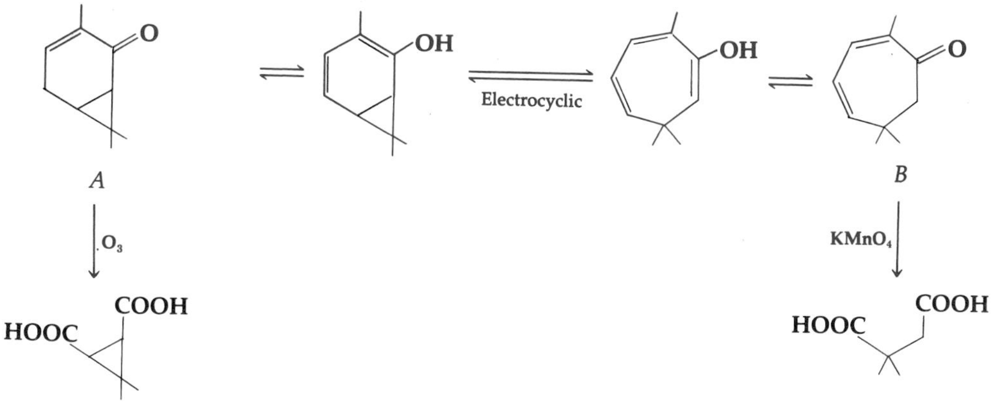

Sesquiterpenes (C$_{15}$ Compounds)

Compounds of this group are composed of three isoprene units, which are arranged in enough different ways to provide open-chain, monocyclic, bicyclic, and even tricyclic structures (Fig. 27-3). Elucidation of the structures of these substances has led not only to the discovery of new organic reactions but even to new aromatic systems (e.g., azulene). Dehydrogenation of partheniol (and other terpenes of like skeleton) with hot sulfur or palladium affords guaiazulene, a blue hydrocarbon also found naturally in geranium oil. Such dehydrogenations are widely used in structure determinations.

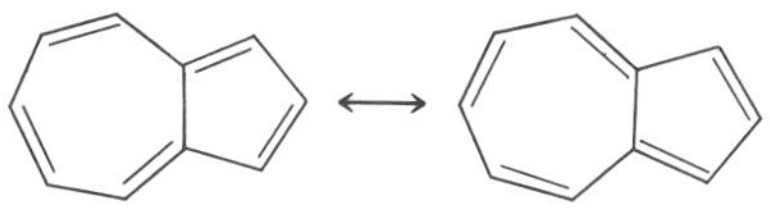

Azulene

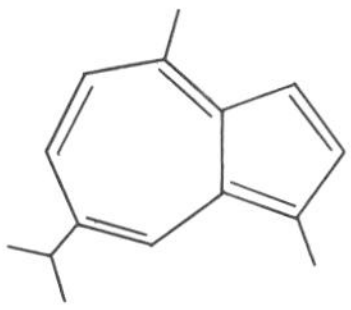

Guaiazulene

Diterpenes (C$_{20}$ Compounds)

The diterpenes (four isoprene units) can have either cyclic or acyclic structures. Phytol occurs as an ester of the porphyrin portion of the chlorophyll molecule (page 967). Vitamin A, which contains one ring and an isoprenoid side chain, is a fat-soluble substance found in fish oils, particularly in shark-liver oil. The substance is required for normal eyesight and for the growth of mammals. The light-sensitive pigments that function in photoreception in the retina of the eye are synthesized in the body from this vitamin.

Vitamin A was first synthesized in 1947 and is currently produced for use in vitamin preparations both by synthesis and by extraction from shark liver. The synthesis was given in detail earlier (page 1073) as an illustration of the repeated application of condensation reactions in a complex commercial process.

The most abundant constituent of pine rosin is abietic acid, apparently produced in the rosin from primary diterpenes like dextropimaric acid (Fig. 27-4), via methyl migration. Vigorous oxidation of optically active abietic acid yielded the optically inactive triacid shown. Since this acid is meso, there are only two possible configurations for it. The correct one, with trans carboxyl groups, was established by a comparison of its pK_a behavior with that of synthetic model acids. This simple study at once established the relative stereochemistry of three out of the four asymmetric centers in abietic acid.

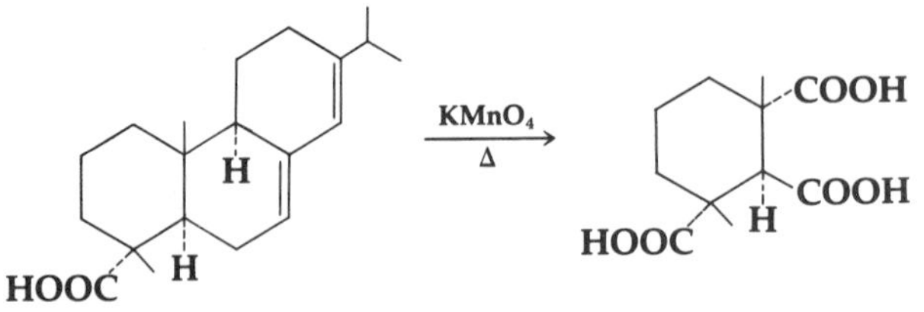

Abietic acid
$[\alpha]_D = -110°$

Meso triacid
$[\alpha]_D = 0°$

Higher Terpenes

Triterpenes are the most common terpenes in plants, usually as solids with pentacyclic structures like those of amyrin and lupeol (Fig. 27-4). All of these compounds are formed in the living cell by cyclizations of squalene epoxide, just as with steroid biosynthesis. Lanosterol (page 1074) is the common triterpene of animal origin and the source of the steroids. The different triterpenes are formed according to the conformation in which the squalene epoxide is folded when it undergoes the sequential additions that form the several rings.

The most common examples of the tetraterpenes are the carotenoids, which are pigments widely distributed in vegetables and animal fats. These substances contain long conjugated systems of double bonds, which are responsible for their color. Lycopene, which contains no rings, is the red coloring matter of tomatoes and other fruit. This substance is structurally related to the carotenes (Fig. 27-4), which provide the carrot with its characteristic orange color and are also found in green leaves.

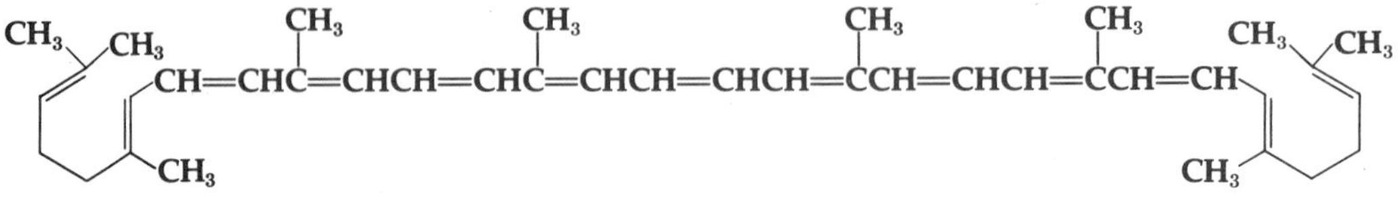

Lycopene (red)

Steroids

Steroids are a family of compounds that contain the perhydro-1,2-cyclopentanophenanthrene ring system with a side chain of varying length on the *D* ring.

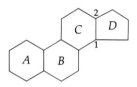

Perhydro-1,2-cyclopentanophenanthrene
ring system

The compounds of this group are widely distributed in plants and animals and are among the most important natural products, having first been isolated and studied in the early nineteenth century. As a consequence, the chemistry of the steroids has been so thoroughly explored that almost all the principles of the science could be illustrated with this class of compound as a vehicle. To this family belong the sterols, bile acids, sex hormones, the hormones of the adrenal cortex, and the cardiac aglycones.

Cholesterol, the most common steroid, is found in almost all tissues of animals, particularly in the brain, in the spinal column, and in gallstones. Derivatives of the substance are deposited in the arteries of human beings, causing high blood pressure and hardening of the arteries. The human body contains about half a pound of cholesterol, far less of other steroids. Gallstones are practically pure crystalline "stones" of cholesterol.

Cholesterol has eight asymmetric centers and thus is one of $2^8 = 256$ possible stereoisomers. When the double bond is saturated the skeleton is a rigid collection of chair cyclohexane rings. For this reason the whole theoretical area of conformational analysis (Chap. 6) was developed largely in the chemistry of steroid derivatives. The steroids generally possess the same absolute stereochemistry as cholesterol, the most common variation being at C-5 in saturated steroids and the hydroxyl configuration at C-3. The two possible dihydrocholesterols are cholestanol, with the hydrogen at C-5 below the ring plane and thus a trans-*A*/*B* ring juncture, and coprostanol with the hydrogen "up" at C-5 and a cis ring junction. Substituents on the periphery of the steroid nucleus are labeled α or β if they lie below or above the ring plane, respectively. The hydroxyl at C-3 in cholesterol and the side chain at C-17 are both β-substituents.

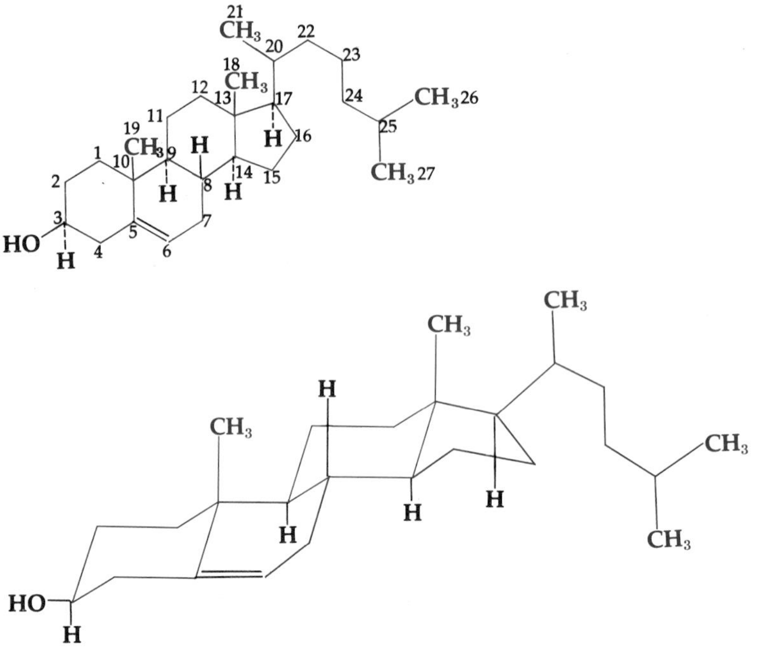

Cholesterol

Establishment of the actual relative stereochemistry at these various asymmetric centers constituted a major part of the total steroid structure problem. The most potent chemical technique involved the formation of rings to demonstrate cis placement of substituents. As a typical example, cholesterol was converted to a diacid derivative with a saturated C-5. The configuration of C-5 relative to that of the hydroxyl at C-3 is demonstrated to be cis by lactone formation and confirmed by showing that the alternative configurations at either C-3 or C-5 (trans) afford no lactone.

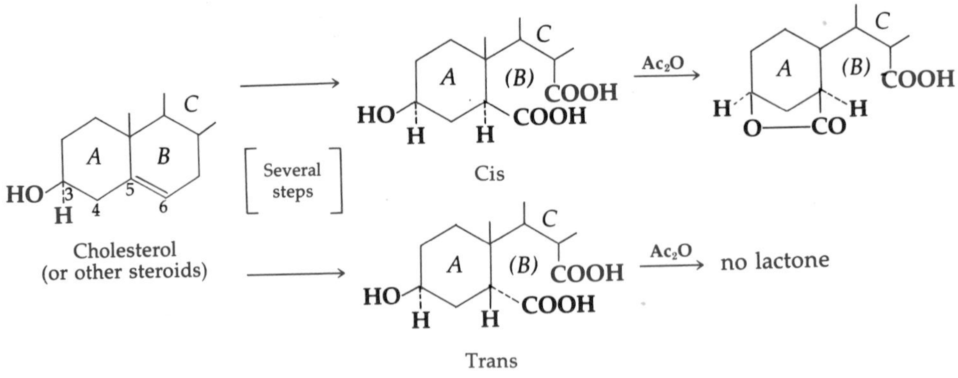

A second example involves the nature of the ring junction between rings C and D. In a multistep degradation by oxidation the D-ring portion of the molecule is isolated as a diacid without loss of the configuration at atoms C-13 and C-14. This diacid on vigorous pyrolysis, slowly yields an anhydride, which on hydrolysis affords an epimeric diacid. This second diacid is reconverted to the same anhydride very easily on heating. Hence the second diacid must have cis carboxyl groups so that the original diacid, and also the steroid C/D ring junction, must have been trans. The proof is completely analogous to that of the maleic and fumaric acids on page 181. In the slow pyrolysis of the trans diacid, the lower carboxyl is first enolized and so epimerized by its own acid catalysis on heating.

Various steroids   Trans   Cis

The steroids mediate a number of bodily functions although these are not generally understood in any chemical detail. One group of steroids is the **bile acids**. Bile is an emulsifying agent generated by mammals, which aids the absorption of fats and other lipids into the body fluids. The substance is a mixture of amides derived from bile acids and the amino acids glycine ($H_2NCH_2CO_2H$) and taurine ($H_2NCH_2CH_2SO_3H$). Hydrolysis of these amides produces the bile acids, which contain a carboxyl group in the side chain of the steroid nucleus and varying numbers of hydroxyl groups in the 3-, 7-, and 12-positions of the ring system. These hydroxyl groups are invariably oriented α (below the plane of the page), and the A/B rings are fused cis. Cholic and deoxycholic acids are the most abundant bile acids, and possess the coprostanol (β-**H** at **C-5**) skeleton. The α-**OH** at **C-7** in cholic acid provides an axial substituent for elimination to the C-6,7 double bond and then oxidation to the diacid needed for the proof of C-5 configuration described above.

Cholic acid   Deoxycholic acid

Conformational analysis explains why acetylation occurs more easily with hydroxyl groups at C-3 of cholic acid than with those at C-7 and C-12. Equatorially oriented hydroxyl groups are less hindered than those in axial positions and are therefore more easily attacked by acylating agents. Only the hydroxyl on **C-3** of cholic acid occupies an equatorial position. Furthermore, the hydroxyl at **C-7** is very much sterically hindered, because the cis fusion of rings A and B places one of the methylene groups of ring A axial and cis to this function.

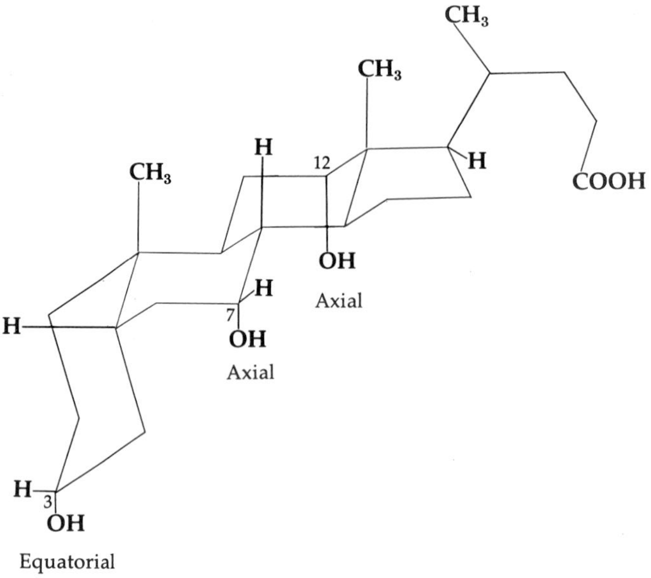

Cholic acid

Sex hormones are steroids generated by the gonads (ovaries or testes) and subsequently liberated into the blood stream. These steroid hormones are responsible for development of sex characteristics and for sexual responses of male and female mammals. Female hormones are called **estrogens** and male hormones **androgens**. Both estrogens and androgens are produced by the adrenal glands of each species.

Of the estrogens, estradiol is the primary hormone, and is distinguished by an aromatic *A* ring. The isolation of estradiol represents one of the feats of organic chemistry. Less than 12 mg of the compound was isolated from about 18,000 pounds of hog ovaries! One of the agents used in contemporary anti-fertility pills is a synthetic steroid, ethinylestradiol, which is closely related.

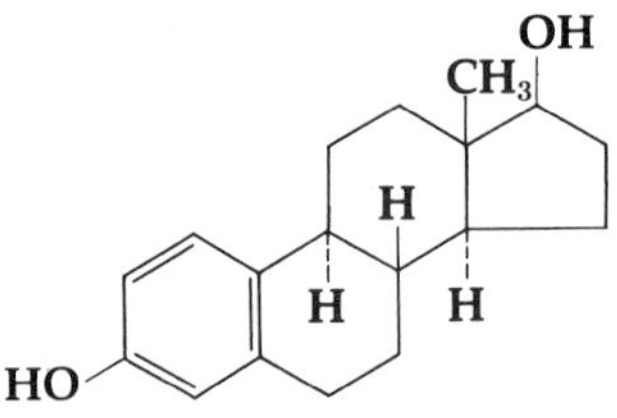

Estradiol 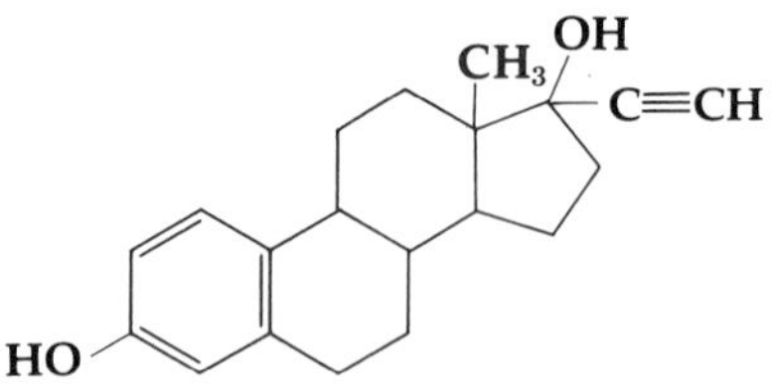   Ethinyl-estradiol

A second type of hormone, called progesterone, is secreted by the *corpus luteum*, a tissue of the ovary. This hormone is involved in the preparation for and maintenance of pregnancy. Unlike the physiological activity of progesterone, which is limited to a few steroid molecules, estrogenic activity is found in many other types of compounds, among which are substituted stilbenes such as stilbestrol.

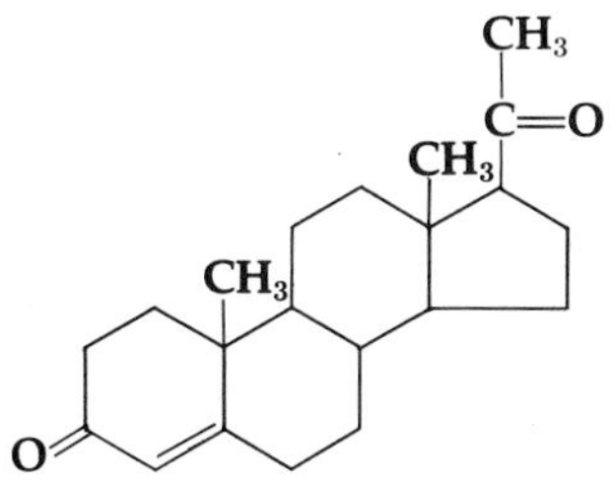

Progesterone

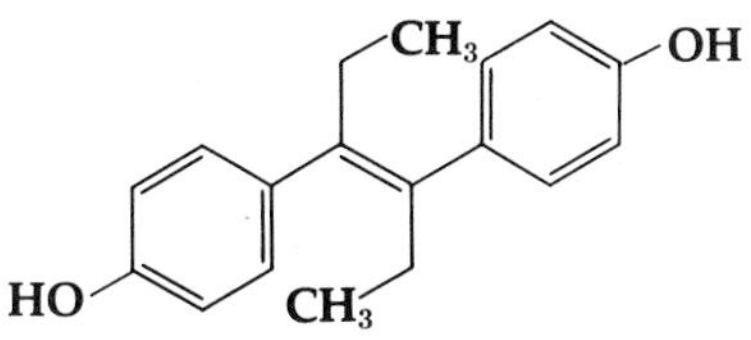

Stilbestrol

Male sex hormones (androgens) bear a structural resemblance to progesterone with respect to the character of the A and B rings and to the estrogens with regard to substituents at C-17. Testosterone and androsterone are the prominent members of the androgen group.

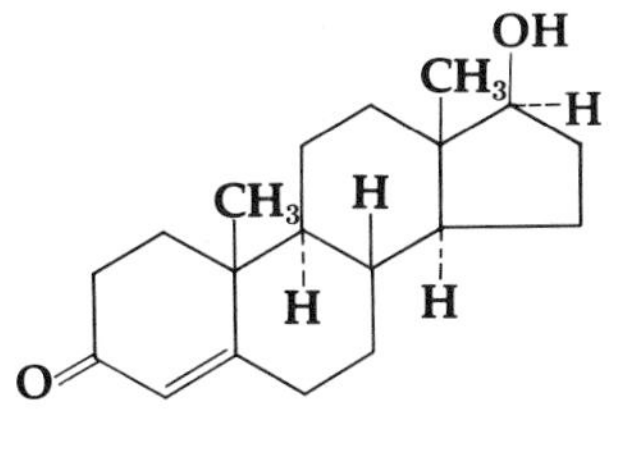

Testosterone

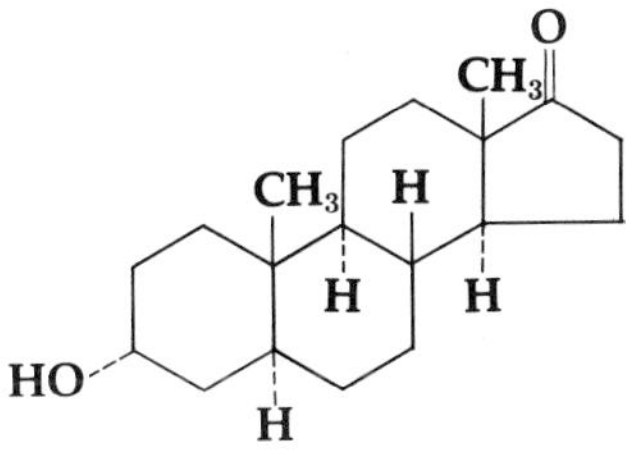

Androsterone

Testosterone is excreted by the testes and is the primary hormone. Androsterone is produced from the primary hormone and is excreted in the urine. About 10 mg of testosterone was isolated from 100 kg of testes tissue of bulls. The total synthesis of epiandrosterone is outlined on page 923.

PROBLEM 27-15

Eucarvone forms a monobenzylidene derivative, $C_{17}H_{18}O$, on treatment with benzaldehyde and base. The derivative shows two unsplit vinyl protons, five aromatic protons, and three unsplit methyl singlets in the NMR spectrum. What is its structure?

PROBLEM 27-16

Santonin (Fig. 27-3) on treatment with strong acid yields desmotroposantonin, an isomeric but *phenolic* substance which exhibits only one aromatic proton in the NMR spectrum as well as bands at 2.8 and 5.65 μ (1770 and 3600 cm^{-1}) in the IR spectrum.

PROBLEM 27-17

Medically valuable steroids are usually obtained by transformation of plentiful ones, such as the bile acids (obtained from the bile fluids of cattle, via the slaughterhouse). The following such transformation converts a derivative of deoxycholic acid to cortisone; only the side chain operations are shown. Fill in the missing information.

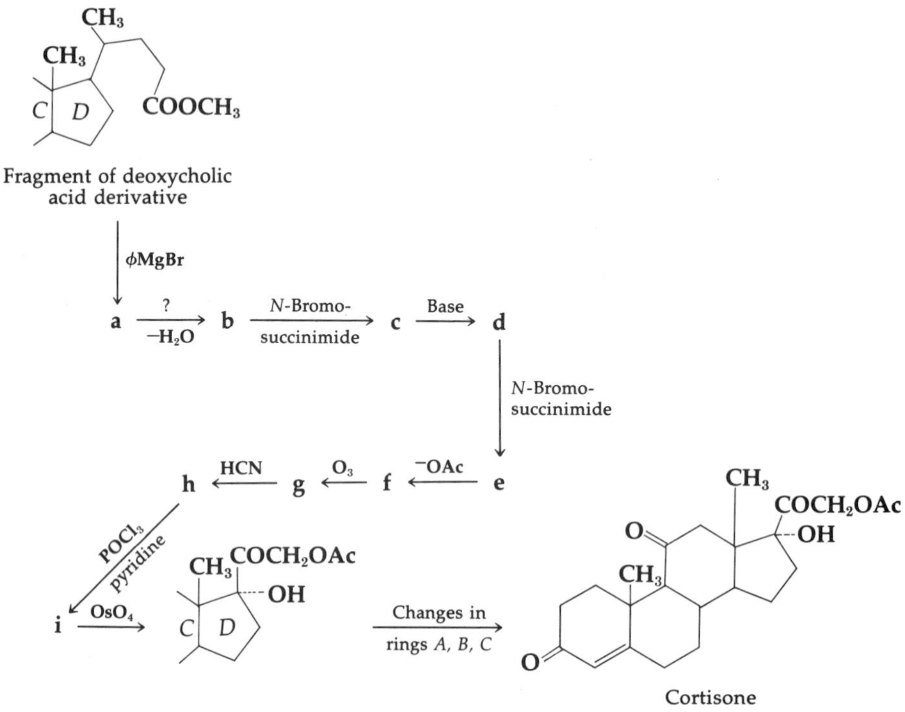

Fragment of deoxycholic
acid derivative

Cortisone

27-6   ALKALOIDS

Alkaloids are nitrogenous bases (usually heterocyclic) widely distributed in plants. Many of these substances have marked physiological effects, a fact discovered by many ancient peoples long before organic chemistry developed. For example, cinchona alkaloids occur in the bark of the *Cinchona* species, indigenous to the high eastern slopes of the Andes. Quinine, whose structure is indicated in Fig. 27-6, is one of the chief constituents and active principles of the bark extract, which was found in 1639 to be an effective antimalarial medicine. As a result, cinchona trees were cultivated in the Dutch East Indies to serve as a commercial source of the compound.

The striking characteristic of alkaloid chemistry is the extraordinary variety of structural features exhibited. (See Fig. 27-6.) The challenge presented by the elaborate ring systems, from the point of view both of elucidation of structure and of synthesis, has absorbed the efforts of some of the most talented research chemists of the past century.

Reserpine is one of a family of *Rauwolfia* alkaloids whose remarkable physiological properties have led to its extensive use in treatment of nervous and mental disorders. Elucidation of the structure of quinine took about 25 years, and its synthesis was achieved only after a period of about 50 years. In contrast, reserpine, an alkaloid of comparable complexity, was isolated in 1952. Its structure was determined, and the compound was synthesized within a span of 5 years; the synthesis is discussed on page 934.

Coniine and Tropine

Many alkaloids are simpler than quinine and reserpine and many have histories, at least as crude plant extracts, that are much interwoven in human affairs. The hemlock of Socrates' poison cup contains chiefly coniine, an alkaloid with the modest constitution $C_8H_{17}N$, the structure of which is readily deduced. Coniine cannot be hydrogenated catalytically, but dehydrogenation with zinc dust yields $C_8H_{11}N$, a pyridine derivative yielding pyridine-2-carboxylic acid on permanganate oxidation. Chromic acid oxidizes coiine itself to *n*-butyric acid.

This information requires that coniine be formulated as a saturated six-membered ring containing nitrogen† and bearing on the adjacent carbon the remaining three required carbons as a linear side chain. The latter is converted to *n*-butyric acid when chromic acid attacks the nitrogen.

Coniine

Among the tropane alkaloids are many well-known medicinal substances such as cocaine (Fig. 27-6), scopolamine (the "truth serum"), and the constituents of such plants as belladonna, henbane, and deadly nightshade. The 24-hour continuous-action capsules sold in drugstores for relief from symptoms of the common cold contain tropane alkaloids sealed in tiny pellets with coatings

---

†The index of hydrogen deficiency is 1. The saturated formula is $C_nH_{2n+3}N$, or $C_8H_{19}N$, and this is two hydrogens more than those in coniine ($C_8H_{17}N$).

designed to dissolve in the stomach at various rates, thus continuously releasing the alkaloids into the system over a long period of time.

Many of these alkaloids are esters of tropine, $C_8H_{15}NO$. The structure proof for tropine is rendered relatively simple by the important fact that it is optically inactive and hence must be a meso compound (or else have no asymmetric centers). The molecule has a secondary hydroxyl, oxidizable to a ketone, a tertiary amine with N—CH_3, and no double bonds (by hydrogenation). The ketone tropinone ($C_8H_{13}NO$), produced by oxidation of tropine, possesses two α-methylene groups since with benzaldehyde and alkali tropinone is converted to a bis-benzylidene derivative ($C_{22}H_{21}NO = C_8H_{13}NO + 2C_6H_5CHO - 2H_2O$). This accounts for all the functionality and demonstrates that there are two rings since the index of hydrogen deficiency is 2 (saturated = $C_8H_{19}NO$; $C_8H_{19}NO - C_8H_{15}NO = H_4$).

The meso molecule must have a plane of symmetry. Hence we can build up the molecule by adding the structural pieces equally on each side of this mirror plane. The secondary $>$CHOH and $>$NCH$_3$ are each present as only one of a kind and hence must be placed *on* the plane of symmetry. There are two CH_2 groups, one on each side of the $>$CHOH (which is oxidized to $>$C=O), and the remaining four carbons and six hydrogens must also be placed in pairs across the plane, as shown below.

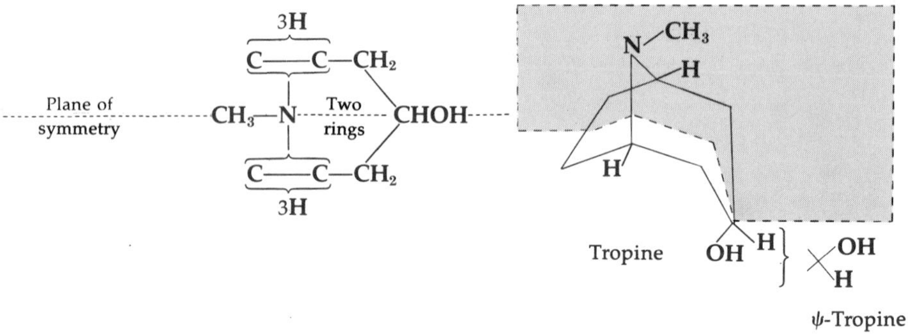

Tropine ψ-Tropine

Ignoring unstable three-membered rings this analysis allows only two structures for tropinone. Each of these in turn corresponds to two stereostructures for tropine since the hydroxyl (lying on the symmetry plane) can be up or down (cis or trans) with respect to the ring-junction hydrogens. The correct formulas for tropine and its hydroxyl epimer, ψ-tropine, are shown above, with the plane of symmetry in color.

In 1917 Robinson developed his ideas for alkaloid biogenesis via the Mannich reaction (page 1075). He envisioned tropinone as arising in the plant by a symmetrical two-part Mannich reaction from succinic dialdehyde, methylamine, and acetone. To test the possibility of this notion he mixed these three compounds in water at room temperature and was able to isolate authentic tropinone from the mixture. This feat was especially astonishing at the time since a previous synthesis of tropine had required a laborious 18-step sequence from cycloheptanone!

Physical Methods in Structure Proof

The development of spectroscopic methods (Chap. 7) has revolutionized the elucidation of structure of these complex alkaloids, since most of the extensive and time-consuming chemical degradation formerly necessary is replaced by analysis of spectra. An example of such a structure proof may be found in the alkaloid vindoline, one of a complex of indole alkaloids found in the periwinkle plant and used in the treatment of certain cancers. A variety of modern techniques is exemplified in this elucidation.

Vindoline is a white crystalline substance, mp 155°, with a constitution of $C_{25}H_{32}O_6$. The UV spectrum is practically identical with that of 6-methoxy-dihydroindole and the IR spectrum implied hydroxyl (2.8 μ) and ester (5.75 μ) functions. The ester functions were defined by lithium aluminum hydride reduction, and by saponification and reacetylation, to be $-COOCH_3$ and $-OCOCH_3$. A single unconjugated double bond was also detected by NMR and by hydrogenation.

6-Methoxy-dihydroindole

$$C_{25}H_{32}N_2O_6 \equiv C_{20}H_{23}N_2$$

Vindoline
(index of hydrogen
deficiency = 11)

$-COOCH_3$
$-OCOCH_3$
$-OH$
$-OCH_3$

The index of hydrogen deficiency is 11, composed of two for the esters, three for the unsaturation in the benzene ring, one for the double bond, and hence five rings. Two of these rings must be those of a 6-methoxy-dihydroindole skeleton to account for the UV spectrum.

The indole skeleton suggested that the alkaloid was formed in the plant from tryptophan and this was confirmed by growing the plant in a medium containing radioactive tryptophan, then isolating radioactive vindoline. Following common biosynthetic experience the dihydroindole skeleton was therefore extended by $-CH_2CH_2N$ in the β-position as it is in tryptophan after decarboxylation.

The NMR spectrum is tabulated in Table 27-3. This spectrum distinguishes and accounts for all 32 protons on the skeleton of vindoline, and these are identified in the table. In examining the table, note that each coupled hydrogen finds its coupling constant duplicated in the signal of the other hydrogen which couples with it. A large part of the carbon skeleton can be assembled from this data as shown at the bottom of the table. In fact all but three carbons are incorporated in the partial skeleton plus its pendant func-

TABLE 27-3   **NMR Spectrum of Vindoline with Assignments of Signals**

τ		
τ 9.52	3H (t) J = 7.5 cps	$-CH_2-$ H_3
8.65	2H (q) J = 7.5	$C-CH_2-CH_3$
7.93	3H (s)	$-O-CO-CH_3$
7.0–7.9	4H (multiplet)	$C-CH_2-CH_2-N$
7.35	1H (s)	$N-C-H$
7.32	3H (s)	$N-CH_3$
6.60	2H (dd) J = 2, 5	$N-CH_2-CH=CH-$
6.25	1H (s)	See text
6.20	6H (s)	$2O-CH_3$
4.77	1H (d) J = 10	$-CH=CH-C$
4.57	1H (s)	$CH_3CO-O-C-H$
4.12	1H (ddd) J = 2, 5, 10	$CH_2-CH=CH-$

3.92   1H (d) J = 2.5
3.70   1H (dd) J = 2.5, 8
3.09   1H (d) J = 8

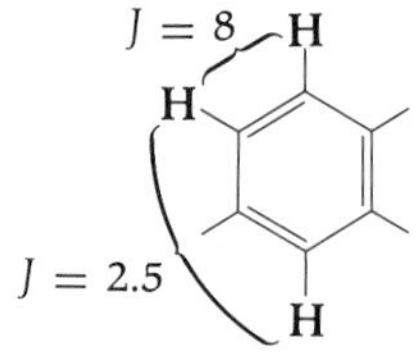

J = 8

J = 2.5

1.00   1H (s) $C-OH$

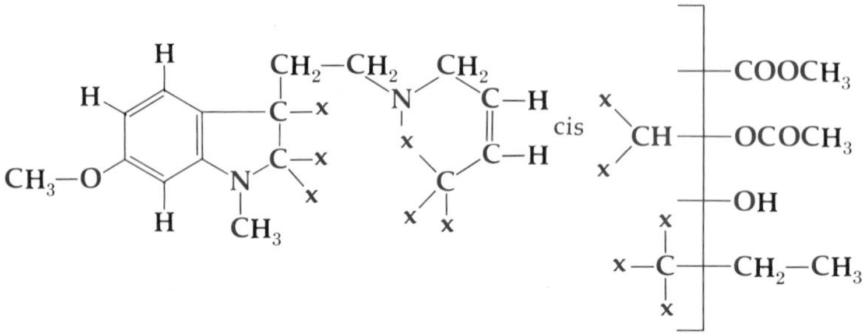

Each peak is reported with chemical shift position in τ units, number of hydrogens involved, the splitting and splitting constant(s), J:

(s) = singlet
(d) = doublet
(dd) = pair of doublets
(t) = triplet

The hydrogens responsible for the signal are shown in color. The neighboring hydrogens which cause splitting are in black. There can be *no more* hydrogens on adjacent atoms than those shown in black, owing to the observed splittings.

tionalities. These several attached functions (on the bracket) must be linked to the three skeletal carbons still unplaced.

Heating dihydrovindoline with acid yields a ketone (IR: $5.85\,\mu$ = six-membered ring) with the same UV spectrum as vindoline but with the IR showing disappearance of the hydroxyl and *both* ester functions. The new ketone shows two α-hydrogens in the NMR as a doublet signal, split by one adjacent hydrogen at $\tau\,7.5$. When treated with base in deuteromethanol (i.e., CH_3ONa/CH_3OD) the two α-hydrogens are exchanged for deuterium via enolization.† The split signal of the adjacent hydrogen then goes to a singlet (still at $\tau\,7.5$) since the two vicinal deuterium atoms now cause no splitting.

The chemical change occurring in the acid reaction may be deduced on mechanistic grounds. With the limited skeleton and groups allowed by the foregoing analysis there is only one way to unite them into a single molecular unit which will account for the observed chemical change in acid.

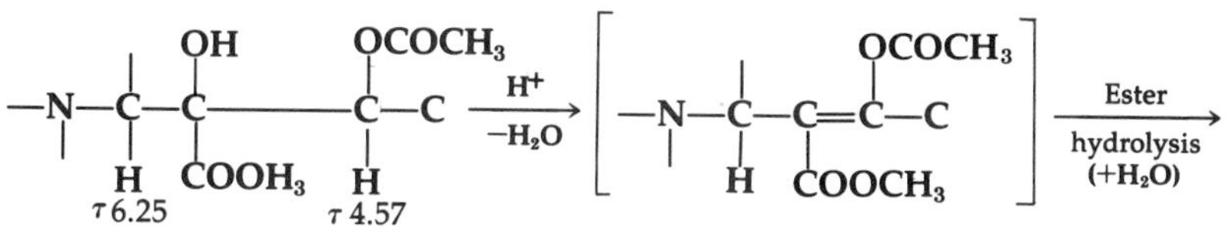

Vindoline

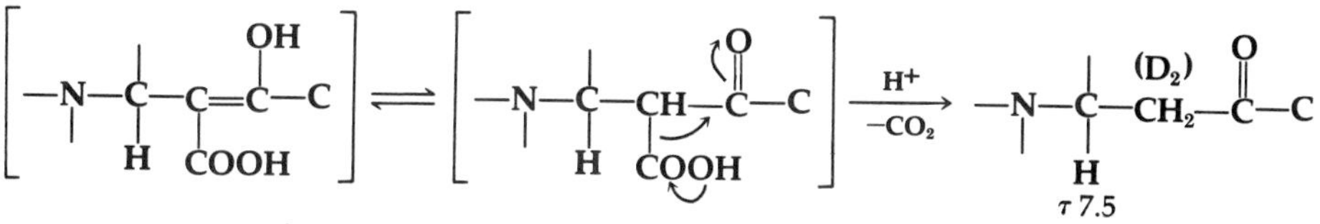

Ketone product

This molecular subunit contains all the remaining carbons and must be fitted together with the partial skeleton in Table 27-2 in order to deduce a full formula for vindoline. The nitrogen of the subunit is one of the two nitrogens of the partial skeleton and the linking up must be done so as to create three new rings (and the ketone ring six-membered) and not place any hydrogens in a new vicinal relationship which would imply NMR splittings not actually observed.

The structural logic involved offers an interesting challenge which we can examine by selecting arbitrarily one nitrogen to be common to the two formulas, as shown below. The other nitrogen is therefore attached to the hitherto unassigned —CH [$\tau\,7.35$ (s)] which in turn must be connected to atoms not bearing hydrogen since it has a singlet NMR signal. What remains is the connection of the open (colored) bonds to form three rings. Note that neither the —CH_2CH_3 nor the —CH—$OCOCH_3$ may be attached to the N—CH— (nor

---

† The fact that exactly two hydrogens are exchanged for deuterium is discerned from the mass spectrum, which shows a new parent peak two units higher.

to each other) since this would require new splittings of vicinal hydrogens, which are not actually observed. The combination shown can *only* lead to one formula for vindoline; the final bond connections are shown in color.

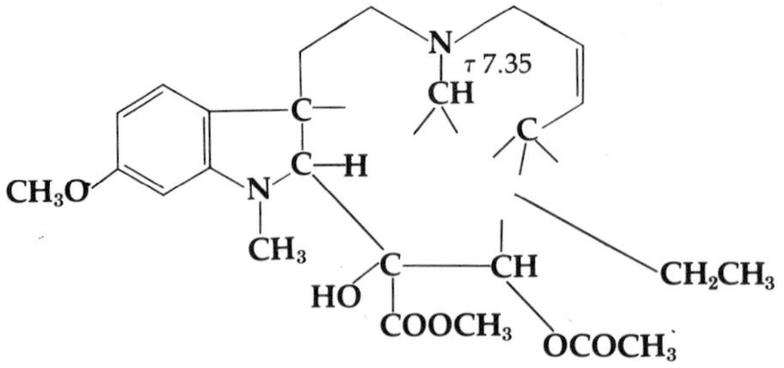

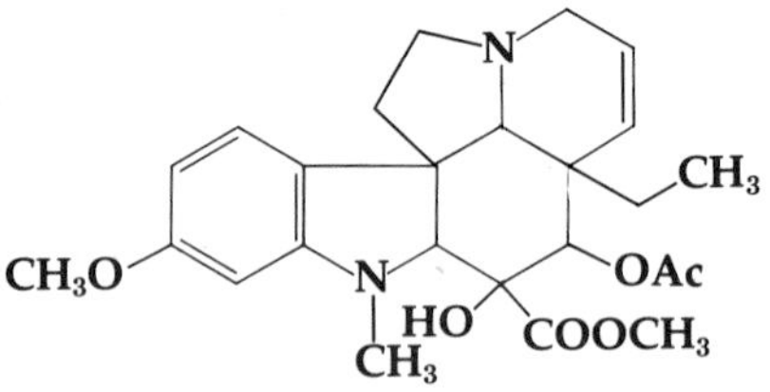

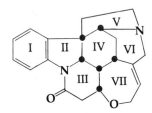

Vindoline

Strychnine Stereochemistry

The establishment of the relative stereochemistry at several asymmetric carbons in a natural product can often be a very difficult task, as the case of the steroids (page 1106) implied. However, this task can also be relatively simple if the asymmetric centers are interlocked in a rigid polycyclic bridged ring skeleton. The infamous poison strychnine provides a happy example. Crystalline strychnine was first isolated in 1817, but the elucidation of its structure was not complete until 1948. Chemists were engaged in this effort throughout nearly the entire history of organic chemistry.

Strychnine

The two-dimensional structure contains six asymmetric carbons (colored dots), theoretically capable of $2^6 = 64$ possible stereoisomers. Yet so interdependent are the configurations at these six carbons that we may deduce the full relative stereochemistry and conformation without further experiment!

Start with an examination of ring IV in a chair conformation. Ring VI is fused across this ring at 1,3-positions, which can only be bridged from cis-diaxial bonds (refer to Fig. 6-14). This in turn requires one bond of ring V to be axial (the bond to N) so that the other bond of ring V must be cis and equatorial. This arises as a consequence of the impossibility of bridging 1,2-trans-diaxial positions with an ordinary-sized ring.

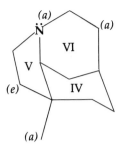

Continuing to move about ring IV, we find that its bond to the aromatic ring is now necessarily axial. As with ring V above, ring II must also therefore be cis-fused, thus obliging the bond to the lactam nitrogen to be equatorial. Working in the other direction from the cis-diaxial fusion of ring VI, we find that ring VII must be cis-fused since one bond in that ring is already axial. The other then must be cis and hence equatorial. This leaves the juncture of rings III and IV trans and defines the ring system as shown below.

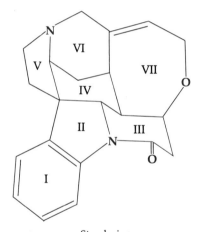

Strychnine

Tracer Studies of Biosynthesis

Illumination of the routes utilized by living organisms to synthesize the molecules of natural products is usually achieved through tracer experiments. Radioactive precursors are fed to the organism and radioactive product isolated and degraded to find where the radioactive, or tracer, atoms have gone. In studying the biosynthesis of alkaloids from amino acids, the technique involves the following steps:

1 Synthesis of the amino acid with isotopic atoms in known positions of the molecule
2 Metabolism of the labeled molecule by the plant
3 Isolation of the alkaloid in question
4 Determination of the isotope level in the alkaloid and degradation of the molecule
   to show the exact positions of the tracer

The principles involved in the use of this technique will be illustrated in their application to the biosynthesis of morphine.

Morphine is one of a group of alkaloids that occur in poppy seeds and their crude extract, known as opium. The compound and its phenolic methyl ether, codeine, are both powerful analgesics and have been used as painkillers for several centuries. The drugs are highly toxic in any but small doses and are habit-forming. Considerable research effort has been devoted to the synthesis of a molecule with the analgesic but not the addictive properties of morphine.

The structure of morphine suggests that the molecule might be mainly derived in the plant from two molecules of phenylalanine or tyrosine, as outlined on page 1076. This hypothesis was tested by the following experiments.

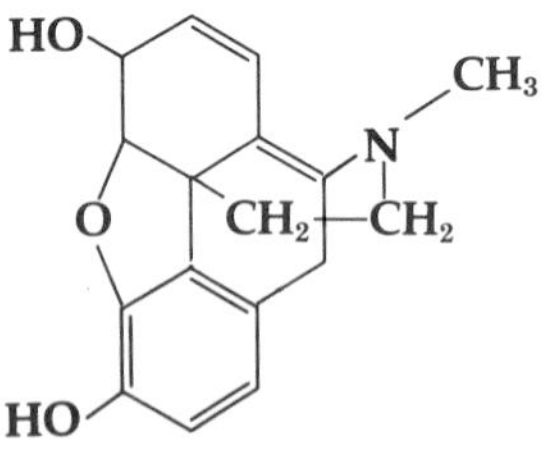

Morphine 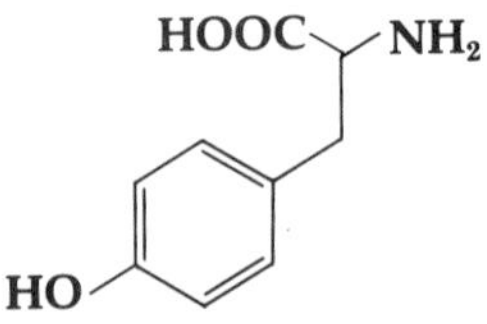   Tyrosine 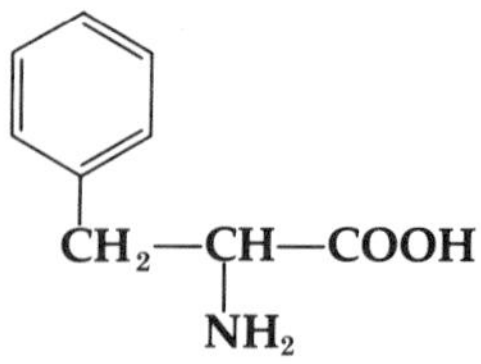   Phenylalanine

Phenylalanine labeled with C^{14} (radioactive isotope of carbon) on the carbon attached to nitrogen was prepared as indicated from malonic ester labeled in the α-position. Labeled tyrosine was prepared by a similar route, except that p-acetoxybenzyl bromide was employed as the alkylating agent in place of benzyl bromide. The p-acetyl protective group was readily removed by hydrolysis. In the synthesis outlined and in subsequent formulas, the presence of the C^{14} label is indicated with a heavy dot and use of color.

$$Br_2 + \overset{\bullet}{C}H_2(COOC_2H_5)_2 \longrightarrow Br\overset{\bullet}{C}H(COOC_2H_5)_2$$

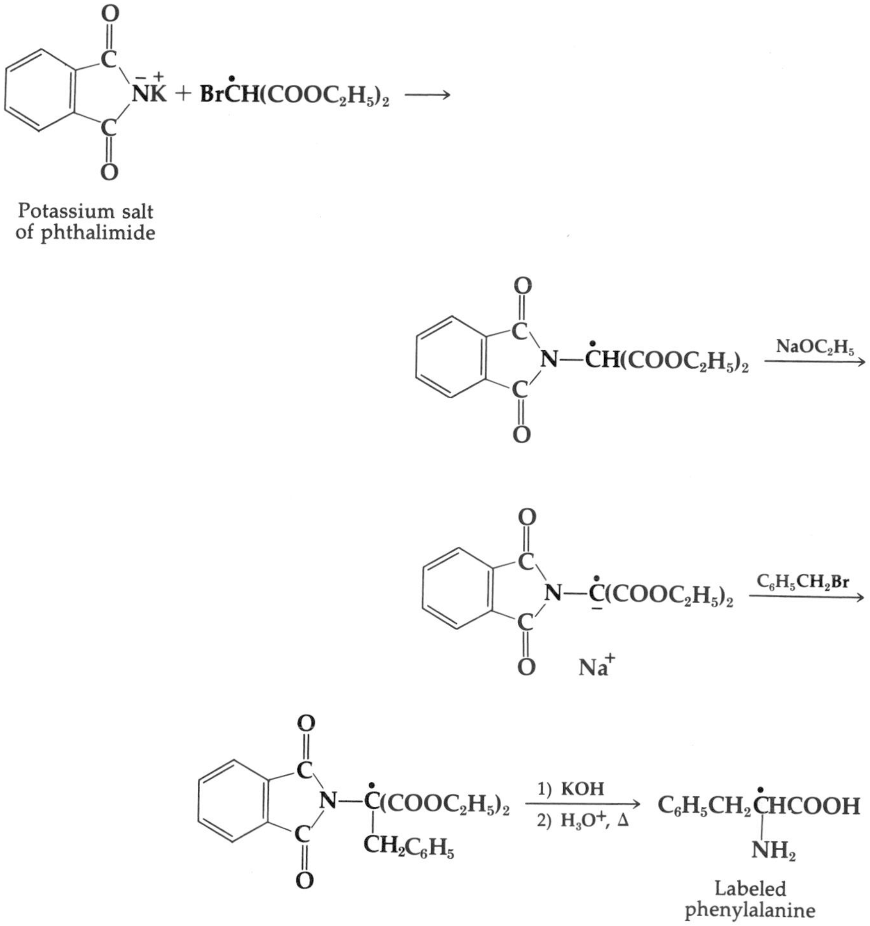

Potassium salt
of phthalimide

+ $Br\overset{\bullet}{C}H(COOC_2H_5)_2 \longrightarrow$

$\xrightarrow{NaOC_2H_5}$

$\xrightarrow{C_6H_5CH_2Br}$

$\xrightarrow[\text{2) } H_3O^+, \Delta]{\text{1) KOH}}$ $C_6H_5CH_2\overset{\bullet}{C}HCOOH$
$\overset{|}{N}H_2$

Labeled
phenylalanine

Opium poppies were grown with labeled phenylalanine added as nutrient in one experiment and labeled tyrosine in a second experiment. Morphine was then isolated from the seeds of the plant and found to contain the C^{14} label in both experiments. Thus, both amino acids appear to be precursors of morphine in the biosynthesis of the alkaloid. The next step was the location of the labels in morphine to see if they corresponded to the positions expected if two moles of amino acid were utilized per mole of morphine. The morphine isolated from poppy plant fed with labeled phenylalanine was degraded as follows:

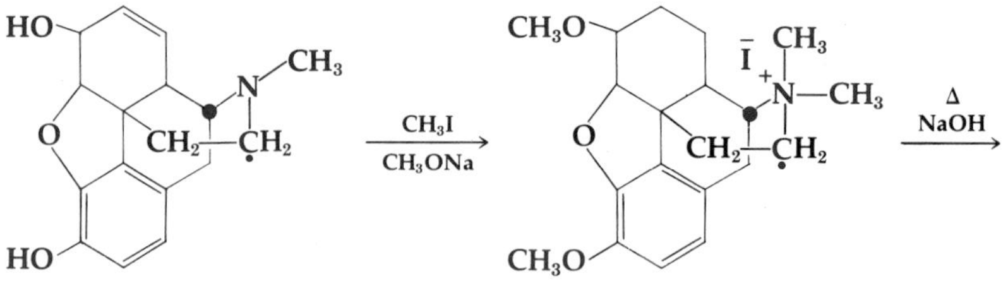

Morphine from
labeled-phenylalanine
feeding experiments

$\xrightarrow[CH_3ONa]{CH_3I}$

Methiodide

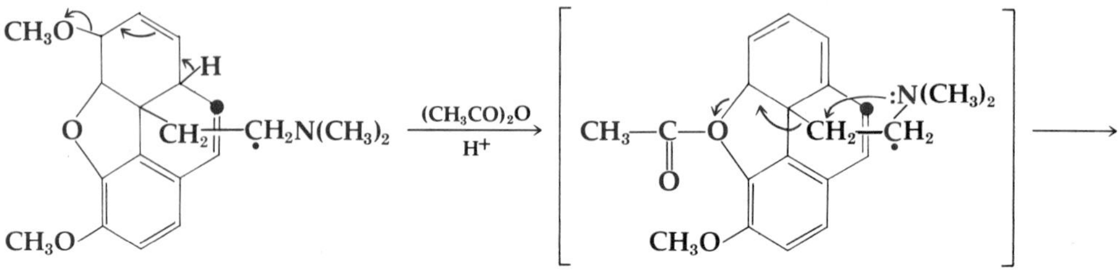

$\xrightarrow[NaOH]{\Delta}$

Tertiary
amine

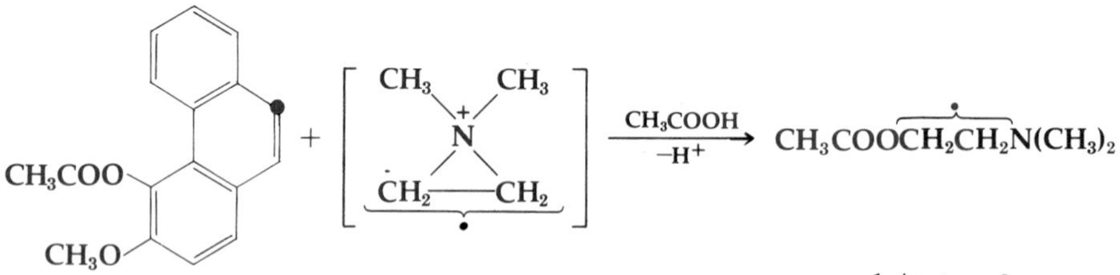

$\xrightarrow[H^+]{(CH_3CO)_2O}$

1-Acetoxy-2-methoxy-
phenanthrene (half
original C^{14} in this
molecule)

$\xrightarrow[-H^+]{CH_3COOH}$ $CH_3COOCH_2CH_2N(CH_3)_2$

1-Acetoxy-2-
dimethylamino-
ethane (half
original C^{14}
in this molecule)

In the first step, morphine was alkylated both at nitrogen and at oxygen with methyl iodide. The methiodide product, when heated with base, underwent a Hofmann elimination reaction to produce a new carbon-carbon double bond and a tertiary amino group. The elimination reaction occurred in this direction because the proton α to the phenyl group was the most acidic (Table 8-2). Application of the Hofmann elimination reaction to alkaloid degradation has played a very important role in determining the structure of this class of compounds, and the combination of methylation with the elimination reaction is often referred to as **Hofmann degradation**.

The tertiary amine, when heated with acetic anhydride in the presence of a trace of acid, underwent eliminations as indicated to give 1-acetoxy-2-methoxyphenanthrene and 1-acetoxy-2-dimethylaminoethane. Each molecule contained half the radioactive carbon originally present in the morphine. The driving force for this last reaction derives from the facts that an aromatic ring was produced, a mole of methanol eliminated, and the breaking of the carbon-

carbon bond was aided by participation of the neighboring dimethylamino group of the side chain.

The position of the label in the phenanthrene nucleus was further delineated by oxidation of the 1-acetoxy-2-methoxyphenanthrene to give phthalic acid, which was converted to phthalic anhydride. All the radioactive label present in the starting material was retained in the product. This oxidation reaction illustrates the general principle that benzene rings substituted with electron-donating substituents are more subject to oxidation than other benzene rings. Retention of all the label in the phthalic anhydride indicates that none of the label could have been at those carbon atoms lost in oxidation. When submitted to a Schmidt rearrangement (page 707), the phthalic anhydride gave anthranilic acid, which contained only half the label present in the phthalic anhydride. Clearly, the label must be restricted to the carbonyl groups of the anhydride.

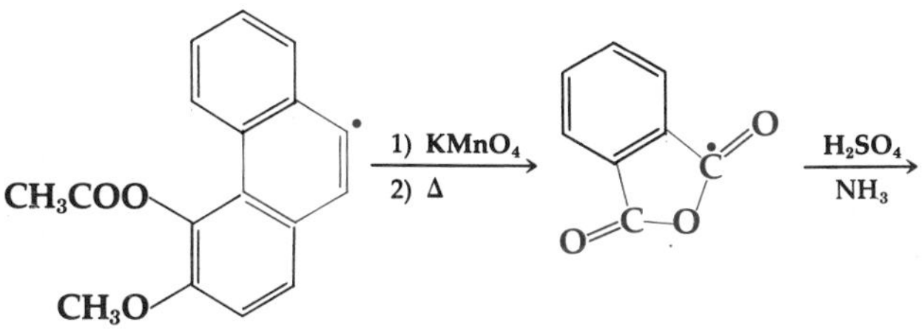

1-Acetoxy-2-methoxy-
phenanthrene (contains
half the label present
in original morphine)

Phthalic anhydride
(contains all the
label present in
phenanthrene
compound)

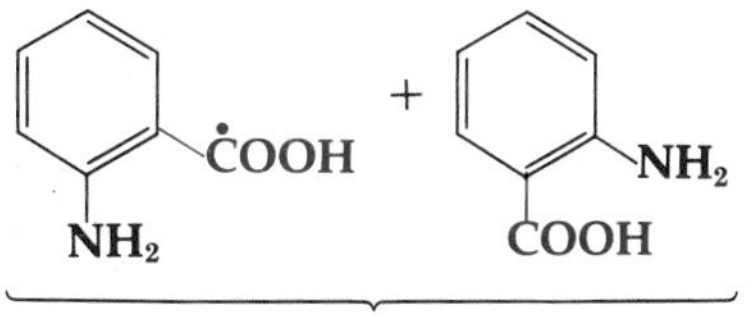

Anthranilic acid
(contains half the
label present in
phenanthrene
compound)

In a second set of experiments, the morphine grown with tyrosine as nutrient was carried through the first two steps of the above procedure to yield the tertiary amine with the same level of activity as the original morphine. The two nonaromatic double bonds of this amine were catalytically reduced, and the resulting substance was submitted to a second Hofmann degradation. The trimethylamine produced was free of label, which demonstrates that the methyl group originally in morphine was free of C^{14}.

The olefin produced in the Hofmann elimination reaction was hydroxylated with osmium tetroxide (page 635), and the resulting 1,2-diol was cleaved with periodic acid (page 761). The formaldehyde produced was converted to a crystalline derivative with the β-diketone, dimedone. The reaction is an aldol condensation followed by Michael addition. The derivative contained half the total label originally present in morphine. The other aldehyde fragment from the periodic acid reaction contained the other half.

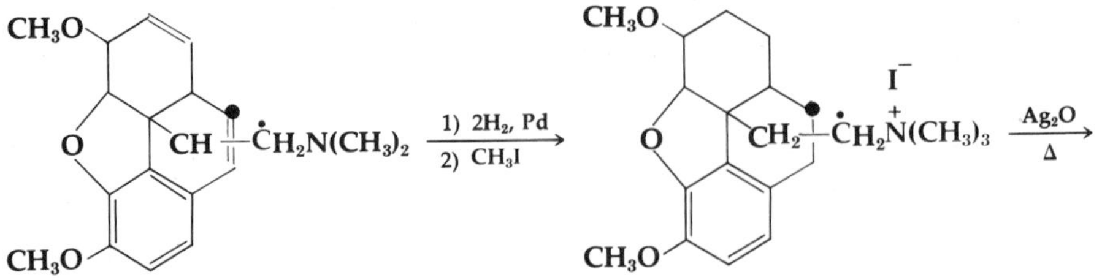

Tertiary amine derived
from morphine extracted
from labeled-tyrosine-fed plants

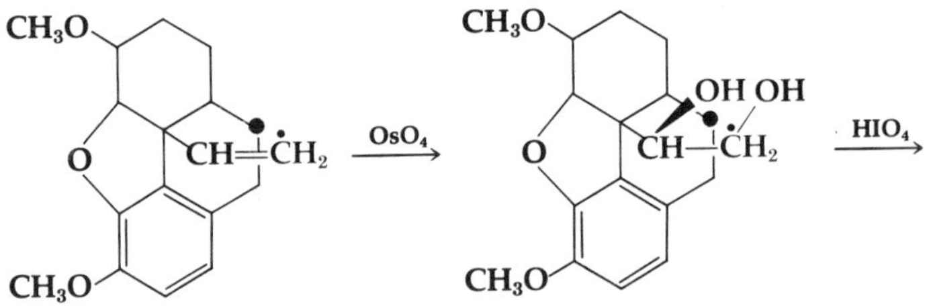

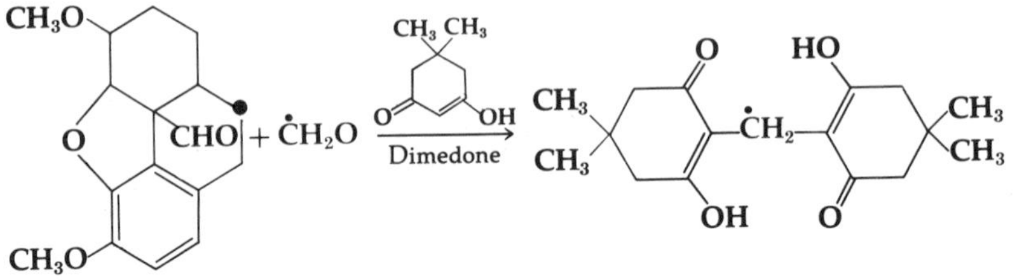

Aldehyde (contains
half of label present
in original morphine)

Dimedone derivative
of formaldehyde
(contains half of label
present in original
morphine)

PROBLEMS

27-18 Papaverine, one of the alkaloids in opium, has the formula $C_{20}H_{21}NO_4$. Hot permanganate oxidation yielded the following three acids, as proved by synthesis. What is the structure of papaverine? (*Note:* Acid *A* is not the source of acid *B* under these conditions.)

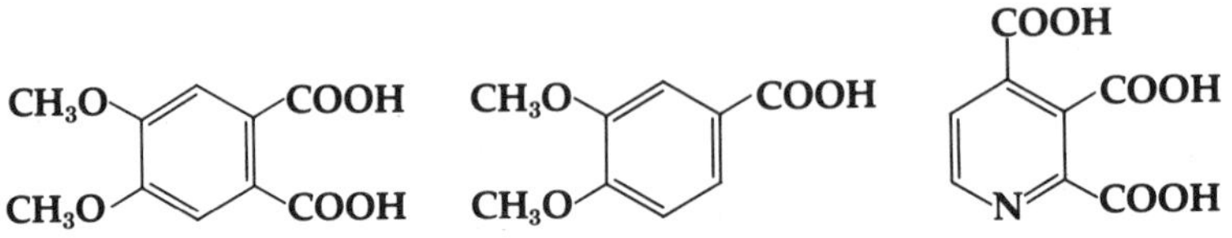

A   *B*   *C*

27-19 The structural reasoning for tropine as shown in the text implies a second possible structure. What is it and how would you distinguish between the two structures so derived for tropine? Show both a physical and a chemical method of distinguishing them.

27-20 The structural reasoning for vindoline proceeded to its final conclusion from an arbitrary assignment of one nitrogen atom common to both partial structures. Deduce one or more other complete structures for vindoline assuming the *other* nitrogen to be common to both partial structures. How would you distinguish your new structure(s) from the correct one given for vindoline in the text?

27-21 In the original assignment of stereochemistry to reserpine, "isoreserpinol" was treated with toluenesulfonyl chloride in pyridine and yielded a quaternary ammonium salt.

    a Show what reaction occurred and what stereochemical conclusions may be drawn from the result.

    b Write the reaction sequence used to prepare "isoreserpinol" from reserpine.

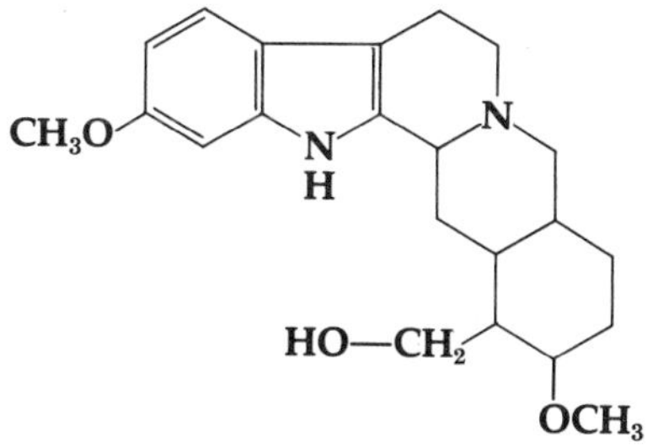

Isoreserpinol ·

27-22 In structure studies on the alkaloid mitraphylline, pyrolysis led to recovery of mitraphylline as well as a diastereomer named isomitraphylline. A third product obtained was $C_{10}H_9NO$. The arrows drawn on the structure of mitraphylline constitute what is believed to be the first reaction in pyrolysis, i.e., a tautomerism and, with the reverse reaction, an equilibrium.

    a How many diastereomers of mitraphylline could be produced in principle by this equilibrium?

    b What is the structure of $C_{10}H_9NO$?

    c What IR and NMR spectra do you expect for $C_{10}H_9NO$?

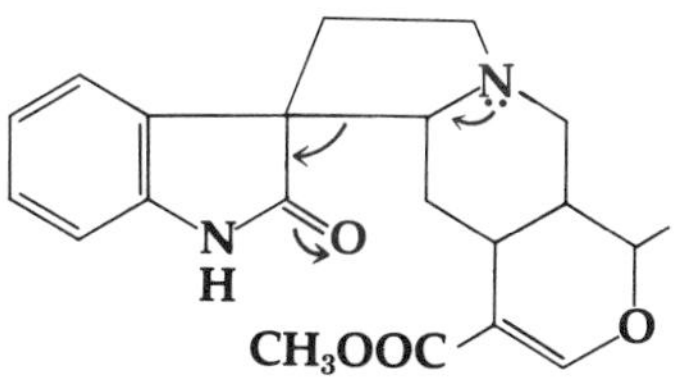

Mitraphylline

27-23 Outline a biosynthetic pathway to fisetin (Fig. 27-7).

27-24 In the early structure work on quinine (Fig. 27-6), the alkaloid was first dehydrated and the olefin (i.e., enamine) product cleaved by acid hydrolysis to form 6-methoxy-4-methyl-quinoline and an amino acid. Considering the quinoline ring, protonated in acid, as capable of ketonic behavior (Fig. 24-1), write a mechanism for the cleavage and a structure for the amino acid formed.

27-25 In biosynthetic studies the **Kuhn-Roth oxidation** is a very potent degradative tool for the isolation of small parts of the molecule after labeling. In this oxidation a compound is oxidized with hot chromic acid, which destroys all of the carbon skeleton except for methyl groups attached to carbon, which are converted to acetic acid nearly quantitatively. The acetic acid is then quantitatively determined by distillation and titration. The method is also called a **C-methyl determination**.

$$-\overset{|}{\underset{|}{C}}-CH_3 \xrightarrow[\Delta]{H_2Cr_2O_7} HOOC-CH_3$$

a In squalene biosynthesis from $C^{14}H_3COOH$, what proportion of the total radioactivity of the squalene is found in the acetic acid isolated by Kuhn-Roth oxidation?

b In cholesterol biosynthesis from $C^{14}H_3COOH$, what is the proportion?

c In cholesterol biosynthesis from $CH_3C^{14}COOH$, what is the proportion?

d Similarly, examine the following compounds from the chapter, using either of the two labeled acetates for feeding experiments: Menthol, vitamin A, β-amyrin, griseofulvin.

27-26 An alkaloid of unknown structure possessed a molecular formula $C_8H_{17}NO$. The substance, when acetylated, formed a monoacetate and in a C-methyl determination gave 0.7 mole of acetic acid (Prob. 27-25). The alkaloid also gave an iodoform test. When oxidized under controlled conditions, the alkaloid produced $C_6H_{11}NO$, a neutral substance that gave no acetic acid in a C-methyl determination. Hydrolysis in acid of $C_6H_{11}NO$ gave an amino acid, $C_6H_{13}NO_2$, which gave no nitrogen gas on treatment with HNO_2. With correct structures, trace the above interconversions.

27-27 Design a series of experiments that would differentiate L-arabinose from the other L-aldopentoses.

27-28 Devise synthetic schemes for the following:

a Preparation of mannonic lactone from glucose.

b Preparation of the pentaacetate of galactonic acid from galactose (simple acetylation of galactonic acid leaves one hydroxyl unacetylated owing to the lactone).

c Preparation of 3,5,6-triacetylglucose from glucose.

d Preparation of phenyl β-D-glucopyranoside from glucose and phenol.

e Conversion of D-gulose into

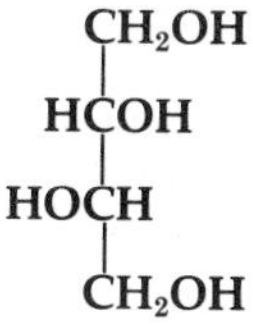

$$
\begin{array}{c}
\mathrm{CH_2OH} \\
\mathrm{H\overset{|}{C}OH} \\
\mathrm{HO\overset{|}{C}H} \\
\mathrm{\overset{|}{C}H_2OH}
\end{array}
$$

f Conversion of D-allose into

$$
\begin{array}{c}
\mathrm{CH_3} \\
| \\
\vdash \\
\vdash \\
\vdash \\
| \\
\mathrm{CH_2OH}
\end{array}
$$

g Conversion of D-mannose into

$$
\begin{array}{c}
\mathrm{CHO} \\
\dashv \\
\dashv \\
\vdash \\
| \\
\mathrm{CH_3}
\end{array}
$$

h Preparation of 3-*O*-methyl-D-glucose from glucose.
i Preparation of 6-methylcoumarin from *p*-methylphenol.
j Synthesis of the following from resorcinol and anisole:

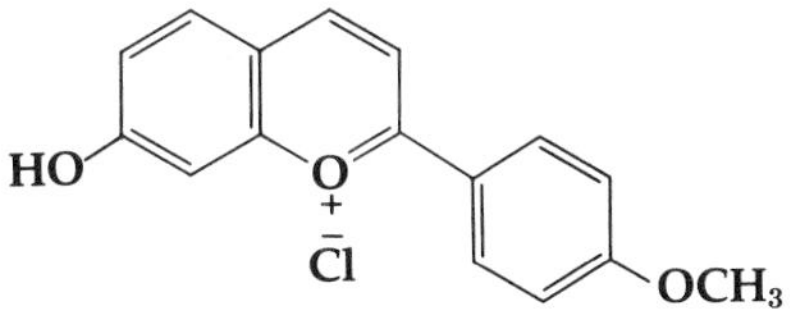

k Synthesis of the following flavanone from resorcinol and phenol:

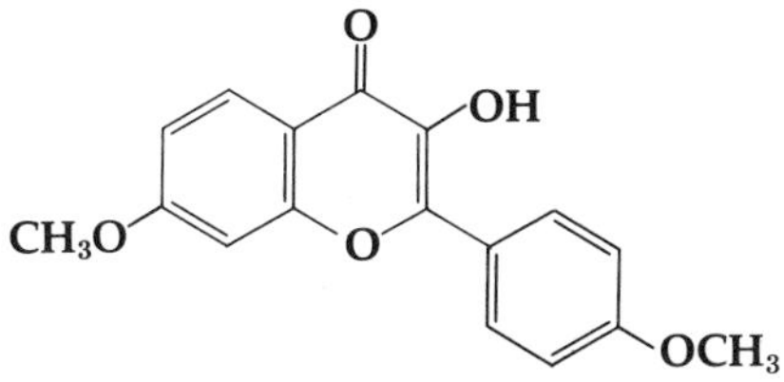

27-29 Interchange the functions on the ends of the eight D-aldohexoses without otherwise disturbing the configurations of the asymmetric centers, and name the products of this operation.

27-30 A ketohexose of D, but otherwise unknown, configuration upon oxidation gave a mixture of (+)- and (−)-tartaric acid. What is the total structure of the ketohexose?

27-31 Two different aldohexoses gave the same glycaric acid (six-carbon dicarboxylic acid) when oxidized. Degradation of the aldohexoses to their aldopentoses and oxidation of these substances gave two different five-carbon dicarboxylic acids, of which one was optically active, the other inactive. Both aldohexoses were converted to their methyl glycopyranosides, which when oxidized with periodic acid gave the same compound as that obtained from similar treatment of methyl α-D-glucopyranoside. With the correct formulas, trace the above reactions.

27-32 Indicate the steric structure of the product you would expect to obtain from the following reactions:

    a Epoxidation of a compound that possesses the structure of cholestanol except for the presence of a double bond in the 9,11-position.

    b Hydrolysis in strong acid of the above epoxide.

    c Bromination of a compound that contains the structure of coprostanol except for a double bond in the 9,11-position.

27-33   **a** Develop a synthesis for mevalonic acid labeled with C^{14} at the carbon α to the carboxyl group. Presume that $C^{14}H_3CO_2H$ was one of the available starting materials.

    b Presume that mevalonic acid labeled at the carbon α to the carboxyl group was fed to an organism that produced squalene and cholesterol. Indicate the positions of the labels in these compounds.

27-34 A neutral, optically active solid isolated from plant material gave the NMR spectrum illustrated and an analysis for $C_{19}H_{22}O_6$. The IR spectrum exhibited no carbonyl band and the UV spectrum showed only weak ($\epsilon \sim 1000$) absorptions around 270–280 nm, typical of polyalkoxybenzenes. Mild oxidation yielded $C_{19}H_{20}O_6$ with a new IR band at 5.85 μ (1710 cm^{-1}). Derive a structure for this natural product and assign it to a structural family of natural products.

27-35 A neutral solid was isolated from the leaves of *Lomatium columbianin* and purified. This optically active natural product, named columbianetin, yielded an analysis for $C_{14}H_{14}O_4$ and the accompanying NMR spectrum (Fig. P27-35). The UV spectrum was essentially identical to that of 7-methoxy-8-methyl-coumarin and the IR showed no other carbonyl absorption. Acid-catalyzed dehydration yielded an optically inactive compound, $C_{14}H_{12}O_3$. Derive a structure for columbianetin and deduce some features of its biosynthesis if possible.

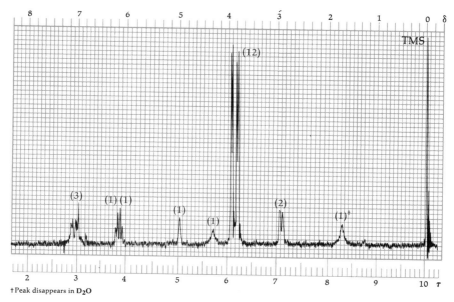

† Peak disappears in D_2O

FIGURE P27-34

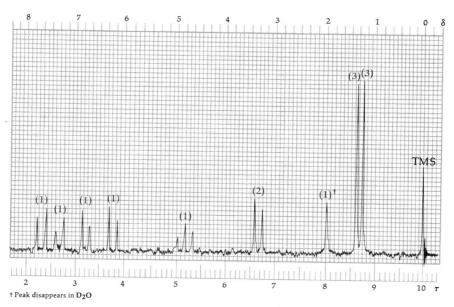

† Peak disappears in D_2O

FIGURE P27-35

27-36 A pleasant-smelling, optically active oil, $C_{10}H_{14}O$, can be isolated from caraway
seeds. The IR spectrum shows no bands below about 3.2 μ (above 3100 cm^{-1})
but shows two strong bands at 5.95 and 6.08 μ (1640 and 1680 cm^{-1}). The UV
spectrum, λ_{max} 236 nm (log ϵ 4.3), is unchanged on catalytic hydrogenation to
the dihydro compound but disappears in the tetrahydro compound. The NMR
is illustrated (Fig. P27-36). What is the structure of the compound (and its two
hydrogenation products)? Is it recognizable as a member of a family of natural
product structures?

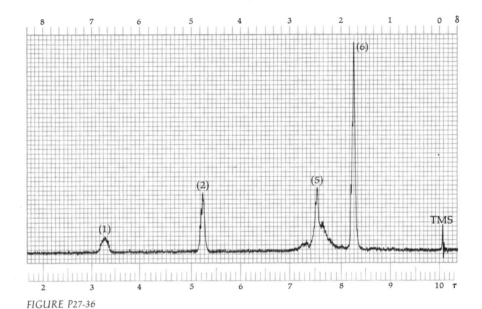

FIGURE P27-36

27-37 Articulone, $C_{15}H_{22}O$, is an oil isolated from a West African plant (*Cyperus
articulatus*). Catalytic hydrogenation yielded successively a dihydroarticulone
and a tetrahydroarticulone, and ozonolysis afforded acetone, isolated as its
2,4-dinitrophenylhydrazone. Dehydrogenation studies were carried out by
heating with palladium both the crude product from **LiAlH$_4$** and that from
CH$_3$MgI treatment of articulone. The first yielded 1,6-dimethyl-4-isopropyl-
naphthalene and the second produced 1,6,8-trimethyl-4-isopropylnaphthalene.
Relevant spectral data are shown below. Deduce a structure for articulone and
classify it in a family of natural products.

	IR	UV	NMR
Articulone	5.95, 6.12 μ (1630, 1680 cm^{-1})	240 nm (log ϵ 4.0)	One vinyl **H**
Tetrahydroarticulone	5.83 μ (1715 cm^{-1})	None with $\epsilon > 100$	No vinyl **H**

27-38 Construct reasonable biosynthetic pathways to 7-hydroxycoumarin and 6,7-dimethoxycoumarin from tyrosine.

27-39 Comment on the biosynthesis of helenalin, a constituent of "sneeze weed."

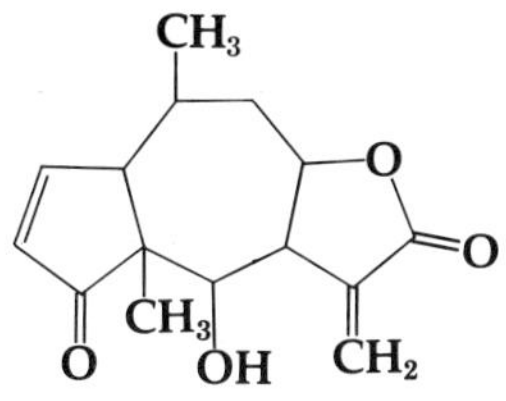

Helenalin

READING REFERENCES

Hendrickson, J. B., "The Molecules of Nature," W. A. Benjamin, Inc., New York, 1965.
Yates, P., "Structure Determination," W. A. Benjamin, Inc., New York, 1966.

THE LITERATURE OF ORGANIC CHEMISTRY

THE development of organic chemistry has created a vast and sprawling literature, which reaches into physics on the one side and biology on the other. In spite of the many languages, different systems of nomenclature, and millions of organic compounds known, a good library readily supplies the organic chemist with available information he needs on a given subject. It is important that he be able to locate previous work which relates to research he has in progress. It is more important that he not undertake unwittingly a project that has already been done. He must be able to find recipes for preparing compounds he needs which have previously been made. Types of source material include the following.

1 *Advanced general textbooks* provide a general picture of the state of development of the field as a whole.
2 *Reference books* provide specific information regarding special topics, synthetic procedures, and properties of organic compounds.
3 *Reviews* of research literature summarize advances made in particular fields of investigation.
4 *Handbooks and dictionaries* provide physical constants for particular organic compounds, and in some cases references to the original literature.
5 *Encyclopedias* are exhaustive collections of information regarding organic compounds. They frequently include physical data and references to the original literature.
6 *Abstracts* of research literature summarize the content of individual articles printed in all significant journals.
7 *Research journals* themselves are the primary source of all scientific information. They consist of articles setting forth research results and detailed laboratory accounts of the experiments involved.

With the accelerating increase in research support and numbers of practicing research chemists there has been a nearly explosive increase in the volume of primary literature in research journals. New journals are being founded each year, and older ones expanded, to accommodate this outpouring of new information. In consequence it has become increasingly difficult for chemists both to read (or even scan) all the new work relevant to their own and also to locate specific details of information they require from the traditional format of indexes, compendia, and reviews. Accordingly, new modes of information retrieval are being vigorously examined.

This chapter is designed as an introduction to the chemical literature in its traditional format of the seven categories above as well as some of the newer retrieval techniques. A large and representative collection of book titles on special topics is included (Sec. 28-1) but is not intended to be comprehensive for even such a list would be beyond reasonable lengths. Books written in English are stressed.

28-1   REFERENCE AND REVIEW BOOKS

A number of the books available in organic chemistry are advanced or specialized textbooks. The paperbacks and short texts in this category are listed as reading references at the ends of the appropriate chapters. Others which are more general in nature and do not specifically pertain to only a single chapter are included in the lists below. One of the most exhaustive general reference books written in English is that edited by Rodd. This work presents the whole field of organic chemistry in an integrated and systematic way and has been recently updated in a new edition.

"Chemistry of Carbon Compounds," E. H. Rodd, 2d ed., edited by S. Coffey, Elsevier, Amsterdam, 1964–1970.

The greater number of books on organic chemistry are in the nature of reviews and discussions of particular topics and serve the research chemist as reference books. Many of these monographs are collections of separate-topic chapters written by different specialists under the overall editorship of one person. A selection of these books is collected into the several lists below. In most cases the titles are self-explanatory. A number of works are in several volumes, and some are so organized as to be continuing series, with new volumes appearing on a roughly regular basis every year or two. These works are indicated by the notation "(continuing)."

Physical and Theoretical Organic Texts

These books are concerned with such topics as resonance, reaction mechanisms, stereochemical principles, steric effects, and correlations between structure and reactivity and between structure and physical properties.

"Mechanism and Structure in Organic Chemistry," E. S. Gould, Holt, New York, 1959.
"Physical Organic Chemistry," J. Hine, McGraw-Hill, New York, 1962.
"Rates and Equilibria of Organic Reactions," J. E. Leffler and E. M. Grunwald, Wiley, New York, 1963.
"Advanced Organic Chemistry: Reactions, Mechanisms, and Structure," J. March, McGraw-Hill, New York, 1968.
"Physical Organic Chemistry," L. P. Hammett, McGraw-Hill, New York, 1940.

"Advanced Organic Chemistry," G. W. Wheland, Wiley, New York, 1960.

"Physical Organic Chemistry," K. B. Wiberg, Wiley, New York, 1964.

"Introduction to Physical Organic Chemistry," E. W. Kosower, Wiley, New York, 1968.

"Resonance in Organic Chemistry," G. W. Wheland, Wiley, New York, 1955.

"Steric Effects in Organic Chemistry," edited by M. S. Newman, Wiley, New York, 1956.

"Molecular Orbital Theory for Organic Chemists," A. Streitwieser, Wiley, New York, 1961.

"The Molecular Orbital Theory of Organic Chemistry," M. J. S. Dewar, McGraw-Hill, New York, 1969.

Analytical and Other Techniques

A large number of books have been published on techniques of organic chemistry, which range all the way from simple laboratory methods to the use of spectrophotometers and the interpretation of spectra.

"Techniques of Organic Chemistry," 13 vols., edited by A. Weissberger, Interscience-Wiley, New York, (continuing), 1956–   . (Physical methods of Organic Chemistry. Catalytic, Photochemical and Electrolytic Reactions. Separation and Purification. Laboratory Engineering. Distillation. Adsorption and Chromatography. Micro and Semimicro Methods. Organic Solvents. Investigation of Rates and Mechanisms of Reactions. Chemical Applications of Spectroscopy. Fundamentals of Chromatography. Elucidation of Structures by Physical and Chemical Methods. Thin-layer Chromatography. Gas Chromatography.)

"The Systematic Identification of Organic Compounds, R. L. Shriner, R. C. Fuson, and D. Y. Curtin, Wiley, New York, 1964.

"Quantitative Analysis Via Functional Groups," S. Siggia, Wiley, New York, 1963.

"Infrared Absorption Spectroscopy," K. Nakanishi, Holden-Day, Inc., Publisher, San Francisco, 1962.

"The Infrared Spectra of Complex Molecules," L. J. Bellamy, 2d ed., Wiley, New York, 1958.

"Theory and Application of Ultraviolet Spectroscopy," H. H. Jaffe and M. Orchin, Wiley, New York, 1962.

"Interpretation of the Ultra Violet Spectra of Natural Products," A. I. Scott, Pergamon, New York, 1963.

"Handbook of Ultraviolet and Visible Absorption Spectra of Organic Compounds," K. Hirayama, Plenum, New York, 1967.

"Organic Electronic Spectra Data," 4 vols., Interscience-Wiley, 1946–1959.

"Applications of Nuclear Magnetic Resonance Spectroscopy in Organic Chemistry," L. M. Jackman, Pergamon, New York, 1960. 2d ed., 1969.

"Applications of NMR Spectroscopy in Organic Chemistry: Illustrations from the Steroid Field," N. S. Bhacca, and D. H. Williams, Holden-Day, Inc., Publisher, San Francisco, 1964.

"High Resolution Nuclear Magnetic Resonance Spectroscopy," 2 vols., J. W. Emsley, J. Feeney, and L. H. Sutcliffe, Pergamon, New York, 1966.

"Mass Spectrometry: Applications to Organic Chemistry," K. Biemann, McGraw-Hill, New York, 1962.

"Mass Spectrometry of Organic Compounds," H. Budzikiewicz et al., Holden-Day, Inc., Publisher, San Francisco, 1967.

"Applications of Mass Spectrometry to Organic Chemistry," R. I. Reed, Academic, New York, 1966.

"The Mass Spectra of Organic Molecules," J. H. Beynon, R. A. Saunders, and A. E. Williams, American Elsevier, New York, 1968.

"Structure Elucidation of Natural Products by Mass Spectrometry (Vol. I. Alkaloids)," H. Budzikiewicz, C. Djerassi, and D. H. Williams, Holden-Day, Inc., Publisher, San Francisco, 1964.

"Structure Elucidation of Natural Products by Mass Spectrometry. (Volume II. Steroids, Terpenoids, Sugars, and Miscellaneous Classes)," H. Budzikiewicz, C. Djerassi, and D. H. Williams, Holden-Day, Inc., Publisher, San Francisco, 1964.

"Optical Rotary Dispersion: Applications to Organic Chemistry," C. Djerassi, McGraw-Hill, New York, 1960.

"Optical Rotatory Dispersion and Circular Dichroism in Organic Chemistry," P. Crabbe, Holden-Day, Inc., Publisher, San Francisco, 1965.

Organic Reactions and Functional Groups

The discussions in the following books are organized either around functional-group types and their reactions or around types of reactions. "Organic Reactions" is the central collection of studies on specific reaction types. A new volume of "Organic Reactions" appears every 18 months. It summarizes the facts concerning five or six different general organic reactions, such as the Friedel-Crafts alkylation reaction or the Mannich synthesis. Each chapter is written by a different author familiar with the field and contains discussions of scope and limitations of reactions, sample procedures, tables of examples, and numerous references.

"Organic Reactions," various editors, Wiley, New York, (continuing) 1942– .

"Chemistry of Functional Groups," Interscience-Wiley, New York, (continuing) 1964– .

"Oxidation Mechanisms: Applications to Organic Chemistry," R. Stewart, W. A. Benjamin, Inc., New York, 1964.

"Mechanisms of Oxidation of Organic Compounds," W. A. Waters, Barnes & Noble, New York, 1964.

"Oxidation in Organic Chemistry," K. B. Wiberg, Academic, New York, 1965.

"Carbanion Chemistry," D. J. Cram, Academic, New York, 1965.

"1,4-Cycloaddition Reactions," J. Hamer, Academic, New York, 1967.

"Cyclobutadiene and Related Compounds," M. P. Cava and J. J. Mitchell, Academic, New York, 1967.

"Steroid Reactions: An Outline for Organic Chemists," C. Djerassi, Holden-Day, Inc., Publisher, San Francisco, 1963.

"Organic Compounds with Nitrogen-Nitrogen Bonds," C. G. Overberger, J-P. Anselme, and J. G. Lombardino, Ronald Press, New York, 1966.

"The Chemistry of Organic Fluorine Compounds," M. Hudlicky, Pergamon, New York, 1961.

"Organic Sulfur Compounds," 4 vols., N. Kharasch and C. Meyers, Pergamon, New York, 1961–1970.

"Hydroboration," H. C. Brown, W. A. Benjamin, Inc., New York, 1962.

"The Chemistry of Open-Chain Organic Nitrogen Compounds," 2 vols., P. A. S. Smith, W. A. Benjamin, Inc., New York, 1965–1966.

"Oxidation Mechanisms," R. Stewart, W. A. Benjamin, Inc., New York, 1964.

Organic Synthesis

The works below contain collections of synthetic methods as well as exhaustive compendia, discussed below.

"Organic Syntheses," collective vols. I–IV, and subsequent single volumes, various editors, Wiley, New York, (continuing), 1948–   .

"Newer Methods of Preparative Organic Chemistry" (English trans.), edited originally by W. Foerst, Interscience-Wiley, New York, (continuing), 1948–   .

"Methods of Organic Chemistry" (in German, referred to as Houben-Weil), 14 vols. published, edited by E. Müller, Georg Thieme Verlag, (continuing), 1952–   .

"Synthetic Methods of Organic Chemistry," W. Theilheimer, S. Karger, Basel, and Interscience-Wiley, (continuing), 1946–   .

"Reagents for Organic Syntheses," L. F. Fieser and M. Fieser, Wiley, New York, 1967.

"Modern Synthetic Reactions," H. O. House, W. A. Benjamin, Inc., New York, 1965.

In "Organic Syntheses," detailed and verified procedures for preparations of particular organic compounds are set forth. A new volume, edited by a different author, is published almost every year, and every 10 years collective volumes appear. Great emphasis is placed on reproducibility of experimental results in these volumes, and a representative group of reactions is selected.

In the multivolume "Synthetic Methods of Organic Chemistry" (frequently referred to as "Theilheimer"), a systematic classification of organic transformations was developed and into this are fitted organic reactions taken from the literature each year. An elaborate general index aids in locating procedures for the preparation of particular classes of compounds, irrespective of starting material. This compendium contains the most comprehensive and up-to-date listing of organic synthetic methods.

The single volume by the Fiesers is an especially valuable collection of modern synthetic usages arranged in an alphabetical order of the reagents used; nearly 1,200 reagents are discussed, with leading references.

Heterocycles and Natural Products

A number of the central works on natural products and heterocycles are listed here, while the main lists of naturally occurring compounds are included in Sec. 28-2. Such biochemistry texts as are cited are listed at the end of Chap. 26, but no effort has been made here to provide any coverage of the vast volume of published work in biochemistry.

"Tetracyclic Triterpenes," G. Ourisson, P. Crabbe, and O. Rodig, Holden-Day, Inc., Publisher, San Francisco, 1964.

"Steroids," L. F. Fieser and M. Fieser, Reinhold, New York, 1959.

"Indole Alkaloids," W. I. Taylor, Pergamon, New York, 1966.

"Vitamins and Coenzymes," A. F. Wagner and K. Folkers, Interscience-Wiley, New York, 1964.

"The Chemistry of Flavonoid Compounds," edited by T. A. Geissman, Macmillan, New York, 1962.

"The Terpenes," 5 vols., edited by Sir John Simonsen, Cambridge, 1947–1957.

"The Alkaloids," edited by R. H. F. Manske, Academic, New York, (continuing), 1950– .

"Ergebnisse der Alkaloid-Chemie bis 1960," H. G. Boit, Akademie Verlag, Berlin, 1961.

"The Biosynthesis of Terpenes, Steroids and Acetogenins," J. H. Richards and J. B. Hendrickson, W. A. Benjamin, Inc., New York, 1964.

"Medicinal Chemistry," A. Burger, Interscience-Wiley, New York, 1960.

"The Chemistry of Heterocyclic Compounds," edited by A. Weissberger, Interscience-Wiley, New York, (continuing), 1950– .

"Heterocyclic Compounds," edited by R. C. Elderfield, 9 vols., Wiley, New York, 1950–1965.

"Physical Methods in Heterocyclic Chemistry," edited by A. R. Katritzky, Academic, New York, 1963.

Special Topics

Those works which supplement the latter chapters of this textbook have been selected.

"Non-benzenoid Aromatic Compounds," edited by D. Ginsburg, Interscience-Wiley, New York, 1959.

"Grignard Reactions of Nonmetallic Substances," M. S. Kharasch and O. Reinmuth, Prentice-Hall, New Jersey, 1954.

"Reduction with Complex Metal Hydrides," N. G. Gaylord, Interscience-Wiley, New York, 1956.

"Organic Photochemistry," O. L. Chapman, Marcel Dekker, Inc., New York, 1967.

"Mechanistic Organic Photochemistry," D. Neckers, Reinhold, New York, 1967.

"Organic Photochemistry," R. T. Kan, McGraw-Hill, New York, 1966.

"Molecular Photochemistry," N. J. Turro, W. A. Benjamin, Inc., New York, 1965.

"Macromolecular Syntheses," 2 vols.: vol. 1 edited by C. G. Overberger, 1963; vol. 2 edited by J. R. Elliott, Wiley, New York, 1966.

"Bioorganic Mechanisms," 2 vols., T. C. Bruice, and S. J. Benkovic, W. A. Benjamin, Inc., New York, 1966.

"Carbene Chemistry," W. Kirmse, Academic, New York, 1964.

"Structure and Mechanism in Organophosphorus Chemistry," R. F. Hudson, Academic, New York, 1965.

"The Organic Chemistry of Phosphorus," A. J. Kirby and S. G. Warren, American Elsevier, New York, 1964.

"Divalent Carbon," J. Hine, Ronald, New York, 1964.

"Organo-Sulfur Chemistry," M. J. Jansseen, Interscience-Wiley, New York, 1967.

"Vistas in Free-Radical Chemistry," edited by W. A. Waters, Pergamon, New York, 1959.

"Free Radicals," W. Pryor, McGraw-Hill, New York, 1966.

Problems

Several collections of problems on synthesis, mechanism, structure determination, etc., have recently been compiled from the literature. Many of these problems are suited to the level of students using this textbook, and the intellectual challenge of these puzzles provides pleasure as well as instruction.

"Seminar Problems in Advanced Organic Chemistry," T. Goto, Y. Hirata, and G. H. Stout, Holden-Day, Inc., Publisher, New York, 1968.

"Problems in Organic Structure Determination," A. Ault, McGraw-Hill, New York, 1967.

"Fascinating Problems in Organic Reaction Mechanisms," S. Ranganathan, Holden-Day, Inc., Publisher, San Francisco, 1967.

"Problems in Spectroscopy: Organic Structure Determination by NMR, IR, UV, and Mass Spectra," B. M. Trost, W. A. Benjamin, Inc., New York, 1967.

Reviews of Research Literature

A number of journals and books are published each year by publishers, and by chemical societies of various countries, with the purpose of reviewing the research literature. *Chemical Reviews* is a journal published every two months in which review articles appear covering topics that range over the whole field of chemistry, and not restricted to recent work. *Annual Reports* is a series of books published once a year (since the beginning of the century) which reviews the chemical research literature of the prior year. *Quarterly Reviews*, which is published in England four times a year, resembles *Chemical Reviews*. *Angewandte Chemie*, published in Germany monthly, contains valuable review articles as well as short communications; an International Edition is published in English.

Chemical Reviews, published by the American Chemical Society.

Annual Reports, published by the Chemical Society (London).

Quarterly Reviews, published by the Chemical Society (London).

Angewandte Chemie, International Edition in English, published by the Gesellschaft Deutscher Chemiker.

Special reviews on the current state of understanding in various specialized areas are to be found in a variety of review volumes which appear on a regular periodic basis, commonly annual. The following list includes the chief review series in organic chemistry.

Advances in Alicyclic Chemistry
Advances in Carbohydrate Chemistry
Advances in Free-radical Chemistry
Advances in Heterocyclic Chemistry
Advances in Magnetic Resonance
Advances in Organic Chemistry
Advances in Organometallic Chemistry
Advances in Photochemistry
Advances in Physical Organic Chemistry
Advances in Protein Chemistry
Annual Surveys of Organometallic Chemistry
Progress in the Chemistry of Organic Natural Products
Progress in Organic Chemistry
Progress in Stereochemistry
Progress in Physical Organic Chemistry
Progress in Polymer Science
Progress in the Chemistry of Fats and Other Lipids
Topics in Phosphorus Chemistry
Topics in Stereochemistry

With such a vast literature just of review articles on the primary literature, it is necessary to have available an index of the topics to be found in review articles. These indexes of reviews are provided in the following books.

"Index to Reviews, Symposia Volumes and Monographs in Organic Chemistry," (1940–1960), N. Kharasch and W. Wolf, and supplements, Pergamon, New York, 1962.

Bibliography of Chemical Reviews (Chemical Abstracts Service), annually from 1958.

28-2 COMPENDIA OF ORGANIC COMPOUNDS

Handbooks and Dictionaries

Handbooks and dictionaries containing lists of organic compounds are useful for looking up the physical constants of common substances. A number of these are available in single volumes, as listed below, and most organic chemists own one or more of these. The two "Handbooks of Chemistry" also contain extensive lists of tables of constants and physical properties, conversion and mathematical tables, and equations of value. The "Ring Index" lists all possible ring systems, carbocyclic, heterocyclic, and polycyclic, of which examples are known, with references. The "Buyers Guide" ("Chem Sources") lists the thousands of commercially available organic compounds with their suppliers. The last three works on the list are comprehensive lists of specific natural products, Karrer's containing, for example, the 2669 nonalkaloid plant products characterized by 1958.

"Ring Index," 2d ed. and supplements, A. M. Patterson, American Chemical Society, Washington, D. C., 1960.

"Composition Tables: Data for Compounds Containing C, H, N, O, S," G. H. Stout, W. A. Benjamin, Inc., New York, 1963.

"Handbook of Organic Structural Analysis," Y. Yukawa, W. A. Benjamin, Inc., New York, 1965.

"Chem Sources Chemical Buyers Guide," Directories Publishing, 1968.

"Handbook of Chemistry," N. A. Lange, McGraw-Hill, New York, 1966.

"Handbook of Chemistry and Physics," Chemical Rubber Company, Cleveland, new edition annually.

"The Merck Index of Chemicals and Drugs," 8th edition, Merck Sharp & Dohme Research Laboratories, 1968.

"Tables for Identification of Organic Compounds," Chemical Rubber Company, Cleveland, 1962.

"Konstitution und Vorkommen der organischen Pflanzenstoffe (exclusive Alkaloide)," W. Karrer, Birkhäuser, Basel, 1958.

"The Pfizer Handbook of Microbial Metabolites," M. W. Miller, McGraw-Hill, New York, 1961.

"Physical Data on Indole and Dihydroindole Alkaloids," Eli Lilly Research Laboratories, 1960, and later supplements.

Two more ambitious works run to several volumes with more comprehensive lists. The Elsevier volumes are especially valuable for collections of complex polycyclic compounds such as steroids.

"Dictionary of Organic Compounds," 4th ed., 5 vols. with supplements, Oxford, New York, 1963–   .

"The Encyclopedia of Organic Chemistry," Elsevier, New York, (continuing), 1946–

Beilstein

With the number of organic compounds that have been described rising into the millions, the need for a catalog of all known compounds is obviously imperative. In the early part of the century an effort was begun in Germany to fulfill this need. The result was the "Handbuch der organischen Chemie," commonly referred to as "Beilstein," after its first editor. The Beilstein Institute continues this important cataloguing but it is increasingly difficult to keep up with the rate of publication of new compounds. However, in approximately one hundred volumes Beilstein lists all the organic compounds known through 1929 and is usually considered the only source necessary for searching compounds reported previous to that date.

Beilstein consists of three sets of volumes. The *main series* covers the literature through 1909, the *first supplementary series* from 1909 through 1919, and the *second supplementary series* from 1919 through 1929. A *third supplementary series*, covering the 1930–1949 period, is currently being prepared. Each supple-

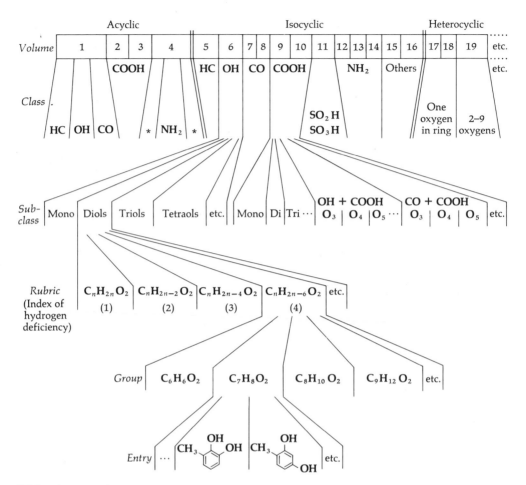

*Other classes not shown

FIGURE 28-1 **Schematic diagram of the "Beilstein" system**

mentary series contains the same number of volumes as the main series, and the arrangement of material in each supplementary series *parallels* that of the main series. For example, page 341 of volume 3 of the *main series* lists 3,3-dimethyl-2-hydroxybutanoic acid. Volume 3 of the *first supplementary* series lists this same compound on a page that has two numbers. The conventional page number appears at the top right (e.g., page 125). In the top middle appears the number 341. As soon as a compound is located in any series, it can quickly be found in any other series.

Three procedures are available for locating a particular compound in "Beilstein." The simplest method involves use of the cumulative formula indexes. Here compounds are listed by name in order of increasing complexity of their molecular formulas. Structural isomers are differentiated on the basis of names of the compounds.

A second cumulative index exists in which names of compounds are arranged in alphabetical order. The requisites for using this index are a knowledge of German nomenclature and some knowledge of the organization of "Beilstein." Use of this method for finding a particular compound is perhaps the least satisfactory.

The third method of finding a compound in "Beilstein" requires familiarity with the system of organization used in the volumes.

The system is a general and logical arrangement of compounds by structure such that any compound will have a *place* in the order whether it is known or not. In order to find this place for a given structure, we need to understand the system.

A book has been written ("A Brief Introduction to the Use of Beilstein's Handbuch der organischen Chemie," E. H. Huntress, Wiley, New York, 1938) to describe the organization of this compendium and facilitate its use. Another aid is the "Programmed Guide to Beilstein's Handbuch," O. A. Runquist, Burgess, Minneapolis, 1966. The categories of arrangement follow this outline, which is diagrammed schematically in Fig. 28-1. The categories are listed in the order in which they appear.

I Divisions

Acyclic compounds	Volumes 1–4
Isocyclic (carbocyclic) compounds	Volumes 5–16
Heterocyclic compounds	Volumes 17–27
Subdivided by number and kind of heteroatoms in the rings	
Indexes	Volumes 28–29
Natural products not otherwise classified	Volumes 30–31

II Classes (within each division)

Arrangement by functional groups; most of the entries are found in the starred classes.

*1 Hydrocarbons
*2 Hydroxy ("oxy") compounds
*3 Carbonyl ("oxo") compounds
*4 Carboxylic acids
 5 Sulfinic acids
 6 Sulfonic acids
 7 Selenium compounds
*8 Amines
 9 Hydroxylamines
10 Hydrazines
11 Azo compounds
12–28 Other relatively rare groups mostly involved with compounds containing elements other than C, H, O, S, N, X

III Subclasses (within each class)

Arrangement by number of groups present or, for hydrocarbons, by increasing unsaturation.

IV Rubrics (within each subclass)

Arrangement by increasing unsaturation, i.e., increasing index of hydrogen deficiency. The order for oxygenated compounds, for example, would be the following (with the index of hydrogen deficiency in parentheses):

$C_nH_{2n+2}O_m$ (0); $C_nH_{2n}O_m$ (1); $C_nH_{2n-2}O_m$ (2); $C_nH_{2n-4}O_m$ (3); etc.

V Groups (within each rubric)

Arrangement by increasing number of carbons. Thus monocarboxylic acids of the formula $C_nH_{2n-4}O_2$ will be arranged: $C_4H_4O_2$, $C_5H_6O_2$, $C_6H_8O_2$, $C_7H_{10}O_2$, etc. Individual isomers are listed in each group.

In order to find a compound in Beilstein one must locate the volume and page(s) bearing the desired functionality class, with the correct index of hydrogen deficiency and molecular size. The tables of contents in each volume greatly assist the rapid scanning of the system for the correct location.

Polyfunctional molecules are found *under that class which appears last in the above list.* For example, hydroxy-carboxylic acids appear under carboxylic acids, and amino-carboxylic acids under amines. This organization is called the "principle of latest position."

Compounds that contain groups not found among the functional classes of compounds above are considered *derivatives of these functional classes.* These derivatives are listed immediately after the parent compounds and in the order shown. For instance, after propionic acid are listed its amide, ester, etc. These derivatives are arranged in the following order:

1 *Functional derivatives,* or compounds that can be hydrolyzed to the parent compound.
2 *Substitution derivatives,* or compounds that contain halogen, nitroso, nitro, or azido ($-N_3$) groups, replacing hydrogens on the carbon skeleton. These derivatives are offered in the order: **F, Cl, Br, I, NO, NO$_2$, N$_3$**.
3 *Sulfur* (and *selenium*) *compounds,* which are listed as derived from the corresponding oxygen compounds (except for separate *classes 5 to 7,* above).

When a functional derivative can be hydrolyzed into two or more organic molecules, the substance is listed as a derivative of the hydrolytic product appearing latest in the list of basic classes of compounds, i.e., the principle of latest position. For example, esters are listed as derivatives of carboxylic acids. Ethers listed as derivatives of the parent alcohol, and secondary and tertiary amines as derivatives of the parent primary amine. The parent alcohol or primary amine in each case is the one of latest position in the system. With *polysubstituted derivatives,* the principle of latest position also applies, as it does in any ambiguous situation.

Heterocyclic compounds are defined as any compounds with one or more atoms other than carbon in the ring. Even cyclic anhydrides and lactones appear as heterocycles although they are always cross-referenced with their corresponding acyclic acids.

Each *entry* for a given compound is arranged in this order, with literature references running through it as required:

1 Structure and configuration
2 Occurrence, formation, preparation
3 Physical properties
4 Chemical reactions
   a Physical means (e.g., heat and light)
   b Inorganic reagents
   c Organic reagents
5 Physiological action and uses
6 Analytical methods for determination
7 Addition compounds and salts
8 Transformation products of unknown structure

Although the system appears complicated at first sight, it will be clear that some complexity of definition is necessary to obtain a rigorous logical organization for all the possible variations of organic structure. In fact Beilstein is easier to use than to describe, as a little practice, with the assistance of the generous tables of contents in each volume, will quickly demonstrate.

The following examples illustrate the procedure.

$$CH_3CHBrCONH_2$$

$$C_6H_5CH\!\!=\!\!CHCH\genfrac{}{}{0pt}{}{COOH}{NHC_6H_5}$$

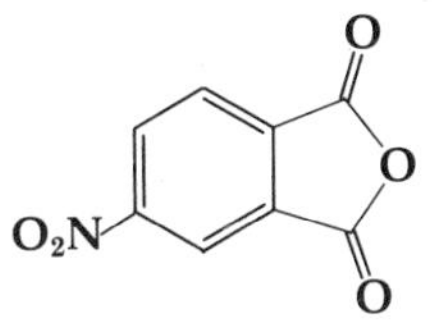

       A   *B*   *C*

A The structure above is a functional derivative (the amide) of propionic acid and a substitution (bromo) derivative of that. The order of various derivatives of propionic acid appearing under the propionic acid entry will be this:

$$CH_3CH_2COOCH_3, \ CH_3CH_2CONH_2, \ CH_3CHClCOOH, \ CH_3CHBrCOOH,$$
$$CH_3CHBrCONH_2, \ etc.$$

Propionic acid will be found in the acyclic division under the carboxylic acid class, monoacid subclass, $C_nH_{2n}O_2$ rubric, and $C_3H_6O_2$ group. The physical location of the entry will be in volume 2 of the main series (see Fig. 28-1), and the table of contents quickly guides us to pages 234 to 264. Scanning these pages turns up the entry on page 256. Information subsequent to the year 1909 (main series) is now found by examining the three supplementary series volumes 2 for **II-256** marked at top center of the page.

B This compound is a nitro substitution derivative of phthalic anhydride listed in the heterocyclic division under heterocycles containing one oxygen atom in the ring. It is then classed as a *dioxo* (carbonyl) compound of the $C_nH_{2n-12}O_3$ rubric, and found in volume 17, page 486.

C This amino acid will appear under amines, as the function of latest position. The particular parent primary amine could be either $C_6H_5NH_2$ or $C_6H_5CH{=}CHCH{-}$ (COOH)—NH_2 and the latter is chosen for reasons of latest position in the system. The latter primary amino acid is listed in the isocyclic division after monoamines, diamines, amino alcohols, etc., as a monoamine-monoacid in volume 14. It is listed as an amino derivative of the parent acid, $C_6H_5CH{=}CH{-}CH_2COOH$, or $C_{10}H_{10}O_2$- ($C_nH_{2n-10}O_2$), in the contents of volume 14, as page 525. Both the parent amino acid and the N-phenyl derivative are found together on that page. Had we required the ethyl ester we should not find it in the main series, nor under pages marked XIV-525, in the supplements. However, we know its place in the system. When it does not appear there, we know that the compound had not been prepared prior to 1929.

28-3 ABSTRACTS AND INDEXES

The American Chemical Society publishes an abstract journal known as *Chemical Abstracts*, which summarizes, in a paragraph or so each, all original articles on chemistry that appear in journals in all countries of the world. This journal is published twice a month, and each year exhaustive author, subject, and formula indexes appear. Since 1960 these indexes have been assembled every six months owing to the vast, and growing, content of Chemical Abstracts. Every 10 years a decennial index is published, and since 1957 these collective indexes have been reduced to 5-year instead of the previous 10-year (decennial) collections. These collective indexes include separate author, subject, and formula indexes, just like the annual indexes.

In 1951 a cumulative formula index appeared which covers the period from 1920 to 1946. Unfortunately, large numbers of organic compounds were not included in the index prior to 1945, and therefore the cumulative formula index should not be considered exhaustive. The decennial and collective indexes take up formula indexing following the 1920–1946 volume.

Chemisches Zentralblatt is the German counterpart of *Chemical Abstracts* and serves as a second source of abstracts of research articles. The formula index of *Chemisches Zentralblatt* is very useful. Collective indexes (formula, subject, and author) have appeared every five years up to 1939. Before 1939, the coverage of this journal was more complete than that of *Chemical Abstracts.*

28-4 THE PRIMARY LITERATURE

Research journals are fundamental to science. Almost every country in which chemical research is conducted publishes at least one research journal. The following list includes the major chemical journals of the world. In recent years many new journals have been founded to deal only with growing publication in specific areas of interest, such as steroids or photochemistry. These journals are not included in the list below. The boldface portions of the names indicate how the names of these journals are abbreviated.

Journal of the **Americal Chemical Society**, in English (USA), all branches of chemistry.

Journal of **Organic Chem**istry, in English (USA), organic chemistry.

Journal of **Biological Chem**istry, in English (USA), biological chemistry.

Journal of the **Chemical Society**, in English (British), all branches of chemistry.

Journal of the **Indian Chem**ical Society, in English, all branches of chemistry.

Biochemical Journal, in English (British), biological and natural-product chemistry.

Canadian Journal of **Chem**istry, in English, all branches of chemistry.

Chemische **Ber**ichte, in German, all branches of chemistry.

Annalen der **Chem**ie, in German, organic chemistry.

Zeitschrift für **physiol**ogische **Chem**ie, in German, organic and biological chemistry.

Helvetica Chimica Acta, in German, French, Italian, and English (Swiss), all branches of chemistry.

Monatshefte für **Chem**ie, in German (Austrian), all branches of chemistry.

Bulletin de la **société chim**ique de **France**, in French, all branches of chemistry.

Bulletin des **sociétés chim**iques **Belges** (Belgian), in French, English, and Dutch, all branches of chemistry.

Recueil des **travaux chim**iques des Pays-Bas, in English, French, German, and Dutch (Dutch), all branches of chemistry.

Acta Chemica **Scand**inavica, in English, German, and French, all branches of chemistry.

Arkiv för **Kemi**, in English, German, and French (Swedish), all branches of chemistry.

Gazzetta **chim**ica **ital**iana, in Italian, all branches of chemistry.

Collection of **Czechoslov**ak **Chem**ical **Comm**unications, in English, Russian, and German, all branches of chemistry.

Bulletin of the **Chem**ical **Society** of **Japan**, in Japanese, English, German, and French, all branches of chemistry.

Bulletin of the **Acad**emy of **Science** of the **U.S.S.R.**, in Russian (English translation available), all branches of chemistry.

Journal of **General Chem**istry (U.S.S.R.), or **Zhur**nal **Obshcheĭ Khim**ii, in Russian, (English translation available), all branches of chemistry.

Tetrahedron, various languages, international journal of organic chemistry.

When a chemist has completed a piece of research, he writes his experiments and results into a research paper which he sends to one of these journals. The editor in turn forwards the manuscript to one or two other chemists with like research interests and these *referees* review the paper and advise on its acceptability for publication. Following its acceptance (in original form or with recommended revisions) the paper is then published in the journal. The common format for organic chemical papers consists of an abstract, or one-paragraph summary of results; a brief introduction placing the work in the context of current understanding; a discussion of the experiments, their results and the authors' conclusions; and lastly, an experimental section devoted to detailed recipes for all the experiments and the new compounds prepared.

The publication process can consume a number of months. For this reason there exist several journals devoted solely to the quicker publication of "preliminary communications," more concise and without experimental sections. A number of the important journals of this kind are listed below (Some of the major journals above also carry a section for communications.)

Angewandte **Chem**ie (German, and international edition in English)
Chemical **Comm**unications (British)
Experientia (Swiss)
Nature (British)
Science (USA)
Tetrahedron Letters (international)
The Journal of **Chem**ical **Ed**ucation (USA) carries articles of interest to teachers and
    students.

28-5 INFORMATION RETRIEVAL

In order to keep abreast of current publication in his areas of interest the chemist has several options. One of these consists in scanning the relevant abstracts of papers in *Chemical Abstracts*. In order to locate certain subjects or concepts rapidly in the abstracts, a *Keyword Index* is provided. The Keyword Index is an alphabetical listing of the key words in the titles of every paper. The full title is oriented for each entry such that the key word alphabetized is centered on the page and printed in boldface. This provides an alphabetical center column of words easily scanned and located. An example is found in Fig. 28-2.

Besides *Chemical Abstracts* there are several publications which simply publish the titles or tables of contents of current research journals, in order to make the subject matter of current journals rapidly available to view. These publications include:

Chemical Titles, (USA), from Chemical Abstracts Service, includes keyword index.
Current Chemical Papers (British), from The Chemical Society (London).
Current Contents (USA), photoreproduced tables of contents from journals.

In order to be placed in context a scientific paper always contains many references to previous papers and reviews in the same area. The *Science Citation Index* provides an index of references cited in current papers. Hence knowledge of an older paper allows one to find all the papers published more recently which cited this older paper as a reference. Advantage can be taken of these references to locate papers published recently on specific matters for which only older references are known. Thus, if one wishes to know what work has been done this year to extend the results in a paper of 5 or 10 years ago, he can quickly find all the papers this year which cited the known, older paper.

Ultimately, the process of gathering, storing, and sorting chemical information will have to be done by computer. Indeed great strides have already been taken in this direction, but there are special difficulties. Computers take in and put out information in simple linear form, as in a sequence of letters and numbers. Hence a system is necessary for transforming three-dimensional

Keywords   *Journal-vol-page*

```
ALITY.=                 ORGANIC MATTER AND FINISHED WATER QU  JAWWA5-0061-0107
GEN COMPOSITION OF SOIL ORGANIC MATTER EXTRACTS.=+ THE NITRO  SOSCAK-0107-0108
MINO OXIDASE INHIBITOR+ ORGANIC NITRATE EXPLOSIVES AS MONO A  AEHLAU-0018-0311
ILATOR EFFECTIVENESS OF ORGANIC NITRATES.=          +AND THE VASOO  JPETAB-0165-0286
ERIAL ACTION OF IRRADI+ ORGANIC PER OXIDES AND THE ANTI-BACT  RAREAE-0037-0531
CT PRECIPITATION OF THE ORGANIC PHASE.=            +AMINE AND DIRE  ICBEAJ-32-15-775
OXYGEN RE+INTRACELLULAR ORGANIC PHOSPHATES AS REGULATORS OF   NATUAS-0221-0618
G OF THE PERSISTENCE OF ORGANIC PHOSPHORS.=             +DAMPIN  OPSPAM-0026-0092
ROBLEMS IN THE STUDY OF ORGANIC PHOTOCHEMICAL REACTIONS.=  P  ICBEAJ-32-15-042
AK TYPE CZECHOSLOVAKIAN ORGANIC POLYMER IN THIN LAYER CHROMA  CHPUA4-0019-0074
ERMINATION OF OXYGEN IN ORGANIC PRODUCTS.=       +ANALYTICAL DET  ICBEAJ-32-15-198
. REACTION OF POTASSI+ ORGANIC REACTIONS OF METAL CARBONYLS  BCSJA8-0041-2990
ULTIVALENT ELEMENTS AND ORGANIC REAGENTS.  INTERACTION OF     ZAKHA8-0024-0044
RATING OXYGEN SENSITIVE ORGANIC REDOX SYSTEMS WITH REDUCING   ANBCA2-0027-0536
LET LASER GENERATION IN ORGANIC SCINTILLATOR MOLECULES.=+VIO  PZETAB-0009-0015
PHOTO ELECTRIC STUDY OF ORGANIC SEMICONDUCTOR IN VIDICON      BBPCAX-0073-0086
QUEOUS SOLUTION AND THE ORGANIC SOLVENT.=        +BETWEEN THE A  ZAKHA8-0024-0015
OF LUMI FL+FLAVINES IN ORGANIC SOLVENTS.  SPECTRAL BEHAVIOR  PHCBAP-0009-0045
F CARBONYL COMPOUNDS IN ORGANIC SOLVENTS.=       +OF TRACES O  ZACFAU-0244-0244
M(II) BIS(DI OXIMES) IN ORGANIC SOLVENTS.=+ OF SOME PALLADIU  IASKA6-1969-0045
TION SPECTRA OF COMPLEX ORGANIC SUBSTANCES.=    +THE ABSORP  ZFKHA9-0043-0039
THE OXIDATION OF POLAR ORGANIC SUBSTANCES.=     +THE MEDIUM IN  ZFKHA9-0043-0115
IN-PILE IRRADIATION IN ORGANIC SULFUR COMPOUNDS IN AQUEOUS  JINCA0-0031-0605
ELECTROLYTIC GENERATIO+ ORGANIC SYNTHESIS BY ELECTROLYSIS.    TELEAY-1969-0687
ADICAL YIELDS IN FROZEN ORGANIC SYSTEMS.=+ TRANSFER ON THE R  BAPCAQ-0016-0527

LING FILTER MEDIA.=     ORGANICS REMOVAL BY A SELECTED TRICK  WWAEA2-06-01-22A
M EXCHANGE REACTIONS OF ORGANICS. LABELING OF SYNTHETIC ESTR  JLCAAI-0005-0021
ITS DISTRIBUTION IN THE ORGANISM.=     +ACTION OF CARBINE AND  FATOA0-0032-0100
DNA TURNOVER IN HIGHER ORGANISMS.=         +IMPLICATIONS OF  AHSUAV-0008-0041
CONTAMINATION OF MARINE ORGANISMS.=      +OF RUTHENIUM ON THE  CHDOAT-0268-0976
ION OF COBALTOCENE WITH ORGANO (BENZYL, ALLYL, AND PROPARGYL  JORCAI-0016-0301
EAR SILOXANE BONDS WITH ORGANO BORON COMPOUNDS.=     +OR LIN  VYSAAF-0011-0426
RED SPECTRA OF AROMATIC ORGANO BORON COMPOUNDS.=       INFRA  ZFKHA9-0043-6022
TION IN THE PRESENCE OF ORGANO COBALT COMPLEX CATALYSTS.=+LU  MGKFA3-0075-0126
OF TRI ETHYL INDIUM WI+ ORGANO INDIUM COMPOUNDS.  REACTIONS  JORCAI-0016-0215
ZATION OF PIPERYLENE BY ORGANO LITHIUM COMPOUNDS IN SOLUTION  KCRZAE-28-02-004
E BOND OF A+ADDITION OF ORGANO LITHIUM REAGENTS TO THE DOUBL  TELEAY-1969-0707
YL AND ALLY+REACTION OF ORGANO MAGNESIUM COMPOUNDS WITH BENZ  CHDCAQ-0268-0547
ESIS OF AZIRIDINES FROM ORGANO MAGNESIUM COMPOUNDS AND KET   TELEAY-1969-0759
ARACTERISTICS OF 8-DI + ORGANO METAL COMPLEXES.  AROMATIC CH  ANCEA0-0081-0150B
RING EXPA+REACTIVITY OF ORGANO METAL COMPLEXES.  SOLVOLYTIC  ANCEA0-0081-0153
                        ORGANO METALLIC CATALYSIS.=          KETIAL-0081-0020
OF COBAL+REACTIVITY OF ORGANO METALLIC COMPLEXES.  REACTION  JORCAI-0016-0301
I+REACTIONS OF GROUP IV ORGANO METALLIC COMPOUNDS.  BENZOYL  JSOOAX-1969-0701
ILIC S+POLY FLUORO ARYL ORGANO METALLIC COMPOUNDS.  NUCLEOPH  TETRAB-0025-0565
OF DI FERROCENYL KETO+ ORGANO METALLIC SERIES.  PROTONATION  JORCAI-0016-0283
TION OF FERROCENYL CAR+ ORGANO METALLIC STUDIES.  DE-PROTONA  TELEAY-1969-0817
D METABOLISM STUDIES ON ORGANO PHOSPHATE TERATOGENS AND THEI  BCPCA6-0018-0373
```

FIGURE 28-2   **Sample of a section of a keyword index**

organic structures into some linear sequence of information bits. Several systems for doing this have been worked out, but they are very complex to describe and to use.

Nevertheless, the Chemical Abstracts Service is currently coding all the published literature with one of these systems and filing all compounds in this fashion into computer memory. They can also retrieve on request any known structure and its attendant references. In principle the computer is ideally suited to store and quickly retrieve references for any given structure. We may hope that every library will soon have a machine which will offer a series of journal references every time a structure is written onto its sensor pad!

Meanwhile, the most rapid way of finding all original references to a given organic compound is the following:

1 Calculate the molecular formula for the compound, and name the substance in two or three ways.

2 Consult the cumulative formula index of "Beilstein," and differentiate among possible structural isomers by the name of the compound. Alternatively, find the location of the entry by means of the system of organization (Fig. 28-1). The first procedure is faster but the second is more certain should the formula index not produce the compound.

3 Consult the collective formula indexes of *Chemical Abstracts*, which cover the years 1920 through 1956.

4 The years 1956 to the present are covered by the yearly formula indexes of *Chemical Abstracts*.

5 Should the preceding steps not succeed in finding the compound, then it must be looked up by name in the collective and annual indexes of *Chemical Abstracts* after 1929.

Practice in this searching procedure and in the use of "Beilstein" can best be obtained by picking compounds at random from the previous chapters in this text and proceeding as indicated to locate the original research papers which describe them.

READING REFERENCES

Crane, E. J., A. M. Patterson, and E. B. Marr, "A Guide to the Literature of Chemistry," Wiley, New York, 1957.

Mellon, M. G., "Chemical Publications: Their Nature and Use," McGraw-Hill, New York, 1965.

"Searching the Chemical Literature," American Chemical Society, Washington, D.C., 1961.

NOMENCLATURE OF ORGANIC COMPOUNDS

DURING the first half of the nineteenth century, scientists working in organic chemistry discovered many new compounds and bestowed upon them names with little or no structural significance. Such *common* or *trivial* names might reflect the source of the material (e.g., formic acid means ant acid), the method of preparation of the substance (e.g., dichloride of ethylene), or the name of a person (e.g., barbituric acid is said to perpetuate the name of a certain Barbara). The confusion was not helped by the fact that chemists frequently used their classical education by choosing Latin or Greek roots for the names they invented. While such a system of nomenclature has a certain charm, the increasing number of compounds and names began to turn organic chemistry into Babel. Development of a structural theory in the 1860s was gratefully received for many reasons, principal among which was the prospect of using interrelated names that reflected structural relationships among compounds. Many such names were coined during the next few decades. However, many individualists continued to create new common names; in many instances common names were necessary to permit discussion of newly discovered compounds of *unknown structure*. Furthermore, it became obvious that several systematic names could be given to most compounds because of the multiplicity of structural relationships. For example, $ClCH_2CH_2Cl$ obviously *is* both the dichloride of ethylene and dichloroethane.

An obvious answer was to draw up rules for systematic nomenclature to be used by all chemists. In 1889, the International Chemical Congress appointed a committee to formulate definitive rules. The committee found that the multiplicity of both chemical structures and logical structural relationships made progress slow. However, a definitive report was made and accepted at Geneva in 1892. Although the report covered much ground very well, it was admittedly incomplete. However, organized work on nomenclature was not resumed until the project was taken over by the International Union of Chemistry (IUC) in 1919. The first report of the IUC committee was made in 1930 (Liége Rules). In 1947, a new Commission on Nomenclature was appointed by the union, which had become the *International Union of Pure and Applied Chemistry (IUPAC)*. The commission has issued periodic reports since that date.

Many problems of nomenclature arise for reasons that are not immediately obvious. For example, organic chemistry is reported and discussed by people who use many languages and a number of alphabets. Committees on nomenclature have wisely refrained from attempts to strike all common names from systems. Such an act would render future chemists unable to read the invaluable literature of the past century and a half unless they learned two chemical languages. Furthermore, judicious use of common names provides a convenient goup of parent names from which large numbers of derived names can be built. For example, no one has been seriously tempted to rename naphthalene as bicyclo[4.4.0]2,4,6,7,9-decapentaene, a name that would be both cumbersome and misleading because it would imply fixed positions of double bonds in the compound.

One of the greatest problems of nomenclature is the variety of purposes that names must serve. Names that are most valuable for the preparation of indices to the chemical literature may seem cumbersome and stilted if they are used repeatedly in other forms of literature such as research papers or textbooks. Similarly, names that are acceptable and useful in a textbook may be insufficiently graphic or too long for use in lectures or conversation.

The upshot is that *all chemists use several kinds of nomenclature.* They should all be familiar with the basic rules of the IUPAC system and should have access to the definitive reports issued by the international committee. The most recent report appeared in *The Journal of the American Chemical Society,* **82,** 5545 (1960), and contains references to the earlier reports. In addition, an organic chemist must know a considerable number of common names and understand the general principles of nomenclature.

This chapter is organized with emphasis on types of nomenclature. Consequently, nomenclature of the members of the various functional classes of compounds is not integrated. The following guide will assist readers who wish to compare systems for nomenclature as applied to a particular class of compounds.

Compound class	*Pages*
Acid (carboxylic)	1151, 1156, 1157, 1166, 1167, 1172, 1173, 1174
Acid (other)	1157, 1173
Acid halide and anhydride	1160, 1161, 1172
Alcohol	1158, 1165, 1166, 1172, 1174
Aldehyde	1157, 1166, 1172, 1173, 1174
Amide	1157, 1166, 1172, 1173, 1174
Amine	1159, 1166, 1172, 1174
Azo, azoxy	1151, 1165, 1167, 1174
Ester	1166, 1173
Ether	1159, 1166, 1172, 1174, 1176
Halide	1158, 1168, 1172, 1174
Ketone	1160, 1166, 1172, 1174
Mercaptan (thiol)	1151, 1160, 1165, 1174
Nitrile	1157, 1166, 1172, 1174
Nitro, nitroso	1165, 1172, 1174

General Principles

There are, broadly speaking, seven kinds of names in organic nomenclature:

1 *Functional names,* which are designed to specify the principal functional group and the degree of unsaturation of the molecule.

2 *Substitutive names,* which enumerate the substituents attached to some simpler parent structure. The parent structure is given a functional name.

3 *Additive names,* in which compounds are considered to be formed by *addition* of one or more atoms to a parent structure, e.g., ethylene oxide.

4 *Replacement names,* in which a compound is considered to be fabricated by replacement of one or more atoms in some related structure, e.g., C_6H_5SH, thiophenol.

5 *Conjunctive names,* which are formed from the names of two molecules joined together either directly or through some designated group, e.g., $C_{10}H_7CH_2COOH$, naphthaleneacetic acid; $C_6H_5N=NC(C_6H_5)_3$, benzeneazotriphenylmethane; $C_6H_5—C_6H_5$, biphenyl.†

6 *Radicofunctional names,* which couple the name of a radical (group) and the name of a functional class, e.g., acetyl chloride, ethyl alcohol.

7 *Common names.*

All systems of nomenclature attempt to define a relatively small number of parent structures and name all other compounds as derivatives of these parent structures through substitution names, replacement names, conjunctive names, etc. The IUPAC Commission uses a special definition of the term *systematic name*; it maintains that a systematic name is one in which every syllable is especially selected to indicate a structural feature (e.g., pentane, cyclopentanol). All names in which only some syllables are especially coined are called *semisystematic* (e.g., meth*ane*, cholester*ol*). The distinction is obviously one of terminology and does not imply that no system is involved in the formation of other kinds of names.

The utility of various names for the same compound may be illustrated by the following compound, which is known variously as pivalic acid, trimethylacetic acid, and 2,2-dimethylpropionic acid.

$(CH_3)_3CCOOH$

The common name, pivalic acid, is well entrenched in the chemical literature; it is the shortest of the available names and would surely be the name of choice if one were writing or speaking for an audience very familiar with the names

---

† The 1957 IUPAC report does not classify *replicate names,* such as biphenyl, as conjunctive; in fact, the names are not classified at all. The classification is slightly artificial but useful.

and structures of the fatty acids. The name trimethylacetic acid indicates the limited symmetry of the molecule in a way that would be immediately recognizable to any reader familiar with organic chemistry. The IUPAC name, 2,2-dimethylpropanoic acid, is the most cumbersome of the three; however, since the name is a part of the most extensive system of names, it has many uses. Although *Chemical Abstracts* uses many IUPAC names, the common names of simple carboxylic acids are retained. For example, one is certain to find an entry under propionic acid, 2,2-dimethyl.†

Both hydrocarbon and functional groups are given names so that they may be designated as substituents in the formation of substitutive names.

Functional Names

Functional names contain two or three parts: a *root*, which indicates the number of carbon atoms in the molecule; an *ending*, which shows the nature of the functional group and the degree of unsaturation; sometimes a prefix (such as *cyclo-* or *iso-*), which specifies a particular skeletal arrangement. The names of hydrocarbons and carboxylic acids provide the roots used to form functional names of other classes of compounds.

Hydrocarbons. Since the names of hydrocarbons are basic to a large portion of organic nomenclature, a large number of hydrocarbons are named in Table A-1.

The uses of several prefixes and suffixes are illustrated in the table. The suffix *-ane* indicates a hydrocarbon structure containing no double or triple bonds‡; the ending *-ene* indicates the presence of a double bond. Use of the ending *-yne* to indicate the presence of triple bonds is restricted almost exclusively to IUPAC names (page 1169). The prefix *n-* (normal) designates an unbranched chain, and the prefix *iso-* indicates the presence of a $(CH_3)_2CH-$ group at the end of a major chain.§ The prefix *neo-* indicates a $(CH_3)_3C-$ group; *cyclo-* shows the presence of a ring; *spiro-* indicates that two rings have a single carbon atom in common. The use of *sym-* and *unsym-*, meaning symmetrical and unsymmetrical, differentiates between isomers. The prefixes *o-* (*ortho*), *m-* (*meta*) and *p-* (*para*) indicate, respectively, 1,2-, 1-3-, and 1,4- disposition of two substituents attached to a benzene nucleus. The latter terms are most often used in substitutive names, e.g., *m*-chlorotoluene, but may also be used as a part of common names, e.g., *m*-xylene.

---

† It is interesting to note that the entry is "see pivalic acid." It is also noteworthy that IUPAC names can be used to locate subjects in *Chemical Abstracts*. The entry under "propanoic acid" in the index is "see propionic acid."

‡ Note that the ending does not indicate the degree of hydrogen deficiency, since the presence of cyclic structures is indicated by a prefix.

§ There are exceptions to this generalization. The prefix *iso-* sometimes merely indicates that the compound is an isomer of some other compound that has a well-recognized common name, e.g., leucine, $(CH_3)_2CHCH_2CH(NH_2)COOH$, and isoleucine, $CH_3CH_2CH(CH_3)CH(NH_2)COOH$.

TABLE A-1 **Names of Hydrocarbons**

Formula	*Name*	*Formula*	*Name*
	Alkanes		
CH_4	Methane	$CH_3(CH_2)_{10}CH_3$	Dodecane
CH_3CH_3	Ethane	$CH_3(CH_2)_{11}CH_3$	Tridecane
$CH_3CH_2CH_3$	Propane	$CH_3(CH_2)_{12}CH_3$	Tetradecane
$CH_3(CH_2)_2CH_3$	Butane	$CH_3(CH_2)_{13}CH_3$	Pentadecane
$(CH_3)_3CH$	Isobutane	$CH_3(CH_2)_{14}CH_3$	Hexadecane
$CH_3(CH_2)_3CH_3$	Pentane	$CH_3(CH_2)_{15}CH_3$	Heptadecane
$(CH_3)_2CHCH_2CH_3$	Isopentane	$CH_3(CH_2)_{16}CH_3$	Octadecane
$(CH_3)_4C$	Neopentane	$CH_3(CH_2)_{17}CH_3$	Nonadecane
$CH_3(CH_2)_4CH_3$	Hexane	$CH_3(CH_2)_{18}CH_3$	Eicosane
$CH_3(CH_2)_5CH_3$	Heptane	$CH_3(CH_2)_{19}CH_3$	Heneicosane
$CH_3(CH_2)_6CH_3$	Octane	$CH_3(CH_2)_{20}CH_3$	Docosane
$CH_3(CH_2)_7CH_3$	Nonane	$CH_3(CH_2)_{28}CH_3$	Triacontane
$CH_3(CH_2)_8CH_3$	Decane	$CH_3(CH_2)_{33}CH_3$	Pentatriacontane
$CH_3(CH_2)_9CH_3$	Undecane	$CH_3(CH_2)_{38}CH_3$	Tetracontane
	Alkenes		
$CH_2{=}CH_2$	Ethylene	$CH_3CH_2CH_2CH{=}CH_2$	α-Amylene†
$CH_3CH{=}CH_2$	Propylene	$CH_3CH_2CH{=}CHCH_3$	β-Amylene†
$CH_3CH_2CH{=}CH_2$	*unsym*-Butene	$(CH_3)_2CHCH{=}CH_2$	Isopentene
$CH_3CH{=}CHCH_3$	*sym*-Butene	$(CH_3)_2C{=}CHCH_3$	Trimethylethylene
$(CH_3)_2C{=}CH_2$	Isobutene	$(CH_3)_2C{=}C(CH_3)_2$	Tetramethylethylene
$CH_2{=}CH{-}CH{=}CH_2$	Butadiene	$CH_2{=}\underset{CH_3}{C}{-}CH{=}CH_2$	Isoprene
$CH_2{=}C{=}CH_2$	Allene		

Cycloalkanes and cycloalkenes

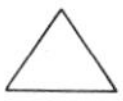

Cyclopropane   Cyclopentane   Cyclohexene

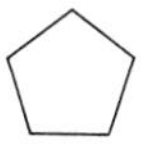

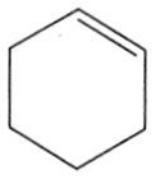

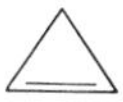

Cyclopropene   Cyclopentene   Spiropentane

Cyclobutane   Cyclopentadiene   1,3-Cyclohexadiene

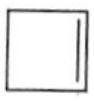

Cyclobutene   Cyclohexane   Decalin

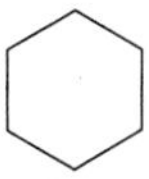

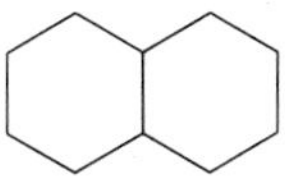

† IUPAC names 1-pentene and 2-pentene are now used almost universally, but the root *amyl* persists in some names, e.g., amyl alcohol.

TABLE A-1 **Names of Hydrocarbons** (*Continued*)

Formula	Name	Formula	Name

Aromatic hydrocarbons

	Benzene	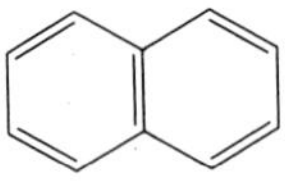	Naphthalene
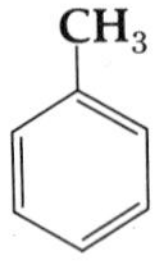	Toluene	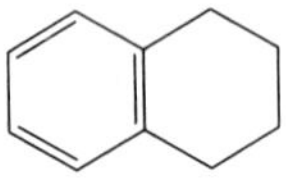	Tetralin
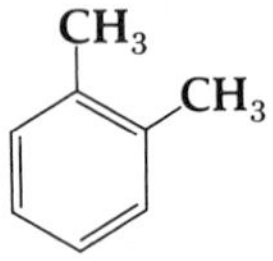	o-Xylene	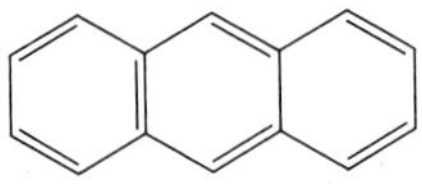	Anthracene
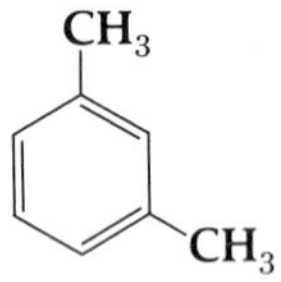	m-Xylene	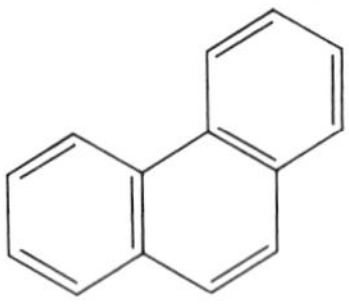	Phenanthrene
	p-Xylene	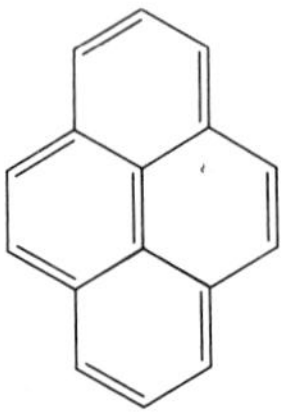	Pyrene
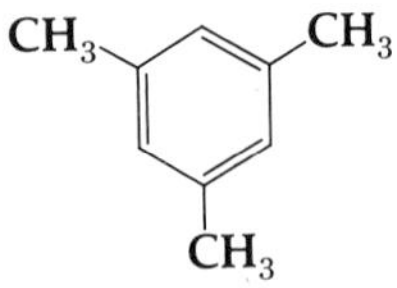	Mesitylene	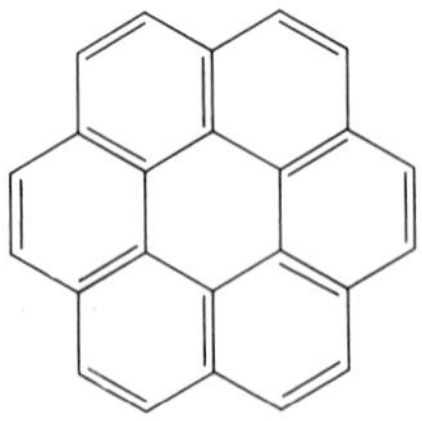	Coronene
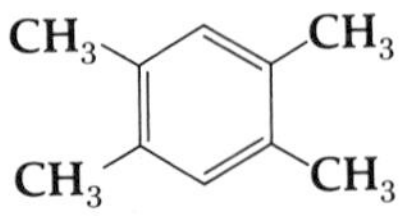	Durene		
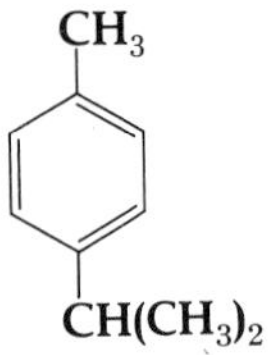	p-Cymene	$C_6H_5CH{=}CH_2$	Styrene
		$C_6H_5CH{=}CHC_6H_5$	Stilbene

Hydrocarbon Groups. Hydrocarbon groups (sometimes called radicals), formed by removing single hydrogen atoms from hydrocarbon structures, are usually named by replacing the suffix *-ane* by *-yl*, *-enyl* and *-ynyl*. The two groups that can be derived from propane are distinguished by the prefixes *n-* and *iso-*.

$CH_3CH_2CH_2$— $(CH_3)_2CH$—

n-Propyl   Isopropyl

Four groups are derived from butane and isobutane. The discriminants *sec-* and *tert-* stand for "secondary" and "tertiary." A secondary group always has the part structure $(C)_2CH$—, and a tertiary group has the part structure $(C)_3C$—. The terms "secondary" and "tertiary" lead to unique structural designations only with the simpler groups and are useful in specific nomenclature only when used without ambiguity.

$$CH_3CH_2CH_2CH_2- \qquad CH_3CH_2\overset{\overset{\displaystyle CH_3}{|}}{C}H- \qquad CH_3\overset{\overset{\displaystyle CH_3}{|}}{C}HCH_2- \qquad (CH_3)_3C-$$

n-Butyl   *sec*-Butyl   Isobutyl   *tert*-Butyl

Only four of the eight hydrocarbon groups that contain five carbon atoms can be named uniquely, and the prefix *sec-* as applied to this system is useless.

$$CH_3(CH_2)_3CH_2- \qquad (CH_3)_2CHCH_2CH_2- \qquad CH_3CH_2\overset{\overset{\displaystyle CH_3}{|}}{\underset{\underset{\displaystyle CH_3}{|}}{C}}- \qquad (CH_3)_3CCH_2-$$

n-Pentyl   Isopentyl   *tert*-Pentyl   Neopentyl
(*n*-amyl)   (isoamyl)   (*tert*-amyl)

Simple unsaturated hydrocarbon groups have very important names of more or less obscure origin. The more important of these are listed below.

$CH_2{=}CH$— $CH_2{=}CHCH_2$— $CH_3CH{=}CHCH_2$— $HC{\equiv}CCH_2$—

Vinyl   Allyl   Crotyl   Propargyl

$p\text{-}CH_3C_6H_4$— $C_6H_5CH_2$— $C_6H_5CH{=}$ $C_6H_5C{\equiv}$

p-Tolyl   Benzyl   Benzal   Benzo
                              (benzylidene)

C_6H_5—   —C_6H_4—

Phenyl   Phenylene   α-Naphthyl   β-Naphthyl

Saturated hydrocarbon groups are referred to as *alkyl groups* and are designated by the general symbol **R**. Unsaturated hydrocarbon groups are called *alkenyl groups*, while the aromatic hydrocarbon groups are called *aryl groups* and are represented by the general symbol **Ar**.

TABLE A-2 **Names of Common Carboxylic Acids**

Structure	Name	Structure	Name
Acids		**Related acyl or aroyl groups**	
HCOOH	Formic acid	HCO—	Formyl
CH_3COOH	Acetic acid	$CH_3CO—$	Acetyl
CH_3CH_2COOH	Propionic acid	$CH_3CH_2CO—$	Propionyl
$CH_3(CH_2)_2COOH$	n-Butyric acid	$CH_3CH_2CH_2CO—$	n-Butyryl
$(CH_3)_2CHCOOH$	Isobutyric acid	$(CH_3)_2CHCO—$	Isobutyryl
$CH_3(CH_2)_3COOH$	Valeric acid	$CH_3(CH_2)_3CO—$	n-Valeryl
$(CH_3)_2CHCH_2COOH$	Isovaleric acid	$(CH_3)_2CHCH_2CO—$	Isovaleryl
$(CH_3)_3CCOOH$	Pivalic acid	$(CH_3)_3CCO—$	Pivalyl
$CH_2{=}CHCOOH$	Acrylic acid	$CH_2{=}CHCO—$	Acryloyl
$CH_3CH{=}CHCOOH$	Crotonic acid	$CH_3CH{=}CHCO—$	Crotonyl
C_6H_5COOH	Benzoic acid	$C_6H_5CO—$	Benzoyl
$C_6H_5CH{=}CHCOOH$	Cinnamic acid	$C_6H_5CH{=}CHCO—$	Cinnamoyl
$C_6H_5CH_2COOH$	Phenylacetic acid	$C_6H_5CH_2CO—$	Phenylacetyl

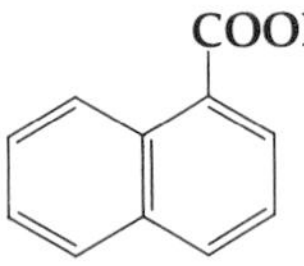

COOH

α-Naphthoic acid $\alpha\text{-}C_{10}H_7CO—$ α-Naphthoyl

COOH

β-Naphthoic acid $\beta\text{-}C_{10}H_7CO—$ β-Naphthoyl

Diacids

Structure	Name	Structure	Name
HOOCCOOH	Oxalic acid		Maleic acid
$HOOCCH_2COOH$	Malonic acid		
$HOOC(CH_2)_2COOH$	Succinic acid		Fumaric acid
$HOOC(CH_2)_3COOH$	Glutaric acid		
$HOOC(CH_2)_4COOH$	Adipic acid	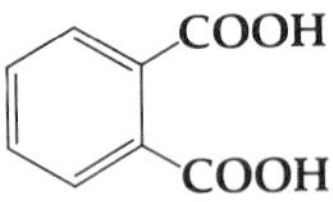	
$HOOC(CH_2)_5COOH$	Pimelic acid		Phthalic acid
$HOOC(CH_2)_6COOH$	Suberic acid		
$HOOC(CH_2)_7COOH$	Azelaic acid	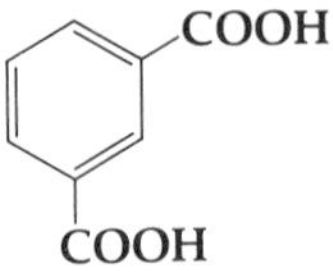	Isophthalic acid
$HOOC(CH_2)_8COOH$	Sebacic acid		

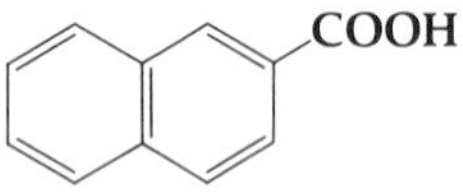

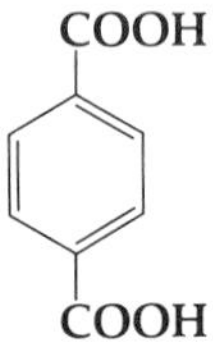

Maleic acid:

H—C—COOH
‖
H—C—COOH

Fumaric acid:

HOOC—C—H
‖
H—C—COOH

Phthalic acid:

Isophthalic acid:

Terephthalic acid:

Carboxylic Acids.   A second group of enumerative roots are derived from the names of carboxylic acids. For purposes of nomenclature, the acids are considered parents of most compounds containing acyl (RCO—) and aroyl (ArCO—) groups. Table A-2 gives the names of monoacids and the acyl or aroyl groups that they contain. The names of a few dibasic acids are also included. Groups having names derived from those of dibasic acids are bifunctional, e.g., —COCH$_2$CO—, malonyl.

Other Acids.   Acids such as sulfonic and phosphonic acids are given functional names by coupling the name of the parent hydrocarbon to the class name. The same system is occasionally applied to carboxylic acids.

CH_3SO_3H $p\text{-}CH_3C_6H_4SO_3H$ $C_6H_5PO_3H_2$

Methanesulfonic acid p-Toluenesulfonic acid   Benzenephosphonic acid

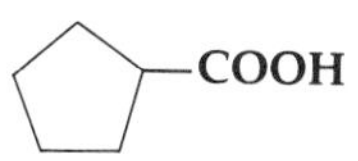

COOH

Cyclopentanecarboxylic acid α-Naphthalenecarboxylic acid

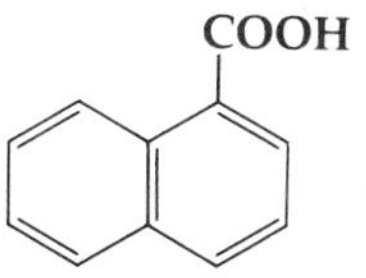

Other Functional Names.   Several classes of compounds whose names are derived from those of carboxylic acids are given functional names. These classes include aldehydes, amides, and nitriles.

Aldehydes:

CH_3CHO $(CH_3)_3CCHO$ C_6H_5CHO $OHC(CH_2)_4CHO$

Acetaldehyde   Pivalaldehyde   Benzaldehyde   Adipaldehyde

Amides:

$C_6H_5CH_2CONH_2$ $CH_3(CH_2)_2CONH_2$ $C_6H_5CONH_2$ $H_2NOCCH_2CH_2CONH_2$

Phenylacetamide   Butyramide   Benzamide   Succinamide

Nitriles:

$(CH_3)_2CHCH_2CN$ $CH_2{=}CHCN$ $CH_3CH{=}CHCN$ $\beta\text{-}C_{10}H_7CN$

Isovaleronitrile   Acrylonitrile   Crotononitrile β-Naphthonitrile

Acid halides and anhydrides, which are also given names derived from those of carboxylic acids, are given radicofunctional names (next section). Some of the other common classes of compounds are given functional names under the IUPAC system. In other systems of nomenclature, they are usually given radicofunctional names.

Radicofunctional Names

A radicofunctional name is formed from the name of a group (radical) and the name of a class. The parts of the name are separated if the class name is not the name of the parent compound of the series. Thus, ethyl alcohol (C_2H_5—OH) is written as two words because "alcohol" is not the name of the parent structure (H—OH). The name "triphenylcarbinol" is run together because "carbinol" is a little-used name for methyl alcohol, the parent of the series. Names of amines are run together, e.g., trimethylamine, since "amine" is a contraction of the name of the parent, ammonia.

Alcohols. Alcohols are named by two systems. The system for designation as *alkyl alcohols* is straightforward as long as the alkyl groups are easily named.

CH_3OH   **$(CH_3)_3COH$**

Methyl   *tert*-Butyl
alcohol   alcohol

Alcohols are also given radicofunctional names in which they are considered to be derivatives of methanol (the functional unit). However, the word "carbinol" is substituted for the words "methyl alcohol."

$(C_6H_5)_3COH$ **CH_2=$CHCH(OH)CH_3$**

Triphenylcarbinol   Methylvinylcarbinol

Phenols. Very familiar common names such as phenol,† naphthol, and cresol are always used. Other phenols are named by the IUPAC system in which the suffix -*ol* is added to the name of the parent hydrocarbon (or the root of the name).

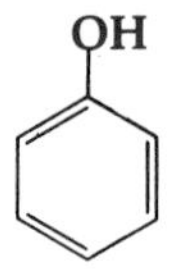

Phenol 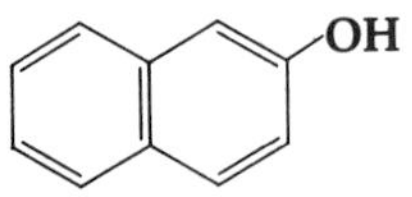   β-Naphthol 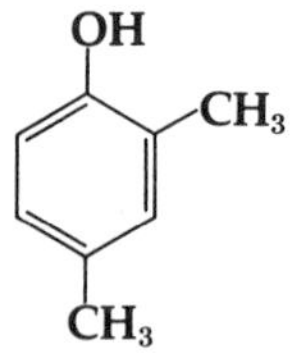   2,4-Xylenol 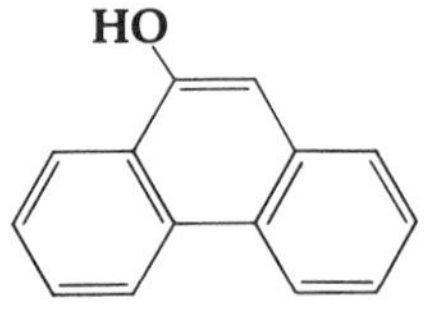   9-Phenanthrol

Halides. Halides are given names closely related to those of the corresponding alcohols. Most simple aliphatic halides are simply named as alkyl halides.

CH_3CH_2Br   **CH_2=CHI**   **$C_6H_5CHCl_2$**

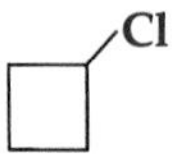

Ethyl bromide   Cyclobutyl   Vinyl   Benzylidene
                chloride   iodide   chloride
                                            (benzal chloride)

† Note that phenol is really a contraction of phenyl alcohol.

Halides that contain complex alkyl groups are sometimes named as *carbinyl halides*. The use of the suffix *-carbinyl* corresponds exactly to the use of *-carbinol* in the nomenclature of alcohols.

$$CH_2{=}CHCH_2\underset{\underset{Cl}{|}}{C}(CH_3)_2$$

Allyldimethylcarbinyl chloride

Ethers. Attempts to make the nomenclature of ethers systematic has rendered the systems more complex than seems necessary. Formerly ethers were given radicofunctional names in which the names of both groups attached to the oxygen atom were designated. A few aromatic ethers have common names.

$CH_3CH_2OCH_2CH_3$ $C_6H_5OCH_3$

    Diethyl ether   Methyl phenyl ether
                      (anisole)

At the present time *Chemical Abstracts* omits the prefix *di-* from the names of symmetrical ethers and indexes unsymmetrical ethers under the name of the alcohol corresponding to the more complex of the two groups. The latter system generates names that are unusable in ordinary writing or conversation.

$CH_3CH_2OCH_2CH_3$ $CH_3CH_2CH_2CH_3OCH_3$

    Ethyl ether   Butyl methyl ether

To make things worse, the IUPAC system (page 1174) names ethers as alkoxy and aryloxy substituents of hydrocarbons.

Amines. The radicofunctional names of amines are really based upon consideration of the compounds as derivatives of ammonia. Many arylamines and heterocyclic amines are always known by common names. Some secondary and tertiary amines are named as derivatives of well-known parents, such as aniline, and a capital *N*, rather than a number, is used to locate the substituents. When *sec-* and *tert-* appear in the names of amines, they are *always* part of the names of hydrocarbon groups and *never* give an indication of the classification of the amine.

$$CH_3CH_2\underset{\underset{NH_2}{|}}{C}HCH_3 \qquad CH_3CH_2CH_2N(CH_3)_2$$

sec-Butylamine   *N, N*-Dimethyl-*n*-propylamine

p-Toluidine 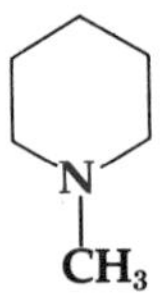   *N*-Methylpiperidine

Ketones. Ketones are named by naming the groups attached to the carbonyl group and adding "ketone." The functional names assigned to ketones under the IUPAC system (page 1172) are being used more and more commonly in ordinary writing and conversation. Phenyl and naphthyl ketones have common names, such as acetophenone, benzophenone, and acetonaphthone.†

$CH_3CH_2COCH_2CH_3$ $(CH_3)_2CHCH_2COCH_3$

   Diethyl ketone   Isobutyl methyl ketone

$C_6H_5COCH_3$ $(C_6H_5)_2CO$

   Acetophenone   Benzophenone   β-Acetonaphthone
                                   (methyl β-naphthyl ketone)

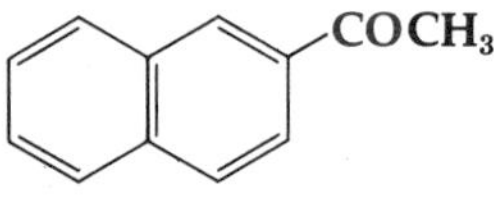

Sulfur and Phosphorus Compounds. Compounds of sulfur and phosphorus, except for the oxy acids, are usually given radicofunctional names.

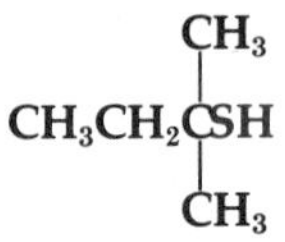

   tert-Amyl   Diethyl   Dimethyl   Dimethyl   Diphenyl
   mercaptan   sulfide   disulfide   sulfoxide   sulfone
               (ethyl sulfide)   (methyl   (methyl   (phenyl sulfone)
                               disulfide)   sulfoxide)

Phosphorus compounds:

$(C_6H_5)_3P$ $CH\,CH_2PHCH$ $(C_6H_5)\overset{+}{P}{-}\overset{-}{O}$

Triphenyl-   Ethylmethyl-   Triphenylphosphine   phosphine   phosphine oxide

As with ethers, *Chemical Abstracts* uses radicofunctional names of sulfides, disulfides, sulfoxides, and sulfones but omits the prefix *di-* before the names of symmetrical species.

Acid Halides. Acid chlorides are always given radicofunctional names using the names of acyl groups (Table A-2).

---

† The names are derived from the old names of the acyl groups plus hydrocarbon names plus the ending *-one*. Note the derivation from *phene,* an old name for benzene.

CH$_3$COCl

Acetyl chloride   Benzoyl   Isophthaloyl
                   chloride   chloride

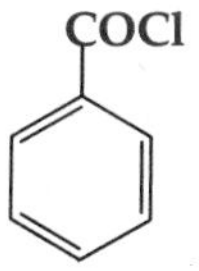

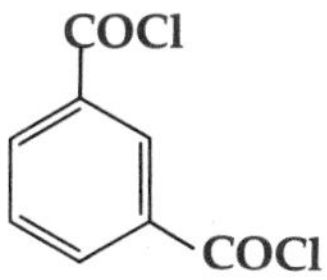

Acid Anhydrides. Symmetrical anhydrides are simply named as anhydrides of the parent acids.

(CH$_3$CO)$_2$O   (C$_6$H$_5$CO)$_2$O

Acetic anhydride   Benzoic   Phthalic anhydride
                       anhydride

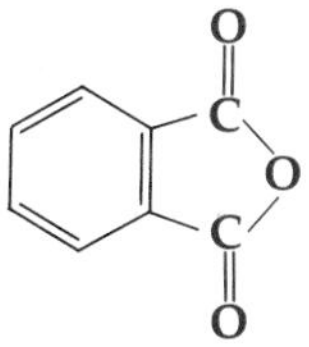

No student of nomenclature seems to have made a careful study of the nomenclature of mixed anhydrides. Relatively few of these compounds have been characterized so the need for a nomenclature system has not been pressing. Various haphazardly chosen ways are used to designate the two acyl groups in mixed anhydrides; *Chemical Abstracts* double-indexes mixed anhydrides under both the parent acids.

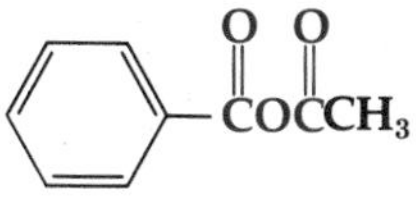

Acetic benzoic anhydride
(acetyl benzoyl anhydride)
(mixed anhydride of benzoic and acetic acids)
(acetic acid—anhydride with benzoic acid)
(benzoic acid—anhydride with acetic acid)

The chaos existing in the nomenclature of a rare group of compounds such as mixed anhydrides illustrates the need for commonly accepted systems of nomenclature.

Carbonium Ions, Carbanions, and Free Radicals.   No consistent scheme has been devised for the nomenclature of these species, which usually occur only as transient reaction intermediates. It is, therefore, necessary to distinguish carefully between desirable nomenclature and actual practice. Free radicals are commonly given radicofunctional names. The name of the group is followed by "radical." Groups are frequently named as derivatives of the methyl group.

CH$_3$CH$_2$·   (C$_6$H$_5$)$_3$C·
Ethyl radical   Phenyl radical   Triphenylmethyl radical

Common nomenclature of cations and anions is not as desirable. Logically, carbonium ions should be given names analogous to those of radicals in which the charge type of the group is indicated by the words "anion" and "cation."

$CH_3CH_2{:}^-$ $(CH_3)_2CH^+$

Ethyl anion   Isopropyl cation

The words "carbonium ion" and "carbanion" could then be reserved for names in which the species are considered as derivatives of methyl cation (*carbonium ion*) and methyl anion (*carbanion*). The methyl anion can also be conveniently designated as *methide ion* in substitutive names.

$$CH_2\!-\!\overset{\displaystyle CH_3}{\underset{\displaystyle CH_3}{\overset{|}{\underset{|}{C}}}}{}^+$$ $C_6H_5CH_2{}^+$

Trimethylcarbonium ion   Benzyl cation
(*tert*-butyl cation)

$CH_2{:}^-$ $C_6H_5{:}^-$

β-Naphthylcarbanion   Phenyl anion
(β-naphthylmethide ion)

Unfortunately, the words "carbonium ion" and "carbanion" are often used in the way that cation and anion are used in the above examples; thus, $(CH_3)_3C^+$ is often called "*tert*-butyl carbonium ion."† Use of the compound term "carbonium ion" in place of the simple name "cation" seems unnecessary, and the opportunity is lost to use the former name in a useful way for naming complicated branched cations.

Substitutive Names

Substitutive names, sometimes called *derived names*, are assigned to the majority of organic compounds. Compounds are named as derivatives of simpler parent molecules. The system is very useful, especially in chemical discussion, as a means of assigning picturesque names to convey a clear visual image of structural formulas to a reader or listener. The best substitutive names for oral communication are those which make maximum use of symmetry or use very familiar compounds as parents. The IUPAC system (page 1149) is fundamentally a carefully devised system for the choice of parent structures for the formation of substitutive names.

Compounds that contain two or more functional groups are almost always given substitutive names in which the parent structure is defined to include

---

†Note the divided spelling.

one of the functional groups and the others are designated as substituents. Occasionally, all the functional groups are considered as substituents on a hydrocarbon parent. The only rigid rules for assignment of substitutive names are those of the IUPAC system.

Hydrocarbons. Hydrocarbons provide good illustrations of the use of substitutive names. Methane, ethane, ethylene, acetylene, cycloalkanes, cyclo-alkenes, benzene, and other aromatic nuclei are often chosen as parent structures. Carbon atoms in the parent skeleton are assigned numbers or Greek letters to show the location of substituents. The following hydrocarbon could be given a variety of names:

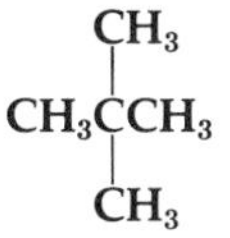

$$CH_3CCH_3$$
with CH_3 above and CH_3 below

Neopentane
(tetramethylmethane)
(2,2-dimethylpropane)
(1,1,1-trimethylethane)

The common name, neopentane, is sufficiently familiar to be the name of choice for most purposes. However, *tetramethylmethane* would be instantly recognizable to anyone familiar with organic chemistry because of the simplicity of the units used in the name. In *Chemical Abstracts* the compound is indexed under the IUPAC name, 2,2-dimethylpropane. It is hard to see any advantage at all in the name in which ethane is chosen as the parent. The following provide further examples of useful substitutive names:

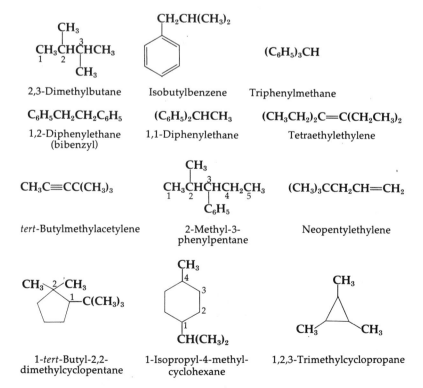

2,3-Dimethylbutane   Isobutylbenzene   Triphenylmethane

$C_6H_5CH_2CH_2C_6H_5$ $(C_6H_5)_2CHCH_3$ $(CH_3CH_2)_2C{=}C(CH_2CH_3)_2$

1,2-Diphenylethane   1,1-Diphenylethane   Tetraethylethylene
(bibenzyl)

$CH_3C{\equiv}CC(CH_3)_3$

tert-Butylmethylacetylene   2-Methyl-3-   Neopentylethylene
                                    phenylpentane

1-*tert*-Butyl-2,2-   1-Isopropyl-4-methyl-   1,2,3-Trimethylcyclopropane
dimethylcyclopentane   cyclohexane

The terms *ortho, meta,* and *para* are used as prefixes to designate relative positions of substituents on disubstituted benzenes. The words are also used as adjectives and are sometimes employed to indicate substituent relationships independent of a particular name. An alternative method uses numbers, as in the following examples:

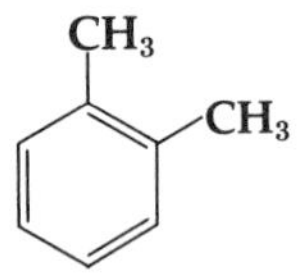

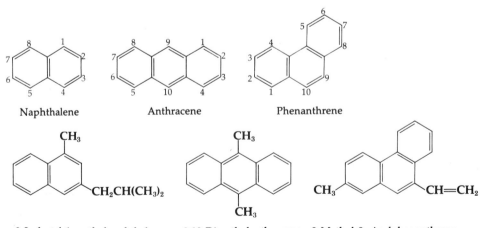

o-Xylene
(*o*-dimethylbenzene)
(1,2-dimethylbenzene)

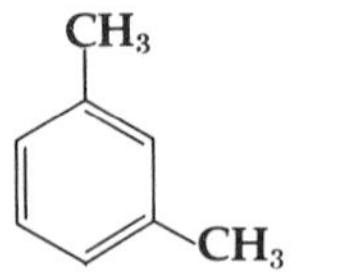

m-Xylene
(*m*-dimethylbenzene)
(1,3-dimethylbenzene)

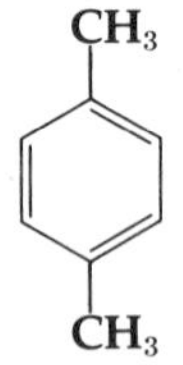

p-Xylene
(*p*-dimethylbenzene)
(1,4-dimethylbenzene)

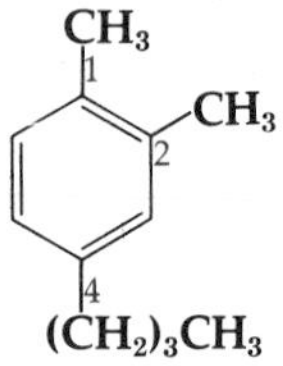

1,2-Dimethyl-4-
n-butylbenzene†

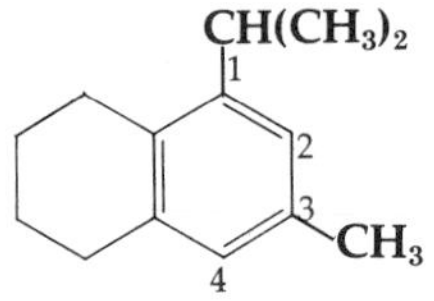

1-Isopropyl-3-
methyltetralin

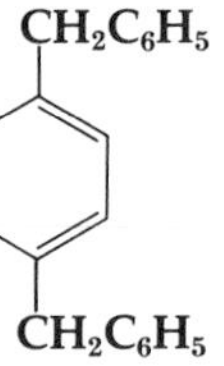

p-Dibenzylbenzene
(1,4-dibenzylbenzene)

Polycyclic aromatic rings are numbered according to rigidly fixed rules, as illustrated with naphthalene, phenanthrene, and anthracene. Within the limitations of these fixed assignments, the numbers used in names are chosen in such a way as to have the lowest possible values. The 1- and 2-positions of naphthalene are also designated as α and β, respectively, and 1- and 8-positions in the same molecule are referred to as the *peri* positions.

Naphthalene   Anthracene   Phenanthrene

3-Isobutyl-1-methylnaphthalene   9,10-Dimethylanthracene   2-Methyl-9-vinylphenanthrene

---

† This name is identical to the IUPAC name except for inclusion of *n*- in the name of the four-carbon substituent. Although the discriminant is actually superfluous, it is often included when a name is used in a context that does not make clear what system of nomenclature is being used.

If alkenes other than ethylene are regarded as parent compounds, the double bond must be located in the carbon skeleton by number. The longest hydrocarbon chain that contains the double bond is numbered from the end closest to the double bond. Numbers are assigned in such a way as to place the double bond between two *consecutively* numbered carbon atoms, and the parent name contains the lower number to locate the position of the double bond. The following examples illustrate these conventions.

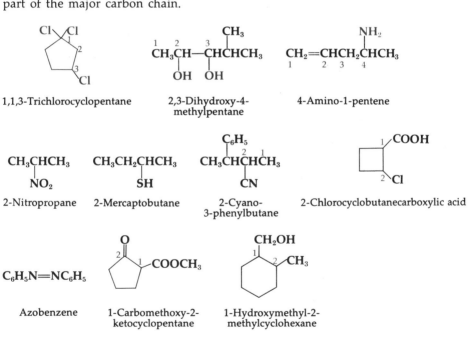

5-Methyl-2-hexene   4-Phenyl-1,3-pentadiene   3-Allylcyclobutene

Compounds Containing Functional Groups. Compounds that contain functional groups can frequently be named as derivatives of hydrocarbons, and the function can be located on the parent carbon skeleton by number. Such a system works particularly well with functional groups that are not themselves part of the major carbon chain.

1,1,3-Trichlorocyclopentane   2,3-Dihydroxy-4-methylpentane   4-Amino-1-pentene

2-Nitropropane   2-Mercaptobutane   2-Cyano-3-phenylbutane   2-Chlorocyclobutanecarboxylic acid

Azobenzene   1-Carbomethoxy-2-ketocyclopentane   1-Hydroxymethyl-2-methylcyclohexane

In an alternative system, a parent structure containing a functional group is chosen and numbered in such a way as to give the functional group the lowest possible number. This system works best when a functional group itself is part of a major carbon chain, as with many aldehydes, acids, and ketones.

$HOCH_2CH_2COOH$

$$CH_3$$
$$CH_3CCHO$$
$$CH_3$$

3-Hydroxypropionic
acid

α,α-Dimethylpropionaldehyde
(2,2-dimethylpropionaldehyde)

$BrCH_2CH_2OCH_2CH_3$

$$(CH_3)_3CCC(CH_3)_3$$
$$O$$

2-Bromodiethyl
ether

Hexamethyl
acetone

Use of Greek Letters.   Greek letters are often used in place of numbers to form substitutive names. When the parent structure contains a principal functional group that has no carbon atom, the first carbon atom, i.e., the one bearing the functional group, is α, the next β, etc. If the principal functional group contains carbon (i.e., aldehyde, ketone, carboxylic acid, ester, nitrile, or amide), *enumeration starts with an adjacent carbon atom.* The carbon atom in the functional group is not assigned a letter. The letter ω always designates the last atom in a chain irrespective of its length.

C—C—C—C—OH
δ γ β α
(ω)

C—C—C—C—COOH
δ γ β α
(ω)

$$CH_3CH_2CHOH$$
$$C_6H_5$$

α-Phenylpropyl alcohol

$$CH_3CH_2CHCOOH$$
$$NH_2$$

α-Aminobutyric acid

$ClCH_2CH_2COCH_2CH_3$

β-Chloroethyl ethyl ketone

$HO(CH_2)_7NH_2$

ω-Hydroxyheptylamine

Multiple Numbering and Lettering.   Two or more branches of *symmetrical* parent structures may be numbered or lettered independently by the use of prime signs.

β α α' β'
$CH_3CH(NH_2)COCH_2CH_2Br$

α-Amino-β'-bromodiethyl ketone
(α-aminoethyl β-bromoethyl ketone)

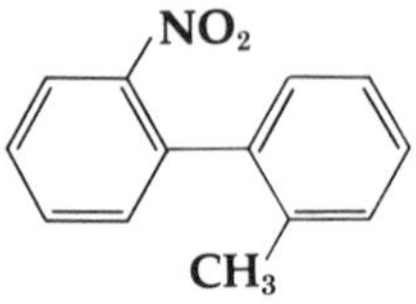

2-Methyl-2-nitrobiphenyl

Conjunctive Names

Sometimes compounds are broken into two parent structures. Each is named, and the names are joined. It is understood that the parents are joined with the loss of hydrogen from each.

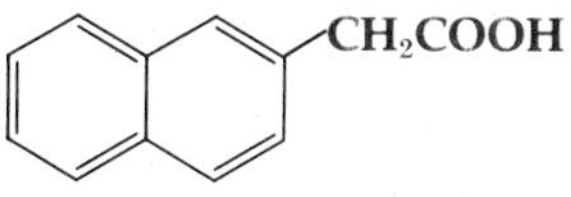

β-Naphthaleneacetic acid

Azo, Hydrazo, and Azoxy Compounds. These are all given essentially conjunctive names. However, the name of the nitrogen function is placed before the name of the parent structure in the case of symmetrical systems and between the names of the two parents if the molecule is unsymmetrical.

$$CH_3N{=}NCH_3 \qquad\quad C_6H_5N{=}NC(C_6H_5)_3$$

Azomethane   Benzene azotriphenylmethane

$$\overset{\displaystyle O^-}{\underset{\displaystyle |}{C_6H_5\overset{+}{N}}}{=}NC_6H_5 \qquad C_6H_5CH_2NHNHCH_2C_6H_5$$

Azoxybenzene ω, ω'-Hydrazotoluene
                      (N,N'-dibenzylhydrazine)

Replicate Names

Symmetrical compounds may be named by dividing the compound into two equivalent groups and naming the substance as a dimer of the radical. The prefix *bi-* (rather than *di-*) is used to indicate the dimeric structure.

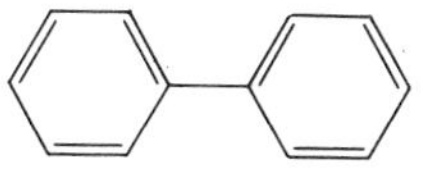

Biphenyl 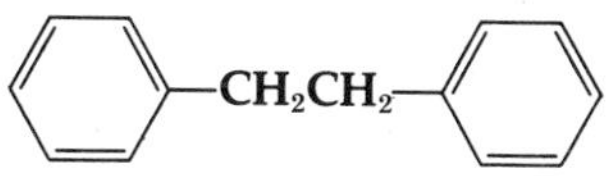   Bibenzyl

Larger assemblies may be named by an extension of the system. The Latin prefixes, *ter-, quater-, quinque-,* etc., are used to show the number of times the fundamental unit is repeated. The concept of repeating units is no longer perfect since the structure includes bifunctional internal units and terminal monofunctional groups. Names of hydrocarbons, rather than radicals, are usually used to form replicate names, although well-established names, such as terphenyl, are retained.

| *p*-Terphenyl | *m*-Terphenyl | Quatercyclopropane |

Additive Names

Compounds that can be considered as derived from bivalent radicals are sometimes given additive names. The commonest examples are compounds actually formed by addition to carbon-carbon double bonds, e.g., ethylene oxide and ethylene dibromide. However, the system is sometimes extended to include other symmetrically substituted systems.

$BrCH_2CH_2Br$ $CH_2\!-\!CH_2$
                         $\diagdown O \diagup$

Ethylene dibromide   Ethylene oxide

Tetramethylene oxide $ClCH_2CH_2CH_2CH_2Cl$
(tetrahydrofuran, preferred)   Tetramethylene dichloride

1,4-Dihydronaphthalene   Decahydronaphthalene
                                    (decalin)

Subtractive Names

Compounds that can be considered as formed from well-known structures by removal of groups or atoms are frequently named accordingly. The commonest of such subtractive operations is removal of a pair of hydrogen atoms, indicated by the prefix *dehydro*-.

1,2-Dehydrobenzene   5,6-Dehydrocamphor
      (benzyne)   (1,7,7-trimethylbicyclo[2.2.1]
                              hept-5-ene-2-one)

Fusion Names

Some bicyclic compounds are considered, for purposes of nomenclature, as formed by the fusion of a pair of cyclic systems. Such fusions always involve sharing of two adjacent ring members. The names of the two parent structures are mentioned in sequence; the letter "o" is introduced after the name of the first ring system. The first name is often shortened, e.g., *benzo-* instead of *benzeno-*.

Benzofuran   2,3-Cyclopentanophenanthrene

The IUPAC System

The system of nomenclature developed by the International Union of Pure and Applied Chemistry is based upon the following principles for assignment of substitutive names:

1 The longest sequence of carbon atoms containing the principal function is the parent structure of acyclic compounds.

2 The common names of familiar cyclics are frequently chosen as parent structures of compounds containing cyclic structures.

3 An order of preference is assigned to various functional groups. The groups higher in the order are given preference in the definition of parent structures. Some functional groups are *always* considered as substituents.

Only the highlights of the IUPAC rules can be presented here. The definitive reports should be consulted for further details.†

Aliphatic Hydrocarbons. The names of aliphatic hydrocarbons illustrate many of the principles of IUPAC nomenclature. The names of the unbranched members of the group are the same in all common systems and are found in Table A-1. Since all branched compounds are given substitutive names, prefixes such as *n-* and *iso-* are never used to modify the names of principal chains. The endings *-ane*, *-ene*, and *-yne* mean "saturated," "double bond," and "triple bond," respectively. The name of a parent structure is chosen so as to include the multiple linkage when one is present and to include as many as possible when there is more than one such group. Numbers are assigned so that the numbers of the carbon atoms involved in multiple bonding are as small as possible.

† (1) International Union of Chemistry Definitive Report, *J. Am. Chem. Soc.,* **55,** 3905 (1933); (2) International Union of Pure and Applied Chemistry Reports, *J. Am. Chem. Soc.,* **82,** 5545 (1960).

$$CH_3CH_2CH_2CH_3 \qquad CH_3\overset{\overset{\displaystyle CH_3}{|}}{C}HCH_3 \qquad CH_3CH_2\overset{\overset{\displaystyle CH_3}{|}}{C}HCHCH_2CH(CH_3)_2$$
$$\underset{\displaystyle CH_2CH_3}{|}$$

Butane   2-Methylpropane   4-Ethyl-2,5-dimethylheptane

$$CH_3CH_2CH_2\overset{\overset{\displaystyle CH(CH_3)_2}{|}}{C}HCHCH_2CH_3 \qquad CH_3CH{=}CHCH_3 \qquad CH_3CH{=}\overset{\overset{\displaystyle CH_3}{|}}{C}CH_2CH_3$$
$$\underset{\displaystyle CH_3}{|}$$

4-Isopropyl-
3-methylheptane   2-Butene   3-Methyl-2-pentene

$$CH_3\overset{\overset{\displaystyle CH_3}{|}}{C}H\overset{\overset{\displaystyle CH_2CH_2CH_3}{|}}{C}HCHCH{=}CH_2 \qquad CH_3CH{=}CHCH_2CH{=}CH_2 \qquad CH_3CH_2\overset{\overset{\displaystyle CH_3}{|}}{C}HC{\equiv}CH$$
$$\underset{\displaystyle CH_3}{|}$$

4,5-Dimethyl-3-
propyl-1-hexene   1,4-Hexadiene   3-Methyl-1-pentyne

Preferences. If a molecule contains both double and triple bonds, no distinction is made between *-ene* and *-yne* structures in seeking the lowest number combination as long as assignment can be made unambiguously.

$$CH_3CH{=}CHC{\equiv}CH \qquad CH_3C{\equiv}CCH{=}CH_2$$

     3-Penten-1-yne   1-Penten-3-yne

However, if two methods for assignment will give the same set of numbers, preference is given to the name that assigns the lowest numbers to *-ene* structures.

$$HC{\equiv}CCH{=}CHCH{=}CH_2$$

    1,3-Hexadien-5-yne
    (*not* 3,5-hexadien-1-yne)

Group Names. Univalent groups (radicals) derived by removal of hydrogen from a terminal carbon atom are named by replacing the suffix *-ane* of the name of the hydrocarbon by *-yl*, e.g., $CH_3(CH_2)_4{-}$, pentyl. Univalent branched groups are named using the longest branch as a parent structure and assigning the number 1 to the carbon atom having a free valence. Shorter branches are included as substituents. Note that the longest sequence must *begin* with the point of attachment.

$$CH_3CH_2CH_2CH_2\overset{\overset{\displaystyle }{|}}{C}HCH_3 \qquad CH_3(CH_2)_3\overset{\overset{\displaystyle }{|}}{C}HCH_3$$
$$CH_3(CH_2)_5\overset{\overset{\displaystyle }{|}}{C}H(CH_2)_5CH_3$$

    1-Methylpentyl   7-(1-Methylpentyl)tridecane

Common names of a number of groups are retained; these include the following:

Isopropyl	$(CH_3)_2CH-$
Isobutyl	$(CH_3)_2CHCH_2-$
sec-Butyl	$CH_3\overset{\mid}{C}HC_2H_5$
tert-Butyl	$(CH_3)_3C-$
Neopentyl	$(CH_3)_3CCH_2-$
tert-Pentyl	$C_2H_5\overset{\mid}{C}(CH_3)_2$
Isohexyl	$(CH_3)_2CHCH_2CH_2CH_2-$
Benzyl	$C_6H_5CH_2-$
Allyl	$CH_2=CHCH_2-$
Vinyl	$CH_2=CH-$
Phenyl	C_6H_5-

Order of Substituents.   The order in which substituents are listed is either (1) alphabetical or (2) in order of increasing complexity. *Chemical Abstracts* always uses alphabetical listing, as has been done throughout this book. Because spelling varies in different languages, alphabetical listing is not entirely desirable for a universal system of nomenclature. However, establishment of a system for unequivocal determination of group complexity is complicated. The IUPAC committee has completed a system for ranking hydrocarbon groups (1957 Rules) and will presumably address itself to the problem of ranking other substituents.

Monofunctional Compounds.   The IUPAC Rules define systematic (or semisystematic) names for many functional classes. The endings *-ane, -ene,* and *-yne* of hydrocarbon names are shortened to *-an-, -en-,* and *-yn-,* and an additional ending is added to indicate the nature of the functional group. Other functional groups are always indicated as substituents. Table A-3 summarizes the devices to indicate the presence of the various functional groups. If both a prefix and a suffix are included in the table, the former is ordinarily used only in naming polyfunctional compounds. Sometimes groups such as carboxyl and cyano are treated as substituents in naming monofunctional compounds if the systematic name is unduly complex; for example, monofunctional derivatives of cycloalkanes and cycloalkenes are often named in this manner. The following examples illustrate the use of Table A-3 for the formation of IUPAC names.

Systematic names:

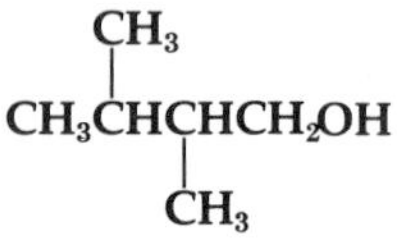

$$\underset{\underset{CH_3}{|}}{\overset{\overset{CH_3}{|}}{CH_3CHCHCH_2OH}}$$

$$\underset{\overset{\|}{O}}{CH_3CH\!=\!CHCCH_3}$$

$$\underset{\underset{OH}{|}}{\overset{\overset{CH_3}{|}}{CH_3CCH_2CH_2CH_2COOH}}$$

2,3-Dimethyl-1-butanol 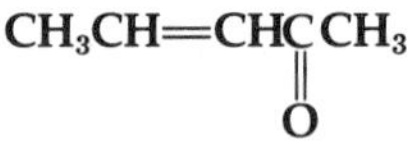   3-Penten-2-one 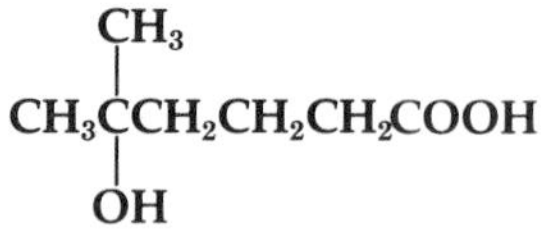   5-Hydroxy-5-methylhexanoic acid

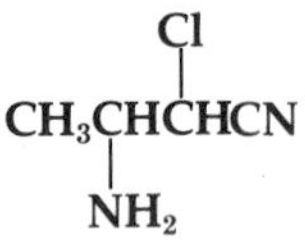

$$\underset{\underset{NH_2}{|}}{\overset{\overset{Cl}{|}}{CH_3CHCHCN}}$$

$$CH_3O\,CH_2CH_2CHO$$

$$HOOC(CH_2)_6COOH$$

3-Amino-2-chlorobutanonitrile   3-Methoxypropanal   Octanedioic acid

$$\underset{\underset{OH}{|}}{\overset{\overset{CH_2CH_2OH}{|}}{CH_3CHCH_2CH_2CONH_2}}$$

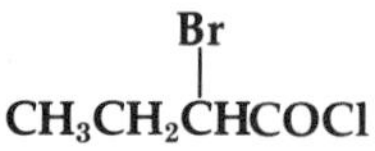

$$\overset{\overset{Br}{|}}{CH_3CH_2CHCOCl}$$

$$HC\!\equiv\!CC\!\equiv\!CCH_2COOH$$

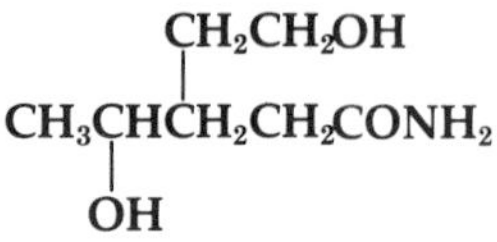

4-Hydroxy-3-[2-hydroxyethyl]   2-Bromobutanoyl   Hexa-3,5-diynoic
pentanoic amide   chloride   acid

Substitutive names:

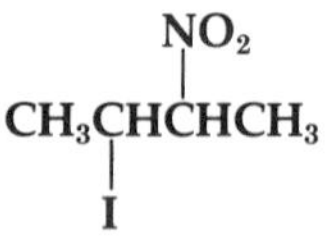

$$\underset{\underset{I}{|}}{\overset{\overset{NO_2}{|}}{CH_3CHCHCH_3}}$$

$$(CH_3)_2CHOCH(CH_3)_2$$

$$N_3CH_2\overset{\overset{ONO_2}{|}}{CHCH_3}$$

2-Iodo-3-nitrobutane   2-Isopropoxypropane   1-Azido-2-nitrato-
propane

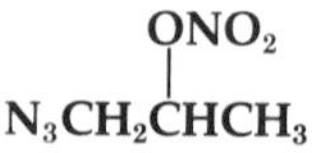

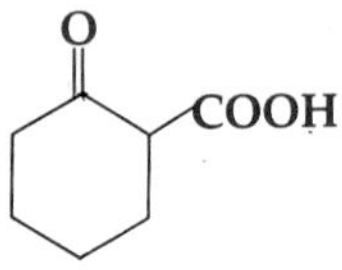

COOH

$$\underset{O}{\overset{}{CH_2\!-\!CHCH_2Cl}}$$

$$CH_3SO_2C_2H_5$$

2-Carboxycyclohexanone   3-Chloro-1,2-   Methylsulfonylethane
(not preferred, see below)   epoxy propane

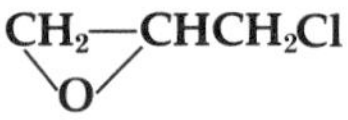

Acids.   If inclusion of a carboxyl or related group in the parent structure is clearly inconvenient, the attached group is given a parent name and the name of the functional group is *added* to the name of the parent. The system is preferred to the one illustrated above in which carboxy, etc., are given substituent names. The suffixes formed in this manner must *always* contain some indication of the presence of carbon atoms not found in the parent structure. Names such as acid, amide, and aldehyde are class names and are not sufficient to show the presence of carbon atoms.

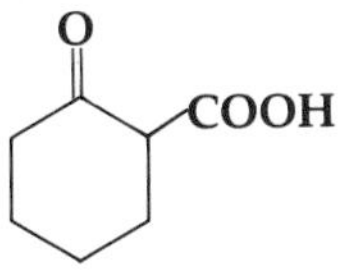

2-Oxocyclohexanecarboxylic
acid

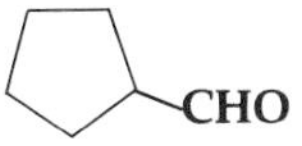

Cyclopentanecarboxaldehyde
(*not* cyclopentanealdehyde)

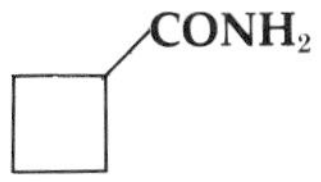

Cyclobutanecarboxamide

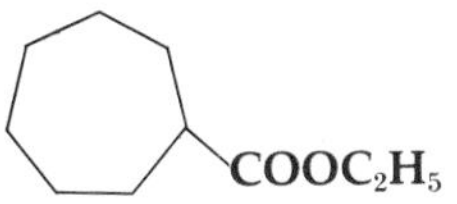

Ethyl cycloheptanecarboxylate

The same system is always used for the nomenclature of other acids, such as sulfonic, sulfinic, sulfenic, and phosphonic acids.

CH_3SO_3H $CH_3CH_2\underset{\underset{SO_2H}{|}}{CH}CH_3$

Methanesulfonic acid   2-Butanesulfinic acid

$CH_3\underset{\underset{CH_3}{|}}{\overset{\overset{CH_3}{|}}{C}}SOH$

2-Methylbutane-2-sulfenic
acid

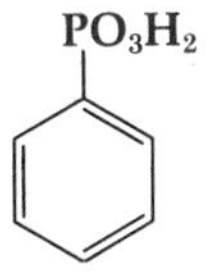

Benzenephosphonic
acid

Polyfunctional Compounds. When two or more different functional groups for which suffixes are given in Table A-3 occur within the same molecule, *only one suffix can be used to form a systematic name.*† The other functional groups must be considered as substituents. For example, the compound $HOCH_2CH_2COCH_3$ must be named 4-hydroxy-2-butanone, *not* 4-butanol-2-one. The name 3-oxo-1-butanol does not violate the rule against multiple suffixes but is not preferred because the hydroxyl group is designated as the principal function. An order of preference is established for the assignment of the role of principal function.

1 First preference is given to acidic groups and to other groups that by their nature must terminate chains. Although variations occur, the following order is usually followed for choice among members of the class: $COOH > SO_3H$, SO_2NH_2, $SO_2R > CONH_2$, CN, $COOR > CHO$.

2 The nonterminal groups, keto, amino, and hydroxyl, are given last preference, with the order among them usually being $CO > N > OH$.

3 All other groups are listed as substituents.

† As the table is formulated, double and triple bonds are exceptions. The designations *-ene* and *-yne* are considered to be part of the root of the names of compounds containing other functional groups.

The following examples illustrate the rules:

$$\underset{\substack{\text{7-Aminoheptanoic}\\\text{acid}}}{\text{NH}_2(\text{CH}_2)_6\text{COOH}}$$

$$\overset{\displaystyle \underset{|}{\text{COCH}_3}}{\underset{\text{4-Acetylheptanoic acid}}{\text{CH}_3\text{CH}_2\text{CH}_2\text{CHCH}_2\text{CH}_2\text{COOH}}}$$

$$\underset{\text{Ethyl 6,8-dioxooctanoate}}{\text{HCOCH}_2\text{CO}(\text{CH}_2)_4\text{COOC}_2\text{H}_5}$$

$$\overset{\displaystyle \underset{|}{\text{SC}_6\text{H}_5}}{\underset{\substack{N,N\text{-Dimethyl-2-phenylthio-}\\\text{heptanoic amide}}}{\text{CH}_3(\text{CH}_2)_4\text{CHCON}(\text{CH}_3)_2}}$$

$$\underset{\text{1-Amino-2-heptanone}}{\text{CH}_3(\text{CH}_2)_4\text{COCH}_2\text{NH}_2}$$

$$\overset{\displaystyle \underset{|}{\text{OH}}}{\underset{\text{4-Hydroxy-1,7-heptanediamine}}{\text{H}_2\text{N}(\text{CH}_2)_3\text{CH}(\text{CH}_2)_3\text{NH}_2}}$$

TABLE A-3 **IUPAC Nomenclature of Principal Functional Groups**

Functional class	Prefix	Suffix
Acid (carboxylic)	carboxy-	-oic acid, -carboxylic acid
Alcohol	hydroxy-	-ol
Aldehyde	oxo-, aldo-, or formyl-	-al; -carboxaldehyde
Amide	carboxamido-	-amide, -carboxamide
Amine	amino-	-amine
Azo	azo-	
Azoxy	azoxy-	
Double bond		-ene
Ether	alkoxy- or aryloxy-	
Ethylene oxide	epoxy-	
Halide	halo-	
Hydrazine	hydrazino- or hydrazo- **(RNHNHR)**	
Ketone	oxo- or keto-	-one
Mercaptan	mercapto-	-thiol
Nitrile	cyano-	-nitrile, -carbonitrile
Nitro	nitro-	
Nitroso	nitroso-	
Quaternary nitrogen		-onium, -inium
Sulfide	alkylthio- or arylthio	
Sulfone	alkylsulfonyl- or arylsulfonyl-	
Sulfoxide	alkylsulfinyl- or arylsulfinyl-	
Triple bond		-yne
Urea	ureido-	

Bicyclic Compounds.   Polycyclic compounds, in which two or more carbon atoms are common to two or more rings, take the name of an open-chain compound having the same number of carbon atoms, and suffixes such as bicyclo- and tricyclo- indicate the number of rings. The points of fusion of the rings are indicated by listing the numbers of carbon atoms in each of the bridges, i.e., the number of atoms between the fusion points (bridgeheads). The bridges are listed, within brackets, in order of decreasing length.

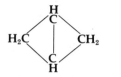

Bicyclo[1.1.0]butane   Bicyclo[3.2.1]octane

Numbers are assigned to the carbon atoms by beginning at a bridgehead and moving along the bridges. The longest bridge is numbered first, the second largest next, etc.

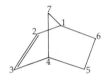

Bicyclo[2.2.1]hept-2-ene

A polycyclic system is considered to contain a number of rings equal to the number of bond scissions necessary to convert the skeleton to an open chain. Tricyclic and tetracyclic systems are named by first specifying a bicyclic system containing the three principal (longest) bridges; the location of additional bridges is then indicated by a superscript number.

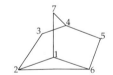

Tricyclo[2.2.1.0^{2,6}] heptane

In the above name the superscript numbers show that the zero-membered bridge connects ring members 2 and 6.

Replacement Names.   Compounds containing hetero atoms are sometimes given names based upon those of hydrocarbons that have the same total skeleton with the hetero atom replaced by carbon. The presence of a hetero atom and its nature is indicated by a characteristic prefix. The prefixes characteristic of common hetero atoms are listed in Table A-4. Since the prefixes all end in *a*, the system has been referred to as *a nomenclature*. The *a* is sometimes elided if the name of the parent skeleton begins with a vowel.

$CH_3CH_2CH_2NHCH_2CH_2CH_3$ $CH_3OCH_2CH_2OCH_2CH_2OCH_3$

4-Azaheptane   2,4,6-Trioxanonane

Heterocyclic Compounds. A vast number of heterocyclic compounds are known, including many important substances. Rules for their nomenclature are too extensive for presentation here; only a brief synopsis can be given. The common names of many ring systems have been accepted by the IUPAC system; the 1957 Report lists 71 such names. In addition, the *Hantzsch-Widman system* for monocyclic rings containing hetero atoms is recommended for use in modified form. The system combines prefixes, which indicate the nature of the hetero atom present, with stems, which indicate the size of the ring. The prefixes are the same as those listed in Table A-4 for use in substitution names, except the terminal *a* is usually elided since the stems, which are listed in Table A-5, all begin with vowels. The prefixes listed in columns 2 and 4 under "unsaturation" are used for naming rings that contain the maximum number of conjugated double bonds. The names listed in the "saturation" columns are used only for compounds containing no double bonds. Compounds of intermediate degrees of saturation are named by using the prefixes dihydro-, tetrahydro-, etc., with the unsaturated name. The following are representative names:

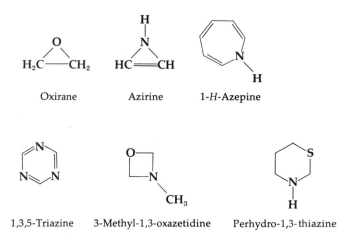

Oxirane   Azirine   1-*H*-Azepine

1,3,5-Triazine   3-Methyl-1,3-oxazetidine   Perhydro-1,3-thiazine

TABLE A-4 **Characteristic Prefixes for Replacement Names**

Element	Prefix
Oxygen	oxa-
Nitrogen	aza-
Sulfur	thia-
Phosphorus	phospha-
Silicon	sila-

TABLE A-5 **Stems for Use in Nomenclature of Heterocyclic Compounds**

No. of ring members	*Rings containing nitrogen*		*Rings not containing nitrogen*	
	Unsaturation	*Saturation*	*Unsaturation*	*Saturation*
3	-irine	-iridine	-irene	-irane
4	-ete	-etidine	-ete	-etane
5	-ole	-olidine	-ole	-olane
6	-ine	†	-in	-ane
7	-epine	†	-epin	-epane
8	-ocine	†	-ocin	-ocane
9	-onine	†	-onin	-onane
10	-ecine	†	-ecin	-ecane

† Use *perhydro* with the unsaturated name

PROBLEMS

1 Assign an IUPAC name and, if possible, at least one other name to the following compounds:

a $(CH_3)_2CHCOCH_3$

b 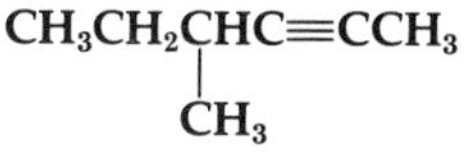

c 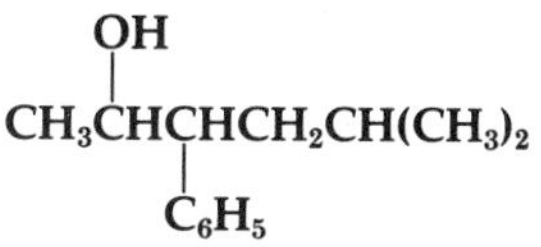

d 

e $C_6H_5CH_2OC_2H_5$

f $(CH_3)_2C{=}CHCHO$

g 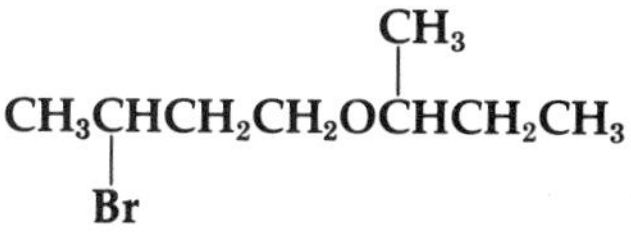

h 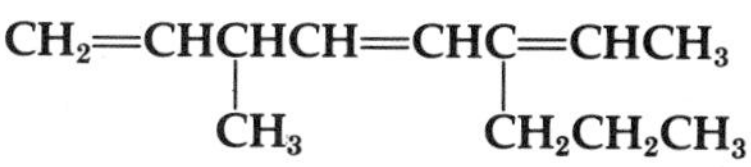

i $BrCH_2CH_2SO_2CH_3$

j $OHC(CH_2)_3COOH$

k 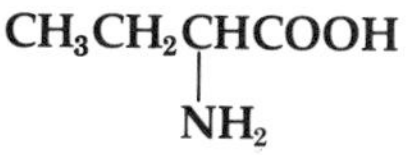

l $C_6H_5SSC_6H_5$

m $(CH_3)_3CCH_2CH_2NHCH_3$

n 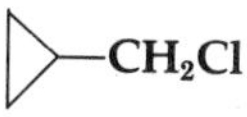

o 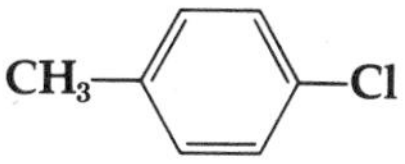

p 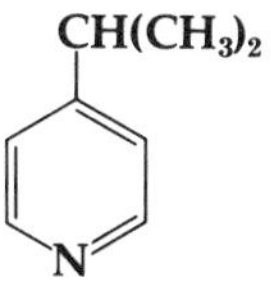

q 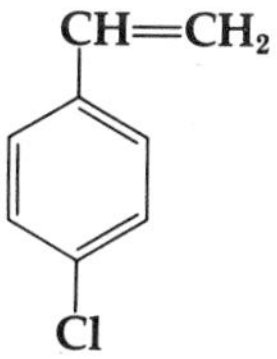

r 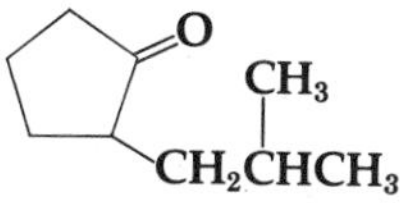

s (naphthalene with OH and NO$_2$ substituents)

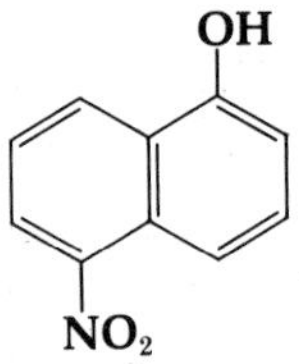

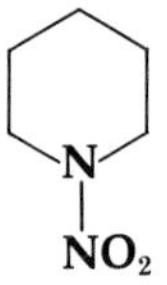

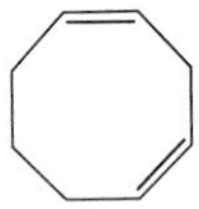

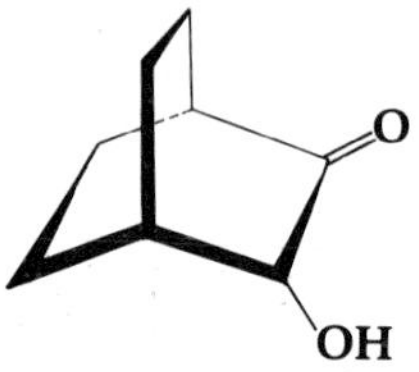

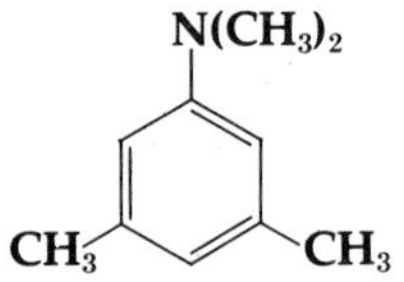

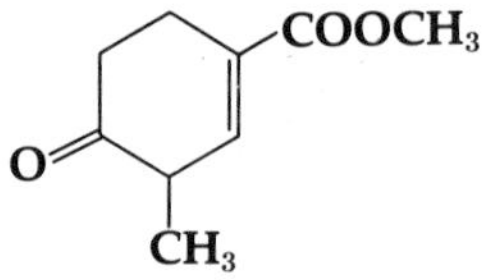

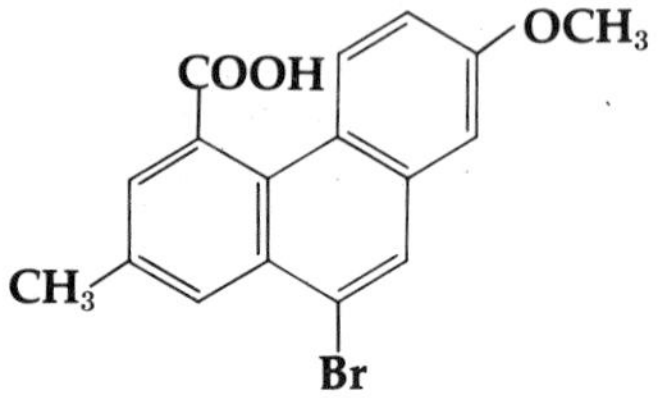

t (piperidine N–NO$_2$)

u (cyclooctadiene)

v (bicyclic ketone with O and OH)

w (benzene with N(CH$_3$)$_2$ and two CH$_3$)

x (cyclohexenone with COOCH$_3$, O, CH$_3$)

y (phenanthrene with COOH, OCH$_3$, CH$_3$, Br)

z (quinoline with OH, CH$_3$O, CH$_2$C$_6$H$_5$)

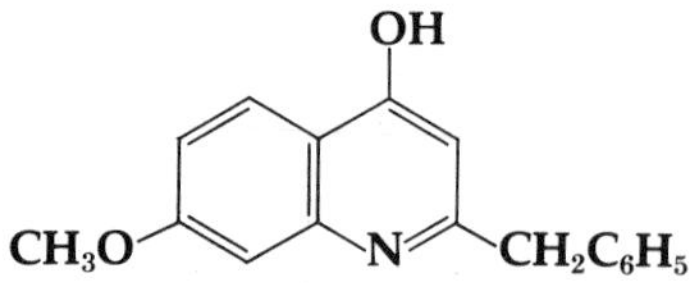

2 Write structural formulas for the following compounds. Give an IUPAC name to each compound not already designated by that system.

a Pivalic acid

b *tert*-Amyl bromide

c α,β'-Dibromodiethyl sulfoxide

d 2,4-Dinitro-5-methylbenzoic acid

e α-Naphthyl β-naphthoate

f 2,2'-Bipyridyl

g Trieicosylcarbonium ion

h Acetylacetone

i 4-Methyl-2-cyclohexenone

j Tetrabutylammonium bromide

k Allyl benzyl ether

l 2-Methyl-4-methoxy-2-hexene

m N-Methyl-N-acetylaniline

n 3-Hexene-1-carboxylic acid

o β-Phenylbutyramide

p *m*-[*sec*-Butyl]-benzoic acid

q Divinylcarbinol

r Ethylmethylphenylcarbanion

s 2-Carbomethoxycyclobutanone

t α-Phenylethanol

u 9,10-Dihydroxy-1-methyl-4-nitroanthracene

v Isobutyl isobutyrate

w [2.2.1]Bicycloheptane

x 4-Oxanonane

y 4,8-Dithiaundecane

z 2,4-Dinitrophenylhydrazone of acetophenone

3 Criticize the following names, and where possible write the correct name and structure for the compound.

a [2.2.2]Bicycloheptane

b 5-Methyl-α-chlorobutanoic acid

c 3-Carboxy-1-butanol

d 4-Propylhexanal

e 2-Methyl-4-chlorobutanoic amide

f 1-Cyclohexanone

g 4-Isopropyl-5-octanone

h 4-Methyl-8-nitronaphthalene

i Methyl propyl cation

j Triphenylmethyl carbonium ion [for $(C_6H_5)_3C^+$]

k Bibutyl

l Tetraphenylcarbinol

m *sec*-Valeric acid

n *trans*-Maleic acid

o Dibenzoylmethane

p Methyl 3-nitrophthalate

q 3-Naphthoic acid

r α-Methylglutaric acid

s 3-Methyl-3-ketohexanoic acid

t 2-Vinylhexanoic acid

4 Give the page reference for the entry of each of the following compounds in the subject index of *Chemical Abstracts* for 1959:

a $C_2H_5OC_2H_5$

b

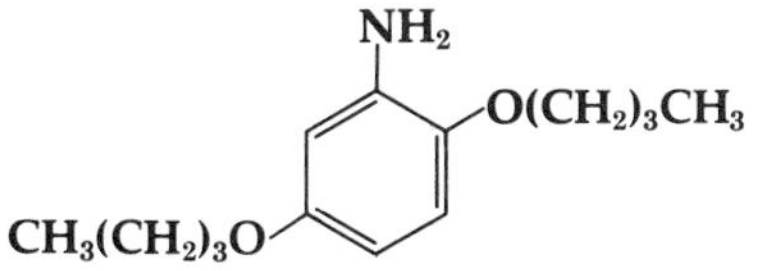

c

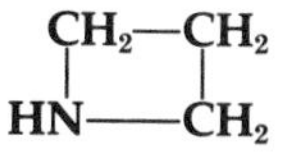

d $CH_2{=}CHO(CH_2)_4OCH{=}CH_2$

e

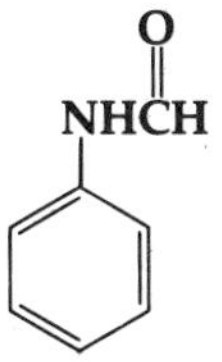

f $CH_3(CH_2)_{16}NH_2$

g

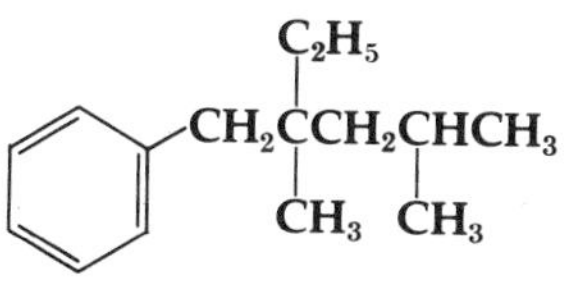

h

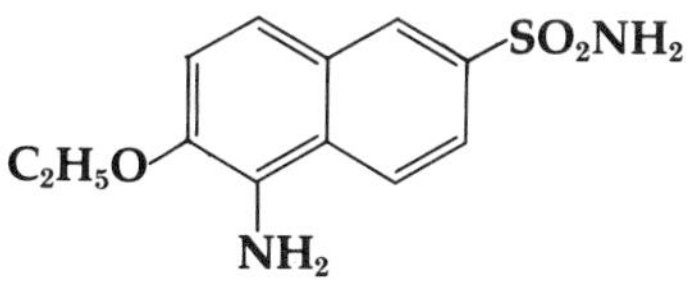

i

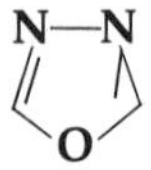

$$\begin{array}{c} N\!-\!N \\ \diagup\ \ \ \diagdown \\ O \end{array}$$

j $C_6H_5COCHCN$
         $|$
       $CH_2CH_2CH_3$

k

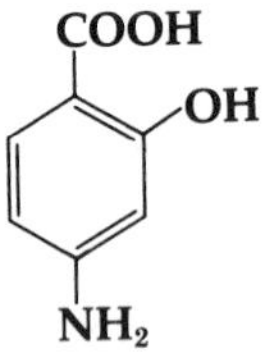

COOH
OH

NH$_2$

l

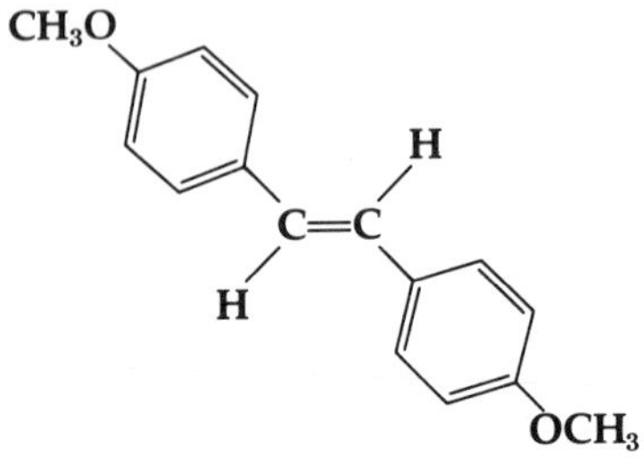

CH$_3$O

$$C=C$$
H   H
             H

OCH$_3$

m

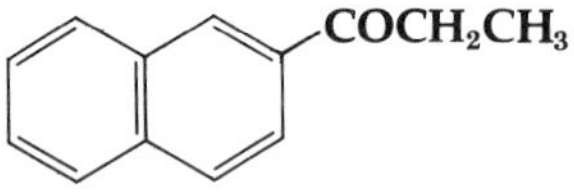

COCH$_2$CH$_3$

n

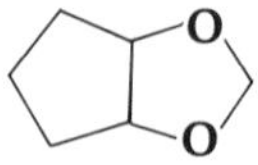

O
O

o

OH
(CH$_3$)$_3$C
          CH$_3$
NH$_2$

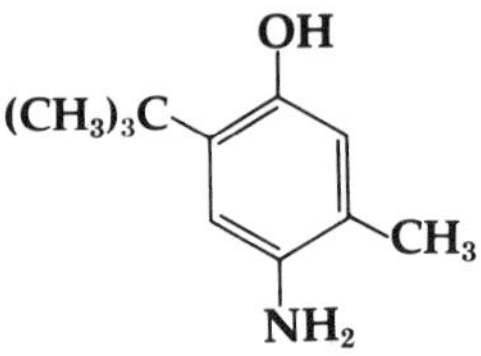

ANSWERS

CHAPTER 1

1–2 Each peak represents a pure substance, the intensity (area) of the peak equivalent to the relative quantity. B and C have the same three compounds in different quantities, A has those three and more; when heated the first substance in the A trace is converted to the third.

1–7   a $87.8 + 12.2 = 100$, hence no elements besides C and H; C_4H_6, C_8H_{12}, $C_{12}H_{18}$, etc.

       c   Another element; assuming oxygen $= 100.0 - (73.1 + 7.4) = 19.5\%$: $(C_5H_6O)_n$

1–8 C_5H_6O

1–11   a $C_8H_8O_5$ b $C_8H_{10}O_2$ d C_4H_5NO

1–12 $A = C_{10}H_{16}$, $B = C_{10}H_{18}$, $C = C_{10}H_{14}O$

1–13 $(CH_2)_n$

CHAPTER 2

2–3   a $CH_3-\ddot{S}-CH_3$, $CH_3-CH_2-\ddot{S}-H$

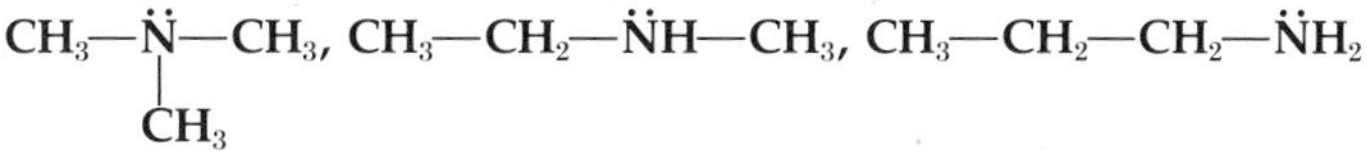

       c $CH_3-\overset{\displaystyle |}{\underset{\displaystyle CH_3}{\ddot{N}}}-CH_3$, $CH_3-CH_2-\ddot{N}H-CH_3$, $CH_3-CH_2-CH_2-\ddot{N}H_2$

       f $CH_3-\underset{\displaystyle :\ddot{O}H}{CH}-\underset{\displaystyle :\ddot{O}H}{CH_2}$, $CH_3-CH_2-\underset{\displaystyle :\ddot{O}H}{CH}-\ddot{O}H$, $CH_3-\ddot{O}-CH_2-CH_2-\ddot{O}H$,

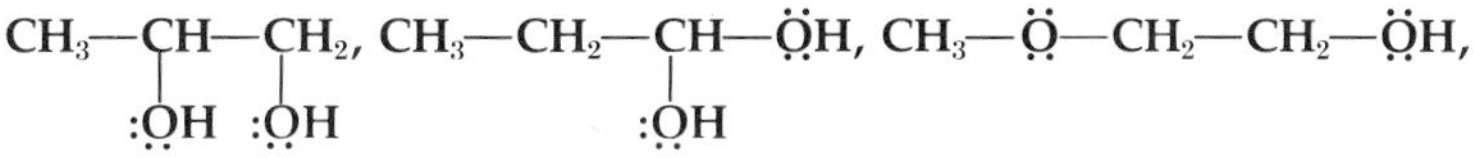

         and six more

       h $CH_3-CH_2-CH_2-CH_2-CH_2-\ddot{F}:$, $CH_3-\overset{\displaystyle |}{\underset{\displaystyle CH_3}{CH}}-CH_2-CH_2-\ddot{F}:$,

         $CH_3-\overset{\displaystyle CH_3}{\overset{\displaystyle |}{\underset{\displaystyle |}{\underset{\displaystyle :\ddot{F}:}{C}}}}-CH_2-CH_3$, and five more

       j $CH_3-CH_2-CH=\ddot{O}$, $CH_3-\overset{\displaystyle :O:}{\overset{\displaystyle \|}{C}}-CH_3$, $CH_2=CH-CH_2-\ddot{O}H$,

         $CH_2=CH-\ddot{O}-CH_3$, $\underset{\displaystyle CH_2-\ddot{O}:}{\overset{\displaystyle CH_2-CH_2}{|}}$, $CH_3-\overset{\displaystyle :O:}{\underset{\displaystyle \diagdown\diagup}{CH-CH_2}}$,

         $H\ddot{O}-CH=CH-CH_3$, $CH_2=\overset{\displaystyle |}{\underset{\displaystyle :\ddot{O}H}{C}}-CH_3$

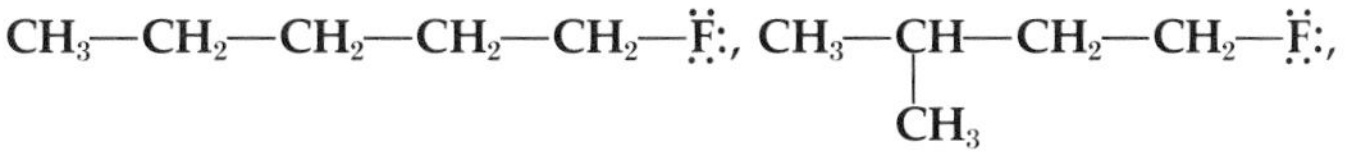

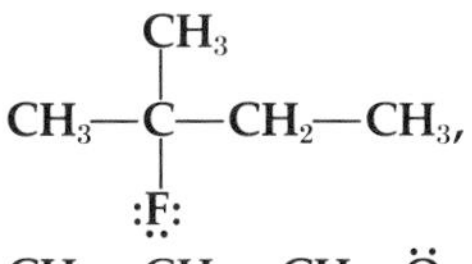

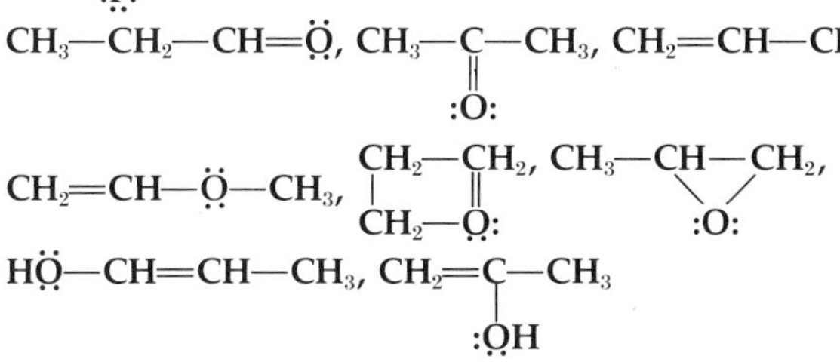

2–4 a

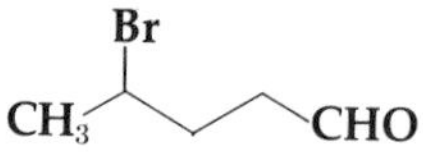

CH_3—$\overset{Br}{\underset{}{}}$ ... CHO **or** Br ... O

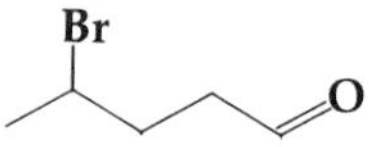

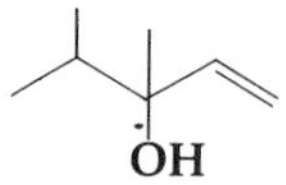

b $\overset{}{\underset{OH}{}}$ (structure)

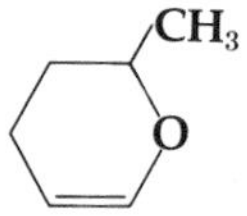

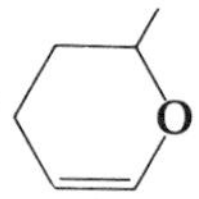

d (structure with CH_3, O ring) **or** (structure with CH_3, O ring)

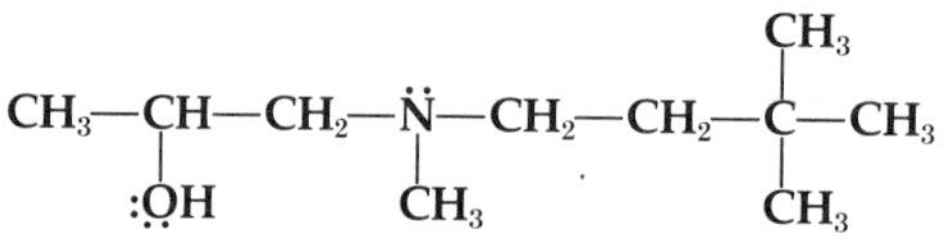

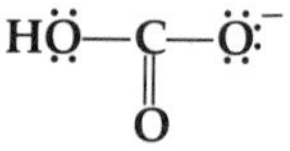

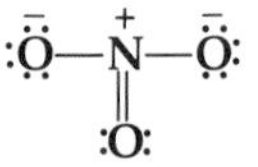

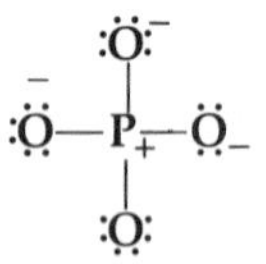

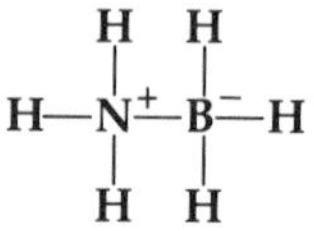

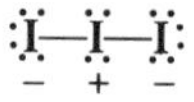

f CH_2=CH—CH=N—CH$_2$—$\overset{CH_3}{\underset{Br}{C}}$—CH$_2$—OH

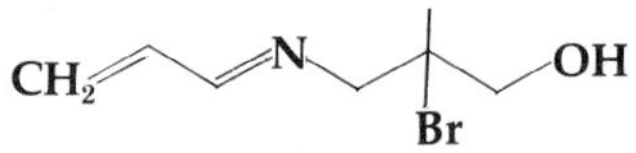

2–5 b $\overset{CH_3}{\underset{CH_3}{C}}$=CH—CH$_2$—$\ddot{O}$H

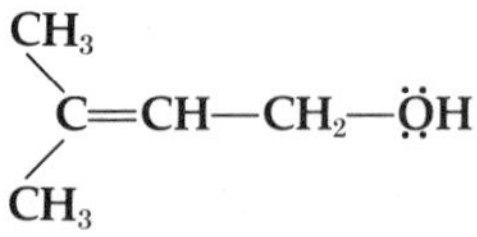

d

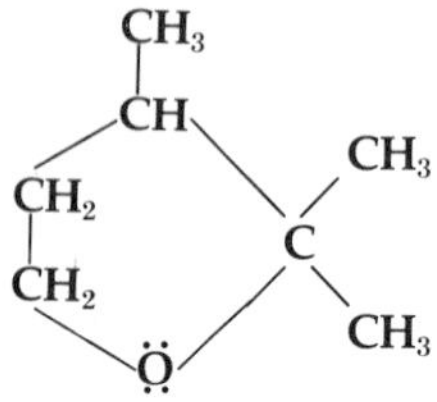

$\overset{CH_3}{\underset{}{CH}}$

CH_2 $\qquad$ CH_3

CH_2 $\quad$ C

$\ddot{O}$ $\qquad$ CH_3

f CH_3—$\overset{}{\underset{:\ddot{O}H}{CH}}$—CH$_2$—$\overset{}{\underset{CH_3}{\ddot{N}}}$—CH$_2$—CH$_2$—$\overset{CH_3}{\underset{CH_3}{C}}$—CH$_3$

k CH_2=CH—$\overset{CH_3}{\underset{CH_3}{C}}$=C—$\overset{}{\underset{:\ddot{O}:}{C}}$—CH$_3$

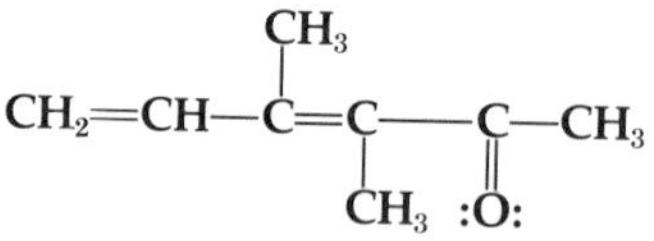

2–7 a $\ddot{O}$=N—$\ddot{O}$:$^-$

b $H\ddot{O}$—C—$\ddot{O}$:$^-$
$\qquad\quad\overset{}{\underset{O}{}}$

c :$\overset{-}{\ddot{O}}$—$\overset{+}{N}$—$\overset{-}{\ddot{O}}$:
$\qquad\quad:\overset{}{O}:$

d :$\overset{:\ddot{O}:^-}{\underset{:\overset{}{O}:}{\ddot{O}}}$—$\overset{+}{P}$—$\overset{-}{\ddot{O}}$_

f :$\overset{-}{C}$≡$\overset{+}{O}$:

h H—$\overset{H}{\underset{H}{N}}$$^+$—$\overset{H}{\underset{H}{B}}$$^-$—$H$

i :$\overset{-}{I}$—$\overset{+}{I}$—$\overset{-}{I}$:

k :$\overset{-}{\ddot{O}}$—$\overset{}{\underset{:\overset{}{O}:}{B}}$—$\overset{-}{\ddot{O}}$:

l :$\overset{:\ddot{C}l:}{\underset{:\ddot{C}l:}{\ddot{C}l}}$—$Al$$^-$—$\ddot{C}l$:

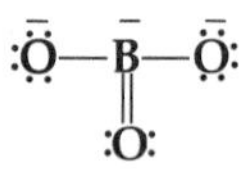

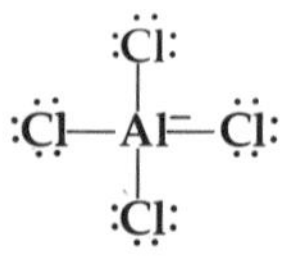

2–8 a $(CH_3)_3\overset{+}{N}H$ $\qquad$ **f** CH_2=$\overset{-}{\ddot{C}}$—CH$_3$, CH_2=$\overset{+}{\ddot{O}}$—CH$_3$

c CF_4, $\overset{+}{N}F_4$

2–12 a 4 C—H bonds = 4 × 99 = 396 kcal/mole

d 6 C—H + 2 C—C + 1 C=O = (6 × 99) + (2 × 83) + 179 = 939

g 6 C—H + 2 C—O + 1 C=C = (6 × 99) + (2 × 86) + 146 = 912

2-14 a HO— (naphthalenone structure with O double bond)

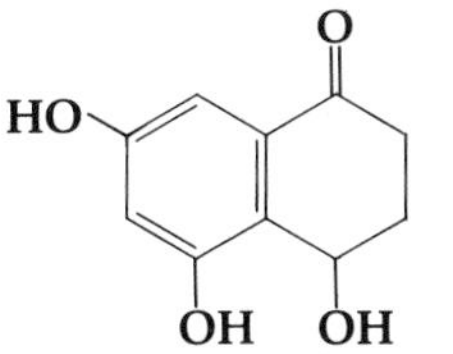

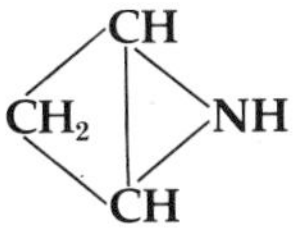

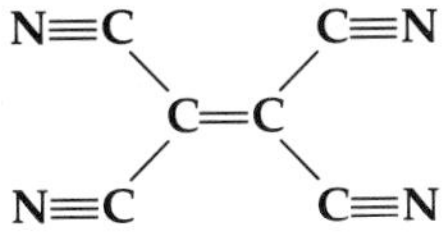

f CH_2 (bicyclic structure) CH / NH / CH

e (structure with OH, OH groups)

$N\equiv C$, $C\equiv N$
$C=C$
$N\equiv C$, $C\equiv N$

g CH_2 CH_2 / $\overset{+}{N}$ / CH_2 CH_2

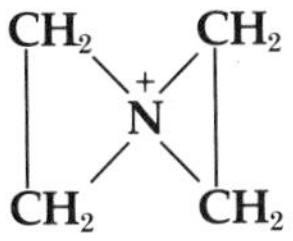

2-15 $\ddot{O}=C=\ddot{O}$, $\ddot{O}=\overset{+}{N}=\ddot{O}$, $\ddot{O}=\overset{-}{B}=\ddot{O}$, $\bar{N}=\overset{+}{N}=\ddot{O}$, $\bar{N}=\overset{+}{N}=\bar{N}$, $\ddot{O}=C=\bar{N}$,

$\bar{\ddot{N}}=C=\bar{\ddot{N}}$, etc.

2-18 a Bonds broken: C—C 83   Bonds formed: C—H 99
                O—H 111   C=O $\underline{179}$
                C—O $\underline{86}$ 278
                     280

Endothermic by 2 kcal/mole

b Bonds broken: H—Br 88   Bonds formed: C—H   99
                C=C $\underline{146}$ C—Br 68
                     234   C—C $\underline{83}$
                                                       250

Exothermic by 16 kcal/mole

2-19 Stability = energy to cleave weakest bond: **d > e > b > c > a**

2-20 a CH_2—$\dot{N}$ and $CH_2=\overset{+}{N}=\bar{\ddot{N}}$
       N

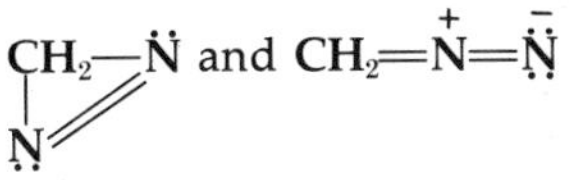

b H—$\ddot{O}$—C$\equiv$N: and H—C$\equiv\overset{+}{N}$—$\bar{\ddot{O}}$:

f CH_3—$\ddot{O}$—$\ddot{S}$—CH_3 and
       CH_3
       $\overset{+}{S}$—$\bar{\ddot{O}}$:
       CH_3

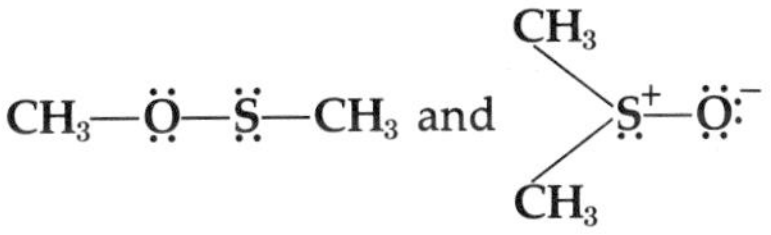

2-21 c

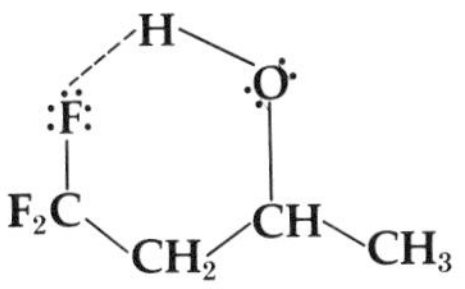

(cyclic hydrogen-bonded structure with F_2C, CH_2, CH, CH_3, :$\ddot{F}$:, $\dot{\ddot{O}}$, H)

2-22 d :O:
       ‖
       —C—$\dot{N}H_2$

h :$\bar{\ddot{O}}$:
       |
       —$\overset{++}{S}$—$\ddot{O}$—H
       |
       :$\underset{-}{\ddot{O}}$:

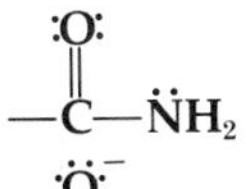

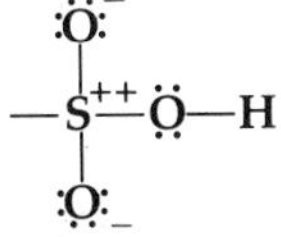

CHAPTER 3

3–1 **a** 1   **c** 3   **e** 4
   b 2   **d** 2   **f** 6

   Compare: **a** ⬠ and ⌒⌒⌒

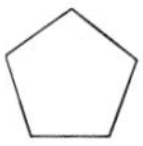

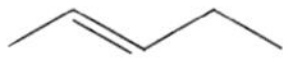

   d  ◁ and $CH_3C\equiv CH$ or $CH_2=C=CH_2$

3–5 $CH_3-\overset{\overset{\displaystyle O}{\|}}{C}-CH_2-OH \longrightarrow CH_3-\overset{\overset{\displaystyle N-OH}{\|}}{C}-CH_2-OH$ but not

   $CH_3-CH_2-\overset{\overset{\displaystyle O}{\|}}{C}-OH \longrightarrow CH_3-CH_2-\overset{\overset{\displaystyle N-OH}{\|}}{C}-OH$; there are two more possible

   $C_3H_6O_2$ compounds which would give oximes

3–6 Ethers and alcohols

3–7 **a** $C_6H_{13}NO_2 + H_2O \longrightarrow C_4H_8O_2 + CH_3CH_2NH_2$ and

   $C_4H_8O_2 \equiv C_3H_7-\overset{\overset{\displaystyle O}{\|}}{C}-OH$; compound A is

   $CH_3CH_2CH_2-\overset{\overset{\displaystyle O}{\|}}{C}-NH-CH_2CH_3$ or $(CH_3)_2CH-\overset{\overset{\displaystyle O}{\|}}{C}-NH-CH_2CH_3$

   d [7-membered ring lactam with C=O] $+ H_2O \longrightarrow$ [7-membered ring with COOH and NH$_2$]

3–8 [benzene ring] $+ 3O_3 \longrightarrow 3O=CH-CH=O$

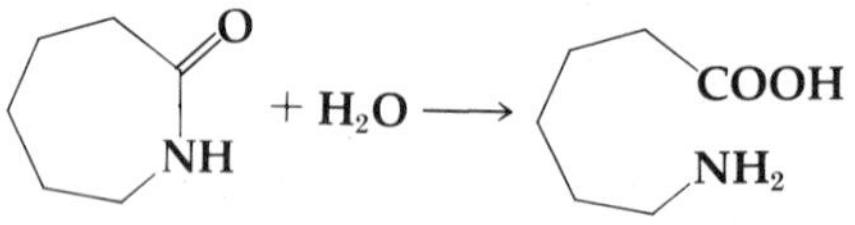

3–10 [cyclopentene with CH$_3$] $+ O_3 \longrightarrow CH_3COCH_2CH_2CH_2CHO$

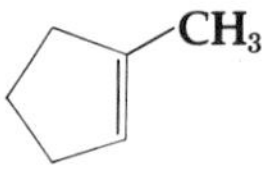

3–12 [cyclohexanone derivative] or [cyclopentanone derivative] etc.

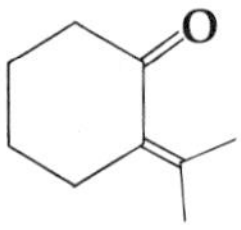

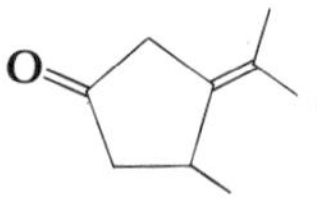

3–13 **a** 2-Methyl-1-butanol
   c 2-Methyl-2-butenoic acid
   g Methyl 2-hydroxy-2-methyl-3-bromobutanoate
   h 2-Chloromethyl-3-ethylamino-2-butenal
   i 2-Methyl-2-propyl-3-buten-1-ol
   l 2-Chloro-4-cyano-6-propyl-cyclohexanone
   n 1-Cyclopropyl-4-hydroxy-1-butanone

3–14 **b** $(CH_3)_3CCH_2CONHCH_3$
   d $CH_3C\equiv CCH=CHCH=CHCHO$

g

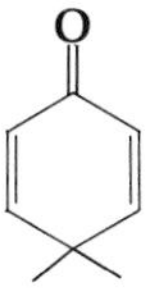

h

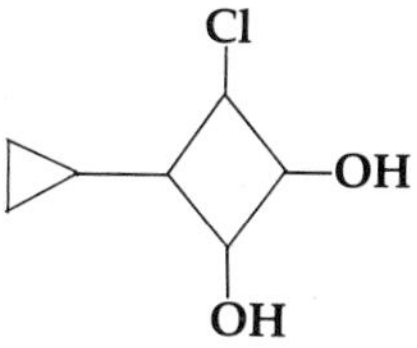

3–17   Indexes: **a** 1   **c** 2   **e** 2
                 b 4   **d** 8   **f** 2

Compare **d** 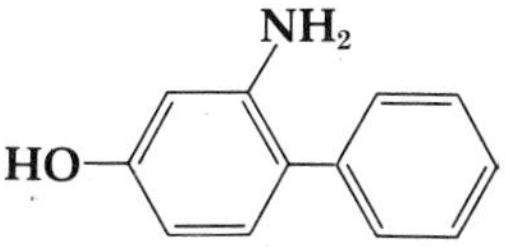   named as 1-hydroxy-3-amino-4-phenyl-benzene

3–19 $CH_3COCH_2CH_2CHO$

3–21

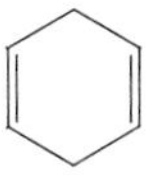

CHAPTER 4

4–2   **a**   2,6-Dimethyl-4-isopropyl-octane
      d   1,3-Dimethylcyclohexane
      e   1,1-Dimethyl-1-phenyl-ethane or *t*-butyl-benzene
      h   1-Vinyl-4-methyl-cyclohexane

4–3   **b**   **f** $(CH_3)_2CHCH_2C{\equiv}CCH_3$

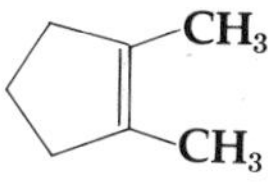

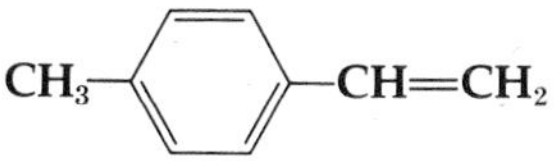

                                         h

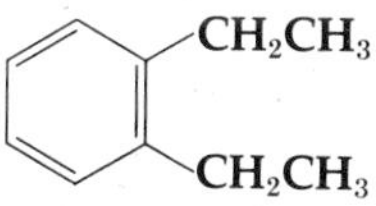

      e CH_3— —$CH{=}CH_2$

4–6   **b**   **g**

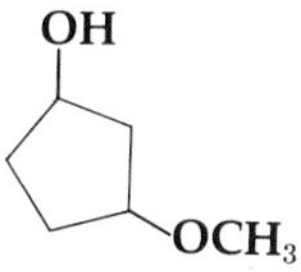

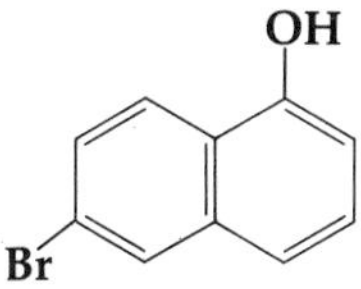

      d

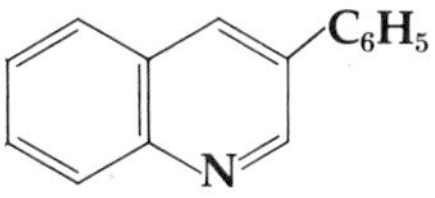

4–10 $CH_3{-}Li + H_2O \longrightarrow CH_4 + Li^+ + OH^-$

4–11 a 2   c 6   e 6
          b 1   d 3   f 6

4–14 a Cl—⟨benzene ring⟩—COOH   f ⟨cyclobutene⟩—COOC$_2$H$_5$

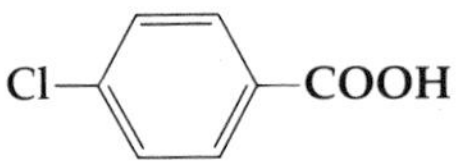

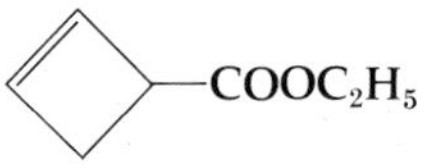

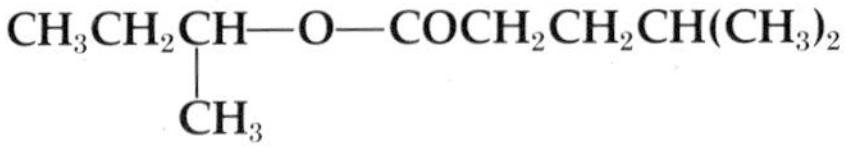

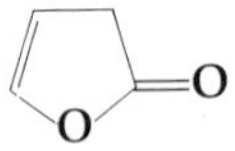

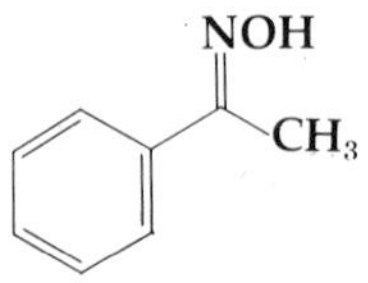

          e CH$_3$CH$_2$CH—O—COCH$_2$CH$_2$CH(CH$_3$)$_2$ i CH$_3$CO—N—C$_6$H$_5$
                         | |
                         CH$_3$ CH$_3$

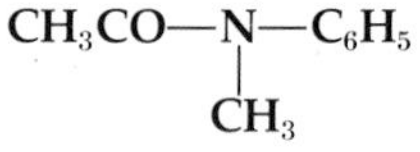

4–15 a ⟨furanone⟩=O   g ⟨structure with NOH and benzene ring, CH$_3$⟩

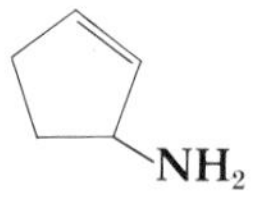

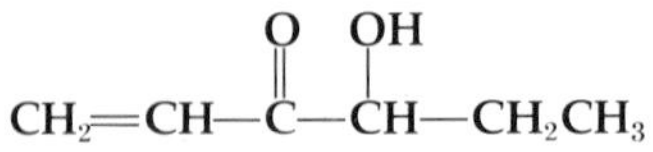

          b ⟨maleic anhydride structure⟩   h CH$_3$COCH$_2$CH$_2$COCH$_3$
                                                 l CH$_3$CH=NNHCONH$_2$

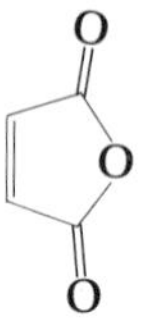

4–17 b HC≡C—C(CH$_2$CH$_3$)$_2$
                     |
                     CH$_3$

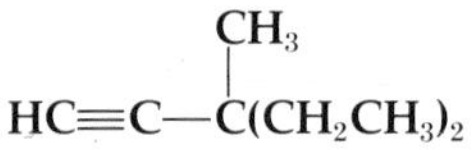

          f ⟨cyclopentene ring⟩—NH$_2$

          i CH$_2$=CH—C—CH—CH$_2$CH$_3$
                       ‖ |
                       O   OH

          m CH$_3$ ⟨structure⟩ CH$_3$ better named as 3,4-dimethyl-4-ethyl-5-hepten-2-one

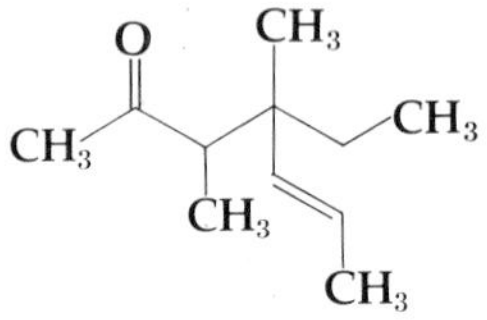

4–18 c 4,4-Dimethyl-2-amino-pentanoic acid
          f 2-Methyl-5-amino-2-cyclohexenone
          g 2-Methyl-2-ethyl-4-ethylamino-butanal   or   2-methyl-2,*N*-diethyl-4-amino-butanal
          i 3-Hydroxy-3-(2-hydroxyethyl-)-pentanoamide

CHAPTER 5

5–2 b CH$_3$—N—Ö:$^-$ ⟷ CH$_3$—N=Ö
                ‖ |
               :O: :Ö:$^-$

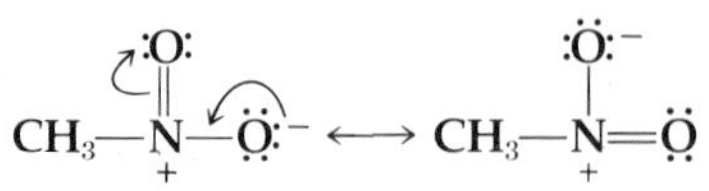

         c R—C—NH$_2$ ⟷ R—C=NH$_2$
               ‖ |
              $^+$NH$_2$:NH$_2$

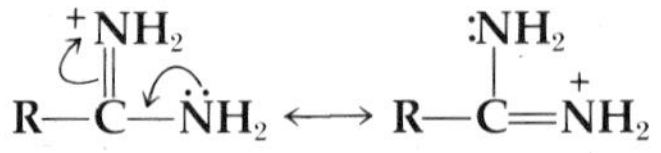

e 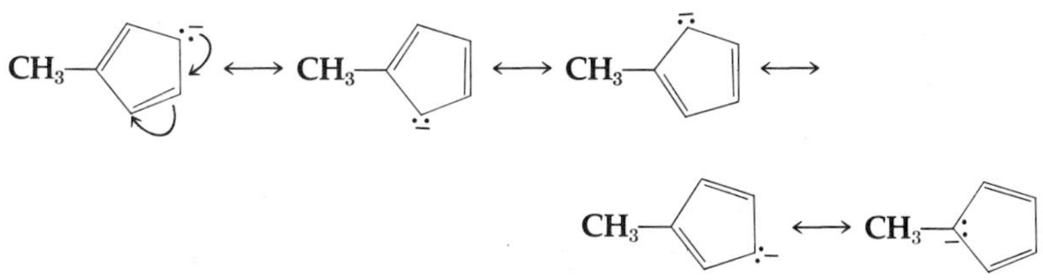

5–3 a 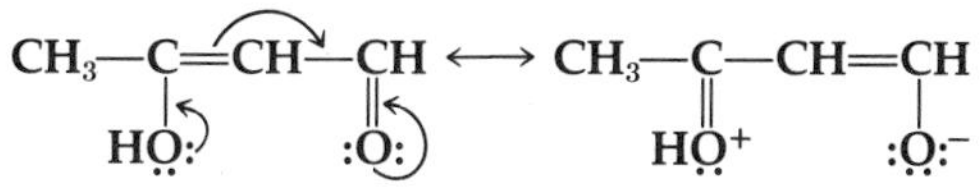

c 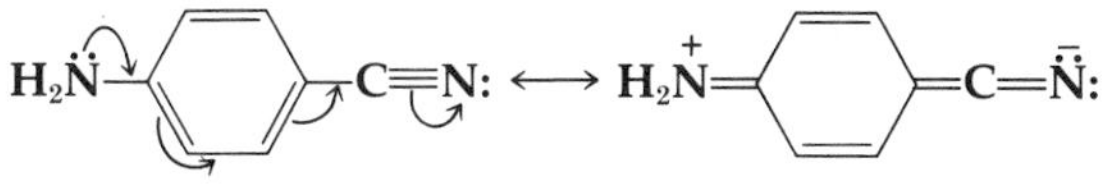

5–4 Only **a** is a resonance form

5–6 Aromatic = **a, c, e, f, h, i, k, l**; others are antiaromatic
Compare:

c 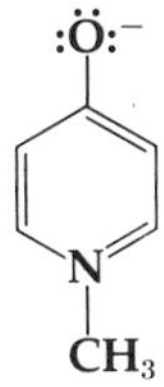

g 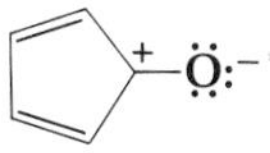

f 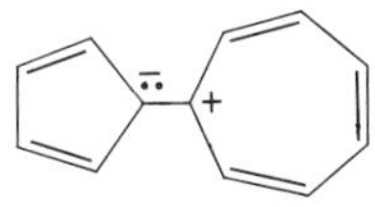

k 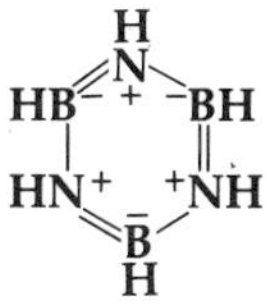

5–7 a 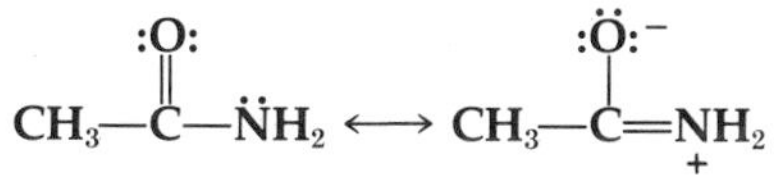

k 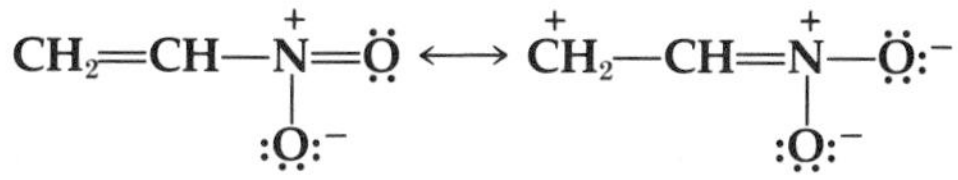

v 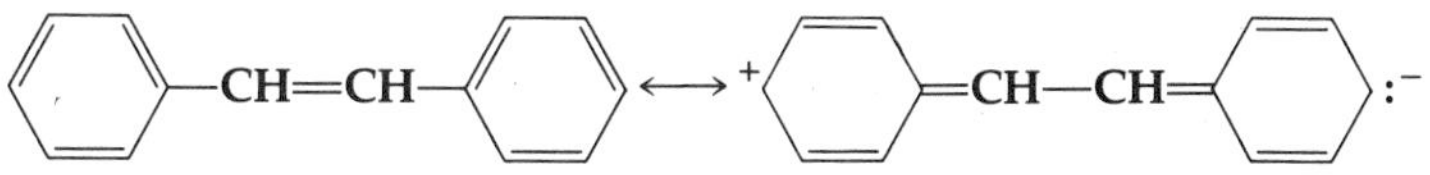

5–9 b 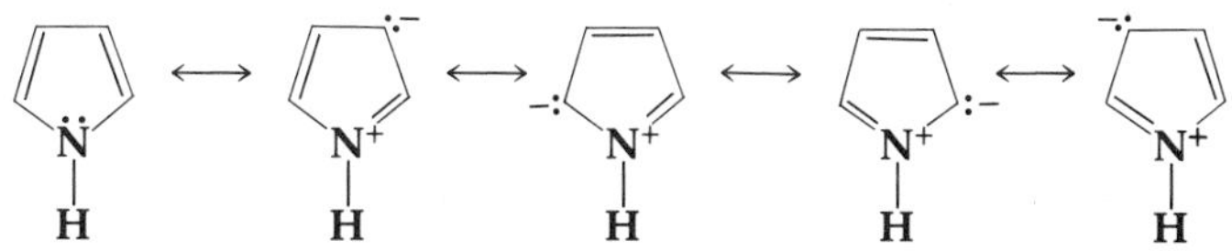

5–11 a $\xrightarrow{-H^+}$ $\xrightarrow{+H^+}$

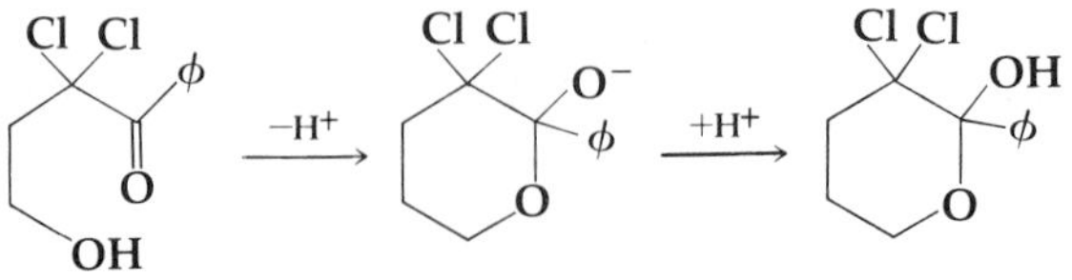

d HO—⟨⟩—N=O $\xrightarrow{-H^+}$

$$\left[\bar{O}-\!\!\bigcirc\!\!-N=O \longleftrightarrow O=\!\!\bigcirc\!\!=N-\bar{O} \right] \xrightarrow{+H^+}$$

$$O=\!\!\bigcirc\!\!=N-OH$$

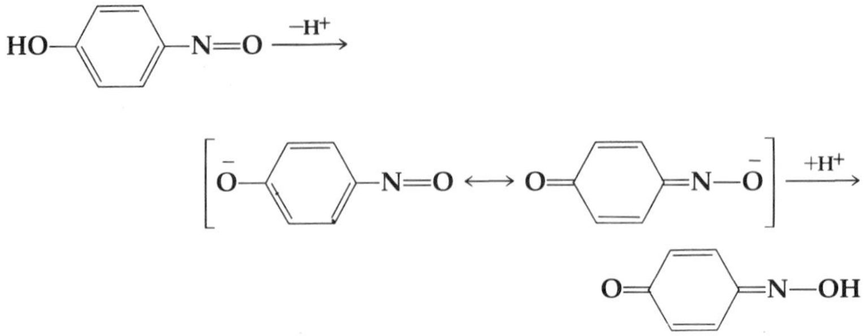

e

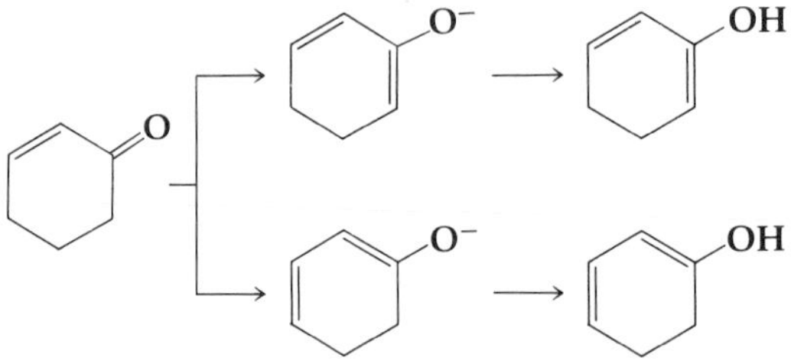

5–14 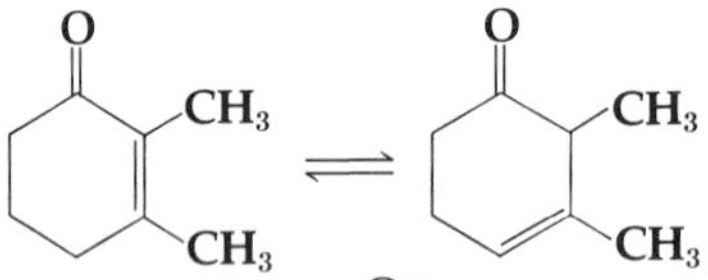 ⇌ interconvert via formation of interme-

diate enolate

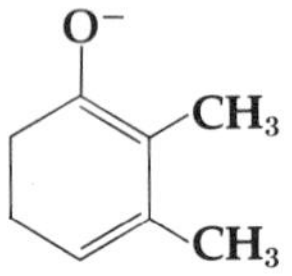

5–16 a $C_6H_5CH{=}CH{-}CH_2^+$ more stable because of added resonance of conjugated phenyl

b Only ⟨⟩ has aromatic 6 π-electron resonance stabilization

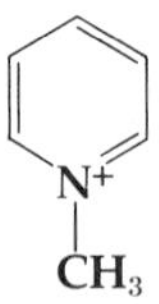

N$^+$

CH$_3$

d ⟨⟩ has aromatic stabilization

f Identical (only resonance hybrids, not different ions)

CHAPTER 6

6–1 a Trans **b** Cis **c** Trans **e** Cis

6–3   The conformer with fewest axial groups is more stable in each pair; in **a**, **d**, and **e** there can be no distinction. Stability: $f = c > b > e = d > a$.

6–5   **d**   [cyclohexane ring with —OH, OH, OH]

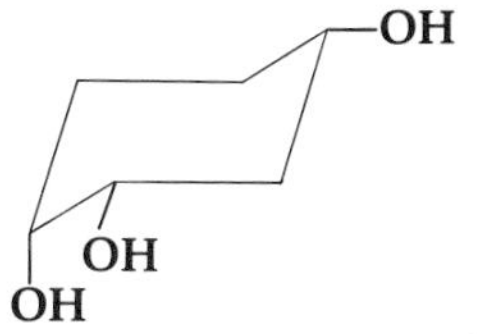

   i   [cyclohexane ring with —CH$_3$, HO—, COOCH$_3$]

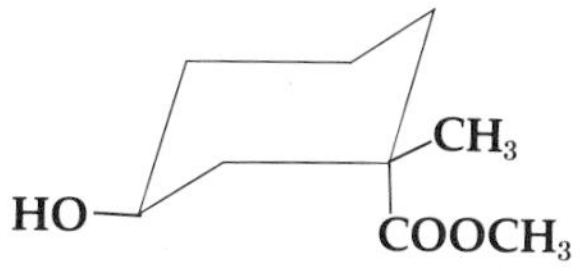

   f   Cl— [cyclohexane ring with —COOH, CH$_3$]

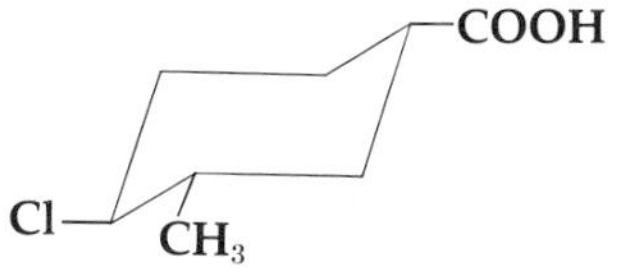

6–6   **a**   Stability: $1 > 2 > 5 = 6 > 3 = 4$

6–7 $CH_2{=}CH{-}\overset{\overset{\textstyle OH}{|}}{\underset{\underset{\textstyle CH_3}{|}}{C}}{-}CH_2CH_3$ (unique answer)

(with * on central C)

6–8   **a** R   **d** R   **f** S

6–13   **a**   3 asymmetric centers $= 2^3 = 8$ stereoisomers, or 4 pairs of enantiomers;

    most stable $=$ [cyclohexane ring with —COOH, CH$_3$, OH]

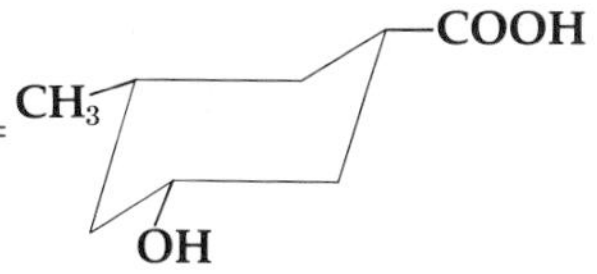

6–14   Only **c** and **g** can be asymmetric; **d** would be asymmetric if it did not rotate freely

6–22   **a**   s-Lactic acid (s-α-hydroxy-propionic or s-2-hydroxy-propanoic acid)

    d   R,R-2,3-Diphenylbutane

6–25   **a, b** $2^3 = 8$ stereoisomers or 4 diastereomeric pairs of enantiomers (4 diastereomeric racemates)

     c   All substituents equatorial:

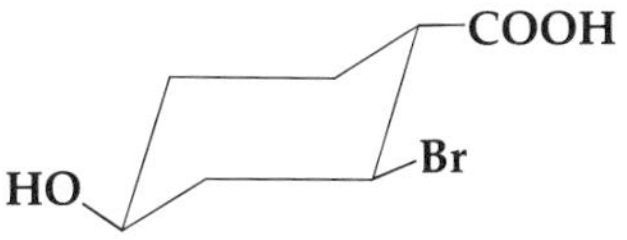

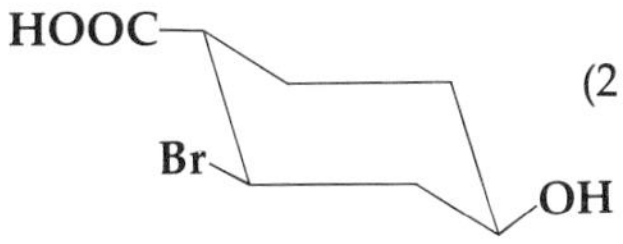

     [cyclohexane ring with —COOH, Br, HO]   [cyclohexane ring with HOOC—, Br, OH]   (2 enantiomers)

     d, e   2-Epimer $=$ [cyclohexane ring with —COOH, HO, Br] $\rightleftharpoons$ [cyclohexane ring with OH, Br, COOH]

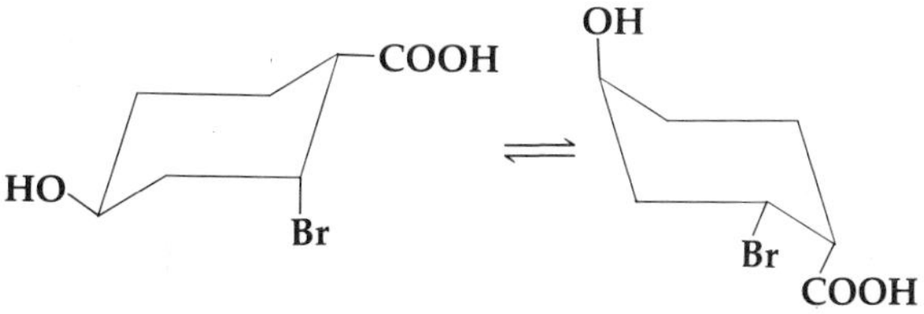

         (HO, Br trans; Br, COOH cis; HO, COOH trans)

6-27

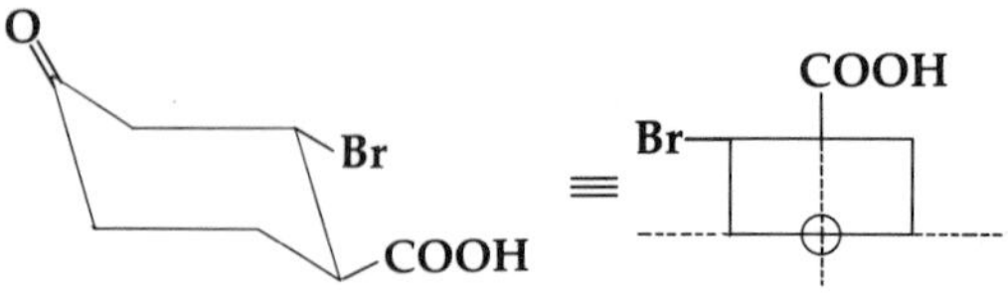

Positive Cotton effect; other enantiomer = negative

6-29

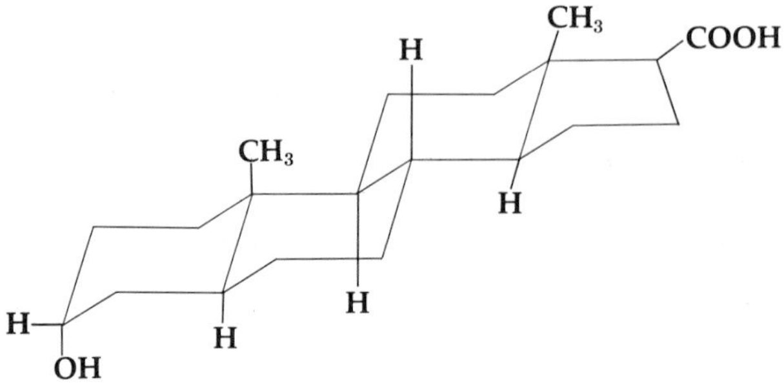

This is the only comformation since it includes rigid *trans*-decalin systems. The hydroxyl is axial, hence less stable than the equatorial hydroxyl epimer. Number of asymmetric carbons = 8; $2^8 = 256$ stereoisomers.

6-33

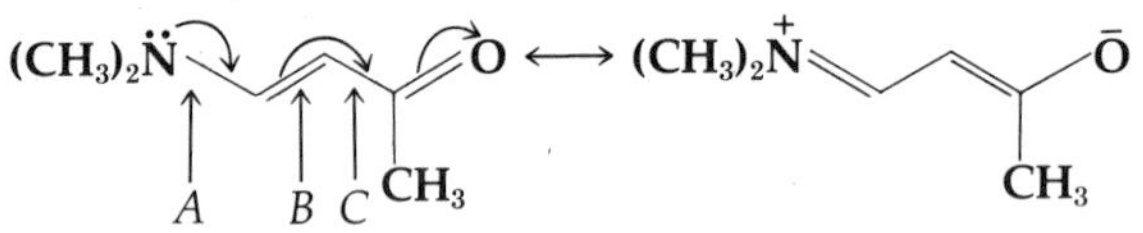

Owing to resonance there is partial double bond character in bonds A and C but less than in bond B which is a double bond in the predominant (uncharged) resonance form. If the charged form accounts for ~25% of the resonance character, then $A, C \approx \frac{1}{4}(60) = 15$ kcal/mole and $B \approx 45$ kcal/mole.

CHAPTER 7

7-2 Dipole moments in **a, d, i,** and **j**

7-3 Solubility in H_2O: $f > e > a > d > b > c$

7-5 $HOCH_2CH_2OH$

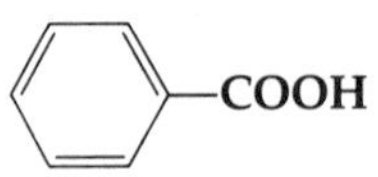

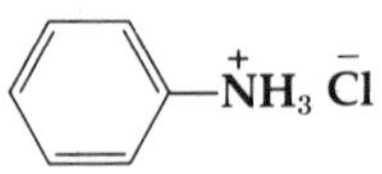

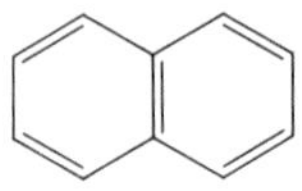

      A B C D

7-6 **b** $248 - 215 = 33$ nm $= 12 + 12 + 10 = \beta + \beta + \alpha$; index of hydrogen deficiency $= 2$, which requires acyclic unsaturated ketone:

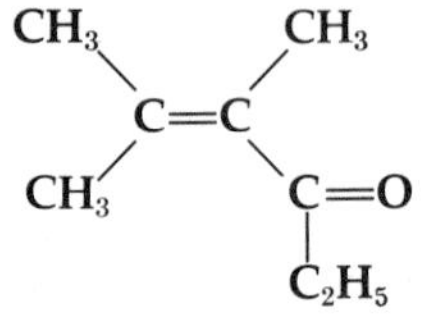

7-7 Thebaine =

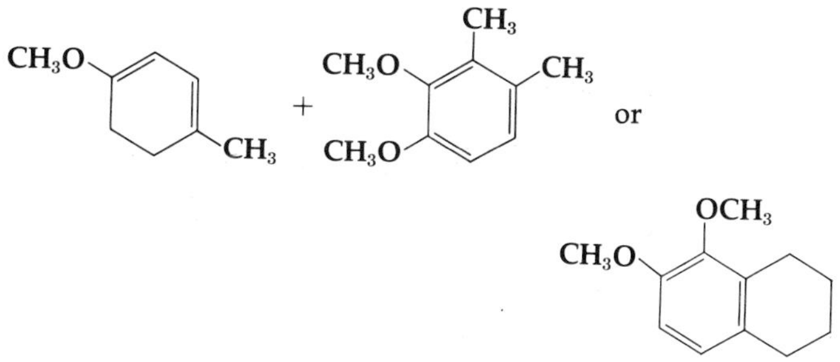

7-8 **b** 2.8, 6.0–6.1, 6.2–6.3 μ (~1600, 1650, 3500 cm^{-1})

  g 4.7, 5.70 μ (1750, 2100 cm^{-1})

  h ~3.0, 5.70, 5.75, 5.95 μ (1690, 1740, 1750, ~3300 cm^{-1})

  i 2.8, 5.65 μ (1770, 3500 cm^{-1})

7-10 **a** Must be saturated ether, index = 1;

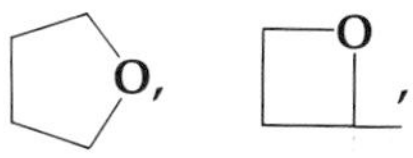

, etc.

  c Two triple-bond bands with only one nitrogen must be C≡C and C≡N. Only carbonyl could be ketone or ester (not amide; nitrogen already placed) and not in a small ring. Index = 5; the three functions account for it, hence no rings or more unsaturation, and four of the carbons are placed, an ester accounts for the fifth and both oxygens. Unique solution is N≡C—C≡C—COOCH$_3$.

7-11 **a** C$_6$H$_5$COCH$_2$CH$_2$COOH: phenyl = 5H ~ 2 τ; methylenes = 4H ~ 7.6 τ; acid = 1H ~ 1 τ

  d CH$_2$(COOC$_2$H$_5$)$_2$: methylene = 2H ~ 7 τ; —OCH$_2$CH$_3$ = 4H ~ 6.5 τ and 6 H ~ 8.8 τ

  h CH$_2$=CH—CH(OCH$_3$)$_2$: 3 vinyl peaks (1H each) from 3.5–5 τ; 1 acetal H ~ 5.5 τ; 6 methoxyl H (one peak) ~ 6.6 τ

7-12 **a** First signal = CHO without adjacent CH; second two are aromatic and split only by each other with a large J = 9, hence two identical pairs of ortho H, implying a para-disubstituted benzene; last is a methyl without adjacent CH; hence the second aromatic substituent is CH$_3$, COCH$_3$, or OCH$_3$; chemical shift best for CH$_3$ or COCH$_3$:

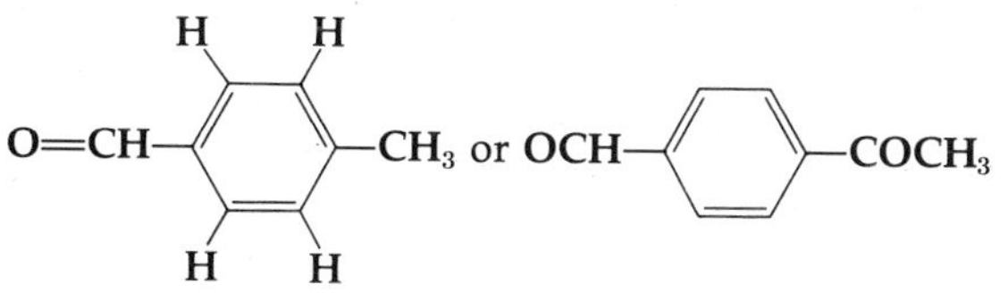

IR ≈ 5.95 μ (1680 cm^{-1}) in either case

7-15 UV = 250 − 215 = 35 ≈ 10 + 12 + 12 or α,β,β-trisubstituted unsaturated ketone of index = 2, hence acyclic: (CH$_3$)$_2$C=C(CH$_3$)COCH$_3$. Alternative = acyclic diene (+ ether or —OH) with 35/5 = 7 substituents = impossible.

7–17 Index = 2; UV = no conjugation; IR = ester or 5-ring ketone, no OH;

NMR= $\diagdown$ CH—$\underline{CH_3}$ (8.85 τ), —CO—$\underline{CH_2}$—CH_2— (7.65 τ chemical shift for

CH_3 next to CO, splitting for 2 adjacent H), and 3H more at O—CH positions. Hence if ester, must be ring (index = 2) and $\geq$ 6-membered; with C_5 this leaves no $CHCH_3$ possible. Therefore, 5-ring ketone + ether. The *isomer*

has 2H and 6H without adjacent H; the latter is $\diagdown$ $C(CH_3)_2$; IR = 4-ring

lactone.

a

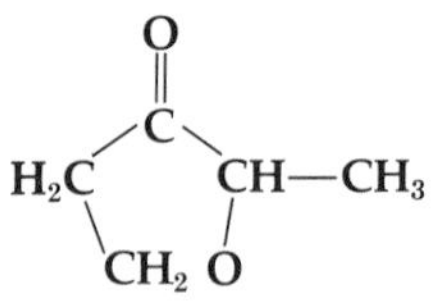

$$
\begin{array}{c}
\text{O} \\
\parallel \\
\diagup\text{C}\diagdown \\
H_2C \qquad CH—CH_3 \\
\diagdown CH_2 \; O \diagup
\end{array}
$$

b

$$
\begin{array}{ccc}
\text{O} & & \text{O} \\
\parallel & & \parallel \\
\text{C}——\text{O} & \text{or} & \text{C}—\text{O} \\
| \qquad | & & | \quad | \\
CH_2—C(CH_3)_2 & & (CH_3)_2C—CH_2
\end{array}
$$

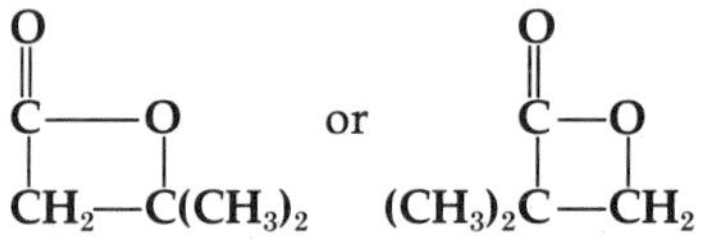

7–21 Index = 6; UV = no conjugation, probable aromatic; IR = OH(or NH), normal ketone, amide. NMR: 2.82 and 3.20 = pair of ortho H doublets = para-disubsituted benzene; 5.66 = 2H adjacent only to NH or OH and also to something else (ϕ, CO, etc.) for low τ; 6.24 = —OCH_3 and 6.63 = CH_2 between *two* unsaturated groups (ϕ, CO, etc.); 7.79 = —$COCH_3$. Total functional groups = NO_3 = CO + CO—N + OCH_3; hence CONHR and no OH.

Summary:

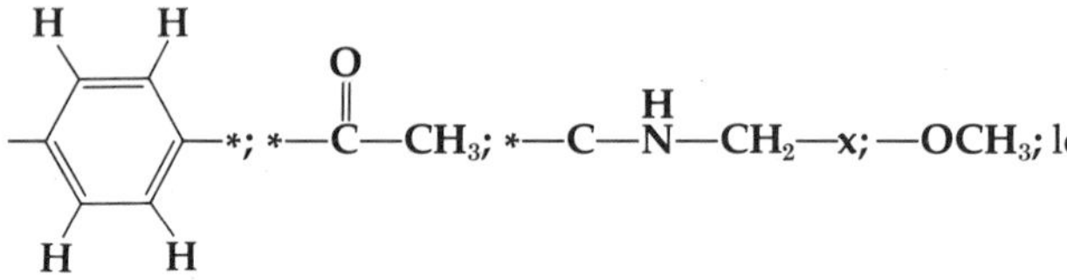

$$
\begin{array}{c}
\text{H} \qquad \text{H} \\
\diagdown \diagup \\
\bigcirc —*; \; *—\overset{\displaystyle \overset{\text{O}}{\parallel}}{C}—CH_3; \; *—C—\overset{\displaystyle \overset{\text{H}}{}}{N}—CH_2—x; —OCH_3; \text{ leaves} \\
\diagup \diagdown \\
\text{H} \qquad \text{H}
\end{array}
$$

no rings or other unsaturation and CH_2 to be placed between any two groups marked *

Structures: CH_3O—⬡—$CH_2CONHCH_2COCH_3$;

CH_3O—⬡—$CH_2NHCOCH_2COCH_3$;

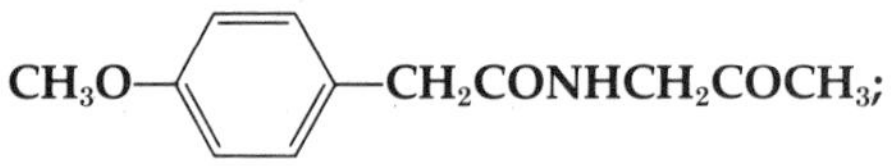

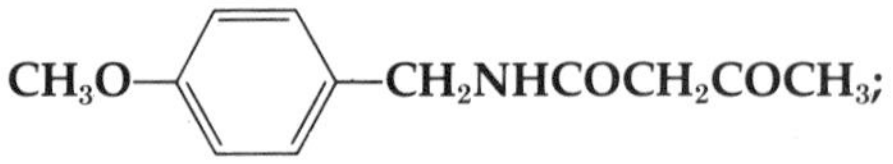

or interchange CH_3O— and —$COCH_3$.

7–22 **a** Mol. wt. = 102; τ 2.6 with 5 H implies C_6H_5—, mol. wt. 77; difference = 102 − 77 = 25 = C_2H. Only $C_6H_5C{\equiv}CH$ fits.

b Index = 2; no H has neighbors; no C=C—H or —O—CH; 12 H = $[(CH_3)_2C\diagup]_2$; 4 H = $[x—CH_2—x]_2$ (in each case these are two *identical* molecular units); 2 H = x—CH$_2$—x; total = $C_3H_6 \times 2 + CH_2 \times 3 = C_9H_{18}$, leaving C=O *and one ring* (index = 2):

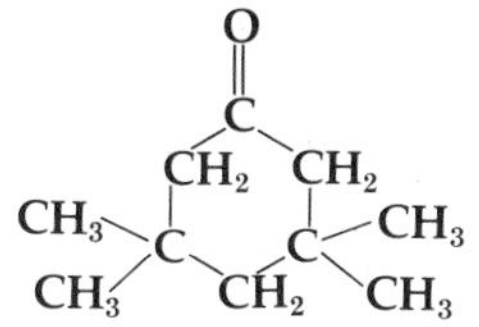

7-24 Mol. wt. = 131 ÷ 14 ≈ 9 first-row atoms (**C, N, O**); NMR shows typical —O—CH$_2$CH$_3$ as low quartet (CH$_2$), high triplet (CH$_3$); IR implies ester, i.e., —COOC$_2$H$_5$, mol. wt. = 73; difference = 131 − 73 = 58; N = 14, and 58 − 14 = 44 or about 3 carbons left; NMR shows N(CH$_3$)$_2$ and x—CH$_2$—x. (CH$_3$)$_2$NCH$_2$COOCH$_2$CH$_3$ is the compound.

CHAPTER 8

8-1 Base must have equal or higher pK_a than H in compound
  a C$_6$H$_5$NH$_2$ = only ϕ_3C:$^-$

  e CH$_3$—⟨benzene⟩—OH; all except CH$_3$COO$^-$

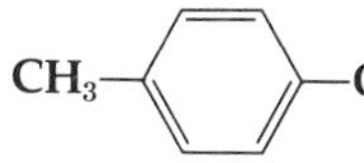

8-2 **b** CH$_3$O$^-$ + CH$_3$COCH$_2$COOCH$_3$ ⇌ CH$_3$OH + CH$_3$COC̄HCOOCH$_3$;
    pK_c = 10.2− 16 = −5.8 K_c = 10$^{5.8}$ ≈ 630,000 (essentially goes to completion)
  c CO$_3^{--}$ + HCN ⇌ HCO$_3^-$ + CN$^-$; pK_c = 9.1 − 10.2 = −1.1; K_c = 10$^{1.1}$ ≈ 12
    or ~70% completion

8-4 **a** Resonance stabilizes conjugate base more than acid and is more effective in para isomer

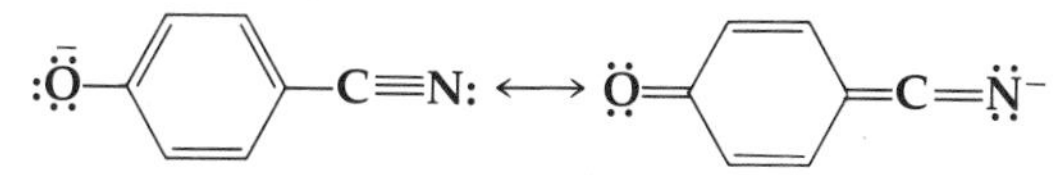

    p-cyanophenol is stronger acid (lower pK_a)
  b Ketone conjugate base more stable since e^- pair is *sp*^{2}; protonated ketone is stronger acid

  h Resonance in 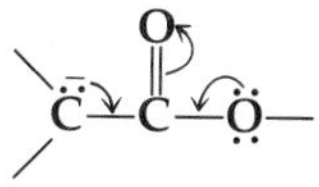 competes with cross-conjugation from oxygen, hence is weaker in ester than ketone; ketone = stronger acid

8-10 Most basic site is e^- pair on carbonyl (*sp*^{2}):

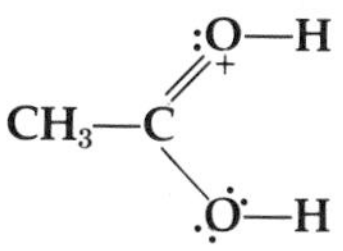

8–13 Initial protonation at most basic carbonyl e^- pair; *protonated β-diketone* enol then has more resonance stabilization and is weaker acid. Similarly, enolate anion after proton removal is more stabilized and β-diketone is a stronger acid.

8–16 **a** ϕ—CH—$\overset{\displaystyle O}{\overset{\|}{C}}$—OR $\rightleftharpoons$ ϕ—$\overset{\displaystyle OH}{\underset{}{C}}$=$\overset{}{C}$—OR; resonance favors enol more
$\quad\quad\quad\;\;$ $\underset{\text{COOR}}{|}$ $\quad\quad\quad\quad$ $\underset{\text{COOR}}{|}$

than in simple ester but not enough to overcome bond-energy preference of keto over enol

d From bond energies triketone more stable than aromatic triol; actually, triol appears to be preferred, however

i Two main tautomers identical

8–19 **a** $\overset{\displaystyle O}{\overset{\|}{C}}$—CH + t-BuO$^-$ $\rightleftharpoons$ $\overset{\displaystyle O^-}{\overset{|}{C}}$=C + t-**BuOH**; pK_c = 20 − 19 = 1; K_c = 0.1 = $\dfrac{x^2}{1-x^2}$; x = 0.24 moles enolate present

b Weaker base, hence less enolate present to do second reaction

c No enolate adjacent to **C**—**Cl** can occur in three cases; these are slower, but in the phenyl-substituted compound, the **CH** in the 2-position is more acidic and more easily removed so that reaction is faster

8–20 Nonsense; no neutral oxygen functional group is basic enough to be observed in water (pK_a 0–12); see Table 8–2

8–21 **a** Index = 4 like pyridine, pK_a 5; e.g.,

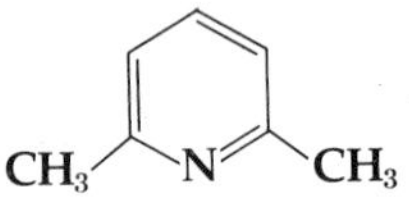

c Normal saturated acyclic (index = 0) amine with asymmetric carbon:

$$\text{CH}_3\text{CH}_2\overset{\overset{\displaystyle \text{CH}_3}{|}}{\text{CH}}\text{—NH}_2$$

g Index = 6, allowing aromatic; phenols = pK_a ~ 10; nitrogen neutral; e.g.,

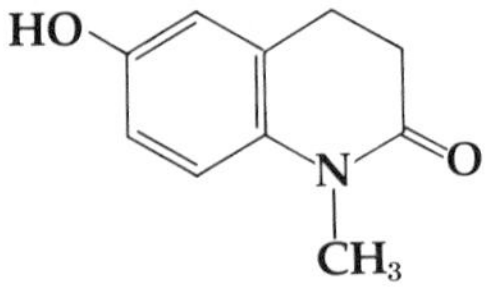

8–22 **a** Acid, pK_a < 7 = **RCOOH**; index = 2 = $\overset{\displaystyle O}{\overset{\|}{C}}$ + one ring (no double bond); e.g.,

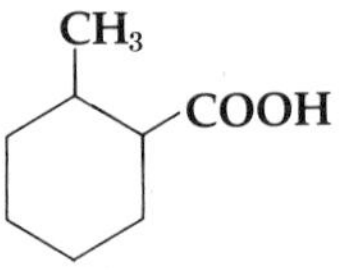

c CH$_3$CH$_2$CN nonbasic; HC≡C—CH$_2$NH$_2$ basic, dissolves in **HCl**

8–23 Nitrogen is nonbasic in amides owing to resonance $\overset{\frown}{N}\!\!-\!\!C\!=\!\!\overset{\frown}{O}$ which creates much double-bond character in the **C—N** bond; the rigid structure of *B* does not allow double-bond character at the bridgehead (Bredt's rule) and prohibits this resonance

CHAPTER 9

9–1 **a** Substitution   **b** Elimination   **d** Addition

9–4 **a**
$$CH_2\!\!-\!\!\overset{\overset{\displaystyle O}{\|}}{C}\!\!-\!\!C_6H_5 \qquad \mathbf{c} \quad CH_3\!\!-\!\!\overset{\overset{\displaystyle Br}{|}}{CH}\!\!-\!\!\overset{\underset{\displaystyle \rightarrow H}{|}}{CH}\!\!-\!\!\overset{\overset{\displaystyle O}{\|}}{C}\!\!-\!\!OCH_3$$
$$\underset{\displaystyle H \leftarrow}{|}$$

9–6 **a** Concentration of *B* at equilibrium $= \dfrac{1{,}000}{30{,}000} = 0.033$;

$$K = \dfrac{0.033}{(1 - 0.033)} = 3.4 \times 10^{-2}$$

9–7 **a** Rates $= CH_3O > CH_3 > CH_3CO$: Fig. 9–7 effects *D*, standard, *B*, respectively

9–9 **a** Nucleophilic substitution
   i Nucleophilic addition, base-catalyzed so that active nucleophile is
$$CN$$
   probably CN^-, giving $CH_3\overset{\overset{\displaystyle CN}{|}}{CH}\!\!-\!\!O^-$ which is finally protonated

9–13 The rate-controlling step is the first ionization and $CH_3CO_2^-$ is not involved in this; hence its concentration does not affect rate

9–15 **a** First reaction faster owing to ring-strain release; effect *C*, Fig. 9–7
   d Protonation of —**OH** and ionization to H_2O and carbonium ion are the initial and rate-controlling steps; the secondary, allylic carbonium ion is more stable, hence faster: Fig. 9–7, effect *B*

9–16 **c** Second equilibrium more favorable since product has added aromatic resonance stablization

CHAPTER 10

10–2 **a** Racemic 1-phenyl-1-acetoxybutane via S_N1; s-1-phenyl-1-acetoxybutane via S_N2

10–4 **a** Racemization via S_N1 solvolysis of the tertiary alcohol to its formate ester
   b S_N2 reaction with inversion produces the enantiomer, which is another S_N2 of identical rate returns to initial enantiomer; net result is ultimate racemization

10-5 **b** **AgNO$_3$:** $(CH_3)CCl > (CH_3)_2CHBr > C_2H_5Br$ (no reaction at room temperature) **NaI:** $C_2H_5Br > (CH_3)_2CHBr$ and no reaction with $(CH_3)_3CCl$

10-6 **b** First reaction slower (S_N2) owing to steric hindrance; products = **CN** replacing **I**

c $C_6H_5-\underset{\underset{CH_3}{|}}{\overset{\overset{CH_3}{|}}{C}}-OH \xrightarrow[Br^-]{H^+} C_6H_5-\underset{\underset{CH_3}{|}}{\overset{\overset{CH_3}{|}}{C}}-Br$ and $\xrightarrow[Br^-]{H^+}$

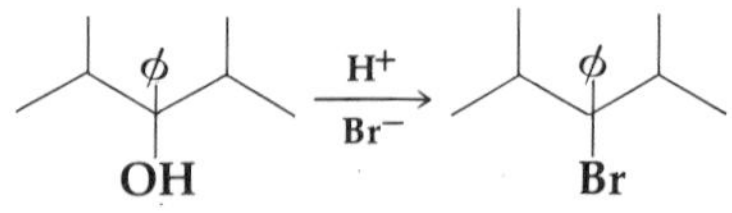

Second is slower (S_N1) because of steric hindrance to forming planar carbonium ion with coplanar phenyl in resonance

10-10 **b** Cis $\longrightarrow$ two cyclohexeneacetic acids; trans $\longrightarrow$ δ-lactone

c Cis $\longrightarrow$ two cyclohexeneacetic acids; trans $\longrightarrow$ γ-lactone

10-12 **a** Triphenylethylene; 1,1,2-triphenylethyl formate

c 1-Methylcyclopentene; 3-methylcyclopentene

10-13 **c** R-2-Phenylpentanonitrile by S_N2/inversion + some 1-phenyl-1-butene

d Racemic 1-phenyl-1-acetoxybutane by S_N1 (racemization) + some excess inverted R isomer + some racemic 1-phenyl-1-ethoxybutane and some 1-phenyl-1-butene

h $C_6H_5CH_2-S-\underset{\underset{S^-}{|}}{\overset{\overset{S}{\|}}{C}}$ and $(C_6H_5CH_2-S)_2C=S$ by S_N2

10-14

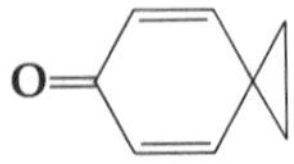

10-16 Cis 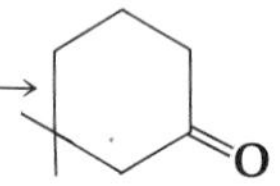 $\longrightarrow$; trans $\longrightarrow$

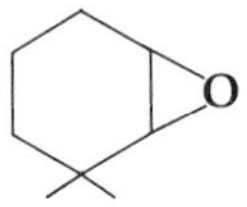

10-19 **a** Carbonium-ion order: $C_{10}H_7CH_2Cl$, $C_6H_5CH_2Cl$, $(CH_3)_2CHCl$, $CH_3CH_2CH_2Cl$

10-22 (1) **HBr**; two diastereomers possible from either cis or trans starting ether owing to S_N1 giving both epimers; (2) propionic acid solvolysis, S_N1, to distinguish two bromines best; (3) **AgBF$_4$** in **CH$_3$OH**; **CH$_3$O$^-$** will lead largely to elimination; both choices are bad in this way

10-23 **a**

c ϕ_3CCH_2OH

e $CH_3CH_2CH=CHCH_2Br$

g

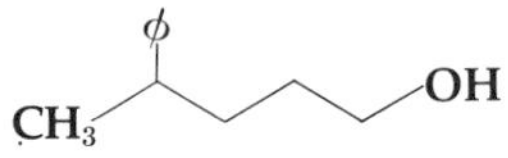

10–26 **b** S_N1 reaction via carbonium ion slower when ion adjacent to positive carbon of $C{=}O$ dipole and so destablized

e Steric crowding released as three bulky groups go from 109.5° to 120° angles (sp^3 to sp^2); this release favors the carbonium ion

h Compression of larger bicyclic system to a flat carbonium ion is easier the larger the ring

CHAPTER 11

11–1 **b** $HC{\equiv}CNa + CH_3I$

d $(CH_3)_2CHONa + CH_3CH_2CH_2I$

h $HC{\equiv}CNa + CH_3CH_2CH_2CH_2Br \longrightarrow$

$$HC{\equiv}C(CH_2)_3CH_3 + 2H_2/Pt \longrightarrow n\text{-hexane}$$

11–3 **c** $Na^+ \bar{:}CH(COOC_2H_5)_2;\ CH_3{-}CH(COOC_2H_5)_2;\ CH_3\bar{C}(COOC_2H_5)_2\ Na^+;$

$$CH_3 \underset{\overset{|}{CH_3}}{\overset{\displaystyle COOC_2H_5}{\underset{\displaystyle COOC_2H_5}{\diagdown}}}$$

11–4 **c** $CH_3CO\overset{\bar{}}{C}HCOOCH_3Na^+ + (CH_3)_2CHI \longrightarrow$

$$(CH_3)_2CH{-}\underset{\overset{|}{\displaystyle COOCH_3}}{CHCOOCH_3} \xrightarrow{CH_3O^-} \xrightarrow{CH_3I} \xrightarrow[H^+]{H_2O} \text{product}$$

f $CH_3COCH_2COOCH_3 \xrightarrow[CH_3(CH_2)_5Br]{OCH_3^-} CH_3(CH_2)_5\underset{\overset{|}{\displaystyle COCH_3}}{CHCOOCH_3} \xrightarrow[CH_3I]{OCH_3^-}$

$$CH_3(CH_2)_5\underset{\overset{\displaystyle COCH_3}{|}}{\overset{\overset{\displaystyle COCH_3}{|}}{C}}COOCH_3 \xrightarrow[\Delta]{^-OH} \text{product}$$
$$\underset{CH_3}{}$$

11–5 **b** $(CH_3)_2C{=}CHCH_3;\ (CH_3)_2\underset{\overset{|}{\displaystyle OCH_3}}{C}{-}CH_2CH_3;\ (CH_3)_3CCH_2OCH_3 \text{ (very little)}$

c $CH_3CH{=}CHCH_3;\ CH_3CH_2CHOHCH_3;\ CH_3CH_2CH{=}CH_2;$ $(CH_3)_2C{=}CH_2;\ (CH_3)_3COH$

11–6 **a** $C_2H_5Br + CN^-$

e $\overset{\displaystyle \overset{O}{\triangle}}{CH_2{-}CH_2} + CN^- \longrightarrow HOCH_2CH_2CN \xrightarrow{TsCl} TsOCH_2CH_2CN \xrightarrow{CN^-}$

$$NCCH_2CH_2CN \xrightarrow{OH^-} HOOCCH_2CH_2COOH$$

l $HC{\equiv}CNa + CH_3I \longrightarrow HC{\equiv}CCH_3 \xrightarrow[\text{cat.}]{H_2} CH_2{=}CHCH_3 \xrightarrow{H_2O}$

$$CH_3CHOHCH_3 \xrightarrow{HCl} CH_3\underset{\overset{|}{\displaystyle Cl}}{CHCH_3}$$

n $CN^- + ClCH_2COOCH_3 \longrightarrow NCCH_2COOCH_3 \xrightarrow[CH_3O^-]{2CH_3I}$

$NCC(CH_3)_2COOCH_3 \xrightarrow[H^+]{H_2O} (CH_3)_2CHCN$

q $HC{\equiv}CNa + \overset{O}{\overset{\triangle}{CH_2{-}CH_2}} \longrightarrow HC{\equiv}CCH_2CH_2OH \longrightarrow$

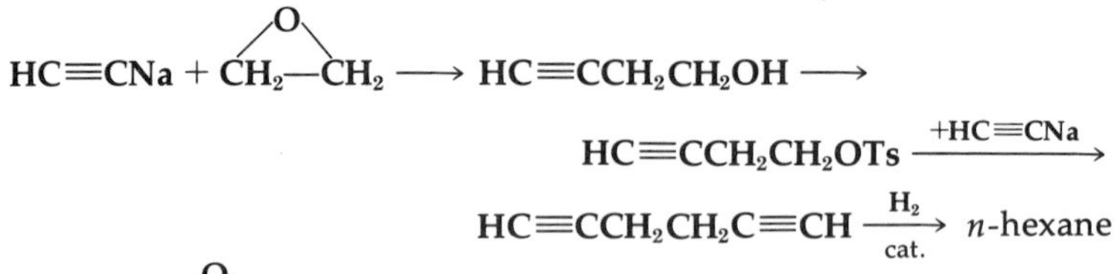

$HC{\equiv}CCH_2CH_2OTs \xrightarrow{+HC{\equiv}CNa}$

$HC{\equiv}CCH_2CH_2C{\equiv}CH \xrightarrow[cat.]{H_2} n\text{-hexane}$

u $CH_3Li + \overset{O}{\overset{\triangle}{CH_2{-}CH_2}} \longrightarrow CH_3CH_2CH_2OH \longrightarrow$

$CH_3CH_2CH_2OTs \xrightarrow[-OC_2H_5]{CH_3CH(COOC_2H_5)_2} CH_3CH_2CH_2\underset{CH_3}{\overset{|}{C}}(COOC_2H_5)_2 \xrightarrow[H^+]{H_2O}$

$CH_3CH_2CH_2\underset{CH_3}{\overset{|}{C}}HCOOH$

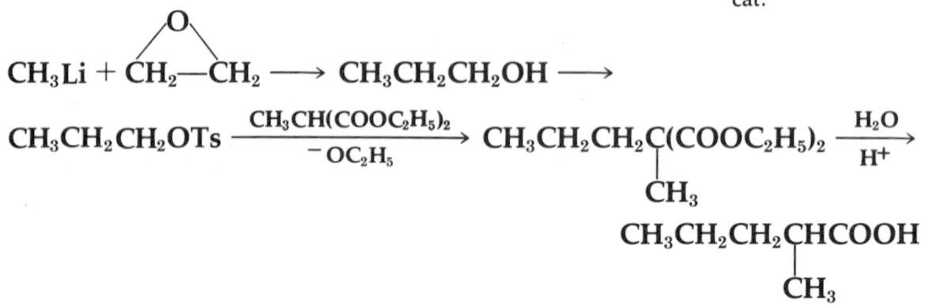

11–9   **a**   The reaction will go well but some amines may arise from S_N2 reaction of NH_3; use another base such as **NaH**. Reaction as written also yields NH_4Cl.

b

c   No cyclization owing to linear acetylene; polymers will form

d   Cis cyclizes, trans polymerizes; cis $\longrightarrow$ $\underset{+}{N}(C_2H_5)_2\ Br^-$

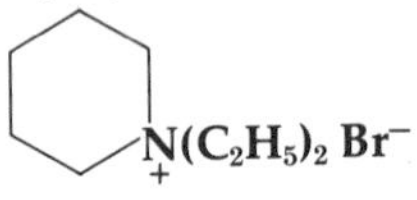

11–11   **c** $\phi\overset{O}{\overset{\triangle}{CH{-}CH_2}} \xrightarrow[H^+]{CH_3OH} \phi\underset{OCH_3}{\overset{|}{C}}HCH_2OH \xrightarrow{TsCl} \phi\underset{OCH_3}{\overset{|}{C}}HCH_2OTs \xrightarrow{NaI} \phi\underset{OCH_3}{\overset{|}{C}}HCH_2I$

$\Big\downarrow LiAlH_4$

$\phi\underset{OH}{\overset{|}{C}}H{-}CH_3 \xrightarrow{H^+} \phi CH{=}CH_2 \xrightarrow[cat.]{H_2} \phi CH_2CH_3$

d $CH_3COCH_2COOCH_3 + CH_3\overset{O}{\overset{\triangle}{CH{-}CH_2}} \xrightarrow{-OCH_3}$

$CH_3CO\underset{CH_2CHOHCH_3}{\overset{|}{C}}HCOOCH_3 \xrightarrow[CH_3I]{-OCH_3} CH_3CO\underset{CH_2CHOHCH_3}{\overset{CH_3}{\overset{|}{\underset{|}{C}}}}COOCH_3 \xrightarrow[H^+]{H_2O}$

$CH_3CO\underset{OH}{\overset{CH_3}{\overset{|}{\underset{|}{C}}}}CH_2CHCH_3$

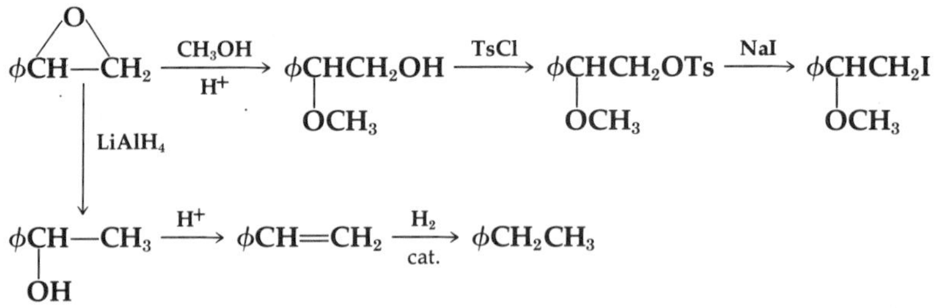

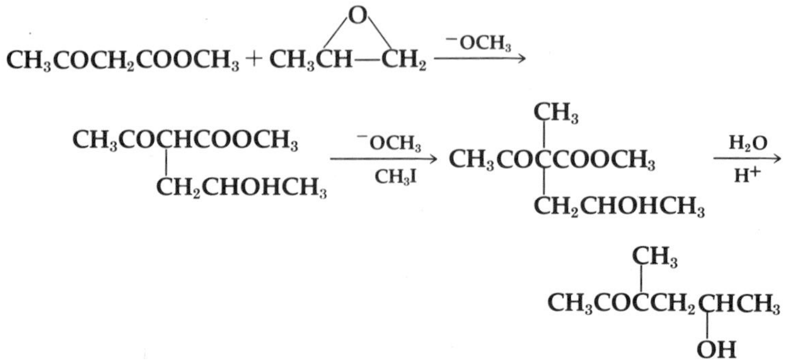

11–12 b

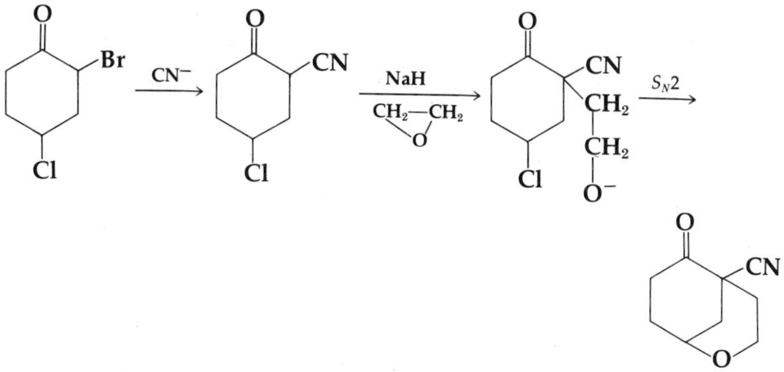

Ignoring epimerization α to **CO**, the bridge from ethylene oxide must form trans to the —**Cl** for S_N2 displacement; hence **CN** is cis to **Cl** and original **Br** was trans to **Cl** since S_N2 displacement by **CN**$^-$ goes with inversion

11–14 IR = ketone (5.85 μ) + ester (5.75 μ); NMR shows two kinds of —CH_2CH_3, one at 9.25 (CH_3) and 8.1 (CH_2), the other at lower fields, 8.75 (CH_3) and 5.8 (CH_2), the latter being —O—CH_2CH_3 of the ester; the former ethyl signal is twice as intense, therefore $\diagup\!\!\!\overset{\diagdown}{C}(C_2H_5)_2$; CH_3CO also from 3H singlet at 7.9 τ

$$CH_3CO—\overset{\displaystyle CH_2CH_3}{\underset{\displaystyle CH_2CH_3}{C}}—COOCH_2CH_3 \text{ (mol. wt. 186)}$$

Mass spectrum shows loss of CH_3 (186 − 15 = 171), loss of $CH_2{=}CH_2$ (182 − 28 = 158), and units $COOC_2H_5$ (m/e 73) and CH_3CO^+ (m/e 43).

CHAPTER 12

12–2 a $CH_3COCH_2CH_2CHOHCN$

c

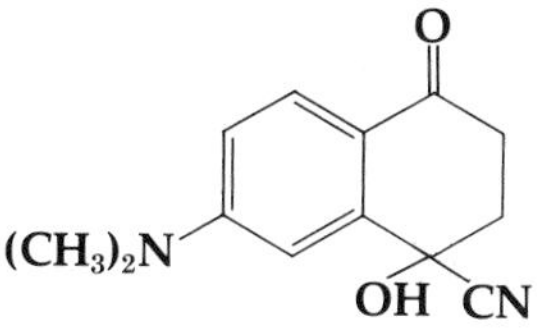

; other **CO** is conjugated and less reactive

e

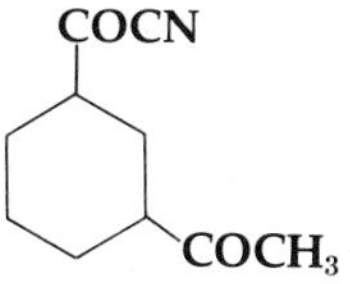

12–3 c *cis*-OCHCH=CHCHO + CH_3OH + H^+

12–4 a 4-Cyano-2-pentanol
   b 1-Chloro-3-pentanol; 3-pentanol
   e Cyclohexanol + acetone

12-6 a $CH_3CH_2CHO + C_2H_5MgBr$

d $CH_3COCH_3 + CH_3Li \longrightarrow (CH_3)_3COH \xrightarrow{HBr} (CH_3)_3CBr$

h $\phi MgBr + CH_3CH_2CHO \longrightarrow \phi CHOHCH_2CH_3 \xrightarrow{H+}$

$\phi CH{=}CHCH_3 \xrightarrow[\text{cat.}]{H_2} \phi CH_2CH_2CH_3$

12-7 b $(\phi CH_2)_2CO + H_2NOCH_3/\text{pyridine}$

d

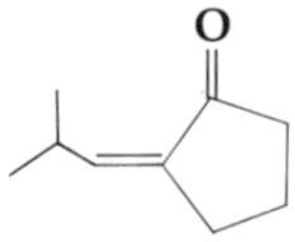

$/OH^-$

i $\phi\underset{\underset{NO_2}{|}}{C}H-\overset{\overset{OH}{|}}{C}(CH_3)_2/OH^-$

m p-Bromobenzaldehyde$/CN^-$

t $CH_3CH{=}P\phi_3 + O{=}$ ${=}O/\text{no catalyst}$

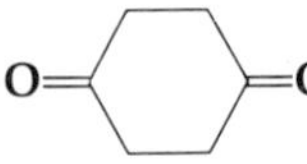

12-9 a 1,3-Cyclopentanedione $\pm$ 3-penten-2-one

d Cyanide ion + 2-cyclopentenone

f $NH_3 + CH_2{=}CHCN$

h $(C_2H_5)_2NH + HC{\equiv}CCOCH_3$

12-11 a $C_6H_5CH_2CN + C_2H_5MgBr$

f $C_6H_5CH_2CN + CH_2{=}CHCOOR$, followed by hydrolysis of **RCN** to **RCONH₂**, which cyclizes spontaneously

12-13 a Owing to repulsion between the two positive carbons in α-dicarbonyl compounds, the carbonyls are less stable and more reactive than normal

d RMgX acts as a base to remove readily ionizable protons, as in R_2NH, ROH, and β-dicarbonyl compounds:

$CH_3COCH_2COOC_2H_5 + RMgX \longrightarrow CH_3CO\underset{\underset{MgX}{|}}{C}HCOOC_2H_5 + RH\uparrow$

12-15

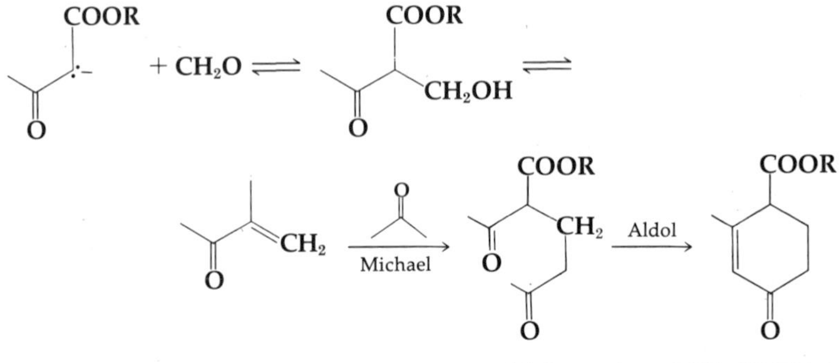

The mechanism can also be written via prior aldol reaction of $CH_3COCH_3 + CH_2O$ to $CH_3COCH{=}CH_2$ and Michael addition of acetoacetate

12–16 a O_3 **b** OH^-

12–17 a

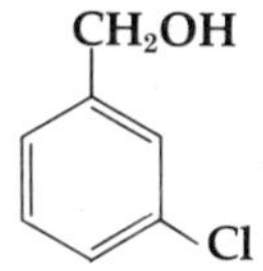

; ϕCH_2CH_2OH; $\phi$$CH_2OH$

b + ; ϕ ;

$$\phi CH=C(CH_3)COOH + \phi CH=C(CH_3)CH_2OH$$

g ; as in **b**; ϕ

k ; $\phi CH_2CHOHCN$;

$$\phi-\underset{\underset{CN}{|}}{C}H-\underset{\underset{CH_3}{|}}{C}H-CHO \text{ (or its cyanohydrin)}$$

p   All three $RCHO \longrightarrow RCH=CHCN$

12–18 a $CH_3Li + CH_3CHO$

c $a + TsCl \longrightarrow (CH_3)_2CHOTs \xrightarrow{CN^-} (CH_3)_2CHCN \xrightarrow[cat.]{H_2}$

$$(CH_3)_2CHCH_2NH_2$$

h $+ CN^- \longrightarrow HOCH_2CH_2CN \xrightarrow{H^+} CH_2=CHCN \xrightarrow{CH_3SH}$

$$CH_3SCH_2CH_2CN$$

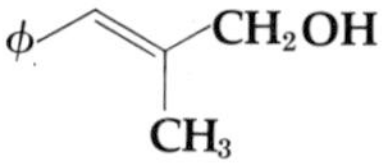

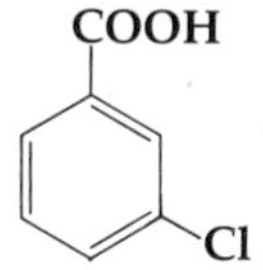

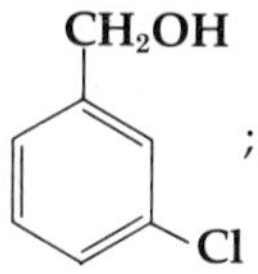

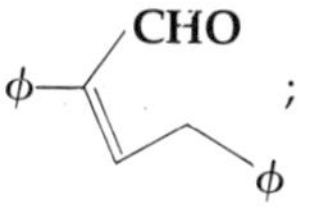

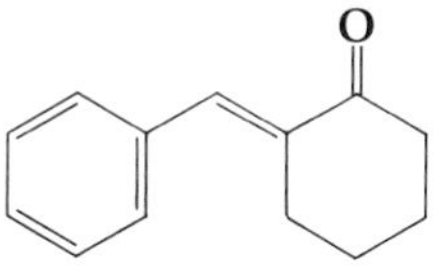

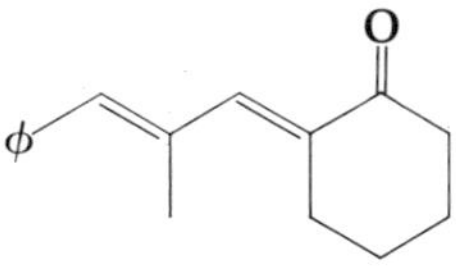

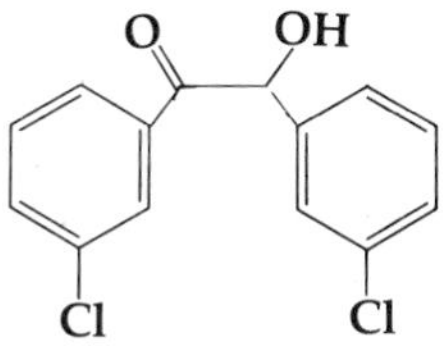

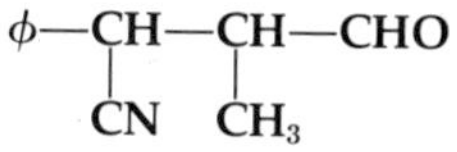

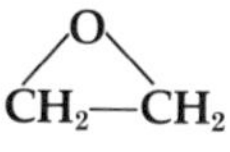

l $CH_3CN + C_2H_5MgI \xrightarrow[H^+]{H_2O} CH_3COCH_2CH_3 \xrightarrow[H^+]{+HOCH_2CH_2OH} \text{product}$

12–19 a $\phi_3P=CH_2 + 4\text{-heptanone}$ **b**   Perkin or Doebner   **e**   Stubbe

12–23 a   Aldol condensation of ϕCHO at 2-position and racemization of asymmetric carbon at 4-position by base-catalyzed enolization:

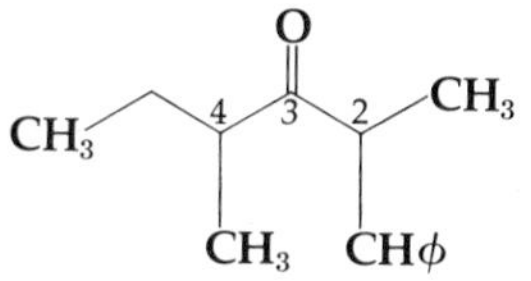

d HO⁻ + CH₂=CH—CN ⟶ HO—CH₂—CH—CN $\xrightarrow{\phi CHO}$

ϕ—CH—CH—CH₂OH $\xrightarrow[2)\ -H_2O]{1)\ H^+}$ ϕCH=C—CH₂OH
  | | |
  O⁻   CN   CN

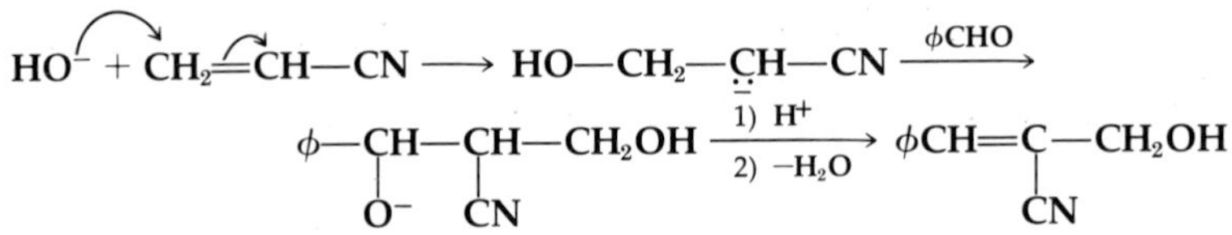

f

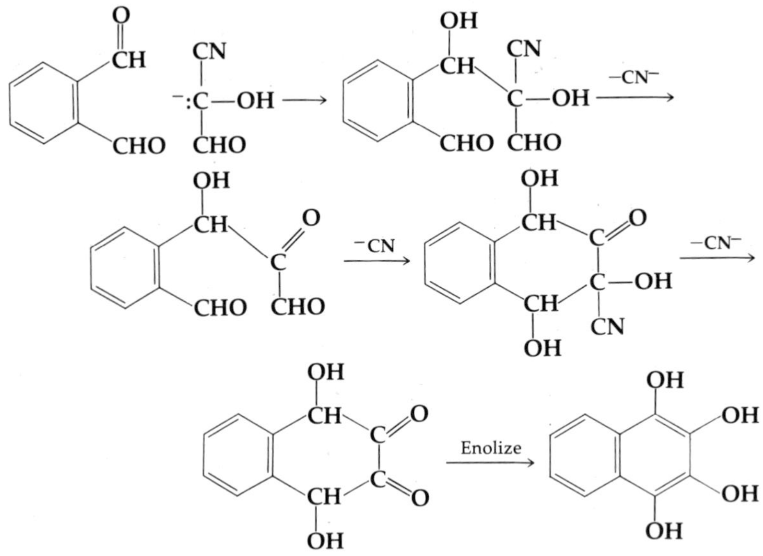

12–26 Index = 5; ketone = 1, remaining 4 suggests aromatic; NMR shows two doublets in aromatic region as for para-disubstituted benzenes; remaining protons are three methyl groups: two different O—CH₃ groups, one from the oxime ether, R—C=N—OCH₃, formed in the reaction, one CH₃
                                                              |
                                                              R'

on an unsaturated carbon:

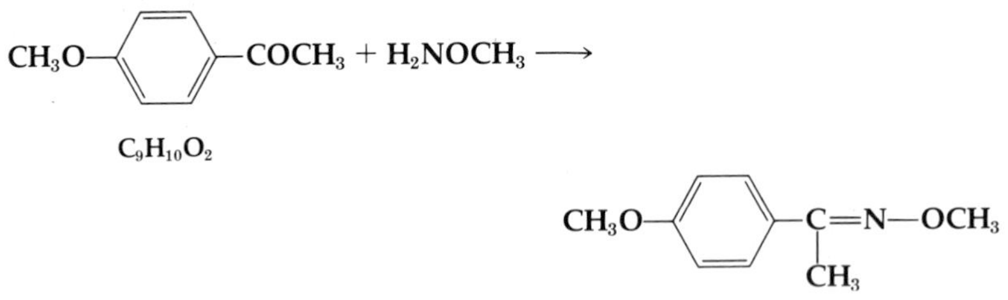

CH₃O—⟨ ⟩—COCH₃ + H₂NOCH₃ ⟶

C₉H₁₀O₂

CH₃O—⟨ ⟩—C=N—OCH₃
                         |
                        CH₃

12–28 C₅H₉NO₂; index = 2; NMR: 5.47 = C=C—H with no adjacent C—H; 6.38 = OCH₃; 8.08 = C—CH₃ with no adjacent C—H; other 2H are probably —NH₂ from reacted NH₃ and exchange in D₂O

Reaction:
$$\text{C}-\overset{\overset{\text{O}}{\|}}{\text{C}}-(\overset{|}{\text{C}}\text{H}) + \text{NH}_3 \longrightarrow$$
original ketone

Summary:
$$\left. \begin{array}{l} \text{CH}_3\text{O}- \\ \text{CH}_3-\overset{|}{\underset{|}{\text{C}}}- \\ \text{H}_2\text{N}- \\ \underset{|}{\overset{/}{\text{C}}}=\overset{|}{\text{C}}-\text{H} \end{array} \right\} +10$$

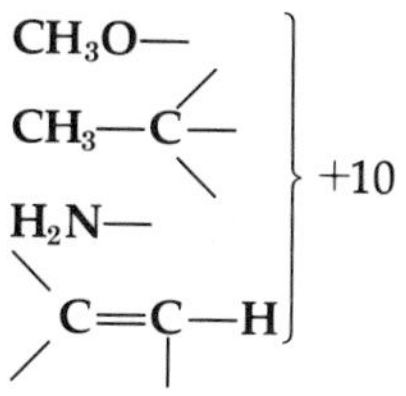

$$\underset{\text{NH}_2}{\overset{\overset{\text{NH}}{\|}}{\text{C}-\text{C}-(\overset{|}{\text{C}}\text{H})}} \rightleftharpoons \underset{}{\overset{\overset{\text{NH}_2}{|}}{\text{C}-\text{C}=\text{C}-}}$$

Required = $(\text{C})-\underline{\text{C}=\text{CH}}-$ $\left. \begin{array}{l} -\text{OCH}_3 \\ -\text{CH}_3 \\ + \text{C} + \text{O} \end{array} \right\}$ $\text{C}_3\text{H}_6\text{O}_2$

$\underset{\text{C}_2\text{NH}_3}{}$

$\Sigma = \text{C}_5\text{H}_9\text{NO}_2$

$$\underset{}{\overset{\overset{\text{NH}_2}{|}}{\text{CH}_3-\text{C}}}=\text{CH}-\overset{\overset{\text{O}}{\|}}{\text{C}}-\text{OCH}_3 \text{ from } \text{CH}_3\text{COCH}_2\text{COOCH}_3$$

$$(\text{CH}_3\text{O}-\text{CO}-\underset{}{\overset{\overset{\text{NH}_2}{|}}{\text{C}}}=\underline{\text{CH}-\text{CH}_3}: \text{CH}-\text{CH}_3 \text{ impossible for singlet } \text{CH}_3)$$

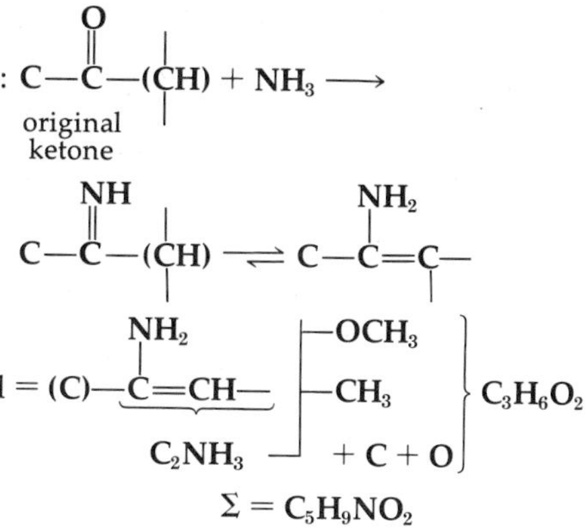

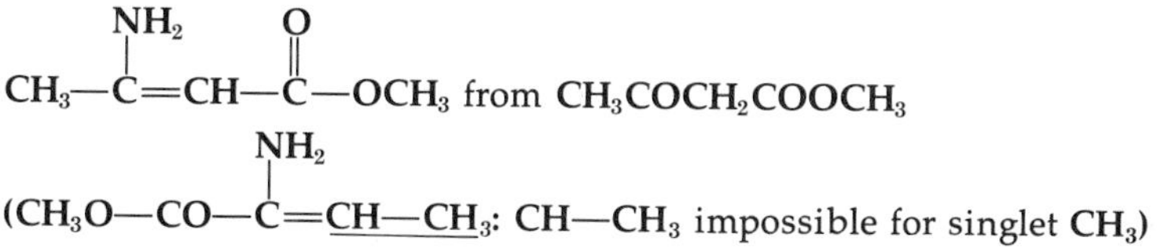

CHAPTER 13

13–1 **a** $\text{HOOCCH}_2\text{CH}_2\text{COOCH}_3$ **i** $\text{CH}_3\text{COOCH}(\text{CH}_3)_2 + \text{C}_6\text{H}_5\text{COOH}$

   c $\text{C}_6\text{H}_5\text{COOCH}_3$ **k** $\text{HOCH}_2\text{CH}_2\text{CH}_2\text{CONHOH}$

   f $\text{CH}_3\text{CH}_2\text{CN}$

13-3

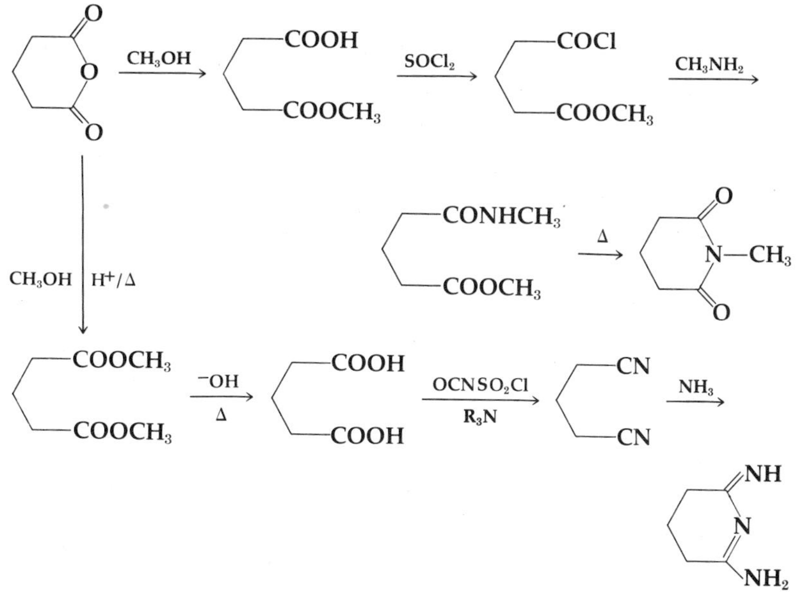

13–4 **a** $\text{CH}_3\text{CH}_2\text{COCHCOOC}_2\text{H}_5$ **f** $\phi\text{COCHCOCH}(\text{CH}_3)_2$
              $\underset{\text{CH}_3}{|}$ $\underset{\text{CO}\phi}{|}$

   b No change   **i**

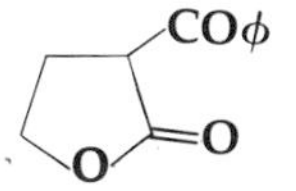

13-6 **a** $NCCH_2COOCH_3 + \phi CH_2Br \xrightarrow{Base} NCCHCOOCH_3 \xrightarrow[\text{base}]{CH_3I}$
$\qquad\qquad\qquad\qquad\qquad\qquad\qquad\qquad\qquad | $
$\qquad\qquad\qquad\qquad\qquad\qquad\qquad\qquad\quad CH_2\phi$

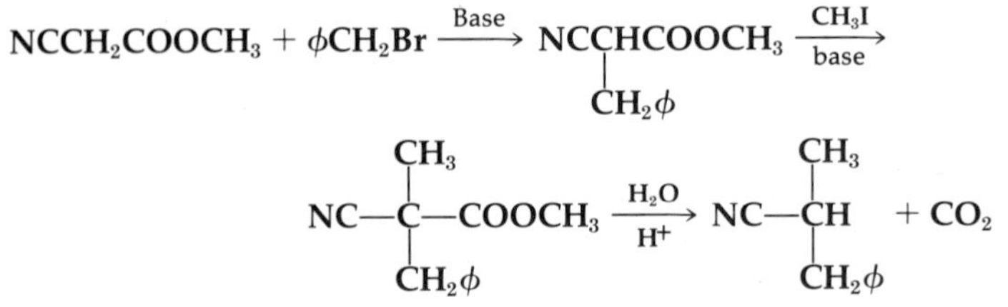

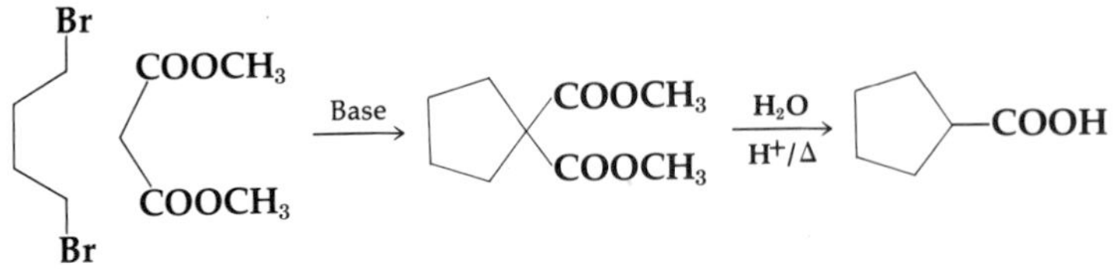

$$NC-\underset{\underset{CH_2\phi}{|}}{\overset{\overset{CH_3}{|}}{C}}-COOCH_3 \xrightarrow[H^+]{H_2O} NC-\underset{\underset{CH_2\phi}{|}}{\overset{\overset{CH_3}{|}}{CH}} + CO_2$$

b [structure: 1,4-dibromobutane + dimethyl malonate] $\xrightarrow{Base}$ [cyclopentane with COOCH3, COOCH3] $\xrightarrow[H^+/\Delta]{H_2O}$ [cyclopentane-COOH]

13-8 [piperidine N—COOH] > [2-oxocyclohexane-COOH] > [2,4-dihydroxybenzoic acid] > [salicylic acid] > [3,5-dihydroxybenzoic acid]

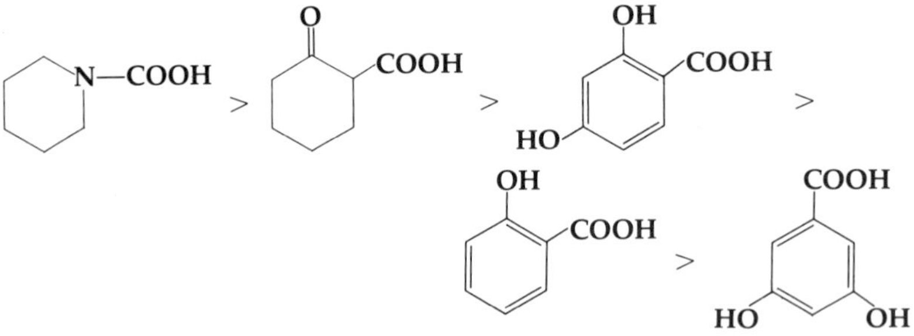

13-10 **d** $COCl_2$ or $CO_2 + LiAlH_4$

e $CH_3OH + HBr \longrightarrow CH_3Br \xrightarrow[\text{(see j.)}]{^-CN} CH_3CN \xrightarrow[H^+]{CH_3OH} CH_3COOCH_3$

f $COCl_2 + CH_3OH$

j $HCOOCH_3 + NH_3 \xrightarrow{\Delta} HCONH_2 \xrightarrow{SOCl_2} HCN$

m $CH_3COOLi + CH_3Li$

q $CH_3COOC_2H_5 \xrightarrow{OC_2H_5} CH_3COCH_2COOC_2H_5 + (CH_3)_2CHI$
$\qquad\qquad\qquad\qquad\qquad\qquad\qquad\qquad \xrightarrow{^-OC_2H_5} product$

u $CH_2(COOCH_3)_2 + CH_2-CH_2(epoxide) \xrightarrow{^-OCH_3} HOCH_2CH_2CH(COOCH_3)_2 \xrightarrow[H^+]{H_2O}$

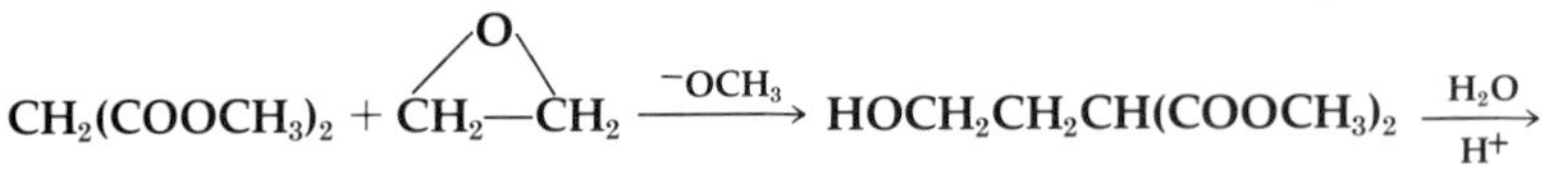

$HOCH_2CH_2CH_2COOH$ and distill $(-H_2O)$; $CH_2{=}CHCOCH_3 +$

$\qquad\qquad\qquad CH_3COCH_2COOCH_3 \xrightarrow[\text{Michael}]{\text{Weak base}} \xrightarrow{^-OH}$

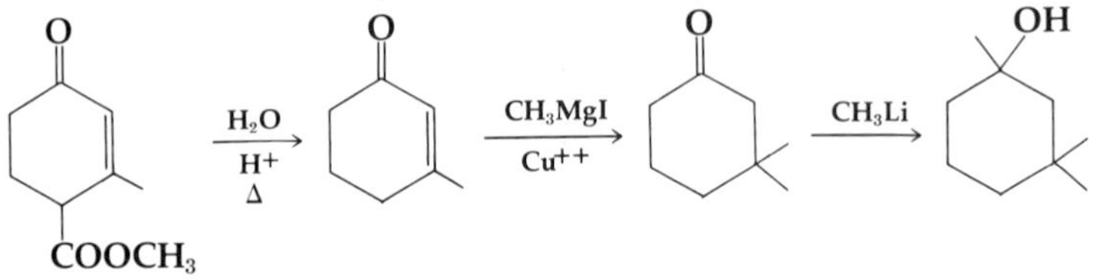

[cyclohexenone with COOCH3] $\xrightarrow[\Delta]{H_2O/H^+}$ [methylcyclohexenone] $\xrightarrow{CH_3MgI / Cu^{++}}$ [dimethylcyclohexanone] $\xrightarrow{CH_3Li}$ [trimethylcyclohexanol]

13–11 c and 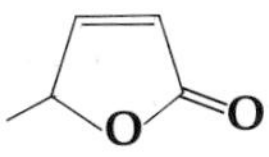 IR 5.60 and 5.70 μ (1790 and 1750 cm^{-1}), respectively

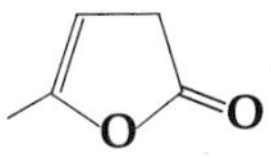

13–13 b

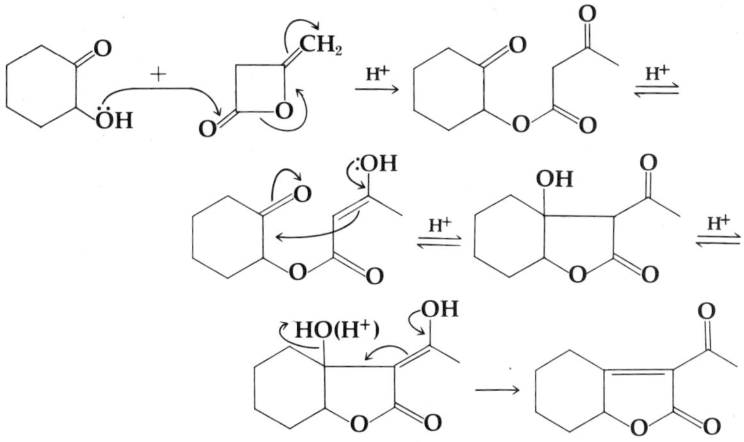

d From part b:

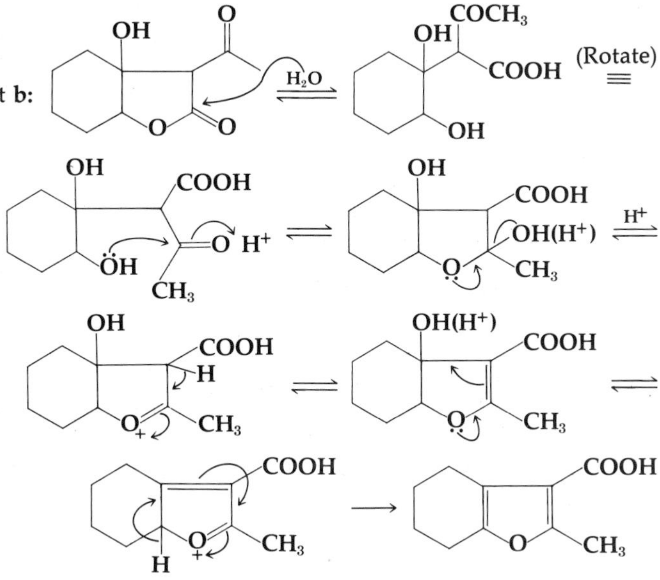

13–15 a Claisen condensation; LiAlH$_4$

c + CO(OC$_2$H$_5$)$_2$ $\xrightarrow{\text{}^-\text{OC}_2\text{H}_5}$ COOC$_2$H$_5$ $\xrightarrow{\text{CH}_3\text{I}}$

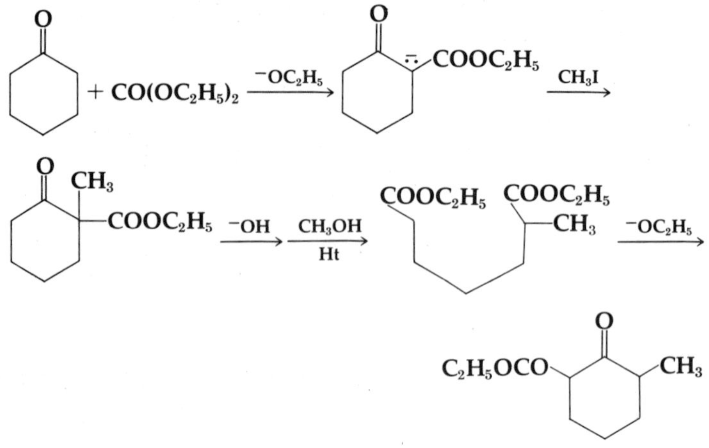

CH$_3$, COOC$_2$H$_5$ $\xrightarrow[\text{Ht}]{\text{}^-\text{OH} \quad \text{CH}_3\text{OH}}$ COOC$_2$H$_5$ COOC$_2$H$_5$ CH$_3$ $\xrightarrow{\text{}^-\text{OC}_2\text{H}_5}$

C$_2$H$_5$OCO CH$_3$

i CN + ⟶ CN $\xrightarrow{\text{}^-\text{OMe}}$ NH

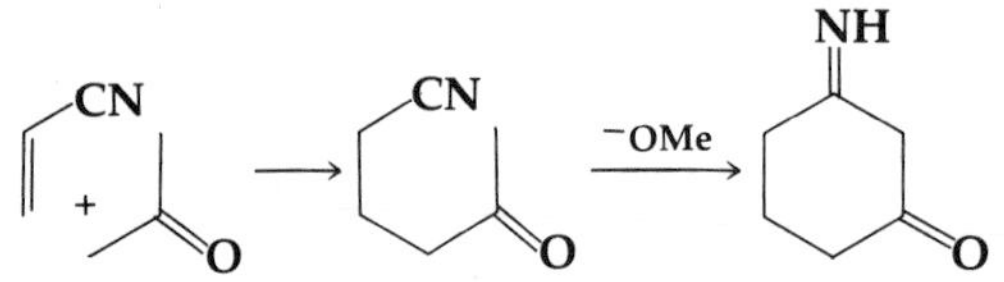

13–16 d CH$_3$COCN + CN$^-$ ⟶ CH$_3$–C–O$^-$ (CN, CN) $\rightarrow$ $\rightarrow$

CH$_3$–C–O–CCH$_3$ + CN$^-$ (CN, CN, O)

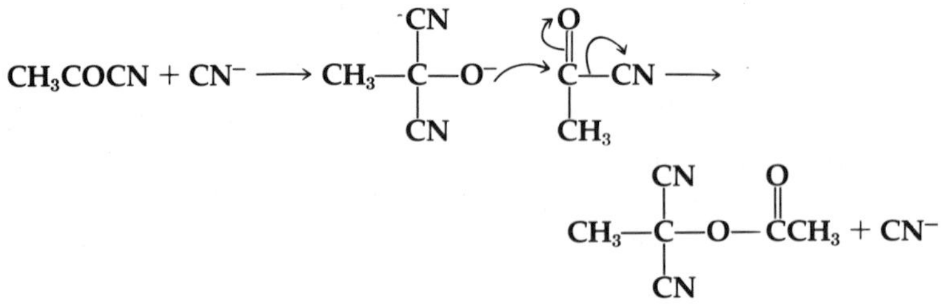

e

(More stable product anion)

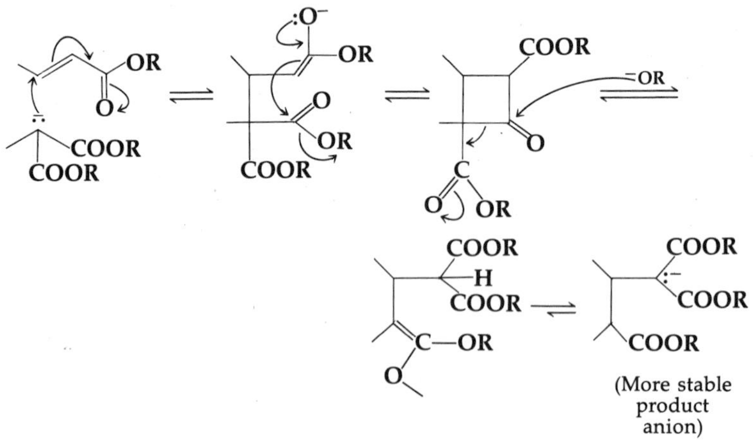

13–18 a Succinonitrile + C$_2$H$_5$MgX ⟶ $\xrightarrow[\text{}]{\text{OR}^- \quad \text{H}_2\text{O}}$

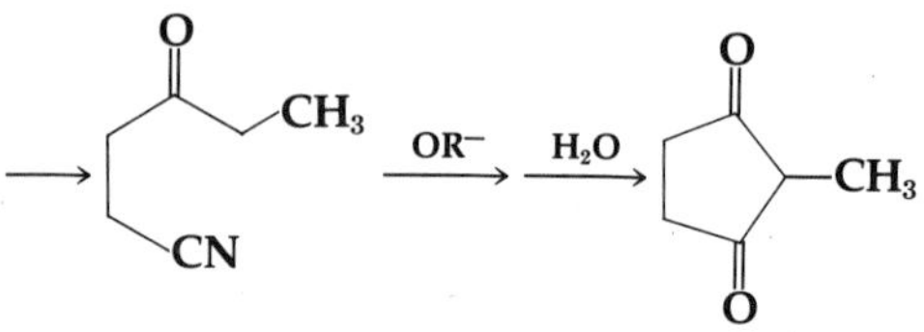

c   Glutaric diester/acyloin reaction

f   Methyl vinyl ketone + 2-methylcyclopentane-1,3-dione (Robinson annellation)

13–19   a, b(d)   [cyclopentanone] $\xrightarrow[-OR]{+(COOR)_2}$ [2-carbethoxycyclopentanone, COOR] $\xrightarrow[\text{base}]{CH_3I}$

[2-methyl-2-carbethoxy-cyclopentanone, COOR] $\xrightarrow{-OH}$ [2-methylcyclopentanone]

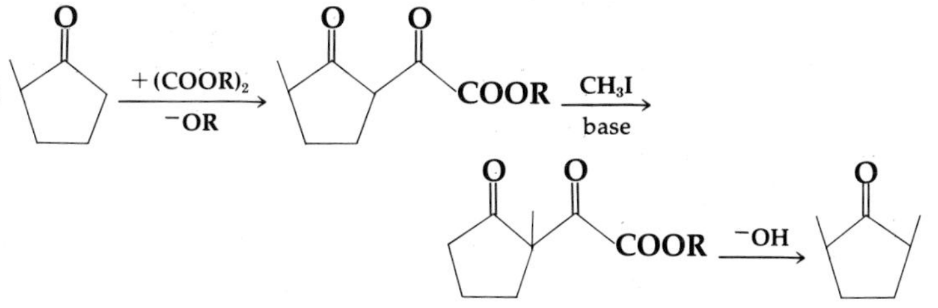

e $HCOOR/OR^- \longrightarrow$ [=CHOH derivative] $\xrightarrow{CH_2N_2}$ [=CHOCH$_3$ derivative]

$\xrightarrow[CH_3I]{KNH_2}$ [dimethyl =CHOCH$_3$ derivative]

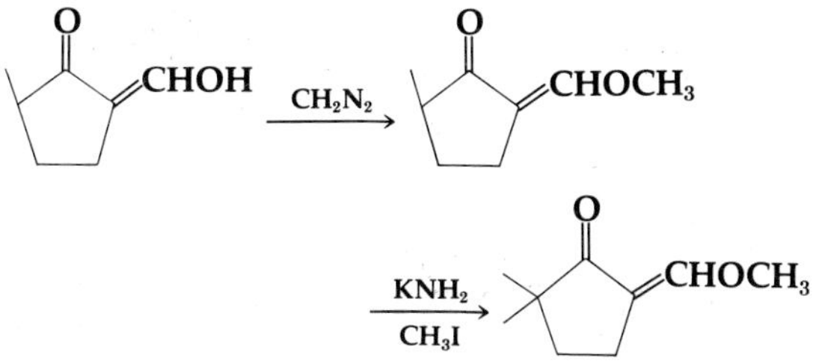

f $NaBH_4;\ \phi_3PBr_2;\ Mg/Et_2O;\ CH_2\!\!-\!\!CH_2$ (epoxide, O)

13–20   a   Enamine + $ClCOOCH_3$; CN^-

b   Michael addition of malonate to 1-cyclohexene carboxylate ester

c CH_3Li; HBr; Li/Et_2O; CH_3COOLi; CH_3COOR/OR^-; OH^- on 2-carbethoxycyclohexanone; $SOCl_2$; $(C_2H_5)_2Cd$; $\phi_3P\!=\!CH_2$

f $\phi_3P\!=\!CHCH_2CH_2CH(OCH_3)_2$; H_2O/H^+

k $NCCH_2COOH/pyridine$; CN^-; $H_2O(H^+ \text{ or } OH^-)$; $(-H_2O)$

n   Block one α-position with $HCOOR/OR^- + CH_2N_2$; dimethylate; remove $=CHOCH_3$ and reduce ketone.

13–21   a   [bicyclic COOR/OR lactone structure] $\xrightarrow[\substack{\text{cleaves}\\ \text{strained}\\ \text{ring}}]{:OR}$ [bracketed intermediate with C=O, OR] $\longrightarrow$ [cyclohexenone, COOR] $\xrightarrow{H_3O^+}$ [dimethyl cyclohexenone, O]

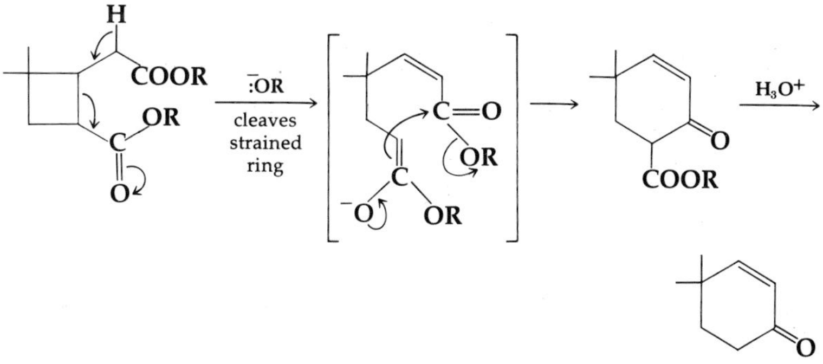

c

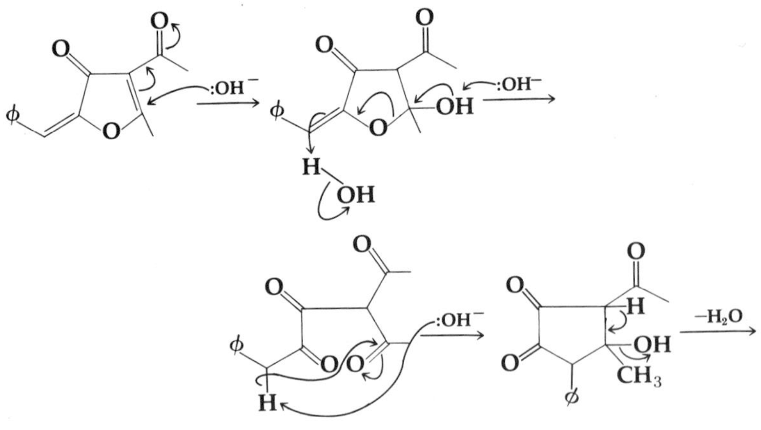

13–23

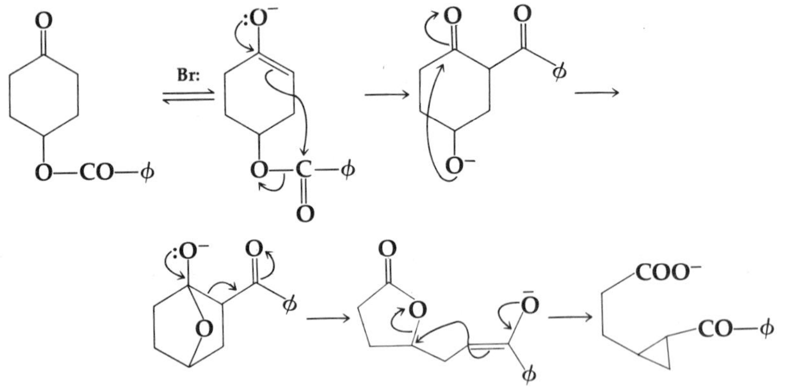

13–24

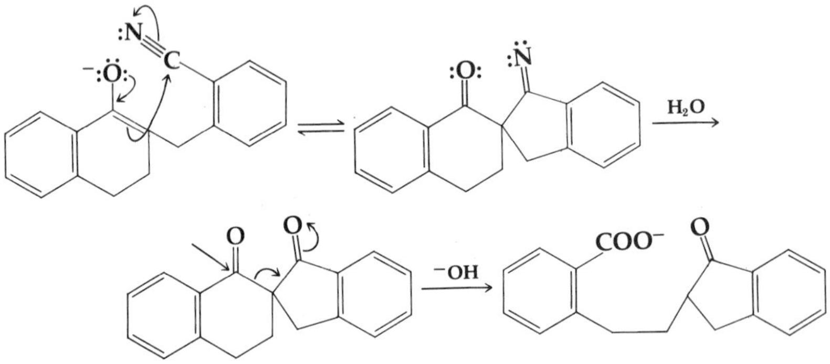

13–26 **a** Internal Claisen condensation of ketone enolate on lactone carbonyl and enolization of cyclic β-diketone product to a di-phenol

c Enol attacks HO—N=O to place —N=O on α-carbon, then cleavage of O=N—C—C=O as with β-diketones

CHAPTER 14

14-1 1,3-elimination is an S_N2 reaction forming a 3-membered ring

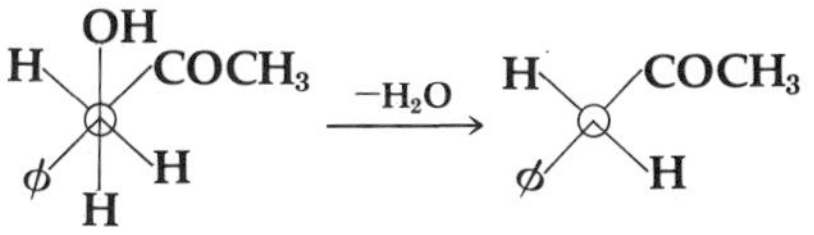

Trans

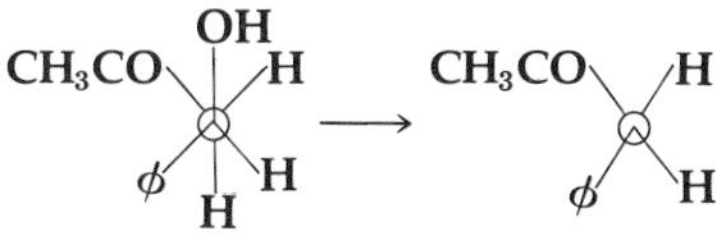

Cis

higher-energy starting transiton state for cis formation owing to CH_3CO- and $\phi-$ crowding

14-4 **a** Trans 1,2-isomer gives epoxide; cis 1,2-isomer has **H** and **Br** trans and gives enol by elimination of **H** on the **C—OH** side; enol then yields cyclohexanone; 1,3-isomers give allylic alcohol on elimination

  c Lactone cannot enolize (Bredt's rule) and no **H** is trans to **Br**

  f Fastest proton removal is α to $C{=}O$ (enolization), hence fastest elimination

14-5 **a** $\phi CH{=}CHCH_3$, probably trans (see Prob. 14–3); conjugation and most substituted double bond

  d Least substituted is $CH_3CH_2\underset{\underset{CH_2}{\|}}{C}CH_2CH_2CH_3$

  e Both products about equally likely

  g Four diastereomers (d, l, and two meso):

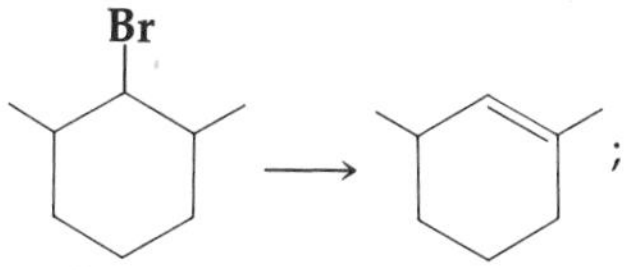

Meso   Racemic 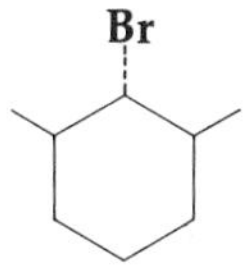   Meso

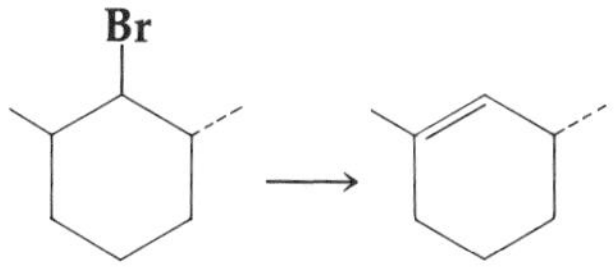

d, l

Each enantiomer yields one enantiomer

14-6 **b** H_2/cat.; CH_3I, excess; OH^-/Δ

  c CH_3I to tertiary amine; H_2O_2; Δ

14-7 **a** $BrCH_2CHOHCHOHCH_2Br$

  c β-Bromobutyrolactone

  d 1,3-Propanediol monomethyl ether $+\ H_2SO_4$

e 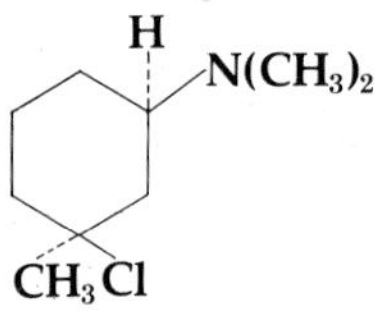   f

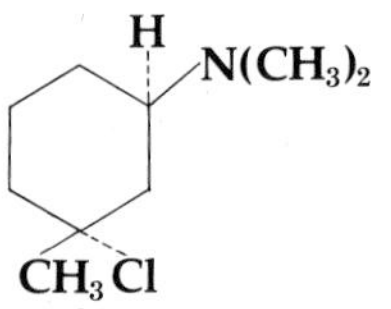

14-9 **a**

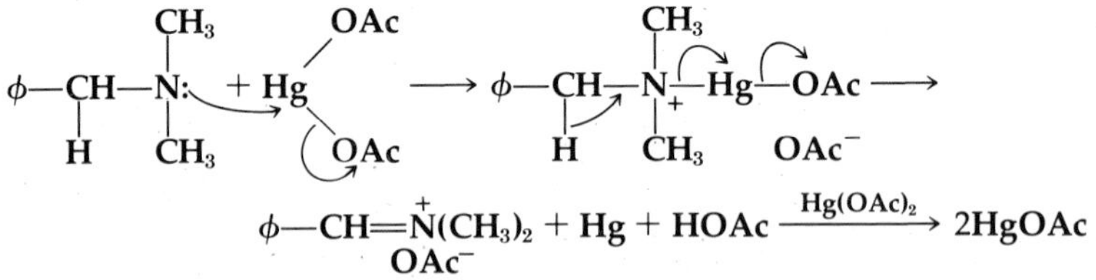

$$\phi-CH=\overset{+}{N}(CH_3)_2 + Hg + HOAc \xrightarrow{Hg(OAc)_2} 2HgOAc$$
$$\overset{|}{OAc^-}$$

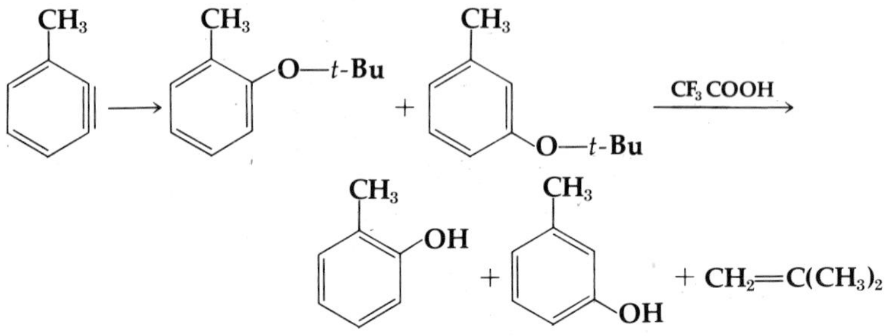

c

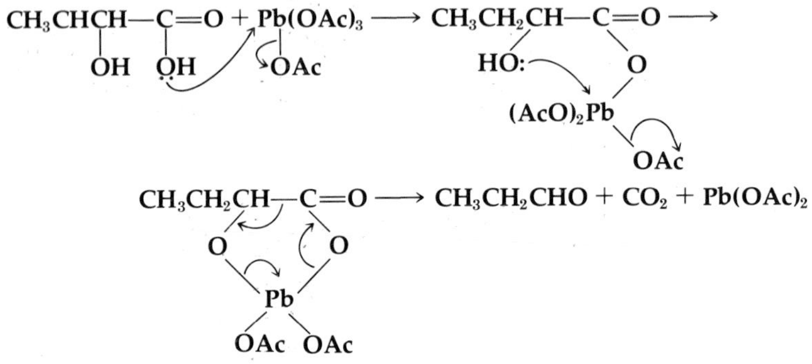

14-10 **a** $\xrightarrow{CF_3COOH}$

$+ CH_2=C(CH_3)_2$

b $t\text{-Bu}-OCOCH_2CH_2\overset{}{\underset{\underset{N-OH}{||}}{C}}COO-t\text{-Bu} \xrightarrow[\Delta]{H^+} HOOCCH_2CH_2CN +$

$$2CH_2=C(CH_3)_2 + CO_2 + H_2O$$

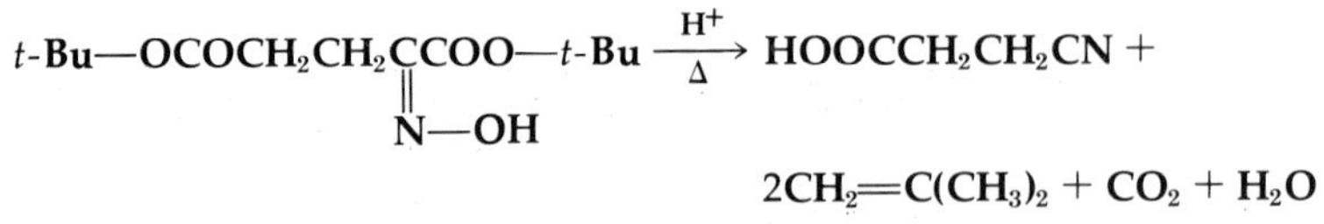

f Cyclododecyne
i 2-Hexyne

14-12 **a** 2-Methyl-2-pentene
         b 2-Methyl-1-pentene

f

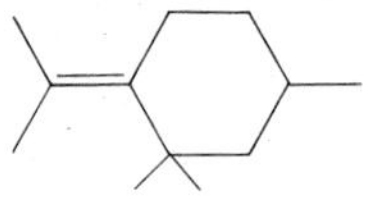

h 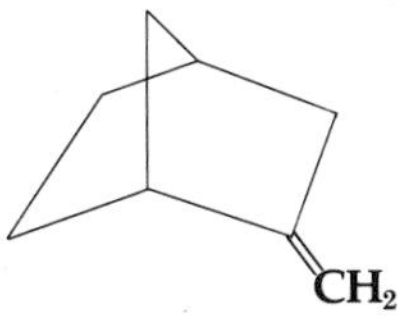

since **H** and **Br** cannot become trans periplanar to eliminate into ring, even though that product is more substituted

14–13 c $H_2SO_4 \longrightarrow RCH{=}CH_2$; Br_2; OH^-/Δ

d $\phi_2CO + CH_3CH_2CH{=}P\phi_3$

h NH_2OH; $SOCl_2$

m CH_3I, excess; OH^-/Δ; CH_3I; excess; OH^-/Δ

q Br_2; OH^-/Δ

14–14 a 1,3-Diol fragmentation on 1,2,3-triphenyl-1,3-butanediol

c 3,5-diacetoxy-3,5-dimethylcyclohexanone

14–16 b $\phi{-}CH(Br){-}CH_2{-}CH_2{-}CH(Br){-}\phi \longrightarrow 2\phi{-}CH{=}CH_2 + ZnBr_2$ (Br ←:Zn)

14–18 Convenient to use partial chair cyclohexane forms.

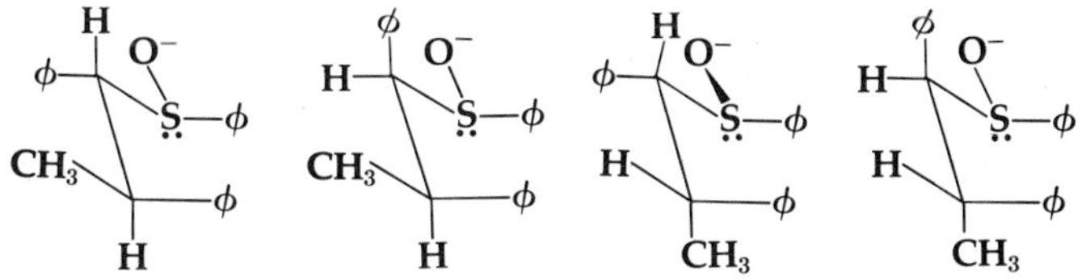

14–19 $CH_2{=}CH{-}C(OH)(CH_3){-}CH_2COCH_3 \xrightarrow{SOCl_2} CH_2{=}CH{-}C(CH_3){=}CH{-}COCH_3$

(258 nm) + $CH_2{=}CH{-}C({=}CH_2){-}CH_2COCH_3$ (220 nm)

14–22 Ketone cannot enolize toward epoxide, 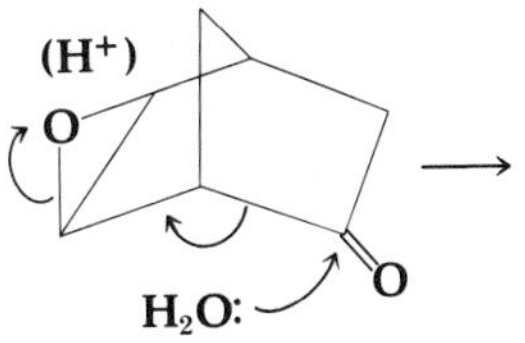

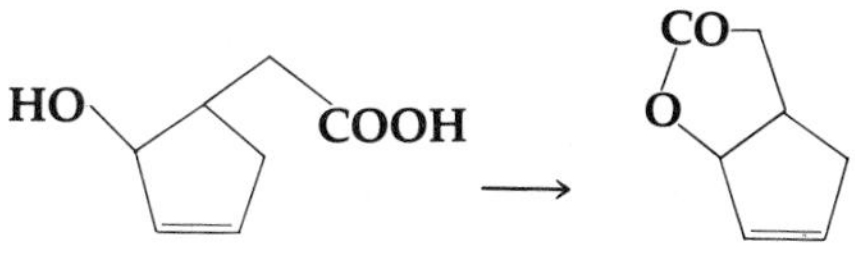

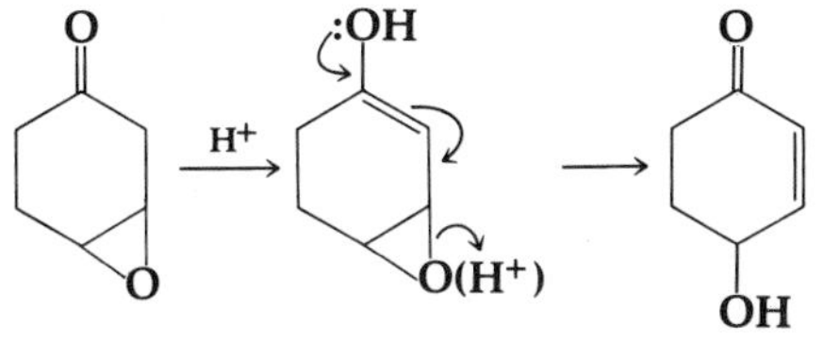

14–24 b

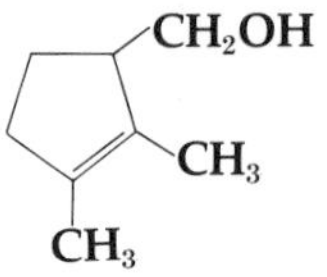

CH_2OH

CH_3

CH_3

CHAPTER 15

15–1   a   Trans addition yields only one diastereomeric racemate; starting material is not optically active so products cannot be; other diastereomer comes from cis addition

d

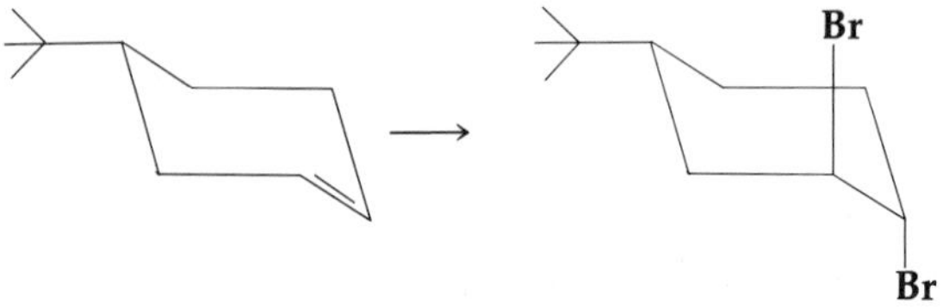

$\longrightarrow$

Br

Br

15–2   a

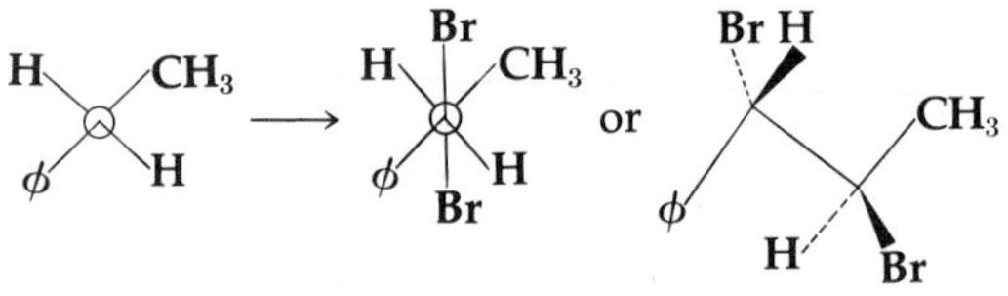

H CH_3 H   Br CH_3 Br H
                                                      CH_3
ϕ H ϕ H   or ϕ
                      Br   H   Br

i

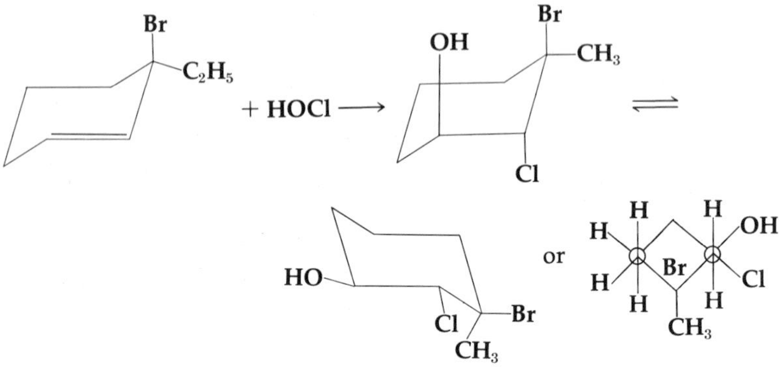

Br

C_2H_5

+ HOCl $\longrightarrow$

Br

OH CH_3

Cl

$\rightleftharpoons$

HO

Cl   Br
   CH_3

or

H   H   OH
H
   Br
H   Cl
H   H
   CH_3

k

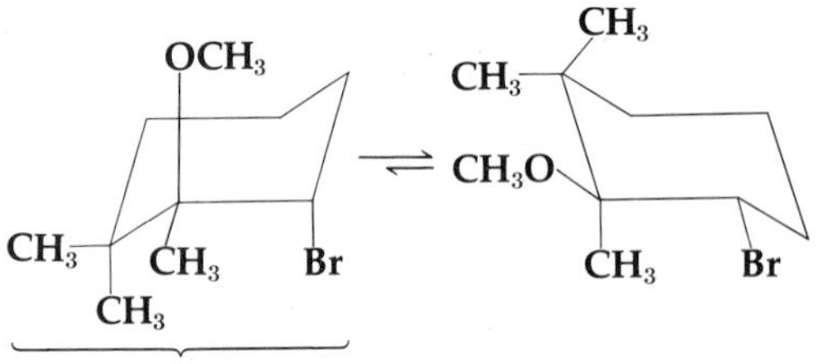

OCH_3

CH_3 CH_3 Br
   CH_3

$\rightleftharpoons$ CH_3O

CH_3

CH_3

CH_3 Br

1,3-Diaxial

15–3   a $CH_3CH-\overset{\displaystyle Cl}{\underset{\displaystyle Cl\ \ OCH_3}{C}}-CH_2CH=CHCH_3$

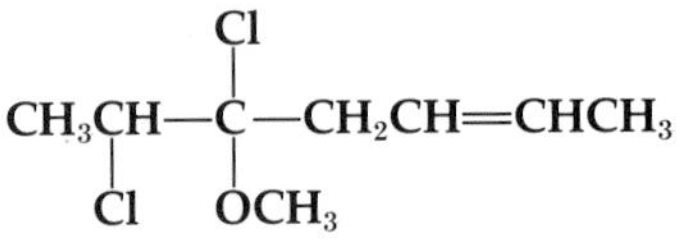

b $ClCH_2CHClCH_2CH_2CH=CHCOOH$

f [structure: cyclohexene fused to cyclobutane with two Cl substituents] (angle strain released: $sp^3 \longrightarrow sp^2$)

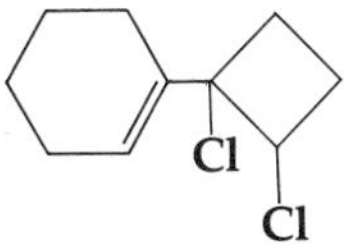

15–4 **a** $\phi CH{=}CHCH_3 + CH_3OH/H^+$

      c $CH_3CH_2CH_2C{\equiv}C\!\!:\!\!^-\ Na + CH_3I;\ HgSO_4/H_2SO_4;\ CH_3Li$

      e 4-t-Butylcyclohexene $+ AgOAc + Br_2 + HOAc/H_2O/\Delta$

15–6 **a** *trans*-2-Methylcyclopentanol

      d *cis*-1,3-Diphenylpropene

      f [structure: bicyclic ring with —CH₃, —Cl, and Cl substituents]

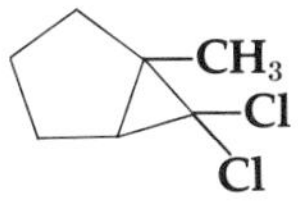

      h Replace double bond by cyclopropane trans to CH_3

15–8 **a** O_3/H_2O_2; CH_2N_2 to octadiendioic acid diester; **Na** (acyloin reaction; $NaBH_4$)

      c O_3; OH^-; $NaBH_4$; $TsCl$; $LiAlH_4$

      e $HgSO_4$ or B_2H_6/H_2O_2

      f $NaBH_4$; $TsCl$; OH^-; $CH_3COCl/AlCl_3$; H_2/cat.

15–10 **a** O_3; OH^-

       b B_2H_6/H_2O_2; CrO_3; $\phi_3P{=}CH_2$

       d Br_2; OH^-/Δ; H_2O/H^+

       g O_3/H_2O_2; Leuckart ($NH_4^+\ HCOO^-$); CH_3I, excess; OH^-/Δ

15–12 *A* (index = 3) has 2 double bonds, one ring; O_3 cleaves $CH_2{=}C{<}$ and $C{-}CH{=}C{<}$ to CH_2O + acyclic ketone-dialdehyde (index = 3) which gains O_2 with $H_2Cr_2O_7$ to keto-diacid *C*; this is β-ketoacid $-CO_2/\Delta$ to ketoacid *D*, which reduces to hydroxy-acid *E*, $C_5H_{10}O_3$, with asymmetric center, $-CHOH-$; *D* cannot be $CH_3CH_2COCH_2COOH$ which would lose more CO_2/Δ. Only 2 structures for *D*:

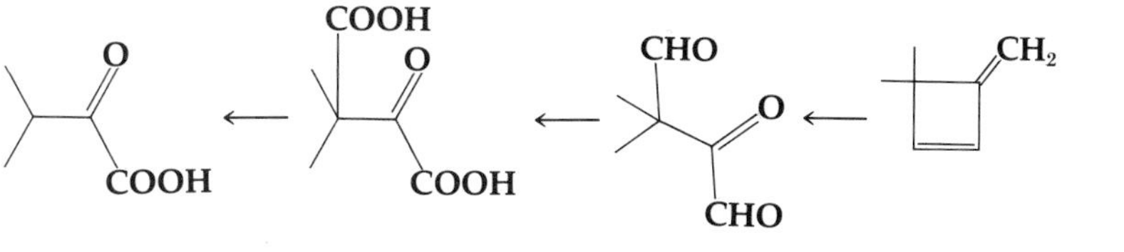

[structures shown left to right:]

D   *C*   *B*   *A*

[second row of structures with arrows]

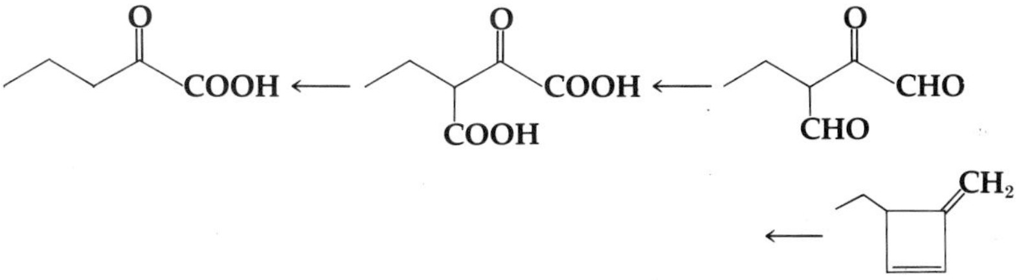

15–15

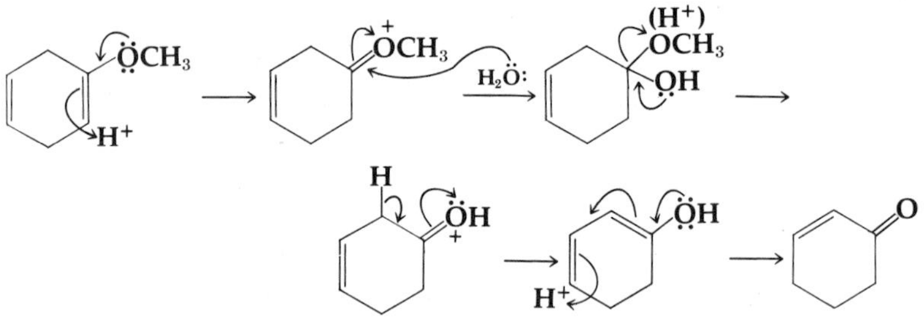

UV = 227 nm

15–17

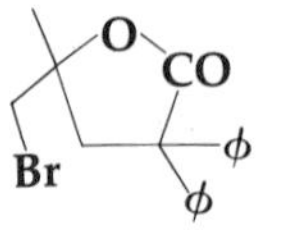

CHAPTER 16

16–1 $A > E > C > D > B$

16–2

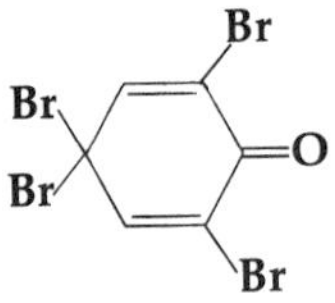

16–4   a   Much 4-nitro-3-methylphenol, little 2-nitro-5 methyl-phenol

b   Same two with **CN** replacing **CH$_3$**

e   2,4-Dinitroacetanilide

16–6

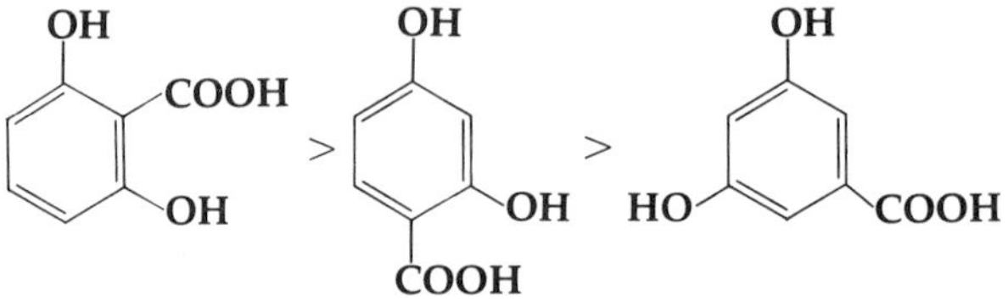

OH must be ortho or para with ortho preferred:

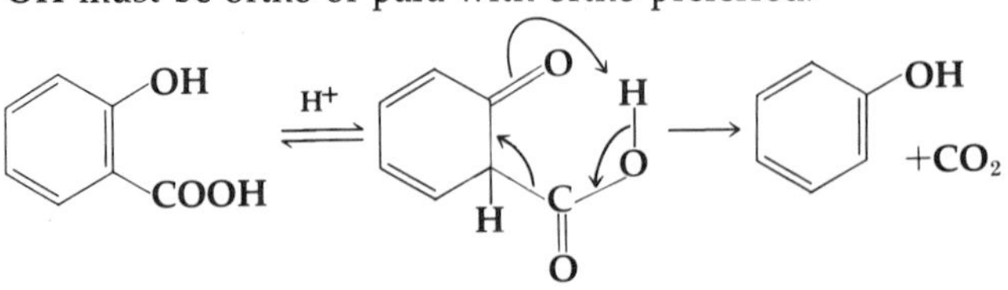

Reaction in base is an equilibrium which can be forced in the direction of carboxylation by using a large amount of **CO$_2$**

16-8 a

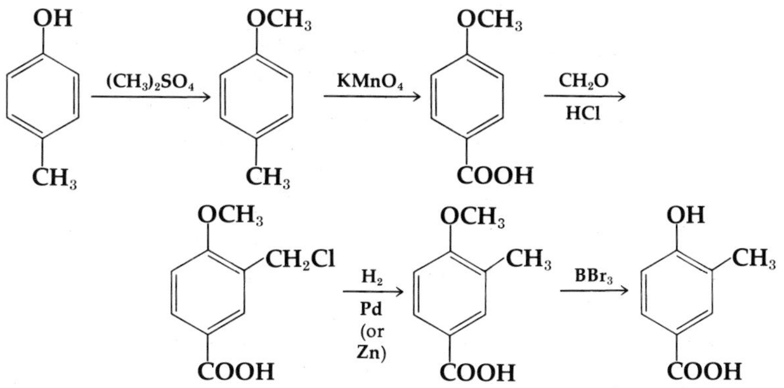

$OH \xrightarrow{(CH_3)_2SO_4} OCH_3 \xrightarrow{KMnO_4} OCH_3 \xrightarrow[HCl]{CH_2O}$

d

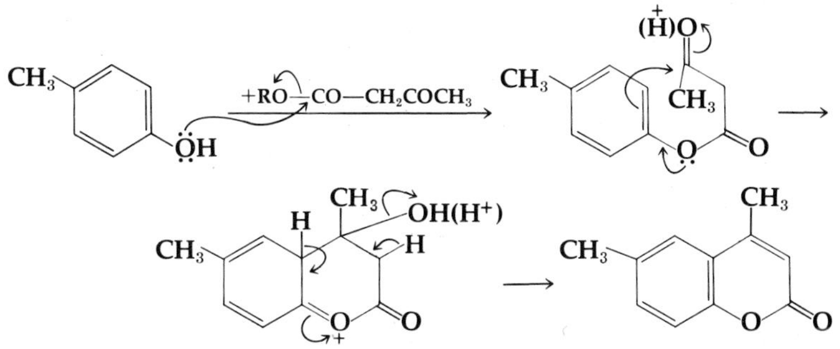

16-11 *Pictet-Spengler:*

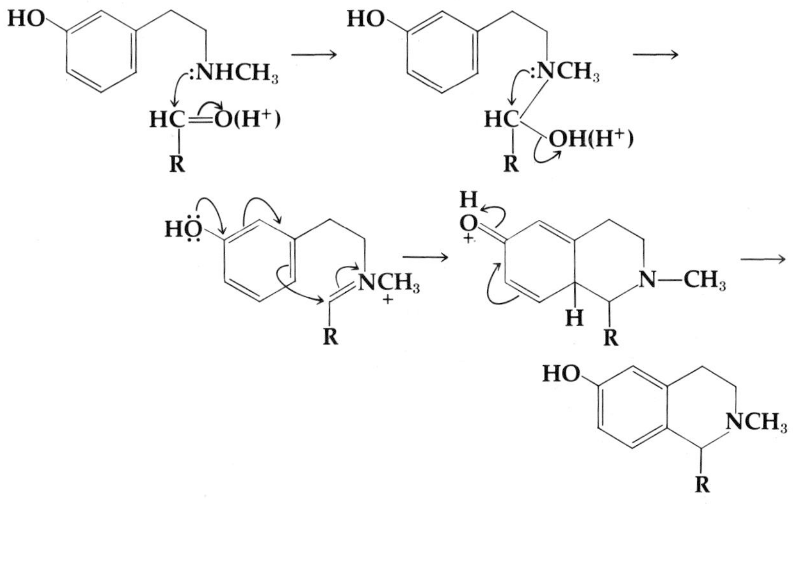

Mannich:

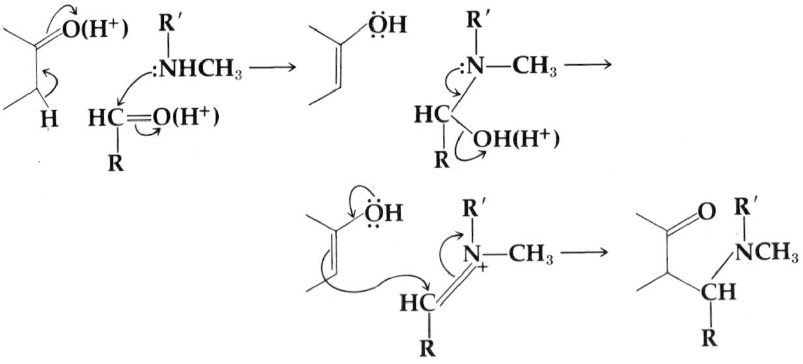

R'
$:NHCH_3$
O(H$^+$)
H   HC=O(H$^+$)
R
→
R'
:N—CH$_3$
ÖH
HC
R   OH(H$^+$)
→

ÖH
R'
N—CH$_3$
HC
R
→
O R'
NCH$_3$
CH
R

16–12 $\xrightarrow{-OH}$

COO$^-$
OH
CH$_3$
HO
OH
CH$_3$
CH$_3$
O
⇌

COO$^-$
:ÖH H
CH$_3$
CH$_3$
CH$_3$
O
OH
O
→ CH$_3$CHO ↑ →

CH$_3$CH
NO$_2$
N—NH
NO$_2$
+
COO$^-$
CH$_3$
CH$_3$
HO
O
OH

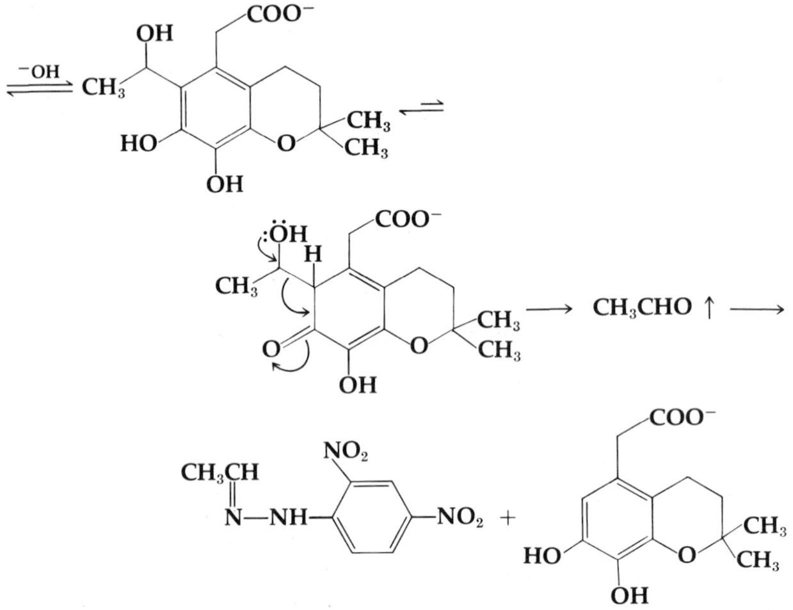

16–13 b Steric hindrance

e

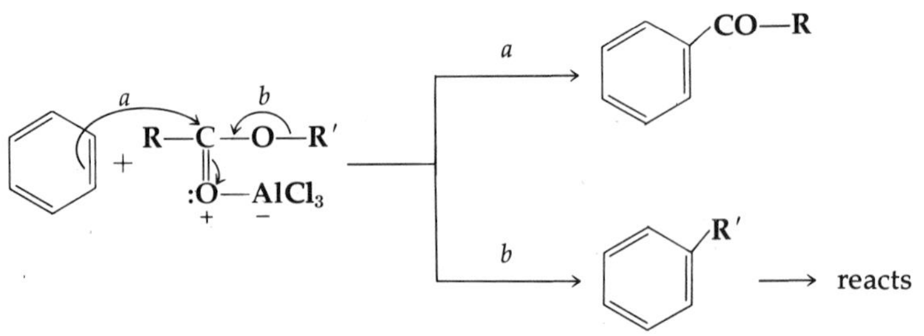

+ R—C—O—R'
:O—AlCl$_3$
+   —

a →   CO—R

b →   R'   → reacts
again, faster than benzene

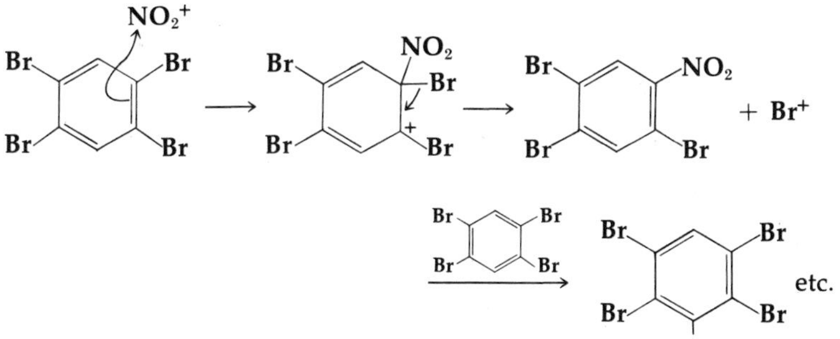

16–14 a HNO_3/H_2SO_4; $KMnO_4$; Zn/HCl

  c $Br_2/FeBr_3$; $KMnO_4$; $LiAlH_4$; $TsCl$; Na^+ $:CH(COOR)_2$; H_2O/H^+

  g HNO_3/H_2SO_4; $Cl_2/AlCl_3$; $SnCl_2$; $Ac_2O \longrightarrow$

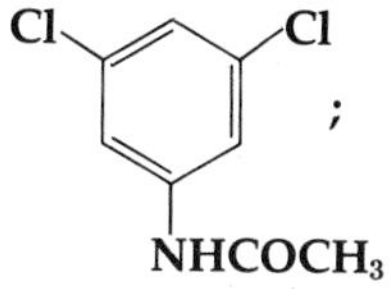

HNO_3/H_2SO_4; OH^-; HNO_2; $NaBH_4$

16–15 a

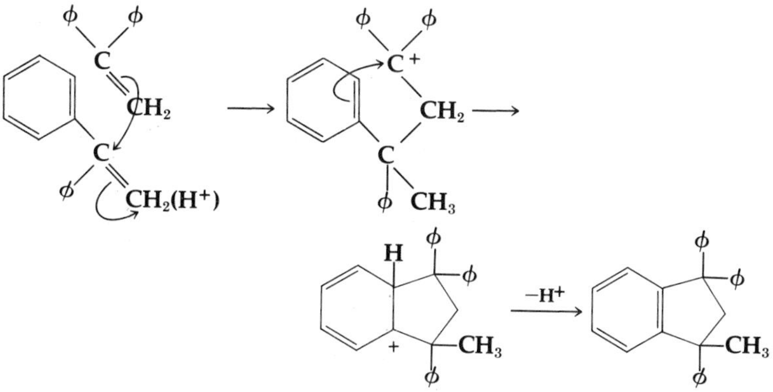

  c

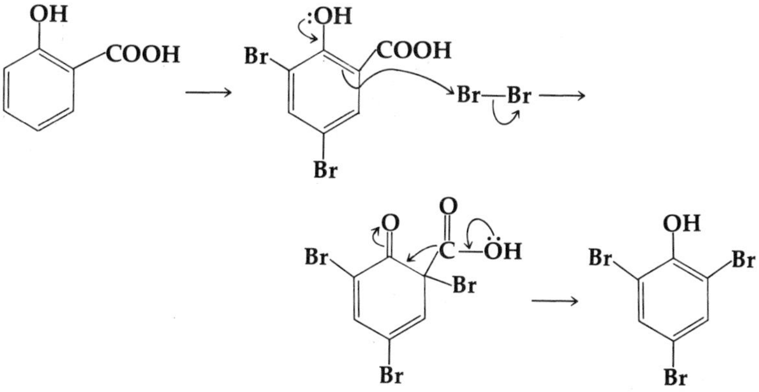

i

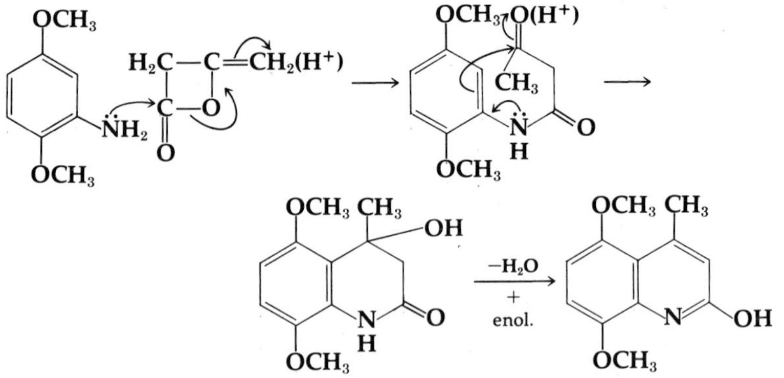

OCH_3 $H_2C-C=CH_2(H^+)$

$\overset{|}{C}-O$

$\overset{\|}{O}$

NH_2

OCH_3

$\longrightarrow$

OCH_3 $O(H^+)$

CH_3

N O
H

OCH_3

$\longrightarrow$

OCH_3 CH_3 OH

N O
H

OCH_3

$\xrightarrow[+\ \text{enol.}]{-H_2O}$

OCH_3 CH_3

N OH

OCH_3

16–16 **a**

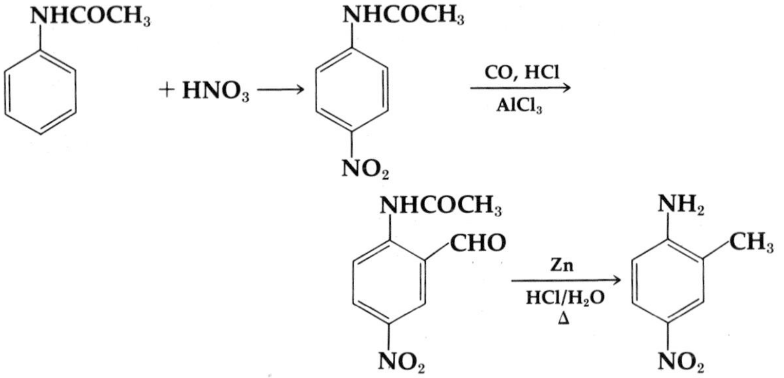

$NHCOCH_3$

$+ HNO_3 \longrightarrow$

$NHCOCH_3$

NO_2

$\xrightarrow[AlCl_3]{CO, HCl}$

$NHCOCH_3$
CHO

NO_2

$\xrightarrow[\substack{HCl/H_2O \\ \Delta}]{Zn}$

NH_2
CH_3

NO_2

d

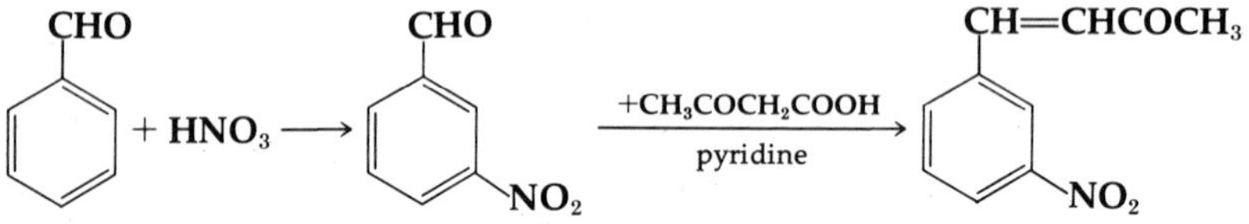

CHO

$+ HNO_3 \longrightarrow$

CHO

NO_2

$\xrightarrow[\text{pyridine}]{+CH_3COCH_2COOH}$

$CH=CHCOCH_3$

NO_2

CH_3MgI; H_2/Pt

f

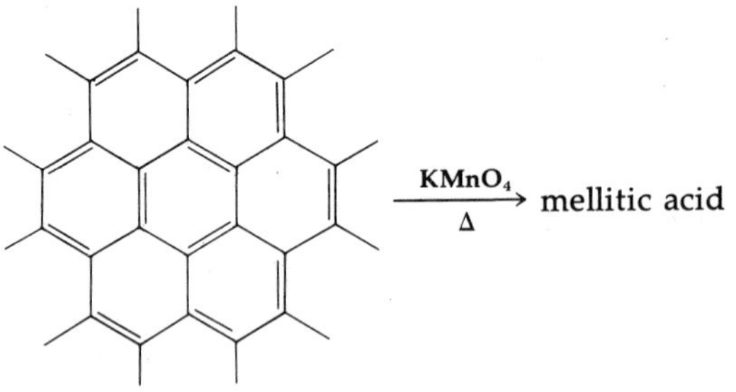

Graphite

$\xrightarrow[\Delta]{KMnO_4}$ mellitic acid

16–17 **a** $D =$

OH

ϕ

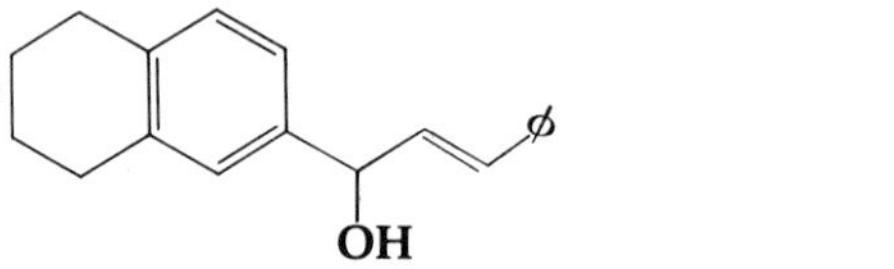

b $E = \phi COCH_2$—⟨benzene ring⟩—CH_2CH_2—⟨benzene ring⟩—$CH_2CO\phi$

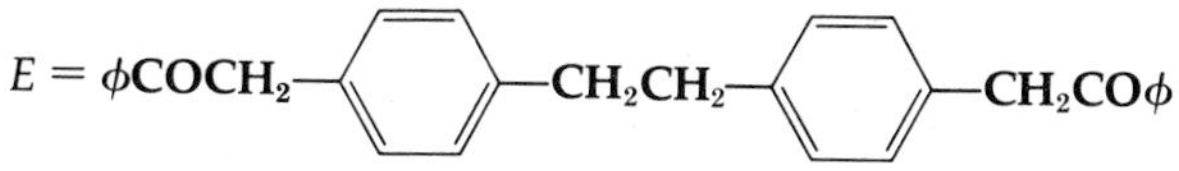

d $E =$

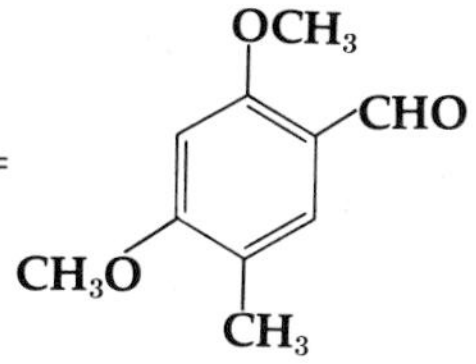

16–20 **a** $C_6H_6 + HNO_3/H_2SO_4 \longrightarrow C_6H_5NO_2$; Sn/HCl; HNO_2; H_2O/Δ

d $C_6H_6 + Br_2/FeBr_3 \longrightarrow C_6H_5Br \longrightarrow C_6H_5MgBr \xrightarrow{CO_2} C_6H_5COOH \longrightarrow$
C_6H_5CN; HNO_3/H_2SO_4; $SnCl_2$; Ac_2O; $\phi NO_2 + \phi COCl/AlCl_3 \longrightarrow m$-nitrobenzophenone; $SNCl_2$; HNO_2; Cu_2Br_2

h $p\text{-}H_2NC_6H_4NO_2 \longrightarrow p$-diaminobenzene; HNO_2; $Cu_2(CN)_2$

m $\phi OH + SO_3 \longrightarrow p\text{-}HOC_6H_4SO_3H$; $1Br_2$; $1Cl_2$; $H_2SO_4/H_2O(-SO_3)$

CHAPTER 17

17–1 **a** $(CH_3)CCHBrCH_3$; $(CH_3)_3CCH-O-CHC(CH_3)_3$; $(CH_3)_3CCH=CH_2$;
with CH_3 and CH_3 substituents

$CH_2=C-CH(CH_3)_2$; $(CH_3)_2\overset{X}{C}CH(CH_3)_2$ (X = HO, RO, Br);
with CH_3 substituent

$(CH_3)_2C=C(CH_3)_2$

17–2 **a**

c $\phi CH=C(CH_3)C_2H_5$; 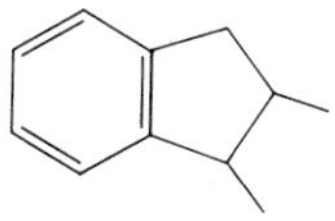 ; $\phi CH_2CH(CH_3)CH=CH_2$;

$\phi CH_2C(CH_3)=CHCH_3$

17–4 **a** D = cyclopentylamine

c

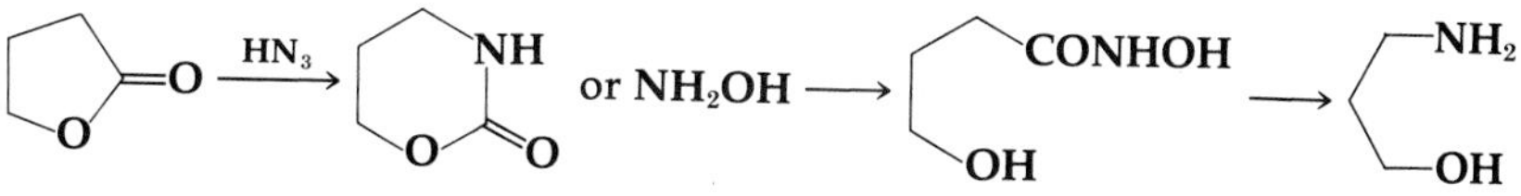

g $SOCl_2$; NaN_3; $/t\text{-}BuOH$; CF_3COOH

17–5 **a** $\phi COOH + HN_3/H_2SO_4$

g 1,2-Diacetyl-4-chlorobenzene + CF_3CO_3H, then OH^-

h $SOCl_2$; CH_2N_2; $Ag_2O/CH_3OH \longrightarrow CH_3OCO(CH_2)_4COOCH_3$; OCH_3^-
(Dieckmann); H_2O/H^+

17-6 b $C_6H_5NHCOCH_3$— phenyl better migrating group via phenonium ion

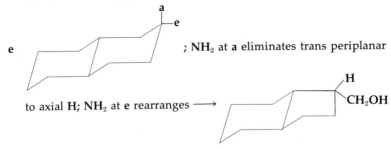

; NH_2 at **a** eliminates trans periplanar

to axial **H**; NH_2 at **e** rearranges $\longrightarrow$

    g Product is *cis*-3,4-dimethyl cyclopentanone (meso)

17-7 a $\phi COCH_2\overset{+}{N}(CH_3)_2CH_2CH{=}CH_2 \xrightarrow{-OH} \phi COCH{-}N(CH_3)_2 \xrightarrow{\phi CH_2Br}$
$$\underset{CH_2CH=CH_2}{}$$

$$\xrightarrow{OH^-} \phi{-}CO{-}\underset{CH-CH=CH_2}{\overset{}{CH}} + \phi CH_2N(CH_3)_2$$

(245 nm)

    c Favorski rearrangement to α-bromo-β-methylsuccinic acid mono-methyl ester and elimination

17-10 a Benzilic acid rearrangement to

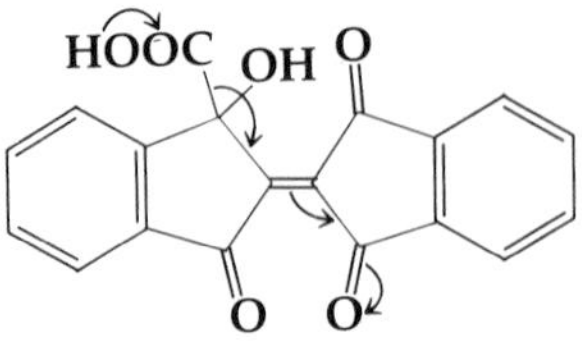

and

decarboxylation

    b Diacetylate and enolize to

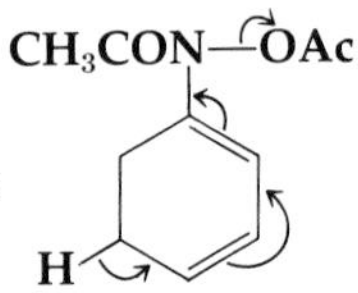

17-11 d $\phi MgBr + CO_2$, then two Arndt-Eistert extensions

    f $\phi H \longrightarrow \phi NO_2 \longrightarrow \phi NH_2 \longrightarrow \phi OH \xrightarrow{H_2}$ cyclohexanone, then CH_2N_2

    i Benzidine rearrangement, HNO_2, Cu_2Cl_2

17-12 b *cis*-3-methyl cyclopentanol   **f** R-2-aminobutane

17-14 c Stable anion forms at —CH_2— and displaces —OTs:

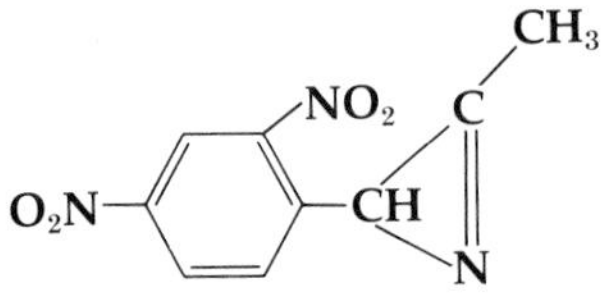

17–16

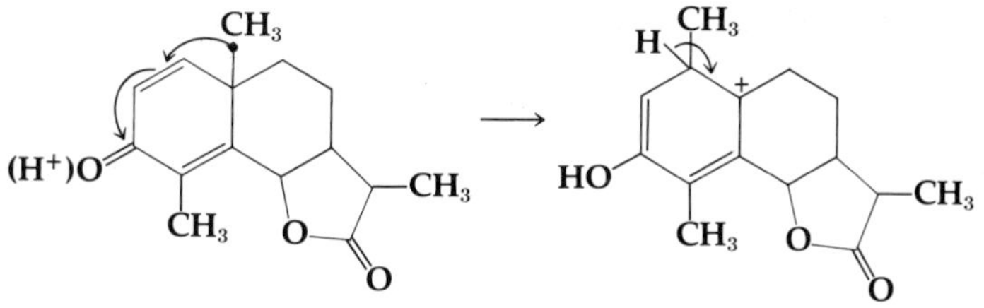

17–17

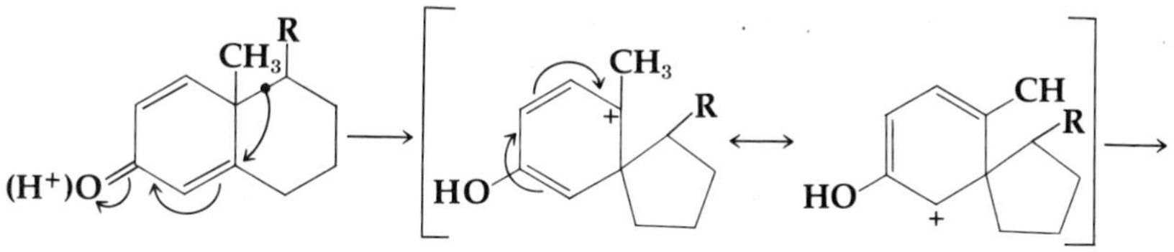

either cyclopentane bond can rearrange a second time

17–19   Enolization is impossible (Bredt's rule):

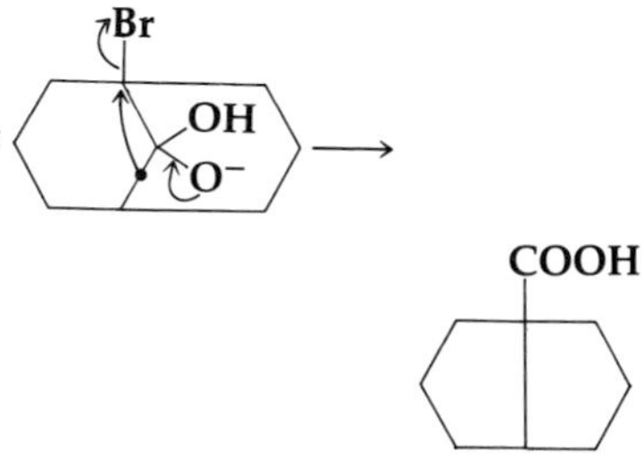

17–20

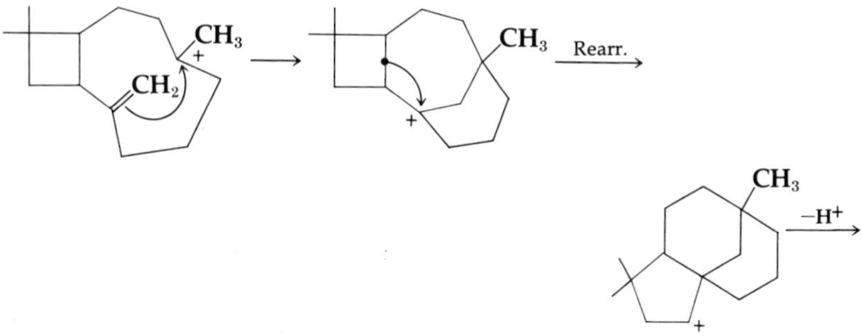

17–21

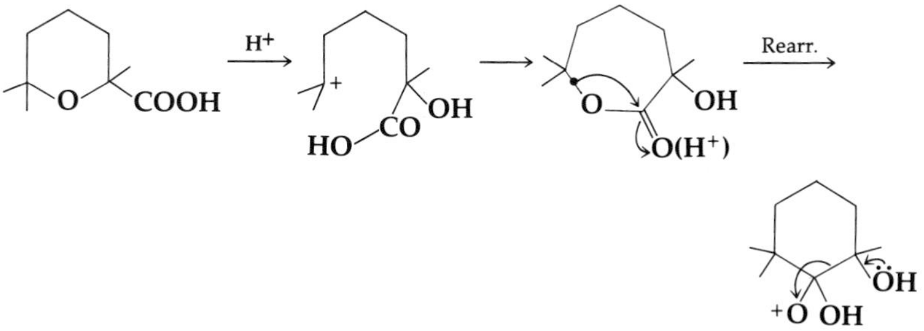

CHAPTER 18

18–2 **a** One carbon oxidized, one reduced; no overall change

b Overall change = one oxidation step, presumably by O_2; overall oxidation

18-3 c $\quad\begin{array}{c}\overset{+3}{-COOC_2H_5}\\[4pt]\overset{+3}{-COOC_2H_5}\end{array}$ + 4Na $\xrightarrow[(\pm4)]{(+H_2O)}$

$\Sigma = +6$

$\begin{array}{c}\overset{0}{-CH-OH}\\[4pt]\overset{}{-C=O}\\[2pt]{\scriptstyle +2}\end{array}$ + 4$\overset{+1}{Na^+}$ + 2C$_2$H$_5$OH + 4OH$^-$

$\Sigma = +2$

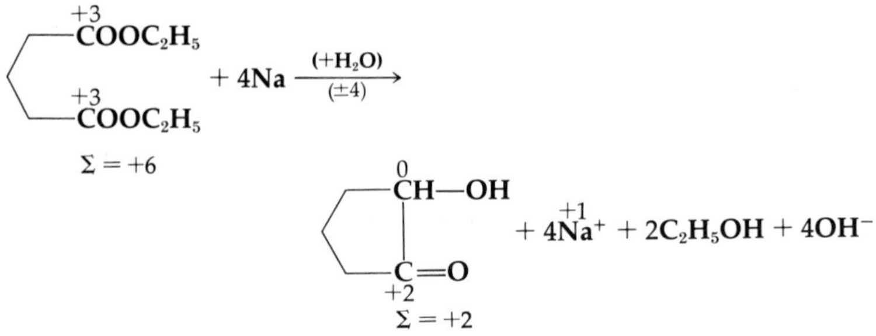

g $\quad$ HOOC$\overset{+1}{-CH_2}\overset{-3}{-NH_2}$ + HO$\overset{+3}{-N}=$O $\xrightarrow[\pm3]{(H_2O)}$ HOOC$\overset{+1}{-CH_2}-$OH + $\overset{0}{N_2}$

18-4 a $\quad$ R$-$COCH$_2-$R$'$ + $\phi-$N$=$O $\xrightarrow[\substack{or\\base\\(via\ enol)}]{Acid}$ R$-$CO$\overset{\overset{\displaystyle N-\phi}{\parallel}}{-C}-R'$ $\xrightarrow{H_2O}$

R$-$CO$-$CO$-$R$'$

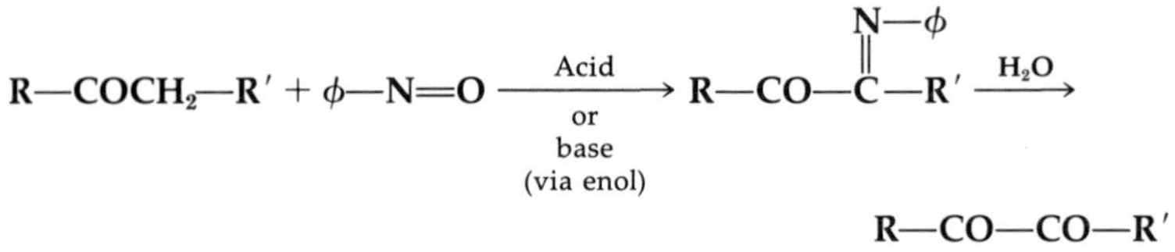

18-5 b $\quad$ R$\overset{}{-CH-}\overset{}{C}-$OH + NBS $\longrightarrow$ $\left[\text{R}-\text{CH}-\text{C}-\overset{\cdot\cdot}{\underset{\cdot\cdot}{O}}\text{H}\right]$ $\longrightarrow$

$\underset{OH}{}\ \underset{O}{}$

R$-$CHO + CO$_2$ + $\begin{array}{c}\diagup CO\\ NH\\ \diagdown CO\end{array}$

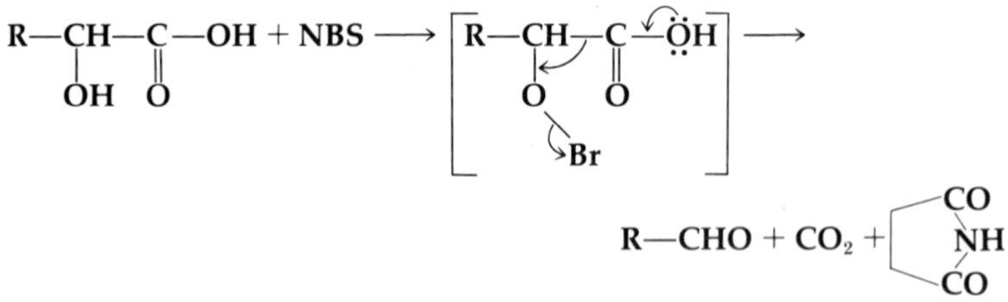

18-6 a $\quad$ NaBH$_4$; H$_2$/Pd; Δ/KMnO$_4$

d $\quad$ ϕCOCH$_3$ + SeO$_2$ + OH$^-$; NaIO$_4$ or Pb(OAc)$_4$

18-8 $\quad$ NBS; HCO$_3$H; OH$^-$; (CH$_3$)$_2$SO/BF$_3$; NaIO$_4$ or Pb(OAc)$_4$; CH$_2$N$_2$; Et$_3$N; OH$^-$; HN$_3$/H$_2$O; CH$_2$(COOCH$_3$)$_2$; H$_3$O$^+$ and H$_2$/Pt

18-9 a $\quad$ Cyclopentane $\qquad$ b $\quad$ Toluene or methylcyclohexane

18-10 a $\quad$ Unlikely $\qquad$ b $\quad$ Unlikely; probably 3-methyl-2-ethyl-benzaldehyde first

18-12 a $\quad\longrightarrow$ $\longrightarrow$

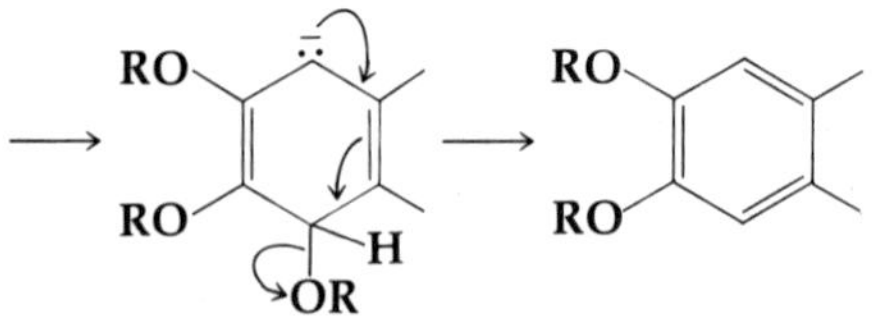

18-14 a $\quad$ 5 + 6MnO$_4^-$ + 18H$^+$ $\longrightarrow$ 5 + 6Mn^{++} + 9H$_2$O

$\Delta = -5$ (over COCH$_3$), $+7$; $+2$; $+3$

$\Sigma = -1 \xrightarrow{\Delta = +6} \Sigma = +5$

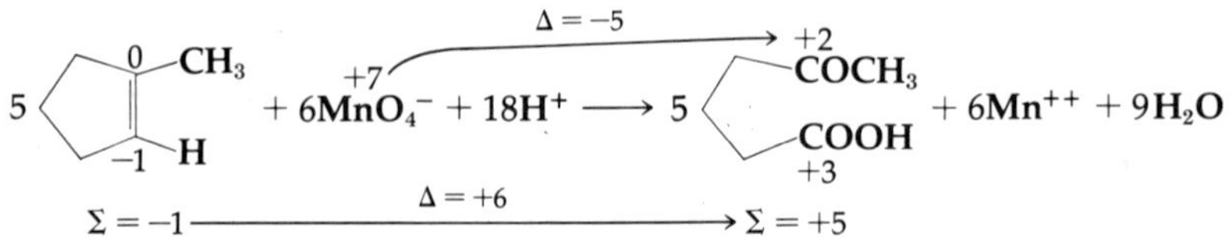

b

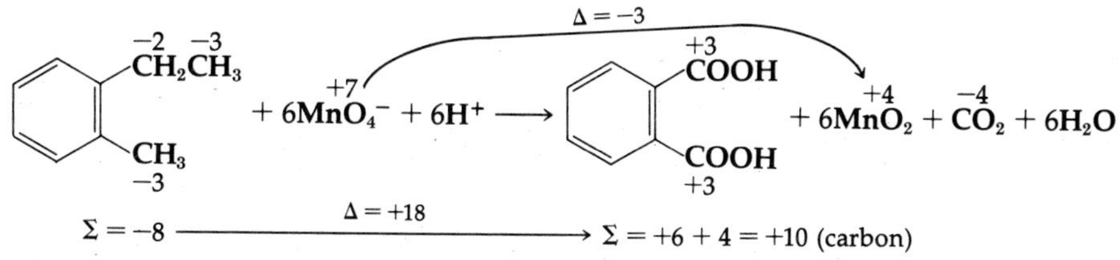

$$+ 6MnO_4^- + 6H^+ \longrightarrow$$

$$\overset{\Delta = -3}{\overbrace{\qquad}}$$

$$+ 6MnO_2 + \overset{+4}{C}\overset{-4}{O_2} + 6H_2O$$

$$\Sigma = -8 \quad\xrightarrow{\quad\Delta = +18\quad}\quad \Sigma = +6 + 4 = +10 \text{ (carbon)}$$

g

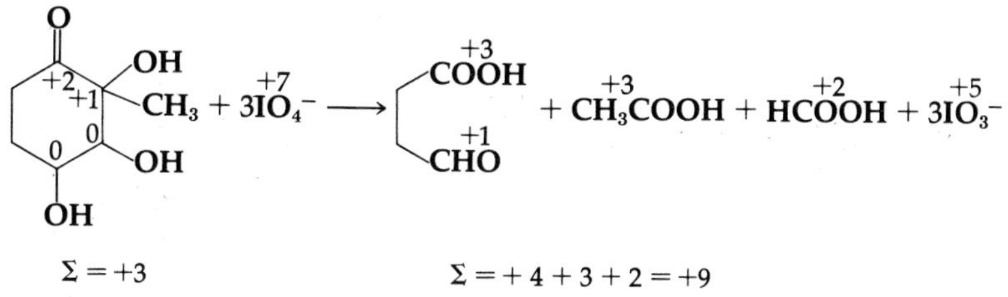

$$+ 3IO_4^- \longrightarrow$$

$$+ CH_3COOH + HCOOH + 3IO_3^-$$

$$\Sigma = +3 \qquad\qquad \Sigma = +4 + 3 + 2 = +9$$

j

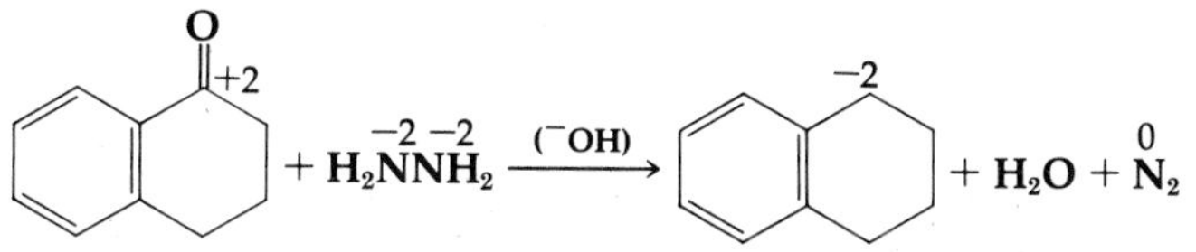

$$+ H_2\overset{-2}{N}\overset{-2}{N}H_2 \xrightarrow{(^-OH)}$$

$$+ H_2O + \overset{0}{N_2}$$

18–15 a Phenol $+ H_2$

c $\phi MgBr + CH_2\!\!-\!\!CH_2;\ CrO_3$

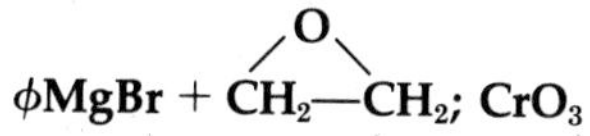

g

$$+ Li/HNEt_2 \longrightarrow \qquad \xrightarrow{+O_3}$$

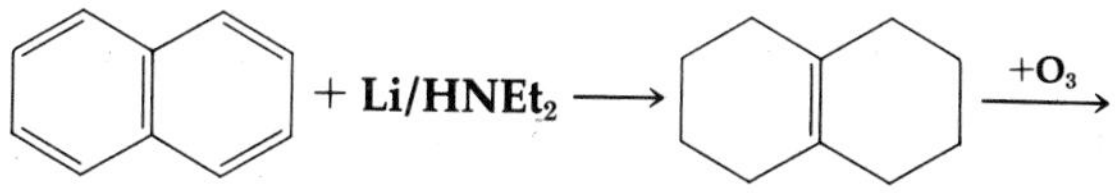

j $C_6H_6 + CH_3COCl/AlCl_3 \longrightarrow \phi COCH_3 \xrightarrow[H_2SO_4]{HNO_3}$

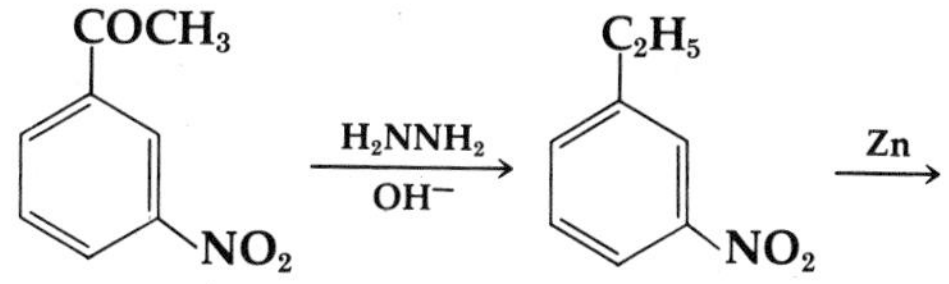

$$\xrightarrow[OH^-]{H_2NNH_2} \qquad \xrightarrow{Zn}$$

n

$$+ Na^+\ :CH(COOR)_2 \longrightarrow \qquad \xrightarrow{H_2O_2}$$

$$\xrightarrow[\Delta]{H_3O^+}$$

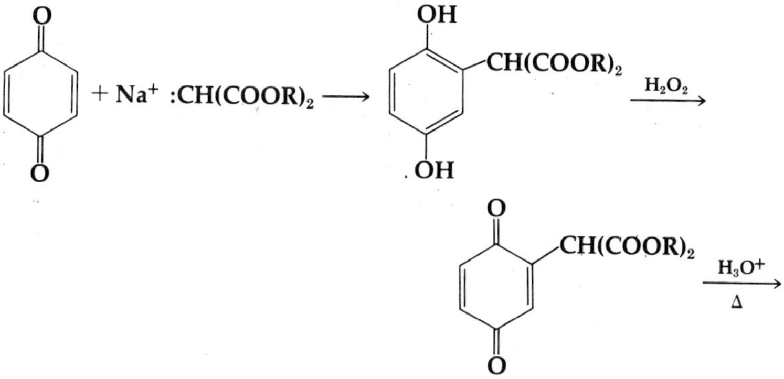

r $S^{--} + CH_3(CH_2)_4Br; H_2O_2$

18–17

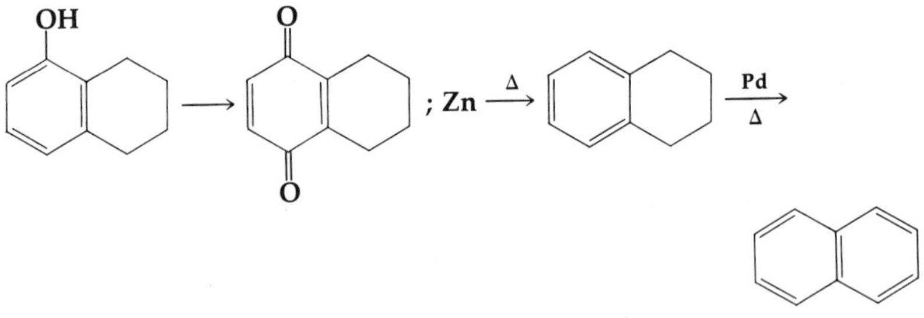

18–19 b $R_3{}'C{-}N{-}NH{-}SO_2\phi \longrightarrow R_3{}'C{-}N{=}N{-}H + \phi SO_2{}^-: \longrightarrow$
$\quad\quad\quad\quad\quad\quad$:OR$^-$

$\quad\quad\quad\quad\quad\quad\quad\quad\quad\quad\quad\quad\quad\quad\quad\quad R_3{}'CH + N_2$

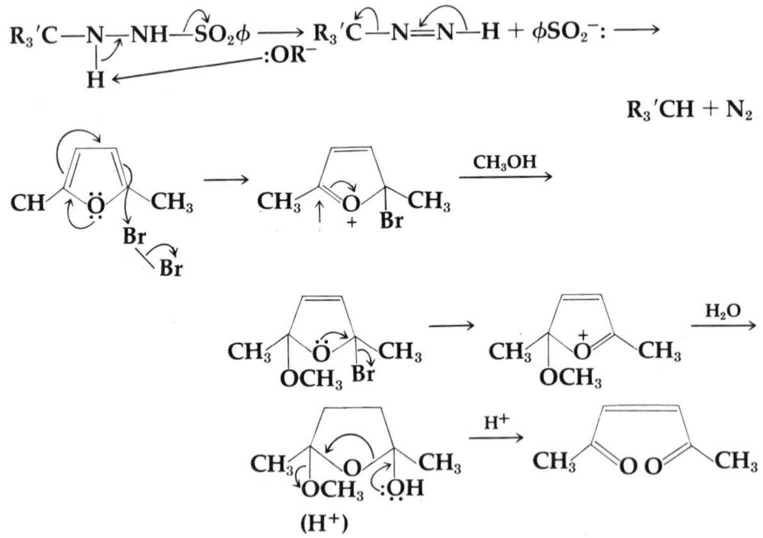

18–21   Index = 5; $RCOOC_2H_5 \xrightarrow{LiAlH_4} RCH_2OH + C_2H_5OH$; $C_{12}H_{16}O_3$ has benzene (index = 4) + ester and an asymmetric carbon; the asymmetric carbon must be

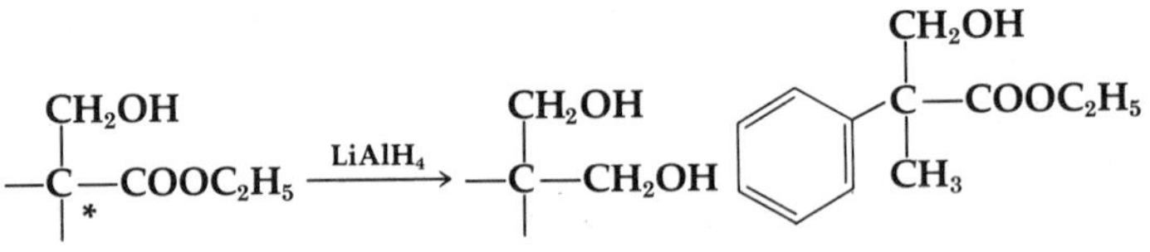

(meso)

since UV shows only one carbon on ring (as in toluene)

18–22

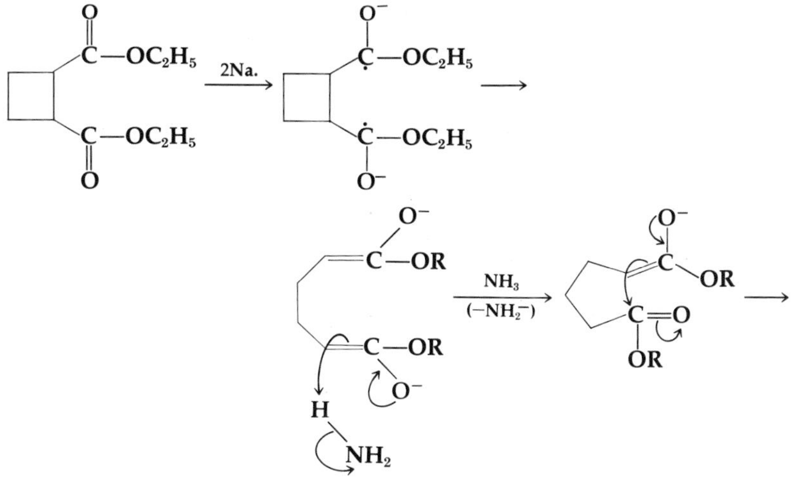

CHAPTER 19

19-1 **a** Anion stabilization in —:CCl$_3$ by back-donation, not possible in

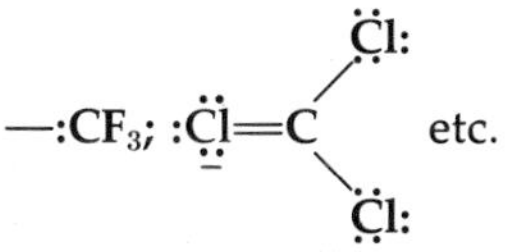

—:CF$_3$; :Cl—C etc. (with Cl: groups)

19-2 **b** Si—O bonds shorter than Si—Si bonds (large second-row atom) and resonance via back-donation:

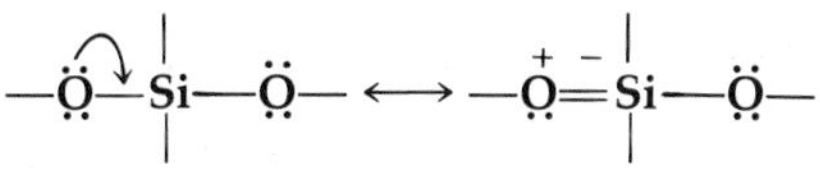

$$—\ddot{O}—Si—\ddot{O}— \longleftrightarrow —\overset{+}{\ddot{O}}=Si—\overset{-}{\ddot{O}}—$$

d The pentacovalent phosphorus has a bond angle of 90° available so that 4-ring has no angle strain at phosphorus

19-4 **b** C$_6$H$_5$—$\overset{+}{\ddot{I}}$—$\overset{-}{\ddot{O}}$: $\longrightarrow$ C$_6$H$_5$—$\ddot{I}$=$\ddot{O}$ (back-donation important to collapse charge separation, as in —SO$_2^-$, $\equiv$PO, etc.) I$^-$ (−1); ϕI$^+$ (+1); ϕIO (+3); ϕIO$_2$ (+5)

19-5 **a** R—C=O $\xrightarrow{\text{:P}\phi_3}$ R—C—O—$\overset{+}{\text{P}}\phi_3$ $\longrightarrow$ R—C$\equiv$N: +N$_2$ $\uparrow$ + ϕ_3PO

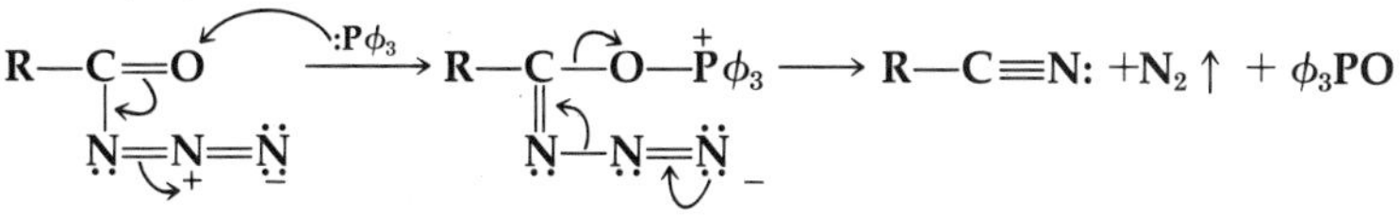

(with N=N=N and N—N=N groups below)

d Cl—CCl$_2$—C—N(CH$_3$)$_2$ $\xrightarrow{\text{:P}\phi_3}$ Cl$_2$C=C (with O—Pϕ_3 and N(CH$_3$)$_2$) $\longrightarrow$

Cl$_2$C=C—$\overset{+}{\text{N}}$(CH$_3$) $\longrightarrow$ Cl$_2$C=C—$\ddot{\text{N}}$(CH$_3$)$_2$

Cl$^-$ + ϕ_3PO   Cl

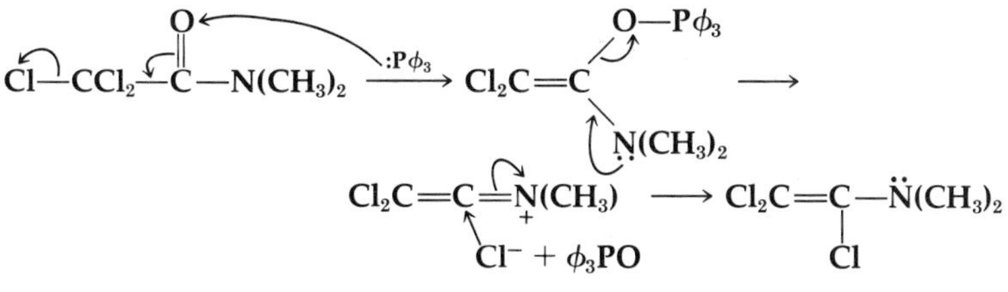

g $(CH_3O)_3P: \quad CH_2=CH-C=O$ with OH $\longrightarrow$ $(CH_3O)_2\overset{+}{P}$ ... $C-OH \longrightarrow$ with CH_2-CH

$(CH_3O)_2POCH_2CH_2COOCH_3$

19-8 c $2CH_3MgX + PCl_3 \longrightarrow (CH_3)_2PCl \xrightarrow{CH_3OH} (CH_3)_2POCH_3$

h $PCl_3 + CH_3OH \longrightarrow P(OCH_3)_3 + RCOCl \longrightarrow$

$$RCO-\overset{OCH_3}{\underset{O-CH_3 \quad Cl^-}{\overset{+}{P}}}-OCH_3 \longrightarrow R-\underset{O \quad O}{C-P}(OCH_3)_2$$

i $\phi N_3 + P\phi_3 \longrightarrow \phi-\overset{..}{\underset{..}{N}}-\overset{+}{P}\phi_3 + N_2 \uparrow$

19-10

$$\begin{array}{c} HOCH_2CH_2 \quad CH_3 \\ \diagdown \diagup \\ C \\ \diagup \diagdown \\ HOOCCH_2 \quad H\overset{..}{O}: \end{array} + HO-\underset{OH}{\overset{O}{P}}-O-\underset{OH}{\overset{O}{P}}-O-\underset{OH}{\overset{O}{P}}-O-R \longrightarrow$$

Good leaving group

$$\begin{array}{c} HOCH_2CH_2 \quad CH_3 \\ \diagdown \diagup \\ C \\ \diagup \diagdown \\ H\overset{..}{O}-\underset{O}{C}-CH_2 \quad O-\underset{OH}{\overset{O}{P}}-O-\underset{OH}{\overset{O}{P}}-OH \end{array} \xrightarrow{-CO_2}$$

$$\begin{array}{c} HOCH_2CH_2 \\ \diagdown \\ C-CH_3 \\ \diagup \\ CH_2 \end{array} \xrightarrow{\text{Same reagent}}$$

$$\begin{array}{c} CH_3 \\ | \\ CH_2=C-CH_2CH_2-O-\underset{OH}{\overset{O}{P}}-O-\underset{OH}{\overset{O}{P}}-OH \end{array}$$

19-11 a Slow step (rate-determining) is internal displacement, hence first order and 3 > 4-ring:

$$R-\overset{..}{S}: \begin{array}{c} (CH_2)_{n=1,2} \\ | \\ CH_2-Cl \end{array} \xrightarrow[\text{Slow}]{} R-\overset{+}{\overset{..}{S}} \begin{array}{c} (CH_2)_n \\ | \\ CH_2 \quad Cl^- \end{array} \xrightarrow[\text{(solvent)}]{\text{Fast} \\ \text{SOH}} R-\overset{..}{\overset{..}{S}}-(CH_2)_n \, CH_2-OS$$

:Nu

b

$$\triangle\!\!O \xrightarrow{SCN^-} \xleftarrow[]{} \begin{array}{c} S \\ \diagdown \\ C \\ | \\ O \quad N \end{array} \rightleftharpoons \begin{array}{c} S \\ \diagdown \\ C=N^- \\ | \\ O \end{array} \rightleftharpoons$$

$$\begin{array}{c} S^- \\ | \\ O-C\equiv N \end{array} \longrightarrow \triangle\!\!S + OCN^-$$

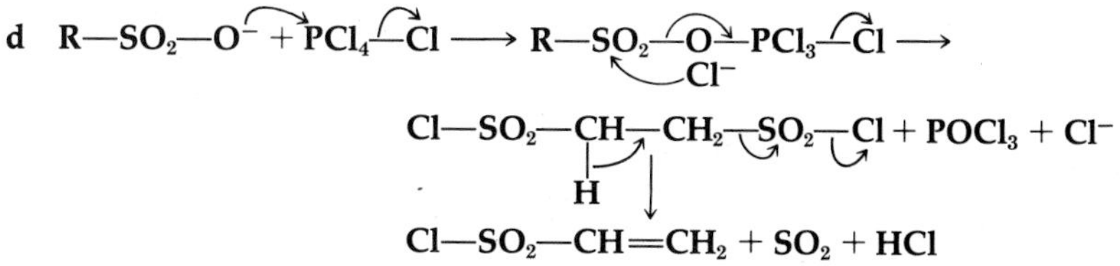

d $R-SO_2-O^- + PCl_4-Cl \longrightarrow R-SO_2-O-PCl_3-Cl \longrightarrow$
$$Cl^-$$

$$Cl-SO_2-CH-CH_2-SO_2-Cl + POCl_3 + Cl^-$$
$$H$$

$$Cl-SO_2-CH=CH_2 + SO_2 + HCl$$

19–13 a, c

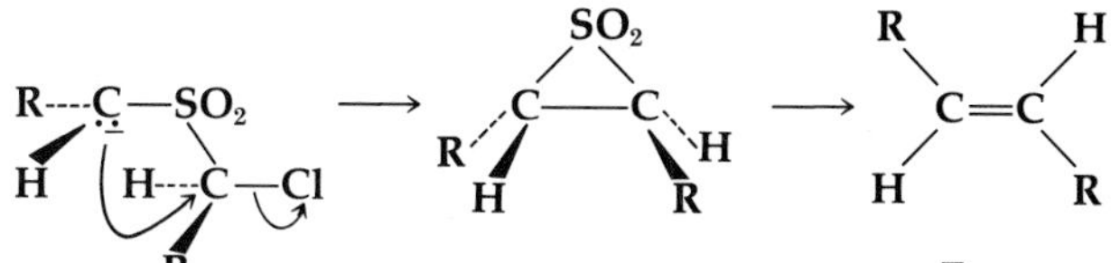

Trans

b $RCH_2OTs + R'SO_2{:}^-$

19–14 a $\phi SO_2H \longrightarrow \phi SO_2CH_3$

c $\longrightarrow p\text{-}CH_3OC_6H_4CON-NH-SO_2C_6H_5 \longrightarrow$
$$H\frown:B$$

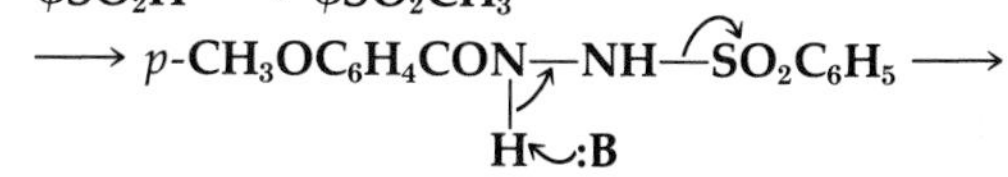

$$[CH_3OC_6H_4CON=NH] + C_6H_5SO_2^- \xrightarrow{-N_2} CH_3OC_6H_4CHO \longrightarrow$$
$$p\text{-methoxytoluene}$$

$$CH_3OC_6H_4CH=N-NHSO_2C_6H_5 \xrightarrow{NaBH_4} p\text{-methoxytoluene}$$

19–15 a $S_8 + 8O_2 \longrightarrow 8SO_2 \xrightarrow{CH_3MgI} CH_3SO_2{:}^- MgI \xrightarrow{Cl_2} CH_3SO_2Cl;$

e, f $\phi MgBr \longrightarrow \phi SO_2{:}^-$ then $LiAlH_4$ or Cl_2, then CH_3SH

19–17
$$H$$
$$RC-O-NH-SO_2C_7H_7 \longrightarrow RCOOH + C_7H_7SO_2NH_2$$
$$OH$$

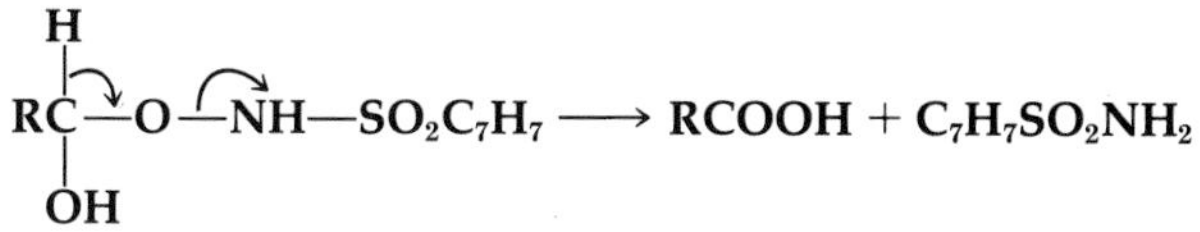

19–18

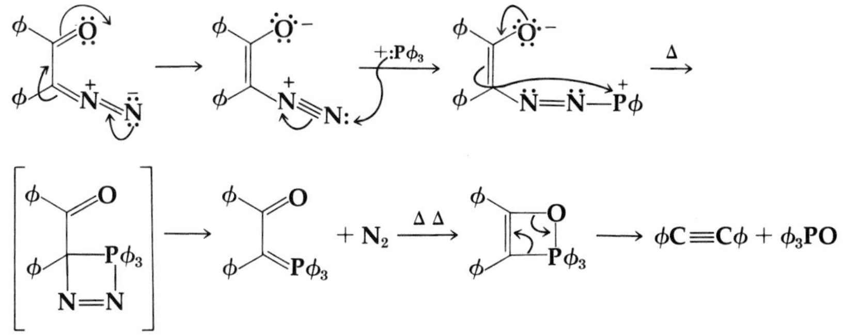

19–20 a

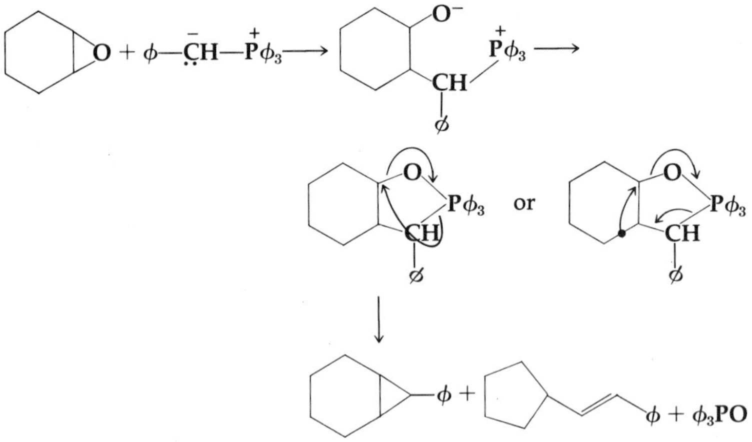

$$\text{(cyclohexene oxide)} + \phi\!-\!\overset{-}{\underset{\cdot\cdot}{C}}H\!-\!\overset{+}{P}\phi_3 \longrightarrow \text{(intermediate with } O^- \text{, } \overset{+}{P}\phi_3\text{)} \longrightarrow$$

(betaine/oxaphosphetane intermediates) or (diradical intermediate)

↓

(bicyclic cyclopropane)$-\phi$ + (cyclopentyl)$-CH=CH-\phi$ + ϕ_3PO

CHAPTER 20

20–1 a $(CH_3)_2CO + (CH_3)_2CHOH \xrightarrow[\text{(3000 Å)}]{h\nu} (CH_3)_2\underset{\underset{OH}{|}}{C}\!-\!\underset{\underset{OH}{|}}{C}(CH_3)_2 \xrightarrow{H_2SO_4}$

$$(CH_3)_3CCOCH_3 \xrightarrow[\text{2) } H_3O^+]{\text{1) NaOI}} (CH_3)_3CCOOH$$

d $CH_3CH=CH_2 \xrightarrow[h\nu]{Cl_2} 2ClCH_2CH=CH_2 \xrightarrow[-NaCl]{Na}$

$$CH_2=CHCH_2CH_2CH=CH_2$$

g $\underset{\underset{COOH}{|}}{\overset{\overset{COOH}{|}}{(CH_2)_3}} \xrightarrow[-H_2O]{\Delta} (CH_2)_3 \text{(glutaric anhydride)} \xrightarrow[\text{CH}_3\text{OH}]{\text{1 mole}} \underset{\underset{COOH}{|}}{\overset{\overset{COOCH_3}{|}}{(CH_2)_3}} \xrightarrow[\text{to pH 7}]{NaOH}$

$$\underset{\underset{COONa}{|}}{\overset{\overset{COOCH_3}{|}}{(CH_2)_3}} \xrightarrow{\text{Electrolysis}} CH_3OOC(CH_2)_6COOCH_3$$

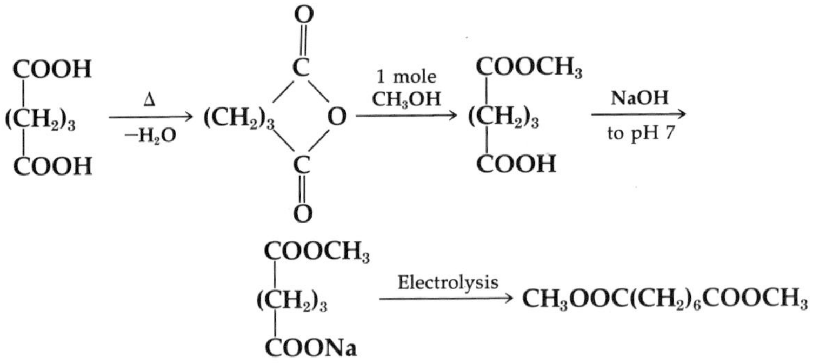

m $C_6H_5CH_3 \xrightarrow[h\nu]{Cl_2} C_6H_5CH_2Cl \xrightarrow{KCN} C_6H_5CH_2CN \xrightarrow[\text{2) } H_3O^-]{\text{1) LiAlH}_4}$

$$C_6H_5CH_2CH_2NH_2$$

q $(CH_3)_2CO \xrightarrow{(CH_3COO)_2} CH_3COCH_2CH_2COCH_3 \xrightarrow[\text{> 2 moles}]{CH_3MgI}$

$$(CH_3)_2COHCH_2CH_2COH(CH_3)_2 \xrightarrow[Al_2O_3]{\Delta} (CH_3)_2C=CHCH=C(CH_3)_2$$

(Separate from some
isomeric alkenes
which also form)

20-2 **a**

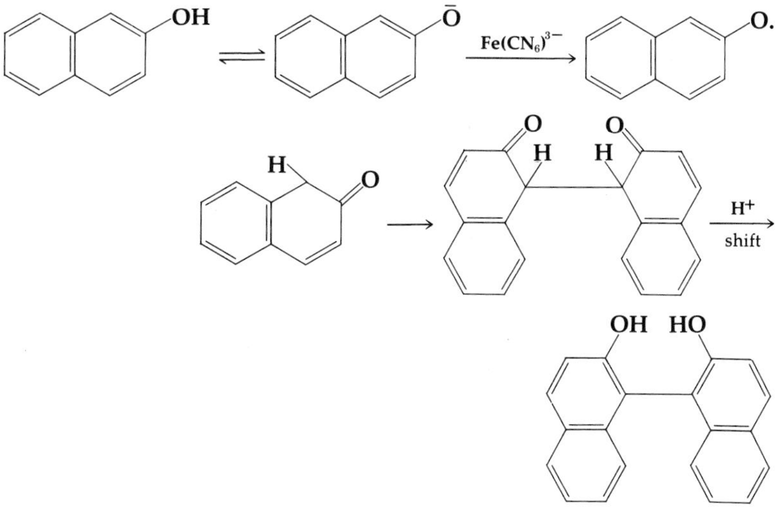

d (CH₃)₂C—N=N—C(CH₃)₂ ⟶ N₂ + (CH₃)₂ĊCN ⟷

$$(CH_3)_2C-N=N-C(CH_3)_2 \longrightarrow N_2 + (CH_3)_2\dot{C}CN \longleftrightarrow$$

$$(CH_3)_2C=C=\dot{N}$$

$$(CH_3)_2\dot{C}CN + (CH_3)_2C=C=N\cdot \longrightarrow (CH_3)_2C=C=N-\underset{\underset{CN}{|}}{C}(CH_3)_2$$

$$2(CH_3)_2\dot{C}CN \longrightarrow \begin{matrix}(CH_3)_2CCN\\(CH_3)_2CCN\end{matrix}$$

Apparently the second compound is *thermodynamically* more stable than the first compound, so that it is the principal product after a length of time.

f $N_2O_4 \overset{h\nu}{\rightleftharpoons} 2NO_2\cdot$

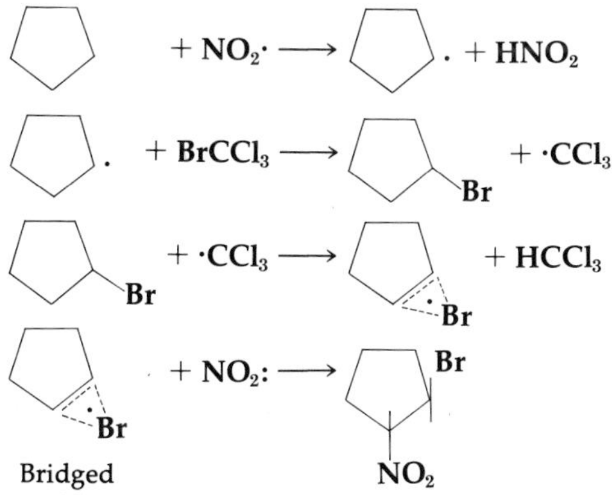

Bridged
radical

One need *not* invoke a bridged radical, but instead can use a free radical and the steric effect of **Br** and **NO₂** to explain the trans product, thus:

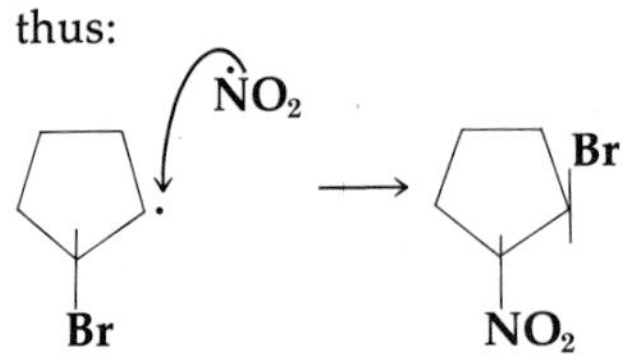

(Trans attack for steric reasons)

j $CCl_3Br \xrightarrow{h\nu} \cdot CCl_3 + Br\cdot$

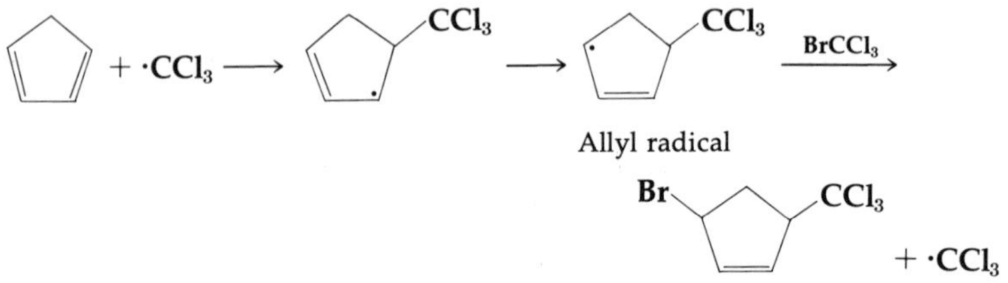

Allyl radical

$+ \cdot CCl_3$

Presumably no 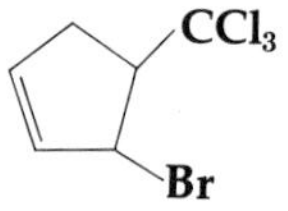 forms for steric reasons: both CCl_3 and **Br** are large

m $(C_6H_5)_3C-O-O-C(C_6H_5)_3 \xrightarrow{\Delta} 2(C_6H_5)_3C-O\cdot \xrightarrow[\text{migrates}]{C_6H_5}$

$2(C_6H)_2\dot{C}-O-C_6H_5 \longrightarrow$

20–4 **a** $+ (C_6H_5)_3CH$

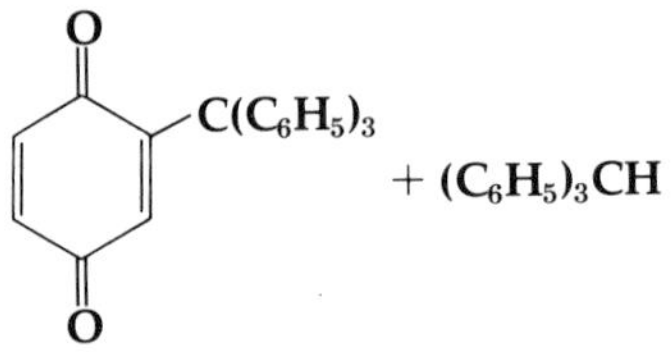

c

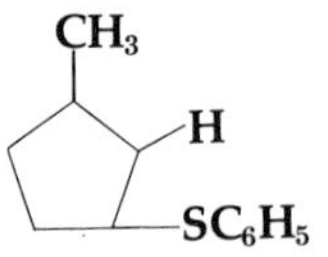

f

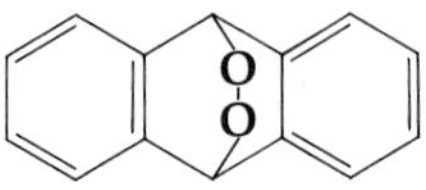

20–5 The compound which solvolyzes very rapidly must be a *tertiary* chloride. The one which reacts very slowly with ethanol must be primary. After drawing out all possible compounds which have C_6H_{14} as the molecular formula, one arrives at the following:

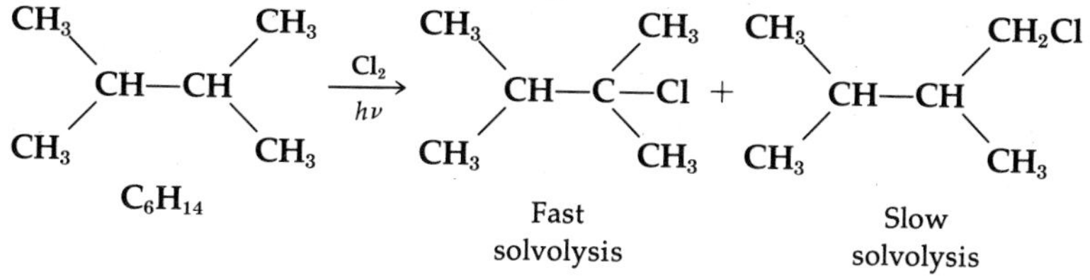

$$
\underset{C_6H_{14}}{\overset{\displaystyle CH_3 \quad\quad CH_3}{\underset{\displaystyle CH_3 \quad\quad CH_3}{CH-CH}}}
\xrightarrow[h\nu]{Cl_2}
\underset{\displaystyle CH_3 \quad\quad CH_3}{\overset{\displaystyle CH_3 \quad\quad CH_3}{CH-C-Cl}}
\;+\;
\underset{\displaystyle CH_3 \quad\quad CH_3}{\overset{\displaystyle CH_3 \quad\quad CH_2Cl}{CH-CH}}
$$

Fast solvolysis Slow solvolysis

CHAPTER 21

21–1 a

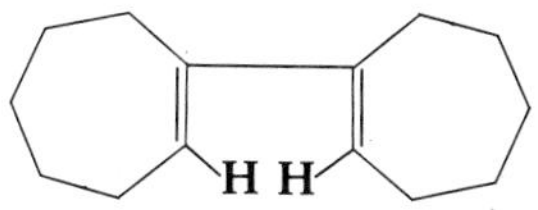

b AcO and AcO

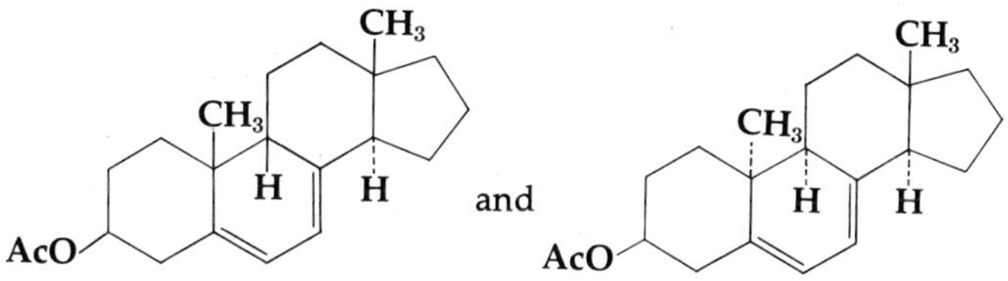

21–2 a

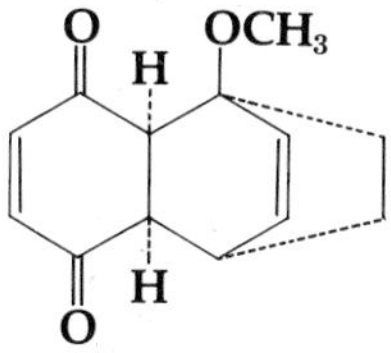

c

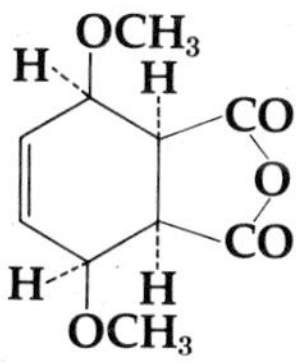

f

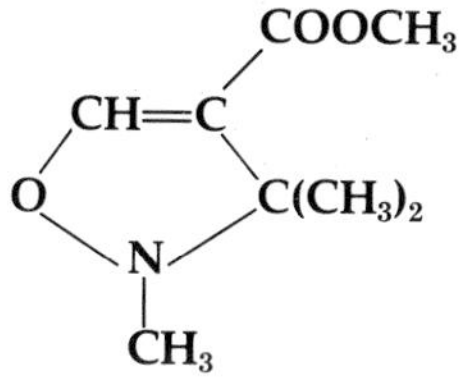

21–3

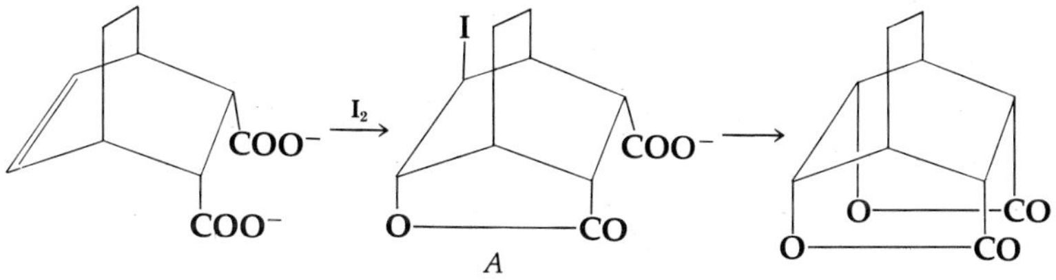

21-4 a $CH_3CH{=}CHCH{-}$ (2,6-dimethyl-4-hydroxyphenyl) with CH_3 groups and OH

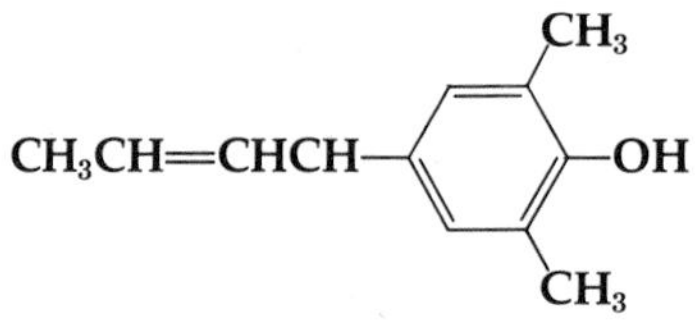

c

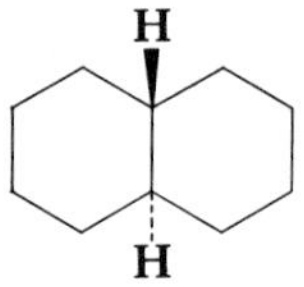

21-6 Conrotatory: $A \xrightarrow{\Delta}$ [intermediate with H, ϕ] $+$ maleic anhydride $\longrightarrow$ cycloadduct

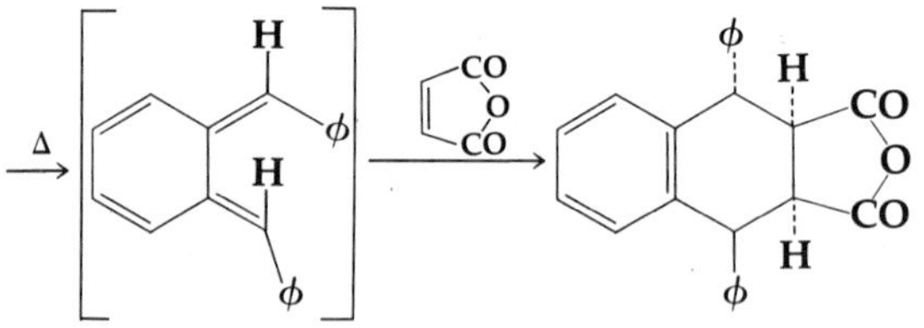

and similarly $B \longrightarrow$ cycloadduct

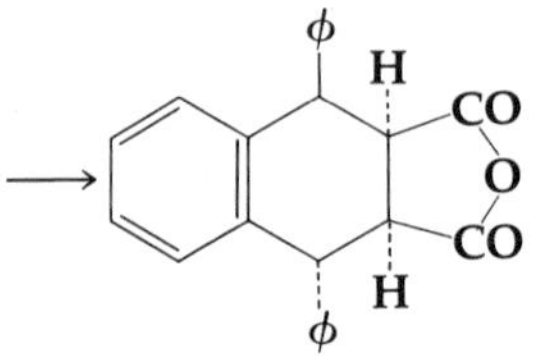

21-9 a Diels-Alder with 1,3-butadiene; $LiAlH_4$; $C_6H_5CO_3H$; H_2O/H^+
b Diels-Alder dimer $+$ $KMnO_4$ (or O_3)

21-10 furan (R on both α-positions) $+$ $CH_3OCOC{\equiv}CCOOCH_3 \longrightarrow$ bicyclic adduct (with CH_3OCO $COOCH_3$) $\xrightarrow{H_2}$

reduced bicyclic adduct (CH_3OCO $COOCH_3$) $\xrightarrow{\Delta}$ substituted furan (CH_3OCO $COOCH_3$) $+$ $CH_2{=}CH_2$

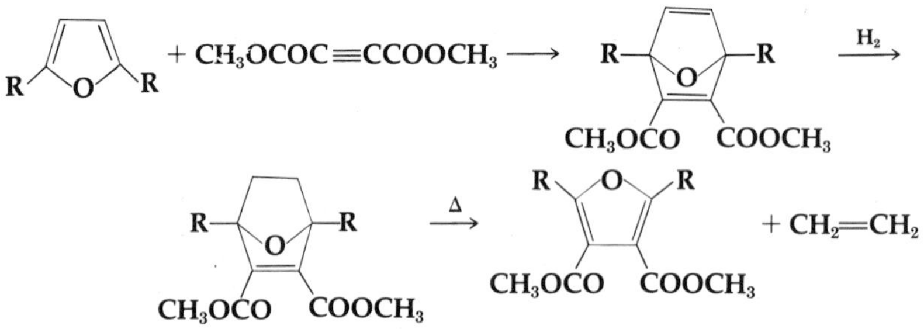

21-11 Compound B has nitrogen electrons partly drawn off in cross-conjugation to carbonyl, hence less available for stabilizing aromatic resonance which prevents cycloaddition in A

21-13 $[2 + 14]$-cycloaddition thermally forbidden if suprafacial (normal) but is thermally allowed if antarafacial; this case is antrafacial (note trans hydrogens)

21-15 cycloheptatriene derivative (CH_3, CH_3, CH_3) $\xrightarrow[\text{allowed}]{\text{Disrotatory}}$ bicyclic intermediate (CH_3, CH_3, CH_3) $\xrightarrow[\text{shift}]{[3,3]\text{-sigmatropic}}$

cyclopropane-fused ring (CH_3, CH_3, CH_3) $\longrightarrow$ cycloheptatriene (CH_3, CH_3, CH_3)

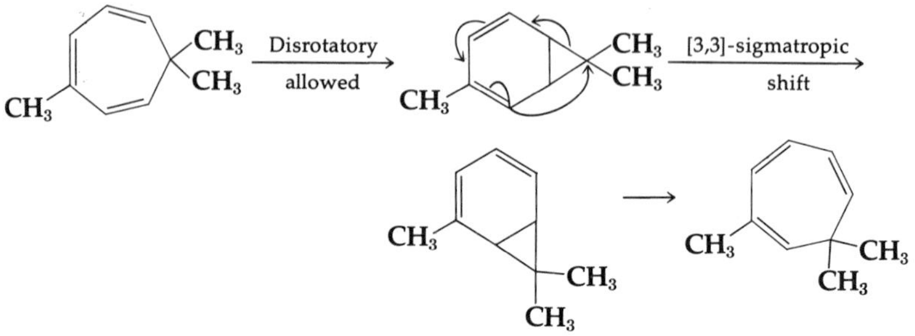

CHAPTER 22

22-1 a

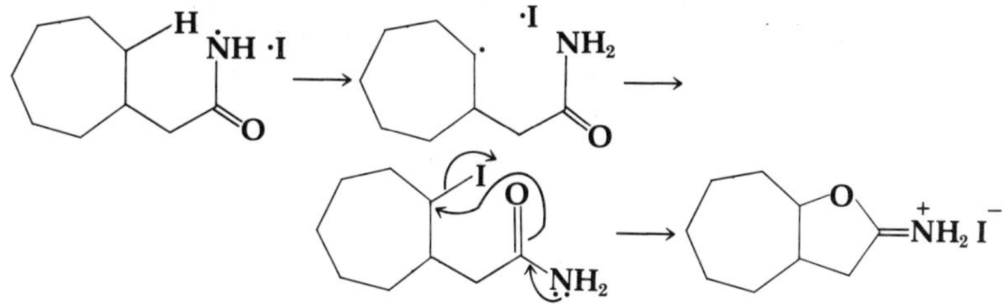

b

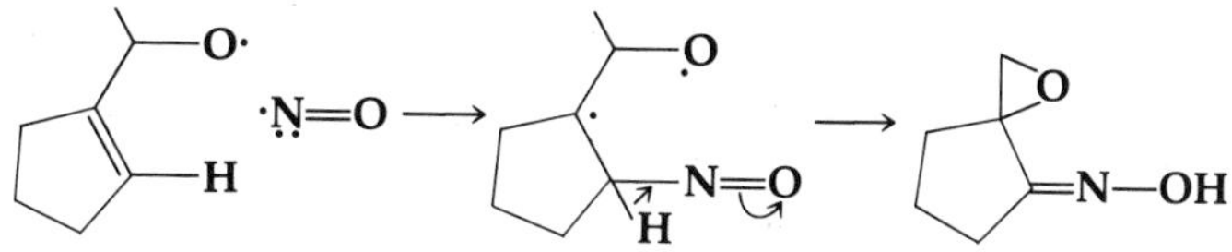

22-2 b

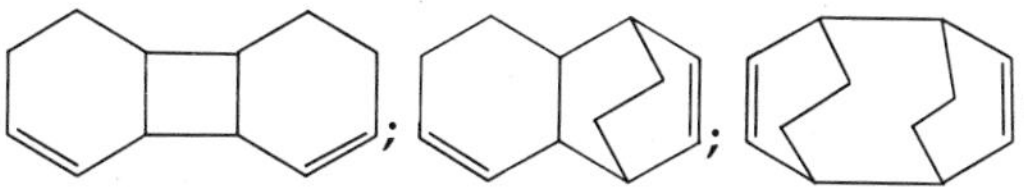

; ; unlikely

c $CH_3(CH_2)_4CH{=}C{=}O$; $CH_2{=}CH(CH_2)_4CHO$

22-4 a

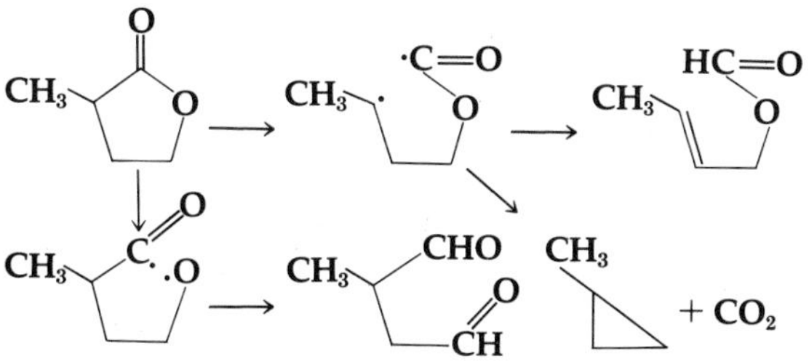

also

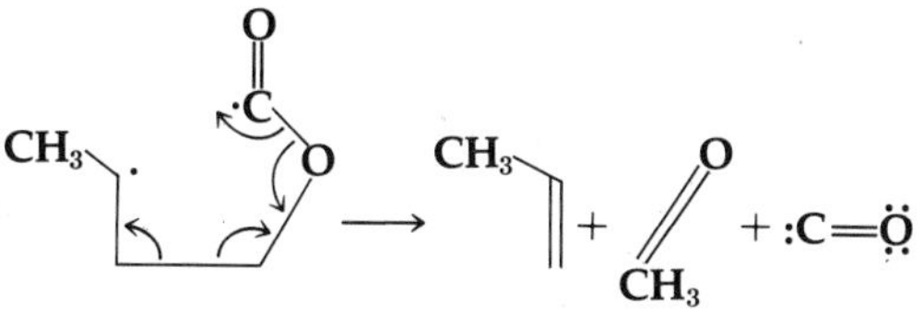

c

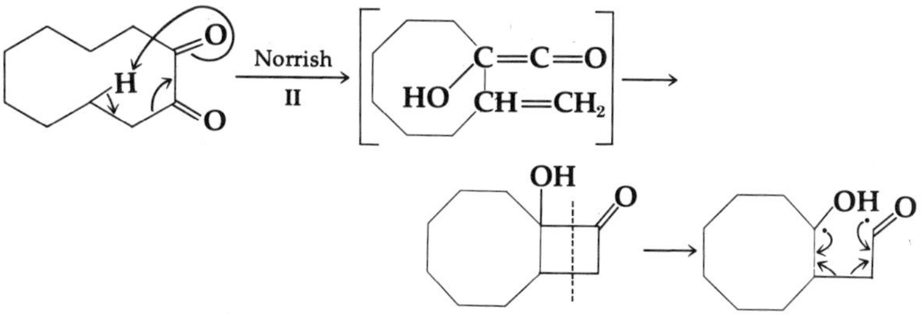

d

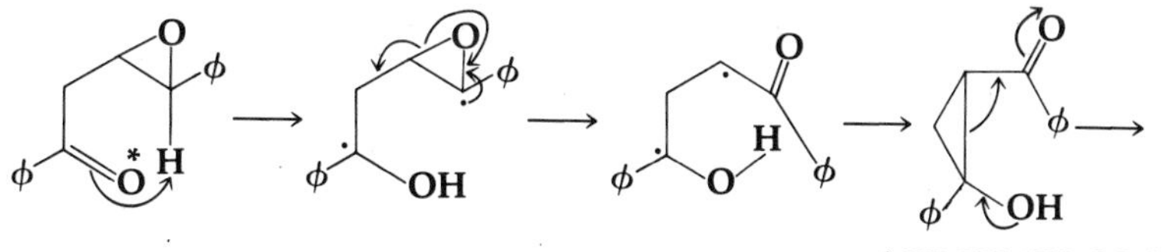

$\phi COCH_2CH_2CO\phi$

CHAPTER 23

23-1   *F* diol ⟶ ditosylate; *E2* elimination with base ⟶

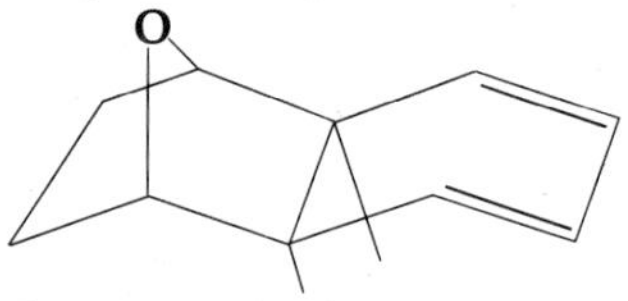

+ **KMnO₄ (or O₃)** ⟶ diacid;

or *F* $\xrightarrow{\text{CrO}_3}$ dione; φCHO/OH⁻ ⟶ *bis*-benzylidene; **KMnO₄ or O₃**

23-2   **a**   CH₃CH₂COOCH₃ + OCH₃⁻ ⟶ CH₃CH₂CO̅CCOOCH₃ + CH₃OH
                              |
                             CH₃

b   Aldol on φCHO

c   Wittig on acetylcyclopentonone

d   Friedel-Crafts cyclization of *m*-methoxyphenylbutyraldehyde

e   Diels-Alder

f   **CH₃MgI** on ester

g   Claisen rearrangement (sigmatropic shift)

h   pinacol reduction

23-3   **a**   No good; convert —CHO to acetal first, then reduce

c   No; elimination more likely; use **CH₃COO⁻/DMF**, then saponification

e   No; **KMnO₄** reacts with amines; —NH₂ must first be acylated

23-4   **c**

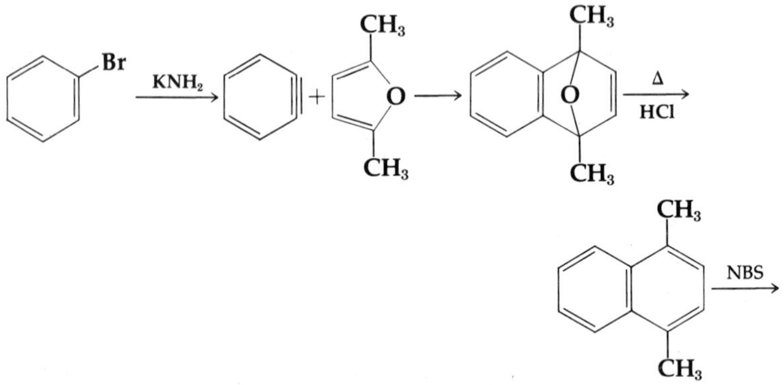

e   Use HC≡C:⁻ to obtain diacetylene for cis reduction

g   CH₃COCH₂COOR/base + CH₃I, C₂H₅I, then LiAlH₄

j   Symmetry implies Diels-Alder: benzoquinone + 2 cyclopentadiene; reduction of ketones; O₃; LiAlH₄; φ₃PBr₂; LiAlH₄

23-6   Require syntheses of small molecules and aromatic rings first:

$Fe + C/\Delta + HCl \longrightarrow HC\equiv CH$; $CO_2 + LiAlH_4 \longrightarrow CH_3OH$; $HC\equiv C: +$
$CH_3I \longrightarrow HC\equiv CCH_3 \longrightarrow (CH_3)_2CO$; $CH_3OH + CrO_3 \longrightarrow H_2CO$;
$(CH_3)_2CO + CH_2O \longrightarrow CH_3COCH=CH_2$; $CH_3COCH_3 + CH_2=CHCOCH_3$
(Robinson annellation)

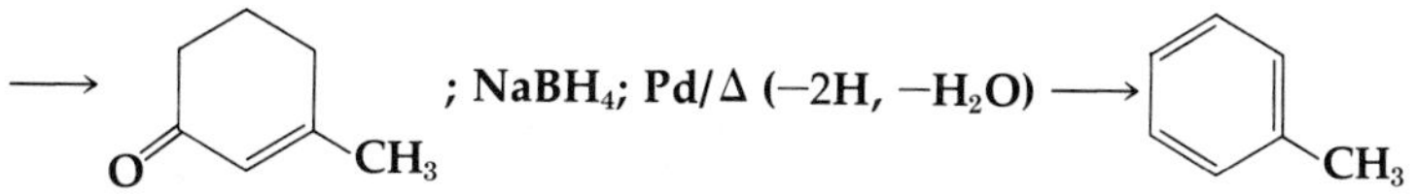

a

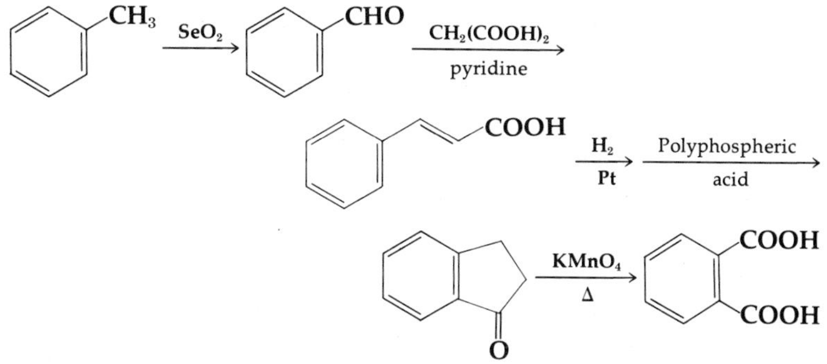

b $HNO_3 \longrightarrow$

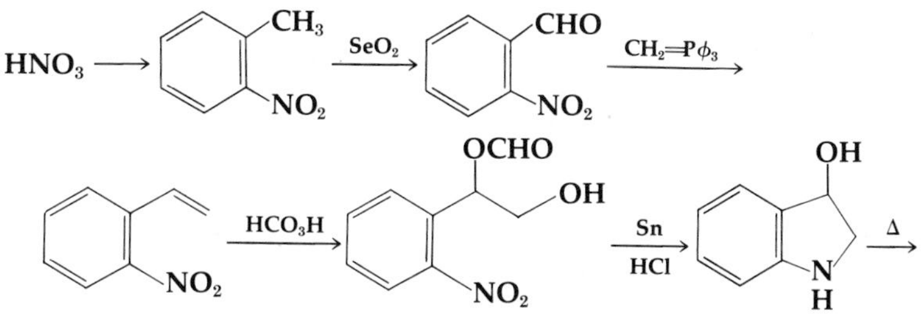

f $CH_3COCH=CH_2 + CH_3COCH_2COOCH_3 \longrightarrow$

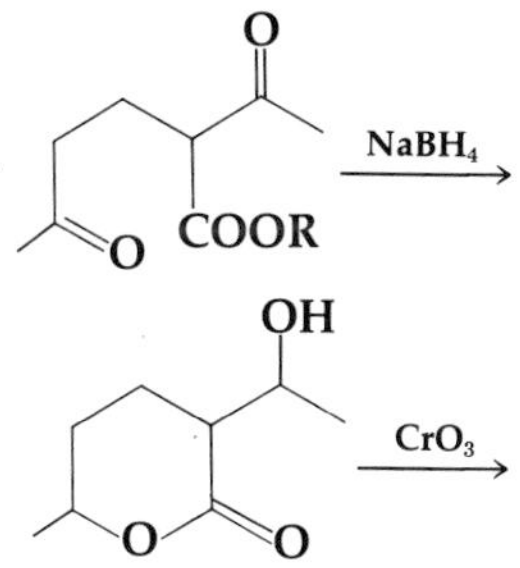

23-7   a $SOCl_2$; CH_2N_2; Ag_2O/CH_3OH (Arndt-Eistert)

d Mg/Et_2O; $CH_2\!-\!CH_2$ (epoxide O); H_2CrO_7

h Cl_2; OH^- (Favorski) $\longrightarrow$ cyclobutane carboxylic acid; HN_3

23–8 *o*-Methoxybenzaldehyde offers a 6-ring with adjacent —O and —C, hence ring *A* of product; hence cyclize a 5-ring to the benzaldehyde first, then reduce it, as in this possible synthesis:

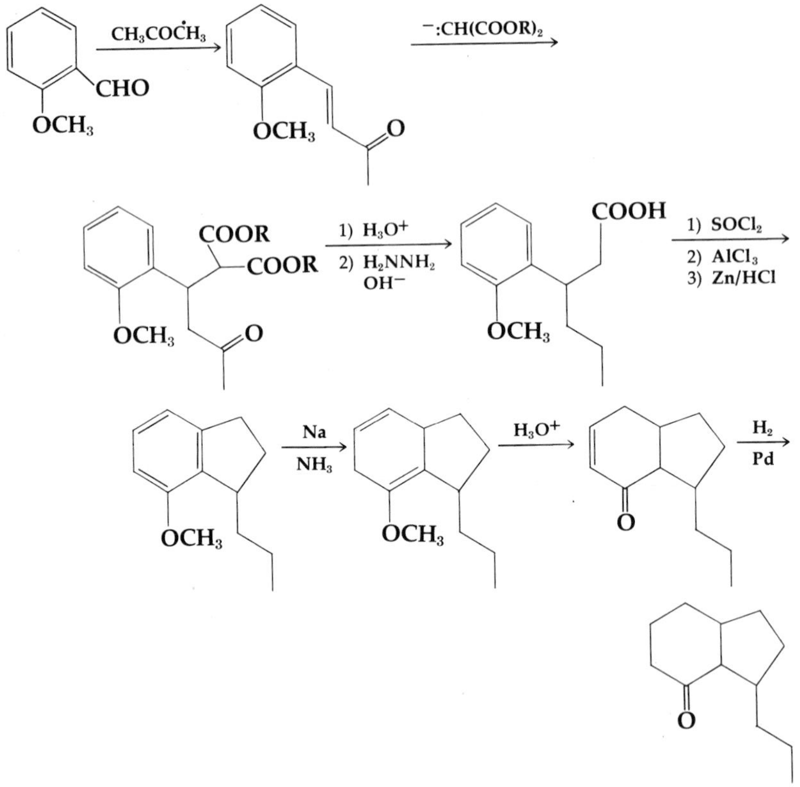

CH$_3$COCH$_3$

CHO

OCH$_3$

$^-$:CH(COOR)$_2$

OCH$_3$

O

COOR

COOR

OCH$_3$

O

1) H$_3$O$^+$
2) H$_2$NNH$_2$
OH$^-$

COOH

OCH$_3$

1) SOCl$_2$
2) AlCl$_3$
3) Zn/HCl

OCH$_3$

Na
NH$_3$

OCH$_3$

H$_3$O$^+$

O

H$_2$
Pd

O

Isomers = trans ring junction by enol equilibration in H$_3$O$^+$; third center should appear as both epimers, uncontrolled

23–11 Acyloin condensation on diester; LiAlH$_4$; H$_2$/Pt to hydrogenolyze ϕCH$_2$—; HIO$_4$ cleavage was followed by spontaneous internal Mannich reaction:

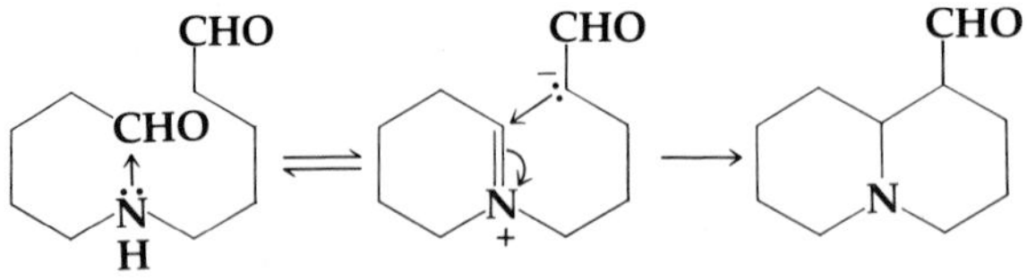

CHO

CHO

N
H

CHO

N
+

CHO

N

23–12   HBr; t-**BuO**$^-$ **K**$^+$; **R$_3$P**=**CH$_2$**; **CH$_2$**=**CH**—**CHO**/100°, Diels-Alder; **LiAlH$_4$**, **CH$_3$SO$_2$Cl**/pyridine, **LiAlH$_4$**; **OsO$_4$**, **C$_7$H$_7$SO$_2$Cl**/pyridine; rearrangement below; ϕ_3**P**=**CH$_2$**

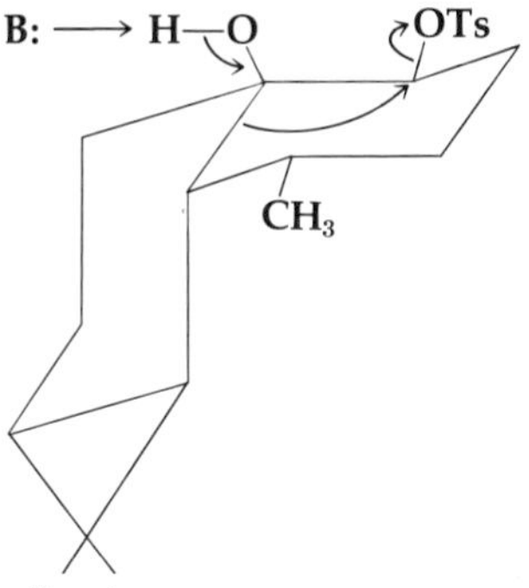

B: ⟶ H—O   OTs

CH$_3$

Toeylate
(trans periplanar orientation of migrating bond and leaving group)

23–14   CH$_3$OOC ⟋⟍⟋ COOCH$_3$ + ⟋⟍⟋ COOCH$_3$

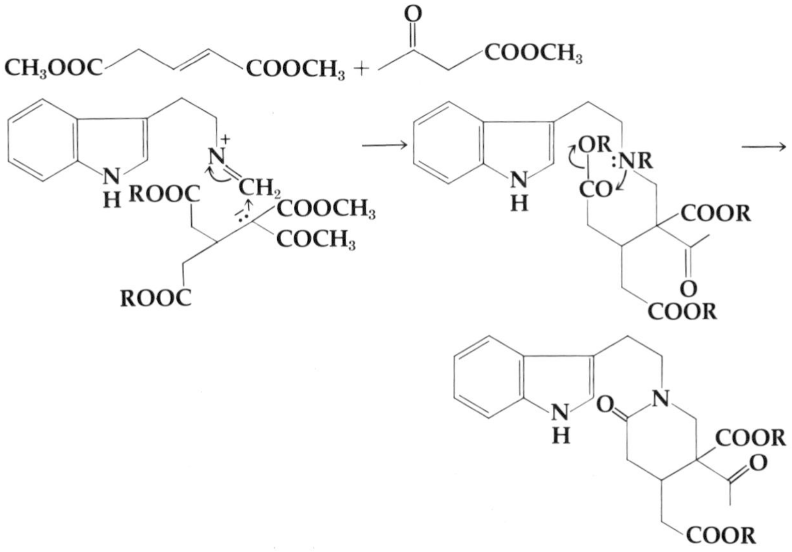

POCl$_3$, **NaBH$_4$**; **HCOOCH$_3$**/ϕ_3**C:**$^-$ **Na**$^+$; **CH$_3$OH/H**$^+$ opens lactone to methyl ester, rotation of **HO**—**CH**=**C** and tautomerism to aldehyde allows internal hemlacetal formation followed by β-elimination of **H$_2$O**

23–16   ; HNO₃; Sn/HCl; HNO₂; H₂O/Δ; (CH₃)₂SO₄ ⟶

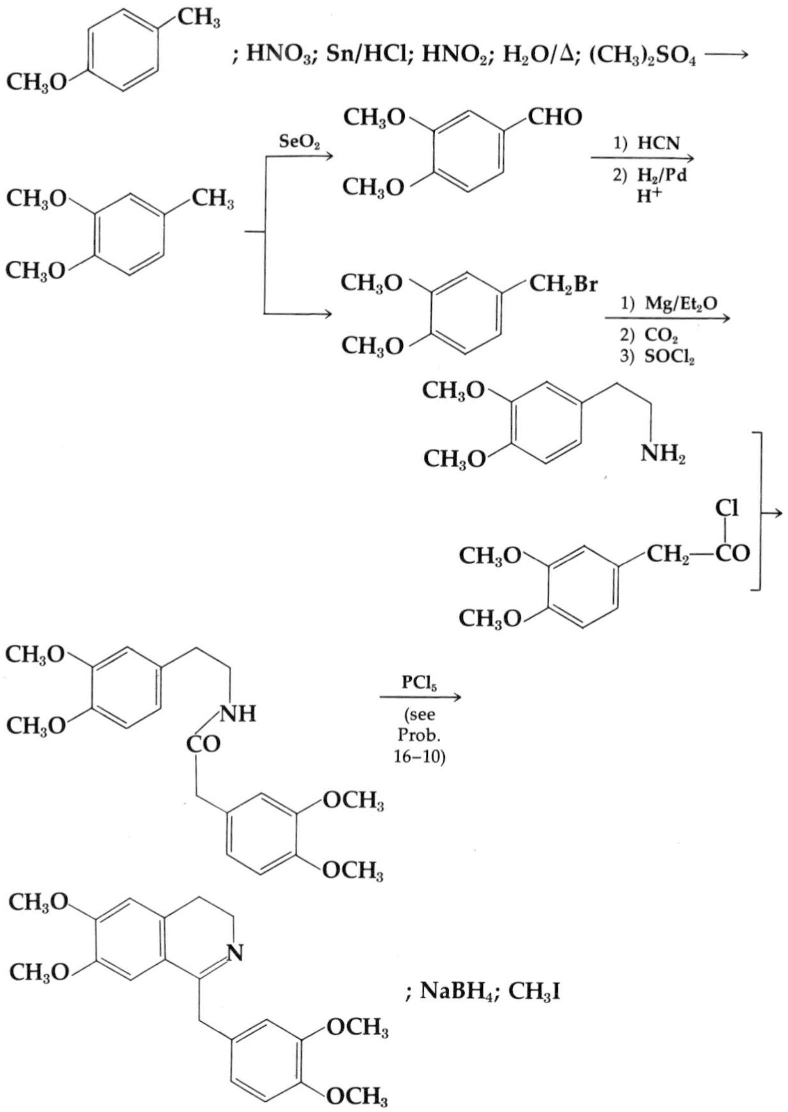

CHAPTER 24

24–1   a   5-Methyl-isoxazole or 5-methyl-1,2-oxazole

c   3-Chloro-1,2,4-triazine-5-carboxylic acid

e

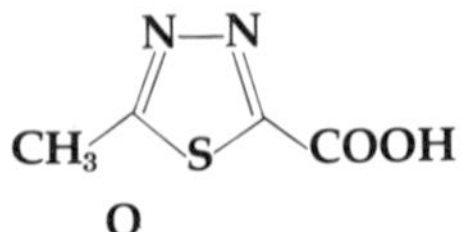

h

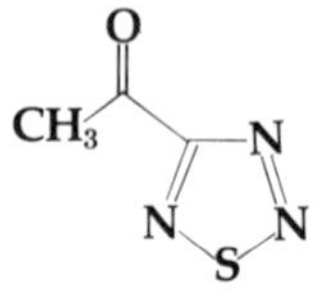

24-3 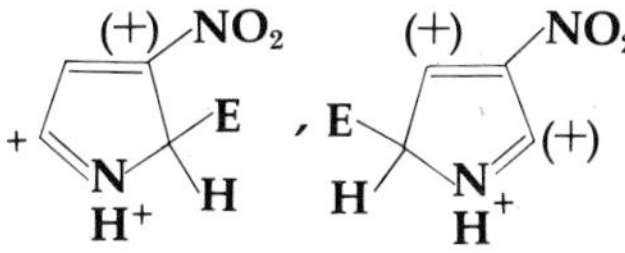 ; second is preferred since no (+) next to —NO$_2$

24-4

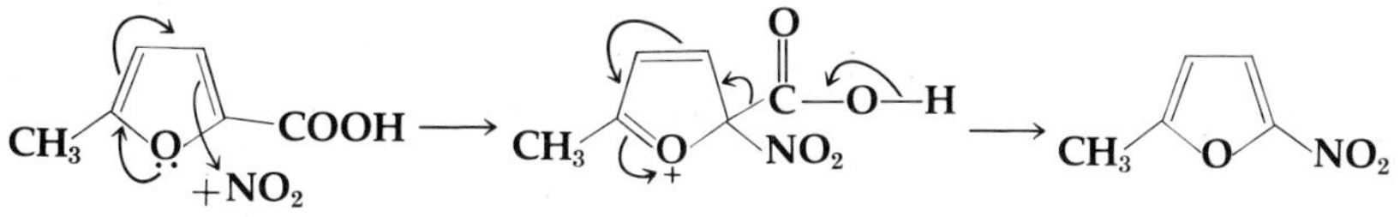

24-5

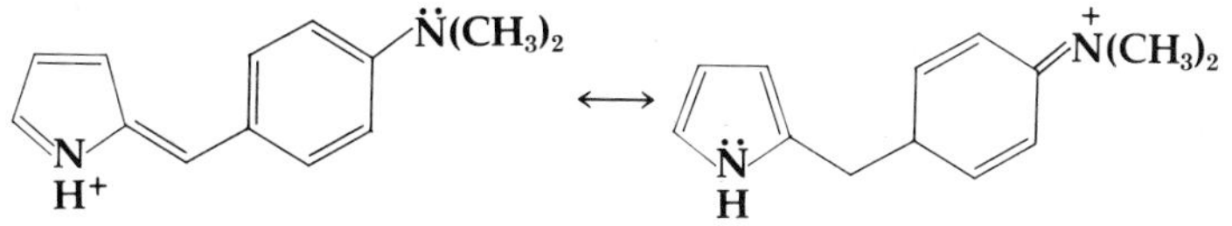

24-6 **b** 2-Nitro-   **c** 5-Nitro-   **g** 3-Nitro-

24-8

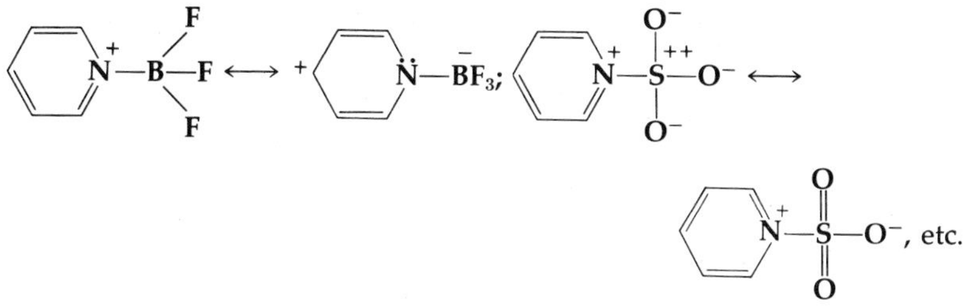

24-10 **a**

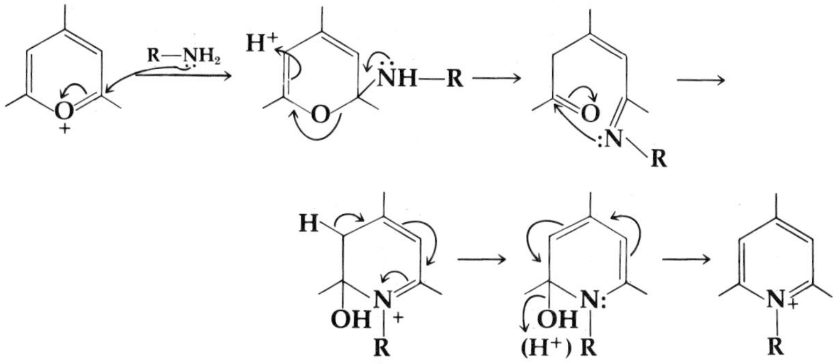

24-11 2- and 4-methyl- and 2- and 4-propyl-pyridines

24-13

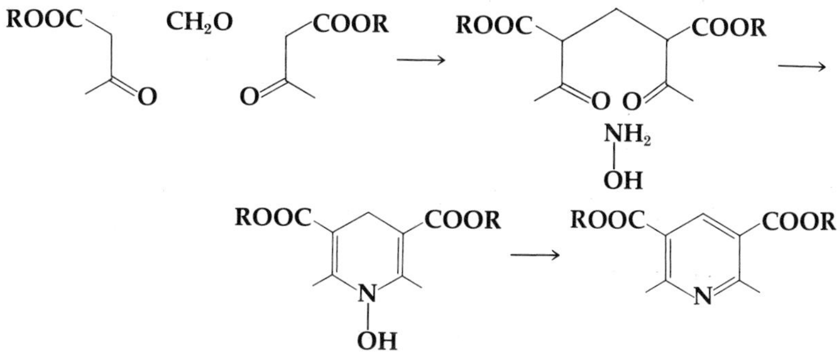

24–15

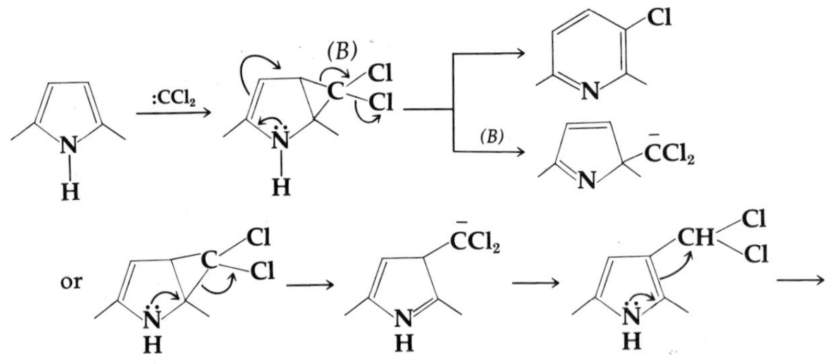

easy solvolysis to aldehyde

24–17 **a** $C_6H_5NHNH_2 + CH_3CH_2COCOCH_2CH_3 \longrightarrow$

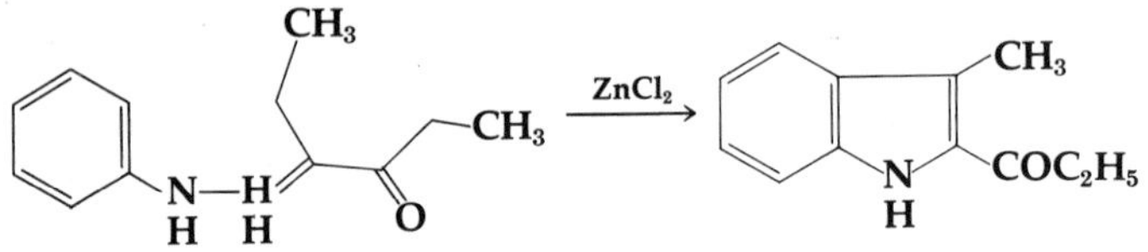

b $C_6H_5NH_2 + CH_3COCH_2COCH_3 \longrightarrow$

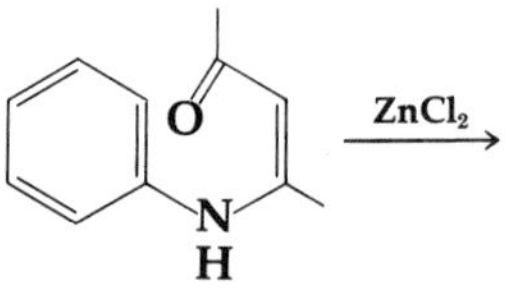

f $\phi COCHBrCH_3 + \phi CSNH_2/\Delta$

24–19

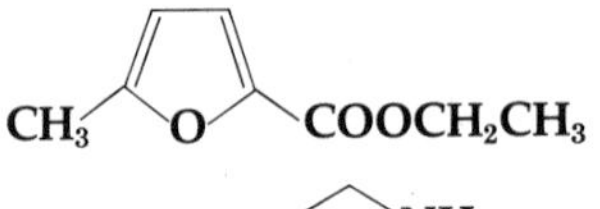

24–21 **a**

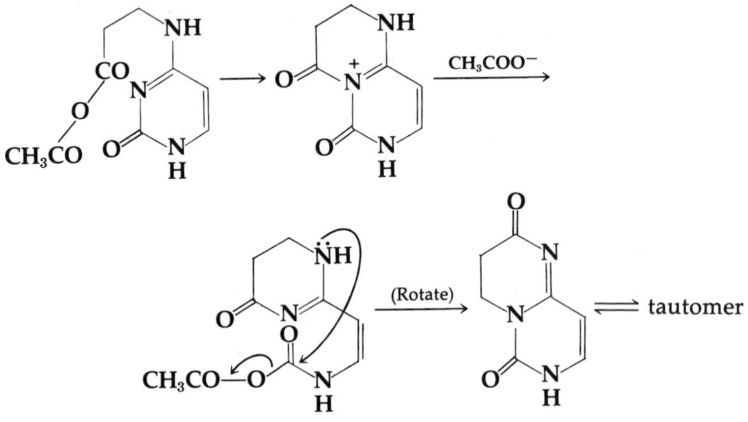

c

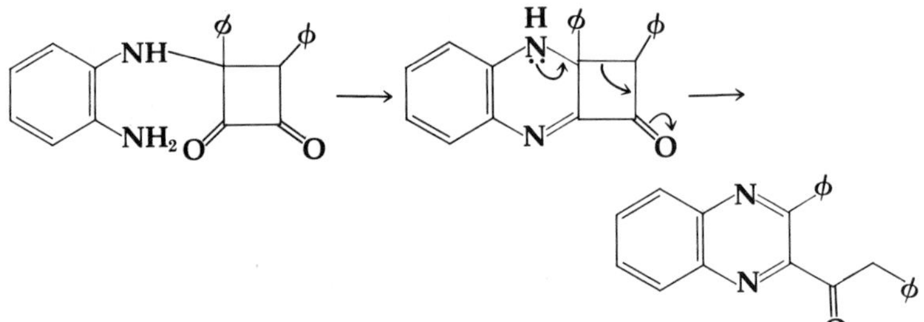

CHAPTER 25

25-1 **a** $CH_2{=}CCl{-}CH{=}CH_2$ via free-radical initiators, e.g., $(t\text{-}BuO)_2$

 d $CH_2{=}CHCN$ via bases, e.g., $NaNH_2$

25-3 **a**

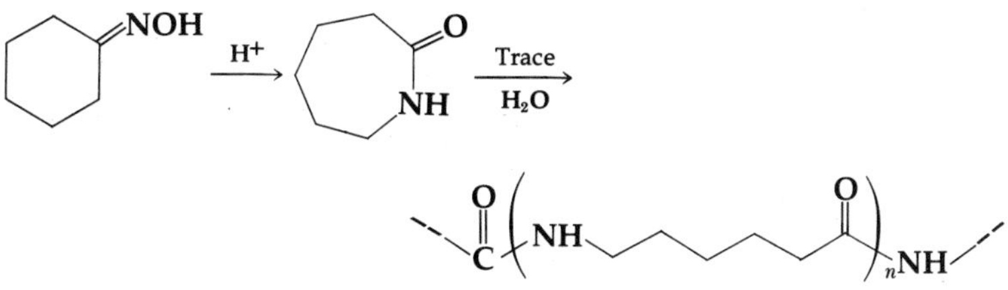

25-4

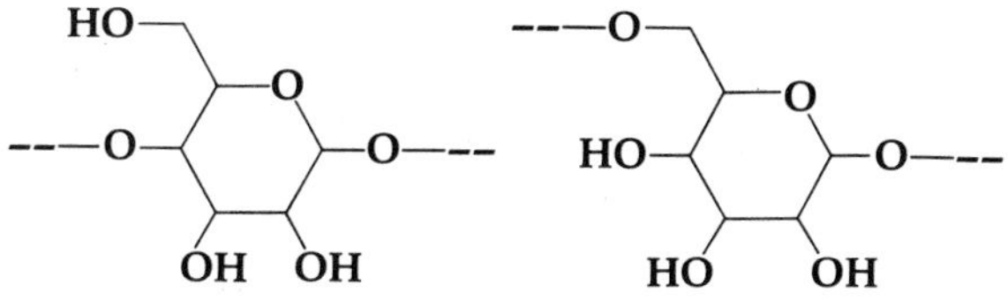

| 1,4-linkage reacts with one HIO_4 | 1,6-linkage reacts with two HIO_4 |

25-7 Cysteine + H_2/Raney nickel $\longrightarrow$ alanine

25-8

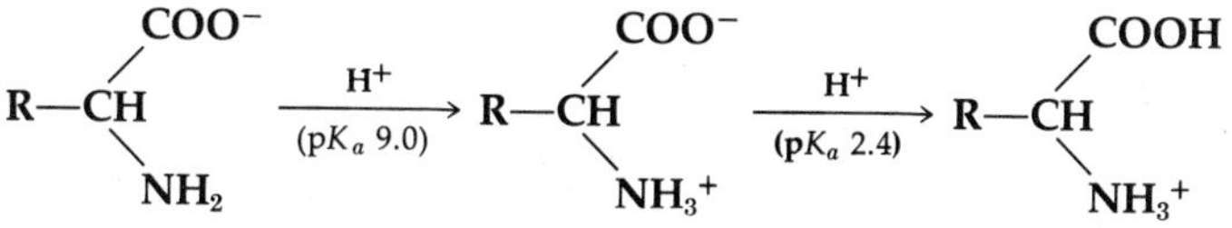

| At pH 12 | At pH 7 | At pH 1 |

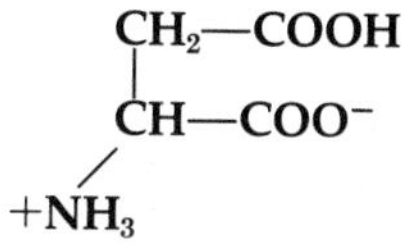

Aspartic acid

above pH 2–3; loss of next H (COOH) resisted by internal H bond, hence pK_a above normal (> 5)

25-11 Cytosine $\rightleftharpoons$

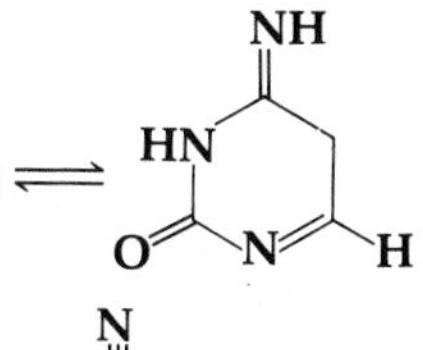

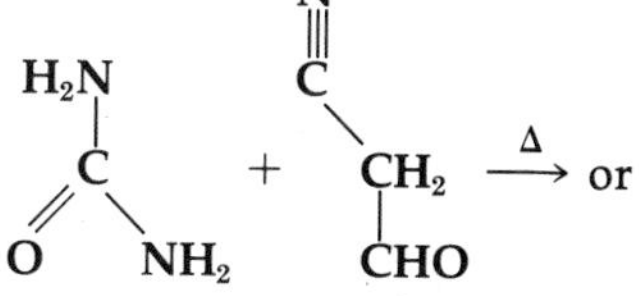

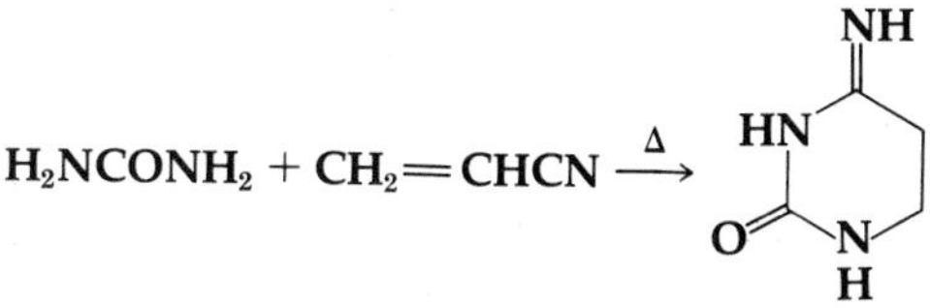

$H_2NCONH_2 + CH_2{=}CHCN \xrightarrow{\Delta}$

and then dehydrogenate/Δ/**Pd**

25-13 a $CH_3CONH\bar{C}(COOC_2H_5)_2 + C_6H_5CH_2Br$, then hyrolysis

b Pyrrole + ClCOOR, then hydrolysis and hydrogenation, or
$CH_3CH_2CH_2CH(NH_2)COOCH_3$ + chlorination and light (Sec. 22-4),
or cyclopentanone oxime + $H_2SO_4 + H_2O \longrightarrow$
$H_2NCH_2CH_2CH_2CH_2COOH + Br_2/PBr_3$

g

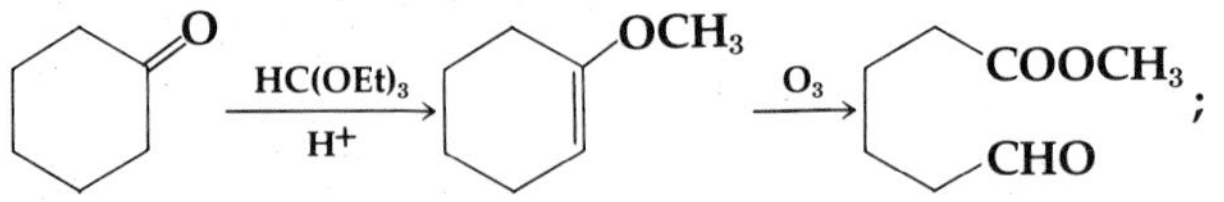

$C_6H_5NHNH_2$; $ZnCl_2$ (Fischer indole synthesis)

25-14 $CH_2{=}CHCOOCH_3 + Br_2 \longrightarrow BrCH_2{-}CHBrCOOCH_3 \xrightarrow{^-OCH_3}$

$CH_2{=}CBrCOOCH_3 \xrightarrow{^*CN^-} NCCH_2CHBrCOOH \xrightarrow[H_2O]{\Delta/NH_3}$

$HOO\overset{*}{C}CH_2CH(NH_2)COOH$

25-16 $-CO{-}\underset{|}{C}H{-}NH{-}CO{-}$

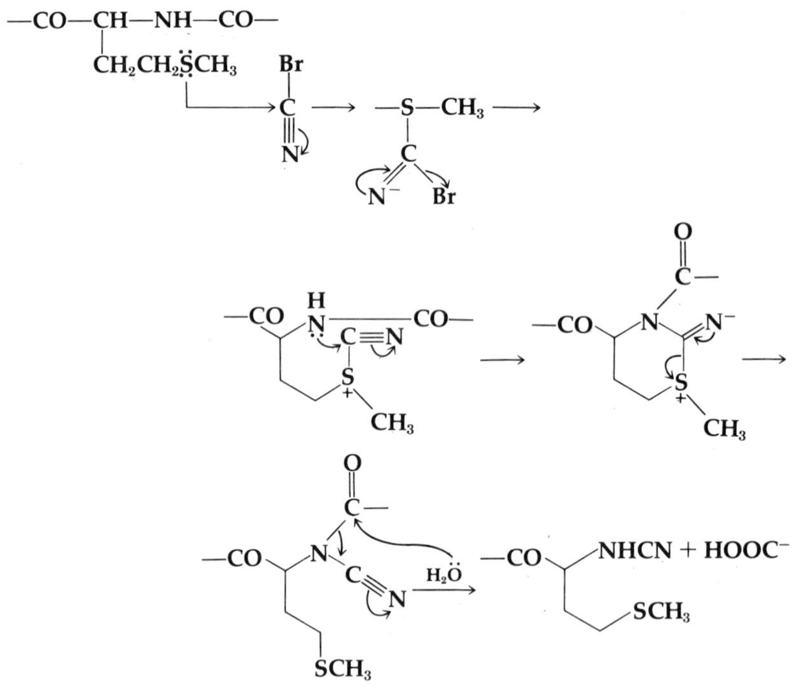

25-19 $ROH + R'NCO \longrightarrow RO{-}CO{-}NHR'$; initial compounds like

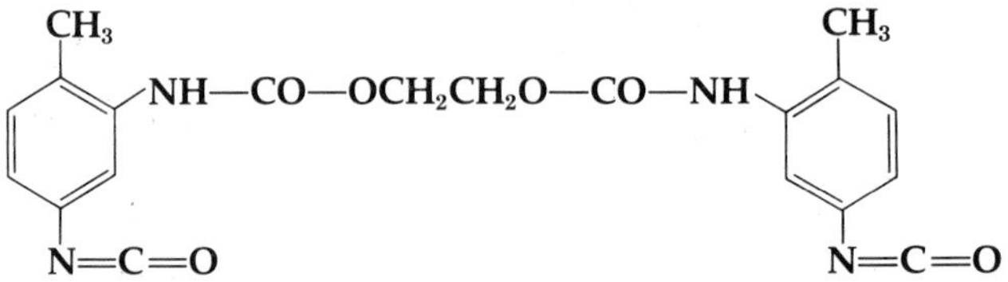

Reaction with H_2O: $R{-}N{=}C{=}O \longrightarrow [R{-}NH{-}CO{-}OH] \longrightarrow$

$R{-}NH_2 + CO_2\uparrow \xrightarrow{+RNCO} R{-}NH{-}CO{-}NH{-}R$ CO_2 loss affords gas
and half of $R{-}NCO$ becomes $R{-}NH_2$ to link with remaining $R{-}NCO$

25–22 **b** Diels-Alder $\longrightarrow$

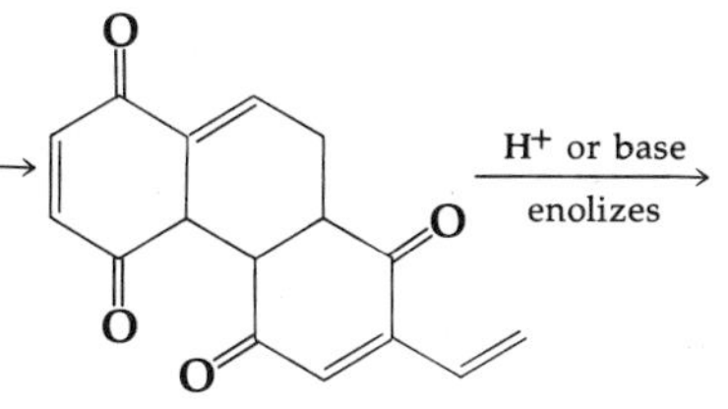

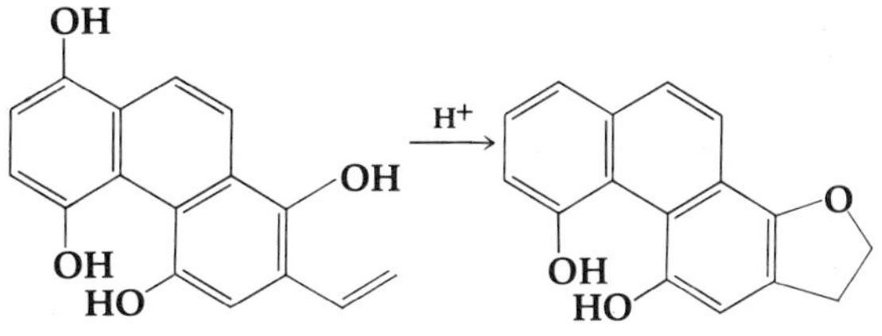

$\xrightarrow{\text{H}^+ \text{ or base}}$ enolizes

$\xrightarrow{\text{H}^+}$

d Alanine + $COCl_2$; $H_2O \longrightarrow [HOCONHCH(CH_3)COOH] \xrightarrow{-CO_2}$

$H_2NCH(CH_3)COOH$

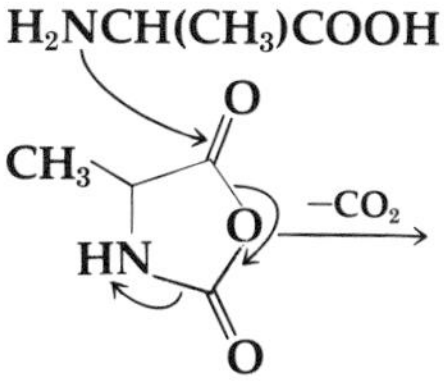

$\xrightarrow{-CO_2} H_2NCH(CH_3)CONHCH(CH_3)COOH$, etc.

25–24 IR: NH (or OH), COOR, CON $\diagdown$; NMR: 3.40 = same NH (or OH); 4.81 =

$\diagdown$CH—NH (or OH), attached to CO; 5.75 and 8.70 = (—O—CH_2—CH_3)$_2$;

7.92 = $COCH_3$; $CH_3CONHCH(COOC_2H_5)_2$ is the starting material

CHAPTER 26

26–1 $HOC\overset{-1}{H}_2$—$(\overset{0}{C}HOH)_4$—$\overset{+1}{C}HO \longrightarrow \overset{+4}{C}O_2$; = +5, 4 (+4), + 3 = 24 or 24/2 = 12

2-e^- oxidation steps. Average energy = 684/12 = 57 kcal/mole per step;

$5HIO_4 \longrightarrow 5CH_2O + 1HCOOH \xrightarrow{CrO_3} 6HCOOH \xrightarrow{KMnO_4} 6CO_2.$

26–2 "Natural fit" has cavities for $C_6H_5CH_2$—, H—, RCONH— in L isomer;
D isomer has two reversed so one cannot fit in a cavity (i.e., no benzyl in
H— cavity). Other compounds show that **RCONH**— cavity is not satisfied
by H— only (slower rate = 15) and benzyl is not satisfied by methyl
(rate = 2), while methyl is too large for **H**— cavity (rate ~ 0). If —**CONH**—
is tied back to phenyl ring, however, it can occupy **H**— cavity very well
cannot occupy normal **RCONH**— cavity effectively.

26–3 Resolve enantiomers of deutero-lactic acid and determine absolute stereochemistry by reaction with ammonia on the ester-tosylates to the two
alanine enantiomers by inversion (S_N^2). Now see which one is formed from
pyruvate by NAD stereoselectively.

26-4 Key reaction = $HO-CH_2-CH-N\!\!=\!\!\!<$ etc. $\longrightarrow$ CH_2O + glycine deriva-

$\qquad\qquad\qquad\qquad\quad$ |
$\qquad\qquad\qquad\qquad\;\,$ COOH

tive; cannot occur in peptide (no NH_2)

26-5

$$\begin{array}{c} COOH \quad O \\ | \qquad\quad \| \\ CH-O-P-OH \\ \| \qquad\quad | \\ CH_2 \qquad OH \\ \qquad C\!\!=\!\!O \text{ or } HC\!\!=\!\!O \\ \qquad \| \qquad\quad | \\ \qquad O \qquad\quad R \end{array} \longrightarrow \begin{array}{c} COOH \quad O \\ | \qquad\quad\; \| \\ CH\!\!=\!\!O\;\;P-OH \\ | \qquad\quad | \\ CH_2 \qquad OH \\ | \\ COOH\,(CHOH) \\ \qquad\qquad | \\ \qquad\qquad R \end{array} \xrightarrow{H_2O}$$

$$\begin{array}{c} COOH \qquad\qquad\quad O \\ | \qquad\qquad\qquad\quad \| \\ C\!\!=\!\!=\!\!=\!\!O + HO-P-OH \\ | \qquad\qquad\qquad\quad | \\ CH_2 \qquad\qquad\qquad OH \\ | \\ COOH \end{array}$$

26-6 **a** 2-C^{14}-pyruvate: $C^{14}O_2$ comes out in step (6), first cycle; 3-C^{14}-pyruvate: $C^{14}O_2$ comes out in step (5), *second* cycle around

b Fumarate with C-3 of pyruvate intact (as **COOH**) but C-2 of pyruvate gone in step (6)

c The cycle is stopped at reaction (4) and pyruvate only goes to isocitrate but (labeled) glutamate is converted to α-ketoglutarate thus allowing more cycle to continue from there, pick up pyruvate at step (1) and proceed to isocitrate; the four lower carbons of α-ketoglutarate become the lower four in isocitrate and would be labeled from labeled glutamate

26-9 $R-CH_2CH_2COOH \xrightarrow{CoSH} R-CH_2CH_2COSCoA \xrightarrow[(-2H)]{NAD}$

$\qquad\quad RCH\!\!=\!\!CHCOSCoA \xrightarrow{H_2O} RCHOHCH_2COSCoA \xrightarrow[(-2H)]{NAD}$

$\qquad\qquad\quad RCOCH_2COSCoA \xrightarrow{H_2O} RCOOH + CH_3COSCoA$

26-10 **a** 12-unit oligonucleotide + all four mononucleotides + polymerase enzyme: mononucleotides bind to counterparts on 12-unit chain and are linked by enzyme to form the complementary 12-unit strand. Continue on each to develop more of both 12-unit complementary strands, each synthesized using the other as template.

b consider building 3 particular 12-units:

then add mononucleotides and polymerase to get two complementary strands of 24-units each, etc.

26-12 **a** 20 specific enzymes to link the amino acids to their counterpart *t*-RNA carriers, one coupling enzyme to form peptide bond (Fig. 26-13)

26-13 One DNA $\longrightarrow$ one enzyme protein to catalyze one necessary metabolic reaction. Hence estimate number of required metabolic reactions to produce energy and synthesize components, e.g., amino acids and nucleotides. Assume glucose as only initial carbon input.

26–15 Enzyme—S—CO

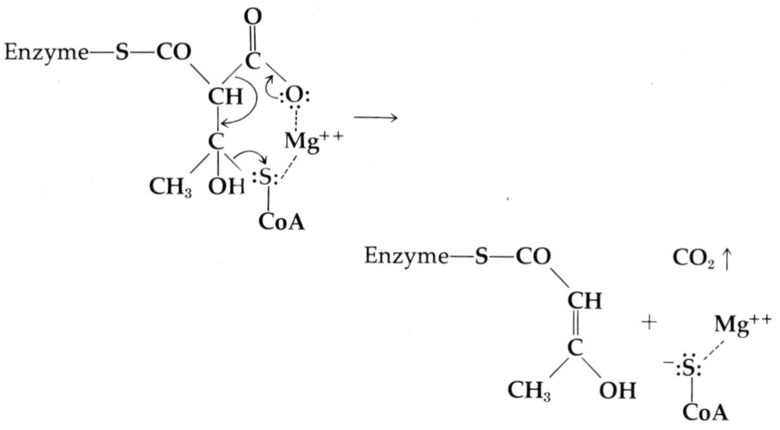

CH

C

CH₃ OH

Mg⁺⁺

:S:

CoA

⟶

Enzyme—S—CO   CO₂ ↑

CH

C

CH₃   OH

+   Mg⁺⁺

-:S:-

CoA

26–17 2CH₃C—SCoA $\xrightarrow[\text{condensation}]{\text{"Claisen"}}$ CH₃—C—CH₂—CO—SCoA $\xrightarrow[\text{(aldol)}]{\text{"Base"}}$

+ CH₃—COSCoA

(or HOOC—CH₂—COSCoA)

OH

CH₃—C—CH₂—COSCoA

CH₂—COSCoA

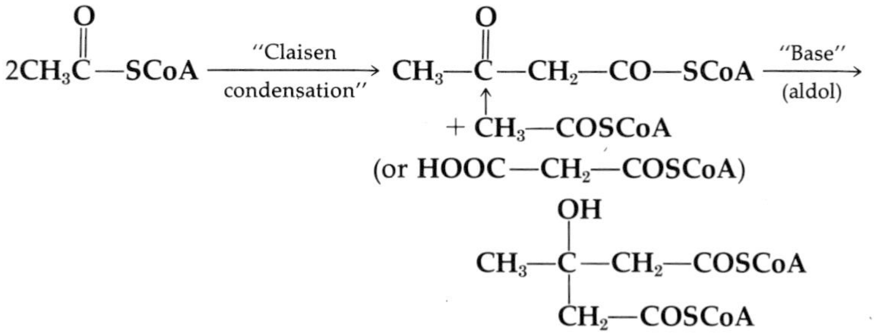

CHAPTER 27

27–1 IR = much absorption ~3μ; also 5.8μ; no UV; NMR = 2H (s) ~ −1.0τ (disappear/D₂O); one other peak disappears/D₂O; complex 1H peak or **CHOH** around 6τ; complex 1H peak ~7.5τ (**CH—COOH**); 2H (*t*) peak ~8τ and a multiplet (or double quartet) of 2H a little higher and a 3H triplet at highest shift (~9τ)

27–2 b Two diastereomers form on **NaBH₄** reduction

27–3 a 5CH₃COSCoA ⟶

$\xrightarrow{\text{Methylation}}$

SCoA

CH₃

O

SCoA
CH₃

O CH₃

⟶

CH₃

HO CH₃

CH₃ SCoA

⟶

CH₃

O

O CH₃

CH₃

⟶

CH₃

HO

OH

O CH₃

CH₃

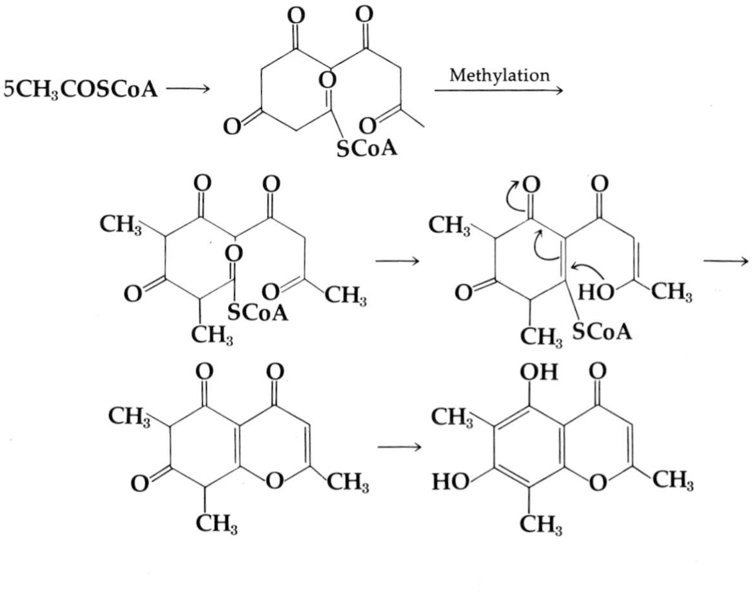

c

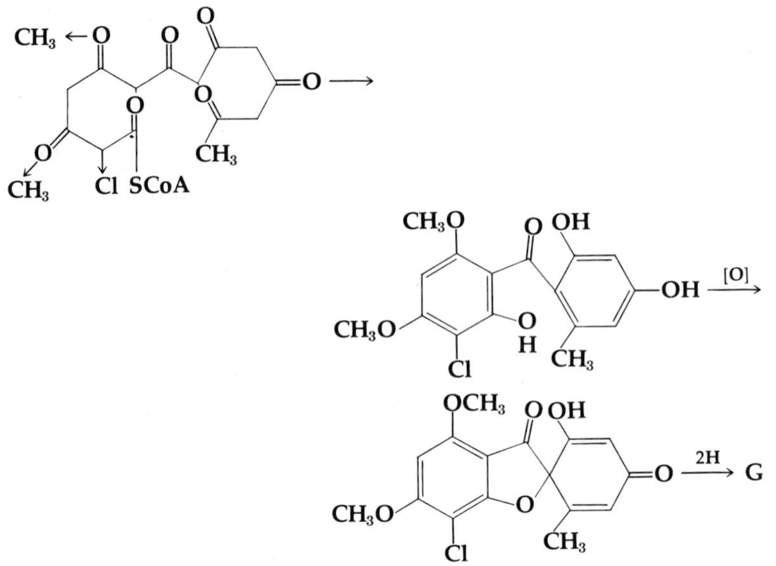

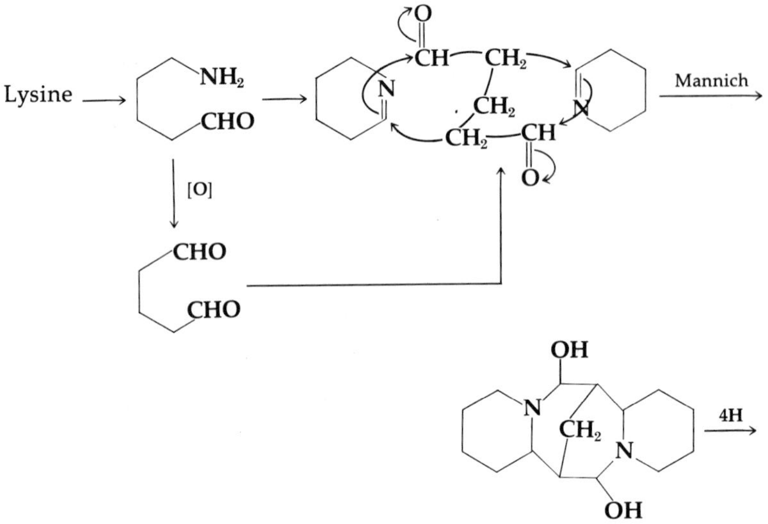

$CH_3 \leftarrow O$ Cl SCoA CH_3

CH_3O OH   OH $\xrightarrow{[O]}$

CH_3O OH CH_3 Cl

OCH_3 OH

CH_3O O $\xrightarrow{2H}$ G

Cl CH_3

Lysine $\longrightarrow$ NH_2 CHO $\longrightarrow$ $\xrightarrow{Mannich}$

$\downarrow$ [O]

CHO   CHO

OH   N CH_2 N $\xrightarrow{4H}$ OH

27–4

HO   OH

CH_3 O $\xrightarrow{Via}$ CH_3

O CH_3

(COOH)

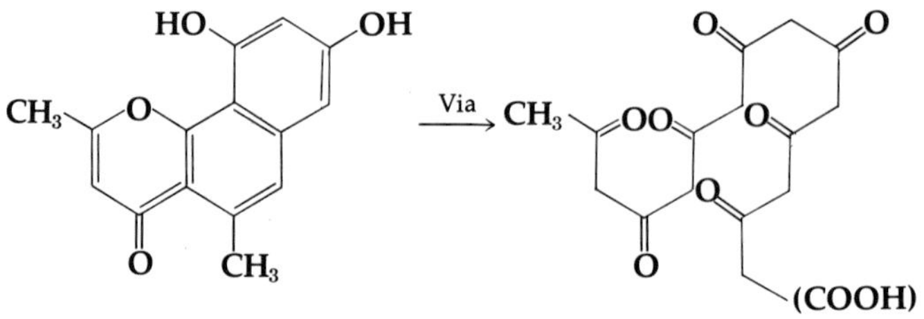

27-7 a 2 moles $HIO_4 \longrightarrow$ 1HCOOH   **e** 3 moles $HIO_4 \longrightarrow$ 2HCOOH

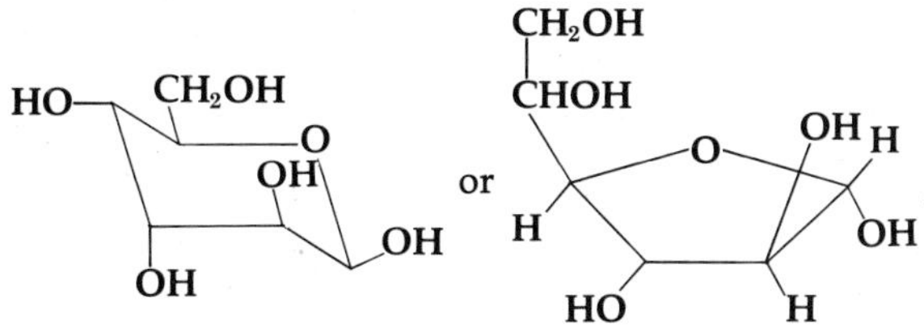

; favors furanoside

2 axial —OH   All trans substituents

27-10

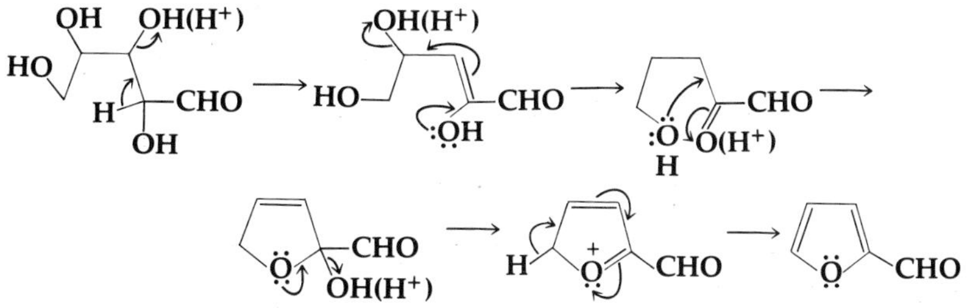

27-13 $(X,Y = H$ or $OH)$

$$\xrightarrow{\begin{array}{c} 1)\ (CH_3)_2SO_4 \\ 2)\ OH^-/\Delta \end{array}}$$

$X,Y = \begin{cases} H \\ OCH_3 \end{cases}$

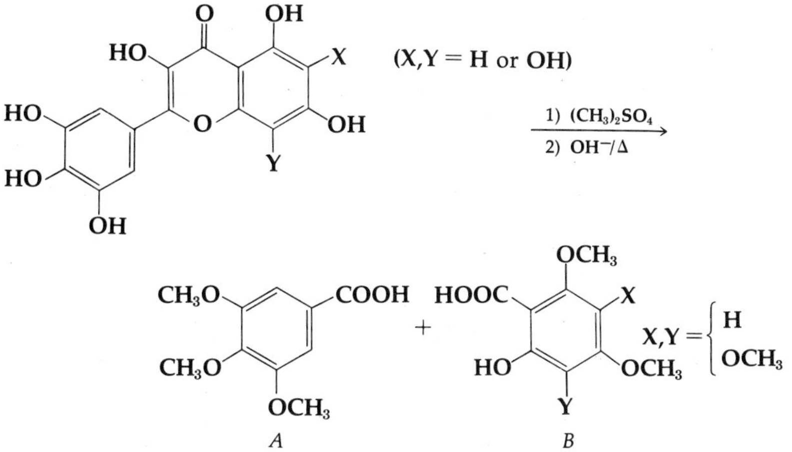

A   B

27-16

$(H^+)O$ HO

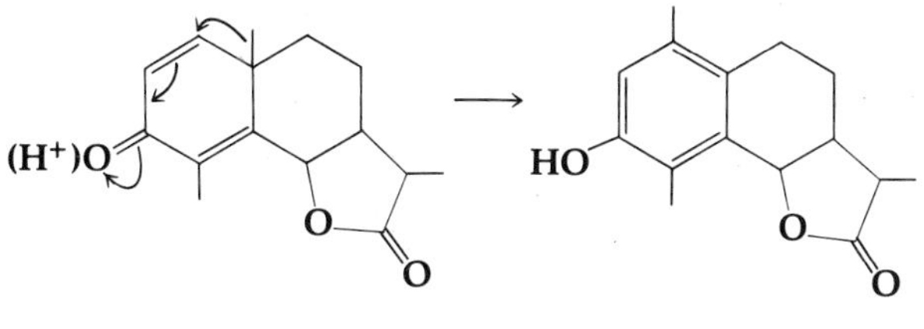

27–17

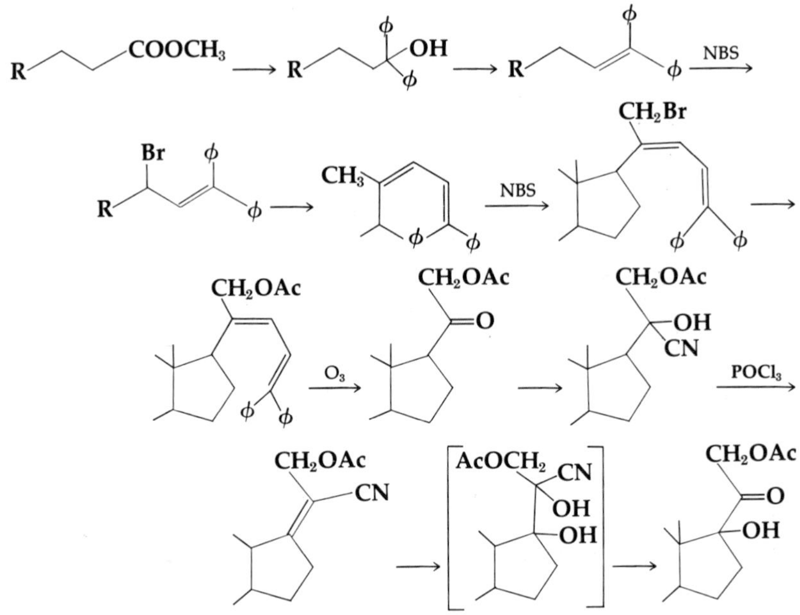

27–18

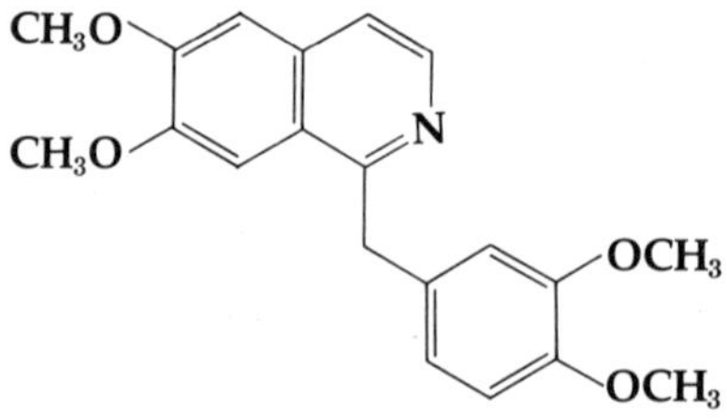

27–21   a

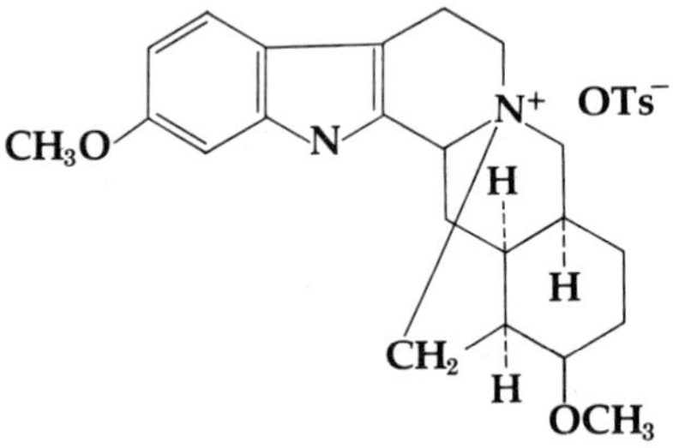

Proves labeled hydrogens must all be cis for such a cyclization to be physically possible

27–24

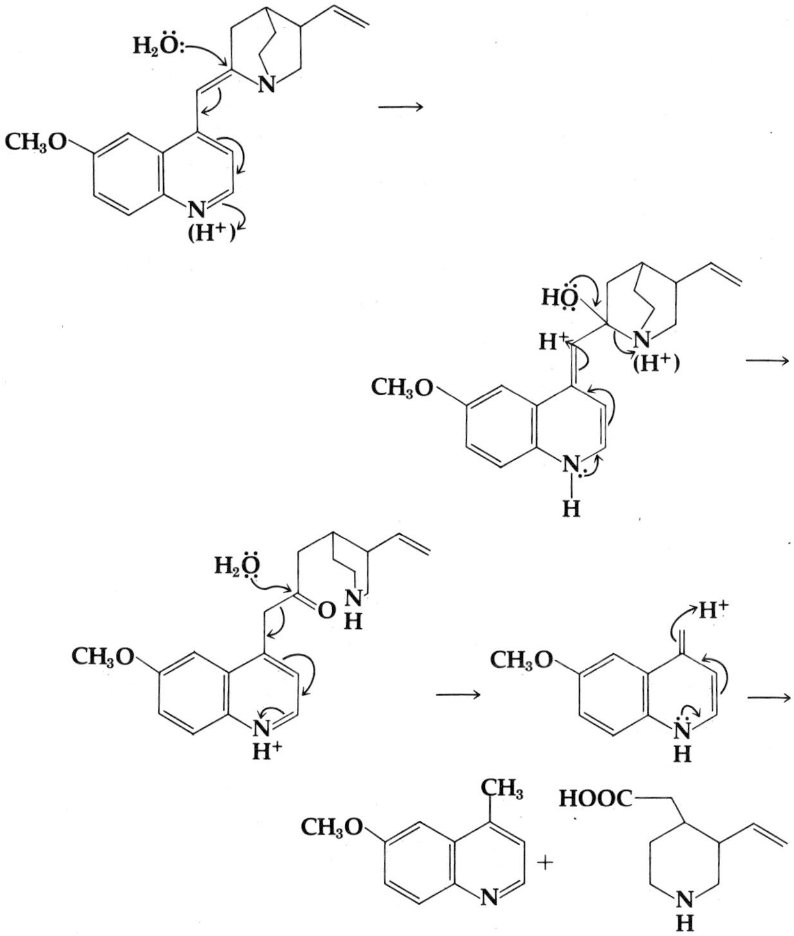

$H_2\ddot{O}:$

CH_3O-

$\longrightarrow$

$H\ddot{O}$

H^+

(H^+)

CH_3O-

$\longrightarrow$

$H_2\ddot{O}$

CH_3O-

H^+

$\longrightarrow$

H^+

CH_3O-

$\longrightarrow$

CH_3O- (4-methylquinoline) $+$ $HOOC-$ (3-vinylpiperidine-4-carboxylic acid)

27–25 a

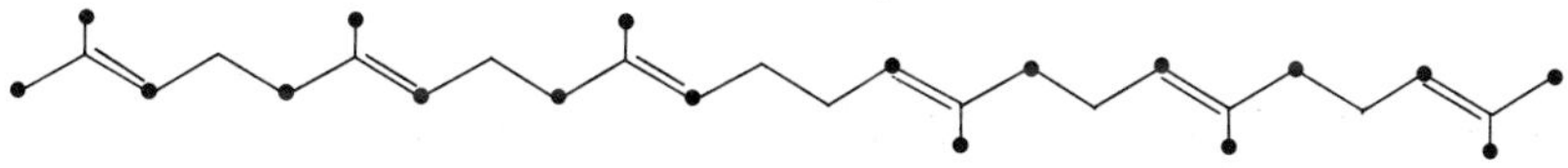

$C^{14} = \bullet$; total $= 18$; $C - C^{14}H_3 = 8$

b 5 out of 15 (18 − 15 = 3CH₃ lost from squalene to cholesterol)

27–26 $C_8H_{17}NO \xrightarrow[-2C]{[O]} C_6H_{11}NO$

Index = 1   Index = 4

—COCH₃ or   neutral =

—CHOHCH₃

$\underset{\;}{-\overset{|}{N}-C=O}$

∴ one ring or C=C

No C—CH₃ $\xrightarrow{H_3O^+}$ $C_6H_{13}NO_2$

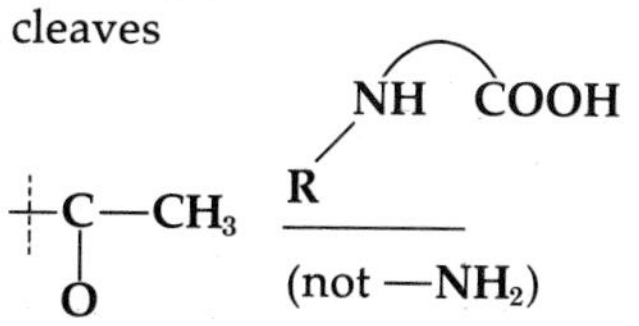

means [O]

cleaves

$+C-CH_3$

$\underset{O}{\overset{|}{\|}}$

$\underset{(not -NH_2)}{\underline{R\quad\overbrace{NH\quad COOH}}}$

Requires

a

ring

Possible: (N-methyl-2-(1-hydroxyethyl)piperidine) $\longrightarrow$ (N-methyl-2-piperidone) Also, (pyrrolidine derivative) etc.

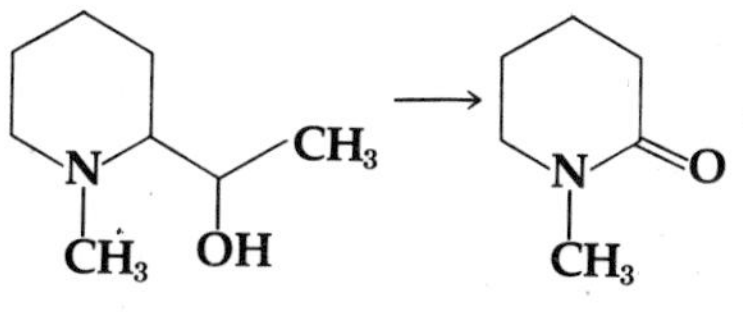

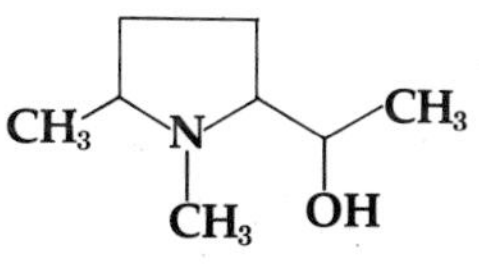

27-30

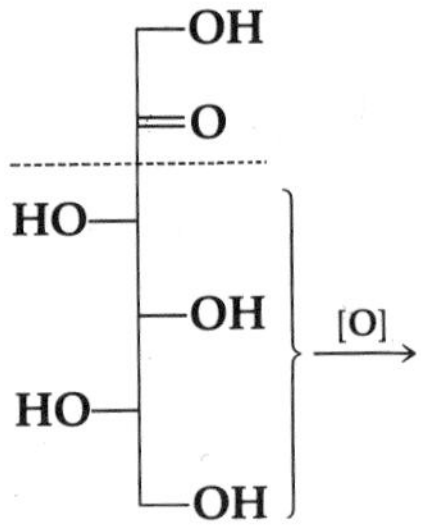

$\xrightarrow{[O]}$ one tartaric acid enantiomer; the other from carbons 2, 3, 4, 5

27-32

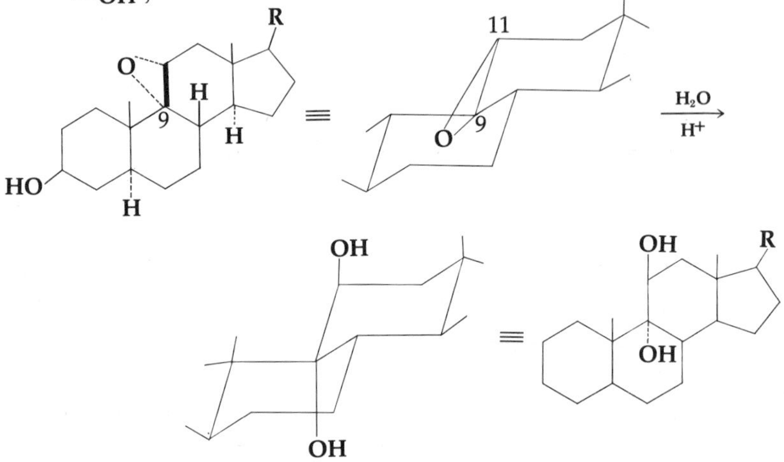

$\xrightarrow[\text{H}^+]{\text{H}_2\text{O}}$

27-34

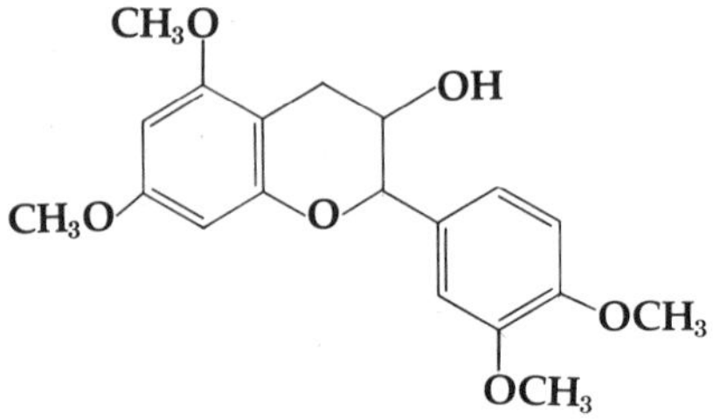

(Flavone type)

27-35

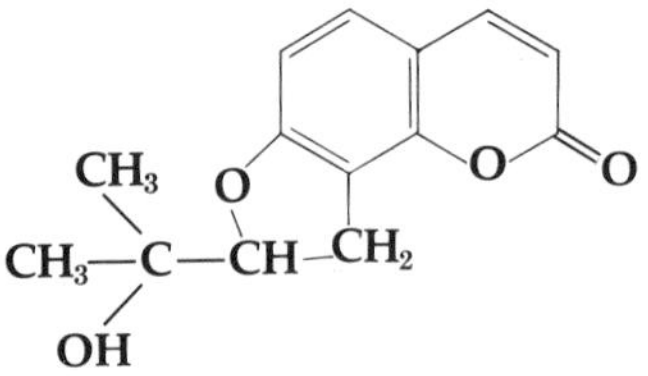

(Coumarin + C_5 terpene side chain)

27-36

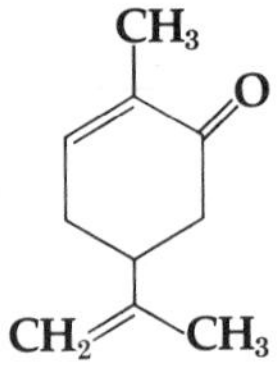

(Monoterpene)

27-37 $C_{15}H_{22}O$ $\xrightarrow[H:^-]{CH_3:^- \text{ or}}$ $\xrightarrow[[-H]]{Pd}$

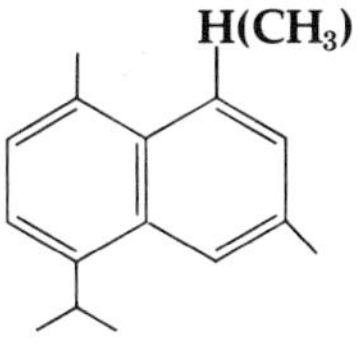

Index = 5

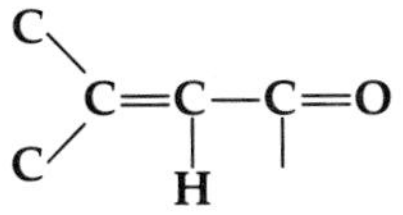

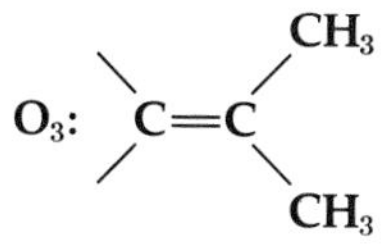

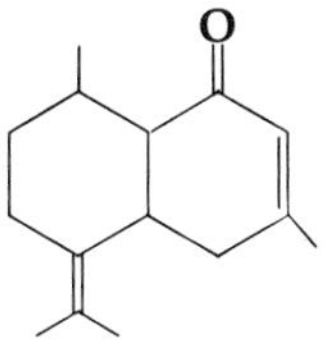

H(CH₃)

C=C—C=O with C, C on left and H below

O_3: $C=C$ with CH₃, CH₃

2 rings

O

Articulone
(sesquiterpene)

27-39 Cannot be constructed from head-to-tail isoprene units but arises by later rearrangement in biosynthesis:

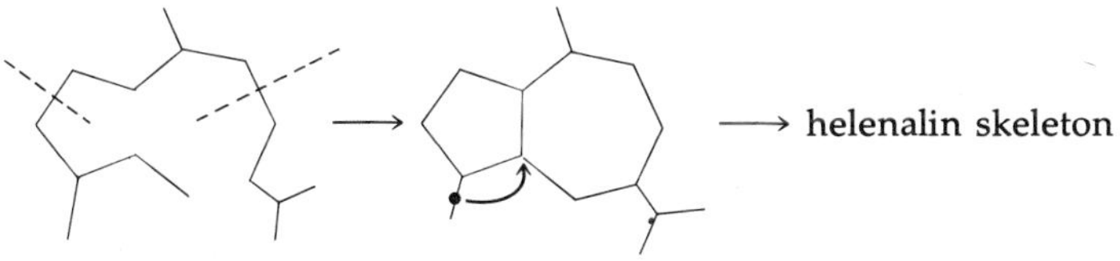

⟶ helenalin skeleton

INDEX

TABLE 2-1   Normal Bond Lengths (angstroms)

H—C	1.09 Å	C=C	1.35 Å	C≡C	1.20 Å
H—N	1.00	C=N	1.30	C≡N	1.16
H—O	0.96	C=O	1.22		
C—C	1.54				
C—N	1.47				
C—O	1.43				
C—Cl	1.76				
C—Br	1.94				
C—I	2.14				

TABLE 2-3   Electronegativities of Atoms

Kernel charge:	+1	+4	+5	+6	+7
	H	C	N	O	F
	2.2	2.5	3.0	3.5	4.0
		Si	P	S	Cl
		1.9	2.2	2.5	3.0
					Br
					2.8
					I
					2.5

TABLE 6-2   Axial-Equatorial Energy Differences
for Substituents on Chair Cyclohexane

Substituent	ΔF, kcal/mole	Substituent	ΔF, kcal/mole
—CH$_3$	1.7	Cl, Br, I	0.5
—CH$_2$CH$_3$	1.8	OH, OR	0.7
—C(CH$_3$)$_3$	Very large	COOR(H)	1.1
—C$_6$H$_5$	3.1	CN	0.2